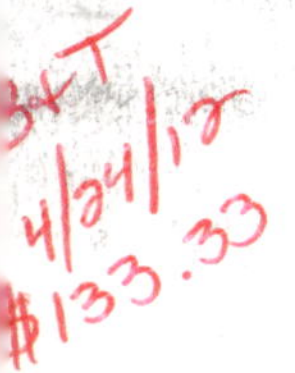

Letter to Readers

Our goal in writing *Fundamentals of HVACR* is to encourage a high level of professionalism for future and current technicians in the industry by producing a text that is both accessible and thorough. To accomplish this, we followed a few guiding principles:

- Explain why things work.
- Organize material so that it progresses from fundamental concepts to specific applications of those concepts.
- Organize the material so that it progresses in difficulty from easily understood ideas to more challenging concepts and applications.
- Write a text that is accessible to readers while still providing the detailed information necessary to be successful.
- Write a book that HVACR students will want to keep for reference when they become professionals.

The "HVACR Science" section lays the foundation for understanding virtually all the concepts and applications introduced later in the book. Taking the time to understand the science behind HVACR systems is an investment in your future. The specific details of individual units change over time, but the science that governs how they operate stays the same.

The most difficult part of writing a comprehensive text is deciding what material to include. Everything there is to know simply won't fit into a single book. We focused on one simple question: Does the material help the reader understand how to install, maintain, and service HVACR systems?

The presentation of the material is equally important. There is always more than one way to describe or explain something. We chose the most transparent and easily understood explanations possible without leaving out crucial detail. Concepts and applications are reinforced with specific, real-world examples. Our writing style is engaging to draw the students in, as there is really no reason a textbook can't be written in a style that encourages the reader to read out of interest rather than obligation.

HVACR students and technicians are our target audience. Most technicians are visually oriented learners. A good picture or illustration can often explain more than several pages of text, so we have included many pictures, illustrations, graphs, and diagrams to show everything from basic concepts to operational sequences.

One of the great things about teaching is sharing in other people's success. There is nothing more gratifying than seeing students succeed. We hope that this book launches you on a successful HVACR career and finds a permanent place in your library or service truck.

Fundamentals of HVACR

Second Edition

CARTER STANFIELD
Athens Technical College

DAVID SKAVES
Maine Maritime Academy

Boston Columbus Indianapolis New York San Francisco Upper Saddle River
Amsterdam Cape Town Dubai London Madrid Milan Munich Paris Montreal Toronto
Delhi Mexico City São Paulo Sydney Hong Kong Seoul Singapore Taipei Tokyo

Editorial Director: Vernon R. Anthony
Acquisitions Editor: David Ploskonka
Editorial Assistant: Nancy Kesterson
Director of Marketing: David Gesell
Executive Marketing Manager: Derril Trakalo
Senior Marketing Coordinator: Alicia Wozniak
Marketing Assistant: Les Roberts
Senior Managing Editor: JoEllen Gohr
Associate Managing Editor: Alexandrina Benedicto Wolf
Production Project Manager: Maren L. Miller
Senior Operations Supervisor: Pat Tonneman
Operations Specialist: Deidra Skahill
Senior Art Director: Diane Y. Ernsberger
Cover Designer: Bryan Huber
Cover Art: Shutterstock, Sergey Mironov
Image Permission Coordinator: Mike Lackey
Rights and Permissions Research: Jaime Jankowski and Martha Hall, PreMediaGlobal
Media Editor: Susan Watkins
Lead Media Project Manager: Karen Bretz
Full-Service Project Management: Peggy Kellar, Aptara®, Inc.
Composition: Aptara®, Inc.
Printer/Binder: R. R. Donnelly, Willard
Cover Printer: Lehigh-Phoenix Color/Hagerstown
Text Font: Vector LH

Credits and acknowledgments borrowed from other sources and reproduced, with permission, in this textbook appear on the appropriate page within the text.

Library of Congress Cataloging-in-Publication Data available upon request.

10 9 8 7 6 5 4 3 2 1

ISBN 13: 978-0-13-285961-5
ISBN 10: 0-13-285961-0

Take the Guided Tour

PEARSON'S CURRICULUM SOLUTION FOR HVACR

We didn't just revise this textbook. We rethought the way you use it and the way you teach with it. We revised the entire line of materials for your HVACR program. Every item, from the supplements to the PowerPoints and the media program, was evaluated so that they would all seamlessly cover your entire curriculum. We wanted to save you time and headaches while giving your students a consistent and engaging learning experience for their entire degree or certificate.

PEARSON CLASS MASTER

Your program requires a lot of material—textbook, videos, online media, PowerPoint slides, lecture notes, and more. You have to juggle and then customize them for your teaching. That is a lot of work. But *not anymore.*

With **Class Master** you just build your calendar, select your topics, and instantly you have your syllabus. Not just a syllabus—it tailors our lecture notes to your schedule. What's more, you now know exactly what student and instructor tools are needed and available for each day.

Finally, **Class Master** will include a list of learning objectives in your instructor syllabus mapped to industry standards like AHRI's ICE and HVAC Excellence to assist you with accreditation and planning sessions.

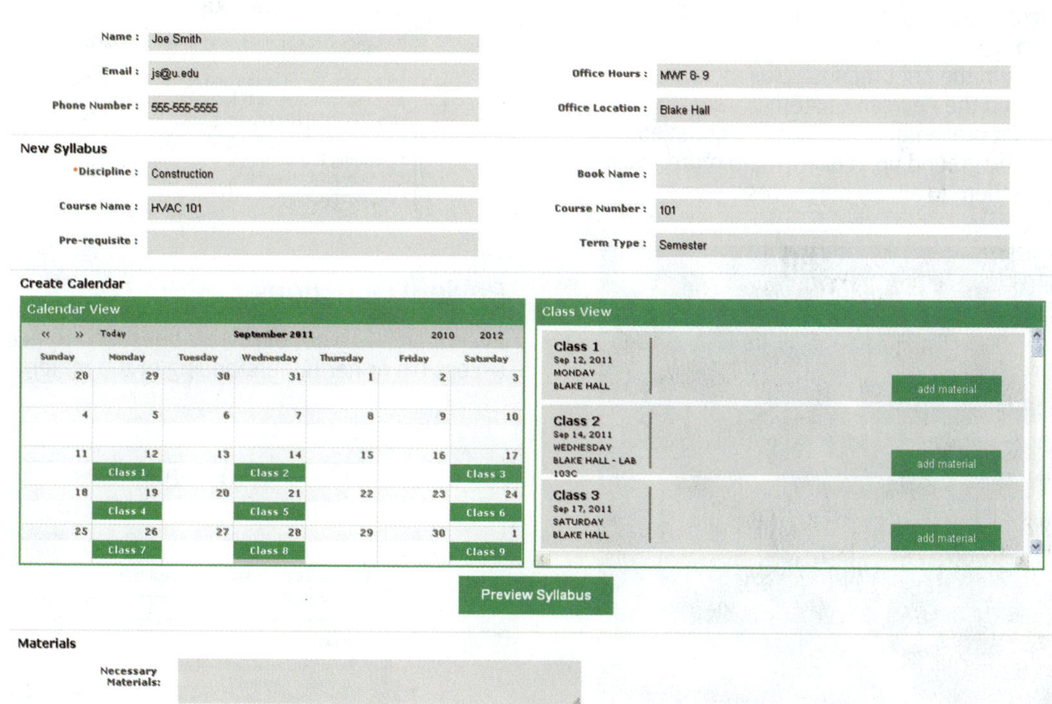

the videos
ons, eBook,
Lab is back
r **fifty new**
never been

Instructor Supplements

Comprehensive and broad in scope, we've left nothing to chance.

The supplements have been revised from the ground up, are integrated into **Class Master**, and are mapped to industry standards.

- Robust instructor's manual with new detailed lecture notes customized to your needs
- Expanded test questions for every learning objective in your program
- Additional labs in the print manual, plus more in MyHVACLab and the Pearson Custom Library
- The PowerPoint presentations have been completely revised from the ground up, making them more striking and matching them to all key objectives

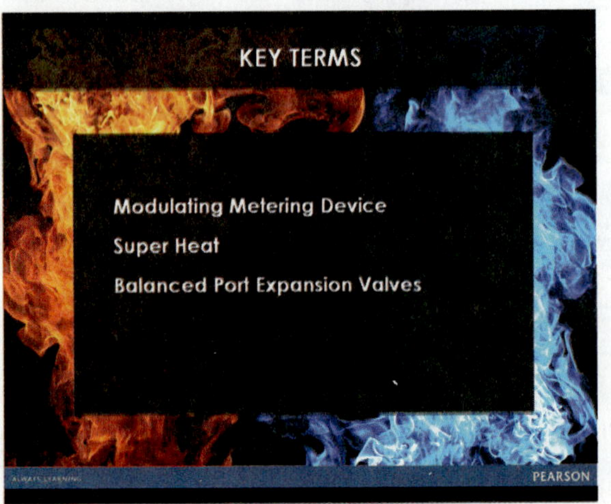

Fundamentals of HVACR, 2nd Edition

Created with a clear-cut vision of what students need, this groundbreaking text provides comprehensive coverage of heating, ventilating, air conditioning, and refrigeration. This fully updated edition includes additional coverage of electrical, commercial, codes, and sustainability.

Learning Objectives

Each unit begins with clearly stated objectives that enable you to focus on what you should achieve by the end of the unit.

OBJECTIVES

After completing this unit, you will be able to:

1. define *refrigerant*.
2. identify the type of refrigerant by its number designation.
3. list the different types of refrigerant chemical composition.
4. explain the difference between compounds, zeotropes, and azeotropes.
5. describe the pressure-temperature relationship of saturated refrigerant.
6. list the different types of refrigerant contaminants.
7. explain the refrigerant safety rating system.
8. match refrigerants with the proper refrigeration oil.

Unit Introductions and Unit Summaries

These pull together the main points of the unit to prepare and remind students of what they should remember.

19.1 INTRODUCTION

Any study of the refrigeration cycle would be incomplete without a look at the refrigerant in the system. The refrigerant is the first component of a refrigeration system that must be determined, since its chemical and physical properties will play a large role in selecting all other components in a refrigeration system. A number of materials have physical and thermodynamic properties that make them suitable for use as refrigerants. In fact, anything that can boil at a low temperature and condense at

UNIT 19—SUMMARY

Refrigerant is the fluid used in a refrigeration system to transfer heat. It does this by absorbing heat while evaporating at a low pressure and temperature and condensing at a high pressure and temperature. Refrigerants may be classified by many characteristics, including the following:

Operating Pressure

Very high pressure
High pressure

Review Questions

Every unit has a set of review questions to help the reader assess his or her understanding of the material.

UNIT 19—REVIEW QUESTIONS

1. Define *refrigerant*.
2. What organization has established the number designation for refrigerants?
3. What are the chemical components of CFC refrigerant?
4. What are the chemical components of HCFC refrigerant?
5. What are the chemical components of HFC refrigerant?
6. What is ozone?
7. Which chemical family of refrigerants has the highest ozone-depletion potential?

Caution Tips and Safety Tips

These tips contain information students should know to operate equipment properly and protect themselves from harm.

CAUTION

Do not use pliers to tighten or loosen brass fittings. No matter how delicately you use pliers on these fittings, the fitting surface will be damaged. This damage can prevent the next technician from using the proper wrench on the damaged fitting.

SAFETY TIP

Some portable electric tools provide for double-insulated casings and do not need to be grounded. This type of tool may only have a two-prong plug rather than three. If a tool plug has three prongs, never use an adapter to make it fit a two-prong receptacle unless the adapter is properly grounded.

Code Tips

These tips help students provide additional information about codes that technicians will need to know.

CODE TIP

On December 7, 2009, the EPA declared that the current and projected atmospheric concentrations of the six key, well-mixed greenhouse gasses, including HFCs, "threaten the public health and welfare of current and future generations." While no regulations are currently in place in the United States to regulate or phase out HFC refrigerants, the statement does establish the justification for future HFC regulations.

Tech Tips and Service Tips

These tips provide extra detail and information for students who want to go beyond the basics and get practical applications for the information in the unit.

TECH TIP

Air-conditioning and refrigeration hand tools do not get as dirty as most mechanics' tools will. However, it is a good idea to wipe them down with a clean, dry rag after each use. In addition, from time to time they should be rubbed with a lightly oiled cloth to protect them from rusting.

SERVICE TIP

Wrenches should always fit securely over the bolt head or nut before force is applied. If the wrench slips, you will jam your fingers. It is always best to try to pull rather than push on a wrench (Figure 4-39). You will have more control and will be less likely to jam your fingers or rap your knuckles. If you must push, then use an open palm so you're your fingers do not become jammed between the wrench and the work as the wrench begins to turn (Figure 4-40).

Green Tips

These tips provide important details about sustainable practices in the HVAC industry.

GREEN TIP

Blower door readings are useful when performing house tightening. They provide an objective measurement of the degree of improvement achieved.

COMPREHENSIVE TEACHING AND LEARNING PACKAGE

For the Instructor

To access supplementary materials, go online to www.pearsonhighered.com/irc and register. Within forty-eight hours you will receive a code via e-mail and instructions. Within forty-eight hours of registering, you will receive a confirming e-mail including an instructor access code. Once you have received your code, locate your text in the online catalog and click on the Instructor Resources button on the left side of the catalog product page. Select a supplement, and a login page will appear. Once you have logged in, you can access instructor material for all Prentice Hall textbooks. If you have any difficulties accessing the site or downloading a supplement, please contact Customer Service at http://247pearsoned.custhelp.com/. *Note:* The supplements are also housed in MyHVACLab at www.MyHVACLab.com.

Class Master

You can gain access to the Class Master Web site through the supplements pages referenced above and through MyHVACLab. Once you are at Class Master, you will not need a new password. There is a walk-through to guide you through the system.

HVACR Blog

For teaching tips and more, visit Carter Stanfield's HVACR blog at http://hvacrfundamentals.blogspot.com/

Instructor's Manual with Lecture Notes and Correlation Guides ISBN-10: 0-13-287932-8

Even though you can customize your own instructor's manual through Class Master, you can also download a PDF of it as well. We've also created downloadable correlation guides to show you how our content matches other books and industry standards like AHRI's ICE and HVAC Excellence.

PowerPoint Slides ISBN-10: 0-13-288088-1

These comprehensive, colorful PowerPoint slides provide a powerful lecture or study tool.

Lab Manual ISBN-10: 0-13-287974-3

The print lab manual covers the basics, providing you and your students with labs for every key topic in the book. There are approximately three hundred labs in the print version, and four hundred more are waiting for you at www.pearsoncustomlibrary.com, where you can choose all the labs you want for your courses. You can build a lab manual for every course to your specifications, choosing from all of our HVACR labs. It will give you a custom ISBN, and your bookstore can order it just like it would a traditional book.

MyTest ISBN-10: 0-13-286113-5

MyTest is a comprehensive set of test questions matching key objectives for all of your courses.

For the Student and Instructor

PEARSON **myhvaclab**

With a **new and intuitive interface, MyHVACLab** provides learning management tools to save you time as well as media and assessment to engage and track your students. With over fifty new service-call simulations, and more than three hundred minutes of video, your students will have a learning experience like none other.
Each unit includes

- Objectives and outlines
- eBook
- Homework questions
- PowerPoint slides
- Digital flashcards
- Animations of key concepts
- Interactive exercises
- Concept spotlights: multimedia cliff notes for the book
- Video clips
- A pretest and a posttest
  - This supercharged study guide allows students to take a pretest to determine what they don't already know. It will then create a customized study plan based on those learning needs. A posttest follows to make sure they have learned everything they need to know.

AHRI and HVAC Excellence Correlation Guides

Standards for these two associations have been mapped to our materials.

Supplemental Text

Guide to the NATE/Ice Certification Exams by Featherstone and Riojas, order ISBN-10: 0-13-231970-5

ACKNOWLEDGMENTS

We could not produce learning experiences like this without instructors like you. Thanks to everyone listed below, and anyone who joined the team after the book went to print.

An asterisk denotes contributors to more than one area of the program.

REVIEWERS

Bruce Bowman
Davidson County Community College

Jerry Britt
Horry Georgetown Technical College

Michael Brock
Florida State College

Douglas Broughman
Augusta Technical College

Danny Burris
Eastfield College of Dallas County Community College District

Thomas Bush
South Florida CC

Victor Cafarchia
El Camino College

Gabriel Cioffi
LA Trade Tech

Hugh Cole
Certification & Training Services

Peter J. Correa
TCI College of Technology

Jonathan Darling
Des Moines Area CC

David DeRoche
St. Clair College

Daniel Foust
Austin Community College

Lani Greenway
Illinois Central College

Nick Griewahn
Northern Michigan University

Kevin Harmon
Jefferson College

Patrick Heeb
Long Beach City College

Jeffrey Hess
Waubonsee Community College

John Holley
Calhoun CC

Randy Hughes
Wallace CC

Timothy Hummel
Southeast Technical Institute

Torry Jeranek
Winona Area Institute

Tom Kissell
Terra Community College

Keith Klix
Texas State Technical College, Waco

Jim Kroll
Virginia Highlands Community College

Dan Leathers
College of DuPage

Joe Marchese
Community College of Allegheny County

Rick Marks
Cisco Junior College

Scott McClure
Vernon College

Richard McDonald
Santa Fe Community College

Alan R. Mercurio
Oil Tech Talk

Michael Mutarelli
Lehigh Carbon Community College

Joel Owen
Alabama Power/HVAC Training Center

Joe Owens
Antelope Valley College

William J. Parlapiano III
President of BP Consulting, serves on the NATE, ARI-ICE and BPI Technical Committees

Michael Partyka
College of the Albemarle

Whit Perry
Northwest Mississippi Community College

Jevaris Pettis
Midwest Technical Institute

Roger Raffaelo
Daytona State College - Main

Terry M. Rogers
Midlands Technical College

Robert Rossell
Pittsburg Technical Institute

Manuel Sanchez
Imperial Valley College

David Shehadeh
UALR, Morrilton College

Dalton W. Thacker
Augusta Technical College

Monty Timm
Ivy Tech

S. Shane Todd
Ogeechee Technical College

Roger Tomfohrde
Minnesota State College–Southeast Tech

Mark VanDoren
Ivy Tech

Juan Villela
Saint Phillips College

Wayne Whitfield
Fitchburg State College

Freddie Williams
Lanier Technical College

Clifford Wilson
Nunez CC

INDUSTRY STANDARDS CONSULTANTS

Jeffrey Hess
Waubonsee Community College

Keith Klix*
Texas State Technical College, Waco

Johnny McDonald*
Middle Georgia Technical College

Jevaris Pettis*
Midwest Technical Institute

LAB CONTRIBUTORS

Chris R. Maaks
Midwest Technical Institute

Johnny McDonald*
Middle Georgia Technical College

Glen McNamara*
Branford Hall Career Institute

Robert Polchinski
New York City College of Technology

Donald Steeby*
Grand Rapids Community College

ASSESSMENT CONTRIBUTORS

Michael Garrity
Branford Hall Career Institute

Elwin Hunt
San Joaquin Valley College

Glen Martin
Branford Hall Career Institute

Patrick Monahan
Branford Hall Career Institute

Tom Owen
Sullivan College of Technology & Design

Michael Patton*
Branford Hall Career Institute

Edward Rosenberg
Branford Hall Career Institute

POWERPOINT SLIDES AND LECTURE NOTES CONTRIBUTORS

Bruce Bowman
Davidson County College

Jerry Britt
Horry Georgetown Technical College

Michael Brock
Florida State College

James Chadwick
Kaplan University

Gabriel Cioffi
LA Trade Tech

Clint Cooper
Chattahoochee Technical College

Mike Falvey
Copiah-Lincoln Community College

Rick Marks
Cisco College

Joe Owens*
Antelope Valley College

Kevin Pulley
Career Institute of Technology

Roger Raffaelo
Daytona State College, Main

Doug Sallade
Cypress College

Donald Steeby
Grand Rapids Community College

Monty Timm
Ivy Tech Community College

S. Shane Todd
Ogeechee Technical College

ADVISORY BOARD

Doug Broughman
Augusta Technical College

Fred Brown
Career & Tech Center

Whit Perry
NW MS Community Coll

Patrick Tebbe*
Minnesota State University

We would also like to acknowledge the following individuals, companies, and colleges for allowing photographs to be taken of their equipment and facilities for use in this text.

PICTORIAL SUPPORT

Jerry Markley and Lynn Darnell,
Maine Maritime Academy, Castine, ME

Charlie Veilleux, Rick Gomm, and Jim Peary,
Eastern Maine Community College, Bangor, ME

Tom Kissell,
Terra Community College, Fremont, OH

Glenn Carlson,
Hannaford Bros. Co., Scarborough, ME

David Kuchta,
The Jackson Laboratory, Bar Harbor, ME

Jeff Vose,
Allen's Blueberry Freezer, Inc., Ellsworth, ME

Robin Tannenbaum LEED AP and Phil Kaplan AIA,
LEED AP, Kaplan Thompson Architects,

Portland, ME Keith Collins, M.D.,
BrightBuilt Barn, Rockport, ME

Bob Morse and Mike Hudson,
Getchell Bros. Inc. Brewer, ME

About the Authors

Carter Stanfield is program director of the Air Conditioning Technology Department at Athens Technical College, where he has taught since 1976. His industry credentials include both an RSES CM and NATE certification and a State of Georgia Unrestricted Conditioned Air Contracting license. He graduated from the University of Georgia magna cum laude in 1995 with a bachelor of science degree in education. Mr. Stanfield believes that successful educational programs are focused on what the students do. Students start with a strong background in fundamental concepts and theory and then actively apply them to solve real problems. Practice and active application are the keys to students building both confidence and competence. For teaching tips and more, see his HVACR blog at http://hvacrfundamentals.blogspot.com/.

David Skaves, P.E., has been a faculty member at the Maine Maritime Academy since 1986 and received the Teaching Excellence award at the college in 2006. His career background includes employment as a marine engineer on supertankers in the merchant marine, a production planner at Maine's Bath Iron Works Shipbuilding, and an engineering consultant for combined cycle power plant performance testing throughout the United States as well as in Mexico and South America. In addition to his MBA from the University of Maine at Orono, Professor Skaves is a registered professional engineer, licensed first-class stationary engineer, and licensed marine chief engineer. He is currently a member of ASHRAE, AFE, and the Maine State Board of Boiler and Pressure Vessels.

Freddie Williams is the author of **MyHVACLab** and the program chair of the Air Conditioning Technology Department at Lanier Technical College, with over twenty-three years of professional experience. Mr. Williams is a graduate of Athens Technical College and has been a faculty member at Lanier Tech since 2002, teaching air conditioning and industrial systems. He was awarded Instructor of the Year at Lanier Tech in 2010 and the Master Teacher award in Hall County, Georgia, in 2010. Mr. Williams is a member of ASHRAE and RSES and holds degrees in management and technical studies. He has extensive experience with electrical and mechanical systems in industrial, commercial, and military environments.

Contents

SECTION 6

Air Conditioning Systems

SECTION 7

Heating Systems

SECTION 8

Heat Pump Systems

SECTION 9

System Design, Sizing, and Layout

SECTION 10

Commercial Environmental Systems

SECTION 11

Commercial Refrigeration Systems

SECTION 12

Installation, Maintenance, Service and Troubleshooting

UNIT 1

Introduction to Heating, Ventilation, Air Conditioning, and Refrigeration

OBJECTIVES

After completing this unit, you will be able to:

1. give a brief history of HVACR.
2. define environmental heating and air conditioning.
3. give the advantages of freezing foods quickly.
4. explain the importance of having a clean background.
5. list the various types of HVACR jobs and explain what they might do.
6. list the HVACR professional organizations.

1.1 INTRODUCTION

The abbreviation HVACR is certainly a mouthful, and so it is not unusual to ask the question, "What does this mean, and how does it impact me?" However, the answer is not so simple, and a standard definition may not explain very much. This is because the HVACR industry is a complex network that our entire society relies on more today than ever before. Just think how your world would change without refrigeration for your food or drinks and without air conditioning in your car or classroom. Try to visualize how this would affect the greater population, from food distribution networks, to hospital care, to housing for the elderly. As a trained and skilled HVACR technician, you can make a positive impact on society. You can contribute to this growing industry to ensure that systems work efficiently and safely and are environmentally friendly (Figure 1-1).

1.2 HISTORY AND OVERVIEW OF HVACR

Heating

In an attempt to better understand HVACR, let's break it down component by component. The *H* for *heating* seems easy. The history of heating a space by burning wood starts in our earliest times and continues to the present. Elaborate systems using firewood heated Roman buildings. Channels were built underneath the floors to draw heat from a fire, thus warming the building and creating the first central heating systems (Figure 1-2).

Wood, peat, and coal remained the primary heating fuels for centuries. Many early buildings had open fireplaces. But fireplaces are an inefficient way of heating because too much of the heat produced is drawn up the chimney. Although early seventeenth-century European masonry-type stoves burned wood safely at high efficiency, the next

Figure 1-1 Think green! New innovative technologies will allow some HVACR systems to operate on power supplied by wind turbines.

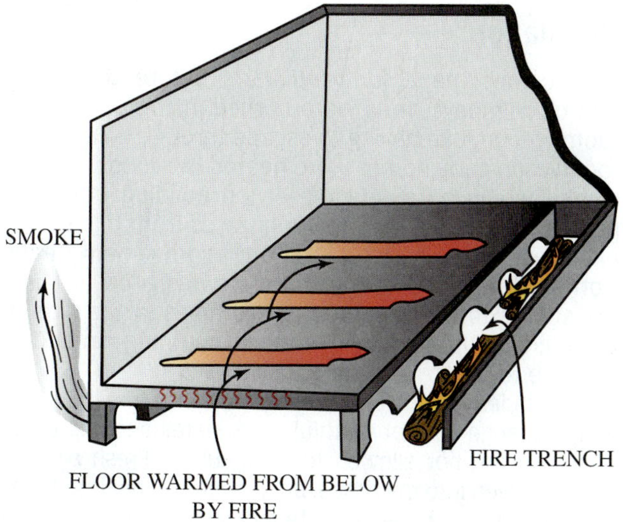

Figure 1-2 Romans used fires channeled below floors as early heating systems.

Figure 1-3 Woodstove.

major step in heating technology in America was the metal stove. Benjamin Franklin is credited with inventing a cast-iron stove that was several times more efficient than any other stove at that time. Many people still use decorative, efficient stoves to provide much, if not all, of their heating needs (Figure 1-3).

However, wood heat is only one alternative, because today there are many more choices for heating. Gas heat, oil heat, electric heat, and solar heating systems are common. Heat pumps that use a refrigeration system for heating can be very efficient. Geothermal heating systems that utilize the heat from within the earth are becoming more popular. New, environmentally friendly ideas and efficient designs are continually being developed, tested, operated, and maintained by people just like you entering the industry. So you can see that just the *H* alone is a large and important sector.

Ventilation

Next comes the *V* for *ventilation*. Before the invention of chimneys, fires were burned in the center of a room with smoke having to escape through holes in the roof. When early homes were heated by wood fires, the smoke would permeate the entire building. Although people were warm, the health hazards from this smoke exposure were harmful. As an improvement, early Norman fireplaces in England were designed to allow the smoke to escape through two holes in the side of the building. It was obvious that something needed to be done to improve the air quality.

A properly ventilated building allows for the air to flow and exchange so that harmful particulates such as those in smoke are not allowed to accumulate. Fresh air also brings oxygen into the space, but it becomes depleted over time. A simple ventilation system can consist of only a fan and some minor ductwork for transporting the air. More complex systems circulate air throughout entire buildings through a vast network of ducts and blowers.

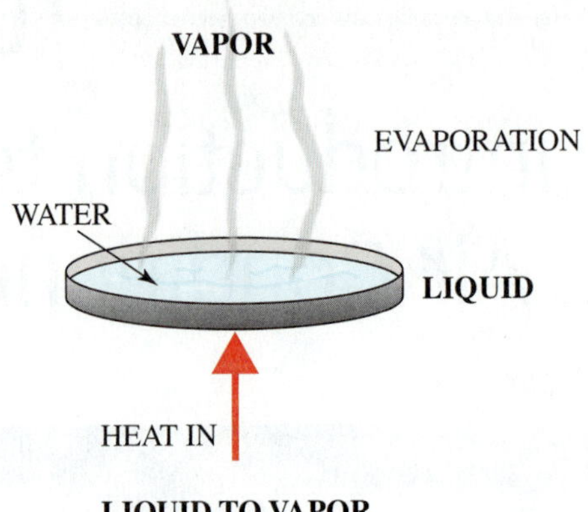

Figure 1-4 When water evaporates, heat is absorbed. This change of state is also referred to as a phase change.

Air Conditioning

The *AC* stands for *air conditioning*. Generally this is considered by most people to be a way to cool a space, but as you will learn, this term encompasses much more. Artificially cooling the air in a living space dates back to the earliest centuries. In ancient Greece, large wet woven tapestries were hung in natural drafts so that the air flowing through and around the tapestries was cooled by the evaporating water. As the water evaporated it would remove heat, just like when you perspire to remain cool (Figure 1-4). Some manufacturers sprayed water in factories for cooling as early as the 1720s. Evaporative cooling is still used extensively in residences and businesses throughout the southwestern United States, where typical summer conditions are very hot and dry.

Ice was the primary means of cooling air for many years. The Romans packed ice and snow between double walls in the emperor's palaces. John Gorrie patented the first mechanical air-conditioning system in 1844. His system was used to cool sick rooms in hospitals in Florida. The United States capitol building in Washington, DC, was first air conditioned using ice in 1909. Rumor has it that when the legislators got really involved in controversial debates, more ice was required to keep the building cool. The phrase "tons of air conditioning" we use today came from this era in history, when tons of ice were used for cooling (Figure 1-5).

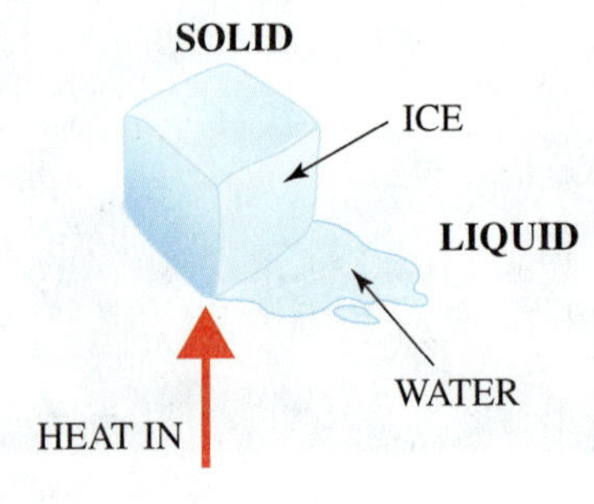

Figure 1-5 When ice melts, heat is absorbed.

TECH TIP

Refrigerant capacity is measured in tons. One ton of capacity is equivalent to the amount of heat that 2,000 lb of ice can absorb in one day. The amount of latent heat required to change 1 ton of ice into 1 ton of water is 288,000 BTU. If this amount is divided by 24 hr per day, the equivalent is 12,000 BTU/hr.

Refrigeration

Finally, the *R* stands for *refrigeration*, which is a necessary component for most air-conditioning systems; however, refrigeration systems are more commonly considered to be used for keeping food cold. That is why very often you may see the abbreviation HVAC, which implies air conditioning only. The broader term HVACR includes both air conditioning and refrigeration systems.

The first use of refrigeration was for the preservation of food. Ice was harvested from frozen lakes and stored for later use. Sometimes it could be kept all summer long in ice houses. Ice harvesting remained a flourishing industry well into the twentieth century.

Archeologists have discovered that the first evidence of man making ice appeared more than 3,000 years ago, about 1,000 BC. Peoples living in northern Egypt, the Middle East, Pakistan, and India made ice using evaporation. Archeological excavations in these regions have discovered ice-producing fields that covered several acres. The ice was produced in shallow clay plates, about the size of a saucer. The water in these clay plates wept through the clay. This water dampened the small straw mats holding the clay plates in racks a few feet above the ground (Figure 1-6). The straw aided evaporative cooling of the water. Under the right conditions of temperature and humidity, a thin film of ice would form overnight on each clay plate.

Producing ice in this way is also the principle behind modern snow-making equipment. A snow-producing machine like the one in Figure 1-7 can make snow by evaporative cooling even when the temperatures on the ski slopes are above freezing.

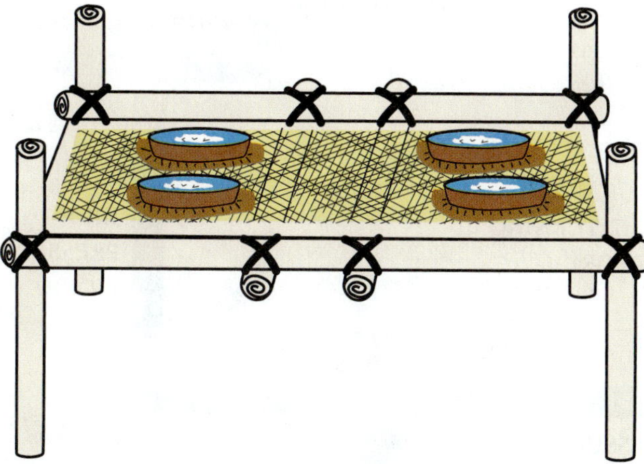

Figure 1-6 Ice was first artificially produced to be used for food preservation more than 3,000 years ago.

Figure 1-7 Snowblowers can produce artificial snow by evaporative cooling. *(Courtesy of Red River Ski Area)*

Today, a majority of refrigeration systems use what is referred to as mechanical vapor compression. The mechanical process of compressing a gas to produce cooling can be traced back to coal mines in England. Large steam-driven or water-powered compressors were used to force air into the deepest mines so miners could work in a safe atmosphere. Over long hours of operation, miners observed the formation of ice around the air nozzles (Figure 1-8). This ice was collected and used for food preservation. The construction of steam-powered compressed-air plants that produced ice soon followed. The first maritime refrigeration units were made by putting steam-powered compressors on sailing ships to make it possible for beef to be shipped from Australia to England, starting in 1876.

HVACR and the Refrigeration Cycle Now that you have a better understanding of what HVACR means, it is easy to see that it encompasses a broad spectrum of needs and applications. Although the methods for heating can vary considerably, the majority of cooling applications are based on the refrigeration cycle. When ice changes to water, heat

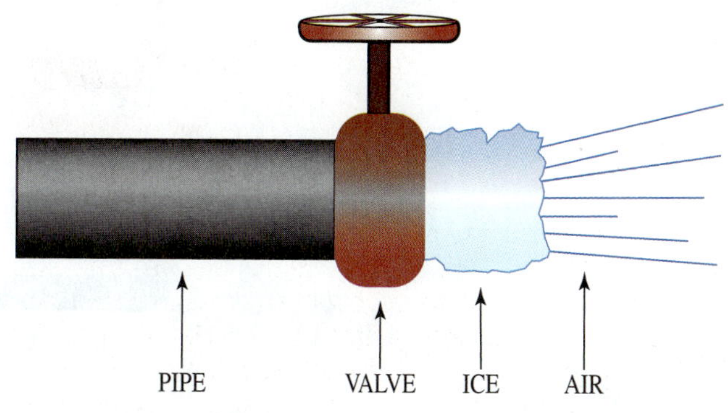

Figure 1-8 Ice forming around an air nozzle.

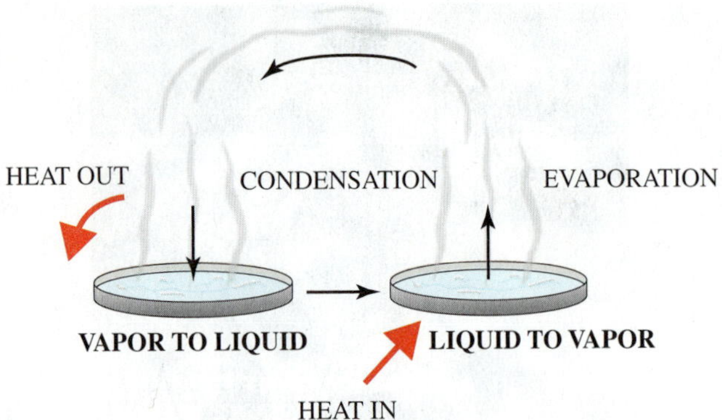

Figure 1-9 Water evaporates to vapor and absorbs heat, and then the vapor is condensed back to water to release its heat.

is absorbed, which makes ice a viable refrigerant. But ice is hard to store and takes up a lot of space. Water is easier to use because it can be pumped and doesn't need the insulation that ice requires. When water evaporates to vapor it also absorbs heat, but then the water needs to be replaced, and this uses up a lot of water over time.

If the vapor can be recovered and turned back into water, then this cycle reduces the total amount of water needed (Figure 1-9). Even so, the major disadvantage with this type of evaporative cooling is that the lowest temperature that can be reached is dependent on the properties of water.

Notice that with both ice and water, it is their change of state that allows for heat to be absorbed. It is this important principle that serves as the basis for most refrigeration systems today, but instead of using water, other fluids with different properties and lower boiling points, called refrigerants, are now used. This allows for much colder temperatures, far below freezing. The "refrigeration cycle" therefore continually evaporates and condenses refrigerants to absorb and then throw away the heat.

A compressor is used like a pump to raise the pressure and circulate the refrigerant through the system (Figure 1-10). A condenser is used to remove heat from the refrigerant as it turns into a liquid. An expansion device drops the pressure to allow the refrigerant to change back from liquid to vapor in the evaporator. Heat is absorbed in the evaporator and then thrown away in the condenser. The refrigerant does not wear out and circulates around and around during operation. Most refrigeration systems in use today operate using this type of cycle.

1.3 TODAY'S HEATING, AIR CONDITIONING, AND REFRIGERATION

"Environmental heating and air conditioning" refers to the control of a space's air temperature, humidity, circulation, cleanliness, and freshness, and it is used to promote the comfort, health, and/or productivity of the inhabitants. Homes, offices, schools, colleges, factories, sporting arenas, hotels, cars, trucks, and other vehicles such as aircraft and spacecraft are heated and cooled. The main purpose of environmental heating or cooling is to help maintain the body temperature within its normal range. Generally, the term *air conditioning* is used when the space temperature is above 60°F (15°C), and *refrigeration* is the term used when the space temperature is below 60°F (15°C).

Figure 1-10 The basic refrigeration cycle consists of four major components: compressor, condenser, expansion device, and evaporator.

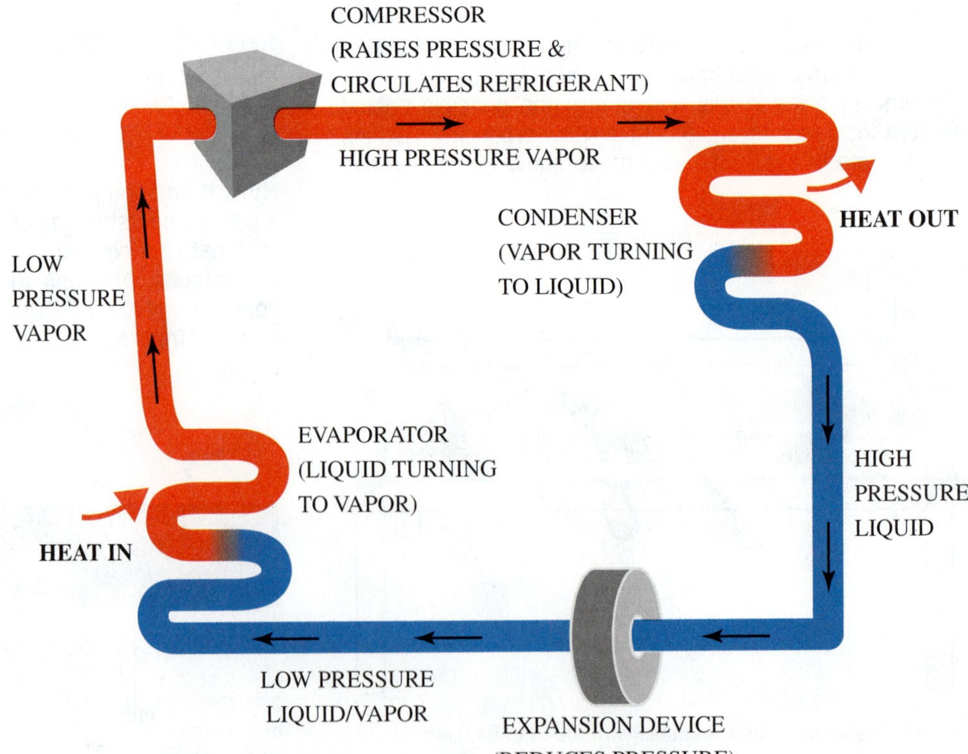

TECH TIP

Without our ability to control the environment, it would be impossible for us to explore space or the bottom of the ocean or even to enjoy the comfort of a transcontinental jet ride at 35,000 ft. So our ability to control our environment has served both to improve the quality of life and to enhance our scientific endeavors.

Process heating and cooling are used to aid in manufacturing or to keep equipment at a desired temperature. An area used to process meat or vegetables may be cooled to help preserve the product. Computer rooms are cooled so the equipment lasts longer and is able to stay online due to the heat being removed from the space. Computers would not operate properly if heat was not absorbed from the space. Remote pumping stations may be heated to prevent pipes from freezing. The main purpose of process heating or cooling is to maintain the temperature of things or processes within their required range.

TECH TIP

An operating room is cooled to aid with the surgery as well as for the comfort of the patient or surgeon. Therefore, an operating room is an example of process cooling even though it may be within the normal air-conditioning temperature range.

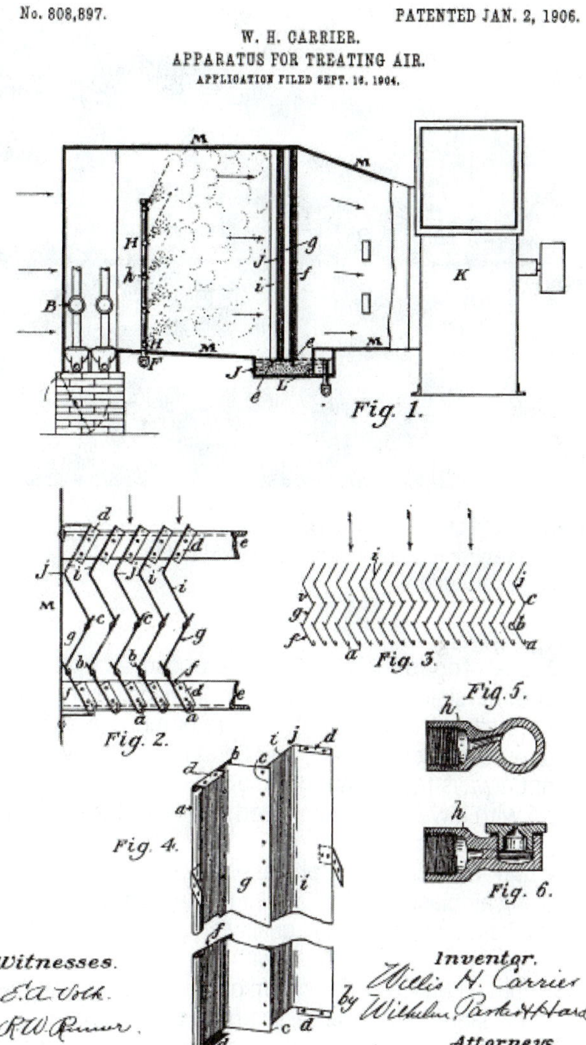

Figure 1-11 The patent for the first apparatus for cooling air, invented by Willis Carrier.

Modern Heating

Central heating of homes and businesses dates back to ancient times, but the first commercial warm-air fan-driven system was marketed in the 1860s. By the 1900s a number of different central warm-air systems were available for residential and commercial applications, and in 1908, the essential elements for heating, cooling, humidifying, dehumidifying, and filtering air were defined.

Today central heating systems can use warm air, hot water, steam, electric resistance, or a reverse refrigeration cycle (heat pump). The basic theory for the heat pump dates back to 1852.

Modern Air Conditioning

The development of modern air conditioning is often credited to Dr. Willis Carrier. Dr. Carrier, an engineer, was confronted with a problem facing printers. As paper was printed with one color, the dampness in the ink caused the paper to stretch slightly, and it was nearly impossible for the second color to be printed without being misaligned. Dr. Carrier determined that a means for controlling the humidity was necessary and developed the first air-conditioning system for the printing industry. His invention, called an "Apparatus for Treating Air," was patented in 1906 (Figure 1-11). His invention quickly found favor not only for dehumidifying but also for cooling. Through the 1940s and 1950s, businesses would proudly display signs reading "Air Conditioned." Dr. Carrier designed the psychrometric chart as we know it today. (This chart displays the properties of air, such as temperature, humidity, and volume, and is commonly used for many HVAC applications.)

Mass air conditioning of homes began in the late 1950s with window air conditioners. Central residential air conditioning started to become popular in the mid-1960s. Today most of us cannot imagine living in a home anywhere in the country that does not have air conditioning.

Modern Refrigeration

Clarence Birdseye made another major contribution to the industry. He developed the process of freezing foods in 1922. Today, supermarket freezer displays provide us with a variety of food products that would not be possible to preserve any other way (Figure 1-12). In 2006, a new era in eating occurred when the American public purchased more heat-to-eat and thaw-to-eat foods than any other type of food.

Figure 1-12 Modern refrigeration display cases provide us with a variety of food products that would not be available without refrigeration.

- **Frozen foods** Before Clarence Birdseye began commercially freezing food, people had allowed food to freeze naturally during the winter months as a way of preserving it for later use. Food frozen this way did not always taste that good, so the trick was to come up with a way of freezing food and having it still taste good when it was thawed.
- **Quick freezing** The process of rapidly freezing food using air blast, contact, and/or immersion freezing was the key to improving the quality and taste of thawed frozen foods. The problem with freezing food slowly is that when ice crystals form over time, they become much larger. These large, sharp ice crystals grow through the cell walls of the food, and when the food thaws, all of the nutrients in the food are allowed to drain away. Quick freezing causes the ice crystals to be very small and less likely to penetrate cell walls, so the food retains nutrients and flavor when it is thawed.

1.4 EMPLOYMENT OPPORTUNITIES

The HVACR industry represents one of the largest employment occupations in the country. Our industry, for example, is one of the largest consumers of electric and gas utilities in the nation. More electricity and natural gas is consumed producing heating and cooling than for any other single use. The size of the industry has been growing steadily since the late 1960s, when residential central systems became popular. The installation and servicing of HVACR systems will always be an expanding occupation. No one builds a home or business without some type of heating and/or cooling system, which requires designing, installing, and servicing by skilled and trained technicians.

Residential Air Conditioning and Heating

Most residential heating systems have a heating capacity of 50,000 to 150,000 BTU/hr. The majority of residential air-conditioning systems are 5 tons or less. Both of these sizes will obviously vary greatly, depending on the region of the country you are working in. In addition, there are many very large homes being built, requiring systems that could easily be classified as light commercial because of their size and/or complexity.

TECH TIP

To protect the public from potentially dangerous individuals, some businesses and/or local and state governments require criminal background checks on anyone involved with in-home service work. These checks may go back 25 years or more into an individual's past. Check with your local or state governmental department that regulates in-home service work if you feel there is something in your past that might affect your ability to work in residential service. In most states these checks are only required for in-home service work, so you may still be able to work in new construction or in the commercial or industrial areas.

Commercial Air Conditioning and Heating

The term *commercial* is used to refer to any system that is used in commercial buildings (for business) that provides cooling or heating. These systems may be as small as a fraction of a ton in size to several thousand tons in cooling capacity and/or from 1,000 BTU/hr to hundreds of thousands of BTU/hr.

Commercial systems may be operated independently of any other system or be integrated with a building automation system. Because of the vast differences in the types of equipment and system complexity, commercial technicians often specialize in a single type of system or group of systems.

Commercial and Industrial Refrigeration

The terms *commercial refrigeration* and *industrial refrigeration* are applied to retail food and cold-storage equipment and facilities. Examples of commercial equipment and systems include refrigeration equipment found in supermarkets, convenience stores, restaurants, and other food service establishments. Industrial refrigeration can include long-term storage either as cold storage or medium or low-temperature refrigeration systems that are generally larger-scale operations.

Types of Jobs

There are a variety of occupational specialties offered within the HVACR industry. These occupations range from the basic entry-level helper to the systems designer. Although the work involved with heating, air conditioning, or refrigeration equipment and systems is similar in theory, there is a significant difference between the work done in the areas of residential, light commercial, commercial, and industrial. These areas of heating, air conditioning, and refrigeration generally relate to the size (capacity) and complexity of the system. However, technicians may find the exact same equipment used in one home being used in a commercial shop or factory. In these cases the distinguishing factor is whether you are working in someone's home or in a business.

- **Entry-level helper** The entry-level helper (first-year apprentice) provides the senior technician with assistance installing and servicing equipment. Most medium and large mechanical contracting companies use a number of helpers to assist with the installation and service of residential and commercial systems. A helper may be expected to assist in lifting, carrying, or placing equipment or components. He or she may also run errands to pick up parts and clean up the area following installation or service. Helpers receive basic safety training, and if they will be driving, they must have good driving records.
- **Rough-in installer** The initial installation process is referred to as rough-in. In this process the technician (first- through third-year apprentice) will install the refrigerant lines, electrical lines, thermostat and control lines, duct boots, and duct run and set the indoor and outdoor units. The rough-in technician must have an understanding of duct layout, blueprint reading, and basic hand tools and good brazing skills.
- **Start-up technician** Once the system has been installed and all of the components are ready for operation, a start-up technician (fourth- and fifth-year apprentice) will go through the manufacturer's recommended procedures to initially start a system. Because much of the HVAC system has been field installed, this checkout procedure is essential to ensure safe and efficient operation. The start-up technician records all of the information requested by the manufacturer's warranty. Start-up technicians must be skilled with electrical troubleshooting and refrigerant charging and have good reading comprehension and writing skills.
- **Service technician** The service technician (fourth-year apprentice to journeyman) is the individual who provides the system owner with repair and maintenance. Service technicians are the people who must be able to diagnose system problems and make the necessary repairs. Service technicians must be skilled in diagnosing electrical problems, refrigerant problems, and air-distribution problems.

TECH TIP

Technology has enabled the field tech to stay in close contact with his service manager. This allows the highly experienced service managers to provide assistance to technicians as they come upon new problems. The technician can also call upon the office to research unique problems to determine the best, most efficient way of making the repair.

- **Sales** HVACR sales are divided into two major categories: inside sales and outside sales. Inside sales deal primarily with system sales to other air-conditioning contractors. Outside sales may be to both contractors and end users. Working in outside sales or consumer sales requires the technician to have a good understanding of cost and value of equipment so that the owner can make an informed choice.
- **Equipment operator** Equipment operators are required by local ordinance and state law to be present anytime large central heating and air-conditioning plants are in operation. Their primary responsibility is to ensure the safe and efficient operation of these large systems. They must have a good working knowledge of the system's mechanical, electrical, and computer control systems to carry out their job. They sometimes need to hold a city or state license to become an operator. Equipment operators generally work by themselves or as part of a small crew. They often are required to have good computer skills when buildings have computerized building-management systems.
- **Facilities maintenance personnel** Facilities-maintenance personnel are responsible for planned maintenance and routine service on systems. They may work at a single location or have responsibilities for multiple locations, such as school systems. Facilities-maintenance personnel typically maintain systems and provide planned maintenance. They may work alone or as part of a crew, depending on the size of the facility. Maintenance personnel may from time to time have duties and responsibilities outside of the HVACR trades, such as doing minor electrical plumbing and carpentry projects for the upkeep of the building.
- **Service manager** A service manager is typically a skilled HVACR technician with several years of experience. This individual oversees the operation of a company or maintenance department. He or she must have good management skills, communication skills, and technical expertise. Service managers typically assign jobs to other technicians and employees. They must then oversee these individuals' jobs.
- **Systems designer** For small buildings, contractors normally size and select HVAC systems and equipment. There are many industry-standard sizing and design guides available from trade associations such as the ACCA (Air Conditioning Contractors of America). For larger buildings, mechanical, architectural, or building services engineers may be required by law to design and specify the HVAC systems. Specialty mechanical contractors will work with the design plans to build and commission these systems.

1.5 TRADE ASSOCIATIONS

With the rapid growth and variety of interests, trade associations naturally evolved to represent specific groups. The list includes manufacturers, wholesalers, contractors, sheet metal workers, and service organizations. Each is important and makes a valuable contribution to the field. Space does not permit a detailed examination of all of these organizations, or all of their activities, but throughout the book many of these associations will be acknowledged as specific subjects are covered.

Certifications

Many trade associations offer training programs and competency examinations for the industry to help ensure a workforce of qualified technicians. In addition, the 1990 Clean

Air Act passed by the United States Congress requires that anyone who performs maintenance, service, repair, or disposal that could be reasonably expected to release refrigerants must be certified. To become certified, technicians are required to pass an Environmental Protection Agency (EPA)–approved test given by an EPA-approved certifying organization. Four different types of certifications have been developed to address different types of equipment. A person meeting the requirements for all four types is issued a universal certification. This is certification process is further described in Unit 28, Refrigerant Management.

Air-Conditioning, Heating, and Refrigeration Institute (AHRI)

The Air-Conditioning, Heating, and Refrigeration Institute (AHRI) is a national trade association representing manufacturers of over 90 percent of U.S.-produced central air-conditioning, gas appliances, and commercial refrigeration equipment. AHRI was formed in 2007/2008 when ARI (Air Conditioning and Refrigeration Institute) merged with GAMA (Gas Appliance Manufacturers Association). ARI, now AHRI, was originally formed in 1954 through a merger of two related trade associations and traces its history back to 1903 when it started as the Ice Machine Builders Association of the United States. Today AHRI has over 180 companies as members.

Many services are provided by AHRI to assist HVACR technicians. Some of these services, which would supplement this text, are listed below:

- ICE is an industry competency exam. This test is made available to students of educational institutions to test their knowledge of fundamental and basic skills necessary for entry-level HVACR technician positions. The information in this text covers the topics in the AHRI curriculum guide and would assist the student in taking this examination. A directory of those who pass the examination is published nationally to assist prospective employers in identifying job candidates.
- Equipment donations to schools participating in the ICE competency exam. AHRI contacts industry sources having no-cost or low-cost equipment available to supply a school's laboratory needs.
- Technician certification program. In accordance with EPA's enforcement of the Clean Air Act, the sale of refrigerants is made only to those technicians who have been certified. AHRI is among the many approved by EPA to administer the test for certification. In addition, AHRI provides study material to prepare for the test.
- Reclaimer certification program. EPA also requires certification of any processor of recovered refrigerant for resale. AHRI is among those assigned by EPA to carry out a certification program for companies that seek to reclaim refrigerants. Technicians handling reclaimed refrigerant should become familiar with the *Directory of Certified Reclaimed Refrigerants*, published every March and September by AHRI.
- Certification program for equipment used to recover and recycle refrigerant. AHRI is one of the companies approved by EPA to certify equipment used to recover and recycle refrigerants. Technicians should become familiar with the *Directory of Certified Refrigerant Recovery/Recycling Equipment*, published every March and September by AHRI.
- HVACR equipment certification program. AHRI maintains a certification service, which tests a wide variety of equipment and products to verify the performance described by the manufacturer. Certified directories for various products are published semiannually and annually.

AHRI has a full program of educational activities geared toward helping the nation's vocational and technical schools improve and expand their education and training programs. Under the direction of AHRI's education director and its Education and Training Committee, AHRI serves as a resource for manufacturers, school instructors, department heads, and guidance counselors. In addition to this textbook and its companion materials, AHRI produces the *Bibliography of Training Aids*, a career brochure, and a promotional video for schools to use to recruit students into HVACR programs. Many schools around the country have adopted the ICE competency exams as final exams for their programs. AHRI's most recent efforts involve participation in developing national HVACR competency standards.

Having students pass the ICE competency exams and training toward national competency standards will improve the quality of installation and service. New HVACR technicians will be better prepared, resulting in three basic advantages:

- Limited training required for contractors
- Limited rework or repeat calls due to error
- Limited warranty/replacement for manufacturers

The cost of repeat service calls, which is borne by contractors, may be reduced substantially by employing properly trained technicians. Every new technician receives training and serves as an apprentice for a period of time. That is essentially a period where contractors pay two people to do one job. A properly trained technician will generally require less training time and function sooner than a poorly trained technician.

In co-sponsorship with AHRI, ASHRAE holds an annual international Air Conditioning Heating Refrigeration Exposition, which may draw 30,000 to 50,000 people in the field. Product exhibits, technical displays, and business seminars highlight the event.

American Society of Heating, Refrigeration, and Air-Conditioning Engineers (ASHRAE)

The American Society of Heating, Refrigeration, and Air-Conditioning Engineers (ASHRAE) is an organization started in 1904 as the American Society of Refrigeration Engineers (ASRE) with seventy members. Today its membership is composed of thousands of professional engineers and technicians from all phases of the HVACR industry. ASHRAE also creates equipment standards for the industry. Its most important contribution probably has been a series of our books that have become the reference books of the industry: *HVAC Applications, Refrigeration, Fundamentals,* and *HVAC Systems* and *Equipment.*

TECH TIP

Becoming an active participating member in a professional trade association will provide you with an opportunity to continue your HVACR education. The HVACR field is such a dynamic and evolving industry that to stay competitive you must continually attend seminars and take classes. This is a field where your success will depend on your continued education.

American Society of Mechanical Engineers (ASME)

The American Society of Mechanical Engineers is an organization composed of engineers in a wide variety of industries. Among other functions, ASME writes standards related to safety aspects of pressure vessels.

Air Conditioning Contractors of America (ACCA)

The Air Conditioning Contractors of America is a service contractor's association concerned with the education of technicians and service managers with business-improvement techniques. ACCA provides technician EPA certification.

Refrigeration Service Engineers Society (RSES)

The Refrigeration Service Engineers Society is the international professional association for all HVACR workers and is dedicated to education and certification of technicians in the HVACR industry. RSES offers Specialist Certification for senior technicians in eight HVACR areas and has a technician EPA certification program. RSES chapters conduct classroom training in technical areas and are a source for educational printed material and books.

HVAC Excellence

HVAC Excellence is a not-for-profit organization that has been serving the HVACR industry since 1994. The organization's goal is to improve competency through validation of the technical education process by offering progressive levels of technician certification through its HVAC Excellence programs.

SERVICE TIP

The AHRI list of certified equipment is available to anyone through the Internet. This material is very helpful when trying to make a determination of the best equipment to recommend for customers and their specific application needs. On the Web, very often all of the various pieces of equipment are available.

UNIT 1—SUMMARY

Since the beginning of time, people have had a desire to control their environment to live and work more comfortably. That trend will not stop, and that is the good news for anyone entering this ever-growing, financially rewarding, and personally satisfying field. HVACR technicians are required to understand the theories behind designing, installing, and servicing a wide range of systems. This diversity ensures that each day on the job will be new and unique, ever changing, and challenging.

UNIT 1—REVIEW QUESTIONS

1. List some of the different ways that homes and buildings may be heated.
2. What were some of the primary heating fuels that early civilizations used?
3. When is it believed that ice was first artificially made for food storage?
4. How did early man make ice?
5. Why did some manufacturers spray water in factories in the early 1700s?
6. How did early Romans cool palaces?
7. What do the terms *environmental heating* and *air conditioning* refer to?
8. What does the term *process heating and cooling* refer to?
9. When did central warm-air systems for residential and commercial applications become well defined?
10. Who developed what is referred to as modern air conditioning?
11. When did mass air conditioning of homes with window units begin?
12. Why is it important to freeze foods quickly?
13. Why do some businesses and/or local and state governments require criminal background checks for HVACR workers?
14. What size range might a commercial air conditioner fit into?
15. Give an example of some of the types of equipment that a commercial refrigeration technician might work on.
16. What type of things might an entry-level helper do?
17. Whose job is it to do the initial installation process, such as install the refrigerant lines, electrical lines, thermostat and control lines, duct boots, and duct run, as well as setting the indoor and outdoor units?
18. What skills must a service technician have?
19. What are some of the things that a service manager must be able to do?
20. What is the ICE exam, and who might take it?
21. What are some of the RSES's activities?

UNIT 2

Being a Professional HVACR Technician

OBJECTIVES

After completing this unit, you will be able to:

1. list some of the most popular HVACR publications.
2. explain the importance of professional certifications.
3. list the seven specialty areas of NATE specialty certification.
4. explain the value of taking the Industry Competency Exam (ICE).
5. list the items that help make for a professional appearance while on the job.
6. describe how to develop good communication with the customer.

2.1 INTRODUCTION

The air-conditioning and refrigeration industry has more professional organizations, trade associations, publications, and other related organizations than most any other technical field. As a student, you should consider becoming involved with a student organization such as the student clubs of ACCA, RSES, or ASHRAE. These will give you an opportunity to begin developing extremely important business and technical contacts in the local HVACR industry, and these contacts will serve you very well as you enter the profession.

As a professional in the trade, it is to your advantage to maintain a close relationship with one or more of these professional organizations. Each group provides its members with the latest trends and most current technical information. This gives members a significant edge. Most of the organizations provide ongoing technical and business training classes. Many of the classes cover the latest trends in equipment, regulations, codes and standards, local building regulations, and business practices. Being able as a member to participate in these ongoing educational opportunities will keep you at the leading edge of your new profession.

Each of these professional organizations has publications, and many provide the industry with codes and standards. These publications would be an excellent addition to your technical library. Having an up-to-date library will help you provide your employer and customers with the best possible service while making you significantly more valuable as an employee. HVACR is an ongoing learning process for even the most skilled technician.

Figure 2-1 lists a number of these professional associations. Most have Web sites, and many have local, regional, and state chapters that you can become affiliated with.

Many of the HVACR professional organizations have industrial trade shows. These shows provide excellent opportunities for you to see the various manufacturers' latest equipment, tools, supplies, and services. Some of the trade shows are local, others may be regional, some are national, and a few are international (Figure 2-2).

TECH TIP

Once you enter the profession as an HVACR technician, all costs associated with taking additional classes, purchasing books, and membership dues may be tax deductible.

2.2 PUBLICATIONS

Another excellent way of keeping up with the latest information in the HVACR field is by subscribing to one or more of the HVACR publications (Figure 2-3). Some of these publications are weekly, while others are monthly. They all contain well-written articles specifically addressing HVACR industry concerns. Many of them are written to teach their readers troubleshooting skills. Their articles are very valuable even to the skilled technician.

Some of the professional organizations have their own newsletters that are published and provided to their members. Some local and state chapters of these organizations have additional newsletters that are provided to their members.

2.3 PROFESSIONAL CERTIFICATION

Every HVACR technician must become certified under the EPA Section 608 regulations. Compliance with these regulations regarding the management of refrigerants is mandatory for everyone in the trade. Following your successful completion of any and all of the appropriate levels, it remains your responsibility to stay current with any changes in these regulations . As unfair as it may seem, you can be fined significantly for violating an EPA regulation pertaining to refrigerants even if that regulation took effect after your successful completion of the certification. In addition, it is your sole responsibility to remember and follow all of the

ACCA—Air Conditioning Contractors of America
AFEAS—Alternative Fluorocarbons Environmental Acceptability Study
AGA—American Gas Association
AHAM—Association of Home Appliance Manufacturers
AMCA—Air Movement & Control Association
ANSI—American National Standards Institute
AHRI—Air-Conditioning, Heating, and Refrigeration Institute
ARWI—Air-Conditioning & Refrigeration Wholesalers International
ASAE—American Society of Association Executives
ASHRAE—American Society of Heating, Refrigerating, and Air-Conditioning Engineers
ABC—Associated Builders & Contractors
BOMA International—Building Owners and Managers Association
COBRA—The Association of Cogeneration
CDA—Copper Development Association
COSA—Carbon Monoxide Safety Association
EEI—Edison Electric Institute
EPRI—Electrical Power Research Institute
EHCC—Eastern Heating & Cooling Council
Envirosense Consortium Inc.
EPEE—European Partnership for Energy and the Environment
FMI—The Food Marketing Institute
GEO—Geothermal Exchange Organization
GMA—Grocery Manufacturers of America
Green Mechanical Council
HARDI—Heating, Airconditioning, and Refrigeration Distributors International
HPBA—Hearth, Patio and Barbeque Association
HI—Hydraulic Institute
HRAI—Heating, Refrigerating, & Air-Conditioning Institute of Canada
HVAC Excellence
IDDBA—International Dairy, Deli, Bakery Association
IFPA—International Fresh-Cut Produce Association
IGSHPA—International Ground Source Heat Pump Association
IIAR—International Institute of Ammonia Refrigeration
IHACI—Institute of Heating and Air Conditioning Industries
ISA—The Instrumentation, Systems, and Automation Society
MCAA—Mechanical Contractors Association of America
MSCA—Mechanical Service Contractors of America
NACS—National Association of Convenience Stores
NADCA—National Air Duct Cleaners Association
NAHB—National Association of Home Builders
NAFEM—National Association of Food Equipment Manufacturers
NAM—National Association of Manufacturers
NATE—North American Technician Excellence Program
NRA—National Restaurant Association
NEMA—National Electrical Manufacturers Association
NFFS—Non-Ferrous Founders' Society
NIPC—National Inhalant Prevention Coalition
PHCC—Plumbing Heating Cooling Contractors Association
PIMA—Polyisocyanurate Insulation Manufacturers Association
PMA—Produce Marketing Association
RACCA—Refrigeration & Air Conditioning Contractors Association
RSES—Refrigeration Service Engineers Society
SMACNA—Sheet Metal and Air Conditioning Contractors' National Association
SMWIA—Sheet Metal Workers International Association
UL—Underwriters Laboratories Inc.
UA—United Association

Figure 2-1 U.S. organizations.

CANMET—Canadian Centre for Mineral and Energy Technology
EPEE—European Partnership for Energy and the Environment
EUROVENT—European Committee of Air Handling & Refrigerating Equipment
ICARMA—International Council of Air-Conditioning and Refrigeration Manufacturers' Association
globalEDGE
IEA—International Energy Agency
IIAR—International Institute of Ammonia Refrigeration
IIR—International Institute of Refrigeration
JRAIA—Japan Refrigeration and Air-Conditioning Industry Association
LATCO's Tools of the Trade (Latin American international trade sites)
Trade Compass
UNEP—United Nations Environment Programme
USA*Engage
World Bank
WTPF—World Trade Point Federation

Figure 2-2 International Organizations.

The Air-Conditioning, Heating & Refrigeration (ACHR) News
American School & University (AS&U)
APPLIANCE magazine
Appliance Manufacturer
ASHRAE Journal
Buildings
Building Design and Construction
Consulting-Specifying Engineer
Contracting Business
Contractor magazine
Energy User News
Engineered Systems
Facilities Net
Heating/Piping/AirConditioning (HPAC)
Japan Air Conditioning, Heating & Refrigeration News (JARN)
Plant Engineering
RSES Journal
SchoolDesigns.com
Skylines
Supply House Times
Western HVACR News

Figure 2-3 Publications.

EPA regulations pertaining to refrigerant management. For that reason, it would be a good business practice to occasionally take a refresher course in EPA rules and regulations.

2.4 INDUSTRY COMPETENCY EXAM (ICE)

The Air-Conditioning, Heating, and Refrigeration Institute (AHRI); the Air Conditioning Contractors of America (ACCA); Heating, Airconditioning, and Refrigeration Distributors International (HARDI); the Plumbing, Heating, Cooling Contractors Association (PHCC); North American Technician Excellence (NATE); the Partnership for Air-Conditioning, Heating, Refrigeration Accreditation (PAHRA); and the Refrigeration Service Engineers Society (RSES) have established a competency examination that is designed for students who have completed or nearly completed a technical training program. This examination is voluntary, but it does provide students leaving a training program, whether from high school, trade school, or community college, with an opportunity to evaluate their knowledge with an industry standardized test.

The ICE has been developed over the years with input from manufacturers, trade associations, instructors, and other industry experts. This exam can also provide your institution with an overall evaluation of its training program. Upon your successful completion of the ICE, your name, along with your school's name, is published and made available to area contractors who might be looking for new skilled employees. In short, the successful completion of the ICE can put you well ahead of other graduates from programs not participating in the ICE. The ICE is in three parts—Residential Heating and Air Conditioning, Light Commercial Heating and Air Conditioning, and Commercial Refrigeration. A good student will make testing and certification achievement a challenge for him- or herself, always setting goals high.

AHRI and its affiliates provide training institutions with incentives as encouragement to participate in the ICE by directing many of its manufacturing members' equipment donation programs toward the schools, institutes, and colleges that participate in the ICE program. These equipment donations can become an excellent source of the latest equipment you will be seeing in the field.

2.5 SKILLS USA

Skills USA is a vocational industrial club for students in high schools, trade schools, and community colleges. Skills USA clubs are open to students in all areas of specialty, including HVACR. The national organization provides local chapters and students with many opportunities to develop leadership, citizenship, and interpersonal skills that are invaluable to the success of individuals in any profession. The Skills USA logo is shown in Figure 2-4a.

In addition to the opportunities for individual professional growth, Skills USA sponsors regional, state, national, and international skills competitions. The contestants for the HVACR competitions will be tested in the following areas: written

(a)

(b)

(c)

Figure 2-4 (a) Skills USA logo; (b,c) Students competing in the National Skills Olympics.

exam, brazing skills, refrigerant component service, air measurement and troubleshooting, refrigerant recovery, and electrical troubleshooting. Various industry equipment used may include ice machines, refrigerated display cases, small package HVAC units, furnaces, and split-system air-conditioning units. These skills competitions are available on the secondary and post-secondary levels. By participating in the competition, students are given the opportunity to demonstrate their troubleshooting skills in diagnosing real-world problems under the supervision of highly skilled individuals serving as judges.

The winners of local, state, and national competitions move on to the international skills competition held in a different country each year. At the international competition, the best and brightest students from around the world compete to see who has the greatest knowledge and expertise in each of the technical areas, including HVACR. Every student who participates in these skills Olympics at any level receives recognition. Students involved in this program are shown in Figure 2-4b, c. This recognition is invaluable to the students because prospective employers value such recognition.

2.6 COUNCIL OF AIR CONDITIONING AND REFRIGERATION EDUCATORS (CARE)

The Council of Air Conditioning and Refrigeration Educators (CARE) is an organization that was founded in the late 1990s by a group of air-conditioning educators, counselors, and administrators responsible for the various aspects of HVACR training. This group is made up of individuals from secondary schools, post-secondary schools, and colleges, representing institutions from various regions of the country.

The purpose of CARE is to educate to meet or exceed the needs of the industry.

CARE membership is open to instructors, counselors, and administrators who are involved in some aspect of HVACR training. Through this organization, individuals can come together to learn and share experiences for the betterment of the HVACR students and program.

2.7 NORTH AMERICAN TECHNICAL EXCELLENCE (NATE)

NATE is an independent, third-party certification body formed in 1997 as a result of a concern expressed by many in the industry that there was not a way of distinguishing quality, highly skilled HVACR technicians from every other person working in the field. NATE is supported by all of the major equipment manufacturers, major component manufacturers, professional and trade associations, and the National Skills Standard Board. Figure 2-5 shows a technician with a patch showing NATE certification working on a system. You can earn installation, service or senior certification in one or more of the following specialty areas:

1. Air conditioning
2. Air distribution
3. Air-to-air heat pumps
4. Gas furnaces
5. Oil furnaces
6. Hydronics gas
7. Hydronics oil
8. Light commercial refrigeration
9. Commercial refrigeration
10. Ground source heat pump loop installer
11. HVAC efficiency analyst (Senior Level)

Installation certification is primarily designed for the technician who is involved with the installation or removal of HVAC equipment. Installers assemble the system and fabricate the necessary connections to complete an efficient system. They also set up the operational controls under the supervision of a service technician. After the system is started, the installation technician records the readings for temperature, pressure, voltage, current, and any other

Figure 2-5 NATE-certified air-conditioning technician checking the refrigerant charge in a residential system.

measurements required by the manufacturer or service company for the completion of the warranty paperwork.

The service technician must have all of the skills of the installation technician plus be able to work independently. Service technicians must be able to perform field diagnostics to determine the cause of system failures and to make the needed repairs.

In addition to having the technical skills required to pass the various areas of specialization, each technician must take a core exam as part of the exam process. The core test covers basic math, customer relations, comfort, heat transfer, and the fundamentals of electricity. A well-trained, experienced technician should be able to take and pass both the core and area specialty exams. A number of organizations provide NATE pretest tutorial classes to bring technicians' skill levels up to pass the exam. The exam passing score is 70 percent or better. Certification lasts for five years, after which time a technician must recertify. There are many classes that qualify for credits that can be used toward recertification. If a technician receives 60 hours of credit, the technician does not have to take a test at five years. If the technician obtains 30 credit hours, he or she will only be required to take a fifty-question specialty test instead of a hundred questions.

NATE recommends that you have at least one year of field experience before taking an installation series test and at least two years of experience before taking a service series test. It is further recommended that you have some instruction from an educational institution or trade association.

NATE has provided technicians with Knowledge Areas of Technician Expertise (KATEs), which give detailed outlines of all of the material that a technician can expect to be questioned about on an exam. These KATEs represent all of the knowledge and skills that a quality HVACR technician should have. The KATEs also help the prospective test candidate to focus on those areas that the industry feels are most important.

2.8 HVAC EXCELLENCE

HVAC Excellence established the national standards for HVACR programs in the summer of 1999, becoming the industry's first accrediting body. Programmatic accreditation is an independent third-party review of an educational program. This begins with experienced auditors conducting an on-site visit at each school applying for accreditation. The process validates that established standards of excellence for HVACR educational programs are met. Standards require a thorough examination of mission of the program, administrative responsibilities, finances and funds, curriculum, plan of instructions, facilities, equipment and tools, cooperative training, and instructor's qualifications.

2.9 CODES AND STANDARDS

There are many organizations and agencies that provide codes that are used throughout the HVACR industry, and Figure 2-6 lists many of them. In addition to codes, these groups provide standards. The difference between a code and a standard is that codes often carry with them the force of law, and standards do not. Organizations providing standards to the HVACR industry are listed in Figure 2-7.

International Association of Plumbing and Mechanical Officials (IAMPO)
International Code Council (ICC)
International Fire Code Institute (IFCI)
National Conference of States on Building Codes and Standards (NCS/BCS)
National Fire Protection Association (NFPA)

Figure 2-6 Code groups.

AFNOR—Association Francaise de Normalisation
ANSI—American National Standards Institute
BSI—British Standards Institution
CSA International—Canadian Standards Association
DIN—Deutches Institut für Normunge. V.
CEN—The European Committee for Standardization
CENELEC—The European Committee for Electrotechnical Standardization
EU—European Union
IEC—International Electrotechnical Commission
ISO—International Organization for Standardization
JIS—Japanese Industrial Standards
NSSN—National Standards Systems Network
SASO—Saudi Arabian Standards Organization
SES—Standards Engineering Society

Figure 2-7 Standards associations.

DOE-EERE—Department of Energy, Energy Efficiency and Renewable Energy
FCIC—Federal Citizen Information Center
NATE—North American Technician Excellence Program
CEE—Consortium for Energy Efficiency

Figure 2-8 Consumer information.

Many consumer groups over the years have worked with HVACR industry leaders to help the industry provide consumers with the most efficient and effective service. Some of these consumer information groups are listed in Figure 2-8.

2.10 THE PROFESSIONAL TECHNICIAN'S APPEARANCE

As a service technician you are seen by the customer as a representative of your company. Customers often assume the appearance and professionalism of the technician is a reflection on the technician and company's technical skills. It is therefore important that you present yourself professionally, in a very clean and neat manner, to the customer. In some cases you might find during the course of the day that your uniform becomes soiled. It is therefore a good idea to carry at least an extra clean shirt in the service van so that you can change if necessary.

If your company does not provide uniforms, you should dress appropriately. In some cases, blue jeans and a jersey or denim shirt are acceptable. In other cases, where you might be working in an office building, slacks and a shirt would be appropriate. Check with your employer to see what the company dress code is. In addition to your clothing, you must have a clean and neat personal appearance. That means clean, well-kept hair and either being clean shaven or having a well-kept beard for men, and clean hands for everyone.

You must keep your service vehicle clean and neat. It is a rolling billboard for your company, and it is important that it look sharp. A clean service van provides a better, more efficient work area, making it easier to find tools and supplies, and it also makes a better impression on the customer. Unless you have permission from residential customers, you should not park your service vehicle in their driveway. Always make it a habit to park your vehicle on the street. You do not want to be responsible for cleaning up an oil spill, and your vehicle can obstruct their access, which to some customers is very aggravating.

When you present yourself to a customer's door, you should have your hands in plain sight either at your side or holding your tools and clipboard. When the customer opens the door, if you take one step back, it will give the customer a greater sense of comfort. Many companies provide photo ID badges, and you must have yours clearly displayed. If it is necessary for you to enter a dwelling for service, ask permission from the homeowner before entering.

2.11 WORKING NEAT

Some companies provide technicians with paper shoe covers to prevent tracking dirt into residences. If you suspect your shoes are dirty, either remove them or use the covers. You and your company can be responsible for cleaning the carpet if it becomes soiled. When working on an indoor furnace, place a drop cloth on the floor in front of the furnace so that any debris will be contained. When you are finished working in the furnace area, a small battery-powered vacuum cleaner can be used to pick up loose dirt and debris in the area. Use a damp rag to wipe down any fingerprints that are on the equipment or that may be on the door or woodwork.

TECH TIP

Cleaning up the equipment area in a residence is an excellent PR move. You should vacuum up all of the debris in the area. Many customers judge your work by appearance, and all they may understand is neat and clean. When changing out a unit, electricians or plumbers may also be involved. You should clean up any mess even if you did not make it. Leaving behind a mess will reflect on you, since the final repair is still your responsibility.

In many residences, the air handler is located in an attic. If access to the unit is through a pulldown staircase, make certain that any dirt or debris that falls from the stairs when it is pulled down is cleaned up. If you are going to be going in and out of the attic a number of times during the service, place a drop cloth on the floor below the stairs to catch any insulation or other debris that might fall on the floor. Do not attempt to carry a large number of tools up and down the stairs. This can cause you to be caught off balance and possibly drop your toolbox. In addition, most attic stairs have a weight limit that could be exceeded by you and a large toolbox. Attic stair treads, like any ladder, are strongest next to the side rail. For that reason, when you ascend and descend the stairs, place your foot as close as possible to the side next to the side rail to reduce the possibility of breaking the stairs. Always be careful in an attic so that you do not step through the homeowner's ceiling.

Many customers have extensively landscaped their homes, including the area around an outdoor condensing unit. Even though their landscaping may encroach within the manufacturer's recommended free air space around the unit, do not remove this landscaping. Notify the homeowner as to how it should be trimmed so that they or their landscaper can remove the vegetation. When working around vegetation, it is important that you be as careful as possible so as not to damage any plants. On very soft, wet ground your repeated trips to the service van can wear a path. To avoid this, each time you have to cross soft, wet ground, take a slightly different route when possible.

It is considered bad practice to leave packaging and boxes left over from the installation of new equipment in the customer's trash receptacles. This material should be taken with you so as not to overload their receptacle.

TECH TIP

Working in air conditioning and refrigeration is often a very hot job. From time to time customers may offer you a glass of water, tea, or cold drink. Use your discretion on accepting these offers. However, it is never appropriate for you to accept a beer or other alcoholic beverage from a customer. Even if you do not open the drink but take it with you, you have significantly damaged your credibility with the customer and your boss. If you take a beer, even though you do not drink it, and there is a problem with the service you provided the customer, the first thing the customer is going to tell your boss and everyone else is that you were drinking on the job, in spite of the fact that they gave you the beer. *Do not take any alcoholic beverages that are offered by customers.*

2.12 TECHNICIAN AND CUSTOMER COMMUNICATION

A large part of the technician's job is to educate the customer as to the problem that was found and the available options for its repair. Tell the customer what failed, why it failed, and the options to fix the problem. Do not simply tell the customer that they need a part and that you are going to replace it.

Many customers would like an estimate of the job's cost. In some municipalities you are obligated under consumer protection laws to provide customers with such a quote. In some cases the quote must be in writing. If, however, as part of your service you uncover a situation that could not have been foreseen or was not visible in your initial evaluation that will require additional work, immediately stop and inform the customer of the new problem. Do not simply make the repairs and expect the customer to accept the higher charge at the conclusion of the job.

TECH TIP

If you locate an additional problem or pending problem with a customer's system and, after notifying the customer of your concerns, they choose not to have you provide that repair, you should note that on the customer's invoice as part of your service call recordkeeping and get them to sign the acknowledgment.

Under no circumstance should you tell a customer that the previous technician messed up their system when they did the last service. All that will result from such statements is the loss of faith the customer may have in you and your professionalism; but, more important, you might find out that one of your colleagues who works for your company did the last service job.

TECH TIP

Good professional technicians never knock the competition. Their skill and knowledge will set them apart from everyone else without the need to brag.

One of the primary complaints customers have on any service call is punctuality. It is not always possible to be at a job exactly when you anticipated being there, as earlier service calls may take more time than you initially estimated. However, as soon as you realize you are not going to make the schedule, and as early as possible, let the customer know and give them an opportunity, if they so choose, to reschedule. If you do reschedule, you must be at the next scheduled appointment on time.

It is also extremely important that you provide a very clear, clean, and legible invoice or bill to the customer. If it is necessary for you to get prices, look up information, or check your spelling, you should do all of that in your service vehicle and not in front of the customer. Even though you may not be a literary expert, customers do expect you to provide them with clear, concise, well-written statements. If you have a problem writing clear and concise statements, you may want to invest in a PDA (personal digital assistant) device. There are many of these devices on the market, and many can have spell-check and writing programs installed.

TECH TIP

Some air-conditioning and refrigeration work in the summer requires that the technician work in relatively hot environments such as attics or buildings without working air-conditioning systems. To prevent heat stress injury, it is important that you guard against dehydration by drinking large quantities of water or sports drinks. You must drink enough so that you have to use the restroom at least once every couple of hours throughout the day. In addition to dehydration, you may also develop kidney and bladder problems if you do not drink enough fluids while working in hot environments.

Carbonated beverages, milk, fruit juice, beer, and many other beverages do not replace the body's electrolytes. Without replacing the electrolytes, you can feel fatigue and may experience cramps. For that reason, only sports drinks with the essential electrolytes and water are recommended as your primary drinks when working in the heat.

UNIT 2—SUMMARY

You should become familiar with the various professional organizations and trade associations that support the air-conditioning and refrigeration industry and consider joining those that are suitable to your career goals. Read the many available publications so that you may keep up-to-date with the latest changes in industry practices and standards.

As a refrigerant technician, you must become certified under EPA Section 608 regulations. To further demonstrate your ability as a service technician, you may also consider taking an industry standardized test such as the Industry Competency Exam (ICE). There are also certification programs such as North American Technical Excellence (NATE) that allow you to focus on specialty areas.

Always remember to present yourself to the customer as a neat and clean professional. Carry an extra set of clothes in the event you need to change because your uniform becomes soiled. Think of using paper shoe covers rather than tracking dirt across the customer's carpet. Finally, always be polite to the customer and make sure that you communicate clearly. A large part of your job is to provide guidance for the customer by explaining the nature of the equipment-related problem. Be prepared to tell the customer what failed, why it failed, and the options to fix the problem.

UNIT 2—REVIEW QUESTIONS

1. List three student organization clubs.
2. What does RSES stand for?
3. What does ACCA stand for?
4. What does AHRI stand for?
5. What does ASHRAE stand for?
6. What does NATE stand for?
7. What section of the EPA regulations requires technician certification?
8. The ____________ has been developed over the years with input from manufacturers, trade associations, instructors, and other industry experts.
9. What is the name of the vocational industrial club for students in high schools, trade schools, and community colleges?
10. What organization was founded in the late 1990s by a group of air-conditioning educators, counselors, and administrators responsible for the various aspects of HVACR training?
11. List the seven specialty areas for the NATE service technician certification program.
12. The core exam for the NATE certification program covers what topics?
13. What is the difference between a code and a standard?
14. What precaution should be taken if you find that during the course of the day your uniform becomes soiled?
15. When might it be necessary for you to use a drop cloth?
16. When making a repair, what information should you relay to the customer?
17. If you locate an additional problem or pending problem with a customer's system and they choose not to have you provide that repair, what should you do?
18. What is one of the primary complaints that customers have regarding service calls?
19. Why are many common drinks, such as sodas, not adequate for your protection against heat stress injury?
20. When would it be appropriate to tell the customer that the previous technician messed up their system when performing the last service call?

UNIT 3
Safety

OBJECTIVES

After completing this unit, you will be able to:

1. explain how to work safely to avoid accidents.
2. discuss the material that appears on all Material Safety Data Sheets (MSDS).
3. discuss how to safely use hand and power tools.
4. discuss how to practice safety in the shop.
5. describe four types of fire extinguishers.
6. demonstrate a safe method of lifting heavy objects.
7. discuss safe welding and cutting practices.
8. discuss electrical safety rules.
9. discuss the safe use of refrigerants, their storage, and proper disposal.
10. tell how to safely handle refrigerant cylinders.
11. discuss refrigerant system safety.
12. name three major hazards of pressure vessels.
13. discuss how a technician's driving record can affect employability.
14. tell what steps should be followed in case of an accident.

3.1 INTRODUCTION

In every trade, safety is a major concern, and safety is everyone's first job responsibility. Accidents, no matter how minor, can cost the technician, the company, and the customer unnecessary losses of time and money. There is no reason to feel that accidents are inevitable and something you must just accept. Good working habits, good tools, and being vigilant to potential hazards can virtually eliminate accidents. Never do anything you feel is unsafe. When working with new equipment or tools, read all the safety instructions and follow them.

Most companies require some type of safety training for all their employees. In addition, some of the businesses where you may be asked to do HVACR work at may have their own safety program that you must pass before beginning work at their site.

Most accidents are caused by carelessness, as well as lack of awareness of proper safety procedures. This unit deals with some of the basic safety tips and procedures the installer and service technician should follow—whether on the job site or at related locations where hazards could exist; additional specific safety facts are covered in each unit. Read and follow all safety rules.

3.2 PERSONAL PROTECTIVE EQUIPMENT

Personal protection equipment (PPE) is designed to reduce your exposure to hazards that cannot be eliminated or controlled. PPE may include equipment or devices to protect your head, face, eyes, ears, respiratory system, hands, and feet. Some devices, such as safety glasses, are commonly used, while others, such as respiratory protection, may be less frequently used.

If PPE is required for any job, OSHA (Occupational Safety and Health Administration) recommends that all employees be trained. This training may be as short as a few minutes or as long as several hours or more. The length of the training time depends on the level of hazard and the complexity of the PPE to be used.

Keeping all PPE clean and in good working condition is essential to ensure that, when needed, it will work properly.

Head Protection

An approved hard hat (Figure 3-1) should be worn whenever there is a danger of things dropping on the head or where the head may be bumped. On a construction site, proper safety head gear is a must.

Eye and Face Protection

The majority of eye injuries are the result of flying or falling objects. Most of these objects are smaller than the head of a pin but can cause serious injury. Approved eye or face protectors (Figure 3-2) must be worn whenever there is

Figure 3-1 Hard hats are required to be worn on many job sites.

Figure 3-2 Eye protection equipment: (a) these safety goggles can be worn over glasses; (b) safety glasses with side protection.

a danger of objects striking the eyes or face. Side shields must be part of any safety glasses worn, even if they are prescription eyeglasses. Safety glasses or goggles must be worn over prescription eyeglasses if they do not have side shields. Eye and face protectors come in various shapes and sizes, and some of them are very specialized.

Special eye protectors must be worn when arc welding, spot welding, and burning to cut out harmful light radiation. These special face visors come with various shades of viewing eyepieces that filter out the harmful emissions. Take time to identify the right one for the job. For example, never wear oxyacetylene welding goggles when an arc welding face shield is needed.

Confine long hair and loose clothing before operating rotating equipment.

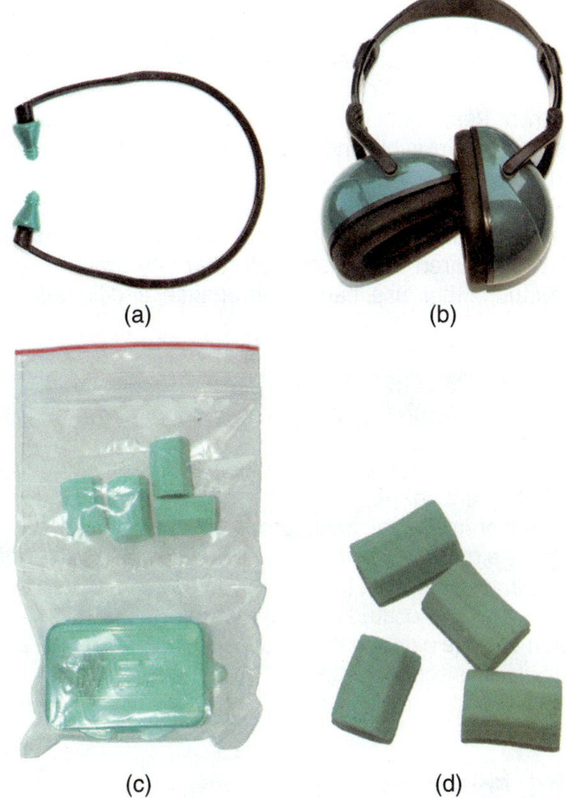

Figure 3-3 Ear protection equipment: (a) disposable earplugs on a lanyard (that fits around the neck) allow for easy removal and reuse; (b) headphones protect both ears and hearing; (c, d) disposable, one-time-use earplugs.

TECH TIP

To control insurance costs, many air-conditioning companies have adopted very stringent policies on personal protection equipment. Unlike your shop teacher, who may have reminded you each day about safety glasses, ear protection, and so on, many employers may terminate you with a single safety infraction. Others may warn you once or twice about wearing proper safety equipment while working. But no HVACR companies are going to give you unlimited warnings before your employment with them is terminated. These policies are good for you and the company, because they reduce the likelihood of your being injured on the job.

Ear Protection

Hearing protection devices (Figure 3-3) must be worn whenever there is exposure to high noise levels of any duration. These devices are of two types: (1) ear plugs, which are inserted in the ear, and (2) headphones, which cover the ear. Either one must be properly selected on the basis of how much protection is required.

Respiratory Protection

There are two main types of respirators, as shown in Figure 3-4: (1) *air-purifying respirators* are ones that purify the air by filtering out harmful dust, mist, metal, fumes, gas, and vapor; and (2) *atmosphere-supplying respirators* are

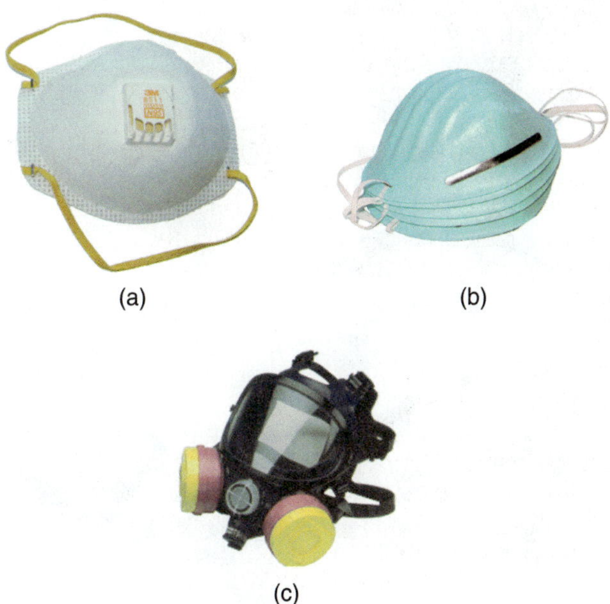

Figure 3-4 Filtration masks: (a, b) light-duty filter; (c) respirator with replaceable filters.

ones that supply clean breathing air from a compressed air source. The second type should always be worn when working in a confined space where concentrations of harmful substances are very high or where the concentration is unknown. Remember that most refrigerants are odorless, tasteless, and invisible and can cause asphyxiation in a very short time.

Respirators must fit tightly against the skin so that there is no leakage from the outside into the face. Workers who are required to use respirators at any time must be instructed in their use, care, maintenance, and limitations.

TECH TIP

Respirators are required to be located in all equipment rooms where that equipment contains large quantities of refrigerant. These respirators are provided in case there is a massive refrigerant leak. If you work in one of these areas, you must familiarize yourself with where the respirators are located and how to quickly put them on. You may have only a matter of seconds once a refrigerant leak alarm is sounded to safely put on this equipment.

Hand, Foot, and Back Protection

There are many different kinds of gloves used for hand protection, as shown in Figure 3-5. Some are made for special uses, such as gloves of steel mesh or Kevlar to protect against cuts and puncture wounds. Different glove materials are needed to protect against a variety of different chemicals. Choose the right kind from a dependable supplier who can supply information to lead you to the right gloves. Discard any damaged ones.

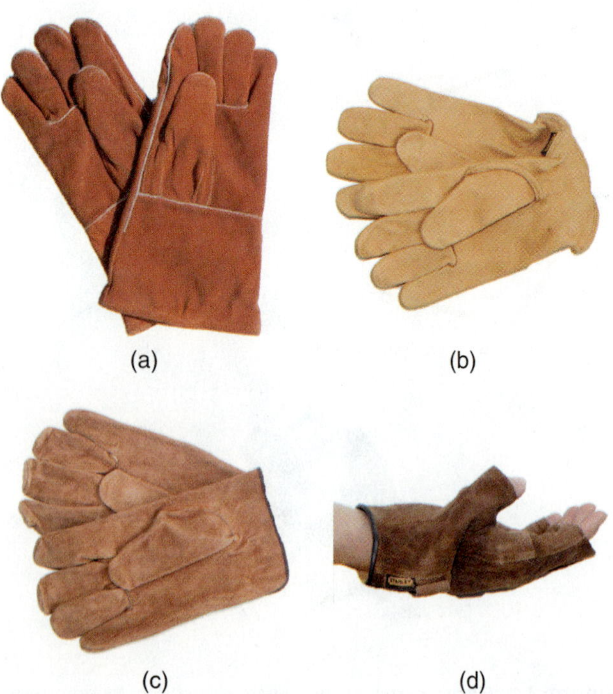

Figure 3-5 (a) Gauntlet-type work gloves; (b) work gloves; (c) welding gloves; (d) open-tipped gloves.

Figure 3-6 High-top work boots.

Lifting heavy objects improperly can lead to back injury. Always bend at the knees and lift straight upward rather than bending over, which will place undue stress on the spine. There are a number of different types of back-support belts available. These should be used if heavy lifting is expected to be performed on a regular basis.

When choosing foot protection, as shown in Figure 3-6, use the following guidelines:

1. All footwear must be well constructed to support the foot and to provide secure footing.
2. Where there is danger of injury to the toes or top of the foot or from electrical shock, the proper shoe or boot, such as steel toe, must have Construction Safety Approval (CSA) indicated.
3. Where there is danger of injury to the ankle, footwear must cover the ankle and have a built-in protective element/support.
4. If there is danger of harmful liquids spilling on the foot, the top of the shoe must be completely covered with an impervious material or treated to keep the dripped substance from contacting the skin.

3.3 LADDERS AND SCAFFOLDS

Access equipment refers to ladders and scaffolds that are used to reach locations not accessible by other means. The following precautions should be practiced in the use of ladders.

1. Only use CSA- or ANSI-approved ladders. Maintain ladders in good condition, and inspect ladders before each use. Discard ladders needing frequent repairs or showing signs of deterioration.
2. All portable ladders must have no-slip feet.
3. Place ladders on a firm footing, no farther out from the wall than one quarter of the height required, as shown in Figure 3-7.
4. Ladders must be tied, blocked, or otherwise secured at the top where the ladder meets the building to prevent them from slipping sideways.

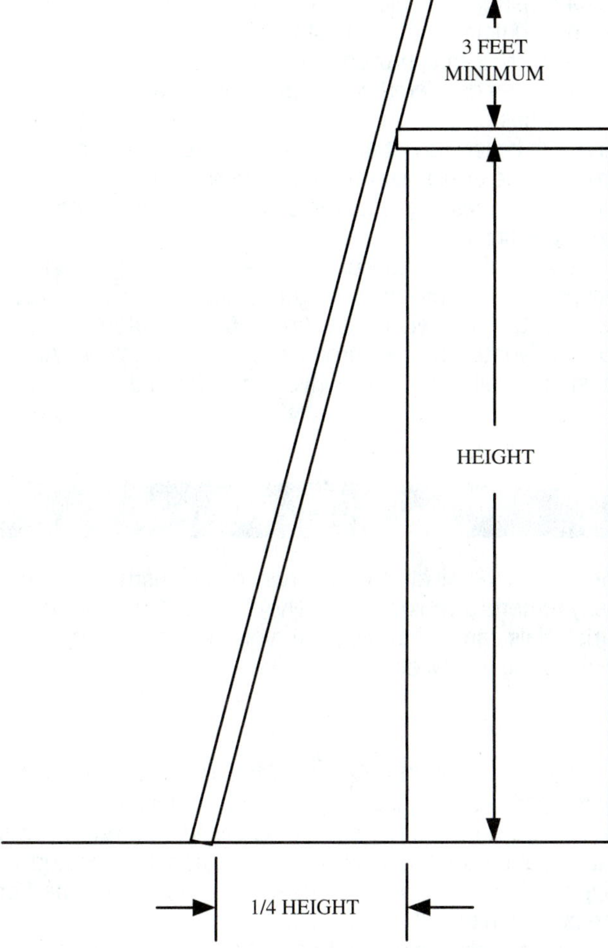

Figure 3-7 A ladder must be placed so the top is at least 3 ft above the roof and at an angle in which the distance from the building is one-fourth the height of the building.

5. Never overload a ladder. Follow the maximum carrying capacity of the ladder, including the person and equipment. The American National Standards Institute (ANSI) sets the standard for ladders.
6. Only one person should be on a ladder, unless the ladder is designed to carry more people. Follow maximum load rating.
7. Never use a broken ladder. Never place a ladder for use on top of scaffolding, and never use a borrowed ladder on someone's property. Always use your own ladder even if you need to leave the job site to go get it.
8. Always face the ladder and use both hands when climbing or descending a ladder.
9. Use fiberglass or wood ladders when doing any work around electrical lines (Figure 3-8).
10. Ladders should be long enough so you can perform the work comfortably, without leaning or having to go beyond the two rungs below the top rung safety barrier.
11. Stepladders should only be used in the fully open position.

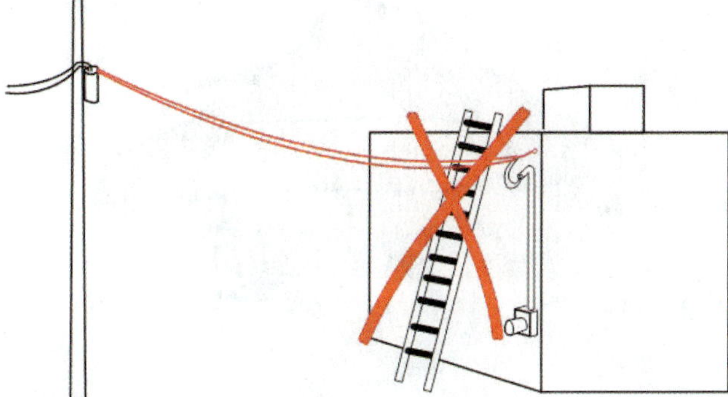

Figure 3-8 Never use a metal ladder near electrical wires!

CAUTION

Ladders must be inspected from time to time to ensure their safety. Some companies will require that a company safety official inspect the ladders you carry on your service van. Damaged or worn ladders must be repaired or removed from service.

The following recommendations apply to scaffolds.

1. Scaffolds must be supported by solid footings.
2. A scaffold having a height exceeding three times its base dimension must be secured to the structure.
3. When rolling scaffolds are used, the wheels must be locked when there are workers on the scaffold.
4. No worker is to remain on the scaffold while it is being moved. All equipment should also be removed before moving a scaffold.
5. Access to the work platform must be a fixed vertical ladder or other approved means.

3.4 FALL PROTECTION

Two types of equipment for preventing injury from falling are (1) fall-prevention equipment and (2) fall-arresting equipment. Either of these methods is required when working at heights over 10 ft above grade when no other means has been provided for preventing falls. Figure 3-9 illustrates a safety belt.

In fall prevention, a worker is prevented from getting into a situation where a fall can occur. For example, a safety belt attached to a securely anchored lanyard will limit the distance a worker can move.

In fall arresting, the worker must wear a safety harness attached to a securely anchored lanyard, which will limit the fall to a safe distance above impact. The harness helps prevent the worker from suffering internal damage. Belts should not be used to arrest a fall because they do not provide the measure of safety that harnesses do. Where a fall-arresting system is not practicable, a safety net should be suspended below the work activity. The worker should be secured separately from the tools and equipment.

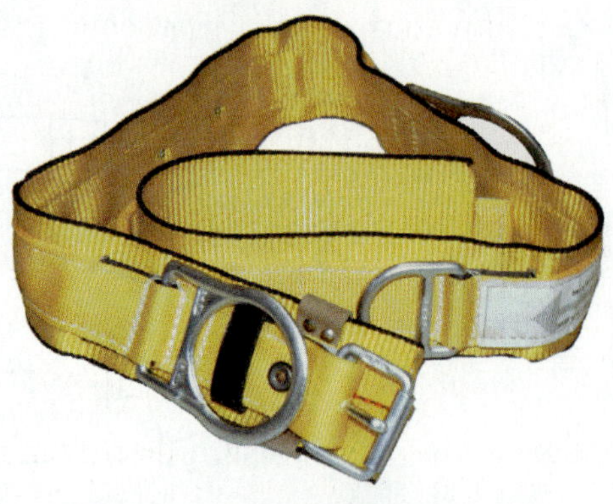

Figure 3-9 Safety equipment for heights over 10 ft above grade—a safety lanyard would be attached to this safety belt.

CAUTION

Fall-protection harnesses are designed to suspend you in a vertical position if you accidentally slip and fall from a height. These harnesses, however, are not designed to suspend you for long periods of time. In recent years, workers have survived a fall, only to die in the safety harness. The safety harness can constrict blood flow to your legs as you dangle at the end of the safety line. The restriction of blood flow to your legs can cause enough blood to pool in your legs so that you might pass out or even die if allowed to dangle motionless for a long period of time. If you are the victim of a fall and are suspended on your safety line, you should remember to move your legs to help keep the blood flowing until you are rescued.

3.5 HOT AND COLD

HVACR technicians often work with extremely hot or extremely cold vapors, liquids, and solid objects. Similar PPE is required for both extremely hot and extremely cold work because of the potential for burns. Burns can occur on your skin from accidentally contacting either extremely hot or extremely cold material. It is easy to see how a hot material can cause a burn but more difficult to see how something that is cold can burn your skin.

When you touch something that is extremely cold, your body heat is drawn out so quickly that it causes the surface of your skin to burn and blister. Remember that refrigerant will cause frostbite, so be careful.

3.6 HARMFUL SUBSTANCES

Workers in the mechanical trades can be exposed to a variety of harmful substances, such as dust, asbestos, carbon monoxide, refrigerants, resins, adhesives, and solvents.

All dust can be harmful. Where dust cannot be controlled by engineering methods, an approved respirator designed to filter out specific dust must be worn.

When asbestos-containing material (insulation) is being cut or shaped, the particles must be removed by a ventilation system that discharges the particulate matter through a high-efficiency particulate air (HEPA) filter. All waste materials that contain asbestos must be placed in impervious bags for transfer to an approved disposal site. These fibers, when inhaled, are considered carcinogenic. If you are asked to remove asbestos from, for instance, a piece of equipment such as an old boiler, do not attempt this. You do not want to put anyone's life at risk, so do not be forced into improperly removing asbestos. Improper removal of asbestos can also lead to serious fines.

Engine-driven mobile equipment operating in an enclosed area can produce dangerous levels of carbon monoxide (CO). Oil-fired or gas-fired space heaters without suitable vents can also produce carbon monoxide. Areas must be well ventilated while being heated with these devices.

TECH TIP

Read and follow all label safety and user instructions on any materials you use in the HVACR field. Some of these materials can be hazardous to your health if you do not follow the label's directions.

Some refrigerants are more dangerous than others, but all refrigerants are dangerous if they are allowed to replace the oxygen in the air. Even the so-called safe refrigerants can produce a poisonous phosgene gas when heated to high temperatures, and refrigerants sprayed on any part of the body can quickly freeze tissue.

Resins, adhesives, and solvents can be dangerous if not properly handled. Ensure that the workspace is continuously ventilated with large amounts of fresh air.

Never use carbon tetrachloride for any purpose, because it is extremely toxic, either inhaled or on the skin. Even slight encounters with it can cause chronic problems.

To provide workers and health-care professionals with specific reactions and treatments for exposure to materials on the job, all manufacturers must provide, on request, a Material Safety Data Sheet (MSDS) on all of their products. It is your responsibility to request a MSDS for all of the products you work with or carry on your service vehicle. Some site safety officers, building inspectors, or job managers may even ask to see the MSDS before you are allowed to start or continue work.

3.7 MATERIAL SAFETY DATA SHEETS (MSDS)

Material safety data sheets are required by law and have specific important information listed in specific areas so that emergency personnel can easily read them. If an area does not apply to the product, the manufacturer must mark the space as being nonapplicable. No blank spaces are allowed on a MSDS. This is done so that there will not be any confusion regarding the safety or reactions to any products. You should read the MSDS on any material before you use it so you know how to use it properly and safely and know what to do if there is an accident involving the material.

Figure 3-10 is an example of an MSDS for a coil cleaner. The following material will appear on every MSDS:

1. **Identity.** This section gives the name of the product.
2. **Section I.** This section gives the manufacturer's name and address and contains the emergency contact phone number and company phone number.
3. **Section II: Hazardous Ingredients/Identity Information.** A list of the hazardous ingredients by chemical name is given here. If a material has a secret ingredient, the company must either list it or provide it to a health worker on request. The item listed as OSHA PEL indicates the *personal exposure limit*. This is the maximum safe amount of contact or exposure time that is allowed. The ACGIH TLV is the time weighted average exposure over an 8-hr period of time. It is the maximum amount of contact or exposure allowed during an 8-hr working day. Any other specific limitations are given in this section.
4. **HMIS Information.** The information in this section refers to the hazardous material identification system (HMIS). This information provides health-care workers with a relative number according to how significantly the material will affect health, how reactive it is, and its flammability.
5. **Section III: Physical/Chemical Characteristics.** This section gives the properties of the material, such as its boiling point, vapor pressure, solubility in water, specific gravity, melting point, evaporation rate, and its appearance or odor.
6. **Section IV: Fire and Explosion Hazard Data.** Gives the flammability of the material and lists its flash point, flammability limits, extinguishing media, any special firefighting procedures, and any unusual fire or explosion hazards.

In addition to the information on the MSDS, you must read and follow all manufacturers' listed instructions for

Material Safety Data Sheet
May be used to comply with
OSHA's Hazard Communication Standard,
29 CFR 1910.1200. Standard must be
consulted for specific requirements.

U.S. Department of Labor
Occupational Safety and Health Administration
(Non-Mandatory Form)
Form Approved
OMB No. 1218-0072

(1) IDENTITY *(As Used on Label and List)* Coil Master Non Acid Coil Cleaner CMD-2A

Note: Blank spaces are not permitted. If any item is not applicable, or no information is available, the space must be marked to indicate that.

(2) **Section I**

Manufacturer's Name	Emergency Telephone Number
Crow Marketing & Distribution, Inc.	800-255-3924
Address *(Number, Street, City, State, and ZIP Code)* P. O. Box 29171, Dallas, TX 75229	Telephone Number for Information 214-241-3049
	Date Prepared 08-23-95 reviewed 12/01/99
	Signature of Preparer *(Optional)*

(3) **Section II – Hazardous Ingredients/Identity Information**

Hazardous Components (Specific Chemical Identity; Common Name(s))	OSHA PEL	ACGIH TLV	Other Limits Recomended	% (optional)
NaOH Sodium Hydroxide 50%	$2 mg/m^3$	$2 mg/m^3$		
CAS#1310-73-2 NA1824 Corrosive				

(4) **HMIS INFORMATION**

Health	2
Reactivity	1
Flammability	0

(5) **Section III – Physical/Chemical Characteristics**

Boiling Point	180°F	Specific Gravity (H_2O = 1)	1.01
Vapor Pressure (mm Hg.)	760	Melting Point	N/A
Vapor Density (AIR = 1)	1	Evaporation Rate (Butyl Acetate = 1)	N/A

Solubility in Water
Complete - 100%

Appearance and Odor
Yellow to gold liquid, characteristic odor

(6) **Section IV – Fire and Explosion Hazard Data**

Flash Point (Method Used)	Flammable Limits	LEL	UEL
None	None		

Extinguishing Media
N/A

Special Fire Fighting Procedures
Will react with metals to produce flammable hydrogen gas

Unusual Fire and Explosion Hazards
None known. Contents under pressure. Exposure to temperatures above 120°F may cause bursting.

Figure 3-10 Sample of a completed Material Safety Data Sheet (MSDS).

proper use and handling of every material you use on the job. Many companies have policies that require that you be fired if you do not follow these instructions. These policies are both for your protection and for the protection of the company and the customers. Handle all material properly according to the manufacturer's instructions.

SERVICE TIP

It is a good work practice to request a MSDS on all products that you get from the supply house. Some job site safety officials require that you have a MSDS on all of the products you will be using at their job site. An easy way of maintaining these documents is to put them all in a large manila envelope that is labeled MSDS and keep this envelope in your service vehicle.

3.8 SAFE WORK PRACTICES

The installer and service technician works in many areas: in the shop, in various types of buildings, in equipment rooms, on rooftops, under houses, and on the ground outside buildings. Each location requires different activities where safe performance is essential.

In addition, the worker deals with many potentially dangerous conditions, such as handling pressurized liquids and gases, moving equipment and machines, working with electricity and chemicals, and being exposed to heat and cold. It is important, therefore, that the technician practice good safety procedures wherever or whatever the work is.

Confined Space Entry

A confined space is an area that has been closed off from any outside source of ventilation and is a space large enough for a person to enter and perform work. Entering a storage tank through a manhole opening would be an example of this. Never enter such a space unless the atmosphere has been checked to ensure that no hazardous vapors exist. Even without the presence of hazardous vapors, the space could be potentially deadly due to a lack of oxygen. A person entering an oxygen-deficient space may not realize it until too late and suddenly pass out unconscious. Never enter a confined space until the air quality in the space has been verified as acceptable. Even then, always leave someone standing outside the space with emergency breathing apparatus available and ready.

3.9 TOOLS

All tools come with safety instructions. Do not assume you know how to use a tool; read and follow the manufacturer's safety instructions. These instructions have been developed and tested to ensure that, when followed, you are not likely to be injured.

Part of the manufacturer's literature may include instructions on how to service or repair the tool. Make sure to follow these instructions so that your tools will last longer and you can use them safely.

Hand Tools

1. Keep all hand tools sharp, clean, and in safe working order.
2. Defective tools should be repaired or replaced.
3. Use correct, proper fitting wrenches for nuts, bolts, and objects to be turned or held.
4. Do not work in the dark; use plenty of light.
5. Do not leave tools on the floor. Keep them properly put away neatly in your vehicle.

Power Tools

1. Only use power tools that are properly grounded.
2. Stand on dry, nonconductive surfaces when using electrical tools.
3. Use only properly sized electrical cords in good condition with a GFI or GFCI.
4. Turn on the power only after checking to see that there is no obstruction to proper operation.
5. Disconnect the power from an electrical tool (or motor) before performing any maintenance.
6. Disconnect the power supply when equipment is not in use.

CAUTION

Never use portable electrical equipment unless it is connected to a ground fault circuit interrupter (GFCI) or ground fault interrupter (GFI). GFI is the term that is sometimes used when referring to a GFCI. A GFCI is a safety device that cuts off electric current to a circuit when a short or ground is detected. These devices are designed to prevent burns and electrocution if a portable electric tool is being used. OSHA requires that all extension cords used for outside service work be equipped with GFCIs. Some extension cords now come with these devices built into the plug. If your extension cords do not have GFCIs built in, one can be purchased and added to the end.

3.10 SHOP SAFETY

1. Keep the shop or laboratory floor clear of scraps, litter, and spilled liquid.
2. Store oily shop towels or oily waste in metal containers in an open, airy place.
3. Clean the chips from a machine with a brush; do not use a towel, bare hands, or compressed air.
4. Keep safety glasses and gloves in a prominent location adjacent to machinery used for grinding, buffing, or hammering and where material with sharp edges is handled.
5. Establish cleaning periods regularly, at least on a daily basis. Make sure everyone is clear when using compressed air to clean.

Maintaining an Orderly Shop

Proper organization of machinery and equipment in the shop helps to promote safety. In addition, there will be less wasted time searching for the required tools or materials. Organizing and preparing is often as important as doing the job itself.

1. Arrange machinery and equipment to permit safe, efficient work practices and ease in cleaning.
2. Materials, supplies, tools, and accessories should be safely stored in cabinets, on racks, or in other readily available locations.
3. Working areas and workbenches should be clear. Floors should be cleaned on a daily or more frequent basis. Keep aisles, traffic areas, and exits free of materials and obstructions.
4. Combustible materials should be properly disposed of or stored in approved containers.
5. Drinking fountains and wash facilities should be clean and in good working order at all times.
6. Eyewash stations should be periodically tested to make sure that they work properly, and they should not be blocked and should always be readily accessible.

Fire Extinguishers

The danger of fire is always present. Rags soaked in oil, grease, or paint can ignite spontaneously. Keep used rags in tightly closed approved metal containers.

Sparks, open flames, and hot metal can ignite many materials. Always have a fire extinguisher close at hand when welding or burning.

Extreme caution should be taken with highly flammable and volatile solvents. Due to its low flash point (the temperature at which vapors will ignite), gasoline should never be used as a cleaning solvent.

Fire extinguishers should be readily accessible, properly maintained, regularly inspected, and promptly refilled after use. Fire extinguishers are classified according to their capacity for handling specific types of fires (Figure 3-11).

1. **Class A extinguishers.** These are used for fires involving ordinary combustible materials such as wood, paper, and textiles, where a quenching, cooling action is required.

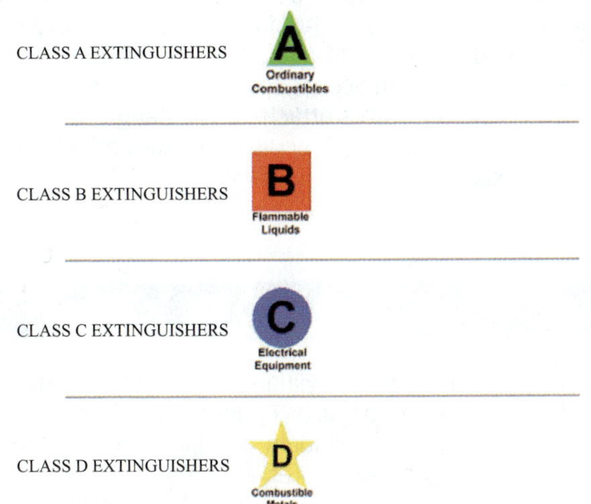

Figure 3-11 Fire extinguisher classification symbols.

2. **Class B extinguishers.** These are for flammable liquid and gas fires involving oil, gas, paint, and grease, where oxygen exclusion or flame interruption is essential.
3. **Class C extinguishers.** These are for fires involving electrical wiring and equipment where nonconductivity of the extinguishing agent is critical. This type of extinguisher should be present whenever functional testing and system energizing take place.
4. **Class D extinguishers.** Some metals, such as magnesium, can actually catch fire. Class D extinguishers are used to put out combustible metal fires.

CAUTION

Having the proper fire extinguisher is not enough if it is not large enough to do the job. Make sure that your fire extinguisher is large enough to handle the size of fire that might occur in your work area. If there are a lot of combustible materials, you need a bigger fire extinguisher. In some cases you may want to have more than one fire extinguisher available.

Material Handling

Use mechanical lifting devices whenever possible. Use a hoist when lifting tools or equipment to a roof. If you are required to lift a heavy object, get help. To avoid straining your back, the following procedures should be observed when lifting heavy objects.

1. Bend your knees and pick up the object, keeping your back straight.
2. Gradually lift the weight using your leg muscles, continuing to keep your back vertical.

3.11 WELDING AND CUTTING

Welding and cutting are specialized skills that require special training. Many refrigeration and air-conditioning technicians require this training due to the need to perform some of these operations as part of their work. It must be recognized, strictly from a safety standpoint, that this work should not be attempted without adequate knowledge and instruction.

Acetylene

Acetylene gas under high pressure is unstable, so to make it safe to use it is absorbed in acetone inside the acetylene cylinder. However, an acetylene cylinder can explode if it is struck or jarred severely, so keep the cylinders secured so they cannot be accidentally knocked over or fall. Because acetylene can explode at high pressures, it is illegal to operate a torch with an acetylene pressure greater than 15 psi. For HVACR work, we generally use pressure less than 5 psi. Most acetylene gauges have a red mark indicating 15 psi (Figure 3-12).

Figure 3-12 Acetylene regulator with gauges.

Figure 3-13 Portable oxyacetylene welding and cutting rig. *(Courtesy of Victor, a Brand of Thermadyne Industries, Inc.)*

CAUTION

The acetone used to stabilize the acetylene is absent in the neck of the tank. For this reason, never store or transport an acetylene cylinder in the horizontal position. Storage in this manner even for short periods is *very* dangerous.

Oxygen

Oxygen is provided in highly pressurized cylinders. A full oxygen cylinder contains approximately 2,000 lb/psi of oxygen. Because the oxygen is under such great pressure, care must be taken to ensure that the oxygen cylinder valve is not damaged or knocked off. If the oxygen cylinder valve were to be broken off of a full oxygen cylinder, the cylinder could fly around the room much like a child's balloon. To prevent possible damage to the valves and regulators on oxygen and acetylene cylinders, the cylinders should be securely attached to a frame or cart when they are in use (Figure 3-13).

CAUTION

Certain materials react explosively if exposed to high-pressure pure oxygen. Never use oil or grease on any part of the oxygen or acetylene regulator, hoses, or torch and tips. Ensure that all fittings are clean and dry when assembling. Repair using only manufacturer-approved replacement parts.

Air acetylene torches are used by HVACR technicians to produce sufficient heat for silver (hard) soldering brazing. These torches use a mixture of air and acetylene as a fuel.

Oxyacetylene torches are used for brazing, welding, and cutting. These torches mix oxygen and acetylene gases to produce a very intense, hot flame.

The following safety rules should be practiced when using this equipment.

1. Always use a welding regulator on acetylene and oxygen cylinders.
2. Never use oil or grease on a regulator, as this could cause a violent explosion with compressed oxygen.
3. Always secure the cylinder to something solid to prevent it from being accidentally knocked over.
4. Wear the proper colored safety glasses.
5. Open the valve on the acetylene cylinder only one and one-quarter turns.
6. Light the torch with a striker.

SAFETY TIP

Always remove regulators from oxygen and acetylene cylinders before you put them back in your service vehicle. In many states and cities it is against the law to transport oxygen and acetylene cylinders in a motor vehicle with the regulators attached. The danger is that in an accident the regulators could be broken off, causing a fire or explosion.

CAUTION

All oxygen and acetylene cylinders should be stored and used in secure racks or carts. Federal safety regulations (OSHA) require 55-ft^3 cylinders and larger to be safety chained to carts or structures anytime the cylinder is being used with its safety cap removed. There are no exceptions to these OSHA regulations regarding cylinders.

3.12 FIRST AID

Refrigeration and air-conditioning workers are advised to enroll in an approved first-aid course. Some vocational programs and technical colleges offer first aid as an elective course. Be sure to talk to your adviser about taking a first-aid course. Prompt and correct treatment of injuries not only reduces pain but could also save lives. A classification of accidents that occur to HVACR personnel, related to the hazards described, include the following:

1. Injuries due to mechanical causes
2. Injuries due to electrical shocks
3. Injuries due to high pressure
4. Injuries due to burns and scalds
5. Injuries due to explosions
6. Injuries due to breathing toxic gases
7. Injuries due to negligence when working with electricity (Always pay strict attention to the work, and never allow any horseplay, as this could cause you or someone you are with to receive serious injury or die.)
8. Heat exhaustion or heat stress

3.13 ELECTRICITY SAFETY

All possible precautions must be practiced to prevent electrical shock—that is, current passing through the body. Very few realize the damage that can be done by even a small amount of current.

The following information applies to low-voltage circuits where current is measured in milliamps (mA). One amp (A) is equal to 1,000 mA.

The illustration in Figure 3-14 indicates the effect on the body when various amounts of current pass through the body at 100 mA or less.

Noncontact voltage (NCV) testers should be a standard tool for all HVACR service technicians. Some digital multimeters incorporate an NCV feature.

Electrical Safety Rules

1. Check all circuits for voltage before doing any service work. Tag and lock out all electrical disconnects when working on live circuits.
2. Stand on dry, nonconductive surfaces when working on live circuits.
3. Work on live circuits only when absolutely necessary.
4. Use only properly insulated tools to work on electrical circuits.
5. Never bypass an electrical protective device.
6. Properly fuse all electrical lines.
7. Properly insulate all electrical wiring.
8. Use ground fault circuit interrupters whenever using power tools.

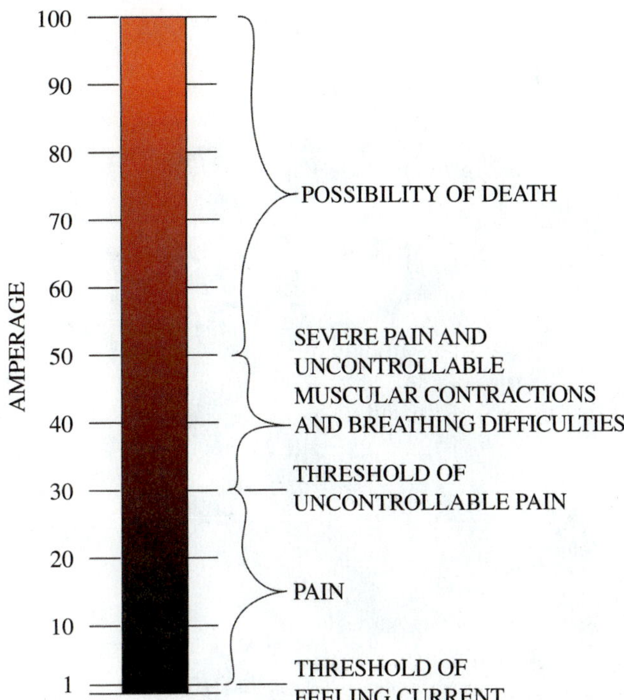

Figure 3-14 Amperage rating of electric current creating various shock effects from 1 mA up to 100 mA. *(© WorkSafeBC. Used with permission from* Working Safely Around Electricity.*)*

3.14 REFRIGERATION SAFETY

The hazards associated with refrigeration service are principally associated with the proper use of refrigerants and their storage in closed containers and systems. A large improvement was made when the HVAC industry started using the so-called safe refrigerants (Class I, or fluorocarbons), which were nontoxic and nonflammable. Dangers now relate to the use of pressurized gas or liquid and the fact that these chemicals, when released accidentally, can replace oxygen in a confined space without sensory detection.

CAUTION

From time to time it may be necessary to warm a refrigerant cylinder so that system charging can proceed. Never heat a refrigerant cylinder with an open flame. This could cause two problems: the open flame could cause a rapid rise in cylinder pressure above the rupture point of the cylinder, and the high heat can cause the refrigerant to decompose. Three safe ways of warming cylinders are to place the cylinder in a warm bath of water, place the cylinder in the warm discharge air from the condenser fan, or wrap an electrical cylinder heater around the cylinder (Figure 3-15).

Used Refrigerants and Refrigerant Oils

The EPA classifies all used refrigerants and used refrigerant oils as hazardous waste that must be disposed of properly. They must be taken to a state or locally approved recycler that specializes in used refrigerants and used refrigerant oils. Often these recyclers have local or state certifications that you can obtain copies of for your records. It is a good idea to keep accurate and complete records showing how and where all of the used refrigerant and used oil you collect was disposed. Although these records are not required

Figure 3-15 An electrical cylinder heater can be used to safely raise cylinder temperature. *(Courtesy of Airserco Manufacturing Company)*

by the EPA, you may be asked to prove that you are complying with all disposal requirements. Having good documentation can help show the EPA, if your records are audited, that you are complying with all the laws and regulations regarding disposal of hazardous waste.

Used refrigerant oil and used automotive oils are not compatible and cannot be mixed for disposal. That is why used refrigerant oil cannot be disposed of at an automotive oil recycler unless they have special containers just for used refrigerant oil.

CAUTION

It is illegal to sell any used CFC (chlorofluorocarbon) or HCFC (hydrochlorofluorocarbon) refrigerants that have not been reclaimed by an EPA-certified reclaimer.

Handling Refrigerant Cylinders

1. Do not fill a cylinder with liquid refrigerant to more than 80 percent of its volume. Heat can expand the liquid enough so that it fills the cylinder. Once that occurs, the cylinder will rupture. Space must be available inside the cylinder for proper expansion of the liquid to take place. In recovering refrigerants, this is particularly important. Special cylinders with an automatic volume-limiting device have been designed for recovery.
2. In using a cylinder or transporting it, the cylinder must be secured with a chain or a rope in an upright position. Do not drop a cylinder.
3. Mixing refrigerants is dangerous. Cylinders are color coded to help identify each refrigerant. Each system has an identifying label. Do not mix refrigerants. Maintain the identification system.
4. Never apply a torch to a system containing refrigerant. If heat is needed to vaporize refrigerant, use hot water at a temperature not to exceed 125°F (52°C).
5. Do not refill disposable refrigerant cylinders.
6. Replace the cylinder cap when not using a cylinder. The cap protects the valve. Do not lift or carry a cylinder by the valve.

SERVICE TIP

Most refrigerant cylinders come in cardboard boxes. These boxes contain important safety information. Some industrial plant safety officers will require that you maintain these cardboard containers with the safety material listed whenever you are working on their job site.

Refrigerant System Safety

1. Never use oxygen or acetylene to pressurize a system, because an explosion can occur. Use dry nitrogen or carbon dioxide from a tank properly fitted with a pressure regulator.
2. When isolating a section of piping or component of a system, exercise caution to prevent damage and potential hazard from liquid expansion.
3. Always charge refrigerant vapor into the low side of the system. Liquid refrigerant entering the compressor could damage the compressor or cause it to burst.
4. Never service a refrigeration system where an open flame is present. The flame must be enclosed and vented outdoors. If a fluorocarbon refrigerant comes in contact with intense heat, it can produce poisonous phosgene gas.

3.15 PRESSURE VESSEL SAFETY

Pressure vessels pose three major hazards if they rupture: The blast from the sudden explosion can cause serious internal injury. The pressurized liquid can cause severe burns. And fragments of metal thrown from the exploding vessel can cause lacerations and punctures. All pressure vessels pose some degree of hazard to individuals working in the area, so they must be inspected for flaws that might weaken their integrity.

There are a number of pressure vessels used in the HVACR field. These include accumulators, receivers, refrigerant cylinders, low-pressure boilers, high-pressure boilers, and hot water tanks. Both low- and high-pressure boilers and many commercial hot-water tanks are required to be inspected periodically. The type and frequency of inspection varies depending on the type of pressure vessel. There must be a placard displayed on or near the pressure vessel showing when it was inspected and giving the date when it needs to be reinspected.

Some pressure vessels, such as accumulators and receivers, may not have to be inspected. It is, however, a good idea to inspect them for rust, corrosion, cracks, and other signs of physical damage that may render them unfit for service.

Refillable refrigerant cylinders, such as those that are used for refrigerant recovery, must be inspected every five years. The inspection date must be stamped on the top of the cylinder. Cylinders that have not been inspected within five years cannot be used.

3.16 DRIVING SAFETY

A good driving record is essential for any job that requires that you drive. Almost all HVACR work requires that you do some driving as part of your job. Outside service technicians may spend much of their day driving from one job to another.

Your driving record will affect the insurance rate the company must pay. In some cases a very bad driving record could make you unemployable due to the high insurance rates your employer would have to pay so you could drive their service vans.

You cannot speed or drive recklessly or aggressively in a service vehicle. To do so could result in bad public relations for your company. In addition, with almost everyone having a cell phone, the chance that someone would call your boss and report your actions is very high. Even worse would be if they use their cell phone to take a picture or video of your driving and send that to your boss. You might be the best technician the company has, but you can still lose your job because of your actions behind the service vehicle's wheel.

If you are involved in an accident, make sure to get all of the other driver's information. You should also get any contact information from anyone who may have witnessed the accident.

TECH TIP

Carrying a disposable camera in the vehicle glove box will let you take photos of the accident and of any cars in the area. Make sure to get photos of the license plates of potential witnesses to the accident. A few photos taken shortly after the accident can help you prove your side of the story, if necessary.

3.17 STEPS TO BE FOLLOWED IN CASE OF AN ACCIDENT

1. If the injuries appear to be serious, call 911 or your local emergency number.
2. First aid should be administered, if needed, only by those qualified to do so.
3. All accidents, injuries, and illnesses should be reported to your boss or supervisor no matter how minor the injuries may seem.
4. An accident report form should be filled out and turned in to your company office.
5. An investigation of the accident may be done to determine the cause of the accident.
6. Clean up the area before resuming work.

UNIT 3—SUMMARY

Schools provide a safe environment in which to *learn*, but it is your responsibility to always *work* safely. You must follow all of the rules and standards, because electricity, compressed gases, and other potential hazards do not know you are a student; you may not get to make a safety mistake more than once.

If you have not been trained on the safe operation of any tool, equipment, or device, do not attempt to use it.

Safe working habits can last a lifetime. Start now and learn everything you can to make your HVACR career a long and safe one.

UNIT 3—REVIEW QUESTIONS

1. What causes accidents, and how they can be eliminated on the job site?
2. What is PPE?
3. Discuss ladder safety.
4. Discuss scaffold safety.
5. What is the difference between fall-prevention equipment and fall-arresting equipment?
6. Describe how cold can cause a burn on skin.
7. List harmful respiratory substances.
8. Explain some of the dangers of refrigerants.
9. Explain what material appears on every Material Safety Data Sheets (MSDS).
10. Discuss how to safely use hand tools.
11. Discuss how to safely use power tools.
12. Discuss how to practice safety in the shop.
13. List four types of fire extinguishers.
14. Tell how to safely lift heavy objects.
15. Discuss safe welding cylinder practices.
16. Discuss electrical safety rules.
17. When a refrigerant cylinder needs to be warmed, how should it be done?
18. How should used refrigerants and refrigerant oils be disposed of?
19. Discuss refrigerant system safety.
20. List three major hazards of pressure vessels.
21. Discuss how a technician's driving record can affect employability.
22. Tell what steps should be followed in case of an accident.

UNIT 4
Hand and Power Tools

OBJECTIVES

After completing this unit, you will be able to:

1. identify the major tools used in HVACR work.
2. describe how the major HVACR tools are used.
3. explain the safety procedures required for using different types of tools.
4. demonstrate how to care for tools.

4.1 INTRODUCTION

The care and use of tools are important considerations for the technician. The customer often judges a technician on the appearance of his tools. They often assume that a technician that has a dirty, poorly maintained, and disorderly tool bag will provide the same type of service. To a large extent this is true, because it is harder to work with poorly maintained tools, and a tool bag that is not well organized makes it take longer to locate the tools for the job. The way you maintain your tools reflects on your quality of workmanship. In addition, injury can frequently be traced to a lack of, or improper use of, hand tools. Clean, sharp tools are better to work with and safer to use. This unit describes the purpose and use of some of the most basic hand and power tools needed by the HVACR technician. Specialty instruments, electrical meters, gauge manifold sets, and tubing/piping tools are described in other follow-on units of the text.

4.2 TOOL KITS

Tool manufacturers package tools in kits (see Figure 4-1), and tool kits are available for most types of HVACR jobs. They are easier to store and have greater convenience than loose tools. Kits come in many sizes and price ranges. A complete kit includes wrenches, hex keys, screwdrivers, nut drivers, pliers, knives and cutters, measurement tools, electrical repair tools, flaring/swaging/cutting tools, a multimeter, a gauge manifold set, and a clamp-on ammeter.

The basic tool kit should include varying sizes of combination, adjustable, refrigeration service, and hex key (Allen) wrenches. There should be an assortment of slotted and Phillips screwdrivers along with nut drivers. Pliers are an essential tool for the technician, and different types—such as lineman's side cutting pliers, needle-nose (long-nose) pliers, diagonal cutting pliers, tongue-and-groove pliers, slip-joint pliers, and multipurpose wire strippers—will be necessary.

Specialty tools can be added to the tool kit over time, and because all tools are expensive, they should be maintained properly. It is recommended that you keep your tool kit locked and placed in a secure location when not in use. Keep an inventory list so that you can check it after each job. Good quality tools will provide for many years of service if you take care of them.

Figure 4-1 Prepackaged tool kit.

4.3 TOOL CARE AND USE

Keep tools in a safe place. Never carry tools in your pockets or leave them lying around. A ripped car seat caused by a screwdriver left in your back pocket spoils a good day on the job. A round screwdriver lying on the ground is an accident waiting to happen because someone will step on it, slip, and fall. Other accidents can occur from using damaged tools. Pieces of metal will fly from a chisel that has a mushroomed head when struck with a hammer (Figure 4-2). That is why safety glasses must be worn whenever using hammers and chisels. Don't keep damaged tools. Throw them away and make sure to dispose of them properly.

You not only need to have the right tool, but it must be used properly. As an example, pliers are not made for tightening or loosening nuts (Figure 4-3). This will round off the corners of the nut and make it increasingly more difficult to

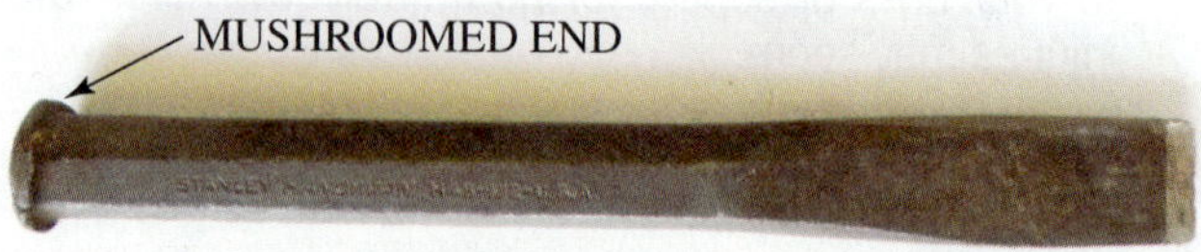

Figure 4-2 Chisel with mushroomed end.

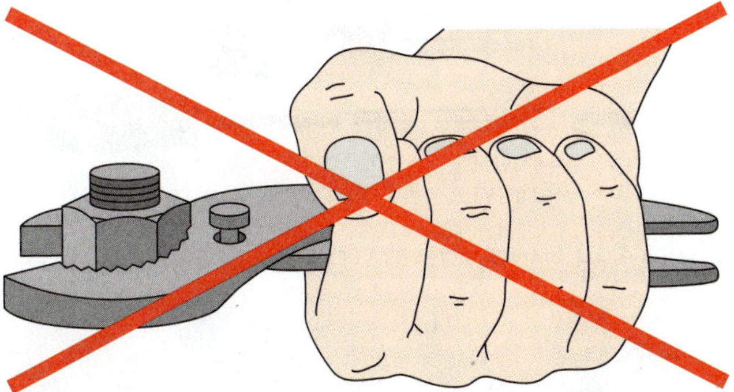

Figure 4-3 Pliers are not made for tightening or loosening nuts.

remove. Use the proper wrench or socket that fits even if it takes a few extra minutes to find it in your toolbox.

TECH TIP

Air-conditioning and refrigeration hand tools do not get as dirty as most mechanics' tools will. However, it is a good idea to wipe them down with a clean, dry rag after each use. In addition, from time to time they should be rubbed with a lightly oiled cloth to protect them from rusting.

4.4 SCREWDRIVERS

Screwdrivers are commonly required for almost every type of HVACR work. These are available in many different shapes, sizes, and materials. They are used for driving or removing screws or bolts with slotted, recessed, or special heads. Never use a screwdriver for prying, punching, chiseling, scoring, or scraping, and they should never be pounded with a hammer! Do not use a screwdriver near a live electrical wire or circuit.

Screwdrivers have a handle connected to a blade (sometimes called a shank) by a ferrule as shown in Figure 4-4. Flat and Phillips tips are the most common types of screwdrivers (Figure 4-5). A flat-tip screwdriver is usually called a flat blade. However, the length of the blade from the handle to the tip may be round or square. A heavy-duty screwdriver may have a square blade so that a wrench can be used to turn it (Figure 4-6). A screwdriver should be chosen to match the head of the screw (Figure 4-7). It is important to have a tight fit between the screw and screwdriver

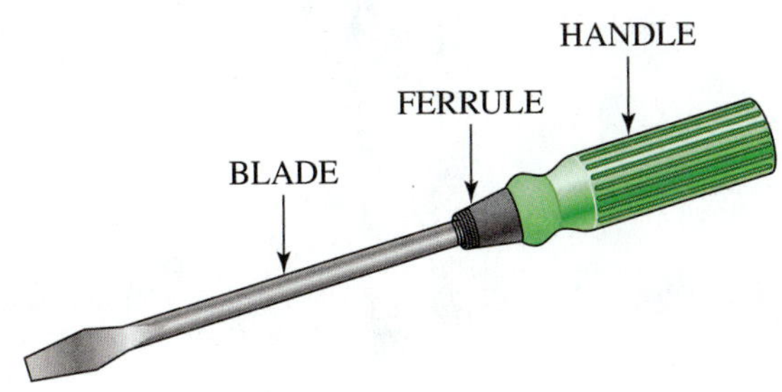

Figure 4-4 Parts of a screwdriver.

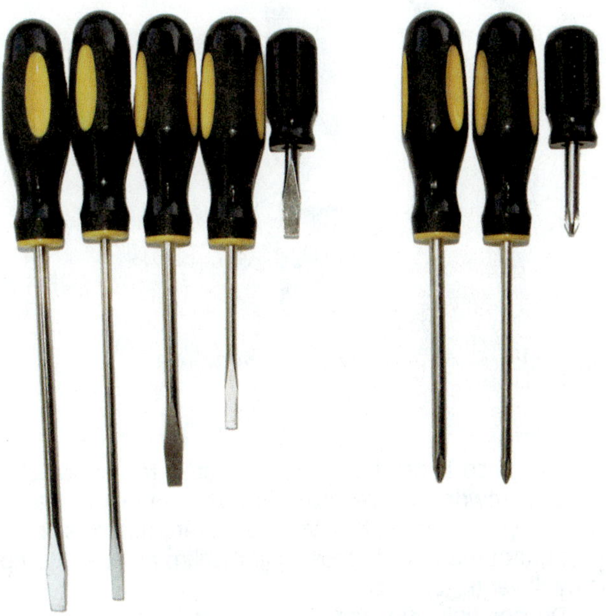

Figure 4-5 Screwdrivers come in a variety of sizes and lengths.

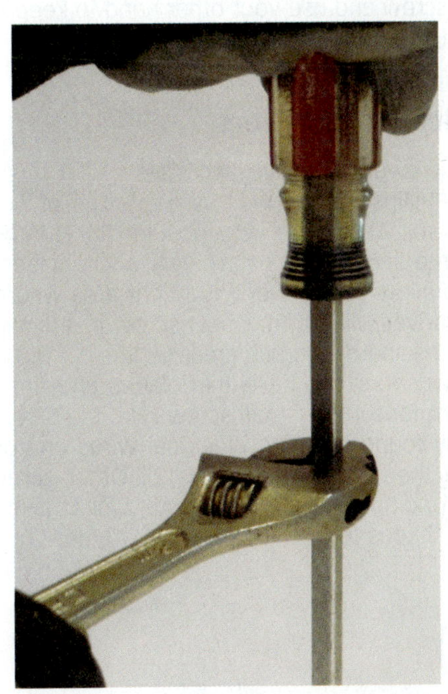

Figure 4-6 Using a wrench to assist in turning a square blade screwdriver.

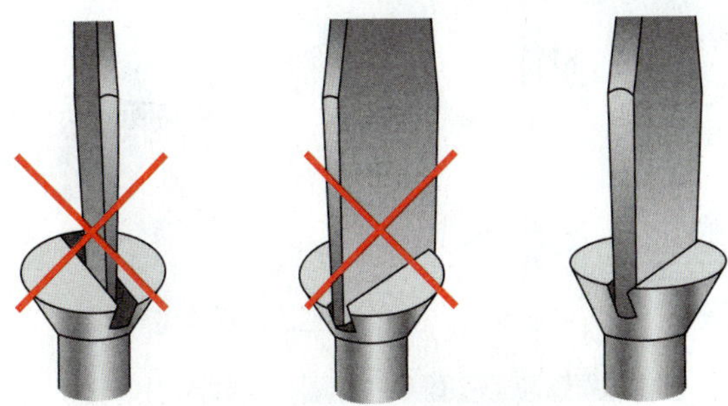

Figure 4-7 The screwdriver should match the head of the screw.

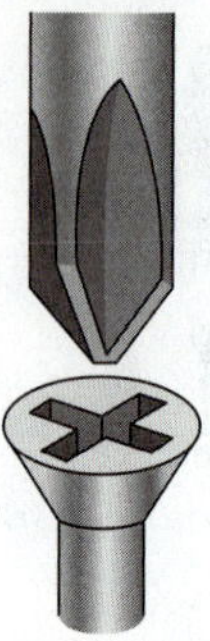

Figure 4-8 Cross-point Phillips screwdriver tip.

tip, otherwise it can be possible to strip the screw slot. Phillips screwdrivers are designed with a tip that is shaped like a cross. Therefore, this type is also referred to as cross-tip (blunt at the tip) and cross-point (sharp point at the tip) screwdriver (Figure 4-8).

Do not hold the work in one hand while using the screwdriver with the other. If the screwdriver slips out of the slot, it may gash your hand. Keep the screwdriver in line with the screw and use your other hand to keep the blade steady (Figure 4-9).

Specialty Screwdrivers

Six-way screwdrivers have two sizes of flat tips and two sizes of Phillips tips as well as two sizes of nut drivers (Figure 4-10). These are very popular for HVACR service work. Some screwdrivers have very short blades and are stubby to allow you to work in a tight area where a regular screwdriver will not fit. Offset screwdrivers can also be used in these hard-to-reach areas (Figure 4-11).

Screwdrivers can have thin blades and tips that are convenient for turning small screws like the types that are commonly found securing electrical wires on control circuits and thermostats (Figure 4-12). Other screwdrivers, such as Torx, Spline Drive, and Bristol types, have different blade tip shapes (Figure 4-13). These are designed so that

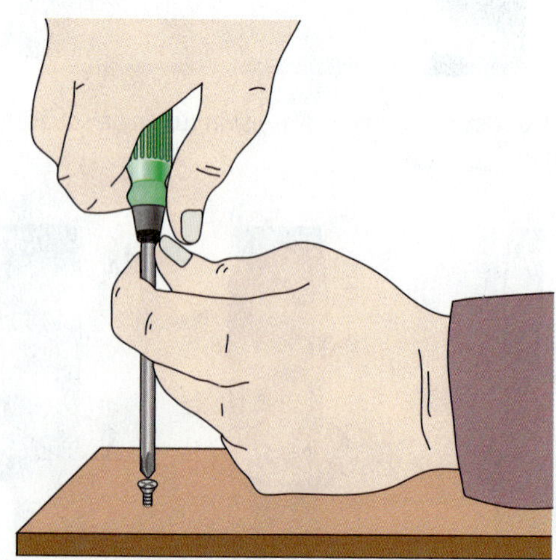

Figure 4-9 Proper use of a screwdriver.

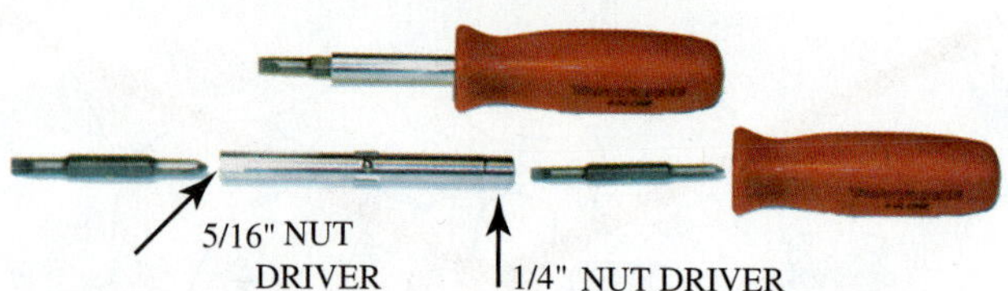

Figure 4-10 Six-way screwdriver.

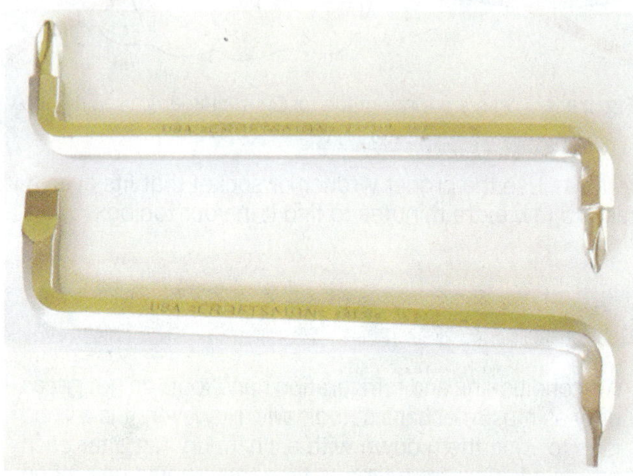

Figure 4-11 Offset screwdrivers.

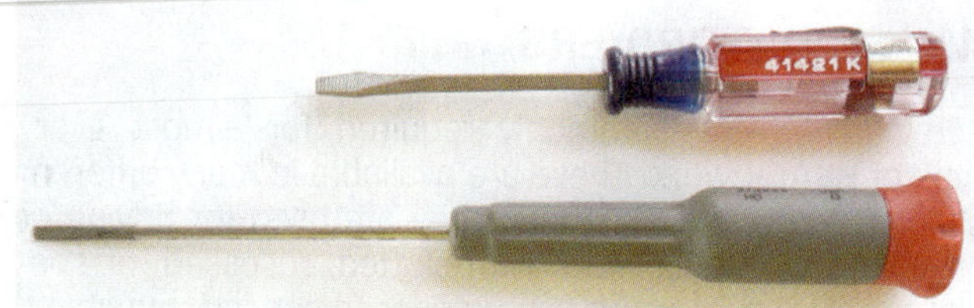

Figure 4-12 Screwdrivers used for smaller screws such as those found on thermostats.

Figure 4-13 Screwdrivers with specialty blade tips.

Figure 4-14 Some specialty blade tips provide greater contact and less chance for slippage.

there is less chance for the screwdriver to slip out of the slot (Figure 4-14).

Nut Drivers

The nut driver is useful in tightening or removing hex-head sheet-metal screws or machine screws that hold equipment panels in place or fasten control box covers. They look like screwdrivers with a handle, ferrule, and blade (Figure 4-15). However they differ from screwdrivers in that the tip has a drive socket to fit over the nut or screw hex head (Figure 4-16). Some nut drivers have magnetic ends that hold the screw head in the nut-driver end.

(a)

(b)

Figure 4-15 Nut driver set: (a) side view; (b) end view; often the ends of the handles on nut drivers are color coded and identified with the size to make it easier to pick them out of your tool bag.

(a)

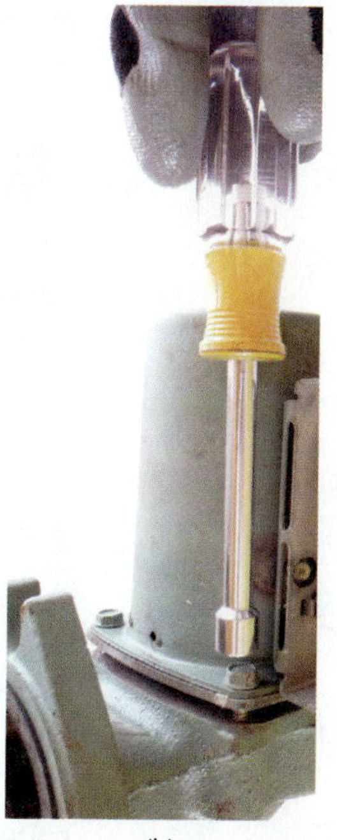

(b)

(c)

Figure 4-16 Using a nut driver: (a) nut driver and screw head; (b) positioning nut driver; (c) nut driver in place and ready to turn.

4.5 PLIERS FOR MECHANICAL WORK

There are a number of different types of pliers used in HVACR work. Slip-joint, tongue-and-groove, and locking pliers are primarily used for mechanical work.

Slip-Joint Pliers

Slip-joint pliers have serrated (grooved) jaws and are handy for general use and for holding hot parts or grabbing something that is being removed (Figure 4-17). The pivot is used to adjust the jaw opening to handle large or small objects (Figure 4-18).

Tongue-and-Groove Pliers

Tongue-and-groove pliers (multitrack) are used like slip-joint pliers except they open wider to hold or grab larger items, such as pipe (Figure 4-19). They come in different sizes with various jaw configurations, such as straight jaw, smooth jaw, and narrow jaw.

Figure 4-17 Using pliers to grab a cotter pin.

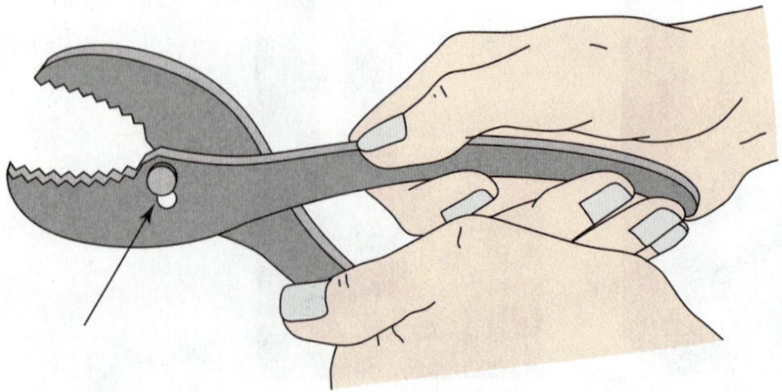

Figure 4-18 Adjusting slip-joint pliers to make them smaller or larger.

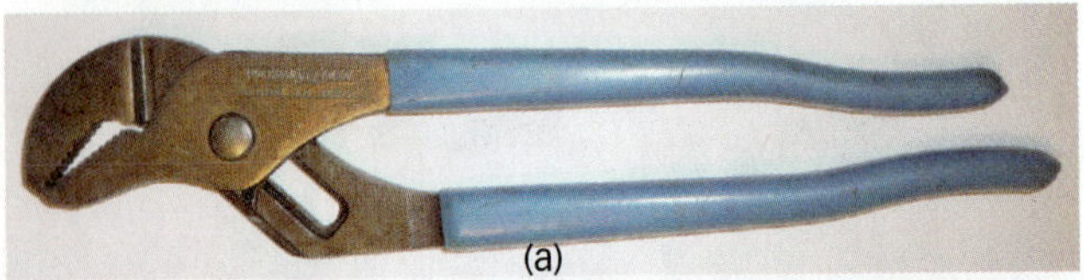

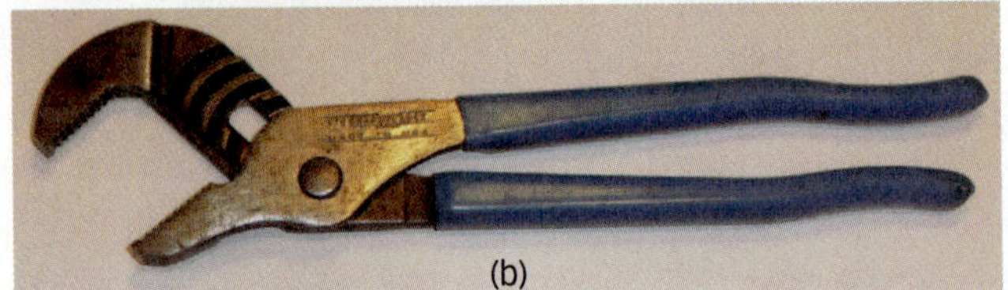

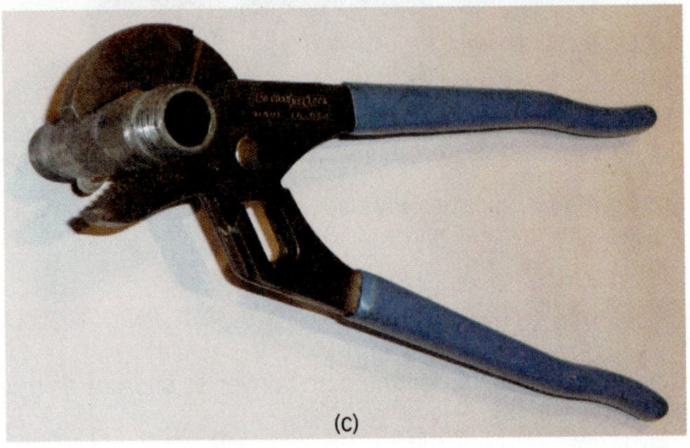

Figure 4-19 Tongue-and-groove pliers: (a) adjusted to minimum size; (b) adjusted to maximum size; (c) adjusted to hold pipe.

Locking, or Vise-Grip, Pliers

Locking, or vise-grip, pliers also come in different sizes with different jaw styles (Figure 4-20). They clamp and lock to hold parts securely. Their locking-clamp action is beneficial because once the pliers are clamped in place you can let go of them, which frees your hand (Figure 4-21).

CAUTION

Do not use pliers to tighten or loosen brass fittings. No matter how delicately you use pliers on these fittings, the fitting surface will be damaged. This damage can prevent the next technician from using the proper wrench on the damaged fitting.

4.6 PLIERS FOR ELECTRICAL WORK

Diagonal cutters, needle-nose, and lineman's pliers are primarily used for electrical work (Figure 4-22). These types of pliers can be damaged at the cutting surface if they are used for cutting materials other than copper wire. Cutting a hard material such as a nail can put a notch in the cutter and/or dull the cutting edge. In either case, damaging the cutter will render your pliers useless for HVACR work. If you have to cut nails, use a hacksaw or an old pair of pliers, not your good electrical pliers.

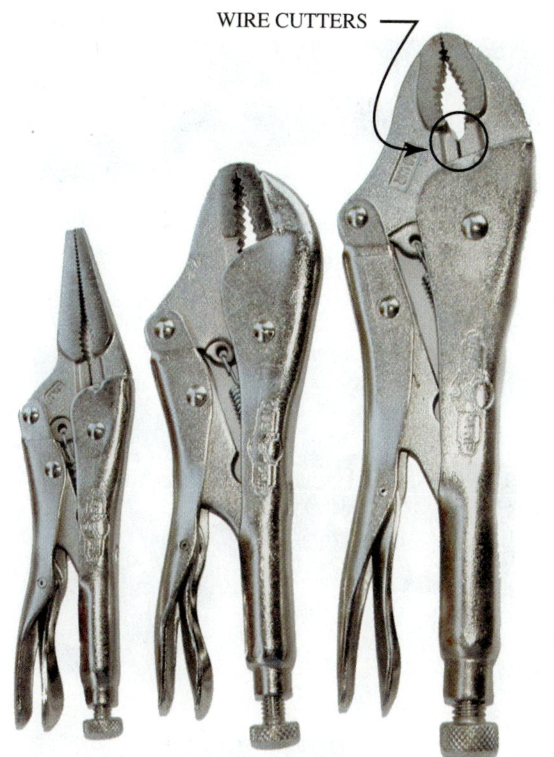

Figure 4-20 Locking pliers.

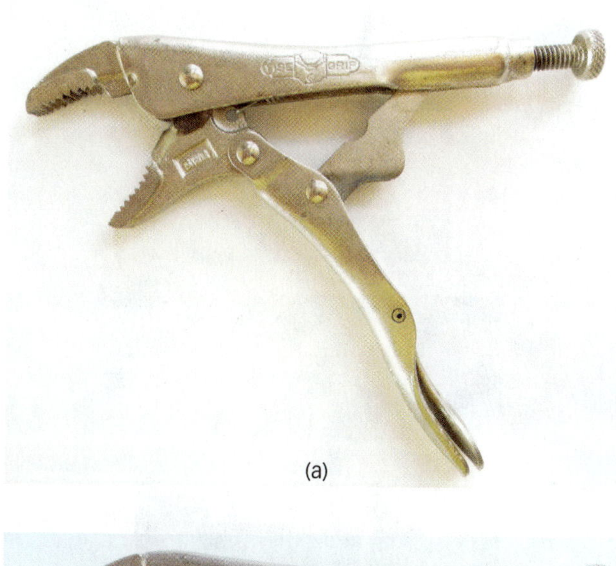

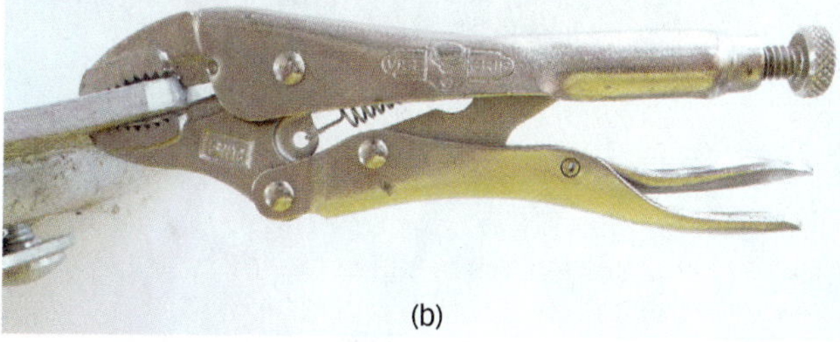

Figure 4-21 (a) Locking pliers before clamping; (b) locking pliers clamped.

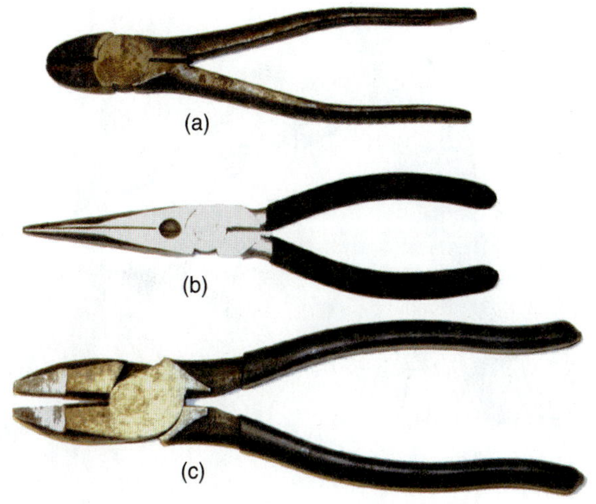

Figure 4-22 (a) Wire cutters; (b) needle-nose pliers; (c) lineman's pliers.

Lineman's Side-Cutting Pliers

Lineman's side-cutting pliers have serrated jaws, a rod-gripping section, side cutters, and a wire cropper. The flat jaws are used to bend sheet metal and twist electrical wire (Figure 4-23). The rod-gripping section is used to hold and

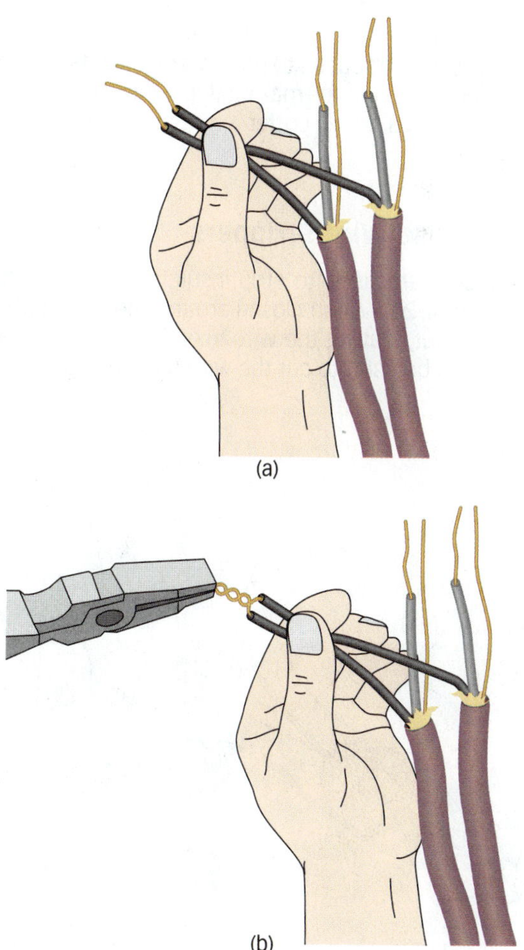

Figure 4-23 Lineman's pliers: (a) preparing electrical wires; (b) twisting electrical wires.

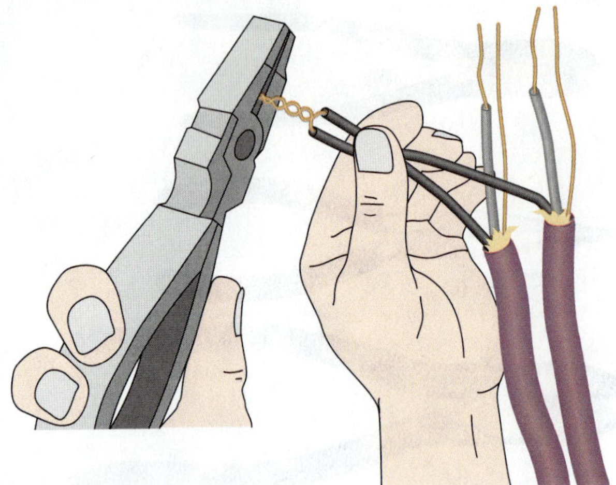

Figure 4-24 Cutting electrical wires with lineman's side-cutting pliers.

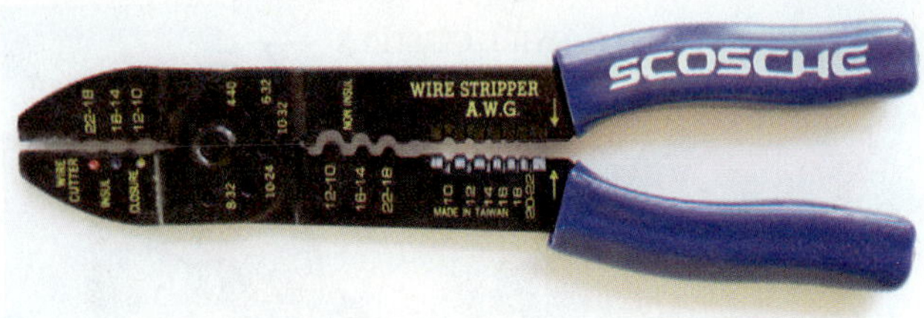

Figure 4-26 Multipurpose wire stripper.

bend small rods. The side cutters permit sharp flush cuts on electrical wire (Figure 4-24). A pair of croppers located above the pivot is used to shear larger wire.

Diagonal Cutting Pliers

The jaws on diagonal cutting pliers are offset by about 15 degrees. They are not used to hold or grip objects. Diagonal cutting pliers are used for making flush cuts on small, light materials such as wire, cotter pins, and similar objects (Figure 4-25).

Multipurpose Wire Strippers

Wire strippers are used to strip insulation from electrical wires (Figure 4-26). When closed around the wire, only the insulation is cut, leaving the wire itself intact (Figure 4-27). They can also be used to cut the wire (Figure 4-28).

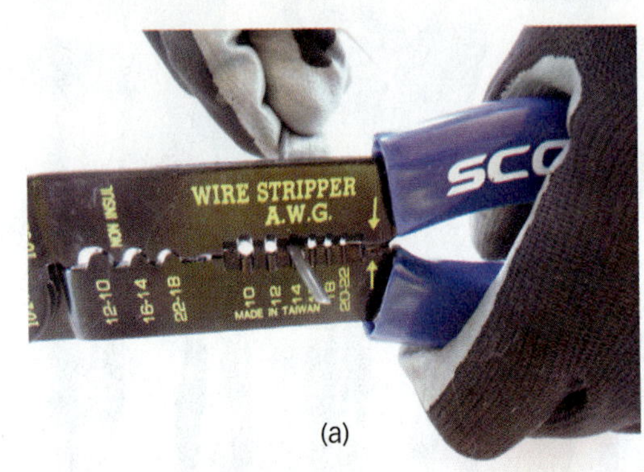

(a)

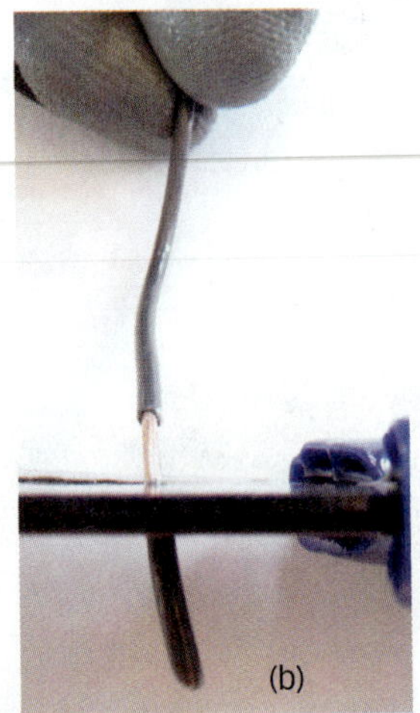

(b)

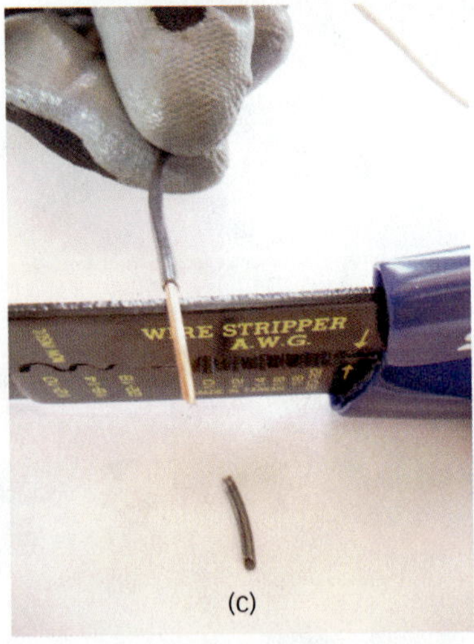

(c)

Figure 4-27 (a) Preparing to strip 14-gauge electrical wire; (b) squeeze handles and pull wire; (c) wire is stripped.

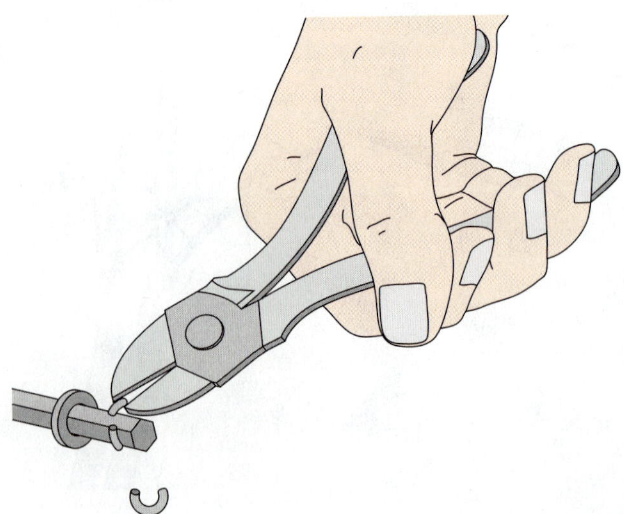

Figure 4-25 Snipping cotter pin with diagonal cutting pliers.

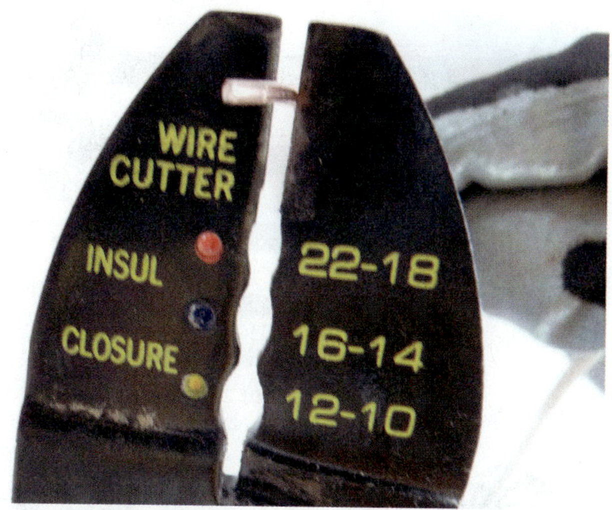

Figure 4-28 Cutting stripped electrical wire to length.

4.7 WRENCHES

The term *wrench* is used to describe tools that grip and turn threaded parts such as nuts, bolts, valves, and pipes. Some wrenches are adjustable, so they will fit a range of sizes of nuts, bolts, or pipes, as shown in Figure 4-29. Other wrenches have fixed sizes and come as sets to fit common sizes, as shown in Figure 4-30. Some wrenches are specifically designed for HVACR jobs, such as the service valve wrench as shown in Figure 4-31.

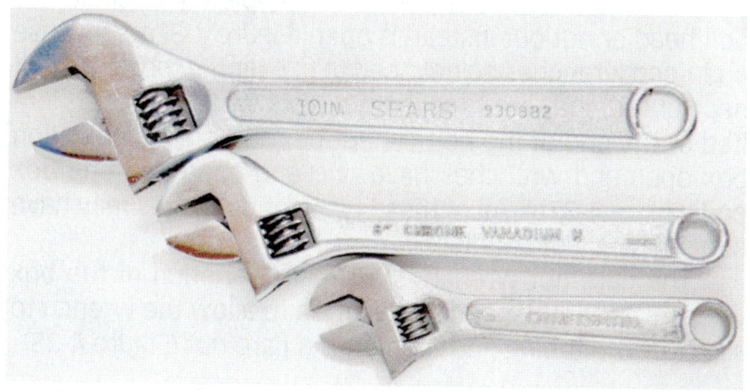

Figure 4-29 Adjustable wrenches come in a variety of sizes.

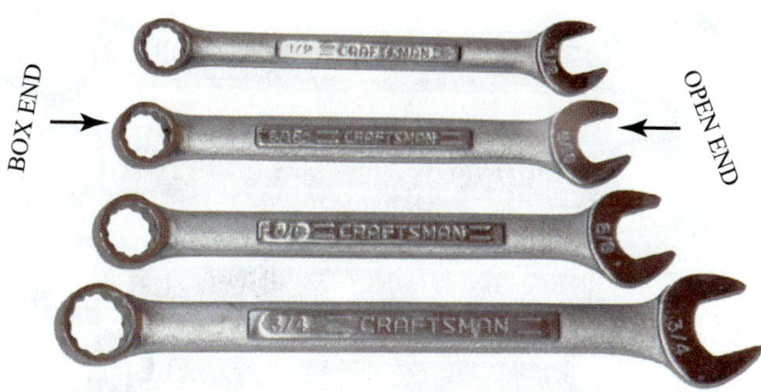

Figure 4-30 Combination wrenches: one end is open and the other end is boxed.

Figure 4-31 Refrigerant service valve wrenches have square openings.

Service Valve Wrenches

Service valves on some compressors and condensers are ¼-in (6 mm) square stems and require a special service valve wrench. Service valve wrenches come in several styles and shapes. Some service valve wrenches use a ratchet mechanism. The ratchet may be fixed in one direction so the wrench has to be removed and turned over to change directions. Other service valve wrenches may have a lever for reversing the rotation of the wrench as required for either opening or closing the valve (Figure 4-32). Reversible-ratchet service valve wrenches can have up to four fixed sizes of square openings, ranging from $^{3}/_{16}$ in, ¼ in, $^{5}/_{16}$ in, and $^{3}/_{8}$ in (4 mm, 6 mm, 8 mm, and 9 mm) (Figure 4-33). These service valve wrenches look similar to the ratchet wrenches used by auto mechanics, but those wrenches have hex openings to fit the heads of standard nuts and bolts.

The handles on service valve wrenches can be flat or offset. The offset gives your hand a little more clearance in tight places (Figure 4-34). Most HVACR technicians carry

Figure 4-32 Ratchet service valve wrench.

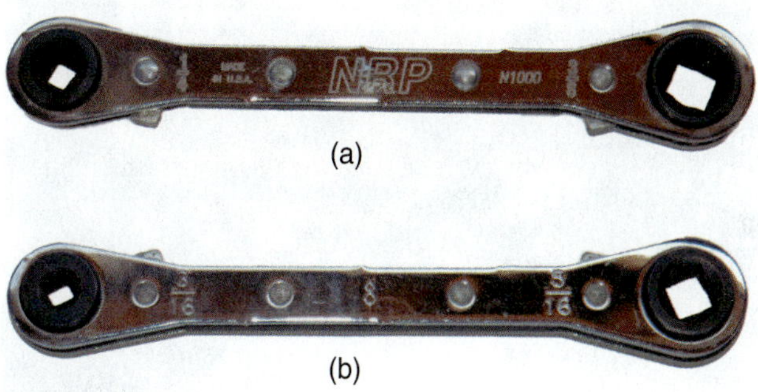

Figure 4-33 Some ratcheting refrigerant service valve wrenches provide four different-sized openings by having two different sizes on each end: (a) ¼ and ⅜ in sizes; (b) $^{3}/_{16}$ and $^{5}/_{16}$ in sizes.

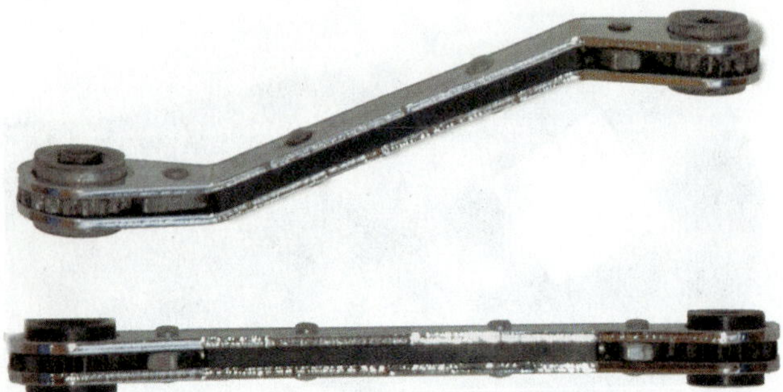

Figure 4-34 Refrigerant service valve wrenches are available with offsets to provide some hand clearance when using the wrench.

more than one type of service valve wrench to meet the varying job needs.

Box-End, Open-End, Combination, and Flare-Nut Wrenches

Box-end wrenches completely encircle the bolt head or nut and are strong and are resistant to slipping. They are ideal for working on long-shaft bolts (Figure 4-35) or tight spaces where a socket would not easily fit (Figure 4-36). Another very common wrench, the open-end wrench, is appropriately named because it does not completely encircle the bolt head or nut but instead is open (Figure 4-37a). Because open-end wrenches do not need to fit over the top of the bolt head or nut, they are excellent for work where access is limited or the end of the bolt cannot be reached. Combination box open-end wrenches have one end open and one box end (Figure 4-37b). All of these types of wrenches may have straight or offset handles.

The flare-nut wrench is a special variation of the box wrench in that the heads are slotted to allow the wrench to slip over the tubing and then onto a flare nut (Figure 4-38).

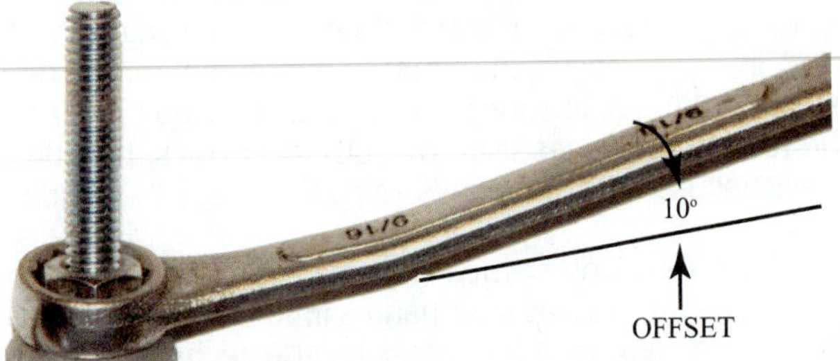

Figure 4-35 Offset box-end wrench.

Figure 4-36 A box wrench is used because the compressor process tube would interfere with a socket wrench.

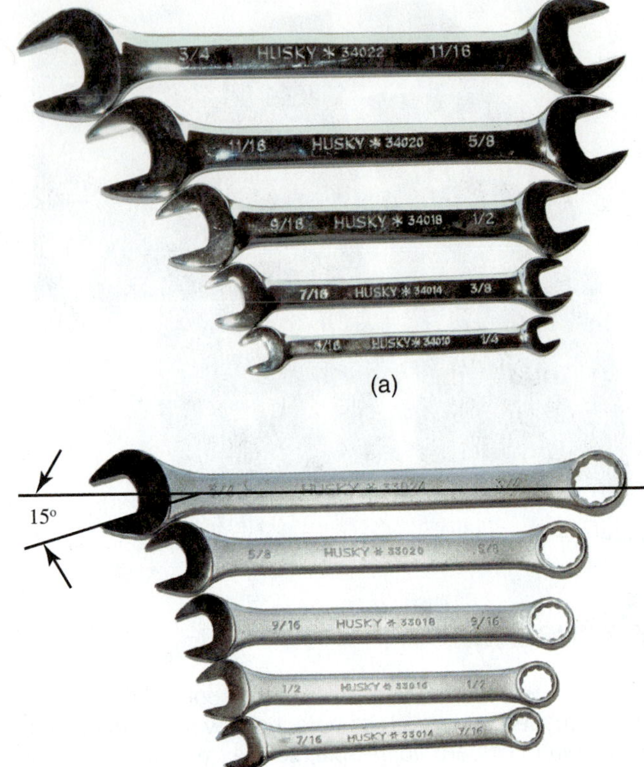

Figure 4-37 (a) Open-end wrench with 15-degree offset openings, both ends; (b) combination box open-end wrenches.

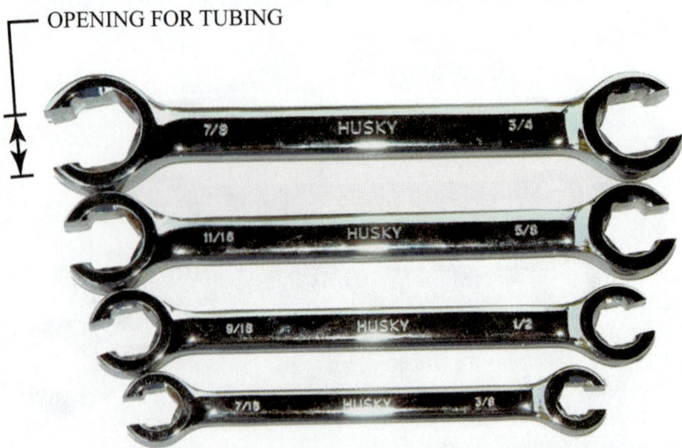

Figure 4-38 Tubing wrenches.

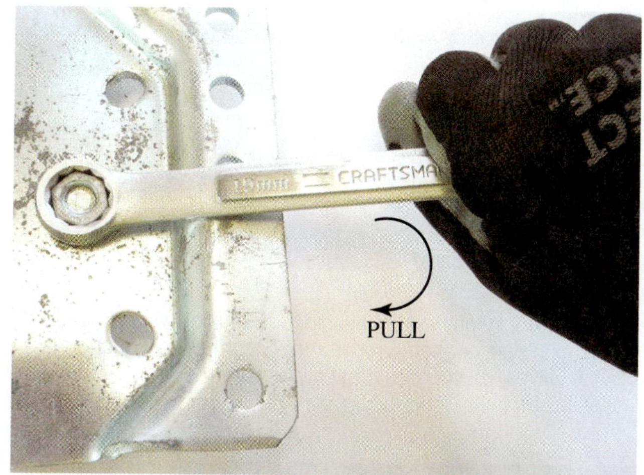

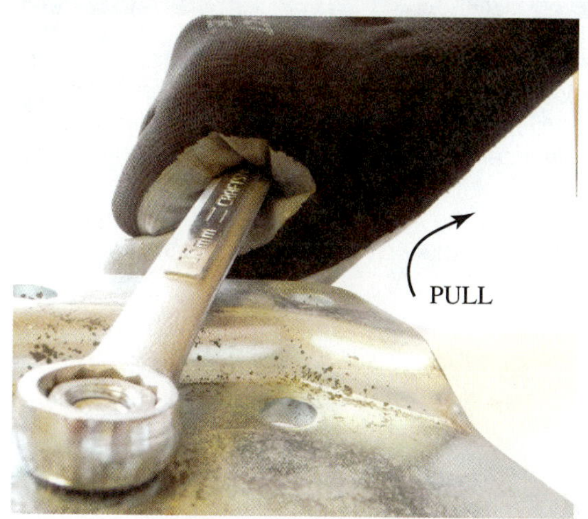

Figure 4-39 Pulling the wrench provides for better control.

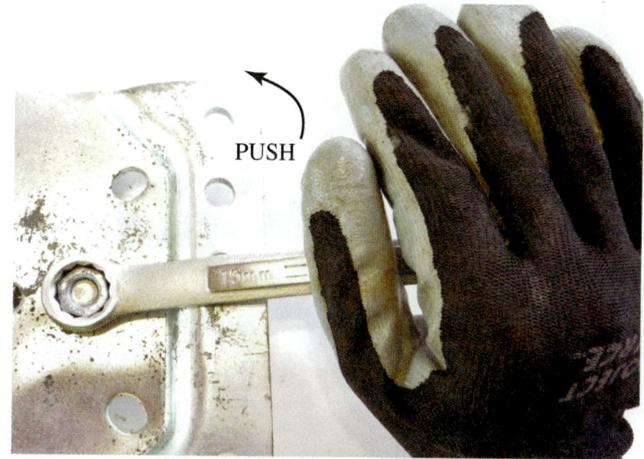

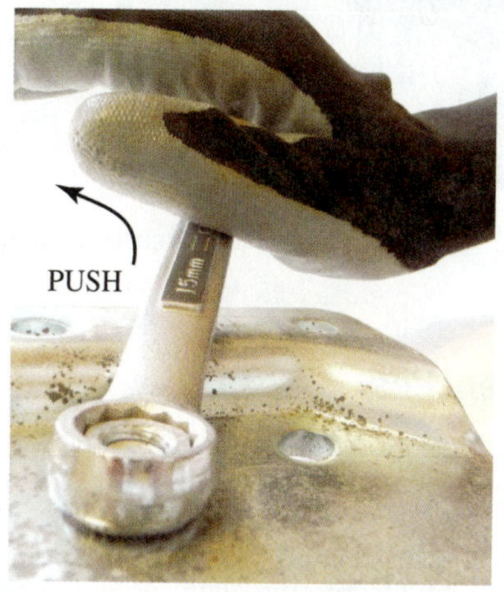

Figure 4-40 Push the wrench with your palm to protect your knuckles.

SERVICE TIP

Wrenches should always fit securely over the bolt head or nut before force is applied. If the wrench slips, you will jam your fingers. It is always best to try to pull rather than push on a wrench (Figure 4-39). You will have more control and will be less likely to jam your fingers or rap your knuckles. If you must push, then use an open palm so you're your fingers do not become jammed between the wrench and the work as the wrench begins to turn (Figure 4-40).

Adjustable Wrenches

Crescent Tool Company was the first to introduce the adjustable wrench (Figure 4-41). The adjusting wheel permits fitting the flat to any size object within the maximum and minimum opening. Always use this type of wrench in a manner such that the force is in a down or counter-clockwise direction with the movable jaw on the bottom when loosening a bolt. This keeps the force against the fixed head so the wrench is less likely to suddenly slip and injure you (Figure 4-42). Adjustable wrenches are available in sizes from 4 in to 12 in (102 mm to 305 mm).

Figure 4-41 Adjustable wrench.

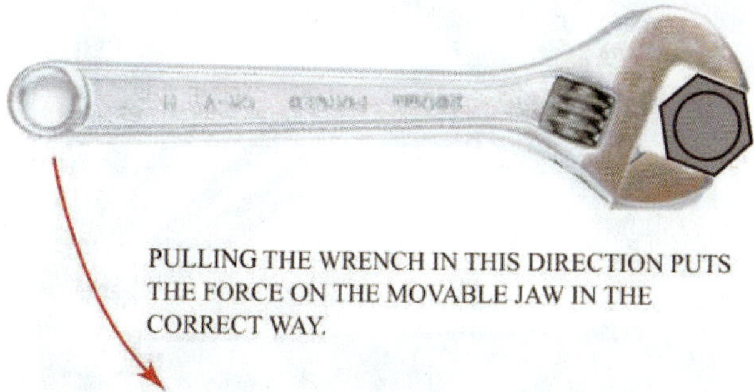

Figure 4-42 Always pull an adjustable wrench so that the force is correctly applied to the movable jaw and the wrench is less likely to slip.

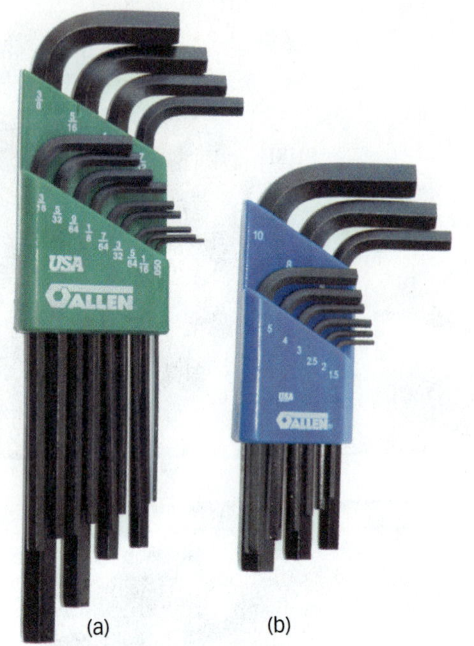

Figure 4-43 (a) Standard size Allen wrenches; (b) metric Allen wrenches.

Allen Wrenches

Allen wrenches often come in sets. Some are short, referred to as standard Allen wrenches, and others are long. The sets should be marked as to whether they are standard or metric sizes (Figure 4-43). The hexagonal shaft of the Allen wrench fits into the Allen screw typically used on fan pulleys, fan blade hubs, and other components. Allen wrenches are also available in a pocket set (Figure 4-44). There are some Allen wrenches that have balled ends that allow the wrench to be at an angle and still go inside the set screw. Another specialty Allen wrench is the one used with the refrigerant ratchet wrench to access the service valves on most condensing units (Figure 4-45).

Figure 4-44 Pocket Allen wrench set.

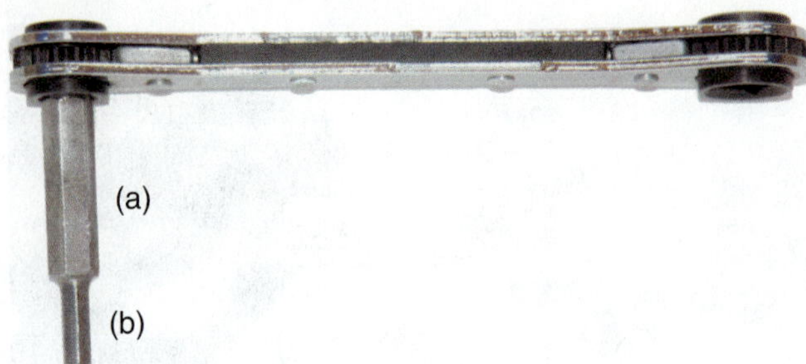

Figure 4-45 Allen wrench designed to fit in service ratchet wrench: (a) part of Allen wrench for accessing most suction-line service valves; (b) section of Allen wrench for accessing liquid-line service valves.

SERVICE TIP

Allen wrenches are often used to loosen and tighten pulley set screws. These screws are often very tight and difficult to remove. Make certain that the Allen wrench tip is fully inserted to the complete depth in the set screw before trying to tighten it or loosen it (Figure 4-46). Failure to insert the Allen wrench to its full depth can result in damage to the wrench or to the set screw.

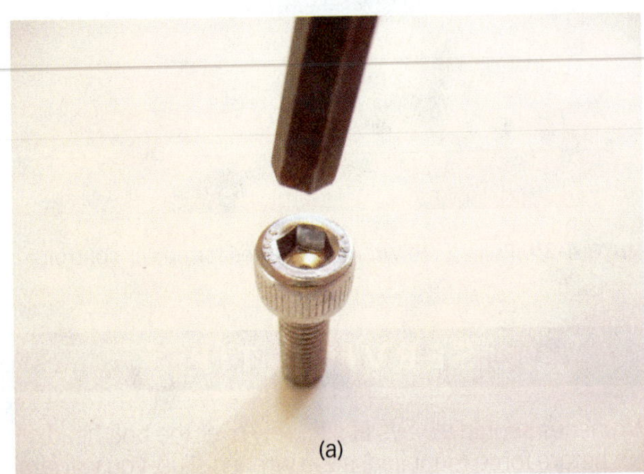

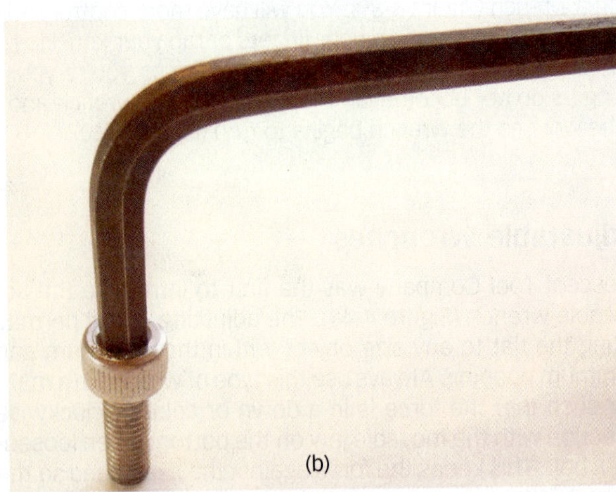

Figure 4-46 (a) Allen wrench and set screw; (b) the Allen wrench must be fully inserted into the set screw before turning.

4.8 SOCKET WRENCHES

Socket wrenches are made to slip over the heads of bolts and nuts (Figure 4-47). They are made of tool steel and usually come as sets in either standard or metric sizes.

Although standard sockets are available in sizes ranging from $\frac{1}{16}$ in up to several inches in size, the most common set would have sockets sizes starting with ¼ in up to 1 in by ⅛ in or $\frac{1}{16}$ in increments (Figure 4-48). Common metric socket sets range from 6 mm to 25 mm.

Both standard and metric socket sets for standard hexagonal nuts or bolts can be either six point or twelve point (Figure 4-49). The six-point sockets are stronger, but the twelve-point sockets allow easier alignment and shorter swing for tight locations. Swivel sockets and universal joints are also useful for reaching bolts that are hard to get at (Figure 4-50).

Socket Handles

A number of designs of socket wrench handles, referred to as socket handles, are available. Ratchet socket wrench handles are the most common type of handles because they can be used to quickly loosen and remove or install and tighten nuts and bolts. Some ratchets have swivel heads that are helpful for working in tight places (Figure 4-51). There are two dimensions used to refer to socket handles: length and size. The length refers to the distance from the tip of the handle to the head. The size refers to the dimension of the square shank that fits into the socket. The three common sizes for this shank are ⅛ in, ¼ in, and ⅜ in. Adapters are available to allow larger or smaller shank sizes to be used with different sized socket sets (Figure 4-52).

Breaker-bar socket wrench handles are used when a great deal of force is required to remove a stuck or stubborn bolt or nut. They are designed to take significantly more force than ratchets, but never use a cheater bar or pipe to extend the handle for more leverage, because this could damage the wrench.

Figure 4-47 The socket slides over the heads of nuts and bolts.

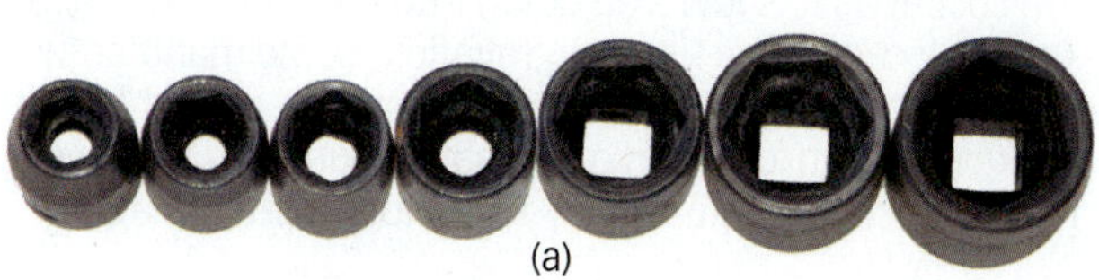

(a)

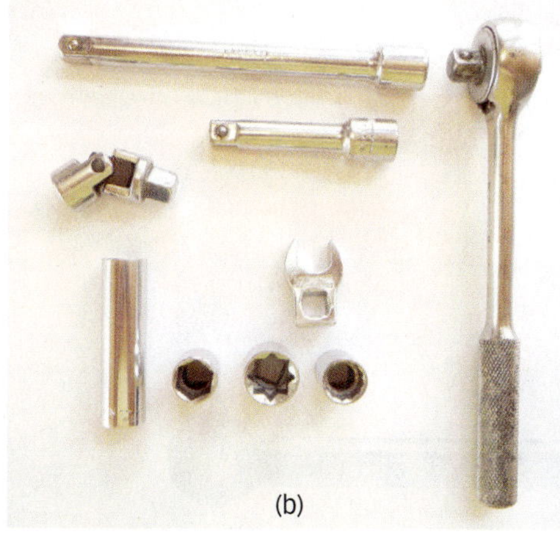

(b)

Figure 4-48 (a) Socket set. (b) Sockets with ratchet and extensions.

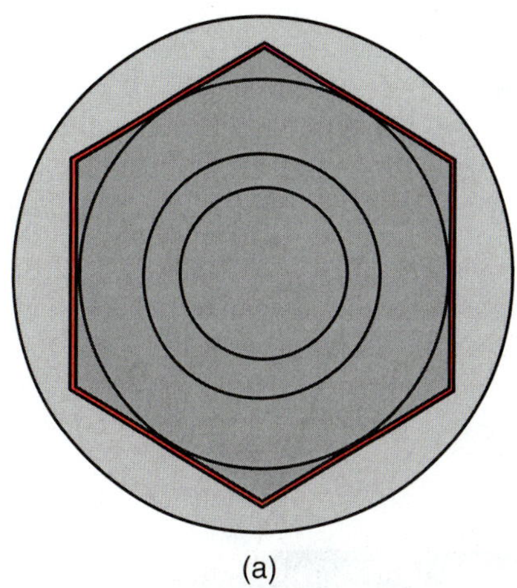

(a)

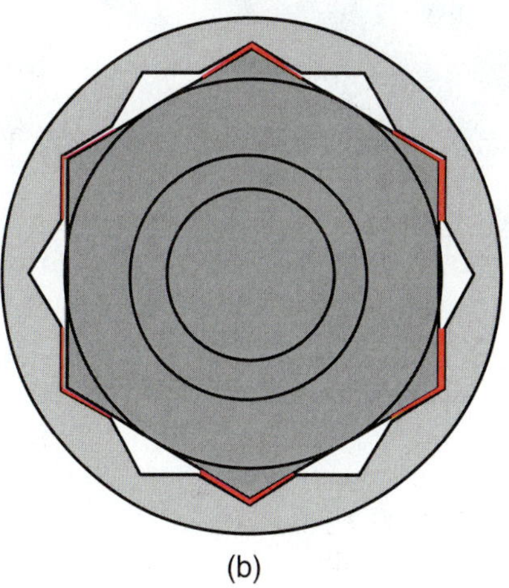

(b)

AREA OF CONTACT BETWEEN SOCKET AND BOLT

Figure 4-49 (a) Six-point sockets provide greater contact with the head of the nut or bolt than (b) twelve-point sockets.

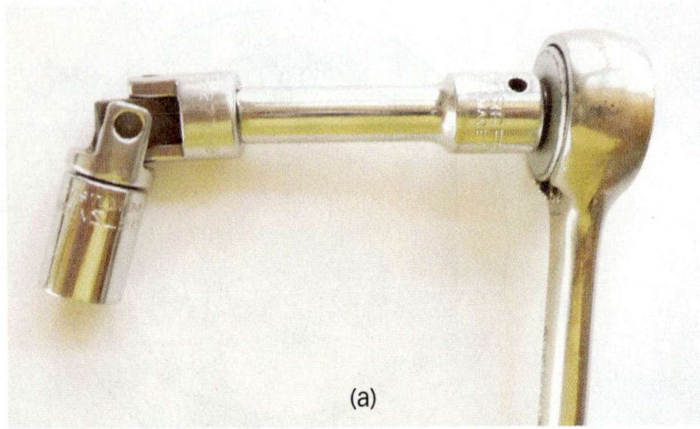

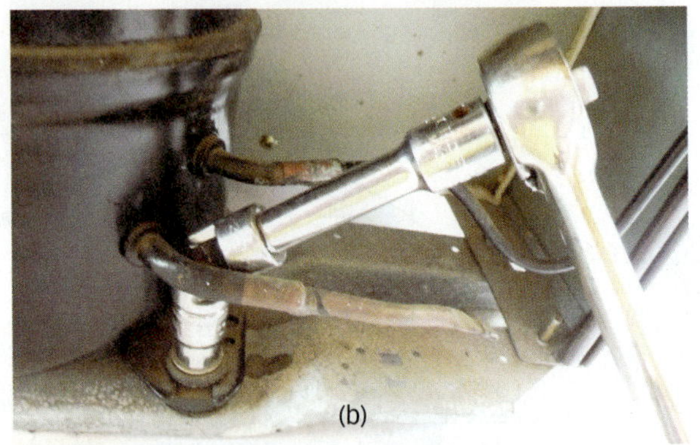

Figure 4-50 (a) Universal joint; (b) the universal joint allows the extension to be used at a slight angle.

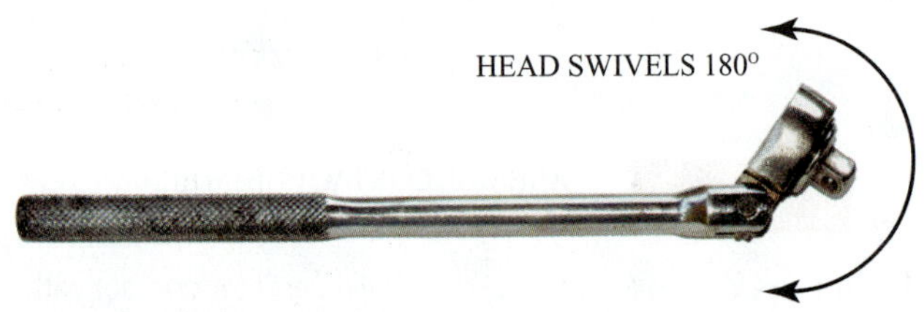

Figure 4-51 Swivel-head ratchet handle.

Figure 4-52 Socket wrench adaptors are available that permit more than one size socket to be used with different-sized wrenches.

SERVICE TIP

From time to time it is necessary to use an extension on a socket to reach way down inside a piece of equipment to remove or install a bolt or a nut. For example, the mounting bolts on a compressor inside of a residential condensing unit may require a foot-long extension or more. A problem can exist when you try to reach back in this tight space to reinstall the bolts. There may not be room for your hand and the wrench. However, if you use a small piece of paper, possibly folded one or two times, and place it over the head of the bolt before pushing the head into the socket, this can be used to hold the bolt or nut in the socket as it is lowered down into the confined space. This trick can save a lot of aggravation. Some manufacturers do provide small magnets that can be placed in the sockets, which can serve the same purpose. However, the paper trick works when you do not have the magnets.

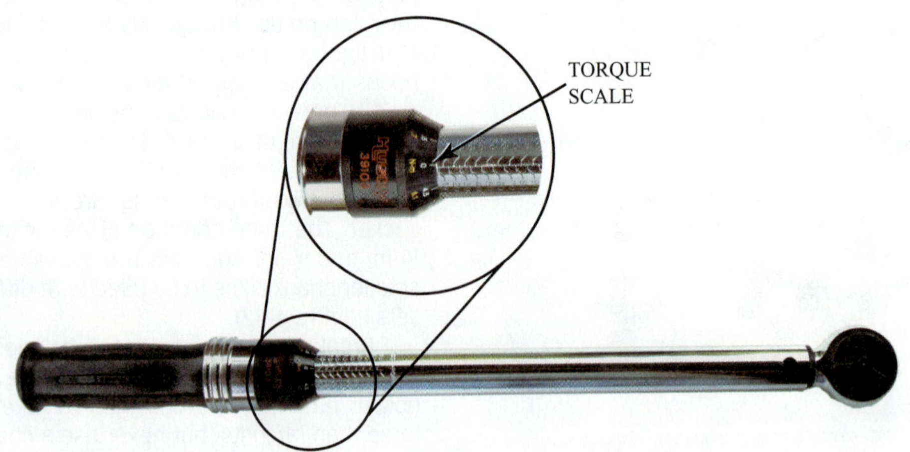

Figure 4-53 Torque wrench.

4.9 TORQUE WRENCHES

Torque wrenches have a gauge to measure the force being applied. The dial on the wrench indicates the amount of pressure being applied to turn the bolt (Figure 4-53). Torque is a twisting action. Torque is measured in inch pounds (in lb), foot pounds (ft lb), or newton meters (Nm) as applied to the wrench handle. Making sure the same force is used on nuts and bolts is important, especially when tightening the heads on compressors. Uneven tightening can result in the head being warped or bolts being stripped or broken. Either case can render the compressor unusable or more expensive to repair. Typical specifications will give both the torque and the tightening sequence (Figure 4-54). A good torque practice is to first apply only a light torque to each bolt in the recommended sequence. Then repeat the tightening sequence and each time increase the torque slightly until the required torque is reached.

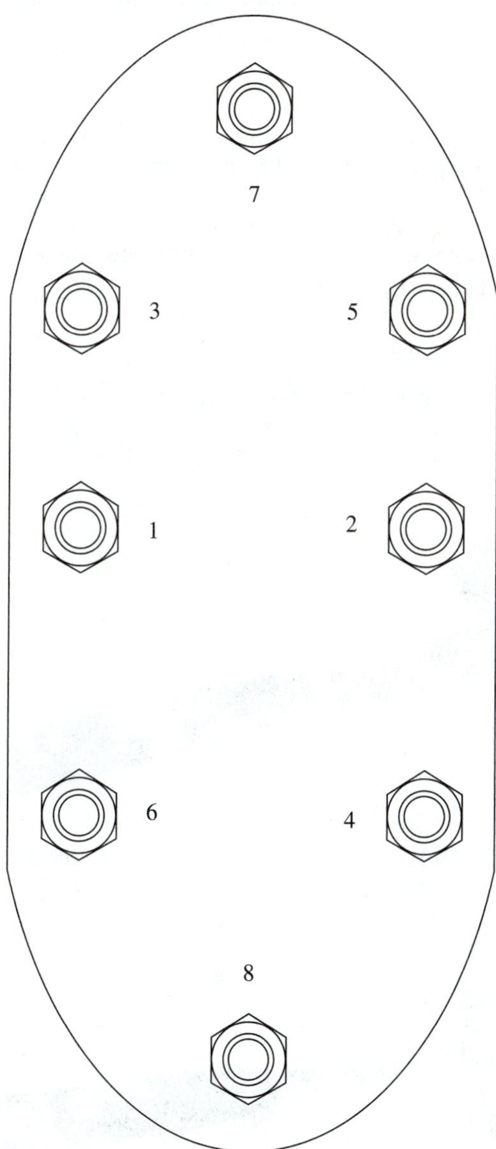

Figure 4-54 Typical head-torquing sequence—check with the manufacturer for specific torquing sequences and torque specifications.

Figure 4-55 Pipe wrench.

4.10 PIPE WRENCHES

The pipe wrench is used in refrigeration installation and service work to assemble or disassemble threaded pipe (Figure 4-55). At least two sizes are recommended. An 8 in (203 mm) wrench can handle up to 1 in (25 mm) diameter pipe, and a 14 in (356 mm) size can handle up to 2 in (50 mm) diameter pipe (Figure 4-56). Some have replaceable jaw inserts to extend the life of the tool. The chain wrench is another form of adjustable pipe wrench. This wrench can make work easier in a confined area or on round, square, or irregular shapes.

4.11 HAMMERS, PUNCHES, AND CHISELS

There are a wide variety of hammers and mallets used in the HVACR trade. Some of the more common hammers used are ball-peen hammers, sheet-metal hammers, claw

Figure 4-56 10-in, 14-in, and 24-in pipe wrenches.

Figure 4-57 The proper way to hold a hammer is near the end of the handle.

hammers, mallets, and sledge hammers. The proper way to hold any hammer is near the end of the handle (Figure 4-57). The wrist and arm motion that you use will depend on how hard you must strike the object. A light tap is almost entirely wrist motion. Heavy blows come from the wrist, forearm, and shoulder.

Ball-Peen Hammer

The ball-peen hammer is sometimes referred to as a machinist's hammer. These hammers are used for some sheet-metal work (Figure 4-58) and can be used when striking metal surfaces.

Sheet-Metal Hammers

The flat end of a sheet-metal hammer is used to close Pittsburgh joints and drive in a drive cleat, bend the end of a drive cleat, and other similar tasks. The tapered end is sometimes used to open up the folded edge of a drive cleat or to unbend a seam (Figure 4-59). These hammers can be used to strike metal surfaces.

Claw Hammer

The claw hammer is commonly used in carpentry work and should never be used to strike hard metal surfaces. The hammer may be damaged or the face may be chipped if used to strike metal. The claws are used for pulling nails (Figure 4-60).

Mallets

Mallets are generally larger in diameter than hammers and may or may not have hard faces. Some mallets are made of wood, rawhide, plastic, rubber, or other similar material (Figure 4-61). Mallets are used when it is important that the surface not be damaged, such as driving shafts, tapping large pieces of equipment into place, or other similar tasks.

Sledge Hammers

Sledge hammers may have long or short handles and are used when heavier strikes are required to dislodge stuck large motor shafts, break up concrete, and so on (Figure 4-62).

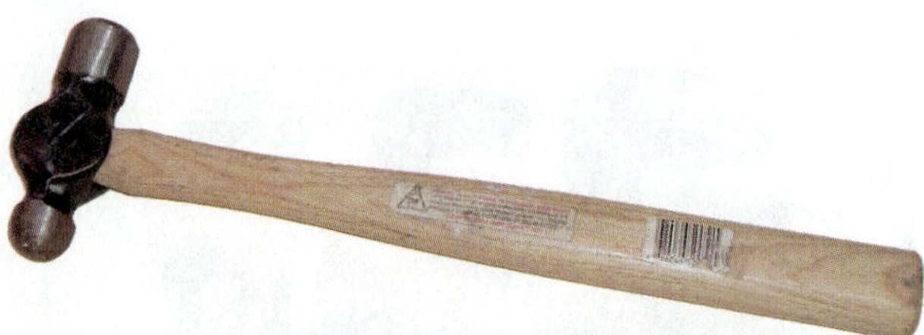

Figure 4-58 Ball-peen hammer. *(Courtesy of Terra Community College HVAC Program, Fremont, Ohio)*

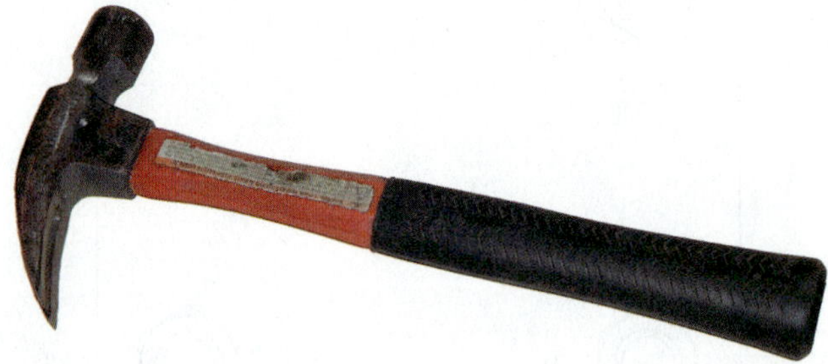

Figure 4-60 Claw hammer. *(Courtesy of Terra Community College HVAC Program, Fremont, Ohio)*

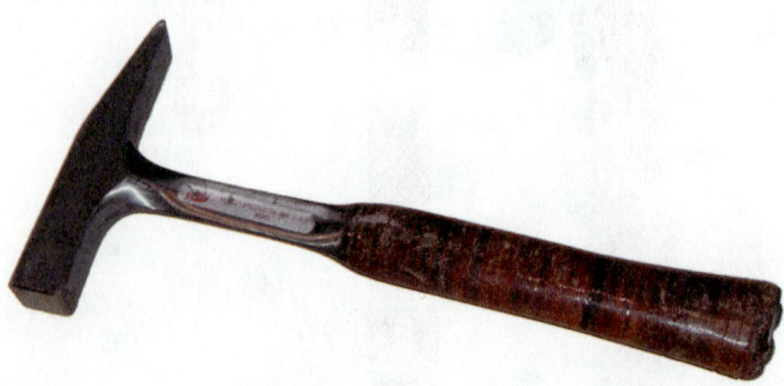

Figure 4-59 Sheet-metal hammer. *(Courtesy of Terra Community College HVAC Program, Fremont, Ohio)*

Figure 4-61 Wood and rubber mallet. *(Courtesy of Terra Community College HVAC Program, Fremont, Ohio)*

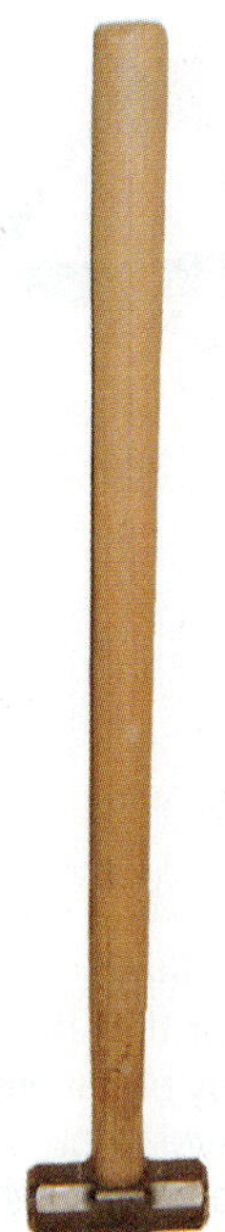

Figure 4-62 Sledge hammer. *(Courtesy of Terra Community College HVAC Program, Fremont, Ohio)*

Chisels and Punches

Chisels and punches are usually struck with a hammer. Strike the tool squarely and on center to prevent the hammer from glancing off. There are a large variety of chisels. Machinist's chisels are designed to cut and shape cold metal. The cape chisel is used for cutting keyways or slots in metal. The rivet buster is used for cutting rivets. The end of the chisel that is struck by the hammer may become mushroomed over time. When this happens the chisel should be replaced because as the end mushrooms, pieces of metal may fly off when the chisel is struck by the hammer.

Punches are used to mark metal, drive pins, and align holes. Drift punches are generally used to remove shafts, pins, and rivets and to align small parts. An alignment punch helps to line up mating parts for assembly. The drive punch has a flat tip for driving nails or pins, while the prick punch has a pointed tip to mark metal (Figure 4-63).

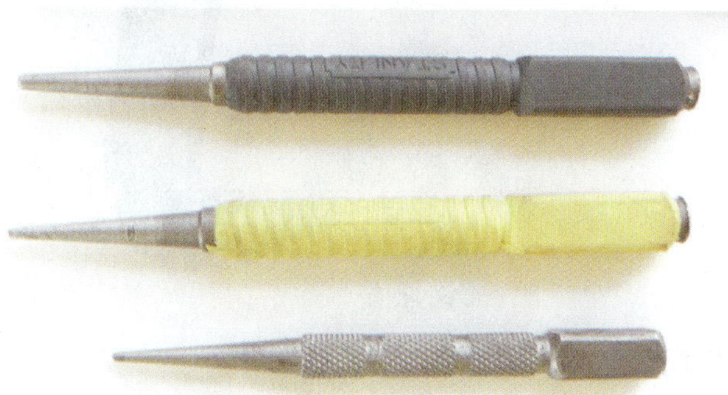

Figure 4-63 Handheld drive and prick punches.

4.12 HACKSAWS AND FILES

The hacksaw is used to cut metal. There are two types of blades: hard and flexible. The blade teeth are designed to cut in one direction only, and the blade will be marked to show which end is connected near the saw handle (Figure 4-64). Cutting is accomplished as you push the saw, not when you pull back (Figure 4-65). Some hacksaws are designed for tight spaces and hold the blade differently than conventional hacksaws (Figure 4-66).

Files come in several shapes: flat or rectangular, round, half round, triangular, square, and so on. In refrigeration work, the common flat file and the half round are used in preparing tubing for soldering by squaring the end or for removing burrs.

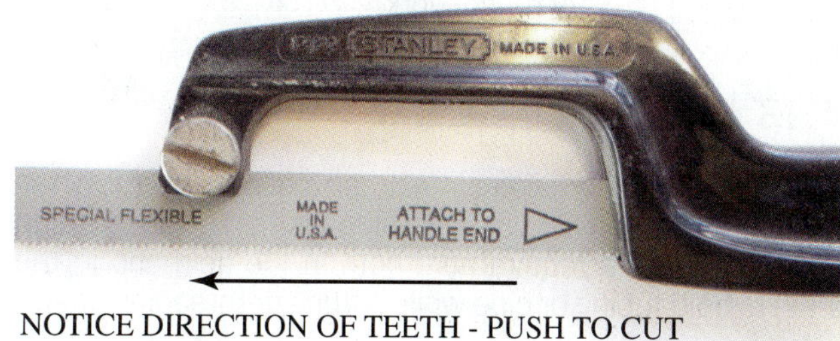

Figure 4-64 Install a hacksaw blade with the teeth pointed in the proper direction.

Figure 4-65 The hacksaw will cut as you push.

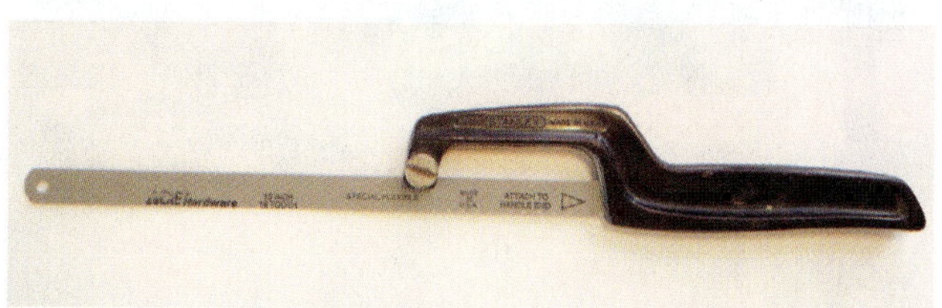

Figure 4-66 Smaller hacksaws can cut in areas where larger hacksaws cannot reach.

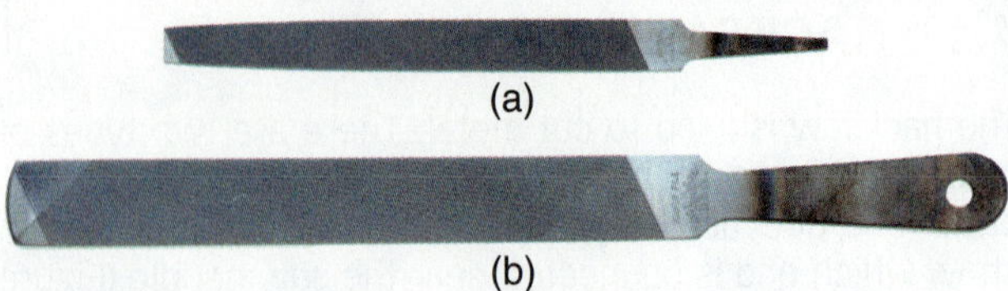

Figure 4-67 (a) This single-cut file will require its handle to be attached before safe use; (b) This cross-cut file has its handle formed as part of the file.

Files can be either single cut or cross cut (Figure 4-67). A single-cut file is used for finishing a surface, such as in preparing copper pipe for soldering. A cross-cut file is coarser and would be used where deeper and faster metal removal is needed. A rasp is an extremely coarse cross-cut file intended for very rough work.

4.13 VISES AND CLAMPS

A machinist's vise (Figure 4-68) can be very useful when mounted in a service van. The pipe vise on a tripod stand (Figure 4-69) is a requirement both in the field and in the shop to hold tubing or pipe while cutting or threading operations are taking place. C-clamps are used to hold work that cannot be held in a vise or that has to be held for extended periods (Figure 4-70). They are available in a variety of sizes. The threads and the swivel should be periodically cleaned with a rag and lubricated with a light coat of oil.

Figure 4-69 Universal base with chain pipe clamp. *(Courtesy Reed Manufacturing Company)*

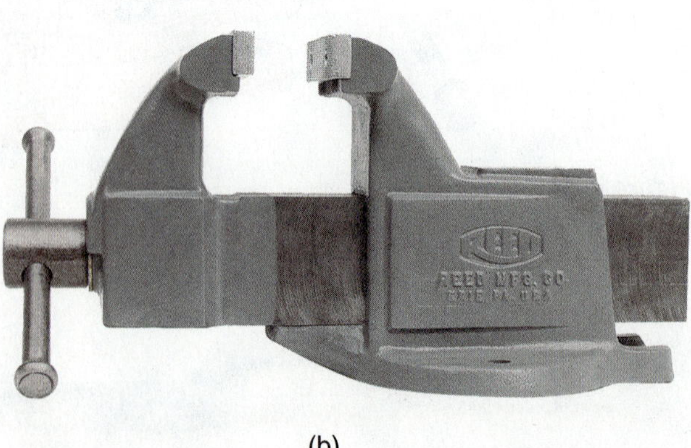

Figure 4-68 (a) This vise is designed to be allowed to swivel; (b) this vise is designed to be fixed in place. *(Courtesy Reed Manufacturing Company)*

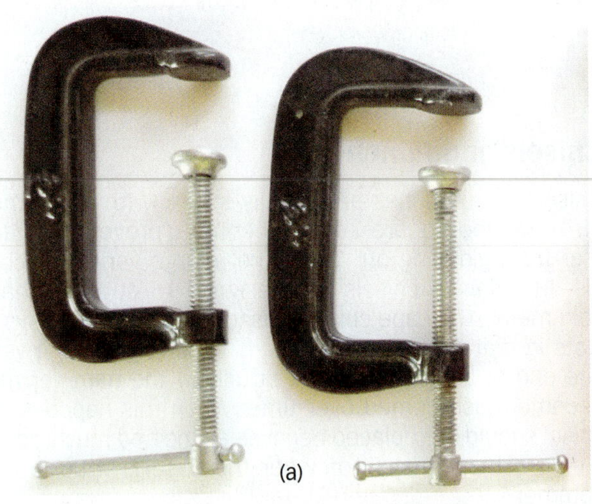

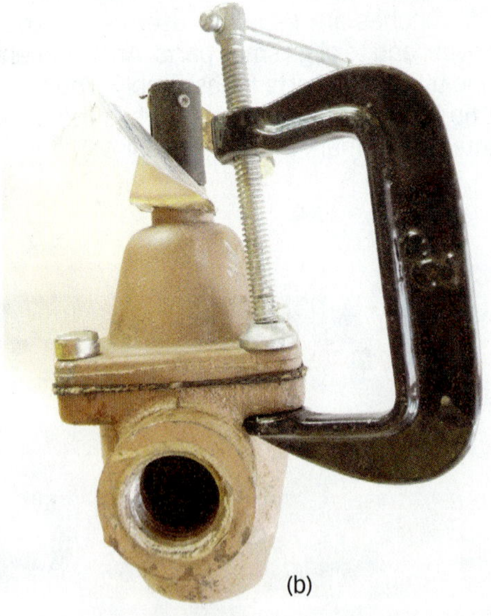

Figure 4-70 (a) C-clamps; (b) C-clamp holding bonnet of valve.

CAUTION

When holding copper or other soft metal, be sure that the jaws are soft so as not to mar the surface and that the clamping pressure does not squeeze the tubing out of round.

4.14 KNIVES AND SCRAPERS

Knives are universal tools that have many functions. There will always be times when you need to cut twine or open a cardboard box. A handy tool is the combination blade and razor knife (Figure 4-71). The razor held at one end of the knife is replaceable. When using a knife, always cut away from your body, not toward it. When not in use, the knife should be folded or put in a sheath. Never carry an open knife in your pocket.

Flat-blade scrapers or putty knives are useful for removing gaskets and cleaning flange faces (Figure 4-72). Be careful when scraping a surface not to slip. A sharp scraper edge is less likely to slip. Be careful not to scratch the surface of a flange face when scraping off an old gasket. When a scraper is not in use, it is a good practice to coat the blade with a film of light oil to prevent it from rusting.

4.15 TAPS AND DIES

Taps and dies are used to cut threads in metal, plastics, or hard rubber. Taps are for the internal (female) threads, and dies are for the external (male) threads (Figure 4-73). They are made of hardened steel and must be placed squarely over the stock or in the hole to prevent cross-threading. Rethreading dies (thread chasers) often come in handy to restore threads that are rusty or have been nicked or damaged. When using a thread chaser, make sure the tool turns on hand tight to start to prevent cross-threading. Once taps and dies become dull they cannot be resharpened and therefore must be replaced.

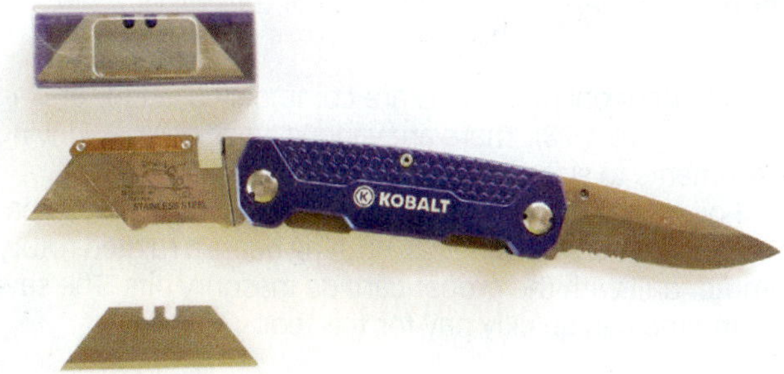

Figure 4-71 Combination razor and blade folding knife.

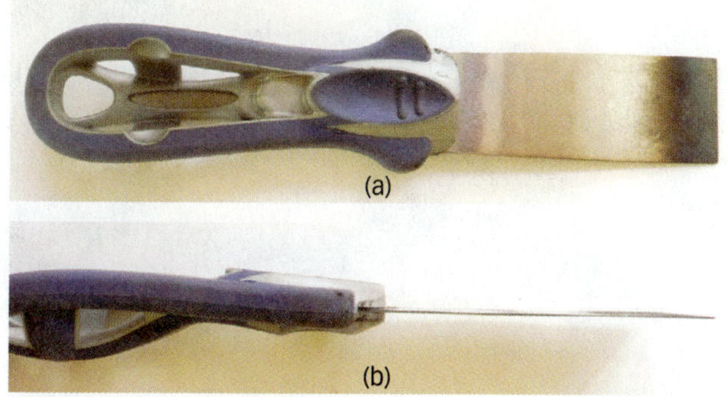

Figure 4-72 (a) Flat-blade scraper; (b) side view.

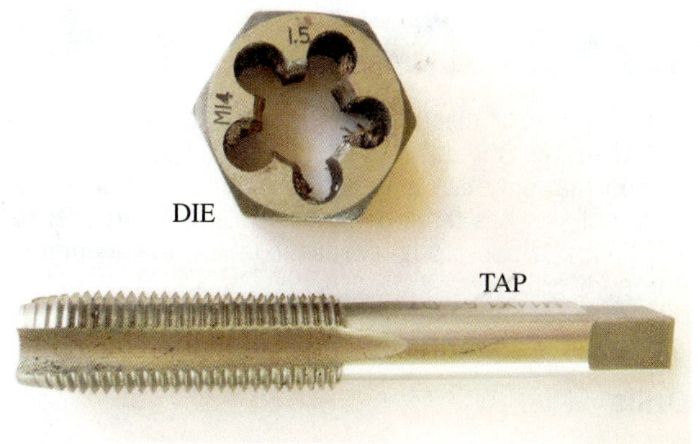

Figure 4-73 Tap and die.

4.16 POWER TOOLS

Safety

Never operate any power equipment unless you are completely familiar with its control and features. Inspect all portable power tools before using them. See that they are clean and in good condition with no frayed or cut wires. Make sure there is plenty of light in the work area. It is important to ensure that your tools are properly grounded. This means that the plug on the tool should have three prongs. Always make sure the tool is in the OFF position before connecting it to a power source.

Many power tools have built-in safety features. Take, for example, a battery-powered staple gun. You would not want the staples to accidently fly out in every direction if the trigger were accidently pressed. Most powered staple guns have a safety switch that will not allow the gun to operate unless it is held firmly against the work (Figure 4-74).

Figure 4-74 The power stapler safety switch must be depressed with stapler held in place on the work before the stapler will operate.

SAFETY TIP

Some portable electric tools provide for double-insulated casings and do not need to be grounded. This type of tool may only have a two-prong plug rather than three. If a tool plug has three prongs, never use an adapter to make it fit a two-prong receptacle unless the adapter is properly grounded.

Drills

Drills are frequently used to both drill holes and drive screws. Cordless drills with interchangeable batteries are ideal for HVACR work (Figure 4-75). A useful drill should have a variable speed control plus a reversing switch to back out stuck bits. Bit selection will depend, of course, on the nature of the material, but at a minimum, a set of high-speed alloy steel bits for metal is a must (Figure 4-76). Such bits can also be used for drilling wood and plastic. However, masonry bits and wood-boring bits may be added for more intensive job requirements. Drill guides are useful tools to quickly measure the diameter of a drill bit (Figure 4-77). In addition to drilling holes, battery-operated drills are commonly used for driving screws (Figure 4-78). There are various types of screwdriving attachments to suit many particular jobs (Figure 4-79).

For more than occasional drilling of concrete or masonry for anchors and shields, nothing beats a rugged rotary hammer drill with the proper carbide masonry bits. The savings in time can quickly pay for the tool.

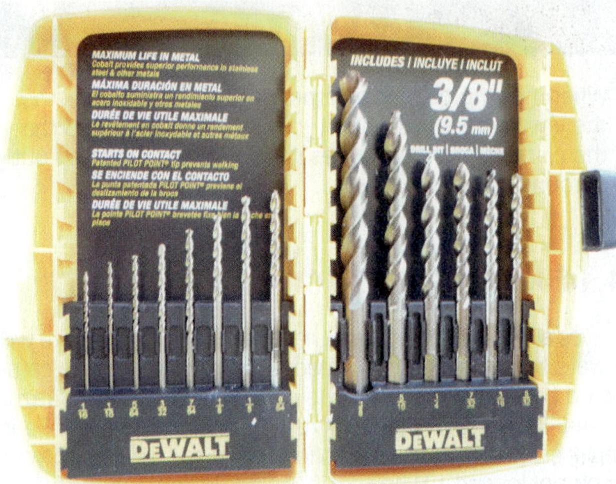

Figure 4-76 High-speed alloy steel drill bit set.

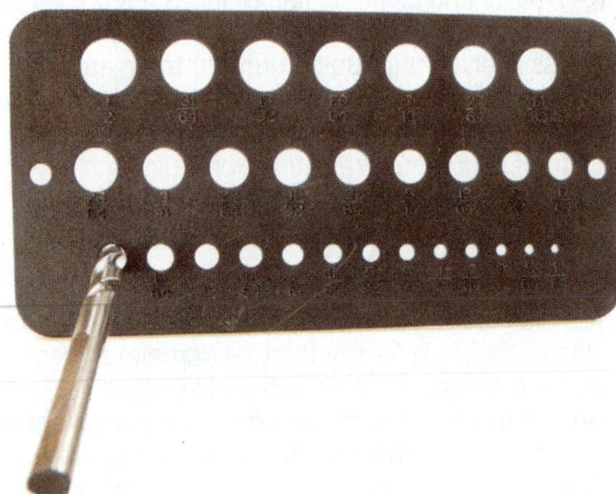

Figure 4-77 Measuring drill bit diameter with guide.

(a)

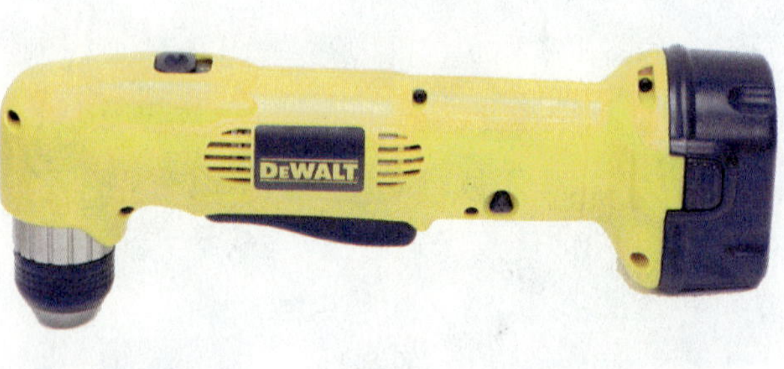

(b)

Figure 4-75 (a) Battery-operated electric drill; (b) right-angle battery-operated drill.

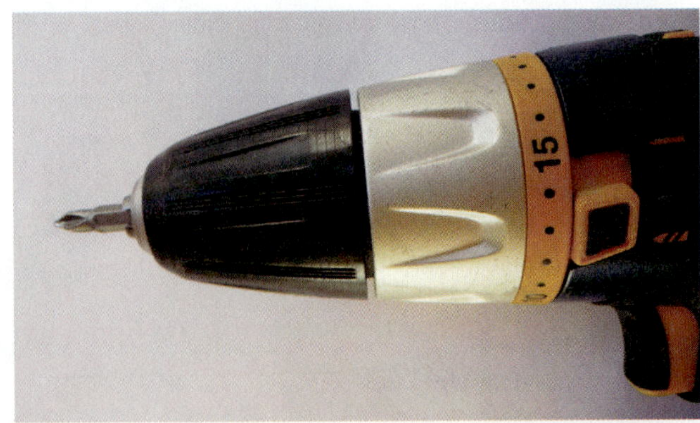

Figure 4-78 Battery-operated drill with Phillips-head driving attachment.

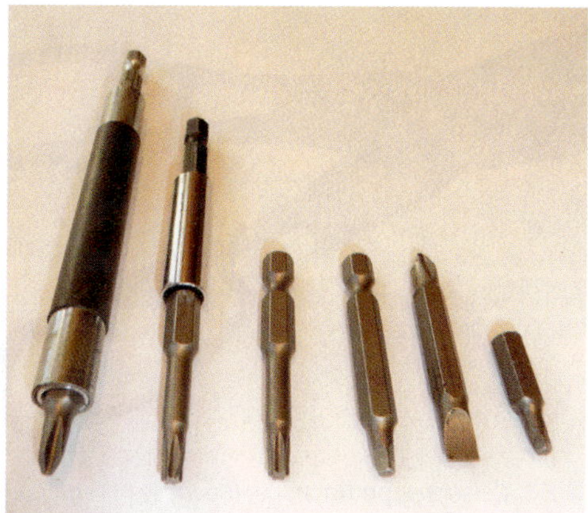

Figure 4-79 Example of different driving attachments for a drill.

TECH TIP

Power converters are devices that plug into a vehicle's 12V outlet. They convert 12V DC to 120V AC. Power converters are excellent additions to your service vehicle tools because they allow standard portable drill battery chargers to be used during the workday to keep your spare battery charged.

Bench and Hand Grinders

Bench grinders are used to sharpen tools, dress screwdrivers, and shape and smooth metal stock. A bench grinder will have two different types of grinding wheels: one is fine and the other is coarse (Figure 4-80). Aluminum, brass, and copper should not be used on the bench grinder unless the grinding stone is made for those materials. Bench grinders are equipped with Plexiglas safety shields and tool rests. The space between the grinding wheel and the tool rest is approximately 1/16 in, but never over 1/8 in (Figure 4-81). Always be sure that the wheel guards, tool rests, and shields are properly positioned before applying power. Stand to the side of the grinder when turning it on, and allow the machine to run for one minute before engaging the wheel with work. Never use a glazed, worn, or uneven wheel.

Figure 4-80 Bench grinder.

Figure 4-81 Always ensure the proper spacing between the grinding wheel and the tool rest.

Handheld grinders must be firmly held with both hands (Figure 4-82). For added safety, they are equipped with a spring-loaded switch that cuts off the power when the hand grip is released. Always use a vise or clamp to hold the work, and never hold the object to grind by hand. Eye protection must always be worn when using grinders.

Figure 4-82 Hold a hand grinder firmly with both hands.

Figure 4-83 The handy jigsaw can cut wood, plastic, and light metal.

Power Saws

There are a variety of different types of portable power saws. The most familiar may be the circular saw, which is good for new construction when cutting studs and other framing members. Also used in construction and renovation is the reciprocating saw with a single straight blade, which can be used to saw out holes in walls and remove existing framework. The portable electric band saw can be used for sawing PVC tubing and light-duty pipe. The handy jigsaw can be used to cut wood, plastic, and metal, depending on the type of blade inserted (Figure 4-83). Its small size and scrolling features for cutting irregular shapes can sometimes be useful in new installations or retrofits.

4.17 SHEET-METAL TOOLS

Often it is necessary for technicians to cut sheet-metal ductwork during installation or service. There are a variety of tools available that can make this much easier.

CAUTION

Always wear gloves when cutting or handling sheet metal.

Tin and Aviation Snips

Tin snips are used for making straight or circular cuts in sheet metal. The scissor-type tin snips have been used for years (Figure 4-84), but they have been replaced in recent years with the aviation-type tin snips (Figure 4-85). The aviation snip's hinged joint and serrated cutting edge have greatly improved the tin snip's use.

Aviation snips are available for right hand, left hand, or straight cuts. For example, the right-handed snips are designed so that the scrap material is bent up and away from the cut, leaving the right-hand side of the sheet metal flat and smooth. They can also be used to cut right-hand curves or circles. The left-hand snips are used for cuts in the opposite direction. Straight snips may bend the edges of the cut sheet metal slightly upward and downward.

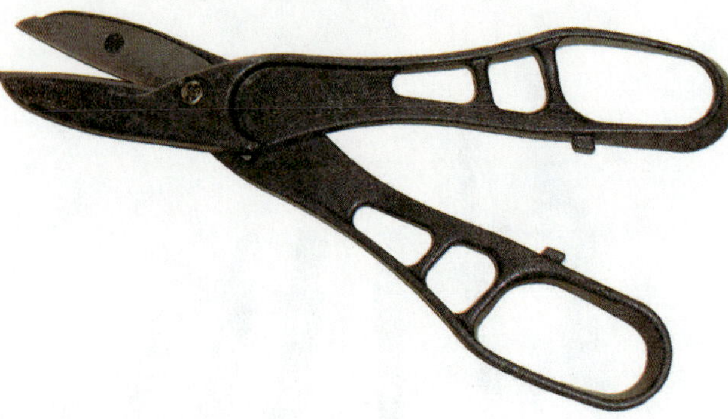

Figure 4-84 Scissor-type tin snips. *(Courtesy of Terra Community College HVAC Program, Fremont, Ohio)*

Figure 4-85 Aviation-type tin snips. *(Courtesy of Terra Community College HVAC Program, Fremont, Ohio)*

TECH TIP

Often the right-handed aviation snip's handles are green, the left-handed aviation snip's handles are red, and the straight aviation snip's handles are black. But these are not standardized colors and may vary from one tool manufacturer to another.

Offset-handled aviation snips are designed to keep your hand as far away from the sharp cut edge of the sheet metal as possible (Figure 4-86).

Nibblers

The nibbler is a powered hand tool that uses a circular punch and die to remove small crescent-shaped pieces of sheet

Figure 4-86 Offset-handled aviation tin snips. *(Courtesy of Klenk Tools, A Division of Everhard Products, Inc.)*

metal to produce a cut. Nibblers are available with fixed or swivel heads. The fixed head tool can be used for making straight cuts or can be guided around a curve or circle. A swivel head allows for easy following of curved lines or circles.

The nibbler leaves a very flat edge on the cut sheet metal, but it produces a small kerf of removed metal. A kerf is the gap being formed by the cutting tool. It is important that you provide for kerf width when laying out the dimensions for a cut to ensure that the part does not wind up being too small.

A nibbler cut must be started on an edge of the metal. To start a nibbler cut in the middle of a surface, a hole must first be drilled, so that the punch and die can be started on the edge of the hole.

CAUTION

Large quantities of sharp crescent-shaped chips are produced by a nibbler. Use a vacuum cleaner or brush to clean up the chips in your work area, because they can get stuck in your shoes and clothing. Be careful that chips stuck in your shoes do not damage your customer's floor. Be sure that chips stuck in your clothes do not scratch you.

Figure 4-87 Hand seamers. *(Courtesy of Terra Community College HVAC Program, Fremont, Ohio)*

Hand Seamers

Hand seamers are used to bend the edge of sheet metal (Figure 4-87). The edge of sheet metal is often bent to form a smooth edge, to make a slip joint, and/or to form a flange.

Duct Stretcher

A duct stretcher is used to pull a slip joint together when assembling rectangular sheet-metal duct. The tool's rollers (wheels) fit into the sheet-metal bends that form the slip joint and allow you to leverage them together so the drive cleat can be installed (Figure 4-88).

4.18 PIPING AND TUBING TOOLS

Unit 24, Piping and Tubing, describes the procedures and tools used for bending, cutting, flaring, and swaging in detail. Therefore, only a brief description of the tools is provided in this unit.

Tubing and Pipe Cutters

Soft copper tubing comes in rolls, while hard copper tubing comes in straight lengths. Tubing cutters can be used for either type of tubing and use a hardened steel rolling

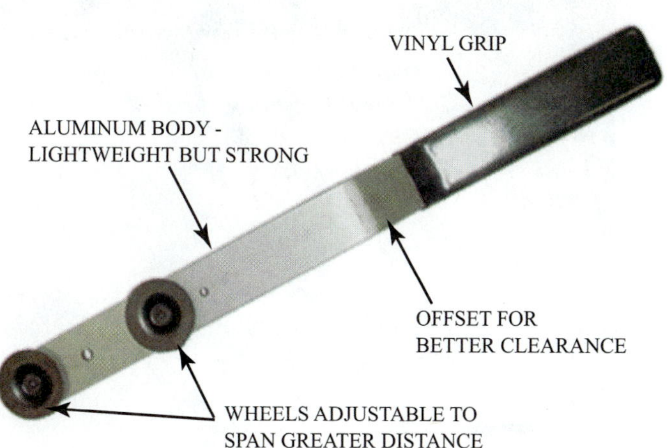

Figure 4-88 Duct stretcher. *(Courtesy of Klenk Tools, A Division of Everhard Products, Inc.)*

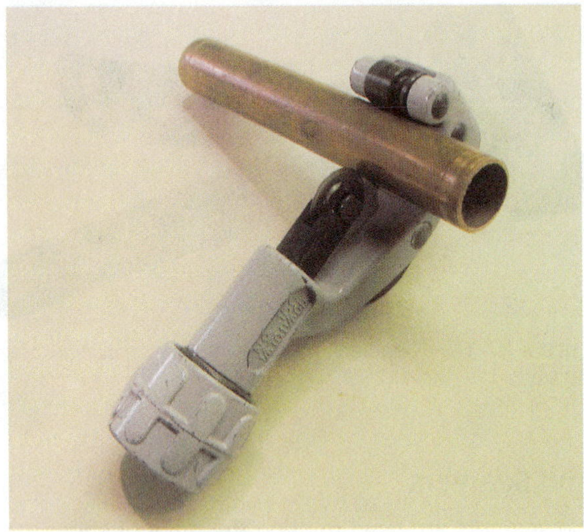

Figure 4-89 Tubing-cutter. *(Courtesy of Terra Community College HVAC Program, Fremont, Ohio)*

cutter (Figure 4-89). Pipe cutters leave a small burr inside the cut end of pipe and tubes. As the cutter is rotated around the pipe, the cutting tool does not remove metal but simply forces an ever-deepening groove into the pipe or tube surface until it has been cut through. This burr must be removed, and there are deburring tools made for this purpose.

There are a number of different types and sizes of tubing cutters. Standard tubing cutters will cut tubing ranging in size from ⅛ in to 1 in diameter. Larger tubing-cutting tools are available to cut tubing up to a 4-in diameter. Larger diameter tubing is usually cut with a portable band saw or other such power tool.

There are small diameter and compact cutters that can be used in tight or limited spaces (Figure 4-90). These cutters are a little more difficult to use, but they work very well

Figure 4-90 Compact tubing cutters 1/8 in to 1 1/8 in OD and 1/8 in to 5/8 in OD.

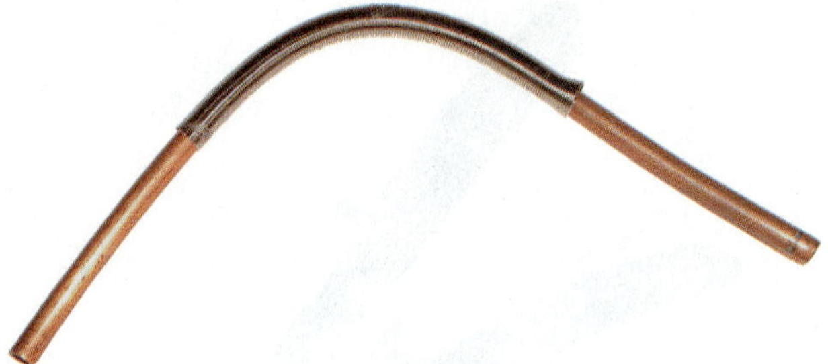

Figure 4-91 Spring-type tubing bender. *(Courtesy of Terra Community College HVAC Program, Fremont, Ohio)*

when cutting tubing that is close to other tubes or otherwise has restricted access. Another type of tubing cutter sits around the tubing, and a ratcheting mechanism turns the cutter around as the handle is moved back and forth.

Spring and Hand Benders

A spring bender is made up of a coil steel spring. It slips over the tube being bent and provides lateral support to the side of the copper tube as it is bent. This lateral support prevents the tubing from crimping by keeping the tube relatively round (Figure 4-91). Different-sized spring-bending tools are available to fit various-sized refrigeration tubing. Spring-bending tools are limited to smaller tubing sizes.

Hand benders are tools that can be used to make accurate short radius bends in copper tubing (Figure 4-92). These tools are available for various diameters of copper tubing and in several different radiuses.

Flaring Tool

The end of a tube may be flared outward so that it can be secured using a flare nut and fitting. The two most popular flare fittings used in HVACR work are the National Pipe (NP) and the Society of Automotive Engineers (SAE) standard thread fittings. For these fittings to be used, a flare must be formed at the end of the tubing. To form the flare end of

Figure 4-92 Copper tubing hand bender. *(Courtesy of Terra Community College HVAC Program, Fremont, Ohio)*

Figure 4-93 Flaring block.

the tubing, a flaring block, shown in Figure 4-93, is placed around the end of the tube to be flared. The flaring block has various-sized openings for each tubing size to be flared.

Swaging Tool

Swaging is sometimes used to join copper tubing, but it is not as popular as flaring. It is the process of joining two pieces of copper tubing together by brazing or soldering without the use of fittings. The end of one piece of copper tubing is expanded so that the other piece will fit into it. Hand-punch swaging tools of different sizes are shown in Figure 4-94.

Tubing Brushes

It is often necessary to clean oxides, oil, or dirt off of copper tubing so it can be soldered or brazed. Tubing brushes (Figure 4-95a) come in a variety of sizes for cleaning out the inside of fittings. Figure 4-95b shows three sizes of tubing

Figure 4-94 Swaging tool set. *(Courtesy of Northway's Machinery, Inc.)*

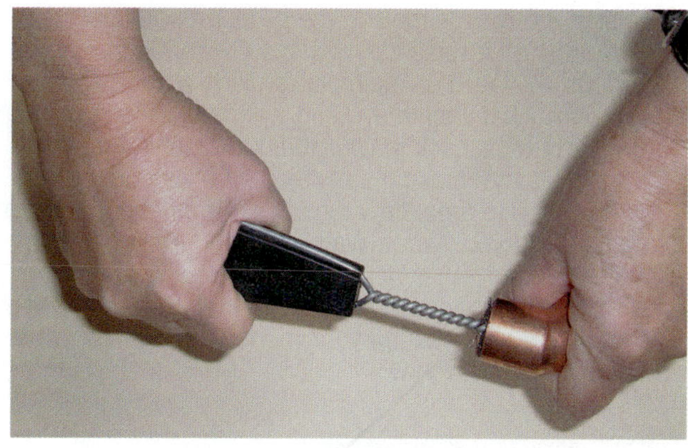

(a)

(b)

Figure 4-95 (a) Tubing brush; (b) three different-sized tubing brushes. *(Courtesy of Terra Community College HVAC Program, Fremont, Ohio)*

brushes. One end is designed to clean the outside of the pipe, and the other end is designed to clean the inside of the fitting. These are available in sizes to fit most standard air-conditioning and refrigeration tubing.

UNIT 4—SUMMARY

All HVACR installation and service technicians use a wide variety of tools daily in the course of doing their jobs. Tools are a significant investment, and keeping them in good working condition is important. These tools must be cared for as prescribed by their manufacturer's instructions so they work properly and provide years of dependable service.

Using the proper tool for the proper job is important, because misusing tools can damage the tool and may be hazardous to you. For example, striking a mushroomed chisel with a hammer may send fragments flying. Avoid putting dirty tools back in your tool kit. Cleaning the tools after each use will make the next job go smoother.

From time to time, new tools are introduced that may help speed up your work. Watch trade journals and supply houses for advertisements for these new items. Talking with other technicians to see if any of them have used the new tools before investing your money in them can save you both time and money.

UNIT 4—REVIEW QUESTIONS

1. Why is it important for an HVACR technician's tools to be clean and organized?
2. Describe some of the features available on service valve wrenches.
3. What are the most common sizes of sockets?
4. Explain how socket handles are sized.
5. When should a breaker-bar socket wrench handle be used?
6. Why is it important that the same force be used when tightening nuts and bolts on compressors?
7. What is a good practice to follow when tightening bolts with a torque wrench?
8. How does a flare-nut wrench differ from a box wrench?
9. Explain the technique that should be used when using an adjustable wrench.
10. How does the tubing cutter cut?
11. What is a deburring tool used for?
12. Can you sometimes use a screwdriver as a chisel and pound it with a hammer?
13. Should pliers be used to tighten or loosen brass fittings? Why or why not?
14. Why are tubing brushes used?
15. List some of the safety precautions to be observed when using power tools.
16. What is the proper way to hold a hammer?
17. What two sizes of pipe wrenches are recommended for refrigeration installation and service work?
18. What type of wrench should typically be used on the screws on fan pulleys and fan blade hubs?
19. When would a nut driver typically be used?
20. What types of pliers are primarily used for mechanical HVACR work?
21. Electrical pliers should only be used for cutting what material?
22. How is cutting accomplished when using a hacksaw?
23. How are files and vises used in refrigeration work?
24. What are hand seamers, and when are they used?
25. What are the three different types of aviation snips?
26. How does a nibbler work?

UNIT 5

Screws, Rivets, Staples, and Other Fasteners

OBJECTIVES

After completing this unit, you will be able to:

1. explain how threaded fasteners are used in the HVACR industry.
2. explain size and length measurement as used with screws.
3. identify screw types and heads.
4. explain NC and NF screw threads.
5. explain how and why rivets are used.
6. demonstrate the use of HVACR duct staples.

5.1 INTRODUCTION

Assembling HVACR systems, components, and equipment requires technicians to use a variety of fastening devices. You might use a sheet-metal screw on ductwork or a condenser panel or a bolt to hold a head on a compressor. The range of items is extensive, and this unit does not cover all of the fasteners you may encounter in your career. It does cover the most commonly used fasteners, the ones you are most likely to see on a regular basis.

5.2 SCREWS AND BOLTS

Screws and bolts are both types of threaded fastener. Screws are typically smaller in diameter and length than bolts. In addition, screws may or may not thread into a nut. Typical types of screws are sheet-metal screws, wood screws, and machine screws. Sheet-metal screws are the most commonly used threaded fasteners in the HVAC industry. Wood screws are sometimes used in HVAC work to attach hanging straps to wooden rafters. Machine screws have standard threads and screw into nuts or prethreaded holes. Machine screw threads are available with coarse or fine threads in both standard and metric diameters (Table 5-1).

5.3 SHEET-METAL SCREWS

A typical specification for a sheet-metal screw might be ½-in #8 hex sheet-metal screw (half-inch #8 hexagonal-head sheet-metal screw) (Figure 5-1). The length of a screw is given from the tip of the screw to the base of the head of the screw or level with the top for flat-head screws (Figure 5-2). The most commonly used lengths of screws in HVAC are ½ and ¾ in (13 and 20 mm) in length. The diameter of a screw is given as a standard number.

Table 5-1 Comparison of Thread Sizes Between Standard and Metric for Both Coarse and Fine Threads

Standard Threads			Metric Threads		
Number or Diameter	Coarse Threads per Inch	Fine Threads per Inch	Diameter in mm	Coarse Threads per mm	Fine Threads per mm
#6	32	40			
			4	0.7	0.5
#8	32	36			
#10	24	32			
			5	0.8	0.5
#12	24	28			
			6	1	0.5
1/4	20	28			
			7	1	.05
5/16	18	24			
			8	1.25	0.5
			9	1.25	0.75
3/8	16	24			
			10	1.5	0.75
			11	1.5	0.75
7/16	14	20			
			12	1.75	1
1/2	13	20			
			14	2	1
9/16	12	18			
			15	—	1
5/8	11	18			
			16	2	1
			17	—	1
			18	2.5	1
3/4	10	16			
			20	2.5	1
			22	2.5	1
7/8	9	14			
			24	3	1
			25	—	1
1	8	12			

Figure 5-1 Number 8, ½-in, hex-head sheet-metal screw.

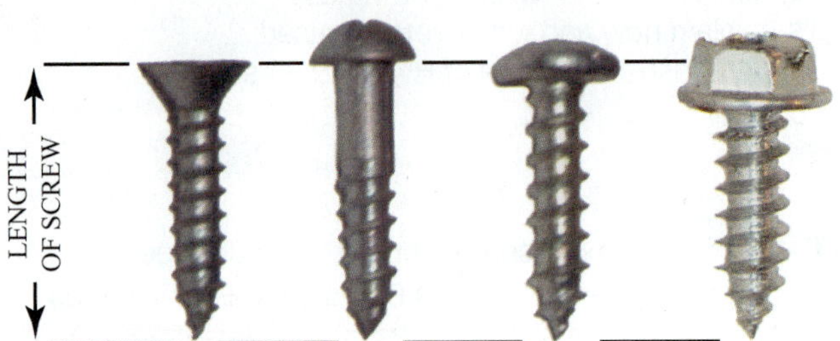

Figure 5-2 Screw length.

Figure 5-3 Sheet-metal hex-head screws: (a) with flat-blade slot; (b) without flat-blade slot.

Number 8 and number 10 screws are the most commonly used diameter sizes in the HVAC trade.

The hex-head sheet-metal screw shown in Figure 5-3a,b with (a) or without (b) a screwdriver slot are commonly used in the HVAC industry. The ¼-in and 5⁄16-in (6 mm and 8 mm) hex-head sheet-metal screws are the most commonly used sizes in the field.

Pointed

Pointed sheet-metal screws, shown in Figure 5-4a, are often referred to as self-starting. This is because they can be spun into sheet metal without having to predrill a hole. Self-starting sheet-metal screws can be used on sheet metal ranging from 30 gauge through 18 gauge.

Self-Threading

Self-threading sheet-metal screws must have a pilot hole drilled into the metal before these screws are used. A small notch at the tip of the screw acts as a thread tap to cut the thread into the metal as the screw is being installed. Self-threading sheet-metal screws can be used in sheet metal ranging from 16 gauge to ⅛ in thickness. The threads on the self-threading screw of Figure 5-4b can tap a hole in 16-gauge sheet metal. The threads of the self-threading sheet-metal screw in Figure 5-4c are more closely spaced, and this screw could thread a hole in ⅛-in sheet metal.

Round-Tipped Sheet-Metal Screws

Round-tipped sheet-metal screws like the one shown in Figure 5-4d are often used on electrical panels or within electrical panels to hold components in place. The tip of these screws is rounded and the end is not threaded. This protects against accidental stripping or shorting of wires within the electrical panel.

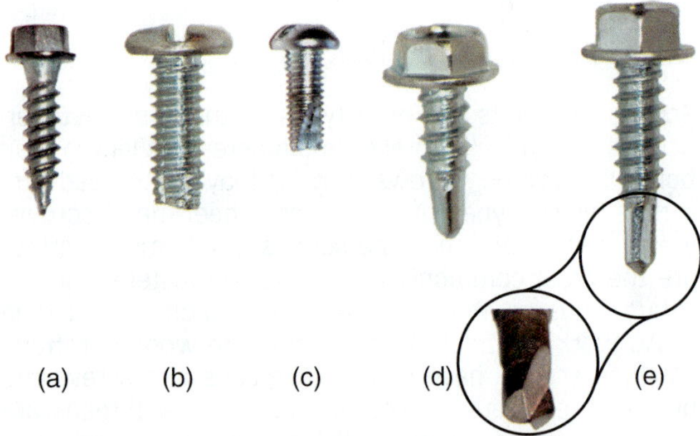

Figure 5-4 Common types of screws used with HVAC work: (a) self-starting; (b) self-threading; (c) self-tapping; (d) round tipped; (e) self-drilling.

Self-Drilling

Self-drilling sheet-metal screws have a tip built like a small drill bit, as can be seen in the inset in Figure 5-4e. Self-drilling sheet-metal screws predrill the hole as the screw goes in. Self-drilling sheet-metal screws can be used in metal thicknesses from 30 gauge to 16 gauge.

TECH TIP

The purpose of using round-tipped sheet-metal screws is to eliminate the possibility of a sharp point damaging electrical wire insulation.

An electric drill with a magnetic nut driver can be used to spin in self-starting and self-tapping sheet-metal screws. Self-starting sheet-metal screws are commonly used for on-site sheet-metal duct fabrication and installation. When using an electric drill with self-starting or self-tapping sheet-metal screws, the technician must be careful not to strip out the screw by overtightening it. Practice is required to develop the skill required to stop the drill the moment the screw is tight without overtightening. Some electric drills have torque adjustments, such as the dial on the drill in Figure 5-5, that allow the operator to set the drill so that it will apply enough force to sink the screw into the sheet metal but not strip it out.

SERVICE TIP

When using a magnetic nut driver and a drill, be certain that the head of the screw is completely in the end of the nut driver. If the nut is not completely in or is in at a slight angle, the screw head can slip inside the nut driver. This will quickly hollow out the end of the nut driver, rendering it useless.

Flat-blade screwdrivers and Phillips-head screwdriver tips may be used on some sheet-metal screws. The flat-blade screwdriver tip works well when manually removing or inserting screws, but it is very difficult to keep the bit aligned in the slot if a drill on a power screwdriver is being used. Phillips-head screws can be used with a Phillips bit chucked in a drill or power driver more successfully than the flat-head-type bit. Unfortunately, when the screwdriver bit begins to wear, it is more difficult to tighten and loosen screws.

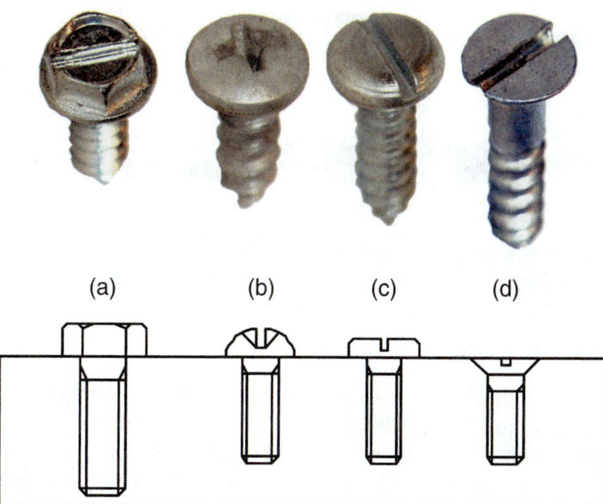

Figure 5-6 Common types of screw heads: (a) hex; (b) round; (c) pan; (d) flat; the bottom half of the figure shows how each screw will be represented in a drawing.

Screw Head Design

Screws can have several different head designs, shown in Figure 5-6, including hex (a), round (b), pan (c), and flat (d). Hex-head screws are the most common type used in HVACR work. They are used for almost everything inside and outside of the units. The round-head screw and the pan-head screw may be found inside electrical boxes and on electrical terminals. Round- and pan-head screws can have flat or Phillips slots, and some even have combination flat/Phillips slots, as shown in Figure 5-7. Phillips- and slot-head screws can be used in drills or power screwdrivers

Figure 5-5 Cordless electric drill with adjustable torque chuck.

Figure 5-7 Combination Phillips and slot-headed screw.

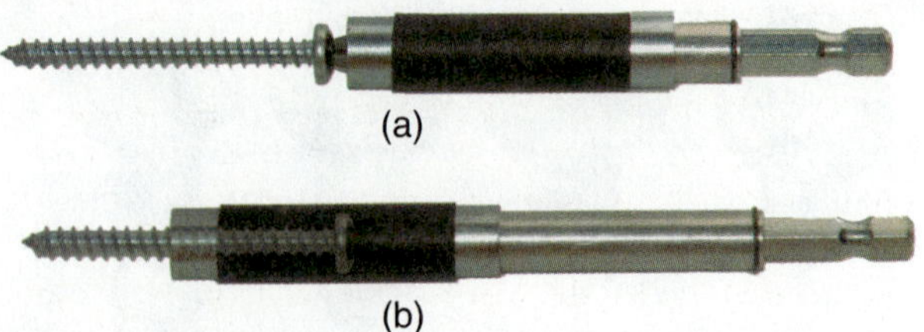

Figure 5-8 Screwdriving adapter: (a) with collar retracted to put Phillips bit into screw head; (b) with collar extended to keep contact between Phillips bit and screw head.

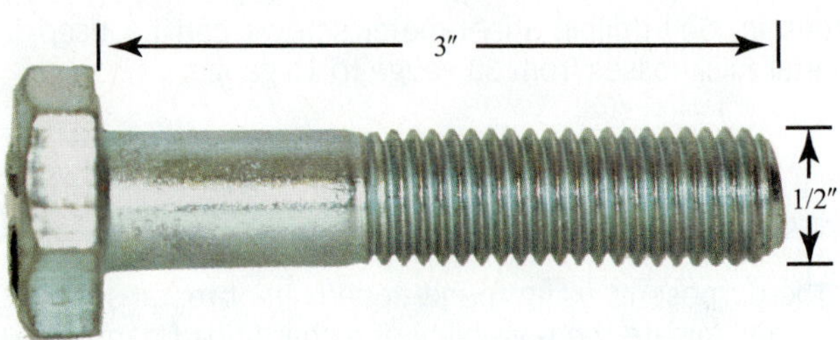

Figure 5-10 Bolt, 3″ × ½″ NC-13 hex head.

with an adaptor (Figure 5-8). The adaptor has a magnetic screwdriver bit inside a collar. The collar slides down over the screw so it can be held straight as it is driven in with a power screwdriver or drill. Wood screws are the most common type of flat-head screws. They are used to attach strapping to beams and joists.

Wood and Sheetrock Screws

Wood screws and sheetrock screws (Figure 5-9a,b) are often used in the HVACR industry to attach hanging straps to wooden structures. Wood screws for attaching hanging straps (a) are most often flat-head screws. Sheetrock screws are almost exclusively Phillips heads. The major difference in appearance between wood screws and sheetrock screws is that the wood screw is shiny and has a portion of the shank that does not have any thread. Sheetrock screws are almost always black, and their thread extends the full length of the screw.

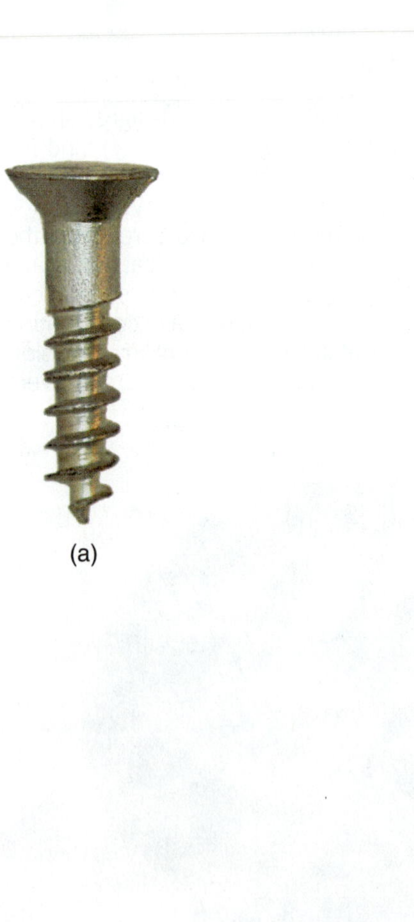

Figure 5-9 (a) Wood screw; (b) sheetrock screw.

5.4 BOLTS AND NUTS

Bolts and nuts have machined threads. These threads will only work when matched parts are used. A typical specification for a bolt would be 3″ × ½″ NC-13 hex head, which is shown in Figure 5-10. The first dimension given, 3″, is the length of the bolt. The second dimension, ½″, is the outside or major diameter of the bolt. The letters NC refer to the fact that this a national coarse threaded bolt. It has 13 threads per inch, and the bolt has a hex head. The length of the bolt is measured from the base of the head to the tip. The diameter of a bolt is measured across the threads. The type of thread is listed as either NC for national coarse or NF for national fine. Each standard-diameter bolt has a uniform number of threads per inch, depending on whether it is coarse or fine. Although most of the bolts used are hex-head bolts, other head designs include carriage bolt, square, and Allen wrench. Square-head and Allen wrench bolts are typically used on fans to hold the blades to the rotating motor shaft.

5.5 NAILS

Nails come in a variety of sizes, lengths, and materials and with different types of heads. The most commonly used nail material for HVACR work is steel, and they may be zinc coated to resist rusting.

Nail Types

Figure 5-11a,b illustrates some of the common nail types and sizes used in HVACR work.

- **Common nails** Large diameter, flat-headed nails used for general wood frame construction.
- **Box nails** Similar to common nails in shape and use but they are thinner in diameter.
- **Finishing nails** Small diameter, small-headed nails used to attach trim boards to door casings and floor molding. These nails are generally designed to be countersunk or recessed for a finished appearance.

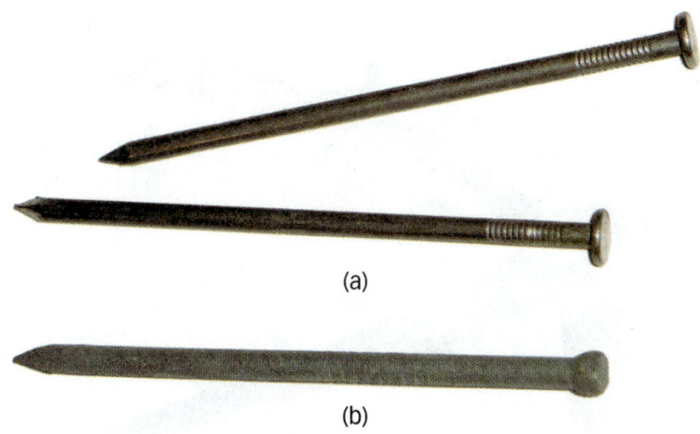

Figure 5-11 (a) Common nails; (b) finishing nail. *(Courtesy of HVAC Department, Terra Community College, Fremont, Ohio)*

Nail Sizes

Penny is the term commonly used for nail sizes. It is abbreviated as "d" and is used with the wire gauge number. (Note: This notation is actually a carryover from English unit weight measurement. Even the penny coin was abbreviated "d.") The wire gauge number refers to the size of the wire used to make the nail. The larger the number, the smaller the nail size will be. Some nails are sold with both their penny size and length, which can make choosing the correct nail size easier.

5.6 CONCRETE FASTENERS AND ANCHORS

The terms *concrete fastener* and *concrete anchor* are often used interchangeably; however, *fasteners* are nonstructural while *anchors* can be used for both nonstructural and structural attachments. Concrete fasteners can be grouped into two general categories: driven-in and predrilled. Anchors are almost always predrilled, but some can be preplaced before the concrete is poured.

Driven-in concrete nails can be installed using a hammer or with a gun and ballistic charge. The hammered-in concrete nails are usually short and work best when driven in with the fewest hammer blows. Repeatedly hammering of the nail can loosen it in the concrete.

Ballistic concrete nail guns may be trigger activated or hammer-strike activated. In both cases they use a small charge that resembles a .22-caliber blank cartridge to drive the nail in quickly. The nail lengths and cartridge power must be matched to the job so that the nail is driven into its desired depth.

CAUTION

Always wear proper eye and face protection when nailing, and read and follow all manufacturer's safety instructions. Special safety protection should always be used when driving ballistic concrete nails in anticipation of flying chips of cracking or exploding concrete.

Early concrete anchors used lead sleeves placed in a drilled hole, and these type of anchors are still used today. One drawback, however, is that the hole drilled into the concrete has to be a lot larger in diameter than the anchor. This means that the holes have to be accurately predrilled so they align with the part being attached. Newer types of anchors can fit into the same diameter hole as the anchor itself. This allows the preplacement of the part being attached, so hole alignment is not as much a problem. Figure 5-12a–c shows concrete anchors, drill bits, and the drilling through concrete for an anchor hole.

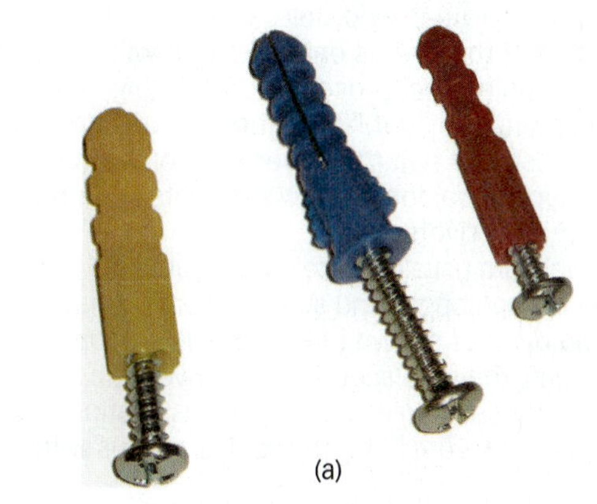

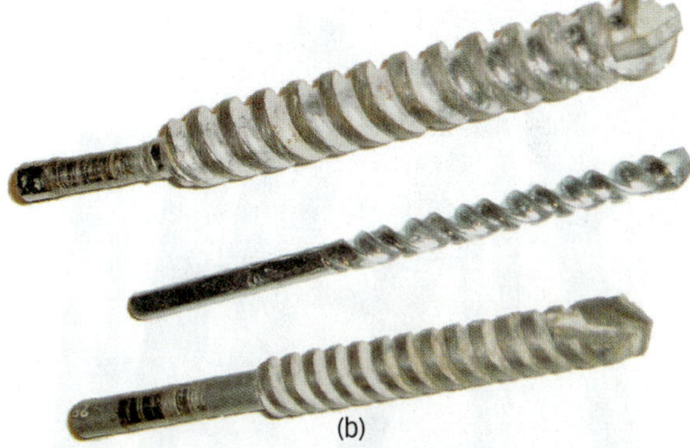

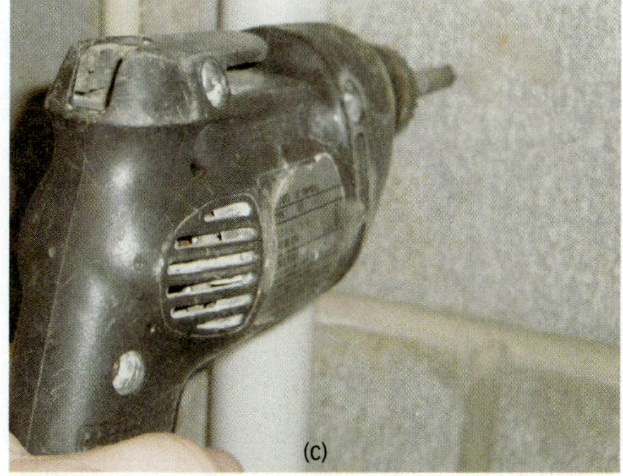

Figure 5-12 (a) Concrete anchors; (b) concrete drill bits; (c) drilling concrete anchor hole. *(Courtesy of HVAC Department, Terra Community College, Fremont, Ohio)*

TECH TIP

Concrete is the most widely used material on Earth. It is used for all types of construction. Masonry is a type of construction using bricks, stones, blocks, and so on. Masonry construction is held together with mortar or similar types of materials. Concrete fasteners can be used on most masonry construction.

5.7 RIVETS

A rivet is a small, unthreaded piece of metal that is installed through a predrilled hole. Once it is through the hole, the end of the rivet is enlarged so it will not fit back through the hole. The process of enlarging the end is called "upsetting the rivet." Over the centuries, rivets that had to be upset by hammers were in common use, but today this type is seldom used. Pop rivets have replaced the older style of rivets.

Pop rivets are usually made out of aluminum, but other metals, including copper and stainless, are available. A variety of pop rivets is shown in Figure 5-13. A pop rivet has a central shaft that is placed in a pop rivet gun, shown in Figure 5-14. The sequence showing the installation of a rivet is represented in Figure 5-15. The central shaft is withdrawn back through the center of the hollow rivet shaft. As the bead on the end of the center shaft is pulled through the rivet sleeve, the sleeve deforms outward. When the bead at the end of the central shaft is pulled snugly against the back side of the drilled metal, it pops off so that the nail-like shaft can be removed and discarded.

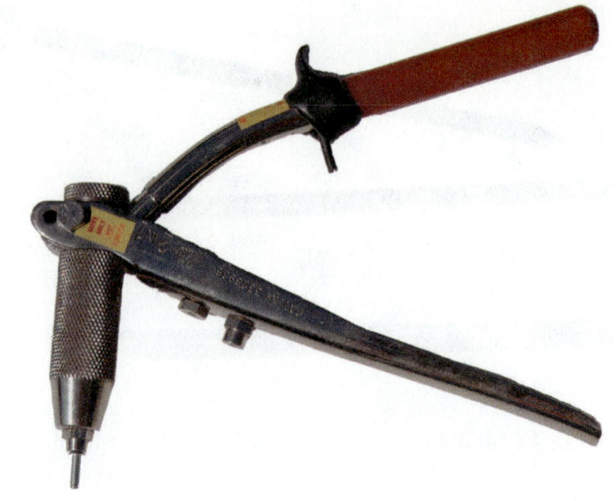

Figure 5-14 Pop rivet gun.

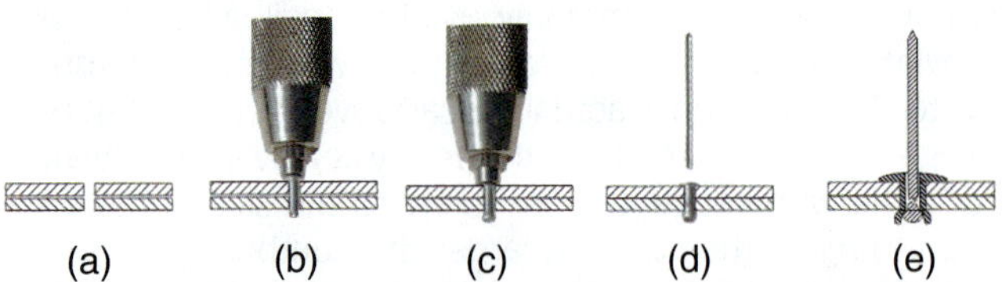

Figure 5-15 Steps in pop riveting: (a) drill a hole; (b) insert a pop rivet into the hole; (c) depress the pop rivet gun handle to upset the rivet; (d) rivet stem pops off when blades are pulled tight and rivet is seated; (e) cross-section of pop rivet.

Figure 5-13 Pop rivets come in a variety of types and sizes.

Pop rivets come in a variety of sizes, ranging from the most popular 1/8-in diameter up to 1/4-in diameter. The length of a pop rivet is determined by the distance from the tip to the head of the pop rivet. You should select a pop rivet that will extend through the joint materials by at least 1½ times the diameter of the pop rivet. For example, a 1/8-in diameter pop rivet should extend through the material being joined by at least 3/16 in, and a 1/4-in diameter pop rivet should extend 3/8 in through the material.

SERVICE TIP

Usually rivets have limited field applications in the HVAC industry because of the time required to predrill the holes. However, a 1/8-in pop rivet will fit into a #8 sheet-metal screw hole. Placing one pop rivet in the corner of a panel will reduce the possibility of unauthorized people tampering with the system.

The best way to remove rivets without damaging the base metal is to drill them out. Pop rivets can be easily drilled

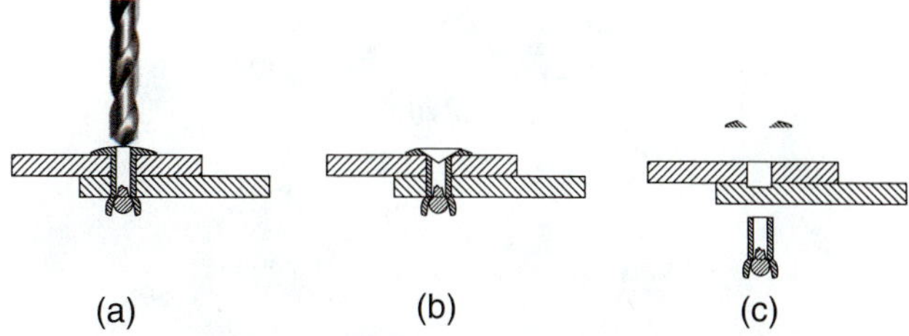

Figure 5-16 Steps in removing a pop rivet using a drill: (a) the drill bit should be slightly larger than the original pilot hole; (b) the drill only cuts out the rivet head; (c) the head and the shaft fall away.

using a bit slightly larger than the original shaft of the hole. As the drill cuts through the rivet head, the shaft of the rivet will be removed once the head has been cut free. Figure 5-16 shows the sequence of the drill bit next to the rivet (a), drilling out just the rivet head (b), and the pop rivet, with its head drilled out, falling away (c). Take care not to drill through and increase the size of the rivet hole.

SERVICE TIP

Sometimes the drill will grab the rivet and the rivet will spin. If this happens, you cannot drill out the rivet head. But by angling the drill bit slightly to one side, the bit will begin cutting again. Only angle the bit enough for it to start cutting. Too much angle may cause the bit to slip off of the head, damaging the finish of the part.

5.8 STAPLES

Staples hold by either driving their points into a soft material such as wood or by having the legs folded, as with a paper stapler. Staples come in one standard width (Figure 5-17). Staples can be purchased with a variety of lengths of leg, from ¼ in to 15/16 in.

Specially designed staples are used to attach fabric or vinyl-backed duct insulation (Figure 5-18a). Duct staplers are designed so that the tips of the staple are bent outward as the staple is driven through the fabric (Figure 5-18b). Duct staples are the most commonly used device for attaching external insulation to sheet-metal ducts.

Figure 5-17 Staples come as preassembled strips.

(a)

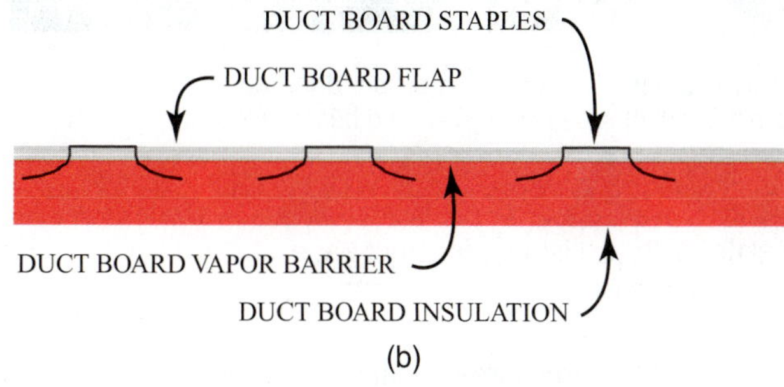

(b)

Figure 5-18 Staple gun used to (a) attach duct insulation or (b) seal duct board flaps.

5.9 TAPE

Two of the most commonly used types of tape in the HVACR industry are duct tape and electrical tape. Tape is manufactured to a variety of specifications, but new codes require that all duct tape meet UL 181 standards. The quality of tape used will directly affect the length of time it stays in place to seal the duct. For example, some vinyl gray duct tape is sold at deep discounts because it does not contain fabric reinforcement. There may even be a pattern embossed in the plastic to give it the illusion of having cloth reinforcement while no reinforcement exists.

Because of the heat in an attic, vinyl or cloth duct tape mastic will dry out over time and become hard and brittle. When this happens, no tackiness is left in the tape, and it simply comes loose (Figure 5-19). It is not a matter of *whether* the tape will dry out and fail but *when* it will dry out and fail. For this reason most codes require UL 181 for duct-closing systems. There are several different types of tape under the UL 181 listing. Each of these tapes has a different purpose (Table 5-2).

Aluminum pressure-sensitive tapes are easily formed to the surface, creating a tight fit. They have a rubber or acrylic adhesive that is protected on the roll with a paper backing. The paper backing is removed when the tape is being used (Figure 5-20). Cloth tapes have a Mylar coating and are designed for application on flex ducts (Figure 5-21). All approved UL 181 tapes must be identified with lettering on the face of the tape.

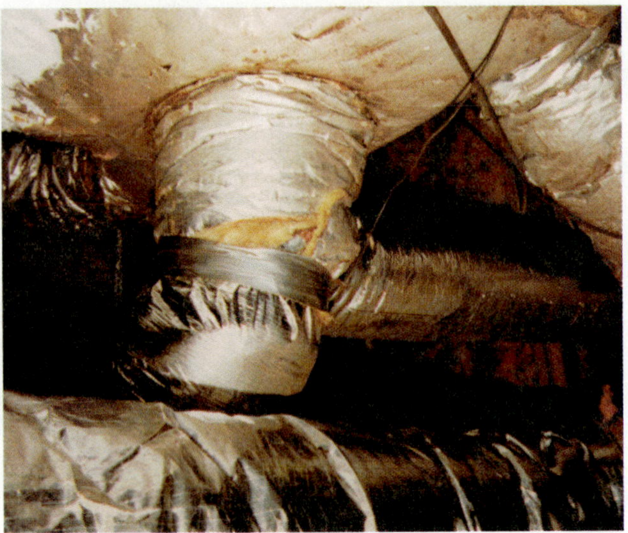

Figure 5-19 Always use the proper duct tape for the job. Improper duct tape will dry out and become loose as shown.

Table 5-2 UL Listings for Duct Tape

Flex duct
UL-181B-FX: Pressure-sensitive tape, for flexible ducts
Duct board
UL 181A-P: Pressure-sensitive tape, for duct board
UL 181A-H: Heat-activated tape, for duct board
Metal duct connectors to flex or duct board
UL 181 A-P or UL181 B-FX

The adhesive on foil tapes is very tacky. If it is accidentally touched to any surface before it is in place it is difficult if not impossible to remove it. Foil tape comes with a paper backing to keep it from being stuck permanently to itself in the roll. This paper must be peeled back as the tape is applied to the surface. Once the tape is in place

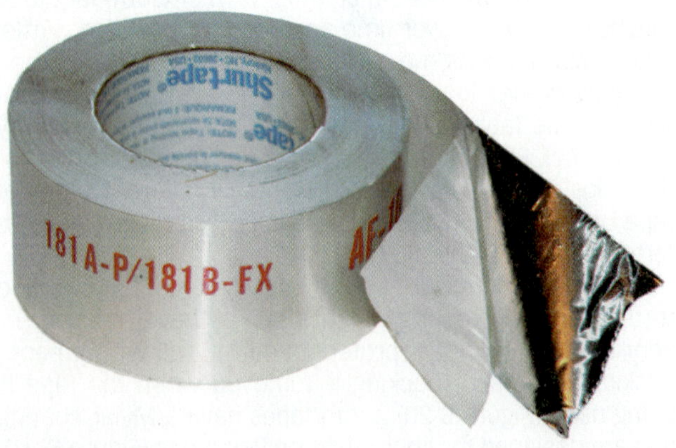

Figure 5-20 Aluminum pressure-sensitive duct tape.

Figure 5-21 Mylar-coated pressure-sensitive duct tape.

it should be smoothed down securely to ensure a proper seal. Failure to wipe the tape down can result in air gaps beneath the tape, which will allow its mastic to dry out. Keeping duct tape securely in place over time is the only way to ensure that conditioned air does not leak out of the duct system.

Vinyl electrical tape is often referred to as electrical tape or simply black tape. As with duct tape, there is a range of quality in the tape available. A good vinyl electrical tape should have some stretch so it can form itself around the wires or connection. Too much stretch and the tape cannot be pulled tightly around the parts. The tape must resist heat without loosening or softening.

5.10 ADHESIVES

There are a number of adhesives used in HVACR work. The three general groupings of adhesives are mastics, glues, and caulk. Generally in HVACR work, mastics are used to seal large areas such as duct joints, glues are used to hold items together, and caulks are used to seal cracks. All three types of material are classified as adhesives. *Adhesion* is their ability to stick to a surface, and *cohesion* is their ability to stick to themselves.

When using any adhesive, it is important that the surface is clean, dry, and free of loose material. Any of these surface conditions can prevent adhesion. Some adhesives have reinforcing fibers to add to their cohesive strength, while others do not have reinforcing fibers.

Adhesives may be water based or use another type of volatile organic compound (VOC) as their solvent. Some VOCs are flammable, and others can produce hazardous fumes. Adequate ventilation must be provided to prevent the buildup of hazardous or explosive fumes in the work area.

Some adhesives are paintable, and some are not. Check to see whether the product you plan to use on any interior or exterior surfaces can be painted before you use it. Not being able to paint the caulk can be a problem, especially when it is used around public areas, such as the trim around a supply grill and the like.

CAUTION

Always follow all manufacturer's instructions when using adhesives. Wear all personal protection equipment suggested by the manufacturer.

TECH TIP

If you pull a glued component apart and the glue comes off of the surface, that is an adhesion failure. If the glue separates itself, it is a cohesive failure.

5.11 TIE STRAPS (DRAW BANDS)

Tie straps, or draw bands, are usually made out of nylon and are widely used in the HVACR industry. Large sizes are used to secure flexible duct to start collars or boots, shown in the installation sequence in Figure 5-22a–d. These tie straps must be resistant to ultraviolet light (UV) and have a tensile strength of 150 lb and a service temperature of 165°F. When nylon straps are used on duct installations, a special tool called a strap tensioner must be used. The strap-tensioning tool pulls the strap tight, reducing the likelihood of air leaks and mechanically connecting the duct to the sheet metal boot or collar securely. Once the proper tension has been applied, a lever on the strap-tensioning tool extends a razor knife to cut off the unused nylon strap end.

Smaller ties are used to secure wires into bundles inside of units. When the nylon straps are used around electrical wires, they are simply pulled tight by hand or with a pair of pliers, and the unused portion is cut off, as shown in the installation sequence in Figure 5-23a–d.

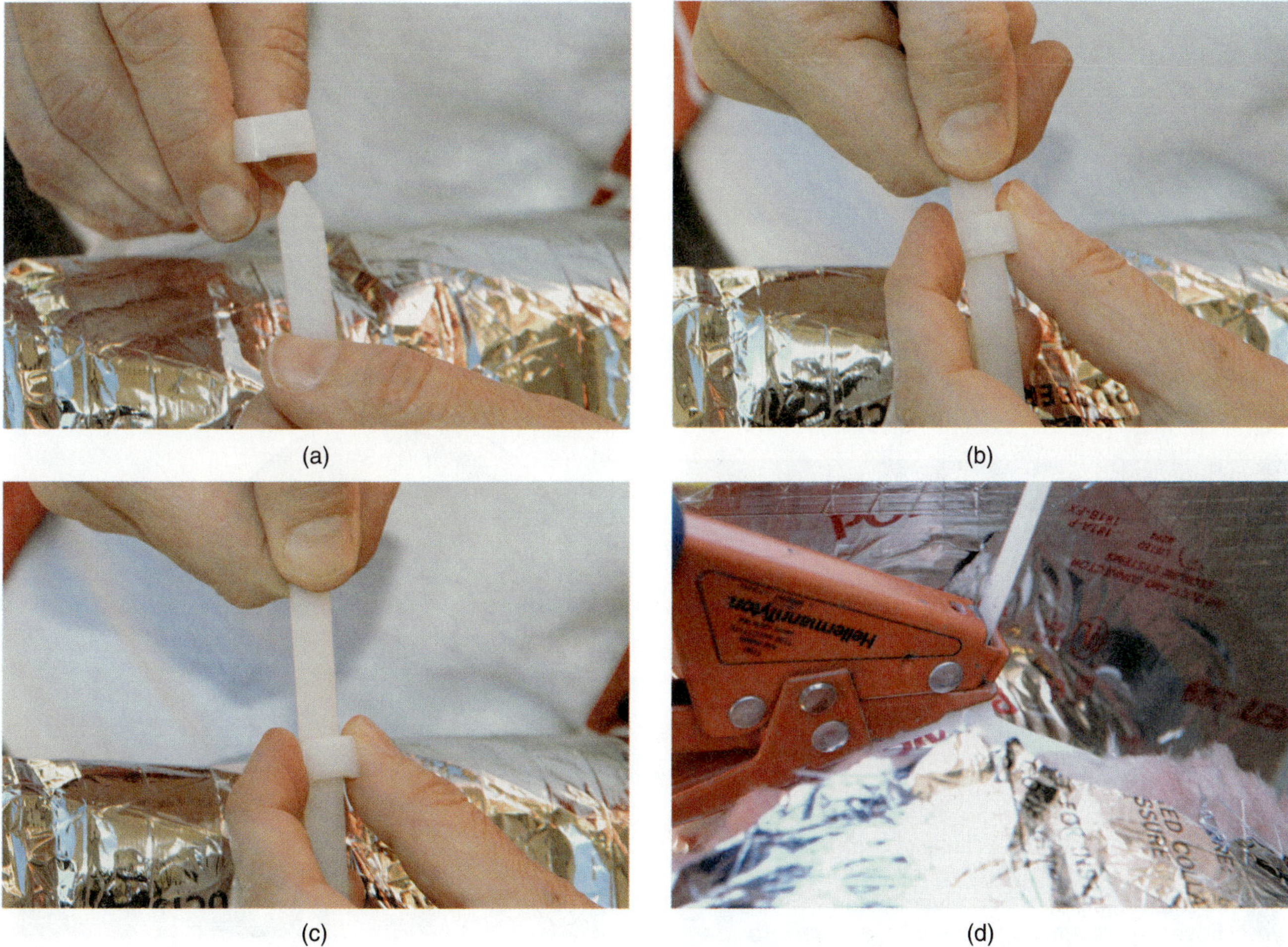

(a) (b) (c) (d)

Figure 5-22 (a) Insert the end of the tie strap through the ratchet mechanism; (b) use your hand to pull the strap through; (c) pull the strap with your hand as tight as possible; (d) use a duct strap tightening tool to finish tightening, and cut the end of the strap.

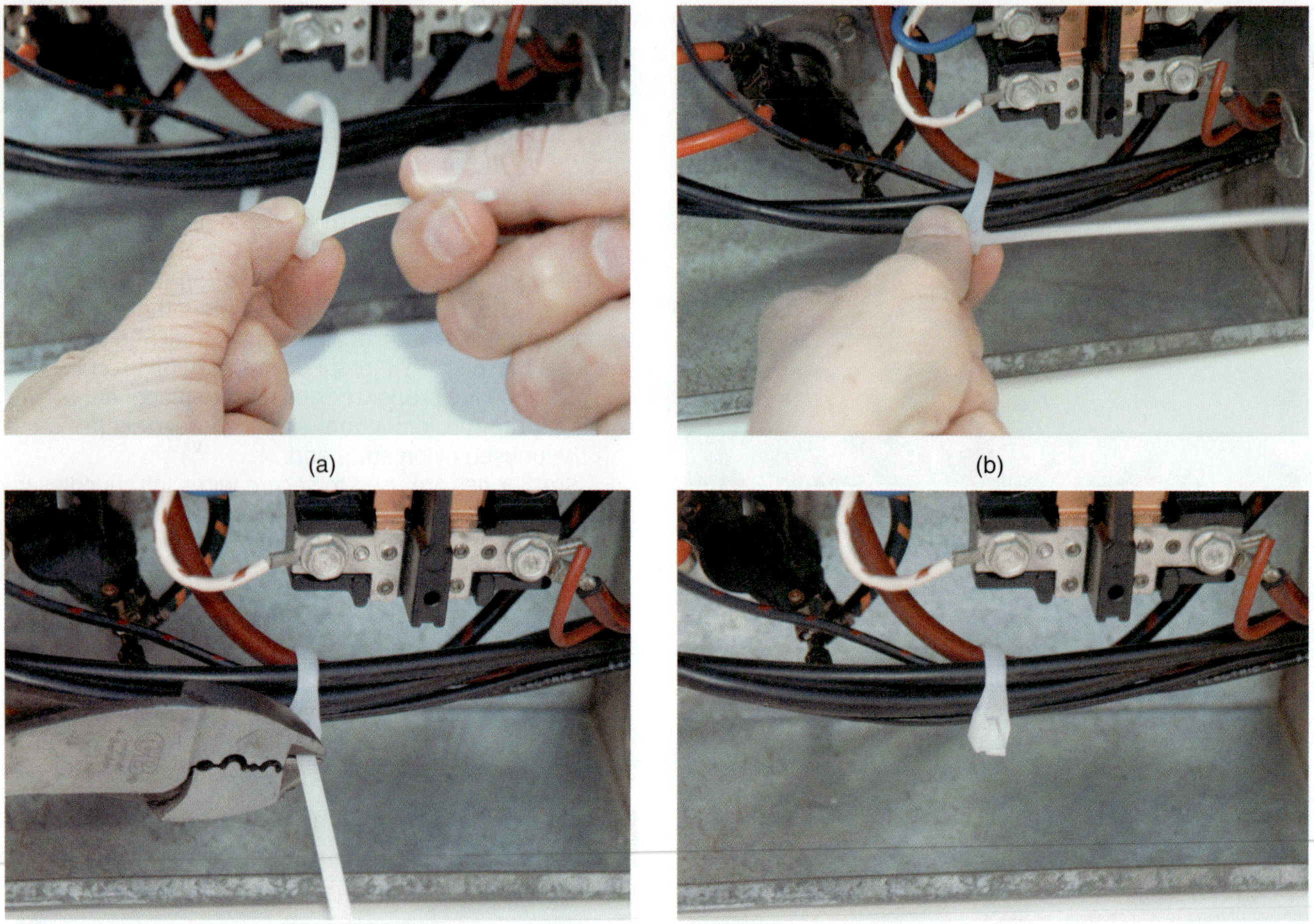

(a) (b)

(c) (d)

Figure 5-23 (a) Small wire tie straps; (b) pull tight by hand; (c) clip the strap; (d) finally, with the end cut off, the job looks neater.

SERVICE TIP

From time to time a tie strap may be put in place and only later is it found that it has to be removed. Ties may need to be removed when a wire must be added or removed from the bundle or when a duct needs to be cut shorter or straightened. Tie straps are expensive and must not be wasted. They can be removed easily, if the end has not been cut off, by placing the tip of a small screwdriver under the ratchet mechanism (Figure 5-24a,b). By putting just a little force on the corner of the screwdriver blade, the tie will be able to slip through the mechanism. Do not force the screwdriver tip under the mechanism. You can damage the ratchet, or the screwdriver could slip and stick your finger. As a precaution, you might want to use pliers to hold the strip the first few times you try this. Once the ratchet is released, use your thumb to slide the end up and off to remove the tie strap.

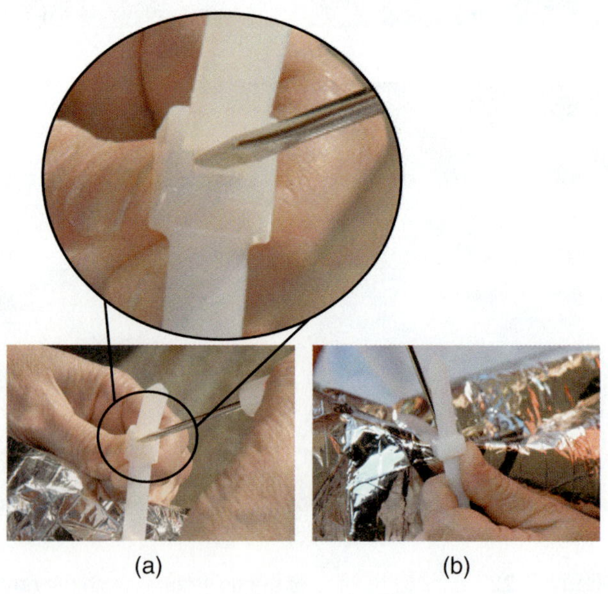

(a) (b)

Figure 5-24 Use a screwdriver to release the ratchet mechanism so that a duct strap can be removed and reused.

UNIT 5—SUMMARY

Thread gauges and screw and bolt size gauges are available to help you find the correct size of threaded fastener for the job. Never force a nut, bolt, or screw if it does not fit correctly. You can cross-thread the part and strip the threads, which can result in a significantly longer time to repair. You may also break off the bolt or screw in the part, which will require a lot of work to remove.

Although there are many different types and sizes of fasteners that can be used in HVACR work, you will find that the number and type you use on a regular basis is relatively small. The commonly used ones are the ones you will have in your tool bag and stock in your service vehicle.

Most HVACR parts houses carry the commonly used fasteners. You may need to go to a nut and bolt supply store to locate the specialty fasteners you need for some jobs.

UNIT 5—REVIEW QUESTIONS

1. What are the typical types of screws used on HVACR jobs?
2. How are the length and diameters of screws given?
3. List the common tip shapes found on sheet-metal screws.
4. What is the difference between the way a self-threading and a self-drilling sheet-metal screw is used?
5. Why are rounded-tipped sheet-metal screws used?
6. What is a common use for self-starting sheet-metal screws?
7. Why are flat-blade screwdrivers usually used manually to remove or install slotted-head screws?
8. List the common screw head designs.
9. What is a common use for wood screws and sheetrock screws in HVACR work?
10. Explain the significance of each part of this bolt specification: 2″ × ¼″ NC-20 hex head.
11. List three common types of nails.
12. What is the difference between a concrete fastener and a concrete anchor?
13. What drives a ballistic concrete nail in?
14. What is the term used to describe the enlarging of the end of a rivet?
15. What metals are commonly used to make pop rivets?
16. What is the best way of removing a pop rivet?
17. How does the tip of an installed duct staple differ from other types of staples?
18. Duct tape must meet what standard?
19. What is the problem with using vinyl or cloth duct tape in an attic?
20. What are two qualities that a good vinyl electrical tape must have?
21. List the three common types of adhesives used in HVACR work, and tell how they are used.
22. Why is it better to use water-based adhesives as opposed to VOC-based solvent adhesives?
23. Why must a tensioner tool be used on duct tie straps?

UNIT 6

Measurements

OBJECTIVES

After completing this unit, you will be able to:

1. list the four physical properties that are commonly measured.
2. list common units used in the inch-pound system.
3. explain the basic concept of the SI system.
4. explain the concept of rate measurement.
5. calculate the area of a surface.
6. calculate the volume of a space.
7. estimate length and area based on standard building material dimensions.

6.1 INTRODUCTION

Standard measures are necessary for people to communicate the quantity and size of objects, areas, substances, energy forms, and just about everything that we have to deal with in life. Without standard measures, the only way for someone to get an idea of something's size would be to see it. Measurement standards are crucial to any form of contracting work, including heating, air conditioning, and refrigeration, so technicians must be familiar with many types of measurement. There are four physical properties that must often be measured: distance, area, volume, and weight. Technicians also measure things that are not physical properties, such as water flow or air flow. Flow measurements measure a quantity over time. In the United States we currently use a mix of two measurement systems: customary units based on the inch-pound system and SI units based on the metric system.

6.2 CUSTOMARY MEASUREMENTS IN THE UNITED STATES

The United States is the last industrialized nation still using traditional customary measurements for most of our day-to-day work. Our customary system is not really a system at all but a collection of different units passed down from several different cultural traditions. The British have been the largest influence, but measures they passed on were adopted from several different cultures, including Celtic, Roman, Saxon, and Norse. Our system is sometimes called the inch-pound system after the units we use for small linear measurement and weight. Table 6-1 lists many of the most common customary measures used today in the United States.

Table 6-1 Customary U.S. Measurements

Property	Unit	Relationship to other Units	Example
Weight	Grain	7,000 Grains in a pound	The weight of water vapor in a pound of air is measured in grains
Weight	Ounce	16 ounces in a pound	Refrigerant charge in refrigerators is in ounces
Weight	Pound	2,000 pounds in a ton	Refrigerant charge in large units is in pounds
Weight	Ton	2,000 pounds in a ton	A 1/2 ton pickup can carry 1,000 pounds
Length	Inch	12 inches in a foot	Filter dimensions are given in inches
Length	Foot	3 feet in a yard	Duct length is given in feet
Length	Yard	1,760 yards in a mile	A run in football is measured in yards
Length	Mile	5,280 feet in a mile	Distance to the job is given in miles
Area	Square Inch	144 square inches in a square foot	The area of a grille is measured in square inches
Area	Square Foot	9 square feet in a square yard	The area of a room is measured in square feet
Area	Acre	43,560 square feet in an acre	Land area is measured in acres
Dry Volume	Cubic Inch	1,728 cubic inches in a cubic foot	Compressor displacement is measured in cubic inches
Dry Volume	Cubic Foot	27 cubic feet in a cubic yard	The space in a room is measured in cubic feet
Dry Volume	Yard (Cubic Yard)	Space occupied by a cube 1 yard in all dimensions	Sand, gravel, and concrete are sold by the cubic yard

6.3 LINEAR MEASUREMENT

Linear measurements are used to indicate distance. Linear means that we are measuring a line. Remember, the shortest distance between any two points is a straight line. When we say that the ceiling is 8 feet high, we are stating that a line drawn from the floor to the ceiling would be 8 feet long. Anything that can be measured using a line can be expressed as a linear measurement. Typically, distance is measured in yards, feet, and inches using the traditional English system of measurements. There are 3 feet in 1 yard and 12 inches in 1 foot (Figure 6-1). Longer distances are measured in miles. The term *mile* derives from the Roman Legion. "Mille passuum" in Latin means a thousand paces. A mile was the distance they would cover in a thousand paces. Today, a statute mile is equal to 5,280 feet.

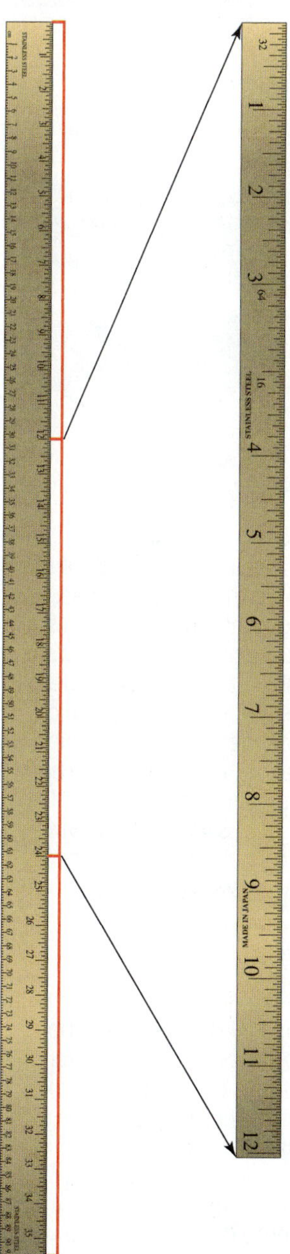

Figure 6-1 One yard = 3 feet; one foot = 12 inches.

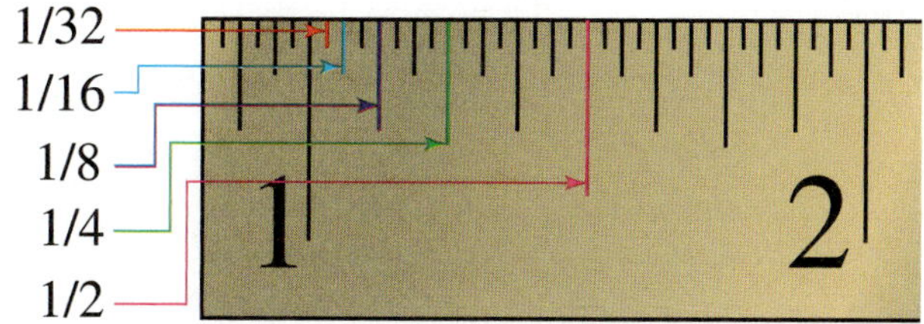

Figure 6-2 Close-up of ruler with dimension lines showing different heights.

Small measurements are taken in fractions of an inch. Each division is half of the previous division, producing common units of 1/2, 1/4, 1/8, 1/16, and 1/32. Extremely small dimensions used in machining are given in thousandths of an inch instead of fractional sizes. Traditional rulers often show the relationship of the divisions by making the lines for each division progressively smaller to make recognizing the different units easier (Figure 6-2).

TECH TIP

Many people find the relationship of fractions with different denominators confusing. To compare two fractional sizes, multiply the bottom number (denominator) of the fraction with the smaller number on the bottom by 2 until it equals the denominator of the other fraction. Then multiply the top number (numerator) by 2 the same number of times. For example, comparing 3/4 inch and 11/16 inch: multiply 4 by 2 twice to get 16 ($2 \times 4 = 8$, $8 \times 2 = 16$). Then multiply 3 by 2 twice to get 12 ($3 \times 2 = 6$, $6 \times 2 = 12$). Now you can easily see that 12/16 is larger than 11/16.

6.4 AREA MEASUREMENT

Area is used to express two-dimensional measurement of a surface. Examples of items that are measured by area would be a desktop, a plot of land, and the floor of a building. Area measurements require two dimensions. The term *square* is used when describing an area because area measurements are broken down into several squares of a certain size. For example, a square foot represents the area of a square with sides of 1 foot. An area of 10 square feet would be equal to the area covered by ten of these 1-x-1-foot squares (Figure 6-3). Technicians sometimes need to convert between square feet and square inches. It might be logical to assume that since there are 12 inches in a foot, that there are 12 square inches in a square foot. However, this is incorrect. Indeed there are 12 inches in a foot, and since a square foot is 1 foot wide and 1 foot long; a square foot can be divided into 12 rows of boxes with 12 boxes in each row. Multiplying 12×12 you arrive at the factor for converting square feet to square inches: 144 (Figure 6-4). To convert any square measurement to another, multiply the area by the linear conversion twice, once for each dimension. For instance, to convert square yards to square feet, the number of square yards would be multiplied by 3 twice, since there are 3 feet in a yard. 5 square

1 FOOT

1 FOOT

1	2
3	4
5	6
7	8
9	10

Figure 6-3 Drawing showing 10 sq ft made up of 10 squares.

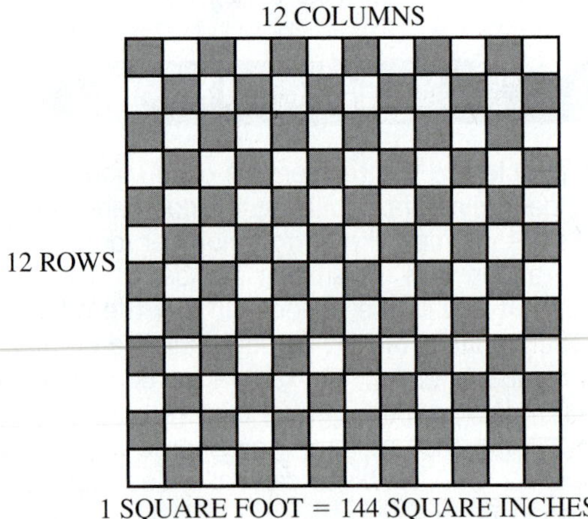

Figure 6-4 Drawing showing 144 sq inches in sq ft.

yards can be converted to square feet by multiplying 5 × 3 × 3 = 45 square feet. Of course you could also multiply 5 times 3 squared and get the same result. When we talk about a number squared, we just mean the number multiplied by itself. In the case of 3, 3 squared equals 9, so 5 square yards × 9 = 45 square feet.

Common Uses

The floor space in your home is measured in area. Floor space is generally measured in square feet, while filter, grille free, and duct work cross-sectional areas are normally given in square inches. To convert from square inches to square feet, divide by 144. For example, a 16 × 16 filter has an area of 256 square inches (16 × 16 = 256). To convert this to square feet: 256/144 = 1.78 square feet (Figure 6-5).

6.5 VOLUME MEASUREMENT

Volume is another physical measurement you are likely to deal with. Linear measure is used for lines, area for surfaces, and volume for things that take up space. All objects exist in three dimensions, one more dimension than can be expressed by area measurement. To find volume, you need length, width, and height. For area we use square measurement; for volume we use cubic measurement. Remember, square measurement subdivides a surface into several equally sized square sections, such as square feet or square inches. Cubic measurement subdivides an object or space into equally sized cubes, such as cubic feet or cubic inches (Figure 6-6). To find the volume of anything, multiply the area of its base by its height. For example, a box with a base of 4 × 6 inches and a height of 10 inches would have a base area of 24 square inches (4 × 6 = 24). Multiplying the 24-square-inch base area times its height of 10 inches would yield a volume of 240 cubic inches (24 × 10 = 240).

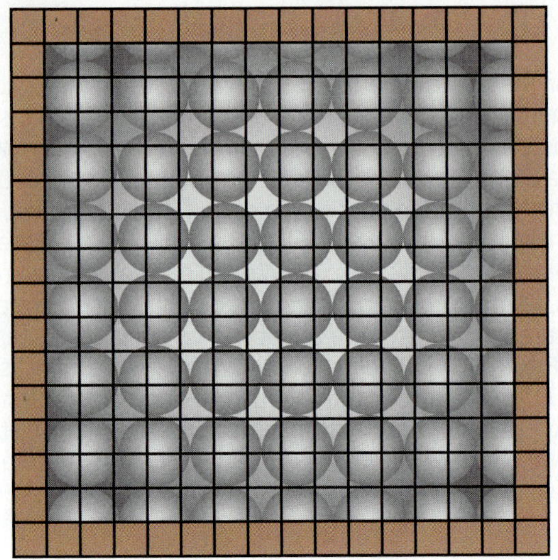

Figure 6-5 There are 256 square inches in a 16 × 16 filter.

Figure 6-6 Drawing showing little cubes for cubic measurement.

Common Uses

A cubic yard is a volume equal to the volume of a box that is 1 yard (3 feet) in all dimensions. Cubic yards are typically used to measure large volumes of material, such as dirt, sand, and concrete. Often, the "cubic" prefix is dropped, so that contractors order concrete by the yard. In air-conditioning work, cubic measurement is used to express airflow. Airflow is measured in cubic feet per minute, CFM. When air is flowing at the rate of 1 CFM, the air will fill up a 1-cubic-foot box in 1 minute (Figure 6-7). The volume inside electrical connection and junction boxes is given in cubic inches (Figure 6-8).

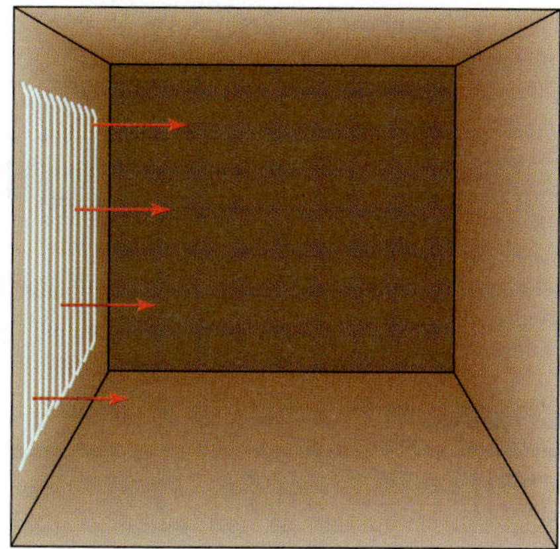

Figure 6-7 Air filling up a 1-cubic-ft box in one minute equals 1 CFM.

Electrical Box Specifications	
Dimensions in inches	**Volume in cubic inches**
3×2×1.5	7.5
3×2×2	10.0
3×2×2.25	10.5
3×2×2.5	12.5
3×2×3.75	14.0
3×2×3.5	18.0

Figure 6-8 Spec sheet for electrical box showing volume in cubic inches.

6.6 LIQUID VOLUME MEASUREMENT

Liquid volumes are more commonly measured in units designed specifically for measuring them. In the United States we use gallons, quarts, pints, cups, and ounces. Note that these units are not "cubed" because they are inherently a cubic measurement. A cubic foot is equal to 7.48 U.S. gallons, making a U.S. gallon approximately 1/8 of a cubic foot. Gallons are divided into quarts, with 4 quarts equal to a gallon. Quarts are divided into pints, with 2 pints equal to a quart. And pints are divided into cups, with 2 cups equaling a pint. Finally, cups are divided into ounces, with 8 ounces equaling a cup. Note that fluid ounces are a measure of volume, not weight. Table 6-2 shows the relationship of all these fluid volume measurements.

Table 6-2 Customary Liquid Volume Measurement

Unit	Relationship to Other Units	Example
Ounce	8 ounces in a cup	Compressor oil is measured in ounces
Cup	2 cups in a pint	Liquid measure in recipes are in cups
Pint	2 pints in a quart	PVC cement is sold in pints
Quart	4 quarts in a gallon	Motor oil is sold in quarts
Gallon	128 ounces in a gallon	Refrigeration oil is sold in gallons

Common Uses

In air conditioning we normally measure oil containers in quarts or gallons, while compressor oil quantities are often stated in fluid ounces. Water flow is normally stated in gallons, as in gallons per minute, GPM.

SERVICE TIP

To determine how many quarts of oil are required for an oil charge stated in fluid ounces, divide the quantity of fluid ounces by 32. For example, an oil charge of 64 ounces would require 2 quarts: 64 fluid ounces/32 = 2 quarts.

6.7 AREA AND VOLUME OF ROUND OBJECTS

The discussion so far has been about areas and volumes for rectangular objects. The formulas that work for rectangles do not apply to circles.

Area of a Circle

The area of a circle is found by using the formula: area $= \pi r^2$ where $\pi = 3.14$ and r = circle radius (Figure 6-9). The radius of a circle is the distance from its center to the outside of the circle. This amounts to half of the diameter, which is the distance across the circle.
Example: Calculate the area of a circle with a 12-in diameter. The formula is $3.14 \times 6^2 = 3.14 \times 6 \times 6 = 3.14 \times 36 = 113$ square inches (approximately). The area of a 10-in diameter circle would be $3.14 \times {}^2 = 3.14 \times 5 \times 5 = 3.14 \times 25 = 78.5$ square inches (approximately). Notice how much smaller the area of the 10-inch circle is than the area of the 12-inch circle.

Circumference of a Circle

Occasionally it is necessary to determine the circumference of a circle. The circumference is the distance all the way around the outside of the circle (Figure 6-9). The formula for

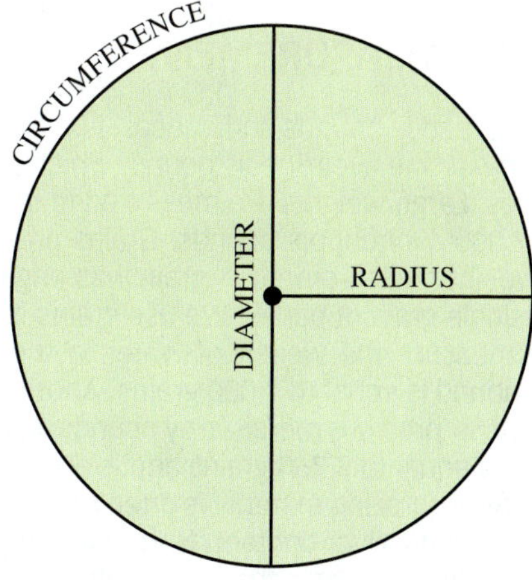

Figure 6-9 Drawing showing radius, diameter, and circumference of circle.

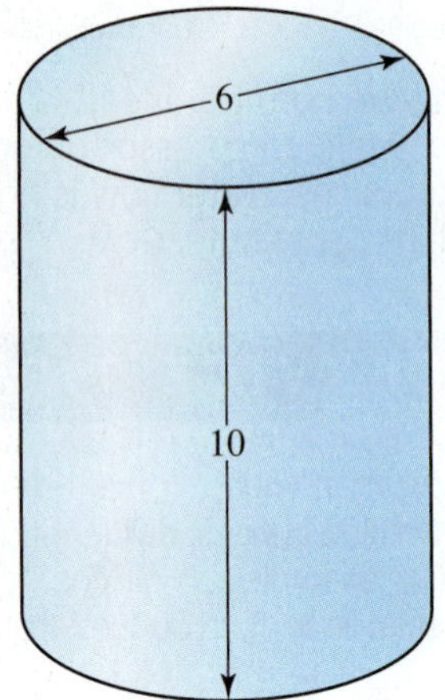

Figure 6-10 A cylinder with 6-in diameter and 10-in height.

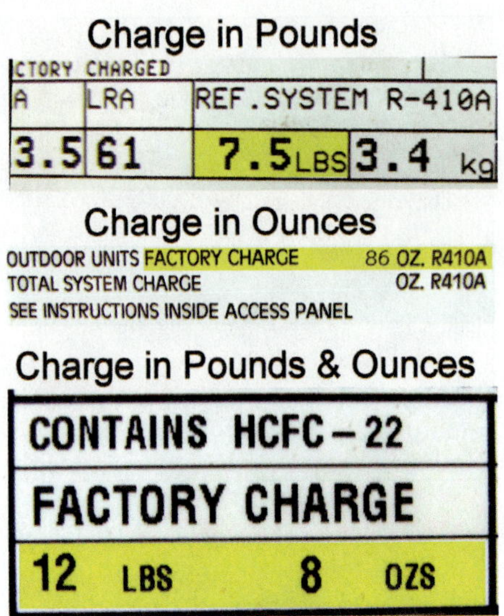

Figure 6-11 Refrigerant charge can be listed in pounds, ounces, or pounds and ounces.

the circumference of a circle is πd, where $\pi = 3.14$ and d = the circle diameter. The circumference of a 12-inch circle would be $3.14 \times 12 = 37.7$ inches (approximately).

Volume of a Cylinder

Calculating the volume and surface area of round objects is somewhat more complicated than calculating the volume of boxes. Round objects include spheres, cones, and cylinders. For our purposes, we only discuss cylinders. The volume of any regular solid can be determined by multiplying the area of its base times its height. A regular solid is one that keeps the same shape all the way up. Boxes and cylinders both fit into this category, while spheres and cones do not. Therefore, to find the volume of a cylinder, first find the area of the circle that forms its base. Then multiply this area by the height of the cylinder.

Example: Calculate the volume of a cylinder that has a diameter of 6 inches and a height of 10 inches. First, calculate the area of the circle that forms the base: $3.14 \times 3^2 = 3.14 \times 3 \times 3 = 3.14 \times 9 = 28.3$ square inches (approximately). Next, multiply this area by the height: 28.3 square inches $\times$ 10 inches = 283 cubic inches. (approximately) (Figure 6-10).

6.7 WEIGHT MEASUREMENT

Weight is measured in pounds and ounces. For most measurements, we use the avoirdupois pound, which is divided into 16 ounces. Large weights are measured in tons. A ton is equal to 2,000 avoirdupois pounds. Grains are used for very small weight measurement. A grain was originally the weight of a single grain of barley. We use grains in air conditioning to measure the weight of water in the air. The avoirdupois pound is equal to 7,000 grains. Another weight system is used for precious metals: troy pounds and ounces. A troy pound is equal to 5,760 grains and is divided into 12 troy ounces. Silver brazing material is often sold by the troy ounce because of its silver content. Note that a troy ounce is 480 grains, which is larger than a "regular" avoirdupois ounce of 437.5 grains. However, a troy pound is smaller than a "regular" avoirdupois pound.

TECH TIP

Early refrigerated food preservation was done with ice. Refrigeration was measured by the amount of ice required. In the United States, refrigeration capacity is still stated in tons. A ton of refrigeration is the amount of cooling accomplished by melting 2,000 pounds (a ton) of ice over a 24-hour period. This is how a measurement for weight also became a measurement of refrigeration capacity.

The quantity of refrigerant in a system can be given in ounces, pounds and ounces, or decimal pounds. Most charging scales do not allow technicians to enter ounce quantities greater than 16, so charges specified in ounces must be converted to pounds and ounces if the charge exceeds 16 ounces. Divide the number of ounces by 16 to arrive at the weight in pounds. For example, a charge of 76 ounces would be 76 ounces/16 = 4.75 pounds. This would work for a scale that reads in decimal pounds. For a scale that requires the weight in pounds and ounces, convert the decimal portion of the answer back to ounces by multiplying by 16: $0.75 \times 16 = 2$ ounces. This charge can be stated three ways: 76 ounces, 4.75 pounds, or 4 pounds, 12 ounces (Figure 6-11).

6.8 COMPOUND MEASUREMENTS

Distance, area, volume, and weight are all quantitative measurements; they measure the quantity or amount of something. How long, how high, the amount of space something occupies, or how much it weighs are all quantities. Volume and weight are often used to measure the amount of a substance. But neither measurement by itself tells the whole story. This is illustrated by an old joke: "Which is heavier, a pound of lead or a pound of feathers?" Of course, they weigh the same since the question stated "a pound" as the quantity. The joke is that many people instinctively answer that the lead is heavier. Of course, lead is heavier than feathers if you compare an equal volume of each. However, the question was comparing an equal weight. If the question is rephrased, "Which weighs more, a cubic foot of feathers or a cubic foot of lead?" the lead would be heavier. A compound measurement measures two or more properties and compares them to each other. It is useful to compare the weight and volume of substances when you need to know "how much" you have. Three comparisons are used for this purpose: density, specific volume, and specific gravity. Density compares weight to volume, specific volume compares volume to weight, and specific gravity compares the density of a substance to the density of air or water. Unit 7, Properties of Matter, discusses these in more detail.

Pressure

Pressure is a compound measurement. It compares force to the area supporting the force. The pressure on any surface can be determined by dividing the total amount of applied force by the area. An object that weighs 100 pounds and has a base of 100 square inches will exert a force of 1 pound per square inch: 100 pounds/100 square inches = 1 pound per square inch. We abbreviate this as 1 psi. If the object weighed 100 pounds but had a base area of only 50 square inches, the pressure would be 2 psi because the same weight is supported by a smaller area: 100 pounds/50 square inches = 2 psi. If the same 100 pounds was supported by a larger base of 200 square inches, the pressure would be only 0.5 psi: 100 pounds/200 square inches = 0.5 psi. To calculate pressure, divide the force by the area. Unit 11, Pressure and Vacuum, discusses pressure in more detail. In air conditioning, pressure is most commonly measured in psi. Figure 6-12 shows a typical pressure gauge.

6.9 RATE MEASUREMENTS

All the measurements discussed up to now have been static measurements. That is, what is being measured does not change. However, air conditioners are dynamic devices. Many performance characteristics that technicians need to measure are moving. Measuring something that is moving is done by rate. A rate measurement is a compound measurement that compares a quantity with time. In air-conditioning systems, airflow, water flow, and even a system's heating or cooling capacity are stated as a rate. The air flowing through an air-conditioning unit is a perfect example. The output of a fan is measured in cubic feet per minute, describing the volume of air in cubic feet that the fan moves every minute. A common rate measurement for water flow is gallons per minute, describing the number of gallons a pump can move in a minute. System capacity is stated as a rate of heat flow in BTUs per hour. This is the amount of heat produced or rejected in 1 hour of operation. Two common electrical measurements are also rates: amperes and watts. Amperes measure the flow of electrons; an amp is the flow of 1 coulomb of electrical charge per second. Watts measure the rate of electrical power; a watt is the transfer of electrical energy at the rate of 1 joule per second.

Figure 6-12 A typical pressure gauge.

Water Flow

Rate measurements require both a quantity measurement and a time measurement. When measuring water flow, gallons represent the quantity while minutes or hours are used for time. A reasonably accurate measurement of water flow can be obtained by collecting water in a container for a minute. The number of gallons collected in a minute represents the rate in gallons per minute. For a more accurate estimate, collect water for a longer period of time and divide the gallons of water collected by the number of minutes. Water-flow gauges are available that allow technicians to read the water flow just by looking at the gauge (Figure 6-13).

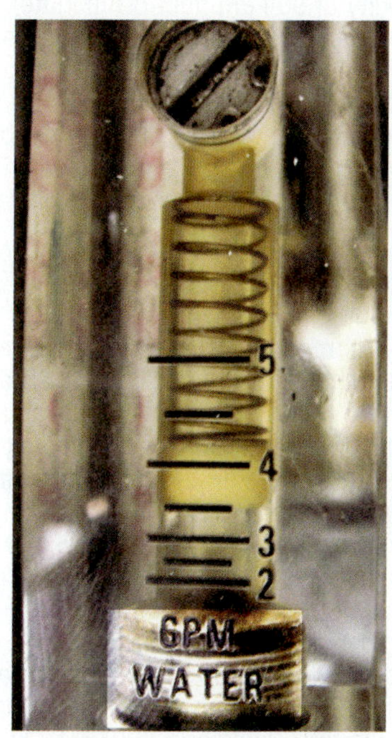

Figure 6-13 A water-flow gauge.

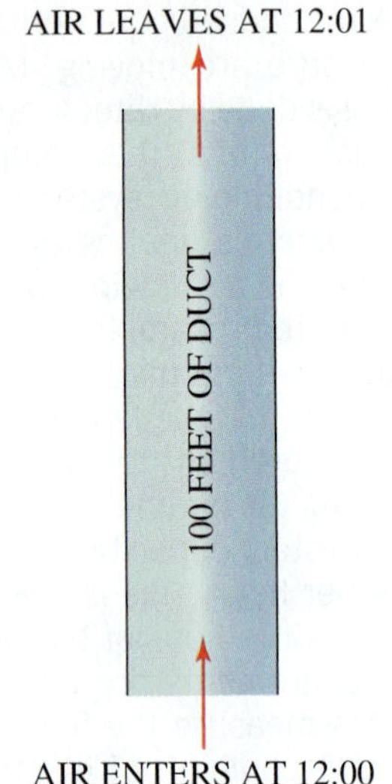

Figure 6-14 Air moving at 100 FPM takes exactly one minute to travel 100 feet.

Airflow

You can't really "collect" the air to check the rate of airflow. However, it is possible to check the speed of the air and use it to determine the amount of airflow. Speed is typically stated as a rate. A speed of 50 miles per hour indicates that a car traveling at that speed for an hour will cover a distance of 50 miles. Airspeed, or velocity, is measured in feet per minute. Feet per minute (FPM) describes how far the air will travel in a minute (Figure 6-14). This still does not really measure how much airflow there is, just how fast it is moving. To determine the volume of the air, multiply the velocity of the air times the area of the hole it is leaving. Air traveling at a rate of 100 feet per minute through a 1-square-foot opening will make a column of air with a volume of 100 cubic feet in 1 minute. Example: Calculate the airflow if air is leaving a 24 × 36–inch hole at a velocity of 50 FPM (feet per minute). The area of the hole is calculated by multiplying 24 inches × 36 inches = 864 square inches. Convert square inches to square feet by dividing by 144: 864 square inches/144 = 6 square feet. Alternately, you can convert the dimensions to feet before multiplying: 24/12 = 2 ft, 36/12 = 3 ft, 2ft × 3ft = 6 square feet. The airflow in CFM (cubic feet per minute) is calculated: 50 FPM × 6 square feet = 300 CFM.

Heat Flow

Heat quantity is measured in British thermal units, abbreviated BTU. A BTU is defined as the amount of heat required to raise 1 pound of water 1 degree Fahrenheit at atmospheric pressure. BTUs are calculated by multiplying the weight of water by the temperature change in degrees Fahrenheit. For example, raising 100 pounds of water 50 degrees would require 5,000 BTUs: 100 pounds × 50 degrees = 5,000 BTUs. BTUs are a quantity, not a rate. The amount of time required to raise the water temperature is also needed to specify the rate. If this is accomplished over the period of an hour the rate would be 5,000 BTUs per hour. If the same amount of temperature rise took 2 hours, the rate would be 5,000 BTUs/2 hours = 2,500 BTUs per hour. Unit 8, Types of Energy and Their Properties; Unit 9, Temperature Measurement and Conversion; and Unit 10, Thermodynamics, discuss temperature, heat, and heat flow measurements in greater detail.

TECH TIP

Temperature is not a measurement of heat quantity. Temperature alone cannot measure the amount of heat in a substance. Instead, temperature measures heat intensity. It measures the average kinetic energy of the molecules in a substance; it is not affected by the amount of the substance. A 16-oz glass of water and an 8-oz glass of water can be the same temperature even though the 16-oz glass contains more heat due to the greater volume of water.

6.10 SI MEASUREMENTS

The International System of Units, abbreviated SI, has become the accepted international system of measurement. All science is done in SI units, and most international trade is done using SI units. Any company that wants to market its products outside of the United States offers unit specifications in SI (Figure 6-15). SI is the implementation of the metric system, which was first adopted in France in 1791. The system was developed around three core ideas: there should be only one unit for each type of measurement, the units should logically relate to each other, and all subdivisions are in base 10. Smaller and larger quantities are expressed by applying prefixes that stand for different powers of 10. Only a few prefixes are commonly used today. For small measurements, *centi-* is used for 1/100th and *milli-* for 1/1,000th. For example, a centimeter is 1/100th of a meter: there are 100 centimeters in a meter. A milliliter is 1/1,000th of a liter: there are 1,000 milliliters in one liter. For large measurements, *kilo-* is used to indicate 1,000 of something and *mega-* is used to indicate a million of something. For example, a kilometer is equal to 1,000 meters and a megaohm is equal to 1 million ohms.

Base Units

The base units originally defined in the metric system were the meter for length, the gram for weight, and the liter for liquid volume. The milliliter was defined as the volume of 1 cubic centimeter, and the gram was defined as the weight of that same quantity of water. This tied distance, volume, and weight all neatly together. Today, SI uses seven defined base units:

- Meter for length
- Kilogram for mass
- Second for time

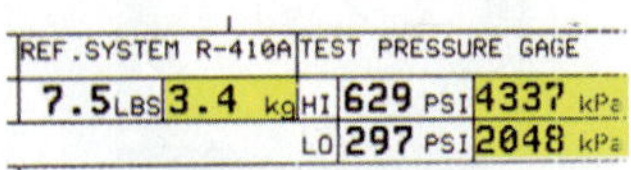

Figure 6-15 The SI specifications for this unit are highlighted in yellow.

Table 6-3 Common SI Prefixes

Prefix	Represents	Units	Example
micro	One millionth	microamps	Flame rod output is measure in microamps
milli	One thousandth	millivolt	Thermocouple output is measure in millivolts
kilo	One thousand	kilowatt	Power is measured in kilowatts
		kilogram	Refrigerant charge is measured in kilograms

- Ampere for electric current
- Kelvin for temperature
- Candela for luminous intensity
- Mole for the amount of substance

Liters are still commonly used, although they are not officially part of the SI system. One liter of water weighs 1 kilogram and takes up 1,000 cubic centimeters of space.

Derived Units

Many things cannot be measured in the base units. Derived units are measurements that require more than one measurement and some type of mathematical relationship. Area and volume are good examples because they require multiple dimensions that are multiplied by each other. Area is expressed in square meters and volume in cubic meters. Pressure is a derived unit because it measures the force applied over an area. Pressure is calculated by dividing the applied force by the area. The SI unit for pressure is the pascal, which is the force of 1 newton applied over an area of 1 square meter. Pascals are used to measure the pressure in ductwork, but they are too small a unit for measuring refrigerant pressure. Kilopascals are used when SI units are used to measure the pressure in a refrigeration system. Figure 6-16 shows a gauge calibrated in kilopascals on the outside scale and psi on the inside scale.

Figure 6-16 A gauge calibrated in kilopascals on the outside scale and psi on the inside scale.

Powers of 10

One of the most important innovations of the metric system was making all subdivisions powers of 10. This makes converting between prefixes as easy as moving the decimal place. For example, to convert centimeters to millimeters, just move the decimal one place to the right. A measurement of 20.4 centimeters becomes 204 millimeters. To change 20.4 centimeters to meters, move the decimal two places to the left to get 0.204 meters. The key to working with SI units is in understanding the prefixes. Some prefixes for smaller measurements are *deci-* for 1/10, *centi-* for 1/100, and *milli-* for 1/1,000. Some prefixes for larger measurements are *deca-* for 10x, *hecto-* for 100x, and *kilo-* for 1,000x. Only *milli-*, *centi-*, *kilo-*, and *mega-* are used regularly in everyday language. Table 6-3 lists the prefixes most commonly used in HVAC/R.

6.11 USING A METER STICK

SI measurements of length for common everyday use are done with a meter stick, or a metric rule (Figure 6-17). A meter is slightly longer than a yard. Meter sticks are typically marked in centimeters, with smaller millimeter lines between each centimeter line and larger lines every 10 centimeters (Figure 6-18). To take a measurement in centimeters, read the closest centimeter and count how many millimeter lines are past the nearest centimeter marking. The centimeters make up the whole number, and the millimeters go in the first place to the right of the decimal. For example, a reading of 27 centimeters and 4 millimeters would be 27.4 centimeters (Figure 6-19). To change this to millimeters, just move the decimal over one place, and 27.4 centimeters becomes 274 millimeters. To change the reading to meters, move the decimal two places to the left, and 27.4 centimeters becomes 0.274 meters.

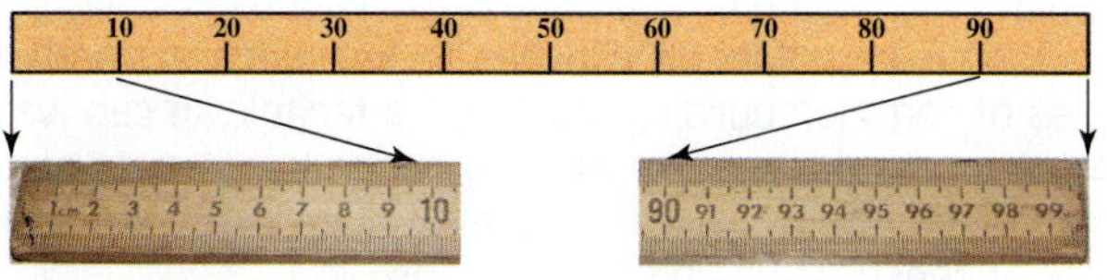

Figure 6-17 A meter stick is marked with 100 cm.

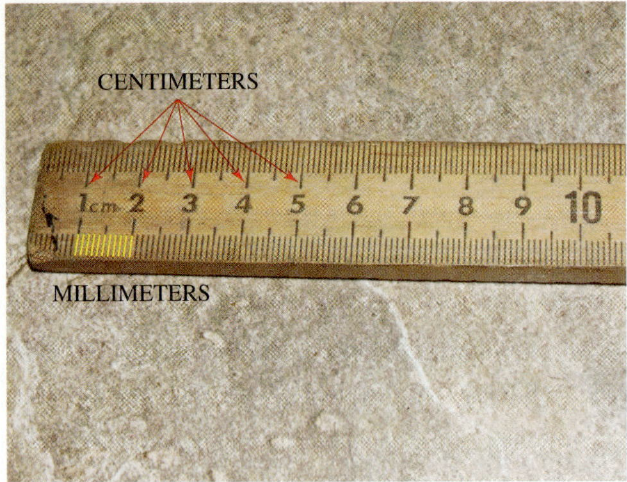

Figure 6-18 The numbers on a meter stick are centimeters, the lines between the centimeters are millimeters.

Figure 6-19 Measurement of 27.4 cm or 274 mm.

TECH TIP

Because the meter stick is based on a decimal base 10 system instead of a fractional system, many people who have difficulty reading rulers calibrated in inches find using a meter stick considerably easier.

6.12 ESTIMATING DIMENSIONS USING COMMON BUILDING MATERIAL DIMENSIONS

Construction in the United States is still done exclusively in traditional units of feet and inches. By knowing the standard sizes of common building materials, a technician can walk into a room and give a reasonably accurate estimate of its size. Most all buildings use repetitive patterns of some basic type of construction unit.

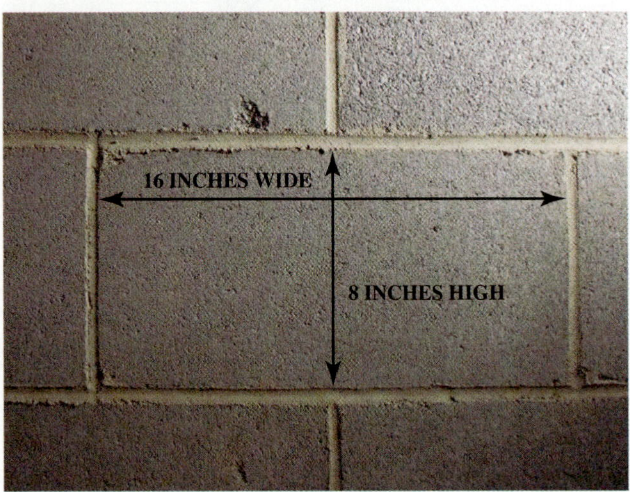

Figure 6-20 A concrete block has face dimensions of approximately 8 in high and 16 in wide.

Concrete Blocks

The walls in a building built with concrete blocks consist of rows of concrete blocks stacked on top of one another. Since each block is essentially the same size, you can estimate the length of any wall simply by counting the number of blocks in the wall and multiplying the number of blocks times the length of each block. Concrete blocks come in many sizes, but the most common size for the face of the block is 16 inches long by 8 inches high. In reality, the actual block is a little smaller, but with the addition of mortar joints, the overall construction unit size is 8 by 16 inches (Figure 6-20).

Example: Estimate the length and height of a wall that is 39 blocks long and 12 blocks high. You could multiply 39 blocks times 16 inches per block and then divide the answer by 12 inches per foot to arrive at an answer. However, an easier way to estimate length using small units like this is to group them into larger units. Grouping the blocks in threes makes the problem much easier. Three 16-inch units equal one 48-inch unit, or 4 feet. So working the problem this way, 39 blocks would be 13 groups of three. Multiplying 13 groups times 4 feet yields a length of 52 feet. Use a similar grouping technique for the height. Since the blocks are 8 inches high, each group of three blocks takes up 2 feet (8 inches $\times$ 3 = 24 inches). Twelve blocks high would be 4 groups of 3, and 4 groups times 2 feet per group gives a height of 8 feet. Figure 6-21 illustrates how this grouping works.

Framing

Floor joists, ceiling joists, roof rafters, and wall studs are normally spaced either 16 inches or 24 inches apart. This distance is measured center to center. The center of each framing unit is either 16 inches or 24 inches from the center of the previous framing unit. Like concrete blocks, the 16-inch framing units can be grouped by threes to form 4-foot counting units. Do not count the first stud, joist, or rafter when counting framing units because you are really

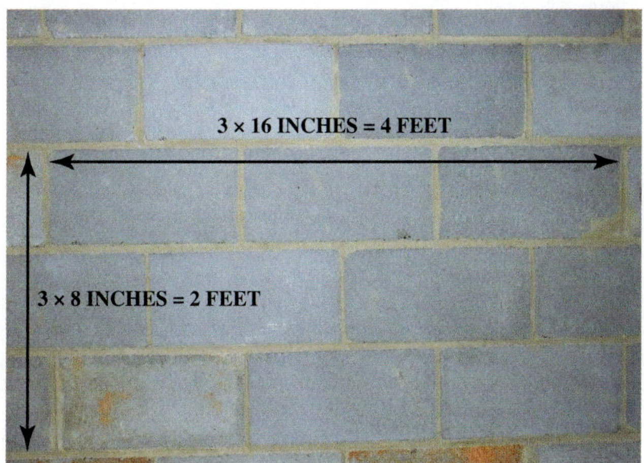

Figure 6-21 Three blocks high equals 2 ft. Three blocks long equals 4 ft.

counting the spaces between them (Figure 6-22). So a wall that has 10 studs would be counted as 9 stud spaces, or three groups.

Example: Estimate the length of a wall that has 31 studs 16 inches on center. After subtracting for the first stud, this wall would have 30 stud spaces, or 10 groups of three. The length would be 10 groups × 4 feet = 40 feet.

Example: Estimate the length of a building that has 35 roof rafters 24 inches on center. After subtracting for the first rafter, this wall would have 34 rafter spaces. The length would be 34 spaces × 2 feet = 68 feet.

Drop Ceilings

Drop ceilings work well for estimating room size. Ceiling tiles for drop ceilings are usually either 2 feet by 2 feet or 2 feet by 4 feet. For length: count the tiles and multiply by either 2 or 4, depending on the type of tile. For area: count the tiles and multiply by either 4 or 8, depending on the tile dimensions.

Example: Estimate the dimensions of a room with a 2 × 2 drop ceiling that is 10 tiles wide and 15 tiles long. The dimensions are found by multiplying the number of tiles by 2: 10 tiles × 2 feet = 20 feet, and 15 tiles × 2 feet = 30 feet.

Example: Estimate the area of the same room. There are two solutions. One would be to calculate the dimensions (as above) and then multiply the dimensions: 20 feet × 30 feet = 600 square feet. Another solution would be to calculate the number of tiles and multiply by 4: 10 tiles × 15 tiles = 150 tiles, and 150 tiles × 4 square feet = 600 square feet.

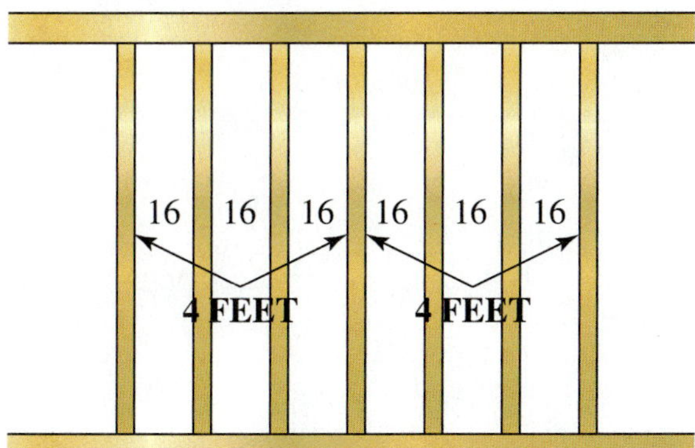

Figure 6-22 Because the center to center distance between studs is 16 inches, three stud spaces covers approximately 4 feet.

Sheetrock, Sheathing, and Plywood

Sheetrock, sheathing, and plywood are easy to use for estimation because they all commonly are 4 feet by 8 feet. Some exterior sheathing and siding products also come in 9-foot and 10-foot lengths. Sheetrock is often run horizontally rather than vertically. When run vertically on the wall, the 4 foot width can be used for estimating distance. Count the number of pieces wide on a wall, roof, or floor and multiply by 4 to get an estimate of length.

Example: Estimate the length of a wall that is covered by 12 pieces of sheathing. Multiply 12 × 4 feet = 48 feet in length.

UNIT 6—SUMMARY

The four physical properties measured are distance, area, volume, and weight. Compound measurements, such as density or pressure, compare more than one physical measurement. Flow measurements measure a quantity over time. Examples of flow measurements used in HVAC/R include airflow, measured in cubic feet per minute (CFM), and water flow, measured in gallons per minute (GPM). In the United States, we currently use a mix of customary units based on the inch-pound system and SI units. Distances are measured in yards, feet, and inches. Small measurements are done in fractions of an inch. Area measurements require two dimensions and are stated in square units, such as square inches. Volume measurements require three dimensions and are stated in cubic units, such as cubic feet. Everyday weight measurement is done in avoirdupois pounds and ounces, with 16 ounces in a pound. Very small weight measurement is done in grains: there are 7,000 grains in an avoirdupois pound. Silver brazing alloy is sold in troy pounds and ounces, with 12 ounces in a pound. SI measurement is the standard system for all science work and most international trade. SI measurements are all base 10, using decimals instead of fractions. Distance is measured in meters, weight in grams, and liquid volume in liters. Prefixes are used to indicate smaller or larger quantities in multiples of 10. The most common prefixes are *milli-* for 1/1,000th, *centi-* for 1/100th, and *kilo-* for 1,000. SI units can be converted from one to another just by moving the decimal. Measurements can be estimated using groupings of common construction materials. Materials with 16-inch dimensions, such as concrete blocks or wall studs, can be grouped in threes to form 4-foot groups. Materials with 24-inch dimensions, such as drop ceiling tiles, can be counted and doubled.

UNIT 6—REVIEW QUESTIONS

1. What are four physical properties of objects that can be measured?
2. What units are used to measure area?
3. Name a common use of area in HVACR.
4. Calculate the area of a grille that is 6 in high and 30 in long.
5. What units are used to measure volume?
6. Calculate the volume of a room that is 12 ft wide and 15 ft long with a 10-ft ceiling.
7. What is a compound measurement?
8. List some examples of compound measurements.
9. What is the difference between a quantity measurement and a rate measurement?
10. List common rate measurements used in HVACR.
11. What is the water flow rate of a pump that moves 72 gallons of water in 6 minutes?
12. Why are the lines different heights on a traditional ruler?
13. Convert 57 ounces to pounds and ounces.
14. What is the difference between a troy ounce and an avoirdupois ounce?
15. List the units used to measure liquid volume.
16. What do grams measure?
17. What are the three fundamental concepts that were used in developing the metric system?
18. Convert 569 millimeters to centimeters.
19. What is the weight in grams of 262 milliliters of water?
20. List the most common prefixes used in the SI system, and explain their meaning.
21. What SI unit is used for measuring refrigerant pressure?
22. What is the approximate width and height of a concrete block wall that is 27 blocks long and 15 blocks high?
23. What is the length of a roof with 22 rafters spaced 16 in on center?
24. Calculate the area of a circle with an 8-in diameter.
25. Calculate the volume of a cylinder with a diameter of 8 centimeters and a height of 10 centimeters.

UNIT 7

Properties of Matter

OBJECTIVES

After completing this unit, you will be able to:

1. explain why the properties of matter are important to the HVACR field.
2. explain the law of conservation of matter and give an example.
3. discuss the unique characteristics of the three states of matter.
4. name the physical and thermal properties of matter, and tell why it is important for the HVACR worker to understand them.
5. explain the difference between weight and mass.
6. explain the importance of testing standards.
7. demonstrate how to calculate the density, specific volume, and specific gravity of a material, and discuss how this information is used.
8. identify the common temperature reference points.

7.1 INTRODUCTION

Matter is all around us; matter is what makes up the earth and everything in and on the earth. Understanding the states and properties of matter is important to the HVACR technician because all heating and cooling requires some change to matter. During both the heating and cooling process, matter may expand or contract, melt or freeze, or even appear to vanish through evaporation or combustion.

Some types of matter are easily heated or cooled, while other types of matter take more heat to get a temperature change. Different forms of matter respond differently when heated or cooled. To better understand what happens when things are heated and cooled and why it happens, you must understand matter and its properties.

Matter can be identified by its properties because each type keeps its own unique set of properties regardless of the sample size. A single drop of water reacts and responds to pressure and temperature just like an ocean of pure water would.

7.2 MATTER

Matter is anything that has mass and occupies space. Here on earth, matter may be thought of as anything that has weight and occupies space. Water and air are example of matter; light and sound are not matter. Only a limited amount of water or air can be put in a container until it is full and there is no more space inside. However, a container will not "fill up" with light or sound because they do not occupy space.

The law of conservation of matter states that we cannot create or destroy matter. Matter can change its form, but it cannot disappear. We can grind, crush, or burn matter, but it still exists in the same quantity, just in a different form. When a log burns in a fire, its matter is not destroyed; it simply changes from one form to another. The log is mainly made up of carbon and hydrogen atoms. When it burns up in a fire, much of it is changed from a solid to a gas. The carbon and hydrogen combine with oxygen to form carbon dioxide, carbon monoxide, and water. Some unburned carbon leaves as smoke and some minerals and carbon are left as ash. If you could capture all of the products of combustion and weigh them, they would weigh the same as the wood and oxygen you started with.

7.3 THE STRUCTURE OF MATTER

Everything on earth consists of different combinations of ninety-four naturally occurring chemicals called elements. We are familiar with some elements because they are used in their pure form. For example, metals like gold, platinum, lead, mercury, iron, copper, and aluminum can be used in their pure element form. Some gases, like argon and nitrogen, are used as pure elements.

Atoms

The smallest piece of an element that can exist and still retain the same chemical properties is the atom. Atoms are made of small particles called protons, neutrons, and electrons. Protons are positively charged, electrons are negatively charged, and neutrons have no charge. Atoms have a nucleus composed of protons and neutrons. Clouds of negatively charged electrons orbit around this nucleus (Figure 7-1). However, most matter does not exist as an element but as a compound made up of more than one atom.

Molecules

The smallest part of anything that is not an element is a molecule. When the atoms of an element join together with themselves or with other types of atoms, they form

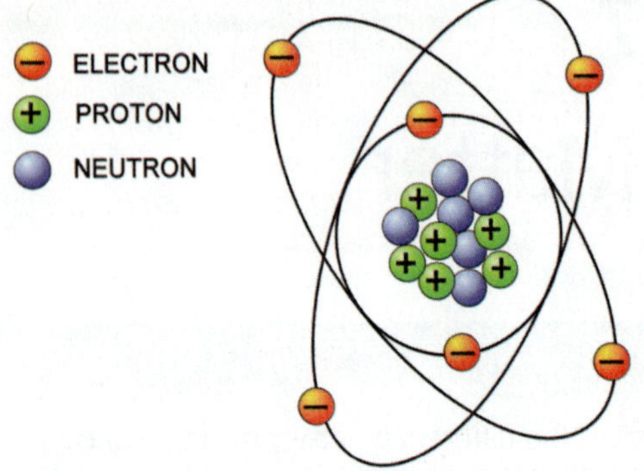

Figure 7-1 The carbon atom has six protons and six electrons.

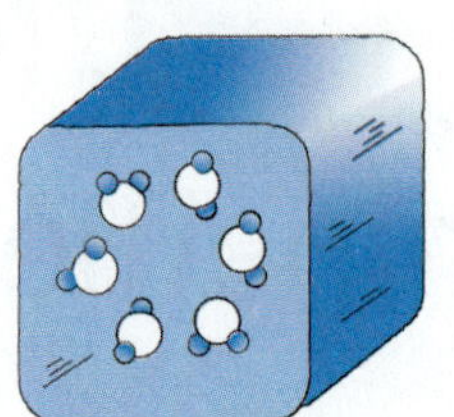
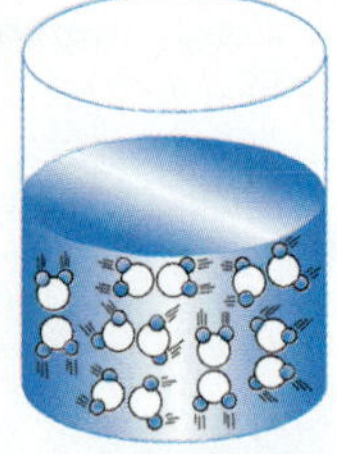
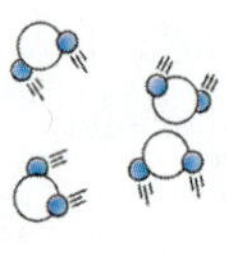

Figure 7-3 Three states of matter.

molecules. Energy binds atoms to each other to form molecules (Figure 7-2). Because of their atomic structure, some materials have a predisposition to combine with other material. Oxygen is an example of an element that is normally only found in molecules made up of the same type of atoms. Two oxygen atoms join to form one molecule of O_2 (pronounced *O-two*), the air we breathe.

Different types of atoms can also join to form different types of matter. Water (H_2O), for example, is formed when two hydrogen (H + H) atoms and one oxygen (O) atom unite. The amount of energy holding the atoms together changes when they combine to form molecules. This change absorbs or releases energy. When hydrogen combines with oxygen, a large amount of heat is released. Chemical energy is the energy change that takes place when the atoms of materials rearrange themselves to form entirely new materials from the same atoms. It is important to remember that no matter or energy is "lost"; it just changes.

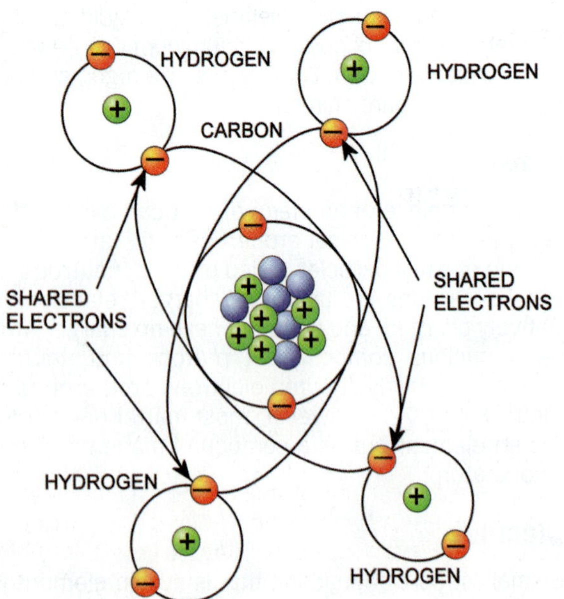

Figure 7-2 The methane molecule is composed of one atom of carbon and four atoms of hydrogen.

7.4 STATES OF MATTER

In air-conditioning and refrigeration work, you will encounter matter in three states: solid, liquid, and gas, as shown in Figure 7-3. The state that a substance is in depends upon its physical properties, the pressure applied to it, and its temperature. Water is an example of matter that is commonly found at standard atmospheric pressure in three states. It is a solid (ice) at 32°F (0°C), a liquid between the temperature of 32°F (0°C) and 212°F (100°C), and a vapor or gas at 212°F (100°C) and above. These three different states of matter each represent a different level of molecular energy. The energy levels from lowest to highest are solid, liquid, and gas.

7.5 SOLID

Solids have the lowest energy level of these three physical states. The molecules of a solid remain basically fixed in place. They do vibrate, but they do not move about. Solids have a definite shape and they will hold their shape under some stress. Gravitational force on solids creates downward pressure on the surface that the solids rests on. Small changes in the volume of a solid occur from temperature change. The volume increases slightly when the temperature is increased and decreases slightly when the temperature is decreased. Examples of solids are steel, bricks, rocks, ice, wood, and paper.

7.6 LIQUID

The molecules of a liquid have a higher energy level than the molecules of a solid. They move freely about while still maintaining close contact with each other. Liquids take the shape of their container and must be supported on the sides as well as the bottom. When liquid fills an open container, it exerts both a horizontal and a downward pressure on the container. This pressure is due to gravity and is the greatest at the bottom and decreases to zero on top of the surface. Examples of liquids are water, oil, gasoline, paint, alcohol, milk, honey, and mercury.

Small changes in the volume of a liquid can occur with changes in temperature. The volume increases slightly when the temperature is increased and decreases slightly when the temperature is decreased.

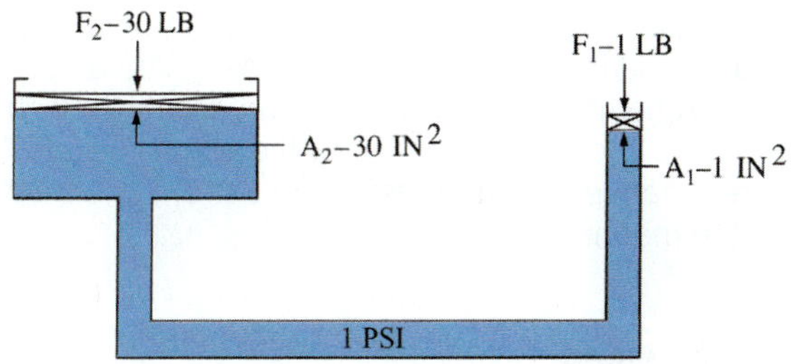

Figure 7-4 Pressure is exerted evenly on all surfaces in a hydraulic system.

Liquids are noncompressible, meaning their volume will not change under pressure. In a closed container that is completely filled with liquid, any pressure applied to the liquid pushes out evenly throughout the container. This is the principle behind the application of hydraulics.

Figure 7-4 shows a hydraulic system containing a liquid such as oil. The system has a small cylinder and a large cylinder connected by a pipe, with tight-fitting pistons in each cylinder. The cross-sectional area of the small piston is 1 in^2, and the area of the large piston is 30 in^2. A force of 1 lb when applied to the smaller piston will create a pressure of 1 psi throughout the system. This pressure will support a weight of 30 lb on the larger piston because the force of 1 psi is applied to every square inch of surface: 30 $in^2 \times 1$ psi $= 30$ lb.

7.7 GAS

The molecules of a gas have a higher energy level than the molecules of a liquid. They move freely about and avoid close contact with each other. A gas or vapor has no fixed shape or volume. Gases must be contained on all sides in a closed container or they will escape into the atmosphere. A gas fills its container space completely, and its pressure is dependant on its mass and temperature. That means that the more gas you put into the same size container, the higher the pressure. A gas exerts its pressure on its container uniformly in all directions (Figure 7-5). Examples of gases are air, steam, refrigerant, CO_2, natural gas, and helium.

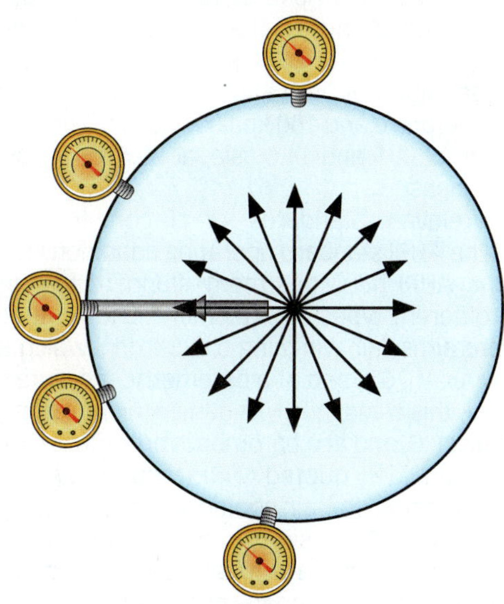

Figure 7-5 Gas exerts pressure evenly on all surfaces containing it.

As with both solids and liquids, changes in the volume of a gas can occur with changes in temperature. The volume increases when the temperature is increased and decreases when the temperature is decreased. Unlike solids and liquids, the volume of a gas can be changed dramatically by simply reducing the size of its container. Reducing the volume of a gas by reducing the size of its container is called compression. Gases are the most variable of the three states because all of their physical characteristics can be easily manipulated.

TECH TIP

A fluid is anything that flows. Most people think of liquid as being the only fluid, but gases are also fluids. The wind blowing outside is acting as a flowing fluid, in this case air. Flowing air is much like running water in a stream—flowing air exerts force on objects. In the same way, refrigerant vapor is a fluid flowing through the system, and its inertia applies force. When it makes a turn as a result of a 90°-angle in a pipe, the inertia of the fluid is affected by the 90°-turn.

7.8 CRITICAL POINT

As a liquid-vapor mixture is heated in a closed space, the pressure and temperature of both increase. The liquid and vapor become more alike as the temperature and pressure increase. The density of the liquid decreases while the density of the gas increases. At a high enough temperature and pressure, there is no longer a distinction between liquid and vapor, and all of the substance behaves like a vapor. The temperature and pressure where the liquid state no longer exists is called the critical point. The temperature, pressure, and specific volume at the critical point are called the critical temperature, critical pressure, and critical volume.

7.9 TRIPLE POINT

The state of a substance is dependent on both the pressure and temperature. There is only one specific pressure and temperature for each substance where it can exist as a solid, liquid, and gas simultaneously in thermal equilibrium. This specific pressure and temperature is called the triple point. This means that all three states are present and they are not in the process of changing from one state to another. A slight change in either the pressure or temperature will cause one of the three states to disappear. The triple point of most substances is less than atmospheric pressure. The triple point of pure water is at 0.01°C at 7.58 mm of mercury absolute pressure, which is the same as 32.02°F at 0.09 psia. Psia measures absolute pressure—pressure above a perfect vacuum. The pressure of 0.09 psia is very small. By comparison, atmospheric pressure is 14.7 psia.

TECH TIP

Since the triple point will only occur at a very specific pressure and temperature for each substance, the triple points of several common materials are used to precisely calibrate thermometers.

7.10 PROPERTIES OF MATTER

The properties of matter can be divided into two areas: chemical properties and physical properties. A substance's chemical properties describe how it reacts to other chemicals. These properties determine whether a chemical will react with other chemicals to form new compounds and what type of compounds will be formed. Some of matter's chemical properties include heat of combustion, reactivity, and pH. Chemical properties of matter are covered as required in the appropriate units in this book.

A substance's physical properties help determine its size, shape, form, and state. Some of matter's physical properties include weight, mass, density, and thermal properties. The thermal properties of matter are particularly important because most HVACR work relates to the addition or removal of heat from matter.

7.11 WEIGHT AND MASS

Mass is a way of measuring "how much" matter an object has. The definition of *mass* is the fundamental measure of all the matter in the object as a measure of its inertia. Inertia is the tendency of things that are at rest to stay at rest and for things that are in motion to stay in motion. Items with a larger mass have more inertia than items with less mass. Heavier objects are harder to start moving than lighter objects because of their inertia. Heavier objects are also harder to stop moving once they start due to their inertia.

On earth, weight is used to measure mass because the weight of an object is proportional to its mass. The definition of *weight* is the force of gravity acting on an object. This means that objects with more mass weigh more. The weight of an object will change when the gravitational forces on it change. The mass of an object does not change and is not dependent on gravity. Although the *Apollo* astronauts weighed less on the moon than they did on earth, their mass remained the same. Figure 7-6 compares the weight and mass of an astronaut on earth and on the moon.

Figure 7-6 An astronaut weighing 180 lb on earth weighs only 30 lb on the moon. *(Courtesy of NASA)*

7.12 VOLUME

Volume is a measurement of the space taken up by matter. In traditional units, this is usually cubic feet (ft^3); in SI units, it is usually liters (L). The volume of a substance is not fixed; it changes with changes in temperature. In general, all matter expands when it is heated and contracts when it is cooled. The volume of a solid changes only slightly with temperature change. Liquids change more in volume than solids with temperature change, and gas volumes change a great deal with temperature change.

Solids and liquids do not change in volume with pressure changes, but gases do. A gas can be compressed into a much smaller volume or expanded into a much larger volume.

7.13 STANDARD CONDITIONS

The conditions under which a test is performed can affect the results of the test. The pressure and temperature conditions that a material is tested at can affect its volume. This will have an impact on density, specific volume, and specific gravity calculations. It is important when comparing system performance or performing calculations on material properties that the tests and calculations be done at the same conditions. Standard conditions are established by industry organizations to make it easier to compare the results of tests run by different people.

Unfortunately, since there are many types of tests and many organizations, there are many standards. It is not enough to state that you are using "standard conditions"; you need to specify whose standard you are using. For most chemistry and physics, the Standard Temperature and Pressure, or STP, is used. These were established by the International Union of Pure and Applied Chemistry in 1997 as 0°C temperature and 100 kpa (kilopascals) pressure. This translates into 32°F and 14.5 psia, or 29.5 in Hg absolute in traditional measure.

More relevant standards for air-conditioning technicians are the AHRI standard operating conditions for testing equipment. AHRI publishes many standard testing conditions for different types of equipment. The AHRI definition of *standard air* is "air weighing 0.075 lb/ft^3, which approximates dry air at 70°F and at a barometric pressure of 29.92 in Hg." In SI, this translates to a density of 1.2 kg/m^3, a temperature of 21°C, and at a barometric pressure of 101.3 kpa.

The most widely quoted AHRI standard is probably the "A" test condition for air-cooled air-conditioning equipment. This is 95°F outside ambient temperature, 80°F indoor dry bulb temperature, and 68°F indoor wet bulb temperature. When a company publishes the performance of their air-conditioning equipment, they typically reference this standard.

7.14 DENSITY

Density compares the weight of a substance to its volume. To find the density of something, divide its weight by its volume. In the United States, we typically use pounds for the weight and cubic feet for the volume. Thus, the density of something tells us how much it weighs per cubic foot. "Heavier" substances, such as lead, will have a high density, while "lighter" substances, such as wood, will have a low density.

A material's density is not fixed. Since a substance's volume changes with pressure and temperature, its density also changes with pressure and temperature. This is especially true for gas because gas volume is subject to far greater changes than either solid or liquid. Gas can be compressed, or squeezed, to a smaller volume. The amount that a solid or liquid can be squeezed is very slight. Gas also expands and contracts with temperature changes much more than either solid or liquid. Since volume can change with temperature and pressure, a reference temperature and pressure are needed when determining density. Density is meaningless without a reference temperature and pressure.

In air conditioning, one practical impact of density changes in a gas is the fact that hot air rises. Hot air is less dense than is cold air because hot air expands. This means that the same weight of air takes a greater volume. Another way to look at this is that there are fewer air molecules in the same volume, making the hot air lighter. The less dense hot air rises above the denser cooler air. This is why the temperature of the air near the ceiling is usually a few degrees warmer than the temperature of the air near the floor (Figure 7-7).

Figure 7-7 Warm air rises to the top of a room because of its lower density.

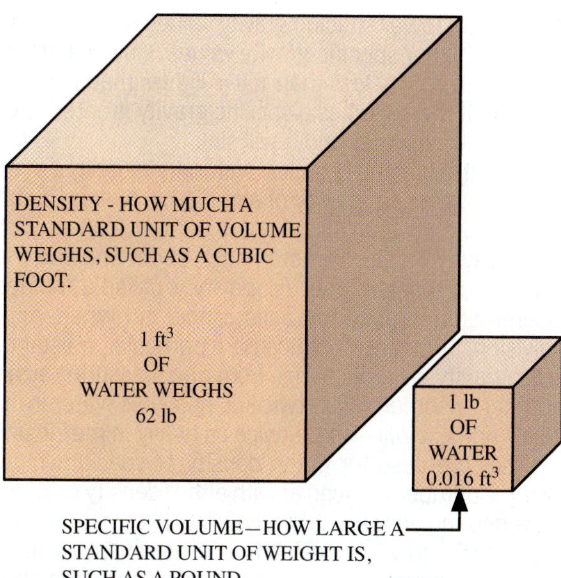

Figure 7-8 Density and specific volume.

7.15 SPECIFIC VOLUME

Specific volume is the weight of a standard unit of measure for a material. Specific volume uses the same information that density does, only backward. Specific volume compares the volume of a substance to its weight (Figure 7-8). To find the specific volume of a substance, divide its volume by its weight. Using traditional measures, specific volume tells us how much space a pound of something will take up. Like density, a specific-volume specification should have a reference temperature and pressure because a substance's specific volume will also change with pressure and temperature.

TECH TIP

The specific volume of a refrigerant can be used to help size refrigeration compressors. The amount of refrigerant required for any particular job is determined by dividing the total amount of refrigeration needed by the refrigeration capacity of 1 lb of refrigerant. This yields the number of pounds of refrigerant the compressor must move. However, compressors are rated by the volume of gas they can move, not by the weight of refrigerant. Multiplying the refrigerant's specific volume times the weight of refrigerant yields the volume of refrigerant the compressor must move.

7.16 SPECIFIC GRAVITY

Specific gravity is a way of relating a material's mass and volume to a constant. It is the ratio of the density of a material to the density of a common material. Solids and liquids are compared to water; gases are compared to air. The constant used has a specific gravity value of 1 because that is always the result when any number is divided by itself.

Materials with lower specific-gravity values will float on materials with higher specific-gravity values. If a solid material has a specific gravity less than 1, it is lighter than water and it will float. If the solid has a specific gravity greater than 1, it is heavier than water and it will sink.

The same is true of gases. Natural gas is lighter than air and has a specific gravity of about 0.60, making it about 40 percent lighter than air. LP gas is heavier than air with a specific gravity of approximately 1.52, about one and a half times heavier than air. Specific gravity is called a "unitless" measurement because the units cancel out when you do the division. This is useful because a particular substance's specific gravity will not change from one measuring system to another. If something is twice as heavy as water in "traditional" units, it will also be twice as heavy in metric units. This is not the case for either density or specific volume. The actual numbers arrived at with either density or specific volume depend very much on the units used. On the other hand, specific gravity remains the same regardless of the units used to arrive at it. It is important to compare apples to apples. If you state water's density in pounds per cubic foot, you also must state whatever you are measuring in pounds per cubic foot.

SERVICE TIP

The specific gravity of a fuel gas is important when using pipe sizing charts. A pipe's resistance to flow changes with the specific gravity of the gas traveling through it. Gas piping offers more resistance to gases with higher specific gravities than to gases with lower specific gravities. For this reason, gas piping charts specify the specific gravity of the gas they are designed for.

7.17 EXAMPLE CALCULATIONS

The AHRI definition of *standard air* will be used as the reference temperature and pressure for all our examples. Take the example of a gas with a weight of 4 lb and a volume of 64 ft^3. To calculate density, divide 4 lb by 64 ft^3 to arrive at a density of 0.0625 lb/ft^3. To calculate specific volume, divide 64 ft^3 by 4 lb to get a specific volume of 16 ft^3/lb. You can easily convert between density and specific volume by dividing them into 1. In our above example, dividing 16 into 1 yields 0.0625, and dividing 0.0625 into 1 yields 16.

For another example, examine a solid that weighs 300 lb and takes up 3 ft^3. To calculate density, divide 300 lb by 3 ft^3 and arrive at a density of 100 lb/ft^3. To calculate specific volume, divide 3 ft^3 by 300 lb to get a specific volume of 0.01 ft^3/lb. You can easily convert between density and specific volume by dividing them into 1. In our above example, dividing 0.01 into 1 yields 100; dividing 100 into 1 yields 0.01.

Notice that in each case density and specific volume are "opposites." If something has a high density, it will have a low specific volume. Likewise, if it has a low density, it will have a high specific volume. In general, solids and liquids have high densities and low specific volumes. Gases are just the opposite; they tend to have high specific volumes and low densities.

To calculate specific gravity of a solid or liquid, divide its density by the density of water. You only need to remember to compare apples to apples. If you state water's density in pounds per cubic foot, you also must state whatever you are measuring in pounds per cubic foot.

The density of water in traditional units is usually stated as 62.4 lb/ft^3 at standard conditions. Applying this to the solid in the second example, the specific gravity of the solid would be 100 divided by 62.4, or 1.6 (rounded). This substance would be much heavier than water and would sink.

For gases, the density of the gas is compared to the density of air. The density of air at standard conditions is usually stated as 0.075 lb/ft^3. The specific gravity of the gas used in the first example would be 0.0625 lb/ft^3 divided by 0.075 lb/ft^3, or 0.83 (rounded). This substance is lighter than air and would rise.

SI Calculations

SI calculations for density and specific volume are performed in the same way as for traditional units, just using SI units. For example, density can be calculated in kilograms per liter, while specific volume would be liters per kilogram. Take the example of a liquid with a weight of 3 kg and a volume of 4 L. To calculate density, divide 3 kg by 4 L to arrive at a density of 0.75 kg/L. To calculate specific volume, divide 4 L by 3 kg to get a specific volume of 1.3 (rounded) L/kg.

Metric units provide a shortcut when determining specific gravity. Since 1 L of water weighs 1 kg, the density of water is 1 kg/L. This means that densities calculated in kilograms per liter also represent the specific gravity. In the example above, the specific gravity of the liquid is 0.75 because it is equal to the density. This also works for the smaller units of grams and milliliters. A substance's density in grams per milliliter is also its specific gravity.

7.18 MELTING AND FREEZING POINTS

The melting point of a material is the temperature at which the material begins to change from a solid state to a liquid state when heat is added. When a substance changes from a solid to a liquid, the potential energy of the molecules increases, but the temperature stays the same. This means that when a solid melts, the liquid being formed is the same temperature as the solid it came from.

The freezing point of a material is the temperature at which the material begins to change from a liquid state to a solid state when heat is removed. The melting point and freezing point are the same except for the direction of heat flow. If heat is flowing into the material, it is melting, and if it is flowing away from the material, it is freezing. Figure 7-9 lists a single temperature for both the melting and freezing temperature for the materials shown.

Mixtures of two or more substances can create a solution with a substantially lower freezing point than either substance alone. Ethylene glycol and water is a good example. Pure ethylene glycol, the primary component in antifreeze, freezes at around 8°F, and water freezes at 32°F. Systems exposed to cold winter temperatures easily freeze at these temperatures. But a solution of 60 percent

Material	F°	C°
Ethanol	−173°F	−114°C
Mercury	−38°F	−39°C
Beeswax	145°F	65°C
Copper	1,984°F	1,085°C
Aluminum	1,220°F	660°C
Ethylene glycol	8.6°F	−13°C

Figure 7-9 Melting points for several materials.

ethylene glycol and 40 percent water freezes at a much lower −48°C (Figure 7-10).

7.19 EVAPORATION

Evaporation is gradual change of state from liquid to gas that occurs when the liquid is below the boiling point. Only the molecules at the surface of the liquid can evaporate. The molecules in the liquid are in constant motion but traveling at different speeds. Liquid molecules are frequently colliding with each other. If a molecule at the surface is struck hard enough by another molecule, it may gain enough kinetic energy for it to leave the liquid and become a gas molecule even though the temperature of the liquid is below the boiling point.

Since higher temperatures mean higher average molecular speed, evaporation increases as the temperature of the liquid increases. Increased surface area also increases evaporation because evaporation only occurs at the surface. A third factor is the concentration of the gas molecules around the liquid. Water evaporates more slowly when the air around it is at a high relative humidity.

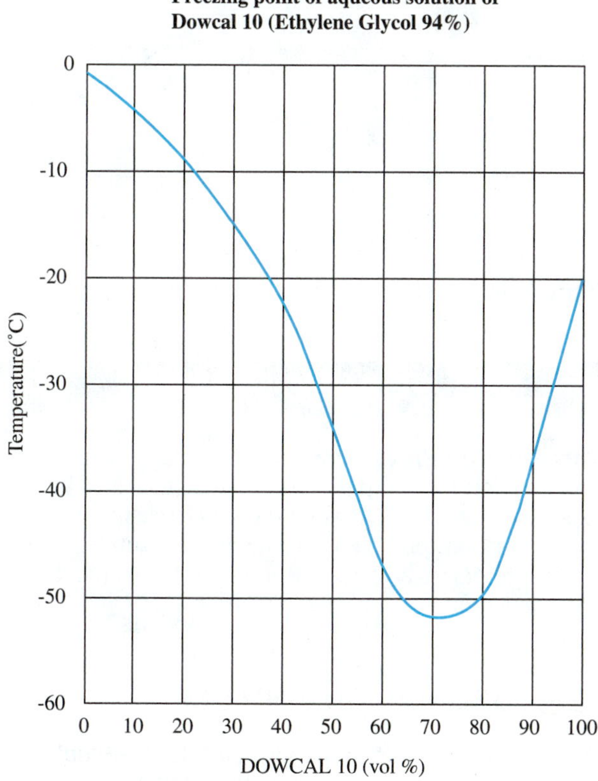

Figure 7-10 The freezing point of water/glycol solutions changes depending on the percentage of glycol.

TECH TIP

Evaporation is very important to everyday life. If water did not evaporate at temperatures below the boiling point, after mopping a floor the water would remain on the floor indefinitely.

Some liquids, like oils, do not appear to evaporate. If you spill some on the counter today it will still be there tomorrow. This is because their molecules do not strike each other hard enough or at the correct angle to eject the surface molecules from the liquid.

7.20 BOILING AND CONDENSING POINTS

Boiling describes a rapid change from a liquid state to a gaseous state. The boiling point of a liquid is the temperature at which the saturated vapor pressure of the liquid equals the surrounding atmospheric pressure. The difference between evaporation and boiling is that when a liquid is boiling, molecules are changing throughout the liquid and forming gaseous bubbles, while evaporation takes place only on the surface. Heat is required to make a substance boil. When a liquid boils, the potential energy of the molecules increases, but the temperature stays the same. This means that when water boils, the water and the steam coming off the water are both 212°F (Figure 7-11).

The condensing point is the temperature at which gaseous molecules are releasing heat and joining together to form a liquid. Condensation of a vapor occurs when the molecules of gas slow down as heat is removed. The molecules begin to clump together and form a liquid. Condensing releases the potential energy that was absorbed when the

Figure 7-11 Both the water and the steam are 212°F.

Figure 7-12 The water in a pressure cooker boils at 257°F under a pressure of 15 psig.

Figure 7-13 The cap on a car radiator increases the boiling point of the engine coolant by increasing the pressure in the radiator.

molecules changed from a liquid to a gas. The difference between the boiling point and the condensing point is which direction heat is flowing.

7.21 EFFECT OF PRESSURE ON BOILING POINT AND CONDENSING POINT

The boiling and condensing points of any liquid are controlled by the pressure. Remember that the boiling point occurs when the vapor pressure of the gas bubbles in the liquid equal the pressure surrounding the liquid. When that pressure is increased, the liquid molecules have to gain more kinetic energy to reach escape velocity to become a gas. Thus, increasing the pressure on a liquid increases the boiling point.

For example, the boiling point of water at atmospheric pressure is 212°F. When the pressure on the water is increased to 15 psi above atmospheric pressure, the boiling point is increased to 257°F. This is the secret of pressure cookers. By increasing the pressure, the water in the pressure cooker heats to a temperature higher than 212°F (Figure 7-12).

This same principle is used in car radiators. The radiator cap maintains a pressure above atmospheric pressure in the engine's cooling system, effectively raising the boiling point of the engine coolant to prevent it from boiling (Figure 7-13).

A lower pressure will result in a lower boiling point. Reducing the pressure below atmospheric with a vacuum pump will lower the boiling temperature of water. Figure 7-14 shows the boiling point of water at different pressures.

The refrigerant cycle is based on changing the pressure of a liquid to change its boiling and condensing temperatures. By lowering the pressure, the boiling point of the refrigerant is lowered enough so it can boil and pick up heat at a lower temperature. Raising the pressure on the refrigerant vapor makes it possible to condense the vapor back into a liquid at a higher temperature by removing heat.

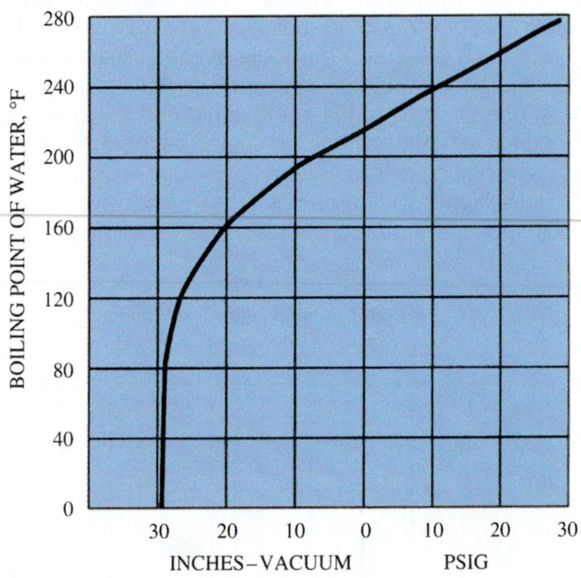

Figure 7-14 Pressure-temperature curve for water.

SERVICE TIP

The purpose of a vacuum pump is to ensure that the refrigeration system has no air or moisture in it before charging it with refrigerant. The low pressure created by the vacuum pump causes any water droplets in the system to boil to a gas so the water can be removed by the vacuum pump.

7.22 SUBLIMATION AND DEPOSITION

When a solid changes to the gaseous state without forming a liquid, it is called sublimation. Under some condition, molecules at the surface of the solid can change from a solid to a gas. An example of sublimation is ice that disappears in a frost-free refrigerator. The ice has not truly disappeared; it has sublimated to a gas. The water vapor is then deposited on the evaporator as frost, changing from a gas back to a solid. When a gas changes to a solid without going through the liquid phase, that process is called deposition.

SERVICE TIP

"Freezer burn" is caused by sublimation. The water in frozen food slowly sublimes out, leaving the food dry and tough. Freezer burn can be reduced by using heavy-duty plastic wrap and containers designed to reduce the amount of water that can pass through the container.

UNIT 7—SUMMARY

Matter exists in three forms: solids, liquids, and gases. Adding heat can change a solid to a liquid and a liquid to a gas. Solids and liquids do not change size by adding increased pressure. Gases do change size with changes in pressure. An increase in pressure can increase the boiling point of a liquid, or the temperature at which a liquid turns to a gas. This is the basic principle on which all HVACR systems work.

UNIT 7—REVIEW QUESTIONS

1. What is matter?
2. What is the difference between an atom and a molecule?
3. List the three states of matter in order of their energy level, from lowest to highest.
4. Explain the difference between the weight of an object and the mass of an object.
5. Calculate the density, specific volume, and specific gravity of a gas that occupies 20 ft^3 and weighs 2 lb.
6. If the gas in question 5 were released in a room, would it be more likely to be found near the ceiling or near the floor?
7. Calculate the density, specific volume, and specific gravity of a solid that occupies 3 ft^3 and weighs 9 lb.
8. If the solid in question 7 were placed in water, would it sink or float?
9. Explain the relationship between specific volume and density.
10. What is the AHRI definition of *standard air*?
11. Why are testing standards important when specifying density and specific volume?
12. What is the difference between the melting and the freezing point?
13. What effect does pressure have on the boiling temperature of a liquid?
14. What is the temperature of the water formed when ice melts?
15. Give a practical application in the HVACR industry for specific gravity.
16. State the law of conservation of matter.
17. What is a substance's critical point?
18. What are the two general categories for properties of matter?
19. What physical change of state is the opposite of boiling?
20. What physical change of state is the opposite of sublimation?

UNIT 8

Types of Energy and Their Properties

OBJECTIVES

After completing this unit, you will be able to:

1. explain the difference between potential and kinetic energy.
2. list common forms of energy and the units used to measure them.
3. explain how the structure of molecules and atoms affects chemical and electrical energy.
4. explain the relationship among energy, work, and power.
5. calculate energy conversions using watts, BTU, foot-pounds, and joules.
6. diagram the electrical distribution system.

8.1 INTRODUCTION

At its most fundamental level, heating and air conditioning is about controlling energy. Many tasks performed by air-conditioning technicians involve controlling and measuring energy. Heating is the process of adding thermal energy; air conditioning and refrigeration are processes that remove thermal energy. Other types of energy are often used when transferring thermal energy. Understanding energy and how it affects our world is crucial to understanding heating, air conditioning, and refrigeration.

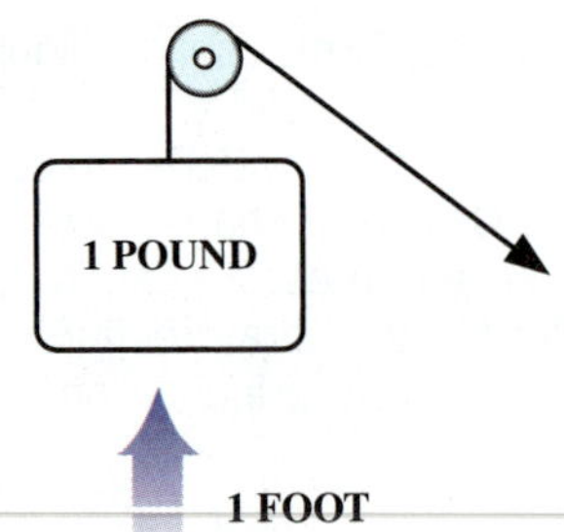

Figure 8-1 A foot-pound of work is moving 1 lb 1 ft.

8.2 ENERGY, WORK, AND POWER

The world and everything in it can be divided into two large categories: matter and energy. On earth, matter has weight and takes up space; everything around us is made of matter. Energy affects matter, but energy has no weight and does not take up any space.

Energy acts on matter to perform work. Work is done when energy is transferred to an object. For example, the mechanical energy used to compress a spring is transferred to the spring. The spring contains more energy than it did before it was compressed. When energy is transferred into an object, the object is changed in some way. Energy makes matter move, changes its temperature, and even changes its physical state.

The relationship among energy, work, and power is typically explained using the example of a moving object. Mechanical work is defined as force multiplied by distance. A foot-pound of mechanical work is performed when 1 lb is moved a distance of 1 ft (Figure 8-1). Moving 5 lb a distance of 2 ft would represent 10 ft lb of work. This describes the amount of work done, but it does not describe how long it took. Horsepower measures the rate of mechanical work. One horsepower (hp) is a rate of work equal to 33,000 ft lb/min. Moving 33,000 lb a distance of 1 ft over the course of 1 min would require 1 hp (Figure 8-2). Therefore, 1 horsepower (hp) is a rate of work equal to 33,000 ft lb/min. The relationship among energy, work, and power can be stated as follows: energy is the ability to do work, work is the result of transferring energy, and power is the rate at which the work is accomplished.

TECH TIP

James Watt designed and sold steam engines. He needed a way to compare the output of his machines to the primary means of power being used at the time: horses. He studied the work rate of horses and developed his formula for horsepower. He determined that on average, a horse could work at a rate of 33,000 ft/lb per minute. He rated his machines by horsepower so that people buying the machines had an idea of how much work they could accomplish.

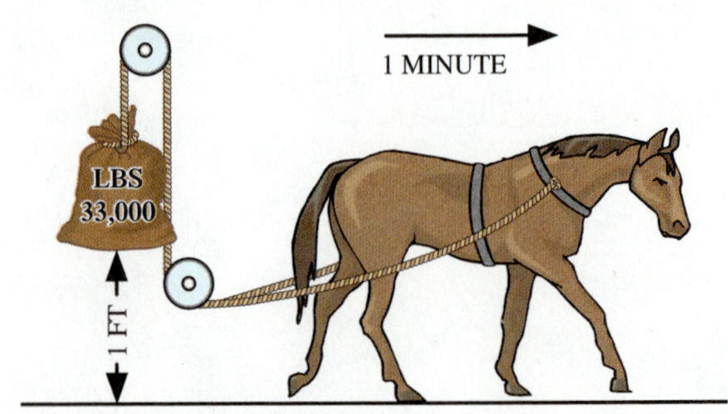

Figure 8-2 One horsepower is moving 33,000 lb 1 ft in 1 min.

8.3 POTENTIAL AND KINETIC ENERGY

Something that has potential has ability. Potential energy is energy that is stored, waiting to perform work. Something that is kinetic is in motion. Kinetic energy is energy in motion performing work. A battery is a good example of potential energy. The battery has the potential to create a flow of electrical current, but current does not flow until a circuit is created. When a path is formed, electrons flow from one pole to the other, creating a flow of electrical current. This converts some of the potential energy in the battery to kinetic energy because the electrons in the circuit are in motion. Other examples of potential energy include a compressed spring or water held in a reservoir behind a dam. Releasing the spring releases the mechanical energy by putting the spring in motion, converting potential energy to kinetic energy. Releasing water through the spillway of the dam converts the water's potential energy into kinetic energy because the water is in motion (Figure 8-3).

Potential and kinetic energy are also present at the molecular level. At temperatures above absolute zero, all molecules move or vibrate. The vibrations and movements are examples of kinetic energy because they are associated with motion. As the temperature of the molecules increases, the motion and kinetic energy also increase.

The physical state of the matter determines the molecule's potential energy. It takes energy to change the state of matter. The molecules of a liquid have less attraction for each other than the molecules of a solid. They move about freely because they have a higher level of potential energy than the molecules of a solid. Similarly, the molecules of a gas have an even higher amount of potential energy than a liquid. The energy that goes into changing the state of a substance is potential energy because it does not affect the motion of the molecules, only their positioning relative to each other. Large amounts of energy are required to change the state of a substance, but its temperature remains the same during this change of state. At the molecular level, energy that is associated with molecular motion and temperature is kinetic energy, while energy that is associated with a physical change of state is potential energy.

Figure 8-3 Flowing water is an example of kinetic energy. *(Colin Stitt/Shutterstock.com)*

8.4 FORMS OF ENERGY

We all use energy hundreds of times a day while going about our daily routine. Driving a car, heating a home, operating a computer, and even playing music all use energy. Energy exists in many forms, and some common forms of energy used in heating and air conditioning are the following.

- Chemical
- Thermal (heat)
- Radiant (light)
- Electrical
- Magnetic
- Mechanical

This is by no means a complete list of energy forms, just the forms of energy most often involved in the air-conditioning trade.

8.5 CHEMICAL ENERGY

The smallest part of anything that is not an element is a molecule. Energy binds atoms to each other to form molecules (Figure 8-4). Because of their atomic structure, some materials have a predisposition to combine with other materials. The amount of energy holding atoms and molecules together changes when atoms combine with each other to form new molecules. This change absorbs or releases energy. Chemical energy is the energy change that takes place when the atoms of materials rearrange themselves to form entirely new materials from the same atoms. It is important to remember that no matter or energy is "lost"; it just changes.

One common use of chemical energy is for the generation of electricity. Batteries are an example of chemical

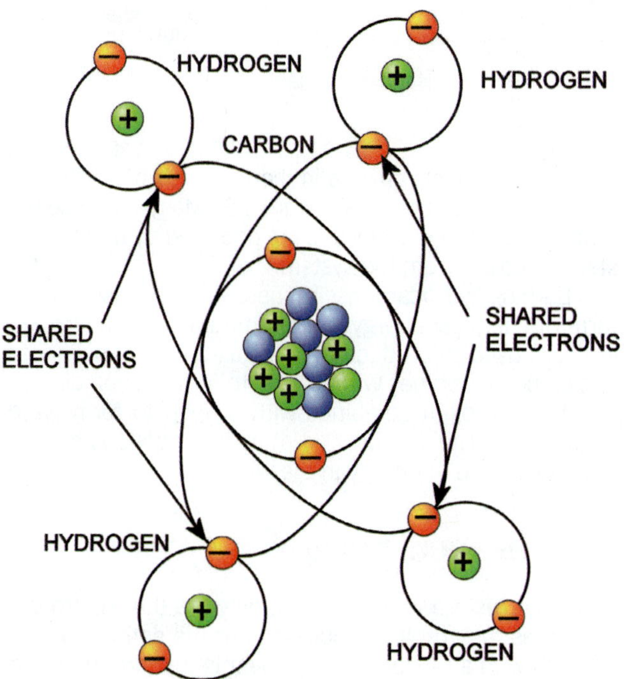

Figure 8-4 Methane molecule.

Figure 8-5 Batteries convert chemical energy into electrical energy.

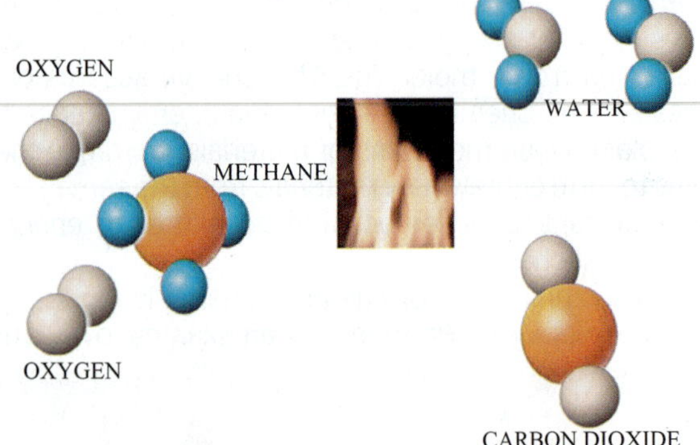

Figure 8-6 Combustion turns chemical energy into heat energy.

energy. A chemical change in the materials of the battery creates electrical potential (Figure 8-5). Many tools use batteries, and some thermostats are operated by batteries instead of power from the system.

Burning hydrocarbons, like natural gas, is also another form of chemical energy. Large amounts of chemical energy are used to produce heat in gas-burning appliances. The carbon combines with oxygen to form carbon dioxide, and the hydrogen combines with oxygen to form water (Figure 8-6). The combustion of hydrocarbons is still the primary source of energy in the world.

8.6 THERMAL ENERGY—HEAT

All molecules are in motion. Not only are the electrons in motion as they orbit the nucleus, but the entire molecule vibrates with a random motion. Heat is the form of energy that gives molecules their motion. For the same state, more heat equals faster molecular motion, while less heat equals slower molecular motion. Temperature increases or decreases as molecular motion increases or decreases. A change in thermal energy that results in a change in temperature is called a sensible heat change because the temperature change can be sensed. But not all changes in thermal energy result in a temperature change. Some changes result in a physical state change.

The arrangement and energy level of the molecules in a substance determine its physical state: solid, liquid, or gas. Solids are at the lower end of the energy spectrum. The molecules of a solid vibrate but remain more or less in the same spot. As their energy level increases, the temperature of the substance increases. When they reach the melting point, the molecules cannot stand still any longer and begin to freely move about. Now they are forming a liquid. As a liquid, all the molecules can move about, but they still maintain close contact with each other. As more energy is added, the molecules move about faster and faster and their temperature increases. At the boiling point, their energy level is too high to tolerate each other, and they separate from each other as much as possible. Now they are forming a gas (Figure 8-7).

Each of these state changes requires an enormous amount of energy. The energy added to a material at the

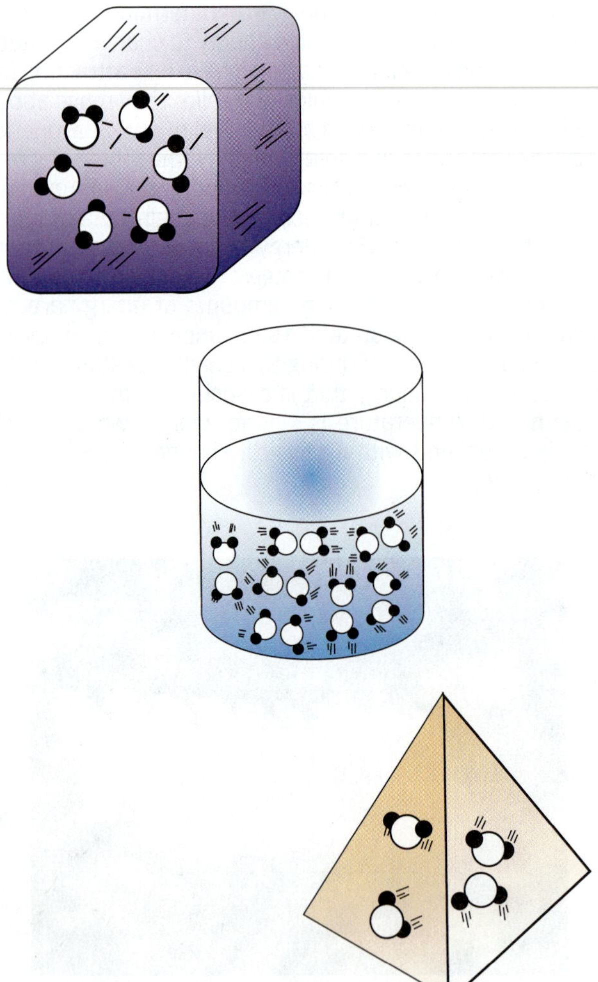

Figure 8-7 Matter can exist as a solid, liquid, or gas.

Figure 8-8 The water and ice are both 32°F.

melting point or boiling point does not result in a temperature change because all the energy is used to change the state of the substance. This is a crucial concept: when matter is changing state, its temperature remains the same. When ice melts, both the ice and the water melting from the ice are 32° (Figure 8-8). Once all the material has changed state, then the temperature can start to change.

A change in heat that results in a change in state is called latent heat, meaning hidden heat. This is because there is no temperature change accompanying the heat change. Of course, it is possible to see the steam coming off of boiling water, but it is not possible to measure a temperature difference between the water and the steam, because the boiling water and the steam are the same temperature.

Controlling thermal energy is usually the desired end result in heating and air conditioning. The end product of operating a furnace is supplying heat to the house. The end product of operating an air conditioner is removing heat from the house. Heat is also involved in the generation of most of the world's electrical power. Chemical energy is used to make heat, the heat is used to make motion, and motion and magnetic energy are used to make electrical energy.

8.7 RADIANT ENERGY

Radiant energy is electromagnetic energy that travels in waves. Heat, light, radio waves, and x-rays are examples of radiant energy. Radiant energy is measured by its wavelength and frequency. Wavelength refers to the distance between the peaks of the waves; frequency refers to the number of complete waveforms per second. Lower frequencies have longer wavelengths; higher frequencies have shorter wavelengths. Figure 8-9 illustrates the relationship between frequency and wavelength. The electromagnetic energy spectrum extends from radio waves at the lowest energy level to gamma rays at the highest energy level.

8.8 ELECTRICAL ENERGY

Electrons are the subatomic particles whizzing around the outside of an atom. Electricity is basically the flow of electrons. It is not like pouring electrons in one end of a wire and dumping them out the other end. Rather, electricity is more like a string of dominoes. One bumps the next and the energy flows from one end to the other, but no individual domino moves very far. With electricity, an electron moving from one atom to another creates an imbalance in that atom, which causes it to reject an electron. This electron then has a similar effect on a neighboring atom. This chain reaction carries through the wire to provide an electrical current.

Electricity is the mainstay of all heating and air-conditioning systems. The controls are all electrical, and many of the systems use electricity exclusively for their power source. While it is true that hydrocarbon

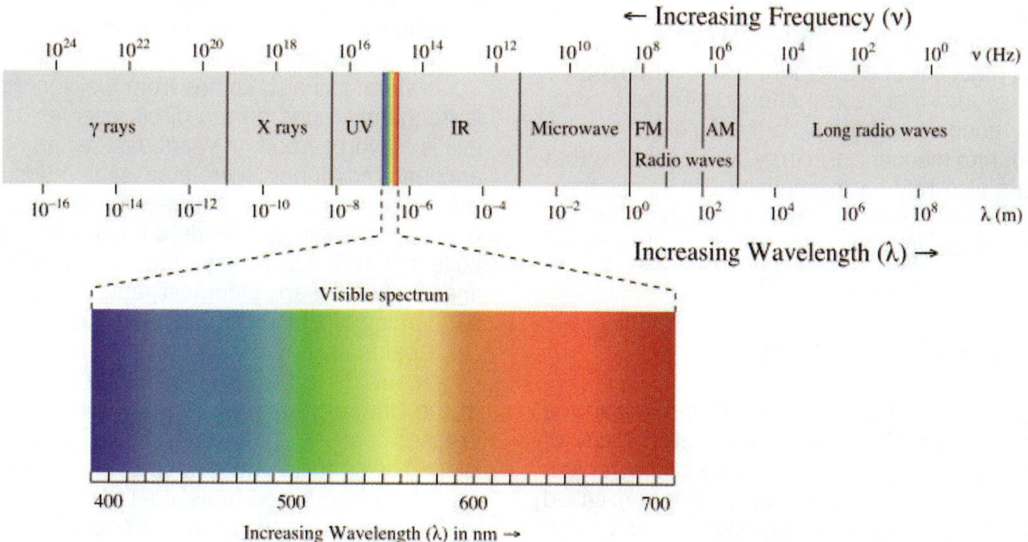

Figure 8-9 As the frequency increases, the wavelength gets shorter. *(Courtesy of NASA)*

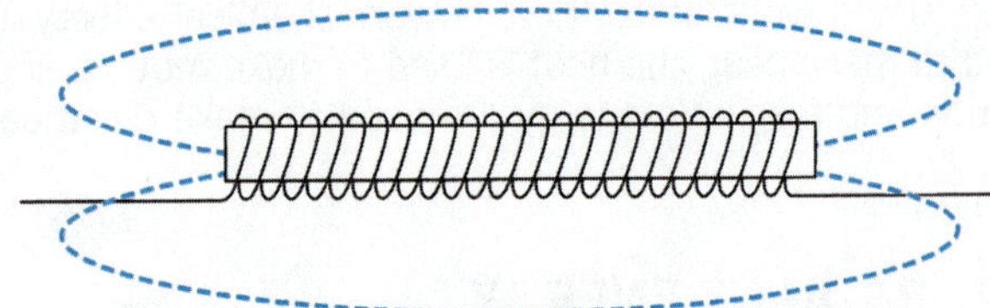

Figure 8-10 Electromagnetism produced by current flowing through a conductor.

combustion powers the world, a large percentage of that heat energy is used to generate electricity.

8.9 MAGNETIC ENERGY

Magnetic energy is created by the alignment of electric charges. This occurs when all the molecules of a polar material are aligned so that the positive ends of all the molecules face the same direction in the material. This alignment gives the entire object a polarity and sets up an invisible magnetic field. Some materials, like iron, are attracted to this field. A magnetic field is also created around a wire that has an electrical current flowing through it (Figure 8-10). Conversely, moving a magnetic field past a wire will "pull" electrons in the wire in the direction of the magnetic field. This creates an electric current in the wire. This relationship between magnetic fields and electrical flow is the foundation of electrical generators and motors.

It might appear that making electricity by moving a magnet past a wire is "free" energy because it is difficult to see anything happening, but turning a generator by hand will dispel this misconception. When the generator is not connected to an operating circuit, it is relatively easy to turn. When an electric circuit with a small light is connected, the generator gets considerably harder to turn. If the light is replaced with a larger one, the difficulty increases. With a large enough electrical load, the generator becomes nearly impossible to turn. However, if the load is turned off, the generator turns easily again. This is just another illustration of the law of conservation: you cannot get energy for free!

Magnetic energy is usually not an end result but a way of transferring mechanical energy into electrical energy or electrical energy into mechanical energy. All electric motors used in air conditioning use a magnetic field to turn an electric current into mechanical energy. Many controls that operate motors also rely on magnetism to turn electrical energy into mechanical energy. Although magnetic energy is seldom the end result, it has a part in nearly all electrical generation.

8.10 SOURCES OF ENERGY

We depend on a variety of energy sources to maintain the lifestyle enjoyed by people in the United States. Many of the energy sources we currently use cannot be replenished; after we use them up, they are gone. These are called nonrenewable energy sources. Other energy sources can be replenished, and so they are called renewable energy sources.

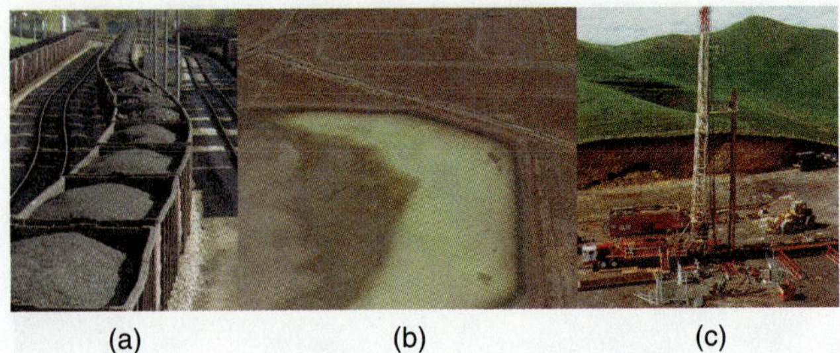

Figure 8-11 Non-renewable energy sources (a) Coal cars *(Courtesy of Indiana Office of Energy)* (b) Uranium mining in the Navajo Nation *(Courtesy of Sprol.com)* (c) Drilling for petroleum based energy *(Courtesy of OSHA)*

TECH TIP

Conservation of the limited supply of nonrenewable natural resources is important. There is a great deal of interest by citizens' groups and the government behind the HVACR industry's push for higher efficiency equipment. Heating and air conditioning accounts for approximately 50 percent of the energy used in most homes. Selling, properly installing, and properly servicing high-efficiency equipment is good for the environment. Properly working high-efficiency equipment will result in our using less of these valuable nonrenewable sources of energy. Properly working high-efficiency equipment is also good for customers, because their utility bills will be lower as a result of consuming less energy.

Nonrenewable Energy Sources

Nonrenewable energy sources have a limited supply. Currently, most of the world's energy is supplied by nonrenewable energy sources. Oil, natural gas, coal, and uranium are examples of nonrenewable energy sources (Figure 8-11).

Crude oil is recovered from wells (Figure 8-12) and distilled into a large number of hydrocarbon fuels, including fuel oil and gasoline. These refined products are used for heating, electricity generation, and transportation.

Natural gas also comes from wells in the ground. Gas is frequently found on top of oil reserves. Like oil, natural gas is a hydrocarbon. In years past, when natural gas was encountered during petroleum exploration, the gas was burned off to allow the petroleum companies to get to the oil. Today the gas is considered as valuable as the oil. It is collected, refined, and distributed nationally through pipelines for heating and electrical generation. Gas is the cleanest burning of all the carbon-based fuels.

Coal is essentially carbon. It must be mined by digging it out of the ground (Figure 8-13). Coal is still plentiful in the United States, but many problems come with increased coal use. Coal is the dirtiest burning of the carbon fuels. The sulfur content in coal contributes to acid rain. Like all carbon-based fuels, coal produces carbon dioxide when burned. However, since coal is nearly all carbon, it produces more carbon dioxide than either gas or oil. Because the United States has such large coal reserves,

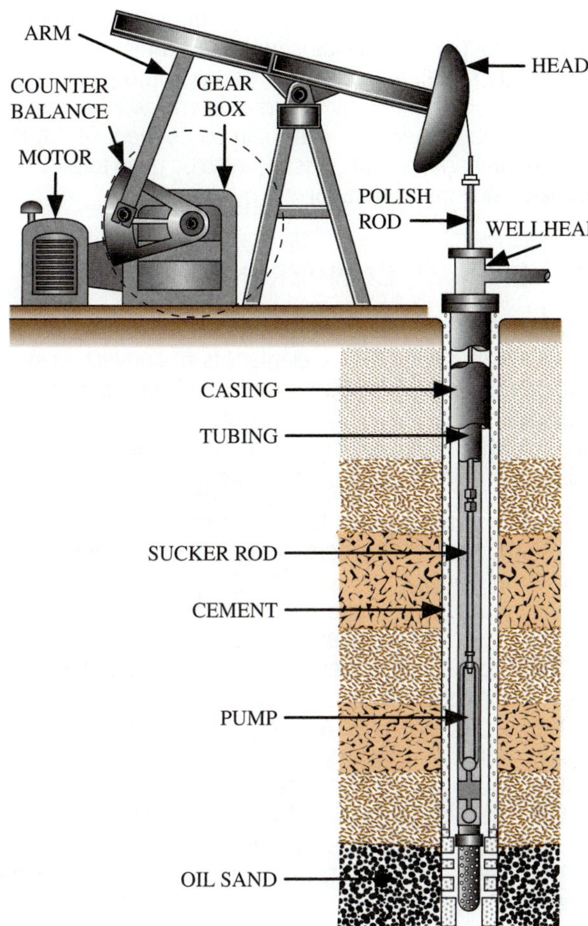

Figure 8-12 Oil well cross-section. *(Source: California Department of Conservation)*

many studies are underway to find cleaner uses of coal. The primary use for coal today is generating electricity.

Uranium is a radioactive element used today in nuclear power plants. It is possible to split uranium atoms to form two smaller atoms and release vast amounts of energy. The energy released from the fission, or splitting, of the uranium atom is used to heat water to steam to generate electricity. Nuclear power holds great promise, but public support has weakened because of the concern for safety. This is because there is a great potential for catastrophe, as was demonstrated at Chernobyl in Russia and during the 2011 earthquake in Japan, both of which caused nuclear emergencies. Another concern with nuclear power is managing the radioactive material left over from operating the plant.

Figure 8-13 Surface coal mine. *(abutyrin/Shutterstock.com)*

The world's growing energy demands have strained the current methods of retrieving and distributing these nonrenewable resources. The increasing energy demands of the world cannot be supported indefinitely, as we are using up finite sources of energy at an ever-increasing rate.

Renewable Energy Sources

Renewable energy sources can be replenished. Interest in renewable energy sources is increasing as researchers look for long-term solutions to the world's energy appetite. Some of the renewable energy sources that have been explored include solar, wind, hydro, ocean, geothermal, and biomass.

Solar energy can be used to generate electricity using photovoltaic cells, or it can be converted into thermal energy for heating water and buildings. A few large-scale solar-generating stations have been constructed in the southwestern United States that generate electrical power and distribute it over the existing electrical grid (Figure 8-14).

Wind energy is primarily used to generate electricity. Wind generators are in use from small windmills to generate electricity in homes to large-scale projects that harvest wind power with hundreds of large, strategically placed windmills (Figure 8-15).

Figure 8-14 Large-scale solar power–generating plant. *(Courtesy of the Office of Energy Efficiency and Renewable Energy, U.S. Department of Energy)*

Figure 8-15 Wind farms use multiple large wind turbines to generate electrical power. *(Stephen Bures/Shutterstock.com)*

Hydroelectric power has been used for years. Projects like the Tennessee Valley Authority (TVA) have sought to turn the raw power of large rivers into electricity. However, the potential is somewhat limited, and environmental concerns have been raised regarding the drastic changes in the ecosystem that a large-scale hydroelectric plant brings. Even in the TVA, one of the world's most ambitious uses of hydroelectric power, only 10 percent of the electricity is generated by hydroelectric power.

Ocean energy can be harnessed from the waves or from the rising and falling tide. The wave motion has been used to turn generators. The tidal change can be used in coastal areas that have a high degree of tidal change between high and low tide. Gates are constructed that control the flow of sea water into and out of large areas, trapping water at high tide. The water is then released through a turbine to generate electricity.

Geothermal energy is heat from inside the earth. Hot water and steam flowing out of the ground can provide heat in areas with active thermal springs or geysers. Geothermal heat pumps can extract heat from the ground, even at relatively low temperatures. A geothermal heat pump can move three times as much heat as it consumes.

Biomass energy refers to energy derived from plants and animals. This can be as simple as burning plants for heat or as complex as converting garbage into gas and oil. Decaying trash, plants, animals, and animal wastes can be used to produce methane, the main component of natural gas. Plant and animal materials can also be used to produce fuels like biodiesel or ethanol.

8.11 ENERGY CONVERSION

Energy can be converted from one form to another. The purpose of an electrical power plant is to convert existing forms of potential energy, such as coal or natural gas, into electrical energy. (Figure 8-16) shows a modern electrical power plant. The energy conversions required to operate an air conditioner include the following:

1. Coal or natural gas is burned at the generating plant to produce steam, converting chemical energy into thermal energy.
2. The steam is used to turn a turbine, converting thermal energy into mechanical energy.
3. The turbine turns a generator, converting mechanical and magnetic energy into electrical energy.
4. The electricity is used to operate electric motors, turning electric energy into magnetic energy.
5. The electric motors use the spinning magnetic field to turn the compressor and fans, converting magnetic energy into mechanical energy.
6. The mechanical energy is used to transfer thermal energy, heat, from where it is not wanted to where it is unobjectionable .

Note that the whole process began by turning some matter, coal and oxygen, into other forms of matter, water and carbon dioxide. When hydrocarbons are burned, the carbon is combined with oxygen to form carbon dioxide, while the hydrogen is combined with oxygen to form water. The coal does not disappear; it is changed into carbon dioxide gas and water vapor. When the entire process is through, there is just as much matter as before. Also, there is just as much energy as before; it is just converted into heat.

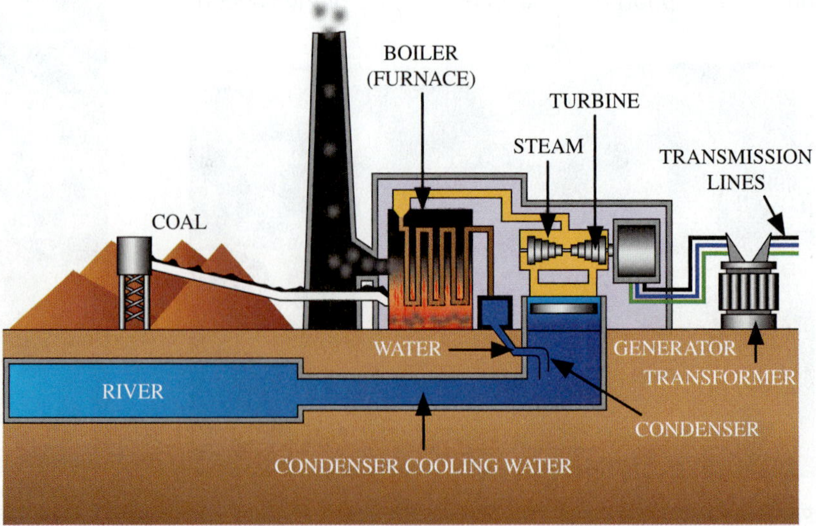

Figure 8-16 Diagram of a coal-fired electric-generating plant.

These conversions are not perfect. Some energy is "lost" with each conversion. But energy is not really "lost." Some of it is not directed toward the desired outcome. When the coal is burned, some of the heat leaves with the hot combustion gases. This energy is not lost in the sense that it no longer exists. It still exists, but some of the heat energy was not captured when it was converted. Likewise, all the heat energy that is captured is not used to turn the turbine; there is still a considerable amount of energy in the steam after it has passed through the turbine. Once again, all the energy was not completely captured, but the "missed" energy still exists. In virtually any energy conversion there is "lost" energy. However, it is important to understand that when energy is "lost" it does not disappear; it just does not go where it is supposed to go.

Common Energy Conversions

It would be accurate to say that the modern lifestyle we enjoy depends upon energy conversions taking place all around us that we take for granted. Operating any appliance, heating and cooling our homes, and transporting people and goods all depend upon energy conversion. Even enjoying your favorite music on your MP3 player requires converting chemical energy into electrical energy to operate, converting that electrical energy into magnetic energy to drive a speaker or headset, converting that magnetic energy into mechanical energy to make the speaker cone move, and finally converting the mechanical movement of the speaker into sound energy so that it can be heard. Some common energy conversions are shown in Figure 8-17.

Energy Conversion Is Reversible

Not only can energy be converted from one form to another, but the conversion can be reversed. A car battery is a good example of a practical use for reversible energy conversion. A car battery is basically a tank full of acid with two different types of metal plates dipped in the acid. The chemical reaction between the acid and one set of metal plates produces an excess of electrons on the metal plate. The reaction between the acid and the other set of metal plates produces a lack of electrons. When an electrical circuit is connected between the two plates, electrons run from the plate with an electron surplus to the plate with an electron deficit, providing electricity. This is a chemical-to-electrical conversion. As electricity is used, the chemicals that supply this reaction are changed. This is how a battery is "used up." The chemicals that react to provide the electron difference are changed to chemicals that do not support this chemical reaction. If an electrical current is applied across the set of metal plates, the chemical reaction is reversed and the chemicals are restored to their original state. This is how a battery is recharged. Most energy conversions are reversible.

	Energy Conversion	
Type of Product/Action	**From**	**To**
Gasoline in a car	Chemical	Mechanical
Electricity going through a lightbulb	Electrical	Radiant
Water turning a turbine	Mechanical	Electrical
Heat energy producing steam to turn a turbine	Thermal	Mechanical
Chemical in a battery	Chemical	Electrical
Burning coal moving the wheels on a steam engine	Chemical	Mechanical
Windmill	Mechanical	Electrical
Electricity in a heater	Electrical	Thermal
Natural gas burning	Chemical	Thermal
Light on a photocell	Radiant	Electrical

Figure 8-17 Examples of energy conversion.

Table 8-1 Power and Energy Conversions

Conversion	Formula
Electricity to heat	Watt-hours × 3.41 = BTU
Electric power to mechanical power	Watts/746 = horsepower
Electricity consumed by electric motor	(horsepower × 746)/motor efficiency*

*Efficiency should be expressed as a decimal.

Conversion Formulas

The most common energy-conversion calculations are electricity to heat and electricity to mechanical energy. Table 8-1 lists the formulas for some of the most common energy and power conversions.

For example, a 5 kW electric heater can produce 17,050 BTU of heat if operated for an hour: 5,000 W × 3.41 BTU/W = 17,050 BTU. A 3 hp electric motor will produce the equivalent of 2,238 W of work: 3 hp × 746 W/hp = 2,238 W. If the motor is 75 percent efficient, it will consume 2,984 W to produce 3 hp of work: 2,238 W/0.75 = 2,984 W.

8.12 ENERGY CONSERVATION

The subject of energy conversion brings up an important concept in dealing with matter and energy: matter and energy cannot be created or destroyed. This is the first law of thermodynamics, often referred to as the law of conservation. It is not possible to increase the amount of heat somewhere by "making heat." It is necessary to convert another form of energy into heat or move heat from somewhere else to where it is needed. Likewise, it is not possible to simply "destroy" the heat to reduce the amount of heat somewhere. It is necessary to convert the heat to another form of energy or move it somewhere else.

The operation of an electric heater is an example of how the law of conservation works. The heat produced by the heater represents the same amount of energy as the electricity used to produce the heat.

8.13 ENERGY DISTRIBUTION

Most of the energy we use is produced in a central location and distributed throughout the country. The two forms of energy used most in heating and cooling are electricity and natural gas.

Electrical Distribution

The laws of physics work against distributing large amounts of electrical power over long distances. According to Ohm's law, the voltage lost through a conductor will equal the resistance of that conductor multiplied by the current traveling through the wire. Power lines have resistance. The longer the wire is, the higher its electrical resistance. To reduce the loss through the line it is necessary to reduce the current traveling through it. Power-distribution systems accomplish this by using high voltages. Electrical work, measured in watts, is a product of volts times amps. A comparison of two 1-kW electrical systems can illustrate the point (Table 8-2).

Note that the higher voltage in Table 8-2 delivers 99.5 percent of the original power while the lower voltage loses half of its power through the wire.

In practice, the voltage is changed to a very high voltage for transmission over long distances and then lowered to a safer level for use. Alternating current is used for our power distribution because it is much easier and cheaper to accomplish the required voltage changes. The voltage is changed, or transformed, through a device called a transformer. Transformers can either increase or decrease voltage. The steps taken from the plant to your house are shown in Figure 8-18.

Table 8-2 Comparison of Two 1-kW Electrical Systems on an 8ohm (Ω) Conductor Ohms

	1 Amps at 1,000 Volts	10 Amps at 100 Volts
Voltage loss	5 volts (1 amp × 5 ohms)	50 volts (10 amps × 5 ohms)
Lost power	5 watts (5 volts × 1 amp)	500 watts (50 volts × 10 amps)
Delivered power	995 watts	500 watts
Percent of power delivered	99.5%	50%

1. Electricity is generated at a high voltage: 20,000–33,000 V.
2. It is transformed to a higher voltage: 69,000–500,000 V.

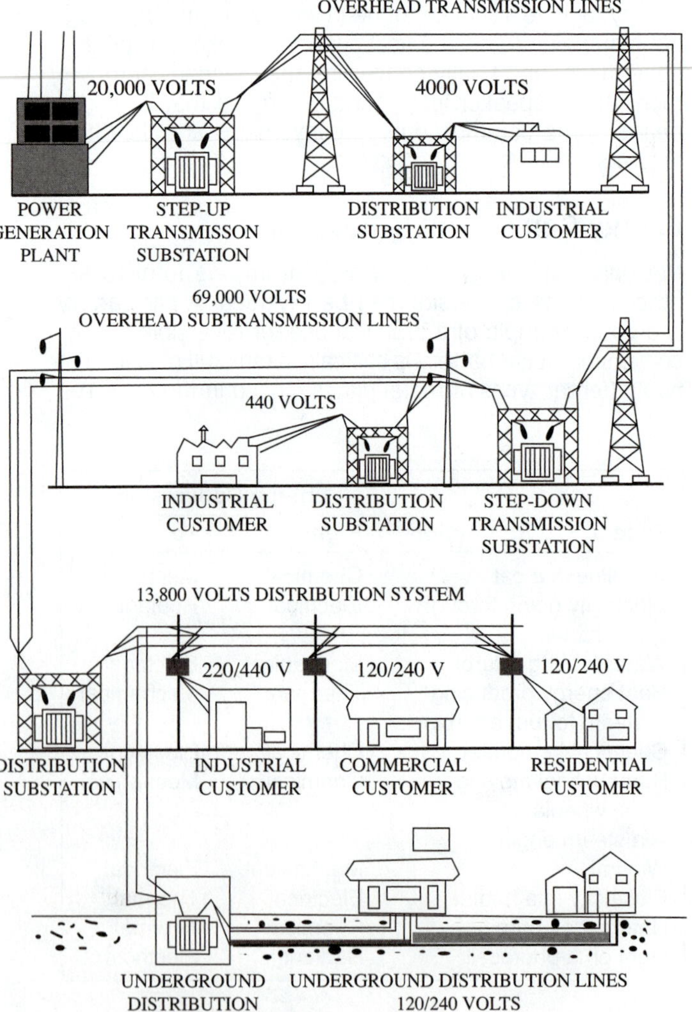

Figure 8-18 Electrical power–distribution system. *(Courtesy of OSHA)*

3. It is transmitted long distances at this voltage.
4. It is transformed to subtransmission voltages at a substation: 34,500–69,000 V.
5. It is transmitted to regional substations at this voltage.
6. The voltage is transformed to a distribution voltage: 4,800–13,800 V.
7. It is distributed to pole transformers.
8. The pole transformer drops the voltage to the end user: 120–240 V.

Natural Gas

Natural gas is collected, refined, and distributed through a nationwide network of pipelines. Figure 8-19 shows a map of the major gas pipelines in the United States. Compressors are used to reduce the volume and raise the pressure of the gas in the large interstate pipelines. These large interstate pipelines typically have a 16–48-in diameter and operate at pressures ranging from 200 to 1,500 psig. Compressor stations (Figure 8-20) are located at 40- to 100-mi intervals along the pipeline to maintain the pipeline pressure. Lateral pipelines deliver the gas from the main pipelines to local distribution networks.

While large gas consumers like electric utilities and industrial plants may get their gas directly from the main pipeline, most users rely on local distribution networks. The gas pressure is reduced before entering the local distribution system. Line pressure may be as low as 3 psig because of the reduced volume of gas. Compressor stations may still be required for large local distribution systems, but they are much smaller than the stations used in the interstate pipelines. Finally, the gas passes through a regulator at each customer's location, and its pressure is reduced to very low pressures. Delivery pressure for residential natural gas is typically 6 inches water column. The cost of building and maintaining these local distribution networks represents almost half of the price of natural gas to residential consumers.

8.14 STANDARD AND SI UNITS OF WORK

Mechanical Work—Traditional Units

Mechanical work is measured by the weight of an object and how far the object is moved. In traditional English units, this is foot-pounds. A foot-pound is the product of multiplying the weight in pounds times the distance in feet. Moving a 50-lb weight a distance of 2 ft requires 2 ft × 50 lb = 100 ft lb of mechanical work. The torque wrench is commonly used in many mechanical fields to measure the mechanical force used when tightening bolts. Standard torque wrenches are calibrated in foot-pounds (Figure 8-21). The United States is one of the few countries still using traditional (standard) units of weight and measure.

Figure 8-20 Typical natural gas compressor system.

Mechanical Work—SI Units

Mechanical force in the SI system is measured in newton (N) meters. This is the force of 1 N times the distance of 1 m. Metric torque wrenches are calibrated in newton meters (see Figure 8-21).

Thermal Work

When asked for a measurement of heat, most people respond with a temperature. In fact, they are partly correct. Temperature is a measure of the concentration of heat. But if a quantity is needed, temperature will not work. Temperature is a comparative measurement; it compares the concentration, or level, of heat against known standards. It does not measure the amount of heat energy in a

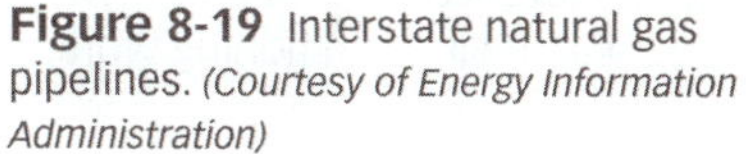

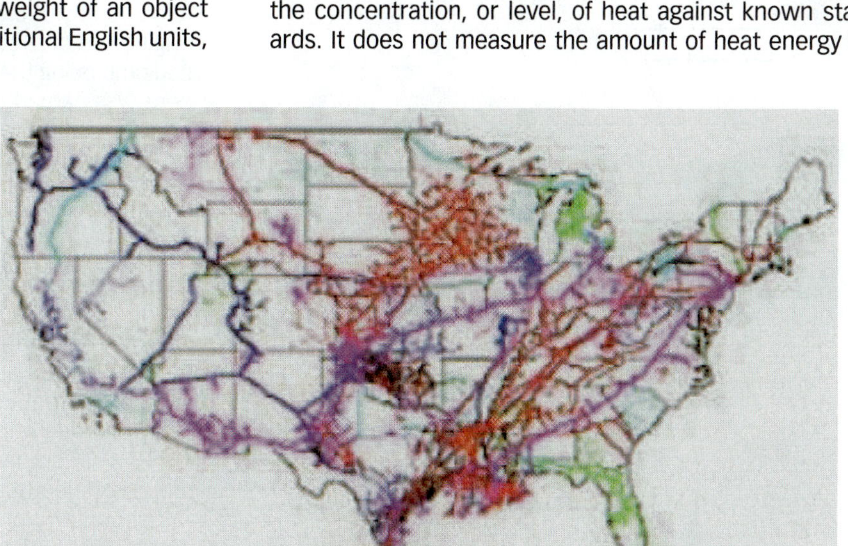

Figure 8-19 Interstate natural gas pipelines. *(Courtesy of Energy Information Administration)*

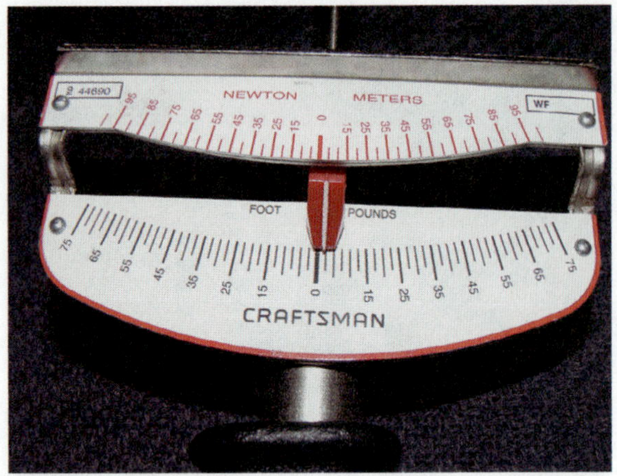

Figure 8-21 Torque wrench calibrated in foot-pounds in black and newton meters in red.

substance. There is a difference between heat intensity and heat quantity and how each is measured. Heat intensity describes how concentrated heat is. Heat quantity describes the amount of heat in something.

Thermal work is measured in British thermal units (BTUs), calories, or joules (Figure 8-22). The definitions for both BTUs and calories are similar. A BTU is the amount of heat required to raise 1 lb of water 1°F. A calorie is the amount of heat required to raise 1 g of water 1°C.

None of these units of heat can be directly measured; they all require calculation. Calculating the amount of heat transferred in water is easy because both BTUs and calories are defined by a temperature change in water. The formulas below show how to calculate heat quantity based on the temperature change of water.

BTU = (water temp difference in °F) × (water weight in pounds)

calories = (water temperature difference in °C) × (water weight in grams)

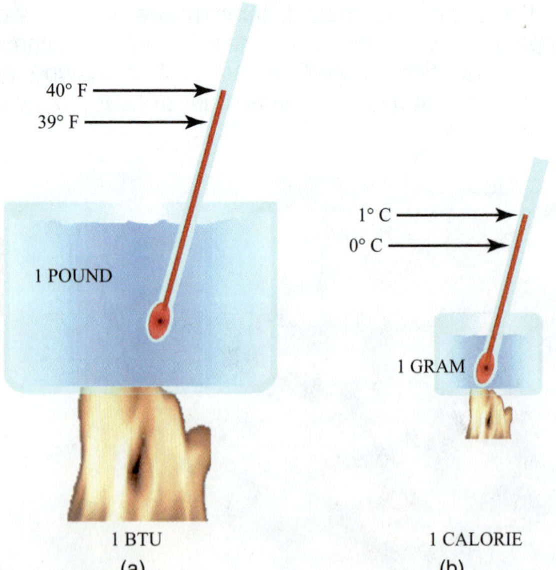

Figure 8-22 (a) 1 BTU raises the temperature of 1 lb of water 1°F; (b) 1 calorie of heat raises 1 g of water 1°C.

Both BTUs and calories represent relatively small amounts of heat. One BTU is roughly equivalent to the heat given off by a single wooden kitchen match. A calorie is even smaller: it takes approximately 250 cal to equal 1 BTU! Larger units of heat quantity have been derived for many real-world applications.

In the United States we use the therm to count very large quantities of heat. A therm is equal to 100,000 BTU. Natural gas is sold by the therm. For metric heat measurement, the kilocalorie (kcal) is used. The kilocalorie is equivalent to 1,000 cal. Frequently kilocalories are referred to as Calories. The capital *C* differentiates the large Calorie from the small calorie. So 1,000 calories equals 1 Calorie.

TECH TIP

Kilocalories, Calories with a capital *C*, are the same Calories people watch when they are dieting! The phrase "burning Calories" is actually quite accurate. Your body is oxidizing the energy stores and converting chemical energy into heat energy.

Electrical Work

Since electricity is the flow of electrons, a logical unit of electricity would measure the flow of electrons. This unit is the ampere, abbreviated amp or A as a unit. One amp is a flow of 1 coulomb in 1 second. But electron flow is only part of the picture. There must be a difference in electrical charge from one side of the circuit to the other before current will flow. This difference is called potential difference and is measured in volts. A volt is defined as the potential difference across a conductor when a flow of 1 Ampere dissipates 1 Watt of electrical power. The watt is the unit used to measure the rate of electrical work.

TECH TIP

A coulomb is the SI unit of electric charge. This amounts to a quantity of 6.2415×10^{18} electrons. The ampere was originally a derived unit, based on the coulomb with the coulomb being the original SI base electrical unit. The ampere was named the SI base electrical unit in 1960. The ampere is a commonly used unit in air-conditioning work.

Since the ampere is a rate and not a quantity, the watt is also a rate. To measure the quantity of electrical work, it is necessary to state how long work is being done at the rate of 1 Watt. The watt-hour is the amount of electrical work done at the rate of 1 W for 1 hr. Watt-hours is the product of watts and hours. Operating a 100-W light for a period of 2 h requires 100 W × 2 h = 200 W hr.

In practical terms, the watt-hour is a relatively small unit. Power companies use kilowatt-hours to measure the amount of electrical work used during the month. Figure 8-23 shows a typical kilowatt-hour meter. This is calculated by measuring the kilowatt use over time. It can be expressed using the formula V × A × 0.001 × hr = kWh.

Figure 8-23 Typical kilowatt hour meter.

Operating ten 100-W lightbulbs for 1 hr would require 10 × 100 W × 1 hr × 0.001 = 1 kWh.

Why is something with "hour" in its name a quantity and not a rate? Remember that amperes is a rate of electron flow. The "hour" in *kilowatt-hour* makes the kWh a quantity by giving a specific amount of time to work at that rate. A comparison with water would be to specify the amount of water pumped at a rate of 5 gal/min (GPM) for the period of 1 hr. The 5-GPM measure just tells the rate of water flow but does not tell how much water is pumped altogether. By specifying a time, the total quantity of water pumped can be determined to be 300 gal: 5 GPM × 60 min.

Joules—SI Unit of Energy

One problem with units of measurement based on the temperature change of water is that the specific heat of water changes slightly as its temperature changes. So the amount of heat represented by a temperature change in water depends upon the starting temperature. This has led to the development of many different definitions of both calories and BTUs, each based on a different starting temperature and representing a slightly different heat value. For this reason, joules have replaced calories in SI measurements of thermal energy.

Joules are a heat equivalent for mechanical energy but are used today in the SI system for all forms of energy. One joule is the work done by a force of 1 Newton moving an object 1 meter. Consuming electrical power at the rate of 1 W for 1 s is the electrical equivalent of 1 joule. Today, the joule represents the standard SI measurement of energy for all types of energy.

8.15 STANDARD AND SI UNITS OF POWER

Most of the terms used to describe the capacity of air-conditioning systems and components are rate measurements. Rate measurements always measure a quantity over time. Power is a measure of the rate of work; it measures a quantity of work over time. Although any form of energy can produce power, in air conditioning the three primary types of power measured are mechanical power, thermal power, and electrical power.

Mechanical Power

Horsepower is a rate of doing mechanical work. Like foot-pounds, horsepower involves distance and weight. Unlike foot-pounds, horsepower also uses a time component because it is a rate, not a quantity. One horsepower is defined as producing 33,000 ft lb of mechanical work in 1 min. Horsepower is literally supposed to be the amount of work a horse can do, as shown earlier in the unit in Figure 8-2. Motors are still rated by this somewhat archaic term. For electric motors, 1 hp is defined as 746 W. This does not mean that a 1-hp motor consumes 746 W; it means that the mechanical work performed by the motor is equivalent to the electrical work rate of 746 W. An electric motor consumes more than 746 W per hp because electric motors are not 100 percent efficient.

Thermal Power

A very common rate used to specify the capacity of heating and cooling systems is BTU per hour. A heating system with a rating of 40,000 BTU/hr would provide 40,000 BTU of heat energy for every hour of operation (Figure 8-24). A common rating for refrigeration equipment is the ton. A ton

AIR TEMP RISE		MAX EXTERNAL STATIC PRESSURE			
20-50F 11.1-27.8C		0.5WC/0.12KPA			
ALTITUDE IN FEET	GAS ORIFICE		INPUT MAX	OUTPUT CAP	THERMAL EFFICIENCY	EQUIPED FOR USE WITH
0 - 2000	NO. 44	BTU/HR	40000	32800	81.9 %	NATURA GAS
	SIZE 0.086	KW	11.71	9.59		
2000 - 4500	NO. 49	BTU/HR	33300	27300	81.9 %	
	SIZE 0.073	KW	9.75	7.98		
GAS SUPPLY PRESSURE		13.0WC/3.2KPA MAX	4WC/0.99KPA MIN			
MANIFOLD PRESSURE		3.5INWC/0.87KPA				

Figure 8-24 BTU/hr input and output rating of gas-fired heating equipment.

of cooling is defined as the amount of cooling accomplished by melting a ton of ice over a 24-hr period. Notice that this specifies both a quantity, the cooling done by 1 ton of ice, and a time, 24 hr. This quantity of cooling is 288,000 BTU because it takes 144 BTU to melt 1 lb of ice: 144 BTU × 2,000 lb = 288,000 BTU. This time period is not usually the most convenient. Dividing both the quantity and time by 24 gives an hourly rate of 12,000 BTU/hr. This is what is more commonly thought of as a ton. Do not forget the per-hour part. A ton is not 12,000 BTU but 12,000 BTU/hr. A ton can also be stated as 200 BTU/min: 12,000 divided by 60. All three of these rates are identical; the only difference is whether you prefer to measure by the minute, the hour, or the day.

Electrical Power

The ability of electricity to do work depends on two factors: voltage and current. Voltage is a measure of the potential difference. Potential difference provides the electrical pressure, which makes current flow possible. Without a difference in voltage from one end of the wire to the other, there will be no current flow. Watts is a measure of the rate at which electrical work is done. Wattage depends upon both voltage and current. A watt is defined as 1 A of current flowing at a potential difference of 1 V. Wattage can be calculated using the formula V × A = W. Wattage measures the rate at which work is accomplished. Wattage is a rate because it is calculated using amps, a measure of electrical current flow. Flows are always measured as a rate.

Wattage is a reasonable measure for small appliances and lightbulbs but for jobs like heating a house—something is needed that works better for large measurement. Kilowatts, abbreviated kW, are used to specify the energy use of heating and air-conditioning equipment. A kilowatt is equal to 1,000 W. The formula is simply W/1,000 = kW.

Watts—SI Unit of Power

Although the watt was originally conceived as a unit of electrical power, it is used today as the standard SI unit of power for all types of power. The watt is a rate of power equal to 1 joule of energy per second. In SI measurement systems, watts are used to measure the output of engines, electric motors, and furnaces.

8.16 ENERGY-EFFICIENCY MEASUREMENTS

A rating called the energy efficiency ratio (EER) is used to compare the efficiency of air-conditioning units. The EER is the number of cooling BTUs per hour a unit produces for every watt of electricity it uses. It is determined by dividing a unit's cooling capacity in BTUs per hour by the energy use in watts. The higher a system's EER is, the more efficient it is. The EER is calculated at the AHRI standard rating condition. The seasonal energy efficiency ratio (SEER) is adjusted for operation through a cooling season. It also represents the number of cooling BTUs per hour a unit produces for every watt of electricity it uses, but at conditions adjusted to represent an entire cooling season. A unit's SEER is normally higher than its EER.

UNIT 8—SUMMARY

Heating, air conditioning, and refrigeration are about controlling energy. Energy that is stored is called potential energy, while energy that is in use is called kinetic energy. There are many forms of energy, including chemical, thermal, radiant, electrical, magnetic, and mechanical. Energy can be converted from one form to another, but it cannot be created or destroyed. There is a definite relationship among energy, work, and power. Energy is the ability to do work, work is the amount of energy transferred, and power is the rate of energy transfer. The difference between work and power is that work is a quantity while power is a rate.

UNIT 8—REVIEW QUESTIONS

1. List the forms of energy used in the heating, air-conditioning, and refrigeration industry.
2. What units are used to measure heat quantity?
3. What units are used to measure heat intensity?
4. Explain the relationship among energy, work, and power.
5. How many BTUs will a 5-kW heater produce if operated for 1 hr?
6. How many BTUs are required to raise the temperature of 15 lb of water from 100°F to 120°F?
7. A heat-transfer process results in 20 kcal of heat being absorbed by 1 L (1 kg) of water. What temperature will the water be if it started at a temperature of 25°C?
8. Describe chemical energy, and give an example.
9. How much mechanical force is required to lift a 250-lb compressor 2 ft?
10. What energy unit is used today in the SI system for all forms of energy?
11. How many joules of heat does a 100-W heater produce if it is operated for 5 min?
12. Use a block diagram to outline a common electrical-distribution system from the power plant to your house.
13. What units of measurement are used in the United States to rate the cooling capacity of air-conditioning equipment?
14. Arrange the following heat quantities from smallest to largest.
    a. BTU
    b. joule
    c. calorie
    d. kilocalorie
    e. therm
15. Explain the difference between a renewable energy resource and a nonrenewable energy resource.
16. Explain the difference between potential and kinetic energy.
17. Describe the natural gas–distribution system.
18. Explain the relationship between watts and joules in the SI system.
19. What is the required input in watts to operate an 80-percent efficient 5-hp electric motor?
20. Explain the units used to measure torque.

UNIT 9

Temperature Measurement and Conversion

OBJECTIVES

After completing this unit, you will be able to:

1. define *temperature* and explain the effect it has on a substance.
2. discuss the concept of absolute zero.
3. compare various temperature-measuring devices.
4. explain the concept of absolute temperature scales.
5. explain the four types of temperature scales: Fahrenheit, Celsius, Rankine, and Kelvin.
6. demonstrate how to convert temperature from one scale to another.

9.1 INTRODUCTION

Temperature control is the primary function of most air-conditioning and refrigeration systems. Temperature measurements are among the most important measurements an air-conditioning technician will make. Taking accurate temperature readings is crucial in analyzing system performance; incorrect readings can lead to incorrect system diagnosis. This unit discusses what temperature is, the different methods of measuring temperature, and how to take accurate temperature readings.

9.2 WHAT IS TEMPERATURE?

Temperature is the measure of the average kinetic energy in a substance. Kinetic energy makes molecules move: the molecules of solids vibrate in place; the molecules of liquids and gases move about. All the molecules are not moving at exactly the same speed, and this is why temperature is defined as a measure of the average kinetic energy of a substance. The higher the temperature, the faster the molecules move. The lower the temperature, the slower the molecules move (Figure 9-1).

Temperature does not measure the amount of heat in a substance, because molecules have both potential and kinetic energy. Temperature can only measure kinetic energy; it cannot measure potential energy. A 40°F gas will have more heat in it than a 100°F liquid because the gas has a much higher potential energy than the liquid.

9.3 TEMPERATURE AND RADIANT ENERGY

Solids both absorb and emit radiant energy. They emit radiant energy at any temperature above absolute zero. The frequency of the radiant energy wave increases as the temperature increases. At around 1,000°F, the radiant energy is in the red spectrum of visible light and the object glows a dull red. As the temperature increases, the frequency of the light increases, producing shifts in the color of the visible light emitted by the solid. Figure 9-2 shows the visible light spectrum associated with different temperatures.

Below 1,000°F, materials are not hot enough to produce visible light. They still produce radiant energy but at a lower frequency. This unseen light is called infrared light because it has a wavelength that is longer than red light.

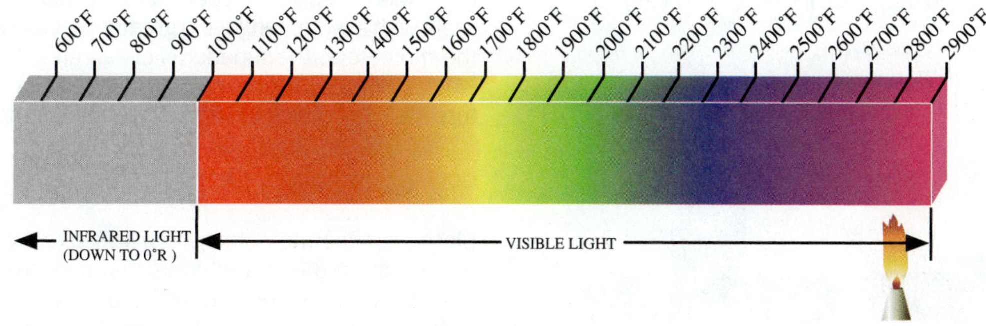

Figure 9-1 Above 1,000°F, the vibrations are fast enough to begin emitting a dull red visible light. As the temperature continues to rise, the color of the light changes. When it is very hot, it glows white.

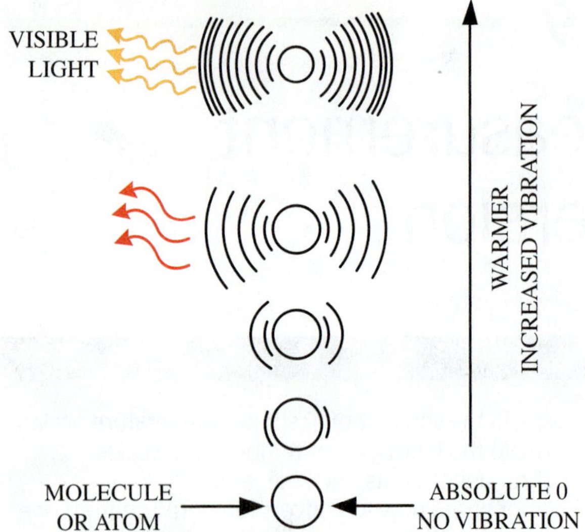

Figure 9-2 As the temperature rises above absolute zero, the molecules begin to vibrate. The hotter the material, the faster the vibrations.

This is the light that an infrared thermometer sees when reading the temperature of a material's surface from several feet away. As the temperature increases, the color of the object changes because the wavelength of the radiation it is emitting gets shorter and shorter.

9.4 EARLY THERMOMETERS

Scientists have been devising methods of measuring temperature for nearly two thousand years. In the first century, Hero of Alexandria published his studies on pneumatics in which he described a tube with the top closed and the bottom open. The bottom is submerged in a container of water and the tube is filled with air at the top and water at the bottom. When the air that is trapped in the top of the tube was heated, the expanded air would push the water down the tube. When the air cooled, the air would contract and water would rise in the tube. This made a sort of backwards thermometer (Figure 9-3). Later, in the eleventh century, Abu Ali ibn Sina developed air thermometers based on this same principle. In the early seventeenth century, Galileo experimented with a device called a thermoscope (Figure 9-4). It consisted of a column of liquid with several glass balls partially filled with water to give them different densities. The density of the fluid they were immersed in changed with temperature, causing them to rise or fall. The position of the glass balls indicated the relative temperature.

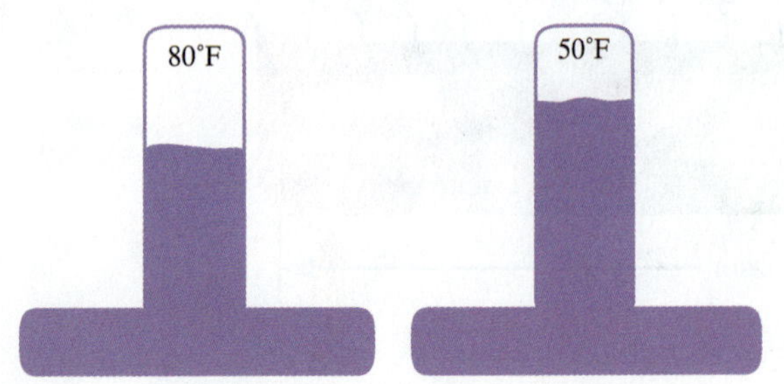

Figure 9-3 The volume of the trapped air changes with temperature, changing the level of the water.

Figure 9-4 In a thermoscope, the position of the balls in the fluid changes with temperature. *(Courtesy of G.W. Schleidt, Inc. GalileoShop.com.)*

All of these early instruments were affected by atmospheric pressure changes as well as temperature changes. In 1654, Ferdinando II de Medici made the first modern-style thermometer that was not affected by changes in atmospheric pressure. His thermometer used sealed tubes that were partially filled with alcohol. Many scientists experimented with different styles of thermometers using different liquids. Unfortunately, each thermometer and scale was unique, and no accepted standard existed. In 1724, Daniel Fahrenheit began producing thermometers that used mercury. Mercury's large coefficient of expansion allowed a scale with a wider range and greater precision than previous thermometers. The superiority of Fahrenheit's thermometer led to its wide adoption, resulting in the Fahrenheit temperature scale becoming the first widely used temperature scale.

9.5 GLASS-STEM THERMOMETERS

Glass-stem thermometers are still used today. An advantage of glass-stem thermometers is that they are highly reliable because they rely on properties of physics, not on a mechanical or electronic device. The temperature is measured on a scale as compared to the tip of the expanding or contracting liquid inside the thin tube. When the liquid is heated,

Figure 9-5 Glass-stem thermometer.

it expands and pushes farther up the glass tube. When it is cooled, it contracts, and the liquid level falls. Any liquid can be used, but mercury or alcohol are the most commonly used liquids. Mercury looks metallic in color, and alcohol is usually dyed red (Figure 9-5). The thinner the size of the hollow glass tube, the greater the movement per degree of temperature change. Longer, more accurate models for laboratory work are available, as are pocket sizes. Breakage is a problem, since glass thermometers are fragile.

9.6 DIAL THERMOMETERS

Dial thermometers indicate temperature by a pointer moving over a circular scale (Figure 9-6). The pointer moves because it is attached to a bimetal spring. The spring may be wound into a spiral or twisted. A bimetal is made by joining together two metals with different rates of expansion. When the bimetal is heated, one metal expands more than the other and the bimetal spring opens; when it is cooled, the spring tightens. This creates a turning motion that is used to move the pointer on the thermometer.

An advantage of dial thermometers is that they are more rugged than glass-stem thermometers and can be used for a wide variety of applications. They are also relatively inexpensive. Service technicians have used dial-type pocket thermometers for years. Figure 9-7 shows a typical dial-type pocket thermometer. The disadvantage of dial thermometers is that they are not very accurate compared to other types of thermometers.

Figure 9-6 Each increment on this dial thermometer represents 2°F—this thermometer is reading 84°F.

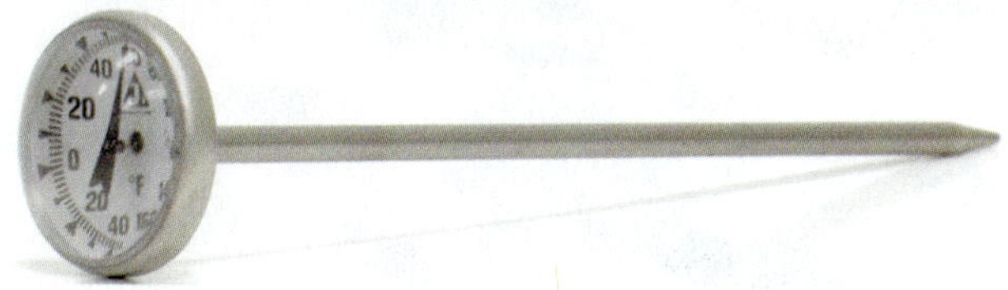

Figure 9-7 Analog pocket thermometer.

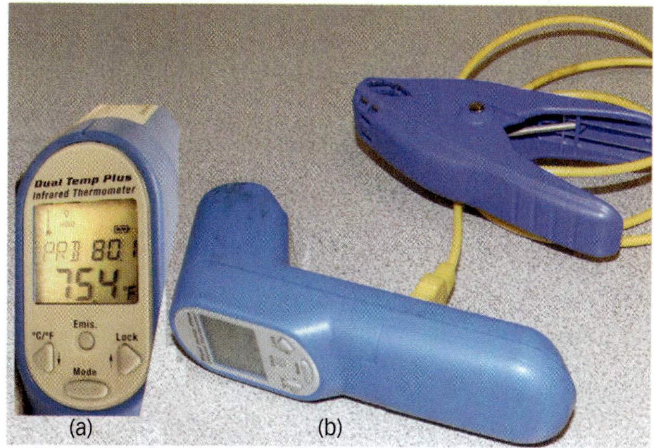

Figure 9-8 (a) Infrared thermometer; (b) infrared thermometer with a temperature contact probe.

9.7 INFRARED THERMOMETERS

Infrared thermometers can "see" the temperature of a surface from a distance (Figure 9-8). What they are seeing is the color of the invisible infrared (IR) light given off by the object. A detector is located inside the instrument that senses the infrared light waves. A change in the IR light's color occurs as a result of temperature changes, which results in a change in the electrical output of the detector. A small integrated circuit converts that electrical output into a displayed temperature. The distance the detector is from the object, the surface color, and its reflectivity all affect the accuracy of the temperature reading. The most accurate reading is obtained when the object whose temperature is being read is a black, dull surface. IR thermometers are generally not useful for measuring air, liquids, shiny metal objects, or small objects.

One difference between different IR thermometers is their distance to spot ratio. This ratio tells how large the target area is compared to the distance (Figure 9-9). An instrument with a 10 to 1 ratio will measure an area of 1 in diameter when the target is 10 in away. Inexpensive tools have ratios of 4:1, while more expensive models have ratios of 12:1. More expensive models also have adjustable emissivity, so that they can be used on different surfaces.

Not all HVACR field work requires extreme accuracy. Technicians are often looking for trends in temperature change. For example, during an air-balancing job in a

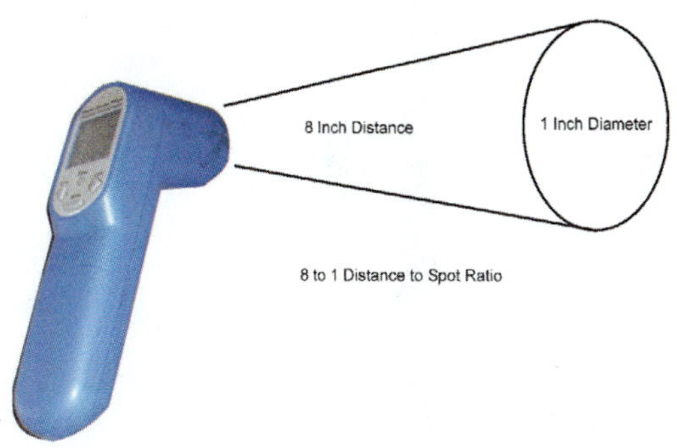

Figure 9-9 Distance to spot ratio.

residence, the technician may need to know if the air coming out of the duct is colder or warmer than it was before the dampers were adjusted.

SERVICE TIP

A disadvantage of IR thermometers is that they cannot detect air temperatures; however, they can detect the temperature of the grille or register that the air is passing through. To obtain the most accurate estimate of the actual duct air temperature, point the detector into the louvers so it is pointed at the inside of the duct boot. Some IR thermometers have a laser pointer to make this easier. Move the instrument around and watch the temperature display. Watch for the highest or lowest temperature (highest for heating and lowest for AC) indicated on the display. That temperature is going to be the closest to the actual air temperature. Practice will help develop your skill in obtaining accurate temperature readings with an IR thermometer.

There are many advantages to using infrared thermometers. The fact that a temperature reading can be made from a distance may mean that you do not need to climb a ladder to get air duct readings. It may also mean that you can get a reading from the back of a hard-to-reach space. The temperature is displayed instantaneously, which speeds up your work.

9.8 DIGITAL THERMOMETERS

Digital thermometers with compact sensing elements are popular due to their speed in sensing a change in temperature and the ease in reading the alphanumeric display instead of a dial or a scale. Figure 9-10 shows a digital pocket thermometer. Most digital thermometers require batteries or a source of electrical power to produce the digital display.

Digital thermometers can perform many functions that analog instruments cannot. These include recording temperature differences between two or more sensing elements, remembering the highest and lowest temperatures from several readings, and averaging the temperature of several readings. The display on most digital thermometers can easily be changed to show the temperature reading as Fahrenheit or Celsius.

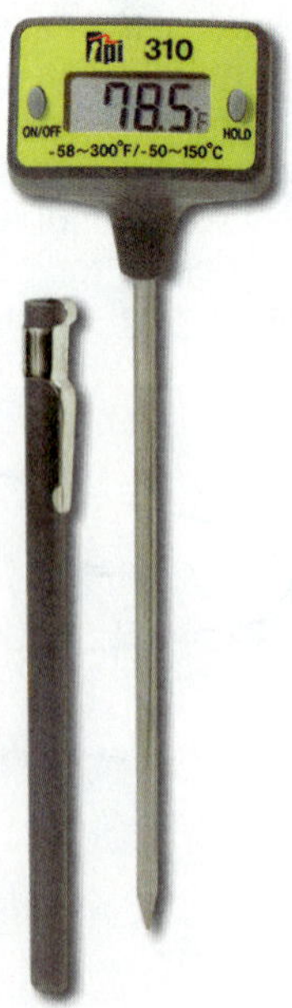

Figure 9-10 Digital pocket thermometer.

9.9 ELECTRONIC TEMPERATURE SENSORS

Digital thermometers use three different types of temperature sensors: wire-wound elements, thermistors, or thermocouples.

Wire-Wound Sensors

A wire-wound temperature sensor changes resistance with temperature change. The resistance of most conductors increases with an increase in temperature. Normally this change is very small. This resistance change can be magnified by winding many turns of small-diameter wire into a coil that is used as a sensor. Wire-wound sensors are very accurate, but they are also the most expensive of the three types of electronic sensors.

Thermistors

A thermistor is a semiconductor that changes resistance with temperature. Thermistors are inexpensive and fairly rugged, but they are the least accurate of the three types of digital temperature measurement because their resistance change is not linear. Thermistors are frequently used as temperature-sensing devices on air-conditioning equipment. Figure 9-11 shows a thermistor used to measure the outdoor coil temperature on a heat pump.

Thermocouples

Thermocouple temperature sensors can be used to detect temperatures that are displayed on analog or digital

Figure 9-11 Outdoor air thermistor.

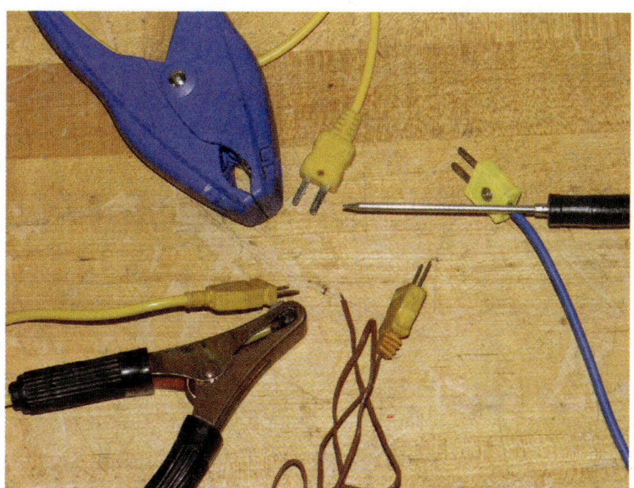

Figure 9-12 Variety of type J thermocouple probes.

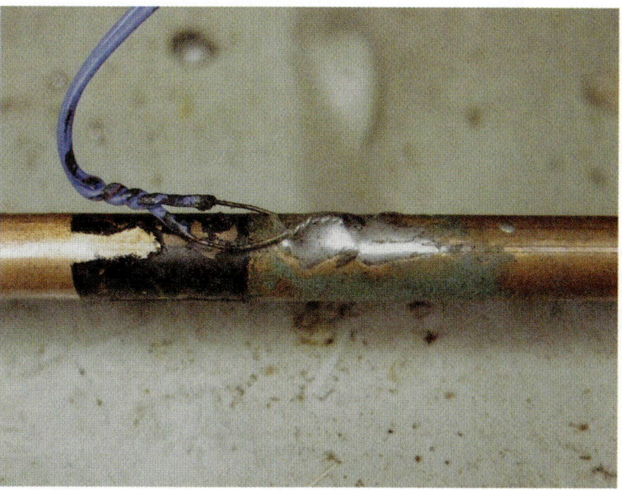

Figure 9-13 Thermocouple soldered to line in unit.

instruments. The thermocouple produces a small voltage when two dissimilar metal wires are joined, forming a tip or junction that can be heated or cooled. The two most common thermocouple types are chromel and alumel (type K) and iron and constantan (type J). Type J sensors have been used for years and are still very popular. Type K is gaining in popularity because of its lower cost. Any manufacturer's thermocouple will work with any manufacturer's digital thermometer as long as they are both the same type. This standardization has led to the development of a wide array of different probes for many specialized applications (Figure 9-12).

Thermocouples are less accurate than wire-wound sensors but more accurate than thermistors. They may provide a temperature that is 2°F higher or lower than the actual temperature. You can test the accuracy of a thermocouple by checking the temperature of ice, which should read 32°F. Once you know the temperature error factor, you can mark it on the thermocouple. That way you can add or subtract a degree or two as needed to keep the error factor within an acceptable range for most HVACR work.

Thermocouples are relatively inexpensive and flexible. They can be attached to several locations inside of the equipment, and the equipment cover panels can then be replaced so the equipment can operate normally (Figure 9-13). With the thermocouple leads run so that they are outside the cabinet, the equipment can be operated normally. This allows technicians to diagnose what is happening to temperatures inside operating equipment.

TECH TIP

Sometimes people confuse thermometers and thermostats. The difference between a thermometer and a thermostat is that a thermometer simply indicates the temperature. A thermostat controls equipment based on the temperature it reads.

9.10 COMMON TEMPERATURE SCALES

Everyday temperatures are indicated using either the Fahrenheit or Celsius temperature scales. The Fahrenheit scale is still predominant in the United States but is seldom used anywhere else. The Celsius scale is part of the SI system of measurement and is used throughout the world. Figure 9-14 shows a comparison of the two scales.

Fahrenheit

By the early eighteenth century, many different temperature scales were proposed and used, but no clear standard had emerged. Daniel Fahrenheit produced the Fahrenheit scale in 1724. The Fahrenheit scale was based on three temperatures. The temperature of 0° was set at the lowest

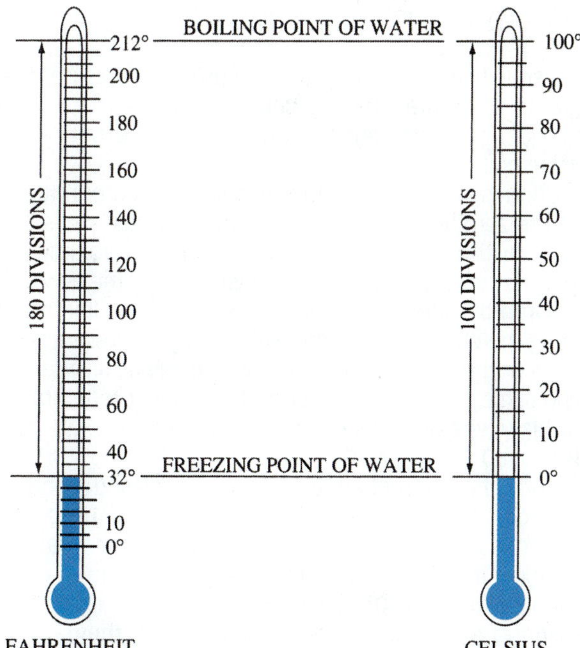

Figure 9-14 Comparison of the Fahrenheit and Celsius temperature scales.

temperature he could produce in his lab using ice and salt; the freezing point of water was set at 32°; and the human body temperature was set at 96°. The boiling point of water on the Fahrenheit scale is 212° at sea level. Horse body temperature was set at 100° and was considered more stable than human body temperature.

Celsius

In 1742, Anders Celsius proposed a temperature scale with 100° as the freezing point of water and 0° as the boiling point of water. Linnaeus reversed this, placing 0° at freezing and 100° at boiling. This scale was widely adopted by scientists and was known as the Centigrade scale because of its division into 100° from freezing to boiling. The name was changed to Celsius in 1948 to honor the scientist who first conceived the scale.

TECH TIP

The Fahrenheit scale was initially set to read from the lowest temperature obtained in the lab (0°F) to horse body temperature (100°F). The Celsius scale has been set to read from the freezing point of water (0°C) to the boiling point of water (100°C). Using the Celsius scale, every degree is equal to 1/100 of the interval between freezing and boiling points. Using the Fahrenheit scale, every degree is equal to 1/180 of the interval between freezing and boiling (212°F − 32°F = 180°F). Due to the increased number of degree intervals and the lower freezing temperature, the Fahrenheit scale requires less decimals and less negative signs as compared to the Celsius scale.

9.11 ABSOLUTE TEMPERATURE

Once scientists had a way to measure temperature, they began to speculate what the coldest temperature possible would be. Most agreed that there must be a bottom, or starting point.

Charles' law was used to determine the temperature of absolute zero. In theory, if the volume of a gas keeps contracting with temperature, at some point the volume and temperature will be nothing. Scientists plotted the decrease in volume of different gases and extended the line representing the change in volume with each gas. Regardless of the gas or the starting quantity of gas, the lines all intersect at the same point, absolute zero. The coldest temperature possible where all molecular motion stops is absolute zero. This is −460°F, or −273°C.

Both the Fahrenheit and Celsius scales depend on reference points. What 5°C really means is that the temperature is 5°C warmer than the freezing point of water. A more exacting place to start measuring temperature would be at absolute zero (0°). There are two scales that do start measuring temperature from absolute zero; they are the Rankine and Kelvin scales (Figure 9-15). There are no negative temperatures in either Rankine or Kelvin because they start at absolute zero.

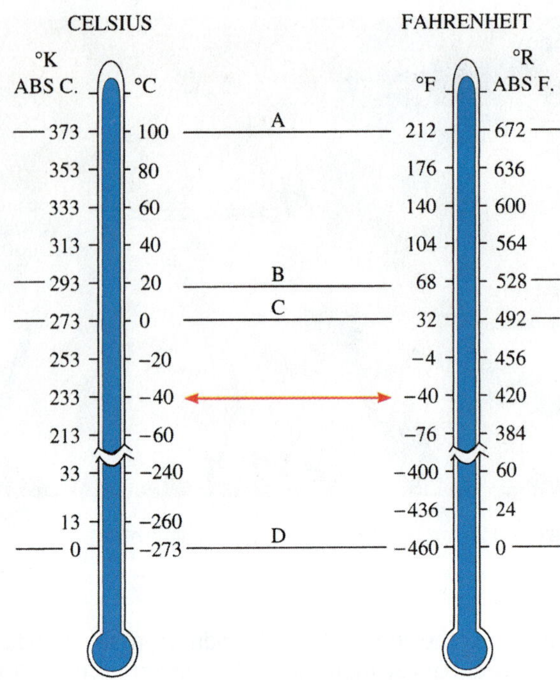

Figure 9-15 Fahrenheit, Celsius, Rankine, and Kelvin temperature scales: (A) boiling temperature of water; (B) standard temperature; (C) freezing temperature of water; (D) absolute zero; note that −40°C (red arrow) equals −40°F.

Both Rankine and Kelvin scales are primarily used for scientific measurements. All gas-law problems require absolute temperatures. The lack of negative temperatures makes absolute temperature scales work correctly with the ratios set up in the ideal gas law. If measurements are taken in Fahrenheit or Celsius they must be converted to Rankine or Kelvin before being used in a gas-law problem.

Rankine

The Rankine temperature scale is named after William Rankine, who proposed it in 1859. Rankine is the absolute temperature scale for Fahrenheit and has the same spacing between degrees as does the Fahrenheit scale. The numbers are just shifted so that 0° starts at absolute zero. Just as there are 180°F between freezing and boiling on the Fahrenheit scale, there is a 180°R difference between freezing and boiling on the Rankine scale as well.

Because their degree spacing is the same, conversion between Fahrenheit and Rankine is simple. Adding 460° to Fahrenheit yields Rankine; subtracting 460° from Rankine yields Fahrenheit.

Kelvin

The Kelvin temperature scale is named in honor of William Thomson, the first Baron Kelvin. Kelvin is the absolute temperature scale for Celsius and is the official temperature scale for the SI system. It has the same spacing between degrees as does the Celsius scale, just shifted so that 0° starts at absolute zero.

Because their degree spacing is the same, conversion between Celsius and Kelvin is simple. Adding 273° to Celsius yields Kelvin; subtracting 273° from Kelvin yields Celsius.

TECH TIP

Temperature scales on thermometers can go higher or lower depending on the use of the instrument. Outdoor thermometers usually have scales from −60°F to 120°F, and indoor thermometers usually range from 50°F to 90°F.

9.12 TEMPERATURE CONVERSION

As long as the United States continues to measure temperature in degrees Fahrenheit while the rest of the world uses Celsius, we will need to make conversions from one system to the other. Two primary differences in how the scales are established must be addressed. The reference point for each system is different, and the amount of temperature change represented by a degree is different.

There are 180° on the Fahrenheit scale between the freezing and boiling points of water, while the Celsius scale only has 100°C between freezing and boiling. This can be converted using a ratio. 180 divided by 100 gives us a ratio of 1.8°F for every 1°C. For people who prefer fractions, 9/5 does the trick. This would be all we needed if 0 represented the same temperature on both scales, but it does not. Since water freezes at 32°F and 0°C, we can adjust for this by adding 32. The traditional formula for converting from Celsius to Fahrenheit is

$$(\text{Celsius} \times 1.8) + 32 = \text{Fahrenheit}$$

or if you prefer fractions

$$(\text{Celsius} \times 9/5) + 32 = \text{Fahrenheit}$$

This can be rearranged algebraically to get the formula for converting from Fahrenheit to Celsius.

$$(\text{Fahrenheit} - 32)/1.8 = \text{Celsius}$$

or if you prefer fractions

$$(\text{Fahrenheit} - 32) \times 5/9 = \text{Celsius}$$

Notice that when converting Celsius to Fahrenheit, you multiply first and then add; but when converting Fahrenheit to Celsius, you subtract first and then divide. This standard conversion process is based on comparing the freezing points of the two scales. Unfortunately, the subtle variation in execution has vexed many an aspiring technician.

9.13 CONVERSION EXAMPLES

To illustrate how this formula works, we will convert 212°F to Celsius, a temperature for which we know the answer.

The basic formula is

$$(F - 32)/1.8 = C$$

Substituting the temperature we want to convert

$$(212°F - 32)/1.8 = C$$

That works out to

$$180°F/1.8 = 100°C$$

Next we will convert 100°C to Fahrenheit.

The basic formula is

$$(C \times 1.8) + 32 = F$$

Substituting the temperature we want to convert

$$(100°C \times 1.8) + 32 = F$$

That works out to

$$180 + 32 = 212°F$$

Now we will do a couple of examples where the answer is not as obvious. It would be useful to know what temperature Celsius is equivalent to a normal Fahrenheit temperature, so we will convert 75°F to Celsius.

The basic formula is

$$(F - 32)/1.8 = C$$

Substituting the temperature we want to convert

$$(75°F - 32)/1.8 = C$$

That works out to

$$43°F/1.8 = 26.9°C$$

Next we will convert 40°C to Fahrenheit.

The basic formula is

$$(C \times 1.8) + 32 = F$$

Substituting the temperature we want to convert

$$(40°C \times 1.8) + 32 = F$$

That works out to

$$72 + 32 = 104°F$$

SERVICE TIP

While the two scales do not correspond at freezing, there is a temperature where the two scales align: −40°. Forty degrees below zero Fahrenheit is exactly the same temperature as 40° below zero Celsius. We can use this coincidence to build an easier system. The system has three short steps. The first and last steps are always the same regardless of which direction you are going. Only the middle step changes, and it is relatively easy to remember.

1. Add 40.
2. Multiply or divide by 1.8.
   Multiply by 1.8 when converting from Celsius to Fahrenheit.
   Divide by 1.8 when converting from Fahrenheit to Celsius.
3. Subtract 40.

9.14 TEMPERATURE CONVERSION TABLES

The easiest way of making the conversion is to use a conversion table, like the one on the back cover. Regardless of whether the starting temperature is Fahrenheit or Celsius, you start by finding the temperature you are converting in the center column. Look across to the number in the right column if you are converting from Fahrenheit to Celsius. If you are converting from Celsius to Fahrenheit, look at the number in the left column. To convert 75°F to Celsius, find 75 in the middle column. Looking at the right-hand column you can see that 77°F is equal to 25°C. To convert 25°C to Fahrenheit, find 25 in the center column. Looking at the left-hand column you can see that 25°C is equal to 77°F. Referring to the back cover, it is interesting to note that −40°F is equivalent to −40°C. This is the only place where the two scales coincide.

9.15 ABSOLUTE TEMPERATURE CONVERSION

The most useful conversions to absolute temperatures will be between each regular scale and its corresponding absolute scale: Fahrenheit to Rankine or Celsius to Kelvin. Fortunately these are very easy conversions to make. Fahrenheit can be converted to Rankine by adding 460; Rankine can be converted to Fahrenheit by subtracting 460. Celsius and Kelvin work in a similar manner. Adding 273 to Celsius produces Kelvin; subtracting 273 from Kelvin produces Celsius.

Since Rankine and Kelvin both start at absolute zero, they can be converted from one to the other by either multiplying or dividing by 1.8. There is no need to adjust the scales as is done with Fahrenheit and Celsius. This method produces results that are off by a little less than a degree because the temperatures for absolute zero are approximated at −460°F and −273°C.

9.16 TAKING ACCURATE TEMPERATURE READINGS

Getting accurate temperature readings requires the right equipment and the right techniques. Common temperatures read while working on HVACR equipment include air, water, and refrigerant line temperatures.

Air temperature can be read with dial-type pocket thermometers, digital pocket thermometers, or electronic temperature testers. Infrared thermometers are less desirable for air temperature readings, because they must read the temperature of objects. However, an infrared thermometer can be used to measure air by targeting the grille.

Water temperature readings can also be read with dial-type pocket thermometers, digital pocket thermometers, or electronic temperature testers. If possible, the thermometer should be inserted directly into the water stream. When using electronic thermometers, be sure that the sensor is rated for liquid immersion. Infrared thermometers do not work well for measuring liquid. An infrared thermometer can be used if the water is contacting an object that the infrared can target.

Figure 9-16 Thermocouple clamp on pipe.

Pipe temperatures are best taken with a thermocouple-type temperature tester. Probes are available with clamps that hold the thermocouple against the pipe (Figure 9-16). Bead-type thermocouples can be held against the pipe with a Velcro strap for an accurate reading (Figure 9-17). Pocket-type thermometers can be used only if the tip can be held securely against the pipe and insulated from the surrounding air. A Velcro strap or Permagum can be used to hold the tip of the thermometer against the pipe and insulate it from the surrounding air. Most infrared thermometers are not suited for measuring pipe temperature because of their distance-to-spot ratio. The emissivity of the pipe can also be a problem for shiny materials like copper. Infrared thermometers with a spot ratio of 10:1 or greater will work fine with dull pipes larger than ½ in.

Some systems are equipped with thermometer wells. These are designed for use with thermometers that have metal stems, like pocket thermometers (Figure 9-18). The

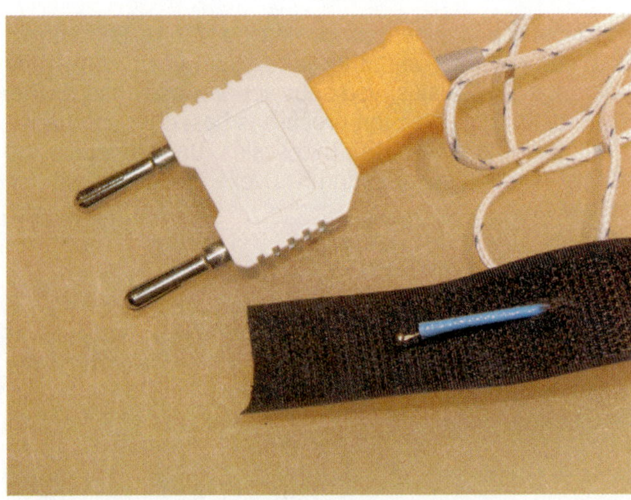

Figure 9-17 Thermocouple bead with Velcro.

Figure 9-18 Thermometer well with stem thermometer in it.

Figure 9-19 Thermometer reading supply air temperature.

well should be filled with mineral oil to ensure good thermal contact between the well and the thermometer.

9.17 AIR TEMPERATURES

Three air temperatures are commonly read when working on air-conditioning equipment: outdoor ambient temperature, return air temperature, and supply air temperature. Ambient temperature is the temperature around the unit, typically outside. Return air is the air returning to the system from the conditioned space. Supply air is the conditioned air being supplied to the conditioned space.

Outdoor Ambient Temperature Readings

Outdoor ambient temperature readings should be taken in areas that are free from discharge air from condensing units or dryer vents and out of direct sunlight. Holding the thermometer in front of the air intake grille of the outdoor unit normally works well.

Return Air Temperature

Accurate return air temperatures can be obtained by holding a pocket or digital thermometer in front of the return air grille. Infrared thermometers can be used with return air grilles by targeting the grille.

Supply Air Temperature

Supply air temperatures can be more difficult. The unit should operate long enough for supply air temperatures to stabilize. If possible, the thermometer should be inserted inside the supply air register (Figure 9-19). When the supply air leaves the register, it mixes with room air, affecting the temperature of the supply air. Infrared thermometers can be used with supply air registers by targeting the supply air register. When measuring supply air temperature with an infrared thermometer, the unit must run long enough for the register to be the same temperature as the supply air.

Duct Temperatures

The temperatures read at the registers will be different than the temperature at the unit because of duct loss and gain. When checking a system performance specification, such as furnace temperature rise, it is best to check temperatures at the unit. Drilling a small hole in the ductwork allows temperature readings of the air just as it enters and leaves the unit. Holes should generally not be drilled in the actual unit or in flex duct. A hole in the wrong place in the unit can damage the unit. Flex ducts should not be pierced; poking holes in a flex duct allows air to pass between the inner and outer jacket. This increases duct loss and can lead to condensation on and inside the duct.

The return air plenum is the best location for checking return air temperatures, especially if there are several return ducts. The thermometer will read the mixed air temperature, which may be different from the temperature in individual returns.

The supply air plenum can be used on systems with a radial duct design using flex duct. The hole in the plenum should not be too close to where the plenum connects to the unit. If there are any components installed in or on the plenum, be careful to avoid them. Airflow from the return air through a bypass humidifier or bypass damper will be a very different temperature than the supply air in the plenum.

When possible, take supply air temperatures in a trunk duct leaving the plenum (Figure 9-20). This avoids the problems of radiant heat from furnace heat exchangers or return air being bypassed through a humidifier or bypass damper.

Figure 9-20 Thermometer placed in hole in supply duct.

SAFETY TIP

When drilling holes in ductwork to measure temperature, make certain you know what is on the other side of the duct; hopefully just air. Contacting a set of electric heat strips with a metal thermometer that is sticking through a metal duct can cause a direct short on a high-current device and give you a serious shock!

UNIT 9—SUMMARY

Understanding temperature measurements and conversions is essential to every aspect of HVACR work. Most HVACR jobs require temperature measurement. Temperature is the measure of the average molecular kinetic energy of a substance. The faster the molecules move, the higher the temperature. A solid emits radiant energy that is proportional to its temperature. Different types of thermometers include glass stem, dial, digital, and infrared. Thermocouples are the most common type of electronic temperature sensors. They produce a small DC voltage in response to temperature. There are four widely used temperature scales: Fahrenheit, Celsius, Rankine, and Kelvin. Fahrenheit is primarily used in the United States, while Celsius is used throughout the world. Rankine and Kelvin are absolute temperature scales whose zero point is absolute zero. Getting accurate temperature readings depends upon using the correct equipment and technique.

UNIT 9—REVIEW QUESTIONS

1. What is temperature?
2. What is the first color of light that can be seen when a piece of metal is being heated?
3. How does a glass-stem thermometer measure temperature?
4. What is an advantage of a glass-stem thermometer?
5. What causes the pointer to move on a dial-type thermometer?
6. What is an advantage of digital thermometers?
7. Which type of thermometer can see the temperature of a surface from a distance?
8. What is a disadvantage of an IR thermometer?
9. What two metals are joined to create a type K thermocouple?
10. What are the two points used to create the Celsius temperature scale?
11. The Rankine temperature scale uses the same spacing between degrees as which other system?
12. What would an indoor thermometer range usually be?
13. Convert the following Celsius temperatures to Fahrenheit.
    a. 50°C
    b. 25°C
    c. 150°C
    d. 350°C
14. Convert the following Fahrenheit temperatures to Celsius.
    a. 50°F
    b. 104°F
    c. 158°F
    d. 356°F
15. Convert the following temperatures.
    a. –60°F to Rankine
    b. 30°C to Kelvin
    c. 630°R to Fahrenheit
    d. 200°K to Celsius
16. Convert the following temperatures.
    a. 125°K to Rankine
    b. 300°K to Rankine
    c. 540°R to Kelvin
    d. 378°R to Kelvin

UNIT 10

Thermodynamics—The Study of Heat

OBJECTIVES

After completing this unit, you will be able to:

1. state the first and second laws of thermodynamics.
2. compare and contrast the three basic methods of heat transfer.
3. explain the factors that affect the rate that heat transfers through various materials.
4. explain the difference between heat energy and temperature.
5. compare sensible heat to latent heat.
6. calculate heat change.
7. differentiate among saturated, superheated, and subcooled refrigerant.

10.1 INTRODUCTION

Almost everything we do in HVACR involves adding or moving thermal (heat) energy. For example, in heating we may be converting chemical energy, such as a natural gas flame, to thermal energy and distributing it through the building. In refrigeration and air conditioning we are removing heat from inside the refrigerator or house and sending it outside.

It is important to understand thermal transfer. The reason we have to heat or cool areas is because heat transfers through the walls. In the heating season it transfers out, and in the cooling season it transfers in. Put simply, HVACR technicians work with thermal energy. In this unit, we cover thermodynamics and then look at heat transfer more deeply.

10.2 FIRST LAW OF THERMODYNAMICS

Thermodynamics is the branch of science dealing with heat and the movement of energy. There are four laws of thermodynamics that describe how heat and energy behave. Two of these are of particular interest in HVACR applications. The first law of thermodynamics states that "energy can neither be created nor destroyed." Heat cannot be "made," but other forms of energy can be converted to heat because different forms of energy can be converted from one form to another. Energy itself is defined as the ability to do work, and heat is the transfer of energy due to temperature difference. Other common forms of energy are mechanical, electrical, and chemical, which may be converted easily from one form to another. The steam-driven turbine generator of a power plant is a device that converts heat energy into electrical energy. Chemical energy may be converted into electrical energy by the use of a battery. Electrical energy is converted into mechanical energy through the use of an electromagnetic coil to produce a push-pull motion or the use of an electric motor to create rotary motion. Electrical energy may be changed directly to heat energy by passing electrical current through wires called resistance heaters, like those used in an electric toaster, grill, or furnace. In all of these transformations, energy is neither created or destroyed; it is just changed from one form of energy to another.

10.3 SECOND LAW OF THERMODYNAMICS

The second law of thermodynamics states that "to cause heat energy to travel, a temperature difference must be established and maintained." Heat energy travels downward on the intensity scale. Heat from a higher temperature (intensity) material will travel to a lower temperature (intensity) material, and this process will continue as long as the temperature difference exists. The rate of travel varies directly with the temperature difference. The higher the temperature difference, commonly called the delta temperature, or ΔT, the greater the rate of heat travel. The lower the ΔT, the lower the rate of heat travel.

10.4 METHODS OF HEAT TRANSFER

There are three principal ways that heat is transferred:

1. Conduction
2. Convection
3. Radiation

Most refrigeration systems utilize all three methods. These three methods of heat transfer are shown in Figure 10-1.

TECH TIP

Heat is always on the move. It is moving constantly from a warmer body to a cooler body. This movement can be slowed with insulation, but no matter how thick the insulation is, heat will continue to move through it until both bodies have reached equilibrium and are at the same temperature. For this reason it is very difficult to store heat for long periods of time without significant loss in the quantity of heat being stored.

10.5 CONDUCTION

Conduction is described as the transfer of heat between the closely packed molecules of a substance, or between substances that are in good contact with one another. When

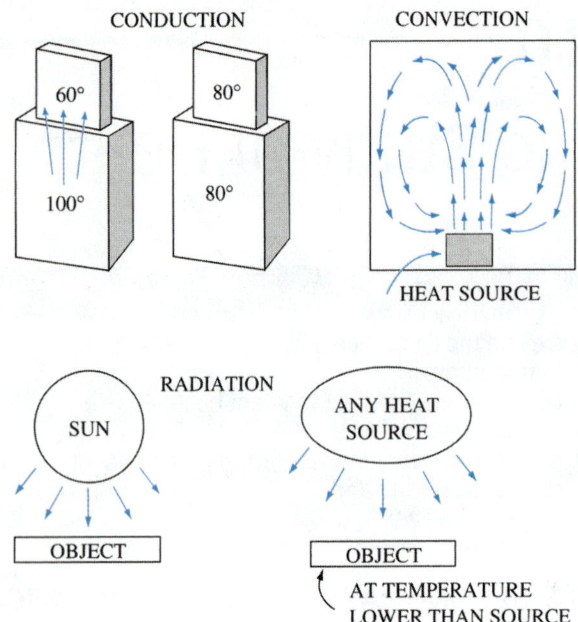

Figure 10-1 Three methods for transferring heat.

heat transfers by conduction in a single substance, such as a metal rod, movement of heat continues until there is a temperature balance throughout the length of the rod. With one end of the rod in a flame and the other end in the air, heat will travel from the end in the flame to the end in the air. Heat will continue to travel through the rod as long as one end is hotter than the other. Even after the rod is removed from the flame, heat will continue to travel from the hotter end to the cooler end until both ends are the same temperature.

If the rod is immersed in water, the rapidly moving molecules on the surface of the rod will transmit some heat to the molecules of water, and still another transfer of heat by conduction takes place. As the outer surface of the rod cools off, there is still some heat within the rod, and this will continue to transfer to the outer surfaces of the rod and then to the water, until a temperature balance is reached.

The rate of heat transfer will vary according to the ability of the material to conduct heat. Solids, on the whole, are much better conductors than liquids; in turn, liquids conduct heat better than gases or vapors.

Most metals, such as silver, copper, steel, and iron, conduct heat fairly rapidly. Many other solids, such as glass, wood, or other building materials, transfer heat at a much slower rate and therefore are used as insulators.

TECH TIP

Materials that are excellent electrical conductors also tend to be excellent thermal conductors. Similarly, materials that are good electrical insulators also tend to be good thermal insulators. Of course there are exceptions. Materials such as ceramics can be designed to be either an insulator or a conductor of electricity or heat. One particular type of ceramic is actually considered to be a superconductor, meaning it has extremely low electrical and heat resistance.

Material	Conductivity *k*
Plywood	0.80
Glass fiber—organic bonded	0.25
Expanded polystyrene insulation	0.25
Expanded polyurethane insulation	0.16
Cement mortar	5.0
Stucco	5.0
Brick (common)	5.0
Hardwood (maple, oak)	1.10
Softwood (fir, pine)	0.80
Gypsum plaster (sand aggregate)	5.6

Figure 10-2 Conductivities for common building and insulating materials; *k* values expressed in BTU/hr/ft^2/°F/in thickness of material.

Copper is an excellent conductor of heat, as is aluminum. These substances are ordinarily used in the construction of heat-transfer coils like evaporators and condensers. Steel and iron are also used for some types of heat-transfer devices.

The rate at which heat may be conducted through any material is dependent on factors like the thickness of the material, its cross-sectional area, the temperature difference between the two sides of the material, the heat conductivity (*k* factor) of the material, and the time duration of the heat flow. The table in Figure 10-2 gives the heat conductivities (*k* factors) of some common materials.

Note that the *k* factors are given in BTU/hr/ft^2/°F/in of thickness of the material. The formula for heat transfer using these factors is

$$\text{Heat in } BTUH = \frac{A \times \Delta T \times k}{X} \quad (10.1)$$

where

A = cross-sectional area, ft^2
k = heat conductivity, BTU/hr/ft^2/°F/in
ΔT = temperature difference between the two sides, °F
X = thickness of material, in

Metals with a high conductivity are used within the refrigeration system itself because it is desirable that rapid heat transfer occur in both evaporator and condenser. The evaporator is where heat is removed from whatever is being cooled; the condenser dissipates this heat to another medium or space.

In the case of the evaporator, the product or air is at a higher temperature than the refrigerant within the tubing, and there is a transfer of heat from the warm air into the cool refrigerant in the evaporator. In the condenser, however, the refrigerant vapor is at a higher temperature than the cooling medium traveling through the condenser, and heat transfers out of the warm refrigerant in the condenser into the cooler surrounding air.

Plain tubing, whether copper, aluminum, or another metal, will transfer heat according to its conductivity, or *k* factor, but this heat transfer can be increased through the addition of fins on the tubing. They will increase the area of heat-transfer surface, thereby increasing the overall efficiency of the system. If the addition of fins doubles the surface area, it can be shown by the use of Equation 10-1 that the overall heat transfer will be doubled compared to that of plain tubing.

10.6 CONVECTION

Another means of heat transfer is by motion of the heated material itself. Convection is limited to heat transfer within a liquid or a gas. When a material is heated, convection currents are set up within it, and the warmer portions of it rise, since heat brings about the decrease of a fluid's density and an increase in its specific volume.

SERVICE TIP

Homeowners and building operators sometimes like to put up their own thermometers in a room. That thermometer reading sometimes is the one you will hear them talking about when they complain about room temperature. There are several factors that will affect the thermometer reading that may result in it being different from the room thermostat. First, the room thermostat is usually close to a height of five feet and should be located on an inside wall. If their thermometer is higher or lower than that, it will read a different temperature. Second, if their thermometer is near an outside wall, it will read higher or lower depending on the season of the year. Certainly a thermometer that is in a window can have wide swings in temperature. One common thermometer people use gives indoor and outdoor temperature readings. By its design, it must be located on an outside wall so that its probe can go to the outside of the building. Those temperature readings will be significantly different most of the time from the room thermostat.

Natural Convection

Natural convection occurs when fluid circulates because of the density changes that are the result of the fluid heating up and cooling down. Systems that rely entirely on natural convection are called passive systems. Air within a refrigerator is a prime example of the results of convection currents (Figure 10-3). The air in contact with the cooling coil of a refrigerator becomes cool and therefore denser and begins to fall to the bottom of the refrigerator. In doing so, it absorbs heat from the food and the walls of the refrigerator, which, through conduction, has picked up heat from the room. After heat has been absorbed by the air it expands, becoming lighter, and rises until it again reaches the cooling coil where heat is removed from it. The convection cycle repeats as long as there is a temperature difference between the air and the coil. In commercial units, baffles may be constructed within the box to direct the air in the desired airflow patterns.

Water heated in a pan will be affected by the convection currents set up within it through the application of heat. The water nearest the heat source absorbs heat, becomes warmer, and expands. As it becomes less dense, it rises and is replaced by the other water, which is cooler and denser.

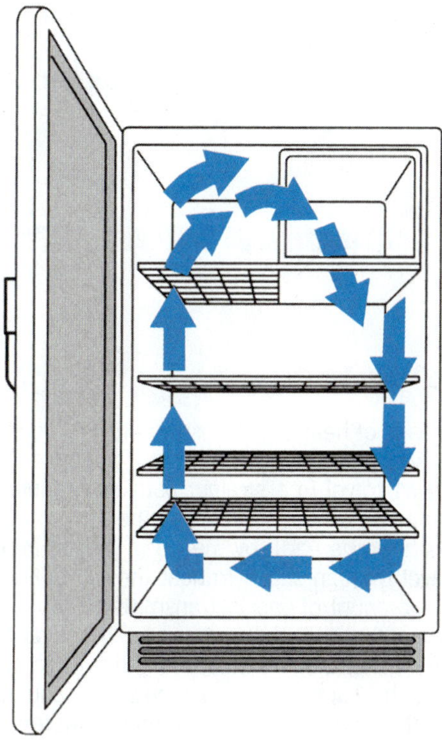

Figure 10-3 Convection currents caused by temperature differential.

Forced Convection

Forced convection is the process of moving a fluid using mechanical means to increase heat transfer. Fans are used to move gases; pumps are used to move liquids. Systems that use fans or pumps to increase convection are called active systems.

Natural convection currents in air or water distribute the temperature only to a certain extent. Natural convection currents will not keep a room's temperature consistent from floor to ceiling. A temperature difference of 10°F from floor to ceiling in a room with a wood stove is common. When there is a large temperature difference from the floor to the ceiling, additional air circulation is needed to break up the stratified air temperature. Ceiling fans like the one shown in Figure 10-4 can help distribute the heated air more evenly.

Figure 10-4 Ceiling fans in vaulted ceilings can help redistribute warm air that has collected at the top of the ceiling.

Figure 10-5 Correlation between the absolute temperature of an object and its color.

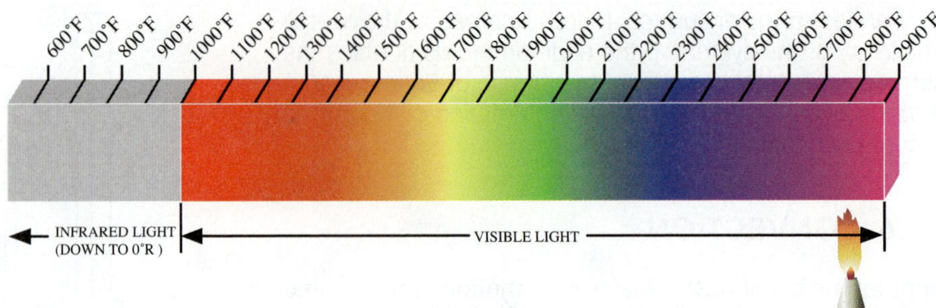

10.7 RADIATION

A third means of heat transfer is through radiation. Radiant heat energy travels by waves similar to light or sound waves. These waves travel in a straight path and require no medium between the heat source and the heated object. The sun's rays heat the earth by means of radiant heat waves, which travel through space to heat the earth. Four factors control the amount of energy transmitted between a radiation source and an object: the frequency of the radiation, the absolute temperature of the source, the surface area of the source, and the distance between the source and the object.

Radiant energy waves exist at many different frequencies. The frequency is related to the distance between the peaks of the waves. The shorter the distance, the higher the frequency. Higher frequency radiation is more powerful than lower frequency radiation.

The total amount of radiant energy emitted by an object increases dramatically as its absolute temperature increases. Doubling the absolute temperature increases the emitted energy by a factor of 16. The amount of radiant energy emitted by a radiation source is also directly related to its surface area. Twice as much surface area produces twice as much radiant energy.

Heat transfer through radiation does not result in an even distribution of heat. The amount of radiant energy transferred decreases with the square of the distance. This means that doubling the distance decreases the amount of energy transmitted to a fourth of the original amount. A fireplace heats through radiation. The further you get away from the fireplace, the colder it is. Next to the fireplace is very warm, but a few feet away can feel much cooler.

Radiant energy travels in straight lines and heats objects, not air. While facing the fireplace in a cold room, your front is warm but your back is cold because it is turned away from the fire. If someone steps between you and the fireplace, your source of heat is cut off and you feel instantly cold.

All objects absorb radiant energy to some extent. When radiant energy strikes matter, it can be absorbed, reflected, or transmitted. The amount of energy absorbed versus the amount of energy reflected and transmitted depends on the characteristics of the material. Solids absorb more radiant energy than liquids; liquids absorb more than gases. Dark-colored solids absorb more energy than they reflect, while light-colored solids reflect more than they absorb. The name for an object that is a perfect absorber of radiation is a black body.

All objects whose temperature is above absolute zero also emit radiation. The amount of radiant energy emitted equals the amount of energy absorbed by the object. Black bodies are also perfect radiation emitters. Radiant objects emit many frequencies of radiation, but they tend to emit more radiation of one particular frequency. This primary frequency changes with the absolute temperature of the object. When the primary frequency is within the visible light spectrum, the object appears that color. Figure 10-5 shows the correlation between the absolute temperature of an object and its color.

TECH TIP

Since radiant energy has little effect on gases, radiant heaters tend to heat the objects in a room without heating the air in the room. This has the advantage of heating people quickly. It also can mean that people can be comfortable at lower room temperatures. The air temperature does rise because the air is heated by the objects in the room, but the primary purpose of radiant heat is not to heat the air but the objects.

10.8 HEAT-TRANSFER PROCESSES

Heat is the transfer of energy because of a difference in temperature. The reaction of matter to the addition or deletion of heat is described as either a sensible or a latent process. A sensible process results in a temperature change, while a latent process results in a physical state change. Notice that these terms describe how heat affects substances; they are not names for different types of heat.

10.9 SENSIBLE HEAT

Heat that can be felt or measured is called sensible heat because it can be sensed. A sensible heat process causes a change in the temperature of a substance but not a change in state. For example, hot air travels across a cold evaporator coil and is reduced in temperature. This is called a sensible heat process because the air temperature was changed when heat was removed from it. Anytime a heat transfer results in a temperature change, that process can be called a sensible heat process.

10.10 SPECIFIC HEAT CAPACITY

Specific heat capacity is a factor that allows technicians to calculate the exact amount of heat required to produce a specific temperature rise in a substance. Calculating how much heat it takes to change the temperature of water is

Water	1.00
Ice	0.50
Air (dry)	0.24
Steam	0.48
Aluminum	0.22
Brass	0.09
Lead	0.03
Iron	0.10
Mercury	0.03
Copper	0.09
Alcohol	0.60
Kerosene	0.50
Olive oil	0.47
Glass	0.20
Pine	0.67
Marble	0.21

Figure 10-6 Specific heats of common substances (BTU/lb/°F).

relatively easy. In either the traditional or SI measurement systems, the weight of the water is multiplied by the difference in temperature. This is because the units of heat are defined by a temperature change in water.

However, everything does not take the same amount of heat to change temperature. For example, it takes far less heat to change the temperature of metal than glass. Each substance has its own specific heat. The specific heat is also different for different states of the same substance. For example, the specific heat of water is 1.0 BTU/lb, while the specific heats of both ice and steam are approximately 0.5 BTU/lb. Figure 10-6 shows the specific heat capacity of several common substances. It is important to use the correct specific heat value when calculating heat change in a substance.

In traditional units, the specific heat capacity is the number of BTUs required to change the temperature of 1 lb of a substance 1°F. Water has a specific heat capacity of 1, since the definition of the BTU is based on a 1°F temperature change in water. A substance that requires only half as much heat as water would have a specific heat capacity of 0.5 BTU/lb, while something that takes twice as much heat would have a specific heat capacity of 2.0 BTU/lb. Things that are easily heated have low specific heat capacities, while things that are slow to heat have high specific heat capacities.

The formula for changing the temperature of anything is

$$\text{heat quantity} = \text{weight} \times \text{temperature difference} \times \text{specific heat}$$

In traditional units, the weight is measured in pounds and the temperature in degrees Fahrenheit. The formula becomes

$$\text{BTU} = \text{lb} \times {}^\circ\text{F}\ \Delta T \times \text{specific heat capacity in BTU/lb/}{}^\circ\text{F}$$

In the SI system, specific heat capacity is commonly given as the number of kilojoules (kJ) required to change 1 kg of a substance 1°K. The formula becomes

$$\text{kJ} = \text{kg} \times {}^\circ\text{K}\Delta T \times \text{specific heat capacity in kJ/kg/}{}^\circ\text{K}$$

10.11 LATENT HEAT

The word *latent* means hidden. Latent heat is hidden heat because there is no accompanying temperature change to sense. Of course the heat exists—it goes into the molecules to change their state. Latent heat refers to a heat transfer that causes a change in state but not a change in temperature. For example, the refrigerant in the evaporator coil of an air conditioner boils from a liquid to a gas while it is absorbing heat from the warm air passing over the coil. Since the refrigerant is changing state and not temperature, this heat-transfer process is called a latent heat process because the result of the heat transfer is a change in state, not a change in temperature. Latent heat is always involved in a change of state. There are several possible state changes, each with its own latent heat value. Each of these changes has a name that identifies the state change. Figure 10-7 lists the different types of state changes.

It takes a lot more energy to change the state of a substance than it does to change its temperature. It takes 144 BTU to change 1 lb of ice to water, but it only takes 0.5 BTU to change the temperature 1°F. The state changes involving gas require even more heat. It takes 970 BTU to boil a pound of water, but it only takes 180 BTU to raise the temperature of 1 lb of water from 32°F to 212°F. The formula for calculating a state change is

$$\text{BTU} = \text{weight} \times \text{latent heat factor}$$

The latent heat factor is different for different substances. It is also different for different state changes of the same substance. Figure 10-8 lists the latent heat factors for a few common substances.

Process	Common Name	Description	Result	Example
Fusion	Freezing	Liquid to solid	Releases heat	Water freezing
Liquefaction	Melting	Solid to liquid	Absorbs heat	Ice melting
Vaporization	Boiling	Liquid to gas	Absorbs heat	Water boiling
Condensation	Condensing	Gas to liquid	Releases heat	Dew
Sublimation		Solid to gas	Absorbs heat	Dry ice sublimating
Deposition		Gas to solid	Releases heat	Frost forming

Figure 10-7 Physical changes of state.

Substance	Melting Point Fahrenheit°	Heat of Fusion BTU/lb	Boiling Point Fahrenheit	Heat of Vaporization BTU/lb
Water	32	144	212	970
Copper	1,981	91	5,301	2,176
Mercury	−38	5	674	226
Paraffin	133			
Ethyl alcohol	−179	45	172	369
Oxygen	−362	6	−298	92
Carnauba wax	120	54	320	
Beeswax	144	75	650	

Figure 10-8 Table of latent heat values.

10.12 APPLICATION OF HEAT-TRANSFER PROCESSES

Simply stated, a sensible heat process causes a change in temperature; a latent heat process causes a change in state. A substance can either change temperature or change state, but it cannot change both at the same time. Water will not boil until it reaches 212°F, but once it does, its temperature will not change. A heat-transfer process requires a heat source to supply the heat and a heat sink to absorb the heat. One can undergo a sensible change while the other undergoes a latent change. In the evaporator coil shown in Figure 10-9, the refrigerant is boiling from a liquid to a gas and absorbing heat. This is a latent change. The fact that the refrigerant is changing state keeps it cold even as it is absorbing heat. The air blowing over the evaporator is changing temperature from warm to cool. This is a sensible change. The sensible change in air temperature is the goal.

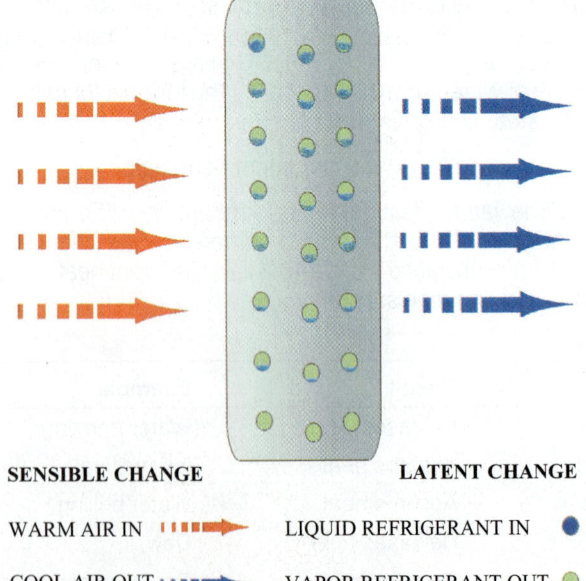

Figure 10-9 The refrigerant in this coil is undergoing a latent change from liquid to vapor while the air is undergoing a sensible change from warm to cool.

10.13 CALCULATING HEAT CHANGE

The specific heat formula is used to calculate the amount of heat required to change the temperature of a substance; the latent heat formula is used to calculate the amount of heat required to change the state of a substance. Both formulas must be used if the change will involve both a temperature change and a state change. Something cannot change state and temperature at the same time, but it can change temperature and then change state. The process of changing water from below-freezing ice to superheated steam is illustrated in Figure 10-10.

The line from A to B in Figure 10-10 shows the temperature change in the ice from 10°F to 32°F. The line from B to C shows the 32°F ice melting to 32°F water. Notice that the temperature of the water is the same as the ice. The line from C to D shows the water temperature rising from 32°F to 212°F. The line from D to E shows the 212°F water boiling to 212°F steam. Notice that the steam temperature is the same as the water temperature. Finally, the line from E to F represents raising the steam temperature from 212°F to 230°F. Each of these steps is solved independently, and the results are summed to solve for the overall amount of heat required.

- The heat required to increase the temperature of 1 lb of ice from 10°F to 32°F is calculated using the specific heat formula:

  BTU = lb × °FΔT × specific heat capacity in BTU/lb/°F
  BTU = 1 lb × 22 × 0.5 = 11 BTU

- The heat required to melt 1 lb of ice is calculated using the latent heat formula:

  BTU = lb × latent heat factor in BTU/lb
  BTU = 1 × 144 = 144 BTU

- The heat required to increase the temperature of 1 lb of water from 32°F to 212°F is calculated using the specific heat formula:

  BTU = lb × °F ΔT × specific heat capacity in BTU/lb/°F
  BTU = 1 lb × 180 × 1 = 180 BTU

- The heat required to boil 1 lb of water is calculated using the latent heat formula:

  BTU = lb × latent heat factor
  BTU = 1 × 970 = 970 BTU

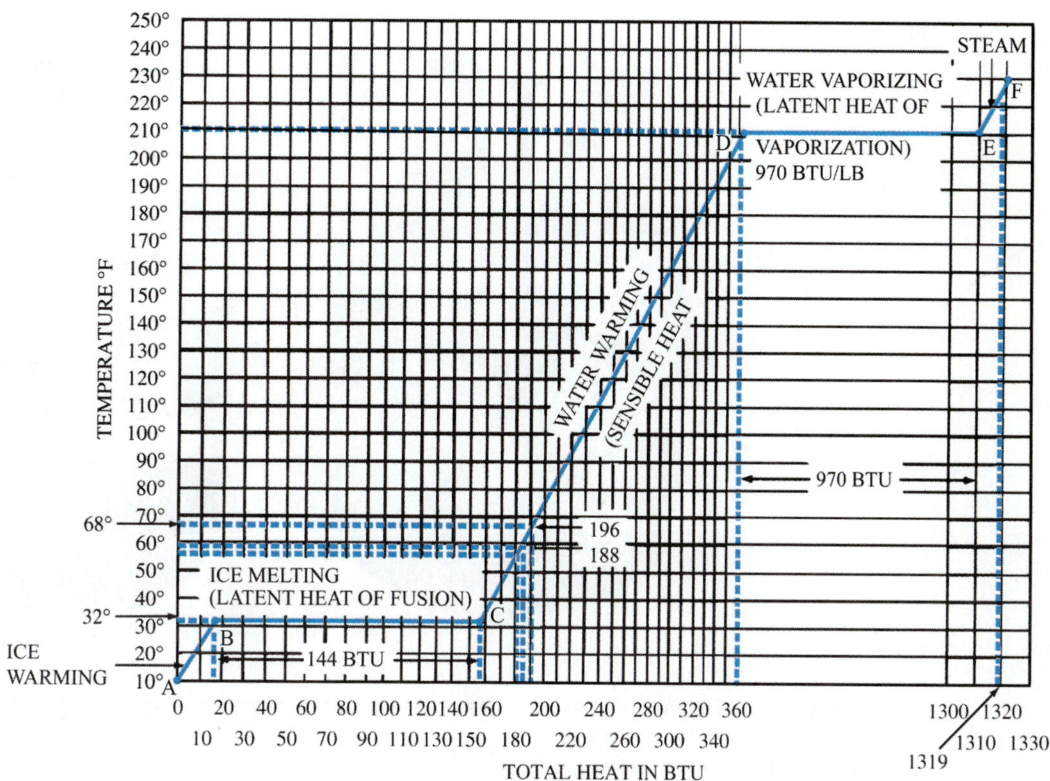

Figure 10-10 Chart demonstrating sensible and latent heat relationships in melting ice, changing ice to water and water to steam.

- The heat required to increase the temperature of a pound of steam from 212°F to 230°F is calculated using the specific heat formula:

  BTU = lb × °F ΔT × specific heat capacity in BTU/lb
  BTU = 1 lb × 18 × 0.5 = 19 BTU

It is evident that changes of state require much more heat than changes in temperature. The total amount of heat required for the temperature change from 10°F to 230°F is 210 BTU, while the total amount of heat required to change state from ice to water and from water to steam is 1,114 BTU. The latent heat is the turbocharger of the refrigeration cycle. Much greater amounts of heat can be moved by changing the refrigerant's state than by changing its temperature.

10.14 EXAMPLE HEAT CALCULATION

Calculate the amount of heat needed to change 10 lb of ice at –20°F to steam at 250°F.

- The heat required to increase the temperature of 10 lb of ice from –20°F to 32°F is calculated using the specific heat formula:

  BTU = lb × °F ΔT × specific heat capacity in BTU/lb/°F
  BTU = 10 lb × 52 × 0.5 = 260 BTU

- The heat required to melt 10 lb of ice is calculated using the latent heat formula:

  BTU = lb × latent heat factor
  BTU = 10 × 144 = 1,440 BTU

- The heat required to increase the temperature of 10 lb of water from 32°F to 212°F is calculated using the specific heat formula:

  BTU = lb × °F ΔT × specific heat capacity in BTU/lb/°F
  BTU = 10 lb × 180 × 1 = 1,800 BTU

- The heat required to boil 10 lb of water is calculated using the latent heat formula:

  BTU = lb × latent heat factor
  BTU = 10 × 970 = 9,700 BTU

- The heat required to increase the temperature of a pound of steam from 212°F to 250°F is calculated using the specific heat formula:

  BTU = lb × °F ΔT × specific heat capacity in BTU/lb/°F
  BTU = 10 lb × 38 × 0.5 = 190 BTU

- Finally, sum the results

  260 BTU + 1,440 BTU + 1,800 BTU + 9,700 BTU + 190 BTU = 13,390 BTU

10.15 SATURATED LIQUID-GAS MIXTURES

The word *saturated* means completely full. When a sponge is saturated with water, it is holding all the water it can possibly hold. If you try to add more water, some water will spill out because there is nowhere for it to go. When a liquid is saturated with heat, it has all the heat is can hold and still remain a liquid. When you add additional heat to a saturated liquid, some of the liquid changes to a gas. The gas that boils off of a saturated liquid is called a saturated gas. If heat

Figure 10-11 Even though these cylinders have different amounts of liquid refrigerant in them, they are the same pressure because they are the same temperature.

is removed from a saturated gas, some of the gas will condense to a liquid. The saturation point is the boiling point. As long as the liquid-gas mix remains at a constant pressure and some liquid remains, the temperature will remain the same. Additional heat does not change the temperature; it simply makes more saturated gas. This concept of saturation helps explain why a liquid's temperature remains stable when it boils. After you reach the boiling point, any extra heat goes into turning some of the liquid to gas rather than raising the temperature of the liquid.

While it is technically possible to have a saturated gas and not have any saturated liquid along with it, in practice, where there is saturated gas there is also saturated liquid. For this reason, many texts refer to saturated mixtures when discussing either saturated gases or saturated liquids.

Saturated gases do not follow the ideal gas laws. The liquid in a saturated mixture will continue to vaporize until the gas pressure reaches the saturation point for the temperature of the mixture. The pressure of a cylinder of saturated refrigerant with 5 lb of refrigerant will have the same pressure as a cylinder with 20 lb of refrigerant if they are both at the same temperature. For saturated mixtures, the pressure is determined entirely by the temperature and not by the quantity of refrigerant (Figure 10-11).

SERVICE TIP

A drop in cylinder pressure is a common problem encountered when charging refrigerant vapor from a cylinder into a system. This does not happen because there is less refrigerant in the cylinder, but because the remaining liquid in the cylinder cools off as the vapor leaves. When vapor boils off of the liquid, it removes heat from the liquid, dropping its temperature. When the temperature of a saturated mixture drops, so does the pressure. Heating the cylinder with warm water can offset this. When the liquid temperature is raised, the pressure comes up as well.

10.16 USING PRESSURE-TEMPERATURE CHARTS

Remember that the boiling point, or saturation temperature, is controlled by pressure. The higher the pressure is, the higher the saturation temperature will be. The pressure and temperature of any saturated mixture go up and down together. If you know one, you can find the other. Charts called pressure-temperature charts, or PT charts for short, are commonly used in refrigeration (Figure 10-12). These charts correlate the saturation temperature and pressure for commonly used refrigerants. Most of these charts list temperature in a column on the left-hand side of the chart and corresponding pressures in columns on the right-hand side of the chart. To use the chart, find the saturation temperature in the left column and follow across that same row to the column under the refrigerant being used. That lists the pressure of the refrigerant that corresponds to that particular saturation temperature. In Figure 10-12, for example, 40°F in the left-hand column corresponds to 68.5 psig in the column under R-22.

Of course the chart can be used the other way also. Find the pressure in the column under the refrigerant and then follow across that same row to the temperature column on the outside. This lists the saturation temperature for the refrigerant at that particular pressure. Again in Figure 10-12, for example, 124.1 psig under the R-134a column corresponds to 100°F in the temperature column on the left-hand side.

10.17 SUPERHEATED GAS

Remember that when heat is added to a saturated mix, the temperature does not change. More liquid turns to gas, but the temperature remains the same. However, what happens if enough heat is added to a saturated mixture until all the liquid has turned to gas? Now, if more heat is added, the gas temperature will rise because there is no more liquid to boil. When a gas is heated above its boiling point, it is

VAPOR PRESSURE, PSIG

Temp (°F)	11	12	22	113	114	500	502	134a	123
5	23.9	11.8	28.2	27.9	16.2	16.4	35.9	9.1	25.3
10	23.1	14.6	32.8	27.6	14.4	19.7	41.0	11.9	24.6
15	22.1	17.7	37.7	27.2	12.4	23.3	46.5	15.0	23.7
20	21.1	21.0	43.0	26.8	10.2	27.2	52.5	18.4	22.8
25	19.9	24.6	48.7	26.3	7.8	31.5	58.8	22.1	21.8
30	18.6	28.4	54.9	25.8	5.2	36.0	65.6	26.0	20.7
35	17.2	32.5	61.5	25.2	2.3	40.8	72.8	30.3	19.5
40	15.6	36.9	68.5	24.5	0.4	46.0	80.5	35.0	18.1
45	13.9	41.6	76.0	23.8	2.0	51.6	88.7	40.0	16.6
50	12.0	46.7	84.0	22.9	3.8	57.5	97.4	45.4	15.0
55	10.0	52.0	92.5	22.2	5.8	63.9	106.6	51.1	13.1
60	7.8	57.7	101.6	21.0	7.9	70.6	116.4	57.3	11.2
65	5.4	63.7	111.2	19.9	10.1	77.8	126.7	63.9	9.0
70	2.7	70.2	121.4	18.7	12.6	85.4	137.6	71.0	6.6
75	0.0	76.9	132.2	17.3	15.2	93.4	149.1	78.6	4.0
80	1.5	84.1	143.6	15.8	18.0	101.9	161.2	86.6	1.2
85	3.2	91.7	155.7	14.3	20.9	111.0	174.0	95.1	0.9
90	4.9	99.7	168.4	12.5	24.1	120.5	187.4	104.2	2.5
95	6.8	108.2	181.8	10.6	27.5	130.5	201.4	113.8	4.2
100	8.8	117.1	195.9	8.6	31.1	141.1	216.2	124.1	6.1
105	10.9	126.5	210.7	6.4	35.0	152.2	231.7	134.9	8.1
110	13.2	136.4	226.3	4.0	39.1	164.0	247.9	146.3	10.3
115	15.6	146.7	242.7	1.4	43.4	176.3	264.9	158.4	12.6
120	18.3	157.6	259.9	0.7	48.0	189.2	282.7	171.1	15.1

PRESSURE/TEMPERATURE CHART

Temp. Deg. F	R-408A (FX-10) Liquid Pressure	R-404A (FX-70) Liquid Pressure	R-409A (FX-56) Liquid Pressure	R-409A (FX-56) Vapor Pressure	R-407C Liquid Pressure	R-407C Vapor Pressure	R-410A Liquid Pressure
5	34.2	38.6	18.1	9.7	33.0	22.9	55.2
10	39.3	44.0	21.7	12.5	38.0	27.3	62.3
15	44.8	49.9	25.5	15.4	43.5	32.0	70.0
20	50.7	56.2	29.6	18.7	49.3	37.2	78.3
25	57.0	63.0	34.0	22.2	55.7	42.7	87.3
30	63.7	70.3	38.7	26.0	62.5	48.7	96.8
35	71.0	78.1	43.8	30.1	69.8	55.2	107.0
40	78.7	86.4	49.2	34.5	77.6	62.1	118.0
45	87.0	95.2	54.9	39.2	86.0	69.5	129.7
50	95.8	104.7	61.0	44.3	94.9	77.5	142.2
55	105.1	114.7	67.6	49.8	104.5	86.0	155.2
60	115.1	125.3	74.5	55.6	114.6	95.1	169.6
65	125.6	136.6	81.8	61.9	125.4	104.8	184.6
70	136.8	148.6	89.5	68.6	136.9	115.2	200.6
75	148.7	161.2	97.7	75.8	149.1	126.2	217.4
80	161.2	174.6	106.4	83.4	162.1	137.8	235.3
85	174.4	188.8	115.5	91.5	175.8	150.2	254.1
90	188.4	203.7	125.2	100.2	190.2	163.4	274.1
95	203.1	219.4	135.3	109.4	205.5	177.4	295.1
100	218.7	235.9	146.0	119.2	221.6	192.1	317.2
105	235.0	253.4	157.2	129.6	238.5	207.8	340.5
110	252.1	271.7	169.0	140.6	256.4	224.4	365.0
115	270.2	290.9	181.4	152.3	275.1	241.9	390.7
120	289.1	311.1	194.4	164.7	294.7	260.5	417.7

Figure 10-12 The pressures of common refrigerants at 40°F saturation and 100°F saturation are shown in the highlighted bands. *(Courtesy of Arkema Inc.)*

called a superheated gas. In theory it is possible to have a gas that has no liquid in it and is still at the saturation point. However, in practice, if there is only gas present, it is usually superheated. What happens when heat is removed from a superheated gas? The temperature of a superheated gas drops as it is cooled until it reaches the saturation point. If heat continues to be removed after the gas reaches the saturation point, some of the gas will turn back to liquid. When the saturated gas begins to change to liquid, the temperature remains the same.

10.18 SUBCOOLED LIQUID

As long as there is some gas in a saturated mix, the removal of heat will result in some of the gas changing to liquid at the same temperature. After all the gas has turned to liquid, if more heat is removed, the liquid temperature will drop because there is no more gas to turn into liquid. When a liquid is cooled below its boiling point it is called a subcooled liquid.

10.19 DISTINGUISHING AMONG SATURATED, SUPERHEATED, AND SUBCOOLED

Refrigerant can be found in a refrigeration system as a superheated gas, a saturated mixture, or a subcooled liquid. Determining the condition of the refrigerant is an important step in working on refrigeration systems. It is not really difficult to determine if the refrigerant in a part of a refrigeration system is saturated, superheated, or subcooled. All that is needed are the folowing:

- An accurate temperature reading
- An accurate pressure reading
- A pressure-temperature chart for the type of refrigerant involved

Take accurate pressure and temperature readings where the state of the refrigerant is to be identified. Use a saturated pressure-temperature chart to look up the saturation temperature that corresponds to the pressure reading.

Saturated Mix

If the actual temperature reading is within 3° of the temperature on the chart, the refrigerant is most likely saturated. Ideally it should be dead on, but a slight measurement error on either the pressure or temperature reading can make it difficult to get a perfect reading.

Superheated Vapor

If the actual temperature reading is more than 3° above the temperature on the chart, the gas is superheated.

Figure 10-13 Comparison of a saturated mixture of liquid and vapor, a superheated vapor, and a subcooled liquid.

Subcooled Liquid

If the actual temperature reading is more than 3° degrees below the saturated temperature on the chart, the liquid is subcooled.

This procedure is summarized below.

1. Measure the refrigerant pressure and temperature.
2. Look up the saturation temperature for the refrigerant pressure on a PT chart.
3. Compare the chart temperature to the actual temperature.

Figure 10-13 illustrates the difference among a saturated mix, a superheated vapor, and a subcooled liquid.

UNIT 10—SUMMARY

Thermodynamics is the study of heat. The first law of thermodynamics states that energy cannot be created or destroyed; the second law of thermodynamics states that heat travels from hot to cold. Heat can travel through conduction, convection, or radiation. Conduction is the transfer of heat through a material by physical contact, convection is heat transfer by fluid circulation, and radiation is heat transfer by waves of radiant energy.

A sensible heat-transfer process results in a temperature change; a latent process results in a change of state. It takes more heat to change the state of matter than to change its temperature. Heat changes that involve both temperature changes and state changes must be broken down into steps before solving.

A saturated mixture is a mixture of liquid and vapor at the boiling point. The pressure or temperature of a saturated mixture can be predicted using a pressure-temperature chart. A gas whose temperature is above the saturation temperature contains no liquid and is called a superheated gas. A liquid whose temperature is below the saturation temperature contains no gas and is called a subcooled liquid.

UNIT 10—REVIEW QUESTIONS

1. Summarize the first and second laws of thermodynamics.
2. How does temperature difference affect the rate that heat travels?
3. List the three methods of heat transfer, and give an example of each.
4. What is the relationship between electrical conductivity and thermal conductivity?
5. List two materials that are considered to be good thermal conductors.
6. According to their *k* values, which material is a better insulator: expanded polystyrene or expanded polyurethane?
7. Explain why the addition of fins on the tubing will increase the rate of heat transfer.
8. Describe the process of natural convection.
9. What factors determine the amount of radiant energy that is transferred between two objects?
10. Explain the relationship between radiant energy and color.
11. Can a substance change temperature and state at the same time? Explain.
12. Describe the following terms: *fusion, liquefaction, vaporization, condensation, deposition*, and *sublimation*.
13. Define *sensible heat*.
14. Define *latent heat of fusion*.
15. What is a saturation pressure-temperature chart?
16. The refrigerant R-22 in a system component has a pressure of 70 psig and a temperature of 50°F. Is it saturated, superheated, or subcooled?
17. The refrigerant R-134a in a system component has a pressure of 35 psig and a temperature of 40°F. Is it saturated, superheated, or subcooled?
18. The refrigerant R-410a in a system component has a pressure of 340 psig and a temperature of 95°F. Is it saturated, superheated, or subcooled?
19. Calculate the amount of heat in BTUs required to change the temperature of 6 lb of copper from 75°F to 100°F.
20. Calculate the amount of heat in BTUs required to turn 5 lb of 10°F ice to 242°F steam.

UNIT 11

Pressure and Vacuum

OBJECTIVES

After completing this unit, you will be able to:

1. explain the relationships between atmospheric pressure, gauge pressure, and absolute pressure.
2. convert atmospheric, absolute, gauge, and vacuum pressures to different scales.
3. define *head pressure* and tell how it is used.
4. explain how a manometer measures pressure.
5. explain how a refrigeration gauge measures pressure.
6. describe how gas responds to changes in temperature, volume, and pressure.
7. solve for new pressure, temperature, or volume using the ideal gas law.

11.1 INTRODUCTION

Pressure is a fundamental physical property that is measured when servicing HVACR equipment. Hydronic, pneumatic, refrigeration, and air systems all depend upon pressure differences to create flow, and HVACR technicians must be familiar with pressure and vacuum readings so they can properly install, service, and repair these systems. The operation of HVACR equipment depends on controlling system pressures. This unit covers the different types of pressure readings and how to convert one type of reading into another.

11.2 PRESSURE

Pressure is a comparative measurement. It compares force pushing on a surface to the area supporting the force. Pressure is uniformly distributed over the surface it is pushing on. The pressure on any surface can be determined by dividing the force applied by the area it is pushing against.

Different states of matter exert pressure in different ways. A solid exerts pressure downward because of gravity. Liquids exert pressure downward and outward because of gravity. Gases exert pressure in all directions against the walls of their container (Figure 11-1).

The force is normally expressed as weight. The pressure created by an object that weighs 100 lb and has a base of 100 in^2 is 100 lb/100 in^2 = 1 pound per square inch. We abbreviate this as 100 psi. If the object weighed 100 lb but had a base area of only 50 in^2, the pressure would be 100 lb/50 in^2 = 2 psi, because the same weight is supported by a smaller area. If the same 100 lb was supported by a larger base of 200 in^2, the pressure would be 100 lb/200 in^2 = 0.5 psi. All that is necessary to calculate pressure is to divide the force by the area. In standard units, pressure is measured in pounds per square inch (psi).

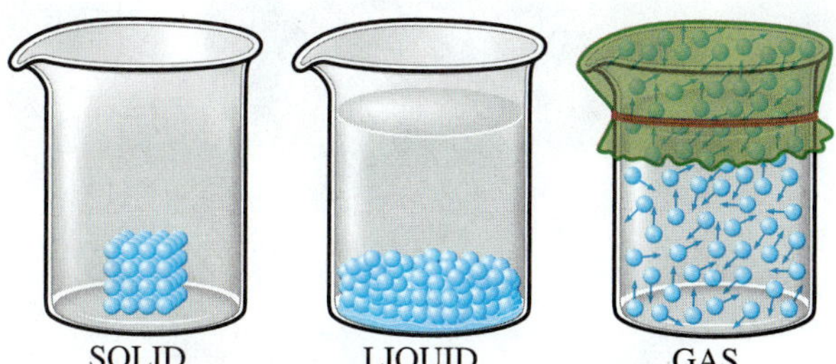

Figure 11-1 A solid exerts pressure down, a liquid exerts pressure down and to the sides, and a gas exerts pressure in all directions.

11.3 LIQUID COLUMN PRESSURE

Another common way to measure pressure is to measure how high a column of liquid the pressure will support. In the case of atmospheric pressure, the atmosphere will support a column of mercury about 30 inches high. What this means is that a column of mercury 30 inches high weighs the same as the column of air pushing on it.

Higher pressures are measured in inches of mercury because mercury is heavy. Lower pressures are measured in inches of water column, because water is much lighter. While atmospheric pressure will support a column of mercury 30 inches high, it will support a column of water 34 feet high! In air conditioning, we use mercury columns to measure vacuum. A vacuum is a pressure less than atmospheric pressure. Since standard atmospheric pressure measures 29.92 inches of mercury, inches of mercury can be used to measure vacuum by comparing the vacuum to atmospheric pressure. A 10-inch mercury vacuum will only support a mercury column 19.92 inches high, 10 inches less than the atmospheric pressure of 29.92 inches mercury. Water columns are used to measure the air pressure in ductwork and the gas pressure in gas furnaces because the pressures in an air duct and in a gas line are much smaller pressures.

11.4 ATMOSPHERIC PRESSURE

The air around us exerts a pressure because it has weight. The weight of all the air sitting on top of each square inch of surface area is about 15 lb at sea level. This varies depending upon the elevation of your location. The higher the elevation, the lower the atmospheric pressure. Since higher elevations have less air above them, the column of air weighs less, making the air pressure lower. Standard atmospheric pressure at sea level with a temperature of 59°F is 14.7 psi, or 101.325 kilopascals (kPa) for SI. In mercury column measurement, that is 29.92 inches of mercury (in Hg) or 760 millimeters of mercury (mm Hg).

The atmospheric pressure at any location on earth will also change with the weather. Both temperature and humidity changes affect the atmospheric pressure. Higher temperature and higher relative humidity both produce lower atmospheric pressure. Higher temperature causes the air to expand, reducing its density and weight, which reduces the atmospheric pressure. Water molecules are lighter than either nitrogen or oxygen molecules. Air with a high relative humidity has more water molecules and fewer nitrogen and oxygen molecules than dry air, making high relative-humidity air lighter than dry air.

11.5 BAROMETRIC PRESSURE

Barometric pressure is a term that is used for the current atmospheric pressure. Meteorologists study barometric pressure to predict weather patterns. A barometer is a device designed to measure the atmospheric pressure. Atmospheric pressure was first demonstrated by the type of simple barometer shown in Figure 11-2.

Early experimenters used a glass tube about 36 in long and closed at one end, an open bowl, and a supply of mercury. They filled the tube with mercury and inverted it in the bowl of mercury, holding a finger at the open end to keep the mercury from spilling out while the tube was inverted. Upon removal of the finger, the level of the mercury in the tube dropped somewhat, leaving a vacuum at the closed top of the tube. The atmospheric pressure bearing down on the open bowl of mercury forced the mercury in the tube to stand up to a height determined by the air pressure. Standard air pressure at sea level and 59°F is 29.92 in Hg. That is 760 mm in SI. That is how barometric pressure came to be measured in inches of mercury (in Hg).

SAFETY TIP

Mercury is considered to be a hazardous chemical, and it must be disposed of properly when an instrument containing mercury is to be disposed of. Mercury can contaminate an area, resulting in an expensive cleanup operation if it is carelessly discarded on the floor of an office, school, hospital, or any other public area. Mercury vapors have been associated with a number of health issues. Therefore, anytime mercury is spilled, you must notify the proper authorities so that its potential threat can be determined and an appropriate method of removing the contamination can be done. Be very careful when instruments are used that contain mercury to avoid any accidental spillage.

Although early scientists did touch mercury with their hands, this is not safe, because mercury is toxic. Handling of mercury should be avoided.

One problem with this type of instrument is that it is not very portable. The aneroid barometer was developed as a more portable instrument. Its operating mechanism is an evacuated bellows called an aneroid cell (Figure 11-3). All the air is removed from the bellows, but it has a spring inside that balances standard atmospheric pressure and prevents the bellows from collapsing. An increase in the atmospheric

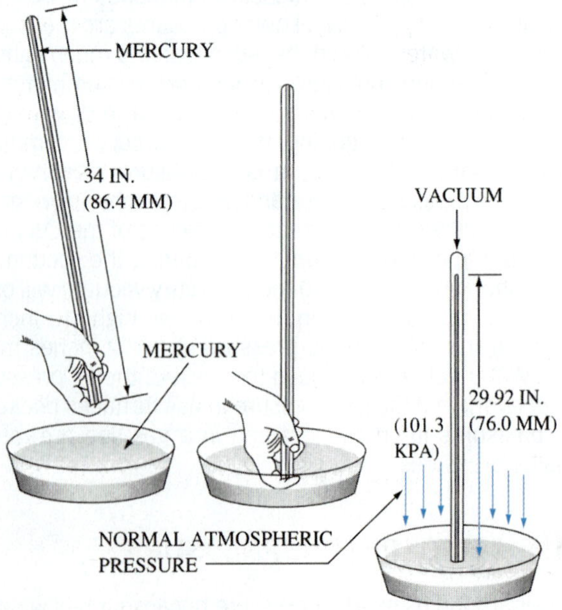

Figure 11-2 Column of mercury supported by normal atmospheric pressure.

Figure 11-3 Aneroid cell barometer.

pressure causes the bellows to contract; a decrease in pressure causes it to expand. The movement of the bellows moves a needle that indicates the atmospheric pressure.

11.6 LIQUID COLUMN PRESSURE MEASUREMENT

A manometer is one type of device used in the refrigeration and air-conditioning field for the measurement of pressure. This type of pressure gauge uses a liquid, usually mercury, water, or gauge oil, as an indicator of the amount of pressure involved. The water column manometer is customarily used when measuring pressures like gas pressures or pressures in air ducts because the low density of water makes it suitable for measuring small amounts of pressure.

SERVICE TIP

There are electronic devices that can measure air pressure so accurately that they can tell the difference in altitude of a distance as little as 5 ft. These devices use small electronic sensors to determine the difference in pressure. Most of these instruments use a piezo crystal as their sensor. A piezo crystal is a crystal that changes its electrical resistivity as a result of external force. As the cost of piezo crystal sensors decreases, their use in air conditioning will increase.

A simple open-arm manometer is shown in Figure 11-4. The U-shaped glass tube is partially filled with water and is open at both ends. The water is at the same level in both arms of the manometer, because both arms are open to the atmosphere, and no external pressure is being exerted on them.

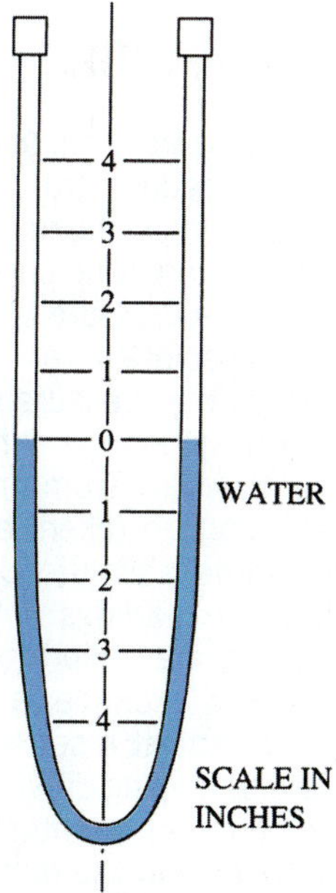

Figure 11-4 U-tube manometer.

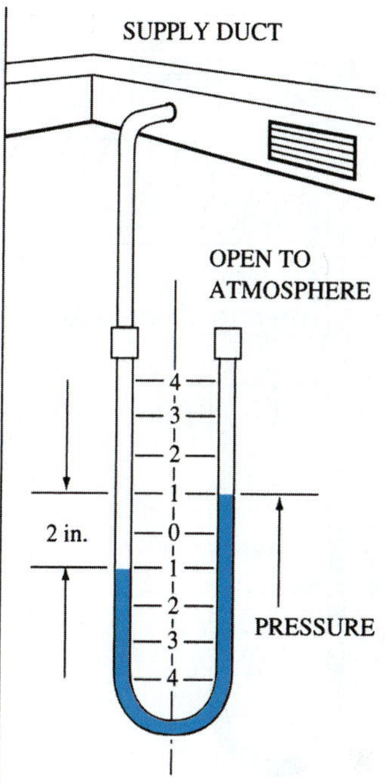

Figure 11-5 When pressure is applied to one side of the manometer, the liquid in that side is forced down and the other side is forced up.

Figure 11-5 shows the manometer in use with one arm connected to a source of positive pressure that is being measured. The amount of pressure being applied is shown by the difference in fluid level from one side to the other. In this case, the pressure is 2 in water column.

A space that is void, or lacking any pressure, is described as having a perfect vacuum. If the space has pressure less than atmospheric pressure, it is defined as being a partial vacuum. It is customary to express this partial vacuum in inches of mercury vacuum, abbreviated as in Hg vacuum.

If a vacuum pump pulls a partial vacuum on the left arm of the manometer, as shown in Figure 11-6, atmospheric pressure will push the mercury in the right arm lower and the mercury on the left higher. The measured vacuum is the difference between the two sides of the manometer, in this case 2 in Hg vacuum.

Liquid column measurement in SI uses millimeters of mercury. Atmospheric pressure is equal to 760 mm Hg.

11.7 GAUGE PRESSURE

The gauges used in the HVACR field typically measure pressure compared to atmospheric pressure. Atmospheric pressure is used as the starting point, so the gauges are calibrated to read 0 at atmospheric pressure. Pressures above atmospheric pressure register as a positive pressure, while pressures below atmospheric pressure register as a vacuum. A "g" is added to the end of psi to indicate gauge pressure. A pressure of 20 psig would mean 20 psi gauge, or 20 psi above atmospheric pressure.

Pressure gauges most commonly used in the field by service technicians, to determine pressure within the refrigeration system, are of the Bourdon tube type. As is shown

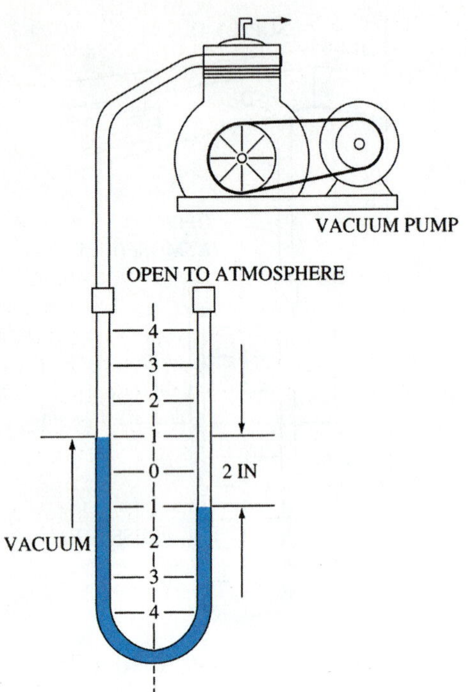

Figure 11-6 When a vacuum is applied to one side of the manometer, the liquid in that side is forced up and the other side is forced down.

in Figure 11-7, an internal view, the essential element of this type of gauge is the Bourdon tube. This oval metal tube is curved along its length and forms an almost complete circle. One end of the tube is closed, and the other end is connected to the equipment or component being tested.

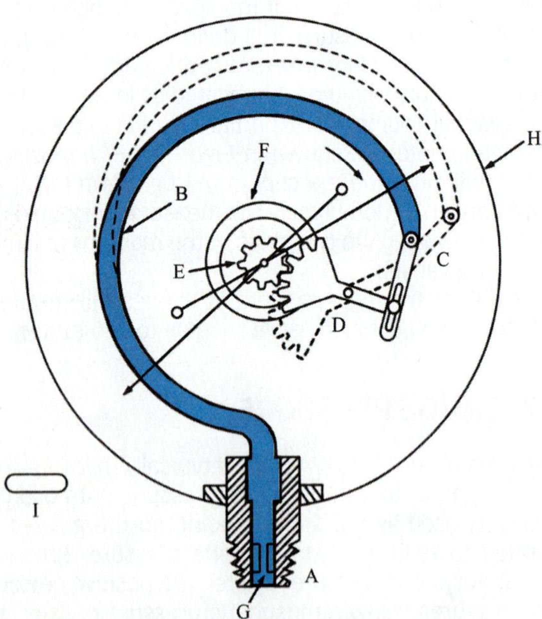

Figure 11-7 Internal construction of a pressure gauge: (A) adapter fitting, usually a ⅛-in pipe thread; (B) Bourdon tube; (C) link; (D) gear sector; (E) pinter shaft gear; (F) calibrating spring; (G) restricter; (H) case; (I) cross-section of the Bourdon tube; the dashed lines indicate how the pressure in the Bourdon tube causes it to straighten and operate the gauge.

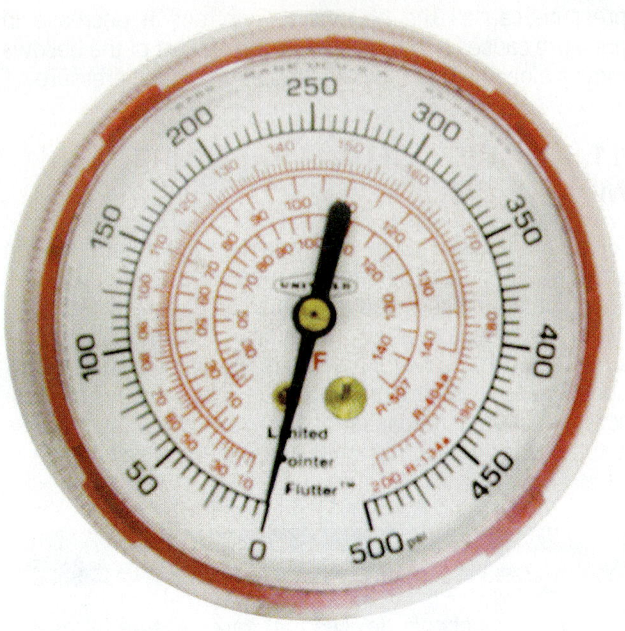

Figure 11-8 A refrigeration pressure gauge measures pressure above atmosphere pressure.

High-pressure gauges, as shown in Figure 11-8, are preset at 0 lb, which represents atmospheric pressure of 14.7 psi. Any additional pressure applied when the gauge is connected to a piece of equipment will tend to straighten out the Bourdon tube, thereby moving the needle or pointer and its mechanical linkage, thus indicating the amount of pressure being applied. High-pressure gauges can only indicate pressures above atmospheric pressure.

Compound gauges are used to measure pressures above and below atmospheric pressure (Figure 11-9). Pressures below atmospheric are customarily expressed in units of in Hg. There is an indication of the range between 0 gauge and 30 in Hg on the compound gauge.

11.8 ABSOLUTE PRESSURE

Of course atmospheric pressure is not 0; it is approximately 15 psi. Some applications require true, or absolute, pressure. Absolute pressure is simply the real pressure. In absolute pressure, 0 means 0. There is no pressure lower than 0 absolute pressure. Absolute pressure is normally stated in pounds per square inch absolute (psia).

Absolute pressure is the pressure without regard to the current atmospheric pressure. Where atmospheric and gauge pressures may vary from time to time based on when and where they are taken, absolute pressure never changes. The same absolute pressure can give different gauge pressure readings if the gauges used to measure the pressure are calibrated to different atmospheric pressures. For example, a cylinder containing a quantity of refrigerant at a set temperature might have a gauge pressure of 85 psig down at the coast. That same cylinder at the same temperature might only have a gauge pressure of 87 psig in the mountains. However, the cylinder's absolute pressure at both locations would be 100 psia.

Figure 11-9 A refrigeration compound gauge measures pressure above and below atmospheric pressure. *(Courtesy of Ritchie Engineering Company, Inc., Yellow Jacket Products Division)*

11.9 VACUUM

A vacuum is any pressure less than atmospheric pressure. Mercury columns can be used to measure vacuum by measuring how much pressure has been removed. A pressure of 20 in Hg vacuum means that the pressure is 20 in of mercury less than atmospheric pressure. Since atmospheric pressure is 29.92 in of mercury, a perfect vacuum would be 29.92 in Hg vacuum. This would indicate that all the pressure had been removed. There is no vacuum using absolute pressure because absolute pressure starts with 0 equal to a perfect vacuum. It is impossible to have less pressure than no pressure at all.

Absolute pressure can be used to measure vacuums by measuring how much pressure is left. In the case of a pressure of 20 in Hg vacuum, the absolute pressure would be 9.92 in Hg absolute because 9.92 in of the original 29.92 in of pressure is left. A perfect vacuum using absolute measurement would be 0 in of mercury absolute. This indicates that no pressure remains.

11.10 SI PRESSURE MEASUREMENT

The SI system has a unit made specifically to measure pressure: the pascal. A pascal is equal to a newton per square meter. This is too small a pressure for air-conditioning purposes, so many gauges that read pressure using SI units use kilopascals (kPa). A kilopascal is equal to 1000 pascals. A kilopascal is still a much smaller unit of pressure than psi. A pressure of 100 kilopascals is roughly equivalent to 15 psi. That means that 1 kPa is about one-seventh of 1 psi.

Some metric gauges use bars. A bar is equivalent to 100 kPa and is roughly equal to atmospheric pressure. The bar and barometric pressure both get their names from the same Greek root, *baros*, meaning "weighty."

11.11 CONVERTING GAUGE AND ABSOLUTE PRESSURES

There are times when it will be necessary to convert between gauge pressure and absolute pressure. Calculating compression ratio is an instance when a service technician will need to convert gauge pressure to absolute pressure. A set of refrigerant gauges shows the system pressure in gauge pressure, but absolute pressures are needed to calculate compression ratio.

Absolute pressure equals gauge pressure plus atmospheric pressure. Use 14.7 if the actual local atmospheric pressure is not known. For most air-conditioning work, the atmospheric pressure can be approximated at 15 psi.

$$\text{psia} = \text{psig} + 14.7 \quad \text{or} \quad \text{psia} = \text{psig} + 15$$
$$\text{psig} = \text{psia} - 14.7 \quad \text{or} \quad \text{psig} = \text{psia} - 15$$

For example, to convert the low side pressure gauge reading of 65 psig to absolute pressure:

$$\text{psia} = \text{psig} + 14.7$$
$$\text{psia} = 65 + 14.7$$
$$\text{psia} = 79.7$$

or rounded to

$$\text{psia} = 80$$

To convert a chart reading of 105 psia to the gauge reading you would see in gauge pressure:

$$\text{psig} = \text{psia} - 14.7$$
$$\text{psig} = 105 - 14.7$$
$$\text{psig} = 90.3$$

or rounded to

$$\text{psig} = 90$$

SERVICE TIP

Many refrigerant pressure charts and psychrometric charts are based on sea-level readings. You must use the correct chart when working at altitudes that are significantly above sea level. The pressure-temperature chart used by a technician in Denver, Colorado, is going to be different from one used by a technician in Dallas, Texas. If you look at the bottom of the chart, there is often a small note stating the barometric pressure, altitude, or other indications as to the chart's intended use.

11.12 CONVERTING MERCURY COLUMN PRESSURE TO PSI

Standard atmospheric pressure is stated as 14.7 psia or 29.92 in Hg. This means there are approximately 2 in Hg for every psi; 2.04, to be more precise. To convert from inches

Figure 11-10 Mercury manometer.

of mercury to psi, divide inches of mercury by 2.04, or simply use 2 for an easy estimate:

$$\text{psi} = \text{in Hg}/2.04 \text{ or psi} = \text{Hg}/2 \text{ for an easy estimate}$$

For example, to estimate the psi equivalence for 18 in of mercury, divide 18 by 2 to get 9 psi (psi = 18 in Hg/2 = 9 psi). For a more precise conversion, divide 18 by 2.04 (18/2.04 = 8.8 psi).

To convert from psi to inches of mercury, simply multiply by 2.04 for precision, or by 2 for a quick estimate. For example, 8 psi would be 16.32 in of mercury ($8 \times 2.04 = 16.32$).

Most positive mercury column measurement is absolute. One end of the tube that the mercury is in is sealed, and the space between the mercury and the sealed end of the tube is in a vacuum (Figure 11-10). When converting gauge pressure, psig, to mercury column pressure absolute, the gauge pressure should first be converted to absolute pressure, psia.

11.13 VACUUM CONVERSIONS

Sometimes vacuums are measure by how much pressure has been removed, as in inches of mercury vacuum, and sometimes the vacuum is measured by how much pressure is left, as in inches of mercury absolute. To convert between the two, subtract the value you want to convert from 29.92. You can use 30 for a very close estimate. For example, an 18 in Hg vacuum is approximately 12 in Hg absolute. Subtracting either one from 30 will produce the other: 30 − 18 in Hg vacuum = 12 in Hg absolute; 30 − 12 in Hg absolute = 18 in Hg vacuum.

SERVICE TIP

Parts of low-pressure refrigeration systems operate in a vacuum. Some manufacturers specify their pressures in psia, while others specify their pressures in inches Hg vacuum. For a close estimate of psia from inches Hg vacuum, first subtract the inches of Hg vacuum from 30, and then divide the answer by 2. In the case of 16 in Hg vacuum:

$$30 - 16 = 14$$

$$14/2 = 7 \text{ psia}$$

For more precision, use 29.92 and 2.04:

$$29.92 - 16 = 13.92$$

$$13.92/2.04 = 6.8 \text{ psia}$$

11.14 FLUID PRESSURE

A fluid is a substance that deforms and flows under pressure. Fluids change shape and move from one point to another because of pressure difference. Both liquids and gases are considered fluids because both can take the shape of their container and can flow from one place to another because of pressure difference. Like all matter, fluids have weight—even gases.

The weight of any solid material acts as a downward force on whatever is supporting it. The force of a solid object is the overall weight of the object, which is distributed over the area on which it lies. The weight of a given volume of water, however, acts not only as a force downward on the bottom of the container holding it but also as a force laterally on the sides of the container. If a hole is made in the side of the container below the water level, as in Figure-11-11, the water above the hole will be forced out because of its force acting downward and sideways.

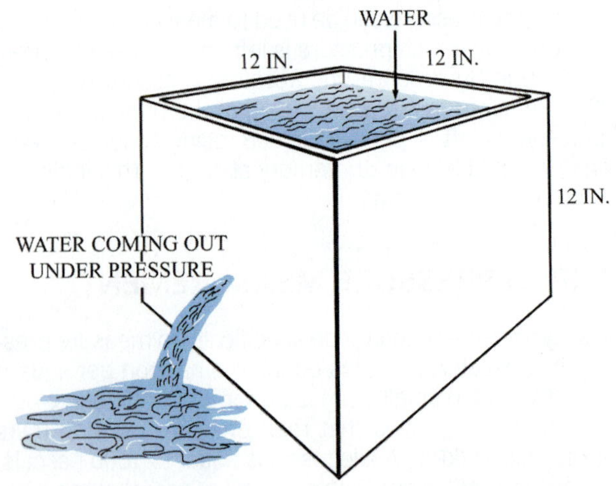

Figure 11-11 Water exerts pressure on the sides and bottom of its container.

TECH TIP

The most commonly overlooked fluid that affects air-conditioning performance is air. Many people do not realize that air is a fluid. However, either a liquid or a gas can be a fluid because a fluid is anything that can flow. That makes air a fluid with all of the associated characteristics of weight, density, inertia, and pressure.

Fluid pressure is the force per unit area that is exerted by a gas or a liquid. Fluid pressure varies directly with the density and the depth of the fluid. At the same depth below the surface, the pressure is equal in all directions. Notice the difference between the terms used: *force* and *pressure*. Force means the total weight of the substance; pressure means the unit force or pressure per square inch.

The tank in Figure 11-11 measures 1 ft in all dimensions and holds exactly 1 ft^3 of water. The weight of 1 ft^3 of water is approximately 62.4 lb; therefore, the total force being exerted on the bottom of the tank is 62.4 lb. The force is spread over an area of 144 in^2 (12 in $\times$ 12 in $=$ 144 in^2). Using the equation

$$\text{Pressure} = \frac{\text{Force}}{\text{Area}}$$

a pressure of 0.433 psi is exerted on the bottom of the tank (62.4/144 = 0.433).

11.15 HEAD PRESSURE

Pressure and depth have a close relationship when a fluid is involved. The head is the height of water expressed in feet and is used to express water pressure in a system. Water pressure varies directly with its depth. As an example, if the tank in Figure 11-11 was 2 ft high and filled with water, it would contain a volume of 2 ft^3 of water and would weigh 2 $\times$ 62.4 lb, or 124.8 lb. Now the force of the water on the bottom of the tank would still be distributed over 144 in^2 and the unit pressure would be 0.866 psi (124.8 lb $\div$ 144 in^2). This is twice the amount of pressure that was exerted when the head of water was only 1 ft. Therefore, in an open-top container, the pressure of the water will equal 0.433 psi for each foot of head.

SERVICE TIP

A pump only moves fluid; it does not create pressure. The resistance to the flow of a fluid is what results in pressure. A pump that has a fully unrestricted outlet will have zero or nearly zero pressure reading. It is only when a pipe is attached to the outlet and the water directed through the pipe or up some distance that the pressure in the system is created.

The same relationship follows when pumping water in a system. The head is the height difference from the lowest point to the highest point. An increase in the height that a pump must move the water increases the pressure. The head pressure can be determined by multiplying 0.433 times the height. If the height is 10 ft, the pressure will be 0.433 $\times$ 10, or 4.33 psi.

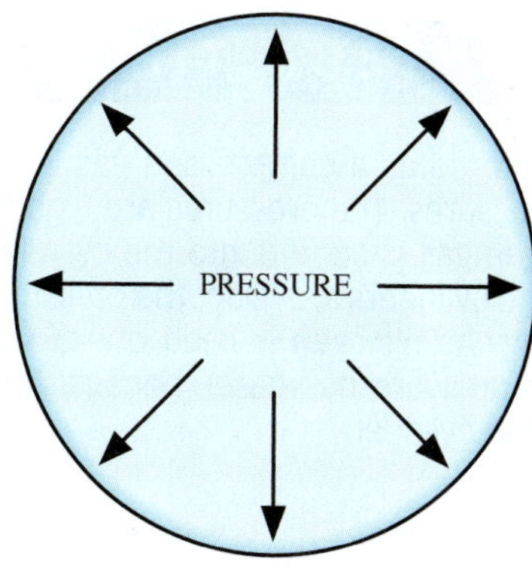

Figure 11-12 Gas exerts force evenly against all sides of its container.

The tank in Figure 11-11 has an area of 1 ft^2 with 1 ft of head; therefore, the pressure on the bottom of the tank is 0.433 psi. If there is a fish pond covering an area of 50 ft^2, and the depth of water in it is 1 ft, the pressure on the bottom of the pond will still be just 0.433 psi, even though there is a larger total volume of water. The pressure is determined by the depth of the water, not the volume.

11.16 GAS PRESSURE

An ideal gas is one in which the molecules of the gas have no attraction for each other. An important characteristic of an ideal gas is that it will expand to take up the entire volume that it is contained in. When gas is confined, it pushes against the sides of its container, creating pressure (Figure 11-12). This pressure is created by the kinetic energy of the gas molecules as they strike the surface of the container. If more molecules are put into the same container, the pressure goes up because now there are more molecules to strike the surface (Figure 11-13).

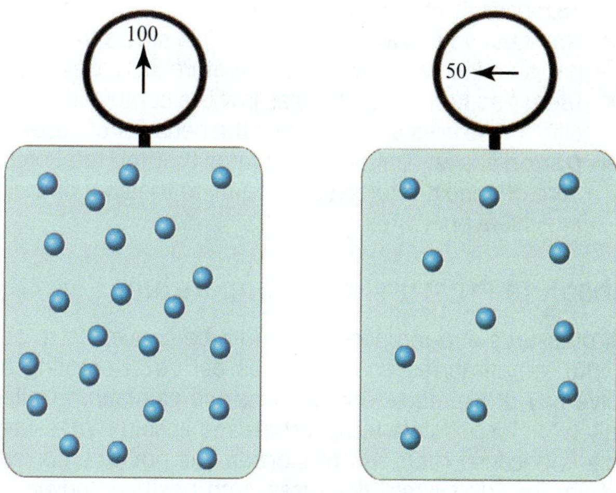

Figure 11-13 Increasing the number of gas molecules in the container increases the pressure on the sides of the container.

TECH TIP

Nitrogen is stored as a compressed gas in cylinders at very high pressures. The pressures are high because so much nitrogen gas is stuffed into the cylinder. Immediately after the cylinders are filled, the pressure is as high as 2,400 psig. As nitrogen is used out of the cylinder, the cylinder pressure decreases because there is less nitrogen in the cylinder.

If the temperature of the gas is increased, the increased speed and kinetic energy of the molecules increases the number of times molecules strike the surface and increases the pressure. If the volume of the container is reduced, the pressure increases because the molecules are packed in more closely and strike the surface more often. Therefore, the three ways that gas pressure can be increased are by increasing the quantity of gas in the container, increasing the temperature of the gas, or decreasing the size of the container. Naturally, these all work in reverse as well: gas pressure can be reduced by removing some of the gas from the container, decreasing the gas temperature, or increasing the size of the container.

11.17 GAS LAWS

Many scientists have studied the behavior of gases and contributed to our knowledge of gas laws. Several scientists have published their findings demonstrating a particular aspect of gas behavior. The math formulas describing the particular gas behavior they studied have become known as the gas laws. Each law is named after the scientist credited with creating it. The laws governing gas behavior are summarized in the list below.

- **Pascal's law** Pascal's law states that gas pressure is exerted uniformly everywhere on its container.
- **Boyle's law** Boyle's law states that pressure increases as volume decreases.
- **Charles's law** Charles's law states that volume increases as temperature increases.
- **Gay-Lussac's law** Gay-Lussac's law states that the pressure increases as the temperature increases.
- **Ideal gas law** The ideal gas law is a combination of all of these laws and describes the behavior of gases.
- **Dalton's law** Dalton's law states that the total pressure of a mixture of gases is equal to the sum of their individual pressures.

Absolute Pressure and Temperature

All pressures and temperatures must be converted to absolute temperatures and pressures before attempting to solve any of the equations because the gas law formulas only work for absolute temperature and absolute pressure. This is because ratio and proportion will not give correct results for any system that uses both positive and negative values. Absolute temperature and absolute pressure have no negative numbers. The coldest temperature on an absolute scale is 0°; there are no negative temperatures. The lowest pressure on an absolute pressure scale is 0; there is no vacuum.

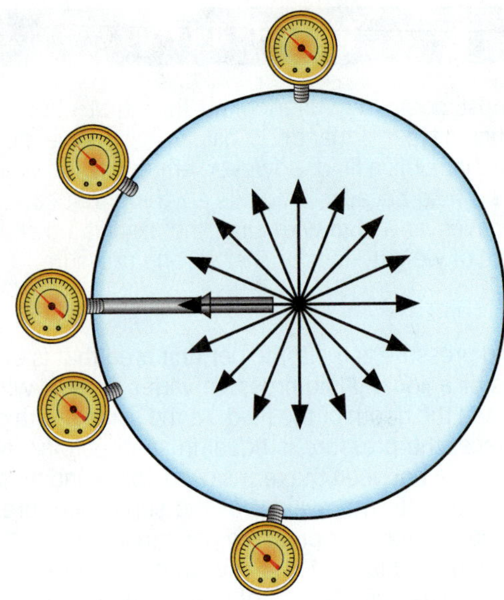

Figure 11-14 Pascal's law states that gas pressure is applied evenly on all surfaces of the container holding the gas.

11.18 PASCAL'S LAW

Pascal's law describes how a gas reacts in a confined space. Pascal's law states that when a gas is confined, it exerts equal pressure in all directions. This means that if the pressure were measured from the outside or inside or from the top or bottom, the pressure readings would all be the same (Figure 11-14).

11.19 BOYLE'S LAW

Boyle's law describes the relationship between the volume and pressure of a gas. Boyle's law states that the volume of a gas varies inversely with its pressure if the temperature of the gas remains constant. This means that if the volume increases, the pressure will decrease. For example, if the volume of a gas is doubled, the pressure will be halved. This also works the other way: when the volume decreases, the pressure increases. If the volume is halved, the pressure is doubled. This is the operating principle of compressors. Figure 11-15 shows a fixed quantity of gas before and after being compressed.

11.20 CHARLES'S LAW

Charles's law describes how a gas volume reacts to changes in temperature. Charles's law states that the volume of a gas is in direct proportion to its absolute temperature, provided that the pressure is kept constant. This means that a gas will expand in volume with an increase in temperature and contract in volume with a decrease in temperature. Figure 11-16 shows a balloon at two different temperatures. When the

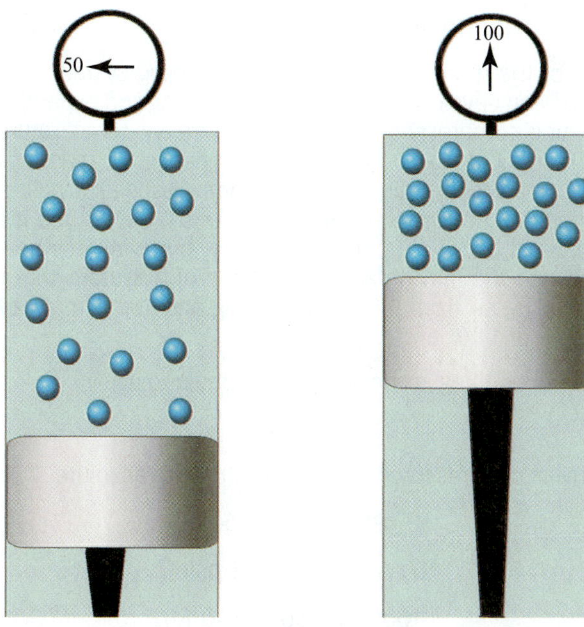

Figure 11-15 Confining the same number of gas molecules in a smaller space increases the gas pressure.

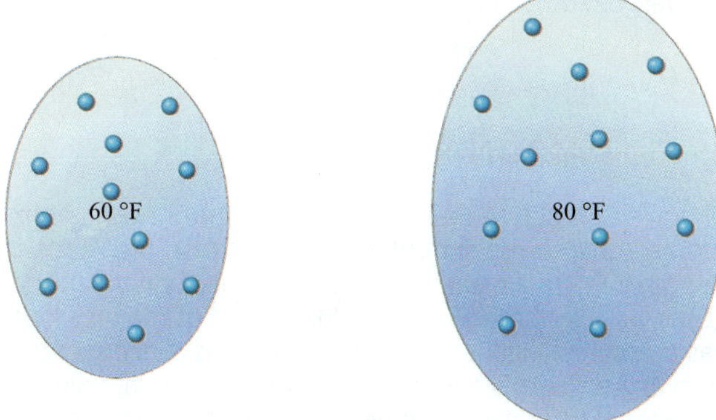

Figure 11-16 The same number of gas molecules take up more space at a higher temperature.

gas in the balloon is at a higher temperature, the gas volume expands, making the balloon larger.

TECH TIP

The increase in gas volume that accompanies an increase in gas temperature is the operating principle of hot-air balloons. The heated air inside the balloon expands, making it less dense than the surrounding air.

11.21 GAY-LUSSAC'S LAW

Gay-Lussac's law describes the relationship between gas pressure and gas temperature: the pressure of a gas is directly proportional to its temperature with the volume remaining constant. This means that increasing the temperature increases the pressure, or decreasing the temperature

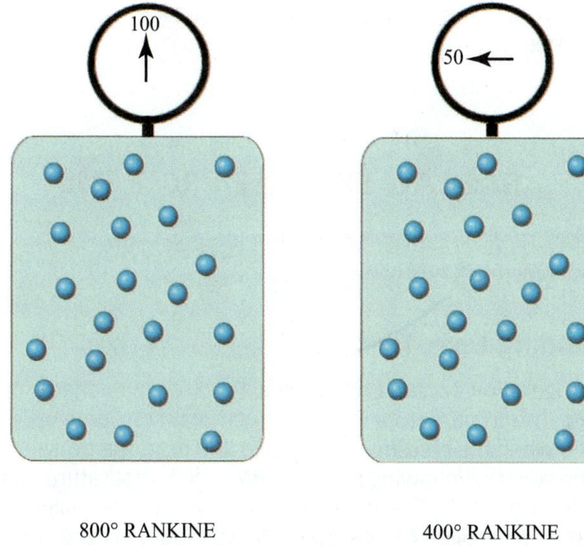

Figure 11-17 With the same number of molecules and the same amount of space, the container with the higher temperature will have a higher pressure.

will decrease the pressure. Figure 11-17 shows two identical volumes of gas at different temperatures. The volume at a higher temperature is also at a higher pressure.

11.22 IDEAL GAS LAW

A combination of all of the above laws is called the ideal gas law. All of the previous relationships assumed that one of the variables remained constant. In reality, this is seldom the case. The ideal gas law makes predictions of gas behavior possible even when all of the variables are changing. Mathematically, the ideal gas law may be written as

$$P_1 V_1/T_1 = P_2V_2/T_2$$

where

V_1 = the starting volume
V_2 = the ending volume
P_1 = the starting pressure
P_2 = the ending pressure
T_1 = the starting temperature
T_2 = the ending temperature

The basic formula can be rearranged to show the solution for new volume, new pressure, or new temperature:

New volume

$$V_2 = (T_2 \times V_1 \times P_1) \div (T_1 \times P_2)$$

New pressure

$$P_2 = (T_2 \times V_1 \times P_1) \div (T_1 \times V_2)$$

New temperature

$$T_2 = (T_1 \times V_2 \times P_2) \div (V_1 \times P_1)$$

Just plug in the relevant numbers to solve for new volume, new pressure, or new temperature. When working a problem that does not use all three variables, just leave off the variable you are not using. For example, to solve for new pressure when the volume remains constant, the formula becomes

$$P_2 = (T_2 \times P_1) \div T_1$$

11.23 GAS LAW EXAMPLE PROBLEMS

A few practical applications of the ideal gas law will illustrate how these formulas work.

Standing Leak Test

A refrigeration system is charged with nitrogen to check for leaks. The temperature of the system and surroundings is 100°F when the system is charged to 100 psig. The pressure is checked the following morning after the temperature has dropped to 50°F. The pressure will have dropped slightly even if no nitrogen has escaped. What should the pressure be if no nitrogen has escaped?

The formula to use is

$$P_2 = (T_2 \times V_1 \times P_1) \div (T_1 \times V_2)$$

Since the space in the system, the volume, is staying the same, this can be shortened to

$$P_2 = (T_2 \times P_1) \div T_1$$

Convert the pressures and temperatures to absolute.

$$100 \text{ psig} + 15 = 115 \text{ psia}$$

$$100^\circ\text{F} + 460 = 560^\circ\text{R}$$

$$50^\circ\text{F} + 460 = 510^\circ\text{R}$$

Plug the numbers into the formula.

$$(510^\circ\text{R} \times 115 \text{ psia}) \div 560^\circ\text{R} = 105 \text{ psia}$$

Convert 105 psia to psig.

$$105 \text{ psia} - 15 = 90 \text{ psig}$$

Compressed Gas

What would be the final volume of 4 ft^3 of gas that is compressed from 60°F at 70 psig to 150°F at 210 psig?

Since everything is changing, the formula to use is

$$V_2 = (T_2 \times V_1 \times P_1) \div (T_1 \times P_2)$$

Convert the pressures and temperatures to absolute.

$$70 \text{ psig} + 15 = 85 \text{ psia}$$

$$210 \text{ psig} + 15 = 225 \text{ psia}$$

$$60^\circ\text{F} + 460 = 520^\circ\text{R}$$

$$150^\circ\text{F} + 460 = 610^\circ\text{R}$$

Plug the numbers into the formula.

$$V_2 = (610^\circ\text{R} \times 4 \text{ ft}^3 \times 85 \text{ psia}) \div (520^\circ\text{R} \times 225 \text{ psia})$$

$$V_2 = 1.8 \text{ ft}^3$$

Compressed Air Cooling

One of the more interesting air-conditioning systems that have been developed works by compressing and expanding air. Air is compressed, raising its pressure and temperature. It is then cooled to outdoor ambient temperature by passing it through a cooling coil. Finally, the compressed air is passed through an orifice where its pressure is dropped and it is expanded back to its original volume, but with less heat. Calculate the leaving air temperature of a system that is expanding air from 5 ft^3/lb at 30 psig and 100°F to a new volume of 14 ft^3 at 0 psig.

Since everything is changing, the formula to use is

$$T_2 = (T_1 \times V_2 \times P_2) \div (V_1 \times P_1)$$

Convert the pressures and temperatures to absolute.

$$30 \text{ psig} + 15 = 45 \text{ psia}$$

$$0 \text{ psig} + 15 = 15 \text{ psia}$$

$$100^\circ\text{F} + 460 = 560^\circ\text{R}$$

Plug the numbers into the formula.

$$T_2 = (560^\circ\text{R} \times 14 \text{ ft}^3 \times 15 \text{ psia}) \div (5 \text{ ft}^3 \times 45 \text{ psia}) = 523^\circ\text{R}$$

Convert 523°R to °F.

$$523^\circ\text{R} - 460 = 63^\circ\text{F}$$

11.24 DALTON'S LAW

Dalton's law states that the total pressure of a gas mixture is equal to the sum of all the gas partial pressures in the mixture. This means the total pressure of a mixture of gases can be determined by adding each of the individual gas pressures (Figure 11-18). Dalton's law is a logical extension of Avogadro's law, which states that equal volumes of gases, at the same temperature and pressure, contain the same number of atoms or molecules. The type of the gas has no effect on the gas pressure, only the number of molecules in the container.

TECH TIP

In the field, you can easily check to see whether a cylinder of refrigerant is contaminated with air. If a cylinder of refrigerant is contaminated with air, the pressure gauge reading will be equal to the reading for the refrigerant and the air added together. Take the cylinder's pressure and temperature, and compare the results to a refrigerant chart. If the pressure you obtained is higher, the cylinder contains something other than pure refrigerant, and that is usually air.

11.25 SATURATED GAS

So far, all the discussion has related to the behavior of an ideal gas in which the molecules have no attraction for one another. However, when a gas is at the saturation point,

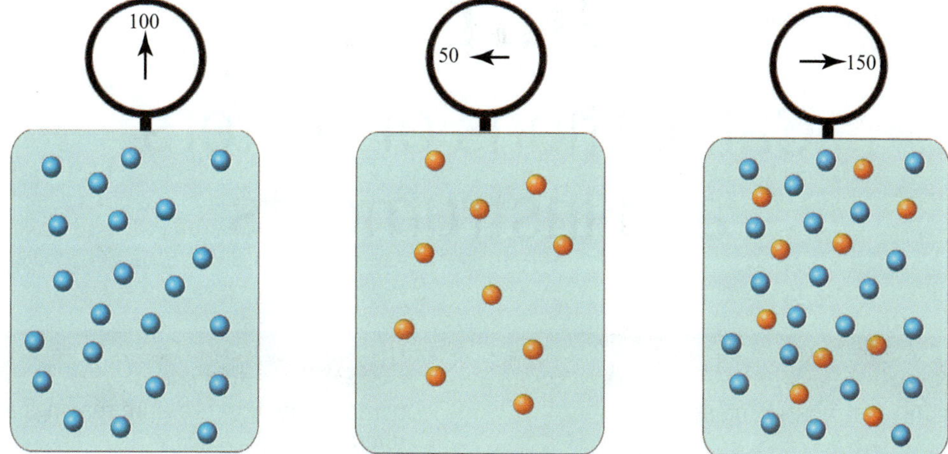

Figure 11-18 When two separate gases are put in the same container, the total pressure in the container is the sum of the individual pressures.

the molecules have enough attraction for each other that removing heat from them will cause them to condense to a liquid. A saturated gas behaves very differently from an ideal gas because the gas can change state at the saturation point. For this reason, saturated gases do not follow the ideal gas laws.

The temperature and pressure changes of a saturated mixture are far greater than those of an ideal gas. Adding heat to a saturated mixture in a container creates more gas molecules while leaving the volume the same. More gas molecules create more gas pressure. This increase in pressure is much more dramatic than the increase in pressure due to simply increasing the temperature of the molecules that are already there, as in an ideal gas.

UNIT 11—SUMMARY

Hydronic, pneumatic, refrigeration, and air systems all depend upon pressure differences to create flow. That is why HVACR technicians must be familiar with pressure and vacuum readings so they can properly install, service, and repair these systems.

Absolute pressure is the only pressure reading that is not dependant on the surrounding atmospheric pressure. Absolute pressure is not often used by the HVACR field technician, but it is important to know about and may be a factor in some refrigeration applications. For example, low-temperature refrigeration often operates in a vacuum. As you now know, the total pressure difference between an absolute vacuum and atmospheric pressure is 14.7 psi. That is not very much pressure to work with as compared to air conditioning that might have 300–400 psi. Therefore, in some low-temperature refrigeration service, technicians may have to convert gauge reading to absolute to know exactly how the system is working and at what temperature.

UNIT 11—REVIEW QUESTIONS

1. How does atmospheric pressure affect a gauge pressure reading?
2. What is a fluid?
3. Define *pressure*.
4. What creates atmospheric pressure?
5. What is the difference between absolute pressure and gauge pressure?
6. What causes atmospheric pressure to change?
7. What is the relationship between atmospheric pressure and barometric pressure?
8. The reading on a low side pressure gauge is 65 psig. What is the absolute pressure?
9. Convert a chart reading of 100 psia to the reading you would see in gauge pressure.
10. Convert a barometric pressure reading of 26 in Hg to psia.
11. Convert a reading of 20 in Hg vacuum to psia.
12. Define a vacuum, and tell how it is measured.
13. Why are some vacuum pressures expressed in inches of water and others in inches of mercury?
14. What causes fluid pressure?
15. What is the pressure at the bottom of a square tank that has a cross-section of 5 ft^2 and is filled with water to a depth of 4 ft?
16. What is the pressure in psi required to lift water 15 ft?
17. Explain how a simple open-arm manometer measures pressure.
18. Explain how a Bourdon tube gauge works.
19. Briefly summarize Dalton's law.
20. What temperature in degrees Fahrenheit would a fixed volume of gas need to be heated to in order to double its pressure if it starts out at 25°F?
21. A volume of 4 ft^3 of gas is compressed from 50°F at 70 psig to 2 ft^3 with the resulting temperature of 100°F. What is the new pressure?
22. Explain the bar pressure measurement system.
23. Which is a larger unit of pressure: 1 kPa or 1 psi?
24. Why do saturated gases not follow the ideal gas law?
25. Explain Pascal's law.

UNIT 12

Calibrations of Meters and Instruments

OBJECTIVES

After completing this unit, you will be able to:

1. use the correct terminology when talking about instrument calibration.
2. perform simple procedures to verify the calibration of temperature-measuring instruments.
3. demonstrate that pressure readings are within tolerance.
4. verify the proper operation and accuracy of leak-detection instruments.
5. perform calibration-verification checks of flue-gas analysis instruments.

12.1 INTRODUCTION

Instruments are calibrated for several reasons, including ensuring improved troubleshooting accuracy and that a system is working according to design specifications. If the instruments the technicians use are not calibrated and are not providing accurate information, it will be difficult to make a determination of actual equipment performance. Many commercial buildings and some residential homes are now being commissioned. Commissioning is the process of testing all the building's systems and certifying to the owner that they are all operating according the manufacturer's specifications. To commission a system or building, technicians must be using instruments that are calibrated.

12.2 CALIBRATION

Calibration is a baseline testing and adjustment procedure performed on instruments to ensure that the measured indication or output is an accurate representation of the process. The technician may calibrate some instruments, and others may need to be sent to a meter shop to be calibrated.

The instrument manufacturer may specify how often an instrument must be calibrated. In some cases, local codes may require some critical instruments to be calibrated and recertified on a regular schedule. The instrument used to certify the calibration of your instrument will have its calibration traced back to the National Institute of Standards and Technology. Certified calibrated equipment will have a sticker listing who performed the calibration, the date of the calibration, the date the calibration expires, and any error factor that must be used to correct the instrument's reading (Figure 12-1).

Not all instruments can be calibrated. Some electronic instruments cannot be calibrated but must be tested periodically to ensure that they are working accurately. If they are found to be out of calibration beyond the acceptable limits, they must be replaced.

SERVICE TIP

The meter movement on many analog meters is very sensitive. In some cases the weight of the instrument needle can affect the reading. The needle's weight is less when the meter is laying flat as compared to when it is set upright.

The faceplate is typically made of plastic (not glass) to provide a more robust instrument capable of handling the rough environment and conditions to which it may be exposed. Static electricity on the plastic faceplate can affect the reading. To test for static buildup on the plastic face, slowly wave your hand over the top of the faceplate without actually touching it. If the indicator needle moves, static is present. The static electricity can be removed using an anti-static spray available from either the manufacturer or a supply house.

Calibration Terminology

Accuracy Accuracy is how exact a device is when making a measurement. An example is a meter measuring a conductor carrying 100 V. If the meter is 100 percent accurate, it indicates exactly 100 V with no tolerance. Highly accurate instruments are those used in a laboratory environment.

Range The range is the difference between the smallest indication on the scale and the largest indication on the scale. When talking about an instrument with a 0 percent reading of 4 mA and a 100 percent reading of 20 mA, the range is 16 mA.

Span The span is the value that represents 100 percent of the scale of an instrument. When talking about an instrument with a range of 4 mA to 20 mA, the span is 20 mA.

Tolerance The tolerance is the amount of variation allowed from a standard. Typically tolerance is expressed in percent (%). Your digital volt-ohmmeter

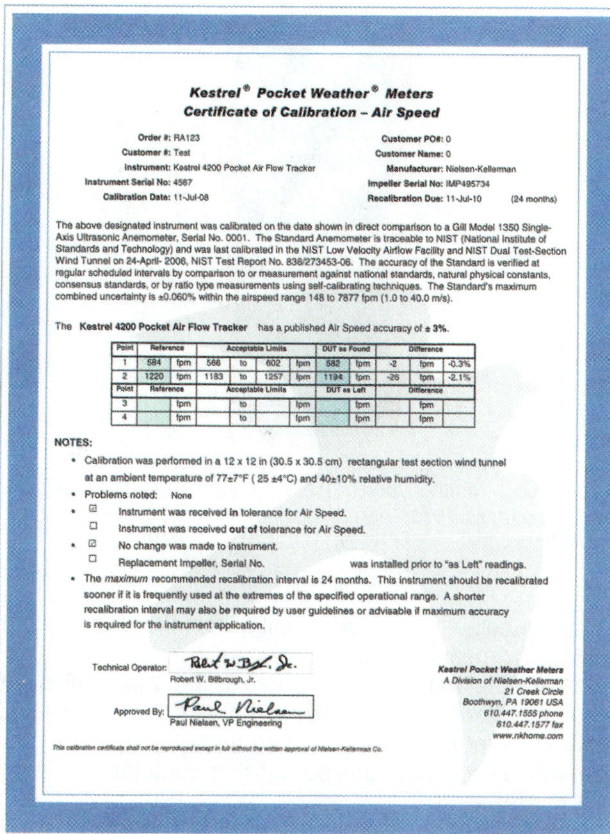

Kestrel® Pocket Weather® Meters
Certificate of Calibration – Air Speed

Order #: RA123
Customer #: Test
Instrument: Kestrel 4200 Pocket Air Flow Tracker
Instrument Serial No: 4567
Calibration Date: 11-Jul-08

Customer PO#: 0
Customer Name: 0
Manufacturer: Nielsen-Kellerman
Impeller Serial No: IMP495734
Recalibration Due: 11-Jul-10 (24 months)

The above designated instrument was calibrated on the date shown in direct comparison to a Gill Model 1350 Single-Axis Ultrasonic Anemometer, Serial No. 0001. The Standard Anemometer is traceable to NIST (National Institute of Standards and Technology) and was last calibrated in the NIST Low Velocity Airflow Facility and NIST Dual Test-Section Wind Tunnel on 24-April- 2006, NIST Test Report No. 836/273453-06. The accuracy of the Standard is verified at regular scheduled intervals by comparison to or measurement against national standards, natural physical constants, consensus standards, or by ratio type measurements using self-calibrating techniques. The Standard's maximum combined uncertainty is ±0.060% within the airspeed range 148 to 7877 fpm (1.0 to 40.0 m/s).

The **Kestrel 4200 Pocket Air Flow Tracker** has a published Air Speed accuracy of **± 3%.**

Point	Reference		Acceptable Limits				DUT as Found		Difference		
1	584	fpm	566	to	602	fpm	582	fpm	-2	fpm	-0.3%
2	1220	fpm	1183	to	1257	fpm	1194	fpm	-26	fpm	-2.1%
Point	**Reference**		**Acceptable Limits**				**DUT as Left**		**Difference**		
3		fpm		to		fpm		fpm		fpm	
4		fpm		to		fpm		fpm		fpm	

NOTES:

- Calibration was performed in a 12 x 12 in (30.5 x 30.5 cm) rectangular test section wind tunnel at an ambient temperature of 77±7°F (25 ±4°C) and 40±10% relative humidity.
- Problems noted: None
- ☑ Instrument was received **in** tolerance for Air Speed.
  ☐ Instrument was received **out of** tolerance for Air Speed.
- ☑ No change was made to instrument.
  ☐ Replacement Impeller, Serial No. was installed prior to "as Left" readings.
- The *maximum* recommended recalibration interval is 24 months. This instrument should be recalibrated sooner if it is frequently used at the extremes of the specified operational range. A shorter recalibration interval may also be required by user guidelines or advisable if maximum accuracy is required for the instrument application.

Technical Operator: Robert W. Billbrough, Jr.

Approved By: Paul Nielsen, VP Engineering

Kestrel Pocket Weather Meters
A Division of Nielsen-Kellerman
21 Creek Circle
Boothwyn, PA 19061 USA
610.447.1555 phone
610.447.1577 fax
www.nkhome.com

This calibration certificate shall not be reproduced except in full without the written approval of Nielsen-Kellerman Co.

Figure 12-1 Instrument-calibration sticker with test data, expiration date, and error factor. *(Courtesy of Nielsen-Kellerman)*

may have a tolerance of +/−1 percent. What this means is that if a conductor is carrying 100 V, the meter may indicate anything from 99 V to 101 V and still be within tolerance. If the meter indicates 98.9–101.1 V, it is out of tolerance and requires calibration.

Zero Zero is the value that represents 0 percent of the range of an instrument. It is not necessarily 0. Zero, when talking about an instrument with a range of 4 mA to 20 mA, is 4 mA. As an example, a refrigerant gauge may read zero, but that only means the gauge pressure is zero; it does not mean there is no atmospheric pressure.

12.3 TEMPERATURE-MEASURING INSTRUMENTS

There are a number of ways that temperature-measuring instruments may sense temperature. Some, like resistance thermal devices (RTDs), change their electrical resistance with changes in temperature. Thermocouples provide varying voltages with temperature changes. In contrast, a bimetallic device produces mechanical changes that are translated to a dial.

TECH TIP

"Standard atmospheric pressure" is based upon conditions at sea level, and this is typically taken to be 14.7 psi. However, depending on your geographic location, atmospheric pressure will be different. In Denver, Colorado, often referred to as being a "mile high," the atmospheric pressure is only slightly above 12 psi.

To perform an operational check on any temperature-sensing device, it must be subjected to a known temperature. An easy way to generate a known temperature is to use water and apply heat. Water boils at 212°F and freezes at 32°F at standard atmospheric pressure. These are two easily duplicated temperatures. If the heat-sensing portion (tip) of the temperature-sensing device is placed in boiling water, it should generate a reading of 212°F at standard atmospheric pressure, and in an ice/water bath it should read between 32°F and 33°F. Ice water is never exactly 32°F due to the heat flow from a warmer substance, the water, to a cooler substance, the ice. For most shop calibrations of temperature-sensing devices, this is within the acceptable level of accuracy.

Not all equipment can be immersed in water, and sometimes temperatures higher than 212°F are required for calibration. For these instruments, a sand bath can be used. A sand bath is a container full of sand. It is surrounded by an electric heating element. To increase the temperature, the electricity to the heating element is increased. To decrease the temperature, the electricity going to the heating element is decreased. Air is slowly forced up from the bottom of the container to ensure that the temperature is constant in the bath. As you might expect, the time required to perform this type of calibration is prolonged. It takes a while for the mass of the sand in the container to both heat up and cool down. A certified test instrument measures the temperature of the sand.

A third method to provide heat is the use of an oven. Varying the electricity applied to the heating elements varies the temperature in the oven. A certified instrument indicates the temperature inside the oven.

Resistance Thermal Devices (RTDs)

Resistance thermal devices provide an output in units of ohms or resistance. These are devices that change their resistance according to how much temperature they are sensing. They are manufactured to measure within specific temperature ranges. To verify their correct operation, we need the manufacturer's chart and an ohmmeter. If we subject the RTD to a known temperature, we can refer to the manufacturer's chart and see how much resistance the RTD should be generating. You can then measure the resistance and determine if the RTD is operating correctly.

To perform a calibration check of the RTD, it must be subjected to known temperature values at 0%, 25%, 50%, 75%, and 100% of its range. By measuring the resistance generated at these values and comparing them to the information provided on the manufacturer's chart, you can determine whether the RTD is working properly. The technician records the resistance-to-temperature relationships on an instrument data sheet. The technician cannot adjust an RTD. If the unit is out of tolerance, it cannot be used, and a new one will have to be obtained. Even if a new RTD is being used, its accuracy must be checked/verified using the previously mentioned procedure.

RTDs come in two types: negative temperature coefficient (NTC) and positive temperature coefficient (PTC). In the case of an NTC, the resistance increases as the temperature decreases. In the case of a PTC, the resistance increases as the temperature increases. These types of devices are often used in residential refrigerators and freezers and as temperature-measuring instruments and sensors installed in equipment utilizing an electronic circuit for control.

Thermocouples

Thermocouples generate a millivoltage (mV) of electrical power when heated. The millivoltage generated for each degree of temperature differs depending on the makeup of the thermocouple wires. Refer to the manufacturer's data sheet to see what the millivoltage output should be for the thermocouple you are testing. Due to length of thermocouple wire or other factors, thermocouple readings may vary slightly. Checks should be made at 0%, 25%, 50%, 75%, and 100% of the required range. The millivoltage output of the thermocouple at these points is recorded on the instrument data sheet and compared to the required tolerance as specified by the manufacturer.

The thermocouple reading may deviate by a degree or two. Often technicians will write the degree of deviation on the thermocouple lead, instructing the user to correct the reading they obtain by adding or subtracting 1° or 2° from that reading.

The hot junction is typically the point where thermocouples fail. The connection comes apart. Once the connection is broken, it cannot be soldered back together. Therefore, we cannot repair thermocouples. Devices such as these can be measured for accuracy using a multifunction process calibrator, as shown in Figure 12-2.

Bimetallic Devices

Bimetallic devices provide a visual indication of temperature, such as an analog thermostat connected to a residential heating/cooling system. A scale is labeled to indicate the temperature that the device is sensing.

Factors that affect the accuracy of this device are the graduations of the scale and the consistency of the expansion/contraction rate of the metals used. Both of these factors are a function of their initial construction at the manufacturer's facility and are beyond the control of the technician. On spiral-type bimetallic springs, a small screwdriver can be used to adjust the temperature pointer upward or downward if its reading is not correct. On conical bimetallic-type instruments, like a dial pocket thermometer, a small wrench can be used to hold the stem while the dial is turned so the pointer is directed at the correct temperature. Repeated rotating or twisting of these conical, helical, or coil-wound devices will distort them and render the reading meaningless.

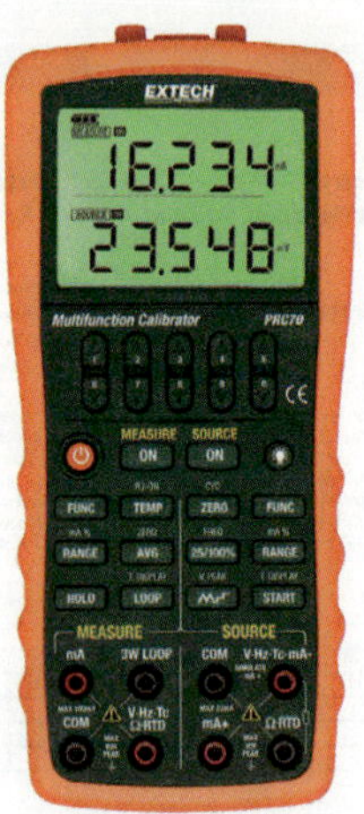

Figure 12-2 Multifunction process calibrator used as a precision source and measure for thermocouples, RTD, mA, mV, V, Hz, and Ohm devices. *(Courtesy of Extech Instruments, A FLIR Company)*

A reading that is off by as much as 5 to 7 percent is not uncommon for these types of instruments. For this reason, most temperature-measuring instruments today are digital. Figure 12-3 shows a digital pocket thermometer.

12.4 PRESSURE-MEASURING INSTRUMENTS

Pressure gauges can be electronic or mechanical. In most cases, electronic pressure sensors cannot be calibrated. They can be tested against a known pressure, and, if their reading falls within the acceptable range, they can be used, but if they are found to be out of tolerance, they must be replaced.

The mechanical gauges like the ones on a refrigerant manifold test set may have a small adjusting screw on the gauge face (Figure 12-4). There are two ways this screw can be adjusted. One is to adjust the needle to the 0 psi point. The other method is to attach the gauge to a known pressure source and adjust the needle to point at that pressure. One possible pressure source would be a source attached to a certified pressure gauge. A certified source like this can be adjusted to make sure the gauge reads accurately all the way across the scale. A second way to set the gauge is to attach it to a nearly full new refrigerant cylinder that has a stable measured temperature. Compare the temperature of the refrigerant to the gauge reading and adjust the needle as required.

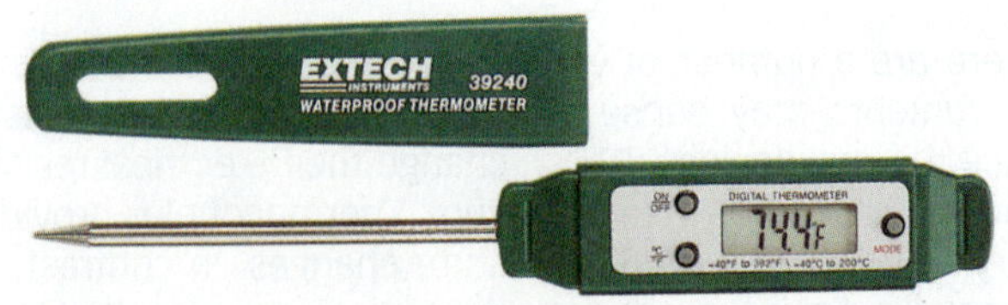

Figure 12-3 Digital pocket thermometer. *(Courtesy of Extech Instruments, A FLIR Company)*

Figure 12-4 Gauge manifold adjustment screw. *(Terra State Community College HVAC Program, Fremont, Ohio)*

SERVICE TIP

A full or nearly full refrigerant cylinder is required because attaching the gauge can cause a significant pressure drop in cylinders containing small quantities of refrigerant.

12.5 VACUUM-MEASURING INSTRUMENTS

U-shaped vacuum instruments have a moveable scale that when attached looks much like a ruler (Figure 12-5). This scale is adjustable in that it has the ability to slide up and down in relationship to the tube. This scale reads in inches of water, sometimes expressed as WC (water column) or inches of mercury, expressed as inches Hg. The scale has a 0 in the middle and goes to plus reading (positive pressure) and a minus reading (vacuum).

Typical values are plus 10 inches to minus 10 inches, but this can vary depending on the range of the instrument.

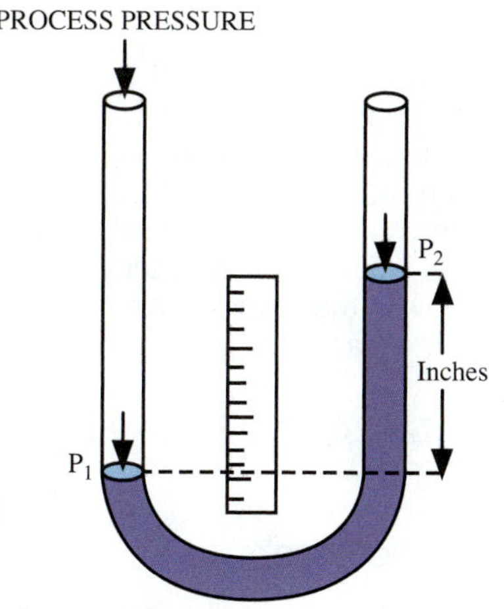

Figure 12-5 U-tube manometer. *(Courtesy of RadonAway Inc.)*

Water-filled U-shaped vacuum instruments use distilled water. Since distilled water is clear and hard to see against a scale, a dye is added to change the clear water to a green, blue, or red liquid. This makes it much easier to see. Once the colored water is poured into the tube, the calibration process can be performed. Simply slide the scale up or down until the 0 on the scale is perfectly lined up with the water level in the tubes. That is all the technician can do.

SERVICE TIP

The distance the water moves when vacuum is applied is a function of the water's specific gravity, and the indication on the scale is a function of the accuracy of the scale's manufacturer. Since a fixed specific gravity of the water is necessary for accurate readings, it is important to use only the special dye designed for this purpose. If you were to use food coloring, for instance, it would change the specific gravity of the water, resulting in inaccurate readings.

A variation of this instrument is a tube filled with mercury. Its operating principle and method of calibration are identical to the water-filled tube. This instrument reads out in inches or millimeters of mercury (in/mm Hg). This instrument is not used much in the field because of the chance of spilling the mercury, which is a dangerous (and expensive) consequence. Mercury spillage/contamination is an Environmental Protection Agency (EPA) reportable situation, and extreme care should be exercised in its use and disposal (Figure 12-6).

Analog Electronic Vacuum Instruments

Analog electronic vacuum instruments are both easy to use and to calibrate. They are powered by electrical energy and convert vacuum to a reading on a scale through the use of a transducer. The units on scale are microns.

The unit has a scale and a switch, and the switch has two positions. One position is labeled ON or READ, and the other position is labeled CALIBRATE. The calibration procedure requires a number of steps. First, take a close look at the transducer. It is a metal tubular device with an electrical cord coming out of one end and a threaded connector on the other end. The electrical cord is connected to the main body of the instrument. The threaded connector is used to facilitate the connection of a hose that is connected to the system on which vacuum readings are being taken (Figure 12-7).

As you perform a closer visual inspection of the transducer, you will notice a number stamped into its case. Although there is no standard number, a typical number would be 30.25. Before connecting the transducer to the system being measured, place the unit in the ON position. Now observe the number stamped into the transducer. Say it is 30.25. Place the switch into the CALIBRATE position. There will be an adjustment knob on the main body of the instrument. Adjust this knob until the needle indicates the value stamped into the transducer, in this case 30.25. Place the switch into the READ position. The instrument is now ready for use.

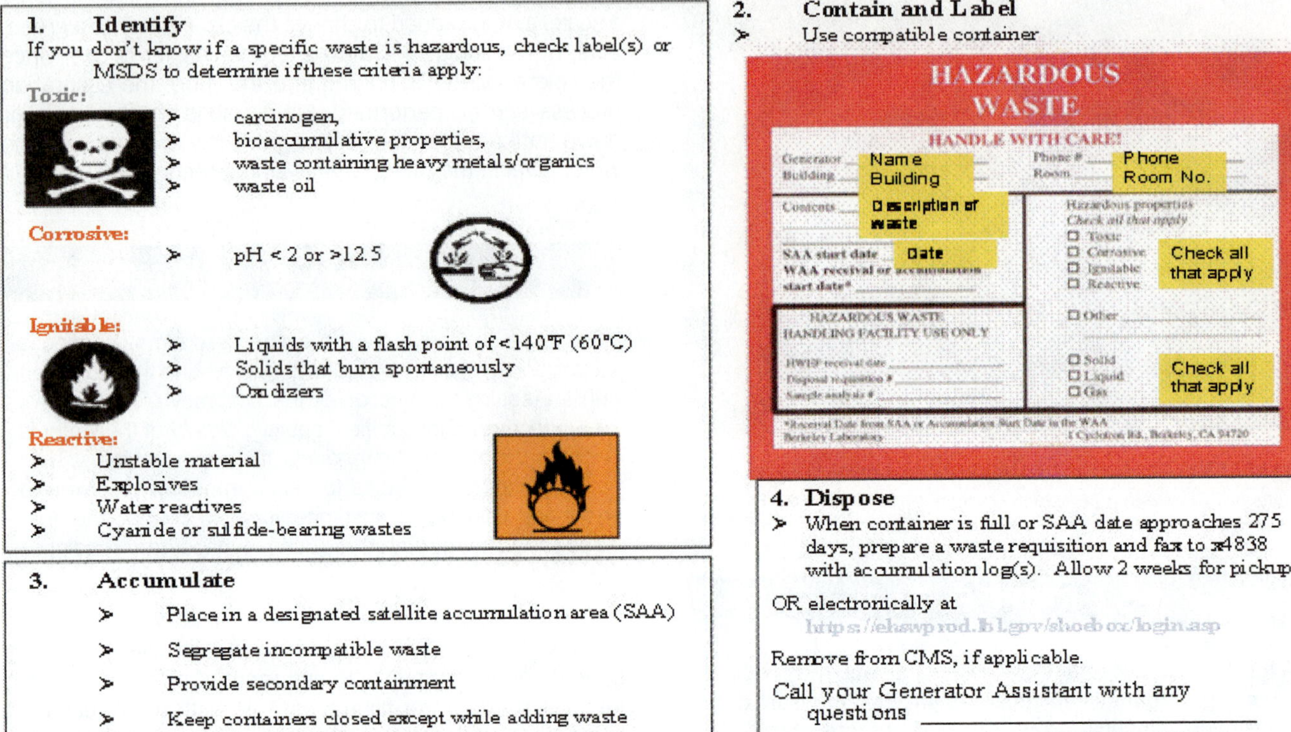

Managing Hazardous Waste

1. Identify

If you don't know if a specific waste is hazardous, check label(s) or MSDS to determine if these criteria apply:

Toxic:
- carcinogen,
- bioaccumulative properties,
- waste containing heavy metals/organics
- waste oil

Corrosive:
- pH < 2 or >12.5

Ignitable:
- Liquids with a flash point of <140°F (60°C)
- Solids that burn spontaneously
- Oxidizers

Reactive:
- Unstable material
- Explosives
- Water reactives
- Cyanide or sulfide-bearing wastes

2. Contain and Label
- Use compatible container

HAZARDOUS WASTE

HANDLE WITH CARE!

Generator: Name — Phone #: Phone

Building: Building — Room: Room No.

Contents: Description of waste

SAA start date: Date

WAA receival or accumulation start date*

Hazardous properties — Check all that apply: Toxic, Corrosive, Ignitable, Reactive, Other — Check all that apply

HAZARDOUS WASTE HANDLING FACILITY USE ONLY

Solid, Liquid, Gas — Check all that apply

3. Accumulate
- Place in a designated satellite accumulation area (SAA)
- Segregate incompatible waste
- Provide secondary containment
- Keep containers closed except while adding waste
- Use an accumulation log

4. Dispose
- When container is full or SAA date approaches 275 days, prepare a waste requisition and fax to x4838 with accumulation log(s). Allow 2 weeks for pickup

OR electronically at https://ehswprod.lbl.gov/shoebox/login.asp

Remove from CMS, if applicable.

Call your Generator Assistant with any questions ______________

Name and Phone

Figure 12-6 A mercury spill is a hazardous waste reportable situation. *(© 2009 The Regents of the University of California)*

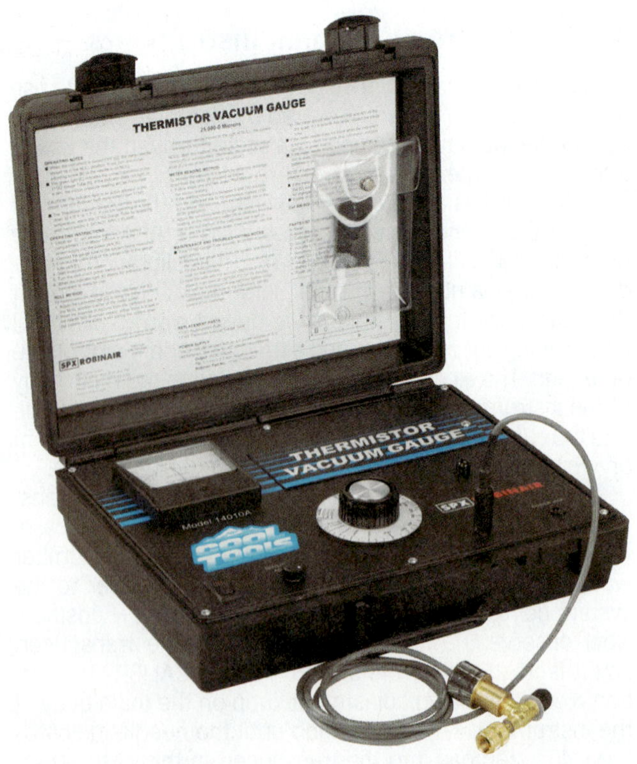

Figure 12-7 Analog electronic vacuum instrument transducer. *(Courtesy of SPX Corporation)*

There is one consideration with this type of instrument. Be careful not to expose the transducer itself to oil or liquid refrigerant. This contamination will render the transducer useless and another one will have to be obtained.

Digital Electronic Vacuum Instruments

There is no user calibration required on digital electronic vacuum instruments. Calibration has been performed by the manufacturer and will last the life of the instrument.

12.6 ELECTRICITY-MEASURING INSTRUMENTS

Electronic instruments such as volt meters, ohm meters, and ammeters, along with wattmeters, power meters, and phase angle meters, may be out of calibration. If the readings provided are not within tolerance, these instruments must be returned to the manufacturer for calibration. There are no user level adjustments that can be made, and, typically, even removing the cover to gain access to the circuit board will void the warranty.

12.7 REFRIGERANT LEAK–DETECTION INSTRUMENTS

Analog Leak Detectors

Analog detectors have a small sensing wand connected to the main body of the instrument. This sensing wand is used to "sniff" along the system tubing/piping in its entirety. There is a small pump in the body of the instrument. There is usually a small filter at the end of the sensing tube that should be changed before calibration begins.

It is important to note that the sensing tips of these wands should never come in contact with refrigerant. If they are exposed to refrigerant, they must be replaced. There is little that can be adjusted other than using a small screwdriver to adjust the meter movement for a zero reading. Calibration of the most complex leak-detection instruments is beyond the scope of the technician. These instruments must be returned to the manufacturer, where the calibration can be performed in an instrument-calibration laboratory.

To perform a calibration-verification procedure, test gases of 0 and 100 percent concentration must be purchased. These are available from the manufacturer and some supply houses. First, the sensor is exposed to a gas with no refrigerant or zero concentration. At this point, the minimum or zero indication is adjusted mechanically by slightly rotating a limit screw on the face of the meter or electronically by carefully adjusting a potentiometer on the circuit board to a value specified in the manufacturer's calibration procedures. Next, the sensor is exposed to a gas of 100 percent concentration. At this point there are no mechanical adjustments, electronic only. Access to the circuitry inside is required, and a potentiometer must be adjusted to provide the voltage recommended by the manufacturer or until a 100 percent maximum (span) reading is indicated on the meter. With all mechanical 0 percent (zero) and electronic 100 percent (span) adjustments, one adjustment affects the other. This requires you to go back and check 0 percent again because it will be off just a bit. Then recheck the 100 percent adjustment. Continue checking both the minimum and maximum indications until both are accurate. Electronic adjustments are not so interrelated, but it is good practice to check 0 percent and 100 percent several times to ensure both are correct.

Digital Leak Detectors

These detectors have some of the same limitations as the analog models. Exposing the wand sensing tips to refrigerant renders them useless. These types of instruments have removable/replaceable tips that can be installed if needed. They operate on a slightly different principle from the analog units. Digital instruments do not draw in an air sample. They measure the conductivity between two sensor plates. Clean air provides less conductivity as compared to air with refrigerant in it. The amount of difference in conductivity is the input, which is then converted into audible signals via microprocessors or integrated circuits.

Given this principle of operation, it should be understandable that these instruments are more effective in enclosed spaces as compared to operating on the roof of a building with the wind blowing. There is no calibration on these instruments that the technician can perform. You can verify operation by first purposely exposing the wand tip to a refrigerant source and listening for the audible indication. Next, proceed to a location where there is no refrigerant, and no audible detection signal should be heard.

12.8 FLUE GAS ANALYSIS INSTRUMENTS

The newer types of instruments have a small pump built into them. Like their predecessors, they must be inserted into the flue pipe, but its location is not as critical. When the sensing tip is inserted, the technician initiates the test. The small internal pump takes an air sample for X amount of seconds and electronically performs an analysis of the gases it has drawn in. A digital display is provided, although attachments can be purchased that provide a printout of the results. The newer types of instruments perform many tests on the exact same sample of flue gas. This helps eliminate an inaccurate reading caused by the improper placement of the sensing tip. The calibration of this type of instrumentation varies according to functionality.

Analog Flue Gas Instruments

For analog flue gas instruments, compensation must be made for the condition of the batteries in the unit. When batteries are new, they provide more voltage and current than older batteries. This can affect instrument readings. You must also compensate for the temperature of the batteries. Even with new batteries, the amount of voltage and current provided varies with the temperature. Colder batteries provide less voltage and current than warmer batteries. Placing a selector switch to the BATT (or CAL) position and adjusting a knob for the desired indication on the faceplate will compensate for both of these battery conditions.

Digital Flue Gas Instruments

Unlike their analog counterparts, digital displays are not affected by static electricity. The condition of the batteries is not as important either. Usually, digital display models will continue to operate with the advertised accuracy until a LOW BATT icon is displayed. There is also no zeroing of the display required.

Calibration Verification of Flue Gas Instruments

The calibration verification of both types of these instruments is accomplished in the same way. Sample gases must be purchased from the manufacturer of the instrument. The sensing tip is then exposed to these gases and an indication provided.

If the indication is within tolerance, calibration verification is complete and no further action is required. If the indication provided is not within tolerance, the instrument must be returned to the manufacturer for calibration. There are no user level adjustments that can be made, and typically even removing the cover to gain access to the circuit board will void the warranty.

UNIT 12—SUMMARY

Technicians use a variety of electronic instruments to provide an accurate picture of exactly how a specific HVACR system is performing. Most of these parameters cannot be visually observed, so the use of instrumentation is required.

It is the responsibility of the technician to determine whether the electronic instruments are operating correctly. If the instruments are not operating correctly, the faulty measurement could waste the technician's valuable time when troubleshooting a condition that does not exist or, even worse, cause the technician to miss a condition that needs to be rectified.

Manufacturer-level calibration is available for a fee, and expedited work for an increased fee is available from some manufacturers. This service is one that should be considered when initially purchasing your particular instrument.

UNIT 12—REVIEW QUESTIONS

1. What is calibration?
2. Why is it important for your instrumentation to give you accurate readings?
3. If your voltmeter was 100 percent accurate and you applied 68.7 V direct current (DC) to it, what would you expect your voltmeter reading to be?
4. If your voltmeter had an tolerance of +/−2 percent and you applied 68.7 V alternate current (AC) to it, any reading between _____ and _____ would be considered within tolerance.
5. What is the range of an instrument whose 0 percent output is 3 psig and whose 100 percent output is 15 psig?
6. Resistance thermal devices come in two basic types. Name them.
7. If the RTD you are using is a negative temperature coefficient type, and the temperature is increasing, the resistance output would be ________.
8. What is an easy way for you to generate a temperature reading of 212°F?
9. What type of instrument would you use to generate temperature readings in excess of 212°F?
10. The output signal a thermocouple generates is in units called ________.
11. In a thermocouple, the point at which the two dissimilar wires are connected to each other is called the ________.
12. Briefly describe the construction of a bimetallic device.
13. The two types of bimetallic devices are ________ and ________.
14. A sand bath is used to calibrate conical bimetallic devices. TRUE or FALSE?
15. Units of vacuum are expressed in what?
16. Why is it important to use only the special dyes made to color the water in a U-tube manometer?
17. Mercury spillage/contamination is reportable to ________.
18. What information is located on the transducer of the micron gauge?
19. To perform a calibration of a leak detector, ________ of ________ must be purchased.
20. How do digital refrigerant leak detectors differ from analog refrigerant leak detectors?

UNIT 13

Types of Refrigeration Systems

OBJECTIVES

After completing this unit, you will be able to:

1. define *refrigeration*.
2. list the different types of cooling systems.
3. explain the fundamental principles of evaporative cooling.
4. explain the basic operation of thermoelectric refrigeration.
5. explain the fundamental principles behind the compression-refrigeration cycle.
6. discuss the difference between the compression cycle and the absorption cycle.

13.1 INTRODUCTION

The purpose of any refrigeration system is to move heat. Typically the heat is being moved from a place where it is not wanted to a place where it is unobjectionable. Most often this involves moving heat from a cool place to a warmer place. Pushing heat uphill like this requires energy. There are many types of cooling systems, including absorption, evaporative, thermoelectric, and mechanical compression. The different types of refrigeration systems differ in how they accomplish this task, but they all involve using energy to move heat. The mechanical compression-refrigeration cycle is by far the most common. The mechanical compression-refrigeration cycle is applied in air conditioning, commercial refrigeration, and industrial process refrigeration. Air-conditioning systems cool people. They use refrigeration to provide comfort cooling and dehumidification in residential and commercial buildings (Figure 13-1). Commercial refrigeration systems are chiefly concerned with cooling products. Commercial refrigeration systems can be found in grocery stores, restaurants, refrigerated warehouses, and science labs (Figure 13-2). Industrial process refrigeration is used to cool equipment and machinery in large industrial manufacturing plants (Figure 13-3).

Figure 13-1 Residential air conditioner. *(Image courtesy of Bryant Heating & Cooling Systems)*

Figure 13-2 Commercial refrigeration unit.

GREEN TIP

The heat being moved can often be used. For example, heat rejected by an air-conditioning system can be used to heat hot water. Using the heat that is removed from the area being cooled to heat domestic hot water reduces the energy required to heat the water, improving the overall system efficiency.

Figure 13-3 Industrial process chiller.

13.2 COMPRESSION-CYCLE OPERATING PRINCIPLES

The transfer of heat in the compression-refrigeration cycle is performed by a refrigerant operating in a closed system. Most machines that cool or refrigerate something use the same mechanical refrigeration cycle. This cycle depends upon a few basic physical principles. Some of these principles are intuitive, such as that heat travels from hot to cold. Other principles are less intuitive, such as understanding that boiling is a cooling process. Understanding the following handful of basic physical principles involved makes understanding the mechanical refrigeration cycle much easier.

- "Cold" is not a substance or an energy form; it is the lack of heat.
- Heat travels from hot to cold.
- Liquids absorb large amounts of heat when they boil off to a gas.
- Gases give off large amounts of heat when they condense to a liquid.
- When something is boiling or condensing, its temperature remains the same.
- The temperature at which a liquid boils is controlled by its pressure.

Refrigeration machines do not pump cold in; they pump heat out. Imagine that your air conditioner is like a pump on the bottom of a ship. The purpose of the pump is to pump water out, preferably faster than it leaks in. An air conditioner has a similar purpose, to pump heat out faster than it leaks into your house. The problem is that heat only wants to travel from hot to cold, and when you need your air conditioner it is usually hotter outside than inside. So the challenge is to pump heat from inside, where it is relatively cool, to outside, where it is relatively hot.

Remember that boiling is a cooling process! When a liquid boils, it absorbs large amounts of heat in much the same way that a sponge absorbs water. When you place a dry sponge on a puddle, the sponge turns soggy. The puddle ends up drier, but the sponge ends up wetter. A boiling liquid acts like a sponge soaking up heat. As the liquid boils it takes heat away from whatever it is touching, leaving the surrounding area cooler. When water is poured on a hot frying pan, clouds of steam billow forth. The energy comes from the frying pan to produce the steam, cooling off the frying pan. The reason people think of boiling as hot is because water boils at 212°F under normal atmospheric pressure, and that is hot. But suppose water boiled at 50°F instead? People might pour boiling water over themselves in the summer to help cool off. It actually is possible to boil water at 50°F! Remember that a liquid's boiling temperature is controlled by its pressure. Water will boil at 50°F if placed under a low enough vacuum (Figure 13-4).

Refrigerant, like water, can boil at almost any temperature by controlling its pressure. The refrigerant in an air conditioner boils in the cooling coil at a low pressure. The low pressure keeps the refrigerant boiling temperature below the temperature of the air being cooled. The refrigerant temperature stays low even as it absorbs heat because the temperature of a boiling liquid remains stable. This is how heat is absorbed from the space being cooled, by boiling the refrigerant at a low pressure and temperature. During this process, the refrigerant changes from a liquid to a gas. This is done in a component called an evaporator because the refrigerant is evaporating. Heat flows from the warm air traveling over the evaporator to the cold refrigerant boiling inside the evaporator, as shown in Figure 13-5.

The compressor pumps gas from the low-pressure evaporator to the high-pressure condenser (Figure 13-6). The word *compress* means to squeeze. It increases the gas pressure by squeezing the gas into a smaller space. It is important to note that the compressor is designed to pump only gas. Liquid does not squeeze very well. Putting liquid through the compressor will damage it.

When the gas is squeezed into a smaller space, both the pressure and temperature rise. The high pressure raises the refrigerant temperature above the hot outdoor air temperature. Now the refrigerant is sent outside to be condensed at a high temperature and pressure. The heat that was absorbed in the evaporator is expelled outside by condensing the refrigerant back to a liquid. The temperature of the refrigerant stays high even as the refrigerant loses heat because temperature remains constant during condensation. Heat travels from the hot refrigerant to the outdoor

Figure 13-4 Water boils at 50°F under a vacuum.

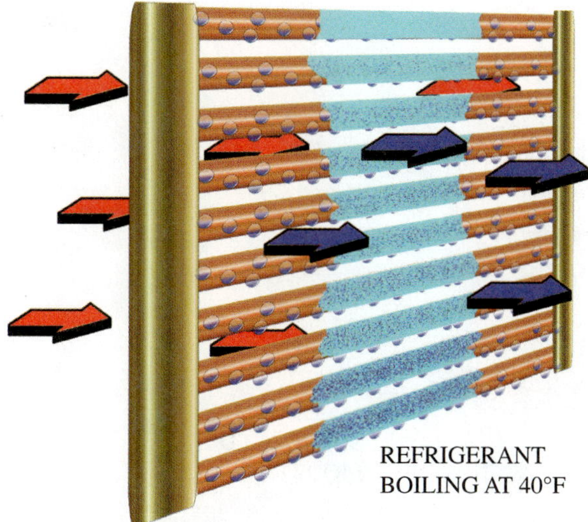

Figure 13-5 The evaporator is cold because refrigerant boils inside the evaporator.

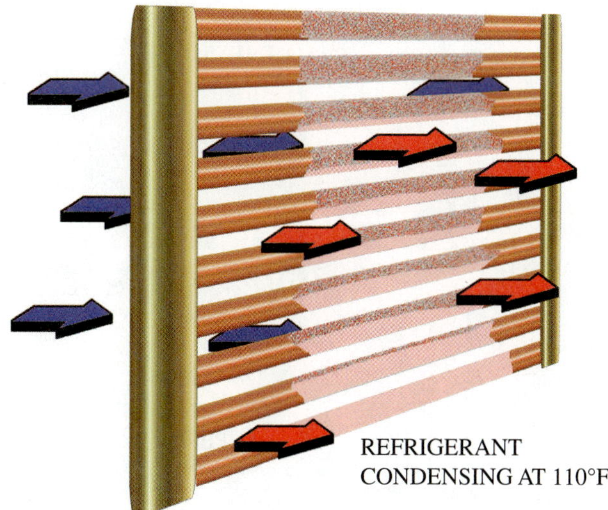

Figure 13-7 The condenser expels heat from the refrigerant by condensing it from a gas to a liquid.

Figure 13-6 The compressor draws in refrigerant vapor at a low pressure and compresses it to increase the refrigerant pressure and temperature.

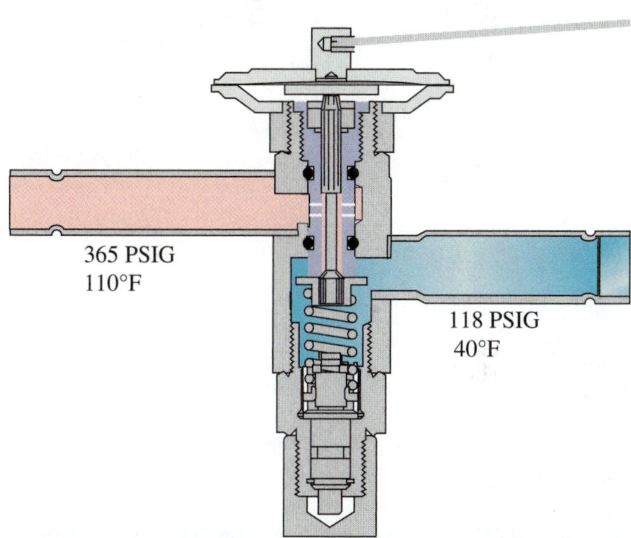

Figure 13-8 The metering device drops the pressure and temperature of the refrigerant by restricting the refrigerant flow.

air because the air is cooler than the refrigerant. During this process, the refrigerant changes from a gas to a liquid. This is done in a component called a condenser because the refrigerant is condensing (Figure 13-7). Notice that in both the evaporator and condenser, heat is going from hot to cold. The trick is keeping the refrigerant pressures low in the evaporator where heat is absorbed and high in the condenser where heat is expelled.

The high pressure and temperature of the refrigerant must be reduced before sending it into the evaporator to boil at a low pressure and temperature. The metering device does this (Figure 13-8). The metering device is basically a small restriction that limits the amount of refrigerant entering the evaporator. Limiting the flow of refrigerant increases the pressure before the metering device and decreases the pressure after the metering device. The drop in pressure causes the refrigerant to also drop in temperature. This cold refrigerant is now ready to absorb heat in the evaporator.

13.3 SYSTEM HIGH SIDE AND LOW SIDE

For all practical purposes there are two pressures in the system: the low-side pressure and the high-side pressure. The compressor and the metering device work in partnership to maintain this pressure difference. Figure 13-9 shows how the compressor and metering device, the expansion valve, divide the system into the high- and low-pressure sides.

The low side contains the low-pressure liquid and vapor refrigerant and is the side that absorbs heat. The metering device controls the flow into the evaporator; the expansion of the refrigerant causes a pressure drop. The low side of the system starts at the metering device and extends through the evaporator and the suction line up to the compressor inlet.

The high side contains the high-pressure vapor and liquid refrigerant and is the part of the system that rejects

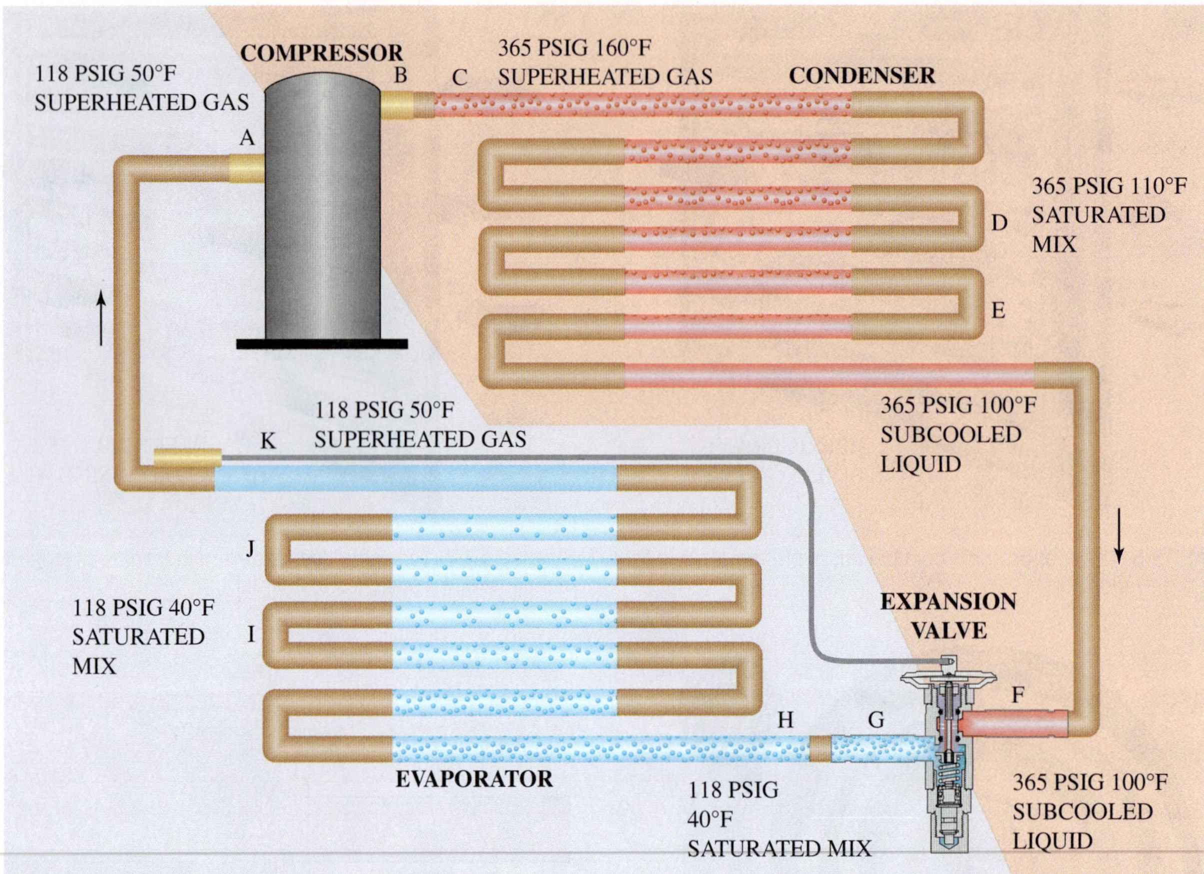

Figure 13-9 The compressor and metering device divide the system into a high-pressure side and a low-pressure side.

heat. The compressor pumps the refrigerant into the condenser and maintains the high pressure. The high side of the system starts at the compressor outlet and continues through the discharge line, condenser, liquid line, and up to the metering device.

TECH TIP

The body is cooled by evaporation of the perspiration on the skin. The evaporation is reduced when the air is at a higher relative humidity, so the body's cooling mechanism does not work as well. That is why you feel more comfortable at 80° with 40 percent relative humidity as compared to 75° and 90 percent relative humidity. To be effective, air-conditioning systems must remove both heat and humidity.

13.4 REFRIGERATION SYSTEM RELATIONSHIPS

There are key relationships within the refrigeration cycle. Understanding these will improve your understanding of the refrigeration system. The most obvious is the link between pressure and temperature. If the refrigerant is a high pressure, it is also a high temperature. If it is a low pressure, it is also a low temperature. Another important link is between the refrigerant state and its heat content. Gas always has a relatively high heat content, and liquid always has a relatively low heat content. The heat content of the refrigerant is affected much more by the state of the refrigerant than by its temperature. This is why the refrigerant can have a low temperature and a high heat content leaving the evaporator. Refrigerant will be found in the system in one of three conditions: superheated, saturated, or subcooled. It will be superheated anywhere it is all gas, subcooled anywhere it is all liquid, and saturated anywhere it is changing from one to the other. The refrigerant is superheated leaving the evaporator, through the compressor, and up to the condenser. It is sub-cooled leaving the condenser up to the expansion device. It is saturated in both the condenser, where it is changing from gas to liquid, and in the evaporator, where it is changing from liquid back to gas.

Each of the four main components has an opposite. The compressor and expansion device are opposites. The compressor changes a low-pressure and temperature gas into a high-pressure and temperature gas by reducing its volume. The expansion device changes a high-pressure and temperature liquid into a low-pressure and temperature liquid by expanding its volume. Together they maintain the pressure difference in the system. The condenser and evaporator are opposites. The condenser changes a high-pressure and temperature gas into a high-pressure and temperature liquid, losing a large amount of heat in the process. The evaporator turns a low-pressure and temperature liquid into a low-pressure and temperature gas, gaining a lot of heat in the process.

13.5 HEAT FLOW IN THE REFRIGERATION CYCLE

Heat is added to the refrigerant in the evaporator. Heat from the product load is transferred to the refrigerant as it boils in the evaporator. This is the majority of heat absorbed by the refrigerant. A small heat gain occurs in the piping from the evaporator up to where the refrigerant enters the compressor. Heat gain in the suction line is not desirable and can be reduced by insulating the suction line. The compressor adds a sizable quantity of heat to the refrigerant. This heat is equivalent to the work done in compressing the refrigerant. In a suction gas cooled semi-hermetic or hermetic motor compressor unit, the motor heat is also transferred to the refrigerant.

The heat added in the evaporator and by the compressor is removed in the condenser. The heat is transferred from the refrigerant in the condenser to the air or water flowing over the condenser. The heat balance of the overall system is shown in the following formula: Heat added in the evaporator + heat of compression = heat rejected in the condenser.

13.6 ABSORPTION REFRIGERATION SYSTEMS

Absorption refrigeration systems use heat and the process of absorption instead of a compressor to move the refrigerant. The absorption process takes advantage of the fact that some chemicals have an affinity for other chemicals. One of the most common combinations is to use ammonia as the refrigerant and water as the absorbent. This combination is used in smaller absorption systems. Large absorption chillers use water as the refrigerant and lithium bromide as the absorbent. Absorption systems really have two cycles: a refrigerant cycle and an absorbent cycle. Some systems use a third gas, typically hydrogen, to help the refrigerant move from the low side of the system back to the high side without using any mechanical devices. Other systems use a solution pump to move solution from the low side to the high side. This discussion will focus on ammonia-water absorption systems used for residential chillers. Figure 13-10 shows a simplified drawing of an ammonia-water absorption cycle.

SAFETY TIP

Ammonia is a poisonous and somewhat flammable gas! Do not attempt to service an ammonia system unless you have received specific training in the safe handling of ammonia refrigerant.

The transfer of heat is essentially the same in the absorption system as in the traditional mechanical compression cycle. Heat is absorbed in the evaporator and rejected in the condenser. Under a high pressure, the ammonia refrigerant condenses from a gas to a liquid in the condenser, giving up its latent heat in the process. The high-pressure liquid now passes through a refrigerant restrictor, where its pressure and temperature are reduced. This low-pressure liquid boils inside the evaporator to a gas, absorbing latent heat in the process and cooling off the water passing over the evaporator coil. So far, the compression and absorption systems are identical. The difference is in how the low and high pressures are created.

In a compression system, the compressor provides the force to create both the high- and low-side pressures. In an

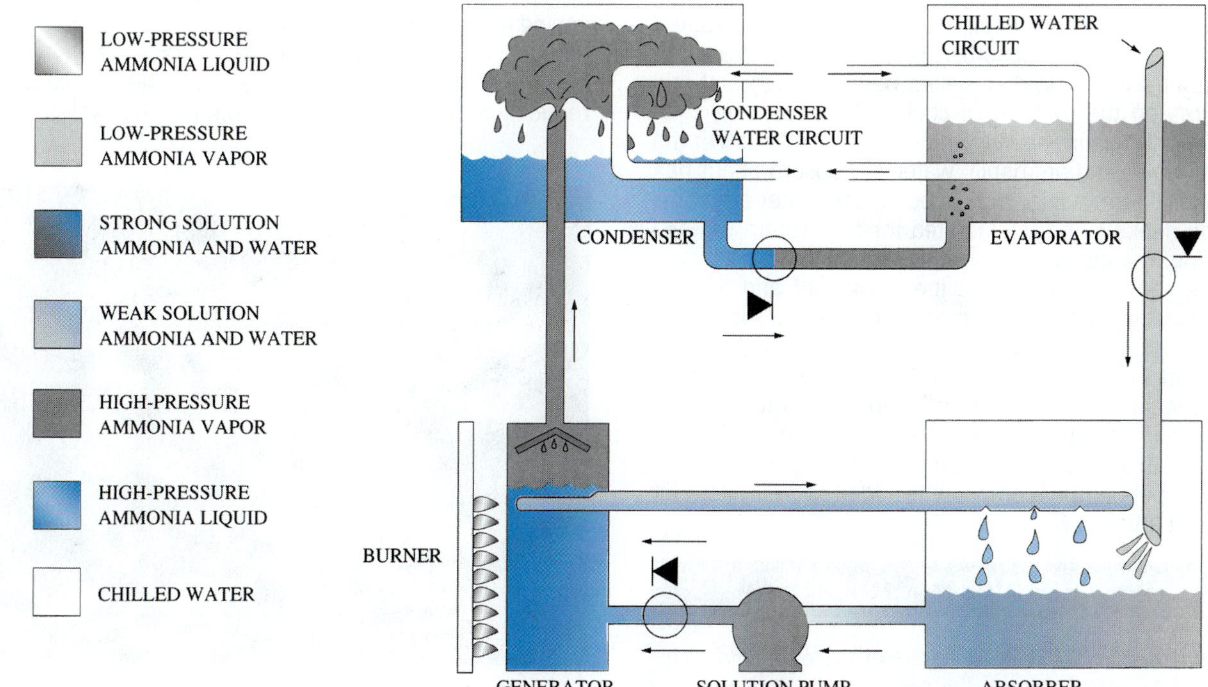

Figure 13-10 Simplified ammonia-water absorption cycle. The solution cycle is on the left; the refrigeration cycle is on the right.

absorption system, the compressor is replaced by three components: the generator, the absorber, and the solution pump.

One component that absorption systems have that is not found on compression systems is the absorber. A second fluid that has an affinity for the refrigerant absorbs the refrigerant in the absorber. This has the effect of "sucking" the refrigerant out of the evaporator and creating a low refrigerant pressure in the evaporator. The absorption process replaces the suction stroke of the compressor. When the vapor refrigerant is absorbed into the solution, it condenses back into a liquid. This gives off a great deal of heat, normally called the heat of absorption. If the heat of absorption is not removed from the solution, the absorption process will stop. Some units actually have two absorbers: a solution-cooled absorber and an air-cooled absorber.

The solution-cooled absorber is where the refrigerant and weak solution are first mixed. The solution-cooled absorber cannot cool the solution adequately enough to remove all the heat of absorption; therefore, much of the refrigerant remains in a vapor state inside the solution. To remove the heat of absorption, the solution passes through the air-cooled absorber, where the heat of absorption is removed and the refrigerant vapor is condensed. A solution pump is used to pump the solution from the low-pressure absorber to the high-pressure generator.

The solution pump is the closest thing to a compressor in concept. The solution pump used in many units is a diaphragm pump and pumps somewhat like a heart, in pulses of pressure that are directed by check valves. On the suction stroke, the valve to the absorber opens, the valve to the generator closes, and solution is drawn into the pump from the air-cooled absorber. On the pressure stroke, the valve to the generator opens, the valve to the absorber closes, and the solution is forced into the generator. The pulsing action of the solution pump causes the solution level in the solution-cooled absorber to rise and fall. This helps entrain refrigerant vapor in the weak solution before it is circulated through the air-cooled absorber. The solution pump and the restrictors are the dividing points between high and low pressures on the system. Of course, the absorbing solution can only absorb a limited amount of refrigerant. In the case of water and ammonia, water will absorb about half its weight in ammonia. At this point, the water-ammonia solution needs to be regenerated for the system to keep operating. The generator does this.

The generator separates the refrigerant and the absorbent by boiling the refrigerant out of the absorbent. This creates a high pressure on the high side of the system. The refrigerant vapor is sent on its way toward the condenser and the weak solution is sent through a restrictor, back to the solution-cooled absorber. The solution cycle is now complete. In addition to the components mentioned, there are a number of other components added to the system for increased efficiency.

SERVICE TIP

Never use standard brass gauges to service an ammonia system! Ammonia attacks the brass. Ammonia systems require gauges made specifically for ammonia.

13.7 EVAPORATIVE COOLING SYSTEMS

Evaporative cooling reduces the air temperature by evaporating water. The heat required to evaporate the water comes from the air, reducing the temperature of the air. Air that has been cooled through direct evaporative cooling still has the same amount of heat, but its temperature is lower and its humidity is higher. This method works well in dry climates, such as in the southwestern United States. Evaporative coolers can drop the air temperature by as much as 20°F if the entering air is dry. Evaporative cooling efficiency increases as the air temperature increases because the water evaporates more readily. Evaporative cooling does not work as well in humid climates, such as coastal areas or the southeastern United States. The high humidity in these areas reduces the effectiveness of evaporative cooling because the water evaporates slowly. The installation and operation costs for an evaporative cooler are much lower than conventional air conditioning using a refrigeration cycle. This makes evaporative cooling popular in large factories and plants, where traditional refrigerated air conditioning would be prohibitively expensive.

Outdoor misters are an increasingly popular form of direct evaporative cooling. Nozzles that spray out micro-droplets of water cool the outside air through evaporation. Amusement parks now often use outdoor misters to provide outdoor cool spaces during the summer. Mist systems can also be applied to residential outdoor spaces such as patios (Figure 13-11).

A typical evaporative cooled air conditioner, also called a "swamp cooler," is shown in Figure 13-12. The unit consists of a water sump, a pump, wetted media to increase the surface area for evaporating water, and a blower to move air. The pump pumps the water in the sump at the bottom to the top of the media where it runs down the media. A float controls the water level in the sump by allowing makeup water to fill the sump when the water level drops. The blower moves outside air across the media and into the house. Water evaporates from the wetted media, reducing the temperature of the air. For this system to op-

Figure 13-11 Mist system applied to cooling an outdoor patio. *(Courtesy of MicroCool)*

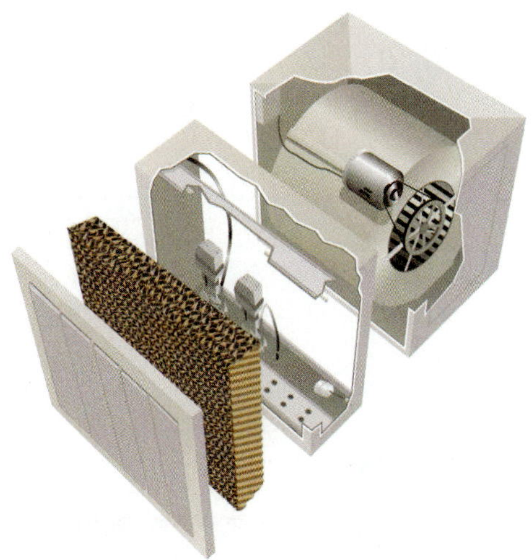

Figure 13-12 Evaporative cooler. *(Impco Air Coolers)*

erate correctly, air in the house must be allowed to escape. This is normally done by opening the windows. Evaporative coolers are constantly replacing the air in the house with air from outside that has been cooled by evaporation. This type of evaporative cooling is called direct evaporative cooling because the water is directly exposed to the air entering the building.

Indirect evaporative cooling uses a heat exchanger and two airstreams. The heat exchanger keeps the airstreams separate while allowing one airstream to cool the other. Figure 13-13 shows an indirect evaporative cooler. The water evaporation takes place in the airstream that passes through the heat exchanger without entering the building. The airstream entering the building is cooled by the heat exchanger without adding any water to it. Indirect evaporative cooling can be used to precool air before it passes through a traditional compression-cycle conditioning system. This saves energy by reducing the amount of cooling that the compression system must provide (Figure 13-14).

The effectiveness of evaporative cooling can be increased using multiple stages. One manufacturer, Coolerado, makes a twenty-stage indirect evaporative

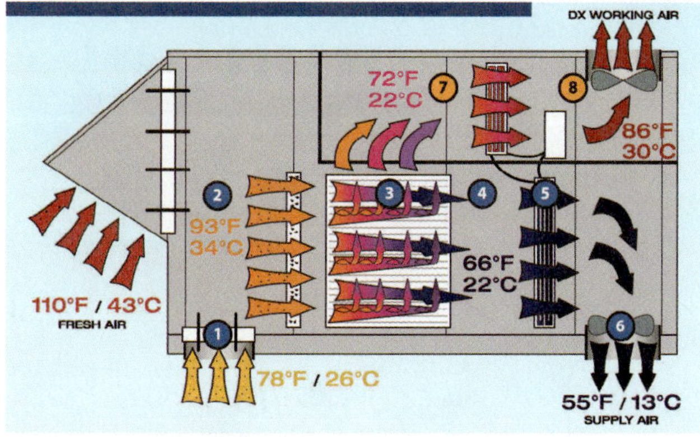

Figure 13-14 Indirect cooling can be used to precool the air entering a traditional compression system. *(Courtesy of Coolerado Corp.)*

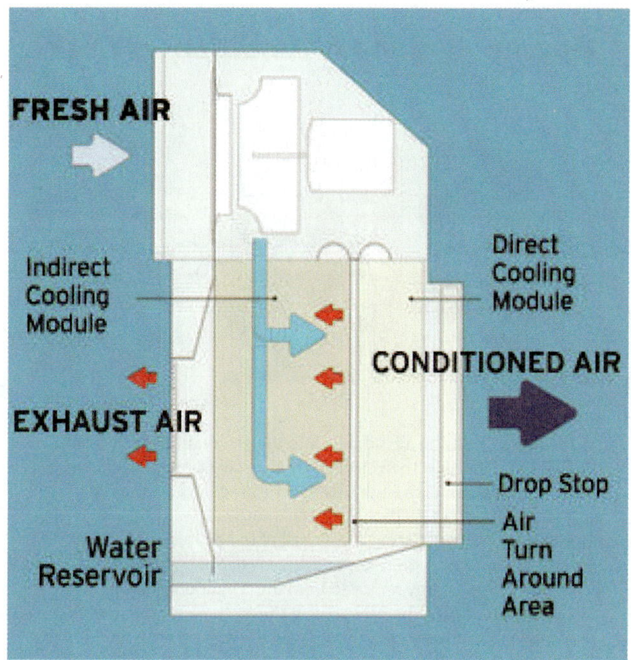

Figure 13-13 Indirect evaporative cooling process.

Figure 13-15 Indirect, multistage evaporative air conditioner. *(Courtesy of Coolerado Corp.)*

cooler using the Maisotsenko Cycle (Figure 13-15). The system uses a unique heat and mass exchanger (HMX) (Figure 13-16) that consists of several plates of a special plastic that is designed to wick water evenly on one side and transfer heat through the other side. The plates are stacked on each other, separated by channel guides. The channel guides divide the incoming air stream into product air and working air. The product air will be delivered into the house; the working air will be expelled to the outside. Product air is always separate from the working air so that the air entering the house does not pick up moisture. Air that is cooled by the first set of plates is used to cool the next set of plates. This process occurs multiple times in a short physical space within the exchanger, resulting in progressively colder product and working air temperatures.

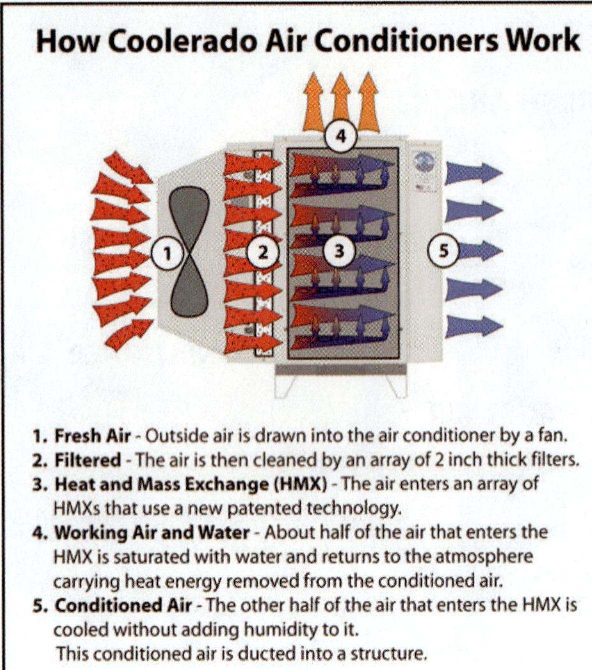

Figure 13-16 The Maisotsenko Cycle. *(Courtesy of Coolerado Corp.)*

13.8 THERMOELECTRIC REFRIGERATION

A thermocouple is a device composed of two dissimilar metal wires joined on both ends but separated or insulated from each other in between. Thomas Seebeck discovered that small amounts of electrical current flow through the wires if the two junction points are at different temperatures (Figure 13-17). This effect has been used for years as a flame safety device in gas-burning appliances. While studying the Seebeck effect in 1834, Jean Pelletier discovered that if a current is imposed on the thermocouple, one end will heat up and the other will cool off. Reversing the direction of the current will swap the hot and cold junctions. This is called the Pelletier effect (Figure 13-18). This characteristic is exploited in thermoelectric refrigeration.

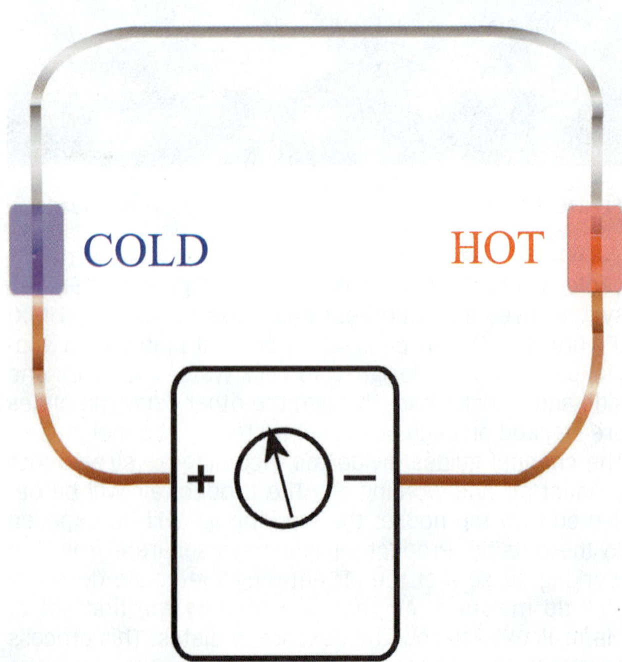

Figure 13-17 Temperature difference between the two junctions creates a DC current flow.

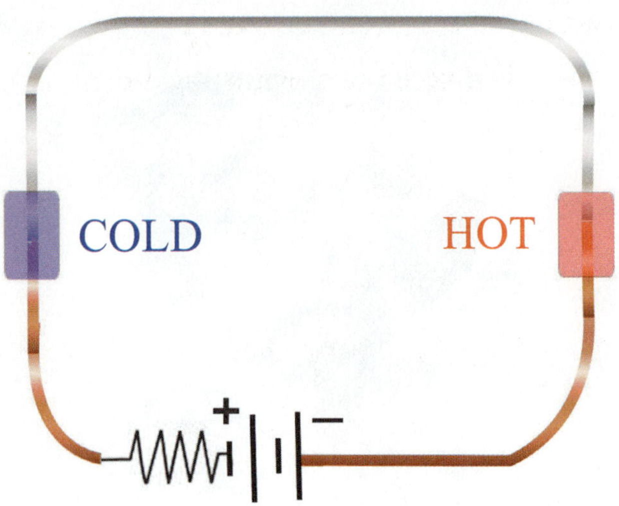

Figure 13-18 A DC current imposed across the two junctions creates a temperature difference.

Today's thermoelectric modules typically use semiconductors rather than metal. Figure 13-19 shows the construction of a typical thermoelectric module. These modules can be sandwiched between two heat sinks to produce a unit capable of moving heat. Figure 13-20 shows a complete thermoelectric cooling module, including the heat sinks. Thermoelectric refrigeration is not very energy efficient and is best suited for small loads. Thermoelectric cooling does provide a small, lightweight cooling system for areas that would be difficult to cool with a traditional refrigeration system. Thermoelectric cooling has been applied to electronic systems, spaceships, and picnic coolers.

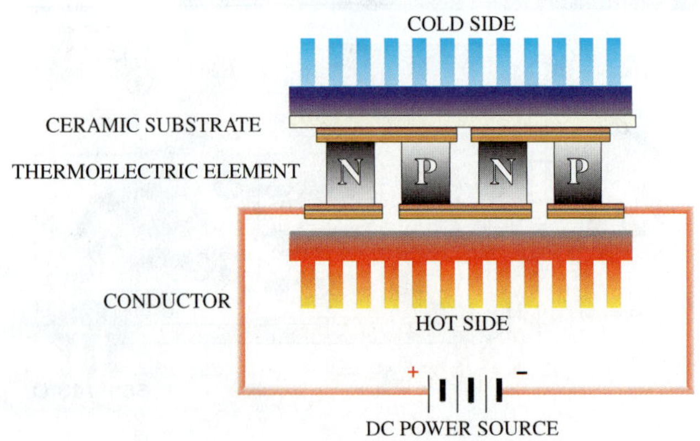

Figure 13-19 Modern thermoelectric modules use semiconductors to create the Pelletier effect.

Figure 13-20 Complete thermoelectric cooling module, including heat sinks.

UNIT 13—SUMMARY

The purpose of a refrigeration system is to move heat from where it is not wanted to where it is unobjectionable. There are several kinds of cooling systems, including absorption, evaporative, thermoelectric, and mechanical compression. The mechanical-compression refrigeration cycle is the most common. The four main components of the mechanical-compression refrigeration cycle are the compressor, the condenser, the metering device, and the evaporator. The refrigeration cycle is made possible by maintaining a pressure difference between the low side, which absorbs heat when the refrigerant boils in the evaporator, and the high side, which rejects heat when the refrigerant condenses in the condenser. This pressure difference is maintained by the compressor and metering device. The compressor increases the refrigerant pressure by squeezing it; the metering device reduces the refrigerant pressure by restricting its flow into the evaporator.

Other cooling mechanisms include evaporative coolers, which cool the air by evaporating water; the absorption cycle, which uses heat and an absorbent fluid to create the needed system pressure difference; and thermoelectric cooling, which transfers heat using electric current and the Pelletier effect.

UNIT 13—REVIEW QUESTIONS

1. List the different types of cooling mechanisms.
2. Which type of cooling mechanism is used in the most applications?
3. Briefly explain the fundamental principles that make the compression-refrigeration cycle work.
4. What does an air conditioner do with the heat in your house?
5. How does a compression-cycle refrigeration system move heat from a relatively cold temperature to a relatively warm temperature?
6. Explain why boiling is considered a cooling process.
7. List the four major components of the compression refrigeration cycle in order.
8. Why do both the compression cycle and absorption cycle have a high-pressure side and a low-pressure side?
9. What components are responsible for maintaining the high and low pressures in a compression refrigeration system?
10. What are the three refrigerant conditions inside a compression-cycle system?
11. What components are responsible for maintaining the high and low pressures in an absorption refrigeration system?
12. What are the two cycles that make an absorption system work?
13. Explain how an evaporative cooling system works.
14. Where is evaporative cooling the most effective?
15. What is the advantage of the Maisotsenko cycle?
16. What acts as the refrigerant in a thermoelectric refrigeration system?
17. What is the Pelletier effect?
18. Explain how a thermoelectric refrigeration system works.
19. What is the purpose of refrigeration?

UNIT 14
The Refrigeration Cycle

OBJECTIVES

After completing this unit, you will be able to:

1. explain the fundamental principles behind the refrigeration cycle.
2. identify the four major components of the compression cycle.
3. name the three main refrigeration-system interconnecting lines.
4. identify where in the refrigeration system the refrigerant is saturated, superheated, or subcooled.
5. draw a basic compression-refrigeration cycle and identify the pressure, temperature, state, and heat content in and out of each component.

14.1 INTRODUCTION

The mechanical vapor-compression–refrigeration cycle is used in many types of HVACR systems. It can be found in all sizes of systems, from small dormitory refrigerators to large industrial chillers. The compression cycle has been used for a diverse range of applications, including air conditioning, commercial refrigeration, and industrial process refrigeration. A thorough understanding of the refrigeration cycle is absolutely essential to working in the HVACR field. Understanding exactly what is supposed to take place at any point in the cycle is necessary for proper system installation, servicing, and troubleshooting. Because we only discuss the mechanical vapor-compression–refrigeration cycle throughout this unit, we refer to it as simply the refrigeration cycle for the rest of the unit.

14.2 REFRIGERATION CYCLE FUNDAMENTALS

The refrigeration cycle is a practical application of the gas laws discussed in Unit 11. The system consists of four components: the compressor, condenser, expansion device, and evaporator. All four are visible in the air conditioner shown in Figure 14-1. The names tell what the components do. The compressor compresses, or squeezes, the gas into a smaller volume (Figure 14-2). This raises the gas pressure and temperature. The high-pressure, high-temperature, superheated gas travels to the condenser. This is where the heat is rejected from the refrigeration system. The condenser condenses the refrigerant from a high-pressure, high-temperature, superheated gas to a high-pressure, warm, subcooled liquid (Figure 14-3). The high-pressure, warm, subcooled liquid travels to the expansion device, also called the metering device. The expansion device drops the pressure and temperature of the refrigerant by restricting its flow (Figure 14-4). At the same time, the refrigerant is passing into a larger space and expanding. The combination of pressure drop and expansion dramatically drops the temperature of the refrigerant and it starts to flash off to a vapor. The refrigerant leaving the expansion device is a low-pressure, low-temperature, low-heat saturated mixture. This low-temperature saturated mixture enters the evaporator, where it evaporates from a liquid to a gas, absorbing heat. The evaporator is the component that cools by absorbing heat from its surroundings (Figure 14-5). The temperature of the refrigerant stays the same through most of the evaporator because the temperature of a boiling liquid is set by its pressure, not by the amount of heat in it.

Figure 14-1 The four refrigeration-cycle components can be seen in this packaged air-conditioning unit.

Figure 14-2 The compressor raises the gas pressure and temperature by squeezing the gas to a smaller volume.

Figure 14-3 The condenser removes heat by condensing gas to a liquid at a high pressure.

After all the liquid has evaporated, the gas temperature increases. The refrigerant leaves the evaporator as a low-pressure, low-temperature, superheated vapor and returns to the compressor.

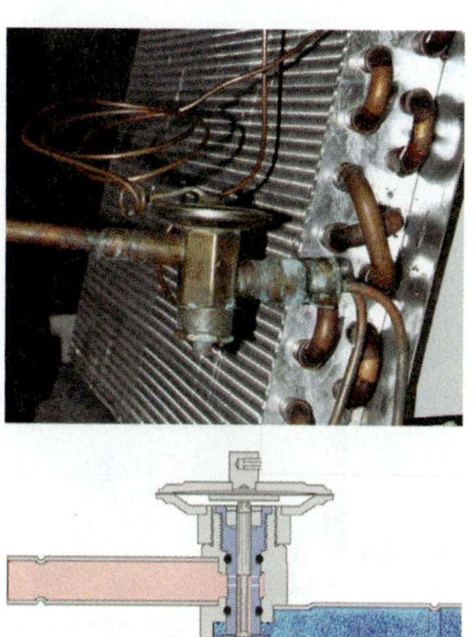

Figure 14-4 The expansion device drops the refrigerant pressure and temperature by restricting its flow.

Figure 14-5 The evaporator cools by evaporating refrigerant at a low pressure and temperature.

TECH TIP

The refrigerant temperature actually drops slightly as the refrigerant travels through the evaporator. The small pressure drop through the evaporator causes a small but measurable temperature drop from the beginning of the evaporator to the end of the portion that contains saturated refrigerant. Once all the liquid has boiled off, the temperature starts to rise. Normally this occurs in the last 10 percent of the evaporator.

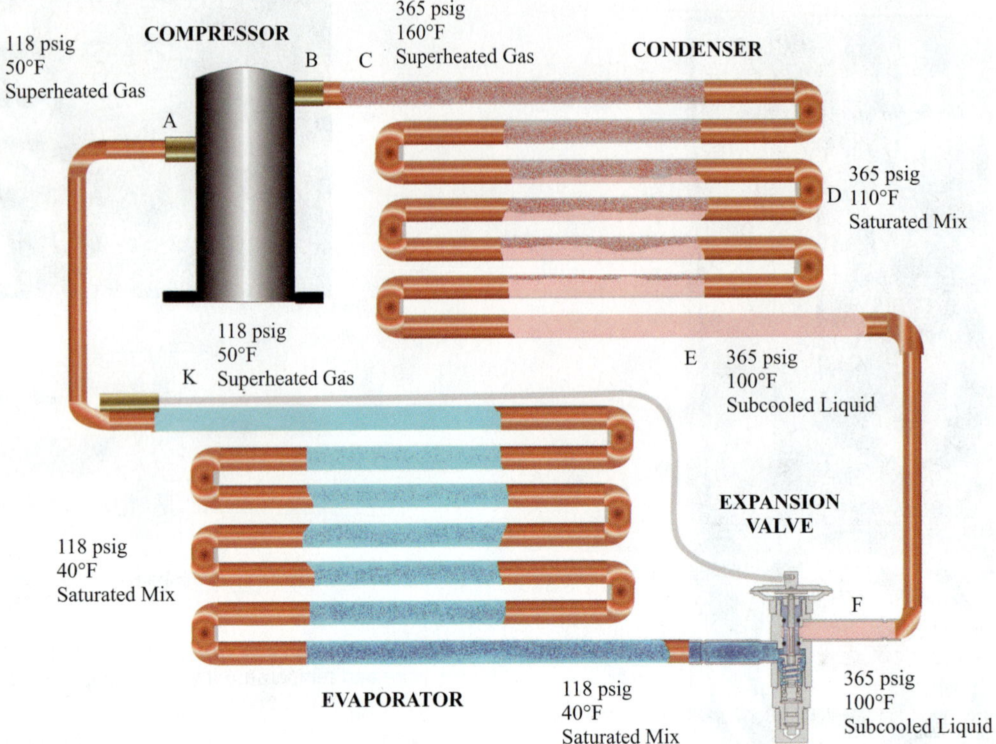

Figure 14-6 Basic refrigeration cycle of an R-410A air conditioning system.

14.3 REFRIGERANT CHANGES THROUGH THE CYCLE

We examine the specific changes in the refrigerant through a typical air-conditioning system. Our example system is a residential air-conditioning system using HFC 410a refrigerant. The refrigeration cycle is a never-ending circle; there is really not a starting or stopping point. This discussion starts with the compressor and works through the cycle. Refer to Figure 14-6 to follow the refrigerant changes through the cycle.

A. **Compressor Suction 118 psig 50°F Superheated Gas** The refrigerant enters the compressor as a low-temperature, low-pressure, superheated vapor. The refrigerant has picked up some superheat in the final evaporator circuit. Superheating is the process of continuing to heat the refrigerant after sufficient latent heat has been added to vaporize all the liquid. Superheating ensures that no liquid slugs will reach the compressor and cause damage to the valves and pistons. The state of the refrigerant is 100 percent vapor.
B. **Compressor Discharge 365 psig 160°F Superheated Gas** The refrigerant leaves the compressor as a high-pressure, high-temperature, superheated vapor. The refrigerant contains both the heat absorbed in the evaporator and the heat of compression from the compressor. The state of the refrigerant is 100 percent vapor.
C. **Beginning of Condenser 365 psig 160°F Superheated Gas** The refrigerant is superheated as it enters the condenser. The first portion of the condenser is used to reduce the temperature of the refrigerant to the saturation temperature. This portion of the condenser is often referred to as the de-superheating section. The state of the refrigerant is 100 percent vapor.
D. **Center of Condenser 365 psig 110°F 50 Percent Liquid/50 Percent Gas Saturated Mix** As additional latent heat is removed, the vapor condenses. The amount of liquid increases and the amount of vapor decreases as the refrigerant travels through the condenser. During this time, the refrigerant temperature remains stable. The state of the refrigerant is 50 percent liquid and 50 percent vapor saturated mix.
E. **End of Condenser 365 psig 100°F Subcooled Liquid** At the lower portion of the condenser, the refrigerant is all liquid. The temperature of the liquid refrigerant is reduced as heat continues to be expelled to the surrounding air. This portion of the condenser is called the subcooling section. Adequate subcooling will prevent the refrigerant from starting to boil as it experiences small pressure drops through the piping or components. Such boiling, called flash gas, can reduce the system capacity. It is desirable to subcool the liquid refrigerant either in the condenser or in the liquid line before the metering device. Subcooling the liquid refrigerant reduces flash gas and increases

mass flow. The state of the refrigerant is 100 percent liquid.

F. **Metering Device Inlet 365 psig 100°F Subcooled Liquid** The refrigerant is a highpressure, warm, subcooled liquid at the inlet of the metering device. The state of the refrigerant is 100 percent liquid.

G. **Metering Device Out 118 psig 40°F 75 Percent Liquid/25 Percent Gas Saturated Mix** The pressure of the liquid refrigerant drops as it goes through the small opening of the metering device, eventually reaching a saturation point at a low pressure and temperature. In passing through the metering device to the low pressure evaporator, some refrigerant is evaporated, cooling the remaining liquid. The refrigerant is typically a mixture of 75 percent liquid and 25 percent vapor at this point.

H. **Evaporator in 118 psig 40°F 75 Percent Liquid/25 Percent Gas Saturated Mix** As the refrigerant enters the evaporator, the air (or liquid, or whatever is being cooled) gives off heat to the refrigerant. The refrigerant takes in this heat as latent heat of vaporization. The refrigerant is about 25 percent vapor and 75 percent liquid at this point.

I. **Evaporator Center 118 psig 40°F 50 Percent Liquid/50 Percent Gas Saturated Mix** Heat from the air or the product being cooled in the evaporator is absorbed by the liquid refrigerant and causes the refrigerant to boil or vaporize. As the compressor draws the vaporized gas from the evaporator, the metering device admits more refrigerant, continuing the process. The refrigerant is about 50 percent vapor and 50 percent liquid at this point.

J. **Nearing the End of Evaporator 118 psig 40°F 25 Percent Liquid/75 Percent Gas Saturated Mix** As the refrigerant continues through the evaporator, the mixture becomes more saturated gas than saturated liquid. However, while there is still liquid in the tube, the temperature remains at the saturation temperature of the refrigerant. The refrigerant is a saturated mixture of liquid and gas.

K. **Leaving the Evaporator 118 psig 50°F 100 Percent Superheated Gas** After all the saturated liquid has boiled to a vapor, the refrigerant vapor starts to increase in temperature in the final evaporator circuit. Superheating is the process of continuing to heat the refrigerant after sufficient latent heat has been added to vaporize all the liquid. Superheating ensures that no liquid slugs will reach the compressor and cause damage to the valves and pistons. The state of the refrigerant is 100 percent gas.

Figure 14-7 Piston compressor (reciprocating).

Figure 14-8 Cutaway of a rotary compressor.

14.4 THE COMPRESSOR

The compressor is a mechanical device for pumping refrigerant vapor from the low-pressure evaporator to the high-pressure condenser. The compressor increases the refrigerant pressure and temperature by decreasing the gas volume. The main types of compressors are reciprocating (piston), rotary, centrifugal, screw, and scroll, shown in Figures 14-7 to 14-11.

The compressor type describes the mechanical operation of the compressor. In the reciprocating compressor, a piston travels back and forth (reciprocates) in a cylinder. Gas is compressed in a reciprocating compressor by squeezing it between the piston and a valve plate at the top of the cylinder. The rotary compressor has a roller that rotates in an orbital motion within a cylinder. Gas is compressed by squeezing it between the roller and

Figure 14-9 Cutaway of a centrifugal compressor. *(Photo courtesy of McQuay International)*

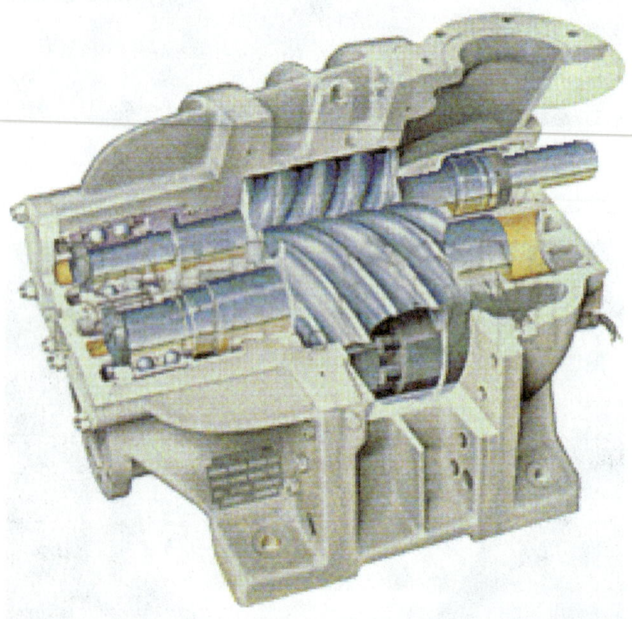

Figure 14-10 Screw compressor.

a spring-loaded vane. The centrifugal compressor has a very-high-speed centrifugal impeller that spins within a housing. The gas enters the impeller in the center and is thrown outward by centrifugal force, compressing the gas. The screw compressor uses two rotating screws shaped like augers. Gas is compressed by squeezing it between the two screws. The scroll compressor has a stationary scroll and an orbiting scroll that moves within the stationary scroll. Gas is compressed by squeezing it between the two scrolls.

Figure 14-11 Scroll compressor.

TECH TIP

Air-conditioning and refrigeration compressors have very large motors. Most of these motors are located inside of the refrigerant system so that the refrigerant can keep the motor cooler. Some large compressors actually have the cold suction vapor come into the compressor at the motor end to provide even greater motor cooling.

14.5 HEAT TRANSFER AND COIL DESIGN

Heat is always trying to reach a state of balance by flowing from a warmer object to a cooler object. Heat only flows in one direction, from warmer to cooler. Temperature difference (TD) is what allows heat to flow from one object to another. The greater the temperature difference, the more rapid the heat flow. For the high side of a refrigeration unit to reject heat, its temperature must be above the ambient, or surrounding, temperature. Also, for the evaporator to absorb heat, its temperature must be below the surrounding ambient temperature (see Figure 14-5).

The two factors that affect the quantity of heat transferred between two objects are the temperature difference and the mass of the two objects. The greater the temperature difference between the refrigerant coil and

Figure 14-12 These two condensing units have the same cooling capacity; the larger unit is a higher-efficiency unit.

Figure 14-13 These two evaporators have the same cooling capacity; the larger coil is a new higher-efficiency evaporator coil.

the surrounding air, the more rapid the heat transfer. Increasing the coil size and the amount of refrigerant in the coil also increases the rate of heat transfer. Engineers can either design coils to have high temperature differences or larger areas to increase the heat transfer rate.

New high SEER systems are designed with larger coils to increase energy efficiency. Larger coils can operate with a lower temperature difference between the coil and the surrounding air. Less energy is required to produce the lower condensing temperature, improving system efficiency. Manufacturers of new high-efficiency air-conditioning systems use this principle. That is why the newer high SEER outdoor condensing units are significantly larger than older models having the same capacity (Figure 14-12).

The same principle has been applied to the evaporator coils of new high SEER systems. The temperature difference between the evaporator coil and the entering air is less than on older systems. Higher-temperature evaporator coils can pick up the same amount of heat as lower-temperature coils if the coil has greater surface area. The expanded area allows more refrigerant exposure to the airstream for absorbing heat (Figure 14-13).

SERVICE TIP

The larger size of the new high-efficiency equipment may make replacing older equipment more difficult. The increased size of newer equipment may make it impossible to fit it in the same location as an older, smaller, low-efficiency system. Check with the manufacturer for the size of new equipment before starting to change out older equipment.

Air-conditioning and refrigeration design engineers must take into consideration a variety of factors when designing systems for higher efficiency. For example, the higher evaporative coil temperature may produce less dehumidification. In humid climates, dehumidification can be an important part of the total air conditioning. Manufacturers spend thousands of hours and tens of thousands of dollars researching the effective energy efficiency of systems. These tests are carried out in large calibration rooms where the condenser operates at specific temperature and humidity conditions in one area and the evaporator operates under separate conditions in another area.

The results of manufacturing research are incorporated within the manufacturer's technical data sheets provided to the technician during installation. This material may also be found in AHRI's certified equipment guides.

14.6 THE CONDENSER

The condenser is a device for removing heat from the refrigeration system. In the condenser, the vapor at high temperature and high pressure transfers heat through the condenser tubes to the surrounding medium, usually air or water. The first portion of the condenser cools off the superheated gas and lowers its temperature to the saturation point. This part of the condenser is called the de-superheating section. When the temperature of the vapor reaches the saturation temperature, the additional latent heat removed causes condensation of the refrigerant, producing liquid refrigerant. The refrigerant temperature remains the same while it is changing state. The end of the condenser subcools the refrigerant, lowering the refrigerant temperature below the saturation point. This is called the subcooling section of the condenser. There are three types of condensers: air cooled, water cooled, and evaporative (Figure 14-14). The air-cooled condenser uses air as the condensing medium, the water-cooled condenser uses water as the condensing medium, and the evaporative condenser uses both air and water.

Air-Cooled Condensers

Air-cooled condensers consist of three types: natural draft, forced draft, and induced draft. Natural-draft condensers do

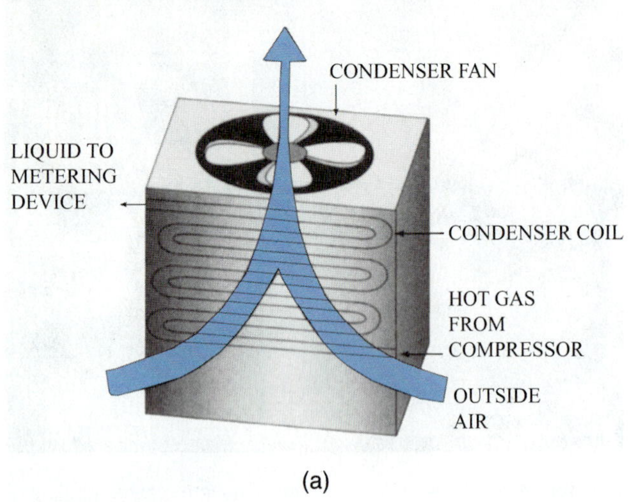

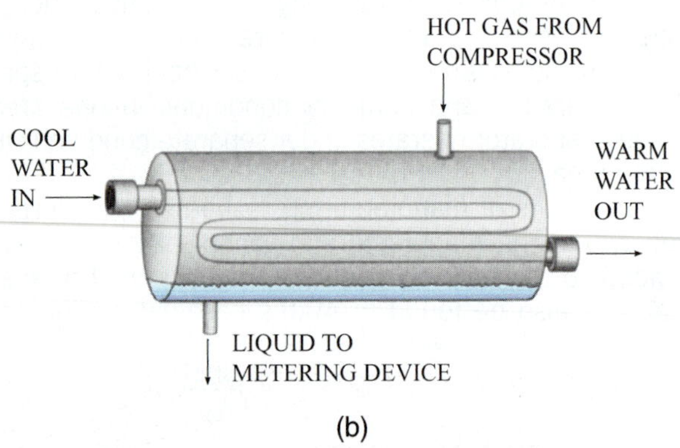

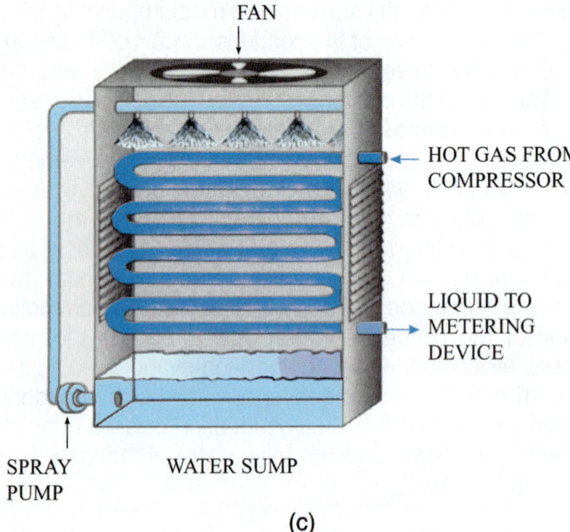

Figure 14-14 Condensers: (a) air cooled; (b) water cooled; (c) evaporative condenser, also called a sump.

Figure 14-15 Natural-draft condenser.

not have fans; they rely on natural air currents for air movement across the condenser (Figure 14-15). Forced-draft condensers use a fan to force air across the condenser. In a forced-draft condenser, the cups of the fan blade will be facing toward the coil (Figure 14-16). Induced-draft condensers use a fan to suck air across the condenser. In an induced-draft condenser, the cups of the fan blade will be facing away from the coil (Figure 14-17).

Figure 14-16 The cups of the fan blades face the coil on a forced-draft condenser.

Figure 14-17 The cups of the fan blades face away from the coil on an induced-draft condenser.

Figure 14-18 The tube-in-tube single-circuit condenser.

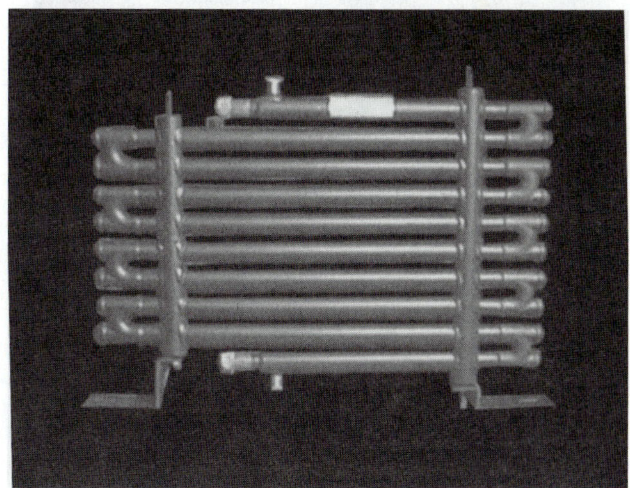

Figure 14-19 Tube-in-tube multiple-circuit condenser. *(Courtesy of Doucette Industries, Inc.)*

SERVICE TIP

Many units with induced-draft condensers require the cabinet panels to be in place for proper airflow. Without the panels in place, air can enter without going across the condenser, reducing the airflow across the condenser.

Water-Cooled Condensers

Water-cooled condensers typically operate at a lower condensing temperature than air-cooled condensers. This produces a lower high-side pressure on water-cooled systems. There are four types of water-cooled condensers: double pipe, open vertical shell and tube, horizontal shell and tube, and shell and coil.

Double-pipe condensers can be a single continuous loop, as shown in Figure 14-18 , or several parallel pipes, as shown in Figure 14-19. In either case, the water and refrigerant should flow through the condenser in opposite directions. This is called counterflow (Figure 14-20). This is more efficient because the coldest water contacts the coldest refrigerant and the hottest water contacts the hottest refrigerant. Single-circuit double-pipe condensers can be constructed using a tube in tube, as shown in Figure 14-21, or two tubes soldered together and wrapped in a protective outer tube (Figure 14-22).

Evaporative Condensers

Evaporative condensers use water evaporation to help cool the refrigerant. Evaporation of the water running over the condenser surface cools the condenser faster than air alone can. Evaporative coolers require makeup water to replace the water that evaporates. Evaporative condensers also require more maintenance than air-cooled systems because the warm open water sump is a perfect place to grow bacteria, mold, and fungus. Also, evaporation causes mineral scale to build up. If the scale is not removed, the condenser's efficiency will drop off.

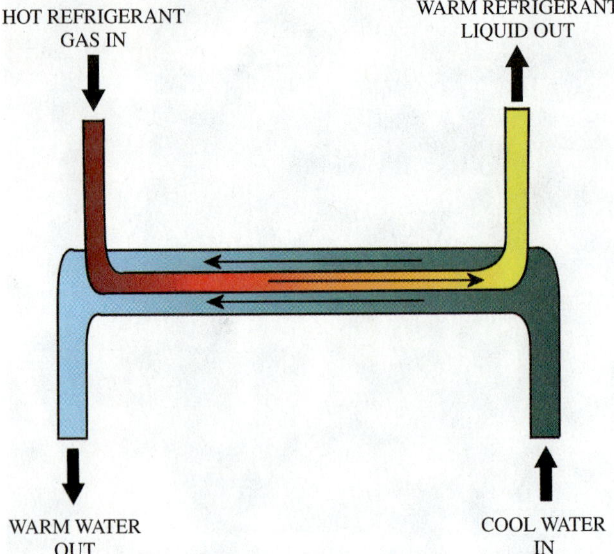

Figure 14-20 The water and refrigerant travel in opposite direction in a counterflow heat exchange.

Figure 14-21 Cross-section of tube-in-tube coil.

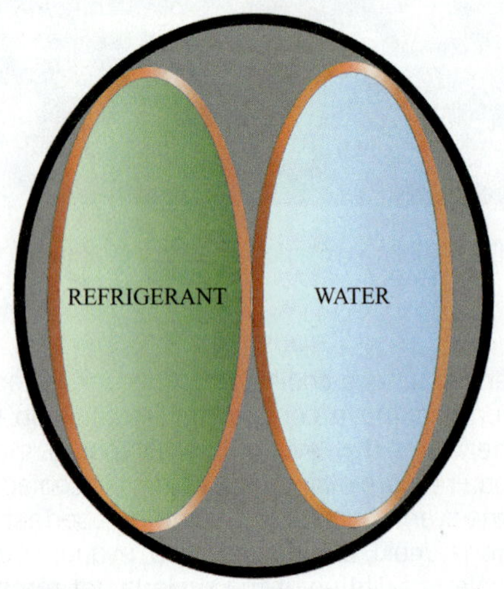

Figure 14-22 Schematic cross-section of tube-by-tube coil.

14.7 THE METERING DEVICE

A metering device controls the flow of refrigerant to the evaporator. It separates the high- and low-pressure sides of the system. High-pressure, high-temperature subcooled liquid enters the metering device and exits as low-pressure, low-temperature saturated mixture. The pressure is low because the compressor is continuously pumping vapor from the evaporator. Two actions occur in the metering device: (1) the pressure of the refrigerant is reduced to a pressure corresponding to the evaporator temperature at the saturated condition, and (2) the refrigerant liquid is cooled to the evaporator temperature by actual evaporation of some of the liquid refrigerant. The saturated mix leaving the metering device is typically 25 percent vapor and 75 percent liquid. There are two general types of metering devices commonly used in modern refrigeration systems: fixed-bore expansion devices and thermostatic expansion valves.

Fixed-Bore Expansion Devices

Fixed-bore expansion devices work by limiting the amount of refrigerant that can pass through the restriction and are best suited for constant heat loads because they cannot adjust to changing load conditions. The pressure difference across a fixed-bore expansion device affects the refrigerant flow through it. A higher pressure difference produces more refrigerant flow; lower pressure differences produce less refrigerant flow. Two types of fixed-bore expansion devices are used: capillary tubes and orifices (Figure 14-23 and Figure 14-24). The capillary tube's length and inside diameter determine the correct amount of liquid flow into the evaporator and the correct pressure drop.

Thermostatic Expansion Valves

A thermostatic expansion valve, TEV, can adjust to varying load conditions and is designed to maintain a constant superheat. The valve measures the suction line temperature with a refrigerant-filled sensing bulb that installs on the suction line leaving the evaporator. The pressure from the sensing bulb is balanced by the evaporator pressure and the adjustment spring pressure. The

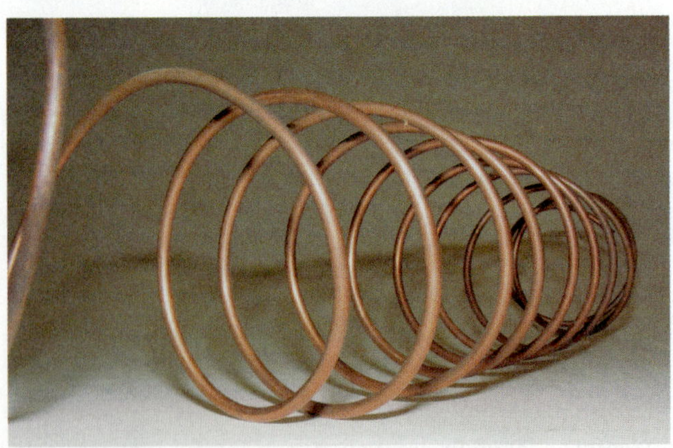

Figure 14-23 Capillary tube metering device.

Figure 14-24 Orifice-type metering device.

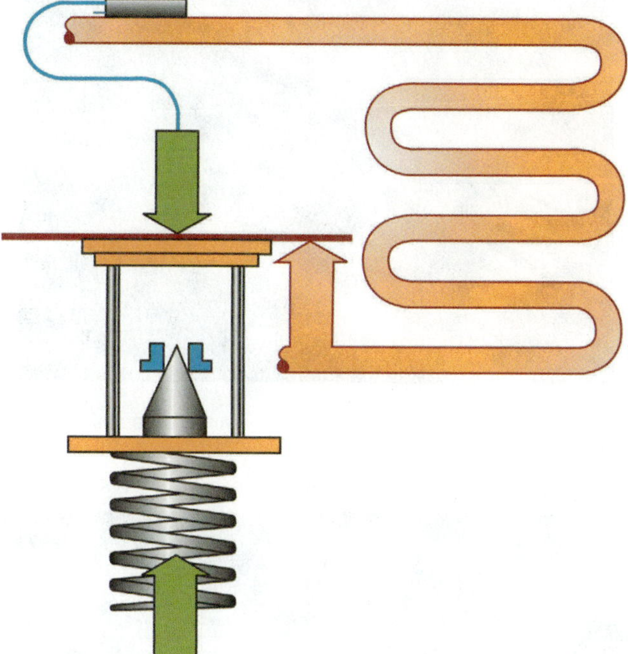

Figure 14-25 The TEV bulb pressure is balanced by the evaporator pressure and spring pressure.

sensing bulb pressure tends to open the valve as the suction line temperature increases; the evaporator pressure tends to close the valve as the evaporator pressure increases (Figure 14-25).

TEVs are used on systems with varying loads because of their ability to adjust to varying load conditions. They are also used frequently on high-SEER systems because of their ability to allow a fully active evaporator under all load conditions (Figure 14-26).

14.8 THE EVAPORATOR

The evaporator is a device for absorbing heat into the refrigeration system. In the evaporator, the saturated refrigerant absorbs heat from its surroundings and boils into a low-pressure vapor. Refrigerant enters the evaporator from the metering device as a saturated mixture. The saturated liquid turns to saturated vapor as the refrigerant travels through the evaporator. The refrigerant is superheated in

Figure 14-26 Thermostatic expansion valve on high-efficiency coil.

the end of the evaporator after all the liquid has boiled off. Some superheat is desirable to ensure that no liquid enters the compressor because liquid refrigerant could damage the compressor.

TECH TIP

Because it takes heat to vaporize water, removing humidity from air also removes heat from the air. The more humidity there is in air, the more heat the air contains. Condensing water on an evaporator coil uses system cooling capacity and reduces the amount of cooling capacity available for changing the air temperature. As the relative humidity of the return air increases, the temperature drop across the evaporator coil decreases.

The evaporator or cooling coil is fabricated from metals such as copper or aluminum or both. These metals are selected because of their good thermal conductivity. Although there are many variations and modifications of evaporators, there are three basic types of construction: bare pipe, finned tube, and plate (Figure 14-27). On finned-tube evaporators, the tubing is interconnected by aluminum fins that serve to both direct the airflow through the coil and increase heat transfer by increasing the surface area of the evaporator (Figure 14-28). Like condensers, evaporators can be forced air or natural draft.

Most evaporators operate below the dew point of the air being cooled. This causes water to condense on them.

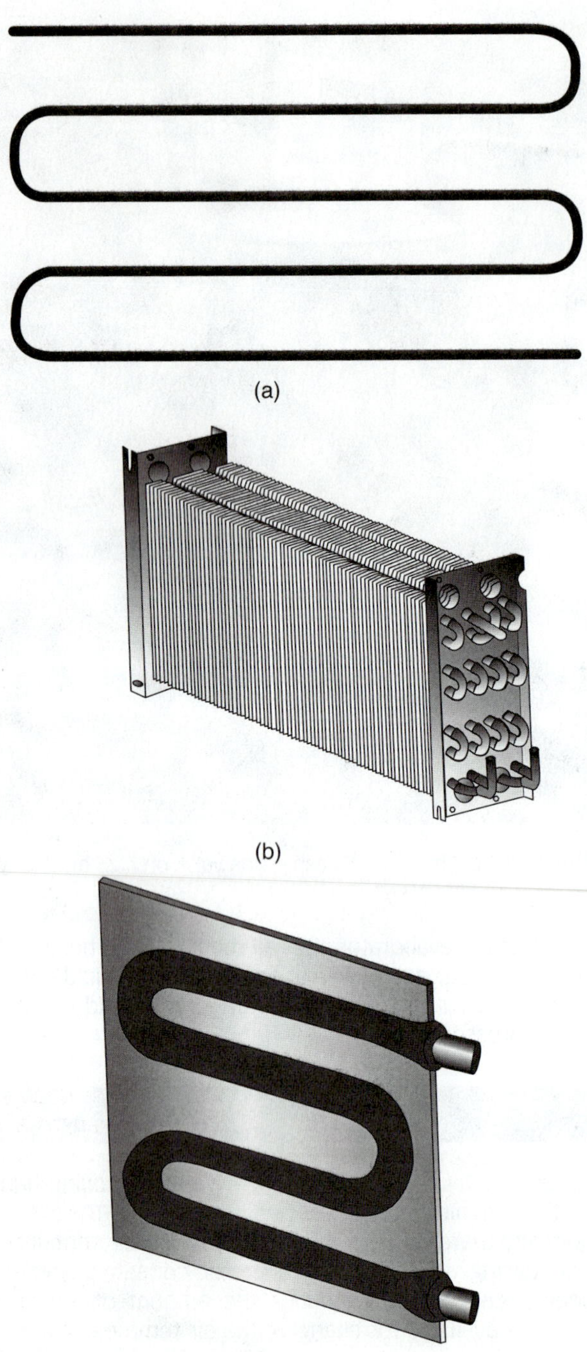

Figure 14-27 Different types of evaporators; (a) bare pipe; (b) finned tube; (c) plate.

Figure 14-28 Cross-section of typical finned-tube coil construction.

Figure 14-29 Evaporator-coil drain pan.

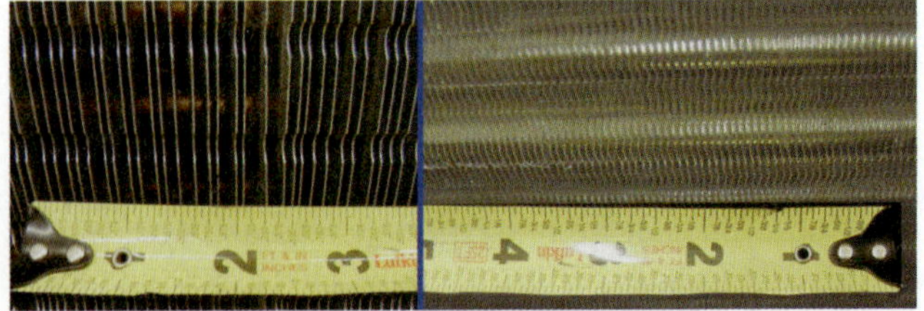

Figure 14-30 The fins are spaced wider apart on coils that operate below freezing.

The condensed water vapor drains into a collection pan and is routed to a drain by tubing (Figure 14-29).

The fin spacing on an evaporator is partly determined by its intended operating temperature. Evaporators that constantly operate below freezing have a wider fin spacing, so frost accumulation does not interfere as much with airflow. Evaporators that typically operate above freezing can use a closer fin spacing because frost does not accumulate. Figure 14-30 shows the difference in fin spacing between a frosting and nonfrosting evaporator coil.

Figure 14-31 The suction line entering the compressor is the large line; the discharge line leaving the compressor is the small line.

Figure 14-32 Suction and liquid lines on a residential split-system air conditioner.

14.9 REFRIGERANT LINES

The components of the refrigeration system are connected by three refrigerant lines: the discharge line, the liquid line, and the suction line. The discharge line is the smaller of the two lines connected to the compressor (Figure 14-31). The discharge line carries the high-pressure, hot, superheated gas from the compressor to the condenser. It is called the discharge line because refrigerant is discharged from the compressor into it. The liquid line carries the high-pressure, warm, subcooled liquid from the condenser to the metering device. The suction line is the larger of the two lines connected to the compressor (Figure 14-31). The suction line carries the low-pressure, cool, superheated vapor from the evaporator coil to the compressor inlet.

Units with all the refrigeration components in one piece of equipment are called packaged units. The discharge, liquid, and suction lines of a packaged unit are contained within the unit. A split system is a unit with parts of the refrigeration system located in different pieces of equipment. Most split-system refrigeration units have the evaporator and metering device located inside and the condenser and compressor located outside. The piece of equipment located outside is called a condensing unit. The two lines connecting these systems are the large suction line and the small liquid line (Figure 14-32). The discharge line is not actually an external line in these systems but part of the outdoor condensing unit. A few split systems locate all the components except the condenser in one piece of equipment and locate the condenser in a separate piece of equipment. These systems also have two lines connecting them together, but they are the lines that go to and from the condenser: the discharge line and the liquid line. On these systems the suction line is an internal line.

Split-system heat pumps have two lines. The large line's function changes depending upon the operating cycle of the unit. During cooling, the large line is the suction line and the small line is the liquid line. During heating, the large line is the discharge line and the small line is still the liquid line. To avoid confusion, the large line on split-system heat pumps is often called the gas line because that is what travels through it in either cycle.

14.10 REFRIGERANT STORAGE

The capacity of many refrigeration systems varies depending upon their operating conditions. Capacity changes also produce different refrigerant flows. An increase in refrigeration capacity circulates more refrigerant, while a decrease in capacity circulates less refrigerant. This can make it difficult for systems operating under a broad range of temperature conditions to maintain the desired level of subcooling and superheat. The extra refrigerant needs a place to sit when it is not circulating. Liquid receivers and suction accumulators provide a place for the surplus refrigerant to sit until it is needed. Systems do not usually have both receivers and accumulators. Typically, systems using thermostatic expansion valves use liquid receivers, and systems using fixed bore expansion devices use accumulators.

Liquid Receivers

Liquid receivers are located at the end of the condenser outlet to collect liquid refrigerant (Figure 14-33). The liquid receiver allows the liquid to flow into the receiver and any vapor collected in the receiver to flow back into the condenser to be converted back into a liquid. The line connecting the receiver to the condenser is called the condensate line and must be large enough in diameter to allow liquid to flow into the receiver and vapor to flow back into the condenser. The condensate line must also have a slope

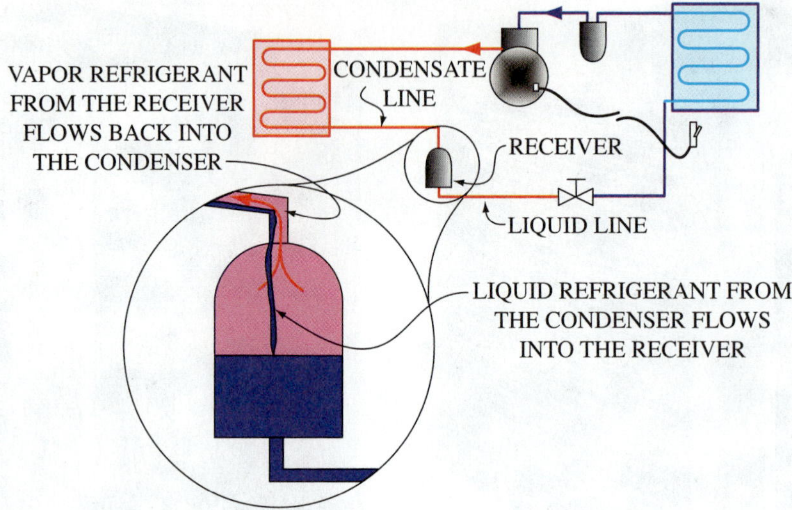

Figure 14-33 Liquid refrigerant drains into the receiver, and vapor flows back to the condenser through the condensate line.

toward the receiver to allow liquid refrigerant to freely flow from the condenser into the receiver. The outlet side of the receiver is located at the bottom, where 100 percent liquid can flow out of the receiver into the liquid line.

Receivers should be sized so that 100 percent of the refrigerant charge can be stored in the receiver. When the receiver is properly sized, it can be used to store the refrigerant during a pump-down cycle or for some types of approved system service.

Some refrigeration-condensing units come with receivers built into the base of the condensing unit (Figure 14-34).

Accumulators

An accumulator is a device located at the end of the evaporator that allows liquid refrigerant to be collected in the bottom of the accumulator and remain there as the vapor refrigerant is returned to the compressor. The inlet side of the accumulator is connected to the evaporator, where any liquid refrigerant and vapor flow in. The outlet of the accumulator draws vapor through a U-shaped tube or chamber (Figure 14-35). There is a small port at the bottom of the U-shaped tube or chamber that allows liquid refrigerant and oil to be drawn into the suction line. Without this small port, refrigerant oil would collect in the accumulator and not return to the compressor. The small port does allow some liquid refrigerant to enter the suction line. However, it is such a small amount of liquid refrigerant that it boils off rapidly, so there is no danger of liquid refrigerant flowing into the compressor.

Accumulators are often found on heat pumps. During the changeover cycle, liquid refrigerant can flow back out of the outdoor coil. This liquid refrigerant could cause compressor damage if it was not for the accumulator, which blocks its return.

Figure 14-34 Receiver tank as part of a refrigeration condensing unit.

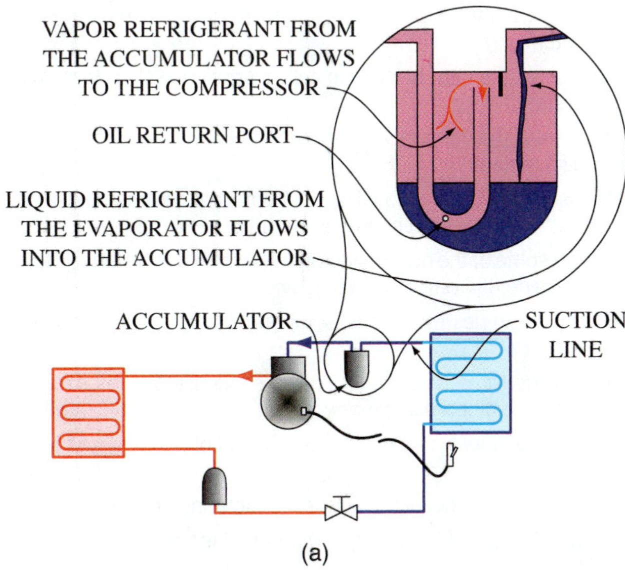

Figure 14-35 Unevaporated liquid refrigerant flows into the accumulator; vapor refrigerant is drawn off the top of the accumulator. A screen covers the oil return port to prevent debris in the system from plugging the port; (a) diagram; (b) accumulator.

14.11 OTHER REFRIGERATION SYSTEM COMPONENTS

All mechanical refrigeration systems have a compressor, a condenser, a metering device, and an evaporator, but many systems have additional components that help the system perform its job. Other refrigeration system components that may be found on some refrigeration systems include discharge-line muffler, oil separator, liquid receiver, liquid line filter drier, sight glass, liquid line solenoid, suction to liquid heat exchanger, evaporator pressure regulator, suction line filter, suction accumulator, crankcase pressure regulator, and oil-level control. Some of these components improve the efficiency or reliability of the refrigeration system, while others are essential for specialized systems to operate correctly.

Optional High-Side Components

The discharge-line muffler dampens pulsations and noise created by the compressor. The oil separator removes the oil entrained in the discharge gas and returns it to the compressor crankcase. The liquid line filter drier removes moisture and impurities from the liquid refrigerant and helps protect the system, especially the metering device. A sight glass allows the technician to visually see the condition of the refrigerant in the line. Many sight glasses also have a moisture indicator that changes color when exposed to moisture. A liquid line solenoid is an electrically operated valve used to control the flow of liquid refrigerant. With a normally closed solenoid valve, refrigerant will only flow through when the solenoid is energized. When the solenoid is de-energized, refrigerant does not flow through the solenoid valve. The suction to liquid heat exchanger exchanges heat between the warm liquid line and the cool suction line. This increases both the suction gas superheat and the liquid line subcooling.

Optional Low-Side Components

The evaporator pressure regulator, EPR, controls the pressure in the evaporator. It will not allow the evaporator pressure to drop below the EPR setpoint. EPR valves are used on systems with multiple evaporators operating at different pressures and temperatures. The evaporators share a common suction line but are able to maintain separate evaporator pressures because of the evaporator pressure regulators. Suction line filters are used to protect the compressor from contaminants. Typically, suction line filters are used when replacement compressors are installed. The filter is removed after the system has operated long enough for the filter to do its job. Suction line accumulators are used to ensure that no liquid enters the compressor. They are commonly used on heat pumps. Crankcase pressure regulators, CPR, are valves used to keep the suction pressure to the compressor from rising above the setpoint of the CPR valve. CPR valves are commonly used on commercial freezers. They keep the compressor from being overloaded when the evaporator temperature is above its normal operating temperature. Oil-level controls are used on systems with multiple compressors that are piped into common discharge and suction headers.

UNIT 14—SUMMARY

The mechanical vapor-compression–refrigeration cycle is the most common of all refrigeration cycles. The four main components are the compressor, the condenser, the metering device, and the evaporator. The refrigeration cycle is made possible by the pressure difference, which is maintained by the compressor and metering device. The compressor increases the refrigerant pressure by squeezing it; the metering device reduces the refrigerant pressure by restricting its flow into the evaporator. The low side absorbs heat when the refrigerant boils in the evaporator; the high side rejects heat when the refrigerant condenses in the condenser. The refrigerant is a superheated gas leaving the evaporator, through the compressor, and going up to the condenser. The refrigerant is saturated in the condenser, where it changes state from a gas to a liquid. The refrigerant is subcooled in the last portion of the condenser. It remains a subcooled liquid, leaving the condenser up to the expansion device. It is saturated, leaving the metering device due to a drop in pressure. The refrigerant is saturated throughout the evaporator until the last 10 percent, where it picks up superheat before returning to the compressor.

UNIT 14—REVIEW QUESTIONS

1. Briefly explain the fundamental principles that make the compression-refrigeration cycle work.
2. List the four major components of the compression-refrigeration cycle in order.
3. What components are responsible for maintaining the high and low pressures in a compression-refrigeration system?
4. Where in the compression-refrigeration cycle is the refrigerant superheated?
5. Why is superheating desirable?
6. Where in the compression-refrigeration cycle is the refrigerant subcooled?
7. Why is subcooling desirable?
8. Where in the compression-refrigeration cycle is the refrigerant saturated?
9. Explain the heat balance in a compression-cycle refrigeration system.
10. What is the difference between a liquid receiver and a suction accumulator?
11. List some of the optional components that can be found on the high side of a refrigeration system.
12. List some of the optional components that can be found on the low side of a refrigeration system.
13. List the types of compressors, and briefly explain how they operate.
14. List the three types of air-cooled condensers, and briefly explain the difference between them.
15. List the different type of water-cooled condensers, and give a description of each type.
16. What are the two large classes of metering devices?
17. Which type of metering device provides the most efficient operation?
18. Compare coil design between older, less-efficient air-conditioning systems and newer, high-SEER air-conditioning systems.
19. Name the three common refrigerant lines found on compression systems.
20. Which line is the larger of the two lines going to the compressor?
21. Which two lines connect the inside and outside units of a split system?
22. Why is the refrigeration cycle called a cycle?
23. Draw the basic refrigeration cycle. Label the four main components and the refrigerant lines. List the pressure, temperature, and state going into and out of each major component. Be sure to indicate where the refrigerant is superheated, saturated, or subcooled. Note: The pressure and temperature can simply be "high" or "low." Specific numbers are not required.

UNIT 15
Compressors

OBJECTIVES

After completing this unit, you will be able to:

1. identify the major types of compressors.
2. explain how each type of compressor works.
3. give examples of applications for each type of compressor.
4. explain horsepower and compressor capacity.
5. explain common compressor problems.

15.1 INTRODUCTION

The compressor is a mechanical device for pumping refrigerant vapor from the low-pressure side in the evaporator to the high-pressure side in the condenser. The word *compress* means to squeeze. The compressor reduces the volume of the refrigerant gas by squeezing it. Since gas volume, pressure, and temperature are related, reducing the volume causes an increase in both pressure and temperature.

Compressors are identified by the mechanical parts that perform the actual pumping of the refrigerant vapor. In the reciprocating compressor, a piston travels back and forth in a cylinder (Figure 15-1). The scroll compressor has a stationary scroll and an orbiting scroll that moves within the stationary scroll (Figure 15-2). The centrifugal compressor uses an impeller rotating within a housing at very high speed, as shown in Figure 15-3. The screw compressor uses a rotating screw within a tapered housing (Figure 15-4). The rotary compressor uses an offset rolling piston that orbits within a cylinder (Figure 15-5).

The type of compressor used depends on the application. Application factors include the type of refrigerant, the size (tonnage), the pressure difference, the evaporator and condenser saturation temperatures, and the type of cooling available for the compressor. Figure 15-6 indicates the various applications for the different types of compressors.

15.2 COMPRESSOR TYPES

Compressors can be classified by their physical and operational characteristics. They are classified by general operating principle, physical enclosure, mechanical design, cooling mechanism, and evaporator temperature. The different classifications are listed below.

General Operating Principle

- Positive displacement
- Nonpositive displacement

Physical Enclosure

- Open
- Hermetic
- Semihermetic

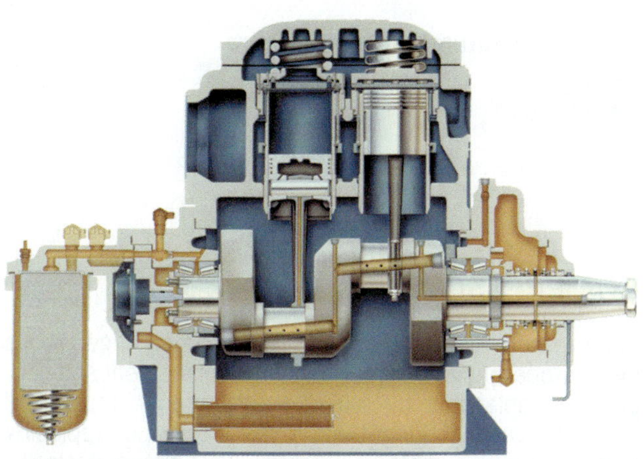

Figure 15-1 Open-drive reciprocating compressor. *(Emerson Climate Technologies)*

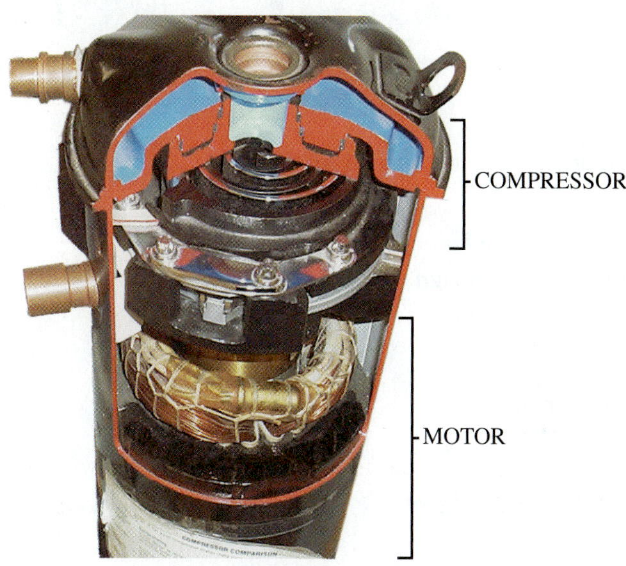

Figure 15-2 Scroll compressor.

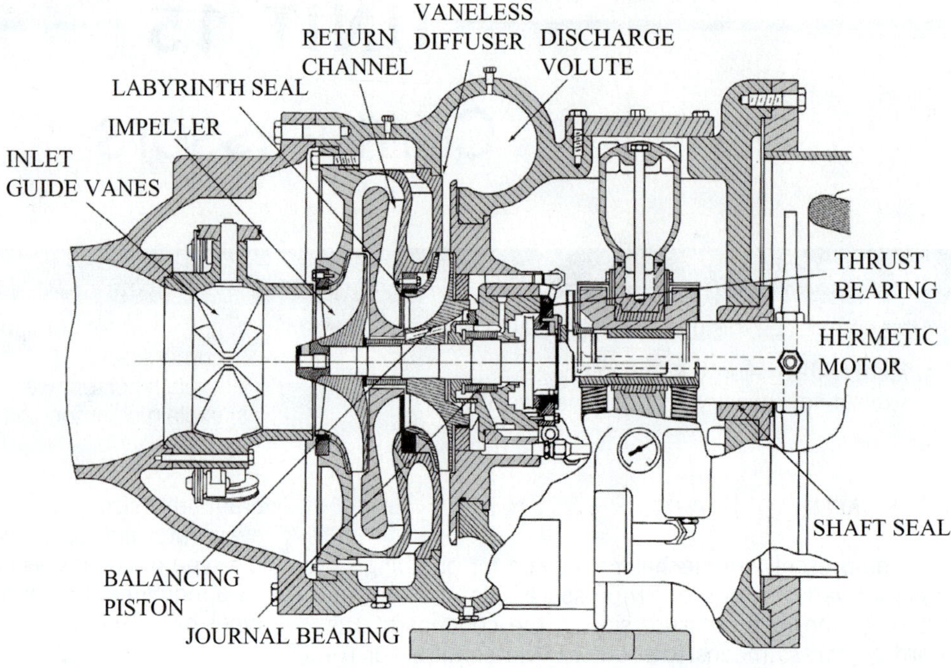

Figure 15-3 Centrifugal compressor.

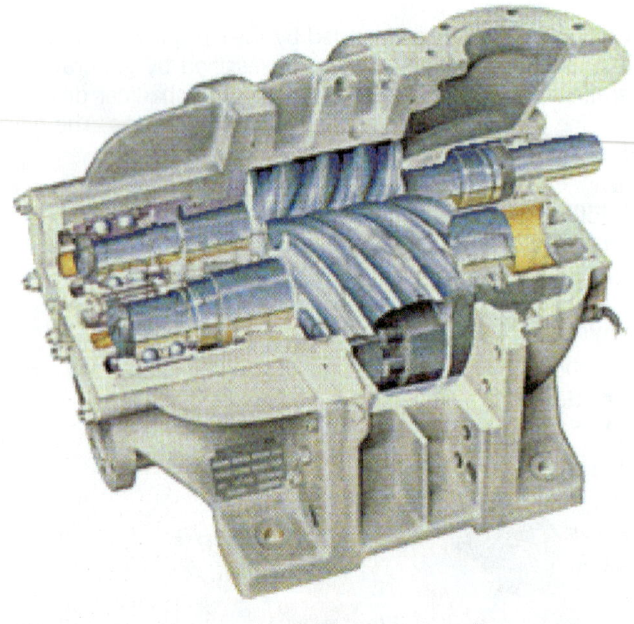

Figure 15-4 Screw compressor.

Figure 15-5 Rotary compressor.

Mechanical Design

- Reciprocating
- Rotary
- Scroll
- Screw
- Centrifugal

Cooling Mechanism

- Air
- Water
- Refrigerant

Evaporator Saturation Temperature

- High
- Medium
- Low
- Ultralow

15.3 POSITIVE-DISPLACEMENT COMPRESSORS

Compressors are commonly divided into two large groups by their operating principle: positive displacement and non-positive displacement. Positive-displacement compressors move the gas by taking it into a space and then replacing that space with something physical, like a piston. Basically, whatever is in the space is positively shoved out by the mechanism. Blocking the discharge port of a positive-

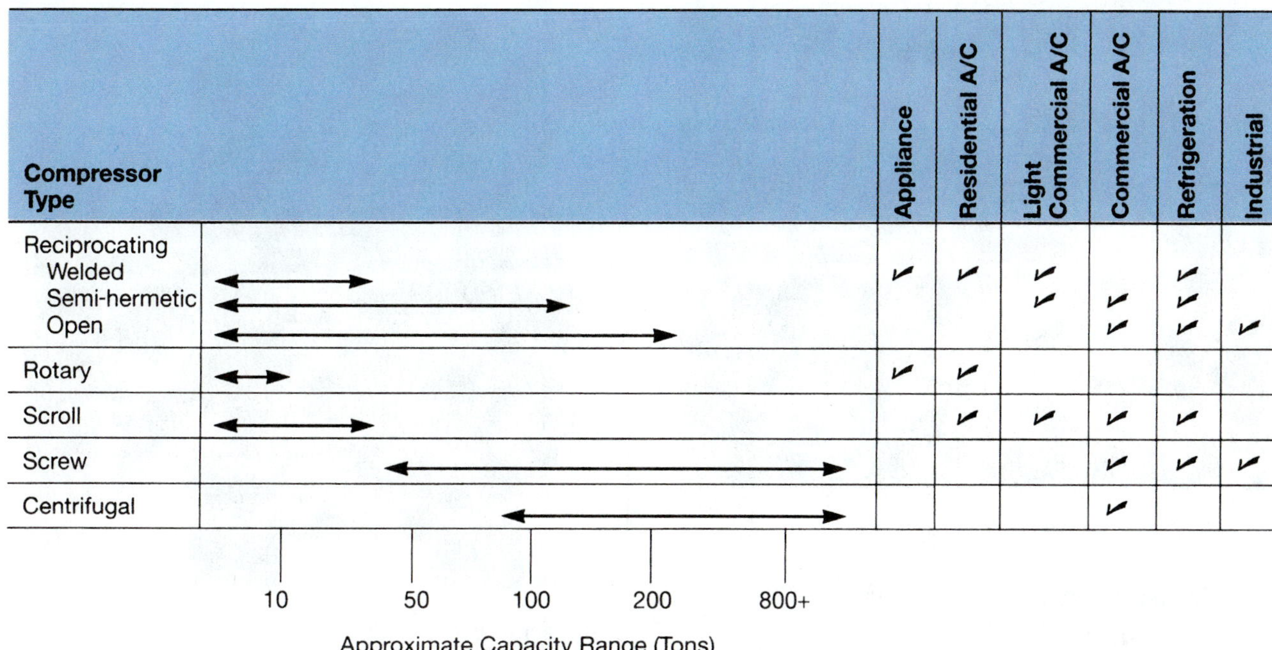

Compressor Type	Appliance	Residential A/C	Light Commercial A/C	Commercial A/C	Refrigeration	Industrial
Reciprocating						
Welded	✓	✓	✓		✓	
Semi-hermetic			✓	✓	✓	
Open				✓	✓	✓
Rotary	✓	✓				
Scroll		✓	✓	✓	✓	
Screw				✓	✓	✓
Centrifugal				✓		

Figure 15-6 Compressor application by types.

displacement compressor can cause compressor damage, because the gas must go somewhere.

SAFETY TIP

Closing off or blocking the discharge valve on a positive-displacement compressor while it is operating can be dangerous. With no place to go, the discharge pressure increases rapidly, creating extremely high pressures.

15.4 NONPOSITIVE-DISPLACEMENT COMPRESSORS

Nonpositive-displacement compressors do not fill the gas space with something else. These machines operate like giant fans or turbines turning at very high speeds. They basically throw the refrigerant outward, creating an increase in pressure and temperature. This outward force is like the centrifugal force you feel on those spinning pieces of playground equipment, merry-go-rounds. These compressors are called centrifugal compressors after the physical force that makes them work. The discharge of a centrifugal compressor should not be blocked because it can set up a condition known as surging, which can damage the compressor.

Figure 15-7 Open compressor.

15.5 OPEN-DRIVE COMPRESSORS

The oldest compressor design is the open-type compressor shown in Figure 15-7. Open compressors may be driven by means of a pulley on a shaft that extends through a seal. The pulley is driven by a separate motor or engine. Open compressors have performed reliably over the decades but have been largely superseded by newer designs. Open machines (unlike hermetic) can be used with ammonia (NH_3) refrigerant as long as all components and accessories are made of iron or steel (or other suitable materials). Open compressors range in size from less than a ton capacity to several hundred tons in capacity. They are often used for refrigeration and industrial applications where rugged,

Figure 15-8 Industrial-duty open-type compressor.

serviceable machinery is required. Figure 15-8 shows an industrial-duty open compressor.

All open compressors use a shaft seal to prevent refrigerant from leaking around the crankshaft. These seals are subject to forming leaks and need to be checked during planned maintenance or service.

In direct-drive compressors (Figure 15-9), the motor shaft is coupled to the compressor shaft and driven at motor speed.

Belt-driven machines offer the flexibility of selecting a compressor speed to match the load. The belts require additional space and a protective guard. They require increased maintenance and increased power due to operational losses. An example of a belt-driven compressor is shown in Figure 15-10.

One advantage of an open-drive compressor is that the motor and compressor are separate units so that the motor or compressor can be repaired or replaced without having to buy a complete system. Another advantage is that the motor's heat is not absorbed in the refrigerant, so it does not have to be rejected in the condenser like most other compressor types.

The most common use of open compressors is for automobile air conditioning. In automobiles, the air-conditioning compressor is driven by a pulley and belt off of the main vehicle engine. Open compressors are also used for transport

Figure 15-9 Direct-drive compressor. *(Emerson Climate Technologies)*

Figure 15-10 Open-type direct-drive compressor.

refrigeration on trucks, tractor trailers, and shipping containers. These applications generally use a separate engine to drive the compressor.

15.6 HERMETIC COMPRESSORS

Hermetic compressors have their mechanical components and electric motor enclosed in a single welded shell (Figure 15-11). The compressor crankshaft extends into the rotor of the electric motor, eliminating the need for drive components. The benefits of hermetic compressors include:

- Reliable operation due to the direct connection between the motor and compressor with no belts or couplings to wear or break
- No shaft seal to leak because the motor and compressor are sealed in a welded steel shell
- Very compact in size, permitting increased refrigerated storage space
- Lower sound level, ideal for domestic appliances

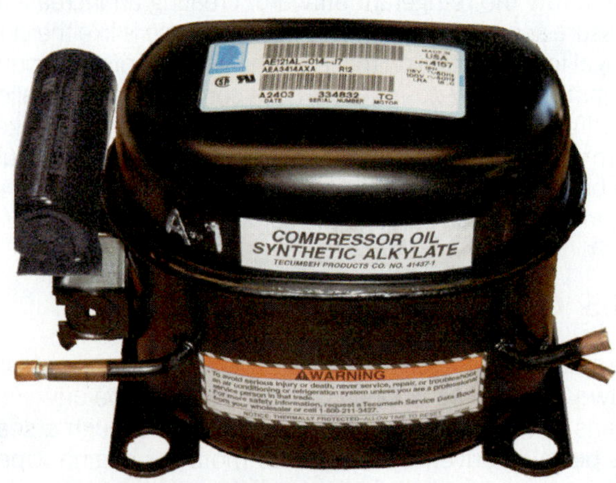

Figure 15-11 Hermetic compressor.

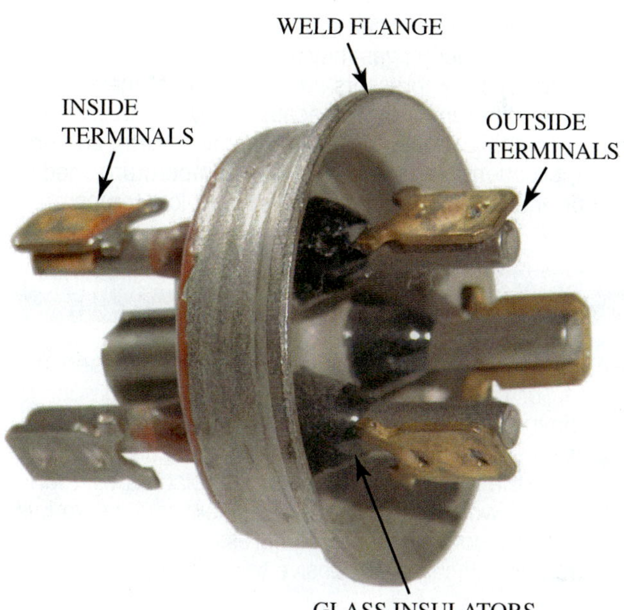

Figure 15-12 A glass-to-metal connection is used to isolate the electrical terminals from the metal compressor shell.

One challenge with hermetic compressor design is to pass electrical connections through the welded steel shell without grounding out to the shell. This is accomplished by passing the electrical connections through glass or ceramic insulators. High-strength glass insulates the terminals from the metal that the terminals pass through (Figure 15-12).

SAFETY TIP

The terminal cover should always be in place when operating a hermetic compressor. Abnormally high current can create enough heat to crack the glass holding the terminals, and the electrical terminals can shoot out under high pressure. This is usually followed by ignition of the refrigerant oil. This acts like a bullet followed by a flamethrower. This is known as terminal venting, a rare but potentially deadly occurrence.

The shells on reciprocating hermetic compressors are called low-side shells because the suction gas enters the shell. The motors in hermetic compressors are cooled by the refrigerant vapor traveling over them. Operating a hermetic compressor on a low charge can cause the motor to overheat because of poor motor cooling. Many hermetic compressor manufacturers do not like their compressors operating in a vacuum. The motor varnishes used in the motor winding lose some of their insulating ability under deep vacuums, and motors can arc out very quickly, even before they have time to overheat. Keeping the refrigerant clean is especially important in a system with a hermetic compressor because the refrigerant and oil flow directly over the motor and all internal electrical connections.

Hermetic compressor units (Figure 15-13) are made in a variety of sizes, from tiny fractional horsepower units meant for small appliances to larger units up to about 20 tons for

Figure 15-13 Hermetically sealed electric motor compressor units. *(Courtesy Danfoss Inc.)*

air-conditioning use. They are sometimes called welded hermetics, full hermetics, or sealed hermetic compressors.

The welded steel shell prevents any field service access, so there are no service procedures to replace damaged internal components such as motors, bearings, valves, and so on. If damaged or defective, the entire hermetic compressor is replaced. Reciprocating hermetic compressors are usually internally spring isolated to reduce the inherent vibration caused by the reciprocating action of the pistons (Figure 15-14).

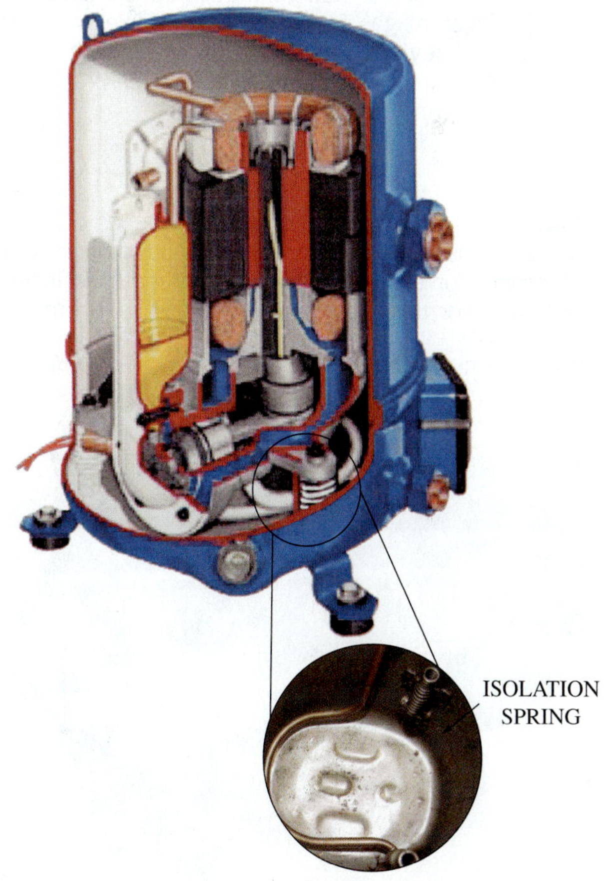

Figure 15-14 Cutaway view of hermetic compressor showing internal isolation spring inset.

15.7 SEMIHERMETIC COMPRESSORS

Semihermetic compressors also have the motor and mechanics in one enclosure, but the enclosure has some form of access, like bolted plates and gaskets. The semihermetic motor compressor unit, which is field serviceable by virtue of its bolted construction, has evolved over the past fifty to sixty years. It may have a bolted cast-iron construction or a bolted, flanged, drawn steel shell. An advantage of semihermetic compressors is that they provide the advantages of hermetic compressors but are still field or factory repairable. Parts such as valve reeds, gaskets, bearing inserts, or motor stators may be replaced on various units. These machines are known by a variety of names: semihermetic, serviceable hermetic, accessible hermetic, and bolted hermetic. The units are available in sizes that range from approximately 0.25 to 125 tons cooling capacity. Figure 15-15 shows a semihermetic compressor used for medium and low-temperature refrigeration.

Cooling

Semihermetic compressors can be classified by their means of motor cooling: air cooled or refrigerant cooled. Air-cooled semihermetic compressors use airflow over the motor portion of the compressor to cool the compressor motor. The refrigerant goes straight into the cylinders on these compressors. Maintaining a minimum superheat of 20°F at the compressor is critical for air-cooled semihermetic compressors. Refrigerant-cooled compressors use the refrigerant gas for cooling, much like hermetic compressors. They can operate safely at a much lower superheat because the refrigerant goes over the motor first before entering the cylinders.

Duty

The size and type of motor a compressor needs depends upon the job it is required to do. Hermetic and semihermetic compressors come with a motor already matched to the compressor. For this reason, hermetic and semihermetic compressors are classified by the saturation temperature of the returning suction gas they are designed to handle. Two semihermetic compressors can be identical mechanically but have different motors because they are intended for different applications. In general, semihermetic compressor applications are classified as high temperature, medium temperature, low temperature, and ultra-low temperature.

Figure 15-15 Semihermetic compressor used for medium- and low-temperature refrigeration.

SERVICE TIP

Although semihermetic compressors are field serviceable, for cost reasons many are removed and sent to companies that specialize in remanufacturing compressors. These facilities are set up to thoroughly clean and inspect the entire compressor and replace those parts that have worn beyond tolerance. Field service work is commonly done for large compressors that would be extremely difficult or expensive to remove.

15.8 RECIPROCATING COMPRESSORS

Reciprocating compressors have been used in refrigeration service for a long time. The reciprocating compressor is used in the majority of domestic, small commercial, and industrial condensing unit applications. Reciprocating compressors are available in open, hermetic, and semihermetic designs.

In a reciprocating compressor the piston is driven up and down in the cylinder by the connecting rod and crankshaft (Figure 15-16). Besides the compressor body, the

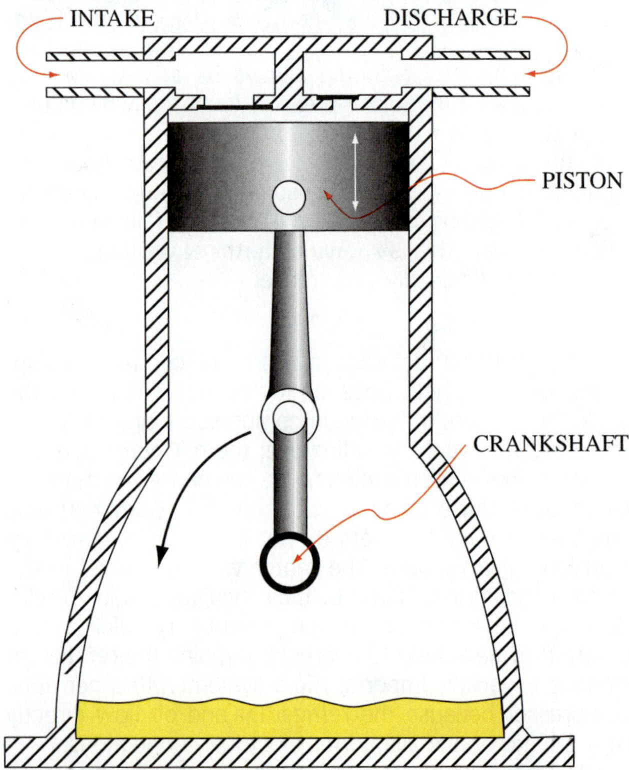

Figure 15-16 The crankshaft rotates while the piston moves up and down.

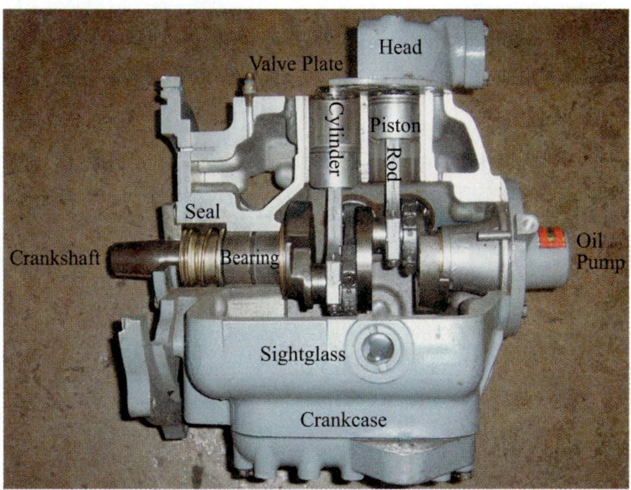

Figure 15-17 Reciprocating compressor parts.

Figure 15-19 Eccentric crankshaft.

basic mechanical components of a reciprocating compressor are the crankshaft, connecting rods, piston, wrist pin, valves, and valve plate (Figure 15-17).

The crankshaft operates like a rotating lever. Two types of crankshafts are the crankthrow and the eccentric. Crankthrow crankshafts look like the crankshaft on an engine; they have offset sections where the rods attach (Figure 15-18). The ends of the rods bolt around the crankshaft on a crankthrow crankshaft. Eccentric crankshafts look like the cam on an engine. An eccentric crankshaft is a straight shaft with offset lobes (Figure 15-19). The connecting rods slide over these lobes. The connecting rods used with eccentric crankshafts are one piece; they slide over the crankshaft when the compressor is assembled. An eccentric crankshaft provides smoother operation than a crankthrow crankshaft.

The connecting rods are either two piece for crankthrow crankshafts or one piece for eccentric crankshafts (Figure 15-20). The shaft on some connecting rods has a hole that carries oil up the piston wrist pin. Other systems use oil in the refrigerant vapor inside the compressor to lubricate the piston wrist pin.

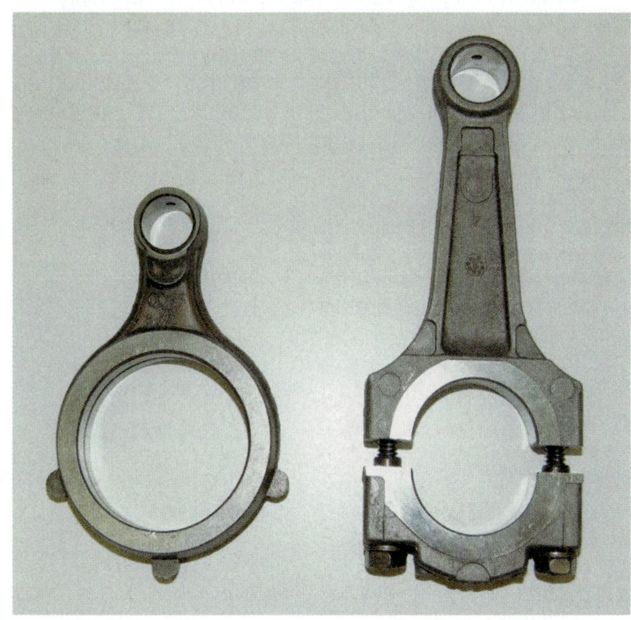

Figure 15-20 One-piece and two-piece connecting rods.

Figure 15-18 Crankthrow crankshaft.

The pistons are usually manufactured slightly smaller in diameter than the cylinder they will operate in. Piston rings are used on large compressors to fill this gap (Figure 15-21). Small compressors do not have rings but "oil grooves" where oil collects to help block gases from "blowing by" the pistons during the compression stroke.

The valve plate serves as the top of the cylinder and holds the valves. Typically, the discharge valves are located on top of the valve plate and the suction valves are located on the underside of the valve plate. The valves are self-operating: they open and close due to pressure difference. They are sometimes spring loaded, but the principle is the same. Note that this is quite different from an automobile engine, where the valves are mechanically opened and closed by a camshaft synchronized to the crankshaft.

Figure 15-22 shows the different types of reciprocating cylinder valves: (a) the flexing reed valve; (b) the floating

Figure 15-21 Compressor piston with rings.

reed valve; (c) the ring valve; and (d) the reduced-clearance poppet valve used for the discus compressor.

Pistons within the compressors may have the suction valve located in the top of the piston; this is classified as a valve-in-head type, or the piston may have a solid head, with the suction and discharge valves located in a valve plate or cylinder head.

Figure 15-23 shows the complete compression cycle of a reciprocating piston compressor. Position (a) shows the beginning of the intake cycle; the suction valve has not opened because the pressure in the cylinder is still above the system suction pressure. As the piston strokes down in the cylinder, the pressure will drop to a point where it is below the pressure in the suction manifold. At (b), the higher pressure in the suction line forces the suction valve open, and cold (but superheated) low-pressure suction gas from the evaporator will flow into the cylinder. At (c), the piston continues downward, drawing in refrigerant vapor. At (d), the refrigerant pressure above the rising piston forces the suction valve closed and the discharge valve open. At (e), the rising piston pushes the high pressure and temperature refrigerant out the discharge valve. At position (f), the suction valve remains closed and the discharge valve also closes because the system discharge pressure is greater than the pressure left in the cylinder.

15.9 CYLINDER GAS REEXPANSION

The clearance volume is the space between the top of the piston and the cylinder head. During the compression stroke, gas leaves the cylinder and passes into the discharge manifold until the pressure in the cylinder is the same as

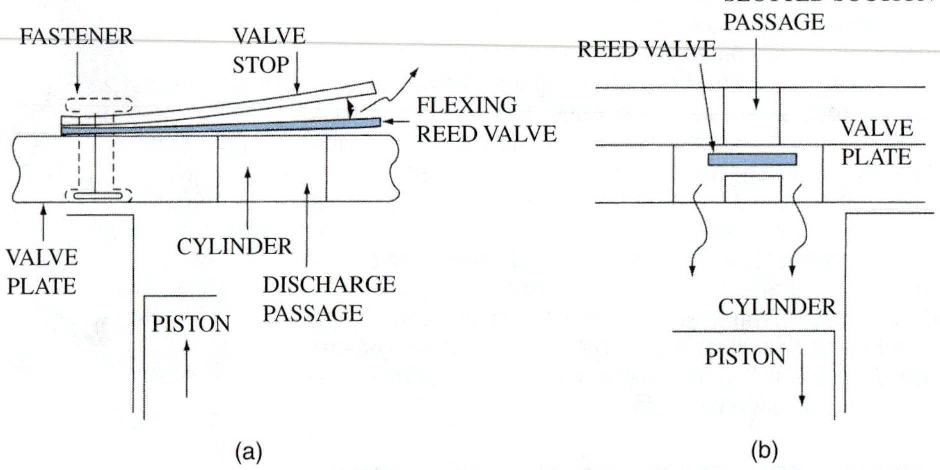

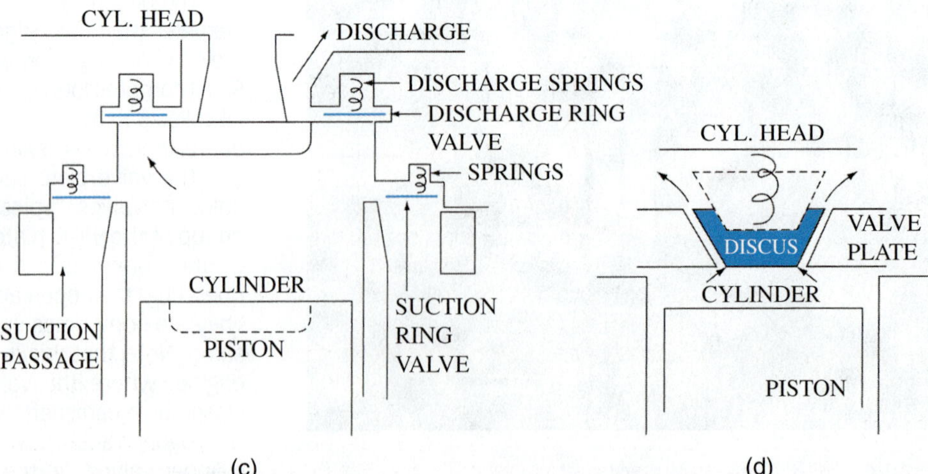

Figure 15-22 Typical reciprocating compressor cylinder valves: (a) flexing reed valve; (b) floating reed valve; (c) ring valves; (d) reduced-clearance poppet valve, discus compressor.

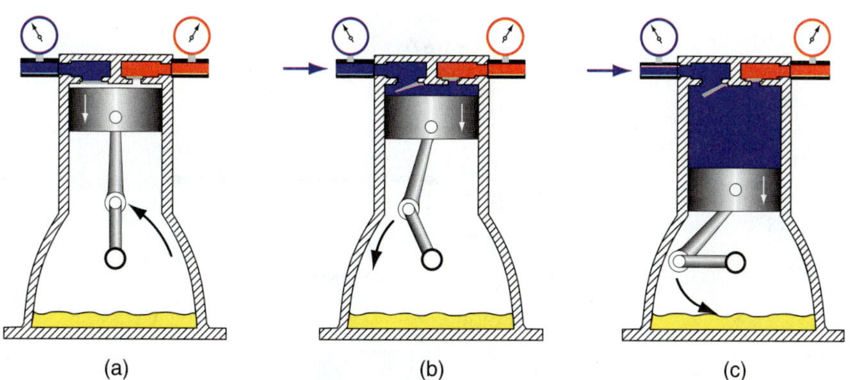

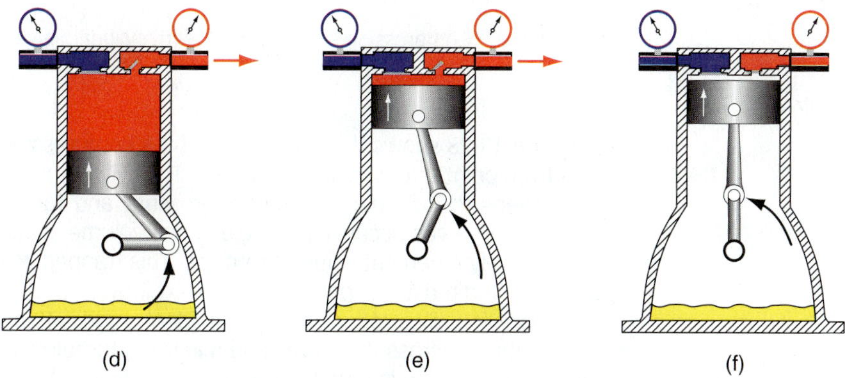

Figure 15-23 Complete compression cycle of a reciprocating piston compressor.

the discharge manifold pressure. At the completion of the compression stroke, this space is filled with compressed gas that is not going anywhere. This gas will have to expand on the downstroke before the pressure can drop to a point to admit fresh suction gas. This reexpansion represents wasted work and lost capacity. The compressor designer strives to minimize the clearance volume while at the same time leaving adequate room for thermal expansion and for unwanted incompressible slugs of oil and liquid refrigerant. These slugs can enter the cylinder, and they possess the ability to do great damage to the valves, piston, and cylinder head. Theoretically, 0 percent clearance volume would be good; practically, it would be disastrous.

Figure 15-24 shows a pressure-volume plot that is used to study reciprocating compressor performance. The horizontal axis represents the cylinder volume, and the vertical axis represents the cylinder pressure. A piston, cylinder, and valves have been added to graphically illustrate the stroke. Figure 15-25 uses this type of graph to show the reexpansion volume at the beginning of the downstroke. Point A represents top dead center of the piston stroke and shows the clearance volume and discharge pressure. As the piston travels down, the trapped compressed clearance volume gas has to reexpand until it reaches suction pressure at point B. Figure 15-26 illustrates the useful portion of the intake stroke. As the downstroke continues from point B to point C, the cylinder pressure is below the suction manifold pressure, and the pressure difference opens

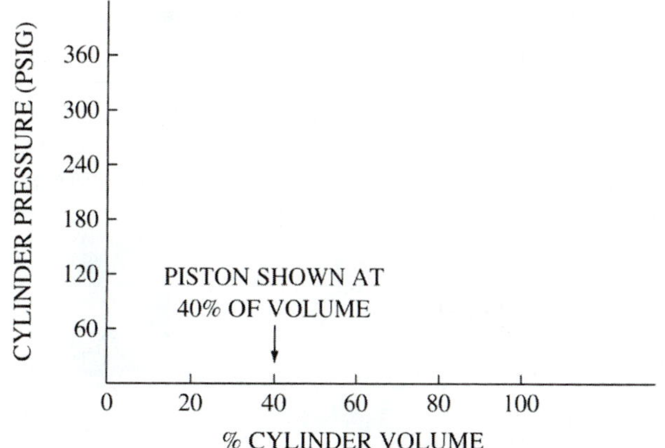

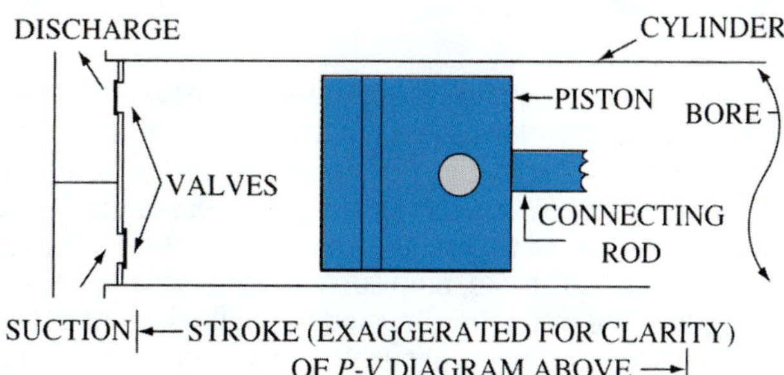

Figure 15-24 The two elements of a pressure-volume plot, used in the study of reciprocating compressor performance.

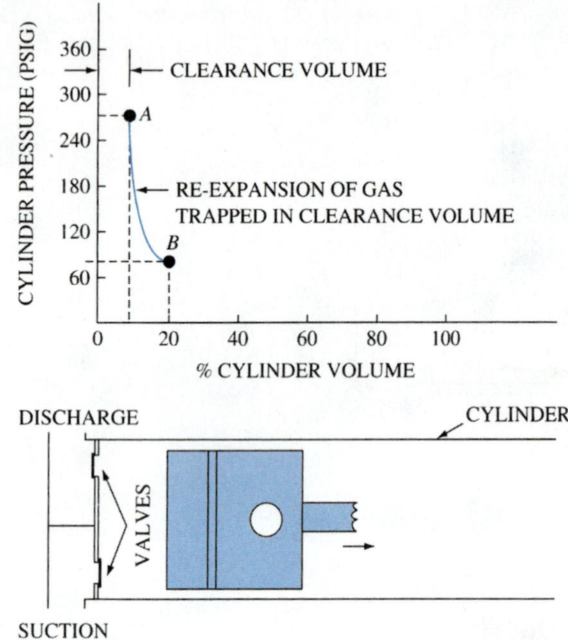

Figure 15-25 Beginning of the intake stroke; *A*: Top dead center; *B*: Suction valve ready to open.

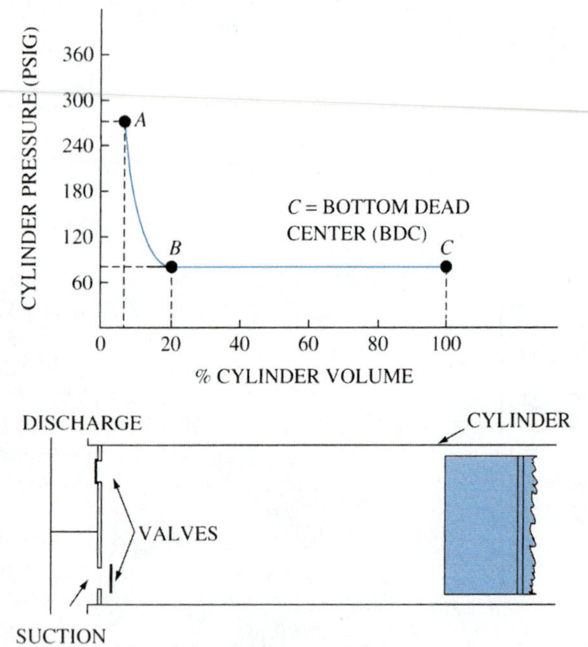

Figure 15-26 Useful portion of intake stroke, *B* to *C*.

the suction valve and fills the cylinder with new refrigerant vapor. The suction valve closes at the bottom of the stroke.

Figure 15-27 shows the compression stroke from bottom dead center (C) to point D. This compresses the gas to a point about equal to discharge manifold pressure. During the compression stroke, both suction and discharge valves are closed. The piston has traveled from 100 percent of cylinder volume to a point about 20 percent of cylinder volume. On air-conditioning R-410A applications, the suction pressure would be about 118 psig (40°F saturation) and the discharge pressure about 417 psig (120°F saturation).

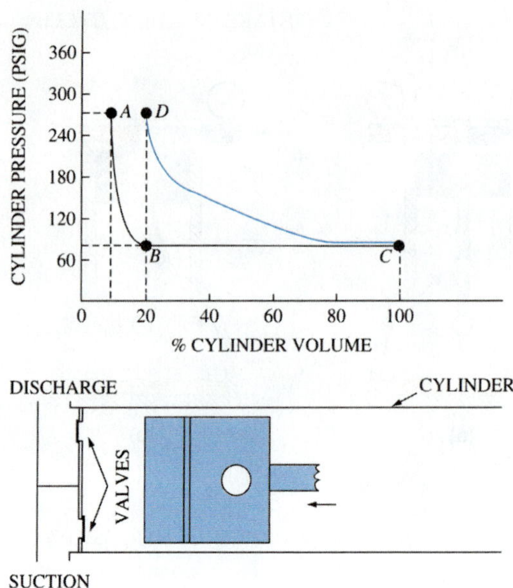

Figure 15-27 Compression stroke, *C* to *D*, with cylinder valves closed.

Figure 15-28 shows the completion of the compression stroke from point D to A (top dead center). The cylinder pressure exceeds the discharge manifold pressure and forces the gas out. The gas occupying the clearance volume space (approximately 4 percent volume) remains. This trapped gas will remain and must reexpand with the next intake stroke.

The previous illustrations assume an ideal world with no pressure losses across the valves and minimum turbulence. A more realistic view is found in Figure 15-29, which shows pressure losses. The area enclosed by points A, B, C, and D on the pressure-volume plot represents the work done by the compressor on the gas passing through the cylinder.

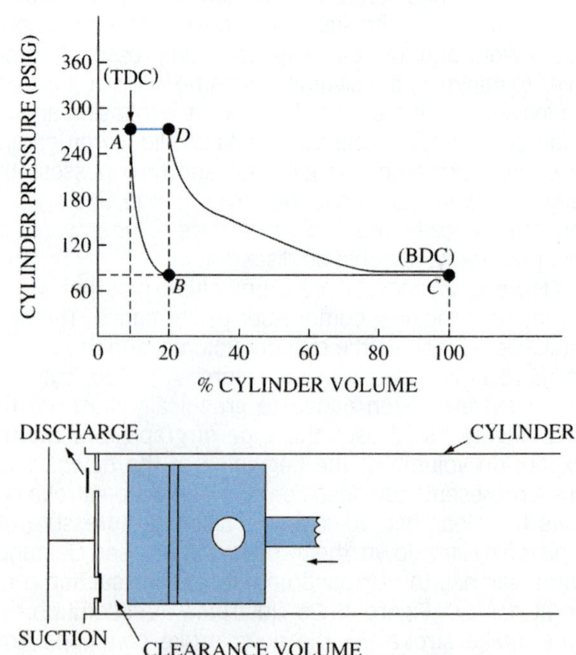

Figure 15-28 Discharge portion of compression stroke, *D* to *A*.

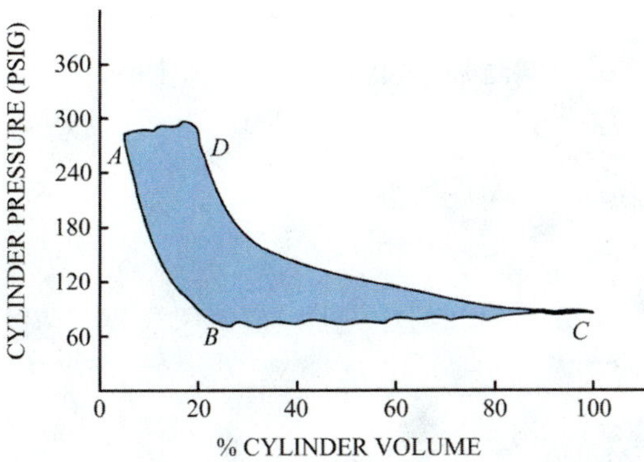

Figure 15-29 Complete *P–V* diagram for a one-cycle compressor cylinder.

Cylinder gas reexpansion causes inefficiency because part of the intake stroke is not used to bring in new gas. Some other types of positive-displacement compressors (scroll, rotary, and screw) do not have trapped reexpansion gas to deal with and thus have inherently higher efficiencies.

Increasing the discharge pressure and/or reducing the suction pressure reduces the output of the compressor. Increased pressure difference increases the amount of cylinder gas reexpansion because the distance the piston must travel before the suction valve opens increases. Other things being equal, a compressor will produce the greatest output by running at the highest suction temperature and the lowest discharge temperature practical: the most work at the lowest energy rate.

TECH TIP

Discus valves in compressors make a significantly different sound than other valve mechanisms. Occasionally, technicians who are not familiar with the sound may feel that the compressor has a bad component or is about to fail. Discus compressors are identified on the nameplate. Check the nameplate before you jump to the wrong conclusion.

15.10 ROTARY COMPRESSORS

Rotary compressors are widely used in small welded hermetic sizes to power small refrigerated appliances, window air conditioners, packaged terminal air conditioners, and ductless split systems. They are efficient and run smoothly and quietly. Two common designs for rotary compressors

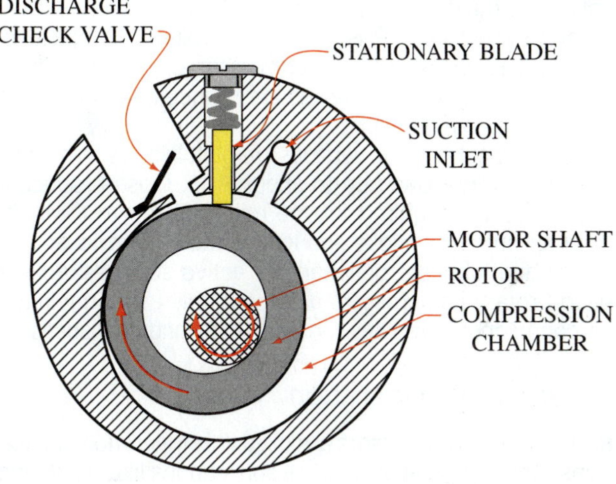

Figure 15-30 Stationary-blade (rolling piston) rotary compressor.

are rotary blade and stationary blade (rolling piston) (Figure 15-30). They are not internally spring isolated, so external vibration isolators must be used on the mounting feet.

Both types of rotary compressors use a high side shell. The refrigerant inside the shell is the discharge gas, not the suction gas. This means that the motor is in a very hot vapor as compared to other compressors. Most other compressors use a low side shell, where their motors are operating in a much lower temperature environment. Rotary compressors are found in window AC units, window heat pumps, and other small units because they are higher efficiency and lighter than reciprocating compressors.

Rolling Piston—Stationary Blade Operation

Rolling piston compressors are also called stationary blade compressors. They use an offset rolling piston that seals against a blade that remains stationary and moves in and out of its slot to seal against the rolling piston. The rolling piston is a rotating off-center cam or lobe that sweeps a path inside a round cylinder (see Figure 15-30). The drive shaft is centered in the cylinder. A close-fitting spring-loaded sliding blade follows the piston and separates the cylinder openings for suction (inlet) gas and discharge (outlet) gas.

As the rolling piston rotates, gas will flow in and fill the suction cavity due to pressure difference, much the same as in a reciprocating compressor (Figure 15-31). The piston continues to rotate, closes off, and passes the suction port and compresses the gas until its pressure is high enough to flow out through the discharge (reed)

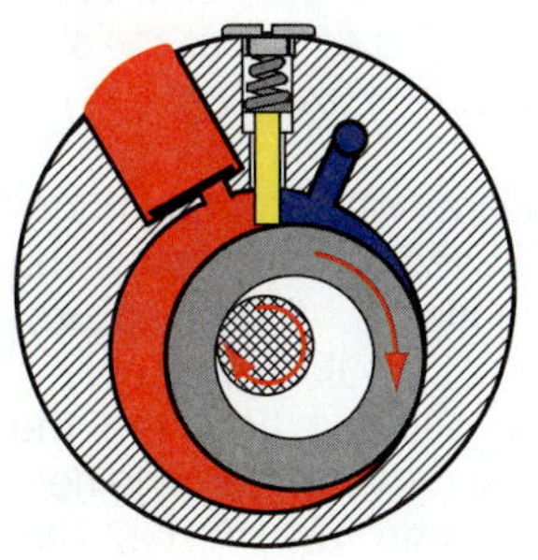
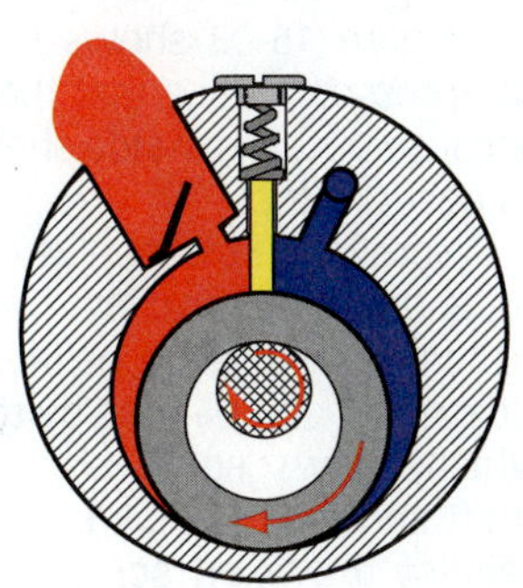
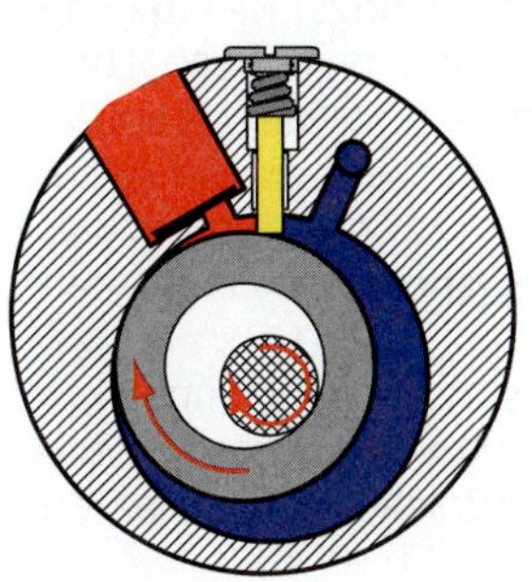

Figure 15-31 Compression cycle for a rotary compressor.

valve into the discharge manifold. No suction valve with its inherent pressure loss is needed, so volumetric efficiency is high. Because of the rotating rather than reciprocating piston motion, vibration levels are reduced and are easier to dampen (externally).

Compact hermetic versions are built in sizes from small fractional horsepower up to about 5 hp. Close production tolerances are required for the rolling piston to clear the cylinder but come close enough for the oil film to provide a seal and lubricate between the active surfaces. Rolling-piston rotary compressors are commonly used in window air conditioners, package terminal units, and appliances.

Rotary Blade Operation

Rotary blade compressors are used in refrigeration applications. One type is used in small applications like residential refrigerators and freezers, while another is used primarily in larger commercial refrigeration applications.

A large off-center rotating shaft carries multiple blades in slots, and they have a reciprocating action as they move in and out, following the cylinder walls. The moving cavities (cells) formed between adjacent vanes fill with gas as they sweep by the suction port. The gas trapped in the cell is then compressed as the moving space between the vanes, cylinder wall, and shaft is reduced. Finally, the compressed gas is released through a discharge port and the process repeats. An oil-flooded design provides sealing and lubrication. Large rotary-vane compressors usually are open-type compressors with external drive shafts. Typically, these machines are used as booster compressors for low-temperature ammonia systems as well as for single-stage compressors for industrial and process applications.

Rotary Compressor Operational Characteristics

Rotary compressors are somewhat tolerant of small amounts of liquid. The spring-loaded blade can lift off the rolling piston and let liquid scoot under it if liquid is trapped and cannot escape through the discharge valve. Rotary compressors are also more efficient than reciprocating compressors because they do not have any clearance volume.

Because rotary compressors use a high-side shell, the refrigerant absorbed by the oil in a rotary compressor takes somewhat longer to leave than with low-side-shell hermetic compressors. Systems with a small refrigerant charge will operate as if they are undercharged for several minutes until all the refrigerant leaves the oil.

Rotary compressors are constructed of precisely machined metal parts with no gaskets or rings. Maintaining clean systems and proper lubrication is critical. Long refrigerant lines can cause problems with rotary compressors because they can lose their oil in the lines and seize. Small impurities can also cause them to seize. Even a hair placed between the rolling piston and the cylinder can cause it to seize.

Keeping condensers clean is especially important with rotary compressors because they are cooled by the hot discharge gas. Dirty condensers lead to overheated compressors.

Figure 15-32 A movable scroll orbits inside a fixed scroll.

15.11 SCROLL COMPRESSORS

Scroll compressors compress gas between two close fitted spiral scroll members (Figure 15-32). One is fixed and the other moves (but does not rotate) in an orbital path. The only rotating part in a scroll compressor is the motor shaft. Progressively reducing cavities compresses the refrigerant gas with little or no vibration.

Scroll compressors are positive-displacement machines with high volumetric efficiencies, currently available in sizes from about 1 hp to 30 hp. Because they are approximately 10 percent more efficient than a comparable reciprocating compressor, they have experienced extraordinary growth in the world markets of residential and commercial air conditioning and refrigeration. Scroll compressors are built in a welded hermetic design.

Scroll Compressor Operation

The sides of the scroll shape are called flanks; the edge of the scroll is called the scroll tip. For a scroll machine to work well, it must be finished to very close tolerances so the contact between the flanks and tips of the scroll members are very snug. Suction vapor is captured in pockets at the outside of the scroll members. The pockets close as the scroll orbits, trapping a small quantity of refrigerant vapor in each pocket. The trapped pocket of vapor is forced around the scroll toward the center in an ever-decreasing pocket size as the scroll continues to orbit. The refrigerant's pressure and temperature increase as the pocket size decreases. The compressed gas is discharged at the center through an opening in the fixed scroll.

Figure 15-33 shows the key components of a scroll compressor. Notice that the suction gas enters at the side because of the low-side shell design. The discharge gas exits near the top.

Scrolls are classed as compliant or noncompliant designs depending on the method used to accomplish the critical task of gas sealing. Compliant designs can provide radial compliance by the use of an Oldham coupling, which allows some side-to-side slipping between the shaft and the moveable scroll. The movable scroll is held against the fixed scroll through centrifugal force. This

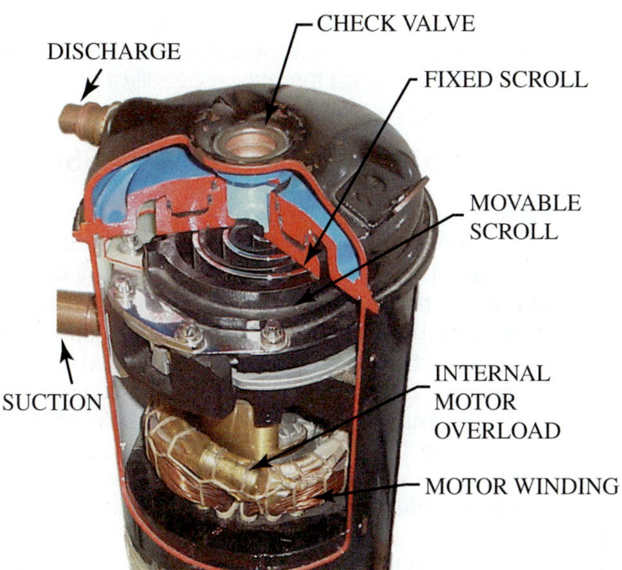

Figure 15-33 Scroll compressor parts shown in a cutaway.

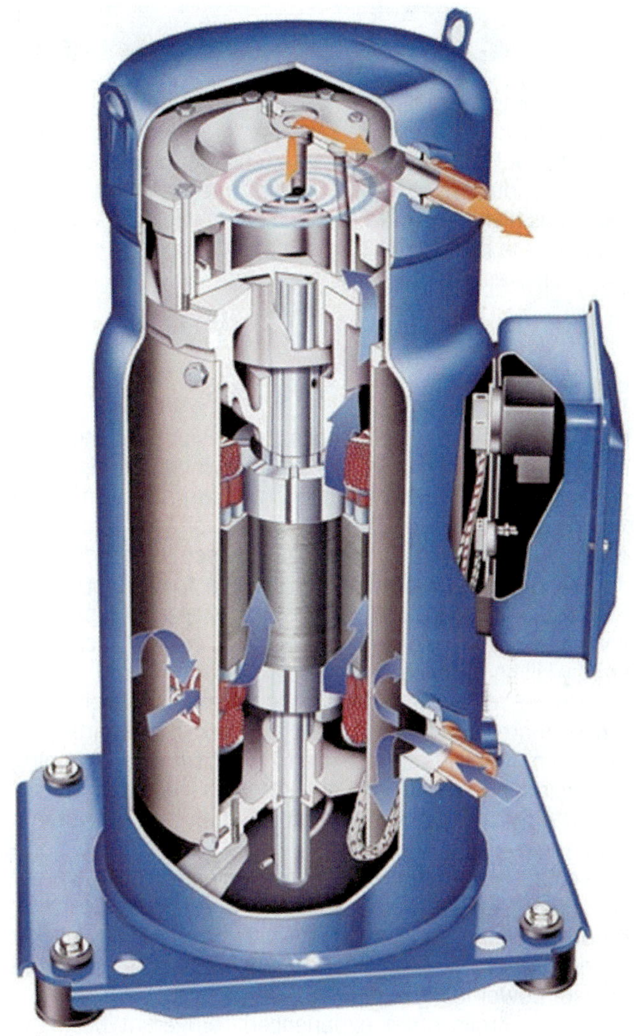

Figure 15-34 Partial cutaway of a scroll compressor. *(Courtesy Danfoss Inc.)*

allows the scroll flanks to maintain close contact, and yet the orbiting scroll can separate and unload to pass a small slug of liquid. Axial compliance is maintained by an adjustable force applied to maintain the seal between the scroll tips during compressor operation. This force is provided by the discharge gas and is released on shutdown. A compliant scroll compressor can handle small amounts of liquid refrigerant without damage. However, large amounts of liquid will create enough pressure on the flanks of the scroll to snap them off.

Figure 15-34 shows a cross section of a scroll compressor. The Oldham coupling is used to drive the orbiting scroll providing radial compliance, permitting the compressor to handle liquid slugs.

Figure 15-35 shows the scroll compression sequence in diagram form. One pair of cavities is shown in the compression sequence.

Scroll compressors have very few moving parts as compared to reciprocating compressors. Scroll compressor motors are pressed to fit into the compressor housing. The compressor is not spring isolated inside the welded shell. Vibration isolation is only external. As a result, scroll compressors can sometimes produce an unusual high-pitched sound. Some manufacturers enclose these compressors in soundproofing boxes to keep the sound level (db) low enough to meet standards.

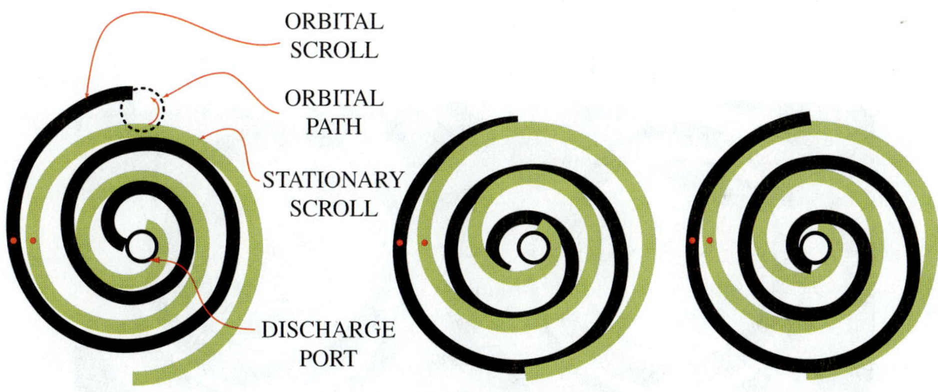

Figure 15-35 Scroll compressor movement; notice that the red dots on the scrolls stay very close together through the compression cycle.

TECH TIP

Scroll compressors may be noisy on startup and shutdown, making a very loud rattle. This rattle is caused by the movable scroll rattling until the discharge pressure is high enough to seal the scroll tips and the compressor speed is high enough to hold the moveable scroll tightly against the fixed scroll. It does not indicate a problem in the compressor.

Scroll Compressor Precautions

Most scrolls are low-side shells. The majority of the shell is low side, but the top of the scroll is where the high-side gas leaves. Do not allow the suction pressure to drop below 25 psig during charging or below 7 psig for even a few seconds. Low pressures can overheat the scrolls and damage the compressor. Do not run a unit with a scroll compressor with the suction valve closed. The compressor can pull into a vacuum and cause internal electrical terminal arcing, instantly killing the compressor.

Unlike most reciprocating compressors, scroll compressors are directional. They only work in one direction. Scroll compressors are extremely noisy when running backward. They also will not compress when operated backward. Check for proper rotation when starting systems with three-phase scroll compressors. This can be done with gauge manifolds. If normal suction and discharge pressures are seen soon after startup, the rotation is correct. If normal suction and discharge pressures do not quickly develop, shut off the machine and reverse any two-compressor power leads. Then recheck.

Early scrolls often experienced refrigerant backflow every time they shut down. The high-pressure discharge gas would push through the compressor to the suction side and make the compressor spin backward. Momentary power interruptions would occasionally cause the compressor to operate backward because of this. Today, systems using scroll compressors typically use anti–short cycle controls to avoid immediate restarts. On today's scroll compressors, a discharge line check valve prevents continued backflow through the scroll when shutting down. A momentary sound will be heard for one or two seconds as the internal pressures equalize backward through the scroll.

It is possible for a scroll to seal some refrigerant in it when it shuts down. Always be sure to vent both the high and the low sides before working on the lines, especially if brazing.

15.12 TWO-STEP SCROLL COMPRESSORS

The Copeland two-step scroll compressor operates much like the standard scroll. The difference is that by activating an internal solenoid, the compressor's capacity can be changed from 67 percent to 100 percent instantly. For example, a 36,000 BTUH compressor (BTUH is the commonly used unit referring to BTU/hr in the industry) would provide approximately 24,000 BTUH on the low step of cooling.

A rotating internal plate opens and closes two drilled ports in the scroll (Figure 15-36). When these two ports are open, refrigerant exits the scroll in the first compression chamber. This balances the pressure at this point in the compression cycle. By moving the actual starting point of the compression cycle a little closer to the center discharge, the compressor is partially unloaded. The stepping between full and partial load capacity is done while the compressor is running.

The term *partially unloaded* means that the compressor capacity has been reduced from its full load capacity. Large commercial refrigeration compressors have used unloading techniques for many years to save energy. This two-step scroll was designed for that same energy-saving purpose. The unloading allows the compressor to run longer when the residence is under a lighter load, such as summer nights and early mornings or warm spring or fall days. Longer run times can increase the dehumidification part of the air-conditioning process.

The two-step scroll compressor can be used in heat pumps. Typically, most homes require more heating capacity than cooling capacity. Using a system with too much capacity in cooling produces inefficient operation and poor dehumidification. The two-step scroll allows a heat pump to have a high capacity in heating and a lower capacity in cooling, improving the efficiency in both heating and cooling.

Two-Step Scroll Compressors Operation

The second step of cooling is controlled by an electronic module (Figure 15-37). This control module receives a 24 V AC signal from the thermostat's second stage. Within the module, the 24 V AC is rectified to DC, which operates the internal unloading solenoid coil. The module has an

Figure 15-36 The two-step scroll compressor activates a solenoid to go from 67 to 100 percent capacity.

Figure 15-37 Control module for two-step scroll compressor.

operating input voltage of 18–28 V AC at 20 V AC. The internal coil has an operating input voltage range of 12–27 V DC.

CAUTION

The unloading solenoid coil operates on 24 V DC and will be damaged if connected to 24 V AC. If the internal coil is damaged, the compressor will only operate on the first step of cooling.

15.13 DIGITAL SCROLL COMPRESSORS

Digital scroll compressors get their name from the fact that they only operate under two conditions: either at 100 percent capacity or 0 percent capacity. It is the ability to switch very quickly between full-on capacity to full-off capacity that allows digital scrolls to produce an infinite range of capacity. The compressor motor operates at a constant speed. The scroll's capacity is controlled by an external solenoid. This solenoid can be energized and de-energized at a rate fast enough to change the effective compressor capacity from 10 to 100 percent as needed.

When the scroll is unloaded, the current draw is reduced. This allows the compressor to provide exactly what is needed for the application and to provide outstanding energy savings.

The 10 percent minimum capacity is designed to ensure that the compressor gets proper lubrication. If the solenoid were energized all of the time, the compressor would have no refrigerant flowing, and it would not have any lubrication. Under these conditions, the compressor would quickly destroy itself.

The digital scroll compressor works much like turning a light off and on very quickly. If you are fast enough with the switch, you could make the light range from a slight glow to 100 percent bright. That is how fast the solenoid turns the scroll off and on.

Digital Scroll Compressors Operation

The external solenoid valve controls the downward refrigerant pressure on the top scroll. When this solenoid is energized, the pressure on top of the scroll is released into the suction line (Figure 15-38). With the pressure reduced on the top scroll plate, it is allowed to lift ever so slightly from the orbiting lower scroll plate. This allows gas to scoot under the scroll, effectively stopping compression. It is like sticking a pin in a balloon: it takes less than a blink of an eye for the pressure to be released. Once the solenoid is de-energized, the compressor instantly begins to compress refrigerant again.

A digital compressor controller turns the solenoid on and off. The controller is connected to a thermostat that

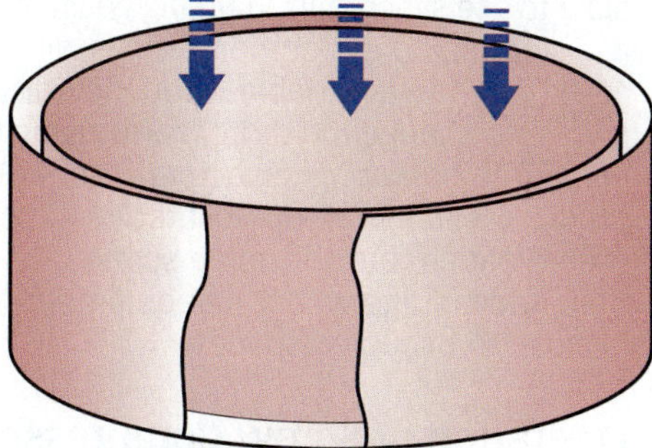

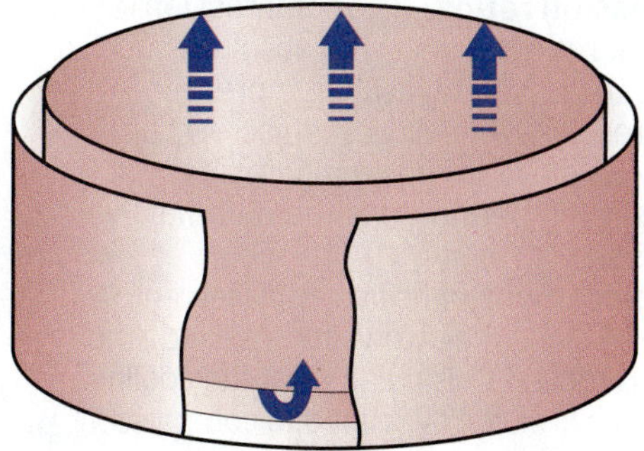

Figure 15-38 Digital scroll operation.

tells it how much cooling is needed. In addition, there are a number of safety devices and sensors that can be attached to this controller.

One major safety/control device that can be added to a digital scroll compressor is the optional refrigerant-injection system. As discussed later in this unit, overheating of all compressors occurs as they are used for lower and lower evaporator temperatures. The refrigerant-injection system on the digital scroll and the scroll's ability to change its effective compression ratio through unloading has allowed it to be successfully used in low-temperature applications. The discharge temperature thermistor and optional suction transducer work through the compressor controller to keep the internal operating temperature below the critical temperature.

Digital Scroll Applications

Digital scroll compressors are being used for refrigeration and heat pump applications under operating conditions that would be detrimental to most compressors. The controller, along with its safety and control sensors, helps to keep the compressor within its operating limits.

Digital scroll compressors are available in sizes ranging up to over 10 tons. The large compressors have been used economically in heat pump applications with outdoor temperatures below 0°F. One large-scale test used a 6 hp R-22 digital scroll compressor to heat a four-story apartment building in Beijing, China. The heat pump system kept the building heated with a discharge air temperature over 100°F even with the outdoor temperature below 15°F.

15.14 SCREW COMPRESSORS

Rotary screw compressors are positive-displacement machines made for large-tonnage applications. Rotary screw compressors are available in single-, twin-, and triple-rotor types. They are relatively simple in concept but demand advanced manufacturing to very close tolerances. Some of the advantages of screw compressors are as follows.

- **High volumetric efficiency** Volumetric efficiency expresses the rate of internal refrigerant leaking past the internal compressor parts. Very little of the refrigerant that enters the compressor at the suction port is allowed to slip back by the screws. Almost all of it forced out at the discharge port.
- **High compression ratios** The compression ratio is expressed as a ratio between the absolute suction pressure and the absolute discharge pressure. A high compression ratio allows the screw to produce very cold evaporator temperatures and high condenser temperatures without damaging the compressor.
- **Positive displacement** Positive displacement means that every time the screws turn, a fixed amount of refrigerant is moved. This means that the cooling capacity changes are directly related to compressor speed. This makes it much easier for screw compressors to have a variable capacity.
- **High-pressure refrigerants** High-pressure refrigerants tend to more easily leak past internal compressor parts. However, the screws' oil coating seals small gaps, thus keeping a high volumetric efficiency.
- **Large tonnage** Screw compressors are available in capacity up to 700 tons.
- **Variable capacity** There are a variety of ways that screw compressors can vary their capacity to meet the demands of changing loads.
- **Refrigeration or air-conditioning applications** Screws are capable of producing low-temperature refrigeration down to −40°F or air-conditioning temperatures ranging up to 45°F.

Figure 15-39 Single-screw open compressor. *(Courtesy of Emerson Climate Technologies)*

Screw compressors are available in open, semihermetic, or hermetic designs (Figure 15-39). Hermetic designs are available from 35 to 175 hp and are often used in multiples for larger capacity equipment.

A single-screw design using gate rotors is available. Single-screw compressors use a single helical main rotor and either one or two gate rotors.

The compression volume is formed by the rotor grooves, the cylinder wall, and by the meshing gate-rotor teeth. The rotor groove fills with gas, turns, and compresses the gas as the volume is reduced by the meshing action with the gate rotor. Oil flooding is used for sealing as well as lubrication and cooling.

The twin-screw models are used for larger refrigeration and air-conditioning (usually water chilling) applications (Figure 15-40). They often consist of a driven male rotor and mating female rotor carried on bearings and rotating within a stationary housing containing gas inlet and discharge ports. Either rotor can be the driven component, although it is possible to use synchronized timing gears to drive both rotors.

A triple-screw compressor is also available (Figure 15-41). It has a variable-frequency drive motor that allows it to change the motor rpm, giving the triple screw a very large operating capacity range. Both the single- and twin-type screw compressors get much louder as they are unloaded, but the triple screw gets quieter. This is a major advantage when the equipment room in a building is close to occupied spaces.

Operation of Screw Compressors

All three types of screws operate similarly. They draw the refrigerant gas in through the inlet port into voids created as the rotors turn, as shown in Figure 15-40. The entire length

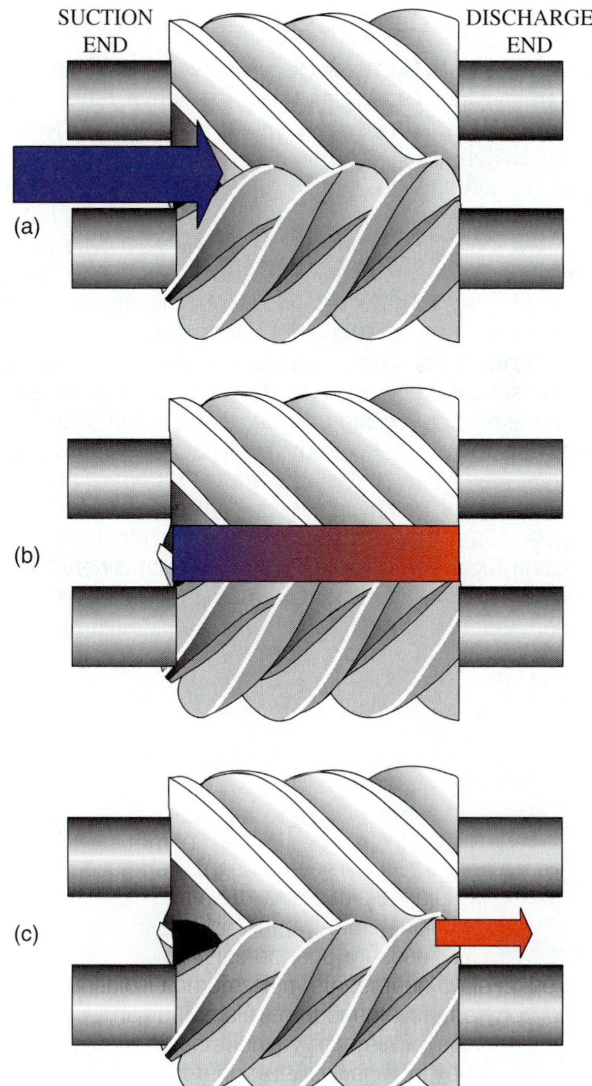

Figure 15-40 Cross-section view of a twin screw showing the gas flow through the compressors.

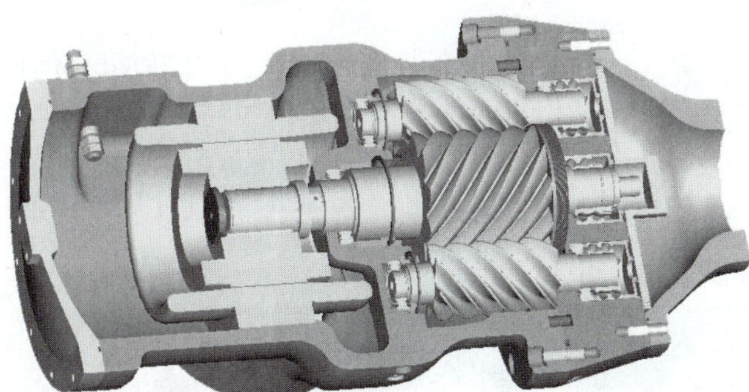

Figure 15-41 Triple-screw compressor. *(Courtesy of Carrier Corporation)*

of the rotor space will fill with gas. As the rotors continue to rotate, they reengage one another on the suction end. This progressively compresses the trapped gas as they mesh, moving it toward the discharge end of the machine. The compressed gas is released through the discharge port as the rotors turn and uncover the port. The cycle repeats as interlobe spaces are created, fill with gas, diminish (compressing the gas), and vent gas to the discharge. The individual discharges blend into a smooth flow of gas with little pulsation compared to a comparable reciprocating compressor.

Rotor bearings may be sleeve or magnetic, referred to as antifriction, depending on machine design and size. The oil separator is an important component in oil-flooded machines. The compressor could quickly run out of oil without the oil separator returning oil to the compressor. An oil separator may be built as an integral part of the design.

Capacity control is possible by speed adjustment or valves. Valve types include slot valves, lift valves, or slide valves. The most commonly used device is the slide valve. It is a simple arrangement that bypasses part of the gas back to the suction and regulates the discharge port. It permits satisfactory unloading over a wide capacity range and is an uncomplicated and reliable mechanism.

15.15 CENTRIFUGAL COMPRESSORS

Centrifugal compressors are widely used in single-stage, multi-stage, open, and semihermetic configurations to chill water for building air conditioning in capacities from 100 to 8,000 tons and higher. Centrifugal compressors are used for moving large volumes of gas against low-pressure differences. One example of a centrifugal compressor is shown in Figure 15-42. Currently, centrifugal capacities are limited to less than 1,000 tons. Larger chiller units utilize multiple compressors. These machines compete with the screw machines for large-capacity chilled-water cooling plants.

Most centrifugal systems are extremely large and used as part of a chiller system in large commercial buildings. These systems are so large that they fall under local, state, or federal guidelines for equipment room operators. An individual must be present at all times when these systems are running. Most run twenty-four hours a day, seven days a week. The individual who operates these units may be referred to as an equipment room operator or an operating engineer.

Centrifugal compressors are grouped into low-pressure and high-pressure machines. Low-pressure centrifugals work

Figure 15-42 Multistage centrifugal compressor.

at pressures ranging from an 18 in. Hg vacuum on the low side to 10 psig on the high side. Older low-pressure centrifugal machines used R-11, a CFC. Newer low-pressure centrifugals use R-123, an HCFC. High-pressure centrifugal compressors are available in both open and semihermetic configurations. They may use refrigerants such as R-134a or R-22.

TECH TIP

Most centrifugal chiller systems cannot produce more than an 18°F temperature difference in the chilled water. If a greater temperature drop in chilled water temperature is needed, two centrifugal chillers may be installed in series. In a series installation, the chilled water from the first unit is passed through the second unit, where it is cooled to a lower temperature. Two centrifugal chillers in series could produce a 36°F temperature difference in the chilled water.

Operation of Centrifugal Compressors

The impeller acts like a fan. It can move large volumes of refrigerant vapor as long as there is relatively little pressure difference between the evaporator and the condenser. Refrigerant enters in the middle of a centrifugal impeller and is thrown outward by centrifugal force (Figure 15-43). Figure 15-44 shows a precision die-cast centrifugal impeller. A two-pole 3,600 rpm motor uses a gear-type speed increaser to drive the impeller at speeds of 8,000 to 10,000 rpm and higher. To reduce sound levels, the impeller vanes are asymmetrically arranged but fully balanced.

Centrifugal compressors are not positive-displacement compressors. If there is too great a pressure difference between the evaporator and the condenser, the impeller may not be able to move refrigerant. If this happens, it may begin to surge, which can destroy the impeller. Capacity control can be accomplished by inlet vane controls or variable motor speed control. Inlet vane controls are the most common.

Centrifugal compressors were once considered low-pressure machines. Today there are many centrifugal compressors that use high-pressure refrigerants. High-pressure

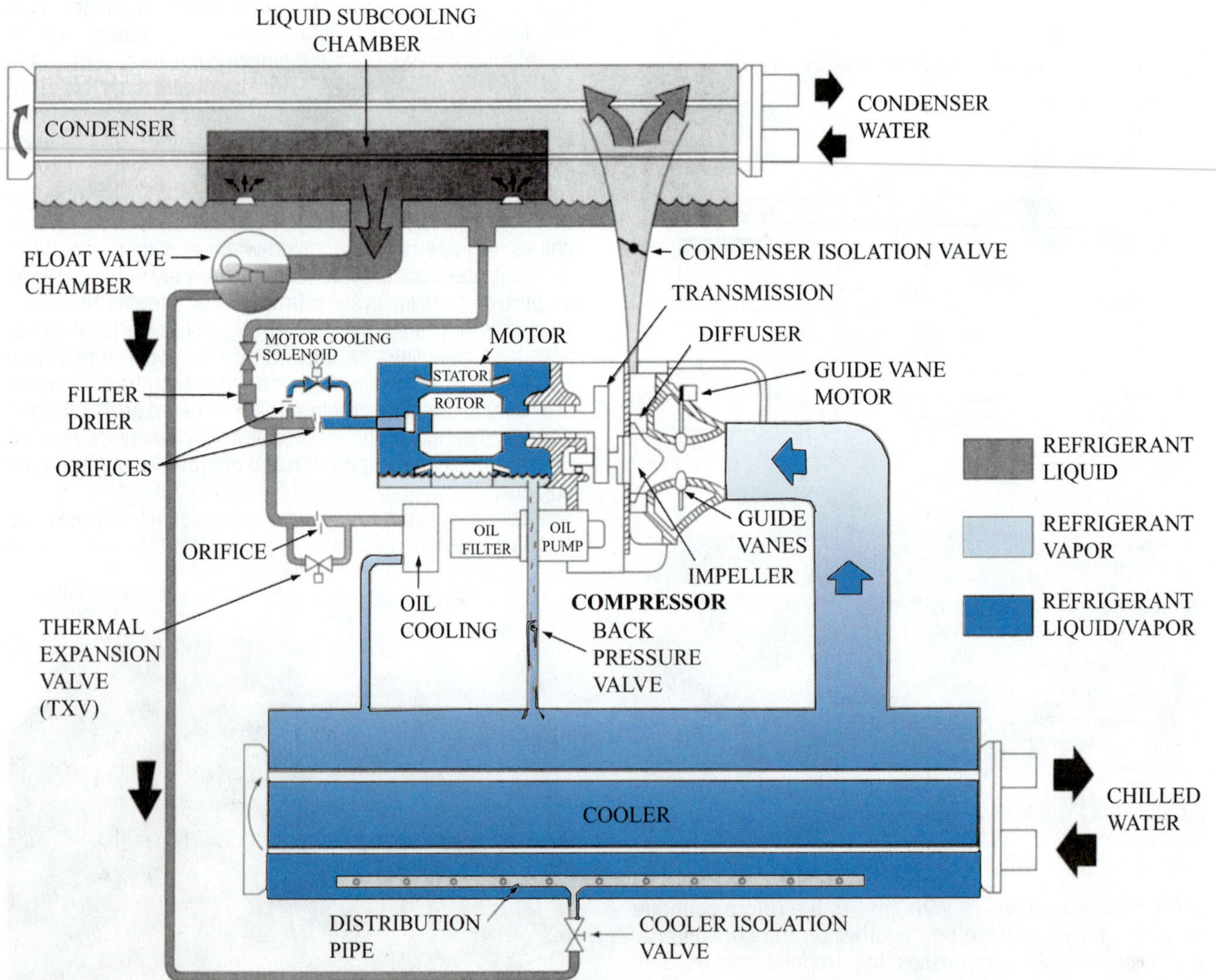

Figure 15-43 The gas enters the impeller in the center and is thrown outward by centrifugal force.

Figure 15-44 Lightweight, high-speed centrifugal impeller.

centrifugal chillers are smaller than low-pressure chillers for the same capacity.

15.16 ENGINE-DRIVEN COMPRESSORS

Not all compressors are operated by electric motors. Some compressors use gas or diesel engines to drive them. The most common application of engine-driven compressors is found in transport refrigeration, where diesel engines are mated to open compressors (Figure 15-45). Engines are used out of necessity in transportation refrigeration because the system is traveling down the road.

Some large stationary chiller applications also use engine-driven compressors. These typically use natural gas engines. These systems allow efficient cooling in large applications where reducing electrical usage is a priority. The engine speed can be varied to achieve variable capacity.

Figure 15-45 Transport refrigeration compressor mounted to diesel engine.

Figure 15-46 Oil slinger on small semihermetic compressor.

15.17 COMPRESSOR LUBRICATION

All refrigeration compressors require lubrication of moving surfaces. Delivery of lubricant can be by simple splash systems or a separate oil pump. A common splash system uses a slinger that slings the oil around the inside of the compressor (Figure 15-46). A cup catches the oil and it drains by gravity into the crankshaft. The crankshaft throws the oil out by centrifugal force through holes in the crankshaft (Figure 15-47).

Pressure lubrication is used with larger compressors. Typically the oil pump is mounted on the compressor end of a semihermetic compressor (Figure 15-48). The oil pump picks oil out of the crankcase pump and forces it down the crankshaft. Holes in the crankshaft deliver the oil to the rods and bearings. Typically the oil pump is mounted on the compressor end of a semihermetic compressor.

Open and semihermetic compressors often have an oil sight glass to view the compressor oil level. The oil level should be judged with the compressor running after it has run long enough for the pressures to stabilize and the crankcase to become warm. The correct oil level will depend on the model and manufacturer, but the oil level should be above the bottom of the sight glass on most compressors. The maximum oil level will vary between three-eighths and

Figure 15-47 Oil is delivered through the holes in the crankshaft.

Figure 15-48 Compressor with oil pump mounted.

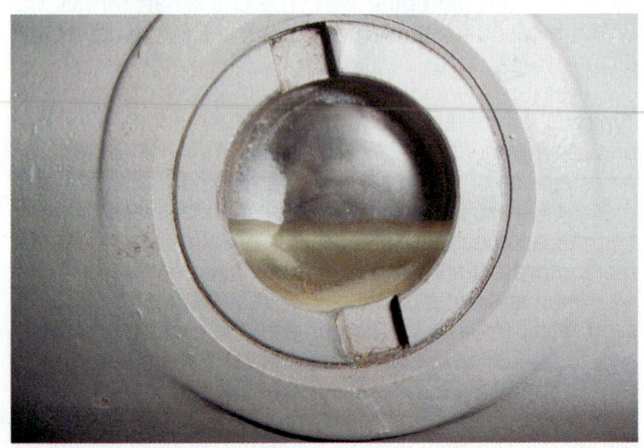

Figure 15-49 Oil sight glass three-eighths full.

one-half of the way up the glass on most compressors. Figure 15-49 shows an oil sight glass with the correct level for that compressor.

Some oil leaves with the discharge gas. The refrigerant must be able to carry the oil back. The oil should be miscible with the refrigerant for this to happen. System piping is also critical. The refrigerant must travel at a minimum speed to keep the oil moving with it.

Some machines are designed to have high oil circulation rates and are always provided with discharge-oil separators. Oils used in hermetic and semihermetic systems have to be compatible with the electrical components and be good insulators. Mineral oils should be wax free. All oils should have a low moisture content to ensure long, corrosion-free system life and freedom from electrical insulation deterioration and freeze-ups.

CAUTION

Refrigerant oil does not evaporate from a system. If you have a compressor that is operating on low oil, the oil is trapped somewhere in the system. Determine where the oil is and what problem has caused the oil to be entrapped in the system away from the compressor, and resolve that problem.

HFC chlorine-free refrigerants require use of special polyol ester (POE) synthetic oils. Older CFC machines used mineral oil–based lubricants. Current recommendations for HCFC refrigerant installation and conversions include mineral oil, alkyl-benzene refrigeration oil, or a combination of the two. The choices have become more complicated. POE refrigeration oil is more expensive but can be used with all halocarbon refrigerants. It must be used with HFC refrigerants as specified.

Pressure lubricated compressors are usually equipped with an oil pressure switch to shut the compressor down should lubrication fail. Oil pressure switches measure the differential pressure between the oil pressure and the suction pressure to measure actual oil pressure. They include a time delay to allow for system startup and to avoid nuisance trips. A safety lockout circuit prevents the system from restarting after shutdown. Refer to manufacturer data for minimum pressure settings, normal operating pressures, and high relief valve settings.

Oil and refrigerant liquid are miscible, so it is possible to have refrigerant migrate during long off cycles to a compressor in a cold location. The oil and refrigerant mixture raises the oil level above the oil sight glass (if so equipped),

Figure 15-50 The oil level must be visible in the sight glass.

as shown in Figure 15-50. On startup, the crankcase pressure drops, the liquid refrigerant flashes to gas, and the mixture turns to foam. The oil pump does not pump the foam effectively, and the foam is a poor bearing lubricant. Once the system has started and operated until the oil is warm, the problem clears. Where such problems are encountered, crankcase heaters can be used as well as timed pump-down cycles and piping changes. A crankcase heater can be added to the outside of hermetic compressors to raise the temperature of the oil to prevent this migration from occurring, as shown in Figure 15-51. A flat crankcase heater can be attached to the bottom of semihermetic compressors for the same purpose, as shown in Figure 15-52.

15.18 COMPRESSION RATIO

Compression ratio is the ratio of the absolute discharge pressure over the absolute suction pressure. High compression ratios can cause overheating. Compression ratio

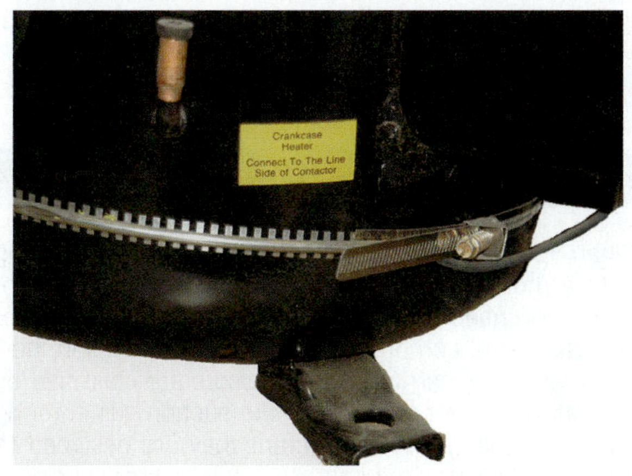

Figure 15-51 External crankcase heater for hermetic compressors.

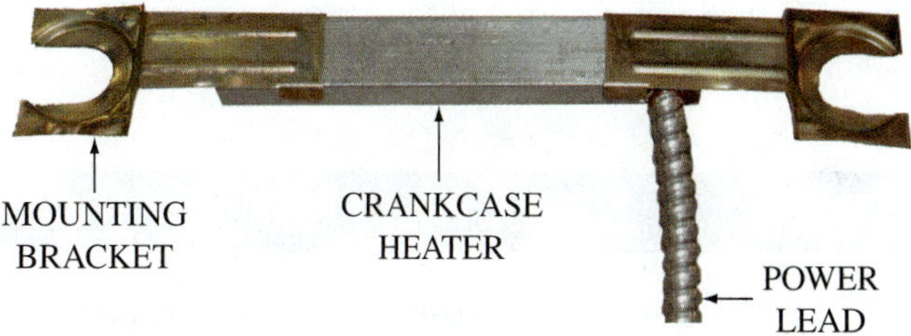

Figure 15-52 External crankcase heater for semihermetic compressors.

is calculated by dividing the absolute discharge pressure by the absolute suction pressure. For example, to find the compression ratio for a system with a 210 psia discharge pressure and a 70 psia suction pressure:

$$210 \text{ psia}/70 \text{ psia} = 3/1 \text{ compression ratio}$$

It is important to remember that compression ratio is calculated using absolute pressure, but your gauges read gauge pressure. The system pressures must be converted from gauge pressure to absolute pressure to calculate the compression ratio of an operating compressor. A reasonable estimate of absolute pressure can usually be obtained by adding 15 to gauge pressure. In the previous example, the discharge pressure would be 195 psig and the suction pressure 55 psig. When starting with gauge pressures, the calculation looks like this:

$$(195 \text{ psig} + 15)/(55 \text{ psig} + 15) = 210 \text{ psia}/70 \text{ psia}$$
$$= 3/1 \text{ compression ratio}$$

System efficiency drops as compression ratio increases. Compressors also increase in temperature as the compression ratio increases. The compression ratio should be kept as low as possible to increase efficiency and to extend life. Factors that help in keeping the compression ratio as low as possible are keeping the condenser temperature as low as possible and keeping the evaporator pressure/temperature as high as possible. However, the suction pressure is more critical than the discharge pressure.

Consider a comparison of the effects of increased head pressure and decreased suction pressure on a system operating at 240 psig discharge pressure and 70 psig suction pressure. The compression ratio is

$$(240 \text{ psig} + 15)/(70 \text{ psig} + 15) = 255 \text{ psia}/85 \text{ psia}$$
$$= 3/1 \text{ compression ratio}$$

If the head pressure is increased by 60 psi while the suction remains the same, the compression ratio becomes

$$(300 \text{ psig} + 15)/(70 \text{ psig} + 15) = 315 \text{ psia}/85 \text{ psia}$$
$$= 3.7/1 \text{ compression ratio}$$

If the head pressure stays the same while the suction pressure drops to 50 psig, the compression ratio becomes

$$(240 \text{ psig} + 15)/(50 \text{ psig} + 15) = 255 \text{ psia}/65 \text{ psia}$$
$$= 3.9/1 \text{ compression ratio}$$

Lowering the suction pressure 20 psi increased the compression ratio more than raising the discharge pressure 60 psi. Dirty filters, iced coils, partially restricted filter driers, and misadjusted expansion valves can all cause small drops in suction pressure that can have a large effect on the compression ratio and system performance.

TECH TIP

It may be cost-effective to use a multistage compressor system or a cascade system when a low-temperature refrigeration system would have too high a compression ratio. The added cost of the more expensive compressors may be offset by the cost of repeatedly replacing overheated and failed compressors.

15.19 COMPRESSOR CAPACITY AND HORSEPOWER

It is common to refer to compressors by a fixed capacity—that is, 1 ton, 2 ton, and so on. However, a compressor's capacity is not a fixed value. There are many factors that affect a compressor's BTUH capacity per horsepower of input. These factors include evaporator temperature, condenser temperature, refrigerant properties, and compressor rpm.

Low Evaporator Temperatures

Less refrigerant enters the cylinder for each compression cycle at lower evaporator temperatures. Less cooling occurs when less refrigerant flows through the system. This requires more horsepower for each BTU of cooling. From 1 to 5 hp may be needed for a ton of cooling, depending on the box temperature. Examples of low-evaporator-temperature systems include medium- and low-temperature refrigeration units and heat pumps.

High Evaporator Temperatures

Higher temperatures increase the BTUH capacity, requiring less horsepower for each BTU of cooling, for example, normal and high-efficiency air conditioning.

Low Condenser Temperatures

The cooler the air or water entering the condenser, the greater its heat-removal capacity. This results in lower condenser pressures, and the compressor has to do less work. This reduces the horsepower required for each BTU of cooling. Air conditioners that are used to cool equipment rooms even in the fall or spring provide higher cooling capacities with less energy or horsepower used.

High Condenser Temperature

The higher the condenser temperature, the harder a compressor has to work to move the refrigerant vapor out of the evaporator into the compressor. This requires more horsepower for each BTU. It is much like driving up a steep hill requires more horsepower for a given speed than does driving at the same speed on nearly level roads. Air conditioners in the southwestern states experience the high summer desert temperatures. But rooftop air conditioners may experience similar temperatures through much of the United States.

Refrigerant Properties

Each refrigerant type has unique chemical properties that affect a compressor's capacity.

Compressor RPM

Most compressors operate on 60 or 50 cycle incoming power. The frequency of the power supply affects the compressor's rpm and therefore its pumping capacity.

Horsepower versus Capacity

There are a few generalizations that can be given regarding horsepower and BTUH capacity. For normal condenser operating temperature ranges, the lower the evaporator temperature, the more horsepower required to produce the same BTUH capacity. The following general rule of thumb can be helpful as a starting point.

- **Air-conditioning temperatures** 1 hp per ton of cooling
- **Medium-temperature refrigeration** 1 hp per two-thirds ton of cooling
- **Low-temperature refrigeration** 1 hp per one-third ton of cooling

Low-temperature refrigeration may take 3 or more horsepower to produce 1 ton of cooling depending on the box temperature and outside air temperature. Refrigeration compressors can be very large compared to air-conditioning compressors with the same ton capacity.

The intake ports on low- and medium-temperature compressors are larger in size than the intake ports on air-conditioning compressors. They have to be larger to allow enough low-density, low-pressure refrigerant vapor into the compression cylinder. If a compressor is used on a higher-temperature system it will take in too much refrigerant for each compression cycle. This will result in the compressor overheating and failing—that is, a low-temperature compressor cannot be used on a medium-temperature or air-conditioning application, nor can a medium-temperature compressor be used on an air-conditioning application.

The smaller intake ports on air-conditioning or medium-temperature compressors will not allow enough refrigerant to enter the compressor cylinder if they are used on a low-temperature application. Too little cool refrigerant entering the cylinder will cause the compressor to overheat, too.

TECH TIP

When you buy a hermetic compressor you are buying both a motor and a compressor. The size and type of motor you need will change depending upon what you are asking the compressor to do. For this reason, hermetic and semihermetic compressors are classified by the saturation temperature of the suction gas entering them and the type of refrigerant they are designed to handle. Never use a compressor for an application for which it is not intended to be used.

15.20 CAPACITY-CONTROL FACTORS

Compressor capacity control refers to controlling how much work the compressor can perform. The idea of capacity control is to make the compressor capacity match the load. The energy required to operate the compressor is minimized by matching the compressor capacity to the load because an unloaded compressor uses less power than a fully loaded compressor. A compressor that runs continuously at reduced capacity will draw fewer total watts of power than one that starts and stops while running at full capacity.

Capacity control also improves building comfort. Many large buildings have a latent load that must be maintained even when there is a light load on the building. By providing this unloading capacity, the system can run in what is sometimes referred to as background so that the air is dehumidified without significantly cooling it.

There are a variety of ways of changing compressor capacity to match system load:

1. Cycling multiple compressors on and off
2. Cylinder unloading
3. Hot-gas bypass
4. Speed control
   a. Two speed
   b. Variable-frequency drives (VFD)

Multiple compressors with either common or independent refrigerant circuits are sometimes used. A twin compressor installation would give the choice of 0, 50, and 100 percent capacity steps (Figure 15-53).

Figure 15-53 Capacity control by using two compressors on a common refrigerant circuit.

Figure 15-54 The standard head is on the left; the head with the unloader is on the right.

Cylinder unloading is available on reciprocating compressors with multiple heads. Compressor unloaders can be actuated by gas, by oil pressure, or electrically. They can unload one or more cylinders by holding the suction valves open, bypassing the cylinder discharge to the suction manifold, or blocking off the suction inlet.

Figure 15-54 compares a normal compressor head and a head with an unloader. When the unloader is energized, gas passes from the high side of the head to the low side of the head, eliminating any work from that head.

Depending on the compressor, multiple stages of control can be provided and power savings achieved. The number of steps depends upon the number of heads. A three-head compressor can have steps of 33, 66, and 100 percent.

Hot-gas bypass is sometimes used on small machines where cylinder unloading is unavailable or on larger machines when additional capacity reduction is required. With hot-gas bypass, some of the compressor discharge gas is piped back into the suction of the compressor. Care must be taken not to overheat the compressor. It can be a practical solution in dealing with light loads. This process is not energy efficient.

Some compressors are available with two-speed motors. This technique (two-speed control) is not widely used at present. Single-phase compressors change speed by switching from a four-pole motor on low speed to a two-pole motor on high speed. Three-phase motors change speed by changing from a wye arrangement on low speed to a delta arrangement on high speed. Wye arrangements join the three windings together in the center to form a Y shape. Delta arrangements join the three windings together to form a triangle, or delta shape. Typically, an arrangement of contactors controls the speed of the compressor motor. These types of compressors must shut down to change speeds.

Variable frequency drives (VFD) enable continuous variable speed (capacity) control on many machines. These have been applied to large centrifugal compressors as well as to smaller comfort cooling applications. Variable speed control works very well and is likely to be increasingly used as the cost of VFDs (also called inverters) continues to drop. It can be very energy efficient.

Brushless DC motors offer a very efficient way of controlling compressor motor speed and system capacity. Two types are currently used: electronically commutated (ECM) and switched reluctance. Currently only a few very small fractional horsepower hermetic compressors use ECM technology. Compressors up to 5 hp are available with switched reluctance drive . These motors do not have a starting current surge but ramp up slowly to avoid inrush current.

Slide valves are commonly used on screw compressors to regulate capacity. The valves work well and save energy. They function based on diverting gas to the suction manifold before it is compressed.

Inlet vane capacity controls are commonly used on centrifugal chillers. These vanes can control down to 10 to 25 percent capacity. They function by giving the entering gas a pre-rotation swirl.

15.21 TROUBLESHOOTING THE COMPRESSOR

Compressors built today are expected to provide many years of constant, trouble-free, quiet operation. In many applications, the compressor is required to run twenty-four hours per day, 365 days a year. Continuous operation is not as hard on a compressor as is cycling operation, because oil temperature and viscosity constantly change during cycling operation.

CAUTION

Some INCORRECT procedures used in the past for testing compressors recommended that the suction line service valve be closed to see how deep a vacuum the compressor would pull. DO NOT use this procedure! It can result in deep vacuum, which can allow internal arcing to occur in the electrical windings of the compressor. This practice is NOT recommended by any refrigerant compressor manufacturer and must not be performed. It can result in compressor failure.

The compressor must not only be designed to withstand normal operating conditions but also occasional abnormal conditions such as liquid slugging and excessive discharge pressure. Compressors have been designed to take extra punishment and yet function properly. Most compressor failures are caused by system faults and not from operating fatigue. The degree of skill used by the technician to install, operate, and maintain the equipment will ultimately determine the actual life of the system, particularly the compressors. It is therefore helpful to review some of the factors that shorten the life of a compressor.

Compressor failures can be grouped into two general categories: mechanical failure and electrical failure. Sometimes a mechanical failure will lock the compressor, causing it to burn out the motor, thus looking like an electrical failure. This may cause a technician to overlook the true cause of this mechanically/electrically failed compressor. The result can be that the uncorrected mechanical problem will cause replacement compressors to fail in a similar manner. Only the mechanical failure problems will be covered in this unit.

The most common mechanical failures include liquid refrigerant floodback, liquid refrigerant slugging, oil slugging, flooded starts, high discharge temperature, and loss of lubrication. Many of these mechanical failures can be diagnosed by disassembling and inspecting the failed compressors. System operational diagnosis can often locate the cause of these problems. A thorough check of the system operation should always be performed after replacing a failed compressor.

SERVICE TIP

If a compressor had both a mechanical and electrical failure, you can bet the mechanical failure occurred first. There is virtually no way that an internal mechanical failure can occur once a compressor fails electrically. If you have both types of failures, look for the cause of the mechanical failure. The electrical failure is probably only a secondary failure resulting from the mechanical problem.

15.22 LIQUID REFRIGERANT FLOODBACK

Liquid refrigerant floodback occurs while the system is operating. A system operating with a suction line superheat of 5° or less is probably experiencing flooding. This can be caused by dirty air filters, dirty or iced evaporator coils, inoperative evaporator fan motors, overfeeding expansion valves, or overcharged systems.

Oil Dilution

Liquid returning to the compressor does damage, even if the compressor does not immediately drop dead. Small amounts of liquid dilute the oil. This oil-refrigerant mixture is not as good a lubricant as the oil alone. This causes premature bearing failure. Any liquid that reaches any lubricated part of the compressor effectively washes off the lubrication and causes excess wear. Figure 15-55 shows the result of oil dilution.

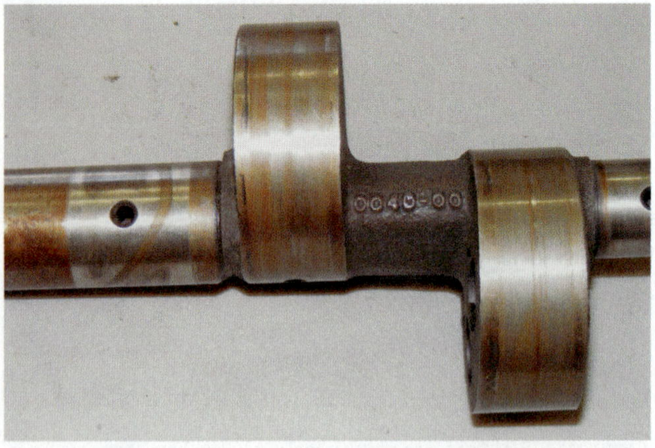

Figure 15-55 Bearing surface is worn due to oil dilution.

Figure 15-56 Liquid slugging can break internal compressor parts.

Figure 15-57 Compressor heads should not be discolored; this head shows the effect of high discharge temperature.

Liquid Slugging

If large amounts of liquid are returning in the suction line, some of it may make it into the cylinder space. Liquid does not compress. If the liquid cannot leave quickly, the mechanical force is quickly converted into thousands of pounds of hydraulic pressure, and something will break. Broken valves, valve plates, rods, or crankshafts are all symptoms of liquid slugging. Figure 15-56 is an example of the damage that can be done with liquid slugging.

Oil Slugging

Oil slugging does the same type of damage that liquid refrigerant slugging does. It is caused by the sudden return of large amounts of refrigerant oil. This usually indicates a refrigerant piping problem or system issue. Suction lines with large traps can fill with oil and then send it back in one burst. Systems operating at reduced capacity for long periods of time are more likely to trap out oil in the lines.

15.23 FLOODED STARTS

Refrigerant dissolves in the oil during the off cycle. System arrangement and location can play a part in allowing large quantities of refrigerant to migrate into the oil in the compressor while it is shut down. When the compressor starts, the refrigerant boils out violently, creating a foamy refrigerant-oil mixture. This is similar to what happens when you shake a carbonated beverage and then open it. This foam can fill the crankcase and make its way into the cylinder. Enough foam in the cylinder will cause hydraulic pressure and break parts. Off-cycle migration can be minimized with crankcase heaters and pump-down cycles.

15.24 COMPRESSOR OVERHEATING

All compressors get hot while working. The lower the saturation temperature of the suction gas, the hotter the compressor gets. One reason for this phenomenon is the amount of energy required to move heat at a low temperature level up to a much higher temperature level. It takes a lot of mechanical energy to do this work.

There is a limit to how hot a temperature the internal parts of a compressor can withstand. Internal compressor parts can be damaged due to the high temperatures. The oil used to lubricate the compressors starts to fail around 300°F. It is critical to keep the internal compressor temperature below 300°F to avoid certain damage. The compressor internal temperature is approximately 50–75° higher than the discharge line temperature. A compressor operating at excessive temperatures can be diagnosed by measuring the discharge line. The maximum discharge line temperature should be between 180°F and 220°F. A compressor operating at or above these temperatures is living on borrowed time. Varnished oil deposits on the head and valve plate indicate that this system has been operating at consistently high temperatures (Figure 15-57).

Excessive operating temperatures are caused by having too little refrigerant entering the compressor. This can be because of system charge, a refrigerant restriction in the system, or a misadjusted metering device.

15.25 OIL FAILURE

Oil failure can occur when internal compressor parts reach temperatures in excess of 300°F. At temperatures ranging from 315°F to 325°F, many refrigerant oils do not provide adequate lubrication. At a temperature of 350°F and above, most refrigerant oils begin to break down chemically. *Carbonization* is the term used to describe the formation of free carbon in refrigerant oil. Free carbon in overheated oil is what makes the oil look black in color. When oils carbonize they lose most of their ability to lubricate moving parts.

SAFETY TIP

Compressor oil may contain acid. This is especially true of oil that has been overheated. Care should be taken when handling the oil from a failed compressor, because the acid can cause serious burns.

TECH TIP

Polyol ester oils (POE) are clear but over time may turn from a light straw color to a darker brown. This does not mean the oil is carbonizing but simply that the oil's protective additives are becoming activated.

15.26 LOSS OF EFFICIENCY

The loss of efficiency of a compressor is usually an indication that the compressor is being subjected to system problems that are wearing some of the component parts. For a reciprocating machine, this can result from a number of conditions.

Leaking Valves

A reciprocating compressor with leaking valves cannot maintain the desired pressure difference. A high suction pressure and low discharge pressure are the most common symptoms. The compressor gets hot because the amount of refrigerant circulated is less. The amp draw is low because the compressor is doing less work. The system superheat is high and the system subcooling low.

Blown Head Gasket

Liquid in the clearance space can force its way from the high side to the low side by making a new path through the head gasket. This has the same effect as leaking valves.

Worn Wrist Pins, Rods, and Bearings

Systems that have leaking discharge valves or blown head gaskets will eventually develop problems in their piston wrist pins and rods. Gas leaking by the discharge valves puts continual downward pressure on the pistons. This prevents oil from getting in the spaces at the top of the bearing surfaces, so these surfaces wear. The holes elongate, causing slop in the piston action (Figure 15-58). When a compressor with this problem starts, it is very noisy until the head pressure builds up. Worn bearings, especially loose connecting rods and wrist pins, prevent the pistons from rising as far as they should on the compression stroke. This has the effect of increasing the clearance volume and results in excessive reexpansion of refrigerant vapor in the cylinder.

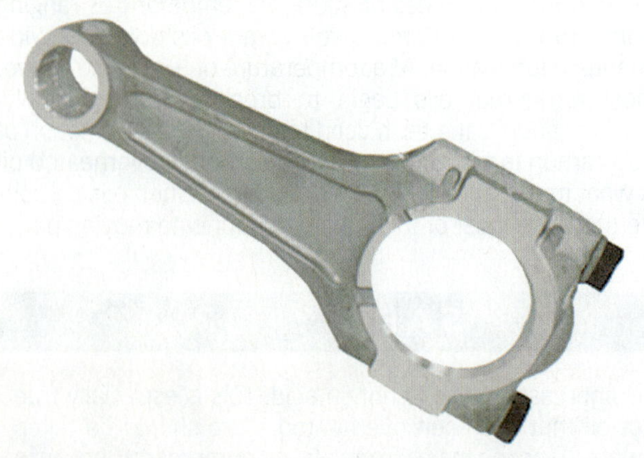

Figure 15-58 Elongated rod was caused by constant pressure on top of the piston because of leaking discharge valves.

15.27 MOTOR LOADING

When the compressor is not performing satisfactorily, the motor load sometimes provides a clue to the trouble. Either an exceptionally high or exceptionally low motor load is an indication of improper operation. Mechanical problems such as loose pistons, improper suction valve operation, or excessive clearance volume usually lead to reduction in motor load but an increase in motor temperature. Another common problem is a restricted suction chamber or inlet screen (caused by system contaminants). The result is much lower actual pressure in the cylinders at the end of the suction stroke than the pressure in the suction line as registered on the suction gauge. If so, an abnormally low motor load will result.

Improper discharge valve operation, partially restricting ports in the valve plate (which do not show up on the discharge pressure gauge), and tight pistons will usually be accompanied by high motor load. Abnormally high suction temperatures created by an excess heat load will cause a high motor load. Abnormally high condensing temperatures, created by problems associated with the condenser, will also lead to high motor load. Low voltage at the compressor will cause high amp draw and overheated motors.

15.28 COMPRESSOR CHANGEOUT

Compressor changeout is considered one of the more important and difficult jobs a service technician will perform. This is because the compressor is the most expensive component and because so many electrical and refrigeration skills are required in order to change a compressor. There are several steps involved in replacing a hermetic compressor that involve procedures covered in other units. References to these procedures will be included. For safety, you should first disconnect and lock out power to the unit (Unit 32). Use an electric meter to test for voltage at the unit to be sure the power is off (Unit 35). It is also a good idea to discharge the compressor run capacitor since the run and start wires to the compressor are connected to the start and run terminals of the compressor (Unit37). Recover the system refrigerant (Unit 28).

On larger systems with both suction and discharge service valves, the valves can be front seated to isolate the compressor so that you only need to recover refrigerant from the compressor. After the refrigerant has been recovered, the piping connections to the compressor are either cut out or unbrazed. Cutting the lines to the compressor is recommended because it is both safer and cleaner. Use a tubing cutter to avoid creating chips that can fall into the lines. Try to cut as closely to the compressor as possible. Oxides that can contaminate the new compressor are formed inside the tubing when the compressor is unbrazed. Technicians have been severely burned by unbrazing a compressor that still had a refrigerant charge. Even a compressor without a refrigerant charge will push out a small flame spout when the tubing is pulled free.

Next, physically remove the compressor and mount the new compressor. Remove the caps or plugs holding the nitrogen charge in the new compressor and connecting the system tubing. To avoid allowing air and moisture into the new compressor, the caps should not be removed until just before the new connections are made. If the replacement compressor is not an exact match, you may need to fabricate some tubing to make the connections between the compressor and the existing tubing. The connections should be nitrogen brazed to reduce oxide formation (Unit 25). Ideally, driers should be installed on both the suction and liquid lines. On severely contaminated systems, the lines may need to be flushed first. (See Unit 91 for more details on system cleanup.) After the compressor and filter driers are connected, the system should be leak tested (Unit 29), evacuated to below 500 microns (Unit 30), and charged according to the manufacturer's recommendations (Unit 31).

UNIT 15—SUMMARY

Compressors are vapor pumps. The compressor's job is to raise the pressure and temperature of the refrigerant so that it can be condensed. Together, the compressor and the metering device create the high side and low side of the refrigeration system. Compressors may be classified by their general operating principle, physical enclosure, mechanical design, cooling mechanism, and evaporator saturation temperature. The different compressor mechanical designs include reciprocating, rotary, scroll, screw, and centrifugal.

All types of compressors work by reducing the gas volume. Reciprocating compressors use pistons moving up and down in cylinders, rotary compressors use a rolling piston orbiting in a cylinder, scroll compressors use a movable scroll orbiting inside a fixed scroll, screw compressors use helical screws meshing together, and centrifugal compressors use a high-speed impeller that throws the refrigerant outward using centrifugal force.

Compressor enclosure types include open, hermetic, and semihermetic. Open compressors are driven by an external motor and have a shaft seal. Both the compressor and the motor are enclosed in hermetic and semihermetic compressors. Hermetic compressor shells are welded together, while semihermetic compressor housings are bolted together. Hermetic compressors are cooled by the refrigerant; semihermetic compressors are cooled by either refrigerant or air.

Common mechanical compressor problems include liquid refrigerant floodback, liquid refrigerant slugging, oil slugging, flooded starts, high discharge temperature, and loss of lubrication. Most compressor problems are caused by system problems. If the system problem that caused the initial failure is not corrected, repeat failures are certain. Technicians should analyze failed compressors to determine the cause and then check the system to eliminate the cause of the compressor failure.

WORK ORDERS

Service Ticket 1501

Customer Request: Compressor Replacement

Equipment Type: R-410a Split Refrigeration System

A customer has a 3-ton scroll compressor that was almost new when it was removed from the building's air-conditioning system during remodeling last year. The compressor has been sealed and kept dry. They want you to install it on their walk-in cooler. The cooler is operated at 30°F, and the condenser is located in the storage room, which is kept at a constant 80°F. You check the box heat load and see that it was calculated to be 35,000 BTUH.

The customer looked up the compressor data on the Internet. They point out to you that the compressor has slightly more capacity with a 30°F evaporator temperature and with an 80°F condensing temperature than the box load. The customer then asks you, "Can you use the compressor?"

You tell them no. You explain that the evaporator temperature for a walk-in cooler will have to be 10°F to 20°F below the box temperature. That would mean that the evaporator temperature would have to be somewhere between 10°F to 20°F, and according to the chart, it would produce between 23,200 to 28,800 BTUH—not enough to keep the walk-in at 30°F.

But most important, you tell them that the scroll they have is for air conditioning and they need a refrigeration compressor. The compressor looks like the old one on the walk-in, but internally it is much different. It would not last very long if installed in a refrigeration application.

Service Ticket 1502

Customer Complaint: Compressor Shutting Off

Equipment Type: R-22 Commercial Refrigeration System with a Semihermetic Air-Cooled Compressor

The customer tells you that the compressor keeps tripping on the low-pressure cutout. They tell you that they know it is working just fine because the evaporator keeps making a good layer of ice between defrost cycles.

You check the system pressures and see that the evaporator is 20 psi below the designed operating pressure/temperature. On investigation you see that one of the evaporator fans is out. You change the fan, and the system pressure returns to its normal reading.

After learning from the customer that they had been pushing the low pressure cutout reset button for months before calling you, you tell them you want to check the refrigerant oil. The customer does not understand why that is necessary because the level is OK in the sight glass.

You explain that you are concerned that the compressor may have been overheated and that overheating can result from too low an evaporator temperature. After a brief explanation of compression ratio and overheating, they agree to let you do the additional work.

Upon investigation, you find the oil to be black and slightly acidic. You follow the manufacturer's recommendation to change the oil and deacidify the system. Several months later, while back on a different system service call, you see this compressor is still running well. You know your oil change and deacidification kept it going.

Service Ticket 1503

Customer Complaint: Poor Cooling Capacity

Equipment Type: R-22 Air-Conditioning System with a Hermetic Reciprocating Compressor

You are called to look at a residential air-conditioning system that is not cooling properly. The airflow across the evaporator is very low. The customer suggests that the system may need a charge because another technician added refrigerant two months ago when the evaporator coil was freezing up. The evaporator is not frozen, but the coil is so packed with dirt that the fins are not visible. You clean the evaporator coil and change the air filter. With proper airflow, the temperature drop across the evaporator is only 8°F. The suction pressure is 95 psig, the head pressure is 150 psig, the compressor body is hot, and the amp draw is low. You conclude that the compressor valves are damaged, reducing the pumping capacity of the compressor. Most likely, the damage was caused by liquid slugging as a result of poor airflow and overcharging.

Service Ticket 1504

Situation: New System Startup

Equipment Type: R-410a Rooftop Three-Phase System with a Scroll Compressor

You are asked to start a new R-410a rooftop air conditioner. The system operates on three-phase power. On initial startup, the compressor makes a loud rattling noise. You notice that the pressures are still equalized and the condenser fans appear to be running backward. You shut down the system to avoid compressor damage, swap L1 and L2 at the power lug on the unit, and restart the system. The compressor noise has disappeared, the pressures look normal, and the fans are running the right way. You conclude that two of the three-phase power legs were swapped, causing the three-phase motors to run backwards. Since scroll compressors only work in one direction, the compressor could not pump running backwards.

UNIT 15—REVIEW QUESTIONS

1. What is the only thing all compressors must pump?
2. List the major types of compressors by mechanical design.
3. Explain the difference between a positive-displacement compressor and a nonpositive-displacement compressor.
4. What is the function of the compressor within the refrigeration cycle?
5. What factors affect compressor capacity?
6. Why are the intake ports on low- and medium-temperature compressors larger in size than the intake ports on air-conditioning compressors?
7. Why are hermetic compressors designed for specific applications?
8. Why is more horsepower required to produce a ton of cooling at an evaporator saturation temperature of 0°F than at 40°F?
9. What is the purpose of the shaft seals on an open compressor?
10. List two advantages of open-drive compressors.
11. What is a hermetic compressor?
12. What is an advantage of the semihermetic compressor design?
13. What opens and closes the valves in a reciprocating compressor?
14. What is the clearance volume of a reciprocating compressor?
15. What are some common applications for rotary compressors?
16. How does a rolling-piston rotary compressor work?
17. How does a rotary-vane compressor work?
18. How does a scroll compressor work?
19. What rotates in a scroll compressor?
20. Explain what compliance means for scroll compressors.
21. Why should hermetic compressors not be operated in a vacuum?
22. What is the purpose of compressor capacity control?
23. How does a two-step scroll compressor achieve capacity reduction?
24. What controls the internal unloading solenoid coil of the two-step scroll?
25. At what two capacities can a digital scroll compressor operate?
26. What gives the digital scroll the appearance of varying its capacity?
27. What is the lowest percentage of full capacity at which a digital scroll can be operated, and why?
28. What are the three different screw compressor types?
29. How are the small gaps on the screw compressor sealed?
30. Why is the oil separator important to screw compressors?
31. How does a centrifugal compressor work?
32. What type of a system are most centrifugal compressors used on?
33. How can oil be delivered to internal moving compressor parts?
34. What is the purpose of a crankcase heater on a compressor?
35. List the ways to change the refrigeration capacity to match the system load.
36. Above what temperature do most refrigerant oils begin to break down chemically?
37. What is the compression ratio of a compressor with a suction pressure of 110 psig and a discharge pressure of 360 psig?
38. Why is it incorrect to close the suction line service valve to test a compressor?
39. What problem can be caused if liquid refrigerant enters the compressor?
40. What is cylinder gas reexpansion, and why is it a problem?
41. How can a technician tell if a failed compressor has been operating at excessive discharge temperatures?
42. How can a technician determine if an operating compressor is experiencing floodback?

UNIT 16
Condensers

OBJECTIVES

After completing this unit, you will be able to:

1. explain how air-cooled and water-cooled condensers work.
2. list the types of air-cooled condensers.
3. list the types of water-cooled condensers.
4. explain why it is important to keep the condenser clean.
5. compare the operation of an air-cooled condenser to a water-cooled condenser.
6. tell how to keep a cooling tower working efficiently.
7. calculate the heat rejected by a cooling tower.

16.1 INTRODUCTION

The purpose of a condenser coil is to reject unwanted heat from a refrigeration or air-conditioning system. It does this by condensing the high-pressure, high-temperature gas to a liquid. That is where the name *condenser* comes from; it describes what is happening inside it. For residential and light commercial split systems, the condenser coil is usually a part of the unit containing the compressor. This combination of compressor and condenser coil is referred to as a condensing unit.

16.2 CONDENSER OPERATION

The condenser is located on the discharge side of the compressor, as shown in Figure 16-1. The superheated refrigerant vapor enters the top of the condenser from the compressor and leaves the condenser as subcooled liquid refrigerant.

The function of the condenser is to transfer heat that has been absorbed by the system to air or water. In an air-cooled condenser, the outside air passing over the condenser surface dissipates the heat to the atmosphere. In water-cooled condensers, water is used to cool the condenser. The water may either be wasted or recirculated. Because of environmental concerns regarding wasting water, most of the cooling water today is recirculated by pumping it to a cooling tower, where the heat is transferred to the atmosphere by means of evaporation.

The pressure-enthalpy diagram shown in Figure 16-2 illustrates the action performed by the condenser. The hot discharge gas from the compressor is superheated when it enters the condenser. The superheat must be removed first before the refrigerant can condense. This occurs in the first

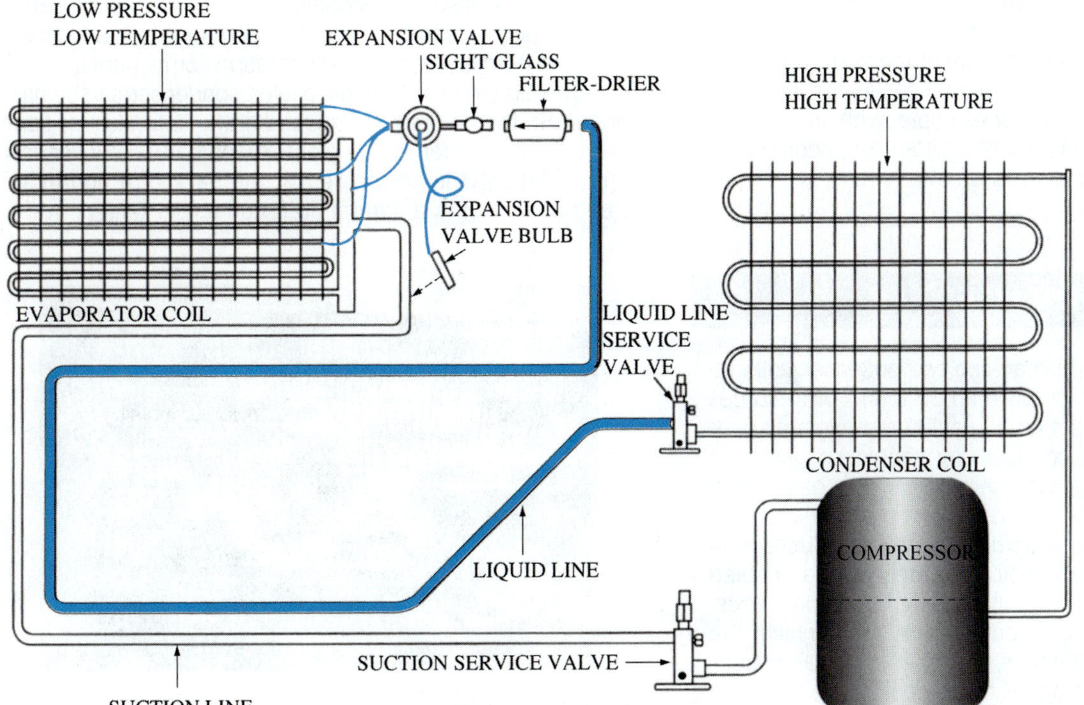

Figure 16-1 The condenser is the first major component after the compressor.

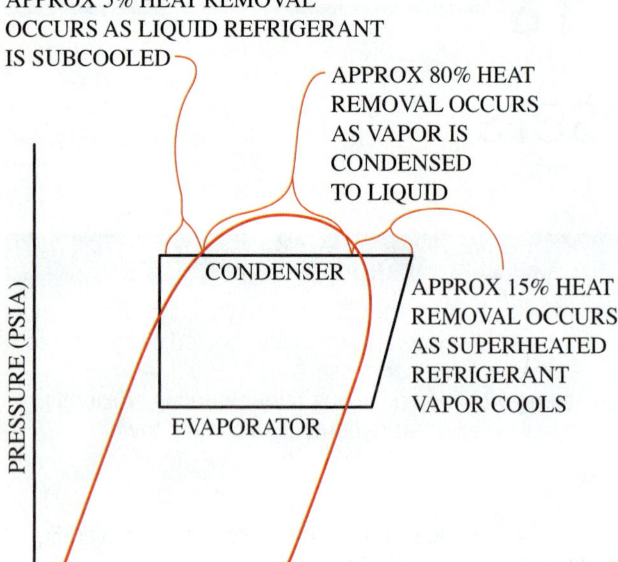

Figure 16-2 Heat removed from the refrigerant in the condenser.

portion of the condenser, known as the desuperheating section. Approximately 15 percent of the heat is removed here. Then the vapor is condensed throughout most of the condenser. Approximately 80 percent of heat is rejected through condensation. After all the vapor is condensed to liquid, the liquid is subcooled in the last portion of the condenser. Approximately 5 percent of the heat is rejected during subcooling.

Even though the subcooling accounts for only a small amount of the total heat rejection, it is important for two reasons:

- It ensures that a solid stream of liquid will enter the metering device.
- It adds to the cooling capacity of the system at a rate of about 0.5 percent of the total cooling capacity per degree of subcooling. For example: with 10°F of subcooling, 5 percent (0.5%/°F × 10°F = 5%) additional capacity is added to the system.

SERVICE TIP

It is very important that air-cooled condenser coils be kept clean, as dirt does build up on them. Dirt reduces the airflow and can act as an insulator, preventing heat transfer. Condenser coil cleaning should be part of a seasonal startup program. The coil must be cleaned with an appropriate solution to remove the dirt and debris. Failure to use a manufacturer-approved solution can result in damage to the coil. Extremely dirty air-cooled condenser coils can result in very high head pressures. If the head pressure and temperature get too high, the refrigerant can decompose, forming acids. Dirty coils result in high utility usage and less cooling.

16.3 TYPES OF CONDENSERS

Three types of condensers are used with HVAC/R systems:

- Air cooled
- Water cooled
- Evaporative, which is a combination air and water cooled

A condenser is a very useful application of a heat exchanger. The condenser has to be able to dissipate or expel all of the heat picked up in the evaporator plus heat from the compressor. Heat exchangers are made of metal to permit fast and efficient heat transfer. The hot refrigerant vapor is in contact with one side of the heat exchanger surface, and the transfer medium, such as air or water, is on the other side.

The application of an air-cooled condenser is usually the simplest arrangement, particularly if the condenser is located outside. Although the water-cooled condenser is more efficient, it is more costly to install. The examples that follow demonstrate some of the common uses of condensers.

16.4 AIR-COOLED CONDENSERS

Air-cooled condensers are commonly used for residential air-conditioning systems up to 5 tons and commercial systems up to about 50 tons in capacity. Air-cooled condensers give off heat to the air and are normally located outdoors. However, the condenser is part of the appliance on small systems like refrigerators and some medium-sized commercial refrigeration systems. These systems expel the heat to the air surrounding them. At normal peak load conditions, the temperature of the refrigerant in the condenser is 25°–30°F higher than ambient temperature. This means that on a 95°F day, the condensing temperature is between 120° and 125°F.

Air-cooled condensers can be part of a packaged system that contains all the refrigeration components in one cabinet; they can be part of a condensing unit that contains the compressor, condenser, and condenser fan in one unit located outside, or they can be a separate component in a location separate from all the other refrigeration components.

The airflow across an air-cooled condenser can be natural draft, forced draft, or induced draft. Natural-draft condensers have no fan and rely entirely on natural convection and radiation. The condenser on the back of a non-frost-free refrigerator is a natural draft condenser (Figure 16-3).

Figure 16-3 Example of a natural-draft air-cooled condenser.

Figure 16-4 The fan pushes air across a forced-draft condenser.

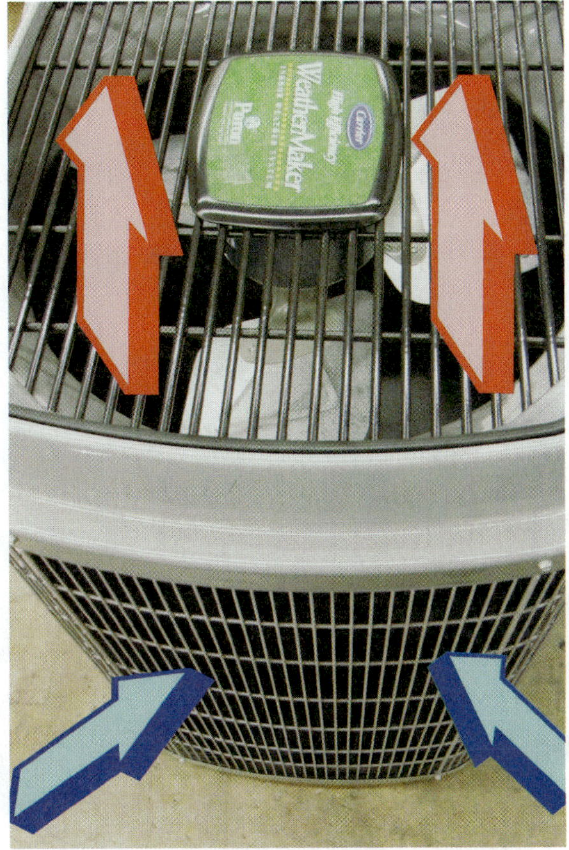

Figure 16-5 The fan pulls the air across an induced-draft condenser.

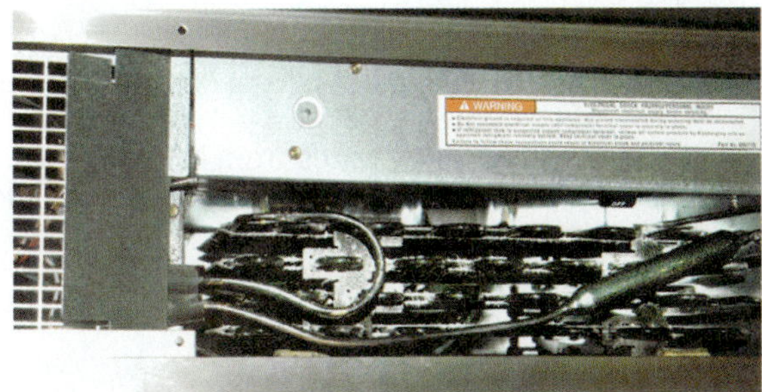

Figure 16-6 Condenser on a domestic refrigerator.

Figure 16-7 Condenser on a residential packaged air-conditioning system.

Forced-draft condensers use a fan to push air across the condenser. The cup of the fan blade faces the condenser on a forced-draft condenser (Figure 16-4). Induced-draft condensers use a fan to pull air across the condenser. The cup of the fan on an induced-draft condenser faces away from the coil (Figure 16-5).

16.5 AIR-COOLED CONDENSERS IN PACKAGED UNITS

One common use of the air-cooled condenser is for domestic refrigerators and upright freezers, as shown in Figure 16-6. The condensers are located in the lower part of the cabinet, and a small fan is used to pull room air over the surface of the condenser coil and exhaust it back out into the room. Some refrigerators use a natural-draft-type condenser coil on the back of the cabinet.

Packaged room coolers and through-the-wall air conditioners both use air-cooled condensers. These units are manufactured in sizes ranging from ½ to 3½ tons of cooling capacity. They are commonly used in houses, apartments, townhouses, condominiums, offices, schools, and motels.

Another application of the air-cooled condenser is in the residential and commercial packaged air-conditioning unit (Figure 16-7). Residential package units are available

Figure 16-8 A condensing unit consists of a compressor, a condenser, a fan, and controls.

(a)

(b)

Figure 16-9 Typical remote air-cooled condenser application.

for central air conditioning in sizes from 1½ to 5 tons of cooling capacity. Commercial packaged units can be much larger sizes, ranging up to 300 tons.

16.6 AIR-COOLED CONDENSERS IN CONDENSING UNITS

The residential condensing unit is a popular use of air-cooled condensers. This unit includes the compressor, condenser, fan, and outdoor circuitry, as shown in Figure 16-8. Refrigerant lines connect to an evaporator coil located on the top of a forced-air furnace indoors or to a fan coil unit. They use ambient air to remove heat from the condenser.

These units are called split systems because they consist of two parts: an outdoor condensing unit and an indoor blower coil.

Commercial field-assembled air-conditioning systems often have a remote air-cooled condensing unit. The portion of this system inside the building is primarily an air-handling unit, including the evaporator, the metering device, and a blower-filter unit. The portion on the outside includes the compressor, condenser, and condenser fan. This type of system can be labeled a split system because it is similar to the residential split systems, except larger.

16.7 REMOTE AIR-COOLED CONDENSERS

The air-cooled condenser can be installed as a separate component in a remote location, away from all the other refrigeration components (Figure 16-9). With this arrangement, the compressor is placed inside the building out of the weather, where it will last longer. This is popular with ice machines. Ice machines with remote condensers avoid exhausting more heat into the kitchen, which is already a very warm place. The two lines going from the system to the condenser are the discharge line and the liquid line. They are both high-pressure, high-temperature lines and are normally not insulated. Occasionally the liquid line is insulated if it is run a long distance through a very hot space.

16.8 AIR-COOLED CONDENSER CONSTRUCTION

Air-cooled condensers are made of tubing with fins added to the tubing to increase the surface area in contact with the air. This is done to help accelerate the heat transfer rate

Figure 16-10 A copper tube and aluminum fin coil is shown on the left; a spine-fin aluminum coil is shown in the middle, and a micro-channel coil on the right.

out of the tubes when cooling the refrigerant. Air is not a good a conductor of heat. By adding tight-fitting aluminum fins to the refrigerant circuit's tubes, the air side surface and heat transfer can be greatly increased. The three most common constructions are copper tubing with aluminum fin sheets, aluminum tubing with aluminum "spine fins" that are wrapped around the tubing, and microchannel aluminum tubing with aluminum fins. Figure 16-10 shows an example of each type.

Copper Tubing with Aluminum Fin Coil Construction

To ensure good heat transfer, the fins must be tightly mechanically bonded to the tubes. The tighter the bond, the better the heat can be conducted to the fins from the refrigerant tubing and refrigerant. The fins are pre-stamped from thin sheets of metal, usually aluminum. They may be left flat or embossed, forming curves or waves in their surfaces (Figure 16-11). The embossed fins create turbulence in the air as it travels over the coil. This improves heat transfer by increasing the contact between the air and the coil. The fin sheets are stacked with even spacing on the copper tubing. The tubes are then mechanically or hydraulically expanded into the tube collar. Figure 16-12 shows the manufacturing process of expanding the tubes within the fin sheets.

Air-cooled finned condenser coils can be flat, U-shaped, square, or round. The coils are made flat and then formed into the desirable shape with large hydraulic presses. Some fin coils are actually two separate coils tightly formed together.

Figure 16-11 The coil on the left uses straight fins; the coil on the right uses wavy fins.

Spine-Fin Coil Construction

The other major system for manufacturing coils is to wrap aluminum fins around aluminum tubing. This produces a coil that has a prickly, or spine-like, look—thus their name, spine-fin coils. The aluminum fins for these coils are long, thin strips of aluminum with slits. This material is wrapped around the aluminum tubing to produce tubing with spine fins (Figure 16-13). This tubing is then formed into round or square coils that resemble a slinky. Several of these are stacked on top of each other to produce multisection coils, and the coils are glued together in the corners (Figure 16-14).

Micochannel Coil Construction

Microchannel coils have been used for many years in the automotive industry, but they are relatively new in HVAC systems. They are made from ribbons of aluminum, with several small passages through them called microchannels. Fins zig-zag between the ribbons. Figure 16-15 shows the construction of a microchannel coil. Because microchannel coils are more efficient than traditional tube-and-fin coils, microchannel condensers are considerably smaller than tube-and-fin condensers of the same capacity (Figure 16-16).

SERVICE TIP

Microchannel coil condensers hold less refrigerant than comparable capacity tube-and-fin coils because they have much less internal volume. The refrigerant charge for a microchannel system can be half the charge of a similar tube-and-fin system.

16.9 HIGH-EFFICIENCY CONDENSERS

Due to changes in federal law regarding the minimum SEER and EER for air-conditioning systems, manufacturers are being required to design higher-efficiency condensers. There are several ways that an air-cooled condenser can be made more efficient. Three common methods are as follows:

- Increase the coil size to increase the heat-transfer area.
- Increase the airflow rate to remove more heat.
- Decrease the condenser tube size so that more tubing can be added, which increases the heat-transfer rate.

In most cases, the higher-efficiency condensers will be larger in size than a comparably sized lower-efficiency condenser. Figure 16-17 shows a 2-ton 10-SEER air-cooled condenser and a 2-ton 13-SEER condenser side by side. This size increase must be considered when replacing an existing system. Make sure the new condenser will fit in the old condenser's spot. Some manufacturers have switched to

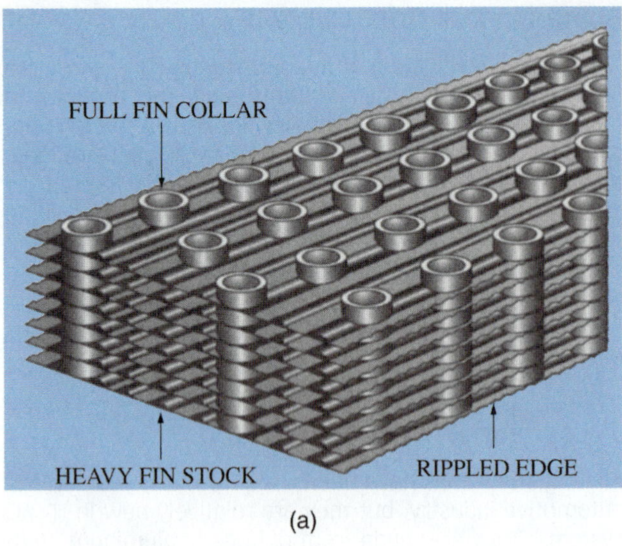

(a)

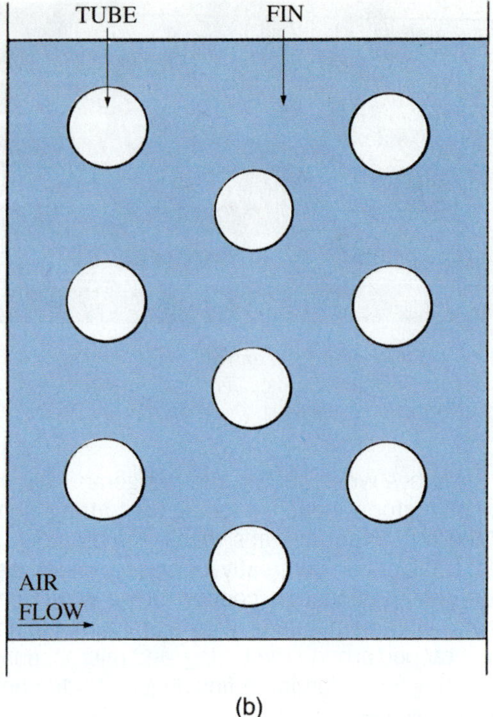

(b)

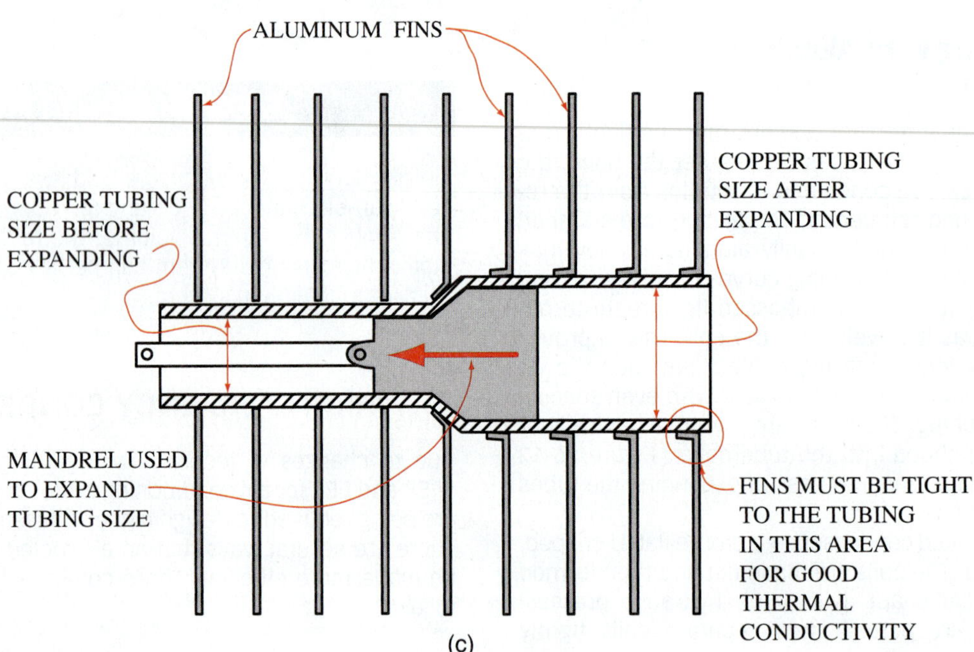

(c)

Figure 16-12 Plate-fin coil details: (a) four-row coil construction; (b) staggered tubes; (c) cross-section of finned tube assembly.

microchannel coils to keep the condenser sizes smaller on high-efficiency units.

16.10 LOW AMBIENT CONTROL

In the case of both commercial refrigeration and commercial air conditioning, low ambient operation is an important requirement. Many air-conditioning and refrigeration systems must operate when the outdoor temperature is low because of building heat loads. Restaurants, computer labs, hospitals, and office buildings all have excessive indoor heat loads, which require cooling even in cold weather. High internal heat loads from people, lights, and electronic equipment may demand cooling even when outside temperatures go down to 35°F. Low ambient controls are essential for these systems.

Operating an air-cooled condenser at low ambient temperatures can drastically reduce the system high-side pressure. If the drop in pressure is too severe, the metering

Figure 16-13 Aluminum strips are wrapped around tubing to create spine-fin coils.

Figure 16-14 Spine-fin coils are glued together at the corners.

Figure 16-15 Microchannel coil construction. *(Terra State Community College HVAC Program, Fremont, Ohio)*

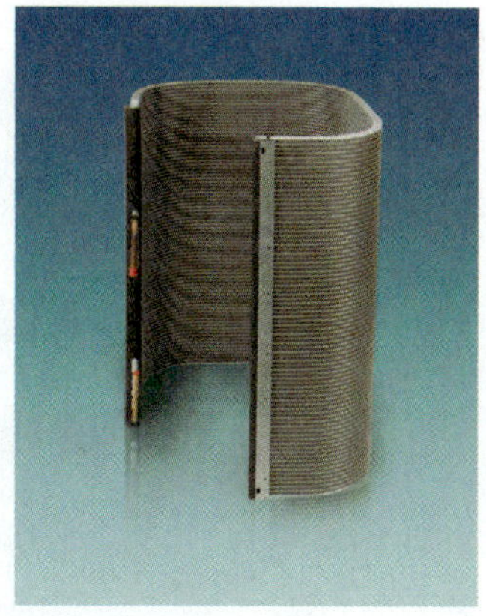

Figure 16-16 The smaller microchannel coil is the same capacity and SEER as the larger traditional tube-and-fin coil. *(Delphi Thermal Systems)*

Figure 16-17 The 13-SEER condenser on the right is considerably larger than the 10-SEER condenser on the left, even though they are the same capacity. *(Courtesy of INTERNACHI)*

device will not feed correctly. Low ambient operation can cause a drop in system capacity. Most metering devices will feed too little refrigerant when the pressure drop across the metering device is lower than the system design pressure differential. In a normal air-cooled condenser operating between 80°F and 115°F ambient, condensing pressures are sufficiently high. But in winter, condensing pressures can drop 100 psi or more, thus the pressure across the expansion device may be insufficient to maintain control of liquid flow. Evaporator operation becomes erratic.

Systems with fixed-restriction-type metering devices are most affected by low head pressure because they have no means of adjusting to changing conditions. When the pressure difference falls, the flow of refrigerant is severely reduced. At outdoor temperatures of 65°F and lower, systems with fixed restrictions and no head pressure control cannot operate reliably. Even systems with standard thermal expansion valves can adjust only so far. Low ambient temperature can cause thermal expansion valves to underfeed as well.

The solution to the problem of low ambient operation is to maintain a minimum head pressure to ensure proper feeding of refrigerant to the evaporator. There are several different ways of doing this. The most common are fan cycling, fan speed control, damper control, and condenser flooding.

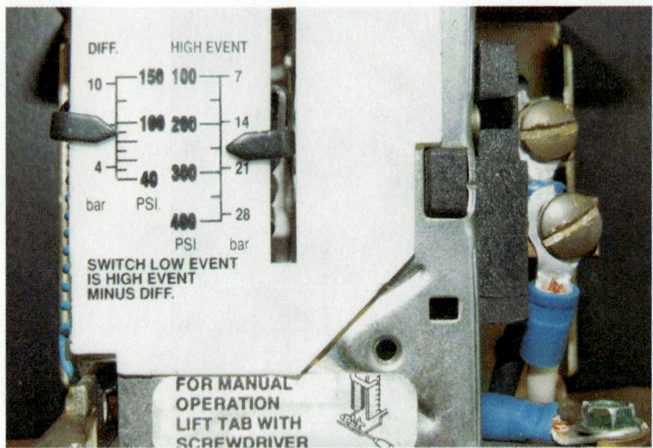

Figure 16-18 A "close on rise of high pressure" switch with an adjustable differential can be used to cycle condenser fans.

The use of improved expansion devices is reducing the need for head pressure control in some systems. Head pressure controls force the system to work as if it is 95°F outside all the time. Improved metering devices that can regulate with very low pressure drop reduce the need for head pressure control and save energy by allowing the system to operate against lower head pressures.

16.11 FAN-CYCLING LOW AMBIENT CONTROL

Where multiple condenser fans are employed on a single coil, the control system of the condensing unit can be equipped with devices to switch or cycle off the fans in stages. These controls are usually "close on rise" thermostats that measure outdoor ambient temperature or "close on rise" pressure controls that sense actual head pressure (Figure 16-18). The airflow across the coil is reduced as fans are turned off. Condensing temperatures rise with the reduced airflow. Under very low ambient conditions, all of the fans can be turned off.

One disadvantage of fan cycling is that the condenser pressure can swing rapidly between the low and high set points of the fan control. Although the system works with these swings in pressure, they can result in a loss of efficiency. Another disadvantage is that on windy days, even with the fans off, there may be too much cooling of the condenser to maintain minimum high-side pressure for the system to work properly.

Systems with a single condenser fan are generally not good candidates for fan cycling because of undesirable pressure swings. Fan cycling should never be used with air-cooled compressors if there is only one fan. Air-cooled compressors depend on the airflow across them to stay cool, so shutting off the fan, even at low ambient temperatures, can overheat an air-cooled compressor.

16.12 FAN SPEED CONTROL

Where only one condenser fan is used, air capacity can be reduced by employing a two-speed motor or an electronic speed control. The electronic speed controls designed for

Figure 16-19 This control will vary the speed of a single-phase fan motor to keep the head pressure from dropping below its set point.

head pressure control work by chopping part of the sine wave out, reducing the amount of current delivered to the motor (Figure 16-19). This is the same type of control often used for ceiling fans. These controls can maintain a very steady head pressure. Speed controls should not be used with condenser fan motors that have sleeve bearings. Sleeve bearings depend on a minimum speed to maintain lubrication, and turning too slowly will grind up their bearings.

16.13 DAMPER LOW AMBIENT CONTROL

Another technique of restricting airflow across the condenser coil is the use of dampers on the fan discharge. These are used on nonoverloading centrifugal-type fans, not propeller fans. Blocking the discharge of a centrifugal fan will cause a reduction in amp draw, but blocking the discharge of a propeller fan will increase the amp draw. Dampers modulate from a head pressure controller to some minimum position, at which time the fan motor is shut off. This type of head pressure control is generally only used on larger systems.

One common and important characteristic of the airflow-restriction methods is that the full charge of refrigerant and entrained oil is in motion at all times, ensuring positive motor cooling and oil lubrication to the compressor.

16.14 FLOODED-CONDENSER LOW AMBIENT CONTROL

Condenser-flooding pressure control systems are common on commercial refrigeration systems and rare on air-conditioning systems. This is because condenser-flooding controls normally require a liquid receiver. Receivers are common in commercial refrigeration but unusual in air conditioning.

Flooding controls artificially raise the head pressure by backing the liquid refrigerant up into the condenser tubes. An "open on rise of inlet pressure" control valve, or ORI, is used to accomplish this. The ORI control is placed in the outlet of the condenser, which closes when the condenser pressure starts to drop. This restricts drainage of the liquid refrigerant out of the condenser, flooding it. The pressure in the flooded condenser increases because less condenser area is available for condensing refrigerant. When the condenser pressure is at or above the valve's setpoint, the ORI control opens and allows liquid refrigerant to leave the condenser. Used by itself, this valve would keep the condenser pressure at the required level, but the liquid pressure after the condenser would actually drop even further when the valve is closed.

An "open on rise of differential pressure" control, or ORD, keeps the receiver pressure at the required setting. The ORD valve bypasses hot gas from the beginning of the condenser to the top of the liquid receiver. The purpose of this valve is to ensure enough pressure on the liquid receiver to keep the liquid in the liquid line at the minimum required pressure. The receiver is necessary because there must be a supply of liquid to the system while the condenser is being flooded. These two valves combined, the ORI and ORD, provide effective head pressure control under a wide range of operating conditions (Figure 16-20).

The functions of these two valves have now been incorporated into a single control, called an "open on rise of outdoor ambient," or OROA. This single, non-adjustable control simplifies selection and application of condenser-flooding head pressure controls (Figure 16-21). However, it is available only for certain refrigerants. It floods the condenser and bypasses gas to the receiver based on the ambient temperature to maintain a minimum pressure.

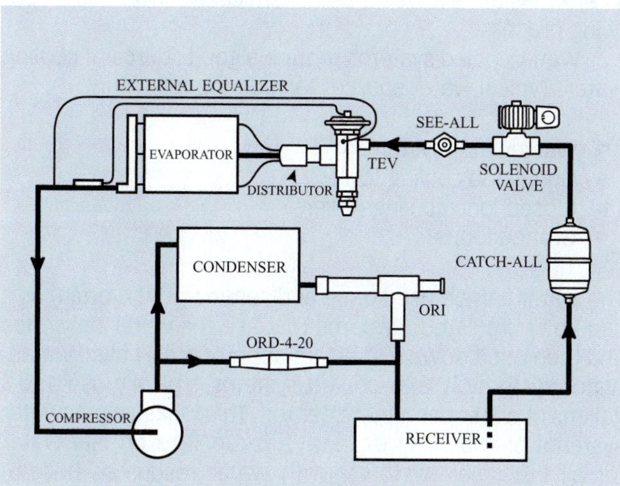

Figure 16-20 Piping diagram for application of ORI and ORD valves to control head pressure.

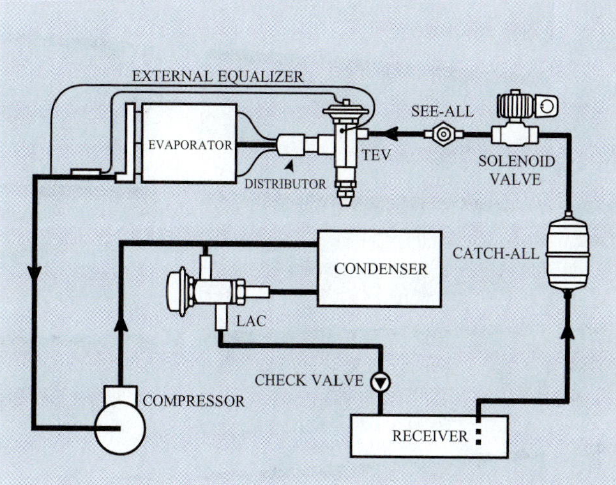

Figure 16-21 Piping diagram for application of an OROA valve to control head pressure.

Condenser-flooding head pressure controls do not suffer from pressure swings the way fan-cycling controls do, but they are more difficult to install in the field since they are piped into the refrigeration system.

TECH TIP

It is an unacceptable practice to overcharge a condenser during winter months to artificially increase the head pressure by flooding the condenser. This practice is not recommended by any manufacturer, because the first warm summer day can result in compressor failure.

16.15 SERVICING AIR-COOLED CONDENSERS

Most condenser problems are related to some type of restriction of the air. When a new condenser is installed, the operating pressures and temperature need to be recorded. These values can be referred to each time the system is serviced. A change in the observed operating conditions is a good indication that additional service work is needed.

CAUTION

If the condenser becomes dirty, its condensing pressure increases. This can be a serious problem when high-pressure refrigerants like R-410A are used. R-410A operates at normal head pressure of around 400 psig, but a dirty condenser can cause these pressures to greatly increase. Condensers are designed to withstand some amount of pressure beyond normal operating pressures. However, over time, the condenser can get progressively more dirty, resulting in elevated pressures. Catastrophic failure of the refrigerant circuit can result from excessively high pressures. It is very important that condensers containing high-pressure refrigerant, such as R-410A, be thoroughly cleaned on a regular basis.

Figure 16-22 Accumulated dirt can restrict airflow and raise the head pressure.

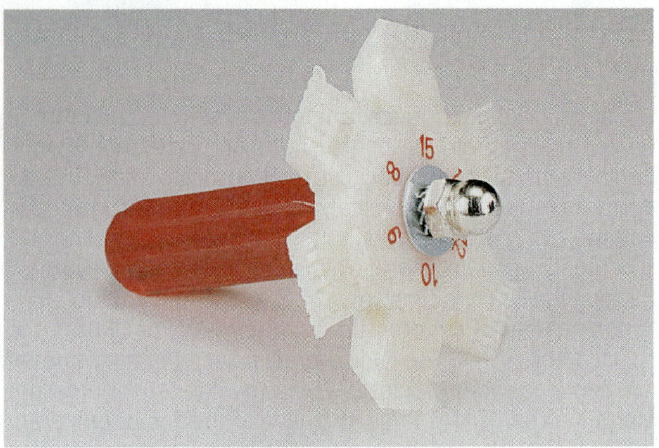

Figure 16-23 Damaged fins can be straightened using a fin comb. *(Courtesy of Ritchie Engineering Company, Inc.—YELLOW JACKET Products Division)*

Accumulation of dirt on the condenser will reduce the heat transfer rate, and the compressor head pressure can rise to damaging levels (Figure 16-22). High head pressures will result in decreased system capacity, increased operating cost, and possible compressor damage. Air-cooled condenser coils will naturally collect dirt over a period of time. Annual cleaning is sufficient under normal operating conditions, but more frequent cleaning may be necessary in particularly dirty environments. Light surface accumulation can be cleaned with a stiff bristle brush and vacuum cleaner. Grease or heavy accumulations will require coil-cleaning chemicals. The equipment manufacturer should be consulted regarding acceptable cleaning products. Not all condenser coils can be chemically cleaned. Read and follow both the coil manufacturer's literature and the coil-cleaning chemical manufacturer's instructions before starting. The wrong chemical cleaner can attack the condenser and do more harm than good.

SAFETY TIP

Proper personal protection equipment should always be used when applying chemical cleaners. Typically this includes safety goggles for eye protection, gloves to protect your hands, and a long-sleeve shirt to protect your arms.

Coils may need repairing because they have been damaged mechanically or as a result of corrosion. Mechanical damage can be weather related, such as from ice and sleet, or from vandalism. Corrosion can result from environmental pollution, sprinklers, or pets.

Mechanical damage of the fins can be straightened with a fin comb (Figure 16-23). If more than 20 percent of the fins have been damaged, the condenser's efficiency will be so low that under heavy summer loads, the compressor is likely to fail. Light corrosion of the condenser fins can be removed with an approved coil cleaner. Severe corrosive damage to condenser fins may not be reparable, so the condenser coil must be replaced.

16.16 WATER-COOLED CONDENSERS

Water is a more efficient heat-transfer medium than air. The temperature difference between the refrigerant and the cooling medium is lower with a water-cooled condenser than with an air-cooled condenser. The temperature differential between the refrigerant and the leaving water is typically 10°F. Water-cooled condensers normally operate with a 15°F lower compressor discharge temperature than air-cooled units. This allows the compressor to operate at a lower discharge pressure, increasing its capacity and lowering the power requirements. This makes water-cooled units potentially more energy efficient than air-cooled units. Water-cooled condensers are also smaller and more compact than comparable air-cooled condensers.

Water-cooled systems are more expensive to install and require more maintenance than air-cooled condensers. The water must be tested and treated regularly to prevent corrosion, fouling, and scaling. While most air-cooled condensers only need an annual inspection and cleaning, the water in a water-cooled condenser may need to be checked as often as once a week, and chemical treatment is an ongoing process.

Water cooled systems require a good source of cooling water. Typical water sources include the following:

- Wastewater systems
- Open-loop cooling towers
- Closed-loop cooling towers
- Ground loops

The water can be used once and sent down the drain, or it can be recirculated. Systems that use the water only once and then put it down a drain are appropriately called wastewater systems. Water-cooled systems typically use 1.5–3 gallons of water per minute per ton. This makes wastewater systems prohibitively expensive to operate on city water for all but small systems. With water resources becoming a precious commodity in many parts of the country, wastewater systems appear to be an extravagant use of a valuable resource. Recirculating systems must find a way to cool the water before returning it to the condenser. The most common method is to use a water tower that cools the majority of the water by evaporating a small portion of it. Cooling towers can be either open loop or closed loop. Open-loop cooling towers are called open loop because the water in the cooling tower is open to the air and the cooling tower water is the same water that circulates through the condenser. Closed-loop cooling towers use built-in heat exchangers to separate the cooling tower water from the recirculating condenser water. Another method is to use a loop of pipe buried in the ground. The water is cooled by the ground and returned to the system. These systems are called closed loops because they are not open to the atmosphere.

16.17 TYPES OF WATER-COOLED CONDENSERS

There are three basic types of water-cooled condenser construction:

- Tube in tube (coaxial)
- Shell and coil
- Shell and tube

16.18 TUBE-IN-TUBE CONDENSERS

Double-pipe condensers can be a single continuous loop, as shown in Figure 16-24, or several parallel pipes, as shown in Figure 16-25. In either case, the water and refrigerant should flow through the condenser in opposite directions. This is called counterflow. Counterflow is more efficient because the coldest water contacts the coldest refrigerant and the hottest water contacts the hottest refrigerant (Figure 16-26). Single-circuit double-pipe condensers can be constructed using a tube in tube, as shown in Figure 16-27, or two tubes soldered together and wrapped in a protective

Figure 16-24 Continuous-loop coaxial water-cooled condenser.

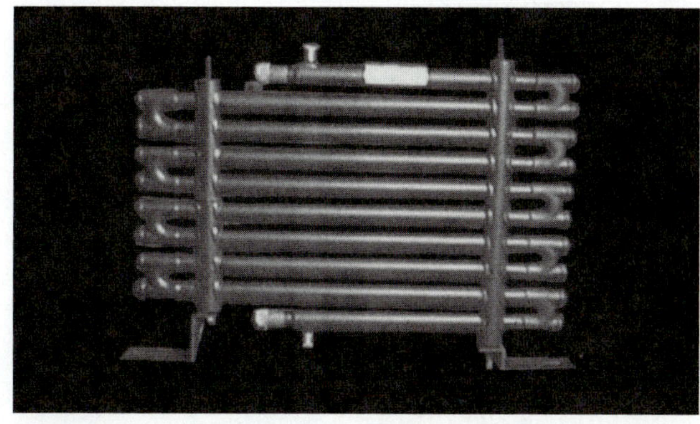

Figure 16-25 Parallel-flow double-tube water-cooled condenser. *(Courtesy of Doucette Industries, Inc.)*

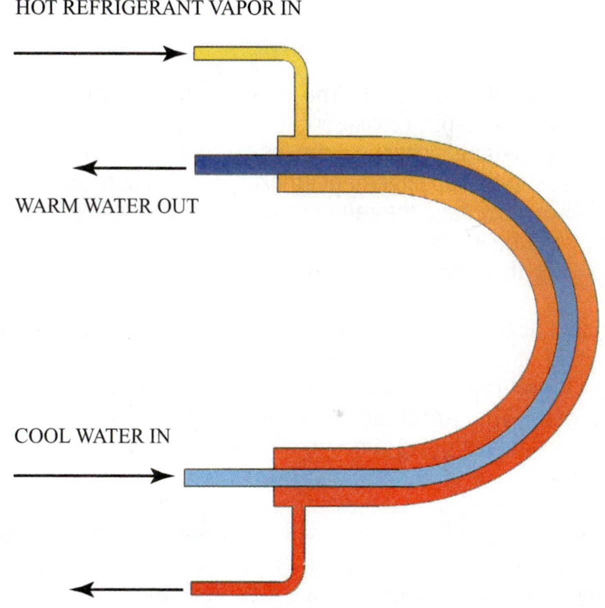

Figure 16-26 The water and refrigerant flow in opposite directions to improve efficiency.

Figure 16-27 Cross-section of a tube in tube coaxial water-cooled condenser.

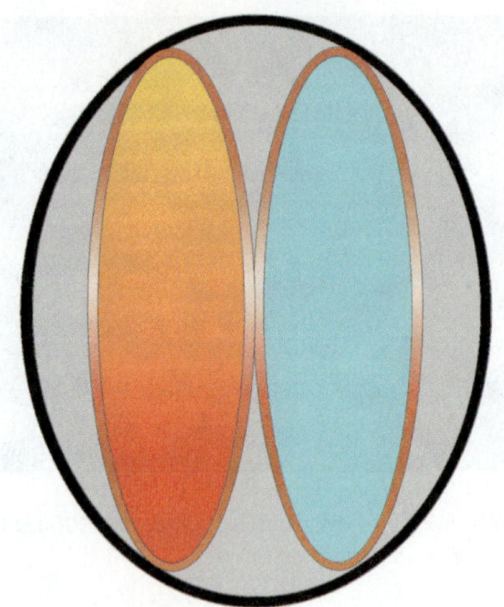

Figure 16-28 Cross-section of a double-walled water-cooled heat exchanger.

outer tube (Figure 16-28). The tube-in-tube construction is more efficient, but cross-contamination between refrigerant and water is possible if the inside tube leaks. The tube-beside-a-tube arrangement makes cross-contamination less likely because the refrigerant and water are separated by two tube walls.

The usual arrangement for tube-in-tube coils is for a smaller tube to be placed inside a larger tube, which is sealed at the end. Water travels through the smaller tube and refrigerant travels through the larger tube.

When multiple tubes are used, they connect to the headers. The water flows through the inner tube and the refrigerant flows through the angular space between the tubes. The inner tube may have a spiral surface (Figure 16-29). This agitates the water as it flows through the tube so that it does not stratify with the cooler water flowing in the center of the tube. An advantage of this design is that it can be wrapped into a shape to fit the space available. For example, the packaged condensing unit shown in Figure 16-30 uses the tube-in-tube condenser.

Figure 16-29 The twisted tubing shape improves heat transfer by creating turbulence in the water.

Figure 16-30 Typical use of a tube-in-tube condenser designed for in-the-room installation.

Single-circuit tube-in-tube condensers are used on most water-source heat pumps. They are generally not accessible and can only be cleaned and descaled chemically. Parallel-circuit condensers with headers can usually be mechanically cleaned by removing the header access plates on each end.

The water tubes of a double-pipe condenser are typically made of copper, while the refrigerant pipe is made of copper or steel. The pipes can be constructed of stainless steel or cupronickel if more resistance to corrosion is required. Cupronickel coils are made of an alloy of 90 percent copper and 10 percent nickel that is more resistant to corrosion than copper.

CAUTION

Care is required when recovering refrigerant from a water-cooled condenser. During the recovery process, liquid refrigerant in the water-cooled condenser can vaporize and draw the temperature down to below the freezing temperature of water. Ice formed inside any sealed container such as a water-cooled condenser can rupture the vessel or damage the internal piping. To avoid this, liquid refrigerant should be recovered first to avoid boiling the refrigerant at a low pressure and temperature. The water must either be continuously circulated or else completely drained before recovering the refrigerant.

16.19 SHELL-AND-COIL CONDENSERS

Shell-and-coil condensers are typically used in larger systems. Shell-and-coil condensers are constructed with a welded steel shell and an internal finned copper tubing coil. Water flows through the finned coil, and the refrigerant in the shell condenses on the coil. The shell can run vertically for better heat transfer through convection or horizontally to meet possible space requirements. Shell-and-coil condensers also act like liquid receivers.

In a vertical shell-and-coil condenser, the water enters at the bottom of the coil and travels up. The refrigerant enters at the top of the shell, condenses, and leaves at the bottom of the shell.

Shell-and-coil condensers are sometimes used with water-source heat pumps and can be combined with the compressor to form a condensing unit package. These units are usually limited to 20 tons or less. Vertical-packaged air-conditioning units from 20 to 60 tons use shell-and-coil condensers due to their compact dimensions.

Shell-and-coil condensers are difficult to service in the field because of their construction and cannot be cleaned mechanically. They can only be cleaned and descaled chemically.

16.20 SHELL-AND-TUBE CONDENSER

The shell-and-tube type condenser (Figure 16-31), is used in the largest condensers. Capacities extend to 1,000 tons. Water flows through the tubes (the tube side of the condenser), and refrigerant flows on the outside of the tubes (the shell side of the condenser). These condensers have long, straight finned tubes, connected to a steel plate (tube sheet) at each end. At each end, water manifolds, called heads, are bolted to the shell (Figure 16-32). These heads direct the water to make from one to eight passes, depending on the size and design of the condenser. The heads can be removed to permit cleaning the individual tubes. Rubber gaskets provide a watertight seal.

The entering water in recirculated systems is typically warmer than with wastewater systems. Single pass and double pass designs are used for water tower systems to keep the difference between the condenser saturation and incoming water temperatures as low as possible, improving the efficiency of the refrigeration system. A higher water flow rate is required with fewer water passes because the condenser is operating at a lower water temperature rise. Multiple-pass designs are used to minimize water use by increasing the amount of heat picked up by each pound of water. Multiple-pass configurations operate with a higher temperature difference between the condenser saturation and incoming water temperatures and are used when water is scarce or expensive, such as for city wastewater systems (Figure 16-33).

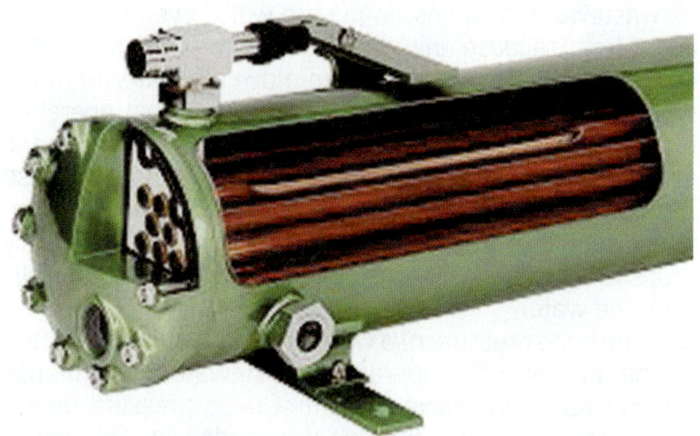

Figure 16-31 The water travels through the tubes and the refrigerant travels through the shell in a shell and tube condenser. *(Image Courtesy of Alfa Laval)*

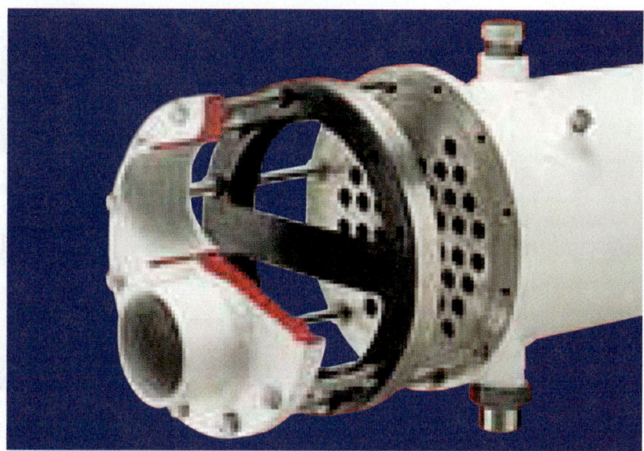

Figure 16-32 Shell-and-tube condenser header configuration. *(Image Courtesy of Alfa Laval)*

In a typical water condenser, the hot gas inlet and purge valve are on the top, the liquid outlet valve is on the bottom, and the pressure-relief valve is on the side. Many of these shell-and-tube condensers are used in large, water-cooled condensing units ranging in size from 5 to 150 tons. A common use of the shell-and-tube condenser is as a component of a water chiller, as shown in Figure 16-34.

16.21 WASTEWATER SYSTEMS

In earlier applications of water-cooled condensers to refrigeration and air conditioning, it was common practice to tap the water supply and then waste the discharge water to a drain connection as shown in Figure 16-35. An adjustable automatic water-regulating valve was placed in the line of the flow of incoming water to control the condenser operating head pressure and temperature. It got the system pressure through a pressure tap.

The temperature of the incoming water would naturally affect the condenser performance and flow rate of any heat load. Depending on the geographic location, water temperatures in city water mains rarely rise above 60°F in summer and frequently drop to much lower temperatures in winter.

Condensers piped for tap water flow were always arranged for series flow and circuited for several water passes to achieve maximum heat rejection to the water, which was then wasted (see Figure 16-33). Condensers drawing on city water used only 1 to 1½ gpm (gallons per minute) per ton of refrigeration. The multipass circuit created high water pressure drops ($P_1 - P_2$) of 20 psi or more; however, most

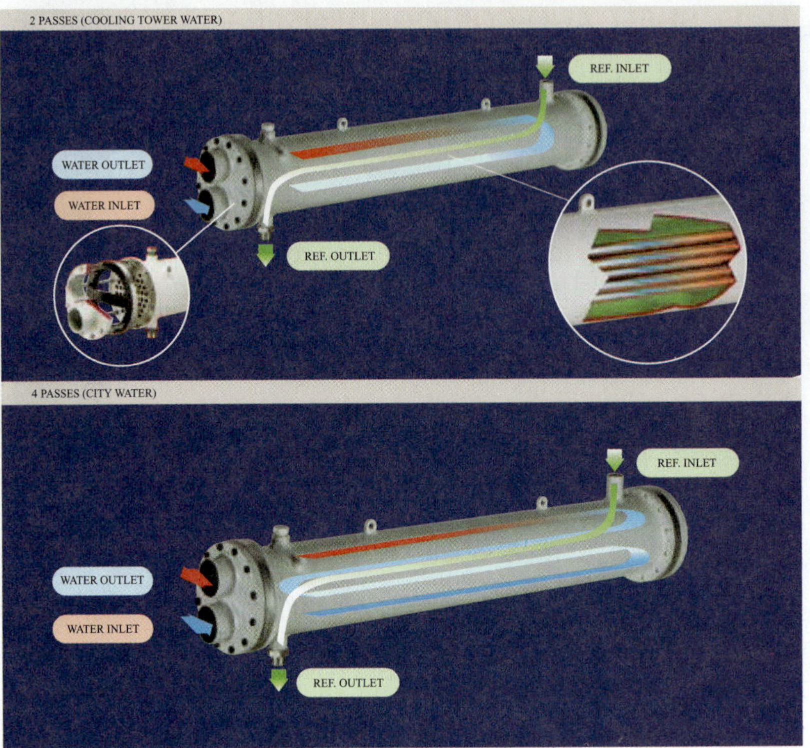

Figure 16-33 Water flow through a shell-and-tube condenser. *(Image Courtesy of Alfa Laval)*

Figure 16-34 Typical packaged liquid chiller unit with a shell-in-tube condenser.

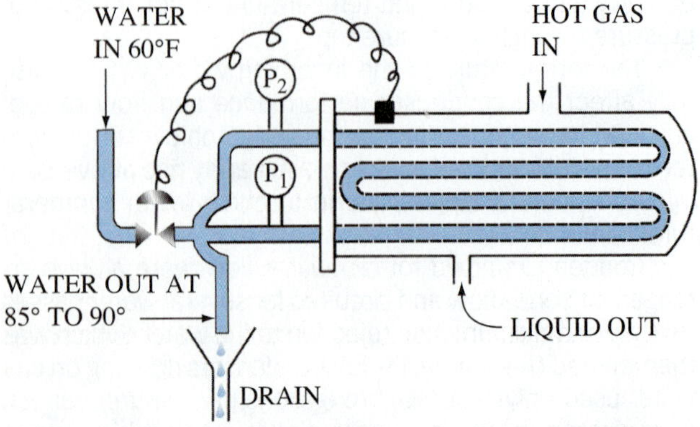

Figure 16-35 Wastewater system using water-regulating valve.

city pressures were able to supply the minimum pressure requirement (usually 25 psig).

Because of water shortages experienced in many areas of the country, water-cooled condensers that waste water are not recommended. The cost and scarcity of water (unless drawn from a lake or wells and returned) has become prohibitive and is even outlawed for refrigeration and air-conditioning use by many local city codes. Ordinances have restricted the use of water to the point where such installations have been forced to use all air-cooled equipment. Some water regulations are so restrictive that they may even require existing wastewater-cooled systems to be replaced. Most system operators have switched to water-saving devices such as the evaporative condenser or water tower.

Residential Well Systems

Many residential water-cooled systems have been installed as wastewater systems on private wells. Even well water is not free, because energy is required to pump the water. Operating a ½-hp water pump to move the water from the well to the unit can be a significant additional operating expense. Wastewater systems are more prone to mineral scaling than other types of water-cooled systems because the continual supply of fresh water also provides a continual supply of fresh minerals. The condenser efficiency drops off as scale builds inside it, insulating the tube walls from the water.

The head pressure of a water-cooled system is affected by the inlet water temperature. Wastewater systems use a water-regulating valve to control head pressure by adjusting the water flow through the condenser. The water-regulating valve measures the head pressure, increases the water flow when the head pressure increases, and decreases the water flow when the head pressure decreases.

16.22 COOLING-TOWER SYSTEMS

In most installations using water-cooled condensers, water is supplied to the condenser from a cooling tower. The heated water from the condenser is returned to the tower to dissipate the heat picked up from the refrigerant, as shown in Figure 16-36). There are four steps in the process:

- Heat is transferred from the refrigerant to the water in the condenser.
- The water is pumped from the indoor condenser to the cooling tower outdoors.
- At the tower, the heat is rejected to the outdoor air, cooling the water.
- The cooled water is returned to the condenser to pick up more heat, making a continuous process.

Typical operating conditions for a water tower system are as follows:

- The refrigerant condensing temperature is about 105°F.
- The water entering the condenser is approximately 85°, which is 20°F lower than the condensing temperature.
- The water temperature rises about 10°F in passing through the condenser, leaving the condenser at about 95°F.
- The leaving water is only 10°F below the saturated refrigerant temperature.

When the water gets to the cooling tower, it is at 95°F. The warm water drops over the cooling tower's wetted decking, while outside air moves past it in the opposite direction, causing evaporation. Each pound of water that is evaporated removes 970 BTU from the remaining water, cooling it. A 10°F drop in temperature occurs by the time the water reaches the bottom of the tower. This 85°F water is returned to the condenser to absorb more heat. Water is lost by evaporation, drift, and blowdown. Drift is atomized water that has not vaporized but is caught by airstreams and carried away. Blowdown is the periodic intentional release of water to reduce the mineral concentration in the water. This water is replaced by makeup water, which is regulated by a float valve in the sump of the tower. See Unit 82, Cooling Towers, for more information on cooling-tower systems.

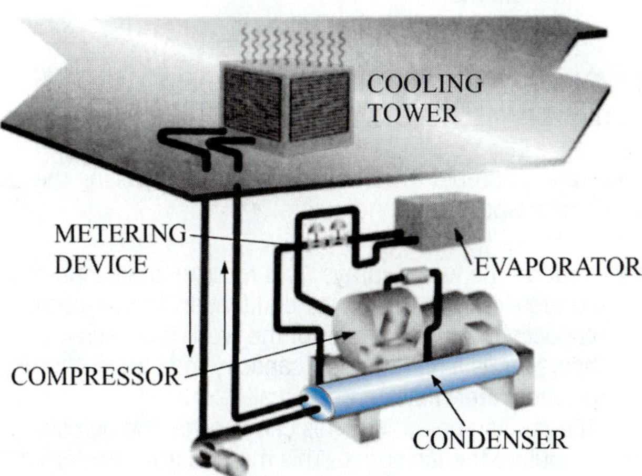

Figure 16-36 The cooling tower cools off the water from the water-cooled condenser and returns it to the condenser.

TECH TIP

Water-cooled condensers are more efficient than air-cooled condensers. This does not mean that a homeowner should put a sprinkler on an air-cooled condenser to increase its efficiency. Continuous watering of an air-cooled condenser coil will result in mineral buildup on the coil, which will ultimately reduce the coil's efficiency. In addition, many minerals are corrosive to the fin material, so that over time the aluminum fins will be corroded away.

16.23 CONDENSER WATER TREATMENT

When a water-cooled condenser is used, a constant supply of water is needed. If the water is being recirculated through a cooling tower, it must be treated to prevent corrosion, scaling, and the formation of algae. If the water treatment is not maintained, the tower will become fouled and stop working. Regular testing of the condenser water is the key to maintaining a healthy water-cooled system. Three different aspects of water conditioning must be monitored:

- Water pH
- Mineral content
- Biological growth

Water pH Control

The pH scale is a measure of the concentration of hydrogen ions in solution. It runs from 0 to 14. Lower numbers represent acidic conditions, and higher numbers represent alkaline conditions. A neutral solution is 7 on the pH scale. Typically, cooling tower water is maintained at a pH between 6.5 and 9. The circulating water can become corrosive if it is too acidic. A pH lower than 6 will aggressively attack metal surfaces, and high pH will encourage mineral deposits. Biological growth tends to raise the water pH level, leading to scaling. Most tower water will be alkaline, high pH, if left alone. Small amounts of acid are added to keep the pH down.

Scaling

Scaling occurs when minerals dissolved in the water precipitate out on the condenser and cooling-tower surfaces (Figure 16-37). The hard scale deposits act like insulation. Heat-exchange surfaces that have 1/16 in of scale have a 50 percent reduction in capacity. Scaling can be controlled by pH control, the addition of inhibitors that discourage mineral precipitation, and with water bleed-off.

The concentration of minerals in the cooling tower increases as the tower is operated. When water evaporates it leaves its mineral behind, effectively increasing

Figure 16-37 Scale deposits on cooling-tower components reduce the tower's cooling capacity. *(Courtesy of ChemFresh)*

the mineral concentration of the remaining water. A small amount of water is intentionally drained off of the bottom of the cooling tower sump to keep mineral concentration under control. The high-mineral-concentration water is replaced with relatively low-mineral-concentration water, reducing the overall mineral concentration in the sump.

Biological Fouling

Biocides are added to the water to prevent growth of organisms in the cooling-tower water. Water is a fundamental requirement for life. Leave some water sitting around in the open, and soon it will be teeming with life. The warm, untreated cooling-tower water becomes a petri dish, and an impressive array of bio-organisms springs forth, including bacteria, fungi, algae, and even protozoa organisms. As the organisms flourish, they attach themselves to all the tower surfaces, creating a coating called a biofilm. A well-established biofilm can be difficult to remove. Biological fouling is considered by many to be the root of most cooling-loop water-treatment problems. Problems created by biological fouling include reduction in heat-transfer efficiency, clogged cooling-tower fill, water-flow blockages, corrosion, and hazards to human health, like *Legionella*.

SAFETY TIP

Legionella was discovered because of an outbreak in 1976 at an American Legion convention, where 221 people became ill and 34 people died. Investigation revealed *Legionella* was growing in cooling-tower water and was spread through the mist created by the cooling tower. Failure to control its growth can lead to illness and death of people who come in contact with the spray from the tower.

Water Control in Open Systems

Water treatment is very different in open systems and closed systems. Open systems are exposed to more potential contaminants. Typically chemicals are added to maintain the proper pH, to inhibit scale formation, and to act as a biocide. Chemical treatment is usually done continuously using automatic chemical feed devices. Even with an automated system, the water should be tested and monitored on a weekly basis. The types and amount of chemicals used may need to be adjusted as conditions change.

Water Control in Closed Systems

Water control is easier in closed systems because the water is not exposed to the atmosphere, and makeup water is not continuously bringing in new minerals. A single chemical charge is usually sufficient when the loop is filled. The water should be tested quarterly because the water conditions will change over time, even in a closed system.

16.28 EVAPORATIVE CONDENSERS

Evaporative condensers are similar to closed-loop cooling towers, except the coil is a refrigeration coil instead of a water coil. They work by evaporating water to cool the refrigeration coil. Hot refrigerant vapor travels through the inside of the copper coil as cold water is sprayed over the exterior of the coil. Fresh air is pulled up over the exterior of the coil while cold water is sprayed down over the coil. Heat from the refrigerant is transferred to the spray water, causing the refrigerant to condense from a gas to a liquid. Heat from the spray water is discharged to the atmosphere through evaporation of the spray water. Evaporative condensers provide lower condensing temperatures and can offer significant horsepower savings over conventional air-cooled and water-cooled condensing systems.

The condensing coil rejects heat through both evaporative cooling using the fresh airstream and through sensible cooling using the precooled recirculating spray water. The recirculating spray water falls from the coil to a fill surface section, where it is cooled by a second fresh airstream using both evaporative and sensible heat-transfer processes. The cooled water increases the temperature differential between the water and the refrigerant. This cool water is then pumped back up to the spray heads to absorb more heat from the condenser.

16.29 CONTROLLING EVAPORATIVE CONDENSER CAPACITY

There are a number of arrangements for controlling the capacity of evaporative condensers:

- Shut off the water sprays. This reduces the evaporative condenser to an air-cooled condenser. The air-cooled capacity is 50–60 percent of the wetted capacity. This, however, is a huge step in capacity reduction. Shutting off the water may increase scaling.
- The fan can be cycled. This can shorten the belt life.
- Modulate the fan speed. This method gives better control with less likelihood of increased scaling.
- The airflow through the condenser can be controlled using dampers.

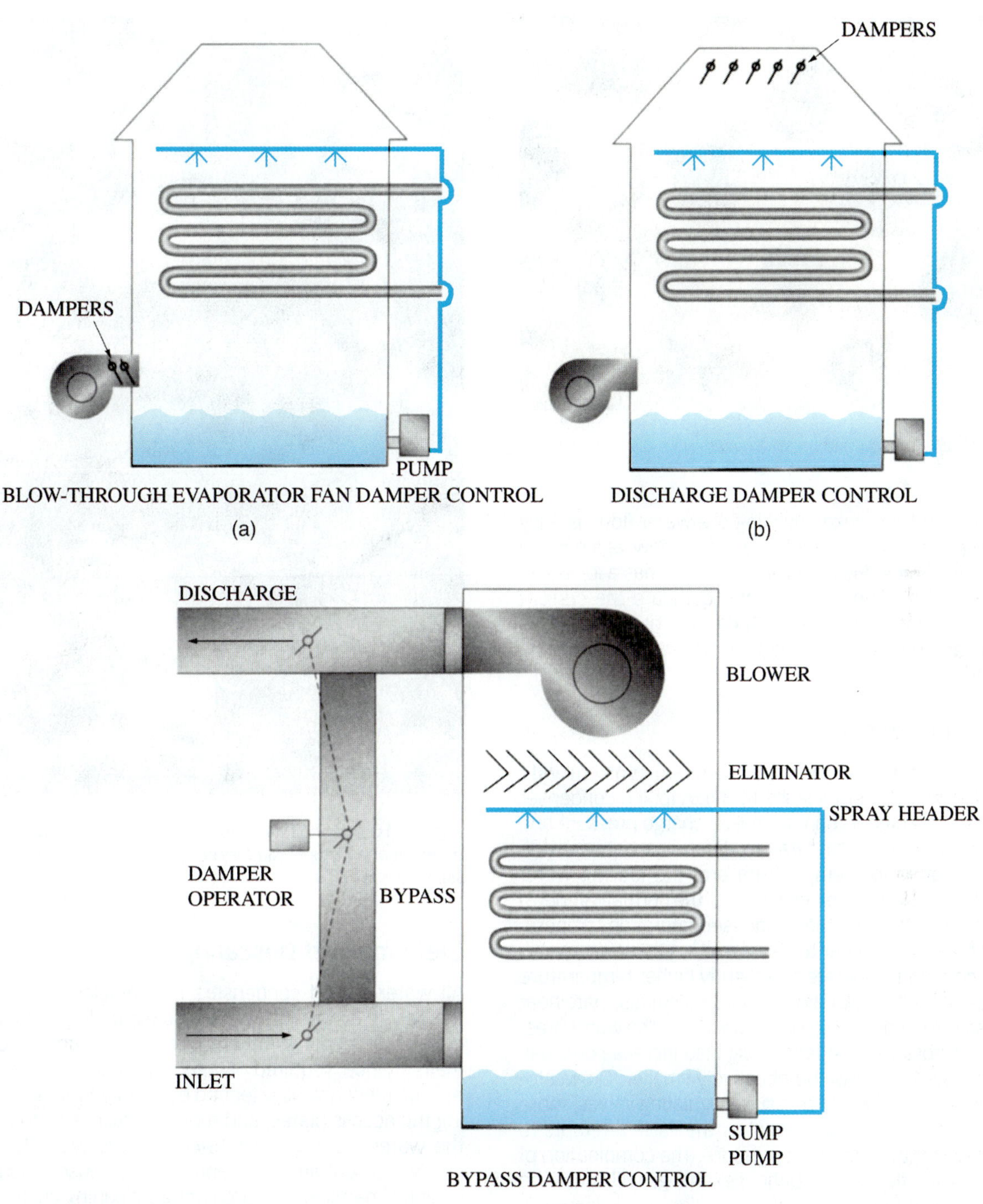

Figure 16-38 Three ways to control the airflow on the evaporative condenser.

Airflow can be controlled using dampers in the outlet of the blower, in the outlet of the tower, or in a bypass duct. Warm humid air from the condenser discharge can be mixed with outdoor air by the use of bypass dampers and ductwork. This increases the relative humidity of the air entering the unit, reducing evaporation and, therefore, reducing the condenser capacity. Figure 16-38 shows three ways to control the capacity of an evaporative condenser by controlling the airflow.

16.30 SERVICING WATER-COOLED CONDENSERS

The first step in servicing water-cooled equipment is to check the water flow. All other checks on a water-cooled system are useless if the water flow and temperature are not right! The water flow rate can be checked by checking the pressure drop through the unit and comparing it to the manufacturer's specifications. Some systems have water

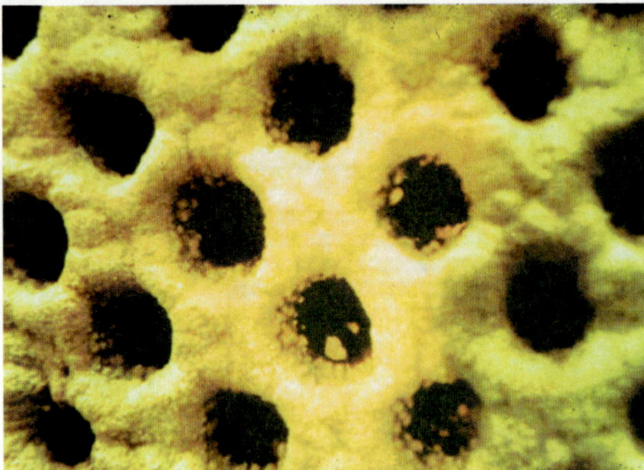

Figure 16-39 Fouled condenser. *(Courtesy of www.tubetech.com)*

flow indicators that make checking the water flow as easy as looking at a gauge. Check the water flow against the manufacturer's specifications. If the system has a log book, check the data in the log book and compare the current water flow against the baseline. Once the water flow is correct you can proceed to check the other system operating characteristics.

Fouled Condenser

Fouled condensers are the most common problem with water-cooled equipment (Figure 16-39). A fouled condenser is indicated by a higher than normal discharge pressure and a higher than normal temperature difference between the condenser saturation temperature and the leaving water temperature. In cooling tower systems, the normal temperature difference between the condenser saturation temperature and the leaving water temperature is 10°F. Wastewater systems commonly operate at a slightly higher temperature difference of 15°F. Fouled condensers do not dissipate heat as well, so the condenser temperature rises. The water pressure drop across the condenser will also increase because the condenser tubing will be effectively smaller. On wastewater systems with water-pressure-regulating valves, more water will be flowing than normal as the valve attempts to compensate for the high head pressure. The combination of increased water flow and poor heat exchange will cause the temperature difference across the condenser to decrease. The normal temperature difference across the condenser on cooling-tower systems is 10°F. Wastewater systems using city water can vary widely because of large temperature differences in the incoming water.

Low Water Flow

Reduced condenser water flow will cause an increase in head pressure and increase the temperature difference across the condenser. The temperature difference between the condenser saturation temperature and the leaving water temperature will remain about the same. Low water flow is often caused by fouled water-handling components. Plugged pump inlet screens or scaled water pipes can reduce the amount of water the pump can deliver.

Figure 16-40 Shell-and-tube condenser with head off having tubes cleaned. *(Photo courtesy of Goodway Technologies Corp., Stamford, CT)*

Cleaning and Descaling

All water-cooled condensers lose efficiency over time as the result of internal water-tubing fouling. Proper water maintenance will greatly reduce the amount of condenser maintenance required. Large tube-and-shell condensers may be mechanically cleaned by draining the water, removing the access plates, and running brushes or rods through the water tubes (Figure 16-40). This is typically done annually. Coaxial and shell-and-coil condensers can only be cleaned chemically by circulating a cleaning chemical agent to dissolve scale and oxides. The water system should be drained and flushed following a chemical cleaning. Follow the manufacturer's cleaning instructions, and read and follow all of the cleaning chemicals' safety instructions.

16.31 HEAT RECLAIMING

All refrigeration and air-conditioning systems produce waste heat when they are making something cool. That heat from either cooling process is by definition being moved from a place where it is not wanted to a place where it does not matter. Heat reclaiming is a process that makes use of the waste heat from a cooling process by moving it to a place where it is wanted.

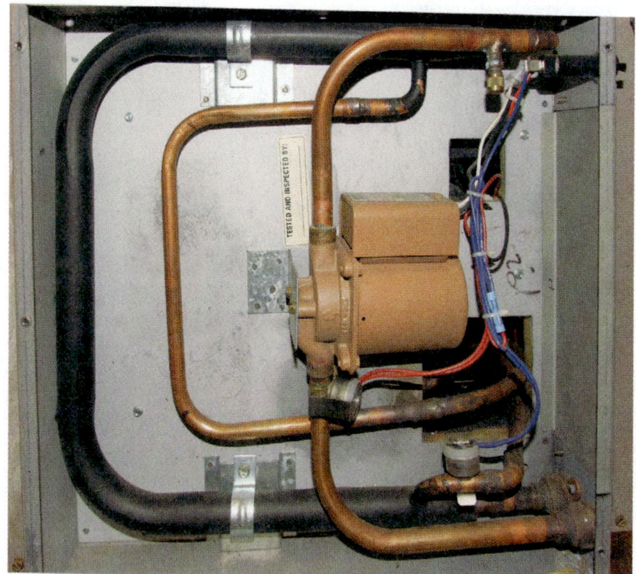

Figure 16-41 This unit has a heat exchanger built into it for heating domestic hot water.

Everyone needs some hot water even on the hottest day in the summer. Too often we use expensive energy sources like natural gas or electricity to produce that hot water. On the coldest day of the winter, restaurants, supermarkets, schools, and hospitals have something they are keeping cold in a walk-in freezer or refrigerator.

A desuperheating coil added to the hot gas line of any air conditioner or refrigeration unit can produce the heat to make hot water. The waste heat from these cooling systems is always free and in some cases can reduce the high-side pressure and temperature on the condenser to save even more money. Desuperheaters can be added to existing systems or built into the system at the factory (Figure 16-41).

CAUTION

Desuperheaters must have thermostats and safety switches to protect the water heater from being overheated. Desuperheaters may even boil the water in the heater without these controls.

The core of a large building often does not need heat even on cold winter days. The heat from lights, equipment, and people is often more than is needed to keep this area warm. In fact, sometimes some mechanical cooling is required on all but the coldest days. The waste heat from this core cooling can be redirected to the perimeter of the building, which probably needs some heat on these cold days.

Waste building heat can be recycled through the building's forced-air or hydronic system. Forced-air systems are the easiest to set up to use waste heat. In most cases, all that is needed is to add a condenser coil in the supply or return air ducts. For a hydronic system, the heat can be collected with a desuperheater to be added to the hydronic heating system.

Some supermarkets use heat-reclaiming condensers to capture the heat from operating their refrigerated cases in the winter. The refrigerated cases operate year round, but the supermarket typically needs heat in the winter. The discharge gas from the compressor is directed to a condenser in the store's duct system, where it can be used to heat the store.

UNIT 16—SUMMARY

The job of the condenser is to reject heat from the refrigerant by condensing a hot refrigerant vapor to a warm liquid. The refrigerant is desuperheated, then condensed to a liquid, and finally the liquid is subcooled. Approximately 80 percent of the heat removal occurs because of the condensing process.

WORK ORDERS

Service Ticket 1601

Customer Complaint: Poor Cooling and High Power Bills

Equipment Type: R-22 Split-System Air Conditioner with an Air-Cooled Condenser

The customer lives on a heavily traveled dirt road. He complains that the system runs more than it did when it was new and does not seem to cool as well. He also notes that his power bill is considerably higher than when the system was first installed. He asks the technician to add some refrigerant to make the system cool better. The technician notes that the system is five years old and has never been serviced. The technician shows the customer that the high-side pressure is 325 psig, almost 100 psi higher than the pressure specified on the system charging chart located on the inside panel of the equipment. The technician explains that adding refrigerant would only make matters worse and shows the customer the dirt caked on the condenser. The customer agrees to have the condenser cleaned. After cleaning, the system operating pressures return to normal, the compressor cools off, and the unit amp draw decreases.

Service Ticket 1602

Customer Complaint: System Not Operating

Equipment Type: R-22 Split-System Air Conditioner with an Air-Cooled Condenser

A customer complains that her one-year-old system is not operating. The customer is especially upset that she paid to have the system recharged just last month. The technician sees a large white area on the condenser fins and uses a volt meter to determine that the low-pressure switch is open and preventing the system from operating. A pressure check reveals that there is no refrigerant in the system at all. A large, overly friendly dog approaches, slobbers on the technician, lifts his leg, and wets the condenser coil. The technician adds a nitrogen charge to look for leaks and hears a hissing sound from the white area of the condenser. The technician informs the customer that the condensing coil leaks because of corrosion due to the acidic nature of the dog urine and recommends replacing the condenser and moving the dog to an area away from the condenser.

Service Ticket 1603

Customer Complaint: High Process Water Temperature

Equipment Type: Process Water Chiller Connected to Open Water Tower

A manufacturing company uses a water chiller to cool the water used to cool the machines in their plant. The company has noticed that the chiller is unable to maintain the desired process water temperature. The technician notes that the temperature of the water entering the condenser is 100°F and appears to be climbing. The temperature leaving the condenser is 106°, only 6°F warmer than the entering temperature, indicating reduced system capacity. The condenser saturation temperature is 120°F. The water in the cooling tower appears murky, and parts of the tower are heavily coated with scale. The technician shows the plant manager the cooling tower and recommends a thorough cleaning of the tower and condenser, followed by a routine water-treatment regimen to prevent the problem from recurring. The plant manager agrees to the cleaning but defers the decision on water treatment to the home office, noting that they had discontinued regular water-treatment service about a year ago to reduce operating cost. The technician points out that the increased electrical use due to inefficient operation probably exceeded the cost of regular maintenance, so neglecting maintenance effectively increased operating cost.

Service Ticket 1604

Customer Complaint: System Cuts Off Periodically

Equipment Type: R-22 Water-Source Heat Pump Unit with a Tube-in-Tube Condenser on a Wastewater Well System

A customer complains that his ten-year-old water-source heat pump periodically shuts itself off. He has figured out that if he pushes a red button marked RESET, the system will often start and run for a while before stopping again. The button is a manual reset high-pressure switch. The technician measures the system operating pressures and temperatures. The high-side pressure starts high and continually rises until it reaches 375 psig, when the technician shuts the system down. The pressure drop across the coil is higher than normal, and the temperature rise in the condenser water is only 5°F. The leaving water temperature is 90°F. The technician informs the customer that his water coil is fouled, causing high head pressure, which trips the high-pressure switch. The technician recommends either replacing the coil or the entire unit because this type of coil cannot be mechanically cleaned.

UNIT 16—REVIEW QUESTIONS

1. What is the purpose of a condenser?
2. Describe the changes in the refrigerant as it passes through the condenser.
3. What do environmental concerns have to do with water-cooled condensers?
4. Why is subcooling important?
5. How much additional cooling capacity would 20°F of subcooling have on a refrigerant system?
6. List the three types of condensers used with HVACR systems.
7. How much heat must a condenser be able to dissipate?
8. What would the expected condensing temperature be for an air-cooled condenser on a 100°F day?
9. What advantages are there to using a remote air-cooled condenser?
10. What are some ways that an air-cooled condenser can be made more efficient?
11. Why would you need air conditioning if the outdoor temperature was 35°F?
12. List the different types of low ambient controls for air-cooled condensers.
13. List the different types of water-cooled condensers.
14. What can happen to a water-cooled condenser if the refrigerant is recovered improperly?
15. What can be placed in the water line on a wastewater system to keep the head pressure constant?
16. What does a cooling tower do?
17. What water conditions must be monitored to keep a cooling-tower system working properly?
18. What can happen to a cooling tower if the water is not treated?
19. Compare the discharge pressure and operational cost of air-cooled condensers versus water-cooled condensers.
20. List some of the factors that can cause the condensing temperature to be too high on a water-cooled system.
21. List some of the factors that can cause the condensing temperature to be too high on an air-cooled system.
22. List the methods of reclaiming condenser heat.
23. What is the purpose of the fins on finned tube coils?
24. Explain how an evaporative condenser works.
25. What is the difference between open and closed cooling towers?
26. What type of water-cooled condenser is easiest to clean?
27. Which requires more maintenance: air-cooled condensers or water-cooled condensers?
28. What problems can be created if the system high-side pressure is too low?
29. What is the most important operating characteristic to check whenever working on a water-cooled system?

UNIT 17

Metering Devices

OBJECTIVES

After completing this unit, you will be able to:

1. list the different types of metering devices.
2. explain the difference in operation between fixed and modulating metering devices.
3. describe how to measure superheat.
4. explain the purpose of liquid distributors.
5. explain the difference in operation between conventional and balanced-port expansion valves.
6. discuss common metering device problems.

17.1 INTRODUCTION

A metering device controls the flow of refrigerant to the evaporator. It separates the high-pressure side from the low-pressure side. The evaporator pressure is low because the compressor is continuously pumping vapor from the evaporator while the metering device is restricting the flow of refrigerant into the evaporator. Refrigerant enters the metering device as a subcooled liquid and leaves it as a saturated mixture that is approximately 75 percent liquid and 25 percent vapor. The pressure and temperature at the outlet of the metering device correspond to the saturated condition required to establish the desired evaporator temperature.

17.2 TYPES OF METERING DEVICES

Metering devices are selected by application. They can be generally classified by operation as belonging to one of two groups: fixed metering devices and modulating metering devices. Fixed metering devices are simpler and less expensive than modulating metering devices. They are basically an engineered hole, designed for a specific set of operating conditions. The capillary tube and the orifice (piston) are examples of fixed-type metering devices.

Modulating metering devices have the ability to respond to changes in system operation. This allows modulating-type metering devices to be used on a wider range of applications. Modulating metering devices will usually operate more efficiently than fixed metering devices, especially if the refrigeration load conditions vary. The low-side float, high-side float, automatic expansion valve, thermostatic expansion valve (TEV), and electronic expansion valve are examples of modulating metering devices.

TECH TIP

A type of metering device that is seldom used today is the hand-operated expansion valve. It is a fixed valve whose capacity is adjusted by hand. The obvious disadvantage of the hand expansion valve is that it has to be monitored and adjusted by hand to match the load. These valves were used in the past on applications where the load was fairly constant and an operator was present to manually make adjustments when necessary. The hand expansion valve was often used on ammonia systems that had a nearly constant load and where an operator was available to make adjustments when needed. It is seldom used today, except in laboratory tests to explore optimum flow rates.

17.3 METERING DEVICE OPERATION

A metering device is a type of restrictor placed in the liquid line between the condenser and the evaporator to produce a difference in pressure between the high side and the low side of a refrigeration system by regulating the flow of refrigerant. The amount of the restriction must balance the work of the compressor to maintain the correct condensing and evaporating temperatures.

The temperature of the warm refrigerant liquid must be reduced before it can absorb heat in the evaporator. The pressure of the refrigerant is reduced as it flows through the metering device. The drop in pressure causes a small portion of the liquid refrigerant to vaporize, cooling the remaining liquid.

Refrigerant that evaporates as the result of a pressure drop is called flash gas because the liquid turns to vapor instantly, "in a flash" (Figure 17-1). Heat is required

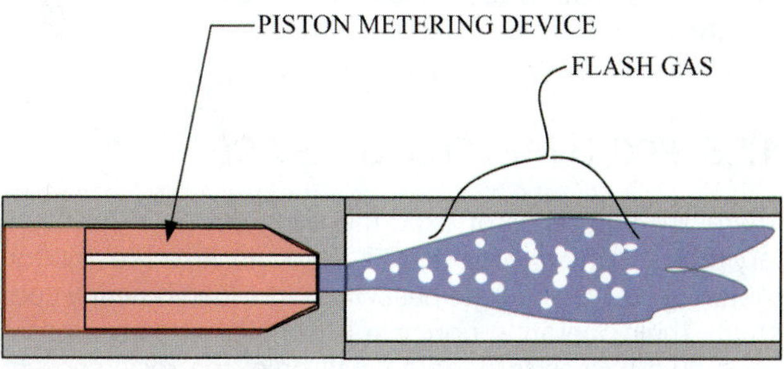

Figure 17-1 How flash gas is formed when the liquid refrigerant flows through the metering device.

	Inlet	Outlet
Volume	28 in^3/lb	268 in^3/lb
Pressure	418 psig	118 psig
Saturation temperature	120°F	40°F
Actual temperature	105°F	40°F
State	Subcooled liquid 100% liquid	Saturated mixture 30% vapor/70% liquid
Specific enthalpy	55 BTU/lb	55 BTU/lb

Figure 17-2 Typical values for 410a air-conditioning system at ARI rating condition.

to make the liquid refrigerant boil. Since no heat is being added, the heat must come from the remaining liquid refrigerant. The flashing refrigerant absorbs heat from the remaining liquid refrigerant, reducing the temperature of the remaining liquid. Between 25 and 30 percent of the liquid typically flashes to vapor as the refrigerant flows through the metering device. Expansion that occurs without adding any heat from outside the process is called adiabatic expansion.

Referring to Figure 17-2, the volume of the refrigerant increases from around 28 in^3/lb entering to approximately 268 in^3/lb leaving. This expansion creates a large pressure drop from 418 psig to 118 psig. The pressure drop causes a corresponding temperature drop from 105°F to 40°F. The specific enthalpy is unchanged because the heat required to vaporize the flash gas came from the remaining liquid.

Figure 17-3 The metering device comes installed as part of an air conditioning evaporator.

17.4 METERING DEVICE LOCATION

The metering device is located between the condenser and the evaporator. Typically the metering device is actually connected to the evaporator. The metering device is supplied with most air-conditioning evaporator coils. Air-conditioning evaporators are typically manufactured with the metering device as an integral component (Figure 17-3). Ductless mini-split systems are a notable exception. Many of these systems have the metering device located in the outdoor condensing unit (Figure 17-4). The liquid line on these systems must be insulated because they carry cold, saturated liquid rather than a warm, subcooled liquid. The metering device is more commonly field installed in commercial refrigeration. It is still attached to the evaporator, but the installation technician must install it in the field (Figure 17-5).

17.5 FIXED METERING DEVICES

There are two types of fixed metering devices: the capillary tube and the fixed orifice (Figure 17-6). Fixed-restriction metering devices operate best where the load is nearly constant. Their operating principle is simple: when a pressurized fluid passes through a small hole, the resistance to flow through the hole causes a drop in pressure. While the operating concept is simple, the design and application of fixed-restriction metering devices is anything but simple. Since a fixed hole cannot adjust itself to pressure changes, improper sizing can easily lead to flooding or starving the evaporator, both of which may be harmful to the compressor. The size of that opening and the amount of charge are critical to correct operation.

Many technicians expect a fixed amount of refrigerant to pass through a fixed metering device since the device cannot adjust. However, changes in pressure difference across a fixed metering device will cause changes in flow precisely because the device cannot adjust. The reason refrigerant goes through a restriction at all is because of a difference in pressure across it. Both the low- and high-side pressures tend to be influenced by factors, such as the outdoor

Figure 17-4 The metering device on this mini–split system is located in the outside unit.

Figure 17-5 Commercial refrigeration evaporators typically do not come with a metering device already installed; this expansion valve was installed in the field.

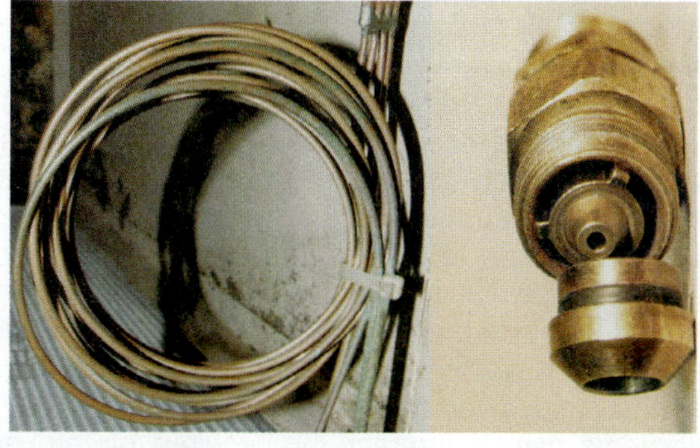

Figure 17-6 The capillary tube on the left and the orifice on the right are the primary types of fixed-restriction metering devices.

temperature, that are beyond the control of the machine. In an air-conditioning system, the head pressure increases as the outdoor temperature rises. This increase in head pressure causes an increase in the pressure difference across the orifice. As the pressure difference across the restriction increases, more refrigerant will go through it. This increased flow will cause a rise in suction pressure, but the suction pressure will not increase quite as much as the head pressure that caused the increased flow. The increased flow will also have the effect of reducing superheat, since more refrigerant is now being delivered to the evaporator. A check of the compressor motor amperage will reveal a relatively higher current draw from an increased system load.

Lower outdoor ambient temperatures will produce lower head pressures, reducing the pressure drop across the fixed restriction. This reduced pressure difference will result in reduced refrigerant flow to the evaporator. Decreased flow will cause a drop in suction pressure and an increase in superheat, since less refrigerant is now being delivered to the evaporator. A check of the compressor motor amperage will reveal a decrease in the system load. Figure 17-7 shows the operating characteristics of an air conditioner with a fixed restriction at different outdoor ambient temperatures.

Changes in temperature at the evaporator have an opposite effect from changes in condenser temperature. An increase in temperature at the evaporator causes an increase in evaporator pressure. This has the effect of reducing the pressure difference across the metering device, which reduces the refrigerant flow to the evaporator and raises the superheat. Decreased temperature at the evaporator decreases the evaporator pressure, increasing the refrigerant flow through the metering device and decreasing the superheat. Figure 17-7 shows the effect of indoor wet bulb temperature on system operation.

Because the flow through a fixed restriction changes depending upon the operating conditions, the only time a system with a fixed metering device produces its full capacity is at design conditions. The AHRI standard design conditions for air conditioning are 95°F outdoor ambient temperature, 80°F indoor dry bulb temperature, and 67°F indoor wet bulb temperature. A 24,000 BTU, AHRI-rated air conditioner with a fixed expansion device will not produce 24,000 BTU worth of cooling at an outdoor ambient of 75°F.

This change in capacity shows up in the system superheat. Most manufacturers using orifices or capillary tubes publish superheat charts that specify what the system superheat should be at any operating condition that falls within the operating range of the unit. A look at one of these charts will show that as the outdoor ambient temperature goes up, the system superheat goes down; and conversely, as outdoor ambient temperature goes down, the system superheat goes up. In fact, most air conditioners with fixed-restriction metering devices are operating under starving conditions at ambient temperatures less than 95°F. An orifice-equipped unit can have a superheat as high as 45°F at its lowest operating limit. This superheat is what determines the system's lowest operating temperature limit. The compressor on a system using a fixed restriction can overheat due to lack of cooling for the compressor motor when operating at low ambient conditions. That's correct: low ambient operation can cause overheated compressors!

Outdoor Temperature	Indoor Wet-Bulb Temperature	Temperature Drop across Evaporator	Suction Pressure	Discharge Pressure	Superheat	Amp Draw
100	77	10	145	530	24	15.3
	72	14	142	523	20	14.9
	67	17	135	417	15	14.5
	62	21	130	410	4	14.1
95	77	10	144	402	23	15.6
	72	14	135	396	20	14.5
	67	17	132	390	17	14.1
	62	21	128	382	6	13.7
90	77	11	140	378	25	14.2
	72	14	134	372	24	13.8
	67	18	130	365	19	13.5
	62	20	120	355	8	13.1
85	77	11	139	346	27	13.3
	72	15	132	340	24	13
	67	18	128	334	21	12.6
	62	21	119	328	10	12.2
80	77	11	135	335	26	12.4
	72	15	131	329	24	12.0
	67	18	126	323	23	11.6
	62	22	119	317	14	11.2
75	77	12	134	320	25	11.7
	72	15	130	314	25	11.2
	67	18	120	308	25	10.8
	62	22	118	300	17	10.5

Figure 17-7 Operational characteristics of R-410a air conditioner with fixed restriction.

Adding charge at the low ambient condition will produce a more fully active evaporator and reduce the superheat, but then the system will flood back at higher ambient temperatures when the superheat naturally drops.

A characteristic of systems using fixed-restriction metering devices is that the system high-side and the low-side pressures equalize when the compressor shuts off. Since the metering device is simply a hole, refrigerant will continue to flow through it until the pressures on both sides of it are the same. This can be an advantage in system design because the compressor does not have to start against a pressure differential. This permits the use of a low starting torque compressor motor. This can also be a disadvantage, since the evaporator can be filled with liquid on shutdown. On the next startup, the liquid can flood back to the compressor and cause damage. System equalization also encourages refrigerant migration to the compressor crankcase. When the system starts, the oil and refrigerant mixture foams violently, reducing lubrication. Neither of these problems is pronounced on small systems with limited charge, but they can become serious on larger systems. Crankcase heaters and suction line accumulators can be employed to mitigate these problems on larger systems.

Systems using fixed-restriction metering devices do not use liquid receivers. Without a receiver, if refrigerant is condensing faster than it is flowing through the metering device, liquid starts to pile up in the bottom of the condenser. This raises the head pressure and helps force more refrigerant through the metering device. Any refrigerant storage must be accomplished on the low side of the system with a suction accumulator.

Systems can counteract the problem of refrigerant floodback by using a suction line accumulator at the entrance to the compressor, as shown in Figure 17-8. This

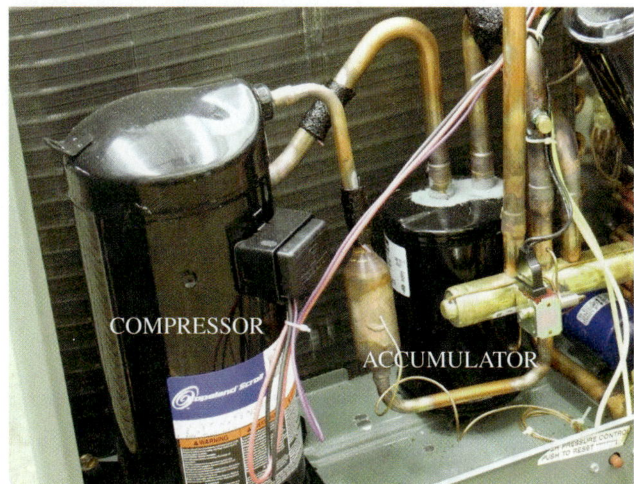

Figure 17-8 A suction line accumulator can trap liquid refrigerant before it reaches the compressor.

captures the liquid refrigerant and returns it to the system as vapor. Because liquid refrigerant is evaporated from an accumulator, these devices get very cold and often sweat. This condensation can cause accumulators to rust over time, as in Figure 17-9. On older units, the rust can weaken the accumulator to the point where it can fail.

Figure 17-9 Rust formed on the accumulator due to condensation.

SERVICE TIP

It is easy to overcharge an air-conditioning system with a fixed restriction in mild weather. For example, suppose a technician is checking an air conditioner on a 75°F day. The unit operates with a moderately high superheat of 30°F. The suction line feels cool, but not really cold. Since the evaporator is obviously starved, the technician adds refrigerant until the suction line is cold and the superheat is 10°F. The unit begins cooling a little better. The customer is happy until the temperature climbs to 95°F outside. As the outdoor temperature rises, the superheat drops. Since the unit has been charged to operate at a low superheat under a 75°F ambient, the evaporator will flood at a 95° ambient. The flooded evaporator does not cool as well as it should, so the capacity of the unit is reduced just when it is needed the most. Since the unit is operating at reduced capacity and increased pressure, it uses more electricity, costing the customer more money. Because the evaporator is flooded, liquid will be returned to the compressor and damage it, further reducing the capacity of the system. The moral: air-conditioning systems with fixed-restriction metering devices should operate at relatively high superheats in moderate weather. Do not overcharge them.

17.6 CAPILLARY TUBES

The capillary tube is probably the simplest of all metering devices. It is constructed of a seamless copper tube with an inside diameter ranging in size from 0.026 in to 0.090 in (Figure 17-10). Capillary tubes may be found on older models of central air conditioners but are seldom used today in residential air-conditioning systems. Capillary tubes are still commonly used as a metering device on domestic refrigerators, window units, and other small appliances. In these appliances, the capillary tube doubles as a metering device and the liquid line; the refrigerant travels from the condenser to the evaporator through the capillary tube (Figure 17-11). Capillary tubes are frequently longer than the distance they must travel and are sometimes coiled to conserve space, as in Figure 17-12.

Small systems like refrigerators typically use a single capillary tube for their metering device. The evaporator coil on these systems is a single circuit coil: refrigerant enters at the start of the coil and exits at the end of the coil. Larger units using capillary tubes may use multiple capillary tubes feeding multicircuit coils. A multicircuit coil has several entry points and exits. Essentially, a multicircuit coil is like several small coils manufactured together. Each circuit has its own capillary tube (Figure 17-13). Systems with multiple capillary tubes use a small distributor tube to feed all of them (Figure 17-14).

The size and the length of the tube are carefully selected to match the pumping capacity of the compressor at full load. Capillary tubes depend on their length as well as their diameter to determine their total restriction. Increasing the diameter or decreasing the length of the capillary tube decreases the pressure drop through it, increasing its capacity. Decreasing the diameter or increasing the length of the capillary tube increases the pressure drop through it and decreases its capacity. The pressure of the liquid refrigerant passing through the capillary tube drops slowly in the first 25 percent of the capillary tube length. When the pressure has dropped to the saturation pressure for the refrigerant temperature, the liquid starts to flash. The flash gas causes

Figure 17-10 Typical capillary-tube metering device.

Figure 17-11 The capillary tube in a window unit runs from the outlet of the condenser to the inlet of the evaporator.

Figure 17-12 The capillary tube is coiled when it is longer than the distance it must travel.

Figure 17-13 The coil has four circuits, each fed by its own capillary tube.

Figure 17-14 The four capillary tubes for this multicircuit coil are fed by a capillary-tube distributor.

a rapid drop in pressure starting at that point. This point on the capillary tube can be felt. The first part is still warm to the touch; the part containing flash gas is cool to the touch.

Systems using capillary-tube metering devices have no provision for refrigerant storage. The charge in these systems is critical because the capillary tube cannot adjust, the systems have no provision for refrigerant storage, and a small change in charge can have a large impact on system pressures. Most systems that use a capillary tube as their metering device are packaged units such as domestic refrigerators, freezers, and window air conditioners. These systems are evacuated and charged at the factory.

17.7 INSTALLING CAPILLARY TUBES

Capillary tubes are not generally field-selected components. However, capillary tubes may occasionally need replacing. The best sizing procedure is to replace the capillary tube with exactly the same size and length. Even the coiling makes a difference. Many manufacturers sell replacement capillary tubes that are the exact replacement, including the shaping of the capillary tube. If a capillary tube must be cut during installation, it should be cut with a small triangle file (Figure 17-15). File a groove all the way around the tube and then bend the tube back and forth until it breaks at that point without creating a burr. Capillary tubes are too small for most tubing cutters. Even if a tubing cutter manages to cut a capillary tube, it leaves a large burr, which adds additional restriction. This additional restriction will create an excessive pressure drop. Wire-cutting pliers will actually pinch the tubing shut. Figure 17-16 compares a capillary tube that has been cut using wire pliers with a capillary tube that has been cut with a small file. When installing a capillary tube, it should be inserted at least an inch past the point it enters to prevent brazing material from plugging its end (Figure 17-17).

Figure 17-15 A small triangular file like this one can be used to cut capillary tubing. *(Photo used courtesy of Snap-on Incorporated. © 2011 Snap-on Incorporated)*

Figure 17-16 The capillary tube on the left was cut with a file; the capillary tube on the right was cut with wire pliers.

Figure 17-17 The capillary tube should be inserted far enough into the coil so that the end will not get plugged with brazing alloy.

Due to the small size of the tube, it can be easily plugged. Metal shavings, dirt, moisture, filings, flux, or oxides from brazing can all enter the system during manufacture and installation. Every precaution should be taken to keep these contaminants from entering the system during the installation process. Many systems use a built-in liquid line filter at the entrance of the tube (Figure 17-18).

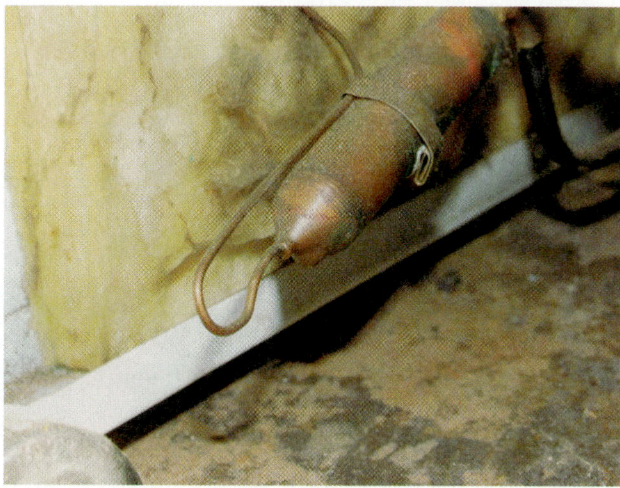

Figure 17-18 This filter prevents debris from clogging the capillary tube.

17.8 TROUBLESHOOTING CAPILLARY TUBES

A restricted capillary tube is the primary problem encountered on capillary-tube metering devices. A restricted capillary tube will operate with a low suction pressure, a high superheat, and a high subcooling. The solution depends upon the reason for the restriction. The restriction can be from solid debris like copper shavings and copper oxide scale, or it can be from water. If the capillary tube is plugged with solid debris, it will have to be changed. Sometimes the strainer ahead of the capillary tube is restricted, not the tube itself. Normally the outlet of the filter should be warm because the refrigerant has not yet begun to flash. A restricted filter will be cool at its outlet. In this case, changing the strainer or filter should fix the problem.

Water in low-temperature systems can freeze up at the capillary tube, creating a restriction. The difference is that the water will thaw and the restriction will clear if the unit is allowed to thaw. This does not fix the problem; it only identifies it. A system with moisture will operate correctly until the moisture hits the metering device and freezes. Then the suction pressure will drop, the superheat will rise, and the unit will lose its cooling ability. The solution is to recover the refrigerant, replace the filter, pull a good vacuum on the system, and charge the system with clean refrigerant.

17.9 FIXED ORIFICES

The fixed orifice is a newer type of fixed metering device (Figure 17-19). *Orifice* is just a three-syllable word for "hole." Although many names have been used for the small brass piece that has the hole, the most common name is piston. Its operating principle is similar to the capillary tube: a pressurized fluid passes through a small hole, and the resistance to flow through the hole causes a drop in pressure. Since the entire pressure drop occurs at one opening, the size of that opening is all the more critical. Since the pressure drop occurs to some extent throughout the entire length of a capillary tube, its sizing is much less critical than that of an orifice.

The orifice is like the capillary tube in a number of ways:

- It must be carefully selected to match the load.
- The system must be critically charged.
- It permits refrigerant migration into the evaporator during the off cycle and requires all of the same protective accessories as the capillary tube.

Many orifices come in some form of mechanical fitting. On these systems, the orifice is contained in the male fitting on the unit where the liquid line connects (Figure 17-20). The orifice looks like a small, solid brass piston with a small hole. The two chief advantages of these metering devices are economy and flexibility. Since these devices are installed like flare couplings, they can be added after the coil-manufacturing process instead of during manufacture. They can also easily be changed to match specific system requirements. Split-system evaporators that use orifices typically come with an installed orifice that matches the normal condenser match for the evaporator. However, there are usually several correct matches for any condenser. Many split-system manufacturers ship a selection of orifices with the outdoor unit and a sizing chart to help the installing technician select the proper orifice for the application. The sizing charts give the flow-check device size for just about any combination of condensing unit and evaporator coil imaginable. Matching a 3-ton evaporator with a 2-ton condensing unit is possible by simply changing the flow-check piston. This flexibility helps manufacturers and distributors reduce inventory and still cover most applications.

17.10 ORIFICE METERING DEVICES AND HEAT PUMPS

The orifice has the advantage for heat pump applications of operating as both a metering device and a check valve. It meters refrigerant in one direction and allows unrestricted flow in the other direction. When refrigerant is flowing into the coil, toward the piston, the piston will seat and refrigerant will be forced to go through the orifice in the piston (Figure 17-21). When refrigerant is leaving the coil, the piston will unseat, allowing unrestricted flow around the piston (Figure 17-22). Prior to the development of the piston-type metering devices, air-conditioning and heat pump indoor coils were different. Heat pump coils contained a check valve that allowed refrigerant to bypass the metering device in heating. Today most manufacturers use the same indoor coils for both air-conditioning and heat pump applications.

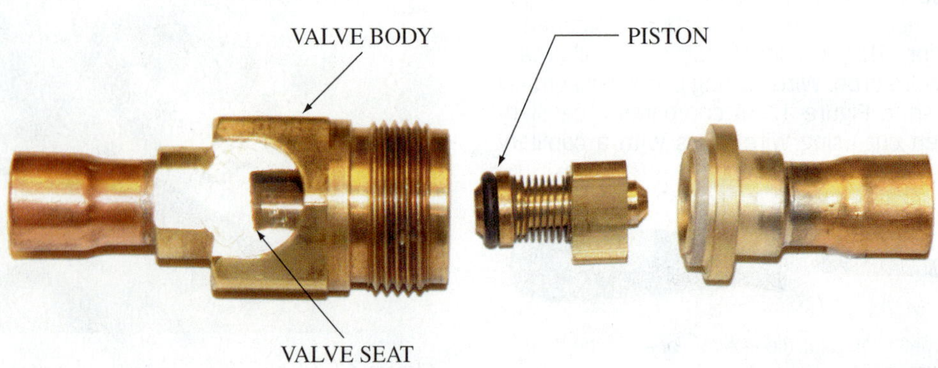

Figure 17-19 Construction of a fixed-orifice-type metering device.

Figure 17-20 The metering device for this evaporator coil is an orifice located inside the liquid line connection.

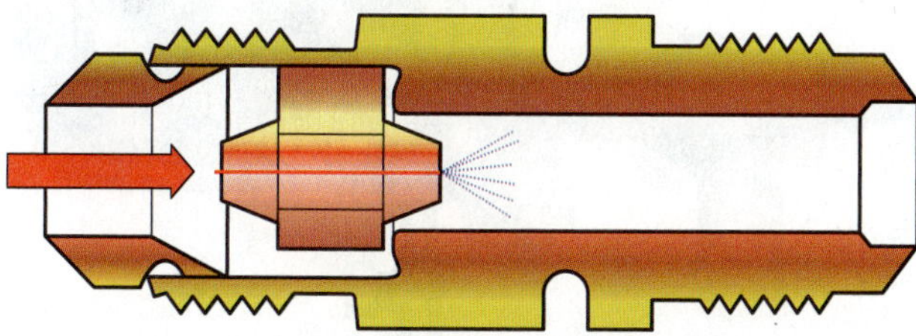

Figure 17-21 In the forward direction, the piston seats and acts like a metering device.

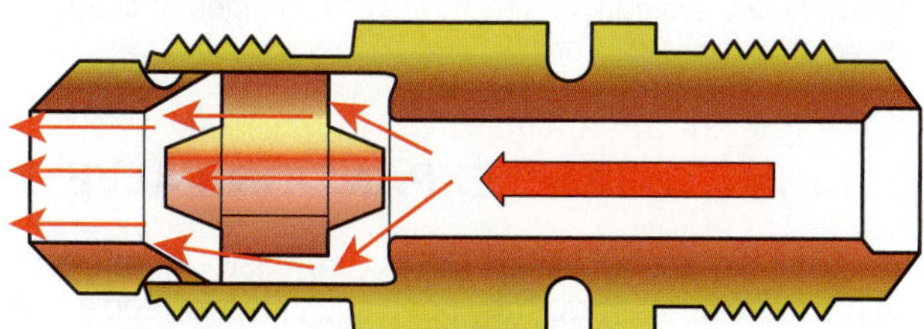

Figure 17-22 In the reverse direction, refrigeration can flow around the piston for unrestricted flow.

17.11 TROUBLESHOOTING FIXED ORIFICES

Like capillary tubes, fixed orifices can become clogged. Since all the flow restriction is occurring at one point, a very small amount of debris buildup will cause a problem. Like capillary tubes, a clogged orifice will operate with a low suction pressure, a high superheat, and a high subcooling. Unlike capillary tubes, clogged orifices are fairly easy to change. Debris can also cause orifices to seat improperly and allow too much refrigerant to pass. In this case, the low-side pressure will be high and the superheat and subcooling will both be low.

SERVICE TIP

Many assemblies that hold the orifice and distributor have a screen inside the copper tube leading to the orifice and distributor tubes. This screen is not generally visible without removing the whole assembly. Brazing oxides and contamination from compressor burnouts can clog this screen, making it difficult to locate refrigerant restriction.

Orifices are more likely to be sized incorrectly than capillary tubes because they are more likely to have been installed in the field. An oversized orifice will overfeed, producing high suction pressure, low superheat, and low subcooling. An undersized orifice will cause low suction pressure, high superheat, and high subcooling.

Any orifice problem can usually be solved by recovering the refrigerant; installing a new filter drier; replacing the orifice with a new, correctly sized orifice; evacuating the system; and charging the system with clean refrigerant.

SERVICE TIP

Do not try to clean or resize a metering-device orifice by drilling it out. The orifice in a fixed-restriction metering device should never be drilled out. The orifices are actually conical-shaped holes that you do not have the capability to drill in the field. Drilling an orifice is much more likely to create even worse system problems than existed before the operation. A drilled orifice can easily overfeed, leading to flooding and dead compressors. Simply changing the piston takes less time and is much safer for the system.

17.12 MODULATING METERING DEVICES

A modulating metering device can adjust to changing operating conditions. Modulating metering devices are preferred for systems that have significant load variation. The ability to adjust to changing operating conditions makes these metering devices more efficient than fixed metering devices. Types of modulating metering devices include the following:

- Low-side float (LSF)
- High-side float (HSF)
- Automatic expansion valve (AEV)
- Thermostatic expansion valve (TEV)
- Electronic expansion valve (EEV)

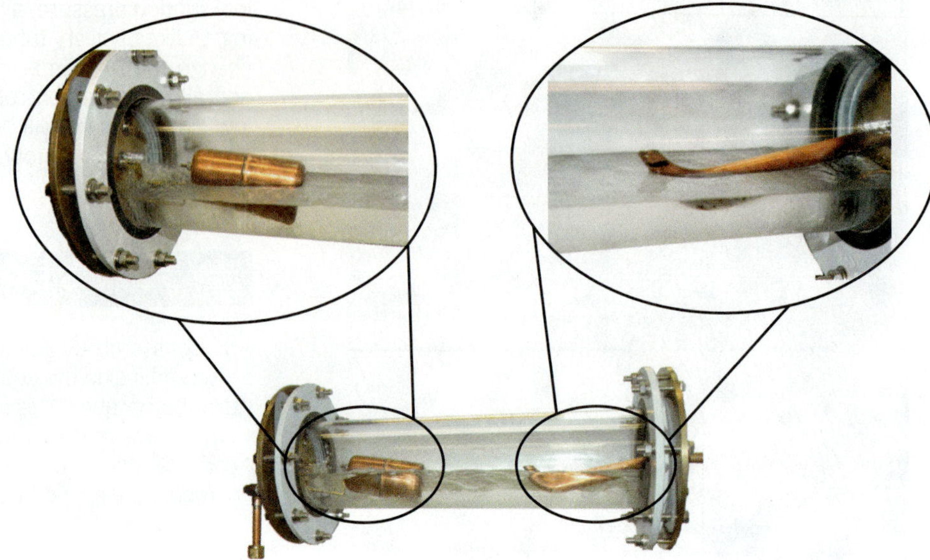

Figure 17-23 (Left inset): Float opens when the level of refrigerant lowers; (right inset): the outlet has a small opening that lets some refrigerant and oil flow back to the compressor to ensure proper oil return to the compressor during operation.

17.13 LOW SIDE FLOAT

In the low-side-float (LSF) metering device, the evaporator is flooded with refrigerant liquid and the level in the evaporator is maintained by a float (Figure 17-23). Their operation is similar to the floats used in toilets. If the load is increased, more refrigerant boils away. This causes the liquid level to drop, and the float lowers and the valve opens wider, allowing more refrigerant into the coil. If the load is reduced, less refrigerant boils away. The liquid level rises as refrigerant continues to enter until the float rises enough to close the valve. The low-side float can be installed in the evaporator or in a separate float chamber.

One advantage of this arrangement is that the heat-transfer rate from a flooded coil is higher than with a mixture of liquid and vapor. One disadvantage is the possibility of oil collecting in the evaporator under light loads and not returning to the compressor.

The low-side float has been rather commonly used on ammonia systems that use flooded evaporators. However, low-side floats are not used on typical air-conditioning or commercial refrigeration applications.

17.14 HIGH-SIDE FLOAT

The high-side float (HSF) modulates refrigerant flow to the evaporator based on liquid level. Unlike the low-side float, this float assembly is located on the high-pressure side of the metering device's orifice. As the load increases, more refrigerant is condensed in the condenser and flows into the high-side float assembly. The valve opens when the liquid level rises in the chamber, permitting greater flow into the evaporator. The pressure-reducing valve, or weight valve, is used in a long liquid line to prevent evaporation before the liquid reaches the evaporator. If the load is decreased, the float is lowered, reducing the flow of refrigerant. The most common application of the high-side float is in the flooded cooler of a centrifugal chiller. Figure 17-24 shows a high-side float controlling refrigerant flow to a flooded chiller evaporator.

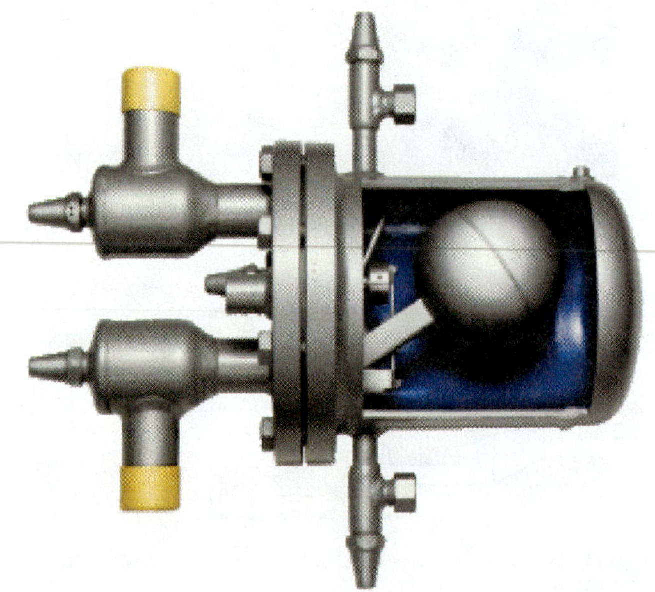

Figure 17-24 Diagram of a high-side float refrigerant control. *(WITT cold Maschinenfabrik GmbH)*

17.15 AUTOMATIC EXPANSION VALVE

Automatic expansion valves (AEVs) are basically refrigeration pressure regulators. They open and close to keep a constant pressure at the valve outlet. This type of valve consists of a diaphragm, a control spring, the valve needle (or ball), and seat (Figure 17-25). The evaporator pressure exerts a force against the bottom of the diaphragm. An adjustable spring exerts a pressure on the top of the diaphragm. When the spring pressure is higher than the evaporator pressure, the diaphragm flexes down, moving the valve pin off the seat. As the evaporator pressure increases, it overcomes the spring pressure and moves the diaphragm up, thus closing the valve.

The valve has an adjustment at the top for setting the evaporator pressure to produce a desired evaporating

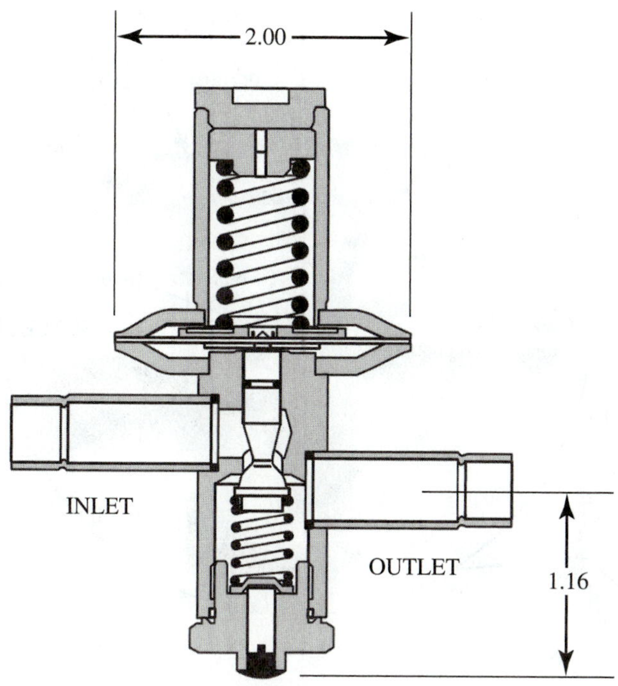

Figure 17-25 Automatic expansion valve.

temperature. Turning the adjustment clockwise increases the evaporator pressure; turning the adjustment counterclockwise decreases the evaporator pressure (Figure 17-26).

AEVs do not allow system pressures to equalize on the off cycle. When the compressor shuts off, the evaporator pressure builds up and overcomes the spring pressure, closing the valve. The low-side pressure quickly reduces when the compressor starts operating. The valve does not open until the evaporator pressure is lower than the spring pressure.

AEVs are good for maintaining a specific temperature because they keep the saturation pressure in the evaporator steady. One of the most common uses for AEVs today

Figure 17-26 Turning the adjustment clockwise increases the evaporator pressure; turning the adjustment counterclockwise decreases the evaporator pressure.

Figure 17-27 Automatic expansion valves are often used on refrigerated air driers.

is with refrigerated air driers (Figure 17-27). A refrigerated air drier condenses water out of compressed air. These systems run continuously, even when no air is traveling through them. AEVs allow the evaporator to be as cold as possible without freezing, even when there is no air flowing over the evaporator.

17.16 AUTOMATIC EXPANSION VALVE RESPONSE TO LOAD CHANGES

AEVs respond to changes in the evaporator load, but not in a way that is normally useful. If the load is light, less liquid will boil in the evaporator, causing the evaporator pressure to fall. The valve will open to try and keep the pressure up. This excess refrigerant floods the evaporator, creating a potential hazard for the compressor. Then as the load increases, more liquid refrigerant will boil, raising the evaporator pressure. The valve will close in response to this increased pressure, reducing the flow of refrigerant just when the evaporator could use more refrigerant. This causes the refrigerant vapor going into the compressor to have an increased amount of superheat. The increased suction superheat can cause the compressor to overheat, resulting in high discharge temperature and pressure, oil breakdown, carbonizing of the valves, and poor efficiency.

17.17 THERMOSTATIC EXPANSION VALVE

The thermostatic expansion valve (TEV) is widely used for nearly all types of air-conditioning and refrigeration systems, except for small appliances. The TEV is shown in Figure 17-28. The advantage of the TEV is its ability to automatically adjust to match the load, as it is designed to sense changes in the refrigeration load and respond to them. The TEV is not designed to maintain a constant temperature or pressure. Instead, it is designed to maintain a constant superheat. Imagine a system with an AEV that was monitored by a system operator. The operator could tell how much heat load was currently on the evaporator by monitoring the suction line pressure and temperature leaving the evaporator and calculating the superheat. If the superheat increased, the operator would

Figure 17-28 Construction of a TEV.

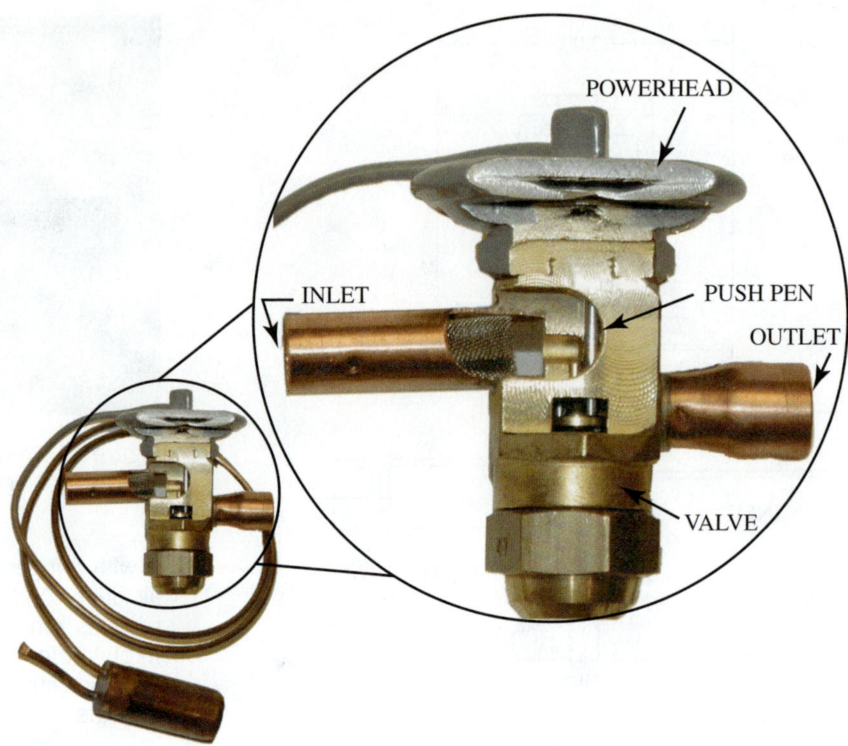

increase the spring pressure on the AEV to increase the flow of refrigerant to the evaporator. If the load and superheat decreased, the operator would decrease the spring pressure on the AEV to decrease the flow of refrigerant to the evaporator. The system would be operating at higher pressures and temperatures during periods of high load and lower pressures and temperatures during periods of low loads, but the superheat would remain fairly stable. This is essentially how a TEV works, except the valve, instead of a human operator, monitors the superheat.

Like the AEV, the TEV uses a diaphragm to control the flow of refrigerant. However, the TEV uses a remote bulb instead of spring pressure for the opening force on top of the diaphragm. The position of the valve pin is controlled by three pressures:

1. The evaporator pressure acting on the bottom of the diaphragm to close the valve
2. The spring pressure acting on the bottom of the diaphragm to close the valve
3. The bulb pressure opposing these two pressures and acting on the top of the diaphragm to open the valve (Figure 17-29 shows the relationship among these three pressures.)

The remote bulb is mounted on the suction line, sensing the suction line temperature. The bulb is filled with a volatile fluid whose pressure increases when the bulb temperature increases. On traditional expansion valves, the fluid in the bulb has the same type of refrigerant as that in the system. The bulb pressure is transmitted to the top of the diaphragm via a capillary tube (Figure 17-30). When the suction line temperature increases, the bulb pressure increases the opening force on the valve. When the suction line temperature decreases, the bulb pressure decreases the opening force on the valve. The evaporator pressure is

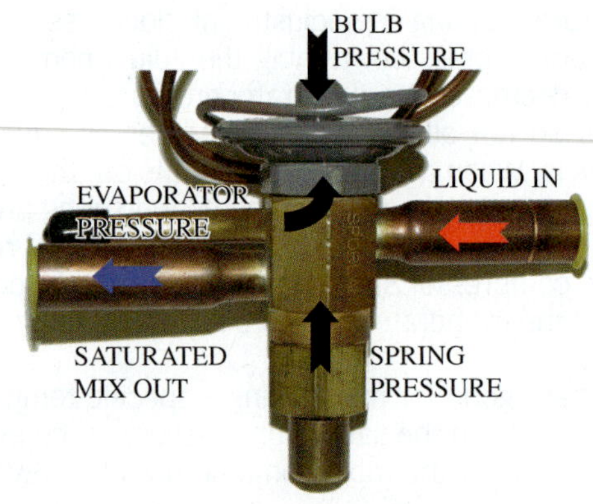

Figure 17-29 Basic TEV operation can be expressed as bulb pressure = evaporator pressure + spring pressure.

Figure 17-30 The power head of a TEV, including diaphragm, bulb, and connecting capillary tube.

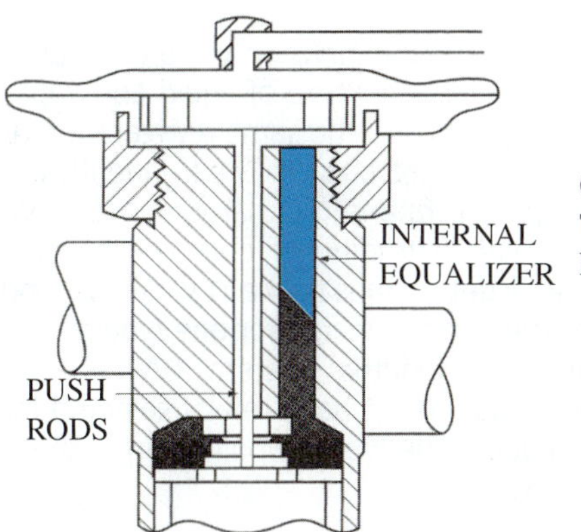

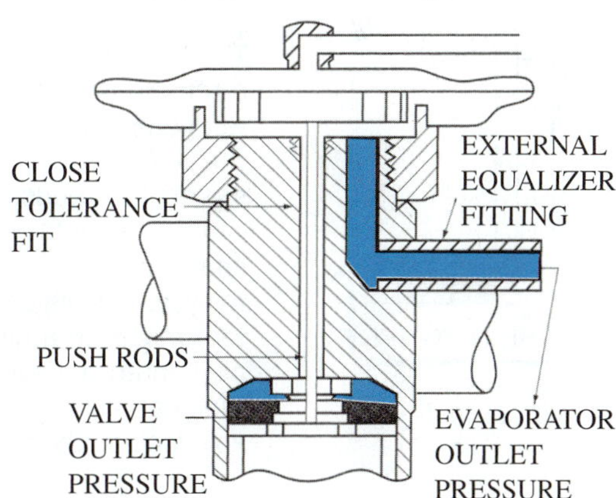

Figure 17-31 Evaporator pressure must reach the underside of the diaphragm for the TEV to operate.

supplied to the underside of the diaphragm on the TEV by either an internal passageway or an external line (Figure 17-31). An increase in evaporator pressure increases the pressure on the underside of the diaphragm, increasing the closing force on the valve. A decrease in evaporator pressure decreases the pressure on the underside of the diaphragm, decreasing the closing force on the valve. The addition of a spring on the underside of the diaphragm adds additional closing pressure, producing superheat.

The system is in equilibrium when its operating superheat equals the valve setting and the flow of refrigerant through the valve matches the load. At this condition, the pressure on the bottom of the diaphragm exactly balances the bulb pressure on the top of the diaphragm. For example, an R-410a system, with a suction pressure of 118 psig, a bulb temperature of 50°F, and a superheat setting of 10°F, would be in equilibrium. To calculate the superheat, look up the saturation temperature for 118 psig R-410a: it should be 40°F. Subtract the 40°F saturation temperature from the 50°F actual suction line temperature for the superheat: 50°F − 40°F = 10°F.

17.18 INTERNALLY EQUALIZED VALVES

Thermostatic expansion valves (TEVs) must be able to sense the evaporator pressure to operate correctly. The port or line that transmits the evaporator pressure to the underside of the diaphragm is called an equalizer line. On internally equalized valves, this is done with a passageway through the valve from the valve outlet to the underside of the diaphragm. Often the passageway is nothing more than a clearance around the push rods (Figure 17-32). This type of valve is called an internally equalized valve because the equalizer port is inside the valve. Internally equalized valves are easier to install than externally equalized valves because they do not require any connection to the suction line. However, since internally equalized valves measure the evaporator pressure at the beginning of the evaporator, they will not work well with evaporators that have a large pressure drop. The pressure drop causes the saturation temperature to drop also. The actual saturation temperature at the end of the evaporator will be considerably lower than the saturation temperature the valve sees at the beginning of the coil. This is why internally equalized valves are normally limited to single-circuit evaporators with a maximum pressure drop that is equivalent to a 2°F saturated temperature drop. Figure 17-33 lists the maximum recommended pressure drop for several refrigerants and evaporator temperatures. Notice that less pressure drop can be tolerated at lower evaporator temperatures than at higher evaporator temperatures.

Figure 17-32 Evaporator pressure passes through the holes around the triangular push rods on this internally equalized TEV.

	Evaporating Temperature F°				
	40°F	20°F	0°F	−20°F	−40°F
Refrigerant	**Pressure Drop in psi**				
R-134a	2	1.5	1	0.75	NR
R-22	3	2	1.5	1.0	0.75
R-404a, R-507	3	2.5	1.75	1.25	1.0

Figure 17-33 Maximum recommended pressure drop for R-134a, R-22, R-404a, and R-507.

17.19 EXTERNALLY EQUALIZED VALVES

It is necessary to use an externally equalized thermal expansion valve to obtain an accurate superheat setting when the pressure drop in the evaporator (including the distributor) is

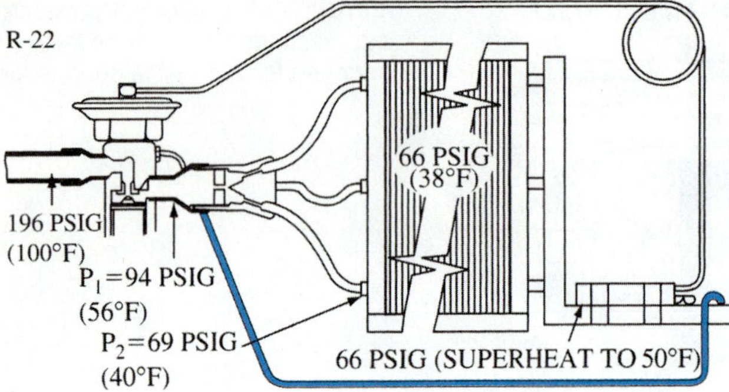

Figure 17-34 Effect of an external equalized TEV on the operation of the system.

sizable (Figure 17-34). Externally equalized valves measure the evaporator pressure through a separate line connected to the suction line. The underside of the diaphragm is isolated from the outlet of the valve so that the only pressure influencing the operation of the valve is the suction pressure coming in at the external equalizer line. To ensure that the valve outlet is isolated from the evaporator side of the diaphragm, the push rods are tightly fitted (Figure 17-35). Externally equalized valves are named for the line external to the valve that provides the evaporator pressure that balances the bulb pressure. Thus, the difference between an internally equalized valve and an externally equalized valve is where the evaporator pressure is sensed. If the evaporator pressure reaches the underside of the diaphragm via a passageway inside the valve, it is an internally equalized valve. If the evaporator pressure reaches the underside of the diaphragm through a separate line external to the valve, it is externally equalized (see Figure 17-31).

External equalizers must also be used when a distributor is used. Distributors are devices that divide the refrigerant evenly between several different refrigerant circuits in a multiple-circuit evaporator. A pressure drop is created across the distributor. This prevents the use of an internally equalized valve because the pressure at the valve outlet will not be the same as the pressure at the evaporator inlet.

Figure 17-35 The push rods on this externally equalized expansion valve fit tightly in the guide holes; evaporator pressure is provided by a separate, external tube and passes through the hole shown in the picture.

17.20 BULB CHARGES

The conventional thermostatic expansion valve (TEV) uses a remote sensing bulb with a mixture of liquid and vapor refrigerant in it. They typically use the same refrigerant that the system uses. However, bulb charges are frequently designed to produce specific operating characteristics that cannot be obtained with a traditional charge. The bulb charge is what primarily determines how a TEV will respond to system changes. The perfect charge for one type of system might be a disaster in another system. TEVs are not plug-and-play devices; they must be carefully selected for their particular application. The types of bulb charges available include the following:

- Liquid charge
- Liquid cross-charge
- Gas charge
- Gas cross-charge
- Adsorption charge

Liquid Charge

Liquid-charged bulbs are the standard. They are charged with the same type of refrigerant as is in the system. Enough charge is put in the bulb so that there is always liquid in the bulb. A liquid-charged bulb will exert an opening force equal to the saturation pressure for bulb temperature. An advantage to this type of charge is that the sensing bulb will never lose control over the valve as can happen with some other types of charge. A disadvantage of this type of charge is that it will cause high suction pressures on system startup and will also cause slow pull-down times.

Liquid Cross-Charge

A cross-charge contains a liquid that is different from the system refrigerant, usually a mixture of refrigerants. The components in the bulb charge are selected to produce a pressure response that is different from the saturation curve of the refrigerant in the system. If the bulb response and the refrigerant saturation pressure were graphed, they would cross each other at some point; thus the name *cross-charge*.

A valve with a traditional liquid charge requires more superheat to open at low pressures than at higher pressures. Liquid cross-charges help address this by flattening out the valve response. Valves with liquid cross-charges require about the same amount of superheat to open at low temperatures as at higher temperatures. Liquid cross-charges are often used on commercial refrigeration systems.

Gas Charge

A standard gas charge uses the same refrigerant as is in the system but does not have enough liquid refrigerant in the bulb to ensure liquid at all temperatures. At normal operating temperatures, the gas-charged bulb will respond more quickly than a liquid-charged bulb because of the limited charge in the bulb. The liquid will all be vaporized at or above the maximum operating pressure of the valve. Additional temperature rise will produce very little pressure rise because there is no more liquid in the bulb to vaporize. This is used to produce a valve with a maximum operating pressure.

A potential problem with this charge is that the liquid refrigerant can migrate from the bulb to the diaphragm if the valve body becomes colder than the sensing bulb. Once the liquid has migrated to the diaphragm, the sensing bulb no longer controls the valve operation.

Gas Cross-Charge

A gas cross-charge uses a refrigerant or mixture of refrigerants that are different from the system refrigerant. There is limited charge in the bulb, giving this type of charge a maximum-operating-pressure feature. The pressure response of a gas cross-charge is different from a traditional gas charge because the fluid in the bulb is different. One application of a gas cross-charge is to produce a flat response at normal operating pressures and temperatures and still have a maximum operating pressure.

Gas cross-charges are subject to charge migration just like traditional gas charges. It is important that the bulb remain cooler than the valve body.

Adsorption Charge

The adsorption charge uses an adsorbent material in the bulb and a noncondensable gas charge. The amount of gas the adsorbent material will adsorb depends upon its temperature. It tends to adsorb more gas at lower temperatures and release gas at higher temperatures. The variation in the amount of gas adsorbed changes the gas pressure in the power element. One advantage of this type of charge is that it will not lose control due to charge migration because there is no adsorbent in the diaphragm. However, the adsorption charge will not provide a maximum operating pressure.

17.21 MAXIMUM OPERATING PRESSURE (MOP) VALVES

The conventional sensing bulb can cause operating problems during the off cycle as well as on startup. This type of charge can cause excess pressure buildup in the evaporator during the off cycle, causing compressor overloading on startup. When the system starts, the valve opens wide because the suction line is warm, creating high suction pressures that can overload the compressor. This can be prevented by using a maximum operating pressure (MOP) valve. As the name implies, the valve has a maximum pressure that it will let through, regardless of the superheat. MOP valves have a gas-charged sensing bulb. The charge in the bulb is a mixture of liquid and gas at normal operating conditions but turns to all gas at the maximum operating pressure point (Figure 17-36). An increase in temperature of a vapor past the normal operating temperature range produces only a slight increase in pressure. Nearly all air-conditioning systems using TEVs limit the amount of pressure that can develop in the evaporator.

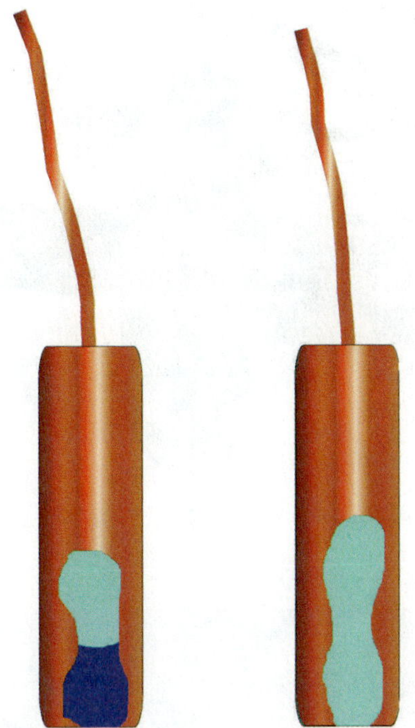

Figure 17-36 There is a small amount of liquid at normal operating temperatures in a gas-charged valve with MOP. When the temperature reaches the MOP point, all the liquid has vaporized, so the pressure levels off.

17.22 VALVE SUPERHEAT SETTING

All systems do not operate at the same superheat setting. The "correct" superheat for any system can only be determined by consulting the equipment manufacturer. However, there are standards that are widely used throughout the industry. First, a lower superheat setting delivers a higher capacity by using more of the evaporator coil to boil refrigerant. All systems should operate at the lowest possible superheat that will ensure dry vapor returning to the compressor. Next, the desired operating superheat decreases as the temperature difference between the evaporator and the load decreases. The evaporators in low-temperature refrigeration systems operate as little as 5°F lower than the freezer they are cooling, while the evaporators in air-conditioning systems can be 40°F cooler than the air they are cooling. Low-temperature freezers often operate with superheat as low as 4°F, while air-conditioning systems typically have superheat settings as high as 15°F. In general, the lower the operating temperature, the lower the normal operating superheat. As always, the manufacturer's recommendations should be the deciding factor in what a superheat setting should be. One manufacturer recommends 3–5°F superheat for low-temperature systems, 5–9°F superheat for medium-temperature systems, and 10–15°F superheat for high-temperature systems. An all-purpose superheat for generic refrigeration systems would be somewhere around 10°F.

Many manufacturers put nonadjustable TEVs on their systems (Figure 17-37). These valves have no provision for adjusting the superheat because the manufacturer has them made for one particular application. Most over-the-counter valves are adjustable. They typically have an adjustment stem, as in Figure 17-38, or an adjustment screw, as in Figure 17-39. They usually come from the factory set between 8°F and 12°F. These valves may need setting when they are installed because they are not manufactured for any particular system.

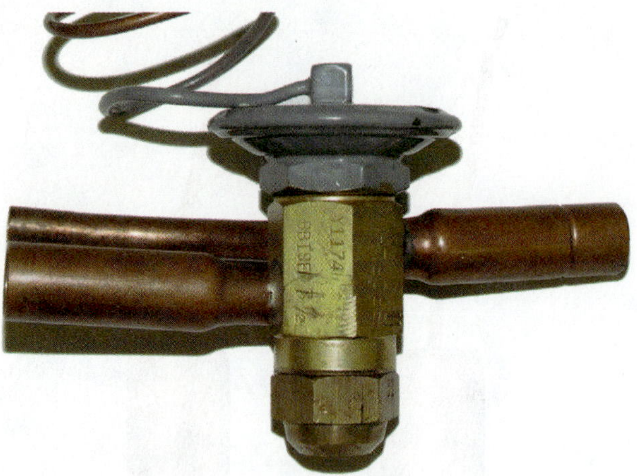

Figure 17-37 Some TEVs are not field adjustable.

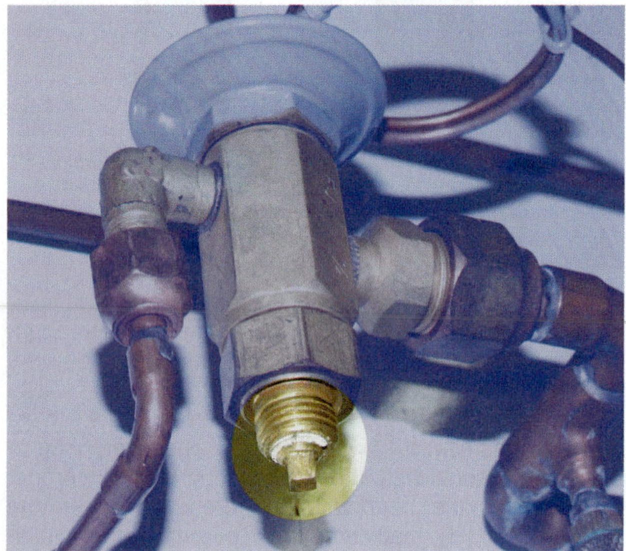

Figure 17-38 Most adjustable TEVs have square adjustment stems that must be turned with a refrigeration wrench.

17.23 MEASURING SUPERHEAT

Operate the system long enough for the system pressures to stabilize before trying to read the superheat. This can take 15–30 min. To adjust a TEV accurately, it is first necessary to measure its operating superheat. The basic formula for determining superheat is

superheat = (suction line temperature) − (evaporator saturation temperature).

This is illustrated in Figure 17-40.

Suction line temperature can usually be obtained quite easily by measuring the temperature of the suction line at or near the expansion valve bulb. The temperature should be measured at the bulb since that is the point at which the valve is measuring the suction line temperature. An accurate electronic thermometer is recommended (Figure 17-41). Infrared thermometers are usually not accurate for suction line temperature measurements. The line is usually smaller than the smallest point they can focus on, and the surface emissivity of the copper tubing is not close to the 0.95 that most infrared thermometers expect.

Figure 17-39 Some adjustable TEVs have adjustments that are turned with a screwdriver.

Evaporator saturation temperature is not as easily measured. To begin with, just getting to a spot where the

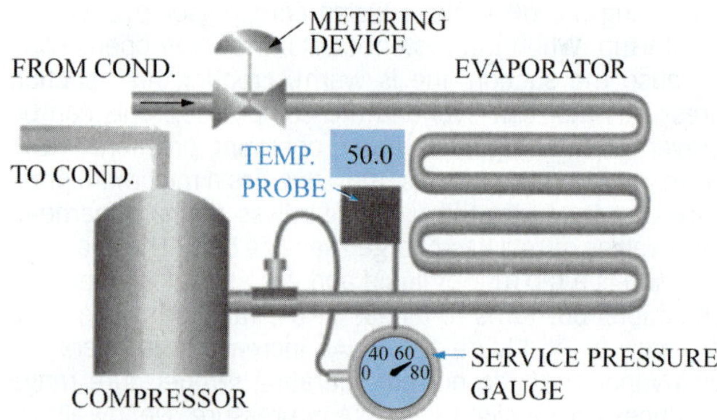

TO MEASURE SUPERHEAT:

1. FIND SUCTION PRESSURE
2. FIND MATCHING SATURATION TEMPERATURE
3. READ TEMPERATURE LEAVING EVAPORATOR
4. SUPERHEAT = TEMP. LEAVING — SATURATION TEMP.

Figure 17-40 Superheat is measured by subtracting the evaporator saturation temperature from suction line temperature.

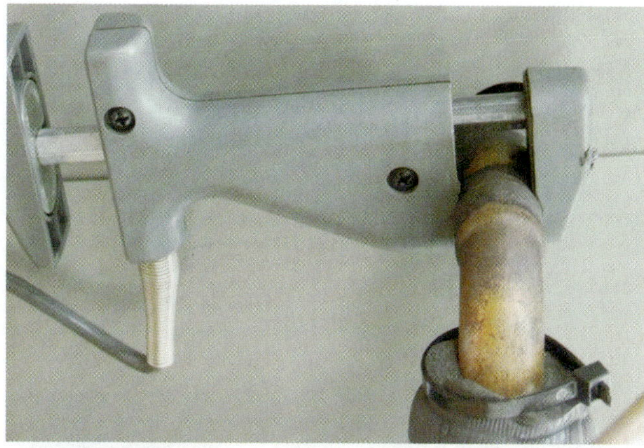

Figure 17-41 Clamp-on thermocouple temperature probes can be used to get an accurate suction line temperature measurement.

evaporator saturation temperature can be measured is usually not practical. Therefore, the saturation temperature must be approximated since it cannot be directly measured. Second, the saturation temperature does not remain constant throughout the evaporator because the pressure does not remain the same. The most accurate way to determine the saturation temperature of the evaporator is to measure the suction pressure as close to the outlet of the evaporator as possible. If appreciable pressure drop exists between the evaporator outlet and where you actually measure the suction line pressure, this pressure drop must be added to the actual pressure reading to obtain an accurate evaporator saturation pressure reading.

Next, refer to a pressure/temperature chart to determine the evaporator saturation temperature (Figure 17-42). Finally, subtract the evaporator saturation temperature from the suction line temperature: the difference between the two is the operating superheat of the valve.

The obvious limitation is knowing the pressure drop between the outlet of the evaporator and the nearest suction line port. On split systems, this is usually outside at the condensing unit. Standard line sizing practice is to design for a 2°F saturated temperature drop in the suction line. The amount of pressure would vary depending upon the evaporator temperature and type of refrigerant. The pressures in Figure 17-33 will apply here as well. Some manufacturers do provide a pressure tap at the indoor coil, making the process much easier (Figure 17-43).

Temperature	R-134a	R-22	R-410a
-25	7" Hg	7	23
-20	4" Hg	10	27
-15	0	13	32
-10	2	17	37
-5	4	20	43
0	7	24	49
5	9	28	55
10	12	33	62
15	15	38	70
20	18	43	78
25	22	49	87
30	26	55	97
35	30	62	107
40	35	69	118
45	40	76	130
50	45	84	142
55	51	93	155
60	57	102	170
65	64	111	185
70	71	121	201
75	79	132	217
80	87	144	235
85	95	156	254
90	104	168	274
95	114	182	295
100	124	196	317
105	135	211	340
110	146	226	365
115	158	243	391
120	171	260	418
125	185	278	446
130	199	297	476
135	214	317	507
140	229	337	539
145	246	359	573
150	263	382	608

Figure 17-42 Saturated refrigerant pressure-temperature chart.

17.24 ADJUSTING VALVE SUPERHEAT

Before attempting to adjust the superheat on an expansion valve, make sure that some other system problem is not causing the incorrect superheat. Adjusting a valve that does not need adjusting will just create another problem to solve once you have solved the original problem. For example, a high superheat can be caused by a system undercharge or a liquid line restriction. The valve cannot feed more refrigerant if it is not available. On the other hand, a system with a very low load may have a low superheat.

The superheat is adjusted by turning the valve stem in small increments to change spring tension. Increasing the spring tension increases the superheat setting; decreasing the spring tension decreases the superheat setting. The adjustment on most valves is clockwise (cw) to increase superheat and counterclockwise (ccw) to decrease superheat (Figure 17-44). Small adjustments can make big changes, especially if the operating superheat is within a few degrees of the desired superheat. Usually a quarter turn at a time is sufficient. A half turn might be warranted if the setting is off

Figure 17-43 The coil has a Schrader valve connection that can be used to take an accurate evaporator saturation-pressure reading.

Figure 17-44 Turning the adjustment clockwise increases the superheat; turning the adjustment counterclockwise decreases the superheat.

by 10°F or more. Time is required after each adjustment for the system pressures and temperatures to stabilize. This is an operation where being patient and taking your time will save time in the long run.

17.25 LIQUID DISTRIBUTORS

The refrigerant leaving the expansion valve must be distributed evenly to each circuit of multiple-circuit coils. This cannot be done using a common header because the refrigerant leaving the metering device is a mixture of liquid and gas. Since the liquid and gas are different densities, they will not flow through the header the same way . The result will be that some circuits are flooded while others

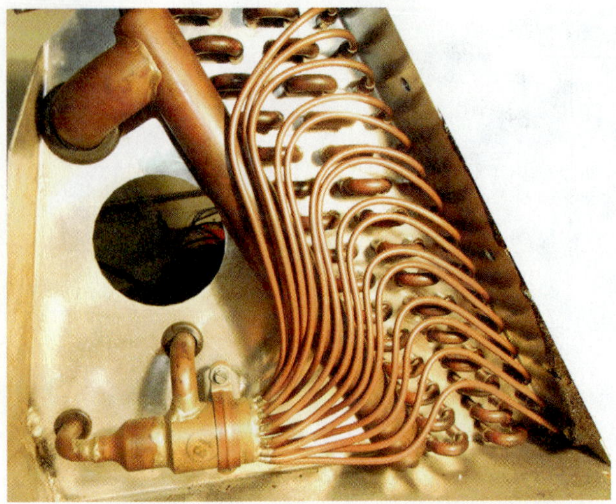

Figure 17-45 The distributor allows the TEV to feed multiple circuits evenly.

Figure 17-46 The distributor has a conical point that spreads the refrigerant evenly to all its circuits.

are starved. Distributors are used to equally distribute the refrigerant to each circuit on a multiple-circuit coil. They are placed between the expansion valve and the evaporator, as shown in Figure 17-45. Distributors are usually supplied by the coil manufacturer and are used where one metering device serves from two to forty evaporator circuits, using connecting tubes ranging from $^{5}/_{32}$ in OD to $^{3}/_{8}$ in OD. The distributor consists of a nozzle, a conical divider, and multiple holes evenly spaced around the divider cone (Figure 17-46). The liquid and gas are evenly mixed when they pass through the nozzle. This homogenous mixture is then split evenly by the divider cone, which sends equal amounts to each of the holes around it.

There is a pressure drop across the distributor. The effect of the distributor pressure drop is to reduce the amount of pressure drop across the expansion valve. Expansion valves with external equalizers must be used when a distributor is used because the pressure leaving the valve is higher than the pressure entering the evaporator coil.

17.26 LOW LOAD LIMITS

Both distributor nozzles and TEVs have low load limitations. A distributor will usually operate properly between 50 and 200 percent of its design capacity. When the flow through the nozzle drops below 50 percent, the gas and liquid refrigerant are not equally mixed. The superheat in the top circuits of the evaporator can be excessive due to liquid starving. The liquid floods back to the compressor from the lower circuits. This can easily be checked by measuring the temperature difference between the top and the bottom circuits of the coil. This problem can sometimes be solved by replacing the nozzle with the next smaller size.

The minimum load for most traditional TEVs is 35–50 percent of their capacity. When the flow through a TEV drops below 30 percent, the valve will not hold constant settings. The valve will alternately overfeed the coil and then underfeed the coil. This condition is called hunting. Under these conditions, superheat values fluctuate. Using a smaller expansion valve can sometimes correct this condition.

Systems that must operate over a wide capacity range can solve the problem of underloading the expansion valve and distributor by using two expansion valves and distributors. Each valve would be sized for approximately 50 percent of the coil capacity. The control system can operate a solenoid to cut off one section of the evaporator when the load drops below 50 percent capacity.

17.27 BALANCED-PORT EXPANSION VALVES

The primary reason for the operating limits of traditional TEVs is the design of the port and pin in the valve. Typically, the valve pin seats in a hole to oppose flow through the seat. The pressure on one side of the seat is much higher than on the other. This pressure difference acts as another opening force, lifting the pin off the seat. The problem is that this extra opening force changes as the system pressure drop changes. When the head pressure decreases, the pressure drop across the port decreases, and the lifting force decreases. The result is less refrigerant flow and a starved evaporator. When the evaporator pressure decreases, the pressure drop increases, increasing this lifting force. The result is a flooded evaporator.

A solution to these problems is to design the port in the expansion valve to minimize the effect of pressure drop across the valve. These valves are called "balanced port" or "double ported" valves. Although there are several designs of balanced-port valves from different manufacturers, the operating concept remains the same: refrigerant inlet pressure is applied at more than one point on the modulating assembly so that it counterbalances itself. Balanced-port valves can operate within a much wider range of operating conditions, including both low-ambient and low-load operation. Balanced-port valves can typically operate properly on air-cooled condensers down to 35°F without the need for condenser-head pressure control. They can modulate correctly down to 15 percent of full evaporator load.

17.28 BI-FLOW EXPANSION VALVES

For years, expansion valves have been one-way devices: they were made to only operate in one direction. Most still are designed this way. However, TEVs are now available that are specifically designed to allow refrigerant flow in both directions. Some of these valves meter the refrigerant in both directions, while others have built-in check valves that allow unrestricted flow in the "reverse" direction.

Figure 17-47 Bi-flow expansion valves are used in water-source heat pumps because they meter refrigerant in both directions.

Valves that meter in both directions are used in close-coupled systems, like water-source heat pumps (Figure 17-47). This way the system can be manufactured with a single metering device. Since these valves meter in both directions, they are not directional. These valves must be externally equalized, and the equalizer line must run to a common suction that is suction in both heating and cooling. Likewise, the sensing bulb must be attached to a suction location that is always on the low side, regardless of the operating cycle.

The valves with built-in check valves are typically used with split-system heat pumps (Figure 17-48). This eliminates an external check valve that would normally be used to bypass flow around the expansion valve in the reverse direction. Even though these are also called bi-flow valves, they are directional because they meter refrigerant in one direction and let it flow unrestricted in the other direction.

17.29 THERMOSTATIC EXPANSION VALVE SIZING

Although most TEVs are labeled with a nominal capacity, the operating conditions used to produce this rating may not match the actual operating conditions for the system the valve will be installed in. Extended capacity charts give the valve ratings after considering all the factors that affect the valve capacity (Figure 17-49). The factors that affect the valve capacity are as follows:

- The type of refrigerant
- The evaporator temperature
- The pressure drop across the valve
- The actual temperature of the liquid entering the valve

Refrigerant Type

Selection charts are typically organized by refrigerant type, with a different selection chart for each refrigerant type.

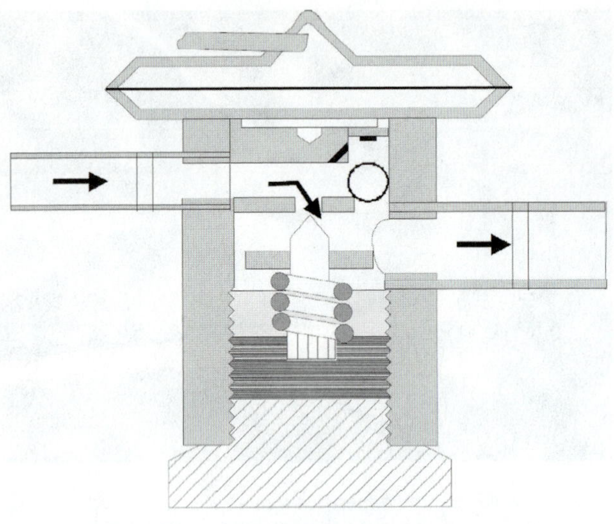

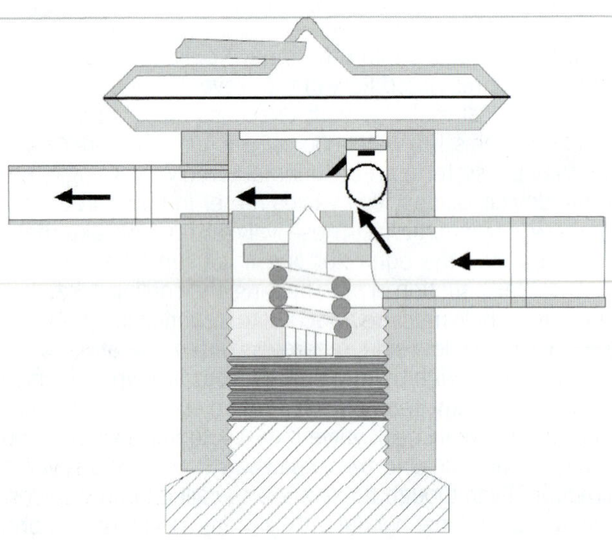

Figure 17-48 Expansion valves with built-in check relief valves are used on air-source heat pumps at both the indoor coil and outdoor unit.

The heading of the table in Figure 17-49 specifies that this table is for R-410a valves. Sometimes refrigerants with similar operating characteristics are grouped together on the same selection chart.

Evaporator Temperature

A range of possible operating evaporator saturation temperatures is listed under each type of refrigerant. A typical selection chart has columns for temperatures from 50°F to –40°F. Select the column that is closest to the evaporator saturation temperature for the system. In general, the valve capacity will decrease as the evaporator saturation temperature decreases.

Pressure Drop Across the Valve

A range of possible operating pressure drops across the valve is listed under the evaporator temperature column. In Figure 17-49 there are six pressure drops to choose from under each temperature column. The highlighted 160 psi column under the 40°F temperature column is the nominal rating condition for these valves. To determine what the pressure drop at the valve will be, start by determining the condenser and evaporator saturation temperatures for the system. Convert these saturation temperatures to pressures using a saturation pressure/temperature chart. Subtract the evaporator pressure from the condenser pressure. Next, subtract any additional pressure drops that do not occur at the expansion valve, including the following:

- Pressure drop through the refrigerant lines
- Pressure drop across any accessories in the liquid line, such as filter driers
- Pressure drop due to vertical lift in the liquid line
- Pressure drop through the distributor if a distributor is used

Looking Up the Correct Nominal Size

Look under the pressure-drop column that most closely matches the actual pressure drop across the valve. Go down the column until you find an actual capacity that is close to the capacity needed. Frequently the desired capacity falls between two valve ratings. In general, a slightly smaller valve will control better than a slightly larger valve. Look to the left to find the nominal size of the valve that

R 410A Thermostatic Expansion Valve Extended Capacity Table												
Valve Capacity in Tons of Refrigeration												
	40°F Evaporator						20°F Evaporator					
Nominal	Pressure Drop in psi						Pressure Drop in psi					
Capacity	120	160	200	240	280	320	120	160	200	240	280	320
1.5	1.3	1.5	1.7	1.8	2.0	2.1	1.3	1.5	1.6	1.8	1.9	2.1
3	2.6	3	3.4	3.4	4.0	4.2	2.6	2.9	3.3	3.6	3.9	4.2
5	4.3	5	5.6	6.1	6.6	7.0	4.2	4.9	5.5	6.0	6.5	6.9
7	6.0	7	7.8	8.6	9.3	9.9	5.9	6.9	7.7	8.4	9.1	9.7
9	7.8	9	10.0	11.0	11.9	12.7	7.6	8.8	9.9	10.8	12.5	12.5

Figure 17-49 TEV extended-capacity selection chart.

Liquid Refrigerant Temperature Correction Factors								
Liquid Line	70°F	80°F	90°F	100°F	110°F	120°F	130°F	140°F
R-134a	1.21	1.11	1.07	1.00	0.93	0.87	0.81	0.71
R-22	1.18	1.12	1.06	1.00	0.94	0.88	0.82	0.77
R-410A	1.09	1.06	1.03	1.00	0.97	0.93	0.89	0.84

Figure 17-50 Valve-capacity correction factors for entering liquid temperature.

delivers the correct capacity at the system's specific operating conditions. The extended sizing tables can also be used to determine what the actual capacity of any particular valve is at any given condition.

Actual Liquid Temperature

Most valve extended-capacity charts assume a liquid temperature of 100°F. Correction factors are applied to the capacities listed in the extended-capacity chart for liquid temperatures other than 100°F (Figure 17-50). Take the case of a 3-ton nominal-capacity R-410a valve with an evaporator saturation temperature of 20°, a pressure drop across the valve of 200 psi, and an entering liquid temperature of 80°F. In Figure 17-49, the 3-ton nominal-capacity valve has an actual capacity of 3.3 tons at 20°F with a pressure drop across the valve of 200 psig. The liquid correction factor shown in Figure 17-50 for 80°F R-410a liquid is 1.06. The capacity of the valve would be $3.3 \times 1.06 = 3.5$ tons. On the other hand, if the liquid were 120°F, the correction factor from Figure 17-50 would be 0.93. Then the valve capacity would be $3.3 \times 0.93 = 3$ tons.

17.30 INSTALLING THERMOSTATIC EXPANSION VALVES

A TEV should be mounted as close to the evaporator as is practical. When using a distributor, it should be mounted directly to the expansion valve outlet. The valve body itself can be in any position, but it should be in a position that allows access to the adjusting stem in case the valve needs setting. Be careful when soldering or brazing in an expansion valve. The valve body can be wrapped with wet rags or heat-absorbing paste to keep it cool. In particular, the power head and bulb should not be overheated.

SERVICE TIP

Many air-conditioning manufacturers are now supplying evaporator coils for high-efficiency units with TEVs installed inside the case of the coil. The thermal bulb is already attached to the suction line inside the coil. It is easy to destroy the expansion valve when brazing in the suction line on these coils. The heat conducts through the suction piping to the bulb, which is fastened to the suction line on the other side of the coil casing. The pressure created from the extreme heat destroys the valve. When possible, remove the bulb from the suction line before brazing the suction line, and then reattach it when the line cools.

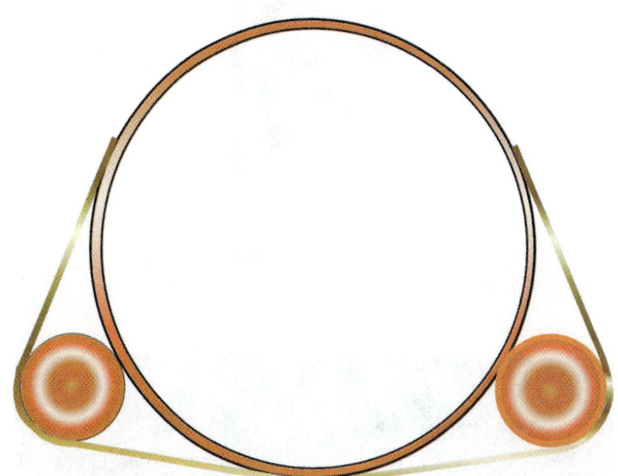

Figure 17-51 The sensing bulb should be mounted at the four- or eight-o'clock position on lines that are $^7/_8$ in outside diameter or larger.

Bulb Location

The thermal sensing bulb should ideally be located on a clean, horizontal section of suction line as close to the evaporator as possible. The sensing bulb should never be located on the bottom of the suction line. Liquid and oil can puddle at the bottom of the line and affect the temperature on the bottom of the line. The bulb is normally mounted at the four- or eight-o'clock positions on suction lines that are $^7/_8$ in OD or larger (Figure 17-51). The bulb may be mounted anywhere except on the bottom on lines that are smaller than $^7/_8$ in OD.

If a vertical bulb installation cannot be avoided, the bulb may be installed on a descending vertical line with the capillary-tube end at the top (Figure 17-52). The bulb should never be located in a trap or downstream from a trap. Liquid collecting in the trap can throw off the temperature reading at that point.

Good thermal contact is essential for the valve to operate correctly. The bulb should be fastened to the line with copper straps that will not only hold the bulb securely but will also help improve the thermal contact between the line and the bulb. Finally, the bulb should be insulated from the surrounding air so that the temperature the bulb is sensing is the suction line temperature, not the air temperature.

External Equalizer Connection

If a valve has an external equalizer, it must be connected; an externally equalized valve cannot operate without its equalizer line being connected. The equalizer connection

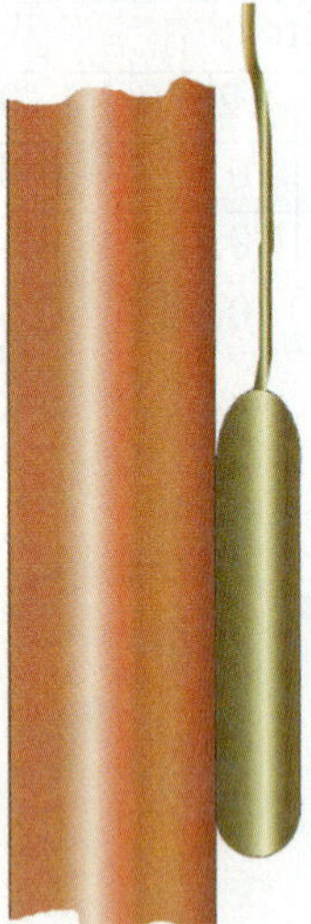

Figure 17-52 The capillary tube should come off the top if the bulb must be mounted on a vertical line.

Figure 17-53 The external equalizer connection should come off the top of the suction line, downstream from the bulb.

normally penetrates the suction line 6 or 8 in downstream from the sensing bulb, unless the manufacturer's instructions advise differently. The equalizer line should come off the top of the suction line, never off the bottom (Figure 17-53). Piping the equalizer off the bottom creates a potential trap that can interfere with the proper operation of the valve.

SERVICE TIP

The external equalizer line should never be upstream of the bulb. Although refrigerant is not supposed to flow through the equalizer line from the expansion valve, small internal leaks in the valve can allow a small flow of liquid through the equalizer line. If the equalizer line enters upstream of the bulb, this small flow of liquid can fool the bulb into thinking there is liquid leaving the evaporator. This will cause the valve to underfeed and starve the evaporator.

17.31 TROUBLESHOOTING THERMOSTATIC EXPANSION VALVES

TEV problems fall into three general categories:

- The valve overfeeds, flooding the evaporator.
- The valve underfeeds, starving the evaporator.
- The valve hunts, alternately overfeeding and underfeeding.

Valve Overfeeding

Valve overfeeding is most often caused by incorrect thermal bulb installation. The valve overfeeds because it cannot accurately sense the suction line temperature. The thermal bulb should be securely fastened to a straight, clean, horizontal section of suction line. The bulb should also be insulated from the surrounding air. The bulb should sense the suction line temperature, not the air temperature. If a bulb is loose or poorly insulated, it will be reading a temperature that is higher than the suction line temperature, and the valve will overfeed. Overfeeding can also be caused by improper valve sizing, incorrect application, or improper adjustment. In the case of valves that are assembled in the field from a kit of components, improper component selection can easily result in overfeeding. However, before investigating any of these possibilities, make sure the sensing bulb is correctly installed and insulated.

Valve Underfeeding

Check the system charge before assuming that the expansion valve is underfeeding. A valve cannot supply refrigerant that is not there. Also check for restrictions in the liquid line such as a plugged filter drier. Typically, a restriction in the liquid line will cause a temperature drop across the restriction. This can be assessed by checking the temperature difference across the suspected restriction.

A common cause of underfeeding on valves in systems with HFC refrigerant and POE lubricant, such as R-410a systems, is clogged screens and valves. Many valves have screens at their inlet to prevent foreign debris from entering the valve. A system with scale inside the lines caused by brazing can become clogged with this scale when the refrigerant and oil scrub it off the inside of the refrigerant lines and deposit it at the expansion valve. If the valve does not have a screen, the debris may become lodged in the valve itself. This usually requires changing the valve. One equipment manufacturer has had so many problems with debris-clogged valves that they now require the liquid line filter to be installed as close to the indoor unit as possible.

A TEV that has lost the charge in its bulb will underfeed severely. Often, systems that have valves with dead power elements will pull a vacuum on the low side. One way to diagnose a dead power element is to remove the bulb from the suction line and hold it in your hand during system operation. The suction pressure should rise. If there is no change, the power element is probably dead. Some valves have replaceable power elements. The parts on the valve shown in Figure 17-54 can be replaced. The entire valve must be replaced on one-piece valves (Figure 17-55).

Figure 17-54 The power head on some expansion valves can be replaced.

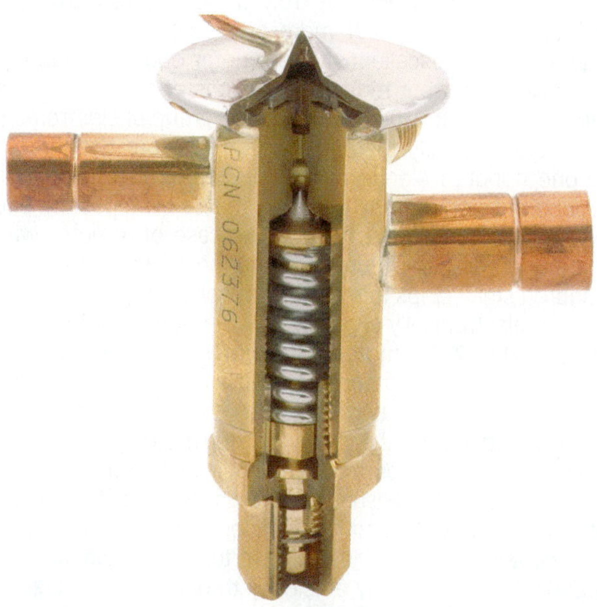

Figure 17-55 Some TEVs are constructed with a one-piece body; if the power head dies, the entire valve must be replaced. *(Courtesy Emerson Climate Technologies)*

Valve Hunting

A valve that is too large for the load it is trying to control will alternately overfeed and underfeed the evaporator. This happens when the smallest amount that the valve can open lets too much refrigerant through. When the valve opens to minimum position, it floods the evaporator. It will sense this and then shut. This causes the evaporator to starve. It will then sense this and open again. The low-side pressure will actually rise and fall on systems with severe hunting problems. Hunting is usually the result of an oversized valve or a lower than normal evaporator load.

SERVICE TIP

Manufacturers that sell systems with orifice metering devices often offer high-efficiency upgrade kits that can be field applied to their units. These kits are essentially a TEV with all the necessary fittings to connect it to the system. Replacing the orifice with a TEV increases the efficiency of most units by about 1 point on the SEER rating: a 12-SEER unit with a fixed restriction becomes a 13-SEER unit with an expansion valve. Remember that the charging chart on the units are for the orifice, not the TEV that was just installed. The charging charts on an orifice unit will not work with a TEV. Typically orifice charging charts specify the operating superheat at different conditions. Since TEVs try to maintain a stable superheat, these charts will not work with a TEV. Normally there is a statement to this effect in small print somewhere beside the charging chart. A few manufacturers provide a new chart to paste on the panel, but most do not.

17.32 ELECTRONICALLY CONTROLLED EXPANSION VALVES (EEV)

A number of mechanisms have been developed to control refrigerant flow with electric valves. Electronically controlled valves sense the evaporator saturation temperature and the suction line temperature using thermistors (Figure 17-56). A thermistor is a device that changes resistance when its temperature changes. The thermistors are wired to electronic controls that control the operation of the valves. One advantage of using electronic temperature sensing is being able to have the valve and the temperature-sensing device in different locations. These valves are all controlled by electronic signals, but the mechanisms used are quite different from one another. Types of electronically controlled valves that have been successfully applied include the following:

- Pulse-width modulated solenoid valves
- Magnetically controlled analog valves
- Heat motor valves
- Stepper motor valves

Pulse-Width Modulated Solenoid

A pulse-width modulated solenoid valve is shown in Figure 17-57. Unlike the TEV, which can modulate to positions between fully open and fully closed, solenoid valves are either

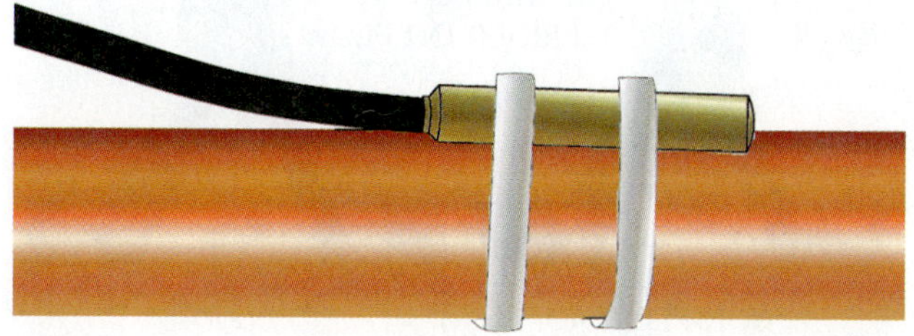

Figure 17-56 Thermistors are used to sense the line temperature for electronically controlled valves.

Figure 17-57 Pulse-width modulated solenoid valve. *(Courtesy of Danfoss, Inc.)*

open or closed. Pulse-modulated valves control flow by increasing or decreasing open time during each cycle. When coupled with the proper electronic control system, a solenoid valve can pulse rapidly, opening and closing quickly in response to the cooling load. The illustration on the left in Figure 17-58 shows the valve closed, and the solenoid valve is de-energized; the view on the right shows the energized condition. This valve also serves the function of a liquid line solenoid valve, blocking flow to the evaporator during the off cycle and at other times when refrigerant flow is not required.

Magnetic Analog Valves

The magnetic analog valve is really a specially designed solenoid valve that can modulate. It is controlled by a variable

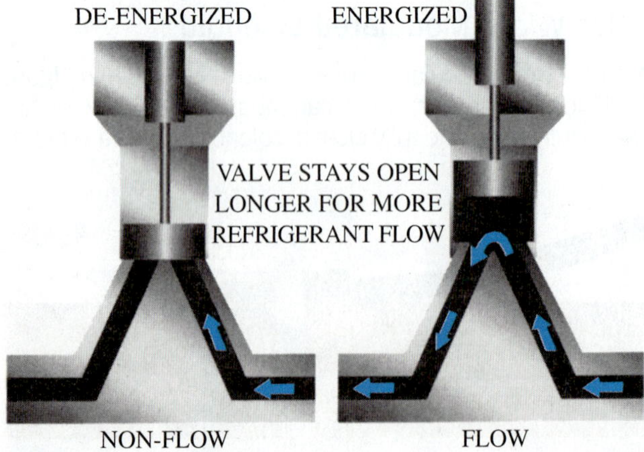

Figure 17-58 Pulsating solenoid valve, both energized and de-energized.

magnetic field. These valves have been used successfully in transport refrigeration, but they are not common. They are difficult to control, and their repeatability is poor.

Heat Motor Valves

Heat motor valves operate somewhat like traditional TEVs. They are filled with a fluid that expands when it is heated and contracts when it cools. A small heater that is controlled by an electronic circuit is immersed in the fluid. When the heater heats the fluid, it expands and increases the valve's opening force. When less heat is produced and the fluid cools, it contracts, and the opening force is reduced.

Step Motor Valves

The pulse-modulated solenoid valve, the magnetic analog valve, and the heat motor valve all have been successfully applied, but they all share some limitations. Getting consistent control from these is generally difficult. Their repeatability is poor. That is, they do not always behave the same way under the same conditions. All three are essentially fairly old electromechanical technologies being controlled by sophisticated controls.

Step motor valves are truly a more advanced technology. They can deliver more precise control, and their repeatability is excellent. A step motor does not spin like a regular motor but moves in fractions of a revolution, or steps. The motor winding is actually a group of electromagnet pairs. Typically there are anywhere from twenty-four to one hundred electromagnet pairs arranged in a circle around the step motor. The motor moves in steps from one pair of magnets to the next. In the case of a motor with one hundred pairs of magnets, the motor would take one hundred small steps to make one revolution.

An electronically controlled step motor valve is shown in Figure 17-59. The motor shaft rotates in tiny steps to control a set of gears that position a pin in a seat. The operation of step motor valves is governed by the software and electronics used to drive the valve. The step motor gets its signal from an electronic control panel that is attached to an electronic sensor that measures refrigerant superheat. Since the step motor valve is controlled independently of pressure, this valve is able to provide safe startup, shutdown, and operation and high energy efficiency through its full range of operating conditions. One particular step motor valve has 1,596 steps in a 0.125-in stroke and can operate at a rate of 200 steps per second. Not only can this provide very precise control, but the control points are very repeatable. The

Figure 17-59 Step motor expansion valve.

position of the valve pin at five hundred steps will always be the same. As electronic control becomes more popular, more electronically controlled valves will be put into use.

UNIT 17—SUMMARY

The metering device reduces the pressure and temperature of the refrigerant entering the evaporator by restricting how much refrigerant can pass through the metering device. All metering devices have some amount of flash gas. The evaporation of the flash gas helps cool the remaining liquid. The two general types of metering devices are fixed and modulating. Modulating devices can adjust their flow in response to changes in the system; fixed metering devices cannot. The flow through a fixed metering device is not fixed but varies with both the condenser and evaporator temperature. Two types of fixed metering devices are the capillary tube and the orifice. The capillary tube is a small-diameter, seamless copper tube. The orifice is a small, calibrated hole.

Modulating metering devices include the low-side float, the high-side float, the automatic expansion valve, the thermostatic expansion valve (TEV), and the electronically controlled step motor expansion valve. The TEV controls the refrigerant flow to maintain a constant superheat. The three operating pressures affecting TEV flow are bulb pressure, evaporator pressure, and spring pressure. The bulb pressure is balanced by the spring and evaporator pressure. An electronically controlled step motor valve uses a step motor that operates by moving in tiny incremental steps. The step motor operates gears that position the valve pin. Step motor valves are not affected by pressure variables and can provide control with much more precision than traditional TEVs. The electronic controls that operate step motors use thermistors to sense system temperatures.

WORK ORDERS

Service Ticket 1701

Customer Complaint: Poor Cooling

Equipment Type: Recently Installed R-22 Split System with an Orifice Metering Device

A customer has recently installed a new split system. The system does not seem to be cooling well. The technician notices that the high-side pressure is a little low, the low-side pressure is a little high, and both the subcooling and superheat are low. The service technician calls in and asks the installation mechanics if they remember what orifice they used. The installation mechanics reply that they do not know what number the orifice is; they simply left in the orifice that came with the coil. The service mechanic looks up the match between the coil and the condensing unit because the nominal coil rating is 12,000 BTU/hr larger than the nominal rating of the condensing unit. The coil is an AHRI match for the condenser, but the manufacturer specifies a smaller orifice for the match than is normally shipped with the coil. The technician closes the liquid line valve, pumps the system refrigerant into the condenser, and then closes the suction line valve. The orifice is replaced with the correct size, the coil and line set are evacuated, and the system valves are opened. The unit now operates within manufacturer's specifications.

Service Ticket 1702

Customer Complaint: Refrigerator Stops Cooling

Equipment Type: Household Refrigerator with a Recently Replaced Compressor

A customer has just had the compressor replaced in their refrigerator. The refrigerator starts and begins to cool, but it stops cooling a little while after starting. The technician starts the refrigerator and monitors its pressure. The pressures appear normal and the temperature begins to drop. However, after a few minutes the suction pressure pulls into a vacuum and the refrigerator begins to warm up again. The technician notices that although the compressor has been changed, the filter drier appears to be the original. The technician explains that the cause of the problem is most likely moisture freezing at the capillary tube as a result of not getting all the moisture out of the system when the compressor was changed. The refrigerant is recovered, the filter drier replaced, and the system is evacuated and charged with new, clean refrigerant. The refrigerator now stays cold after pulling down.

Service Ticket 1703

Customer Complaint: Air Conditioner Not Cooling

Equipment Type: Newly Installed Split System with a Factory-Installed TEV in the Evaporator Coil

A customer has just had a new high-efficiency air-conditioning system installed. The system uses a thermostatic expansion valve (TEV) that is factory installed inside the evaporator coil. The technician notices that the suction pressure is very low, and both the superheat and subcooling are very high. The system is doing very little cooling. The technician takes the front cover off of the evaporator and removes the TEV bulb from the suction line. The evaporator pressure does not change, even while holding the bulb. The technician notices that the paint on the coil casing is burned around the suction line and that the TEV bulb is installed just on the other side of the coil casing. The technician determines that the TEV power element was destroyed during installation from overheating when the suction line was brazed in. The technician recovers the system refrigerant, replaces the TEV, and evacuates and recharges the system. The suction pressure rises to the factory-specified level, and the system begins cooling.

Service Ticket 1704

Customer Complaint: Poor Cooling and Noisy Compressor

Equipment Type: Walk-in Cooler with a TEV

A new technician changed the TEV on a walk-in cooler last week. The system owner noticed that the compressor seemed noisier than usual and also noticed that the system was not cooling well. The company's senior technician is sent to see what is going on. There is no superheat, and the evaporator is flooding back to the compressor, causing the noise and poor cooling. The technician decides to check the bulb installation and finds that insulating gum is between the bulb and the suction line, keeping the bulb from sensing the suction line temperature. The senior technician asks the new technician about the gum, and he replies that he was instructed to insulate the bulb well, so he wrapped insulating gum all the way around the bulb. The senior technician explains that the purpose of the insulation is to ensure that the bulb senses the suction line temperature, so insulation between the bulb and suction line is not a good idea. The senior technician corrects the bulb installation, and the system operates with an 8°F superheat and a much quieter compressor.

UNIT 17—REVIEW QUESTIONS

1. What happens to the refrigerant as it passes through the metering device?
2. Explain the difference between a fixed metering device and a modulating metering device.
3. What are the two types of fixed metering devices?
4. Explain how an air conditioner with a fixed metering device reacts to changes in outdoor temperature.
5. Explain how an air conditioner with a fixed metering device reacts to changes in indoor temperature.
6. List the advantages of the capillary tube.
7. What advantages are there to orifice-type metering devices?
8. List types of modulating metering devices.
9. What is the most common application of the high-side float?
10. What is the difference between the low-side float and the high-side float?
11. Explain the basic operation of the automatic expansion valve.
12. Why is the automatic expansion valve not suited for systems with varying loads?
13. Which way would the adjustment on top of an automatic expansion valve be turned to increase the evaporator pressure?
14. What does a TEV try to regulate and keep stable?
15. What is the difference between internally and externally equalized TEVs?
16. How does a TEV with MOP work?
17. What is the difference between a traditional TEV and a balanced-port TEV?
18. What is the difference between a traditional liquid-charged TEV bulb and a liquid cross-charged TEV bulb?
19. What is the purpose of a liquid distributor?
20. The electronic expansion valve is activated by a(n) _______.
21. How do pulse-modulated valves control flow?
22. How do electronically controlled step motor valves work?
23. An R-22 TEV is set for a 12°F superheat. The system is operating with a suction pressure of 76 psig and a suction line temperature at the bulb of 50°F. What is the actual operating superheat?
24. Which way would the adjusting stem on an expansion valve be turned to increase the superheat?
25. Which way would the adjusting stem on an expansion valve be turned to decrease superheat?
26. What types of problems could cause a TEV system to operate at too high a superheat?
27. Why must technicians be careful when brazing lines to evaporator coils that have TEVs?
28. How can a capillary tube that is restricted with debris be differentiated from a capillary tube that is restricted with ice?
29. What is a bi-flow expansion valve?
30. An R-134a TEV is set for an 8°F superheat. The system is operating with a suction pressure of 22 psig and a suction line temperature at the bulb of 50°F. What is the actual operating superheat?
31. What system operating characteristics would you expect from a system that has a TEV with a dead power element?
32. What system operating characteristics would you expect from a system that has a TEV with a loose sensing bulb?
33. What system operating characteristics would you expect from a system that has a restricted-orifice-type metering device?
34. What system operating characteristics would you expect from a system that has an oversized-orifice-type metering device?

UNIT 18

Evaporators

OBJECTIVES

After completing this unit, you will be able to:

1. identify the four broad categories of evaporator application.
2. discuss the concepts of sensible and latent cooling as they relate to evaporators.
3. explain how bypass air and dehumidification work.
4. list the types of water-cooling evaporators.
5. explain the difference between flooded and direct-expansion evaporators.
6. explain the purpose of the condensate drain trap in air-conditioning evaporators.
7. list the three most common types of defrost in commercial refrigeration systems.
8. discuss the importance of using an evaporator coil that is an AHRI match to the condensing unit with which it is used.

18.1 INTRODUCTION

The evaporator is a device for absorbing heat into the refrigeration system and is named for the refrigerant evaporation taking place inside it. In the evaporator, the refrigerant absorbs heat from its surrounding load as it evaporates from a low-pressure saturated liquid to a low-pressure vapor. The evaporator is a heat exchanger with the refrigerant contained within tubes, passages, or a vessel. The load or product is separated from the refrigerant by the heat exchanger walls or shell. The heat flows into an evaporator because the temperature of the refrigerant is lower than the temperature of the product being cooled. The refrigerant temperature is set by its pressure. The evaporator pressure is maintained by the metering device.

18.2 EVAPORATOR LOADS

In refrigeration, the term *product* refers to anything that is being cooled or frozen, such as food. But it can be used to describe things being manufactured, such as plastics and electronics, health products such as medicines and medical labs, and products that help in manufacturing, such as solvent baths, machining coolants, and hydraulic fluids. It is not normally used to describe air, water, or brine, although they can also be considered products.

The evaporator heat load is the quantity of heat the refrigeration system must remove from the product. The product can be anything that contains a quantity of heat. An evaporator can be used to cool gases, liquids, or solids. Air is the most common gas cooled, water and brine are the most common liquids, and ice or other frozen products are the most common solids.

When evaporator coils are used to cool air, the heat load can be both sensible and latent. The sensible load is the amount of temperature change in the air as it passes through the coil. Sensible heat change results in a temperature change that can be measured by a thermometer.

Most air contains some water vapor. That water vapor will condense to a liquid as it passes over the evaporator if the evaporator temperature is below the dew point. As the water vapor changes from a vapor state to a liquid state, it releases its latent heat to the refrigerant. Latent heat reduces the amount of water vapor in the air. This requires refrigeration capacity, but it does not result in a temperature change in the air. A cooling process that is entirely latent will result in no temperature change in the air. In air conditioning, the load is typically both sensible and latent. Figure 18-1 shows how an increase in latent capacity corresponds to a decrease in sensible capacity.

Dehumidifying the air is the process of removing water vapor from the air. In mechanical refrigeration, water vapor is condensed to form liquid water. Dehumidifying takes a lot

Indoor Wet Bulb Temperature	Total Capacity	Latent Capacity	Sensible Capacity	Temperature Drop
62°F WB	31,000 BTU/hr	0 BTU/hr	31,000 BTU/hr	21°F
67°F WB	34,000 BTU/hr	6,000 BTU/hr	28,000 BTU/hr	19°F
72°F WB	37,000 BTU/hr	16,000 BTU/hr	21,000 BTU/hr	14°F

Figure 18-1 Evaporator sensible and latent capacity vs. entering-air wet bulb.

of energy and can be a major load on a cooling coil. It takes approximately 970 BTU of heat to vaporize 1 lb of water. A pint weighs about 1 lb, so it takes 970 BTU to change it all to vapor. This has the effect of adding 970 BTU of heat to the air. The same quantity of heat, 970 BTU, has to be removed when that water vapor condenses on the evaporator. The cooling capacity that is used to condense water vapor out of the air does not change the air temperature. In humid climates like the southeastern United States, dehumidification is just as important as temperature reduction.

18.3 PRESSURE-ENTHALPY DIAGRAM AND THE EVAPORATOR

A plot of a refrigeration cycle using R410A is shown in Figure 18-2. The line representing the evaporator is shown in green; the lines representing the other three components are shown in gray. Refrigerant enters the evaporator as a saturated mixture at the point indicated by the blue arrow. At this point, the refrigerant is a 30 percent vapor saturated mix with a pressure of 133 psia, a temperature of 40°F, with a heat content of approximately 58 BTU/lb. As the mixture travels through the evaporator, the vapor percentage increases as the liquid percentage decreases. This causes an increase in enthalpy, but the temperature remains 40°F and the pressure remains 133 psia. Eventually, the refrigerant is 100 percent vapor near the end of the evaporator at the point indicated by the purple arrow. At this point the refrigerant is still 133 psia, and 40°F, but its enthalpy has risen to approximately 120 BTU/lb. The last part of the evaporator is used to superheat the refrigerant. By the time the refrigerant leaves the evaporator at the point indicated by the red arrow, the refrigerant has changed to a superheated gas with a pressure of 133 psia, a temperature of 60°F, and a heat content of 135 BTU/lb.

18.4 TYPES OF EVAPORATORS

Evaporators are made for many diverse applications. They can be classified by their application, general operating principle, physical construction, and normal operating temperature range. Evaporators can be grouped by application into four broad categories:

- Domestic refrigeration
- Commercial refrigeration
- Air conditioning
- Chillers

Domestic refrigeration includes household appliances such as refrigerators, freezers, and dehumidifiers. Figure 18-3 shows evaporators for two domestic refrigeration appliances. The top picture is the evaporator for a dehumidifier; the bottom picture is the evaporator for a frost-free refrigerator. Notice that they are both constructed of aluminum.

Commercial refrigeration would be the large refrigerated cases and freezers in grocery stores and restaurants (Figure 18-4). Air-conditioning coils would include residential and light commercial air conditioners and heat pumps. Chillers are systems that cool water or water and antifreeze solutions (Figure 18-5). Each of these groups can be broken down further by such characteristics as physical construction, operating principle, or normal temperature range.

Domestic and commercial refrigeration evaporators can be listed both by their construction and operating

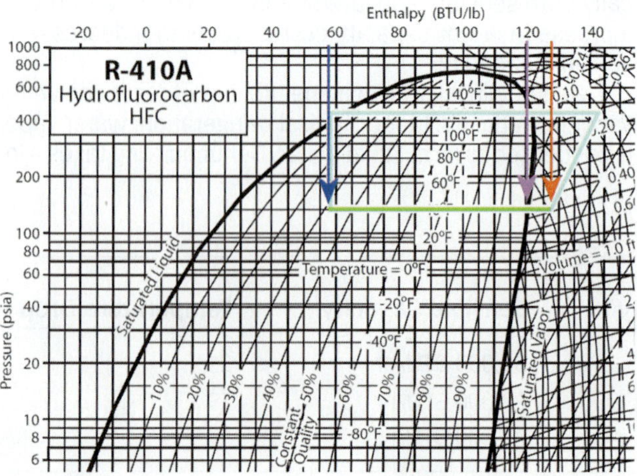

Figure 18-2 Enthalpy change through the evaporator.

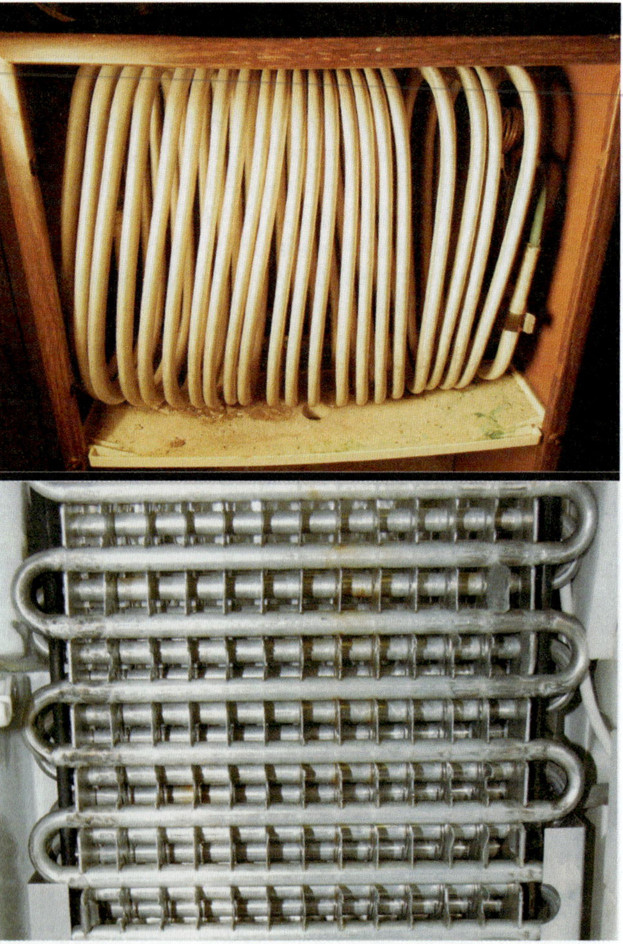

Figure 18-3 Two aluminum domestic refrigeration evaporators.

Figure 18-4 Commercial refrigeration evaporator for walk-in cooler. *(© 2011 Heatcraft Refrigeration Products LLC)*

temperature range. By construction, the types of domestic refrigeration evaporators are as follows:

- Bare pipe (Figure 18-6)
- Plate (Figure 18-7)
- Tube and fin (Figure 18-8)

Listed by their temperature range, they can be classified as the following:

- High temperature, 47°F to 60°F
- Medium temperature, 28°F to 40°F
- Low temperature, –20°F to 32°F

High-temperature applications are generally for storage of such items as flowers, candy, and dry goods. Medium-temperature applications are typically used for perishable fresh foods that must stay above freezing, such as eggs, produce, and fresh meat. Meat cases actually operate below 32°F, but the meat does not begin to freeze until around 28°F. Low-temperature applications are for ice and frozen food. Everything does not freeze at the same temperature. Ice cream needs to be –20°F to be frozen hard, but ice only has to be 32°F.

Air-conditioning evaporator coils are all tube-and-fin coils operating at about the same temperature range. However, there are many variations of shape and thickness

Figure 18-5 Large-tonage chiller.

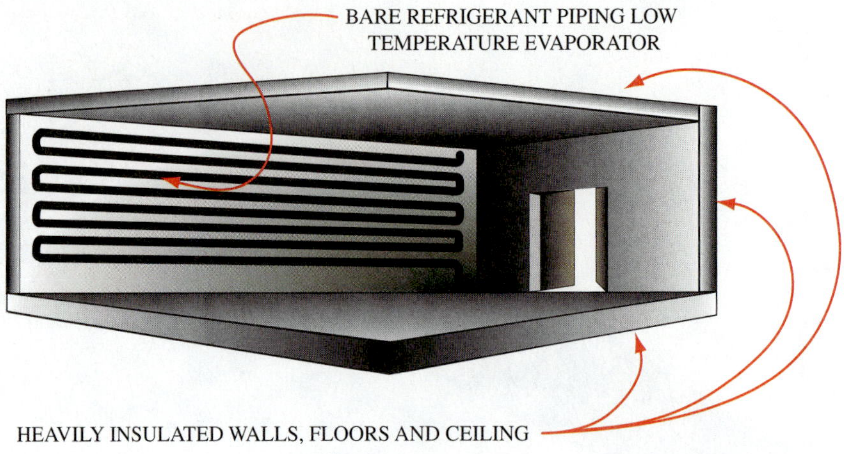

Figure 18-6 Bare-pipe evaporator coil with gravity air circulation.

in tube-and-fin air-conditioning coils. Figure 18-9 shows the four most common types of air-conditioning coil shapes:

- Slab coils (top left)
- Slant coils (top right)
- "A" coils (bottom right)
- "N" or "M" coils (bottom left)

Air-conditioning coils can also be listed by the direction that air flows through them. Listed by airflow, they can be as follows:

- Upflow: the air blows up through the coil.
- Downflow: the air blows down through the coil.
- Horizontal: the air blows horizontally through the coil.
- Multi-position: this combines one or more of the other positions.

Figure 18-10 shows the airflow direction for each of these types of coils.

Chiller evaporators can be listed both by their general operation and by their construction. By construction they can be as follows:

- Tube in tube (Figure 18-11)
- Shell and coil (Figure 18-12)
- Shell and tube (Figure 18-13)

Figure 18-7 Plate-type evaporator.

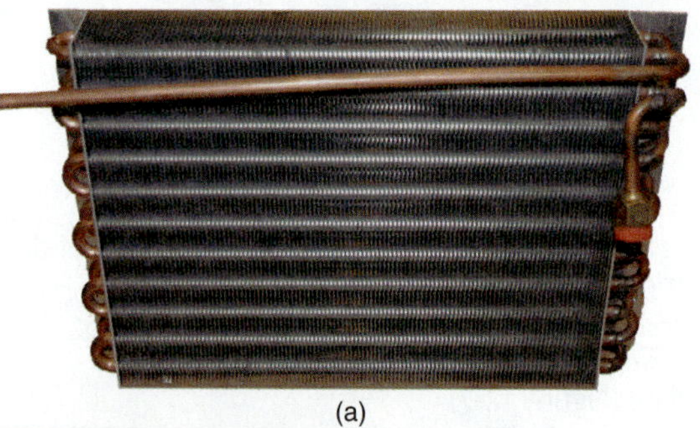

(a)

(b)

Figure 18-8 Tube-and-fin evaporator.

Figure 18-9 Four styles of air-conditioning coils, from top left clockwise: slab, slant, "A," and "M."

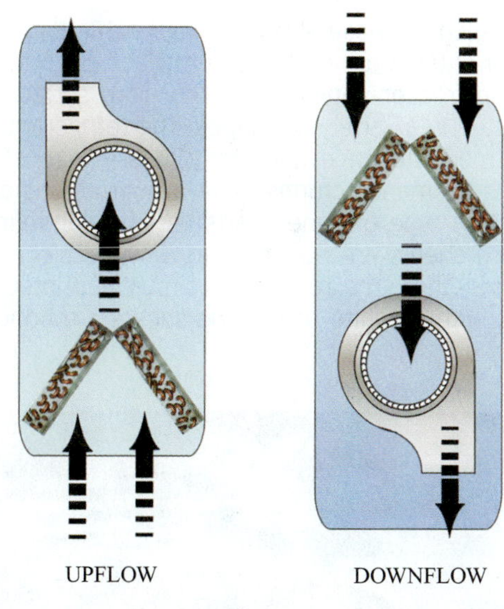

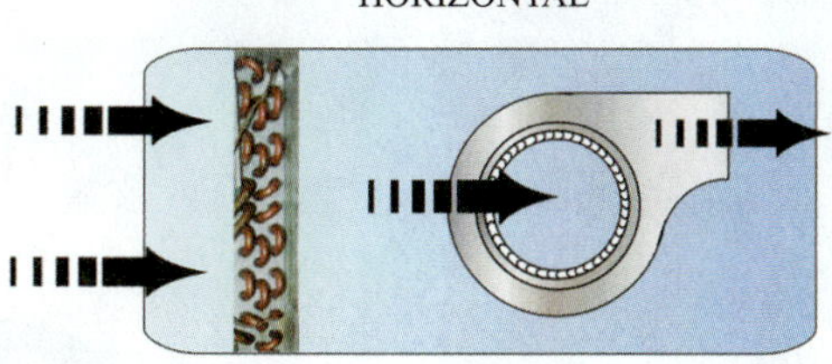

Figure 18-10 Three directions of airflow through an evaporator coil.

Figure 18-11 Cross-section of a tube-in-tube evaporator; the water travels through the inside copper tube, and the refrigerant travels through the steel tube in the opposite direction.

Figure 18-12 The water in the tank is cooled by the refrigerant in the tubing, which is wrapped around the tank.

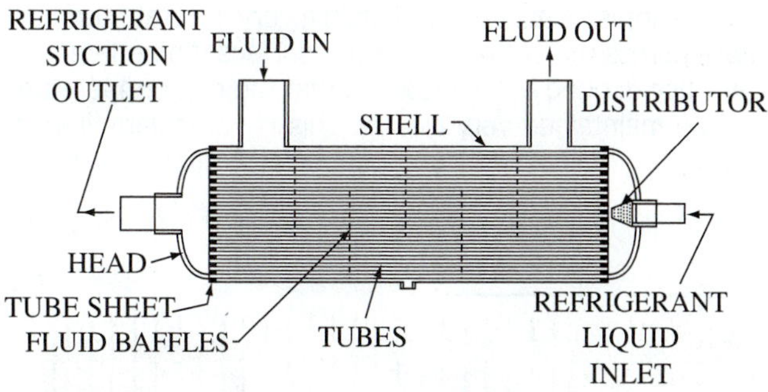

Figure 18-13 Direct-expansion chiller; refrigerant tubes, water in the shell.

By operating principle, chillers can be either direct expansion or flooded.

Evaporator design is specific to the intended application. An evaporator that performs perfectly in one application might not work at all in a different application.

18.5 FLOODED AND DIRECT-EXPANSION EVAPORATORS

Two general operating principles are used in evaporator design: direct-expansion evaporators and flooded evaporators.

Direct-Expansion Evaporators

Most air-conditioning and refrigeration systems use direct-expansion evaporators, in which the refrigerant travels through tubing. It enters as a saturated mixture and leaves as a superheated vapor. The refrigerant entering a direct-expansion evaporator is a saturated mixture of liquid and vapor. Generally 25–30 percent of the refrigerant entering a direct-expansion evaporator is vapor. This refrigerant flashed to vapor as it was expanding in the metering device. All the refrigerant is a vapor by the time it reaches the end of a direct-expansion evaporator. The very last portion of a direct-expansion evaporator is used for superheating the refrigerant (Figure 18-14).

Flooded Evaporators

The refrigerant enters a flooded evaporator as all saturated liquid, and there is saturated liquid throughout the evaporator. If an evaporator does not have to superheat refrigerant vapor, it can produce more cooling capacity. Flooded evaporators can circulate more pounds of refrigerant (more cooling capacity) per square foot of heat-transfer surface than direct-expansion evaporators because no evaporator surface needs to be used to superheat the suction vapor. On small systems the difference is negligible, but on very large systems, the increase in evaporator performance can be significant.

Large water chillers use flooded evaporators. (A flooded evaporator is shown in the example in Figure 18-43.) The evaporator in a flooded chiller is basically a large tank full of low-pressure, low-temperature liquid refrigerant with water lines submerged underneath the liquid.

Flooded evaporators typically use a float mechanism as their metering device. One of the advantages of a flooded evaporator is its ability to maintain a very accurate temperature across the entire evaporator surface. Some large ice rinks use flooded evaporators so that the rink temperature can be maintained very closely. This is particularly important for Olympic skating competitions. Skaters can tell the difference in ice temperature of a few tenths of a degree. Very cold ice sticks to the skate and is more difficult to glide across than slightly warmer ice.

It is very important to ensure that the saturated refrigerant flowing to the compressor does not contain quantities of liquid that could cause mechanical damage. Flooded chiller evaporators have eliminators whose function is to catch droplets of liquid in the returning suction vapor and prevent them from entering the compressor.

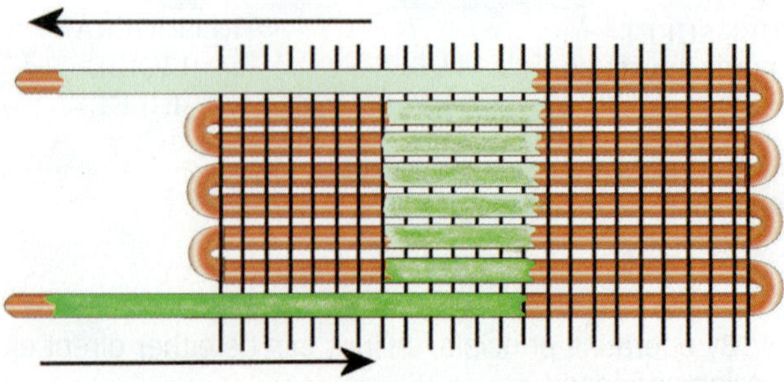

Figure 18-14 Refrigerant enters this direct-expansion evaporator at the bottom as a saturated liquid/vapor mixture and leaves at thetop as a superheated vapor.

18.6 EVAPORATOR CONSTRUCTION

The three most common types of evaporator construction are the following:

- Bare-pipe evaporators
- Plate evaporators
- Tube-and-fin coil evaporators

Bare-Pipe Evaporators

Bare-pipe evaporators are used in domestic upright freezers. The evaporator tubing also serves as the freezer shelves. In commercial refrigeration, bare-pipe evaporators may be used for low-temperature refrigeration, liquid cooling, ice skating rinks, and thermal storage applications. Because air is not blown across low-temperature bare-pipe evaporators, ice buildup on them will not significantly reduce the evaporator's efficiency, unless the ice becomes very thick.

Plate Evaporators

Plates such as those shown in Figure 18-15 are a special form of extended heat-transfer surface used in refrigeration and freezer applications. The plate surfaces may be fabricated in a variety of shapes. The refrigerant passages are formed into the evaporator plates. The most common process for manufacturing plate evaporators is called roll-bond. This type of panel starts as two aluminum sheets. The two sheets are bonded into a single piece by rolling them together. The refrigerant circuit is printed onto the sheets with graphite before the sheets are bonded. After

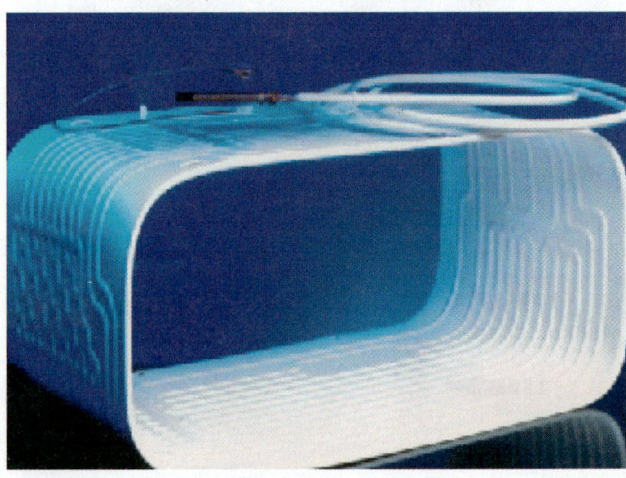

Figure 18-15 Roll-bond refrigerator evaporator.

Figure 18-16 Plate evaporator with tubes brazed to a galvanized plate.

rolling, the channels are created by pressurizing the panel at between 1,500 and 2,250 psig.

Plate evaporators can also be made by brazing tubing to the plate (Figure 18-16). Either way works well as long as the mechanical contact between the plate and the refrigerant passages is tight. A tight connection is needed for good thermal conductivity of heat from the plate to the refrigerant.

SERVICE TIP

Many plate evaporators are used in small residential refrigerators. Under-the-counter refrigerators do not defrost automatically. When manually defrosting these units, individuals often puncture the evaporator refrigerant passages with a sharp metal object. These punctures can be repaired using a two-part epoxy specifically designed for repairing holes in plate evaporators. Follow the manufacturer's instructions on using the epoxy, and only use those epoxies specifically designed and approved for this application. Some epoxies may have toxic substances that would not be appropriate to have in a domestic refrigerator freezer where food is stored.

Tube-and-Fin Coils

The most common evaporator for cooling air is the tube-and-fin coil. This applies to small refrigerated reach-in and walk-in boxes, as well as to small and large comfort air-conditioning units. The most common tube material is copper.

Fins are added to the copper tubing to increase the surface area. This is done to help accelerate the heat transfer rate into refrigerant evaporator tubes when cooling air. Air is not good as a conductor of heat. By adding tight-fitting aluminum fins to the refrigerant circuit's copper tubes, the air-side surface and corresponding heat transfer can be greatly increased.

18.7 TUBE-AND-FIN COIL CONSTRUCTION

To ensure good heat transfer, the fins must be tightly mechanically bonded to the tubes. The tighter the bond, the better the heat can be conducted from the fins into the refrigerant tubing and refrigerant. The fins are pre-stamped from thin sheets of metal, usually aluminum. They may be left flat or embossed, forming curves or waves in their surfaces. The fin sheets are stacked with even spacing on the copper tubing. The tubes are then mechanically or hydraulically expanded into the tube collar.

The fins are most often formed out of aluminum, but other materials can be used to increase the fins' corrosion resistance. Light protection can be provided with a baked-on powdered coating that is applied to the fin sheets before assembly. The baking process melts the powder, and it flows evenly over the fin surface and bonds much like paint.

It is possible to order coils with special protective coatings for industrial use where ordinary fins may be attacked by corrosion. The heaviest protection is provided when the finished coil is dipped in a thin epoxy. The epoxy cures and covers all of the fins, tubing, and coil headers. These coils provide the longest service in industrial applications where corrosive fumes are present, but they may reduce capacity.

For sprayed-coil dehumidifiers, where water sprays on the coil fins to maintain close temperature and humidity conditions, copper fins would be recommended. Because both the fins and tubing are made from copper, galvanic corrosion is eliminated. This is also important for marine or coastal applications.

Fin Spacing

The distance between fins is called "fin spacing" or "fin pitch" and is expressed as the number of fins per inch. The fin spacing on low-temperature coils can be 1–4 fins per inch to reduce ice bridging. *Ice bridging* is the term for what happens when the ice on one fin joins up with the ice formed on the adjoining fin. As shown in Figure 18-17, it restricts all airflow through the coil.

For air-conditioning applications, fin spacing ranges from 8 to 14 fins per inch. Figure 18-18 compares the fin spacing on a low-temperature refrigeration coil, a medium-temperature refrigeration coil, and an air-conditioning coil. The closer fin spacing produces increased heat transfer per square foot of coil, but the additional fins require more fan energy to move the air through them. This type of tradeoff in design is not unusual.

Tube-and-Fin Coil Tubing

The number of refrigerant tube rows in each refrigerant circuit and their spacing has a great influence on heat transfer in fin coils. The number of refrigerant tubing rows can range from one to ten. Two to four rows are most common for residential and light commercial air conditioning, while commercial air-conditioning coils may have ten or more rows.

Liquid refrigerant absorbs more heat than does refrigerant vapor. There are improvements in the tubing designs to increase the area wetted by liquid, including creating internal tubing grooves. These grooves churn the liquid up

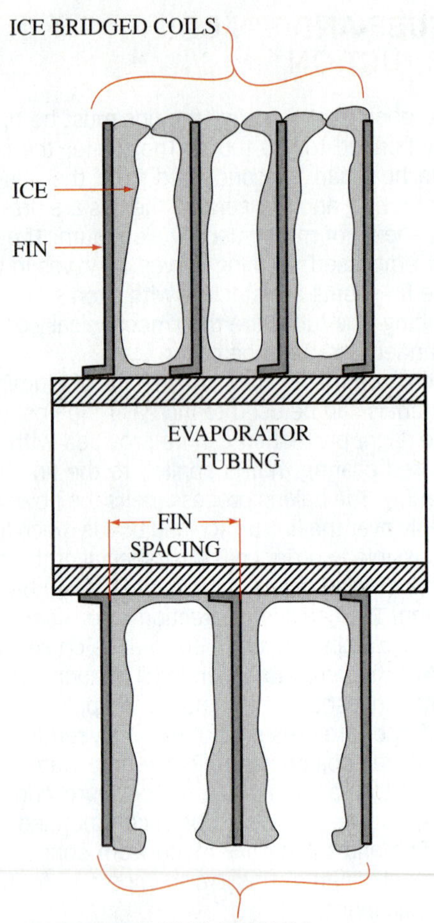

Figure 18-17 Ice bridging between fins on a tube-and-fin coil.

Figure 18-19 The grooves, or rifling, on the inside of this tubing increase heat transfer by creating turbulent flow.

onto the tubing sides as it flows through the evaporator (Figure 18-19). This increases the wetting of the internal tubing walls for better heat transfer. Another improvement has been to make the tubing oval shaped. This, too, increases the area inside the tubing that is exposed to liquid refrigerant.

Tube-and-Fin Coil Thickness

The more rows a coil has, the thicker it is. The thickness of fin coils greatly affects the coil's heat transfer and air resistance. Fin coils range in thickness from 1 in to about 1 ft. Residential and light commercial air-conditioning coils range from 1 in to 4 in thick. Commercial refrigeration coils may be up to 12 in thick.

Thin coils offer less air resistance and provide mostly sensible cooling. Many high-efficiency air-conditioning coils are built today with only one row of tubing. This provides the maximum possible heat transfer between each tube row and the air. Single-row coils operate at a slightly higher saturation temperature than multi-row coils, so they frequently operate above the dew point. Thin coils offer great economy of operation in arid areas where little or no dehumidification is required.

Thicker multi-row coils are advantageous for nearly all other areas where some degree of dehumidification is required. Multi-row coils use multiple rows of copper tubing held together by aluminum fins (Figure 18-20). The tubing in multi-row coils is staggered so that the front row does not block the airflow to the succeeding rows. Multi-row coils typically operate below the dew point. The air temperature is decreased further because the air that has already been cooled by one tube row is cooled even further by the following tube rows. Multi-row coils offer greater resistance to airflow and require more material to manufacture because their heat-transfer efficiency is less than single-row coils.

Figure 18-18 Comparison of fin spacing on coils for three different applications; the low temperature is on the left, medium temperature is in the middle, and is air conditioning on the right.

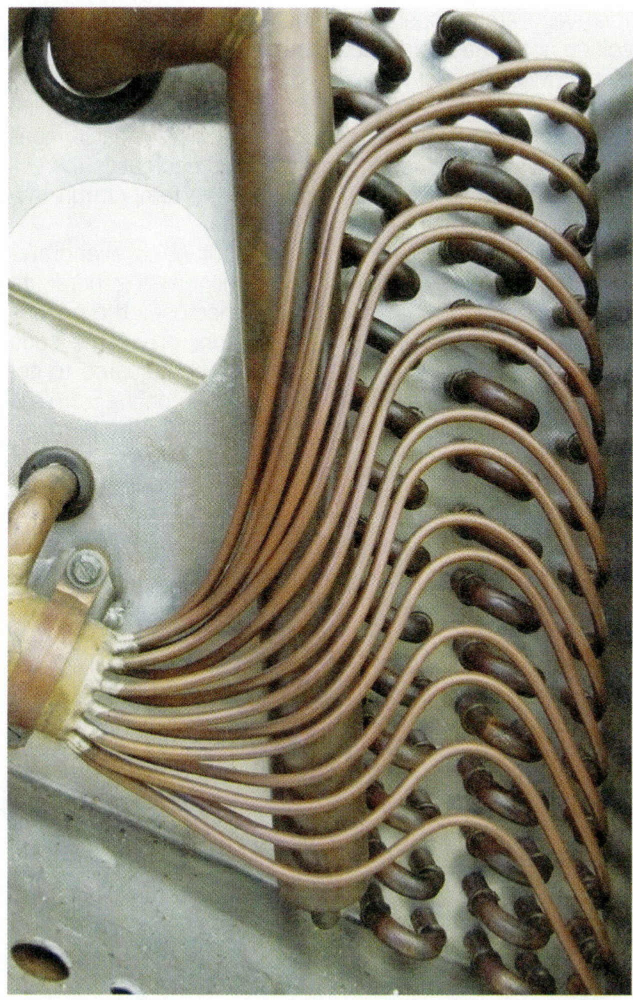

Figure 18-20 Multi-row coil used in commercial refrigeration application.

Figure 18-21 Slant evaporator used in a heat pump blower coil.

18.8 AIR-CONDITIONING COIL DESIGN

Air-conditioning evaporators use tube-and-fin coils and are manufactured in many shapes, including slab coils, slant coils, "A" coils, and "N" or "M" coils. Slab coils are rectangular in shape and are usually installed vertically or at a slight angle. The airflow through them is horizontal. Slab coils are usually multi-row coils to keep their size down. Slab coils are common in air-conditioning packaged units.

Slant coils are a variation of a slab coil. They are essentially a slab coil installed diagonally (Figure 18-21). Slant coils are usually provided with drain pans that allow both vertical and horizontal installation of the equipment. Slant coils are common in heat pump blower coils.

"A" coils are shaped like the letter *A*. They look like two slab coils joined at the top (Figure 18-22). "A" coils are popular with split-system air conditioners and heat pumps. Most "A" coils are designed to be installed in the upflow position, with air coming up through the wide opening. Some have been designed for use in counterflow or horizontal positions as well. Most "A" coils are multiple-row coils, but there are many being manufactured today that are single-row coils. These are typically taller because each side must be larger.

Figure 18-22 Air-conditioning "A" coil.

Figure 18-23 Air-conditioning "M" coil.

Coils shaped like the letter *N* or *M* use many smaller slabs to get more coil surface into a smaller space (Figure 18-23). These coils are typically one- or two-row coils. They are trying to achieve the efficiency improvement of a single-row coil and keep the physical size down.

18.9 TYPICAL AIR-CONDITIONING EVAPORATOR OPERATION

The saturation temperature in an air-conditioning coil is typically 30–40°F below the return air temperature entering the coil. The air temperature drop across the coil is usually between 10°F and 20°F.

For years the standard design evaporator saturation temperature was 40°F at AHRI rating conditions. The standard AHRI rating condition includes 80°F dry bulb and 67°F wet bulb return air. Thus, the evaporators were designed to operate 40°F below the return air temperature at the rating condition. Now many manufacturers are designing the evaporator to operate at 45°F under the same conditions.

The evaporator saturation temperature and the temperature drop across the coil are largely affected by the entering-air wet bulb temperature. A high wet bulb indicates a large amount of water vapor in the air. The condensing of the water vapor on the coil reduces the system's sensible capacity, so the temperature difference across the coil is decreased. The evaporator temperature is increased because the air and coil do not get as cold since more of the cooling capacity is going toward condensing water on the coil. Lower entering-air wet bulb temperatures produce the opposite effect. With less latent cooling to do, more of the cooling capacity goes into reducing the air's temperature. This decreases the coil saturation temperature and increases the temperature differential across the coil.

Standard airflow across an air-conditioning evaporator is 400 CFM (cubic feet per minute) per ton. In general, increasing the airflow across the coil increases the cooling capacity and decreasing the airflow decreases the cooling capacity. However, the airflow can be adjusted to get the correct amount of sensible and latent cooling from the evaporator. Decreasing the airflow increases the percentage of latent cooling capacity and decreases the percentage of sensible cooling capacity. Although both sensible and latent cooling will decrease with decreased airflow, a greater percentage of the cooling will be latent cooling. Increasing the airflow does just the opposite; it increases the percentage of sensible cooling and decreases the percentage of latent cooling. In areas of high humidity, it may be advantageous to sacrifice a small amount of total cooling capacity to increase the percentage of latent cooling capacity in order to control humidity. Equipment manufacturers provide system performance data that show the system's latent and sensible capacities at different airflow rates (Figure 18-24).

18.10 AIR-CONDITIONING COIL-FACE AREA AND FACE VELOCITY

The face area of a coil limits the face velocity of air that can pass through the coil without causing the water that is condensed on the coil to blow off. *Condensate blowoff* and *moisture carryover* are terms that refer to the small droplets of water that are blown free from the coil and travel a short distance down the duct. Coil manufacturers list the maximum volume in CFM and/or the maximum face velocity in feet per minute (fpm) for each coil they produce. Typically, the air face velocity for coils ranges from 300 fpm to 500 fpm. The face velocity should not exceed 550 fpm. If the condensate blowoff carries past the drain pan, it can leak from the air-handler cabinet and may cause expensive

PERFORMANCE DATA

COOLING CAPACITIES (MBH) - PURON® REFRIGERANT

CNPV UNIT SIZE	INDOOR COIL AIR		SATURATED TEMPERATURE LEAVING EVAPORATOR (°F)														
			30			35			40			45			50		
	CFM	EWB	TC	SHC	BF	TC	SHC	BF	TC	SHC	BF	TC	SHC	BF	TC	SHC	BF
1814	450	72	31.00	15.20	0.00	28.60	13.90	0.00	25.80	12.50	0.00	22.80	11.10	0.00	19.30	9.60	0.00
		67	26.00	15.70	0.00	23.50	14.30	0.01	20.70	12.80	0.01	17.60	11.30	0.01	14.10	9.80	0.01
		62	21.50	16.00	0.01	18.90	14.50	0.01	16.10	13.00	0.01	13.00	11.50	0.01	10.10	10.10	0.03
	600	72	38.30	18.70	0.00	35.30	17.10	0.00	32.00	15.50	0.00	28.20	13.80	0.00	23.90	12.00	0.01
		67	32.30	19.50	0.01	29.20	17.80	0.02	25.70	16.10	0.02	21.80	14.30	0.02	17.50	12.40	0.02
		62	26.70	20.10	0.02	23.50	18.40	0.02	20.00	16.60	0.02	16.20	14.80	0.02	12.90	12.90	0.06
	750	72	44.30	21.50	0.00	40.90	19.90	0.00	37.00	18.10	0.00	32.70	16.10	0.02	27.70	14.00	0.02
		67	37.40	22.80	0.03	33.90	20.90	0.03	29.90	19.00	0.03	25.40	16.90	0.03	20.40	14.70	0.04
		62	31.10	23.80	0.04	27.40	21.80	0.04	23.40	19.80	0.04	19.10	17.70	0.04	15.60	15.60	0.10

Figure 18-24 Evaporator sensible and latent cooling capacities.

water damage to the building or its contents. It is important to stay within the manufacturer's published airflow limits to avoid condensate blowoff.

18.11 BYPASS FACTOR

Not all of the air that passes through a coil comes in contact with the coil, even though the fins are very closely spaced. *Bypass factor* is a term used to quantify the air that gets through the coil without touching anything. The bypass factor for a coil is listed as a percentage of the total air volume. Manufacturers often list an evaporator's bypass factor in the performance data (see Figure 18-24).

The bypass factor of a coil affects the coil's ability to cool and dehumidify the air. No moisture is removed from the bypass air that passes through the coil without coming in contact with a cold surface.

There are a number of items that affect a coil's bypass factor, including the following:

- **Fin spacing** The closer the fins are spaced, the lower the bypass factor; the further they are spaced apart, the higher the bypass factor.
- **Fin shape** The greater the fin-embossed curves or waves, the lower the bypass factor; the flatter or straighter, the higher the bypass factor.
- **Coil thickness** The thicker the coil, the lower the bypass factor; the thinner the coil, the higher the bypass factor.
- **Rows of tubing** The number of rows of refrigerant tubing is closely associated with coil thickness. So the more rows, the thicker the coil and the lower the bypass factor; the fewer the rows, the higher the bypass factor.
- **Air velocity** The slower the air speed through the coil, the lower the bypass factor; the faster the air, the higher the bypass factor.

18.12 DEHUMIDIFYING THE AIR

Dehumidifying the air is second to cooling the air as the most important job of tube-and-fin cooling coils. If cooling coils did not remove some of the moisture in the air, our homes and businesses would be cold and damp. It would be like living and working in a cave with mold growing everywhere.

A coil's ability to dehumidify the air is closely associated with a coil's bypass factor. The higher the bypass factor, the lower the dehumidification, and the lower the bypass factor, the greater the dehumidification. The items listed above that affect the bypass factor of a coil, therefore, affect the dehumidification ability of a coil. There are other items beyond those for a coil's bypass factor that affect the coil's dehumidification ability, including the following:

- **Coil temperature** The colder the coil, the more dehumidification; the warmer the coil, the less the dehumidification.
- **Longer run times** The longer the system operates, the greater the dehumidification; the more the system cycles, the less the dehumidification.
- **Parallel air and refrigerant flow** Air and refrigerant can flow through a coil in the same direction—this is called parallel flow. When they flow parallel, the coldest refrigerant is in contact with the warmest air, and the coldest air is in contact with the warmest refrigerant. This results in less dehumidification.
- **Counter air and refrigerant flow** When the air and refrigerant flow in opposite directions through the coil, it is called counterflow. When they flow in opposite directions, the warmest refrigerant is in contact with the warmest air, and the coldest air and refrigerant are in contact. This results in the greatest dehumidification (Figure 18-25).
- **Number of refrigerant circuits** The larger the number of paths refrigerant has through the coil, the higher the dehumidification; the fewer the number of refrigerant paths, the lower the dehumidification.

There are three common ways that manufacturers list a coil's ability to dehumidify the air. One is to list the sensible and latent cooling values for the coil. The second is to list the total cooling capacity and the sensible cooling capacity, with the latent capacity being the difference. The third way is to give the total cooling capacity and give the latent as a percentage of the total. For example, a coil with a nominal rating of 30,000 BTU per hour might have a total capacity of 30,000 BTU, a sensible capacity of 22,000 BTU, and a latent capacity of 8,000 BTU. This would mean that 22,000 BTU are being used to reduce the temperature of the air and 8,000 BTU are being used to condense water on the coil. In this case, the latent capacity is 27 percent, or 0.27 of total capacity.

18.13 PRESSURE DROP ACROSS THE COIL

The term *pressure drop* when applied to a fin coil refers to the difference in static pressure from one side of a coil to the other side. The high side of the incline manometer reads the static pressure of the air entering the coil, while the low side of the manometer reads the static pressure of the air leaving the coil (Figure 18-26). Coil manufacturers publish coil data sheets with this information. The pressure drop across a coil depends on the design of the coil and the amount of air moving over it. All coils produce a higher pressure drop as more air is moved across them. The static pressure drop across an air-conditioning coil is usually the largest single pressure loss in the entire duct system.

The pressure drop, or ΔP (pronounced *delta P*), across the coil is measured in inches of water column. The pressure drop across residential air-conditioning coils is typically less than a half inch of water column. To measure this small amount of a static pressure, you will need an instrument such as an incline manometer or a magnehelic gauge (Figure 18-27). Most manufacturers provide the static pressure for both "wet" and "dry" coil conditions. The term *wet coil* means that the coil is condensing water. The water droplets on the coil fins add to the air resistance and cause a larger static drop across the coil. When taking "wet" coil readings, let the system run for 10–15 min to make sure it

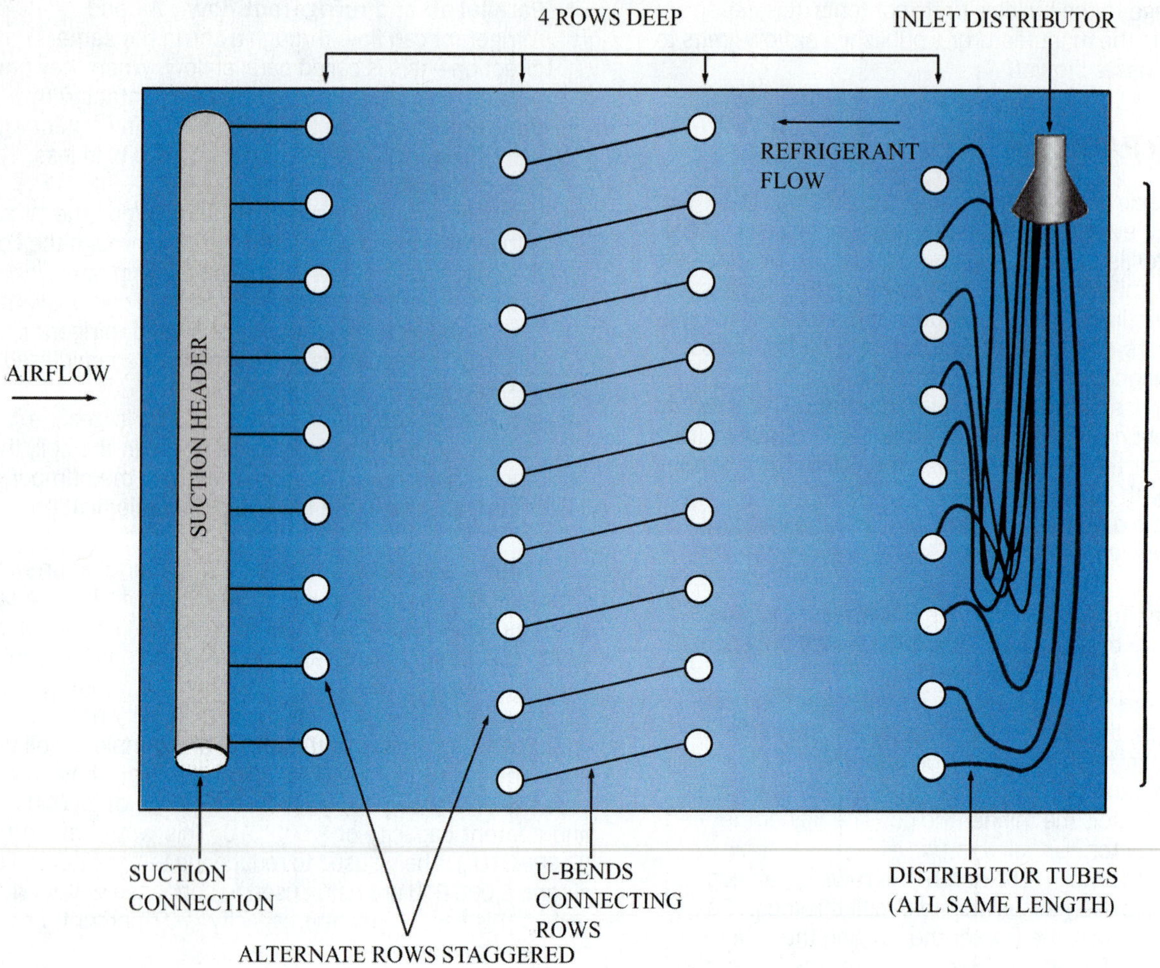

Figure 18-25 Direct-expansion evaporator coil showing counterflow airflow.

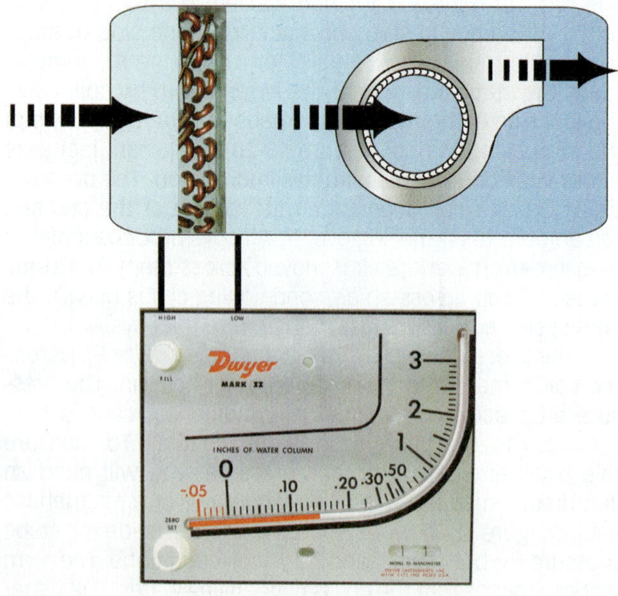

Figure 18-26 The incline manometer is measuring 0.15 in WC static pressure drop across the evaporator coil.

is "wet" before reading the static pressure drop across the coil.

Once you have the pressure drop, you can compare your readings to the manufacturer's data sheet to determine the CFM passing over the coil (Figure 18-28).

SERVICE TIP

The coil may be dirty if the pressure drop across the coil exceeds the manufacturer's specifications. Check the system airflow by checking the pressure drop across the blower and comparing it to the blower data. If the system airflow is not excessive, the extra pressure drop across the coil is caused by restriction in the coil itself. This usually can be corrected by cleaning the coil.

18.14 AIR-CONDITIONING CONDENSATE

Direct-expansion coils operating below the dew point of the entering airstream will cool the water vapor (humidity) in the air and condense it as droplets on the coil. The droplets

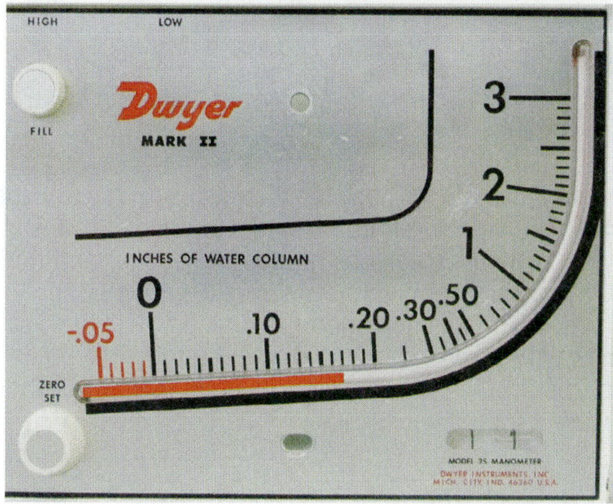

Figure 18-27 Static pressure drop can be read using eithera magnehelic gauge, shown on top, or an incline manometer, shown below; both are reading 0.17 in WC here.

Figure 18-29 The sides of this drain pan slant in and toward the drain outlet to reduce the amount of standing water in the drain pan.

drain by gravity into the drain pan and flow through a condensate drain line to a drain or sink disposal. The drain pans in older coils were often simple rectangular pans under the coil. The drain pans in modern coils are designed to minimize the amount of standing water in the coil by sloping the condensate toward the drain outlet (Figure 18-29). The drain line should be provided with a water seal trap and should never be connected directly to a sanitary drain.

The condensate drain must be trapped to allow the condensate to drain freely. Even when the static pressure in the coil cabinet is positive, the velocity of the air blowing past the drain outlet can create a negative pressure at the drain that tends to hold the water in the drain. This effect is even more pronounced in draw-through coils where the pressure inside the coil portion of the cabinet is negative to start with.

A condensate drain trap is used to offset the negative pressure at the drain outlet. Standard practice for residential air conditioning includes a 3-in-deep trap. The outlet of the trap should be lower than the outlet of the evaporator drain, and the line leaving the trap should slope toward an open drain or sump. It should not be directly connected to a sewer line. Figure 18-30 illustrates proper condensate drain construction. This design lets the condensate water create a siphon suction pressure equal to the unit's negative static pressure to offset the system's negative pressure so that the condensate water is drawn out of the condensate pan.

SERVICE TIP

Failure to properly design a draw-through condensate trap will allow the negative pressure in the coil drain to stop the condensate from draining. Excessive condensate will then overflow the drain pan, which can cause damage to the structure.

PERFORMANCE DATA (CONT.)

COIL STATIC PRESSURE DROP (IN. WC.) PURON® AND R-22 REFRIGERANTS

UNIT SIZE	Standard CFM																		
	400	500	600	700	800	900	1000	1100	1200	1300	1400	1500	1600	1700	1800	1900	2000	2100	2200
1814	Dry																		
	0.078	0.114	0.156	0.198	0.253														
	Wet																		
	0.096	0.138	0.183	0.213	0.277														

Figure 18-28 Evaporator coil static pressure drop (in WC).

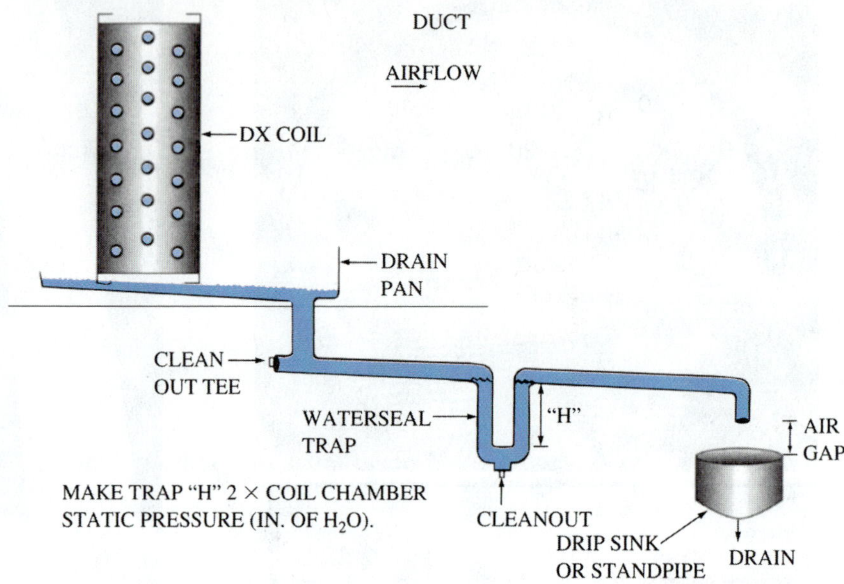

Figure 18-30 Correct condensate drain-line piping includes a properly sized trap and clean-outs.

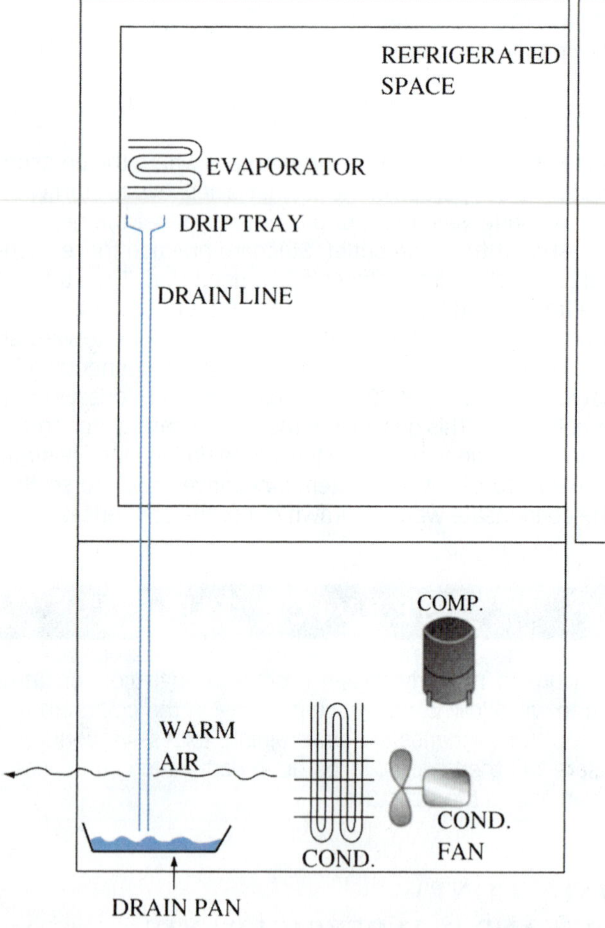

Figure 18-31 Condensate disposal for a refrigerator.

Condensate on refrigerator-freezers can be drained to a tray under the appliance and evaporated by warm air from the air-cooled condenser, as shown in Figure 18-31.

Condensate on package terminal air conditioners (PTAC) and window air conditioners is drained to a sump, picked up by a slinger on the condenser fan, and hurled onto the condenser coil to evaporate (Figure 18-32).

When condensate cannot be drained by gravity, small pumps with a reservoir and float-operated switch can be used to pump the condensate to a nearby drain, as shown in Figure 18-33.

18.15 AIR-CONDITIONING CAPACITY CONTROL

Evaporator coils can be arranged in multiple circuits and the refrigerant flow controlled by solenoid valves in larger systems. Each circuit in a dual-circuited evaporator is fed by a separate expansion valve and distributor. The circuits can be in separate rows, or the coil face can be circuited in two sections. Figure 18-34 illustrates how row and face controls work. Face control is usually preferred for better humidity control (dehumidification). When unloading face control coils, the lower coil section should always be the first on and last off to ensure that condensate will not drain over an inactive section of coil and re-evaporate.

TECH TIP

Capacity control adds efficiency to systems, but the fan speed cannot simply be slowed down as a means of capacity control. This is because as the fan speed slows, less heat is picked up by the evaporator. Therefore the evaporator coil temperature can decrease below the freezing point. Capacity control can be achieved on evaporators by using solenoids to shut down part of the refrigerant circuit within a single evaporator or in an evaporator when there are multiple evaporators in the system.

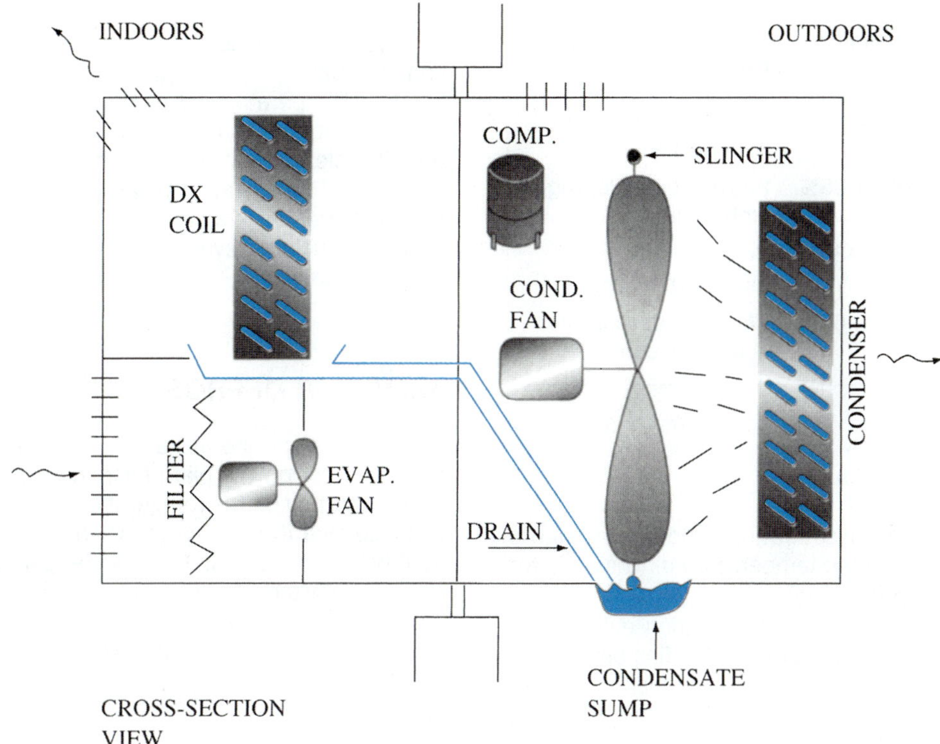

Figure 18-32 Condensate disposal for window air conditioner.

Figure 18-33 Evaporator condensate pump.

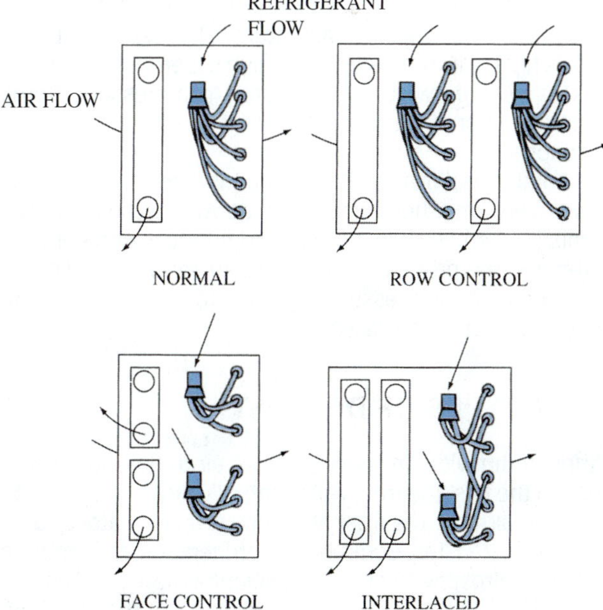

Figure 18-34 Direct expansion coil, row, and face control.

18.16 COMMERCIAL REFRIGERATION EVAPORATOR OPERATION

Although most commercial refrigeration evaporators are also tube-and-fin coils, their operating characteristics are quite different from air-conditioning coils. The saturation temperature for commercial refrigeration coils is much closer to the temperature of the air they are cooling. Unlike air conditioning, where one of the objects is to reduce the relative humidity of the space, commercial refrigeration typically tries to maintain a high a relative humidity to preserve the food. The evaporator saturation temperature is only 7–22°F colder than the return air entering the evaporator. The higher the desired relative humidity, the lower the temperature difference between

Temperature Differential	Relative Humidity	Product
7–9°F	90%	Vegetables, produce, flowers, ice
10–12°F	80%–85%	Packaged meats, vegetables, fruit
12–16°F	65%–80%	Potatoes, onions, tough-skin fruits
17–22°F	50%–65%	Warehouses, cold storage, prep rooms

Figure 18-35 Typical coil operating parameters for commercial refrigeration applications.

the box and the evaporator saturation temperature. Figure 18-35 shows recommended temperature differentials for typical commercial refrigeration applications.

Frost forms on commercial refrigeration evaporators because they operate below freezing. The fins on a refrigeration coil are spaced much wider apart than on an air-conditioning coil to allow for some frost buildup without too much airflow restriction.

18.17 HOT PULLDOWN

Refrigeration evaporators are sized for the normal operating temperatures of the system. When a freezer or walk-in cooler is first started, the temperature in the box is much higher than the design temperature. This extra heat load causes high operating pressures, high superheat, high compressor amp draw, and overheated compressors. Some equipment manufacturers actually list operating characteristics for a hot pulldown.

Commercial freezers have what amounts to a hot pulldown every time they come out of defrost because the coil is much warmer than normal. Commercial refrigeration systems often protect the compressor from hot pulldown using a crankcase pressure regulator to limit the pressure returning to the compressor.

18.18 TYPES OF DEFROST

Although the wide fin spacing allows air to flow with some frost on the fins, the frost will eventually build up enough to block the airflow. The frost also acts as an insulator and inhibits heat transfer. Commercial refrigeration coils require periodic defrosting to ensure sufficient airflow and cooling.

The amount of frost buildup depends on the condition of the air and evaporator temperature. Open cases build up the most frost because they have a continual supply of warm, moist air. Closed cases that are only opened infrequently build up frost more slowly because new air and moisture only enters the case when the door is opened.

Medium-temperature cases frequently defrost naturally on the off cycle. Their evaporators operate below freezing and build up frost when the compressor is running. However, the box temperature is above freezing, so the coil defrosts naturally when the compressor shuts off. This works fine unless the operating cycle is extended beyond normal.

Low-temperature cases must have defrost cycles because the frost remains on their evaporators even during the off cycle. The three most common means of defrosting commercial refrigeration evaporators are electric defrost, hot gas defrost, and air defrost. Other less common means of defrost include reverse-cycle defrost and heated water or glycol sprays.

18.19 AIR DEFROST

Coolers that operate at 28–30°F suction temperature will build up frost on the plate fins. They can be defrosted by simply shutting off the refrigeration and letting uncooled air circulate over the coil, as shown in Figure 18-36. Air at 35–39°F will melt the frost. Periodic defrosting can be initiated by a time switch. Air defrosting may go slower than other means, but it uses less energy. Refrigeration units running below 28°F with box temperatures below freezing (32°F) will not air-defrost.

(a)

(b)

Figure 18-36 (a) Air-defrost evaporator unit; (b) the air moved over the coil during the off cycle melts the ice.

Figure 18-37 Electric defrost evaporator unit: (a) end cover removed; (b) bottom cover removed; (c) electric resistance pan heater to prevent ice from forming during the defrost cycle.

18.20 ELECTRIC DEFROST

The electric defrost cycle is simple and widely used because of its simplicity and effectiveness. Electric defrost uses electric resistance heaters to melt ice off the coil, the drain pan, and the drain line, as shown in Figure 18-37.

Defrost is usually initiated by a time switch (Figure 18-38), although pressure or temperature sensors have also been used. When defrost is initiated, both the liquid refrigeration solenoid valve and evaporator fan motor are de-energized. The solenoid stops flow through the liquid line and the compressor pumps the system down. The low-pressure switch opens, de-energizing the compressor. At the same time this is happening, the coil heater and drain pan defrost heaters activate. The cycle is usually temperature terminated with a backup-time termination provision to prevent the box from overheating should the temperature control fail for any reason. When the coil reaches the termination temperature, the defrost termination thermostat energizes the termination solenoid in the defrost timer. The evaporator fan is wired through the defrost termination thermostat, which is a single-pole, double-throw switch. The termination thermostat opens the circuit to the evaporator fan when it terminates the defrost cycle. This allows the coil to cool down before the fan comes on. The liquid line solenoid is energized, allowing flow into the evaporator. The pressure increases, closing the low-pressure switch and starting the compressor. Once the coil is cold again, the termination thermostat will close the circuit to the evaporator fan.

18.21 HOT-GAS DEFROST

Using hot gas available from the refrigeration cycle can be an attractive means of defrost. Hot gas is common in large systems with multiple evaporators. A correctly designed hot-gas defrost system is faster than most electric defrost systems. The basic concept of a hot-gas defrost system is

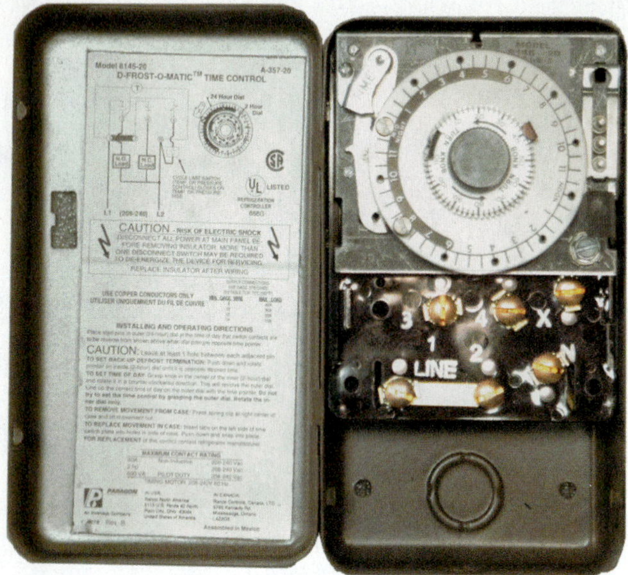

Figure 18-38 Typical commercial refrigeration defrost timer.

that the condenser is bypassed, sending hot gas directly to the evaporator. Returning liquid slugs to the compressor is a potential pitfall with hot-gas defrost because the hot gas can easily condense in the frozen evaporator. This problem is usually addressed in one of two ways:

- The pressure in the evaporator is limited, preventing the gas from condensing.
- The refrigerant is allowed to condense and is then returned to a liquid receiver.

Limited Evaporator Pressure

On single-evaporator systems, hot-gas defrost sets up a short cycle between the compressor and the evaporator. To prevent liquid refrigerant from returning to the compressor, it is necessary to keep the refrigerant from condensing in the evaporator. If the pressure is maintained below the saturation point, the refrigerant will not condense. This type of system feeds hot gas through a side port in the refrigerant distributor, so there is an appreciable pressure drop through the distributor nozzle. This system uses only the sensible heat from the compressor.

Liquid Receiver Return

In a multiple-evaporator system, the evaporator that is defrosting temporarily becomes a condenser. The liquid that condenses in the evaporator is returned to the receiver via the liquid line. This type of system uses an electrically controlled evaporator pressure regulator and a defrost solenoid valve for each coil. When an evaporator coil is being defrosted, the solenoid on the electrically controlled pressure regulator is de-energized, closing the connection to the suction line. The defrost solenoid is energized at the same time, sending hot gas into the evaporator through the suction line. Liquid is condensed in the evaporator and returned through a solenoid to the liquid receiver. A check valve must be used to allow the liquid to bypass the TEV. Alternately, a TEV with a built-in check valve can be used.

18.22 CONDENSATE DRAINAGE IN COMMERCIAL REFRIGERATION

All defrost systems must have provisions to allow the melted condensate to drain clear of the coil before the system restarts. Otherwise, the condensate can freeze. This can block drain lines and create slabs of ice in the bottom of the evaporator. A time delay between the end of the defrost cycle and the restarting of the compressor and refrigerant cycle helps ensure that all condensate is drained and not refrozen.

Drain lines from condensate drain pans in freezers must be free and clear to handle the drainage during the defrost cycle. The entire drain should be continuously heated by an electric heater cable. They should have a steep pitch (4-in drop per 12-in run) for good drainage and be insulated. Line sizes are typically a minimum of $\frac{7}{8}$-in OD tubing, with cleanout tees provided for maintenance.

Heater cable capacity should be sized in accordance with supplier recommendations. A typical capacity is about 6 W per lineal foot. It is also possible to run a hot-gas line strapped to the drain line as a heat source.

SERVICE TIP

Some coils that are used in medium-temperature applications where the space temperature is above 32°F may still form ice because the evaporator temperature is below 32°F. These coils may require some form of defrost either using air or other artificial heating systems, such as electric or hot gas. Coils that ice over without defrost systems may result in a compressor tripping on low pressure. The problem for the technician is when they arrive on the job, the ice has melted and there is no evidence of a problem. For that reason, it may be necessary to temporarily install a temperature recorder on the system to ascertain if the system at times is dropping down below freezing and forming ice. Temperature plotters are available as electronic recorders or paper tape.

18.23 EVAPORATOR EFFICIENCY

The physical construction of the coil affects its efficiency. Decreasing the coil's bypass factor improves the efficiency because more of the air comes into contact with the coil. Decreasing the number of tube rows increases the heat-transfer efficiency of each row because all of the rows are exposed to the warm air.

The closer to the end of the evaporator that the last bit of refrigerant evaporates, the more of the coil is being used to absorb heat. That makes the evaporator perform the most cooling for the highest level of efficiency. However, if there is too little superheat added to the refrigerant, some liquid refrigerant may return to the compressor. *Floodback* is the term used to describe liquid refrigerant flowing back to the compressor. Floodback can cause compressor damage.

Direct-expansion evaporators are commonly fed refrigerant into the lower tubing circuits, and the vapor is drawn out of the top tubing circuits. This reduces the possibility of liquid leaving the evaporator, helping to prevent floodback while still operating at a high level of efficiency.

The overall system efficiency is improved by maintaining as high an evaporator saturation temperature as possible. Higher evaporator temperatures mean higher suction pressures and a lower system pressure differential. The compressor can more easily move the refrigerant in a system with a lower pressure differential. This is especially crucial in low-temperature applications. A freezer operating at −20°F can consume 50 percent more energy than a similar freezer operating at 0°F. Setting controls to force refrigeration systems to operate at the lowest possible temperature increases the required system compression ratio and consumes energy needlessly.

18.24 WATER-COOLING EVAPORATORS

Refrigerated liquid coolers can range from small drinking fountains to huge water chillers capable of cooling large buildings. Water-cooling evaporators are made in a variety of sizes and designs. Small units can be of a tube-in-tube coaxial design, as shown in Figure 18-39. A common construction is copper water tube and steel refrigerant tube. Water flows through the inner tube, and refrigerant flows through the outer tube. The tubing is typically twisted or rifled to create turbulent flow. Turbulent flow improves heat transfer by ensuring that all the water passing through touches the surface of the tubing. The water and refrigerant flow in opposite directions. This counterflow design improves efficiency. Coaxial tube in tube evaporators are normally not used for systems above 5 tons in capacity.

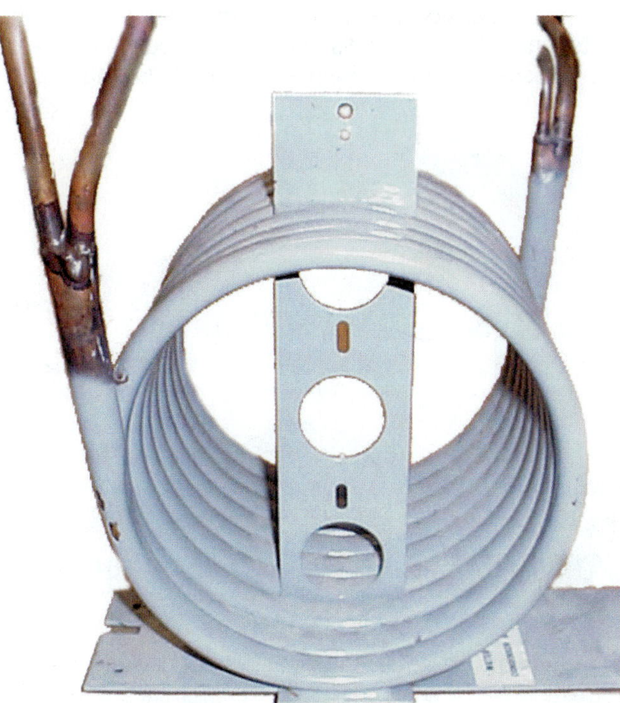

Figure 18-39 Coaxial-tube heat exchanger.

Larger chillers use shell-and-tube designs. Both direct-expansion and flooded shell-and-tube chillers are available. The shell in a direct-expansion chiller is filled with water, while the shell in a flooded chiller is filled with liquid refrigerant. The largest chillers usually use flooded evaporators.

SERVICE TIP

It is important that chiller evaporator temperatures stay above freezing to avoid damaging the chiller. It is possible to let the refrigerant pressure and temperature drop below freezing when adding or removing refrigerant from the chiller. Evacuated chillers should be charged with vapor until the vapor pressure in the chiller corresponds to a 36°F saturation temperature. Liquid refrigerant can then be added without freezing the water in the chiller. Liquid refrigerant should be recovered first when a chiller is being evacuated. After all the liquid has been removed, vapor recovery may proceed.

18.25 DRINKING WATER FOUNTAINS

Water fountain evaporators are shell-and-coil evaporators. The evaporator consists of a small storage tank cooled by an external coil of tubing wrapped around the tank and insulated with a Styrofoam cover. This is called double-wall construction, and it is required for potable water for safety. Refrigerant and/or oil would have to pass through both the wall of the refrigerant coil and the water tank to get into the water. Incoming water can be precooled by the cold water draining from the fountain to enhance capacity. Figure 18-40 shows the evaporator on a typical water cooler.

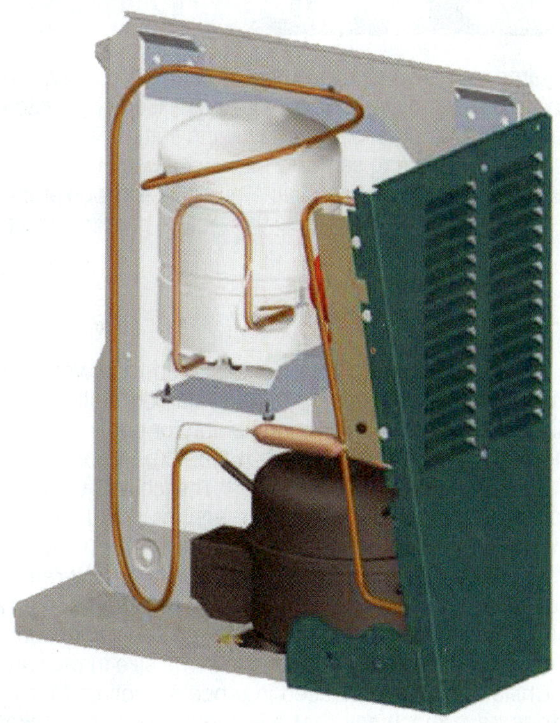

Figure 18-40 The highlighted area shows the evaporator in this water cooler.

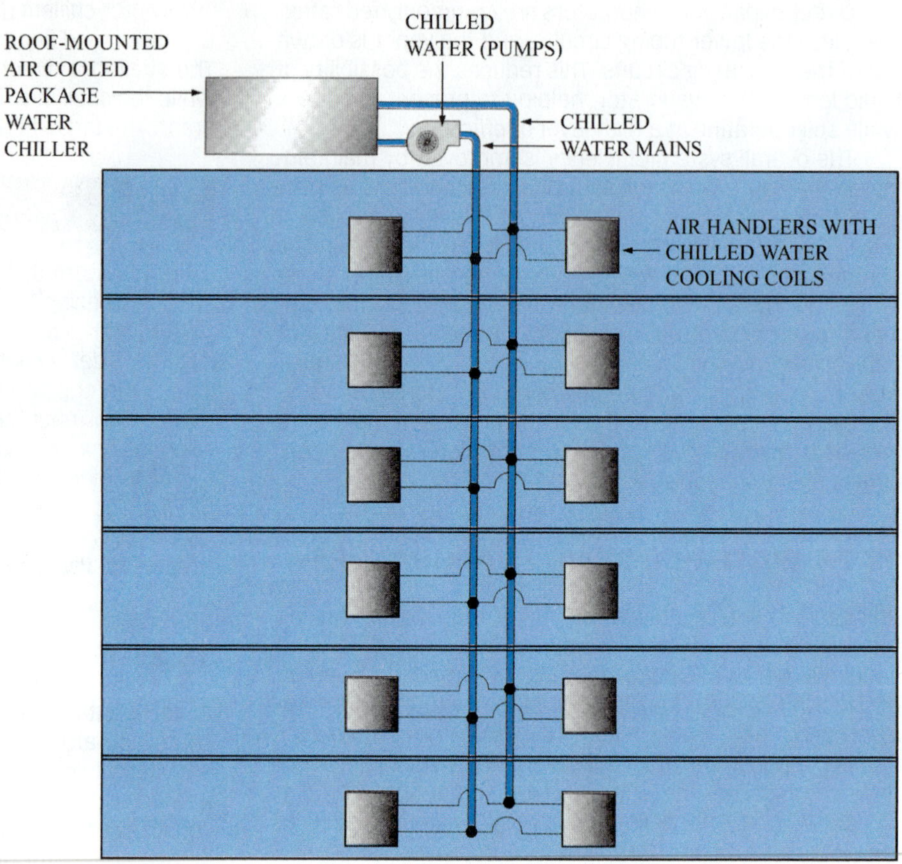

Figure 18-41 Roof mounted, air-cooled packaged water chiller.

SERVICE TIP

Anytime you are working around systems that have potable water (drinking water), you must keep the system clean. It is a violation of many local and state health codes to return a system to service that has not been thoroughly sanitized. Refer to the manufacturer's guidelines on proper system cleanup before putting the system back in service.

18.26 WATER CHILLERS

The most common application for shell-and-tube water chillers is to provide a secondary coolant, namely chilled water. Chilled water is used in air conditioning for cooling multiple remote air-handling units where it would be impractical to run long, multiple refrigerant lines. The chilled water can be pumped wherever required through insulated piping, as shown in the roof-mounted unit in Figure 18-41. The air handlers have water coils instead of direct-expansion refrigerant coils. Water coils are usually several rows deeper than an equivalent direct-expansion coil to get an equivalent cooling performance. They are normally the same size in face area.

Chilled water is also used in process cooling. Many industries have machines that are cooled by water jackets. Chillers are used to remove the heat from the water that is circulated through the machines.

18.27 DIRECT-EXPANSION CHILLERS

In direct-expansion water chillers, the shell is filled with water and refrigerant flows through the tubes. A direct-expansion shell and tube design is shown in Figure 18-42. The

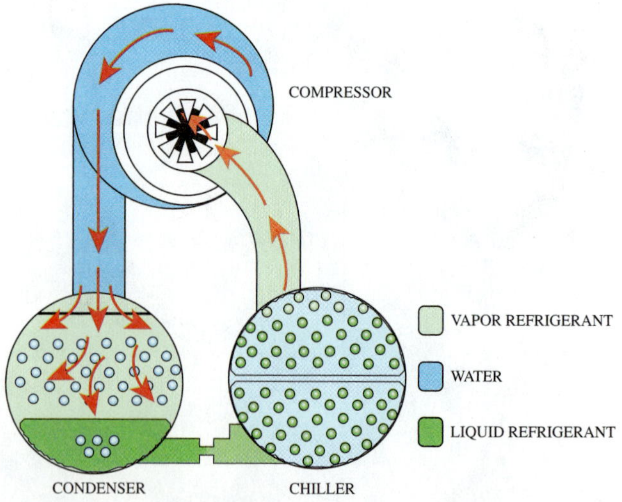

Figure 18-42 Water is in the shell of this direct-expansion chiller; liquid refrigerant flows through the tubes, which are submerged in the water; the refrigerant is superheated before leaving the chiller to prevent any liquid from returning to the compressor.

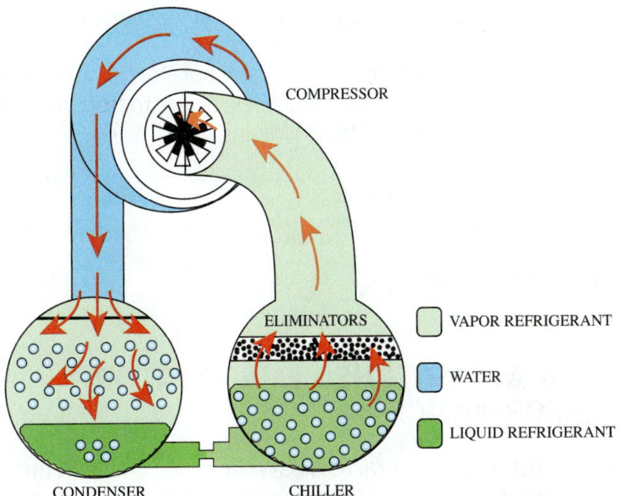

Figure 18-43 Liquid refrigerant is in the shell of this flooded chiller; the water flows through the tubes that are submerged in the liquid refrigerant; the eliminators prevent any liquid from returning to the compressor.

Figure 18-44 Packaged brine cooler.

shell is usually steel, although brass pipe has been used in some smaller diameters. The refrigerant tubes are copper. Some have an integral rolled fin to increase the heat-transfer surface. Some designs utilize metal inserts or turbulators in the tubes to enhance heat transfer on the refrigerant side. The shell holds the water, and baffles provide for even water flow and loading of all circuits. Like all direct-expansion evaporators, the refrigerant is superheated before leaving the chiller.

18.28 CHILLERS—FLOODED

For larger-capacity systems, several hundred tons and up, it is common to use flooded chillers (Figure 18-43), instead of direct-expansion chillers. Flooded chillers also use a shell-and-tube construction. However, the fluids are reversed from the direct-expansion chiller construction. The refrigerant fills the shell and water (or brine) and passes through the tubes. They use enhanced surface tubing for improved refrigerant heat transfer. The tubes are submerged in liquid refrigerant and fill the lower portion of the shell. Refrigerant is controlled by a float valve or orifice, and the shell can act as a surge chamber, permitting separation of refrigerant vapor from liquid. Closely spaced chevron-type eliminator baffles or mesh screens are normally provided to ensure satisfactory separation of dry vapor from liquid. Flooded chillers do not superheat suction vapor and provide very efficient heat transfer.

18.29 CHILLER CAPACITY CONTROL

Controlling the chiller cooling capacity is important. Excess capacity can freeze the water circulating through the chiller. The capacity of the system load is controlled by controlling the water flow. Water flow to small and large cooling coils is controlled by modulating valves controlled by room thermostats or other system controls. The chiller capacity can exceed the load when the valves on the water coils start throttling. It is better to avoid starting and stopping the chiller compressor. Instead, the chiller capacity can be reduced by unloading the compressor. Chillers can usually be unloaded down to 25 percent capacity. Beyond that, the water circulating in the system provides a reserve load to be cooled and provides enough cool water permitting satisfactory operation during light load conditions without excessive on/off cycling.

18.30 BRINE COOLERS

Chillers are also used to cool brine. Brine is a solution of salt and water. The saltwater solution can be cooled below 32°F without freezing. Other secondary coolants such as ethylene glycol and propylene glycol and water (antifreeze) are used to make water solutions that can be cooled below 32°F. A brine cooler is shown in Figure 18-44. These are found in skating rink applications and can also be used to produce ice for thermal storage.

Brine often also contains corrosion inhibitors. The same chemical companies that manufacture antifreeze for automobiles manufacture antifreeze for chillers, but the two are not the same. The corrosion inhibitors added to automobile antifreeze are not the same as the inhibitors needed for chiller systems. These chemicals can actually create problems in the water-circulation system.

SERVICE TIP

Check to make sure the antifreeze you intend to use with a chiller is approved for that use. Normally the equipment manufacturer can provide a list of recommended products. Antifreeze solutions will normally identify their intended use on the container.

18.31 SERVICING CHILLERS

When servicing chillers, care should be taken to avoid possible damaging freeze-ups. Water expands when it freezes, destroying the chiller tubes. Chillers normally have flow interlocks that will not let the chiller operate if the water is not flowing. Make sure the flow interlocks are working, so the chiller cannot operate until water flow is established. Check to see that the chilled water thermostat is correctly set. The usual minimum exiting water temperature is about 36°F. Glycol is normally used when chiller temperatures will approach freezing.

Chillers have many safeties, including freeze-stats, low-suction pressure cutouts, and low-refrigerant temperature cutouts. Make certain they are set to the manufacturer's specifications, wired correctly, and operating. Safeties should never be jumped or bypassed.

TECH TIP

A freeze-stat is a thermostat that monitors the chiller water temperature and shuts down the chiller before the water can freeze. Allowing the water in the chiller to freeze can ruin a chiller because the ice expands and ruptures the chiller tubes.

18.32 OVERSIZED EVAPORATOR AND HIGH LOAD

The effect of an oversized evaporator and a system operating with a high load are similar. In both cases there will be more heat entering the refrigerant, and the evaporator temperature will be high. Compressors rely on the returning cool vapor to keep the motor from overheating. A high heat load will cause the compressor to work harder and may cause it to overheat. The temperature of the vapor returning to the compressor will be increased, so it will be less effective in cooling the compressor.

The space temperature will rise, and if the system is on a medium- or low-temperature walk-in, the food product may spoil.

18.33 UNDERSIZED EVAPORATOR AND LOW LOAD

The effect of an undersized evaporator and a system operating with a low load are similar. In both cases there is less heat being transferred to the refrigerant, and the evaporator pressure and temperature will be low. The evaporator pressure and temperature will become lower and lower until a new balance point is reached. There will be little or no superheat, and liquid refrigerant may flood back to the compressor.

If the system is designed for air conditioning, the evaporator temperature can drop below freezing, and the coil will begin to ice up. The formation of ice on the coil can further reduce the temperature, causing more ice to form. If the evaporator temperature drops too low, the compressor can experience an excessively high compression ratio. Excessively high compression ratios are associated with compressor damage and failure.

Some of the reasons that an evaporator temperature and pressure will drop are listed below. Restricted air flow can be caused by any of the following:

- A dirty evaporator coil
- A dirty air filter
- A bad blower motor
- Ice on the coil

18.34 MISMATCHED EVAPORATOR PERFORMANCE

In the past, many contractors have routinely changed out a condensing unit without changing the existing coil. Typically, the SEER rating of the new condensing unit would be mentioned when the new unit was quoted without informing the customer that the SEER only applies to a matched system. There is no practical way of determining the result from matching a new condenser to an older, unmatched coil that may have been designed for a different SEER condensing unit. The actual performance of these mismatched systems seldom lives up to the rating.

It is important to understand that a system's SEER rating only applies to matched systems. Coils do not have to be manufactured by the same company that manufactured the condensing unit, but they should be tested and listed with that unit as an AHRI match. Just matching components by their nominal rating does not ensure correct operation. AHRI maintains an online directory that allows anyone to check a system match, at www.ahridirectory.org. Certificates verifying the system rating and performance can be printed from this site as well (Figure 18-45).

In January 2006, 13 SEER became the minimum efficiency for new equipment. There is a significant difference in system design between 13 SEER and the previous 10-SEER minimum. If a high-efficiency (SEER) air-conditioning condensing unit is installed and the old evaporator coil is left in service, the coil may be inadequate. The newer units are designed to run at higher evaporating temperatures and need more evaporator coil surface to achieve a satisfactory balance. Also, the older, fixed-orifice metering devices are designed to operate with a higher pressure difference. These metering devices will not operate properly when left installed with newer, higher-efficiency condensing units, as the liquid pressure in these units is lower. Manufacturers recommend changing the evaporator coil when a new high-efficiency unit is installed, to ensure a proper equipment match.

TECH TIP

One major equipment manufacturer installed a new 13-SEER condenser with a new 10-SEER coil and took accurate measurements of the system performance. Using the factory system charge, the system capacity went down and the overall SEER dropped from 10 to 8. Installing a new, higher-efficiency condenser on an old, lower-efficiency coil actually dropped the system efficiency.

Certificate of ARI-Certified Performance

The following

RCU-A-C

Outdoor Unit Model Number: AY018MA322

Manufactured by: YORK, UNITARY PRODUCTS GROUP

combined with

Indoor Unit Model Number: HD47436+TD

Manufactured by: ADVANCED DISTRIBUTOR PRODUCTS

under the Trade/Brand name: ADP

has been rated in accordance with

ARI Standard 210/240-2005 for UNITARY AIR-CONDITIONING AND AIR-SOURCE HEAT PUMP EQUIPMENT

and is certified by the Air-Conditioning and Refrigeration Institute to meet

the following product performance ratings:

Cooling Capacity (Btuh):	18000
EER Rating (Cooling):	12.00
SEER Rating (Cooling):	14.00

* Voluntarily revised, unless accompanied with a WAS in which case the change is involuntary.

ARI Reference #:	**1402949**
Today's Date:	**2/5/2008**
Status:	**Active**

CERTIFIED RATINGS ARE VALID ONLY FOR THE PARTICULAR COMBINATION OF INDOOR AND OUTDOOR UNITS LISTED IN THE AIR-CONDITIONING AND REFRIGERATION INSTITUTE'S DIRECTORY OF CERTIFIED EQUIPMENT. VISIT WWW.AHRIDIRECTORY.ORG TO VERIFY THAT THIS COMBINATION IS AN ACTIVE LISTING AND THE DATA LISTED ON THIS CERTIFICATE IS ACCURATE. SEARCH ON THE ARI REFERENCE # TO QUICKLY LOCATE THIS COMBINATION IN THE DIRECTORY

TERMS AND CONDITIONS

This Certificate shall be used for individual, personal, and confidential reference purposes only, and may be used only pursuant to the terms and conditions listed. This Certificate and the contents hereof are proprietary products of ARI. The contents of this Certficate may not, in whole or in part, be reproduced;copied; disseminated; entered into a computer database; or otherwise utilized, in any form or manner or by any means, except for the user's individual, personal and confidential reference. Contained herein are product information and certified ratings. ARI does not endorse the product(s) listed in this Certificate and makes no representations, warranties or guarantees as to, and assumes no responsibility for, the product(s) listed in this Certificate ARI expressly disclaims all liability for damages of any kind arising out of the use or performance of the product(s), or the unauthorized alteration of data, listed in this Certificate.

Figure 18-45 This certificate shows that these two pieces of equipment will work properly together to give the results listed.

UNIT 18—SUMMARY

Evaporators are manufactured for four broad types of application: domestic refrigeration, commercial refrigeration, air conditioning, and chillers. Air-cooling evaporators can be listed by their construction as bare pipe, plate, and tube-and-fin coil. Evaporators may be divided by general operating principle as either direct expansion or flooded. A small percentage of the refrigerant entering a direct-expansion chiller is already gas. The refrigerant leaving a direct-expansion chiller is superheated. The refrigerant entering a flooded chiller is all liquid, and the refrigerant leaving is a saturated gas.

Domestic and commercial refrigeration evaporators may be further listed by their operating temperature: high, medium, and low. The fin spacing on medium-temperature coils is closer together than on low-temperature coils. Most refrigeration evaporators typically operate below freezing and require defrosting. The most common forms of defrost in commercial refrigeration systems are air, electric, and hot-gas defrost. When they are coming out of defrost, the high evaporator temperature can create high pressures, temperatures, and superheat. This is known as hot pulldown.

Air-conditioning evaporators are tube-and-fin type. They may be single-row coils or multi-row coils. Single-row coils produce more sensible cooling and less latent cooling than multi-row coils do. Air-conditioning coils may be listed by their shape as slab, slant, "A," "N," or "M." By airflow, air-conditioning coils can be upflow, downflow, horizontal, or multi-position. The evaporator pressure and temperature are affected by the return-air wet bulb temperature. High wet bulb temperatures produce high evaporator saturation temperatures. The drain on air-conditioning evaporators should include a trap that is at least 3 in deep to keep negative air pressure from preventing the condensate from draining.

Chillers can be constructed as tube in tube, shell and coil, or shell and tube. Tube-in-tube evaporators are used for small applications up to 5 tons. The water flows in the inner copper tube while the refrigerant flows in the opposite direction through the outer steel tube. Shell-and-coil chillers are used for domestic water coolers. The evaporator is coiled around the outside of the water tank, providing double-wall protection for the potable water in the cooler. Large chillers up to several hundred tons are typically shell-and-tube construction. They may be either direct expansion or flooded. Direct-expansion chillers have the refrigerant in the tubes and the water in the shell, while flooded chillers have the refrigerant in the shell and the water in the tubes.

WORK ORDERS

Service Ticket 1801

Customer Complaint: Poor Cooling and Iced Suction Line

Equipment Type: R-22 Split Air-Conditioning System

A customer has called because her system is not cooling well.

She mentions that there is ice visible on the unit outside. The service technician arrives and sees ice on the suction line and all over the compressor. The customer is surprised that the system can make ice but not cool the house.

The technician explains that ice on an air conditioner does not mean it is working well. In fact, ice is a bad sign in an air-conditioning system. When the technician opens the blower door, the air filter is completely plugged up with dust, and the coil is completely iced over. The technician explains that the dirty filter restricted the airflow over the evaporator, causing its temperature to drop below freezing. Once the ice started forming, it also caused airflow restriction. Liquid refrigerant then slowly worked its way down the suction line to the compressor.

The technician turns system off and turns the fan to ON to melt the ice. This way, the ice melts slowly enough for the drain pan to handle the water. When the ice is clear, the technician restarts the system with a clean air filter and measures the airflow. After determining that the airflow is correct, the system pressure, superheat, and subcooling are all checked. The system is operating within the manufacturer's specifications with the correct airflow.

Service Ticket 1802

Customer Complaint: Chiller Not Working

Equipment Type: R-134a Flooded Chiller Used for Process Cooling

A customer calls to say that his process chiller is not working. Occasionally the plant maintenance crew had found the chiller off on the freeze-stat. When they pushed it, the chiller would work for a while until it went off again on the freeze-stat. Eventually, they jumped out the freeze-stat to keep the chiller operating. This worked for about a day, but now it does not work at all. The technician checks system refrigerant pressure and notices that the pressure is very low. Suspecting ruptured chiller tubes, the technician drains some water from the chiller loop and notices that it is murky and oily. The chiller is drained, the tubes are pressure tested, and the rupture is confirmed.

The technician explains that the freeze-stat was probably tripping because the exiting water temperature was getting too close to freezing. The freeze-stat tripping was not the problem but a symptom. Jumping the safety only forced the chiller to operate at a dangerous condition, freezing the water in the chiller and rupturing the tubes. The chiller has been ruined. The technician warns that the original problem may still exist if something in the water loop outside the chiller was causing low water flow.

Service Ticket 1803

Customer Complaint: Poor Cooling and High Electric Bills

Equipment Type: R-22 Split System with New Condenser and Old Evaporator

A customer calls a company to look at a new air-conditioning unit that is not cooling properly. The unit was installed by the customer's brother-in-law while he was visiting from out of state. The compressor in the old system died and the brother-in-law changed out the entire condensing unit with a new 13-SEER model. The new condensing unit was connected to the twenty-year-old capillary-tube evaporator coil. The technician checks the system airflow and refrigerant pressures. The airflow is normal, but both pressures are low and the superheat is high. The technician suspects that the condenser and coil are not an AHRI match. A quick check of the AHRI online directory confirms that the system is not an approved match. The technician explains to the customer that the problem is that condenser and evaporator are not matched and suggests that the customer have a matched coil installed.

Service Ticket 1804

Customer Complaint: Water Leaking on Ceiling Under Unit

Equipment Type: R-410a Heat Pump with the Blower Coil Installed in the Attic

A customer has called about water leaking on the ceiling under a heat pump blower coil in a new house. The technician notices that the blower coil is installed directly on the ceiling joists with no secondary drain pan. There is very little fall in the condensate line, which also runs across the ceiling joists until it exits. Further, there is no trap in the condensate line. Water is dripping out from the cracks in the bottom panels, and the ceiling insulation under the unit is wet. The technician informs the customer that the blower coil should be elevated about 6 in so that a 3-in trap can be installed in the drain and the drain can slope toward the exit point. The technician also points out that a secondary drain pan should be installed under the unit to avoid similar problems in the future.

Service Ticket 1805

Customer Complaint: Compressor Shutting off on Overload After Defrost Cycle Is Completed

Equipment Type: R-404a Open-Display Commercial Refrigeration Freezer

The compressor on a new frozen food open display case sometimes shuts off on overload when it starts after the completion of a defrost cycle. The technician monitors the case operation during the cooling cycle and finds the pressures, temperatures, and current draws to be within the manufacturer's specifications. The case goes into defrost and defrosts successfully. When the case comes out of defrost, the suction pressure and compressor amp draw are both high. After running a few minutes at the elevated temperatures and pressures, the compressor overload opens and the compressor shuts down. The technician waits for the overload to cool and reset and then restarts the compressor. This time the technician throttles the suction pressure to the compressor by running the suction service valve toward the front seat position and watching the compressor amp draw. The compressor continues operating. As the coil temperature drops, the suction pressure and amp draw also drop. After fifteen minutes the compressor operates normally without throttling.

The technician explains to the customer that the compressor amp draw is too high on restart after defrost because it is trying to pull down the hot evaporator coil and is being overloaded. The technician explains that this can be remedied by installing a crank case pressure regulator, which will prevent the returning pressure from exceeding the maximum safe operating pressure for the system.

UNIT 18—REVIEW QUESTIONS

1. What is the purpose of the evaporator in the refrigeration system?
2. Why does the heat flow into the evaporator?
3. What is the difference between sensible and latent cooling?
4. Explain how an evaporator can dehumidify the air.
5. Define *product* as it relates to refrigeration.
6. List the four broad types of evaporator application.
7. List the types of water-cooling evaporator construction.
8. Where are plate-type evaporators used?
9. Why are fins added to the copper tubing on tube-and-fin coils?
10. How is the mechanical tight bond created between the refrigerant tube and the fin-on-fin coils?
11. What are two ways of adding corrosion protection to the aluminum fins on a fin coil?
12. Explain the difference between the fin pitch on air-conditioning coils and commercial refrigeration coils.
13. Explain the difference in operation between a single-row evaporator and a multi-row evaporator.
14. In what shapes are air-conditioning tube-and-fin coils available?
15. Define *condensate blowoff*.
16. What is the maximum velocity for air through a coil to avoid blowoff?
17. Define *bypass factor*.
18. In what two ways should an air-conditioning evaporator condition the air?
19. List the items that can affect a coil's dehumidification ability.
20. What are the common ways that manufacturers list a coil's ability to dehumidify the air?
21. Explain the relationship between the static pressure drop across an evaporator coil and the airflow going over the coil.
22. Why do coil manufacturers specify both wet and dry coil static pressure drops?
23. How does a coil being dirty affect its static drop?
24. What is a direct-expansion evaporator?
25. What is a flooded evaporator?
26. What type of a system uses a flooded evaporator?
27. What is the purpose of the trap in a condensate drain?
28. Why should the bottom coil be first on, last off when using a face control to unload a coil?
29. List the common types of defrost for commercial refrigeration evaporators.
30. What is brine?
31. Why are direct-expansion evaporators typically fed at the bottom and drawn from the top?
32. Explain the difference between a direct-expansion chiller and a flooded chiller.
33. List the three general temperature ranges for commercial refrigeration evaporators, and give examples of each.
34. What is meant by an AHRI match?

UNIT 19

Refrigerants and Their Properties

OBJECTIVES

After completing this unit, you will be able to:

1. define *refrigerant*.
2. identify the type of refrigerant by its number designation.
3. list the different types of refrigerant chemical composition.
4. explain the difference between compounds, zeotropes, and azeotropes.
5. describe the pressure-temperature relationship of saturated refrigerant.
6. list the different types of refrigerant contaminants.
7. explain the refrigerant safety rating system.
8. match refrigerants with the proper refrigeration oil.

19.1 INTRODUCTION

Any study of the refrigeration cycle would be incomplete without a look at the refrigerant in the system. The refrigerant is the first component of a refrigeration system that must be determined, since its chemical and physical properties will play a large role in selecting all other components in a refrigeration system. A number of materials have physical and thermodynamic properties that make them suitable for use as refrigerants. In fact, anything that can boil at a low temperature and condense at a higher temperature can be used as a refrigerant. Water can even be used as a refrigerant under the right conditions. Although water boils at 212°F at standard atmospheric pressure, it boils at 40°F in a vacuum of 29.75 in Hg (Figure 19-1).

19.2 DEFINITION OF A REFRIGERANT

Refrigerant is the fluid used in a refrigeration system for transferring heat. Gases or liquids can do this by increasing in temperature. Cooling operations that depend on a temperature change in the refrigerant use the sensible heat properties of the refrigerant. However, the effectiveness of this approach is limited to the specific heat capabilities of the refrigerant and the volume of refrigerant that can be moved. Refrigeration processes that depend only on sensible cooling must move large amounts of refrigerant.

A much larger amount of heat can be absorbed and transferred when a liquid is vaporized to a gas. This method takes advantage of the latent heat characteristics of the refrigerant and results in much higher operating efficiency due to the reduced quantity of refrigerant that must be circulated. Refrigerants in mechanical refrigeration systems do this by changing their physical state. They absorb heat by evaporating from a liquid to a vapor; they release heat by condensing from a vapor back to a liquid. The refrigerant's temperature is controlled by its pressure. Lower pressures yield lower temperatures, while higher pressures yield higher temperatures.

TECH TIP

A comparison between the specific and latent heat properties of R-134a can illustrate how much more heat can be absorbed through changes in state. The specific heat of R-134a vapor is 0.19 BTU/lb. A pound of vapor can only absorb 19 BTU while changing in temperature 100°F. Fifty-three pounds of R-134a would have to be circulated to absorb 1,000 BTU, even if the refrigerant were to rise 100°F in temperature (1,000/0.19). Even more refrigerant, 530 lb, would be required if the refrigerant were to rise in temperature only 10°F. By comparison, the latent heat of vaporization of R-134a is 92.8 BTU/lb. Only 11 lb are required to absorb the same 1,000 BTU by boiling from a liquid to a vapor.

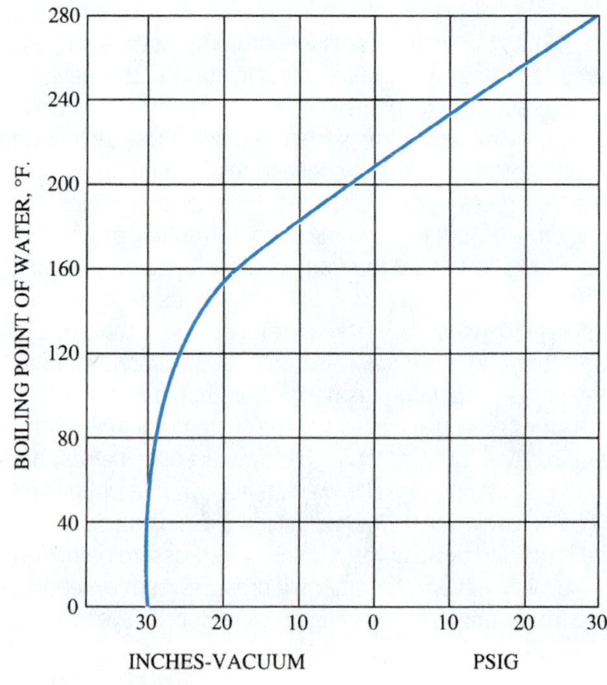

Figure 19-1 Pressure-temperature curve for water.

19.3 TEMPERATURE-PRESSURE RELATIONSHIPS

The pressure inside a refrigerant cylinder, or inside a system, is created by vapor. If all the vapor inside a closed cylinder condenses, there is no pressure. The boiling temperature of any liquid is controlled by the vapor pressure exerted on the liquid. The pressure and temperature of any saturated liquid tend to go up and down together. If you know one, you can find the other.

Charts called pressure-temperature charts, or PT charts for short, are commonly used in refrigeration. These charts correlate the saturation temperature and pressure of commonly used refrigerants (Figure 19-2). The vapor pressure of most common refrigerants can be found by consulting a PT chart. If the temperature of a saturated refrigerant can be determined, its pressure can be determined. Most of these charts list temperature in a column on the left-hand side of the chart. To use the chart, find the saturation temperature in the left-hand column and follow across that same row until reaching the column of the refrigerant being used. That gives the pressure of that refrigerant corresponding to that particular saturation temperature. In Figure 19-3, at a temperature of 40°F, saturated R-134a will have a pressure of 35 psig. Of course the chart can also be used the other way. Find the pressure in the column under the refrigerant being used and then follow across that same row to the temperature column on the outside. This gives the saturation temperature of the refrigerant at that particular pressure. In Figure 19-4, saturated R-22 with a pressure of 43 psig will have a temperature of 20°F.

There is an important point that must remembered: these charts are specifically designed for saturated refrigerant only. They do not work with superheated gas or subcooled liquid. To be saturated, the refrigerant must be ready to evaporate or condense. In practical terms, saturated refrigerant is usually a mixture of liquid and vapor, not all vapor or all liquid. It is true that PT charts are used to calculate superheat and subcooling, but PT charts cannot be used to predict the pressure or temperature of a superheated gas or a subcooled liquid. Rather, superheat and subcooling are calculated by comparing the saturation temperature on the PT chart to the actual refrigerant temperature.

VAPOR PRESSURE, PSIG

Temp (°F)	11	12	22	113	114	500	502	134a	123
-50	**28.9**	**15.4**	**6.2**	–	**27.1**	**12.8**	**0.2**	**18.7**	**29.2**
-45	**28.7**	**13.3**	**2.7**	–	**26.6**	**10.3**	1.9	**16.9**	**29.0**
-40	**28.4**	**11.0**	0.5	–	**26.0**	**7.6**	4.1	**14.8**	**28.9**
-35	**28.1**	**8.4**	2.6	–	**25.4**	**4.6**	6.5	**12.5**	**28.7**
-30	**27.8**	**5.5**	4.9	**29.3**	**24.6**	**1.2**	9.2	**9.8**	**28.4**
-25	**27.4**	**2.3**	7.4	**29.2**	**23.8**	1.2	12.1	**6.9**	**28.1**
-20	**27.0**	0.6	10.1	**29.1**	**22.9**	3.2	15.3	**3.7**	**27.8**
-15	**26.5**	2.4	13.2	**28.9**	**21.8**	5.4	18.8	**0.1**	**27.4**
-10	**26.0**	4.5	16.5	**28.7**	**20.6**	7.8	22.6	1.9	**27.0**
-5	**25.4**	6.7	20.0	**28.5**	**19.3**	10.4	26.7	4.1	**26.5**
0	**24.7**	9.1	23.9	**28.2**	**17.8**	13.3	31.1	6.5	**25.9**
5	**23.9**	11.8	28.2	**27.9**	**16.2**	16.4	35.9	9.1	**25.3**
10	**23.1**	14.6	32.8	**27.6**	**14.4**	19.7	41.0	11.9	**24.6**
15	**22.1**	17.7	37.7	**27.2**	**12.4**	23.3	46.5	15.0	**23.7**
20	**21.1**	21.0	43.0	**26.8**	**10.2**	27.2	52.5	18.4	**22.8**
25	**19.9**	24.6	48.7	**26.3**	**7.8**	31.5	58.8	22.1	**21.8**
30	**18.6**	28.4	54.9	**25.8**	**5.2**	36.0	65.6	26.0	**20.7**
35	**17.2**	32.5	61.5	**25.2**	**2.3**	40.8	72.8	30.3	**19.5**
40	**15.6**	36.9	68.5	**24.5**	0.4	46.0	80.5	35.0	**18.1**
45	**13.9**	41.6	76.0	**23.8**	2.0	51.6	88.7	40.0	**16.6**
50	**12.0**	46.7	84.0	**22.9**	3.8	57.5	97.4	45.4	**15.0**
55	**10.0**	52.0	92.5	**22.2**	5.8	63.9	106.6	51.1	**13.1**
60	**7.8**	57.7	101.6	**21.0**	7.9	70.6	116.4	57.3	**11.2**
65	**5.4**	63.7	111.2	**19.9**	10.1	77.8	126.7	63.9	**9.0**
70	**2.7**	70.2	121.4	**18.7**	12.6	85.4	137.6	71.0	**6.6**
75	0.0	76.9	132.2	**17.3**	15.2	93.4	149.1	78.6	**4.0**
80	1.5	84.1	143.6	**15.8**	18.0	101.9	161.2	86.6	**1.2**
85	3.2	91.7	155.7	**14.3**	20.9	111.0	174.0	95.1	0.9
90	4.9	99.7	168.4	**12.5**	24.1	120.5	187.4	104.2	2.5
95	6.8	108.2	181.8	**10.6**	27.5	130.5	201.4	113.8	4.2
100	8.8	117.1	195.9	**8.6**	31.1	141.1	216.2	124.1	6.1
105	10.9	126.5	210.7	**6.4**	35.0	152.2	231.7	134.9	8.1
110	13.2	136.4	226.3	**4.0**	39.1	164.0	247.9	146.3	10.3
115	15.6	146.7	242.7	**1.4**	43.4	176.3	264.9	158.4	12.6
120	18.3	157.6	259.9	0.7	48.0	189.2	282.7	171.1	15.1
125	21.0	169.0	277.9	2.2	52.8	202.8	301.4	184.5	17.7
130	24.0	180.9	296.8	3.7	58.0	217.0	320.8	198.7	20.6
135	27.1	193.5	316.5	5.4	63.4	231.9	341.2	213.6	23.6
140	30.4	206.5	337.2	7.2	69.0	247.4	362.6	229.3	26.8
145	34.0	220.2	358.8	9.2	75.0	263.7	385.0	245.7	30.2
150	37.7	234.5	381.5	11.2	81.3	280.7	408.4	263.0	33.8

Bold Numbers - Inches Hg. Below 1 ATM

PRESSURE/TEMPERATURE CHART

Temp. Deg. F	R-408A (FX-10) Liquid Pressure	R-404A (FX-70) Liquid Pressure	R-409A (FX-56) Liquid Pressure	R-409A (FX-56) Vapor Pressure	R-407C Liquid Pressure	R-407C Vapor Pressure	R-410A Liquid Pressure
-50	**1.6**	0.6	**12.4**	**17.2**	**2.9**	**11.4**	3.5
-45	1.1	2.7	**9.7**	**15.2**	0.4	**8.5**	8.5
-40	3.3	5.0	**6.8**	**13.1**	2.5	**5.2**	11.6
-35	5.6	7.6	**3.5**	**10.7**	4.8	**1.5**	14.9
-30	8.2	10.4	0.0	**8.1**	7.3	1.3	18.5
-25	11.0	13.4	2.0	**5.1**	10.1	3.6	22.5
-20	14.1	16.8	4.1	**1.9**	13.1	6.1	26.9
-15	17.5	20.5	6.5	0.8	16.5	8.8	31.7
-10	21.2	24.5	9.0	2.8	20.1	11.9	36.8
-5	25.2	28.8	11.8	4.9	24.0	15.2	42.5
0	29.5	33.5	14.8	7.2	28.3	18.9	48.6
5	34.2	38.6	18.1	9.7	33.0	22.9	55.2
10	39.3	44.0	21.7	12.5	38.0	27.3	62.3
15	44.8	49.9	25.5	15.4	43.5	32.0	70.0
20	50.7	56.2	29.6	18.7	49.3	37.2	78.3
25	57.0	63.0	34.0	22.2	55.7	42.7	87.3
30	63.7	70.3	38.7	26.0	62.5	48.7	96.8
35	71.0	78.1	43.8	30.1	69.8	55.2	107.0
40	78.7	86.4	49.2	34.5	77.6	62.1	118.0
45	87.0	95.2	54.9	39.2	86.0	69.5	129.7
50	95.8	104.7	61.0	44.3	94.9	77.5	142.2
55	105.1	114.7	67.6	49.8	104.5	86.0	155.2
60	115.1	125.3	74.5	55.6	114.6	95.1	169.6
65	125.6	136.6	81.8	61.9	125.4	104.8	184.6
70	136.8	148.6	89.5	68.6	136.9	115.2	200.6
75	148.7	161.2	97.7	75.8	149.1	126.2	217.4
80	161.2	174.6	106.4	83.4	162.1	137.8	235.3
85	174.4	188.8	115.5	91.5	175.8	150.2	254.1
90	188.4	203.7	125.2	100.2	190.2	163.4	274.1
95	203.1	219.4	135.3	109.4	205.5	177.4	295.1
100	218.7	235.9	146.0	119.2	221.6	192.1	317.2
105	235.0	253.4	157.2	129.6	238.5	207.8	340.5
110	252.1	271.7	169.0	140.6	256.4	224.4	365.0
115	270.2	290.9	181.4	152.3	275.1	241.9	390.7
120	289.1	311.1	194.4	164.7	294.7	260.5	417.7
125	308.9	332.3	208.0	177.8	315.2	280.1	445.9
130	329.7	354.5	222.3	191.6	336.7	300.9	475.6
135	351.5	377.8	237.2	206.3	359.2	322.9	506.5
140	374.3	402.2	252.9	221.8	382.6	346.2	539.0
145	398.1	427.7	269.3	238.2	407.0	370.8	572.8
150	423.0	454.4	286.4	255.5	432.4	396.9	608.1

Bold Numbers - Inches Hg. Below 1 ATM

Figure 19-2 Saturated refrigerant pressure-temperature chart. *(Courtesy of Arkema Inc.)*

VAPOR PRESSURE, PSIG									
Temp (°F)	11	12	22	113	114	500	502	134a	123
	23.9	11.8	28.2	**27.9**	**16.2**	16.4	35.9	9.1	**25.**
10	**23.1**	14.6	32.8	**27.6**	**14.4**	19.7	41.0	11.9	**24.**
15	**22.1**	17.7	37.7	**27.2**	**12.4**	23.3	46.5	15.0	**23.**
20	**21.1**	21.0	43.0	**26.8**	**10.2**	27.2	52.5	18.4	**22.**
25	**19.9**	24.6	48.7	**26.3**	**7.8**	31.5	58.8	22.1	**21.**
30	**18.6**	28.4	54.9	**25.8**	**5.2**	36.0	65.6	26.0	**20.**
35	**17.2**	32.5	61.5	**25.2**	**2.3**	40.8	72.8	30.3	**19.**
40	**15.6**	36.9	68.5	**24.5**	0.4	46.0	80.5	35.0	**18.**
45	**13.9**	41.6	76.0	**23.8**	2.0	51.6	88.7	40.0	**16.**

Figure 19-3 Saturated R-134a at 40°F has a pressure of 35 psig. *(Courtesy of Arkema Inc.)*

VAPOR PRESSURE, PSIG									
Temp (°F)	11	12	22	113	114	500	502	134a	123
	23.9	11.8	28.2	**27.9**	**16.2**	16.4	35.9	9.1	**25.**
10	**23.1**	14.6	32.8	**27.6**	**14.4**	19.7	41.0	11.9	**24.**
15	**22.1**	17.7	37.7	**27.2**	**12.4**	23.3	46.5	15.0	**23.**
20	**21.1**	21.0	43.0	**26.8**	**10.2**	27.2	52.5	18.4	**22.**
25	**19.9**	24.6	48.7	**26.3**	**7.8**	31.5	58.8	22.1	**21.**
30	**18.6**	28.4	54.9	**25.8**	**5.2**	36.0	65.6	26.0	**20.**
35	**17.2**	32.5	61.5	**25.2**	**2.3**	40.8	72.8	30.3	**19.**
40	**15.6**	36.9	68.5	**24.5**	0.4	46.0	80.5	35.0	**18.**
45	**13.9**	41.6	76.0	**23.8**	2.0	51.6	88.7	40.0	**16.**

Figure 19-4 Saturated R-22 with a pressure of 43 psig has a temperature of 20°F. *(Courtesy of Arkema Inc.)*

PRES	
Temp. Deg. F	R-410A Liquid Pressure
40	118.0
45	129.7
50	142.2
55	155.2
60	169.6
65	184.6
100	317.2
105	340.5
110	365.0
115	390.7
120	417.7
125	445.9

Figure 19-5 The saturation temperature of the evaporator and condenser can be determined using the pressure-temperature chart. *(Courtesy of Arkema Inc.)*

Saturated PT charts are very useful for determining the actual evaporating and condensing temperatures of the refrigerant in the system. Generally, the PT charts will yield a much more accurate evaporating or condensing temperature than attempts to measure them directly with a thermometer. For example, an R-410a system operating at pressures of 118 psig on the low side and 365 psig on the high side would have an evaporator saturation temperature of 40°F and a condenser saturation temperature of 110°F (Figure 19-5).

19.4 REFRIGERANT CLASSIFICATION

Refrigerants have many physical and chemical characteristics that determine their suitability for any particular system. There are many groupings of refrigerants based on these characteristics, with the most common characteristics as follows:

- Operating pressure
- Chemical composition
- Formulation
- Toxicity
- Flammability
- Ozone-depletion potential
- Global-warming potential

19.5 OPERATING PRESSURE

The EPA regulations for preventing the release of ozone-depleting substances divide refrigerants into three categories of operating pressures: low pressure, high pressure, and very high pressure. The recovery methods and levels specified for each refrigerant depend upon the refrigerant's operating pressure range. The operating pressure range is determined by the boiling point of the refrigerant at atmospheric pressure. Low-pressure refrigerants have a boiling point at atmospheric pressure above 10°C (50°F). High-pressure refrigerants have a boiling point at atmospheric pressure between −50°C (−58°F) and 10°C (50°F). Very-high-pressure refrigerants have a boiling point at atmospheric pressure below −50°C (−58°F). Generally speaking, the lower the refrigerant's operating pressure, the higher the evaporator saturation temperature. Systems using low-pressure refrigerant typically have an evaporator saturation temperature of 40°F or higher. Systems using very-high-pressure refrigerant typically have evaporator saturation temperatures of −40°F and lower. High-pressure refrigerants are used in the widest variety of applications. Systems using high-pressure refrigerant may have an evaporator saturation temperature between −20°F and 50°F. Figure 19-6 shows examples of refrigerants according to their operating pressure range.

Operating Pressure	Refrigerants
Low pressure	R-11, R-123
High pressure	R-12, R-22, R-134, R-502, R-507
Very high pressure	R-503, R-410A

Figure 19-6 Refrigerants listed by operating pressure.

19.6 REFRIGERANT CHEMISTRY

Refrigerants can be divided into five major chemical categories based on their chemical composition:

- Hydrocarbons
- Chlorofluorocarbons
- Hydrofluorocarbons
- Hydrochlorofluorocarbons
- Natural refrigerants

Refrigerants are now frequently referred to by their chemical family using the first letters of each chemical component in the refrigerant to produce abbreviations (HC, CFC, etc.). Common refrigerant chemical families are shown in Figure 19-7.

Compounds

Until recently, most refrigerants were compounds. Water is a good example of a compound. Take a highly flammable gas, hydrogen, combine it chemically with oxygen, a gas that is necessary for combustion, and the result is water, a chemical that does not burn and is widely used to extinguish flames. Two chemicals have been combined to make a new chemical with distinct chemical properties that are completely separate from the properties of the original chemicals. Further, these chemicals cannot be separated by physical means. The hydrogen cannot be distilled from the oxygen by boiling the water. Any pure chemical or compound has a specific saturation temperature for any given pressure. The boiling temperature can be precisely determined if the pressure on the liquid is known. Examples of refrigerants that are compounds include some of the old standards: R-11, R-12, and R-22. These are all based on the same molecule, methane (Figure 19-8).

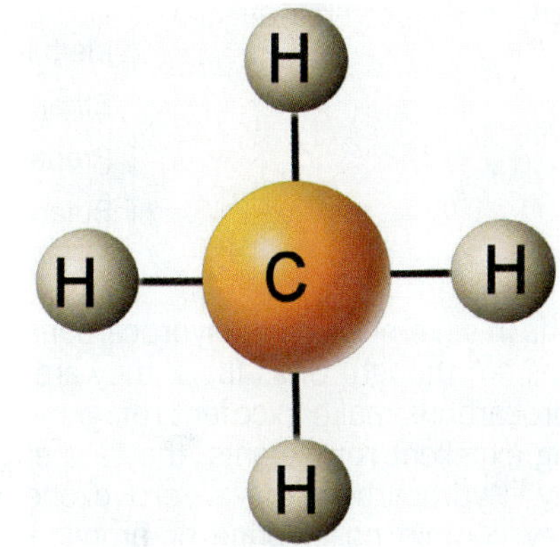

Figure 19-8 Methane molecule.

19.7 HC REFRIGERANTS

Hydrocarbons are compounds consisting of carbon atoms tied to each other in chains surrounded by hydrogen atoms. Figure 19-9 shows the arrangement of hydrogen and

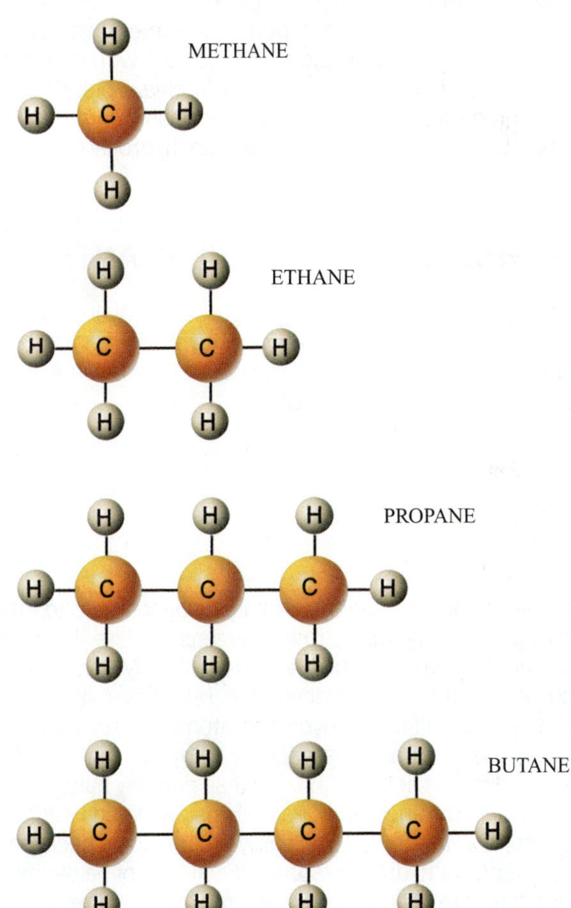

Figure 19-9 Hydrocarbon refrigerants.

Figure 19-7 Refrigerant chemical families.

Symbol	Refrigerant Family	Chemical Components
HCs	Hydrocarbons	Hydrogen and carbon
CFCs	Chlorofluorocarbons	Chlorine, fluorine, carbon
HFCs	Hydrofluorocarbons	Hydrogen, fluorine, carbon
HCFCs	Hydrochlorofluorocarbons	Hydrogen, chlorine, fluorine, carbon
CO_2, NH_3	Natural refrigerants	Carbon, oxygen, nitrogen, hydrogen

Figure 19-10 Hydrocarbon refrigerants.

Methane	R-50	4 hydrogen atoms	1 carbon atom
Ethane	R-170	6 hydrogen atoms	2 carbon atoms
Propane	R-290	8 hydrogen atoms	3 carbon atoms
Butane	R-600	10 hydrogen atoms	4 carbon atoms

carbon atoms in several common hydrocarbons. Normally, hydrocarbons are thought of as fuels; they are burned for energy. Hydrocarbons make excellent refrigerants. In addition to being excellent refrigerants, they are environmentally friendly. Hydrocarbons have zero ozone depletion because they contain no chlorine or bromine, and their global-warming potential is very low compared to halogenated refrigerants. Unfortunately, hydrocarbons are not just flammable but explosive. Hydrocarbons are only approved for limited applications in the United States. They are more widely used in other places, including Australia, Canada, China, and Europe. Hydrocarbons are sometimes used in small quantities in zeotropic blends to help with oil return. The amount of hydrocarbons used in these blends is so low that the mixture retains a nonflammable safety rating. Figure 19-10 shows a list of common hydrocarbon compounds and their refrigerant number.

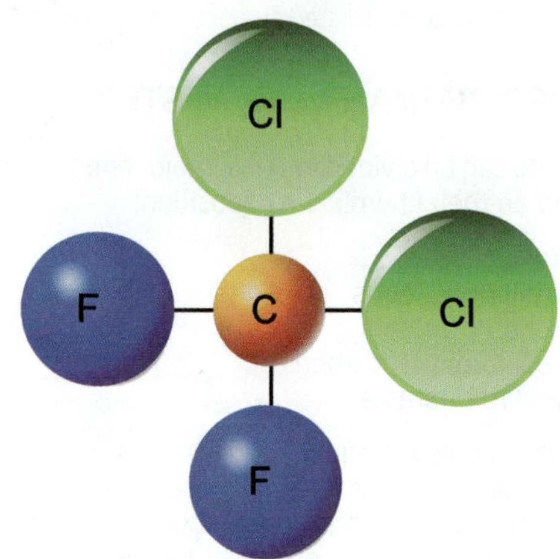

Figure 19-11 CFC-12 is based on the methane molecule.

19.8 HALOGENATED REFRIGERANTS

All halogenated refrigerants are based on hydrocarbons, like methane. Halogenated refrigerants are formed by combining a hydrocarbon with halogens. Halogens are a group of five highly reactive chemicals:

- Chlorine
- Fluorine
- Bromine
- Iodine
- Astatine

Chlorine and fluorine are the primary halogens used to make halogenated refrigerants. The halogens replace hydrogen atoms in the hydrocarbon molecules. If only some of the hydrogen atoms are replaced, the refrigerant is partially halogenated. If all of the hydrogen atoms are replaced, the refrigerant is fully halogenated. Fully halogenated CFC refrigerants are more chemically stable than partially halogenated HCFC and HFC refrigerants. Partially halogenated HCFC and HFC refrigerants can be more easily broken down because of the presence of hydrogen. Some advantages of halogenated refrigerants are that they are generally nonflammable, low in toxicity, and do not chemically react with many materials. This is primarily because they are very chemically stable molecules. The chlorine in halogenated refrigerants and the bromine in halon fire suppressant are the halogens responsible for ozone depletion. Chemicals that contain either chlorine or bromine contribute to ozone depletion.

19.9 CFC REFRIGERANTS

CFC is short for *chlorofluorocarbon*. CFC refrigerants contain chlorine, fluorine, and carbon. They are fully halogenated refrigerants with chlorine and fluorine replacing all the hydrogen atoms of a hydrocarbon compound. Most CFC refrigerants are built around the methane or ethane molecules. Figure 19-11 shows the chemical structure of R-12, which is built on a methane molecule.

CFC refrigerants were once the most common type of refrigerants in use. However, CFC refrigerants are the worst offenders in terms of ozone depletion; they have the highest ozone-depletion potentials of any group of refrigerants. CFCs also have a relatively high global-warming index and a long atmospheric lifetime. CFCs make good refrigerants because they are very chemically stable. This stability is also what makes them an environmental liability. CFCs do not break down in lower earth atmosphere. They find their way up to the stratosphere, where the high-intensity ultraviolet light breaks the CFCs apart, releasing the chlorine in the stratosphere.

It has been illegal to intentionally vent CFC refrigerants into the atmosphere sine July, 1, 1992, and 1995 was the last year that CFCs were allowed to be manufactured or imported into the United States, so the supply of CFCs is very low. Some CFC refrigerant is still available at very high prices, but there is now very little demand. Figure 19-12 lists some of the most common CFC refrigerants.

19.10 HCFC REFRIGERANTS

HCFC is short for *hydrochlorfluorocarbon*. HCFCs contain hydrogen, chlorine, fluorine, and carbon. Most HCFC refrigerants are built around the methane or ethane molecules. Figure 19-13 shows the chemical structure of R-22, which is built on a methane molecule.

HCFCs are only partially halogenated because they still have some hydrogen atoms. This makes them less stable than CFC refrigerants and more environmentally friendly.

Refrigerant Number	Refrigerant Name	No. Carbon Atoms	No. Chlorine Atoms	No. Fluorine Atoms
R-11	Trichlorofluoromethane	1	3	1
R-12	Dichlorodifluoromethane	1	2	2
R-13	Chlorotrifluoromethane	1	1	3

Figure 19-12 CFC refrigerants.

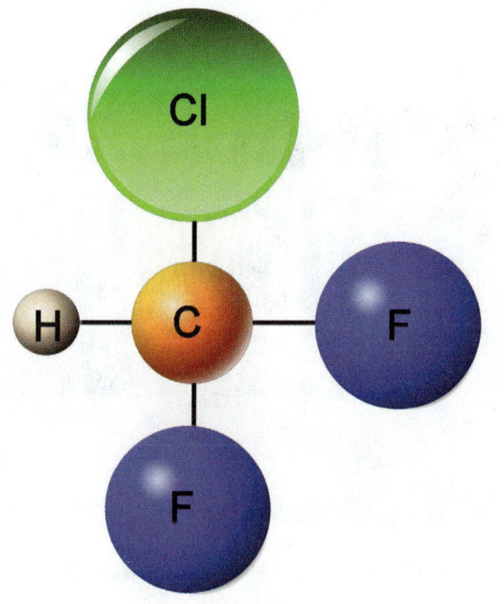

Figure 19-13 HCFC-22 is based on the methane molecule.

HCFCs still have an ozone-depletion potential because they still have chlorine. Their ozone-depletion potential is far less than the ozone-depletion potential of CFC refrigerants because HCFCs are more likely to break down in the lower atmosphere. Once the chlorine is released in the lower atmosphere it finds plenty of chemicals to attack, so it will not reach the stratosphere.

The most common HCFC refrigerant is R-22. Since January 1, 2010, new equipment charged with R-22 cannot be manufactured or imported, and R-22 is scheduled for a total phaseout on January 1, 2020. All HCFC refrigerants are scheduled to be phased out on January 1, 2030. R-22 has been the least expensive halogenated refrigerant available, but the price of R-22 is rising quickly because its production has been cut back. Figure 19-14 lists some of the most common HCFC refrigerants.

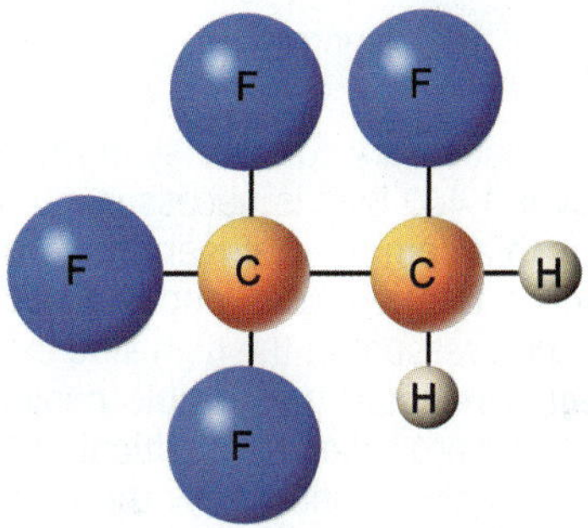

Figure 19-15 HFC 134a is based on the ethane molecule.

19.11 HFC REFRIGERANTS

HFC is short for *hydrofluorocarbon*. HFCs contain hydrogen, fluorine, and carbon. Most HFC refrigerants are built around the ethane molecule. Figure 19-15 shows the chemical structure of R-134a, which is built on an ethane molecule.

HFC refrigerants are only partially halogenated because they still have some hydrogen atoms. This makes them less stable than CFC refrigerants. They are more environmentally friendly than either CFC or HCFC refrigerants because they contain no chlorine and therefore have no ozone-depletion potential. However, HFC refrigerants do still have a global-warming potential. It has been illegal to intentionally vent HFC refrigerants into the atmosphere since November 15, 1995. Figure 19-16 lists some of the most common HFC refrigerants.

19.12 CO_2 REFRIGERANT

Interest in using CO_2 as a refrigerant has been increasing because it is nontoxic and nonflammable, does not deplete the ozone layer, has a very low global-warming potential, and is inexpensive. The biggest challenge is the extremely

Refrigerant Number	Refrigerant Name	No. Carbon Atoms	No. Hydrogen Atoms	No. Chlorine Atoms	No. Fluorine Atoms
R-22	Monochlorodifluoromethane	1	1	1	2
R-123	Dichlorotrifluoroethane	1	1	2	3
R-124	Chlorotetrafluoroethane	2	1	1	4
R-142b	Chlorodifluoroethane	2	3	1	2

Figure 19-14 HCFC refrigerants.

Refrigerant Number	Refrigerant Name	No. Carbon Atoms	No. Hydrogen Atoms	No. Fluorine Atoms
R-125	Pentafluoroethane	2	1	5
R-134a	Tetrafluoroethane	2	2	4
R-23	Trifluoromethane	1	1	3
R-32	Difluoromethane	1	1	2
R-143a	Trifluoroethane	2	3	3
R-152a	Difluoroethane	2	4	2

Figure 19-16 HFC refrigerants.

high pressures necessary for CO_2 systems to operate. A pressure of over 1,400 psig is necessary for an air-cooled system similar to a common refrigeration system. CO_2 will not condense at temperatures above 88°F because the temperature and pressure in the high side exceed the critical point. Heat is rejected by sensible cooling of the gas. These CO_2 systems are called transcritical systems because they operate both above and below the critical point. The high side of a transcritical system operates above the critical point, and the low side operates below the critical point. CO_2 can also be applied as the low-temperature portion of a cascade system, where the CO_2 condenser is cooled by the evaporator of another refrigeration system that uses a more traditional refrigerant. By keeping the condenser at a low temperature, the CO_2 system operates like a "normal" refrigeration system, below the critical point.

Figure 19-17 R-410A is an example of a zeotropic refrigerant.

19.13 FORMULATION

All refrigerants are combinations of two or more chemicals, but there are differences in how these chemicals are combined. The chemicals used to form a refrigerant can be combined in one of three ways:

- Compounds
- Zeotropes
- Azeotropes

TECH TIP

The terms *zeotropic* and *azeotropic* derive from Greek. *Zeo* means "to boil," and *trope* means "to turn." Thus, zeotropic refrigerants turn, or change, as they boil. The mixture changes (trope) as it boils (zeo). The prefix *A-* means "not." Placing the prefix *A* in front of a word effectively turns the word upside down. For example, an amoral person is one who does not have morals. An azeotropic refrigerant does not change as it boils.

19.14 ZEOTROPES

One result of the demise of CFC refrigerants has been the proliferation of refrigerant mixtures. Unlike compounds, these are combinations of chemicals that do not form a new chemical and whose physical properties are a sort of blending of the physical properties of the individual components. Zeotropic refrigerants change in percentage mixture as they change state. A more common term that is widely used for zeotropic refrigerants is *blends*. Sometimes the number of components in the blend is mentioned. Ternary blends are mixtures of three refrigerants, while binary blends are made up of two refrigerants. All refrigerants whose number is in the 400 series are zeotropes (Figure 19-17).

The behavior of a zeotropic refrigerant can be compared to a mixture of alcohol and water. The alcohol can be separated from the water through distillation. The alcohol tends to boil before the water, leaving more water and less alcohol in the remaining liquid. The boiling point of the mixture changes as the percentage of the two components in the mixture changes. The mixture changes as it boils.

From the standpoint of refrigeration, this change in composition is significant because the boiling temperature of the remaining liquid also changes as the mixture percentage changes. This is not like a normal refrigerant that maintains a fixed boiling temperature for any particular pressure. Zeotropic refrigerants will boil within a range of temperatures for any given pressure.

Bubble Point and Dew Point

When heating a saturated liquid mixture, the temperature at which small vapor bubbles first begin to form is known as the bubble point. Coming the other way, the temperature at which small liquid droplets first start to form when cooling a saturated vapor is known as the dew point. In compounds these two points are the same for any pressure. With zeotropes, however, the dew point is typically a higher temperature than the bubble point at any particular pressure. The difference between these two temperatures is known as glide. For example, the bubble point and dew point for R-12 at a pressure of 28 psig is 30°F. For R-401a, an R-12 substitute, the bubble point is 20°F and the dew point is 30°F at 28 psig. So while an evaporator with R-12 will always be 30°F when boiling at 28 psig, the evaporator temperature of an R-401a system can be anywhere between 20°F and 30°F at the same 28 psig saturation pressure.

Not all zeotropic blends have such a large glide. For example, the dew point and bubble point for R-410a are typically within 1°F. In fact, most PT charts do not even bother listing both the dew point and bubble point for R-410a.

TECH TIP

Many confusing names have been coined in a marketing effort to distinguish between high-glide zeotropes, like 401a, and low-glide zeotropes, like 410a. Low-glide zeotropes have been referred to as NARMS (for "nearly azeotropic refrigerant mixtures"), "near azeotropic blends," or, worst of all, "azeotropic mixtures." If you de-Greek the word *azeotropic* it means "does not change while boiling." Saying something "nearly does not change when it boils" really makes no sense. Even more confusing is the term *azeotropic mixture*. Since azeotropes are mixtures by definition, this term really confuses the issue. A few simple rules will keep the difference between zeotropes and azeotropes clear. If the refrigerant number is in the 400 series, it is a zeotrope. If the refrigerant must be removed from the cylinder as a liquid, it is a zeotrope. Azeotropes have numbers in the 500 series and may be removed from the cylinder as either a vapor or a liquid.

Pressure-Temperature Charts for Zeotropes

Simple PT charts work for pure compounds, like R-134a, but they do not work for zeotropic refrigerants since the boiling point shifts some as they boil. Therefore, charts for high-glide blends include two temperatures for each pressure: a dew point and a bubble point (Figure 19-18). Low-glide zeotropes like R-410a are commonly still listed as a single point because the difference between the bubble point and the dew point is typically less than 1°F.

Fractionation

The separation of the components in a zeotropic blend is called fractionation. The refrigerant is essentially being distilled by removing a fraction of its mixture. Zeotropic refrigerants should only be removed from the cylinder as a liquid to avoid fractionation.

Fractionation can also occur due to a leak, with one component leaking out at a faster rate than the other components. Fractionation is more likely if the leak is in the

Temperature (°F)	Bubble Point psig Pressure (in Hg vacuum)	Dew Point psig Pressure (in Hg vacuum)
−60	*(9.1)*	*(15.9)*
−55	*(5.9)*	*(13.5)*
−50	*(2.4)*	*(10.9)*
−45	0.7	*(7.9)*
−40	2.9	*(4.5)*
−35	5.2	*(0.7)*
−30	7.9	1.7
−25	10.7	4.0
−20	13.9	6.5
−15	17.3	9.3
−10	21.1	12.4
−5	25.2	15.8
0	29.6	19.5
5	34.4	23.6
10	39.6	28.0
15	45.2	32.7
20	51.3	37.9
25	57.8	43.6
30	64.7	49.6
35	72.2	56.2
40	80.2	63.2
45	88.7	70.7
50	97.8	78.8
55	107.5	87.5
60	117.7	96.8
65	128.7	106.7
70	140.2	117.2
75	152.5	128.4
80	165.5	140.4
85	179.2	153.1
90	193.6	166.5
95	208.8	180.8
100	224.9	195.8
105	241.8	211.8
110	259.6	228.7
115	278.2	246.5
120	297.8	265.3
125	318.3	285.2
130	339.9	306.1
135	362.4	328.2
140	386.0	351.4
145	410.7	375.9
150	436.5	401.7

Figure 19-18 Pressure-temperature chart for R-407C has columns for the temperature, bubble point, and dew point.

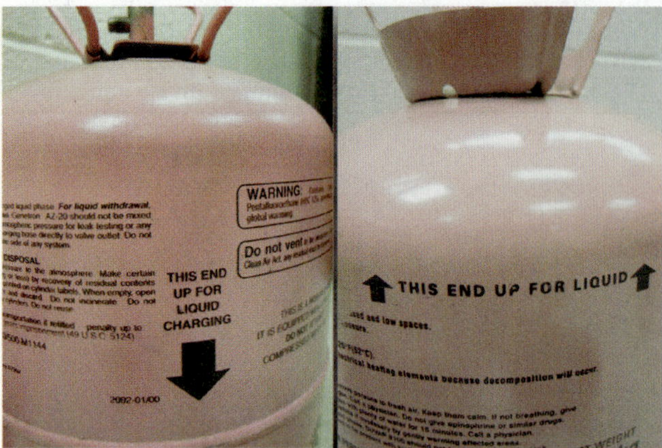

Figure 19-19 Follow the instructions on the cylinder to ensure that R-410A leaves the cylinder as a liquid.

evaporator or condenser, because fractionation occurs during a change of state. Because of this possibility, early advice for recharging systems using zeotropic refrigerants was to recover the remaining system charge and recharge the system with new refrigerant. This was based on the assumption that the refrigerant would fractionate as it leaked out. Practical experience has shown that concerns about fractionation due to leaks are largely unfounded. However, the absolute best way to charge a system that uses a zeotropic refrigerant is to recover the remaining refrigerant and recharge the system with a full charge of new refrigerant. Most manufacturers making systems that use low-glide blends now state that it is acceptable to top off a system charge without recovering the residual system refrigerant.

Handling Zeotropes

The primary difference between handling zeotropes and other refrigerants is how the refrigerant is removed from the cylinder. The instructions for working with zeotropic refrigerants are typically on the refrigerant cylinder (Figure 19-19). All zeotropes are handled the same way: they must leave the cylinder as a liquid to stay mixed in the correct ratio. If vapor is taken out of the cylinder, the liquid will boil to replace the vapor being removed. This causes a separation of the components in the mixture. Not only will the gas going into the system be the wrong mix, but all the remaining liquid left in the cylinder is ruined as well! Remember that all 400 series refrigerants are zeotropes and must leave the cylinder as a liquid.

SERVICE TIP

Some cylinders contain dip tubes so that liquid comes out of the valve on top of the cylinder when the cylinder is in the upright position. Other cylinders do not have the dip tube. How do you know which is which? Read the instructions. The cylinders have arrows painted on them to inform technicians exactly how to use the product.

19.15 AZEOTROPES

Occasionally some binary mixtures of refrigerants will have one specific mixture ratio where the bubble point and dew point temperatures are the same. In short, the mixture behaves like a compound. These mixtures cannot be physically separated since the components are evaporating in the same ratio as the mixture. These are called azeotropic refrigerants, for "not" zeotropic, because they do not change their boiling point as they boil. Azeotropic refrigerants are still mixtures because they are not chemically bonded, but they cannot be separated physically. For example, R-507 is 50 percent R-125 and 50 percent 143a. When R-507 boils, the R-125 and the R-143a boil in the same 50/50 ratio. It is not possible to separate the two original refrigerants by boiling the mixture. Azeotropes can be handled just like compounds, or "pure" refrigerants, since they will not separate when they change state. All azeotropes have a 500 series number.

19.16 SAFETY

ASHRAE Standard 34, Safety Code for Mechanical Refrigeration, addresses refrigerant safety. Refrigerants are grouped by their toxicity and flammability. Each refrigerant is assigned a safety rating consisting of a letter and a number. Letter *A* identifies refrigerants with lower toxicity, and letter *B* identifies refrigerants with higher toxicity. A number indicating the flammability of the refrigerant follows the letter. The lower the number, the lower the flammability.

Toxicity

Almost anything can be toxic in high enough concentrations. For example, the air we breathe is approximately 78 percent nitrogen. However, breathing 100 percent nitrogen will cause asphyxiation. The maximum safe exposure level in parts per million is called the threshold limit value (TLV). The TLV is not for a one-time exposure but is based on an exposure for an eight-hour day, forty-hour week. A refrigerant's TLV is used to rate its toxicity level. *A* is used to designate low-toxicity refrigerants having a TLV of 400 or more. *B* designates high-toxicity refrigerants having a TLV less than 400. Type B refrigerants are prohibited in many applications by building and safety codes.

Most common CFC, HCFC, and HFC refrigerants are listed as type A refrigerants. One notable exception is R-123. R-123 is a low-pressure HCFC refrigerant that is used in large, low-pressure chillers. It is considered a type B refrigerant. Many older refrigerants are type B, including ammonia, sulfur dioxide, and methyl chloride. Note that this rating system does not distinguish between refrigerants like these that will kill you in a few minutes from refrigerants like HCFC-123, which will cause some physical distress when you are exposed for eight hours a day, forty hours a week.

Flammability

A refrigerant's flammability is listed as type 1, nonflammable; type 2L, lower flammability with a maximum burning velocity less than 10 cm/s; type 2, lower flammability; or type 3, higher flammability. Nonflammable refrigerants will not ignite and will not support combustion. Refrigerants listed as type 2L or 2 will ignite when exposed to a flame but will not continue to burn when the flame is removed. Refrigerants with a type 2 flammability rating do not have a flash point; they will not explode. Refrigerants with a flammability rating of 3 will ignite, they will support combustion, and they do have a flash point. R-12 is an example of a refrigerant with a flammability rating of 1. HFO 1234yf is an example of a refrigerant with a 2L flammability; ammonia, R-717, is an example of a flammability 2 refrigerant; and propane, R-290, is an example of a flammability 3 rating.

Combined Rating

A refrigerant's safety rating is a combination of the toxicity letter and the flammability number. A refrigerant that is nontoxic and nonflammable would have an A1 rating. A refrigerant that is toxic and very flammable would have a B3 rating. Figure 19-20 shows the combined ratings for several refrigerants.

Safety Rating for Zeotropes

Zeotropic refrigerants can fractionate when they boil. This raises the possibility that a mixture's characteristics can change. For this reason, zeotropic refrigerants are assigned a dual rating. The first rating is for the characteristics of the original mixture; the second rating is for the characteristics in a worst-case fractionation. For example, a zeotropic refrigerant could have a rating of A1/A2. This would mean that the original mixture would be nonflammable and nontoxic but that the result of a worst-case fractionation would be nontoxic and somewhat flammable. Most commonly used zeotropic refrigerants have a safety rating of A1/A1, but there are a few with a rating of A1/A2. This indicates that the mixture can be ignited if the refrigerant has been fractionated. Azeotropic refrigerants are rated just like pure compounds because they do not separate when they boil.

Chemical Interactions

There are safety concerns that arise from the interaction of refrigerants and their surroundings. For example, CFCs and HCFCs are not flammable. However, they chemically decompose when exposed to flames or very high temperatures, releasing their chlorine and fluorine. This creates highly irritating and potentially toxic gases, including chlorine gas, fluorine gas, and phosgene gas. HCFCs and HFCs can react violently with some metals. HCFCs should never be exposed to magnesium. HFCs should not be allowed to come into contact with freshly abraded aluminum.

	Toxicity Group	
	A	B
Flammability Group	Lower Toxicity	Higher Toxicity
3	A3	B3
Higher flammability	R-50	R-1140
	R-290	
	R-600	
2	A2	B2
Lower flammability	R-142b	R-717
	R-143a	R-30
	R-152a	R-611
	A1	B1
2L	A2L	B2L
Lower flammability	R-1234yf	
Flame speed < 10 cm/s		
1	R-11	R-10
No flame propagation	R-12	R-123
	R-22	R-764
	R-113	
	R-114	
	R-134a	
	R-500	
	R-502	

Figure 19-20 Refrigerant safety classification.

Aluminum that has been recently ground up can react violently with HFCs. Some HFCs are flammable under high pressure with large concentrations of oxygen. Of course you should never put oxygen or compressed air into any refrigeration system. The oxygen under pressure can cause an explosion when mixed with the refrigerant oil.

Asphyxiation

Most refrigerants are heavier than air and will displace the air in a room. Releasing large quantities of nontoxic refrigerant into a confined space can cause asphyxiation. This can sneak up on technicians because the refrigerant does not attack the body. Nobody would hang around long enough to asphyxiate from a large release of ammonia; if you are able to leave, you will. On the other hand, R-22 does not attack the body. Victims just get lightheaded from lack of oxygen. This can be very dangerous in confined spaces. So not only will preventing large releases of CFCs help preserve the ozone layer, but it helps preserve technicians as well! In the event of a large release of refrigerant, leave! Ventilation, particularly in low-lying places, will help clear the room. The mechanical room for large systems containing a class B refrigerant should have a self-contained breathing apparatus. However, the best course of action is to get out of the room.

Frostbite

Of course, all refrigerants can be cold. Letting liquid refrigerant come into contact with exposed skin can cause "burns." Immediately after the incident, the skin will just look irritated and red. In one hour after the exposure, blisters will start to appear and the skin will start to swell and puff up. In bad cases, the outer skin comes off and the area is treated like a third-degree burn. Technicians should wear safety goggles to protect their eyes and neoprene-lined gloves to protect their hands when handling refrigerants.

SAFETY TIP

HFC-410 boils at −62°F at atmospheric pressure. Skin exposed to a direct liquid stream for only a minute can receive severe frostbite that acts like a second- or third-degree burn. Keep your hands out of streams of liquid refrigerant. Refrigerant can be replaced, but your hands cannot!

19.17 OZONE DEPLETION

M. J. Molina and F. S. Rowland published a laboratory study in 1974 demonstrating the ability of CFCs to break down ozone in the presence of ultraviolet light. Observations of the ozone layer have since demonstrated that the concentration of ozone in the stratosphere is decreasing. The most severe ozone loss was discovered over Antarctica. The loss of ozone over Antarctica is referred to as the "ozone hole." Many people once believed that ozone depletion could only occur at the poles, but reductions in ozone have also been observed in northern middle latitudes.

Ozone Layer

Scientists studying the earth's atmosphere generally divide it into six regions: troposphere, stratosphere, mesosphere, thermosphere, exosphere, and ionosphere (Figure 19-21). The troposphere extends from the earth's surface to about 6 miles above the earth. The stratosphere starts at 6 miles above the earth and extends to 30 miles above the earth. The ozone layer is really just a concentration of ozone molecules within the stratosphere (Figure 19-22). Ozone is a form of oxygen.

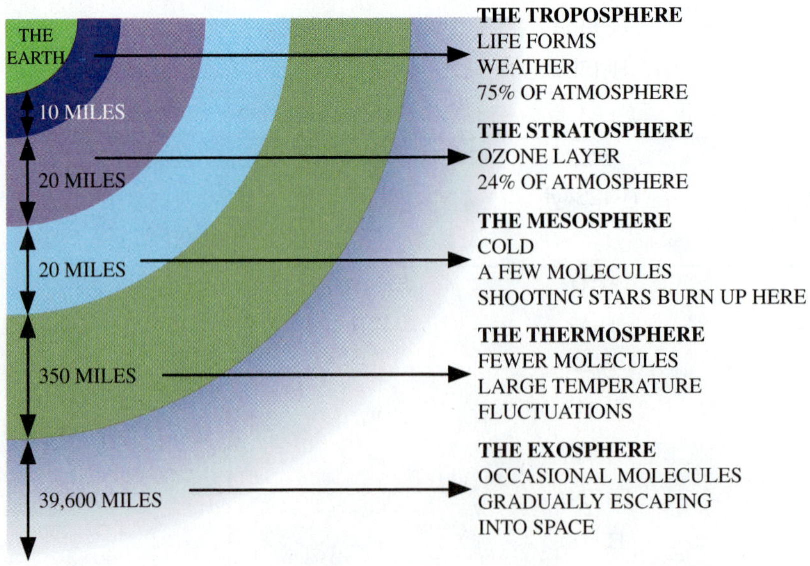

Figure 19-21 Layers of the earth's atmosphere.

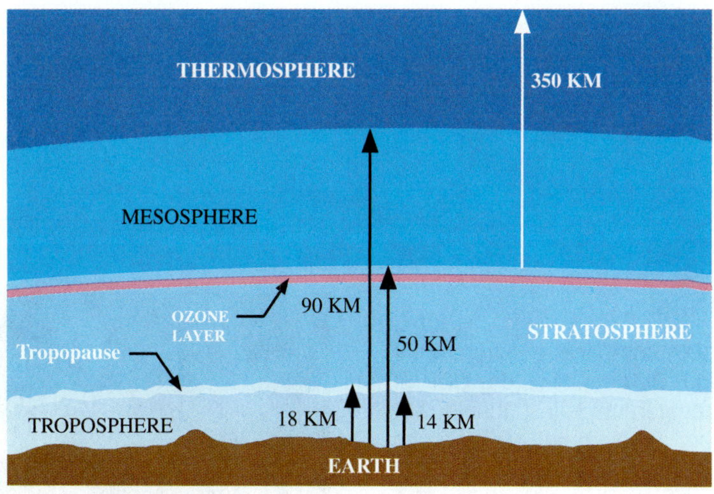

Figure 19-22 The ozone layer is inside the stratosphere. *(NASA)*

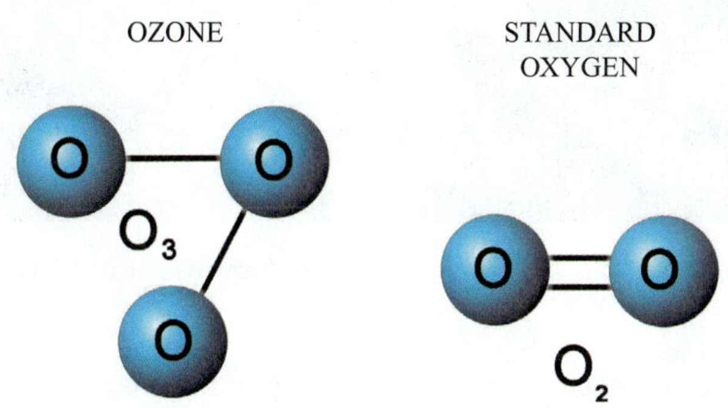

Figure 19-23 Ozone molecules are composed of three oxygen atoms, while the standard oxygen molecule is composed of two oxygen atoms.

The oxygen we breathe, O_2, is formed from two oxygen atoms. Ozone molecules are formed from three oxygen atoms, O_3. Figure 19-23 shows normal O_2 and ozone, O_3. Ozone is a relatively unstable molecule that normally has a short lifetime. Ozone is constantly produced and destroyed in a natural cycle. However, the overall amount of ozone has remained essentially stable as a result of a natural balance between ozone creation and destruction. Large increases in stratospheric chlorine and bromine have upset that balance by adding to the destruction side of the equation.

Ozone-Depletion Chemistry

The ozone-depletion process begins when ozone-depleting substances, including CFCs and HCFCs, are released into the atmosphere. Winds efficiently mix all the different gases in the part of the atmosphere closest to earth, the troposphere. CFCs are extremely stable and do not break down in the atmosphere. After a period of several years, CFC molecules reach the stratosphere, the atmospheric layer above the troposphere. Strong ultraviolet light breaks apart the CFC and HCFC molecules. CFCs and HCFCs release chlorine atoms when they are broken apart. It is these chlorine atoms that actually destroy ozone, not the intact CFC or HCFC molecules. The free chlorine pulls an oxygen atom off of the O_3 ozone molecule to form O_2 and chlorine monoxide, ClO. Chlorine monoxide then pulls an oxygen atom off of another O_3 ozone molecule to form two O_2 molecules, and once again the chlorine is free. It is estimated that one chlorine atom can destroy over 100,000 ozone molecules before it is removed from the stratosphere. Figure 19-24 illustrates this process.

Since ozone filters out harmful ultraviolet B (UVB) radiation, less ozone means higher UVB levels on the ground. UVB has been linked to skin cancer, cataracts, and harm to certain crops and marine organisms. Although some UVB reaches the surface even without ozone depletion, the harmful effects of UVB will increase as the concentration of stratospheric ozone decreases.

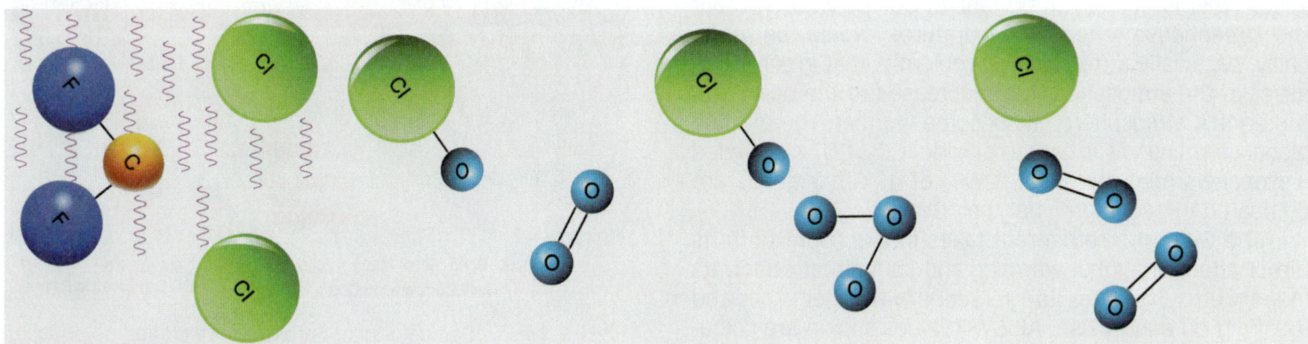

Figure 19-24 Stages of ozone depletion caused by a CFC = 12 molecule.

Figure 19-25 Classes of ozone-depleting substances.

Refrigerant Class	Ozone Depletion Potential	Refrigerants
Class I	ODP > 0.2	CFC-11, CFC-12, R-500, R-502
Class II	ODP < 0.2	HCFC-22, HCFC-123

Ozone-Depletion Potential

Ozone-depletion potential, ODP, compares a refrigerant's ozone-depleting effect to R-11. By definition, the ODP of R-11 is 1. The ODP of most refrigerants is less than 1, but there are other substances, such as halons, with ODPs much greater than 1. Note that refrigerants that do not contribute to ozone depletion are not classified in this system. The EPA divides ozone-depleting refrigerants into two classes based on their ozone-depletion potential.

- **Class I** refrigerants have an ODP greater than 0.2.
- **Class II** refrigerants have an ODP less than 0.2.

This classification is becoming a historical artifact because nearly all Class I refrigerants are CFCs, and CFC production and importation have been banned since the end of 1995. It is still possible to purchase CFC refrigerants from a supplier who still has refrigerant left over from 1995, or from a refrigerant reclaimer. Class II refrigerants will be around for some time to come, but their time is also limited. HCFC refrigerants are scheduled to be phased out completely by 2030. Figure 19-25 shows refrigerant classes according to their ozone-depletion potential.

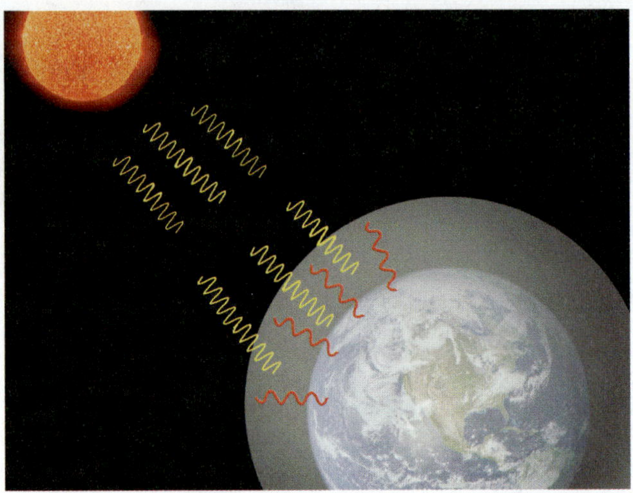

Figure 19-26 Illustration of the greenhouse effect and global warming. *(NASA)*

19.18 GLOBAL WARMING

Global warming is the gradual rise in the earth's temperature because of an increase in greenhouse gases. The glass in greenhouse works by allowing most wavelengths of light energy in and blocking the escape of infrared energy. Visible light has short wavelengths and can penetrate the glass. The light strikes surfaces and some of the visible light energy is converted into heat energy. The longer-wavelength infrared waves cannot penetrate the glass. This has the effect of trapping in heat. Some gases in the earth's atmosphere behave like the glass in a greenhouse, trapping heat in (Figure 19-26). These gases are referred to as greenhouse gases. This is actually important to our survival. Without the greenhouse effect, temperatures would be quite chilly, especially at night. The problem is that greenhouse gases in the atmosphere have increased to the point that the earth's atmosphere has become warmer. The primary global-warming gas is carbon dioxide, CO_2. CO_2 is assigned a global-warming potential, GWP, of 1. Other gases are compared to CO_2 to arrive at their GWP.

The GWP of a refrigerant takes into account both its direct effect on global warming and its indirect effect. Its indirect effect includes its interaction with other gases and its effect on energy use. All GWP values shown are calculated over a 100-year time horizon. All halogenated refrigerants contribute to global warming. In fact, the GWP of HFCs is generally worse than for the HCFCs they replace. Figure 19-27 shows the relative ozone-depletion and global-warming potential for several common refrigerants.

CODE TIP

On December 7, 2009, the EPA declared that the current and projected atmospheric concentrations of the six key, well-mixed greenhouse gasses, including HFCs, "threaten the public health and welfare of current and future generations." While no regulations are currently in place in the United States to regulate or phase out HFC refrigerants, the statement does establish the justification for future HFC regulations.

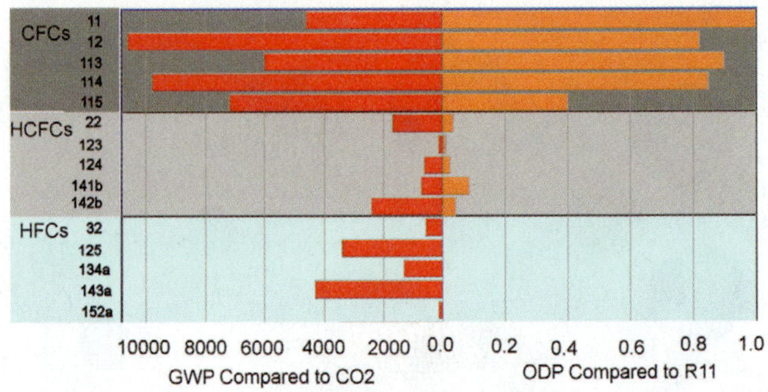

Figure 19-27 Comparison of both ozone-depletion potential and global-warming potential for several halogenated refrigerants.

CODE TIP

A European Union regulation in place since January 2011, known as the mobile air conditioner directive, or MAC, requires the refrigerant in new car air-conditioning systems to use refrigerant with a GWP of 150 or less. This requires auto manufacturers to find a replacement refrigerant for 134a. Many auto manufacturers are electing to use a new fluorinated refrigerant with an extremely low GWP of 4 called HFO 1234yf. It is a near drop-in replacement for 134a, so it requires very little system redesign. This new refrigerant is moving to the United States as well. General Motors has announced that they will be using HFO 1234yf in their new cars beginning with the 2013 models.

19.19 NAMING REFRIGERANTS

All modern refrigerants are named using a system of chemical nomenclature. The name describes a compound's chemicals and their arrangement. For example, monochlorodifluoromethane is the chemical name for R-22. This tells the reader that the chemical is composed of one chlorine (mono chloro) and two fluorines (di fluoro) attached to a methane molecule. However, these names are more useful to chemists than to air-conditioning technicians. Some simpler refrigerant naming conventions exist that help identify the makeup of some types of refrigerants in a more user-friendly manner.

Refrigerant Pseudonyms

Halogenated refrigerants are identified by a series of letters indentifying the chemical element in the compound. The letters *C*, *F*, and *H* are used to represent chlorine, fluorine, and hydrogen. A final *C* is used to represent carbon.

These letters combine to identify the chemical components of the refrigerant.

- **CFC** contains chlorine, fluorine, and carbon.
- **HFC** contains hydrogen, fluorine, and carbon.
- **HCFC** contains hydrogen, chlorine, fluorine, and carbon.

Trade Names

Some manufacturers have created trade names to help promote their products. One of the most successful is Freon. Freon is DuPont's trade name for their line of CFC and HCFC refrigerants. Freon does not refer to a single refrigerant but a family of refrigerants. Another popular trade name is Puron. Puron is the Carrier corporation's trade name for R-410a. Puron does refer to a single refrigerant.

19.20 NUMBER DESIGNATION

Refrigerants are identified by number, preceded by the letter *R*, for "refrigerant." Some refrigerant numbering conventions exist that help identify the makeup of some types of refrigerants. These numbering designations were originally developed by DuPont; have been established by the American Society of Heating, Refrigeration and Air Conditioning Engineers (ASHRAE); and are used throughout the industry.

Decoding Halogenated Refrigerant Numbering

The refrigerant number assigned to single-component halogenated refrigerants describes the number of each type of atom in the refrigerant molecule. Adding 90 to the refrigerant number yields a three-digit number, with each digit representing the number of a particular type of atom. The first digit on the left represents the number of carbon atoms, the second digit represents the number of hydrogen atoms, and the last digit on the right represents the number of fluorine atoms. Look at HCFC-22, for example: $22 + 90 = 112$.

The number shows that HCFC-22 contains one carbon, one hydrogen, and two fluorine atoms.

The number of chlorine atoms can be calculated using the information about the other components. The number of bonds available on a carbon-based molecule is twice the number of carbon atoms plus 2. For HCFC-22, which has one carbon atom, there are four bonds. Chlorine atoms occupy any remaining bonds after the fluorine and hydrogen atoms. Since HCFC-22 has four bonds and three of them are occupied, HCFC-22 contains one chlorine atom.

Let's look at another example: HFC-134a.

$$134 + 90 = 224$$

This indicates that R-134 contains two carbon atoms, two hydrogen atoms, and four fluorine atoms.

There are six bonds: $2 \times 2 + 2$. There are no bonds left over, so there are no chlorine atoms.

Isomers

Some halogenated refrigerants have a lowercase letter after their number, such as HFC-134a. The *a* at the end describes how these atoms are arranged. Isomers of a given compound contain the same atoms but are arranged differently. The atoms in compounds containing more than one carbon atom can usually be arranged in more than one way. Isomers usually have different properties from each other. Simply moving the components around changes the properties of the chemical. The letter following the number helps identify which isomer of the compound is being used.

The most symmetrical arrangement of atoms does not have a letter after the refrigerant number, as in HFC-134. The second most symmetrical arrangement is followed by the letter *a*, as in HFC-134a. If there is a third isomer, the third most symmetrical is followed by a *b*. Tetrafluoroethane has only two isomers: HFC-134 and HFC-134a, shown in Figure 19-28.

Zeotropic and Azeotropic Refrigerant Numbers

Zeotropic refrigerants all receive a 400 series number. The numbers following the 4 indicate the order that the refrigerant number was assigned by ASHRAE. The uppercase letter

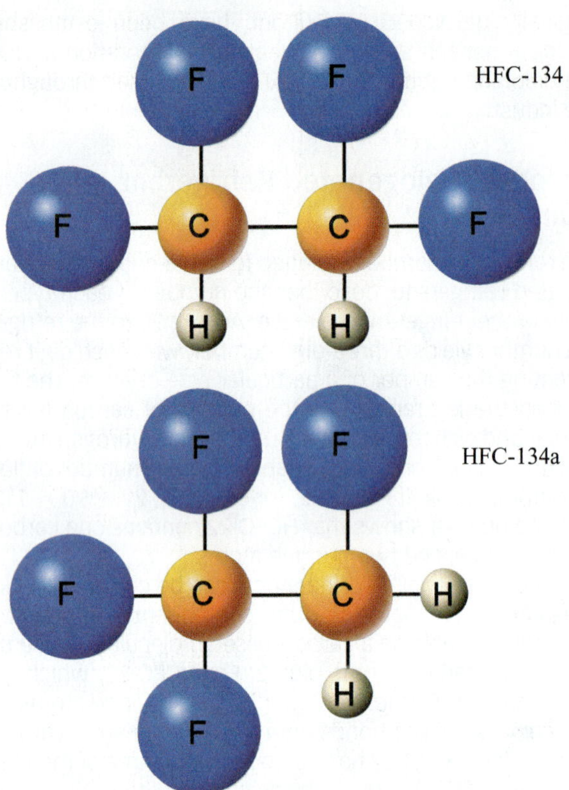

Figure 19-28 Comparison of tetrafluoroethane isomers 134 and 134a.

after the number tells which percentage mix of the particular chemicals is used. For example, R-401A was the first blend assigned a number and the first percentage mix of the chemicals in R-401A. R-401B is composed of the same chemicals but with a different percentage of each chemical.

Azeotropic refrigerants all receive a 500 series number. Like zeotropes, the following two numbers identify the order in which the refrigerant was approved. Unlike zeotropes, azeotropes have no letter after the number because there is only one percentage mix where the components will behave like an azeotrope. For example, R-507 was the seventh azeotropic refrigerant to receive a number.

19.21 COMPATIBILITY

Refrigerants, like any chemical, can affect materials in detrimental ways. The effect of the refrigerant on the materials used in a system must be known. Preferably, the refrigerant should have no chemical or physical effect on the materials that the refrigeration system is built of. The refrigerant must not react with or deteriorate materials it comes in contact with during the operation. These include metallic compressor parts, gaskets, O-rings, seals, motor insulation and windings, piping, and condenser and evaporator heat-transfer surfaces.

Refrigerants will often cause swelling in plastics and elastomers. Petroleum-based refrigerants may dissolve some petroleum-based plastics. When designing plastic insulation, O-rings, and seals, this becomes important. However, this has seldom been any concern to the refrigeration technician. With the introduction of new HFC refrigerants, the effect of a refrigerant on elastomers is suddenly of great concern. These new compounds are generally more chemically reactive than the older refrigerants they are designed to replace. This makes retrofitting an old system with new refrigerant difficult without technical information about the materials used inside the system. Systems should not be field retrofitted without the advice and consent of the equipment manufacturer.

Most refrigerants, including halogenated refrigerants, have no affect on any type of metal. There are some notable exceptions. Ammonia, still in wide use, chemically acts upon copper, brass, tin, and zinc. Therefore, all these metals are restricted from use with an ammonia system. Also, while pure ammonia is essentially inert to aluminum, ammonia and water mixtures will eat aluminum away. Aluminum is not a recommended metal for use with ammonia because of this.

Halogenated refrigerants are generally acceptable with most any common metal, but they should not be used with aluminum alloys containing more than 2 percent magnesium or zinc since both magnesium and zinc are reactive with a combination of halogenated refrigerants and water. HFC-410a is reactive with abraded aluminum. Abraded aluminum is aluminum that has been roughed up or ground up into pieces.

19.22 CONTAMINANTS IN REFRIGERANT

The only two chemicals that should be inside a refrigeration system are refrigerant and refrigeration oil. However, contaminants are often introduced through poor installation and service practices or through compressor failure. Contaminants that are frequently found in refrigerant include water vapor, acid, noncondensables, carbon, and particulates. When a refrigerant is recycled, many of these impurities can be removed. To resell the refrigerant, it must be reclaimed by a certified source and brought back to the original level of purity indicated by AHRI Standard 700-2006, which gives maximum allowable contamination levels for both new and reclaimed fluorocarbon refrigerants.

Water is undesirable in all but a very few refrigeration systems; notably, those that use water as either a refrigerant or an absorbent in an absorption-cycle machine. Water may cause numerous problems, such as the following:

- Freezing up at the metering device can cause a loss of system performance.
- Water can promote oxidation in the system.
- Water may hydrolyze the refrigerant and form powerful acids that will corrode the system.

Water is a great catalyst for chemical reaction; it accelerates many chemical reactions, such as oxidation. Since chemical reactions are very undesirable in a refrigeration system, water is most unwelcome. Most of the metallic parts inside a refrigeration system are not made to withstand corrosion since they are, in theory, exposed only to refrigerant and oil. Neither of these substances will corrode the metals used in the system, but water will corrode most any type of metal. Since nonresistant metals are used, a little water in the system can wreak havoc. While water will not cause any direct chemical reaction with halogenated refrigerants, water will break down halogenated refrigerants in the presence of extreme heat. The temperatures on the discharge side of a compressor can be high enough to begin this process. Since halogenated refrigerants contain chlorine and fluorine, hydrochloric and hydrofluoric acids can form. These acids will eat away the metals in the system and attack the insulation of motor windings in hermetic compressors. Water can also be present in refrigerant cylinders. Using only clean, new refrigerant from a reliable source can effectively prevent this particular hazard.

TECH TIP

Noncondensables in a refrigerant cylinder or in a system will increase the pressure above the normal saturation pressure. A refrigerant cylinder with noncondensables will have a higher pressure than is indicated on a PT chart for its temperature.

19.23 REFRIGERATION OILS AND THEIR APPLICATIONS

Refrigerants need to be chemically compatible with the oil used in the system. Chemical reactions occurring between the system refrigerant and the oil are undesirable. Refrigeration oils and refrigerant must also be physically compatible. Most refrigeration systems depend on the refrigerant to carry the oil back to the compressor. In the best of all possible worlds, all the oil would remain in the compressor. However, in the real world it is impossible to keep all the oil in the compressor. Once the oil leaves the compressor, the only way to get it back is for the refrigerant to bring it back. For this to happen, the oil must be miscible with the refrigerant. Miscibility means that the oil will dissolve in the refrigerant in any proportion.

Mineral Oil

For years the primary oil in refrigeration systems has been mineral oil. Most CFCs and HCFCs are compatible with mineral oil. One of the primary challenges posed by HFC refrigerants is their incompatibility with mineral oil. Mineral oil is not the least bit miscible with HFC refrigerants. This means that once the oil has left the compressor, it does not return. Early work with HFC refrigerants and mineral oil showed that the oil tended to drop out at the first available opportunity, in the condenser. Normally, oil is more apt to drop out in the evaporator where the low temperatures thicken up the oil and make it more prone to separate. This tendency of oil to drop out in the condenser was dubbed "reverse solubility." This necessitated the development of new oils.

Polyalkylene Glycol

The first oil used with R-134a automotive systems was polyalkylene glycol, PAG. PAG is a cousin of the polyethylene glycol used as antifreeze in cars. Anyone who has handled the glycol used for auto antifreeze knows that it is extremely slick and seems impossible to clean off. PAG oils are like a thick antifreeze. Glycol-based oils offer some challenges. They are chemically incompatible with CFCs, HCFCs, and mineral oil. Early attempts to convert R-12 systems to R-134a were frequently plagued by the formation of a sticky goo due to the incompatibility of PAG with both mineral oil and CFC refrigerant. PAG works well in new systems, but it proved too problematic for retrofit.

Like its antifreeze cousin, the glycol oils are very friendly with water. PAG sucks it up like a sponge, and this is not desirable for a refrigeration system. Materials like the glycols, which have an affinity for water, are called hygroscopic. PAG oil must be packaged in metal containers because it can literally suck water out of the air through plastic containers. PAG containers should not sit around on the shelf after they are opened, so PAG is best purchased in small quantities that are only opened immediately before use.

Polyol Ester

Polyol ester (POE) is the predominant oil of choice today. It is compatible with both old and new refrigerants and is also somewhat compatible with mineral oil. Ester oils are aggressive solvents, making material compatibility an issue. Ester-based oils can sometimes be used in retrofit situations, but they can be incompatible with the plastics and varnishes used in older systems. POE is also hygroscopic, and once mixed with water, it hydrolyzes to form acid. All systems containing POE must have a filter drier to avoid water contamination. Like PAG, it is shipped in metal containers. Also like PAG, it should be purchased in small quantities and only opened immediately before use.

Alkyl Benzene

Alkyl benzene is an older synthetic oil. It was designed to be used where mineral oil is used. Alkyl benzene can be used with zeotropic blends that contain a mixture of HCFCs and HFCs. Material compatibility is not as big an issue with alkylbenzene because it was developed for older systems.

Polyvinylether

A promising new refrigerant lubricant is polyvinylether, or PVE. It is compatible with all common refrigerants, including HFCs. It is not hygroscopic but instead is hydrophobic: it does not absorb or mix with water. It can be used in systems with small quantities of residual mineral oil or POE. Figure 19-29 shows refrigerant and oil compatibility.

19.24 REPLACEMENT REFRIGERANTS

All replacement refrigerants must be evaluated and approved by the EPA. The program that evaluates new refrigerants is called the Significant New Alternatives Policy, or SNAP. Substitutes are reviewed on the basis of ozone-depletion potential, global-warming potential, toxicity, flammability, and exposure potential. Lists of acceptable and unacceptable substitutes are updated several times each year.

CFCs	HCFCs	HFCs	Blends with HFCs and HCFCs
Mineral oil	Mineral oil	Polyalkylene glycol	Alkyl benzene
Alkyl benzene	Alkyl benzene	Polyol ester	Polyol ester
Polyol ester	Polyol ester	Polyvinylether	Polyvinylether
	Polyvinylether		

Figure 19-29 Refrigerant and oil compatibility.

TECH TIP

A "drop in" refrigerant can replace the original system refrigerant without any changes to the system, such as compressor oil changes, metering device adjustments, or seal changes. Although there are many advertised "drop in" replacement refrigerants, the EPA does not recognize any refrigerants as being true drop-in refrigerants requiring no system modifications or servicing.

The EPA lists alternative refrigerants as suitable for retrofit, new application, or unacceptable. Alternatives listed as retrofit refrigerants can be used to replace the refrigerant in an existing system. This does not mean that they will work with no system alterations, but they can be made to work in an existing system. Refrigerants listed as new alternatives will only work in new applications that accomplish the same thing. Sometimes these new alternatives are completely different approaches to cooling. One example is evaporative cooling, which is listed as an alternative for many types of systems. Evaporative cooling is a cooling system that reduces air temperature by evaporating water. Obviously, an evaporative system is a completely different type of system from a traditional mechanical refrigeration system. It is only a replacement in the sense that it accomplishes roughly the same thing as a mechanical refrigeration system if humidity control is not an issue.

There are some refrigerants that are specifically designed to replace CFCs or HCFCs in new equipment. HFC-134a replaces CFC-12 in new equipment but is not generally considered a retrofit refrigerant. HFC-407C is a very close replacement for HCFC-22 in retrofit situations. HFC-410A is the replacement for HCFC-22 in new equipment. It has pressures 40–60 percent higher than R-22, so it cannot be used as a retrofit. It is a replacement only in the sense that it is used in the same general types of cooling applications where HCFC-22 is used.

CODE TIP

There are many refrigerants available for retrofitting HCFC 22 systems, including R-407A, R-407C, R-417A, R-421A, R-422B, R-422C, R-424A, R-427A, and R-437A. However, using a new chemical in a refrigeration system will generally void the equipment's third-party certification, such as UL or CSA, because the system was not tested by the testing organization with the new replacement refrigerant.

19.25 CHARACTERISTICS AND APPLICATIONS FOR REFRIGERANTS

The following information provides descriptions and tables for the properties of common refrigerants. The refrigerants are arranged by common application.

Refrigerants Primarily Used in Air Conditioning

R-22: monochlorodifluoromethane

Chemical family	HCFC Formulation Compound $CHCl_2F$
Safety rating	A1
Oil compatibility	Mineral oil, alkyl benzene, polyol ester
ODP	0.034
GWP	1,700

R-22 has been the refrigerant used in nearly all air-conditioning systems manufactured for several decades, from window units up to 25-ton commercial direct-expansion systems. It was used extensively in commercial refrigeration after the phaseout of CFCs, but low-temperature systems using R-22 often experience high compressor temperatures. It has a low ozone-depletion potential and a moderate global-warming potential. Because R-22 is an HCFC, it cannot be used in new equipment since January 1, 2010. It will no longer be manufactured or imported beginning January 1, 2020. The price of R-22 has risen dramatically in recent years as its production is being phased out.

R-407C: 23% pentafluoroethane, 25% difluoromethane, 52% tetrafluoroethane

Chemical family	HFC
Formulation	Zeotrope 23% R-125, 25% R-32, 52% R-134a
Safety rating	A1/A1
Oil compatibility	Polyol ester, poly alkylene glycol
ODP	0
GWP	1,700

HFC-407C is a high-glide zeotrope with a glide around 10°F. It is intended as a long-term replacement for HCFC-22. Its operating pressures are very close to those of HCFC-22. Its operating efficiency is less than HCFC-22. HFC-407C can be used as a retrofit refrigerant in many systems with a change in refrigeration oil. It must leave the cylinder as a liquid because it is zeotropic.

R-410A: 50% pentafluoroethane, 50% difluoromethane

Chemical family	HFC
Formulation	Zeotrope 50% R-125, 50% R-32
Safety rating	A1/A1
Oil compatibility	Polyol ester, poly alkylene glycol
ODP	0
GWP	2,000

HFC-410A is a low-glide zeotrope with a glide of 1°F or less. It has no ozone-depletion and a moderate global-warming potential. HFC-410A is the refrigerant that is replacing HCFC-22 in new air-conditioning applications. However, its pressure is 40–60 percent higher than HCFC-22, so it cannot be used to replace HCFC-22 in existing systems. The operating efficiency of HFC-410A tends to be higher than similar HCFC-22 systems. HFC-410A requires service tools designed for it because of its high pressures. Since it is zeotropic, HFC-410A must leave the cylinder as a liquid.

SAFETY TIP

In the air-conditioning/refrigeration industry there are no distinctions between the service valve connections for the different refrigerants used. This differs from the automotive industry, where different refrigerants have different-sized access ports to prevent the accidental use of an improper gauge set on an unapproved refrigerant. In air conditioning it is strictly up to the technician to make certain before attaching the gauge set to a system that it is the proper gauge set. This is significant when working around systems that may contain R-410a, because it has significantly higher pressure than most refrigerants used in the industry. Although the gauges will not explode, they can be overstressed to the point that they will no longer work properly.

Refrigerants Primarily Used in Commercial Refrigeration

R-404A: 44% pentafluoroethane, 52% trifluoroethane, 4% tetrafluoroethane

Chemical family	HFC
Formulation	Zeotrope 44% R-125, 52% R-143a, 4% R-134a
Safety rating	A1/A1
Oil compatibility	Polyol ester, poly alkylene glycol
ODP	0
GWP	3,800

HFC-404A is a long-term replacement for the CFC-502 in new systems. HFC-404A is a low-glide zeotrope with a glide of 2°F or less. It has no ozone-depletion and a moderately high global-warming potential. It is primarily used in medium- and low-temperature systems, including commercial freezers, ice machines, and transportation refrigeration. HFC-404A is becoming the refrigerant of choice in low-temperature commercial refrigeration applications. Since it is zeotropic, it must be removed from the cylinder as a liquid.

R-507: 50% pentafluoroethane, 50% trifluoroethane

Chemical family	HFC
Formulation	Azeotrope 50% R-125, 50% R-143a
Safety rating	A1
Oil compatibility	Polyol ester, poly alkylene glycol
ODP	0
GWP	3,900

HFC-507 is a true azeotrope designed to replace CFC-502 in new commercial refrigeration applications. As an azeotrope, it may be removed from the cylinder as either a gas or a liquid.

Refrigerants Primarily Used in Domestic Refrigeration

R-12: dichlorodifluoromethane

Chemical family	CFC
Formulation	Compound CCl_2F_2
Safety rating	A1
Oil compatibility	Mineral oil, alkyl benzene, polyol ester
ODP	0.82
GWP	10,600

CFC-12 was widely used for years in domestic refrigerators, automobiles, medium-temperature commercial refrigeration, and ice machines. It has both a high ozone-depletion potential and a high global-warming potential. It has been illegal to manufacture or import CFC-12 since January 1, 1996. Because of its previous widespread use, some systems are still in operation that use CFC-12.

R-134a: 1,1,1,2-tetrafluoroethane

Chemical family	HFC
Formulation	Compound CH_2FCF_3
Safety rating	A1
Oil compatibility	Polyol ester, poly alkylene glycol
ODP	0
GWP	1,300

This is a chlorine-free fluorinated refrigerant designed to replace R-12 in new appliances. It is generally not considered a retrofit refrigerant. It has an ODP of 0.0 and a GWP of 1,300. It is suitable for use in domestic refrigerators, automotive air conditioning, and medium- and high-temperature commercial applications. R 134a is also now widely used in high-pressure industrial chillers. It is a suitable replacement for R-12 wherever the evaporation temperature is −10°F or higher.

Refrigerants Primarily Used in Large Industrial Systems

R-11: trichlorofluoromethane

Chemical family	CFC
Formulation	Compound CCl_3F
Safety rating	A1
Oil compatibility	Mineral oil, alkyl benzene, polyol ester
ODP	1
GWP	4600

CFC-11 was widely used in low-pressure centrifugal chillers. It has a high ozone-depletion potential and a moderately high global-warming potential. It has been illegal to manufacture or import CFC-11 since January 1, 1996. Because of its previous widespread use and the high cost of large centrifugal chillers, some systems are still in operation that use CFC-11.

R-123: 2,2-dichloro-1,1,1-trifluoroethane

Chemical family	HCFC
Formulation	Compound $CHCl_2CF_3$
Safety rating	B1
Oil compatibility	Mineral oil, alkyl benzene, polyol ester
ODP	0.012
GWP	120

This refrigerant was designed to replace R-11 in low-pressure chillers. The pressure-temperature curves show close performance characteristics between these two refrigerants. R-123 has a safety group classification of B1, making it objectionable from the standpoint of toxicity. As a result, some service companies refuse to use it. As yet, no new universally acceptable refrigerant has been developed to replace R-11. R-123 has a small ODP and a small GWP. Because R-123 is an HCFC, it will no longer be used in new equipment starting in January 2020. It will no longer be manufactured or imported beginning January 1, 2030.

R-717: ammonia

Chemical family	NH_3
Formulation	Compound
Safety rating	B2
Oil compatibility	Mineral oil
ODP	0
GWP	0

Ammonia is one of the oldest refrigerants in use today. It has a much higher latent heat of evaporation than the other common refrigerants. This means that smaller piping can be used. It is corrosive to copper but not to iron, steel, or aluminum. It therefore requires an all-iron, steel, or aluminum system, including the compressor, condenser, evaporator, controls, and piping. It is not destructive to the ozone layer. Its greatest defect is its toxicity and flammability. It has a safety rating of B2.

Due to ammonia's safety hazards, it is not found in appliances or normal comfort cooling applications. Typically it is used in large commercial or industrial applications where its operating efficiencies (lower horsepower per ton) are important and where plant engineers are available to operate the system. Examples would be dairies, ice cream plants, and large cold-storage facilities.

Due to its toxicity, an operator must take special precautions to limit the quantity inhaled. Leaks are detected by use of litmus paper, which changes color in the presence of ammonia, or a sulfur candle, which creates smoke in contact with ammonia. Technicians must also take care to avoid ammonia contact with the skin.

SERVICE TIP

In addition to the gloves and safety glasses that technicians should use with all refrigerants, there must be a self-contained breathing apparatus available for each worker in the area when working around systems that contain ammonia refrigerant. It is not possible to simply hold your breath and run from an area if an ammonia leak occurs. Part of the problem is the reaction the lungs have when a small quantity of ammonia is inhaled. Even a small amount of concentrated ammonia can damage lungs, and exposure to concentrated ammonia can temporarily blind you. The only way to prevent these problems is to have the appropriate equipment ready and available for emergency use.

UNIT 19—SUMMARY

Refrigerant is the fluid used in a refrigeration system to transfer heat. It does this by absorbing heat while evaporating at a low pressure and temperature and condensing at a high pressure and temperature. Refrigerants may be classified by many characteristics, including the following:

Operating Pressure
- Very high pressure
- High pressure
- Low pressure

Chemical Composition
- CFCs
- HCFCs
- HFCs

Formulation
- Compounds
- Zeotropes
- Azeotropes

Toxicity
- A: lower toxicity
- B: higher toxicity

Flammability
- 1: nonflammable
- 2: somewhat flammable
- 3: highly flammable

Ozone-Depletion Potential
- 0 ODP: no ozone depletion
- 1 ODP: high ozone depletion (CFC-11)

Global-Warming Potential
- 0 GWP: no global-warming potential
- 4,000 GWP or greater: high global-warming potential

The majority of modern halogenated refrigerants are built around the hydrocarbon molecules methane and ethane. Halogens replace some or all of the hydrogen atoms in the hydrocarbon molecules. CFCs and HCFCs have ozone-depletion potential because they contain chlorine. HFCs have no ozone-depletion potential because they contain no chlorine.

Refrigerant is available in disposable and refillable cylinders. Disposable cylinders should not be refilled by anyone with anything. Refillable cylinders are returned and refilled by the refrigerant manufacturer but should not be refilled by field technicians. Recovery cylinders are used in the field to recover refrigerant. They should not be refilled more than 80 percent full by volume because liquid expansion can cause hydrostatic pressures that can rupture the cylinders if they are filled completely with liquid.

UNIT 19—REVIEW QUESTIONS

1. Define *refrigerant*.
2. What organization has established the number designation for refrigerants?
3. What are the chemical components of CFC refrigerant?
4. What are the chemical components of HCFC refrigerant?
5. What are the chemical components of HFC refrigerant?
6. What is ozone?
7. Which chemical family of refrigerants has the highest ozone-depletion potential?
8. What is the primary safety concern when working with R-410A?
9. Define the term *glide* when referring to zeotropic refrigerants.
10. Explain the ASHRAE refrigerant safety designation regarding refrigerant toxicity and flammability.
11. What is the difference between a disposable refrigerant cylinder and a refillable refrigerant cylinder?
12. What is the difference between a compound and a zeotrope?
13. What is the difference between a zeotrope and an azeotrope?
14. What contaminants can be found in refrigerant?
15. State two reasons ammonia has limited use.
16. What refrigerant is being used in small appliances in place of R-12?
17. What refrigerant is being used in new air-conditioning systems in place of R-22?
18. When working with ammonia, what safety equipment must be available?
19. How can the presence of a noncondensable material in a refrigerant cylinder be detected?
20. Describe the difference between latent heat and sensible heat, and explain why these concepts are important for a refrigerant.
21. What is the difference between a low-pressure refrigerant and a high-pressure refrigerant?
22. Which type of refrigerants are listed as Class I ozone-depleting substances by the EPA?
23. What chemical family is the basic building block for most modern refrigerants?
24. What is a halogenated refrigerant?
25. What effects does water have when introduced in a system with refrigerant?
26. What differences are there between handling zeotropes and azeotropes?
27. Explain the difference between a bubble point and a dew point.
28. What is fractionation, and how does it affect the way zeotropic refrigerants are handled?
29. What is the difference between a high-glide blend and a low-glide blend?
30. Why do CFCs deplete the ozone layer?
31. What is the ozone layer?
32. What benefit is the ozone layer to us?
33. Why are HCFCs not as harmful to the atmosphere as CFCs?
34. What is the difference between ozone depletion and global warming?
35. What gas is used as the standard for comparison when assigning a refrigerant's global-warming potential?
36. What has been used for years as a refrigeration lubricant with CFCs and HCFCs?
37. What lubricants can be used with HFCs?
38. Why should a refrigeration lubricant be miscible with the system refrigerant?
39. What is a hygroscopic lubricant?

UNIT 20

Special Refrigeration Components

OBJECTIVES

After completing this unit, you will be able to:

1. Identify each special refrigeration system component.
2. explain the purpose of each special refrigeration system component.
3. locate special refrigeration components on a refrigeration system.
4. discuss the need for filter driers in a refrigeration system.

20.1 INTRODUCTION

There are a variety of refrigeration components that can be added to a basic refrigeration system to improve the operation of the system, increase system efficiency, enhance serviceability, provide system protection, or ensure safe operation. Service and installation technicians need to understand the operation and function of these special components when working on a refrigeration system. Some are necessary for the proper operation of the system. In this unit we discuss the special refrigeration system components and their location in the refrigeration cycle, beginning at the compressor. Figure 20-1 shows the location of many common types of special refrigeration components in the refrigeration system.

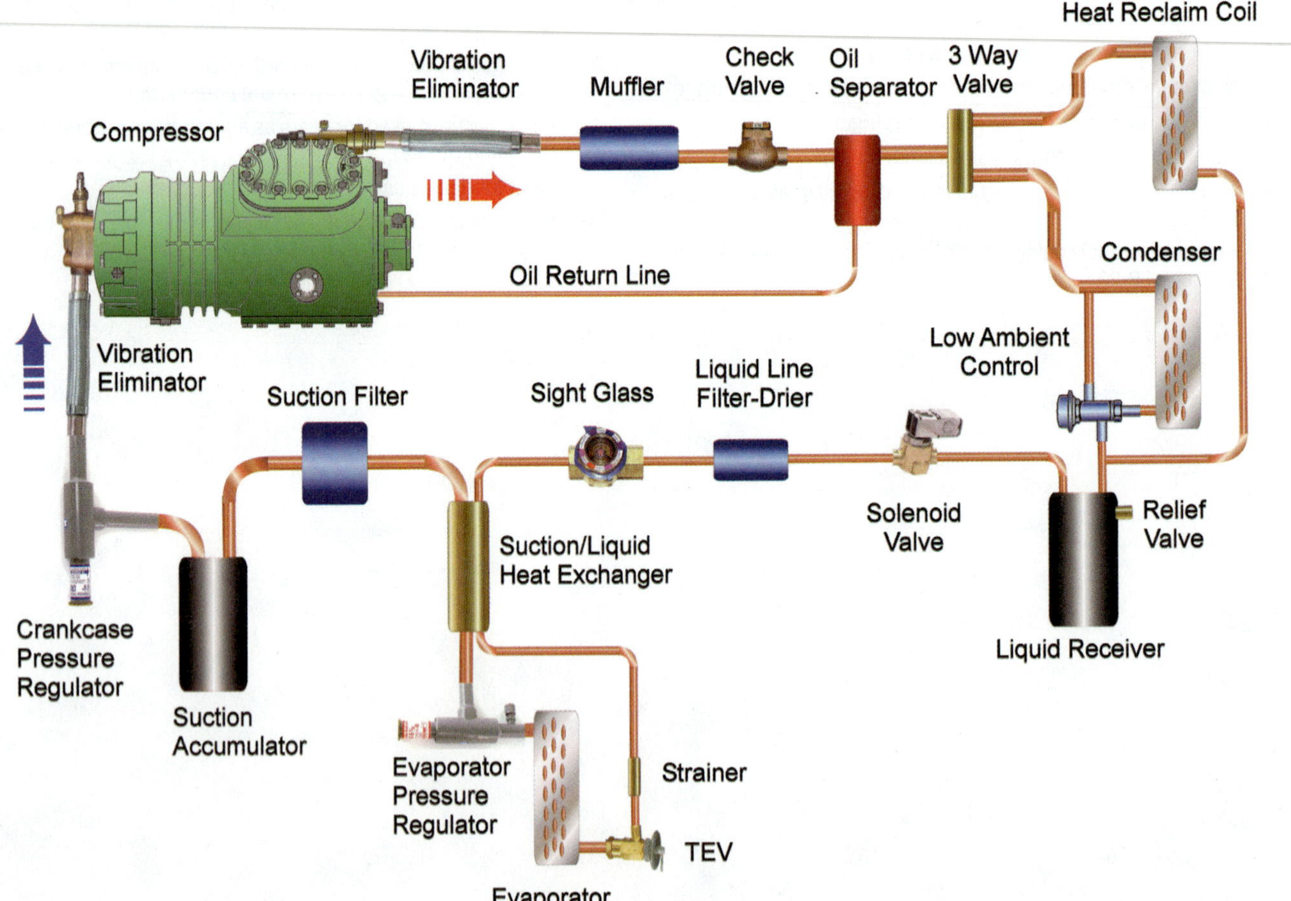

Figure 20-1 The location of common types of special refrigeration components in the refrigeration system.

SAFETY TIP

Refrigeration equipment contains various safety devices to protect equipment and people. Never bypass a safety device because a dangerous situation results that can lead to personal injury or equipment damage. Notify the system owner if a safety device is bypassed, and make the safety device operational.

20.2 VIBRATION ELIMINATORS

As a compressor operates, it produces some vibrations in the suction line and discharge line. On larger systems the vibration could cause abnormal stress on the lines, which could lead to refrigerant leaks on those lines. Vibration eliminators are used to absorb some of the vibration created by the compressor. They can be located at either the inlet or the outlet of the compressors (see Figure 20-1). Typically they are made of flexible bronze tubing that expands and contracts like an accordion. Braided bronze strips with copper absorb the vibrations and prevent overstressing the copper lines leading to and from the compressor. Figure 20-2 shows a vibration eliminator.

20.3 HOT GAS LINE MUFFLERS

Mufflers are used on systems to reduce the noise that can be produced by the gas pulsations of the refrigerant being compressed and pumped by a compressor. Mufflers are normally installed on the discharge line of the compressor (see Figure 20-1). A typical discharge-line muffler is shown in Figure 20-3 . They are made of a brazed or welded cylinder with baffle plates mounted on the inside. Compressor manufacturers can also install a muffler inside the compressor. Most hermetic reciprocating compressors have discharge mufflers inside their shell. When a muffler is added to a system, it will commonly be placed in the discharge line as close to the compressor as possible.

Figure 20-2 Suction-line vibration eliminator

Figure 20-3 Hot gas line muffler.

20.4 CHECK VALVES

Check valves are used on some larger refrigeration systems to ensure that the refrigerant flows in one direction only. They are located on the discharge line, usually near the compressor (see Figure 20-1). Figure 20-4 shows a refrigeration check valve used to prevent liquid refrigerant from flowing back to the compressor from the condenser during the off cycle of the system. Liquid refrigerants returning to the compressor during the off cycle can seriously harm the compressor on startup. These valves are also used on systems with two evaporators operating at different temperatures and controlled by a single condensing unit. A check valve is placed in the suction line of the lower-temperature evaporator to prevent the suction vapor from the higher-temperature evaporator from entering the lower-temperature evaporator.

20.5 OIL SEPARATORS

Reciprocating and rotary compressors will discharge some oil with the refrigerant as it leaves the compressor. This oil will then travel with the refrigerant throughout the system

Figure 20-4 Refrigerant check valve.

Figure 20-5 Oil separators used to remove oil from the discharge gas. *(Courtesy Danfoss Inc.)*

and eventually return to the compressor's crankcase. For the oil to travel with the refrigerant throughout the system, good piping practices must be employed. Poor piping practices can lead to oil being trapped out in the system, preventing it from returning to the compressor. If enough oil is trapped out in the system, the compressor could become starved for oil and bearing damage could occur.

Oil separators are used to separate and collect some of the oil in the discharge line as it leaves the compressor and return it to the compressor's crankcase. They are located in the discharge line (see Figure 20-1). Figure 20-5 shows an oil separator. The refrigerant/oil mixture discharged from the compressor enters the oil separator; the velocity of this mixture is slowed down from the use of internal baffles and impingement screens. This slowing down of the refrigerant/oil mixture causes a major portion of the oil to drop out of the mixture and fall to the bottom of the oil separator.

At the bottom of the oil separator is a float assembly connected directly back to the crankcase of the compressor. As the oil level at the bottom of the oil separator increases, the float will cause a valve to open, and, due to the pressure difference between the oil separator and the compressor's crankcase, some of the oil will be returned to the compressor's crankcase.

Although oil separators can be quite efficient (some will be as high as 98 percent efficient), they do not separate all of the oil from the refrigerant. Some quantity of oil will always travel with the refrigerant throughout the system. Good piping practices must always be followed while installing a system. Even with an oil separator, poor piping practices will eventually lead to oil being trapped out in the system, causing a deficiency of oil in the compressor's crankcase.

Figure 20-6 Typical three-way valve on a system.

Oil separators are often used on supermarket rack systems with multiple compressors connected together to a common manifold. Oil management is necessary when operating several compressors on the same system to ensure that each compressor receives proper lubrication.

20.6 THREE-WAY VALVES

Some larger refrigeration systems will reclaim the heat from the discharge vapor leaving a compressor and use this heat to either assist the heating of the building or assist in the heating of the domestic hot water for the building. The three-way valve directs the flow of the refrigerant to either the condenser or the heat-reclaim coil. Three-way valves used for heat reclaim are located in the discharge line between the compressor and the condenser (see Figure 20-1). Figure 20-6 shows a typical three-way valve on a system.

20.7 HEAT-RECLAIM COILS

When it is possible to reclaim some of the heat from the hot discharge gas, the three-way valve sends it to another heat exchanger, commonly referred to as a heat-reclaim coil, shown in Figure 20-7. This heat exchanger will extract some useful heat from this discharge vapor and then the cooler discharge vapor will be piped to the system's condenser. Heat-reclaim coils used for heating air are located in the air-handling equipment. They are used to provide heat in the winter in supermarkets or for reheating the discharge air during dehumidification.

GREEN TIP

Heat-reclaim coils improve a building's overall efficiency by using waste heat that would normally be thrown away to heat the building or hot water.

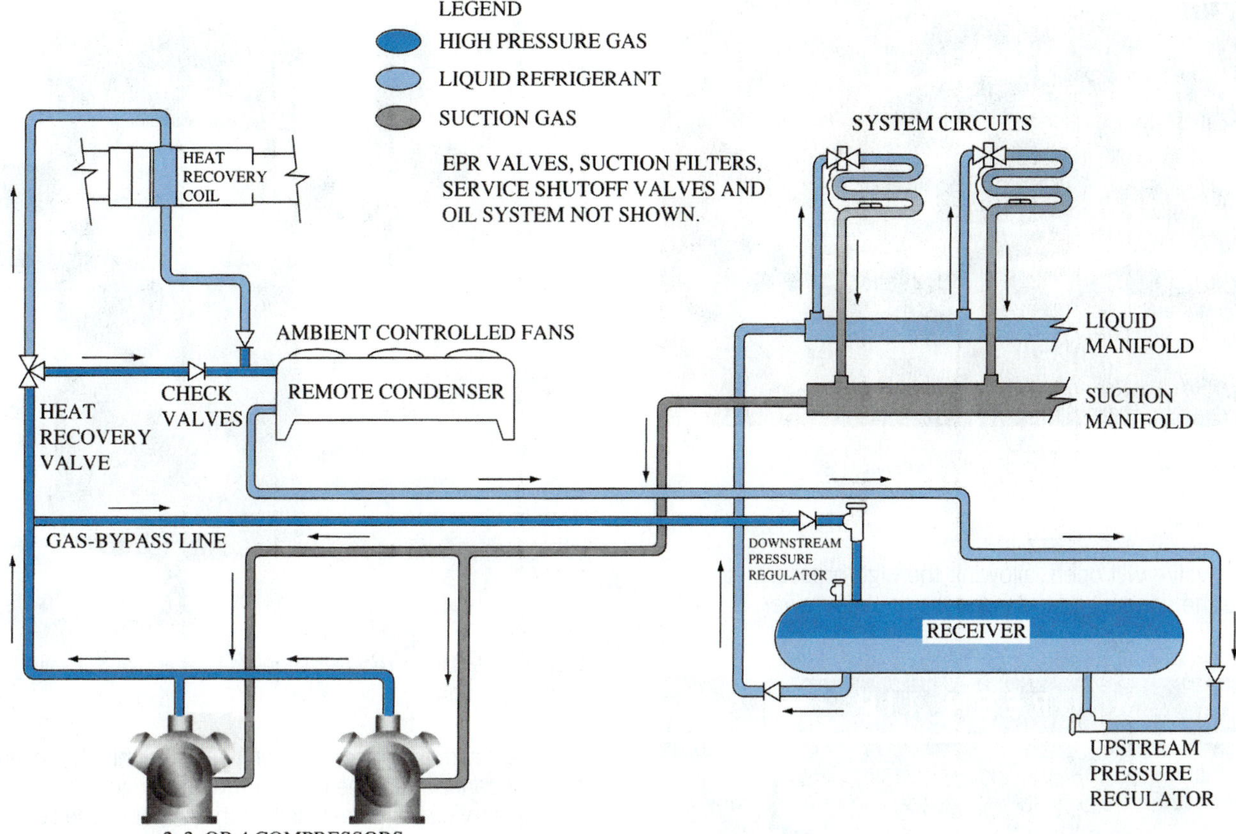

Figure 20-7 Heat-recovery system for supermarket installation.

20.8 LOW AMBIENT CONTROLS

Many large refrigeration systems are designed with outdoor air-cooled condensers. Air-cooled condensers can provide free subcooling and lower condensing pressures during low ambient conditions. However, there is a downside to low ambient operation: if the condensing pressure falls too low, it will cause the metering devices at the refrigerated cases to underfeed. Typically, thermostatic expansion valves (TEV) are used as the metering device on these systems. A standard TEV requires a minimum pressure difference across its port to properly feed refrigerant into the evaporator. If the pressure difference across the TEV drops below a minimum value, the valve will not feed enough refrigerant into the evaporator, starving it. Low ambient controls are designed to maintain a minimum liquid pressure during low ambient conditions.

One method of elevating the system liquid pressure is to control the operation of the condenser fan motors. A system can be designed to either cycle the condenser fan motor(s) on and off as needed, or the speed of the condenser fan motor(s) can be controlled to maintain a minimum condensing pressure. As the condensing pressure falls, the condenser fan motor(s) are either cycled off or slowed down to reduce the amount of air flowing across its coils. This reduces the condenser's ability to reject the heat from the refrigerant, and its pressure increases. Fan cycling is often used for small systems with refrigerant-cooled compressors because of its simplicity.

TECH TIP

Fan cycling should never be used on air-cooled compressors. Air-cooled compressors depend on the airflow from the condenser passing over the compressor body to cool the compressor. Shutting off the condenser fan during cold weather can actually cause the compressor to overheat.

Many commercial refrigeration systems maintain a minimum liquid pressure using condenser flooding controls. They work by controlling the flow of refrigerant through the condenser using an OROA pressure regulating valve (Figure 20-8). OROA valves "open on rise of outlet" pressure. They are preset, so they must be carefully selected to match the refrigerant and the system in which they will be used. This control is installed at the outlet of the condenser, with a line going back to the condenser inlet, as shown in Figure 20-1. When the liquid pressure falls below a preset minimum, the regulator will close down, causing the refrigerant to back up in the condenser, flooding it with liquid refrigerant. This reduces the condensing surface, causing the condensing temperature and pressure to increase. While the flow leaving the condenser is restricted, the OROA valve will bypass a portion of the discharge vapor directly to the receiver to maintain a minimum liquid pressure in the receiver. When the receiver pressure increases above the valve setting, the

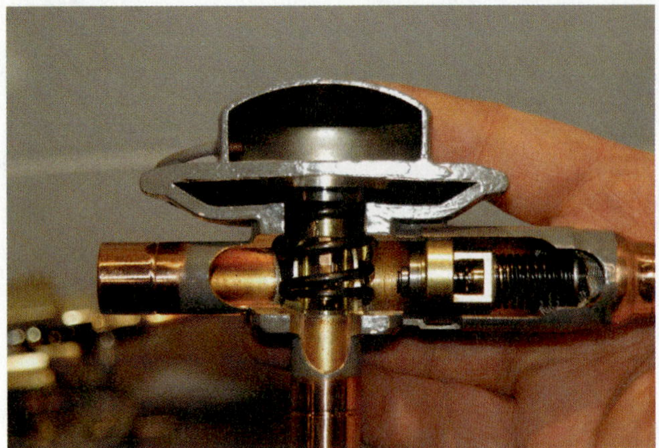

Figure 20-8 Open on rise of outlet pressure head pressure-control valve.

OROA valve will open, allowing the high-pressure liquid to leave the condenser and enter the liquid receiver.

GREEN TIP

Low ambient controls force the compressor to work harder than it normally would need to in cold ambient temperatures. Improved refrigerant metering devices are available that allow the system to operate at much lower pressure drops across the metering device, reducing or eliminating the need for low ambient controls. Since compressors use less energy when operating at lower discharge pressures, using advanced refrigerant metering devices to reduce the need for low ambient control improves system efficiency.

20.9 RECEIVERS

A receiver is a vessel that allows a system to store excess refrigerant that it does not currently need. A typical receiver is shown in Figure 20-9. Liquid receivers are normally used on systems with modulating metering devices, such as TEVs. When the TEV throttles, the receiver provides a place for the excess refrigerant to go. When the TEV opens, the stored refrigerant can be used. Only systems that have the ability to regulate the flow of refrigerant and are exposed to varying heat loads should have a receiver installed. Liquid receivers are not used with fixed restrictions such as orifices or capillary tubes.

Receivers are normally installed in the liquid line of a system after the condenser and before the metering device (see Figure 20-1). The inlet connection of the receiver will allow refrigerant to enter the top of the receiver. The outlet of the receiver will be connected to an internal dip tube connected to the bottom of the receiver. This ensures that only liquid refrigerant leaves the receiver and travels to the system's metering device.

Receivers are typically designed to hold 20 percent more than the entire system's refrigerant charge. This additional

Figure 20-9 Liquid receiver showing inlet and outlet valves and safety relief.

20 percent allows a vapor cushion to exist within the receiver so that it cannot become overfilled with refrigerant.

The service valve installed at the outlet of the receiver is referred to as the king valve. This valve is very useful for a service technician. It allows a technician to read the system's pressure at the receiver as well as trap the system's refrigerant charge in the condenser and receiver. This allows a technician to work on any section of the system from the outlet of the king valve to the inlet of the compressor without having to recover the refrigerant charge.

20.10 RELIEF VALVES

Relief valves are used to release abnormally high pressure inside a vessel before the pressure causes it to erupt. They are normally located on a system's condenser or receiver, but they can be located on any other vessel on a system. There are two basic types of relief valves: automatic relief valves and one-time relief valves.

Automatic relief valves are spring-loaded valves normally encased in a brass body with a neoprene seat. They are designed to automatically reset once the pressure inside the vessel reaches a safe level. They will be located on a section of the vessel where refrigerant vapor is located. This allows only the vapor to be released from the vessel rather than any liquid refrigerant. Some automatic relief valves will also have a thread connection on the top to allow piping to be attached to it so the released refrigerant can be vented out of the building or mechanical room where the vessel is located.

A one-time relief valve, sometimes called a fusible plug, is a valve that cannot be reset once it releases the pressure from a vessel. It is normally constructed from a fitting with a drilled hole filled in with a low-temperature solder. At a specific temperature the solder will soften and pressure within the vessel will cause the solder plug to blow out and release the refrigerant within the vessel. Once the solder plug blows out of the fitting, the relief valve must be replaced.

20.11 SOLENOID VALVES

A solenoid valve is a flow-control device that stops or allows the flow of a refrigerant in a refrigeration system. It can be installed in any of the refrigerant lines. Most commonly, it is installed in the liquid line to stop or allow the flow of liquid refrigerant into the metering device and the evaporator (see Figure 20-1).

Solenoid valves are available as normally open or normally closed. When voltage is applied to the coil of a normally closed valve, it picks up the plunger and allows the flow of refrigerant through it. When voltage is no longer applied to the coil, the plunger will close due to an internal spring and system pressures. When voltage is applied to the coil of a normally open valve, it will close and stop the flow of refrigerant through it. When voltage is no longer applied to the coil, the valve will again open and allow refrigerant to flow through it.

The direction in which a solenoid valve is installed is important to its proper operation. Normally there will be an arrow stamped on the body of the valve to indicate the direction of refrigerant flow (Figure 20-10). If installed incorrectly, the solenoid will not properly shut down and stop the flow of refrigerant when needed.

A typical solenoid valve consists of two parts: the valve body and its electrical coil. Figure 20-11 shows a picture of a solenoid valve. When purchasing a solenoid valve, both parts are normally ordered separately, allowing a technician to use the same valve body with different coil voltages.

Figure 20-10 The solenoid must be installed with the arrow pointing in the direction of refrigerant flow.

Figure 20-11 Solenoid valve.

20.12 LIQUID-LINE FILTER DRIERS

The filter drier's job is to keep the system clean and dry. Filter driers are designed to remove three types of contaminants: particles, moisture, and acid. Filter driers normally incorporate a screen to filter out any debris in the system that could potentially clog the small orifice in the metering device. Particle debris found in refrigeration systems includes dirt, oxide, carbon, flux, and metallic particles. Other contaminants would include sludge and wax formed by oil breakdown. This debris can damage cylinder walls and bearings and plug metering devices and screens.

Filter driers also have desiccant inside that absorbs water vapor and acid in the refrigerant. Water vapor is very harmful to a refrigeration system, and even a very small amount of moisture can cause serious system problems. Excessive amounts of water vapor can cause two problems: (1) it can freeze at the metering device, causing a restriction, or (2) it can react with heat and oil inside the compressor to produce sludge, wax, and acids. Acid formation starts in systems with POE (polyol ester) lubricant with water concentrations as low as 75 ppm (parts per million).

Filter driers must also be able to remove acids from the system. Chemical interactions between the refrigerant, the lubricants, and moisture can form acid in the system. This is particularly true for systems that use POE lubricant. POE is made by combining alcohol and acid to form polyol ester and water. Adding water and heat can reverse the process, forming alcohol and acid inside the system. This is known as hydrolysis.

Desiccants are the materials used to remove water from the refrigerant. The two desiccants most often used in filter driers are molecular sieve and activated alumina. Molecular sieve is superior at moisture removal but is less effective on acids. Activated alumina is effective on both organic and inorganic acids but is less effective at moisture removal. Activated charcoal is often added to help remove hydrocarbon residues and wax. The mix of desiccant materials in the drier depends upon its intended use. A drier that is used on new equipment to help with dehydration will have more molecular sieve, while a drier that is used for cleaning a system after a compressor burnout may have more activated alumina.

Two methods are used to hold the desiccant materials in the driers. One method mixes different types of desiccant material and forms it into a solid molded core (Figure 20-12). The other uses individual desiccant beads that are compressed together inside the shell (Figure 20-13). Standard filter driers have a definite inlet and outlet with arrows indicating the direction of refrigerant flow (Figure 20-14). Piping the drier in backwards can decrease its ability to filter out particles. Driers intended for use with heat-pump systems are designed to allow flow in either direction. These are called bi-flow driers. Their ability to allow flow in either direction is normally indicated with arrows pointing in both directions (Figure 20-15).

Figure 20-12 Solid-core filter drier.

Figure 20-13 Cutaway view of a beaded-core filter drier.

Figure 20-14 This filter must be installed with the refrigerant flowing in the direction of the arrow.

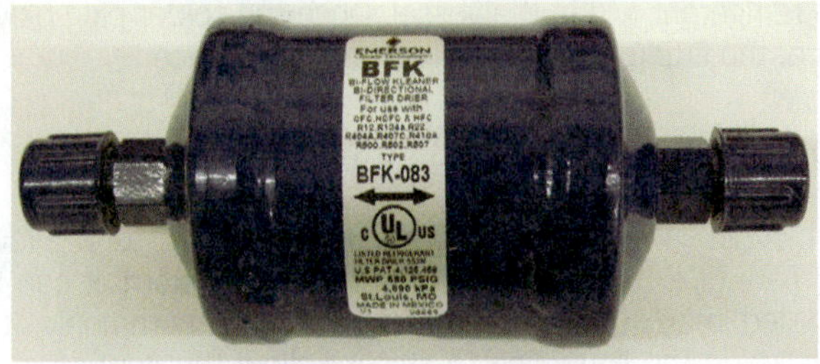

Figure 20-15 The arrows pointing in either direction indicate that this is a bi-flow drier.

SERVICE TIP

One advantage of driers with flare connections is that they are easier to replace.

Small driers for residential systems are one piece inside a welded steel shell (Figure 20-16). The entire drier shell and its desiccant must be removed and discarded when the drier is changed. These driers are available with either brazed or flared connections (Figure 20-17). Replaceable core driers

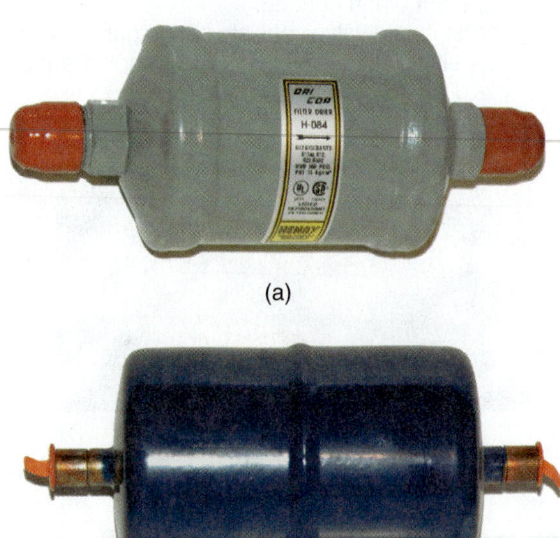

(a)

(b)

Figure 20-16 Small welded-shell filter driers.

(a)

(b)

Figure 20-17 Filter driers are available with flare ends or sweat ends.

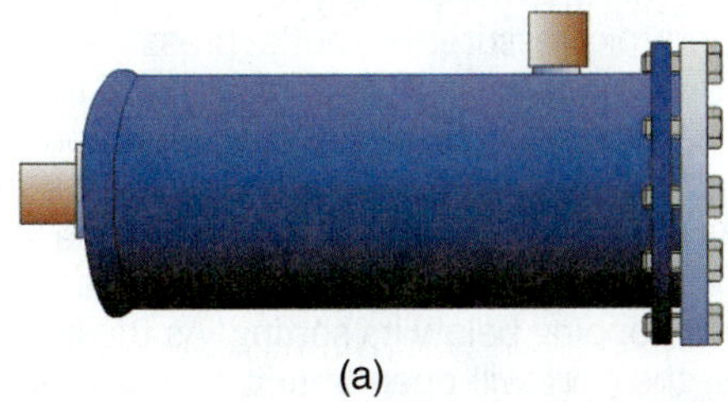

(a)

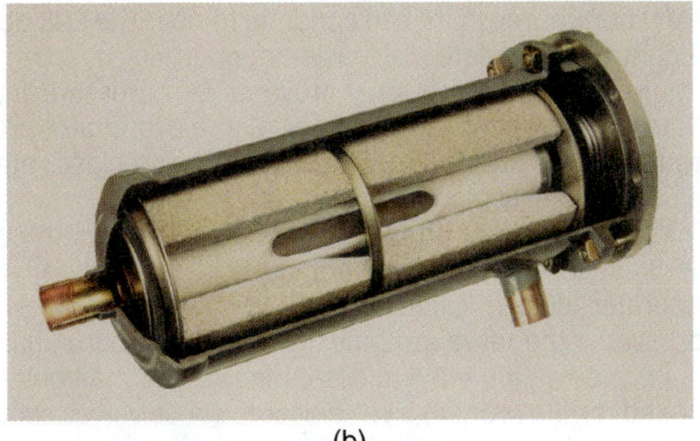

(b)

Figure 20-18 Replaceable-core filter drier.

are available for larger systems. For these, the shell remains in place and the drier cores are replaced (Figure 20-18).

SAFETY TIP

Always make sure that the filter drier you are installing is rated for the refrigerant the system uses. For example, HCFC 410A systems operate at substantially higher pressures than HCFC 22 systems. Using a drier intended only for HCFC 22 on an HFC 410a system could result in the drier shell rupturing.

A filter drier should be installed with all new refrigeration systems that use POE lubricants, because POE is such a hygroscopic material. Even a deep evacuation will not remove moisture from POE. The filter drier should be replaced anytime a refrigeration system is opened for service, regardless of refrigerant type. The pressure drop across a filter will increase when the filter is clogged. Normally, a filter drier with a pressure drop exceeding 2 psig should be replaced. Driers should not be removed by debrazing. Heating the drier with a torch will cause the desiccant to release some of the contaminants back into the system. Instead, brazed-in driers should be cut out to ensure that the contaminants remain in the drier.

SAFETY TIP

Always make sure the refrigerant has been recovered or the system pumped down before removing a liquid-line filter drier.

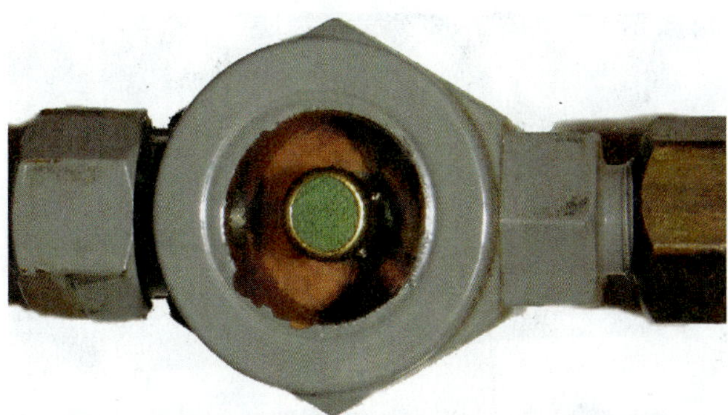

Figure 20-19 Liquid-line sight glass moisture indicator.

20.13 SIGHT GLASSES/MOISTURE INDICATORS

A sight glass can be added to the liquid line to allow a technician to view the condition of the refrigerant in the liquid line at the point it is located. A typical liquid-line sight glass is shown in Figure 20-19. On a properly operating system, the liquid line should contain only liquid refrigerant. A sight glass will show this condition, as only clear liquid will be seen flowing through the sight glass. If vapor exists along with the liquid refrigerant, it will be seen flowing through the sight glass along with the liquid refrigerant.

An indicator can also be added to a sight glass to help indicate the moisture content of the refrigerant. The indicator is a porous filter paper impregnated with chemical salt that is sensitive to moisture. A technician can view the indicator in the sight glass and the color of the paper will show the moisture content of the refrigerant. A common color-coding system is green for a dry system and yellow for a wet system. Always check with the sight glass manufacturer to verify their color-coding system.

20.14 STRAINERS

Strainers are similar to filter driers except there is no desiccant inside to absorb the moisture in the system. Strainers are simply filters used to prevent solid particles from entering metering devices and causing a restriction. They are normally located just ahead of the metering device (see Figure 20-1). Capillary tube systems often have strainers between the liquid line and the capillary tube (Figure 20-20).

20.15 EVAPORATOR PRESSURE REGULATORS (EPR)

An evaporator pressure regulator (EPR), such as that shown in Figure 20-21, is a refrigerant flow-control device that prevents the refrigerant pressure in the evaporator from operating below a minimum value. It is normally installed in the suction line just after the outlet of the evaporator (see Figure 20-1). EPR valves are normally used on systems with multiple evaporators operating at different

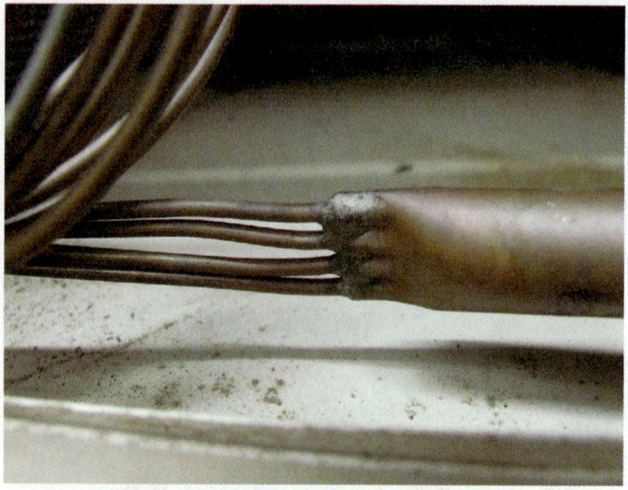

Figure 20-20 Capillary-tube systems often have strainers just before the capillary tubes.

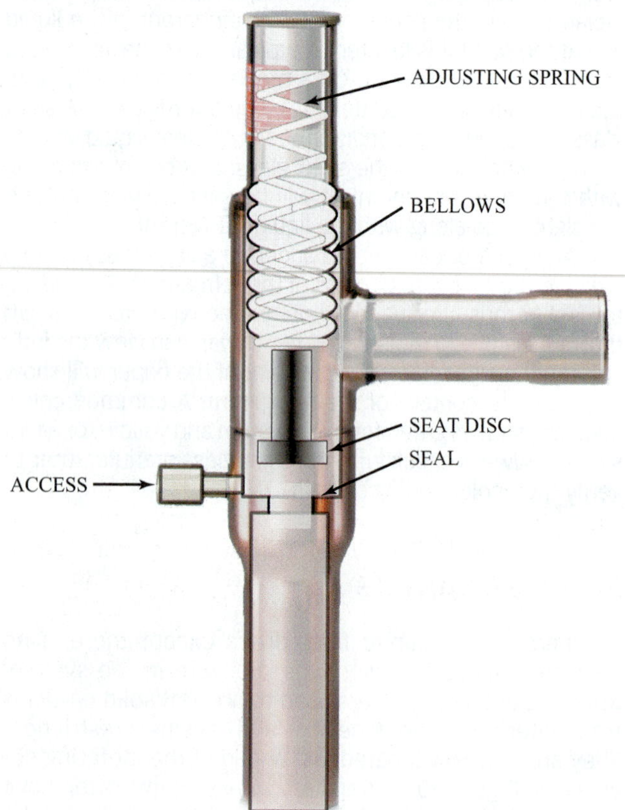

Figure 20-21 Evaporator pressure regulating (EPR) valve.

evaporating pressures. Each of the evaporators operating at a pressure higher than the lowest on the system will have an EPR installed at its outlet. The EPR will prevent these evaporators from operating at the lower operating pressure. An EPR can also be used as a safety, such as on a water chiller, to prevent the evaporator pressure (temperature) from dropping too low and freezing the water in the chiller barrel.

An EPR is an "open on rise of inlet" pressure regulator. The two forces that regulate the port opening on this type of valve are an internal spring pressure and the regulator inlet evaporator pressure. The outlet pressure of the regulator has no factor in the opening or closing of this regulator. The internal spring pressure is the closing force, and the inlet evaporating pressure is the opening force. As the inlet pressure drops, the internal spring pressure will start to close down the port, preventing the inlet (evaporator pressure) from dropping below its setting. As the inlet pressure increases, the port will open until it reaches its maximum port size. Once its maximum port size has been reached, the evaporator pressure can continue to rise. The EPR does not control or regulate the maximum operating evaporator pressure of a system; it only prevents this pressure from dropping below a minimum setting. Some evaporator pressure regulators are designed to pop open and close rather than continuously modulating.

The selection of an EPR is based on the system conditions, such as the refrigerant used in the system, the evaporator design capacity, and the available pressure drop across the valve at its design load. Check the manufacturer's selection table to determine what conditions are required for the proper valve selection. These valves should not be selected based simply upon on the suction line size, since a manufacturer may use the same line size for many different capacities.

Adjusting an Evaporator Pressure Regulator

To adjust the control point of an EPR, first attach a pressure gauge on the inlet side of the regulator. Turn its adjustment screw clockwise to increase the valve's control point and counterclockwise to decrease its control point. It may take up to thirty minutes for the new balance to take place after an adjustment is made. If the valve is being adjusted to a lower setting, an immediate response to an adjustment should be noticed. Adjustments to an EPR should be made when the evaporator is under a minimum load condition.

20.16 SUCTION-LINE FILTERS

Suction-line filters are installed in the suction line before the compressor (see Figure 20-1) and are used to protect the compressor from contaminants in the refrigeration system. Suction-line filters are often added after a compressor burnout to protect the replacement compressor. Suction filters contain the same desiccants that are found in liquid-line filter driers, but often in mixtures that enhance the removal of acids and particulates. Suction filters must have a large cross-sectional area because gas travels through them, not liquid. The larger area helps reduce pressure drop through the filter since filter pressure drop adversely affects the system efficiency. Suction filters are usually removed after the system has been cleaned, or if the filter becomes so loaded with contaminants that there is a significant pressure drop across the filter. Most suction-line filters have Schrader ports that allow technicians to check the pressure drop

Figure 20-22 Suction-line filter drier.

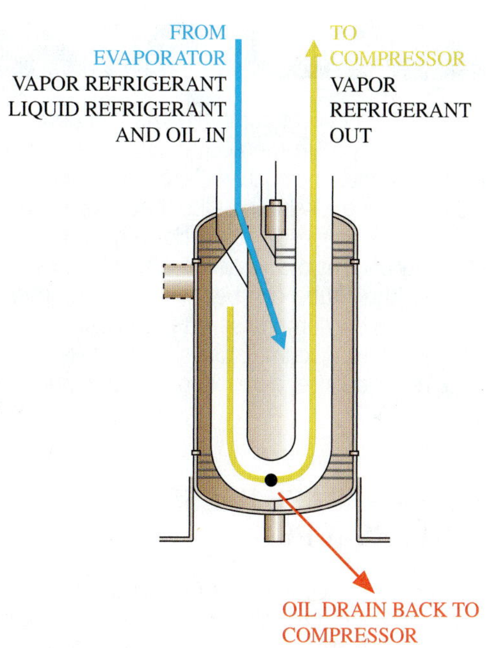

Figure 20-23 Cutaway view of a suction-line accumulator.

across the filter (Figure 20-22). In large systems, replaceable core shells can be used in the suction line.

20.16 SUCTION-LINE ACCUMULATOR

The suction accumulator is located in the suction line just before the compressor (see Figure 20-1). The suction-line accumulator stores refrigerant and helps prevent liquid refrigerant from returning to the compressor. Small quantities of liquid refrigerant returning to a compressor can cause bearing washout. The liquid refrigerant returning to the compressor mixes with the oil and is then used to lubricate the bearings within the compressor. This mixture is a very poor lubricant and will lead to premature bearing failure. Large quantities can cause liquid slugging if liquid gets in the compressor cylinders. When the compressor tries to compress the liquid, pressures increase to thousands of psi and parts break.

While suction accumulators can be applied to most types of systems, they are often used on systems that cannot use liquid receivers but have a need to store refrigerant. Heat pumps are a very common example. They cannot use a liquid receiver because the coil serving as the condenser changes depending upon the cycle. The amount of refrigerant circulating through the system ranges from less than half the charge during cold-weather heating to all the charge in hot-weather cooling. The accumulator gives the extra charge a place to sit without adversely affecting system operation.

The basic design of a suction-line accumulator is a U-shaped tube enclosed in a large-volume vessel. A cutaway view of a suction-line accumulator is shown in Figure 20-23. One end of the U-shaped tube is connected to the suction line leading back to the compressor, and the opposite end is open to the tank. The suction line from the evaporator is then connected to the top of the shell of the tank. If any liquid refrigerant enters the accumulator, it will be exposed to a large volume of the tank, causing it to evaporate and return only refrigerant vapor to the compressor. There are also two small holes drilled into this U-shaped tube. One small hole is drilled at the bottom bend of the U-shaped tube. This allows a small amount of oil and liquid refrigerant, which collects on the bottom of the tank, to be safely metered back to the compressor. Another small hole is drilled at the top end of the U-shaped tube and is referred to as a pressure-equalization orifice. When the system shuts down, it is possible for liquid refrigerant to collect in the bottom of the suction-line accumulator, and because of the small hole drilled in the bottom bend of the U-shaped tube, liquid refrigerant will also collect in this tube. On startup, liquid refrigerant could then be sucked out of the U-shaped tube and return to the compressor. To prevent this from occurring, a small hole is drilled at the top end of the U-shaped tube. The small hole equalizes the pressure on both sides of the liquid in the U-shaped tube and prevents the liquid refrigerant from being sucked out of the tube on startup.

Some suction-line accumulators will have a section of the liquid line run through its bottom section. This warm liquid refrigerant helps to evaporate any liquid refrigerant that collects at the bottom of the accumulator. Another benefit to this design is that the liquid refrigerant in the liquid line will gain additional subcooling, increasing the net refrigeration effect of the system.

SERVICE TIP

A suction-line accumulator should not be used to solve a system problem that is causing liquid refrigerant to return to a compressor. If liquid refrigerant is returning to a compressor, the root cause should be identified and repaired.

SERVICE TIP

Accumulators will collect contaminants and debris. Particles become trapped in the accumulator after any major system contamination, such as a compressor burnout. Accumulators have a J-shaped tube that pulls in gas at the top, dips down into the oil at the bottom, and sucks in a small quantity of oil as the gas leaves the accumulator. The small hole in the J tube can become clogged, interfering with oil return to the compressor. This can lead to a second compressor failure. Accumulators should be removed and flushed or replaced after a major compressor burnout.

20.17 CRANKCASE PRESSURE REGULATORS (CPR)

Crankcase pressure regulators (CPR) are commonly added to many low-temperature refrigeration applications, such as walk-in and reach-in freezers. A typical CPR is shown in Figure 20-24. They prevent the compressor's motor from overloading when the crankcase pressure rises above its designed level. On many low-temperature applications, this can occur during or after a defrost cycle or upon restarting after a normal shutdown period. Compressor overload is likely when starting, especially if the cooler/freezer is at ambient temperature. The CPR is an outlet pressure regulator and will not allow the crankcase pressure to rise above a predetermined level.

The CPR is installed in the suction line just before the compressor (see Figure 20-1). Normally there are no other components installed downstream between the outlet of the CPR and the compressor. This is to ensure that the outlet of the CPR senses the true crankcase pressure of the compressor.

It is not recommended that this type of regulator be used on a system that also uses a maximum operating pressure (MOP)–type of expansion valve. The use of both of these valves on the same system may cause longer pull-down times. This, however, may be overcome if the pressure settings of both valves are sufficiently spread apart. When using a CPR on a system that is also using a discharge bypass valve for capacity control, the pressure setting of the CPR must be higher than that of the bypass valve.

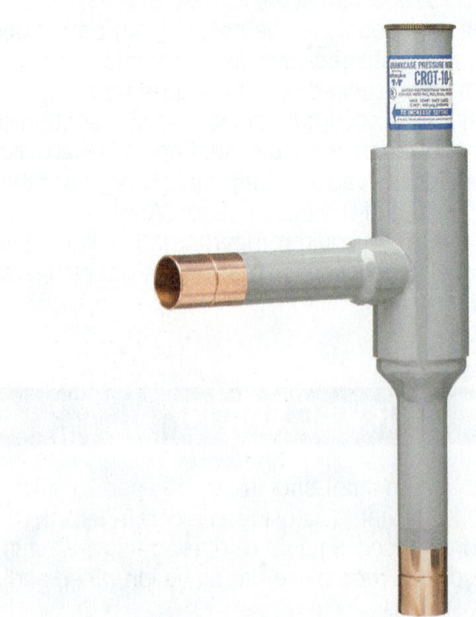

Figure 20-24 Crankcase pressure regulator (CPR).

The typical CPR is a close-on-rise regulator. The outlet pressure (crankcase pressure) of the valve is the closing force within it. The opposing opening force within the valve is an interior spring. These two forces oppose each other to regulate the port size within the valve. The interior spring is adjustable to apply a maximum opening force. If the pressure applied at the outlet of the valve (crankcase pressure) is above the spring force, the valve will close down. The inlet pressure of the CPR has no effect of the operation of the valve. The inlet pressure is applied equally to both the underside of its bellows and the top of its valve seat disc, canceling out the effect of the inlet pressure.

The selection of a CPR is based on five basic system conditions:

1. Refrigerant
2. Refrigeration capacity of the system
3. Design suction pressure of the system
4. Maximum crankcase pressure recommended by the compressor manufacturer
5. Pressure drop across the valve at design load conditions

Once these conditions are known, a valve can be chosen from the manufacturer's selection table. These valves should not be selected based on the line size of suction line, since a manufacturer may use the same line size for many different capacities.

Servicing Crankcase Pressure Regulators

The control setpoint of a CPR will need to be adjusted on the startup of a new system or anytime the valve is replaced. Always refer to the instructions provided by the manufacturer when adjusting these valves. If the instructions are not available, the following guidelines can be helpful:

- Allow the system to be off long enough to allow the pressures to stabilize and the evaporator pressure to reach a high level—one that would simulate the pressure during or after a defrost period or during the initial startup of the case.
- Turn the adjustment screw on the valve all the way out (normally this is done by turning the screw counterclockwise) so that the valve is set to an initial low setting.
- Start the system and observe both the crankcase pressure and the amperage draw of the compressor. Slowly turn the adjustment screw in (normally this is done by turning the screw clockwise) until the amperage draw reaches the maximum allowed by the compressor manufacturer.

This should allow a technician to achieve an acceptable setpoint for the valve. Some systems may have two CPRs that are piped in parallel. When adjusting these valves, both will need to be adjusted at the same time and at the same

Figure 20-25 A suction-to-liquid heat exchanger transfers heat from the warm liquid passing through the small, coiled tubing to the cool suction gas passing through the shell.

rate. This will ensure that the load is divided equally across both valves.

20.18 SUCTION-TO-LIQUID HEAT EXCHANGER

A suction-to-liquid line heat exchanger can be added to improve system efficiency and help prevent liquid from returning to the compressor. Normally, the low-pressure suction line and the high-pressure liquid line are piped next to each other as they travel between the condensing unit outside and the evaporator inside. The suction-to-liquid heat exchanger is installed in both the suction line and the liquid line (see Figure 20-1). The heat exchanger is a shell with a coil of tubing inside (Figure 20-25). The suction gas passes through the shell of the heat exchanger and the liquid passes through the tubing coiled inside the shell. Heat is transferred between the cool vapor in the suction line returning to the compressor and the warm liquid refrigerant passing through the tubing on its way to the metering device. This warms the cool suction gas and cools the warm liquid. The effect is to increase both suction gas superheat and liquid subcooling. The increased subcooling adds system efficiency by preventing flash gas in the liquid line and reducing the amount of flash gas formed in the metering device.

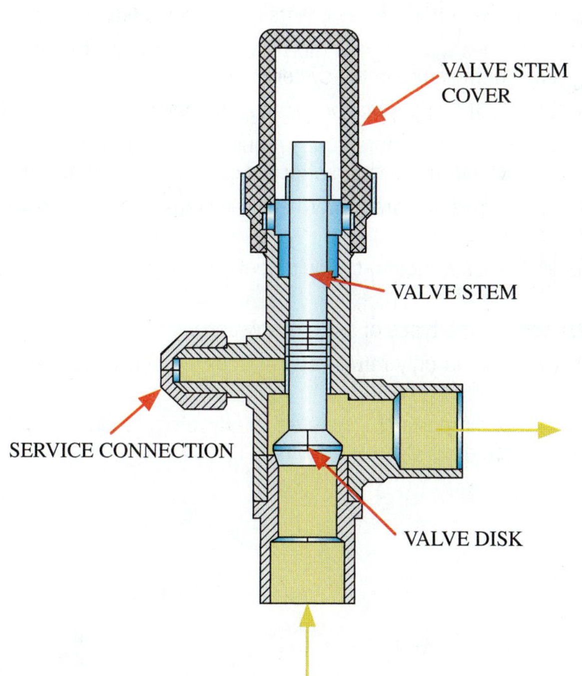

Figure 20-26 Refrigeration service valve.

20.19 ISOLATION VALVES

Isolation valves are used to isolate system components when servicing a system. Isolation valves can be stem-type refrigeration service valves (Figure 20-26), diaphragm valves (Figure 20-27), or ball valves (Figure 20-28). Ball valves are often preferred over other valves because they cause less restriction to the flow of refrigerant. It is advantageous to service technicians to be able to isolate a component when servicing systems. Once a component is isolated from the rest of the system, only the refrigerant charge for that component needs to be removed to service that component, not the entire system charge. When working on systems with a large refrigerant charge, having isolation valves installed can save much time for a technician.

SAFETY TIP

Only valves specifically designed for refrigeration equipment should be used for isolation valves in a refrigeration system! Valves designed for lower pressures, such as water valves, can rupture when exposed to refrigeration pressures. Valves should also be rated to withstand pressures normally generated by the refrigerant in the system. Valves rated for HCFC 22 may not be safe if installed in an HFC 410A system.

Figure 20-27 Refrigeration diaphragm valve.

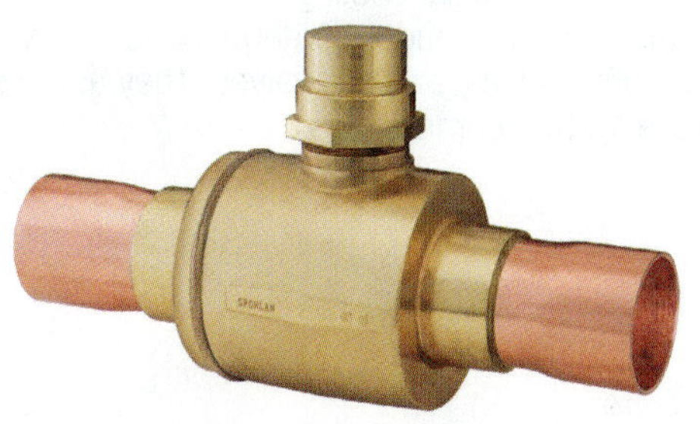

Figure 20-28 Refrigeration ball valve.

UNIT 20—SUMMARY

There are many components that could be added to the basic refrigeration system. Some of these components are designed to improve the operation of a system, while others are safety components that protect the major system components and equipment owner. Special refrigeration components discussed in this unit include the following:

- Vibration eliminators prevent compressor vibration from being transmitted to the lines.
- Mufflers reduce the noise produced by the gas pulsations of the refrigerant being compressed and pumped by the compressor.
- Check valves ensure that the refrigerant flows in one direction only.
- Oil separators remove oil in the discharge gas and return it to the compressor's crankcase.
- Three-way valves direct hot gas to either the condenser or a heat-reclaim coil.
- Heat-reclaim coils take waste heat in the discharge gas and use it to heat air or water.
- Low ambient controls maintain a minimum liquid pressure during low ambient conditions.
- A receiver allows TEV systems to operate properly by storing excess refrigerant.
- Relief valves release excess pressure inside the receiver before the pressure causes it to burst.
- Solenoid valves control the flow of a refrigerant by using an electromagnetic coil to open or close the valve.
- A liquid-line filter drier has a screen inside that filters out debris and desiccant that absorbs water vapor in the system.
- A sight glass allows a technician to view the condition of the refrigerant in the liquid line at the point it is located.
- An evaporator pressure regulator (EPR) prevents the evaporator pressure from dropping below the setpoint of the EPR valve.
- Strainers filter solid particles out of the refrigerant entering the metering devices.
- Suction filters protect replacement compressors by removing contaminants from the suction gas.
- Suction-line accumulators prevent liquid refrigerant from returning to the compressor.
- A crankcase pressure regulator (CPR) prevents compressor overload by keeping the crankcase pressure below the CPR setpoint.
- A suction-to-liquid line heat exchanger improves efficiency, increases the system superheat, and increases system subcooling.
- Isolation valves include refrigeration service valves, diaphragm valves, and ball valves. They are used to isolate system components.

UNIT 20—REVIEW QUESTIONS

1. What is an evaporator pressure regulator?
2. What systems use crankcase pressure regulators?
3. How does a CPR work?
4. What is the purpose of a solenoid valve?
5. How can a technician tell which direction a solenoid valve should be installed?
6. Explain the different methods used to maintain a minimum high-side pressure on outdoor air-cooled condensers.
7. What are the two basic types of relief valves used on refrigeration systems?
8. What is the function of a suction-line accumulator in a refrigeration system?
9. What is the function of an oil separator in a refrigeration system?
10. What is the function of a receiver in a refrigeration system?
11. Why are suction-to-liquid line heat exchangers used on refrigeration systems?
12. What device can be installed on a system to indicate the moisture content of the refrigerant?
13. Where are vibration eliminators installed in a system?
14. What is the purpose of a hot-gas muffler?
15. Why do some systems have discharge check valves?
16. What is the purpose of the three-way valve in a system with a heat-reclaim coil?
17. How can a system with a heat-reclaim coil improve a building's overall efficiency?
18. What is the difference between a liquid receiver and a suction-line accumulator?
19. Explain the operation of a solenoid valve.
20. List the types of contaminants that filter driers must remove.
21. Why are filter driers required when installing a new system that uses POE lubricant?
22. What is a desiccant?
23. What two ways are desiccants held inside a filter drier?
24. How does a sight glass moisture indicator indicate the presence of moisture in the system?
25. What is the difference between a strainer and a filter drier?
26. Why do some manufacturers require that suction-line filters be removed from a system after the system has been cleaned?
27. What types of systems normally use crankcase pressure regulators?
28. How does a suction-to-liquid heat exchanger improve system efficiency?
29. Name three types of system isolation valves.
30. Why should only refrigeration-rated isolation valves be used on a refrigeration system?
31. Why is it important to ensure that any refrigeration component used on an HFC 410A system be rated for HFC 410A use?

UNIT 21

Plotting the Refrigeration Cycle

OBJECTIVES

After completing this unit, you will be able to:

1. explain how refrigerant heat, pressure, and temperature change as refrigerant flows through a refrigeration system.
2. identify the lines and scales on a PH (pressure-enthalpy) diagram.
3. draw a PH diagram of a basic refrigeration cycle.
4. calculate the pounds of refrigerant flowing through a system for a given set of operating conditions and capacity.
5. explain the difference in PH diagrams for compounds and zeotropes.

21.1 INTRODUCTION

Pressure-enthalpy (PH) diagrams are primarily used for designing refrigeration systems. All aspects of system performance can be predicted using a PH diagram. A system's refrigeration capacity, the amount of refrigerant that must be circulated, the amount of heat absorbed by the evaporator, the amount of heat rejected by the condenser, and the required compressor capacity can all be determined using a PH diagram. Technicians can use PH diagrams to troubleshoot systems by comparing a plot of what a system is doing to what it should be doing.

21.2 PRESSURE-ENTHALPY DIAGRAMS

Pressure-enthalpy diagrams are referred to as PH diagrams. Enthalpy is the amount of heat in a substance, represented by the uppercase letter *H*. Specific enthalpy is the amount of heat per pound, represented by the lowercase *h*. PH diagrams show the relationships between a refrigerant's pressure, heat, temperature, volume, and state. The lines and scales used on PH diagrams represent the same thing on every PH diagram; each refrigerant just has different values. So once you understand the R-410A PH diagram shown in Figure 21-1, you can work with any other refrigerant's PH diagram.

21.3 PH DIAGRAM LINES AND SCALES

The key to understanding a PH diagram is learning what the different lines represent, the direction they run, and what units of measurement are used for each line.

Saturation Curve

The large hump in the middle of the PH chart is the saturation curve (Figure 21-2). Every point on or inside the saturation curve is saturated. The critical point is at the very top of the hump. The critical point is the pressure and temperature at which liquid and vapor are indistinguishable; refrigerant cannot be saturated above the critical point. The left side of the curve is the saturated liquid line, and everything to the left of the saturation curve is subcooled. The right side of the curve is the saturated vapor line, and everything to the right of the curve is superheated. Everything inside the curve is saturated.

Constant Quality

The constant-quality area is the area inside the dome shape outlined by the saturated liquid and saturated vapor lines. The constant-quality lines start at the bottom and extend upward toward the critical point located in top center of the dome (Figure 21-3). These lines represent the percentage of vapor in the saturated refrigerant. Any point along the line marked 20 percent would have 20 percent vapor and 80 percent liquid. The lines run from 0 percent vapor at the saturated liquid line on the left, to 100 percent vapor at the saturated vapor line on the right. As you move from left to the right across these lines, there is more vapor and less liquid refrigerant. Moving from the right to the left increases the percent of liquid as compared to percent of vapor.

During the evaporation process in the evaporator, the percentage of vapor will increase and the percentage of liquid will decrease. This is shown as a left-to-right movement across the area. During the condensation process in the condenser, the percentage of vapor will decrease and the percentage of liquid will increase. This is shown as a right-to-left movement across the area (Figure 21-4).

Pressure

The pressure lines run horizontally from the right side to the left side of the diagram (Figure 21-5). The line spacing becomes closer together as they go up from the bottom. Each line is one-half the space of the previous line. If all the lines were drawn with the same scale units, by the time they got to the top of the diagram, they would be so close that it would appear as a single line. To make the pressure scale readable, the unit spacing changes from the bottom to the top.

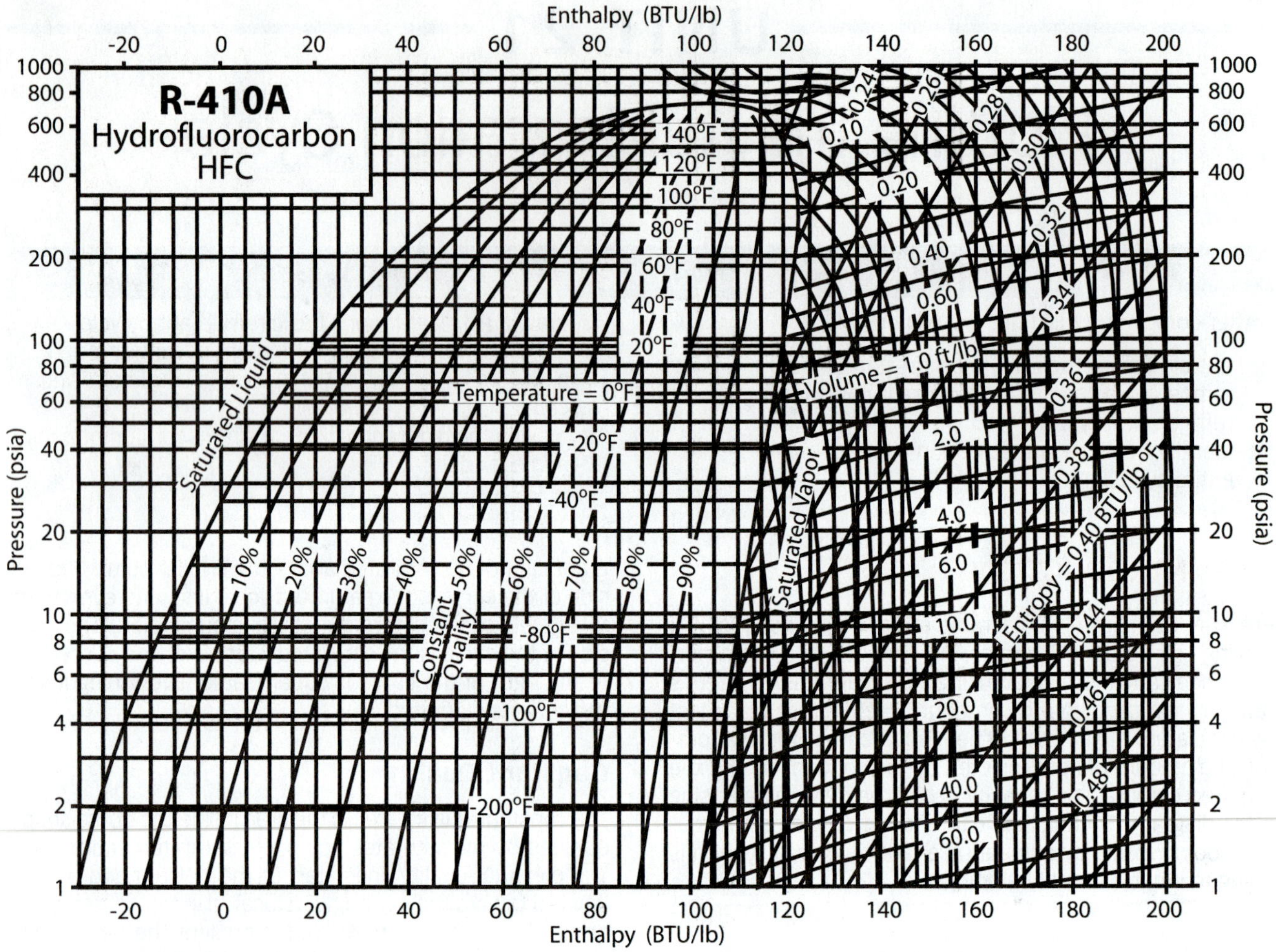

Figure 21-1 Pressure-enthalpy diagram for refrigerant R-410A.

Figure 21-2 The saturation curve.

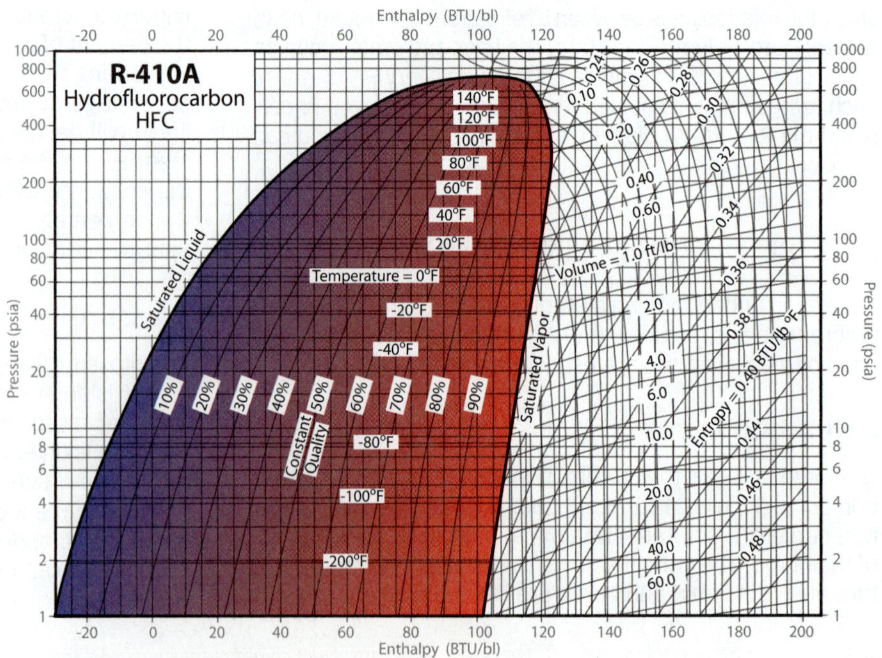

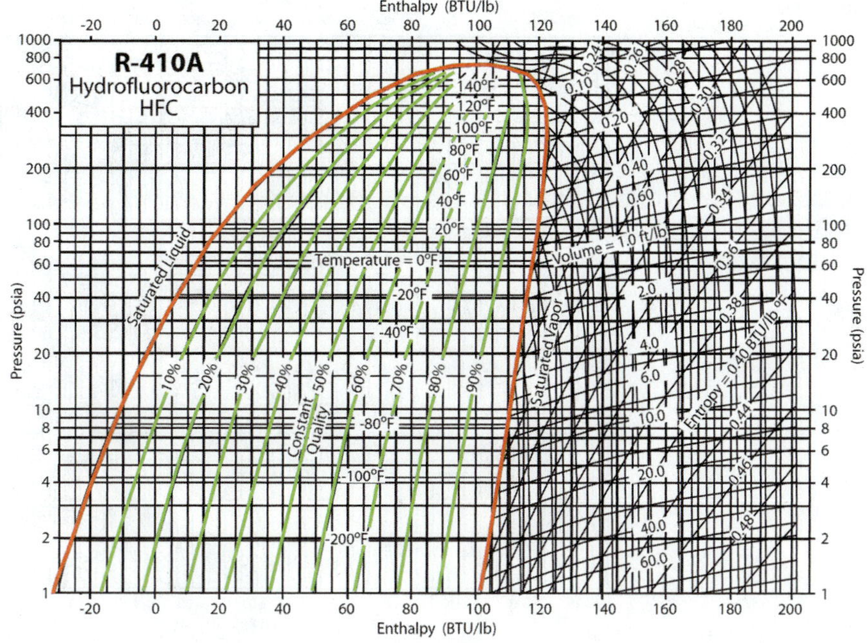

Figure 21-3 The constant quality lines inside the saturation curve.

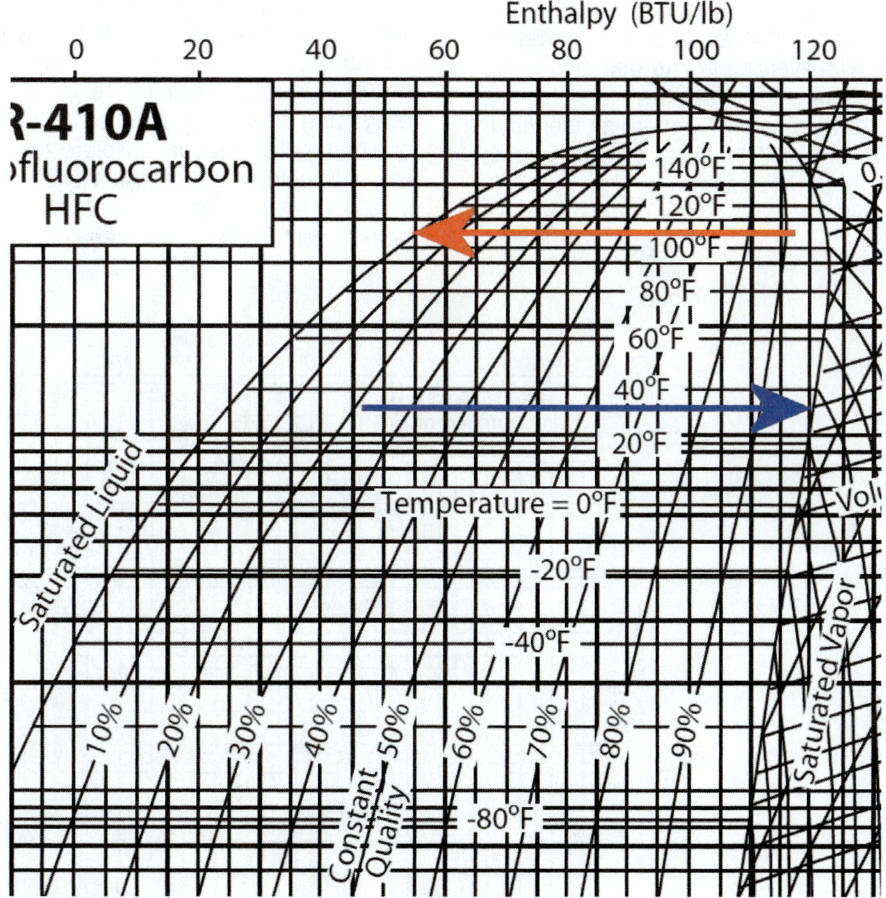

Figure 21-4 The percentage of vapor refrigerant increases as the process moves from left to right, (blue arrow); the percentage of vapor refrigerant decreases as the process moves from right to left (red arrow).

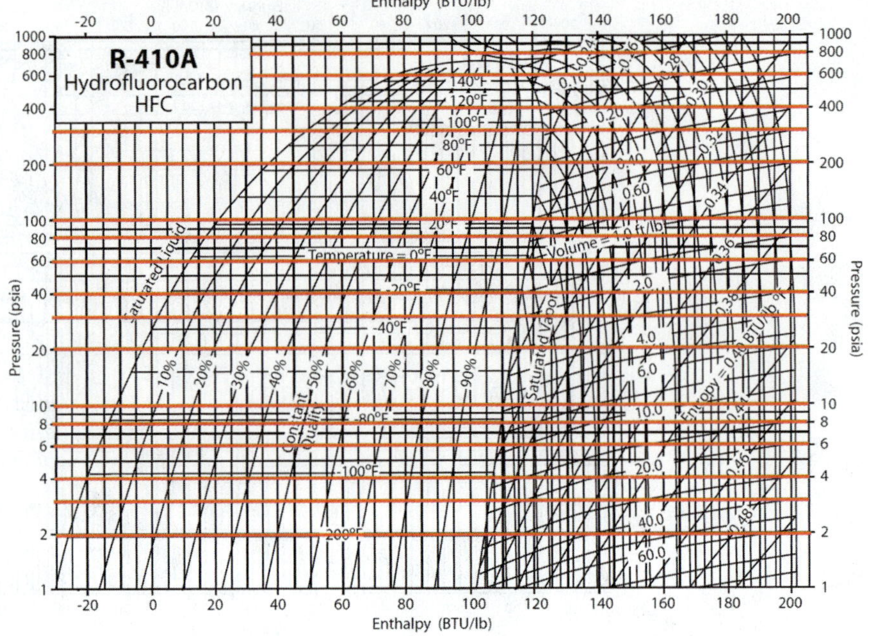

Figure 21-5 The absolute pressure lines run horizontally.

The pressure scales shown on the right and left sides of PH diagrams are given in pounds per square inch absolute, psia. Notice that the scale changes three times from the bottom to the top. The bottom lines are drawn to represent increments of 1 psia, with every other line identified as 1, 2, 4, 6, 8, and 10 psia. The middle scale units are given in increments of 10 psia, with every other line identified as 10, 20, 40, 60, 80, and 100 psia. At the top, the scale changes to increments of 200 psia, with every other line identified as 200, 400, 600, 800, and 1,000. The pressure increments increase as you move up the PH diagram.

Absolute pressures are used so that the diagram's values can be universally used without regard to the current weather or altitude. Subtract approximately 14.7 from these absolute pressures to get gauge pressure.

Enthalpy

The specific enthalpy lines are vertical lines going from top to bottom; the scale is shown along the top and bottom of the PH diagram. The units are given in BTU per pound of refrigerant, BTU/lb (Figure 21-6). Although the true enthalpy for a material starts from absolute zero, PH diagrams define their zero point to be the enthalpy of −40°F saturated liquid. Thus, it is possible to have a negative specific enthalpy on the PH chart. The specific enthalpy increases as you move

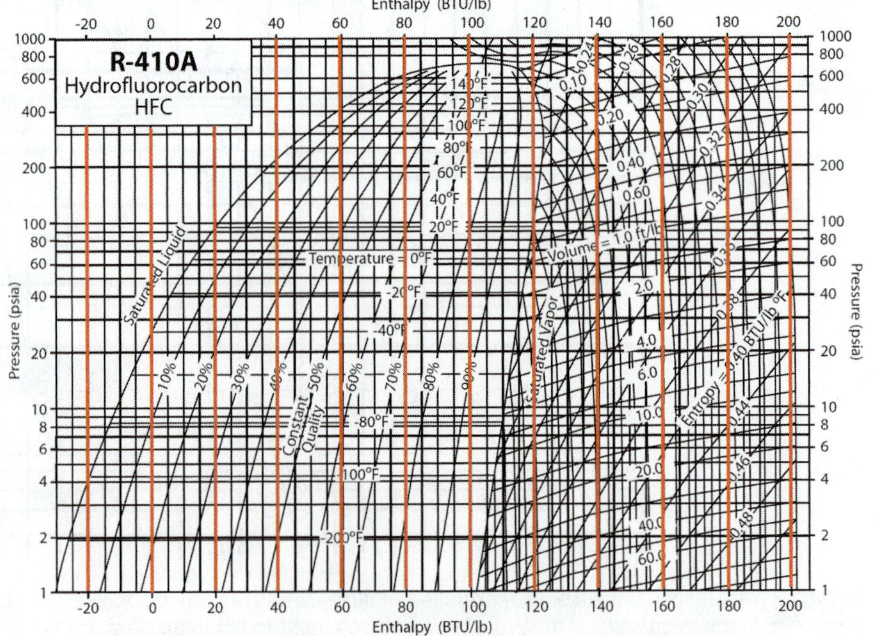

Figure 21-6 The specific enthalpy lines run vertically.

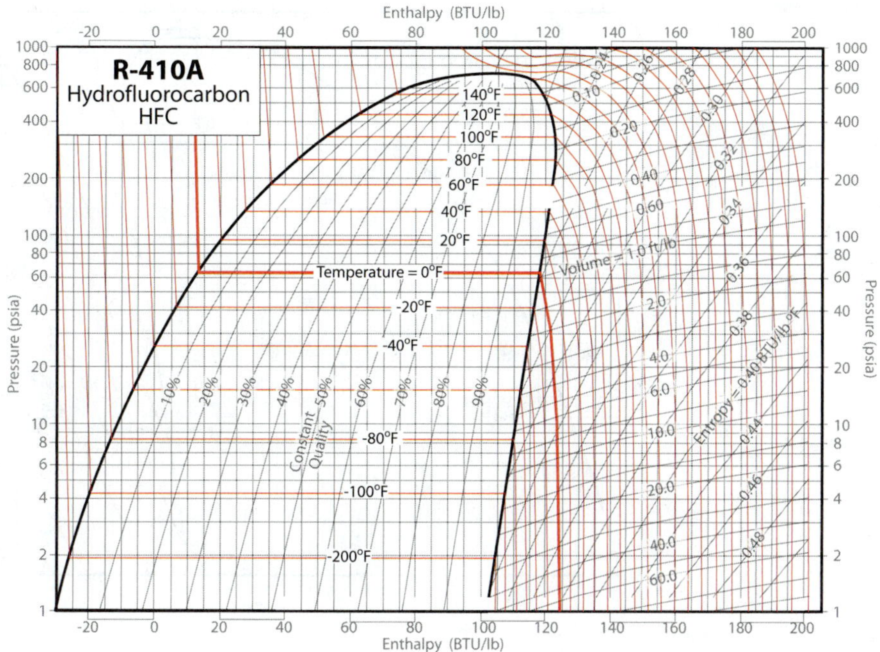

Figure 21-7 The constant temperature lines are not straight; they are semivertical in the liquid and vapor areas and horizontal in the saturated area.

from the left to the right and decreases moving from right to left.

Constant Temperature

The constant temperature lines start along the left top of the diagram and drop nearly vertically to a point where they intersect the saturated liquid line (Figure 21-7). At this point they travel horizontally across to the saturated vapor line, where they drop nearly vertically again. The temperature lines run horizontally inside the saturated region because the temperature of a saturated mixture stays the same as it changes state. All of the constant temperature lines run parallel as they make this zigzag path across the diagram. The temperature at any point on a constant temperature line is the same. For example, in Figure 21-7, you can follow the same 0°F temperature along the line down through the saturated liquid, across the constant quality area, and down through the saturated vapor.

> **TECH TIP**
>
> Not all PH diagrams show the temperature lines in the saturated liquid area. Many PH diagrams' temperatures look more like Figure 21-1 than like Figure 21-7.

Constant Entropy

In refrigeration, entropy is defined as the ratio of the heat content of the refrigerant to its absolute temperature in degrees Rankine. On a PH diagram, entropy is expressed in terms of BTU per pound per degree (BTU/lb/R°). The constant entropy lines start at the bottom of the PH diagram and slope upward to the right (Figure 21-8). Although constant entropy lines can be shown across the entire PH diagram, many diagrams only show the entropy lines in the superheated vapor area. The refrigerant follows the constant entropy lines as it is compressed by the compressor.

Constant Volume

The constant volume lines are nearly horizontal lines that start at the saturated vapor line and gradually slope up to the right (Figure 21-9). These lines represent the change in density of the refrigerant at a constant pressure due to the increase in enthalpy and temperature. Follow the 60 psia line to the right starting at the saturated vapor, where it is about 0°F with about 120 BTU/lb enthalpy. At this point, 1 lb of refrigerant occupies about 1 ft^3. At the right side of the PH diagram, following the same pressure line to about 360°F, this is at a point around 200 BTU/lb enthalpy. The same 1 ft^3 volume now contains around 2 lb of refrigerant.

21.4 REFRIGERANT BLENDS

All refrigerants with a three-digit number whose number starts with 4 are zeotropic blends. A zeotropic blend is a mixture of two or more refrigerants that will physically separate when the refrigerant boils, changing the composition of the blend. Zeotropic refrigerant blends do not condense and evaporate at a single temperature. At any given pressure they have a bubble point and a dew point. The bubble point is the temperature where evaporation begins. It is when small gas bubbles start to form in the liquid. The dew point is the temperature where condensation begins. It is where small droplets of liquid first start to condense out of

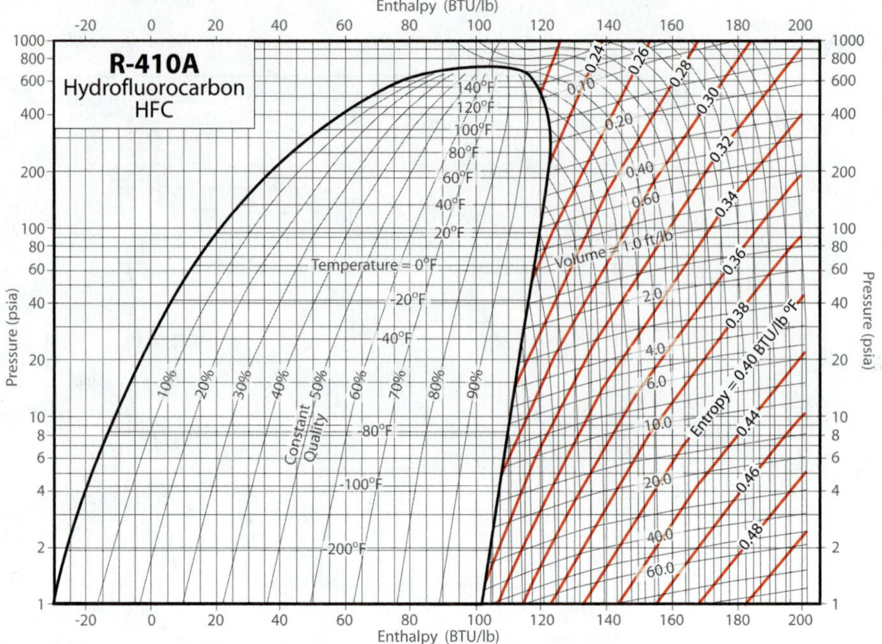

Figure 21-8 The constant entropy lines are diagonal lines in the vapor area.

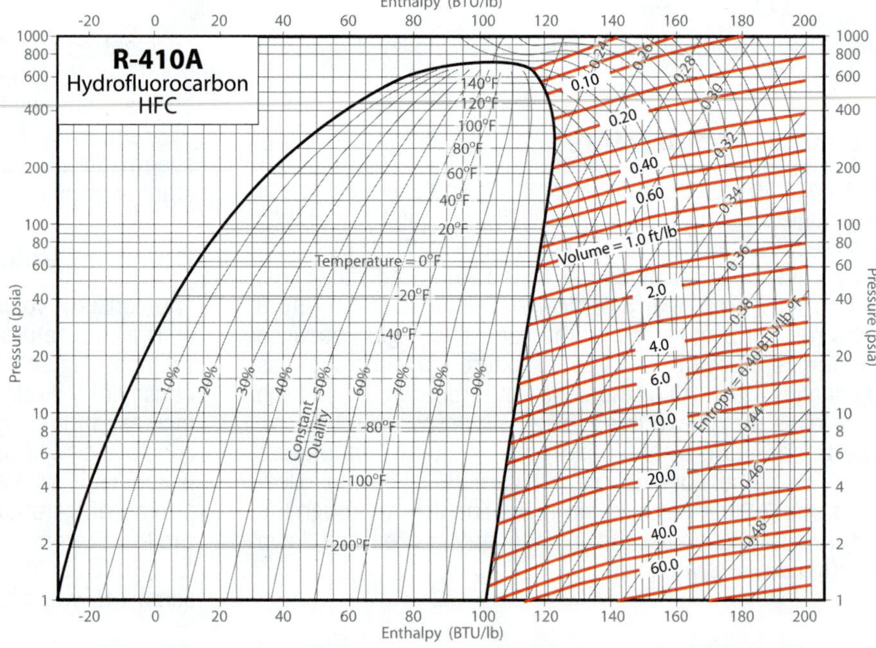

Figure 21-9 Constant volume lines.

a gas. *Temperature glide* is the term used to describe the difference between the bubble point and dew point. Some zeotropes have a glide of 2°F or less, while others have a glide as high as 10°F. Figure 21-10 is a PH diagram for a high-glide blend, R-407C. The temperature lines (highlighted in blue) are higher on the left side of the constant-quality area than on the right side because evaporation starts at a slightly higher pressure with 100 percent liquid than condensation does with 100 percent vapor. The line passing through the saturated area represents the average temperature where a phase change occurs for the saturated mixture.

21.5 THERMODYNAMIC TABLES

One limitation of pressure-enthalpy diagrams is that obtaining precise data is difficult. PH diagrams allow a good overview of what is happening, but getting precise figures is difficult. For example, the pressure lines in Figure 21-1 go from 100 psia to 200 psia. Any pressures between those two must be interpolated. Refrigerant manufacturers publish tables of thermodynamic properties for each of the refrigerants that they sell. These allow much more precision when determining specific values. Figure 21-11 gives the

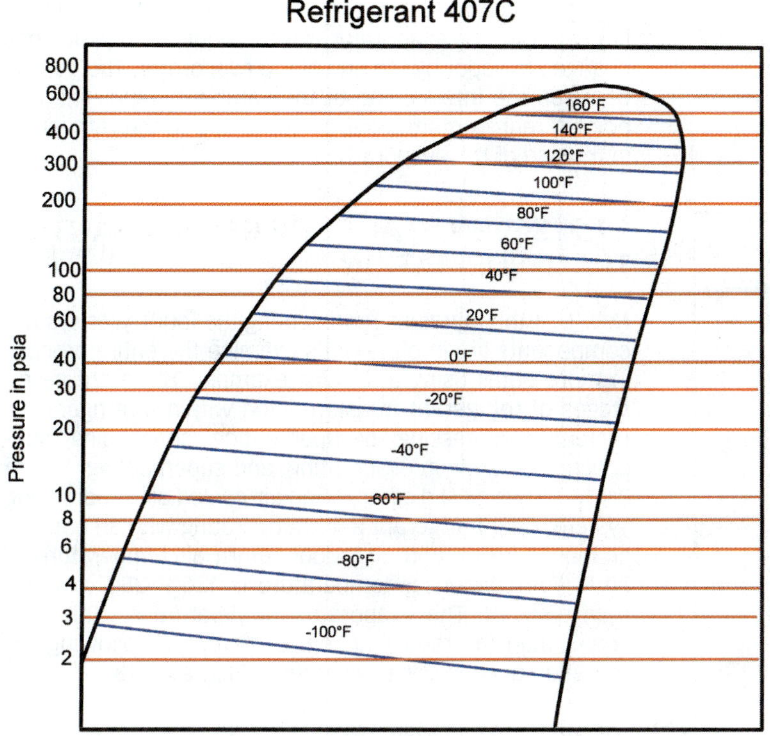

Figure 21-10 The temperature lines are slanted from left to right in the saturated area of a zeotropic blend.

thermodynamic properties of saturated refrigerant R-410A, including pressure, temperature, enthalpy, and entropy for both liquid and vapor states. It is important to recognize that this chart is accurate only for saturated refrigerant. Charts are also published for superheated and subcooled refrigerant. Figure 21-11 can be used to find the pressure for any particular saturation temperature. Two pressures are listed—a liquid pressure and a vapor pressure—because R-410A is zeotropic. Since R-410A is a low-glide zoetrope, the pressure difference is slight. For example, the pressure of 10°F saturated refrigerant is approximately 76 psig. At 10°F, the saturated vapor has an enthalpy of 119.8 BTU/lb, while saturated liquid has an enthalpy of 16.2 BTU/lb.

21.6 REFRIGERATION-CYCLE PH DIAGRAM

PH diagrams can be used to show what happens in a refrigeration cycle, and a basic refrigeration cycle is plotted on the PH diagram in Figure 21-12. A refrigeration-cycle PH diagram contains four lines, each representing one of the four major components. Each line shows how the refrigerant changes in temperature, pressure, heat content, and state as it travels through the cycle.

Compressor

The line between points A and B in Figure 21-12 represents the compression process. The compressor line follows the entropy lines up. This line travels up because the refrigerant pressure is increased in the compressor and to the right because of the heat added by the compressor. Heat of compression is the amount of heat energy added by the compressor due to mechanical work. The discharge temperature predicted by the enthalpy diagram is the theoretical discharge temperature based solely on heat added by mechanical work. In actual operation, the discharge temperature may be 20–35°F higher because of the heat added by the compressor motor in systems with hermetic or semi-hermetic compressors.

Condenser

The line between points B and E in Figure 21-12 represents the condenser. The line is horizontal because the pressure remains the same through all the condenser processes. The portion of the line from point B to the saturation curve represents the desuperheating portion of the condenser. Here, the refrigerant temperature drops until it reaches the saturation point. The refrigerant temperature remains the same from the intersection with the saturated vapor line until the intersection of the saturated liquid line. This portion of the process is called isothermal because the temperature remains the same while the refrigerant changes state from a gas to a liquid. At the saturated liquid line, the refrigerant is all liquid. The portion of the line between the intersection of the saturated liquid line and point E is the subcooling portion of the condenser. The liquid temperature drops in the subcooling section.

Metering Device

The line between point E and point G in Figure 21-12 represents the metering device. This line drops straight down because no heat is gained or lost in the metering device, but the pressure drops dramatically. The refrigerant increases in volume through the metering device. This is why one of the most common metering devices is called an expansion valve. This type of expansion without adding heat is

Thermodynamic Properties of Forane® 410A - Saturation

Saturation Temperature (°F)	Liquid Pressure (psia)	Vapor Pressure (psia)	Liquid Density (lb/ft³)	Vapor Density (lb/ft³)	Liquid Enthalpy (BTU/lb)	Vapor Enthalpy (BTU/lb)	Liquid Entropy (BTU/lb °R)	Vapor Entropy (BTU/lb °R)
-50	20.04	20.01	82.82	0.342	-3.1	111.8	-0.007	0.273
-45	22.72	22.68	82.30	0.385	-1.5	112.5	-0.004	0.272
-40	25.68	25.63	81.78	0.431	0.0	113.2	0.000	0.270
-35	28.94	28.88	81.24	0.482	1.5	113.9	0.004	0.269
-30	32.53	32.46	80.70	0.538	3.1	114.6	0.007	0.267
-25	36.46	36.38	80.15	0.599	4.7	115.3	0.011	0.266
-20	40.77	40.67	79.59	0.666	6.3	116.0	0.015	0.264
-15	45.48	45.36	79.02	0.738	7.9	116.6	0.018	0.263
-10	50.61	50.47	78.45	0.817	9.5	117.3	0.022	0.262
-5	56.18	56.03	77.86	0.902	11.2	117.9	0.025	0.261
0	62.24	62.06	77.26	0.995	12.8	118.6	0.029	0.259
5	68.80	68.60	76.66	1.095	14.5	119.2	0.033	0.258
10	75.89	75.67	76.04	1.204	16.2	119.8	0.036	0.257
15	83.54	83.29	75.41	1.322	18.0	120.3	0.040	0.256
20	91.79	91.51	74.76	1.449	19.7	120.9	0.043	0.255
25	100.65	100.35	74.11	1.586	21.5	121.4	0.047	0.254
30	110.18	109.85	73.44	1.735	23.3	121.9	0.051	0.253
35	120.39	120.03	72.75	1.895	25.1	122.4	0.054	0.251
40	131.32	130.93	72.05	2.067	27.0	122.9	0.058	0.250
45	143.00	142.58	71.33	2.254	28.8	123.3	0.062	0.249
50	155.46	155.01	70.60	2.455	30.8	123.8	0.065	0.248
55	168.75	168.27	69.85	2.671	32.7	124.1	0.069	0.247
60	182.89	182.38	69.07	2.905	34.7	124.5	0.073	0.246
65	197.92	197.38	68.28	3.157	36.7	124.8	0.077	0.245
70	213.88	213.31	67.46	3.429	38.8	125.1	0.080	0.244
75	230.80	230.21	66.61	3.724	40.9	125.3	0.084	0.242
80	248.72	248.10	65.74	4.042	43.1	125.5	0.088	0.241
85	267.67	267.03	64.84	4.387	45.3	125.6	0.092	0.240
90	287.69	287.04	63.90	4.762	47.5	125.7	0.096	0.238
95	308.82	308.16	62.92	5.169	49.9	125.7	0.100	0.237
100	331.10	330.43	61.91	5.613	52.3	125.7	0.104	0.236
105	354.56	353.89	60.84	6.098	54.8	125.5	0.109	0.234
110	379.25	378.58	59.72	6.630	57.3	125.3	0.113	0.232
115	405.19	404.53	58.54	7.216	60.0	125.0	0.117	0.231
120	432.44	431.79	57.29	7.867	62.8	124.6	0.122	0.229
125	461.02	460.40	55.94	8.593	65.7	124.0	0.127	0.227
130	490.97	490.40	54.49	9.412	68.7	123.2	0.132	0.224
135	522.34	521.81	52.91	10.348	72.0	122.3	0.137	0.222
140	555.16	554.70	51.15	11.439	75.4	121.0	0.143	0.219
145	589.47	589.08	49.14	12.747	79.2	119.4	0.149	0.215
150	625.31	625.01	46.75	14.403	83.5	117.2	0.155	0.211

Figure 21-11 Thermodynamic properties of Forane R-410A saturation table. *(Copyright © Arkema Inc. Used by permission.)*

called adiabatic expansion. *Adiabatic* means without heat. The metering device line starts in the subcooled region and ends up in the saturated region of the PH diagram.

Evaporator

The line between point G and point A in Figure 21-12 represents the evaporator. It travels horizontally because the pressure remains the same in the evaporator. The line between point G and the intersection with the saturated vapor line represents the portion of the evaporator where refrigerant is changing state from a liquid to a gas. The temperature remains the same from point G to the intersection of the saturated vapor line because evaporation is an isothermal process. This is where the vast majority of the heat is gained in the evaporator. All the liquid refrigerant has changed to vapor by the end of the evaporator. The portion between the intersection of the saturated vapor line and point A represents the superheat added to the refrigerant in the end of the evaporator.

21.7 SEVEN STAGES OF REFRIGERANT TRANSFORMATION

Now that the four lines representing the four major system components are in place, let's examine the entire process in more detail using a specific example. There are seven stages of refrigerant transformation within a refrigeration system: compressing, desuperheating, condensing, subcooling, expanding, evaporating, and superheating. Figure 21-13 shows a PH diagram of an air-cooled air-conditioning system using refrigerant R-410A operating with an outdoor ambient of 95°F and an indoor return air temperature of 80°F. The condenser temperature is 120°F–25°F warmer than ambient. The evaporator temperature is 40°F–40°F cooler than the return air temperature. The following explanation of each stage will refer to Figure 21-13.

Stage 1: Compression

At point A, the cool refrigerant vapor enters the compressor suction valve at 133 psia and 60°F. The compressor raises its temperature and pressure to approximately 175°F and 433 psia at point B. The temperature and pressure are raised by reducing the gas volume from 0.5 ft³/lb to 0.17 ft³/lb. The compressor line follows the constant entropy lines. Vertical lines drawn from points A and B to the constant enthalpy line show the heat added to the refrigerant by the compressor (Figure 21-14). The difference between the enthalpy at point A and the enthalpy at point B is the heat of compression. In this case, the heat of compression is 145 BTU/lb − 127 BTU/lb = 18 BTU/lb.

Stage 2: Desuperheating

Desuperheating, cooling off of the hot vapor refrigerant, occurs along the line between points B and C. Desuperheating decreases the sensible temperature and heat, but the pressure remains the same. Vertical lines drawn from points B and C to the constant enthalpy line show the heat removed from the refrigerant during desuperheating (Figure 21-15). The amount of desuperheating accomplished is the difference between the enthalpy of the vapor refrigerant at points B and C. In this case, 145 BTU/lb −122 BTU/lb = 23 BTU/lb of heat removed by desuperheating the vapor refrigerant.

Stage 3: Condensing

Condensation begins after all the refrigerant superheat has been removed. Condensation of the refrigerant vapor into a liquid begins at point C, where the condenser line crosses the saturated vapor line. The temperature stops dropping; it is now 120°F. That is the saturated vapor-liquid temperature for 433 psia R-410A. Condensing continues to take place with no change in pressure or temperature until the

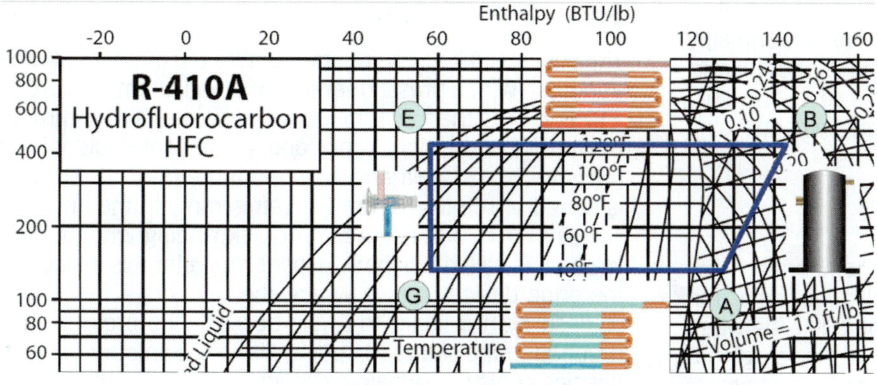

Figure 21-12 Pressure-enthalpy (PH) diagram of the operation of the basic refrigeration cycle.

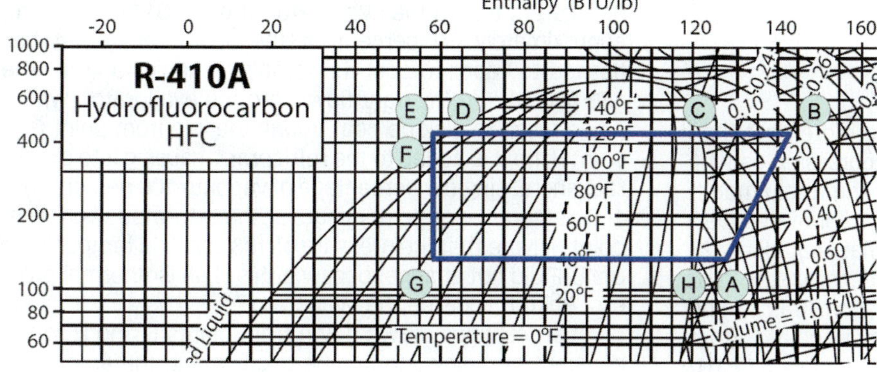

Figure 21-13 PH diagram of the operation of the basic refrigeration cycle showing stages of refrigerant transformation.

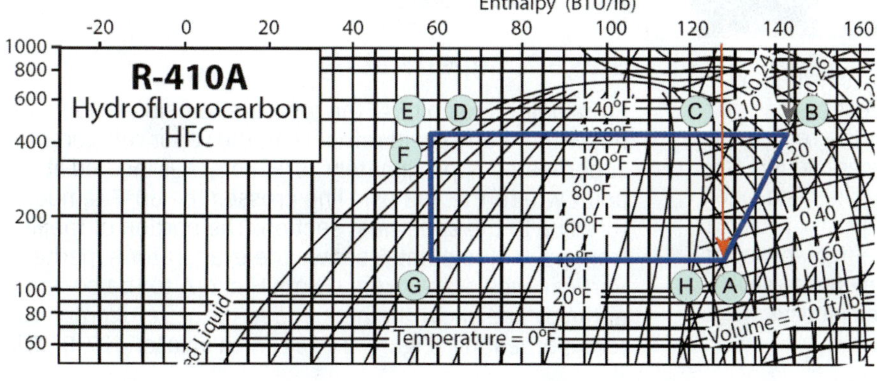

Figure 21-14 The heat of compression is the difference in the enthalpy at points A and B.

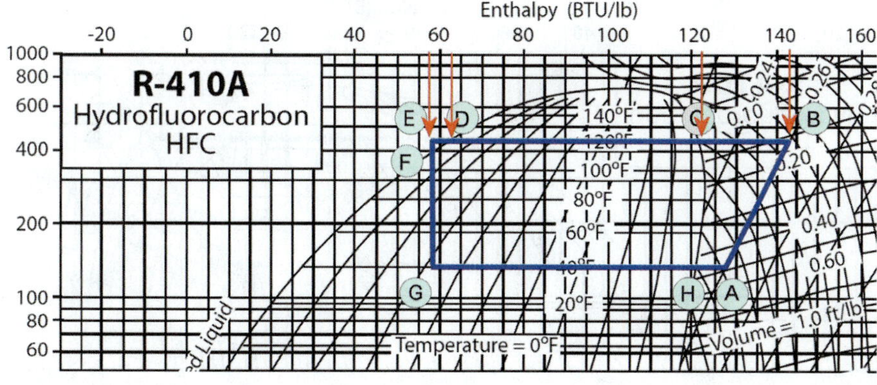

Figure 21-15 The condenser desuperheats the refrigerant between point B and points C, condenses the refrigerant between point C and D, and subcools the refrigerant between points D and E.

refrigerant reaches point D on the condenser line. Vertical lines drawn from points C and D to the constant enthalpy line would show the heat removed from the refrigerant during condensation (see Figure 21-15). In this case, 122 BTU/lb – 62 BTU/lb = 60 BTU/lb of heat removed by condensing the refrigerant.

Stage 4: Subcooling

Subcooling is the lowering of a liquid's temperature below its saturated liquid pressure. Subcooling begins as the condenser line crosses the saturated liquid line between points D and E. If 10°F of subcooling occurs, the refrigerant temperature lowers to 110°F but the pressure remains at 433 psia. Vertical lines drawn from points D and E to the constant enthalpy line show the heat removed from the refrigerant during subcooling (see Figure 21-15). In this case, 62 BTU/lb – 57 BTU/lb = 5 BTU/lb.

Total Condenser Heat Rejection

Notice that the condenser must dissipate more heat than the evaporator absorbs—the heat of compression as well as the heat picked up in the evaporator. The total amount of heat rejected by the condenser is the total of desuperheating, condensing, and subcooling. That would be 23 BTU/lb desuperheating + 60 BTU/lb condensing + 5 BTU/lb subcooling = 88 BTU/lb total heat rejection. This can be found directly by subtracting the enthalpy at point E from the enthalpy of point B. In this case, 145 BTU/lb – 57 BTU/lb = 88 BTU/lb.

TECH TIP

Although subcooling accounts for a very small amount of heat compared to the rest of the process, it is still very important. Subcooling ensures that the refrigerant entering the metering device is 100 percent liquid. Any pressure drop in the liquid line between the condenser and the metering device will cause a saturated liquid to flash, creating gas bubbles. Bubbles of refrigerant vapor decrease the amount of refrigerant the metering device feeds, reducing the system's operating efficiency.

Stage 5: Expansion

The subcooled liquid refrigerant enters the metering device at point E with a pressure of 433 psia and a temperature of 110°F. It exits the metering device at point G as a saturated mixture of 30 percent vapor and 70 percent liquid with a pressure of 133 psia and a temperature of 40°F. No heat is added or removed from the refrigerant, so the enthalpy remains 57 BTU/lb (Figure 21-16). Expanding without adding heat is called adiabatic expansion. Point F is inside the metering device. This is where the refrigerant first starts to flash from a liquid to a gas because the pressure has dropped to the saturation point. The small amount of liquid flashing cools the remaining liquid.

Stage 6: Evaporating

Refrigerant enters the evaporator at point G. As this point, approximately 30 percent of the liquid refrigerant has flashed to vapor to cool the remaining liquid to 40°F. The 133 psia 40°F refrigerant flows into the evaporator, where it can begin to pick up heat. Along the line from point G to point H, heat is added to the refrigerant, causing it to evaporate, increasing the percentage of vapor to 100 percent. The 100 percent vapor point occurs at the saturated vapor line, point H. Note that the pressure of 133 psia and temperature of 40°F did not change, because this is an isothermal process. Point H is inside the evaporator.

Vertical lines drawn down from points G and H to the constant enthalpy line can be used to find the amount of heat that was added to the refrigerant as it changed from approximately 30 percent vapor to 100 percent vapor. In this case, 122 BTU/lb – 57 BTU/lb = 65 BTU/lb heat gain through evaporation (Figure 21-17).

Stage 7: Superheating

From point H to point A, the pressure remains at 133 psia, but without any liquid refrigerant, the vapor refrigerant is superheated approximately 20°F. Point A is located at the point where the 133 psia line crosses the 60°F temperature line in the superheat section. The portion of the line from point H to point A shows the evaporator superheat. Vertical lines drawn from points H and A to the constant enthalpy line show the heat added to the refrigerant due to superheating (Figure 21-18). In this case, 127 BTU/lb

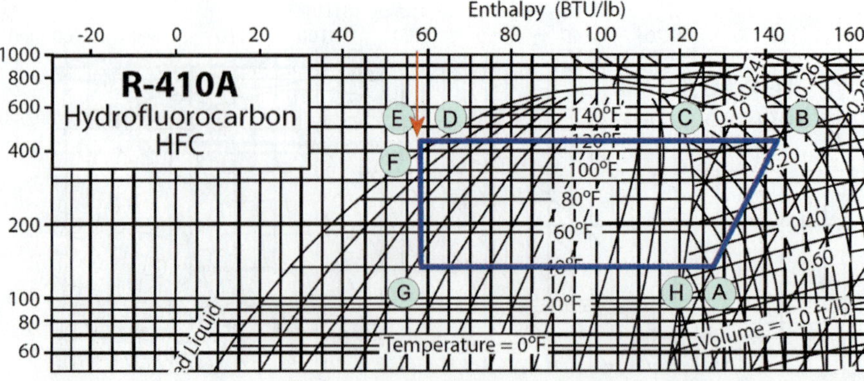

Figure 21-16 The pressure and temperature drops in the metering device are shown between points E and G.

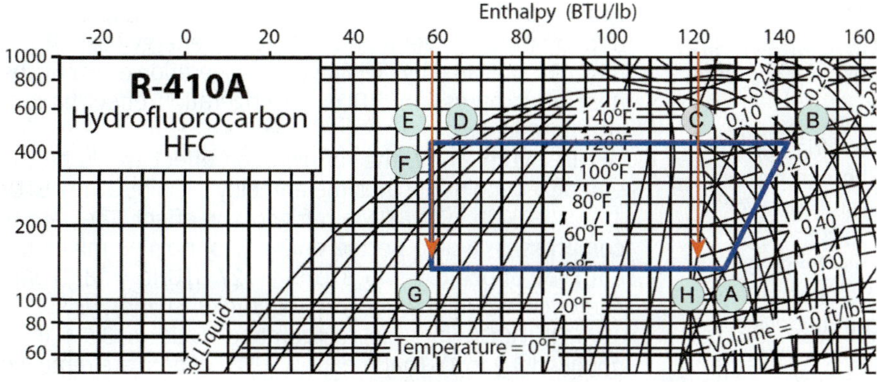

Figure 21-17 The refrigerant change of state in the evaporator from 30 percent to 100 percent vapor takes place between points G and H.

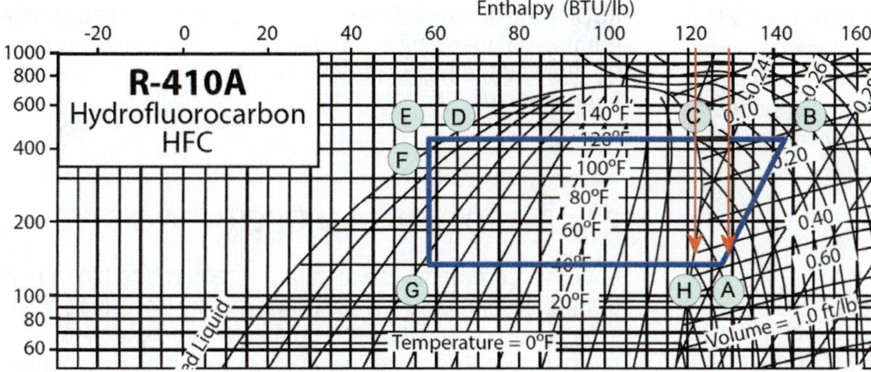

Figure 21-18 The evaporator superheat is shown between points H and A.

– 122 BTU/lb = 5 BTU/lb heat added due to evaporator superheat.

21.8 CALCULATING REFRIGERATION EFFECT

The refrigeration effect is the difference in the enthalpy of the refrigerant entering the evaporator and the enthalpy of the refrigerant leaving the evaporator. The refrigeration effect is rated in BTU per pound of refrigerant (BTU/lb). Referring to Figure 21-19, the enthalpy of the refrigerant at point G, where the refrigerant enters the evaporator, is 57 BTU/lb, and the enthalpy of the refrigerant leaving the evaporator at point A is 127 BTU/lb. The refrigeration effect is then 127 BTU/lb – 57 BTU/lb = 70 BTU/lb.

The refrigeration effect can be used to determine the amount of refrigerant that needs to be circulated to meet any particular load. This can then be used to determine the required compressor capacity.

21.9 REFRIGERATION SYSTEM CAPACITY

Once we know how many BTUs a pound of refrigerant can pick up each time it passes through the evaporator, we can determine the pounds of refrigerant that must flow through the system to meet the system's load requirements. The amount of refrigerant that must be circulated can be calculated by dividing the system load in BTU per minute by the refrigeration effect. One ton of refrigeration is a rate of 200 BTU/min. For a refrigeration capacity of 1 ton, this system would need to circulate (200 BTU/min)/(70 BTU/lb) = 2.9 lb/min. The required capacity per ton can be determined by dividing 200 BTU/min by the system's refrigeration effect as determined by the PH diagram.

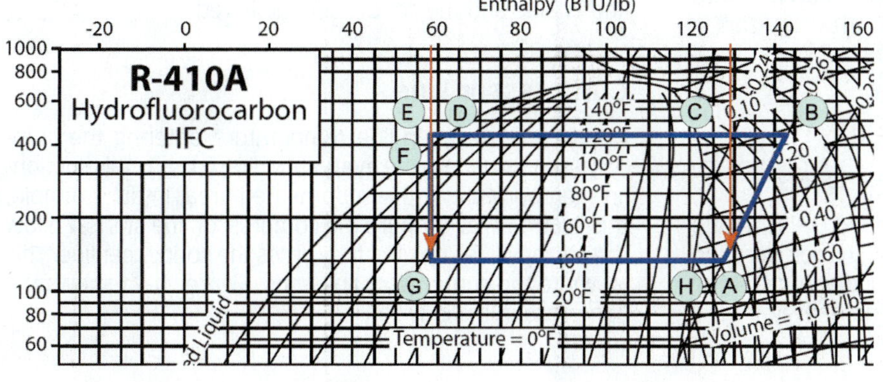

Figure 21-19 The refrigeration effect is the difference in enthalpy between points G and A.

21.10 NET REFRIGERATION EFFECT

In their equipment performance standards, AHRI defines *net refrigeration effect* as

> the rate of total heat absorption by the refrigerant, at stated evaporator conditions, of the complete refrigeration system. This effect is equal to the product of the refrigerant mass flow rate through the system and the enthalpy difference between the refrigerant vapor leaving the evaporator and the refrigerant liquid entering the liquid control device of the evaporator, BTU/h.

The refrigeration effect in the example was determined to be 70 BTU/lb, and the mass flow required 2.9 lb/min. Therefore, the net refrigeration effect is 70 BTU/lb × 2.9 lbs/min = 200 BTU/min. The system's net refrigeration effect should match the heat load. (Note: The result of multiplying 2.9 × 70 is actually 203. The result should be 200, but compound error from rounding both the mass flow and enthalpy numbers yields the slight discrepancy.)

21.11 COMPRESSOR CAPACITY

For design conditions to be maintained within a refrigeration circuit, there must be a balance between the requirements of the evaporator coil and the capacity of the compressor. The compressor must have enough capacity to remove all the refrigerant that has vaporized in the evaporator and send the same weight of refrigerant vapor on to the condenser. If the compressor is too small, it will be unable to move all the vaporized refrigerant, and some of the vapor will remain in the evaporator. This will cause an increase in evaporator pressure and temperature and a decrease in system capacity.

Compressors are rated by the volume of gas they can move at particular low- and high-side pressures. The pressures the compressor must operate at are determined by the evaporating and condensing temperatures. The volume of gas to be moved is determined by the system capacity calculation. In this example, the amount of refrigerant to be circulated is 2.9 lb/min. The units of pounds per minute (lb/min) need to be turned into a volume: cubic feet per minute (ft^3/min). The enthalpy diagram shows that the volume of the refrigerant where it enters the compressor at point A is 0.5 ft^3/lb. Multiplying 0.5 ft^3/lb × 2.9 lb/min gives the required total gas volume of approximately 1.5 ft^3/min. The compressor in this example must be able to move 1.5 ft^3/min at a suction pressure of 133 psia and a head pressure of 433 psia to produce a capacity of 200 BTU/min. This is equivalent to 1 ton of refrigeration.

21.12 REFRIGERANT COEFFICIENT OF PERFORMANCE

The refrigerant coefficient of performance (RCOP) is a ratio of the efficiency of a specific refrigerant in the refrigeration cycle comparing the utilization of expended energy during the compression process in a ratio to the energy that is absorbed in the evaporation process. RCOP is expressed as a ratio of the refrigeration effect to the energy used. RCOP may be calculated by dividing the refrigeration effect by the heat of compression: RCOP = refrigeration effect/heat of compression.

In Figure 21-19, the refrigeration effect would be the enthalpy of point A minus the enthalpy of point G: 127 BTU/lb − 57 BTU/lb = 70 BTU/lb refrigeration effect. The heat of compression would be the enthalpy of point B minus the enthalpy of point A: 145 BTU/lb − 127 BTU/lb = 18 BTU/lb. The RCOP would then be (70 BTU/lb)/(18 BTU/lb) = 3.9.

The RCOP for this refrigeration cycle is 3.9:1 (pronounced 3.9 to 1). That means 3.9 BTU of heat are removed for every BTU of heat added by the compressor. The less energy expended in the compression process, the larger the RCOP ratio of the system would be. When comparing refrigerants and operating conditions, the refrigerant and condition having the highest RCOP would be the best choice for energy efficiency.

21.13 PLOTTING AN OPERATING SYSTEM

Four measurements are all that are required to plot an operating system on a PH diagram: low-side pressure, high-side pressure, suction-line temperature, and liquid-line temperature. The system should be operating long enough for these readings to be stable before taking them. This example will use a system with R-410A, operating with a suction pressure of 116 psig, a head pressure of 417 psig, a suction-line temperature of 40°F, and a liquid-line temperature of 110°F.

Evaporator and Condenser Lines

The first step in plotting a refrigeration cycle on a PH diagram is to establish the condensing and evaporating lines. Take the system operating pressures and convert them to absolute pressure by adding 15. These pressures will be used to establish the evaporator and condenser lines. For example, with an R-410 system operating with pressures of 116 psig on the low side and 417 psig on the high side, the absolute pressures would be 131 psia and 432 psia. Draw a horizontal line across the saturation curve at 131 psia for the evaporator line and at 432 psia for the condenser line. Do not worry about exactly where the lines should start and stop; just make sure they go all the way across the saturation curve on both sides. Figure 21-20 shows the evaporator line in blue and the condenser line in red.

Compressor Line

Measure the suction-line temperature entering the compressor. Find where the evaporator line crosses this suction-line temperature in the superheated area. In our example, this is 60°F. This is where the compressor line starts. Follow the entropy lines up until they cross the condenser line. This is where the compressor line ends. Figure 21-21 shows the compressor line in green.

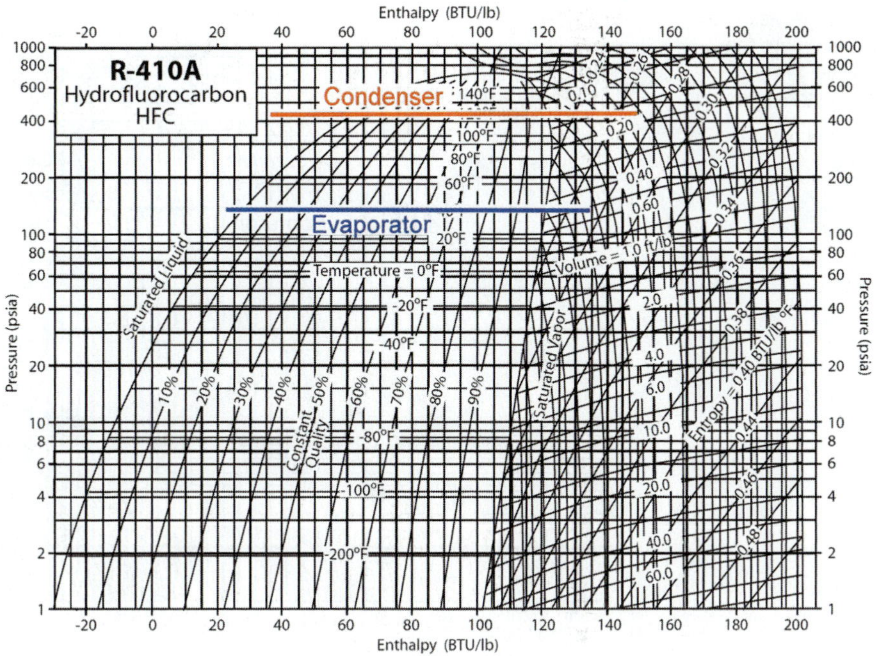

Figure 21-20 The evaporator line is drawn in blue at 131 psia; the condenser line is drawn in red at 432 psia.

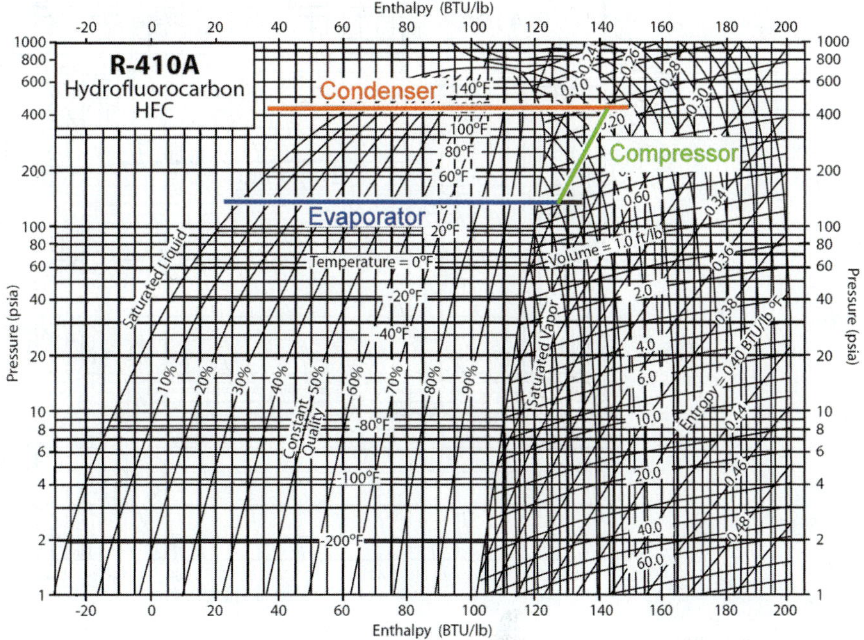

Figure 21-21 The compressor line, drawn in green, starts where the evaporator line intersects the 60°F temperature line and follows the entropy lines up to the condenser line.

Metering Device Line

Measure the temperature of the liquid line entering the metering device. Follow the condenser line until it crosses this liquid-line temperature in the subcooled area. In our example, this is 110°F. This is where the metering device lines starts. Draw a line straight down until it crosses the evaporator line. This is where the metering device line ends. Figure 21-22 shows the metering device line in purple. Figure 21-23 shows the completed plot of the cycle with the evaporator and condenser lines cleaned up.

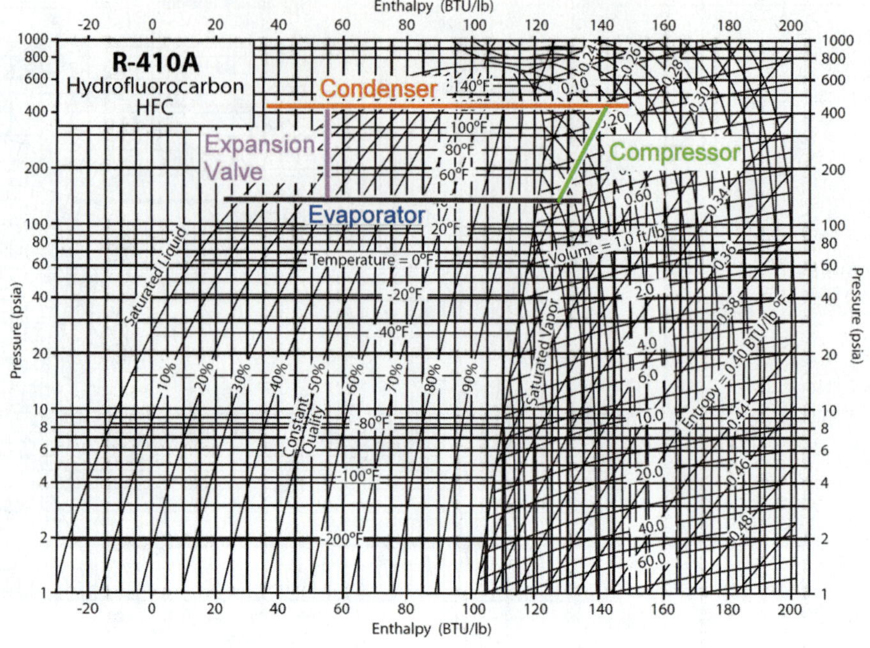

Figure 21-22 The metering device line, shown in purple, starts where the condenser line intersects the 110°F temperature line and drops straight down to the evaporator line.

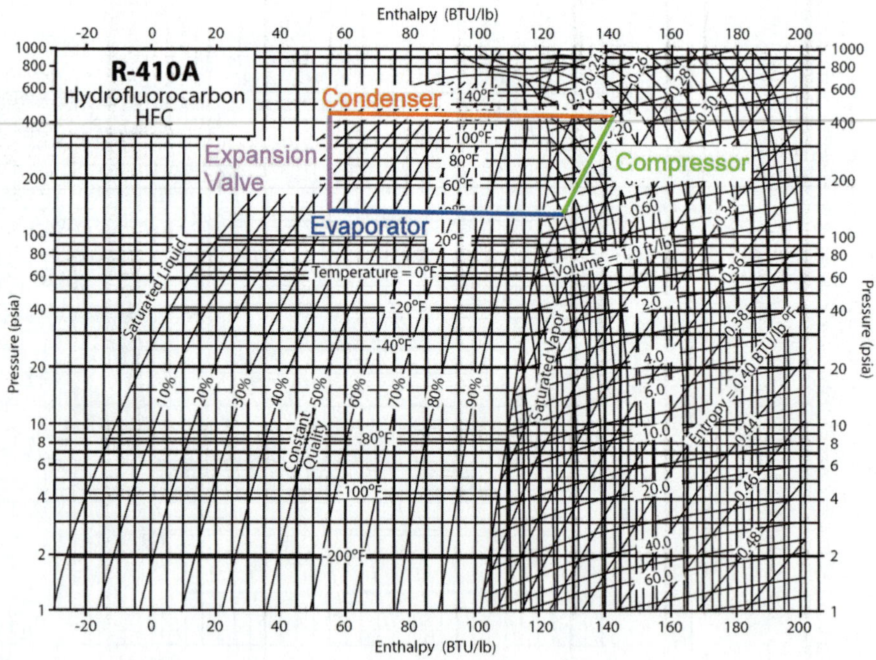

Figure 21-23 Completed plot of refrigeration cycle.

UNIT 21—SUMMARY

A refrigerant pressure-enthalpy diagram, called a PH diagram, shows the relationship of all the physical characteristics of a refrigerant, including pressure, temperature, heat, volume, and state. PH diagrams can be used to determine a system's refrigeration capacity, the amount of refrigerant that must be circulated, the amount of heat absorbed by the evaporator, the amount of heat rejected by the condenser, or the required compressor capacity. PH diagrams have a large hump in the middle called the saturation curve. Everything inside the curve is a saturated mixture of liquid and gas. Everything to the right of the curve is superheated gas, and everything to the left of the curve is subcooled liquid. Enthalpy is the amount of heat in something; specific enthalpy is the amount of heat per pound. The specific enthalpy lines are vertical and marked in BTU/lb. The pressure lines are horizontal lines that are marked in absolute pressure, psia. The temperature lines are nearly vertical in the vapor and liquid areas but horizontal in the saturated area, giving them a zigzag appearance. The temperature remains constant when heat is added inside the saturation curve. This is called an isothermal process. The refrigeration cycle is shown with four lines. A horizontal line at the bottom shows the evaporator, and a horizontal line at the top shows the condenser. They are connected by the compressor line, which follows the entropy lines up. Entropy is the amount of heat per pound per degree in the refrigerant and is measured in BTU/lb/°F. No heat is added or lost when the refrigerant passes through the metering device. This is why the metering device is shown as a vertical line dropping straight down form the condenser line to the evaporator line.

UNIT 21—REVIEW QUESTIONS

1. List the refrigerant characteristics that are shown on a pressure-enthalpy diagram (PH).
2. What is the saturation curve?
3. Where are the pressure lines, and how are they marked?
4. Define *enthalpy*.
5. Explain the difference between enthalpy and specific enthalpy.
6. Where are the specific enthalpy lines, and how are they marked?
7. Why are the temperature lines horizontal inside the saturated curve?
8. What are the constant quality lines?
9. As you move to the right or left along a constant pressure line in the saturated liquid area, what happens to the temperature?
10. As you move to the right or left along a constant pressure line, what happens to the specific enthalpy?
11. As you move up or down along a constant enthalpy line, what happens to the pressure?
12. As you move up along a constant enthalpy line in the constant quality area, what happens to the temperature?
13. In refrigeration, how do we define *entropy*?
14. Why does some of the refrigerant liquid flash-vaporize in the metering device?
15. At what point in the refrigeration cycle does the last of the liquid vaporize?
16. What happens to the refrigerant pressure during desuperheating?
17. When does liquid refrigerant first begin forming in the condenser?
18. What is subcooling?
19. How much heat enters the refrigerant in the metering device when some of it flashes to a vapor, cooling the remaining liquid to the evaporator temperature?
20. What can bubbles of refrigerant vapor do to the metering device's operating efficiency?
21. When does a refrigerant pick up most of the heat as it circulates through a system?
22. Define *adiabatic*.
23. Define *isothermal*.
24. Explain how a system's refrigeration effect can be determined.
25. Determine the refrigeration effect of the following system: the liquid entering the evaporator has an enthalpy of 64 BTU/lb, and the vapor leaving the evaporator has an enthalpy of 121 BTU/lb.
26. What is the net refrigeration effect for the system in question 25 if the system is circulating 4 lb of refrigerant each minute?
27. When drawing a PH diagram of an operating system, how is placement of the condenser and evaporator lines determined?
28. When drawing a PH diagram of an operating system, how is placement of the compressor line determined?
29. When drawing a PH diagram of an operating system, how is placement of the metering device line determined?
30. Why is the condenser line wider than the evaporator line in a PH diagram of a refrigeration cycle?
31. How can a PH diagram be used to select the system compressor?
32. How can a PH diagram be used to determine the heat of compression?
33. How can a PH diagram be used to determine the RCOP for a particular system?
34. How many pounds of refrigerant must flow through a 4-ton evaporator each minute if each pound has an enthalpy of 50 BTU/lb?
35. What will happen in the evaporator if the compressor capacity is not large enough to remove all of the vapor?
36. On a blank sheet of paper, sketch a typical PH diagram for a refrigeration system. You do not need to duplicate all the scales on the PH diagram. Just show the saturation curve and the four cycle lines.

UNIT 22
Refrigerant Safety

OBJECTIVES

After completing this unit, you will be able to:

1. demonstrate how to obtain and read a refrigerant MSDS.
2. apply proper first aid in the event of a refrigeration accident.
3. explain the refrigerant safety rating system.
4. list the safety hazards associated with handling refrigerants.
5. explain the difference among nonrefillable, refillable, and recovery refrigerant cylinders.
6. list the personal protection equipment (PPE) for handling refrigerant.
7. properly dispose of a DOT 39 refrigerant cylinder.
8. explain the difference among a safety relief valve, a rupture disk, and a fusible plug.
9. describe the operation of machinery room refrigerant sensors.

22.1 INTRODUCTION

The refrigeration machines of the early 1900s were the forerunners for today's advanced food processing, product storage, and air-conditioning systems. Unfortunately, these first machines used refrigerants that were flammable, toxic, and highly reactive. In the late 1920s, Thomas Midgley began work to develop nontoxic, nonflammable refrigerants referred to as fluorocarbons. Little did anyone realize at the time that most of these "safe" refrigerants also contained chlorine (CFCs and HCFCs) and were slowly damaging the earth's protective ozone layer.

To reverse this trend, the phase-out of these chlorinated refrigerants has led to the development of totally new refrigerants as well as the increased usage of longstanding refrigerants such as ammonia. This has resulted in a potentially large field of refrigerants that you must become familiar with. Fortunately, refrigerant-specific information is readily available. The specific hazards of each type of refrigerant can be determined by referring to its MSDS. Also, refrigerant safety ratings designated by ANSI/ASHRAE Standard 34-2010 help to provide a simple breakdown for toxicity and flammability.

22.2 MATERIAL SAFETY DATA SHEETS (MSDS) AND PERSONAL PROTECTION EQUIPMENT (PPE)

The first place to begin with any new refrigerant is to review the MSDS (Material Safety Data Sheet). An MSDS will accompany a newly purchased refrigeration cylinder if requested. Another way to access an MSDS is to download it from the Internet, as these are often made available by refrigerant manufacturers. It is advisable to have a reference copy available and on hand for each refrigerant type being used.

Material Safety Data Sheets

Although the refrigerant MSDSs will vary slightly depending on the manufacturer, they all contain similar information. The first section of the MSDS, Chemical Product and Company Identification, will have an emergency overview and call number. Section 3, Hazards Identification, provides a listing of potential health hazards to the skin and eyes and from inhalation. Section 4, First Aid Measures, includes first-aid measures for the skin, eyes, inhalation, and ingestion, along with advice for the physician. This information will be important if an accident does occur and may need to be referred to quickly. This is why it is important to review the MSDS beforehand so that you are not scrambling for information at the time of an accident.

Just as important in taking the right steps when an accident occurs is trying to prevent an accident in the first place. Section 8, Exposure Controls/Personal Protection, lists the exposure controls and personal protection equipment required for handling refrigerants. Items such as gloves for skin protection, goggles for eye protection, and breathing apparatus for respiratory protection are listed, along with any additional recommended special equipment.

Other important sections include 5, 6, and 7, Firefighting Measures, Accidental Release Measures, and Handling and Storage. Section 9, Physical and Chemical Properties, lists information such as vapor pressure and boiling point. The MSDS is a valuable reference that should be made available to anyone who is handling refrigerants or servicing refrigeration systems.

Personal Protection Equipment

It is important to protect yourself whenever working with refrigerants. Hoses sometimes fail and fittings sometimes leak, which could cause refrigerant to come in contact with your skin or eyes. Approved safety glasses, marked Z87 on the frame or lens, are impact and shatter resistant and should be worn whenever working with refrigerants. Chemical splash goggles provide for even further protection, and some are designed to be worn over prescription glasses. Face shields can be worn with safety glasses, or goggles and can be used when maximum protection is required.

Protective gloves should be worn whenever handling refrigerants. They should not only protect against exposure but also resist physical hazards such as abrasion and cuts. The thickness and therefore the amount of dexterity the glove allows is important because a bulky glove may be inadequate for the job required. Glove size and length should also be considered, as longer gloves will protect more of the arm. Proper-fitting gloves are important because tight gloves cause fatigue and large gloves have loose finger ends, which make work more difficult. Generally, leather gloves will provide adequate protection for common refrigeration work. If prolonged contact with liquid or gas is anticipated, then insulated gloves constructed of neoprene, nitrile, Viton, butyl, or polyvinyl alcohol (PVA) should be worn. Some manufacturers offer glove compatibility charts to help you select the best glove for the refrigerant you will be handling.

A self-contained breathing apparatus (SCBA) is a device worn by rescue workers, firefighters, and others to provide breathable air in an "immediate danger to life and health" (IDLH) atmosphere. The term *self-contained* means that there is a high-pressure tank and regulator, and therefore a long supply hose from an external source is not required. An SCBA is very similar to an underwater scuba (self-contained underwater breathing apparatus) tank. Machinery spaces that have the potential for large refrigerant spills must meet special mechanical ventilation requirements because the refrigerant will displace the oxygen in the space. If a refrigerant spill occurs, then a Mine Safety and Health Administration/National Institute for Occupational Safety and Health (MSHA/NIOSH)–approved SCBA must be worn upon entering the space whenever the refrigerant concentration levels are suspected to be above permissible limits.

22.3 FIRST AID

The first steps taken to help someone in distress can often be the most important to ensure his or her safety. If an accident does occur, time will pass before trained medical personnel arrive on site. By understanding the hazards involved when working with refrigerants, you will be better prepared to provide basic medical care until help arrives.

Frostbite

Of course, all refrigerants can be cold. Letting liquid refrigerant come into contact with exposed skin can cause "burns." Immediately after the incident, the skin will just look irritated and red. In one hour after the exposure, blisters will start to appear and the skin will start to swell and puff up. In bad cases, the outer skin comes off and the area is treated like a third-degree burn. Technicians should wear safety goggles to protect their eyes and neoprene-lined gloves to protect their hands when handling refrigerants.

If refrigerant comes into contact with the skin, the area should be flushed with water. If there is evidence of frostbite, then bathe (do not rub) with lukewarm (not hot) water. If water is not available, cover the affected area with a clean, soft cloth. If refrigerant comes into contact with the eyes, immediately flush with large amounts of water for at least fifteen minutes with warm (not hot) water.

SAFETY TIP

HFC R410a boils at –62.9°F at atmospheric pressure. Skin exposed to a direct liquid stream for only a minute can receive severe frostbite that acts like a second- or third-degree burn. Keep your hands out of streams of liquid refrigerant. Refrigerant can be replaced, but your hands cannot!

Inhalation

If refrigerant has been inhaled, then immediately remove the victim from the space to fresh air. If breathing has stopped, give artificial respiration. Use oxygen as required, provided there is someone qualified to administer it. At high levels of inhalation, cardiac arrhythmia may occur. Because of the possible disturbances of cardiac rhythm, catecholamine drugs, such as epinephrine (adrenaline), should not be given.

Asphyxiation

Most refrigerants are heavier than air and will displace the oxygen in a room. Because of this, releasing large quantities of nontoxic refrigerant into a confined machinery space can cause asphyxiation (suffocation). Most of these refrigerants are odorless, so there may be no outward signs other than the victim becoming lightheaded. Under such circumstances, unconsciousness can occur quickly and with little to no warning. Coming to someone's aid in such circumstances requires good judgment. You do not want to rush into the space to rescue someone and become the second victim. First start the ventilation system and use a MSHA/NIOSH-approved SCBA to enter the space if the refrigerant concentration levels are still excessive.

TECH TIP

All commonly used refrigerants except ammonia (R-717) and water (R-718) are heavier than air.

22.4 AMMONIA

More than 80 percent of all ammonia produced is used for agricultural purposes. Less than 2 percent is used for refrigeration, and it must be 99.98 percent pure dry (anhydrous) ammonia to prevent freeze-up. Ammonia attacks copper, zinc, and alloys of copper and zinc. However, ammonia can be safely used as a refrigerant provided the system is properly designed, constructed, operated, and maintained. Unlike most refrigerants, ammonia is lighter than air and rises. Pure ammonia vapors are a fire and explosion hazard at concentration ranges of approximately 15–28 percent. When mixed with lubricating oils, its flammable range is increased even further. Table 22.1 provides information on the exposure limits and health risks for ammonia.

Table 22-1

Concentration	Effect of Ammonia
5 ppm	Odor threshold
25 ppm	TWA exposure limit and noticeable odor
35 ppm	Short-term exposure limit
50 ppm	Permissible exposure limit (PEL), detectable by most people
140 ppm	Eyes irritated, strong odor
300 ppm	Immediately dangerous to life and health (IDLH)
400 ppm	Major throat irritation
700 ppm	Coughing, severe eye irritation may lead to loss of sight
1,700 ppm	Convulsive coughing, serious lung damage, possible death
2,000 ppm	Skin blisters and burns within seconds, no exposure permissible
5,000 ppm	Respiratory spasms, rapidly fatal

First-aid treatment for ammonia exposure is similar to that for most refrigerants. If the possibility of exposure above 250 ppm exists, use a MSHA/NIOSH-approved SCBA with a full face piece operated in a pressure-demand or other positive-pressure mode. Bring the victim to fresh air and flush the affected parts with warm water. Eyes affected by ammonia close involuntarily, so the eyelids must be held open. Clothes saturated with ammonia may freeze to the skin. While still clothed, the victim should immediately get under a shower. Remove clothes only after they are thawed out and can be freely removed from the frozen areas.

TECH TIP

Ammonia is widely used in large industrial refrigeration systems because it has a high net refrigeration effect and low impact on the environment. However, most fatal refrigeration accidents have resulted from the sudden release of ammonia. OSHA statistics show that from 1984 through 2006, there were 224 accidents and 50 fatalities involving the release of ammonia in the United States.

22.5 ANSI/ASHRAE STANDARD 34-2010

ANSI/ASHRAE Standard 34-2010, Designation and Safety Classifications of Refrigerants, addresses refrigerant safety. Refrigerants are grouped by their toxicity and flammability. Each refrigerant is assigned a safety rating consisting of a letter and a number. The letter *A* identifies refrigerants with lower toxicity, and the letter *B* identifies refrigerants with higher toxicity. The numbers 1–3 that follow the letters are used to indicate the flammability of the refrigerant. The number 1 would indicate a low flammability, while the number 3 would indicate a high flammability. As an example, A1 would indicate low toxicity and low flammability. A3 would indicate low toxicity and high flammability. B3 would indicate high toxicity and high flammability (Table 22.2).

CODES AND STANDARDS

ANSI/ASHRAE Standard 34-2010, Designation and Safety Classification of Refrigerants, describes a shorthand way of naming refrigerants and assigns safety classifications and refrigerant concentration limits based on toxicity and flammability data.

Toxicity

Almost anything can be toxic in high enough concentrations. For example, the air we breathe is approximately 78 percent nitrogen. However, breathing 100 percent nitrogen will cause asphyxiation. *Toxicity* is defined as the ability of a refrigerant to be harmful or lethal due to acute or chronic exposure by contact, inhalation, or ingestion. (Section 11 of a refrigerant's MSDS provides toxicological information.)

Table 22-2

	Safety Group	
Higher Flammability ↑	A3	B3 Toxic & Flammable
Lower Flammability ↑	A2 A2L	B2 B2L
No Flame Propagation ↑	A1 Non-toxic & Non-flammable	B3
	Increasing Toxicity →	

The maximum safe exposure level in parts per million by volume is called the threshold limit value (TLV). The time weighted average (TWA) concentration for a normal eight-hour day, forty-hour week that is safe without any adverse effect is referred to as the threshold limit value–time weighted average (TLV-TWA). This is often commonly expressed as the occupational exposure limit (OEL).

TECH TIP

Parts per million (ppm) can sound confusing because it is difficult to comprehend a number as large as 1 million. If you started counting nonstop from 1 to 1 million, it would take you over 11½ days to finish. Ammonia has an occupational exposure limit of 25 ppm, which is equivalent to less than ½ minute of an 11½-day count. Another way to look at it is that 25 ppm would be equivalent to .0025 percent.

A highly toxic refrigerant would require a small amount as measured in ppm to reach the OEL, while a nontoxic refrigerant would require a much higher value. As an example, nontoxic R-134a has an OEL of 1,000 ppm, while toxic ammonia has an OEL of 25 ppm. Only a small amount of ammonia needs to be released to reach the threshold limit. Since this value is based on an average person, a small percentage of people may experience discomfort at concentrations below the TLV depending on their health and medical background.

The permissible exposure level (PEL), which is similar to OEL, is based upon a TWA as set by the Occupational Safety and Health Administration (OSHA). The safety group rating "A" is used to designate low-toxicity refrigerants having a PEL of 400 ppm by volume or greater. Safety group rating "B" designates high-toxicity refrigerants having a PEL of less than 400 ppm by volume. Type "B" refrigerants are prohibited in many applications by building and safety codes.

Most common CFC, HCFC, and HFC refrigerants are listed as type "A" refrigerants. One notable exception is R-123, which is a low-pressure HCFC refrigerant used in large, low-pressure chillers. It is considered a type "B" refrigerant. Many older refrigerants are type "B," including ammonia, sulfur dioxide, and methyl chloride. An example of different refrigerants and their safety group classifications are provided in Table 22.3. Note that this rating system does not distinguish between refrigerants that will kill you in a few minutes and refrigerants like HCFC-123 that will cause some physical distress when you are exposed for eight hours a day, forty hours a week.

TECH TIP

Refrigerant toxicity safety ratings are normally based on the permissible exposure level (PEL) set by OSHA. Chemical manufacturers may also publish similar recommendations, such as occupational exposure limit (OEL), acceptable exposure level (AEL), and industrial exposure limit (IEL), generally for substances for which PEL has not been established.

Table 22-3

		Safety Group
Methane Series		
R-10	Tetrachloromethane	B1
R-11	Trichlorofluoromethane	A1
R-12	Dichlorodifluoromethane	A1
R-22	Chlorodifluoromethane	A1
R-30	Dichloromethane (methylene chloride)	B2
R-50	Methane	A3
Ethane Series		
R-113	1,1,2-trichloro-1,2,2-trifluorethane	A1
R-114	1,2-dichloro-1,1,2,2-tetrafluoroethane	A1
R-123	2,2-dichloro-1,1,1-trifluoroethane	B1
R-134a	1,1,1,2-tetrafluorethane	A1
R-142b	1-chloro-1,1-difluoroethane	A2
R-143a	1,1,1-trifluoroethane	A2
R-152a	1,1-difluoroethane	A2
Propane		
R-290	Propane	A3
Hydrocarbons		
R-600	Butane	A3
Oxygen Compounds		
R-611	Methyl formate	B2
Inorganic Compounds		
R-717	Ammonia	B2
R-764	Sulfur dioxide	B1
Unsaturated Organic Compounds		
R-1140	1-chloroethene (vinyl chloride)	B3
Zeotropes		
R-404A	R-125/143a/134a (44.0/52.0/4.0)	A1
R-407C	R-32/125/134a (23.0/25.0/52.0)	A1
R-410A	R-32/125 (50.0/50.0)	A1
Azeotropes		
R-500	R-12/152a (73.8/26.2)	A1
R-502	R-22/115 (48.4/51.2)	A1
R-507	R-125/143a (50.0/50.0)	A1

Flammability

A refrigerant's flammability is listed as type 1, nonflammable; type 2, somewhat flammable; or type 3, very flammable. There is also one optional subclass, type 2L. These ratings are based on the refrigerant's lower flammability limit (LFL), the heat of combustion, and an optional burning velocity measurement. Refrigerant flammability for single-compound refrigerants is based on their worst case of formulation for flammability (WCF). The rating for refrigerant blends is based on the worst case of fractionation for flammability (WCFF). (Sections 5 and 9 of a refrigerant's MSDS provide information on the refrigerant's flammable properties.)

The LFL is the minimum concentration in ppm of flammable liquid vapor in air that will support the propagation of flame upon contact with an ignition source. A liquid vapor/air mixture can be too lean (not enough, low ppm) to burn below the LFL or too rich (too much, high ppm) to burn above the upper flammable limit (UFL). As an example, ammonia is flammable between the concentration limits of 15 and 28 percent. Below 15 percent, the ammonia mixture is too lean to burn, and above 28 percent the mixture is to rich to burn.

Nonflammable type 1 refrigerants will not ignite and will not support combustion, which means they do not have an LFL limit. An example of a single-compound refrigerant type 1 rating is R-134a, based on its WCF. The UFL is not considered for flammability ratings.

Refrigerants listed as type 2 are somewhat flammable and in the proper concentrations (not too lean or too rich) will ignite when exposed to a flame. However, they will not continue to burn when the flame is removed. An example of a single-compound refrigerant type 2 rating is R-717, ammonia. It has an LFL of 167,000 ppm by volume. Concentrations of ammonia above this limit but below the UFL can be ignited.

A refrigerant with a 2L rating would have a slower burning velocity, meaning the flame would not burn as fast as a refrigerant classified with a type 2 rating. Some refrigerant manufacturers may choose to apply for this optional category, which would indicate a better safety rating than type 2.

Refrigerants with a flammability rating of 3 will ignite and support combustion. An example of a single-compound refrigerant type 3 rating is R-290, propane. It has an LFL of 21,000 ppm per volume. As compared to ammonia (167,000 ppm by volume), it would take far less propane, almost 8 times less, to support a flame.

Safety Rating for Zeotropes (Blends)

Azeotropic refrigerants are rated just like pure compounds because they do not separate when they boil. However, zeotropic refrigerants can fractionate (separate) when they boil. This raises the possibility that a mixture's characteristics can change. For this reason, zeotropic refrigerants are assigned a safety group classification based on the worst case of fractionation.

The worst case of fractionation for flammability (WCFF) is defined as the composition during fractionation that results in the highest concentration of the flammable component(s) in the vapor or liquid phase. So if one part of a three-part blend vaporizes, and that part is the most flammable, the refrigerant will be assigned a rating based on it even though the other two parts may be less flammable or maybe not even flammable at all.

The worst case of fractionation (WCF) for toxicity is defined as the composition during fractionation that results in the highest concentration of the component(s) in the vapor or liquid phase for which the TLV-TWA is less than 400 ppm by volume. Remember, more toxic refrigerants have lower allowable ppm concentrations, therefore less than 400 ppm means more toxic.

22.6 CHEMICAL INTERACTIONS

There are safety concerns that arise from the interaction of refrigerants and their surroundings. (Section 10 of a refrigerant's MSDS provides information on the refrigerant's stability and reactivity.) For example, chlorofluorocarbons and hydrochlorofluorocarbons (CFCs and HCFCs) are not flammable. However, they chemically decompose when exposed to flames or very high temperatures, releasing their chlorine and fluorine. This creates highly irritating and potentially toxic gases, including chlorine gas, fluorine gas, and phosgene gas.

SAFETY TIP

Never braze tubing or do any hot work on systems that contain HCFCs and CFCs without first evacuating the lines, because these refrigerants will chemically decompose and produce toxic gases.

HCFCs and HFCs can react violently with some metals. HCFCs should never be exposed to magnesium. HFCs should not be allowed to come into contact with freshly abraded aluminum. Aluminum that has been recently ground up can react violently with HFCs. Some HFCs are flammable under high pressure with large concentrations of oxygen. Of course you should never put oxygen or compressed air into any refrigeration system. The oxygen under pressure can cause an explosion when mixed with the refrigerant oil.

22.7 REFRIGERANT STORAGE CYLINDERS

Refrigerants need to be transported and stored safely. Metal cylinders are specifically designed for this purpose. They come in different sizes and are often color coded. The cylinders will contain both liquid and vapor refrigerant under pressure, so they will be equipped with some form of burst protection. Due to the potential hazards encountered in the storage and handling of refrigerants, cylinders must meet regulatory requirements. The three basic types of refrigerant cylinders are as follows:

- Nonrefillable refrigerant cylinders
- Refillable refrigerant cylinders
- Recovery cylinders

A typical recovery cylinder is shown in Figure 22-1.

CODES AND STANDARDS

The regulations for nonrefillable refrigerant cylinders are specified in Title 49, Code of Federal Regulations (CFR), Section 178.65, Specification 39.

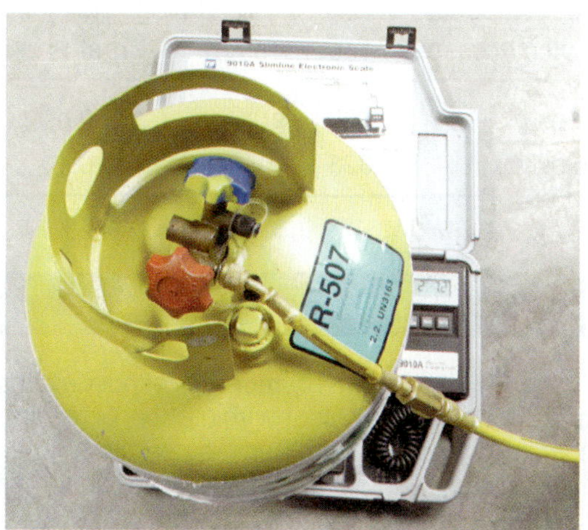

Figure 22-1 Refrigerant recovery cylinder in use.

22.8 DOT 39 NONREFILLABLE CYLINDERS (DISPOSABLE)

As the name suggests, the nonrefillable refrigerant cylinder is only filled one time, by the refrigerant manufacturer (Figure 22-2). The regulations for these cylinders are specified in Title 49, Code of Federal Regulations (CFR), Section 178.65, Specification 39. It should never be refilled by anyone with anything, not even the refrigerant manufacturer. Disposable cylinders come with check valves built in that make refilling the cylinders nearly impossible, because the check valves prevent flow-back into the cylinder (Figure 22-3).

Figure 22-2 Nonrefillable refrigerant.

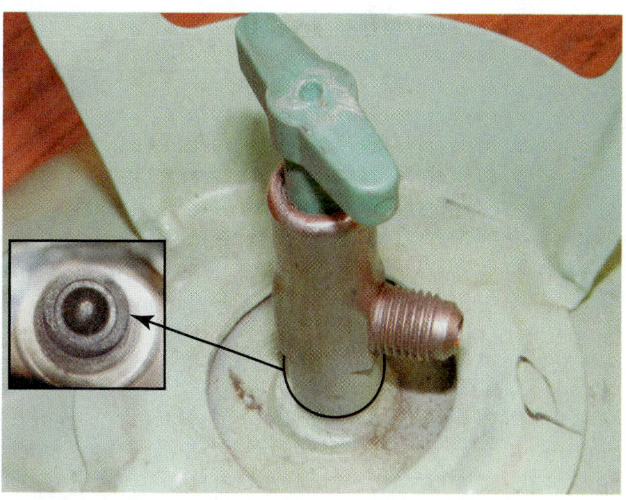

Figure 22-3 Nonrefillable cylinders have a check valve in the stem to prevent flow-back into the cylinder.

These cylinders are nonrefillable for a reason: they are not built to last as long as refillable cylinders and cannot withstand the same rigors as an approved recovery cylinder. The number-one reason for not refilling disposable cylinders is to keep all your body parts firmly attached. The number-two reason not to refill a disposable cylinder is that the potential penalty for refilling and transporting a nonrefillable cylinder is now up to a $500,000 fine and five years in prison (Figure 22-4). This regulation is a Department of Transportation rule. It is designed to keep the highways safe, and it is enforced. (Section 14 of a refrigerant's MSDS provides transport information.)

To prevent disaster, disposable cylinders should not be stored or used in temperatures exceeding 125°F. All refrigerant cylinders are rated for a maximum recommended service pressure. Most DOT 39 disposable cylinders have a service pressure of 260 psig. Disposable cylinders for use

Figure 22-4 The penalty for refilling and transporting a nonrefillable cylinder is a $500,000 fine and/or five years imprisonment.

Figure 22-5 Pressure ratings of DOT 39 cylinders.

Refrigerant	Service Pressure	Test Pressure	Burst Pressure
R-410A	400 psig	500 psig	1,000 psig
All others R-134a, R-22, etc.	260 psig	325 psig	650 psig

with R-410A have a service pressure of 400 psig. These pressures correspond to a temperature of 120°F for saturated refrigerant. At the maximum recommended cylinder temperature of 125°F, these pressures increase to 278 psig and 447 psig, respectively. They are tested at the factory at 125 percent of their rated service pressure and are supposed to have a minimum burst pressure of 250 percent of their rated service pressure. Figure 22-5 lists the service pressure, test pressure, and burst pressure for both types of DOT 39 disposable cylinders.

DOT 39 disposable cylinders have a frangible disk built into them to prevent pressures from reaching the bursting point of the cylinder (Figure 22-6). These disks look like knockouts in electrical boxes. A restriction on the other side of the disk limits the amount of flow out of the hole to prevent turning the cylinder into a missile (Figure 22-7).

Figure 22-6 Frangible disk on nonrefillable DOT 39 cylinder.

Figure 22-7 Flow-limiting orifice on the inside of the cylinder underneath the frangible disk

SAFETY TIP

The frangible disk cannot protect the cylinder against all hazards. Heating the cylinder with a torch, or leaving the cylinder in the weather for extended periods of time, can create new weak spots that can lead to a violent cylinder rupture.

Disposable Cylinder Recycling

The cylinder cannot be reused, but the metal may be recycled. Any remaining refrigerant cannot be vented to atmosphere but instead must be recovered using proper recovery equipment and procedures. After all remaining refrigerant has been removed, the cylinder valve may be opened. A metal chisel can be placed on the inside serration of the rupture disk located on the shoulder of the tank. A hammer can be used to lightly tap the chisel to pierce the rupture disk open. After the rupture disk has been completely removed, a large circle is to be drawn with permanent marker around the open rupture disk and labeled EMPTY. The tank can then be recycled along with other steel recyclables at a local steel recycling center.

22.9 REFILLABLE REFRIGERANT CYLINDERS

The name says it all: refillable cylinders are intended to be refilled. However, only the owners of the cylinders or their agent may refill them. Refillable cylinders are large cylinders with water capacities of 123 lb or more. Refrigerant can be purchased in a large cylinder as shown in Figure 22-8. The customer generally pays a deposit on this cylinder when purchasing the refrigerant. When the cylinder is empty it is returned to the wholesaler. It is important to note that the refrigerant company still owns the cylinder. This means that the field service technician is not authorized to refill it.

CODES AND STANDARDS

The regulations for refrigerant recovery cylinders are specified in Title 49, Code of Federal Regulations (CFR). Recovery cylinders of the DOT 4BA type are covered in Section 178.51.

22.10 RECOVERY CYLINDERS

The primary differences between a refillable cylinder and a recovery cylinder are size and ownership. Recovery cylinders are normally available in water capacities of 27 lb and

Figure 22-8 150-lb net weight cylinder of CFC R-114. Note the tag lists a deposit fee for the cylinder.

47 lb. Refillable cylinders and recovery cylinders are regulated by the DOT in Title 49, Code of Federal Regulations (CFR), covering transport of hazardous materials. This is an extensive set of rules covering many things, and refrigerant cylinders are only a very small part of this regulation. The primary hazard refrigerant cylinders pose is rupture.

Recovery Cylinder Service Pressure

Cylinders are marked with the particular DOT specifications for that cylinder. The numbers and letters immediately following the letters DOT describe how the cylinder is made while the last group of numbers describes the cylinder service pressure. For example, a DOT-4BA-350 recovery cylinder is a welded or brazed steel cylinder with a recommended service pressure of 350 psig (Figure 22-9). This

Figure 22-9 This specification shows that the service pressure for this recovery cylinder is 350 psig.

Figure 22-10 Refrigerant recovery cylinder cut in half showing the liquid valve tube.

is the most common specification for refrigerant recovery cylinders. There are also some DOT-4BA-260 and DOT 4BA-400 cylinders in use. These cylinders have recommended service pressures of 260 psig and 400 psig, respectively. Figure 22-10 shows the inside of a recovery cylinder that has been cut in half.

Pressure-Relief Valve

Recovery cylinders have an automatic pressure-relief valve built in that opens at 150 percent of the rated cylinder service pressure (Figure 22-11). The outlet of the relief valve should never be plugged or obstructed by anything. These valves reset once the pressure in the cylinder has dropped to a safe level.

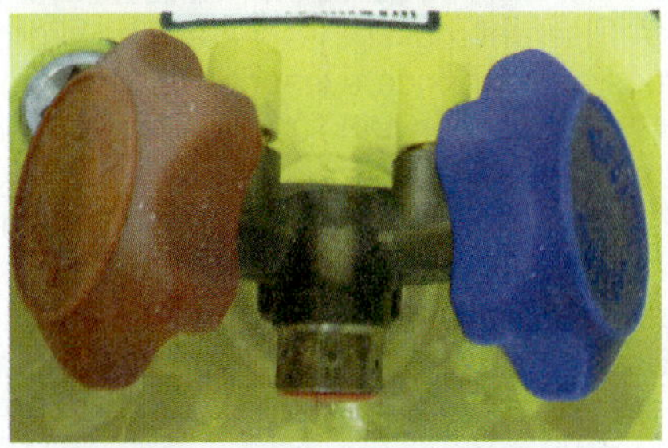

Figure 22-11 Pressure-relief valve on recovery cylinder.

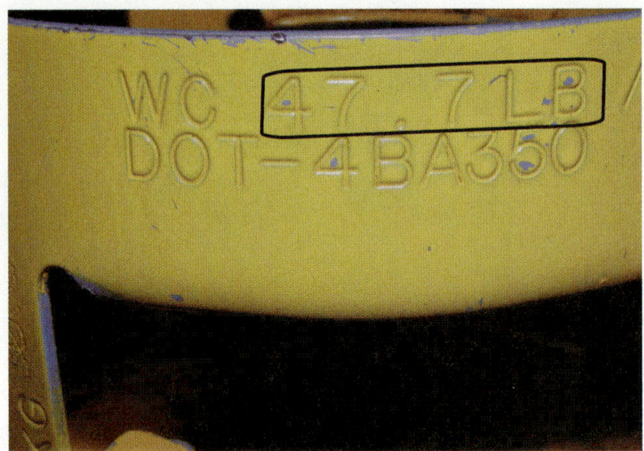

Figure 22-12 This recovery cylinder can hold 47.7 lb of water.

Recovery Cylinder Markings

In addition to the DOT specifications and service pressure markings, there are other crucial markings stamped on the recovery cylinder, including the following:

- Water capacity
- Tare weight
- Date of manufacture
- First retest date
- Most recent test date
- Last allowable retest date

The water capacity is equal to the weight of an equivalent amount of water required to completely fill the cylinder (Figure 22-12). The tare weight is the weight of the empty cylinder (Figure 22-13). These pieces of information are used to calculate how much refrigerant the cylinder can safely hold. This topic will be covered in detail in Unit 28, Refrigerant Management.

CODES AND STANDARDS

The regulations for requalification of refrigerant recovery cylinders are specified in Title 49, Code of Federal Regulations (CFR), Section 180.205.

Recovery Cylinder Requalification (Retest)

AHRI Guideline K-2009 requires that refillable cylinders used to recover refrigerant must be inspected and hydrostatically tested a minimum of once every five years in accordance with Title 49 CFR, Section 180.201. The date of manufacture and the first retest date are stamped onto every cylinder (Figure 22-14). The first retest date gives the month and year that a cylinder must be requalified. If the first requalification date has not yet arrived, the cylinder should be still on its first five years. Note that the five years starts when the cylinder is made, not when it is purchased. A cylinder that was made two years ago can only be used for another three years before getting requalified, even if it has been

Figure 22-13 This recovery cylinder weighs 28.1 lb when empty.

sitting on the shelf in the store. If the first retest date has already passed, the cylinder should not be used unless it was retested. If it was retested, the date that it was retested should be stamped on the cylinder. If the date was less than five years ago, the cylinder should be safe to use.

Some cylinders also have a last permissible use date (Figure 22-15). These are commonly known as 5 + 5 cylinders. They can only be requalified one time. These cylinders may not be used past the last permissible use date.

CODES AND STANDARDS

AHRI Guideline N-2008 and Guideline K-2009 address suggested colors for most refrigerant containers and recovery cylinders.

Cylinder Color and Labeling

AHRI Guideline N-2008, Guideline for Assignment of Refrigerant Container Colors, addresses suggested colors for most refrigerant containers except for recovery cylinders (Figure 22-16). AHRI Guideline K-2009, Guideline for Containers for Recovered Non-Flammable Fluorocarbon

Figure 22-14 This recovery cylinder was manufactured July 1998.

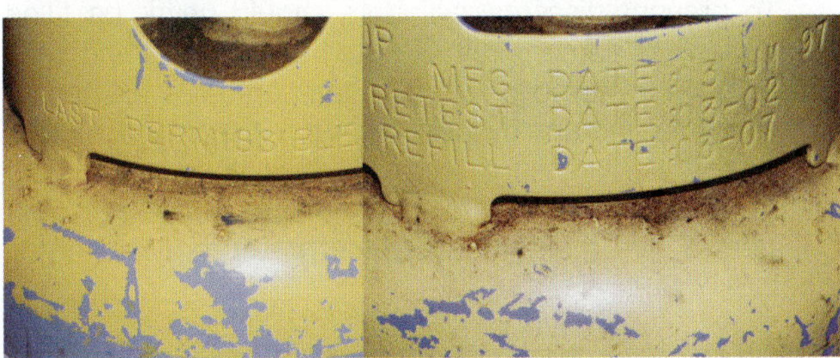

Figure 22-15 This 5 + 5 cylinder was manufactured in March 1997 and retested in March 2002, and the last permissible refill date was March 2007.

Refrigerants, specifies that recovery cylinders all be painted gray with a yellow top regardless of the type of refrigerant in the cylinder (Figure 22-17). This allows the recovery cylinders to be fairly generic and used with a variety of different types of refrigerants. Of course these cylinders must be clearly labeled with the type of refrigerant they contain since cylinder color cannot be relied upon to identify them. The only thing the technician knows for sure by the color is that refrigerant in a gray and yellow cylinder is not new.

The Environmental Protection Agency (EPA) requires that any cylinder containing a recovered refrigerant that is classified as a Class I (CFC) or Class II (HCFC) must display a warning statement: *WARNING: Contains [insert refrigerant type, such as R-22 etc.], which harms public health and environment by destroying ozone in the upper atmosphere.*

Figure 22-16 The suggested refrigerant cylinder colors in AHRI Guideline N.

Figure 22-17 Recovery cylinders are gray with a yellow top. *(Courtesy of National Refrigerants, Inc.)*

Figure 22-18 DOT nonflammable gas label.

Each container should be labeled with the filler's name, address, and date filled. The cylinder should be labeled with a DOT green, 4 × 4–in, diamond-shaped nonflammable gas label (Figure 22-18). Each container should display a precautionary label prepared in accordance with ANSI Z129.1, American National Standard for Hazardous Industrial Chemicals—Precautionary Labeling. In addition, Federal Law 49, CFR Sections 172.301 and 172.302, must be followed. An example, the labeling that is standard on a DOT 39 disposable cylinder and its cardboard shipping carton, is shown in Figure 22-19. Federal law requires the label to include the following:

- Product identity
- Instruction in case of fire, spill, or leak
- Instruction in case of contact or exposure
- Signal word (such as DANGER!, WARNING!, or CAUTION!)
- Statement of hazards
- Instructions for container handling and storage
- Antidotes

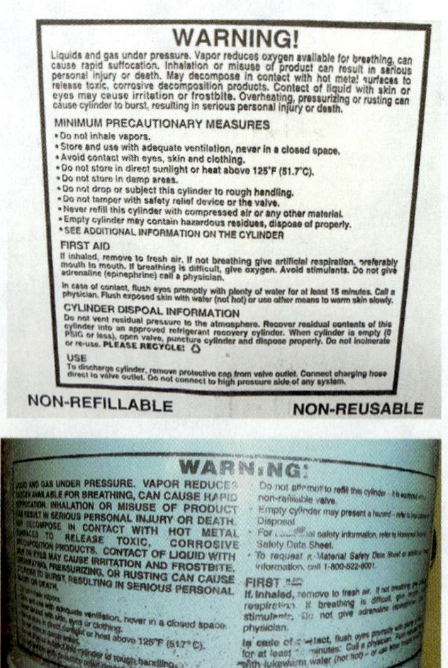

Figure 22-19 Warnings listed on DOT 39 cylinder and its cardboard shipping carton.

- Precautionary measures
- Notes to physicians

TECH TIP

Refrigerants should never be mixed. This is why it is important to clearly mark the recovery cylinder to identify the type of refrigerant it contains. Only use the recovery cylinder for one type of refrigerant even if it has been emptied and then refilled again.

22.11 CYLINDER FILL LEVEL

The contents of any pressurized cylinder will increase in pressure if heated. As long as the cylinder still has some gas in it, the gas will compress and the rise in pressure will normally be slow enough that the built-in safety devices can relieve the pressure when it exceeds 150 percent of the cylinder service pressure. However, if the cylinder is filled completely with liquid, a relatively small rise in temperature can create a very quick rise in pressure. This is because liquid is not compressible. A cylinder that is filled with liquid can go from 100 psig to over 1,000 psig simply by warming up 10°F (Figure 22-20). This can cause a violent explosion. Neither the relief disk on disposable cylinders nor the relief valve on recovery cylinders can relieve pressure quickly enough if a cylinder full of liquid should rise in temperature by 10°F in a few minutes. This type of dramatic pressure increase due to liquid expansion is called hydrostatic pressure.

To prevent disaster, cylinders should never be filled more than 80 percent full of liquid by volume, and they should not be stored or used in temperatures exceeding 125°F. All refrigerant cylinders have a maximum recommended service pressure. This provides a good idea of how much pressure is too much. Do not pressurize recovery cylinders above their rated service pressure. More detail

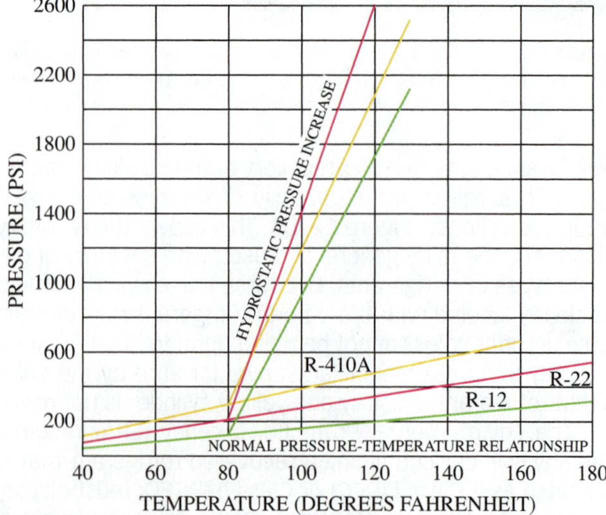

Figure 22-20 This chart shows the rapid rise in cylinder pressure when completely filled with liquid.

on safely filling recovery cylinders is provided in Unit 28, Refrigerant Management.

CODES AND STANDARDS

ANSI/ASHRAE Standard 15-2010, Safety Standard for Refrigeration Systems, is directed toward the safety of persons on or near the premises where refrigeration facilities are located. The standard established safeguards for life, limb, health, and property and prescribes safety requirements.

22.12 ANSI/ASHRAE STANDARD 15-2010

ANSI/ASHRAE Standard 15-2010, Safety Standard for Refrigeration Systems, addresses the design, construction, operation, and testing of refrigeration systems. Mechanical considerations include the rupture or fracture of a component resulting in a release of refrigerant, fire, and explosion. Also addressed is the prevention of personal injury such as suffocation (asphyxiation), cardiac arrhythmia, toxic effects, corrosive attack, and freezing of tissue.

The potential risk of exposure depends on whether a refrigerant leak can enter an occupied space. These occupancy categories are listed in Table 22.4. Systems are classified as low probability or high probability within an occupancy area. The refrigeration system located in a high-probability area must be sized so that a complete discharge of its refrigerant does not exceed the concentration limits set by ANSI/ASHRAE Standard 34-2010. This does not apply to equipment containing less than 6.6 lb., certain laboratories, and industrial occupancies and refrigerated rooms as long as appropriate design conditions are met.

Table 22-4

Occupancy Classification	Examples
Institutional	Hospitals, nursing homes, asylums, prisons
Public assembly	Auditoriums, classrooms, passenger depots, restaurants, theaters
Residential	Dormitories, hotels, apartment buildings, private residences
Commercial	Offices, professional buildings, markets
Large mercantile	Shopping centers where more than 100 persons congregate
Industrial	Manufacturing where access by authorized persons is controlled
Mixed	Two or more occupancies located in the same building

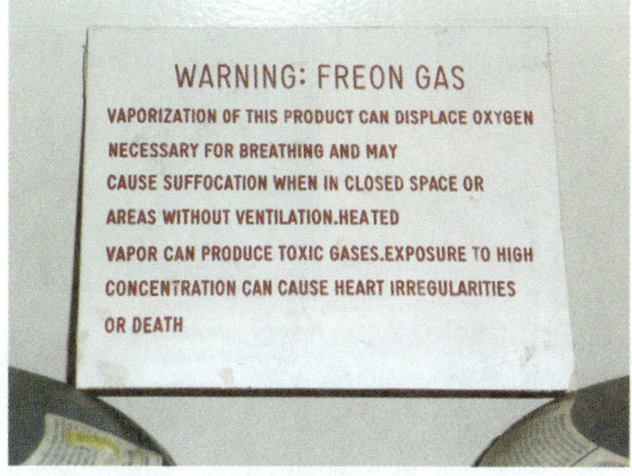

Figure 22-21 Warning sign for a refrigeration machinery room.

Machinery Rooms

Refrigeration machinery rooms located indoors must follow special provisions. Access will be restricted to authorized personnel with clearly marked signs posted at the entrance (Figure 22-21). Doors will be tight fitting, open outward, be self-closing, and be sufficient in number to ensure freedom for persons to escape in an emergency. All machinery rooms are required to have detectors located in an area where refrigerant from a leak will concentrate.

The refrigerant detector will activate an audible and visible alarm and mechanical ventilation within the space at a value not greater than the corresponding TLV-TWA. The alarm must be manually reset, with the reset located inside the machinery room. Do not rely on an automatic detector to announce that the event is over. When entering the space, use a MSHA/NIOSH-approved SCBA if the refrigerant concentration levels are still excessive.

As a special case, ammonia systems require the mechanical ventilation system to be run continuously. Failure of the ventilation system will sound an alarm. Depending on local codes, there may be a provision for a manual emergency discharge of the ammonia refrigerant.

Machinery rooms that use refrigerants classified as flammable (A2, A3, B2, B3) must conform to Class 1, Division 2, of the National Electric Code. In addition, some machinery rooms are required to have a manual shutoff for the refrigeration equipment and a ventilation fan switch located immediately outside the machinery room door.

SAFETY TIP

The total amount of refrigerant stored in a machinery room in all containers not provided with relief valves and piping in accordance with ANSI/ASHRAE Standard 15-2010, Section 9.7, shall not exceed 330 lb.

Pressure Relief

Refrigeration systems need to have some method to relieve excessive pressure that can be caused by fire or other abnormal condition. A large refrigerant receiver contains mainly liquid refrigerant by weight. In the event of a fire, the excessive heat would cause the liquid refrigerant to turn into vapor. The refrigerant volume would greatly increase

Figure 22-22 Compressor internal pressure-relief valves.

while the receiver volume would remain the same. The resulting increase in hydrostatic pressure could rupture the receiver, causing a violent explosion.

Pressure-relief valves, rupture disks, and fusible plugs are installed to prevent this from happening. Pressure-relief valves may be set to relieve back to the low-pressure side of the system or through escape piping safely to atmosphere. Figure 22-22 shows two different internal relief valves found on refrigeration compressors that recirculate the discharge gas back into the suction side. They will reseat when the pressure returns to a safe level. External safety valves are sealed by the manufacturer and are marked with the data required by Section VIII, Division 1, of the ASME Boiler and Pressure Vessel Code.

Rupture disks are one-time failure devices that release the refrigerant upon excessive pressure. Rupture disks are marked with the data required in paragraph UG-129(e) of Section VIII, Division 1, of the ASME Boiler and Pressure Vessel Code. A rupture disk and its label are shown in Figure 22-23. Fusible plugs melt at a set temperature and will release the refrigerant as a one-time failure device. They are marked with the melting temperature in Fahrenheit or Celsius. Both rupture disks and fusible plugs release directly into the atmosphere, so they are normally piped to direct the refrigerant safely outside the building in accordance with ANSI/ASHRAE Standard 15-2010, Section 9.

22.13 REFRIGERANT SENSORS FOR COMMERCIAL APPLICATIONS

ANSI/ASHRAE Standard 15-2010 requires that refrigeration machinery rooms be equipped with refrigerant sensors to sound a visible and audible alarm in the event of a release above the TLV-TWA. However, many building codes across the country are now being modified to meet new environmental regulations. This makes ANSI/ASHRAE Standard 15 increasingly open to interpretation by local jurisdictional codes. The standard is being applied for many different applications:

- Banana and produce rooms
- Beverage plants
- Bakeries
- Chemical plants
- Confined-space entry
- Food processing
- Gas bottling plants
- Ice rinks
- Product coolers
- Rack houses
- Refrigeration systems
- Supermarkets
- Wineries
- Mushroom farms
- Ice cream storage
- Poultry, meat, and fish processing

As an example, a typical walk-in cooler located in a supermarket can be interpreted as a "confined space," creating the need for refrigerant-specific sensing. Ice rinks provide another example. Most ice rinks use R-22 as the refrigerant, but ammonia is becoming more popular in rinks because it does not bring the same environmental issues as halocarbon refrigerants. Refrigerant sensors may need to be placed strategically around the arena.

Figure 22-23 Rupture disk and label.

Sensor Types

There are many different sensor types. Some require pumps and filters to draw a sample of refrigerant into the sensor. These may take periodic samples rather than continuous samples. Other types, such as diffusion-type sensors, have no moving parts and will continuously monitor the space. Some sensors can detect more than one type of refrigerant, while others are designed for one refrigerant type only.

Ceramic metal oxide semiconductor (CMOS) sensors are the earliest versions. The conductivity of sensitized metal oxides will change when exposed to refrigerant gases. More common now are infrared sensors. The operation of this type is based on the principle that most gases absorb infrared energy (light) at a characteristic frequency (wavelength).

Infrared non-dispersive (NDIR) or absorptive infrared systems compare an air sample from the space to a sample of inert gas (usually nitrogen) stored in the monitor. The sample and the inert gas are irradiated with infrared light, and the light absorption of each is measured and compared. With this type of sensor, inaccuracies can occur over time and are referred to as "drift."

Photoacoustic infrared (PIR) sensors eliminate the need to compare results to a reference sample and therefore reduce the possibility of drift. This type of sensor directly measures the changes in pressure that occur when an infrared light is absorbed by the refrigerant in the sample.

Diffusion-design sensors use infrared technology but do not require pumps or filters for sampling. This type of sensor utilizes a broadband infrared source to emit energy at the refrigerants narrow range of frequencies. The sensor contains a pulsed infrared source, a gas sample cell, an optical filter, and a detector. A released refrigerant gas collecting in the sample cell will absorb energy reaching the detector. This change in energy is detected, amplified, and sent to the signal-processing portion of the system. The space is monitored continuously, unlike a sample-draw system. This type of sensor can be used for most of the common refrigerants, including ammonia. The standard refrigerant detection range is 0–3,000 ppm, but units can be rescaled in the field if lower trip points are required. Ammonia can be detected over a range of 0 to 2 percent (20,000 ppm).

Electrochemical (EC) sensors can be used to detect concentrations in the low-ppm range and are often used for specific harmful gases such as carbon monoxide, chlorine and ozone. This type of sensor can also be desirable for toxic refrigerants such as ammonia. A pair of polarized electrodes is located within the sensor and isolated from the space by a gas-permeable membrane. When released ammonia passes through the membrane, a measured redox (oxidation reduction) reaction occurs, which is then amplified and sent to the signal-processing portion of the system.

Photoionization detectors (PIDs) can also be used for ammonia systems instead of EC sensors. Since EC sensors detect low ppm levels, they may not have the range required for determining high levels of concentration. Ammonia has an ionization potential of 10.18eV and can be readily measured with a PID containing a standard 10.6eV lamp. PIDs may have a range of up to 10,000 ppm, which is helpful when making appropriate PPE decisions.

SAFETY TIP

These are the recommended personal protection guidelines for an ammonia release:

Over 35 ppm for 15 minutes: put on a respirator.

Over 250 to 300 ppm: put on a SCBA.

Over 250 to 5,000 ppm: put on Level A suit.

Over the LEL (lower explosion limit): evacuate the area.

Sensor Location and Maintenance

A single sensor takes a sample at the location where it is mounted. Some sensors are equipped with multipoint sampling that allows samples to be taken at more than one location by a single sensor. A good rule of thumb is that there should be a sensor or sample point between each chiller located in a machinery space. As an example, for two chillers, sample between them; for three chillers, sample between each pair of chillers; etc. For refrigerated product storage areas (cold rooms), a sensor should be installed for each 20,000–30,000 cu ft of room volume. As an example, a banana storage room that is 20 ft high, 40 ft long, and 35 ft wide would take up 28,000 cu ft of space.

TECH TIP

1 lb of refrigerant will evaporate to occupy about 3 to 4 cu ft of volume. To put this in perspective, consider that a basketball takes up about 1 cu ft of space, so about 3 to 4 basketballs. This 1 lb of refrigerant would spread out if released into a 30,000–40,000 cu ft space. The average concentration resulting from 1 lb of refrigerant released into the space would be about 100 ppm.

The sensor should be located where the leak is most likely to occur. CFC/HCFC/HFC vapors are heavier than air and will settle to a low area. In this case, sensors should be mounted about 1–2 ft from the floor, close to the potential leak source. Because ammonia is lighter than air, this type of sensor would be mounted near the ceiling. If the primary application for the sensor is for personal protection (TLV-TWA) then mount the sensor at a height in the breathing zone of the employees, typically 5 ft off the floor.

Airflow patterns in the space must be taken into account so that the sensor is mounted downstream from the potential leak source. Place it in a location that is easily accessible for calibration and maintenance. Do not mount the sensor over a door. Sensors may also be located in refrigerant safety vent lines to protect against and accidental loss of refrigerant due to a slowly leaking relief valve, rupture disk, or fusible plug.

The sensor should not be placed in locations of excessive humidity or in a water wash-down area unless the sensor is specifically designed for such use. A sensor utilizing a sample-draw system could have ice form in the draw tube, which would block the lines and disable the system. There are some sensors that are specifically built to resist dust and water for mounting in harsh locations. They can operate in conditions of 100 percent relative humidity and at ambient temperatures as low as –40 to –50°F.

For routine maintenance, sensors should be exposed to a refrigerant sample on a monthly basis to ensure that they are operating correctly. This will also provide a check for the visible and audible alarms. Infrared sensors should have signal voltages measured and logged on a regular basis. Calibration should be performed with certified calibration gas every six months. Calibration kits to perform these tests can be obtained from the equipment manufacturers. All tests and calibrations should be logged.

UNIT 22—SUMMARY

Safety in the workplace is always the number-one priority. The hazards that can result from refrigerant exposure and misuse fall into a number of different categories: toxicity, frostbite, flammability, explosion, decomposition, and pressure. Varying levels of toxicity can result from short-term exposure to high concentrations or long-term exposure to lower concentrations. Acceptable levels are referred to as threshold limit values (TLV). Frostbite results from direct exposure and can be avoided by wearing the correct PPE. An explosion could result from a flammable refrigerant, which is different as compared to a cylinder rupturing because of excessive pressure. Both cases can lead to serious injury or even death. Always refer to the refrigerant MSDS, because even some nontoxic refrigerants can produce poisonous gases if exposed to a flame. To protect yourself and the customer, always take proper precautions, wear the appropriate PPE, and review the refrigerant MSDS before servicing any refrigeration system.

UNIT 22—REVIEW QUESTIONS

1. Who was the person instrumental in developing the first CFCs?
2. Why it is important to review the MSDS before an accident happens?
3. What should be done to treat frostbite to the skin?
4. What should be done if refrigerant comes into contact with the eyes?
5. High levels of refrigerant inhalation may lead to what?
6. How do you treat a person who has inhaled large amounts of refrigerant?
7. Can you physically feel that there is a lack of oxygen in a confined space?
8. What steps would you take if a person fell unconscious due to a lack of oxygen in a confined space?
9. Most fatal refrigeration accidents have occurred due to which type of refrigerant?
10. How would a refrigerant safety rating of A1 compare to B3?
11. What is the maximum safe exposure level in parts per million by volume called?
12. What is the difference between TLV, OEL, and PEL?
13. What is meant by lower flammable limit (LFL)?
14. How would the LFL of propane compare to ammonia?
15. What happens to CFCs and HCFCs that are exposed to a flame?
16. Can a DOT 39 cylinder be refilled? Why or why not?
17. What is the maximum temperature that a DOT 39 cylinder can be exposed to?
18. What is the purpose of a frangible disk on a nonrefillable refrigerant cylinder?
19. Can a DOT 39 cylinder be recycled?
20. How often must a recovery cylinder be retested?
21. Do recovery cylinders need to be labeled?
22. Can a recovery cylinder be filled to 100 percent capacity?
23. What happens when a machinery room refrigerant sensor detects above-normal concentration levels of refrigerant?
24. What is the difference between pressure-relief valves, rupture disks, and fusible plugs?
25. How are machinery space refrigerant sensors maintained?

UNIT 23

Refrigerant System Servicing and Testing Equipment

OBJECTIVES

After completing this unit, you will be able to:

1. list the different types of thermometers.
2. explain the advantages and disadvantages of each type of thermometer.
3. list common air temperature measurements technicians make.
4. list the different types of gauges that can be used to measure pressure.
5. explain how gauge accuracy is stated.
6. explain the operation and use of a manifold gauge set.
7. list the types of positive-shutoff low-loss fittings available.
8. explain why a micron vacuum gauge is used to measure vacuum instead of a compound gauge.
9. list the different types of refrigerant leak detectors.

23.1 INTRODUCTION

Measuring and testing is the only way an HVACR technician can verify correct system installation and operation. Technicians engaged in installing, servicing, and troubleshooting HVACR systems need to measure the performance of the refrigeration system. During installation, technicians measure the vacuum level using vacuum gauges. System pressures and temperatures are measured to verify system operation. The temperatures and pressures within a refrigeration system are key pieces of information that all HVACR service technicians need to diagnose refrigeration problems. Instruments that test temperature, pressure, and vacuum and check for refrigerant leaks are all essential to HVACR work. Testing and measuring to verify correct operation is the only way to ensure that high-efficiency equipment will perform as designed.

Most of the instruments used are digital, although many analog instruments are still in use. The introduction of digital electronics to temperature and pressure sensing has greatly increased both accuracy and speed. Today's test instruments can provide readings to one or more decimal points. That level of accuracy is often important to ensure that the system is not only running but also operating at the highest efficiency.

TECH TIP

Sometimes people confuse thermometers and thermostats. The difference between a thermometer and a thermostat is that a thermometer simply indicates the temperature. A thermostat controls equipment based on the temperature it reads.

23.2 TEMPERATURE MEASUREMENT

HVACR technicians measure temperature daily, and taking accurate temperature readings is an essential part of the job. System performance is affected by the temperature of the air or water flowing through the units. System performance affects the temperature of the final product. Temperature readings allow technicians to monitor system operating conditions and system performance. The performance of particular refrigeration system components can often be determined by taking temperature readings. Part of taking accurate temperature readings is using the right thermometer. There are many types of thermometers available, including glass stem, dial, digital, and infrared. Each type has advantages and disadvantages. Some types of thermometers are better suited for a particular job than others are.

23.3 GLASS-STEM THERMOMETERS

Glass-stem thermometers once were commonly used by refrigeration service people. The temperature is measured on a scale by comparing the tip of the liquid inside the thin tube to a scale on the side of the tube. Any liquid can be used, but mercury and alcohol are the most commonly used. Mercury looks metallic in color, and alcohol is usually dyed red (Figure 23-1). The thinner the size of the hollow glass tube, the greater the movement per degree of temperature change. Glass thermometers are available from pocket-sized models to longer, more accurate models for laboratory work. Breakage is a problem since glass thermometers are fragile. Many technicians would consider the slow response time of this type of thermometer to be a disadvantage.

An advantage of the glass-stem thermometer is that they are highly reliable. They do not rely on any mechanical

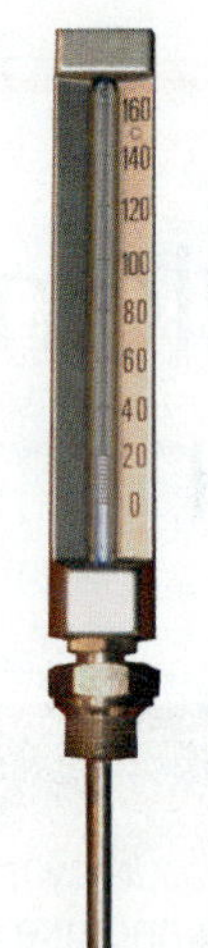

Figure 23-1 Glass-stem thermometer.

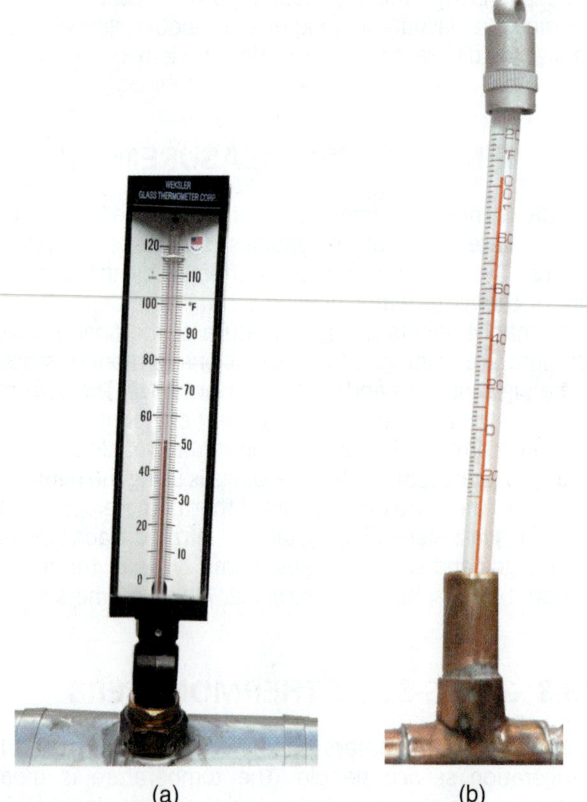

Figure 23-2 Glass-stem thermometers: (a) in protective case; (b) with glass tube exposed.

or electronic device that might stick or quit working (Figure 23-2). Their reliability and relative low price are reasons they are extensively used on large central-heating and cooling plants used in large buildings or factories.

23.4 DIAL THERMOMETERS

Dial thermometers indicate temperature by a pointer moving over a circular scale (Figure 23-3). The pointer moves because it is attached to a piece of metal inside the thermometer that

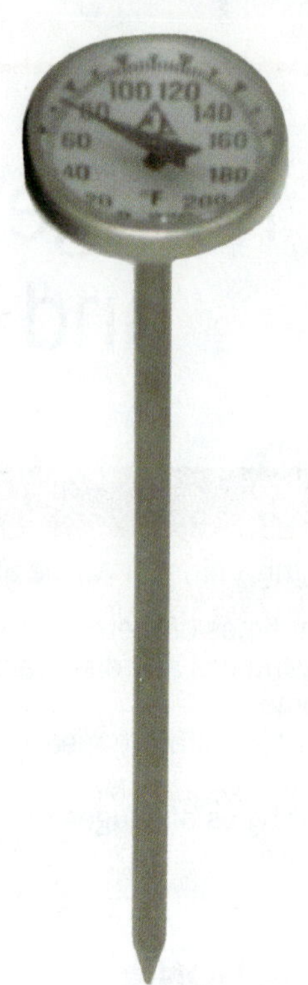

Figure 23-3 Dial-type pocket thermometer.

has a high rate of thermal expansion and contraction with temperature changes. The metal inside the thermometer can be straight or it may be wound into a spiral or spring. If it is straight, it expands linearly; if it is in a spiral or spring, it rotates (Figure 23-4). Spiral or spring-wound metal strips are constructed from two different types of metal attached together. Because each strip of metal has a different rate of thermal expansion, the twisting or bending motion is greater than it would be if a single type of metal were used. An advantage of dial thermometers is that they are more rugged than glass-stem thermometers and can be used for a wide variety of applications. A disadvantage is that they are not as accurate as either glass-stem thermometers or digital thermometers.

23.5 DIGITAL THERMOMETERS

Digital thermometers with compact sensing elements are popular because of their speed in sensing a change in temperature and the ease in reading the alphanumeric display instead of a dial or a scale (Figure 23-5). Most digital thermometers require batteries or a source of electrical power to produce the digital display. The display can be easily changed to display the temperature reading in the Fahrenheit or Celsius scales.

Advantages of digital thermometers are the many features available on some models, including temperature differences between two or more sensing elements, highest and lowest temperatures, average temperature, and lighted displays. They are useful in measuring surface temperatures, such as suction superheat temperatures when adjusting an expansion valve.

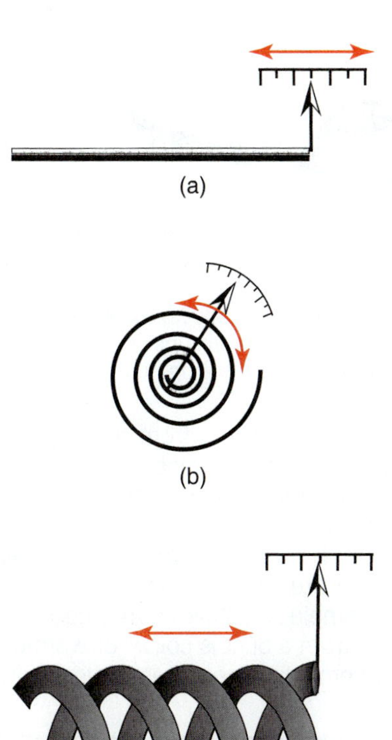

Figure 23-4 Spiral or spring-wound dial thermometer operation.

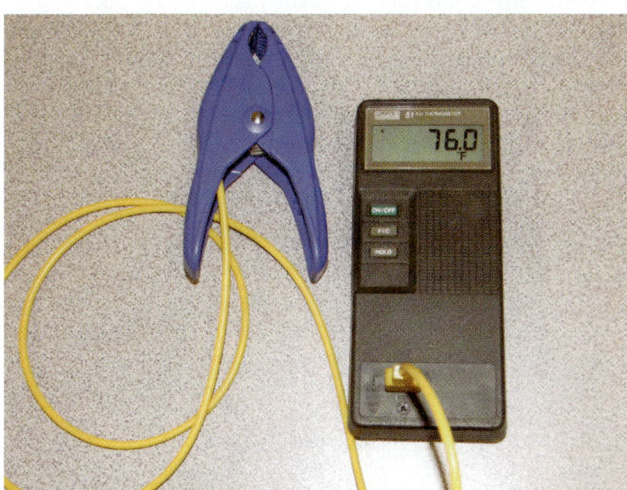

Figure 23-5 Digital thermometer.

23.6 THERMOCOUPLE TEMPERATURE SENSORS

Thermocouple temperature sensors can be used to detect temperatures that are displayed on analog or digital instruments. The thermocouple produces a small voltage when two dissimilar metal wires are joined, forming a tip or junction that can be heated or cooled. A number of dissimilar metals, such as chromel and alumel or iron and constantan, are used in commercially available thermocouple wires. Type J (iron and constantan) sensors have been used for years but are being replaced with the more popular type K (chromel and alumel) because of its lower cost. One problem with thermocouple sensors is their accuracy. They may provide a temperature that is 2°F higher or lower than the actual temperature. You can test the accuracy of a thermocouple by comparing its temperature reading with another type of thermometer. Once you know the temperature error factor, you can mark it on the thermocouple. That way, you can add or subtract a degree or two as needed to keep the error factor within an acceptable range for most HVACR work.

Some of the advantages of thermocouples are their relative low cost, which makes them readily available, and their flexibility. They can be attached to several locations inside the equipment, and the equipment cover panels can then be reinstalled. With the thermocouple leads run so they are outside the cabinet, the equipment can be operated normally. Now the technician can better diagnose what is happening to temperatures inside the operating equipment.

When taking the temperature of a refrigerant line, make sure the tip of the thermocouple is tightly pressed against the line. You should also insulate the tip, because thermocouples are very sensitive to their surrounding temperature. Some meter manufacturers supply a short strip of Velcro to wrap around the thermocouple tip to hold the tip securely against the refrigerant line and insulate it.

To make it easier to get accurate refrigerant line temperatures, some manufacturers produce line-clamping thermocouple heads (Figure 23-6). These clamping thermocouples provide both a tight connection and insulation from ambient air, Figure 23-7.

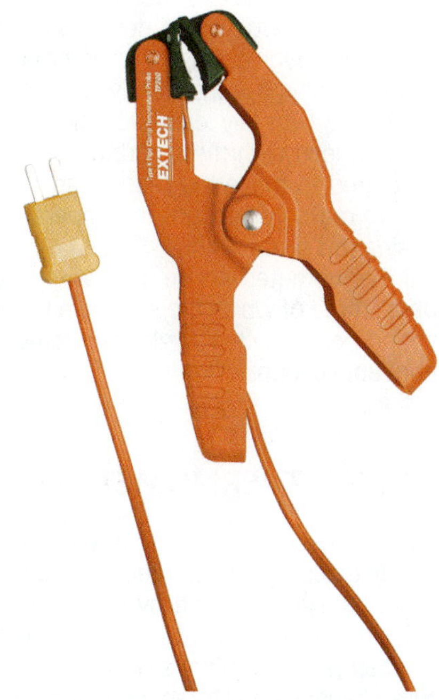

Figure 23-6 Thermocouple clamp. *(Courtesy of Extech Instruments, A FLIR Company)*

Figure 23-7 Line-clamping thermocouple head. *(Courtesy of HVAC Department Terra Community College, Freemont, Ohio)*

TECH TIP

Sometimes the thermocouple leads used with digital temperature gauges may have a slight error in the temperature reading. For this reason, it is a good idea to test a new set of temperature probes against each other or against another temperature-measuring instrument. This comparison will allow you to determine the accuracy and any variance in readings that you may receive from these leads. If you consistently find a variance greater than 1° between any lead and other test instruments, mark that lead with the amount of temperature difference required to make the reading accurate. This will allow you to make corrections in temperature readings when using this probe.

A digital thermometer can be used with a wide variety of probes, three of which are shown in Figure 23-8. The different probes allow the single thermometer to be used for more applications.

23.7 INFRARED THERMOMETERS

An infrared thermometer can "see" the temperature of a surface from a distance. What it is seeing is the color of the invisible infrared (IR) light given off by the object. A detector is located inside the instrument that senses the infrared light waves. A change in the IR light's color occurs as a result of temperature changes, which results in a change in the electrical output of the detector. A small integrated circuit converts that electrical output into a displayed temperature.

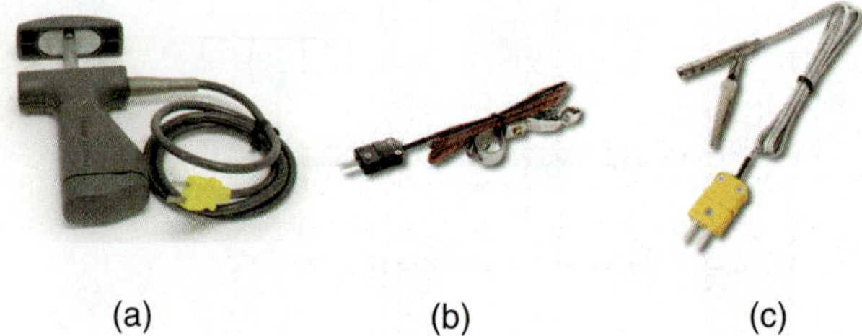

(a) (b) (c)

Figure 23-8 Temperature sensor probes: (a) temperature sensor for refrigerant piping; (b) ambient temperature sensor with clamp; (c) flame-sensing temperature probe.

The distance the detector is from the object, the surface color, and its reflectance or reflectivity all affect the accuracy of the temperature read. The most accurate reading obtained is when the detector is closest to a black, dull surface. However, not all HVACR field work requires extreme accuracy. Technicians are often looking for trends in temperature change. For example, a technician may need to determine if the air coming from a duct is colder or warmer than it was before the system adjustment was made (Figure 23-9).

SERVICE TIP

A disadvantage of IR thermometers is that they cannot detect air temperatures; however, they can detect the temperature of the grille or register that the air is passing through. To obtain the most accurate estimate of the actual duct air temperature, point the detector into the louvers so it is pointed at the inside of the duct boot. Some IR thermometers have a laser pointer to make this easier. Move the instrument around and watch the temperature display. Watch for the highest or lowest temperature (highest for heating and lowest for AC) indicated on the display. That temperature is going to be the closest to the actual air temperature. Practice will help develop your skill in obtaining accurate temperature readings with an IR thermometer.

Figure 23-9 Using an infrared thermometer for measuring air damper temperature.

There are many advantages of infrared thermometers. The fact that a temperature reading can be made from a distance may mean that you do not need to climb a ladder to get air duct readings. It may also mean that you can get a reading from the back of a hard-to-reach space. The temperature is displayed instantaneously, which speeds up your work.

23.8 TAKING ACCURATE TEMPERATURE READINGS

Getting accurate temperature readings requires the right equipment and the right techniques. Common temperatures read while working on HVACR equipment include air, water, and refrigerant line temperatures.

Air temperature can be read with dial-type pocket thermometers, digital pocket thermometers, or electronic temperature testers. Infrared thermometers are less desirable for air temperature readings, because they must read the temperature of objects. However, an infrared thermometer can be used to measure air by targeting the register where the air is leaving.

Water temperature readings can also be read with dial-type pocket thermometers, digital pocket thermometers, or electronic temperature testers. If possible, the thermometer should be inserted directly into the water stream. When using electronic thermometers, be sure that the sensor is rated for liquid immersion. Infrared thermometers do not work well for measuring liquid. An infrared thermometer can be used if the water is contacting an object that the infrared can target.

Pipe temperatures are best taken with a thermocouple temperature tester. Probes are available with clamps that hold the thermocouple against the pipe (Figure 23-10). Bead-type thermocouples can be held against the pipe with a Velcro strap for an accurate reading. Pocket-type thermometers can be used only if the tip can be held securely against the pipe and insulated from the surrounding air. A Velcro strap or Permagum can be used to hold the tip of the thermometer against the pipe and insulate it from the surrounding air. Most infrared thermometers are not suited for measuring pipe temperature because of their distance to spot ratio. Infrared thermometers with a spot ratio of 10:1 or greater will work fine with dull pipes larger than ½ in. The emissivity of the pipe can also be a problem for shiny materials like copper.

Figure 23-10 Thermocouple clamp on pipe.

Figure 23-11 Thermometer well with stem thermometer in it. *(Courtesy of Ritchie Engineering Company, Inc.-Yellow Jacket Products Division)*

Some systems are equipped with thermometer wells. These are designed for use with thermometers that have metal stems, like pocket thermometers (Figure 23-11). The well should be filled with mineral oil to ensure good thermal contact between the well and the thermometer.

23.9 AIR TEMPERATURE READINGS

Three air temperatures are commonly read when working on air-conditioning equipment: the outdoor ambient temperature, the return air temperature, and the supply air temperature. Ambient temperature is the temperature around the unit, typically outside. Return air is the air returning to the system from the conditioned space. Supply air is the conditioned air being supplied to the conditioned space.

Outdoor Ambient Temperature Readings

Outdoor ambient temperature readings should be taken in areas that are free from discharge air from condensing units or dryer vents and out of direct sunlight. Holding the thermometer in front of the air intake grille of the outdoor unit normally works well.

Return-Air Temperature

Accurate return-air temperatures can be obtained by holding a pocket or digital thermometer in front of the return air grille. Infrared thermometers can be used with return-air grilles by targeting the grille.

Supply-Air Temperature

Supply-air temperatures can be more difficult. The unit should operate long enough for supply-air temperatures to

Figure 23-12 Thermometer reading supply-air temperature.

stabilize. If possible, the thermometer should be inserted inside the supply-air register (Figure 23-12). When the supply air leaves the register, it mixes with room air, affecting the temperature of the supply air. Infrared thermometers can be used with supply-air registers by targeting the supply-air register. When measuring supply-air temperature with an infrared thermometer, the unit must run long enough for the register to be the same temperature as the supply air.

Duct Temperatures

The temperatures read at the registers will be different from the temperature at the unit because of duct loss and gain. When checking a system performance specification, such as furnace temperature rise, it is best to check temperatures at the unit. Drilling a small hole in the ductwork allows temperature readings of the air just as it enters and leaves the unit. Holes should generally not be drilled in the actual unit or in flex duct. A hole in the wrong place in the unit can damage the unit. Flex ducts should not be pierced; poking holes in a flex duct allows air to pass between the inner and outer jacket. This increases duct loss and can lead to condensation on and inside the duct.

The return-air plenum is the best location for checking return-air temperatures, especially if there are several return ducts. The thermometer will read the mixed air temperature, which may be different from the temperature in individual returns. If there are any components installed in or on the plenum, be careful to avoid them. The temperature of air flowing to the return air through a bypass humidifier or bypass damper will be a very different temperature than the rest of the return air.

The supply-air plenum can be used on systems with a radial duct design using flex duct. The hole in the plenum should not be too close to where the plenum connects to the unit. When possible, take supply-air temperatures in a trunk duct leaving the plenum (Figure 23-13). This avoids the problems of radiant heat from furnace heat exchangers or interference from components installed inside the plenum.

Figure 23-13 Thermometer placed in hole in supply duct.

SAFETY TIP

When drilling holes in ductwork to measure temperature, make certain you know what is on the other side of the duct; hopefully just air. Contacting a set of electric heat strips with a metal thermometer that is sticking through a metal duct can cause a direct short on a high-current device and give you a serious shock!

23.10 PRESSURE MEASUREMENT

The proper operation of refrigeration equipment depends upon maintaining the correct pressures throughout the system. HVACR technicians must be able to check system operating pressures to verify correct operation. Pressure readings are often necessary when diagnosing system problems. The first step in obtaining accurate pressure readings is having equipment that matches the job. Gauges must be designed to work with the particular refrigerant and pressures being measured. Technicians must be familiar with the range of pressures that can be expected inside the equipment. They must also know what equipment is required for those pressure measurements. Using the wrong tool can not only lead to incorrect readings, but it can also be dangerous.

CAUTION

Do not use any refrigerant gauge sets on R-410a that are not rated for its higher pressures. It is dangerous to use older gauge sets designed for R-22 because they can be damaged by the higher R-410a system pressures.

Figure 23-14 Example of a bellows-style refrigeration gauge. *(REFCO Manufacturing (US) Inc.)*

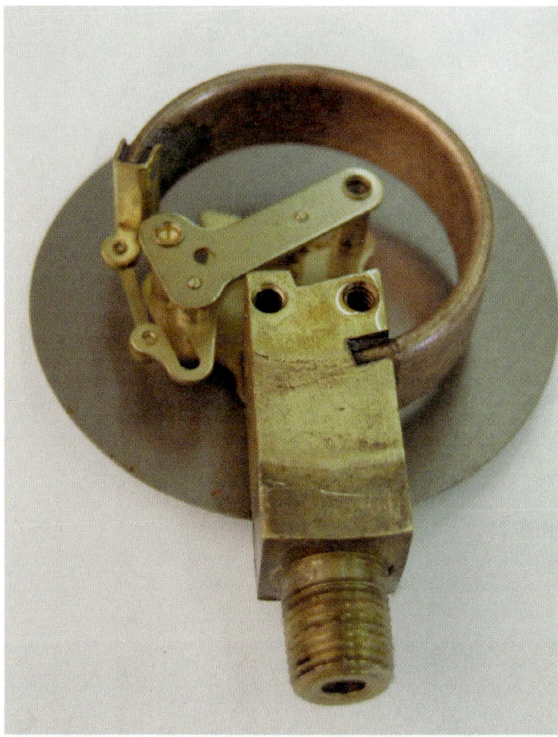

Figure 23-15 Bourdon tube.

23.11 REFRIGERATION GAUGES

Mechanical gauges require a physical device to translate pressure changes into movement that can be used to move a dial. Three types of devices are used in mechanical gauges: diaphragm, bellows, and Bourdon tube. Diaphragm gauges are generally used for lower pressures within a limited range and are usually not used in refrigeration. Bellows gauges are used in refrigeration gauges, but they are not as common as Bourdon tube gauges. The bellows expand with an increase in pressure and contract with a decrease in pressure. Figure 23-14 shows a bellows gauge used in refrigeration.

Most refrigeration gauges use a Bourdon tube as the operating element. The Bourdon tube is a flattened metal tube that is sealed at one end, curved, and soldered to the gauge fitting at the other end (Figure 23-15). Figure 23-16 illustrates how a Bourdon tube operates. The blue shaded section shows the tube at rest at atmospheric pressure. A rise in pressure inside the tube tends to make the Bourdon tube mechanism straighten. The position the Bourdon tube will move to when pressure is applied is shown in Figure 23-16 in the dotted outline. This movement will pull on the link, which will turn the gear sector counterclockwise. The pointer shaft will move clockwise to move the needle. On a decrease in pressure, the Bourdon tube moves clockwise toward its original position and the pointer moves counterclockwise to indicate a decrease in pressure.

Pressure Gauges

The high-pressure gauge has a single continuous scale. Gauge sets made for refrigerants such as R-134a and R-22

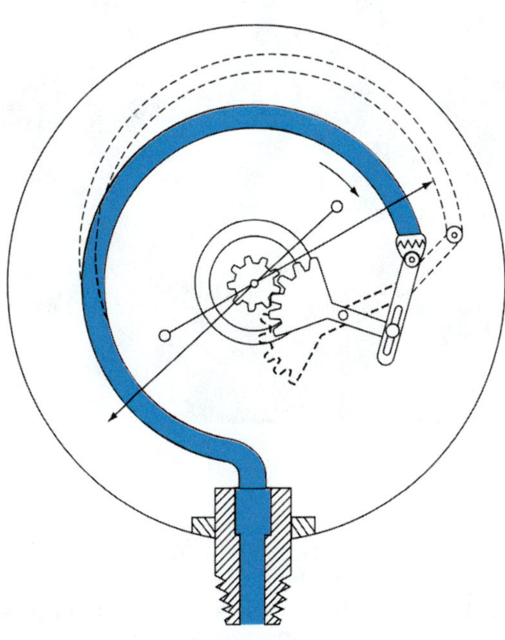

Figure 23-16 The dashed line shows how the Bourdon tube responds to a pressure increase.

are usually calibrated (marked off) to read 0 to 500 psi. Figure 23-17 shows a high-pressure gauge. The scale is usually marked in 5-lb increments. On most gauges, the outer scale is the pressure scale and the inner scales indicate the saturation temperature of different refrigerants at the indicated pressure. For example, if the gauge pointer indicated 200 psi pressure for R-22, the saturation temperature of the refrigerant would be approximately 101°F.

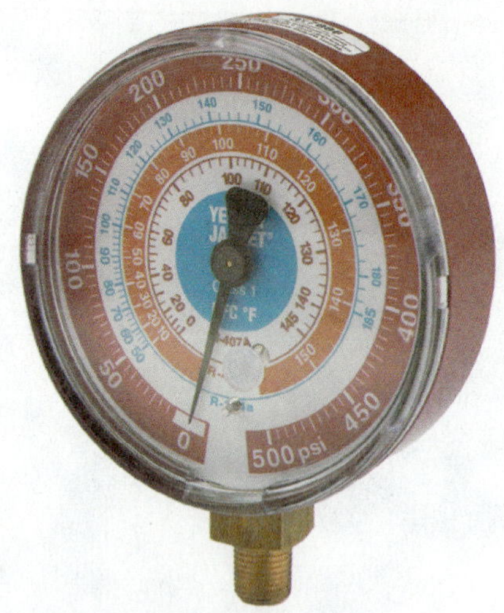

Figure 23-17 High-side pressure gauge for most common refrigerants, except R-410a. *(Courtesy of Ritchie Engineering Company, Inc., Yellow Jacket Products Division)*

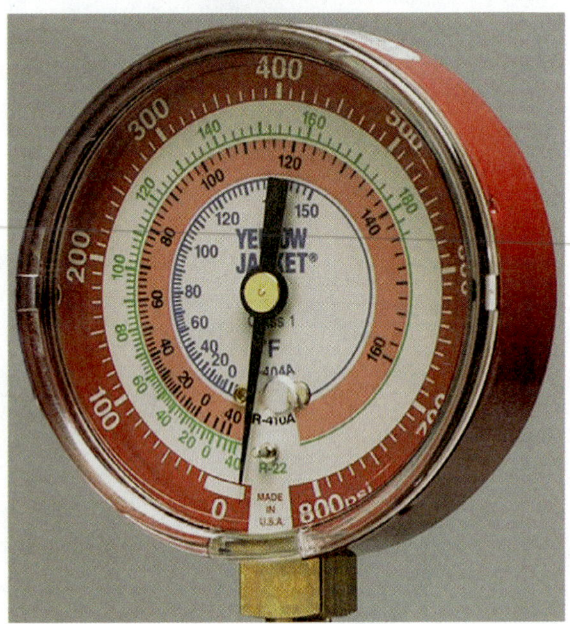

Figure 23-18 High-side pressure gauge for R-410a. *(Courtesy of Ritchie Engineering Company, Inc., Yellow Jacket Products Division)*

Pressure gauges made for R-410a have a higher pressure range because of the higher pressures of R-410 systems. R-410a pressure gauges can typically read pressures up to 800 psig. On an R-410a pressure gauge, each mark on the pressure scale represents 10 psig pressure. Figure 23-18 shows a high-pressure gauge for R-410a systems.

Compound Gauges

The low-side gauge is a compound gauge that measures both pressure and vacuum (Figure 23-19). Gauge sets made for refrigerants such as R-134a and R-22 are usually calibrated from 0 to 30 in of mercury vacuum and from 0 to 120 psig pressure. On a typical compound gauge, each mark on the vacuum scale represents 2 in Hg vacuum, and each mark on the pressure scale represents 1 psig pressure.

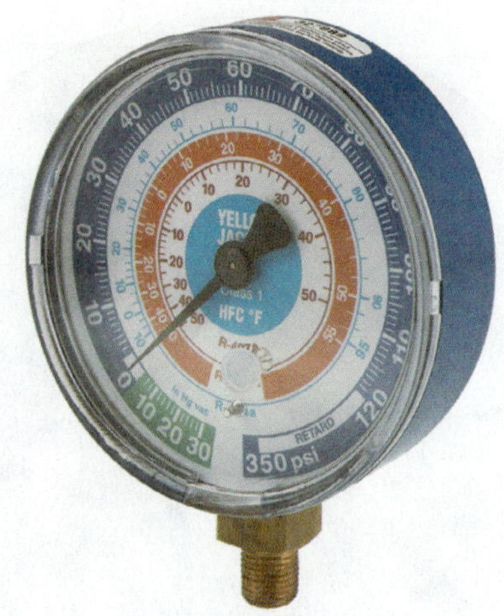

Figure 23-19 Compound gauge for most common refrigerants, except R-410a. *(Courtesy of Ritchie Engineering Company, Inc., Yellow Jacket Products Division)*

It is common for the standing pressure on the low side of a system to exceed the normal range of the compound gauge. The retard range protects the gauges from damage when a pressure above their normal range of operation is connected to them. Gauges for R-134a and R-22 typically have a retard range of up to 350 psig.

Compound gauges made for R-410a have a higher pressure range because of the higher operating pressures of R-410a systems. R-410a gauges can typically read pressures up to 350 psig and have a retard range up to 500 psig. On an R-410a compound gauge, each mark on the pressure scale represents 5 psig pressure. Each mark on the vacuum scale represents 5 in Hg vacuum. On the gauge shown in Figure 23-20, each mark on the vacuum scale represents 10 in Hg vacuum.

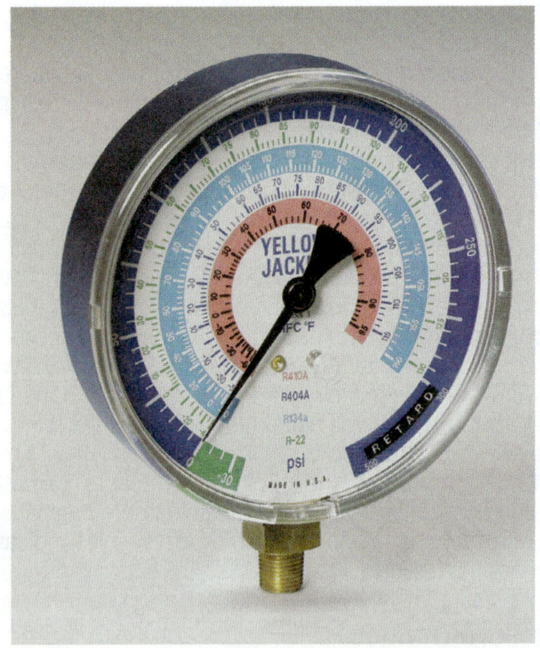

Figure 23-20 Compound gauge for R-410a. *(Courtesy of Ritchie Engineering Company, Inc., Yellow Jacket Products Division)*

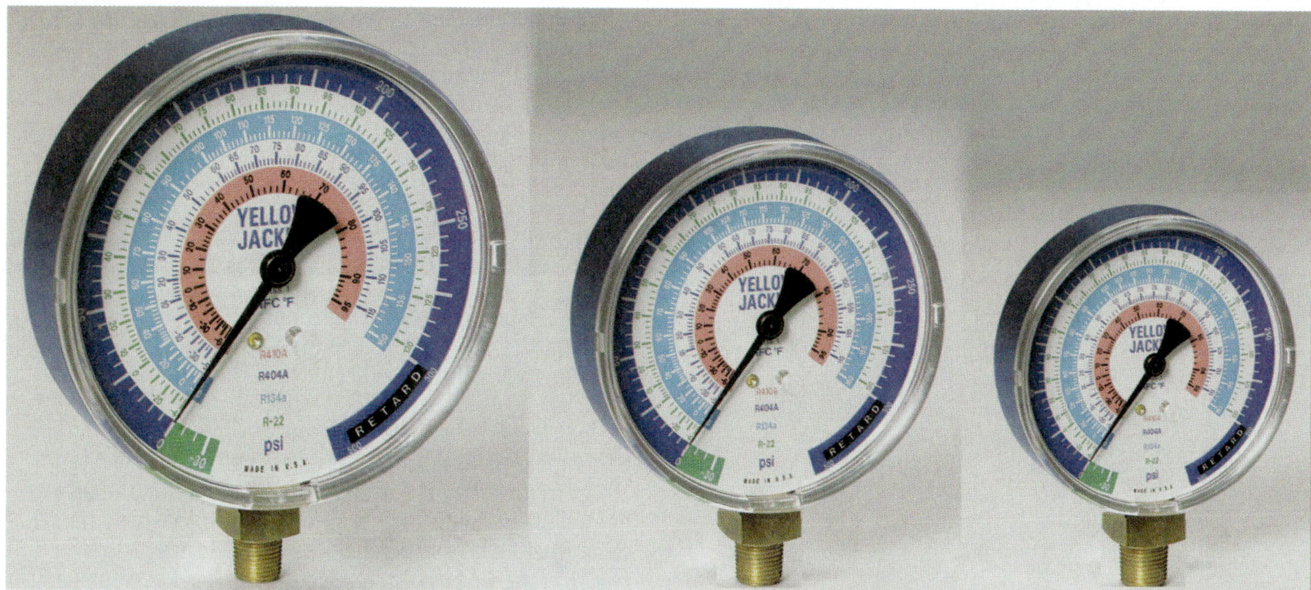

Figure 23-21 Comparison of 2½-in, 3⅛-in, and 4-in mechanical gauges. *(Courtesy of Ritchie Engineering Company, Inc., Yellow Jacket Products Division)*

Like the high-pressure gauge, the compound gauge also has scales calibrated to read saturation temperatures of various refrigerants, such as R-134a, R-22, and R-410a. With these scales, it is not necessary to refer to pressure-temperature tables or curves to calculate pressure-temperature relationships. Figure 23-20 shows a compound gauge for an R-410a system.

Gauge Accuracy

Gauges are classified by their percentage accuracy. Class 1 gauges are accurate to ±1% of the reading throughout their range. This means that 100 psig can read anywhere between 99 and 101. Class 1 gauges are the most accurate mechanical gauges used in refrigeration. Class 1.6 and class 2 gauges are accurate to 1.6 percent and 2 percent, respectively. Class 2-3-2 gauges are accurate to 2 percent for the first third of their scale, 3 percent for the middle third, and 2 percent for the final third of their scale. Class 3-2-3 gauges are accurate to 3 percent for the first third of their scale, 2 percent for the middle third, and 3 percent for the final third of their scale.

Figure 23-22 High-pressure needle vibrating due to compressor pulsing.

Gauge Size

Gauges are available in different sizes. The most common sizes are 2½-in, 3⅛-in, and 4-in diameter. Figure 23-21 shows all three common sizes of gauges. Larger gauges are easier to read than smaller gauges. Their larger circumference produces increments that are spaced further apart, making them easier to read and more accurate. This is especially important for gauges that are used over a wider range of pressures (e.g., gauges designed for use with R-410a).

Liquid-Filled Gauges

During operation of an AC or refrigeration system, the high-side pressure may rapidly pulsate as the compressor valves open and close. This pulsation can cause the high-side gauge needle to flutter or vibrate (Figure 23-22). Liquid-filled analog gauges are available that reduce needle flutter. They contain glycerin to dampen pulsations, lengthen service life, and improve accuracy (Figure 23-23).

Digital Gauges

The most advanced gauges use digital pressure transducers to read pressure instead of the Bourdon tube gauges (Figure 23-24). They reduce reading error by displaying the pressure directly in numbers. Digital gauges can show pressure changes in increments down to 1/10 lb while analog gauges do well to show changes smaller than 5 psig. Digital gauges

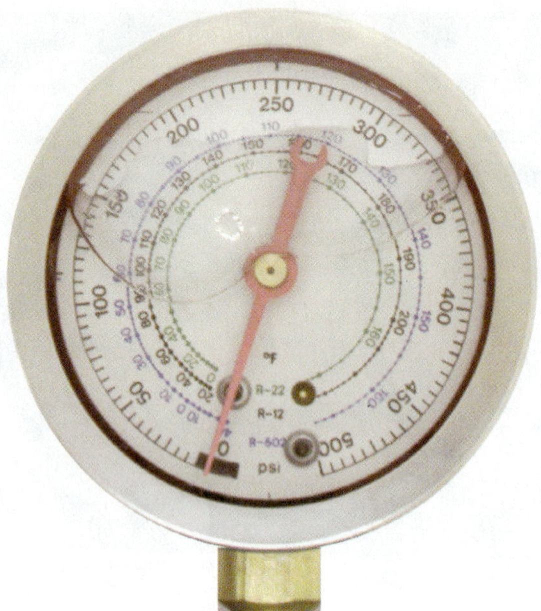

Figure 23-23 Liquid-filled gauges dampen needle vibration.

Figure 23-24 Digital manifold gauge set. *(Courtesy of Ritchie Engineering Company Inc.-Yellow Jacket Products Division)*

also provide accurate built-in saturation temperature correlation for several refrigerants. The digital gauge shown in Figure 23-24 can simultaneously display many important system operating characteristics, including the following:

- Evaporator pressure
- Evaporator saturation temperature
- Suction-line temperature
- Suction-line superheat
- Condenser pressure
- Condenser saturation temperature
- Liquid-line temperature
- Liquid-line subcooling

SERVICE TIP

Refrigeration gauges have a small adjustment screw that allows the gauge to be zeroed. The gauge is adjusted to zero with the hoses open to atmospheric pressure. However, it is possible for a gauge to zero out but not read other pressures accurately. From time to time, the technician should attach the gauges to a new cylinder of refrigerant at a known temperature and compare the pressure reading to the temperature-pressure chart so that the gauge can be checked for accuracy. Gauges that are zeroed and are still inaccurate should be replaced.

23.12 READING REFRIGERANT SATURATION TEMPERATURES

Refrigeration gauges not only read system pressure but also show the refrigerant saturation temperature of common refrigerants. The saturation temperature of a refrigerant may be determined by observing the colored scale for that particular refrigerant. The gauge in Figure 23-25 is showing 150 psig for an R-410a system. The needle also points toward its corresponding saturation temperature, 53°F. The high-pressure gauge shown in Figure 23-26 indicates a pressure of 365 psig and a corresponding temperature for R-410a of 110°F.

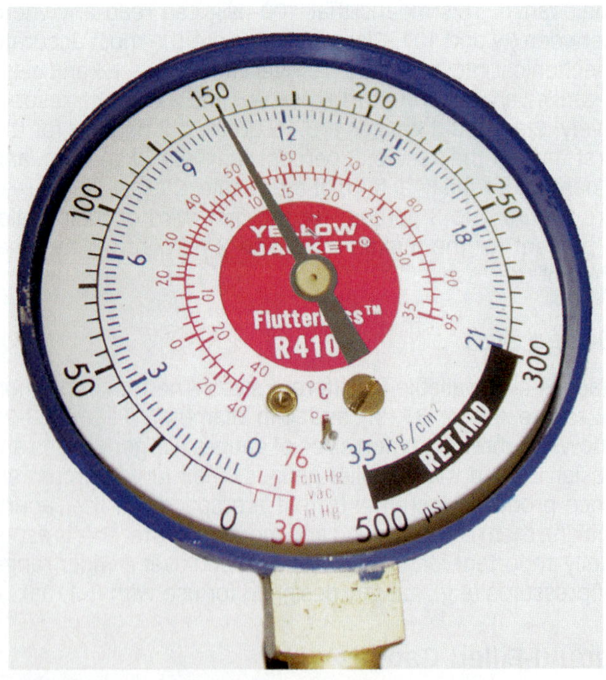

Figure 23-25 This gauge is reading 150 psig with a saturation temperature of 53°F for R-410a.

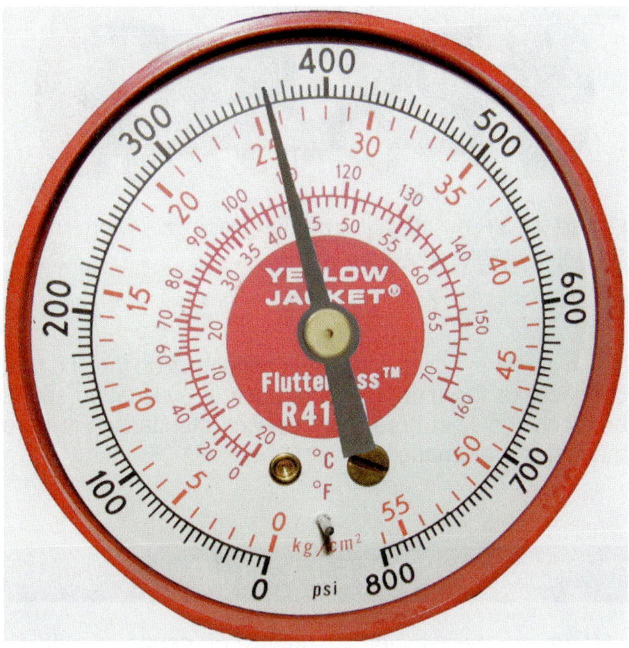

Figure 23-26 This gauge is reading 365 psig with a saturation temperature of 110°F for R-410a.

Figure 23-27 Gauge manifold set.

Digital gauges often have built-in saturation temperature tables for multiple refrigerants. The refrigerant saturation temperature is displayed by pressing a button on the gauge. The saturation temperature display of digital gauges is considerably more accurate than on analog gauges. They can also hold data for many types of refrigerant, compared to the two or three types that analog gauges can display.

23.13 GAUGE MANIFOLD SET

Technicians use the gauge manifold to diagnose trouble in refrigeration systems. Gauges allow the operator to watch both gauges simultaneously during evacuation or charging operations and save time on almost any work that must be done on the system. The most common gauge manifold test sets contains two shut-off valves, three external connections, and two pressure gauges, Figure 23-27. Manifolds are also available with extra ports, valves, and hoses. The gauges and the flexible hoses that connect to the manifold to connect it to the system are color coded; blue is the low side of the system, while red is the high side. The left-hand gauge is called a compound or suction pressure gauge. The blue hose connects to the low side of the system, allowing the blue compound gauge to read the system low-side pressure. The red hose connects to the high side of the system, allowing the red high-pressure gauge to read the system high-side pressure. The center yellow hose is useful for connecting to the charging refrigerant cylinder, vacuum pump, or recovery machine. Figure 23-28 shows a gauge set with color-coded gauges and hoses. Standard hoses have ¼-in inside diameter and are designed to seal on a ¼-in male flare end. Typically, one end is straight and connects to the manifold and the other end is angled to allow easier connection to service valves.

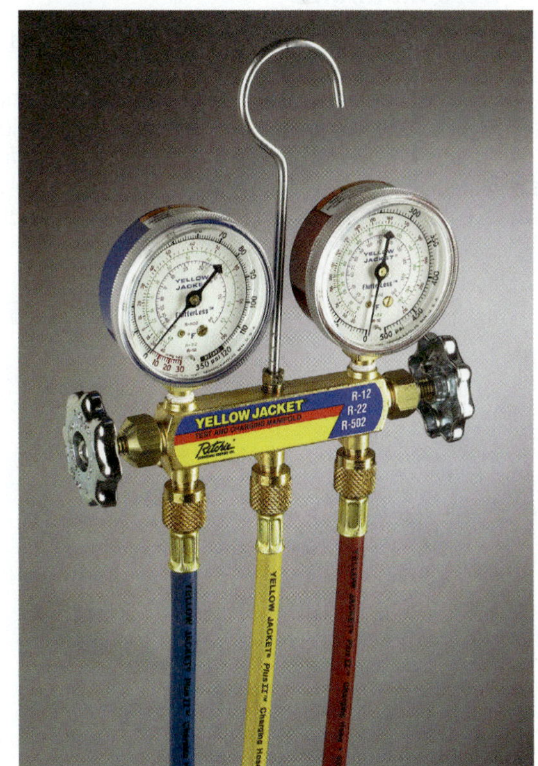

Figure 23-28 This gauge set is color coded: blue for the low-side gauge and hose, red for the high-side gauge and hose. *(Courtesy of Ritchie Engineering Company Inc.-Yellow Jacket Products Division)*

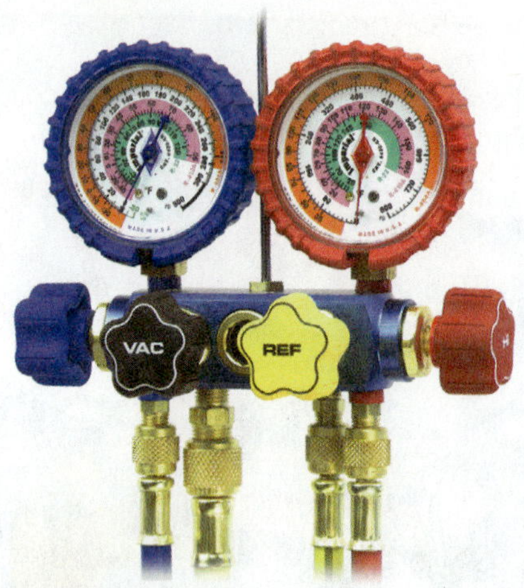

Figure 23-29 Four-valve manifold gauge set. *(Courtesy Stride Tool)*

Four-Port Manifolds

More advanced gauge manifolds are available that have four hose connections and four shutoff valves (Figure 23-29). Besides the low-side and high-side connections, there are two center connections, each with its own shutoff valve. Typically, one of the middle connections allows the use of a ⅜-in hose for faster evacuation. The extra shutoff valves give the technician more control over the flow through the manifold.

Large-Bore Manifolds

The hole through the middle of the manifold is called the bore. Standard-gauge manifolds have a bore of ¼ in, the same as the typical hoses used on standard gauge manifolds. The ¼-in bore becomes a bottleneck and slows down evacuation and recovery through the gauge set. Manifolds are available with a larger 3/8-in bore (Figure 23-30) and ½-in bore (Figure 23-31). The ½-in bore manifold set uses

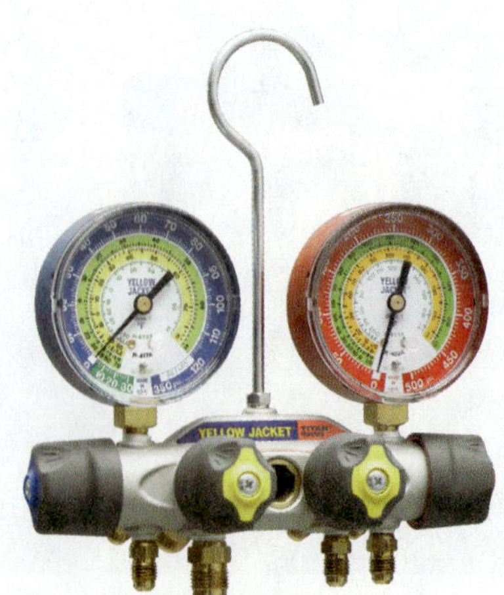

Figure 23-30 Manifold with 3/8-in bore. *(Courtesy of Ritchie Engineering Company, Inc., Yellow Jacket Products Division)*

Figure 23-31 The Megaflow manifold has a 1/2-in bore and several hose connections.

3/8-in and ½-in hoses with ¼-in connectors on the end that connects to the unit. The difference in evacuation speed between the ½-in bore manifold and standard ¼-in bore manifolds is dramatic.

SAFETY TIP

There are two major groupings of gauge manifold sets. One group is designed to be used with R-12, R-22, R-134a, R-408A, R-407C, and R-404A. The other groups of gauge manifold sets are designed to be used with R-410A only. Do not use gauges designed for the lower-pressure refrigerants on the very-high-pressure R-410A system.

23.14 REFRIGERATION HOSES

The refrigeration hoses are an important part of a gauge manifold set. They are what actually connects the gauges to the system. Charging and vacuum hoses are available in sizes of ¼ in, ⅜ in, ½ in, and ⅝ in inside diameter. The most common size is ¼ in. Refrigeration hoses are available in many colors (Figure 23-32). Many gauge sets use hoses that are color coded to the gauges: red for the high side, blue for the low side, and yellow for the center hose. Many four-port gauge manifolds have a ⅜-in port for vacuum and come equipped with a ⅜-in hose. Typical charging hoses have a straight end that connects to the manifold and an angled end that connects to the system. The angled end contains the core depressor for Schrader valves (Figure 23-33).

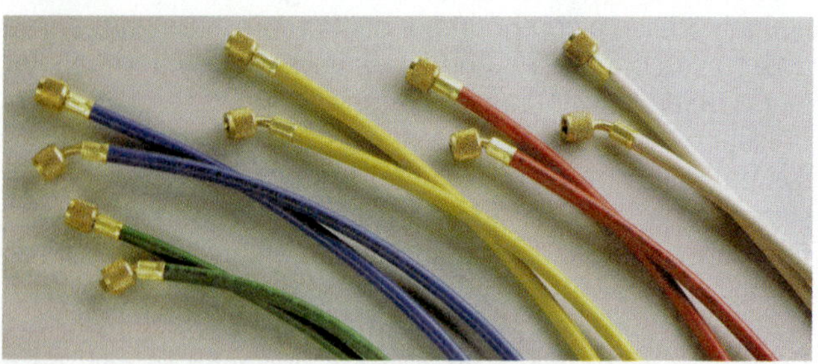

Figure 23-32 Refrigeration hoses are available in many colors. *(Courtesy of Ritchie Engineering Company, Inc., Yellow Jacket Products Division)*

Figure 23-33 The angled end contains a core depressor for Schrader valves.

TECH TIP

Manifold gauge sets have ¼-in male flare connectors for attaching the system end of the hoses when the hoses are not in use. These fittings are there to keep dirt out of the hose end. Any dirt that gets into the hose end will be pushed into the next system you access. It is important to keep the hose ends clean by keeping them on these connectors when not in use.

Refrigeration hoses are not simply a single tube; they are built in layers. Figure 23-34 illustrates the construction of a refrigeration hose. The rubber seen on the outside simply provides physical protection for the inner layers. The woven fabric layer just under the outer rubber layer provides strength. Some refrigeration hoses have two more inner layers, while others have only a single inner layer. The innermost layer actually contains the refrigerant. Hoses with two inner layers have extra permeation protection.

Hoses designed for use with refrigerants that operate at pressures lower than R-410a typically have a working pressure of 500 psig and a minimum burst pressure of 2,500 psig. Hoses designed for use with R-410a typically have a working pressure of 800 psig and a minimum burst pressure of 4,000 psig.

Figure 23-34 Diagram of refrigeration hose construction.

TECH TIP

The small O-ring in the end of the service valve hose is a replaceable part. These O-rings will become damaged over time and will need to be replaced to prevent refrigerant and vacuum leaks from occurring at the service valve connections. It is a good idea to carry a small package of these replacement seals with your tools because from time to time the O-rings will need to be replaced and occasionally one may fall out. When the O-ring falls out of the hose end, the hose cannot be used until a new O-ring has been installed.

23.15 POSITIVE SHUTOFF FITTINGS

Whenever gauges are connected to a charged system, the hoses are filled with refrigerant. With standard hose ends, the refrigerant in the hoses comes out of them when they are disconnected. Positive shutoff fittings allow technicians to control the flow of refrigerant leaving the hoses. They may be built into the hose, or they may be part of an adapter that connects to a standard hose. Positive shutoff fittings reduce refrigerant loss when connecting and disconnecting refrigeration hoses. They also protect the technician by reducing the amount of refrigerant spray that occurs when disconnecting refrigerant hoses. Positive shutoff fittings are very helpful when connecting and disconnecting to the high side of the refrigeration system, particularly on the liquid line. They are less important for the low side. Positive shutoff fittings are actually an annoyance for the middle hose because they create a restriction to flow and will not allow refrigerant purging out of the end of the middle hose. Some types of positive shutoff fittings include check valves built into the end of the refrigeration hose (Figure 23-35), check valve adapters, and short hoses with a ball valve (Figure 23-36).

TECH TIP

Consider the purchase of "low-loss" fittings when purchasing a manifold set.

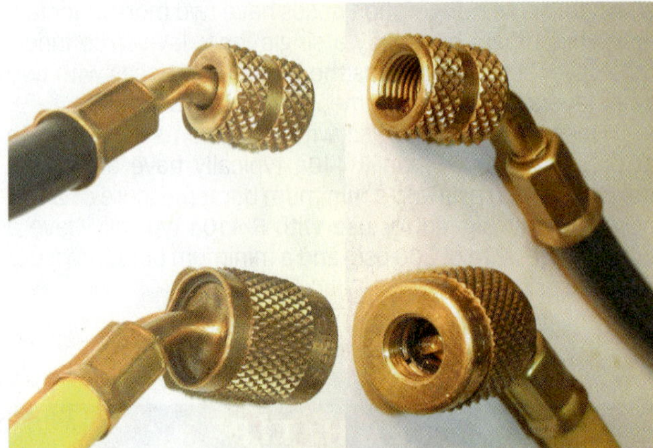

Figure 23-35 A standard hose end, shown on top, compared to a check valve hose end, shown on bottom.

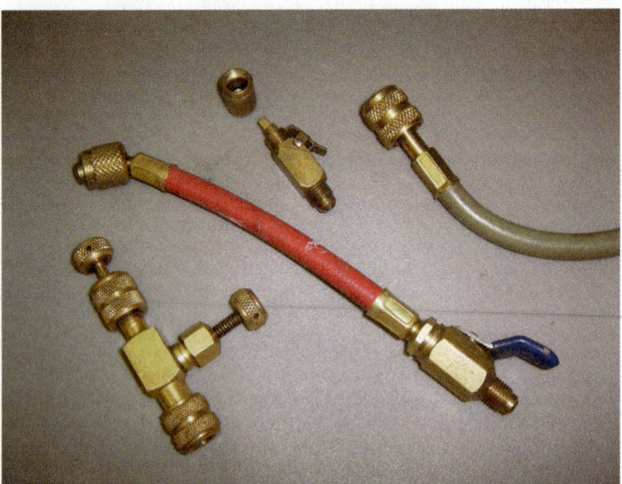

Figure 23-36 Refrigerant hose extensions with mechanical shutoff valve.

23.16 GAUGE MANIFOLD OPERATION

It is important to understand how the valves on a gauge manifold operate. Figure 23-37a shows a diagram of a standard set of manifold gauges. The valves do not control the flow of refrigerant to the gauges. Instead, they control the flow of refrigerant to the center port. The low-side gauge will always read pressure whenever the blue hose is connected to pressure. The high-side gauge will always read pressure whenever the red hose is connected to pressure. It is not necessary to open the hand wheels to read pressure. In fact, opening both valves will cause the two pressures to equalize through the manifold, making the readings inaccurate.

The hand wheels that control the valves on the end of the manifold can be set in one of three positions: fully closed, fully open, or partially opened. When both hand wheels are closed and the hoses are connected to the system, the low-side gauge will indicate the system low-side pressure and the high side will indicate the system high-side pressure (Figure 23-37b).

When the low-side hand wheel is turned open slightly, a partially open path around the valve and valve seat is opened from the low side to the center hose (Figure 23-37c). This is the way a system can be charged with refrigerant. The slight opening allows the technician to control the flow of refrigerant into the system. As the valve is fully opened, free flow between the low side and the center hose can occur (Figure 23-37d). This is the position that would be used to recover only vapor refrigerant from the low side of the system.

When the high-side hand wheel is turned open slightly, a partially open path around the valve and valve seat is opened from the high side to the center hose (Figure 23-37e). This is how the valve is operated when a little liquid refrigerant is being removed from the system. The slight opening allows the technician to control the flow of refrigerant out of the system. As the valve is fully opened, free flow between the high side and the center hose can occur (Figure 23-37f). This is the position that would be used to recover liquid refrigerant from the high side of the system.

When both hand wheels are fully opened, a free-flowing path between the low side, high side, and center port is provided (Figure 23-37g). This is the position for the valves when a vacuum is being pulled on the system or when the entire refrigerant charge is being recovered.

Gauge manifolds designed especially for evacuating and dehydrating a system have larger hose connections and use larger-diameter hoses (Figure 23-38). The larger, usually ⅜-in, inside diameter of the hoses and valves reduces the pressure drop that occurs when pulling a vacuum. Reducing even the very small pressure drop that occurs in normal gauge manifolds can cut the evacuation time by up to 90 percent.

Some gauge manifolds have multiple openings on the utility connection to accommodate various devices being used simultaneously. The additional connections allow you to change from evacuation to charging without losing any of the system vacuum or refrigerant.

SERVICE TIP

The needle on the high side will sometimes start to flutter, making it hard to read. Normally, it does not flutter when the gauges are first attached. That is because the needle flutters when the high-side hose and valve get full of liquid refrigerant. Liquid refrigerant is drawn to the cooler hose as vapor condenses in the manifold and hose. Liquids are noncompressible, so the pulsing pressure in the high-side line is transmitted directly to the gauge when the hose fills with liquid (Figure 23-39a). First, close the cylinder valve so none of the system refrigerant charge will be lost. Now slightly open the high-side valve to let the liquid flow into the center hose (Figure 23-39b). This will dampen the needle fluttering for a moment, allowing you to get a pressure reading. After you have the pressure reading, close the high-side valve. Next, slightly open the low-side valve to let the liquid refrigerant that collected in the hose be drawn back into the system through the low-side service valve (Figure 23-39c). By putting the refrigerant back into the system, the refrigerant charge is not changed.

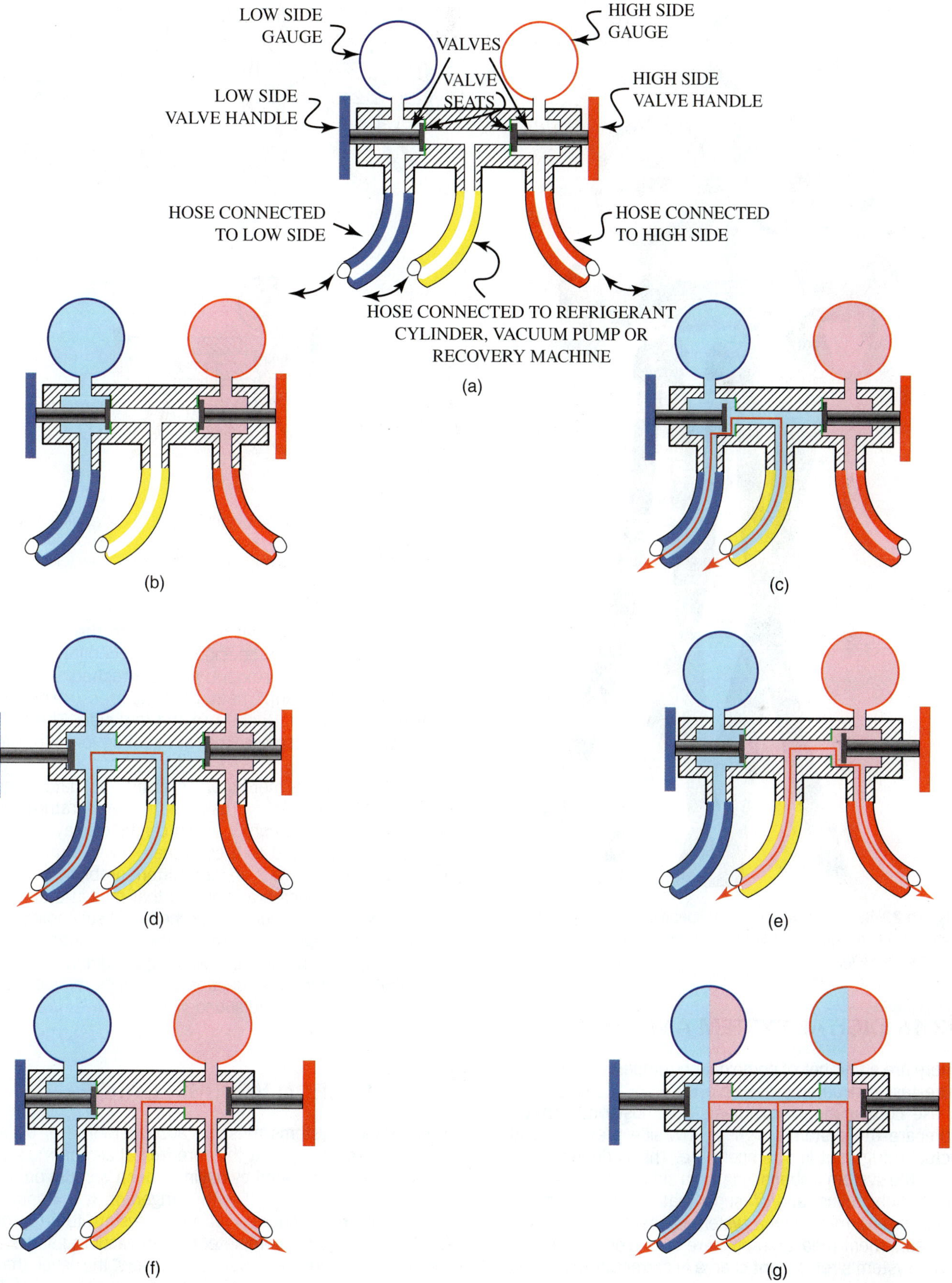

Figure 23-37 Function of the gauge manifold set: (a) parts of the gauge; (b) area pressurized with both valves closed; (c) area pressurized with low-side valve slightly open; (d) area having full flow with low-side valve open; (e) area pressurized with high-side valve slightly open; (f) area pressurized with high-side valve completely open; (g) area pressurized with both low- and high-side valves open.

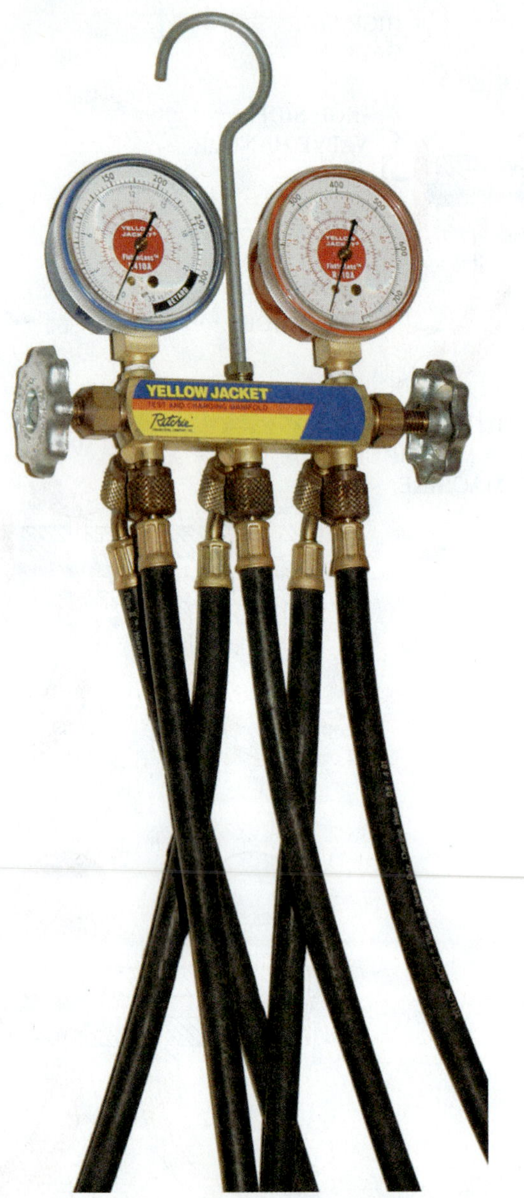

Figure 23-38 Gauge manifold set specifically designed for vacuum. *(Courtesy of Ritchie Engineering Company Inc.-Yellow Jacket Products Division)*

23.16 DIGITAL SYSTEM ANALYZERS

There are a number of combination temperature and pressure test instruments available for HVACR service work. These combination instruments are most often used to compare the system's high- or low-side pressure to the actual refrigerant line temperature. This is done to determine the system's subcooling (high side) or superheat (low side). Subcooling and/or superheat measurements are used to determine how a system is operating. Subcooling and superheat measurements are also used to determine if the system's refrigerant charge is correct or needs to be adjusted.

Subcooling and superheat meters have a pressure port that is attached to the refrigerant's access valve and a thermocouple that is attached to the refrigerant line (Figure 23-40). For checking the subcooling, both the pressure port and thermocouple are attached to the high side. To check the superheat, both the pressure port and thermocouple are attached to the low side. The meter calculates and displays the subcooling or superheat reading.

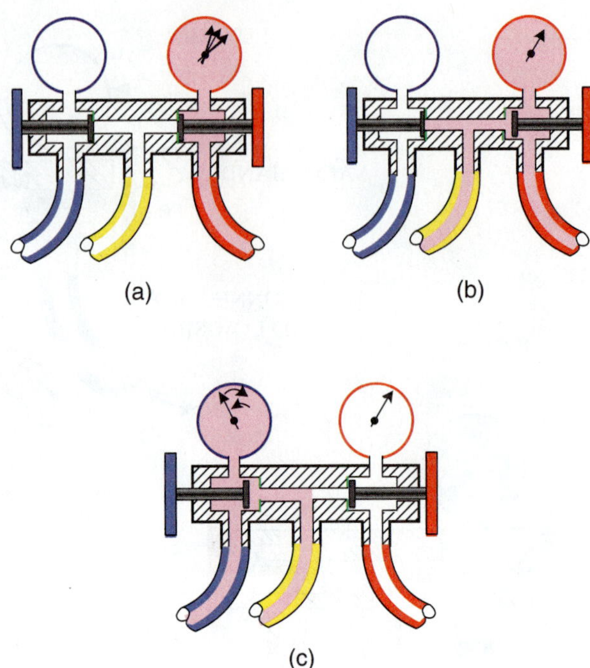

Figure 23-39 Illustration of service tip: (a) high-side gauge needle vibrating due to liquid flooding; (b) gauge needle dampened as liquid is slowly bled into center hose; (c) bleeding liquid from center hose back into system.

Some of these meters are classified as data loggers because they record the pressures and temperatures over a period of time. A graph or report can be displayed on the meter or downloaded to a computer. This feature is very important when there is an intermittent problem.

Manifold gauges are available that measure pressure and temperature, calculate superheat and subcooling, and even provide target pressure, temperature, superheat, or subcooling for a particular operating condition. These instruments combine the function of a gauge manifold with the digital superheat and subcooling meters.

23.17 MICRON VACUUM GAUGE

Refrigeration systems must be absolutely free of air and moisture. To ensure that they are free of air and moisture, systems are dehydrated by pulling a deep vacuum on them before they are charged with refrigerant. It is important to measure the vacuum level on the system to determine when evacuation is complete. Micron vacuum gauges are used to measure deep vacuums below the level that a compound gauge on a manifold gauge set can display. The vacuum scale on a compound gauge cannot show the difference between a terrible vacuum and a great vacuum.

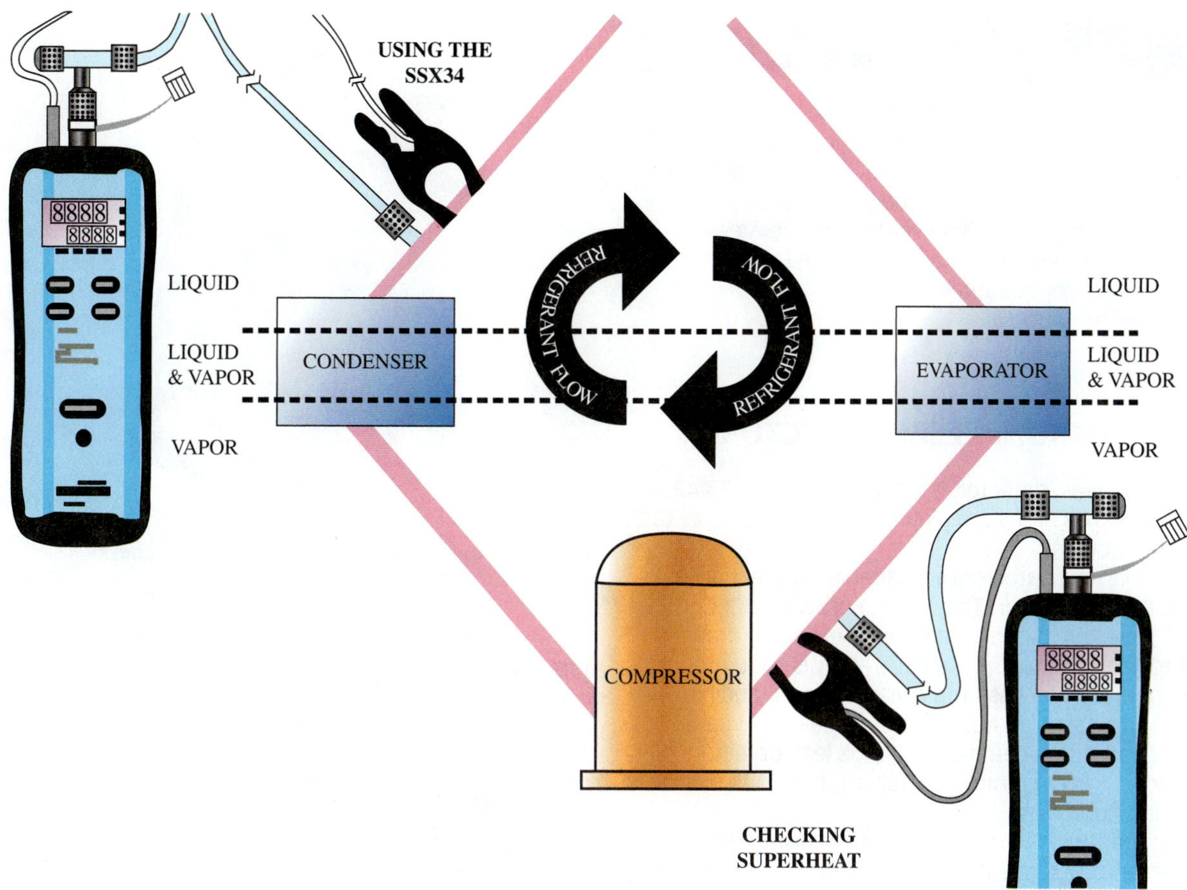

Figure 23-40 Subcooling and superheat meter with pressure port and thermocouple. *(Courtesy of Fieldpiece Instruments Inc.)*

The compound gauge can only show that a vacuum had been pulled down to 29 in Hg. Most manufacturers consider a vacuum level of 500 microns or less the level for a good vacuum. In inches of vacuum, this is 29.9 in Hg. Even if a compound gauge could read this accurately, you could not see it. The needle on most compound gauges is the same thickness as the 1-in Hg vacuum increments. They just cannot show when a good vacuum is achieved. The micron gauge can display vacuums down to around 50 microns. Microns are very small measurements. There are 25,400 microns in 1 inch. The display on micron gauges can show the reading as a number on a numeric display, as in Figure 23-41a, or by lighting an LED next to the micron scale, as in Figure 23-41b. See Unit 30, Evacuation, for more details on using a micron vacuum gauge.

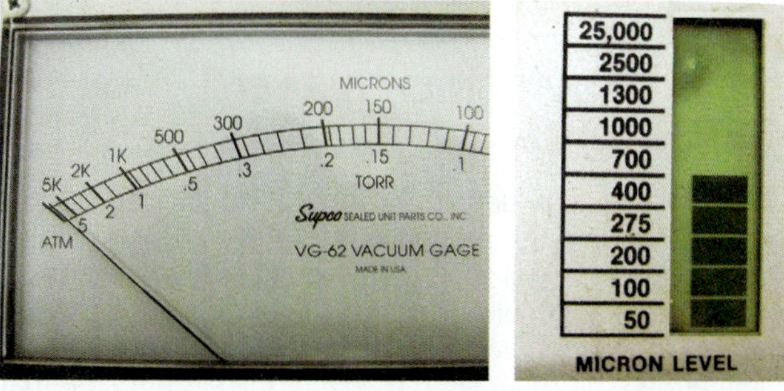

Figure 23-41 (a) Numeric-display micron gauge; (b) LED-display micron gauge.

23.18 LEAK DETECTORS

Ideally, no refrigeration system should leak; however, leaks are a major problem with HVACR systems. It is important that all leaks be located and repaired. Electronic leak-detection instruments have made this process much faster, more accurate, and easier. Some leak detectors work better in some locations than others. You should have more than one leak-detection method available.

Some methods are better for locating the area of a leak and others are better for pinpointing the exact spot of the leak. Applying soap bubbles to a suspected leak is the most common procedure for pinpointing a leak.

TECH TIP

Never use homemade soapy solutions for leak checking. They may be corrosive to the coil or refrigerant piping.

CAUTION

Large leaks of refrigerant in confined spaces are dangerous because they can replace all of the oxygen leading to asphyxiation. If you suspect there is a significant leak, ventilate the area before entering. Equipment rooms with systems that have large refrigerant charges must have self-contained breathing apparatuses available in case of a major refrigerant leak.

23.19 ELECTRONIC LEAK DETECTOR

An electronic leak detector is shown in Figure 23-42. A small pump draws air over a platinum diode located at the end of the test wand.

Electronic leak detectors are capable of detecting leaks as low as 0.4 oz per year. They can be used for HCFC, CFC, and HFC gases. A pump located in the device draws air directly to the sensing tip. No calibration is required. It is battery operated and has both a visual and an audible signal, which increase in frequency as the leak source is approached. Detectors identified as halide leak detectors will only respond to refrigerants containing halogens such as chlorine and fluorine.

If you are looking for a leak in an enclosed space, the leak detector may sense the refrigerant in the air. Calibrate the meter, or adjust its sensitivity to background refrigerant levels, by manually turning the adjustment knob. Some meters calibrate automatically when the technician turns the instrument off and back on.

23.20 ULTRASONIC-TYPE LEAK DETECTOR

An ultrasonic leak detector (Figure 23-43) will detect any gas leaking through an orifice. As refrigerant is escaping from a system, it will generate a sound at a higher than normal frequency. An ultrasonic leak detector can be used to pick up these sound waves. This method works well as long as there are no other sources within the system area that produce sound waves at the same frequency. The ultrasonic leak detector is typically somewhat more expensive as compared to the other leak-detection devices.

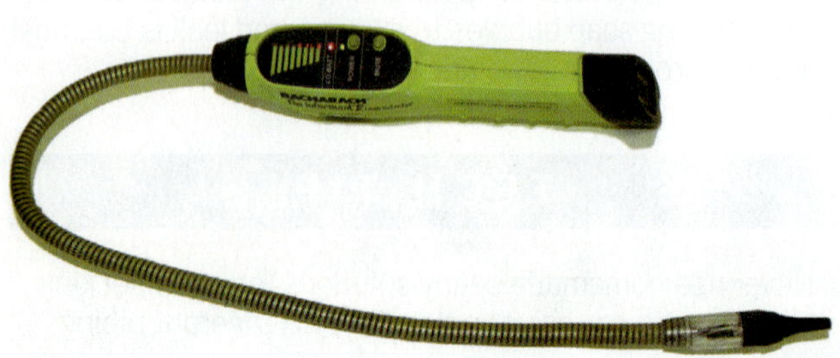

Figure 23-42 Electronic leak detectors.

Figure 23-43 Ultrasonic leak detector. *(Courtesy Robinair)*

The features and specifications for this meter are as follows:

- Detects pressure or vacuum leaks
- Unaffected by windy, rooftop conditions
- Unaffected by background noise
- Detects ultrasonic noise from arcing electrical switchgear
- Can be used for finding leaks in ductwork

23.21 FLUORESCENT DYE LEAK CHECKING

A small quantity of fluorescing oil is put into the system. Use only oil dyes that have been approved by the equipment manufacturer as safe to use in their system. The system is operated for a period of time. As refrigerant leaks, it takes with it a small quantity of oil. Once a UV lamp shines on the system, even the smallest of leaks is easily found.

The advantages of this leak-detection method are that it can pinpoint leaks, and once the dye is in the oil, rechecking for leaks is easy and fast.

23.22 ELECTRONIC SIGHT GLASS

An electronic sight glass is shown in Figure 23-44. It has both visual and audible bubble detection. Transducer clamps fit tubing from ⅛ in to 1⅛ in (3 mm to 32 mm) in diameter. The electronic sight glass uses ultrasonic waves to detect changes in density of the fluid traveling through the pipe. Gas bubbles are much less dense than liquid, so the instrument pings whenever a gas bubble passes in the liquid stream. The electronic sight glass can be used to help diagnose refrigeration problems. The presence of gas bubbles in the liquid line of a refrigeration system can indicate an undercharge. It will also detect liquid droplets in a gas line. Liquid droplets in the low-pressure vapor line can indicate an overcharge.

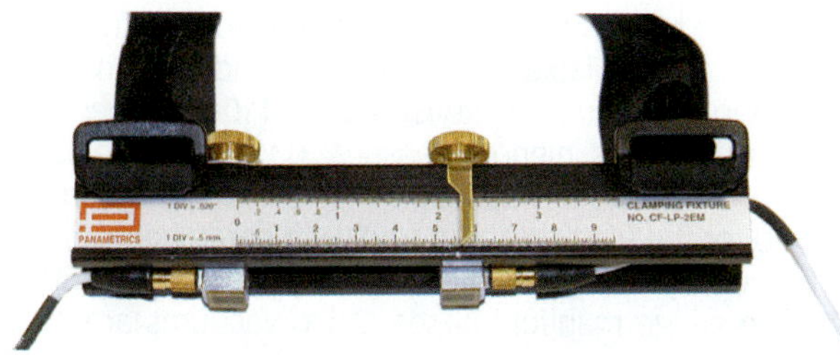

Figure 23-44 Electronic sight glass.

23.23 DIGITAL CHARGING SCALES

Digital charging scales are used to weigh the amount of refrigerant added or removed from a refrigeration system precisely. They are typically accurate to within 0.25 oz. Most can display weight in kilograms, decimal pounds, and pounds and ounces by pushing the Units button to select the desired units (Figure 23-45). Digital scales can be zeroed with weight on them by pushing the Zero button (Figure 23-46). The scale then displays the amount of weight lost or gained after it is zeroed. This is very helpful during charging or recovery. Scales are available in three general forms: one piece with the platform and display built into a single unit (Figure 23-47), with a removable display attached by a cord (Figure 23-48), or with a cordless removable display (Figure 23-49). Programmable scales are also available with a built-in solenoid controlled by the scale. They can be programmed to shut the flow of refrigerant off after a programmed amount of weight change. You should periodically check your scale with a known weight to ensure that it is accurate (Figure 23-50).

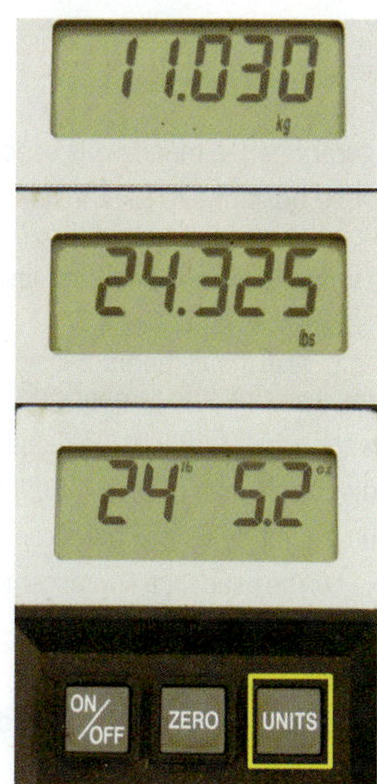

Figure 23-45 Pushing the Units button selects the units displayed.

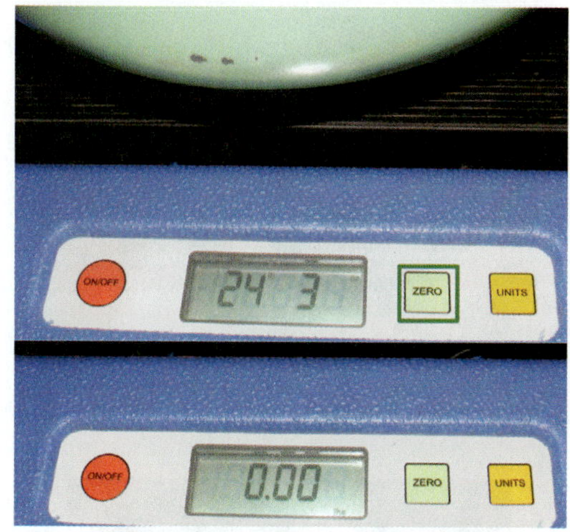

Figure 23-46 Pushing the Zero button zeros the scale.

Figure 23-47 Single-piece scale.

Figure 23-48 Scale with a removable display connected by a cord.

Figure 23-49 Scale with a cordless removable display.

Figure 23-50 This scale is accurate. It is displaying 2 kg with a 2-kg weight on it.

UNIT 23—SUMMARY

Making accurate temperature and pressure measurements is essential to all refrigerant system service. Types of thermometers include glass stem, dial, digital, and infrared. The most common sensors used with digital thermometers are type K and type J thermocouples. Three air temperatures are commonly read when working on air-conditioning equipment: the outdoor ambient temperature, the return-air temperature, and the supply-air temperature. Three types of mechanical pressure gauges are diaphragm, bellows, and Bourdon tube. The Bourdon tube is the most common. The accuracy of mechanical gauges is stated as a percentage. The most accurate mechanical refrigeration gauges are class 1. Digital gauges are more accurate than mechanical gauges. They can show pressures in 1/10 lb. A compound gauge reads both pressure and vacuum; a high-pressure gauge reads only pressure. A gauge manifold set consists of a manifold with a compound and a pressure gauge, valves to control flow through the manifold, and connecting hoses. Manifolds are available with both three and four hoses and either two or four valves. Refrigerant R-410a requires manifolds rated for its higher pressure. Gauge manifolds and hoses are often color coded: blue for the low-pressure side and red for the high-pressure side. Refrigerant spray can be minimized by using positive shutoff adapters and fittings with the gauge manifold hoses. Deep vacuums are measured using a micron vacuum gauge instead of a compound gauge. Common refrigerant leak detectors include soap bubbles, electronic leak detectors, ultrasonic leak detectors, and fluorescent dye.

UNIT 23—REVIEW QUESTIONS

1. Why would a service technician need a system's temperature and pressure?
2. What does a temperature meter measure?
3. What liquids may be used in a glass-stem thermometer?
4. What is an advantage of the glass-stem thermometer?
5. What is the pointer of a dial thermometer attached to that causes it to move?
6. Why are digital thermometers popular?
7. What temperature scales can a digital meter display?
8. What are the two thermocouple temperature sensor types commonly used in HVACR work?
9. What are some advantages of thermocouples?
10. Why must the thermocouple tip be tightly pressed against the line?
11. How does an infrared thermometer measure temperature?
12. What color surface temperature can be read most accurately with infrared thermometers?
13. What units are used to measure pressure in HVACR systems?
14. What is the approximate maximum high-side pressure on an R-22 gauge?
15. What is the approximate maximum high-side pressure on an R-410a gauge?
16. Why is it dangerous to not to use R-410a gauges on a R-410a system?
17. What is a refrigerant gauge manifold is used for?
18. What are the three positions to set the hand wheels that control the valves on the end of the manifold?
19. What happens when the low-side hand wheel is turned open slightly?
20. Why would some test instruments sense both the low-side pressure and the low-side line temperature?
21. Why would a large refrigerant leak be dangerous?
22. How does an electronic leak detector work?
23. How does an ultrasonic-type leak detector work?
24. What type of dye can be used for a fluorescent dye-leak checker?
25. How many microns are in 1 in of mercury?
26. What units can most digital scales display?
27. What refrigeration service procedures are digital scales used for?
28. Why is the ability to zero a scale with weight on it important?

UNIT 24

Piping and Tubing

OBJECTIVES

After completing this unit, you will be able to:

1. list the pipe and tubing materials used for air conditioning and refrigeration.
2. discuss the characteristics of copper tubing.
3. describe the characteristics of ACR copper tubing.
4. discuss the methods of joining copper tubing.
5. discuss the characteristics of iron and steel pipe.
6. describe how to thread iron pipe.
7. describe how to cut and join PVC and CPVC pipe.

24.1 INTRODUCTION

Heating and air-conditioning equipment needs refrigerants, fuels, and fluids to operate. All of these travel through piping or tubing, both inside and outside the equipment. Working with the different piping materials is a key component of installing and servicing air-conditioning and refrigeration equipment. Understanding the material properties and connection techniques is essential knowledge for air-conditioning and refrigeration technicians.

24.2 TYPES OF TUBING AND PIPING MATERIALS

In the air-conditioning and refrigeration industry, the most commonly used materials for piping and tubing are copper, iron, steel, aluminum, PVC, and CPVC. The preferred choice of material depends upon the pressure the tubing must contain, compatibility with the chemical it will be used with, ease of working the material, and the cost of the material.

Pipe Versus Tubing

Pipe is always named by its approximate inside diameter and schedule, or wall thickness. This can be confusing because the actual inside diameter is usually slightly larger than the nominal size. In other words, neither the inside nor the outside diameter of ½-in iron pipe is ½ in. Tubing is usually measured by its outside diameter because the outside diameter of tubing is accurate. This means that ½-in OD tubing actually measures ½ in. However, tubing sold in hardware stores for water piping is commonly sold by its nominal size, which is its approximate inside diameter. Figure 24-1 shows the difference between ½-in steel pipe and ½-in OD copper tubing. This nominal size is arrived at by subtracting ¼ in from the tubing outside diameter. Air-conditioning and refrigeration tubing is always specified by outside diameter. Copper tubing suitable for air-conditioning and refrigeration use, whether rigid or coiled annealed, is dehydrated and marked "ACR."

Figure 24-1 Comparison of ½-in OD copper tubing and ½-in nominal iron pipe.

Refrigeration Lines

Most refrigeration lines are made of copper. However, lines for ammonia systems are made of steel because ammonia chemically reacts with copper. Aluminum is used for internal piping of some air conditioners and appliances, but it is not used for field piping because of the difficulty in joining aluminum tubing.

Fuel Gas Lines

Black iron is used for piping natural gas. Galvanized iron pipe should not be used for natural gas because of the reaction between the odorant in the gas and the zinc in the galvanized coating. The residue generated by this reaction can migrate into the gas valve. Gas piping is being run in new flexible plastic-covered corrugated stainless steel piping (CSST) (Figure 24-2). This material costs more than black iron but is quicker and easier to run. There are limitations on where it can be used, so even systems run with the new material generally have some black iron pipe. Copper is commonly used for liquid petroleum (LP) gas lines, but it should not be used for natural gas with an average of 0.3 grains of hydrogen sulfide per 100 standard cubic feet

Figure 24-2 Flexible corrugated steel tubing used for gas lines.

(scf) of gas. Hydrogen sulfide found in the odorant added to natural gas reacts with copper. Check with the local building code authorities before using copper on natural gas lines. The flexibility of the aluminum tubing has made it the primary material used for piping the pilot gas for standing pilot systems on both natural and LP gas systems.

Water Piping

Galvanized iron, copper, PVC, and CPVC have all been used for water piping. Black iron should not be used with water because it rusts. Copper was the material of choice for water systems before the advent of PVC and CPVC plastic pipe, which have replaced copper for water lines in most new construction because the material is very inexpensive and is easy to join. Drain lines are usually run in PVC but can be run in CPVC or copper if the drain water is high temperature.

24.3 IRON AND STEEL PIPE

Iron and steel pipe are available in different strengths, called schedules. The higher the schedule number, the stronger the pipe. Available strengths range from schedule 5 to schedule 160. Figure 24-3 summarizes the characteristics of standard iron and steel pipe. The most common schedules used for air-conditioning and refrigeration work are schedules 40 and 80. Another system of strength designation simply refers to the pipe as standard, extra strong, or double extra strong.

Iron Pipe Specifications (all dimensions in inches)

Normal Size	Outside Diameter	Inside Diameter	Wall Thickness	Weight Pounds per Foot
Schedule 40, Standard Strength Pipe				
⅛ Std	0.405	0.269	0.068	0.24
¼ Std	0.540	0.364	0.088	0.43
⅜ Std	0.675	0.493	0.091	0.57
½ Std	0.840	0.622	0.109	0.85
¾ Std	1.050	0.824	0.113	1.13
1 Std	1.315	1.049	0.133	1.68
1¼ Std	1.66	1.38	0.140	2.27
1½ Std	1.9	1.61	0.145	2.72
Schedule 80, Extra-Strong Strength Pipe				
⅛ XS	0.405	0.215	0.095	0.31
¼ XS	0.540	0.302	0.119	0.54
⅜ XS	0.675	0.423	0.126	0.74
½ XS	0.840	0.546	0.147	1.09
¾ XS	1.050	0.742	0.154	1.48
1 XS	1.315	0.957	0.179	2.17
1¼ XS	1.660	1.278	0.191	3.00
1½ XS	1.900	1.500	0.200	3.63
Schedule 160 Pipe				
½	0.840	0.464	0.188	1.31
¾	1.050	0.612	0.219	1.95
1	1.315	0.815	0.250	2.85
1 ¼	1.660	1.16	0.250	3.77
1 ½	1.900	1.338	0.281	4.86
Double-Extra-Strong Strength Pipe				
½ XXS	0.840	0.252	0.294	1.72
¾ XXS	1.050	0.434	0.308	2.44
1 XXS	1.315	0.599	0.358	3.66
1¼ XXS	1.660	0.896	0.382	5.22
1 ½ XXS	1.900	1.100	0.400	6.41

Figure 24-3 Pipe specifications.

Standard generally equates to schedule 40, and extra strong equates to schedule 80. Double extra strong is somewhat heavier than the highest schedule pipe, schedule 160. Figure 24-4 shows a cross-section comparison of standard, extra-strong, and double-extra-strong pipe.

TECH TIP

The original idea behind pipe schedules was for the schedule number to have a uniform relationship of 1,000 times the pressure-to-tensile strength ratio in the modified Barlow equation that is commonly used to determine how much pressure a pipe can hold. This proved to be impractical, but the schedule numbers remained as a common designation for different strengths of pipe.

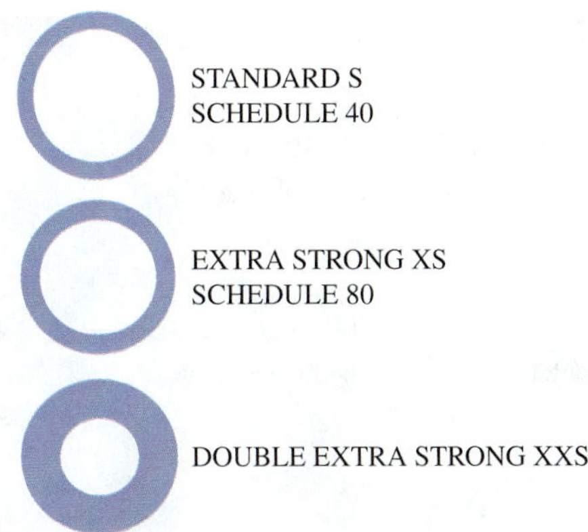

Figure 24-4 Standard, extra-strong, and double-extra-strong pipe are differentiated by their wall thickness.

Figure 24-5 Iron pipe in black and galvanized.

Iron pipe is available in two finishes: black iron and galvanized (Figure 24-5). Black pipe is iron pipe that has its natural finish, while galvanized is iron pipe that has been dipped in molten zinc to make it corrosion resistant. Galvanized has a silver appearance similar to steel. Steel pipe is made from steel, which is made by alloying iron with carbon for added strength. Iron and steel pipe come in 21-ft lengths, threaded on each end. Typically pipe is supplied with a coupling on one end.

Iron pipe is measured by its approximate inside diameter and comes in ⅛-in increments from ⅛ in to ½ in, ¼-in increments from ½ in to 1½ in, and in ½-in increments from 1½ in and up. The nominal size is the approximate inside diameter, but the actual inside diameter is slightly larger than the nominal size. The most consistent measurement is the outside diameter. The outside diameter is approximately ¼ in larger than the nominal size. To determine the size of iron pipe, measure the outside diameter and subtract ¼ in. To measure a pipe fitting, measure the threads and subtract ¼ in. This applies to both internal and external threads. For ⅛-in and ⅜-in pipe, the measurement needs to be to the nearest ⅛ in.

24.4 JOINING IRON AND STEEL PIPING

Iron and steel pipe may be joined by welding, flanges, or threaded connections (Figure 24-6). Flanged and welded connections are only used in large industrial and commercial

Figure 24-6 Examples of welded, flanged, and threaded pipe joints.

applications. Most iron and steel piping used in air conditioning is joined by cutting threads on the end of the pipe and joining the pipe with threaded fittings. Pipe threads are tapered 1⁄32 in per inch of threads; they have a slightly larger outside diameter at the end of the threads than at the beginning (Figure 24-7). The first seven threads are fully formed, but the threads at the back are not. The taper allows pipe threads to tighten and seal. The number of threads per inch decreases as the pipe size increases. Figure 24-8 lists the threads per inch for different sizes of pipe.

Pipe sealant is used when assembling the threads to lubricate the threads and help seal the joint. Check the sealant to make sure it is compatible with the gas or liquid that the pipe will carry. Not all sealants are compatible with all gases and liquids. Some pipe sealants specifically warn against using them with LP gas. Pipe sealant should not be

Figure 24-7 Pipe threads are tapered; the starting threads are smaller in diameter than the last threads.

Nominal Size	Threads per Inch
⅛	27
¼, ⅜	18
½, ¾	14
1, 1 ¼, 1 ½, 2	11 ½
2 ½ up	8

Figure 24-8 National pipe thread specifications.

Figure 24-9 The pipe should be started before thread sealer is applied.

allowed to get inside the pipe when assembling it. Insert the male threads into the female threads for a full turn before applying the pipe sealant (Figure 24-9). This will keep pipe sealant out of the inside of the pipe.

Teflon tape can be used as a pipe sealant, but some mechanical and gas codes prohibit its use. Make sure that Teflon tape is acceptable for your application before using it. Teflon tape is applied by wrapping it around the threads in a direction opposite to the direction the pipe will be turned (Figure 24-10). If it is applied in the same direction, it will bunch up when the joint is tightened. Take care not to have tape all the way to the end of the pipe; the first couple of threads should not have tape on them.

Figure 24-10 Teflon tape should be wrapped opposite to the direction the pipe will be turned, leaving the first threads exposed.

Figure 24-11 Portable tripod pipe vise.

24.5 CUTTING AND THREADING IRON AND STEEL PIPE

When installing a piping system with iron pipe, the pipe must be cut and fitted. The cut ends have to be threaded before they can be joined. The pipe is held in a pipe vise while it is being cut and threaded. The most common vises used in the field are tripod vises that can be transported to the job (Figure 24-11). Vises with regular flat jaws cannot hold pipe properly. The jaws of a pipe vise are semicircular, with ridges designed to hold the pipe (Figure 24-12). Manual pipe-threading tools typically include a pipe cutter, a pipe reamer, a ratchet handle, and several die heads that fit in the ratchet handle. The pipe cutter works just like the tubing cutters used on copper; it is just larger. The reamer is used to cut out the burr left from the pipe cutter. The ratchet handle holds the die heads. A typical ratchet handle can accommodate heads for pipe sizes from ½ in up to 1¼ in. The steps to cut pipe to length and thread it are listed below and shown in Figure 24-13.

1. Mount the pipe in a pipe vise.
2. Cut the pipe to the desired length using the pipe cutter.

Figure 24-12 Pipe vise jaws are designed to hold pipe.

Figure 24-13 Steps in threading iron pipe.

3. Use the pipe reamer to remove the burr left by the pipe cutter.
4. Insert the correct die in the ratchet handle.
5. Apply thread-cutting oil to the pipe.
6. Slide the die over the pipe. Make sure that the end without teeth showing is the end that goes over the pipe first. Also, check the direction of the arrow on the ratchet handle. It should be pointing to the right. If it is not, pull it out and flip it around so that it points to the right.
7. Apply pressure to the die with the heel of your hand while turning the ratchet handle. After one or two full turns it should not be necessary to keep applying pressure.
8. Continue to turn the ratchet handle, stopping every couple of turns to lubricate the threads and die with cutting oil.
9. Continue until the end of the pipe is flush with the teeth on the backside of the die.
10. Flip the arrow on the ratchet handle and turn the ratchet handle in the opposite direction to remove the die head.

SAFETY TIP

Always wear gloves and safety glasses when handling, cutting, and threading pipe. The chips created by the threading process are quite sharp and can easily cut you.

Figure 24-14 Common pipe fittings (from left to right): couplings, cap, union, 90° ell, 45° ell, plug, tee, bell reducer, and bushing.

24.6 PIPE FITTINGS

Iron and steel pipe is most often joined with threaded pipe fittings. Common pipe fittings include pipe couplings, unions, ells, tees, caps, plugs, bushings, and bell reducers (Figure 24-14). Pipe fittings can have either male or female threads; male threads will fit into female threads. To measure pipe fittings, measure the threads to the nearest ¼ in and subtract ¼ in. For male fittings, measure the outside diameter of the threads; for female fittings, measure the inside diameter of the threads (Figure 24-15). The measurement is approximate, that is why you measure to the nearest ¼-in measurement. The two exceptions to this rule are ⅛-in pipe fittings and ⅜-in pipe fittings; they will measure approximately ⅜ in and ⅝ in. Two pieces of pipe can be joined by either a coupling or a union. A coupling is a cylinder with female threads on both ends. The coupling threads onto one pipe and another pipe threads into the coupling. One of the pipes being joined must be able to turn in order to use a coupling (Figure 24-16). A union can connect two pipes when neither can be turned. A union, also called a ground joint union, consists of three pieces:

Figure 24-15 The threads on these ¾-in pipe fittings measure 1 in.

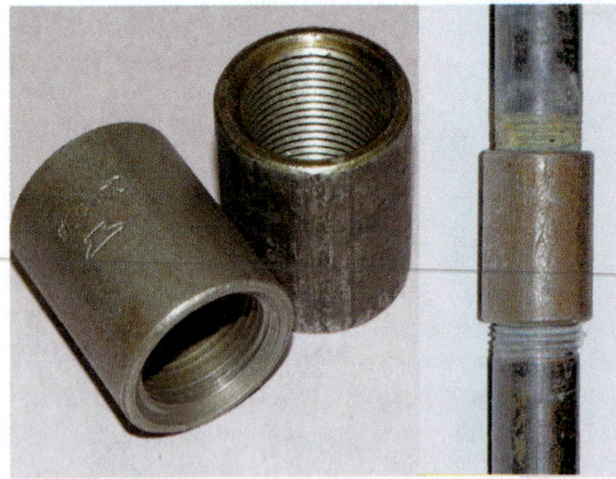

Figure 24-16 A coupling can only be used to join two pipes if one pipe can turn.

two threaded ends that each connect to a pipe and a large nut that pulls the two ends together (Figure 24-17). Elbows, often called ells, are used to change the pipe direction. They are available in 90° and 45° (Figure 24-18). Common ells have female threads on both sides. A street ell has one female and one male end. It is designed to fit into another fitting and allow a pipe to fit into it (Figure 24-19).

Figure 24-17 Because a union has three parts, it can join two pipes even if neither can turn.

Figure 24-18 A 90° elbow and a 45° elbow.

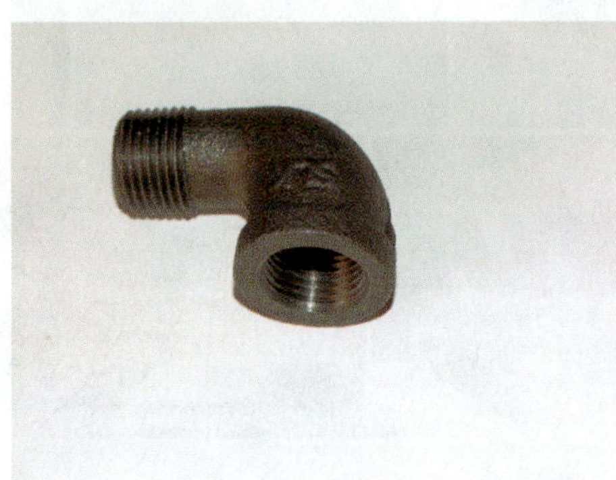

Figure 24-19 A street elbow has a female end an a male end.

TECH TIP

The name "street ell" comes from its use in sewage and waste lines. Because waste lines only flow because of gravity, care is taken to minimize restriction. A street ell has both a female and male end. Pointing the male end toward the street in a waste line offers less restriction because the waste products do not get caught on the edge of the pipe as they do with regular ells.

Tees are used when a line must branch off of another line. They are shaped like the letter *T* and have female threads. Regular tees are the same size at both ends and in the middle. Reducing tees can have different sizes. Reducing tees are specified by the run and branch size. The run is the part straight through the tee; the branch is the side port (Figure 24-20).

Figure 24-20 The run is the part of the tee that goes straight through the tee. The branch is the line coming off the side, like a tree branch.

Use Bell Reducers Instead of Bushings

Two fittings are available for connecting a smaller pipe to a larger pipe: bushings and bell reducers, also called a reducing coupling. A bushing has both male and female threads. It fits inside a larger fitting and allows a smaller pipe to thread into the female threads (Figure 24-21). A bell reducer has female threads on both ends and looks something like a bell because one end is smaller than the other. A large pipe can thread into one end and smaller pipe in the other ends (Figure 24-22).

Caps and Plugs

A cap is used to seal off a threaded pipe. A plug is used to seal the hole in a threaded female fitting (Figure 24-23).

CODE TIP

Most codes require bell reducers rather than bushings for making reductions in size. Bushings increase turbulence and restrict flow.

Figure 24-21 A pipe bushing.

Figure 24-22 A bell reducer.

Figure 24-23 A pipe plug.

24.7 COPPER TUBING

There are several grades of copper tubing. It is available in different wall thicknesses and two hardness. Copper tubing comes in rolls or straight lengths and can be standard or dehydrated.

Types of Copper Tubing

Copper is available in three wall thicknesses: K (heavy), L (medium), and M (light). Type K heavy wall tubing is meant for special use where abnormal conditions of corrosion might be expected. Type M thin wall tubing is not used on pressurized refrigerant lines because it does not have the wall thickness to meet the safety codes. It can be used for water lines and condensate drains. Type L is most frequently used for normal refrigeration applications. Figure 24-24 provides specifications for both type K and type L tubing.

ACR Copper Tubing

Copper tubing intended for refrigeration and air-conditioning work is designated as ACR tubing (Figures 24 and 25). ACR tubing is type L that has been dehydrated. ACR tubing is cleaned, degreased, dehydrated, and sealed to keep its

	Diameter		Wall	Weight
Type	OD (in)	ID (in)	Thickness (in)	per Foot (lb)
K	½	0.402	0.049	0.2691
	⅝	0.527	0.049	0.3437
	¾	0.652	0.049	0.4183
	⅞	0.745	0.065	0.6411
	1⅛	0.995	0.065	0.8390
	1⅜	1.245	0.065	1.037
	1⅝	1.481	0.072	1.362
	2⅛	1.959	0.083	2.064
	2⅝	2.435	0.095	2.927
	3⅛	2.907	0.109	4.003
	3⅝	3.385	0.120	5.122
L	½	0.430	0.035	0.1982
	⅝	0.545	0.040	0.2849
	¾	0.666	0.042	0.3621
	⅞	0.785	0.045	0.4518
	1⅛	1.025	0.050	0.6545
	1⅜	1.265	0.055	0.8840
	1⅝	1.505	0.060	1.143
	2²⁄₈	1.985	0.070	1.752
	2⅝	2.465	0.080	2.479
	3⅛	2.945	0.090	3.326
	3⅝	3.425	0.100	4.292

Figure 24-24 Specification of common copper tubing sizes.

Figure 24-25 ACR tubing is suitable for air-conditioning and refrigeration use.

inside clean. It is purged by the manufacturer with nitrogen gas to seal against air, moisture, and dirt and also to minimize the harmful oxides that are normally formed during brazing. The ends are plugged to keep out air and moisture. These plugs should be replaced after cutting a length of tubing.

Copper Tubing Sizes

ACR tubing is measured by outside diameter. The outside diameter of all three wall thicknesses stays the same; ⅝-in outside diameter copper tubing will be ⅝-in outside diameter regardless of the wall thickness. The inside dimension does change as the wall thickness changes. The approximate inside dimension is given as the nominal size. Since the wall thickness is approximately $\frac{1}{16}$ in, the nominal size is ⅛ in smaller than the outside diameter. Figure 24-26 compares the dimensions of the three types of copper pipe.

ACR tubing is available in ⅛-in increments from ¼ in to ⅞ in and in ¼ in increments from 1⅛ in up. There is an exception: $\frac{5}{16}$-in copper is often used for the refrigerant lines on household refrigerators and occasionally is used as a liquid line on residential air conditioners. To determine the size of ACR tubing, simply measure the outside diameter of the tubing. To find the nominal size, measure the outside diameter and subtract ⅛ in.

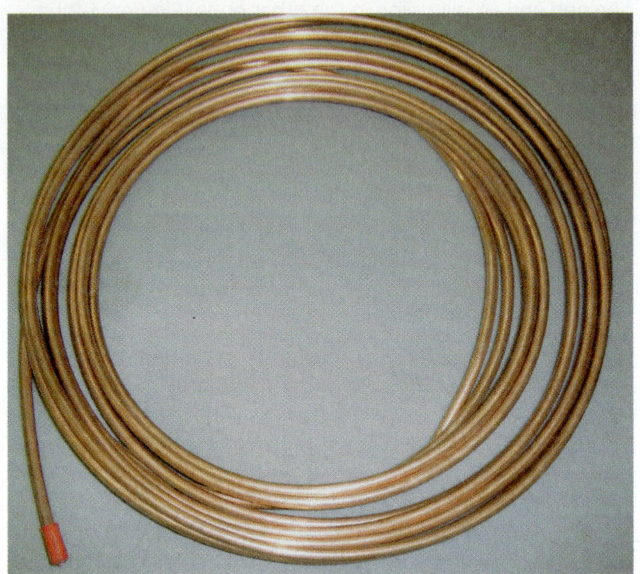

Figure 24-27 Annealed soft copper ACR tubing.

Soft Copper Tubing

ACR copper tubing is available in two tempers: 50-ft rolls of annealed soft copper tubing and 20-ft straight sections of hard-drawn copper tubing. Soft copper tubing is tubing that has been softened by annealing. Annealing is the process of heating the tubing up to a temperature of 1,300°F and letting it cool. This permanently softens the copper and makes it malleable enough to mechanically form bends and flares in the tubing. Soft copper tubing is used extensively for refrigeration lines in residential air-conditioning systems (Figure 24-27).

Soft copper must be unrolled before it can be used. It is much easier to straighten the tubing while it is still on the roll. The end on the outside of the roll should be held firmly against a flat surface and the coil slowly rolled out (Figure 24-28).

As copper is bent, it will become work hardened. The more it is bent, the harder it becomes. If excessive work hardening has occurred, it may crack or buckle when formed. Work-hardened copper can be reannealed by heating it to a dull red and allowing it to cool. If this is done quickly with a large hot flame, a minimum of oxides will form on the copper tubing. Slow heating can cause excessive oxides, which can cause problems with system operation once installation is complete.

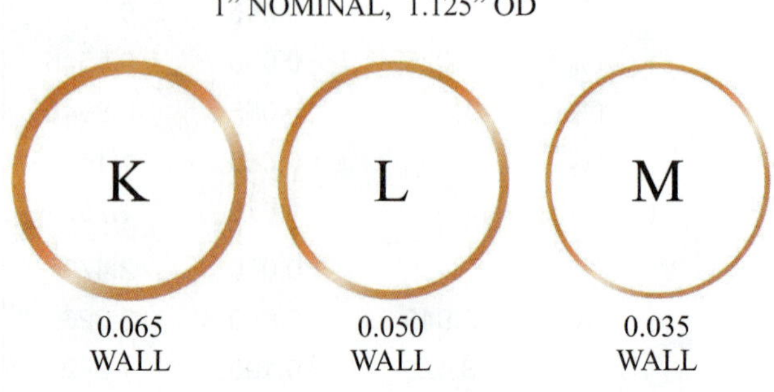

Figure 24-26 The tubing outside diameter is the same for types M, L, and K.

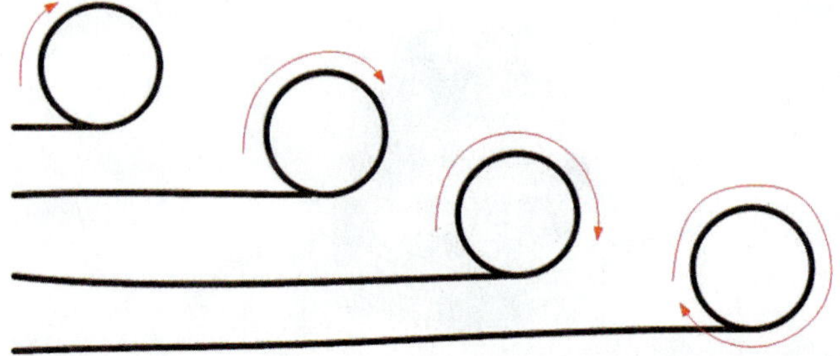

Figure 24-28 To use copper tubing, take the whole roll and unroll what you need from the outside edge across a clean, flat surface.

Figure 24-29 Lengths of hard copper tubing.

Hard Copper Tubing

Hard copper is used extensively in commercial refrigeration and air conditioning systems. Unlike soft tubing, it is hard and rigid and comes in straight lengths (Figure 24-29). Hard copper cannot be mechanically formed; it will crush or buckle rather than bend. It is intended for use with formed fittings to make the necessary bends or changes in direction. Because of its rigid construction, it is more self-supporting and needs fewer supports than soft copper. Sizes range from ¼ in OD to over 6 in OD. Hard-drawn tubing comes in standard lengths of 10 ft and 20 ft. It is dehydrated, charged with nitrogen, and plugged at each end to maintain a clean, moisture-free internal condition. The use of hard-drawn tubing is most frequently associated with large line sizes or where neat appearance is desired.

24.8 CUTTING COPPER TUBING

Copper tubing should be cut using a tubing cutter. Tubing cutters are preferable for refrigeration work because they do not generate chips that can get in the system. Tubing cutters also have the advantage of producing a cut that is perpendicular to the length of the tubing, called a square cut. Tubing cutters are available in different sizes to accommodate different sizes of tubing. A standard tubing cutter typically can cut tubing from ⅛ in OD up to 1⅛ in OD (Figure 24-30). Larger cutters handle from ⅜ in up to 2⅝ in (Figure 24-31). Traditional tubing cutters will not fit everywhere. Smaller cutters are made that make it possible to cut tubing in confined spaces (Figure 24-32).

All these tubing cutters work by rolling a sharp wheel around the outside of the tubing. The wheel scores the tubing as it rolls around it. The cutter must roll around the tubing several times to properly cut; it does not cut through in one pass. The wheel should be adjusted until it contacts the tubing with a slight pressure. Roll the cutter around the tubing, pulling the cutter so that the cutting wheel leads the cut (Figure 24-33). Increase the tension slightly with each rotation by turning the adjusting knob in a clockwise direction. Continue rolling and tightening until the tubing is cut through. Attempting to cut the tubing too quickly will pinch the tubing in and cause constriction. Figure 24-34

Figure 24-30 Standard tubing cutter used for ⅛-in to 1⅛-in copper tubing. *(Courtesy of Ritchie Engineering Company, Inc., Yellow Jacket Products Division)*

Figure 24-31 Large tubing cutters are used for ⅜-in to 2⅝-in copper tubing. *(Courtesy of Ritchie Engineering Company, Inc., Yellow Jacket Products Division)*

Figure 24-32 Small tubing cutters are useful for cutting ⅛-in to ⅞-in tubing in tight spaces. *(Courtesy of Ritchie Engineering Company, Inc., Yellow Jacket Products Division)*

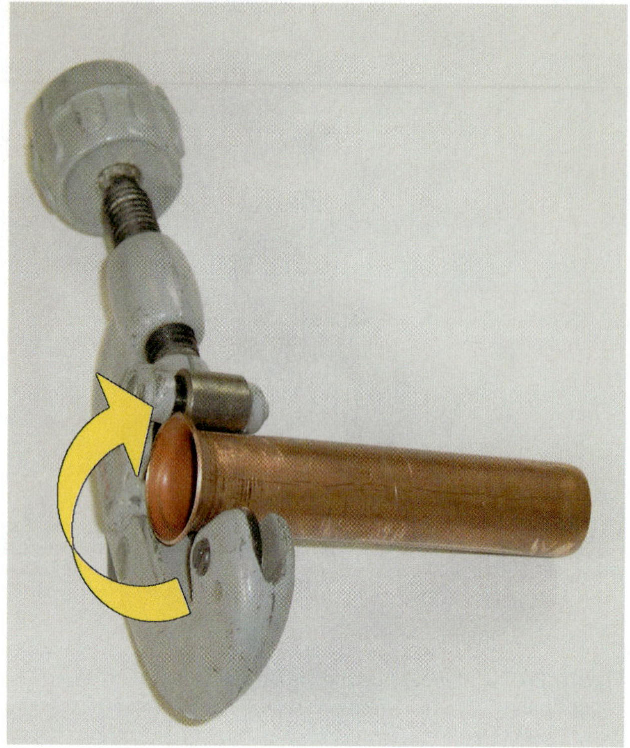

Figure 24-33 Arrow shows direction to pull tubing cutter. *(Courtesy of HVAC Department, Terra Community College, Fremont, Ohio)*

Figure 24-34 Tubing that is cut too quickly will have a much larger burr than tubing that is cut slowly. *(Courtesy of HVAC Department, Terra Community College, Fremont, Ohio)*

compares an improper cut to a correct cut. Even when done correctly, tubing cutters leave a small burr on the inside of the tube. This burr should be removed using the reamer built into the tubing cutter (Figure 24-35).

Most tubing cutters have a groove that can be used for cutting off the flare on a flared piece of tubing (Figure 24-36). This lets the technician cut off a flare with a minimal loss of pipe.

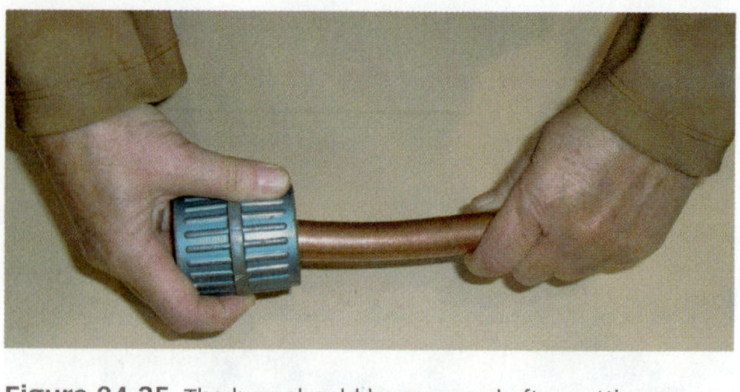

Figure 24-35 The burr should be removed after cutting. *(Courtesy of HVAC Department, Terra Community College, Fremont, Ohio)*

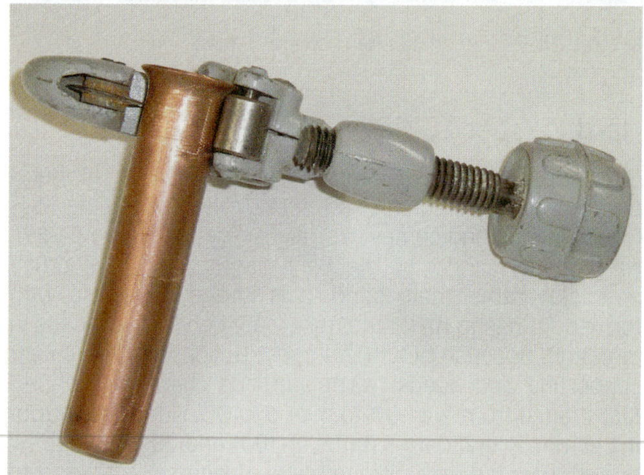

Figure 24-36 The groove in the tubing cutter rollers allows the flare to be cut off. *(Courtesy of HVAC Department, Terra Community College, Fremont, Ohio)*

24.9 BENDING COPPER TUBING

Tubing can be bent with accuracy if the right tools are used and care is taken. It is very important not to crimp the tubing because this can restrict the flow of refrigerant. Never get in a hurry when bending tubing.

When using lever- or wheel-type benders, be sure that the groove in the bender is for the size of tubing you are working with. A tool designed for tubing larger than the tubing you are working with will flatten the tubing (Figure 24-37). The work should be done slowly and steadily to avoid buckling or crimping the tubing.

Most lever-type benders have an "R" mark, which will help you determine the distance from the end of the completed bend to center of the pipe. To use the mark, first measure the distance you want from the end of the pipe to the center of the bend after the bend is completed. This mark lines up on the R mark on the bender. On most benders the distance is measured from the right to the R mark (Figure 24-38). Bend the tubing with a slow, steady pull until the degree mark lines up to the desired angle. The dimension of the completed bend should now be correct.

When making several successive bends on one piece of pipe, it is necessary to calculate the total pipe length needed. To do this, you will need to know the radius of the tubing

Figure 24-37 Using the wrong-sized bender can flatten the tubing.

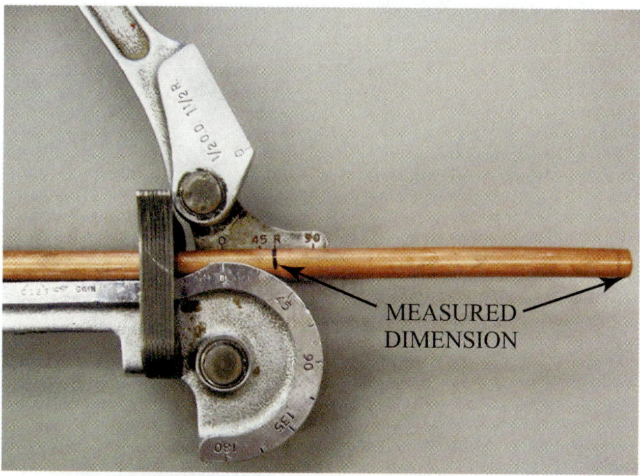

Figure 24-38 The desired dimension is measured from the right to the R mark.

bender you will be using. The radius is the distance from the center of the bending wheel to the outside of the bending wheel and should be marked on the bender. Because of the many sizes and designs of tubing benders, there is no standard set of radii to remember. The technician must look on the actual tool that will be used (Figure 24-39).

Example Bend Calculation

- **Tubing** ½-in OD ACR annealed tubing
- **Bend** 90°
- **Legs** 7 in on one side and 8 in on the other
- **Bender Radius** 1½

Figure 24-39 The tubing-bender radius.

The total length of pipe should be the sum of the two straight lengths plus the length of the 90° arc that connects them. First, calculate the actual straight length of each leg. The actual straight length will be the desired leg length minus the radius of the bender. In this example, the actual straight lengths will be 7 in − 1½ in = 5½ in, and 8 in − 1½ in = 6½ in (Figure 24-40).

To calculate the arc length, first calculate the circumference of a full circle with a radius equal to the radius of the tubing bender. In the example, the circumference would be 1½ in × 2 × 3.14. This equals 3 in × 3.14 = 9.42 in. Since a 90° bend is one-quarter of a circle, the arc length is ¼ of the entire 9.42 in circumference. This is ¼ × 9.42 in = 2.355 in.

The entire length of tubing needed is 5½ in + 6½ in + 2.355 in = 14.355 in. The 0.355 portion must be estimated to the nearest ⅛ in or nearest 1⁄16 in. To estimate to the nearest ⅛ in, divide the decimal remainder by 0.125. Therefore, 0.355 would round off to ⅜ in. To estimate to the nearest 1⁄16 in, divide by 0.0625. Therefore, 0.355 would round to 6⁄16 = ⅜. In this case, there is no difference.

The total length needed is then 5½ in + 6½ in + 2⅜ in = 14⅜ in. A piece of tubing cut to this length, marked correctly and bent with the R mark, will produce a 90° bend with a 7-in leg and an 8-in leg.

Notice that the total required length is actually slightly less than the simple sum of the two center distances, 8 in + 7 in = 15 in. This is because it requires less tubing to make a rounded turn than a square turn. Simply adding the center distance dimensions will yield an approximate length needed to make the bend. If some waste is allowed, the length can be figured in this manner and the excess cut off at the completion of the bend. The estimation method works for bends of 90° or less but does not work for bends over 90° because of the extra length of the arc.

Figure 24-40 Diagram of example bend dimensions.

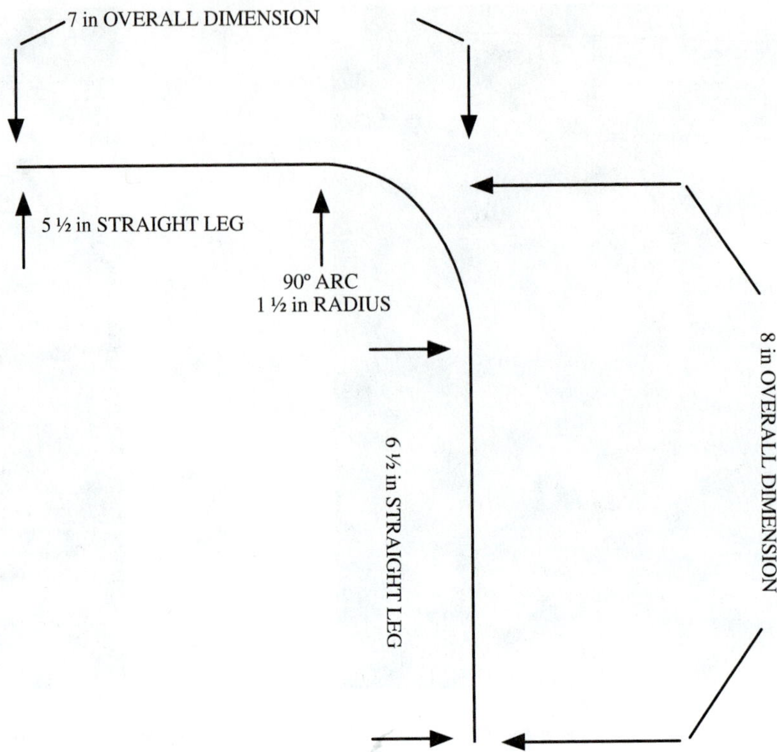

24.10 JOINING COPPER TUBING

Soft copper tubing can be connected in many ways, including compression, flaring, soldering, and brazing. Compression fittings are typically used to connect water lines to icemakers in household refrigerators. They are also used on split-system air conditioners to connect the refrigerant lines to the evaporator and condenser. Soft copper is also connected by flare connections. Flaring is used commonly on refrigeration lines and water lines. Both hard and soft copper may be soldered or brazed together. Soldering and brazing are the only two methods used to connect hard copper tubing.

24.11 SWEAT FITTINGS

Sweat fittings are used with copper tubing when it is soldered or brazed. Common sweat fittings include flare couplings, ells, tees, caps, male adapters, and female adapters (Figure 24-41). Sweat fittings have both male and female ends. The female ends are referred to as solder ends; the male ends are referred to as fitting ends. Solder ends are measured by their inside diameter because they are designed to fit over the tubing. Fitting ends are measured by their outside diameter because they are designed to fit into the solder end of another fitting. A coupling has two solder ends and is used to join two pieces of copper tubing that are the same size (Figure 24-42). A reducing coupling is used to join two different sizes of copper tubing (Figure 24-43). Elbows, or ells, are used to make sharp turns in a copper tubing line. They are available in 90° and 45°. They also are available in both short and long radius. Standard ells have two solder ends. Street ells have one solder end and one fitting end. Figure 24-44 shows different types of

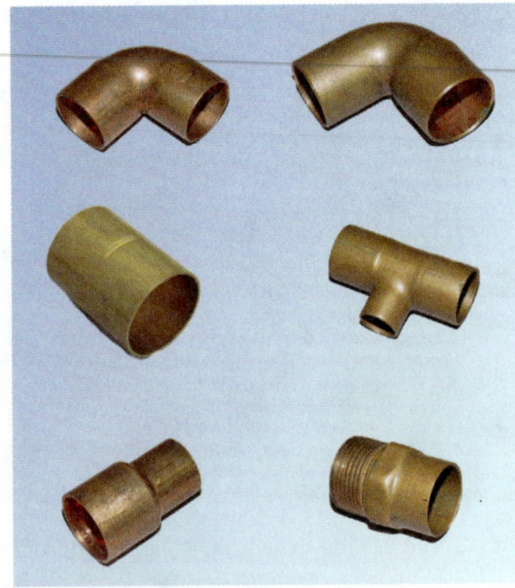

Figure 24-41 Common sweat fittings, including ell, street ell, coupling, tee, reducing coupling, and male pipe thread adapter.

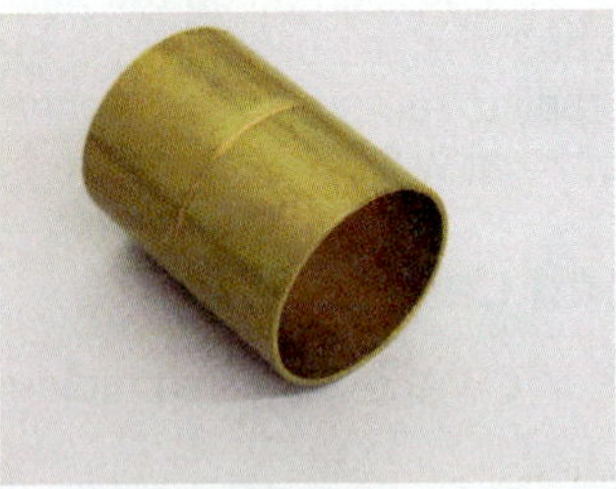

Figure 24-42 Sweat coupling.

Figure 24-43 Sweat reducing coupling.

Figure 24-44 Types of sweat ells, including long radius, short radius, street, 45 degree.

sweat ells. Sweat tees allow one line to branch off of another. They are sized by the run and branch sizes (Figure 24-45). Sweat caps are used to seal off a piece of copper tubing (Figure 24-46). Male and female adapters are used

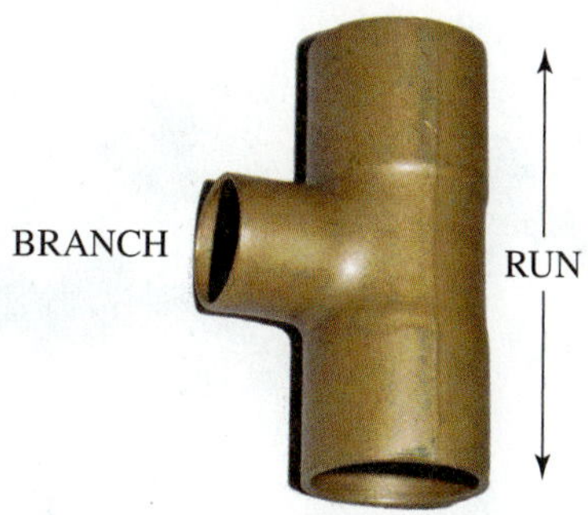

Figure 24-45 Sweat tee (with run and branch identified).

Figure 24-46 Sweat cap.

Figure 24-47 Sweat by male pipe thread and sweat by female pipe thread adapters.

to connect copper tubing to pipe threads. A male adapter has a solder-type sweat fitting on one end and male pipe threads on the other end. A female adapter has a solder end on one end and female pipe threads on the other end (Figure 24-47). The solder end of an adapter is measured by its inside diameter, while the pipe thread end is measured like pipe threads.

24.12 SWAGING TECHNIQUES

A swage joint is used to allow two pieces of copper tubing to be joined by soldering or brazing without any fittings. A swaging tool expands the end of a piece of copper tubing so that it will fit over the end of another tube the same size (Figure 24-48). The depth of the cup on the swage should be equal to the diameter of the pipe being joined. Three types of swaging tools are commonly used: tools that hammer in like a punch, tools that work like a flaring tool, and lever-type tube expanders.

Figure 24-48 Swaged copper tubing. *(Courtesy of HVAC Department, Terra Community College, Fremont, Ohio)*

Figure 24-49 Hammer-style swage tools.

The hammer style is the least expensive and most common. Hammer swages can be sized for each pipe size, or they can be a multi-taper tool that fits many sizes (Figure 24-49). The tubing is held in a flaring block, and it should be high enough in the block that the swage shoulder does not compress against the block. The swage tool is inserted into the tubing and struck lightly with several small taps (Figure 24-50). Large blows will deform and buckle the copper.

The flaring-style swage tool works similarly to flaring but uses a tip that expands the tubing rather than flaring it (Figure 24-51).

The lever-type expander uses cylinders with segments that look like pie slices. When the lever is pulled, the cone slides into the center of the cylinder and the wedges expand outward (Figure 24-52). The lever should be pulled slowly to avoid splitting the copper. Lever tools make consistent swage joints and are easier to use than hammer types, but they are considerably more expensive.

Figure 24-50 The copper is held in a flaring block when using a hammer-type swage tool.

Figure 24-51 This tool swages copper tubing using a spinner with a swage tip. *(Courtesy of HVAC Department, Terra Community College, Fremont, Ohio)*

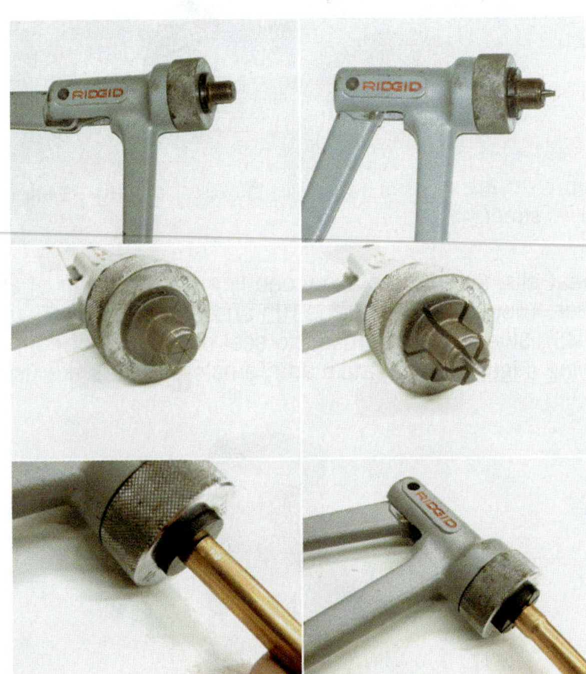

Figure 24-52 Lever-type tubing expander.

24.13 JOINING COPPER TUBING USING COMPRESSION FITTINGS

The compression joint is composed of a compression sleeve, a ferrule, and a compression nut. The copper tubing fits through the nut and the ferrule into the sleeve (Figure 24-53). The nut threads over the sleeve. As the nut is tightened, it squeezes the ferrule around the tubing. The compression of the ferrule around the tubing seals the joint. The sleeve and nut can be reused but the ferrule cannot since it is irrevocably squeezed around the tube (Figure 24-54). To redo the joint, the old ferrule must be cut off and a new

Figure 24-53 Compression fittings, ferrule, and compression nut. *(Courtesy of HVAC Department, Terra Community College, Fremont, Ohio)*

Figure 24-54 Once the ferrule is compressed, it cannot be removed from the tubing. *(Courtesy of HVAC Department, Terra Community College, Fremont, Ohio)*

ferrule used. It is extremely important that the tubing is clean and straight where the joint is to be made and that the nut is not overtightened. It is very easy to assemble a leaky compression joint and very difficult to correct one. Compression nuts and ferrules are measured by the diameter of the hole that the tubing slides through.

SERVICE TIP

Compression fittings for use in refrigeration systems are not interchangeable with the compression fittings used in water systems. You should never use compression fittings designed for plumbing on a refrigeration system. Compression fittings were common on air-conditioning installation valves in the 1980s but are not used today in air conditioning.

Figure 24-55 Flared end on copper tubing.

Figure 24-56 The flared end of the copper tube is squeezed between the male and female flare fittings.

24.14 MAKING FLARE JOINTS

A flared piece of copper tubing has a conical appearance, something like the end of a trumpet (Figure 24-55). The flared end of the tubing is squeezed between two brass fittings, which have complementary 45° angles (Figure 24-56). A flaring tool is used to make the flare on the end of the tubing (Figure 24-57).

A good flare starts with properly cutting and preparing the tubing. The tubing should be cut slowly to minimize the burr left by the tubing cutter. There will still be a small burr, which should be removed with a reamer. The tubing is clamped in the flaring block with approximately ⅛ in above the block (Figure 24-58). The spinner is locked in place over the tubing (Figure 24-59). Put a few drops of refrigerant oil on the spinner before making the flare. The spinner is turned until the flare is formed by pressing the tubing between the spinner and the block. Slide a flare nut to the flared end of the tubing to check the completed flare. The flare should clear the threads of the flare nut but still fill the chamfer (bevel) on the inside of the nut. The flare is too large if it catches on the threads of the nut. The flare is too small if it does not fill the inside chamfer on the nut (Figure 24-60). If

Figure 24-57 Typical flaring tool.

Figure 24-58 Copper tubing loaded in flaring block ready for flaring.

Figure 24-59 The spinner is positioned over the tubing in the flaring block.

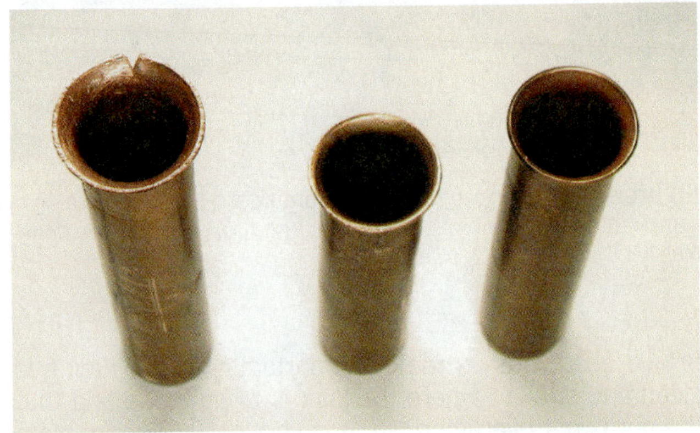

Figure 24-60 The flare on the left is too large, the flare on the right is too small, and the flare in the center is the correct size. *(Courtesy of HVAC Department, Terra Community College, Fremont, Ohio)*

the flare is not the right size, it should be cut off and redone. A good flare does not require thread sealant because the threads are not what is sealing the joint—the flare face is. If the end of the tubing is not properly deburred, the flaring tool will generate a "rolled bead" around the flare, and this will be the only point of flare surface contact (Figure 24-61)

SERVICE TIP

It may be necessary to put the flare nut on the pipe before making the flare. If the pipe has bends, or flares on both ends, it will be impossible to slide the flare nut on after the flare is made. If you forget and flare the tubing before putting the nut on, the only fix is to cut off the flare, put the nut on, and reflare the tubing.

24.15 FLARE FITTINGS

Flare fittings are used to join soft copper tubing that has been flared. Common flare fittings include flare nuts, ells, tees, caps, plugs, unions, and half unions (Figure 24-62). As in pipe and sweat fittings, ells are used to change the tubing direction, unions are used to join tubing, and tees are used to connect three pieces together to form a branch. Flare nuts slip over the outside of the tubing; they are measured by the diameter of the hole in the back of the nut (Figure 24-63). All

Figure 24-61 When the tubing is not correctly cut and deburred, the burr forms a "rolled bead," which holds the flare off of the flare face, creating a leak.

Figure 24-62 Common flare fittings include flare nuts, ells, tees, caps, plugs, unions, and half unions.

Figure 24-63 Flare nuts are measured by the diameter of the hole in the back of the flare nut.

Figure 24-64 Male flare fittings can be measured by measuring the hole in the fitting and adding 1/8 in.

other flare fittings have a male 45° chamfer that matches the angle of the female 45° chamfer in the flare nut. The easiest way to measure a male flare fitting is by finding a flare nut that fits it and measuring the hole in the back of the flare nut. It is also possible to measure the hole in the male flare fitting and add 1/8 in (Figure 24-64). A flare half union is used to connect soft copper to pipe threads. They are called half unions because one side is flare threads and the other pipe threads. The flare side is measured using flare rules; the pipe thread side uses pipe thread rules. A common use for half unions is connecting copper lines to compressor heads and valve bodies. The pipe thread side is threaded into the compressor, and the copper tubing is connected to the flare side (Figure 24-65).

24.16 MAKING FLARE JOINTS FOR R-410A REFRIGERANT

Operating pressures for systems using R-410a refrigerant are 40–60 percent higher than systems using R-22 refrigerant. Flares for systems using R-410a refrigerant are the same 45° angle as regular flares, but the cone shape on the end is larger. Figure 24-66 shows a comparison between a normal flare and a flare that is R-410a compliant. R-410a compliant flaring tools have a larger chamfer and use an offset roller to make more uniform flares (Figure 24-67).

R-410a compliant flare fittings are not interchangeable with traditional flare fittings. Flare fittings designed for R-410a are heavier wall and have a larger chamfer area to accommodate the increased flare cones.

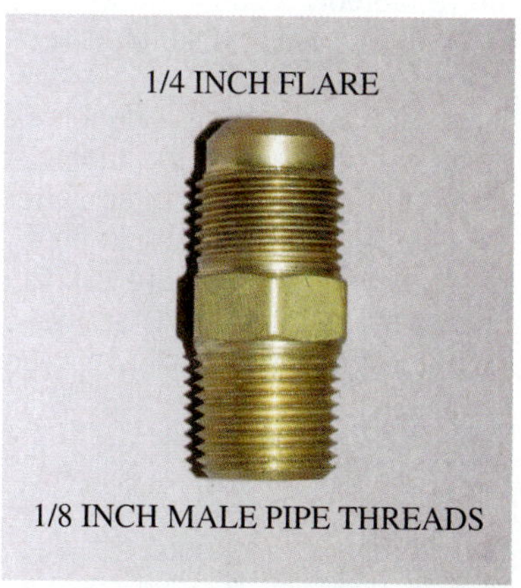

Figure 24-65 Half unions are often used to connect copper tubing to compressor heads.

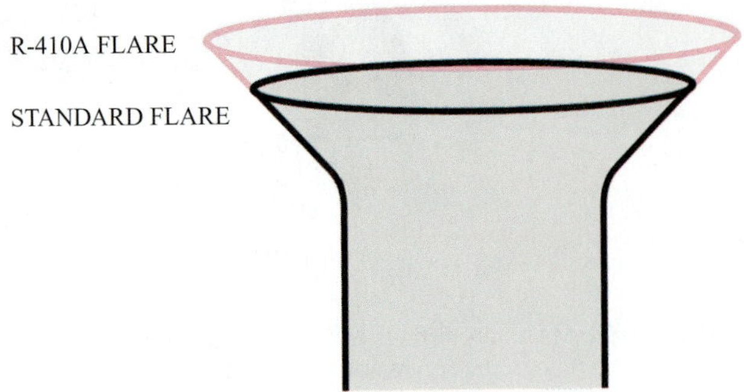

Figure 24-66 The larger pink cone represents the area of an R-410a compliant flare.

Figure 24-67 R-410a compliant flaring tool. *(Courtesy of Ritchie Engineering Company, Inc., Yellow Jacket Products Division)*

24.17 MAKING A DOUBLE-THICKNESS FLARE

A double flare has two thicknesses of metal at the flare face instead of one. This gives the flare more strength. Double flares are more suited for applications where the flared connection will be repeatedly loosened and retightened. The most common application of double flares is on brake lines for cars.

Double flares are made using a double-flare adapter. The adapter folds over the end of the tubing so that there are two layers of copper at the flare face after the flare is formed (Figure 24-68). After the double-flare adapter has been used, it is removed and the flare is finished with the spinner (Figure 24-69). Figure 24-70 shows the steps in making a double-thickness flare.

24.18 PLASTIC PIPE

Plastic pipe is becoming more popular due to its inexpensive price, ease of use, and relative strength. The most common forms of plastic pipe used in air conditioning and refrigeration are PVC, CPVC, and HDPE (Figure 24-71).

Figure 24-68 The tubing is formed into a "mushroom top" by the double-flare adapter. *(Courtesy of HVAC Department, Terra Community College, Fremont, Ohio)*

Figure 24-69 Completed double flare. *(Courtesy of HVAC Department, Terra Community College, Fremont, Ohio)*

Polyvinylchloride, PVC, is normally used for cold-water lines and drainage lines. It can withstand high pressures at low temperatures. It should not be used for hot-water lines. In the air-conditioning industry it is primarily used on condensate drain lines. The PVC is connected to the air-conditioning equipment using adapters that have pipe threads on one side and a PVC socket on the other. A male adapter has male pipe threads, while a female pipe adapter has female pipe threads (Figure 24-72).

PVC pipe comes in 10-ft lengths and ranges in size from ⅜ in to 6 in. PVC and CPVC sizes are nominal sizes, like iron pipe. The nominal size is the approximate inside diameter of the pipe. The size and schedule of PVC and CPVC pipe are stamped all along the pipe length (Figure 24-73). The size of PVC and CPVC fittings is molded into the fitting (Figure 24-74).

Chlorinated polyvinylchloride, CPVC, is more rigid than PVC and can be used for hot-water lines. It can withstand pressures up to 100 psig at 180°F. It is seldom used for drain lines, simply because it is more expensive than PVC.

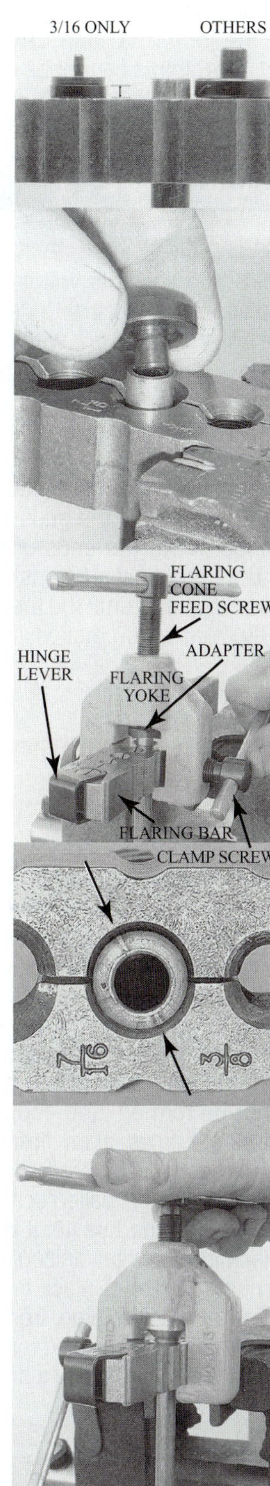

Figure 24-70 The tubing is first folded inward by the double-flare adapter, and the flare is finished with the spinner. *(Images courtesy Ridge Tool Co.)*

High-density polyethylene, HDPE, is used for water-distribution systems, underground gas-distribution systems, and large-scale drainage systems. HDPE tubing is not used much in air conditioning and refrigeration, but it is used for the water loops in ground-source heat pump systems.

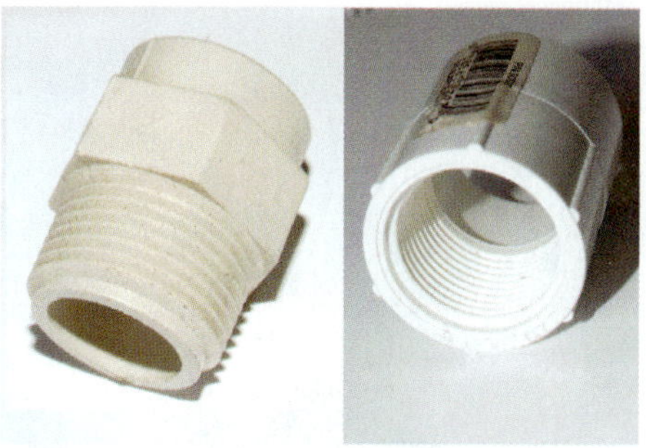

Figure 24-71 Male pipe thread and female pipe thread PVC adapters.

Figure 24-72 Plastic tube comes in many varieties.

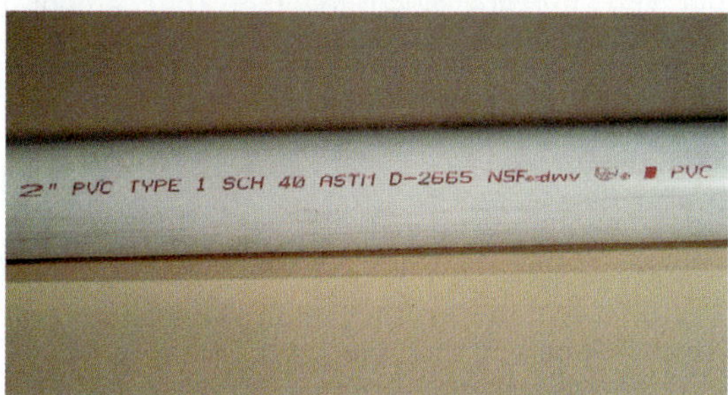

Figure 24-73 PVC pipe specifications. *(Courtesy of HVAC Department, Terra Community College, Fremont, Ohio)*

Figure 24-74 PVC fitting size is molded into fitting. *(Courtesy of HVAC Department, Terra Community College, Fremont, Ohio)*

Figure 24-75 PVC shear produces a clean, square cut; the hacksaw cut produces chips that must be cleaned off before assembly.

24.19 CUTTING AND JOINING PLASTIC PIPE

Plastic pipe can be cut with a fine-tooth saw or a plastic tubing shear. The shear is preferable because it produces a clean, square cut with no chips, shavings, or burrs. Figure 24-75 compares the end of a pipe cut with a hacksaw to one cut with a shear. The shear uses two large, sharp blades to slice through the tubing (Figure 24-76). The shear uses a ratcheting action to produce enough leverage to easily cut through the plastic pipe.

PVC and CPVC are "glued" together. The "cement" used is not a glue in the traditional sense. Instead, it is a solvent that literally dissolves the pipe, allowing it to fuse together.

PVC and CPVC are both very easy to join, and it requires only four simple steps:

1. Clean the pipe to be joined.
2. Prime the pipe.
3. Apply solvent evenly to the primed pipe.
4. Insert the pipe in the fitting and turn it slightly to help spread the cement.

Figure 24-76 Plastic tube being cut by a shear.

If primer is used first to clean the joint and solvent is applied evenly to the joint, PVC and CPVC joints will seldom leak. PVC and CPVC joints cannot be undone. Mistakes must be cut out.

SERVICE TIP

CPVC pipe requires CPVC solvent; PVC solvent will not work with CPVC. Some solvents will work with either, but not all PVC solvents will work with CPVC pipe. Check to make sure the solvent will work with the type of pipe being assembled to avoid embarrassing joint "blowouts" when pressure is applied.

HDPE pipe is most commonly joined by heat fusion. First, a tool is used that squares the ends of the two pieces of tubing to be joined. Next, the ends of the tubing to be joined are heated until the plastic begins to melt. Last, the two pieces are forced together until the melted plastic cools. When done properly, heat fusion makes a very strong joint.

UNIT 24—SUMMARY

The materials most commonly used for piping and tubing in the air-conditioning and refrigeration industry are copper, iron, steel, aluminum, PVC, and CPVC. Copper tubing is available in three wall thicknesses: M, light; L, medium; and K, heavy. Only types L or K may be used for refrigeration lines. ACR tubing is cleaned, dehydrated, and purged with nitrogen. ACR tubing is measured by its outside diameter. Copper tubing is available in two tempers: hard-drawn sticks or annealed rolls. Annealed tubing can be formed; hard-drawn tubing cannot be formed. Soft copper tubing can be joined by compression fittings, flare fittings, soldering, or brazing. Hard-drawn must be soldered or brazed. Iron pipe is available in several strengths, called schedules. The most common strength used is schedule 40. It is available in two finishes: natural black iron and galvanized. Gas lines are run in black iron. The nominal size of black iron pipe is the approximate inside diameter of the pipe. Iron pipe is threaded and assembled with fittings. PVC is used for drain lines on air-conditioning systems. It is joined using a primer and solvent.

WORK ORDERS

Service Ticket 2401

Customer Complaint: Hot Water Pouring out of Broken Joint

Equipment Type: Combination Furnace and Hot Water Heater

A heating and air-conditioning contractor has just installed a new gas-fired combination furnace and hot water heater. The installers had worked with PVC for drain lines but had never performed any hot water work. They knew to use CPVC instead of PVC, but they did not know that PVC solvent will not work with CPVC pipe. After the system had operated for a short time, one of the joints gave way and water started gushing out. The service technician recognized the problem and repaired the system with the correct solvent.

Service Ticket 2402

Customer Complaint: Poor Cooling

Equipment Type: Split-System Air Conditioner

The split-system air conditioner in a new home is performing poorly. The service technician notices that the lines are bent at a sharp angle where they come out of the wall. Both are kinked. The liquid line is cool just past the kink. The technician realizes that the kinked lines are restricting refrigerant flow, causing the liquid to flash at the kink in the liquid line. The technician recovers the system refrigerant, cuts out the kinked sections, and brazes in elbows to make the sharp turn. After the system is evacuated and charged, it operates correctly with the line kinks removed.

Service Ticket 2403

Customer Complaint: LP Gas Leak

Equipment Type: 80 Percent LP Gas Furnace

A customer calls and says that she smells gas near the furnace, which was installed last year. The technician arrives and checks for leaks around the furnace. The threaded joint where the gas line enters the gas valve is leaking. The technician recognizes the color of the pipe sealant and remembers that that product is not listed for use with LP gas. The joint is redone with thread sealant that is listed for LP gas.

Service Ticket 2404

Customer Complaint: Deer Cooler Not Cooling

Equipment Type: Field Assembled Walk-in Cooler

A hunting club calls a refrigeration mechanic to look at their deer cooler. They built the deer cooler themselves and hired someone to install it on the weekend in exchange for hunting privileges. The initial installer has not been able to make the cooler work correctly and has said that the equipment is faulty. Everything is running, but the compressor is hot and the suction pressure is very low. There is frost on the expansion valve, but the evaporator is not even cool. The technician recognizes that there is a restriction at the metering device. Looking at the copper tubing used for the refrigeration lines, the technician notices that it is hard copper, type L, ¾-in nominal size. It does not say that it is ACR tubing. The hunting club members confirm that the copper came from a hardware store that one of them manages. The technician explains that ACR tubing is cleaned and dehydrated, which plumbing tubing is not. There was probably moisture in the tubing that is causing the system to freeze up at the expansion valve. The technician recovers the refrigerant, installs a large filter drier and a moisture indicator in the liquid line, evacuates the system down to 500 microns, and recharges the system with clean refrigerant. The system works for now, but the technician explains that the drier may need to be changed a few more times before all the water is removed.

UNIT 24—REVIEW QUESTIONS

1. What is the difference between pipe and tubing?
2. What materials are used for piping and tubing in heating, air-conditioning, and refrigeration systems?
3. What materials are used for refrigerant piping?
4. What materials are used for fuel gas piping?
5. What materials are used for water piping?
6. What materials are used for drain lines?
7. What is the difference between hard and soft copper?
8. Why should copper tubing be cut with a tubing cutter instead of a hacksaw?
9. How is copper annealed?
10. What methods can be used to join soft copper?
11. What methods can be used to join hard copper?
12. What is the difference between schedule 40 and schedule 80 iron pipe?
13. What is the difference between black iron and galvanized iron pipe?
14. List the steps for threading iron pipe.
15. What is the difference between PVC and CPVC pipe?
16. Why should PVC pipe be cut with a plastic tubing shear instead of a hacksaw?
17. What is the difference between type M, L, and K copper tubing?
18. What is the difference between ACR copper tubing and tubing for water lines?
19. Why are pipe threads tapered?
20. What is the difference between nominal size and outside diameter?
21. What is the inside diameter of schedule 40, ¾-in black iron pipe?
22. How many threads per inch are there on the threaded portion of 1-in iron pipe?
23. Explain how a compression fitting works.
24. List the steps for flaring the end of copper tubing.
25. What is the difference between a flare for an R-410a system and an R-22 system?

UNIT 25

Soldering and Brazing

OBJECTIVES

After completing this unit, you will be able to:

1. describe the proper clothing and PPE that must be worn during soldering and brazing.
2. explain the difference between soldering, brazing, and welding.
3. list the types of torches used for soldering and brazing.
4. list safe handling procedures for acetylene cylinders.
5. list safe handling procedures for oxygen cylinders.
6. explain the correct procedure for soldering.
7. explain the correct procedure for brazing.
8. explain the correct procedure for nitrogen brazing.

25.1 INTRODUCTION

Soldering and brazing are essential skills for HVACR technicians. Soldering and brazing are required for a wide range of service and installation procedures. For example, technicians frequently fabricate piping systems by soldering or brazing the piping components together. Replacing most refrigeration components requires brazing. Soldering and brazing are skills that require practice to become proficient. Making high-quality, leak-free brazed joints is essential, because refrigerant leaks are a problem that often plague HVACR systems. Leaks can often be traced to a field-brazed joint made during installation. Soldering, brazing, and welding all refer to methods of joining metal parts using molten metal. The processes of brazing, soldering, and welding are often misunderstood as all being the same. In the process of welding, the base metal itself is melted as additional metal from the filler is added to the joint. However, during brazing and soldering, only the metal being added to join the base parts is melted. The difference between brazing and soldering is temperature. Soldering takes place at a lower temperature than brazing, but other than that both processes are similar.

25.2 PPE FOR SOLDERING AND BRAZING

Personal protective equipment, PPE, is specialized clothing or equipment worn by workers for protection against health and safety hazards. When using torches, technicians need protection from potential hazards created by the flames: heat, ultraviolet light, sparks, and molten metal. The wrong apparel can increase the hazards to the technician by igniting.

Clothing

The hazards from heat, sparks, and molten metal can be minimized by wearing flame-resistant clothing that covers the technician's exposed skin. This is especially important when welding because of the large quantity of hot sparks produced by most welding processes. One hundred percent cotton treated for flame resistance or leather clothing is the best material to wear while soldering, brazing, or welding. Synthetic materials, such as nylon or polyester, should never be worn while soldering, brazing, welding, or cutting because they burn easily. Remember that plastics are made from petroleum products. Shirts worn while using torches should have long sleeves and either have no shirt pocket or have flaps that cover the pocket to prevent sparks from collecting (Figure 25-1). They should have a collar and be able to be buttoned securely around the neck. One hundred percent cotton blue jeans are ideal pants for working with torches. Leather shoes with high tops are preferable because they resist burns from sparks and molten metal (Figure 25-2). Tennis shoes should never be worn when soldering, brazing, or welding. Properly worn protective clothing will prevent burns from sparks, molten metal, and UV light given off when welding.

HVACR technicians often wear work gloves when soldering or brazing (Figure 25-3a). For welding, leather gauntlet-type welding gloves are preferred (Figure 25-3b). The gauntlet keeps both sparks and welding light from burning your exposed wrists.

Safety Glasses

Safety glasses must be worn at all times when using torches to protect the technician's eyes from harmful radiant energy,

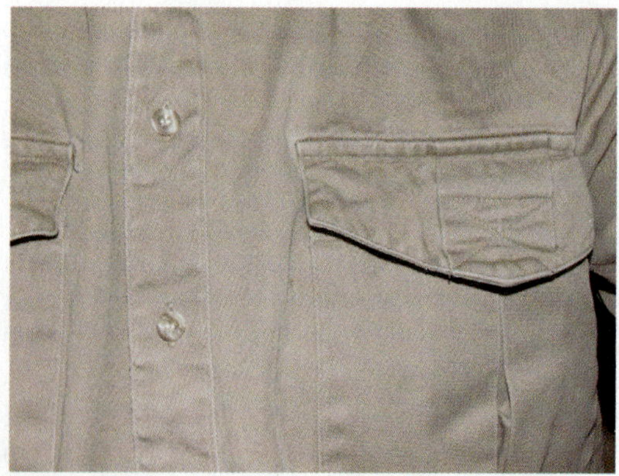

Figure 25-1 Shirts should be 100 percent heavy cotton. Pockets should have flaps to keep out sparks.

Figure 25-2 Welding boots should have smooth toes to prevent sparks from being trapped.

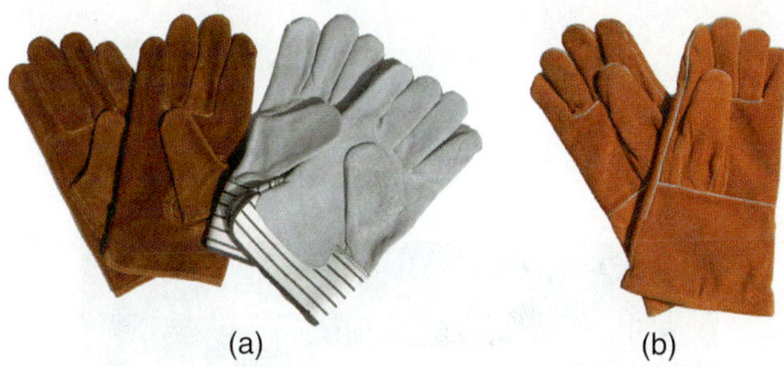

(a) (b)

Figure 25-3 (a) Work gloves may be cloth, leather palm, or all leather; (b) welding gloves are gauntlet-type gloves with high cuffs.

flying sparks, and molten metal. The level of radiant energy produced increases as the process temperatures increase. The most damaging radiation produced is ultraviolet radiation. UV light can cause skin burns similar to sunburn. These burns can occur very quickly to any unprotected skin surface or the eyes because of the intensity of the arc light. Soldering produces very little UV light, while welding produces a great deal of UV light. That is why a welder's safety shield must be worn over the face at all times when arc welding. Safety glasses intended for use with torches come in different shades, from 1.3 for the lightest shade and least protection to 14 for the darkest shade and highest level of protection (Figure 25-4). In general, the highest level of shade that allows the technician to see the work should be chosen. OSHA recommends a minimum shade of 2 for soldering, 3 for brazing, and 5 for gas welding and cutting. Carbon arc welding requires the highest level of protection, level 14. Table 25-1 shows OSHA shade recommendations for different uses.

25.3 TYPES OF TORCHES

A torch is the primary tool required for soldering and brazing. There are many types of torches, each with its own advantages and disadvantages. The two factors affecting torch operation are the type of fuel used and the oxygen source used. Since some types of fuel have higher heat content than others, the type of fuel used will have a bearing on the temperature range in which the torch can operate. The way the fuel is stored and delivered depends on the physical characteristics of the fuel. In addition to the type of fuel, the manner in which oxygen is mixed with the fuel will determine the speed with which the fuel can be burned and, thus, the temperature that the torch can achieve. It is important to understand the torch characteristics to achieve good results in soldering. Generally, torches fall into three categories: air fuel, swirl-tip air fuel, and oxygen fuel.

Air-Fuel Torches

Air-fuel torches aspirate air into the tip utilizing the natural draft created by the flow of fuel as it travels through the tip.

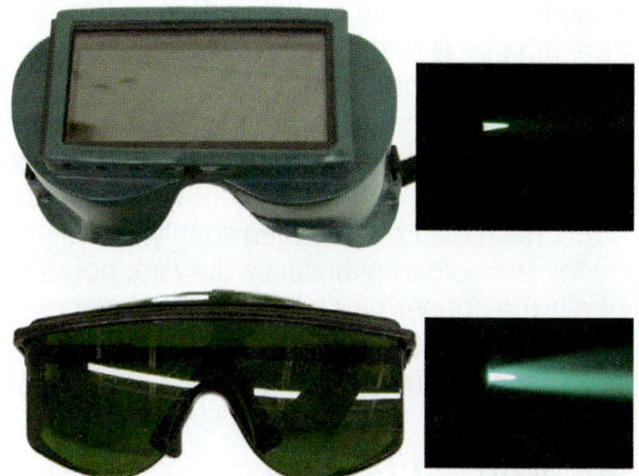

Figure 25-4 Safety glasses are available in different shades. The same flame viewed through two different glasses shows how much more light the darker glasses filter out.

Table 25-1

Operation	Minimum Shade
Torch soldering	2
Torch brazing	3
Gas welding less than 1/8-in thick	4
Gas welding 1/8–½-in thick	5
Gas welding over ½-in thick	6
Torch cutting under 1-in thick	3
Torch cutting under 1–6-in thick	4
Torch cutting over 6-in thick	5
Shielded metal arc welding with less than 3/32-in electrode	7
Shielded metal arc welding with 3/32–5/32-in electrode	8
Shielded metal arc welding with 5/32–8/32-in electrode	10
Shielded metal arc welding with greater than 8/32-in electrode	11
Carbon arc welding	14

The fuel and air are mixed by the venturi action of the tip so that the mixture is ready to burn when it is delivered out the end of the tip. Since the air is only 21 percent oxygen, torches that receive their oxygen by aspirating air are limited in the speed that they can burn the fuel. Therefore, air-fuel torches burn with a relatively cool flame compared to torches that have their own oxygen supply. Air-fuel torches are good for soft soldering because they get hot enough to quickly heat the copper pipe, yet they stay cool enough to prevent oxidizing and burning the pipe. Common fuels used in air-fuel torches include propane, MAPP, and acetylene.

Air-Fuel Swirl-Tip Torches

Standard air-fuel torches will usually not reach temperatures high enough for brazing. Swirl tips create a higher burning temperature for brazing by drawing more air into the tip to mix with the fuel, much like a turbo charger on an engine. While standard air-fuel tips aspirate air near the base of the tip, swirl tips aspirate the air close to the tip (Figure 25-5). Swirl tips are generally available for air-acetylene and air-MAPP torches. Swirl-tip torches usually make a loud whistling sound because of the higher volume of gas and air drawn into the tip. Swirl-tip torches do make brazing possible at a relatively low equipment cost because they do not require a separate oxygen tank and regulator. Figure 25-6 shows a typical swirl-tip torch set for acetylene.

Oxyacetylene Torches

Oxyacetylene torches produce a hotter and more concentrated flame than air-acetylene torches because they use pure oxygen to burn the fuel. Even the coolest oxyacetylene flame is much hotter than an air-acetylene flame. Oxyacetylene torches use an acetylene cylinder, an oxygen cylinder, regulators for each cylinder, connecting hoses, and a torch handle that mixes the acetylene and oxygen (Figure 25-7). They are generally preferred for brazing due to their higher temperature and heating capacity. Oxyacetylene outfits are more expensive and more difficult to transport, but they provide a solution to high-temperature brazing even on larger pipes, where swirl-tip air-acetylene torches won't work. The most common torches used in HVACR service are small portable oxyacetylene brazing torches (Figure 25-8). Larger tanks are typically used for large installations where technicians will be brazing all day long (Figure 25-9).

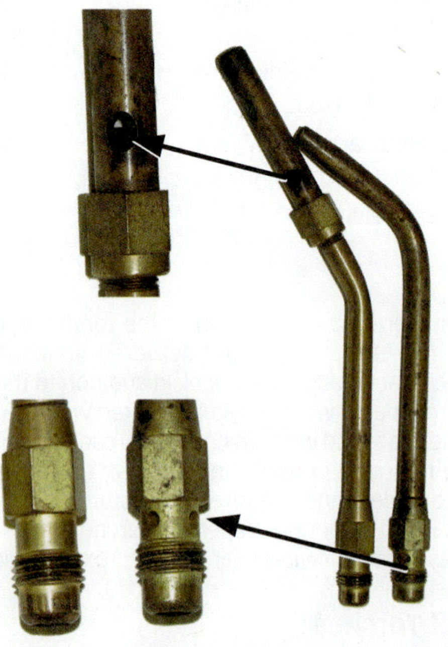

Figure 25-5 Standard air-fuel tips take in air at the base. Swirl tips take in air near the tip.

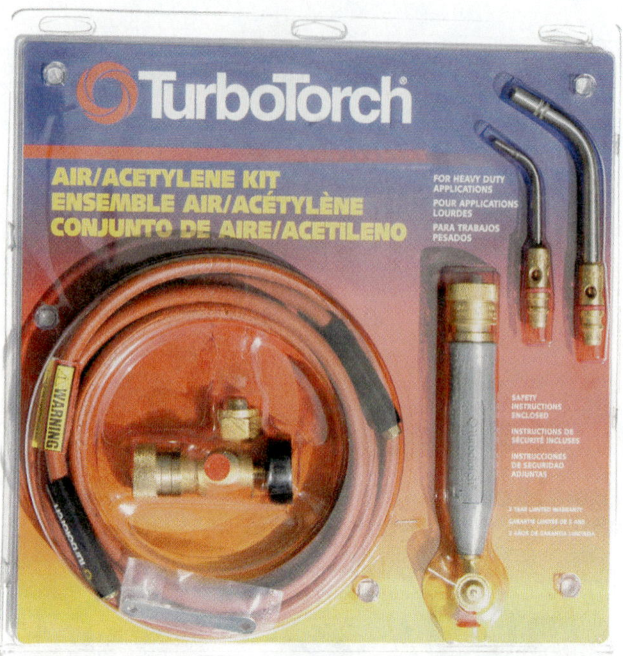

Figure 25-6 Swirl-tip torch set for air acetylene.

Figure 25-7 Oxyacetylene torch outfit.

25.4 ACETYLENE CYLINDERS

To store enough acetylene to make acetylene cylinder storage feasible, acetylene is compressed into the cylinder under a pressure of 250 to 275 psig. However, acetylene gas is unstable under high pressure and can explode just

Figure 25-8 Portable oxyacetylene welding and cutting rig. *(Victor, a Brand of Thermadyne Industries, Inc.)*

Figure 25-9 Heavy-duty oxyacetylene torch with larger cylinders.

from being struck or jarred severely. For this reason, the acetylene gas must be stabilized in the cylinder with liquid acetone. The acetone is capable of absorbing 28 times its own weight in acetylene gas. Acetone reacts with acetylene similarly to carbon dioxide and water. Carbon dioxide can be dissolved in water, forming a carbonated beverage. As the pressure on a carbonated drink bottle is slowly released, you can observe bubbles being formed within the liquid. These bubbles are CO_2 gas being released. The acetone in an acetylene bottle works the same way. To prevent the acetone from simply being poured out accidentally, the acetylene cylinder is filled with an absorbent, coarse material during manufacturing (Figure 25-10). The acetone is stored inside a porous material that soaks it up like a sponge. When the valve is opened, the pressure inside the

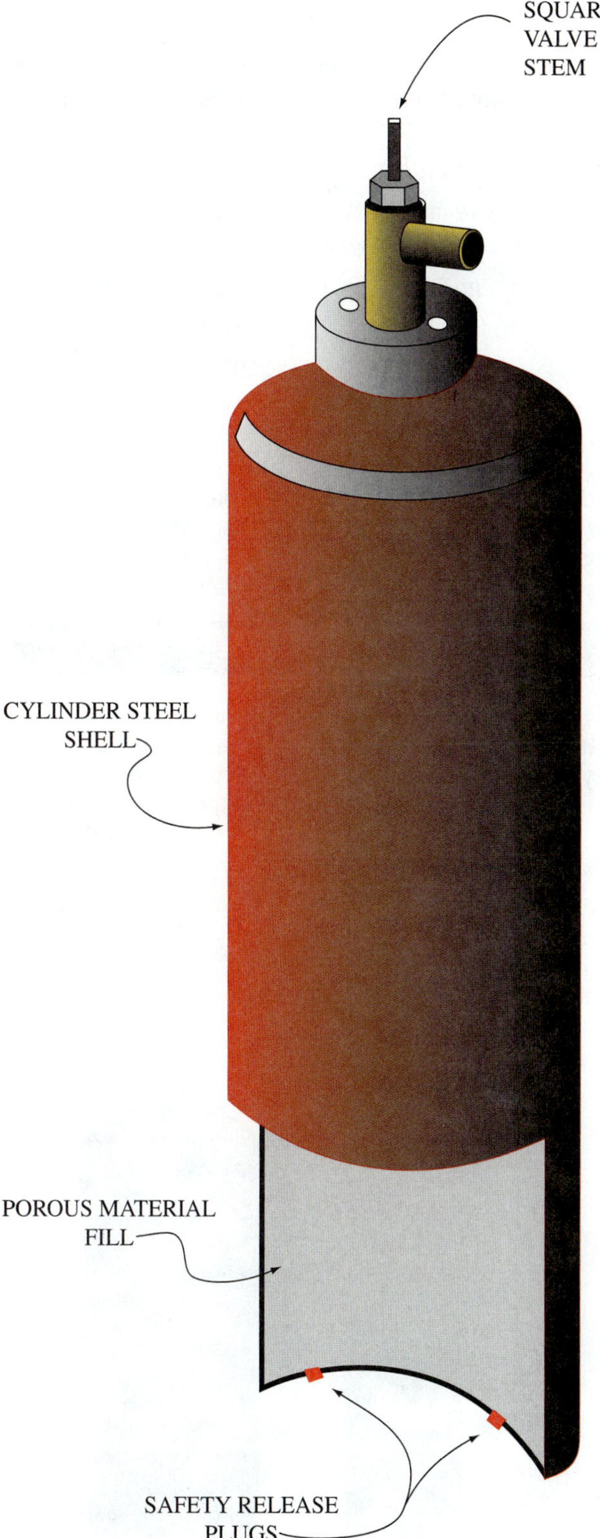

Figure 25-10 Diagram of a cutaway section of a B acetylene tank showing the porous absorbent section.

tank drops and the acetylenes boil out of the acetone. When all the acetylene has boiled out of the acetone, a qualified compressed gas dealer refills the tank. Acetylene tanks come in several sizes, from small tanks that hold 10 ft^3 to large tanks that hold 420 ft^3. Table 25-2 shows the most

Table 25-2 Acetylene Cylinder Specifications

Size	Dimensions	Empty Weight	Full Weight	Cubic Foot Capacity
MC	4 × 12 in	7.5 lb	8.5 lb	10
B	6 × 19 in	22.5 lb	25.5 lb	40
1	7 × 25 in	47 lb	52.5 lb	75
2	8 × 30 in	70 lb	79 lb	130
3	10 × 30 in	100 lb	113 lb	190
4	12 × 36 in	175 lb	197.75 lb	330
5	12 × 39 in	185 lb	209.75 lb	360

common sizes of acetylene tanks available. Types B and MC are the most popular for portable air-acetylene torches and small oxyacetylene torches (Figure 25-11).

CAUTION

Acetylene cylinders must be used only in the upright position to prevent the acetone from being pulled out of the cylinder with the acetylene. If the acetone comes out, a concentrated, sooty flamethrower-type flame leaves the torch tip. The space left inside the cylinder by the lost acetone can fill with acetylene gas, creating an explosion hazard.

(a) (b)

Figure 25-11 (a) The MC acetylene tank holds 10 ft^3; (b) the B acetylene tank holds 40 ft^3.

CAUTION

Acetylene cylinders must be secured during storage, transport, and use. Rough handling and banging the cylinder around can damage the filler inside the tank, creating a dangerous acetylene gas pocket that can lead to an explosion hazard.

CAUTION

Acetylene can become unstable and explode at pressures above 15 psig. Because of acetylene's instability, it is against OSHA regulations to operate a torch with an acetylene pressure greater than 15 psi. Most acetylene gauges have a red warning mark indicating 15 psi (Figure 25-12).

25.5 AIR-ACETYLENE TORCHES

HVACR technicians often use air-acetylene torches for soldering and light brazing. These torch flames are ideal for soldering. An air-acetylene flame burns at a temperature of approximately 4,220°F. Air-acetylene torches work very well for soldering copper pipe from 1/4 in up to 4 inches. Figure 25-13 shows a typical air-acetylene torch.

Figure 25-12 Acetylene regulator with gauges.

Figure 25-13 Typical air-acetylene torch.

Figure 25-14 Before opening the acetylene cylinder, the regulator should be adjusted out counterclockwise to relieve the spring pressure on the diaphragm.

Adjusting the Torch

First, check to see that the regulator is adjusted all the way out counterclockwise (Figure 25-14). The cylinder valve should not be opened with the regulator adjusted in, because the sudden cylinder pressure can damage the regulator. Some cylinders have hand wheels, and others just have square stems that require a wrench, as in Figure 25-15. OSHA requires that a nonadjustable wrench be used on any valve stems that are not equipped with a hand wheel (Figure 25-16). The wrench must be left on the cylinder valve anytime the valve is open. Hold the cylinder steady with one hand while turning the valve counterclockwise one-quarter to one-half turn with the other hand (Figure 25-17). Opening the valve a minimum amount will allow you to shut down the valve quickly in an emergency. Do not leave the torch unattended with the valve turned on. If you must leave the work area, secure the torch out of the way of traffic, close the valve, and take the valve key off the stem.

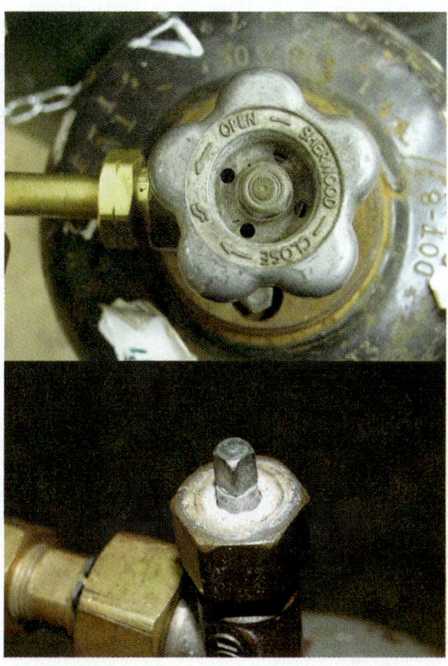

Figure 25-15 Some acetylene cylinders have a hand wheel; others have a square stem.

Figure 25-16 Nonadjustable wrench for acetylene cylinder with a square stem.

Figure 25-17 Support the cylinder with one hand when turning on the acetylene or oxygen valves.

CAUTION

It is against federal regulations (OSHA) to open any acetylene cylinder valve more than two and a half turns. This is to allow you to turn off the cylinder quickly in an emergency. The best practice is to always open the cylinder valve the minimum amount necessary to see a pressure reading on the regulator. Often a half turn will work.

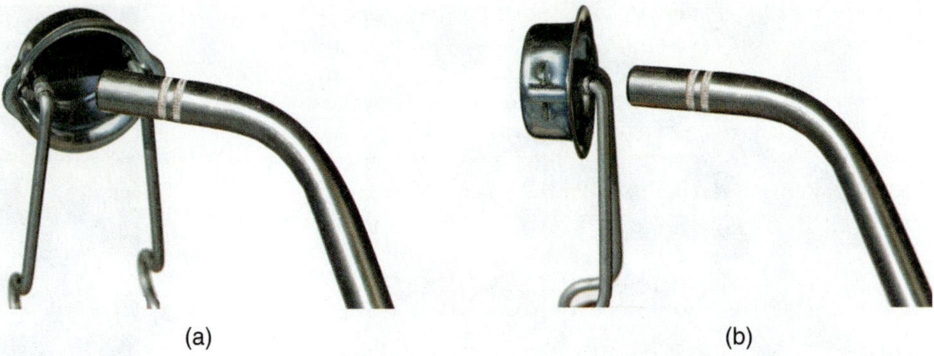

Figure 25-18 (a) Correct position for striker, off to the side of the torch tip; (b) improper position for striker, directly in front of the torch tip.

An air-acetylene torch tip is designed for a specific pressure and flame size. The correct adjustment will depend on the tip being used. Adjust the regulator in clockwise according to the manufacturer's instructions. The fuel delivery should be adjusted at the regulator, not by the valve on the torch handle.

Lighting the Torch

With the pressure adjusted properly according to the manufacturer's recommendations for that torch tip, light the torch by opening the valve on the torch body completely and creating a spark at the torch tip with a torch lighter or sparker. The only safe devices to use to light any torch are those specially designed for that purpose, such as spark lighters or flint lighters. Never use household lighters! When using a flint lighter, hold it slightly off to one side of the torch tip, as in figure 25-18a. Do not hold it directly over the tip, as in Figure 25-18b. To shut off an air-acetylene torch, just turn the valve on the handle clockwise until the gas flow stops.

SERVICE TIP

Holding the striker over the end of the tip may cause a pop when the torch is lit.

CAUTION

Never use a cigarette lighter to light any torch. The torch flame can instantaneously burn through the plastic lighter, causing it to explode. Butane cigarette lighters can explode with the force of a quarter stick of dynamite if they are ignited by the torch. The only thing preventing an explosion is a cheap piece of plastic.

The flame should be a quiet, blue flame (Figure 25-19a). If the flames burn away from the tip or try to light but seem to "blow out," reduce the pressure on the regulator. Gas pressure that is too high will prevent the torch from lighting correctly. If the torch does not light and you can't hear gas escaping from the tip, turn the regulator clockwise a little at a time until you can hear the sound of acetylene escaping from the torch tip.

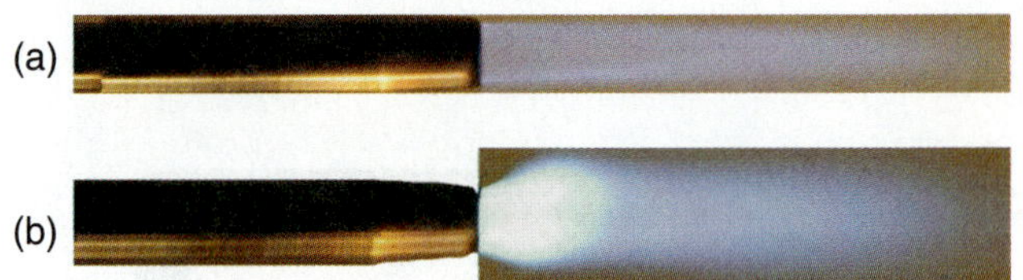

Figure 25-19 (a) Correct appearance of acetylene flame; (b) too much acetylene causes a noisy flame that lifts off the tip.

You can control the size of the flame somewhat by using the regulator to control the gas pressure delivered to the torch tip. However, each tip is designed for a narrow range of operating pressures. If the gas pressure to the tip is too low, the torch tip can become overheated. In some severe cases, the tip may actually begin to glow red hot. Do not adjust the flame size to a point where it lifts off the tip and burns with a loud rushing sound, Figure 25-19(b). The best way to either increase or decrease the amount of heat the torch produces is to change the size of the torch tip. Tip size can be important since the success of a soldered joint depends largely on the manner in which heat is applied to the joint. The air-acetylene tips available from one manufacturer are listed in Table 25-3. Note that the #1 tip is for soldering up to ¼-in tubing, while #6 is needed for soldering 4-in tubing.

Putting up the Torch

When you are finished with the torch, shut the valve on the tank off (clockwise), and remove the valve key. Open the handle valve and bleed the gas out of the handle and regulator. Turn the regulator counterclockwise until you

TABLE 25-3 Air Acetylene Torch Tips

Size	Description	Solders Tubing Up To
BA-1	Fine	¼-in diameter
BA-2	Small	½ in
BA-3	Medium	¾ in
BA-4	Large	1 in
BA-5	X large	1.5 in
BA-6	XX large	4 in

feel the spring tension let up. This prevents a sudden surge of pressure on the regulator diaphragm when the torch is used next. Finally, secure the torch in the designated storage area. Acetylene cylinders and torches should be stored away from flammable materials or oxygen cylinders in a well-ventilated area that is protected from the weather. They should be stored in the upright position and physically restrained to prevent being tipped over.

25.6 SOLDERING

Soldering is used in the HVAC industry to both join electrical wiring and components and for copper pipes when used for water or condensate drains. Soldering is not approved by most manufacturers or codes for use in refrigerant lines or combustible gas lines. Electrical soldering is accomplished using an electric soldering iron. Soldering to join copper pipes is accomplished with an air-acetylene torch, air-MAPP torch, or air-propane torch. Oxyacetylene torches are not recommended for soldering because their flame temperature is so high that the pipe is easily overheated. When overheating occurs, the flux breaks down and excessive oxides are formed so that no bonding can occur.

In both brazing and soldering, the base metal must be heated to a temperature where the bonding phase between the liquid filler and the base metal can occur. This process is called tinning. The soldering alloy must be able to chemically bond with the metals being soldered. The solder forms an alloy with the base metal when the molecules of the solder enter and fill spaces existing between molecules of the base metal. For the bonding or tinning phase to occur between the filler metal and the base metal, often a flux is required (Figure 25-20). A flux is an active compound that removes light surface oxides and promotes the wetting of the base metal with the liquid filler metal. If the metal is oxidized, the solder will not stick because the oxide is already filling the spaces between molecules. Flux prevents the formation of oxides during the heating process, keeping the spaces open for the solder. If the metal being joined is too cool or too hot, tinning cannot occur. Overheating is a major problem when brazing or soldering copper pipe because the flux will burn (become oxidized), stop working, and become a barrier to tinning. In addition, a heavy oxide can be formed on the pipe itself, preventing a bond forming between the filler metal and the pipe surface.

Figure 25-20 Soldering paste flux.

Soldering Alloys

Soft solder is generally a combination of two metals. Older solders were often made of tin and lead. One common alloy, 50 percent tin and 50 percent lead, was used for plumbing. Lead solders are no longer used in plumbing because of the health concerns with lead. An alloy of 95 percent tin and 5 percent antimony is more common now. One problem with 95/5 is that it can become brittle and crack from vibration. A solder alloy called silver-bearing solder is often used where vibration is a concern. It is 96 percent tin and 4 percent silver. Table 25-4 illustrates some of the characteristics of common soldering and brazing alloys.

Preparing the Joint

Begin by cleaning the copper pipe and fitting using abrasive cloths like those shown in Figure 25-21. A wire brush or sand cloth may be used to remove any contaminants, such as oxidation, oil, paint, or dirt. Do not touch the part to be soldered with your hand. Oil from your fingers can prevent solder from flowing completely into a joint. The cleaned end should look like Figure 25-22.

TABLE 25-4 Common Soldering and Brazing Metal and Fluxes Showing the Metals That Can Be Joined

Components	Flow Temp	Flux	Joins
Soldering Alloys			
95% tin 5% antimony	464°F	Paste Acid below 700°F	Copper Brass Steel
96% tin 4% silver	430°F	Paste Acid below 700°F	Copper Brass Steel
85% tin 15% zinc	482°F	Aluminum below 700°F	Aluminum
Brazing Alloys			
93% copper 7% phosphorus	1,475°F	None	Copper
80-90% copper 1–15% silver 5-7% phosphorus	1,450°F to 1,480°F	None	Copper
25-56% silver 20-43% copper 17-33% zinc 0-5% tin 0-2% nickel	1,200°F to 1,435°F More silver gives lower flow temp	Brazing 1,100°F to 1,600°F Work temp	Copper Brass Steel

Figure 25-21 A variety of sanding cloths are available.

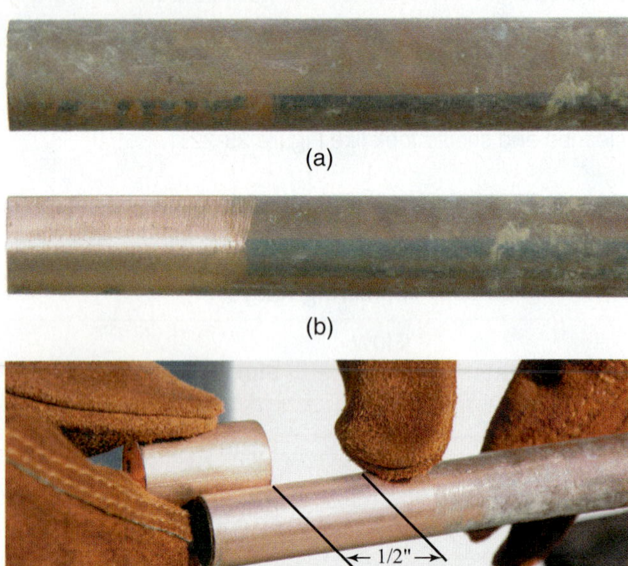

Figure 25-22 (a) Copper pipe before cleaning; (b) copper pipe after cleaning; (c) clean at least ½ in beyond the fitting.

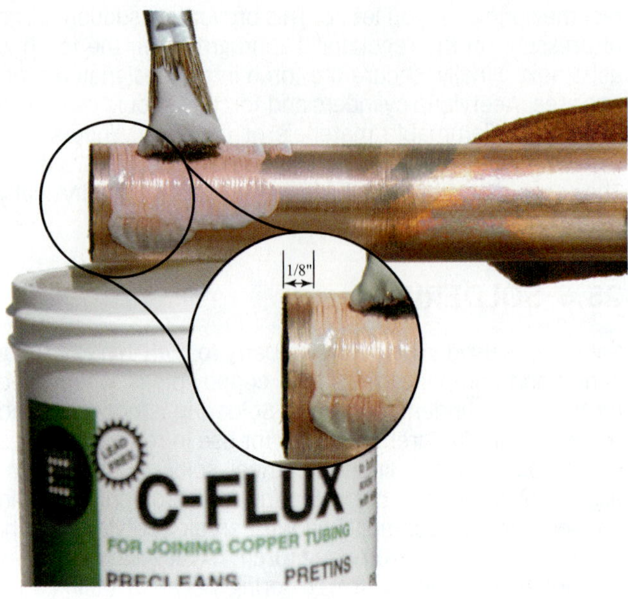

Figure 25-23 Apply the flux up to approximately ⅛ in from the end of the copper pipe.

Figure 25-24 Preheat the pipe before moving the flame over the fitting.

With the joint clean, use a brush to apply the flux. Do not apply flux to the very end of the pipe; you don't want any flux in the pipe after it has been soldered. Apply the flux up to approximately ¹⁄₁₆ in to ⅛ in from the end of the pipe, as shown in Figure 25-23. This will help prevent flux contamination into the system. Insert the flux-covered pipe into the copper fitting and twist the copper pipe inside the copper fitting to spread the flux around inside the joint.

Heating the Joint

During soldering, the molten solder runs toward the heat. Uniform heating of the joint is important to provide uniform solder coverage and completely fill the joint with the solder alloy. It is important to understand that the torch is not used to directly melt the solder. To ensure that both the base metal and the solder are hot enough to bond, the torch should only be used to heat the metal and the metal should melt the solder. Begin heating by heating the pipe first. As the pipe expands, it will make firm mechanical contact with the inside of the copper fitting (Figure 25-24). This mechanical contact will aid in capillary attraction, which pulls the solder into the joint. The mechanical contact also aids in thermal conductivity so that the entire pipe and joint become uniformly heated. As the pipe begins to heat up, move the flame onto the fitting and pipe. Periodically test the backside of the joint with the tip of the solder. Once the solder begins to flow freely, remove the torch. Continue adding the solder until you see a silver ring all the way around the joint (Figure 25-25). Ideally, a small fillet of solder should be left at the joint surface (Figure 25-26).

To prevent overfilling of the joint, before you start, bend the solder to match the diameter of the pipe being soldered. A length of solder equal to the diameter of the pipe being soldered is adequate to make soldered joints in most copper lines smaller than 1 in diameter (Figure 25-27).

Figure 25-25 Add solder until silver ring appears all the way around the joint.

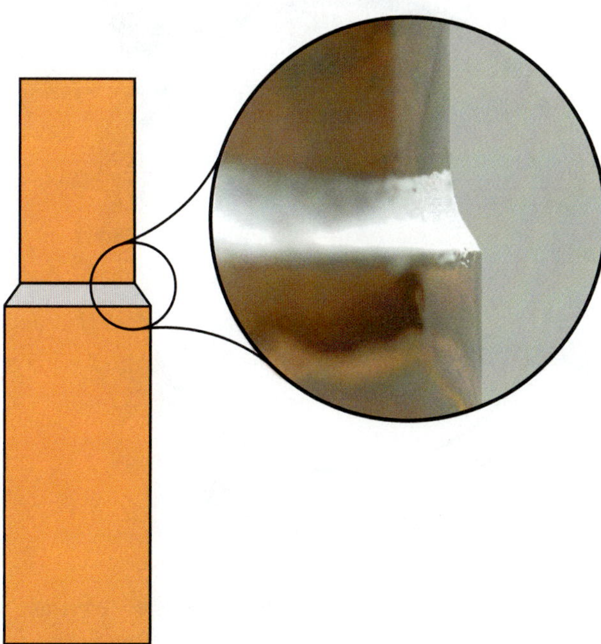

Figure 25-26 Add a small amount of additional solder to create a fillet around the joint.

Solder will not stop flowing into the joint. If you continue to heat the joint and feed more solder, it will continue to go in. Pushing more solder into the joint simply results in solder BBs being formed inside the piping system. In the field, these BBs of solder can circulate back to water pumps where they can cause damage to the pump's impellers. In extreme cases the line can become completely blocked with solder (Figure 25-28).

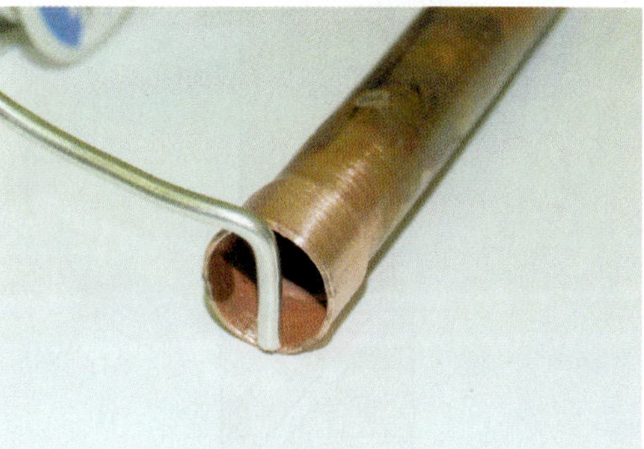

Figure 25-27 A length of solder equal to the pipe diameter is all that is needed.

Figure 25-28 Too much solder was fed into this joint, resulting in solder inside the pipe.

SERVICE TIP

When soldering copper pipe, once the metal has been overheated you must stop to cool and reclean the parts before the joint can be made correctly. If the parts are not disassembled and recleaned, only the outside of the joint will be able to be joined. This will leave a joint that will someday leak.

Testing Soldered Joints

Once the solder joint has been completed, allow it to cool before cutting a slice into the pipe fitting at approximately a 45° angle using a hacksaw (Figure 25-29). Use a slight rocking motion as the cut progresses, so that you are only cutting through the fitting and not the pipe (Figure 25-30). Figure 25-31 shows what the cut should look like when it is done. To release the fitting from the copper pipe, put the tip of a flat-blade screwdriver into the slot, apply pressure, and twist (Figure 25-32). Using a pair of pliers, peel back the copper fitting, exposing the soldered surface. If the surface is smooth and has no large voids, the soldering job was successful (Figure 25-33). Some small flux-filled pockets may remain. These voids are acceptable as long as they do not

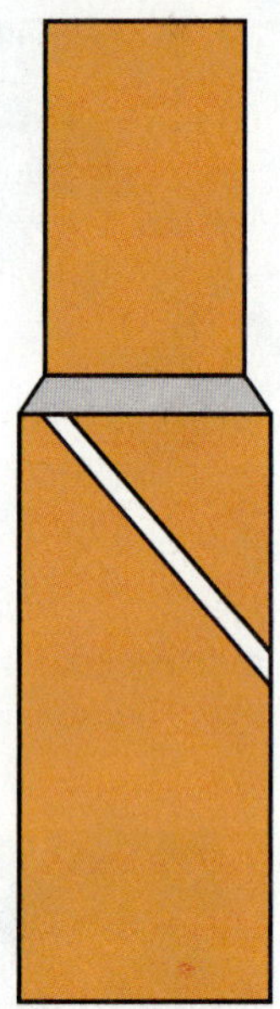

Figure 25-29 The first step in testing a solder joint is to make a hacksaw cut diagonally across the fitting.

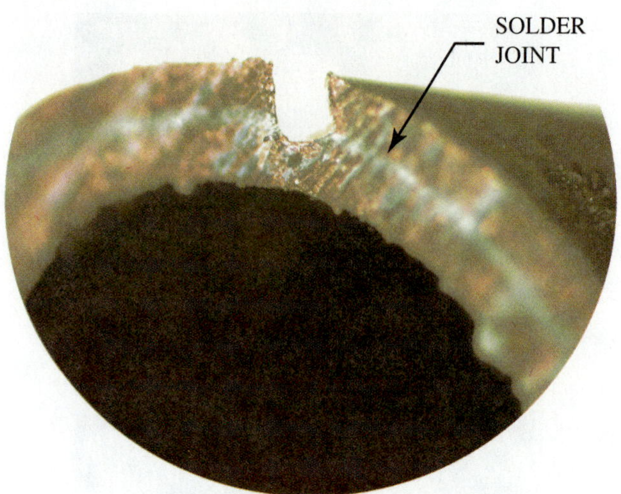

Figure 25-31 The finished cut should completely cut through only the outer fitting. A small cut into a portion of the inner fitting is OK.

(a)

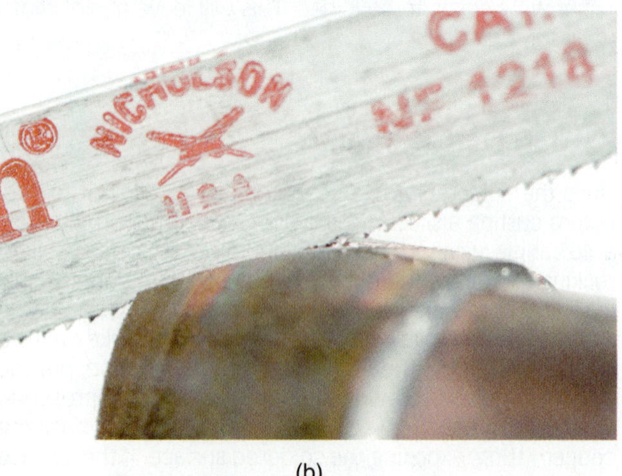

(b)

Figure 25-30 (a) Begin the hacksaw cut flat across the fitting; (b) rock the hacksaw as the cut is made.

Figure 25-32 Place a flathead screwdriver in the hacksaw kerf and twist to begin opening up the joint; use pliers to finish the job.

run from near the inside of the joint to near the outside. Obviously, this test will render the joint unusable. The technician may use this test for the practice and development of brazing skills.

Overheating and under-heating are common problems with soldering. Overheating is indicated by tiny bubbles in the solder (Figure 25-34). The bubbles are formed as the solder boils. Under-heating is indicated if the solder does not flow into the joint. If the joint were heated first instead of the pipe first, then during soldering the pipe would separate

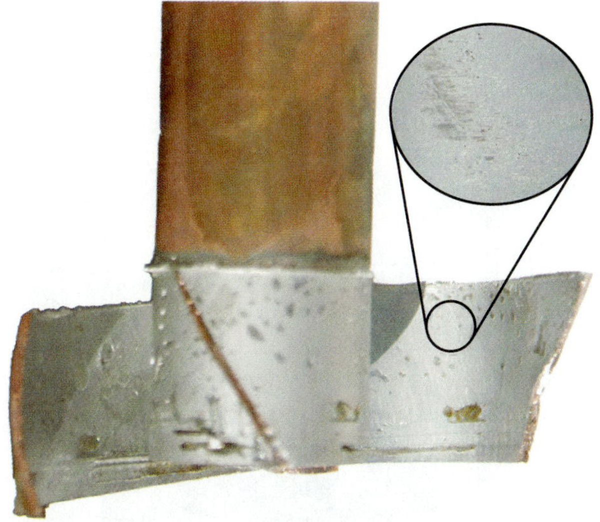

Figure 25-33 An acceptable solder joint is one that will have a minimum amount of small bubbles and flux inclusions.

Figure 25-35 From the outside, this joint looks very good; solder flowed completely across the outside face of the joint, indicating that the fitting was at the proper soldering temperature.

Figure 25-34 Excessive bubbles formed by overheating and boiling solder.

Figure 25-36 The inside of the joint from Figure 25-35; solder did not flow through the joint because the pipe had not reached the soldering temperature due to improper heating technique.

so that heat would not transfer to the pipe from the joint. The solder on the outside of the joint may look great, as in Figure 25-35. In Figure 25-35, the solder not only flowed around the joint but even down the outside. But looking at the inside of the joint in Figure 25-36, we see that the solder penetration was actually very poor. The solder did not flow throughout the joint space and left large unfilled voids. This can be caused by inadequate joint temperature or a swage joint that is uneven.

25.7 OXYGEN CYLINDERS

Oxygen is stored in the cylinder as a compressed gas. Unlike acetylene cylinders, there is nothing inside an oxygen cylinder to absorb the oxygen. Compressing a large volume of oxygen gas into a cylinder creates tremendous internal cylinder pressure. A full oxygen cylinder contains approximately 2,000 psig of oxygen. All high-pressure cylinders that can be transported on public roads must meet federal Department of Transportation (DOT) guidelines. The cylinders must be marked with the cylinder specification, serial number, manufacturing date, owner, retest markings, and tare weight (Figure 25-37).

Since high-pressure, pure oxygen will burn anything that is even marginally combustible, the least piece of trash in the oxygen regulator can cause a fire that can burn up the regulator's diaphragm. Oil and pure oxygen under pressure can spontaneously combust. The ensuing explosion can be lethal. Even small quantities of oil that may get on the oxygen gauge, valve, or hoses can cause an explosion when

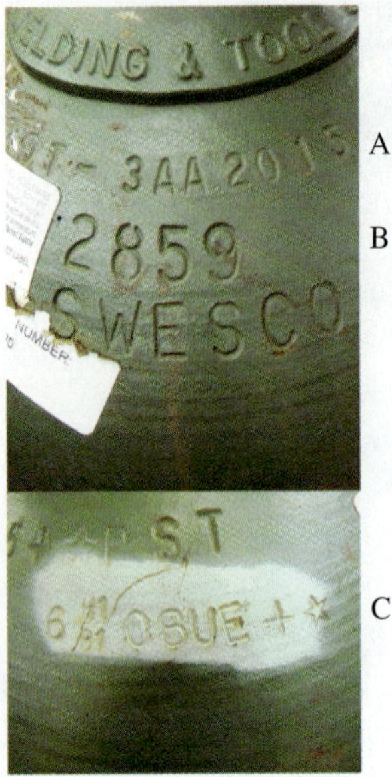

Figure 25-37 Cylinder markings. A: Service pressure is 2015 psig. B: Serial number is 2859. C: Retest marking shows a retest June 2008. The + indicates the cylinder qualifies for 10 percent overfill. The star indicates the cylinder meets requirements for a ten-year retest.

subjected to high-pressure oxygen. It is very important that you never allow oil to come in contact with oxygen and acetylene equipment.

CAUTION

Use no oil or thread sealant on the threads of the oxygen regulator. Go to extreme measures to keep the oxygen-delivery system clean.

Oxygen cylinders should always be stored and transported with their safety caps in place. The safety cap screws over the cylinder valve to prevent physical damage to the valve (Figure 25-38). The regulator and torch assembly must be removed from the cylinder to install the safety cap. Oxygen cylinders are available in sizes for 20 ft^3 to 400 ft^3 (Table 25-5). Two common cylinder sizes used for

Table 25-5 Oxygen Cylinder Sizes

Capacity	Size	Weight
20 ft^3	5 × 14 in	15 lb
40 ft^3	7 × 17 in	25 lb
55 ft^3	5 × 41 in	25 lb
80 ft^3	7 × 36 in	53 lb
125 ft^3	7 × 48 in	79 lb
250 ft^3	9 × 56 in	115 lb
330 ft^3	9.25 × 60 in	143 lb

Figure 25-38 Oxygen cylinder safety cap screws over threaded collar to protect cylinder valve.

Figure 25-39 (a) 20 ft^3 oxygen cylinder; (b) 55 ft^3 oxygen cylinder.

the air-conditioning and refrigeration trades are the 20 ft^3 and the 55 ft^3 cylinders (Figure 25-39).

CAUTION

Federal safety regulations (OSHA) require 55 ft^3 cylinders and larger to be safety chained to a cart or structure anytime the cylinder valve cap is removed. If one of these high-pressure cylinders falls over and the valve breaks off, the cylinder will be propelled through the air like a missile. The force is enough to break through concrete-block walls.

CAUTION

Always remove regulators from oxygen and acetylene cylinders before you put them back in your service vehicle. In many states and cities it is against the law to transport oxygen and acetylene cylinders in a motor vehicle with the regulators attached. The danger is that in an accident the regulators could be broken off, causing a fire or explosion.

CAUTION

Compressed oxygen should never be used in place of nitrogen or compressed air. If oxygen were used to pressurize a refrigeration system and the system were started, there would be a severe explosion of the compressor and compressor shell. This explosion would be hazardous to anyone within the immediate area.

25.8 OXYACETYLENE TORCHES

The oxyacetylene flame is often required to make braze joints in large refrigerant lines and join refrigerant lines to compressor stubs. An oxyacetylene torch requires an acetylene cylinder, an oxygen cylinder, a regulator for each cylinder, a torch handle that mixes the oxygen and acetylene before delivering the combustible mixture to the tip, and a set of hoses to connect the regulators to the handle. Oxyacetylene torches have changeable tips whose size helps determine the temperature of the flame. Figure 25-40 shows a typical oxyacetylene torch kit.

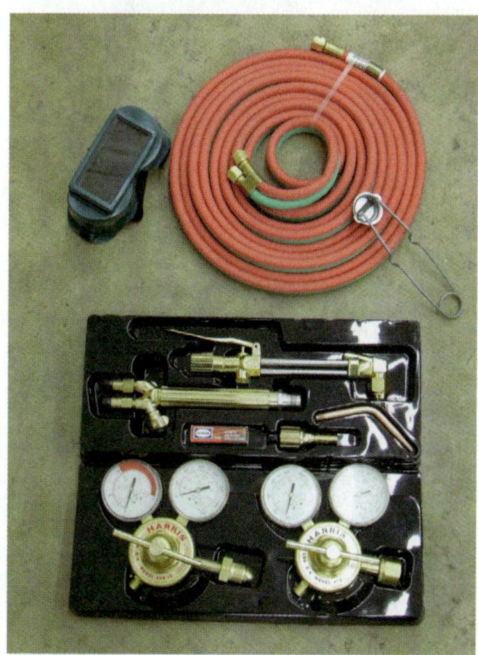

Figure 25-40 Typical large oxyacetylene torch setup.

Figure 25-41 The oxygen regulator connects to the cylinder with a female, right-handed thread.

It is extremely important that the two hoses and regulators not be mixed up and connected to the wrong type of gas. Connecting a regulator that has acetylene in it to a high-pressure oxygen cylinder would cause the regulator to explode. Oxyacetylene equipment is designed to prevent this. The oxygen regulator connects to the cylinder with a female, right-handed thread (Figure 25-41). The oxygen hose is green and connects to both the regulator and torch handle with right-handed threads (Figure 25-42). The acetylene regulator connects to the acetylene cylinder with a male, left-handed thread. The acetylene regulator has hash marks that indicate it has left-handed threads (Figure 25-43). The acetylene hose is red and connects to

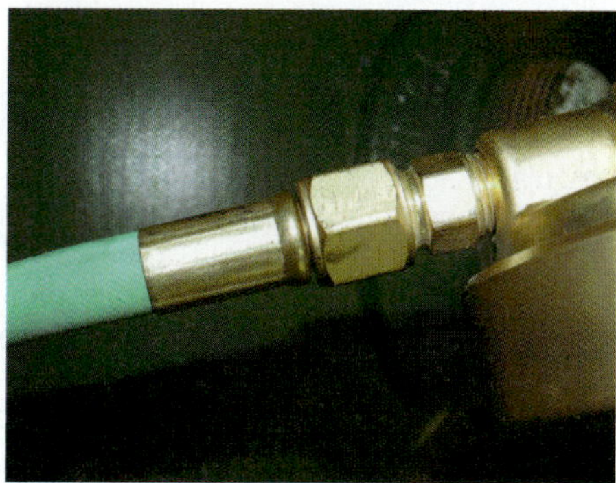

Figure 25-42 The oxygen hose is green and connects using right-handed threads.

Figure 25-43 The acetylene regulator cylinder connects with a male, left-handed thread. Notice the hash mark indicating the left-handed thread.

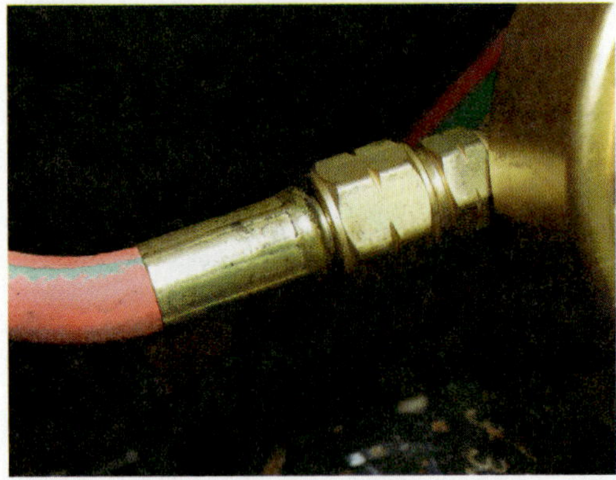

Figure 25-44 The acetylene hose is red and connects using left-handed threads. Notice the hash marks.

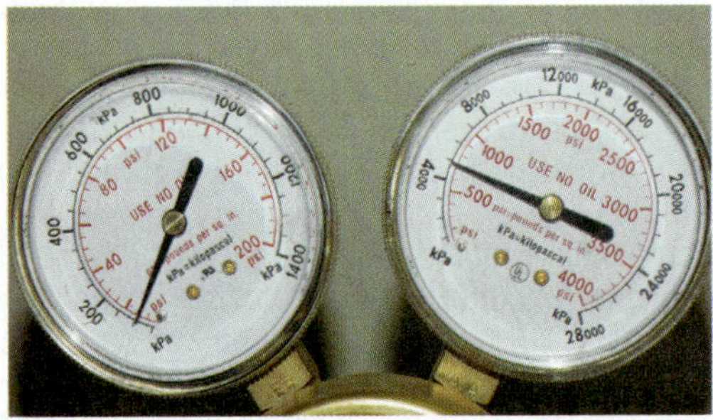

Figure 25-45 The gauge on the right indicates tank pressure; the gauge on the left indicates regulated pressure.

both the regulator and the torch handle with left-handed threads. The left-handed threads are indicated by the hash marks (Figure 25-44).

The regulators have two pressure gauges: one for tank pressure and another for regulated pressure (Figure 25-45). Turn the T handle in clockwise to increase the regulated pressure. Turn the T handle out counterclockwise to decrease the regulated pressure (Figure 25-46).

Figure 25-46 Turn the T handle in clockwise to increase the pressure (red arrow). Turn out counterclockwise to decrease the pressure (blue arrow).

SERVICE TIP

The intense oxyacetylene flame heats up parts faster than air-acetylene flames. This concentrates more heat on the work, allowing braze joints to be completed faster. Spending less time heating the parts reduces the amount of heat conducted to surrounding parts that might be damaged by the brazing heat. Although it sounds counterintuitive, the hotter flame can reduce the amount of heat damage done to connected components if it is used properly.

Adjusting the Torch

First, check to see that the regulators on the acetylene and oxygen cylinders are not adjusted in. They should both be out counterclockwise until there is no spring pressure on the T handle (Figure 25-47). Stand behind the acetylene cylinder on the side opposite the regulator so that you are not in a direct line with the regulator. This reduces the possibility of being struck by the regulator should it fail and come flying apart under pressure. Using the tank valve key, turn the acetylene tank valve counterclockwise the minimum amount required to get a reading on the cylinder pressure gauge. This should never be more than two and a half turns. Make sure to leave the key on the valve while you are using the torch.

Stand behind the oxygen cylinder on the side opposite the regulator so that you are not in a direct line with the regulator and crack open the valve counterclockwise (Figure 25-48).

Figure 25-47 Regulators should be out counterclockwise before opening cylinder valves.

Figure 25-48 Stand behind the oxygen cylinder on the side opposite the regulator when cracking open the oxygen cylinder.

Table 25-6

Oxyacetylene brazing tip chart				
Tip size	**Acetylene Setting**	**Oxygen Setting**	**Acetylene SCFH Use**	**Metal Thickness (inches)**
000	3 psig to 5 psig	3 psig to 5 psig	0.5	1/32″
00	3 psig to 5 psig	3 psig to 5 psig	1.5 – 3	3/64″
0	3 psig to 5 psig	3 psig to 5 psig	2 – 4	5/64″
1	3 psig to 5 psig	3 psig to 5 psig	3 – 6	3/32″
2	3 psig to 5 psig	3 psig to 5 psig	5 – 10	1/8″
3	3 psig to 6 psig	4 psig to 7 psig	8 – 18	3/16″
4	4 psig to 7 psig	5 psig to 10 psig	10 – 25	1/4″
5	5 psig to 8 psig	6 psig to 12 psig	15 – 35	1/2″
6	6 psig to 9 psig	7 psig to 14 psig	25 – 45	3/4″
7	8 psig to 10 psig	8 psig to 16 psig	30 – 60	1.25″
8	9 psig to 12 psig	10 psig to 19 psig	35 – 75	2″
10	12 psig to 15 psig	12 psig to 24 psig	50 – 100	3″

Once you see a reading on the oxygen cylinder pressure gauge, open the oxygen tank valve all the way counterclockwise. This is done to prevent the high cylinder pressure from placing a strain on the cylinder valve threads.

Torch manufacturers have designed their equipment to operate properly at specific acetylene and oxygen pressures. Always consult the manufacturer's tip chart when determining the correct pressures for adjusting an oxyacetylene torch. Table 25-6 shows a typical brazing tip chart. Regulate the acetylene and oxygen pressures by slowly turning the regulators in clockwise and observing the pressure gauges. Brazing seldom requires acetylene pressures greater than 8 psig, and any acetylene pressure adjustment beyond 15 psig is extremely dangerous. Do not exceed the recommended pressure for any given torch and application.

CAUTION

The acetylene and oxygen should never mix anywhere in the torch assembly except in the carburetor and tip. A condition known as reverse flow can occur if unequal acetylene and oxygen pressures are allowed to mix in the hoses, regulators, or cylinders. In a worst-case scenario, reverse flow can result in a catastrophic explosion. To avoid reverse flow, you should keep your equipment in good condition, never use an oxygen cylinder with less than 50 psig in it, always independently bleed each hose before lighting the torch, and never light both gases at once.

Lighting the Torch

To prevent potentially dangerous backfires in the torch, acetylene is always ignited first and turned off last. When lighting an oxyacetylene torch, ignite the acetylene without the oxygen and then add the oxygen to produce a neutral flame. When shutting the torch down, shut off the oxygen first and the acetylene last. To light the torch, open only the torch handle acetylene valve slightly and light the acetylene gas using a spark or flint lighter. This should produce an orange and slightly sooty flame. After the torch lights (Figure 25-49a), increase the flow of acetylene until most of the smoke in the flame has disappeared (Figure 25-49b). Next, slowly open the oxygen valve counterclockwise (Figure 25-49c). Increase the oxygen flow until the center cone is well defined and the feather completely disappears and joins the center cone (Figure 25-49d). This is a neutral flame. Do not increase the oxygen flow rate beyond this point. Further increasing the oxygen will produce an oxidizing flame (Figure 25-49e). Oxidizing flames burn up the work and cause excessive oxides to form on copper during brazing. A properly adjusted oxyacetylene flame will burn with a balanced, or neutral, flame. This means there is not an excess of oxygen or acetylene. It is important to use a balanced or neutral flame so that the flame does not introduce contaminants into the braze joint.

CAUTION

The oxygen cylinder valve should be opened slowly to prevent a sudden compressive force on the regulator from the high cylinder pressure. Sudden high pressure can create heat. If there is an explosive mixture in the regulator, heat from opening the oxygen cylinder quickly can ignite the mixture and cause an explosion.

Adjusting the acetylene flow rate so that most of the smoke disappears will ensure that you have adequate gas flow to keep the tip cool. If there is insufficient gas flow, the tip will overheat. As the tip becomes overheated, it will begin

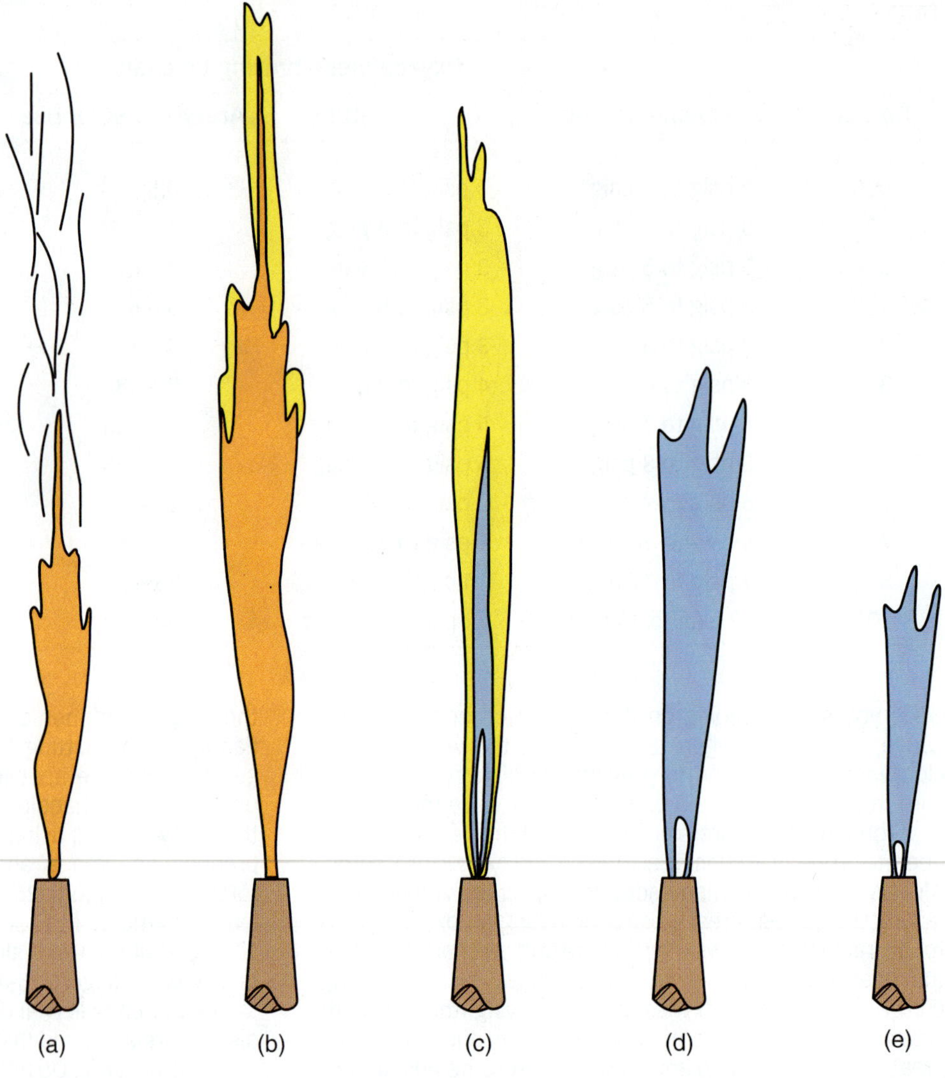

Figure 25-49 (a) When first lit, low-pressure acetylene has smoke; (b) increase acetylene pressure until smoke disappears; (c) turn on oxygen and slowly increase flow; (d) increase oxygen until the outer feather forms into a very smooth uniform cone; (e) excessive oxygen causes the flame to become purple, and the inner cone becomes very small.

to pop. This popping is called a backfire, and backfires can result in damage to the equipment. A backfire can result in a flashback, where flame actually races back through the hoses to the cylinders. If this were to occur, the results could be catastrophic: the system could explode. Flashback arrestors and check valves can be installed on the torch handles to guard against both flashback and reverse flow (Figure 25-50). Incorrect shutoff can also cause the torch to backfire. To safely shut off an oxyacetylene torch and avoid a backfire, close the oxygen valve on the torch handle first and then close the acetylene valve (Figure 25-51).

Putting up the Torch

When you are finished with the torch, close the cylinder valves and release the pressure from the hoses and regulators. Bleed

Figure 25-50 Flashback arrestors and check valves can prevent dangerous reverse-flow conditions.

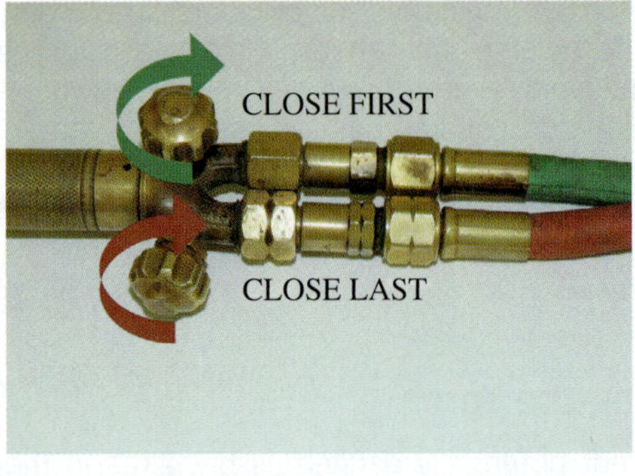

Figure 25-51 Close the oxygen valve first and the acetylene valve last.

the acetylene first and then the oxygen. The acetylene and oxygen should be bled individually, one at a time, to avoid creating a flammable mixture. Once the pressure has been relieved from both the oxygen and acetylene hoses and regulators, back out or loosen the regulator-adjusting valve handle. Make certain you do not over-loosen the valve handle, as in some cases it can completely unscrew and fall out of the regulator body. If this does occur, be very careful not to cross-thread the adjusting handle as you reinstall it. Many manufacturers use nylon threads in the regulator so that oil is never needed for them to operate smoothly. The nylon is easily cross-threaded.

SERVICE TIP

From time to time the oxyacetylene tip becomes dirty. To clean the tip, use a tip cleaner of the proper size and clean out the dirt that has collected in the tip orifice (Figure 25-52). Select a tip cleaner size that will easily fit into the hole in the tip. Tip cleaners are actually small, round files that are accurately calibrated for each tip size. By sliding the tip cleaner in and out several times, you will remove the debris inside the tip, as demonstrated by Figure 25-53.

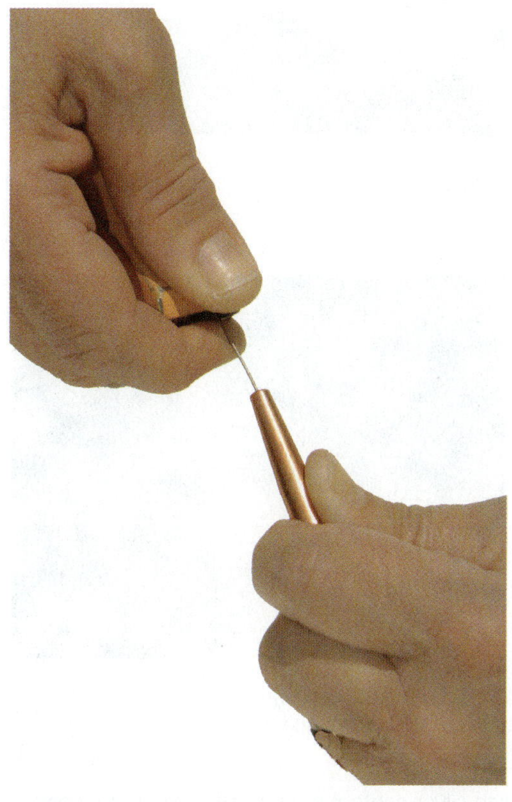

Figure 25-53 Use the round tip cleaner to clean the tip orifice.

25.9 BRAZING

Brazing and soldering are similar processes. The primary difference is that soldering occurs below 840°F and brazing occurs above 840°F. Most manufacturers and codes require that refrigerant line joints be made using brazing alloys. Oxyacetylene torches are preferred for brazing because they can quickly heat up joints to brazing temperatures. The faster a joint is heated, the fewer surface oxides are formed. Heating the joint quickly also lowers the chance that heat will be conducted from the joint to damage surrounding parts. Overheating can damage many refrigerant system valves. The longer that copper is kept at the higher brazing temperature, the greater the formation of copper oxide (Figure 25-54). Copper oxide can become a barrier to the successful completion of the joint as well as becoming a contaminant in the refrigerant circuit.

Figure 25-52 Tip-cleaning set.

Figure 25-54 Copper oxide formed by excessively heating copper pipes and fittings.

Brazing Alloys

Brazing alloys used in HVACR fall into two general categories: copper phosphorus and high-silver-content brazing alloys. Copper-phosphorous alloys typically contain copper, phosphorous, and silver. Copper-phosphorous brazing alloys are typically supplied in flat rods with silver percentages from 0 to 15 percent (Figure 25-55). Copper phosphorous rods are only used for brazing copper to copper joints (Figure 25-56). They do not work on ferrous metals like iron and steel and are usually not recommended for brass. Copper-phosphorous alloys do not require flux; the phosphorous acts as the flux. The phosphorous cleans off the copper oxide deposits at the relatively high temperatures that copper-phosphorus flows (1,200°F –1,600°F). Copper-phosphorous rods are the most commonly used brazing material in HVACR.

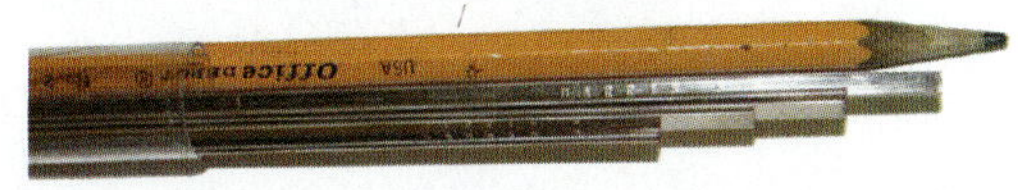

Figure 25-55 Copper-phosphorous brazing rods.

Figure 25-56 Copper-phosphorous rods are only used for copper-to-copper joints.

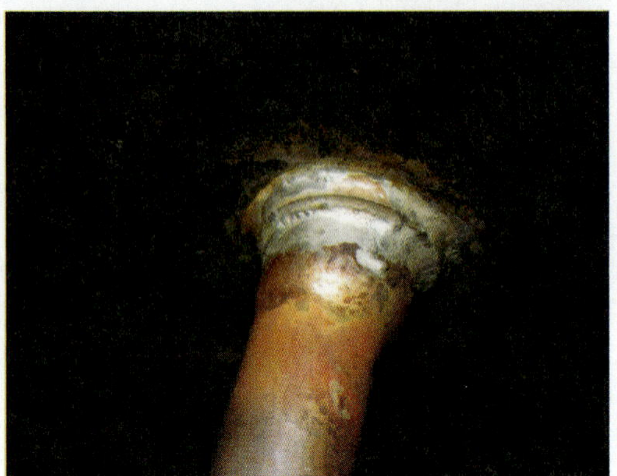

Figure 25-57 High-silver-content silver brazing material was used to braze this copper line to this steel shell.

High-silver-content silver brazing alloys are used for connecting copper lines to brass and steel fittings (Figure 25-57). High-silver-content brazing alloys contain 35–56 percent silver. Most high-silver-content brazing alloys are sold by the troy ounce in spools (Figure 25-58). High-silver-content brazing materials work with copper, ferrous materials such as iron or steel, and brass. They do require cleaning and fluxing with silver brazing flux (Figure 25-59). Some silver brazing rods have flux already applied to the outside of the rod (Figure 25-60). Because of their high cost, high-silver-content brazing rods are normally only used for joining different type of metals.

Figure 25-59 Silver brazing flux.

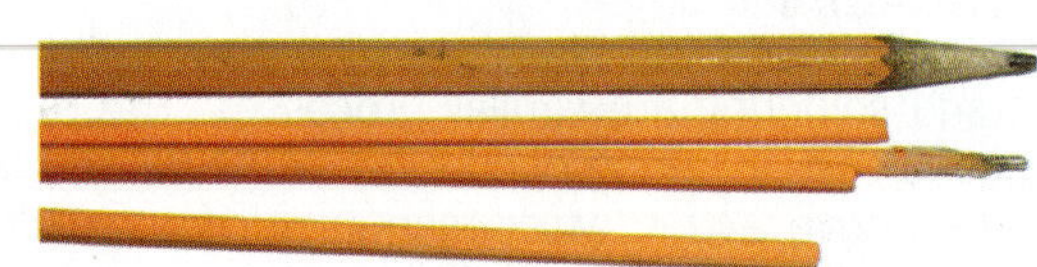

Figure 25-60 High-silver-content silver brazing rod with pre-applied flux.

Preparing a Brazed Joint

To make a proper brazed joint in copper, the copper fittings must be clean. Clean new fittings might only need to be wiped off, but pipe and fittings that have been discolored from oxide must be cleaned mechanically or chemically. Mechanical cleaning is easiest performed using sand cloth (Figure 25-61) or a wire brush (Figure 25-62). Specially designed round wire brushes fit inside of copper fittings to provide better cleaning. These round brushes or small flat brushes can also be used for cleaning the outside of the pipe. Sand cloth or abrasive pads for cleaning copper pipe and fittings are available. Sand cloth should not be confused with emery cloth. Emery cloth is a very hard, abrasive material, and any abrasive grit, if left inside the pipe, can damage a piston or other moving parts. AC-type sand cloth has softer grit that will more easily break down if the grit is accidentally left in the system, reducing the possibility of damage to moving parts.

Figure 25-58 High-silver-content silver brazing alloy is commonly sold by the troy ounce in spools.

The pipe should be cleaned at least ½ in farther back than the pipe fits into the fitting, shown in the series of photos in Figure 25-22. Because copper oxidizes in the air, it is best to clean the fittings immediately prior to producing the brazed joint. Do not to touch the cleaned surfaces with your hands or oily tools. The oil can contaminate the cleaned surfaces and keep the brazing material from sticking.

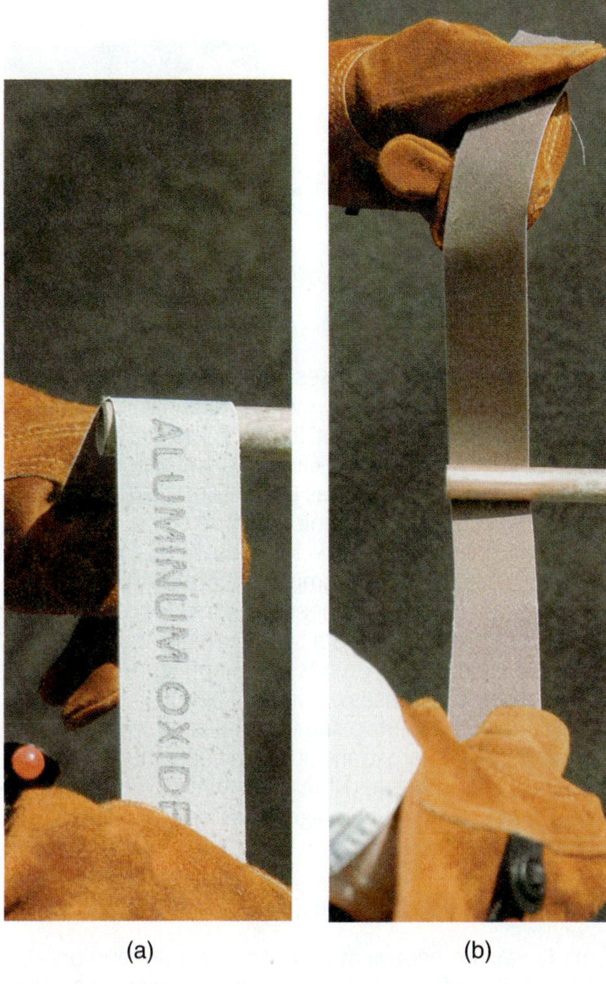

Figure 25-61 (a) Use aluminum oxide sand cloth to clean copper before brazing and soldering; (b) be sure to clean all the way around the fitting.

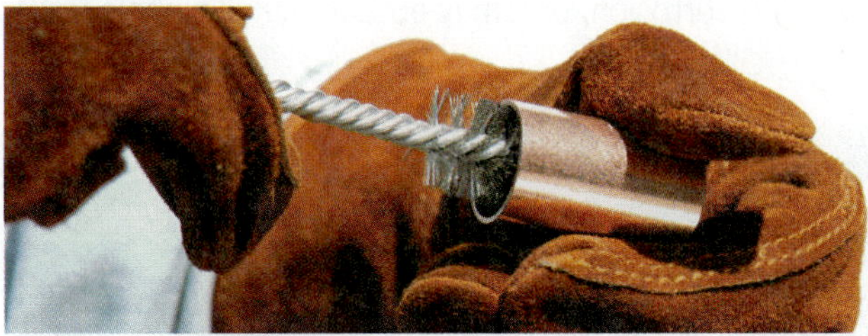

Figure 25-62 Round tubing brushes can be used to clean the inside of pipe and fittings.

SERVICE TIP

Push the cleaned copper pipe as deeply as possible into the copper fitting (Figure 25-63). Twisting the pipe once it is completely inserted can help seat it completely to the bottom stop of the joint (Figure 25-63). On fittings that have a stop all the way around the inside or on swaged fittings, this can prevent braze or solder from flowing into the pipe.

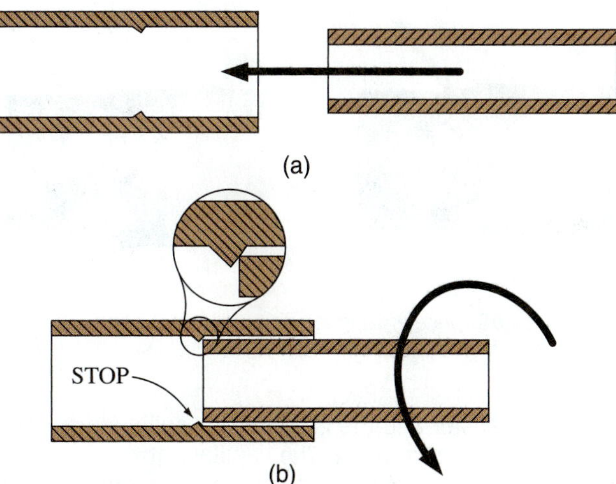

Figure 25-63 (a) Insert the copper pipe into the fitting; (b) twisting the copper pipe against the stop helps seat the fitting.

Heating the Copper Pipe

The order in which you heat the parts is crucial to a quality brazed joint. It is important to heat the pipe first (Figure 25-64a) and then the fitting (Figure 25-64b). As the copper pipe is heated, it will expand. This expansion will tighten the joint space between the fitting and the copper pipe. If the fitting were to be heated first, this space would increase. The tighter the space is, the better

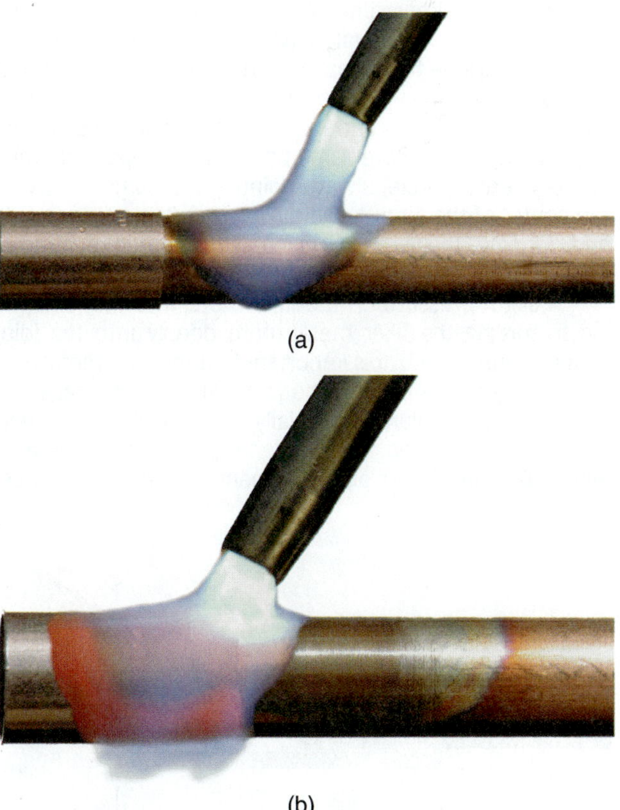

Figure 25-64 (a) Begin heating the pipe near the fitting; (b) once the pipe is hot, move the flame onto the fitting and continue heating.

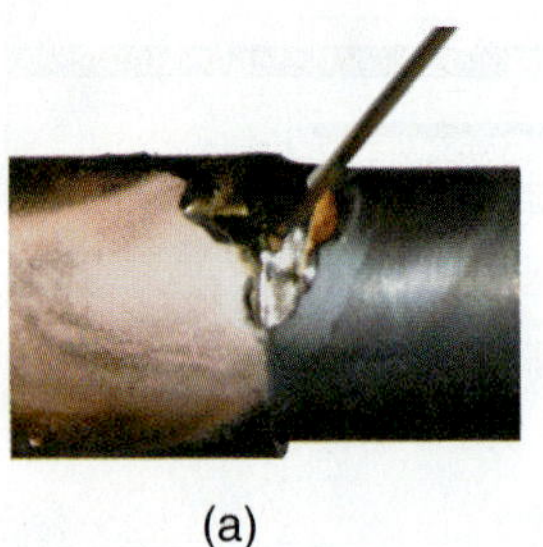
(a)

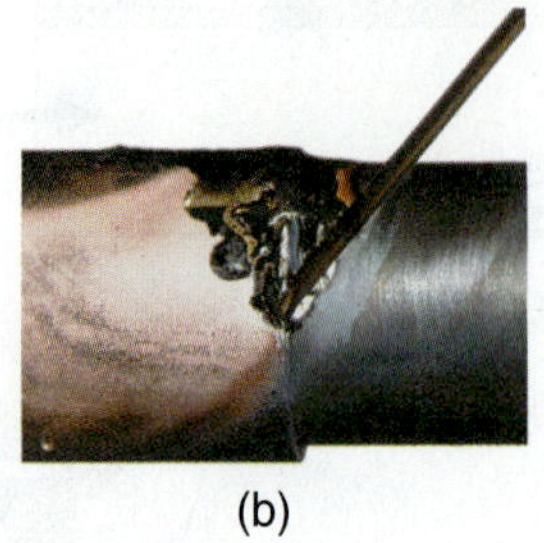
(b)

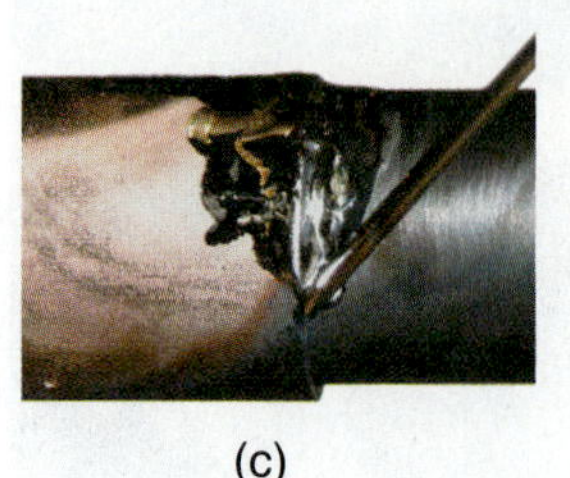
(c)

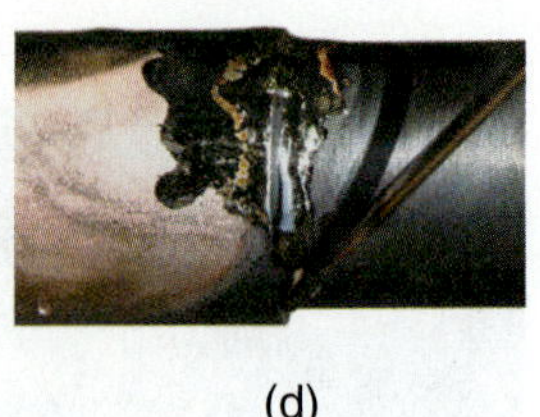
(d)

Figure 25-65 Once brazing temperatures have been reached, use a slight downward pressure on the filler rod at the braze joint and follow the joint around the fitting.

the heat transfer and the greater the capillary attraction for the liquid braze metal. The capillary attraction is the force that draws molten braze metal deep into the joint.

Once the pipe has become hot, approaching a dull red color, move the torch flame onto the fitting so that it envelops both the fitting and the pipe (Figure 25-64b). Continue heating the fitting. Occasionally touch the pipe surface with the tip of the brazing rod as a test of temperature readiness. A small drop of braze material may transfer from the rod to the fitting just before the fitting reaches brazing temperature.

One big difference between soldering and brazing is that you must continue to heat the braze joint even after the braze material starts to flow. (Figure 25-65). You must slowly move the flame while heating to avoid burning through the copper. When the copper gets "cherry red," the copper is very near its melting point of 1,984°F. Some copper-phosphorous brazing alloys do not flow well until the copper is almost cherry red. The temperature of the oxyacetylene torch coupled with the higher pressure of the oxyacetylene flame makes burning holes in the copper very easy. To avoid this, the torch should be in constant motion to avoid concentrating the heat and pressure on any one spot. Move the flame between the pipe and fitting in a concentric circle, always pointing the flame in toward the center of the pipe. Avoid moving the flame side to side and "fanning" the flame. This increases the time required to heat the joint and increases oxide formation.

A slight pressure on the filler metal into the joint will aid in forcing the filler metal more deeply into the joint space (Figure 25-66). It is important that the filler metal flow completely to the base of the joint. Any unfilled gaps can result in joint failure, especially when R-410A refrigerants are being used. These higher-pressure refrigerants will cause the copper pipe to expand slightly as the pressure inside the pipe increases. If the pressure decreases, the pipe will contract. This cyclical expansion can cause a stress fracture if the joint is not completely filled with brazed metal. If you watch carefully, you can see a slight change in the fitting's color as the filler metal flows into the joint space. This change in color is more obvious for larger-diameter fittings. Bringing the torch down on the fitting will help draw the filler metal completely into the joint.

Once the joint gap has been filled, move the torch back up above the joint and add a small amount of braze metal until a fillet of metal surrounds the joint (Figure 25-67). This fillet serves two purposes. First, the fillet makes it easier to visually check the joint to see that it has no leaks. Second, it removes any sharp transition between the two pieces of copper, the fitting and the pipe, which will further reduce stress cracking when high-pressure refrigerants (e.g., R-410A) are being used.

Cleaning the Brazed Joint

Some fluxes are neutral at room temperature and only become active when heated to a specific temperature. Other fluxes are active irrespective of the temperature. All fluxes must be removed from the completed joint. This is particularly true for active fluxes. If flux is allowed to remain on the joint it can cause corrosion. Flux can trap moisture, causing pitted corrosion, or it may obscure small pinhole leaks,

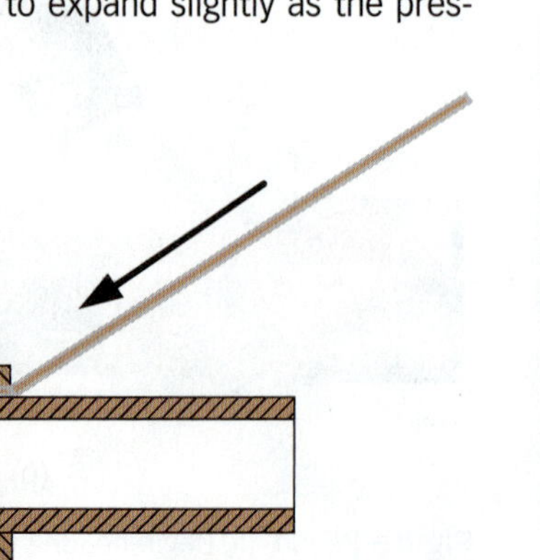

Figure 25-66 Add the filler metal straight into the joint gap.

BRAZE METAL FILLET

Figure 25-67 Add a small amount of braze to the joint so that there is a smooth, uniform fillet to ensure a good joint seal.

Figure 25-68 This horizontal refrigerant line is 1 in above the ground, making access to the bottom of the joint difficult.

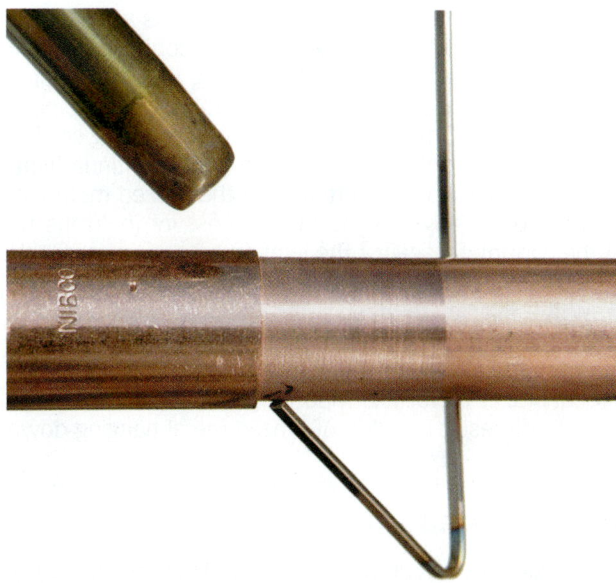

Figure 25-69 Use the flame to heat and bend the brazing rod approximately 2 in from the end to make reaching the bottom of the horizontal joint easier.

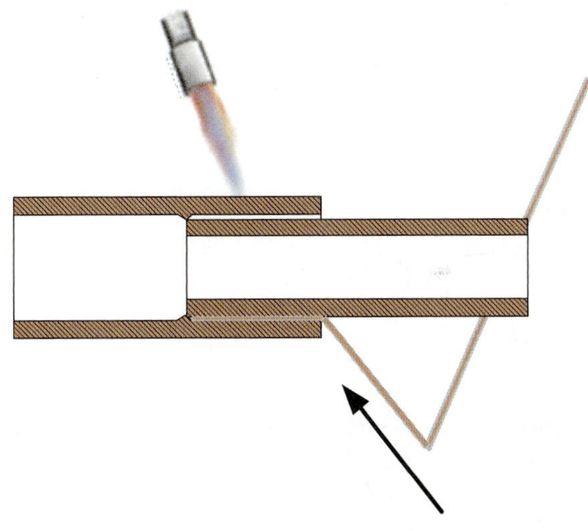

Figure 25-70 When the fitting is completely heated, filler metal can be added from the bottom first; then follow the joint around with the tip of the filler metal ending at the top of the joint.

which could open later, causing system leaks. Active fluxes can significantly damage metal parts if not completely cleaned away and neutralized.

Horizontal Brazed Joint

Horizontal brazed joints are the most common joints experienced in the HVAC field. Horizontal joints are typically found at the junction of the refrigerant line set and the outdoor unit as well as from the refrigerant line set to the evaporative coil. When practicing this joint, it will be beneficial to you later if you will work from just the top side of the joint. Because these joints are typically close to the ground in most installations, you cannot see the bottom of this joint when making it in the field, such as the joint shown in Figure 25-68. Using the torch, heat the filler metal approximately 2 in from the end. When the rod becomes soft, bend it to approximately a 90° angle (Figure 25-69). This will allow you to direct the filler metal into the bottom portion of the joint while your hand and torch are working from above. Again, always heat the copper pipe first. Then move your torch to the fitting. When the pipe and fitting are hot enough, as indicated by the red color shown in Figure 25-64 , add filler metal directly into the groove space (Figure 25-70).

Sometime while you are brazing, you might accidentally touch the tip of the filler metal to the joint. If it becomes stuck, simply move the torch flame over until the rod melts itself free. Watch the filler metal, and do not overfill the joint. Overfilling will result in a drip of brazed metal hanging from the bottom of the joint. When the joint has been completely filled, simply wipe the tip of the brazed metal rod across the bottom joint. This will provide a slight fillet around the joint. On larger-diameter copper pipe and fittings (1 in or larger), it is usually necessary for you to move the torch around so that the flame is directed on the bottom surface. As you move the torch around, pay particular attention so that you do not inadvertently direct the flame toward yourself, others, or combustible materials in your work area. Learning to be aware of the flame direction will help you in the field to avoid burning the paint off of equipment as braze joints are performed.

Vertical Up Joint

Occasionally in the field, a brazed joint must be made in the vertical position with the joint on top. The filler metal must travel uphill to fill the joint. These joints are typically found, for example, when a 90° fitting or coupling is joined in refrigerant line sets. Start by heating the copper pipe just as with the other brazing practices. When the pipe is properly heated, move the torch up on the fitting so that the flame covers both the joint and pipe. Continue heating the copper pipe and fitting until the brazed metal begins to flow. Using a slight inward pressure, hold the tip of the filler metal against the joint and press upward. This slight pressure along with the heat and capillary space will draw the filler metal deeply into the joint (Figure 25-71). When the joint has been completely filled, wipe the tip of the filler metal around the joint groove to produce a soft fillet of brazed metal. It is important not to overfill this joint, as this will result in a drip of brazed metal hanging down on the pipe.

Testing Brazed Joints

In the field, evacuating and charging the system places the brazed joints under both vacuum and pressure conditions.

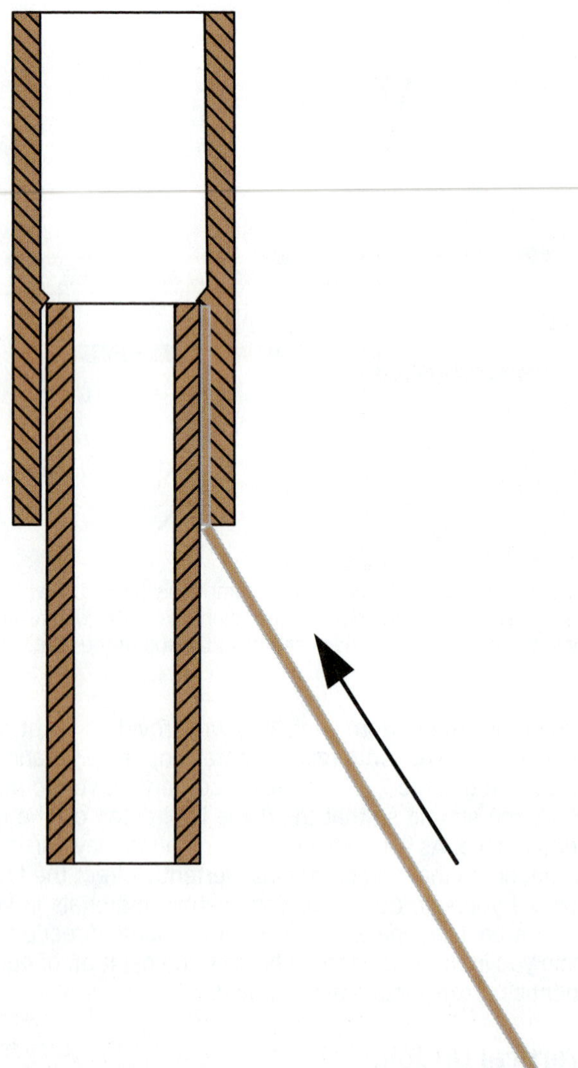

Figure 25-71 Inward pressure of the filler metal helps in filling the joint gap completely with brazes in the vertical position.

Figure 25-72 The first step in braze joint testing is to cut the fitting just beyond where the pipe is seated.

This will serve the purpose of testing the brazed joints. In the refrigeration and air-conditioning lab, you can test the brazed joints in a different manner. It is possible to inspect your brazing by destructively testing the joint. To perform this destructive test, first use a hacksaw to saw off the copper and copper fitting at a point just beyond the depth that the pipe was inserted into the joint (Figure 25-72). Once the fitting has been cut completely apart, clamp the pipe in a vise and cut straight down through the entire joint into the pipe. Rotate the pipe 90° and repeat this process, cutting as shown in the diagram in Figure 25-73. First bend each quarter out with a wrench (Figure 25-74a). Using a hammer and anvil, flatten each of the four corners of the pipe and fittings you just made (Figure 25-74b). As the joint pieces are flattened, it is easy to see areas where 100-percent joint penetration did not occur (Figure 25-75). Repeat the practice until you can successfully pass each of the four various joints without finding voids during testing. Obviously, this test will render the joint unusable. The technician may use this test for the practice and development of brazing skills.

TECH TIP

The stubs on new hermetic compressors and filter driers look like copper. Often, the copper is actually a very thin coating on a steel pipe. The copper coating is put on the stubs to aid in brazing the refrigerant lines to the compressor. This thin copper coating is easily damaged by overheating if the oxyacetylene torch flame is an oxidizing flame or if too much time is taken during brazing. To keep from burning off the copper coating, be sure to use a neutral oxyacetylene flame, and use a torch tip that is big enough to do the job quickly. If the copper coating is damaged, you will have to stop, disassemble the joint, and clean it before restarting. In that case, it will also be necessary to use a silver brazing filler and a brazing flux.

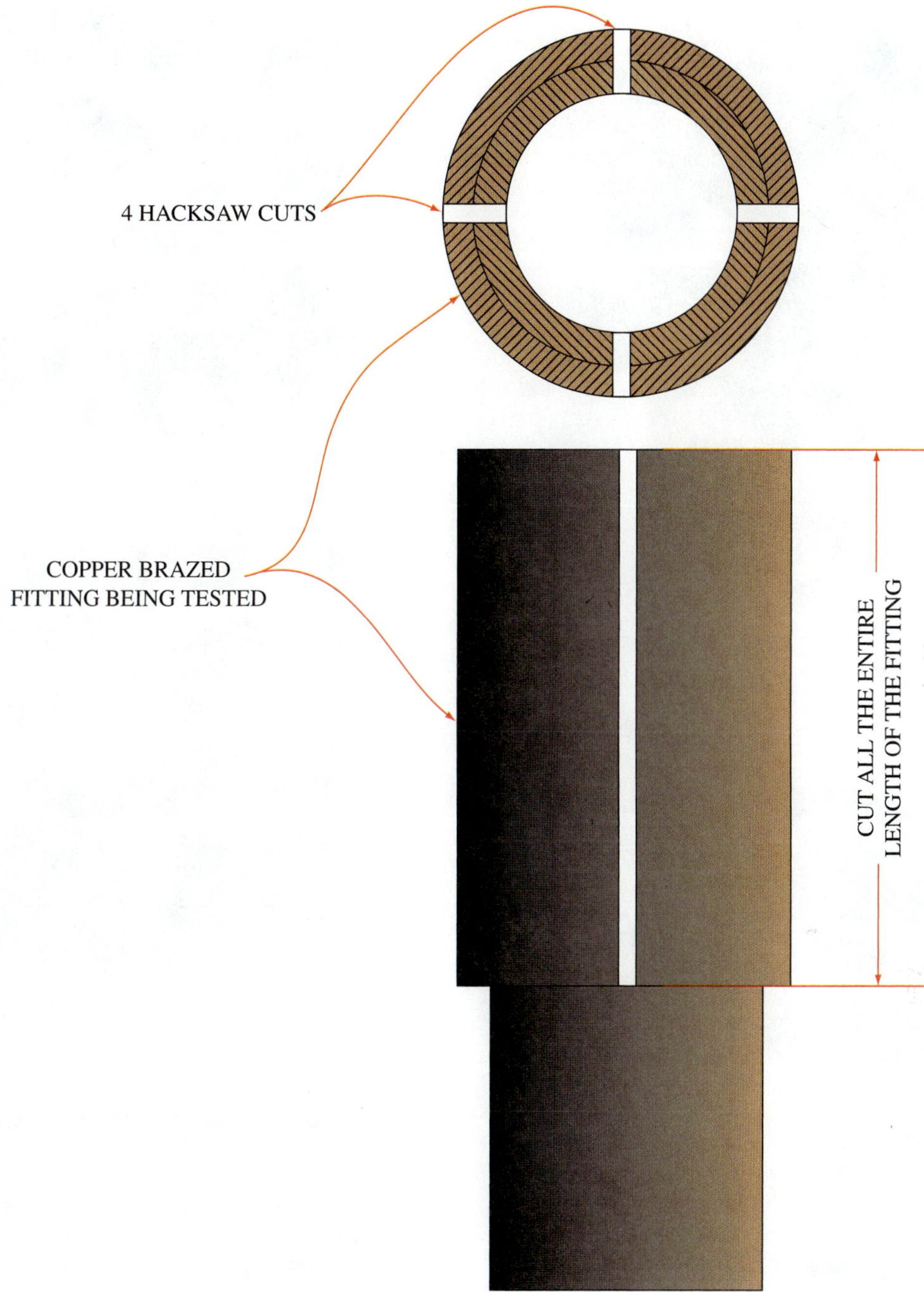

Figure 25-73 Use a hacksaw to make two cuts through the entire length of the fitting.

25.10 NITROGEN BRAZING

Brazing creates metal oxides on all the parts being brazed, both inside and outside the pipes (Figure 25-76). The oxides can come loose and contaminate refrigeration systems. Oxides are formed by a chemical reaction between oxygen and a metal. If there is no oxygen, the oxides cannot form. Refrigeration lines should be purged with nitrogen during brazing to prevent the formation of copper oxides inside the pipe. The nitrogen replaces the air, eliminating the oxygen and preventing the formation of oxides. Figure 25-77 shows a piece of copper after heating while being purged with nitrogen. Notice that it is oxidized on the outside and clean on the inside.

Nitrogen Cylinder Safety

Nitrogen is inherently safe. It does not burn, it does not chemically react with many substances, and it is not poisonous. We breath nitrogen all the time, as air is 78 percent nitrogen. However, nitrogen can be dangerous because it does not irritate our lungs or eyes. You will not have any warning that the closed area you are in is so full of nitrogen that there is no longer enough oxygen. You will just

(a)

(b)

Figure 25-74 (a) Use a pair of pliers to bend each quarter section out; (b) flatten each section using a ball-peen hammer and anvil.

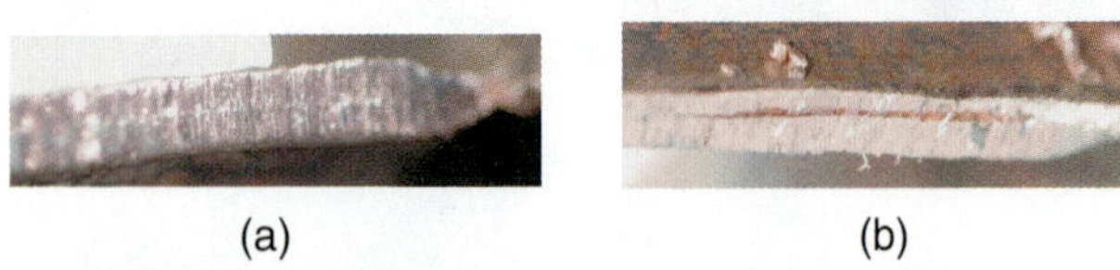

(a) (b)

Figure 25-75 (a) A thin layer of brazed metal can be seen between the pipe and fitting; (b) area not filled with brazed metal will allow the pipe and fitting to slip, slightly revealing a very small shelf between the pipe and fitting; this area was not properly filled with brazing metal.

get dizzy, pass out, and asphyxiate. To avoid asphyxiation, never release large quantities of nitrogen inside a closed space. Nitrogen cylinders can be dangerous because of the very high pressures inside them. They typically have over 2,000 psig of pressure when the cylinder is new. This much pressure on a refrigeration system would cause parts to rupture and explode. You should always use a pressure regulator and a pressure-relief valve when using nitrogen. Figure 25-78 shows a nitrogen regulator with a built-in pressure-relief valve. The pressure regulator lets you safely control the pressure you put into the system. The pressure-relief valve keeps the downstream pressure from reaching full tank pressure should the regulator fail. Most refrigera-

Figure 25-76 Copper oxide formed during brazing.

Figure 25-77 This copper is clean on the inside because it was purged with nitrogen while it was being heated.

Figure 25-78 A nitrogen regulator is used to safely control the nitrogen pressure during nitrogen brazing.

–024–230–01	
F38025	
–410A	DESIGN PRESSURE
IARGE	HI 446 PSIG
OZS	LO 236 PSIG
RATING	NOMINAL VOLTS: 208/230
60 HZ	MIN 197 MAX 253

Figure 25-79 Never exceed system design pressure when pressurizing a system.

tion systems have system design pressures on their data plates (Figure 25-79). These are not the operating pressures but the maximum pressures the system is designed to withstand. You should never exceed the system's design pressure when field testing a system. It can be difficult to ensure that the high side and low side of an installed system will remain segregated, so the safest procedure is to never exceed the low-side design pressure. If the system does not list a design pressure, use 150 psig as the maximum safe test pressure. Hermetic compressor manufacturers list 150 psig as the maximum safe test pressure on the hermetic compressor shell. Normally you should not use pressures anywhere close to system design pressure when nitrogen purging.

Opening the Nitrogen Cylinder

The nitrogen cylinder should be treated with care. To open the nitrogen cylinder, stand behind the cylinder on the side opposite the regulator so that you are not in a direct line with the regulator and slowly crack open the valve counterclockwise. Once you see a reading on the pressure gauge, open the cylinder valve all the way counterclockwise.

SERVICE TIP

Nitrogen brazing is especially important on systems that use HFC refrigerants with polyol ester (POE) lubricant, such as today's new R-410A air-conditioning systems. POE and HFCs are very strong solvents and "clean" the oxides off of the inside of the copper lines. The oxides then end up in the compressor oil and in metering devices, causing major system failures.

Purging Procedure

You cannot braze a system that is pressurized, as the pressure will simply blow a hole in the molten braze material. The nitrogen must be able to leave the system so that pressure does not build up. Introduce nitrogen at one point in the system and release it at another point. Remove the cores from Schrader valves to improve flow and insure the release of the nitrogen (Figure 25-80).

Begin by purging with enough pressure to purge all the existing air from the lines. Adjust the regulator in clockwise until you have a pressure of 50 psig and purge for one minute. This will safely clear most lines (Figure 25-81). The pressure and flow must be reduced before brazing. Turn the regulator out counterclockwise until there is a slight flow of nitrogen leaving the open valve on the system. It should feel similar to blowing on the back of your hand. This pressure is under 2 psig and will not show on most nitrogen gauges (Figure 25-82). Brazing proceeds just as it would normally, except now there will not be any oxides formed inside the lines. Leave the nitrogen flowing after completing the brazing until the lines are cool.

Figure 25-80 Nitrogen cylinder set up to purge the system during nitrogen brazing.

Figure 25-81 Nitrogen cylinder adjusted to 50 psig for initial purge.

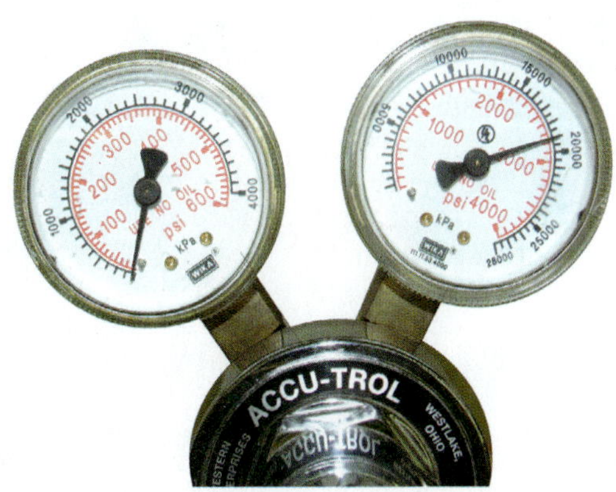

Figure 25-82 Nitrogen cylinder adjusted for purge during brazing.

UNIT 25—SUMMARY

Soldering, brazing, and welding are all methods of joining metal parts using molten metal. During welding, the base metal itself is melted as additional metal filler is added to the joint. During soldering and brazing, only the metal being added to join the base parts is melted. The difference between brazing and soldering is temperature. Soldering takes place below 840°F and brazing takes place above 840°F. Three types of torches used for soldering and brazing are air fuel, swirl-tip air fuel, and oxyacetylene. Air-fuel and swirl-tip air-fuel torches get their oxygen from the air. Oxyacetylene torches use pure oxygen supplied by a highly pressurized oxygen cylinder. Air-fuel torches are mainly used for soldering. Swirl-tip torches get hot enough to braze and do not require a separate oxygen cylinder. Oxyacetylene torches are the hottest burning torches and are preferred for brazing. Oxygen cylinders should always have their safety cap on when being transported. Acetylene cylinders safely store acetylene at high pressures by dissolving it in acetone, which is held inside the cylinder with a sponge-like filler material. Acetylene is unstable and can explode at pressures above 15 psig. It should never be used at pressures above 15 psig.

The first step in soldering and brazing is to clean the joint. Filler alloy will not stick to dirty, oxidized metal. When heating the joint, heat the pipe first and then move to the fitting. The torch should heat the metal and the metal should melt the solder or brazing alloy. Oxidation is created when metal is brazed. This can be avoided by purging the lines with nitrogen during the brazing process.

UNIT 25—REVIEW QUESTIONS

1. Why must proper protective clothing be worn during soldering and brazing?
2. What harmful light is produced during soldering and brazing?
3. What harmful light is produced during welding?
4. What material should clothes worn during soldering and brazing be made of?
5. List the three types of torches used for soldering and brazing.
6. Why must an acetylene cylinder be used in the upright position?
7. List the common sizes of acetylene cylinders.
8. Why should the valve on an acetylene cylinder no be opened more than two and a half turns?
9. What can happen to an air-acetylene torch tip if it is used with too low a flame?
10. What is an advantage of the oxyacetylene flame compared to air acetylene?
11. What can happen to an oxyacetylene torch if it is used with too low a flame?
12. What is the difference between brazing and soldering?
13. What is the purpose of flux?
14. Why is it important to heat braze joints as quickly as possible?
15. Why must the copper fitting be cleaned before brazing it?
16. When making a brazed copper joint, what part is heated first?
17. Why is it important to fill the brazed copper joint completely when R410A refrigerants are used?
18. Why is it important to make the braze on a compressor stub as quickly as possible?
19. What alloys make up the soldering alloy often referred to as silver bearing solder?
20. How does nitrogen purging prevent the formation of oxides?
21. What is "reverse flow" in oxyacetylene torch, and how can it be dangerous?
22. How can reverse flow be avoided?
23. What PPE is required for brazing?
24. What is the minimum OSHA recommended shade for safety glasses used while brazing?
25. List the procedure for adjusting an oxyacetylene torch for brazing.
26. List the procedure for lighting an oxyacetylene torch.
27. Why is nitrogen brazing particularly important when working with systems that use HFC refrigerants?
28. What brazing material is used for connecting copper lines to brass and steel valves?
29. What brazing material is mainly used for connecting copper to copper?
30. What is the purpose of the phosphorous in copper-phosphorous brazing rods?
31. How is the solder melted when soldering?

UNIT 26

Refrigerant System Piping

OBJECTIVES

After completing this unit, you will be able to:

1. explain the importance of minimizing pressure drop through the lines.
2. explain the importance of maintaining minimum gas velocities in the lines.
3. calculate the pressure drop in a liquid line due to vertical lift.
4. list the system modifications required for long-line applications of residential air-conditioning systems.
5. list some of the differences between air-conditioning and commercial refrigeration piping.
6. size refrigeration lines using line sizing charts.

26.1 INTRODUCTION

How does a technician determine what size lines to run when installing a refrigeration system? One common method is to measure the fittings where the refrigerant lines connect on the unit and use that size. If installations are restricted to "standard" installations, this method will often work. However, there are many variables that affect the size of refrigerant lines that cannot be determined without knowing the details of a particular installation. The penalty for incorrectly sizing the refrigerant lines can range from poor system performance to multiple compressor failures. It is in your best interest to know how to size refrigerant lines correctly. Refrigerant line sizing can be important from a service standpoint also, since it affects system performance. Understanding line-sizing concepts will help you solve those elusive problems that other mechanics cannot.

26.2 GOOD REFRIGERANT PIPING PRACTICES

There are good piping practices that should be followed when sizing and installing a refrigerant piping system:

- Keep it clean. Cleanliness is a key factor in the actual installation. Dirt, metal filings, sludge, and moisture will cause breakdown in the system and must be avoided. Neat, clean work will avoid many service difficulties.
- Pressure drop through all sections of piping should be kept to an absolute minimum to maximize system capacity and efficiency. Use as few fittings as possible. Fewer fittings mean less chance for leaks and, more important, less needless pressure drop.
- Each section of the piping system must also be sized to ensure enough velocity for proper oil return. In those installations where oil return and pressure drop are in conflict, proper oil return takes precedence.
- Be careful in making every connection. Use the right material and follow the method recommended by the equipment manufacturer. The lines should be purged with nitrogen during brazing to eliminate the formation of copper oxide inside the lines.
- Pitch horizontal lines in the direction of refrigerant flow (Figure 26-1). To aid in forcing oil to travel through lines that contain vapor (suction line, hot-gas line), horizontal lines should be pitched in the direction of refrigerant flow. To help the oil flow in the right direction, this pitch should be a minimum of ½ in for each 10 ft of run. Pitch also helps to prevent backflow of the oil during shutdown. In piping systems where sufficient return-gas velocity can be ensured at all times, it is satisfactory to run the horizontal suction lines dead level. This may be desirable where headroom is at a premium or where a sloping run will interfere with other piping.

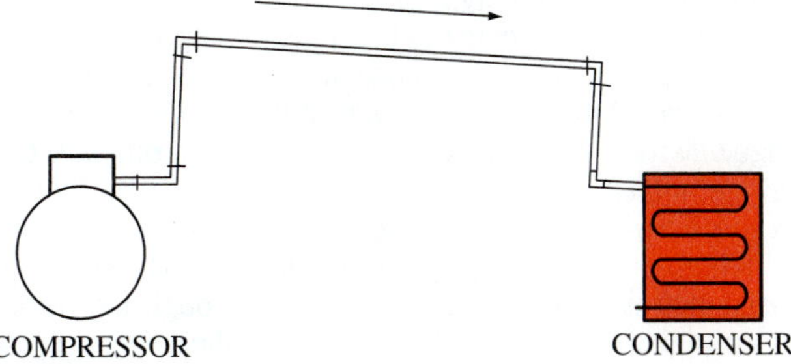

Figure 26-1 Lines should be sloped at least ½ in per 10 ft of run in the direction of refrigerant flow.

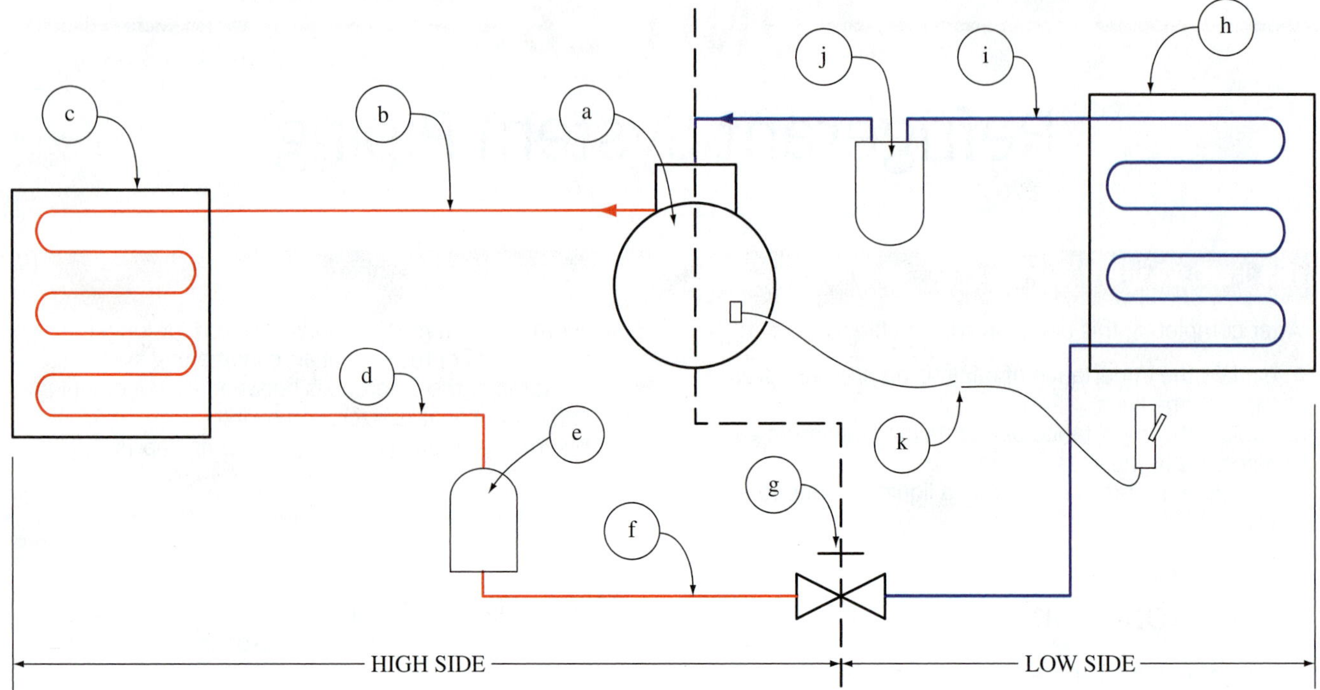

Figure 26-2 Major components of air-conditioning and refrigeration systems: (a) compressor*; (b) hot-gas line; (c) condenser; (d) condensate line; (e) receiver; (f) liquid line*; (g) metering device*; (h) evaporator*; (i) suction (vapor) line; (j) accumulator; (k) operational controls and safeties* (components with asterisks [*] appear on all air-conditioning and refrigeration units).

TECH TIP

Selecting the correct refrigerant lines can be very complex, involving many factors. One factor, however, that should be considered the least is the cost of the materials. Copper line sets are frequently undersized for cost savings. However, undersizing a system's refrigerant lines significantly reduces the system's performance. Reduced performance means extended operation and higher operating costs. The money saved on the initial installation is quickly spent on power bills.

26.3 REFRIGERATION LINE SIZING PRINCIPLES

Figure 26-2 shows a typical refrigeration piping system. The piping that connects the four major components of the system has two major functions: it provides a passageway for the circulation of refrigerant, and it provides a passageway through which lubricating oil carried out with the refrigerant is returned to the compressor. Each section of the piping should fulfill these two requirements with a minimum pressure drop of the refrigerant. Excessive flow resistance will increase pressure drop through the refrigerant piping and reduce the system's efficiency. A pressure drop in the refrigerant lines kills the system capacity and increases operating costs. In general, the larger and straighter the line, the less pressure drop it will have.

However, excessively large piping causes low refrigerant flow velocity, which reduces the amount of oil returning to the compressor. The best way to keep refrigerant velocity up is to make the lines smaller. So piping that is too small causes loss of efficiency, and piping that is too large can prevent refrigerant oil from being returned to the compressor.

26.4 ALLOWABLE PRESSURE DROP

It is desirable to minimize the pressure drop when sizing refrigerant lines. If the lines are too small, the capacity of the equipment is reduced. Standard practice has been to design gas lines for a maximum pressure drop roughly equal to a 2°F drop in saturation temperature. The actual amount of pressure drop this represents depends upon both the type of refrigerant and the temperature of the refrigerant. The acceptable pressure drop in the discharge line is almost twice as much as in the suction line. Systems using R-410a can have a higher pressure drop in the lines than systems using R-134a.

To see how this works, let's examine two systems operating at an evaporator saturation temperature of 40°F: one using R-134a and the other using R-410a. For R-134a, the saturation pressure at 40°F is 35 psig and the saturation pressure of 38°F is 33 psig. So the R-134a system can have a maximum pressure drop in the suction line of 2 psig (Figure 26-3). In the R-410a system, the saturation pressure at 40°F is 118 psig and the saturation pressure at 38°F is 114 psig (Figure 26-3). The R-410a system can have a 4-psig pressure drop for the same 2°F saturation temperature drop. The R-410a system can operate at twice the suction line pressure drop as the R-134a system and still operate efficiently.

Temperature °F	R-134a	R-410a
36°F	31.3	109.9
38°F	33.1	114.3
40°F	35.0	118.8
42°F	37.0	123.4

Figure 26-3 Comparison of pressures for R-134a and R-410a.

TECH TIP

An ideal refrigerant piping system has a pressure differential equivalent to less than 2°F temperature difference between the indoor coil and the outdoor unit. Refrigerant pressures and temperatures are not linear. There is more temperature difference associated with each pound of pressure change at lower temperatures than at higher temperatures. Suction lines for refrigeration applications must be sized for less pressure drop than air-conditioning systems because air-conditioning systems operate at higher evaporator temperatures. Refer to the pressure-temperature chart and the operating temperature range of the system you are working on to determine what pressure difference is related to a 2°F temperature drop at the system's normal operating temperature.

TECH TIP

As the suction pressure drops, it has a greater effect on the compressor's compression ratio. The compression ratio is the absolute suction pressure divided into the absolute discharge pressure. If the suction pressure drops as a result of undersized refrigerant piping, the compression ratio will go up significantly. Compressor discharge temperatures will rise as a result of increased compression ratio, leading to oil breakdown, acid formation, and eventual compressor failure.

26.5 IMPORTANCE OF OIL RETURN

In an ideal system, all the compressor oil would stay in the compressor. However, the compressor surfaces that are moving the refrigerant are lubricated with oil and a small amount of oil is swept out of the compressor with the discharged refrigerant. This results in the unavoidable loss of oil from the compressor. Returning oil to the compressor is crucial for the health of the system. The compressor will eventually be destroyed by lack of oil return if the piping system does not adequately return the compressor oil.

In reciprocating compressors, some of the oil in the crankcase gets on the cylinder walls during the downstroke of the piston and is blown out with the compressed vapor refrigerant through the discharge valve on the compression stroke (Figure 26-4).

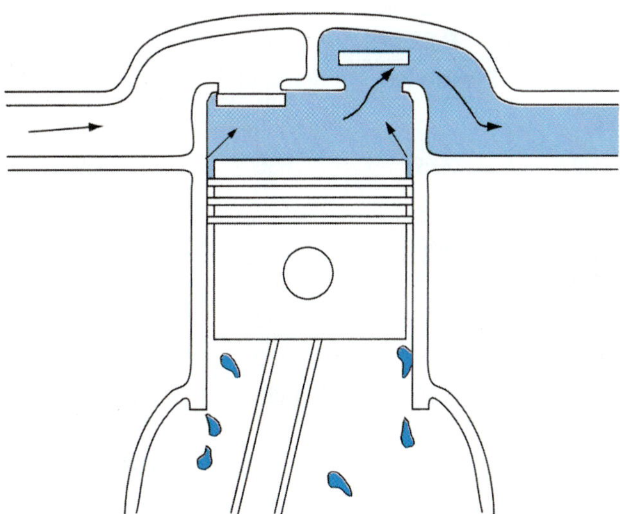

Figure 26-4 Some oil leaves the compressor with the refrigerant.

Some compressors lose much less oil than others, but there is no way to design a compressor that does not lose any oil into the refrigerant piping. This oil serves no other useful purpose in the system except to lubricate the compressor.

The presence of oil in the evaporator and condenser coils can reduce the capacity of the heat-exchange surfaces as much as 20 percent, because the excess oil forms an insulating coating on the interior surfaces of the refrigerant tubes. This, in turn, can cause thermal expansion valves (TEVs) to feed excess amounts of liquid refrigerant. Therefore, the presence of oil in the piping must be taken into consideration when installing piping. A piping system that is not correctly constructed to allow oil return can cause the following problems:

- Seized compressor bearings due to insufficient oil returning to the compressor for lubrication.
- Broken compressor valves, valve plates, pistons, and/or connecting rods due to liquid refrigerant and/or large quantities (slugs) of oil entering the compressor. It is important to understand that the compressor is designed to pump vapor and will not pump liquid.
- Loss of capacity caused by the oil occupying portions of the evaporator, thus reducing the amount of effective surface and the overall system capacity.

26.6 GAS VELOCITY

Oil return in refrigerant piping is accomplished by refrigerant gas velocity. The refrigerant must move fast enough through the lines to carry the oil droplets with it. The minimum required velocity actually varies depending upon the type of refrigerant, the temperature of the refrigerant, and the size of the refrigerant line. In general, large lines require a higher velocity for oil return than smaller lines do. Figure 26-5 shows minimum recommended velocities for oil return based on factors like refrigerant type, evaporator temperature, direction, and line size. Notice that the larger lines require higher velocities than smaller lines.

	R-22 Medium Temp		R-22 Low Temp		R-404A Medium Temp		R-404A Low Temp	
	Horizontal	Vertical	Horizontal	Vertical	Horizontal	Vertical	Horizontal	Vertical
½	280	560	425	850	220	440	330	660
5/8	315	630	475	950	245	490	370	740
7/8	375	750	565	1130	295	590	445	890
1 1/8	430	860	650	1300	335	670	505	1010
1 3/8	480	960	720	1440	375	750	560	1120
1 5/8	520	1040	785	1570	405	810	615	1230
2 1/8	600	1200	905	1810	465	930	705	1410
2 5/8	665	1330	1005	2010	520	1040	785	1570

Figure 26-5 Minimum recommended velocities for oil return (feet per minute, FPM).

Also, lower-temperature lines require higher velocities than higher-temperature lines.

Frequently the complexity of minimum refrigerant line velocity is simplified by quoting a single figure for horizontal velocity and another for vertical line velocity. A safe range of velocities for most applications is 500–750 feet per minute (FPM) for horizontal lines and 1,000–1,500 FPM for risers. Typically the minimum required velocity for a vertical riser is twice the velocity for a similar horizontal section of piping. A quick look at Figure 26-5 shows that velocities considerably less than these can work, depending upon the variables listed in the table. However, it does not hurt for the velocity to be faster than the absolute minimum as long as it does not become excessive. The maximum gas velocity should be kept under 3,000–4,000 FPM to minimize noise. Normally, line sizes that would yield gas velocities in excess of 3,000 FPM would create too high a pressure drop to be acceptable anyway.

26.7 HOT-GAS LINE

In sizing and arranging hot-gas lines, select tubing with a diameter small enough to provide the velocity to carry the hot vaporized oil to the condenser. On the other hand, the diameter must be large enough to prevent excessive pressure drop. In hot-gas lines, a pressure drop equivalent to a 2°F saturated temperature drop is recommended, but a greater loss can be accommodated if conditions demand it.

If a discharge line has excessive pressure drop, the velocity of gas flow through the line can be excessive, causing noise, vibration, and serious reduction in system capacity. There would also be an increase in operating cost due to the higher compressor discharge pressure required.

The velocity needed to move oil up long discharge risers sometimes creates an unacceptable pressure drop. The installation of an oil separator, such as shown in Figure 26-6, is an alternative to moving oil up the discharge riser. The oil

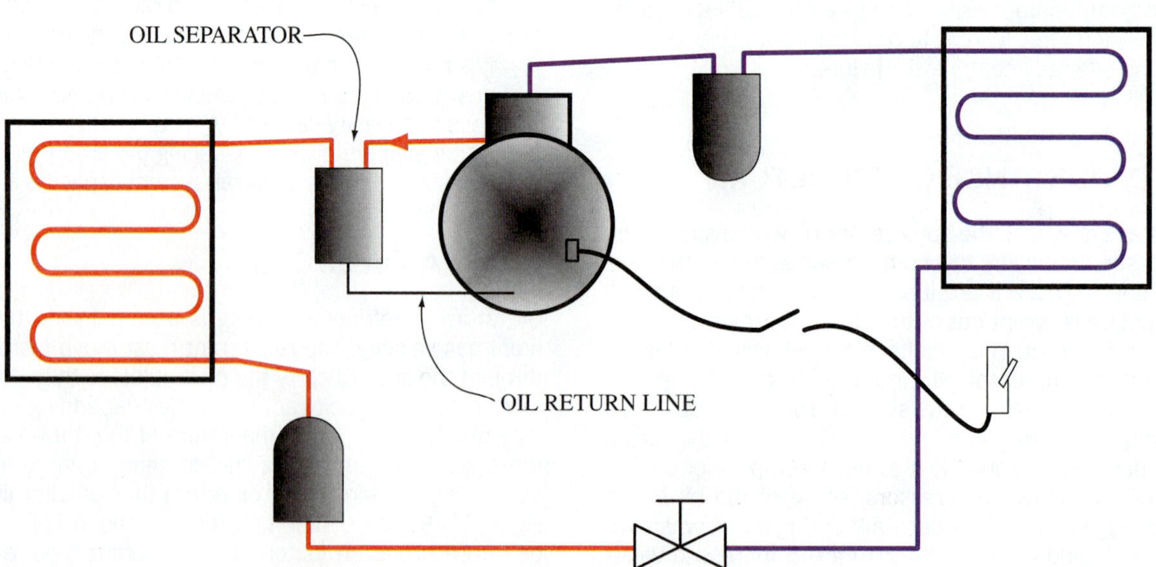

Figure 26-6 Discharge piping with oil separator.

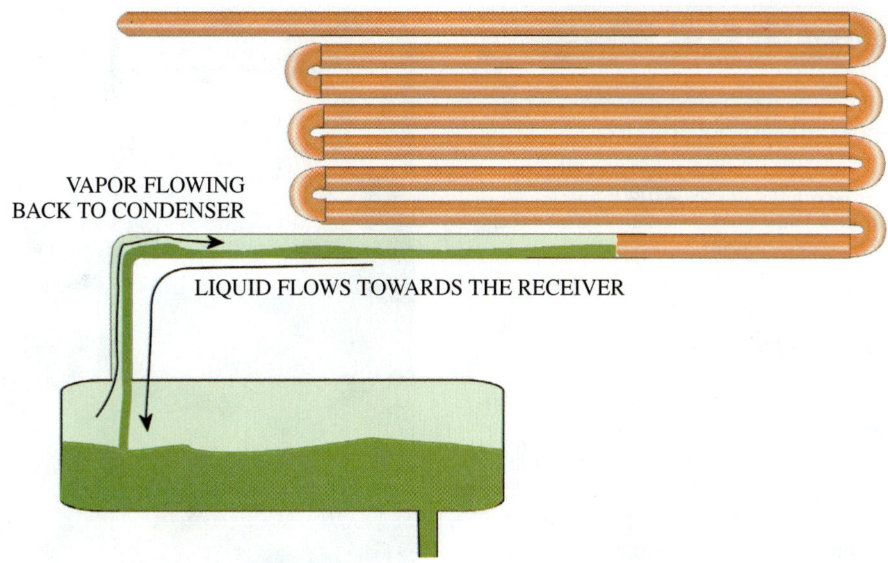

Figure 26-7 The condensate line should be large enough to allow free drainage of liquid from the condenser to the receiver.

from the separator is usually returned to the crankcase of the compressor. With this arrangement, the hot-gas riser can be sized for a lower pressure drop.

26.8 CONDENSER CONDENSATE LINE

The line between a condenser and a liquid receiver is called the condensate line. The objective in sizing condensate lines is to provide ample size, allowing free drainage of the liquid refrigerant to the receiver while the gas refrigerant flows over the liquid in the opposite direction, thus serving as an equalizer line as well as a liquid drain (Figure 26-7). It is almost impossible to oversize the condensate line; 0 psig pressure drop is desirable. Undersizing or installing too long a line should be avoided. An undersized or excessively long line can restrict the flow of refrigerant to the receiver to the extent that some refrigerant will be held back in the condenser. The effective condenser surface is reduced when the condenser begins to fill with liquid, reducing condenser capacity. This causes the head pressure to rise, decreasing the overall system capacity. At the same time, power requirements and operating costs increase.

For proper operation, the length of the line between the condenser and receiver must be kept as short as possible. The top of the receiver should be level with, or preferably below, the bottom of the condenser. This eliminates the need for the liquid to be forced uphill to enter the receiver. The condenser drain line is not usually insulated.

26.9 LIQUID LINE

Liquid lines do not present a problem from an oil-return standpoint because liquid refrigerant and oil mix easily and the oil is carried through the liquid line with ease. Liquid lines, however, are critical for pressure loss both from a pressure drop due to pipe size and, in the case of the vertical upflow line, vertical lift of the refrigerant. Liquid lines are sized for a pressure drop equivalent to a 2°F saturation temperature drop. The pressure drop through the liquid line can be increased if enough subcooling is added to offset the pressure drop. The subcooling should be at least 1°F more than the equivalent saturation temperature drop. If the pressure drop through the liquid line will cause a saturation temperature drop of 4°F, the liquid must be subcooled at least 5°F to prevent flash gas.

Long, large-diameter liquid lines can add a considerable amount of charge to the system. Figure 26-8 shows the amount of refrigerant contained in each foot of liquid line for different diameters of tubing. Line lengths that cause the system charge to exceed the maximum charge limit of the compressor should be avoided. To limit the amount of refrigerant in the system, most residential air-conditioning system manufacturers use ⅜-in liquid lines for split systems using remote condensing units. However, the size of the liquid line will vary with tonnage and line length on commercial systems. Do not change the liquid line size from the original specifications because incorrectly sized liquid lines can seriously reduce system capacity and reduce the equipment life.

Ounces of R-22 per Foot of Line		
Line OD	**Liquid Line**	**Suction Line**
¼ in	0.22 oz	
3/8 in	0.58 oz	
½ in	1.14 oz	
5/8 in	1.86 oz	0.04 oz
¾ in		0.06 oz
7/8 in		0.08 oz
1 1/8 in		0.15 oz
1 3/8 in		0.22 oz

Figure 26-8 Amount of refrigerant charge per foot of line.

TECH TIP

The liquid line is typically not insulated; however, when it runs a great distance through a hot attic, it is advisable to insulate it from the extreme attic heat. In addition, insulating the liquid line will aid in performance of a heat pump. Finally, insulating the liquid line can reduce the noise caused by changeover of heat pumps.

26.10 PRE-EXPANSION FLASH GAS

Liquid leaving the condenser is subcooled, but it is still generally at a higher temperature than the surrounding air. Flash gas generally is not a problem with short liquid lines in normal ambient conditions. However, liquid lines can experience flash gas from pressure drop through the line, pressure drop due to vertical lift, or from passing through very-high-temperature areas. Refrigerant leaving the condenser remains a liquid only as long as its boiling point (condensing temperature) is higher than its actual temperature. Vaporization of some of the liquid can occur before it passes through the metering device if a drop in pressure causes the boiling point to drop below the refrigerant's actual temperature. The vapor formed is called pre-expansion flash gas. Flash gas within the liquid line is undesirable, since it displaces liquid at the port of the expansion valve, greatly reducing capacity. Known as gas binding, it also affects the capacity of capillary tubes, as the ability of the tubes to carry vapor is considerably less than the ability to carry liquid.

Running a liquid line across a metal roof is an example of a situation that will cause pre-expansion flash gas. When the roof is heated by the sun, the roof can easily be hotter than the liquid line. Refrigerant lines should be raised at least 18 in above such a roof to minimize the effects of the sun (Figure 26-9).

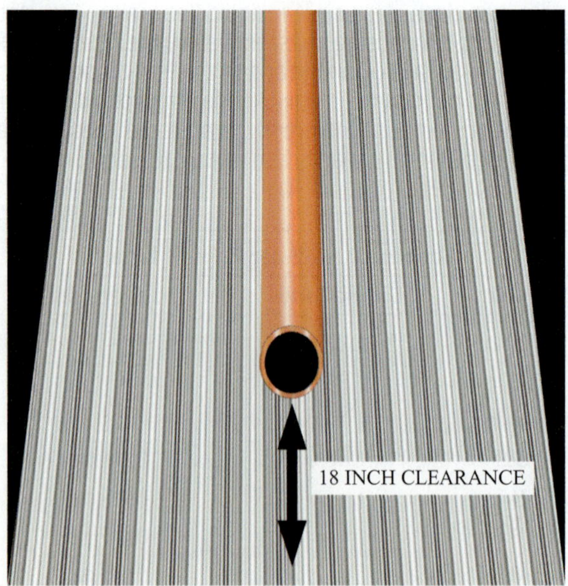

Figure 26-9 Liquid lines should be supported at least 18 in above metal roofs to prevent excessive temperatures that can cause flash gas in the liquid line.

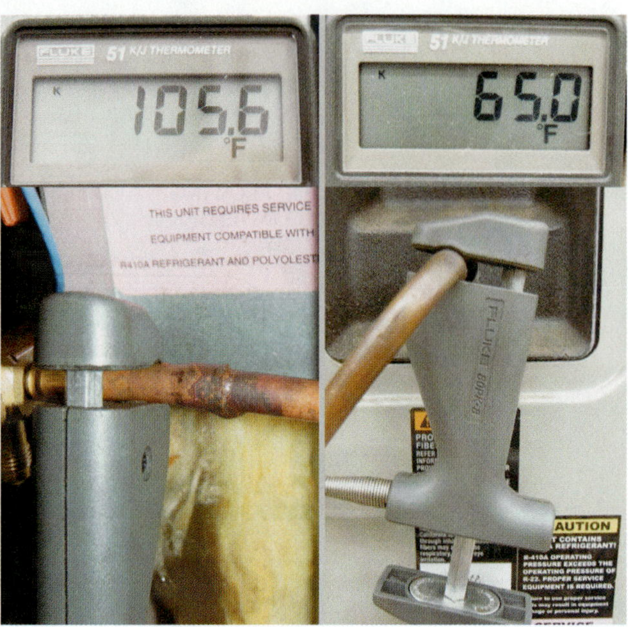

Figure 26-10 The 30°F temperature drop in this liquid line can indicate the presence of flash gas or a restriction in the liquid line.

A method of spotting flash gas is to compare the temperature of the liquid line where a solid stream of liquid exists to the temperature of the liquid line at a point where flash gas is suspected (Figure 26-10). A typical location for the second thermometer would be at the entrance to a TEV valve. If there is a noticeable temperature drop at point 2, flash gas could be present. An exception would be where the presence of gas binding and low liquid flow would cause the liquid line to be warm at the TEV valve and mask the presence of the flash gas.

SERVICE TIP

Liquid-line sight glasses are used in some applications as an indicator of flash gas. Many technicians install the liquid-line sight glass near the condenser. However, the sight glass's usefulness is significantly diminished if it is in the liquid line near the condenser. At that location, it simply tells you there is or is not flash gas there; however, flash gas can be forming in front of your metering device and go undetected. The best location for the sight glass is near the evaporator.

26.11 PRESSURE DROP DUE TO VERTICAL LIFT

Static loss refers to the pressure difference that exists between the bottom and the top of a liquid-filled pipe due to the weight of liquid in the line. Static loss is a frequent cause of the creation of pre-expansion flash gas. Friction and static losses are present in all liquid lines, even properly sized lines. It is important to understand how much pressure loss there will be in the liquid line and offset that by subcooling.

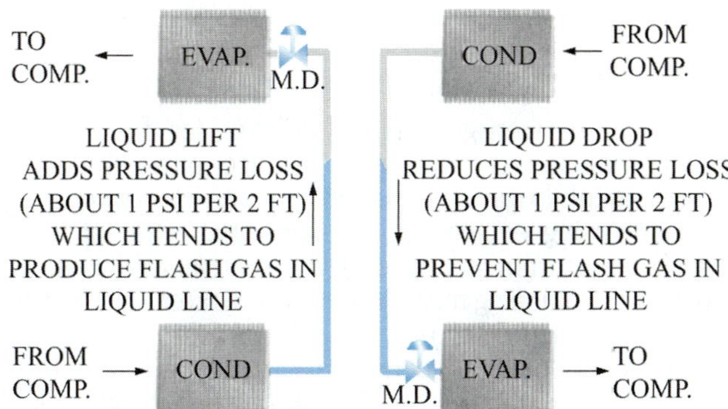

Figure 26-11 Liquid line static head with lift or drop conditions.

Standing columns of liquid exert a force on the bottom of the column due to their weight. Refrigerants are no different. For example, a standing column of liquid R-22 at 100°F and 210 psig exerts a pressure of approximately 0.50 psi for every foot of height of the column due to its weight. R-410A exerts 0.43 psi/ft of height. The pressure at the bottom of a 10-ft column of R-22 is 5 psi greater than the pressure at the top of the column. Conversely, for every 10 ft that R-22 is lifted in a vertical riser, the pressure at the top of the column is reduced by 5 psi. Figure 26-11 shows the effect of vertical lift in liquid lines.

Take the case of an R-22 system with a 30-ft lift, a 6 psi pressure drop in the liquid line, a condensing temperature of 105°F, and 3°F of subcooling. Considering the 3°F of subcooling, the liquid leaves the condenser at 102°F, 211 psig. If the liquid line riser has a 30-ft lift, the pressure at the top of the riser would be 15 psi less due to the weight of the refrigerant. The final pressure would be:

$$211 \text{ psig} - 15 \text{ psig} = 196 \text{ psig}$$
$$196 \text{ psig} - 6 \text{ psig line loss} = 190 \text{ psig}$$

This will produce a boiling point of 98°F. Since the liquid left the condenser at 102°F, enough liquid will vaporize to cool the remaining liquid to 98°F. To avoid the formation of flash gas in this riser, liquid subcooling leaving the condenser of 8°F or more is required to compensate for lift and pressure drop in the line. This can be accomplished by the use of a liquid-to-suction-line heat exchanger.

Again, assume that the same conditions are applied to an air-conditioning unit with 15°F of subcooling. The pressure loss in the liquid line could be much greater before flash gas is formed. Instead of leaving the condenser at 102°F as it did with only 8°F of subcooling, the liquid would leave at 90°F. The saturation pressure for 90°F R-22 is 168 psig. The pressure loss due to vertical lift could be as high as $211 - 6 - 168 = 37$ psi. The vertical lift could increase to 74 ft before flash gas would occur ($74 \times 0.5 = 37$).

This demonstrates the value of a proper amount of subcooling. It not only provides latitude for the designer when laying out the liquid line, but the proper amount of subcooling provides the maximum system capacity at the lowest operating cost.

This also highlights the importance of avoiding vertical lifts in the liquid line. Where possible, avoid installations with large vertical distances between the condensing unit and the evaporator. An example of a problematic installation would be a condensing unit on grade and the air handler in the attic of a multistory dwelling. Flash gas, reduced capacity, and increased operating cost are all likely to be problems.

If you cannot avoid this situation, provide enough liquid subcooling to overcome the pressure loss. Extra subcooling can be obtained by using a suction-to-liquid heat exchanger or stacking the liquid and suction lines together and insulating both lines inside a common wrap to promote heat exchange from the liquid to the suction line.

26.12 SUCTION LINE

Oil circulates throughout the system and must be returned to the compressor to prevent damage to the compressor. The most critical line in performing this function is the suction line. For example, observe the behavior of two common refrigerants used in refrigeration and air conditioning, R-22 and R-410A. In liquid form, these refrigerants will mix with oil and carry it along the piping with ease. Therefore, few, if any, oil problems exist in the liquid line. However, in their gaseous state, the refrigerants are poor carriers of oil.

Oil in the suction line is at a lower temperature than the rest of the system and therefore has higher viscosity, which slows down the flow over pipe surfaces. Also, the refrigerant is in vapor form and has only a mechanical effect on the oil. The vapor does not absorb the oil.

Minimum suction gas velocity is normally between 500–750 FPM for horizontal runs and 1,000–1,500 FPM for vertical runs. This is the minimum velocity needed to bring back the oil to the compressor. If there is a conflict between the size needed to achieve the minimum velocity required for oil return and the size needed to achieve a low-pressure drop, size the suction line to achieve the minimum velocity for oil return. Be sure to check the velocity and size of both the horizontal and vertical sections of pipe, since the minimum velocity for a vertical rise is greater than for a horizontal run. Frequently the best size for the horizontal run will be larger than the best size for the vertical riser (Figure 26-12). It is good practice in these cases to make the vertical piping smaller than the horizontal piping. This keeps pressure drop to a minimum and still maintains the

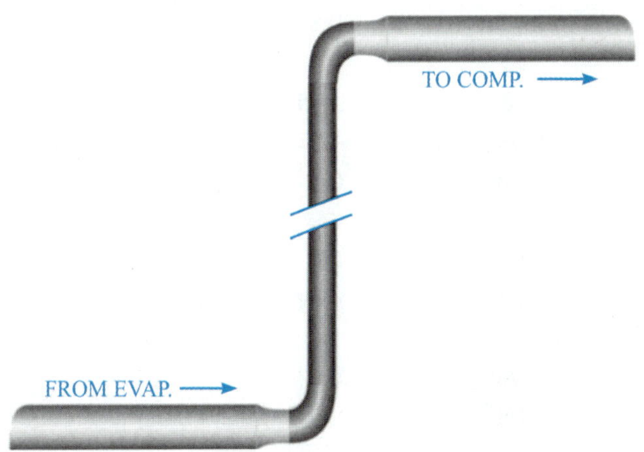

Figure 26-12 Reduced-size suction gas pipe riser.

minimum required velocity for oil return. The suction line, therefore, must be carefully designed to ensure a uniform return of dry refrigerant gas as well as sufficient oil to the compressor.

The maximum recommended saturation temperature drop in the suction line is 2°F. The equivalent pressure drop is different for different refrigerants. The suction pressure drop is important because it has a great effect on the compressor compression ratio. Reducing the suction pressure to the compressor increases the compression ratio. Any increase in this ratio reduces the capacity of the compressor to pump refrigerant vapor and increases the power required.

For example, for R-22 at 0°F, a 10°F temperature drop produces a 7-psi pressure drop from 23.9 psig at 0°F to 16.5 psig at −10°F. With a 195.9 psig, 100°F saturation condenser, the compression ratio at 0°F is calculated as

$$(195.9 + 14.7)/(23.9 + 14.7) = 5.45$$

With a −10°F evaporator the compression ratio becomes

$$(195.9 + 14.7)/(16.5 + 14.7) = 6.75$$

26.13 EQUIVALENT LENGTH OF FITTINGS

There is some pressure loss through every fitting and device in the refrigeration line. This pressure loss must be taken into account when sizing lines. The most common way of accounting for the pressure loss through fittings and devices in refrigeration lines is to compare the loss through each fitting as equivalent to the loss through a length of pipe. Tables give the equivalent length for different sizes and types of fittings. Figure 26-13 gives the equivalent length for common fittings and valves. Note that the equivalent length increases as the pipe size increases.

When looking up line lengths in sizing tables, the length that should be used is the line's total equivalent length. The total equivalent length is the sum of the line's actual length plus the equivalent length of all the fittings and devices in the refrigerant line. It is common for the equivalent length of all the fittings and devices in the line to add up to 50 percent of the line length.

26.14 AIR-CONDITIONING PIPING PRACTICES

What is considered good piping practice for one type of system may be considered poor practice for another type of equipment. In general, the piping recommendations for air-conditioning equipment are somewhat different from the piping recommendations for commercial refrigeration. Most air-conditioning manufacturers do not recommend oil traps, double risers, or expansion loops in their refrigeration lines. In general, air-conditioning piping should be kept as straight and simple as possible. Most air-conditioning manufacturers recommend that refrigerant lines not be run underground. Lengths of underground lines create a large, cool trap to condense significant quantities of refrigerant during the off cycle. This refrigerant is then delivered to the compressor as a liquid slug on startup.

26.15 PREVENTING LIQUID SLUGGING

Slugging is a momentary return of a quantity of liquid to an operating compressor. A slug can be either refrigerant or oil. To learn how to prevent liquid slugging, it is first necessary to understand what happens in a simple system when the compressor stops operating after its cooling requirements are satisfied. The evaporator is still filled with refrigerant—part liquid and part vapor. There will also be some oil present. The liquid refrigerant and oil may drain by gravity to points where, when the compressor starts running again, the liquid will be drawn into the compressor and cause liquid slugging. The piping design must prevent liquid refrigerant or oil from draining to the compressor during shutdown. When the compressor is above the evaporator, this is not a problem. When the compressor is located in an area where the ambient temperature is cooler than the condenser and/or evaporator, off-cycle refrigeration migration can be a problem. Traps in the discharge piping may be required as well as suction-line accumulators and crankcase heaters. It is best to avoid such problematic conditions where possible rather than to try to design fixes for them after the fact.

Copper Pipe Size (in OD)	90° Elbow	45° Elbow	Tee	Gate Valve	Globe Valve	Angle Valve
1/2	0.8	0.4	2.5	0.26	7.0	4.0
3/4	1.0	0.6	2.5	0.3	12.0	6.5
7/8	1.45	0.8	3.6	0.36	17.2	9.5
1 1/8	1.85	1.0	4.6	0.48	22.5	12.0
1 3/8	2.4	1.3	6.4	0.65	32.0	16.0
1 5/8	2.9	1.6	7.2	0.72	36.0	19.5
2 1/8	3.6	2.0	9.6	0.96	48.0	22.5
2 5/8	4.5	2.4	11.2	1.1	56.0	28.0
3 1/8	1.0	2.9	12.8	1.4	72.0	32.0
3 5/8	1.0	3.4	16.0	1.6	80.0	40.0
4 1/8	1.0	4.0	18.6	1.7	100.0	48.0

Figure 26-13 Feet of equivalent length for copper fittings and valves in feet of copper tubing of same size.

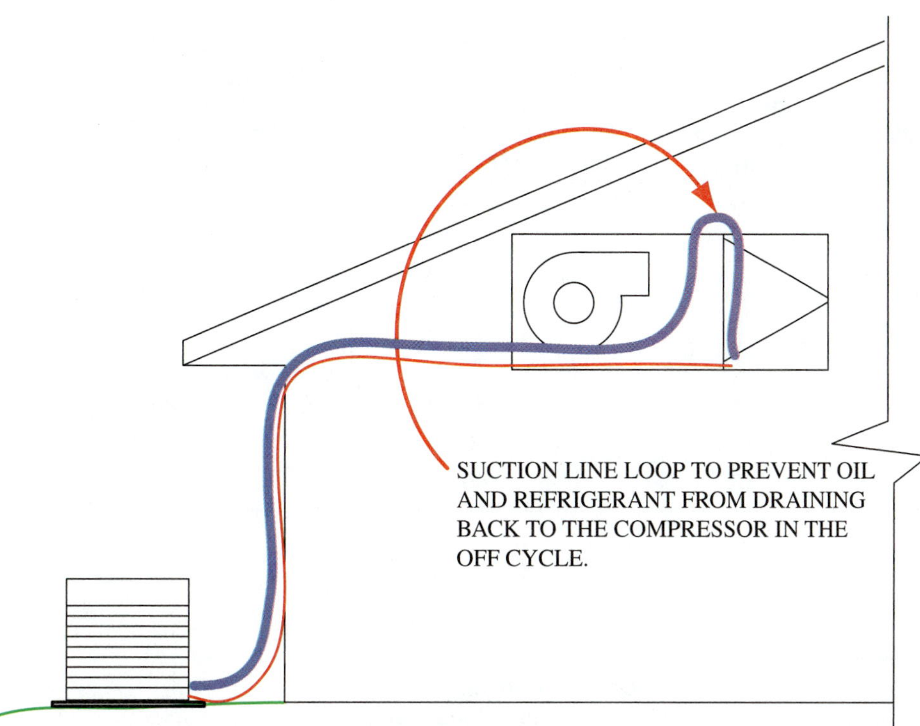

Figure 26-14 Many manufacturers require that the suction line rise above the height of the evaporator when the compressor is installed level with, or below, the evaporator coil.

TECH TIP

Soft refrigeration copper comes in coils, and when properly unrolled it is relatively straight. However, if the copper is simply pulled from the center like a spring, it will have a series of loops. The loops are not significantly deep, but each loop can collect a small amount of refrigerant oil. When the amount of oil in the loops gets large enough, it will begin moving from the first loop to the second, where it picks up that oil and moves to the next. This moving slug of oil then can pass into the compressor and slug it. It moves much in the same way as the last bit of juice is sucked through a child's curlicue straw.

The greater the system refrigerant charge, the more likely it is that compressor failures may be experienced. Common sense dictates close coupled systems with clean, simple piping layouts and minimal refrigerant charges.

Some evaporator coils are designed to feed refrigerant at their top and take suction off the bottom. For top-fed coils, a riser to at least the top of the evaporator must be placed in the suction line if the compressor is below the evaporator, as in Figure 26-14. This inverted loop is to prevent liquid draining from the evaporator into the compressor during shutdown. A hard-shutoff expansion valve will also help prevent refrigerant flow back to the compressor during the off cycle. The sump at the bottom of the riser promotes free drainage of liquid refrigerant away from the TEV bulb, thus permitting the bulb to sense suction-gas superheat instead of evaporating liquid refrigerant. No loop is required for coils that feed at the bottom and take the suction off the top.

26.16 SIZING LINES FOR RESIDENTIAL AIR CONDITIONING

The importance of following the equipment manufacturer's instructions when sizing and installing refrigeration lines cannot be overstressed. Differences in equipment design and application can require different approaches to line sizing. However, there are several design issues that most all residential air-conditioning systems have in common.

Refrigeration lines have a practical limit in terms of both length and size. For safety reasons, hermetic compressors have a system charge limit (Figure 26-15). The charge limit is the maximum amount of refrigerant that the entire system can have in it for safe compressor operation. Extremely long or oversized lines can increase the amount of refrigerant in the system past the safe charge limit of the hermetic compressors used in residential equipment. The "soak out" limit is the maximum charge the oil in the compressor can absorb and not experience liquid slugging on startup.

Further, the compressor has a limited amount of oil. Extra-long or oversized lines can hold so much oil that the compressor becomes low on oil. Adding oil to hermetic compressors is not a viable option because the crankcase of most hermetic compressors is not large enough to hold any extra oil. Residential air-conditioning equipment manufacturers specify the correct line size for their equipment in the installation instructions. Typically, the size for lines less than 50 ft is the same as the condenser stub-out. Lines longer than 50 ft or having a vertical lift more than 20 ft generally require special installation considerations. Figure 26-16 shows one manufacturer's recommendations for line sizes in residential long-line applications.

Compressor Model Number	Compressor BTU/hr Capacity @ AHRI	Refrigerant Oil Charge (Zerol 150T) (pt)	Maximum System r-22 Charge Permissible with Lubricant Zerol 150T Supplied with the Compressor (lb)	Compressor Soak-Out Refrigerant Charge Limits (lb) (Without Crankcase Heater Energized 12 to 16 hr)
H2NG094	87,700	7	15	10.0
H2NG104	101,800	7	17	10.0
H2NG124	119,400	7	20	10.0
H2NG144	144,100	8	24	15.0
H2NG184	183,200	16	30	20.0
H2NG204	212,400	16	35	20.0
H2NG244	252,900	16	40	20.0
H2NG294	287,300	16	45	20.0

Figure 26-15 Hermetic compressor charge limit.

Unit Nominal Size (BTUH)	Acceptable Vapor Line Diameter OD (in)	Cooling Capacity Loss (%) Total Equivalent Line Length (ft.)										
		Standard Application			Long-Line Application Requires Accessories							
		25 ft	50 ft	80 ft	80+ ft	100 ft	125 ft	150 ft	175 ft	200 ft	225 ft	250 ft
18,000	5/8	0	1	1	1	2	3	3	4	5	5	6
R-22 AC	3/4	0	0	0	0	0	1	1	1	1	2	2
24,000	5/8	0	1	3	3	3	5	6	7	8	9	10
R-22 AC	3/4	0	0	0	0	1	1	1	2	2	3	3
	7/8	0	0	0	0	0	0	0	0	1	1	1
30,000	5/8	1	3	5	5	6	8	10	11	13	15	17
R-22 AC	3/4	0	1	1	1	2	3	3	4	5	5	6
	7/8	0	0	0	0	1	1	1	2	2	2	3
36,000	3/4	0	1	2	2	3	4	5	6	7	8	9
R-22 AC	7/8	0	0	1	1	1	2	2	3	3	4	4
42,000	3/4	1	2	3	3	4	5	7	8	9	10	11
R-22 AC	7/8	0	1	1	1	2	2	3	4	4	5	5
	1 1/8	0	0	0	0	0	0	0	1	1	1	1
48,000	3/4	1	2	4	4	5	7	8	10	11	13	14
R-22 AC	7/8	0	1	2	2	2	3	4	5	5	6	7
	1 1/8	0	0	0	0	0	0	1	1	1	1	1
60,000	7/8	1	2	3	3	4	5	7	8	9	10	11
R-22 AC	1 1/8	0	0	1	1	1	1	2	2	2	3	3

Figure 26-16 Vapor line sizing and cooling capacity losses—R-22 air-conditioner applications.

26.17 LONG-LINE AIR-CONDITIONING APPLICATIONS

Another difference between air conditioning and commercial refrigeration is the definition of a long line set. For residential air conditioning, refrigerant line lengths exceeding 80 ft or having a vertical rise of 20 ft are considered "long." A 100-ft line set would not be considered particularly long in a commercial refrigeration system. Consequences of using a long line set include the following:

- Additional refrigerant charge is required.
- Refrigerant migration control is needed.
- Oil return can become a problem.
- Significant system capacity losses are inevitable.
- Metering device adjustments are required.

System modifications that may be required to alleviate problems caused by a long line set include:

- Adding a liquid line solenoid at the outdoor unit. The liquid solenoid closes, preventing refrigerant flow from the condenser to the evaporator during the off cycle.
- Using a hard-shutoff TEV at the evaporator, which prevents migration from the evaporator to the compressor during the off cycle.
- Adding a compressor crankcase heater. The crankcase heater is used to reduce migration to the compressor crankcase during the off cycle by keeping the oil warm.
- Adding a compressor hard-start kit. The hard-start kit is required to get start the compressor against the pressure difference that is created by installing the solenoid and hard-shutoff expansion valve. It will also help the compressor start against the pressure created by a standing column of liquid in the liquid line.

At least one major equipment manufacturer recommends using only a 3/8-in liquid line regardless of the length. The system subcooling will have to be high enough to offset the loss through the liquid line. Even with these modifications, the system capacity will be reduced. System capacity reduction from a long line set can be up to 17 percent (Figure 26-16).

26.18 COMMERCIAL REFRIGERATION PIPING PRACTICES

The refrigerant lines for commercial refrigeration systems are typically longer than the lines in air-conditioning systems. Commercial refrigeration systems frequently have multiple components installed on a common piping system, something that is unusual in air-conditioning systems. Underground lines are avoided in air-conditioning applications but are unavoidable in large supermarket installations. Finally, the evaporator saturation temperatures in commercial refrigeration systems are considerably lower than in air conditioning. These differences create unique challenges for commercial refrigeration piping. Oil traps, double risers, and expansion loops are all typical components of a commercial refrigeration system piping layout.

26.19 OIL TRAPS

Some manufacturers recommend oil traps on vertical risers to ensure oil return up the riser. The oil trap, often called an oil lift, is located at the bottom of a riser. The free oil droplets and oil on the side of the riser are blown upward by the velocity of the refrigerant. But the higher the oil is pushed, the slower it travels. Given enough height, some of the oil will stop moving upward and begin to fall back down. The oil lift will fill with oil. The refrigerant velocity through the trap will increase due to the reduced area for the gas to flow through. This increased velocity helps the vapor entraining the oil carry it up the riser. An inverted trap should be used at the top of the line to keep the oil from draining back down once it has made it up the riser. Figure 26-17 shows how to pipe a vertical suction riser with an oil lift at the bottom and an inverted trap at the top.

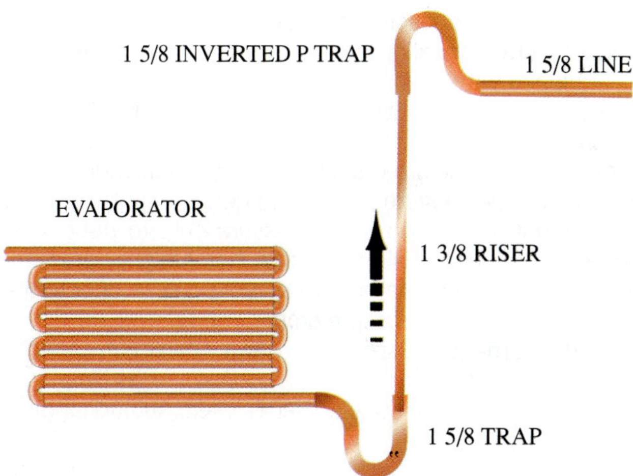

Figure 26-17 Suction risers should have an oil trap at the bottom and an inverted trap at the top.

Where the vertical rise is over 10 ft on either suction or discharge lines, it is recommended that line traps be installed approximately every 20 ft so that the storage and lifting of oil can be done in smaller stages.

SERVICE TIP

It is important to note that not all manufacturers consider oil traps and double risers good piping practice. Instead, they stress sizing for proper velocity and leaving off the traps. They point out that the extra restriction added by the traps increases the pressure drop through the line. Always follow the manufacturer's recommendations for the particular equipment being installed.

26.20 VERTICAL RISERS IN MODULATING SYSTEMS

Riser sizes for both suction and hot-gas lines are critical to permit carrying the oil upward with the force of the flowing gas. Vertical risers require a higher velocity than horizontal lines. The minimum velocity for a vertical riser is typically twice the minimum velocity of a horizontal line the same size. The increased velocity for vertical risers is often accomplished by using a smaller size for the riser than for the rest of the system. Vertical drops (flow down) will return the oil by gravity and are not critical as to size or velocity.

TECH TIP

Commercial systems often operate at varying capacities. That is, the compressor is not always moving the same amount of refrigerant. Compressor capacity is reduced to match the cooling load. Operating at a reduced capacity is called part load operation.

As system capacity reduces, so does the velocity of the refrigerant traveling through the lines. If the compressor is equipped with capacity control, the vertical riser can be sized to return the oil when the system is running at its minimum capacity. Short risers on systems with capacity control will usually be sized smaller than the remainder of the suction line (see Figure 26-12). This helps keep the velocity up without adding too much pressure drop to the overall system. Although this smaller pipe has a higher pressure drop, its short length adds a relatively small amount to the overall suction-line pressure drop.

An alternative to sizing the riser for minimum capacity is to provide a double riser in the suction line, as illustrated in Figure 26-18. If a double riser is used, additional oil is required in the system to fill the trap during periods when the overall capacity of the system is reduced, and only one riser is used. When full capacity is resumed, this extra oil can return to the compressor and overload the oil capacity of the crankcase, causing oil slugging. It is, therefore, desirable to avoid the use of double risers wherever possible.

The smaller riser should be sized for a minimum velocity of 1,000–1,500 FPM at the system's minimum capacity. The larger riser should be sized to carry the remaining system capacity at a minimum velocity of 1,000–1,500 FPM. Referring to Figure 26-18, riser 1 is used to carry the suction gas at minimum load. Riser 1 plus riser 2 are used to carry the suction gas at maximum load. The area of riser 1 plus the area of riser 2 is made equal to the area of the main suction line.

Figure 26-18 Double-suction riser.

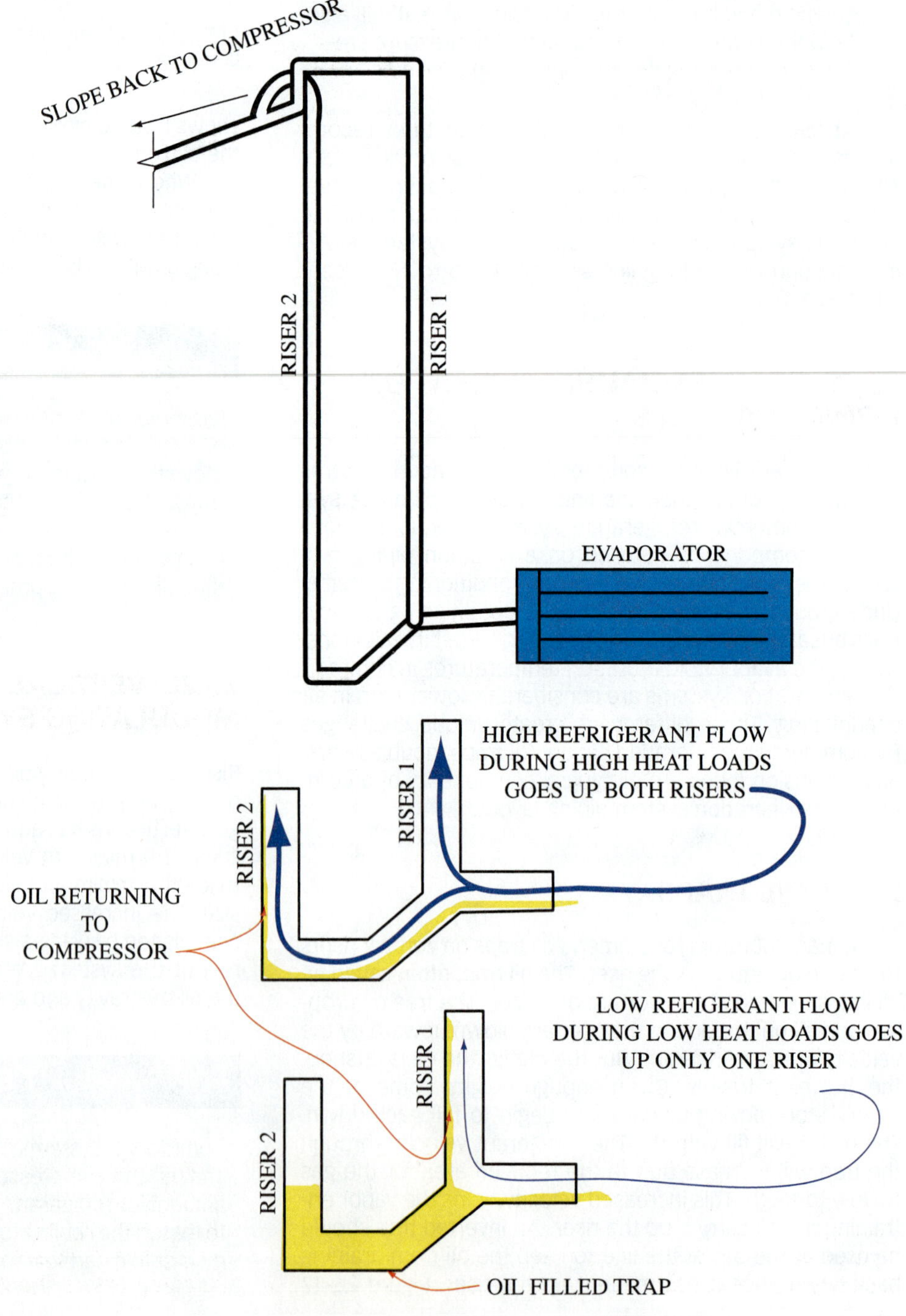

Both risers will carry refrigerant and oil when maximum cooling is required. During part load operation, as the amount of refrigerant being evaporated decreases, the gas velocity will also decrease to a point where it will not carry oil up through the larger riser. Oil will collect in the trap and force the refrigerant to flow up the smaller riser, which is sized for the minimum system capacity (see Figure 26-18). As the system load increases and more refrigerant is passed through the evaporator, this increased pressure will break the oil seal in the trap and carry oil upward through both risers.

SERVICE TIP

Small, high-speed compressors have relatively small crankcase capacities. If double risers with oil seals are to be utilized, it may be necessary to add an auxiliary oil receiver. Be sure the system has sufficient oil to permit proper compression lubrication at full and minimum capacities.

26.21 OFF-CYCLE PROTECTION

When a remote condenser is being used and it is located higher than the compressor, the hot-gas discharge line should not be piped directly up from the compressor. This makes the compressor head an oil and refrigerant trap.

Some oil is always in this line because it is being pushed along with the refrigerant vapor. A vertical riser directly from the compressor would allow oil to drain back to the compressor when it stops. Liquid refrigerant can be formed in the hot-gas line in the winter. In cold weather, once the compressor cycles off, the hot-gas line can be cooled to a temperature below the dew point of the refrigerant. This can result in liquid refrigerant being formed in the hot-gas line. Either oil or liquid refrigerant can break valves on a reciprocating compressor on startup.

To prevent this, a discharge loop can be placed in the hot-gas line with an optional check valve (Figure 26-19).

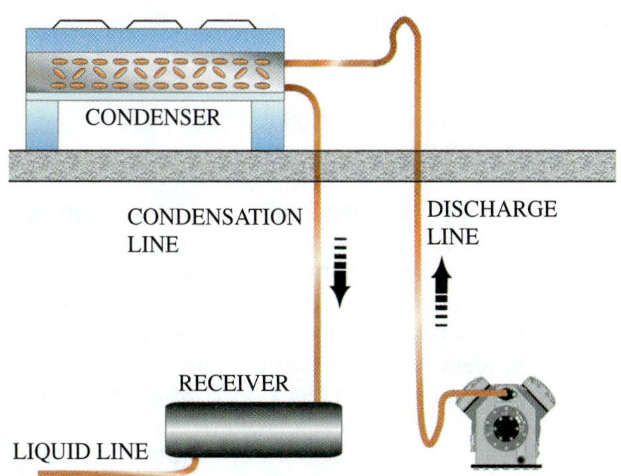

Figure 26-19 Discharge line risers should have a trap at the bottom to prevent liquid accumulation in the compressor head.

Figure 26-20 This discharge check valve keeps liquid refrigerant from draining down on top of the compressor head during the off cycle.

Modest quantities of oil and liquid refrigerant can accumulate in this trap when the compressor shuts down and will be dissipated on startup without damage to the compressor.

A discharge check valve is essential for systems with large quantities of refrigerant and a condenser that is located above the compressor. Figure 26-20 shows a discharge check valve on a transport refrigeration unit.

26.22 OIL RETURN IN PARALLEL-COMPRESSOR SYSTEMS

Parallel compressors are often used to provide systems that can produce different levels of cooling capacity based on system need. When demand is heavy, all of the compressors will be operating; but as the load decreases, one or more of the system compressors can be cycled off.

TECH TIP

Keeping the oil level balanced between parallel compressors can be a problem. Oil-level control systems are used with multiple-compressors systems to ensure proper oil level in each of the compressors. If the oil level is too low, compressor lubrication can be lost and the compressor can be damaged or destroyed. If the oil level is too high, the crankshaft will be pushed through the oil. This will cause excessively high current draw, which can damage or destroy the compressor motor.

In parallel compressors, one of the main problems is returning the oil back to the crankcases of each of the compressors. Without consideration for oil return and equilization, some compressors will end up with too much oil while others have too little oil. Maintaining the oil level in each compressor crankcase is important for reliable operation. The compressors should be mounted on the same level;

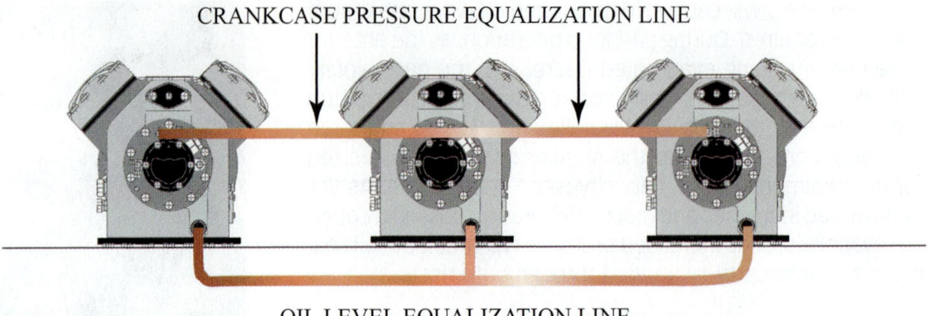

Figure 26-21 Typical oil equilibration piping for multiple compressors.

normally they are mounted on a rack. A compressor rack is a frame constructed from structural steel, angle iron, or square tubing.

Figure 26-21 shows an oil-equalization piping system for multiple compressors. Two separate pipe systems are required to provide for oil-level distribution. One pipe is connected into the bottom of the compressor. This pipe is called the oil-level equalization line. Another pipe is attached to the upper portion of the compressor crankcase, the crankcase pressure equalization line. The lines are to be installed as follows:

- The crankcase oil-equalization lines are screwed into the bottom ports of the crankcase. The lines must be the same diameter as the bottom ports. These lines tie into a crankcase-equalization manifold. The manifold should be mounted below the crankcase to prevent any vapors from entering the manifold. If vapor were to enter the manifold, it would prevent the oil from being equally distributed to the compressors' crankcases.
- The crankcase-pressure equalization line is attached between the compressor and a manifold. The manifold must be higher than the top of the crankcases. This is to prevent any liquid (oil) from entering the manifold, which would also prevent equal distribution of oil to each crankcase.

There should be isolation valves on each of the equalization lines next to each compressor. This is to allow each compressor to be isolated for servicing. The manifold method is the preferred method of oil return because it can transfer oil from one crankcase to the other in case one crankcase has too much oil.

26.23 OIL-LEVEL CONTROL

Oil-level controls can also be used to return oil to the compressor. Figure 26-22 shows piping for an oil-control system using an oil separator and oil-level controls. This is done by first piping the hot-gas line to an oil separator. The outlet of the oil separator is piped to an oil receiver. In the receiver, any liquid refrigerant is vaporized. It is vaporized because there is a line connecting the receiver to the suction line. This lowers the pressure in the receiver. A pressure-differential check valve maintains the pressure in the receiver to a point slightly higher than what the pressure is in the crankcase. This slightly higher pressure allows oil to flow from the receiver to the crankcase level controls. As the oil level drops in the crankcase, the level controls open and allow oil to flow into the crankcase.

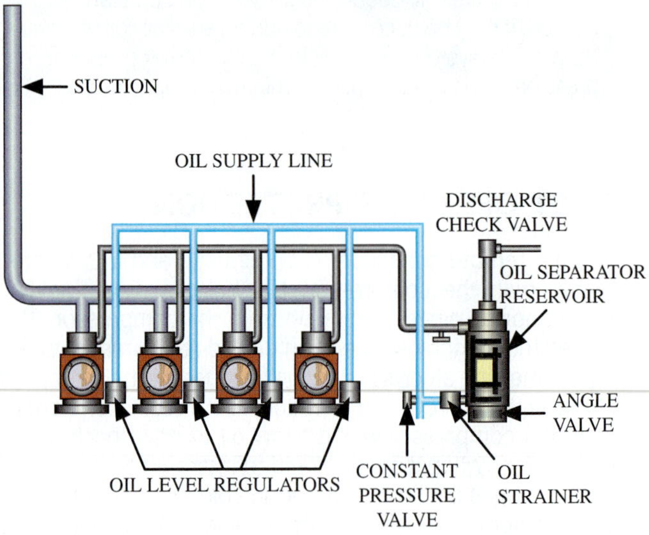

Figure 26-22 Typical oil-control system for multiple compressors.

This oil-level control cannot lower the oil level in a compressor crankcase if the oil level becomes excessively high.

26.24 PIPING FOR MULTIPLE EVAPORATORS

Systems with multiple evaporators have a common liquid line and a common suction line. The evaporators are individually piped to these common lines. Care must be taken to ensure that flow from one evaporator does not affect the operation of other evaporators. When evaporators are stacked vertically, oil and liquid can drain out of the top evaporator into the bottom evaporator, as in Figure 26-23. This can affect the expansion valve bulb, causing the valve to throttle and starve the bottom evaporator. The suction lines should be piped as shown in Figure 26-24 so that liquid cannot drain from the top evaporator into the bottom evaporator.

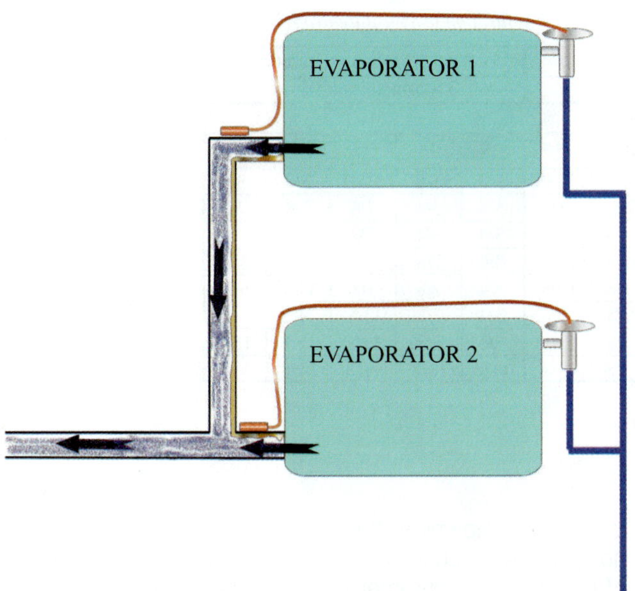

Figure 26-23 Liquid can drain from the top evaporator to the bottom evaporator with incorrectly piped evaporators.

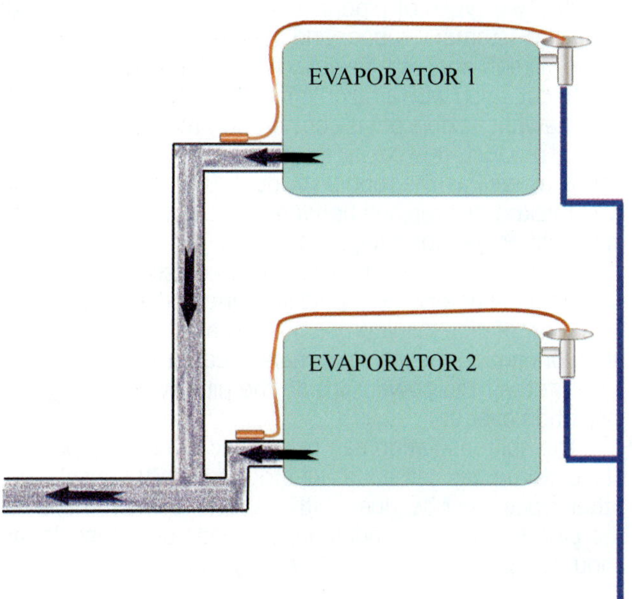

Figure 26-24 Properly piped evaporators prevent liquid from draining into the lower evaporator.

Evaporator pressure regulators are required when multiple evaporators with different temperatures are piped to a common suction line. Each evaporator will have a pressure that corresponds to its saturation temperature. This means that evaporators with different pressures will all be piped into the same suction line. Evaporator pressure regulators keep the pressure in each evaporator from dropping below the correct operating pressure for that evaporator. The common suction line must be at or below the pressure of the lowest temperature evaporator.

26.25 COMMERCIAL REFRIGERATION LINE SIZING

Charts called nomographs are used to size refrigerant lines in commercial refrigeration and air-conditioning applications. Each refrigerant has two nomographs: one for pressure drop and one for velocity. Typically the line size required to meet the maximum pressure-drop requirements is selected and the refrigerant velocity is checked to see that it falls between the minimum and maximum recommended velocities. The five factors needed to use a nomograph are the type of refrigerant, the length of the line in equivalent feet, the capacity of the unit, the condenser temperature, and the allowable pressure drop.

Figure 26-25 shows a simplified velocity nomograph. To use the nomograph, find the system capacity at the top right-hand side of the chart. There are diagonal lines representing different temperature suction lines, the discharge line, and finally the liquid line. Follow the vertical capacity line straight down until it intersects the diagonal line representing the evaporator temperature (see the red line in Figure 26-25). On the left side of the chart are diagonal lines representing different sizes of pipe. Go horizontally across from the intersection until this line intersects the pipe size being checked (see the blue line in Figure 26-25). Drop straight down to the answer at the bottom left side of the chart (see the green line in Figure 26-25). In the case of pressure-drop nomographs, this will be the pressure drop per 100 ft of pipe. In the case

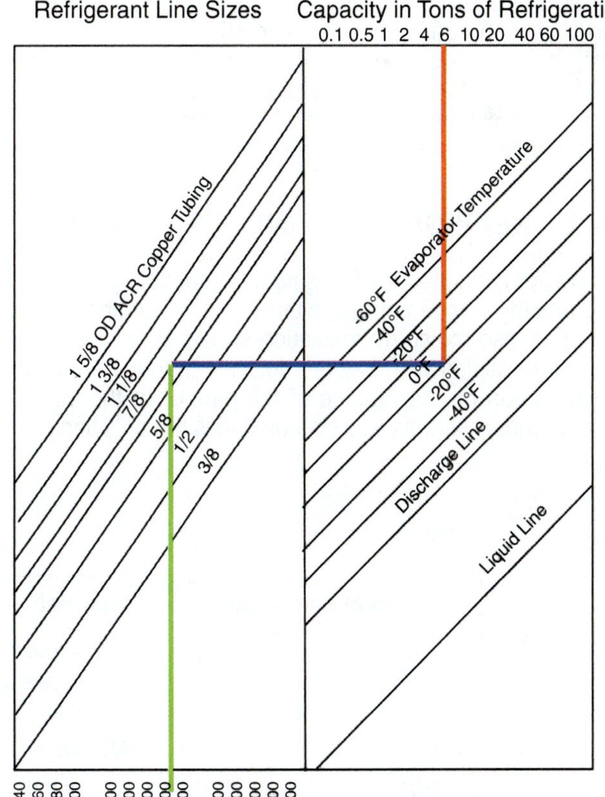

Figure 26-25 Nomographs like this are used to determine the velocity of refrigerant through refrigeration lines.

***Suspection Line Pressure Drop for Refrigerant R 404A** (Sized for 2°F saturation temperature drop)*

Tons	-10° - 9°F Evap Saturation 1.9 psig pressure drop					10° - 29°F Evap Saturation 2.6 psig pressure drop					30° - 50°F Evap Saturation 3.4 psig pressure drop				
	Equivalent Length in feet														
	25	50	75	100	150	25	50	75	100	150	25	50	75	100	150
1.5	3/4	5/8	3/4	3/4	3/4	5/8	3/4	3/4	7/8	7/8	5/8	5/8	3/4	3/4	3/4
2	5/8	3/4	7/8	7/8	7/8	3/4	3/4	7/8	7/8	11/8	5/8	3/4	7/8	7/8	7/8
2.5	3/4	3/4	7/8	7/8	11/8	3/4	7/8	7/8	11/8	11/8	3/4	3/4	7/8	7/8	11/8
3	3/4	7/8	7/8	7/8	11/8	3/4	7/8	11/8	11/8	11/8	3/4	7/8	7/8	7/8	11/8
3.5	3/4	7/8	7/8	11/8	11/8	7/8	11/8	11/8	11/8	13/8	3/4	7/8	7/8	11/8	11/8
4	3/4	7/8	11/8	11/8	11/8	7/8	11/8	11/8	11/8	13/8	3/4	7/8	11/8	11/8	11/8
5	7/8	7/8	11/8	11/8	13/8	7/8	11/8	11/8	13/8	13/8	7/8	7/8	11/8	11/8	13/8
7.5	11/8	11/8	11/8	11/8	13/8	11/8	13/8	13/8	13/8	13/8	11/8	11/8	11/8	11/8	13/8

Figure 26-26 Liquid-line selection using a nomograph for pressure drop.

of the velocity nomograph, this will be the refrigerant velocity in feet per minute.

26.26 QUICK LINE SELECTION TABLES

Some commercial refrigeration equipment manufacturers publish quick refrigeration-line-sizing tables. These tables make assumptions about the type of refrigerant, condenser temperature, evaporator temperature, and line length. They are produced by application engineers who take the assumptions being used for a particular chart and use line-sizing nomographs to determine the correct line size. These are accurate as long as the assumptions made fit your particular application. Figure 26-26 shows a typical quick sizing chart. This chart is for a system using R404a with a condensing temperature of 105°F and evaporator temperatures ranging from −10°F up to 50°F. Notice that the line sizes increase in diameter as the line lengths increase.

26.27 REFRIGERANT LINE INSULATION

It is always good practice to insulate the suction line since it is usually lower in temperature than the surrounding air and can condense moisture. Insulation should be applied and thoroughly sealed with a good moisture barrier to prevent condensation on the outside of the pipe or in the insulation. On most systems, the suction line is the only line that is insulated. However, there are times when the liquid and hot-gas lines need to be insulated, too. The insulation needs of the piping are summarized in Figure 26-27.

All refrigerant-line insulation should be closed cell to reduce the adsorption of water. Closed cell means that the material is made of millions of tiny gas pockets that are each completely sealed. This keeps the material from wicking up moisture and also allows it to act like a moisture barrier. Two types of pipe insulation are commonly used: buna-N elastomeric rubber and polyolefin. The elastomeric rubber is soft and very flexible. Its flexibility makes it difficult to cut, even with a razor knife. The polyolefin insulation is somewhat flexible but is considerably more rigid than the rubber product. It is easier to cut but does not conform to bends as well as the rubber product. Both are available in wall thicknesses ranging between ⅜ in to 1½ in and come in unsplit or split form (Figure 26-28).

For new lines, the unsplit insulation is slid over the end of the tubing before the tubing is connected (Figure 26-29). This form of pipe insulation typically has a fine white powder lubricant on the inside to make it easier to apply. Try to avoid getting the powder inside the pipe while pulling the insulation over it.

The split insulation can be applied after the pipe is in place. Some has self-adhesive strips to join the insulation; other types must be glued with a special cement made for the pipe insulation. In addition, the ends of the insulation should be glued together where they join.

	Type of Insulation	
Refrigerant Line Type	**Thermal***	**Vapor Barrier**
Liquid line	Sometimes on heat pumps in cold climates and on ACs in very hot attics	
Vapor line	Always	Always
Hot-gas line	Sometimes on heat pumps	
Condensate line	Sometimes on rooftop installations	

*Sometimes refrigerant lines are insulated to reduce the transmission of sound from the lines to the occupied area.

Figure 26-27 Refrigerant lines and insulation.

Figure 26-28 Refrigerant-line insulation is available in split and unsplit forms. *(Courtesy of K-Flex USA LLC)*

Figure 26-29 Pipe insulation that is not split is placed on the copper tubing before the pipe is connected to the system.

Suction-Line Insulation

Insulation on the suction line is an absolute requirement. Suction-line insulation eliminates condensation on the line and helps improve system efficiency by preventing the suction line from picking up heat from the surrounding air. Water condensing on the suction line can drip and can cause damage to ceilings, floors, furnishings, and electronics. Insulating the suction line helps keep the suction-gas temperature as low as possible. Hermetic and semihermetic motor compressor assemblies are suction-gas cooled. Lower return gas temperature is better for cooling the motor and keeping the discharge temperature down.

Heat Pump Gas-Line Insulation

Some types of insulation that are acceptable for suction-line insulation are not able to withstand the high temperatures of the gas line on a heat pump when it is operating in heat. Generally speaking, lines on heat pumps are usually insulated with elastomeric rubber insulation; polyolefin insulation material is not used for heat pumps. Make sure that insulation used on the large gas line of a heat pump system is rated for heat pump duty.

TECH TIP

The joints in the insulation on suction lines must be sealed. Often technicians have the misconception that the insulation is simply there to aid in system efficiency. That is only partially true. Because the refrigerant line is below the dew point, moisture can condense on the line. If the insulation vapor barrier is not sealed, water can collect around the copper and draw in through gaps and openings at the joints. The water can then run down the refrigerant line set and weep out, causing water stains and mold growth in the wall cavities. It is important that the seams of the refrigerant-line insulation be sealed properly.

Liquid-Line Insulation

Normally, no insulation is used on the liquid line because the liquid line is usually warmer than its surroundings. Allowing the liquid line to lose heat to the air actually improves performance by increasing subcooling. However, when the liquid line runs through a hot space such as an attic, insulation of the liquid line may be required to prevent boiling from occurring prior to the expansion valve.

Insulation on the liquid line of a computer room air-conditioning system is desirable. The liquid can get quite cold in winter, and lines running through humidified space to the air-conditioning unit will sweat and drip if not covered.

Sometimes the liquid line can be insulated together with the suction line to promote heat exchange between the two lines for long-line applications. The extra subcooling the liquid line receives from the suction line helps to prevent flash gas in the liquid line.

Hot-Gas-Line Insulation

In package units and condensing units with short hot-gas lines between compressor and condenser, no insulation should be used on the hot-gas line.

On remote condensers, insulating the hot-gas line is advisable. If the unit is expected to operate in low outside temperatures, it is possible to reach the condensing temperature of the discharge refrigerant before the refrigerant reaches the condenser. This can cause liquid slugs to fall backward down the hot-gas line into the superheated vapor from the compressor. Violent expansion of the slug vaporizing can cause "steam hammer," resulting in noise and vibration, even to the point of line breakage. Insulating the hot-gas line would prevent this action.

When hot-gas lines are run indoors in machinery rooms, they should be insulated or otherwise protected to prevent accidental burns to operating personnel.

Be sure to use an insulation that can withstand high temperatures if a hot-gas line is insulated.

26.28 PIPING SUPPORTS

All piping must be properly supported (Figure 26-30). The supports must allow for the expansion and contraction of the pipe. The recommended allowance is ¾ in of movement

Figure 26-30 Maximum spacing between supports for type-L copper tubing.

OD Pipe Diameter (in)	5/8	7/8	1 1/8	1 3/8	1 5/8	2 1/8	2 5/8
Maximum Span (ft)	5	6	7	8	9	10	11

Figure 26-31 Proper support for piping.

per 100 ft of pipe. Hangers should be spaced according to the pipe size being supported. Figure 26-31 shows the recommended distances between supports for common sizes of copper pipe.

Piping should always be supported near a bend in the piping, preferably on the longest straight connection to the bend. The hanger must have sufficient width not to crush the insulation on insulated pipe. Extra rigidity must be incorporated into the insulation at the pipe clamp because the insulation will compress under the weight of the pipe. Sheet-metal saddles or wood-block inserts (on large piping) will serve this purpose (Figure 26-32).

26.29 VIBRATION DAMPENERS

It is usually desirable to isolate vibrating equipment to reduce noise and to prevent damage to the piping or other equipment. With soft copper tubing, loops or a coil of tubing can be connected to the moving part. For hard copper tubing, a similar dampening effect can be produced by running the suction and discharge lines 15 times the pipe diameter in each of two or three directions before securing the pipe hanger. This will provide some give to the piping without undue strain. Flexible connectors at the compressor can help isolate the refrigerant piping from compressor vibrations and movement (Figure 26-33).

For larger equipment, rubber-in-sheath or spring vibration-eliminator mounts and flexible piping connections can be provided for the compressor to supply the necessary isolation. Concrete inertia blocks are sometimes incorporated into the base. Piping and electrical lines must be securely anchored beyond the isolators to be effective.

26.30 HOT-GAS MUFFLER

The pulsations from a compressor, usually a reciprocating type, can cause serious vibrations and noise in the hot-gas line. This can be most noticeable where a remote air-cooled

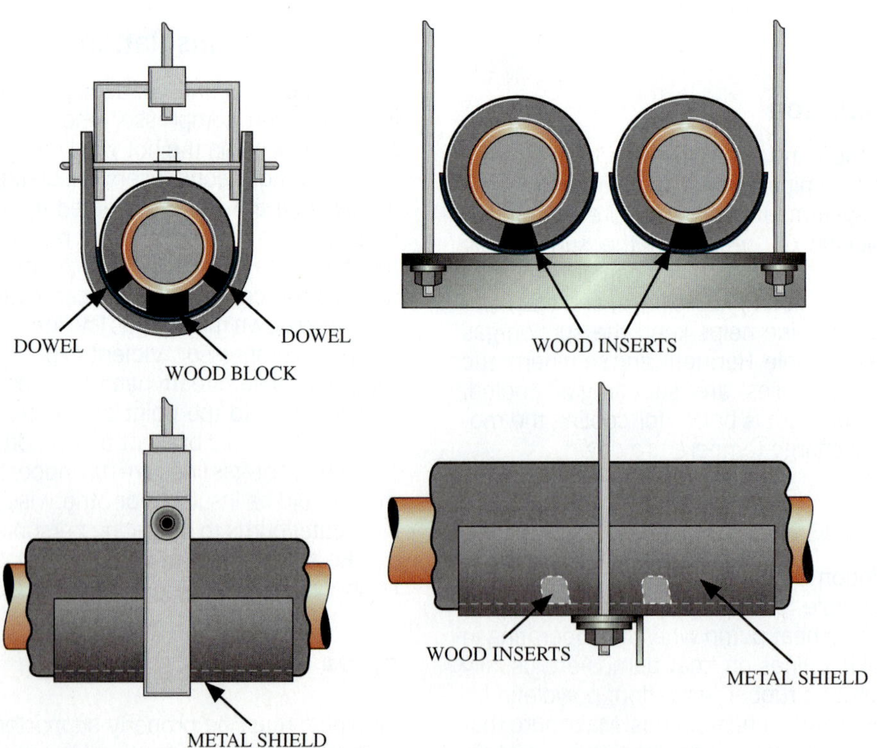

Figure 26-32 Wood pieces are inserted into the insulation to support the pipe at each pipe hanger.

Figure 26-33 This refrigeration piping on this system is isolated from the vibration of this compressor by flexible connectors at the compressor.

condenser requires a long vertical hot-gas line. Usually the larger the compressor, the more noticeable the pulsations, although this is dependent on speed and number of cylinders.

The best way to solve this problem is to install a hot-gas muffler in the compressor discharge line, as shown in Figure 26-34. The hot-gas muffler should be placed in a vertical position (so that it does not trap oil), as close to the compressor as possible, and securely mounted to the compressor. This usually destroys the resonance that the compressor has built up in the hot-gas line. If this does not help, it may be necessary to enlarge the discharge line and relocate it, to destroy the resonance pattern.

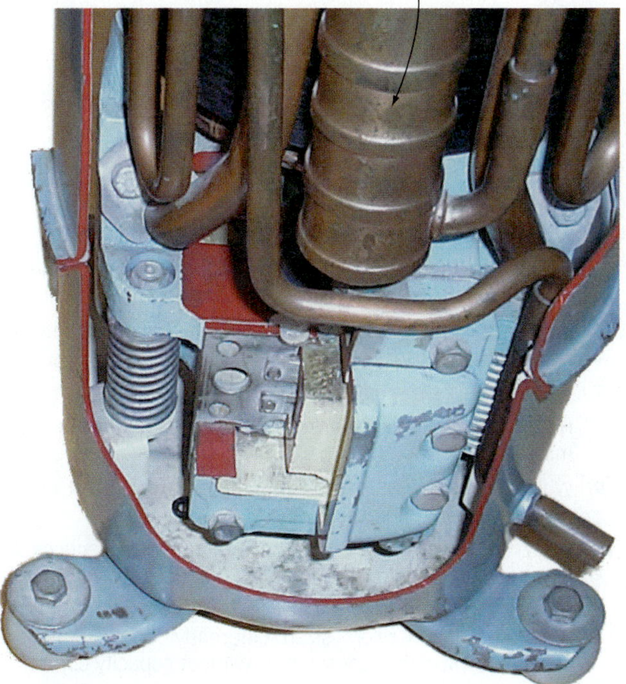

Figure 26-34 This hermetic compressor has a built-in discharge muffler to dampen discharge-gas pulsations.

UNIT 26—SUMMARY

The sizing of the refrigerant lines is important for many reasons. If the lines sizes are too small, the system will lose efficiency. If the line sizes are too large, the system may destroy itself due to slugging or oil starvation.

Pressure drop through the refrigerant lines kills system efficiency. The typical maximum pressure drop through any line is equivalent to a 2°F saturation temperature drop. Refrigeration vapor lines must also be sized to create a high enough velocity for the refrigerant vapor to carry the oil back to the compressor. Typical minimum horizontal line velocities are 500–750 FPM. Typical minimum riser velocities are 1,000–1,500 FPM. Minimizing pressure drop and ensuring oil return are sometimes in conflict. Larger lines minimize pressure drop, but smaller lines increase velocity to ensure oil return. Oil return should always take precedence.

Proper insulation of the various types of refrigerant lines will improve efficiency, and system efficiency is critical for customer satisfaction. A properly insulated piping system can also eliminate operational problems. Piping support is necessary to eliminate potential problems, including undesirable vibration. Vibration can cause both noise complaints and pipe failures.

WORK ORDERS

Service Ticket 2401

Customer Complaint: Compressor Keeps Cutting out on the Low-Oil Safety

Equipment Type: Low-Temperature Walk-in

You arrive at the small butcher shop to find the walk-in warm and the compressor off on oil safety. In talking to the owner you learn that the compressor is on top of the walk-in and the condenser is located on the roof. You also learn that the box had been set to hold the temperature at 40°F up until last month, when it was converted to a low-temperature walk-in with a 0°F setting.

When you inspect the walk-in and compressor, you see that the last technician had made all of the necessary changes and additions to the compressor and evaporator, including adding a defrost control.

Knowing that the compressor oil had to be in the system, you suspect that the evaporator is oil logged. You start the compressor and note that the oil level quickly rises in the compressor-oil sight glass. After a few minutes, the oil level is stable at ½ the oil sight glass. The system has the correct amount of oil.

When you examine the new evaporator, you note that the tech used the original piping. A quick check of the evaporator and compressor technical specification indicates that the low-temperature application has resulted in a lower refrigerant velocity in the suction line. You have determined that the refrigerant line is too large. The flow in the suction line is below 500 FPM.

Because the compressor is only a short distance above the evaporator, you can easily change the suction line with a smaller line to increase the velocity. Once you have replaced the suction line, the walk-in runs properly.

Service Ticket 2402

Customer Complaint: Poor Cooling

Equipment Type: Residential Split System in Third-Story Apartment, Condenser on Ground

A customer complains that their air-conditioning system cannot cool the apartment on a hot day. She notes that her neighbor two stories below has the same size apartment and the same type of unit, and it never has a problem. The lines run up three stories to get from the condenser on the ground to the evaporator in the attic.

The technician notices that even though the system has a vertical rise of 30 ft, the system has not been modified for a long-line application. The technician changes the fixed-restriction metering device to a hard-shutoff expansion valve, adds a liquid-line solenoid at the condenser, adds a hard-start kit, and charges the unit until the subcooling is high enough to overcome the pressure drop from the high vertical lift.

The system cools better, but it still must run longer than the downstairs neighbor's unit because of the added heat load of the upstairs apartment and the reduction in system capacity even after the modifications.

Service Ticket 2403

Customer Complaint: Multiple Compressor Failures

Equipment Type: Residential Split System on Ground Level with Lines Run Underground

A technician is asked to look at a system that has experienced multiple compressor failures. The house has a concrete-slab floor and an exposed-beam vaulted ceiling. The air-conditioner blower coil sits in the middle of the house, and the refrigerant lines run under the slab for about half of the length of the house. The compressor failures are always mechanical.

The technician decides that the compressors are probably dying from liquid slugs due to the long line set run underground. However, there is no alternate path for the lines due to the house construction. The technician suggests treating the system like a long-line application and also adding a suction accumulator to protect the new compressor. The additions of a liquid-line solenoid at the condenser and a hard-shutoff valve on the evaporator reduce the migration during the off cycle. The accumulator keeps the occasional slug from reaching the compressor. A hard-start kit is also added to help the compressor start against pressure. Two years after the modifications, the new compressor is still operating without incident.

Service Ticket 2404

Customer Complaint: Higher than Expected Power Bills

Equipment Type: New High-Efficiency Residential Split-System Change-out

A customer has just purchased a new 2-ton 16-SEER high-efficiency air conditioner to replace his 25-year-old system. To the customer's surprise, it costs about the same to operate as the old unit. The installing contractor changed both the indoor and outdoor sections but used the existing line set. The original line set had a ¼-in liquid line and a ⅝-in suction line. The technician notices that the new system has line stub-outs for a ⅜-in liquid line and a ⅞-in suction line. The installers used sweat reducers to connect the system up to the old line set. The technician checks the manufacturer's literature, and it specifies a ⅜-in liquid line and a ⅞-in suction line. The technician notices that the liquid line is warm leaving the condenser and cool before it enters the evaporator. The technician explains that the small size of the old line set is causing enough pressure drop to create flash gas in the liquid line and reduce the system capacity, which causes it to run longer. The system operates correctly after replacing the old lines with the correct size liquid and suction lines.

UNIT 26—REVIEW QUESTIONS

1. What are the two major functions of refrigerant piping?
2. What effect does oil have on a system's coils?
3. What problems are caused by oversized suction lines?
4. What problems are caused by undersized suction lines?
5. What can be done to prevent refrigerant from being drained or migrating back to the compressor during the off cycle?
6. Why is it important to keep the oil level balanced between parallel compressors?
7. What is liquid slugging?
8. How is the maximum pressure drop for a suction line determined?
9. Which piping pressure drop has the greatest effect on the compressor's compression ratio?
10. What are some differences between piping practices for air-conditioning systems and piping practices for commercial refrigeration systems?
11. Why are suction risers sized differently for systems with capacity control than for systems without capacity control?
12. Why are oil lifts needed in tall vertical refrigerant risers?
13. What is the most common range for minimum velocities in a horizontal line carrying refrigerant vapor?

14. What is the most common range for minimum velocities in a riser carrying refrigerant vapor?
15. Why is refrigerant oil flow through the liquid line not a problem?
16. What two factors affect the hot-gas line size?
17. What are the two common reasons the suction line is insulated?
18. Why may the liquid line need to be insulated?
19. Why might the hot-gas line need to be insulated?
20. How can flash gas be spotted in a liquid line?
21. What would be the pressure drop due to the weight of refrigerant in a 30-ft-tall liquid line containing R-410a refrigerant at 100°F?
22. If the condenser pressure of the above system is 300 psig, how much subcooling will be required to offset the pressure drop created by the vertical lift?
23. According to Figure 26-31, what would be the recommended support spacing for 2⅛-type L copper pipe?
24. Use Figure 26-25 to determine the refrigerant velocity through a ⅞-in suction line on a walk-in cooler with a capacity of 2 tons, an evaporator saturation temperature of 20°F, and a condenser saturation temperature of 100°F.
25. Use Figure 26-13 to determine the total equivalent length of a 50-ft section of 1⅜-in line with four 90° elbows and one angle valve.
26. An engineer wishes to select a suction line that will have a maximum pressure drop that is equivalent to a 2°F saturation temperature drop. What would the pressure drop be for an R-134a system operating with a 0°F saturation temperature?
27. What pressure drop should the engineer use on the nomograph if the equivalent length for the suction line in the above system is 75 ft?
28. What is considered a long-line application for residential air conditioning?
29. List the system modifications that need to be made for a long-line residential air-conditioning application.
30. What is the minimum recommended velocity for a 2⅝-in suction riser in a low-temperature system using R-22?

UNIT 27

Accessing Sealed Refrigeration Systems

OBJECTIVES

After completing this unit, you will be able to:

1. describe the different types of refrigeration service valves.
2. explain the operation of gauge manifold valves.
3. explain how to properly install and remove a gauge manifold set on manual service valves.
4. explain the operation of split-system installation valves.
5. explain how to properly install and remove a gauge manifold set on Schrader valves.
6. describe how to gain access to systems without service valves.

27.1 INTRODUCTION

One of the last things a service technician should do when troubleshooting a system is attach a set of gauges to the system. Each time a sealed system is accessed, there is a chance that some contaminants can be introduced to the system or refrigerant can be lost. However, there are times when a refrigeration system's operating conditions cannot be accurately assessed without accessing the system's refrigerant piping to determine the pressures. Knowing the temperature difference across the coils, the amperage, and the airflow all give the technician vital information; but sometimes a final determination of the problem cannot be accurately made without knowing the system operating pressures.

It is important to attach and remove a gauge manifold set properly. Understanding how to properly manipulate system access valves and install gauge manifolds is vital to the personal safety of the service technician. Improper technique can damage the system or injure the technician. Proper techniques should always be practiced so that they become a habit performed the same way each time.

SAFETY TIP

The proper personal protective equipment (PPE) for installing refrigeration gauges and manipulating valves includes safety glasses and gloves. When liquid refrigerant escapes into the atmosphere, it boils at extremely cold temperatures. Liquid refrigerant sprayed in the eyes can cause blindness, and sprayed onto the skin it can cause frostbite.

27.2 FACTORY-INSTALLED SERVICE VALVES

With the implementation of the Clean Air Act, all manufacturers are required to provide factory-installed service valves on Type II and Type III equipment. Type I equipment is only required to have a process stub. Commercial refrigeration and air-conditioning systems have had factory-installed service valves for years.

SERVICE TIP

You must put service valve caps back on all access ports. This is an EPA refrigerant management requirement. Failure to do so is a violation of EPA rules and regulations. Many service access valve caps have O-rings to seal the system. Be sure these O-rings are in place before installing the cap to ensure a proper seal. Caps that have a metal-to-metal seal must be tightened one-eighth turn with a wrench after they have been finger-tightened.

Factory-installed service valves may be a manually operated stem shutoff valve, as in Figure 27-1, or a Schrader-type valve. Schrader valves are spring-loaded valves similar to the valves used on car tires. They have a core that threads down into them (Figure 27-2). Refrigerant is allowed to flow into and out of the valve when the core is depressed. To

Figure 27-1 Manual three position service valve.

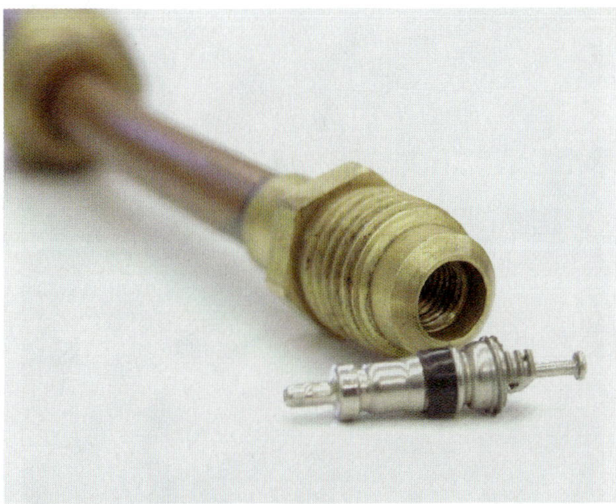

Figure 27-2 Schrader valve and core.

Figure 27-3 Schrader valves should always have their cap on when not in use.

prevent leaks, Schrader valves should always have their valve caps on when not in use (Figure 27-3).

Residential split-system equipment service valves are usually located on the suction line and liquid line (Figure 27-4). The valve on the suction line is used to read the low-side system pressure, and the valve on the liquid line is used to read the high-side system pressure. Most air-conditioning systems use front-seating split-system shutoff valves, as shown in Figure 27-5b. These valves are used to hold the system refrigerant in during shipping. The gauges connect to the Schrader valve that is built into them. These valves are normally opened when the unit is installed and then seldom used after that. It is not necessary to turn them to read pressure because the Schrader valve in them reads line pressure regardless of the valve position.

SAFETY TIP

A few residential split systems use manual stem valves as shown in Figure 27-5a. The gauge connection on these valves is not a Schrader valve. These valves must be opened to read pressure. More important, they must be *closed* before removing the gauges.

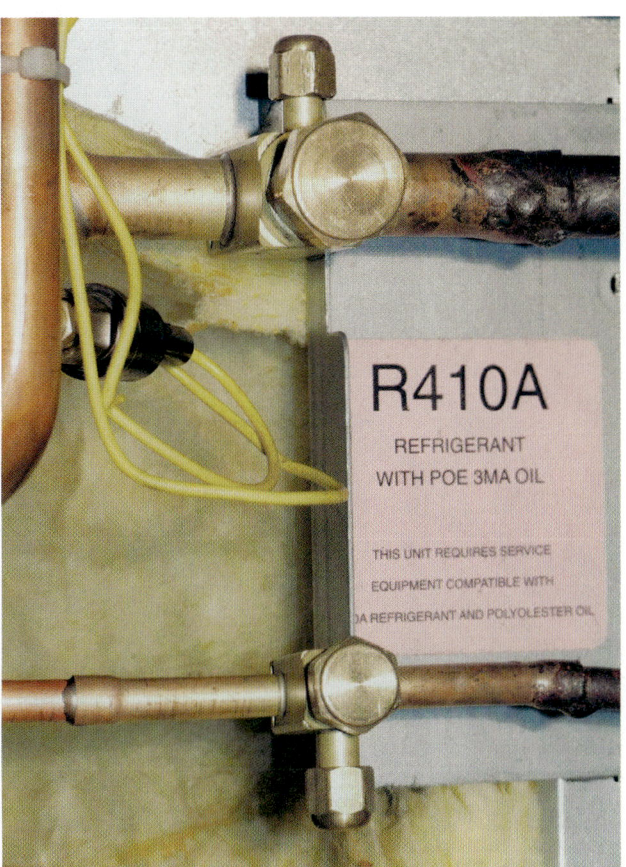

Figure 27-4 Residential split-system installation valves.

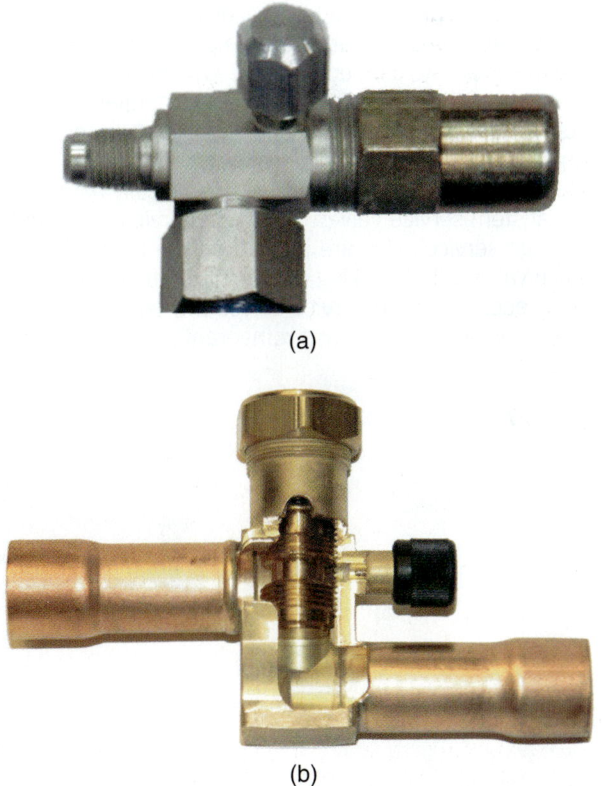

Figure 27-5 Service valves.

Figure 27-6 The arrow points to the port that is always on the low side of the system regardless of the operating cycle.

Figure 27-7 High-side and low-side Schrader valves on a packaged unit.

On heat pump systems, both lines are on the high side during the heating cycle. To allow technicians to read low-side pressures in heat, heat pump systems use a third valve that is always connected to the low side of the system regardless of the cycle (Figure 27-6). Residential packaged units typically have a low-side and high-side Schrader valve mounted on the outside of the equipment (Figure 27-7).

Commercial refrigeration systems typically have three manual stem service valves. A suction service valve and a discharge service valve are located on the compressor. A service valve called the king valve is mounted on the outlet of the receiver. These service valves are equipped with a gauge service port. Operating refrigerant pressures may be observed on the service gauge manifold when hoses are connected to these ports and the valves are cracked open.

27.3 THREE POSITION SERVICE VALVES

Three position service valves have three ports and two seats. The three ports include a middle port that connects to the compressor or system, a line port that connects to the refrigerant line, and a service port that is used for servicing the system (Figure 27-8). The refrigerant has three ways it can take through the valve: it can flow between the middle port to the line port, the middle port to the service port, or

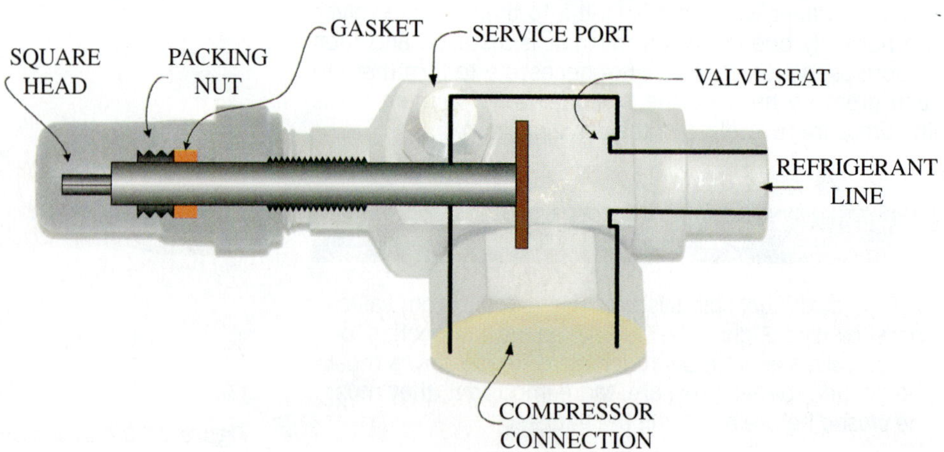

Figure 27-8 Three position system service valve.

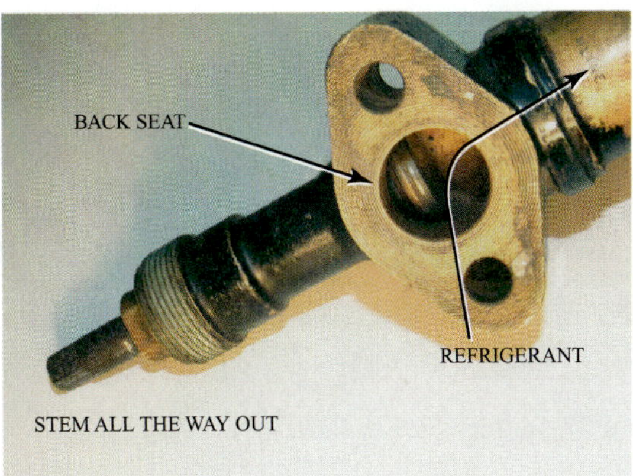

Figure 27-9 Three position service valve in the back-seated position.

between all ports. There is not a closed position on these valves because one of the two exit ports will always be open. Instead, the terms *back-seated* and *front-seated* are used to describe the valve positions.

The valve is back-seated when the stem is turned all the way out counterclockwise. This position is called back-seated because the valve is seated on the back port of the valve. In this position, the service port is closed but the middle port is open to the line port (Figure 27-9). Service valves should always be placed in the back-seated position before installing or removing gauges.

The valve is front-seated when the stem is turned all the way in clockwise. This position is called front-seated because the valve is seated on the front port of the valve. In this position, the line port is closed but the middle port is open to the service port (Figure 27-10). The front-seat position can be used to isolate refrigeration system components.

SAFETY TIP

Be sure that internal pressure in the compressor is relieved by recovery and vacuum procedures before attempting to remove an isolated compressor from the system. Pressure remains in the compressor even after the service valves are front-seated.

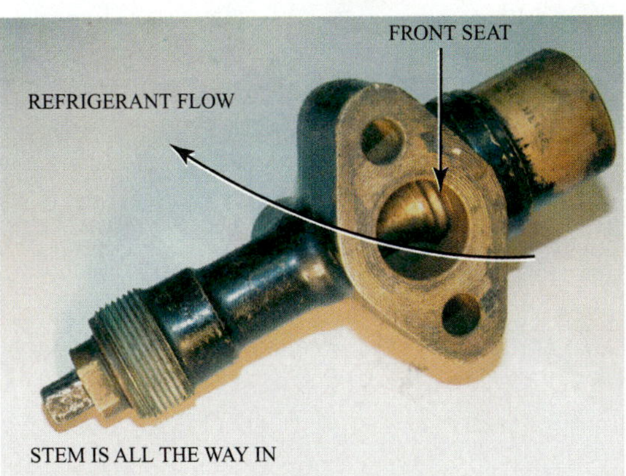

Figure 27-10 Three position service valve in the front-seated position.

Figure 27-11 When a service valve is cracked, the valve is just off of the seat, creating a small opening, or crack.

If the stem is positioned anywhere in between the back-seat and front-seat positions, all three ports will be open. Turning the valve one-half turn off of a fully seated position is called cracking the valve. When checking pressures, the valves are usually in the back-seat cracked position (Figure 27-11).

The stems on manual service valves should only be turned with a valve stem wrench (Figure 27-12). The corners of the valve stems round off easily when they are turned with adjustable wrenches or pliers. Once a stem is rounded it cannot be turned with the correct wrench (Figure 27-13). The cap for the valve stem and the cap for the service port should always be replaced after using manual service valves.

27.4 SCHRADER VALVES

Schrader valves provide a convenient method of checking system pressures or servicing the system where it is not economical or convenient to use manual stem service valves. The Schrader valve core, shown in Figure 27-14, is a spring-loaded device for positive seating. The valve is like those used on automobile tires, but the cores used in refrigeration valves are not the same as the cores used in car tires. The

Figure 27-12 Square-stem service valve and valve wrench.

Figure 27-13 This valve stem has been ruined by a careless technician using the wrong tool.

Figure 27-14 Schrader valve core.

rubber used in tire valves is not compatible with refrigerants and would dissolve if used on a refrigeration system.

The stem must be depressed to force the valve seat open against spring pressure. Refrigerant hoses with built-in core depressors must be used with Schrader valves. When a hose with a core depressor is connected to a Schrader valve, the valve core is pushed in and the valve opens (Figure 27-15). Ideally, the seal in the gauge hose will seat just as the depressor is pushed in and very little

Figure 27-15 The core depressor on the refrigerant hose pushes in the Schrader valve core when the hose is connected.

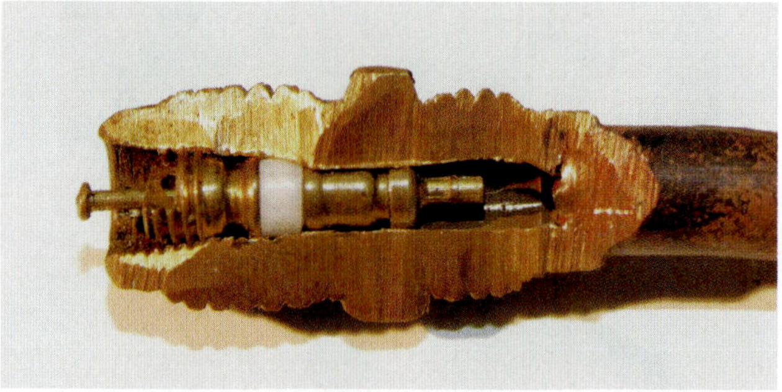

Figure 27-16 This cutaway shows how a Schrader valve works.

refrigerant will escape. Flow through a Schrader valve is limited because the core takes up most of the space in the center of the valve. This leaves only a small passage for refrigerant to go through (Figure 27-16).

If a valve core leaks, it can be replaced by using a core-removal tool to unscrew it. Some tools, such as the one in Figure 27-17, allow this to be done while the system is under pressure. All Schrader valves should have a leakproof cap on them when not in use. The cap can prevent refrigerant loss even if the valve core does leak.

27.5 SPLIT-SYSTEM INSTALLATION VALVES

Most new residential split systems come equipped with unit installation valves, as shown previously in Figure 27-4. These valves are used to hold the refrigerant charge in the

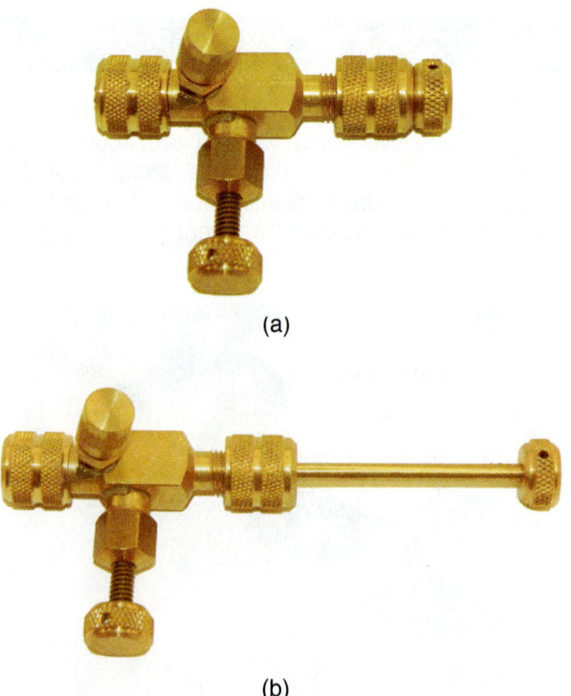

(a)

(b)

Figure 27-17 This tool can be used to change the core in a Schrader valve without losing the refrigerant in the system.

Figure 27-18 The condensing unit is shipped with the valve front-seated to hold in the system refrigerant.

outdoor unit while it is shipped. These valves have three ports: the port to the unit, the line port, and the Schrader valve service port. Even though they have three ports, they do not behave like standard manual service valves. No ports are closed in the back-seat position; the Schrader valve service port is always open to the line port regardless of the position of the valve.

These valves are used primarily for holding the system charge during shipping. The systems are shipped with the valve run all the way down clockwise (Figure 27-18). After connecting the refrigerant lines and indoor coil, the system is evacuated through the service port. The valve is turned counterclockwise to open it and allow the trapped refrigerant to flow throughout the system. Typically the valve is never used again because it does not control refrigerant flow to the service port.

SAFETY TIP

Be careful when opening split-system installation valves. The only thing that keeps the plug in the valve is a small lock ring at the top of the valve. Stop turning when the valve plug approaches the lock ring. The valve does not need to be tightened in the counterclockwise, or up, position. Overtightening the valve can cause the lock ring to pop out, and the plug will fly out under system pressure, with all the system refrigerant behind it.

27.6 GAUGE MANIFOLD VALVES

The valves on a gauge manifold are used to control the flow of refrigerant from either side of the manifold to the center. Figure 27-19 shows a cutaway view of a gauge manifold. It is not necessary to open the valves on the gauge set to read pressure. The parts above and below each valve are interconnected so the gauges will read pressure at all times when connected to the system. The valves open or close the path from each side to the middle port. When both valves are closed (front-seated) the center or utility port is isolated (Figure 27-20a).

Slightly opening a valve is called cracking the valve. Cracking open the low-side valve connects the low side and center hoses (Figure 27-20b). Some refrigerant can now flow from the center hose to the system. Fully opening the low-side valve opens the low-side port to the center port for full flow of refrigerant or for evacuating the system (Figure 27-20c).

With the low-side valve closed, cracking open the high-side valve will allow some refrigerant to flow from the high side to the center hose (Figure 27-20d). Fully opening the

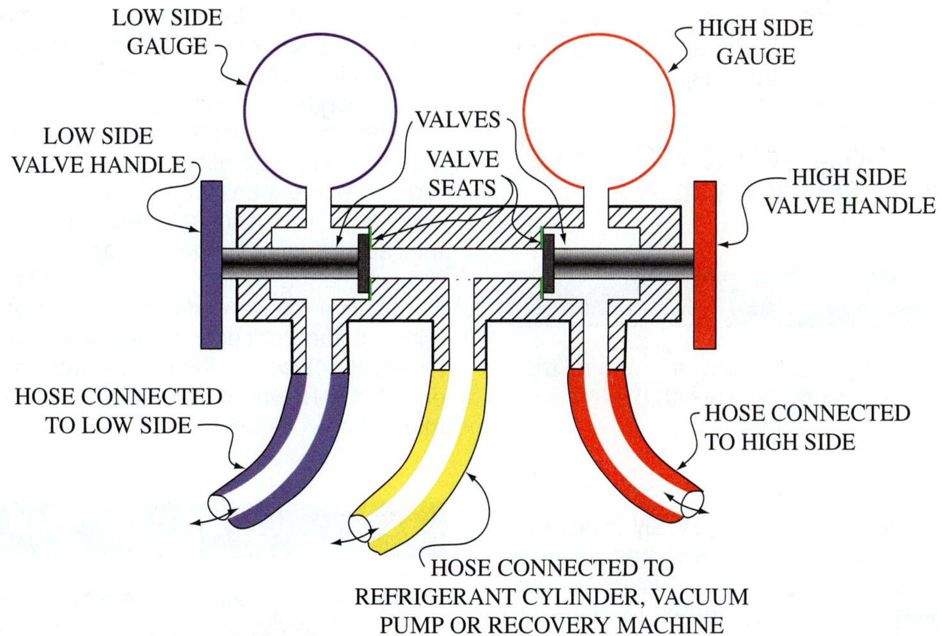

Figure 27-19 Cutaway of a gauge manifold.

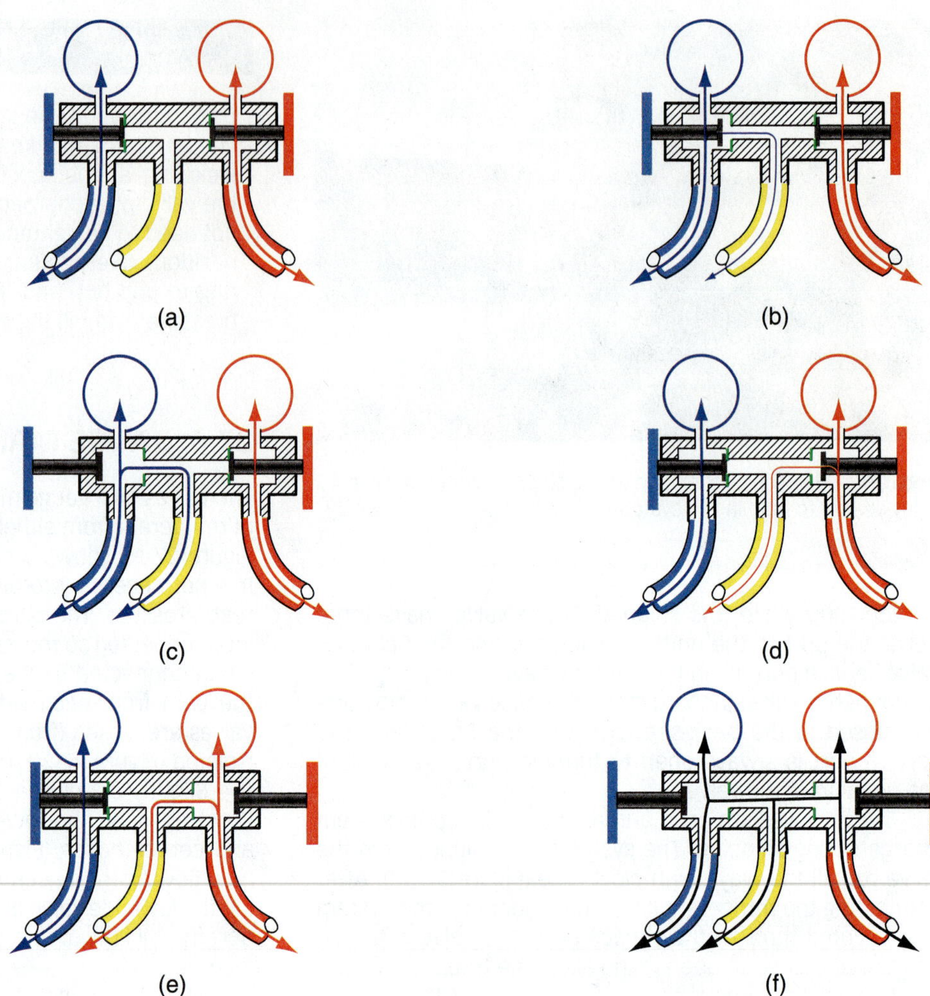

Figure 27-20 Gauge manifold valve positions.

high-side valve opens the high-side port to the center port to remove refrigerant from the system or for system evacuation (Figure 27-20e).

Opening both the low- and high-side valves opens both the low- and high-side ports to the center port. This valve position is used for system evacuation (Figure 27-20f).

27.7 CONNECTING A GAUGE MANIFOLD SET ON MANUAL SERVICE VALVES

SAFETY TIP

Always check the system data plate to confirm the gauge set you are using is compatible with the refrigerant in the system.

Locate the suction service valve. It is usually located on the compressor where the suction line attaches. Check to make sure the service valve stem is back-seated. This will ensure that no pressure is currently applied to the gauge port of the service valve. Remove the gauge port flare cap and attach the hose connected to the compound side of your gauges to the gauge port. Locate the discharge service valve. It is also located on the compressor, but it is the valve connected to the smaller discharge line leaving the compressor. Check to be sure the valve stem is back-seated. Remove the gauge port cap and attach the hose connected to the high side of the manifold to the gauge access port on the discharge service valve.

The gauge manifold and connecting lines should be purged to avoid system contamination from air and moisture. To purge a unit that does not operate in a vacuum, crack both the suction and discharge service valves. Open the gauge manifold valves one at a time and bleed refrigerant out the middle line until all the air is out. This usually just takes a few seconds. This will purge the gauge manifold and the connecting lines. Many technicians simply loosen the hoses at the manifold connections and purge each hose. However, this does not purge the manifold or middle line.

SAFETY TIP

Contact with the gases being purged should be avoided. Oil and liquid refrigerant will sometimes escape during the purging process, and liquid refrigerant can cause frostbite.

The high side is used to purge both sides of the manifold on systems whose low side is likely to be in a vacuum. To purge from the high side only, connect the lines as before, but crack only the high-side service valve. Open the low-side and high-side manifold valves at the same time and loosen the hose connection at the low-side service valve. This will purge everything but the middle hose. It can then be purged separately by loosening the middle hose at the plug connection. This procedure will work on any system with manual service valves but is only necessary when the low side is likely to be in a vacuum.

SERVICE TIP

With most refrigerants, it is common for the standing pressure on the low side of a system to exceed the normal range of the compound gauge. The retard range protects the gauges from damage when a pressure above their normal range of operation is connected to them. Gauges for R-134a and R-22 typically have a retard range of up to 350 psig.

Some technicians purge the hoses and gauges from a separate refrigerant tank. This has the advantage of not losing any of the system charge. It has the disadvantage of requiring a drum of refrigerant even if you are just checking the pressures. To use this method, connect your gauges as before except connect the middle leg to the refrigerant drum. Do not open the service valves. Crack the tank valve and open both manifold valves. Loosen the refrigerant hoses at both the suction and discharge service valves and purge each individually. After both hoses are purged, close the valve on the refrigerant cylinder and close the manifold valves.

27.8 REMOVING A GAUGE MANIFOLD SET FROM MANUAL SERVICE VALVES

The simplest way to remove the gauge set is to back-seat both the service valves. This should prevent any further escape of refrigerant into the gauges. There is still a considerable amount of refrigerant in the gauges and hoses that needs to be carefully released. Generally it is easier to release the refrigerant through the middle leg. Disconnect the middle hose at the plug connection and direct it into an empty can or other safe area. Slowly open each manifold valve, one at a time, until all the refrigerant has been released from the gauges.

There is a way to remove the gauges and avoid losing quite as much refrigerant. Leave the unit running. Instead of back-seating both service valves, back-seat only the discharge service valve. Crack open both manifold valves. This will allow the high-pressure gas in the discharge side of the manifold to pass into the low side of the refrigeration system. The pressure on the gauges will then be equal to the low-side operating pressure. This is the lowest pressure any part of the system will ever have on it. After the pressures in the gauge manifold have come down to the operating low-side pressure, back-seat the suction service valve. The pressure trapped in the gauges can now be bled off through the middle hose, but now the trapped pressure is significantly lower.

SAFETY TIP

If an error or equipment failure causes an unexpected release of refrigerant, keep your hands and body out of the refrigerant spray. Do not try to replace a hose or valve cap on a valve that has refrigerant escaping. Liquid R-410a is approximately –60°F at atmospheric pressure. It does not take long to get a nasty frostbite burn at that temperature.

27.9 CONNECTING A GAUGE MANIFOLD SET TO SCHRADER VALVES

Schrader valves automatically open when a hose with a core depressor is connected to them. The service technician normally does not have control over when the valve opens and closes. When hoses are connected, the valve opens; when hoses are removed, the valve closes.

SERVICE TIP

Typically one end of a refrigeration hose is straight and the other end is angled to allow easier connection to service valves. The straight end connects to the manifold, and the angled end connects to the valves on the unit. Only the angled end contains a Shrader core depressor. When connecting to Shrader valves, the angled end must be connected to the valve to get a reading.

Frequently the valve will open before the seal is made and release refrigerant. If the Schrader valve is on the liquid line, the liquid that comes out will be extremely cold as it flashes off at atmospheric pressure. Occasionally the hose will seat before the core is depressed, and no pressure will be read. Some manufacturers' hoses have adjustable core depressors that can help either of these situations (Figure 27-21).

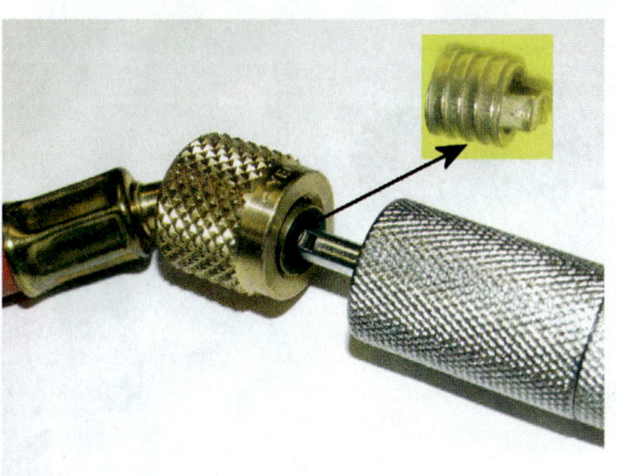

Figure 27-21 Adjustable core depressor.

SAFETY TIP

When servicing a system with R-410a, always use gauges designed for the higher pressures produced by R-410a systems. NEVER connect gauges designed for lower-pressure refrigerants to an R-410a system!

One way of minimizing the risk to the technician is to install the gauges on the Schrader valves with the system off. This reduces the pressures available at the high-side valve when the hose is connected and reduces the chance of liquid refrigerant escaping. Connecting with the system off also makes purging the gauges easier. The gauges may be purged by opening the valves one at a time and releasing the air through the center hose.

Tools are available that give the technician more control over the valve. With control over when the valves open and close, gauges may be safely connected to Schrader valves with the unit in operation. A valve-core-removal tool like the one shown in Figure 27-17 allows gauges to be connected before opening the valve. Then the tool is used to remove the core, opening the system to the gauges. This also has the advantage of improving flow through the valve. This type of core-removal tool can be used to change a Schrader valve core without recovering the system refrigerant.

A simpler tool, the thumbscrew core depressor, also allows the connection of gauges before the valve is opened (Figure 27-22). Turning in the thumbscrew depresses the valve core and opens the valve. This process is reversed to remove the gauges. The thumbscrew is turned out, allowing the valve to close. After the valve is closed, the gauges are removed.

27.10 REMOVING A GAUGE MANIFOLD SET FROM SCHRADER VALVES

When a hose is disconnected from a Schrader valve, the refrigerant that is trapped in the hose and gauges comes back out. This can be a surprising amount of refrigerant when the high side is connected to the liquid line. Positive-shutoff hoses can drastically reduce the amount of refrigerant released when the gauges are disconnected. Positive-shutoff devices include check valves on the end of refrigerant hoses, manually operated ball valves on the ends of hoses, and adapters that can be used with standard hoses. Available adapters include check valves, mechanisms like the quick couplings used on air hoses, and ball valves (Figure 27-23).

Figure 27-22 Thumbscrew core depressor valve. *(Courtesy of Ritchie Engineering Company, Inc., Yellow Jacket Products Division)*

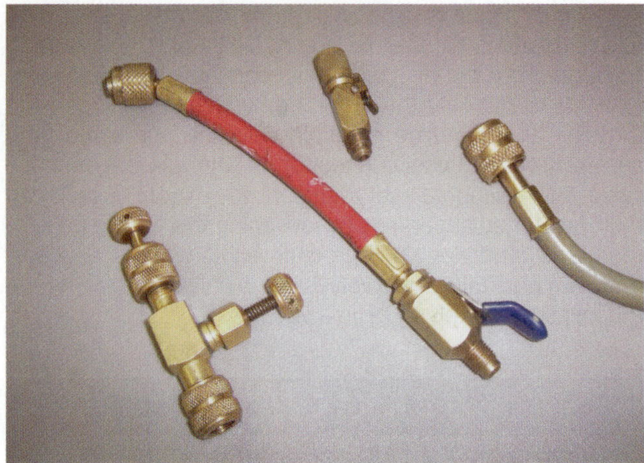

Figure 27-23 Positive-shutoff adapters make connecting to Schrader valves easier.

The check valves built into the end of a hose keep refrigerant from flowing backward out of the hose (Figure 27-24). Refrigerant will be trapped in the hoses after the gauges have been disconnected using a positive-shutoff valve. Most of the trapped refrigerant can be put back in the system if the proper removal procedure is followed.

Using a positive-shutoff hose, disconnect the high side with the unit still operating. Now crack both the gauge manifold valves to allow the trapped high-pressure refrigerant into the low side. Last, disconnect the low side with the unit still operating. The remaining amount of refrigerant in the hoses can be safely and legally vented. Note that the gauges must be purged when they are connected, or part of what is going back into the system will be air and contaminants.

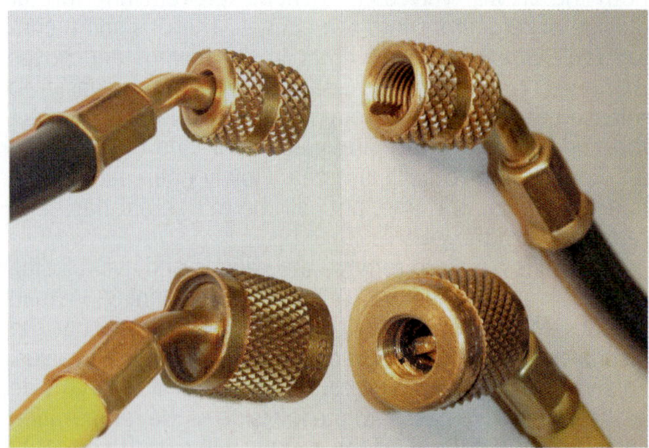

Figure 27-24 The hoses at the top of the picture have a standard Schrader core depressors. The hoses on the bottom of the picture have automatic check valves built in that keep the refrigerant in the hose from coming back out when the hose is removed.

Three simple steps will reduce your exposure to refrigerant spray if positive-shutoff hoses or adapters are not used. First, wear gloves and safety glasses. It is always a good idea to wear gloves and safety glasses when handling refrigerant, but it is especially important when using Schrader valves. Disconnect the low side with the system operating because the low-side pressure is lower with the unit on. Turn the system off and let the system sit for several minutes before disconnecting the high side, because the high-side pressure is lower with the unit off.

27.11 SECURING REFRIGERANT SYSTEMS

Tragically, an increasing number of people have killed themselves by intentionally inhaling refrigerant, a practice called "huffing." The victims are often teenagers looking for a quick thrill from a readily available substance. The "high" is from dizziness due to lack of oxygen and is certainly not worth risking your life. Although the refrigerant in most systems is considered nontoxic, it can still asphyxiate you if you breath in too much. Since refrigerant is heavier than air, once a large quantity gets in your lungs, it can be difficult to displace, and you suffocate. The bottom line is people die.

Safety caps are available that restrict acces to refrigerant using valves caps that require a special tool to remove. The idea is to limit access to the refrigerant to professionals servicing refrigeration systems. Figure 27-25 shows safety caps that are used to secure refrigeration access valves. They can only be removed using the tool shown in Figure 27-26.

CODE TIP

Both the 2009 International Mechanical Code (IMC) 1101.10 and the 2009 International Residential Code (IRC) M1411.6 require tamper-resistant valve caps on refrigerant access ports located outside.

Figure 27-25 Safety locking caps for refrigeration access valves.

Figure 27-26 This tool is required to remove the locking caps.

27.12 ACCESSING SEALED SYSTEMS

Most small appliances, such as refrigerators or window units, do not have any valves installed at the factory. Instead, the systems are evacuated and charged at the factory through process tubes. A process tube, or process stub, is a copper tube connected to the system for the sole purpose of evacuating and charging the unit at the factory. The process tubes can be on the compressor or on one of the system lines (Figure 27-27). After the system is evacuated and charged at the factory, the process tube is pinched off and brazed shut.

The only way to gain access to these refrigeration systems is to use a valve that pierces a hole in the tubing. It is better to avoid entering a sealed system unless it is absolutely necessary, because poking a hole in a sealed system can create problems. The technician could literally be installing a leak. Small systems sometimes only hold a few ounces, and putting gauges on them can remove enough refrigerant to make the system undercharged. It is definitely not a good idea to install piercing valves on small sealed systems just to perform a routine check. However,

Figure 27-27 Process tube on hermetic compressor.

when the refrigeration system must be accessed, a piercing valve or piercing pliers are required.

SAFETY TIP

You must wear eye protection and gloves anytime you access a refrigeration system. Liquid refrigerant accidentally released from a pressurized system can be at an extremely low temperature that can cause blindness and skin burns. Never position your face or head in a direct line with an access port as you are attaching or removing hoses.

27.13 PIERCING VALVES

One of the easiest devices available for sealed-system access is the saddle piercing valve, or tap-a-line. These piercing valves are clamped to the tubing and sealed by a bushing gasket, and then they pierce the tube with a tapered needle. Most contain some sort of shutoff control. The technician should keep in mind that these valves should be used to gain temporary access to a hermetically sealed system for checking system operating pressures or for pressurizing for leak testing. The piercing valve shown in Figure 27-28 allows quick access to system pressures to immediately start diagnosing the refrigeration problem.

SAFETY TIP

Before brazing a Schrader valve onto a existing refrigerant line, all of the refrigerant in the system must be recovered. Any refrigerant that is allowed to enter the torch flame will produce a noxious gas. Breathing this gas can be dangerous and could even cause lung damage. Remove the Schrader core before brazing to keep it from becoming damaged.

Bolt-on piercing valves are bolted on the line, then the piercing needle is run down to pierce the copper tubing (Figure 27-29). Refrigerant flows through the hole pierced by the needle to the service port on the valve.

Figure 27-28 Bolt-on piercing valve.

Figure 27-29 The needle pierces the copper tubing to gain access to the refrigeration system.

Bolt-on piercing valves should be removed once the source of the sealed-system malfunction has been located. The only way to do this is to recover the refrigerant from the system through the piecing valve. After the refrigerant has been recovered, a permanent valve may be brazed on. One way to do this is to install the piercing valve on a process tube. After the refrigerant is recovered, the process tube can be cut and a permanent valve brazed onto the process tube. Figure 27-30 shows Schrader valves made for field application.

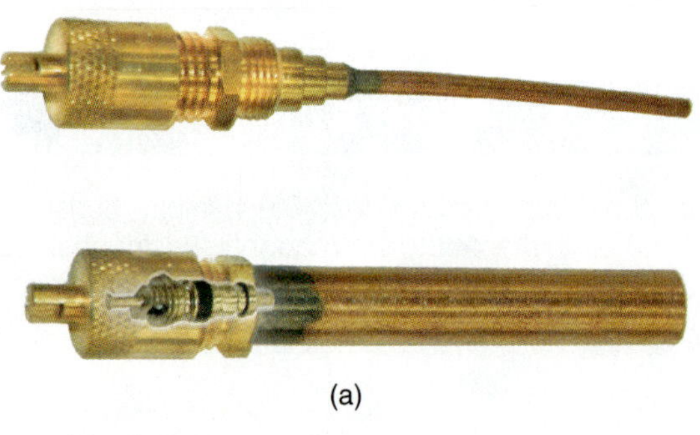

(a)

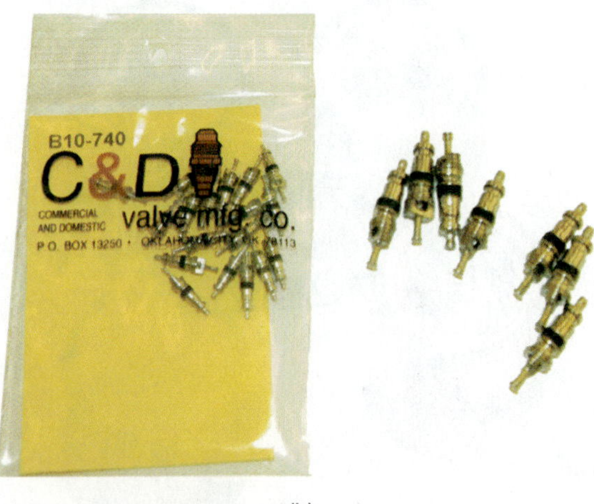

(b)

Figure 27-30 (a) Pigtail-type Schrader valve; (b) Schrader valve replacement cores.

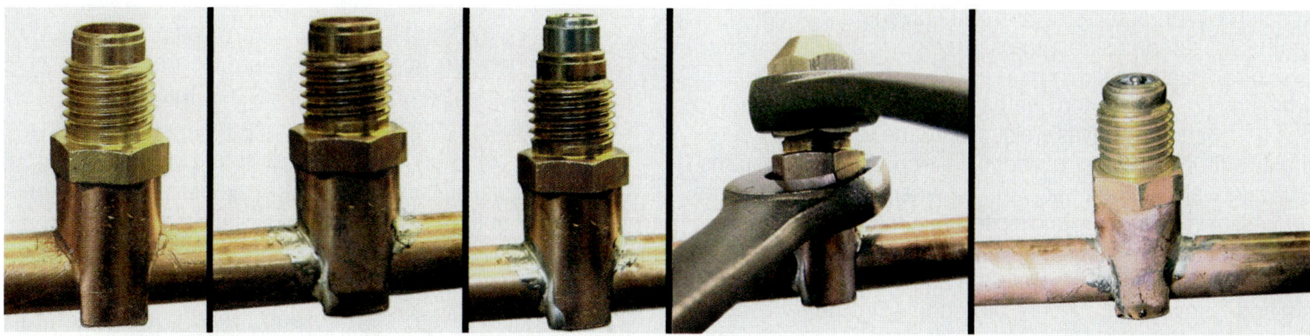

Figure 27-31 Steps in installing a braze-on piercing valve. Braze-on piercing valves can be used as both a means of gaining access to a sealed system as well as a permanent valve.

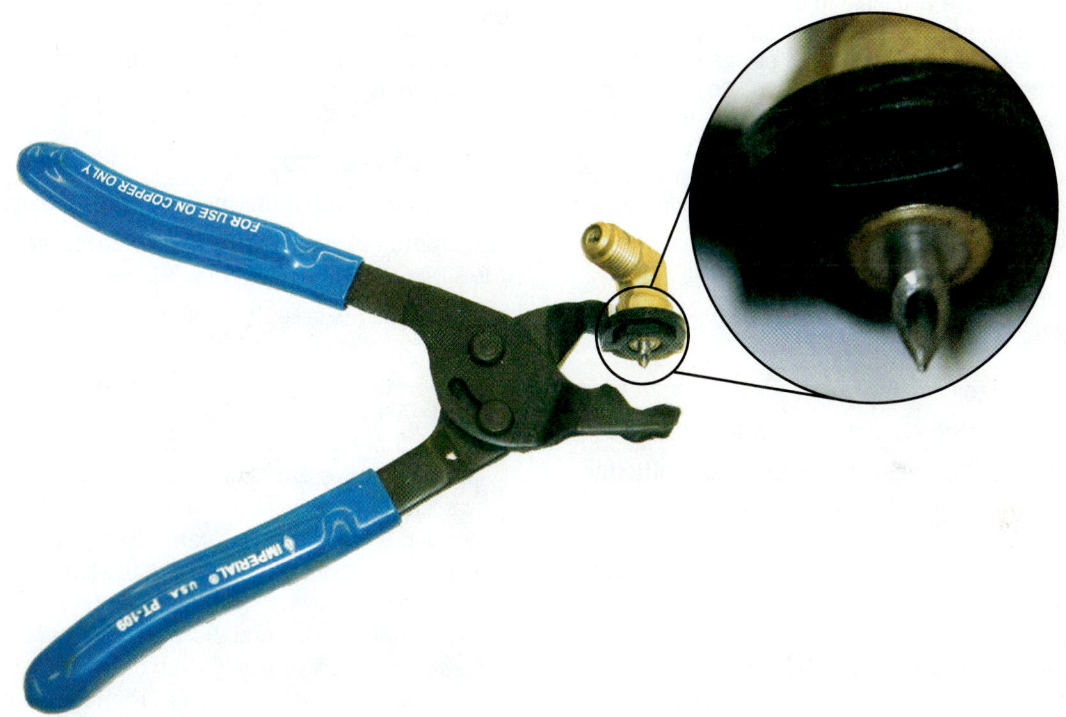

Figure 27-32 These piercing pliers can be used to gain access to sealed systems.

Braze-on piercing valves are available, such as the one in Figure 27-31. They function like a Schrader valve after the tubing is pierced. Braze-on piercing valves can be used for initial access and left on the system. One safety concern with braze-on piercing valves is that they are brazed on while the system is still under pressure. Brazing on a system with pressure still in it is generally considered unsafe. If the tubing is overheated, it can rupture, allowing refrigerant and oil under pressure to blast out unexpectedly. The oil spray can ignite and create a flamethrower effect.

Another tool for accessing sealed systems is a set of piercing pliers (Figure 27-32). These can pierce the tubing and allow temporary access for system diagnosis and refrigerant recovery. Since the pliers obviously must be removed, they can only be used if the technician plans to recover the system refrigerant.

After piercing the system and recovering the refrigerant, a Schrader valve can be installed. Schrader valves are available for field installation in a variety of forms as seen in Figure 27-30. The core should be removed from the valve before brazing to avoid melting the plastic seals.

UNIT 27—SUMMARY

Service valves are used to gain access to the refrigeration system. Type II and Type III systems all have service valves installed at the factory, and Type I systems have process tubes. Commercial refrigeration systems typically use manual stem service valves on the suction and discharge sides of the compressor and on the liquid receiver. The normal operating position of a manual stem service valve is back-seated. In the back-seated position, the service port is closed. System pressures can be read in the back-seat cracked position, in which the valve is turned one-half turn off of the back-seated position. Schrader valves are similar to the valves on car tires and are found on most residential air-conditioning systems. Schrader valves automatically open when a hose with a core depressor is connected and close when the hose is removed.

Type I systems must be accessed using piercing valves. Bolt-on piercing valves should be used for system diagnosis and refrigerant recovery but should not be left on the system. Braze-on piercing valves can be used for access and may be left on the system. Care should be taken when installing a braze-on piercing valve because of the inherent danger of brazing on a system under pressure.

It is important to use gauges designed for the system refrigerant. Gauges are available in a high-pressure range suitable for R-410a and in a lower-pressure range suitable for everything else. Never use gauges designed for lower-pressure refrigerants on R-410a systems. Positive-shutoff hoses or adapters help technicians control the flow of refrigerant in and out of the gauges, especially when using Schrader valves. The saturation temperature of common refrigerants can be read on the gauges by looking at the colored scales.

WORK ORDERS

Service Ticket 2701

Customer Complaint: Air Conditioner Freezes Up

Equipment Type: Packaged Terminal Air Conditioner (PTAC) Using R-22 Refrigerant

A service technician for the Rest Easy Motel chain is asked to look at a PTAC in one of the rooms because it cools poorly and freezes over. Like most PTACs, this one has no service valves of any kind. The technician knows that the two most common causes of air-conditioning evaporators freezing over are poor airflow and refrigerant undercharge. The technician decides to check the airflow first. The air filter is missing and the evaporator is very dirty. This is most likely causing reduced airflow over the evaporator. After cleaning the coil, the unit airflow is noticeably improved and the evaporator no longer freezes. The technician elects not to install a piercing valve and keep the system sealed. Instead, the system operation is checked by examining the system amp draw, airflow, and temperature drop across the evaporator coil.

Service Ticket 2702

Customer Complaint: Unit Operates but Does Not Cool

Equipment Type: Packaged Terminal Air Conditioner (PTAC) Using R-22 Refrigerant

A service technician for the Rest Easy Motel chain is asked to look a PTAC in one of the rooms because it operates but does not cool. Like most PTACs, this one has no service valves of any kind. The system operation is checked by examining the system amp draw, airflow, and temperature drop across the evaporator coil. The amp draw of the compressor is less than half of the normal RLA rating. The coil appears clean, the airflow is normal, but the supply air is the same temperature as the return air; no cooling is taking place. The technician decides to install a bolt-on piercing valve on the low-side process tube. The reading on the low side with the compressor operating is 150 psig. The technician believes the compressor is bad but decides to install a piercing valve on the high-side process tube to be certain. The pressure on the high side with the compressor operating is also 150 psig. The compressor is condemned, the system refrigerant is recovered, the bolt-on piercing valves are removed, Schrader valves are brazed in, a new compressor and liquid line filter are installed, the system is evacuated to 500 microns, and a charge is weighed in.

Service Ticket 2703

Customer Complaint: Unit Does Not Cool Adequately

Equipment Type: Commercial Refrigeration Unit on Walk-in Cooler with R-134a

A technician is called to look at a walk-in cooler whose capacity has noticeably declined. A check of the system pressures at the suction and discharge service valves reveals that the suction pressure is very low and the high-side pressure is just slightly low. Both the system superheat and subcooling are high. The technician looks for system restrictions and finds a sweating liquid-line filter drier. The technician determines that the filter drier is restricted and decides to change the filter drier but does not want to recover all of the refrigerant. Instead, the king valve is front-seated and the system is operated until the filter drier no longer sweats and the pressure on the line leaving the king valve is 0 psig. The technician turns off the unit and waits to make sure the pressure does not go above 0 psig. The old filter drier is replaced with a new filter drier, the system is evacuated from the suction service and king valves to 500 microns, the king valve is back-seated, and the system is returned to service.

Service Ticket 2704

Job: Check Oil for Acidity on Semihermetic Compressor

Equipment Type: Commercial Refrigeration System with Semihermetic Compressor

A-1 Refrigeration changed the compressor in a commercial refrigeration system last week. A technician is sent to collect a small oil sample and perform an oil acid test on the oil to see if the oil is contaminated. The technician does not want to recover all the system refrigerant but knows that the refrigerant pressure must be taken off of the compressor before opening the oil plug, so the technician decides to isolate the compressor using the suction and discharge service valves. First the system is operated long enough to warm up the crankcase and boil out any residual refrigerant from the oil. The system is turned off and locked out. Then gauges are installed on the compressor and both the service valves are front-seated. The compressor is now isolated from the rest of the system. A recovery system is used to recover the small amount of refrigerant in the compressor. The crankcase plug is removed, a small oil sample taken, the plug replaced, and the oil is tested. The oil is clean, so the technician pulls a 500-micron vacuum on the compressor through the suction and discharge service valves. The service valves are back-seated and the system is put back in operation.

UNIT 27—REVIEW QUESTIONS

1. What is the term for slightly opening a valve?
2. List three precautions when handling a gauge manifold set.
3. What can technicians do to reduce refrigerant loss through Schrader vales when they are not being used?
4. Why is it not always a good idea to install a gauge set and check system pressure?
5. What type of units are required to have service valves?
6. What type of units are only required to have process stubs?
7. Explain the difference between the back-seated position and the front-seated position on manual stem service valves.
8. When using manual stem service valves, describe how to position both the service valves and the gauge manifold valves when reading system pressures.
9. Why are positive-shutoff hoses helpful when using Schrader valves?
10. What is the purpose of the saddle or piercing valve?
11. What is the retard range on compound gauges for?
12. What personal protective equipment (PPE) should be worn when using gauge manifolds?
13. What can be done to reduce the risk of exposure to liquid refrigerant spray?
14. What is the difference between the two ends of a refrigeration hose?
15. What service valves are normally found on a commercial refrigeration system?
16. Why do split-system heat pumps have three service valves?
17. Describe how to remove a gauge manifold set from a system and put most of the trapped refrigerant back into the system.
18. Describe how to purge a gauge manifold set using only the high-side service valve.
19. Describe how to purge a gauge manifold set when connecting to Schrader valves.
20. What tool is used to turn the stems on manual stem service valves?
21. Describe how split-system installation valves are used.
22. Why are bolt-on piercing valves not supposed to remain on systems permanently?
23. What is a process tube?
24. Why do both the International Mechanical Code and the International Residential Code require locking valve caps on all outdoor refrigerant access ports?

UNIT 28

Refrigerant Management and the EPA

OBJECTIVES

After completing this unit, you will be able to:

1. discuss the significance of the Montreal Protocol.
2. outline the major provisions of Title VI, section 608, of the 1990 Clean Air Act.
3. determine the EPA-specified recovery level for a given refrigerant and system.
4. discuss the requirements for EPA certification.
5. explain the different types of refrigerant recovery.
6. discuss proper refrigerant-recovery technique.
7. discuss the difference between recovered, recycled, and reclaimed refrigerant.

28.1 INTRODUCTION

Understanding the laws and regulations designed to reduce the environmental impact of refrigerants is now a crucial part of an air-conditioning technician's responsibility. Technicians must now pass a certification exam to demonstrate their understanding of these regulations before they are allowed to handle refrigerant. Reducing refrigerant emissions through good service practices, leak detection, and refrigerant recovery is a big part of any HVACR technician's job.

28.2 THE MONTREAL PROTOCOL

The problems of ozone depletion and global warming affect the whole earth's atmosphere and cannot be addressed by the unilateral action of one country. In 1985, the United Nations sponsored a treaty called the Vienna Convention for the Protection of the Ozone Layer seeking international cooperation to address concerns about ozone depletion. The participants agreed to study the problem, share information, and work toward a solution. In 1987, twenty-nine signatories to the Vienna Convention developed the Montreal Protocol, which established target dates for the phaseout of ozone-depleting substances. To date, 191 countries have signed the Montreal Protocol.

28.3 1990 CLEAN AIR ACT

The United States Congress passed an extensive revision to the Clean Air Act in 1990. The legislation addresses many aspects of air pollution and air quality, including stratospheric ozone depletion. The Clean Air Act is divided into six major sections, called titles. The last section, Title VI, covers Stratospheric Ozone Protection. There are eighteen sections in Title VI, each dealing with specific aspects of the effort to address ozone depletion. These sections are numbered by the title and section. Section 608 is the eighth section of Title VI.

The Clean Air Act establishes the goals to be achieved and sets target dates. It does not set forth specific practices and procedures; those are left to the EPA to establish and enforce. The legislation gives the EPA the authority to regulate ozone-depleting chemicals and their replacements, including refrigerant. Currently, the maximum possible fine for violation of the Clean Air Act is $37,500 per incident per day. Note that multiple violations can add up to a substantial amount of money.

28.4 SECTION 608

Section 608 of Title VI establishes regulations to reduce emissions of ozone-depleting substances from everything except motor vehicle air conditioners, which are covered in section 609. Residential air conditioning, commercial and industrial refrigeration, and even transport refrigeration are covered in section 608. Regulations in section 608 establish:

- A prohibition on venting ozone-depleting refrigerants
- Technician certification
- Restrictions on the sale of ozone-depleting refrigerants
- Reclaimed refrigerant standards
- Certification of recycling and recovery equipment
- Refrigerant recovery levels
- Repair of substantial leaks in equipment with a charge greater than 50 pounds
- Safe disposal of appliances containing ozone-depleting refrigerant
- Recordkeeping

28.5 VENTING PROHIBITION

Effective July 1, 1992, individuals are prohibited from knowingly venting ozone-depleting refrigerants into the atmosphere while maintaining, servicing, repairing, or disposing of air-conditioning or refrigeration equipment or appliances. Four types of releases are permitted under the prohibition:

1. "De minimus" quantities of refrigerant released in the course of making good-faith attempts to recapture and recycle or safely dispose of refrigerant. *De minimus* is

Latin for the minimum, or a little. There is no specific quantity set as de minimus. Instead, de minimus is defined as the amount released while following the manufacturer's instructions for the recovery device being used.

2. Refrigerants emitted in the course of normal operation of air-conditioning and refrigeration equipment. Normal operating releases would include releases due to mechanical purging and small leaks. However, EPA requires the repair of substantial leaks in large equipment.
3. Releases of CFCs or HCFCs that are not used as refrigerants. For instance, mixtures of nitrogen and R-22 that are used as holding charges or as leak test gases may be released, because in these cases, the ozone-depleting compound is not used as a refrigerant. However, a technician may not avoid recovering refrigerant by adding nitrogen to a charged system; before nitrogen is added, the system must be evacuated to the EPA-specified recovery level. Otherwise, the CFC or HCFC vented along with the nitrogen will be considered a refrigerant. Similarly, pure CFCs or HCFCs released from appliances will be presumed to be refrigerants, and their release will be considered a violation of the prohibition on venting.
4. Small releases of refrigerant that result from purging hoses or from connecting or disconnecting hoses to charge or service appliances will not be considered violations of the prohibition on venting. However, recovery and recycling equipment manufactured after November 15, 1993, must be equipped with low-loss fittings.

28.6 TECHNICIAN CERTIFICATION

Since November 14, 1994, anyone who performs maintenance, service, repair, or disposal that could be reasonably expected to release refrigerants into the atmosphere must be certified. Examples of actions that require certification include connecting gauges, adding or removing refrigerant, or opening a sealed refrigeration system. Technicians performing jobs that do not involve the refrigeration system, such as hanging ductwork or wiring units, are not required to be certified. Four types of certification have been developed to address different types of equipment:

- Type I, for servicing small appliances
- Type II, for servicing high-pressure and very-high-pressure appliances
- Type III, for servicing low-pressure appliances
- Universal, for servicing all types of appliances except motor vehicle air conditioners

Type I: Small Appliances

A small appliance is defined as a product that is fully manufactured, charged, and hermetically sealed in a factory with 5 lb or less of refrigerant. Examples of small appliances include domestic refrigerators and freezers, window air conditioners, packaged terminal air conditioners and heat pumps, dehumidifiers, under-the-counter ice makers, vending machines, and drinking water coolers. Technicians working on small appliances must have either Type I or universal certification.

Type II: High-Pressure Appliances

High-pressure appliances use an ozone-depleting refrigerant with a boiling point at atmospheric pressure between –58°F and 50°F. This category covers the widest of variety of appliances, including residential air conditioning, commercial air conditioning, commercial refrigeration, and transportation refrigeration.

Type II: Very-High-Pressure Appliances

Very-high-pressure appliances use an ozone-depleting refrigerant with a boiling point at atmospheric pressure below –58°F and are typically cryogenic scientific equipment operating at extremely low temperatures. These appliances represent a very small portion of the refrigeration industry.

Technicians working on either high-pressure or very-high-pressure appliances must have either Type II or universal certification.

Type III: Low-Pressure Appliances

Low-pressure appliances use an ozone-depleting refrigerant with a boiling point at atmospheric pressure above 50°F. The low-side pressure on these systems usually operates in a vacuum. Large chillers are the only common type of low-pressure systems. Technicians working on Type III low-pressure appliances must have either Type III or universal certification.

MVAC-like Appliances

An MVAC-like appliance is an air conditioner in a non-road vehicle, such as agricultural and construction vehicles. These systems are generally identical to those found in cars and trucks, but the vehicles do not operate on the public road system. A farm tractor with an air-conditioned cab is an example of an MVAC-like appliance (Figure 28-1). Technicians working on MVAC-like appliances may be certified under either section 608 or section 609.

Figure 28-1 Large agricultural machinery, like this cotton picker with an air-conditioned cab, is classified as an MVAC-like appliance. *(Dmitry Kalinovsky/Shutterstock.com)*

Technician Certification Exam

To become certified, technicians are required to pass an EPA-approved test given by an EPA-approved certifying organization. There are four parts to the exam:

- Core
- Type I
- Type II
- Type III

Each section has twenty-five questions. Sections are scored individually, and each section is passed or failed as a section. Technicians must correctly answer eighteen questions to receive credit for a section and must pass the Core and at least one other section to become certified. The number of sections passed determines their certification, and a universal certification is issued to people who pass all four sections.

TECH TIP

Although universal certification implies that technicians are certified to work on anything, universal certification does not cover MVAC systems, motor vehicle air-conditioning systems. A section 609 certification is required to work on motor vehicle air-conditioning systems.

28.7 REFRIGERANT SALES RESTRICTION

Since November 14, 1994, the sale of refrigerant in a container of any size has been restricted to certified technicians. The sales restriction covers refrigerant containing parts as well as refrigerant in cylinders or drums.

What Is Not Covered

Fully assembled appliances containing refrigerant such as household refrigerators, window air conditioners, and packaged air conditioners are not included in the sales restriction. HFC refrigerants are also not covered. Uncertified people may purchase HFC refrigerant such as R-134a. Precharged split systems may also be purchased by uncertified individuals, providing all of the components are sold at one time. Components of precharged systems are still subject to the sales restriction if sold individually.

TECH TIP

While it is legal for uncertified individuals to purchase an entire precharged split system, they may not install the system. Connecting the refrigerant lines requires a certified technician.

28.8 RECOVER, RECYCLE, RECLAIM

Refrigerant that has been removed from a system contains impurities. What is done with refrigerant removed from a system depends upon the level of cleaning the refrigerant undergoes and whether the refrigerant will change ownership. The EPA uses three terms to describe the condition of refrigerant that has been removed from a system: recovered, recycled, and reclaimed. Although these terms all have a similar meaning in general use, they have very specific meanings when used to describe refrigerant in the context of EPA regulations.

Recovered Refrigerant

Recovered refrigerant has been removed in any condition from an appliance and stored in an external container without necessarily testing or processing it in any way. Recovered refrigerant may not change ownership; it may not be removed from one owner's system and then charged into a system owned by someone else. Recovered refrigerant may be charged into the system it came from or another system of the same owner.

Recycled Refrigerant

Recycled refrigerant has been removed from an appliance and cleaned for reuse without meeting all of the requirements for reclamation. In general, recycled refrigerant is refrigerant that is cleaned using oil separation and single or multiple passes through replaceable-core filter driers, which reduce moisture, acidity, and particulate matter. Although recycled refrigerant is safer than recovered refrigerant, it has the same basic restrictions. It may not be removed from one owner's system and then charged into a system owned by someone else. Recycled refrigerant may be charged into the system it came from or another system of the same owner.

Reclaimed Refrigerant

Reclaimed refrigerant has been reprocessed to at least the purity specified in AHRI Standard 700-2007, Specifications for Fluorocarbon Refrigerants. Its purity has been verified using the analytical methodology prescribed in the Standard. Refrigerant reclamation requires specialized machinery not available at a particular job site.

Reclaiming is done by an EPA-approved refrigerant reclaimer at a reprocessing facility. Recovered refrigerant is sent to either a general reclaimer or back to the refrigerant manufacturer for reclaiming. Reclaimed refrigerant may be used anywhere new refrigerant may be used, and it may change ownership.

28.9 RECOVERY EQUIPMENT CERTIFICATION

Refrigerant recovery and recycling equipment manufactured on or after November 15, 1993, must be tested by an EPA-approved testing organization to ensure that it meets EPA requirements. Third-party testing is not required for equipment manufactured before November 15, 1993.

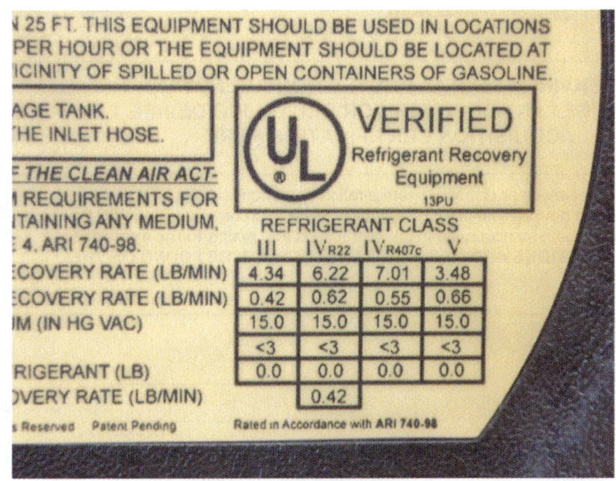

Figure 28-2 Recovery equipment should be certified by AHRI or UL.

Currently, the EPA has approved both the Air-Conditioning, Heating and Refrigeration Institute (AHRI) and Underwriters Laboratories (UL) to certify recycling and recovery equipment. Certified equipment can be identified by a label reading, "This equipment has been certified by AHRI/UL to meet EPA's minimum requirements for recycling and/or recovery equipment. . . ." (Figure 28-2). The intended use for the recovery equipment is listed at the end of this statement.

The standards vary depending on the size and type of air-conditioning or refrigeration equipment being serviced. Recovery equipment for use with Type I small appliances must be able to recover 90 percent of the refrigerant in the small appliance when the small appliance compressor is operating and 80 percent of the refrigerant in the small appliance when the compressor is not operating. Figure 28-3 shows the recovery levels for equipment intended for use with Type II air-conditioning and refrigeration systems. Recovery equipment for Type III, low-pressure systems, must be able to lower the pressure in the system to an absolute pressure of 25 mm of mercury. This is approximately 29 in of mercury vacuum.

Equipment Grandfathering

Equipment manufactured before November 15, 1993, including homemade equipment, may be grandfathered if it meets the standards shown in Figure 28-3.

Companies Must Own Refrigerant-Recovery Equipment

The EPA requires people servicing or disposing of air-conditioning and refrigeration equipment to certify to the appropriate EPA regional office that they have acquired recovery or recycling equipment and that they are complying with the applicable requirements of this rule. This certification must be signed by the owner of the recovery equipment or another responsible officer of the company and sent to the appropriate EPA regional office. Figure 28-4 shows a sample form. Although owners of recycling and recovery equipment are required to list the number of trucks based at their shops, they do not need to have a piece of recycling or recovery equipment for every truck. Owners do not have to send in a new form each time they add recycling or recovery equipment to their inventory.

28.10 FACTORS AFFECTING RECOVERY REQUIREMENTS

Technicians need to know when they have finished recovering the refrigerant from a system. Removing all the refrigerant would require evacuating the system until achieving a perfect vacuum of 29.92 in Hg. Fortunately, the EPA recognized that this would be impractical and in most cases impossible. The EPA uses the term *evacuate* when talking about the use of recovery equipment and the phrase "evacuate to the atmosphere" when describing the use of vacuum pumps. The recommended evacuation levels are based on the following factors.

Type of System

- Type I
- Type II
- Type III

Date Recovery Machine Was Manufactured

- Manufactured before November 15, 1993
- Manufactured on or after November 15, 1993

Type of Refrigerant

- HCFC 22
- Very-high-pressure refrigerant

Type of Appliance	Refrigerant Examples	Manufactured Before 11/15/1993	Manufactured on or After 11/15/1993
Very high pressure	R-13, R-503	0 psig	0 psig
R-22 < 200 lb	R-22	0 psig	0 psig
R-22 ≥ 200 lb	R-22	4 in Hg vacuum	10 in Hg vacuum
Other high pressure < 200 lb	R-12, R-502	4 in Hg vacuum	10 in Hg vacuum
Other high pressure ≥ 200 lb	R-12, R-502	4 in Hg vacuum	15 in Hg vacuum
Low-pressure appliance	R-11, R-123	25 in Hg vacuum	25 mm Hg absolute pressure

Figure 28-3 Type 2 and type 3 recovery levels.

Figure 28-4 Form for registering refrigerant recovery. *(Courtesy of Environmental Protection Agency)*

Form Approved
OMB No. 2060-0256
Expires: 07/31/2010

EPA

ENVIRONMENTAL PROTECTION AGENCY
REFRIGERANT RECOVERY OR RECYCLING DEVICE
ACQUISITION CERTIFICATION FORM

EPA regulations require establishments that service or dispose of refrigeration or air-conditioning equipment to certify that they have acquired recovery or recycling devices that meet EPA standards for such devices. To certify that you have acquired equipment, please complete this form according to the instructions and **mail it to the appropriate EPA Regional Office. BOTH THE INSTRUCTIONS AND MAILING ADDRESSES CAN BE FOUND ON THE REVERSE SIDE OF THIS FORM.**

PART 1: ESTABLISHMENT INFORMATION

Name of Establishment | Street

(Area Code) Telephone Number | City | State | Zip Code

Number of Service Vehicles Based at Establishment | County

PART 2: REGULATORY CLASSIFICATION

Identify the type of work performed by the establishment. **Check all boxes that apply.**

- □ Type A - Service small appliances
- □ Type B - Service refrigeration or air-conditioning equipment other that small appliances
- □ Type C- Dispose of small appliances
- □ Type D - Dispose of refrigeration or air-conditioning equipment other than small appliances

PART 3: DEVICE IDENTIFICATION

	Name of Device(s) Manufacturer	Model Number	Year	Serial Number (if any)	Check Box if Self-Contained
1.					□
2.					□
3.					□
4.					□
5.					□

PART 4: CERTIFICATION SIGNATURE

I certify that the establishment in Part 1 has acquired the refrigerant recovery or recycling device(s) listed in Part 2, that the establishment is complying with Section 608 regulations, and that the information gives is true and correct.

Signature of Owner/Responsible Officer | Date | Name (Please Print) | Title

EPA FORM 7610-31

- Other high-pressure CFCs and HCFCs
- HFCs

Amount of Refrigerant in System

- Under 200 lb
- 200 lb or more

The recovery levels for each of these specific circumstances are discussed in the following sections. Knowing these evacuation levels is extremely important; every section of the EPA certification exam will ask several questions about refrigerant evacuation (recovery) levels.

28.11 RECOVERY-LEVEL EXCEPTIONS

Two exceptions to the EPA-mandated recovery levels are permitted: opening systems for minor repairs and evacuating leaky systems.

Major and Minor Repairs

Systems can be opened for minor repairs after recovering the refrigerant to a level of 0 psig instead of the EPA-mandated evacuation level for that particular system. A minor repair is one that does not involve the compressor, condenser, evaporator, or auxiliary heat-exchange coil. Work on any of these components is considered a major repair and requires the full EPA-mandated evacuation. An example of a minor repair would be replacing a filter or a valve in a refrigerant line.

SERVICE TIP

The practical usefulness of the minor repair exclusion is limited. The EPA does not allow the system to be evacuated to the atmosphere after performing the minor repair. This means that the technician is not allowed to pull a deep vacuum on the part of the system that was opened before placing the system back into service. It is poor refrigeration practice to open systems and not pull a vacuum on them afterward.

Leaky systems may also be evacuated to 0 psig instead of the EPA-mandated recovery level for the system. The system must be evacuated to the lowest practical level, but that level can be no higher than 0 psig. Trying to evacuate a system with a leak below 0 psig can result in sucking air into the system and mixing it with the recovered refrigerant.

28.12 RECOVERY MACHINES: OLD VERSUS NEW

The age of the recovery machine affects the required recovery level in most cases. Specifically, if the recovery equipment was manufactured before November 15, 1993, then the required evacuation levels are not as stringent as the vacuum levels required of machines made on or after November 15, 1993. Older recovery equipment is not required to have low-loss fittings, but equipment manufactured on or after November 15, 1993, must have low-loss fittings on the refrigerant hoses.

TECH TIP

The Clean Air Act was passed in 1990 and specified that it would take effect beginning in 1992. The law did not flesh out all the specific details involved in recovering refrigerant but simply said that it had to be done. Well, 1992 came and went, and the EPA had not completed work on the details, so there was no way to know exactly what would be required, other than the refrigerant could no longer be released to the atmosphere when installing, servicing, or disposing refrigeration equipment. So equipment manufacturers had to design, manufacture, and sell machines that could not possibly be built to any standard, since there was not a standard. These early machines were bulky, expensive, and generally not able to meet the standards that the EPA later published. In fairness to the contractors who purchased these early noncertified machines, the EPA published a less stringent set of recovery levels when they published their first set of regulations. This is why technicians now have two sets of refrigerant-recovery levels to memorize.

28.13 TYPE I EQUIPMENT-RECOVERY LEVELS

When using a recovery device manufactured before November 15, 1993, the recovery level for Type I systems is 80 percent, regardless of the appliance compressor condition. When using a recovery device manufactured on or after November 15, 1993, recovery levels for Type I systems are established based on the condition of their compressor.

- 90 percent for appliances with operational compressors
- 80 percent for appliances with nonoperational compressors

These recovery levels are valid for both system-dependent (passive) and self-contained (active) recovery devices. Self-contained recovery devices may also be certified by achieving a vacuum level of 4 in Hg.

SERVICE TIP

The Type I recovery requirements are unique because they specify what percentage of the system's refrigerant charge must be recovered instead of a vacuum level. The percentage charge that a recovery device can capture is tested using a very specific EPA-prescribed test stand following the device manufacturer's instructions. If the recovery device successfully recovers the desired amount of refrigerant on the EPA test device, it is assumed that the same device will perform similarly in the field when used according to the manufacturer's instructions. It is important to read and follow the instructions provided with the recovery device. Field technicians are not required to prove that they successfully recovered 80 percent of the system's charge; they only have to demonstrate that they followed instructions using an EPA-approved device.

28.14 TYPE II EQUIPMENT-RECOVERY LEVELS

Remembering Type II recovery levels can be quite a challenge, but there are a few organizational tricks that will help.

Very-High-Pressure Refrigerants

The recovery level for systems using a very-high-pressure refrigerant is 0 psig. Neither the age of the recovery machine nor the amount of refrigerant in the system has any effect on this level.

HCFC-22

The recovery level for systems that hold less than 200 lb of R-22 is 0 psig, regardless of when the recovery machine was manufactured. It is important to note that this means 200 lb by weight, not the pressure reading.

The recovery level for a system holding 200 lb or more of R-22 is established by when the recovery machine was manufactured. The recovery level for machines made before November 15, 1993, is 4 in Hg vacuum. The recovery level for machines made on or after November 15, 1993, is 10 in Hg vacuum.

SERVICE TIP

There are very few systems that hold 200 lb of R-22. Typical residential systems hold less than 10 lb, and even 25-ton commercial systems only hold around 50 lb.

Other High-Pressure Refrigerants

High-pressure refrigerants other than R-22 require lower recovery levels than R-22. For systems holding less than 200 lb of refrigerant other than R-22, the recovery level is 4 in Hg vacuum when using a recovery machine made before November 15, 1993, and 10 in Hg vacuum when using a recovery machine made on or after November 15, 1993.

The final set of recovery levels is for systems using 200 lb or more of a high-pressure refrigerant other than R-22. The recovery level when using a recovery machine manufactured before November 15, 1993, is 4 in Hg vacuum. The recovery level when using a recovery machine manufactured on or after November 15, 1993, is 15 in Hg vacuum.

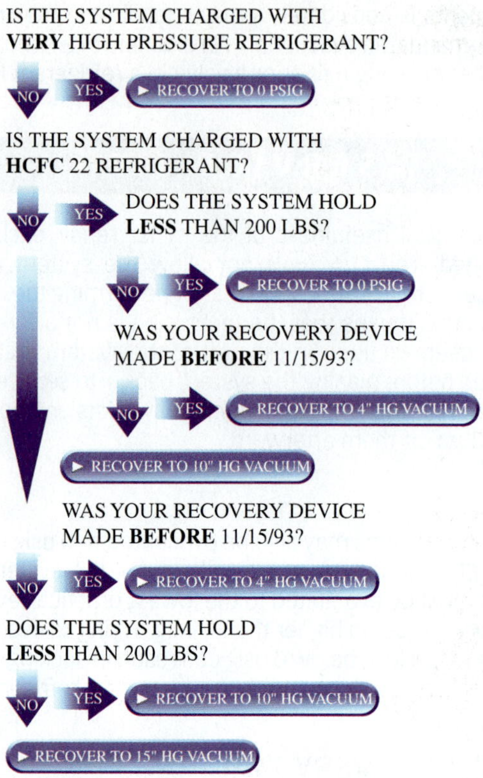

Figure 28-5 Decision flow chart for Type II refrigerant-recovery levels.

Old Recovery Machines

The recovery level for high-pressure refrigerants when using a recovery machine manufactured before November 15, 1993, is never more than 4 in Hg vacuum. This simplifies remembering the recovery level when using an older recovery machine. For very-high-pressure refrigerants and R-22 systems holding less than 200 lb of refrigerant, the recovery level is 0 psig. For every other high-pressure refrigerant situation, when using a recovery machine manufactured before November 15, 1993, the recovery level is 4 in Hg vacuum.

Decision Flow Chart

Figure 28-5 is a decision flow chart that helps organize Type II recovery levels.

28.15 TYPE III EQUIPMENT-RECOVERY LEVELS

Type III recovery levels are easy to remember: the target number is always 25. When using a recovery device that was manufactured before November 15, 1993, the required evacuation level is 25 in Hg vacuum. When using a recovery device manufactured on or after November 15, 1993, the required recovery level is 25 mm Hg absolute pressure. The amount of refrigerant in the system is irrelevant for Type III systems. An easy way to remember the evacuation levels for Type III low-pressure systems is to remember that inches is the old way to measure and millimeters is the new way.

TECH TIP

Although 25 in vacuum and 25 mm absolute look similar, they represent very different evacuation levels. Atmospheric pressure is normally about 29.92 in Hg. A vacuum level of 25 in Hg vacuum means that 25 of the original 29.92 inches have been removed, leaving an absolute pressure of about 4.92 in Hg. This method of establishing vacuum measures how much pressure has been removed from the system. The 25 mm absolute pressure required for newer recovery machines measures how much pressure remains after evacuating the system. Since 25 mm is approximately 1 in, a pressure of 25 mm absolute is roughly equivalent to 29 in Hg vacuum, about 1 in shy of a perfect vacuum.

28.16 SAFE DISPOSAL

Equipment that is typically dismantled onsite before disposal must have the refrigerant recovered in accordance with EPA's requirements for servicing before being dismantled. Examples of equipment requiring refrigerant recovery in the field before disposal include commercial refrigeration equipment, central residential air conditioning, chillers, and industrial process refrigeration. However, equipment that typically enters the waste stream with the charge intact, such as household refrigerators and room air conditioners, is subject to special safe-disposal requirements.

The final person in the disposal chain, such as the scrap metal recycler or landfill owner, is responsible for ensuring that refrigerant is recovered from the equipment before disposal. However, technicians "upstream" can remove the refrigerant and provide documentation to the final person if this is more cost effective. The final person in the disposal chain must have documentation showing who removed the refrigerant if he accepts appliances that no longer hold a refrigerant charge. The documentation must include a signed statement listing the name and address of the person who recovered the refrigerant and the date that the refrigerant was recovered.

The equipment used to recover refrigerant from appliances prior to final disposal must meet the same performance standards as equipment used for servicing, but it does not need to be tested by a laboratory. Technician certification is not required for individuals removing refrigerant from appliances in the waste stream.

28.17 RECORD KEEPING

Technicians, service companies, equipment owners, wholesalers, and refrigerant reclaimers are all required to keep some form of documentation. Technicians must keep proof of their certification (Figure 28-6). Employers are also required to keep copies of their employee's certification. Technicians servicing appliances that contain 50 lb or more of refrigerant must provide the equipment owner with an invoice that indicates the amount of refrigerant added to the appliance. Owners of equipment that contains 50 lb or more of refrigerant must keep servicing records documenting the date and type of service, as well as the quantity of refrigerant added (Figure 28-7). Owners of equipment that contains less than 50 lb of refrigerant are not required to keep records of refrigerant use. Wholesalers who sell CFC and HCFC refrigerants must retain invoices that indicate the name of the purchaser, the date of sale, and the quantity of refrigerant purchased. Reclaimers must maintain records of the names and addresses of people sending them material for reclamation and the quantity of material sent to them for reclamation.

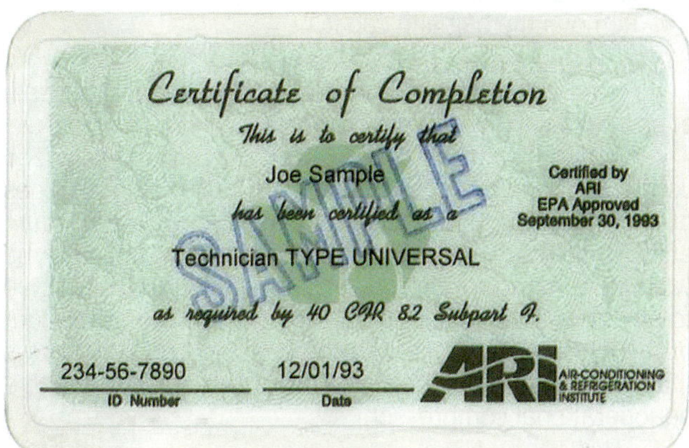

Figure 28-6 EPA certification card.

28.18 HAZARDOUS WASTE

Recycled or reclaimed refrigerants are not considered hazardous under federal law. Used oils contaminated with CFCs are not treated as hazardous on the condition that the oil:

- Is not mixed with other waste
- Has been subjected to CFC recycling or reclamation
- Is not mixed with used oils from other sources

28.19 RECOVERY EQUIPMENT

Recovery equipment is classified as either system dependent or self-contained. System-dependent recovery equipment depends on the system from which the refrigerant is being recovered to help remove the refrigerant. Self-contained recovery equipment does not rely on the system because the recovery equipment has its own means of moving the refrigerant, typically a compressor.

	System Data for Commercial Air-Conditioning Equipment					
	Refrigerant Type			**HCFC 22**		
	Refrigerant Amount			**60 lb**		
				Leak Rate Calculations		
Date	***Service Description***	***Amount Recovered***	***Amount Added***	***% Charge***	***Annual Adjustment***	***Annual Leak Rate***
1/1/2007	Unit placed in service					
6/30/2008	Added refrigerant		1.5 lb	1.5/60 = 2.5%	365/(365 + 182) = 0.0667	2.5% × 0.667 = 1.7%
12/10/2008	Replaced liquid drier	59 lb	60 lb	1/60 = 1.7%	365/163 = 2.2	1.7% × 2.2 = 3.8%
5/30/2009	Added refrigerant		1.5 lb	1.5/60 = 2.5%	365/172 = 2.12	2.5% × 2.12 = 5%

Figure 28-7 A refrigerant data log is used to document the amount of refrigerant added to systems containing 50 lb or more of refrigerant.

System-Dependent Recovery Devices

System-dependent recovery devices may be used for recovering refrigerant from systems containing up to 15 lb of refrigerant. A system-dependent recovery device depends on the system from which the refrigerant is being recovered to help move the refrigerant. The most common system-dependent recovery device is basically a big plastic bag. As refrigerant enters it, its volume expands so that the pressure in the bag does not rise above 0 psig. The EPA refers to these bags as nonpressurized containers.

Whirlpool developed the bags for use with refrigerators. These bags will only hold about 1 lb of refrigerant. To use the bag on a refrigerator with an operating compressor, use the following steps:

- Connect the bag to the high side of the system.
- Let the refrigerant flow into the bag with the unit off for approximately 15 min.
- Start the compressor and let it operate for another 5 min.
- Remove the bag and cap it.

Figure 28-8 shows the correct connection for an appliance with an operational compressor.

To use the bag on a refrigerator with a compressor that will not operate:

- Connect the bag to BOTH the high side and low side of the system.
- Let the refrigerant flow into the bag with the unit off for approximately 15 min.
- Hit the compressor three times. (You read it right: *hit the compressor*.)

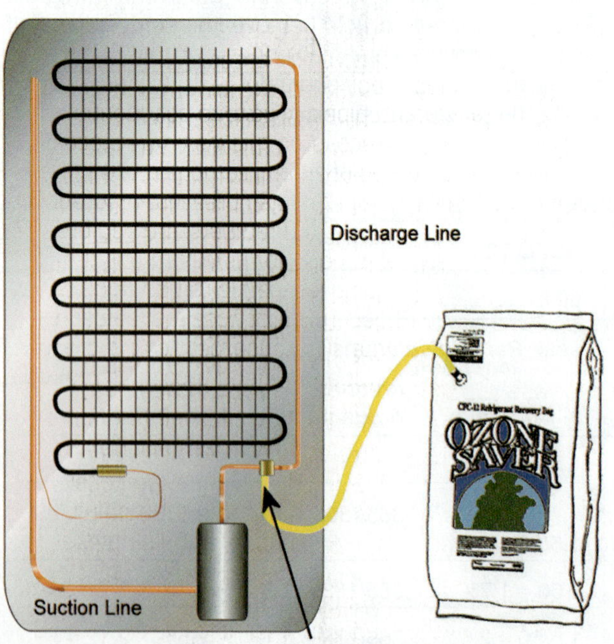

Figure 28-8 Connection of a system-dependent recovery device to a small appliance with an operational compressor.

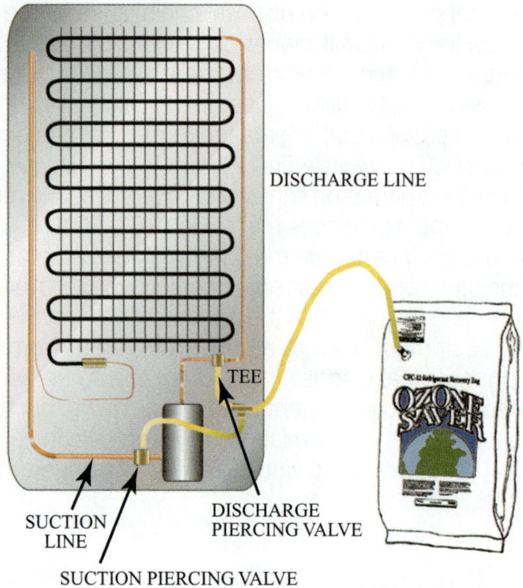

Figure 28-9 Connection of a system-dependent recovery device to a small appliance with a nonoperational compressor.

- Heat the compressor with a hair dryer or heat gun for another 12 min.
- Remove the bag and cap it.

Figure 28-9 shows the correct connection for an appliance with a nonoperational compressor.

A vacuum pump can also be used with a nonpressurized container (bag). Connect the vacuum pump to both sides of the system and connect the discharge of the vacuum pump to the bag. This requires some ingenuity, because the vacuum pump is not designed to have its outlet connected to anything. The vacuum pump will pull the refrigerant out of the system and place it in the bag. Figure 28-10 shows the use of a vacuum pump and bag on an appliance with a nonoperational compressor.

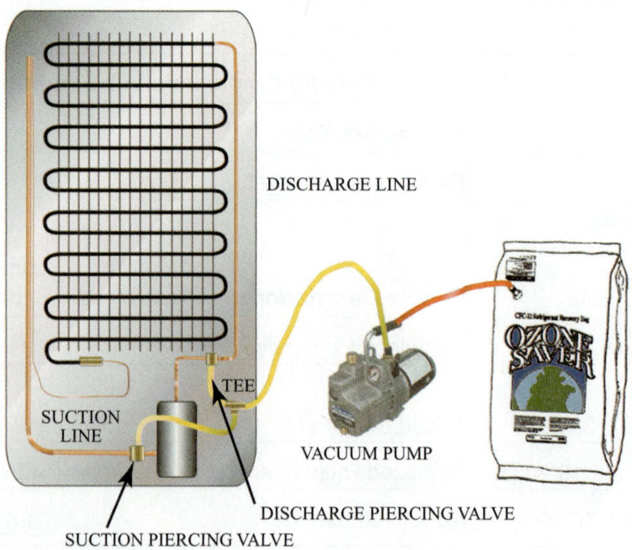

Figure 28-10 Connection of a system-dependent recovery device using a vacuum pump.

Figure 28-11 Self-contained recovery devices.

Self-Contained Recovery Devices

Self-contained recovery devices do not rely on the system the refrigerant is being recovered from for their operation. They typically are small, specialized condensing units (Figure 28-11). They have a compressor, a condenser, and valves that are used to control the refrigerant flow in and out of the recovery machine (Figure 28-12). Most modern self-contained recovery machines use oil-less compressors to reduce cross-contamination between refrigerants when changing from one refrigerant to another. Many also have purge cycles that will pump the refrigerant trapped in their condenser into the recovery cylinder after recovering the system refrigerant. A small amount of refrigerant is left between the outlet of the compressor and the outlet of the recovery machine. It is not a violation to release this small quantity of refrigerant to recover another type of refrigerant.

Figure 28-12 Most self-contained recovery devices consist of a compressor, a condenser, and control valves. *(Courtesy of Ritchie Engineering Company, Inc., Yellow Jacket Products Division)*

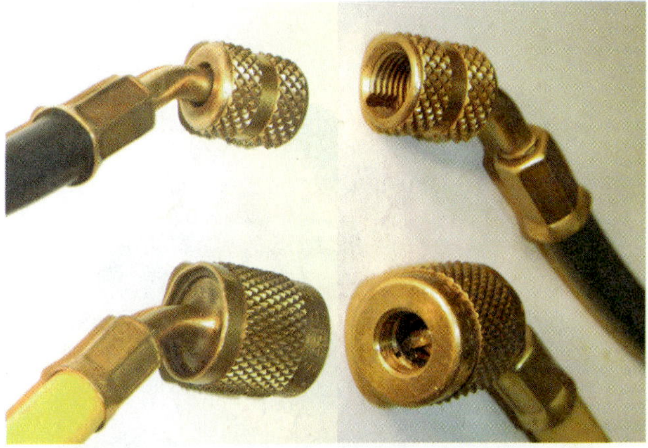

Figure 28-13 A standard hose end, shown on top, compared to a check valve hose end, shown on bottom.

Water-Cooled Recovery Machines

A very few recovery machines are water cooled. The water source for water-cooled recovery machines is city water (tap water).

28.20 LOW-LOSS FITTINGS

Low-loss fittings reduce the amount of refrigerant released when hoses are connected and disconnected by keeping refrigerant from leaving the hose when it is disconnected. The majority of the refrigerant released when disconnecting a hose from a Schrader valve is actually coming out of the hose, not the valve. This can be a significant amount of refrigerant when the hose is filled with liquid. All hoses used with recovery equipment built after November 15, 1993, are required to have low-loss fittings. These can be automatically closing check valves, like the one in Figure 28-13, or they can be manually controlled valves, like the one shown in Figure 28-14.

28.21 RECOVERY CYLINDERS

Nonrefillable cylinders should not be used for refrigerant recovery. Recovery cylinders are normally available in water capacities of 27 and 47 lb. Refillable and recovery cylinders

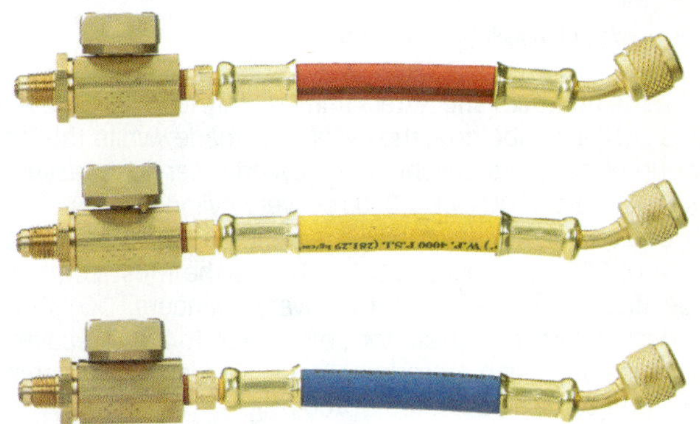

Figure 28-14 Short hose sections with manual shutoff valves.

Figure 28-15 Refrigerant-recovery cylinder.

are regulated by the DOT in Code of Federal Regulations 49, covering the transport of hazardous materials. Figure 28-15 shows a typical refrigerant recovery cylinder.

Cylinder Markings

Cylinders are marked with the particular DOT specifications for that cylinder (Figure 28-16). Some crucial cylinder markings are the following:

- DOT specification
- Water capacity
- Tare weight
- Date of manufacture
- First retest date
- Most recent test date
- Last allowable retest date

The numbers and letters immediately following the letters DOT describe how the cylinder is made, while the last group of numbers describes the cylinder service pressure. For example, a DOT-4BA-350 recovery cylinder is a welded or brazed steel cylinder with a recommended service pressure of 350 psig (Figure 28-17). This is the most common specification for refrigerant-recovery cylinders. DOT 4BA-400 cylinders are used for refrigerant R-410A recovery because of its high pressures. These cylinders have recommended service pressures of 400 psig.

The water capacity is the weight of the water to completely fill the cylinder (Figure 28-18). The tare weight is the

Figure 28-16 The cylinder specifications are stamped into the collar of the cylinder.

Figure 28-17 This cylinder has a service pressure of 350 psig.

Figure 28-18 The weight of the water to completely fill this cylinder is 47.6 lb.

Figure 28-19 This cylinder weighs 28.1 lb empty.

weight of the empty cylinder (Figure 28-19). The tare weight and water capacity are used to calculate how much refrigerant the cylinder can safely hold.

Cylinders with a service pressure under 300 psig must be tested every ten years. Cylinders with a service pressure of 300 psig or more must be tested every five years. The date of manufacture and the first retest date are on every cylinder (Figure 28-20). The first retest date gives the month and year that a cylinder must be requalified. If the first requalification date has not yet arrived, the cylinder should still be on its first five years. Note that the five years starts when the cylinder is made, not when it is purchased. A cylinder that was made two years ago can only be used for another three years before getting requalified, even if it has been sitting on the shelf in the store. If the first retest date has already passed, the cylinder should not be used

Figure 28-20 This cylinder was manufactured in July 1998.

Figure 28-21 The last date that this cylinder may be used is 10 years after its date of manufacture.

unless it was retested. If it was retested, the date that it was retested should be stamped on the cylinder. If the date was less than five years ago, the cylinder should be safe to use.

Some cylinders also have a last permissible use date. These are commonly known as 5 + 5 cylinders. They can only be requalified one time. These cylinders may not be used past the last permissible use date (Figure 28-21).

Cylinder Color

AHRI Guideline K specifies that recovery cylinders all be painted gray with a yellow top regardless of the type of refrigerant in the cylinder (Figure 28-25). Of course, these cylinders must be labeled with the type of refrigerant they contain, since cylinder color cannot be relied upon to identify them. The only thing the technician knows for sure by the color is that refrigerant in a gray and yellow cylinder is not new.

Pressure-Relief Valve

Recovery cylinders have an automatic pressure-relief valve built in that opens at 150 percent of the rated cylinder service pressure (Figure 28-22). The outlet of the relief valve should

Figure 28-22 Recovery cylinder pressure-relief valve.

never be plugged or obstructed by anything. These valves reset once the pressure in the cylinder has dropped to a safe level.

28.22 CYLINDER FILL LEVEL

Cylinders should never be filled more than 80 percent full of liquid by volume, and they should not be stored or used in temperatures exceeding 125°F. All refrigerant cylinders have a maximum recommended service pressure. Do not pressurize recovery cylinders above their rated service pressure.

However, do not rely on pressure to determine the safe fill level. It is possible to fill a cylinder completely up with liquid and remain below the cylinder service pressure. Remember that for saturated mixtures of liquid and gas, the pressure is established by the refrigerant temperature, not the amount of refrigerant in the cylinder. It does not matter if the cylinder is 25 percent full or 95 percent full; the pressure will be the same as long as the temperature is the same. In practice, the cylinder pressure gradually rises as recovery proceeds. This is because the warm refrigerant leaving the recovery machine tends to heat up the cylinder and all the refrigerant in it, raising the cylinder temperature and pressure. If the cylinder is cooled, its pressure will drop. Pressure absolutely cannot be relied upon to tell when a cylinder is full. Most recovery machines have high-pressure safety switches. Remember these are safety switches, not fill-level detectors. In short, do not simply fill the cylinder up until the high-pressure safety device shuts off the recovery machine! Remember that when a safety shuts something off, it generally indicates that there is a problem.

Observing the weight of the cylinder is the most commonly used method for determining when a cylinder is full. All DOT-approved cylinders have their tare weight and water capacity stamped on them. If a cylinder were refilled with water, the maximum safe fill level could be determined by taking 80 percent of the water capacity and adding it to the tare weight. Before blends, all commonly used refrigerants were denser than water. Most of the 400 series refrigerants are less dense, and cylinders cannot safely hold as much of these refrigerants. In general, a cylinder will only hold 75–80 percent as much 400 series refrigerant in the same size cylinder as the older refrigerants used. For example, 30-lb disposable cylinders have been common for years. All refrigerants in them come in the same size cylinders with the same net weight. This does not mean that all these different refrigerants filled the cylinders up equally, but none of them overfilled the cylinders at this level. The familiar 30-lb cylinder can safely hold only 25 lb of R-410a refrigerant. The cylinder is 80 percent full by volume, but it weighs less because the refrigerant is less dense than the older refrigerants. Therefore, it is necessary to calculate the net weight for each specific type of refrigerant to know exactly how much refrigerant a recovery cylinder can safely hold. The formula is

$$\text{tare weight} + (0.8 \times \text{water capacity} \times \text{specific gravity of the refrigerant at } 77^\circ\text{ F})$$

- Tare weight is the weight of the empty cylinder. It is usually stamped as Tare or TW on the cylinder.
- Water capacity, stamped WC, is what the water would weigh if the cylinder were completely filled with water.
- The specific gravity of the refrigerant compares the weight of the refrigerant to the weight of water. This number must be looked up for each refrigerant.

Most refrigerant manufacturers do not give a specific gravity but list the refrigerant's density instead. To calculate the specific gravity, divide the refrigerant saturated liquid density at 77°F by 62.4, the density of water. The formula using the refrigerant density instead of specific gravity is

$$\text{tare weight} + (0.8 \times \text{water capacity} \times (\text{refrigerant density}/62.4))$$

Of course, this only has to be calculated one time for each cylinder. Figure 28-23 gives the safe fill level for the two most common sizes of recovery cylinders for several refrigerants.

Refrigerant	Density @77°	Specific Gravity @77°	11.9 WC 80% Fill	26.1 WC 80% Fill	47.6 WC 80% Fill
12	81.84	1.31	12.47	27.35	49.88
22	76.92	1.23	11.71	25.68	46.84
134a	75.31	1.21	11.52	25.26	46.08
500	72.16	1.16	11.04	24.22	44.17
502	75.95	1.22	11.61	25.47	46.46
401a	74.5	1.19	11.33	24.85	45.32
404	65.17	1.04	9.90	21.72	39.60
407	71.12	1.14	10.85	23.80	43.41
408	66.31	1.06	10.09	22.13	40.36
409	75.91	1.22	11.61	25.47	46.46
410a	67.66	1.08	10.28	22.55	41.13

Figure 28-23 Safe fill level for common refrigerants and cylinder capacities.

Figure 28-24 Some recovery cylinders have a float that shuts off the recovery unit when the cylinder is 80 percent full.

Some recovery machines use float switches that cut the recovery machine off when the cylinder is 80 percent full (Figure 28-24). Others use temperature-sensing devices that sense the liquid level in the cylinder. Both of these are good safety devices, but they should not be used to determine fill level. Floats and switches are mechanical devices and can fail. It is important to always have a good idea how much room is in the cylinder and how much refrigerant will be recovered before starting.

28.23 RECOVERY TECHNIQUES

Remembering a few simple ideas will speed up refrigerant recovery.

- If possible, operate the system before beginning refrigerant recovery.
- Use the large-diameter, short-length hoses.
- Recover liquid first.
- Begin recovery at the lowest available access point.
- Heat the system from which refrigerant is being recovered.
- Cool the cylinder the refrigerant is going into.
- Recover from both sides of the system during vapor recovery.

Operate the System

If possible, operate the system before beginning refrigerant recovery. Refrigerant settles into the oil in the compressor. It takes a long time for recovery machines to vaporize this refrigerant out of the oil. Operating the compressor will pull the refrigerant out of the oil quicker.

Minimize Resistance Through Hoses and Connections

Short, large-diameter connection hoses will reduce the restriction that the recovery equipment must pull the refrigerant through. Reducing the restriction will increase the speed of recovery and decrease the time required to recover the refrigerant.

Recover Liquid First

Remove as much liquid from the system as possible before proceeding with vapor recovery. Liquid recovery is faster than vapor recovery because a pound of vapor takes up hundreds of times more space than a pound of liquid. Vapor recovery requires the liquid to vaporize, dropping its temperature and pressure. Liquid is removed either from the king valve on the liquid receiver or from the liquid service valve at the end of the condenser.

Let Gravity Help

When working on a system that has a difference in height between components, try to start your recovery at the lowest available access point. Rather than trying to pull liquid uphill against gravity, let gravity push the liquid down. For example, liquid recovery is normally done from the king valve on the liquid receiver, but when the condensing unit is located above the evaporator, liquid recovery should start at the liquid line at the evaporator instead (Figure 28-25). Similarly, if the receiver is located above the condenser, start liquid recovery at the outlet of the condenser (Figure 28-26).

Heat the System

One of the basic principles behind the refrigeration cycle is that pressure and temperature move up and down together. Heating the system both before and during recovery will increase the refrigerant pressure, and the refrigerant

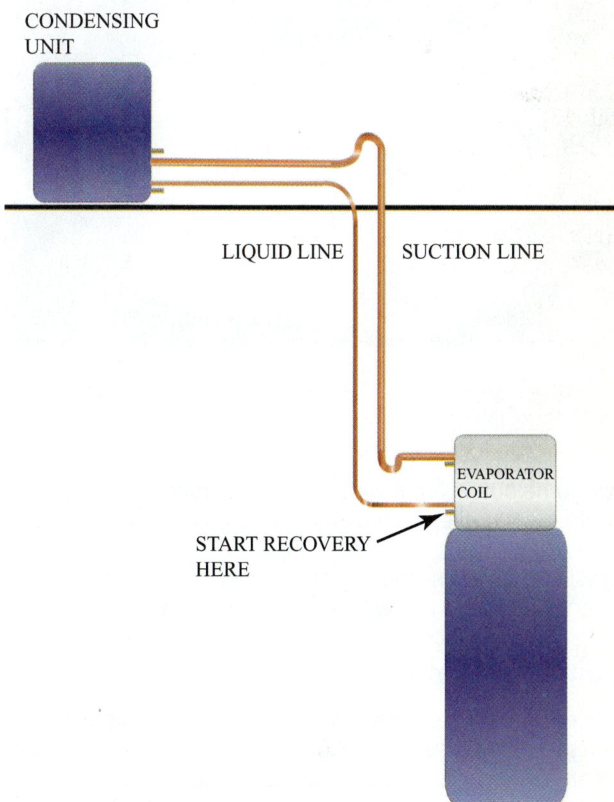

Figure 28-25 Begin recovery at the liquid line entering the evaporator when the condenser is located above the evaporator.

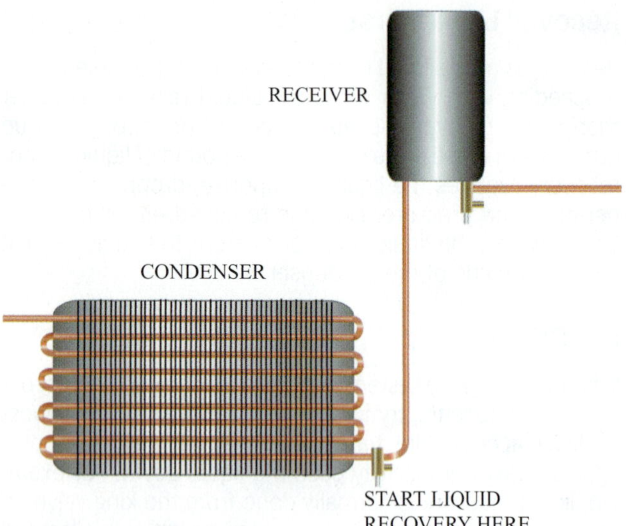

Figure 28-26 Begin recovery at the outlet of the condenser when the condenser is located below the receiver.

Figure 28-27 The frost line on this accumulator indicates there is still liquid sitting in the bottom.

will come out faster. Pay particular attention to spots where liquid can puddle, such as the compressor, liquid receiver, filter driers, and accumulator (Figure 28-27). To prevent excessive pressures, the system should not be heated with torches or steam but with warm air or water.

Cool the Cylinder

The refrigerant-recovery cylinder should be cooled in addition to heating the system. Cooling the cylinder will decrease its pressure and speed the recovery process. Cylinders can be cooled with ice or chemicals or by active cooling (Figure 28-28). Active cooling involves connecting the recovery machine inlet to the vapor port on the cylinder and connecting the recovery machine outlet to the liquid valve on

Figure 28-28 Cooling the recovery cylinder in ice can speed system recovery by keeping the cylinder pressure low.

Figure 28-29 Recovery unit connected to cylinder to perform active cooling on the cylinder.

the cylinder (Figure 28-29). This turns the cylinder into an evaporator, cooling off the liquid and dropping its pressure. Some recovery units have this ability built in. They always connect to the cylinder with two hoses instead of one.

Recovering from Both Sides of the System

Vapor recovery is required to completely evacuate a system down to the EPA-specified level. The recovery machine should be recovering from both sides of the system simultaneously during vapor recovery (Figure 28-30). Attempting vapor recovery through just one port requires some of the gas to travel through either the metering device or the compressor, drastically increasing the time required.

Wait Before Opening the System

After recovering refrigerant to the required level, wait to be sure that the pressure in the system does not rise above the prescribed level. Any refrigerant that is pooled up in pockets, such as in the compressor crankcase, receiver, filter drier, or traps, will vaporize and raise the pressure in the

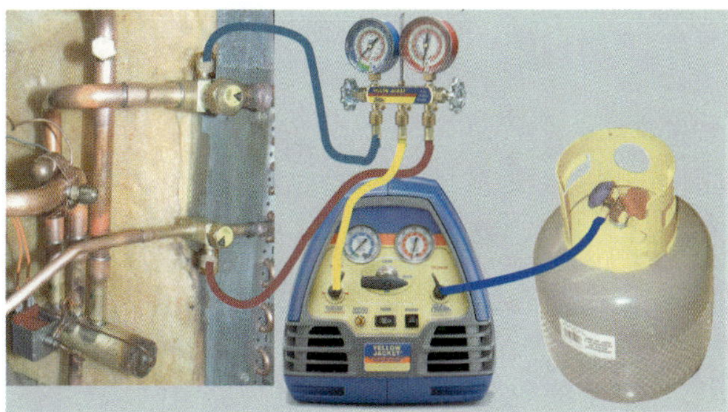

Figure 28-30 Recovery from both sides of the system when performing vapor refrigerant recovery.

system as it heats up. Heating the system at these locations can help combat this. It generally saves time to evacuate beyond the required level, because a slight pressure rise is almost guaranteed.

28.24 TYPE I REFRIGERANT RECOVERY

Some recovery techniques are unique to Type I appliances. Most Type I appliances do not have any type of service valves. The technician will have to add some type of service aperture before any refrigerant can be recovered. The EPA requires all Type I appliances to have a process stub or other "equally effective service aperture." That is what most Type I systems have: a short piece of copper tubing that has been pinched closed on the end and brazed up (Figure 28-31). The only way this can be used is to pierce

Figure 28-31 Process stub on a compressor.

Figure 28-32 Bolt-on piercing valve installed on a system.

it with a piercing valve or piercing pliers. Bolt-on valves are more convenient and are relatively easy to use, but they frequently leak (Figure 28-32). They should not be left on the system for permanent access because of the relatively high probability of future leaks. Braze-on piercing valves are brazed to the line prior to piercing the tubing and are therefore less likely to leak. They can be used for both recovery and permanent access. However, using braze-on piercing valves raises a safety issue. Most equipment manufacturers caution against brazing on a system under pressure, but that is exactly what must be done to install a braze-on piercing valve. Another option for gaining system access is to use a set of piercing pliers. Piercing pliers allow refrigerant recovery from Type I appliances without leaving a valve on the system.

Schrader valves work for permanent access valves but can only be installed on a system that has already been evacuated (Figure 28-33). You should always check Schrader valves for leaks when working on systems using Schrader valves and replace leaky cores when necessary. Replace the valve caps after completing work on the system to avoid leaks.

Refrigerants that Should Not Be Recovered

Always check an appliance's data plate for the type and amount of refrigerant before beginning recovery. This is especially true for small appliances. Some refrigerators built before 1950 contain poisonous chemicals, such as sulfur dioxide, methyl chloride, and ammonia. None of these noxious chemicals are covered under the refrigerant-recovery rules. However, for safety reasons, they should not be casually released to the air. Refer these systems to someone who is trained in handling them.

28.25 TYPE II REFRIGERANT RECOVERY

Type II systems cover the widest variety of applications, so the recovery techniques and equipment used for different Type II applications also varies widely. However, the order

Figure 28-33 Schrader access valve on process stub of small appliance.

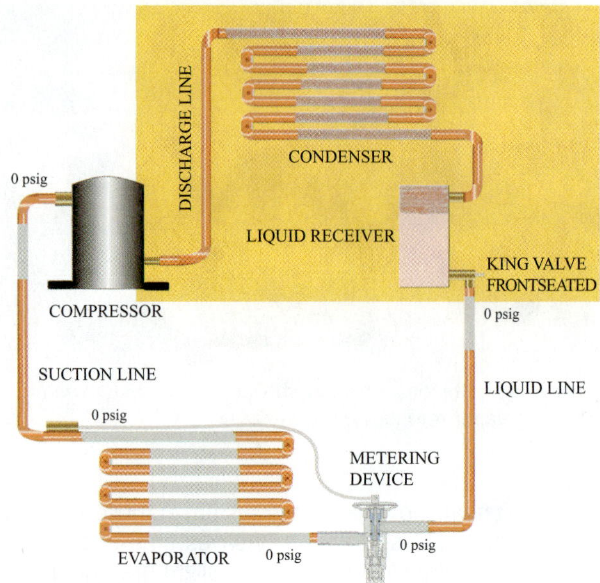

Figure 28-34 All the refrigerant is trapped between the compressor discharge valve and the king valve when the system is pumped down.

Figure 28-35 Front-seating the king valve keeps refrigerants from leaving the liquid receiver.

of preference for speed of recovery from a Type II system is usually as follows:

- Recover into the receiver, also known as pumping a system down
- Push-pull
- Liquid recovery
- Vapor recovery

Recover into the Receiver

On systems with liquid receivers and king valves, time can be saved by pumping the system down into the receiver using the system compressor. Although the EPA uses the phrase "recover the refrigerant into the receiver," this is more commonly known as "pumping the system down" in the field. Pumping all of the refrigerant into the receiver traps it between the discharge valve on the compressor and the king valve on the receiver (Figure 28-34). This is accomplished by front-seating the king valve and operating the system until the pressure on the low side of the system is at the EPA-required recovery level (Figure 28-35). It is much faster to move the refrigerant into the receiver or the condenser than it is to recover all of it. Pumping a system down is only an option if the system compressor operates and the part of the system to be serviced is on the low side of the system or on the liquid line after the king valve. It will not work for repairing anything between the compressor discharge valve and the liquid receiver king valve.

Many systems that do not have receivers can still be pumped down. Most air-conditioning systems have liquid-line service valves and condensers large enough to hold an entire system charge (Figure 28-36). Residential air conditioners are normally shipped with an entire system charge trapped in the condenser. However, if the system has lines over 50 ft, or if it has been overcharged, the condenser may not hold the entire system charge. The compressor discharge pressure should be monitored while pumping down a system. It will become very high if the condenser fills with liquid and there is no more room for vapor. This can cause loud noises from inside the compressor when the internal pressure-relief valve opens.

Push-Pull Recovery

Many recovery-machine manufacturers rate the maximum speed of their recovery unit using the push-pull recovery

Figure 28-36 The liquid-line service valve on residential split systems can be used to pump the system down.

method. The push-pull recovery method pushes the gas discharged from the recovery machine into one part of the refrigerant system while pulling liquid out of the system.

In push-pull mode, either the liquid-line service valve or the king valve is connected to the liquid valve on the recovery cylinder. The vapor valve on the recovery cylinder is connected to the suction side, or inlet, of the recovery machine. The discharge, or outlet, of the recovery machine is connected to the discharge service valve on the compressor. Figure 28-37 shows the correct connections for push-pull liquid refrigerant recovery.

When the recovery machine runs, it will pressurize the condenser by pushing gas into it. The pressure of the recovery cylinder will be reduced because the recovery machine is pulling gas from it. At the same time, pulling gas off of the recovery cylinder forces some of the liquid in the cylinder to boil. This reduces its temperature and pressure. The result is an increased pressure difference between the system and the recovery cylinder. This produces the fastest rate of recovery because the liquid leaves the refrigeration system quickly without having to boil. Not having to boil the liquid in the system keeps the system temperature and pressure higher, helping to maintain the high pressure difference.

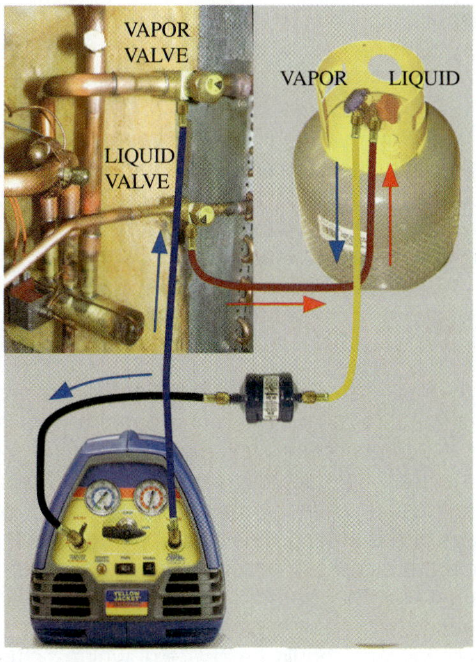

Figure 28-37 Liquid recovery using the push-pull method.

One limitation of the push-pull setup is that all the system charge cannot be recovered this way. After recovering liquid, the recovery machine will have to be reconnected to finish by recovering vapor.

Liquid Recovery Without Push-Pull

Push-pull works well for systems with large amounts of liquid sitting in a receiver. However, it is much less effective with smaller residential air-conditioning equipment and heat pumps.

Most recovery units today have a setting for recovering liquid directly. They do this by having a built-in restriction that acts like a metering device to flash off the liquid before it reaches the recovery unit's compressor. This setting can be used to start the recovery process with liquid and switch over to vapor without moving hoses.

First, check the weight of the recovery cylinder to make sure it has room for the refrigerant in the system. Use the procedure outlined in section 28.22, Cylinder Fill Level, to determine the total weight of the cylinder and refrigerant when filled to the maximum safe fill level of 80 percent. Weigh the cylinder. Subtract the actual cylinder weight from the maximum safe fill-level weight; the difference is the amount of room in the cylinder, which should exceed the amount of charge listed on the equipment data plate.

Figure 28-38 shows a typical setup for recovering liquid refrigerant from a residential split system. The gauges are connected as they normally are, with the high-side manifold

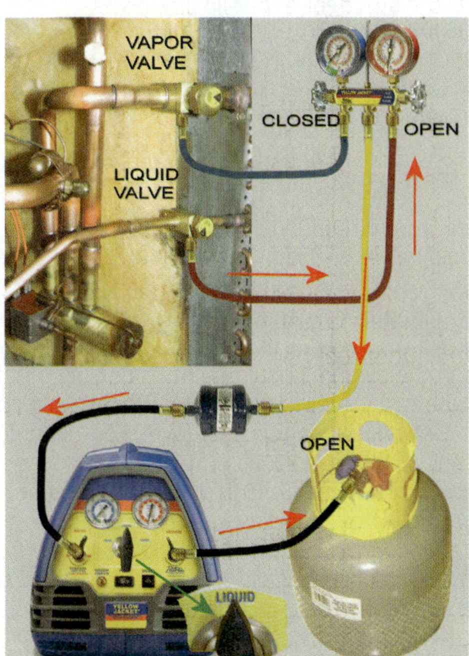

Figure 28-38 Recovery unit connected for direct liquid refrigerant recovery.

Figure 28-39 The recovery unit's valves are positioned for liquid recovery.

Figure 28-40 The recovery unit's valves are positioned for vapor recovery.

hose connected to the liquid-line service valve and the low-side manifold hose connected to the suction service valve. The center manifold hose connects to the inlet of the recovery unit, and the outlet of the recovery unit connects to the vapor port on the recovery cylinder.

All hoses should be purged of air before proceeding with recovery. The high-side manifold gauge valve should be opened, the recovery inlet valve set for liquid recovery, the recovery outlet valve opened, and the refrigerant cylinder vapor valve opened (Figure 28-39). Operate the recovery unit and watch the pressures in the system and on the recovery unit gauges. An operating recovery system should not be left unattended. Most of the liquid will have been removed when the system pressure is substantially less than the saturation pressure corresponding to the system temperature. The gauges and hose connections will feel cold as liquid flashes off coming through them. When all the liquid has been recovered, those parts that were cold will begin to warm up. After all the liquid has been removed, switch to vapor recovery to finish.

Figure 28-41 Recovering to 5 in Hg vacuum will help ensure that the system pressure will stay at or below 0 psig when the recovery unit is shut off.

Vapor Refrigerant Recovery

To switch from direct liquid recovery to vapor, put the recovery inlet valve to vapor (Figure 28-40). This lets the vapor flow into the recovery unit unrestricted and speeds vapor recovery. Open both manifold valves so that the recovery unit pulls from both sides of the system. Any remaining cold pockets in the system should be heated using a heat gun. Monitor the system pressures and let the recovery go a little past the EPA-mandated recovery level. For example, if the system holds less than 200 lb of R-22, the recovery level is 0 psig. Recover down to 5–10 in Hg vacuum instead of stopping at 0 psig. After achieving a vacuum slightly lower than the target vacuum, close the manifold valves and shut off the recovery machine (Figure 28-41). Wait to see if the system pressures are going to stay under the EPA-mandated level. It is normal for the pressure in the system to rise a little; that is why it is a good idea to recover past the required level. If the system pressure is stable and at or below the EPA-mandated level, the recovery is complete. Put the recovery-machine valves in the purge position and operate the recovery machine until it has moved the refrigerant trapped in it into the recovery cylinder (Figure 28-42). This normally only takes a few minutes. Close all valves before breaking down the recovery setup.

28.26 VERY-HIGH-PRESSURE RECOVERY

Very-high-pressure refrigerants have saturated pressures in a range of 250–700 psig at room temperature. They are used in low-temperature systems and cascade systems. Cascade systems are actually a combination of two refrigeration cycles. The evaporator of one is used to cool the condenser of the other. The system operating at the highest temperatures uses a "normal" high-pressure refrigerant such as R-404A. The system with the ultra-low evaporator temperature uses a very-high-pressure refrigerant, such as R-508. Technicians should be aware of which system they are recovering from when working on cascade systems. A

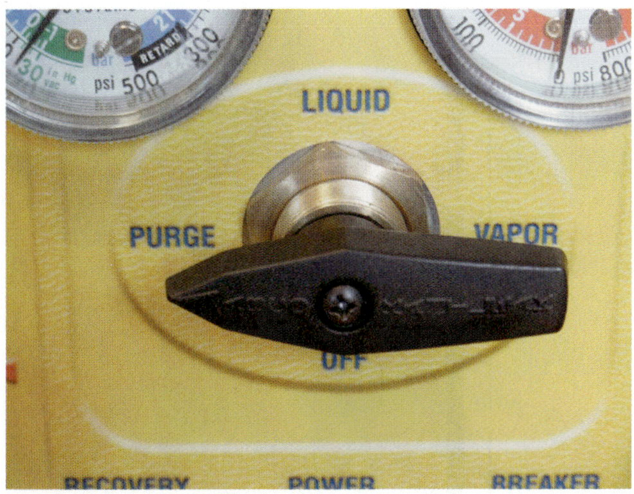

Figure 28-42 The recovery unit valves are positioned to purge its condenser of refrigerant.

recovery system that is certified for R-404A is not suitable for recovering R-508.

Recovery machines and cylinders for very-high-pressure refrigerants are designed for the extreme pressures of these refrigerants. Never use a recovery machine that is not specifically designed for very-high-pressure refrigerant. It is a violation of EPA regulations to use a recovery unit that is not AHRI/UL certified for the type of refrigerant being serviced. Figure 28-43 shows a recovery unit specifically certified for very high pressure refrigerant.

Figure 28-43 Very-high-pressure recovery unit. *(Courtesy of Redi Controls, Inc. Refrigerant Mizer Model RS-503/13-C3)*

SAFETY TIP

Always check the AHRI/UL rating on the recovery device before using it, especially with very-high-pressure refrigerants. Using a "standard" recovery unit on a system with very-high-pressure refrigerant is extremely dangerous because the components in the device are not designed to withstand the extreme pressures of very-high-pressure refrigerant.

28.27 CHILLERS AND WATER-COOLED SYSTEMS

Recovering vapor from a system that still has liquid refrigerant in it forces the liquid to boil to replace the vapor. The liquid temperature and pressure drop as a result. Recovering vapor refrigerant from water-cooled systems or chillers can drop the temperature and pressure of the refrigerant to the point that water in the chiller tubes or condenser shells begins to freeze. Freezing the water in tubes or shells can cause extensive damage to the system when the ice expands. Always recover liquid first from any system with water in either side to prevent damaging the system! Liquid does not have to boil to make vapor during liquid recovery, so the system pressure and temperature will not drop as rapidly. Another trick is to circulate the water while recovering the refrigerant. Running water is far less likely to freeze. It is possible to remove all the water from small systems, but this is normally not practical in larger systems.

28.28 TYPE III REFRIGERANT RECOVERY

The vast majority of Type III systems are chillers, so the procedures in the previous section on chillers also apply to these systems.

Evaporator Operates in Vacuum

The fact that the evaporator on a low-pressure chiller operates in a vacuum presents some unique problems. The recovery levels for low-pressure systems are very low: 25 in Hg of vacuum for recovery machines built before November 15, 1993, and 25 mm absolute pressure for recovery machines built on or after November 15, 1993.

Purge Unit

All low-pressure systems have purge units. These units purge the system of air that sucks into the low side during system operation because of system leaks in the low side. Air sucks in rather than refrigerant leaking out because the air pressure in the room exceeds the pressure in the evaporator, which is in a vacuum. Since air is a noncondensable, it travels to the condenser and stops there. A buildup of noncondensables increases the system high-side pressure and reduces its efficiency. A purge unit gets rid of the gas at the top of the condenser where noncondensables collect. The purge unit on a system that is relatively leak free will not operate very much because noncondensables are not collecting in the system. On the other hand, a leaky system will have a lot of purge operation because noncondensables will constantly be entering the system through the leaks on the low side.

Minor Repair Exclusion

The minor repair exclusion can be used with low-pressure systems. However, instead of recovering refrigerant to reach 0 psig, the low-side pressure must be raised to 0 psig before opening the system for service for a minor repair, since the low side is normally in a vacuum.

Raising the Low-Side Pressure

One common low-pressure refrigerant, CFC-11, has a boiling point of 75°F. This means that the system pressure will be below 0 psig anytime the system temperature is below 75°F. If the low side of a low-pressure system is opened, it will suck in air. This will lead to increased purge operation when the system is put back into operation. Purge operation should be minimized since some refrigerant is always lost when operating the purge unit.

There are two common methods used for raising the pressure of the evaporator on a low-pressure chiller: adding nitrogen until the pressure comes up to 0 psig or circulating warm water through the chiller tubes. Adding nitrogen is better for the system than letting a bunch of air suck in because the air will contain oxygen and water as well as nitrogen. Since both these contaminants can harm the system, they are to be avoided. However, adding nitrogen is discouraged by the EPA because it increases the purge operation and increased purge operation amounts to increased refrigerant release. The EPA endorses circulating warm water through the chiller bundle to raise the evaporator pressure. The increase in temperature will cause a corresponding increase in pressure in the chiller.

SAFETY TIP

Be careful while circulating water in low-pressure chillers! The heat from the pump motors can heat the refrigerant in the chiller enough to build pressures over 15 psig, the point where the rupture disk opens. All the system refrigerant can be lost if the rupture disk breaks. This is an expensive accident for a system that holds thousands of dollars worth of refrigerant.

Rupture Disk

All low-pressure chillers have a rupture disk on the evaporator that opens at 15 psig and vents all the refrigerant out of the evaporator (Figure 28-44). This is a physical safety device that literally breaks, meaning that there is no putting the genie back in the bottle once the disk ruptures. It is considered good practice to limit the pressure on the low side of the system to 10 psig to stay a safe distance away from disaster. Never exceed 10 psig whenever heating the chiller to raise the pressure.

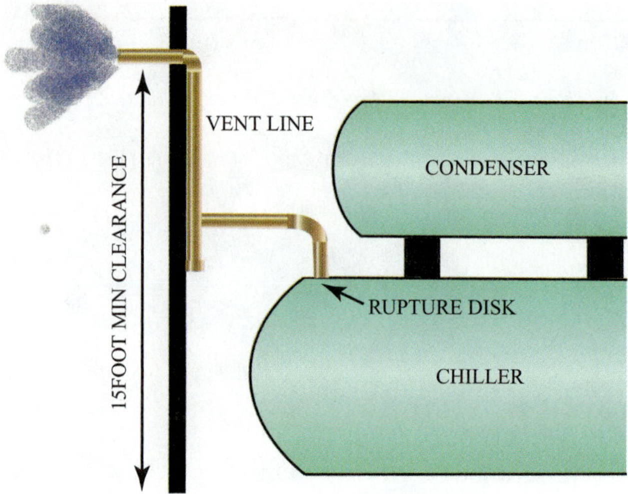

Figure 28-44 Refrigerant escaping through a broken rupture disk on a low-pressure chiller.

Removing Compressor Oil

Compressor oil can contain a large amount of refrigerant. Heat the oil in the compressor to at least 130°F before removing it from the compressor. This will drive out most of the dissolved refrigerant while the oil is still in the system.

28.29 LARGE SCALE

Low-pressure chillers typically hold hundreds of pounds of refrigerant. A 30 lb portable recovery unit will not recover several hundred pounds of refrigerant. Low-pressure chillers usually have separate storage tanks and built-in recovery systems so that refrigerant can be pumped over into the auxiliary holding tank when the system needs to be opened (Figure 28-45). This is a slow process even with large on-site systems. Once the work is completed and the system is evacuated, the refrigerant is transferred back into the chiller. To aid

Figure 28-45 On-site recovery vessel for large low-pressure chiller. *(Reftec International Systems, LLC)*

in refrigerant transfer from the auxiliary storage tank back into the system, many of these storage tanks have built-in heaters that keep the refrigerant pressure up in the storage cylinder when the refrigerant is transferred from them back into the system.

Recovery equipment designed for use with Type III low-pressure systems should have a high-pressure cutoff at 10 psig to avoid over-pressurizing the system and blowing the rupture disk.

28.30 RECOVERING REFRIGERANT FROM SYSTEMS WITH LEAKS

Occasionally, a technician will check the pressure on a system that needs to have refrigerant recovered only to find that the system pressure is 0 psig. It is important to recognize that a pressure of 0 psig does not indicate that the system is completely empty. The correct procedure to follow is determined by the type of system and refrigerant. If it is a Type I system, no further action is necessary. The system may be opened without any further recovery efforts.

If a Type II system has a pressure of 0 psig, the remaining refrigerant must be recovered if the refrigerant type and amount require a vacuum below 0 psig. However, 0 psig may be considered an adequate recovery level for systems with leaks, because the EPA allows recovery to stop at the lowest practical level on systems with leaks. That level cannot be any higher than 0 psig.

The steps for recovering refrigerant from a system with known leaks are as follows:

1. Isolate the leaking part from the rest of the system. Now the rest of the system can be evacuated to the specified level without sucking in any air.
2. Evacuate the rest of the system to the specified level.
3. Recover the leaking part of the system to the lowest practical level. Of course, that must be no higher than 0 psig.

28.31 RECOVERED REFRIGERANT CONTAMINANTS

Recovered refrigerant may contain many contaminants, including refrigerant oil, acids, and water. After recovering the refrigerant, the technician must determine if it qualifies as hazardous waste. It may qualify as a hazardous waste if it is corrosive or contains heavy metals or is carcinogenic or extremely flammable. If it is determined that the recovered refrigerant is indeed hazardous waste, a whole new set of EPA rules for the handling and disposal of hazardous waste must be followed!

Recycled and reclaimed refrigerants are exempt from the hazardous waste rules, but recovered refrigerant is not. This is a very good reason to make certain that the refrigerant passes through a filter when it is recovered. Passing the refrigerant through a filter makes the refrigerant recycled and exempt from hazardous waste regulations.

Refrigerant Oil

Refrigerant oil is also recovered when recovering refrigerant. More oil comes out with liquid recovery than with vapor recovery. This is not a problem if the system will be recharged with the recovered refrigerant, because most of the oil removed will go back into the system with the refrigerant. Oil that is the result of a recycling or reclaiming process is exempt from the hazardous substance rules, provided it is not mixed with other types of oil and is not burned.

Avoid Mixing Refrigerants

Avoid mixing refrigerants when recovering refrigerant. Mixed refrigerants can be very difficult to separate and frequently must be destroyed. Not only does this reduce the available supply of refrigerant, but it is expensive. Most refrigerant reclaimers will take relatively clean refrigerant for a reasonable fee but charge a high disposal fee for mixed or severely contaminated refrigerant. Each recovery cylinder should be used for only one type of refrigerant. Most recovery machines may be used for more than one type of refrigerant, but the refrigerant remaining in the recovery unit must be removed before the recovery unit is used on a different refrigerant. Follow the recovery unit manufacturer's instructions to change the system from one type of refrigerant to another. Typical steps for older systems include changing the oil in the recovery unit compressor, changing the filter, and evacuating the unit with a vacuum pump. Newer oil-less systems do not require an oil change.

28.32 DISPOSING OF RECOVERED REFRIGERANT

Recovered or recycled refrigerant may be reused if it does not change ownership. Refrigerant that has been recovered so that a mechanical repair can be performed is the best candidate for reuse. Refrigerant from compressor burnouts, systems that have experienced multiple compressor failures, or systems that are contaminated with air or moisture are not good candidates for reusing refrigerant. The refrigerant is a minor cost in most systems. Replacing the refrigerant with new, clean refrigerant after performing repairs will increase the system's chances for a long and productive life. However, replacing the refrigerant requires that the technician dispose of the old refrigerant.

The recovered refrigerant is normally returned to a refrigerant reclaimer or manufacturer. This can be accomplished through a local wholesaler, or the refrigerant may often be sent directly to the processor. When refrigerant is shipped, the cylinders must be marked according to both DOT and EPA regulations.

DOT Labeling Requirements

For nonflammable refrigerants, the Code of Federal Regulations (CFR) 49 requires that each cylinder display a DOT diamond (square-on-point) "nonflammable gas" label (Figure 28-46). Each container of a regulated material should be marked with a DOT proper shipping name. This will be a generic chemical name, not a proprietary trade name. Each shipping container should list the consignor's name and address.

Figure 28-46 The DOT requires this marking on refrigerant cylinders while being shipped.

EPA Labeling Requirements

Each cylinder, ton tank, or drum containing a recovered refrigerant designated by the EPA as a Class I (CFC) or Class II (HCFC) substance in the Clean Air Act is required to display a warning statement indicating that the product(s) inside the cylinder or drum harms the earth's ozone layer. The chemical name of the substance may be abbreviated. For example, R-12 may be substituted for dichlorodifluoromethane.

The warning statement is as follows:

"WARNING: Contains [insert name of substance], which harms public health and environment by destroying ozone in the upper atmosphere" (Figure 28-47).

28.33 TIMELINE

The timeline below shows where we have been and where we are going in recognizing and addressing the problem of stratospheric ozone depletion.

1974	Molina-Rowland propose the ozone-depletion theory
1985	Vienna Convention begins global cooperation to study ozone depletion
1987	Montreal Protocol proposes concrete steps to regulate ozone depleting substances
1990	United States passes Clean Air Act amendments
July 1, 1992	Venting ban on ozone-depleting refrigerants from stationary equipment
May 14, 1993	Final ruling outlining required practices
July 13, 1993	Evacuation requirements published in May ruling take effect
November 15, 1993	Refrigerant-recovery machines must be certified
November 14, 1994	Technicians must be certified
November 14, 1994	CFC and HCFC refrigerant sold only to certified technicians
November 15, 1995	Venting ban on HFC refrigerants
January 1, 1996	CFCs may not be manufactured or imported
Jan 1, 2010	No more manufacture of new systems containing R-22
Jan 1, 2020	No more manufacture or importing of R-22 and R-142b
January 2030	No more manufacture or importing of any HCFC refrigerant

Dispos-A-Can® Container
Non Refillable

WARNING: Contains HCFC-22 a substance which harms public health and environment by destroying ozone in the upper atmosphere.

Figure 28-47 This warning is required on all products containing ozone-depleting substances.

28.34 CERTIFICATION OUTLINE

The EPA provides an outline of the topics covered in the different sections of the certification exam. This outline is provided below, along with the section of text that covers each specific piece of information.

CORE

Ozone Depletion

Destruction of ozone by chlorine ***19.17***
Presence of chlorine in CFC and HCFC refrigerants ***19.6, 19.8–19.10***
Identification of CFC, HCFC, and HFC refrigerants ***19.6, 19.8–19.10***
CFCs have higher ODP than HCFCs, which have higher ODP than HFCs ***19.17***
Health and environmental effects of ozone depletion ***19.17***
Evidence of ozone depletion and role of CFCs and HCFCs ***19.17***

Clean Air Act and Montreal Protocol

CFC phaseout date ***19.9***
Venting prohibition at servicing ***28.5***
Venting prohibition at disposal ***28.5, 28.16***
Venting prohibition on substitute refrigerants in November, 1995 ***19.11***
Maximum penalty under CAA ***28.3***
Montreal Protocol ***28.2***

Section 608 Regulations

Definition/identification of high- and low-pressure refrigerants ***19.5***
Definition of system-dependent vs. self-contained recovery/recycling equipment ***28.19***
Identification of equipment covered by the rule ***28.4***
Need for third-party certification of recycling and recovery equipment ***28.9***
Standard for reclaimed refrigerant, AHRI 700 ***19.22***

Substitute Refrigerants and Oils
Absence of "drop-in" replacements ***19.24***
Incompatibility of substitute refrigerants with mineral oil ***19.23***
Fractionation problem in blends ***19.14***

Refrigeration
Refrigerant states and pressures at different points of refrigeration cycle ***14.2–14.3***
Refrigeration gauges (color codes, ranges of different types, proper use) ***23.13***

Three Rs
Definition of *recover* ***28.8***
Definition of *recycle* ***28.8***
Definition of *reclaim* ***28.8***

Recovery Techniques
Need to avoid mixing refrigerants ***28.23***
Factors affecting speed of recovery ***28.23***

Dehydration Evacuation
Need to evacuate system to eliminate air and moisture at the end of service ***30.3, 30.4, 30.7***

Safety
Risks of exposure to refrigerant ***22.3***
Personal protective equipment ***22.2***
Reusable cylinders vs. disposable cylinders ***22.7–22.10***
Risks of filling cylinders more than 80 percent full ***22.11***
Use of nitrogen rather than oxygen or compressed air for leak detection ***29.6, 29.7***
Use of pressure regulator and relief valve with nitrogen ***29.7***

Shipping
Labels required for refrigerant cylinders ***28.32***

TYPE 1 (Small Appliances)

Recovery Requirements
Definition of *small appliance* ***28.6***
Evacuation requirements using equipment made before November 15, 1993 ***28.13***
Evacuation requirements using recovery equipment made after November 15, 1993 ***28.13***

Recovery Techniques
Identify refrigerants and detecting noncondensables ***19.22***
Recovering refrigerant using a system-dependent recovery device ***28.19***
Recover from both high and low side with inoperative compressors ***28.19***
Run system compressor when using a system-dependent recovery device ***28.19***
Should remove solderless access fittings at conclusion of service ***28.24***
R-134a as likely substitute for R-12 ***19.24–19.25***

Safety
Decomposition products of refrigerants at high temperatures (HCl, HFl, etc.) ***22.6***

TYPE 2 (High-Pressure)

Leak Detection
Signs of leakage in high-pressure systems ***29.3***
Need to leak test before charging or recharging equipment ***31.2***
Order of preference for leak test gases ***29.6***

Leak Repair Requirements
Allowable annual leak rate for commercial and industrial process refrigeration ***29. 4***
Allowable annual leak rate for other appliances ***29.4***

Recovery Techniques
Recovering liquid at beginning of recovery process speeds up process ***28.23***
Other methods for speeding recovery ***28.23***
Methods for reducing cross-contamination and emissions ***28.20***
Need to wait a few minutes after reaching required recovery ***28.23***

Recovery Requirements
Evacuation requirements for high-pressure appliances in each of the following situations:
- Disposal ***28.14, 28.16***
- Major vs. non-major repairs ***28.11***
- Leaky vs. non-leaky appliances ***28.11, 28.30***
- Appliance (or component) containing less vs. more than 200 lb ***28.14***
- Recovery/recycling equipment built before vs. after November 15, 1993 ***28.12***
- Definition of "major" repairs ***28.11***
- System-dependent prohibition on systems exceeding 15 lb of refrigerant ***28.19***

Refrigeration
How to identify refrigerant in appliances ***31.3***
Pressure-temperature relationships of common high-pressure refrigerants ***19.3***
Components of high-pressure appliances ***14.2–14.14***

Safety
Shouldn't energize hermetic compressors under vacuum ***15.6***
Equipment room requirements under ASHRAE Standard 15 ***22.12***

TYPE 3 (Low-Pressure)

Leak Detection
Order of preference of leak test pressurization methods ***29.17***
Signs of leakage into a low-pressure system (e.g., excessive purging) ***29.17***
Maximum leak test pressure for low-pressure centrifugal chillers ***29.17***

Leak Repair Requirements
Allowable annual leak rate for commercial and industrial process refrigeration ***29.4***
Allowable annual leak rate for other appliances exceeding 50 lb of refrigerant ***29.4***

Recovery Techniques
Recovering liquid at beginning of recovery process speeds up process ***28.23, 28.27***
Need to recover vapor in addition to liquid ***28.23***
Need to heat oil to 130°F before removing it to minimize refrigerant release ***28.28***
Circulate water in chiller during refrigerant evacuation to prevent freezing ***28.27***
High-pressure cutout level of recovery devices used with low-pressure appliances ***28.29***

Recharging Techniques
Need to introduce vapor before liquid to prevent freezing of water in the tubes ***31.5***
Need to charge centrifugals through evaporator charging valve ***31.5***

Recovery Requirements
Evacuation requirements for low-pressure appliances in each of the following situations:

Disposal ***28.15, 28.16***
Major vs. non-major repairs ***28.11***
Leaky vs. non-leaky appliances ***28.11, 28.30***
Appliance (or component) containing less vs. more than 200 lb ***28.15***
Recovery/recycling equipment built before vs. after November 15, 1993 ***28.25***
Definitions of "major" and "non-major" repairs ***28.11***
Allowable methods for pressurizing a low-pressure system for minor repair ***28.28***
Need to wait a few minutes after reaching required recovery level ***28.23***

Refrigeration
Purpose of purge unit in low-pressure systems ***28.28***
Pressure-temperature relationships of low-pressure refrigerants ***19.3***

Safety
Equipment room requirements under ASHRAE Standard 15 ***22.12***
Need to have equipment room refrigerant sensor for R-123 ***22.12***

UNIT 28—SUMMARY

The Montreal Protocol is an international treaty first passed in 1987 that established specific goals for the reduction of ozone depleting substances. Section 608 of Title VI of the 1990 Clean Air Act established regulations to reduce emissions of ozone-depleting substances from stationary refrigeration and air-conditioning equipment. Important aspects of this regulation include the following:

- A prohibition on venting ozone-depleting refrigerants
- Technician certification requirement starting November 14, 1994
- Restrictions on the sale of ozone-depleting refrigerants starting November 14, 1994
- Reclaimed refrigerant must meet AHRI 700 standard
- Certification of recycling and recovery equipment beginning November 15, 1993
- Refrigerant recovery levels established
- Repair of substantial leaks in equipment with a charge greater than 50 lb
- Safe disposal of appliances containing ozone-depleting refrigerant

Regulations outlining procedures for reducing ozone-depleting refrigerants divide refrigeration systems into three categories:

Type I	Small appliances that are hermetically sealed and contain 5 lb or less refrigerant
Type II	High-pressure appliances whose refrigerant has a boiling point between −58°F and 50°F and very-high-pressure appliances whose refrigerant has a boiling point below −58°F
Type III	Low-pressure appliances that use a refrigerant having a boiling point at atmospheric pressure above 58°F

The EPA certification exam consists of four parts: Core, Type I, Type II, and Type III. To become certified, technicians must pass the Core and at least one of the other sections. Refrigerant that has been removed from a system is classified as one of the following:

Recovered	Removed in any condition from an appliance and stored in external container without necessarily testing or processing it in any way
Recycled	Cleaned using oil separation and single or multiple passes through replaceable core filter driers
Reclaimed	Reprocessed to at least the purity specified in the AHRI Standard 700-1996, Specifications for Fluorocarbon Refrigerants

Recovered and recycled refrigerant may be charged into the system it came from or another system of the same owner. Reclaimed refrigerant may be used anywhere new refrigerant may be used. Refrigerant recovery can be accelerated through proper technique, including the following:

- If possible, operate the system before beginning refrigerant recovery.
- Use the large-diameter, short-length hoses.
- Recover liquid first.
- Begin recovery at the lowest available access point.
- Heat the system from which refrigerant is being recovered.
- Cool the cylinder the refrigerant is going into.
- Recover from both sides of the system during vapor recovery.

Nonrefillable refrigerant cylinders should not be used for refrigerant recovery. Recovery cylinders are gray with a yellow top and have DOT specifications stamped on them identifying the cylinder specifications. They should not be filled over 80 percent by volume.

Refrigeration systems should be leak tested before charging or whenever refrigerant needs to be added. Leak testing techniques include the following:

- Soap bubble solutions
- Halide torches, which work only on chlorinated refrigerants
- Ultrasonic leak detectors, which detect the sound of refrigerant escaping
- Electronic sniffers, which detect CFC, HCFC, and HFC refrigerants
- Dyes that make oil spots more visible
- Standing vacuum test
- Standing pressure test

Significant leaks in systems that hold 50 lb of refrigerant or more must be repaired. A significant leak is defined as an annual leak rate of 35 percent of the charge for commercial and industrial refrigeration or an annual leak rate of 15 percent of the charge for air-conditioning systems and other applications.

WORK ORDERS

Service Ticket 2801

Work to Be Performed: Liquid-Line Filter Replacement

Equipment Type: R22 Split-System Air Conditioner with 30 lb of Refrigerant

Since this is an R-22 system that holds less than 200 lb of refrigerant, the recovery level is 0 psig. The technician decides to pump the system down using the liquid-line service valve to save time. The liquid-line service valve is front-seated and the system is operated. The system is shut down when the liquid-line pressure reaches 0 psig. The technician waits to make sure the pressure will hold. It does. The old filter is removed and the new filter installed. Nitrogen is added to the liquid line to check the connections for leaks. After performing the leak test, the nitrogen is released to the air. A vacuum pump is used to evacuate the liquid line and low side. Then the liquid-line service valve is back-seated and the system returned to operation.

Service Ticket 2802

Work to Be Performed: Compressor Change-out

Equipment Type: R-410A Split-System Air Conditioner with 7 lb of Refrigerant

A technician needs to change a compressor with a bad burnout. The technician selects a recovery system that is rated for R-410A refrigerant. The recovery cylinder is a 4BA-400 to handle the higher pressures. The gauges are connected to the liquid and suction service valves. The center hose connects to an inline filter. Another hose connects to the recovery unit, and a hose with a positive shutoff adapter is connected from the recovery unit to the recovery cylinder. The technician begins liquid recovery from the high side only. Once no more liquid is being removed, the recovery valves are set to vapor recovery, and both manifold gauge valves are open to allow vapor recovery from both sides of the system. The recovery unit runs until the system pressure is 5 in Hg vacuum, slightly lower than the required recovery level of 0 psig. The gauges are closed and the recovery units shut off. The system pressure rises slightly after sitting but remains below 0 psig. The technician replaces the compressor, installs both liquid and suction-line filter driers, and pulls a deep vacuum on the system. The system is recharged with new R-410A. The recovered R-410A is returned to the refrigerant wholesaler, who sends it off for reclaiming.

Service Ticket 2803

Work to Be Performed: Refrigerant Added

Equipment Type: R-22 Split-System Air Conditioner with 50 lb of Refrigerant

A customer asks the local air-conditioning contractor to add charge to his commercial air-conditioning unit, stating that he always needs to add some every spring. The technician checks the system charge and finds that it is indeed low. The system requires 10 lb of charge. The technician informs the customer that the system has a leak that exceeds the allowable rate of 15 percent of the charge per year. The owner says that he really does not want to spend money repairing the system, and the technician informs the owner that the system must be repaired or mothballed, or a plan must be developed to replace the system. (Mothballing requires recovering the system's refrigerant and taking it permanently out of service.) The owner decides to replace the system but wants to operate it for a while before replacing it. The technician tells him that he needs to submit a written plan to the regional EPA office outlining the plan to replace the equipment in no later than a year.

Service Ticket 2804

Work to Be Performed: Dispose of Refrigerator with a Dead Compressor

Equipment Type: Domestic Refrigerator with 12 oz of R-12 Refrigerant

A technician working for a scrap metal dealer is asked to prepare a domestic refrigerator for recycling. The compressor does not operate. The technician intends to use a system-dependent recovery device, so two sets of piercing pliers are used, one on each side of the compressor. The hose from each pair of piercing pliers is run to a tee that is connected to the inlet of the nonpressurized container (bag). The bag begins to increase in volume as the refrigerant travels from the refrigerator to the bag. The compressor is heated with a hair drier and struck three times with a hammer. The pliers and hoses are removed and the bag capped.

UNIT 28—REVIEW QUESTIONS

1. What is the significance of the Montreal Protocol?
2. What is the possible fine for violating the Clean Air Act?
3. List the major provisions of Title VI, section 608, of the 1990 Clean Air Act.
4. List the exceptions to the ban on venting ozone-depleting refrigerant.
5. List the four parts of the EPA certification exam.
6. What is the definition of a small appliance?
7. List the recovery levels for Type I, small appliances.
8. What is a nonpressurized container?
9. What is the difference between system-dependent recovery devices and self-contained recovery devices?
10. List the recovery levels for Type II systems using a recovery system manufactured before November 15, 1993.
11. List the recovery levels for Type II systems using a recovery system manufactured on or after November 15, 1993.
12. List the recovery levels for Type III systems.
13. What is the advantage of beginning with liquid recovery whenever possible?
14. Why should liquid always be recovered first from chillers?
15. Describe the connection setup for push-pull liquid recovery.
16. Describe the connection setup for vapor recovery.
17. List methods of speeding up refrigerant recovery.
18. What is the maximum safe pressure to impose on any low-pressure system?
19. Discuss the difference between recovered, recycled, and reclaimed refrigerant.
20. When was it first illegal to vent ozone-depleting refrigerants from stationary equipment?
21. When were recovery units required to be tested by a third party to prove they perform to EPA specifications?
22. When were technicians required to pass an EPA certification exam to handle refrigerant?
23. When did it become illegal to vent replacement refrigerants?
24. How can a technician identify what refrigerants a particular recovery machine is designed for?
25. Determine the weight of a recovery cylinder when safely filled with the following specifications: tare weight of 20 lb; water capacity of 30 lb; refrigerant specific gravity of 1.1.

UNIT 29

Refrigerant Leak Testing

OBJECTIVES

After completing this unit, you will be able to:

1. list issues to check for in a preliminary visual leak-prevention inspection.
2. explain the trigger rates that require mandatory leak repair.
3. discuss inert gas leak-detection techniques.
4. list types of refrigerant leak detection.
5. discuss proper refrigerant leak-detection techniques.
6. calculate the effect of ambient temperature change on a standing pressure test.
7. discuss leak-testing procedures for low-pressure systems.

29.1 INTRODUCTION

Reducing refrigerant emissions through good service practices, leak detection, and refrigerant recovery is a big part of any HVACR technician's job. Keeping refrigerant inside the system protects the environment, and leaking refrigeration systems are expensive. Not only does replacing the system refrigerant become costly, but the cost of operation also increases because of refrigerant leaks. A system operating undercharged because of a leak uses more energy to produce les cooling, leading to increased operating costs. Customers operating leaky systems may believe they are saving money by not paying for expensive repairs and just topping off the charge. Instead, they are paying for the repairs in the form of increased power bills, but they are not receiving the repairs or comfort they are paying for.

29.2 PREVENTING PROBLEMS

Systems should always be tested for leaks before they are charged. This saves time by eliminating the need to deal with leaking refrigerant, recovering refrigerant from the system to perform repairs, and evacuating the system multiple times. It is good service practice to take steps toward recognizing and fixing potential problems. When inspecting a system for leaks, take the time to do a quick preliminary visual inspection. Make sure refrigerant lines and joints have been assembled properly and are sufficiently braced and supported. Unsupported lines can put extra pressure on joints, leading to eventual failure. Unbraced lines can transmit vibrations that can cause leaks. Look for any metal-to-metal contact points on the refrigerant lines. Any contact with a hard surface can cause leaks at the contact point. Refrigerant lines should not come into direct contact with concrete, as chemicals in the concrete can cause corrosion, leading to leaks.

29.3 SIGNS OF REFRIGERANT LEAKAGE

There are several signs of refrigerant leakage that technicians should be aware of. Refrigerant oil travels with the refrigerant in the system. Trace amounts of oil will escape with the refrigerant leaking out. Over a period of time, the area around the leak will appear oily. Often the oil will attract dust and dirt, making an oily dirt buildup around the leak area. Systems operating with leaks will have a high system superheat and a low system subcooling and low capacity because of the lack of refrigerant. Unit 31, Refrigerant System Charging, discusses how a system charge affects both the superheat and subcooling. The compressor will be hotter than normal because the reduced amount of refrigerant circulating through the system does not cool the compressor as well.

29.4 TRIGGER RATES

Leak testing is an important part of reducing refrigerant emissions. Systems should be leak tested before charging or recharging them with refrigerant. Good service practice demands finding and repairing leaks on all systems. The EPA only requires repairing leaks on systems with a refrigerant charge of 50 lb or more that have leaks meeting EPA established trigger rates. The trigger rate for commercial and industrial process refrigeration is an annual leak rate of 35 percent of the total charge. Comfort cooling and "other" applications are restricted to an annual leak rate of 15 percent of the total charge. Note that the word *rate* includes the speed with which the refrigerant is lost. If a system that holds 100 lb of refrigerant loses 3 lb in a month, the leak rate is 36 lb per year. This leak would require repair, even if only 3 lb had escaped. The system owner is not allowed to wait until 35 percent of the charge has actually been lost before repairing the leak. The EPA uses owner maintenance records to prove that a system does not leak.

When technicians find a significant leak, they are required to inform owners in writing. The owner then has thirty days to repair the leak or submit a plan to the EPA to replace the system. The old unit can remain in service for up to a year if the plan is approved. Another way to put off either repairing or replacing the system is to mothball it. The leaky system does not have to be repaired or replaced if it is mothballed. The system may remain in this state indefinitely.

29.5 SYSTEM PRESSURE

Most leak-detection systems rely on detecting gas escaping from the leak. This assumes that there is still something in the system to escape. It does not matter what technique is used if there is no gas in the system to leak out. The first step should be to check the pressure on the system.

System Design Pressures

Most refrigeration systems have design pressures stamped on them (Figure 29-1). These are not the operating pressures but the pressures used for pressure testing the high and low sides at the factory. The system's design pressures should never be exceeded when field testing a system! Think of the design pressure as the maximum pressure that can be applied to the system components during leak testing. The safest procedure is to never exceed the low-side design pressure, because it is difficult to ensure that the high side and low side of an installed system will remain segregated. If the system does not list a design pressure, use 150 psig as the maximum safe test pressure. Hermetic compressor manufacturers list 150 psig as the maximum safe test pressure on the compressor shell.

29.6 LEAK TESTING WITH INERT GAS

If a system is empty, it can be charged with refrigerant or with an inert gas. Inert gases do not cause any chemical reactions with the materials in the system, and they do not support combustion. Carbon dioxide and nitrogen are the two most common inert gases. It does not make sense to use refrigerant when locating a potentially large leak. Refrigerant is expensive and must be recovered, while inert gases are cheap and do not have to be recovered. They may simply be released to the air. There are two methods of detecting nitrogen leaking from a system: one involves soap bubbles and the other ultrasonic detectors.

CONTAINS HFC-410A		DESIGN PRESSURE	
FACTORY CHARGE		HI 446 PSIG	
12 LBS 0 OZS		LO 236 PSIG	
ELECTRICAL RATING		NOMINAL VOLTS: 208/230	
1 PH	60 HZ	MIN 197	MAX 253
COMPRESSOR		FAN MOTOR	
PH	1	PH	1
RLA	10.3	FLA	2.8

DESIGN PRESSURE - HI SIDE	300 PSIG
DESIGN PRESSURE - LOW SIDE:	150 PSIG
FACTORY SHIPPED CONTAINING HCFC-22 CHARGE:	
FOR AUXILIARY ELECTRIC HEAT PACKAGES SEE UNIT HEA	

Figure 29-1 The system design pressures are shown on the data plate; the top data plate is for an R-410A unit, and the bottom data plate is for an R-22 unit.

Figure 29-2 Large leaks blow bubbles when a soapy solution is used.

SAFETY TIP

Any chemicals used on food service equipment, including soap bubbles, should be approved by NSF International. Using non-approved chemicals can result in sickness or injury to people who consume food prepared in that area.

Soap Bubbles

Soap bubbles and nitrogen can locate any major leaks in field fabrication. Escaping gas blows bubbles when a soapy solution is applied to the leaking area. Larger leaks will blow visible bubbles (Figure 29-2), while tiny leaks will form a white froth of micro-bubbles (Figure 29-3). Wait a few minutes after applying soap bubbles; micro-bubbles take a few minutes to show up. The EPA recommends using soap bubbles as the best method for pinpointing the location of leaks. Concentrate on mechanical joints first and then field-brazed joints.

Figure 29-3 Small leaks create micro-bubbles that collect into a froth after several minutes.

Figure 29-4 This leak detector produces micro-bubbles that froth to show small leaks.

SERVICE TIP

Chemical manufacturers that specialize in HVACR applications make soap bubbles specifically for the HVACR industry. While it is certainly possible to make your own using dishwashing detergent, they will not perform as well. This is particularly true when it comes to micro-bubbles. The leak detector shown in Figure 29-4 is specifically designed to produce micro-bubbles.

Ultrasonic

Ultrasonic leak detectors check for leaks by listening for the high-pitched whistle the gas makes as it escapes through small holes (Figure 29-5). The primary advantage of an ultrasonic detector is that it works for locating very small leaks with any gas, including nitrogen. Ultrasonic detectors usually have a way to "tune out" background noise and concentrate on the frequencies most likely to be caused by gas whistling through a small hole.

29.7 SAFETY WITH CO_2 AND NITROGEN

Nitrogen and CO_2 are inherently safe. They do not burn, they do not chemically react with many substances, and they are not poisonous. However, both can be dangerous because they do not irritate your lungs or eyes. There is little warning that the closed area a technician is in is so full of nitrogen that there is no longer enough oxygen to breathe; the technician will simply get dizzy and pass out.

Nitrogen and CO_2 cylinders can also be dangerous because of the very high pressures inside them. They typically have over 2,000 psig of pressure when the cylinder is new. A pressure regulator and pressure-relief valve should always be used when pressure testing with nitrogen or CO_2 (Figure 29-6). The regulator makes it possible to safely control the pressure on the system. The relief valve provides protection against accidental system overpressurization in the event of a regulator failure. There should be no other devices or valves connected to the outlet of the pressure-relief valve. Having something in the way of the outlet could prevent the valve from operating correctly.

Figure 29-5 Ultrasonic leak detector. *(Courtesy Robinair)*

Never Use Oxygen for Pressure Testing!

Never use oxygen or compressed air to pressurize a refrigeration system. Pure oxygen should be considered a dangerous gas because of the increased risk of fire. Mixtures of refrigeration oil and concentrated oxygen can be explosive. Compressed air contains substances that should not be inside a refrigeration system, including oxygen and water.

Figure 29-6 A pressure regulator and a pressure-relief valve should always be used when using nitrogen for pressure testing.

Figure 29-7 The red-hot copper disk breaks down the refrigerant when it passes over the disk.

Figure 29-8 Most leaks cause a pale green flame color.

29.8 USING R-22 AS A TRACE GAS

Mixtures of nitrogen and up to 10 percent R-22 may be used to check for leaks, using the R-22 as a trace gas. This allows the use of sensitive electronic sniffers that can detect the R-22. Mixing the R-22 with the nitrogen reduces the amount of refrigerant being released. Mixtures of R-22 and nitrogen may be released without being recovered, but nitrogen may not be put in on top of existing R-22. The system must be evacuated first, and then the R-22 and nitrogen mixture is added. From an environmental standpoint, testing with 100 percent nitrogen is preferable to testing with R-22 and nitrogen mixtures.

29.9 LEAK TESTING WITH HALIDE TORCHES

Halide torches were once one the most popular forms of leak-detection devices for chlorinated refrigerants. However, halide torches are disappearing as chlorinated refrigerants are phased out. The flame of a halide torch keeps a copper disk red hot (Figure 29-7) Combustion air is drawn through a small hose, passing over the copper disk. The red-hot disk decomposes any chlorinated refrigerant passing over it, and the flame color changes. Mixtures of CFC or HCFC refrigerant and air turn the flame green (Figure 29-8).

The brightness and intensity of the green color is directly related to the concentration of refrigerant. Higher refrigerant concentrations are indicated by a greener flame. Halide torches work well with CFC and HCFC refrigerants, but they do not work with HFC refrigerants because HFCs contain no chlorine. Of course, a halide torch should never be used to check for leaks of flammable refrigerant. Also be aware that the fumes produced by a halide torch can be very irritating because halogenated refrigerants break down when passed through a flame. This is normally not a very great hazard because the amount of refrigerant passing through the flame is small.

29.10 LEAK TESTING WITH ELECTRONIC SNIFFERS

Electronic leak detectors have largely replaced halide torches (Figure 29-9). They have been around for a long time, but electronic detectors have become the standard in leak detection since the era of the Clean Air Act and refrigerant conservation. Electronic leak detectors are advertised to be able to detect leaks as small as 0.25 oz per year. In practice, hypersensitivity can actually be a detriment to locating the exact location of a leak. This is why the EPA considers electronic sniffers the best tools for locating the

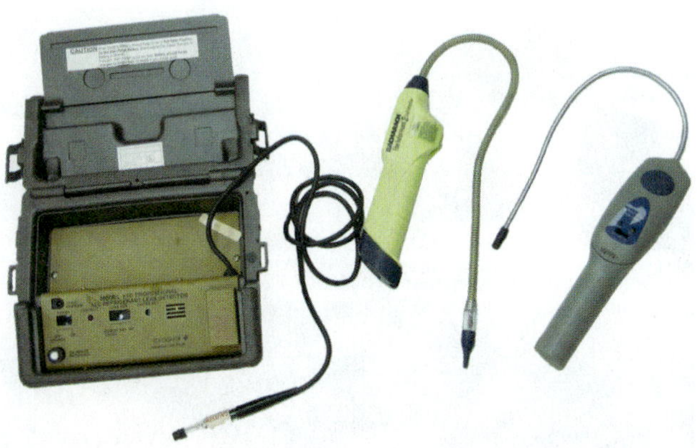

Figure 29-9 Electronic leak detectors.

general area of a leak but soap bubbles as the best tool for pinpointing the leak source.

Important questions to ask when choosing an electronic sniffer would be as follows:

- What types of refrigerants will the detector detect?
- How long will the tip or sensor last before it needs replacing?
- How much does the tip or sensor cost to replace?
- How quickly does the instrument clear itself after finding a leak?
- What types of things can cause false alarms with the instrument?

29.11 CORONA DISCHARGE LEAK DETECTORS

The first electronic leak detectors used corona discharge technology (Figure 29-10). The corona is a high-voltage discharge that ionizes the gas it passes through. Anything that breaks that corona barrier can set off the detector, including dust, moisture, refrigerant, or a good gust of wind. Corona discharge tips tend to become unreliable after as little as 20 hours of use. Fortunately, the tips are normally not expensive, and this type of detector can usually be restored to reliable operation by changing the tip. Most corona discharge units continue to squeal for a long time after they are removed from the leak source, which makes locating exactly where the leak is more difficult. They also give false readings when wind blows on them or when the tip is tapped against something.

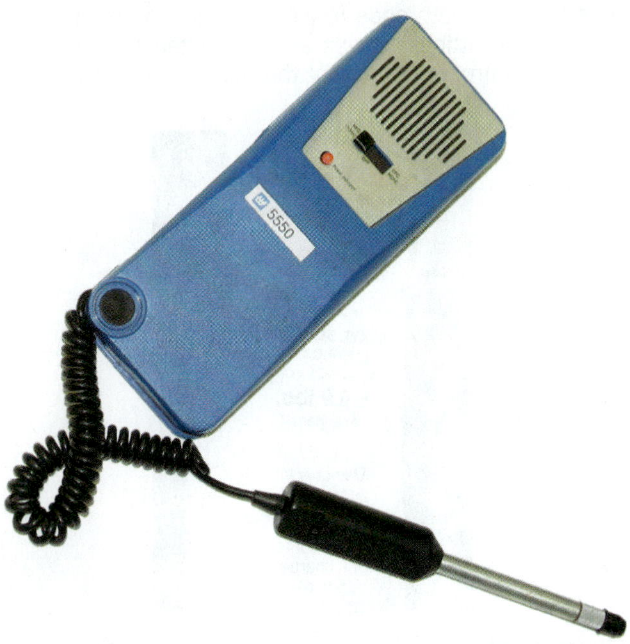

Figure 29-10 Older-style corona discharge electronic detector.

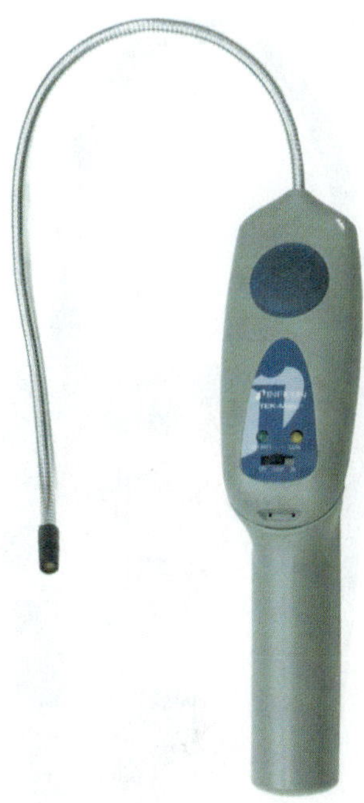

Figure 29-11 Heated diode electronic leak detector.

29.12 HEATED DIODE LEAK DETECTORS

Heated diodes are a newer technology that has proven to be sturdier and more reliable (Figure 29-11). Their tips last much longer than the corona discharge units, but they are more expensive to replace. Heated diode detectors use heated electrochemical sensors. When a CFC, HFC, or HCFC contacts with the hot surface, the chlorine and fluorine atoms are separated from the molecule and ionized. This causes an electrical current flow in the sensor, indicating the leak. Heated diode instruments tend to clear themselves quickly. This aids in tracking down the actual location of the leak and differentiating leaks from ambient contamination. Heated diode detectors are not set off by wind.

29.13 INFRARED LEAK DETECTORS

The newest handheld electronic leak detectors use infrared sensors (Figure 29-12). They are not affected by wind, water, or oil, and the sensor life is measured in years instead of hours. Refrigerant reflects infrared light. Refrigerant is drawn into the tip and travels between an infrared light source and an infrared light sensor. When refrigerant enters

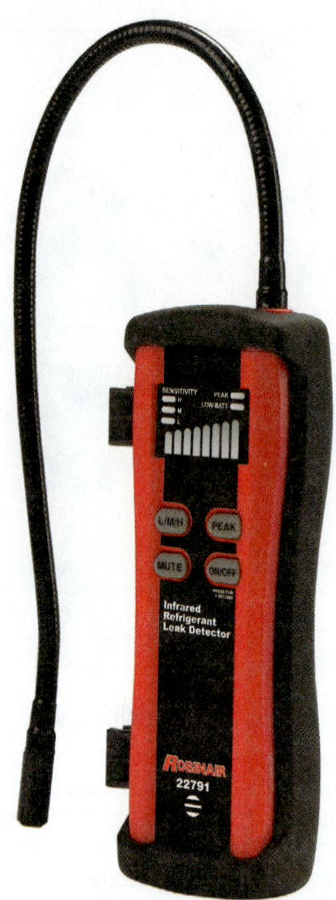

Figure 29-12 This handheld electronic leak detector uses an infrared sensor.

the tip, it reflects the infrared radiation and blocks the sensor from seeing it, much like a cloud can block the sun on a cloudy day. The detector senses the change and indicates a leak. IR sensors detect change. If the level of refrigerant stays the same, the detector stops alarming. Because of this, they are used differently from heated diode detectors. You sweep the detector back and forth, narrowing your target area.

29.14 LEAK TESTING WITH DYES

Refrigerant dyes have been around for many years. The idea is for the escaping refrigerant to leave a visible trail. These dyes do not really dye the refrigerant but the refrigerant oil. Since refrigerant oil travels throughout the system with the refrigerant, oil normally leaks out with the refrigerant. A well-known trick of the trade is to look for oily spots on or around systems when trying to find difficult leaks (Figure 29-13). The dye simply makes this more visible.

SERVICE TIP

It is important to make sure the dye you use is compatible with the oil in the system. Some dyes are designed for specific kinds of refrigerant oil. Check the label on the dye to make sure it is compatible with your system. Figure 29-14 shows the application notes on a container of dye.

Figure 29-13 Oily spots around mechanical or brazed joints can indicate a refrigerant leak.

Two types of dye are available: red and fluorescent. The red dye turns the oil in the compressor a very dark red. Red dye does just what you would expect: it shows leaks by creating a red stain anywhere refrigerant escapes. Fluorescent dye shows up when exposed to ultraviolet light (Figure 29-15). Ultraviolet dye is useful for locating leaks in unusual locations. These "hidden" leaks are often in areas that you would normally not suspect, such as a pinhole in the side of a receiver. Dyes can be an excellent tool for people doing regular maintenance on the same machines. Systems with dye previously installed can be quickly checked using a high-wattage black light.

The dye is available as an additive or loaded in refrigerant cylinders (Figure 29-16). Refrigerant that is already mixed with dye must leave the cylinder as a liquid to keep the dye in solution. The refrigerant charge in the cylinder forces the dye into the system. Dye is most often sold in small cylinders of dye called dye sticks. These have ¼-in flare ports on both ends to allow connection of standard refrigeration fittings. Valves are connected to both ends to allow control of refrigerant through the dye stick. Make

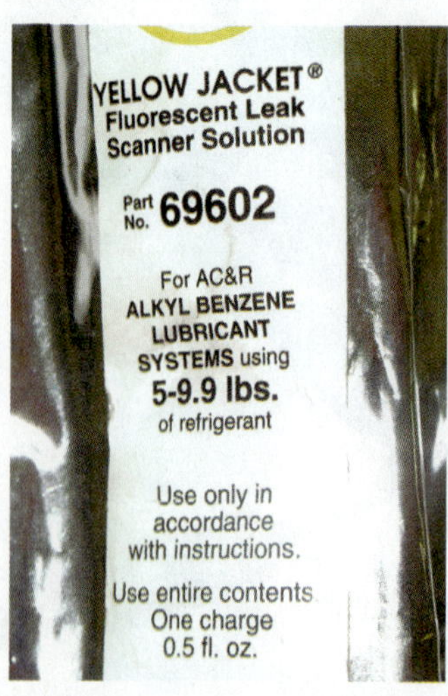

Figure 29-14 The instructions on this dye stick package indicate the type of refrigerant oil it is compatible with.

Figure 29-15 Ultraviolet dye makes small leaks visible that might not be detected using other methods.

Figure 29-16 Red dye additive.

certain the bleed valves are closed, and then connect short refrigeration hoses between the low-side and high-side service valves and the dye stick. The arrow on the dye stick should point toward the low-side service valve (Figure 29-17). Purge the air out of each hose by loosening the hose connection at the bleed valves. Crack the bleed valve on the inlet side of the dye stick and then crack the bleed valve on the outlet side. The system pressure will force the dye into the low side of the system.

Figure 29-17 The dye stick is connected between the high side and low side with the arrow pointing toward the low side.

CAUTION

Do not open the bleed valves all the way. Pushing the dye into the system too quickly can cause a liquid slug and kill the compressor.

Dye is not a one-call method. The dye and oil must have time to circulate in the system before the leaks will show up. A major problem with dyes is that they are messy. Expect a mess every time gauges are connected and disconnected (Figure 29-18). It is necessary to clean up the mess, because it might otherwise lead someone to suspect a leak where the dye has sprayed. Cleaning solutions made specifically for cleaning the dye should be used to clean up before leaving (Figure 29-19). Many equipment manufacturers do not endorse the use of dyes, so check with the equipment manufacturer before adding dye to a system. The use of dyes in some manufacturers' equipment will void the warranty.

29.15 ULTRAVIOLET LEAK-DETECTION LIGHTS

Ultraviolet bulbs that produce UV light with a wavelength of 375 nm are required to find leaks indicated by the fluorescent dye. Older UV lights were large, required 120 volts, and

Figure 29-18 The spray pattern makes a mess when refrigerant gauges are removed from a system containing dye; the top picture shows what the area looks like without the black light; the bottom picture shows the same area with the black light; this should be cleaned up before leaving the job.

used expensive mercury vapor bulbs. Figure 29-20 shows an example of an early model UV light. Newer models use battery-powered LED lights. These are far easier to use and less costly to maintain. Figure 29-21 shows a newer flashlight-style LED UV light for leak detection.

Figure 29-19 This cleaning solution is made specifically for cleaning up dye.

Figure 29-20 Older UV light used to detect fluorescent dyes.

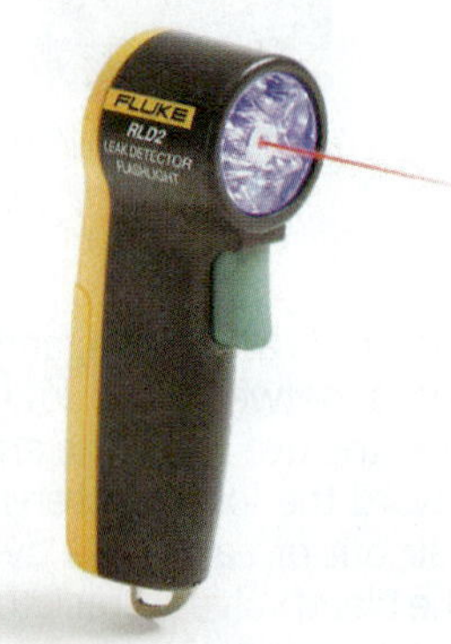

Figure 29-21 Newer LED flashlight-style UV light used to detect fluorescent dyes.

TECH TIP

It is important to get the right UV light. All UV light is not the same wavelength. UV just means the light waves are shorter than normal visible light. Inexpensive lights used for black-light posters will usually not work well because they don't have the right wavelength of light to show the indicator dye.

29.16 PROVING THE SYSTEM IS TIGHT

There are two methods of proving a system is free of leaks that do not involve checking for escaping gas:

- Standing pressure test
- Standing vacuum test

Standing Pressure Test

A standing pressure test involves charging an empty system with an inert gas, recording the pressure, and then checking back later to see if the pressure has dropped. This method is only valid when the system is charged with a gas that cannot possibly condense at the pressure and temperature found in the system because a saturated mixture of liquid and gas refrigerant would change pressure significantly as the ambient temperature changes. A drop in pressure might just reflect a drop in temperature and not a leak. Also, the refrigerant pressure will remain the same at a saturated condition as long as the system temperature remains stable. As long as there is still some liquid in the system, the pressure in the system will be controlled by the temperature, not by the amount of refrigerant in the system. Therefore, a system could have a leak, lose refrigerant, and not see a drop in pressure.

Temperature change will affect the system pressure even if all the gas is in vapor form, but the change will not be as pronounced. The absolute pressure of a gas varies directly with its absolute temperature. The gas laws can be used to predict the effect of temperature. If a system with a pressure of 85 psig and a temperature of 100°F changes to a temperature of 50°F, the pressure would be reduced to 76 psig.

Converting those temperatures to absolute temperature yields a temperature change of 560°R to 510°R. That ratio is 510/560, or approximately 0.91. Multiplying this ratio times an absolute pressure of 100 psia gives 91 psia. 91 psia is 76 psig converted to gauge pressure.

Standing Vacuum Test

It is common practice to wait after pulling a deep vacuum on a system to ensure that the pressure will not rise. A rise in pressure can be interpreted as air entering the system through a leak. Of course, other factors can also cause a rise in system pressure after an evacuation, such as system contamination, slow evaporation of refrigerant dissolved in the compressor oil, or simply reading the vacuum near the vacuum pump instead of near the system.

29.17 LOW-PRESSURE SYSTEMS

Leak testing low-pressure chillers is entirely different from leak testing high-pressure systems. Because the evaporator on a low-pressure system operates in a vacuum, refrigerant does not leak out of a low-side leak. Air leaks in instead. One sign of leaks in a low-pressure system is extended purge operation. When a chiller is leaking, the air sucked in through the leaks forces the purge system to operate more to remove the air. Technicians need to raise the evaporator pressure above 0 psig to detect a refrigerant leak. This can be done by either raising the temperature of the water in the chiller bundle or by adding nitrogen. The preferred method for raising the system pressure is to circulate warm water through the chiller bundle. The advantage of this method is that you are not introducing a noncondensable gas that will have to be removed through purge operation. Either way, it is important to keep the pressure under 10 psig. Chillers have rupture disks that open at 15 psig, releasing the refrigerant in the chiller. This is to prevent the chiller from being pressurized beyond its design limits.

UNIT 29—SUMMARY

Refrigeration systems should be leak tested before charging or whenever refrigerant needs to be added. Leak testing techniques include the following:

- Soap bubble solutions
- Halide torches, which work only on chlorinated refrigerants
- Ultrasonic leak detectors, which detect the sound of refrigerant escaping
- Electronic sniffers, which detect CFC, HCFC, and HFC refrigerants
- Dyes that make oil spots more visible
- Standing vacuum test
- Standing pressure test

Significant leaks in systems that hold 50 lb of refrigerant or more must be repaired. A significant leak is defined as an annual leak rate of 35 percent of the charge for commercial and industrial refrigeration or an annual leak rate of 15 percent of the charge for air-conditioning systems and other applications.

WORK ORDERS

Service Ticket 2901

Work to Be Performed: Run Refrigerant Lines

Equipment Type: R-404A Commercial Refrigeration System

A system installer is asked to run refrigerant lines for a new commercial refrigeration installation. After running the lines, the installer pressurizes the lines with 100 psig of nitrogen to check for leaks. The installer uses soap bubbles at the joints to check for leaks. One joint indicates a leak by forming large bubbles. The nitrogen is released, and the braze joint is redone. The lines are pressurized once more with nitrogen and checked with soap bubbles again. No large bubbles appear on the repaired joint, but a small clump of frothy bubbles appears on another joint. The nitrogen is released, that joint is redone, and the system is checked again. No bubbles or froth appears this time. Because of the multiple leak problems, the installer decides to perform a standing pressure test. The equalized system pressure and temperature are recorded. The next day the installer calculates the expected pressure change due to temperature change and finds the system pressure matches the calculations. The installer releases the nitrogen, pulls a deep vacuum on the system, and charges it to factory specifications.

Service Ticket 2902

Work to Be Performed: Find Elusive Refrigerant Leak

Equipment Type: R-22 Split-System Heat Pump

A split-system heat pump has lost a small amount of refrigerant each year, and it seems to be getting worse. The technician installs gauges to make sure the system has pressure, which it does. Leaving the system off, the technician uses a heated diode electronic leak detector to check inside the evaporator and condenser cabinets for puddled refrigerant vapor near the bottom of the cabinet. The leak detector goes off in the evaporator cabinet. The technician is unable to locate the specific leak location. Fluorescent dye is added to the system and the system charge is brought back up to normal. The system is operated for a week. The technician returns and inspects the evaporator area with an ultraviolet light. Small, bright streaks appear in several locations on the coil where the tubing passes through the endplate. The technician informs the customer that he needs a new evaporator coil.

UNIT 29—REVIEW QUESTIONS

1. What is the purpose of the pressure regulator on the nitrogen cylinder?
2. A refrigeration system data plate states that the low-side test pressure is 150 psig and the high-side test pressure is 450 psig. What does this mean?
3. What is the maximum pressure that should be used for leak testing a refrigeration system with a hermetic compressor?
4. What is the difference between corona discharge and heated diode electronic leak detectors?
5. How does an LED refrigerant leak detector work?
6. What refrigerants can be detected with a halide torch?
7. What is the advantage of an ultrasonic leak detector?
8. How can dye indicate system leaks?
9. Why does increased purge operation indicate a leak in the low side of a low-pressure system?
10. What method is recommended for raising the pressure of the refrigerant in a low-pressure chiller?
11. What is the maximum safe pressure to impose on any low-pressure system?
12. A system in a 90°F room is charged to 100 psig with nitrogen to perform a standing pressure test. When the system pressure is checked later, the room temperature is 70°F. What pressure would you expect to see in the system?
13. How do leaking refrigerant systems cost their owners money?
14. What items should be included in a preliminary visual inspection when looking for refrigerant leaks?
15. What leak-detection method does the EPA consider the best for pinpointing leaks?
16. Compare older UV lights used for fluorescent dye and newer flashlight models.
17. Describe how dye is added to a system.
18. Besides a leak, what can make the vacuum rise on a standing vacuum test?
19. How do soap bubble solutions indicate very small refrigerant leaks?
20. Why should technicians never use oxygen to pressurize refrigeration systems when leak testing?

UNIT 30

Refrigerant System Evacuation

OBJECTIVES

After completing this unit, you will be able to:

1. list the reasons for evacuating a refrigeration system.
2. describe the effects noncondensable gas has on refrigeration systems.
3. describe the effect moisture has on refrigeration systems.
4. explain how to measure a deep vacuum.
5. explain the system of measuring vacuum in microns.
6. describe the ways that the vacuum pump and vacuum gauges can be connected to the system.

30.1 INTRODUCTION

The operating efficiency and life of the compressor both depend heavily on the proper evacuation and charging of the refrigerant circuit. Evacuation to the proper levels is critical to ensure the system has been dehydrated and all noncondensables have been removed. Leaving noncondensable gases in the system increases the system head pressure and compression ratio. These adverse system reactions to noncondensable gases decrease system capacity and efficiency while increasing energy use. Any moisture or noncondensable gas left in the system significantly affects the system's performance and can have a devastating effect on the compressor's longevity. Water can freeze up in the metering device, starving the evaporator and decreasing system capacity. Water can also hydrolyze refrigerant oil to form acid, leading to internal system corrosion.

30.2 LEAK TESTING

Refrigeration systems should always be checked for leaks prior to evacuation. Attempting to evacuate a leaky system will only suck air into it through the leak. Checking for leaks prior to evacuating saves time in the long run, because time spent evacuating a leaky system is wasted. Refer to Unit 29, Refrigerant Leak Testing, for more information on how to test refrigeration systems for leaks.

30.3 THE PURPOSE OF EVACUATION

The only two things that should be circulating inside a refrigeration system are refrigerant and refrigerant oil. Additional gases and/or liquids degrade system performance and reduce equipment life. The purpose of pulling a deep vacuum on a system is to have nothing but refrigeration oil in the system before charging it with refrigerant. Proper evacuation of a unit will remove noncondensables and water. Other contaminants, such as brazing oxides and metal particles, must be removed using a filter drier. When properly done, evacuation also ensures a leak-tight system, because systems with leaks will not pull down into a deep vacuum and hold.

TECH TIP

Evacuation cannot completely dehydrate systems that use hygroscopic lubricants, such as polyol ester (POE) or polyalkylene glycol (PAG) lubricants. Moisture can only be removed from these lubricants using a filter drier. That is why newer HFC systems with POE lubricant require a filter drier in new installations in addition to a thorough evacuation. Note that the filter drier does not eliminate the need for evacuation. The filter can only remove small quantities of moisture and cannot remove noncondensable gas at all.

30.4 SYSTEM CONTAMINANTS

Since a vacuum pump only removes gas, an evacuation can only remove substances that can be in vapor form at normal ambient temperatures under a deep vacuum. Solid contaminants like copper chips or carbon must be filtered out. However, the two most common system contaminants can be removed through evacuation. They are noncondensable gases and moisture.

Noncondensable Gases

Noncondensables are gases that cannot condense at the normal system pressures and temperatures, such as air or nitrogen. The primary problem caused by noncondensable gases is an increase in the system high-side pressure and temperature (Figure 30-1). The pressure of the noncondensable gases is added to the refrigerant pressure, according to Dalton's law. Noncondensable gases tend to collect in the condenser since they do not condense. This has the effect of reducing the amount of condenser surface available for condensing refrigerant, increasing the refrigerant temperature and pressure. The effect of the additional pressure is to increase the compression ratio and decrease the amount of refrigerant circulated. This reduces system capacity, increase energy usage, and leads to compressor overheating.

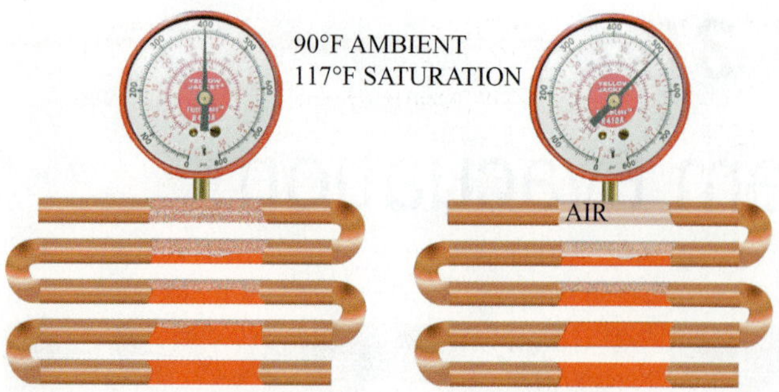

Figure 30-1 Noncondensables in the condenser raise the system high-side pressure.

Air in the system can cause chemical damage as well. Oxygen is one of nature's most chemically active elements. Its rate of chemical reaction with refrigeration oil increases very rapidly with a rise in temperature above 200°F. The rate of chemical reaction between oil and oxygen doubles with every 18°F rise in temperature above 200°F. When this occurs, the oil starts to break down and deposits buildup in the discharge valve area. Further, this breakdown of refrigerant creates an acid condition that leads to compressor motor burnouts. A chain of events with ever-increasing system malignancy develops that leads to system failure.

TECH TIP

The number of systems with large amounts of noncondensables appears to be increasing. Installers that are in too much of a hurry to evacuate systems are leaving large quantities of air in systems. Simply opening the system charging valves after connecting the refrigerant lines traps the air from the lines and indoor coil inside the system. These systems operate with a very high head pressure, a hot compressor, and poor cooling efficiency. The only fix for this once it has been done is to recover all the refrigerant and air in the system into a separate empty recovery cylinder, properly evacuate the entire system, and weigh in the correct charge.

Water

Water is another formidable foe of the refrigeration system. A single drop of water can cause a major blockage of the refrigerant flow within a system by freezing up at the metering device. Water can also cause a breakdown of the refrigerant oils and the refrigerant used in a system. This makes moisture the greatest of all enemies to the inside of the refrigeration system, especially units equipped with hermetic compressors.

The primary ill effects of moisture in the system are:

- Freeze-up at the expansion valve or metering device
- Corrosion of metals
- Valve failure
- Copper plating
- Chemical damage to motor insulation

System Freeze-up

A freeze-up at the metering device occurs when small droplets of water freeze and form ice crystals. This occurs whenever the system temperature drops below the water's freezing point. The flashing of refrigerant at the metering device can cool the water enough to freeze it, creating a restriction.

System Corrosion

Most metals corrode on contact with moisture and oxygen. This can occur inside the refrigeration system when both moisture and air are present; corrosion starts eating away at the metal. This reaction increases with heat. The presence of noncondensables increases the compressor temperature and accelerates the chemical activity of the water and oxygen. Corrosion is not a problem in a properly evacuated and charged system.

Acid Creation

Moisture, when teamed with refrigerant in the refrigeration system, will create acid. Acid in the system can cause a burnout when the refrigerant is exposed to the compressor motor. The acidic reaction increases when heat is added to moisture and refrigerant. An ever-increasing upward spiral of acid level is created, and temperature increases eventually leads to compressor motor burnout.

30.5 MEASURING VACUUM LEVEL

A vacuum is any pressure less than atmospheric pressure. Vacuums can be measured by specifying how much of the atmospheric pressure has been removed or by specifying how much pressure remains. Atmospheric pressure is typically measured by how high a mercury column it will support. Standard atmospheric pressure at sea level is 29.92 inches of mercury, abbreviated in Hg (Figure 30-2). Mercury columns can be used to measure vacuum by measuring how much pressure has been removed. A 20 in Hg vacuum means that the pressure is 20 inches of mercury less than atmospheric pressure (Figure 30-3). Since atmospheric pressure is 29.92 in Hg, a perfect vacuum would be 29.92 in Hg vacuum. This would indicate that all the pressure had been removed.

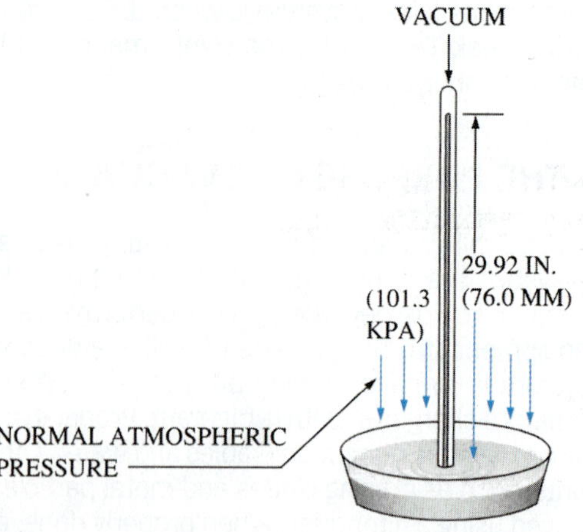

Figure 30-2 Atmospheric pressure will support a mercury column 29.92 in high.

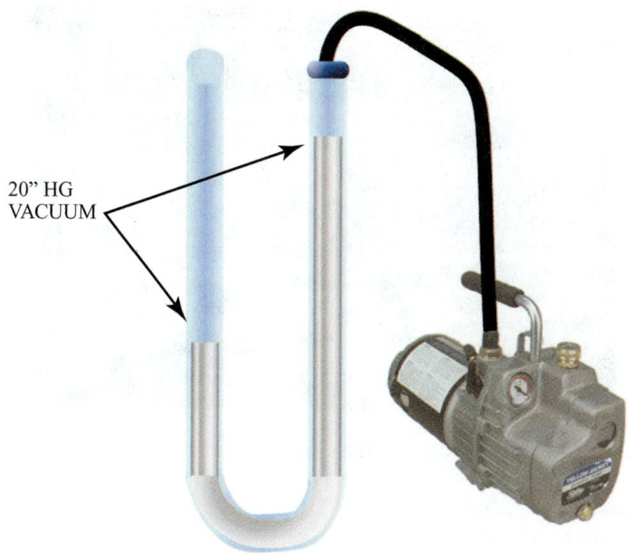

Figure 30-3 The pressure in a 20-in mercury vacuum is 20 in less than atmospheric pressure; this leaves 9.92 in of mercury remaining.

Absolute Pressure

Absolute pressure can be used to measure vacuum by measuring how much pressure is left. In the case of a 20 in Hg vacuum, the absolute pressure would be 9.92 in Hg absolute because 9.92 inches of the original 29.92 inches of pressure are left (see Figure 30-3). A perfect vacuum using absolute measurement would be 0 inches of mercury absolute. This indicates that no pressure remains.

The problem with both of these methods is that an inch is far too large and imprecise a measurement to tell the difference between a really good vacuum and a poor one. Most equipment manufacturers now specify the required vacuum level in microns of mercury absolute pressure, normally shortened to just microns for simplicity.

Microns

A micron is a very small measurement of distance equal to 1/1,000 of a millimeter. There are approximately 25,400 microns in an inch. When vacuum is measured in microns, what is actually being measured is how much pressure remains in the system, not how much pressure has been removed. Absolute pressure is what is really being specified. Enough pressure remains in a 500-micron vacuum to support a column of mercury 500 microns high, or 1/2 mm. To put this number in perspective, full atmospheric pressure can support a mercury column 760,000 microns high. Of the initial 760,000 microns, only 500 are left.

30.6 SYSTEM EVACUATION

Evacuation is the process of removing gases from the refrigeration system. Gases removed from refrigeration systems by a vacuum pump include air, nitrogen, residual refrigerant, and water vapor. The majority of the gases in a refrigeration system can be evacuated in a few minutes. It normally does not take long to remove air and noncondensables, since they are already a gas. Achieving a deep vacuum on a system that is dry takes much less time than on a system that is contaminated with moisture. However, removing air and noncondensables is only one reason for evacuating a refrigeration system. Evacuating a system is also one of the principle means of dehydrating it.

SERVICE TIP

Listening to the sound of a vacuum pump can tell you something about what it is pulling out of the system. Most deep vacuum pumps operate under oil. The gas being evacuated gurgles as it goes through the pump. When the gurgling noise stops, the pump has removed most of the noncondensable gas from the system. This is the point where the pump starts to dehydrate the system. If the gurgling noise never goes away, either the system or the connections are leaking and letting in a constant stream of air.

30.7 SYSTEM DEHYDRATION

Dehydration is the process of removing moisture from a refrigeration system. Moisture in a system may exist as liquid or vapor. Water vapor is removed quickly along with the noncondensables. However, the water in liquid form must be vaporized before it can be removed. The amount of gas created when even a small amount of water is vaporized is quite large. Vaporizing an ounce of water at an absolute pressure of 1,000 microns of mercury creates over 500 ft^3 of water vapor (Figure 30-4). It would take a 5-CFM (cubic feet per meter) vacuum pump an hour and 40 minutes to remove this much vapor! For this reason, the best practice when doing refrigeration work is to avoid letting any water into the system in the first place. That is why all refrigeration components and refrigeration lines should be sealed when not in use.

Water in Compressor Oil

The time estimate in the above example assumes that the water is able to vaporize easily. However, water that collects below the refrigerant oil in the compressor crankcase can be more difficult to remove. Striking the compressor and applying heat to the compressor crankcase can help

Water Vapor Volume		
	Saturation Pressure	
Saturation Temperature (°F)	(Microns)	Volume ft^3 per Ounce
10	1,760	566
20	2,740	354
30	4,180	225
40	6,230	153
50	9,170	106
60	13,250	75
70	18,740	54
80	26,260	40

Figure 30-4 Volume of an ounce of vaporized water at different vacuum levels.

Figure 30-5 The vapor coming off of dry ice (solid carbon dioxide) is an example of sublimation.

remove water from underneath mineral oil. A filter drier is the only reliable way to remove water from hygroscopic refrigerant oil such as polyol ester. Even a deep evacuation will normally not remove water that has been absorbed into polyol ester lubricant.

Sublimation

It is possible to have ice inside the system when systems are assembled in below-freezing conditions. Frozen water can be evaporated through sublimation. Sublimation is the process of turning a solid into a gas without passing through the liquid stage. The vapor that comes off of dry ice is an example of sublimation (Figure 30-5). It is possible to dehydrate a system with ice inside it, but it is a very slow process. Heating any part of the system that is below freezing will speed up evacuation and dehydration if it has any moisture in it.

Speeding up Dehydration

During evacuation, the vacuum level in a system that has moisture in it will appear to get stuck between 1,500–5,000 microns. This is because the vaporizing water continually replaces the gas that is removed by the vacuum pump, preventing the pressure from dropping. If a system is suspected of having liquid water anywhere in it, every effort should be made to remove the water from the system prior to evacuation. Following good refrigeration practices will prevent water from getting into in the system in the first place. Heating any part of the system that is suspected of having moisture in it will speed up evacuation and dehydration by helping the water vaporize quickly. Systems should be evacuated from both the high side and low side of the system using large-diameter hoses that are as short as possible. When possible, connect the vacuum pump to the system or gauge set using multiple hoses. Figure 30-6 shows a vacuum pump that allows multiple connections between the vacuum pump and the system being evacuated. Removing Schrader valve cores using a vacuum-rated valve core tool will also decrease the time required for evacuation.

Figure 30-6 The Appion TEZ8 vacuum pump has multiple connectors, allowing several hoses to be connected at one time. *(Courtesy of Appion Inc.)*

Desired Evacuation Level

The lower the final system evacuation level is, the better. Lower evacuation levels give more assurance of a dry system. However, the actual evacuation level required to vaporize all the water in the system is dependent on temperature. Lower system temperatures require lower levels of evacuation to ensure a dry system. Figure 30-7 shows that at an absolute pressure of 400 microns, water will vaporize at temperatures as low as −20°F. A system pressure as high as 1,100 microns will vaporize water at 0°F. The most common benchmark for a good vacuum is a system pressure no higher than 500 microns. Generally, a system is considered to be dehydrated if the system can hold a pressure of 500 microns or lower. Components with refrigeration oil cannot be evacuated lower than 200 microns, because the oil starts to produce a vapor pressure at 200 microns.

Outgassing

Under deep vacuum, many materials release a small number of molecules that create a slight vapor pressure. This is known as outgassing. Vacuum levels below 500 microns are difficult to maintain due to outgassing of connection hoses and materials in the refrigeration system. Refrigeration hoses do this more when they are new. Although many vacuum pumps are available that can pull down below 50 microns, it is not practical to evacuate most system components to this level in the field due to outgassing.

Keeping Water out of System

It saves a great deal of time to follow good refrigeration practices when assembling systems. This includes keeping system components and piping plugged or capped while awaiting assembly and storing components with a charge of dry nitrogen to discourage the entry of air and moisture (Figure 30-8). One of the principle components of air is moisture; letting in air automatically lets in moisture as well. Allowing air in on a hot, humid day allows a significant amount of moisture in. If the component is stored on a cool location, water that came in as water vapor on a warm day may condense to droplets of liquid water, which will significantly increase the time required for system dehydration.

Temperature°F	Millimeters Hg Absolute Pressure	Microns Hg Absolute Pressure	Inches Hg Absolute Pressure	Inches Hg Vacuum
–70	0.02	20	0.0008	29.9192
–60	0.03	30	0.0012	29.9188
–50	0.07	70	0.0028	29.9172
–40	0.13	130	0.0051	29.9149
–30	0.23	230	0.0091	29.9109
–20	0.4	400	0.0157	29.9043
–10	0.68	680	0.0268	29.8932
0	1.1	1,100	0.0433	29.8767
10	1.76	1,760	0.0693	29.8507
20	2.74	2,740	0.1079	29.8121
30	4.18	4,180	0.1646	29.7554
40	6.23	6,230	0.2453	29.6747
50	9.17	9,170	0.3610	29.5590
60	13.25	13,250	0.5216	29.3984
70	18.74	18,740	0.7378	29.1822
80	26.26	26,260	1.0338	28.8862

Figure 30-7 Water will evaporate at temperatures below freezing in a vacuum.

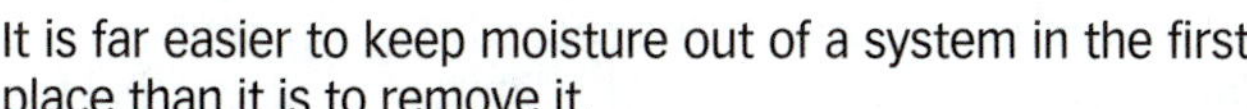

Figure 30-8 Refrigeration components should be sealed when not in use to keep out air and moisture.

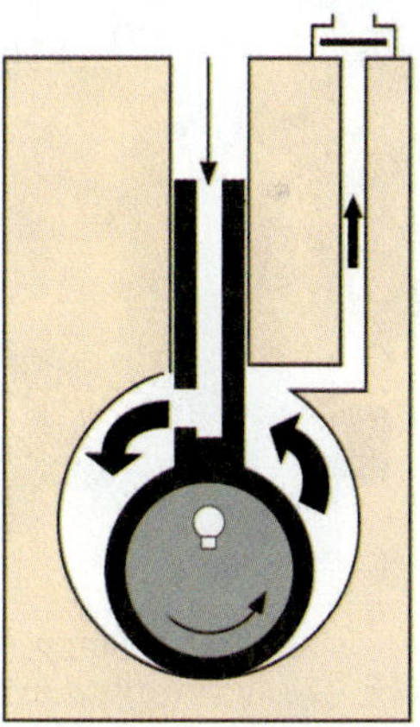

Figure 30-9 Diagram of rolling-piston vacuum pump.

It is far easier to keep moisture out of a system in the first place than it is to remove it.

30.8 VACUUM PUMPS

Two general designs of vacuum pumps are available: rolling piston (Figure 30-9) and rotary vane (Figure 30-10). Reciprocating-piston-type pumps are not used as vacuum pumps. They are incapable of producing a deep vacuum because of the clearance necessary between the piston head and valve plate. The rolling-piston or the rotary-vane designs are more suited for deep vacuum because neither requires any clearance volume. The rolling-piston type is capable of lower pressures and is more rugged. However, these pumps are also much heavier and more expensive

Figure 30-10 Rotary-vane vacuum pump.

Figure 30-11 Rolling-piston vacuum pump.

Figure 30-13 Single-stage vacuum pump.

Figure 30-12 Rotary-vane vacuum pump. *(Courtesy Ritchie Engineering Company, Inc., Yellow Jacket Products Division)*

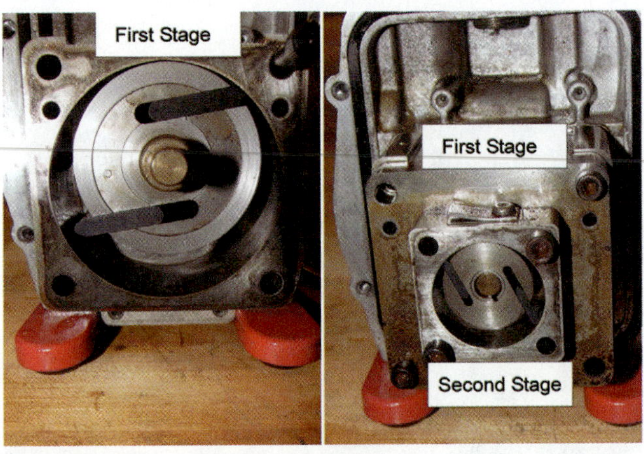

Figure 30-14 Two-stage vacuum pump.

(Figure 30-11). Most refrigeration vacuum pumps are rotary-vane-type pumps (Figure 30-12). These have the advantage of being lightweight and relatively inexpensive.

Vacuum pumps are produced in single-stage and two-stage models. Single-stage pumps discharge directly to the atmosphere (Figure 30-13). Two-stage pumps have a first stage that discharges into the suction side of the second stage (Figure 30-14). Two-stage pumps generally have the best record in the refrigeration industry. They are capable of producing consistently lower pressures and are generally more efficient when removing water vapor.

Gas Ballast Valve

A problem with two-stage pumps is that there is a slight compression of the discharge gas coming out of the first stage and entering the second stage. This can tend to condense the water vapor being removed from the system and leave it in the vacuum pump oil. To prevent this problem, the discharge of the first stage can be directed out to the atmosphere instead of into the second stage through a valve known as a gas ballast valve. This limits the vacuum that the pump can produce, since admission of outside air will raise the absolute pressure of the second stage. The gas ballast valve is typically only open during initial pulldown and is closed once the system pressure has dropped to 20 in Hg vacuum (Figure 30-15).

Isolation Valve

Most vacuum pumps have valves that shut off the pump from the connections to the pump (Figure 30-16). These valves can be used to valve off the vacuum assembly from the vacuum pump. This valve should be shut off before turning the pump off. If the valve is left open, the vacuum in the system can suck the oil in the pump out of the pump and

Figure 30-15 The ballast valve is opened when starting the vacuum pump and closed after the vacuum has pulled down to 20 in Hg.

into the hoses, vacuum gauge, or even the system if left open long enough. The valve should be left closed when the pump is not in use to prevent contamination of the oil in the vacuum pump.

Figure 30-16 Vacuum pump isolation shutoff valve.

System Size in Tons	Pump CFM Capacity
1–10	1.5
10–15	2.0
15–30	4.0
30–45	6.0
45–60	8.0
60 and above	11.0

Figure 30-17 Recommended vacuum pump capacity.

Vacuum Pump Ratings

Vacuum pumps have two ratings: a volume capacity and a blank-off pressure. The volume rating tells how fast the pump can move gas at ideal conditions. Rotary-vane vacuum pumps intended for refrigeration use are available with capacities from 1 CFM up to 12 CFM. The CFM rating tells how many cubic feet of gas the pump can move in 1 minute at ideal conditions. Figure 30-17 shows the suggested maximum system size for different CFM vacuum pumps.

The blank-off pressure tells the lowest pressure the vacuum pump can produce. For refrigeration use, the vacuum pump blank-off pressure should be no higher than 50 microns. The blank-off pressure is more important to pulling a good vacuum than the CFM rating. Once the initial outrush of gas has been accomplished, the speed of evacuation has more to do with the pressure difference between the system and the vacuum pump. The lower the pump's blank-off pressure, the greater this final pressure difference will be. A 1-CFM pump with a 15-micron blank-off pressure will evacuate most systems more quickly than a 10-CFM pump with a 100-micron blank-off pressure.

30.9 VACUUM PUMP OIL

The oil in a vacuum pump acts as a seal as well as a lubricant for the pump. A vacuum pump cannot pull a vacuum any lower than the vapor pressure of its oil. Oil that is not intended for use in vacuum pumps has a much higher vapor pressure than vacuum pump oil and will limit the amount of vacuum the pump can produce. Likewise, dirty oil will do the same thing. All the gases pulled out of a system by the vacuum pump pass through the vacuum pump oil. Refrigerant, water, and acids will dissolve in the oil as they pass through. Contaminants dissolved in the oil will produce a vapor pressure and limit the pump's ability to pull a good vacuum. Clean vacuum pump oil is typically clear (Figure 30-18). Vacuum pump oil that is contaminated with a large amount of water is milky or cloudy in appearance (Figure 30-19). Leaving dirty oil in a vacuum pump also exposes its internal parts to potentially corrosive chemicals, reducing its useful life. To maintain peak performance and prolong equipment life, most vacuum pump manufacturers recommend replacing the oil in the pump after each evacuation. One vacuum pump has a unique design that allows the oil in the vacuum pump to be changed while the pump is operating. The majority of the oil is kept in a clear outboard canister that can be swapped out. The oil is changed by removing the old oil canister and putting in a new one (Figure 30-20).

Figure 30-18 The oil sight glass in a vacuum pump should show clear, clean oil.

Figure 30-19 Cloudy or milky vacuum pump oil is contaminated with water.

Figure 30-20 The oil in the Appion TEZ 8 is held in an outboard canister that can be changed while the pump is operating. *(Courtesy of Appion Inc.)*

Figure 30-21 The vacuum scale on compound gauges is calibrated in Hg vacuum.

The amount of volatile contaminants in the oil can be tested with a vacuum gauge. A vacuum pump should be able to pull down to 50 microns when connected directly to a vacuum gauge if it is in good shape and the oil is clean.

30.10 VACUUM GAUGES

The compound gauge used on most gauge manifolds has a vacuum scale that reads from 0 psig down to 30 in Hg vacuum (Figure 30-21). However, mechanical gauges cannot accurately indicate the difference between a poor vacuum and a great vacuum. The needle on the gauge is typically wider than the space occupied by 1 in Hg vacuum (Figure 30-22). A high vacuum gauge is required for accurate vacuum readings, and the industry has developed electronic instruments to measure high vacuums.

Figure 30-22 The needle thickness is equal to the space occupied by 1 in Hg vacuum on the compound gauge.

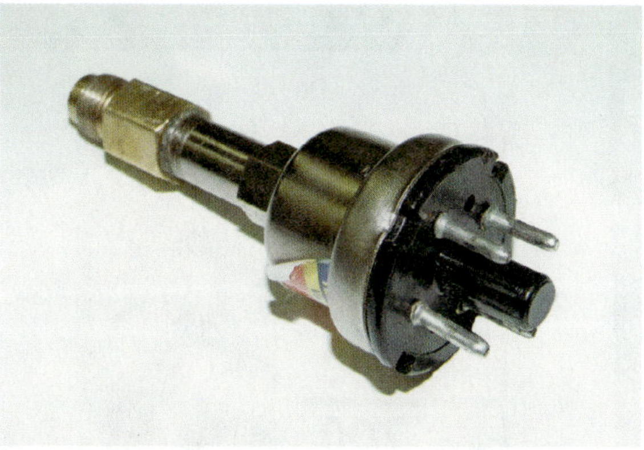

Figure 30-23 Thermistor sensor used with electronic vacuum gauge.

In general, electronic vacuum gauges are heat-sensing devices. The element that is connected to the system generates heat. The rate at which the heat is carried off changes as the surrounding gases and vapors are removed. The output of the sensing element changes as the heat dissipation rate changes. This output is indicated on a meter that is calibrated in microns. The sensors used in these instruments are typically either a thermocouple or thermistor (Figure 30-23). Three general styles of electronic vacuum gauges are available:

- Vacuum gauges with an analog scale and needle (Figure 30-24)
- Vacuum gauges with a digital numeric display (Figure 30-25)
- Vacuum gauges with lights or segments that indicate general vacuum levels (Figure 30-26)

30.11 MAINTAINING VACUUM GAUGE ACCURACY

A deep vacuum gauge is really only useful if it is accurate. To maintain accuracy, the manufacturer's instructions and specification must be followed. Conditions that can lead to inaccurate readings or damaged vacuum sensors include failure to calibrate the vacuum gauge, getting the sensor dirty, and exposing the vacuum gauge to high pressures.

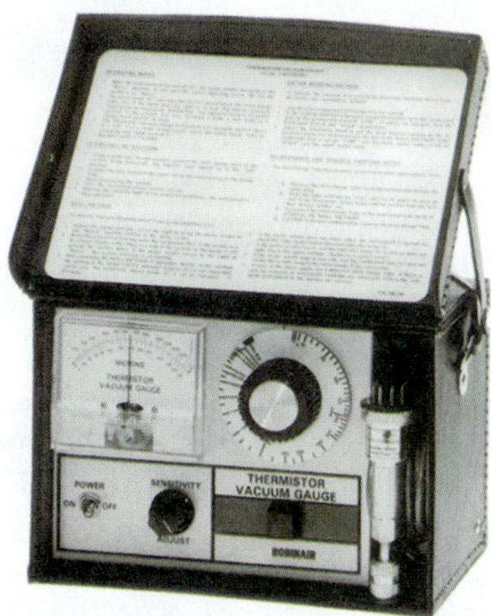

Figure 30-24 Analog electronic vacuum gauge. *(Courtesy Robinair)*

Figure 30-25 Electronic vacuum gauge with digital readout.

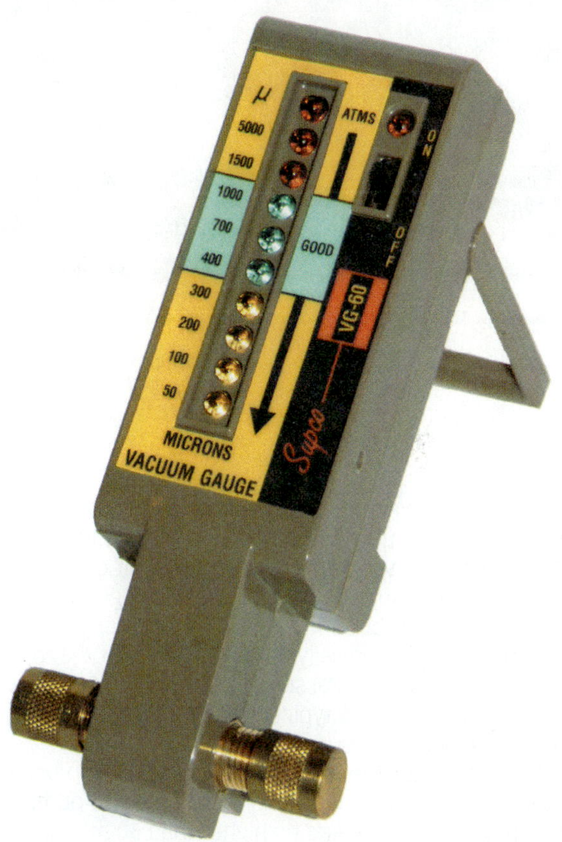

Figure 30-26 Segmented digital vacuum gauge. *(Courtesy of Robinair)*

Vacuum Gauge Calibration

Most analog vacuum gauges require calibrating before using. To calibrate the gauge, turn it on while the sensor is reading atmospheric pressure. Adjust the calibration screw until the needle indicates atmospheric pressure (Figure 30-27).

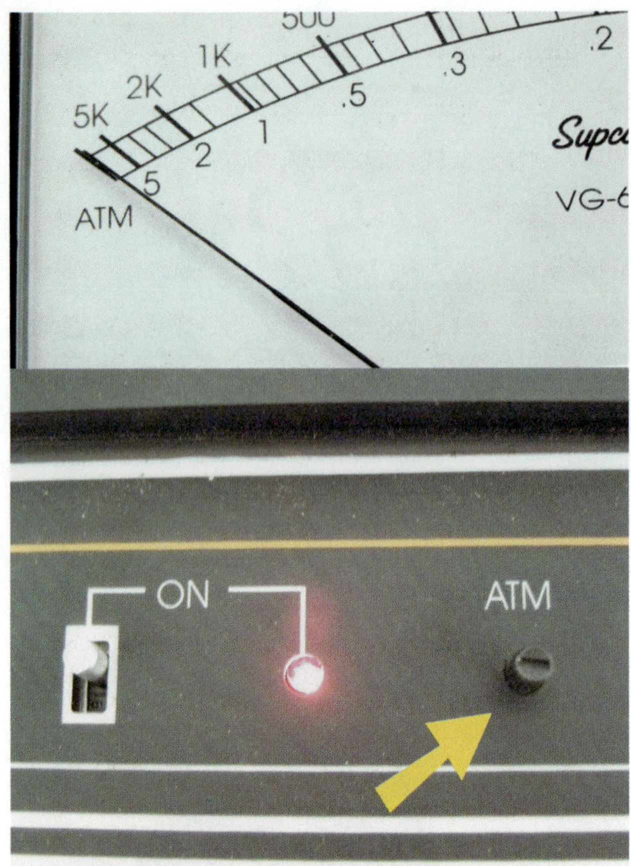

Figure 30-27 Analog vacuum gauges should be calibrated to atmospheric pressure before use.

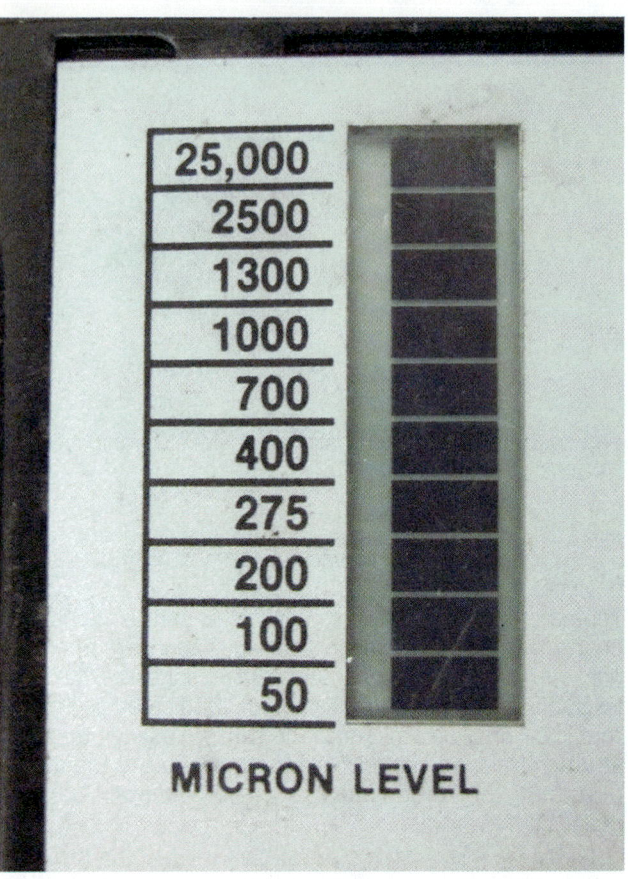

Figure 30-28 Segmented vacuum gauge showing atmospheric pressure.

Most digital vacuum gauges sold today do not require calibration when used. However, a quick way to check the accuracy of a vacuum gauge is to turn it on while it is exposed to atmospheric pressure. The gauge should not indicate a vacuum but should read atmospheric pressure. Segmented gauges typically have a segment at the top that indicates atmospheric pressure (Figure 30-28). Gauges with digital numeric displays typically show a single "1" to indicate atmospheric pressure (Figure 30-29).

Next, connect the gauge to a vacuum pump using as little connection hose as possible. Operate the pump. The vacuum gauge should read no higher than 50 microns. This is also a means of testing the pump. A pump with dirty oil will not pull down to its normal blank-off pressure. These two tests help demonstrate that both your vacuum gauge and pump can perform their jobs.

Figure 30-29 Digital vacuum gauge showing atmospheric pressure.

Cleaning the Vacuum Sensor

Most vacuum sensors will not read accurately if they become dirty or contaminated with refrigeration oil. They may normally be restored to proper working condition by cleaning them using alcohol. After cleaning, turn the gauge upside down and allow the vacuum sensor to dry for 20 to 30 minutes before using the gauge.

SERVICE TIP

Do not attempt to clean the vacuum sensor by sticking Q-tips, wires, small screwdrivers, or any other crude device into the sensor. The chance of destroying the sensor this way is far greater than the chance of improving it.

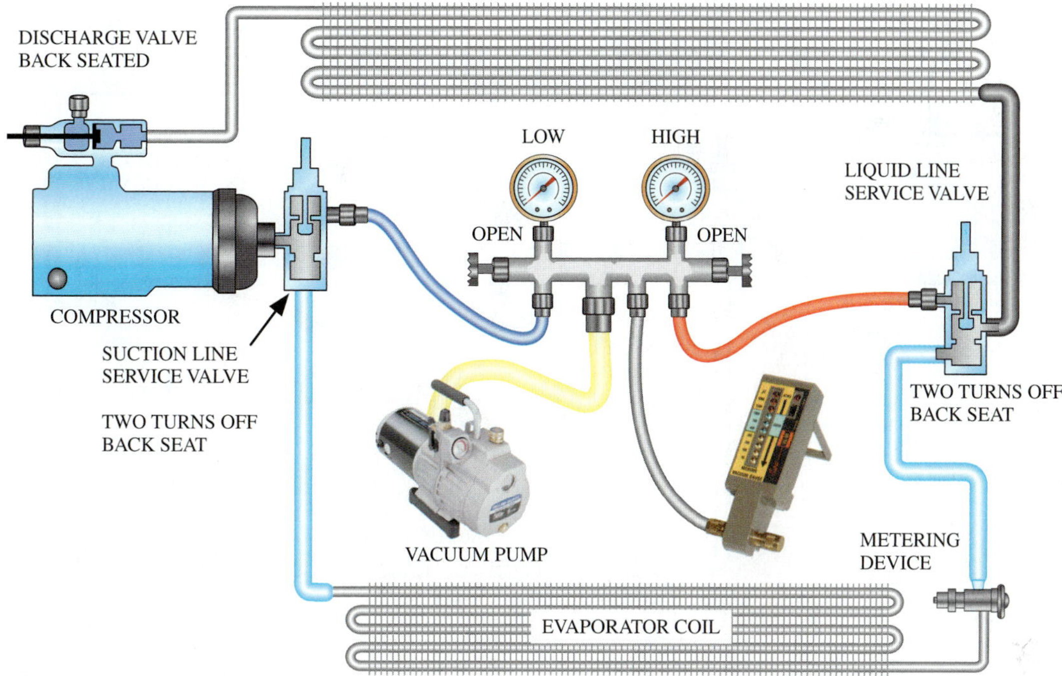

Figure 30-30 The ¼-in center port on a four-port manifold can be used to connect a vacuum gauge.

Maximum Pressure

Never expose the vacuum sensor to pressures higher than its rating. Check with the vacuum gauge manufacturer to determine the sensor's maximum pressure rating. Some vacuum sensors are designed to withstand pressures as high as 400 psig, but the sensors used with some vacuum gauges are not able to withstand high pressures. To avoid damaging these sensors, a shutoff valve should be used between the sensor and the system so that the sensor may be isolated from the system before it is pressurized. This may be accomplished by connecting the vacuum gauge to the ¼-in center port on a four-port manifold gauge set (Figure 30-30).

30.12 CONNECTING VACUUM GAUGES

The best place to connect a vacuum gauge is directly to the system being evacuated. The vacuum in the system will not be quite as low as the vacuum in the pump. Measuring vacuum at the system ensures accuracy. To connect a vacuum gauge directly to the system, either a third system access port is necessary or an adapter that allows connection of both the vacuum sensor and a charging hose to a single port can be used. A short hose or coupler is usually necessary to connect the vacuum sensor to the valve port because the connection on most vacuum gauge sensors is exactly the same as most service valve ports, ¼-in male flare. The ideal connection is a short metal coupler (Figure 30-31).

Commercial refrigeration systems frequently have three service ports: one on the suction side of the compressor, one on the discharge side of the compressor, and one on the liquid receiver. The vacuum gauge can be connected to the discharge service valve port while the vacuum is drawn from both the suction service valve and the king valve (Figure 30-32). Heat pumps typically have a third access port that is always on the suction side of the compressor. The vacuum can be drawn through the common suction port and the liquid-line port while the vacuum gauge is connected to the large gas-line port (Figure 30-33).

Figure 30-31 The vacuum gauge can be connected directly to the system using a short metal coupler or metal hose if the system has more than two gauge ports.

Blank-Off Valves

Many systems have only two service ports. Using a blank-off test valve allows connection of both the vacuum sensor and charging hose directly to the system. The built-in shutoff valves allow the vacuum sensor to be isolated from the system before it is charged.

Vacuum Manifolds

Another way to connect the vacuum gauge to the system is to use a vacuum manifold. A vacuum manifold is a device for connecting a vacuum sensor to both the system and the vacuum pump. Manifolds like the ones shown in Figure 30-34 attach to

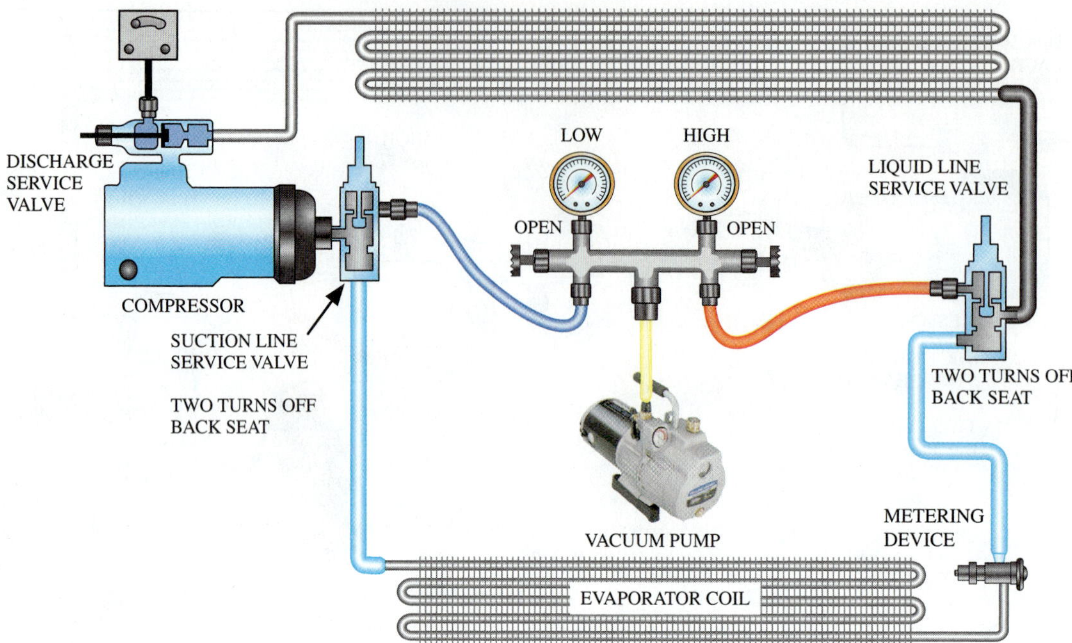

Figure 30-32 The vacuum gauge can be connected to the discharge service value on commercial systems that have king valves.

Figure 30-33 The vacuum gauge can be connected to the vapor-line port on heat pump systems that have a common suction port.

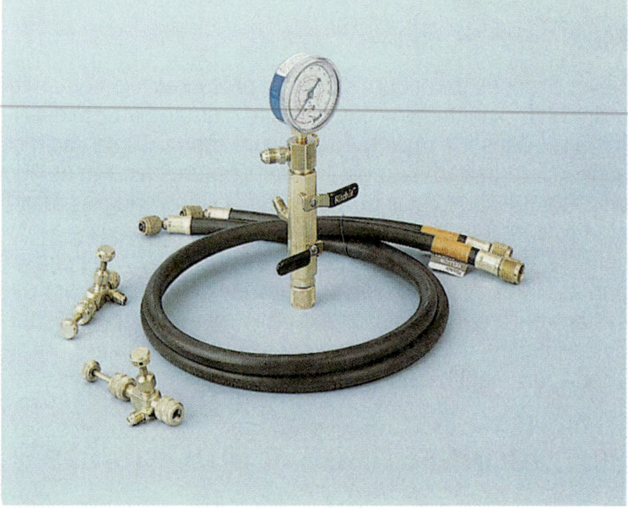

Figure 30-34 A vacuum manifold can be used to allow connection of multiple hoses and a vacuum gauge at the vacuum pump. *(Courtesy of Ritchie Engineering Company, Inc., Yellow Jacket Products Division)*

the vacuum pump and allow the connection of two 3/8-in vacuum hoses as well as connection of the vacuum gauge. One problem with a vacuum manifold that connects to the vacuum pump is that the vacuum is being read at the pump instead of at the system.

Schrader Valves

Most residential systems have Schrader valves for service ports. The core in a Schrader valve is a significant restriction and increases the amount of time required to evacuate and charge a system. Figure 30-35 shows a cutaway of a Schrader valve with a core in it. Valve core removal tools with side ports, like the one shown in Figure 30-36, can be used to remove Schrader cores during evacuation and charging and to replace them when charging is complete. This can cut the evacuation time in half. Be sure to use a vacuum-rated core tool to prevent introducing air into the system during evacuation.

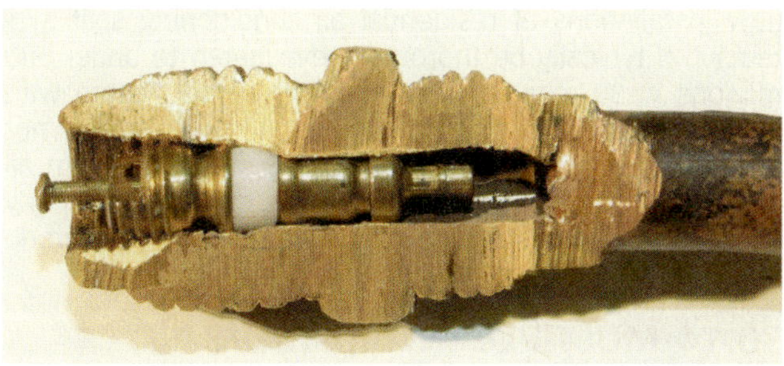

Figure 30-35 Schrader valve cores offer a great deal of restriction to evacuation and charging.

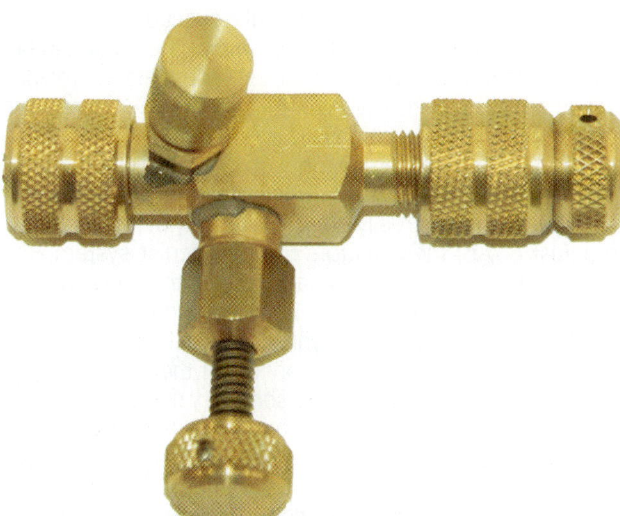

Figure 30-36 A valve core tool with a side port can be used to remove the valve core during evacuation and charging.

Care should be used when removing and replacing valve cores during routine service because Schrader valves are not designed to have their cores removed and replaced frequently. It is possible to damage the Schrader valve, making it necessary to recover the system refrigerant and replace the damaged Schrader valve.

SERVICE TIP

Most manifold gauge sets have an internal opening and hoses with ¼-in diameter. Larger gauges and hoses are available that have a ⅜- or ½-in interior diameter. The larger the interior diameter and the shorter the hose, the faster a system can be evacuated. As the pressure in the system drops, the rate of withdrawal is significantly hampered when working with the smaller-diameter hoses. For most residential applications, this time is not a major problem. However, for larger commercial installations it can become a significant factor in the overall job time. Many manufacturers of gauges provide these larger-diameter hoses, and frequently they are black in color and less flexible than the normal-gauge hoses.

Figure 30-37 The cross-sectional area of a ⅜-in hose is over twice that of a standard ¼-in hose, allowing faster evacuation and charging.

30.13 DEEP EVACUATION

The deep-vacuum method is the most positive method of ensuring a system free of air and water. The equipment required includes a vacuum pump with a blank-off pressure no higher than 50 microns, a reliable electronic vacuum gauge, a gauge manifold or vacuum manifold, and all connecting hoses. Hoses should be as short as possible and connections should be kept to a minimum. Use ⅜-in or larger hoses whenever possible (Figure 30-37).

The actual physical connection used depends very much upon the tools and equipment. The vacuum gauge can be connected directly to the system and the vacuum pump connected to the middle port of a standard three port manifold gauge on systems with three access valves, as in Figures 30-32 and 30-33. Figure 30-38 shows a connection using a standard three-port manifold with the vacuum gauge connected to the heat pump permanent suction service port. Figure 30-30 shows a connection using a four-port manifold with the vacuum gauge connected to the ¼-in center port and the vacuum pump connected to the ⅜-in center port.

These are only a few of the possible connection strategies. Regardless of how the vacuum pump and gauge are connected, the technician should keep in mind a few overall concepts.

- There should be no pressure in the system before connecting a vacuum pump. If there is still refrigerant in the system, it must first be recovered before attempting to pull a vacuum. If the system is pressurized with nitrogen, release the nitrogen before beginning evacuation.
- The vacuum level must be measured with an accurate vacuum gauge. Mechanical gauges are not accurate

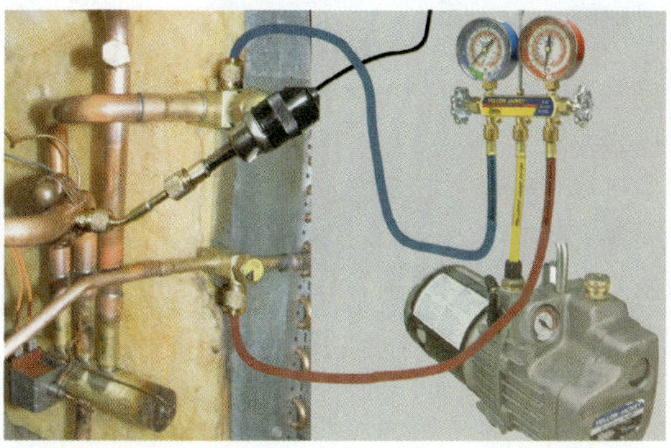

Figure 30-38 Vacuum pump and vacuum gauge connections using standard three-port manifold and a third system gauge port.

enough to determine when a system has been properly evacuated.

- The vacuum integrity of the connections should be checked before evacuating the system.
- The system should be evacuated from both the low side and the high side simultaneously.
- The system should be able to be isolated from the vacuum pump upon completion of the vacuum.
- The vacuum gauge should be able to read the system pressure during a blank-off period at the end of the vacuum.
- The vacuum gauge should be able to be isolated from system pressure during charging.
- After pulling the system down to under 500 microns, valve it off and wait to see if the vacuum level will hold.
- A continuous, rapid rise in pressure indicates leaks either in the system or in the evacuation connections.
- A slow rise that levels off between 1,500 and 5,000 microns indicates system contamination with water.
- If the system leaks, the leaks must be identified and repaired. If the system is contaminated with water, further evacuation should correct the problem.

Whenever possible, check the vacuum integrity of your connections before evacuating the system. This can be done with manual service valves by making all connections and opening all valves except the system service valves. The vacuum pump should be able to pull below 500 microns quickly on the vacuum setup. If it cannot, then you are wasting your time trying to pull the system under 500 microns. To improve vacuum integrity, use as few mechanical connections as possible, and use short, low permeation hoses.

How Long Should It Take to Pull a Deep Vacuum?

The simple answer is that there is no specified period of time that is required for a deep vacuum. A deep vacuum is done when the system pressure will hold below 500 microns (Figure 30-39). A deep vacuum can be pulled on small, dry systems in minutes. The lines and evaporator coil on new installations of residential air-conditioning split systems can typically be thoroughly evacuated to under 500 microns in 30 minutes. However, the same system with water contamination may take several hours. Time is not a replacement for a vacuum gauge. Pulling a vacuum all weekend is not an assurance that the system is properly evacuated. If the system leaks, no amount of time will work.

Figure 30-39 The system should be evacuated to under 500 microns for a deep vacuum.

30.14 MULTIPLE EVACUATION

A multiple evacuation is an alternative to the deep-vacuum procedure. Multiple evacuations can be faster than a single deep evacuation. The multiple-evacuation procedure typically does not require a deep-vacuum gauge. Multiple evacuation, sometimes called triple evacuation, is really several short vacuums in succession that are broken in between with dry nitrogen. The nitrogen absorbs and/or dislodges moisture in the system, making it easier to remove on subsequent evacuations. The number of evacuations, the amount of nitrogen introduced between evacuations, and the time given for the nitrogen to "blot up" the moisture varies. There is really no scientific way of knowing how to adjust any of these factors for a particular system. The most common multiple evacuation is the triple evacuation, in which three successive vacuums are used.

Figure 30-40 shows a typical connection for multiple evacuation using a four-port manifold. If there is nothing in the system, it is advisable to put some dry nitrogen in before the first vacuum. This will yield one extra "blotting" period. Dry nitrogen is nitrogen that is free of water. Nitrogen is available in different grades and can sometimes contain water. Be sure when purchasing nitrogen to request dry nitrogen.

TECH TIP

The term *dry nitrogen* is not very well defined. Suppliers use their own names for the grades of gas products they sell. All compressed gas contains some level of other substances. The amount is typically measured in parts per million (ppm). The grade of the gas is indicated with two numbers that describe the percentage of purity. The first digit tells how many "9" digits are in the percentage, and the second digit identifies the number after the last "9." A grade 5.6 gas is 99.9996% pure. A 3.0 grade, 99.90% pure, is the minimum acceptable for refrigeration purposes. It is important to ask your gas supplier what the gas grade is you are purchasing and what the level of water is in it.

Of course, it is imperative to use a nitrogen regulator and pressure-relief valve when pressurizing the system. Nitrogen tanks hold pressures as high as 2,000 psig, and no refrigeration system will stand up to that kind of pressure. Recommendations on what pressure to use for the blotting period vary from 2 psig to full-test system pressures. For safety, never exceed the low-side system test pressure when adding the nitrogen. Let the nitrogen stand in the system for 10–15 minutes. Blotting times of up to 1 hour are sometimes recommended. However, if you are

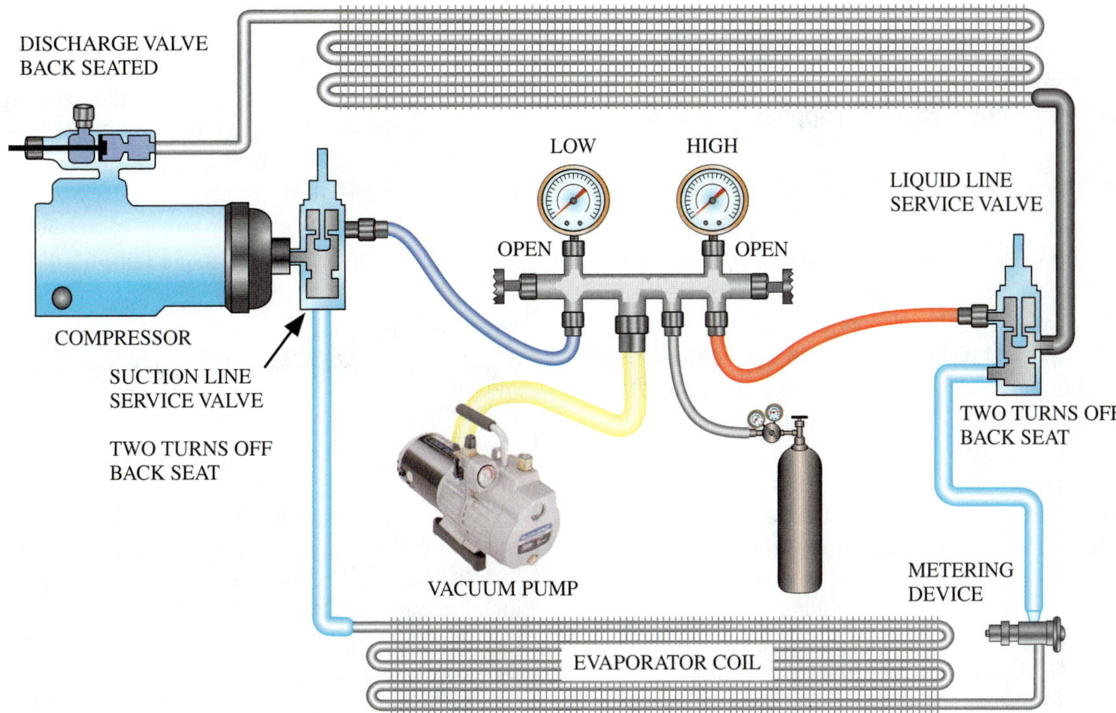

Figure 30-40 Connection for multiple evacuation.

going to invest that kind of time, you might as well pull a deep vacuum. Obviously, longer waiting periods will generally yield better results.

Connect the vacuum pump to the center port of your gauges, open both manifold gauge valves, and pull a vacuum from both sides of the system. Let the pump pull for as long as the compound gauge will give you a visible drop in pressure. In any case, do not shut off the vacuum pump until the compound gauge has bottomed out (Figure 30-41). If an electronic vacuum gauge is used, watch how quickly the system pulls down. There may be no advantage in using the multiple-evacuation method if it is possible to achieve a deep vacuum quickly. Pulling a system down three times to 500 microns is really not any better than pulling it down once if the vacuum holds under 500 microns.

Figure 30-41 The compound gauge should peg out at 30 in Hg vacuum with each evacuation on a multiple evacuation.

SERVICE TIP

Electronic vacuum gauges are generally not used for multiple evacuations, but using one can easily demonstrate the effectiveness of multiple evacuation. Note the speed with which the unit pulls down on each successive evacuation. Generally, the system pulls down faster with each sweep.

UNIT 30—SUMMARY

The only two chemicals that should be circulating in a refrigeration system are refrigerant and oil. Contaminants such as moisture and noncondensable gas reduce system capacity and shorten the life of system components while increasing the system operating cost. Evacuating a system ensures that the system is free of noncondensables and moisture before charging. Deep vacuums are measured in microns using electronic vacuum gauges. The lower the number, the better the vacuum; a deep vacuum should hold under 500 microns. Micron vacuum gauges are used to measure deep vacuum. The vacuum scale on a compound gauge is not accurate enough to read a deep vacuum. The best place to connect the vacuum gauge is directly to the system. Systems should be leak tested before evacuation to avoid introducing contaminants and wasting time. To speed up evacuation use short, large-diameter hoses.

WORK ORDERS

Service Ticket 3001

Customer Complaint: System Is Tripping Main Circuit Breaker

Equipment Type: R-22 Residential Split System

The service technician checks to make sure the thermostat and compressor disconnect are turned off. Then the technician checks to make certain the breaker is good and turns power on to the disconnect. Leaving the disconnect off, the technician checks and determines the compressor is grounded. The technician reports to the customer that the compressor is bad, and the customer wants the compressor replaced.

The technician checks the price book and provides the customer with a cost estimate for the job. The customer accepts the estimate, and the technician calls the office and has the parts department send a replacement compressor.

While the compressor is being delivered, the technician puts on safety glasses and gloves and proceeds to recover the refrigerant. After the refrigerant is recovered to the proper level, the technician replaces the compressor and installs the liquid- and suction-line driers. A leak test is then performed to verify a tight system.

With a leak-free system, the technician then evacuates and dehydrates the system to 500 microns. After determining the system is tight and dry, the refrigerant-charging process is started. To ensure the efficiency and longevity of the system, this process should be performed to the manufacturer's specifications.

After the technician cleans up all of the packaging and other scrap material from the worksite, the customer is provided with a service ticket.

Service Ticket 3002

Customer Complaint: Poor Cooling and High Operating Cost

Equipment Type: R-22 Split System, New Condensing Unit Replacement

A customer complains that his new air conditioner does not seem to cool as well as his old one did and costs more to operate. The technician checks the system pressures and finds that the high-side pressure is 50 psig higher than the manufacturer's specification. The compressor is very hot, the compressor amp draw is high, and the system superheat is 30°F, but the unit calls for 15°F. The technician recognizes that this is not a simple case of overcharge because of the high superheat. The technician asks the homeowner for more information about the installation, and the homeowner confesses that a friend of his changed the unit out one day during an extended lunch break. The technician suspects that the system has noncondensables from not being thoroughly evacuated. All the refrigerant is recovered into a separate cylinder. The system is evacuated to under 500 microns and a new charge of clean refrigerant is weighed in. The technician runs the system until all the pressures and temperatures stabilize and then checks the system pressures and superheat. The system now is operating at the required 15°F superheat with a normal condensing pressure. The compressor is running cooler and the amp draw is down under the data plate current rating.

Service Ticket 3003

Work to Be Performed: New System Startup

Equipment Type: R-410A Split-System Air Conditioner

A technician is called to perform a new system startup on an R-410A split system. The technician notices that the installing technician did not install a liquid-line filter drier. The start-up technician explains that the system has POE lubricant and requires a filter even on new installations because of the hygroscopic nature of POE. He installs a new liquid-line filter drier and tests the lines for leaks using dry nitrogen. There are no leaks, so he releases the nitrogen to the air. The technician connects his gauges to both sides of the system and connects his vacuum pump to the large vacuum hose connected to the 3/8-in port on his gauges. A vacuum gauge is connected to the low side of the system using a vacuum shutoff valve that allows connection of both the low-side hose and the vacuum gauge. The technician turns on the vacuum pump and lets it pull while he completes his visual inspection and other system checks. When he has completed his other checks he looks at the vacuum gauge and sees that the system has pulled down to 350 microns. He valves off the vacuum pump and waits to see if the vacuum will hold. It does, so he isolates the vacuum gauge by closing the valve on the vacuum shutoff valve, closes the gauge manifold valves, and opens the unit-installation valves by turning them counterclockwise. The refrigerant shipped in the condensing unit fills the lines and coil. Last, the technician operates the system and checks the system charge using the manufacturer's charging chart.

Service Ticket 3004

Work to Be Performed: Evacuate System After Repair

Equipment Type: R-134a Commercial Refrigeration Walk-in Cooler

A technician has repaired a leak in a commercial refrigeration system, installed a new filter drier, and checked the system for leaks using dry nitrogen. He is preparing to evacuate the system before recharging it with new, clean refrigerant. Before connecting his gauges, the vacuum pump, and the vacuum gauge to the system, he connects his vacuum gauge directly to his vacuum pump to check the operation of the pump and gauge. The vacuum drops slowly to around 1,500 microns and then levels off. The technician looks in the oil sight glass and notices that the oil is milky. He knows that contaminated oil can prevent a good vacuum, so he changes the oil in his vacuum pump with clean vacuum pump oil. Again he performs the test, and this time the vacuum drops quickly to 100 microns and then slowly drops to 50 microns. He now proceeds to connect his high-side gauge to the king valve, the low-side gauge to the suction service valve, the vacuum gauge to the discharge service valve, and the vacuum pump to the middle hose. He opens both manifold valves and puts the system service valves at mid-position. He operates the vacuum pump until the vacuum on the system reads below 500 microns. He then closes his manifold valves and back-seats both the king valve and the suction service valve to see if the vacuum will hold. It holds, indicating the system is clean and tight. The technician back-seats the discharge service valve to isolate the vacuum gauge, connects a cylinder of refrigerant to the middle hose, and purges just the middle hose. He back-seat cracks the king valve and weighs in the correct refrigerant charge as liquid into the liquid receiver.

UNIT 30—REVIEW QUESTIONS

1. What chemicals should be inside a refrigeration system?
2. How are vacuum pumps rated?
3. What is the purpose of evacuating a refrigeration system?
4. What is a micron?
5. Explain how microns are used to measure vacuum.
6. What happens if noncondensables are left in the system?
7. What happens if moisture is left in the system?
8. Why does the presence of liquid water in a system slow down the evacuation process?
9. What is considered a good vacuum level for a deep vacuum?
10. What are the two most common types of vacuum pumps?
11. How often should the vacuum pump oil be changed?
12. Why should only oil designed for vacuum pump operation be used in vacuum pumps?
13. Where is the best place to connect a vacuum gauge?
14. Why is the vacuum scale on the compound gauge not used to measure deep vacuum?
15. What is a vacuum?
16. What is a perfect vacuum?
17. Why should systems be tested for leaks prior to evacuation?
18. What is indicated if, during evacuation, the vacuum level stalls out around 2,000 microns and quits dropping?
19. What is indicated if, after evacuation, the vacuum level gradually rises to atmospheric pressure?
20. What is indicated by milky vacuum pump oil?
21. What can be done to speed up evacuation?
22. How long does it take to pull a deep vacuum?
23. What is a multiple vacuum?
24. Why is it important to be able to isolate the vacuum gauge from the system after evacuation?
25. Why should the vacuum pump isolation valve be closed before turning off the vacuum pump?

UNIT 31

Refrigerant System Charging

OBJECTIVES

After completing this unit, you will be able to:

1. explain how to determine the correct amount of charge for a split system.
2. discuss the difference between liquid charging and vapor charging.
3. explain how to charge systems that use zeotropic refrigerants.
4. determine the correct superheat for a fixed-restriction air-conditioning system using a superheat charging chart.
5. determine the correct subcooling for a thermostatic expansion valve air-conditioning system using a charging chart.
6. list the variables that affect system operating pressures.
7. define the AHRI standard equipment rating condition.

31.1 INTRODUCTION

There are many outside influences—such as low evaporator airflow due to dirt, improper fan speed, or poor duct design—that can mimic the pressure readings of an undercharged system. Likewise, condenser coil conditions can result in pressure readings that might initially indicate an overcharge. It is, therefore, important that technicians do a thorough job of checking pressures, temperatures, and current draw of systems to determine system charge and performance.

The level of refrigerant charge is a delicate balancing act because both too little and too much refrigerant reduce the system's performance. Overcharging can cause liquid to flood back to the compressor, leading to compressor failure. Overcharging will also cause liquid refrigerant to back up in the condenser, increasing the operating head pressure and reducing system efficiency.

Undercharging can reduce oil return back to the compressor, cause high compression ratios, and lead to compressor overheating. Undercharged systems operate inefficiently due to the increased compression ratio and reduced refrigerant flow.

It is important that the technician understand and employ the evacuation and charging methods recommended by the equipment manufacturer for any piece of equipment. High-efficiency and low-maintenance costs can be ensured by following thorough evacuation and charging procedures.

31.2 LEAK TESTING

Refrigeration systems should always be checked for leaks whenever they need charging. Leak tests should always be performed before evacuating and charging a new system. Checking for leaks prior to evacuation and charging saves time and refrigerant. It is necessary to check for leaks whenever refrigerant must be added to an existing system. Refrigerant does not get used up, so a system that is low on charge has a leak. Just adding refrigerant without finding and repairing the leak costs the customer money, and the system operates inefficiently, using more energy to produce less cooling.

31.3 DETERMINING SYSTEM CHARGE

Whether the system is a new one or an existing one that has been repaired, the final step in putting the system in operation is to charge it with refrigerant. The amount of refrigerant charge is more critical in some systems than in others. Systems that have a receiver and thermostatic expansion valve (TEV) are less critical since extra refrigerant can be stored in the receiver, and the expansion valve feeds the refrigerant into the evaporator as required to match the load.

Any excess refrigerant in systems that do not have a receiver will be stored in some part of the system where it reduces the effectiveness of that part and reduces the capacity of the system. If the system is short of refrigerant, the metering device is not supplied with a solid stream of liquid refrigerant and the evaporator will be starved for refrigerant, reducing the capacity of the system. If the system is overcharged, excessive pressures and floodback to the compressor may damage the compressor. The equipment installation instructions will provide the installer with the correct charging procedure for that piece of equipment. Many equipment manufacturers provide charging information on the equipment access panels (Figure 31-1).

Packaged Units

One way to determine the proper charge is to read it on the name plate, as shown in Figure 31-2. Most manufacturers specify the type and amount of charge on the system data plate. For any one-piece system, such as appliances, window units, or packaged air conditioning systems, the amount on the data plate will be the correct charge.

Split Systems

The amount of charge required for a split system is affected by the length of the refrigerant lines. The length of the refrigerant lines is determined by the conditions of the installation. This is why the correct system charge will be different from one split system to another, even for identical equipment. The condensing unit on a split system has

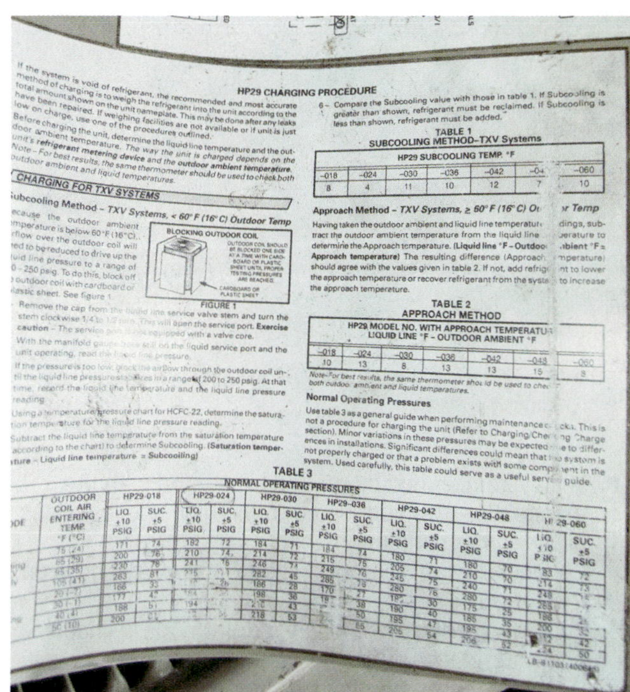

Figure 31-1 Most manufacturers provide charging information inside the unit service panel.

A	LRA	REF.SYSTEM R-410A		TEST PRESSURE		GA
3.5	61	7.5 LBS	3.4 kg	HI	629 PSI	4337
				LO	297 PSI	2048

Figure 31-2 The type and amount of refrigerant is shown on the equipment data plate.

a factory charge amount on the data plate (Figure 31-3). Typically this amount includes enough refrigerant for the condensing unit, the evaporator coil, and some predetermined length of refrigerant line. The assumed line length varies from 15 to 30 ft, depending upon the manufacturer.

SERIAL 1103E05826
PROD 38TDB024---301--
MODEL 38TDB024300
PISTON TXV INDOOR N/A OUTDOOR
FACTORY CHARGED R410A
6.0 LBS 2.72 KG
INDOOR TXV SUB COOLING °F
POWER SUPPLY 208/230 VOLTS AC
1 PH 60 HZ
PERMISSIBLE VOLTAGE AT UNIT
187 MIN

Figure 31-3 The factory charge for this split-system condensing unit is shown on the data plate.

SYSTEM TOTAL CHARGE

INSTALLER
DETERMINE COMPLETE SYSTEM TOTAL CHARGE BY USING PROPER ITEM BELOW. STAMP OR MARK THIS VALUE ON RATING PLATE IN "SYSTEM TOTAL CHARGE" BLOCK.

"OUTDOOR UNIT CHARGE"
ADD PER FT. VALUE SHOWN BELOW FOR EACH FOOT OVER 25 FT. LIQUID LINE TO FACTORY CHARGE. SUBTRACT PER FT. VALUE SHOWN BELOW FOR EACH FOOT UNDER 25 FT. LIQUID LINE FROM FACTORY CHARGE.

.3 OZ. PER FT. FOR 1/4 OD — .4 OZ. PER FT. FOR 5/16 OD
.6 OZ. PER FT. FOR 3/8 OD — 1.2 OZ. PER FT. FOR 1/2 OD
1.9 OZ. PER FT. FOR 5/8 OD

92-23446-64-04

Figure 31-4 Many manufacturers provide details on adjusting the charge for line length inside the system service panels.

Liquid Line Size	R-22 or R-410A (oz/ft)
1/4	0.4
3/8	0.6
1/2	0.8
5/8	1.0

Figure 31-5 Liquid-line allowance table.

The assumed line length used for any particular piece of equipment can be found in the installation instructions or on the inside of the service access panel (Figure 31-4).

The system charge must be adjusted if the actual line length varies from the assumed line length. Most manufacturers only compensate for the liquid line and not the suction line because the amount or refrigerant in the suction line is so slight. Figure 31-5 shows how much refrigerant to add or subtract for every foot of difference in line length. This remains the same regardless of the type of equipment because it is based on the physical volume of the different size liquid lines. For example, a system with a factory charge of 56 oz of R-410A, an assumed line length of 15 ft, and a 3/8 in liquid line that is 35 ft long.

$$\begin{aligned}56 \text{ ounces} + (35 - 15) \times 0.6 \text{ oz/ft} &= 56 \text{ oz} + 12 \text{ oz}\\ &= 68 \text{ oz total system charge}\end{aligned}$$

Repairing Existing Equipment

When the system charge has been recovered to repair an existing system, the total system charge should be weighed in after evacuating the entire system.

New Installations

For new split-system installations, only the additional amount of charge needs to be determined and added because the condensing unit comes precharged. In the above example, only 12 oz needs to be added for a new installation. The procedure for charging a new split system would be as follows:

- Install refrigeration lines and evaporator coil.
- Evacuate the lines and coil below 500 microns.
- Determine the amount of charge adjustment needed for the extra line length.
- Weigh the charge adjustment into the evacuated lines.
- Open the charging valves on the condensing unit.

SERVICE TIP

Too often, technicians in the HVACR field are of the opinion that if a little refrigerant is good, a lot is better, and too much is just about right. Refrigerant is not a magical fluid that adds cooling capacity to a system. The cooling capacity of a system that is overcharged is significantly decreased. The only time a system will operate with its designed peak performance is when the charge is at the manufacturer specified level. It is not an acceptable practice to add just a little more in case the system leaks.

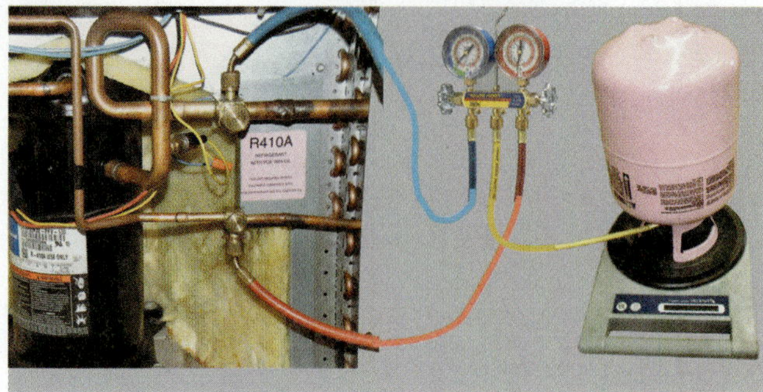

Figure 31-7 Connections for weighing liquid refrigerant into the high side of a system.

31.4 CHARGING BY WEIGHT

The most difficult aspect of refrigerant charging is not putting refrigerant into a system but knowing when to stop putting it in. There are so many variables to take into account that charging to a particular operating condition can be difficult. The easiest way to have confidence in the charge is to weigh in the correct amount of refrigerant into a fully evacuated system using an accurate scale. Scales such as the one shown in Figure 31-6 can be used to accurately weigh the refrigerant into the system. An electronic scale with the ability to be zeroed with the refrigerant cylinder on the scale is preferred.

Most fully evacuated systems will take in the entire charge in liquid form in one shot. Figure 31-7 shows the setup for weighing liquid into the high side of an evacuated system. To weigh in a system charge after evacuating a system:

Figure 31-6 Electronic charging scales like this one can be used to weigh in a refrigerant charge.

- The system should be off and evacuated with the manifold gauges connected to the high and low sides of the system and the manifold valves closed.
- Connect the refrigerant cylinder to the middle hose on the gauges.
- Purge the center hose with vapor from the cylinder while it is in the upright position.
- Do not attempt to purge the other hoses, as they are in a vacuum and do not need purging.
- Invert the cylinder and place it on the scale with the cylinder valve opened.
- Zero the scale with the cylinder on it.
- Open the high-side manifold valve to allow liquid refrigerant to flow from the cylinder into the system.
- The position of the hoses should not be changed during charging because this can change the weight felt by the scale.
- Close the high-side manifold valve when the desired charge is reached.
- Some refrigerant will be left in the hose to the cylinder. To put this in the system: close the cylinder valve, operate the unit, and crack the low side manifold hand valve.

31.5 LIQUID CHARGING

Liquid charging is always much faster than vapor charging. This is because liquid is far denser than vapor and because taking liquid out of the refrigerant cylinder does not affect the cylinder pressure. The primary disadvantage of liquid charging is that it is easy to overcharge a system when adding liquid refrigerant. Liquid is normally charged into the system in one of two ways: into the high side with the system off or by pulling liquid into the liquid line with the system operating and the king valve front seated.

Liquid Charging—System Off

On smaller systems, liquid charging is usually done with the compressor off. Prior to charging, the system must be leak tested and evacuated. Figure 31-7 shows connections for liquid charging in to the high side with the system off. When the charging is started, the system is under vacuum and the refrigerant cylinder is under pressure, so the refrigerant is pushed into the system due to the difference in pressure. After the system has been evacuated, liquid is introduced into the high side with the system off. Since the compressor can be damaged by liquid in the suction line, liquid is normally not charged into the low side of a system.

The system pressure rises as the refrigerant enters the system. Sometimes the system pressure and cylinder pressure will equalize before the full amount of refrigerant has been charged into the system. When this occurs, the charge may have to be completed as a vapor charge with the system operating.

Liquid Charging—System Operating

On larger systems, a king valve located between the condenser and the metering device offers a convenient means of charging the system on the high side, with the compressor running. When the king valve is front-seated (Figure 31-8), flow from the receiver is shut off. As the system operates, the pressure in the liquid line leaving the receiver starts to drop. When the pressure in the refrigerant cylinder exceeds the pressure in the liquid line, refrigerant is drawn into the system from the cylinder (Figure 31-9). This same technique may be used for most residential split systems by front-seating the liquid-line charging valve (Figure 31-10).

SERVICE TIP

It is easy to overcharge a small system using this charging method. When adding liquid refrigerant with the system operating, the gauges will show both low-side and high-side pressures, but these pressures tell the technician nothing about the amount of charge in the system. The amount of refrigerant added should always be checked with a scale to guard against overcharging.

Figure 31-8 The king valve stem is turned in all the way clockwise to front-seat it.

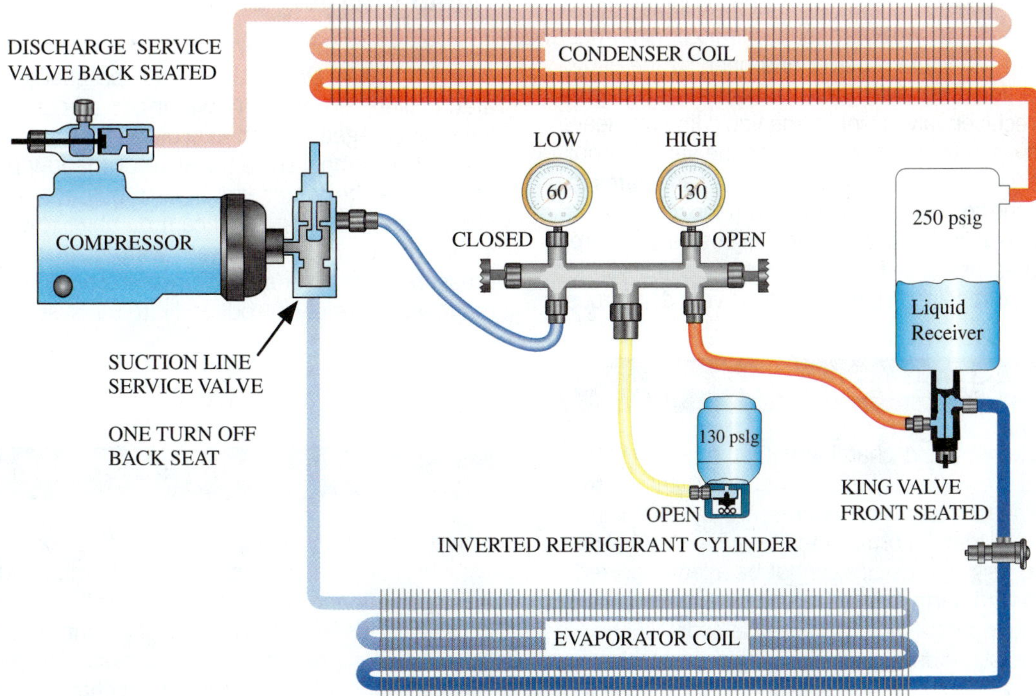

Figure 31-9 With the king valve front-seated and the compressor operating, liquid refrigerant will be drawn into the liquid line.

Figure 31-10 The liquid-line service valve on split-system air-conditioning system can also be used to liquid charge with the compressor operating.

Figure 31-11 King valves that have the gauge port open to the line port when the valve is front-seated can be used for liquid charging with the system operating.

Figure 31-12 King valves that have the gauge port open to the receiver port when the valve is front-seated should NOT be used for liquid charging with the system operating.

This will not work for all systems that have liquid receivers and king valves! To work properly, the liquid line must be isolated from the liquid receiver and open to the gauge port when the liquid receiver valve is front-seated (Figure 31-11). Many receiver valves isolate the liquid line and leave the gauge port open to the receiver when the valve is front-seated (Figure 31-12). These systems are not candidates for this method! If the liquid line is isolated from the receiver and the gauge port is open to the receiver, high-side pressures from the compressor will be forced into the charging cylinder through the liquid receiver. This condition is very dangerous!

SERVICE TIP

On large systems, liquid charging with the compressor operating and the king valve front-seated may cause compressor overheating and/or low-pressure system shutdown. A system that is normally fed by a $\frac{5}{8}$-in liquid line with over 200 psig of pressure cannot be adequately fed by a $\frac{1}{4}$-in line with normal cylinder pressure. The result is that the system operates with a starved evaporator and a low suction pressure during the charging process. Liquid charging with the system operating should be limited to short durations of time to avoid system problems.

Liquid Charging Chillers

Low-pressure chillers are charged through the evaporator charging valve. Liquid should not be charged into evacuated chillers or units with water-cooled condensers. When liquid is charged into an evacuated unit, the liquid boils at a very low temperature because of the low pressure. This can freeze the water sitting in tubes in evaporators and condensers, causing significant damage to the system. To avoid this, charge the system with vapor until the system pressure is at a saturation pressure equal to 36°F. Liquid may then be safely introduced into the system without fear of freezing the water.

SERVICE TIP

On systems that require large quantities of liquid refrigerant to meet the basic charge requirement, liquid can be pumped into the system by using a cylinder heating band. These bands are thermostatically controlled and sense the cylinder pressure with a pressure-sensitive switch that is connected to the refrigerant charging line. Cylinder heaters can significantly reduce the charging time.

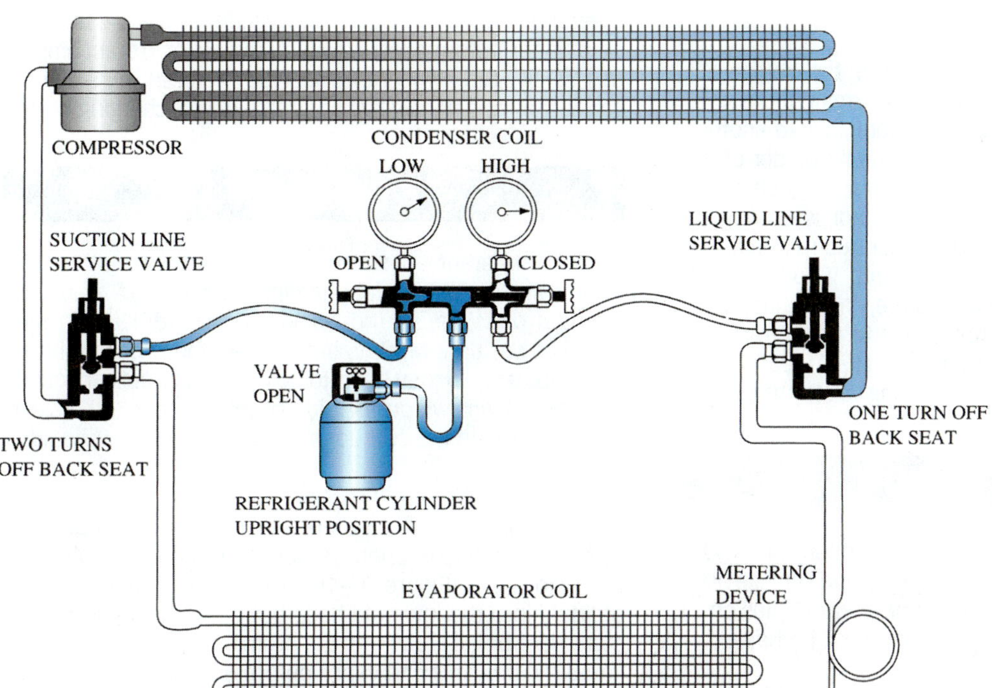

Figure 31-13 Connections for vapor charging a system.

31.6 VAPOR CHARGING

Vapor charging is used to add small amounts of refrigerant slowly because vapor takes up a great deal more space per pound than liquid. When a partial charge is required, vapor charging is usually employed. Figure 31-13 shows a typical connection for vapor charging. The center hose of the gauge manifold is connected to the valve on the refrigerant cylinder with the cylinder in the upright position. The center hose is purged using refrigerant from the cylinder. If the system was evacuated prior to charging, only the center hose needs purging. However, if vapor is being added to a system that already has some charge in it, all gauge hoses should be purged.

SERVICE TIP

Before attempting to vapor charge a system, make sure that the refrigerant to be charged is not a 400 series zeotropic refrigerant, such as R-410A. Zeotropic refrigerants should not be removed from the cylinder as a vapor because they will fractionate and the mixture of refrigerants will be incorrect. Worse, the refrigerant remaining in the cylinder is no longer the correct mixture after fractionation.

Small Appliances

The recommended charging procedure for small appliances whose total system charge is less than a pound is to evacuate the system and introduce vapor into the appliance until the static equalized pressure in the appliance equals the factory specification. This is the most accurate charging method for appliances whose total charge is only a few ounces. This is done after evacuating the appliance with the appliance off.

Refrigerant Cylinder Pressure Drop

The refrigerant cylinder pressure decreases as vapor is drawn out of it, even if there is still plenty of refrigerant left in the cylinder. As the vapor leaves the cylinder, it creates a low-pressure area above the liquid due to the void left by the exiting refrigerant. The liquid in the cylinder boils to fill this void with more vapor. This process continues as long as vapor is being removed from the cylinder. The heat required to vaporize the liquid comes from the liquid refrigerant and the tank itself and leaves with the refrigerant. The result is an ever-decreasing refrigerant temperature. Since the pressure of a saturated refrigerant is determined by its temperature, a drop in temperature also means a drop in pressure. The charging process gradually grinds to a halt because of the drop in cylinder pressure. The frost line on the bottom of the cylinder in Figure 31-14 shows the liquid level of the cylinder. It has been cooled by drawing vapor out of the cylinder.

Figure 31-14 The frost line on this cylinder shows the liquid level; the liquid temperature has dropped because of vapor being removed from the cylinder.

Heating the Refrigerant Cylinder

Heating the refrigerant cylinder will raise the refrigerant temperature and pressure, allowing the charging process to continue. Refrigerant travels from one container to another simply because of pressure difference. A full cylinder of refrigerant with a pressure of 50 psig will not even charge a refrigerator if the pressure in the refrigerator is 50 psig or higher. Remember, the pressure of a saturated refrigerant is controlled by its temperature, not its volume. If both the unit and the cylinder are in a saturated state, there must be a temperature difference to get refrigerant to travel from the refrigerant cylinder into the system. Therefore, heating the refrigerant cylinder is often required during vapor charging.

SAFETY TIP

It is possible to increase the cylinder temperature and pressure to a point where it can rupture and release an awesome amount of energy with one great explosion. Therefore, when a refrigerant cylinder must be heated, these precautions should be taken:

- Wait until some of the refrigerant has been removed from a full cylinder before doing any heating.
- Never use a concentrated heat form, such as a torch, to heat the cylinder. Warm water or air are effective and safe alternatives.
- Limit the maximum temperature exposure of the cylinder to below 125°F.
- Never obstruct or tamper with the safety relief device on the cylinder.

Vapor Charging with Unit Operating

Most vapor charging is done with the compressor operating while charging is taking place. The pressure difference between the tank pressure and the system evaporator pressure will be greater if vapor is introduced into the low side while the compressor is operating (see Figure 31-13). This will delay the problem of cylinder pressure drop since the evaporator pressure is lower while the system is operating. If the amount of refrigerant being added is relatively small, operating the compressor and charging vapor into the suction side of the system generally provides enough temperature and pressure difference to complete the charging job. If large quantities of refrigerant are needed, however, the refrigerant cylinder will still have to be heated.

31.7 CHARGING ZEOTROPIC REFRIGERANTS

Refrigerants that fractionate, such as R-410A, must be charged as liquid to prevent the separation of the refrigerant. Weighing a liquid charge into an evacuated system is the same with zeotropes as with regular refrigerant, since the refrigerant leaves the cylinder as a liquid. Adding small amounts with the system operating is different because the refrigerant should not leave the cylinder as a vapor. These refrigerants must come out of the cylinder as a liquid but must be a vapor before entering the compressor. To accomplish this safely, the refrigerant must be metered either by the gauge manifold hand valve or by an external charging device.

SERVICE TIP

Most refrigerant cylinders have arrows showing the correct cylinder position for charging. Early R-410A cylinders used a dip tube, so the cylinder delivered liquid in the upright position. Most cylinders used today do not have the dip tube, so they need to be inverted to deliver liquid. Check the arrows on the side of the cylinder to be sure which position the cylinder should be in to deliver liquid.

Zeotropes can be metered in with the unit operating by slightly cracking open the low-side manifold valve and then closing it (Figure 31-15). This allows small amounts of liquid into the low-side hose but gives the liquid time to evaporate before reaching the compressor. Be patient. The liquid must flash off to vapor before entering the compressor. Allowing too much liquid in at once can damage the

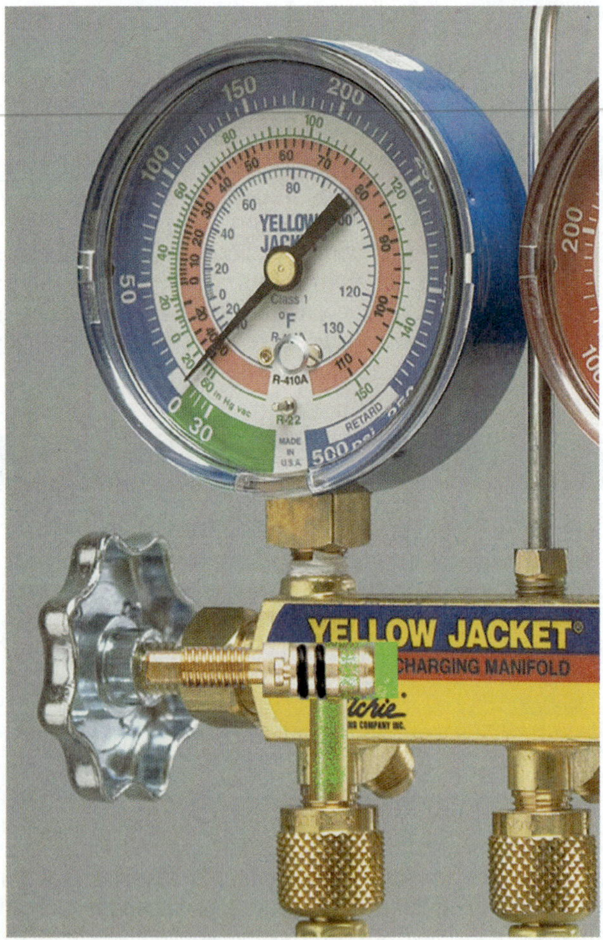

Figure 31-15 Liquid refrigerant can be metered in through the manifold gauges by cracking the hand valve on the suction side of the gauges. *(Courtesy of Ritchie Engineering Company, Inc., Yellow Jacket Products Division)*

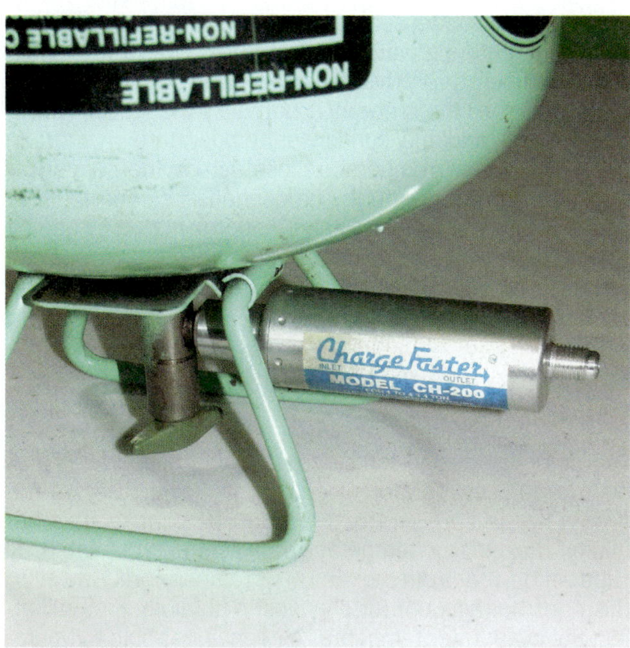

Figure 31-16 This device can be used in the charging line to flash off the liquid refrigerant.

compressor because liquid is not compressible! Charging devices like the one shown in Figure 31-16 increase the safety of low-side liquid charging. They are essentially a restriction that acts as a metering device.

Some gauge manifold sets have sight glasses that make it easy to monitor the condition of the refrigerant (Figure 31-17). For gauges that do not have this, a liquid-line sight glass that has ¼-in flare fittings on both ends can be installed in the low-side refrigerant hose.

31.8 FACTORS AFFECTING SYSTEM PRESSURES

Checking the charge on an operating system involves checking some aspect of system performance. The major difficulty in checking system refrigerant charge by measuring any single aspect of its performance is that there are so many variables that can affect system performance. Getting all variables that affect system performance to conform to the manufacturer's specifications at one time can be quite challenging. Understanding how each of these variables affects system performance is crucial for the technician trying to measure system performance. Some of the most important variables that can affect system performance include the following.

Figure 31-17 The refrigerant condition may be monitored using the sight glass in this manifold.

Outdoor Ambient Temperature for Air-Cooled Units

The high-side pressure on air-cooled systems is largely determined by the outdoor ambient temperature and the amount of air flowing over the condenser. Increased ambient temperature causes increased high-side pressure.

Condenser Airflow for Air-Cooled Units

The airflow across the condenser also affects the high-side pressure. Less condenser airflow means less cooling, causing higher condenser temperatures and pressures.

Inlet Water Temperature for Water-Cooled Units

The high-side pressure on water-cooled systems is largely determined by the condenser inlet water temperature and the water flow. Increased water temperature causes increased high-side pressure; lower water temperature causes lower head pressures.

Condenser Water Flow for Water-Cooled Units

Condenser water flow has a very large impact on high-side pressures in water-cooled units. Reduced water flow means less cooling, causing higher condenser temperatures and pressures. Increased water flow has just the opposite effect.

Return-Air Wet Bulb for Air Conditioners

Return-air wet-bulb (wb) temperature has a big effect on evaporator temperature and pressure in air-conditioning systems. Increased wet-bulb temperatures mean increased heat load on the evaporator, which increases the evaporator temperature and pressure. Low wet-bulb temperatures mean reduced evaporator heat load and reduced evaporator temperature and pressure.

Evaporator Airflow for Air Conditioners

The airflow across the evaporator in an air-conditioning system is critical. Low airflow means low load and lower evaporator temperature and pressure. Reduced airflow also reduces the system sensible cooling (temperature change) while increasing the system latent cooling (removal of water). Increased airflow increases the load on the evaporator and higher evaporator temperature and pressure. Increased airflow increases the system sensible cooling while decreasing the system latent cooling.

Returning Water Temperature for Chillers

Returning water temperature has a big effect on evaporator temperature and pressure in a chiller. Increased water temperatures mean increased heat load on the evaporator, which increases the evaporator temperature and pressure. Lower water temperatures mean reduced evaporator heat load and reduced evaporator temperature and pressure.

Chilled Water Flow for Chillers

The water flow through the evaporator in a chiller is critical. Low water flow means low heat load and lower evaporator temperature and pressure. Reduced water flow can be a hazard for chillers because the water can freeze if its temperature is allowed to drop too low. Increased water flow increases the heat load on the evaporator and increases evaporator temperature and pressure.

31.9 STANDARD INDUSTRY DESIGN AND OPERATION

There are system operating characteristics that have become industry standards through common use. These generic standards should never replace manufacturer information, but understanding what is normal can help a technician understand how all these variables interact with each other.

Standard Design Conditions

The primary rating condition for AHRI-listed unitary air-conditioning equipment is 95°F outdoor ambient and 80°F dry-bulb (db) and 67° wb indoor temperature. Air-conditioning equipment is rated at this condition. Logically, that means this is the ideal operating condition for an air-conditioning system.

Standard Air-Cooled Condenser Temperature

For many years, equipment manufacturers have been designing equipment to operate with a 125°F condenser temperature at the AHRI design condition of 95°F. A reasonably accurate approximation of condenser temperature for older equipment is to add 30°F to the outdoor ambient temperature.

High-Efficiency Air-Cooled Condenser Temperature

Most equipment manufactured today is not designed to operate with a 125°F condenser. Instead, 115°F at 95°F is more common. For newer equipment, condenser temperature can be approximated by adding 20°F to the ambient temperature. It should be noted that there is really an entire range of efficiencies, so the term *approximate* is apt. Some systems now operate as low as 10°F above the ambient temperature.

Water-Cooled Condenser Temperature

Water-cooled condensers typically operate at lower head pressures than air-cooled condensers. Standard design for water-cooled condensers is 105°F condenser temperature with 85°F entering-water temperature and 95°F leaving-water temperature. This provides two important pieces of information: the difference between the leaving-water temperature and the condenser temperature is approximately 10°F. The temperature rise across the condenser is also approximately 10°F. The condenser temperature of a water-cooled condenser can be approximated by adding 10°F to the leaving-water temperature. Its performance can be quickly checked by looking for a 10°F temperature rise in the condenser water temperature.

Evaporator Temperature in Air Conditioners

The common design for air-conditioning evaporators has been a 40°F evaporator when the air flowing across the evaporator is 80°F and 67°F wb (50 percent relative humidity). Evaporator temperature can be approximated by subtracting 40°F from the return-air temperature. However, this does not account for changes in humidity that will affect the evaporator temperature. Higher wet-bulb temperatures will produce higher evaporator temperatures, while lower wet-bulb temperatures will produce lower evaporator temperatures. The dew point temperature at 80°F db, 67°F wb is 60°F. A more accurate approximation would be to subtract 20°F from the dew point of the return air.

High-Efficiency Evaporator Temperatures

New equipment designed for higher efficiency typically operates at a slightly higher evaporator temperature. Many systems are now designed for 45°F evaporators at the 80°F db, 67° wb condition. This would make the evaporator temperature 35°F colder than the return-air temperature, or 15°F colder than the return-air dew point temperature.

Evaporator Airflow

Standard airflow across an air-conditioning coil is 400 CFM per ton. This is a very commonly applied standard. Many installations tweak this to get the desired amount of sensible and latent cooling from a unit, but practically everyone's unit will work with 400 CFM per ton moving across the evaporator coil.

Evaporator Temperature Drop

The temperature drop across an air-conditioning evaporator is typically around 15°F at the AHRI design of 80°F db, 67° wb. As the wet-bulb temperature drops, the temperature difference increases to a maximum of approximately 20°F. Higher wet-bulb temperatures decrease the temperature drop to a minimum of around 10°F.

Thermostatic Expansion Valve Superheat

Expansion valve superheat settings vary with the design and temperature range of the system. In general, the lower the evaporator temperature, the lower the valve superheat setting. Low-temperature systems often operate with a valve superheat of 3–5°F, medium-temperature systems with a valve superheat of 5–9°F, and high-temperature systems and air-conditioning systems with a valve of superheat of 9–12°F.

Fixed-Restriction Superheat

The system superheat of fixed-restriction systems varies widely with system operating conditions. At AHRI design conditions, most fixed-restriction systems operate with a superheat similar to expansion valve systems, around 10°F. However, their superheat increases significantly as the outdoor ambient decreases. Subtracting the ambient temperature from 105 gives an approximation of superheat for a fixed-restriction system operating with return air at AHRI design conditions, 80°F db, 67° wb.

Subcooling

The subcooling on most correctly operating refrigeration systems is between 5°F and 15°F. All systems should have some subcooling. Most split systems will have problems with flash gas in the liquid line if they operate at less than 5°F subcooling. Subcooling values exceeding 15°F can be symptomatic of systems that are either overcharged or have a restriction in the liquid line.

31.10 MANUFACTURER CHARGING CHARTS

Some manufacturers provide charging charts, charging calculators, or performance charts for checking the charge while the system is in operation. Some of these performance charts are intended to be used for both checking and adjusting the charge while others are intended only to check system operation.

Figure 31-18 shows a performance chart for an R-410A packaged unit that correlates suction pressure, discharge pressure, and outdoor ambient temperature. Note that this chart requires the indoor temperature to be at comfort conditions (Figure 31-19). If the indoor temperature is outside of comfort conditions, the chart will not be accurate. The liquid pressure is read on the left side of the chart, and the suction pressure is read across the bottom of the chart. Find the intersection of the liquid and suction pressures. This intersection should fall on or near the diagonal line for the outdoor ambient temperature. If the intersection falls above the ambient temperature line, the system is overcharged. If the intersection falls below the ambient temperature line, the system is undercharged.

Figure 31-20 shows examples of undercharged, overcharged, and correctly charged systems operating at an outdoor ambient of 85°F. The dark blue lines represent a correctly charged system with a liquid pressure of 310 psig and a suction pressure of 140 psig. They intersect exactly on the light blue 85°F diagonal ambient temperature line. The red lines show an example of an overcharged system at the same ambient temperature, 85°F. The liquid pressure of 340 psig and the suction pressure of 160 psig intersect above the light blue 85°F diagonal ambient temperature

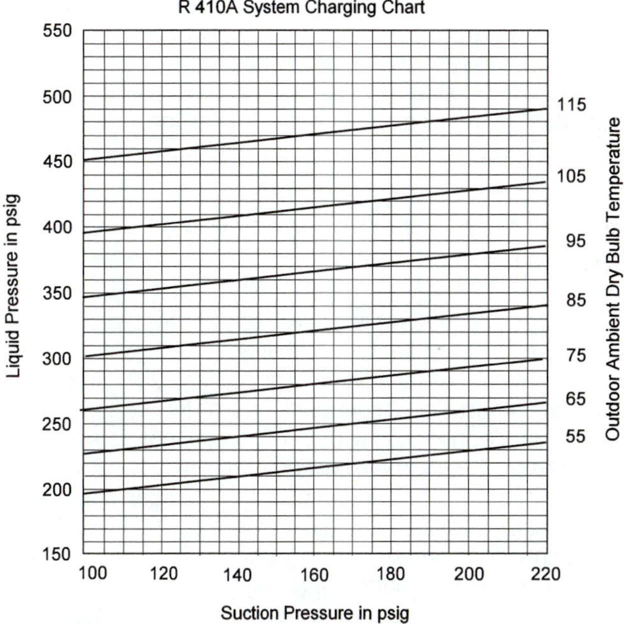

Figure 31-18 Typical manufacturer's charging chart correlating suction pressure, liquid pressure, and outdoor ambient temperature.

Pressure at Suction Service Port (psig)

CAUTION: BEFORE FINAL REFRIGERANT CHECK, INDOOR RETURN AIR TEMPERATURE SHOULD BE AT COMF ORT CONDITIONS FOR MOST ACCURATE RESULTS.

INSTRUCTIONS:

1. CONNECT PRESSURE GAUGES TO SUCTION AND LIQUID PORTS ON UNIT.
2. MEASURE AIR TEMPERATURE TO OUTDOOR COIL.
3. PLACE AN "X" ON THE APPROPRIATE CHART WHERE THE SUCTION AND LIQUID PRESSURES CR OSS.
4. IF "X" IS BELOW AMBIENT TEMPERATURE LINE, ADD CHARGE AND REPEAT STEP 3.
5. IF "X" IS ABOVE AMBIENT TEMPERATURE LINE, RECOVER EXCESS CHARGE AND REPEAT STEP 3.

Figure 31-19 This chart is only accurate if the house is at comfort conditions.

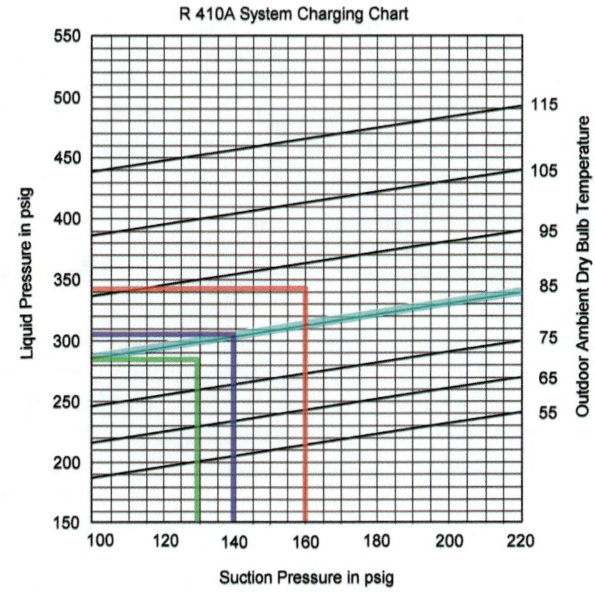

Figure 31-20 The red line shows an example of an overcharge, the blue line shows an example of a correct charge, and the green line shows an example of an undercharge.

line, indicating an overcharge. The green lines show an example of an undercharged system at the same ambient temperature, 85°F. The liquid pressure of 290 psig and the suction pressure of 130 psig intersect below the light blue 85°F diagonal ambient temperature line, indicating an undercharge.

31.11 CHARGING FOR PROPER SUPERHEAT

The superheat method is a very accurate means of checking the refrigerant charge. A change in refrigerant charge of 1 percent can change the superheat 3°F. Superheat charging charts are commonly used for systems with fixed-restriction metering devices. This is not as simple as stating a single superheat because the superheat of systems with fixed restrictions fluctuates with operating conditions. Fixed restrictions cannot adjust to varying load conditions, but a change in pressure drop produces a change in refrigerant flow, which produces a change in superheat. Warmer condenser ambient temperatures reduce system superheat by increasing refrigerant flow; cooler condenser ambient temperatures increase system superheat by decreasing refrigerant flow. Return air wet bulb temperature has just the opposite effect. Warmer return-air wet-bulb temperatures will increase superheat; cooler return-air wet-bulb temperatures reduce superheat. Figure 31-21 graphically shows how system operating conditions affect the superheat of systems with fixed-restriction metering devices.

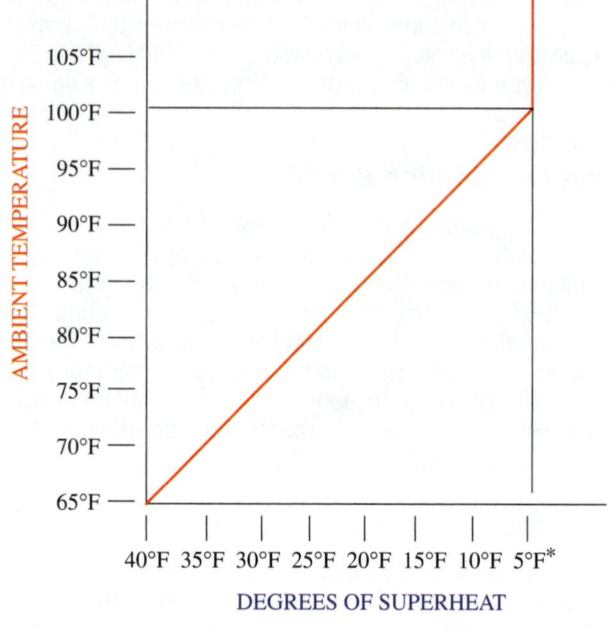

Figure 31-21 The degree of superheat required for a correct charge changes in direct proportion to the ambient temperature:

Charging charts like the one shown in Figure 31-22 give the correct system superheat for any operating condition. This can be used to check the system charge at any particular operating condition by comparing the actual system superheat to the superheat specified by the charging chart.

Both the outdoor ambient temperature and the indoor wet-bulb temperature must be known to use the chart in Figure 31-22. Figure 31-23 shows an example with an indoor wet bulb of 66°F and an outdoor ambient temperature of 90°F. Read across from the indoor wet-bulb temperature on the left (66°F) until you reach the column under the outdoor ambient temperature (90°F). The required superheat at this condition is 11°F.

The superheat of an overcharged system will be lower than specified by the charging chart. This is because liquid will be traveling further through the evaporator before it is all boiled off, leaving less evaporator coil to superheat the refrigerant. In this example, a superheat of 7°F would indicate an overcharge. The superheat of an undercharged system will be higher than specified on the chart. This is because the liquid will be completely boiled off early in the evaporator, leaving more of the evaporator to superheat the refrigerant. In this example, a superheat of 15°F would indicate an undercharge.

Make sure to read all the fine print on any charging chart. Manufacturers attempt to make the charging charts

Figure 31-22 This superheat charging chart gives the correct operating superheat based on both ambient temperature and indoor wet-bulb temperature.

INDOOR WB	OUTDOOR DB°F										
°F[1]	55	60	65	70	75	80	85	90	95	100	105	110
50	9	7										
52	12	10	6									
54	14	12	10	7								
56	17	15	14	10	6							
58	20	18	16	13	9	5						
60	23	21	19	16	12	8	6					
62	26	24	22	19	16	12	8	5				
64	29	27	24	21	18	15	11	9	6			
66	32	31	30	24	23	18	15	11	9	6		
68	35	33	30	27	24	21	19	16	14	12	9	6
70		35	33	30	28	25	22	20	18	15	13	11
72			35	33	30	28	26	24	20	20	17	15
74					34	31	30	27	25	23	22	20
76						35	33	31	29	27	26	25

INDOOR WB °F[1]	OUTDOOR DB°F											
	55	60	65	70	75	80	85	90	95	100	105	110
50	9	7										
52	12	10	6									
54	14	12	10	7								
56	17	15	14	10	6							
58	20	18	16	13	9	5						
60	23	21	19	16	12	8	6					
62	26	24	22	19	16	12	8	5				
64	29	27	24	21	18	15	11	9	6			
66	32	31	30	24	23	18	15	11	9	6		
68	35	33	30	27	24	21	19	16	14	12	9	6
70		35	33	30	28	25	22	20	18	15	13	11
72			35	33	30	28	26	24	20	20	17	15
74					34	31	30	27	25	23	22	20
76						35	33	31	29	27	26	25

Figure 31-23 The correct superheat is found by locating the intersection of the outdoor ambient temperature column and the indoor wet-bulb temperature row.

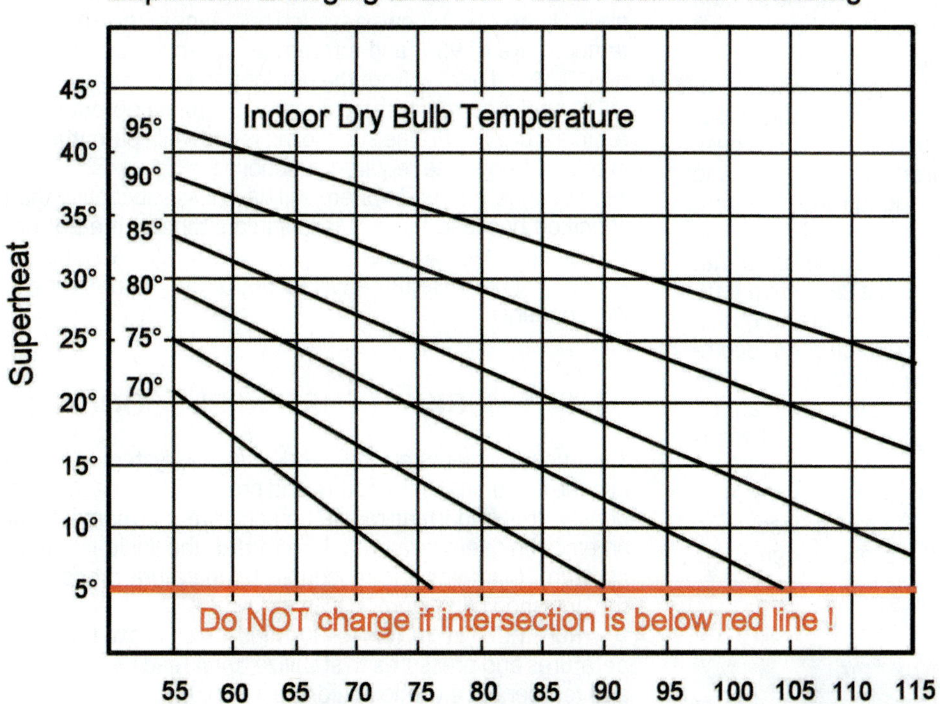

Figure 31-24 The superheat charging chart uses both outdoor and indoor dry-bulb temperature.

as easy to use as possible while still providing accurate information. Most charts assume something. Typical assumptions would include system airflow, indoor temperature, or indoor humidity. Any condition that is assumed for the sake of simplicity is usually listed somewhere on the chart. Rather than asking for indoor wet bulb, some superheat charts ask for a regular dry-bulb temperature both inside and outside (Figure 31-24). They assume the relative humidity to be 50 percent. These charts will not be accurate if the relative humidity is very far off of 50 percent. Other superheat charging charts only specify an outdoor ambient temperature (Figure 31-25). These charts are assuming an

SYSTEM SUPERHEAT

OUTDOOR AMBIENT TEMPERATURE	SYSTEM SUPERHEAT
65°F	45°F–50°F
75°F	30°F–40°F
85°F	20°F–30°F
95°F	8°F–12°F
105°F	3°F–5°F

Figure 31-25 Some superheat charts only require outdoor ambient temperature; they assume an indoor temperature and humidity.

indoor temperature, usually the AHRI design condition of 80°F, 50 percent relative humidity. These charts are only accurate at this one indoor operating condition.

SERVICE TIP

Be sure to check all system variables, especially the evaporator airflow. Other system variables can cause the superheat to be incorrect. Lower than normal evaporator airflow or a dirty evaporator will cause a low superheat. Higher evaporator airflow can cause a high superheat.

31.12 CHECKING SYSTEM SUPERHEAT

Two measurements are required to check system superheat: the suction line temperature and pressure near the compressor. It is important to use an accurate thermometer, or an electronic temperature probe to take the suction line temperature. Clamp on thermocouple probes like the one shown in Figure 31-26 work well.

Run the system for 10–30 minutes to allow the temperatures and pressures to stabilize. Also, record the indoor wet-bulb and outdoor ambient temperatures, since these are required in using the manufacturer's charts. Then read the pressure and temperature of the suction line.

Use a pressure-temperature chart like the one in Figure 31-27 to determine the saturation temperature that matches the suction pressure. Subtract the saturation temperature from the actual vapor line temperature to obtain the superheat. Compare it with the manufacturer's recommendation. Most manufacturers allow a variation of ±3°F. Add more charge if the system superheat is more than 3°F higher than specified; recover charge if the superheat is more than 3°F lower than specified.

The system must operate for at least 10 minutes after any charge adjustment. Then repeat the test procedure to be certain that the charge is within the proper range.

Figure 31-26 Clamp-style thermocouple temperature testers work well for checking superheat and subcooling.

An R-410a system operating with a suction pressure of 118 psig and a suction line temperature of 51°F has a superheat of 11°F. Using the charging chart in Figure 31-22 and the pressure-temperature chart in Figure 31-27, if these readings are obtained with 90°F ambient and 66°F wb, this system is charged correctly.

31.13 CHARGING FOR PROPER SUBCOOLING

Thermostatic expansion valves (TEVs) will regulate the refrigerant flow over a wide range of load and charge conditions. Because TEVs try to maintain a constant superheat, the superheat method is not the best method for testing the charge in an expansion valve system.

Some manufacturers recommend using subcooling to check the charge on expansion valve systems. Normally, under full load conditions, the refrigerant in an air conditioner will be subcooled 5°F to 15°F leaving the condenser. Figure 31-28 shows a typical subcooling charging chart. The example in Figure 31-29 shows a system operating with an ambient temperature of 95°F and a return-air wet-bulb temperature of 67°F. Read across from the outdoor ambient temperature (95°F) until you reach the column with the indoor wet-bulb temperature (67°F). The subcooling is shown in parentheses. In this example, the required subcooling is 5°F.

An undercharged system will have less subcooling than specified because there is less liquid sitting at the bottom of the condenser. An overcharged system will have a higher subcooling because there will be extra refrigerant sitting in the condenser.

31.14 CHECKING SYSTEM SUBCOOLING

Two measurements are required to check system subcooling: the liquid-line temperature and pressure leaving the condenser. It is important to use an accurate thermometer, or an electronic temperature probe to take the liquid-line temperature. Clamp-on thermocouple temperature probes like the one shown in Figure 31-26 work well.

Run the system for 10–30 minutes to allow the temperatures and pressures to stabilize, then read the pressure and temperature of the liquid line.

Use a pressure-temperature chart like the one in Figure 31-27 to determine the saturation temperature that matches the liquid pressure. Subtract the actual liquid temperature from the saturation temperature on the chart to obtain the subcooling. Compare it with the manufacturer's recommendation. Most manufacturers allow a variation of ±1°F. Add more charge if the system subcooling is more than 1°F lower than specified; recover charge if the superheat is more than 1°F higher than specified. The system must operate for at least 10 minutes after any charge adjustment. Then repeat the test procedure to be certain that the charge is within the proper range.

An R-22 system operating with a liquid pressure of 227 psig and a liquid-line temperature of 105°F has a subcooling of 5°F. Using the chart in Figure 31-29, if this system is operating with a 95°F ambient temperature and a 67°F wb temperature, the system is charged correctly.

VAPOR PRESSURE, PSIG									
Temp (°F)	11	12	22	113	114	500	502	134a	123
-50	***28.9***	***15.4***	***6.2***	–	***27.1***	***12.8***	***0.2***	***18.7***	***29.2***
-45	***28.7***	***13.3***	***2.7***	–	***26.6***	***10.3***	1.9	***16.9***	***29.0***
-40	***28.4***	***11.0***	0.5	–	***26.0***	***7.6***	4.1	***14.8***	***28.9***
-35	***28.1***	***8.4***	2.6	–	***25.4***	***4.6***	6.5	***12.5***	***28.7***
-30	***27.8***	***5.5***	4.9	***29.3***	***24.6***	***1.2***	9.2	***9.8***	***28.4***
-25	***27.4***	***2.3***	7.4	***29.2***	***23.8***	1.2	12.1	***6.9***	***28.1***
-20	***27.0***	0.6	10.1	***29.1***	***22.9***	3.2	15.3	***3.7***	***27.8***
-15	***26.5***	2.4	13.2	***28.9***	***21.8***	5.4	18.8	***0.1***	***27.4***
-10	***26.0***	4.5	16.5	***28.7***	***20.6***	7.8	22.6	1.9	***27.0***
-5	***25.4***	6.7	20.0	***28.5***	***19.3***	10.4	26.7	4.1	***26.5***
0	***24.7***	9.1	23.9	***28.2***	***17.8***	13.3	31.1	6.5	***25.9***
5	***23.9***	11.8	28.2	***27.9***	***16.2***	16.4	35.9	9.1	***25.3***
10	***23.1***	14.6	32.8	***27.6***	***14.4***	19.7	41.0	11.9	***24.6***
15	***22.1***	17.7	37.7	***27.2***	***12.4***	23.3	46.5	15.0	***23.7***
20	***21.1***	21.0	43.0	***26.8***	***10.2***	27.2	52.5	18.4	***22.8***
25	***19.9***	24.6	48.7	***26.3***	***7.8***	31.5	58.8	22.1	***21.8***
30	***18.6***	28.4	54.9	***25.8***	***5.2***	36.0	65.6	26.0	***20.7***
35	***17.2***	32.5	61.5	***25.2***	***2.3***	40.8	72.8	30.3	***19.5***
40	***15.6***	36.9	68.5	***24.5***	0.4	46.0	80.5	35.0	***18.1***
45	***13.9***	41.6	76.0	***23.8***	2.0	51.6	88.7	40.0	***16.6***
50	***12.0***	46.7	84.0	***22.9***	3.8	57.5	97.4	45.4	***15.0***
55	***10.0***	52.0	92.5	***22.2***	5.8	63.9	106.6	51.1	***13.1***
60	***7.8***	57.7	101.6	***21.0***	7.9	70.6	116.4	57.3	***11.2***
65	***5.4***	63.7	111.2	***19.9***	10.1	77.8	126.7	63.9	***9.0***
70	***2.7***	70.2	121.4	***18.7***	12.6	85.4	137.6	71.0	***6.6***
75	0.0	76.9	132.2	***17.3***	15.2	93.4	149.1	78.6	***4.0***
80	1.5	84.1	143.6	***15.8***	18.0	101.9	161.2	86.6	***1.2***
85	3.2	91.7	155.7	***14.3***	20.9	111.0	174.0	95.1	0.9
90	4.9	99.7	168.4	***12.5***	24.1	120.5	187.4	104.2	2.5
95	6.8	108.2	181.8	***10.6***	27.5	130.5	201.4	113.8	4.2
100	8.8	117.1	195.9	***8.6***	31.1	141.1	216.2	124.1	6.1
105	10.9	126.5	210.7	***6.4***	35.0	152.2	231.7	134.9	8.1
110	13.2	136.4	226.3	***4.0***	39.1	164.0	247.9	146.3	10.3
115	15.6	146.7	242.7	***1.4***	43.4	176.3	264.9	158.4	12.6
120	18.3	157.6	259.9	0.7	48.0	189.2	282.7	171.1	15.1
125	21.0	169.0	277.9	2.2	52.8	202.8	301.4	184.5	17.7
130	24.0	180.9	296.8	3.7	58.0	217.0	320.8	198.7	20.6
135	27.1	193.5	316.5	5.4	63.4	231.9	341.2	213.6	23.6
140	30.4	206.5	337.2	7.2	69.0	247.4	362.6	229.3	26.8
145	34.0	220.2	358.8	9.2	75.0	263.7	385.0	245.7	30.2
150	37.7	234.5	381.5	11.2	81.3	280.7	408.4	263.0	33.8

Bold Numbers - Inches Hg. Below 1 ATM

PRESSURE/TEMPERATURE CHART							
Temp. Deg. F	R-408A (FX-10) Liquid Pressure	R-404A (FX-70) Liquid Pressure	R-409A (FX-56) Liquid Pressure	R-409A (FX-56) Vapor Pressure	R-407C Liquid Pressure	R-407C Vapor Pressure	R-410A Liquid Pressure
-50	***1.6***	0.6	***12.4***	***17.2***	***2.9***	***11.4***	3.5
-45	1.1	2.7	***9.7***	***15.2***	0.4	***8.5***	8.5
-40	3.3	5.0	***6.8***	***13.1***	2.5	***5.2***	11.6
-35	5.6	7.6	***3.5***	***10.7***	4.8	***1.5***	14.9
-30	8.2	10.4	0.0	***8.1***	7.3	1.3	18.5
-25	11.0	13.4	2.0	***5.1***	10.1	3.6	22.5
-20	14.1	16.8	4.1	***1.9***	13.1	6.1	26.9
-15	17.5	20.5	6.5	0.8	16.5	8.8	31.7
-10	21.2	24.5	9.0	2.8	20.1	11.9	36.8
-5	25.2	28.8	11.8	4.9	24.0	15.2	42.5
0	29.5	33.5	14.8	7.2	28.3	18.9	48.6
5	34.2	38.6	18.1	9.7	33.0	22.9	55.2
10	39.3	44.0	21.7	12.5	38.0	27.3	62.3
15	44.8	49.9	25.5	15.4	43.5	32.0	70.0
20	50.7	56.2	29.6	18.7	49.3	37.2	78.3
25	57.0	63.0	34.0	22.2	55.7	42.7	87.3
30	63.7	70.3	38.7	26.0	62.5	48.7	96.8
35	71.0	78.1	43.8	30.1	69.8	55.2	107.0
40	78.7	86.4	49.2	34.5	77.6	62.1	118.0
45	87.0	95.2	54.9	39.2	86.0	69.5	129.7
50	95.8	104.7	61.0	44.3	94.9	77.5	142.2
55	105.1	114.7	67.6	49.8	104.5	86.0	155.2
60	115.1	125.3	74.5	55.6	114.6	95.1	169.6
65	125.6	136.6	81.8	61.9	125.4	104.8	184.6
70	136.8	148.6	89.5	68.6	136.9	115.2	200.6
75	148.7	161.2	97.7	75.8	149.1	126.2	217.4
80	161.2	174.6	106.4	83.4	162.1	137.8	235.3
85	174.4	188.8	115.5	91.5	175.8	150.2	254.1
90	188.4	203.7	125.2	100.2	190.2	163.4	274.1
95	203.1	219.4	135.3	109.4	205.5	177.4	295.1
100	218.7	235.9	146.0	119.2	221.6	192.1	317.2
105	235.0	253.4	157.2	129.6	238.5	207.8	340.5
110	252.1	271.7	169.0	140.6	256.4	224.4	365.0
115	270.2	290.9	181.4	152.3	275.1	241.9	390.7
120	289.1	311.1	194.4	164.7	294.7	260.5	417.7
125	308.9	332.3	208.0	177.8	315.2	280.1	445.9
130	329.7	354.5	222.3	191.6	336.7	300.9	475.6
135	351.5	377.8	237.2	206.3	359.2	322.9	506.5
140	374.3	402.2	252.9	221.8	382.6	346.2	539.0
145	398.1	427.7	269.3	238.2	407.0	370.8	572.8
150	423.0	454.4	286.4	255.5	432.4	396.9	608.1

Bold Numbers - Inches Hg. Below 1 ATM

Figure 31-27 Standard pressure-temperature chart for saturated refrigerant. *(Courtesy of Arkema Inc.)*

COOLING MODE				
OUTDOOR AMBIENT	INDOOR WET BULB (°F)			
	57	62	67	72
DB (°F)	LIQUID PRESSURE (SUBCOOLING)			
65	136 (3)	138 (4)	140 (4)	143 (4)
70	150 (4)	152 (5)	154 (5)	157 (5)
75	163 (5)	165 (5)	167 (5)	170 (6)
80	178 (5)	180 (5)	182 (5)	185 (6)
85	192 (5)	194 (5)	196 (5)	199 (6)
90	208 (5)	210 (5)	212 (5)	215 (6)
95	223 (5)	225 (5)	227 (5)	230 (6)
100	240 (5)	242 (5)	244 (5)	247 (6)
105	256 (5)	258 (5)	260 (5)	263 (6)
110	274 (5)	276 (5)	278 (5)	281 (6)
115	291 (4)	293 (4)	295 (5)	298 (6)
120	310 (4)	312 (4)	314 (5)	317 (5)
125	328 (3)	330 (3)	332 (4)	335 (4)

Figure 31-28 Typical subcooling chart for checking refrigerant charge.

COOLING MODE				
OUTDOOR AMBIENT	INDOOR WET BULB (°F)			
	57	62	67	72
DB (°F)	LIQUID PRESSURE (SUBCOOLING)			
65	136 (3)	138 (4)	140 (4)	143 (4)
70	150 (4)	152 (5)	164 (5)	157 (5)
75	163 (5)	165 (5)	167 (5)	170 (6)
80	178 (5)	180 (5)	182 (5)	185 (6)
85	192 (5)	194 (5)	196 (5)	199 (6)
90	208 (5)	210 (5)	212 (5)	215 (6)
95	223 (5)	225 (5)	227 (5)	230 (6)
100	240 (5)	242 (5)	244 (5)	247 (6)
105	256 (5)	258 (5)	260 (5)	263 (6)
110	274 (5)	276 (5)	278 (5)	281 (6)
115	291 (4)	293 (4)	295 (5)	298 (6)
120	310 (4)	312 (4)	314 (5)	317 (5)
125	328 (3)	330 (3)	332 (4)	335 (4)

Figure 31-29 The correct system subcooling is found in parentheses at the intersection of the ambient temperature row and the indoor wet-bulb column.

31.15 LIQUID-AMBIENT APPROACH-CHARGING METHOD

The liquid line is always a little warmer than the outdoor ambient temperature. The approach-charging method examines the temperature difference between the liquid line and outdoor ambient temperature. To calculate the approach, measure the liquid-line temperature and outdoor ambient temperature and subtract the outdoor ambient temperature from the liquid-line temperature. It is a good idea to either use the same thermometer to take both readings or to compare the two thermometers used at ambient temperature to make sure they correspond. Just a few degrees of difference can give a different diagnosis. Compare the measured temperature difference to the manufacturer's specification (Figure 31-30). If the temperature difference is higher than specified, the system is undercharged. If the temperature difference is lower than expected, the system is overcharged.

Model No.	Approach Temperature Liquid Line: Outdoor Ambient °F(°C)
10ACB12	7 (3.9)
10ACB18	5 (2.8)
10ACB24	9 (5)
10ACB30	10 (5.6)
10ACB36	12 (6.7)
10ACB42	14 (8)
10ACB48	13 (7.2)
10ACB60	12 (6.7)
10ACB62	12 (6.7)

Figure 31-30 Approach method.

31.16 WATER-COOLED SYSTEMS

The first checks to make on any water-cooled system are water temperature and flow. The pressures of a water-cooled system operating with too little water flow will be higher than normal, making the system look as if it is overcharged. The pressures of a water-cooled system operating with water that is too cold will be lower than expected, making the system appear to be undercharged. The manufacturer's performance data for water-cooled systems typically specify the condenser water flow, inlet water temperature range, and condenser water temperature rise (Figure 31-31). Operating pressure ranges for both the suction and discharge pressures are normally provided as well.

31.17 OIL CHARGING

Very rarely does oil need to be added to a refrigeration system. Unlike engine oil, refrigeration oil does not need to be changed regularly. The initial charge of refrigeration oil should last for the life of the system. A refrigeration system is a closed system, free of air and other contaminants. Further, the temperatures inside a refrigeration system are designed to stay under the temperature at which the oil

ENTERING WATER	FLOW GPM	COOLING PERFORMANCE DATA				HEATING PERFORMANCE DATA			
		WATER TEMP RISE	SUCTION PRESSURE	DISCHARGE PRESSURE	AIR TEMP DROP	WATER TEMP DROP	SUCTION PRESSURE	DISCHARGE PRESSURE	AIR TEMP RISE
35°F	7	NR	NR	NR	NR	5 - 6	38 - 44	163 - 207	19 - 24
	9	NR	NR	NR	NR	4 - 5	40 - 46	165 - 210	19 - 25
45°F	7	13 - 16	66 - 76	122 - 155	21 - 27	6 - 7	47 - 54	176 - 225	22 - 28
	9	10 - 13	66 - 75	118 - 150	21 - 27	5 - 6	48 - 55	178 - 227	22 - 29
55°F	7	13 - 16	67 - 77	133 - 170	21 - 27	7 - 9	55 - 64	192 - 245	25 - 32
	9	10 - 13	66 - 76	129 - 164	21 - 27	6 - 7	57 - 66	194 - 247	26 - 32
65°F	7	12 - 15	70 - 81	156 - 199	21 - 26	8 - 10	68 - 78	213 - 271	29 - 37
	9	10 - 12	70 - 80	151 - 192	21 - 26	7 - 8	69 - 80	215 - 273	30 - 38
75°F	7	12 - 15	72 - 82	172 - 219	20 - 26	9 - 11	74 - 85	223 - 283	32 - 40
	9	10 - 12	71 - 82	166 - 212	20 - 26	7 - 9	76 - 87	225 - 286	32 - 41
85°F	7	12 - 15	73 - 85	200 - 255	19 - 25	10 - 12	83 - 95	236 - 301	34 - 43
	9	9 - 12	73 - 84	194 - 248	19 - 25	8 - 10	84 - 97	238 - 303	34 - 44
95°F	7	11 - 14	75 - 86	227 - 289	19 - 24	NR	NR	NR	NR
	9	9 - 12	75 - 86	220 - 280	19 - 24	NR	NR	NR	NR

NOTE: CHART VALUES ARE ACCURATE ONLY FOR SYSTEMS OPERATING WITHIN DESIGN AIRFLOW LIMITS

Figure 31-31 System-operating data for water-source system.

Figure 31-32 The sight glass is used to check the level of the compressor crankcase oil.

begins to break down. Leaks seldom release enough oil to warrant adding any. A very small amount of oil sprayed out with refrigerant vapor makes a large, oily spot that may appear to be evidence of a large oil leak. When the compressor is low on oil, the oil is usually just somewhere else in the system. However, there are times when it is necessary to remove or add refrigeration oil.

Checking Oil Level

On compressors that have a crankcase sight glass, as in Figures 31-32 and 31-33, the correct oil level can be accurately observed and the quantity adjusted to the level indicated in the sight glass. It is important, however, to observe the oil level after the system has been in operation. Refrigerant will

Figure 31-33 Dissolved refrigerant raises the oil level during the off cycle.

Figure 31-34 The oil level starts out too high, drops too low shortly after startup, and settles at the correct operating level after several minutes of operation.

dissolve in the oil during system shutdown and raise the apparent oil level. This refrigerant boils out when the system starts, carrying some of the oil with it. Most of this oil returns after the system has operated long enough for the pressures to stabilize, but some oil remains in the system all the time. The typical level in an oil sight glass starts out above the normal oil level, drops to the bottom of the sight glass shortly after startup, and then gradually rises to the normal operating level (Figure 31-34). If the oil level is low after the system pressures have stabilized and the compressor crankcase is warm, the system may be low on oil.

Changing Compressor Oil

Oil is a scavenger, and it will end up collecting most of the crud in a system. One of the fastest ways to clean a system that has experienced multiple compressor failures is to check and change the compressor oil. Occasionally, refrigeration oil needs to be changed to make a refrigerant retrofit possible because the new refrigerant does not work well with the old refrigeration oil. Changing refrigeration oil for any reason is only really possible in systems with semihermetic or open compressors.

Figure 31-35 Crankcase plug in small, semihermetic air-cooled compressor.

Hermetic Compressors

It is not practical to remove oil from a hermetic compressor. Hermetic compressors do not have crankcase ports. The only way to remove the oil is to remove the compressor, turn it upside down, and pour the oil out. It is also not possible to check the oil level on hermetic compressors because there is no oil sight glass. It is possible to add oil to a system with a hermetic compressor using a hand pump to force oil into the suction line. However, the technician cannot accurately measure the effect of adding the oil and has no reasonable way of removing the oil once it is in the system.

Semihermetic and Open Compressors

Adding or removing oil is more reasonable with semihermetic and open compressors. Small semihermetic compressors have a crankcase plug above the normal crankcase oil level that may be used for removing or adding oil (Figure 31-35).

Evacuating the Crankcase

Remember that the compressor crankcase is under pressure! The pressure must be removed from the compressor crankcase before removing any crankcase plugs. This may be accomplished by having the compressor pump itself down. Connect a set of gauge manifolds to the suction and discharge service valves on the compressor. Front-seat the suction service valve, energize the compressors, and operate the compressor until the crankcase pressure is below the mandated EPA recovery level for the refrigerant in the system. De-energize the compressor after it has pumped down, and front-seat the discharge service valve. Make sure the compressor cannot be turned back on.

The refrigerant will need to be recovered from systems that cannot or should not pump themselves down. Some compressor manufacturers do not want the compressor operating in a vacuum, even for a short period of time. Figure 31-36 shows the connections and valve positions necessary to recover refrigerant from the compressor. Make sure the compressor is off and cannot be turned on. Connect a set of gauge manifolds to the suction and discharge service valves on the compressor. Front-seat both the suction and discharge service valves, and recover the refrigerant to the EPA-specified level for the refrigerant in the system. Note that the under-200-lb rule may be used because the part that is isolated contains less than 200 lb.

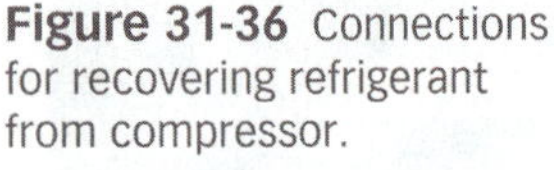

Figure 31-36 Connections for recovering refrigerant from compressor.

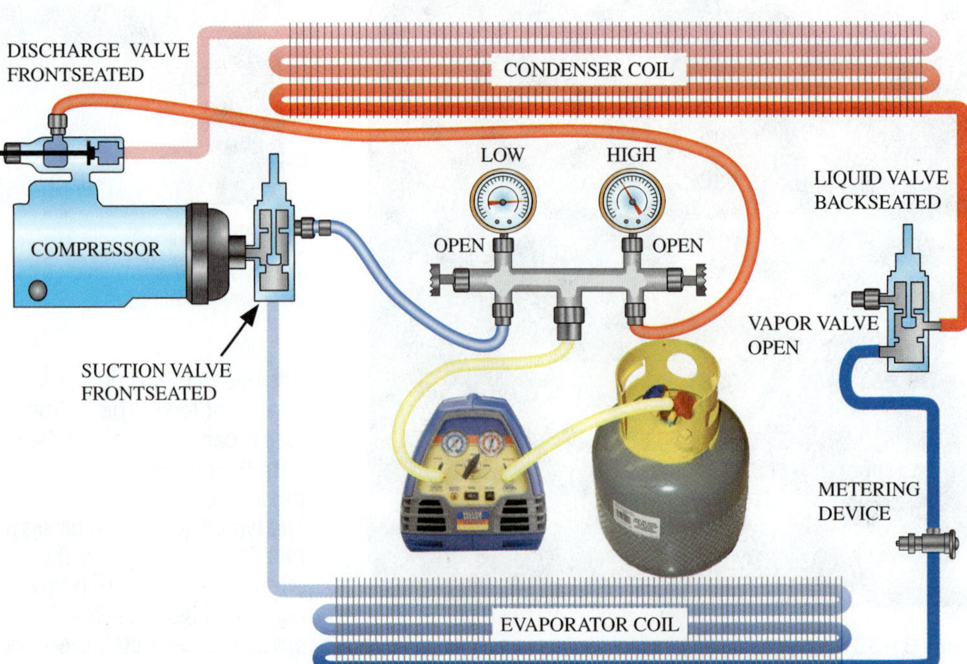

Figure 31-37 Crankcase drain plug on large compressor.

Removing Oil

An adapter can be used to remove oil from small semihermetic compressors. It provides a dip tube to the bottom of the crankcase. Once the adapter is in place, a few pounds of nitrogen introduced through the suction service valve will push the oil out the dip tube in the adapter.

Larger semihermetic and open compressors have crankcase drain plugs. Opening this plug allows oil to flow out of the compressor crankcase (Figure 31-37).

Adding Oil

The oil can be poured into the compressor crankcase after evacuating and opening the compressor crankcase. After the oil is added, the compressor will need to be evacuated. This should not take long, because only the compressor crankcase is being evacuated. After evacuating the compressor, the suction and discharge service valves are back-seated, allowing the system refrigerant back into the compressor.

Adding Oil Without Opening the Compressor Crankcase Using a Vacuum

In the evacuation method, the compressor creates a slight vacuum that is used to suck oil in. The oil line must have a manual valve to control flow through the oil line. One end of the oil hose is connected to the suction service valve, and the other is submerged in the oil container. Purge the air from the oil line by blowing a small amount of refrigerant vapor through it into the oil. Front-seat the suction service valve and operate the compressor long enough to create a slight vacuum. Oil may now be added by opening the control valve in the oil line. Close the oil control valve when the oil level in the compressor crankcase sight glass is at the minimum fill level (see Figure 31-37). Back seat the suction service valve and remove the oil charging hose.

UNIT 31—SUMMARY

A system that has been evacuated is normally charged with liquid into the high side of the system with the system off. The amount of refrigerant added is measured by an electronic scale. The amount and type of refrigerant used in a system can be found by looking at the data plate. This amount must be adjusted for line length on split systems. Common forms of charging charts provided by manufacturers include charts that correlate operating conditions to system pressure, superheat, or subcooling. Superheat charts are used with fixed-restriction systems, while subcooling charts are used with thermostatic expansion valve (TEV) systems.

WORK ORDERS

Service Ticket 3101

Customer Complaint: System Not Cooling Properly

Equipment Type: R-22 Residential Split System

The technician arrives at the residence. The customer states that the system has not been cooling very well for the last few weeks. The technician asks to go in and look at the indoor unit and finds that the fan is running and that there is very little air coming out of the vents. The technician notes that there is a TEV on the evaporator and finds that the outdoor coil is clean and the condenser discharge air temperature is cool.

The technician suspects that the evaporator coil may be freezing. The technician asks to go back in the residence so that the evaporative coil can be defrosted. The technician turns the thermostat fan switch to ON and the system switch to OFF so that the fan will blow air across the coil to defrost it. During the defrost process, the technician watches the condensate drain to be sure that it does not overflow as a result of the large volume of water from the frozen coil.

Once the evaporator coil has been thawed, the technician returns to the outdoor unit to check the charge. The technician puts on his safety glasses and gloves and attaches the refrigerant manifold gauge set to the system. The technician obtains the following pressures and temperatures:

- Suction-line pressure is 50 psi
- Suction-line temperature is 60°F
- Ambient air temperature is 85°F
- High-side pressure is 170 psi

The unit superheat chart calls for a superheat of 20°F, but the actual superheat is 32°F (50 psig = 28°F saturation 60°F − 28°F = 32°F). The technician determines that the system is low on refrigerant because the superheat is high. A leak is found on the Schrader valve. The technician uses a core-removing and -replacement tool to change out the Schrader core without having to remove the charge.

The charge is adjusted so that the proper superheat temperature is obtained. The technician notes the number of pounds of refrigerant that were added to the system. This information is put on the service ticket and recorded on the refrigerant log. The system is now operating satisfactorily.

Service Ticket 3102

Customer Complaint: System Is Tripping Main Circuit Breaker

Equipment Type: R-22 Residential Split System

The service technician checks to make sure the thermostat and compressor disconnect are turned off. Leaving the disconnect off, the technician checks and determines the compressor is grounded. The technician reports to the customer that the compressor is bad, and the customer wants the compressor replaced.

The technician checks the price book and provides the customer with a cost estimate for the job. The customer accepts the estimate, and the technician calls the office and has the parts department send a replacement compressor.

While the compressor is being delivered, the technician puts on safety glasses and gloves and proceeds to recover the refrigerant. After the refrigerant is recovered to the proper level, the technician replaces the compressor and installs the liquid- and suction-line driers. A leak test is then performed to verify a tight system.

With a leak-free system, the technician then evacuates and dehydrates the system to 500 microns. After determining the system is tight and dry, the refrigerant-charging process is started. To ensure the efficiency and longevity of the system, this process should be performed to the manufacturer's specifications.

The manufacturer's specifications dictate an approach-charging method. Using the manufacturer's supplied chart, the technician reaches the appropriate temperature difference between the outdoor ambient temperature and the liquid-line temperature, thus completing the charging process.

After the technician cleans up all of the packaging and other scrap material from the worksite, the customer is provided with a service ticket.

Service Ticket 3103

Customer Complaint: Poor Cooling and High Operating Cost

Equipment Type: R-22 Split System, New Condensing Unit Replacement

A customer complains that her new air conditioner does not seem to cool as well as her old one and it costs more to operate despite the fact that the new system is a 14-SEER unit and the old one was a 10-SEER unit. The technician checks the system charge and finds that it is operating with a low suction pressure, a high superheat, and a high subcooling. The technician tries to adjust the charge, but it seems impossible. To get the superheat down to the recommended level, the high side has to be higher than the charging chart specifies for the ambient temperature. The technician decides to look at the indoor coil to look for clues and notices that the indoor coil is old. The technician uses a Web-capable cell phone to look for an AHRI match between the new condensing unit and the old coil and discovers the two pieces of equipment are not matched. The technician explains that SEER ratings apply to matched systems, not individual components. Since the condenser and evaporator are not an AHRI match, they will not work well together. The technician suggests that the customer have a coil installed that is an AHRI match to solve the problem.

Service Ticket 3104

Work to Be Performed: New System Startup

Equipment Type: R-410A Split-System Air Conditioner

A technician is called to perform a new system startup on an R-410A split system. The technician notices that the manufacturer lists the factory charge as adequate for 15 ft, but the lines on this system are 30 ft. The technician suspects that the system will need a small amount of charge to adjust for the extra line length but decides to check the system operation before adding refrigerant. The technician gets the following readings while checking the system charge: 130 psig suction pressure, 60°F suction-line temperature, 340 psig liquid pressure, 102°F liquid-line temperature. The technician's first impression is that everything looks good. However, the manufacturer calls for an 8°F subcooling, and the unit is operating with only 3°F of subcooling (340 = 105°F saturation). The technician throttles a small amount of liquid refrigerant into the low side with the system running, waits for 10 minutes, then rechecks the figures. The pressures have not changed, but the liquid line is now 100°F, making the subcooling 5°F. The technician throttles in another small amount of liquid, waits, and rechecks the system. The pressures have remained stable, but the liquid-line temperature has dropped to 97°F, providing the correct subcooling.

UNIT 31—REVIEW QUESTIONS

1. Why are systems with fixed-restriction metering devices critical to charge?
2. How does an overcharge of refrigerant affect system efficiency?
3. Why must refrigerants that fractionate, such as R-410A, be charged as a liquid?
4. How does an undercharge affect the superheat of a fixed-restriction system?
5. How does an overcharge affect the superheat of a fixed-restriction system?
6. How does an undercharge affect the subcooling of a TEV system?
7. How does an overcharge affect the subcooling of a TEV system?
8. What is the best method for charging a large system that holds 50 lb of refrigerant?
9. What is the best method of charging a small appliance that holds less than a pound of refrigerant?
10. Using Figure 31-22, what would be the proper superheat for a system with an ambient temperature of 95°F and a return-air wet bulb of 70°F?
11. Using the information in question 10, you measure a suction pressure of 130 psig for an R-410A system. What should the suction line temperature be?
12. Using the information from questions 10 and 11, what charge adjustment should be made if the suction line measures 55°F?
13. Using Figure 31-28, what would be the proper subcooling for a system with an ambient temperature of 115°F and a return-air wet bulb of 62°F?
14. Using the information from question 13, what should the liquid-line temperature be if the liquid pressure is 293 psig for an R-22 system?
15. Using the information from questions 13 and 14, what charge adjustment needs to be made if the liquid line temperature is 110°F?
16. Why should a system be leak tested before it is charged?
17. Outline the procedure for adding oil to a semi-hermetic compressor without opening the compressor.
18. How often should the refrigeration oil in a compressor be changed?
19. Using Figure 31-30, what should the liquid-line temperature be for a model 10ACB30 unit operating in an ambient temperature of 97°F?
20. Using the information from question 19, what charge adjustment needs to be made if the liquid-line temperature is 110°F?
21. Using Figure 31-18, the suction pressure is 140 psig, the liquid pressure is 370 psig, and the ambient temperature is 105°F. What charge adjustment needs to be made?
22. Using Figure 31-18, the suction pressure is 155 psig, the liquid pressure is 360 psig, and the ambient temperature is 95°F. What charge adjustment needs to be made?
23. List the variables that affect system pressures.
24. Give an example of a condition that can make a system appear overcharged even though the charge is correct.
25. Give an example of a condition that can make a system appear undercharged even though the charge is correct.
26. What is the standard AHRI design condition used for rating unitary air-conditioning equipment?
27. Who is the most reliable source of information regarding the charge for any particular piece of equipment?
28. A new split-system air-conditioning system is being installed. The actual line length is 48 ft. The factory charge for the condensing unit is 94 oz. The installation instructions state that the factory charge assumes a line length of 20 ft and gives the per-foot adjustment as 0.6 oz/ft. How much refrigerant should be added to the system after the lines and coil have been evacuated?
29. The compressor in an existing split-system air-conditioning system is being replaced. The actual line length is 35 ft. The factory charge for the condensing unit is 100 oz. The installation instructions state that the factory charge assumes a line length of 15 ft and gives the per foot adjustment as 0.6 oz/ft. How much refrigerant should be added to the system after the system has been evacuated?

UNIT 32

Electrical Safety

OBJECTIVES

After completing this unit, you will be able to:

1. describe the harmful effects of electrical accidents.
2. recognize the significance of proper electrical safety procedures.
3. explain the function of ground fault circuit interrupters (GFCI).
4. determine if a circuit breaker has tripped and how to reset it.
5. lockout and tagout an electrical circuit.
6. test a circuit for voltage to make sure it is de-energized.
7. explain the safety importance of fuse and breaker amperage capacities.
8. describe why electrical wire types and sizes are important to safety.

32.1 INTRODUCTION

It is very likely that during your career you will work with electrical circuits and electrical equipment. You may be involved with new equipment installation and need to perform some finish wiring or retrofit wiring. You will also need to understand wiring and circuits for troubleshooting systems. This can be difficult because electrical theory and wiring are only one part of a HVACR technician's total body of knowledge. It may take years of service experience before you have a complete understanding of electrical systems and circuits. However, that does not mean that you cannot begin working safely beginning on the first day.

There is no excuse for poor safety practices. If the circuit is properly de-energized and locked out, then you will be safe. In the United States, electrical hazards cause more than three hundred deaths and four thousand injuries in the workplace every year. Electrical accidents rank sixth among all causes of work-related deaths. The Electrical Safety Foundation International (ESFI) recently reviewed data for a 5-year period and reported that a total of thirty-nine (15 percent) of the victims were electrocuted while servicing HVACR equipment. You need to understand how to protect yourself and the customer from electrical hazard. This unit will identify the steps that you need to take to ensure safe workplace practices.

CODE

OSHA's general industry electrical safety standards are published in Title 29 Code of Federal Regulations (CFR), part 1910.302–1910.308 (Design Safety Standards for Electrical Systems) and 1910.331–1910.335 (Electrical Safety-Related Work Practices Standards).

32.2 ELECTRICITY FLOW

Electricity needs a path to flow through. Substances such as metals offer very little resistance to electric current flow and are called conductors. This is why electrical wires are normally made of metal such as copper, which not only conducts electricity but bends fairly easily without breaking. Insulation has the opposite effect. It resists the flow of electricity, and this is why electrical wires are not just bare but instead are enclosed by insulation that has a high resistance to electricity flow. A typical electrical cable with insulated wire is shown in Figure 32-1.

Wet and damp areas can often lead to electrical hazard even though pure water itself is a poor conductor. This is because any water or moisture present will not be pure

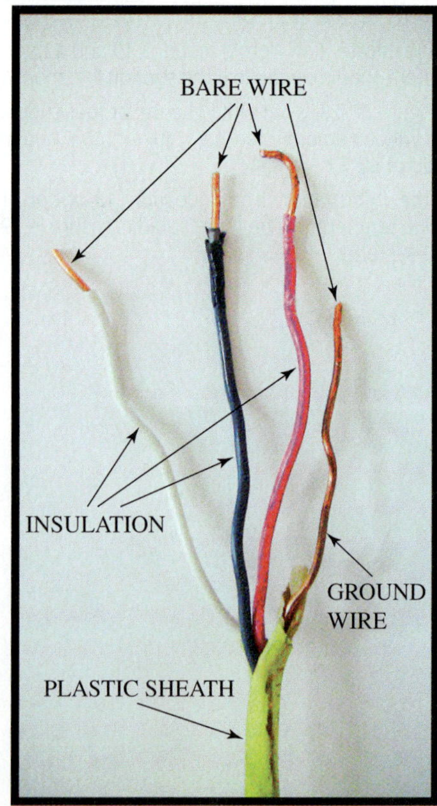

Figure 32-1 Three conductor NM electrical cable with ground.

but instead will most likely hold small amounts of impurities such as salts and minerals, which turn water into a very good conductor. As an example, dry wood is a poor conductor, but when saturated with water, it changes into a good conductor.

Similarly, your skin, when it is dry, also has some resistance to electricity flow. However, normally there will be some perspiration or moisture on your skin, which then turns it into an excellent conductor of electricity. The amount of current in amperes that passes through a body determines the severity of an electrical shock. One milliamp (one one-thousandth of an ampere) will provide a slight tingling sensation. Current flow of over 50 milliamperes will cause severe pain and at higher levels cardiac arrest.

The approximate body electrical resistance underneath the surface layer of your skin, from hand to hand across the body, is 1,000 ohms. Given this resistance, a current flow of 30 milliamperes would be generated with only 30 volts. Fortunately, the outside of your skin would provide for a higher resistance and therefore a lower current flow. However, if your skin was wet or cut, then the resistance would decrease and the current flow could increase to a potentially dangerous level even at this low voltage.

High-voltage systems (600 volts or more) provide a greater chance for fatal shock. However, it is easy to see that with only such a small amount of current flow required, electrical shock can be just as likely to happen when working on a low-voltage system if you come in contact with a conductor.

SAFETY TIP

The majority of electrocution deaths occur at low voltages. The false perception that low voltage is safe to work on may contribute to this statistic.

32.3 ELECTRICAL SHOCK

In an electrical circuit, the electricity will flow from the source through one wire (hot conductor) connected to the load and then back again to make a closed loop or closed circuit. If you touch a live wire, then you can become part of the circuit and the electricity can flow through you. Electricity can flow between parts of your body or through your entire body to the ground, as shown in Figure 32-2. This can lead to burns, electrical shock, and even death.

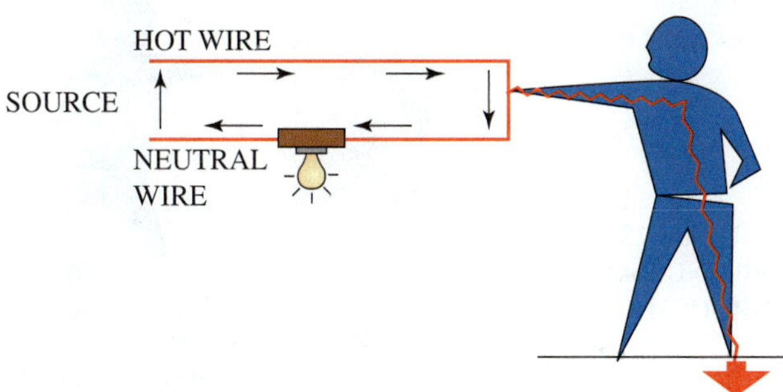

Figure 32-2 Coming into contact with an energized conductor will result in electric shock.

SAFETY TIP

Shocks occur when a person's body completes the current path with both wires of an electric circuit, such as one wire of an energized circuit and the ground; a metal part that accidently becomes energized, such as a break in a wire's insulation; or another conductor carrying current.

A severe shock may cause internal damage and destruction of tissues, nerves, and muscles that may not at first be readily apparent. Also, electrical burns can occur as electricity flows through parts of the body. In the event of severe shock or electrical burns, immediate medical attention is required.

If a person is "frozen" and unable to pull free of the circuit, do not attempt to grab hold of him as then you will also become part of the live circuit. Instead, locate the power source and shut off the switch immediately. If this is not possible, try to use something nonconducting, such as a wooden broom handle, and try to safely push the person away from the contact.

To prevent the possibility of electric shock, the circuit should always be de-energized before it is worked on. Even when the circuit is de-energized, there still may be a residual electrical charge remaining in components such as capacitors. Always assume there is a voltage present when working with circuits having high capacitance, even when the circuit has been disconnected from its power source. Before working on de-energized circuits that have capacitors installed, they must be discharged using a safety shorting probe.

CODE

OSHA's electrical standards are based on the National Fire Protection Association (NFPA) Standards NFPA 70, National Electric Code, and NFPA 70E, Electrical Safety Requirements for Employee Workplaces.

32.4 BURNS

Burns are the most common type of shock-related injury. These can be classified as electrical burns, arc burns, thermal contact burns, or a combination of burns. The most serious are electrical burns. These occur when heat generated by the flow of current through body parts damages tissue. Since the burning is often mostly internal, there may be only a small wound on the surface. Even so, this should not be ignored. An electrical burn requires immediate medical attention.

A flash or explosion near a person's body due to an electric arc can cause severe burns. Protective clothing and gear will reduce the chance of exposure to the high temperatures that can be produced by an arc flash. An electric arc can be produced when there is a short circuit or poor electrical contact between conductors. It is important to note that physical contact with the conductor is not necessary. Even standing several feet away from the source of an arc may lead to severe or even fatal burns.

Bare skin that touches hot wires, hot motors, or other overheated electrical components will experience thermal contact burns. Thermal contact burns can also be the result of an electric arc explosion, as hot flying fragmented metal and splattering molten droplets will burn the skin and catch clothing on fire.

SAFETY TIP

You should always maintain a safe distance from a live circuit. On circuits of less than 300 volts, never contact a live conductor. On live circuits of 300 to 750 volts, you should never approach at a distance closer than 1 ft.

32.5 PROTECTIVE GEAR

Insulated rubber gloves should be used whenever testing live circuits. They should be marked with the appropriate voltage rating and last inspection date. Leather protectors worn over the gloves are used to reduce the chance that the rubber will be punctured. Never wear watches or rings while wearing insulated rubber gloves, and never wear anything conductive. Avoid wearing synthetic-fiber clothing. This type of clothing will burn rapidly and melt, which will increase the severity of the burn. Wear flame-resistant clothing instead.

Safety glasses or face shields should be worn whenever working on electrical circuits. You should always use insulated tools, and make sure that the insulation is not cracked or damaged. Use the proper type of plastic or fiberglass-style fuse puller for installing or removing low-voltage cartridge-type fuses.

CODE

OSHA addresses the use of protective equipment apparel and tools in CFR 1910.335(a). This article is divided into two sections: Personal Protective Equipment and General Protective Equipment and Tools.

32.6 PORTABLE HAND TOOLS

A considerable number of electrical injuries occur from the improper use of portable, electrically operated hand tools. It is important to ensure that your tools are properly grounded. This means that the plug on the tool should have three prongs, as shown in Figure 32-3. For example, a faulty portable electric drill may have an uninsulated metal casing that would provide an electrical shock when turned on. A ground wire from the three-pronged plug will direct the stray current to ground rather than through the person holding the drill. The ground provides an easier path for the electricity to take rather than through your body, as shown in Figure 32-4. This increased flow of current to ground will normally activate the circuit-protection devices, which would shut off the power to the device.

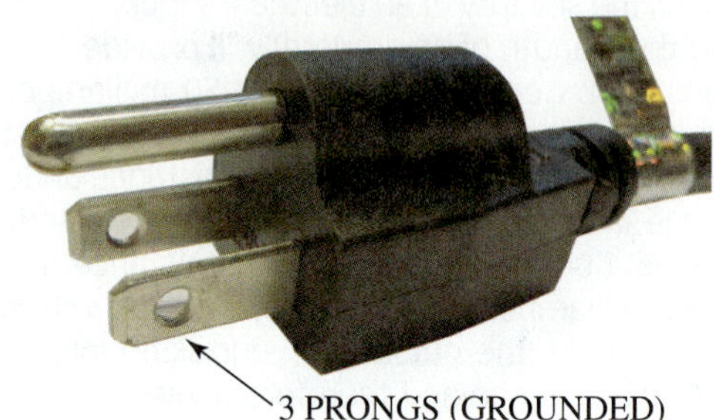

Figure 32-3 Comparison of a three prong and two prong electrical plug.

CODE

If a conductor is connected to the earth or some conducting body that serves in place of the earth, such as a driven ground rod (electrode), the conductor is said to be grounded. Often in residential construction, the only grounding electrode that is available is the metal underground water piping system, as shown in Figure 32-5. The National Electric Code (NEC), section 250.53(D)(2), requires that whenever water piping is used as an electrode it must be supplemented, for example, with a ground rod. Also in NEC section 250.5 (A), the metal underground water pipe must be in direct contact with the earth for 10 ft or more, and the grounding clamp connection must be made within 5 ft of the point where the pipe enters the building. **Caution:** *NEC Section 250.52 (B) requires that metal underground gas piping systems may not be used as a grounding electrode.*

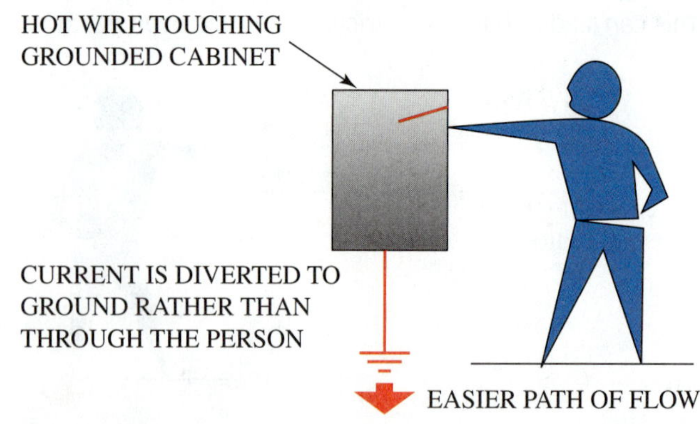

Figure 32-4 A grounded enclosure will protect a person from electric shock.

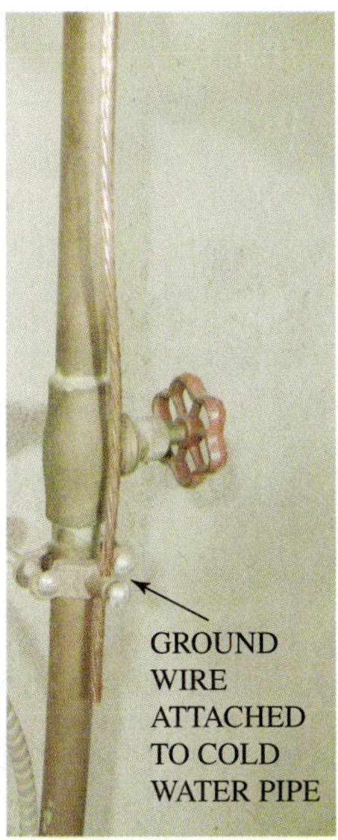

Figure 32-5 A ground wire may be connected to a cold water pipe if attached in accordance with the National Electric Code.

SAFETY TIP

Some portable electric tools provide for double-insulated casings and do not need to be grounded. This type of tool may only have a two-prong plug rather than three. If a tool plug has three prongs, never use an adapter to make it fit a two-prong receptacle unless the adapter is properly grounded. Some adapters, as shown in Figure 32-6, have a grounding connection, but this will not protect you unless it is attached to ground.

32.7 GROUND FAULT CIRCUIT INTERRUPTER (GFCI)

Always use a GFCI (ground fault circuit interrupter) receptacle to plug into if available. A standard GFCI receptacle is shown in Figure 32-7. A GFCI can interrupt the flow of

Figure 32-6 Electrical adapter for three-prong plug to two-prong receptacle with ground wire connection.

Figure 32-7 GFCI receptacle.

electricity in as little as 1/40 of a second to prevent electrocution. It compares the amount of current going into the electric device with the amount of current returning from it, and if the difference exceeds 5 milliamperes, the device automatically shuts down. Take, for example, the wire of an extension cord. If there were a break in the insulation exposing bare wire, and the extension cord was located in a damp or wet area, this could lead to electrical shock for someone using it. However, this type of fault would generate excessive current, and the GFCI receptacle would trip before any further damage was done. As further protection, the GFCI receptacle must be manually reset.

32.8 DETERMINING DE-ENERGIZED CIRCUITS

An energized circuit is normally referred to as being "closed." This would be considered to be a "live" circuit. A de-energized circuit is "open." Think of a typical appliance plug. When it is plugged in to the receptacle, it is energized (closed), and there is power to the device. When it is unplugged from the receptacle, it is de-energized (open), and the device shuts down.

32.9 BREAKERS AND SWITCHES

In electrical circuits, switches and breakers are used to energize and de-energize the circuit. Think of a light switch. The lights are on, the switch is closed, and the circuit is energized. The lights are off, the switch is open, and the circuit is de-energized. Therefore, one way to check a circuit is to check its switch or breaker to determine whether the circuit is energized. In some cases, there may be more than one switch or breaker for the device. Take, for example, a residential service panel, as shown in Figure 32-8. There is a main disconnect switch that will shut off power to the entire building. Then there are separate circuit breakers that supply separate house circuits. Which switches and breakers are to be opened (turned off) will depend on the type of service work to be performed.

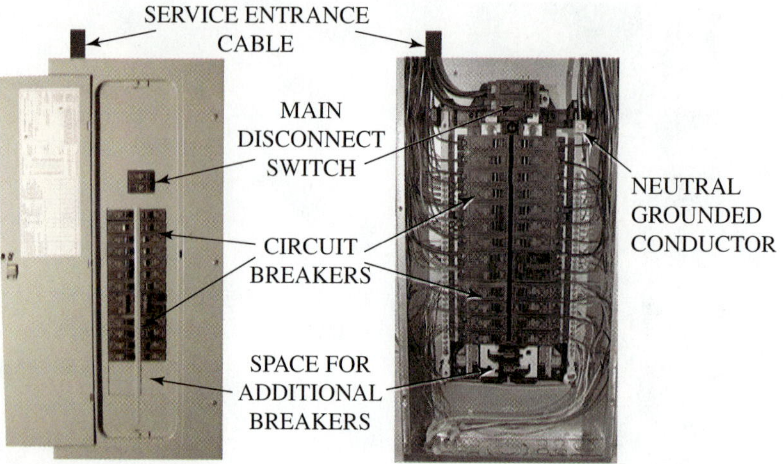

Figure 32-8 Service entrance panel.

SAFETY TIP

There is still power being supplied to a residential service panel even with the main disconnect switch is open (off). This is the power coming from the outside utility pole to the panel. The supply wires of the service entrance cable will still be "live."

32.10 CIRCUIT BREAKERS

Circuit breakers typically have three positions. The two most common positions are when the breaker is on or off. The third position is when the breaker is tripped due to an overcurrent situation. If the breaker is tripped, as shown in Figure 32-9, it is positioned in the middle and is not in the off or on position, so it will need to be reset. Even though the breaker is not in the off position, it is tripped and no power is being supplied to the circuit. The reason for this middle position is to allow you to easily look through all the breakers in a service panel to see if there is a fault somewhere.

Generally, to reset a breaker you will need to push it into the off position, and you should hear a slight click and then push it back into the on position and you will again hear it click as it snaps closed. Generally with breakers there is a "snap" action. Some motive force, such as a spring, will close the contacts. In breakers, the contacts are normally enclosed so they are not visible unless you break them apart.

Circuit breakers can be easily replaced. They simply connect to the bus bar with a clip located on the back side of the breaker, as shown in Figure 32-10. Never replace a circuit breaker until you have tested the circuit to ensure that the power is secured and there is no voltage present. It is not enough to just shut the power off. Always test the circuit for voltage as a double-check that the breaker is safe to remove.

SAFETY TIP

Never reset a circuit breaker until the cause of the fault is determined. The circuit breaker tripped due to a surge of current through the circuit. If you try to reset the breaker and the fault has not been corrected, it will trip again. An overcurrent condition can lead to equipment damage and even fire. Determine why the breaker tripped and correct the condition before resetting the breaker.

Figure 32-9 15-amp circuit breaker.

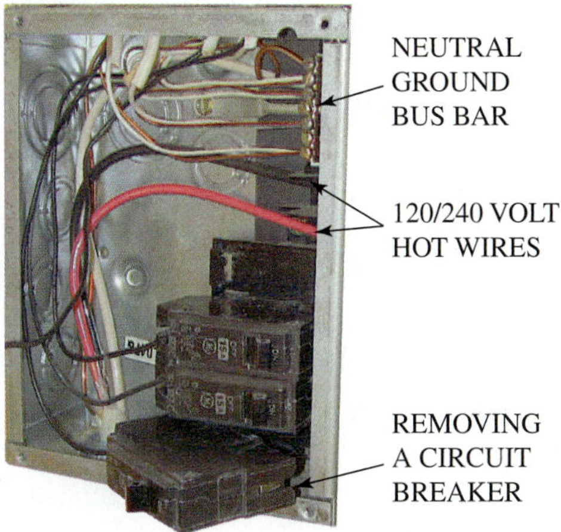

Figure 32-10 Removing a circuit breaker.

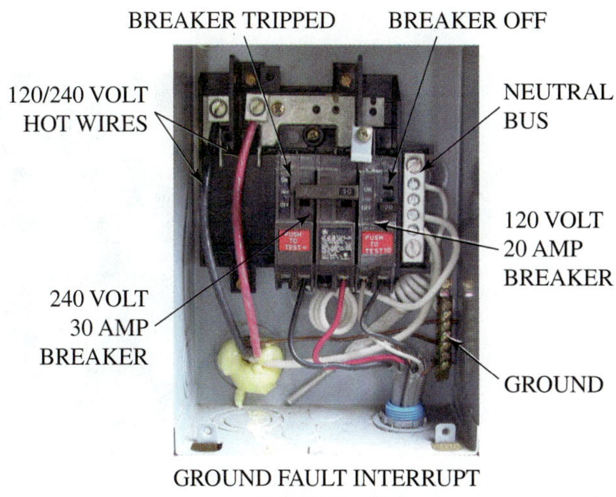

Figure 32-12 Tripped GFCI circuit breaker.

TECH TIP

Electric motors withstand hours and hours of running time and quite a bit of abuse. If a circuit breaker has tripped, it could be due to a shorted electric motor. A motor that runs hot for a prolonged period may experience insulation breakdown. This can lead to a short, where the electricity bypasses its normal path of flow. It is a "shortcut" for the electricity, and this allows for more electricity to flow (current increases). The wires in the circuit will not be large enough to carry this extra current, and they can overheat, melt, and cause a fire. Before this happens, the circuit breaker will trip.

32.11 GFCI BREAKERS

GFCI breakers are often used for equipment to be located outdoors in damp or wet locations. The GFCI breaker, as shown in Figure 32-11, has three positions: on, off, and tripped. It is reset like any other circuit breaker. You will need to push it into the off position, and you should hear a slight click and then push it back into the on position and you will again hear it click as it snaps closed. The GFCI breaker also has a test button. When this button is pushed, the breaker will trip and then must be manually reset. This type of breaker can be periodically tested in this way to make sure that it is functioning properly. A GFCI breaker wired in a circuit is shown in Figure 32-12. From the position of the switch, you can see that the breaker is tripped. GFCI circuit breakers can also be replaced similarly to standard-type circuit breakers, as shown in Figure 32-13.

Figure 32-11 GFCI circuit breaker.

32.12 SWITCHES

General-duty switches are designed for use in residential and commercial applications. These are used for relatively light-load applications such as general air conditioning and appliance loads. There are several different kinds of these switches, and some will have fuses while some others will not. A typical residential safety switch is shown in Figure 32-14. Unlike breakers that have a "snap" action, switches may consist of a throw handle that engages blades and clips, sometimes called a knife switch, as shown in Figure 32-15.

Figure 32-13 Removing a GFCI circuit breaker.

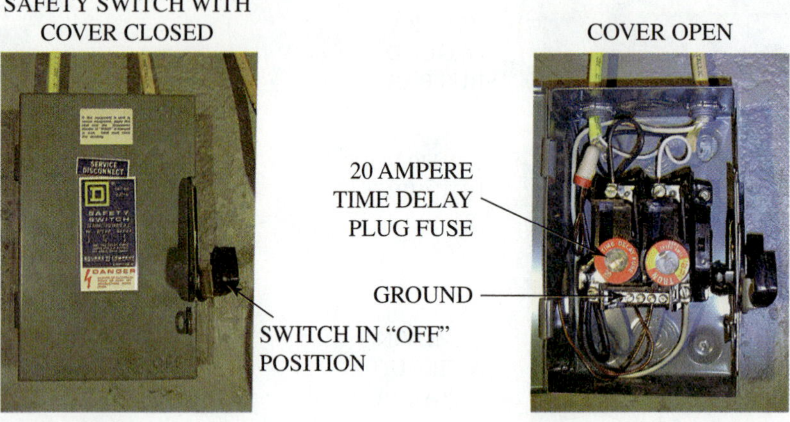

Figure 32-14 Safety switch enclosure.

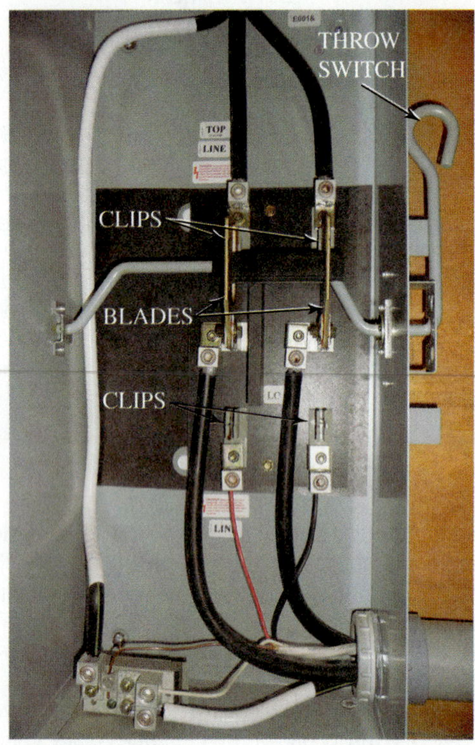

Figure 32-15 Manual switch with blades and clips.

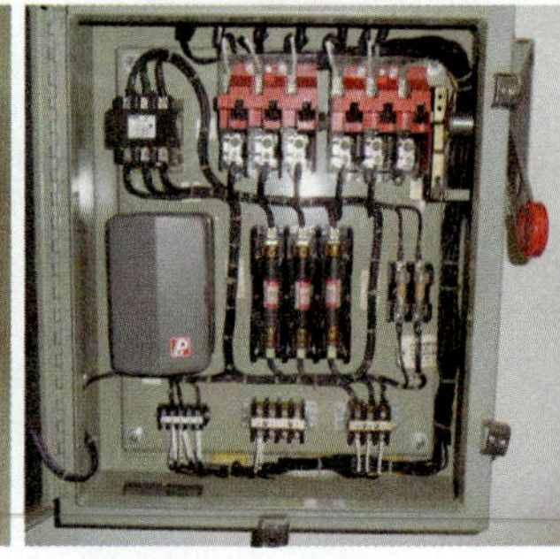

Figure 32-16 Walk-in freezer control cabinet.

On lower-voltage systems, the person throwing the switch manually engages the blades and clips together to energize the circuit. Slow operation of the switch can cause the blades and clips to arc.

Heavy-duty motor-rated switches are commonly found on motor control boxes and feeder boxes for refrigerant controls in larger commercial systems, as shown in Figure 32-16. These will typically have three blades and clips for three-phase AC power, as shown in Figure 32-17. Since arcing is more pronounced on these types of larger switches, a quick-break operation prevents an operator from closing or opening the switch too slowly. This is similar to the "snap" action of a circuit breaker. For safety, when the switch is closed (energized), an interlock switch prevents the door from being opened, as shown in Figure 32-18. The handle must be turned off (de-energized) before the door can be opened. When the door is open, the interlock prevents the handle from being turned on (energized).

32.13 LOCKOUT/TAGOUT

When working in a residential setting, the first step before beginning any electrical work is to turn off the power to the circuit at the service entrance panel. Any other safety switches between the service panel and the component to be serviced should also be shut off. You want to be 100 percent positive that the customer or anyone else working in the area will not turn the power back on to the circuit while you are working on it. **Never assume that they know you are working on the circuit!**

The customer should be informed that you are working on the circuit and the breaker should be physically locked out, as shown in Figure 32-19. You should attach a tag that includes your name, your company, why you have the breaker locked out, and the date. In the event that you do not finish the work and do not return to the residence, this provides information on who shut the power off and when it was secured. It is always preferable to physically lock out the circuit. Disconnect switches normally have a built in method for locking them, as shown in Figure 32-20.

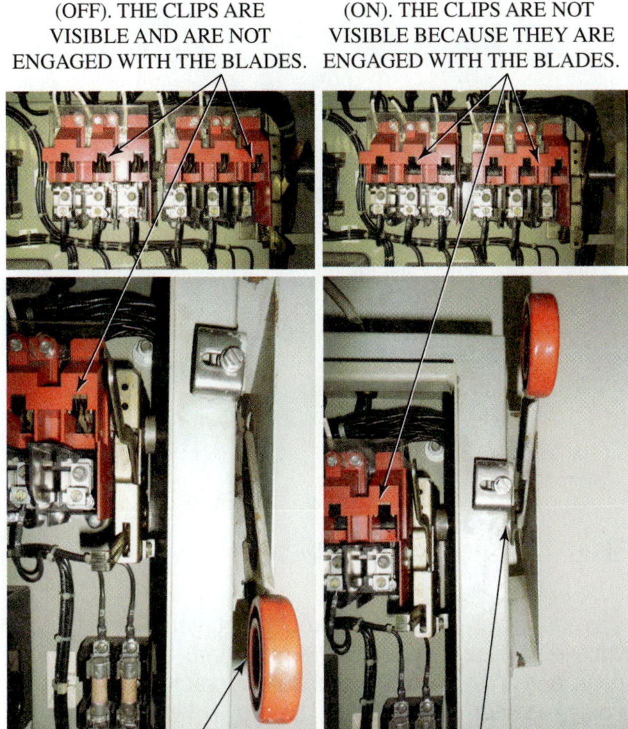

Figure 32-17 Blades and clips for walk-in freezer control cabinet.

Figure 32-19 Breaker locked out on residential service panel.

SAFETY TIP

Always check the component for voltage even after you have shut the power off the circuit. If you temporarily leave the area, always retest the circuit before continuing to work on the component.

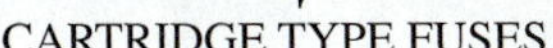

Figure 32-18 Safety interlock for walk-in freezer control cabinet.

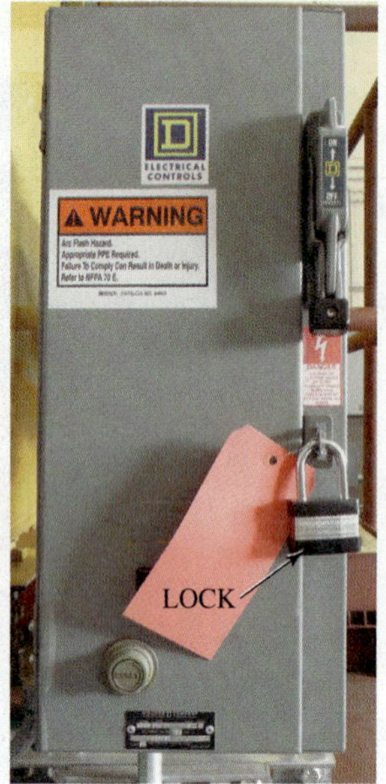

Figure 32-20 Locked-out electrical enclosure.

An apartment or condominium may have a maintenance supervisor overseeing the complex. On this type of job site, there is a greater potential that someone could accidently turn the power on while you are working on the circuit. In this type of setting, it is always preferable to physically lock out the circuit you are working on. The maintenance supervisor should be informed that you are working on the circuit. Never assume that the circuit is safe. Always test the circuit before working on it.

CODE

OSHA's lockout/tagout standards are published in Title 29, Code of Federal Regulations (CFR), part 1910.147, The Control of Hazardous Energy (lockout/tagout), and 1910.147 App A, Typical Minimal Lockout Procedures (when only tagging is possible).

In an industrial setting, there are often lockout/tagout procedures for the facility as shown in Figure 32-21. OSHA has regulations regarding lockout and tagout of equipment while it is being serviced. It is important that you follow these and all other electrical safety practices. You must receive proper authorization before locking and tagging out breakers or removing the lockout/tagout. You must understand and follow the procedures and record the action taken in a job journal, as shown in Figure 32-22. An industrial switchboard is shown in Figure 32-23. A safety placard is attached to the switchboard, as shown in Figure 32-24. The proper way to lockout and tagout a breaker is shown in Figure 32-25.

Figure 32-21 Lockout/tagout instructions.

Figure 32-22 Locks and job journal record book.

Figure 32-23 Industrial electrical switchboard.

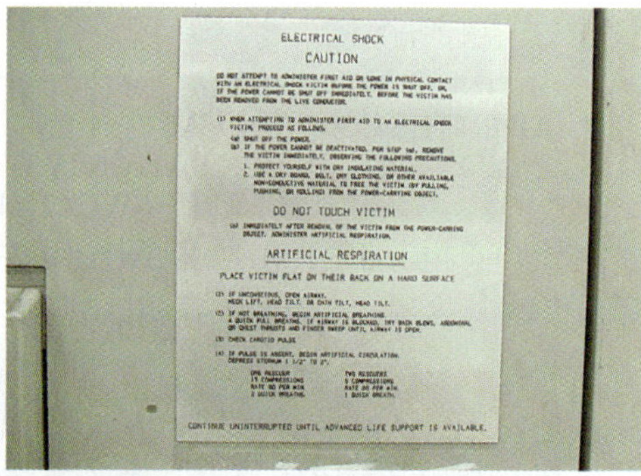

Figure 32-24 Electrical safety placard posted on switchboard.

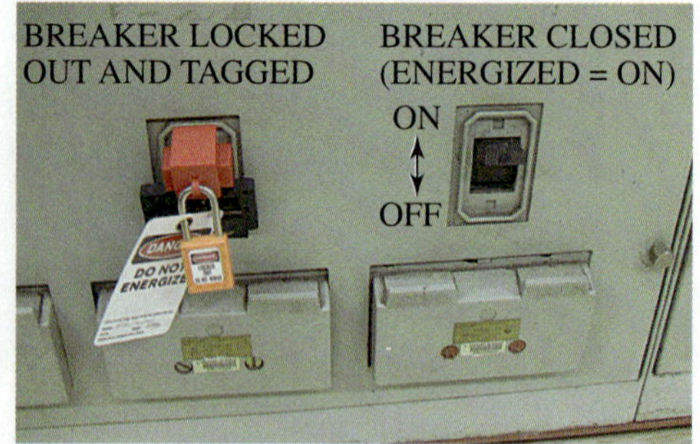

Figure 32-25 Breaker properly locked and tagged.

SAFETY TIP

As a safety precaution, the only person who may remove a lock or tag is the person who placed it there. In some instances there may be multiple locks and multiple tags on one circuit breaker. The circuit cannot be energized until all tags and locks are removed by the individuals who placed them there.

32.14 MULTIMETERS

A volt-ohm-meter (VOM), also known as a multimeter, is often used to test a circuit to determine if it is energized. You should be thoroughly familiar with the meter you are using. A category III multimeter is shown in Figure 32-26. Mulitmeters are explained in further detail in Unit 35, Electrical Measuring and Test Instruments. You should always use the properly rated meter for the circuit you are testing. There are four category ratings for multimeters: CAT I, CAT II, CAT III, and CAT IV. Protected electronic equipment can be tested with a CAT I meter. Appliances, portable tools, and single-phase receptacles can be tested with a CAT II meter. Commercial lighting and equipment would require a CAT III meter. An outside service connection, such as from the utility pole to the power meter, would require a CAT IV meter.

SAFETY TIP

Never use a meter that is rated for a lower category. As an example, a CAT I meter would not provide the proper protection for use on commercial equipment. Also look for a symbol and listing of an independent testing lab, such as UL, CSA, TUV, or other recognized testing organization, to ensure that the meter meets the required standards. The meter test leads should also be certified to a category and voltage as high or higher than the meter.

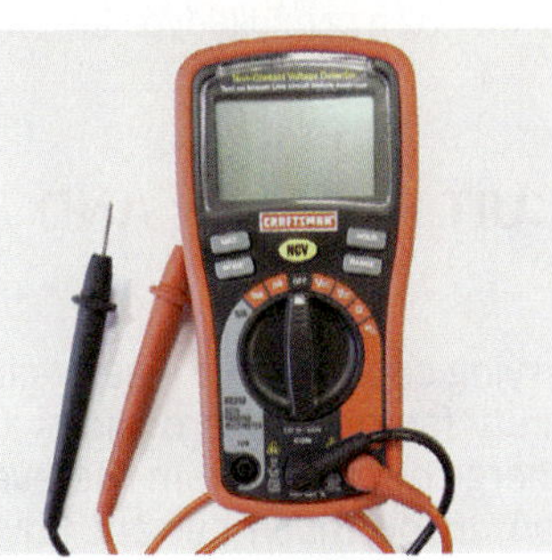

Figure 32-26 Category III multimeter.

32.15 TESTING THE CIRCUIT FOR VOLTAGE WITH A MULTIMETER

Always try to work on de-energized circuits whenever possible, and follow the proper lockout/tagout procedures. Wear protective gear and stand on an insulated mat. Test the circuit in the following manner.

1. Make sure the circuit is off and properly locked out and tagged out.
2. Test the meter on a known voltage source.
3. Test the circuit that should be de-energized (off).
4. Test the meter again on a known source to ensure that you didn't damage your meter or blow a fuse and get a "0" volt reading accidently in step 3.

Try to rest or hang the meter whenever possible rather than holding it in your hands. Connect the ground test lead first and then the hot test lead. Reverse this procedure to do the opposite for removal. Remove the hot test lead first and then the ground test lead. Try to keep one hand in your pocket to lessen the chance of a closed circuit across your chest and through your heart.

To test for voltage, the meter must be set appropriately to read direct current (DC) or alternating current (AC), and also the proper scale must be used. Typical residential service is 120V and 240V. Digital meters are very precise, so the reading may show something slightly higher or lower than the expected reading. When testing a circuit to determine if it is de-energized, a reading of 0 V is required, as shown in Figure 32-27.

32.16 NONCONTACT VOLTAGE TESTERS

Noncontact voltage testers do not require physical contact with the live, current-carrying conductors. This type of voltage tester works by detecting the electromagnetic field that surrounds a "live" conductor. When the noncontact voltage tester is waved close to a conductor, it will provide an audible beep and a light will flash, as shown in Figure 32-28.

When using this type of voltage tester, you must still follow the same four steps outlined for testing a circuit with a multimeter to ensure that the noncontact voltage tester is working properly. Only use a detector that is rated for the level of voltage being measured and is sensitive enough for your application.

Figure 32-27 Testing the voltage of a circuit with a multimeter.

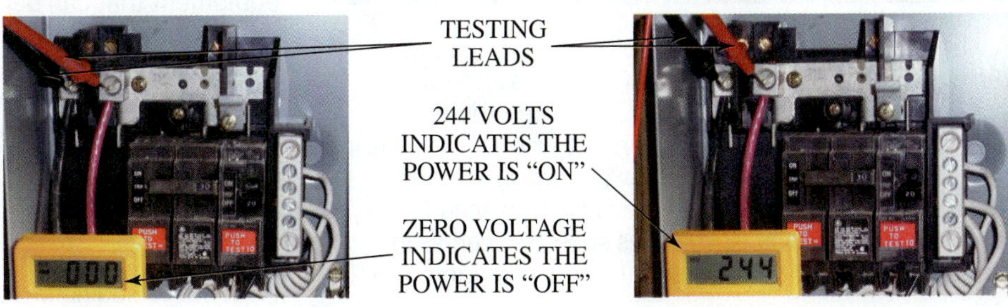

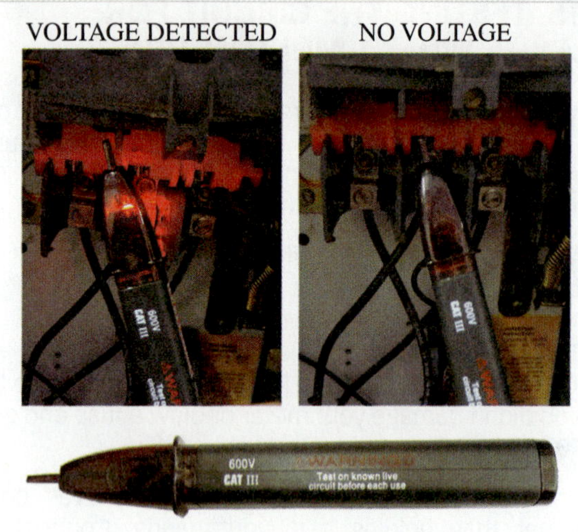

Figure 32-28 Testing the voltage of a circuit with a noncontact voltage tester.

Figure 32-29 Visual indication for the condition of a plug-type fuse.

32.17 FUSES

Circuit breakers are typically designed to trip the circuit and can be reset when the fault condition is corrected. Fuses are often one-time blowout devices that de-energize the circuit in the event of an overcurrent situation. When a fuse is blown, it must be replaced. Plug-type fuses generally have an indicator that will allow you to visually determine if the fuse is blown, as shown in Figure 32-29. If the fuse is of the cartridge type, often the only way to determine its condition may be to test the circuit with a multimeter. However, some cartridge-type fuses have a visual indicator, as shown in Figure 32-30. To replace the fuse, always de-energize the circuit first. Always use a plastic or fiberglass-style fuse puller for removing and installing low-voltage cartridge fuses, as shown in Figure 32-31.

SAFETY TIP

Always de-energize the circuit before replacing fuses. If the circuit is live, then the fuse will act like a knife switch when you try to install it. The blown fuse stopped the flow of current, so as it is removed it will not arc. The new fuse will allow current to flow as soon as it engages into the fuse clips. Since the fuse must be placed firmly into the clips, it will arc severely as soon as it comes into contact with the clips if the circuit is left energized.

Figure 32-30 Visual indication for the condition of a cartridge-type fuse.

Figure 32-31 Removing a cartridge-type fuse with a fuse puller.

32.18 CIRCUIT BREAKER AND FUSE SIZES

The current carrying capacity of a fuse or circuit breaker is an essential safety factor. These devices are designed to open and de-energize the circuit in the event of an overcurrent situation. If too much current flows through the circuit, the wires can become hot and overheat. This can lead to a fire. Excessive current will also cause damage to equipment that can become very expensive. If the breaker is sized too large, then it will not provide for the safety of the circuit. For example, a breaker that is rated at 30 amps installed into a 15-amp circuit would allow for too much current to flow.

It is also incorrect to install too small a fuse or breaker into a circuit, because this will lead to nuisance trips. For example, a breaker rated at 15 amps is installed into a 30-amp circuit. The system might operate normally for a period of time, but as additional components, fans, lights, etc. that are

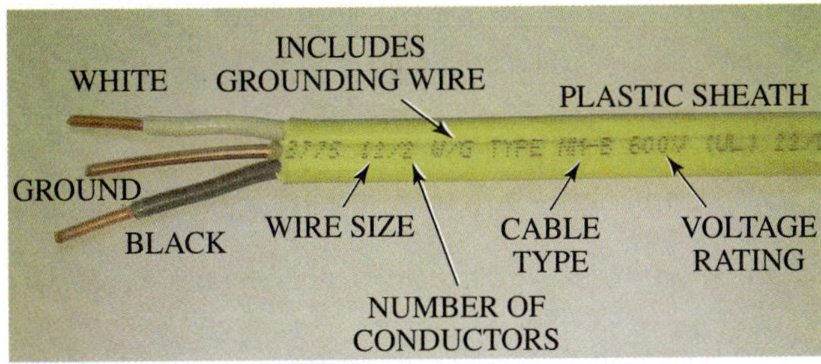

Figure 32-32 Identification of markings found on electrical cable.

part of the circuit begin to turn on, the current will continue to climb. The breaker will trip. If the improper-size breaker was not identified, then this can be difficult to troubleshoot, because there was no actual fault. The circuit was working correctly but the breaker was not sized properly to handle the current flow.

32.19 ELECTRICAL WIRE SIZES AND TYPES

One of the most important considerations in an electrical circuit is the wiring size and type of wiring. National and local electrical codes require specific wiring types and sizes depending on the application. Electrical cable contains two or more insulated wires and may contain a grounding wire. It is classified according to where it will be used for dry, damp, or wet locations and exposure to sunlight and rough use, as shown in Figure 32-32. Nonmetallic sheathed cable (Type NM) is widely used for branch circuits and feeders in residential and commercial systems. NM cable is flame retardant and moisture resistant but suitable for dry locations only. Underground feeder cable (Type UF) is sunlight and moisture resistant, suitable for outside use, and can be buried under the ground. Another common type of cable is THHN, which is an abbreviation that stands for "thermoplastic high heat-resistant nylon-coated" and is suitable for dry and damp locations.

Wire sizing is expressed in American wire gauge (AWG) numbers, ranging from 50 to 0000. The smaller the number (gauge), the larger the diameter the wire will be, as shown in Figure 32-33. Wires larger than 1 gauge are designated as 0, 00, 000, and 0000 (one aught [1/0], two aught [2/0] etc.). A sample of different wire sizes and types is shown in Figure 32-34. While in some cases 14-gauge wire may be suitable for residential lighting circuits, 12-gauge wire is often recommended for residential receptacle outlets. The 12-gauge wire is thicker and harder to work with, but using a wire that is too small in diameter can lead to overheating and potentially fire. Receptacles will draw more current than lights; therefore it is often advisable to use the larger-diameter 12-gauge wire even if it is more difficult to work with. Always refer to the requirements of the local building codes for the location where you are working. Also realize that local codes change from time to time, so you must keep up with the current regulations on a regular basis.

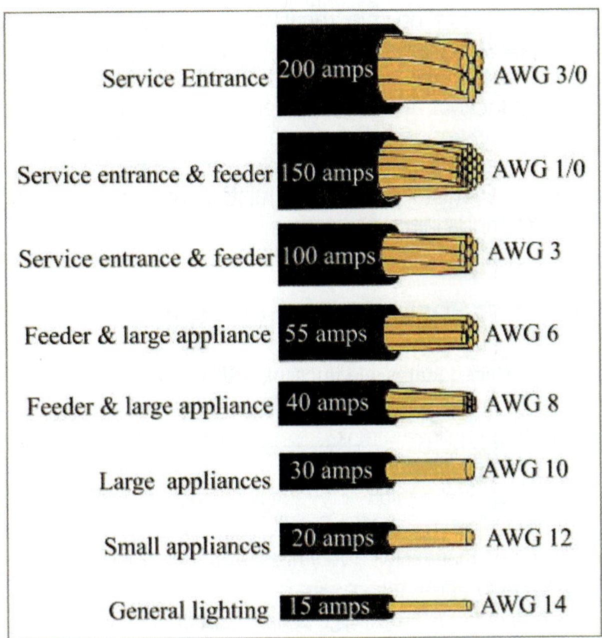

Figure 32-33 Approximate comparison of wires sizes (not to scale).

Figure 32-34 Comparison of different electrical cable types.

UNIT 32—SUMMARY

Electrical work should be performed on systems that are de-energized with all power turned off. It is extremely important that you follow all safety rules and recommendations when working on electrical systems. Circuits should be tagged and locked out before any electrical work is performed. Even after the power is secured, the circuit should be tested for voltage to ensure that it is safe to work on. OSHA has regulations regarding lockout and tagout of equipment while it is being serviced. It is important that you follow these and all other electrical safety practices. Electricity does not provide you with one free mistake. Your first mistake or careless act can result in serious injury or death.

UNIT 32—REVIEW QUESTIONS

1. In the United States, electrical hazards cause more than _____ deaths and ______ injuries in the workplace every year.
2. Electrical accidents rank ______ among all causes of work-related deaths.
3. Materials that exhibit very little resistance to electricity flow are called what?
4. Materials that exhibit a very high resistance to electricity flow are called what?
5. Is it true that a low-voltage system cannot harm you?
6. How would you help a person who is "frozen" and unable to pull free of the circuit?
7. Why would an electric drill have a three-prong plug?
8. How can electrical shock occur?
9. What is meant by the term *grounded* in an electrical circuit?
10. How does a GFCI breaker provide for electrical safety?
11. What is meant by a "closed" breaker as compared to an "open" breaker?
12. How can you determine if a circuit breaker has tripped, and how do you reset it?
13. What steps should you take before resetting a circuit breaker?
14. Who can remove a lock from an electrical circuit that has been locked and tagged?
15. Is lockout/tagout required in a residential setting?
16. What are the four steps to take when testing a circuit for voltage?
17. Why should you never replace a fuse with one that has a higher amperage rating?
18. What electrical wire size is larger in diameter, AWG 14 or AWG 12?
19. Why is consideration of electrical wire size important when wiring an electrical circuit?
20. What type of information can be determined from the markings found on electrical cable?

UNIT 33

Basic Electricity

OBJECTIVES

After completing this unit, you will be able to:

1. explain how the structure of the atom affects electricity.
2. discuss the difference between a conductor and an insulator.
3. explain the difference between direct current and alternating current.
4. list the characteristics of electric current that are commonly measured and the units used to measure them.
5. explain the relationship of potential, current, and resistance in an electrical circuit.
6. discuss the three elements required to make an electrical circuit.
7. explain the difference between series, parallel, and series parallel circuits.
8. use Ohm's law and/or the power formula to calculate values for volts, amps, ohms, and watts.

33.1 INTRODUCTION

Electricity and electrical problems are the most common issues HVACR technicians encounter. Installation and service both require an understanding of the basic electrical characteristics. A fundamental knowledge of electricity is required to truly understand all the electrical components in an air-conditioning system. Understanding electrical circuit characteristics and how to measure them allows technicians to test and analyze circuits in air-conditioning systems. Therefore, it is essential that technicians have a good understanding of basic electrical theory.

33.2 STRUCTURE OF THE ATOM

It is necessary to understand the structure of the atom to understand electricity. The smallest piece of an element that can exist and still retain the same chemical properties is the atom. Atoms are made of small particles called protons, neutrons, and electrons. Protons are positively charged, electrons are negatively charged, and neutrons have no charge. In the center of the atom is a nucleus composed of protons and neutrons. Clouds of negatively charged electrons are arranged in energy levels around this nucleus (Figure 33-1). Usually, the number of electrons and protons are equal, so the atom is neither negatively charged nor positively charged. The negatively charged electrons are held in the atom by their attraction to the positively charged nucleus. The electrons are arranged in energy levels, with fewer electrons in the energy level closest to the nucleus and more electrons in the energy level farthest away from the nucleus. Each energy level has a maximum number of electrons that it will hold. The maximum number of electrons in each energy level is two in the first, eight in the second, eighteen in the third, and thirty-two in the fourth. As atoms become larger with more electrons, some electrons end up farther away from the nucleus. The electrons that are farther away from the nucleus are not as tightly bound to the atom because the attraction is weaker. When just a few electrons are in an outer orbit, they can escape the bonds of their atom. Electrons that have escaped their atomic bonds are called free electrons. Figure 33-2 shows a copper atom. The first three energy levels are filled, leaving one electron in the outermost energy level.

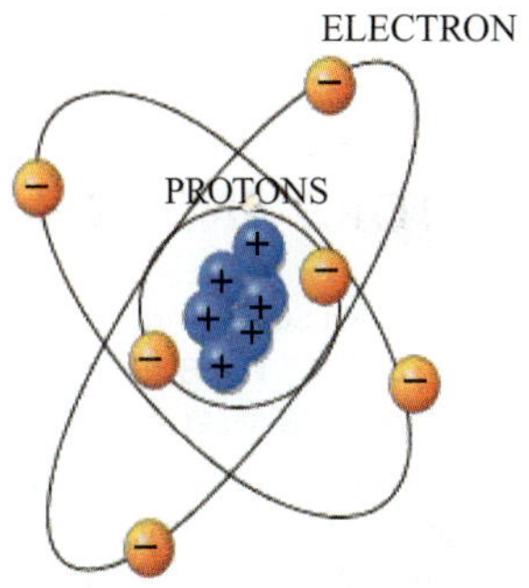

Figure 33-1 A simple carbon atom has a nucleus of positively charged protons surrounded by negatively charged electrons. *(Courtesy of NOAA)*

33.3 FLOW OF ELECTRONS

Electricity is the flow of electrons. Electrical flow is not like pouring electrons in one end of a wire and dumping them out the other. Rather, electricity is more like a string of dominoes. One bumps the next and the energy flows from one end to the other, but no individual domino moves very far. With electricity, an electron moves from one atom to another and causes an electron to leave that atom. This electron then moves to a neighboring atom, which causes it to lose an electron. This chain reaction carries through the wire to provide an electrical current (Figure 33-3). Energy is required to start and maintain this chain reaction and get all the free electrons moving in the same direction.

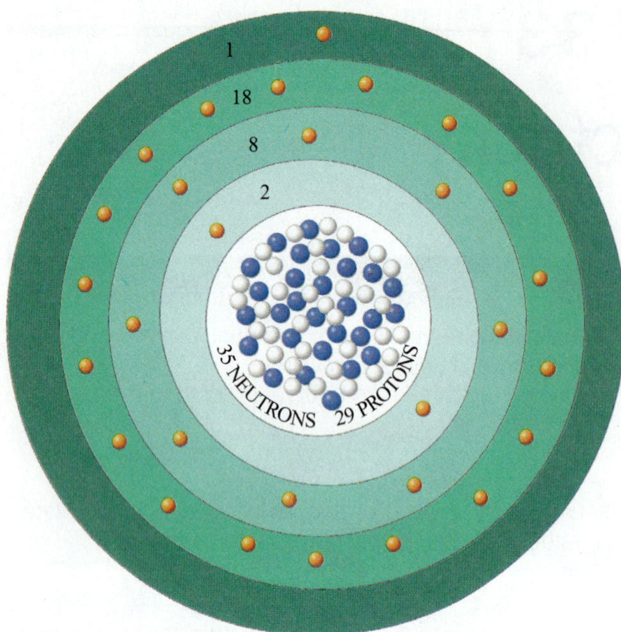

Figure 33-2 The first three electron energy levels of a copper atom are full, but the outer level has only one electron.

33.4 CONDUCTORS AND INSULATORS

Electrons can flow through some materials more easily than other materials. Atoms that have just a few electrons on their outer energy level make good conductors because those electrons can more easily become free electrons (see Figure 33-2). Materials with a large number of free electrons make good electrical conductors. Most metals are good conductors because they have a large number of free electrons that are not bound to any particular atom. Gold, silver, copper, and aluminum are all excellent electrical conductors. This list is arranged in order of the metal's ability to conduct electricity, with gold being the best conductor. Gold and silver are both commonly used as electrical conductors in electronic circuits where the amount of material is small (Figure 33-4). Copper is used most often in wiring because of its lower cost compared to gold and silver and its superiority to aluminum in conductance (Figure 33-5).

Insulators have far fewer free electrons, so they resist the flow of electrons. The outer energy level of an insulator is typically full, making it more difficult for electrons to escape. Ceramic, rubber, and plastic are examples of insulators. Thermoplastics are commonly used for the insulator on wires because a very thin layer of thermoplastic can resist hundreds of volts. A thicker insulation layer is required to contain higher voltages. The voltage feeding sparkplugs in car engines is several thousand volts. Sparkplug wires have a much thicker insulation jacket than house wiring because they need to contain higher voltages. Figure 33-6 shows a comparison of the insulation on a power wire and on a sparkplug wire. Ceramic insulators are used in high-voltage applications that also require physical strength. The insulators on high-voltage power-distribution lines are ceramic.

33.5 SEMICONDUCTORS

The revolution in electronic circuits is based on the discovery and advancement of semiconductors. Semiconductors can behave like a conductor or like an insulator. This behavior allows them to be used like an electronically controlled switch. Silicon, carbon, and germanium can form uniform crystals because they have four electrons in their outer shell. Silicon is the most commonly used semiconductor material. The addition of trace amounts of another chemical into the crystalline structure is called doping. Doping alters the semiconductor's electrical properties. Doping can produce either a positive or negative charge in the semiconductor material, depending on the chemical used. N-type (negative) semiconductors are doped with phosphorus and arsenic; P-type (positive) semiconductors are doped with boron and gallium. The N-type and P-type materials are joined together to make semiconductor devices like diodes, transistors, and rectifiers.

33.6 STATIC ELECTRICITY

Static electricity is a very common phenomenon that we all have experienced at one time or another. You may have walked across the rug and then reached for the doorknob and unexpectedly felt a static shock. The reason for this can be simply explained.

Static versus Dynamic

Electricity can be classified as either static or dynamic. Something that is static is stationary; something that is dynamic is in motion. Static electricity is the result of a standing charge, while dynamic electricity is in motion. Dynamic electricity has a current created by flowing electrons.

Static Charges

Objects can be neutral, positively charged, or negatively charged. If the number of electrons and protons in an object

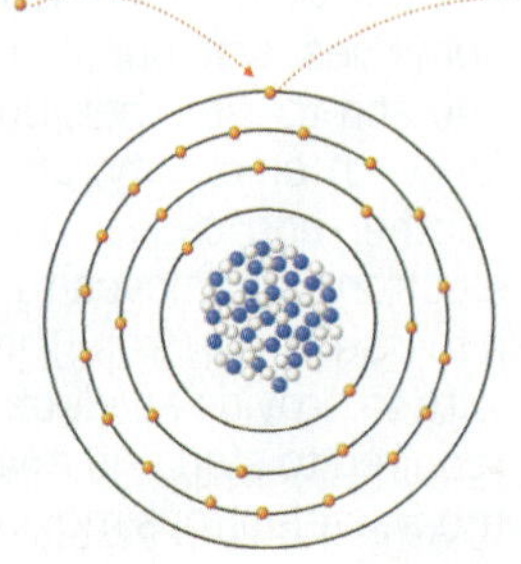
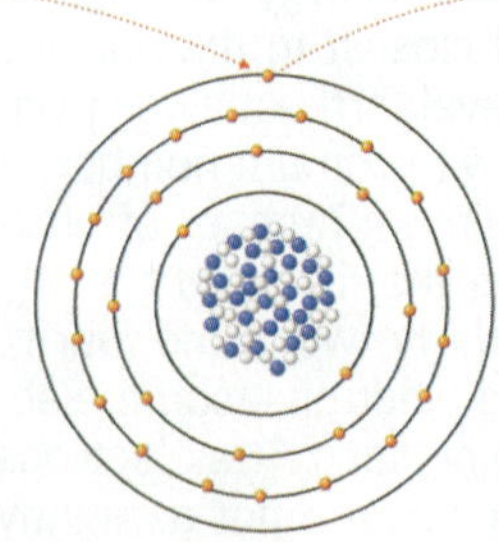
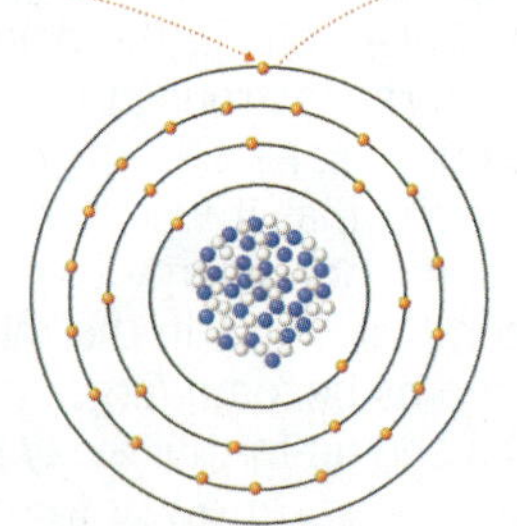

Figure 33-3 Electron flow occurs as a chain reaction.

Figure 33-4 The contacts on the edge connector of this circuit board are made of gold.

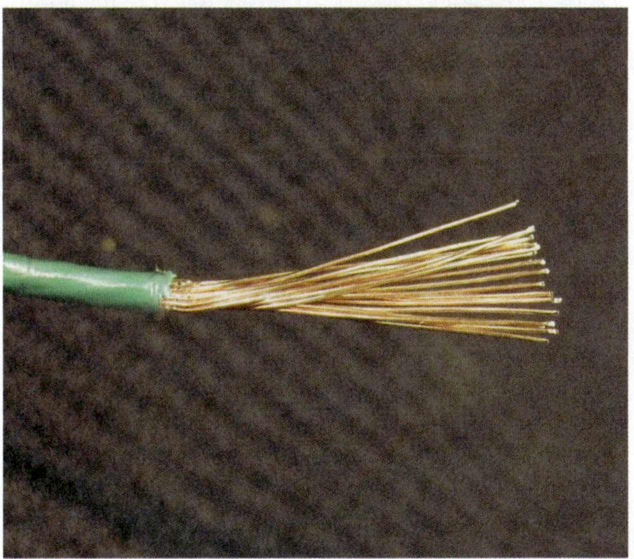

Figure 33-5 This wire is made of copper.

Figure 33-6 The insulation on the wire on the left is designed for up to 600 V; the insulation on the wire on the right is designed for thousands of volts.

Figure 33-7 Lightning is a powerful example of static electricity. *(Courtesy of NOAA Photo Library, NOAA Central Library; OAR/ERL/National Severe Storms Library [NSSL])*

are equal, it will be neutral and will not have an electrical charge. If an object picks up additional electrons, it will become negatively charged because electrons have a negative charge. If an object loses electrons, it will become positively charged. Static electrical charges are the result of friction, where electrons are literally rubbed off from one object to another, creating one positive and one negative charge.

Examples of Static Electricity

If the difference in charge between the two objects is great enough, electrons will jump from the negatively charged object to the positively charged object. Lightning is probably the most common type of static electricity (Figure 33-7). Lightning occurs during a storm when cloud particles acquire large amounts of electric charge as they rub across the ground. Meanwhile, the ground has built up an opposite electric charge. A lightning bolt is created when the difference in charge becomes strong enough and the clouds get close enough to the ground for the electrons to jump. A smaller charge is acquired in dry weather by a person walking along a thick rug, collecting negative charges from the rug as a result of friction. A spark jumps from the person to an object when the person reaches to touch it, because of the static charge he or she has built up.

SERVICE TIP

There are many printed circuit boards currently being used in the HVACR field containing electronic chips that control the functions of the system. These electronic chips can be damaged by an accidental discharge of static electricity. Always discharge yourself by touching the equipment frame when working around equipment with electronic components.

Static electricity has few practical uses, but one rather common device called an electrostatic precipitator uses static electricity to filter the air. Air is passed through a plastic medium that is charged by static electricity from the movement of the air. The static charge is used to attract and trap the particles in the airstream. The process is very effective in removing fine dust particles and pollen from the air.

33.7 PRODUCING DYNAMIC ELECTRICITY

Other forms of energy that can be used to generate electricity include heat, light, chemical, mechanical, and magnetic. They generate electricity by moving free electrons. The two most common forms of energy used to generate electricity are chemical energy and magnetic energy.

Batteries

Batteries generate electricity by turning chemical energy into electrical energy. Two chemical reactions take place in a battery: one gives off free electrons, and the other absorbs free electrons. A car battery is an example of a typical wet-cell battery. Lead and lead oxide are immersed in acid. The negative terminal is connected to the lead because the reaction between the lead and the acid creates free electrons. The positive terminal is connected to the lead oxide because the reaction between it and the acid absorbs free electrons (Figure 33-8). Electrons will flow from the negative terminal to the positive terminal when a path is created between them.

SAFETY TIP

Never create a path from one battery terminal to the other using only a conductor. The large current flow created has enough energy to melt the wire or cause the battery to explode! Any path between the two battery terminals should contain a load, such as a lightbulb, fan motor, or some other device. A load has the ability to resist current flow and turn the electrical energy into useful work.

Generators

Generators use magnetic and mechanical energy to create electricity. When a magnetic field is placed near a conductor, the free electrons in the conductor line up with the

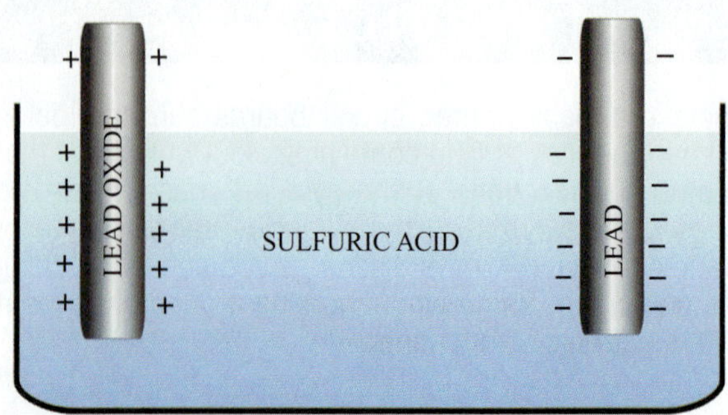

Figure 33-8 In a lead-acid battery, the reaction between the lead and sulfuric acid produces a negative charge while the reaction between the lead oxide and the sulfuric acid produces a positive charge.

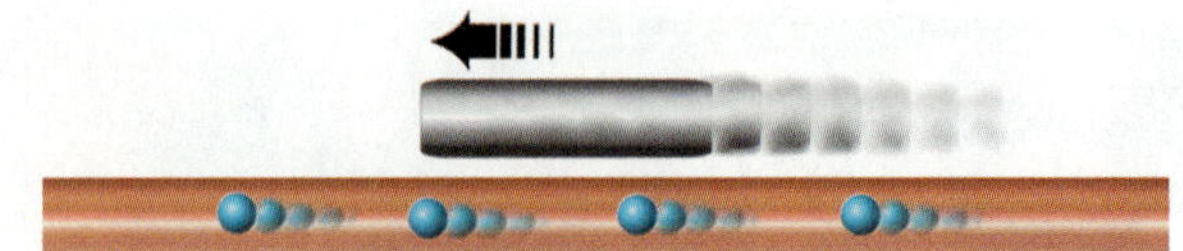

Figure 33-9 Free electrons in a conductor are pulled by the magnetic field moving past the conductor.

magnetic field. Moving the magnetic field pulls the free electrons with it, creating electron movement and current flow in the conductor (Figure 33-9). Generators use this principle to transform magnetic and mechanical energy into electrical energy.

33.8 CURRENT, VOLTAGE, WATTAGE, AND RESISTANCE

Electrical circuit operating characteristics can be measured. The common measurement terms used are *current, voltage, wattage*, and *resistance*. Current refers to the flow of electrons, voltage is a measurement of potential, wattage indicates power, and the opposition to the flow of electricity is called resistance.

Electric Current: Amps

The coulomb is the measure of electrical charge. One coulomb is equal to the charge of 6.241506×10^{18} electrons. Coulombs are seldom used in practical measurement. Instead, the rate of electrical current flow is measured when testing electrical equipment. The rate of electrical current flow is measured in amperes, abbreviated "amps." Electrical current is flowing at the rate of 1 amp when 1 coulomb of electrons passes by in 1 second. Amperes are represented in equations and formulas by the letter I, for "intensity," because amperes are a measure of the intensity of the electric current.

Electric Potential: Volts

The ability to make current flow is called electromotive force or potential. Potential difference provides the electrical pressure that makes current flow possible. Current flows through a conductor because of a difference in potential from one end of the conductor to the other. Voltage is a measure of the potential difference. Without a difference in voltage from one end of the wire to the other, there will be no current flow. Voltage is represented in equations and formulas by the letter E, for "electromotive force."

Electric Work: Watts

The ability of electricity to do work depends on two factors: voltage and current. The watt is a measure of the rate at which electrical work is done. Wattage depends upon both voltage and current. A watt is defined as 1 ampere of current flowing at a potential difference of 1 V. Wattage can be calculated using the formula volts $\times$ amps = watts. Wattage is not truly a quantity but a rate. Wattage does not measure the amount of work accomplished; it measures the rate at which work is accomplished. Wattage is a rate because it is calculated using amps, a measure of electrical current flow. Remember that flows are always measured as a rate. Watts are sometimes represented in equations and formulas by the letter *P*, for "power," and sometimes by the letter *W*, for "watt."

Wattage is a reasonable measure for small appliances and light bulbs, but if you are heating a house you need something that works better for large measurement. Kilowatts, abbreviated kW, are used to specify the energy use of heating and air-conditioning equipment. A kilowatt is equal to 1,000 W. The formula is simply watts/1,000 = kilowatts. Kilowatts are represented in equations and formulas as kW, for "kilowatt."

Watts and kilowatts are really both rates, not quantities. Electrical quantity is measured in kilowatt hours. Power companies use kilowatt hours to measure the amount of electrical work you did during the month. This is calculated by measuring kilowatt use over time. Kilowatt hours are represented in equations and formulas as kWh, for "kilowatt hours."

Resistance: Ohms

To some extent, all conductors and electrical devices resist current flow. The ratio of voltage to current is the resistance, and it is stated in ohms. Ohms are represented in formulas and equations as *R*, for "resistance," or the Greek letter omega, Ω, for "ohms." The resistance of a conductor is affected by the physical properties of the material, the cross-sectional area of the conductor, and the length of the conductor. Smaller-diameter wires have a higher resistance than larger-diameter wires. Long conductors have a higher resistance than shorter conductors.

TECH TIP

The resistance of most conductors increases as they rise in temperature. The low ohm reading of a cold lightbulb increases almost instantly as it heats up and begins to glow, giving off light. This change in resistance occurs with any conductor as it is heated. Hot conductors have higher resistance than cold conductors.

33.9 DIRECT CURRENT

Direct current, DC, is a continuous flow of electrons in one direction (Figure 33-10). All batteries supply DC current. In the HVACR trade, DC current is used primarily for operating cordless tools. Direct current is also used in special applications, such as the control systems in transportation refrigeration equipment, the cells in electronic air cleaners, and operating electronic circuits.

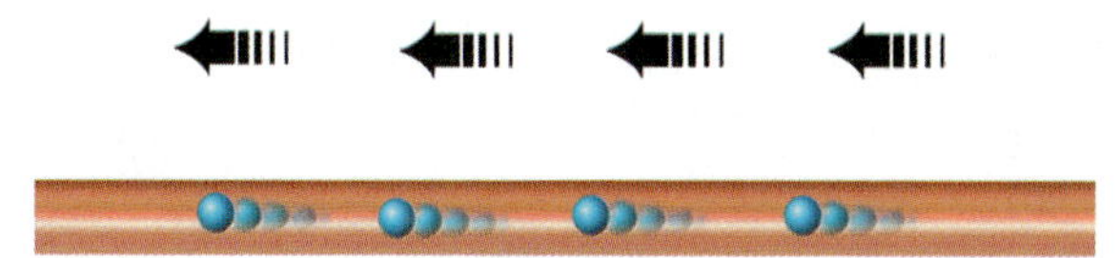

Figure 33-10 Electrons flow in only one direction with direct current.

33.10 ALTERNATING CURRENT

Unlike DC current, which flows continuously in one direction, alternating current alternates its direction of flow through the conductor. It flows in one direction and then the other at regular intervals (Figure 33-11). The AC waveform is a shaped like a sine wave. Half the time the current is positive and half the time it is negative. The period of time required for the voltage to peak in both directions and return to 0 V is called a cycle. Figure 33-12 shows one cycle of alternating current.

Frequency

The frequency of AC current is the number of cycles per second. Frequency is measured in hertz (Hz): 1 Hz is equivalent to one cycle per second. In the United States, 60 Hz power is standard. In Canada and many other parts of the world, the standard is 50 Hz.

Phase

The number of these waveforms occurring at the same time is called the phase. AC is available in single phase and three phase. A single-phase AC circuit requires only two wires. A three-phase circuit requires three wires: one for each phase. Three-phase power is advantageous for operating electric motors and for transferring large amounts of current. Residences use single-phase AC power, while commercial and industrial buildings typically are supplied with three-phase power.

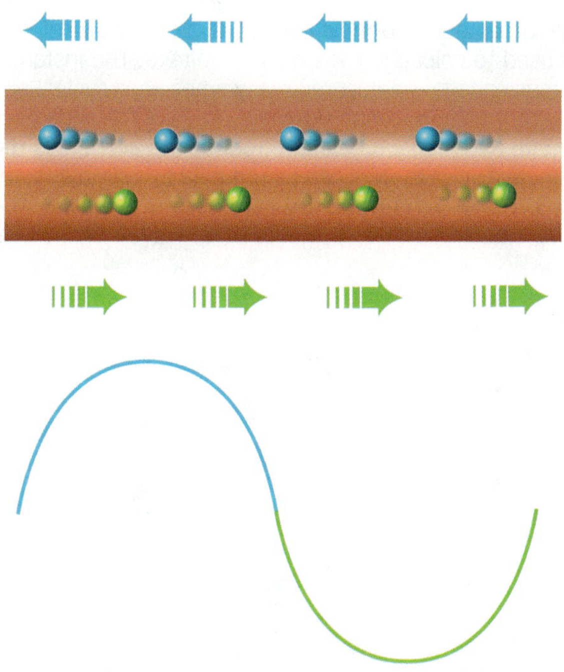

Figure 33-11 In alternating current, the direction of electron flow reverses at regular intervals.

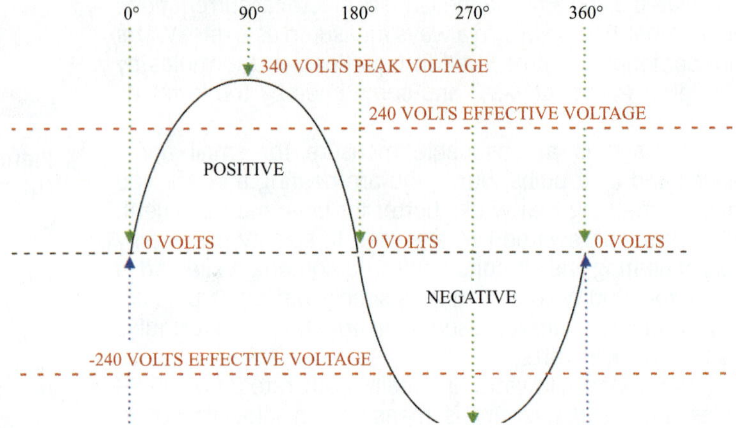

Figure 33-12 One complete cycle of alternating current.

Voltage

The voltage used in an AC circuit changes constantly. The average voltage supplied is 0 because the two halves of the sine wave cancel each other out. Therefore, AC voltages are stated in effective voltage. The DC voltage that would produce the same heating effect in an electric heater is established as the effective AC voltage.

The effective voltage of a pure sine wave is 0.707 times peak voltage. For example, an AC current with a peak voltage of 170 V is rated at 120 V because it produces the same amount of heat in an electric heater as a DC voltage of 120 volts. This is shown mathematically as: 170 peak volts $\times$ 0.707 = 120 effective volts (Figure 33-13).

TECH TIP

A mathematical formula called the "root mean square" is used to calculate the effective voltage. The instantaneous voltage is measured many times each cycle. Each of these values is squared, the average of the squares is found, and finally the square root of the average is the effective voltage. This can be reduced to a simple factor of 0.707 for AC currents that are pure sine waves. The 0.707 factor is not accurate for other waveforms.

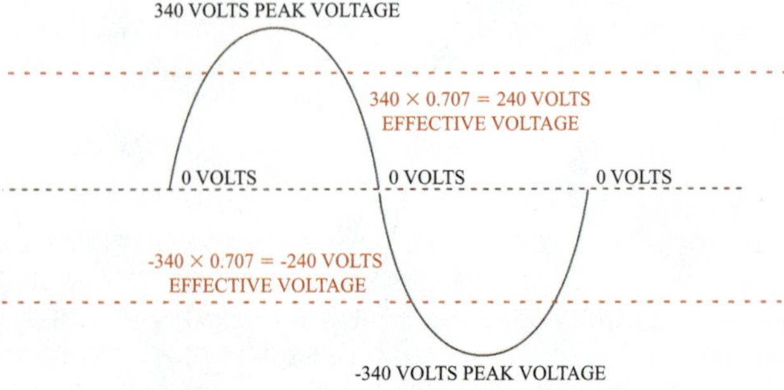

Figure 33-13 The effective voltage is 0.707 times the peak voltage.

33.11 COMPARISON OF AC AND DC POWER

AC power is produced by electrical power-generating plants because it is efficient to generate and can be transmitted over long distances. DC power is commonly found in many low-voltage applications, especially where these are powered by batteries. This is because batteries that are used in many portable electrically operated devices produce only DC power.

Reason for AC Current

Transmission of large amounts of power involves unavoidable line losses due to the resistance of the wires. The power lost in the line is calculated using the formula I^2R. Keeping the current, I, low is the key to reducing line loss. By utilizing high voltages, the power can be transmitted by using relatively low line current, thus minimizing the line losses. The voltage level of AC current can be more easily changed, or transformed, than DC current. This allows AC current to be transformed to extremely high voltages for transmission across long distances and then transformed at the point of use back down to a safe voltage level.

DC Electrical Power Generators

Electrical power generators have a shaft called a rotor that spins in a stationary housing called the stator. DC generators are designed differently than AC generators. A magnetic field induces current to flow in the armature windings located in the rotor for DC generators as compared to the armature windings located in the stator for AC generators. In a DC generator, the magnetic north and south poles are placed in the housing (stator), and the windings located in the rotor generate current as it spins through the magnetic field (Figure 33-14). Drawing this current flow from the rotor, which is spinning very fast, involves a complex arrangement of a commutator and brushes. As the total amount of current and voltage required increases, it becomes more difficult and less practical to draw the current from the spinning rotor.

AC Electrical Power Generators

AC generators spin the north and south poles with the rotor (Figure 33-15). The armature windings are located in the

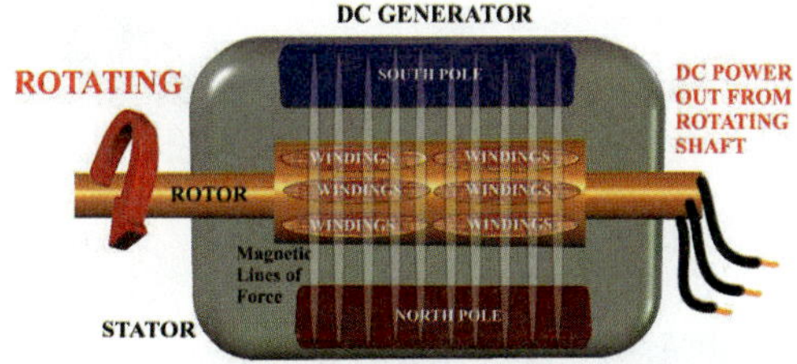

Figure 33-14 Operation of a DC generator.

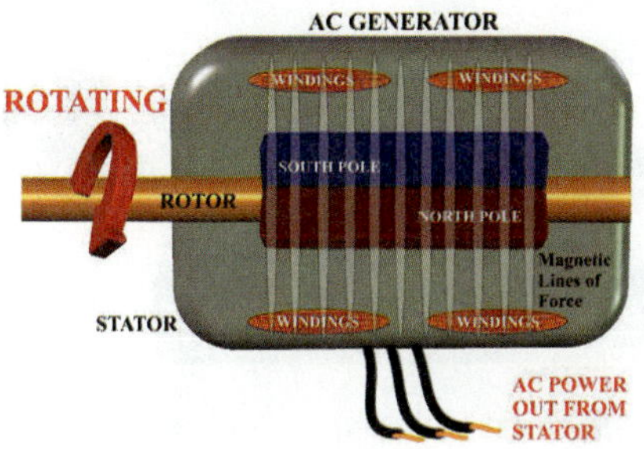

Figure 33-15 Operation of an AC generator.

stator, and the power is tapped directly from them to make for a much simpler arrangement. The distribution wires are larger and the power drawn will be greater as compared to a DC generator. Therefore AC generators can be designed to more easily produce the large amounts of power required of large electrical power plants.

TECH TIP

In an AC generator, each revolution of the rotor causes a change in the direction of current flow. Inside electrical power-generating plants, steam turbines or gas turbines spin the rotor at the high speed of 3,600 rpm. This rotation is equal to 60 times a second, which creates a change in direction of current flow at an AC frequency of 60 cycles, commonly referred to as 60 hertz.

33.12 ELECTRICAL CIRCUITS

Power is supplied to a HVACR unit by the electrical power system. The power systems will have one or more electrical circuits that are used to deliver electrical power to the HVACR equipment. Electrical power systems have three essential requirements and one optional requirement:

- A source of power (could be a transformer)
- An electrical load device
- An electric circuit: a path for the current to flow
- (Optional) A switch to control the flow

Power Source

The electrical power source can be provided by a battery or from the building's central electrical system. Batteries supply DC power, while most buildings will have AC power transmitted from the local utility company. Transformers will step down the high voltage generated at the electrical power-generating plant to the lower voltage required for residential or commercial use. Rectifiers will convert AC power into DC power, while inverters will do just the opposite and convert DC power into AC power.

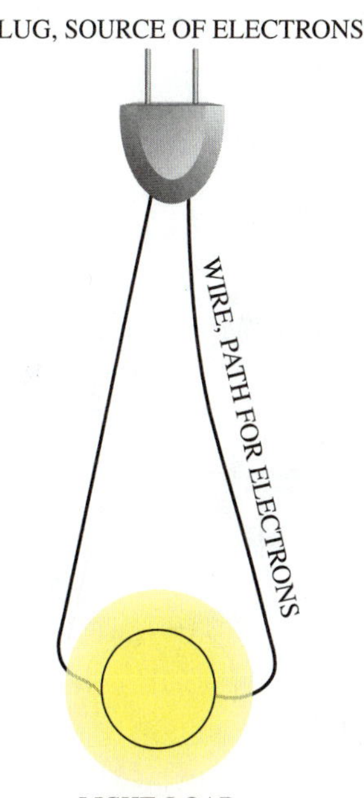

Figure 33-16 Principal components of a simple circuit.

Electrical Loads

A load is any electrical device that requires power to operate. Electric loads convert electricity into useful work. The most common loads for HVACR systems are electric motors. Motors drive the compressors, fans, and pumps. Motors also drive dampers and zone valves. Many other electrical components require power, such as resistance heaters and solenoid valves.

Path for Current to Flow

There must be a path for the current to travel (Figure 33-16). Every electric circuit has at least two wires, often indicated as line terminals L1 and L2. For this illustration, these would be the two prongs on the plug. For there to be a complete circuit, the path of the electricity flows through one wire of the electric circuit, passes through the load, and returns through the other wire of the electric circuit.

Switches

The switch is a device to turn the load on and off. The switch may be manually operated, such as turning on the lights when entering a dark room. Or the switch may be automatic, such as a thermostat that turns a unit on and off in response to the surrounding temperature. Generally a switch will have two positions, open or closed. A switch is referred to as closed when it allows current flow (Figure 33-17). A switch is referred to as

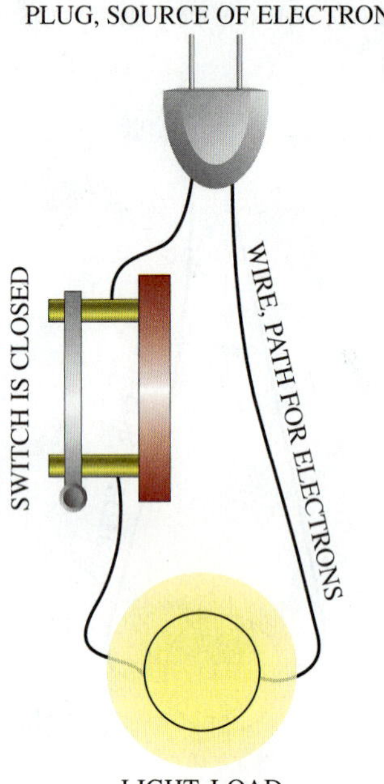

Figure 33-17 A closed circuit with current flowing through the load.

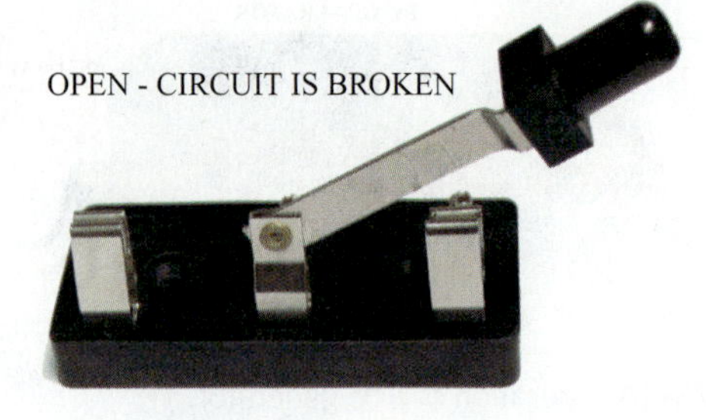

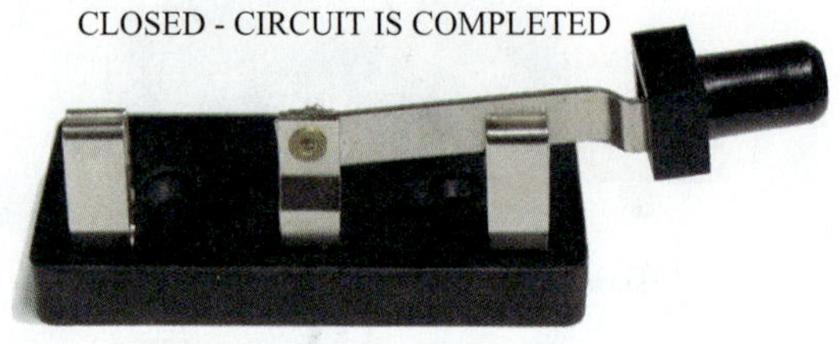

Figure 33-19 The terms *open* for "off" and *closed* for "on" describe the positions of a knife-blade switch; they are used today when describing the position of all types of switches in electrical circuits.

open when it is breaking current flow (Figure 33-18). These terms come from the operation of early knife-blade switches (Figure 33-19).

33.13 ELECTRICAL CIRCUIT OPERATION

Figure 33-20 shows a simple circuit consisting of a source, a path, and a load. A complete circuit has continuity; it has a continuous path from the source to the load and back from the load to the source. A break anywhere in the circuit will keep current from flowing and prevent the load from operating (Figure 33-21). Switches can be added to a circuit to control when the circuit operates and when it does not operate. If the circuit path must pass through a switch, the switch can break the circuit and keep the load from functioning or make a complete circuit and allow the load to function (Figure 33-22).

An open circuit is a circuit without a complete path. Any break, or opening, in the circuit will keep it from operating. A short circuit is a circuit without a load. A short circuit has a power source and a complete path, but no load to do useful work. Without a load, current flow through the circuit is so high that both the source and the circuit conductors overheat.

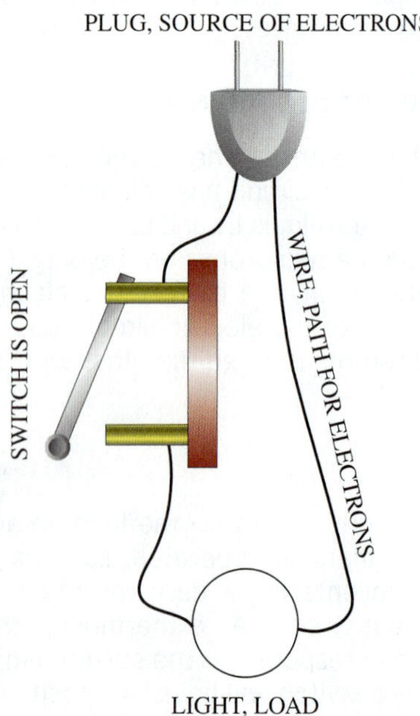

Figure 33-18 No current flows in an open circuit.

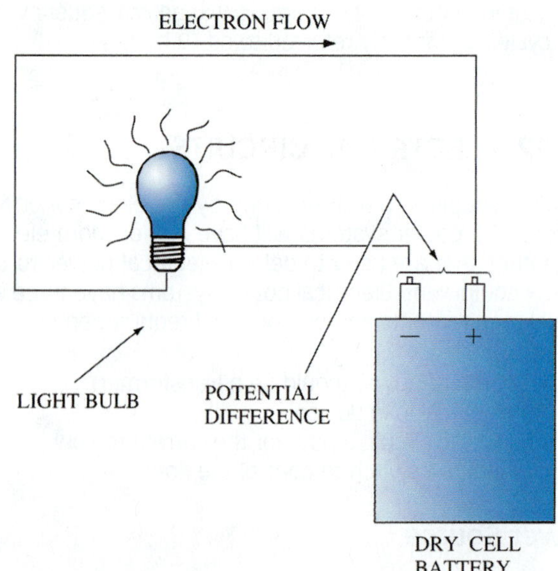

Figure 33-20 A simple electric circuit using a battery power source and a lightbulb for the load.

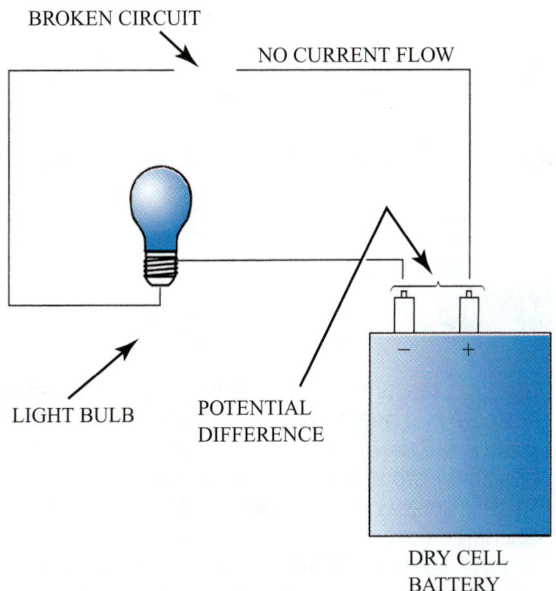

Figure 33-21 A break in the circuit stops electron flow.

TECH TIP

An electrical circuit that contains a source and path without a load is called a short circuit. A circuit that contains a source and load without a path is called an open circuit. Sometimes people refer to any circuit that does not function as having a short. This is an improper use of the term. An open circuit is one that does not work. A short circuit is one that will trip the power breaker or cause something to overheat and possibly catch fire.

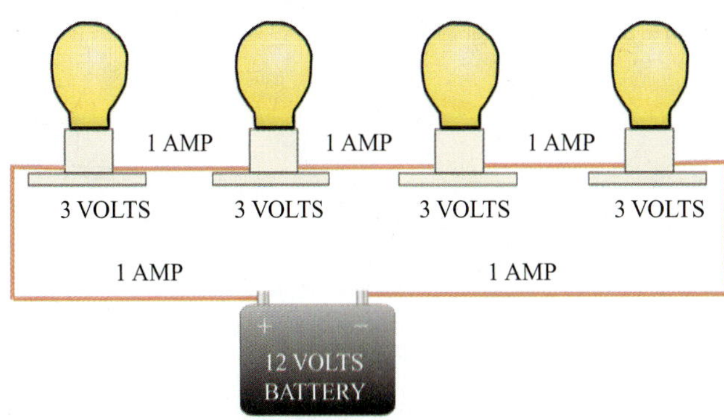

Figure 33-23 Four lights wired in series.

There are three types of path arrangements for circuits:

1. The series circuit, which allows only one path for the current to flow
2. The parallel circuit, which has more than one path
3. The series parallel circuit, which is a combination of series and parallel circuits

33.14 SERIES CIRCUITS

In a series circuit, components are connected along a single path and every device must function for the circuit to be complete. If one lightbulb in a series circuit burns out, then the circuit is broken and none of the lights will come on.

Current Flow in Series Circuits

In a series circuit, there is only one path for the current to follow. The power must pass through each electrical device in succession in that circuit to go from one side of the power supply to the other. An example of a series circuit is shown in Figure 33-23, where four lights are placed end to end in a single circuit. The current in a series circuit is the

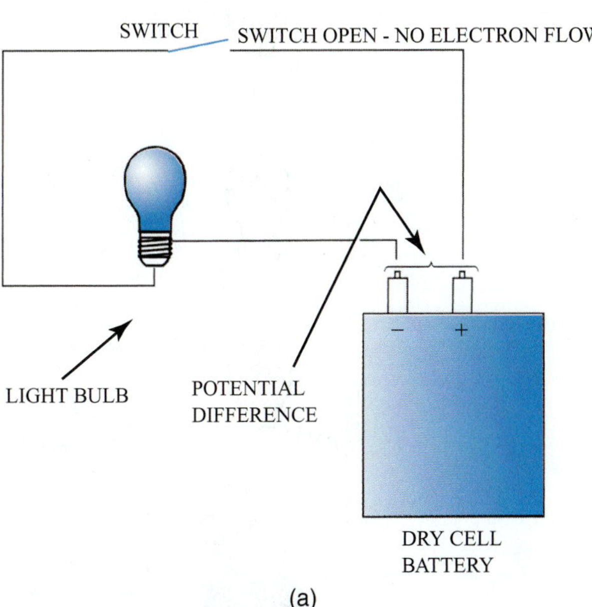

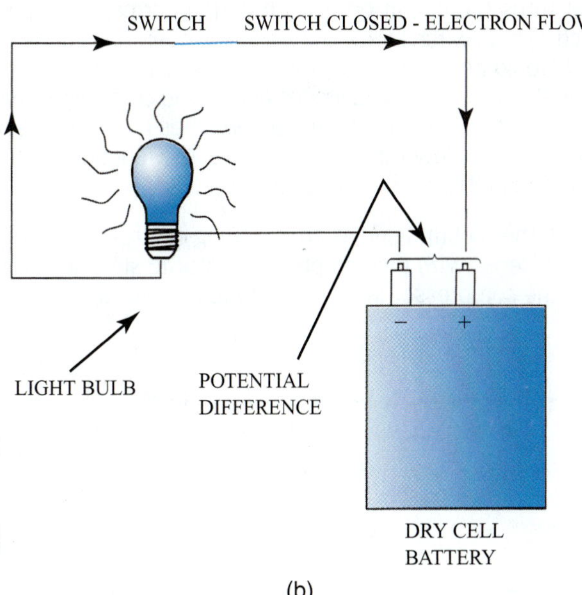

Figure 33-22 A switch is a controlled break, allowing the user to control the flow of electrons; (a) switch open, no electron flow; (b) switch closed, electron flow.

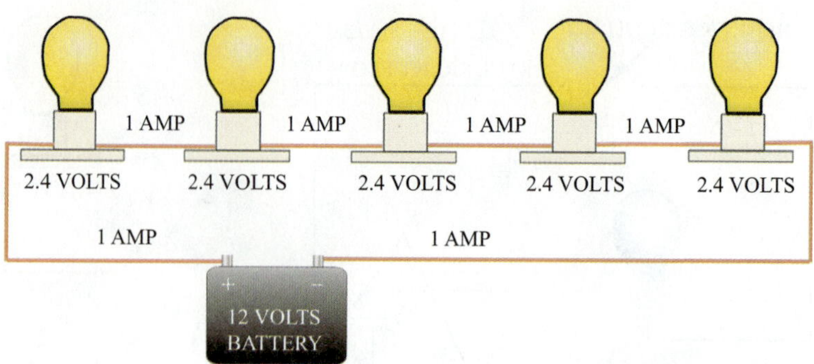

Figure 33-24 Five lights wired in series.

same throughout the entire circuit because there is only one path for the electrons to take. Every device in a series circuit affects the current flow of the entire circuit.

Voltage and Resistance in Series Circuits

The voltage used by each device in the circuit is proportional to its resistance; devices with a higher resistance receive more voltage, while devices with a lower resistance receive less voltage. As an example, a fifth light bulb of identical resistance is placed in the series circuit just described and the current is allowed to remain the same at 1 amp (Figure 33-24). Notice that the voltage at each lightbulb has been reduced from 3 volts to 2.4 volts.

Because of this nonuniform voltage, loads are seldom wired in series with other loads in the HVACR industry. Motors are designed to operate at specific voltages, such as 120 V, and therefore cannot be wired together with other motors in series. Instead, usually there is only one load controlled by a series of switches, as shown in Figure 33-25. In this diagram, the 120 V power supply terminals are indicated with the symbols L1 and L2. The one load is a compressor motor. The switches placed in series with the compressor motor are used to control its operation. The thermostat is the operating control, and the other switches are all safety switches. The operating control is the switch that turns the circuit on and off during normal operation. Each safety switch checks for a specific operation condition that would be dangerous either to the equipment or to the user. If any one of the switches opens, the compressor will shut down. In a series circuit, all switches must be closed for current to flow through the circuit. The functions of the switches in this circuit are as follows:

- A thermostat used as an operating control is placed in series with the compressor motor to start and stop the compressor in response to temperature.
- The low-pressure switch senses compressor suction pressure and opens on a drop in pressure. It is set to cut out at a protective low-pressure limit but remains closed at normal operating pressures.
- The high-pressure switch senses the compressor discharge pressure and opens on a rise in pressure. It is set to open at a protective high-pressure limit but remain closed at normal operating pressure.

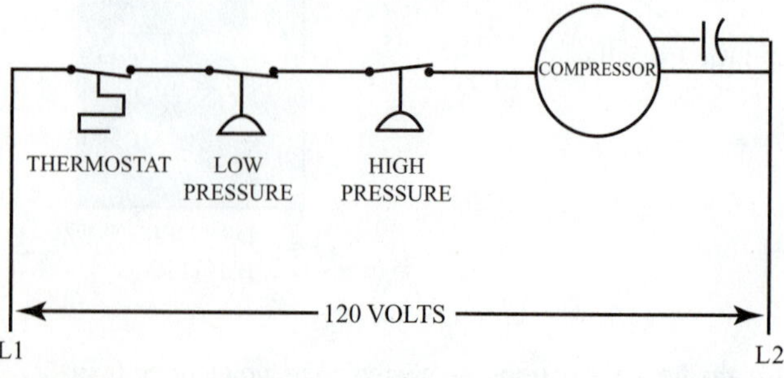

Figure 33-25 Series circuit with three switches and one load.

33.15 PARALLEL CIRCUITS

In a parallel circuit, components are connected so that each has its own circuit. If all the lightbulbs in a parallel circuit are burned out except one, that one good light bulb will still come on.

Current Flow in Parallel Circuits

A parallel circuit has more than one path for current flow. The current for each load depends upon the resistance of that load. Loads with less resistance have more current flow; loads with more resistance have less current flow. Figure 33-26 illustrates a parallel circuit with three loads.

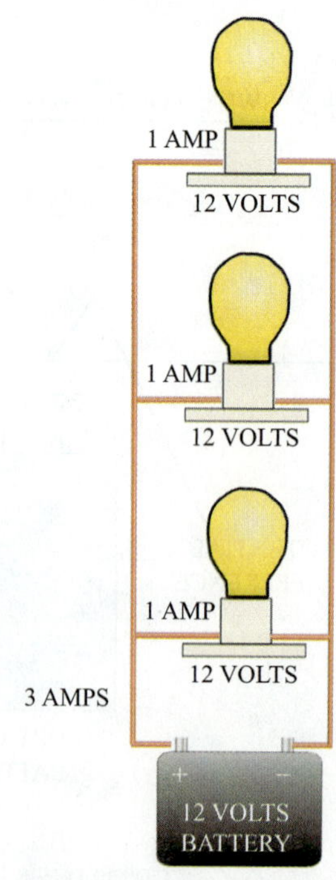

Figure 33-26 Parallel circuit with three loads in parallel.

Notice that the current for each load does not have to pass through any other loads because each load has its own individual path. If one load is broken or removed, the other loads will continue to operate because they have their own paths.

Voltage in Parallel Circuits

The voltage to each load is the same in a parallel circuit because each circuit has its own independent connection to the power source. In HVACR equipment, parallel circuits are used to supply the same voltage to each load.

33.16 COMPARISON OF SERIES AND PARALLEL CIRCUITS

The characteristics of series and parallel circuits are opposite in every respect.

- In a series circuit, the current stays the same.
  - In a parallel circuit, the current changes with each load.
- In a series circuit, the voltage changes with each load according to its resistance.
  - In a parallel circuit, the voltage stays the same.
- In a series circuit, the load with the highest resistance receives the most voltage.
  - In a parallel circuit, the load with the lowest resistance receives the most current.
- In a series circuit, the overall circuit resistance increases as loads are added.
  - In a parallel circuit, the overall circuit resistance decreases as loads are added.
- In a series circuit, the overall circuit current decreases as loads are added.
  - In a parallel circuit, the overall circuit current increases as loads are added.

Roads can be used to illustrate the difference between series and parallel circuits. A series circuit is like a single, lane road with no passing lanes. There is only one way for all vehicles to get from one end of the road to the other, and all vehicles can only travel as fast the slowest vehicle. Adding more distance to the road makes everyone's trip take even longer because there is more road to travel to get from one end to the other.

A parallel circuit is like a road with more than one lane. More vehicles can travel from one end of the road to the other at the same time because there is more than one path. Because there is more than one lane, all vehicles do not have to travel at the same speed. The speed in each lane is controlled by the conditions in that lane. Adding lanes increases the number of vehicles that can travel at one time and reduces the time it takes to move from one end of the road to the other.

33.17 SERIES PARALLEL CIRCUITS

A series parallel circuit, as the name implies, combines both series and a parallel arrangements of electrical devices. A typical series parallel circuit is shown in Figure 33-27. Most equipment is wired with series parallel circuits. The most common use of series parallel circuits is to have several circuits in parallel with each other, with each circuit containing several devices in series. Typically each individual circuit in a series parallel arrangement has one load and one or more switches in series with the load. This allows full voltage to all loads while still maintaining control using the switches that are in series with the loads.

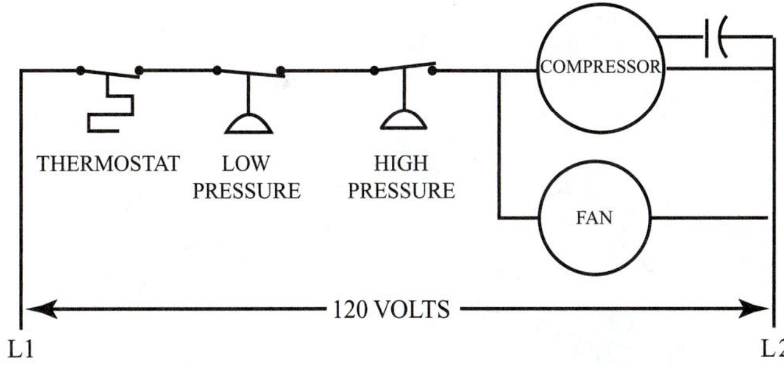

Figure 33-27 In this series parallel circuit, the fan and compressor are in parallel with each other but in series with switches.

33.18 RELATIONSHIP OF VOLTAGE, RESISTANCE, AND CURRENT

It is useful for technicians to understand the relationship between voltage, resistance, and current. A change in one variable will affect the others. The balance shown in Figure 33-28 is a simple way to visualize the relationship between these three variables. Resistance and current are balanced on top of voltage. With voltage remaining the same, if resistance goes up, current will go down (Figure 33-29). On the other hand, if resistance goes down, current will go up (Figure 33-30).

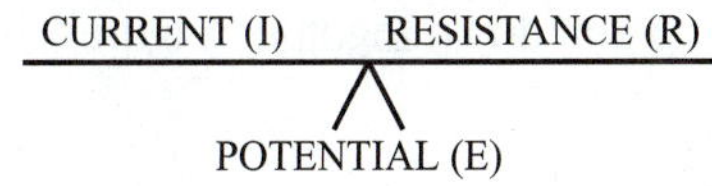

Figure 33-28 The relationship between current and resistance can be visualized as a balance.

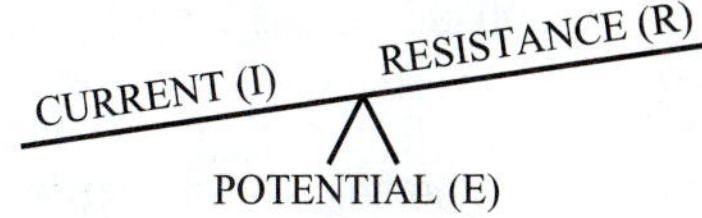

Figure 33-29 When the resistance increases, the current decreases.

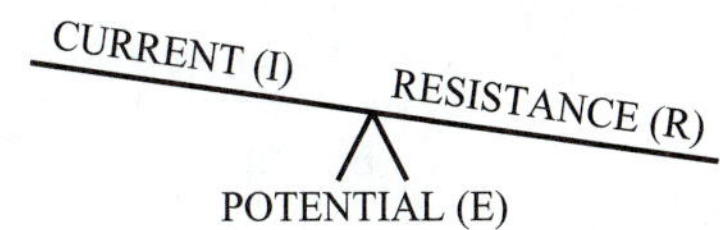

Figure 33-30 When the resistance decreases, the current increases.

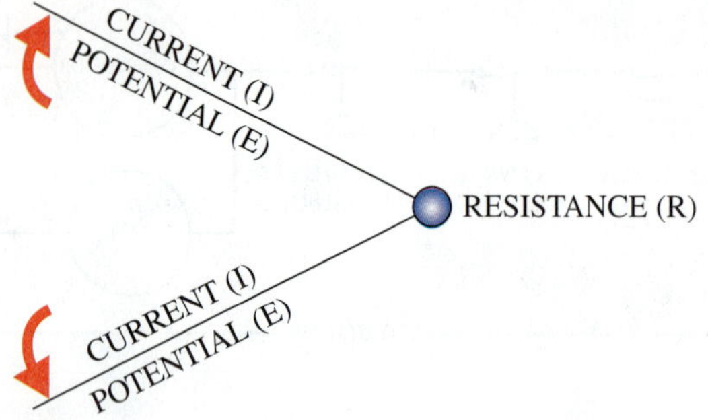

Figure 33-31 If the resistance stays the same, the voltage and current go up and down together.

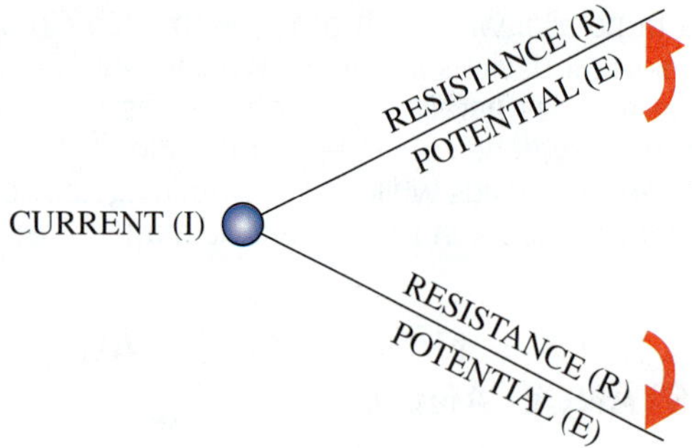

Figure 33-32 For the current to stay the same, the resistance and voltage must go up and down together.

The same relationship is also true if resistance or current remain the same. Using the same principle as the balance beam, if the resistance remains the same, then voltage and current go up and down together (Figure 33-31), and if current remains the same, then voltage and resistance must also go up and down together (Figure 33-32).

33.19 OHM'S LAW

Ohm's law is the relationship between the voltage, current, and resistance in an electrical circuit. Ohm's law stated in simple terms is 1 V applied across a resistance of 1 Ω will produce a current of 1 A. This can be written as the mathematical formula: $E = IR$. If two of the values are known, the remaining value can be calculated. Figure 33-33 shows the three forms Ohm's law can take, depending upon the unknown value.

TECH TIP

It should be noted that Ohm's law does not apply to most AC circuits. The flow of current in AC circuits can be influenced by inductive reactance and capacitive reactance. Magnetic coils of wire produce an induced voltage that opposes current changes in an AC circuit. This has the effect of reducing the current flow through the circuit. As a result, Ohm's law cannot be applied to any AC circuit that contains a capacitor or magnetic coil. The only AC circuits that work correctly with Ohm's law are purely resistive circuits, such as electric heaters. The combined effect of resistance, inductive reactance, and capacitive reactance is called impedance. The Ohm's law formulas will work with AC circuits if the impedance is calculated and substituted in the formulas for resistance.

The following examples illustrate simple applications of Ohm's law.

EXAMPLE 33-1: SOLVING FOR CURRENT

The circuit in Figure 33-34 has a power source of 24 V and a load with a resistance of 8 Ω. Determine the amp draw of the circuit.

$$I = \frac{E}{R}$$

$$I = \frac{24\text{ V}}{8\ \Omega}$$

$$I = 3\text{ A}$$

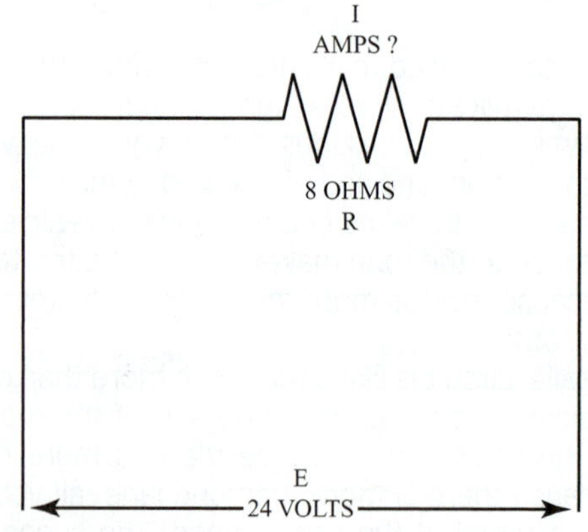

Figure 33-34 Calculating the current using Ohm's law with voltage and resistance known.

Figure 33-33 Ohm's law formulas.

Characteristic	Units	Written Formula	Symbolic Formula
Potential	Volts	voltage = current × resistance	$E = I \times R$
Current	Amps	amps = volts divided by ohms	$I = E/R$
Resistance	Ohms	ohms = volts divided by amps	$R = E/I$

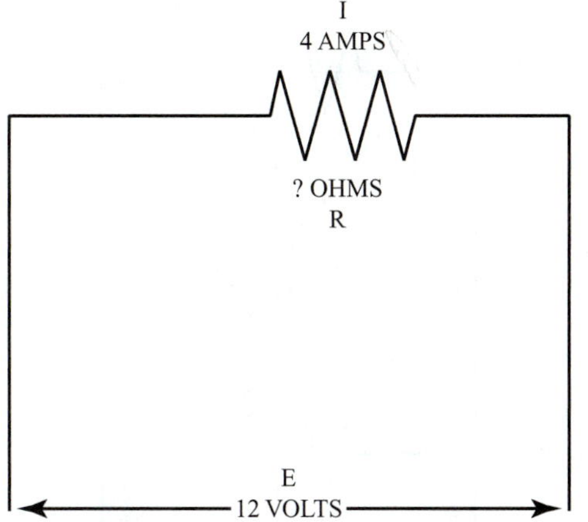

Figure 33-35 Calculating the resistance using Ohm's law with voltage and current known.

EXAMPLE 33-2: SOLVING FOR RESISTANCE

The resistor in Figure 33-35 is drawing a current of 4 A on 12 V. Determine the resistance of the resistor.

$$R = \frac{E}{I}$$

$$R = \frac{12\,V}{4\,A}$$

$$R = 3\,\Omega$$

EXAMPLE 33-3: SOLVING FOR VOLTAGE

In Figure 33-36, the 5 Ω resistor has a current draw of 2 A. Determine the voltage supplied to the circuit.

$$E = IR$$

$$E = 2\,\text{A} \times 5\,\Omega$$

$$E = 10\,\text{V}$$

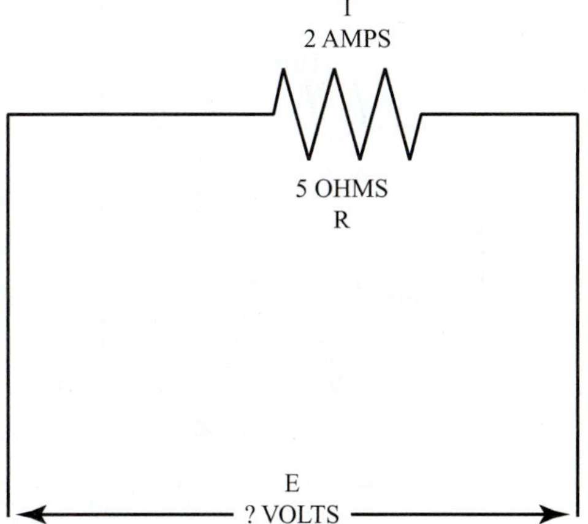

Figure 33-36 Calculating the voltage using Ohm's law with resistance and current known.

33.20 OHM'S LAW CALCULATIONS FOR A SERIES CIRCUIT

The calculations in the previous section used only one load. When a circuit has more than one load, the values can be different for each load in the circuit. The formulas used to calculate the total circuit resistance, current, and voltage depend upon whether the circuit is a series or parallel circuit.

Series Circuit Resistance

The total resistance of a series circuit is the sum of all the individual resistances. This is expressed in symbols as:

$$R_T = R_1 + R_2 + R_3 + R_4$$

For example, the total resistance of a circuit with resistances of 4 Ω, 5 Ω, and 6 Ω is found by adding the three resistances together: 4 Ω + 5 Ω + 6 Ω = 15 Ω total circuit resistance (Figure 33-37).

Series Circuit Current

The current flowing through a series circuit is the same for each load in the circuit. To determine the current in a series circuit, divide the total circuit resistance into the voltage applied to the circuit. For example, if 30 V is applied to a circuit with resistances of 4 Ω, 5 Ω, and 6 Ω, the circuit current is found by adding the resistances and dividing the sum into 30 V: 30/(4 Ω + 5 Ω + 6 Ω) = 2 A. Since the current in a series circuit is the same throughout the circuit, the current for all of the resistances is 2 A (Figure 33-38). This is expressed in symbols as:

$$I_1 = I_2 = I_3 = I_4$$

Series Circuit Voltage

Although the current stays the same throughout a series circuit, the voltage across each load in a series circuit is proportional to its resistance. Loads with different resistances receive different voltages. The sum of the individual voltages equals the total voltage applied to the circuit. The voltage drop across each resistance can be calculated by multiplying the circuit current times the individual resistance. The series circuit with a current draw of 2 A and resistors of 4 Ω,

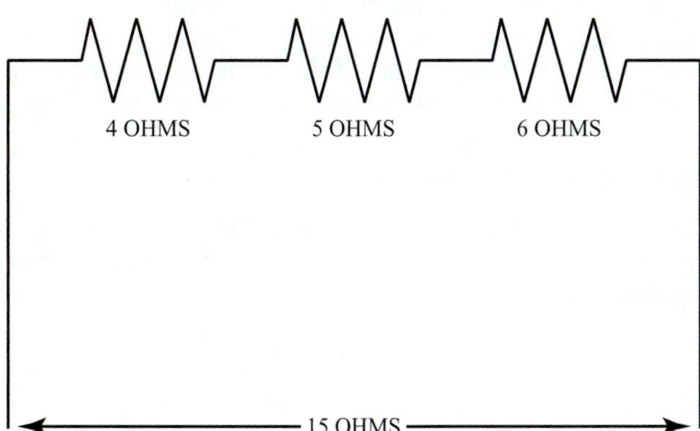

Figure 33-37 The total resistance of a series circuit is the sum of all the individual resistances.

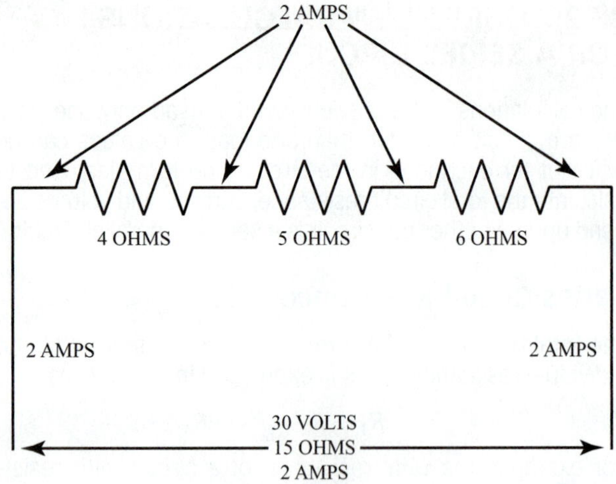

Figure 33-38 The current is the same throughout a series circuit.

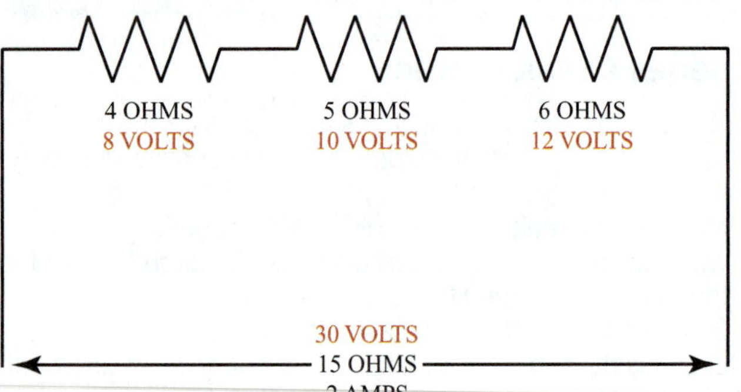

Figure 33-39 The voltage drop across each load in a series circuit is proportional to its resistance.

5 Ω, and 6 Ω would have voltage drops of 8 V, 10 V, and 12 V: 2 A × 4 Ω = 8 V, 2 A × 5 A = 10 V, 2 A × 6 Ω = 12 V. The sum of the individual voltages equals the applied circuit voltage: 8 V + 10 V + 12 V = 30 V (Figure 33-39). This is expressed in symbols as:

$$E_T = E_1 + E_2 + E_3 + E_4$$

33.21 CALCULATIONS FOR A PARALLEL CIRCUIT

Calculations for a parallel circuit are different than those used for series circuits. In a parallel circuit, the voltage across each of the components is the same, and the total current is the sum of the currents through each component.

Parallel Circuit Voltage

The voltage to each load in a parallel circuit is equal to the applied circuit voltage (Figure 33-40). This is expressed in symbols as:

$$E_T = E_1 = E_2 = E_3 \ldots$$

Parallel Circuit Current

The current draw for a parallel circuit is determined for each of its parts. Use the standard $I = E/R$ form of Ohm's law to

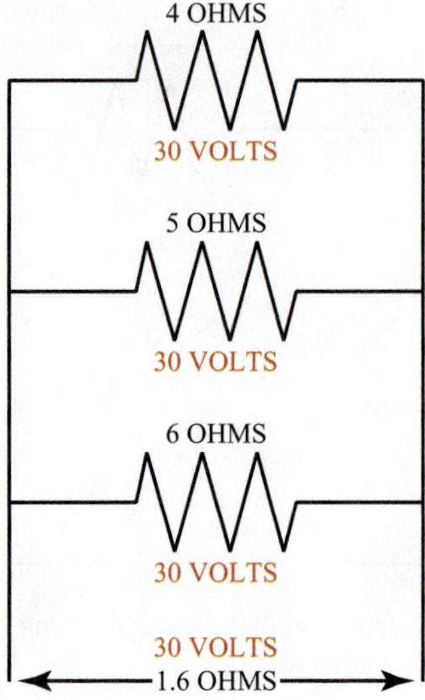

Figure 33-40 The voltage is the same to all loads in a parallel circuit.

determine the current draw through each individual resistance. For example, a parallel circuit with an applied voltage of 30 V and resistances of 4 Ω, 5 Ω, and 6 Ω would have individual current draws of 7.5 A, 6 A, and 5 A: 30 V/4 Ω = 7.5 A, 30 V/5 Ω = 6 A, 30 V/6 Ω = 5 A. The current consumed by the entire parallel system is the sum of the individual currents. 7.5 A + 6 A + 5 A = 18.5 A total circuit current (Figure 33-41). This is expressed in symbols as:

$$I_T = I_1 + I_2 + I_3 \ldots$$

Parallel Circuit Resistance

The total resistance of a parallel circuit gets smaller as more resistances are added. This is because each load in parallel creates another current path, making it easier for electrons to flow. Another way to visualize this is to examine the

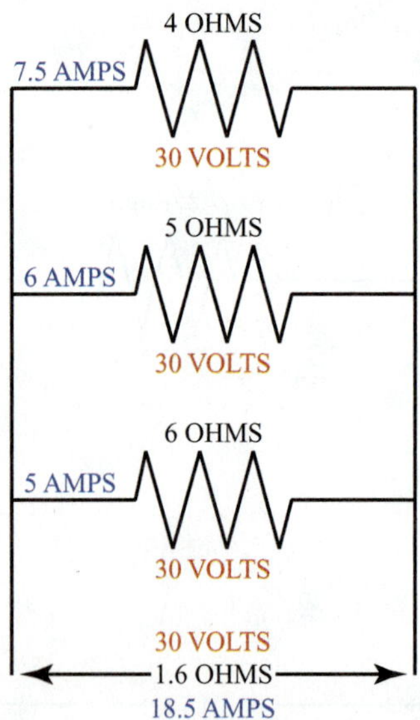

Figure 33-41 The total circuit current in a parallel circuit is the sum of the individual currents.

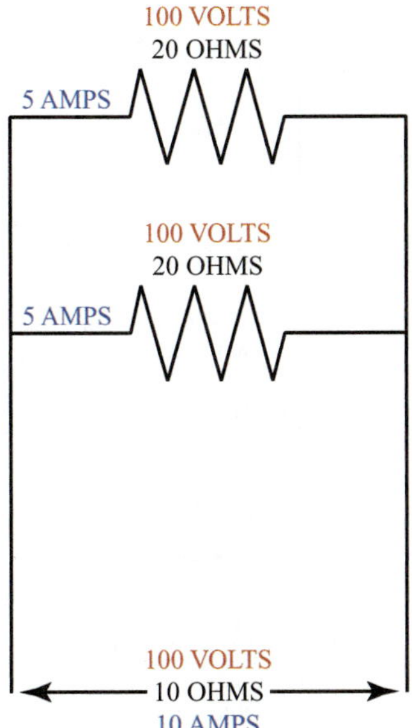

Figure 33-42 The total resistance of a parallel circuit with multiple identical resistances can be found by dividing the value of one resistor by the number of resistors.

relationship between current and resistance. Since voltage = current × resistance, for the current to increase without increasing the voltage, there must be a decrease in resistance. For example, a 20 Ω resistor wired across 100 V would draw 5 A: 100 V ÷ 20 Ω = 5 A. Two 20 Ω resistors wired in parallel across 100 V would each draw 5 A. The total circuit current would be 10 A, 5 + 5. The total circuit resistance can be calculated using Ohm's law as 100 V ÷ 10 A = 10 Ω. In other words, the total resistance for two identical loads is *half* the resistance of one load by itself. Similarly, the total resistance of *any* parallel circuit containing several loads of identical resistance value in parallel can be found by dividing the resistance value for any one load by the number of loads (Figure 33-42).

The formula is more complicated for parallel circuits with different value resistances. If there are only two resistances, the total resistance can be calculated by a formula called "product over the sum":

$$R_T = \frac{(R_1 \times R_2)}{(R_1 + R_2)}$$

For example, to find the total resistance of a parallel circuit with one resistance of 12 Ω and one resistance of 10 Ω:

$$R_T = \frac{(R_1 \times R_2)}{(R_1 + R_2)}$$

$$R_T = \frac{(12\,\Omega \times 10\,\Omega)}{(12\,\Omega + 10\,\Omega)}$$

$$R_T = \frac{120}{22}$$

$$R_T = 5.4\,\Omega$$

If there are more than two resistances, use the following formula and solve for R_T:

$$\frac{1}{R_T} = \frac{1}{R_1} + \frac{1}{R_2} + \frac{1}{R_3}$$

Note that this formula is actually solving for the reciprocal of R_T. The answer must be inverted to find R_T. For example, a parallel circuit with resistances of 4 Ω, 5 Ω, and 6 Ω:

$$\frac{1}{R_T} = \frac{1}{R_1} + \frac{1}{R_2} + \frac{1}{R_3}$$

$$\frac{1}{R_T} = \frac{1}{4} + \frac{1}{5} + \frac{1}{6}$$

$$\frac{1}{R_T} = \frac{15}{60} + \frac{12}{60} + \frac{10}{60}$$

$$\frac{1}{R_T} = \frac{37}{60}$$

$$R_T = \frac{60}{37} = 1.6\,\Omega$$

R_T will always be smaller than the smallest resistor in the problem. In the example, $R_T = 1.6\,\Omega$, which is smaller than 4 Ω, the lowest resistance in the problem.

TECH TIP

Finding a common denominator can get messy when the values are not simple numbers. An easy way to work parallel resistance problems is to use a calculator with a reciprocal key and memory. Enter each resistance followed by the reciprocal key (1/x). Then push the memory plus key (M+) to add the results of each process into memory. Last, push the memory recall key (MR) to display the sum, and push the reciprocal key to show the answer.

33.22 CALCULATIONS FOR A SERIES PARALLEL CIRCUIT

Most circuits are not simple series or parallel circuits but a more complex combination of the two. To analyze a series parallel circuit, solve sections of the circuit that can be solved using the available formulas and then consider those sections as one load. For example, the circuit in Figure 33-43 has two 20 Ω loads in parallel with each other. However, both loads are in series with the 5 Ω load. First calculate the resistance of the two parallel 20 Ω loads, 20 Ω/2 = 10 Ω. The circuit can now be considered a series circuit with one 10 Ω

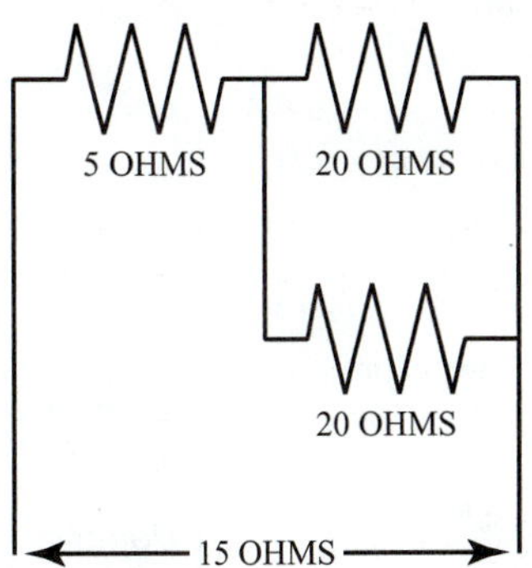

Figure 33-43 Series parallel circuit with a parallel block.

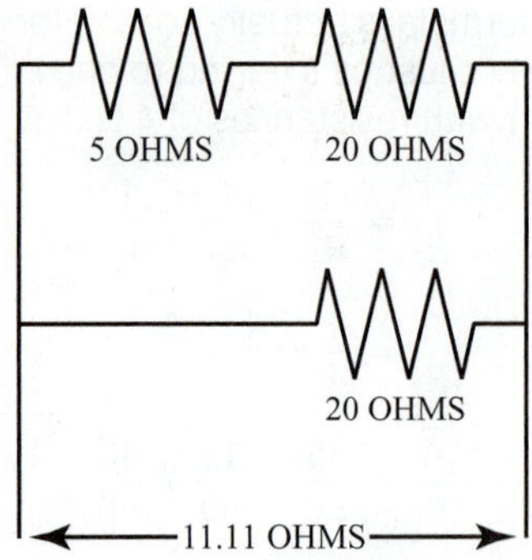

Figure 33-44 Series parallel circuit with a series block.

load and one 5 Ω load. Adding the two together gives you a total resistance of 15 Ω for the circuit. The most expedient solution to this problem was to apply the parallel formula first and then the series formula; however, the formulas used and the order in which they are executed depend upon the particular series parallel circuit being studied.

If the same three components are rearranged as in the diagram in Figure 33-44, the approach and the answer will change. In this diagram, a 20 Ω load and the 5 Ω load are in series with each other and another 20 Ω load is in parallel to both of them as a group. Now it is necessary to use the series formula first to find the total resistance of the 5 Ω and 20 Ω loads, which would be 25 Ω. The problem is now a parallel problem with one 20 Ω load and one 25 Ω load. Using the product over the sum formula: $(25 \times 20)/(25 + 20) = 500/45 = 11.11$ Ω. When circuits advance past two or three loads, several series or parallel blocks may be possible, and there will generally be more than one approach to the solution.

33.23 WATTS

The watt is the unit used to measure the rate of electrical work. The watt is an electrical rate of work equal to 1 joule per second (J/S). A watt is the amount of work done when 1 amp flows with a potential difference of 1 volt. The formula for calculating watts is: watts = volts × amps. This is shown symbolically as $P = IE$. If any two factors are known, the third can be calculated using one of the formulas in Figure 33-45.

EXAMPLE 33-4: SOLVING FOR POWER

In Figure 33-46, find the wattage of an electric heater with a current draw of 8 A on 120 V:

$$P = IE$$
$$P = 8\text{ A} \times 120\text{ V}$$
$$P = 960\text{ W}$$

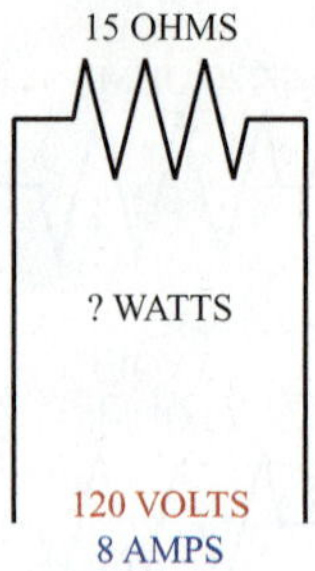

Figure 33-46 Calculating the wattage using the power formula if volts and amps are known.

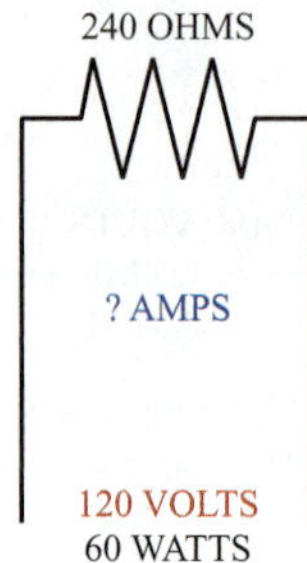

Figure 33-47 Calculating the current using the power formula if volts and watts are known.

EXAMPLE 33-5: SOLVING FOR CURRENT

In Figure 33-47, find the current draw of a 60 W, 120 V heater:

$$I = P/E$$
$$I = 60 \div 120$$
$$I = 0.5\text{ A}$$

EXAMPLE 33-6: SOLVING FOR VOLTAGE

In Figure 33-48, find the voltage applied to an 800 W electric heater drawing 10 A:

$$E = P/I$$
$$P = 800\text{ W}/10\text{ A}$$
$$E = 80\text{ V}$$

33.24 COMBINING OHM'S LAW AND THE POWER FORMULA

Occasionally, the two known values will not work with either Ohm's law or the power formula. To solve this, values for volts, ohms, amps, or watts may be stated in terms of one

Figure 33-45 Power formulas.

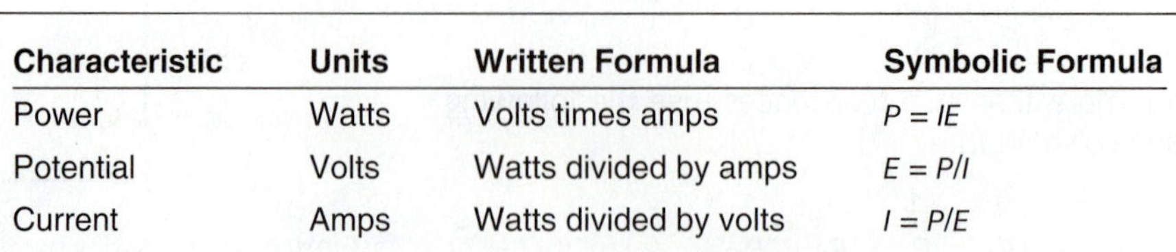

Characteristic	Units	Written Formula	Symbolic Formula
Power	Watts	Volts times amps	$P = IE$
Potential	Volts	Watts divided by amps	$E = P/I$
Current	Amps	Watts divided by volts	$I = P/E$

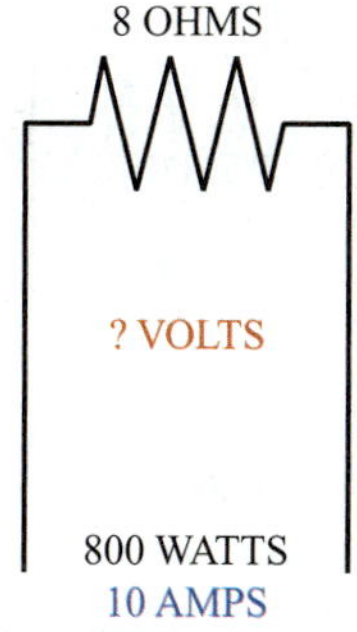

Figure 33-48 Calculating the voltage using the power formula if amps and watts are known.

formula and substituted in another formula. For example, suppose you needed to find the voltage for a device with a known resistance and wattage. These two values will not work in either Ohm's law or the power formula. However, the value P/E can be substituted for I in Ohm's law, so it can be rewritten $E = (P/E)R$. This can be rearranged algebraically as $E = \sqrt{PR}$. Now the value for voltage can be found by multiplying watts times resistance and taking the square root of the product. Figure 33-49 shows all the variations of the Ohm's law and power formulas that can be produced using the two formulas. The formulas that contain either squared values or square roots of values are the result of algebraic combinations of Ohm's law and the power formula.

EXAMPLE 33-7: SOLVING FOR WATTS USING AMPS AND OHMS

In Figure 33-50, find the wattage lost through a conductor with 50 Ω of resistance carrying 3 A:

$$P = I^2 \times R$$
$$P = 3^2 \times 50$$
$$P = 9 \times 50$$
$$P = 450 \text{ W lost through the conductor}$$

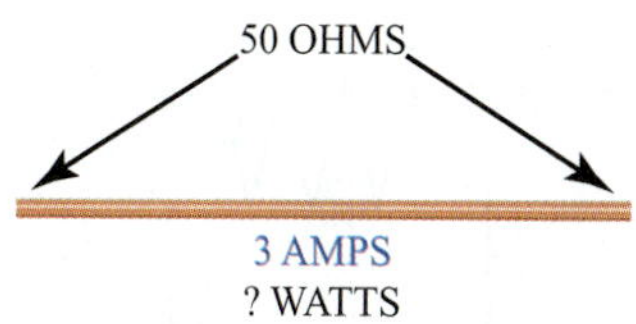

Figure 33-50 Calculating wattage lost in a conductor using both Ohm's law and the power formula when amps and ohms are known.

EXAMPLE 33-8: SOLVING FOR VOLTAGE USING WATTS AND OHMS

In Figure 33-51, find the voltage of a 60 W heater that has 240 Ω resistance:

$$E = \sqrt{W \times R}$$
$$E = \sqrt{60 \times 240}$$
$$E = \sqrt{14{,}400}$$
$$E = 120 \text{ V}$$

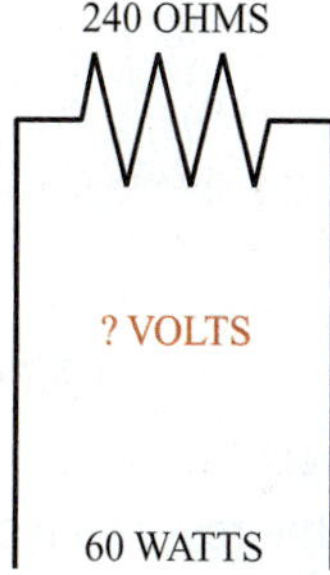

Figure 33-51 Calculating voltage using both Ohm's law and the power formula when watts and ohms are known.

Value	Units	Written Formula	Symbolic Formula
Power	Watts	Amps times volts	$P = IE$
		Amps squared times ohms	$P = I^2R$
		Volts squared divided by ohms	$P = E^2/R$
Potential	Volts	Amps times ohms	$E = IR$
		Watts divided by amps	$E = P/I$
		Square root of (watts times ohms)	$E = \sqrt{PR}$
Current	Amps	Volts divided by ohms	$I = E/R$
		Watts divided by volts	$I = P/E$
		Square root of (watts divided by ohms)	$I = \sqrt{P/R}$
Resistance	Ohms	Volts divided by amps	$R = E/I$
		Watts divided by the square of amps	$R = P/I^2$
		Volts squared divided by watts	$R = E^2/P$

Figure 33-49

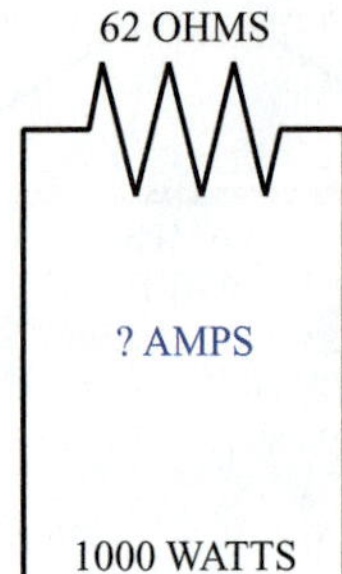

Figure 33-52 Calculating current using both Ohm's law and the power formula when watts and ohms are known.

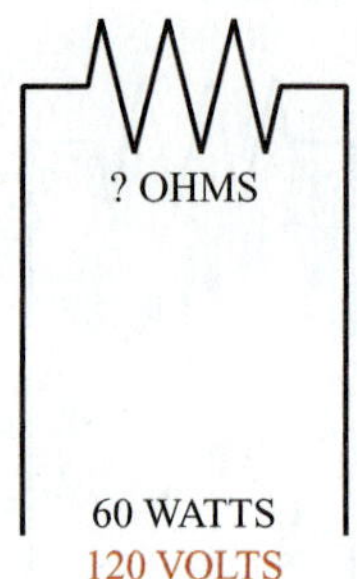

Figure 33-53 Calculating resistance using both Ohm's law and the power formula when watts and volts are known.

EXAMPLE 33-9: SOLVING FOR AMPS USING WATTS AND OHMS

Using Figure 33-52, find the amp draw of a 1,000 W heater with a resistance of 62 Ω:

$$I = \sqrt{P/R}$$
$$I = \sqrt{1{,}000/62}$$
$$I = \sqrt{16}$$
$$I = 4\text{ A}$$

EXAMPLE 33-10: SOLVING FOR RESISTANCE USING VOLTS AND WATTS

In Figure 33-53, find the resistance of a 60 W 120 V heater:

$$R = E^2/P$$
$$R = 120^2/60$$
$$R = (14{,}400)/60$$
$$R = 240\ \Omega$$

TECH TIP

It should be noted that the power formula cannot be directly applied to most AC circuits. The flow of current in AC circuits can be influenced by inductive reactance and capacitive reactance. Magnetic coils of wire produce an induced voltage that opposes current changes in an AC circuit. This has the effect of causing the voltage and current to peak at different times, so that the delivered power is less than the product of volts and amps. Three-phase power has three different currents operating at one time. A measurement of one does not accurately indicate the current used by the device. The only AC circuits that work correctly with the power formula are single phase, purely resistive circuits, such as electric heaters.

UNIT 33—SUMMARY

Electricity is the flow of free electrons. Electrons are negative subatomic particles that exist on the outside of an atom. Electron flow is a chain reaction, with electrons moving from atom to atom. Conductors have many electrons but only a few in their outer orbit. Insulators have several electrons in their outer orbit. Electron flow in one direction is called direct current. This type of current flow is used with battery-powered devices. Alternating current reverses directions at regular intervals. Alternating current is used for power distribution because of its ability to be easily transformed to different voltage levels, allowing for high-voltage transmission and reducing line loss.

Electrical potential is the ability to produce an electrical current; it is measured in volts. Electrical current is the rate of electron flow; it is measured in amps. Resistance opposes the flow of electrons; it is measured in ohms. Electrical work is affected by both the potential and the current flow; it is measured in watts.

A circuit is composed of a source of electrical power, a path for electrons to flow, and a load to convert the electrical energy into useful work. Resistance and current move opposite to one another: when the resistance goes up, the current goes down. A series circuit has only one path for electron flow. The current in a series circuit stays the same throughout the circuit, and the voltage drop across loads in series is proportional to their resistance. The voltage stays the same in a parallel circuit, and the current for each load is determined by that load's resistance, not the entire circuit resistance. Adding loads in series increases the circuit resistance; adding loads in parallel decreases the circuit resistance.

Ohm's law and the power formula can be used to determine any of the commonly measured circuit characteristics: voltage, amps, ohms, or watts. If any two characteristics are known, the others can be calculated using Ohm's law and the power formula. The basic formula for Ohm's law is $E = IR$. The basic power formula is $P = IE$. Ohm's law and the power formula are not applicable to alternating current circuits, which have inductive or capacitive loads.

UNIT 33—REVIEW QUESTIONS

1. Describe the basic structure of the atom.
2. Define *free electron*.
3. What is the difference between a generator and an alternator?
4. What is the difference between static electricity and dynamic electricity?
5. What precautions should be taken before handling printed circuit boards?
6. Give an example of the use of static electricity in the HVACR field.
7. Give the units used to measure the following electrical circuit characteristics:
   a. Potential
   b. Current
   c. Resistance
   d. Power
8. Explain the relationship between current and resistance in an electrical circuit.
9. Explain why copper is a good electrical conductor.
10. Explain the difference between direct current and alternating current.
11. What is meant by the frequency of AC current?
12. What is Ohm's law?
13. What is the current draw of a circuit with a resistance of 6 Ω when 120 V is applied to the circuit?
14. What is the resistance of a circuit operating with a current draw of 5 A and 30 V?
15. What voltage is being supplied to a circuit with a resistance of 25 Ω and a current draw of 5 A?
16. Calculate the electric power for a 6 V DC circuit drawing 2 A.
17. What is the current draw of a 1,500 W electric heater operating on a 120 V power supply?
18. What three elements are required for all electric circuits?
19. Name and describe the three types of path arrangements for circuits.
20. What is the difference between a short circuit and an open circuit?
21. Explain how a switch controls whether a load operates or does not operate.
22. Why are parallel circuits used to supply power for loads in HVACR equipment?
23. Find the total resistance of the complete circuit with 240 V and $R_1 = 6\ \Omega$ and $R_2 = 4\ \Omega$.
24. Draw a series circuit with three loads.
25. Draw a parallel circuit with four loads.
26. Draw a series parallel circuit with one switch and two loads. The loads should be in parallel with each other, but in series with the switch.
27. Explain how to calculate the resistance of a series parallel circuit.

UNIT 34

Alternating Current Fundamentals

OBJECTIVES

After completing this unit, you will be able to:

1. identify the different types and categories of electricity and electrical power systems used in the HVACR industry.
2. explain the basis for magnetic induction in coils and transformers.
3. explain how capacitors work in AC circuits.
4. discuss the uses of single-phase and three-phase power.
5. describe how electrical power is generated.
6. state the importance of overcurrent protection.

34.1 INTRODUCTION

The electrical power provided to residential and commercial customers is alternating current (AC) with a frequency of 60 cycles (hertz). AC power flowing through an electrical circuit has unique characteristics that must be understood by the HVACR technician prior to installing or servicing equipment. It is also important to note that residential service will be very different from commercial service. Residential service requires lower voltage and single-phase power. Commercial systems utilize higher voltages and three-phase power. Therefore, the electrical connections, the types of motors, and the electrical code requirements will not be the same.

This unit shows how resistance, inductance, capacitance, and impedance affect the operation of AC circuits. From this, the HVACR technician can understand the relationship of voltage to current in an AC circuit. Single-phase and three-phase power is also explained and the differences between the two made clear. In addition, there are examples of power-distribution arrangements with the appropriate line voltage connections illustrated for both residential and commercial service. After reviewing this unit, the HVACR technician will have a good understanding of AC circuit behavior and power distribution.

34.2 TYPES OF POWER SOURCES

There are two common types of power:

- Direct current (DC)
- Alternating current (AC)

Many electrical testing instruments operate on DC power. This is because the most common source of direct current is the battery. Batteries allow test instruments to be portable and convenient to use. DC power can also be used for control systems and on solid-state modules for defrost and overcurrent protection. For these applications, the DC power is often converted from AC power through a rectifier instead of using batteries.

Alternating current is the most common source of power for most HVACR systems. AC power is generated by all the power companies. In residences, 120 V AC is used to power most small appliances. Larger appliances such as electric stoves and residential air-conditioning units use 240 V. The power company supplies residential users with 240 V over the incoming lines. A portion of it is tapped to supply the 120 V requirements. Commercial and industrial customers normally use higher voltages in single-phase and three-phase systems.

TECH TIP

In the United States the AC frequency supplied by power companies is normally 60 cycles. Many foreign countries supply power at 50 cycles. It is important to use equipment that is designed to operate at the proper frequency.

34.3 MAGNETISM

The theory of electron flow was discussed in Unit 33. Some materials (conductors) have more free electrons, while others (insulators) have few or no free electrons. These electrons have a spinning motion. When an equal number of electrons spin in opposite directions, they tend to cancel each other out. However, if more electrons spin in one direction than the other, a magnetic field is developed. The materials that exhibit this behavior are called magnets.

A force called a magnetic field transmits outward in all directions from magnetic materials. The invisible lines of force are referred to as magnetic flux and are shown radiating from the magnet in Figure 34-1. Lines of force are always considered to leave the north pole and enter the south pole of a magnet in continuous loops. Like poles repel each other, and opposite poles attract (Figure 34-2).

Since flux will choose the path of least resistance (called reluctance), the field is scattered in Figure 34-1 because air has a high reluctance. However, if an iron bar with

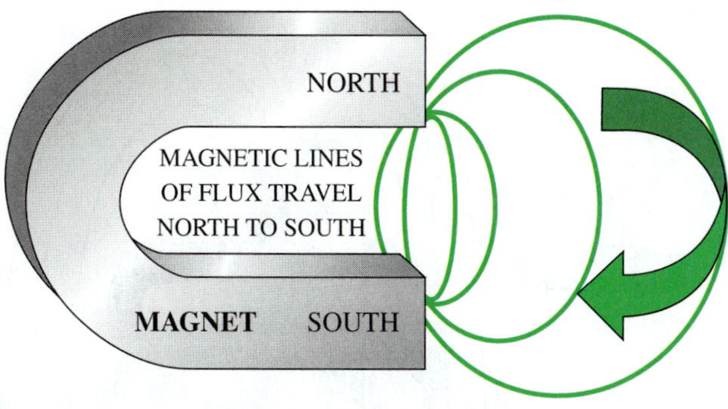

Figure 34-1 Invisible magnetic lines of force.

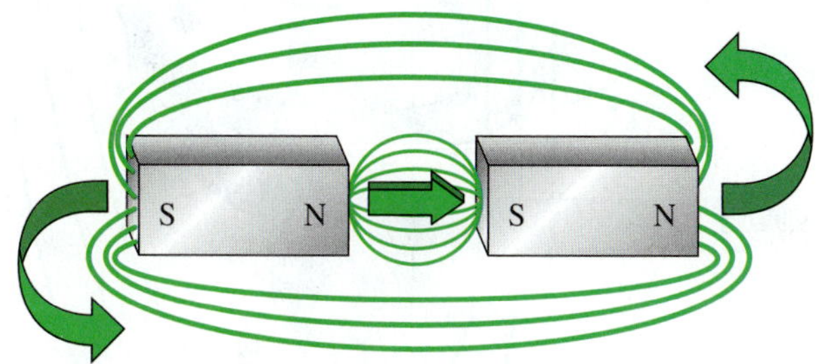

Figure 34-2 Like poles repel each other and opposite poles attract.

a low reluctance is placed between the two ends of the magnet, the flux will concentrate through it, and this force will draw the iron toward the magnet (Figure 34-3).

Electromagnetism and Induced Voltage

When current passes through a conductor, a magnetic field is produced (Figure 34-4). The magnetic field has no poles, but, similar to magnets, there is a continuous loop with a specific direction. The left-hand rule is used to determine the direction of rotation of the flux around the conductor. Imagine the conductor held in your left hand with your thumb pointing in the direction of current flow. Your fingers will align in the direction of the flux (Figure 34-5).

There is another phenomenon that is exhibited with conductors. If the conductor is moved through a magnetic field, a voltage is induced in the conductor (Figure 34-6). The polarity of the induced voltage is affected by which

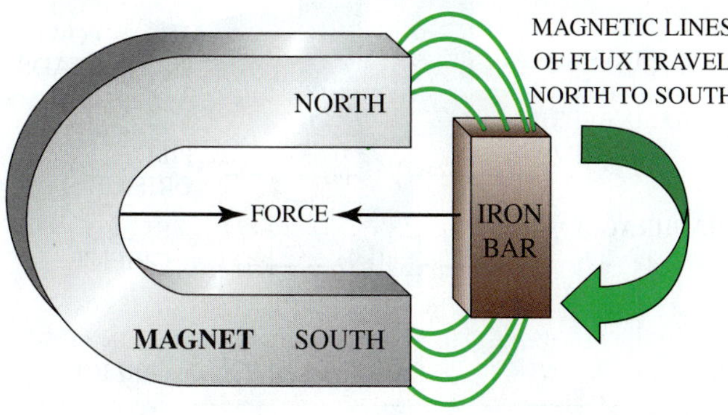

Figure 34-3 Magnetic flux concentrates in the iron core, and this force pulls the iron toward the magnet.

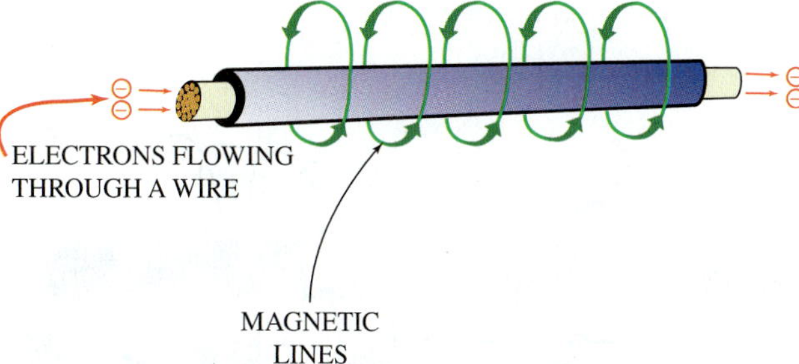

Figure 34-4 Electrons flowing through a straight wire form a weak magnetic field around the wire.

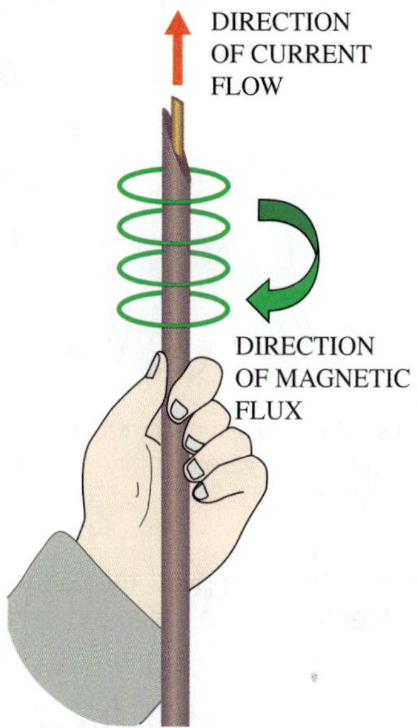

Figure 34-5 Left-hand rule for determining the direction of the flux rotation around a wire.

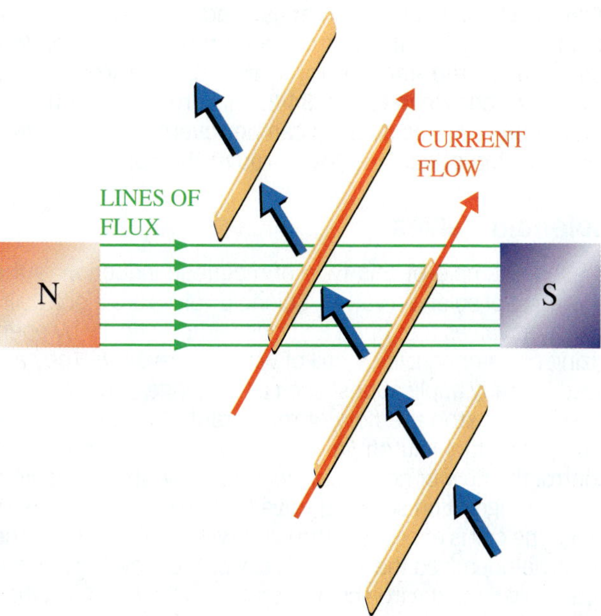

Figure 34-6 Moving a conductor through a magnetic field induces a voltage.

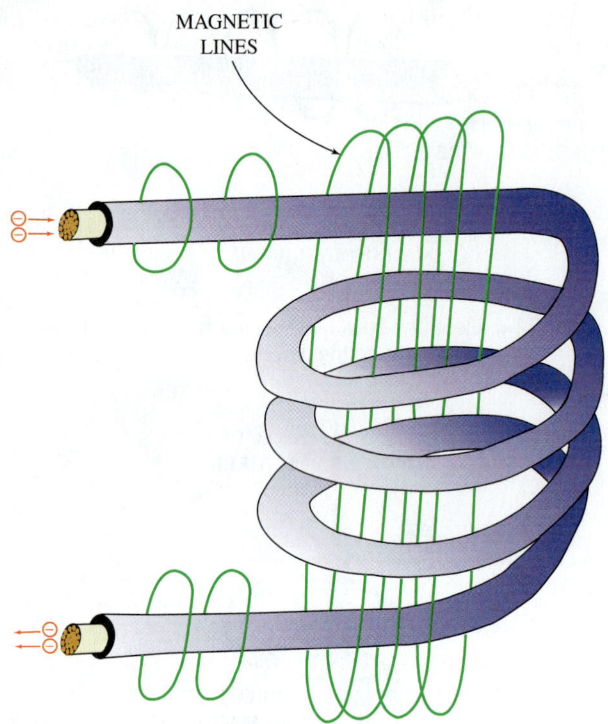

Figure 34-7 When the wire is coiled, the weak magnetic field around each section of wire combines to form a stronger magnetic field around the coil.

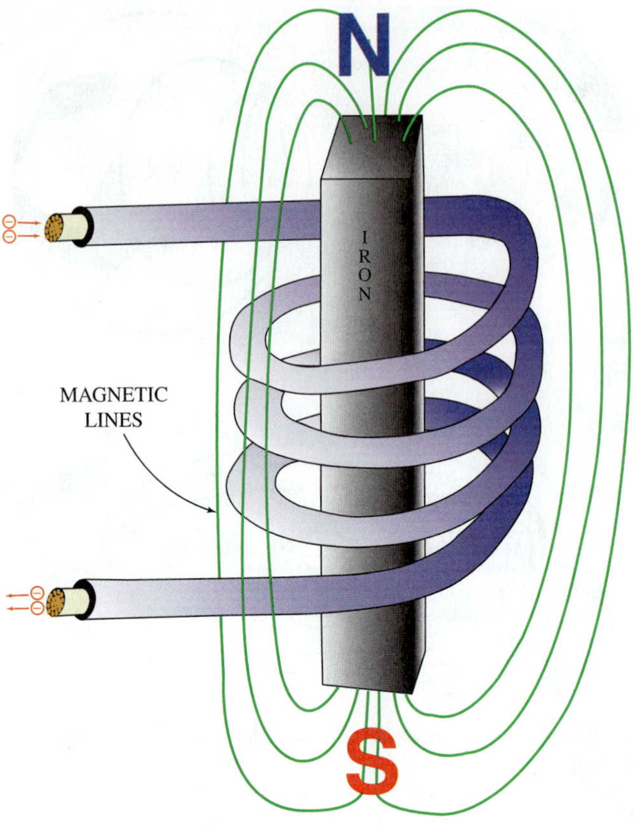

Figure 34-8 The addition of an iron core to the coil concentrates the magnetic lines of force.

direction the conductor is moved through the field (up or down) and the direction of the magnetic field.

34.4 MAGNETIC INDUCTION IN COILS AND TRANSFORMERS

The magnetic field created by current passing through the conductor will intensify if the conductor is coiled (Figure 34-7). If an iron bar is placed inside the coil, as in Figure 34-8, the current flow through the coil will concentrate the lines of force through the iron bar as previously shown in Figure 34-3. This creates a north and south polarity. This polarity can be reversed by changing the direction of current flow through the coil.

Solenoid Valves

A good example of this type of magnetic induction is the operation of solenoid valves. Solenoid valves are electrically operated valves that open and close automatically by energizing or de-energizing a coil of wire (Figure 34-9). They are used in many applications, such as in refrigeration systems to start and stop the flow of refrigerant. They are used as emergency fuel shutoff valves in boilers. They are used to control the flow of hot or cold water in a washing machine.

A refrigeration solenoid valve is shown in Figure 34-10. When the coil is energized, the valve will open, and when the coil is de-energized the valve will close. The movable plunger acts similarly to the iron core described in Figure 34-7. When the coil is energized, the plunger, which has much less reluctance than the air, will provide a path for the flux. The plunger is then drawn upward toward the coil (similar to the

Figure 34-9 A typical coil that would be used for a solenoid valve.

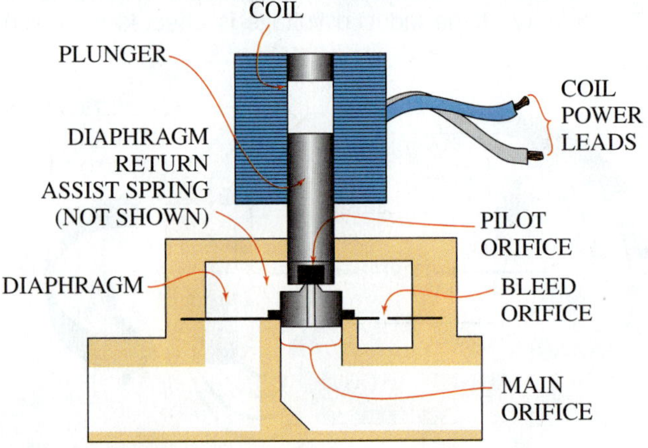

Figure 34-10 The solenoid valve coil is de-energized and the valve is closed.

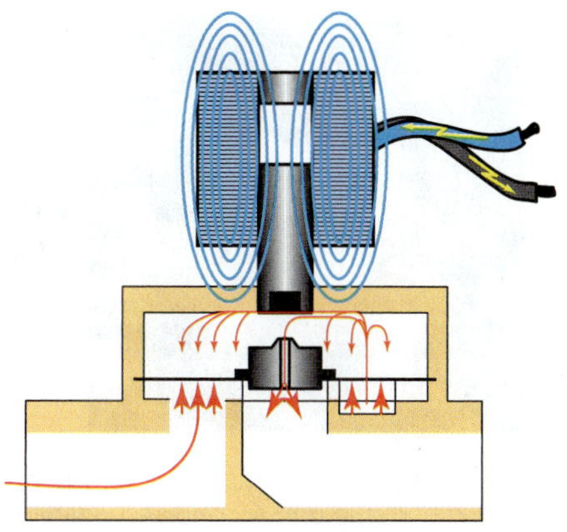

Figure 34-11 The solenoid valve coil is energized and the valve is open.

iron bar being pulled toward the magnet in Figure 34-3), and the valve will open as shown in Figure 34-11. When the coil is de-energized, there is no magnetic force to hold it in the coil and it moves downward to close the valve.

Relays and Contactors

This same principle is applied to the construction of relays and contactors as well. These act as automatic switches to turn loads on and off. In the resting position, gravity keeps the contacts separated. Referring to Figure 34-12a, when current flows through the coil, the iron core is magnetized and is drawn into the coil. This raises the iron core and changes the position of the relay contacts, as shown in Figure 34-12b.

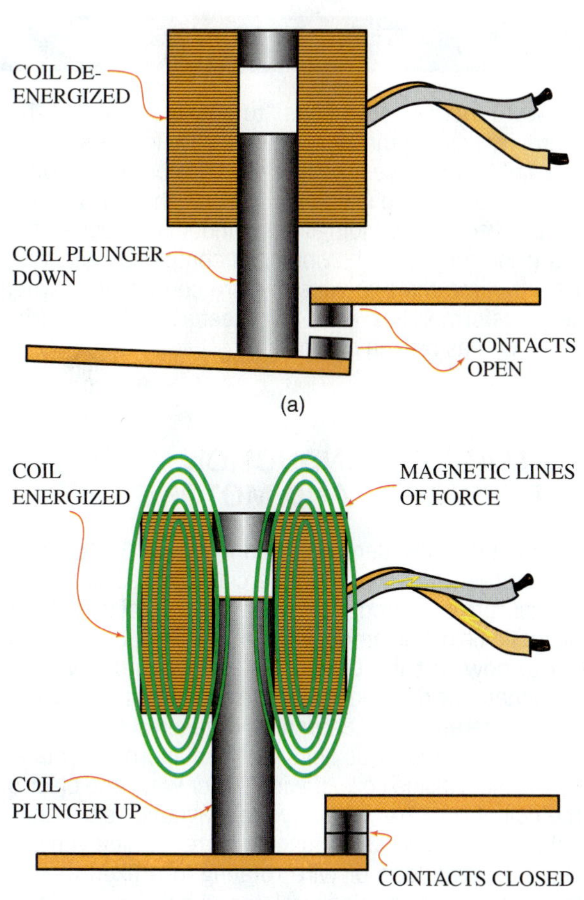

Figure 34-12 (a) Coil de-energized; (b) coil energized.

Transformers

A different application of the induction principle is used in the construction of a transformer, which has no moving parts. Transformers are used to increase or decrease incoming voltages to meet the requirements of the load. Transformers are used to step down the line voltage to 24 V for the controls used in many HVACR units. Transformers contain a single iron core that is wrapped with two separate coiled conductors known as primary and secondary windings (Figure 34-13). When AC voltage is applied to the primary winding, the resulting lines of force are carried through the core. These lines create a current flowing through the secondary winding, inducing a voltage in that winding.

There are two types of transformers: step-down transformers and step-up transformers (Figure 34-14). Step-down transformers are often used to reduce line voltage required for motor starters to lower voltages for the controls. The

(a)

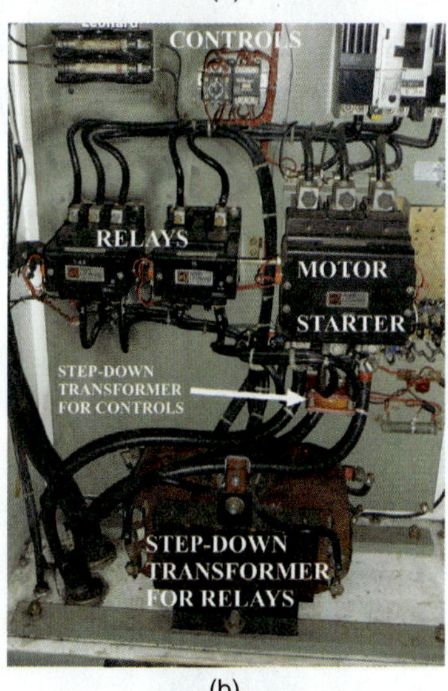

(b)

Figure 34-13 (a) A large transformer for an AC chiller unit; (b) the large transformer steps down line voltage to relay voltage, and the small transformer steps down relay voltage to control voltage; (see next page) (c) a small, low-voltage transformer for a control circuit board; (d) transformer for starting relay for a fan blower motor.

(c)

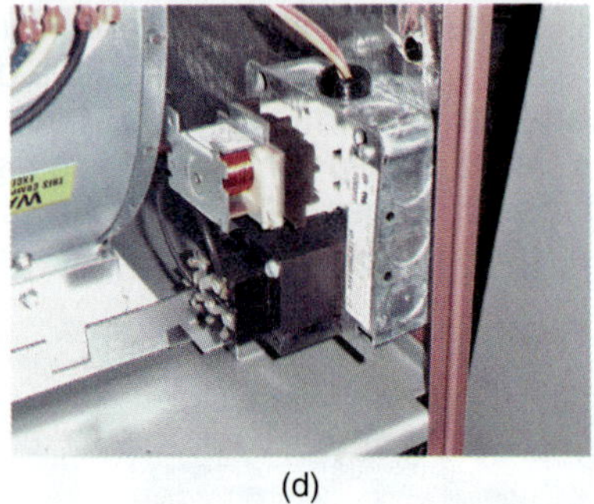

(d)

Figure 34-13 *(Continued)*

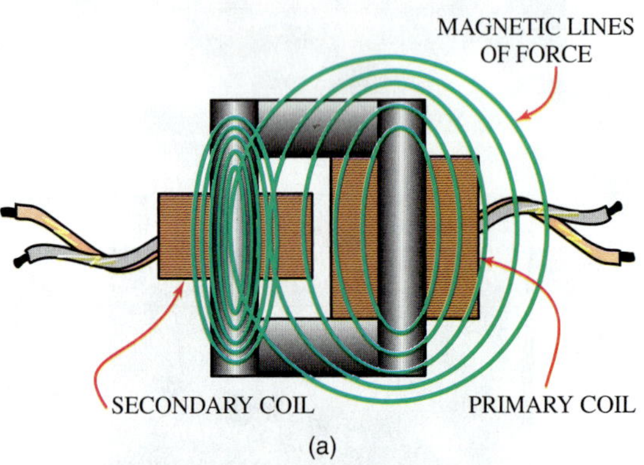

(a)

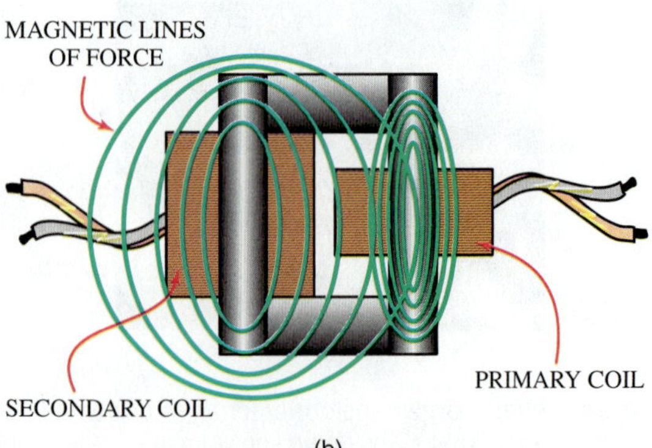

(b)

Figure 34-14 (a) Step-down transformers have more windings on the primary coil; (b) step-up transformers have fewer windings on the primary coil.

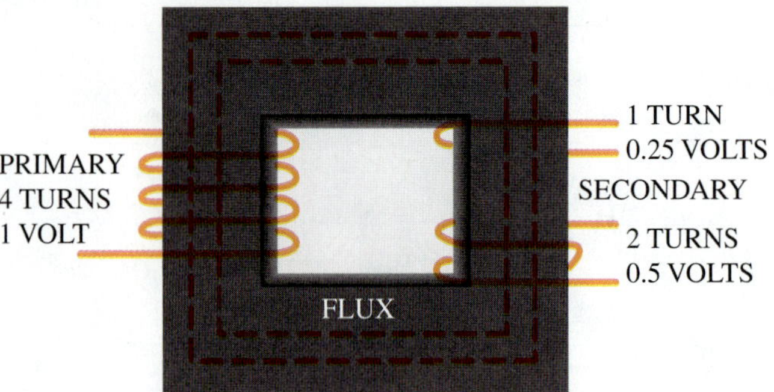

Figure 34-15 The secondary voltage is directly related to the number of turns on both the primary and secondary windings.

440 line voltage supplied to the motor starter shown in Figure 34-13a,b is stepped down by the large transformer to a lower voltage for the relays and then further stepped down by a second smaller transformer to a lower voltage for the controls.

The amount of voltage induced in the secondary winding depends on the ratio of the number of turns in the primary winding to the number of turns in the secondary winding (Figure 34-15). As an example, a primary voltage of 1 volt with four turns would induce a voltage of 0.25 volts to one turn in the secondary or 0.5 volts to two turns in the secondary.

TECH TIP

The word *inductance* means "to cause or create." The electricity flowing through a coil of wire induces a magnetic field that causes another electric current to flow in a parallel coil. There is no electrical-mechanical connection between the windings of an induction device such as a transformer. Each winding is completely separate. The only time the two windings are connected is when the transformer has been overheated and the wiring insulation has broken down.

34.5 MAGNETIC INDUCTION IN GENERATORS AND MOTORS

Magnetic induction, demonstrated in Figure 34-16, is used to generate voltage for commercial and residential use. Electrical current is produced by moving coils of wire through a magnetic field. The amount of voltage produced is proportional to how fast the coils of wire move and how strong the magnetic field is. Electrical current can be produced by either a generator or an alternator. A generator rotates coils of wire inside a magnetic field, while an alternator rotates a magnetic field inside coils of wire. Either way, free electrons in the coil of wire are pulled by the magnetic field, generating electricity. Figure 34-17 shows current being generated through a single loop of wire rotating in a fixed magnetic field. As the loop turns clockwise through the magnetic field, a flow of electrons is established in the wire. Alternating current is generated because the loop cuts the lines of force first

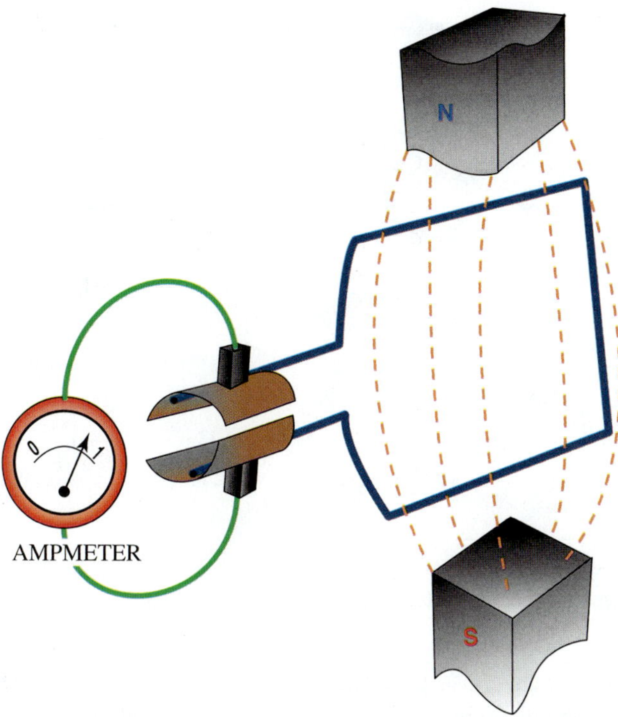

Figure 34-16 Current is induced in a coil of wire rotating in a magnetic field.

in one direction and then in the other as it makes a complete revolution. Notice that the wire near the north pole is always negative and the wire near the south pole is always positive. However, the red half of the wire is near the north pole during the first half cycle and near the south pole during the last half cycle. This means that the red wire is negative during the first half cycle and positive during the last half cycle.

AC Motors

This induction principle is also applied to the construction of AC motors (Figure 34-18). The basic motor has two coils wrapped around stationary cores called stator poles. As AC current flows through the coils, the stator poles are magnetized. The stator field induces an opposing field in the rotor and the principle of attraction and repulsion causes the motor to run.

TECH TIP

Sixty-cycle current can create a low-frequency hum in some motors and transformers. This hum is a result of the frequency of current that passes through the coil. Sometimes this noise can be very disturbing to the residents. Occasionally the sound is amplified because the motor or transformer is not tight on its mounting brackets. Secure the motor tightly to attempt to reduce the sound. If the sound can be tracked to a vibrating sheet-metal component, that component may be secured using a pop rivet or sheet-metal screw to reduce the sound level.

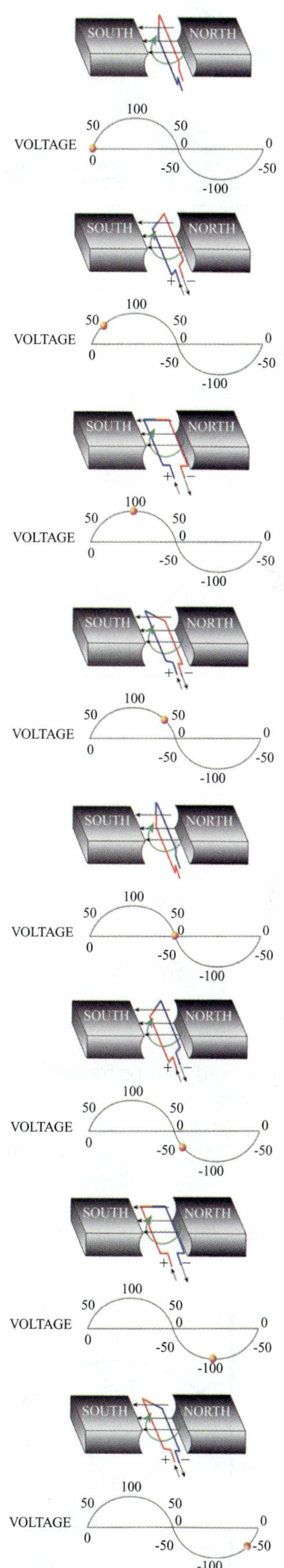

Figure 34-17 Operation of an AC generator.

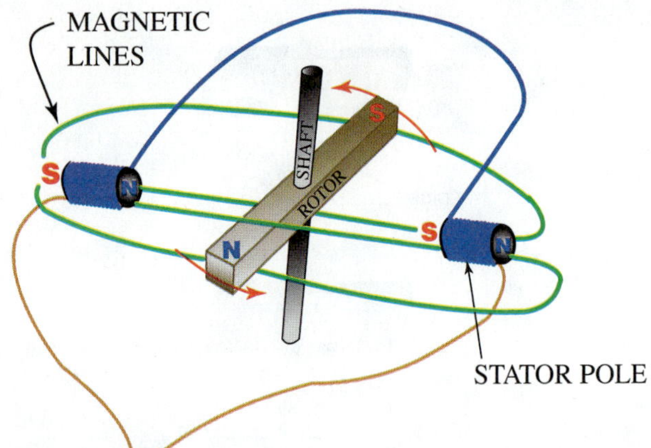

Figure 34-18 When AC power is applied to the coils of a motor, the rotor turns.

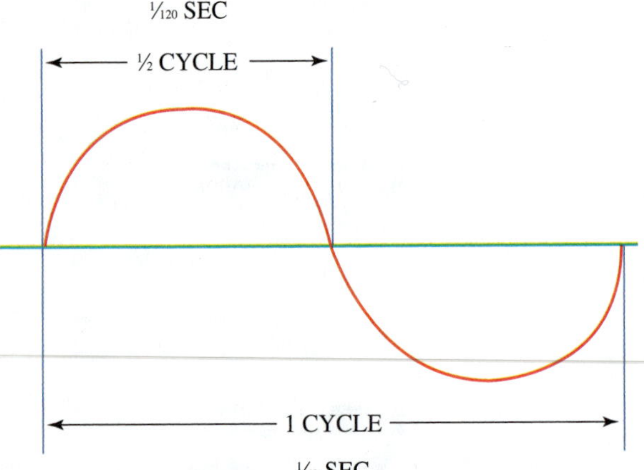

Figure 34-19 One cycle of alternating current.

34.6 SINGLE-PHASE VOLTAGE

Alternating current flow continuously reverses direction to produce a sinusoidal waveform that alternates polarity continuously at a fixed rate or frequency. The number of times per second the polarity is reversed is the frequency expressed in hertz (Hz), or cycles per second. Alternating current is produced by rotating a coil within the magnetic field to produce an output voltage. Figure 34-19 is a diagram of a single cycle that is repeated 60 times per second (60 Hz), on a continuous basis. This is called single-phase voltage.

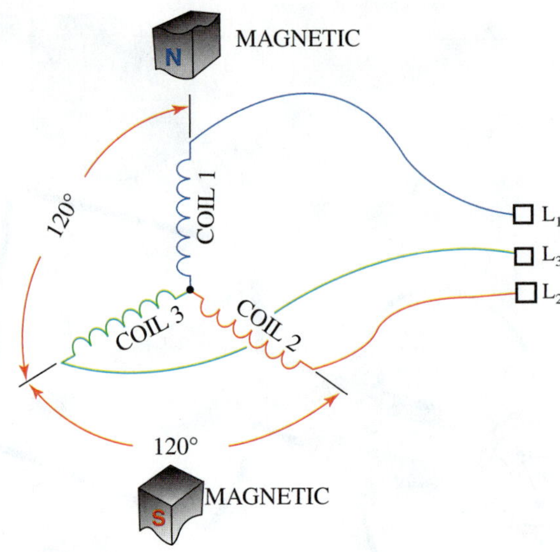

Figure 34-20 Three-phase alternating current is produced when three coils rotate inside of a magnetic field.

Single-phase power is commonly used for residential and small commercial systems. For larger motors, generally 5 hp and above, three-phase power is normally used. This requires the addition of one more conductors to the generator, as shown in Figure 34-20. The three conductors are positioned 120° from each other.

A diagram of the three-phase waveforms is shown in Figure 34-21. They have the same shape, but they are 120° out of phase with each other.

34.7 PHASE SHIFT

In working with AC power, certain characteristics affect the power calculation, $E \times I$, which are known as phase shift factors. The phase shift factors relate to the following:

- Resistive circuits
- Inductive circuits
- Capacitive circuits

34.8 RESISTIVE CIRCUITS

The resistive circuit (Figure 34-22) contains at least one resistive load, such as an electric heater or lamp. The current rises and falls with the voltage and the two are considered

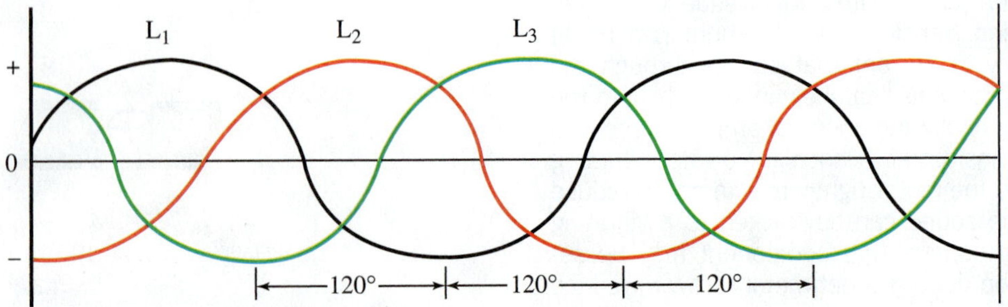

Figure 34-21 The three legs of power produced by a three-phase AC generator.

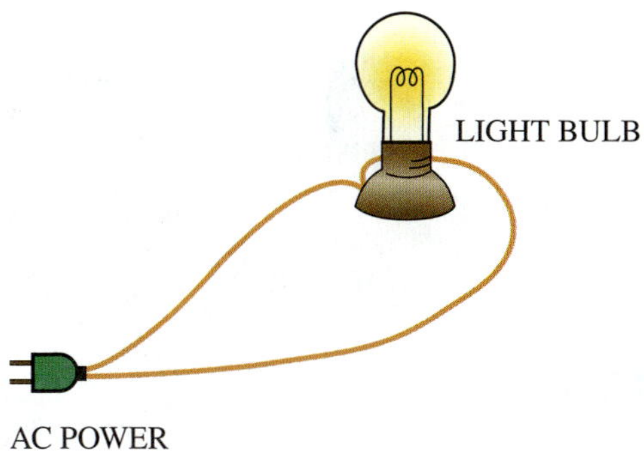

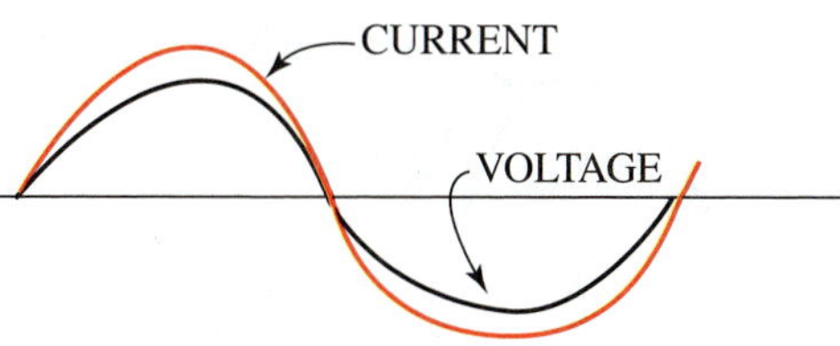

Figure 34-22 The AC voltage and current flow through resistance loads, like lightbulbs, at the same time.

to be synchronized, or "in phase." The maximum voltage occurs at the same time (or same phase angle) as the maximum amperage.

34.9 INDUCTIVE CIRCUITS

Motors, relays, transformers, and some other AC loads are constructed using coils of wire. These coils produce magnetism and are called inductive loads. The voltage and current in circuits containing inductive loads are phase shifted, or out of phase, sometimes as much as 90°.

In the inductive circuit, the current lags, or is out of phase with the voltage, as shown in Figure 34-23. In this circuit, the current waveform peaks 90° after the voltage waveform. Because of this current lag, the measured power in an inductive circuit will always be less than the calculated power ($E \times I$). This is because measured power is an instantaneous reading, and at any particular time, one or the other or both voltage and current readings are not at their peak.

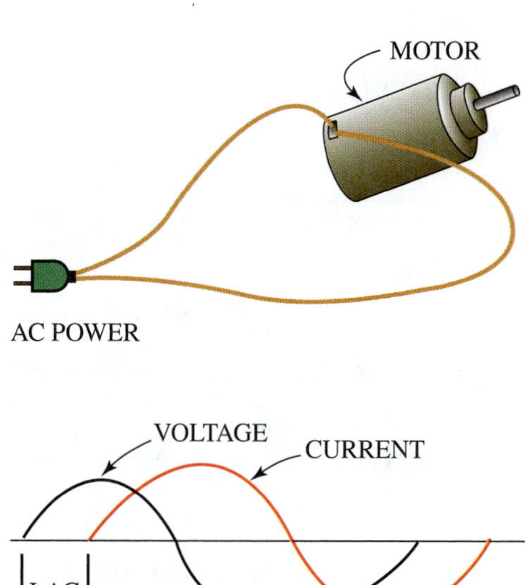

Figure 34-23 The AC voltage leads the current through inductive loads, like motors.

The term *power factor* is used to indicate this difference.

$$\text{power factor} = (\text{true power})/(\text{apparent power})$$

or

$$\text{power factor} = (\text{wattmeter reading})/(E \times I)$$

34.10 CAPACITORS

A capacitor is an electrical device that is used to change the phase relationship between the current and the voltage. This effect can be used to increase the starting power (torque) of an electric motor. A simple capacitor consists of a layer of insulation (a dielectric) placed between two plates of highly conductive metal (Figure 34-24).

Capacitors are broadly categorized as either polarized or nonpolarized and variable (Figure 34-25). The curved line in the symbol will represent the proper lead for some

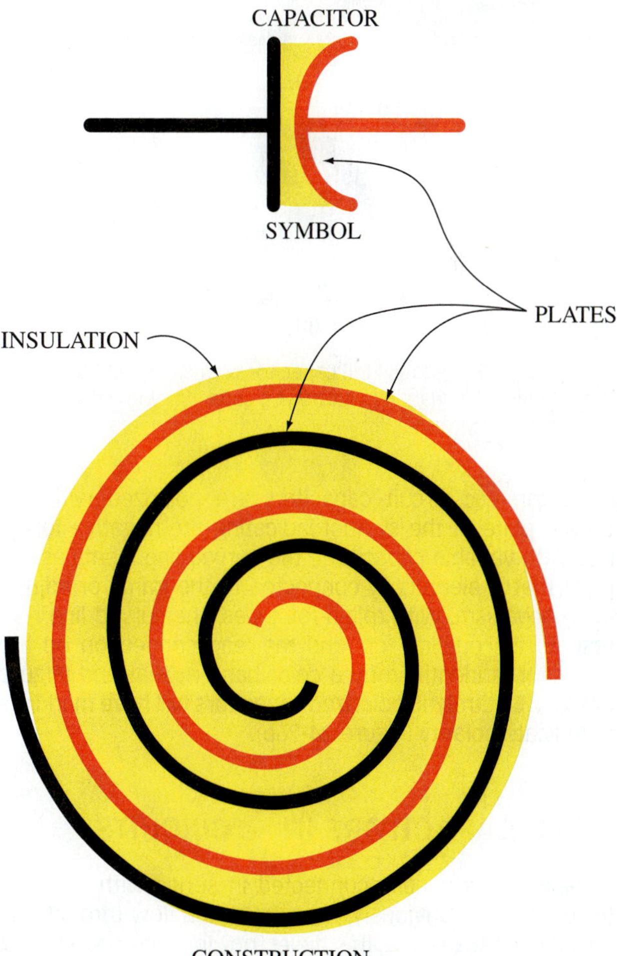

Figure 34-24 Capacitors are made up of two plates, thin metal foil, separated by an insulation paper and rolled up together.

Figure 34-25 Capacitor symbols; (a) polarized; (b) nonpolarized; (c) variable.

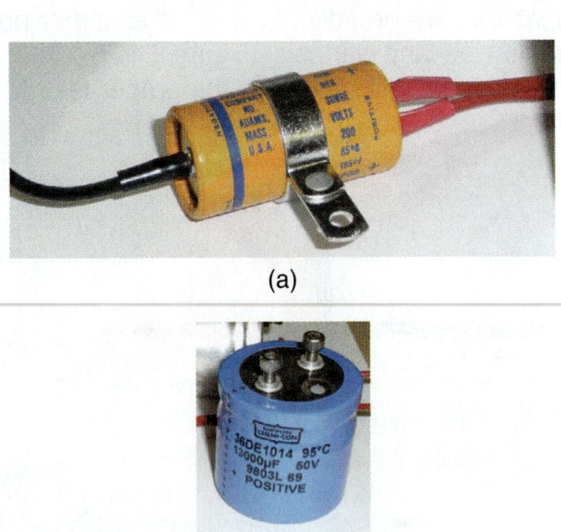

Figure 34-26 (a) Rolled foil capacitor with color band to indicate negative lead; (b) electrolytic capacitor with labeled positive polarity terminal.

capacitors. DC circuit capacitors are polarized with the curved plate of the symbol indicating the negative terminal. With variable capacitors, the curved line identifies the plate that is electrically connected to the frame or adjusting mechanism. With rolled-foil types, the curved line represents the outside foil, and the lead connection on the capacitor is identified by a color band at that end (Figure 34-26a). Electrolytic polarized capacitors will have markings to indicate polarity (Figure 34-26b).

34.11 CAPACITORS IN AC CIRCUITS

A capacitor is usually connected in series with the load (Figure 34-27). Obviously, no current can flow through the capacitor because of the dielectric. Initially, the current does flow through the series circuit (Figure 34-28). When the switch is closed, the supply voltage is applied across the capacitor. At that instant, the electrons flow rapidly from the source to the first plate of the capacitor and from the second plate of the capacitor to the source, causing a current to flow through the load. The capacitor quickly reaches peak current. It is described as charging during this period.

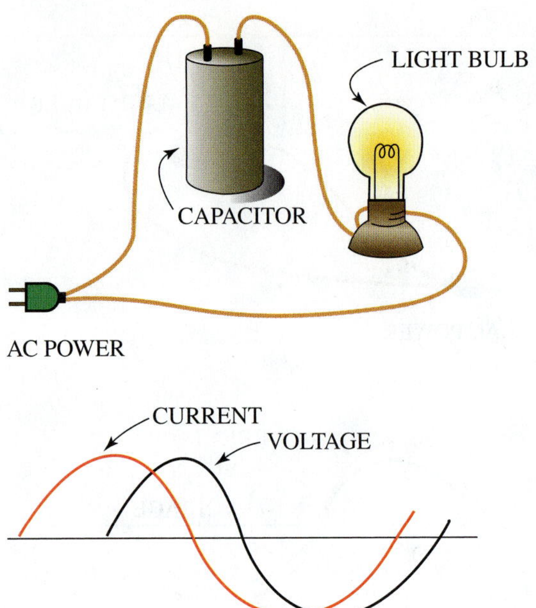

Figure 34-27 A capacitor causes a phase shift, so the current leads the voltage through a capacitance circuit.

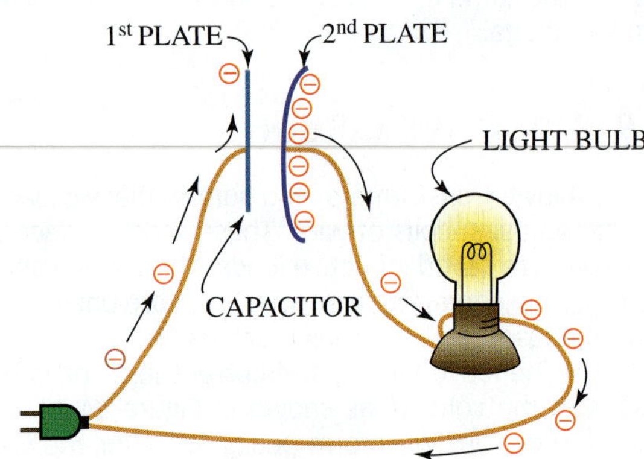

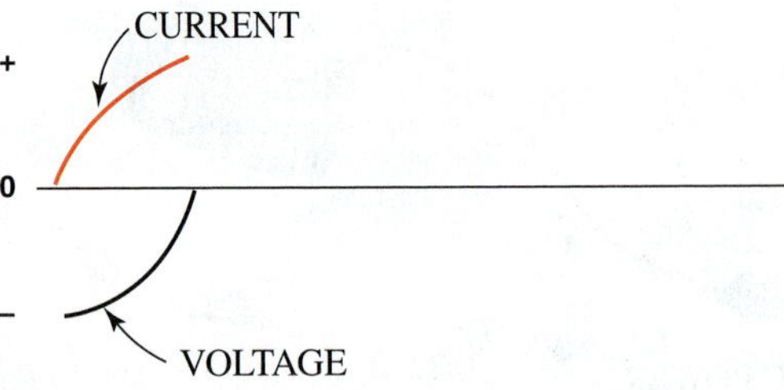

Figure 34-28 The capacitor charges when source voltage is applied. Electrons flow rapidly from the source to the first plate of the capacitor and from the second plate of the capacitor to the source, causing a current to flow through the load.

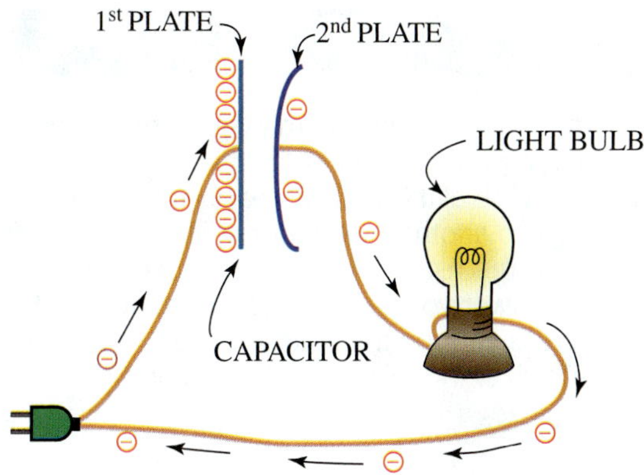

Figure 34-29 As the voltage on the first plate builds to its peak, most of the electrons have flowed off of the second plate.

TECH TIP

DC capacitors have positive (+) and negative (–) terminals and must be connected to the circuit with the proper polarity. When the switch to a capacitor installed in a DC circuit is closed, current initially flows, the capacitor charges, and then current flow stops. Since there is no change in the direction of current, the capacitor remains charged but now would appear to the circuit as an open switch.

Following the initial rapid flow of electrons, the rate of current flow reduces (Figure 34-29). Figure 34-30 shows the resulting downward movement of the current waveform. As electrons leave one plate and accumulate on the other, a potential difference (voltage) begins to develop across the capacitor. This difference is created by current flow and therefore lags the current. In this circuit, which contains a resistive load, current leads the voltage by 45° (Figure 34-30).

As the supply voltage waveform crosses the baseline, the polarity across the capacitor changes and electrons leave the first plate. The capacitor discharges and the current flows through the load in the opposite direction (Figure 34-31). The capacitor continues in this manner as long as the source voltage is being applied.

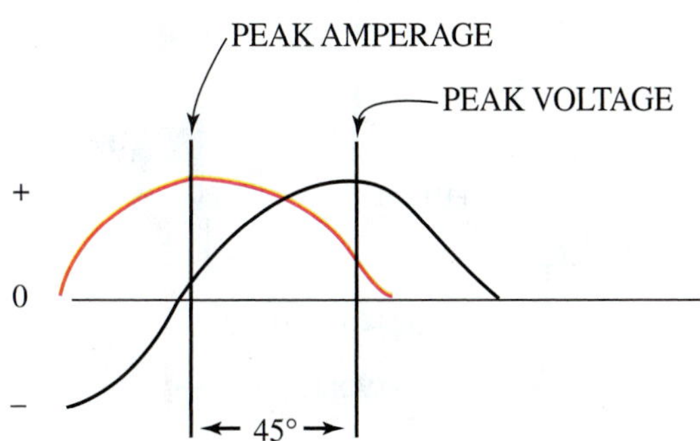

Figure 34-30 Current leads the voltage by 45°.

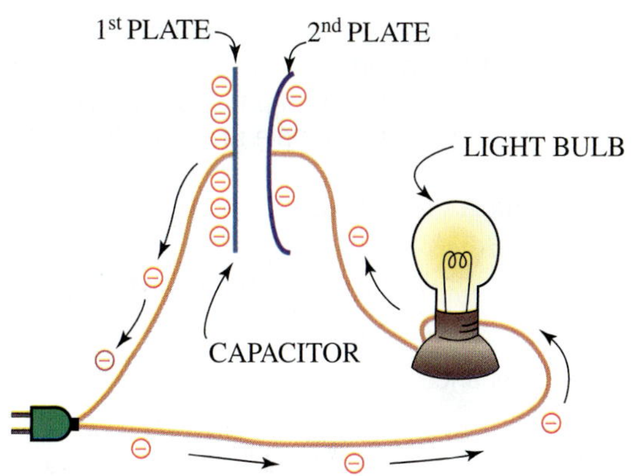

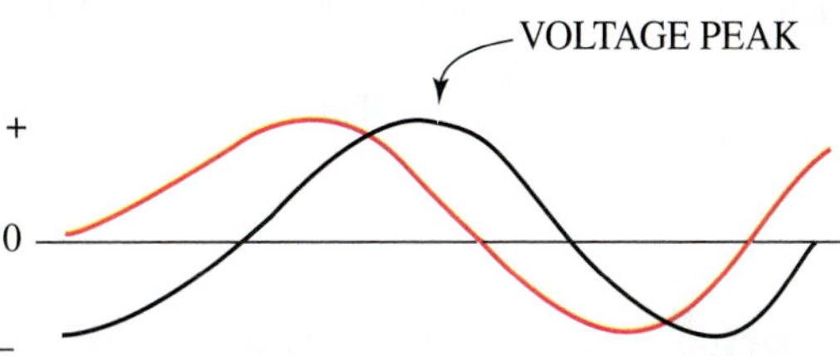

Figure 34-31 As the voltage peaks positive and starts back down toward zero, the electrons that were piled up on the first plate quickly flow off, forming a rush of current out of the capacitor.

TECH TIP

A capacitor provides a phase shift to help an induction electric motor to start. A start capacitor added to an electric motor gives it greater phase shift for greater starting torque. Too much phase shift from an excessively large start capacitor can actually reduce some of the starting torque. Engineers have done extensive studies to determine the most appropriate size of start capacitor to optimize the starting torque. Refer to the manufacturer's literature anytime you are applying a start capacitor to an induction motor.

Capacitive Reactance

Capacitive reactance is a term used to describe a capacitor's opposition to alternating current and is somewhat like resistance that can be used to control the current in a circuit. There are two main factors that affect capacitive reactance: the frequency of the current (and voltage) and the amount of capacitance. Doubling the frequency or capacitance will reduce the reactance by half. Doubling the reactance will reduce the frequency or capacitance by half.

34.12 IMPEDANCE

Impedance is the opposition to the current flow in an AC circuit. Impedance is to an AC circuit what resistance is to a DC circuit; however, multiple impedances in a circuit cannot be added like resistances are added in direct current because the currents through inductive, capacitive, and resistance loads in an AC circuit are out of phase with each other.

The following is an example of calculating impedance in an inductive circuit.

The formula that is used is:

$$Z = E/I$$

where

Z = impedance

EXAMPLE 34-1: SOLVING FOR IMPEDANCE USING VOLTS AND AMPS

What is the impedance of a single-phase inductive circuit having a voltage of 240 V and a current flow of 10 A?

$$Z = \frac{240\text{ V}}{10\text{ A}} = 24\text{ Ohms}$$

34.13 POWER DISTRIBUTION

Almost all of the electrical power used by consumers today is alternating current. It is supplied to substations at voltages as high as 120,000 V. There it is reduced to voltages between 4,800 V and 34,000 V for distribution to areas of commercial or residential users. Most power companies are now using 23,000 V or 34,999 V. Local transformers reduce the voltage levels to suit user requirements, as shown in Figure 34-32.

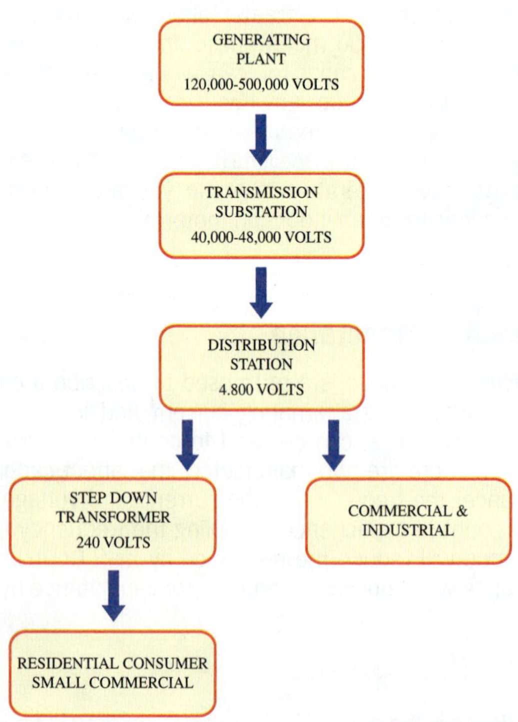

Figure 34-32 Voltage reduction in the power-distribution system.

TECH TIP

The primary reason that alternating current is used as the standard power source in commercial and residential buildings is that it is the most easily transformed from one voltage to another. Without the ability to transform power from the higher voltages used during transmission to the lower voltages used in our homes and businesses, cross-country transmission lines would have to be made with extremely large-diameter wires.

Four common low voltage systems are available to consumers:

- 240 V, single-phase, 60 Hz systems
- 240 V, three-phase, 60 Hz systems
- 208 V, three-phase, 60 Hz systems
- 480 V, three-phase, 60 Hz systems

SERVICE TIP

Although most consumer power is alternating current, DC is still used in many applications. An example is your service van. This DC can be converted to AC with a power inverter that uses an automobile's 12 V battery to produce 110 V of 60 cycle current. A power inverter is a great device to add to your service van because it allows you to run portable drill chargers and other electrical devices that normally must be plugged into a standard outlet.

34.14 ELECTRIC SERVICES

Single-phase current is used for almost all residences. Any electrical appliance that operates on 120 V power is single-phase equipment. The most common service supplied to residential and small commercial users is the 240 V, single-phase, 60 Hz system. The system uses three wires—two hot wires and one grounded neutral. A schematic diagram of this 240 V system is shown in Figure 34-33. Electric utility

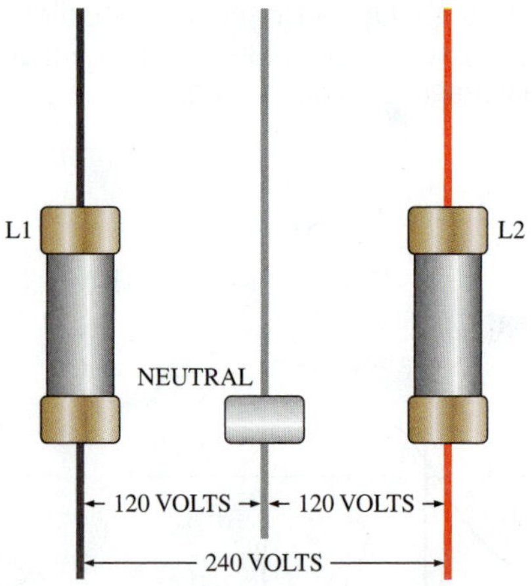

Figure 34-33 Line voltage for a three-wire 240 V single-phase power supply.

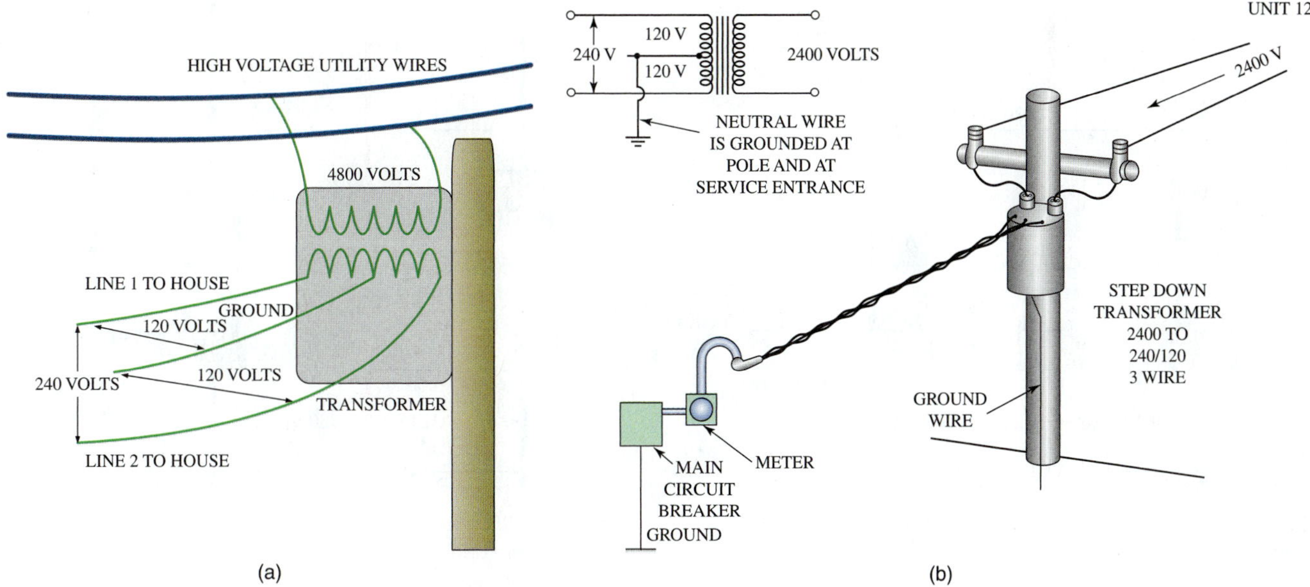

Figure 34-34 Step-down transformer for a 240 V three-wire single-phase power supply.

companies use a transformer to produce this service, as shown in Figure 34-34.

All HVACR equipment is manufactured to operate satisfactorily on voltages of plus or minus 10 percent of the rated voltage unless otherwise specified. For example, if the equipment has a voltage rating of 230 V, the equipment should be able to operate at any voltage between 207 and 253 V. HVACR equipment has a tendency to operate more satisfactorily on maximum voltage than on minimum voltage.

The electric utility attempts to maintain a voltage at the load within this plus or minus (±) 10 percent range. At peak load times, the line voltage may drop to near the permitted minimum. If you suspect any voltage problems, the line voltage at the HVACR load should be measured at these times.

34.15 240 V, THREE-PHASE, 60 Hz DELTA SYSTEMS

Three-phase systems are commonly used for sizable commercial and industrial installations. These transformers have three hot legs of power and one neutral leg, as shown in Figure 34-35. This type of power supply is obtained from a delta transformer secondary hookup, as shown in Figure 34-36.

Three-phase 240 V power is obtained by connecting to the three hot legs. Single-phase 240 V power can be obtained by connecting to any two of the hot legs. Single-phase 120 V power can be obtained by connecting to either of the adjacent hot legs and the midpoint neutral. Single-phase 208 V power can be obtained by connecting to the nonadjacent hot leg and the ground. This is called the wild leg.

34.16 208 V, THREE-PHASE, 60 Hz WYE SYSTEMS

These systems are common in schools, hospitals, and office buildings where 208 V three-phase motors and 120 V single-phase lighting and convenience circuits are required,

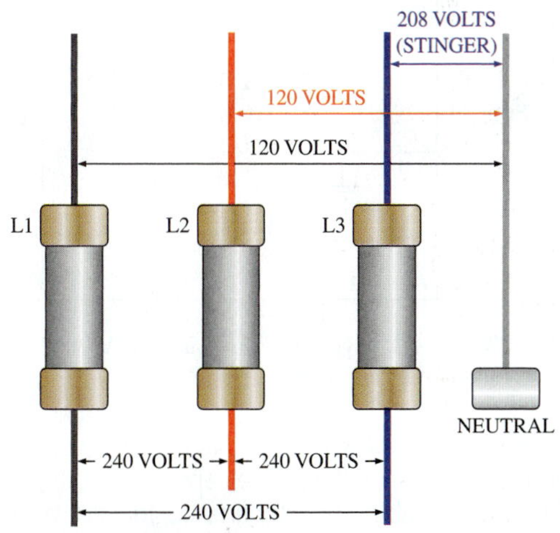

Figure 34-35 Line voltage for a four-wire 240 V three-phase power supply.

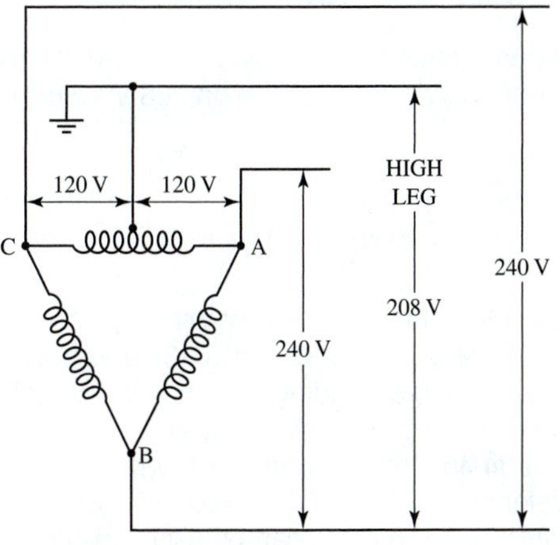

Figure 34-36 Line voltages for a four-wire 240 V three-phase system using a delta transformer secondary.

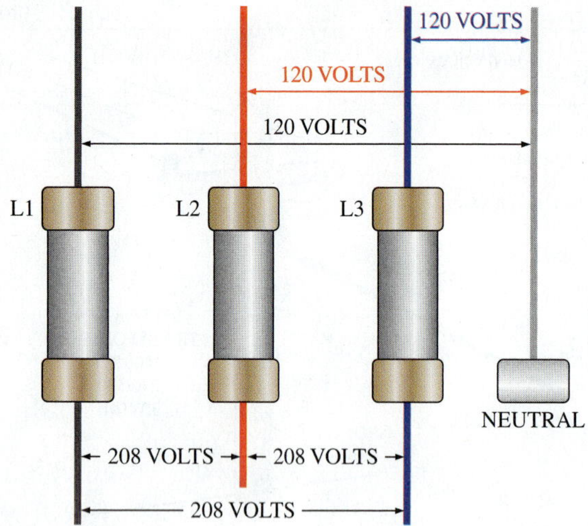

Figure 34-37 Line voltage for a four-wire 208 V three-phase power supply.

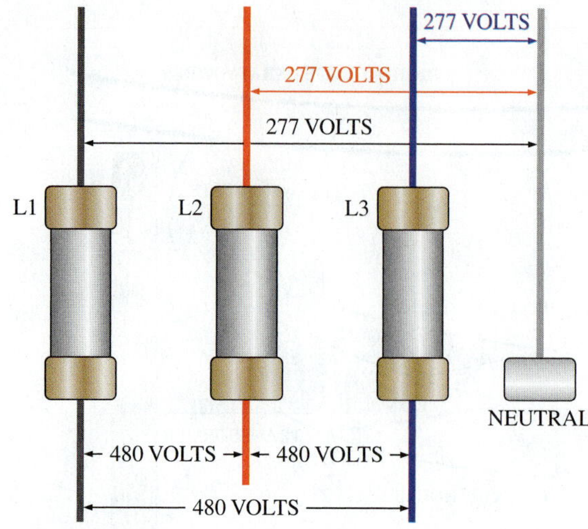

Figure 34-39 Line voltage for a four-wire 227 V/480 V three-phase power supply.

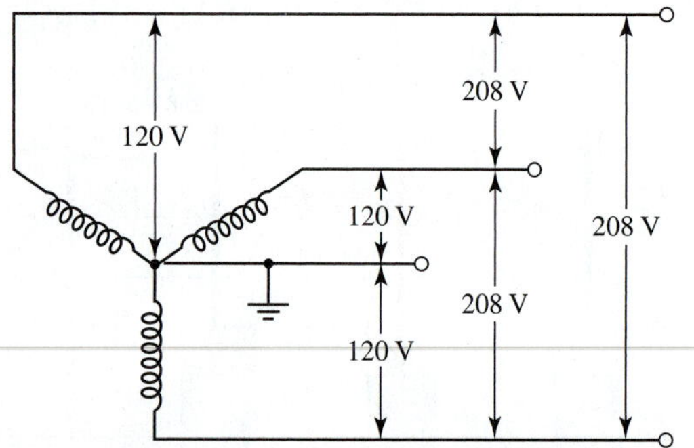

Figure 34-38 Line voltage for a four-wire 208 V three-phase power supply using a wye transformer secondary.

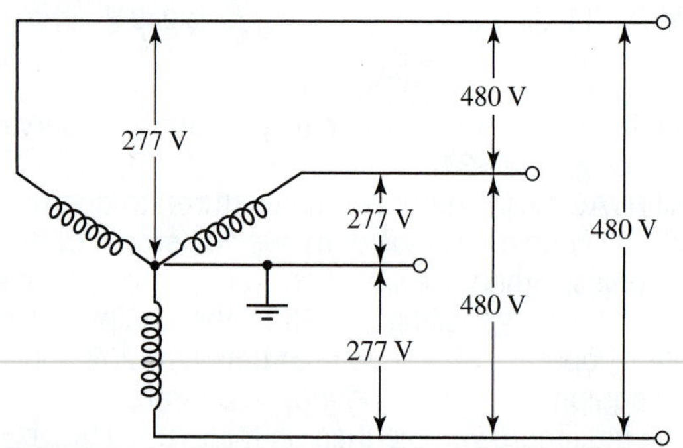

Figure 34-40 Line voltage for a four-wire 227 V/480 V three-phase power supply, using a wye transformer secondary.

as shown in Figure 34-37. From this type of system, 208 V three-phase, 208 V single-phase, and 120 V single-phase services are available. The schematic for the wye transformer secondary hookup is shown in Figure 34-38.

SERVICE TIP

It is recommended that all motors used be rated for 208 V (not 230 V) for operation on 208 V systems.

34.17 480 V SYSTEMS

The 208 V/120 V three-phase four-wire wye-connected system, shown in Figures 34-37 and 34-38, has been generally superseded in large buildings by the 480 V/277 V three-phase four-wire wye-connected system, shown in Figures 34-39 and 34-40. This improvement was made possible by the development of 277 V fluorescent lighting. Standard 460 V three-phase motors can be used on 480 V systems. Convenience outlet circuits at 120 V are provided for by 480 V/208 V or 480 V/120 V step-down transformers. The higher supply voltage permits larger loads to be serviced by smaller wires. In addition, voltage drops are reduced.

TECH TIP

A 460 V three-phase system will draw less current than a 208 V three-phase system. For example, a 50 hp, 460 V three-phase 60 Hz motor would have a full load of 57.5 A, while the same motor at 208 V would have a full load of 122 A.

34.18 SINGLE-PHASE VERSUS THREE-PHASE POWER

Three-phase power is most often used by large HVACR equipment operators. The following are some of the reasons:

- Individual line current for a three-phase load is approximately 58 percent of a similarly sized single-phase load. Since wire is sized by current draw, reducing the line current also reduces the wire size and voltage drop.

- There is less power loss to transformers in using it.
- Three-phase motors are smaller, less expensive, more reliable, and more efficient than single phase. They do not require special capacitors to increase their starting torque.
- Some three-phase motors have an ability to correct the phase shift caused by some other inductive loads. Large commercial consumers of electric power are often faced with a surcharge that is applied to their base electrical rate when their power consumption is out of phase. Three-phase motors can reduce or eliminate this surcharge by correcting the phase angle.

Single-phase power is used for most HVACR residential and light commercial systems. The following are some of the reasons:

- Single-phase power is more readily available than three-phase power.
- Because of the popularity of single-phase power, a larger variety of equipment types and sizes are available.
- Single-phase equipment is more readily available because it is often kept in stock at supply houses.

34.19 OVERCURRENT PROTECTION

The sizing of the electrical power wiring and overcurrent protection are important for a number of reasons, with safety being the most important. If the wiring is not large enough to carry the current, it may overheat and cause a fire. It may also become hot enough to melt the insulation, allowing the wire to come in contact with part of the equipment case, which could cause an electrical shock to anyone touching the charged metal.

The fuse and/or circuit breaker sizing are important. Proper sizing will ensure that in case of an overcurrent event the wiring is protected. This is very important because the wiring often runs through areas of homes and buildings that contain wood and other combustible building materials. If this wiring were to overheat, it could start a fire. All HVACR residential and most commercial units are located inside metal cabinets. As part of the UL certification process, these cabinets are tested to guarantee that they will withstand an internal component fire. This is to confirm their safety should an overcurrent event happen inside the unit.

Many HVACR units have name plates that list the minimum circuit ampacity and the maximum overcurrent protection ratings. If the unit does not have this information, you may locate it in the National Electrical Code or calculate it using one of several similar equations.

34.20 SIZING POWER WIRING

The *minimum circuit ampacity* (MCA) is the amperage load that the conductor must be sized at as a minimum for the equipment load. MCA is calculated using the following equation:

$$MCA = (1.25 \times FLA_1) + FLA_2 + FLA_3 + FLA\ldots$$

where

MCA = minimum circuit ampacity (amperage capacity)

1.25 = a constant that increases the FLA for the motor by 25 percent

FLA_1 = the full load amps (FLA) or rated load amps (RLA) of the largest motor

$FLA_2 + FLA_3 + FLA\ldots$ = the full load amps (FLA) or rated load amps (RLA) of the all the other motors

For example, find the MCA for a rooftop unit that has the following data listed on its name plate: compressor FLA 22, two fans FLA 3.5 each.

$$MCA = (1.25 \times FLA_1) + FLA_2 + FLA_3 + FLA\ldots$$
$$MCA = (1.25 \times 22) + 3.5 + 3.5$$
$$MCA = 27.5 + 3.5 + 3.5$$
$$MCA = 34.5\ A$$

34.21 SIZING CIRCUIT PROTECTION AND DISCONNECTS

The maximum overcurrent protection (MOCP) is the maximum amperage rating for fuses or circuit breakers that are used to protect the system. If wiring in the circuit passes too much current, it will overheat and cause conductor failures or fires. Typically all equipment will have a listed maximum fuse or circuit breaker designation that should always be adhered to. Never use a larger fuse or breaker size than the one originally designated for the equipment. It is also important to note that a motor may be operating in an overloaded condition while the circuit itself is not, and the fuse will not open the circuit. That is why each motor is also protected separately from the circuit from an overload within its own operating range.

SERVICE TIP

HVACR-type fuses or circuit breakers must be used. These are often classified as "slow blow" because they will handle the momentary surge of current associated with starting a motor. If standard circuit breakers or fuses are used, nuisance trips may result.

UNIT 34—SUMMARY

Alternating current (AC) is the most common source of power used for most HVACR systems. The current flow reverses and changes direction at a regular interval referred to as frequency. The typical frequency is 60 times per second, designated as 60 hertz (Hz).

One important characteristic of AC power that needs to be understood is the relationship between the current and voltage in a circuit referred to as phase shift. In a purely resistive circuit, the current rises and falls with the voltage and is considered to be "in phase." However, the current lags the voltage in an inductive circuit, and because of this phase shift, the measured instantaneous power will be less than the calculated power. This difference is referred to as the power factor. In a capacitive circuit, the phase shift is opposite and the current leads the voltage. Another important characteristic of an AC circuit, referred to as impedance, is the opposition to the current flow.

Power-distribution systems can be single or three phase. Single-phase power is typically used for most HVACR residential and light commercial systems. Single-phase power is readily available, and there is a large variety of single-phase equipment to choose from. Three-phase power is more commonly used in commercial applications. Three-phase systems allow for reduced line current, less transformer power loss, and smaller, more efficient motors. Since either of these systems may be encountered on the job site, the HVACR technician needs to understand both single-phase and three-phase circuits.

UNIT 34—REVIEW QUESTIONS

1. What attraction and repulsion behavior do the north and south poles of a magnet exhibit?
2. Explain the left-hand rule for flux direction in conductors.
3. What is the most common source of power for most HVACR systems?
4. What is the purpose of transformers?
5. What is a solenoid valve?
6. Is there an electrical connection between the primary and the secondary windings of a transformer?
7. What could be the cause of a loud hum coming from a transformer?
8. Capacitors are broadly categorized how?
9. Describe how a DC capacitor would differ from an AC capacitor.
10. What is capacitive reactance?
11. What are the two types of transformers?
12. Explain where single-phase power and three-phase power are normally used.
13. The phase shift factors that affect the power calculation $E \times I$ relate to what circuits?
14. What is the impedance of a single-phase inductive circuit having a voltage of 120 V and a current flow of 5 A?
15. What are the four common low-voltage systems available to consumers?
16. Why is three-phase power most often used by large HVACR equipment operators?
17. Why is single-phase power used for most HVACR residential and light commercial systems?
18. What can happen if electrical power wiring is not sized properly?
19. Find the minimum circuit ampacity (MCA) for a rooftop unit that has the following data listed on its name plate: compressor FLA 26, two fans FLA 3.0 each.
20. What can happen if an oversized fuse or circuit breaker is installed and the circuit is allowed to pass too much current?

UNIT 35

Electrical Measuring and Test Instruments

OBJECTIVES

After completing this unit, you will be able to:

1. list the major types of electrical test instruments and explain how they are used.
2. explain the difference between weighted average meters, true RMS meters, and RMS meters.
3. identify the CAT rating on meters and explain its importance.
4. describe the difference between analog and digital meters.
5. demonstrate how voltmeters, ohmmeters, and ammeters are used.
6. take measurements on a de-energized circuit with a test meter.

35.1 INTRODUCTION

The terms *test instrument* and *test meter* are used interchangeably in the HVACR field to describe devices that are used to detect some aspect of an electrical system's operation. Usually they are used to determine how a system is operating or why it is not operating. This information can then be used for servicing and troubleshooting.

The types and availability of test meters have increased to the point that it can be extremely difficult to determine which meter is best for a particular application (Figure 35-1). The flood of inexpensive test meters into the market has also provided what may appear to be a cost-effective way to outfit your toolkit. However, it is important to note that many of these inexpensive test meters will not be appropriate for HVACR use. Test meters should be selected based upon the standards required for their application. Never select a meter that is not rated for the application where it will be used.

This unit provides information on some of the general standards and ratings for test meters, but this information is not intended to replace any operating instructions provided by the meter's manufacturer. Always read and follow all manufacturers' operating and safety instructions; failing to do so can damage the equipment and may result in your being injured.

Figure 35-1 Assortment of analog and digital electrical test meters.

35.2 SAFETY STANDARDS

The two principal government agencies that oversee workplace safety are OSHA (U.S. Occupational Safety and Health Administration) and NIOSH (National Institute for Occupational Safety and Health). OSHA sets and enforces the rules, while NIOSH provides helpful information on workplace safety.

TECH TIP

An eighty-eight-page electrical safety handbook, *Electrical Safety—Safety and Health for Electrical Trades Student Manual*, can be downloaded from the NIOSH Web site at www.cdc.gov/niosh/docs/2009-113/.

The nonprofit National Fire Protection Association (NFPA) has developed standard NFPA 70E, Standard for Electrical Safety in the Workplace. This was written to correlate with the National Electrical Code (NEC), which is applied to many local building codes and regulations.

35.3 CATEGORY RATINGS

Properly designed test meters should not cause a shock, fire, arcing, or explosion. They are also designed to reduce the potential damage that can result from operator error if the meter is connected improperly. Even so, this protection does have its limits! The wrong meter used for the wrong application is extremely dangerous! That is why there are four categories designated for low-voltage test equipment. Be aware that the term *low voltage* is deceiving, because for test meters this is considered to be 1,000 volts or less. Many HVACR technicians would consider any circuit of 480 volts or above to be high voltage.

There are a number of considerations taken into account for meter ratings. These include the safe distance to the power source the technician is working and transient voltage spikes. Motors, capacitors, and power-conversion equipment such as variable speed drives can be prime generators of spikes. Transients can occur on low-voltage power circuits and can reach peak values in the many thousands of volts.

Any time you are working with electricity, there is a possibility of an overvoltage situation occurring. We use surge protectors on our computers to protect them from overvoltage. A momentary surge or spike of voltage can be several times the normal circuit voltage. These surges, called transient voltages, can be caused by electrical storms, even miles away, or they can be caused when a major load is added or dropped from the system. Examples of changes in the load could be an auto accident where a utility pole was damaged, or a large piece of equipment starting or stopping.

Category numbers range from CAT I to CAT IV. A higher CAT number is used for higher voltages and energy transients. A test meter designed to a CAT III standard can resist much higher energy transients than one designed to CAT II standards (Figure 35-2). There are also voltage ratings within categories, such as a CAT III 1,000 V test meter has superior protection when compared to a CAT III 600 V test meter.

- **CAT I** is required for use on low-energy equipment such as protected electronic circuits and on high-voltage/low-energy sources from a high-winding resistance transformer. A typical example would be a photocopier machine.
- **CAT II** is required for single-phase receptacle connected loads such as appliances, portable tools, and other household-type loads. The outlet should be located at least 30 feet from a CAT III source.
- **CAT III** is required for distribution circuits such as three-phase bus and feeder circuits, load centers, and distribution panels. These test meters are also used for permanently installed loads such as three-phase motors, large commercial lighting systems, and heavy appliance outlets with short connections to the service entrance.

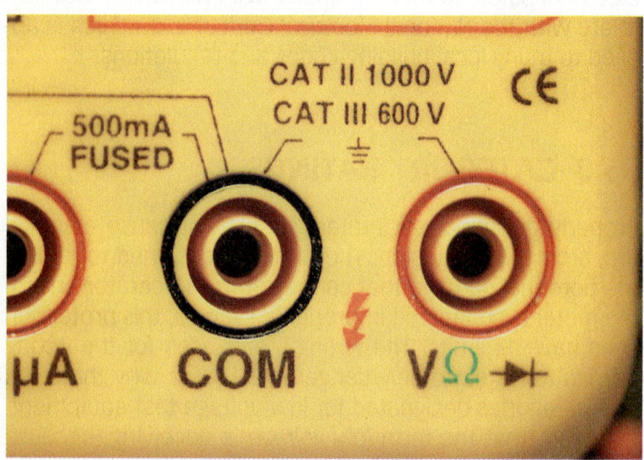

Figure 35-2 Category ratings of two voltage levels: CAT II 1000 V and CAT III 600 V.

- **CAT IV** is required for outside and service entrance use from the pole to the meter and the meter to the service panel. These test meters are also used for outside overhead and underground cable runs because they may be affected by lightning strikes.

Independent Testing Laboratories

A manufacturer may choose to self-certify a meter it produces without any independent verification. These meters may include a statement such as "designed to meet specification." Be aware, however, that only test meters that are stamped with a symbol from a recognizable independent testing laboratory or approval agency can be considered to meet CAT requirements. Typical recognized agencies include the Underwriters Laboratories (UL), CSA, and TUV (Figure 35-3).

Meter Selection and Test Leads

Always choose a test meter rated for the highest category (CAT) you could be working in. A higher-category meter can be used for any lower-category applications. The voltage rating for the meter must also be considered. As an example, a CAT III 600 V meter has superior transient protection compared to a CAT II 1,000 V meter even though the voltage ratings may seem to indicate otherwise.

The test leads for the meter are just as important as the meter ratings. They should be certified to a category and voltage the same or higher than the meter (Figure 35-4). Alligator clips are handy in freeing up your hands from holding the test leads, which allows you more freedom to set the meter for the proper reading (Figure 35-5).

Figure 35-3 Testing laboratory symbol located on test meter.

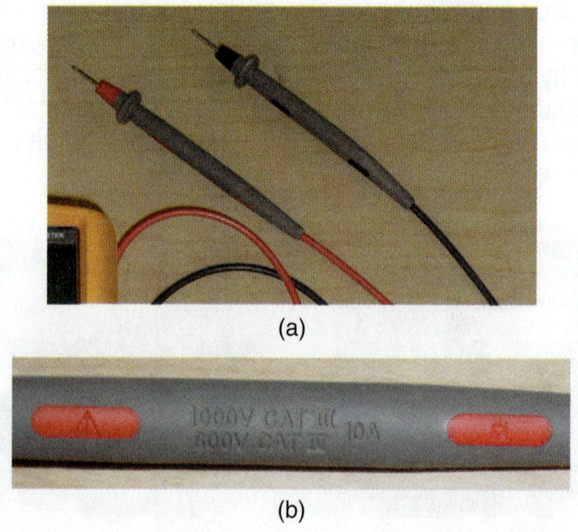

(a)

(b)

Figure 35-4 (a) Meter test leads; (b) CAT rating listed on the test lead.

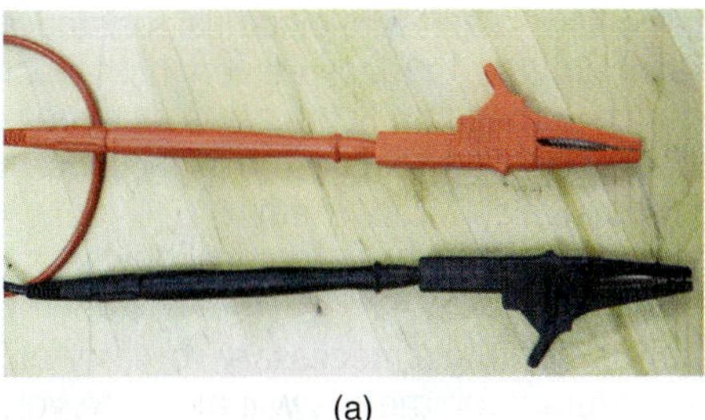

(a)

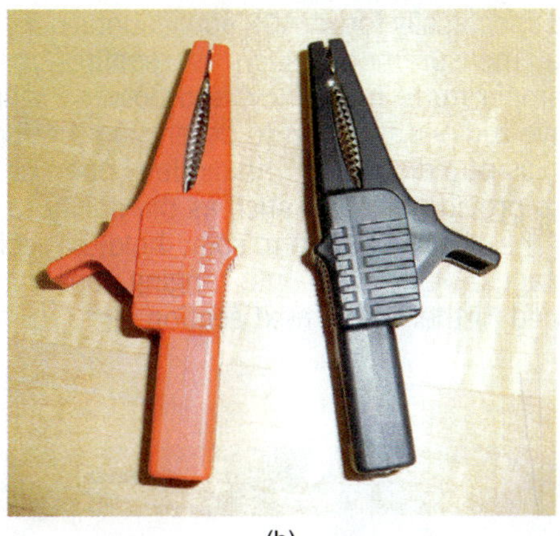

(b)

Figure 35-5 (a) Alligator clips with test lead connected; (b) alligator clips disconnected from test leads.

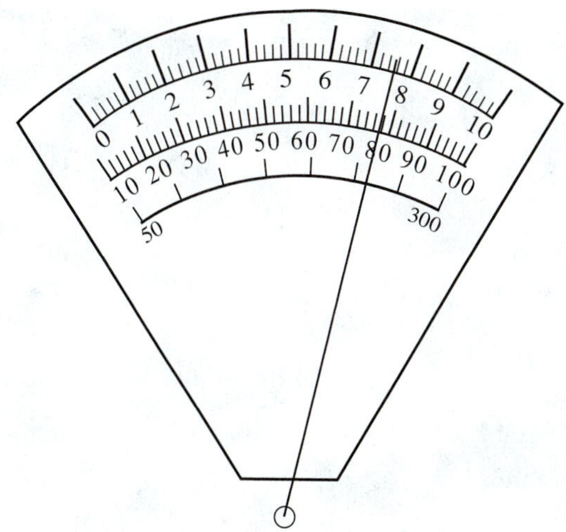

Figure 35-6 An analog multimeter has several scales, so you must know which range you are measuring and read the appropriate scale.

Figure 35-7 Analog electrical test meter.

SAFETY TIP

Older test leads that you have used for many years may not be made with the insulation thickness needed for some of today's electrical systems. Try not to pinch leads in a panel door or kink them excessively in sharp bends. Test leads that are cracked, pinched, or excessively dirty should be replaced with new leads matching the rating of the tester.

35.4 METER TYPES: ANALOG AND DIGITAL

Digital meters are commonly used in the field today and have all but replaced analog-style meters. The following comparisons between them will highlight the advantages and disadvantages of both types.

Analog Meters

Analog meters are seldom used today because digital meters are superior in almost every way. Analog meters have a mechanical needle that swings across the dial to a point over a scale (Figure 35-6). The meter reading is taken by looking directly at the face of the meter dial and comparing the needle location to the appropriate scale. Some technicians prefer this type of display and consider it superior to a digital meter (Figure 35-7).

Digital meters display discrete number values, while analog meters provide a relative reading. An example is a digital watch compared to a watch with minute and second hands. Some people prefer the moving hands. Most cars use a speedometer needle for displaying relative speed rather than a digital number. The type of display then just becomes a matter of preference for the user. Some digital meters offer both types of display. They incorporate an analog sliding scale that appears across the digital meter face so that it is possible for the technician to have both the digital number and an analog reference (Figure 35-8).

Other than the needle display, analog meters offer no other true advantages. They are almost never auto-ranging. If too high a voltage is being tested for the scale setting, then the meter can be damaged. If you prefer to use an analog meter, there are three important characteristics that need to be considered:

- The most accurate reading is at the midpoint of the scale. So whenever the operator has a choice of scales, the one selected should place the pointer in the most favorable (central) position.
- Analog meters periodically need to recalibrated. Most ohmmeters include some type of adjustment and instructions for calibration.

Figure 35-8 A digital meter displaying 7.89 megohms both as a numeric value and on a sliding scale.

- The small coil of wire that forms part of the meter movement is sensitive to excessive current. The meter may be made completely inoperable if subjected to excessive current. In using a meter with multiple scales to choose from, always use the higher scale first and move down to the scale required.

Digital Meters

Digital meters offer a number of advantages.

- They are direct reading, so there is no need to interpret the scale.
- Digital meters can be obtained that will give accurate readings to three decimal places.
- They have no moving parts and are less likely to fail or get out of calibration than analog meters, and they are more rugged.
- They often have automatic scaling features.

SERVICE TIP

Some digital meter displays are susceptible to water or humidity damage. It is important that these meters be kept as dry as possible so that they do not become damaged.

35.5 MULTIFUNCTION METERS (MULTIMETERS)

Having separate single-use meters was the only choice for years, until the introduction of the multifunction meter (multimeter). The different types of single-use meters would include the following:

- A voltmeter to measure electrical potential
- An ammeter to measure electrical current
- An ohmmeter to measure electrical resistance
- A megohm meter to measure very high electrical resistance (a megohm is the same as a mega ohm)
- A capacitor checker to measure capacitance, measured in microfarads (mfd)
- A wattmeter to measure electrical power

Today it is no longer necessary to carry all of these single-use meters. Multimeters will measure voltage, current, and resistance all with one instrument. Multimeters designed specifically for HVACR applications often provide additional measurements such as readings for capacitance, temperature, frequency, and power (Figure 35-9). Some have a diode function to determine the forward or reverse bias of a diode or if the diode is open or shorted. Many also provide features such as the ability to hold the value on the display screen (display hold) or record min/max averages. Many have auto-ranging capability, which is explained further in the next section.

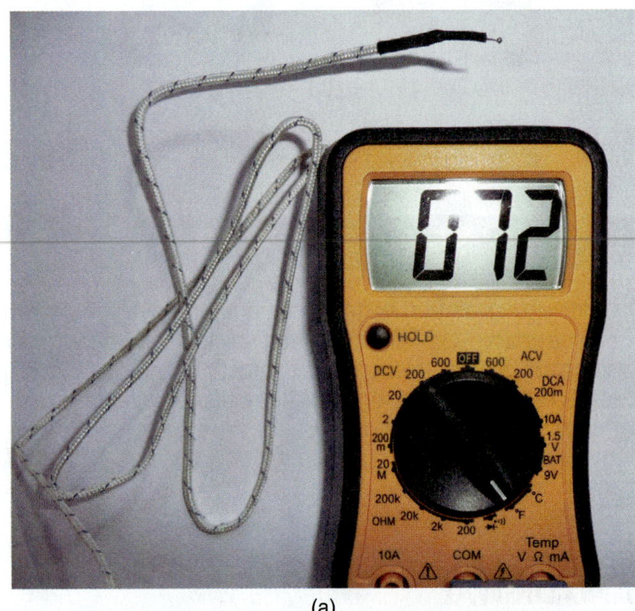

(a)

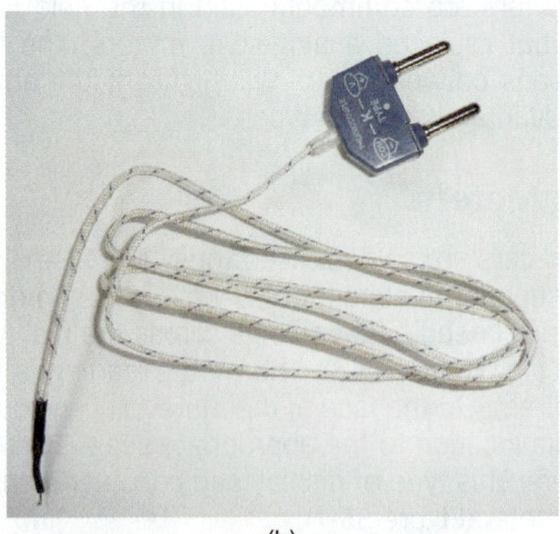

(b)

Figure 35-9 (a) Measuring room temperature at 72°F with a multimeter and a K-type thermocouple; one more turn of the dial would allow for testing diodes; (b) K-type thermocouple.

TECH TIP

Digital meters are expensive and can be damaged if they receive rough treatment. To ensure that your instruments last as long as possible, they should be cleaned after each use and stored in their own individual case or pouch. It is also a good idea to carry your instruments in a separate bag away from your hand tools.

35.6 METER RANGE, RESOLUTION, AND ACCURACY

Range, resolution, and *accuracy* are three terms used to differentiate all meters. Understanding test instrument terminology will help you make the best selection of the instrument for your specific job requirements.

Range The range tells you how low and how high a reading the meter will provide for each function the meter can perform. Many meters today have an auto-ranging function (Figure 35-10). This means the meter sets the range to the best resolution for the input signal. Normally there is also a manual range mode that allows you to set the range yourself.

- Voltmeters should be able to read voltages from a fraction of a volt up to 600 V AC and DC.
- Ohmmeters should be able to read ohms from a fraction of an ohm up to 20,000 (20K) ohms.
- Mega ohm (megohm) meters should be able to read up to 40,000,000 (40M) ohms.
- Amp meters (ammeters) should be able to read AC amperage from 1 A up to 200 A and DC amps from 200 mA to 4,000 mA and from 0 to 20 μA (microamps).
- Capacitance meters should be able to check from 1 μF (microfarad) up to 200 μF.

Resolution The resolution tells you what units of measure are used for each function. On a digital meter specification, it may be expressed as the number of digits or decimal points. This can range from three to four digits and one to three decimal places.

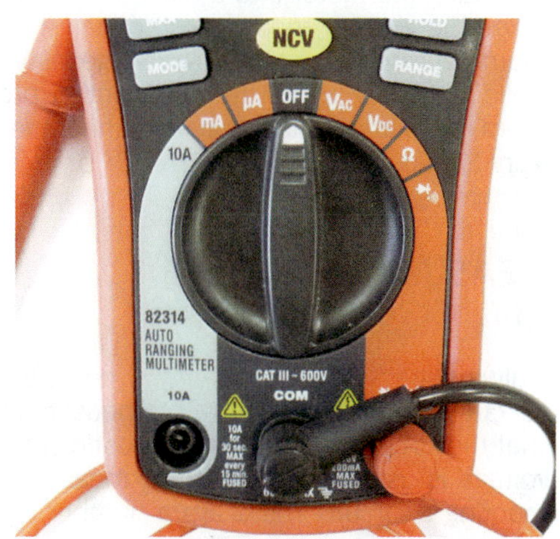

Figure 35-10 Auto-ranging multimeter.

Accuracy The accuracy is how close to the actual value the meter is going to read. These values are usually given as a percentage ± (plus or minus) of the reading. Meters may vary from ±0.5 percent up to ±3 percent, depending on the function and scale.

35.7 TESTING A DE-ENERGIZED CIRCUIT

If you are testing to verify that there is no voltage present in a de-energized circuit before beginning work, there are three different test instruments that may be used. These are a noncontact proximity tester, an electrical tester, and a multimeter.

Testing electrical circuits is an important skill that each technician needs to develop. Although practice and experience are significant, a high degree of success is obtainable by following a proven procedure, such as the following:

1. Know the unit electrically. This means understanding the proper function of each control and the sequence of the control operation. It is as important to know what a meter does not measure as well as what it does.
2. Be able to read schematic wiring diagrams and have them available.
3. Be able to use the proper electrical test instruments. Know the instrument. Read instructions carefully before using.

CODE

NFPA Standard 70 E and NEC Article 110.16 define where and when personal protection equipment must be used. This equipment includes items such as eye protection, insulated hand tools, insulated gloves, and fire-resistant clothing.

Testing Procedures

Always try to work on de-energized circuits whenever possible, and follow the proper lockout/tagout procedures. Wear protective gear and stand on an insulated mat. Test the circuit in the following manner.

1. Make sure the circuit is off and properly locked out and tagged out.
2. Test the meter on a known voltage source.
3. Test the circuit that should be de-energized (off).
4. Test the meter again on a known source to ensure that you didn't damage your meter or blow a fuse and get a "0" volt reading accidently in step 3.

Try to rest or hang the meter whenever possible rather than holding it in your hands. Alligator-clip test leads also allow you to keep your hands free by connecting one at a time and leaving them in place. Connect the ground test lead first and then the hot test lead last. Reverse this procedure to do the opposite for removal. Remove the hot test lead first and the ground test lead last. When connecting test leads, try to keep one hand in your pocket to lessen the chance of a closed circuit across your chest and through your heart.

Figure 35-11 Noncontact voltage detector lights up when voltage is present.

Low-Voltage Proximity or Noncontact Voltage Tester

This type of tester does not have test leads. The tester does not touch the conductor but is placed close (within proximity) to it. The tester will light up if voltage is present (Figure 35-11). These are good for a first test but should always be followed up with a direct-contact meter. This is because proximity tester readings can be inaccurate for a variety of reasons, such as when used inside a metal enclosure. Also, proximity readings will vary by distance and the strength of the voltage field and are therefore subject to some unreliability.

Electrical Testers

There are two general types of voltage testers: solenoid based and electronic. Both these types have test leads that must come in direct contact with the conductors. Solenoid-based testers that vibrate and light up are seldom used anymore (Figure 35-12). This is because there have been a number of safety issues involved with these, including that most are not fused. Another drawback is the sensitivity response of these testers. The circuit measured must have enough energy to move a spring-loaded slug. Therefore, low-voltage control circuits will not normally register on the indicator.

Figure 35-12 Solenoid-based voltage testers such as the one shown are seldom used today due to safety concerns and accuracy at low voltages.

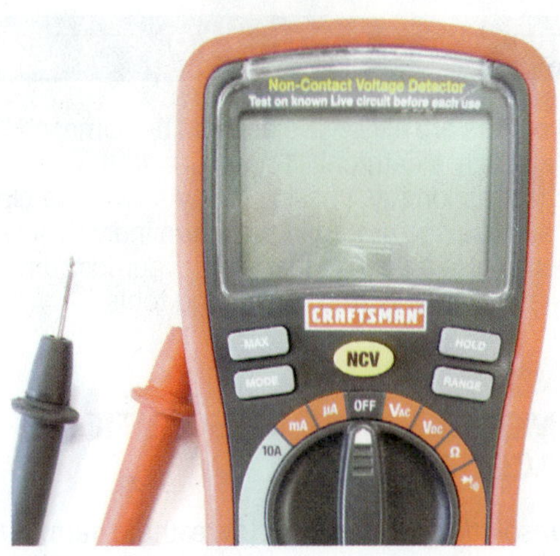

Figure 35-13 Multimeter with built-in noncontact voltage detector.

A better and more popular alternative is the electronic tester. These are fused for transient protection and are much more accurate. They have much higher input impedance so there is less input current required and the instrument remains cooler. Since they work with lower voltages, you are able to use the electronic test meter on a wider range of equipment. They are even designed to vibrate and light up to mimic the solenoid-based type.

Digital Multimeter

Multimeters can be considered the best instrument to test for voltage, but there are a few drawbacks as compared to the more simple voltage testers described previously. One drawback is operator error in turning the multimeter dial to the wrong function, such as amps instead of volts. This cannot happen on a proximity or electronic voltage tester. Another drawback for an older-type meter that is not auto-ranging is putting the meter into a range that is too high, which makes the voltage appear smaller than it really is. Some multimeters are now built with a voltage testing function that allows you to start with a noncontact proximity test and then move to a contact test, with the same instrument (Figure 35-13).

35.8 MEASURING VOLTAGE ON A LIVE CIRCUIT

It is not always possible to test only a completely de-energized circuit. Troubleshooting methods may often require that the circuit tested be "live." Testing for proper supply voltage is usually the first thing measured when troubleshooting a circuit. There are a number of safety considerations for testing a live circuit, such as location and setup.

CODE

OSHA and NFPA 70E, Standard for Electrical Safety in the Workplace, direct workers to de-energize all live parts to which an employee may be exposed, unless live conditions are required for troubleshooting.

Location and Setup

It is never safe to work on high-energy circuits without someone else present. Make sure the test meter is rated high enough for the reading you will be taking. Assess the location before you begin, and make sure the lighting is adequate. Find a location where you will place your meter so that you do not have to hold it. Magnetic hangers allow you to position the meter so that it can be clearly read without needing to be held. Determine if the leads will reach into the electrical cabinet without a need for extenders. Probe extenders keep your hands farther away from the conductors. Test probes with a minimum amount of exposed metal at the probe tips, such as .12 in (4 mm) metal-tip probes, will reduce the risk of accidently shorting between two phases (Figure 35-14).

Attaching the Test Leads

Always use safe testing procedures as outlined in section 35.7. Whenever possible, attach the test leads to the conductors with the circuit de-energized. This can be done with alligator-clip-type test leads. Always connect the neutral test lead first and then the hot test lead last. The circuit may then be energized for the reading to be taken. Once the reading has been obtained, the circuit may once again be de-energized and the test leads removed by disconnecting the hot lead first and the grounded lead last.

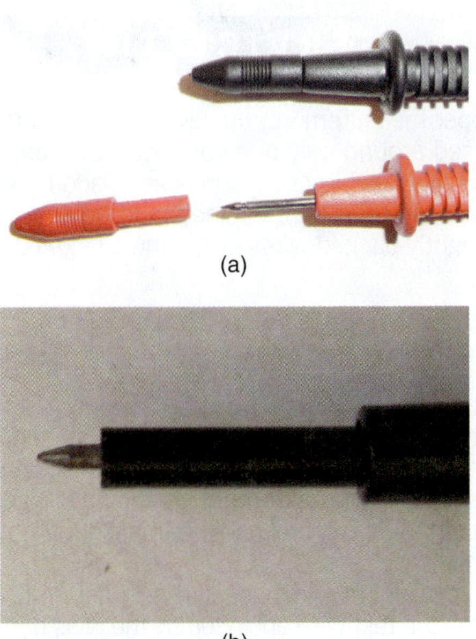

Figure 35-14 (a) Long test lead tip with cover; (b) short test lead tip.

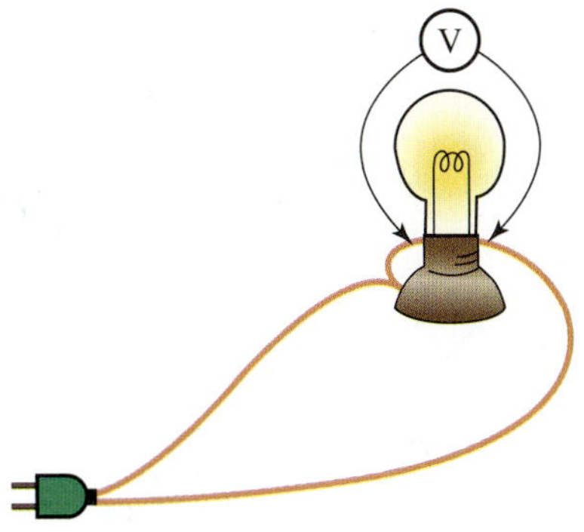

Figure 35-15 A voltmeter is connected parallel to the load.

If you need to probe the circuit, attach the black test lead with an alligator clip and use the red test lead to move from one conductor to the next. Keep one hand in your pocket and out of the panel so as not to offer a closed circuit path for the flow of electricity.

How to Measure Voltage

Voltmeters are connected in parallel with the load to read the voltage drop or potential difference. In testing a circuit for proper voltage, one lead goes on each side of the load, as shown in Figure 35-15. Many test meters today are auto-ranging, which means a specific range of voltage does not need to be selected. However, when using a test meter that is not auto-ranging, always start to measure voltage using the highest range on the meter—for example, 600 V. When the approximate voltage is read, the meter range can then be reduced to the proper range for greater reading accuracy—for example, start at 600 V then move down to 200 V. Also, using the meter to test a higher voltage than the range of the meter could cause burnout or otherwise damage the meter.

Read the instruction manual for the meter you are using. Some meters have function buttons in addition to the selector dial (Figure 35-16). On a DC circuit, the polarity of

Figure 35-16 In addition to a selector switch, some test meters have function buttons.

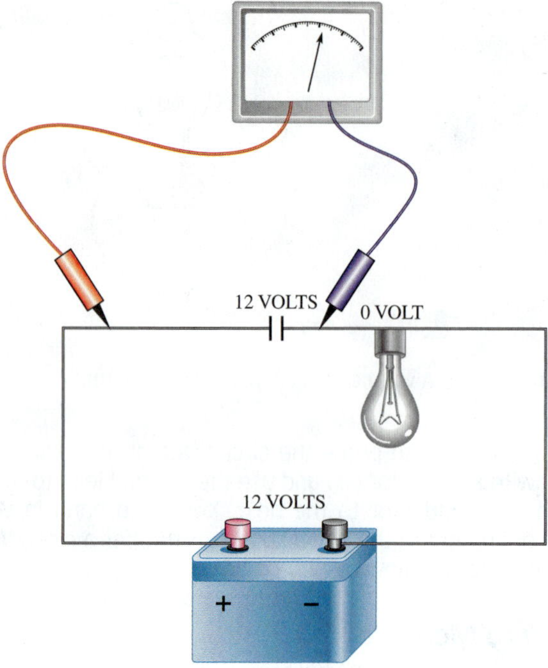

Figure 35-17 Open switch and a positive measured voltage.

the leads must be observed. Therefore, in DC circuits, verify the correct polarity of the probes that are used before connecting the meter to the circuit.

An example of a use for a voltmeter is to determine if a hidden switch is open or closed. This is very helpful in troubleshooting. If there is power in the circuit, and the leads of the voltmeter are placed on each side of the switch, a voltage reading indicates the switch is open (Figure 35-17), and a zero reading indicates the switch is closed (Figure 35-18) (if no other open switches are in the circuit).

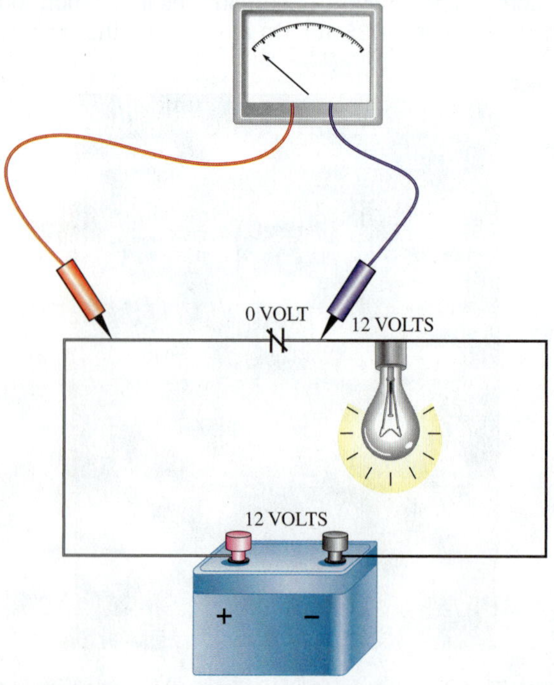

Figure 35-18 Closed switch and no measured voltage.

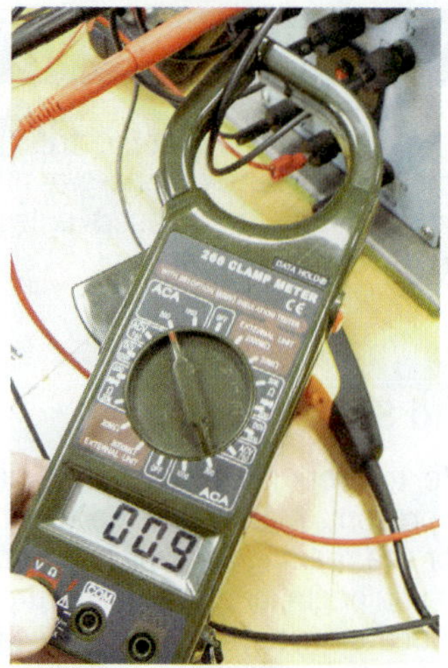

Figure 35-19 Clamp-on ammeter that is also a multimeter.

35.9 AMMETERS

Amp meters (ammeters) are used to measure current flow through a circuit. There are two types of ammeters: clamp-on and inline. An advantage of the clamp-on meter is that it is not necessary to disconnect or make contact with any wires to obtain a reading; this is very convenient. The advantage of an inline meter is that it can provide very accurate current reading because all of the system current flows through the meter. Some of today's clamp-on ammeters are dual use and can serve both as inline (multimeters) and clamp-ons (Figure 35-19). Even so, there are often times when simultaneous readings of both voltage and amperage are required, and most common meters will not be able to read both at the same time. So for troubleshooting purposes, two meters work best, and many technicians will use both a multimeter and a clamp-on ammeter.

SERVICE TIP

For proper measurement, the jaws of a clamp ammeter are placed around only one conductor at a time. Some meters have flexible current probes in addition to solid jaws that can fit in tight cabinets and be squeezed between tightly packed wires or around a large conductor.

Clamp-On Ammeters

The clamp-on ammeter is one of the most useful of the electric meters for HVACR technicians. It is used to measure current flow through a single wire by enclosing the wire within the jaws of the instrument, as shown in Figure 35-20. Some clamp meters are capable of measuring both AC and DC current.

This instrument functions like a transformer. The primary coil is the test wire encircled by the jaws of the instrument. The secondary is a coil of wire within the instrument that is connected to the current-indicating mechanism. The current in the primary wire induces a flow of current in the

Figure 35-20 The most accurate amp measurement is when the wire passes through the center of the meter jaws.

secondary winding, measuring the current flow. The greater the current flow through the test wire, the greater the induced current and the greater the deflection of the needle reading on the scale.

Inline Ammeter

Inline instruments that only measure amperage are not commonly used. Instead, multimeters are the most common electrical measuring instrument for HVACR work because they are able to do so much more. Some multimeters have a removable clamp-on available as an accessory that may be used with the meter (Figure 35-21).

When using a multimeter as an inline meter, the proper location for its connections in a circuit is shown in Figure 35-22. Note that it is connected in series with the circuit being tested. In DC circuits, verify the correct polarity of the ammeter used before energizing the circuit. Never connect an ammeter across a load. It will be destroyed by line current, as there is no load to limit it!

Figure 35-21 Removable clamp-on accessory that can be connected to the multimeter for taking amperage readings.

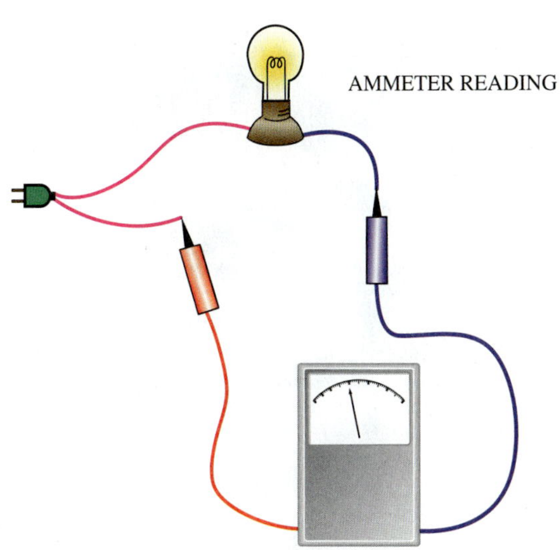

Figure 35-22 An ammeter is connected in series with the load.

35.10 TEST METER FUSES

A somewhat common mistake for someone inexperienced in operating a test meter is to take a current measurement in line with the load and then next try to take a voltage measurement with the test leads remaining in that same position rather than placing them in parallel with the load. This effectively places a short across the voltage source. This could potentially destroy the meter. To prevent this, a fuse is located in series with the meter's test leads.

Not all fuses are the same. Even fuses of the same amperage and voltage ratings can be different from each other. Specially designed "high-energy" fuses have different circuit-interrupt characteristics. Some are filled with sand so at high temperatures the energy will melt the sand, turning it to glass to coat the element for increased protection against a meter explosion and meltdown. Using the wrong fuse or trying to jumper the fuse could result in an accident where a meter explosion could lead to serious burns to your arms, face, and clothing. Always refer to the test meter's manual, or check with the test meter manufacturer to ensure you have the correct fuse.

SAFETY TIP

A meter with a blown fuse should be put aside and not used again until there is a proper replacement fuse available. Never put in a fuse that is not correctly rated for the meter. If a wrong fuse is used, the metal element inside heats up very quickly, which can lead to a small explosion, resulting in the fuse enclosure bursting open and causing serious injury. *Never place a jumper wire around the fuse connections!*

35.11 TRUE RMS METERS

The term *RMS* is an abbreviation for root-mean square, a mathematical formula for calculating the electrical power passing through an alternating current circuit. True RMS

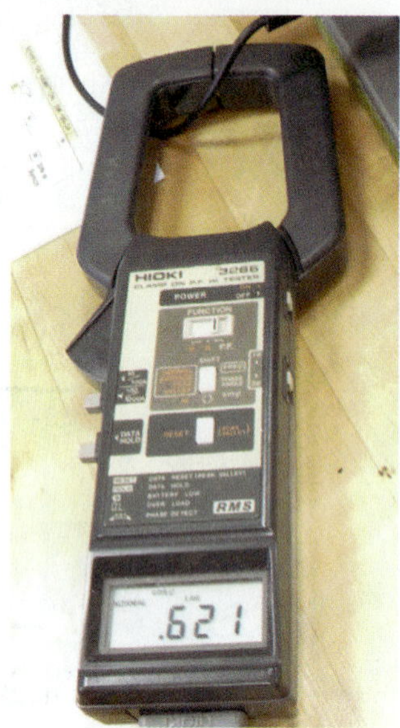

Figure 35-23 True RMS clamp-on ammeter measuring power factor.

meters have sensitive electronic circuits that can accurately measure voltage and amperage (Figure 35-23). Most meters called weighted average meters display the sensed voltage and amperage by averaging the readings over a short time period. However, some weighted-average meters provide RMS readings by applying a correction factor to the sensed voltage or amperage. These RMS meters may be more accurate than standard meters, but they are only averaging the true voltages and amperages.

Why Is True RMS Needed?

In the past, almost all of the equipment serviced by an HVACR technician had linear loads. Linear loads, like transformers, resistant heat strips, and so on, do not affect the power line sine wave (Figure 35-24a). Standard meters work well on linear load circuits. However, more and more HVACR systems and other building loads are using semiconductors in their power and control circuits (Figure 35-24b).

The new semiconductor circuits are faster switchers and more reliable than the ones they are replacing. Their ability to energize and de-energize equipment so quickly causes changes in the power line's smooth waveform. They disrupt this smooth waveform much like a speedboat's wake disrupts the uniform smoothness on a lake. At times the speedboat's wake can make the waves coming ashore larger, and other times they can be smaller. If you took the wave heights over time, you would not perceive any difference in wave power. But the occasional large boat wave comes crashing ashore with higher energy, which could cause damage.

True RMS meters can measure that momentary change in incoming line voltage. If there were only an occasional power peak, no long-term problem would occur. Unfortunately, this is not the case, because equipment such as variable-speed motors so common in HVACR equipment today constantly disrupts the wave power. They use power in very short pulses, which disrupts the

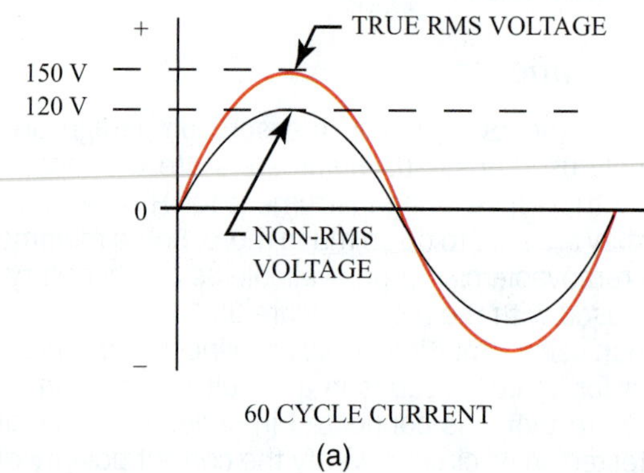

Figure 35-24a True RMS voltage plotted against non-RMS voltage.

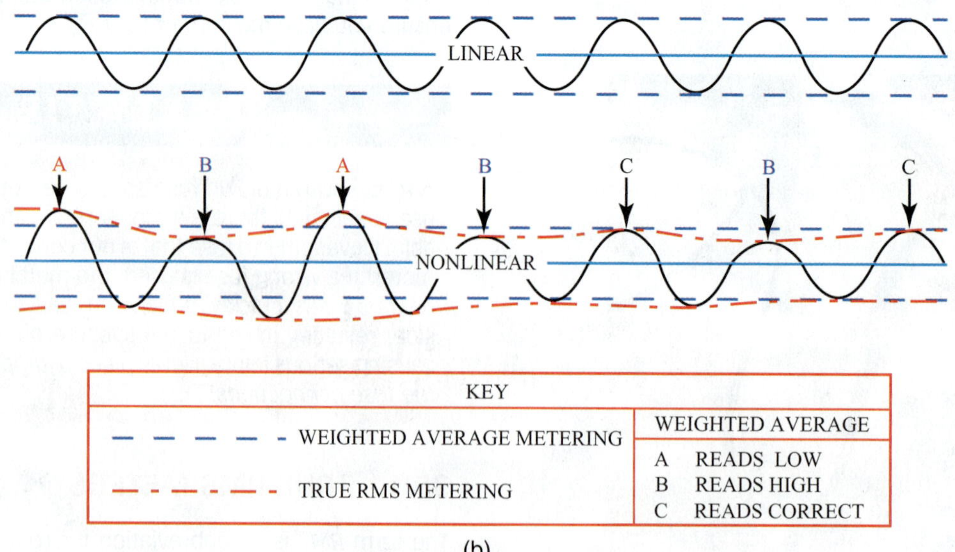

Figure 35-24b Weighted-average metering plotted against true RMS metering.

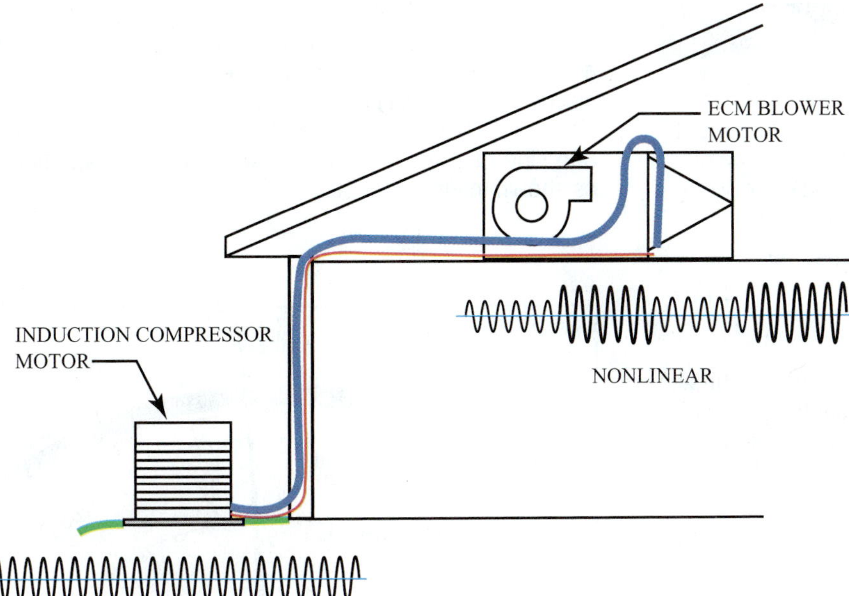

Figure 35-25 Linear vs. nonlinear power supply.

normally smooth linear line current making it rough and choppy (Figure 35-25).

The weighted averaging from non-RMS meters of voltage and amperage can result in readings that are 5 to 40 percent lower than the values that are actually being seen by the equipment. Small changes in operating voltages on induction motors can significantly affect current draw. Lower voltages cause higher amperage draws through the motor, which increases motor heat. A 20°F rise in motor temperature can cut the motor's life in half. Higher voltages can cause excessively high inrush currents at startup, which can also damage motors.

Using a true RMS meter is the only way to accurately measure voltage and amperage in nonlinear power applications.

Figure 35-26 Test leads touching and the meter is reading 0.18 ohms.

35.12 OHMMETERS

Ohmmeters measure the electrical resistance and can be used to check a circuit for continuity. Multimeters will have this function and so will some clamp-on ammeters. *Continuity* means there is a complete electrical path. When the test leads are touching each other there is a complete electrical path (Figure 35-26). The meter is measuring the resistance of the wires of the test leads as 0.18 ohms. When the test leads are separated there is infinite resistance as indicated by the meter reading OL (Figure 35-27).

Not every circuit that tests OK for continuity is actually good. For example, a circuit that is grounded (shorted) to the unit case would show continuity to the case. All ohmmeters can check for continuity, but some have a special position that sounds a beep when there is continuity. The ohmmeter is different from an ammeter or a voltmeter in that it uses a battery as a power supply. The battery furnishes the current needed for resistance measurements.

Figure 35-27 Test leads separated and the meter is reading infinite resistance.

SERVICE TIP

The power to the circuit being tested must be turned off before testing a circuit for resistance. Further, if there are any capacitors in the circuit, they must be discharged before the meter is used. There can be only one source of power to the meter, and that must be from the battery within the meter itself.

Using the ohmmeter to check for open circuits is called continuity testing. Figure 35-28 shows three diagrams representing the three possible responses that the meter can give.

- In Figure 35-28a, the meter is measuring the resistance through a pair of closed contacts in a circuit, and it registers zero. This indicates maximum current flow, or 0 Ω resistance. This is the measurement of a short circuit or a good set of closed contacts.
- In Figure 35-28b, the meter is measuring the resistance of a coil, which has a measurable resistance that is read on the meter.
- In Figure 35-28c, the meter is measuring the resistance of an open circuit, which is read on the meter as infinity. Infinity means that the resistance is so large that it cannot be measured. It means that at this point there is a lack of continuity or no current could flow.

Care must be taken to prevent errors in reading resistance when two or more circuits are connected in parallel, as shown in Figure 35-29. The meter in the illustration is actually reading the combined resistance of two parallel resistances, HTR1 and Blower. To read only one resistance, one side of the component being tested is disconnected, as shown in Figure 35-30.

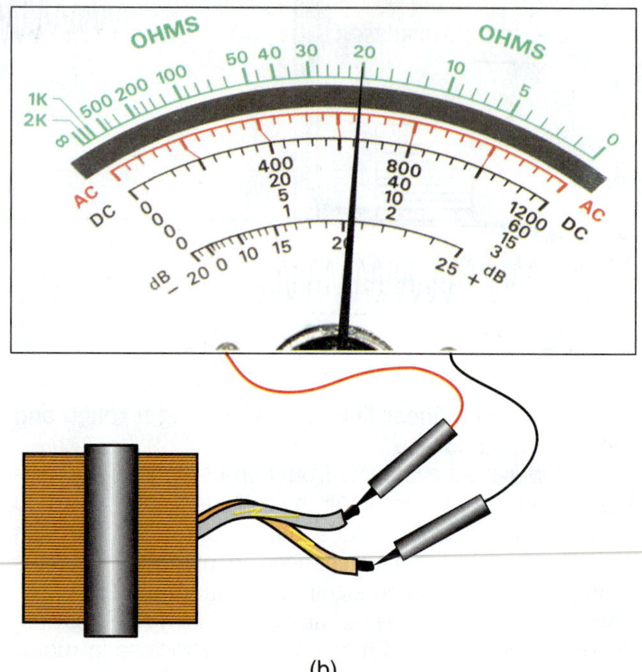

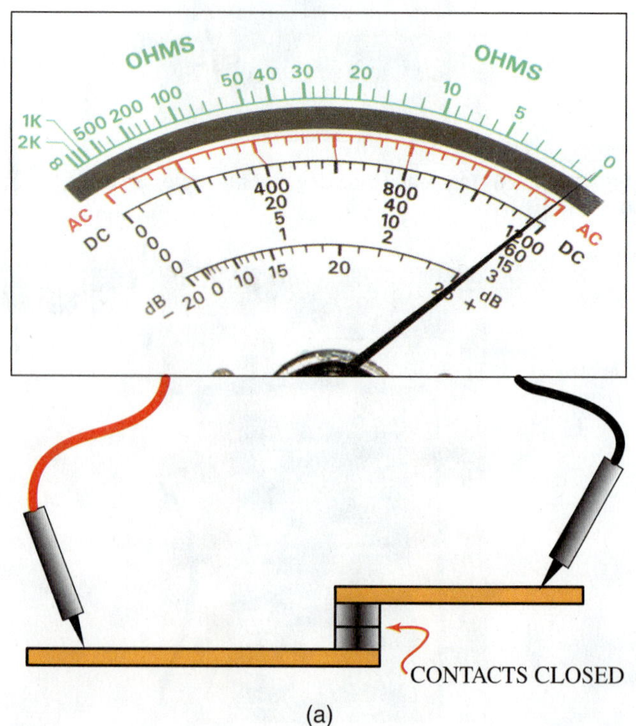

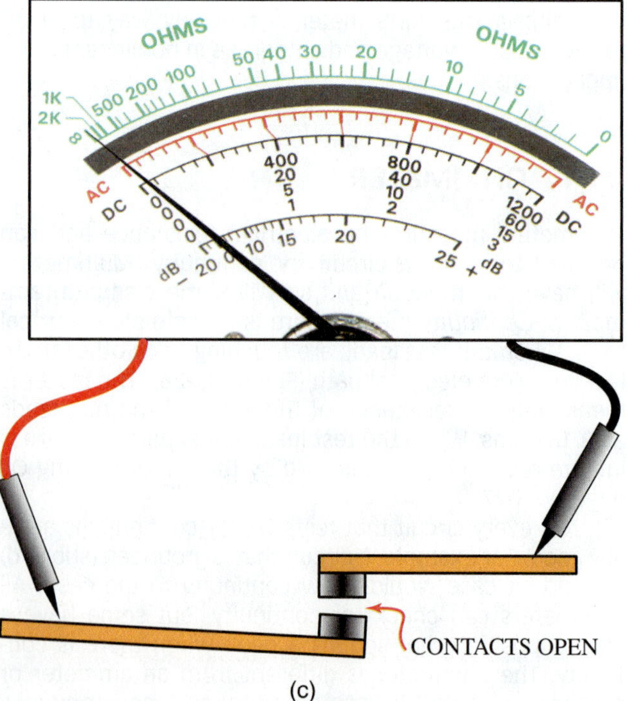

Figure 35-28 (a) Closed contacts or a short will read as 0 Ω; (b) a coil resistor will have a resistance reading between 0 and infinity; (c) an open contact or broken circuit will have an infinite (∞) resistance .

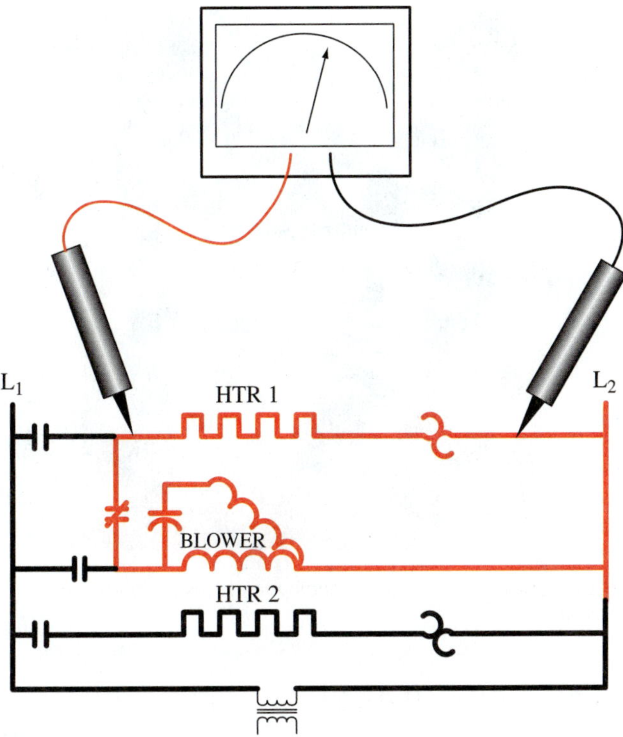

Figure 35-29 The ohm reading for Heater 1 would not be accurate because there are two paths, as shown in red.

Figure 35-31 NEVER use a multimeter to check an integrated circuit board chip.

SERVICE TIP

One caution needs to be followed: do not use an ohmmeter to test a solid state circuit unless the manufacturer specifically allows it. The internal battery voltage of the ohmmeter can damage an integrated circuit chip, as shown in Figure 35-31.

Megohmeters (Meggers)

Some ohmmeters, often referred to as meggers (megohmeters), must be able to read resistances of tens of millions of ohms (megohms) (Figure 35-32). Meggers are designed

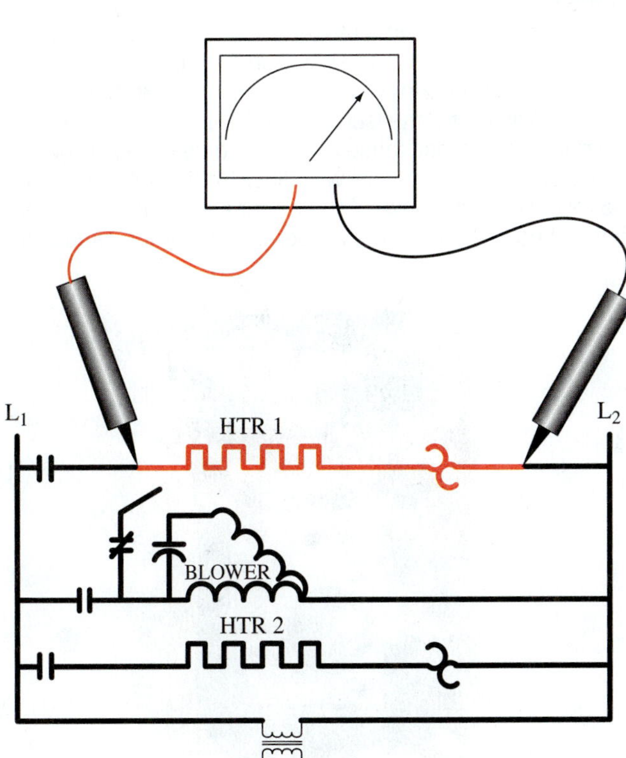

Figure 35-30 The ohm reading would be accurate because the second path is now broken at the normally closed contacts.

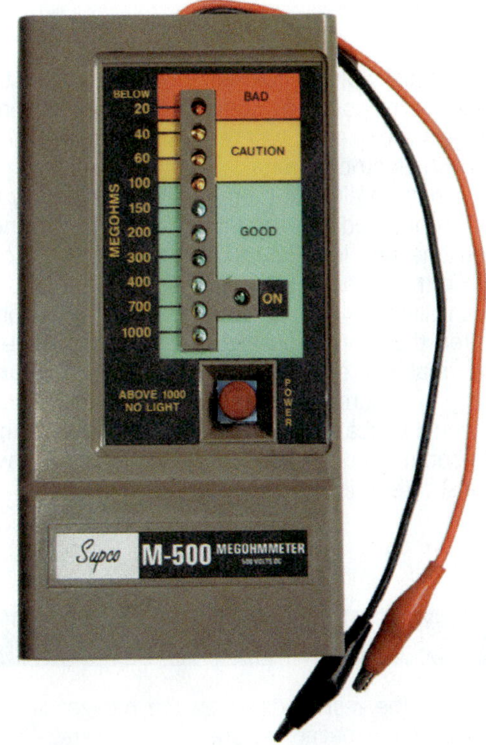

Figure 35-32 Megohmeter.

(a)

(b)

Figure 35-33 (a) Black test lead attached to motor housing; (b) red test lead attached to motor windings and black clip attached to motor housing.

to generate a high voltage across their test leads and therefore require more power as compared to regular ohmmeters. Meggers are used to test the resistance of insulation, particularly in electric motors.

For testing an electric motor, the power supply must first be disconnected and the motor isolated from the circuit. Then one test lead is attached to one phase of the motor while the other test lead is attached to ground, usually with an alligator clip (Figure 35-33). The ground connection is often the motor housing. When attaching the lead to ground, make sure to scrape any paint until bare metal is showing to ensure a good connection. A high megger reading would indicate that the motor winding and insulation are good (Figure 35-34). A low megger reading would indicate a shorted motor winding (Figure 35-35).

SAFETY TIP

Never touch the test leads when the megger is being used, because the megger generates a high voltage across them.

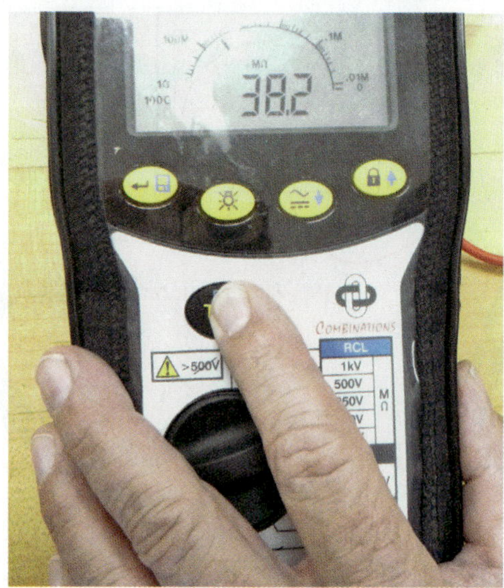

Figure 35-34 High megger reading indicates good insulation resistance.

35.13 CAPACITANCE METERS

Capacitance meters measure in microfarads (Figure 35-36). Most capacitors are within the range for nanofarad to microfarad. One microfarad is equal to one millionth of a farad. Before testing a high-voltage capacitor, always begin by properly discharging the capacitor first. The meter measures capacitance by charging the capacitor with a known current and measuring the resulting time of charging period. From this, the capacitance is then calculated by the meter.

The leads will have a residual capacitance. The known capacitance of the circuit shown in Figure 35-37 is 4 microfarads, as both 2-microfarad switches are in the closed position. The meter measures the circuit at 4.198 microfarads due to the capacitance of the meter leads as shown in Figure 35-38. Most meters will allow for the subtraction of the residual capacitance of the meter and leads. The meter must be connected for the proper polarity of the capacitor.

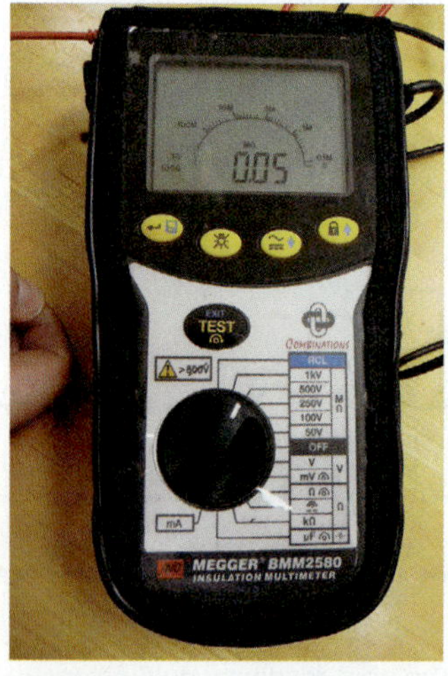

Figure 35-35 Low megger reading indicates shorted winding.

Figure 35-36 Capacitance meter.

Figure 35-38 The meter measures 4.198 microfarads instead of 4.0 due to the capacitance of the meter leads.

TECH TIP

An ohmmeter can be used as a quick check to test for a shorted or open capacitor. Analog meters have been commonly used for this because you can more easily see the meter movement. The test is performed by placing the two leads of the ohmmeter on each of the capacitor terminals for a few seconds and then swapping the leads. What this will do is charge the capacitor with the very low voltage used for ohm checking. In a functioning capacitor, when the leads are reversed, at first there will be a bump in the reading. Then, as the voltage bleeds off the capacitor, the needle will slowly swing back to the infinity (∞) position. If the capacitor is shorted out, there will be no bump and the reading will stay at zero. If there is no needle movement at all, then the capacitor is open.

35.14 WATTMETER

Power that is measured by simply multiplying the current by the voltage is called apparent power. However, this is not a true indication of power. To mathematically calculate power in watts using the measured volts and amps, it is necessary to know the power factor. Wattmeters incorporate the power factor to read true power (Figure 35-39). The meter makes the necessary calculations for power, in accordance with the power formula:

$$\text{watts} = \text{volts} \times \text{amps} \times \text{power factor}$$

Most wattmeters zero themselves when not in use and automatically select the proper range when used. They can be used to measure watts on single, split-phase, and three-phase power sources. Some multimeters and clamp-on ammeters have a power measurement function.

35.15 SPECIAL-PURPOSE METERS

There are many special-purpose meters designed for specific functions. Many digital meters have the capability to freeze and hold the reading or provide for a peak hold, a max hold, or a min/max hold. With some of these meters

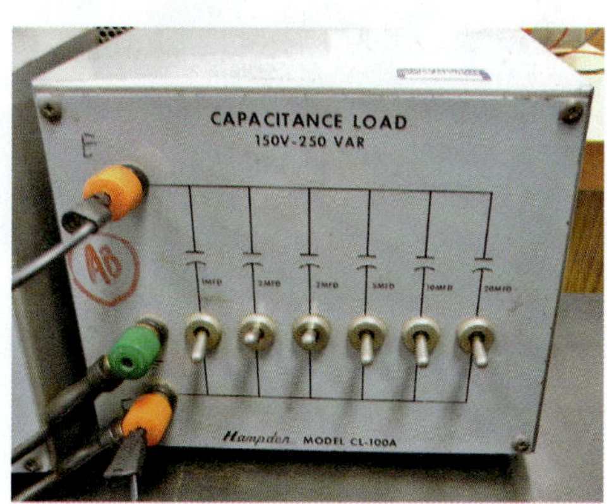

Figure 35-37 4-microfarad circuit.

Figure 35-39 Digital wattmeter.

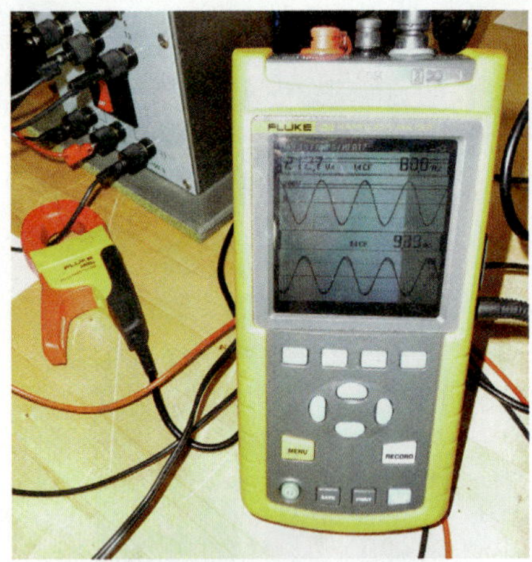

Figure 35-40 Digital power-quality analyzer with clamp-on accessory displaying circuit frequency.

you can see the starting current of a motor during start-up exactly as the circuit protector sees it (Figure 35-40). As the motor starts, the meter begins recording a large number of samples in as little as 100 milliseconds time and processes the samples to display the actual starting current. This type of meter can also be used to resolve intermittent breaker trips, helping to identify whether a nuisance trip is due to a faulty breaker.

Detachable Face

Detachable faces can be found on some multimeters and clamp-on ammeters. Some detachable displays can be separated from the body of the meter by as much as 30 feet to allow you to take remote readings more safely. The detachable face is removed completely and not connected by wires. The connection between the detachable face and the body of the meter is a radio signal.

Data Logging

Data-logging instruments can be set up to record any number of data points from a few seconds to a month or more. These instruments are extremely valuable when dealing with customers' complaints like, "My unit runs all the time," "It's always hot in here," and so on. A data logger can prove or disprove these too common and often frustrating complaints. It can also help to solve troublesome intermittent problems that never seem to happen while you are working on the system.

Some data loggers display the logged data directly on the equipment display panel, while others can download the data to a computer for a comprehensive graphic analysis (Figure 35-41).

The newest line of data-logging instruments prompts the technician for information. They can connect the proper test device to the data logger, where it can collect the information directly from the device. With all of the data collected and the system equipment model number entered, the data logger will provide a comprehensive evaluation of the system's operation and its current level of efficiency. These data are automatically compared to the manufacturer's data, so the report can include specific recommendations if the system is not within the manufacturer's operating specifications.

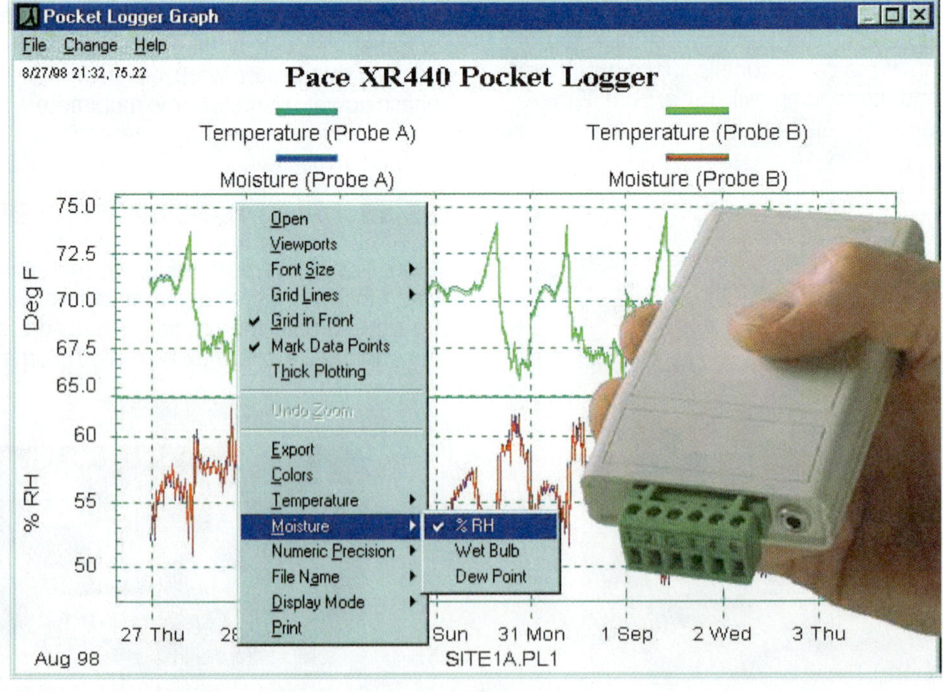

Figure 35-41 Data logger. *(Courtesy of Pace Scientific Inc.)*

UNIT 35—SUMMARY

A number of general principles should be observed in the use of meters, as follows:

- Always use the highest scale on the meter first, and then work down to the appropriate scale. This prevents damaging a meter by applying excessive power. An auto-ranging voltmeter will do this automatically.
- Always check the function of a meter before using it. Do not just assume a meter is working. If it is a battery-operated meter, the batteries may be run down. The meter could be damaged during transportation. Other things could happen to affect the readings.
- In using a clamp-on ammeter, be sure the jaws are around only one wire. If it is around two wires of the same circuit, there will be no reading at all as they will cancel each other out. Start at the high range and work down as described above.
- Always have an extra set of meter fuses on hand for replacement. Sometimes the proper fuses are difficult to obtain when you need them.
- Never use an ohmmeter in a circuit that is powered. An ohmmeter has its own power supply and can be destroyed by connecting to a live power source.
- If the meter uses batteries, have replacement batteries available. Older batteries can affect the readings the instrument provides. Colder batteries provide less voltage and current than warmer batteries. The condition of the batteries is not as important on digital meters, because they can compensate for low battery voltages automatically. Usually, digital meters will continue to operate until a LOW BATT icon is displayed.

UNIT 35—REVIEW QUESTIONS

1. What are the three test instruments that can be used to verify that no voltage is present in a de-energized circuit before beginning work?
2. Which standards reference the personal protection gear required for electrical work?
3. How should meter test leads be connected for testing a circuit?
4. Explain the terms *range, resolution*, and *accuracy* as they relate to meters.
5. What is a transient voltage?
6. Where are transient voltages most dangerous?
7. What are the lowest CAT and highest CAT ratings for meters?
8. What could happen to an analog meter if too high a voltage is tested?
9. Why have some digital meters added an analog sliding scale?
10. What part of an analog scale gives the most accurate reading?
11. List three advantages of digital meters.
12. What do voltmeters test?
13. What can a voltmeter show about a hidden switch?
14. What do ammeters test?
15. How are clamp-on ammeters used?
16. What do the letters RMS stand for?
17. How does the voltage reading made with a true RMS meter differ from a voltage reading taken with an RMS meter?
18. What can cause nonlinear power?
19. What does the beep indicate on some ohmmeters?
20. What does a wattmeter read?
21. What instrument can be used to solve troublesome intermittent problems?

UNIT 36

Electrical Components

OBJECTIVES

After completing this unit, you will be able to:

1. determine the resistance value of a color-banded fixed resistor.
2. provide examples of where and how transformers are used.
3. identify paper and film, electrolytic, ceramic, and mica capacitors.
4. identify the different types of thermostats.
5. explain cut-in, cut-out, and differential on pressure switches.
6. test transformers, capacitors, contactors, and relays.
7. list the different types of fuses and overloads.
8. explain the difference between relay logic and solid-state logic.
9. describe how a silicon rectifier operates.

36.1 INTRODUCTION

Most HVACR service work deals with electrical problems. These problems can be caused by component failure, improper installation, or misuse. It is therefore important that HVACR technicians have a thorough and complete understanding of electrical components and wiring diagrams. Electrical circuits on different pieces of equipment will have similarities. Most circuits will have resistors, capacitors, relays, contactors, switches, and transformers. Understanding the function of each of these components will help you better understand how to troubleshoot a circuit. Electronic circuits with solid-state components are quickly replacing traditional electrical circuits, but many similar operating principles still apply. Therefore, this unit includes not only information on basic circuit components but also introduces the most common solid-state components.

36.2 RESISTORS

Resistors are found in many circuits (Figure 36-1). They are designed to allow for a measured resistance that can affect either voltage or current as calculated by using Ohm's law. As an example, a resistor could be used in an electrical test meter to limit the current flow. Fixed resistors can be made from nickel wire wound on a ceramic tube and then covered with porcelain. Smaller fixed resistors are made from mixtures of powdered carbon and insulating materials molded into a round tubular shape (Figure 36-2). Variable resistors have a tightly wound coil of resistance wire made into a circular shape (Figure 36-3a,b). The resistance value is changed by turning an adjustment that moves the point of contact along the circular coil. Some variable resistors can be controlled by a small knob, while others are adjusted with a screwdriver (Figure 36-3c). Adjustable resistors are often used for electronic circuits.

Figure 36-1 Resistors.

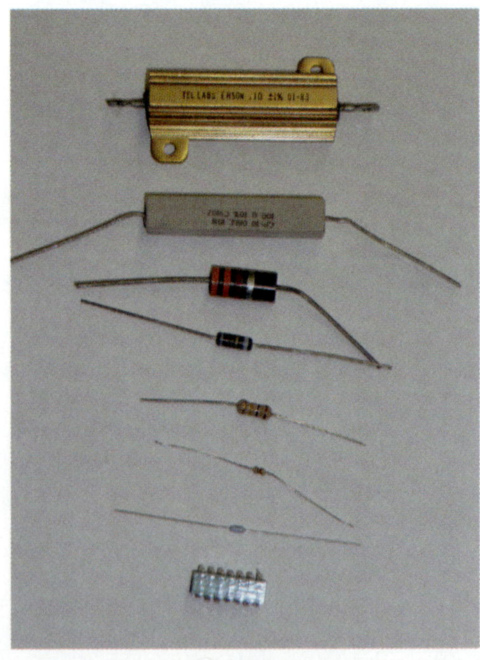

Figure 36-2 Example of different types of fixed resistors.

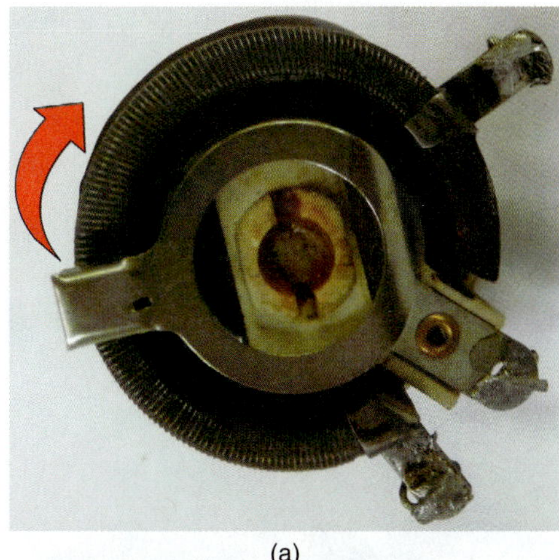

(a)

(b)

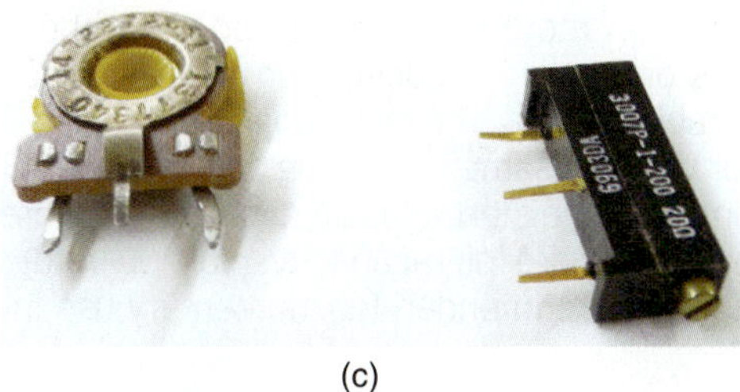

(c)

Figure 36-3 (a) Resistance changes at different points along the resistance coil; (b) assortment of variable resistors; (c) a small, flat-tipped screwdriver would be require to turn the adjustment to change the resistance.

Resistor Color Bands

Markings on resistors can vary. Larger resistors have printed resistance values, while smaller resistors have color-coded bands. To determine the resistance of a color-coded resistor, start from the end opposite the silver or gold band. Use the color code chart from Figure 36-4 to determine the resistance values. The first two bands identify the first and second digits of the resistance value, and the third band indicates the number of zeroes. However, if the third band is silver, this will indicate a 0.01 multiplier. If the third band

0	BLACK
1	BROWN
2	RED
3	ORANGE
4	YELLOW
5	GREEN
6	BLUE
7	VIOLET
8	GREY
9	WHITE
0.1	GOLD
0.01	SILVER
5%	GOLD-TOLERANCE
10%	SILVER-TOLERANCE

Figure 36-4 Resistor color-code chart.

Figure 36-5 The resistor color code is orange, orange, red, silver.

is gold, this will indicate a 0.1 multiplier. The fourth band indicates tolerance. Silver indicates a +/– 10 percent tolerance, and gold indicates a +/– 5 percent tolerance. If there is no fourth band, the resistor tolerance is +/– 20 percent.

Calculate the resistance of the resistor shown in Figure 36-5. The first band is orange, which is listed as number 3 on the chart. The second band is also orange, and the third band is red. The resistance therefore would be 3,300 ohms with a tolerance of 10 percent.

36.3 CAPACITORS

A capacitor will store energy when an electric charge is forced onto its plates from a power source. A capacitor will still retain this charge even after disconnection from the power source. However, it would be impractical to try to discharge the power from the capacitor into a different circuit, as you would do, for example, by placing charged batteries into your radio. Compared to a storage battery, the total amount of energy stored by a capacitor is relatively small. Also, the discharge rate of a capacitor is rapid, so the release of the stored energy only occurs during a short time interval. However, a mishandled capacitor will deliver a shock that can be severe and even fatal, especially for large capacitors charged to a high voltage.

Capacitor Types

Capacitors are rated for a maximum voltage by the manufacturer. This rating is usually expressed as the direct current working voltage (DCWV). Exceeding this voltage will shorten the life of the capacitor.

Capacitors can be used for a number of different applications. As an example, they can be used for tuning,

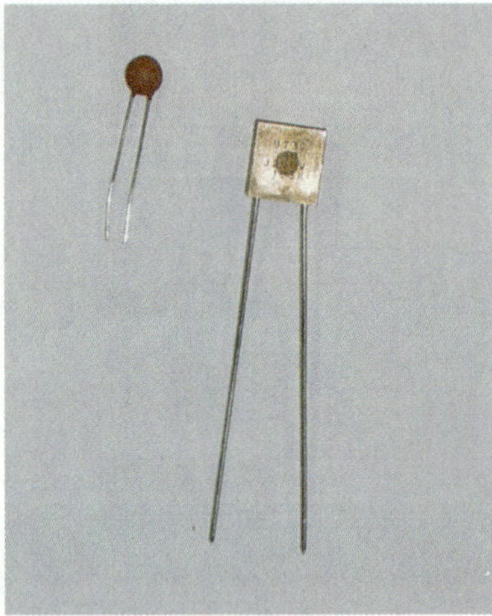

Figure 36-6 Disc ceramic capacitors.

filtering, energy storage, power factor correction, and motor starting. Capacitive filters are used to smooth pulsating DC and separate low-frequency AC from high-frequency AC. Capacitive tuners are used for tuning radios and television sets to the proper channel. Energy-storage capacitors are used in industrial applications such as capacitor discharge welding, where a large amount of stored energy is discharged rapidly. The leading current of a capacitor offsets the lagging current in an inductive load to allow for power factor correction. Many electric motors also utilize capacitors to produce a current phase shift in their windings.

Not all capacitors are made of the same materials. There are paper and film, electrolytic, ceramic, and mica capacitors. Disc ceramic capacitors are commonly found on electronic circuit boards and are typically 0.1 microfarads (mfd) or less (Figure 36-6). Mica capacitors are limited to even lower values than this (Figure 36-7).

For larger-capacitance requirements, paper and film capacitors are used (Figure 36-8). They are constructed using a rolled-foil technique. Once rolled, the capacitor may be dipped into a plastic insulating material. Capacitors of this type used for electronic circuits are rated at generally less than 1 mfd. However, they can also be designed for industrial applications to meet the requirements of several hundred microfarads. In this case, they would be housed in a metal container filled with special insulation oil.

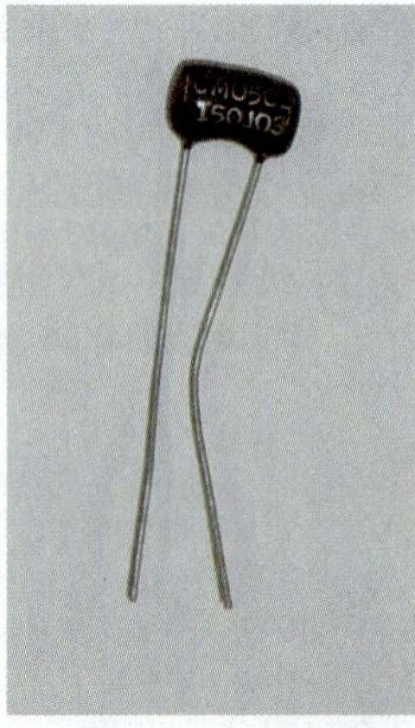

Figure 36-7 Mica capacitor.

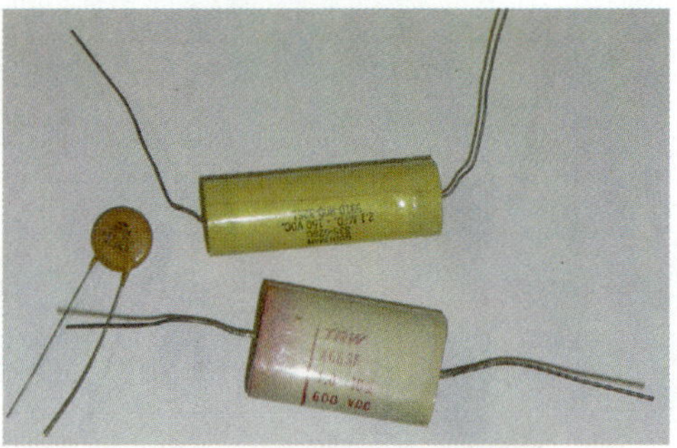

Figure 36-8 Rolled film and paper capacitors.

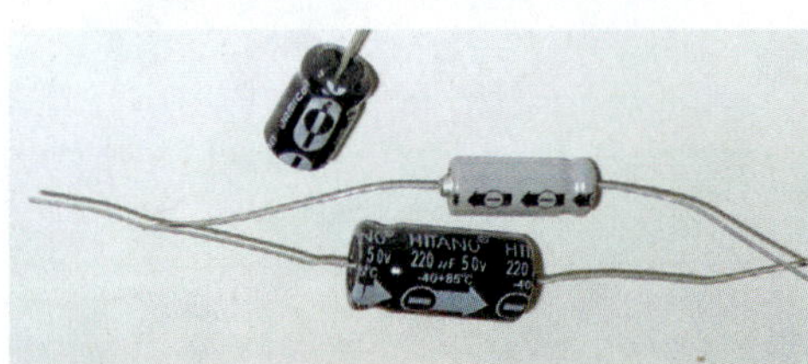

Figure 36-9 Electrolytic capacitors marked for polarity.

The type that provides for the most capacitance in relationship to their size and weight are electrolytic capacitors, which are commonly polarized. The polarity is marked on the body of the capacitor in some manner, as shown in Figure 36-9. Never reverse the polarity on this type of capacitor. This will lead to excessively high current, overheating, and possible explosion of the capacitor. A pop-out hole on some capacitors allows for the insulation to expand if the capacitor is overheated (Figure 36-10). If the hole is ruptured, the capacitor must be replaced.

Motor Capacitors

Capacitors for motors are classified as either starting capacitors or running capacitors. In replacing a capacitor, it is desirable to use an exact replacement. This means a capacitor with the same mfd rating and voltage limit rating. Do not interchange start and run capacitors (Figure 36-11). Start capacitors are high-capacity (50–700 mfd) electrolytic units that are intended for momentary use in starting

Figure 36-10 Motor-start capacitor with pop-out hole.

Figure 36-11 Assortment of both motor-start and motor-run capacitors.

Figure 36-12 159 to 191 mfd motor-start capacitor.

motors (Figure 36-12). They are normally encased in plastic. Run capacitors have much lower capacitance ratings (2–40 mfd) but are made for continuous-duty use. They are normally sealed in a metal can.

Motor-start capacitors are in series with the start switch and starting winding. This allows for a large phase shift to create a good starting torque. Since it is a starting capacitor only, it is not rated for continuous duty and is limited to about twenty starts per hour.

Motor-run capacitors are rated for continuous duty and are commonly used for permanent split-capacitor motors. The capacitor is matched to provide a 90° phase shift between current in the auxiliary and main motor windings at 80 to 100 percent of rated power. It stores and releases an electrical charge in the auxiliary winding to increase the current lag between it and the main winding. This is to balance the effective inductance and inductive reactance of the windings. The capacitor remains in the circuit the entire time the motor is running.

Some run capacitors have some sort of a mark, usually a red dot, as shown in Figure 36-13, to indicate the terminal that should be connected to the run terminal. With

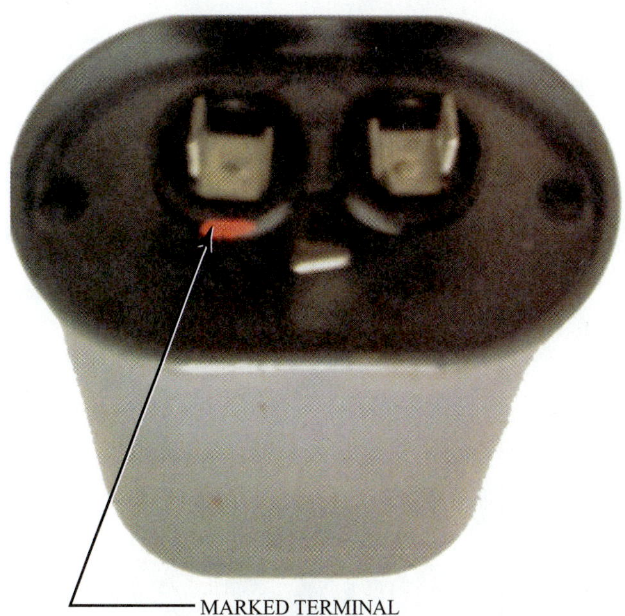

Figure 36-13 The marked terminal on the run capacitor should be connected to the run terminal of the compressor.

this arrangement, an internal short circuit to the capacitor case will blow the system fuses without passing the current through the motor start winding.

36.4 TESTING CAPACITORS

The first operation in testing a capacitor is to discharge it. Do not discharge it by shorting out the terminals, as this can damage the capacitor. To avoid electrical shock, the technician should never place fingers across the terminals before properly discharging the capacitor.

The proper way to discharge a capacitor is to put it in a protective case and connect a 20,000 Ω, 2 W resistor across the terminals, as shown in Figure 36-14. Most start capacitors have a bleed resistor across the terminals. This makes it so the capacitor can be tested with the bleed resistor in place. Even so, it is good practice to make sure the charge has been bled off.

Capacitors can be roughly checked by using an ohmmeter. The ohmmeter used in testing capacitors should be able to read a high resistance and have at least an R × 100 scale. To test the capacitor, disconnect it from the wiring and place the ohmmeter leads on the terminals, as shown in Figure 36-15.

If the capacitor is not shorted, the needle will make a rapid swing toward zero and slowly return to infinity. If the capacitor has an internal short, the needle will stay at zero, indicating that the instrument will not take the charge. What you are actually doing is attempting to charge the capacitor using the battery in the ohmmeter (be sure the battery in the ohmmeter is good). An open capacitor will read high with no dip and no recovery.

The use of a capacitor analyzer is highly recommended (Figure 36-16). This instrument will read the mfd rating and detect any breakdown in the dielectric underload conditions. It will detect any capacitors that have failed to hold

Figure 36-14 Using a bleed resistor on a start capacitor.

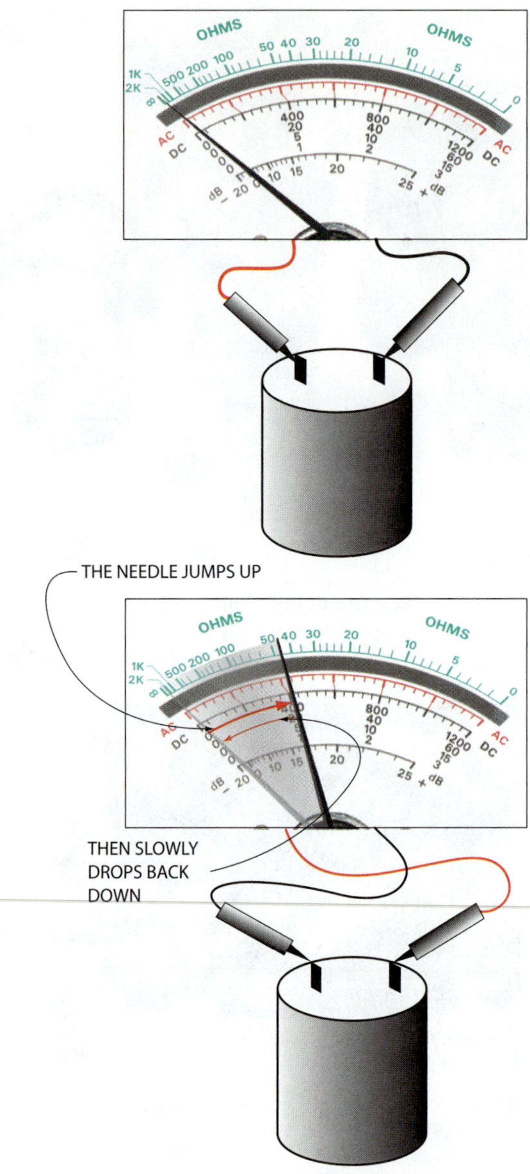

Figure 36-15 Using an ohmmeter to test capacitors.

their ratings. It also is useful in measuring the rating of a capacitor that has an unreadable marking.

Most digital multimeters now have scales for testing capacitors. Set the meter to the capacitance test function and place the leads on the capacitor terminals (Figure 36-17).

36.5 TRANSFORMERS

HVACR equipment often requires more than one voltage. One or more transformers are often used to step down the line voltage to supply load or control requirements. Occasionally a stepup transformer may be used.

Figure 36-16 Capacitor test meter.

TECH TIP

There are several reasons that most HVACR equipment uses 24 V as the control voltage. First, under most state and local guidelines, an electrician's license is not required to install and service these low-voltage wires. Second, most codes do not require that these connections be made at electrical junction boxes. And third, under OSHA regulations, circuits that have less than 80 V fall under less stringent safety requirements.

Figure 36-17 Testing a capacitor using the capacitor test function on a multimeter.

Transformer Design

Transformers are constructed using the induction characteristics of AC power. When current flows through a coil, a magnetic field is produced. When a second coil is placed in the field of the current-carrying coil (primary), electric current can be transferred to the second coil (secondary) (Figure 36-18). The process is made more efficient by wrapping the coils around a common metal core. The voltage transferred is directly in proportion to the ratio of the number of turns on the primary coil to the number of turns on the secondary coil. For example, a control transformer with a 240 V primary winding and a 24 V secondary winding has a 10:1 ratio: 10 turns of wire in the primary for every turn in the secondary.

More than one secondary coil can be used if additional voltages or circuits are required. Likewise, some transformers are made with more than one primary. These multiple windings can then be connected in series or parallel to change the voltage or current capability of the transformer.

Center-tapped transformers allow for a small change in the windings voltage rating by changing taps. Typical residential voltage is 120 V, but this can vary. A center-

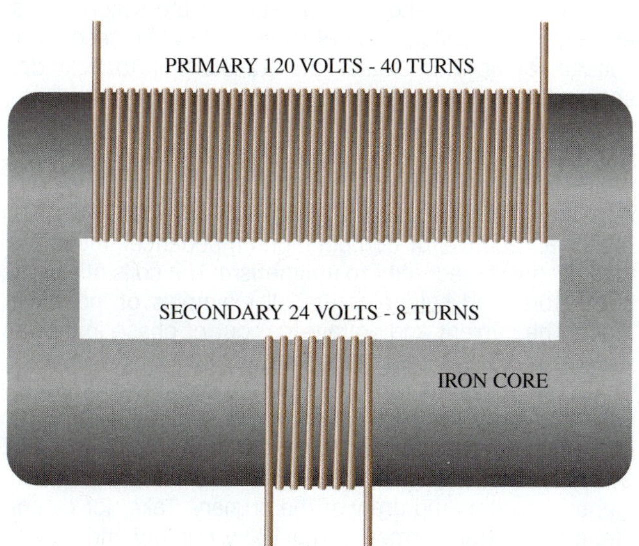

Figure 36-18 Primary and secondary voltages of a step-down transformer.

Figure 36-19 The primary voltage for this transformer depends upon the tap used.

Figure 36-20 Variable autotransformer.

tapped transformer could be connected to meet varying requirements. As an example, a transformer that has multiple primary taps could be connected for 120 V, 208 V, or 240 V, depending on the tap used (Figure 36-19). Variable autotransformers also allow for changing the voltage output (Figure 36-20).

Three-phase transformers are generally used for three-phase power. However, a single-phase transformer for each leg (three single-phase transformers) would produce the same results.

Transformer Application

There are many different types of transformers for many different applications. Large power transformers are designed to operate at electric utility voltages from 115 V to several thousand volts (Figure 36-21). A common use for some smaller transformers is to provide low-voltage AC to

Figure 36-21 Typical line voltage power transformer.

rectifiers for conversion to DC. These would be called rectifier transformers. Air-conditioning systems use low-voltage transformers to provide 24 V AC to control circuits to operate relays and solenoids. These step down the voltage from line voltage to 24 V and are referred to as control transformers (Figure 36-22).

Equipment that is very sensitive to voltage changes use constant voltage transformers. Electronic devices may use the metal chassis where components are mounted as a common conductor. If you touch the metal chassis accidently while there is power, you will receive a shock. Isolation transformers are used to break the circuit and protect the technician.

Figure 36-22 Control voltage transformer.

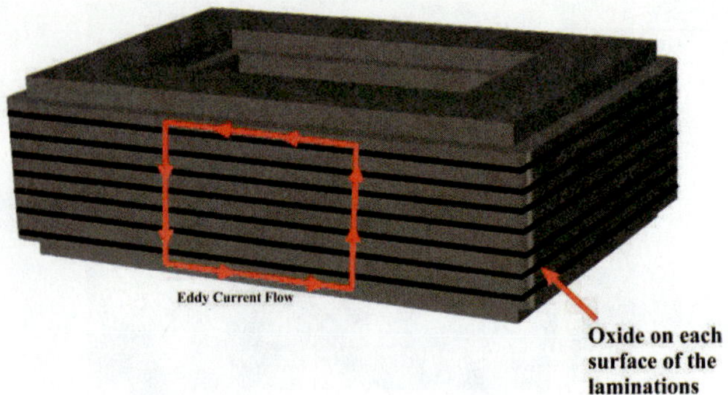

Figure 36-23 Stray eddy current flow through the transformer core is reduced by the oxide surface treatment of the laminations.

Transformer Construction

Transformers will heat up during operation. This production of heat is inefficient. Transformer cores are made of thin sheet-metal strips in the form of a laminate. There are a number of advantages to using thin sheet-metal strips. It is far easier to manufacture thin sheets because they can be stamped, where the thicker material would have to be cut. Also, the voltage induced in the core will cause current, called eddy current, to circulate (Figure 36-23). Each metal strip is insulated with a thin layer of oxide, which resists the flow of eddy currents (Figure 36-24). This reduces unwanted current circulation through the core and excessive heating of the transformer.

Transformer Operation

The voltage ratings for transformers are specified by the manufacturer for both primary and secondary windings. A primary winding operating at above its rated voltage will overheat. Current ratings are often only provided for the secondary winding. This is because the primary winding current capacity cannot be exceeded before the secondary winding. When the current rating of the secondary is exceeded, the voltage output drops below the secondary voltage rating. This causes the transformer to heat up, shortening its life.

Transformers are commonly rated by volt-ampere ratings, abbreviated VA (Figure 36-25). The VA rating is literally the voltage multiplied by the current, amps. VA ratings are used because they apply to any load, whether resistive, reactive, or combination (impedance). Inductive loads convert electricity to magnetism. The coils of relays, contactors, and solenoids are all examples of inductive loads. The current and voltage get out of phase in inductive loads, so the wattage is actually less than the volts times the amps. VA ratings show the combined effect of volts and amps on the transformer regardless of the type of load on the transformer.

Note that the amp draw of the secondary is much higher than the amp draw of the primary. Take, for example, a 48 VA transformer with a 120 V primary and a 24 V secondary. The primary current = 48 VA/120 V = 0.4 amps. The secondary current equals 48 VA/24 V = 2.0 amps. Note that this example ignores the heat given off. This is why the

Figure 36-24 Laminations are clearly visible on most transformers.

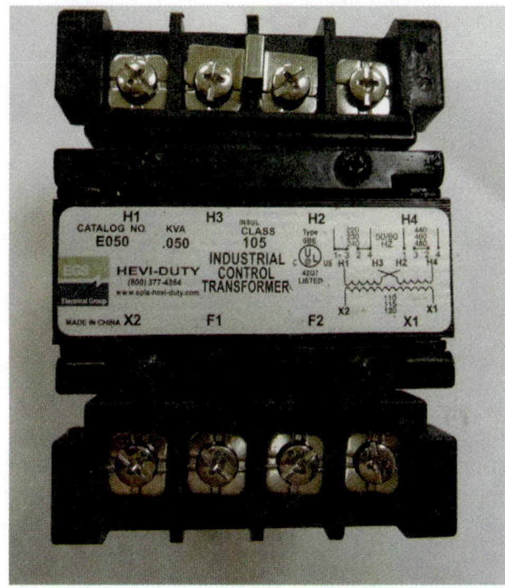

Figure 36-25 Industrial control transformer with a .050 KVA rating.

Figure 36-26 Residential service entrance panel with the access cover open to provide access to the circuit breaker switches.

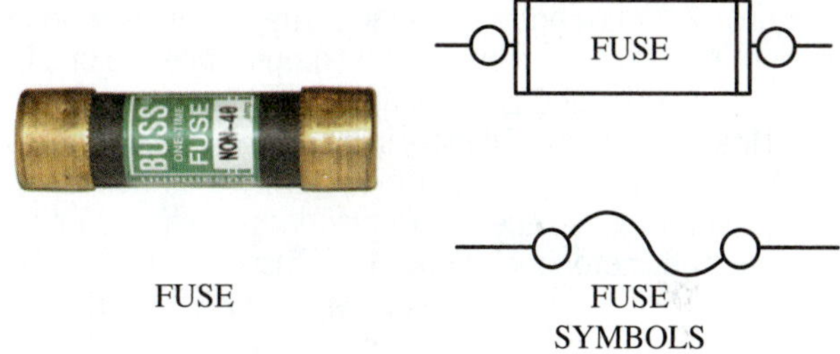

Figure 36-27 A cartridge fuse and its common symbols.

secondary winding is made from heavier-gauge wire than the primary winding.

Efficiencies for transformers are higher when they operate more fully loaded. The primary winding current is mostly inductive, making it nearly as much as 90° out of phase with the voltage. The mostly resistive secondary winding current will offset this at higher transformer loads so that the total current is more in phase with the voltage and therefore more efficient. When operating without a load, all the current the transformer draws is waste. When operating fully loaded, most of the current is going to the load.

36.6 FUSES

Where the power comes into the building, it enters a service entrance panel for distribution to the various electrical loads (Figure 36-26). Each electrical circuit that comes from this panel is electrically protected by either a fuse or a circuit breaker.

A sample fuse and two commonly used fuse symbols are shown in Figure 36-27. A fuse is a special electrical conductor that is placed in series with a load and melts when excessive current flows through it, opening the circuit. Fuses are available in different types and sizes so that they can be selected to match the requirements of specific loads (Figure 36-28). If they are too small, they melt before they should. If they are too large, they do not offer the proper protection. Their selection follows the rules set forth in the electrical code or in the specifications accompanying the load.

Figure 36-28 Fuses located in an air-conditioning console.

Figure 36-29 250 volt fuses: 30 amp on left and 60 amp on right.

Cartridge fuses are the most common in HVACR. They are rated by voltage, current, and maximum instantaneous current. All their values should be matched when replacing a fuse. Fuses are grouped by physical size. Up to 30 amps is one physical size; over 30 amps and up to 60 amps is a larger size (Figure 36-29). Fuses are also rated for voltage. Fuses designed for 600 volts are much larger than fuses designed for 250 volts (Figure 36-30). Fuses must withstand incredibly high amounts of energy when they are subjected to a direct short. Not all fuses have the same ability to withstand the same levels of energy. Less expensive fuses have lower instantaneous current ratings. The two fuses in Figure 36-31 are both 600 V, 30 A fuses, but the one on the right has a rating of only 50,000 amps while the one on the left can withstand 200,000 amps. The fuse on the right would literally explode if it were subjected to 200,000 amps.

Because motors draw four to five times their normal operating current when they start, standard fuses often blow during normal motor operation. Where fuses are used to protect motors in the circuit, a special type of fuse is used called a dual-element time-delay fuse. This type of fuse has a built-in delayed action that will tolerate momentary heavy starting current on motor power-up but functions the rest of the time to protect the motor against excessive running current.

Figure 36-30 Both fuses are rated at 30 amps. The larger one is rated at 600 V, while the smaller one is rated at 250 V.

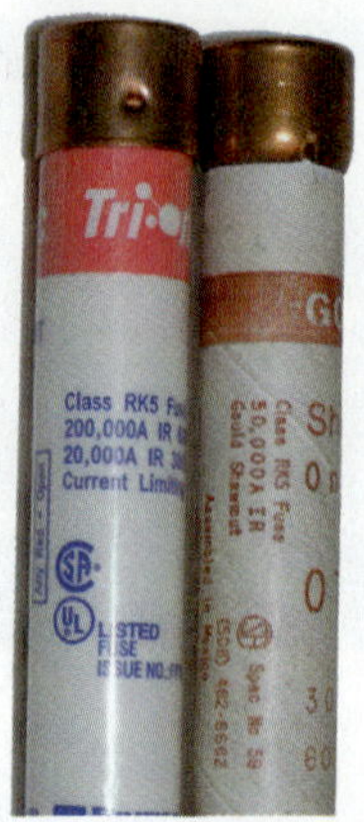

Figure 36-31 Even though both fuses are 600 V 30 amp fuses, the fuse on the left has an IR rating of 200,000 amps, while the fuse on the right has an IR rating of 50,000 amps.

CODE TIP

UL requires a dual-element time-delay fuse to withstand 500 percent of the fuse amp rating for 10 seconds. Dual-element fuses achieve the time delay by using two separate elements: one for short circuit protection and another for overload protection.

36.7 CIRCUIT BREAKERS

All residences have some type of electrical panel, where the electrical service enters the building and is distributed to the circuits in the building. Each circuit has some type of protective device to automatically disconnect the power in case the circuit is overloaded. This protection can either be a fuse as described above or a circuit breaker. Note the symbols used to represent a circuit breaker in an electrical wiring diagram (Figure 36-32). The advantage of a circuit

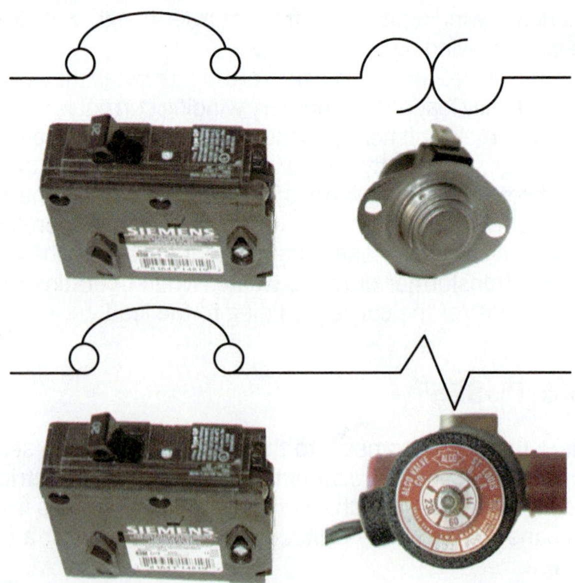

Figure 36-32 A circuit breaker and its common symbol.

Figure 36-33 Small power switch.

breaker over a fuse is that it can be manually reset at the electrical service panel after an overload, rather than replaced. Also, the circuit can be manually opened in case there is a need to perform service on the circuit. Three types of breakers are available: standard, GFCI, and AFCI. Standard breakers protect circuits against too much current, but they cannot detect if current is going where it should not go. In addition to protecting against too much current, ground fault circuit interrupter (GFCI) breakers protect against current that flows somewhere outside of the normal current path. An arc fault circuit interrupter (AFCI) protects against electrical arcs.

36.8 SWITCHES

Switches can come in all shapes and sizes (Figure 36-33). Often they are normally open and when the switch position is changed from OFF to ON, the switch contacts will close and energize a circuit, the same as when you turn on a light. Switches can also be designed to be normally closed, which de-energizes the circuit when the contacts open (Figure 36-34). An example would be a stop switch to shut down a motor. Proximity switches often limit motion and are often operated by the movement of a mechanical device (Figure 36-35). An example would be a switch that stops the motor used for opening air inlet dampers once they have reached a full open position.

General-duty switches are designed for use in residential and commercial applications. These are used for relatively

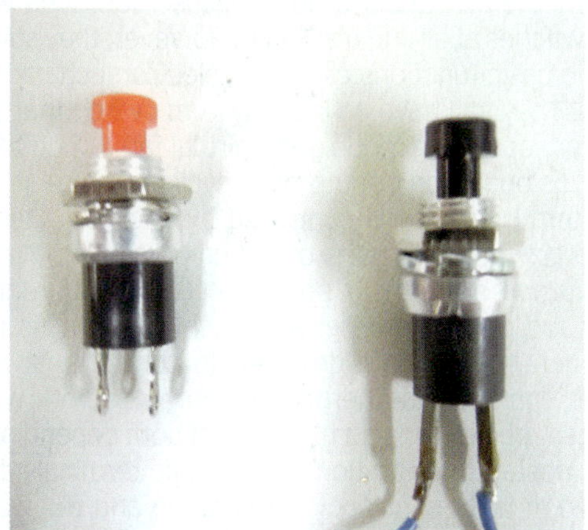

Figure 36-34 The black switch is pushed to start (normally open), while the red switch is pushed to stop (normally closed).

Figure 36-35 This proximity switch is operated by some controlled mechanical movement.

Figure 36-36 Residential safety switch.

light-load applications such as general air-conditioning and appliance loads. There are several different kinds of these switches, and some will have fuses while some others will not. A typical residential safety switch is shown in Figure 36-36. Switches are rated by the voltage and current they can safely switch (Figure 36-37). Operating a switch on a higher voltage or heavier current than its rating is dangerous.

Air-conditioning systems use many switches that open and close based on a particular condition. Examples include thermostats that open and close based on changes in temperature, humidistats that open and close based on changes

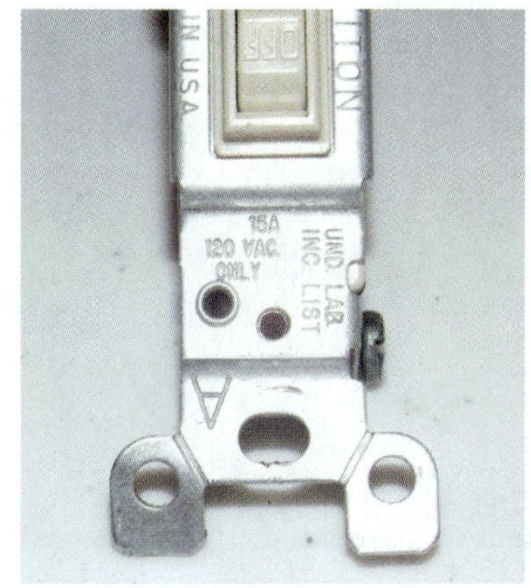

Figure 36-37 This switch is rated for 120 V and 15 amps.

in humidity, and pressure switches that open and close based on changes in pressure. These can be divided into two general categories: line voltage and low voltage. Line-voltage switches are referred to as line duty, while low-voltage switches are referred to as pilot duty.

36.9 LOW-VOLTAGE THERMOSTATS

A thermostat is a switch that is operated based on changes in temperature. Low-voltage thermostats operate on 24 volts and normally do not switch more than 2 amps. Low-voltage thermostats typically control several system functions. They operate like a bunch of switches, each one with a specific function. However, instead of operating the system by manually flipping all the appropriate switches, the thermostat does it automatically for you.

Low-voltage thermostats have two manual switches: a system switch and a fan switch. The system switch controls the operating mode: Off, Cooling, Heating (Figure 36-38). The fan switch controls cycling of the fan. In the ON position, the fan operates all the time, regardless of the thermostat setting. In the AUTO position, the fan cycles with the system (see Figure 36-38).

For many years, thermostats were electromechanical, using bimetal coils to move mercury bulb switches (Figure 36-39). The bimetal element is composed of two different metals bonded together. As the temperature surrounding the element changes, the metals will expand or contract. Since the metals have different coefficients of expansion, one will expand or contract faster than the other. This creates movement in the bimetal such as twisting or turning. A mercury bulb attached to the bimetal acts as a switch. The contacts are enclosed in an airtight glass bulb containing a small amount of mercury. When the bimetal tilts the bulb, mercury in the bulb will roll to one end and complete the electrical circuit.

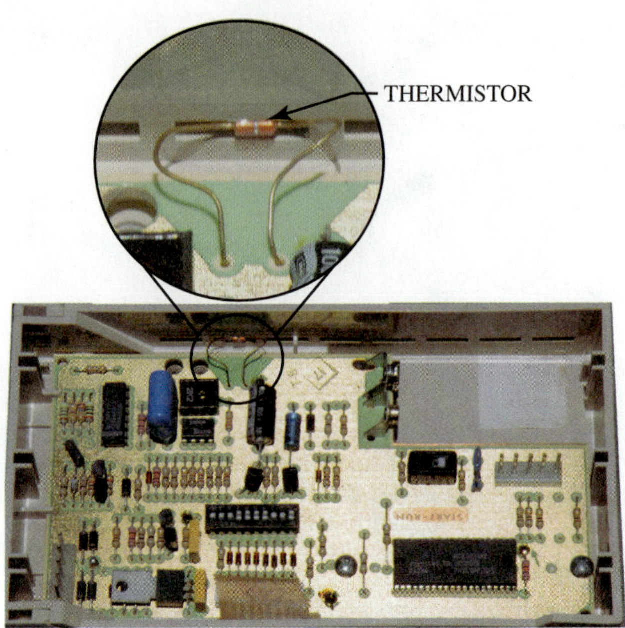

Figure 36-40 Thermistor sensor used in electronic thermostat.

Figure 36-38 The system switch and fan switch on a typical low-voltage thermostat.

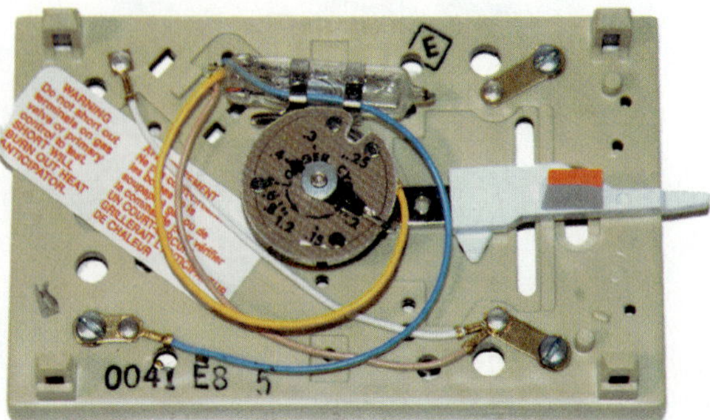

Figure 36-39 Older bimetal element and mercury bulb thermostat.

Newer low-voltage thermostats are now digital, using thermistors to sense the temperature and logic to control the operation of electronic switches or relays inside the thermostat (Figure 36-40). Electronic digital thermostats are considerably more accurate than older electromechanical thermostats. They will control the temperature within 1°F, while the bimetal thermostat controls to an accuracy of about 3°F. Digital thermostats require voltage to operate their logic boards. Some require both sides of the 24 V control power to operate, others steal power from the circuits they are controlling, and many use batteries. Some may work with either 24 volts or batteries.

Many thermostats are now touchscreen units with no actual switches at all (Figure 36-41). However, they still retain the same basic functions of the old electromechanical thermostats. They still have a virtual system switch that selects the system operating mode and a virtual fan switch to select continuous operation or automatic cycling (Figure 36-42).

System operation is controlled by energizing terminals that energize different parts of the unit. Because the operation of furnaces and heat pumps are quite different, thermostats are normally designed to control furnaces and air conditioners, or they are designed to control heat pumps. However, many digital thermostats now will work with furnaces, heat pumps, or a combination of both depending upon the thermostat configuration. The specific terminals that are energized in heating or cooling operation and the way they cycle are determined by the thermostat configuration. Figure 36-43 shows a configuration screen from a touchscreen thermostat that can be configured for many types of systems.

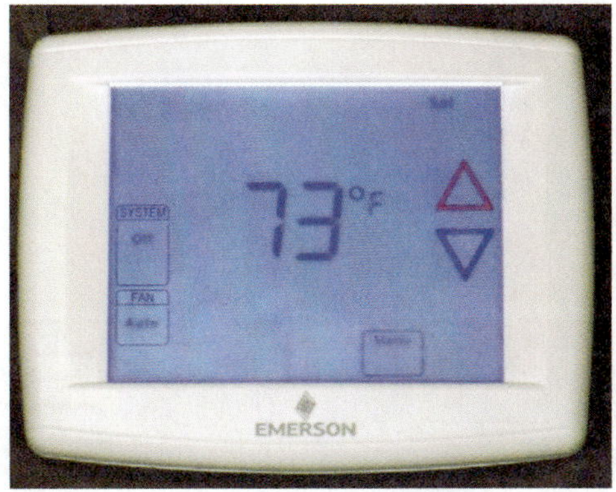

Figure 36-41 Touchscreen thermostats have no physical switches.

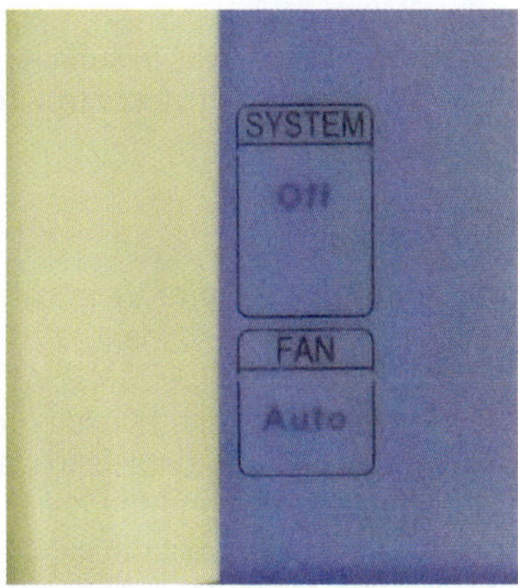

Figure 36-42 Virtual fan switch and system switch on a touchscreen thermostat.

Staging thermostats allow more than one stage of heating or cooling. In general, operating just the amount of heating or cooling necessary saves energy and money. The unit can use less energy by operating on first stage at a lower capacity most of the time. Heat pump systems typically use the second stage of heat to control auxiliary heat. If the first stage cannot meet the demand, a second stage is energized. Many two-stage furnaces are now available with low-fire and high-fire operation. Air conditioners and heat pumps are also commonly available with two capacity compressors. Staging thermostats are required for all these applications. Often, a configurable staging thermostat is used so that the thermostat function can match the system needs. More details on low-voltage thermostats and their circuits can be found in Unit 39, Control Systems.

Electronic programmable thermostats can be set for different temperatures for time periods throughout the week.

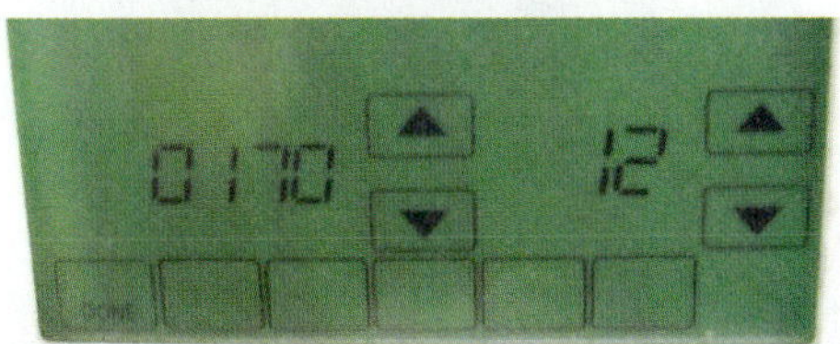

Figure 36-43 Configuration setup screen on a touchscreen thermostat.

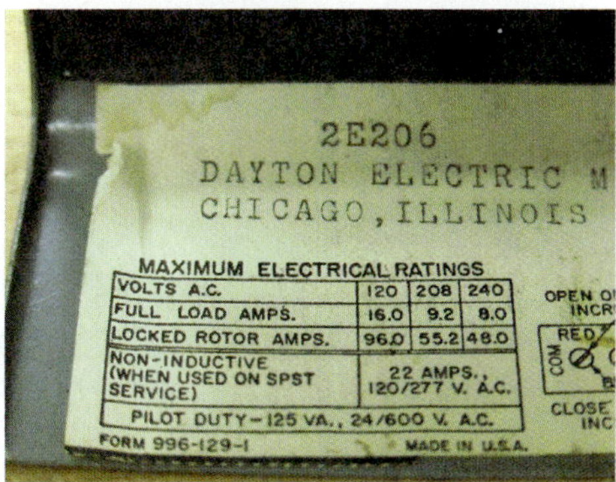

Figure 36-44 The current ratings on a line voltage thermostat.

These settings will take effect automatically. Most programmable thermostats divide the day into four periods: waking, leaving for work, arriving home, and going to bed. They allow you to set a different temperature for each of these four periods. A heating thermostat with night (or unoccupied) setback will automatically reduce the control set point during preset periods when lower than normal temperatures are acceptable. A cooling thermostat will have setup to raise the set point during scheduled hours. The intent of setup and setback is to reduce cooling and heating energy usage. Three levels of programmable thermostats are available: one program, separate weekday and weekend programs, or full 7-day programmability. The least expensive run the same program every day. The weekday-weekend thermostats have one program for the week and another for the weekend. The 7-day programmable allows a different program for every day of the week.

36.10 LINE VOLTAGE THERMOSTATS

Line voltage thermostats are designed to switch circuits that operate on the line voltage supplied to the unit, such as 120 V or 240 V. The switching action is described as open on rise or close on rise. An open on rise thermostat is used to control heating, while a close on rise switch is used to control cooling. They typically can handle 15 to 25 amps at 120 V (Figure 36-44). All line voltage thermostats have a switch, a sensing mechanism, and a mechanical linkage to allow the sensing mechanism to operate the switch (Figure 36-45).

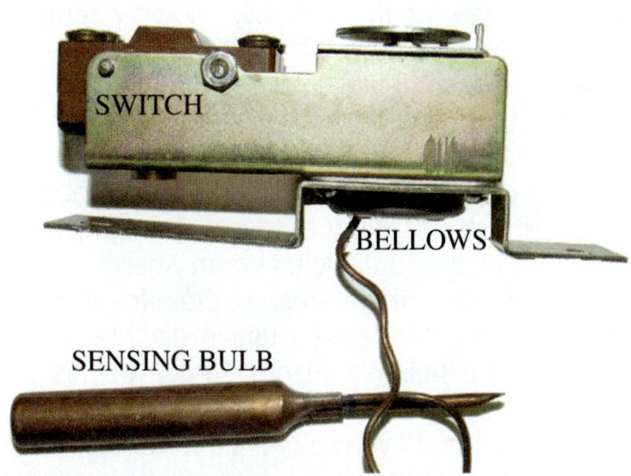

Figure 36-45 The switch, sensing bulb, and bellows on a line voltage thermostat.

Figure 36-46 Remote bulb thermostat.

The most common temperature-sensing mechanisms are charged bulbs and capillary tubes.

Charged bulbs contain a volatile fluid that increases in pressure when its temperature is increased. This pressure pushes on a bellows that operates the switch. An adjustable spring pushes against the bellows, allowing temperature adjustment. Figure 36-46 shows a charged bulb thermostat. This thermostat is sometimes referred to as a remote bulb thermostat. The controlling bulb can be placed in a location away from the body of the thermostat. For example, the bulb for a commercial refrigeration cooler can be placed inside the cooler with the capillary tube run through the wall to the thermostat located outside the room. Thermostat adjustments can be made without entering the refrigerated room.

Capillary-tube thermostats use a diaphragm mechanism instead of a bellows (Figure 36-47). In this control, the diaphragm is completely filled with liquid. The liquid expands and contracts with a change in temperature. The movement of the diaphragm is very slight, but the pressure that can be exerted is tremendous. Because the coefficient of expansion of liquid is small, relatively large-volume bulbs

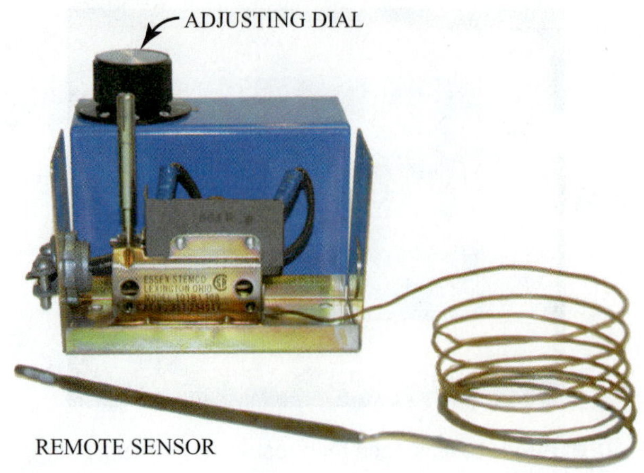

Figure 36-47 Capillary-tube thermostat.

are used on the control. This results in sufficient diaphragm movement and also ensures positive control from the bulb.

36.11 HUMIDISTATS

Humidistats control the operation of their switch based on changes in humidity. Hygroscopic elements are used on these controls, including human hair (Figure 36-48). As the moisture content of the air increases and the humidity rises, the hair expands and allows the electrical contacts to close (or open). As the humidity level decreases and the hair begins to dry, it contracts and once again activates the electrical contacts. This type of control is susceptible to dirt and dust in the air.

A more common form of humidistat uses a nylon element instead of human hair (Figure 36-49). The nylon is bonded to a light metal in the shape of a coil spring. The expanding and contracting of the nylon creates the same effect as that found in the spiral bimetallic strip used in thermostats. Another type uses a thin, treated nylon ribbon as a sensor.

An electronic circuit board can have hygroscopic properties, and lithium salt is used for this purpose. Another arrangement uses carbon particles embedded in a hygroscopic material. In both cases, the sensing element acts as a thermistor. Changes in the humidity affect the resistance of the material and alter the current in the electronic circuit.

Figure 36-48 Humidistat with hair for the sensing element.

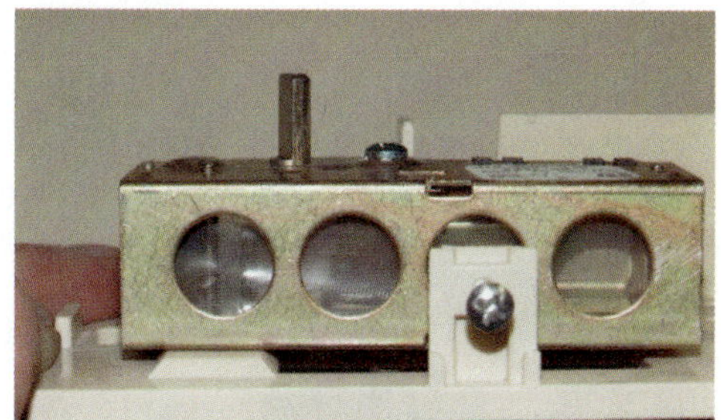

Figure 36-49 Humidistat with a nylon sensing element.

36.12 PRESSURE SWITCHES

Similar to a line voltage thermostat, a pressure switch uses a diaphragm or bellows to operate a switch. The difference is that the pressure operating the pressure switch comes directly from the refrigeration system. The pressure switch is either attached directly to the system, as in Figure 36-50, or it has a capillary tube that connects the pressure switch to the system, as in Figure 36-51.

The switching action of a pressure switch is described as either close on rise or open on rise. A high-pressure switch whose purpose is to shut the unit off when the high-side pressure exceeds a safe level uses an open-on-rise switching action. A low-pressure switch whose purpose is to shut off the system when the low-side pressure drops below a safe level uses a close-on-rise switching action. Because many commercial refrigeration systems use both a close-on-rise low-pressure safety switch and an open-on-rise high-pressure safety switch, dual-pressure switches combine the two switches into one mechanism (Figure 36-52).

Figure 36-50 Pressure switch attached directly to the system.

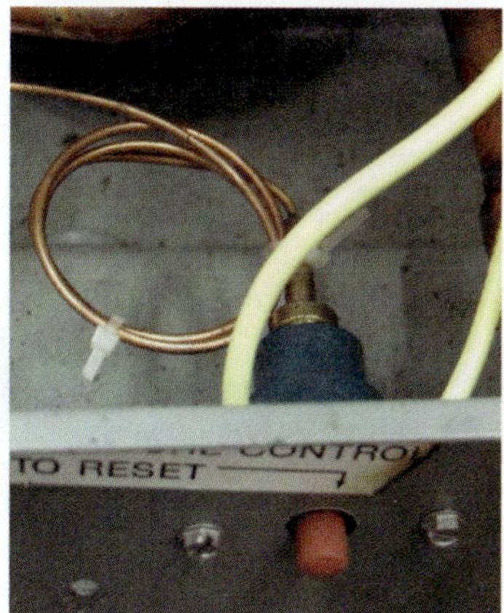

Figure 36-51 Pressure switch connected to the system with a capillary tube.

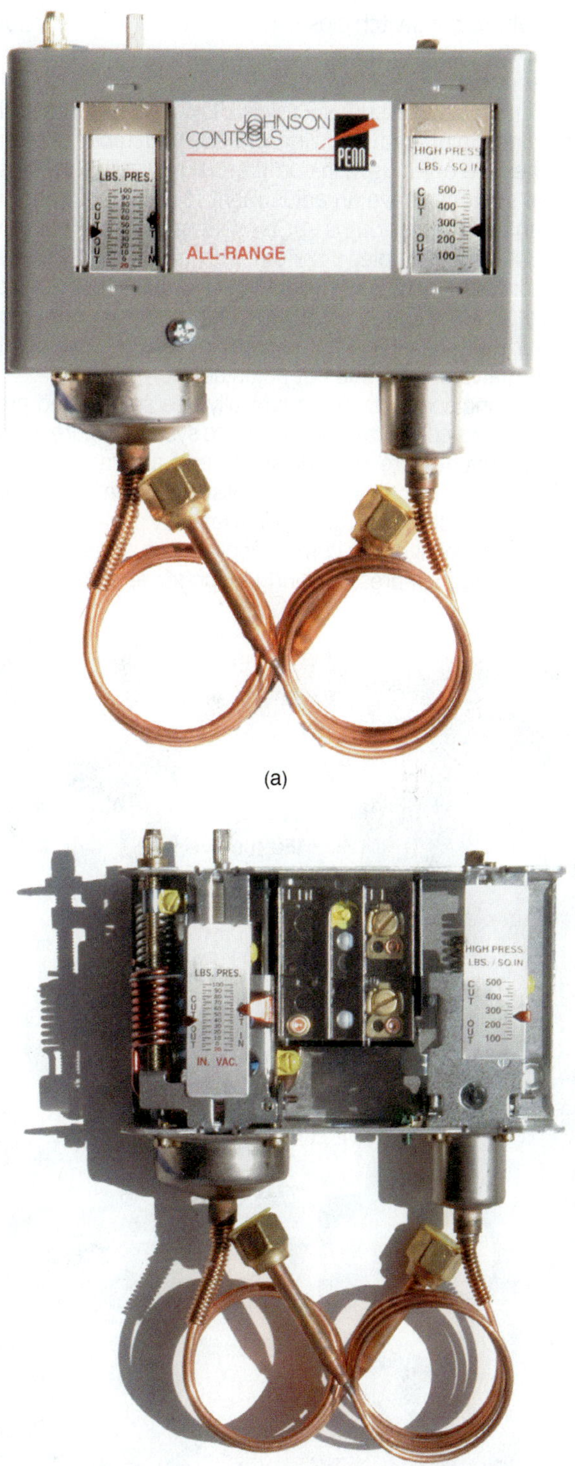

Figure 36-52 (a) Combination high- and low-pressure switch; (b) the high- and low-pressure sides of this control are separated inside the switch.

These have two pressure bellows controlling the same set of electrical contacts. The contacts will open if the high-side pressure exceeds the setting of the high-pressure switch or the low-side pressure falls below the setting of the low-pressure switch.

The pressure at which a switch closes and the pressure at which it opens cannot be the same pressure. There must be a difference between the two for the switch to work. The point where the switch closes is called cut-in, and the point where the switch opens is called cut-out. The difference between the two is differential.

Adjustable pressure switches allow adjustments to the operating point and the differential. Adjustable pressure switches are often used in commercial refrigeration. Close-on-rise switches have an adjustment for the cut-in and the differential. The cut-out is set by setting the cut-in and differential. For example, a switch with a cut-in of 250 psig and a differential of 50 psi would have a cut-out of 200 psig (Figure 36-53). Open-on-rise switches used for safety controls typically only have an adjustment for the cut-out. The differential is pre-set and not adjustable.

Air-conditioning systems typically use small, fixed pressure switches connected directly to the system (Figure 36-54). These switches are not adjustable and are manufactured to open and close at a pre-set pressure. These small, fixed pressure switches are normally pilot-duty switches, rated only for low-voltage, low-current applications. Typical applications include high-pressure and low-pressure safety cut-out switches. They are usually wired in series with the contactor coil. When they open, they break the circuit to the contactor coil and the contactor opens to turn off the compressor.

Figure 36-54 High-pressure switch on residential air conditioner.

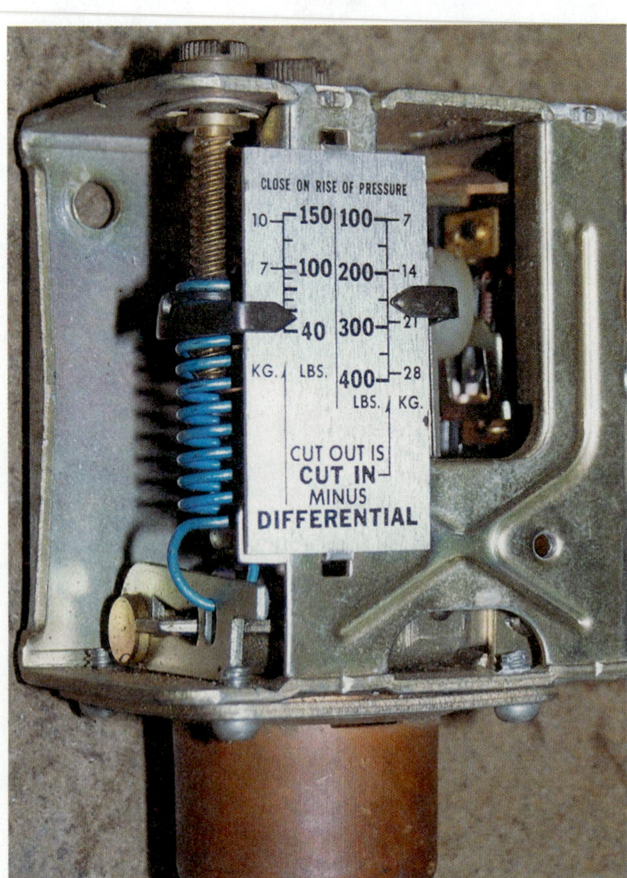

Figure 36-53 The cut-out on this low-pressure switch is 200 psig: 250 psig cut-in and 50 psi differential.

36.13 FLOW SWITCHES

Flow switches open or close based on the movement of fluid. Note that both air and water are fluids. The sail switch, shown in Figure 36-55, is a protective device to prevent the operation of a unit when there is inadequate fluid flow. In an air system, the sail switch is placed in the duct to sense the flow of air. Unless there is an adequate supply of air over the coil, the unit is either not started or shut down. Switches of this type can also be placed in a waterline feed of a water-cooled condenser. If there is an inadequate supply of water, or no water, the unit is prevented from running.

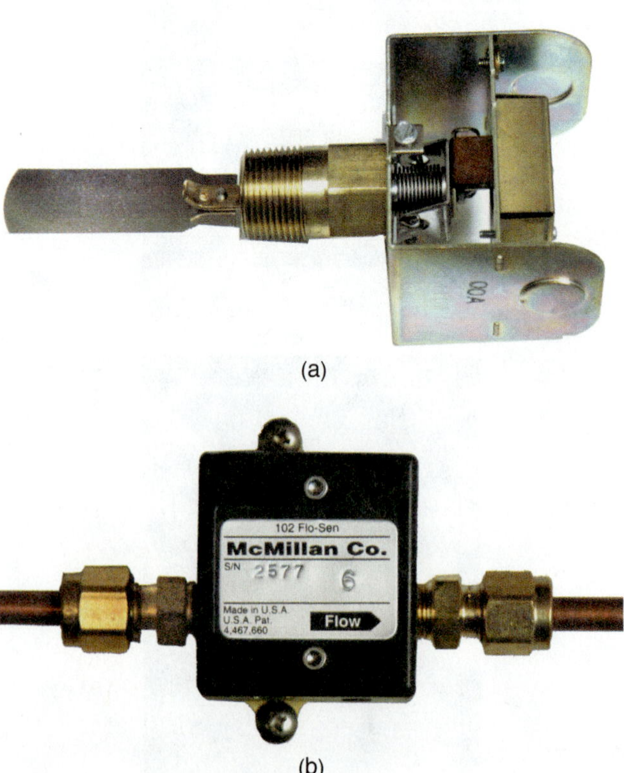

Figure 36-55 (a) Sail switch to detect airflow; (b) electronic flow switch.

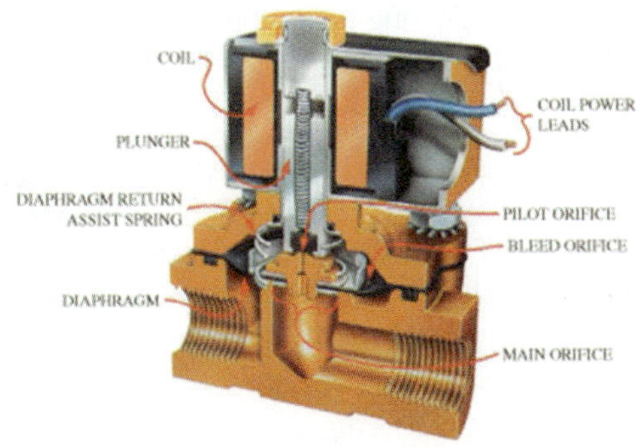

Figure 36-56 Cutaway of a solenoid valve.

36.14 SOLENOID COILS

A solenoid coil is a doughnut-shaped coil of wire that produces a strong magnetic field when energized. Solenoid coils typically have steel plungers that are attracted to the coil when it is energized. The movement of the plunger can be used to control many HVACR electrical devices including solenoid valves, gas valves, relays, and contactors.

Figure 36-56 shows a cutaway of a solenoid valve. A solenoid valve is an electrically operated valve that controls the flow of a fluid by opening or closing when the solenoid coil is energized. They may be either normally open or normally closed. When the coil is energized, the valve will change position. Normally open valves will close; normally closed valves will open.

36.15 RELAYS

An automatic switch requires some method for opening and closing. This is often accomplished through the use of a relay (Figure 36-57). A relay is an electrically operated switch that uses an electromagnet to open or close a set of electrical contacts (Figure 36-58). Normally only a small amount of current is required to energize the electromagnet. This allows for a device with a high current rating, such as an electric heater, to be operated by a control relay operated by a low-current signal. The control wires can be much smaller and separate from the large main supply lines required for the main load.

Figure 36-58 The visible coil of fine wire is the part of the electromagnet arrangement of this relay.

Relays can be designed for normally open switches or normally closed switches. The normal position of the switch is always the position of the switch when the relay coil is de-energized. Figure 36-59a shows a normally closed set of contacts with the relay coil de-energized. A spring is used to hold the contacts together. When the coil is energized (supplied with current) (Figure 36-59b), a magnetic field is set up that attracts the lower contact toward the coil. This will separate the two contacts and open the circuit. As long as the current flows through the coil, the relay is energized, and the switch will remain open. With a normally open

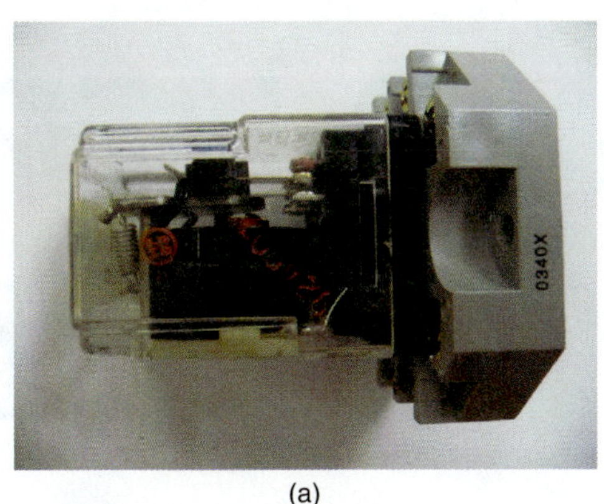

(a)

(b)

Figure 36-57 (a) Control relay; (b) control relay connected in the circuit.

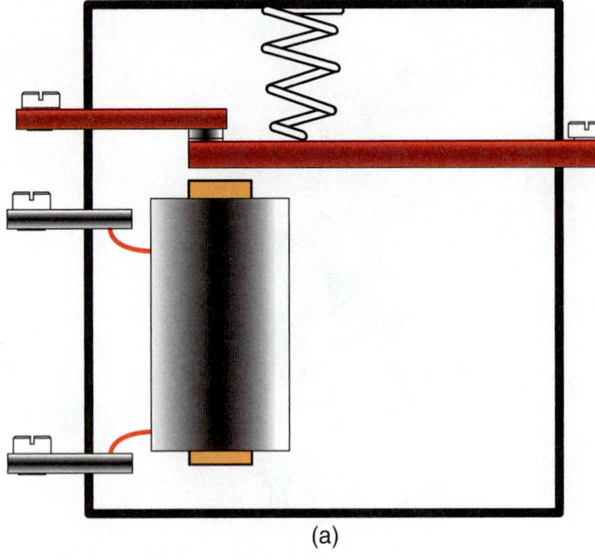

(a)

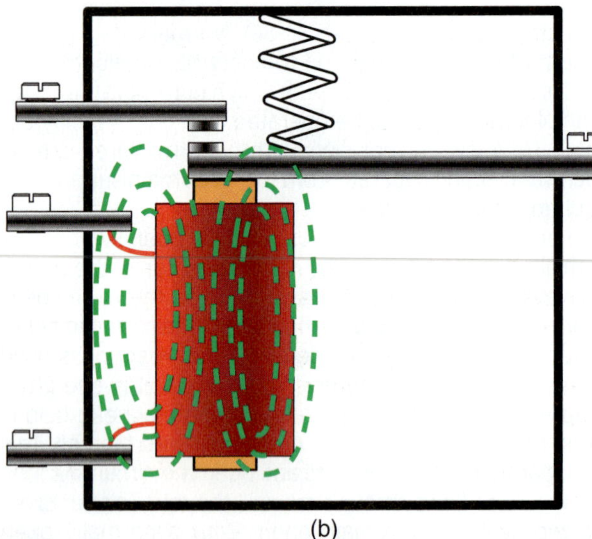

(b)

Figure 36-59 (a) The coil is de-energized and the contacts are held closed by spring force; (b) the coil is energized and the contacts are opened by the magnetic attraction of the coil.

set of contacts, the spring holds the contacts open. When the coil is energized, the magnetic field pulls the contacts together.

Relay coils are available in all common voltages, including 24 V, 120 V, and 230 V. There is no electrical connection between the coil and the contacts. The contacts are separate electrical devices and have their own rating. The contacts are rated for both voltage and amperage. It is common for the contacts to switch higher voltage than the coil voltage.

36.16 TESTING RELAYS

The relay coil can be tested by checking its resistance. The coil should be disconnected from the circuit to ensure that you are only checking the coil. This can be done by disconnecting the wires from the coil. Relay coils should have a measureable resistance: not open and not shorted. A meter reading of infinite Ω (OL on most digital meters) indicates that that the coil is open. A reading of 0 Ω indicates that the coil is shorted. The coil resistance will vary depending upon the design of the relay and its operating voltage. The contacts should ohm as either open or shorted because they are a switch. With the coil de-energized, the normally open contacts should read infinite Ω (OL) and the normally closed contacts should read 0 Ω. When the coil is energized, the readings reverse: the normally open contacts should read 0 Ω and the normally closed contacts should read infinite Ω (OL).They should never have a measurable resistance.

36.17 CONTACTORS AND STARTERS

Relays, contactors, and starters all have contacts that open and close to complete or disconnect a circuit (Figure 36-60). Contacts for control relays are often very small (Figure 36-61). This is because they do not need to be connected to the load current, since they are only relaying the signal. Typically, relay contacts are rated for 15 amps or less.

Contactors that are connected in the load circuit will need to be larger (Figure 36-62). The operation of a contactor is shown in Figure 36-63. There may be one or more sets of contacts located on the armature. Contactor poles are the number of contact sets. A single-pole contactor has one set of contacts, a two-pole has two sets, and a three-pole has three sets. The armature moves up and down in

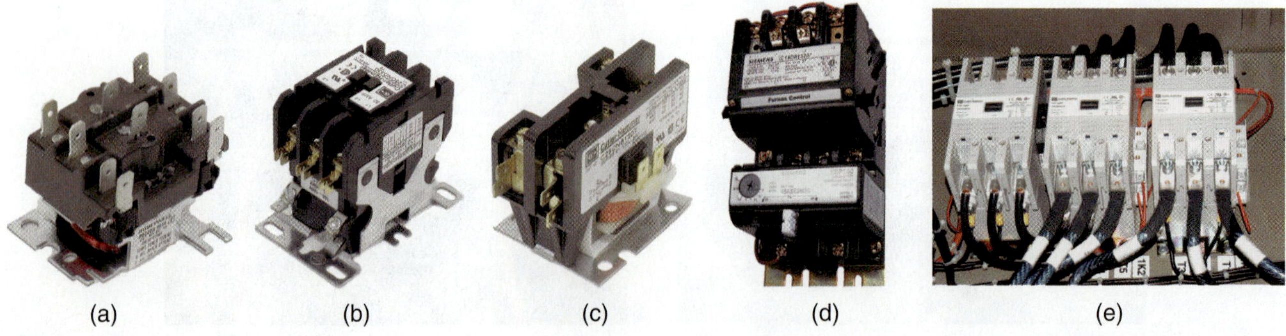

(a) (b) (c) (d) (e)

Figure 36-60 (a) Fan relay; (b) three-phase contactor; (c) single-pole contactor; (d) motor starter; (e) large-amperage systems may use multiple contactors all wired together.

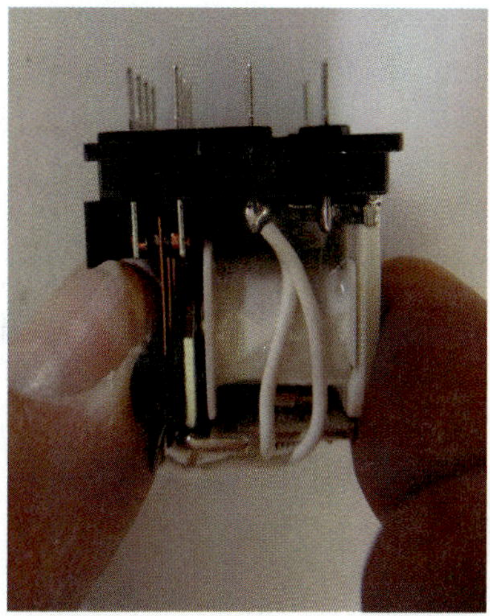

Figure 36-61 The contacts for this control relay are very small.

(a)

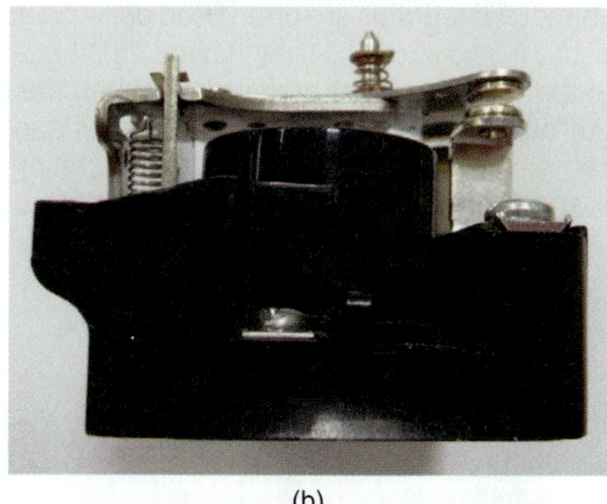

(b)

Figure 36-62 (a) Front view of contacts; (b) side view of contacts.

the holding coil. When the coil is energized the armature moves and closes the contacts. The contacts are considered normally open since they are open when no current is applied to the coil. Contactors may also be designed to have normally closed contacts that are closed when the coil is de-energized.

The contacts are rated by the voltage and current they can switch. There are normally three current ratings: resistive, FLA, and LRA (Figure 36-64). The resistive rating is the amount of current the contacts can switch for a resistive device such as an electric strip heater. The resistive rating is higher than the FLA (full load amps) rating because resistive loads do not have a high starting current. FLA is the current rating for motors. A motor's FLA current is the amount it draws when doing the full amount of work it was designed to do. A contactor's FLA rating is lower than its resistive rating because of the high starting current of motors. The LRA rating stands for locked rotor amps. A motor's LRA is the amount of current the motor draws if it is energized while its rotor is not moving. This current is very high, typically four to five times as high as the FLA operating current.

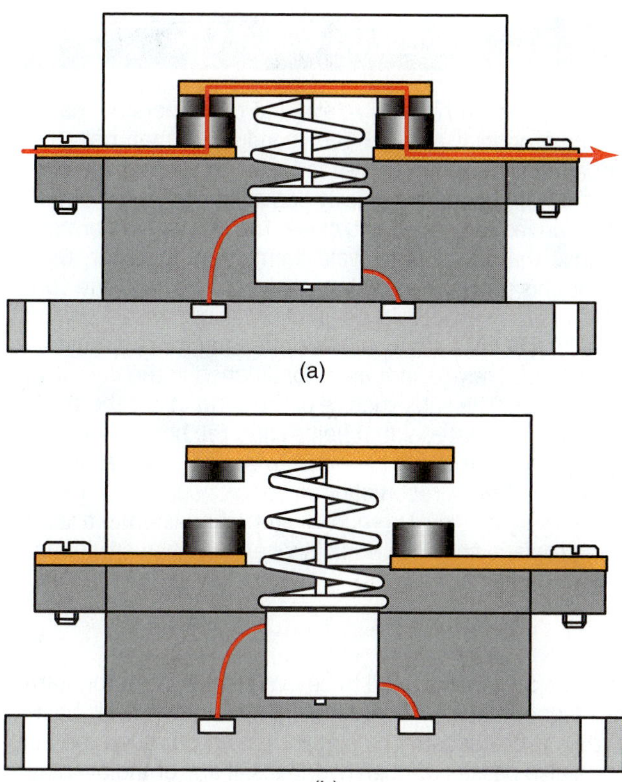

Figure 36-63 (a) Contacts closed to complete the circuit; (b) contacts opened to break the circuit.

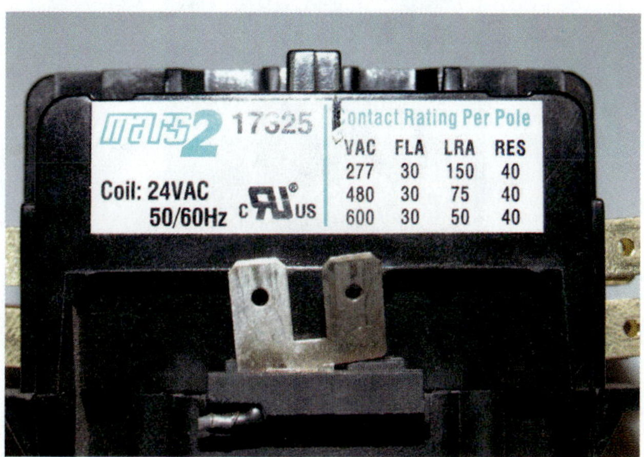

Figure 36-64 The current ratings on a contactor.

SERVICE TIP

The contact surface in relays and contactors is coated with a special alloy of very conductive material. This coating is very thin but effective in preventing the contacts from becoming quickly damaged from the momentary arc as they open and close. This arc would normally cause the contacts to weld themselves together. Over time this protective coating will wear away and the contactors may eventually begin to occasionally stick. When sticking occurs, the contactor or contactor tips must be replaced. They cannot be reconditioned in the field with point files. The substructure of the contact is a thermally conductive material that helps carry the heat away from the coating. It is not resistant to arcing and welding itself together. A reconditioned contact surface will stick again very quickly. Do yourself and your customer a favor and replace sticking contacts whenever they are found.

Starters for motors need to be even larger. A motor starter or magnetic starter is essentially a contactor with built-in overload protection. The current passing through the contacts also passes through overloads. If any of the overloads open, they break the circuit to the starter coil, shutting it down. The control voltage is used to operate the holding coil on a starter as well as for many other control functions. Motor starters work on the same magnetic coil principles as relays and contactors, but the actual contact size will be greater. The contacts should snap open and close with spring assist to reduce arcing across the contacts. A magnetic starter is shown in Figure 36-65.

Motor starts are often used with three-phase motors. Since three-phase motors have three circuits, opening any one leg of power still leaves an energized circuit. The overload in a magnetic starter protects the motor by breaking the circuit to the starter coil, which then opens the contacts, de-energizing all three circuits.

Figure 36-65 Magnetic motor starter.

Figure 36-66 Manual reset overload.

36.18 CURRENT/TEMPERATURE SAFETY COMPONENTS

Electrical overloads provide for protection against excessive current. Overloads can be line duty or pilot duty. Line-duty overloads break the power to the device they are protecting. Pilot-duty overloads operate like relays. They have a separate coil and contacts. Once they have tripped, some overloads reset themselves automatically after cooling down. These are called automatic reset overloads. Others use a mechanical design that requires a button to be pushed to reset it (Figure 36-66).

Line-Duty Overloads

Thermal bimetal overloads used with electric heaters are a good example of a line-duty overload (Figure 36-67). They open to break the circuit to the heater if the strip heater overheats.

The bimetal overloads are often used on small hermetic compressors (Figure 36-68). They have a small heater and a bimetal that controls a set of contacts. When the current passing through the heater exceeds the overload rating, the heater will cause the bimetal to warp, opening up the contacts and breaking the circuit. Many current overload

Figure 36-67 Thermal overload used to protect electric strip heaters from overheating.

Figure 36-68 Bimetal current overload used on small hermetic compressor.

devices sense both the current and temperature of the motor. These devices, like those located within the Bakelite cover of small compressors, must be in their enclosure to function properly. If the Bakelite cap is left off, the overload will only sense the current and not the temperature. This reduces their sensitivity and removes a large portion of the overload safety protection from a motor.

Pilot-Duty Overload Relays

Pilot-duty overloads are called overload relays because they have a coil that senses the overload and a set of contacts. The coil can be either thermal or magnetic. Overload relays are used in conjunction with contactors. When a contactor is used, there is both a control circuit, in which the primary control is inserted, and a load circuit, which is opened and closed by the contactor. When excessive current is drawn in the load circuit, this device will open the control circuit.

The thermal overload relay is shown in (Figure 36-69). The current for the contactor coil passes through the normally closed overload contacts. If the load current passing through the bimetallic element in the overload coil becomes too high, the element bends to the side, forcing the contacts apart. This breaks the control circuit and allows the contactor to open, thereby interrupting the power to the load. The bimetallic element will now cool and return to its original position, but because of the slot in the bottom arm, the contacts will not be remade. The control circuit will remain broken until the reset button is pushed. The reset moves the arm to the right and closes the control-circuit contacts. Because this control must be reset by hand, it is usually referred to as a manual-reset overload. Some overload relays may be field adjusted for manual or auto-reset function.

Magnetic overload relays are another type of pilot-duty overload (Figure 36-70). The advantage of this type of overload is that it is only slightly affected by ambient temperatures, thereby avoiding nuisance trips. The magnetic overload relay is made up of a sealed tube completely filled with a fluid and holding a movable iron core. When an overload occurs, the movable core is drawn into the magnetic field, but the fluid slows its travel. This provides a necessary time delay to allow for momentary high current during motor startup (locked rotor amps) without tripping the overload. When the core approaches the pole piece, the magnetic force increases and the armature is actuated, thus breaking the control circuit.

Figure 36-69 The overload on the bottom of this magnetic starter will break the circuit to the starter coil if the current exceeds its setting.

Figure 36-70 Magnetic overload relay.

On short circuits, or extreme overloads, the movable core is not a factor because the strength of the magnetic field of the coil is sufficient to move the armature without waiting for the core to move. The time-delay characteristics are built into the overload and are a function of the core design and fluid selection.

36.19 ELECTRONICS AND LOGIC CIRCUITS

Electrical circuits can have any number of input devices. Some examples are pushbuttons, mechanical limit switches, pressure switches, and photo cells. This input is then transmitted in a logical fashion to an actuating device. These could be relays, contactors, motor starters, solenoids, or some component that will start a series of event to satisfy the initial input.

Relay Logic

Many systems have been designed to use relay logic. These circuits will make decisions based upon the initial input. Take, for example, that the input is a call for heat by the thermostat. This input signals for contacts to close, which will energize a relay, and this begins a sequence of events to start the furnace. If a flame is detected, the furnace continues to run. If not, the furnace shuts down. This could be seen as an OR function. Stay running OR shutdown.

From this example, it can be seen that with relay logic, relay coils are designed to control other relay coils. Two contacts wired in series would be considered an AND function. First one contact closes AND the second contact closes to energize the load. Contacts in parallel will produce an OR function. Contact 1 OR contact 2 will close.

Relay logic is quickly becoming replaced by electronic systems. One reason is that mechanical switches never truly make a good, clean closing contact. The points tend to bounce against each other several times before fully closing. Sometimes this causes minor sparking. Relays like this cannot be used in an explosive environment unless they are enclosed in an airtight container. Sometimes capacitive switch filters are used to help smooth out this bounce.

Contacts also wear out over time as they continually open and close. They are exposed to any chemicals or dust in the air, and mechanical linkages can eventually stick and the contacts become pitted. In comparison to electronic solid-state logic, relays are slow and heavy.

Solid-State Logic

Unlike relay logic, solid-state components have no moving parts (Figure 36-71). They are sealed from the atmosphere. They take up less space because most of the components are attached to a compact circuit board (Figure 36-72). They generally cost less and require less power. Most important, they can be more than a thousand times faster than relays.

Solid-state logic circuit input signals are used to slightly vary voltage rather than operate contacts. These signals can be designed as low and high. A low voltage (0 V) might be used to open the circuit, while a higher voltage (+5 V) could

Figure 36-71 Solid-state relays.

(a)

(b)

Figure 36-72 (a) Circuit board for gas heater; (b) circuit board for an air-conditioning console.

be used to close it. Since this voltage is low, it consumes little power, but an output amplifier is still required for the actuating device, such as a solenoid. The output amplifier will increase the low-voltage/low-current power to higher-voltage/higher-current output power.

Relay Logic Compared to Solid-State Logic

It is obvious from the descriptions of both systems that electronic solid-state logic systems are faster, cheaper, lighter, and last longer than conventional relay-logic systems. The older conventional relay systems will continue to be replaced. However, relay systems still do have a few distinct advantages. If the circuit is small and simple, they can actually be a cheaper alternative. Outside extraneous noise pickup and signals will not affect relay operation but can raise havoc on some solid-state systems. Solid-state components will generate a lot of heat and need to be cooled, while relays are capable of operating in higher-temperature locations. One last most important consideration is the technician. Many technicians in the field have been trained on relay logic and they fully understand it, which is not the case with many solid-state electronic circuits.

36.20 TRANSDUCERS

Transducers are electrical or electromechanical devices that are often used to provide a control signal (Figure 36-73). They can be used to sense changes in pressure, temperature, light, sound, and vibration. They are most often semiconductors that convert mechanical force into an electrical signal. The signal is used by a control device or microprocessor to stop, start, or adjust a system's operation. For example, a vibration sensor located on a large air handler would stop the motor when excessive vibration occurred. Another example would be a water depth gauge, which would sense water pressure in a container. This signal could be used to stop or start a pump or turn a water valve on or off.

Because transducers are relatively low in cost and have high reliability, they are used frequently to provide remote sensing for HVACR equipment. Transducers have increased the life of systems and improved operating efficiencies by providing constant monitoring.

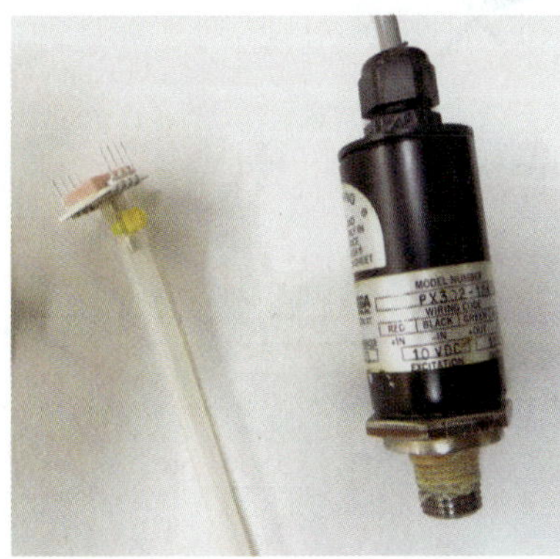

Figure 36-73 Pressure transducer and electrical lead.

36.21 SOLID-STATE COMPONENTS

Circuits using solid-state components are different than conventional mechanical switching devices because the electrons are confined entirely within the solid material. Some common solid-state components include transistors, diodes, and silicon rectifiers.

Semiconductors

Some materials are good conductors of electricity, while others are insulators. A semiconductor is a material that is neither a good conductor nor a good insulator. Semiconductor materials are often made from germanium and silicon. These materials, due to their peculiar crystalline structure, may under certain conditions act as conductors and under other conditions act as insulators. This ability to either conduct current or block current flow can be useful in a control circuit.

Transistors

Transistors were one of the first most popular solid-state devices used and became well known for the mass production of transistor radios. They can be used as a switch or to amplify an electrical signal. As an example, in a HVACR system, these can be used to amplify a signal of low-voltage/low-current power to a higher-voltage/ higher-current power to operate a relay.

Diodes

A diode is a semiconductor that acts similar to a check valve, allowing for one-way flow through an electrical circuit. A diode has an anode and a cathode (Figure 36-74). If the anode is connected to the positive terminal, then the diode is forward biased and current will flow. If the anode is connected to the negative terminal, then the diode is reverse biased and no current will flow. Diodes can vary from the size of a pinhead to much larger sizes for ratings of 500 amperes or more (Figure 36-75).

Silicon Rectifier

Using the principle of the diode, a simple full wave-bridge rectifier can be configured to convert alternating current into direct current. Rectifiers are commonly used because many devices such as meters and control systems often require DC power. Power supplies for battery-operated devices such as cell phones and laptop computers require AC to DC converters. Figure 36-76a shows the current flow to a direct current meter through diode rectifiers 3 and then 2. When the AC current reverses, the meter will still

Figure 36-74 Diode showing anode and cathode.

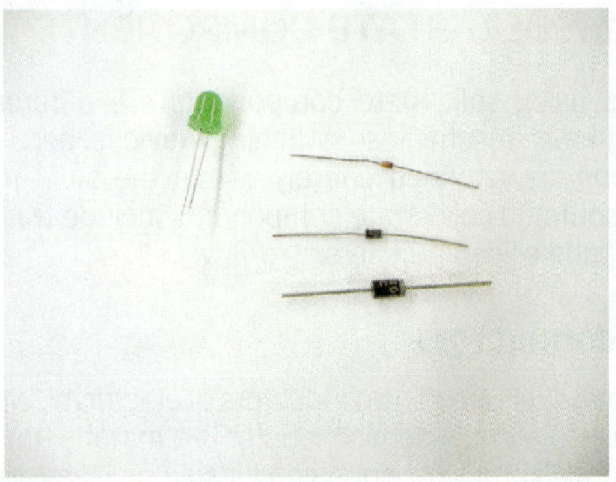

Figure 36-75 Common diodes.

Figure 36-77 Silicon rectifier.

be operating on direct current with current flow through diode rectifiers 1 and 4, as shown in Figure 36-76b. Silicon rectifiers can be designed to operate throughout a wide range of current and voltage levels and are therefore used on many different types of applications (Figure 36-77). They are small, lightweight, and can be made shock resistant. Silicon rectifiers can have efficiencies as high as 99 percent.

Silicon-Controlled Rectifier (SCR)

The increased concern about energy efficiency has led to more systems using variable and controlled power. There are a number of examples, such as lighting, that can be adjusted through a range rather than always at a maximum brightness. Refrigeration compressors and fans use variable-speed motors that can slow down or speed up depending on the demand.

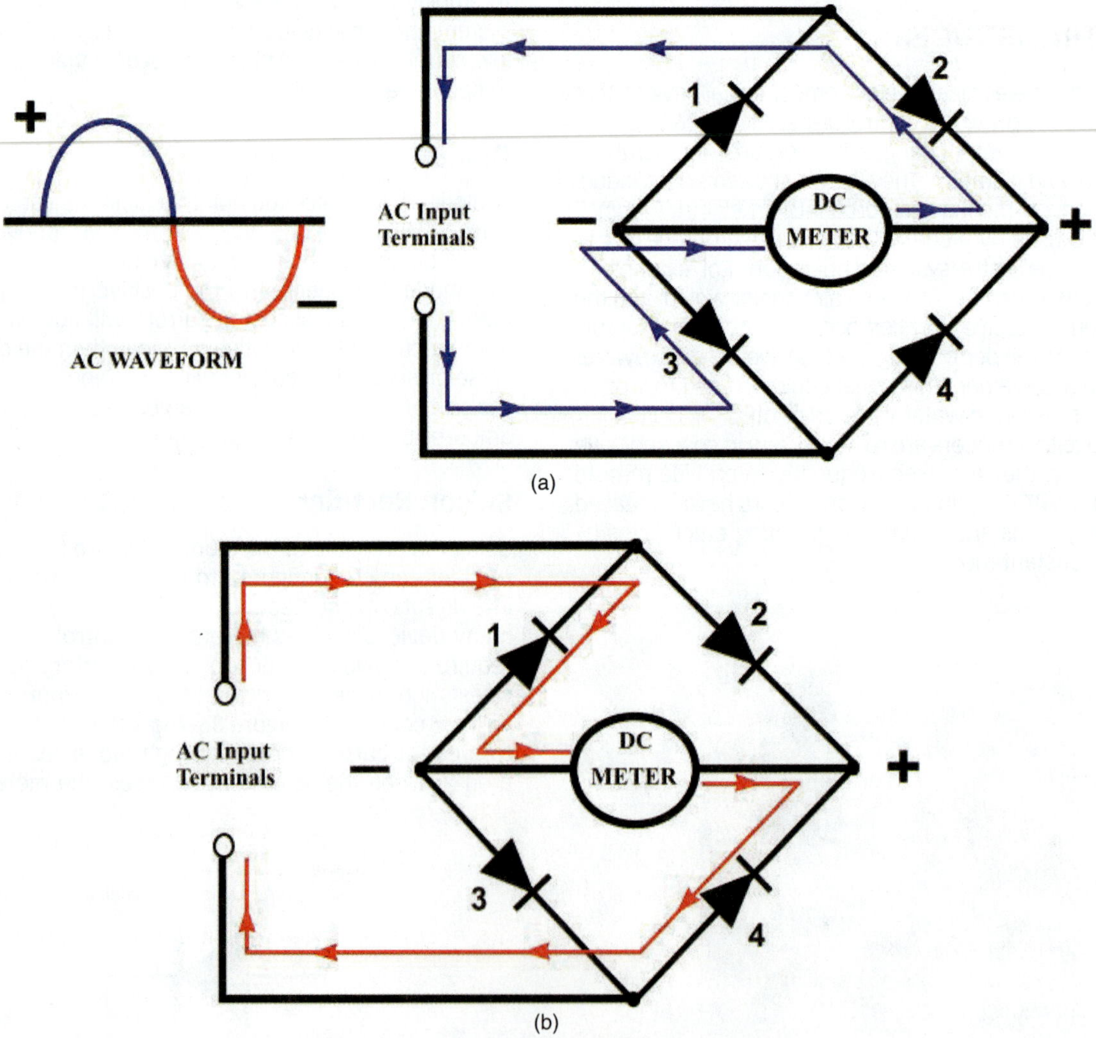

(a)

(b)

Figure 36-76 (a) Full-wave silicon rectifier; (b) full-wave silicon rectifier with opposite current flow.

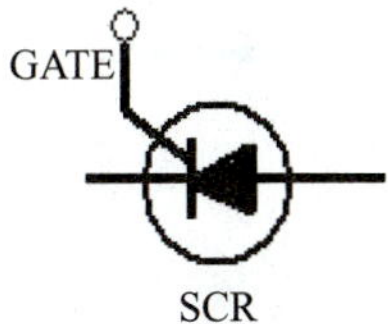

Figure 36-78 Silicon-controlled rectifier with gate control.

Figure 36-79 TRIAC bidirectional device.

Conventional systems use large, expensive transformers that allow for a variable adjustable secondary coil output. These are being replaced by silicon-controlled rectifiers (SCRs) because they are small, relatively inexpensive, require virtually no maintenance, and are very efficient. They can be found in small control circuits as well as in large circuits of several hundred amperes with operating voltages over 1,000 V.

The SCR is similar to a diode in that it will block reverse current. The difference is that the SCR can also block forward current. More important, it can be switched electrically to allow forward current conduction by a pulse of current flowing through its gate (Figure 36-78). When the SCR is triggered by a gate pulse, conduction begins. The important application for an SCR is that it can be turned on and off like a switch without the need for electrical contacts. The speed of switching is also much faster than traditional contacts, and the timing can be varied during operation. A TRIAC, shown in Figure 36-79), is a bidirectional device of two SCRs connected in parallel-opposed with a common gate. A positive voltage applied to the gate will "fire" the TRIAC in either direction.

UNIT 36—SUMMARY

Resistors and capacitors are commonly found in most circuits. Electronic circuits often use variable resistors that allow for adjustment. Capacitors are used for many applications, such as tuning, filtering, energy storage, power factor correction, and motor starting. Transformers have primary and secondary windings that can be configured to step up or step down the voltage. A motor powered by a 440 line voltage will normally require a step-down transformer to lower the voltage for the relays and controls. Fuses and circuit breakers are used to limit the current flow in a circuit to avoid overheating and fire. Switches, relays, and contactors are used to close and open circuits. Relays can be used in combination to start, stop, and engage safety cutouts. This type of relay logic is quickly being replaced by solid-state logic, but many systems still combine some combination of both.

UNIT 36—REVIEW QUESTIONS

1. A fixed resistor has the following color bands: orange, red, brown, gold. What is the rated resistance?
2. A fixed resistor has the following color bands: orange, red, gold, gold. What is the rated resistance?
3. Does a capacitor lose its charge once it is disconnected from the power source?
4. How do ceramic and mica capacitor mfd ratings compare to paper and film capacitors?
5. How do motor-start capacitors differ from motor-run capacitors?
6. Why are some run capacitors marked with a red dot?
7. What is the first step in testing a capacitor?
8. What is the control voltage used by most residential HVACR equipment?
9. What is a center-tapped transformer?
10. What is an isolation transformer?
11. Why are transformer cores laminated?
12. Do transformers operate more or less efficiently when fully loaded? Explain your answer.
13. What is the advantage of a circuit breaker as compared to a fuse?
14. What is a proximity switch, and how does it operate?
15. What type of applications is a close-on-rise thermostat normally used for?
16. What system functions does a low-voltage thermostat normally control?
17. What two manual switches are found on most low-voltage thermostats?
18. What types of systems use staging thermostats?
19. What do modern electronic thermostats use to sense temperature?
20. What types of material are used to sense humidity in humidistats?
21. What three ratings should always be matched when changing fuses?
22. What type of fuses should be used to protect motor circuits?
23. What is a relay?
24. When checking the resistance of the different parts of a relay, what resistance should you expect for the coil, normally open contacts, and normally closed contacts?
25. What is the difference between a relay and a contactor?
26. What is the difference between a contactor and a magnetic starter?
27. What is the cut-out for a close-on-rise pressure switch with a cut-in setting of 45 psig and a differential setting of 20 psi?
28. On a close-on-rise pressure switch, what should the differential be set to for a cut-out of 15 psig if the cut-in setting is 25 psig?
29. Why is the contact surface in relays and contactors coated with a special coating of very conductive material?
30. Consider basic relay logic. What is meant by an AND function? What is meant by an OR function?
31. What are the major differences between relay and solid-state logic?
32. What is a semiconductor material?
33. What is a diode?
34. What is a silicon rectifier?

UNIT 37

Electric Motors

OBJECTIVES

After completing this unit, you will be able to:

1. describe the operation of AC induction motors.
2. explain the importance of torque, speed, and power usage for motors.
3. explain how and why capacitors are often used for single-phase motors.
4. list the different types of single-phase motors.
5. describe the differences between single-phase and three-phase motors.
6. list the different types of motor-protection devices.
7. test a motor circuit for proper operation.

37.1 INTRODUCTION

Electric motors are the most important load device in the various types of HVACR units. They convert large portions of electrical energy to useful work. It is important, therefore, for the technician to understand how they operate and how they can be protected. It is also important to note that not all motors are exactly the same because different motors are designed for various types of service. Even though all motors are not the same, they all share the important characteristics of torque, speed, and power usage.

37.2 AC INDUCTION MOTORTORQUE

Torque is the twisting (or turning) force that must be developed by a motor to turn whatever it is driving. The power required for a motor to drive a fan, compressor, or any other piece of equipment is directly related to the torque and speed required. A greater amount of torque is required to start a motor than to run it. The starting torque requirements for a fan are low. The starting torque requirements for a reciprocating compressor are high. Providing extra starting torque is expensive. Therefore, to keep the cost down the motor is selected with the smallest torque that will adequately perform the work for which it is intended.

TECH TIP

Manufacturers can produce motors with high starting torque, high running torque, or good running efficiency. It is impossible, however, to produce a single motor that has all three characteristics. The performance required of motors will determine what type of motor to use. Motors that frequently start but do not run for long periods of time would benefit from high starting torque. Motors that operate for long periods of time under heavy loads would benefit from having high running torque. Motors that operate for long periods of time under relatively light loads will benefit from good running efficiency. Manufacturers of equipment take these three factors into consideration when specifying which motor is to be used. For that reason, it is recommended that you select a replacement motor within the same category as the one originally supplied with the equipment.

37.3 INDUCTION MOTOR PRINCIPLES

The two principal parts of a motor are the stator and the rotor. The stator is the stationary part, and the rotor (often referred to as the armature) is the rotating part, as shown in Figure 37-1. The stator core is built of slotted

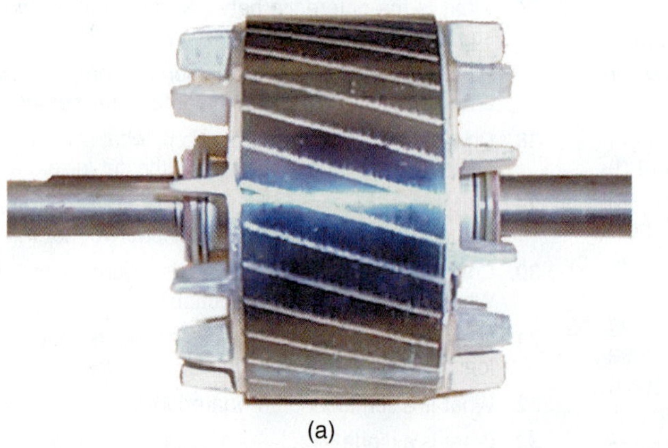

(a)

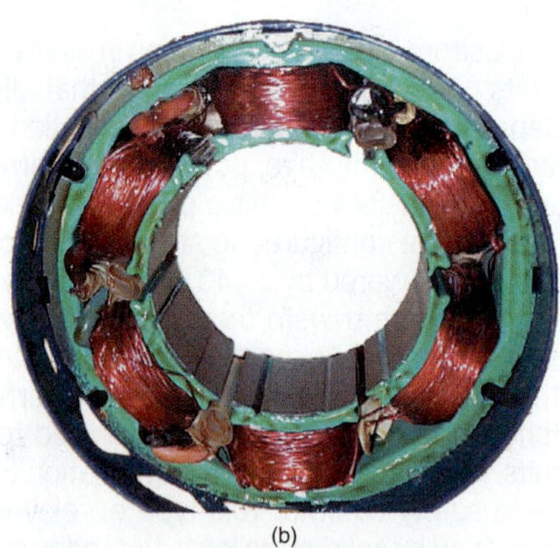

(b)

Figure 37-1 Parts of an electric motor: (a) rotor; (b) stator.

Figure 37-2 Squirrel cage–type rotor.

steel laminations. Coils of wire called windings are spaced in the stator slots. When current is applied to a stator coil, a magnetic field is generated. Depending on the direction of current flow through the coil, either a north or a south pole will be produced. Each winding can be considered a pole.

The rotor is a series of aluminum (or copper) bars mounted on a soft iron core. The core provides a path for the magnetic field of the rotor. The conductor bars are shorted together by an end ring permitting current to flow, as shown in Figure 37-2.

The current passing through the stator creates a powerful magnetic field, as shown in Figure 37-3. No current is actually supplied to the rotor. Instead, the magnetic field of the stator induces current to flow through the rotor. This current induced in the rotor will also produce a magnetic field. However, the current induced in the rotor is in the opposite direction from the stator current. This flow of current in the opposite direction creates an opposing magnetic field in the rotor, which reacts with the stator field. The repulsion and attraction between the poles of the rotor and the stator cause the rotor to turn.

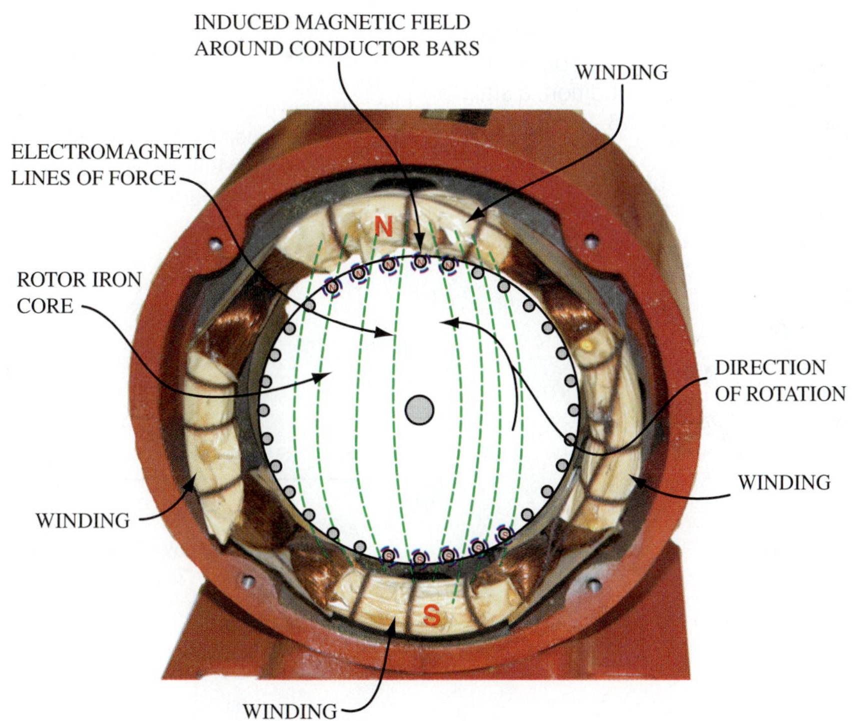

Figure 37-3 The induced magnetic field in the conductor bars is the same as the electromagnetic field; because just as magnetic poles repel each other, the rotor is forced to spin as the fields repel each other.

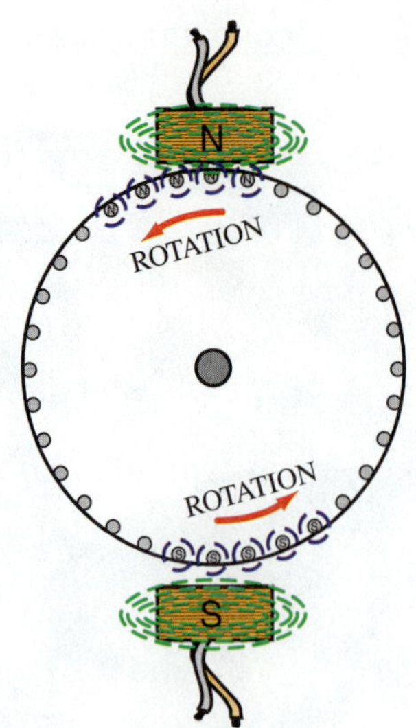

Figure 37-4 The magnetic field induced in the rotor causes it to rotate.

As shown in Figure 37-4, the like poles (north to north and south to south) in the stator coil repel the like poles in the rotor winding, causing rotation. In Figure 37-5, the rotor winding is shown with both north and south poles relatively near each other. The stator coil repels the like poles and attracts the opposite poles.

As the alternating current in the stator coils changes direction, their polarity also changes. The north pole coils now become south poles and vice versa (Figure 37-6). The current flow through the rotor also reverses direction, so the polarity of the rotor changes along with the polarity of the stator. Even though the current is alternating, a continuous rotation is produced.

A problem can occur if the rotor is stopped in the position shown in Figure 37-7. In this position, regardless of polarity, no motion can occur. This position is sometimes described as dead center. To correct this condition, a start winding is added, as shown in Figure 37-8. The original winding is called a run winding. The two windings—the run winding and the start winding—are out of phase with each other, and enough torque is created to start the rotor turning. Motors of this type are called split-phase motors.

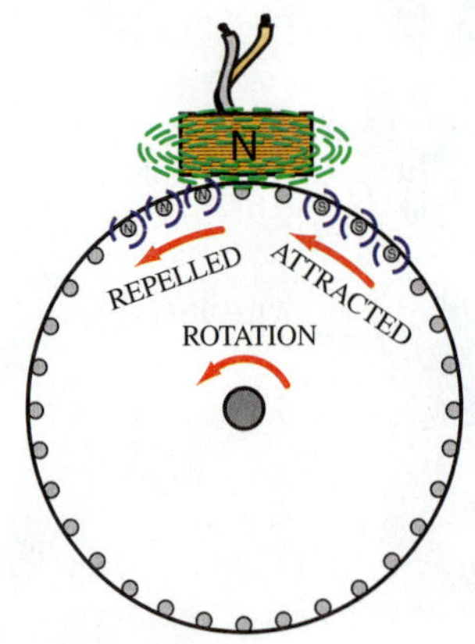

Figure 37-5 On one side the like magnetic poles are being repelled, and on the other side the opposite poles are being attracted.

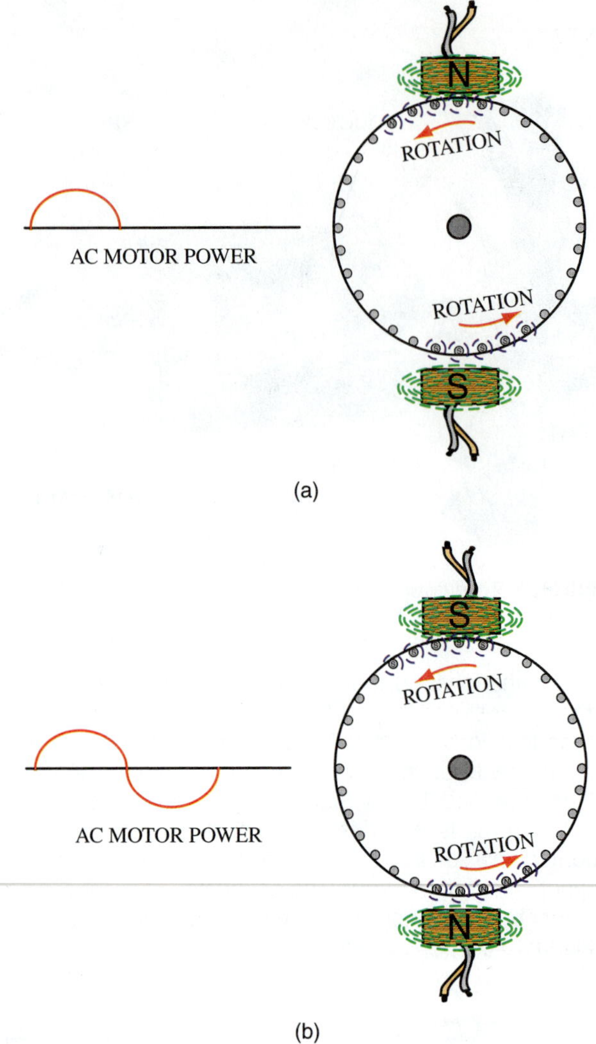

Figure 37-6 (a) In the first half of the AC power cycle, the rotor is pushed away from one side and attracted to the other; (b) in the second half of the AC power cycle, the current direction through the windings changes and the process is repeated; this is what keeps the motor spinning.

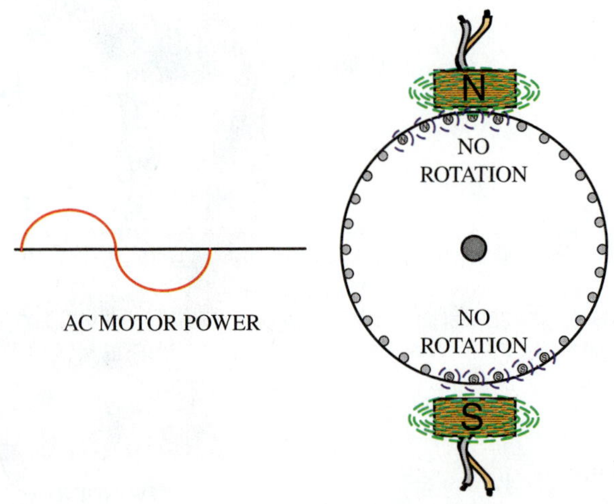

Figure 37-7 A motor cannot start rotating if the magnetic fields are at dead center.

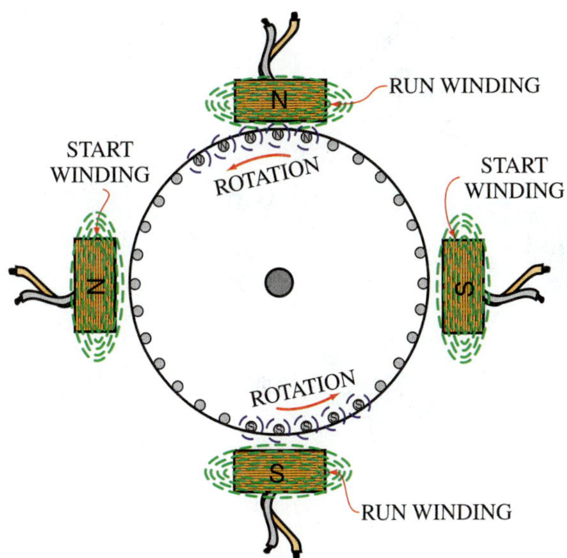

Figure 37-8 Start windings are offset from run windings to get the rotor out of dead center; start windings are often made of finer wire than run windings.

Because of their low starting torque, they are used only on fractional horsepower applications.

37.4 MOTOR SPEED

The speed of a motor is determined by the number of poles (stator coils) and the frequency (hertz) of the alternating current. A greater number of poles results in slower motor speeds. Higher frequency produces faster motor speeds. Speed is measured in revolutions per minute (rpm). The maximum speed of a motor is known as the synchronous speed.

CAUTION

The rpm of induction motors is totally dependent on the number of poles and the frequency of the alternating current. An induction motor will turn at the same rpm even though the voltage supplied to the motor is lowered. However, as the voltage is decreased, the current the motor draws increases. Motors that are operating at a low voltage will quickly overheat and may fail. It is for that reason that rheostat motor controllers should not be used on induction motors.

Motors are not 100 percent efficient. In actual performance, there is some slippage, or inefficiency, in the motor operation. For motors used to power HVACR equipment, the actual speed is usually 95–97 percent of synchronous speed. Figure 37-9 shows two-pole, four-pole, six-pole, and eight-pole motors. The number of poles is used to calculate the synchronous motor speed.

The following is an example of calculating synchronous motor speed. The formula is:

$$\text{rpm} = \frac{\text{Hz} \times 60\ \text{sec/min}}{\frac{1}{2}p} \text{ or } \frac{\text{Hz} \times 120}{p}$$

where

rpm = revolutions per minute
Hz = frequency in cycles/sec
p = number of poles

EXAMPLE 37-1: SOLVING FOR THE SPEED OF A MOTOR

What is the speed of a four-pole motor operating at a frequency of 60 Hz?

$$\text{rpm} = \frac{60 \times 120}{4}$$

$$\text{rpm} = \frac{7{,}200}{4}$$

$$= 1{,}800\ \text{revolutions/min}$$

37.5 CAPACITOR PRINCIPLES

A capacitor is made up of two conductors separated by an insulating material. The conductors are called plates and the insulating material is the dielectric. Referring to Figure 37-10, when voltage is applied, the current through the capacitor resistor circuit will lead the voltage. To provide strong starting torque for a split-phase motor, a start capacitor is placed in series with the start winding (Figure 37-11). When voltage is applied, the magnetism in the start winding will occur earlier than in the run winding and will provide the push needed to start the motor. Figure 37-12 shows the use of the phase shift to advance the magnetic field in the start winding.

When the current direction in the start winding reverses, the polarity of the start winding will also change. By this time, the rotor south magnetic pole has rotated to the point where it is repelled by the new stator south pole, causing continuous rotation.

A capacitor is rated by its capacity and voltage limit. The unit of capacity is the microfarad (mfd and μF are commonly used abbreviations). A high mfd rating is obtained by either using large plates or a small amount of insulation. A low mfd is obtained by using smaller plates or more insulation. The voltage stamped on the outside of the capacitor is the maximum voltage that can be connected safely across the capacitor. If this voltage is exceeded, the capacitor is likely to fail.

SERVICE TIP

There is often a misunderstanding that a capacitor provides the motor with higher voltage. This is because often on the side of the capacitor there is a much higher voltage listed than the actual voltage at which the system operates. For example, 377 is a common voltage listing for capacitors. However, capacitors do not add voltage but merely shift the phase to increase the running efficiency or starting torque. Capacitors in motor circuits are subjected to higher voltages than the voltage powering the unit. The high voltage across the capacitor is the result of the charging and discharging of the capacitor in series with an inductive load.

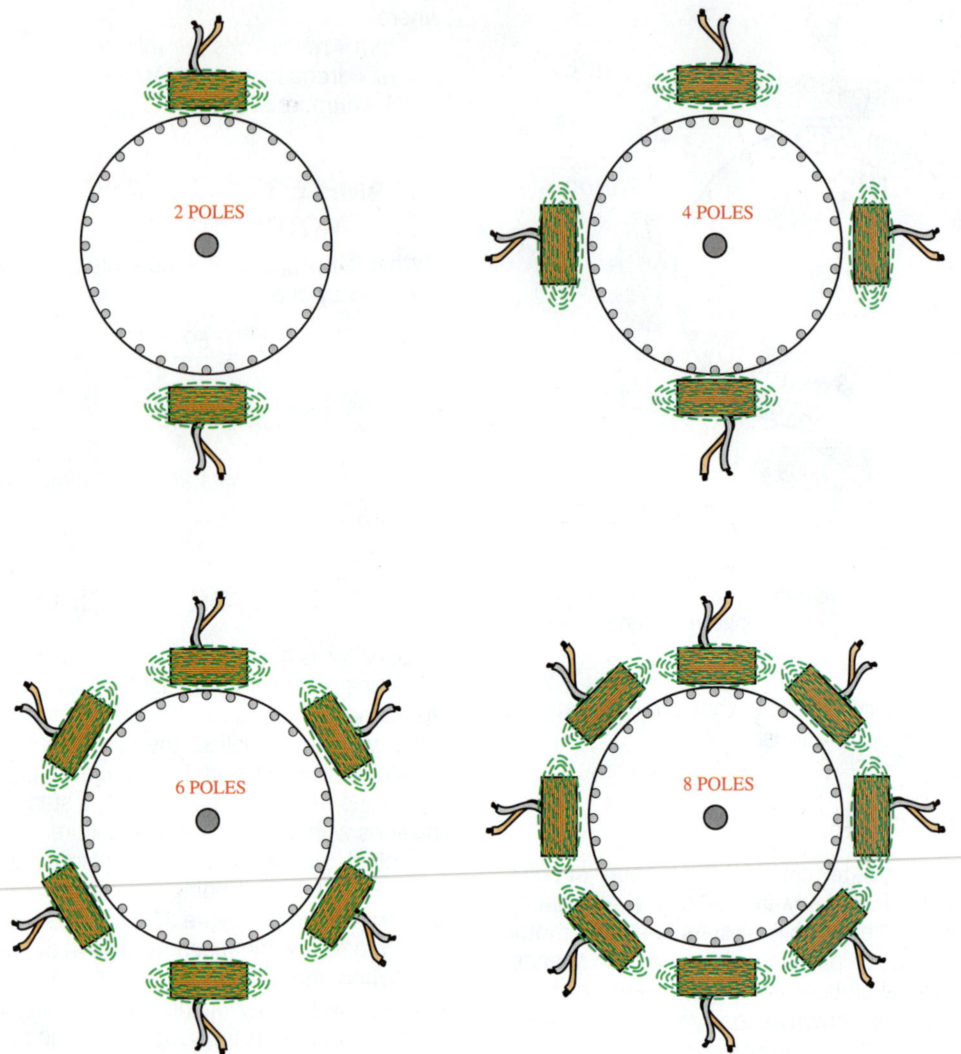

Figure 37-9 Synchronous speed depends on the number of poles.

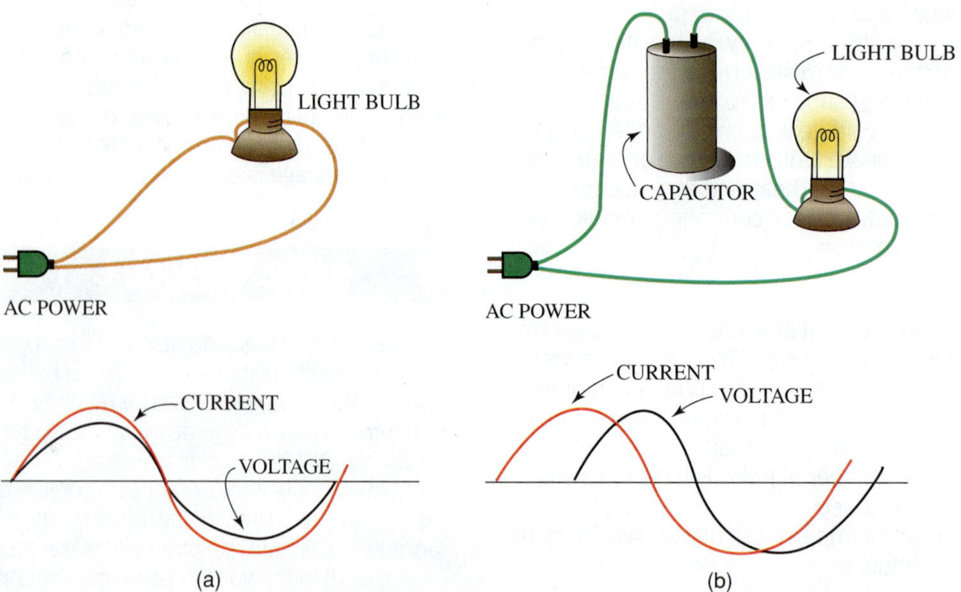

Figure 37-10 (a) The voltage and current flow through a resistor, like the lightbulb load in this circuit, at the same time; (b) however, in the capacitor resistor circuit, the current leads the voltage through the circuit.

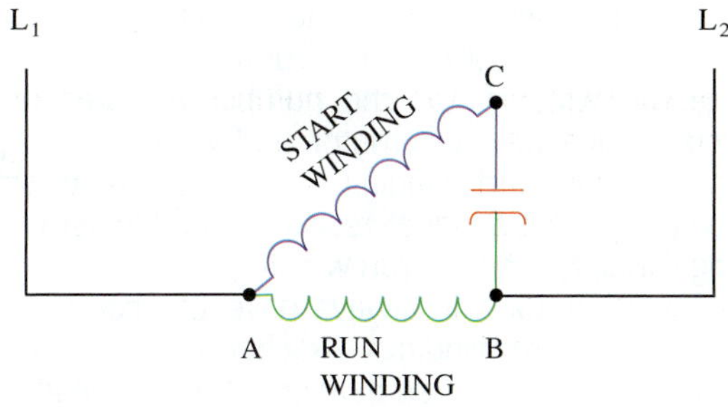

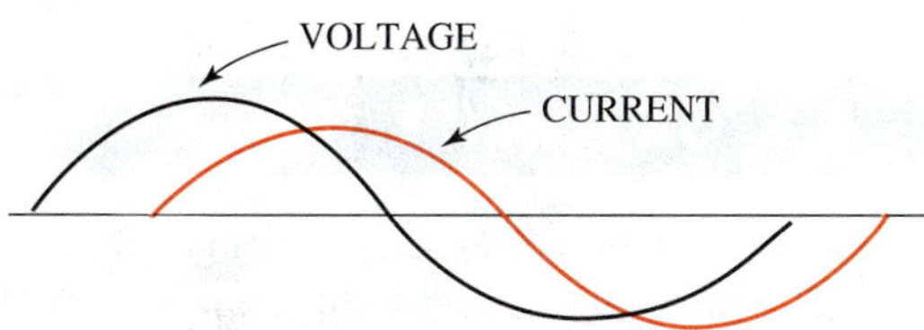

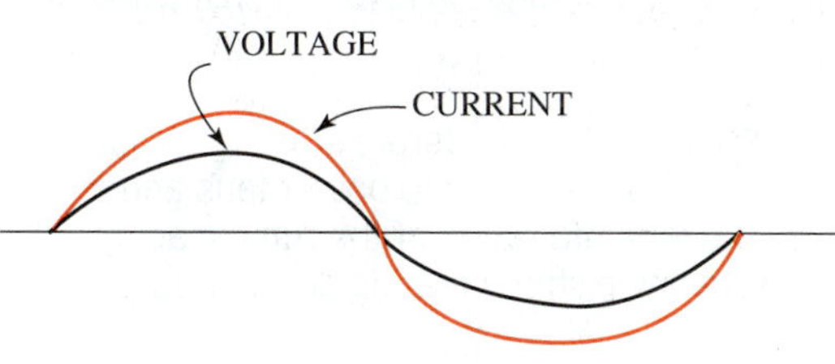

Figure 37-11 The phase shift caused by the capacitor causes the rotor to move, even if it is dead center.

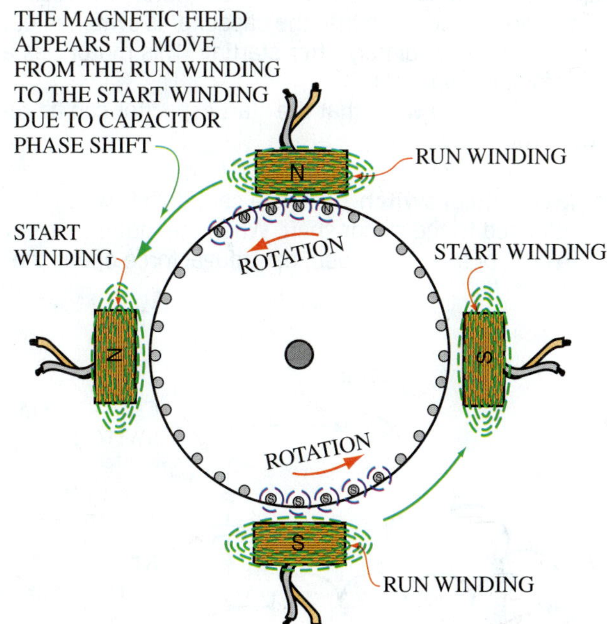

Figure 37-12 Rotating magnetic field due to capacitor phase shift.

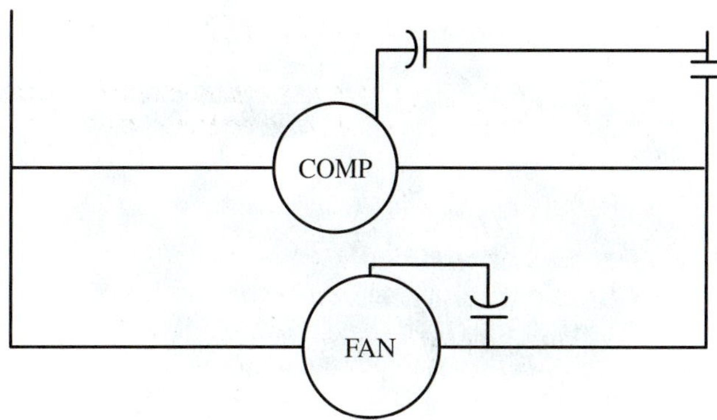

Figure 37-13 Run capacitor connected to the power and start terminals of the fan motor.

Capacitors are used to achieve the desired phase angle shift and to obtain the required current through the series load. Both of these qualities are obtained by selecting the proper mfd rating.

Additionally, whenever a start capacitor and its relay are connected to a compressor motor, the start capacitor and the relay normally closed (NC) switch are in series with the start winding. The start relay coil is connected in parallel with the start winding. With this arrangement, the starting torque of the motor is increased. The relay removes the start capacitor from the circuit as soon as the motor is running.

Referring to Figure 37-13, a run capacitor has been connected to the power and start terminals of the fan motor. The run capacitor stays in the circuit all the time and increases the efficiency of the motor. If the run capacitor should fail, the current draw of the motor would be increased and the motor could overheat. Most PSC motors will not operate if the run capacitor is defective.

Figure 37-14a shows the normal appearance of the run capacitor. The run capacitor, which stays in the circuit continuously, is made of large plates and a large amount of insulation (dielectric) to dissipate the heat. The run capacitor may also be round. The start capacitor does not stay in the circuit long and therefore does not have a heat-dissipation problem. It is typically made in rolled form, sandwiching metal foil and insulating material (Figure 37-14b).

37.6 SINGLE-PHASE MOTORS

There are a number of different types of single-phase motors. They differ from each other mainly by the amount of starting and running torque. The following types are the most commonly encountered:

- Permanent split capacitor (PSC)
- Capacitor start (CS)
- Capacitor start/capacitor run (CSCR)
- Shaded pole
- Split phase

Permanent Split Capacitor (PSC) Motor

The permanent split capacitor (PSC) motor has a run capacitor in series with the start winding, as shown in Figure 37-15. This capacitor stays connected at all times. It starts

(a)

(b)

Figure 37-14 A comparison of the physical shape of start and run capacitors: (a) run capacitor; (b) start capacitor.

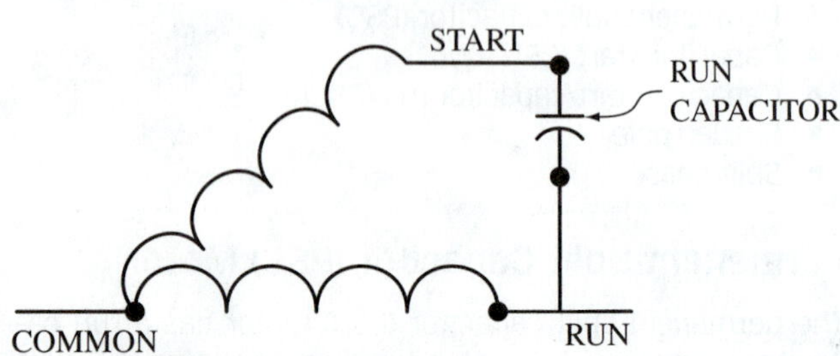

Figure 37-15 Permanent split capacitor (PSC) power.

the motor and then is left in the circuit to improve the efficiency of the motor after it is running.

The run winding has the number of turns of wire required to give the best motor performance at a given line voltage. The start winding has more turns of smaller wire, which gives it a higher resistance and lower current-carrying capacity than the run winding.

The PSC motor will always have the common connection for the two windings attached to power. The run capacitor will be connected between the start winding and the run winding. The run capacitor therefore improves the performance of the motor both in starting and running.

SERVICE TIP

It is important that the proper size run capacitor be used with a PSC motor. If the capacitance changes more than 10 percent from the design capacitance, the motor windings will build up greater heat as they operate. If enough heat builds up, the motor can become damaged. It is therefore very important that the exact capacitor as specified by the manufacturer be installed on any PSC motor.

The PSC motor has moderate starting torque and good running efficiency. It is used to power fans and small compressors. The low mfd rating of the run capacitor results in a small phase angle shift, creating only a moderate starting torque.

Capacitor Start Induction Run (CSIR) Motor

A diagram of the capacitor start induction run (CSIR) motor is shown in Figure 37-16. This motor has a high starting torque but is not as efficient as the PSC motor. The reason for its lower efficiency is that the capacitor is switched out of the circuit immediately after starting. This motor has a high-mfd start capacitor.

There are two ways that the start capacitor can be removed from the circuit:

- **Mechanical switch** This is a centrifugal switch attached to the motor shaft. When the motor reaches ⅔ or ¾ of its rated speed, centrifugal force opens the switch (Figure 37-16).

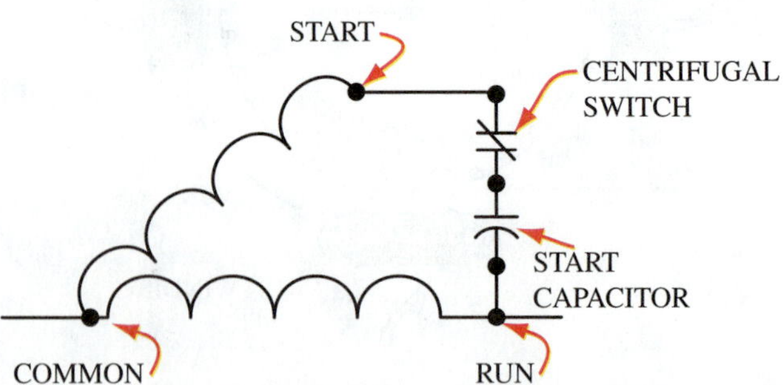

Figure 37-16 Capacitor start induction run (CSIR) motor.

- **Electromagnetic method** A potential relay is placed across the start winding. Its contacts are placed in series with the high-mfd capacitor. When the motor is started, the capacitor produces a high starting torque. As the motor speed increases, the induced voltage across the start relay coil increases until it reaches a preset value higher than the line voltage. The relay coil is energized and the NC switch is opened, removing the start capacitor from the circuit.

Capacitor Start Capacitor Run (CSCR) Motor

The capacitor start capacitor run (CSCR) motor has both a start capacitor and a run capacitor (Figures 37-17 and 37-18). It has excellent starting and running torque but is not as efficient as the PSC motor. It is used to drive most compressors.

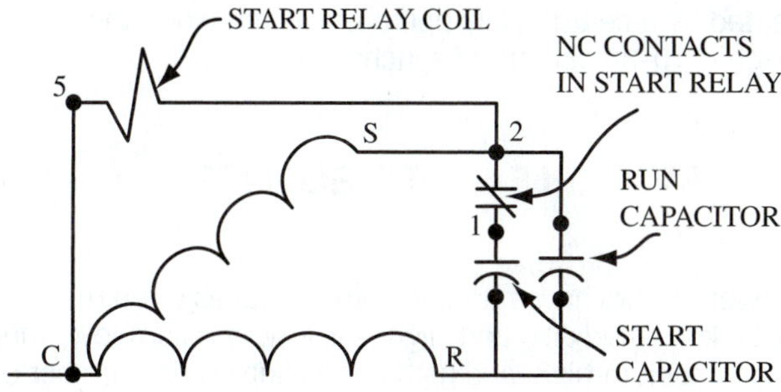

Figure 37-17 Capacitor start capacitor run (CSCR) motor.

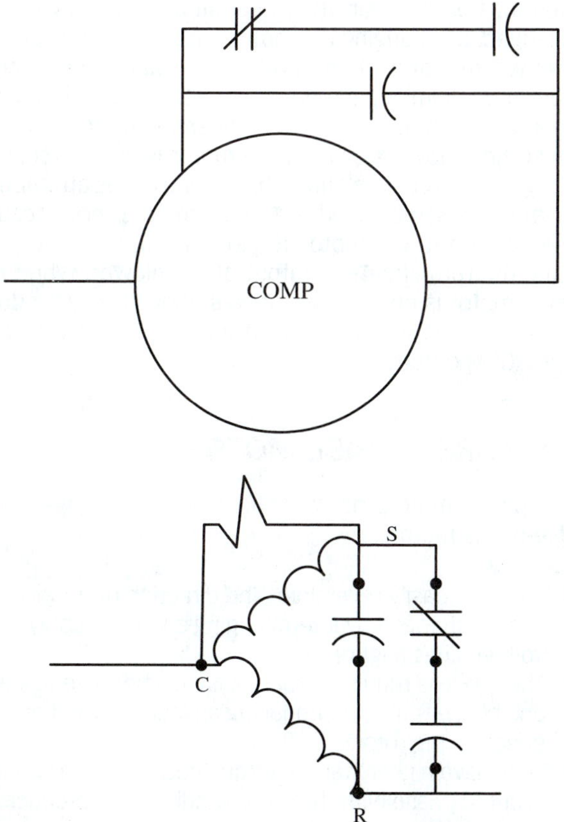

Figure 37-18 CSCR motor diagrams.

Start Kits

Start kits are available for the technician's use whenever it is necessary to improve the starting torque of a motor. When a motor keeps tripping out on overload, the addition of a start kit may solve the problem.

Where a high starting torque is required, a hard-start kit can be installed. Where low voltage or a voltage lag is experienced, a soft-starting kit can be used. A hard-start kit contains a start capacitor and a start relay; a soft start assist uses a PTC device.

SERVICE TIP

Often start capacitors have bleed resistors connected between the two capacitor terminals. The purpose of this bleed resistor is to drain the capacitor's voltage during the motor's off cycle. This is done so that when the motor contactor starts the motor, there is not a capacitor discharge across the contactor's points. Such a discharge causes pitting of the contactor. Over time, this pitting will destroy the contacts. Bleed resistors are available from the supply house to be installed on start capacitors. If one is not installed, the contactor in the unit will become pitted and in some instances weld or stick together.

Positive Temperature Coefficient (PTC) Thermistor

A different soft-starting kit is available to be applied to PSC motors. This kit includes a positive temperature coefficient (PTC) thermistor. The PTC is a temperature-sensitive device whose electrical resistance will increase as its temperature increases. This PTC is placed across the run and start terminals, parallel to the run capacitor of a PSC motor, as shown in Figure 37-19.

At room temperature, the PTC thermistor has a low resistance, about 25 or 50 Ω. When voltage is supplied, an initial surge of high current passes through the start winding. This is because the thermistor is effectively shorting out the capacitor. The surge causes an increased starting torque to start the motor. The temperature increase that results causes the PTC thermistor resistance to rise, removing the short from across the run capacitor. The motor then runs as a normal PSC motor.

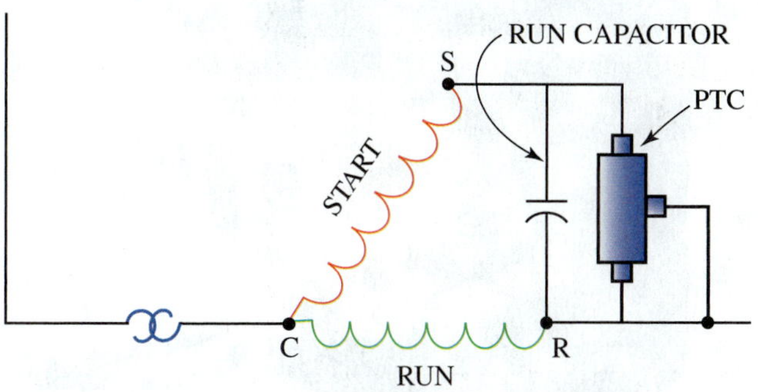

Figure 37-19 PTC start systems are sometimes referred to as soft start because they do not provide as much initial torque as a start capacitor and potential relay.

Shaded Pole Motor

The shaded pole motor (Figure 37-20) has a modified stator pole. A groove separates a small portion of the stator pole from the rest of the pole. A bank of metal is placed around the smaller section of the pole, which provides a phase shift needed to start the motor. Shaded pole motors have a low starting torque, and their speed control under varying load conditions is poor. They offer a low-cost motor for light-duty applications such as running blowers on small air-handling and heating units.

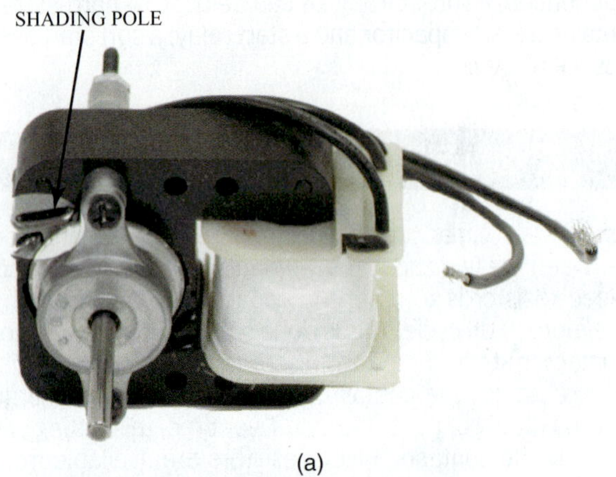

(a)

(b)

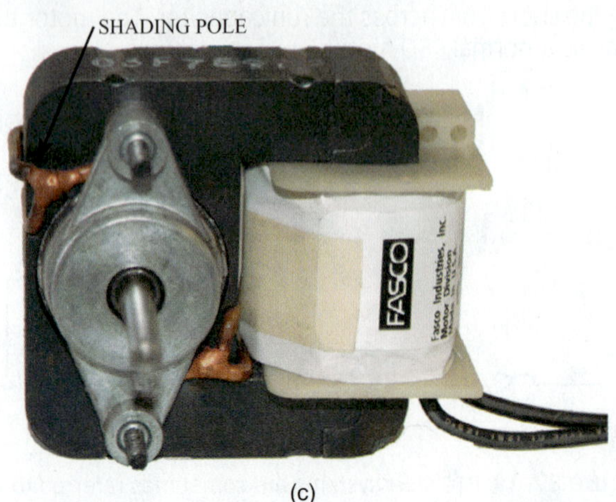

(c)

Figure 37-20 Shading pole on shaded pole motors.

Split Phase

A split-phase motor has very low starting torque and is used for applications that have easily started loads. This motor is seldom used in sizes larger than ⅓ hp. The principle for operation is that two parallel connected motor windings are placed 90° apart. If the windings are identical, then the current through both will lag the voltage by the same phase angle. To generate a revolving magnetic field, one winding has a greater resistance and a lower reactance, so that the currents through the two windings are shifted in time phase. The high-resistance winding is called the starting or auxiliary winding, and the low-resistance winding is called the main winding. The starting winding is usually disconnected by a centrifugally operated switch when the motor reaches 70–80 percent of synchronous speed.

37.7 MULTIPLE-SPEED BLOWER MOTORS

Blower motors have multiple speed settings often referred to as low, medium, and high—or low, medium-low, medium, medium-high, and high depending on the number of speed taps that the motor has. These speed taps do not actually change the motor's rpm. A motor's rpm is based on the number of poles and frequency. By changing the speed taps on a motor, the motor's effective horsepower is changed. By changing the motor output horsepower, the motor will turn at a different rate under different load conditions. For example, a ⅓-hp motor rated at 1,750 rpm will turn at approximately 1,750 rpm when the motor is on the high setting. However, when the motor is on the medium setting, it may produce only ¼ hp. Although it is attempting to turn at the same speed of 1,750 rpm, it cannot because the load causes the motor to slow down, effectively reducing the rpm and CFM output of the blower. When the blower motor is on the low speed setting, it may produce only ⅛ hp. This would result in an even lower rpm and blower CFM output.

37.8 THREE-PHASE MOTORS

Three-phase motors have the following advantages over single-phase motors:

- They are easily reversible. The direction of rotation can be changed by interchanging any two supply voltage lines (Figure 37-21).
- There is less running torque pulsation because at least one phase is always producing an induced rotational effect on the rotor.
- They have higher starting torque because each winding is out of phase with the other windings and produces rotational torque.
- They have higher efficiency.

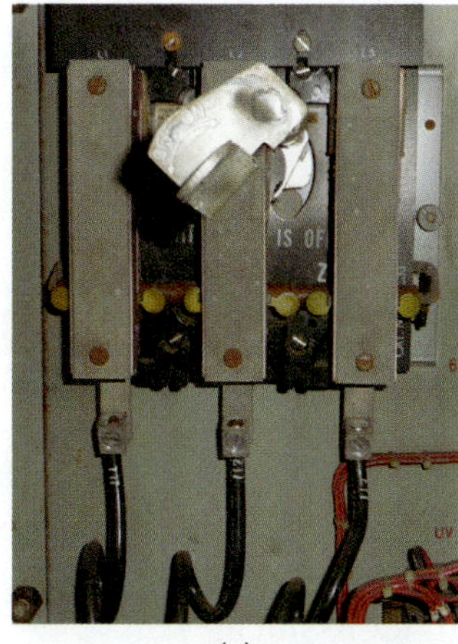

(a)

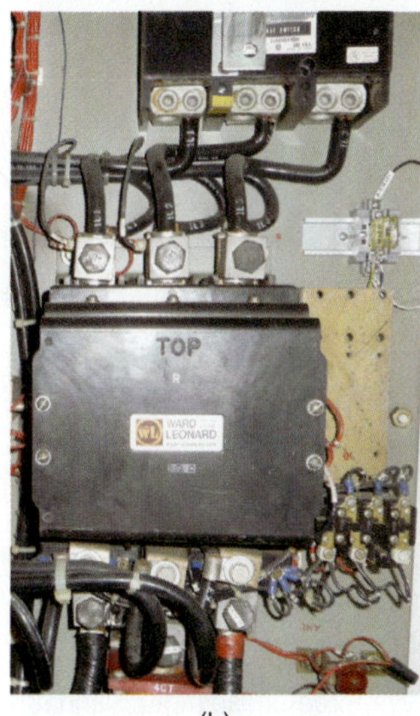

(b)

Figure 37-21 (a) To reverse rotation, switch any two lines; (b) three-phase motor starter.

Figure 37-22 Three sets of wires, one for each phase.

CAUTION

Do not attempt to run a three-phase motor on single-phase power. When wiring the motor, you will find three separate groups of wires for the three different phases, often referred to as "legs" (Figure 37-22). Although the motor may start, it will quickly overheat and damage the windings. The same thing will happen to a three-phase motor if one of its legs of power is lost as the motor is being operated. For that reason, some large motors have single-phase or lost-leg motor protection systems, which will shut the motor down anytime a single leg of power is lost to the motor.

The windings of a three-phase motor are 120° apart, which facilitates starting. Figure 37-23 shows a schematic drawing of a three phase motor. No capacitors, no auxiliary windings, and no switching circuits are required. Three phase motors create their own starting torque.

There are three different phase windings: A, B, and C. Figure 37-24 shows a three-phase, two-pole induction motor, using a wye configuration. The phases are mechanically offset from one another by 120°, and the winding currents are 120° electrically phase shifted. Although poles are represented in the diagram for ease of illustration, actual motor windings are put into slots in the rotor as winding groups. Distributive windings will have a number of windings in a series-connected group. The motor shown in Figure 37-25 has thirty-six slots around its circumference. Two slots are required for each winding, so it provides slots for eighteen windings or six windings connected in series per phase.

Three-phase motors are wired in either a delta configuration (Figure 37-26) or a wye configuration (Figure 37-27). The delta configuration is named because the circuit resembles the triangle shape of the Greek letter delta (Δ). The wye

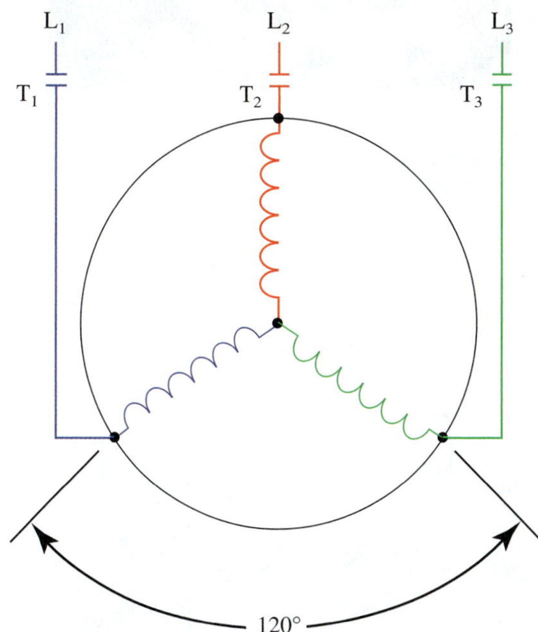

Figure 37-23 The three-phase motor.

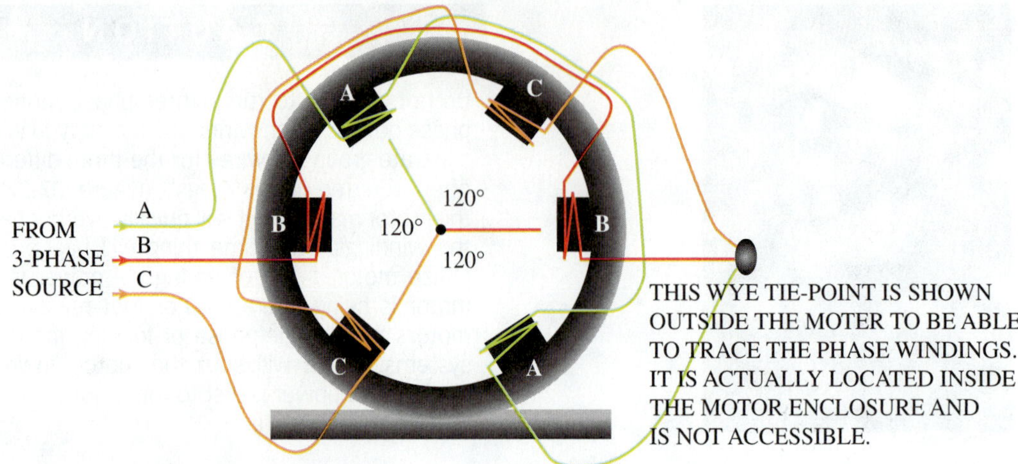

Figure 37-24 Three phase, two-pole induction motor with a wye configuration.

Figure 37-25 (a) Distributive windings; (b) close-up of individual windings and slots.

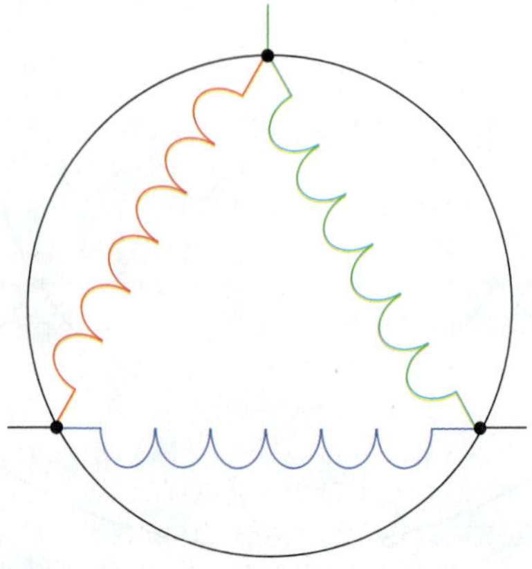

Figure 37-26 Delta three-phase motor windings.

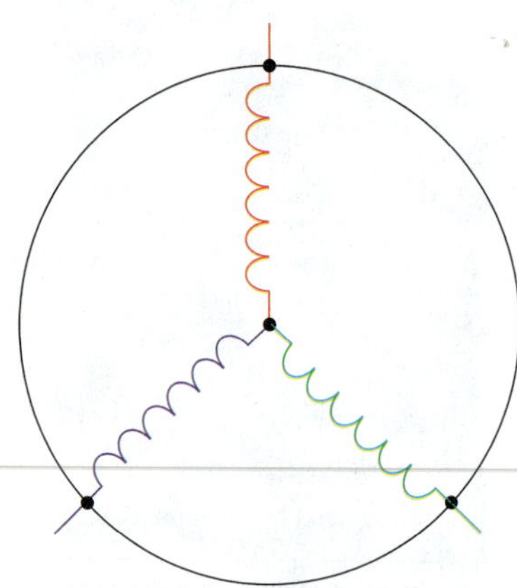

Figure 37-27 Wye three-phase motor windings.

configuration resembles a capital letter Y. In the delta configuration, the line voltages equal the phase voltages, and the line currents are 1.732 times the phase currents. In the wye configuration, the line voltages equal 1.732 times the phase voltages, and the line currents equal the phase currents.

37.9 MOTOR PROTECTION

There are a number of causes of motor failure. The most common problem is excessive heating. Among the causes of excessive heating are the following:

- Defective start relay (on a single phase motor)
- Excessive load
- Loss of refrigerant on a gas-cooled motor
- Excessive time on locked rotor current
- Operation at too high or too low voltage
- Single phasing (on a three-phase motor)

All motor protective devices are designed to cut off the power to the motor before damage has occurred. Repeated cycling on the protective device can damage a motor.

The locked rotor current is the momentary starting current. This current can be 3–5 times the running current of the motor. If the motor is blocked from starting, this excessive current will be prolonged, causing heating and serious

Table 37-1 Voltages Handled by Common Motors

Nameplate	Upper Limit	Lower Limit
208	228	188
230	253	207
460	506	414
575	633	518

damage. A protective device is therefore required to disconnect power on either excessive temperature or excessive current. Fuses or circuit breakers are most often the devices that protect against excessive starting current.

All motors are designed to operate between certain voltage limits. If voltages outside these limits are applied, the life of the motor can be seriously reduced. Table 37-1 shows the limits of voltage that common motors are designed to handle.

Typically motors are designed to be applied within ±10 percent of the nameplate voltage (Figure 37-28). Dual-rated motors such as 208 V/230 V may be rated plus 10 percent, minus 5 percent. The manufacturer's information should be referenced.

On three-phase motors, if the voltage varies more than 2 percent between phases, the life of the motor will be reduced. This is caused by unequal currents and heating.

Single phasing is another serious problem common to only three-phase motors. If the motor loses power in one of the three lines, the motor may continue to operate. This is called single phasing. This will cause sufficient unbalancing to increase the motor temperature and require motor protection.

37.10 TYPES OF MOTOR PROTECTION

There are three types of protective devices that are used to protect the motor against excessive heat:

- Temperature actuated
- Current actuated
- Combination of current and temperature

These devices can be either pilot-duty overload or line-duty overload arrangements, as shown in Figure 37-29.

SAFETY TIP

Motor protection for fractional horsepower compressors is often provided by an external snap disk (Klixon) that clicks onto the compressor case. These devices are wired into the common power lead and are located inside of the terminal cover. They are supposed to be in direct contact with the compressor case to sense its temperature. If the cover is not placed back on the terminal, or if the device is not in direct contact, it will not sense the temperature of the compressor. These devices are designed to open when the temperature and/or current exceeds the compressor's limits. They will not provide the proper protection for the compressor unless they are installed correctly.

(a)

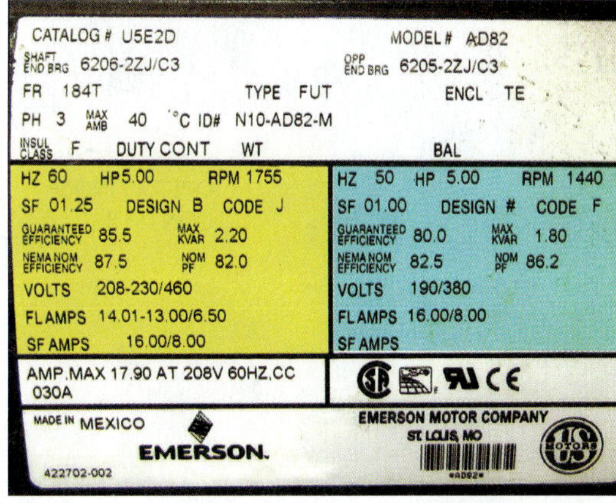

(b)

Figure 37-28 (a) Single-phase fan motor nameplate; (b) notice this motor has nameplate data for both 50 or 60 Hz.

37.11 MOTOR OVERLOAD DEVICES (LINE BREAK)

For a single-phase motor, the overload device must interrupt one of the motor leads. For a three-phase motor, the device must interrupt two or three of the motor leads. On a wye-connected motor with a built-in overload device, this is often where the three windings are connected in common.

In a three-phase motor, a pilot-duty device senses current overload or excessive temperature within the motor and opens the contactor circuit to remove power to the motor. A line-duty device senses current and/or temperature in the motor winding and, if an overload occurs, will disconnect power by directly opening the motor winding circuit.

The line-duty arrangement is commonly found on compressor motors used for domestic service. Most of these reset themselves automatically. However, since they are embedded in the motor winding, as shown in Figure 37-30, it takes some time for the motor to cool down. During this waiting period it is easy to incorrectly diagnose the problem as compressor failure.

Figure 37-29 Two types of motor protection.

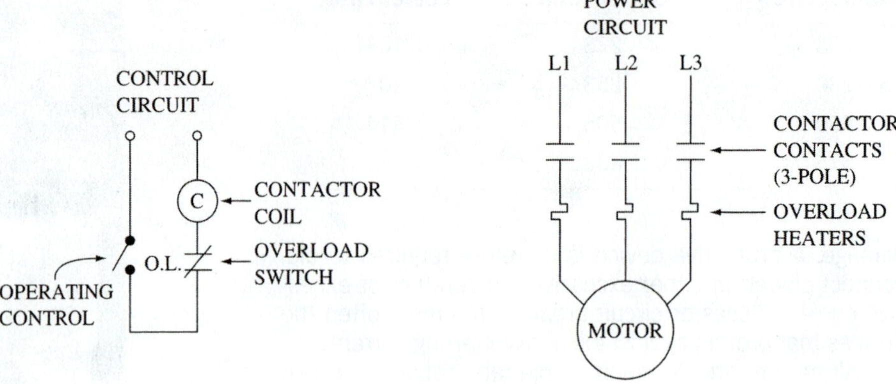

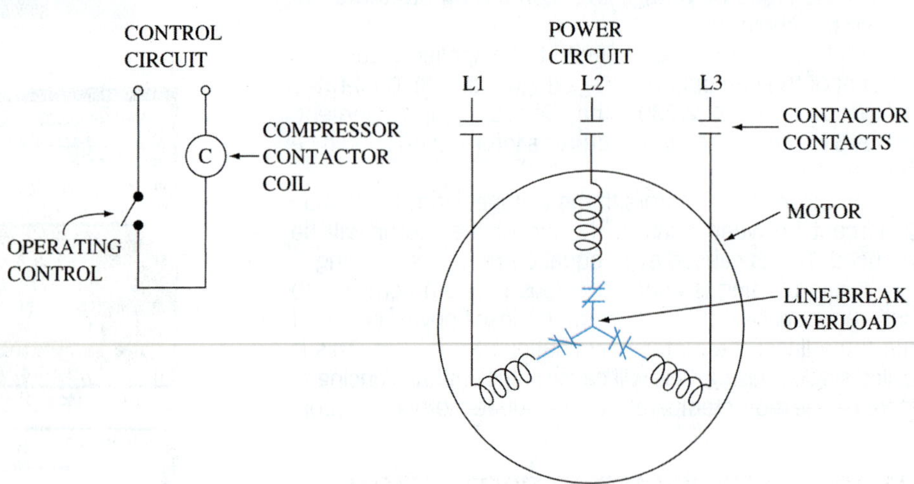

37.12 TYPES OF RESET

After a protective device has opened or tripped out, it needs to be reset. This can be done manually or automatically. The advantage of manual reset is that a service technician has the opportunity to examine the cause of the problem before resetting the protective device. The advantage of the automatic reset is that the unit can automatically go back into service in the case of a nuisance trip-out.

General automatic reset devices are employed only where the time to reset is sufficient to ensure that the motor will not short-cycle and be damaged by the protective device itself. Sometimes the protective device will reset automatically, but the control circuit will require resetting by switching the unit off then on from the thermostat or control switch.

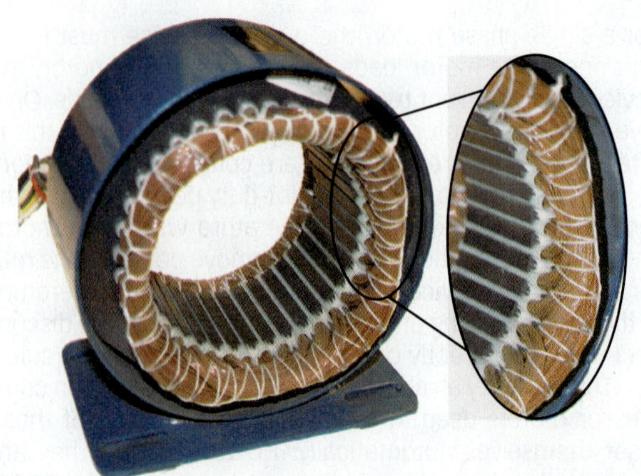

Figure 37-30 It's important to protect the motor windings, because they are made of such small sized wires. *(Courtesy Hampden Engineering Corporation)*

37.13 TYPES OF OVERLOAD DEVICES

The various types of overload devices that will be discussed are as follows:

- Motor starter
- External supplemental overload current or current temperature
- Internal current and temperature overload
- Thermal overload relay
- Heater element current overload

Figure 37-31 Motor starter.

- Magnetic overload (Heinemann)
- General Electric Thermotector
- Three-phase overload

Motor Starter

A motor starter is a contactor with motor overload protection added. Starters made for three-phase motors will have protection on each leg (for a total of three overloads). A motor starter is shown in Figure 37-31.

External Supplemental Overload Current or Current Temperature

External supplemental overload devices are shown in Figure 37-32. These devices provide either current protection or current temperature protection. The bimetallic disk overload protector is widely used in the protection

(b)

(a)

(c)

Figure 37-32 (a) External-type compressor overload; (b) Klixon current and temperature protector can be located inside or outside of the compressor; (c) when it is located outside of the compressor, it is under the motor terminal cover; (d) (see page 602) this motor cover holds it tightly against the compressor so it can sense the motor's temperature.

(d)

Figure 37-32 *(Continued)*

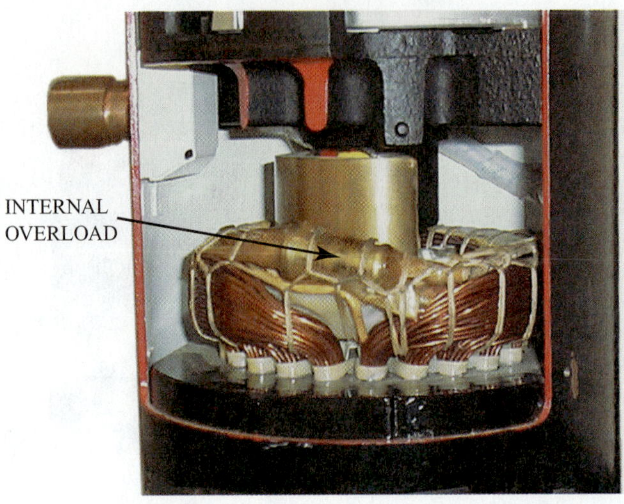

Figure 37-34 Internal motor overload.

of electrical motors. When the temperature is raised, the disk will warp and break the electrical circuit. The external device is mounted on the motor shell and resets automatically when the temperature returns to normal.

Many current overload devices sense both the current and temperature of the motor. These devices, like those located within the Bakelite cover of small compressors, must be in their enclosure to function properly. If the Bakelite cap is left off, then the overload will only sense the current and not the temperature. This reduces their sensitivity and removes a large portion of the overload safety protection from a motor. The current-temperature device is usually connected for line duty, but some versions are used for pilot duty. A three-phase overload device is shown in Figure 37-33.

Internal Current and Temperature Device

Another device that is sensitive to both internal current and temperature and is wound into the motor windings is the internal line-duty overload device (Figure 37-34). As the overload occurs, the bimetal strip within the device warps to open the contacts. The internal device is wound into the motor windings and automatically resets when the motor cools down.

Figure 37-33 Three-phase overload.

CAUTION

Many blower and compressor motors do not have manual reset for thermal overload. Always make sure that the motor has cooled down to allow any internal devices to reset before testing. Testing for an open winding on a motor may lead to inaccurate readings if the internal thermal overload device is tripped.

Thermal Overload Relay

A thermal overload relay is current sensitive with automatic reset and is connected in series with the motor winding. On an increase in heat due to excessive motor current, the upper bimetal strip will warp, opening the pilot-duty contacts wired in series with the contactor. Reset is automatic upon cooling. The lower bimetal strip prevents nuisance trip-outs caused by high ambient temperature.

Heater Element Current Overload

A heater element current overload device for pilot duty uses a bimetal element (spiral or disk type) that responds to heat generated by a heater element. When the bimetal heats sufficiently, the contacts open, disconnecting the contactor. It automatically resets on cooling. The heater elements are interchangeable and calculated for a specific range of current. They should be matched to the motor current.

Another version of this same device uses a solder pot relay instead of a bimetal sensing element. The solder pot relay has a ratchet-and-spindle construction and requires manual reset. It uses interchangeable heaters similar to the bimetal overloads. Heaters are not interchangeable between different makes of equipment.

Magnetic Overload

The magnetic overload is a current sensitive device. A core placed inside a coil will attempt to center itself in the magnetic field created when current flows through the coil. The sensing coil is wrapped around a sealed tube filled with silicone oil containing the core. Excessive current will pull the core toward the armature, causing the contacts that are attached to the armature to open. The silicone has a dampening effect, which prevents operation of the device for temporary overload conditions, such as those that occur on startup. Current rating is fixed and cannot be field adjusted.

General Electric Thermotector

The General Electric Thermotector device is also an internally actuated temperature motor protector. On this device the case and the internal strip expand at different rates. The device will trip early if the temperature rises rapidly. If the temperature rise is gradual it will trip at its normal setting. This means it is a rate-compensated device.

37.14 MOTOR CIRCUIT TESTING

Since motors are so important in HVACR systems, they are carefully protected and can be thoroughly tested to see that they are operating properly. The tests that follow begin with single-phase motors. Since many single-phase motors require capacitors, a word of warning needs to be added about handling capacitors. Capacitors can be dangerous. Safety precautions are required.

SAFETY TIP

Capacitors can hold a high-voltage charge even after the power is turned off. Always discharge capacitors before touching them.

TECH TIP

Capacitors should be discharged before you work on the system; however, the capacitor needs to be discharged with a bleed resistor. If the capacitor is discharged with a screwdriver or other metal device touching the two terminals, the spark that is generated externally can also occur internally, damaging the capacitor.

A variety of instruments can be used for testing motors. The megohmmeter (Figure 37-35) is used to check the resistance of the motor windings. Clamp-on ammeters are used to measure the current in the circuit. In addition to these, voltmeters are used to measure the operating voltages, and capacitor analyzers are used to measure the actual mfd rating of the capacitor.

The first thing to do whenever testing any motor is to make sure that the power supply to the motor has been secured. The motor windings can be checked with a multimeter (Figure 37-36). It may be necessary to remove a guard

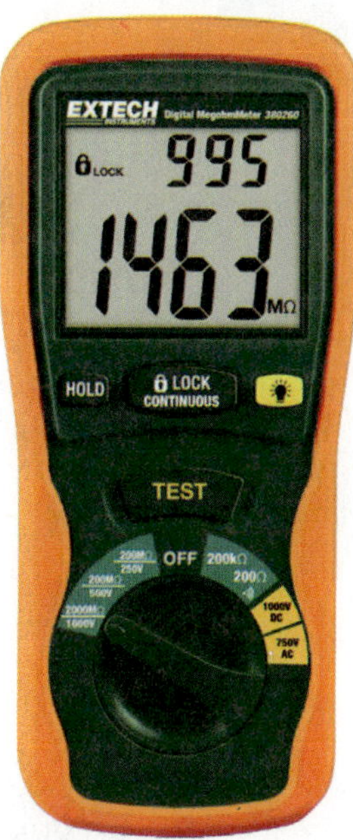

Figure 37-35 Megohmmeter. *(Courtesy of Extech Instruments, A FLIR Company)*

Figure 37-36 To test motor windings, first remove the leads.

that encloses the terminals on a hermetic unit. The terminals should be marked C (common), S (start), and R (run).

CAUTION

Occasionally the terminal has been damaged, and on a pressurized system the terminal can blow out. To avoid possible injury, unless the charge has been removed, use terminal points some distance away from the compressor when checking compressor motor resistance. Be sure that all accessories such as capacitors and relays are disconnected.

Use an ohmmeter to read the resistance in both the run and start winding. Be sure good contact is made with the proper terminals.

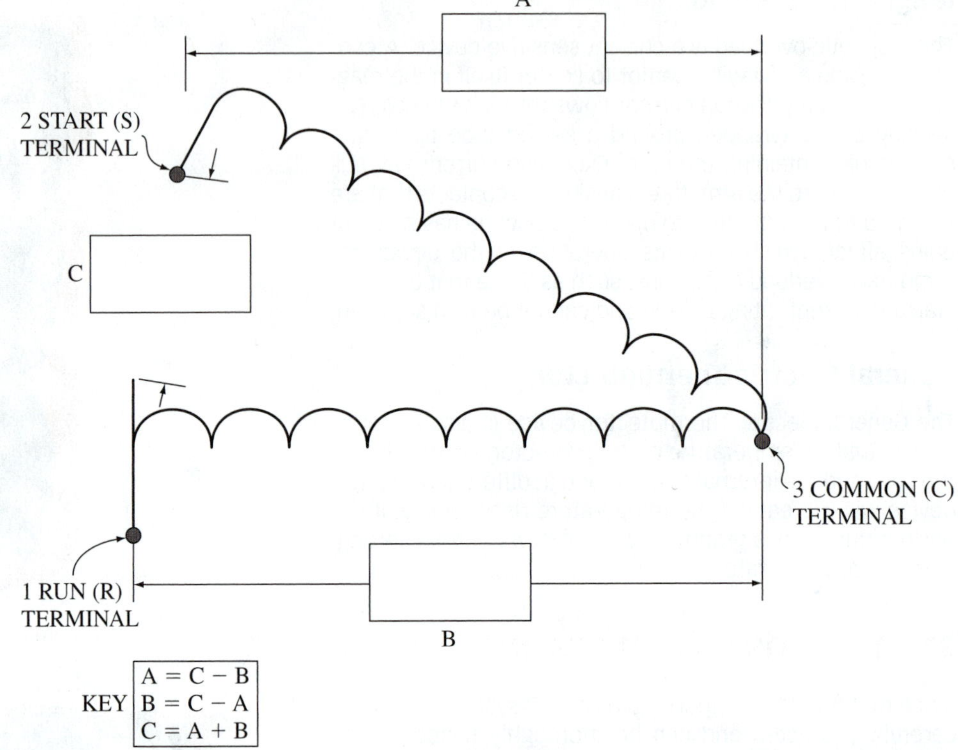

Figure 37-37 Method of determining run, start, and common motor terminals.

If the terminals are not marked, they can be identified by a simple test. First, measure the resistance between each pair of terminals. For example, assume that these readings are 5½, 4, and 1½ Ω. By diagramming them as shown in Figure 37-37, the terminals can be identified. In the example, the greatest reading is between 1 and 2. The common terminal is therefore 3, the one not being touched. So terminal 3 is C. Then, reading from C, the greatest resistance is to 2. Therefore, 2 is S. And, finally, since 2 is S and 3 is C, 1 must be R.

The motor is then tested for an open winding, a broken wire, or a shorted winding, as shown in Figure 37-38. To do this, a good low-range ohmmeter (R × 1) is required. A zero resistance means a shorted winding. Low resistance means the winding is good. An infinity reading means it is an open winding. As a rule, the start winding has a resistance 3–5 times the resistance of the run winding.

Large motors (5–10 hp and higher) have heavy copper windings to carry the motor current. They may therefore indicate a reading very close to a short when winding resistance is read, depending on the ohmmeter.

Figure 37-39 shows the motor being tested for a grounded winding. One lead is placed in contact with bare metal on the compressor casing. For this application, the ohmmeter needs to be capable of measuring very high resistance (R × 100,000). For an ungrounded winding, the resistance is generally 1–3 MΩ (megaohms). This applies to both single-phase motors and three-phase motors.

SERVICE TIP

The temperature of the compressor is important in testing for a partially grounded winding. If the compressor will run, it should be run for about 5 min before testing.

Figure 37-38 Method of testing open, shorted windings and broken wires.

37.15 INSULATION RESISTANCE TESTERS

Electrical insulation is classified by the temperature stability of the materials that are used for the winding. As an example, hermetic compressor windings are generally rated for class H (180°C or 356°F), Table 37-2. Windings of this type are composed of materials that have an acceptable thermal life expectancy at this temperature. Examples of insulating materials for this temperature range are silicone elastomers, mica, glass fibers, and polymers. The insulated windings will often have a varnish or polyester-base

(a)

(b)

Figure 37-39 Testing for a grounded wiring: (a) not grounded; (b) grounded.

Table 37-2 Motor Insulation Temperature Classification

Class	°C	°F
Class 90 (O)	90	194
Class 105 (A)	105	221
Class 130 (B)	130	266
Class 155 (F)	155	311
Class 180 (H)	180	356
Class 220	220	428
Over class 220 (C)	>220	>428

top coat. When exposed to high temperatures over time, the insulation begins to break down. Hairline cracks can form in the top coat, which can absorb dirt and moisture. Insulation testing helps to determine the level of insulation breakdown for motor windings over time. The motor is first tested when it is new and then at regular intervals during its useful life of operation.

These are special testers that are invaluable for testing leakage resistance from motor windings to ground. They are often used to periodically test semihermetic motor insulation. The meters can test leakage at high voltage (500 V for 208–240 V motors and 1,000 V for 480 V motors). They may be battery operated or use a hand-cranked generator and are often referred to as meggers because they must be able to read resistances of tens of millions of ohms (megohms). They may detect insulation faults, where an ordinary multimeter using a few volts DC would show a satisfactory reading.

SERVICE TIP

The wiring that is used for motor windings is insulated with a thin coat of varnish. As the motor is used and the windings are heated, this material will slowly carbonize. As it carbonizes, it will slightly change to a darker tan and eventually to black. When it is completely carbonized, it does not provide any insulating capacity. As a motor heats up, this breakdown process can be accelerated to the extent that in extreme conditions it can carbonize almost immediately when overheated. A megger tests for the degree of breakdown in the insulation. This also provides an indication of the useful life that may remain with a motor or compressor.

SAFETY TIP

When testing for a grounded winding, one test lead is placed on the terminal and the other on the outer shell of the compressor. Be careful to make good contact with the motor shell. A coat of paint or a layer of dirt can hide a grounded winding. The copper refrigerant lines may also be used to check for grounds.

The windings of a three-phase motor all should have the same resistance. Remember when checking a three-phase motor to be sure to reconnect it in the same manner as originally connected. Interchanging any two connections can reverse the rotation of the motor.

37.16 GENERAL ELECTRIC MOTORS (ECM)

ECMs were developed by the General Electric Corporation in the mid-1980s (Figure 37-40). ECM stands for electronically commutated motor. They were originally used as blower motors, but today they are used for every motor need, including combustion air-blower motors, condenser fan motors, and compressor motors. These motors are ultra-high efficiency, programmable, brushless DC motors, 120 V or 240 V AC input. They utilize a permanent magnet rotor and a built-in AC-to-DC inverter to produce the direct current that operates the motor (Figure 37-41).

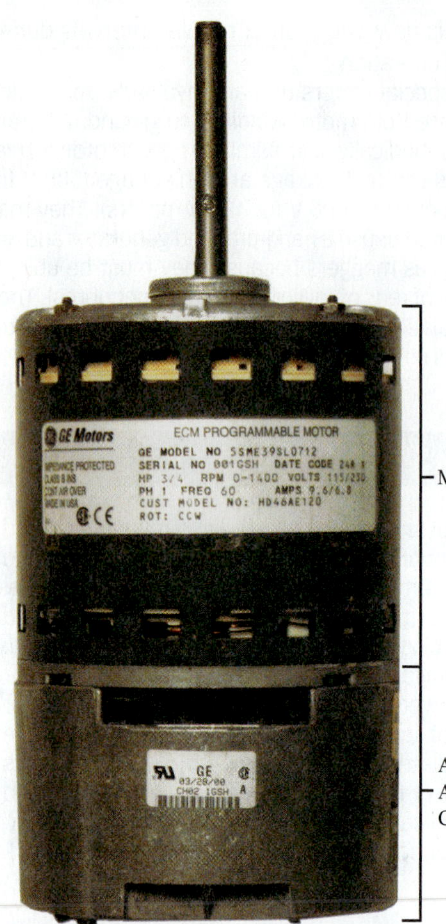

Figure 37-40 GE ECM motor.

Figure 37-41 Cutaway of GE ECM motor.

Figure 37-42 Cutaway of GE ECM motor showing ball bearings.

ECMs have a number of advantages, as follows:

- **High efficiency** ECMs are approximately 70 percent efficient as compared to 45 percent or less for PSC motors.
- **Low maintenance** ECMs use ball bearings as opposed to sleeves, which are used on PSC motors (Figure 37-42). Ball bearings are permanently lubricated as compared to sleeved motors, which must be routinely lubricated.
- **Speed control** ECMs come preset for a specific cfm requirement. Because the rpm of DC motors is more easily adjusted, ECMs can be field adjusted to a specific cfm and with the addition of a controller can be remotely controlled using a central energy-management program (Figure 37-43).
- **Constant CFM** ECMs are designed to provide a constant CFM over a wide range of static pressures.

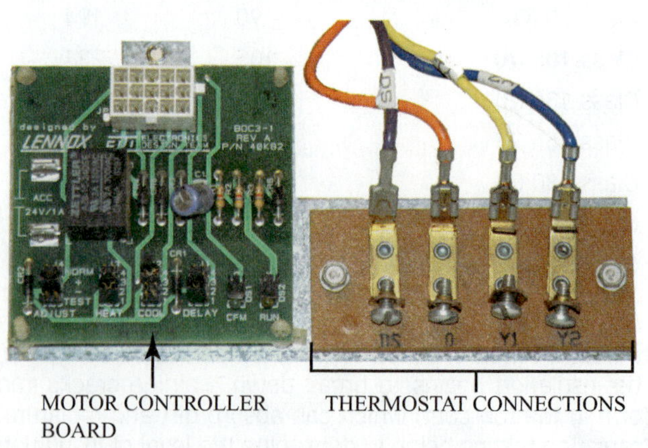

Figure 37-43 ECM motor controller PC board.

The ECM rpm will increase automatically to provide the designed airflow.

- **Heat load** ECMs operate at almost ambient temperatures, as compared to PSC motors, which typically operate from 90° to 150° above ambient.
- **Soft start** ECMs are designed to start at a low rpm and gradually ramp up to the designed speed. This places less stress on the motor fan and other mechanical parts as compared to the almost instant starting of a PSC motor. Soft starting also decreases the sometimes-noticeable blast of air from conventional blowers associated with PSC motor starting.
- **Flexibility in size** Because the ECMs' rpm can so easily be controlled without the loss of efficiency, it is possible for manufacturers to use a limited number of various horsepower-rated ECMs throughout their entire production line. This reduces inventory costs for the manufacturer and for the service company by reducing the number of replacement motors that they must carry on their service vehicles.

ECM X13

This motor is used for indoor blower applications and was developed to meet the 13-SEER mandate for residential air-conditioning units that went into effect in 2006. The X13 offers up to 33 percent greater efficiency than a PSC at rated speed and up to 200 percent in constant fan mode. It allows for variable speed operation from 600 to 1,100 rpm, has up to five selectable constant speed torque settings controlled by 24 V AC inputs, five fixed speeds, and improved high-static CFM performance.

The major difference between the X13 and PSC motors is the ability to deliver constant torque. With standard PSC motors, as the air static pressure increases, both the motor torque and CFM delivered will decrease. The X13 maintains the torque it was programmed for, and although the CFM delivered will still decrease with an increase in static pressure, it will be less as compared to a PSC.

ECM 2.3

The ECM 2.3 is a premium programmable motor that can operate over a wide range of speeds from 200 to 1,300 rpm. This type of motor can deliver constant airflow over a wide range of external static pressures. As an example, if the media filter becomes dirty, this motor will automatically ramp up to ensure that the programmed amount of airflow is still delivered. These are the quietest motors because they ramp up and down slowly. The added benefits for this type of motor also increase its initial cost.

Premium ThinkTank ECM

Premium ThinkTank ECMs have the highest efficiency and advanced features like self-adjusting constant airflow, climate-specific operating profiles, fully variable speed range, and digital communication. This enables true plug-and-play self-recognition software and a user interface that can control system setup and troubleshooting. The OEM control board can automatically recognize the connected units to adjust the required airflow for heating, cooling, and zoning.

37.17 ECM OPERATION

ECMs consist of a motor and a control module (see Figure 37-40). The rotor of the ECM is a permanent magnet, and the stator is wound like a three-phase motor (see Figure 37-41). The module controls the speed and rotation of the motor by energizing the windings in sequence, creating a rotating magnetic field. The permanent magnets in the rotor follow this rotating magnetic field. No more than two windings are energized at one time, leaving one set de-energized. The module checks the speed and rotation of the motor by reading the voltage induced in the windings that are not energized. The motor periodically checks the work load on the motor by not energizing any windings and measuring how much the motor slows down in a fixed period of time. This allows the module to calculate at what speed the motor must turn to provide the amount of airflow it is programmed for. The modules are programmed for the specific blower and unit the motor is matched with. ECM blowers do not have a preset rpm; their speed is variable between 200 rpm to 1050 rpm, with a safety limit of 1,500 rpm. The speed that a motor operates at is determined by its programming, the control signals it receives, and the resistance across the blower.

HVACR equipment uses blower control boards to control ECM operation (Figure 37-44). ECMs are made for three types of controls: pulse-width modulation, 24 V thermostat signals, or a digital serial interface. With pulse-width modulation, the control is not looking for the presence or absence of a particular voltage but is measuring the length, or "width," of a series of voltage pulses. Longer pulses call for increased rpm; shorter pulses call for decreased rpm. Thermostat control looks for 24 volts on pins that represent preprogrammed airflow requirements. Thermostat control is the most common. Digital serial controls are the newest of the three control methods. The unit communicates with the motor control using serial communication.

Figure 37-44 ECM motor control board showing speed jumpers.

Figure 37-45 ECM motor connectors.

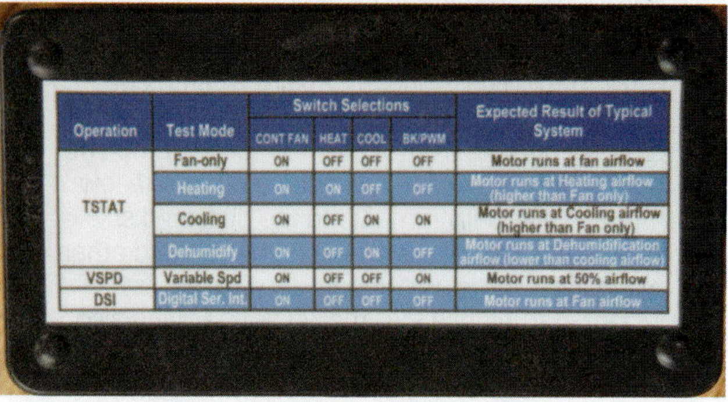

Operation	Test Mode	Switch Selections				Expected Result of Typical System
		CONT FAN	HEAT	COOL	BK/PWM	
TSTAT	Fan-only	ON	OFF	OFF	OFF	Motor runs at fan airflow
	Heating	ON	ON	OFF	OFF	Motor runs at Heating airflow (higher than Fan only)
	Cooling	ON	OFF	ON	ON	Motor runs at Cooling airflow (higher than Fan only)
	Dehumidify	ON	OFF	ON	OFF	Motor runs at Dehumidification airflow (lower than cooling airflow)
VSPD	Variable Spd	ON	OFF	OFF	ON	Motor runs at 50% airflow
DSI	Digital Ser. Int.	ON	OFF	OFF	OFF	Motor runs at Fan airflow

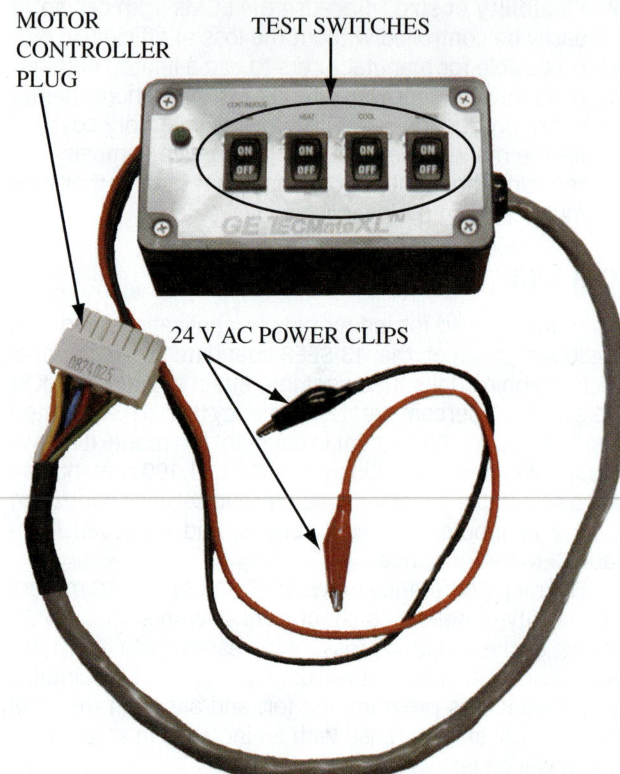

Figure 37-46 ECM motor tester.

37.18 ECM SERVICE

ECMs have two electrical connections: a power connection and a control connection (Figure 37-45). Power is supplied to the power connection all the time, even when the motor is not running. Power to the unit should always be turned off when disconnecting or connecting the power plug. Plugging or unplugging the power plug with power on will cause a high current arc that can damage the motor module. The easiest way to check an ECM is to use a service tool that plugs into the control port to control the motor. Figure 37-46 shows a simple device that can be used to check the operation of an ECM. If the motor will operate using the tester, the problem is in the wiring harness or blower control board. If power is available at the power connector and the motor will not operate using the tester, the problem is most likely the motor control module. Control modules are replaceable without replacing the entire motor (Figure 37-47). The motor windings can be checked by removing the module from the motor and checking their resistance. They should ohm out like a three-phase motor, with all the windings having the same resistance, which should be less than 20 ohms. If the windings are good, the module is bad. Be certain to get the correct module, because the modules are specific for the unit the blower is on.

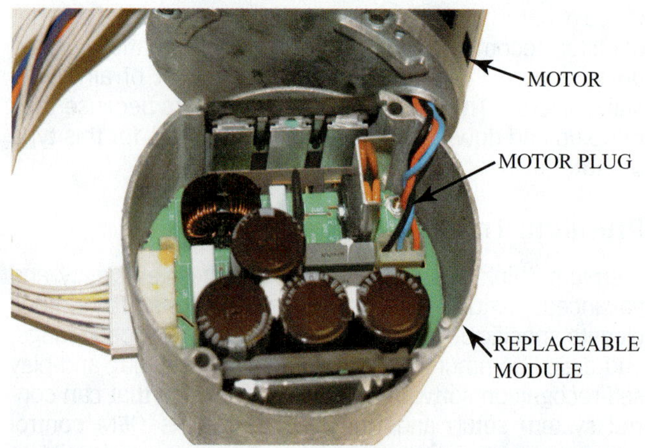

Figure 37-47 ECM motor replaceable module.

37.19 VARIABLE-FREQUENCY DRIVES

The speed of an AC motor is dependent on the number of stator poles and the frequency of the current. Multispeed AC motors designed to operate on constant-frequency (60 Hz)

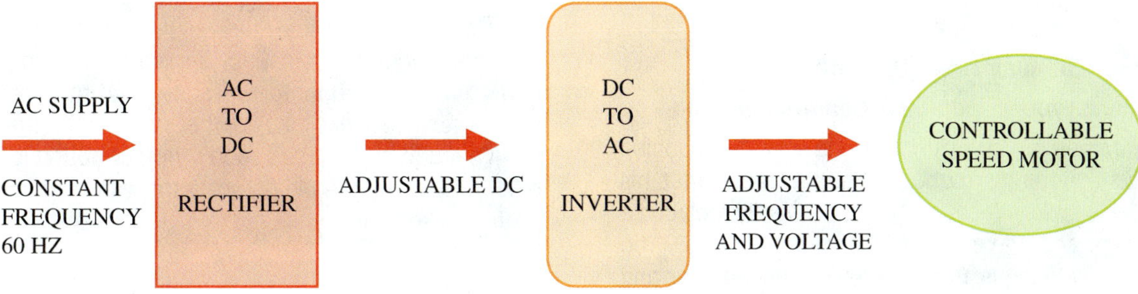

Figure 37-48 Basic configuration of a variable-frequency drive.

systems are provided with stator windings that can be reconnected to form different numbers of poles. A two-speed motor will have a winding that is switched to provide for either full speed or half speed. A four-speed motor will have two windings that each provide for two speeds when switched. This type of motor speed control is suitable for equipment that requires several definite speeds but not necessarily a continuously adjustable speed drive.

A variable-frequency drive (VFD) is a system for controlling the rotational speed of an electric motor by controlling the frequency of the electrical power supplied to it. Drive controllers are solid-state power conversion devices. The usual design first converts AC input power to intermediate power using a rectifier. The DC intermediate power is then converted to quasi-sinusoidal AC power using an inverter switching circuit.

VFDs for AC motor speed control are becoming more commonplace for variable-speed blower and compressor drives. In the past, solid-state motor controls were not cost effective for many applications. However, with the increased focus on efficiency, variable-frequency drives can recover their initial cost over time through reduced electrical operating expenses.

The basic concept of a VFD is shown in Figure 37-48. A solid-state rectifier is used to convert the line frequency AC into DC. The DC power is then converted back into adjustable-frequency AC power through a solid-state inverter. The reason for the VFD is to control the frequency of the AC power. Adjustment of the AC frequency will allow for motor speed control. This speed control is not a step type of control as delivered by changing the configuration of the stator poles. Instead, it is a fully variable speed control that may be continuously adjusted through the use of solid-state devices. Typically this type of drive will generate a considerable amount of heat, and some type of cooling will need to be employed during operation.

UNIT 37—SUMMARY

Motors are used to power compressors and fans that are commonly found in air-conditioning and refrigeration systems. Smaller residential units often use single-phase motors. Many of these utilize start and run capacitors to increase starting torque and to provide better running characteristics. Larger systems often use three-phase motors that do not require capacitors in their circuits. Three-phase motors will draw less current for the same load range as single-phase motors.

It is important to understand how to test a motor circuit. Wiring diagrams are generally supplied by the manufacturer of the unit. These are helpful when testing starting relays. Be sure to always discharge any capacitors before working on a circuit. New electronically commutated motors are growing in popularity because they offer the advantage of variable speeds with higher efficiencies as compared to conventional motors. Motor insulation should be tested on a regular basis with a megger to help determine the extent of insulation breakdown over time. The ability to properly maintain and troubleshoot electric motors is a basic requirement for HVACR technicians.

WORK ORDERS

Service Ticket 3701

Customer Complaint: No Air Flow

Equipment Type: Light Commercial Rooftop Air Conditioner

The building contractor calls the dispatcher to complain about the newly installed rooftop air-conditioning unit. The building is nearing completion, and when the rooftop air-conditioning unit was started, no air flowed from the supply registers. The technician arrives and finds access to the roof and the unit. The technician removes the panel to access the supply fan and secures the panel so that it will not blow from the roof.

The fan is operating, but the technician immediately notices that the fan is spinning in the wrong direction. This is a three-phase motor. The technician secures the power going to the unit at the disconnect switch. The junction box for the motor leads is opened and the technician uses a multimeter to check and make sure that the circuit is not energized. There is no measured power, so the technician switches two of the three main power leads to the motor. After completing this, the disconnect switch is closed and the fan is now running in the proper direction and supplying air to the building.

Service Ticket 3702

Customer Complaint: Not Working

Equipment Type: Window Air-Conditioning Unit

The customer brings a small window air-conditioning unit into the refrigeration shop and explains that his son, who is taking electrical classes, recently replaced a component and now the unit does not work at all. The technician working in the shop looks the unit over and notices that it has a run capacitor mounted on the frame of a hermetic compressor and sees that it appears new. Pulling a technical manual from the shelf, the technician looks up the specifications for the unit. The replaced capacitor is the wrong type. Most likely the motor became overheated and is now damaged. The technician tells the customer that he needs to discharge the capacitor and then check the motor windings with an ohmmeter, but it is suspected that the unit is damaged beyond repair.

Service Ticket 3703

Routine Maintenance

Equipment Type: Light Commercial Walk-in Cooler Three-Phase Motor

The technician is performing a monthly maintenance call at a small family-owned grocery store. One of the items on the checklist is to test the insulation on the compressor motor. The power to the unit was shut off with the circuit breaker located in the electrical panel near to the unit. This is an older nonhermetic unit, so the motor is separately coupled to the compressor. The technician uses a multimeter to double-check that the circuit is deenergized at the motor controller before opening up the junction box. Satisfied that the circuit is de-energized, the technician connects one end of the megger leads to bare metal on the motor frame and then the other end to one of the disconnected motor leads. The three windings are checked in this manner one at a time and the readings recorded. It is obvious in comparing the continually decreasing readings taken over time that the motor will soon need to be replaced. The technician explains to the owner that the motor will need to be replaced soon. Another option is to consider replacing the old compressor with a newer, more efficient semihermetic unit.

UNIT 37—REVIEW QUESTIONS

1. What is meant by motor torque?
2. What are two principal parts of a motor?
3. How is the speed of a motor determined?
4. How are capacitors rated?
5. What is a common application for capacitors in electric motors?
6. List the most common types of single-phase motors.
7. What are the two ways that the start capacitor can be removed from the circuit?
8. What advantages do three-phase motors have over single-phase motors?
9. What is the most common cause of motor failure?
10. According to Table 37-1, what is the upper and lower limit of a motor with the voltage of 460 V?
11. What happens if there is single phasing of a three-phase motor?
12. List the three types of protective devices used to protect the motor against excessive heat.
13. Where is the line-duty device commonly located?
14. What is the advantage of manual reset?
15. What does an external supplemental overload device provide?
16. What is the first thing to do in testing any motor?
17. What are insulation resistance testers?
18. What are ECM motors?
19. What are the principle advantages of ECM motors?
20. What is a VFD, and what is it used for?

UNIT 38

Electrical Diagrams

OBJECTIVES

After completing this unit, you will be able to:

1. identify symbols used for common electrical components.
2. explain the characteristics of different types of electrical diagrams.
3. interpret different types of electrical diagrams.
4. trace an electrical circuit on a ladder diagram.
5. draw a ladder-type diagram.

38.1 INTRODUCTION

Any successful service technician will tell you that understanding electrical circuitry is the single most important aspect of service and that understanding electrical wiring diagrams is the key to understanding electrical circuits. The goal for the technician is to understand how the system operates by studying the system's electrical diagrams.

One difficulty many technicians face when learning to read diagrams is that no two manufacturers use the same format for their wiring diagrams. Terminology and symbols can be confusing. For example, four different symbols are commonly used by manufacturers to represent a relay coil. However, there are enough common practices that learning one manufacturer's diagrams will make reading another manufacturer's diagrams easier. The more diagrams a technician reads, the better he or she will get at reading all diagrams.

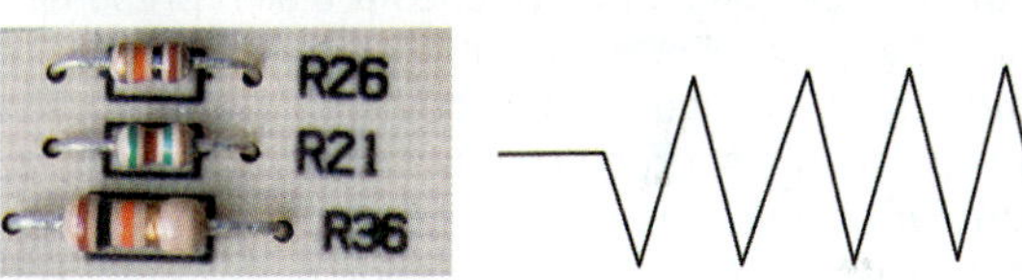

Figure 38-1 Heater symbol.

38.2 SYMBOLS FOR ELECTRIC COMPONENTS

The electrical components used in air conditioning are represented by graphic symbols in wiring diagrams. These symbols may not look very much like the actual physical component they represent. Instead, they show the electrical properties of circuit components. For example, symbols for switches often show that they can break the circuit. The symbol for a common switch is usually drawn open. Symbols for loads usually look continuous, as in the symbol for a heater (Figure 38-1). Many electrical devices consist of a single-circuit component, and so they are represented by a single symbol. A pressure switch is a good example. It only wires into one circuit and is represented by one symbol (Figure 38-2). Other devices contain more than one electrical component and wire in more than one circuit. Because they are part of more than one circuit, multiple symbols are used to represent a single device. Relays are a good example. A relay consists of a magnetic coil and one or more sets of contacts. The magnetic coil is a load and is represented by a symbol for a magnetic coil. The relay contacts are a switch and are represented by a set of parallel

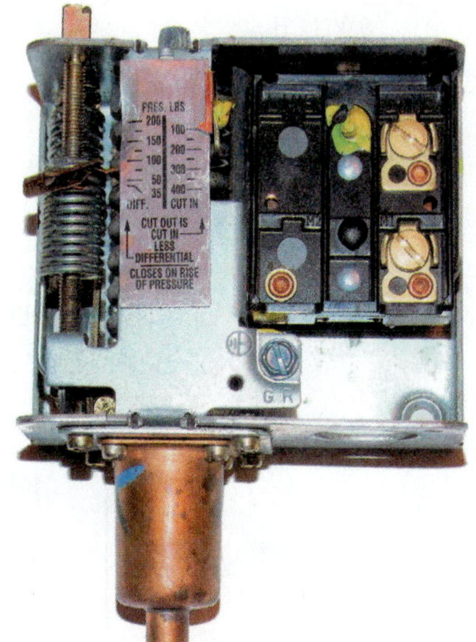

Figure 38-2 Pressure switch symbol.

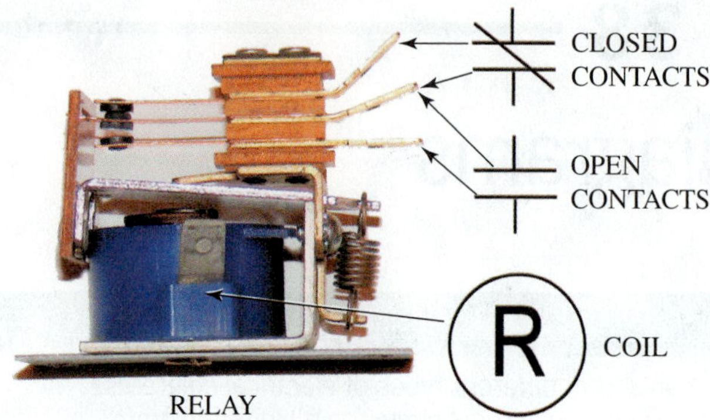

Figure 38-3 Relay and symbols.

lines. Figure 38-3 shows a relay, the symbol for the relay coil, and the symbol for the relay contacts. Even though the relay is a single physical device, it has two electrical components that are in two separate circuits.

38.3 THE POWER SOURCE

A basic electrical circuit is made up of a source of electrons, a path for electron flow, and a load. The source provides the electromotive force to move the electrons through the circuit, the path for the electrons is provided by wires, and the load turns the electrical flow into useful work. Switches are added to control the operation of the circuit. In electrical diagrams, the source, load, and switches are all represented by graphical symbols. The wires that provide the path are represented by lines. Figure 38-4 shows a basic electrical circuit consisting of a source, a path, a load, and a switch.

Electronic circuits and battery-powered devices usually have a direct current (DC) power source. Batteries and DC power sources are drawn to represent a battery's cathode (positive terminal) and the anode (negative terminal) (Figure 38-5). The power source for most HVACR systems is either single-phase alternating current (AC), or three-phase alternating current. The power source is often shown as

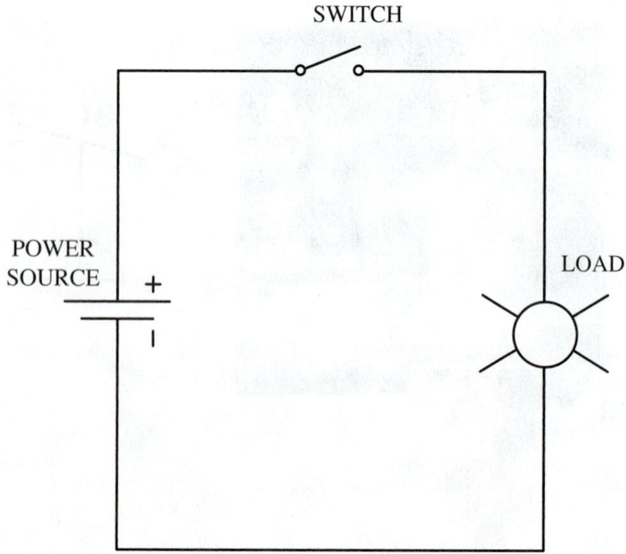

Figure 38-4 Basic circuit with source path and load.

Figure 38-5 Battery symbol with anode and cathode labeled.

an electrical disconnect switch with fuses. The specifications are written by the switch, including voltage, phase, and frequency. A double-pole, single-throw switch is used for single-phase systems; a triple-pole, single-throw switch is used for three-phase systems (Figure 38-6).

SINGLE PHASE DISCONNECT SWITCH

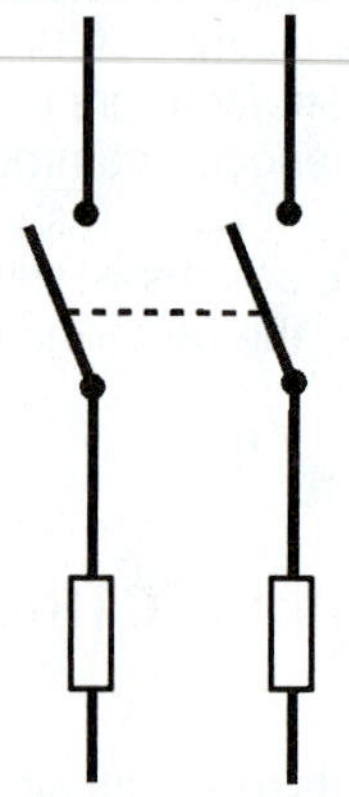

THREE PHASE DISCONNECT SWITCH

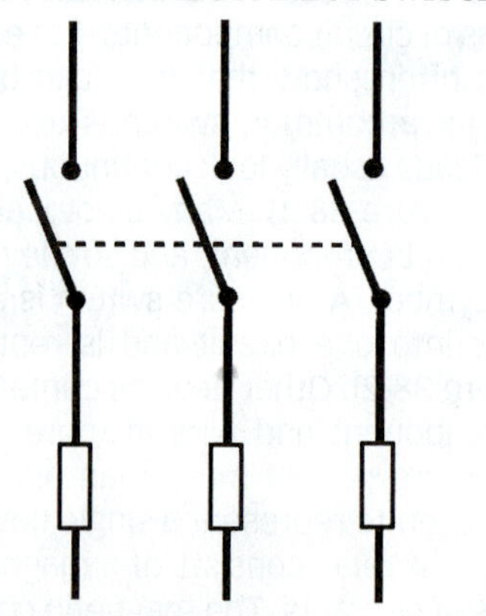

Figure 38-6 Single-phase and three-phase disconnect switch symbol.

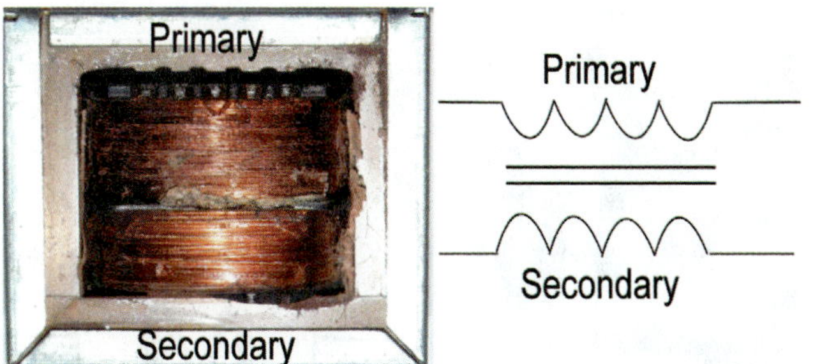

Figure 38-7 Transformer symbol.

Many HVACR system controls operate on a lower voltage than the line voltage. These systems require a transformer to transform, or change, the line voltage to a lower control voltage. Residential equipment uses 24 V AC for control voltage. Commercial equipment may have more than one control voltage. Commercial units frequently have both 24 V AC and 120 V AC controls, requiring two transformers. Transformers are drawn using two parallel lines to represent the iron core of the transformer, with a coil of wire on either side of the parallel lines to represent the primary and secondary windings (Figure 38-7). Transformers can be designed to accommodate different primary voltages by having multiple wires, or taps, on the primary side. These are shown as extra wires coming from the primary winding, with the different primary voltages labeled (Figure 38-8).

CODE BOX

The notes in most diagrams specify that all control wiring comply with NEC Class 2. Class 2 circuits may not be over 30 V or 100 volt-amps, VA. This limits the power of the circuit to a level that is not likely to create fires or cause electrical shock.

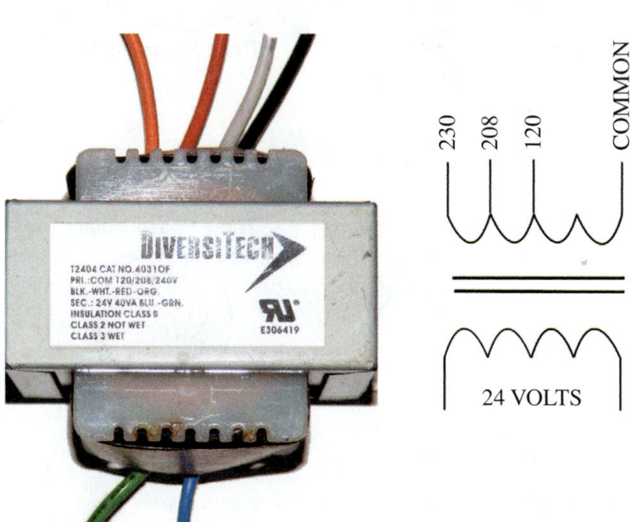

Figure 38-8 Dual-voltage transformer symbol.

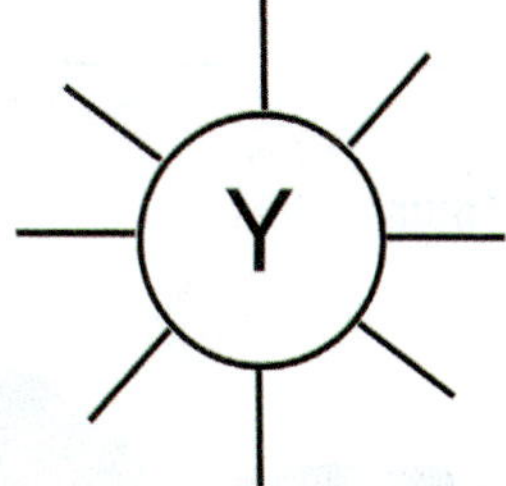

Figure 38-9 Light and symbol.

38.4 SYMBOLS FOR LOADS

A load creates a voltage drop when current passes through it. Loads use the electrical energy to perform work. Common loads are lights, heaters, and motors. A light is represented as a circle with rays radiating from it. Often, a letter inside the circle will represent the light's color (Figure 38-9). Electric heaters operate through resistance. Heat is created when current passes through the electrical resistance, creating a voltage drop and turning the electrical energy into heat. This is why the symbol for a resistor and the symbol for an electric heated are identical (see Figure 38-1). In air-conditioning diagrams, the symbols for heaters often have letters labeling them to help identify their function, such as CH for crankcase heater (Figure 38-10). Electromagnetic coils are another common type of load. They turn electrical energy into magnetism. There are several symbols used for magnetic coils, including a circle, a diagonal line connected to two smaller diagonal lines, a rectangle, and a series of loops (Figure 38-11). Because many electrical devices in HVACR contain magnetic coils, these symbols can be found representing different kinds of components, including solenoid valves, magnetic overload coils, relay and contactor

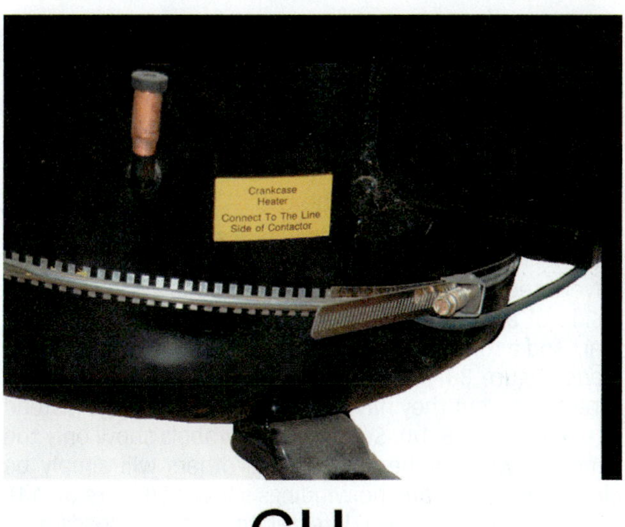

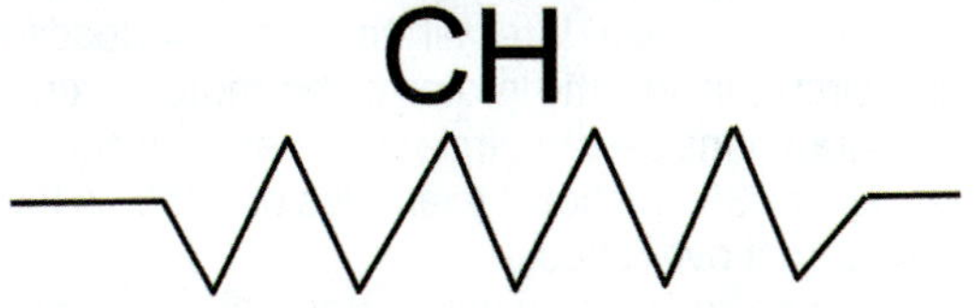

Figure 38-10 Crankcase heater and symbol.

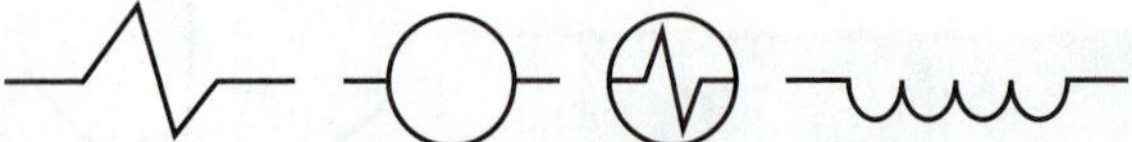

Figure 38-11 Magnetic coil symbols: diagonal, circle, windings.

Figure 38-12 Devices represented by a magnetic coil: solenoid, overload, motor, relay coil, and contactor coil.

coils, and motors. Figure 38-12 shows some different components that can be represented by a magnetic coil.

SAFETY TIP

There should always be a load in any path between L1 and L2. Any path that can be traced from L1 to L2 should pass through a device that uses power, like a motor or heater. A path that does not include a load is called a short circuit.

38.5 MOTORS

Motors are the main loads shown in most air-conditioning diagrams. Because there are a variety of motors used in HVACR, there are a variety of symbols used to represent them. Motors are usually shown as a circle, a coil of windings, or a combination of the two. A simple shaded pole motor is usually shown as a circle with two leads coming out (Figure 38-13). Motors with more than one winding will have three or more leads. They often show the windings inside of a circle. Single-phase motors that have both a start and a run winding will show both windings and three leads (Figure 38-14). Three-phase motors also will show three leads, but they have three internal windings instead of two (Figure 38-14). Some motor symbols show only the windings without the circle, while others will simply be circles with leads and no windings showing (Figure 38-14).

Many motors have internal thermal overloads that break the circuit inside the motor if the motor overheats. These are drawn inside the circle in series with the motor windings. Figure 38-14 shows examples of symbols for motors with internal overloads.

Motors are often labeled with letters representing their function. For example, IFM for indoor fan motor or CFM for condenser fan motor. Compressors use a special type of motor symbol because they have external terminals for connection instead of wire leads. These terminals are usually shown on the symbol (Figure 38-14).

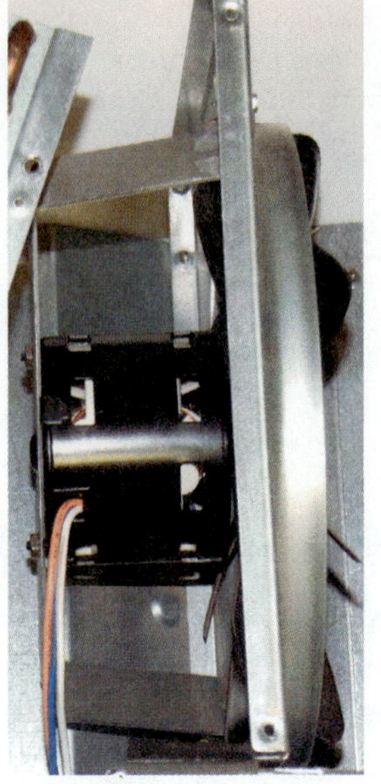

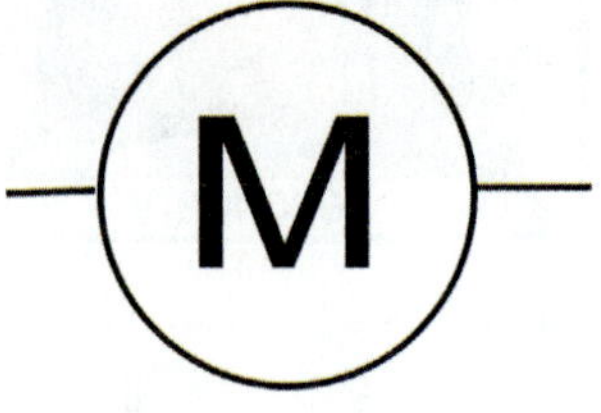

Figure 38-13 Shaded pole motor symbol.

Many single-phase motors require capacitors to operate. These include permanent split capacitor motors (PSC), capacitor start motors (CS), and capacitor start run motors (CSR). Capacitors are always drawn outside the motor. They are usually shown as two parallel lines: one straight and the other curved. Sometimes the capacitors are shown as a circle or an oval with the parallel lines inside the oval (Figure 38-15).

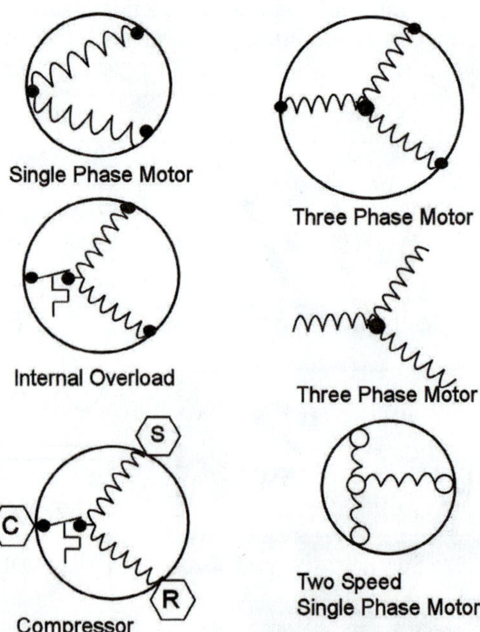

Figure 38-14 From left to right: Motor with start and run winding inside circle; Three-phase motor inside circle; Motor with internal overload; Three-phase motor winding; Compressor motor with common start and run terminals; Two-speed motor.

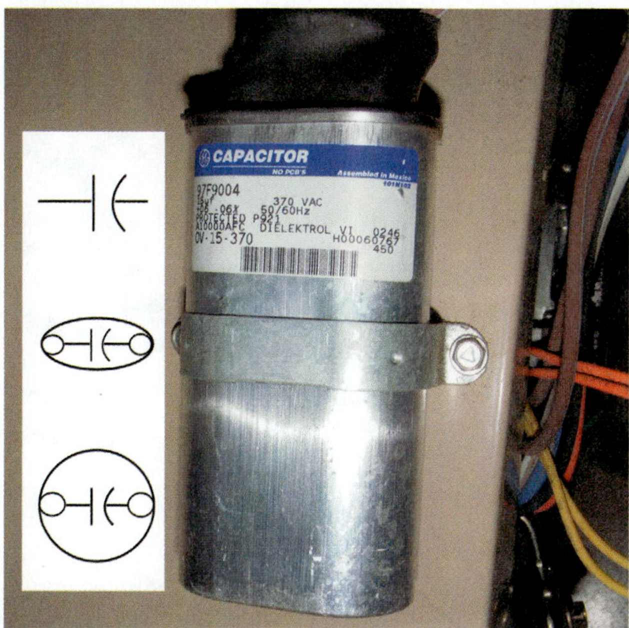

Figure 38-15 Capacitors and symbols.

38.6 SYMBOLS FOR SWITCHES

A switch controls a load by opening or closing the circuit to the load. Switches control loads by making a hole in the circuit. This is why symbols for switches often show how they break the circuit. For example, a simple single-pole toggle switch is drawn to look like the early knife-blade switches (Figure 38-16). When a switch is opened, it breaks the circuit and does not allow current to pass through. When a switch is closed, a path is created for current to flow (Figure 38-16). There should be no voltage drop through a switch when current flows through it.

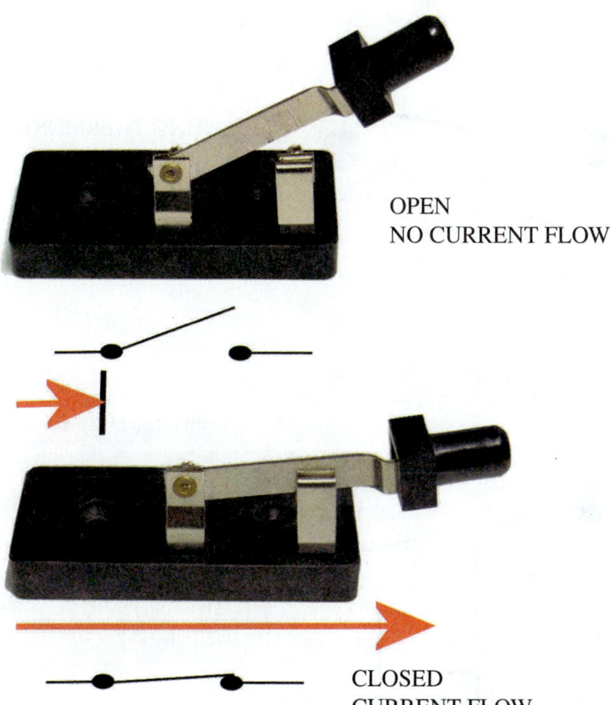

Figure 38-16 Knife-blade switch and symbol; Knife-blade switch in both positions with and without current flow.

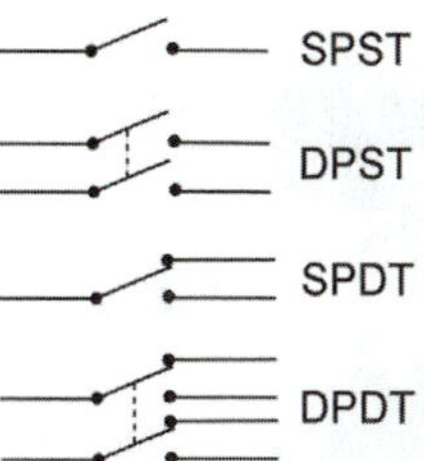

Figure 38-17 Comparison of single- and double-pole switch symbols.

The two most common types of mechanical switches used today are toggle switches and pushbutton momentary switches. A toggle switch alternates between two conditions—for example, off and on. The light switches in your house are toggle switches. Toggle switches are identified by their number of poles and throws. The number of poles identifies how many circuits a switch can handle, while the number of throws identifies and the number of possible paths for each circuit. A switch that can only switch a single circuit is called a single-pole switch, while a switch that can switch two circuits simultaneously is called a double-pole switch (Figure 38-17). A switch that has only one possible path for each circuit is called a single-throw switch. A switch that has two possible paths for each circuit is called a double-throw switch (Figure 38-17). These terms are combined to identify switches as

Single-pole, single-throw (SPST)
Single-pole, double-throw (SPDT)
Double-pole, single-throw (DPST)
Double-pole, double-throw (DPDT)

Figure 38-17 shows the symbols for these switches.

Unlike toggle switches, momentary pushbutton switches only hold their new position while they are being pushed and then return to their original position. The position the switch is in when it is not being pushed is called its normal position. Pushbutton switches can be either normally open or normally closed. A normally open switch starts out open, makes the circuit when it is pushed, and then returns to the open position when it is no longer being pushed. Normally open pushbutton switches are drawn above the line, as in Figure 38-18. A normally closed switch starts out closed, breaks the circuit when it is pushed, and then returns to the closed position when it is no longer being pushed. Normally closed pushbutton switches are drawn below the line, as in Figure 38-18. Start-stop switches and stop, forward, reverse switches are examples of momentary pushbutton switches (Figure 38-19).

38.7 CONTROLS

A control is a switch in an HVACR system that opens and closes in response to a system condition to control the operation of the unit. Controls can regulate the normal operation of the system or monitor system operation for unsafe operating conditions. Controls are drawn like a toggle switch

Figure 38-18 Normally closed (NC) pushbutton switch; Normally open (NO) pushbutton switch.

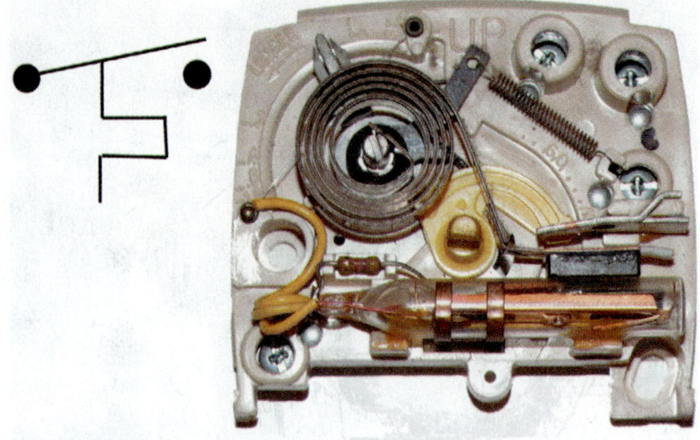

Figure 38-20 Thermostat and symbol.

with a stem under the switch connected to a symbol that represents the sensing mechanism. For example, thermostats look something like a bimetal (Figure 38-20), and pressure switches look like a small bellows (see Figure 38-2).

These switches either close on rise or open on rise. Close-on-rise switches are drawn below the line; open-on-rise switches are drawn above the line. A cooling thermostat that controls an air conditioner would be a close-on-rise switch, closing to energize the circuit when the temperature rises above the thermostat set point. A heating thermostat that controls a furnace would be an open-on-rise switch, opening to de-energize the circuit when the temperature rises above the thermostat set point (Figure 38-21).

Figure 38-22 shows the symbols for several common HVACR controls, including the following:

- Thermostats that open and close in response to temperature change
- Pressure switches that open and close in response to pressure change
- Humidistats that open and close in response to humidity change
- Flow switches that open and close in response to change in air or water flow
- Float switches that open and close in response to change in liquid level

Figure 38-19 Start-stop switch.

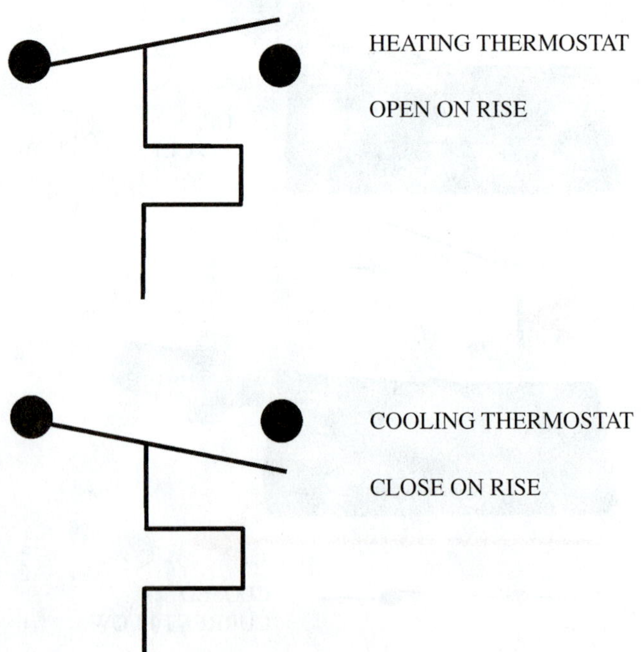

Figure 38-21 Thermostats: close on rise vs. open on rise.

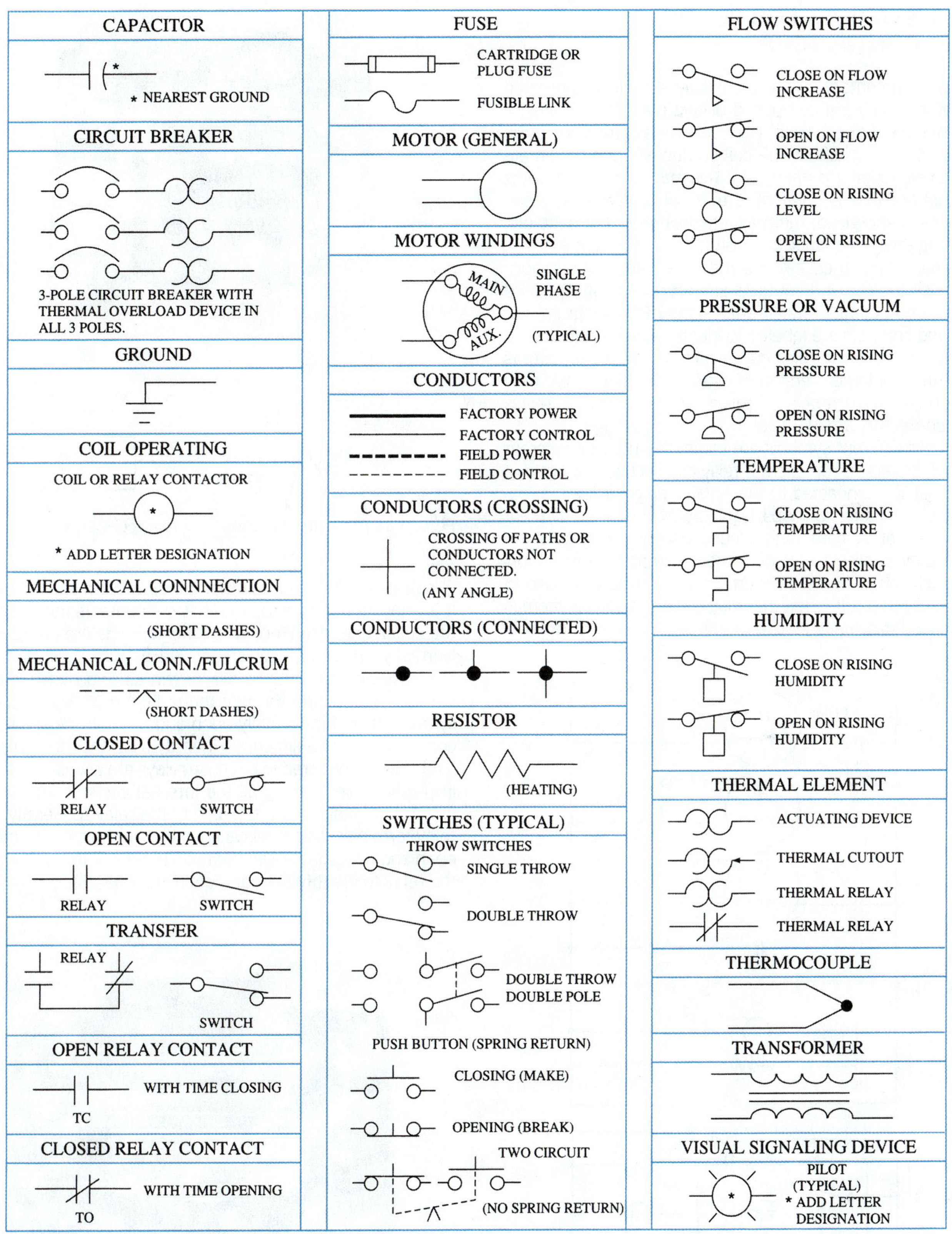

Figure 38-22 Symbols for thermostats, pressure switches, humidistats, float switches, flow controls.

38.8 SYMBOLS FOR RELAYS AND CONTACTORS

Relays and contactors are essentially electrically operated switches. They are composed of two electrical devices: a coil and a set of contacts. The coil is an electrical load; the contacts are switches. The coil produces a powerful magnetic field when it is energized. This magnetic field attracts a steel bar that is connected to a set of contacts. When the coil is energized, the magnetic field pulls the contacts, causing them to change position. The coil and contacts are in separate electrical circuits and have their own symbols. Because the coil and contacts are not drawn together and are often in completely different parts of the diagram, the coil and contacts are labeled to identify which contacts are controlled by which relay coil (Figure 38-23). A contactor is essentially a larger version of a relay. Contactors typically handle larger current than relays. The symbols for relays and contactors are identical.

Four different symbols are commonly used by different manufacturers to represent a relay coil, including a circle, a diagonal line connected to two smaller diagonal lines, a rectangle, and a series of loops. Figure 38-24 shows the different symbols that are commonly used to represent relay coils.

Relay contacts can be either normally open or normally closed. A normally open set of contacts will close to complete a circuit when the relay coil that controls them is energized. Normally closed contacts will open to break a circuit when the relay coil that controls them is energized. It is important to understand that the contacts move when the coil in energized, not when the contacts are energized. Normally open contacts are drawn as two parallel lines. Normally closed contacts are drawn as two parallel lines with a diagonal line through them. Figure 38-3 shows the symbols used for normally open and normally closed relay contacts. It can be difficult to identify what the terminals on a relay are connected to because relays are usually encased in plastic boxes, hiding all the internal parts. Relays often use a small diagram on the side of the relay to identify the relay terminals. Some relays have the symbols for normally open and normally closed contacts molded into the body of the relay to identify the contacts (Figure 38-25).

Figure 38-24 Different symbols for relay coils.

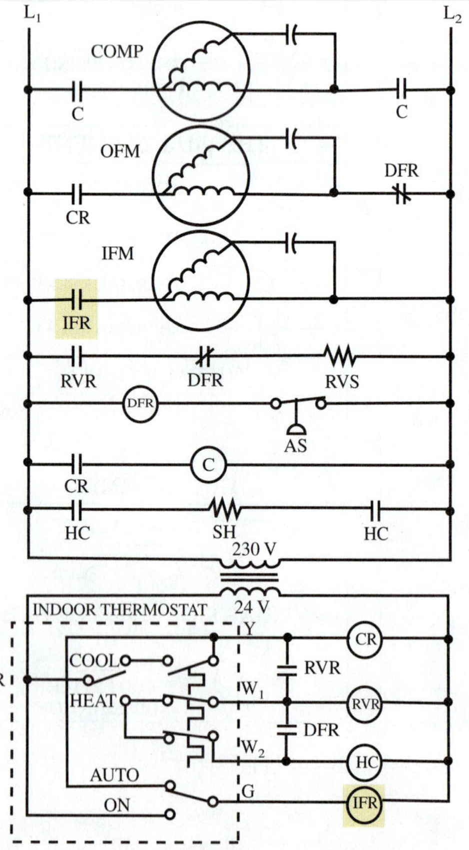

Figure 38-23 The coil of the fan relay is located in the 24 V portion of the diagram, while its contacts are located in the line-voltage portion.

Figure 38-25 Diagram molded into brown relay.

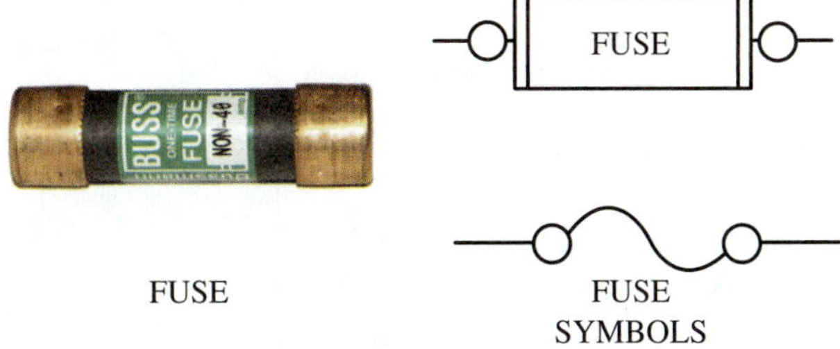

Figure 38-26 Symbols for fuses.

38.9 FUSES AND OVERLOADS

Overloads are controls that open to shut off a circuit if the load is drawing too much current or is overheating. Fuses are the simplest type of overload. They are drawn in series with the circuit they are protecting. Common symbols for fuses include a wavy loop resembling a sine wave and a rectangle that looks like a cartridge fuse (Figure 38-26).

Overloads can be either line duty or pilot duty. A line-duty overload breaks the line voltage circuit to the load it is protecting. When the overload senses too much current or heat it opens, breaking the circuit to the load. A pilot-duty overload has a coil and a set of contacts like a relay. The coil is in series with the load being protected, and the contacts are in series with the coil for the control that normally operates the load, such as a contactor. When the overload senses too much current or heat, the overload contacts break the circuit to the contactor coil. The contactor coil is de-energized, opening the contactor contacts, breaking the circuit to the load. Figure 38-27 shows a diagram of a pilot-duty overload relay.

Overloads sense the current by either heat or magnetism. Heat-sensing overloads are called thermal overloads. They can be drawn as an inverted disk or as a set of hooks. The inverted disk represents a bimetal disk that warps when it heats up. Bimetal overloads are a very common type of thermal overload. Inline thermal overloads can be drawn as a set of normally closed points after the hooks, just the hooks by themselves, or the inverted disk. Magnetic overloads use the same symbol commonly used for solenoid coils. Pilot-duty overload relays use either the hooks or diagonal to represent the coil and a separate set of normally closed contacts. Figure 38-28 shows a variety of overload symbols.

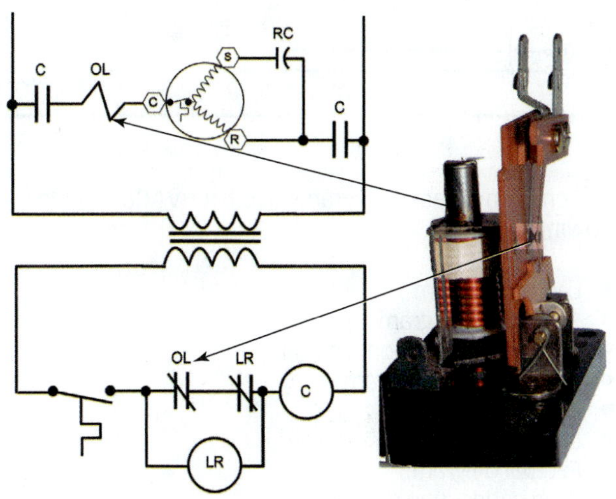

Figure 38-27 Pilot-duty overload.

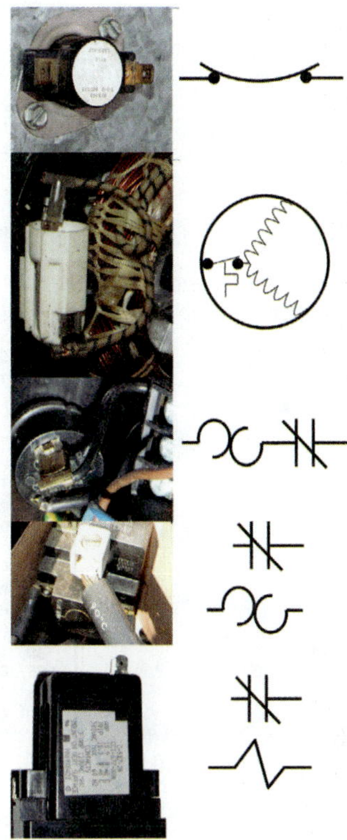

Figure 38-28 Overload symbols: bimetal overload, internal overload, thermal in-line overload, thermal overload relay, magnetic overload relay.

38.10 THE LEGEND

Remembering so many different symbols and variations could definitely be a challenge. Fortunately, diagrams come with a legend that identifies which symbols are used and what they represent. Other information such as abbreviations for wire colors and different line styles are also explained. The first place to begin when reading a new diagram is at the legend. It is the instruction manual for the diagram. Most manufacturers use letter designations to identify the components in a diagram. They typically choose letters that make it easier to identify the components without constantly looking back and forth between the legend and the diagram. For example, the letter *C* is often used to represent a compressor contactor. However, technicians should check the legend first rather than assuming they know what the manufacturer intended. For example, one manufacturer uses the abbreviation IBM to represent an induced draft blower motor while another uses the same three letters to represent an indoor blower motor. Some manufacturers use the letter *M* and a number for all motors and the letter *K* and a number for all relays and contactors. You have to look at the legend to determine which relay is represented by K1. Figure 38-29 shows a schematic and connection diagram with a typical diagram legend.

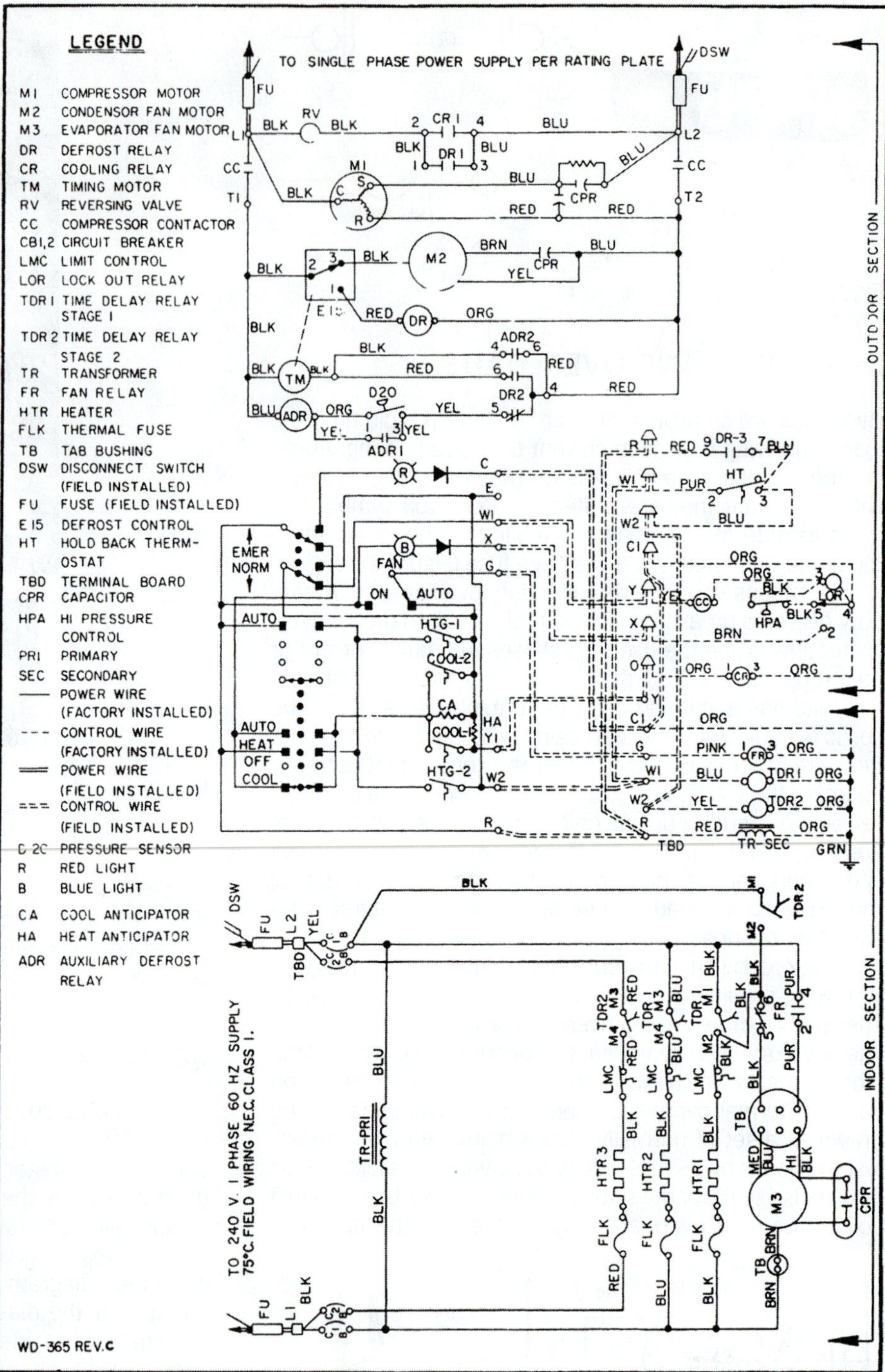

Figure 38-29 The legend on the left identifies the symbols used in the diagram.

38.11 TYPES OF DIAGRAMS

All diagrams do not serve the same purpose. Diagrams are used to show the logic of the electrical circuits, the location of electrical components, the terminal designations on the components, the routing of the actual wires on the unit, the colors and/or labels of the wires used, where the field power wiring for the unit should be connected, and where the field control wiring needs to be connected. Each diagram typically contains more than one type of information, but no single diagram will convey all this information. Manufacturers use multiple diagrams to accomplish this.

Common types of diagrams used in HVACR include the following:

- Ladder diagram
- Schematic diagram
- Component location diagram
- Point-to-point diagram
- Pictorial diagram
- Label diagram
- Factual diagram
- Field connection diagram
- Installation diagram

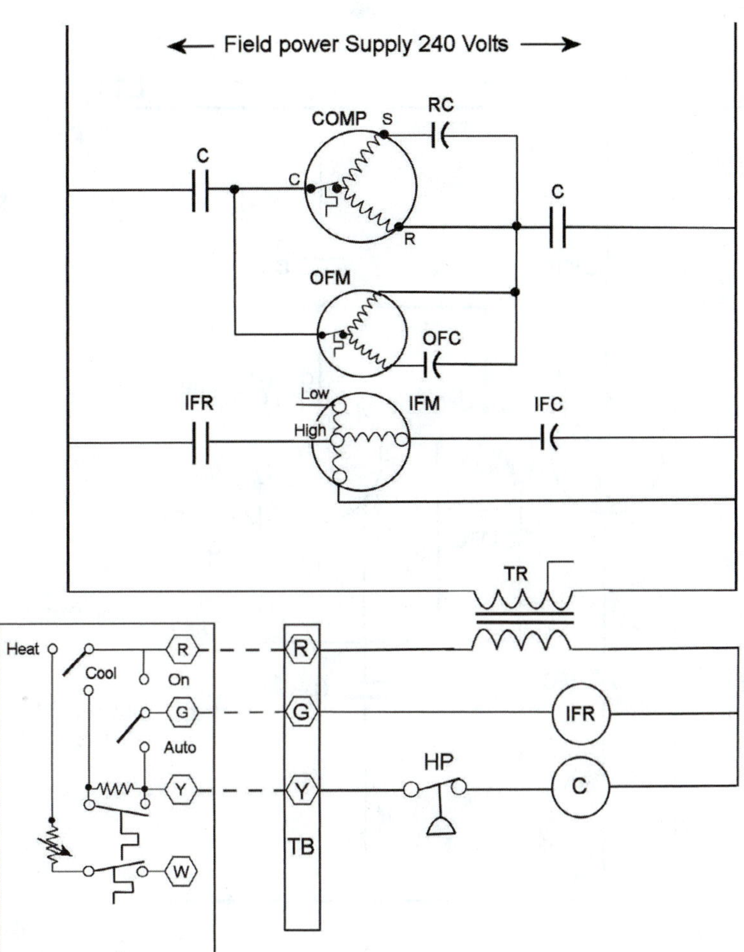

Figure 38-30 Ladder diagram.

38.12 LADDER DIAGRAMS

A ladder diagram is a symbolic representation of circuit logic. Line voltage is represented as rails on a ladder, and individual circuits are shown as rungs (Figure 38-30). Thus, L1 will run down the left side of the diagram, and L2 down the right side. Each circuit is shown as a line running between the two lines representing L1 and L2. Ladder diagrams do not show the actual wire routing, but show a circuit-by-circuit representation of the electrical system. Symbols are placed according to their electrical function, not according to their actual location in the unit. For example, the parts of a relay or contactor are shown in multiple places on the diagram because the coil is in a different circuit from the contacts. The coil and contacts of the contactor are shown in Figure 38-31. Note that the coil is in the 24 V control section of the diagram and the contacts are in the line-voltage portion. When the coil is energized, it closes its contacts, completing a circuit to the compressor motor. Most ladder diagrams have the power supply at the top and locate the largest line voltage loads near the top of the diagram (see Figure 38-30). Small line voltage loads, such as relay coils and solenoids, are usually drawn further down the ladder. For systems with control transformers, the transformer primary winding is the last line voltage load on the ladder. A new ladder is drawn underneath the transformer secondary winding (see Figure 38-30). Although ladder diagrams usually draw the power lines vertically, they can be made with the power lines drawn horizontally, looking like a ladder lying on its side (Figure 38-32).

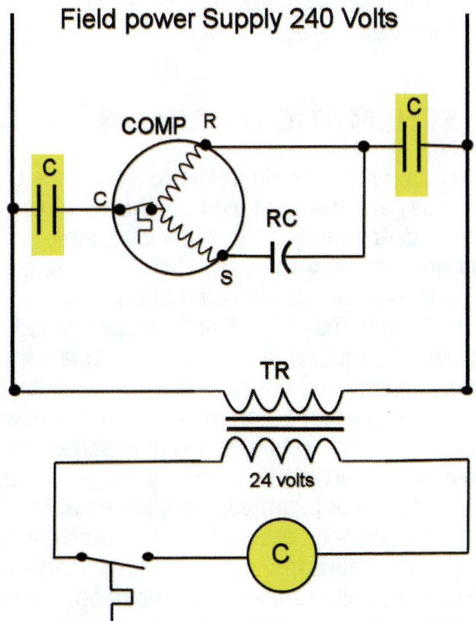

Figure 38-31 Contactor coil and contacts are in different places on the diagram.

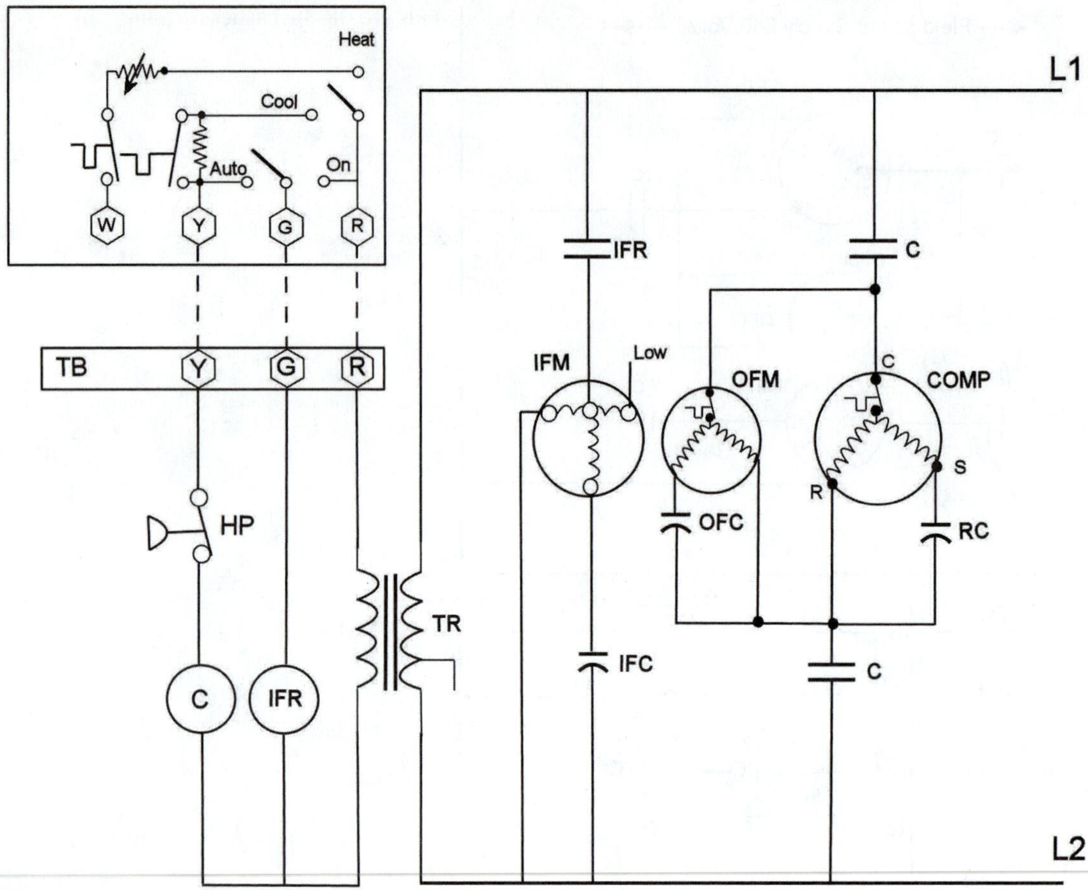

Figure 38-32 Sideways ladder diagram.

SERVICE TIP

Ladder diagrams are very useful for determining the operational sequence of a unit. Technicians can use a ladder diagram to learn how a piece of equipment they have never seen operates.

38.13 SCHEMATIC DIAGRAMS

Schematic diagrams, shown in Figure 38-33, are the most common. They are intended to show the scheme, or logic, of the electrical circuitry. Like ladder diagrams, schematic diagrams do not show the actual wire routing but try to show a circuit-by-circuit representation of the electrical system. Most schematics are at least partial ladder diagrams. However, most manufacturers include some extra information on their schematics. While a pure ladder diagram does not attempt to arrange any of the diagram to represent the physical reality of the unit, schematics commonly have some part of the diagram arranged to represent a part of the unit. For example, many schematics have the thermostat connections or unit terminal board connections drawn in the schematic in the same order and arrangement as in the unit. This allows the schematic to double as a connection diagram, but it breaks the ladder concept because the actual wire connections cannot be shown and keep all the circuits as parallel lines.

Many schematic diagrams vary the style of the lines in the diagram to distinguish between line voltage, low voltage, factory wiring, and field wiring. The heavier lines represent line voltage wires, while the lighter lines represent the low-voltage wires. Solid lines represent factory wiring, and dashed lines represent field wiring. Connections between wires can be represented different ways. In some diagrams, any lines that cross are connected. In these diagrams, when wires cross that are not connected, one line has a half circle loop over the other. Still other diagrams use a small, solid dot where wires connect. Wires that cross without the dot in these diagrams are not connected. Figure 38-34 shows some common line uses.

SERVICE TIP

Wiring diagrams for equipment may change as the manufacturer makes minor modifications to the equipment's components. Because wiring diagrams today are computer generated, it is easy for the manufacturer to make these minor changes without having to redraw the entire diagram. For this reason, many manufacturers' diagrams may be slightly different from one another but look very similar. Make certain that you are using the correct diagram for the piece of equipment you are currently servicing by checking the model number on the diagram with the equipment model number.

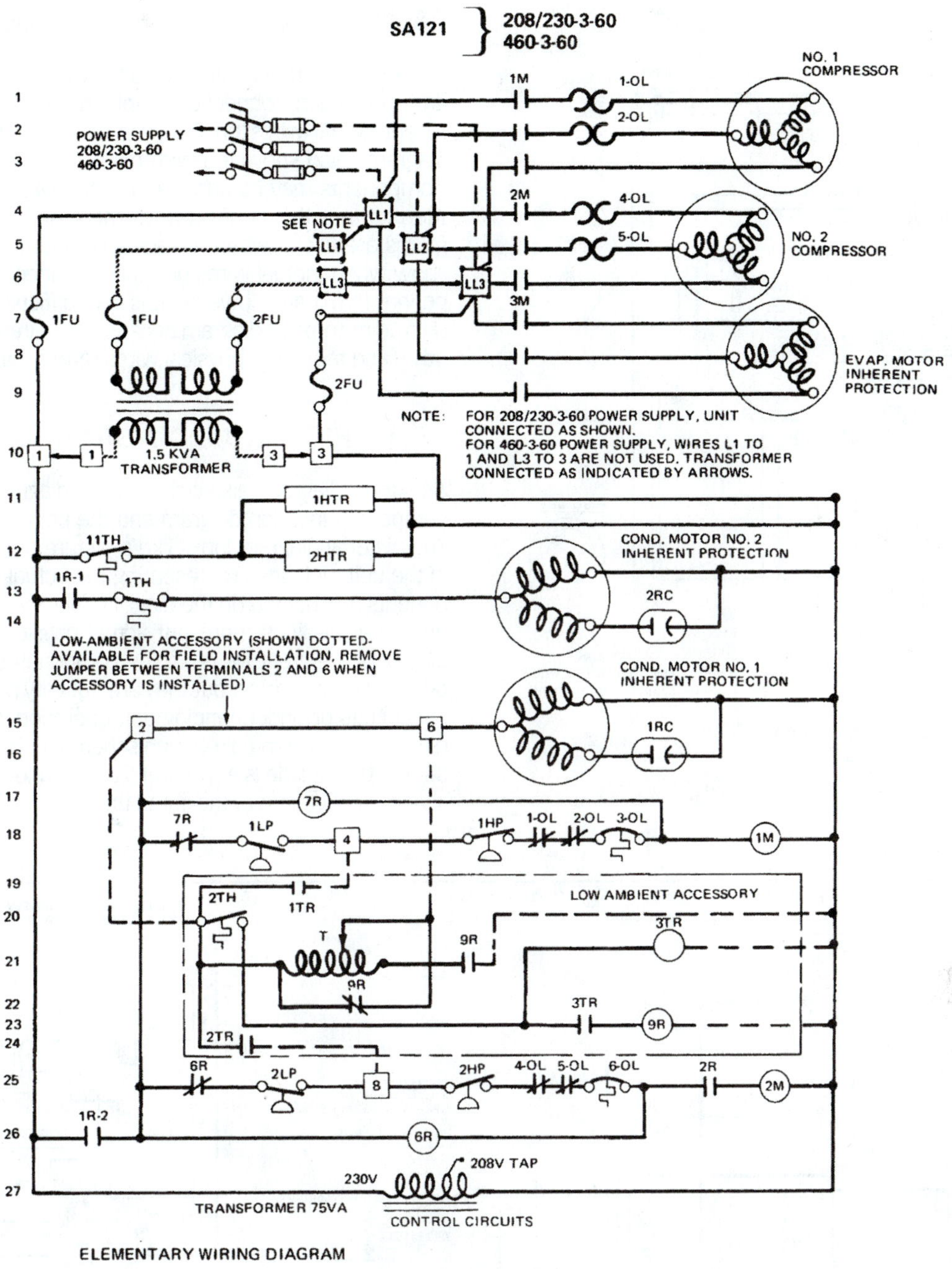

SA121 } 208/230-3-60
460-3-60
NO. 1
COMPRESSOR
POWER SUPPLY
208/230-3-60
460-3-60
SEE NOTE
NO. 2
COMPRESSOR
EVAP. MOTOR
INHERENT
PROTECTION
1.5 KVA
TRANSFORMER
NOTE: FOR 208/230-3-60 POWER SUPPLY, UNIT CONNECTED AS SHOWN.
FOR 460-3-60 POWER SUPPLY, WIRES L1 TO 1 AND L3 TO 3 ARE NOT USED. TRANSFORMER CONNECTED AS INDICATED BY ARROWS.
COND. MOTOR NO. 2
INHERENT PROTECTION
LOW-AMBIENT ACCESSORY (SHOWN DOTTED-AVAILABLE FOR FIELD INSTALLATION. REMOVE JUMPER BETWEEN TERMINALS 2 AND 6 WHEN ACCESSORY IS INSTALLED)
COND. MOTOR NO. 1
INHERENT PROTECTION
LOW AMBIENT ACCESSORY
208V TAP
230V
TRANSFORMER 75VA
CONTROL CIRCUITS
ELEMENTARY WIRING DIAGRAM

Figure 38-33 Schematic diagram.

FACTORY LINE VOLTAGE
FIELD LINE VOLTAGE
FACTORY LOW VOLTAGE
FIELD LOW VOLTAGE
CONNECTED
NOT CONNECTED

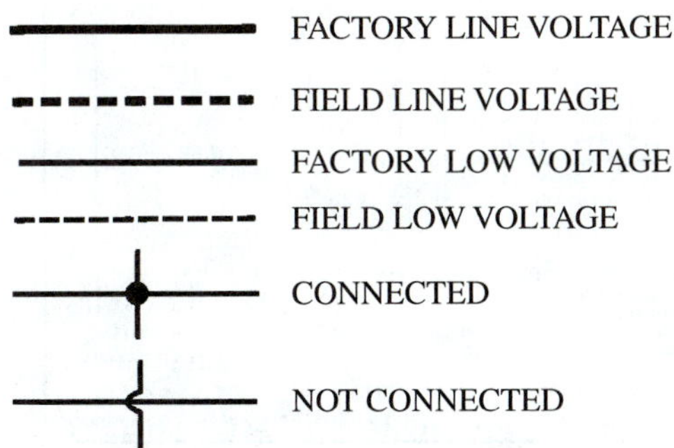

Figure 38-34 Line conventions.

38.14 COMPONENT LOCATION DIAGRAMS

A component location diagram shows where each component is located in the unit (Figure 38-35). This is especially helpful when a unit has multiple components that are physically identical. Component location diagrams usually do not show any wiring or circuits. They are often used in conjunction with ladder or schematic diagrams. The schematic diagram shows how the unit operates and field wiring connections; the component location diagram shows where the parts are actually located in the unit.

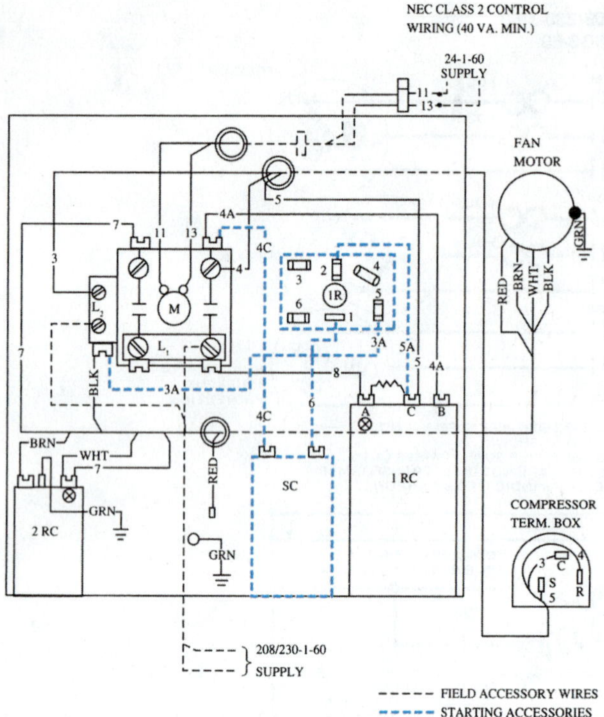

Figure 38-35 Component location diagram.

38.15 POINT-TO-POINT DIAGRAMS

A point-to-point diagram shows how the wires are routed to and from each connection point on the system. Lines represent actual wires on the system. Typically, point-to-point diagrams also show the wire color and any terminal labels. Components may be arranged on the diagram as they are in the unit, but this is not always the case. Point-to-point diagrams are useful when replacing components because they show where actual wires physically connect to specific component terminals. However, it is very difficult to trace circuits on a point-to-point diagram because the wires cross over each other and represent physical wires rather than circuits.

38.16 PICTORIAL DIAGRAMS

A pictorial diagram, also called a label diagram, combines the component location diagram and the point-to-point diagram. The pictorial diagram looks like a picture of the control panel in the unit with lines representing the actual wires. The components are located on the diagram exactly as they are in the unit, and the lines represent actual physical wires. Pictorial diagrams are great for replacing components and identifying original factory wiring, but they are not very much help in tracing circuits or understanding unit operation. Pictorial or label diagrams are often paired with schematic diagrams. In Figure 38-36, the left side is a pictorial diagram and the right side is a schematic diagram of the same unit.

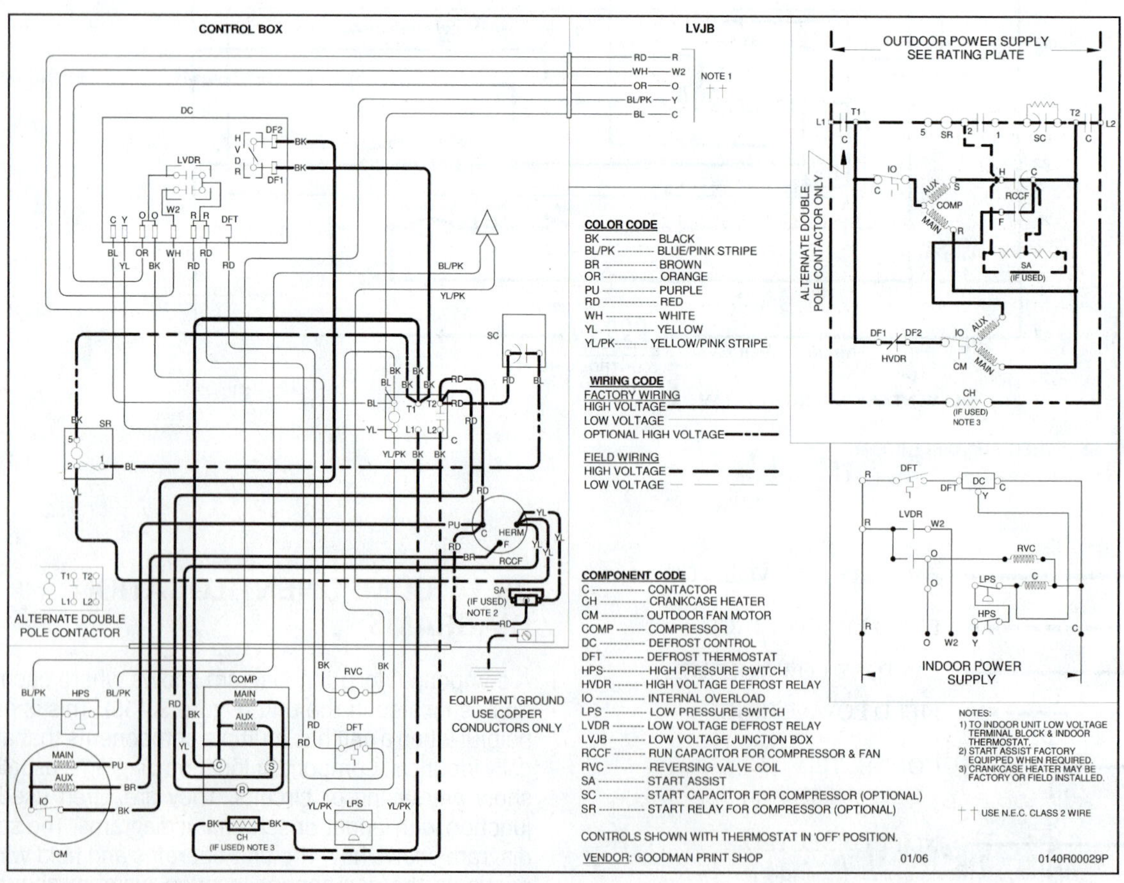

Figure 38-36 Pictorial and schematic together.

SERVICE TIP

Pictorial diagrams are often used in conjunction with ladder or schematic diagrams. The ladder diagram allows the technician to trace circuits and determine the components they want to check, but they don't help locate the actual component and wires that must be checked. Pictorial diagrams can do that. Together they are very powerful.

38.17 FACTUAL DIAGRAMS

A factual diagram is a combination of a ladder diagram and a point-to-point diagram. Factual diagrams are still arranged in a ladder, but references at each connection point show where the actual wires run to and from on the unit. Figure 38-30 shows a ladder diagram of a packaged air conditioner, and Figure 38-37 shows a factual diagram of the same unit. The parallel lines represent specific physical wires. This often results in multiple parallel lines running between the power rails on the side and a component (Figure 38-37). Component terminals are labeled just as they are on the components. Terminals with names or numbers are shown inside a hexagon—for example, the compressor motor terminals in Figure 38-37. Terminals on components that do not have a number or label are shown as small circles—for example, the transformer connections in Figure 38-37. Wires are identified by their color or number at each connection inside the diagram. References along the side rails identify where each wire is headed. For example, in Figure 38-37, there are two wires connected to the L2 contactor terminal: a red wire and an orange wire. The connection on the side rail indicates that the red wire goes to the IFC (indoor fan capacitor). Down lower on the rail, you see

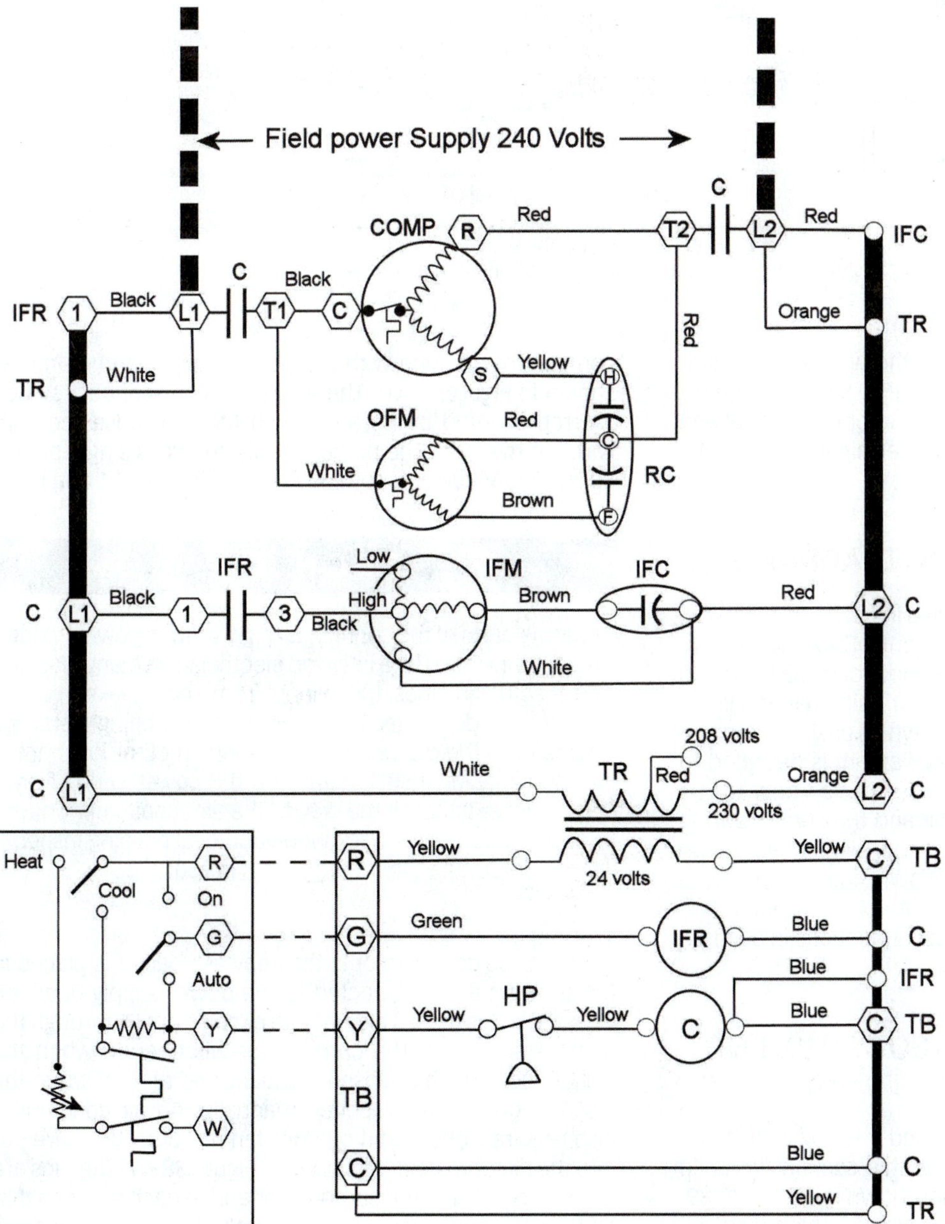

Figure 38-37 Factual diagram.

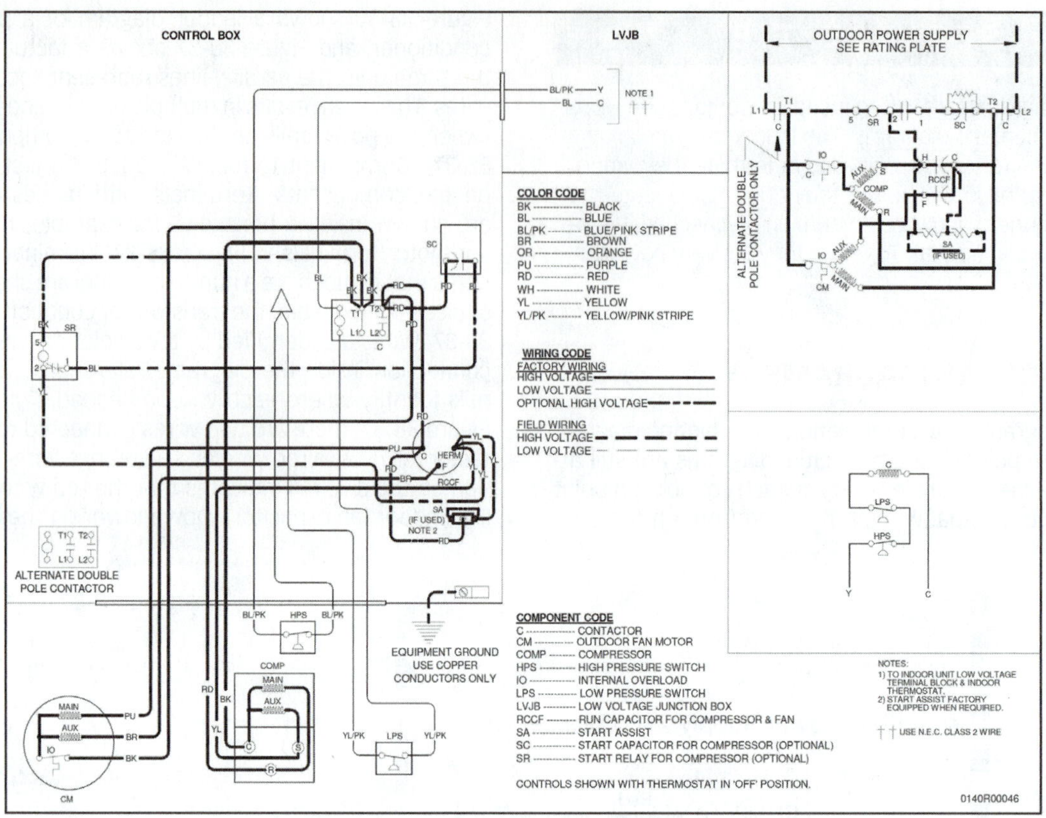

Figure 38-38 The dashed lines on the right show where wires are connected during installation. *(Courtesy Goodman Global Group, Inc.)*

the L2 contactor terminal. Following the wire at that point, you see a red wire connecting to the IFC. Similarly, the orange wire connects between the contactor L2 terminal and the transformer, as indicated by the terminal designations along the power rail.

38.18 FIELD CONNECTION DIAGRAMS

Field connection diagrams, also called installation diagrams, identify the electrical controls in the control box and indicate the necessary field wiring. Field connections are indicated on the right-hand side of Figure 38-38 using dashed lines. Field connections are frequently shown as dashed lines and shaded areas. The field connection diagram is designed to instruct the installing electrician or technician how to run the proper power supply to the unit and the correct wiring between different sections if the unit consists of more than one section. Internal wiring is not shown since the principal use of the field connection diagram is for installation. It is possible for a diagram to show field connections in addition to other details, as in Figure 38-29.

38.19 BUILDING AN AIR-CONDITIONING SCHEMATIC

One of the best ways to understand the logic of schematic diagrams is to draw them. In this section we build the air-conditioning system diagram shown in Figure 38-39. The diagram will be built circuit by circuit, component by component, starting with the power supply to the unit, as shown in Figure 38-40. The wires to the disconnect switch are represented by heavy broken lines because they are wired in the field. The power supply shown is single phase, 60 Hz, 230 V current, coming in through wires L1 and L2.

TECH TIP

In some parts of the country, all high-voltage power wiring must be installed by a licensed electrician. In many places, an electrician does the wiring from the breaker panel to the unit disconnect, and the air-conditioning installer wires from the disconnect to the unit. In other locations, air-conditioning installers may do the power wiring from the breaker panel all the way to the air conditioner under certain conditions. Check with your local building inspector to determine the requirements in your area.

Because the compressor is the heaviest load, it is placed in the diagram first, connected to the power supply between L1 and L2. This makes a complete circuit from L1 through the compressor to L2. The compressor will operate when the disconnect switch is closed manually and turn off when the disconnect switch is opened manually. An air conditioner also requires indoor and outdoor fan motors. These are put into the diagram next, as shown in Figure 38-41. The fans are wired in parallel to the compressor and to each other so that all three motors receive full line voltage. All three motors

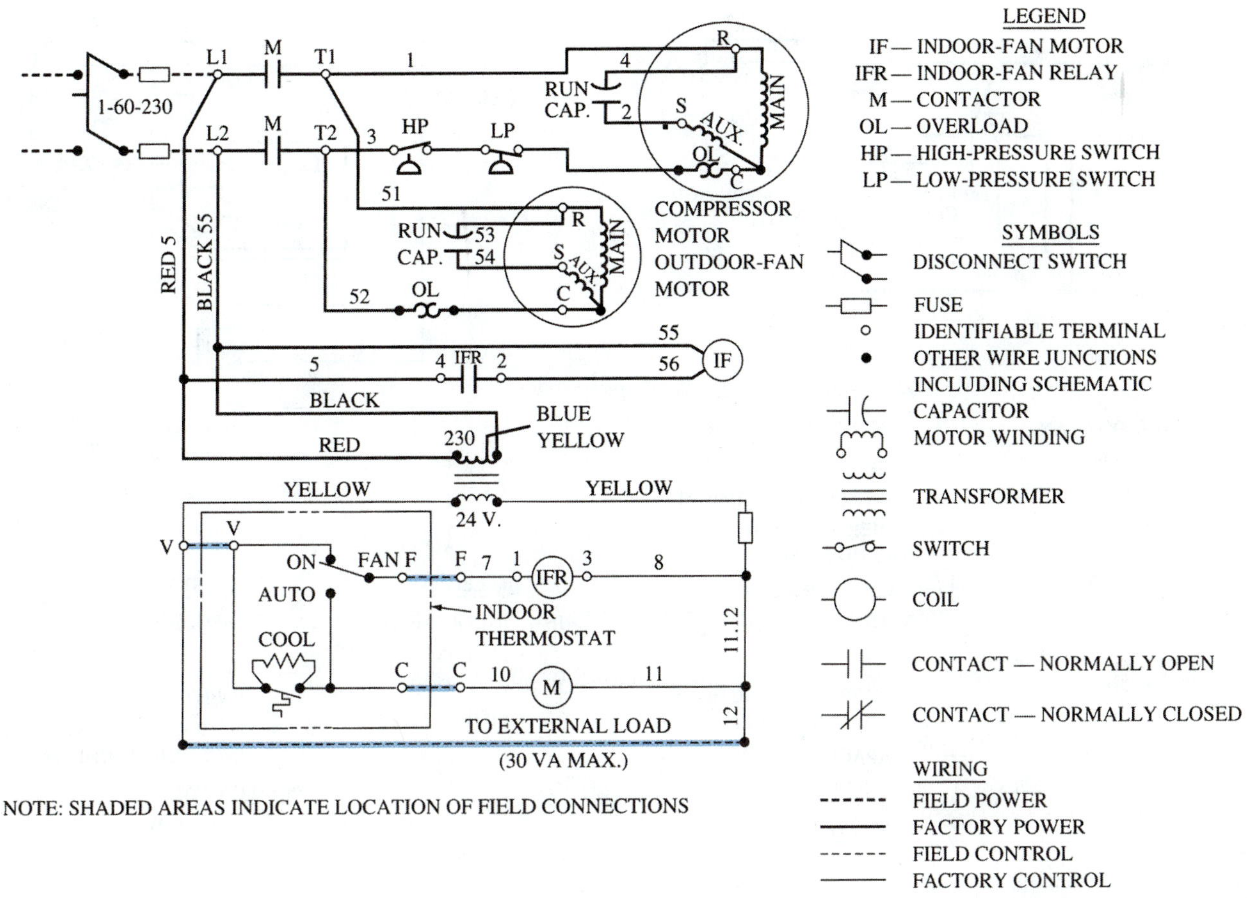

L1
M
T1
1
4
R
RUN CAP.
2
S
AUX.
MAIN
1-60-230
L2
M
T2
3
HP
LP
OL
C
51
COMPRESSOR MOTOR
RUN CAP.
53
54
R
S
AUX.
MAIN
C
OUTDOOR-FAN MOTOR
RED 5
BLACK 55
52
OL
55
5
4
IFR
2
56
IF
BLACK
BLUE
RED
230
YELLOW
YELLOW
YELLOW
24 V.
V
V
ON
FAN F
F
7
1
IFR
3
8
AUTO
INDOOR THERMOSTAT
11.12
COOL
C
C
10
M
11
TO EXTERNAL LOAD
(30 VA MAX.)
12
NOTE: SHADED AREAS INDICATE LOCATION OF FIELD CONNECTIONS
LEGEND
IF — INDOOR-FAN MOTOR
IFR — INDOOR-FAN RELAY
M — CONTACTOR
OL — OVERLOAD
HP — HIGH-PRESSURE SWITCH
LP — LOW-PRESSURE SWITCH
SYMBOLS
DISCONNECT SWITCH
FUSE
IDENTIFIABLE TERMINAL
OTHER WIRE JUNCTIONS INCLUDING SCHEMATIC
CAPACITOR
MOTOR WINDING
TRANSFORMER
SWITCH
COIL
CONTACT — NORMALLY OPEN
CONTACT — NORMALLY CLOSED
WIRING
FIELD POWER
FACTORY POWER
FIELD CONTROL
FACTORY CONTROL

Figure 38-39 Complete wiring diagram for an air-conditioning unit.

are in individual circuits of their own. A path can be traced from L1 to L2 through each motor without going through any other motor. At this point, the system has no controls and does not operate automatically. All three motors will operate whenever the disconnect switch is closed manually. Electrical controls are needed for automatic operation.

L1
1-60-230
L2
RUN CAP.
R
S
AUX.
MAIN
C
COMPRESSOR MOTOR
WIRING
FIELD POWER
FACTORY POWER
SYMBOLS
DISCONNECT SWITCH
FUSE
IDENTIFIABLE TERMINAL
OTHER WIRE JUNCTIONS INCLUDING SCHEMATIC
CAPACITOR
MOTOR WINDING

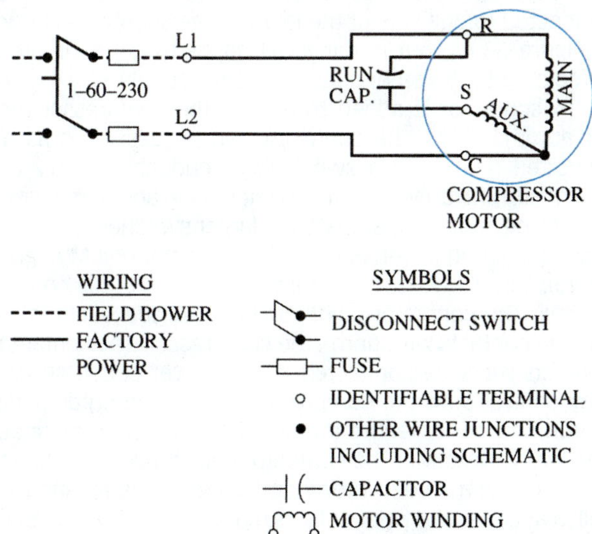

Figure 38-40 Diagram showing the wiring from the disconnect to the compressor motor.

L1
1-60-230
L2
RUN CAP.
R
S
AUX.
MAIN
OL
C
RUN CAP.
OF
COMPRESSOR MOTOR
IF
LEGEND
OF — OUTDOOR-FAN MOTOR
IF — INDOOR-FAN MOTOR
SYMBOLS
DISCONNECT SWITCH
FUSE
IDENTIFIABLE TERMINAL
OTHER WIRE JUNCTIONS INCLUDING SCHEMATIC
CAPACITOR
MOTOR WINDING
WIRING
FIELD POWER
FACTORY POWER

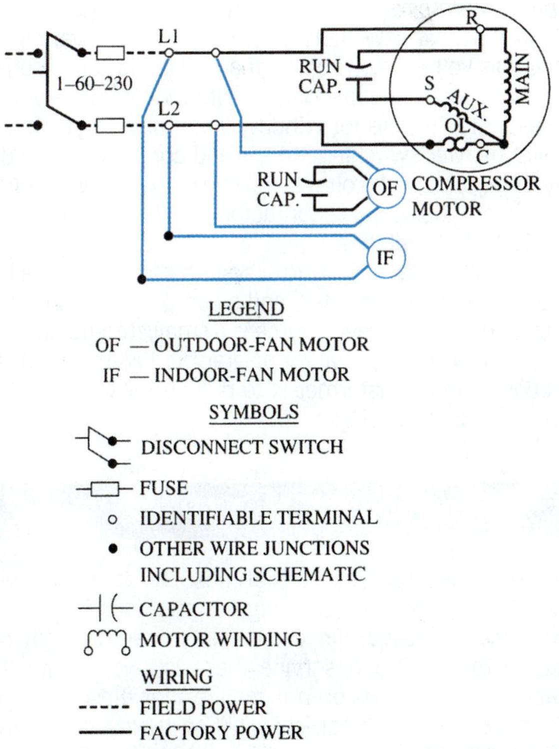

Figure 38-41 Wiring diagram showing the addition of the outside and inside fan motors.

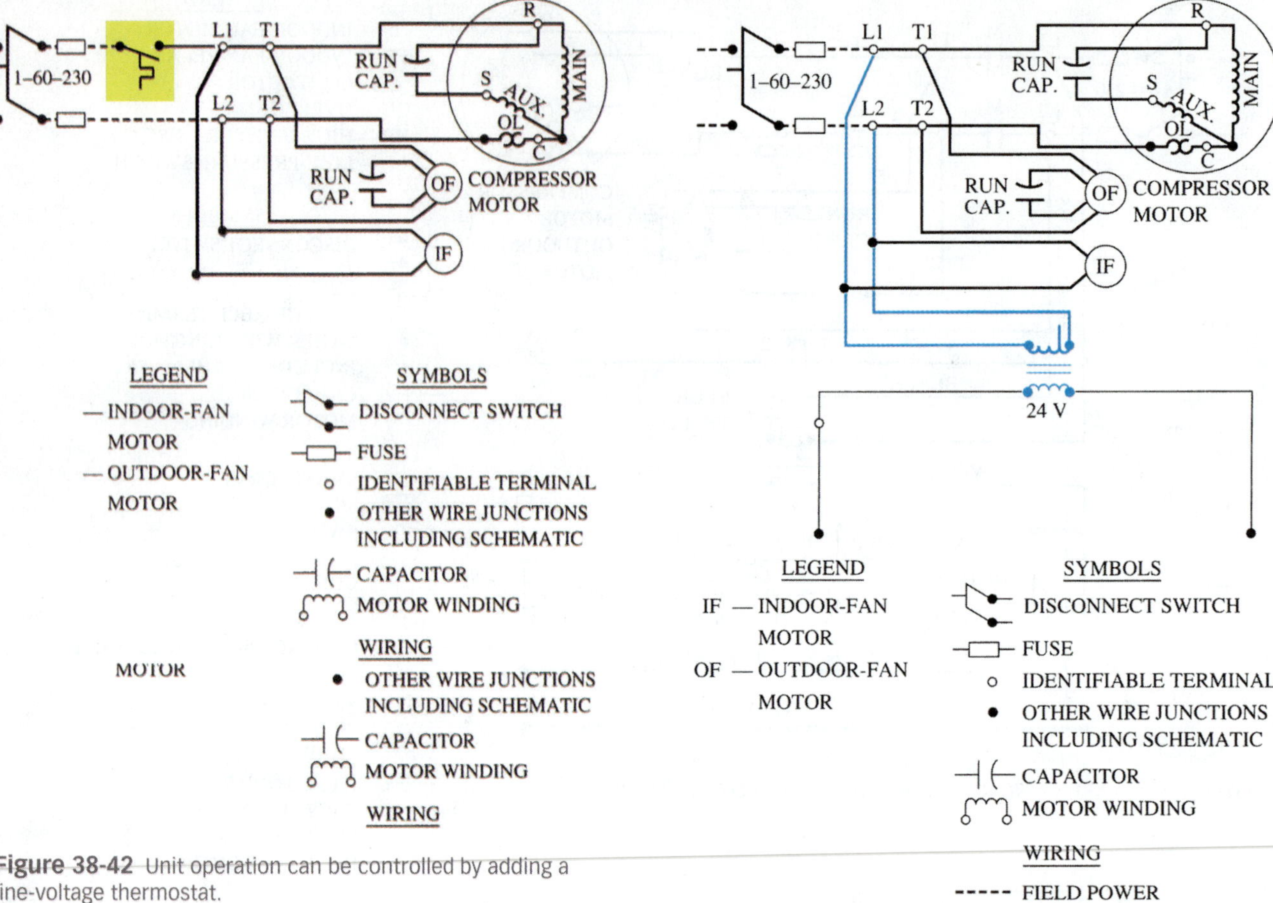

Figure 38-42 Unit operation can be controlled by adding a line-voltage thermostat.

Figure 38-43 Wiring diagram showing the addition of a low-voltage control transformer.

One way of providing automatic operation would be to put a thermostat in series with L1 or L2, (Figure 38-42). This is how window units operate. One disadvantage of using line voltage controls is that switching high current from large loads requires larger components. Line voltage thermostats are fine for window units, but larger systems require heavier switching. Relays and contactors provide a way to switch a high voltage and current with a low voltage and current. Relays and contactors are essentially electrically operated switches.

We will add a 24 V control system consisting of a low-voltage thermostat, a relay, and a contactor to control the motors. The low-voltage source is a small step-down transformer, shown in the wiring diagram in Figure 38-43. The function of the transformer is to reduce the 240 V to 24 V to provide control voltage.

TECH TIP

There are several reasons that most HVACR equipment uses 24 V as the control voltage. First, under most state and local guidelines an electrician's license is not required to install and service these low-voltage wires. Second, most codes do not require that these connections be made at electrical junction boxes. And third, under OSHA regulations, circuits that have less than 80 V fall under less stringent safety requirements.

The thermostat controls system operation by taking the 24 V from the transformer and sending it to the fan relay coil and the contactor coil. Terminal R on the thermostat is connected to one side of the low-voltage supply, as shown in Figure 38-44, but nothing will happen until a circuit is completed to the other side of the 24 V supply.

A contactor is added to control the compressor motor automatically. The contactor can be described as an oversized relay with a switch large enough to carry the heavy current drawn by the compressor and a magnetic coil strong enough to actuate the larger switches. The thermostat is wired in series with the contactor coil M (Figure 38-45). This allows it to control the operation of the contactor coil. The contactor coil M will control the M contacts, and the contacts will control the compressor. The contactor contacts must be connected in the power circuit so that they control both the compressor motor and outdoor fan motor. The indoor fan motor must be able to operate independently. Note that the contactor coil and both contactor contacts are marked M for identification. This means that coil M actuates the two contacts marked M. Since both contacts are normally open, they will close when the coil is energized and open when it is de-energized. Another relay is needed to control the indoor fan motor, IF. A fan relay

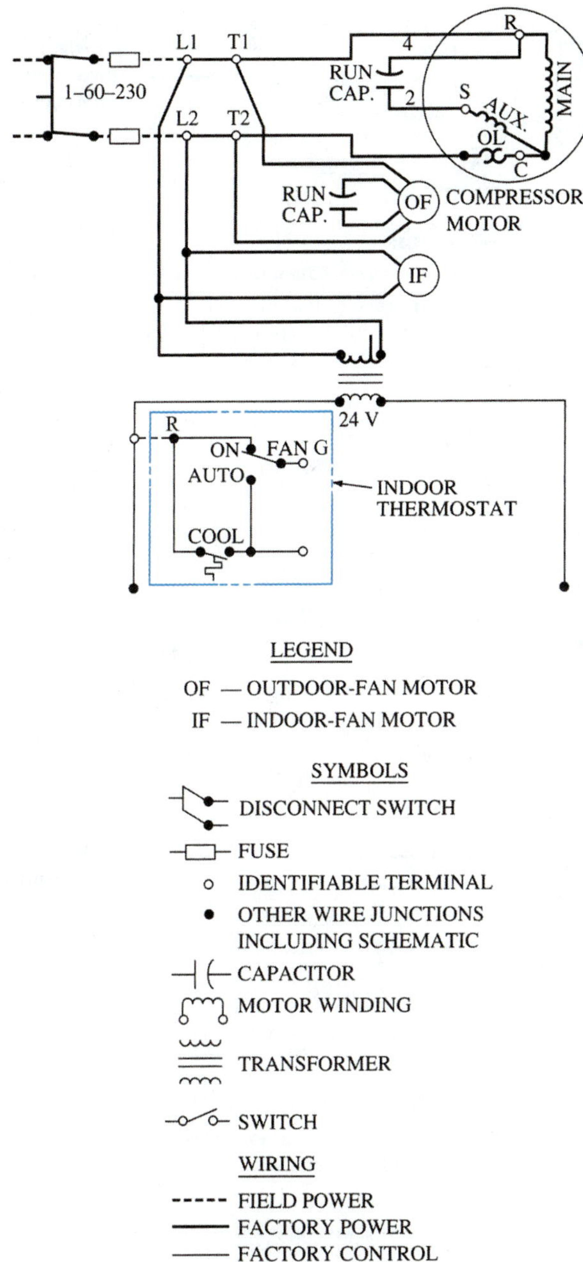

Figure 38-44 Wiring diagram showing the connection of the room thermostat to one side of the low-voltage circuit.

has been added in Figure 38-46. It has a coil and a normally open set of contacts, both marked IFR. Note that the coil is wired in series with the fan switch in the thermostat, and its contacts are wired in series with the indoor fan motor.

Figure 38-47 shows the completed diagram with the unit operating in cooling. On a call for cooling, current flows through the thermostat to both the contactor coil M and the indoor fan relay coil IFR at the same time. This closes both contacts marked M in the power circuit to the compressor and outdoor fan motors, as well as the contacts marked IFR in the power circuit to the indoor fan motor. The compressor and outdoor fan motors are energized and started through the M contacts. At the same time, the indoor fan motor is also started through closed IFR contacts. The entire system is now energized, and all motors are running automatically.

When the thermostat is satisfied, the cooling thermostat opens the circuit to both the contactor and indoor fan relay coils, their contacts return to their normally open positions, and all motors stop—compressor, outdoor fan motor, and indoor fan motor.

Safety controls can be added to protect the system. In Figure 38-48, overloads have been added to protect the compressor and outdoor fan motor windings against excessive current and temperature. Overloads may be installed inside the motor housing, as shown for the compressor motor, or externally, as shown for the outdoor fan motor.

A high-pressure switch and a low-pressure switch have been wired in series with the compressor motor. They will prevent the compressor from operating under abnormal conditions such as excessive discharge pressure or low suction pressure. The high-pressure switch is actuated by system discharge pressure to open and stop the compressor if the discharge pressure exceeds the switch setting. The low-pressure switch is actuated by system suction pressure to open and stop the compressor if the suction pressure falls below the switch setting. A fuse has been installed in series with the transformer secondary winding to protect the transformer and low-voltage control circuit wiring against excessive current in the low-voltage control system.

For closer control of air temperature in the conditioned space, an anticipator has been wired across the terminals of the thermostat cooling switch. It is a fixed, nonadjustable resistor. When the cooling thermostat opens, a new circuit is made with the anticipator in series with the M coil. In a series circuit, the device with the highest resistance uses most of the voltage. The anticipator resistance is several thousand ohms, much higher than the resistance of the contactor coil, so the anticipator uses virtually all the 24 V, and the contactor turns off even though it is still in the circuit. Heat from the anticipator fools the cooling thermostat into closing early, starting the system before the house temperature has risen above the set point. This gives the system the lead time it needs to start delivering cool air.

38.20 SEQUENCE OF OPERATION

Ladder diagrams are used to determine a system's sequence of operation. The sequence of operation is the order, or sequence, in which electrical loads operate and the conditions that will start or stop each load. To determine a unit's sequence of operation, start by identifying any circuits in the line voltage portion that have a complete path from L1 to L2 when power is applied to the unit. In Figure 38-49, the only circuit that is complete when the power is first applied is the circuit to the transformer primary. All the other line voltage circuits have open contacts or switches in the circuit. The transformer will supply 24 volts to the low-voltage controls when the unit is first energized. Next, identify any circuits in the low-voltage portion that have a complete path. In the top portion of Figure 38-49, there are no low-voltage circuits with a complete path. Some of the switches in the thermostat must change position to energize any of the low-voltage control circuits. In most residential control systems, one side of 24 V power goes to the thermostat and the other side of

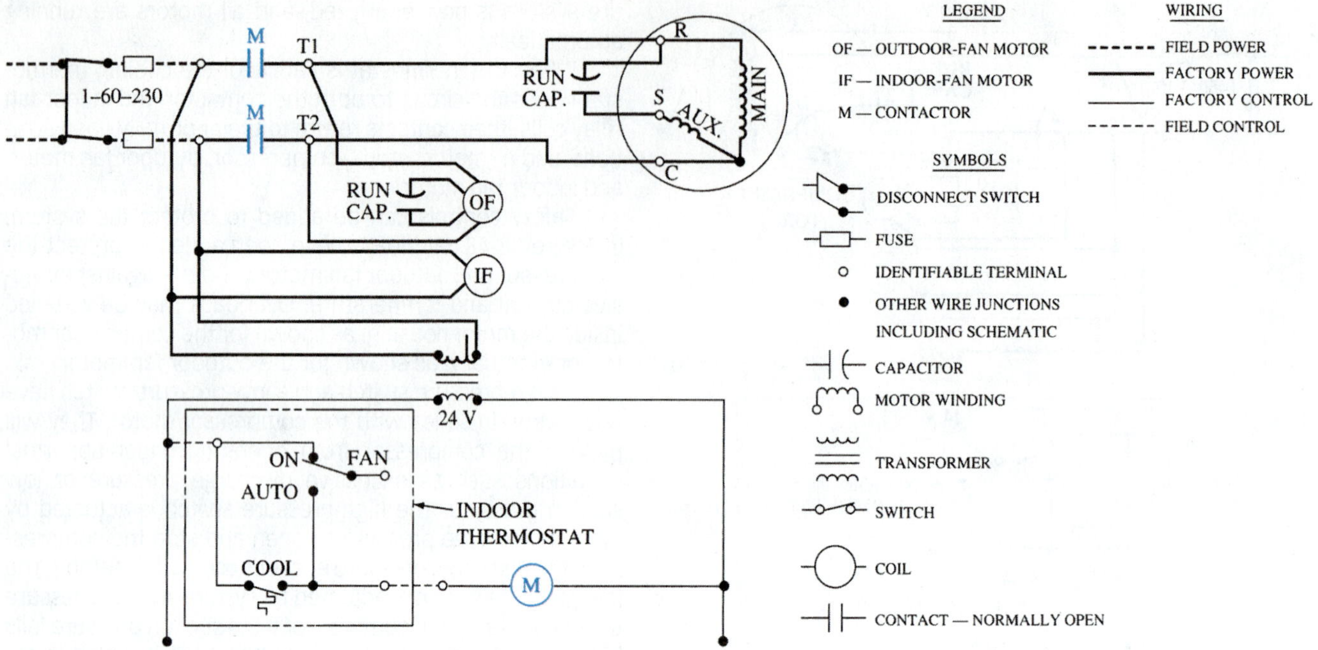

Figure 38-45 Wiring diagram showing the addition of the compressor contactor coil placed in the cooling circuit.

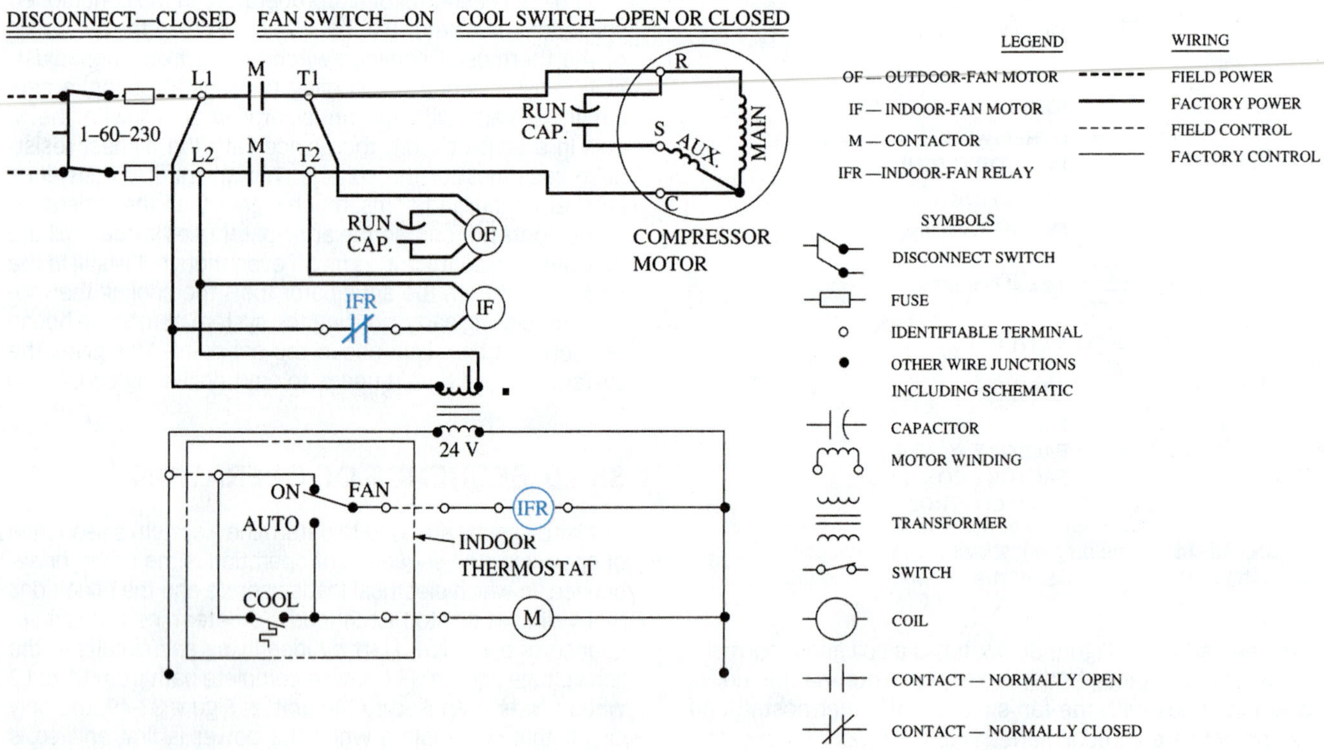

Figure 38-46 Wiring diagram showing the addition of the fan relay operated by the FAN ON switch.

24 V power goes directly to all the low-voltage controls. The thermostat acts like a switch, controlling unit operation by controlling 24 V power to the different low-voltage controls. Most thermostats have two main manual switches: a fan switch with ON and AUTO positions and a system switch with HEAT, OFF, and COOL. The fan switch controls the indoor fan operation. In the ON position, the fan will operate continuously. In the AUTO position, the fan is controlled by the cooling thermostat. When the thermostat calls for cooling, the fan will operate; when the thermostat does not call for cooling, the fan stops. The system switch controls whether the system operates in heating or cooling or is off.

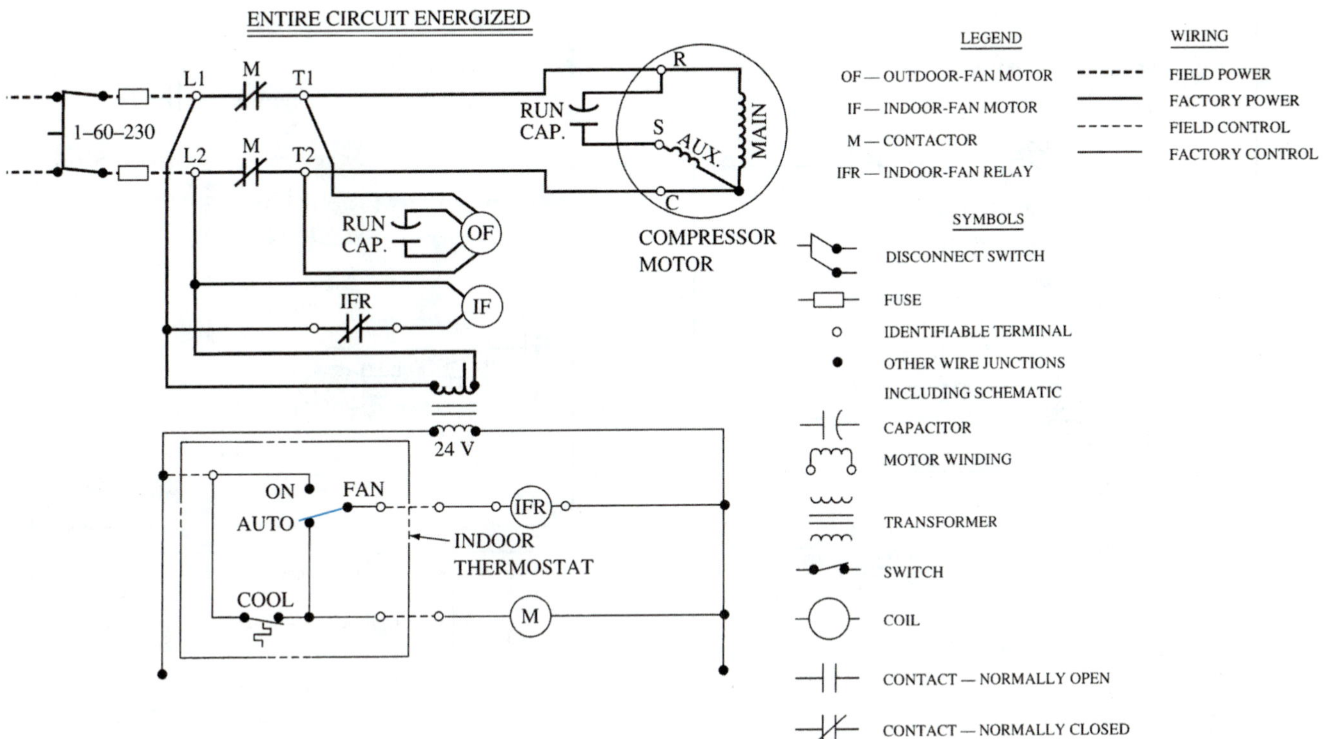

Figure 38-47 Wiring diagram showing fan in AUTO position and thermostat calling for cooling.

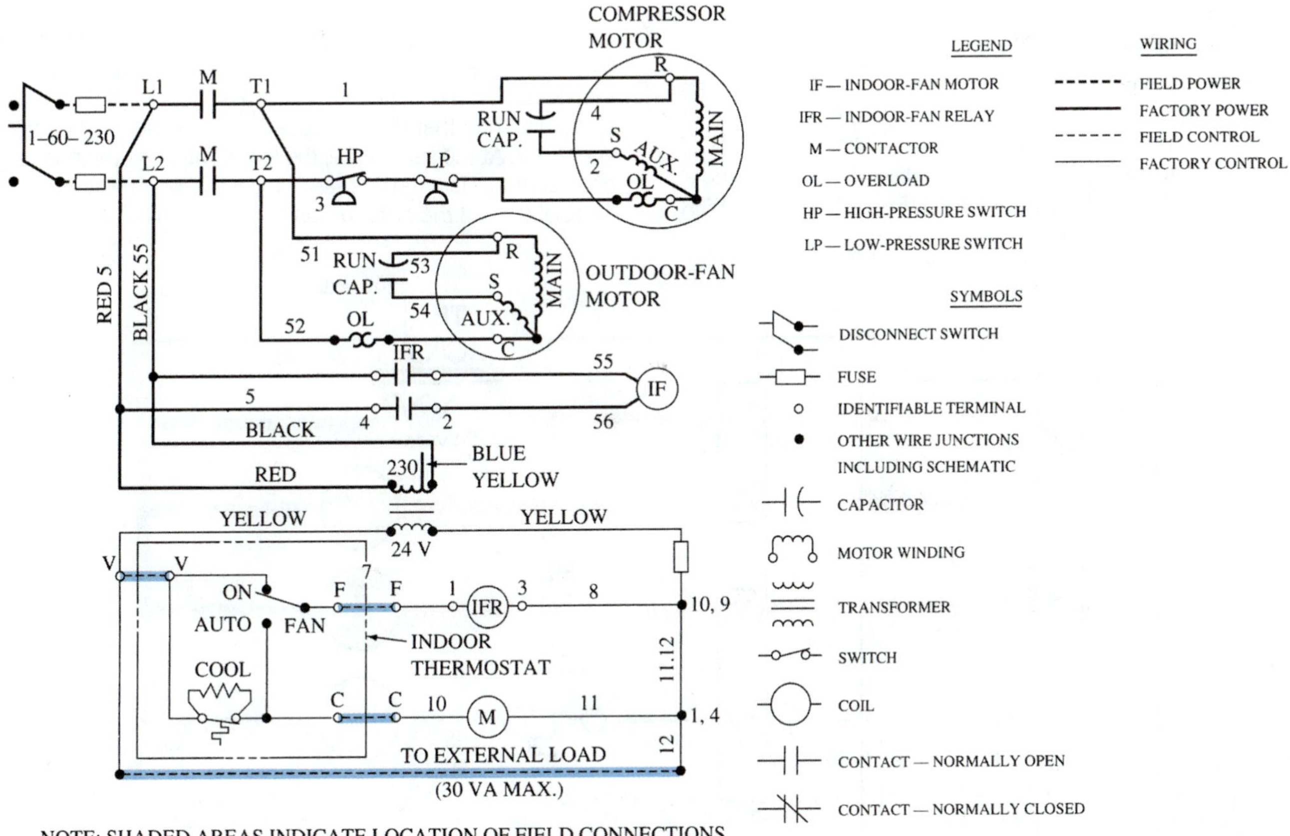

Figure 38-48 Complete ladder diagram for the air-conditioning unit.

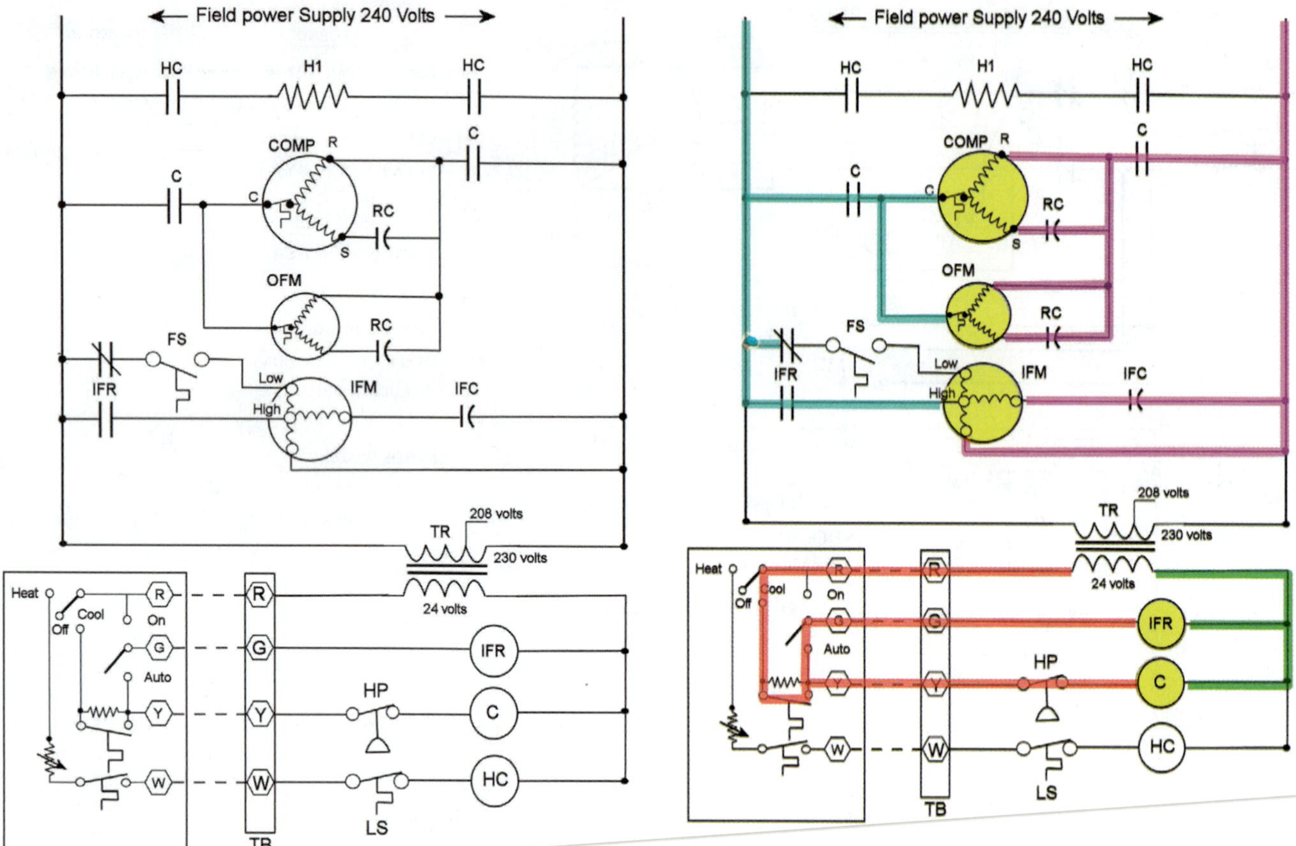

Figure 38-49 Ladder schematic of heating/cooling system.

Figure 38-51 IFR and C contacts energizing compressor, outdoor fan, and indoor fan.

Cooling Operation

With the system switch in the COOL position and the fan switch in the AUTO position, the cooling thermostat will control unit operation. A temperature rise will cause the cooling thermostat to close, completing circuits to both the indoor fan relay (IFR) and the contactor (C) (Figure 38-50). The IFR coil will close the normally open contacts labeled IFR in the line voltage section, completing a path to the high speed of the indoor fan motor (IFM). The normally closed IFR contacts open, ensuring that the low-speed motor winding cannot be energized. At the same time, the C coil will close the normally open contacts labeled C, energizing both the compressor motor (COMP) and the outdoor fan motor (OFM) (Figure 38-51).

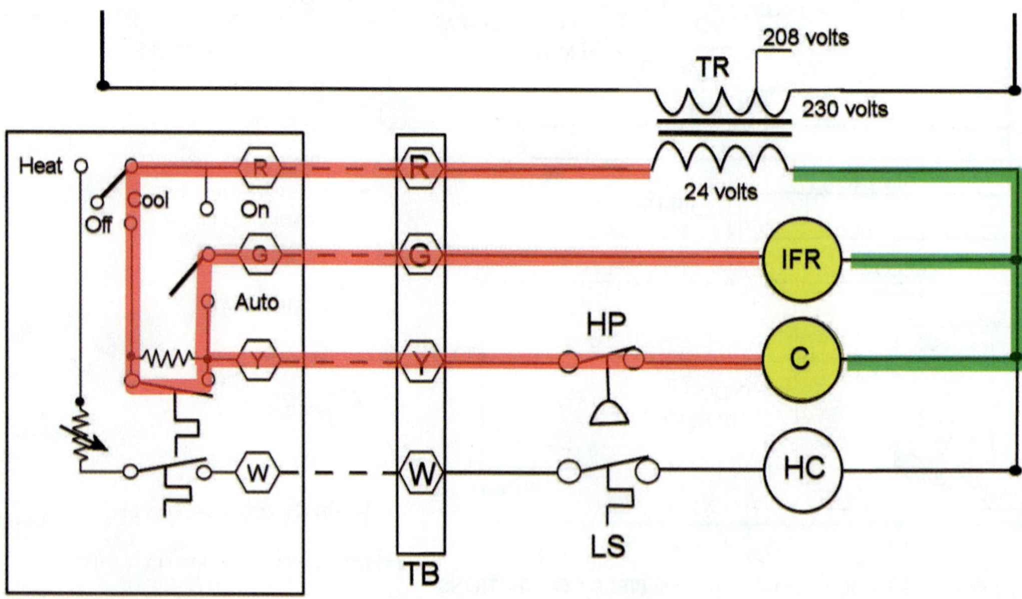

Figure 38-50 Cooling thermostat closing energizing IFR and C.

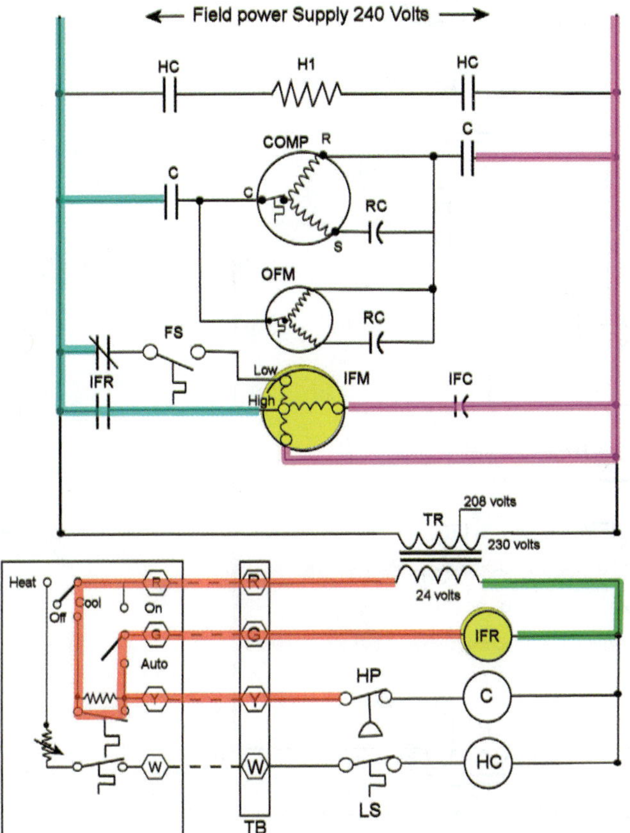

Figure 38-52 HP opening turning off compressor.

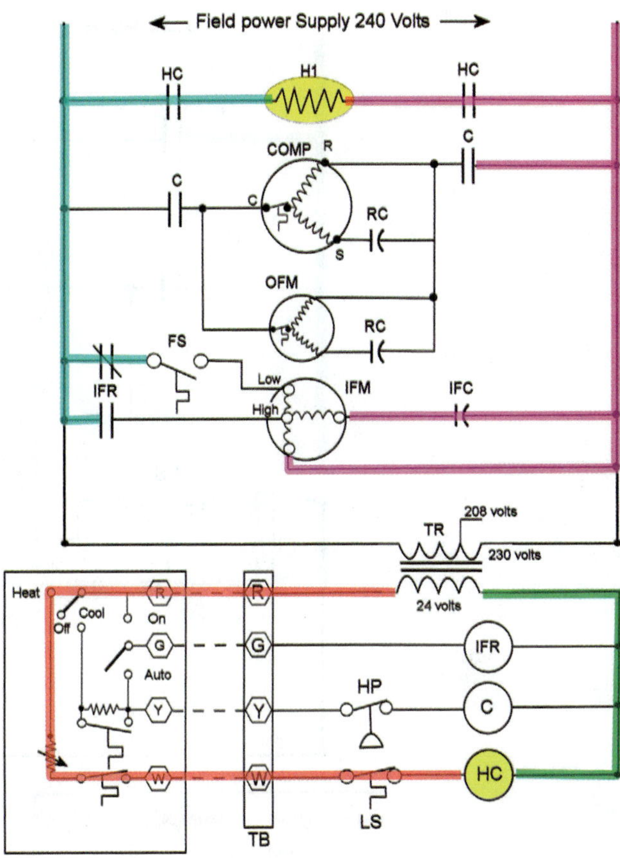

Figure 38-53 Heat strips energized.

If the system pressure rises above a safe operating level, the rise in pressure will open the high-pressure switch (HP). Opening HP will break the circuit to the contactor coil (C), de-energizing it. When the C coil is de-energized, it will open the C contacts in the line voltage section to de-energize the circuits to both the COMP and the OFM (Figure 38-52).

Heating Operation

With the system switch in the HEAT position and the fan switch in the AUTO position, the heating thermostat will control unit operation. A temperature fall will cause the heating thermostat to close, completing a circuit to the heating contactor (HC). The HC coil will close the normally open contacts labeled HC, energizing the electric heater (H1) (Figure 38-53). The fan switch (FS) will close on the rise in temperature created by the heater. The closing of the FS creates a complete circuit through the normally closed IFR contacts to energize the low speed of the IFM (Figure 38-54).

If the heater temperature rises above a safe operating level, the rise in temperature will open the limit switch (LS). Opening LS will break the circuit to the HC, de-energizing it. When the HC coil is de-energized, it will open the HC contacts in the line voltage section to break the circuit to the H1 (Figure 38-55).

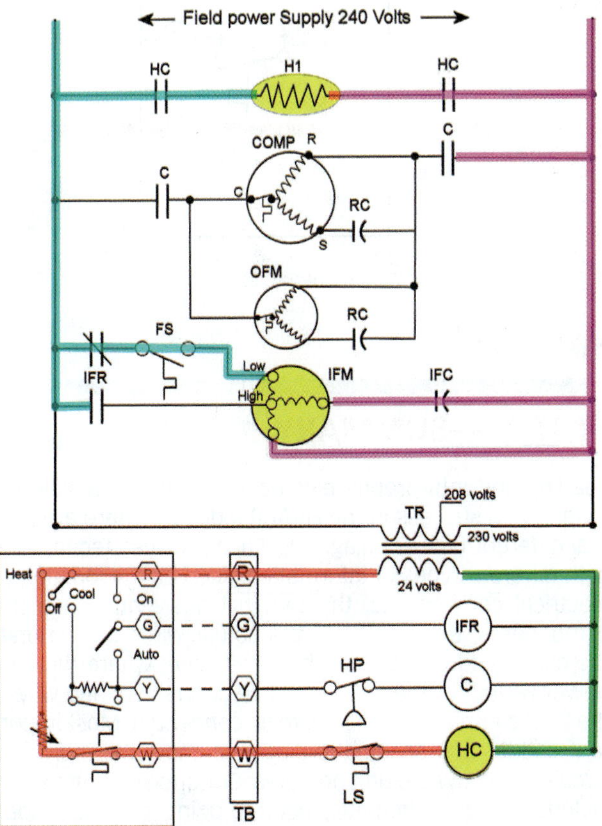

Figure 38-54 Fan switch closing to operate fan in heat.

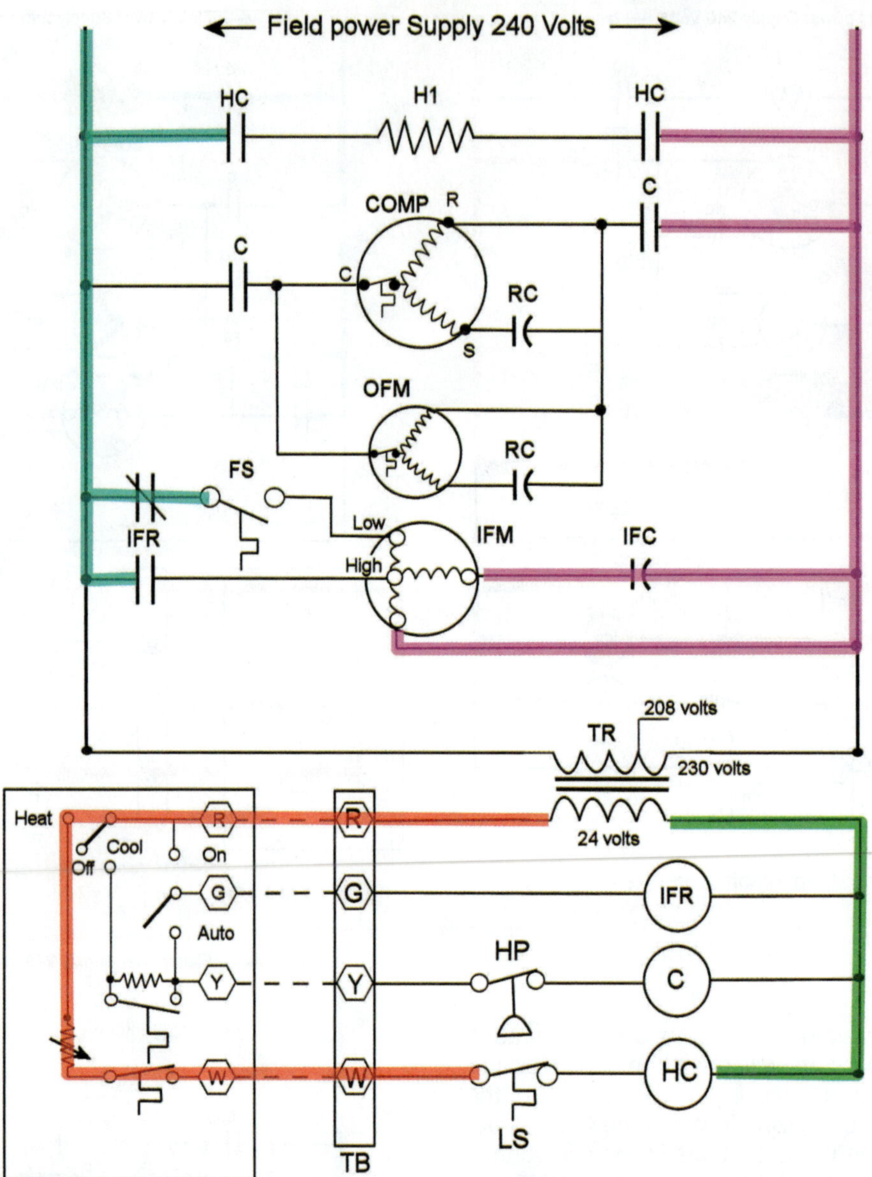

Figure 38-55 Limit switch open to shut off HC; shutting off heater.

UNIT 38—SUMMARY

Reading and interpreting electrical diagrams is absolutely essential to success in the HVACR industry. There are several different types of diagrams. Each provides technicians with different types of information, such as the location of electrical components, the terminal designations on the components, the routing of the actual wires on the unit, the colors and/or labels of the wires used, where the field power wiring for the unit should be connected, and where the field control wiring needs to be connected. Most important, ladder and schematic diagrams show the individual circuits and how the unit operates. Common diagram types include ladder, schematic, point-to-point, pictorial, label, component location, factual, field connection, and installation diagrams. Graphical symbols are used to represent the electrical properties of standard electrical components. Some components wire into a single circuit and can be represented by a single symbol, such a pressure switch. Other components are wired in more than one circuit and are represented by several symbols, such as a relay. Different manufacturers often use different symbols for the same components. The diagram legend identifies the symbols used in any particular diagram.

UNIT 38—REVIEW QUESTIONS

1. List the different types of electrical diagrams used by HVACR manufacturers.
2. Which diagram types are most useful for understanding the operation of the unit?

3. Describe the concept behind ladder diagrams.
4. What information does a point-to-point diagram give the technician?
5. What is the difference between a load and a control?
6. What other diagrams are similar to point-to-point diagrams?
7. Why are relay coils and contacts not shown in the same place on a schematic diagram?
8. What information does a factual diagram give that a ladder diagram does not give?
9. How is field wiring distinguished from factory wiring on most diagrams?
10. How is line voltage wiring distinguished from low-voltage wiring on some diagrams?
11. What is a system's sequence of operation?
12. Which types of diagrams are best for determining a system's sequence of operation?
13. Why do many manufacturers provide both schematic and pictorial wiring diagrams on their equipment?
14. What is the purpose of a component location diagram?
15. Why do most manufacturers supply more than one type of diagram with their equipment?
16. Draw symbols for the following switches:
    a. Close-on-rise thermostat
    b. Open-on-rise thermostat
    c. Close-on-rise pressure switch
    d. Open-on-rise pressure switch
17. What are the two manual switches found on most thermostats?
18. Why are schematic diagrams often not true ladder diagrams?
19. Draw the symbols that are used to represent a(n):
    a. Relay or contactor coil
    b. Normally closed relay contacts
    c. Normally open relay contacts
20. Draw the symbol for a(n):
    a. Thermal overload relay coil
    b. Magnetic overload relay coil
    c. Overload relay contacts
    d. In-line thermal overload
    e. Bimetal thermal overload
    f. Fuse
21. Draw the symbol for a:
    a. Transformer
    b. Capacitor
    c. Red light
    d. Heater
    e. Solenoid coil
22. Draw the symbol for a:
    a. Shaded pole motor
    b. Single-phase compressor motor
    c. Three-phase motor
23. In Figure 38-37, what color wire is connected to the compressor start terminal S?
24. In Figure 38-37, how does the HP switch protect the compressor?
25. In Figure 38-37, what is the purpose of the red wire on the transformer primary?
26. In Figure 38-37, what type of device is the IFC?
27. In Figure 38-37, where is line voltage connected to the unit?
28. In Figure 38-37, what controls the outdoor fan motor (OFM)?
29. In Figure 38-48, what does the M coil control?
30. In Figure 38-48, what type of motor is the indoor fan motor (IFM)?

UNIT 39

Control Systems

OBJECTIVES

After completing this unit, you will be able to:

1. explain the purpose of a control system.
2. list the different types of control systems.
3. explain the difference between open and closed loop control systems.
4. identify standard terminal designations in residential HVACR control systems.
5. use schematic diagrams to describe the operating sequence of standard HVACR equipment.

39.1 INTRODUCTION

The control system is the HVACR nervous system. Like a nervous system, the control system regulates the operation of all the major components to ensure that they function properly together. Effective system troubleshooting requires a thorough understanding of the operation of the control system. Technicians must understand control systems to truly understand HVACR equipment operation. This unit will discuss the major types of control systems and then look at some specific residential relay control systems. Communicating control systems are covered in Unit 40, Communicating Control Systems, and commercial control systems will be discussed further in Unit 78, Commercial Control Systems.

Figure 39-1 Low-voltage thermostat.

39.2 CONTROL SYSTEM FUNDAMENTALS

An HVACR control system regulates the operation of HVACR equipment in response to a monitored condition, such as temperature, pressure, or humidity. At its most basic, a control requires a sensor to monitor a condition, an actuator to cause a change based on what is sensed, and a means of linkage or communication between the sensor and actuator. These elements can be separate devices, or they can all be incorporated into a single device. A room-temperature sensor and an electronic controller connected together by wires in an electronic control system would be an example of multiple devices performing together as a single control. A typical low-voltage wall thermostat would be an example of a control in a single device (Figure 39-1). A control system can consist of a single control, or it can incorporate several controls, operating equipment based on several monitored conditions.

Controls and control systems can be described as either open or closed loop. An open-loop control system does not affect the condition it is sensing. A control that regulates the operation of a piece of equipment based on the outdoor temperature is an open loop since the system cannot control the outdoor temperature. An outdoor thermostat on a heat pump system that will only let the auxiliary heat energize below a particular outdoor temperature is an example of an open-loop control because the system is not controlling the outdoor temperature (Figure 39-2).

A closed-loop control system controls the condition that it is sensing (Figure 39-3). This allows the control to regulate a condition based on the results achieved. Connecting the output back to the input is called feedback. A room thermostat that controls the operation of a heating system based on the room temperature is an example of a closed-loop system because it is controlling the temperature it is sensing. The change in room temperature provides feedback to the thermostat.

Control systems generally control unit operation to maintain a desired condition by one of two methods: two-position control or modulating control. Two-position controls cycle system components on and off, much like a toggle switch. You either have all the light or no light at all. Modulating controls provide levels between on and off; much like a light dimmer controls the light output for

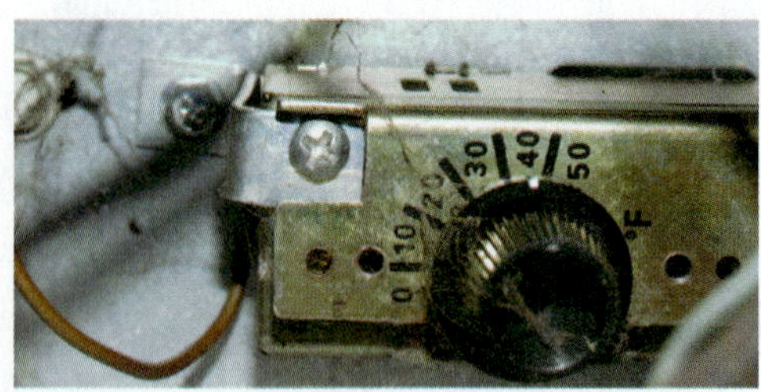

Figure 39-2 Outdoor thermostat.

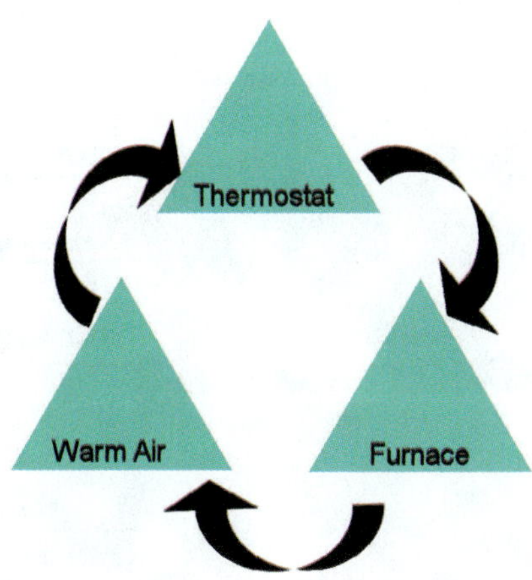

Figure 39-3 A closed-loop control system uses feedback.

light levels between full light and no light. A modulating control varies the amount of conditioning produced to match the need.

39.3 TYPES OF CONTROL SYSTEMS

Control systems can be grouped by their power source and their general operating principle as electromechanical, pneumatic, electronic, and direct digital (Figure 39-4). Most residential units employ electromechanical control systems. Electromechanical control systems are two-position controls that use mechanical switches and magnetically operated devices. Units respond to the presence or absence of 24 V control power. Relays are typical electromechanical

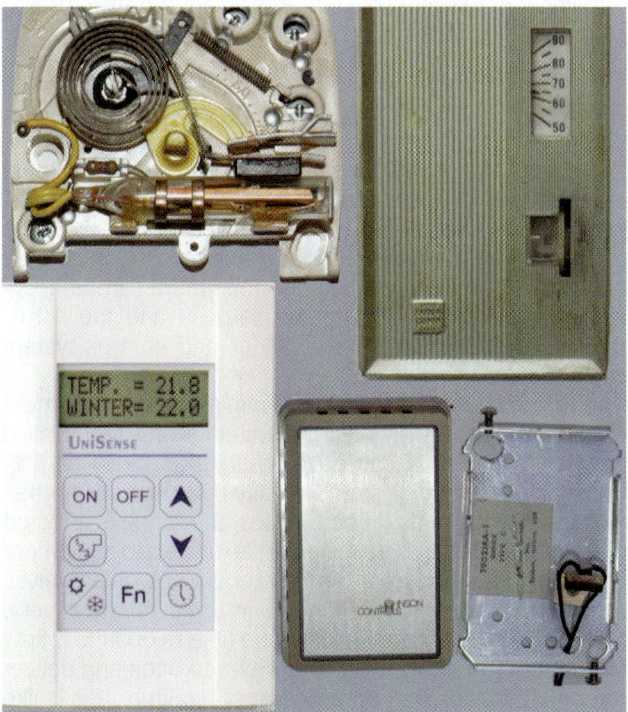

Figure 39-4 Controls shown clockwise from the left: electromechanical, pneumatic, electronic, direct digital.

Figure 39-5 Control board used in a typical air-conditioning system.

controls. When the relay coil receives 24 V, its contacts close to operate a motor or other controls. Most residential units today have electronic boards, but the presence of a board does not make the control system electronic. Although the boards operate electronically, most of them imitate relay functions. Many actually have small relays built into the board. The board either sends voltage to different controls and loads to operate them or does not send voltage to them to turn them off. Figure 39-5 shows a typical electronic board in a relay-type control system.

> **TECH TIP**
>
> Electromechanical control systems are often referred to as relay control systems because of their reliance on relays to control the operation of motors and system components.

Pneumatic controls operate on air pressure. Pneumatic control systems are modulating systems. The output of a pneumatic control is an air pressure that varies in proportion to the condition being sensed. They can position valves and dampers in any position between fully open and fully closed by controlling the amount of air pressure going to the actuator operating the valve or damper. Pneumatic controls were used for many years in commercial applications, and there are still many operating pneumatic systems, but they are seldom used today in new buildings. Figure 39-6 shows a typical pneumatic valve.

Figure 39-6 Pneumatic water valve.

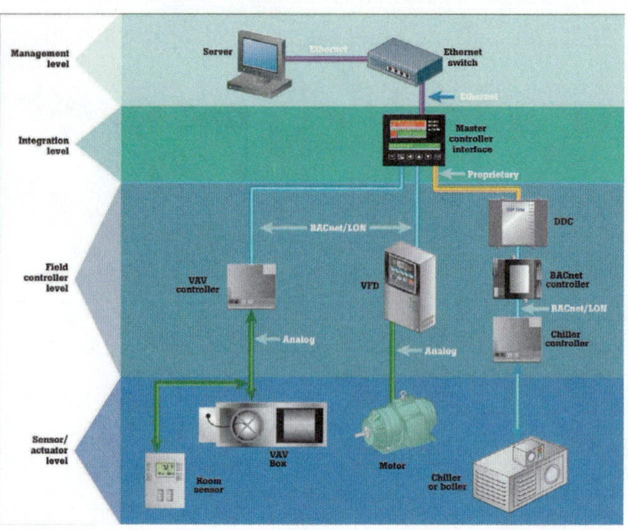

Figure 39-8 DDC control system.

Electronic controls are also modulating controls. The output of an electronic control system is a variable DC voltage or current that is proportional to the condition being sensed. Figure 39-7 shows a typical DC voltage ramp used in an electronic control system. They can position valves and dampers in any position between fully open and fully closed. Electronic controls have the ability to adjust the feedback to the control to maintain a closer tolerance.

The newest controls are called direct digital controls (DDCs; Figure 39-8). Direct digital controls are control systems that have their own built-in microprocessors. DDC systems are similar to computers: they take input signals, process the signals according to a program, and then provide output. The DDC controllers that connect to motors and system sensors typically have both analog and digital input and output. The controllers communicate with each other like a computer network. Information and commands are sent over a network, which allows the components to communicate with each other. DDCs have their own microprocessors, allowing them to respond intelligently to changes. The first DDC systems were exclusively commercial systems. However, there are now many residential communicating systems available in high-end equipment. See Unit 40, Communicating Control Systems, for more information on residential communicating controls. The remainder of this unit concentrates on control systems and diagrams for typical residential relay-type control systems.

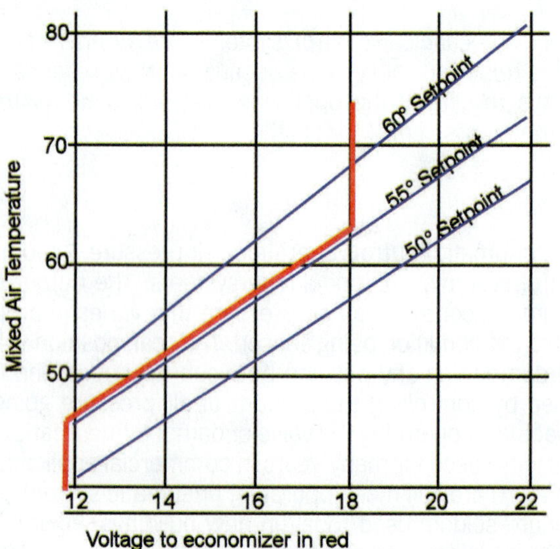

Figure 39-7 Voltage ramp for a DC control system.

39.4 SYSTEM CYCLING

Regardless of the type of control system, the controls must determine when to deliver conditioning to the space and when to stop delivering it. For two-position controls, the resulting on-and-off operation is called cycling. With both two-position and modulating controls, there can be a significant difference between the setpoint and the actual temperature because of equipment and control system characteristics.

The difference between the minimum and maximum room temperatures is called the system swing. Mechanical bimetal thermostats can have a swing as much as 5°F. Proportional control systems usually have less swing than two-position systems. In both cases, the swing is affected by equipment sizing. As a general rule, oversized equipment causes increased cycling and larger swings. Even valves designed for proportional flow have a minimum delivery. If the valve is oversized it will not be able to open to a flow that meets the need. It will either be too open and deliver too much flow or shut off and deliver too little. This is like trying to make a 90° turn into a driveway going 60 miles per hour. You will drive back and forth in front of the driveway but never make the turn. This type of operation is called hunting.

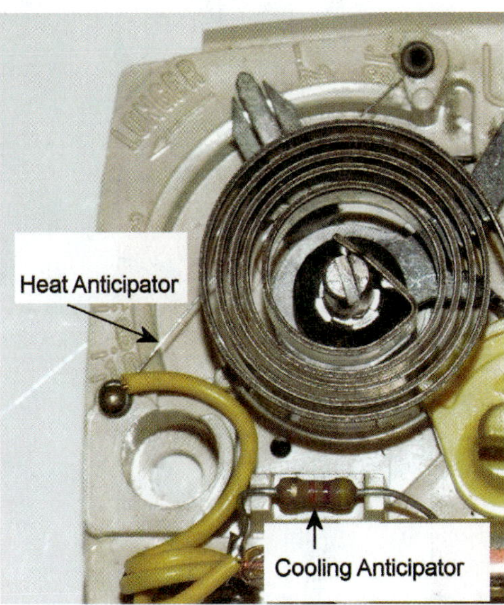

Figure 39-9 Heating and cooling anticipators.

Swing is largely caused by system overshoot and lag. Overshoot describes a condition where the system heats or cools past the setpoint. An example of overshoot is the cool-down cycle in furnaces. Furnace fans typically continue to operate and deliver heat even after the thermostat is satisfied in order to cool the furnace down. If the thermostat does not shut off the call for heat until the room is at the setpoint, the furnace cool-down will heat the room past the setpoint. To avoid overshoot, the call for heat must be terminated before the room reaches setpoint temperature. Mechanical thermostats use heat anticipators to fool the thermostat into shutting off early. A heat anticipator is a small heater with a very low resistance that is wired in series with the heating thermostat (Figure 39-9). It makes a small amount of heat during the heating cycle to fool the thermostat into shutting off early. A heating anticipator reduces the system overshoot in furnaces caused by this cool-down. Figure 39-10 shows the effects of a heat anticipator on a mechanical thermostat.

Lag describes a delayed system response to a change in temperature. Mechanical thermostats use cooling anticipators to fool the thermostat into starting early. A cooling anticipator is a resistor with a high resistance that is wired in parallel with the cooling thermostat (Figure 39-9). They provide a path through the cooling controls during the off cycle, producing a small amount of heat to fool the thermostat into closing early. This reduces system lag by giving the system time to build system pressures and purge warm air out of ductwork before the cool air is needed.

Electronic thermostats have largely replaced the mechanical thermostats. They can sense temperature in tenths of a degree. Electronic thermostats do not use traditional heating or cooling anticipators. Instead, they use improved sensing, timing, and logic rather than false heat to reduce system swing. A well-designed electronic thermostat for a relay control system can keep the temperature swing within a degree.

Even proportional controls have a temperature swing between the minimum and maximum temperature in the room. Proportional controls can use feedback to modify their response to changes in room temperature. Electronic controls have the ability to adjust the feedback to the control to maintain a closer tolerance. One approach is called proportional integral derivative, or PID. A PID control uses programming and math to "learn" the response of the system and reduce the difference between the setpoint and the measured condition. Electronics have made PID control possible, allowing much closer control. DDC controls use PID and advanced programming and computer logic to maintain temperatures within tenths of a degree of setpoint.

39.5 LINE-VOLTAGE CONTROL SYSTEMS

Small appliances and commercial refrigeration systems with fractional horsepower motors often use line-voltage control systems. A line-voltage control system operates on the same voltage supplied to the unit. The switches in a line-voltage control system break the line-voltage power to the components they are controlling. The thermostats, pressure switches, and other controls in a line-voltage system must be able to carry the voltage and current of the motors they are controlling. They are rated by the voltage and current they can safely switch.

39.6 WINDOW-UNIT DIAGRAM

Window-unit air conditioners typically consist of a compressor and fan motor controlled by a selector switch and a thermostat. Figure 39-11 shows a diagram for a window unit. The switch is in series with both the fan motor and the thermostat, while the thermostat is in series with the compressor.

Figure 39-12 shows the selector switch and thermostat from a typical window unit. The switch has connections for incoming line voltage (L); the low (1), medium (2), and high (3) speeds of the fan motor; and the thermostat (4) (Figure 39-13). The unit has two operating modes controlled by the switch: fan only and cooling.

In the low-fan or high-fan modes, the switch completes the circuit to either the low speed or high speed of the fan motor. There is no circuit to the thermostat connection (Figure 39-14). In fan mode, the fan runs continuously and the compressor does not run.

In low cool, medium cool, or high cool, the switch completes a circuit to the low (1), medium (2), or high (3) speed of the fan motor and also to the thermostat connection (4). In cooling mode, the fan runs continuously and the compressor cycles on and off with the thermostat (Figure 39-15).

39.7 WALK-IN FREEZER DIAGRAM

Figure 39-16 shows the diagram for a walk-in freezer. The freezer has four main loads: compressor, condenser fan motor, evaporator fan motor, and defrost heater. The

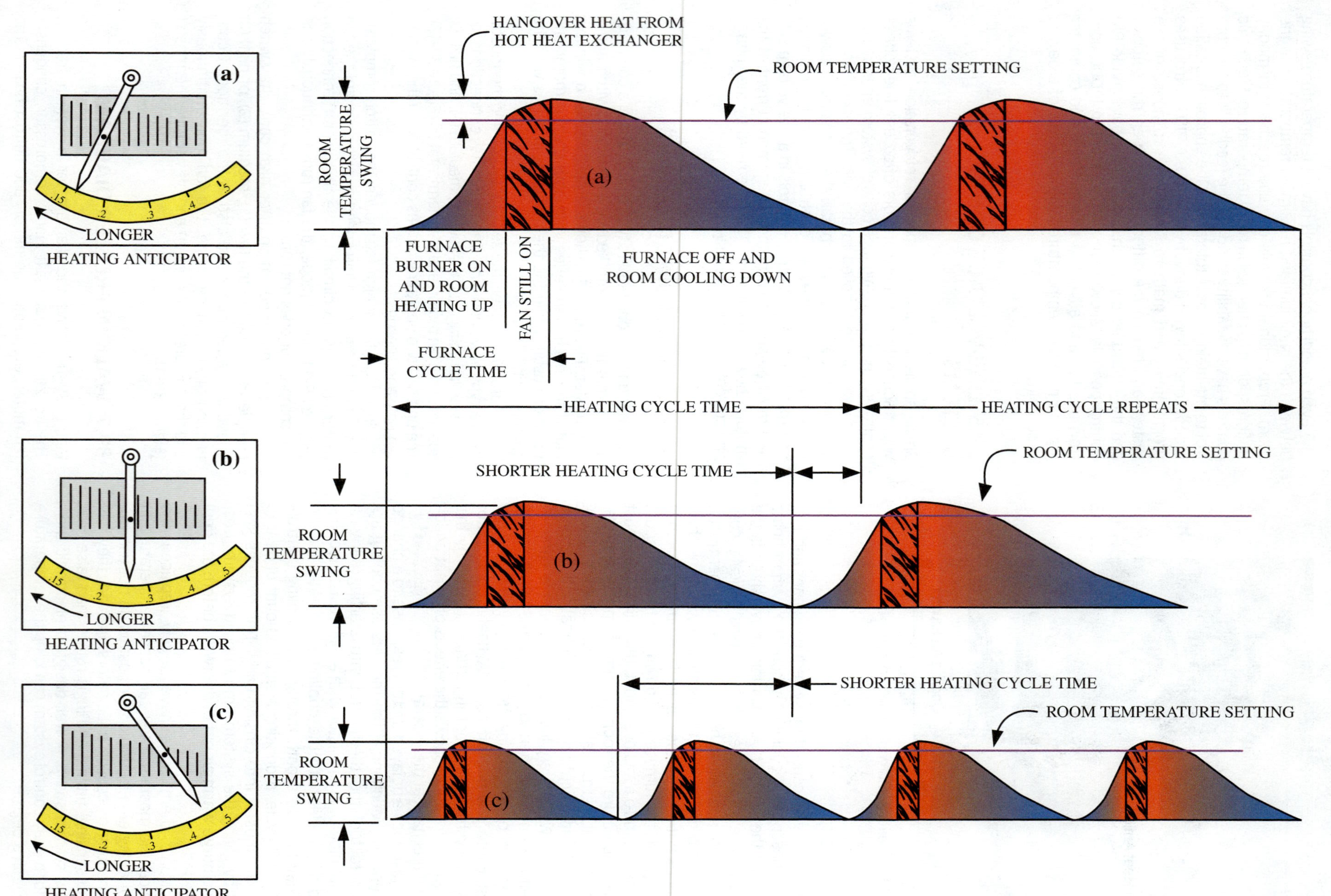

Figure 39-10 Heat cycle with anticipator.

Letters	Component
COMP	Compressor
FM	Fan motor
RC	Run capacitor
S1	Selector switch
TH	Thermostat

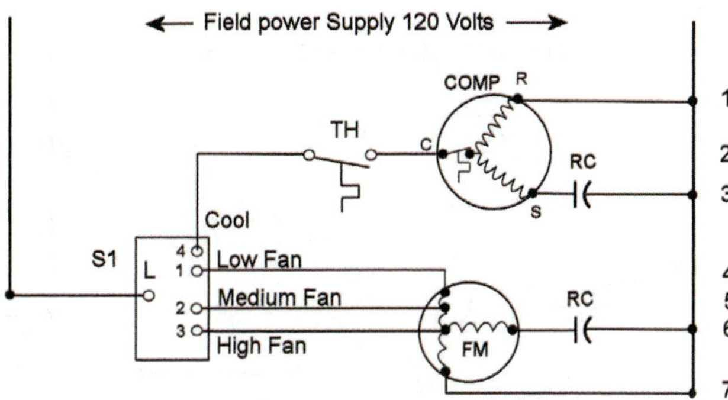

Figure 39-11 Diagram for window unit air conditioner.

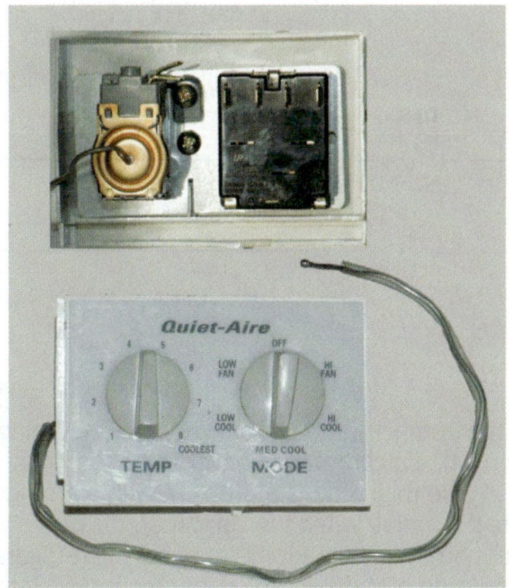

Figure 39-12 Selector switch and thermostat for window unit air conditioner.

compressor and condenser fan motor are cycled by the thermostat. The compressor is protected by both low-pressure and high-pressure switches in series. An additional close-on-rise pressure switch controls the operation of the condenser fan motor. The condenser fan motor does not operate until the head pressure exceeds the cut-in point of the fan cycling pressure switch.

The evaporator fan motor is in series with the open-on-rise contacts of the defrost termination thermostat. It runs all the time, except at the end of a defrost cycle. The defrost timer motor runs all the time. It has two sets of contacts: a normally closed set in series with the compressor and condenser fan motor, and a normally open set in series with

Figure 39-13 The selector switch has connections for the line voltage, three fan speeds, and the thermostat.

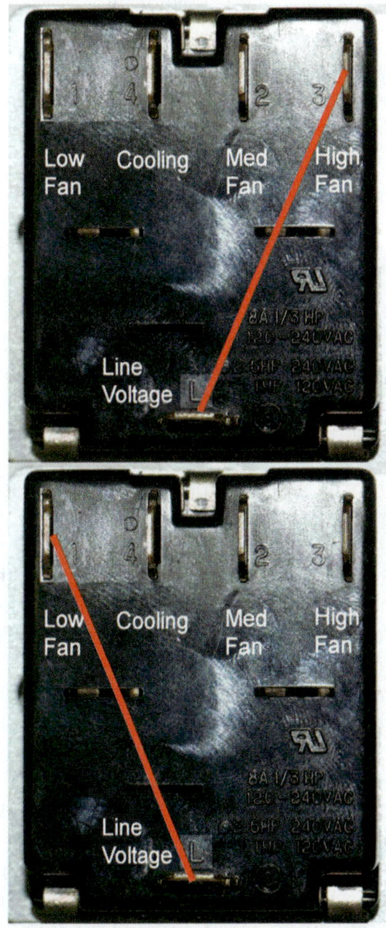

Figure 39-14 Circuits through the selector switch in fan-only operation.

Figure 39-15 Circuits through the selector switch in cooling operation.

Letters	Component
CFM	Condenser fan motor
COMP	Compressor
DFH	Defrost heater
DFT	Defrost timer
DTS	Defrost termination solenoid
DTT	Defrost termination thermostat
EFM	Evaporator fan motor
FCS	Fan cycling switch
FL1	Fuse link
HPS	High-pressure switch
LPS	Low-pressure switch
RC	Run capacitor
TH	Thermostat

Figure 39-16 Diagram for walk-in freezer.

the defrost heater. When the defrost timer indicates that it is time to defrost, the normally closed DFT contacts open, breaking the circuit to the compressor and condenser fan motor. The normally open DFT contacts close, completing the circuit to the defrost heater. When the coil has defrosted, the defrost termination thermostat will switch positions, energizing the defrost termination solenoid and breaking the circuit to the evaporator fan motor. The defrost termination solenoid returns the defrost timer contacts to their normal position, energizing the compressor and breaking the circuit to the defrost heater. When the coil has dropped below freezing once again, the defrost termination thermostat returns to its normally operating position, making a circuit once again to the evaporator fan motor.

39.8 PUMP-DOWN CYCLE DIAGRAM

Refrigeration systems that must operate in cold weather often have a problem with refrigerant migration to the compressor during the off cycle. Refrigerant migration occurs when refrigerant travels to the compressor crankcase and dissolves itself in the compressor oil. When the compressor starts, the refrigerant boils out violently, creating foam similar to when a carbonated beverage is shaken and then opened. While migration is a problem with all refrigeration systems, it is worse when the compressor is cold. One way to prevent off-cycle migration is to pump the refrigerant out of the low side of the system every time the system shuts off so there is no refrigerant to migrate to the compressor. Figure 39-17 shows a diagram for a basic pump-down system.

A pump-down control system requires a normally closed liquid-line solenoid valve and a pressure switch. The control thermostat actually controls the liquid-line solenoid valve instead of the compressor. When the thermostat is satisfied, it opens and de-energizes the solenoid (Figure 39-18). The solenoid shuts, stopping the flow of liquid to the evaporator. The compressor continues to run, but with no more refrigerant flow to the evaporator, the low-side pressure begins to drop as the refrigerant is pumped out of the evaporator. When the low-side pressure reaches the cutout setting on the low-pressure switch, the pressure switch opens and breaks the circuit to the compressor (Figure 39-19). Typically this setting is above 0 psig to ensure that the system does not pull into a vacuum, which could possibly

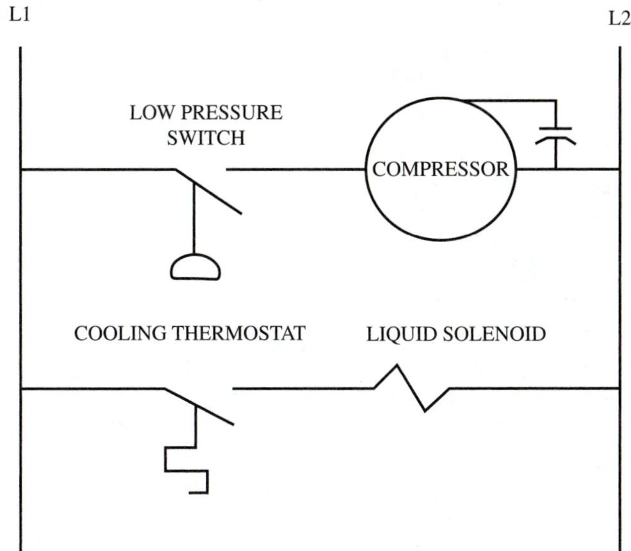

Figure 39-17 Refrigerant pump-down circuit.

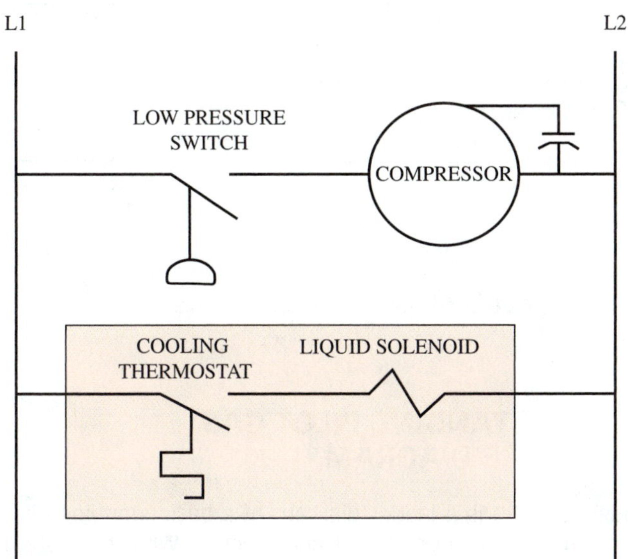

Figure 39-18 Thermostat controlling the pump-down solenoid.

damage the compressor or suck air into the system through any low-side leaks.

The system then waits for the next call for cooling. When the box temperature rises, the thermostat closes, energizing the solenoid. The solenoid opens, allowing refrigerant to flow into the low side of the system. When the pressure reaches the low-pressure switch cut-in setting, the pressure switch closes and starts the compressor.

39.9 LOW-VOLTAGE CONTROL SYSTEMS

Most residential heating and cooling equipment use low-voltage control systems. Low-voltage control systems typically operate on 24 V AC supplied by a control transformer. Control circuits must have a path from one side of the transformer secondary, through a low voltage load, and back to the transformer. The transformer supplies 24 V to a low-voltage

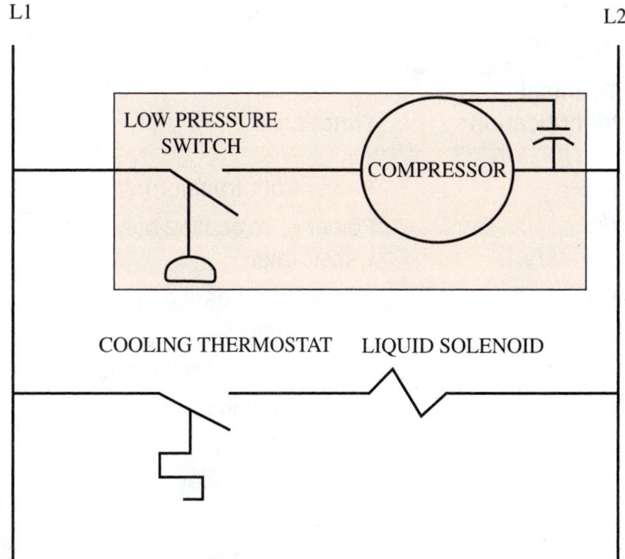

Figure 39-19 Pressure switch controlling the compressor.

thermostat that controls the overall operation of the system. The side of the transformer secondary feeding the thermostat is known as the red side. A low-voltage thermostat does not control the loads directly. Instead, the thermostat controls relays and contactors that then control the motors in the air-conditioning equipment. The relays and contactors are all wired back to the common side of the transformer. It is called the common side because all of the control components have it in common.

39.10 TERMINAL CONNECTIONS

Most manufacturers of thermostats and central systems use a standard terminal identification such as shown in Table 39.1. Standard practice is to use wire colors that match the letter designations: red on R, yellow on Y, and so on. This makes wiring the controls much easier. However, technicians will often find that the thermostat wire that was provided does not contain the exact colors they need.

TECH TIP

Some terminal identification letters are closely associated to the color of wire often used. For example R is often red, W is often white, G is often green, and Y may be yellow; however, not all thermostat control wires contain the same color wires.

Thermostat wire is available in 18- and 20-gauge sizes. The smaller 20-gauge wire may result in a significant voltage drop in the control circuit if the wire run is extremely long. It is therefore recommended that you use 18-gauge wire for any long run. The control wire between the air handler and the condensing unit on a split-system air conditioner is usually a two-wire bundle that contains a red and a white wire. Most heat pumps use thermostat wire with six wires in the bundle: red, white, green, yellow, blue, and

Table 39.1 Standard Terminal Identification for Thermostats

Terminal Identification*	Function
R	Power from transformer
RC	Power from cooling-only transformer
RH	Power from heating-only transformer
C	Common from transformer
X	Same as common from transformer
W	Heat valve or relay
W1	First-stage heat
W2	Second-stage heat
Y	Compressor contactor
Y1	First-stage cooling
Y2	Second-stage cooling
G	Fan relay
B	Heat pump changeover valve to heating Alternate use: common on some systems
E	Emergency heat relay
L	Lamp

orange. For systems requiring larger bundles of conductors in the thermostat wiring, wire manufacturers provide wire bundles with nine and twelve conductors.

TECH TIP

Although the thermostats are often marked with color coded terminals, not all thermostat wiring has the same colors available. It is important when connecting a new thermostat that you reconnect the same terminal identification with the wire previously used for that function. When replacing a thermostat, one way to avoid confusion is to write down the terminal each wire color is connected to before disconnecting anything.

SERVICE TIP

With the thermostat removed from the subbase, you can usually read 24 V from the R terminal on the sub-base to most of the other terminals. That is because the other terminals feed controls that are wired to the common side of 24 V. The meter lead on the B, O, W, Y, and G terminals can read the common side of the transformer through the coils. With no current flowing through the circuit, there is no voltage drop through these controls, so they act just like a wire.

Letters	Component
CC	Compressor contactor
GV	Gas valve
IFM	Indoor fan motor
IFR	Indoor fan relay

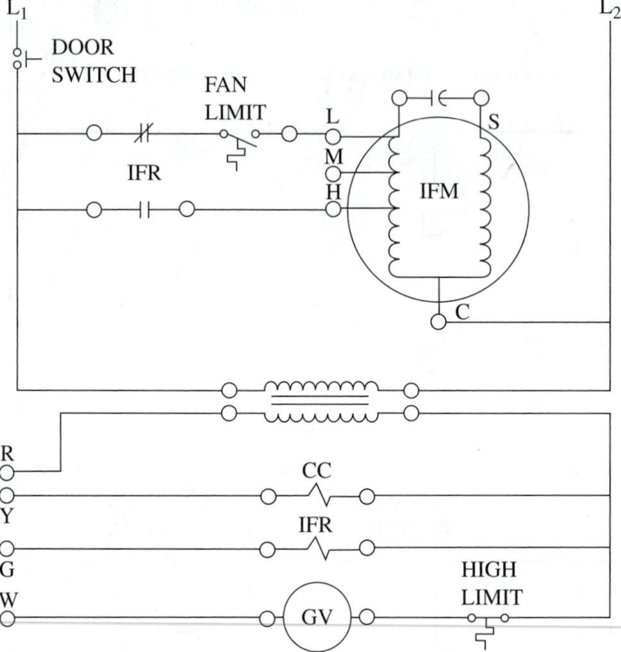

Figure 39-20 Standing pilot gas furnace ladder diagram.

39.11 STANDING PILOT GAS FURNACE DIAGRAM

Figure 39-20 is a ladder diagram of a basic standing pilot gas furnace with central air conditioning. With the system power on and the door switch closed, L1 voltage can be read at the fan switch and at the L1 terminal of the transformer (Figure 39-21). L2 voltage potential can be measured at all of the indoor blower motor (IFM), terminals. It can be read at all of the terminals because without a current flow there is no voltage drop across the motor. L2 voltage can also be measured at the second power lead of the transformer. There is a voltage drop across the primary of the transformer because the transformer is energized any time the system is on and the blower compartment door switch is closed. The transformer provides power to the R terminal on the thermostat. With the thermostat set to off, the transformer common can be read on both sides of the cooling contactor coil, indoor fan relay coil, gas valves, and limit switch. This means that on the thermostat's sub-base a voltage can be read from R to Y, R to G, and R to W.

When the system is in heating, as illustrated in Figure 39-22, L1 voltage can be read on the low, medium, high, and start leads of the indoor blower motor. There is a voltage drop across the motor from the common terminal to all three-speed terminals and the start terminal. A voltage

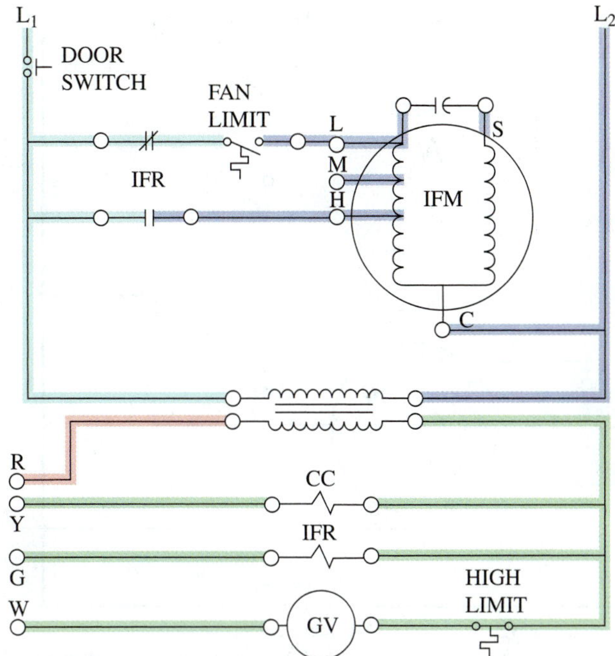

Figure 39-21 Gas furnace with power on but unit not operating.

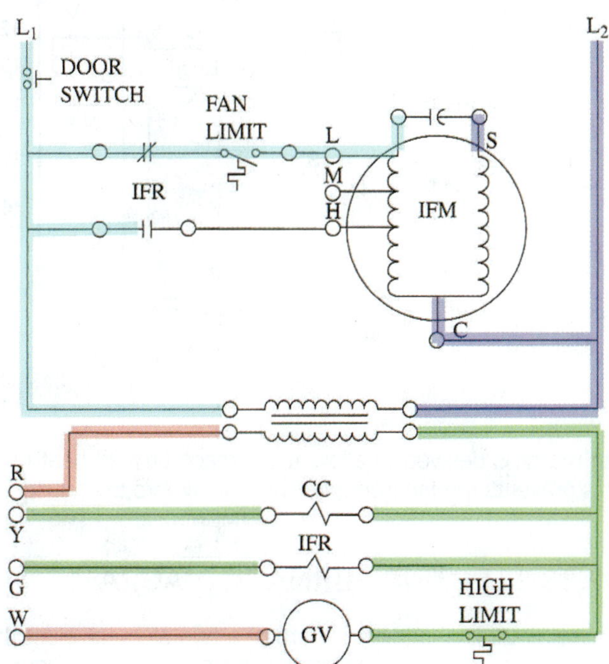

Figure 39-22 Gas furnace operating in heating.

drop can be measured across the primary windings of the control voltage transformer. A voltage drop can be measured across the gas valve, but no voltage drop will occur across the compressor contactor or indoor fan relay (IFR) because those circuits are not energized.

If the high-limit switch opens, a voltage drop would occur across the high-limit terminals and no voltage drop would be measured across the gas valve terminals. There would be no voltage drop on the gas valve because there

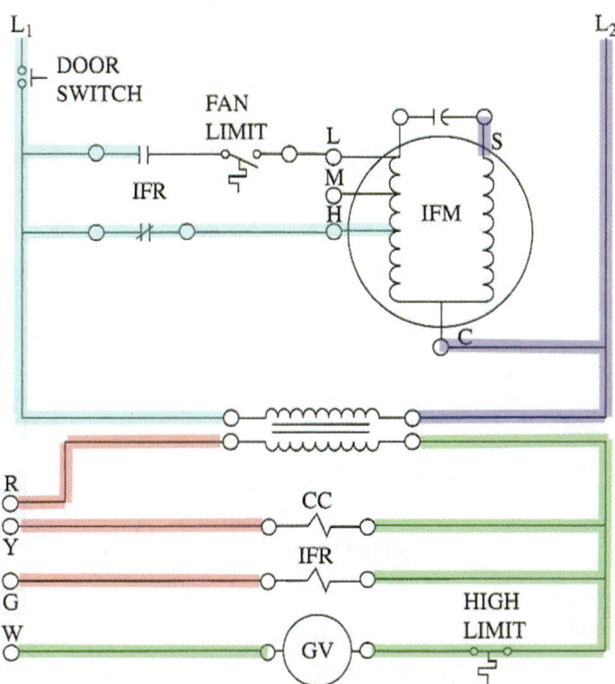

Figure 39-23 Gas furnace operating in cooling.

would not be a current flow through the open high limit. It is important to note that the fan switch would remain closed even though the gas valve is off so that the fan can continue to blow out the excessive heat from the furnace.

When this system is in cooling (Figure 39-23), the indoor blower relay contacts reverse so that the indoor blower motor is operating on high speed. The fan limit switch would remain open since there is no fire in the furnace to cause that temperature switch to close. There would be a voltage drop measurable across the compressor contactor and indoor blower relay. There would not be a voltage drop measurable across the gas valve because the gas valve would not be energized.

39.12 HOT SURFACE IGNITION GAS FURNACE DIAGRAM

Most gas furnaces manufactured today do not use a pilot light. Instead, they ignite the main burners directly using a hot surface igniter. Figure 39-24 shows a ladder diagram for a gas furnace with hot surface ignition. The draft fan relay (DFR), ignition relay (IR), cooling fan relay (CFR), and heating fan relay (HFR) are all part of the ignition control. The ignition control is shown in two places on the diagram because it is involved in both line voltage circuits (lines 1–6) and low-voltage circuits (lines 10–17) . However, the ignition control is one physical device. Splitting it in two preserves the integrity of the ladder diagram.

With the system switch in the heat position, the heating thermostat closes when the room temperature drops. This initiates a call for heat, creating a path from the R side of the transformer to the W terminal on the thermostat, through the limit and roll-out switches to the ignition control (Figure 39-24) (line 14). The ignition control energizes the DFR coil

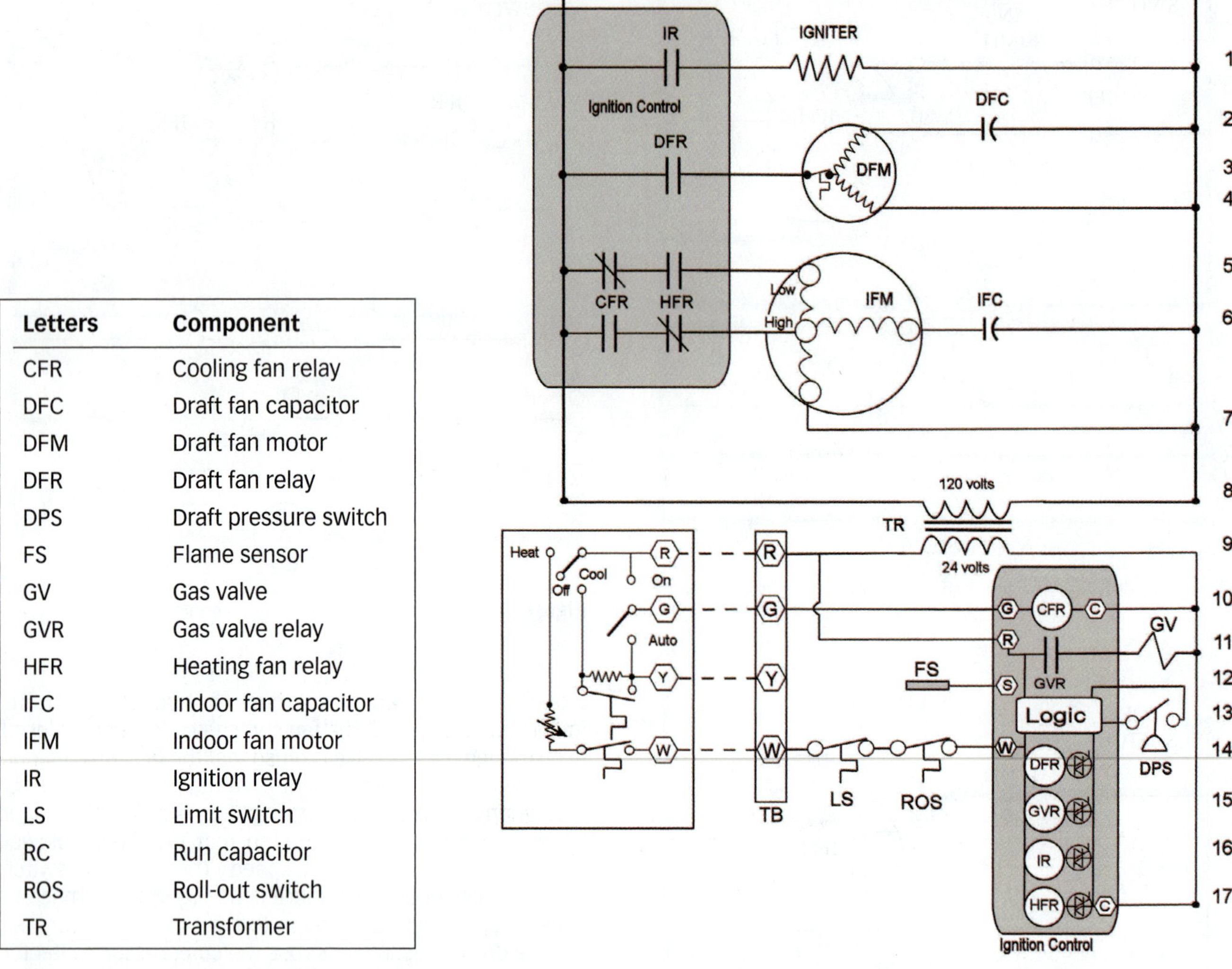

Letters	Component
CFR	Cooling fan relay
DFC	Draft fan capacitor
DFM	Draft fan motor
DFR	Draft fan relay
DPS	Draft pressure switch
FS	Flame sensor
GV	Gas valve
GVR	Gas valve relay
HFR	Heating fan relay
IFC	Indoor fan capacitor
IFM	Indoor fan motor
IR	Ignition relay
LS	Limit switch
RC	Run capacitor
ROS	Roll-out switch
TR	Transformer

Figure 39-24 Diagram for hot surface ignition furnace.

(line 14), closing the DFR contacts, which energize the draft fan motor (DFM) on line 3. The DFM operates for 60 seconds to perform a pre-purge. The pre-purge removes any lingering gas or combustion products from the burner and heat exchanger. Operation of the DFM causes the draft pressure switch (DPS) to close. After the pre-purge, the ignition control checks to see if the DPS is closed (line 13). The ignition process does not continue until the DPS closes. After the ignition control sees that the DPS is closed, it energizes the ignition relay (IR) on line 16, closing its contacts and energizing the igniter (line 1). The igniter is energized for 60 seconds before the gas valve is energized. The gas valve is energized and the flame rectification circuit looks for the presence of flame (line 12). If flame is sensed, the gas valve remains energized and the igniter is de-energized. If flame is not sensed, both the gas valve and igniter are de-energized after 5 seconds. The heating fan relay coil (HFR) on line 17 is energized after a delay of 2 to 3 minutes. The normally open HFR contacts close to energize the heating fan speed and the normally closed HFR contacts open to prevent operation of the cooling fan speed (Figure 39-24) (lines 5 and 6). Starting of the indoor fan motor is delayed to allow the furnace time to heat up. This prevents the fan from blowing cold air into the house.

39.13 ELECTRIC FURNACE DIAGRAM

Figure 39-25 is a ladder diagram of a two-element electric furnace diagram that has central air conditioning. Figure 39-26 shows the part of the circuit that is energized with line 1 and line 2 potential voltages.

Figure 39-27 illustrates all of the circuits that would be energized when the first-stage heating element has come on. Note that there is not a voltage drop across the second-stage heating element at this time. Only the first-stage heat and fan are operating. On the secondary circuit, the sequencer heat motor element has been energized. Figure 39-28 shows the parts of the circuit that would be energized when both the first- and second-stage heating elements have been energized. When the second set of contacts close on the second-stage heating circuit, all of the elements are energized and show a voltage drop.

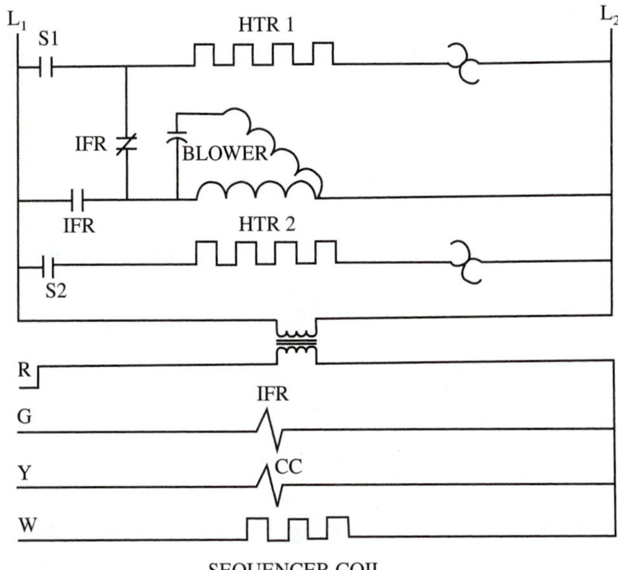

Letters	Component
CC	Compressor contactor
GV	Gas valve
HTR 1, HTR 2	Heater 1, heater 2
IFR	Indoor fan relay
S1, S2	Sequencer contacts

Figure 39-25 Electric furnace ladder diagram.

Figure 39-29 represents those circuits that would be energized when the system is in the cooling mode. Note that the heating elements are not energized and that the indoor fan relay (IFR) has opened the contacts to the first-stage heating element as the normally opened contacts are closed so that the indoor blower motor is operating. On the secondary there is a voltage drop across the indoor blower motor relay and the outdoor compressor contactor.

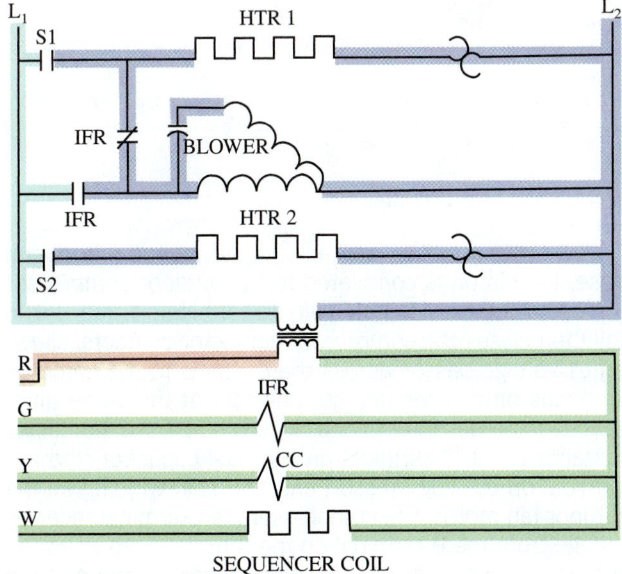

Figure 39-26 Electric furnace with power on and no call for heat.

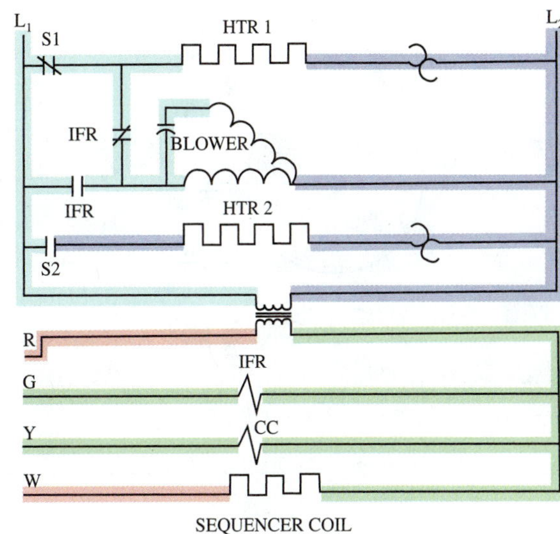

Figure 39-27 First-stage heat energized.

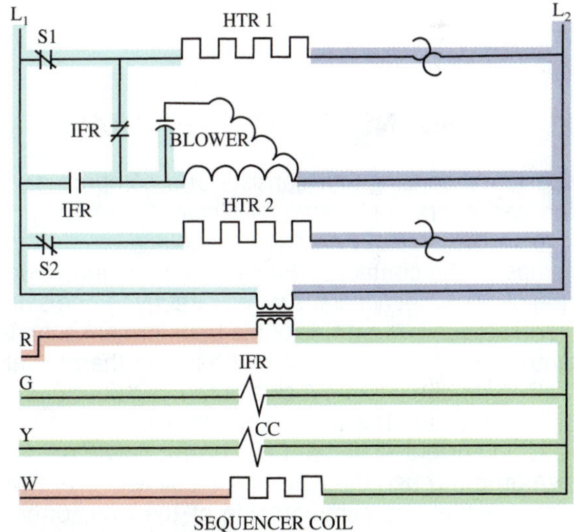

Figure 39-28 Second-stage heat energized.

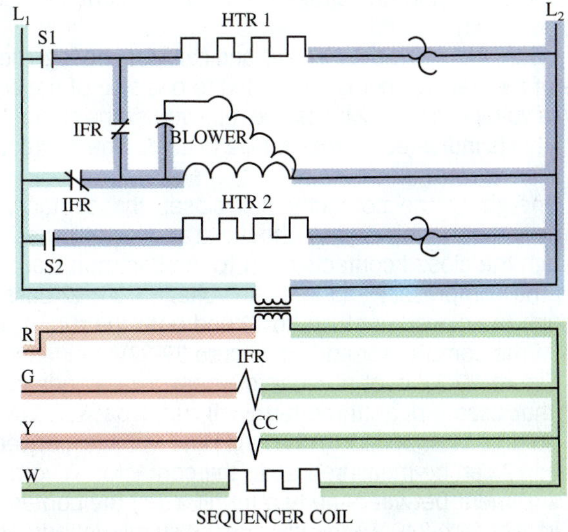

Figure 39-29 Electric furnace operating in cooling.

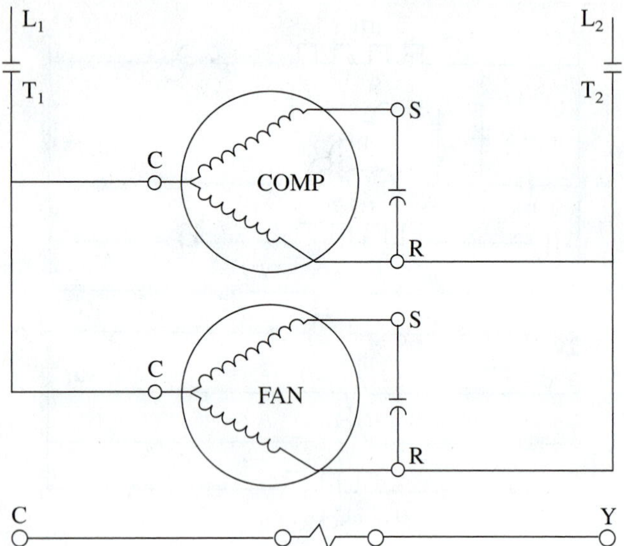

Figure 39-30 Ladder diagram of a condensing unit.

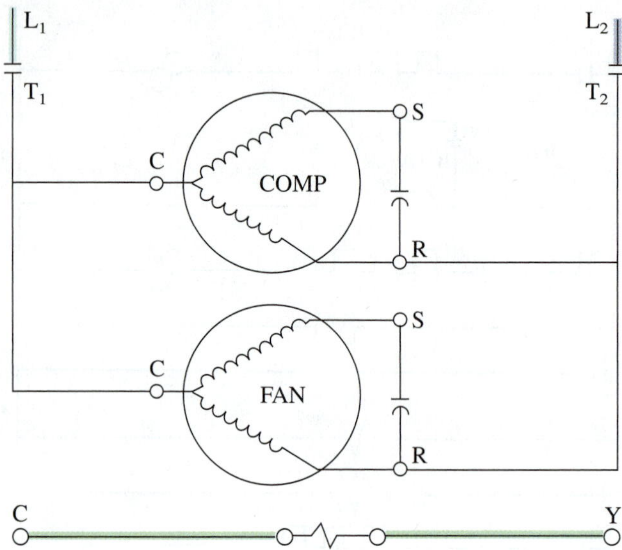

Figure 39-31 Ladder diagram of condensing unit with power on and system off.

39.14 CONDENSING UNIT DIAGRAM

The basic condensing unit consists of a compressor, fan motor, capacitors, and contactor (Figure 39-30). With line voltage to the unit but no call for cooling, the line voltage stops at the contactor. Typically the common side of the transformer secondary is wired directly to one side of the contactor coil. The R side of the transformer secondary is controlled by the thermostat. When the thermostat is not calling for cooling, the R side of the transformer stops at the thermostat. The thermostat breaks the circuit, so there is no current flow. With no current flow there is no voltage drop through the coil, so the coil acts like a wire. This means that the common side of the transformer is available at both sides of the coil and at both control wires to the coil. On the diagram, both wires and both sides of the coil are shaded the same color. The contactor coil is not operating, and no voltage is read at the contactor coil (Figure 39-31).

When the thermostat closes and calls for cooling, the R side of the transformer is connected to one side of the coil. Now a voltage drop of 24 V can be read across the contactor coil. This is indicated by the change in color. The contactor coil is energized.

Energizing the contactor coil closes the contactor's normally open points and the voltage from L1 passes through the closed contacts to T1 to the common leads of both the compressor and fan. The voltage from L2 passes through the closed contacts to T2 and onto the run terminals of the compressor and fan (Figure 39-32).

Figure 39-33 is of a wiring diagram for a condensing unit that uses a dual (three-terminal) run capacitor. When line voltage is supplied to the unit, line voltage between L1 and L2 can be measured up to the contactor. A voltage measurement between the two terminals of the contactor would read 208–240 V depending on the supply voltage. The control voltage to terminals Y and C would read 0 V.

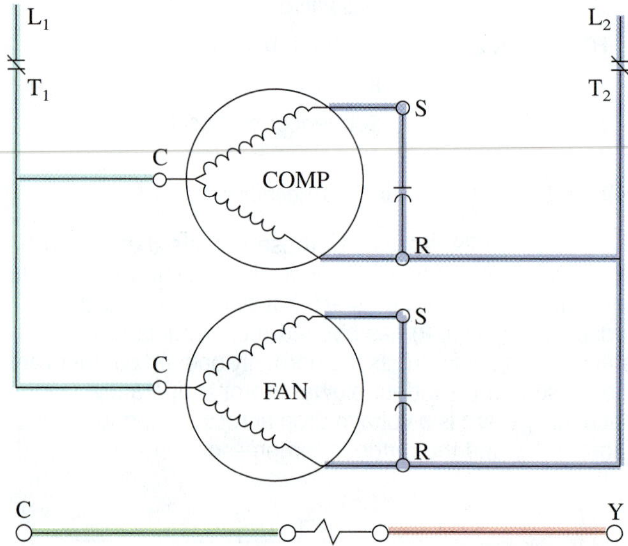

Figure 39-32 Ladder diagram of condensing unit with power on and system on.

When the cooling contacts in the indoor thermostat close, the circuit is completed to the outdoor compressor contactor coil providing 24 volts to terminals Y and C. The coil then closes the compressor contactor contacts, allowing L1 voltage passes across the contacts to the common terminals on the compressor and fan. At the same time, L2 voltage passes through the closed compressor contactor contacts to the C terminal on the dual capacitor, the run terminal on the compressor, and the main winding on the outdoor fan motor. The compressor start terminal receives voltage from the H terminal on the dual capacitor, and the outdoor fan motor auxiliary winding receives voltage from the F terminal on the dual capacitor.

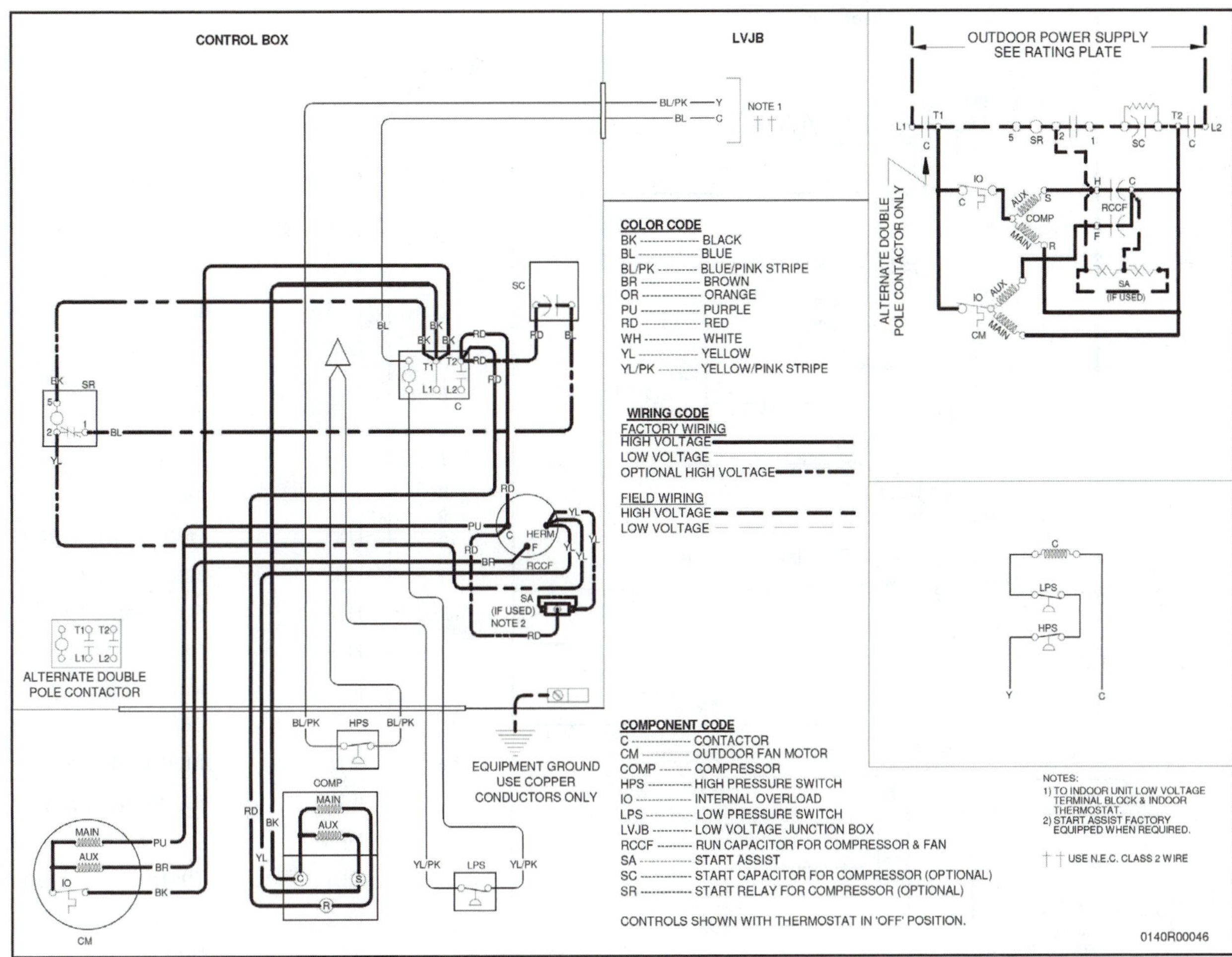

Figure 39-33 Schematic diagram for a dual-capacitor unit. *(Courtesy of Goodman Global Group, Inc.)*

39.15 POTENTIAL RELAY AND START CAPACITOR

Figure 39-33 also has a start capacitor and a potential start relay. The start capacitor improves the compressor motor starting torque. The start relay is used to take the start capacitor out of the circuit after the compressor starts. When this system is on and running, there is a voltage across the coil on the potential start relay (SR) between terminals 5 and 2. This measured voltage is higher than the applied line voltage between L1 and L2. This higher voltage is the result of the back electromotive force created by the compressor motor as the compressor runs. It is this higher voltage, or potential, that energizes the relay coil, opening the normally closed relay contacts located between terminals 1 and 2. This drops the start capacitor from the circuit within a fraction of a second as the compressor starts.

SERVICE TIP

If the contacts in the potential relay do not open, the current flowing through the start capacitor would quickly cause the start capacitor to overheat and explode!

39.16 HEAT PUMP COOLING CYCLE

Figure 39-34 shows a schematic diagram of a traditional air-source heat pump. A heat pump is essentially an air-conditioning system that can reverse the indoor and outdoor coils. The indoor coil is the evaporator in cooling and the condenser in heating. The outdoor coil is the condenser in cooling and the evaporator in heating. This is done by reversing the flow of refrigerant between the two coils using a reversing valve. The reversing valve can be energized in either the heating or cooling cycle, depending upon the manufacturer's design. This particular diagram energizes the reversing valve in the cooling cycle, as do more heat pumps today.

Moving the system switch to COOL sends 24 volts from R to O, energizing the reversing valve solenoid (RVS) on line 17, (Figure 39-34). When the room temperature increases, the cooling thermostat (TSC) closes, sending 24 volts from R to Y, energizing contactor coil C (line 16). If the fan switch is in the AUTO position, TSC also sends 24 volts to G, energizing the indoor fan relay (IFR), on line 15. The contactor coil closes its normally open contacts, energizing the compressor and outdoor fan motor (OFM) on lines 4–9. At the same time, the indoor fan relay (IFR) closes its normally open contacts, energizing the indoor fan motor (IFM) on line 11. All loads necessary to cool are now operating (Figure 39-35).

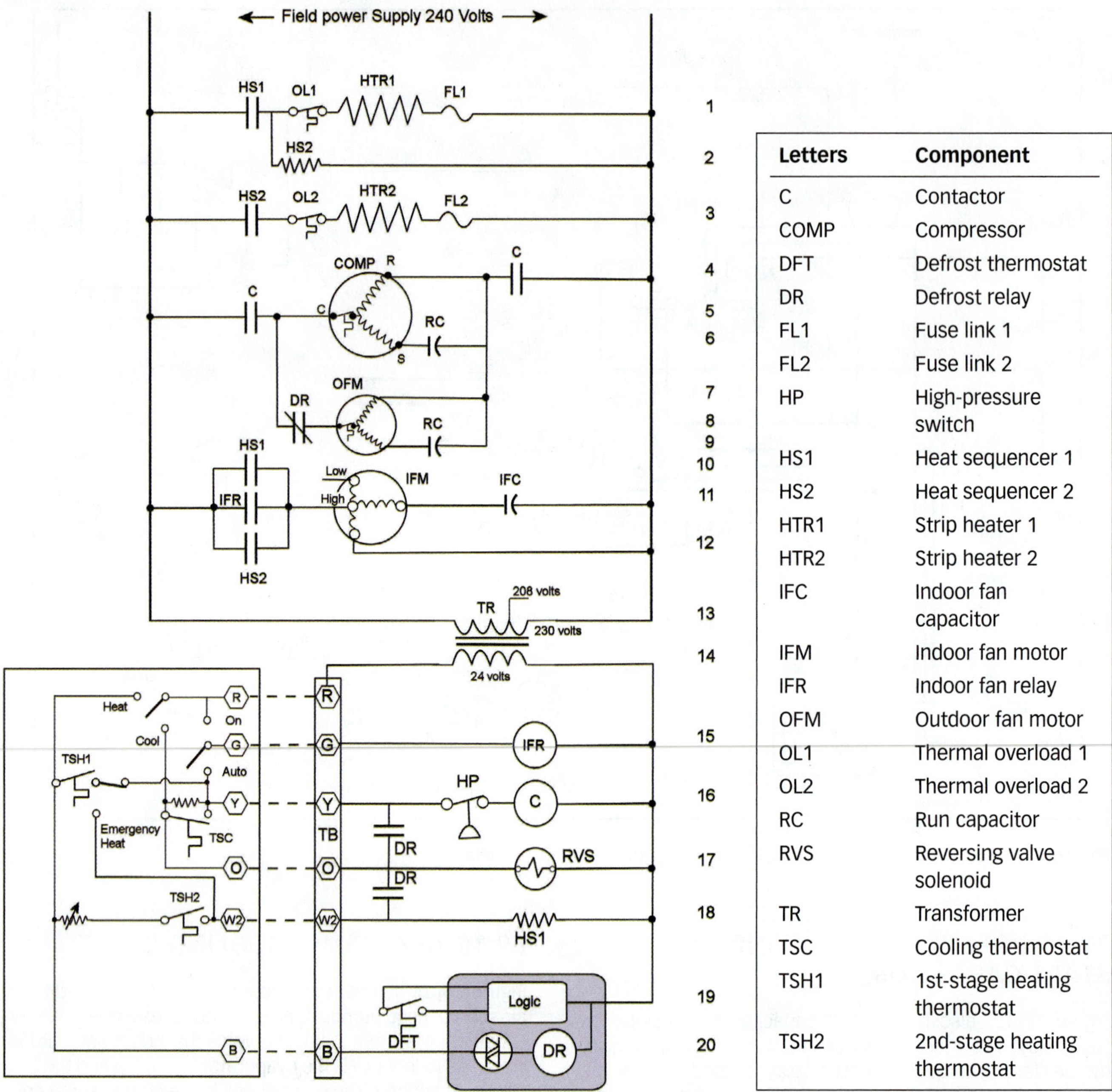

Letters	Component
C	Contactor
COMP	Compressor
DFT	Defrost thermostat
DR	Defrost relay
FL1	Fuse link 1
FL2	Fuse link 2
HP	High-pressure switch
HS1	Heat sequencer 1
HS2	Heat sequencer 2
HTR1	Strip heater 1
HTR2	Strip heater 2
IFC	Indoor fan capacitor
IFM	Indoor fan motor
IFR	Indoor fan relay
OFM	Outdoor fan motor
OL1	Thermal overload 1
OL2	Thermal overload 2
RC	Run capacitor
RVS	Reversing valve solenoid
TR	Transformer
TSC	Cooling thermostat
TSH1	1st-stage heating thermostat
TSH2	2nd-stage heating thermostat

Figure 39-34 Diagram for air-source heat pump.

39.17 HEAT PUMP FIRST-STAGE HEAT CYCLE

The main difference between the cooling cycle and the first-stage heating cycle is the reversing valve. The reversing valve is not energized in heating on this diagram. Moving the system switch to HEAT sends 24 volts from R to B, energizing the defrost board (see Figure 39-34) (line 20). When the room temperature decreases, the first-stage heating thermostat (TSH1) closes, sending 24 volts from R to Y, energizing contactor coil C (line 16). If the fan switch is in the AUTO position, TSH1 also sends 24 volts to G, energizing the IFR (line 15). The contactor coil closes its normally open contacts, energizing the compressor and outdoor fan motor (lines 4–9). At the same time, the indoor fan relay closes its normally open contacts, energizing the indoor fan motor (line 11). All loads necessary for first-stage heat are now operating (Figure 39-36).

SERVICE TIP

Although many systems use terminal B to energize heating-cycle components such as reversing valves or defrost boards, some manufacturers use terminal B as the common terminal. Wiring a thermostat heating terminal B to a heat pump common terminal B will burn out both the thermostat and the transformer when the system switch is moved to the HEAT position. You cannot depend

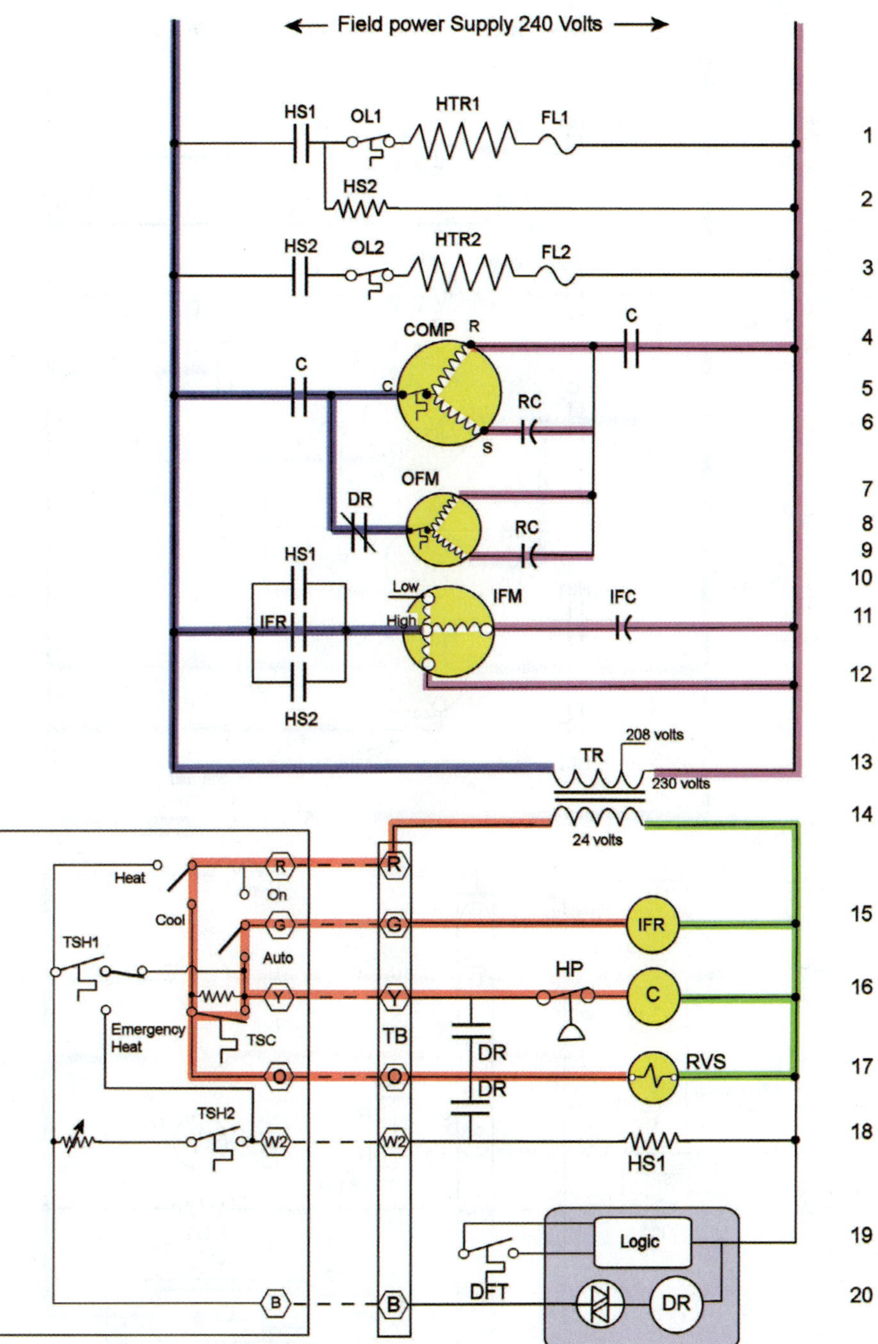

Figure 39-35 Heat pump operating in cooling cycle.

entirely on letters and colors when doing the control wiring. You have to read the instructions and refer to the diagrams on all connected pieces of equipment.

39.18 HEAT PUMP SECOND-STAGE HEAT CYCLE

The second-stage heating thermostat (TSH2) closes if the first-stage heat cannot maintain the room temperature within 1.5 degrees of the thermostat setpoint. TSH2 sends 24 volts from R to W2, energizing heat sequencer 1 (HS1) (see Figure 39-36) (line 18). The HS1 coil is a small heater that warps a bimetal disk to close the contacts after a time delay of 30 seconds. HS1 has two sets of normally open contacts that close after a time delay to prevent the strips from energizing immediately at startup when the compressor and fan motors are drawing high starting amps. The HS1 normally open contacts close, energizing the first heater (HTR1) and the coil to the second heat sequencer (HS2) (lines 1 and 2). HS2 closes its contacts after a time delay to energize the second heater (HTR2) (line 3). Note that HS1 and HS2 normally open contacts are in parallel with the IFR contacts (lines 10 and 12). They ensure that the indoor fan motor will operate whenever the heat strips are operating.

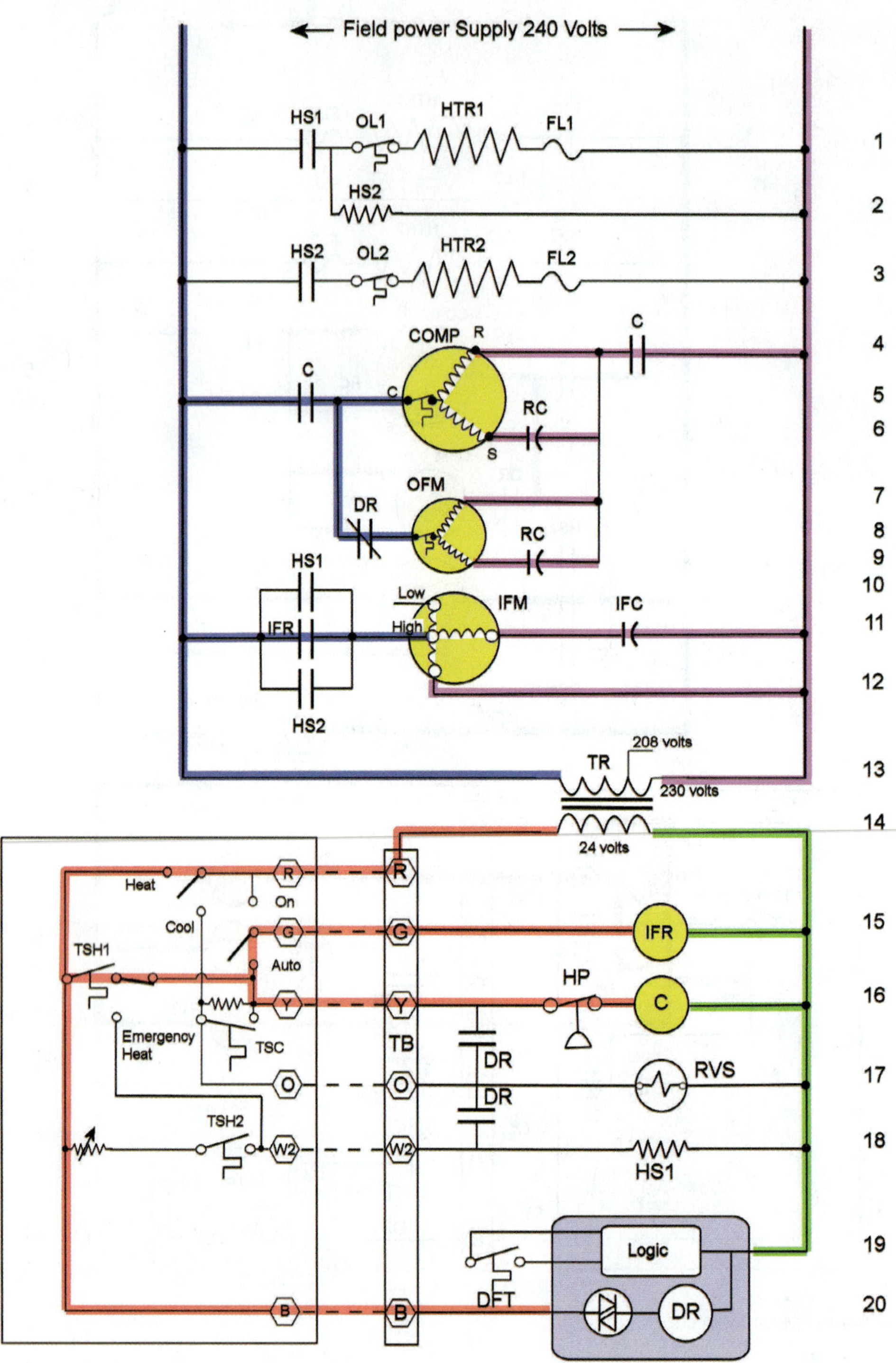

Figure 39-36 Heat pump operating in heating cycle.

All loads necessary for first- and second-stage heat are now operating (Figure 39-37).

39.19 HEAT PUMP EMERGENCY HEAT CYCLE

Air-source heat pump thermostats typically have an EMERGENCY HEAT selection on the system switch. Emergency heat disables the compressor heat and allows the second-stage heat to cycle on the first-stage thermostat. This is useful if the refrigeration circuit is not operating correctly. The thermostat in this diagram accomplishes this through internal circuitry. Moving the system switch to EMERGENCY HEAT sends 24 volts from TSH1 to the W2 terminal instead of the Y terminal (see Figure 39-34) (line 16). When TSH1 closes, it energizes the HS1 coil (line 18). The HS1 contacts close, energizing HTR1 and the HS2 sequencer coil (lines 1 and 2). The second set of HS1 contacts close at the same time to energize the indoor fan motor (line 11). After a delay, the

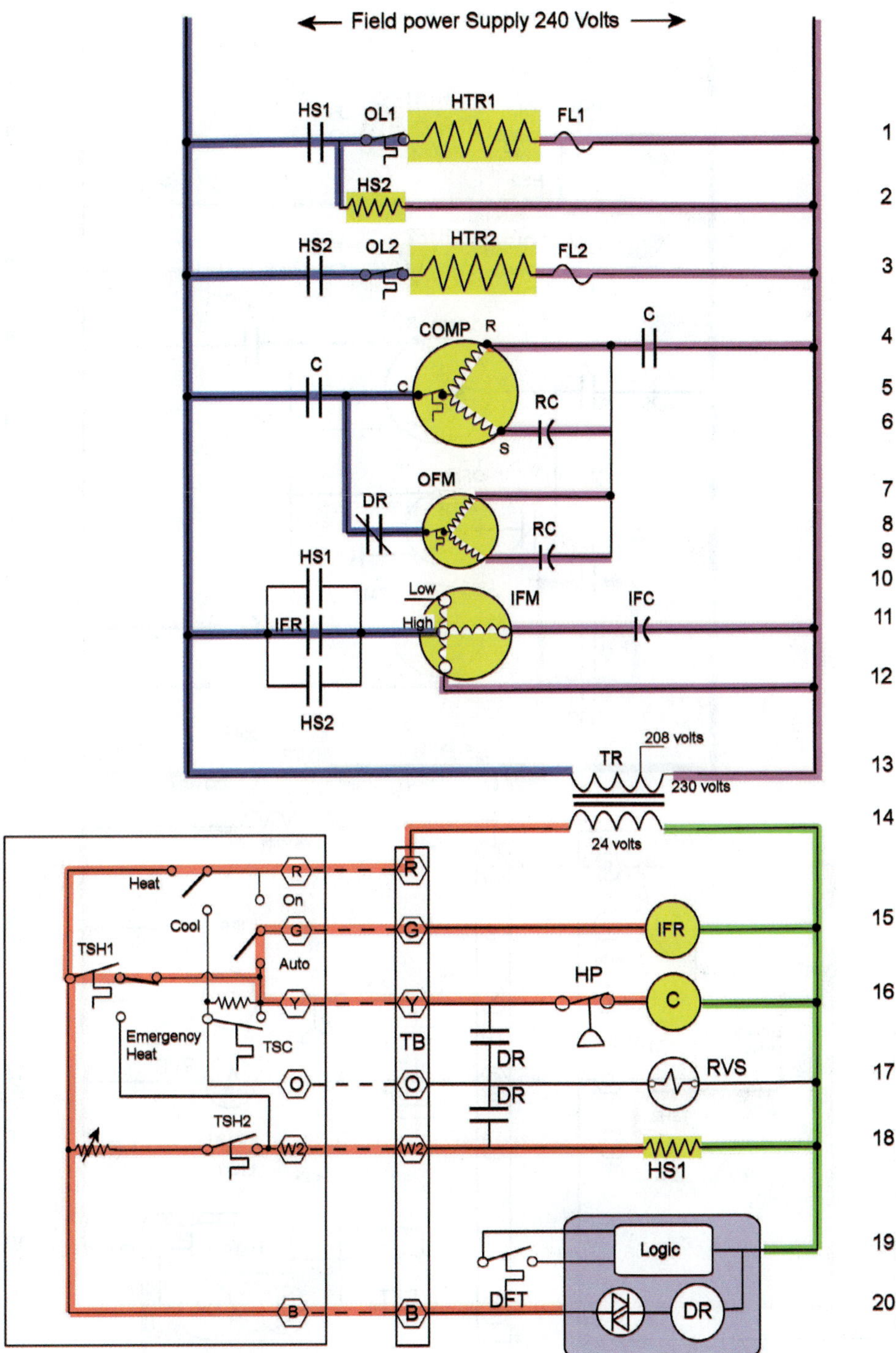

Figure 39-37 Heat pump operating in second-stage heat.

HS2 contacts close, energizing HTR2 (line 3). All loads necessary for emergency heat are now operating (Figure 39-38).

39.20 HEAT PUMP DEFROST DIAGRAM

Most modern heat pump defrost control boards serve as the central nervous system of the heat pump. All control signals pass through the defrost board. Thus the defrost board really controls more than simply the defrost function; it controls all aspects of the heat pump condensing unit operation. Figure 39-39 shows a typical heat pump diagram. The connection diagram on the left shows the terminals on the defrost board, the wire colors, and exactly how the wires are routed on the unit. The schematic on the right shows the electrical circuitry and how the controls in the heat pump operate. The yellow shaded areas in Figure 39-40 represent the defrost board. Notice that the defrost board is shown as one piece in the connection diagram, but its functional parts are split up in the schematic diagram.

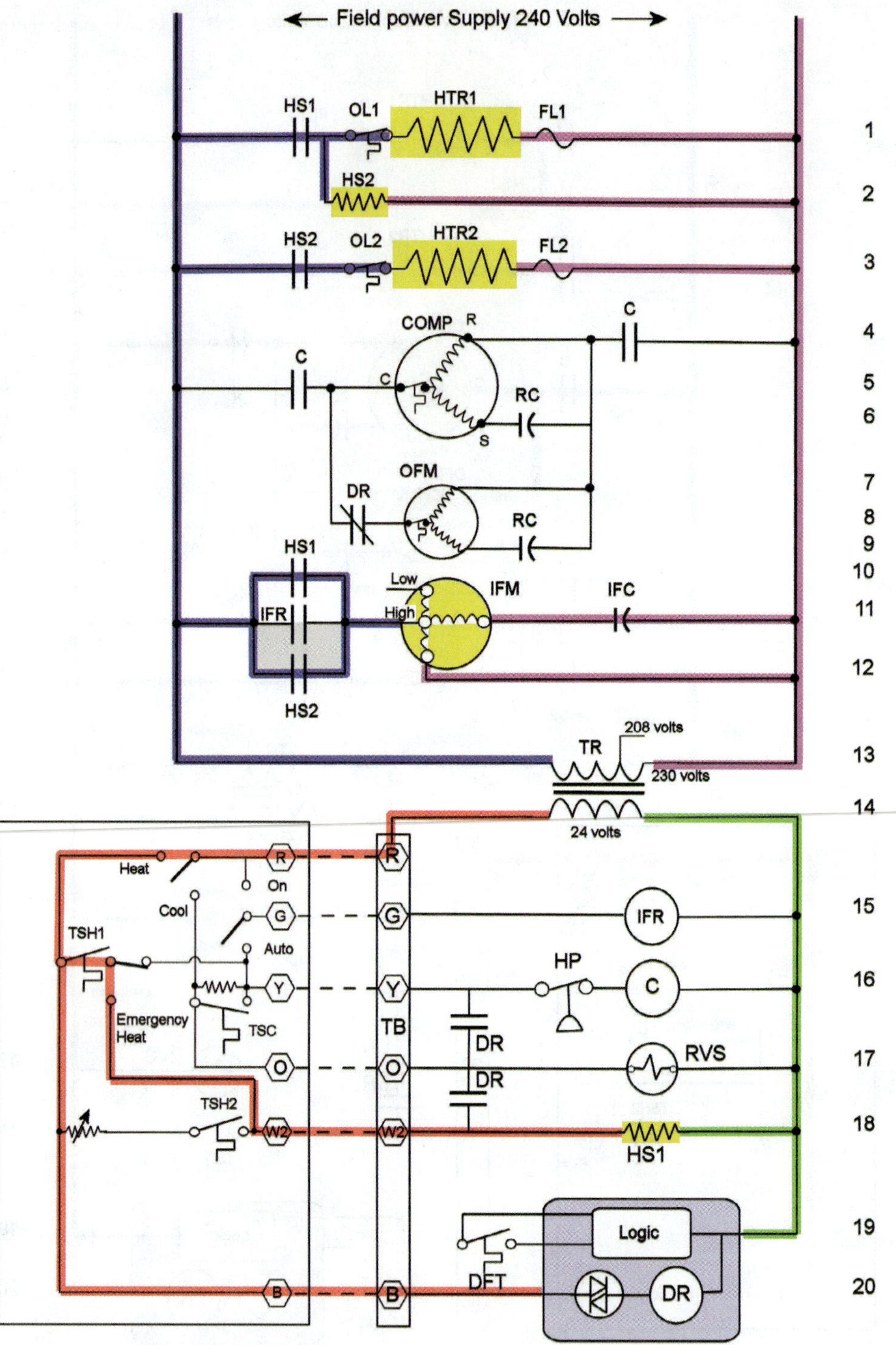

Figure 39-38 Heat pump operating in emergency heat.

To defrost the heat pump, the defrost control must accomplish three things:

- Shift the reversing valve from heating to cooling.
- Turn off the outdoor fan.
- Turn on the auxiliary heat.

The most frequently used method of determining when to initiate a defrost is based on time and temperature. The defrost thermostat, shown in Figure 39-41, senses the outdoor coil temperature. When the coil temperature is below 30°F ±3°F, the defrost thermostat completes a circuit to the logic in the defrost board (Figure 39-42). The defrost board will then start timing. When the timer reaches the elapsed time programmed by the jumper, a defrost cycle is initiated. However, if the defrost thermostat opens before the timing period is reached, the timer is reset. The system must operate for 30, 60, 90, or 120 minutes with the coil

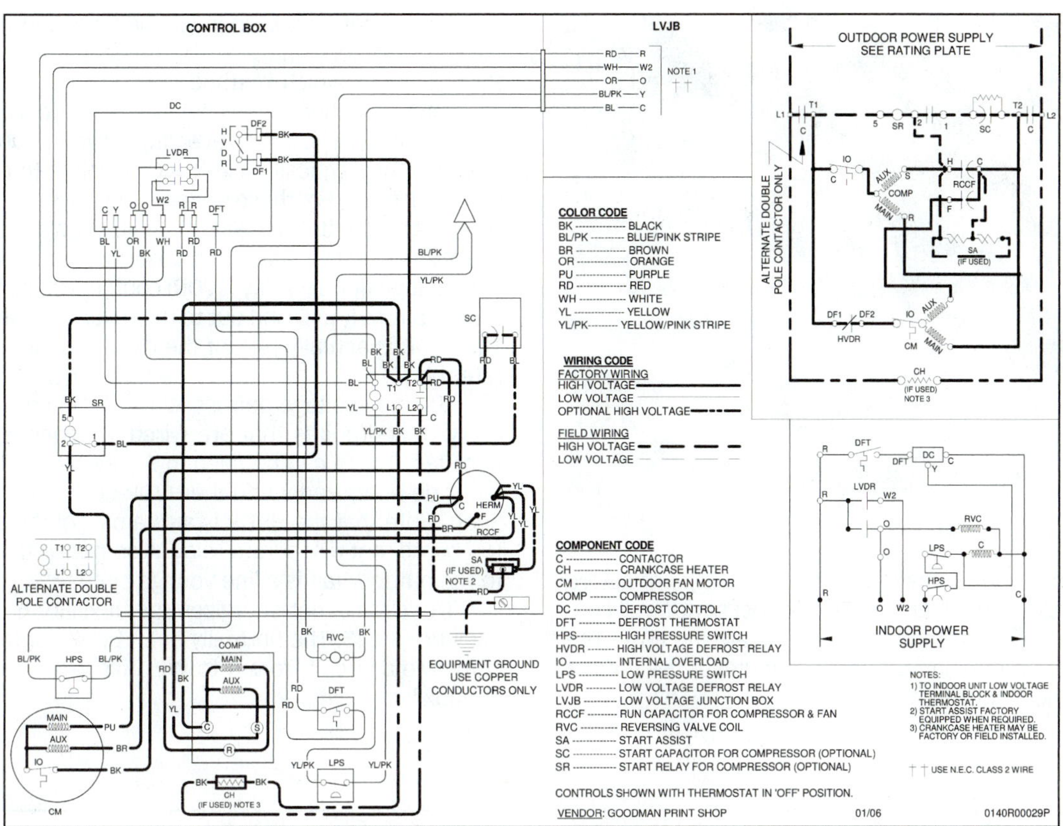

Figure 39-39 Heat pump schematic diagram. *(Courtesy of Goodman Global Group, Inc.)*

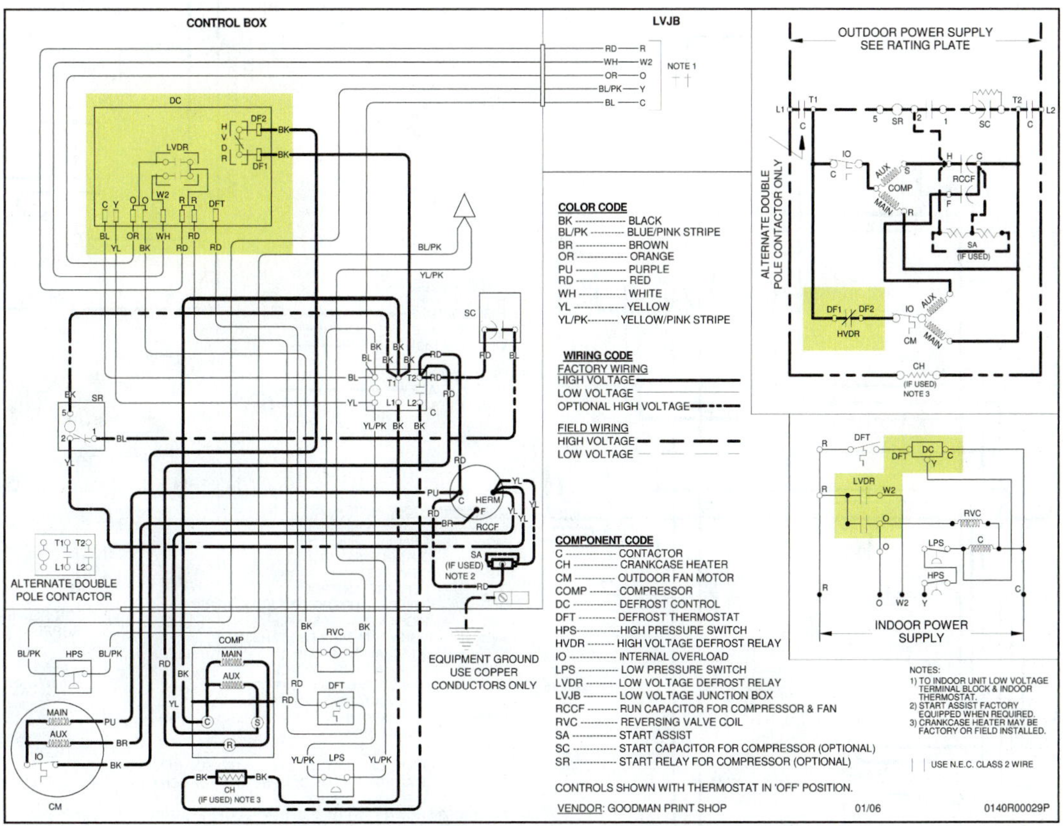

Figure 39-40 Defrost board on heat pump diagram. *(Courtesy of Goodman Global Group, Inc.)*

Figure 39-41 Defrost thermostat.

temperature below the defrost thermostat setpoint before a defrost cycle will be initiated. The defrost period will continue until the defrost thermostat opens around 80°F. The defrost cycle is terminated after 10 minutes if the defrost thermostat does not open.

Some heat pumps energize the reversing valve in heating, while others energize the reversing valve in cooling. Regardless of which method is used, the reversing valve must shift from heating to cooling for a defrost cycle. Today, more units energize the reversing valve in cooling than in heating. They typically use the O terminal to energize the reversing valve. For these systems, the reversing valve must be energized in defrost. Figure 39-43 shows the circuit that energizes the reversing valve in defrost. The normally open low-voltage defrost relay (LVDR) contacts close to complete a circuit from R to O and out to the reversing valve solenoid. The LVDR is actually part of the defrost board, not a separate relay.

The high-voltage defrost relay (HVDR) has a set of normally closed points that are wired in series with the outdoor fan. These can be seen in Figure 39-44. The outdoor fan is de-energized when the HVDR normally closed points open. Note that the HVDR is also a part of the defrost control board, not a separate relay. Also notice that the circuit to the outdoor fan is a line voltage circuit.

The circuit that energizes the auxiliary heat is shown in Figure 39-45. The normally open LVDR contacts close to complete a circuit from R to W2. This provides 24 volts to the auxiliary heat inside.

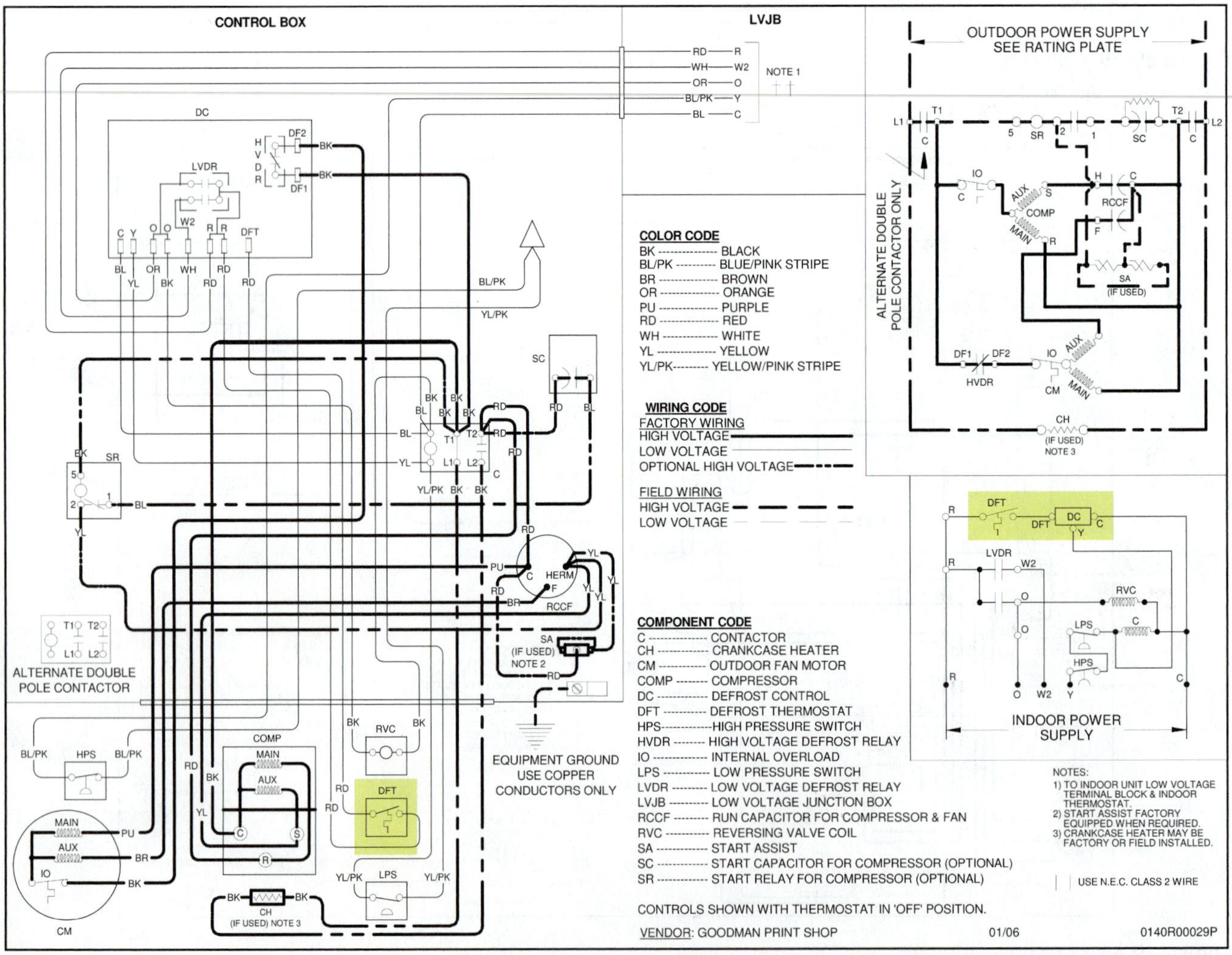

CONTROL BOX

LVJB

OUTDOOR POWER SUPPLY SEE RATING PLATE

COLOR CODE
BK ---------- BLACK
BL/PK ---------- BLUE/PINK STRIPE
BR ---------- BROWN
OR ---------- ORANGE
PU ---------- PURPLE
RD ---------- RED
WH ---------- WHITE
YL ---------- YELLOW
YL/PK ---------- YELLOW/PINK STRIPE

WIRING CODE
FACTORY WIRING
HIGH VOLTAGE
LOW VOLTAGE
OPTIONAL HIGH VOLTAGE
FIELD WIRING
HIGH VOLTAGE
LOW VOLTAGE

COMPONENT CODE
C ---------- CONTACTOR
CH ---------- CRANKCASE HEATER
CM ---------- OUTDOOR FAN MOTOR
COMP ---------- COMPRESSOR
DC ---------- DEFROST CONTROL
DFT ---------- DEFROST THERMOSTAT
HPS ---------- HIGH PRESSURE SWITCH
HVDR ---------- HIGH VOLTAGE DEFROST RELAY
IO ---------- INTERNAL OVERLOAD
LPS ---------- LOW PRESSURE SWITCH
LVDR ---------- LOW VOLTAGE DEFROST RELAY
LVJB ---------- LOW VOLTAGE JUNCTION BOX
RCCF ---------- RUN CAPACITOR FOR COMPRESSOR & FAN
RVC ---------- REVERSING VALVE COIL
SA ---------- START ASSIST
SC ---------- START CAPACITOR FOR COMPRESSOR (OPTIONAL)
SR ---------- START RELAY FOR COMPRESSOR (OPTIONAL)

ALTERNATE DOUBLE POLE CONTACTOR

EQUIPMENT GROUND USE COPPER CONDUCTORS ONLY

INDOOR POWER SUPPLY

NOTES:
1) TO INDOOR UNIT LOW VOLTAGE TERMINAL BLOCK & INDOOR THERMOSTAT.
2) START ASSIST FACTORY EQUIPPED WHEN REQUIRED.
3) CRANKCASE HEATER MAY BE FACTORY OR FIELD INSTALLED.

USE N.E.C. CLASS 2 WIRE

CONTROLS SHOWN WITH THERMOSTAT IN 'OFF' POSITION.

VENDOR: GOODMAN PRINT SHOP 01/06 0140R00029P

Figure 39-42 Defrost thermostat circuit. *(Courtesy of Goodman Global Group, Inc.)*

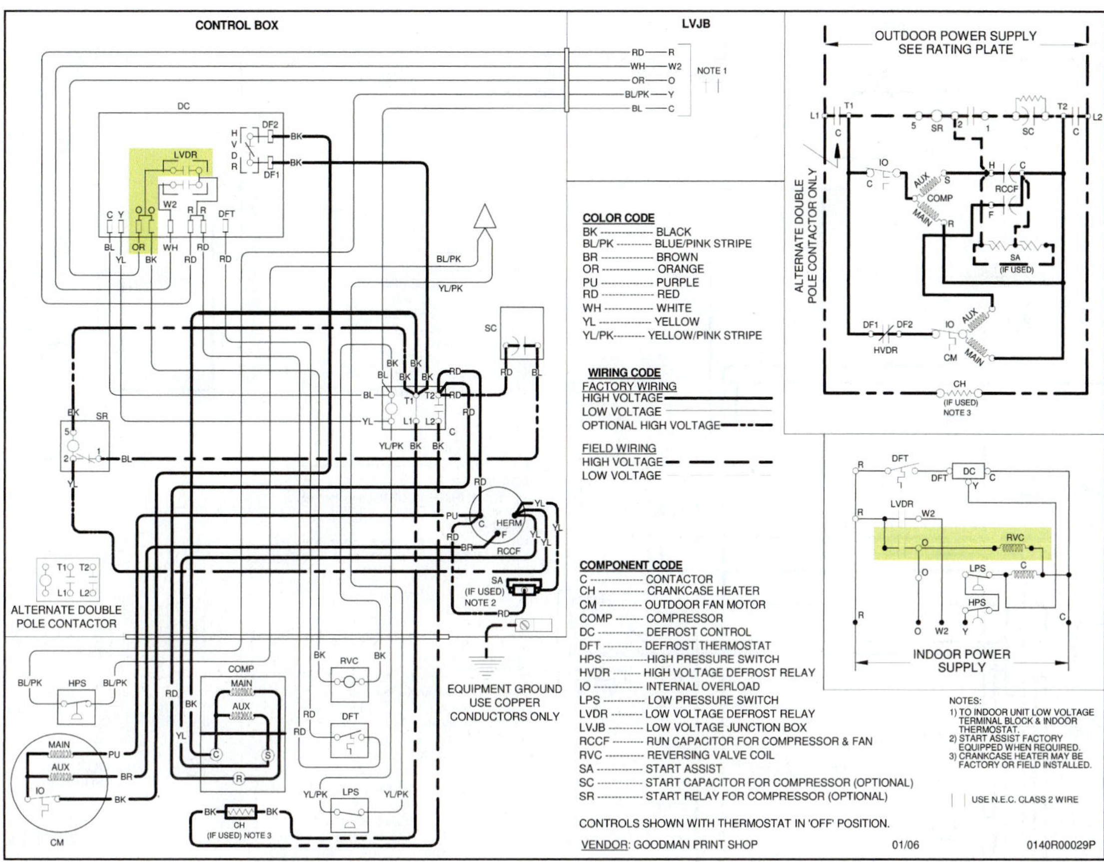

Figure 39-43 Reversing-valve circuit. *(Courtesy of Goodman Global Group, Inc.)*

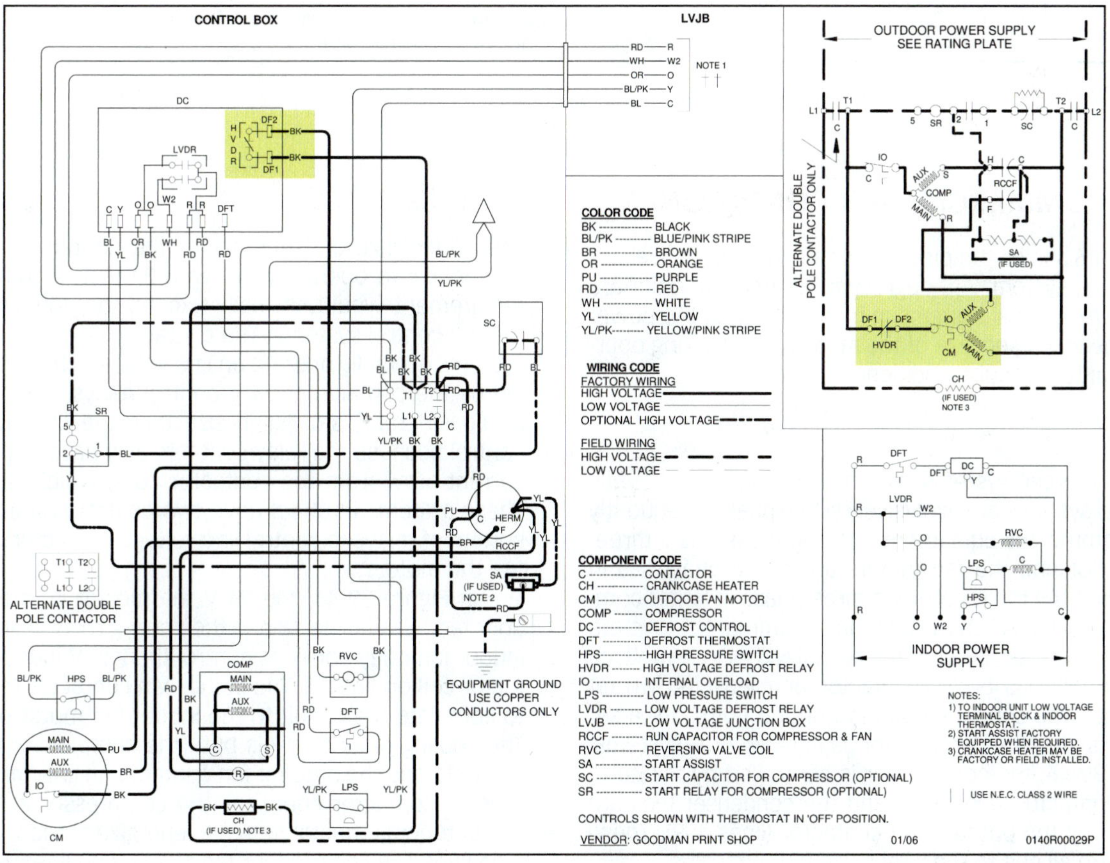

Figure 39-44 Outdoor fan circuit. *(Courtesy of Goodman Global Group, Inc.)*

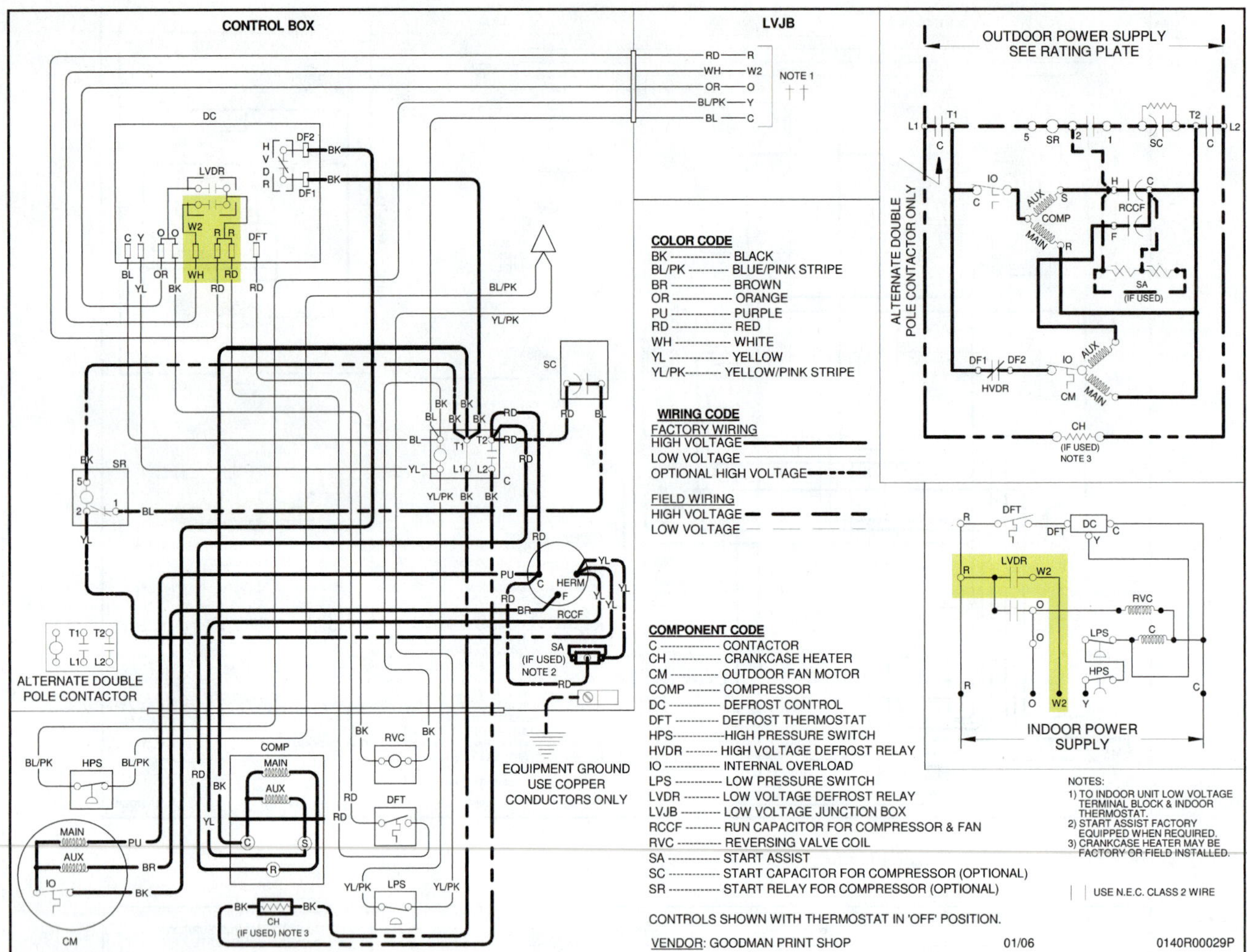

Figure 39-45 Auxiliary heat circuit. *(Courtesy of Goodman Global Group, Inc.)*

39.21 COMMERCIAL AIR CONDITIONER

Figure 39-46 shows the diagram for a light commercial air-conditioning system. Although light commercial units are similar to residential systems in many respects, the power supply, the increased size of the motors, and varying operating conditions require additional controls.

Motors

Most commercial systems use three-phase power, while residential systems are exclusively single phase. Frequently in light commercial equipment, the compressor is a three-phase motor and the fan motors are single phase, as in Figure 39-46. All the motors are three phase in larger commercial systems. All the motors in a commercial unit are normally controlled by contactors instead of relays. Each motor normally has its own contactor, or at least its own set of contacts. The condenser fan motors do not share a set of contacts with the compressor as in residential systems. In Figure 39-46, the compressor is controlled by the compressor contactor (lines 1–3), and the condenser fans are controlled by the outdoor fan contactor (lines 4–9). These changes are all due to the larger size of the motors.

Controls

The added size and number of the contactors makes it impractical to operate them with 24 V. The combined VA requirement would normally exceed the 100 VA limit for Class 2 control systems. The contactor coils in commercial systems normally operate on line voltage. They are controlled by the contacts of 24 V control relays. The relays can be considerably smaller in size than the contactors they control because the relay contacts only have to handle the current draw from the contactor coils, which is very low. These systems in effect have two control systems: a 24 V system of relays to control line voltage contactors that control the motors.

In Figure 39-46, the 24 V thermostat controls the indoor fan relay in line 18 and the cooling relay in line 19. The indoor fan relay contacts control the 230 V indoor fan contactor coil on line 15. The indoor fan contactor then closes to complete the circuit to the indoor fan motor on line 10. The cooling relay controls both the compressor contactor in line 12 and the outdoor fan contactor in line 14. The compressor contactor energizes the compressor in lines 1–3. The outdoor contactor closes to energize outdoor fan motor 1 on lines 4–6 and outdoor fan motor 2 on lines 7–9.

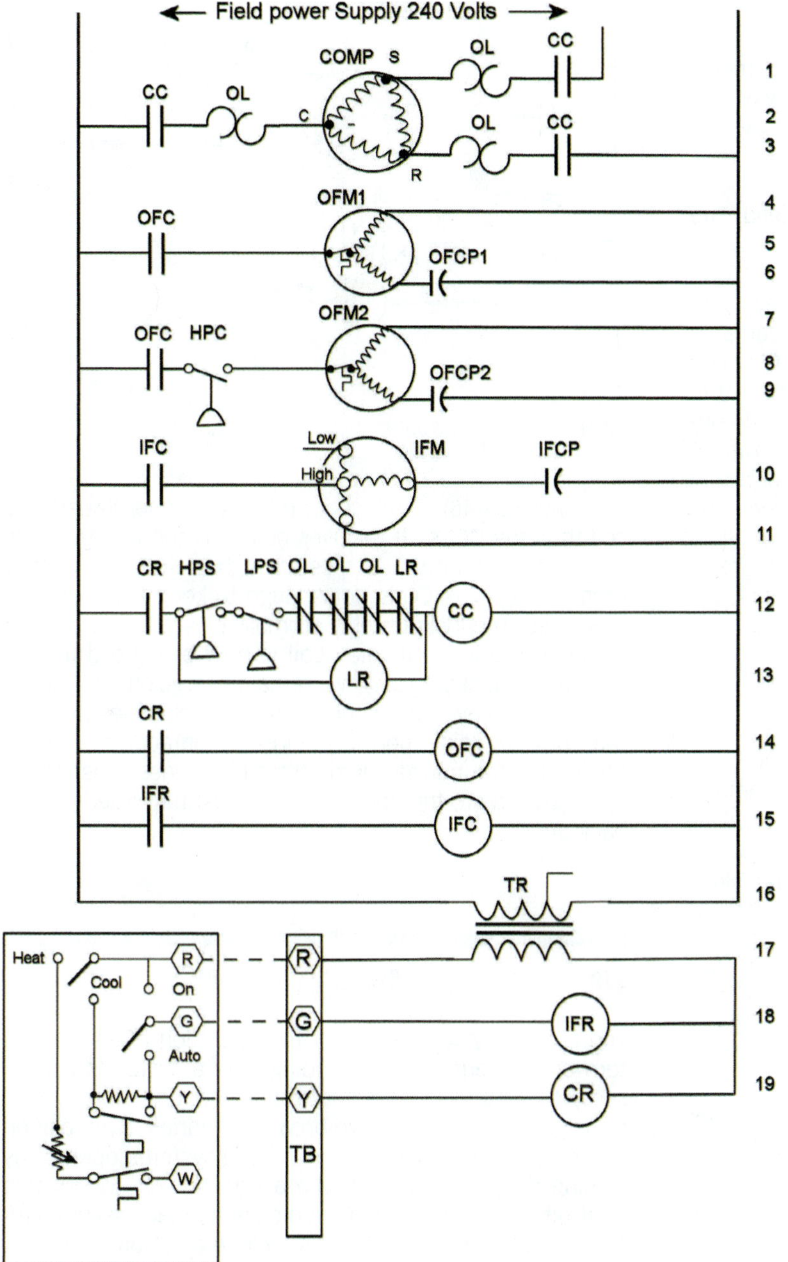

Letters	Component
CC	Contactor
COMP	Compressor
CR	Cooling relay
HPC	High-pressure control
HPS	High-pressure switch
IFC	Indoor fan contactor
IFCP	Indoor fan capacitor
IFM	Indoor fan motor
IFR	Indoor fan relay
LPS	Low-pressure switch
LR	Lockout relay
OFC	Outdoor fan contactor
OFCP1	Outdoor fan capacitor 1
OFCP2	Outdoor fan capacitor 2
OFM1	Outdoor fan motor 1
OFM2	Outdoor fan motor 2
OL	Thermal overload relay
TB	Terminal board
TR	Transformer

Figure 39-46 Diagram for light commercial air conditioner.

Low Ambient Operation

Residential air-conditioning systems seldom need to operate when it is cold outside. Commercial systems, on the other hand, frequently operate in cold weather. Commercial buildings often need cooling even in cold weather because of internal heat loads from lights, machines, and people. Operating air-cooled condensers with air temperatures significantly below design temperature can cause lower than normal high-side pressures, reducing the pressure difference across the refrigerant metering device. Most metering devices need some minimum pressure drop to function properly and will underfeed if the high-side pressure is too low. Commercial systems often use multiple condenser fans with one running whenever the system operates and another running just when needed. In Figure 39-46, the second condenser fan is controlled by a close-on-rise pressure switch on line 8. When the condenser pressure increases enough to require a second condenser fan, the pressure switch closes and energizes the second condenser fan.

Safety Controls

Another big difference is the number of safety controls. Most commercial air-conditioning systems have a full complement of safety controls. The compressors are typically three-phase motors that cannot be protected by inline thermal overloads the way single-phase motors are often protected. Breaking a single line of power to a three-phase motor will cause more harm than good because each leg of power is only in two of the three circuits. One circuit will be left operating, trying to carry the load by itself. This condition is caused single-phasing and quickly leads to motor burnout because the amp draw in the remaining windings doubles, quickly overheating the motor. Overload relays are used instead of inline overloads. An overload relay has a coil that senses the overload and a set of normally closed contacts that open when the coil senses an overload. The coil is in series with the motor and the contacts are in series with the coil for the motor control. In Figure 39-46, the compressor is protected by an overload relay in all three legs of power in lines 1–3. The normally closed overload contacts are in series with the compressor contactor coil (CC) in line 12. When an overload is sensed in any of the lines to the compressor, the overload opens the normally closed contacts in line 12, breaking the circuit to the CC. When the CC is de-energized, it opens its contacts in lines 1–3, breaking all the power to the compressor. Many commercial systems use lockout relays to keep the system from starting again after a safety has tripped. Figure 39-46 has a lockout relay in lines 12 and 13.

39.22 LOCKOUT RELAY

Many safety devices automatically reset after the condition that caused them to open has cleared. A system with an auto-resetting high-pressure switch is a good example. When the pressure switch shuts off the control circuit to the compressor, the pressure in the system will usually fall back down to a pressure that will allow the pressure switch to close again. The system will restart, and the high-pressure switch will trip again. The system will cycle on and off on the safety control, which can be almost as bad for the system as the original issue. A lockout relay prevents the system from restarting after a safety has opened until the control circuit is reset, preventing the system from cycling on and off due to a safety.

Lines 12 and 13 of Figure 39-46 show a lockout relay circuit. Notice that lockout relay coil will not be energized in normal operation because the string of safety controls effectively short the coil out of the circuit. There is no potential difference from one side of the lockout relay coil to the other (Figure 39-47). If any of the safety controls open, a circuit is established with the lockout relay coil in series with the

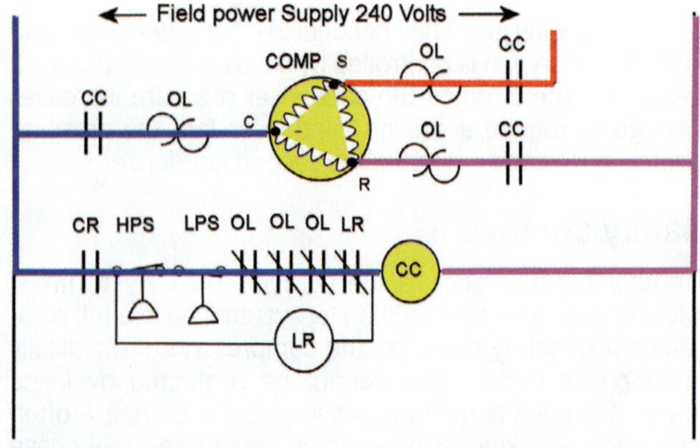

Figure 39-47 Lockout relay with unit operating.

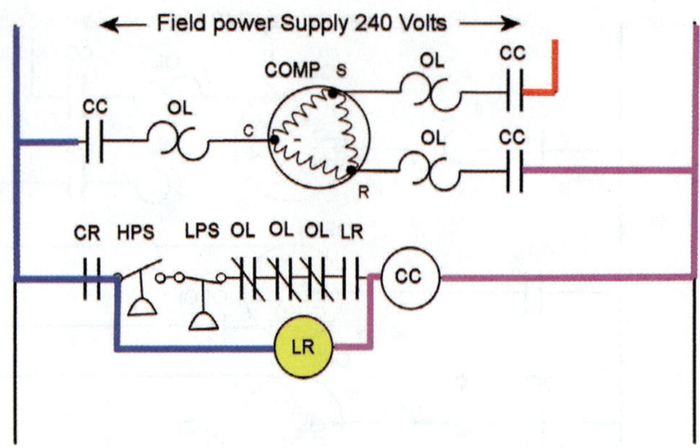

Figure 39-48 Lockout relay safety tripped.

CC (Figure 39-48). The lockout relay is a higher impedance coil than the CC, so it receives nearly all the voltage. Even though the circuit still passes through the CC, the voltage drop through the CC is not enough to keep the contactor energized. The lockout relay normally closed contacts open because the lockout relay coil is now energized (Figure 39-48). The lockout relay will remain energized and the CC de-energized even after the safety control closes because the lockout relay normally closed contacts are open, preventing the normal circuit from being reestablished. To reset the circuit, the control circuit must be turned off and back on.

UNIT 39—SUMMARY

The control system regulates the operation of all the system components. Understanding how a system's control system operates is crucial to troubleshooting. Control systems generally fall into two broad categories: two position and modulating. Two-position control systems operate by turning things off and on, like a light switch. Modulating controls provide levels between on and off, like a light dimmer. Types of control systems include electromechanical, pneumatic, electronic, and direct digital. Electromechanical systems use mechanical switches and relays. Pneumatic control systems are modulating systems that operate by varying the air pressure to dampers and valves. Electronic control systems are modulating systems that use varying DC voltage or current. Direct digital control systems use integrated microprocessors to intelligently control system operation. They normally communicate with each other via a data network that is similar to a computer network.

Control systems can be described as either open loop or closed loop. Open-loop systems do not affect the condition they are sensing. A control system that regulates system operation based on outdoor temperature is an example of an open-loop system. Closed-loop systems receive feedback to maintain setpoint. A furnace with an indoor thermostat that controls furnace operation based on the room temperature is an example of a closed-loop system.

System cycling describes the on-and-off operation of the equipment that occurs as the control system operates the equipment to maintain the setpoint temperature. The difference between the minimum and maximum room temperature is called swing. Two-position controls with mechanical thermostats can have a swing as high as 5°F. They reduce the swing using heating and cooling anticipators. The digital thermostats that are used today in place of mechanical thermostats have far less swing, usually less than a degree. They use more accurate temperature sensing and logic to reduce temperature swing. Proportional systems typically have less swing than two-position systems. Swing is caused by system overshoot and lag. Overshoot is delivering heating or cooling past the setpoint. Lag is a delay between the call for heating or cooling and the delivery of conditioned air.

WORK ORDERS

Service Ticket 3901

Customer Complaint: Unit Running but Not Cooling

Equipment Type: R-22 Commercial Packaged Air Conditioner

Diagram: Figure 39-46

A technician is called to look at a system that is not cooling properly. When he arrives the compressor is off, but both condenser fans are running. The compressor does not try to come back on. The technician looks at the diagram and notices that the compressor is protected by three overloads, a high-pressure switch, and a low-pressure switch in series with the contactor coil. He also notices a lockout relay keeps the system from starting again after a safety trips. The technician checks the voltage drop across each safety, reading 0 V across the high-pressure switch, the low-pressure switch, and all three overloads. He reads 230 V across the LR normally closed contacts, indicating that the lockout relay is keeping the system off. He installs gauges to help determine which safety is tripping and notices that the standing system pressure appears normal. He removes the red wire from the R terminal to reset the lockout relay, places a clamp-on meter around one of the three wires to the compressor, and touches the red wire to the terminal block to restart the system. The compressor hums but does not start, and the amp reading is very high. The tech disconnects power to the unit, removes the wires from the compressor, and ohms the compressor motor. One of the three terminals reads open to the other two, indicating a bad compressor motor.

Service Ticket 3902

Customer Complaint: Furnace Fan Operates Constantly

Equipment Type: Standing Pilot Gas Furnace

Diagram: Figure 39-20

A technician is called to look at a standing pilot furnace whose fan is running constantly. The diagram shows that the fan is controlled by both a fan relay and a fan switch. The technician checks the current of the high-speed fan wire and finds 0 A. The amp draw of the low-speed fan wire reads 3.5 A, indicating that the fan is operating on low speed. The technician checks the settings on the fan switch, turns off the power to the furnace, and checks the continuity of the fan switch. It reads 0 Ω, indicating that it is closed even though the furnace is not hot. The fan switch is replaced and the fan quits operating all the time.

Service Ticket 3903

Customer Complaint: Poor Cooling

Equipment Type: Air-Source Heat Pump

Diagram: Figure 39-34

A technician is called to look at a heat pump that is not cooling well and notices that the outdoor fan is not operating. Consulting the diagram, the technician sees that the outdoor fan motor is controlled by the normally open C contacts and the normally closed DR contacts. A voltage measurement across the C contacts reads 0 V, while a voltage reading across the DR contacts reads 230 V. This reveals that the DR normally closed contacts are open. The technician suspects a faulty defrost board but does further checks to make sure. The diagram indicates that the defrost board is only energized in heating. He checks the voltage at the defrost control between B and common and reads 0 V. Finally he removes the defrost thermostat wires from the board and ohms out the defrost thermostat. It ohms OL, showing that it is open. The technician concludes that the defrost board is indeed bad and replaces it. After replacing the defrost board, the outdoor fan operates normally.

Service Ticket 3904

Customer Complaint: Unit Blows Cold Air in Heating

Equipment Type: Air-Source Heat Pump

Diagram: Figure 39-34

A service technician is called to look at an air-source heat pump that is blowing very cold air during the heating season. He checks the thermostat to ensure that the thermostat is calling for heat and is set correctly. At the outdoor unit, the technician notices that the gas line is cold, indicating the unit is operating in cooling. He checks the diagram and sees the reversing valve is energized in cooling by the O terminal of the thermostat. A voltage measurement at the reversing valve solenoid reads 24 V, indicating that the reversing valve is being energized even though the thermostat is set to heat. Examining the diagram further, the technician also notices that the defrost board can also energize the reversing-valve solenoid. He removes the orange wire from the O terminal and the reversing valve shifts, indicating that the problem is in the thermostat or control wiring. Back inside, the technician removes the thermostat cover to check the control wiring. Everything appears to be wired correctly, but he notices an O B selector switch that determines when the O terminal receives power. It is set to B. He moves the switch to O, replaces the thermostat cover, and turns the system back on. The system operates in heat, and warm air starts coming out of the registers.

Service Ticket 3905

Customer Complaint: Compressor Short-Cycles

Equipment Type: Walk-in Cooler with Pump-Down Circuit

Diagram: Figure 39-16

A technician is called to look at a walk-in cooler whose compressor runs for a minute, remains off for about a minute, and them comes back on. The box temperature is close to the thermostat setpoint, but the customer is concerned about the compressor constantly cycling on and off. The customer confides that he adjusted the differential on the low-pressure switch from 20 psi to 5 psi, thinking it would make the cooler operate at a lower temperature. The technician explains that the differential setting is now so low that the slight pressure rise that occurs naturally after the compressor shuts off is closing the switch before it should. The switch is set back to the correct setting and the compressor stops short-cycling.

UNIT 39—REVIEW QUESTIONS

1. What is the difference between an open-loop and a closed-loop control systems?
2. What is the difference between a two-position control and a modulating control?
3. List the types of control systems.
4. How does a pneumatic control system operate?
5. What is the difference between an electronic control system and a direct digital control system?
6. Why are electromechanical control systems often referred to as relay control systems?
7. Why is there a potential of 24 V between the R terminal and the other terminals on a thermostat sub-base?
8. In Figure 39-34, what is the difference between emergency heat operation and normal second-stage heat operation?
9. In Figure 39-34, what happens when the defrost relay (DR) is energized?
10. In Figure 39-46, list the sequence of events that occur when the cooling relay is energized.
11. In Figure 39-46, what determines whether outdoor fan motor 2 (OFM2) operates?
12. In Figure 39-45, what does the LPS control?
13. In Figure 39-34, when is the reversing valve energized: the heating cycle or the cooling cycle?
14. In Figure 39-46, how does the high-pressure switch (HPS) protect the compressor?
15. In Figure 39-24, what will happen if the limit switch opens while the furnace is operating?
16. In Figure 39-34, what does the field control wire on the thermostat B terminal connect to on the unit?
17. In Figure 39-46, how does the thermostat control the operation of the compressor?
18. In Figure 39-33, what voltage operates the potential start relay coil?
19. In Figure 39-33, what is the purpose of the potential start relay?
20. List the heating operational sequence for the unit in Figure 39-25.
21. In Figure 39-24, what controls the fan operation in heating?
22. List the operational sequence for Figure 39-24.
23. List the operational sequence for the second stage of Figure 39-34. Just list what happens when the W2 terminal is energized.

UNIT 40

Communicating Control Systems

OBJECTIVES

After completing this unit, you will be able to:

1. describe the basic operating concept of a communicating control system.
2. explain the difference between relay controls and communicating controls.
3. discuss the benefits of communicating control systems.
4. list the basic components of a communicating control system.
5. list common communicating system protocols
6. explain how to install a communicating control system.
7. discuss service procedures for communicating controls.

40.1 INTRODUCTION

Regardless of the equipment brand your company sells and services, every technician today needs to understand communicating controls. Just a few years ago, there were not many choices if you wanted a residential communicating system. Today, most major equipment manufacturers offer residential communicating control systems. Communicating systems used to be proprietary, but now there is an open standard to encourage the development of interoperable communicating equipment. The increasing popularity of communicating systems lies in their ability to improve overall system efficiency and improve comfort. The system components can communicate with each other and adjust system operation to match the heat load and operating conditions.

40.2 COMMUNICATING SYSTEMS VERSUS RELAY SYSTEMS

Standard air-conditioning control systems use relay logic or an electronic representation of relay logic. Things are either off or on. These controls work something like a light switch: when the switch is on, the light operates; when the switch is off, the light is off. Thermostats are basically switches that are controlled by temperature. The thermostat closes a set of contacts to complete a circuit to a relay coil, and the relay coil then closes the relay contacts to complete a circuit to a motor. When the thermostat is satisfied, its contacts open, breaking the circuit to the relay coil. The relay opens its contacts, breaking the circuit to the motor. Everything works based on the presence or absence of control voltage. Relay logic control systems have been the basis for HVACR controls for many years. However, this control system needs a separate control wire for each function. Some split-system heat pumps require ten control wires running between the indoor and outdoor section (Figure 40-1). Even with ten wires, the range of control is still somewhat limited. Relay systems cannot use operating data and logic to make control decisions, but communicating systems can.

A communicating control system is more like a computer network. The main controller can be compared to the computer in a home network. The other communicating components are the peripherals connected to the computer. In a communicating control system, the system components communicate over a serial network. Each part has its own unique electronic signature or address, allowing the controller to recognize all the parts and coordinate their operation. Most residential communicating systems use four wires between all components—two for power and two for communication. It does not matter if the component is a furnace, air handler, air conditioner, heat pump, or zone control—everything uses the same four-wire connection (Figure 40-2). This works because the units respond to commands sent over the communication network, not the presence or absence of control voltage. Commands for different components can be sent over the same network, and communication is not just one

Figure 40-1 Typical control wire terminal board for a two-stage heat pump system using 24-volt controls.

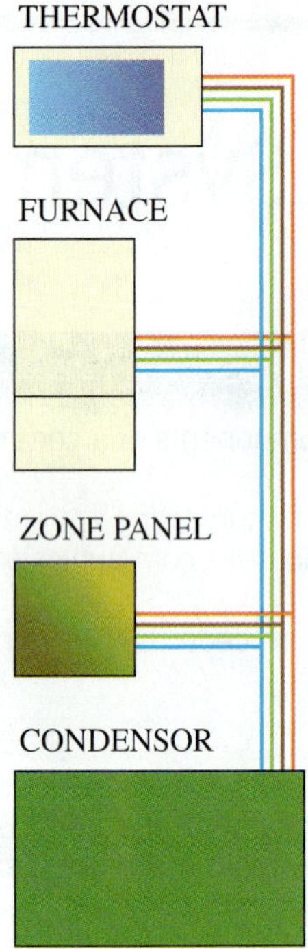

Figure 40-2 Four-wire communicating control system.

way. Communication between components allows each component to know what the other components are doing and adjust accordingly. For example, a communicating zone control knows the total blower airflow in cubic feet per minute (CFM) and the CFM in each zone. Better yet, it can report this to the service technician (Figure 40-3). The system airflow can be ramped up or down to match system capacity. Staged furnaces, air conditioners, and heat pumps give the system the ability to modulate system capacity and airflow as the house load requires, improving efficiency and reducing energy use.

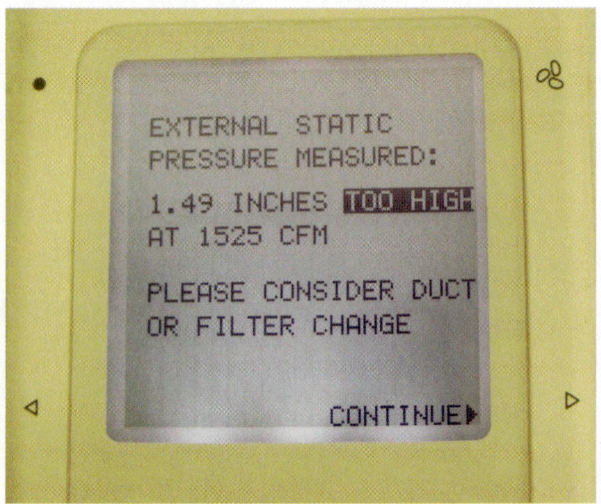

Figure 40-3 Communicating zone controls know how much air the fan is moving and the static pressure it is working against.

40.3 COMMUNICATING APPLICATIONS

Communicating controls are used in a wide range of applications, including ductless mini-split systems, multistage air conditioners and heat pumps, multistage furnaces, multistage split-system air conditioners and heat pumps, water-source heat pumps, and zone control systems. Any application that uses multistage or modulating technology can benefit from communicating controls. Communicating controls can send commands and information over the same network, allowing complex control with just a few wires. Communication makes monitoring and controlling variable-capacity systems much easier. The components send information about their status and operating conditions to the controller, giving the controller real-time information on system operation. Manufacturers are offering more variable-capacity systems because of their increased efficiency. Most of these systems use communicating controls.

Zoning is a strong application for communicating controls. Mini-split systems with multiple indoor coils use communicating controls to match system capacity to the building load using variable refrigerant velocity. These systems have multiple indoor coils connected to a single outdoor condensing unit that has a variable speed compressor. The communicating control system allows the unit to match its capacity to the combined required capacity of the multiple indoor units (Figure 40-4). Communicating zone controls for ducted systems are also available (Figure 40-5). Communication makes it possible for the variable-capacity equipment to match its output to the demand from the different zones. Airflow is also regulated based on the number and size of the operating zones.

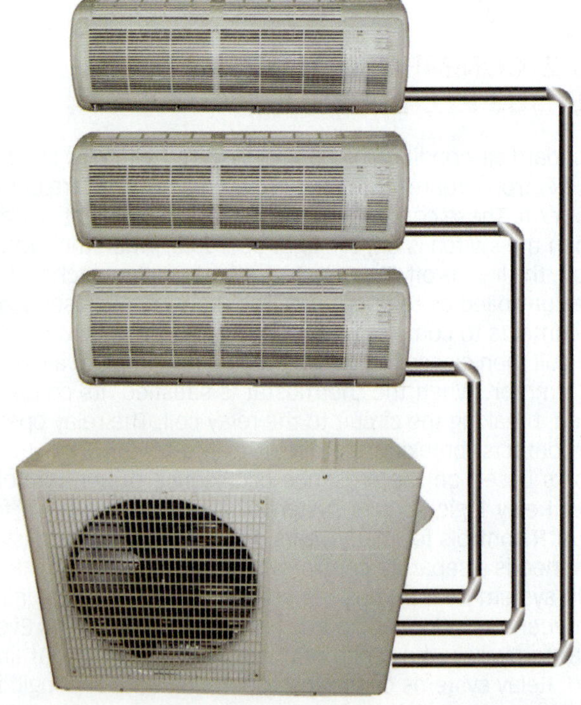

Figure 40-4 Mini-split VRV multiple-blower system.

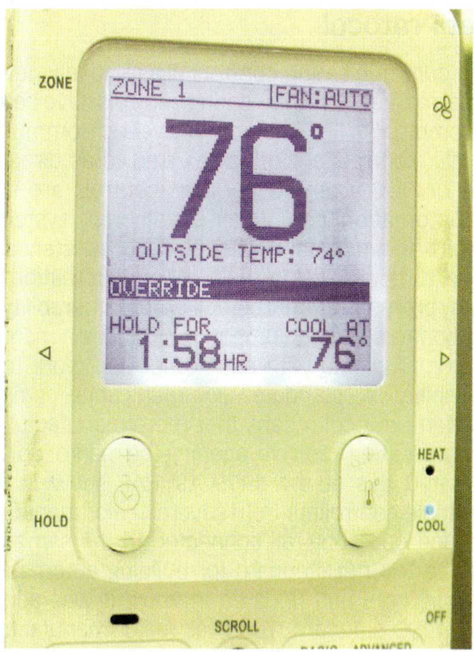

Figure 40-5 Communicating zone control.

GREEN TIP

Communicating variable-refrigerant-velocity mini-split systems can save energy by matching the equipment capacity to the demand. They can also shift capacity from one side of a building to another as the load dictates. This allows for less overall installed capacity and lower energy use without sacrificing comfort.

40.4 HOW COMMUNICATING CONTROLS WORK

Communicating controls send messages over a serial network that is very similar to a computer network. Each communicating component contains a small microprocessor, an electronic address that tells the other components in the network what it is, and the electronics necessary for it to respond to the commands it receives. Communication is accomplished through a series of voltage pulses that represent digital commands and data (Figure 40-6). These pulses are carried over two wires that form a communication bus. The communication bus connects all the communicating components in parallel. Commands sent over the bus are received by all the components. The components simply ignore commands and data that are not addressed to them. This is why communicating systems can accomplish more with four wires than traditional relay logic systems can with twelve wires. With traditional relay logic, each wire can only perform one function. The wires that form the communication bus can deliver commands to all the system's communicating components over the same two-wire communicating bus. Hundreds of commands can be sent over the same two wires.

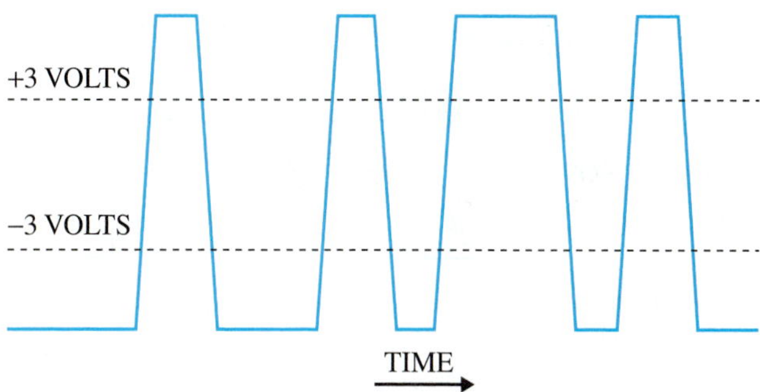

Figure 40-6 Diagram of pulses used to communicate over a communicating network.

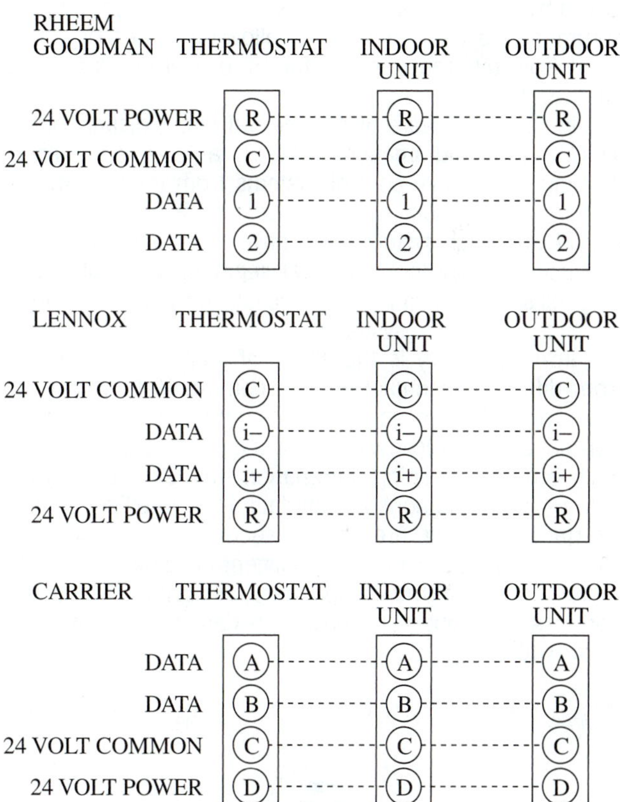

Figure 40-7 Four-wire communicating connections for several manufacturers.

Components are wired to each other using two to four wires, with four wires being the most common. A four-wire connection uses two wires dedicated to control power and two wires strictly for communication (Figure 40-7). Two- and three-wire systems either are powered through the component's primary power source or perform double duty as both a power source and a communication bus (Figure 40-8).

40.5 COMMUNICATING PROTOCOLS

All communicating systems follow some type of protocol. A protocol is an agreed standard set of rules that are used for communication on a network. Proprietary systems follow

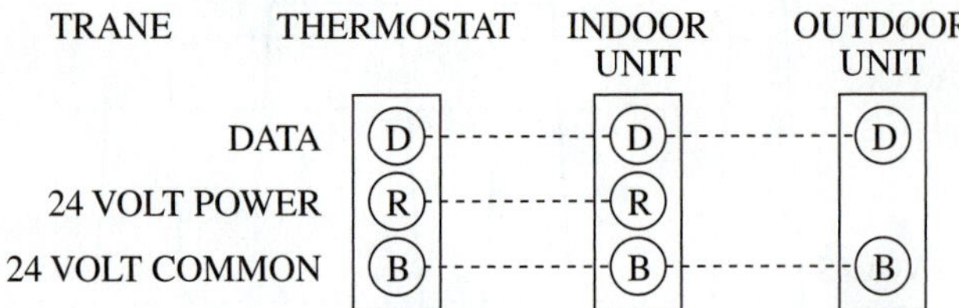

Figure 40-8 Communicating system using a two-wire connection to outdoor unit.

their own unique protocol, which is licensed and controlled by the manufacturer. An open protocol is available to all manufacturers. Open protocols establish rules that are followed by multiple manufacturers so that their systems can communicate with each other, allowing pieces of equipment from different manufacturers to operate as a single intelligent system.

In the early years of commercial direct digital control (DDC) building-automation systems, a large commercial building could have control systems from multiple manufacturers that could not communicate with each other. Individually, each system might have represented the most advanced control for its market segment, but collectively they fell short of their promise because they could not operate as a coordinated system.

Interoperability is the ability of pieces of equipment from different manufacturers to communicate with each other and operate as a single communicating system. Components that follow a standard protocol can work with each other. In building automation, this extends beyond different air-conditioning manufacturers. Intelligent utility interfaces, lighting, fire alarms, security, and HVACR can all tie together, creating an intelligent building automation system where all the components can communicate and optimize their operation based on data from each other (Figure 40-9).

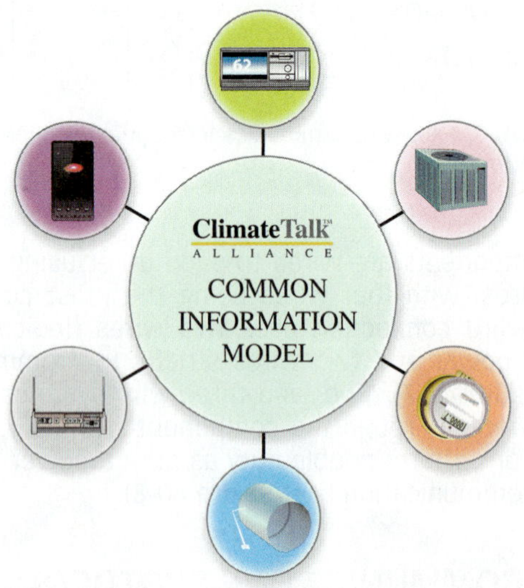

Figure 40-9 ClimateTalk communicating control systems can integrate several types of products with a single system. *(ClimateTalk™ Information Model word mark and artwork are used with the permission of the ClimateTalk Alliance.)*

BACnet Protocol

Although communicating controls are relatively new in the residential market, communicating controls have been common in commercial HVACR for many years. Communicating commercial controls are often referred to as direct digital controls, or DDC. Commercial DDC controls are used for building automation, not just air-conditioning systems.

A building-automation control network standard was developed to help develop the building-automation control industry by promoting interoperability. Interoperability means that components and equipment from different manufacturers can communicate with each other and work together. The availability of "standard" communicating components and an open protocol means that more manufacturers can use communicating controls, opening the market up. BACnet demonstrated this at the 1996 ASHRAE show in Atlanta, when multiple controllers from several major manufacturers were up and operating, all connected to the same BACnet system. The BACnet standard for building automation was published in 1995 and updated in 2001. It was adopted in 2003 as an ISO standard. Today all major manufacturers of large commercial air-conditioning equipment offer BACnet-capable systems.

Climate Talk Protocol

BACnet is primarily a commercial standard. ClimateTalk is an open standard designed to encourage the development of residential communicating equipment. ClimateTalk is a much younger protocol, introduced in 2009. The ClimateTalk Alliance is an organization of companies who are committed to developing a common communication infrastructure for HVAC and smart grid devices, enabling the interoperability of diverse systems. Available ClimateTalk controls include thermostats, furnace control boards, air-conditioning and heat pump control boards, and variable-speed brushless DC motors. Several manufacturers are offering ClimateTalk-capable equipment, including furnaces, air conditioners, and heat pumps.

40.6 COMMUNICATING CONTROL SYSTEM COMPONENTS

The basic communicating control system includes a main controller, a board to control the furnace or heat pump air handler, and a board to control the condensing unit. Many electronically controlled brushless DC variable-speed motors can also send and respond to commands through the communicating control network. System accessories such as zone controls, humidifiers, and air cleaners can connect with the communicating control system for a complete comfort solution that responds intelligently to operating conditions.

TECH TIP

Some manufacturers refer to the component that replaces the thermostat as a communicating thermostat, while others refer to it as a communicating control.

Figure 40-10 Communicating touchscreen thermostat.

Main System Control or Thermostat

The main controller looks like a digital thermostat (Figure 40-10). Like a thermostat, the main controller is responsible for coordinating overall system operation. However, the controller is far more sophisticated. Typically, a single controller model is used for all types of systems, including furnaces and air conditioners, heat pumps, or dual-fuel heat pumps. The controller looks to see what is connected to the network and modifies its operation to fit the system components connected to the network. It not only controls when equipment operates, but it also manages the communication network, controls system capacity to match the load, records system fault history, and provides service data and information to the technician. While thermostats are basically switches, communicating controllers are intelligent and make decisions based on their programming and the information they receive from the rest of the system.

Furnace Control Board

Communicating furnaces are typically multistage or variable capacity. Overall system efficiency is increased by matching the furnace heating capacity to the heat load. Their ability to produce different levels of heat reduces the amount of fuel burned and extends the furnace run time. The furnace board communicates the current furnace status and operating data to the system controller. The board on the furnace is responsible for receiving commands from the controller and coordinating operation of the furnace to produce the level of heat called for by the controller (Figure 40-11). Different firing rates are accomplished by controlling the speed of the combustion air blower and the different stages of the gas valve. Communicating furnaces usually have variable-speed blower motors. The furnace control adjusts the blower speed to match the heat output of the furnace. This ensures that the delivered air stays warm even at reduced firing rates.

Air-Handler Control

Communicating heat pump air handlers and blower coils use variable-speed indoor blowers. The air-handler communicating board controls the indoor blower speed to match the system capacity (Figure 40-12). When the system controller operates the system at reduced capacity, the variable-speed indoor blower also runs slower. This ensures cool air in the cooling cycle and warm air in the heating cycle even at reduced capacity. Many communicating controllers can also control indoor humidity. They can reduce the indoor blower speed during the cooling cycle to increase latent cooling for dehumidification.

Figure 40-11 Communicating furnace board.

Figure 40-12 Communicating blower board.

Figure 40-13 Communicating condensing unit board.

Condensing Unit Controls

Communicating air conditioners and heat pumps typically have two-stage or variable-capacity control. The ability to operate at reduced capacity allows them to operate using less electricity and operate with longer running times. This increases both efficiency and comfort. The condensing unit board controls compressor operation, including staging (Figure 40-13). On heat pump condensing units, the communicating board also controls the defrost cycle. The condensing unit control communicates the condensing unit operating data to the main controller inside, including which stage the compressor is operating in, the outdoor temperature, and information for the defrost cycle on heat pump boards.

Accessories

System accessories can also tie into the communicating network. Zoning boards, humidifiers, and air cleaners are all available for integration in communicating control system (Figure 40-14). Information from the system indoor blower helps all these components perform their jobs better. Communicating controls allow accessories to tailor their operation to the system operation.

40.7 EXTERNAL COMMUNICATION

Communicating systems can also communicate with devices and networks outside of the HVACR network. Most communicating control systems have the ability to connect to the Internet, allowing remote system monitoring and control. Homeowners can "check in" on their HVACR systems while they are away and even change system set points. Internet communication can be done through a traditional Web-connected computer, a wireless laptop, or even a smart phone. Usually, a separate component is added to the system to handle communication with the Internet. However, there are thermostats that are designed to be connected directly to a computer network via a Cat5 Ethernet connection or even a wireless router (Figure 40-15).

40.8 WIRING A COMMUNICATING SYSTEM

Reading the manufacturer's literature and instructions is always important when installing HVACR equipment. With communicating systems, it is an absolute necessity. Even though there are just a few wires, it is still possible to connect them incorrectly. The time spent studying the system and its specific installation instructions before installation will save time after installation by reducing problems.

In addition to the connections for the communicating controls, most communicating furnaces and condensing units have connections for standard 24-volt thermostat control as well as communicating control connections (Figure 40-16). These legacy connections allow the units to be used with standard 24-volt thermostats. However, the system cannot deliver all the benefits of communicating

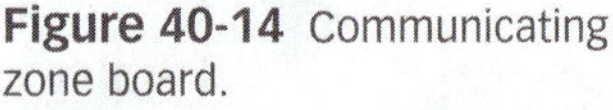

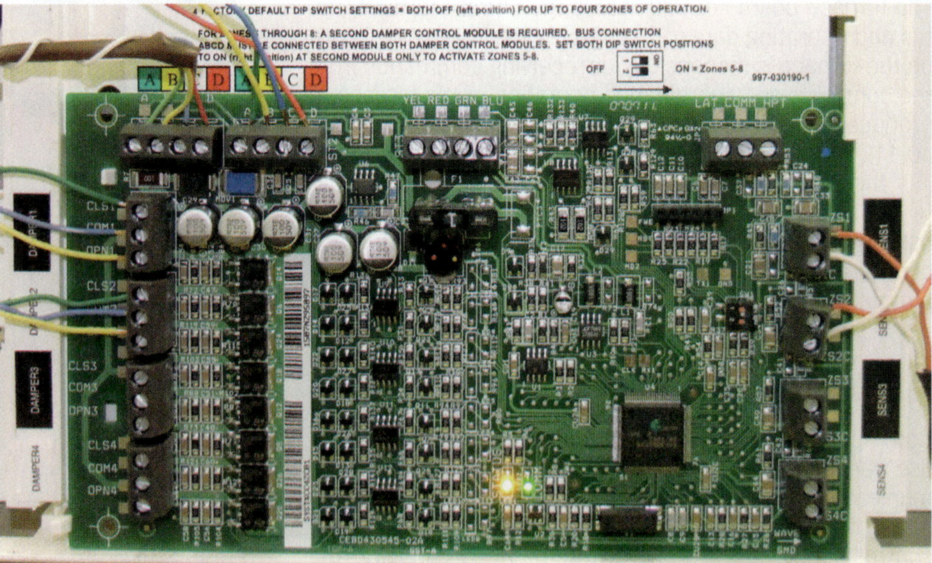

Figure 40-14 Communicating zone board.

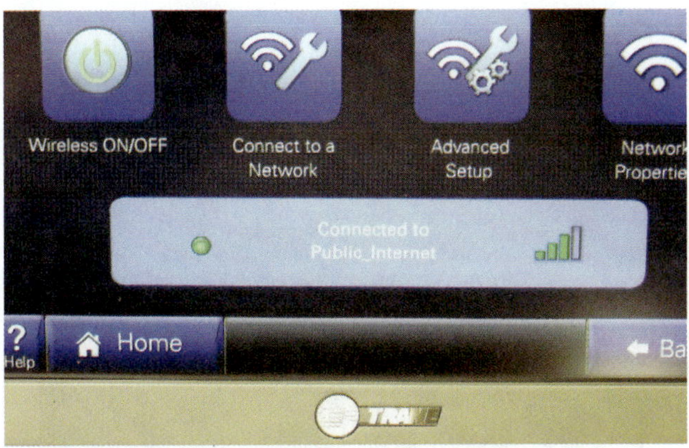

Figure 40-15 This Wi-Fi thermostat can communicate via 802.11b/g wireless.

Figure 40-16 This communicating control board has both legacy and communicating connections.

control when connected using the standard 24-volt controls. It is important for the installer to understand that either the legacy 24-volt connections or the communicating connections should be wired, but not both.

SERVICE TIP

Do not connect both the terminals for the legacy controls and the terminals for the communicating controls. If you are wiring a communicating system, you do not need the standard terminal connections on the control.

Although good connections are crucial for any control system, good electrical connections are even more important with communicating systems. Connections that would work with common relay controls will intermittently fail with communicating systems. Some specific pitfalls to avoid are stranded wires, incorrectly stripped wires, wires improperly placed in the connector, and multiple wires in the same connector (Figure 40-17).

Four-Wire Systems

The most common communicating control systems use four wires to connect the system components. Two wires carry the control voltage and two more wires provide the communication bus. Standard 18-gauge solid thermostat wire is used. Each component uses the same four connections, making connections very easy. The four connections are labeled the same on each control board. Although there are only four wires, it is important not to get the connections confused. The polarity of the power wires needs to be observed. Normally, polarity is not an issue with two wire AC control systems, but it is important with communicating systems. Figure 40-7 shows the wiring for several types of four-wire systems.

Although each communicating board uses four wires, the control board at the furnace or heat pump air handler must connect to two sets of four-wire bundles: one set going to the main controller (the thermostat), and one set going outside to the condensing unit. Some early communicating boards on furnaces and air handlers only had one set of connecting blocks. Although it seems logical to put one wire from each bundle in the connection, this is wrong. Placing multiple wires in the same connector leads to erratic communication problems. To solve this problem, connect a short wire, called a pigtail, to the connector. Then connect the pigtail to one wire from each bundle using small orange wire nuts. This has a decidedly low-tech look for an advanced communicating control system, but it works. Just be careful to make good wire nut connections.

Two-Wire Systems

Four-wire systems still have two more wires than are commonly used for standard air-conditioning condensing units. Two-wire communicating systems make it possible to replace a standard air-conditioning system with a communicating system without pulling new control wires. Some systems use two wires between the furnace or air handler and the outdoor condensing unit, allowing the use of 18-2 thermostat control wire to the condensing unit. The outdoor unit has its own transformer, making it unnecessary to run control power from inside to outside. The two control

Figure 40-17 Typical connection problems from left to right: red wire missing middle of connector clamp; doubling wires in connector; leaving bare wire past the connector.

wires are common and data. It is important not to cross-connect the wires. Although there are only two wires, it is still possible to connect them incorrectly. These systems typically use three or four wires between the furnace or air handler and the comfort control. Figure 40-8 shows the wiring for different types of two-wire systems.

40.9 INITIAL SYSTEM SETUP

When the main controller is first powered up, it searches the network to identify all the components connected to it. Typically, the controller displays what it has found and gives the installer some setup options (Figure 40-18). Many controllers can remind the owner when it is time to change the air filter or perform seasonal service. Dealer contact information can also be programmed in so that it is displayed in service reminders. This type of information is typically entered by the installing technician as part of the system setup.

The controller remembers the equipment connected to the network. In the event that the controller loses power, it will not have to reconfigure when it is powered on. It will simply try to communicate with the equipment it expects to see. If system components are changed, the controller will need to relearn the system. It may give communication errors because it will still be looking for the component that was replaced (Figure 40-19). To solve this issue, go into the service screens. From there you can instruct the controller to rescan the system; it will perform the same network scan that it normally does on initial system power-up. This is similar to rebooting a computer. Remember that a communicating network is essentially a computer network. The controller (thermostat) is the main computer and the units are the peripherals. Unlike computers that automatically detect new components, communicating systems must be told to look for them. The system must be rescanned if new communicating components are added to an existing communicating system.

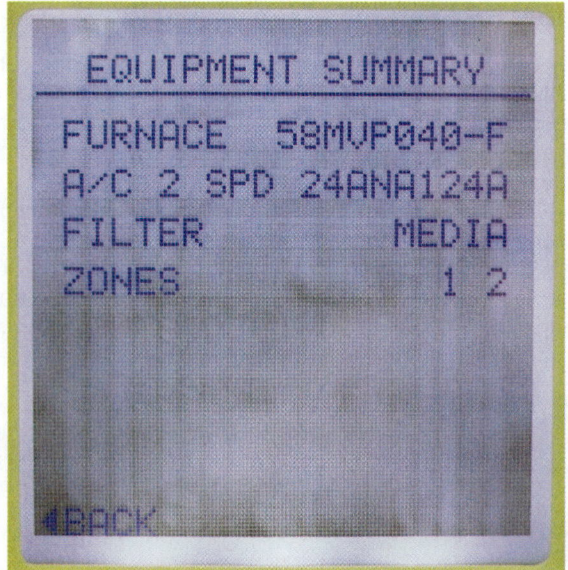

Figure 40-18 After scanning the system, the control displays the components it found on the network.

Figure 40-19 The screen indicates a communication error. The communication LED on the furnace board is out, indicating that the unit is not communicating.

40.10 INSTALLATION AND SERVICE SCREENS

The main controller (thermostat) has installation and service screens to help configure, install, and service the system. These screens typically require key presses or key sequences that are not in the owner's manual provided to the customers. Current system components and their status can be viewed on the main controller in the service screens. Technicians can view the system fault history from the service screens (Figure 40-20). Because the system components communicate with the controller, faults of all connected components can be viewed, not just controller faults. System settings and configuration can also

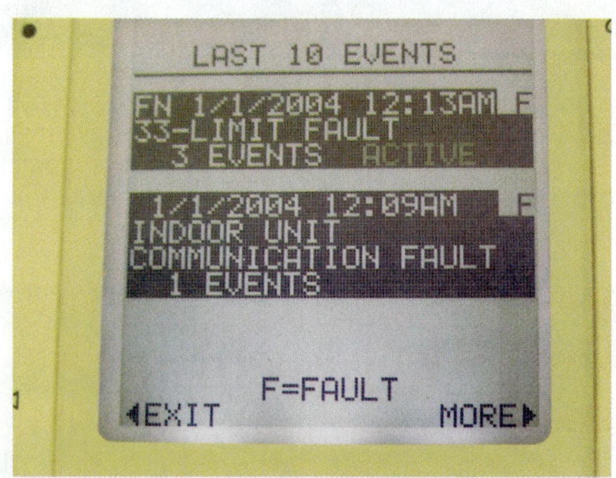

Figure 40-20 The unit's fault history can be displayed on the screen.

Figure 40-21 The top configuration screen is alphanumeric. The bottom one uses number codes.

be changed at the service screens. System configuration varies from an alphanumeric code to a short text message (Figure 40-21). You really need the manufacturer's literature to configure any of these systems, but the systems that use text prompts are definitely easier to set up.

SERVICE TIP

The key presses or touches required to access the service screens on the controller are not published in the user's manual. They are typically only published in the manufacturer's service literature.

40.11 ACCESSING THE INSTALLATION AND SERVICE SCREENS

A lot of valuable service information is available through the controller if you know how to access it. Each system has its own secret combination that unlocks this valuable information.

Amana and Goodman ComfortNet

Figure 40-22 shows a ComfortNet communicating thermostat used for controlling Amana and Goodman ComfortNet products. To view the advanced installer configuration menu, press the Menu button on the bottom right of the touchscreen. A new screen will appear with selections along the bottom. Press and hold the Installer Config button for three seconds. The screen will change. Press and hold the Installer Config button again for three seconds. The screen will change once more and the word *Advanced* will

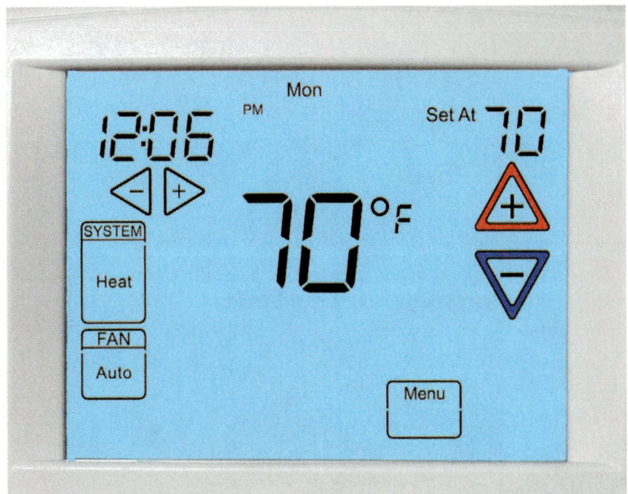

Figure 40-22 Touchscreen communicating thermostat.

appear in the bottom right corner. Figure 40-23 shows this sequence of settings and screens. Any faults will be displayed. You can scroll through the faults using the right and left arrow keys at the top left of the screen. Pressing either of the up or down arrow keys on the right will take you back to the user menu.

Carrier Infinity and Bryant Evolution

Figure 40-24 shows an Infinity communicating controller for Carrier and Bryant communicating products. Carrier's Infinity systems and Bryant's Evolution series are compatible. To view the service screens, press and hold the "advanced" button for ten seconds. The install/service menu will appear. Use the scroll button to scroll down to Service at the bottom of the list. Press the Select button on the right side of the control to display the service screen. Figure 40-25 shows this sequence of settings and screens. You can view the status of any installed equipment and view the system fault history from this screen.

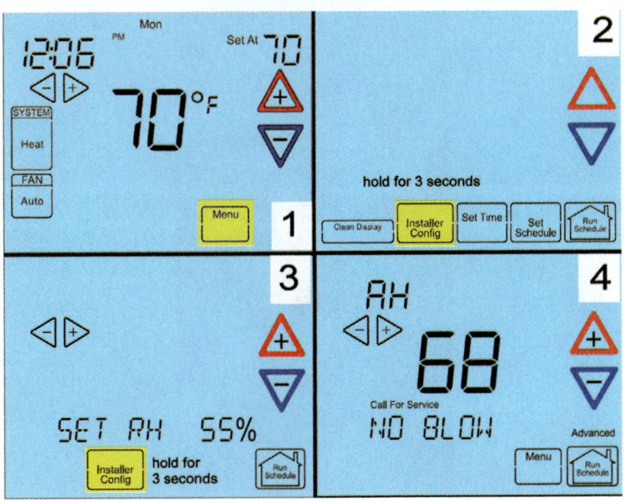

Figure 40-23 Steps to display the service screen on ComfortNet communicating thermostats.

Figure 40-24 Infinity control.

Lennox icomfort

Figure 40-26 shows an icomfort thermostat used to control Lennox icomfort communicating products. The screens are arranged similar to a tabbed browser. Critical alerts are displayed both on the home (user) screen and in the installer screen. Press the Alert tab to view current critical alerts. Minor and moderate alerts are found only in the Installer Alert tab. To access the Installer menu, press the Lennox logo in the bottom right corner of the screen for five seconds (Figure 40-27). A warning will appear: "The following screens are intended for use by qualified Lennox equipment installers only." Press Yes to continue. Press the Diagnostics tab to run diagnostic tests on connected equipment, or press the Alerts tab to view system fault history and current faults.

Figure 40-25 Steps to display the service screen on Infinity control.

Figure 40-26 Lennox icomfort communicating thermostat.

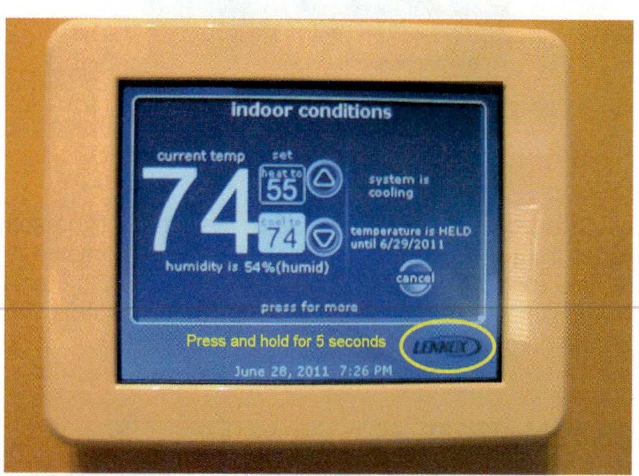

Figure 40-27 Push and hold the Lennox Logo for five seconds to enter the service screen.

Rheem and Ruud Comfort Control

Figure 40-28 shows a Comfort Control thermostat used with Rheem and Ruud Comfort Control communicating products. To view the advanced installer menu, press the left and right buttons simultaneously for three seconds (Figure 40-29). The up and down arrows are used to scroll through the options. Options include Communicating Devices, Heat Pump Disable, Fault Status, Air Handler Lockout, USB Upload, Dual Fuel Setpoint, Thermostat Summary, Heat Cycle Rate, De-Hum Setpoint, Cool Cycle Rate, and Humidity Setpoint. Go

Figure 40-28 Rheem Comfort Control communicating thermostat.

Figure 40-29 Push and hold the two arrow keys simultaneously to go to the service screen.

to Communicating Devices to see details of specific connected devices. Go to Fault Status to read the fault history.

Trane ComfortLink II and American Standard AccuLink

Figure 40-30 shows an XL900 communicating thermostat for Trane ComfortLink II communicating products. Trane ComfortLink II products are compatible with American

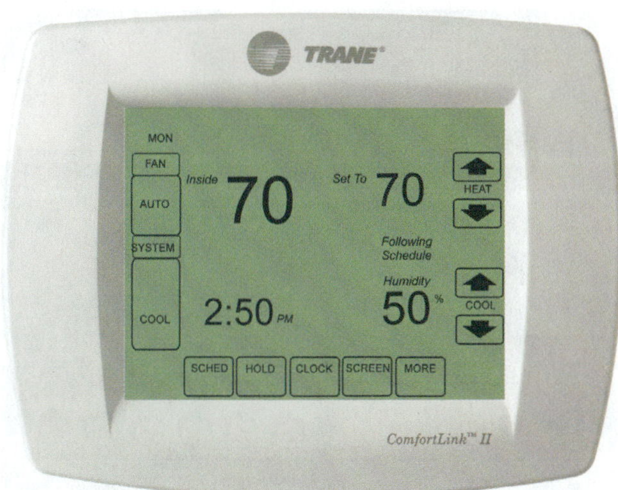

Figure 40-30 Trane XL 900 communicating thermostat.

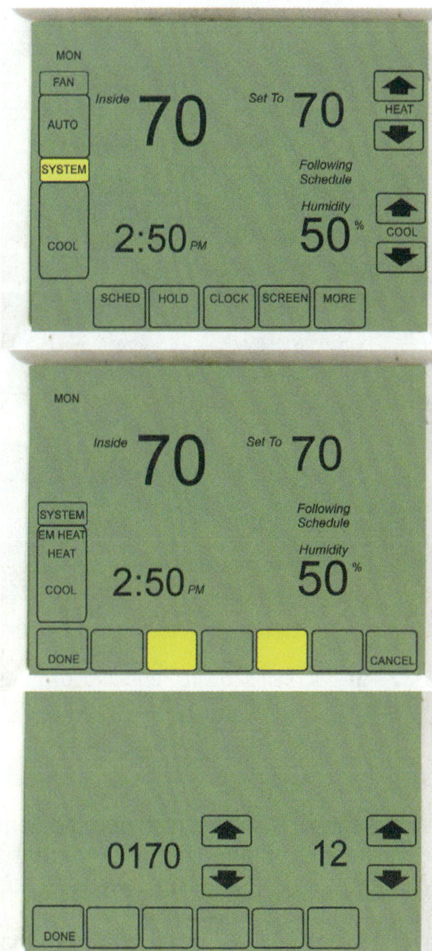

Figure 40-31 From top down: Touch the System button. The screen changes to the second screen. Press and hold the two blank buttons on either side of the middle button for five seconds. The screen changes to the bottom configuration screen.

Standard AccuLink communicating products. To view the Advanced Setup menu, press and release the System button on the middle left of the touchscreen. A row of blank buttons will appear at the bottom of the screen. Press and hold the two blank keys on either side of the center blank key for approximately five seconds until the screen changes, then press and hold the Installer Config button again for three seconds. The screen will change once more, and the word *Advanced* will appear in the bottom right corner. Figure 40-31 shows this sequence of settings and screens. Any faults will be displayed. You can scroll through the faults using the right and left arrow keys at the top left of the screen. Pressing either of the up or down arrow keys on the right will take you back to the user menu.

40.11 TROUBLESHOOTING THE SYSTEM

A big advantage of communicating systems is that they can save and display a fault history for all the connected components, so service technicians can see what has been happening with the system. The first step in troubleshooting a system with communicating controls should be to go

Figure 40-32 Fault screen showing repeated low-pressure-switch fault (lps).

to the service screen. There you can see the components connected to the system and check the current status of each component. Current faults are displayed so that you can see what is currently happening. It is also useful to check the fault history. For example, a system that has repeatedly tripped out on the low-pressure switch will show that on the service history (Figure 40-32). Even if the system is not currently tripped out on low pressure, you can see that it has tripped out on low pressure before. Seeing this, you can look for problems associated with low pressure. After addressing the issue, you can clear the fault history if you want to record only faults that have occurred since your most recent visit.

40.11 TROUBLESHOOTING COMMUNICATION ERRORS

The fault may not be in the operation of a particular component but in communication between components. This will normally show as a communication error on the service screen. The most common error message says, "Communication error with…" and then lists the equipment that it expects to see with which it cannot communicate (see Figure 40-19). This may be a problem with the control or the equipment, but it can also be something simple, such as a loss of power to the equipment or a loose connection.

Check Settings and Power

Like any system, the first step is to make sure the thermostat, or control, is set to call for the system to operate. Some customers are confused by the extra information and touchscreen controls often used with communicating systems and literally do not know how to turn the system on. Customers who are not comfortable with personal electronics may need a little extra time to learn the operation of their system.

SERVICE TIP

Not all customers are tech savvy and may be intimidated by touchscreens and electronics in general. Try to avoid "tech speak" when showing customers how to operate their system.

After establishing that the control is calling for the unit to operate but the system is not operating, you will need to troubleshoot the control system. Again, start with simple checks. Does the equipment have the correct line voltage? Just like relay systems, communicating systems will not operate if they don't have power.

Check Fault Lights and Displays

Communicating control boards have fault lights or displays that show the status of the system. Boards with fault lights often have a light dedicated to indicating communication status (Figure 40-33). This light will change color or flash to indicate that the board is not communicating. Some boards have displays that indicate the status of the system. In either case, checking these fault indicators is relatively easy to do and takes very little time. Checking them early in the troubleshooting process can save a lot of time.

Check Connections

Next, check the connections at the control and the unit. Poor connections are the most common cause of communication errors. Intermittent communication errors frequently can be traced back to the connections. Check these specific items:

- Make sure the correct wire is used—solid, 18-gauge copper.
- Make sure the wire is stripped to the correct length.

Figure 40-33 Communication light on board.

- Do not put multiple wires in the same connector.
- Be sure the wires are not reversed between the control and the unit

Occasionally wires break inside the plastic insulating jacket with the insulation remaining intact. Check near the end of the wire for awkward-looking sharp turns in the wire. You can check for wires with breaks using an ohm meter. Remove the wires from the connectors on both sides, twist two wires at a time together on one end, and ohm out the other end. A reading of a few tenths of an ohm indicates that the wires are OK. If either wire has a break, the reading will be infinite. If any of the wires have a break, you should pull a new set of control wires. This usually takes less time than trying to locate the break. Frequently if one wire is damaged, others are also.

40.12 WIRELESS CONTROLS

Not all communicating controls use wires. Wireless communicating controls use radio signals for interconnection and communication between components. A wireless thermostat can be easily installed in locations that a wired thermostat cannot be installed since they are not limited to locations where wires can be routed. This is especially helpful in renovation and retrofit applications. Currently, wireless systems still must have an interface to translate signals from the wireless system to the air-conditioning components because no air-conditioning systems are currently available with wireless controllers built in. The interface wires into a traditional 24-volt relay control system, allowing the wireless thermostat to control the system. Using wireless systems for zone control makes zone thermostat location easier. This is especially true for a retrofit zoning application. Wireless thermostats can connect to the Internet through a wireless router, similar to a laptop or game console.

UNIT 40—SUMMARY

Standard relay logic air-conditioning control systems work something like a light switch: when the switch is on, the light operates; when the switch is off, the light is off. Everything works based on the presence or absence of control voltage. In a communicating control system, the system components communicate over a serial network. Units respond to commands sent over the communication network, not the presence or absence of control voltage. Communication between components allows each component to know what the other components are doing and adjust to match the operating conditions. The language systems use for communication over a network is called a protocol. Two well-known communicating protocols are BACnet for commercial systems and ClimateTalk for residential systems. The increasing popularity of communicating systems lies in their ability to improve overall system efficiency and improve comfort by matching the system capacity to the heat load. The basic communicating control system includes a main controller, a board to control the furnace or heat pump air handler, and a board to control the condensing unit. Most residential communicating systems use four wires between all components—two for power and two for communication. It does not matter if the component is a furnace, air handler, air conditioner, heat pump, or zone control—everything uses the same four-wire connection. Communicating systems record the system fault history, allowing service technicians to view the problems that have occurred in each of the system components. Fault histories are viewed on service screens on the communicating controller or thermostat. Faults in communication as well as faults in individual components are reported on the service screens. The manufacturer's instructions are critical for understanding and servicing communicating systems. Most communication problems are the result of bad connections.

WORK ORDERS

Service Ticket 4001

Customer Complaint: Unit Not Cooling

Equipment: Four-Wire Communicating System with Furnace and Air Conditioner

A technician is called to look at a communicating system that is not cooling properly. He goes first to the thermostat to check the service screen. The system status screen shows that the condensing unit is off. The fault history shows several low-pressure events. The technician selects the OFF mode on the touchscreen and goes outside to check the system pressure. The equalized pressures look reasonable. He turns off power to the system, disconnects the low-pressure switch, and ohms it out. The reading is OL, indicating that the switch is open even though the equalized pressure is above the cut-out point. The technician determines that the low-pressure switch will have to be replaced.

Service Ticket 4002

Customer Complaint: Unit Not Communicating

Equipment: Four-Wire Communicating Heat Pump Split System

A technician is called to start a new communicating heat pump. On the initial system scan, the control finds the indoor blower but not the outdoor unit. The technician goes to the indoor blower to inspect the connections on the blower board. She finds that the wires to the thermostat and the outdoor unit have both been placed in the same connection because the board only has one set of communicating connections. She removes the wires from the connection block, installs a single set of wires on the terminals, and then connects the thermostat and outdoor unit connections to the pigtail using wire nuts.

Returning to the thermostat, the technician enters the service screens and instructs the system to perform a new scan. This time the controller finds both the indoor and outdoor units.

Service Ticket 4003

Customer Complaint: Communication Error

Equipment: Four-Wire Communicating Variable-Capacity Furnace

A technician is called to start a new communicating furnace. A communication error is reported during the initial system scan. The technician goes to the furnace and checks the communication LED. It is not on. He turns off the power to the furnace, inspects the connections, and records the wire colors used. The connections appear to be fine, so he returns to the thermostat and removes it from the sub-base. While checking the connections he notices that the two wires on the data lines have been swapped between the thermostat and the unit. He swaps the wires at the thermostat and replaces the thermostat on the sub-base. After turning the power back on, the furnace communicating LED lights up. The technician returns to the thermostat, navigates to the service screen, and instructs the thermostat to rescan the system. This time the thermostat recognizes the furnace.

Service Ticket 4004

Customer Complaint: Wireless Thermostat Has Stopped Working

Equipment: Communicating System with Wireless Thermostat and Wireless Receiver

A customer calls because his wireless thermostat has stopped working. He replaced the batteries and it still does not work. The technician goes to the receiver and notices the LED that indicates the presence of a signal from the thermostat is not lit. Next, she goes to the thermostat. The display appears to be working. Even though the customer said he replaced the batteries, the technician removes them and checks the battery voltage. It is 1.45 volts DC. She explains that even though the batteries are rated at 1.5 volts, a good battery will normally read more than 1.5 volts new and with no load. The 1.45 reading indicates weak batteries. She explains that wireless thermostats are very affected by weak batteries because their signal is weakened. The customer says that he just put these in, but they came from the surplus discount center. The technician replaces the batteries with new, high-quality, alkaline batteries. The thermostat now is able to turn on the air-conditioning system. The technician recommends that the customer only use high-quality, in-date alkaline batteries for his wireless thermostat.

UNIT 40—REVIEW QUESTIONS

1. What is a communicating control system?
2. Explain the difference between the way a relay logic control system operates and the way a communicating control system operates.
3. How can a two-stage heat pump be controlled by only four wires?
4. What is a communicating protocol?
5. List two open protocols.
6. What are the four wires used for in a common four-wire communicating system?
7. How do some communicating systems operate with only two wires between the indoor blower and the condensing unit?
8. Why do communicating control boards often have terminals for traditional 24-volt controls?
9. What type of wire should be used for connecting components of a communicating control system?
10. What is the cause of most communication errors?
11. How can a service technician learn what problems a communicating system has been experiencing?
12. Why is the manufacturer's installation and service literature critical when working with communicating systems?
13. How does a communicating thermostat know what type of equipment is connected to it?
14. What applications are wireless communicating thermostats particularly well suited for?
15. How do communicating components determine which commands are for them and which commands are for other devices connected to the network?
16. Why are communicating controls particularly well adapted for modulating capacity systems?
17. How can a technician force a system to scan the system and relearn the components connected to it?
18. Why are communicating controls particularly well adapted for zone control systems?
19. What type of zone control do some communicating ductless mini-split systems use?
20. Explain the concept of interoperability.
21. Discuss the difference between a proprietary system and an open system.

UNIT 41

Fundamentals of Psychrometrics and Airflow

OBJECTIVES

After completing this unit, you will be able to:

1. discuss the factors affecting human comfort.
2. list the ways the body dissipates heat.
3. explain the relationship between dry-bulb, wet-bulb, and dew-point temperatures.
4. plot points on the psychrometric chart given any two variables.
5. locate all unknown air properties on a psychrometric chart from the intersection of two known air properties.
6. calculate the mixed air temperature of two different temperature airstreams.
7. calculate CFM from velocity and area.
8. explain the difference between FPM measurements and CFM measurements.
9. explain the differences between static pressure and velocity pressure.

41.1 INTRODUCTION

Air properties and airflow are the heart of air conditioning. Technicians need to understand air properties and airflow to be able to troubleshoot and maintain an HVACR system. Factors that adversely affect system airflow will also have a negative impact on system efficiency and reliability. When people think about the properties of air and comfort, they think first about temperature. However, temperature is not the only property of air that is important to air conditioning. The properties of air include weight, volume, temperature, water content, and heat content. Humidity is nearly as important to human comfort as temperature. All the properties of air are important for the operation of air-conditioning equipment. The study of air and its properties is called psychrometrics.

41.2 CONDITIONS FOR COMFORT

Comfort is the feeling of physical contentment with the environment. The study of human comfort involves understanding how the area around a person affects the feeling of comfort and how the body adapts to changes in its environment. Many tests have been run to determine what conditions most people consider comfortable in the winter and the summer. The surrounding temperature, relative humidity, and air movement are all influencing factors. In the winter, cooler temperatures can be offset by higher relative humidity and less air movement. In the summer, higher temperatures can be offset by lower relative humidity and increased air movement.

Generalized Comfort Chart

The results of test data have been incorporated in comfort charts, one for summer conditions and one for winter conditions. Figure 41-1 shows the summer comfort conditions in red and the winter comfort conditions in blue. Note that the temperature range for summer comfort is about 5°F higher than the winter temperature comfort range. This assumes the use of warmer clothing in the winter and lighter clothing in the summer.

Body Temperature Regulation

Normal internal human body temperature is 98.6°F, winter and summer. The proper functioning of the body is dependent on constantly maintaining this temperature. The body temperature of warm-blooded animals, including humans, is closely controlled by producing more internal heat than necessary and regulating the release of that heat to the surroundings. The heat is produced by converting food energy into heat. There are four ways the body loses heat to the environment:

- Conduction (Figure 41-2)
- Convection of cool air (Figure 41-3)
- Radiation (Figure 41-4)
- Evaporation (Figure 41-5)

Conduction, radiation, and convention all require surrounding conditions that are cooler than the body temperature to dissipate heat. Conduction cools the body by the transfer of heat from the skin to the air surrounding and touching us. Convection accelerates the cooling process by movement of cool air over the body, increasing the transfer of heat from the body to the air. The wind-chill effect often reported in the winter is due to convection. Cooling by radiation occurs when we get close to a large object that is colder than our body temperature. Heat loss through radiation can be demonstrated by standing next to a cold outside wall in the winter. Heat leaves the body and travels to the cold wall without the person actually touching the wall.

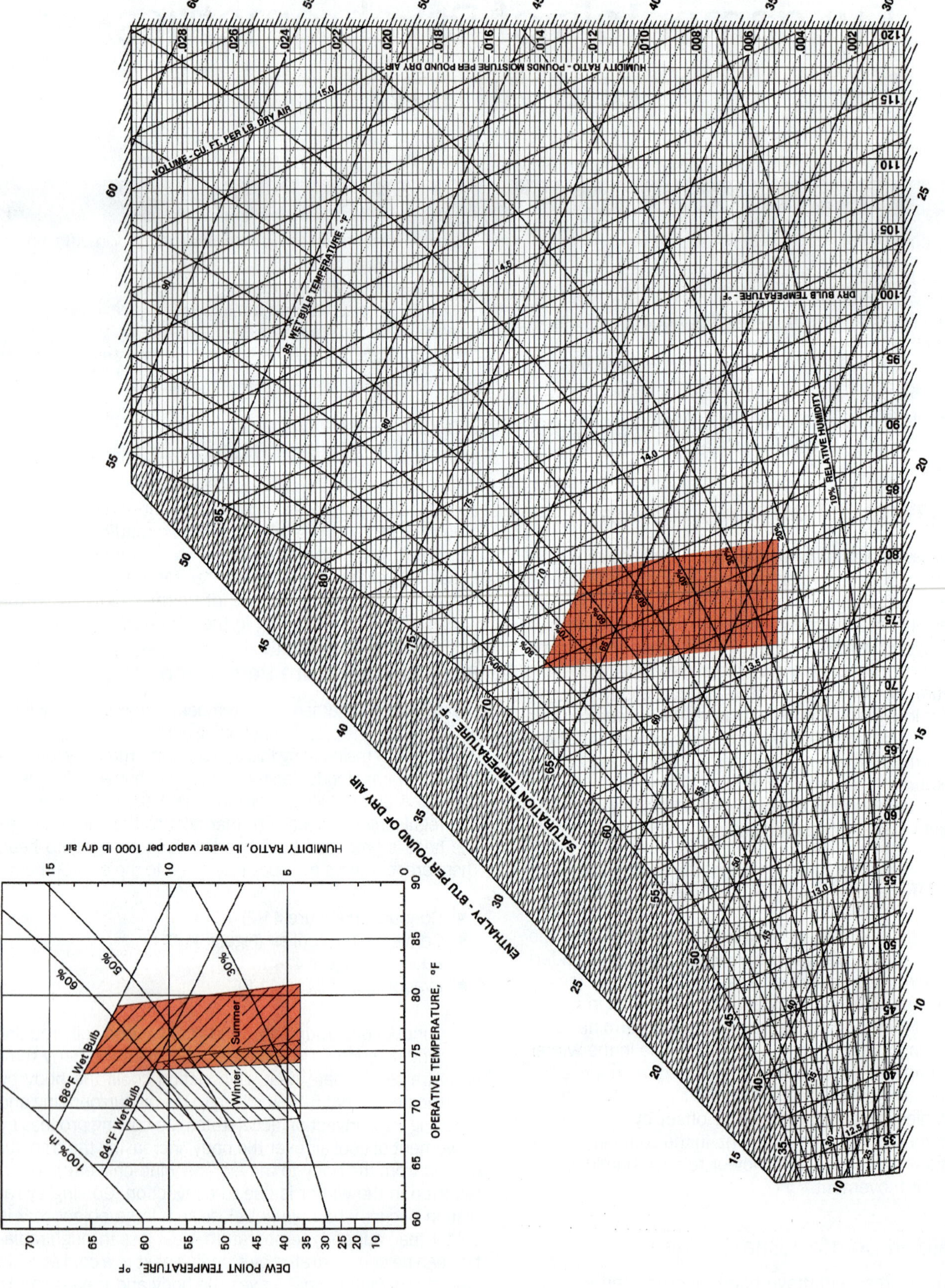

Figure 41-1 Generalized comfort conditions; summer conditions are shown in red and winter conditions are shown in blue.

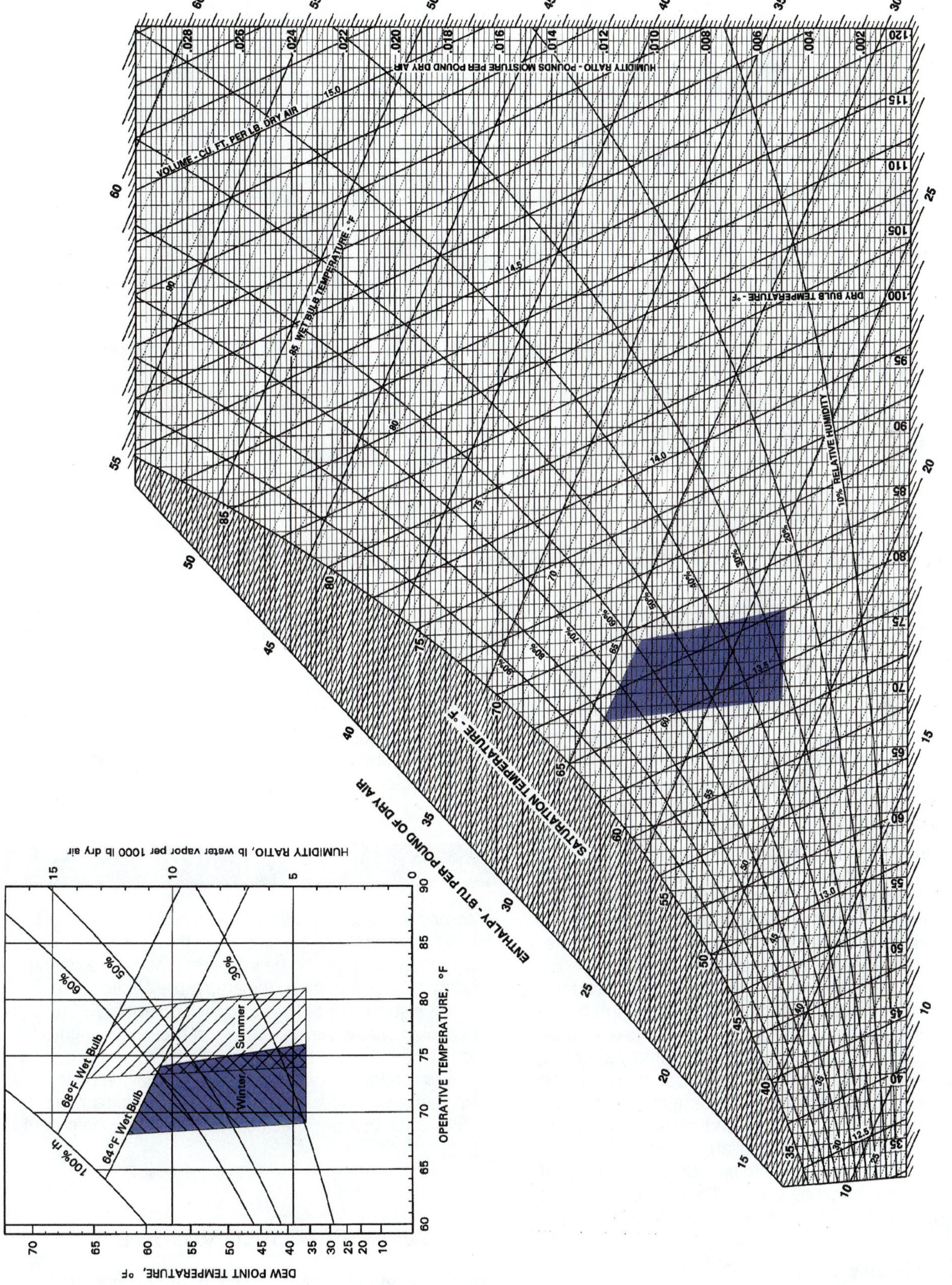

Figure 41-1 *(continued)*

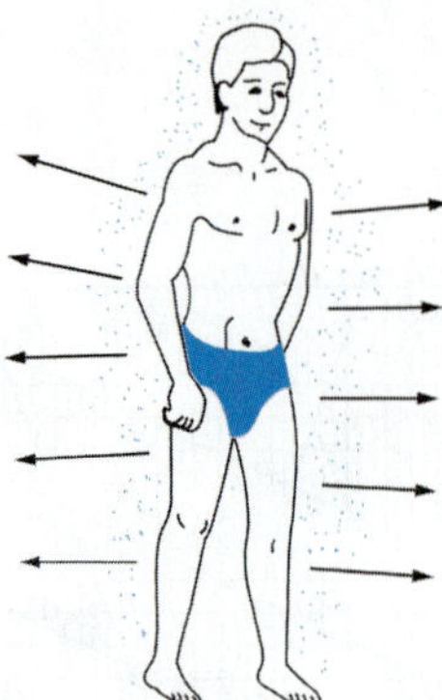

Figure 41-2 Heat is lost from the body by conduction to a film of air in contact with exposed skin.

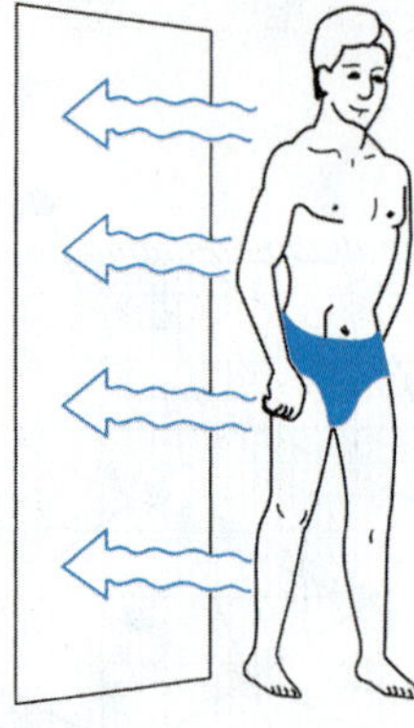

Figure 41-4 The body radiates heat to the surrounding surfaces/objects.

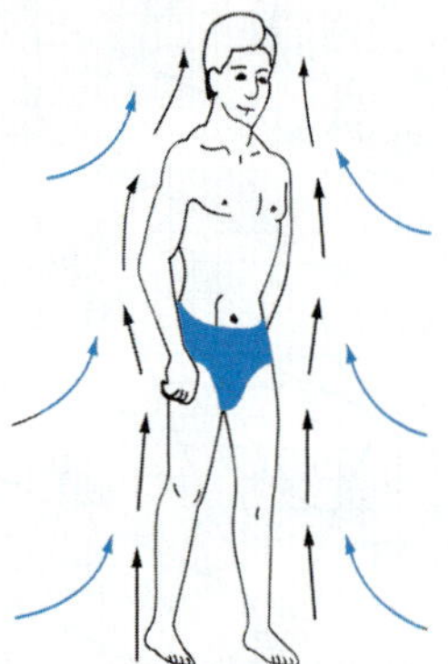

Figure 41-3 Heat is lost from the body by convection of the air around the body.

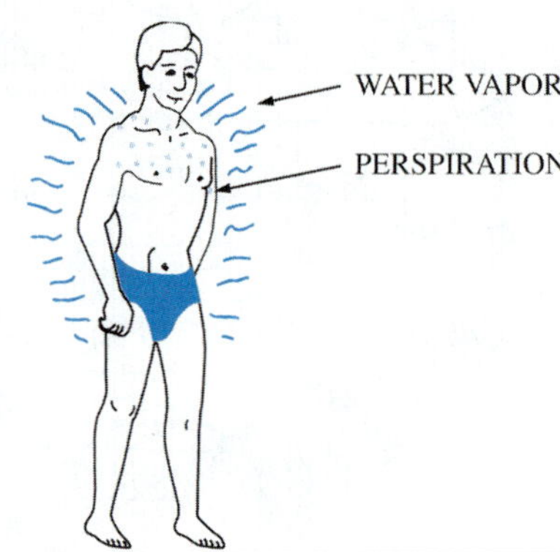

Figure 41-5 The body loses heat by evaporation of water from the skin.

The body cannot be cooled by conduction, convection, and radiation whenever the surrounding temperature is warmer than body temperature. In fact, these processes begin transferring heat in. Evaporation is the only body-cooling process that does not require cooler surroundings. Evaporation only requires that the body produce sweat and that the surrounding air be able to absorb water. Rapid evaporation of perspiration produces a cooling effect. This is why the humidity plays an important role in human comfort. People feel cooler in drier air because their perspiration can readily evaporate. They feel warmer in air with a higher relative humidity because the evaporation process is slower.

The body has a remarkable ability to adjust to temperature change. When a person goes from a warm house into the cold outdoors, some compensation needs to be made to prevent excessive heat loss. Involuntary shivering occurs, providing body heat. When a person goes from an air-conditioned space into an outside temperature of 95°F, an adjustment in the circulatory and respiratory system takes place. The blood vessels dilate to bring the blood closer to the surface of the skin to provide better cooling. If this is not enough, sweating occurs. The sweat evaporates from the skin, producing a large amount of body cooling.

Level of Activity

Comfort is also affected the level of physical activity. Active people feel more comfortable at cooler temperatures than sedentary people. This is because activity creates more body heat, requiring more cooling to maintain comfort. Figure 41-6 shows the comfort temperature ranges for different levels of activity.

TECH TIP

As people age, their circulation tends to decrease. For that reason, elderly residents will often complain about a chill even when the home's system is set well above the norm. One way to make the area more comfortable is to arrange supply-air diffusers so they do not blow into the occupied space. Air movement can cause a slight chill for these individuals even though the room is comfortably warm for the other occupants. It may be necessary to change out the supply-grille faces with longer, curved grille faces that are specifically designed for heat pump applications.

41.3 AIR WEIGHT AND VOLUME

Air has weight and takes up space. Atmospheric pressure is caused by the weight of the air. We are not aware of the air around us or the atmospheric pressure created by

Temperature		
°F	°C	Comfortable Temperature Levels for Different Activities
78	25	Optimal for bathing, showering. Sleep is disturbed.
75	24	People feel warm, lethargic, and sleepy. Optimal for unclothed people.
72	22	Most comfortable year-round indoor temperature for sedentary people.
70	21	Optimum for performance of mental work.
64	18	Physically inactive people begin to shiver. Active people are comfortable.

Figure 41-6 Comfortable temperature levels for different activities.

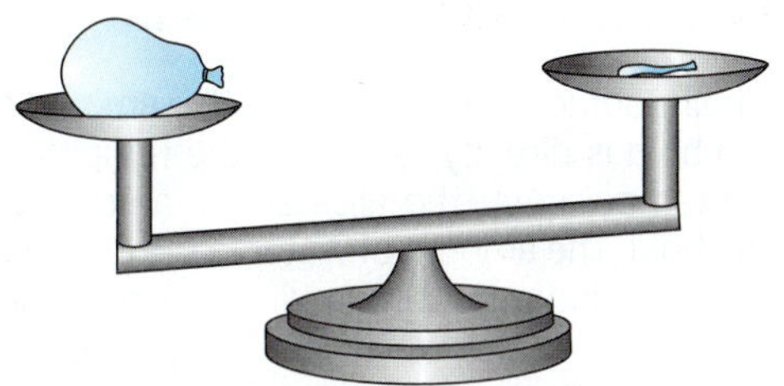

Figure 41-7 The side with the inflated balloon weighs more because of the weight of the air in the balloon.

it because we are continually surrounded by air and air pressure. A simple demonstration can show that air has weight. Place two identical balloons on a balance-beam scale. The scale should show that they are equal in weight. Now inflate one balloon, knot the end, and place it back on the scale. The scale will tilt down on the side with the inflated balloon because of the extra weight of the air (Figure 41-7).

At standard conditions of 70°F and atmospheric pressure 29.92 in Hg, dry air weighs 0.075 lb/ft^3. This is its density. Another way to state this is to say that 1 lb of air at standard conditions occupies 13.33 ft^3. This is its specific volume. However, the weight and volume of air vary with temperature, pressure, and humidity. In general, they affect weight and volume oppositely: something that increases the volume of the air will decrease its weight. Warmer air weighs less than cooler air. Air volume increases as its temperature increases, causing it to weigh less compared to standard air. This is the principle that hot-air balloons use. The hot air inside the balloon is lighter than the surrounding air, so it rises (Figure 41-8).

The weight of air increases as the air pressure increases. Higher pressures stuff more air molecules into the same space, making it weigh more. Standard atmospheric pressure at sea level is 29.92 in Hg. In general, air pressure decreases as altitude increases. Higher elevations have less air above them pushing down and so have lower atmospheric pressures. This is similar to the effect swimmers feel when diving in water. The increased weight of the water above them causes increased pressure (Figure 41-9). The air in Denver, Colorado weighs less than the air in Atlantic City, New Jersey. This affects the density of the

Figure 41-8 Hot-air balloons rise because the heated air in the balloon is less dense than the surrounding air.

air circulated by fans and the amount of horsepower required to move the same volume of air. A fan in Denver does not require as much horsepower to move the same volume of air as an identical fan in Atlantic City because of the difference in the density of the air (Figure 41-10). The amount of oxygen available for combustion in furnaces is also affected. Furnaces are derated at high altitudes because there is less oxygen available in every cubic foot of combustion air (Figure 41-11).

Humid air is lighter than dry air. One pound of moist air occupies a greater space than 1 lb of dry air. This is because water vapor is less dense than air. This can be seen when boiling water. The steam coming off the boiling water rises because water vapor is not as dense as the surrounding air.

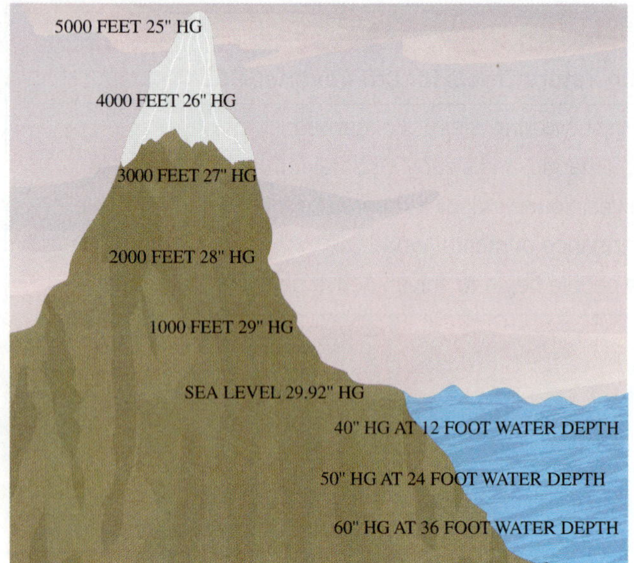

Figure 41-9 Air pressure is affected by elevation; it is higher at lower elevations because of the weight of the air above you; similarly, water pressure increases the further down in the water you go because of the weight of the water above you.

Altitude	Percent of Sea-Level Horsepower
Sea level	1
2,000 ft	0.96
3,000 ft	0.93
4,000 ft	0.89
5,000 ft	0.86
6,000 ft	0.82

Figure 41-10 Fan horsepower vs. altitude.

Altitude	Firing Rate	Percent of Full Capacity
Sea level	100,000 BTU/hr	100%
2,000 ft	96,000 BTU/hr	96%
3,000 ft	94,000 BTU/hr	94%
4,000 ft	92,000 BTU/hr	92%
5,000 ft	90,000 BTU/hr	90%
6,000 ft	88,000 BTU/hr	88%

Figure 41-11 Furnace firing rates are derated based on altitude because both the air and the gas fuel are less dense at higher altitudes.

As the water molecule is absorbed into the air, it is much like putting Styrofoam beads in with sand. Because the Styrofoam beads are much lighter, a box of Styrofoam and sand would become lighter as the proportion of Styrofoam mixed with the sand increases.

SERVICE TIP

Under standard conditions, 1 lb of air occupies approximately 13.33 ft^3. That is a space about the size of a box that measures 28.5 in in all dimensions. The weight of the air in a classroom can be estimated by multiplying the length, width, and height of the classroom to determine the classroom's volume and then dividing that volume by 13.33. For example, a classroom that is 30 ft wide and 40 ft long with a 10 ft ceiling would have 12,000 ft^3 of air; and that air would weigh approximately 900 lb (12,000/13.33).

41.4 WATER VAPOR IN THE AIR

All air contains some water (humidity). The amount of water the air can hold is directly related to the temperature: the higher the temperature, the greater the amount of water that air can hold. The amount of water in the air is measured by both specific and relative humidity. Specific humidity is the actual weight of the water in each pound of dry air. Specific humidity is measured in grains per pound of dry air. A grain is a very small unit of weight; there are 7,000 grains in 1 lb.

Relative humidity compares the amount of water in the air to the amount of water the air can hold at a given temperature. For example, the most water the air can hold at 75°F is 130 grains. A sample of air at 75°F with 65 grains of water is at 50 percent relative humidity because it is holding 50 percent as much water as it can hold at that temperature.

41.5 DRY-BULB AND WET-BULB TEMPERATURES

The air dry-bulb temperature is the temperature measured on an ordinary thermometer. Instruments used to measure wet-bulb temperatures are called psychrometers. The wet-bulb temperature is measured by placing a wick soaked with water around the thermometer bulb and moving it rapidly through the air. Water evaporates from the wick, cooling the bulb and lowering the temperature.

The most common psychrometers are called sling psychrometers because they are swung in circles to help evaporate the water from the wet bulb, Figure 41-12a. There are also a number of electronic psychrometers available, such as the one shown in Figure 41-12b). These instruments do not actually evaporate water but measure the relative humidity using electronic sensors. They typically can display relative humidity, dry-bulb temperature, wet-bulb temperature, and dew-point temperature.

The relation of the wet-bulb temperature to the dry-bulb temperature is a measure of the relative humidity. At 100 percent relative humidity, both the wet-bulb and dry-bulb temperatures will be the same because water will not evaporate from the wet bulb. However, at lower relative humidity levels, the wet-bulb temperature will be considerably lower than the dry-bulb temperature. The relative humidity

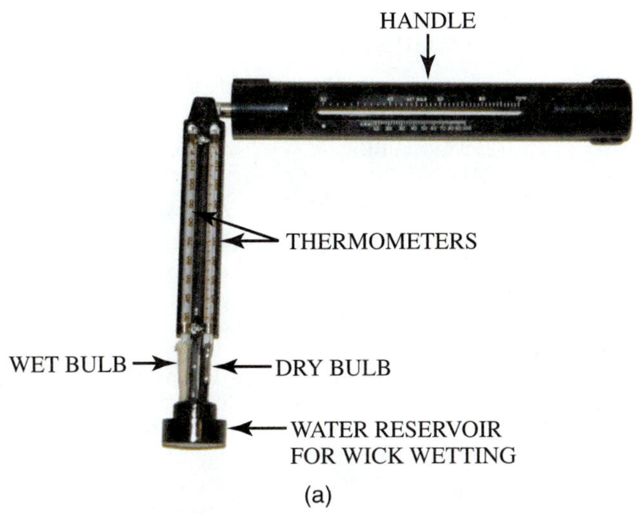

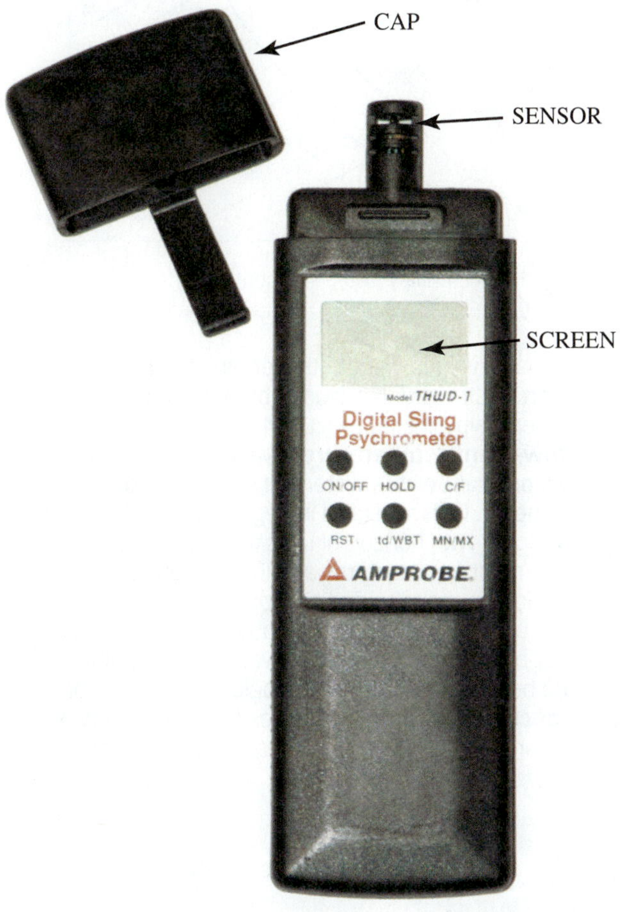

Figure 41-12 (a) Sling psychrometer for measuring dry-bulb and wet-bulb temperatures to determine relative humidity; (b) electronic psychrometer.

value can be obtained by using either a wet-bulb depression chart or a slide rule provided with the psychrometer. For example, air with a dry-bulb temperature of 80°F and a wet-bulb temperature of 67°F would be at 50 percent relative humidity (Figure 41-13).

Figure 41-13 With the arrow pointing to the dry-bulb temperature of 80°F, read under the 67°F wet-bulb temperature to find the relative humidity of 50 percent.

41.6 DEW-POINT TEMPERATURE

Dew point is another key air characteristic that is affected by the amount of water vapor in the air. The dew-point temperature is the temperature at which the water vapor can condense out of the air. In an air-conditioning system, dehumidification takes place when the air passes over a coil whose temperature is below the dew-point temperature of the air.

As shown in Figure 41-14, when air with a 55°F dew-point temperature passes over the 40°F evaporator coil, the water vapor in the air will condense on the coil and drip into the drain pan.

TECH TIP

As the air velocity through an air-conditioning coil increases, the amount of dehumidification decreases. Conversely, as the velocity of the air slows through a coil, greater dehumidification will occur. You can use slight changes in fan speed to help control an area's humidity. This can be very important along the coast or in areas of particularly high relative humidity.

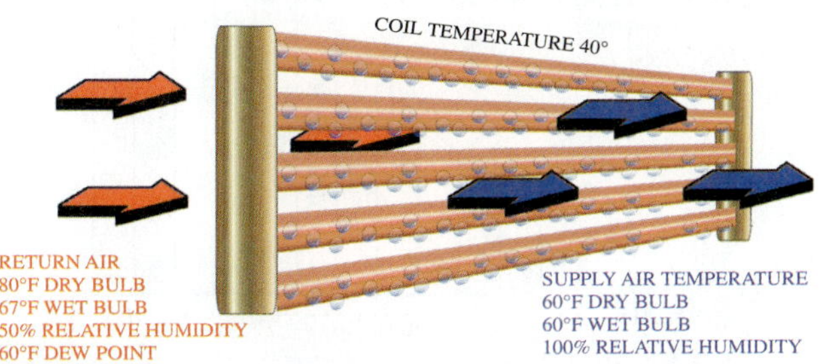

Figure 41-14 Water condenses on the evaporator coil when the coil temperature is below the dew point of the return air.

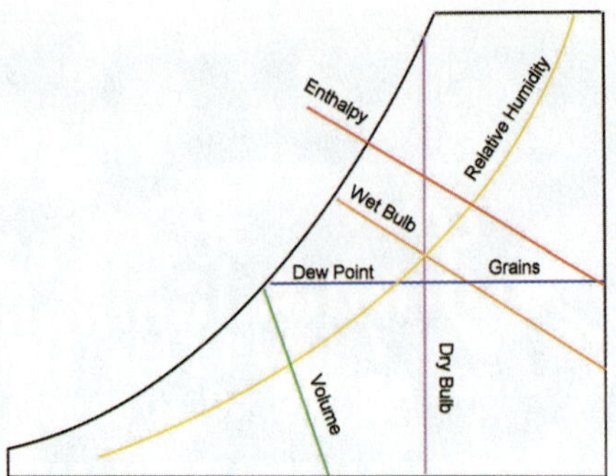

Figure 41-15 The lines on the psychrometric chart represent different air properties.

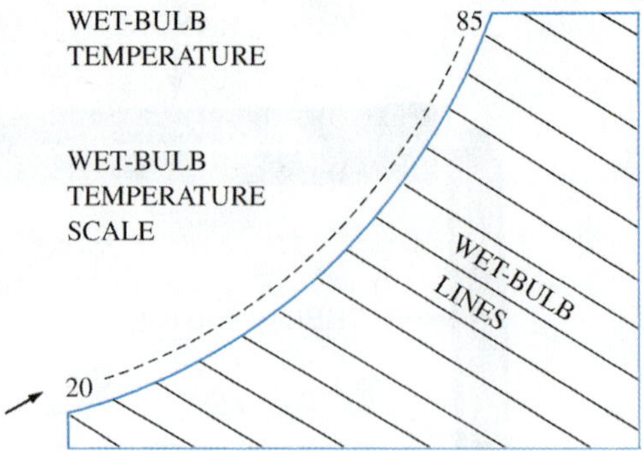

Figure 41-17 Wet-bulb temperature lines on psychrometric chart.

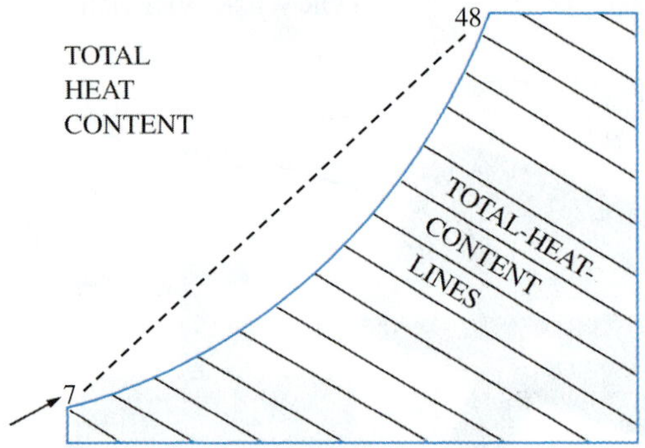

Figure 41-18 Enthalpy (total heat content) lines on psychrometric chart.

41.7 PSYCHROMETRIC CHART

The psychrometric chart is a graphical representation of air properties (Figure 41-15). Psychrometric charts are used by engineers to plot system performance when designing air conditioning systems. A system plot provides a visual picture of the changes taking place in the air passing through the air conditioner.

The psychrometric chart can show the following air properties:

- **Dry-bulb temperature** Measures the normal air temperature in degrees Fahrenheit and shown with vertical lines (Figure 41-16). The dry-bulb temperatures are shown across the bottom of the psychrometric chart.
- **Wet-bulb temperature** Measures the wet-bulb temperature in degrees Fahrenheit and shown with diagonal lines (Figure 41-17). The wet bulb temperatures are shown along the curved portion on the left-hand side of the psychrometric chart.
- **Enthalpy.** Measures the total heat content of the air in BTU per pound of air. The enthalpy lines are diagonal lines that are nearly parallel to the wet-bulb temperature lines (Figure 41-18). The enthalpy values are shown on the left along a straight line that is roughly parallel to the curved line. Instead of having a separate set of diagonal lines, some charts have the enthalpy values on both the left and right side of the chart. A straightedge is used between them to indicate the enthalpy line (see Figure 41-1).
- **Dew-point temperature** Measures the temperature where water will start to condense out of the air. It is measured in degrees Fahrenheit and is shown with horizontal lines (Figure 41-19). The dew point temperatures are shown along the curved portion on the left-hand side of the psychrometric chart. They are the same numbers used for reading wet-bulb temperature. However, the dew point lines are horizontal while the wet-bulb lines are diagonal.
- **Specific humidity** Measures the weight of water vapor in the air in grains and is shown with horizontal

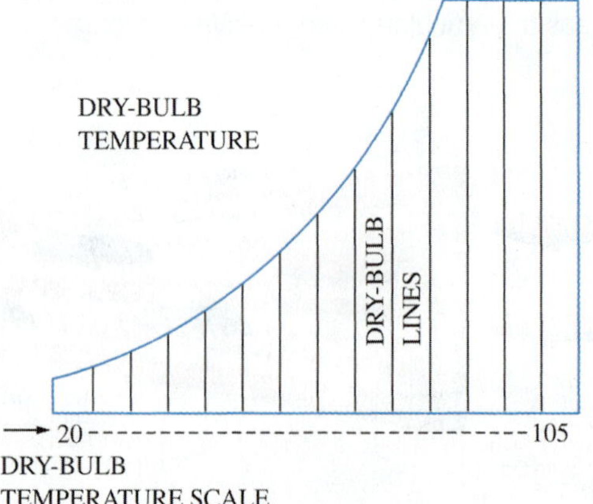

Figure 41-16 Dry-bulb temperature lines on psychrometric chart.

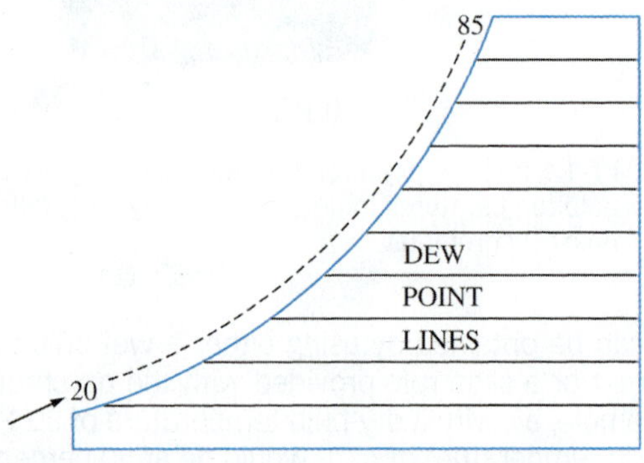

Figure 41-19 Dew-point temperature lines on the psychrometric chart.

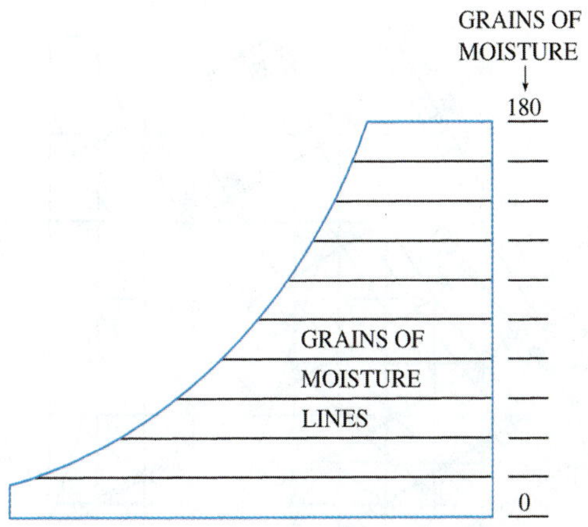

Figure 41-20 Grains of moisture line on the psychrometric chart.

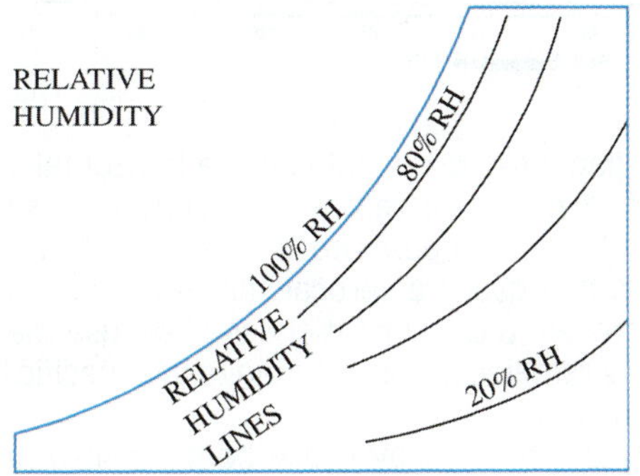

Figure 41-21 Relative humidity lines on the psychrometric chart.

lines (Figure 41-20). The specific humidity readings are shown on the right-hand side of the chart.

- **Relative humidity** Expresses the percentage of water vapor in the air compared to the amount of water vapor the air can hold. Relative humidity is shown by the sweeping curved lines (Figure 41-21). The values are written on the lines.
- **Specific volume** Measures the space in cubic feet occupied by each pound of air. The specific volume lines are diagonal lines with a less severe slope (Figure 41-22). The values are written on the lines.

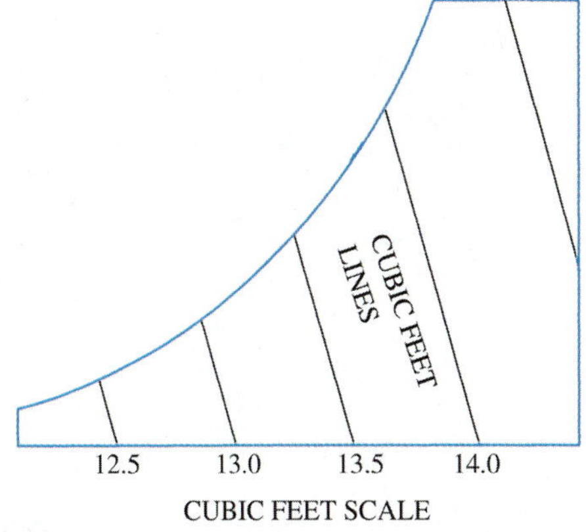

Figure 41-22 Specific volume lines on the psychrometric chart.

41.8 PLOTTING POINTS ON THE PSYCHROMETRIC CHART

Any two air properties that use separate lines can be used to plot a point on the psychrometric chart. All the other properties can be determined once the intersection is found. For example, the intersection of 80°F db (dry bulb) temperature and 67°F wb (wet bulb) temperature is shown by the red dot in Figure 41-23. Blue arrows from that point

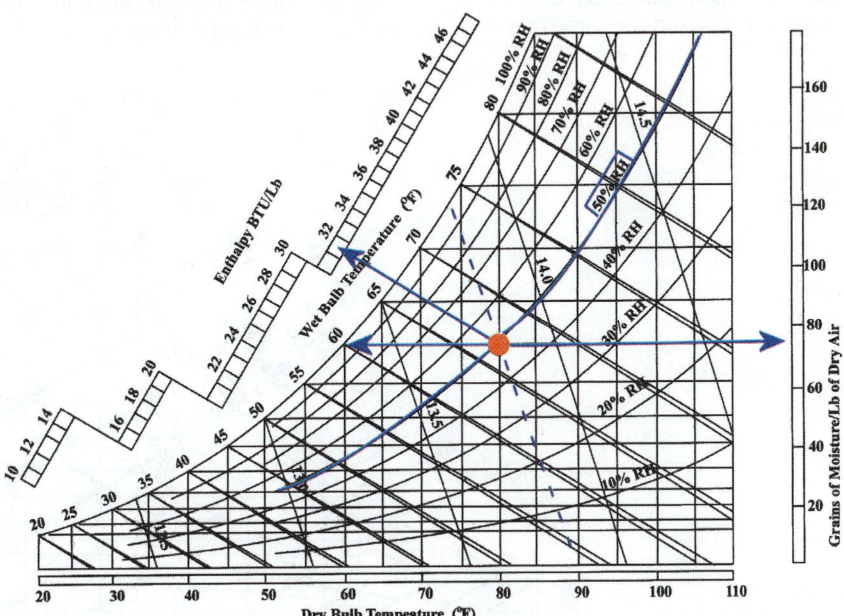

Figure 41-23 Psychrometric chart with 80°F db and 67°F wb temperatures.

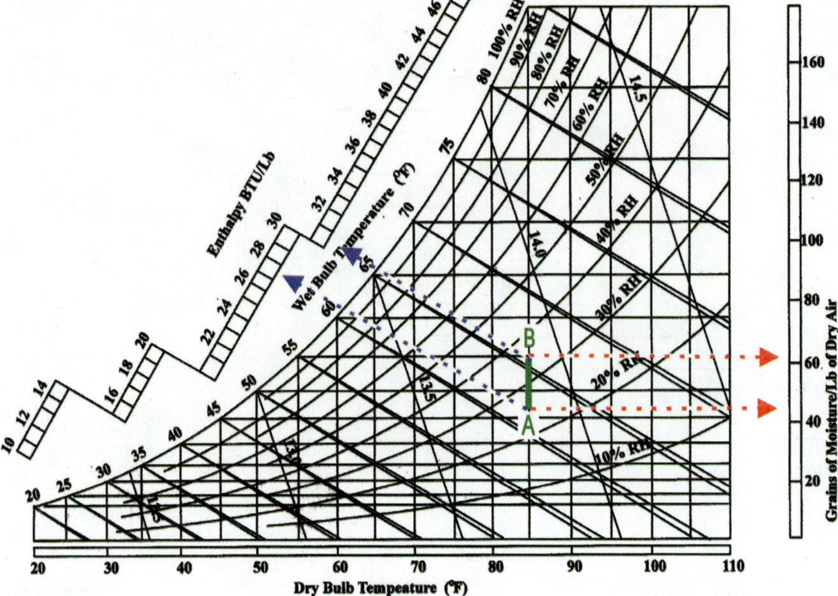

Figure 41-24 Green line between A and B shows a sensible heat change; the red arrows show the temperature change from 70°F to 80°F; the blue arrows show the enthalpy change from 27.5 BTU to 30.5 BTU.

show the values of the other properties at that condition. Summarized, they are:

Dry-bulb temperature	80°F (given)
Wet-bulb temperature	67°F (given)
Dew-point temperature	60°F
Relative humidity	50 percent
Enthalpy	31.4 BTU
Specific humidity	75 grains
Specific volume	13.8 ft^3

41.9 SENSIBLE HEAT AND LATENT HEAT PROCESSES

Nearly all processes will involve a change in sensible heat, latent heat, or both. The exception would be a process whose beginning and ending points lie on the same diagonal enthalpy line. Processes that cause a horizontal shift from the start to the finish represent a sensible heat change because the air temperature is changing (Figure 41-24). Processes that cause a vertical shift from the start to the finish represent a latent heat change because the air temperature is remaining the same while the specific humidity is changing (Figure 41-25).

Many processes may cause both sensible and latent heat changes. It is possible to use the psychrometric chart to determine the amount of sensible heat change and latent heat change for an air-conditioning process. Plot a third point to form a right triangle between the beginning and ending points of the process. The triangle should have one vertical leg and one horizontal leg. Look up the enthalpy of all three points. The sensible change will be the difference in enthalpy from end to end of the horizontal leg. The latent change will be the difference in enthalpy from end to end of the vertical leg (Figure 41-26).

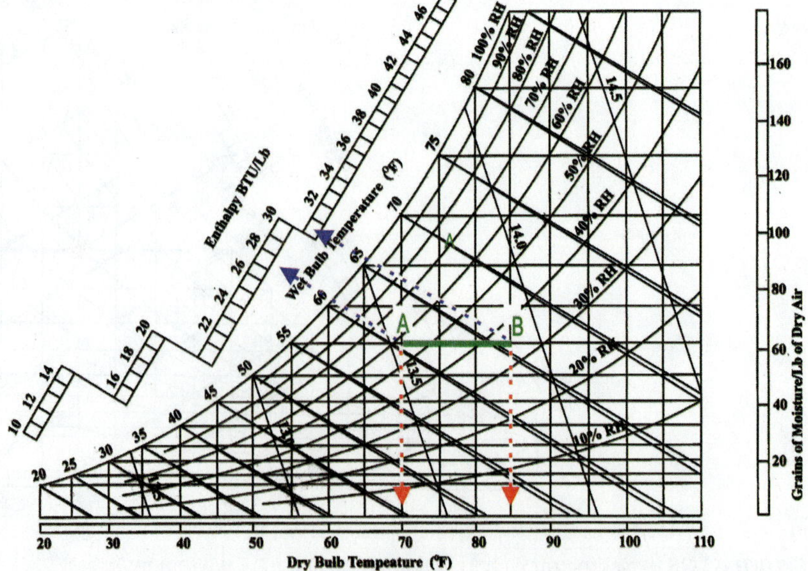

Figure 41-25 Green line between A and B shows a latent heat change; the red arrows show the specific humidity change from 43 grains to 62 grains; the blue arrows show the enthalpy change from 27.5 BTU to 30.5 BTU.

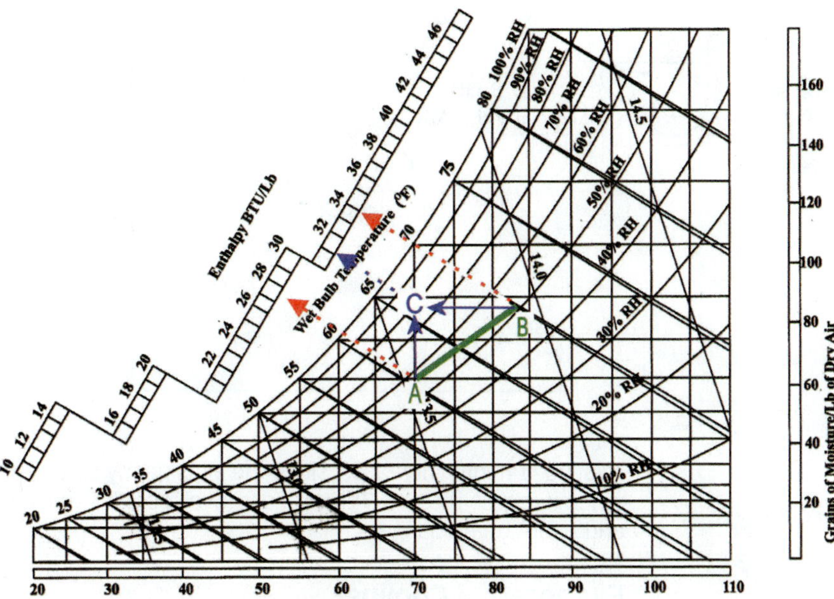

Figure 41-26 B to C shows the sensible change, while A to C shows the latent change.

41.10 PLOTTING COMMON AIR-CONDITIONING PROCESSES

The changes that take place in the properties of the air during an air-conditioning process can be plotted. Examples of plotting common air-conditioning processes are shown next.

Heating

Figure 41-27 shows the addition of sensible heat. The process is shown moving along a horizontal line to the right. The wet-bulb temperature and enthalpy increase and the relative humidity decreases. The dew-point temperature and the water content do not change. This process involves only sensible heat change.

Cooling with a Dry Evaporator Coil

Figure 41-28 shows the reduction of sensible heat. This process represents cooling with a coil whose temperature is above the dew point of the entering air. The process is also shown moving along a horizontal line, but in the opposite direction. The wet-bulb temperature and enthalpy decrease and the relative humidity increases. Again, there is no change in dew point or water content. This process involves only sensible heat change.

Humidifying

Figure 41-29 shows the addition of water. This action is represented by a vertical line moving upward as the specific humidity increases. The only property that does not change is the dry-bulb temperature, which is held constant. Heat is added as evidenced by the increase in enthalpy, but the heat is used to evaporate the water. The wet-bulb temperature, dew point, and relative humidity all increase. This process involves only latent heat change.

Simultaneous Heating and Humidifying

Figure 41-30 shows the simultaneous addition of sensible heat and water. The line representing the operation moves upward to the right. The air picks up both heat and water

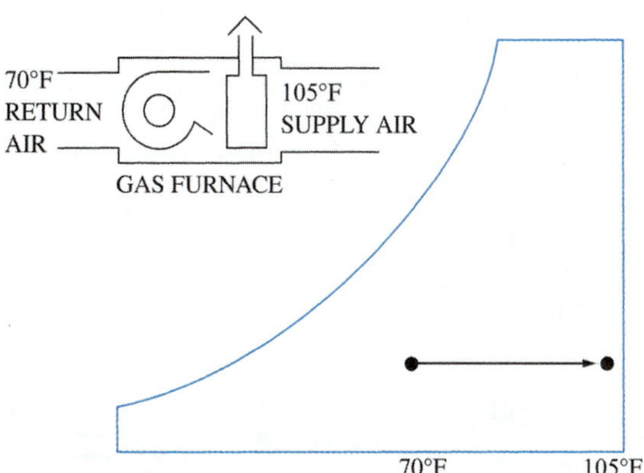

Figure 41-27 Plotting the addition of sensible heat by a furnace on the psychrometric chart.

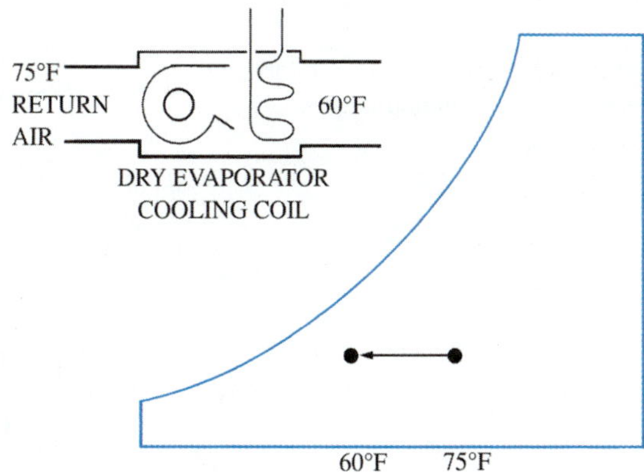

Figure 41-28 Plotting the temperature of air being cooled without dehumidification on the psychrometric chart.

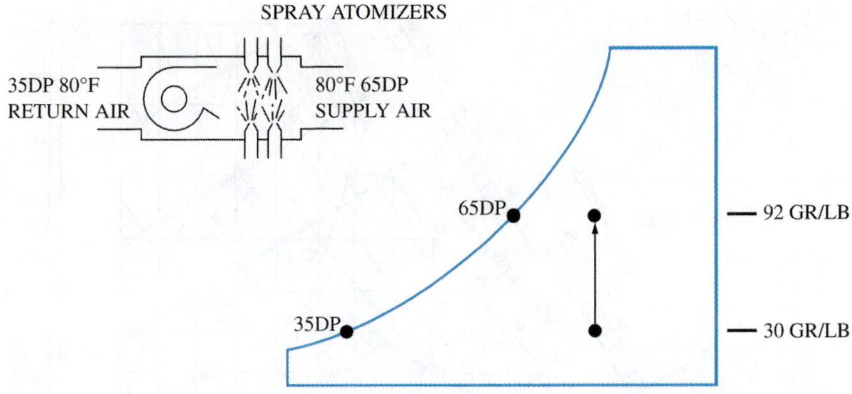

Figure 41-29 Plotting the addition of moisture to the air produced by a humidifier.

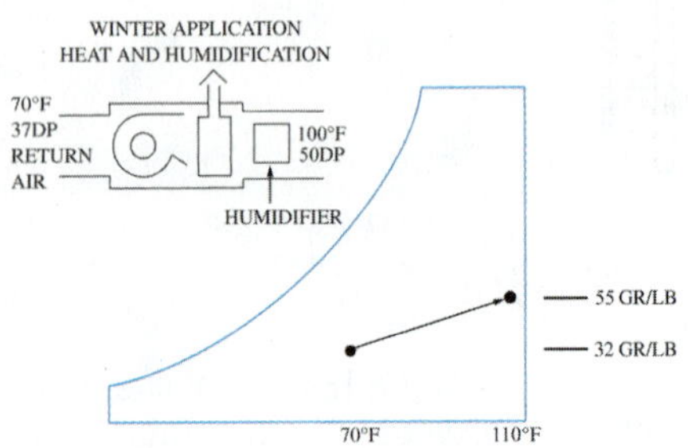

Figure 41-30 Heat and moisture added by a hot-air furnace with a humidifier.

as it moves through the conditioner. The enthalpy, wet-bulb temperature, specific humidity, and dew-point temperature all increase. The relative humidity could increase, remain the same, or decrease, depending on how much water is added. In actual operation, the amount of heat and water added must match the heat and water lost by the structure to maintain the same temperature and humidity in the building. This process involves both sensible and latent heat change.

Cooling and Dehumidification

Figure 41-31 shows the reduction of both sensible heat and water. This is the traditional refrigerated air-conditioning cycle. The line representing the operation moves downward to the left. The cooling coil both cools the air and condenses water as the air passes through it. The enthalpy, wet-bulb temperature, and dew-point temperature all decrease. The relative humidity increases, since the air is nearly saturated when it leaves the coil. This process involves both sensible and latent heat change.

Evaporative Cooling

Figure 41-32 shows the process of removal of sensible heat accompanied by a simultaneous increase in water. The process uses evaporation of water to cool the air. The process follows the wet-bulb line in the chart. The dry-bulb temperature decreases as the water in the air and the relative humidity increase. This process essentially converts sensible heat to latent heat; enthalpy is constant.

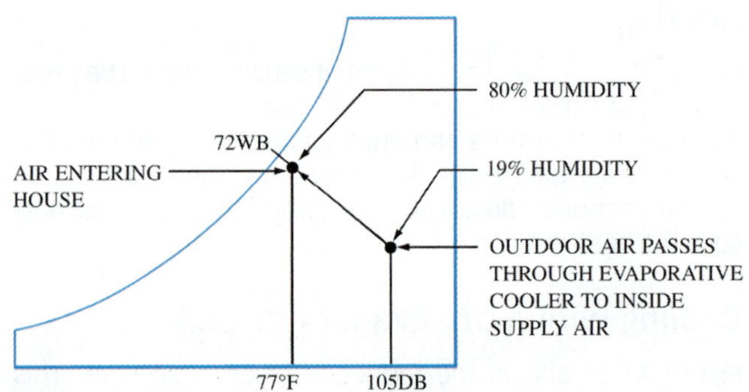

Figure 41-32 Cooling and humidification produced by an evaporative cooler plotted on the psychrometric chart.

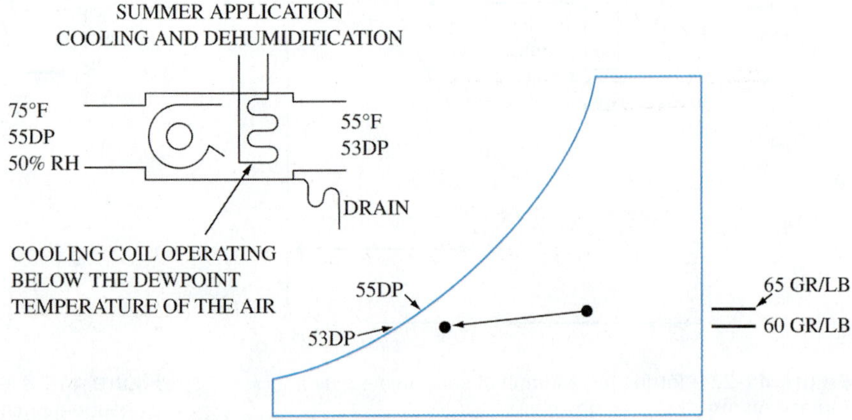

Figure 41-31 Cooling and dehumidification produced by an air conditioner, plotted on the psychrometric chart.

GREEN TIP

The importance of airflow for the proper operation of air conditioning systems cannot be overstated. Poor airflow has become one of the most common problems seen in the field. Systems with inadequate airflow operate inefficiently with a reduced capacity, causing increased energy use. Prolonged operation with inadequate airflow will lead to early liquid floodback and early compressor failure.

41.11 AIRFLOW FUNDAMENTALS

The air in an air-conditioning system moves. Rates are used to describe something that is moving. A rate measurement compares a quantity with a time. The output of a fan is measured in cubic feet per minute (CFM), describing the volume of air in cubic feet that the fan moves every minute. The speed of the air is used to determine the amount of airflow. Speed is also stated as a rate. A speed of 50 miles per hour indicates that a car traveling at that speed for an hour will cover a distance of 50 miles. Airspeed, or velocity, is measured in feet per minute (FPM). Feet per minute describes how far the air will travel in a minute. This still does not really measure how much airflow there is, just how fast it is moving. To determine the volume of the air, multiply the velocity of the air times the area of the hole it is leaving. Air traveling at a rate of 100 FPM through a 1 square foot opening will make a column of air with a volume of 100 CFM.

Example: Calculate the airflow if air is leaving a 24 × 36–in hole at a velocity of 50 FPM. The area of the hole is calculated by multiplying 24 in × 36 in = 864 in^2. Convert square inches to square feet by dividing by 144: 864 in^2/144 = 6 ft^2. Alternately, you can convert the dimensions to feet before multiplying. For example 24/12 = 2ft; 36/12 = 3ft; 2ft × 3ft = 6 ft^2. The airflow in CFM is calculated as 50 FPM × 6 ft^2 = 300 CFM.

TECH TIP

Ventilation hoods work by moving air at high enough speeds to suck in surrounding air and effectively capture the fumes, smoke, grease, or other small particles in the airflow. They have a specification called the capture velocity, which is how fast the air must move to capture particular contaminants and move them out through the vent system. The capture velocity is measured in feet per minute (FPM).

SERVICE TIP

Standard airflow through air-conditioning evaporator coils has been 400 CFM per ton for many years. Most air-conditioning equipment will operate safely at 400 CFM per ton.

Figure 41-33 Digital rotating vane anemometer.

41.12 USING ANEMOMETERS AND FLOWHOODS TO MEASURE AIRFLOW

Anemometers measure air velocity (Figure 41-33.). They are used to measure the velocity of the air leaving registers and grilles to calculate airflow. Taking accurate measurements is complicated by the fact that air does not travel uniformly through all parts of the register or grille. This can be remedied by taking several readings and averaging them. Some anemometers will take a real-time average, making this fairly easy to do (Figure 41-34.). For calculating CFM, another complicating factor comes in: determining the area of the register. A register that measures 6 in by 10 in does not have a free area of 60 in^2 because some of that space is taken up by the metal bars of the grille. The free area can vary a lot depending on the design of the register or grille. The only reliable method is to look up the free area in the manufacturer's literature.

SERVICE TIP

A small pyramid-shaped cone can be used to direct the air through an unobstructed hole with a known free area. A cone with a free area of 1 ft^2 is ideal since the average velocity would also be the CFM.

Figure 41-34 This anemometer can average the velocity reading over the period of time the reading is taken.

Figure 41-35 Flowhood.

Flowhoods solve the problem of not knowing the free area by directing all the air through an opening with a known area (Figure 41-35). They measure the average velocity and calculate the volume using the average velocity. See Unit 73, Testing and Balancing, for more information on anemometers, flowhoods, and airflow measurement.

41.13 TYPES OF DUCT PRESSURE

Air moves through ductwork for only one reason: it travels from an area of high pressure to an area of low pressure. The blower provides the pressure difference to move the air through the ductwork. The amount of air the blower can move and the amount of energy needed to move the air is controlled by the resistance to airflow from the ductwork and all the system components in the airstream. In general, fans have the highest capacity when they are in open air, without any restriction to airflow. This is called free-air in fan ratings. The capacity of the fan decreases when it must move air against a pressure difference. A pressure difference is created when air is constrained by ductwork. There are two kinds of pressure created by moving air through the duct system: static pressure and velocity pressure (Figure 41-36).

Static Pressure

The word *static* means stationary, or not moving. When you blow up a balloon, the air pushes out evenly in all directions, creating static pressure. When a fan blows air into a confined space, it creates static pressure on all the surfaces of the confined space. When a fan moves air through ductwork, it pressurizes the duct system in much the same way a balloon is inflated. Static pressure in the duct pushes in all directions (Figure 41-37). If a large box is placed over the outlet of a blower, pressure will build up in the box and will be distributed evenly in all directions. In air conditioning, this box is called the supply plenum.

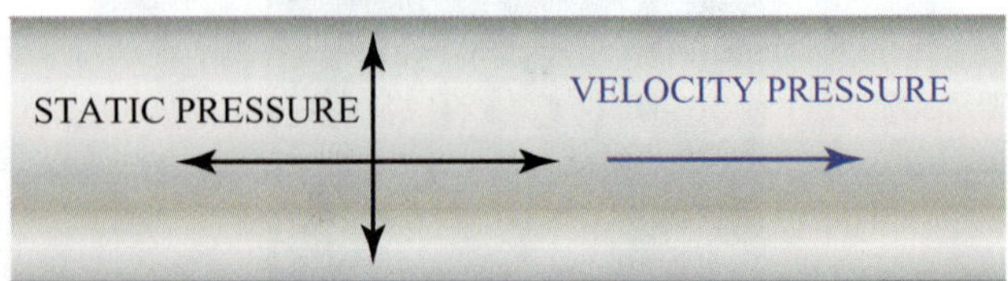

Figure 41-36 The two pressures created by air moving through a duct are static pressure and velocity pressure; total pressure is the sum of these two pressures.

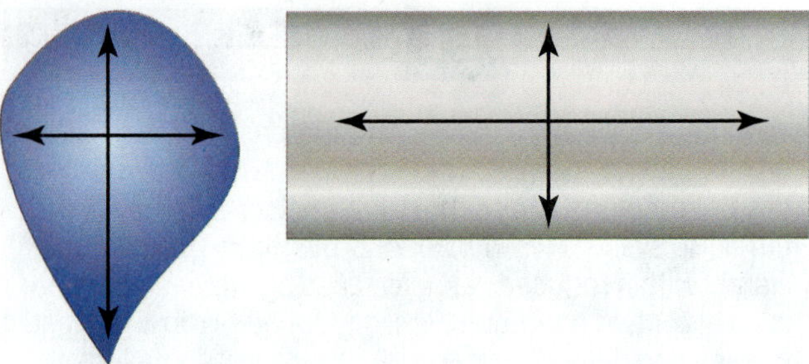

Figure 41-37 Static pressure in a duct like the pressure inside a balloon—pushing out in all directions.

Velocity Pressure

Velocity pressure is pressure created by moving air. Velocity pressure only pushes in the direction of airflow. If the end of an inflated balloon is released, the balloon will fly around the room because the pressure in the balloon goes from the high static pressure inside the balloon to outside of the balloon where the pressure is lower. As this happens, static pressure is converted to velocity pressure. Velocity pressure is the pressure exerted by moving air. The word *velocity* means speed. The air now has velocity because it is moving, creating velocity pressure. If a hole is cut in the supply plenum, the static pressure will force air out of the hole, creating velocity pressure. The pressure in the supply plenum will drop as some of the static pressure is converted to velocity pressure. Total pressure is the combination of static pressure and velocity pressure. The difference in total pressure from one area to another is what pushes the air forward.

Pressure Changes

Static pressure can be transformed into velocity pressure, as in the example of the balloon. Velocity pressure can also be transformed to static pressure. In the example of the cardboard box over the blower, velocity pressure is transformed into static pressure. As the air moves through the blower it achieves significant velocity. When the moving air hits the box, its velocity, and thus its velocity pressure, drops to nothing. However, the energy does not simply disappear; it is transformed into static pressure, pushing out in all directions.

As a general rule, changes in duct geometry affect static and velocity pressure in opposite ways. An example would be a change in duct diameter. If a duct decreases in size, the velocity of the air moving through it will increase because the same amount of air now must pass through a smaller opening. The only way it can do that is to move

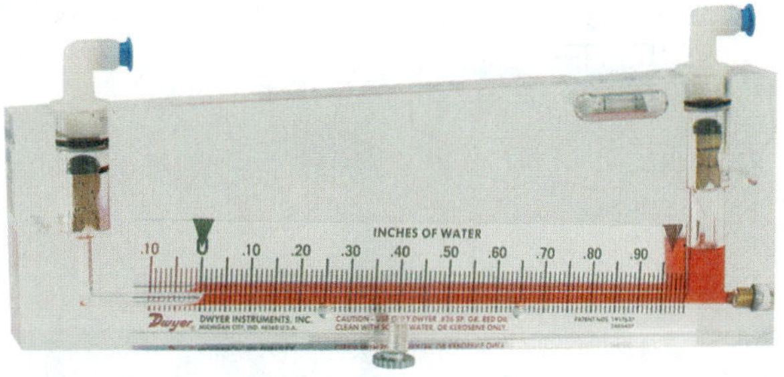

Figure 41-38 Inclined manometer for measuring air pressure. *(Courtesy Dwyer Instruments, Inc.)*

faster. This increases the velocity pressure in the duct. The energy for the increased velocity pressure comes at the expense of static pressure. Since more energy is being used to move the air forward, less is available for pushing out in all directions.

It should be noted that increasing or decreasing the amount of airflow will cause a proportional change in both the static and velocity pressures. Pushing more air through the same size duct will increase both the static and velocity pressures. Another important exception is that when static pressure is reduced because of friction loss, velocity pressure is not increased. In this case, the static pressure is converted to heat, not velocity pressure.

41.14 READING DUCT PRESSURE

Duct pressures are read in inches of water column with an incline manometer (Figure 41-38), a digital manometer (Figure 41-39), or a magnehelic gauge (Figure 41-40). Three types of pressure readings are taken in duct systems: total pressure, static pressure, and velocity pressure.

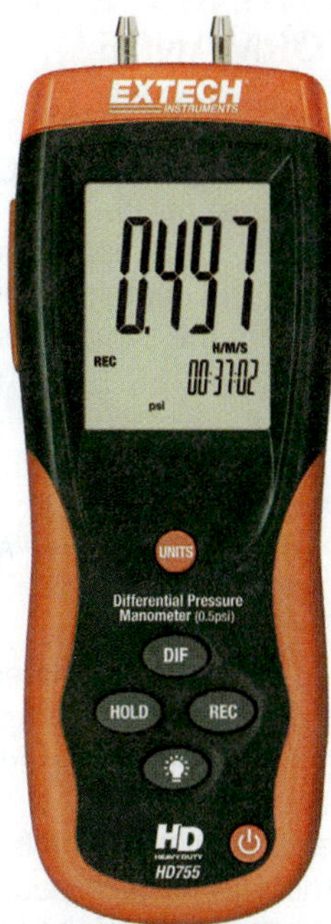

Figure 41-39 Electronic digital manometer used for taking accurate duct pressure readings. *(Courtesy of Extech Instruments, A FLIR Company)*

Figure 41-40 Magnehelic gauge can be used for taking accurate duct pressure readings.

SERVICE TIP

If a manometer reading must be taken in a run of flex duct, cover the area with duct tape before cutting into the flex duct. This will prevent excessive damage to the flex duct.

Measuring Total Pressure

Total pressure is read by placing the manometer sensing tube so that air flows directly into the tube opening, as shown in Figure 41-41. Placing the sensing tube directly into the airstream measures both velocity pressure and static pressure. It senses the velocity pressure because the opening is facing the flow of air; and it senses the static pressure because static pressure pushes out in all directions.

Measuring Static Pressure

The trick to measuring static pressure is to place the probe where it will not pick up any velocity pressure. Placing the probe perpendicular to the airflow is the most common method of reading static pressure (Figure 41-42).

Measuring Velocity Pressure

Velocity pressure cannot be read directly at a single point because static pressure is present anyplace a probe is placed. The velocity pressure is the difference between the total and static pressures and can be read by subtracting the static pressure from the total pressure. Velocity pressure is read by placing one probe in the airstream to

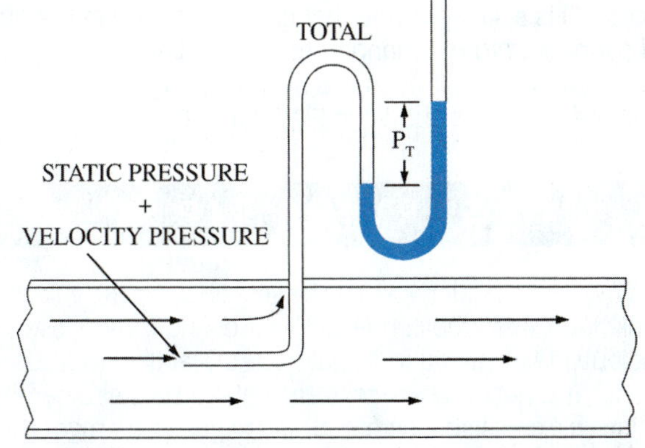

Figure 41-41 Manometer measuring total air pressure, in inches of water column.

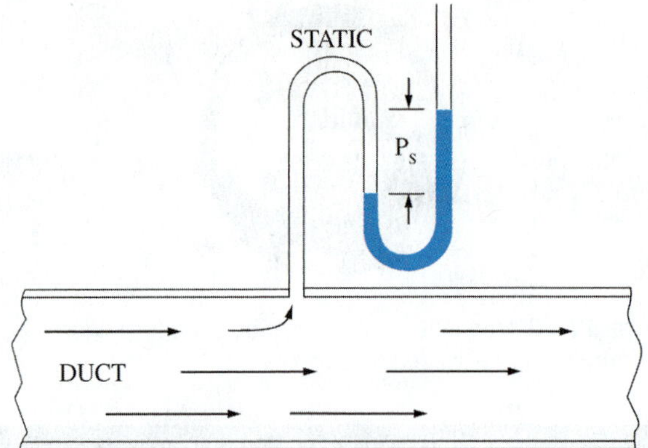

Figure 41-42 Manometer measuring static pressure, in inches of water column.

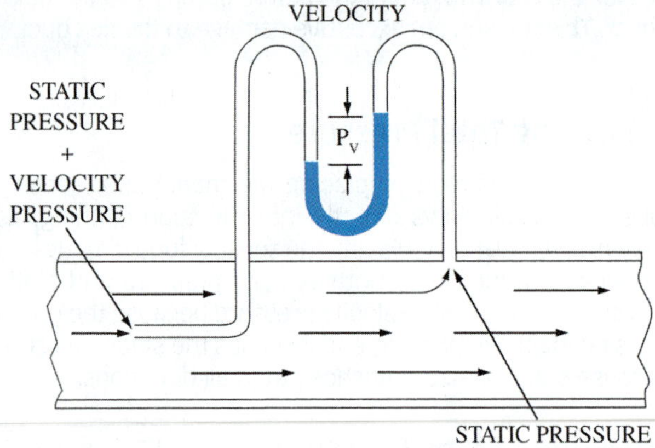

Figure 41-43 Manometer measuring velocity pressure, in inches of water column.

measure total pressure and connecting a probe on the opposite side of the manometer to read static pressure, as shown in Figure 41-43.

The velocity pressure is useful in measuring the velocity of air in a duct. The formula to convert velocity pressure to velocity is

$$4{,}005\sqrt{\text{velocity pressure}}$$

Some magnehelic gauges and inclined manometers are calibrated to read duct velocity directly in feet per minute. Most digital manometers can automatically convert velocity pressure to velocity. The *pitot tube* (pronounced "pea-toe") is a single probe that can be inserted in the duct and connected to the manometer for reading velocity pressure (Figure 41-44). It does this by measuring both total and static pressures, as shown in Figure 41-45.

SERVICE TIP

Airflow through ductwork can be very uneven. The velocity is higher toward the center of the duct and lower close to the outside of the duct. To get accurate airflow measurements, it is necessary to take several readings at different points in a cross-section of the duct to get an average. This is called traversing the duct. This will be discussed in more detail in Unit 73, Testing and Balancing Air Systems.

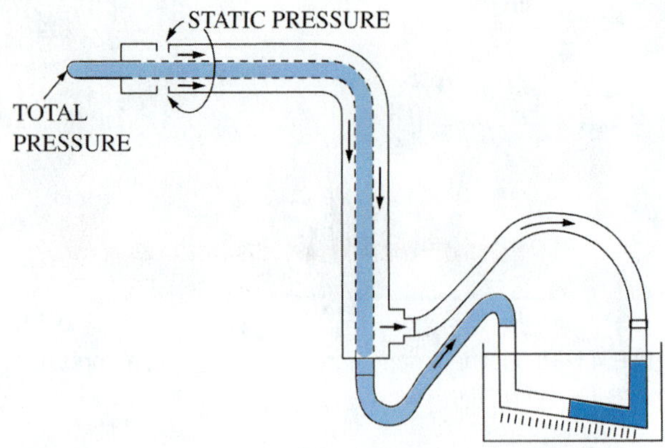

Figure 41-44 A pitot tube measuring velocity pressure by measuring both total pressure and static pressure.

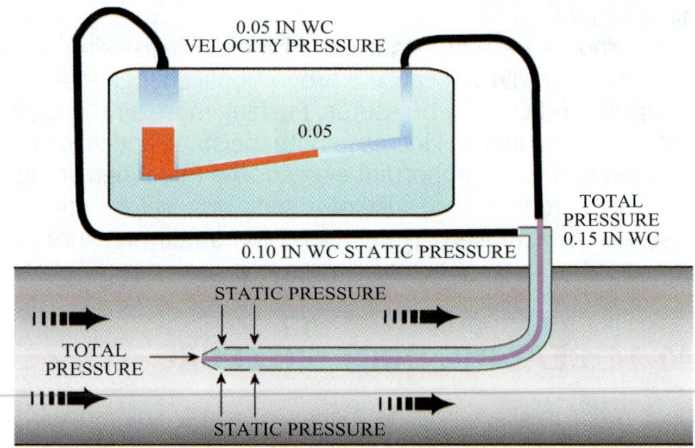

Figure 41-45 A pitot tube reading velocity pressure in a duct.

41.15 CALCULATING AIR-CONDITIONING SYSTEM PERFORMANCE

The psychrometric chart can be useful in checking the performance of air-conditioning equipment, as shown in the following example. Because the wet bulb and enthalpy lines nearly overlap, the wet-bulb temperature is really all that is needed to determine the enthalpy. Four measurements are necessary to calculate the actual cooling capacity of an air-conditioning system:

- Wet-bulb temperature of the entering air
- Wet-bulb temperature of the leaving air
- Dry-bulb temperature of the leaving air
- Airflow volume in CFM
- Entering-air dry bulb (only necessary if sensible and latent values are to be calculated)

The wet-bulb temperatures are used to determine the enthalpy of the entering and leaving air. The difference in the enthalpy between the entering and leaving air will show the amount of heat lost from each pound of air. The dry-bulb and wet-bulb temperatures of the leaving air are used to determine the volume of each pound of air. The CFM and this volume are used to determine the pounds of air per minute. The total cooling capacity in BTU per minute is

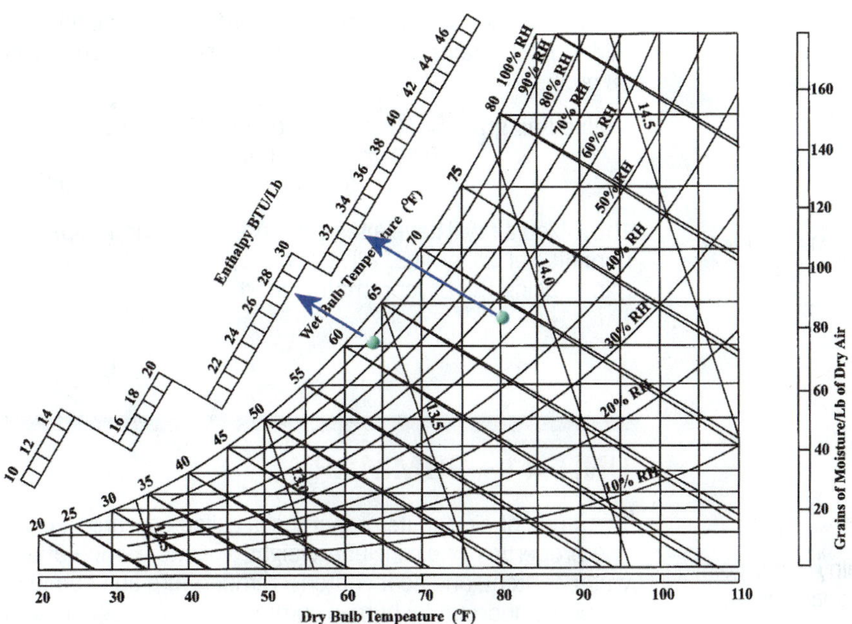

Figure 41-46 The enthalpy of the entering air is 33 BTU/lb, and the enthalpy of the leaving air is 28 BTU/lb.

determined by multiplying the enthalpy difference in BTU times the pounds per minute of air. This value is multiplied by 60 to show the cooling capacity in BTU/hr. This can be summarized by the following formula:

Enthalpy change = enthalpy of air in − enthalpy of air out

Pounds/minute of air = CFM airflow/cubic feet per pound of air

BTU/hr = 60 × pounds of air/minute × BTU/lb enthalpy change

Example Calculation

An air conditioner is operating with a return air temperature of 80°F db, 69°F wb. The supply air temperature is 63°F db and a 61°F wb with an airflow of 850 CFM. Referring to Figure 41-46, the enthalpy of the return air is 33 BTU/lb, the enthalpy of the supply air is 28 BTU/lb and the volume of the supply air is a little less than 13.5 ft^3/lb. To calculate the system capacity:

Enthalpy difference = 33 BTU/lb − 28 BTU/lb
= 5 BTU/lb

Pounds/minute circulated = 850 ft^3/min/13.5 ft^3/lb
= 63 lb/min

Total capacity calculation = 60 × 63 lb/min × 5 BTU/lb
= 18,900 BTU/hr

Sensible and Latent Capacity Calculations

To determine the sensible and latent capacities of this system, a third point is added to form a right triangle (Figure 41-47).

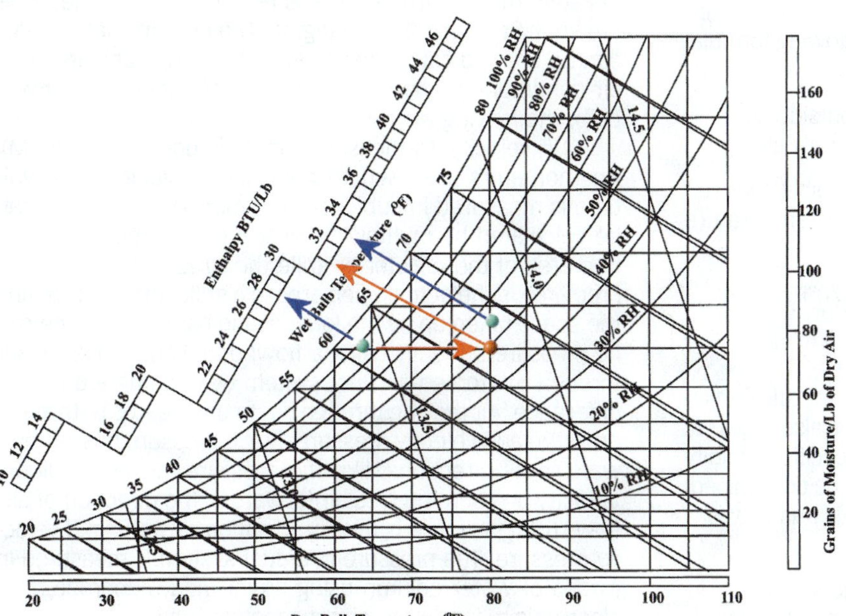

Figure 41-47 The enthalpy of the third point that forms the right triangle is 31 BTU/lb.

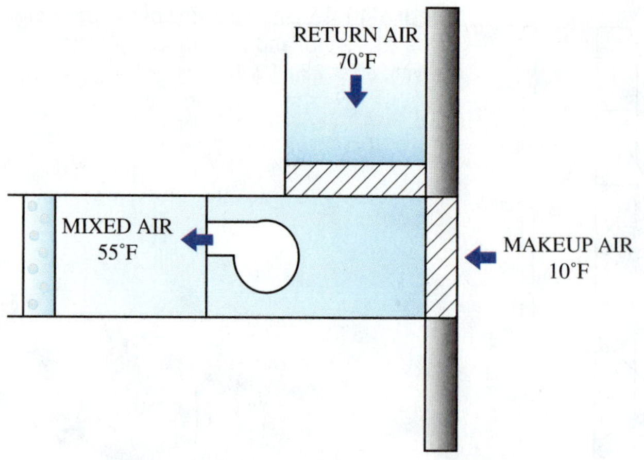

Figure 41-48 Mixed-air example.

This point is at 80°F db and 66°F wb. The enthalpy of this point is 31 BTU/lb. To calculate sensible and latent capacities:

$$\text{Sensible change} = 31\ \text{BTU/lb} - 28\ \text{BTU/lb} = 3\ \text{BTU/lb}$$

$$\begin{aligned}\text{Sensible capacity} &= 60 \times 63\ \text{lb/min} \times 3\ \text{BTU/lb}\\ &= 11{,}340\ \text{BTU/hr}\end{aligned}$$

$$\text{Latent change} = 33\ \text{BTU/lb} - 31\ \text{BTU/lb} = 2\ \text{BTU/lb}$$

$$\begin{aligned}\text{Latent capacity} &= 60 \times 63\ \text{lb/min} \times 3\ \text{BTU/lb}\\ &= 7{,}560\ \text{BTU/hr}\end{aligned}$$

41.16 AIR MIXTURES

It is sometimes necessary to calculate the expected result of mixing two different airstreams. Ventilation air in commercial systems is one example. Most codes require at least 25 percent fresh ventilation air in commercial installations. This means that 25 percent of the air entering the conditioned space came from outside. Typically, the air from outside is called makeup air, the air returning from the building is called return air, and the result of mixing the two airstreams is called mixed air (Figure 41-48). The expected temperature of the mixed air can be determined by the following formula:

$$\begin{aligned}&\text{Mixed-air temperature}\\ &= (\text{makeup-air temperature} \times \text{percent outside air})\\ &+ (\text{return-air temperature} \times \text{percent return air})\end{aligned}$$

For example, a system with 25 percent fresh air with an outdoor temperature of 50°F and a return-air temperature of 70°F would be calculated as follows.

$$\begin{aligned}\text{Mixed-air temperature} &= (50°\text{F} \times 0.25) + (70°\text{F} \times 0.75)\\ &= 12.5°\text{F} + 52.5°\text{F}\\ &= 65°\text{F}\end{aligned}$$

Similar calculations can describe the final specific humidity of the mixed air if the water vapor content is also required. By determining both the temperature and specific humidity of the mixed air, all aspects the mixed air can be determined. The formula to determine the mixed-air specific humidity is:

$$\begin{aligned}&\text{Mixed-air grains}\\ &= (\text{outdoor-air grains} \times \text{percent outdoor air})\\ &+ (\text{return-air grains} \times \text{percent return air})\end{aligned}$$

In the previous example, if the outdoor-air specific humidity is 40 grains and the return-air specific humidity is 28 grains:

$$\begin{aligned}\text{Mixed-air grains} &= (40\ \text{grains} \times 0.25) + (28\ \text{grains} \times 0.75)\\ &= 10\ \text{grains} + 21\ \text{grains}\\ &= 31\ \text{grains}\end{aligned}$$

The mixed air will have a temperature of 65°F and a specific humidity of 21 grains. All other properties can be determined using the psychrometric chart.

UNIT 41—SUMMARY

The study of air and its properties is called psychrometrics. The properties of air include weight, volume, temperature, water content, and heat content. Temperature and humidity are both important for human comfort. The body regulates its internal temperature by dissipating heat to its surroundings through conduction, convection, radiation, and evaporation. The weight and volume of air is affected by its temperature. Increasing the temperature of air increases its volume and decreases its weight. Decreasing its temperature decreases its volume and increases its weight. The amount of water in the air is measured by comparing a regular dry-bulb temperature to a wet-bulb temperature. The closer the two temperatures are to each other, the higher the relative humidity. Specific humidity is the amount of water in the air measured in grains. Dew point is the temperature that water vapor can start to condense out of the air.

The psychrometric chart is a graphical representation of air properties. If two properties are known, the other properties can be determined from the chart. Processes that cause a horizontal shift from the start to the finish represent a sensible heat change because the air temperature is changing. Processes that cause a vertical shift from the start to the finish represent a latent heat change because the air temperature is remaining the same while the specific humidity is changing. The psychrometric chart can be used to calculate the actual cooling capacity of an air-conditioning system by measuring the entering-air wet bulb, the leaving-air wet bulb, and the airflow.

Air velocity is measured in feet per minute (FPM). Anemometers are used to measure air velocity. Air volume is measured in cubic feet per minute (CFM). CFM can be calculated by multiplying air velocity in FPM times the free area of the register or grille the air is moving through. Free area is the actual open area in a grille after subtracting the space taken up by the louvers and bars. Air volume can be measured directly using a flowhood. Fans move the air through the forced-air duct system by creating a pressure difference. Air moving through ductwork creates both static pressure and velocity pressure. Static pressure pushes outward in all directions, like the pressure inside a balloon. Velocity pressure pushes air forward in the direction of airflow. Total pressure is the combination of static and velocity pressure. The pressure in a duct system is measured in inches of water column using an inclined manometer, an electronic manometer, or a magnehelic gauge.

WORK ORDERS

Service Ticket 4101

Customer Complaint: House Feels Stuffy

Equipment Type: High-Efficiency Split-System Air Conditioner 2-Ton Condensing Unit with AHRI Matched 2½-Ton Evaporator Coil

A customer complains that his house feels stuffy since his new high-efficiency system was installed even though the thermostat is at the same setting that he used to keep the old system. The technician notices that there does not seem to be any water coming out of the evaporator drain line. Measurements of the return and supply air show 80°F db, 68°F wb return air, and 64°F db, 63°F wb supply air. Consulting the psychrometric chart, the technician sees that the specific humidity is the same for both readings, indicating that all the cooling is sensible cooling and no water is being removed. The indoor fan ECM speed control board is set to 3 tons at 450 CFM per ton, but the cooling capacity of the condensing unit is only 2 tons. The technician moves the fan speed tap to 2 tons at 350 CFM per ton, reducing the overall airflow from 1,350 CFM to 700 CFM. New readings are taken after operating for 15 minutes at the new fan speed. The return air is about the same, but the supply air temperature has changed to 62°F db and 58°F wb. Water starts coming out of the evaporator drain.

Service Ticket 4102

Customer Complaint: Drafty Air Coming from Supply Registers in Winter

Building: Commercial Office Building with Fresh-Air Ventilation System

The customer complains that the air coming out of the supply registers feels drafty and the building is chilly. Upon arrival, the technician notices that the building has an unusually high positive pressure; the doors are almost standing open. The technician measures the air leaving the supply air registers and finds it is 85°F. Next, the boiler and water pumps are checked. The water flow is correct and the boiler appears to be operating correctly. The temperature rise across the hot water coil is 50°F, but the air going into the coil is only 40°F. The technician measures the outdoor air temperature as 32°F and the return-air temperature as 64°F. A quick calculation confirms that the air mixture is 75 percent outside air. For 25 percent outdoor air, the mixed-air temperature should be 56°F. An adjustment of the minimum air position of the damper control fixes the problem, and the supply registers start delivering 100°F supply air. After the building heats up, the supply air rises to 110°F.

Service Ticket 4103

Customer Complaint: Indoor Swimming Pool Area Is Muggy

Building: Large Residence with Indoor Pool Area that Is Not Air-Conditioned but Is Ventilated

The customer complains that the pool area feels oppressively muggy and that water is running down the walls. The technician takes both a dry-bulb and a wet-bulb temperature and finds that the wet bulb is only 2°F cooler than the dry bulb. The technician suspects that there may be a problem with the ventilation air and checks the ventilation air inlet. There is very little ventilation airflow. The technician finds that the inlet screen to the ventilation blower is clogged with leaves. Once the leaves are cleared, the ventilation air returns. After an hour of proper ventilation, the wet-bulb temperature and relative humidity in the room drop to normal levels.

Service Ticket 4104

Customer Complaint: Finished Basement Is Musty Smelling

Building: Residence with a 1,000-ft^2 Finished Basement

The customer complains that the basement is musty smelling. The temperature is fine, but the basement smells musty and feels dank. The technician takes both a dry-bulb and a wet-bulb reading in the basement. The dry bulb is 70°F and the wet bulb is 65°F. The technician aligns the dry bulb and wet bulb on the scales built into the sling psychrometer and reads a relative humidity of 76 percent. Next, the technician consults a psychrometric chart and finds that the difference in the specific humidity of 76 percent relative humidity and 50 percent relative humidity is about 30 grains. The technician calculates the basement volume at 8,000 ft^3 (1,000 ft^2 × 8 ft), and the weight of air at 588 lb (8,000 ft^3/13.6 ft^3/lb). The house construction is average, consistent with one air change per hour. The technician recommends a dehumidifier with a capacity to remove the water in one air change. The amount of water needing to be removed is 588 lb of air × 30 grains/lb = 17,648 grains = 2.5 lb of water, about 2.5 pints. He recommends a unit with a capacity of 2.5 pints of water per hour, or 60 pints per day.

Service Ticket 4105

Job: Measure Airflow Through New Supply Register

Building: Residence with a Finished Basement

A technician is sent to check the installation of a newly installed register in a finished basement. The plans call for an airflow of 150 CFM through the 8 × 10 register. The technician looks up the register free area and finds that it is 48 in^2. He uses an averaging anemometer to take the average velocity of the air leaving the register. The average velocity measures 450 FPM. He converts 48 in^2

to square feet: 48/144 = 0.33 ft^2. Finally, the free area in square feet is multiplied by the velocity of 450 FPM to arrive at a CFM of 150 CFM.

Service Ticket 4106

Job: Measure Airflow in a Duct

Equipment Type: Residential Split System

A service technician is sent to check the airflow in the trunk duct of a new residential split-system installation to ensure that the air velocity is less than 1,000 FPM. She makes a small ¼-in hole in the duct and inserts a pitot tube. Next, she connects the total pressure hose to the high input of the magnehelic gauge and the static pressure hose to the low input on the magnehelic gauge. The magnehelic reads a velocity pressure of 0.06 in water column (wc). The technician calculates the velocity: $4{,}005 \times \sqrt{0.06} = 981$ FPM.

UNIT 41—REVIEW QUESTIONS

1. What three environmental conditions have the most effect on comfort?
2. List four ways the human body expels heat.
3. Why does the air's relative humidity affect comfort?
4. How is the air volume affected by temperature?
5. What is the relationship between air volume and air weight (density)?
6. What are grains, and what air property is measured in grains?
7. In general, which is heavier: humid air or dry air?
8. What is a wet-bulb temperature reading?
9. How is relative humidity measured?
10. What is the dew-point temperature?
11. What is the difference between specific humidity and relative humidity?
12. What is a psychrometric chart?
13. List the air properties that are plotted on a psychrometric chart, and give a brief description of each.
14. What would the mixed-air temperature be for a system with 20 percent outdoor air if the outdoor temperature is 10°F and the return-air temperature is 75°F?
15. Identify the scales represented by the lines on Figure 41-49.

    A. B.
    C. D.
    E. F.
    G.

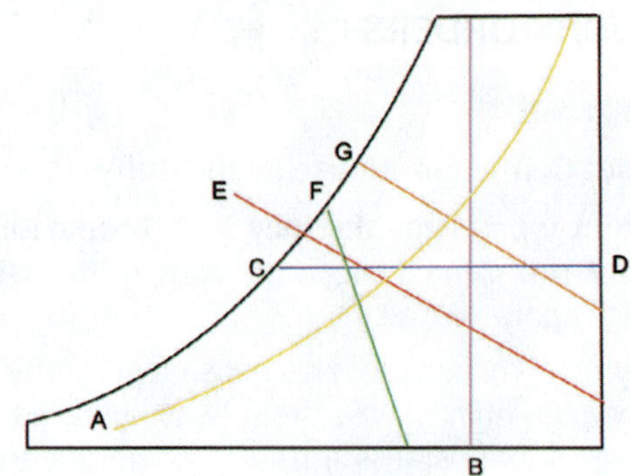

Figure 41-49 Figure for Review Question 15.

16. Use the psychrometric chart in Figure 41-1 to look up the missing data in this table.

Dry Bulb	Wet Bulb	Relative Humidity	Dew Point	Specific Humidity	Volume	Enthalpy
72				59 grains		
	61	45%				
85		60%				
			45			28
85	70					

17. What pressure units are used for measuring duct pressures?
18. What two pressures are created by air moving through the duct system?
19. What instruments are used for measuring duct pressure?
20. Explain how to take a static pressure reading on a duct system.
21. Explain how to take a velocity pressure reading on a duct system.
22. What units are used to measure air velocity?
23. What tools are used to measure air velocity?
24. What units are used to measure air volume?
25. What is the airflow through a register with a free area of 72 in^2 if the air velocity is 300 FPM?
26. How does a flowhood measure airflow?
27. What is the velocity of the air in a duct with a total pressure of 0.15 in wc and a static pressure of 0.1 in wc?
28. What is the airflow in CFM through a 10 × 20–in duct with a total pressure of 0.15 in wc and a static pressure of 0.1 in wc?
29. What is the typical airflow per ton of residential air-conditioning equipment?

UNIT 42

Air Filters

OBJECTIVES

After completing this unit, you will be able to:

1. list the types of air filters.
2. list the types of air contaminants.
3. explain the MERV efficiency rating system.
4. explain how an electronic air cleaner works.
5. discuss the importance of pressure drop and airflow rating when selecting an air filter.

42.1 INTRODUCTION

Total comfort involves more than heating and cooling the air. As the public becomes more aware of indoor air quality and energy efficiency, they also are becoming more aware of air cleanliness. Advanced air cleaners that remove minute particles from the air are gaining in popularity as customers expect their systems to be part of an indoor environmental conditioning system. Installing and servicing advanced filters is part of an air-conditioning technician's job description.

42.2 AIR CONTAMINANTS

Anything in the air besides nitrogen, oxygen, and the normal mix of trace gases can be considered a contaminant. Common air contaminants include dust, smoke, biological contaminants, and chemical contaminants. The size of contaminant particles in the air is measured in microns. A micron is an SI unit of measurement equal to 1/1,000 of a millimeter. There are approximately 25,400 microns in an inch. A human hair is about 100 microns in thickness. The particles suspended in the air vary in size from under 0.3 microns to greater than 10 microns (Figure 42-1). Particles must be 10 microns or larger to be visible with the naked eye. Over 99 percent of the particles in the air are below 1 micron in size.

The chart in Figure 42-1 shows that pollen is relatively easy to remove, but a filter must be capable of removing particles down to 0.3 microns in size to remove airborne bacteria. The smallest particles of tobacco smoke are as small as 0.01 microns. These minute particles can be removed from the air with either a very-high-efficiency mechanical filter or an electronic air cleaner. The common throwaway filters used in many furnaces are only capable of removing particles over about 10 microns in size. They are fine for pollen and sawdust but not very effective on tobacco smoke.

42.3 TYPES OF AIR FILTERS

The most common types of air filters use mechanical media to trap contaminants. Figure 42-2 shows a magnified view of fine fiber filter media. Some filters use electrostatically charged media or a viscous coating to help attract particles in the air. Filters that use an electrostatic charge can be

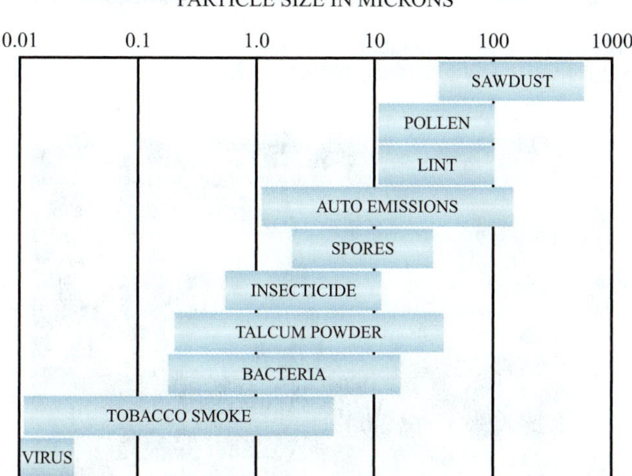

Figure 42-1 Size of common air contaminants.

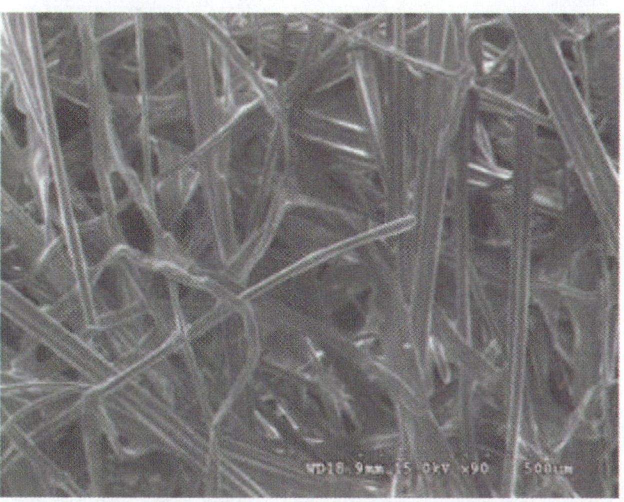

Figure 42-2 Magnified view of a fine fiber media used for filtration. *(Courtesy of NIOSH)*

Figure 42-3 Air filters are available in a wide variety of efficiencies and configurations. *(Courtesy of AAF International)*

divided into mechanical media filters with an electrostatic charge and electronic air cleaners that use an active power supply and charged metal plates. Mechanical filters are available in a wide variety of constructions and efficiencies from simple panel filters to extended surface filters like pleated filters, cube filters, and bag filters (Figure 42-3). They can be generally classified by their efficiency as low-efficiency, medium-efficiency, or high-efficiency particle air filters (HEPA).

Low-Efficiency Panel Filters

Most residential equipment comes with a low-efficiency, 1-in-thick panel filter installed (Figure 42-4). Commercial equipment typically uses similarly constructed 2-in filters. The purpose of these filters is to protect the equipment; they do little for cleaning the air. They are inexpensive and create very little pressure drop across the filter.

Figure 42-4 Most residential air-conditioning equipment is supplied with low-efficiency 1-in panel filters.

Pleated Medium-Efficiency Filters

Pleated 4-in and 5-in medium-efficiency filters are available for both residential and commercial installation (Figure 42-5). These filters do improve the air quality of the building, as well as protect the system. Medium-efficiency filters create a higher pressure drop across the filter, but the pressure drop is still usually within the operating limits of most residential and light commercial equipment.

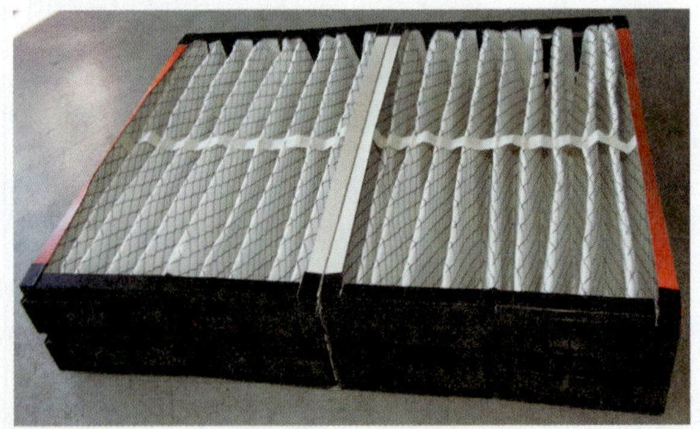

Figure 42-5 Medium-efficiency pleated filter.

HEPA Filters

HEPA stands for high-efficiency particulate air filter. HEPA filters were originally developed to remove radioactive contaminants from the air during the development of the atomic bomb. HEPA filters have a very large extended surface area of submicron media (Figure 42-6). Their surface area is extended by forming long bags or large pleats with deep folds. HEPA filters have a very high efficiency at removing very small particles. There are many different standards for HEPA filters, ranging from 99.97 percent efficiency with particles down to 0.3 microns to 99.99 percent efficiency with particles down to 0.12 microns. HEPA filters typically create too high a pressure drop to be used in residential applications. They are used in commercial applications like clean rooms and surgical suites, where removing airborne contaminants is essential.

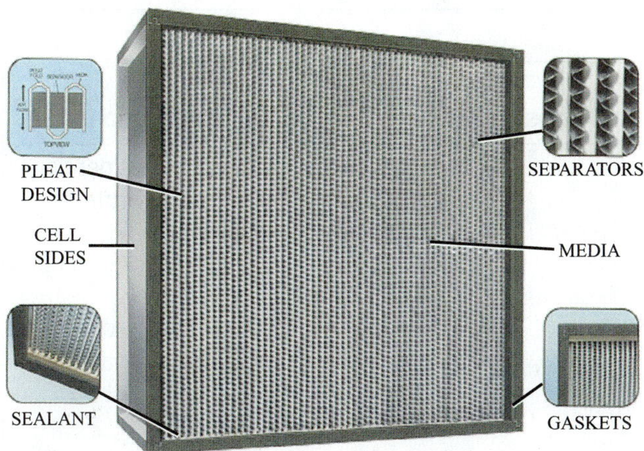

Figure 42-6 HEPA filter. *(Courtesy of Camfil Farr)*

SERVICE TIP

In some commercial applications, a prefilter is used in front of HEPA filters to extend the life of the more expensive HEPA filters. In some applications, this prefilter media can be mounted on spools so that new filter material can be simply unwound from one spool as the dirty filter material is wound on the takeup spool. The dirty spool of filter material is significantly heavier than a clean spool.

42.4 AIR FILTER RATINGS

There are four important factors to be considered when choosing an air filter: the size of the contaminants the filter must remove, its efficiency at removing the contaminants, the airflow the filter is rated to handle, and its resistance at the rated airflow.

Filter Efficiency

Many different methods have been devised to measure air filter efficiency. The two most commonly used methods are called the filter arrestance and the dust spot efficiency. The arrestance method measures the weight of the material captured by the filter and compares it to the weight of the test material that was released. A filter with an efficiency 50 percent arrestance captures half of the material passing through the filter when measured by weight. Special test dust with calibrated particle sizes is used for these tests. Although many filters are advertised by stating an average arrestance percentage, filter efficiency is not as simple as a single percentage.

All filters capture different size particles at different levels of efficiency. A filter that removes 99 percent of the particles larger than 10 microns may only remove 1 percent of the particles smaller than 1 micron. But since large particles weigh more than small particles, the average arrestance efficiency can be quite high even though most of the particles pass through.

The dust spot efficiency test is designed to measure a filter's ability to remove very small particles that can easily pass through many filters and cause staining on walls and surfaces. The dust spot method uses atmospheric dust instead of calibrated test dust. Air is passed over test paper to achieve a stain from deposits of small particles in the air. This is done both before and after the filter. The efficiency is calculated by comparing the amount of "clean" downstream air required to create the same stain as the dirty upstream air.

Filter efficiency ratings are complicated by the fact that the efficiency of all filters changes as they become dirty. Most mechanical filters become more efficient as they fill with particles and the space for particles to pass through becomes smaller. For this reason, many manufacturers advertise a peak arrestance efficiency that is measured when the filter is ready to be changed.

ASHRAE Standard 52.2-2007 addresses these problems by establishing a rating system that measures the arrestance of particles in twelve different ranges of size (Figure 42-7). The minimum efficiency reporting value, or MERV, looks at the efficiency of a filter over each range with different levels of filter loading. "Filter loading" is a term that describes the amount of material the filter has collected. The minimum efficiency for each range is then used to determine the filter's MERV rating. MERV ratings range from 1 for the lowest efficiency filters to 20 for the highest efficiency filters. Figure 42-7 shows the specifications for the MERV ratings. As a filter's MERV rating increases, so does its pressure drop.

One advantage of the MERV rating is that a user can match the MERV rating to the size of contaminant that must be removed. Filters with a relatively low MERV rating can effectively remove pollen, but a high MERV rating is required to remove bacteria.

Airflow

Filters have a maximum airflow rating. Exceeding the airflow rating of a filter will decrease its efficiency and increase the pressure drop across the filter. Operating a filter at decreased airflow levels will increase its efficiency and decrease its pressure drop. The recommended airflow is based on the filter face velocity—that is, the speed of the air passing through the filter. Increased airflow produces increased airspeed, which reduces the effectiveness of the filter.

TECH TIP

It is important to understand that a MERV rating should always be accompanied by a specified air velocity. Air velocities greater than the velocity tested will result in decreased filter performance.

Pressure Drop

Pressure drop is created across an air filter because of its resistance to airflow. This resistance ranges from 0.1 in wc for clean low-efficiency filters to over 1 in wc for some HEPA filters. Filter manufacturers specify the recommended final

MERV	Composite Average Particle Efficiency Range E1 0.3 – 1.0 microns	 Range E2 1.0 – 3.0 microns	 Range E3 3.0 – 10.0 microns	Effective On	Applications
1	Not applicable	Not applicable	E3<20%	Pollen, dust mites, pet hair, carpet and textile fibers	Minimum Residential filtration to protect equipment
2	Not applicable	Not applicable	E3<20%		
3	Not applicable	Not applicable	E3<20%		
4	Not applicable	Not applicable	E3<20%		
5	Not applicable	Not applicable	20%≤E3<35%	Mold spores, cement dust	Most commercial and better residential
6	Not applicable	Not applicable	35%≤E3<50%		
7	Not applicable	Not applicable	50%≤E3<70%		
8	Not applicable	Not applicable	70%≤E3		
9	Not applicable	E2<50%	85%≤E3	Legionella, lead dust, auto emissions	Superior residential and better commercial
10	Not applicable	50%≤E2<65%	85%≤E3		
11	Not applicable	65%≤E2<80%	85%≤E3		
12	Not applicable	80%≤E2	90%≤E3		
13	E1<75%	90%≤E2	90%≤E3	All types of bacteria, tobacco smoke, most forms of smoke	Hospitals, surgical suites, superior commercial buildings
14	75%≤E1<85%	90%≤E2	90%≤E3		
15	85%≤E1<95%	90%≤E2	90%≤E3		
16	95%≤E1	95%≤E2	95%≤E3		
	HEPA Filters				
17	>99.97% of 0.30 micron particles			Viruses, carbon particles	Cleanrooms, pharmaceutical manufacturing
18	>99.99% of 0.30 micron particles				
19	>99.999% of 0.30 micron particles				
20	>99.9999% of 0.10 – 0.20 micron particles				

Figure 42-7 MERV parameters.

pressure drop for their filters. This is the maximum pressure the filter is designed to withstand. Typically, the final pressure is produced when the filter is holding its maximum capacity of dirt. The filter must be changed at this point. However, most filters should be changed before they reach this point. Most residential air-conditioning equipment cannot operate correctly at pressure drops exceeding 0.5 in wc across the blower. All of that cannot be used for the filter—some of that pressure drop is used by coils, ductwork, and registers. Most of these systems will not move enough air if the pressure drop across the air filter starts to exceed 0.2 in wc. An air filter in a residential air-conditioning system with a pressure drop of 0.5 in wc across it is letting very little air through.

The filter's airflow resistance can be measured with an inclined manometer, digital manometer, or magnehelic gauge. It is common practice on large commercial central systems to include a magnehelic gauge or inclined manometer to check the differential pressure across for the filter section (Figure 42-8). This instrument compares the air pressure on the upstream and downstream sides of the filter assembly. The gauge on the instrument indicates when the filters should be changed.

SERVICE TIP

There are 1-in pleated high-efficiency filters for sale whose pressure drop when clean approaches the limit of most residential equipment. These filters cause airflow problems for many systems when they become just a little dirty. Always check a filter's airflow rating and pressure drop, not just its advertised efficiency. There are many filters for sale that simply will not work in most residential systems.

42.5 MECHANICAL FILTRATION

Mechanical air filters work by passing the air and contaminants through a media. The contaminants get stuck in the fibers and the air passes through. A number of materials have been used as the filtering media, including coarse glass fibers, animal hair, vegetable fibers, synthetic fibers, metallic wool, expanded metals and foils, crimped screens, random matted wire, and synthetic open cell foams (Figure 42-9). In

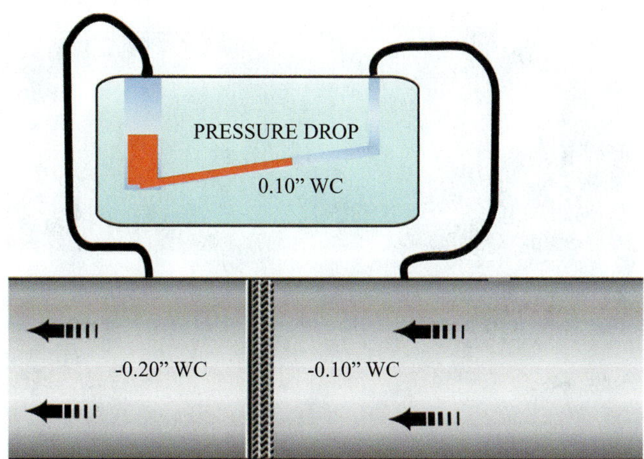

Figure 42-8 Inclined manometer reading pressure drop across air filter.

general, finer fibers produce higher filter efficiency. Filter efficiency also increases as the fiber density increases. However, the air pressure drop across the filter also increases as the filter density increases. The efficiency of mechanical filters increases as the filter becomes dirty because it becomes more and more difficult for particles to pass through the filter. The pressure drop across the filter also increases as the filter gets dirty, reducing the airflow through the filter.

Filter Coatings

The efficiency of coarse fiber media is sometimes increased by coating the media with a viscous substance, such as oil, which acts as an adhesive to the airborne particles coming into contact with it. Another technique is to use materials with an electrostatic charge to attract particles. One problem with these strategies is that they lose their effectiveness as the filter becomes dirty.

Extended-Surface-Area Filters

The area of the filter material exposed to the air can be extended to improve efficiency and minimize the pressure drop created by denser filter material. Extended-area filters are made of random fiber mats or blankets of bonded glass fiber, wool felt, or synthetic materials. Pleating of the media provides a larger filter surface area compared to the face area. This allows for higher filtering efficiency at a reasonable pressure drop. The efficiency of uncoated extended-surface-area filters is usually higher than the viscous coated coarse fiber filters. In addition to their effectiveness, the extended surface filters have a larger dust-holding capacity for longer periods of use.

Figure 42-9 Different filter media; from top left; expanded metal, polyester, foam, hog's hair, light density fiberglass, pleated high density fiberglass.

42.6 PANEL FILTERS

The low-efficiency filters that come as standard equipment on forced-air heating and cooling equipment are panel filters. Both disposable and washable panel filters are available. The dust-holding capacity of panel filters is generally low because of the limited amount of media in them. They need frequent replacement or cleaning; typically every month. These filters protect the equipment but do not do much for cleaning the air.

SERVICE TIP

Panel filters are directional; the airflow is intended to flow through the filter in only one direction. The direction is clearly marked on the side of the filter with an arrow (Figure 42-10). Look for the directional arrow when installing panel filters and install the arrow pointing toward the unit.

Disposable Panel Filters

Disposable panel filters consist of a cardboard frame with a grill to hold the filter material in place (Figure 42-11). The filter media is made of glass fibers that are loosely woven on the entering-air side and more densely packed on the leaving air side. Some makes also have the fiber media coated with an adhesive substance to attract and hold dust and dirt. Filter thickness is usually 1 in for residential equipment and 2 in for commercial equipment. These filters typically have an initial clean resistance of 0.1 in wc with a face velocity of 300 FPM. The maximum face velocity is typically around 300 FPM, and the maximum final pressure drop is typically between 0.3 to 0.5 in wc.

Figure 42-10 The directional arrow on panel filters should point toward the unit.

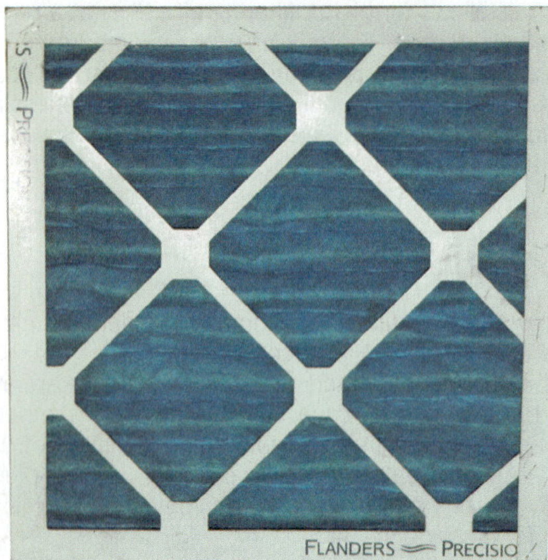

Figure 42-11 Throwaway filter.

Figure 42-12 Reusable filter.

Washable Panel Filters

Permanent washable panel filters are also available. These filters may be removed, cleaned with detergent, dried, and reinstalled. They typically have a metal or plastic frame with a cleanable media (Figure 42-12). The media is typically expanded metal or foam. Another washable filter is the hog's hair filter (Figure 42-13). It is made from hog's hair, bonded together in a mat. These filters have enough rigidity that they do not need a frame, but they do frequently have metal rods running through them to help stiffen them.

SERVICE TIP

Getting the homeowner to keep filters clean is not easy, and dirty filters are probably the main contributor to malfunctioning equipment. As the mechanical filter clogs with dirt, airflow is reduced to a point where the cooling coil will freeze, causing compressor failure. In a heating system, dirty filters can cause overheating and reduce the life of the heat exchanger or cause nuisance tripping of the limit switch. Operating costs increase with loss of efficiency.

Figure 42-13 Hog's hair filter.

Figure 42-14 Pleated replaceable media filter.

42.7 RENEWABLE-MEDIA FILTERS

Renewable-media filters keep the structural components of the filter and replace the media when it becomes dirty. Two common types of renewable-media filters are pleated filters and moving-curtain filters. Renewable-media pleated filters use a frame and combs to support and separate the pleated material (Figure 42-14). When the material becomes dirty, the media is replaced, but the existing frame and combs are reused for the replacement media. Similar filters are available that hold the pleated filter material in a cardboard frame. For these, the entire cardboard box is replaced (Figure 42-15).

Renewable pleated filters are used in residential and light commercial applications. These filters are generally medium efficiency. The increased amount of media gives these filters a much higher dust-holding ability. They do not require changing as often as the disposable panel filters. Most pleated 4-in and 5-in-deep extended-media filters will last between 6 months to a year between changes.

The moving curtain filter is an automatic moving curtain with the random fiber medium, treated with a viscous material and furnished in roll form. The material rolls down

Figure 42-15 Pleated boxed filter.

from the top of the unit. As the exposed area becomes saturated with dirt, a clean section is automatically rolled into place. The used portion is collected in a roll at the bottom and thrown away. A fresh roll is again placed at the top to the unit and the filtering continues.

Figure 42-16 shows a moving curtain renewable-media filter. Moving-curtain filters are used in commercial applications to remove lint and aerosols in textile mills, dry cleaners, and press rooms.

Figure 42-16 Rolling curtain renewable media filter.

42.8 ELECTROSTATIC FILTERS

Did you ever notice how electronic components collect dust? The dust is attracted to them because of their electrostatic charge. Electrostatic filters make use of this principle to increase their efficiency. Some disposable filters are manufactured using materials that have an inherent electrostatic charge. However, this charge tends to dissipate as the filter gets dirty, and the filter efficiency declines. Another type of electrostatic filter uses materials that produce static electricity when air flows across the filter. A static electric charge is developed as the air flows over the material, charging the particles. They are then attracted to the media inside the filter. A filter media, such as foam or polyester fibers, is sandwiched in between the statically charged panels (Figure 42-17). These filters typically have a relatively high initial air pressure drop, which increases as they become dirty. These filters must be cleaned monthly to prevent airflow problems.

SERVICE TIP

It is very important to clean electrostatic filters often. It can be very difficult to clean one of these once it has become clogged with dirt. This is especially true of the filters that use a fuzzy-appearing polyester media in the center. Once one of these becomes excessively dirty, it just becomes an expensive disposable filter.

42.9 ELECTRONIC AIR CLEANERS

The operating efficiency of electronic air cleaners varies with airflow. Each model may be used over a range of airflow quantities (CFM). The stated efficiency will be rated at a nominal CFM. The operating efficiency at other airflows within the filter's operating range will be listed separately. Higher airflow produces lower efficiency. Residential electronic air cleaners come in sizes of 800 CFM to 2,000 CFM. Static pressure drop will run about 0.20 in wc at the nominal

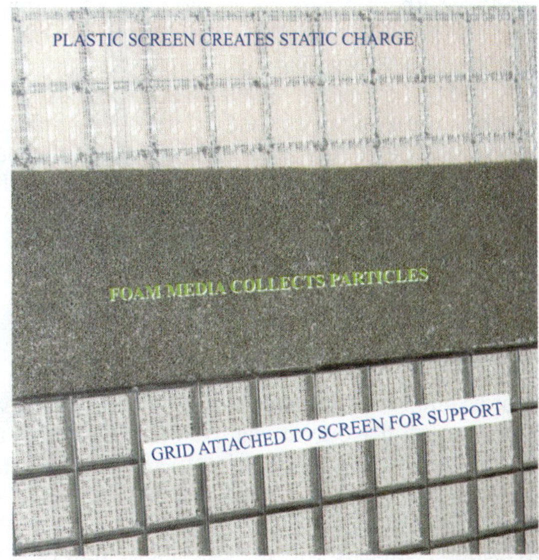

Figure 42-17 Composition of electrostatic air filter.

Figure 42-18 The ionization section of an electrostatic air cleaner consists of ionizing wires alternating with grounded plates.

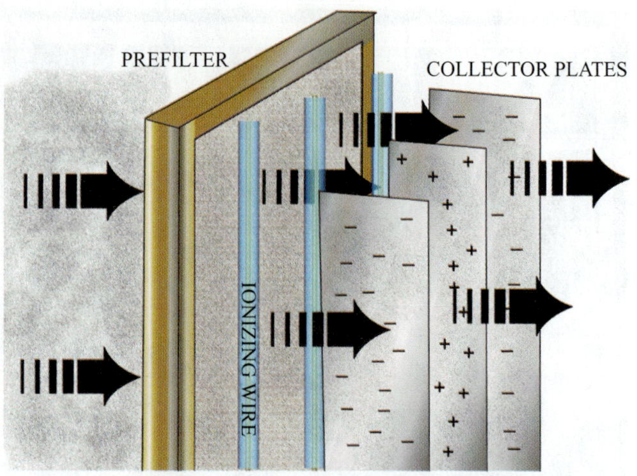

Figure 42-19 Operation of two-stage electrostatic precipitation.

airflow rating point. Ratings are published in accordance with the National Bureau of Standards Dust Spot Test and are certified under AHRI standards. There are three types of units used for commercial service: ionizing plate, charged-media nonionizing, and charged-media ionizing.

Ionizing Plate Air Cleaners

Ionizing plate electronic air cleaners use high-voltage DC current to ionize particles in the air and attract them to oppositely charged plates. A series of thin wires is suspended between a group of grounded plates (Figure 42-18). A high-voltage DC current produces an electric field between negatively charged thin wires with and grounded plates. The dust particles passing through receive a negative charge from the electrical field. This is called the ionizing section because it is ionizing the dust particles. The collection section is a series of plates located just behind the ionizing section. The ionized dust particles are attracted to the oppositely charged collection plates and fasten onto them (Figure 42-19). This process is called two-stage electrostatic precipitation because it occurs in two stages and the dust particles precipitate out of the air.

The voltage is typically 6,000–10,000 V DC. The plates are parallel to the airstream and offer little resistance to the airflow. For best results, the airflow through the plates should be evenly distributed. Conventional-type prefilters are used ahead of the electronic air cleaners to screen out the larger particles in the air (Figure 42-20).

Figure 42-20 The prefilter is seen on the left; the prefilter on the right has been removed.

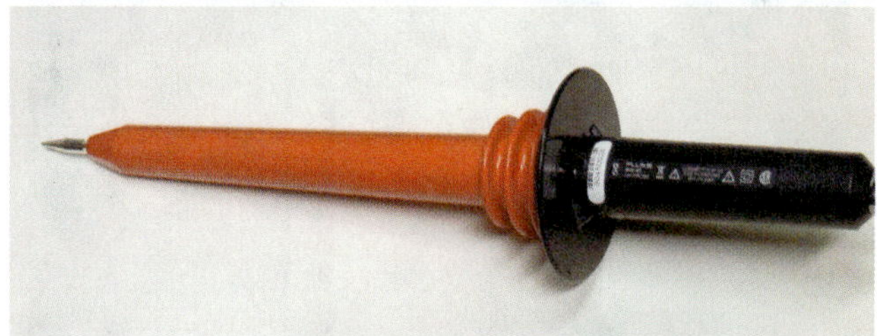

Figure 42-21 This high-voltage probe and a standard digital multimeter can be used to read the high DC voltage on an electronic air cleaner circuit.

SAFETY TIP

Do not attempt to use a standard multimeter and lead set to measure the output voltage on an electronic air cleaner. The voltage is much higher than either the leads or meter is designed for. The voltage can be read using a special high-voltage probe (Figure 42-21). This probe reduces the voltage to a level that can be safely read by a standard multimeter.

The collector plates must be periodically cleaned, as their efficiency declines as they get dirty. Most collector plates can be removed easily and can be placed in a dishwasher for cleaning. Due to the high voltage used by the electronic units, safety switches are provided that will turn off the power when the filter access door is opened. A few filters have built-in cleaning mechanisms. They use water sprayed on the plates in place. Suitable drains are provided in the bottom of the filter compartment. Ionizing plate air cleaners should be cleaned two to four times a year.

SERVICE TIP

Electronic air cleaners do not make dirt disappear; they collect it. Cleaning the plates is essential. If the plates are allowed to get excessively dirty, arcing can occur between the plates, which burns the dust to a fine white ash. This fine white powdery ash then escapes the air cleaner and can pack in the fins of the indoor coil. Adding an electronic air cleaner and not cleaning it is really worse than using low-efficiency filters.

Charged-Media Nonionizing Air Cleaners

The charged-media nonionizing filters are very different in construction. These filters consist of a dielectric filtering medium, usually arranged in pleats, as in typical extended-surface media filters (Figure 42-22). The dielectric medium consists of glass fiber mat, cellulose mat, or similar material supported by a grid consisting of alternately grounded and charged members. The charged members are supplied with 12,000 V DC power. Airborne particles that approach the field are polarized and drawn to the filaments of the fibers of the media. This type of filter offers about 0.10 in wc resistance to the flow of air when clean, with a face velocity of 250 FPM.

Charged-Media Ionizing Air Cleaners

Charged-media ionizing electronic filters combine the effects of the other two designs. Dust is charged in a corona discharge ionizer and collected on a charged-media filter mat. This construction increases the effectiveness of the filter but is more critical to operate successfully.

SERVICE TIP

The high voltage on an electronic air filter can cause discharges when the surface of the collector plate becomes excessively dirty or if a large particle or insect comes in contact with the collector plate. This electrical discharge or spark can be annoying to homeowners because of its sound and interference with television and radio reception. For installations where persistent arcing is a problem, the source of the arcing must be identified and controlled. It may mean more frequent cleaning of the collector plate or possibly the installation of a prefilter to catch or keep out larger particles or insects.

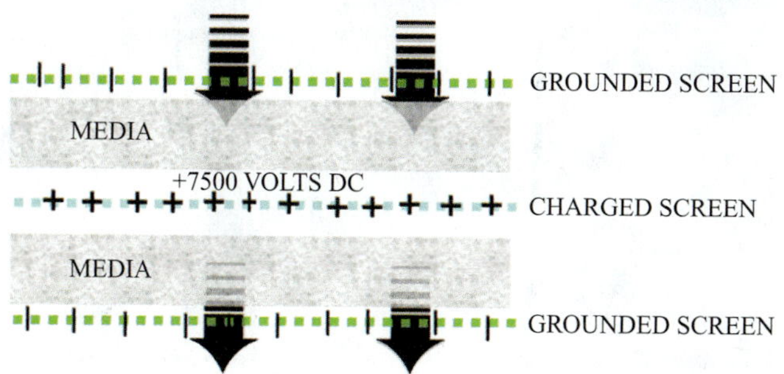

Figure 42-22 Nonionized charged-media filter operation.

Figure 42-23 Charged particles can cause staining if they escape the electronic air cleaner.

Undesirable Operating Conditions

Two operating conditions that can cause problems with electronic air filters are (1) space charge and (2) ozone. The unit needs to be carefully built and installed so that charged dirt particles do not escape into the filtered space (space charge). If they do, they can darken the walls faster than if no cleaning arrangement was used. Figure 42-23 shows the effects of small ionized particles escaping the air cleaner. Proper installation and application are key to preventing this condition. Moving air at speeds exceeding the filter's rating can allow charged particles to escape.

Electronic air cleaners produce a small amount of ozone because of their high voltage. Ozone is a poisonous gas. When the unit is operating correctly, the amount of ozone produced is well within the recommended safe limits. If the unit is continuously arcing, it may yield levels of ozone, which are annoying and even poisonous. High levels of ozone are indicated by a strong bleach-like odor.

42.10 ADVANCED HYBRID AIR CLEANERS

Several manufacturers now offer air cleaners that clean, sterilize, and deodorize the air. These filters offer some combination of HEPA media filter, ultraviolet light, charged-media electronic air cleaner, and activated charcoal (Figure 42-24). The HEPA filter or electronic air cleaner removes the minute particles, the ultraviolet light or charged media sterilizes the air, and the activated charcoal removes odors. They are aimed at the high-end residential market for customers who are concerned about the indoor air quality of their home.

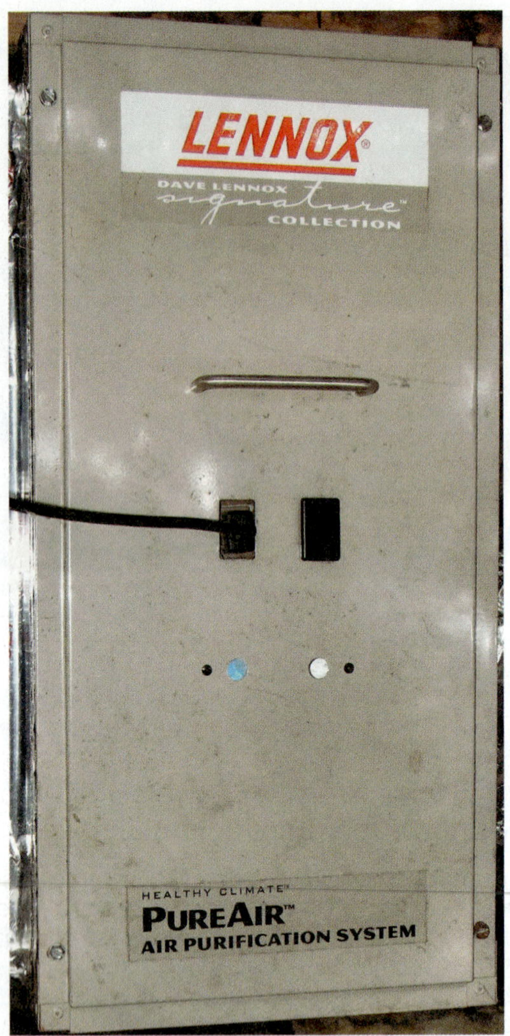

Figure 42-24 This high-efficiency filter cleans and disinfects the air.

42.11 FILTER INSTALLATION

Filters should be placed ahead of heating and cooling coils and other mechanical equipment to protect the system from dust.

Panel Filters

Panel filters have an arrow that indicates the direction of airflow. It should be pointed toward the unit (see Figure 42-10). The published performance of filters is based on straight-through unrestricted airflow. The filter size should be appropriate for the amount of air passing through it. For example, a panel filter rated for a maximum face velocity of 300 FPM should have a face area of 2.67 ft^2 for a 2-ton unit: 400 CFM/ton $\times$ 2 tons = 800 CFM, 800 CFM/300 fpm = 2.67 ft^2, or 384 in^2. A 20-in $\times$ 20-in filter would work (400 in^2); a 16-in $\times$ 20-in would not (320 in^2). It is common to pull return air in from both sides of a unit if the airflow capacity requires more filter face area than can be accommodated easily by a single filter (Figure 42-25).

Figure 42-25 Filters can be installed on both sides of a furnace to increase the filter surface area.

Extended-Media Filters

Extended-media filters require the installation of a filter box (Figure 42-26). This box can be mounted to the unit or in the return duct. A transition should be used if the duct size is smaller than the opening in the filter. In either case, sealing the filter box to the duct or the unit is important to maintain maximum filter efficiency. It does not make sense to install an efficient filter and leave air leaks around the filter that will allow dirty, unfiltered air to enter the system. Service access should be considered when installing the filter box. The filters that must come out are as long as the box. It will

Figure 42-26 Medium-efficiency 5-in filters require a filter box.

Figure 42-27 This airflow sensor turns the electronic air cleaner on when it senses airflow.

be difficult to change the filters if a 25-in-wide filter box is installed with a 20-in service access. The filter airflow capacity should match the system airflow.

Electronic Air Cleaners

Electronic air cleaners are installed like extended-media filters, except that they need a power supply. The air cleaner should operate whenever the blower runs, but it should not operate when the blower is off. Some air cleaners have an airflow sensor that detects airflow and automatically turns the filter on and off with airflow (Figure 42-27). Other air cleaners require a relay or sail switch to energize the filter when the fan runs.

42.12 ULTRAVIOLET (UV) LIGHTS

Ultraviolet lights (UV) are used to sterilize the surfaces inside duct systems. These lights have been popular for years in hospitals, where 100 percent of the air entering an operating room has to be outside air. The air passes through a duct box containing a large number of ultraviolet lights, where the air is sterilized before entering the operating room (Figure 42-28). Ultraviolet lights will kill mold, fungus, bacteria, and any other living organisms if the light is intense enough and the organisms are exposed to it for a long enough period of time.

In recent years, ultraviolet light sterilizing units have become available for the residential and light commercial market. Most of the lights used in residential applications are not strong enough to sterilize the air as it passes by. These lights are typically located in the plenum, where they sterilize the coil, drain pan, or filter. Location of the UV light is critical, because UV lights can only disinfect surfaces that the light can see. Typically, more than one lamp is required to adequately cover the coil (Figure 42-29). Lights that are designed to keep the coil, drain pan, or filter clean operate all the time, even when the system is not running. Lights that attempt to sterilize the air come on during system operation. This is accomplished by tying their circuit into the fan blower relay circuit in the indoor equipment.

Figure 42-28 This bio-wall unit can kill viruses and bacteria in the air in one pass. *(Courtesy of Sanuvox Technologies, Inc.)*

The intensity of the ultraviolet radiation decreases with use, so the lamps in UV lights require replacement every year in most cases. The effectiveness of the UV light drops off as the lamp nears the end of its useful life. Some longer-life lamps can go 2 to 3 years, but their UV output is typically only half of the original output by the end of 3 years.

SERVICE TIP

Some indoor coil drain pans are made of plastic materials that can be degraded by ultraviolet light. Before installing an ultraviolet light in a plenum or near a coil, make certain that the drain pan is UV resistant.

Figure 42-29 Residential application of UV light to keep the coil clean. *(Courtesy of Triatomic Environmental)*

SAFETY TIP

Ultraviolet light at the levels required for air sterilization is dangerous to technicians. The ultraviolet light can result in severe skin burns like those caused by too much exposure to the sun. In addition, burns on the eyes can occur. The intensity of the sterilizing lights is such that it only requires a momentary exposure to cause damage. Ultraviolet light boxes must have an interlock circuit that prohibits the light from coming on when access panels are open.

UNIT 42—SUMMARY

Common air contaminants include dust, smoke, biological contaminants, and chemical contaminants. Contaminant particles are measured in microns. A micron is an SI unit of measurement equal to 1/1,000 of a millimeter. Most of the particles in the air are below 1 micron in size. The most common filters use a media that traps the contaminants and lets the air pass through. These filters get more efficient as they get dirty. Filters have a recommended airflow rating that should not be exceeded. A filter's efficiency is stated as its MERV rating. Higher MERV ratings mean higher efficiency filtration. When choosing a filter, it is important to select a filter that has the correct airflow rating and pressure drop for the application as well as a high efficiency. Electronic air cleaners use high DC voltage to attract dust particles to charged media or charged plates. Ultraviolet lights are used to sterilize air-conditioning coils and the inside of system ducts.

WORK ORDERS

Service Ticket 4201

Customer Complaint: Poor Cooling

Equipment Type: Split-System Air Conditioner with a 1-in Electrostatic Air Filter

A customer calls because her unit runs continuously and does not cool. She also has noticed that very little air is coming out of the supply registers. The technician sees that the suction line at the outdoor condensing unit is iced over. Checking inside, the suction line is iced there as well. The filter is a relatively clean electrostatic air filter that the customer bought at the local home improvement store. The technician shows the homeowner the filter packaging, which says the filter's nominal pressure drop is 0.35 in wc when the filter is clean, and explains that her system cannot move enough air with that much pressure drop across the filter. The low airflow caused the coil to freeze over. The technician replaces the filter with a lower efficiency and lower resistance panel filter and leaves the fan operating without the air conditioner to allow the coil to thaw. Once the coil has thawed out, the airflow returns to normal and the unit operates correctly. The technician advises the customer to consider a 5-in extended-surface media filter if they want a truly high-efficiency filter that will not severely impede the system airflow.

Service Ticket 4202

Customer Complaint: Poor Cooling

Equipment Type: Split-System Air Conditioner with a 1-in Panel Filter

A customer calls because his unit runs continuously and does not cool. He also has noticed that very little air is coming out of the supply registers. The technician sees that the suction line at the outdoor condensing unit is iced over. Checking inside, the suction line is iced there as well. The filter is an inexpensive panel filter that is so dirty that light will not pass through it. The customer is surprised because he remembers changing the filter last year. The technician explains that the panel filters should be changed monthly and recommends cleaning the evaporator coil as well. After cleaning the evaporator coil and replacing the filter, the unit runs properly.

Service Ticket 4203

Customer Complaint: Popping Sound Coming from the System

Equipment Type: Split-System Air Conditioner with an Electronic Air Cleaner

A customer calls because she hears a popping sound coming from her system and is concerned that there may be a loose wire or electrical malfunction. The technician arrives, hears the sound, and realizes that it is coming from the electronic air cleaner. The technician explains that the sound is from the air cleaner arcing due to a large buildup of dirt on the collector plates. The technician removes the cells and places them in the dishwasher to clean while he cleans the prefilter. The technician allows the cells and prefilter to dry, replaces them in the unit, and puts the unit back into operation. The popping sound is now gone.

UNIT 42—REVIEW QUESTIONS

1. How does mechanical filtration work?
2. What is the normal maximum air velocity across the face of a low-efficiency panel filter?
3. How often should a low-efficiency panel filter be changed?
4. How often should a renewable-media pleated surface be changed?
5. What factors should be considered when choosing an air filter?

6. Describe the MERV filter rating system.
7. Why are the highest rated MERV filters not used in residential applications?
8. Why is panel filter media sometimes coated with a viscous substance?
9. What is a HEPA filter?
10. How does increased airflow affect a filter's efficiency?
11. What is the purpose of expanded media area in pleated filters?
12. How does the increased pressure drop across a filter affect system operation?
13. List the three types of electronic air filter units.
14. What is the common operating voltage range on the cells of electronic air cleaners?
15. Describe the operation of an ionizing plate electronic air cleaner.
16. What filter characteristics should be considered when selecting an air filter?
17. Why should a MERV rating always state the air velocity through the filter?
18. How can a high-pressure drop across an air filter adversely affect system performance?
19. What air contaminants are addressed by UV lights?
20. How often do UV lamps need to be replaced?

UNIT 43

Ventilation and Dehumidification

OBJECTIVES

After completing this unit, you will be able to:

1. distinguish between infiltration, exfiltration, ventilation, and exhaust.
2. list the factors that affect building pressure.
3. describe types of ventilation systems.
4. calculate building air changes.
5. discuss the difference between heat-recovery ventilators and energy-recovery ventilators.
6. explain the benefits of dehumidification.
7. discuss the methods of dehumidification.
8. list types of dehumidifiers.

43.1 INTRODUCTION

Ventilation and dehumidification are necessary for a complete comfort system. A total comfort system controls more than temperature. Controlling the amount of air entering the building and the humidity in the building are also important aspects of comfort. This is especially true of today's tight buildings that have limited infiltration. Sealing up the building increases energy efficiency but reduces the air changes to a level where odors and stuffiness can be a problem. Ventilation provides a solution to this problem. Unlike infiltration, ventilation can be controlled. Dehumidification is always important in humid climates like the southeast, but it is especially important with tighter building construction. Normal living activities such as cooking, bathing, and washing produce moisture. Dehumidification keeps the moisture buildup inside the house from becoming hazardous to the building and its occupants.

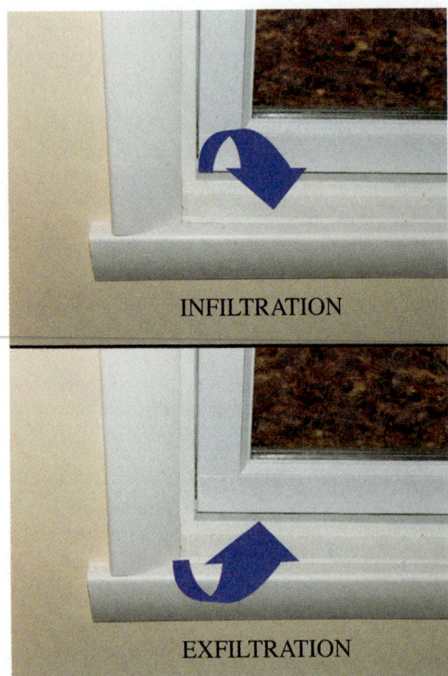

Figure 43-1 Air leaking in is infiltration; air leaking out is exfiltration.

43.2 BUILDING AIRFLOW

Air is constantly moving in or out of most buildings. Some of this air leaks in or out of the building, and some of the air is drawn in or exhausted out by ventilation equipment. Air that leaks in is called infiltration; air that leaks out is called exfiltration (Figure 43-1). The amount of infiltration or exfiltration is affected by building construction, wind velocity, inside combustion appliances, and building pressure. Ventilation and exhaust are the intentional movement of air in or out of the building with fans. Ventilation refers to blowing air in, while exhaust refers to blowing air out (Figure 43-2). Of course, you cannot blow air into the building without letting some of the air that is already in the building leave. Similarly, exhaust fans cannot work if there is no source of makeup air coming in. Makeup air is air that is drawn in to replace air that is exhausted out. In small residential applications like bathroom exhaust fans, the assumption is made that there is enough infiltration to provide makeup air. In larger commercial applications, this simply won't work.

TECH TIP

Many people are not aware of the necessity for makeup air for exhaust systems. Often, problems with an exhaust system not operating properly can be traced to an inoperative makeup-air blower or a blocked makeup-air duct.

43.3 AIR CHANGES

Replacing all the air in a building with air from outside is called an air change. It means that all the air inside has been changed for new air from outside. The amount of infiltration a building has, or the amount of ventilation required, is stated

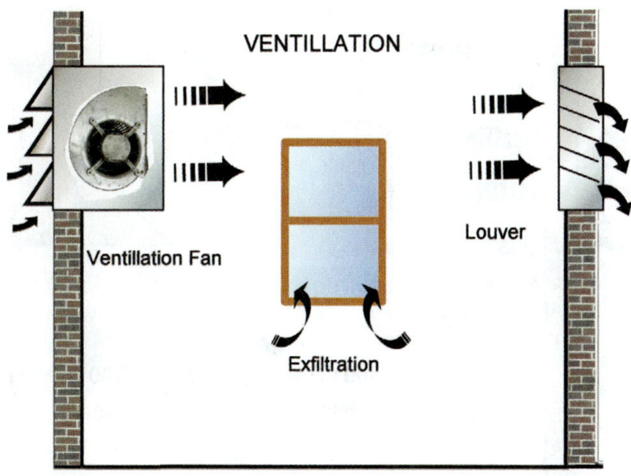

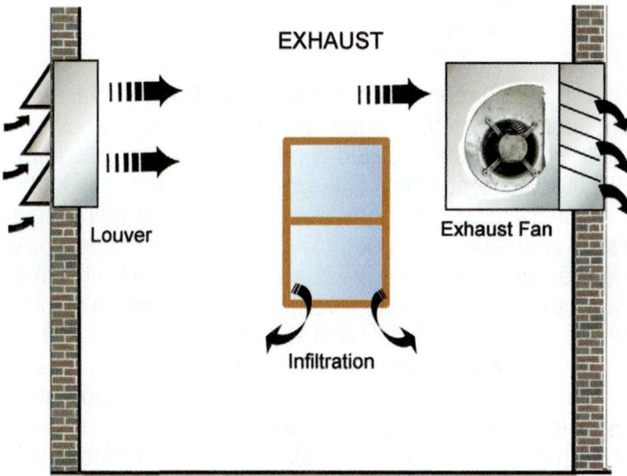

Figure 43-2 Ventilation fans blow air into a building; exhaust fans blow air out of a building.

in air changes per hour. One air change an hour means that it takes an hour for all the air in the building to be replaced with air from the outside. The average air change rate for existing homes is between one and two air changes per hour. However, new homes built today usually have an air change rate between one half to one air change per hour. Extremely tight new construction can achieve air change rates as low as one third air change per hour. Homes with such low air change rates need some form of mechanical ventilation to bring in fresh outside air in to compensate. Commercial buildings use mechanical ventilation to achieve much higher air change rates. The higher air change rates are necessary due to the increased number of occupants and processes giving off carbon dioxide, water, and fumes. Figure 43-3 shows typical air change rates for different commercial applications.

The amount of airflow in cubic feet per minute (CFM) required to produce one air change per hour can be calculated by dividing the building volume by 60:

$$\text{Air change CFM} = (\text{floor area} \times \text{ceiling height})/60$$

Multiply this number times the number of air changes required to get the total ventilation CFM required:

$$\text{Total ventilation CFM} = \text{Air change CFM} \times \text{required air changes}$$

RECOMMENDED AIR CHANGES PER HOUR	
OFFICES	
Business Offices	6-8
Lunch Rooms	7-8
Conference Rooms	8-12
Medical Offices	9-10
Copy Rooms	10-12
Computer Rooms	10-14
RESTAURANTS	
Dining Area	8-10
Food Staging	10-12
Kitchens	14-18
Bars	15-20
PUBLIC BUILDINGS	
Hallways	6-8
Retail Stores	8-10
Churches	8-12
Restrooms	10-12
Auditoriums	12-14
Smoking Areas	15-20

Figure 43-3 Recommended air changes per hour for different types of buildings.

These two formulas can be combined:

$$\text{Total ventilation CFM} = (\text{floor area} \times \text{ceiling height} \times \text{air changes})/60$$

43.4 INFILTRATION MEASUREMENTS

The only truly accurate method of measuring building infiltration is to pressurize the building with a large fan and measure the airflow required to maintain that pressure. The amount of air required to maintain that pressure provides a reference point in describing the infiltration rate of a structure. This measurement can be accomplished with a blower door (Figure 43-4). A blower door contains a large fan, a manometer for measuring the building pressure, and an airflow gauge to measure the amount of air required to maintain a particular pressure. It is installed in an exterior door, the house is pressurized, and the CFM required to maintain that pressure is read.

GREEN TIP

Blower door readings are useful when performing house tightening. They provide an objective measurement of the degree of improvement achieved.

Figure 43-4 Blower door in operation. *(Courtesy of BuilderFish)*

43.5 WHY VENTILATE?

Air that enters a building through infiltration is uncontrolled. It cannot be cleaned, heated, cooled, or dehumidified before it enters the building. The location and amount of infiltration is also largely outside of our control. In the winter, infiltration around windows and doors creates a draft. This increases the heat load and decreases comfort. On the other hand, ventilation air can be controlled. We can control the amount of air entering the building and we can condition it before it enters the occupied space, eliminating drafts. In commercial buildings with many occupants, ventilation provides fresh air and prevents the buildup of carbon dioxide inside the building. Ventilation can also dilute odors and contaminants produced inside the building by people, machines, or processes. Exhaust systems can be located at the source of contaminants and carry them out of the building before they can escape into the rest of the building. Many builders have started to use tighter construction techniques such as sealing penetrations, foaming around windows and doors, and using housewrap. This helps reduce drafts and heat loss but increases stuffiness and odors. Ventilation is really necessary in extremely tightly constructed buildings. Ventilation and exhaust are used to gain control over the air entering and leaving the building and to control building pressure.

43.6 BUILDING PRESSURE

You probably don't think of a building as having a pressure, just as you seldom notice the atmospheric pressure. It is not unusual for the air pressure in a building to be a little lower or higher than the outside air pressure. Air comes through cracks in the walls whenever there is a pressure difference between inside and outside. Commercial structures normally try to maintain a pressure slightly above the outside air pressure, called a positive pressure. Buildings with a positive pressure will experience exfiltration rather than infiltration. This improves overall comfort by reducing drafts. Ventilation is needed to maintain a building at a positive pressure. A positive pressure is created when more air enters the building through ventilation than leaves through exhaust. Pressure builds and the remaining air is forced out through exfiltration (Figure 43-5). A negative pressure is created when the exhaust air exceeds ventilation air. A negative pressure is created and the difference is made up through infiltration (Figure 43-6). Factors that reduce building pressure include any exhaust system, all combustion appliances, and leaks in supply air ducts located in unconditioned space (Figure 43-7). Leaky ductwork is a major contributor to negative building pressure in many older residential systems. Air that leaks out of the supply duct system in unconditioned space can equal all the air leaking through the rest of the structure.

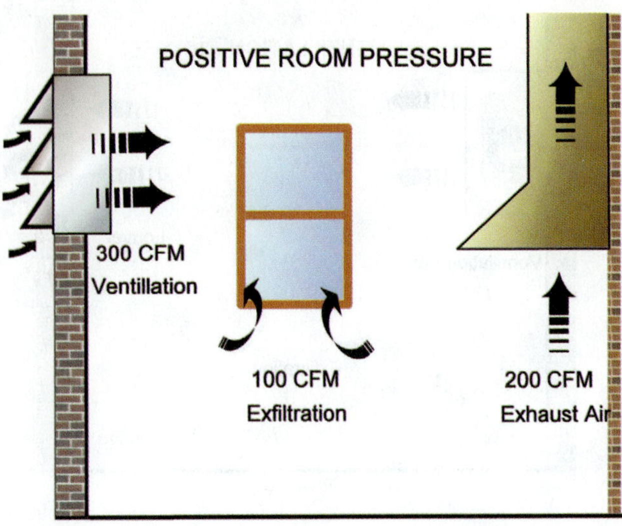

Figure 43-5 Illustration showing positive building pressure.

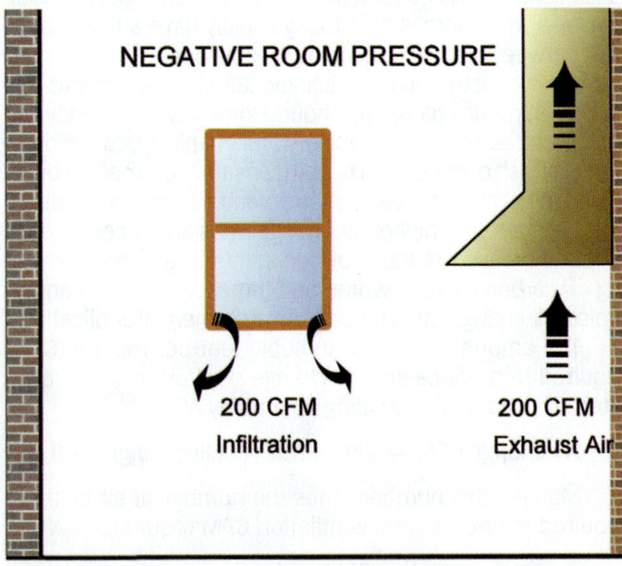

Figure 43-6 Illustration showing negative building pressure.

Figure 43-7 Factors affecting building pressure.

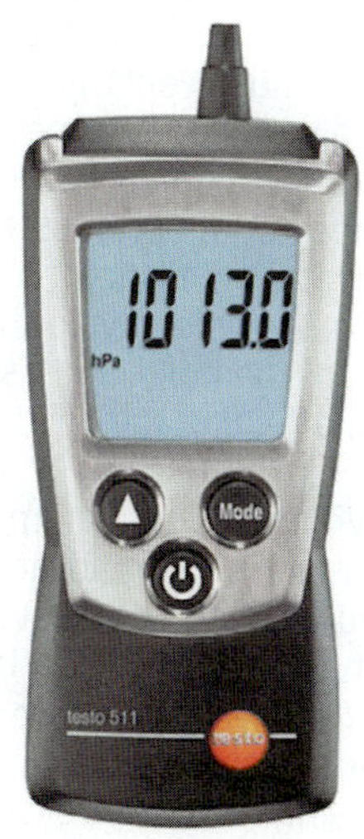

Figure 43-8 This digital manometer can read the absolute pressure in a room. *(Courtesy of Testo, Inc.)*

SAFETY TIP

Combustion appliances should not be located in negative pressure areas. This is why the return air should not be located in the same room as the furnace. The furnace blower can lower the pressure in the equipment room to the point that the vent does not operate correctly and combustion products spill out into the room. This sets up the possibility of circulating combustion products, including carbon monoxide, throughout the house.

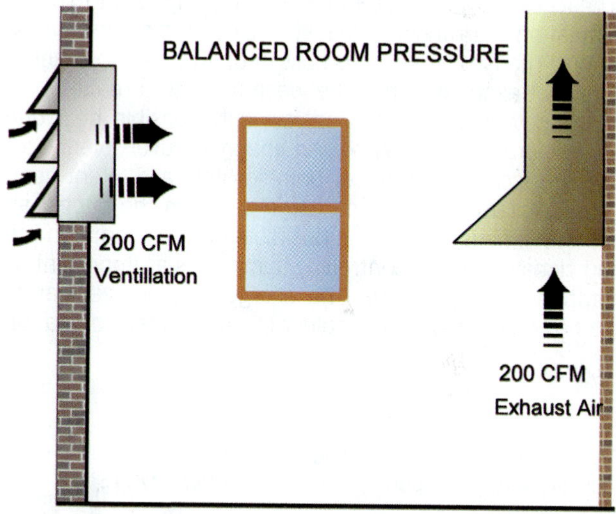

Figure 43-9 Illustration of balanced ventilation system.

Factors that increase building pressure are ventilation systems and leaks in return-air ducts located in unconditioned space. The pressure within a building can vary from room to room. A room with a large return air grille and no supply registers can be negative, while a room next to it with a large supply register and no returns is positive. The pressure difference between inside and outside, or between two rooms, can most easily be measured using a sensitive digital manometer that can read absolute pressure in inches of water or pascals (Figure 43-8).

43.7 TYPES OF VENTILATION AND EXHAUST SYSTEMS

Ventilation systems grouped by their effect on the building are as follows:

- Exhaust: negative pressure (see Figure 43-6)
- Supply: positive pressure (see Figure 43-5)
- Balanced: neutral pressure (Figure 43-9)

An exhaust system uses a fan to force exhaust air out and relies on infiltration or makeup air ducts to supply air from outside. This type of system creates a negative pressure in the entire building so that outside air is sucked in. An example would be an attic fan in a house that sucks air out of the house and blows it into a ventilated attic (Figure 43-10). The entire house is under a negative pressure while the attic becomes pressurized.

Figure 43-10 A whole-house attic fan works by drawing in air through open windows and expelling it through the attic vents.

A supply system blows air into the building and relies on exfiltration and exhaust vents to carry air out of the building. A makeup-air blower in a mechanical room with a large number of combustion appliances would be an example. The mechanical room becomes a positive pressure, aiding in combustion and ventilation.

A balanced pressure system typically uses fans to blow air both into and out of the building. This prevents the ventilation system from affecting the building pressure either way. A slight positive pressure can be maintained with a balanced system by controlling the amount of air leaving compared to the air entering. One advantage of using blowers for both the ventilation and exhaust air is that you have more control. Another advantage of a balanced system is that heat-recovery or energy-recovery ventilators can be used to recover much of the heat from the exhaust air before it leaves, increasing the overall system efficiency.

43.8 HEAT- AND ENERGY-RECOVERY VENTILATORS

A heat-recovery ventilator uses a supply-air blower to move ventilation air in and an exhaust-air blower to move exhaust air out. The two air streams pass through a heat exchanger core that exchanges heat between them (Figure 43-11). A heat-recovery ventilator transfers only sensible heat due to temperature difference. An energy-recovery ventilator uses desiccants to transfer both sensible and latent heat. The desiccant transfers moisture between the supply air and the exhaust air. This is desirable in the summer in humid climates where controlling humidity is as important as controlling temperature. The energy-recovery ventilators are not recommended in colder climates where operation below 25°F for several hours is likely. In colder climates, heat-recovery ventilators should be used instead. Heat-recovery ventilators are available with a defrost feature for very cold climates. When the incoming air is below 23°F, the unit goes into defrost. During the defrost cycle, the incoming outdoor air is shut off by either a motorized damper or by simply shutting off the supply-air blower. Warm air from the house is circulated through the unit until the incoming side of the heat exchanger heats up.

GREEN TIP

Heat and energy-recovery ventilation systems save energy by transferring up to 80 percent of the energy between the ventilation air and the exhaust air while reducing infiltration.

The core in a heat-recovery unit is the part that actually exchanges the heat. There are several different types of heat-recovery cores used in energy-recovery ventilators, including flat plate, rotary wheel, and heat pipe. Flat-plate cores are the most common. They consist of several thin plates spaced about 1/8 in apart. The plates are connected at the ends and edges so the incoming outdoor airstream and the leaving indoor airstream can move through the alternating spaces staying separate from each other (Figure 43-12). Flat-plate cores can be made of plastic, aluminum, or treated paper.

Rotary-wheel cores use a rotating wheel made out of a corrugated plastic material that forms thousands of honeycomb air passages. Heat wheels can be from 1 to 10 ft in

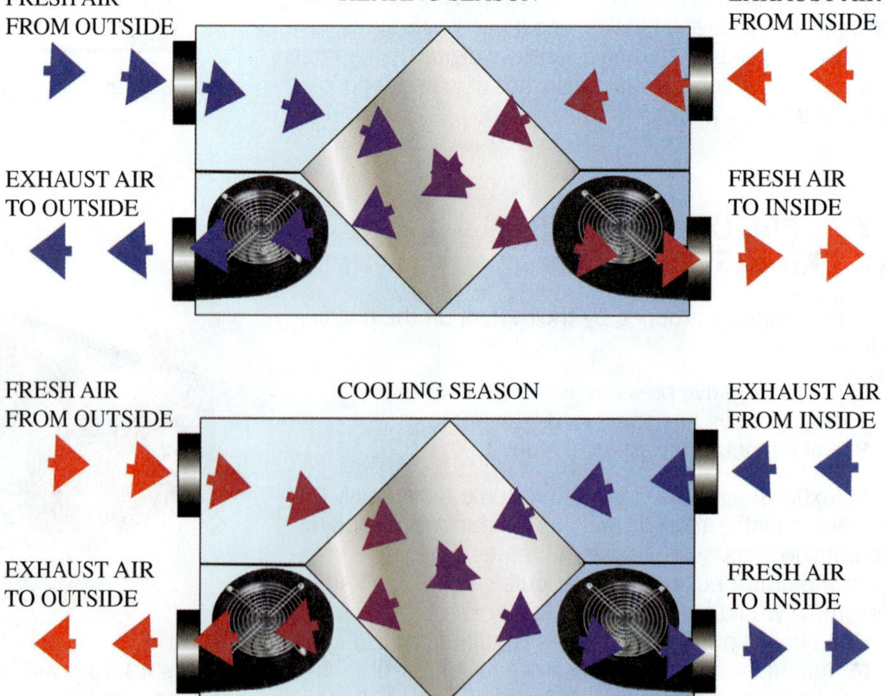

Figure 43-11 Illustration of heat-recovery ventilator operation.

Figure 43-12 Flat-plate heat exchanger used in heat-recovery cores.

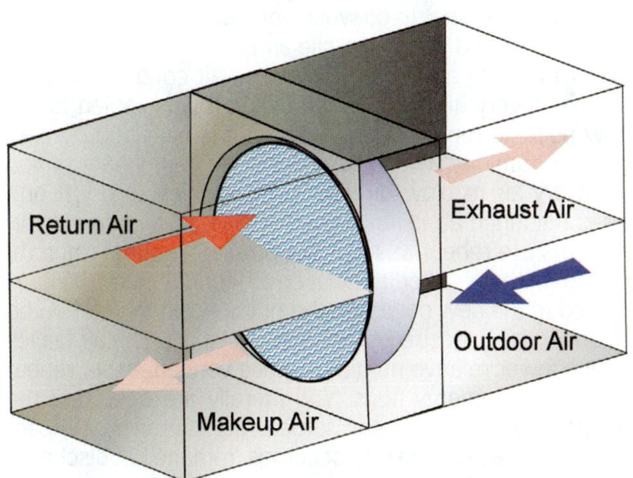

Figure 43-13 Heat-recovery ventilation unit using a heat wheel.

diameter and 6 to 30 in thick. The wheel is mounted with half of the wheel in the exhaust air duct and half in the supply air duct (Figure 43-13). The wheel rotates slowly: one to twenty rounds per minute. The warmer air transfers heat to the wheel, the wheel rotates to the cooler air, and the heat is transferred from the wheel to the cooler air.

A heat pipe is a miniature refrigeration system in a closed tube. A heat pipe is essentially a closed tube installed on a slant with a refrigerant sealed inside. When one side is heated, liquid boils and absorbs heat. The gas rises and goes to the other side because of its lighter density. The gas then condenses on the cool side, giving off the heat. The condensed liquid flows back to the warm side by gravity (Figure 43-14). Heat pipes can be installed so that they transfer heat from one airstream to another (Figure 43-15).

Figure 43-14 Illustration of heat-pipe operation.

Heat-recovery ventilators and energy-recovery ventilators are rated by either sensible recovery efficiency or apparent sensible effectiveness. The sensible recovery efficiency is found by dividing the temperature rise of the outdoor air by the temperature difference between the two airstreams. For example, with 40°F outdoor air, 70° indoor air, and 60°F fresh air entering the house, the sensible recovery efficiency would be (60 − 40)/(70 − 40) = 0.67, or 67%. "Apparent sensible effectiveness" is the term used in CSA standard C439M for testing heat-recovery ventilators. This measurement takes into account other variables, including motor heat gain, cross-leakage gain, and casing gain. It is usually numerically higher than the sensible recovery efficiency of the HRV.

43.9 WHY DEHUMIDIFY?

In humid climates such as the Southeast, removing humidity from the air in warm weather is just is just as important to comfort as reducing temperature. Being warm-blooded, our body normally produces more heat than needed and then regulates our temperature using different cooling mechanisms. The primary cooling mechanism is evaporation of perspiration from our skin. Dry air makes us feel cooler because it accelerates the evaporation from

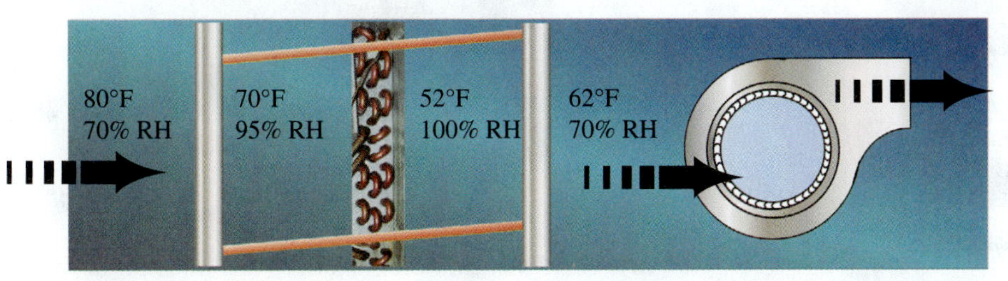

Figure 43-15 Heat pipes installed on an air-conditioner coil.

our skin. Humid air makes us feel warmer because the evaporation process is slower. Dehumidification can be the difference between being comfortable at 78°F and being uncomfortable. Dehumidification can save energy by reducing the amount of sensible cooling required for comfort. Many people with oversized air-conditioning systems essentially overcool their house to be comfortable because their systems do not run long enough in mild weather to reduce humidity, so they do not feel comfortable until they reach temperatures of 70°F in their house. Humidity represents heat because of the latent heat contained in the water vapor. When adding outdoor ventilation air during the cooling season, humid air adds to the heat load on the air-conditioning system.

43.10 AIR CONDITIONING AND DEHUMIDIFICATION

Air-conditioning units are the most common form of dehumidification. The air conditioner's evaporator coil typically operates below the dew point of the air flowing over it, causing water to condense on the coil. This is why air conditioners require a drain. However, there are some limitations on an air conditioner's ability to dehumidify. It takes several minutes for most air-conditioning coils to get cold enough to sweat. An oversized system will often not operate below the dew point for very long before satisfying the thermostat. Since air conditioners are sized for conditions that only occur a small percentage of the time, most air conditioners are oversized most of the time. The key to dehumidifying with an air conditioner is long run times and slow evaporator air flow. A properly sized air-conditioning system will run longer, allowing longer operation with an evaporator operating below the dew point. Systems with ECM blowers and thermidistat controllers have a special dehumidification mode that reduces system airflow for dehumidification (Figure 43-16). This increases the latent system capacity and decreases its sensible capacity. Two-stage cooling systems can help by allowing longer system operation at moderate loads. All of these are a big improvement over the typical oversized single-capacity system with a PSC blower. However, an air conditioner is still not a dehumidifier.

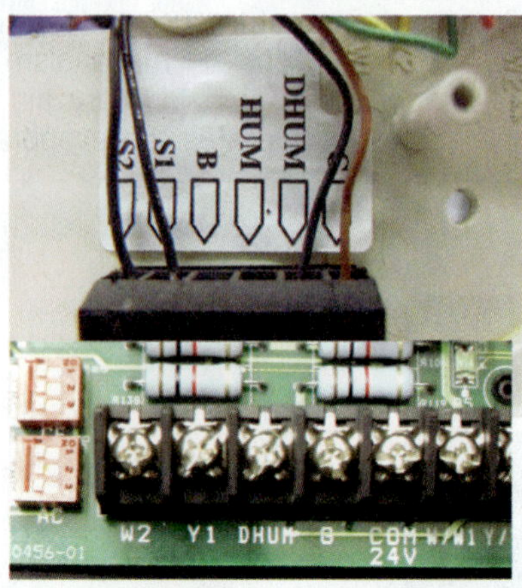

Figure 43-16 Dehum terminal on thermidistat and blower board.

GREEN TIP

Using a whole-house dehumidifier with an air-conditioning system can decrease overall energy use. The dehumidifier takes care of the latent load, allowing the air conditioning load to be mostly sensible cooling. The lower relative humidity allows for a higher thermostat setting, reducing the run time on the air conditioner. When the air conditioner does operate, it is performing mainly sensible cooling. Air-conditioning systems are more energy efficient when providing sensible cooling.

43.11 REHEAT SYSTEMS

The primary problem with using air conditioning for dehumidification is what to do when you need to dehumidify but you do not need to cool? While air conditioners do remove water from the air, the air leaving an air conditioner actually has a very high relative humidity. It was cooled to the dew point to remove the water, so it is near 100 percent relative humidity when it leaves the air conditioner. If you need dry air to blow directly on something to dry it out, air-conditioned air is not a good choice. Many commercial systems use reheat to solve this problem. Reheat refers to reheating the air after it has been air-conditioned. The air is cooled to the dew point to dehumidify it and then reheated to offset the undesired effect of overcooling. The air leaves with a lower relative humidity. While very effective, reheat systems are energy hogs. You literally are operating the heating and cooling at the same time. The energy impact can be mitigated if the reheat comes from the hot discharge gas coming out of the compressor. Units that use hot-gas reheat during dehumidification use a solenoid-operated three-way valve to direct hot gas to the reheat coil (Figure 43-17). Electric-strip heat can also be used for reheat, but it is more expensive to operate. The main advantage to electric reheat is the system's simplicity.

Reheat systems typically are controlled by both a thermostat and a humidistat. The air conditioner will operate if either the temperature or the humidity is too high. The heat will operate whenever the temperature is too low. If the humidistat turns on the air conditioner and the temperature drops below the thermostat heating setpoint, the reheat will energize.

Figure 43-17 Reheat valve and coil on unit with hot-gas reheat dehumidification.

43.12 HEAT PIPES

Heat pipes do not use any energy; they work entirely by temperature difference. A heat-pipe system used to enhance coil dehumidification has one part before the evaporator coil and another after the evaporator coil (see Figure 43-15). Heat pipes improve an air conditioning system's ability to dehumidify by transferring some heat from the air entering the evaporator coil to the air leaving the evaporator coil. They cool the air before it enters the evaporator, so that the evaporator operates at a lower temperature, increasing the dehumidification effect of the coil. The heat that was removed from the air entering the coil is added back to the air leaving the evaporator. In effect, the air is reheated using heat taken from the air itself. Coils that combine the evaporator and heat pipe coils into a single unit make applying the heat pipe easier.

TECH TIP

Heat pipes help improve the dehumidification properties of an air-conditioning coil, increasing the latent cooling capacity and decreasing the sensible cooling capacity. This causes longer operating cycles. Although heat pipes do not use any energy to operate, they do add to the operating cost of the air conditioner by reducing the sensible capacity and extending the operating time.

43.13 DEHUMIDIFIER FUNDAMENTALS

Basically, a compression-cycle dehumidifier is an air conditioner with a single blower that moves air first over the evaporator and then over the condenser. The air first passes over the evaporator, where it is cooled to the dew point to remove water, and then the same air passes over the condenser, where it is reheated to a temperature slightly above its original temperature because of the heat of the compressor and fan motor (Figure 43-18). Thus, the dehumidifier reduces both the amount of water in the air and the relative humidity of the leaving air.

Dehumidifier capacity is listed in pints per day. This represents the amount of water the unit can remove operating continuously for 24 hours at design conditions. ANSI/AHAM DH-1-2003 standard conditions are 60 percent RH, 80°F. Dehumidifier capacity drops as either the air temperature or the relative humidity drops. The capacity can drop by as much as 50 percent when operated at more moderate conditions, such as 70°F and 50 percent RH. Dehumidifier energy use is rated in liters of water per kilowatt-hour of energy use (L/kWh), called the energy factor. Larger capacity dehumidifiers tend to be more energy efficient than smaller ones. The energy factor ranges from around 1.0 for a small, inefficient console model to 3.5 for a more efficient, larger whole-house model.

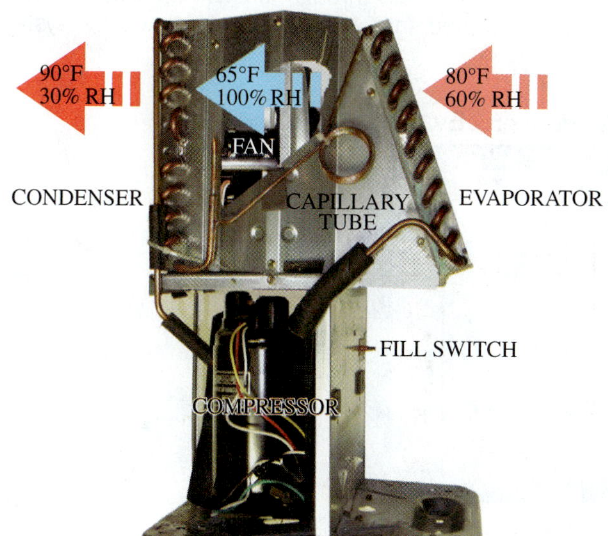

Figure 43-18 Basic dehumidifier operation.

43.14 CONSOLE DEHUMIDIFIERS

Console dehumidifiers are small plug-in electrical appliances that sit in the room they are conditioning. They have capacities ranging from 20–70 pints per day. Console dehumidifiers have a compressor and a fan motor. The compressor is controlled by a humidistat, which cycles it on and off to maintain the humidity setting. The humidistat can be either a mechanical switch controlled by an element that expands and contracts with different humidity levels or an electronic sensor that changes capacitance with different humidity levels (Figure 43-19). The fan motor runs all the time on many models but can cycle with the compressor in some models. Figure 43-18 shows the parts of a typical console dehumidifier. The evaporator is designed to operate just above freezing to maximize dehumidification. When the entering air temperature is below 65°F, the evaporator coil temperature can drop below freezing, and frost will start to form (Figure 43-20). Frost buildup will impede airflow and reduce the unit's effectiveness. Continuing to operate with a frozen evaporator can damage the compressor. A frost-sensing switch shuts off the compressor when the evaporator coil starts to frost (Figure 43-21). The fan continues to operate.

Water from the evaporator coil drains into a plastic bucket that is tilted slightly so that it presses against a switch (Figure 43-22). The switch opens and shuts off the system when the bucket is full. A console unit operated in a basement will typically need to be emptied twice a day.

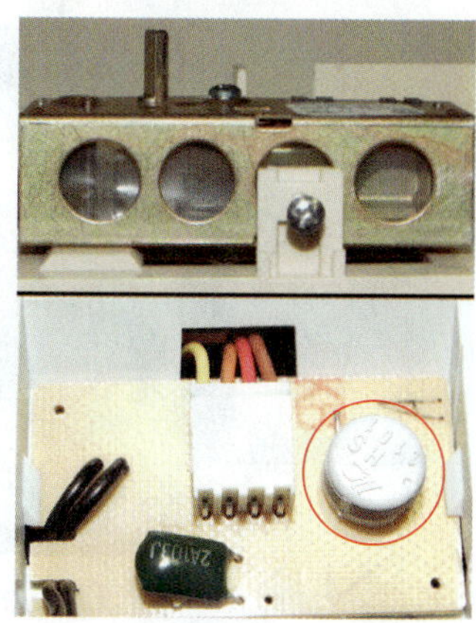

Figure 43-19 Mechanical and electronic humidistats.

Figure 43-20 Frost on dehumidifier evaporator coil.

Figure 43-21 Frost-sensing switch.

Figure 43-22 Condensate bucket and fill switch.

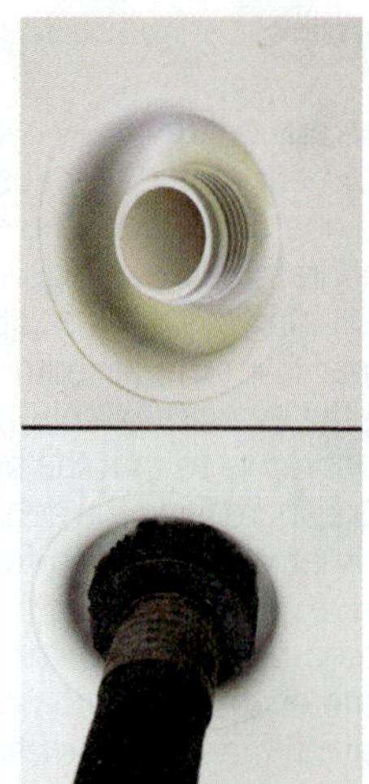

Figure 43-23 Hose connected to back of bucket.

Most units have provisions for attaching a garden hose to the bucket so that the water can be run to the outside or into a floor drain (Figure 43-23). This eliminates the need to empty the bucket.

Console dehumidifiers do have some drawbacks, including relatively low capacity, noise, and the inconvenience of emptying the bucket. These are all related to their design: a small plug-in appliance that sits in the room. They are typically noisy because of the compressor.

43.15 WHOLE-HOUSE DEHUMIDIFIERS

Several companies now offer whole-house dehumidifiers that can be integrated into a complete comfort system for your house. Like the console units, they consist of a complete refrigeration system that has a single airflow over the evaporator and condenser coils (Figure 43-24). Whole-house systems offer many advantages over small console units, including higher capacity, higher energy efficiency, filtration, and ventilation. They have a blower that is strong enough to move air through ductwork, allowing them to be integrated into the heating and cooling system

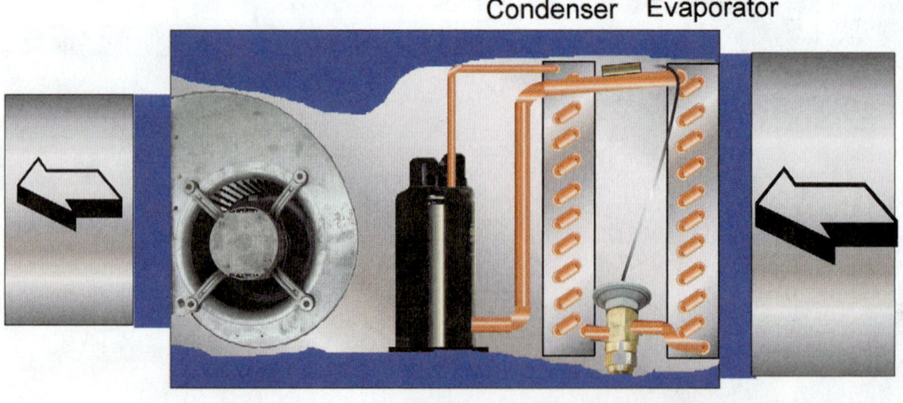

Figure 43-24 Parts in a whole-house dehumidifier.

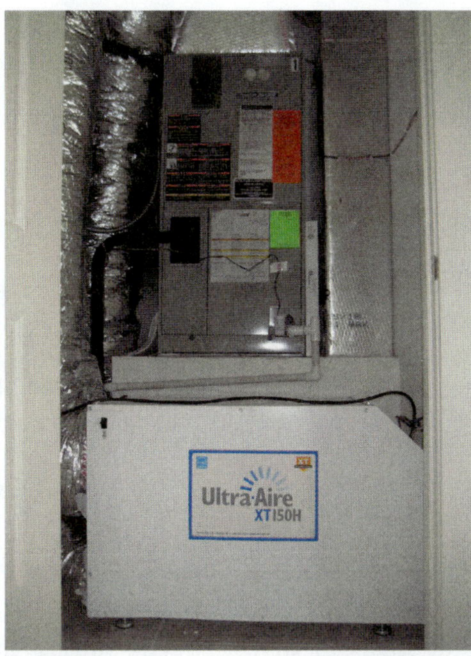

Figure 43-25 Whole-house dehumidifier connected to system ducts. *(Courtesy of Therma-Stor, LLC)*

Figure 43-26 Whole-house dehumidifier operating as a standalone unit. *(Courtesy of Basement Systems)*

ductwork (Figure 43-25). They can also operate as a standalone unit (Figure 43-26). Even when operated as a standalone unit, whole-house systems do not have to be located in the space they are conditioning because they can be connected to ductwork. Most whole-house dehumidifiers also offer high-efficiency air filtering, with filters ranging from MERV 8 to MERV 14. This is especially important when using a model that adds outdoor ventilation. A whole-house dehumidifier with outdoor ventilation and a high-efficiency filter can make a significant improvement to the indoor air quality of most houses.

43.16 DESICCANT WHEEL DEHUMIDIFIERS

A desiccant is a chemical that adsorbs water vapor by chemical attraction. Desiccants will adsorb water vapor from the air. When the desiccants are heated, they release this water

Figure 43-27 Close-up of heat wheel material.

vapor. Desiccants are used in filter driers and in the little bags packed in electronic equipment to keep it dry during storage and shipping. A desiccant wheel dehumidifier uses desiccants imbedded in a rotating wheel made out of a corrugated plastic material. The corrugated material forms thousands of honeycomb air passages (Figure 43-27). The wheel rotates very slowly: ten to twenty rounds per hour. Two separate airstreams pass through the wheel. One airstream is the air to be dehumidified, and the other is the regenerating air. As the wheel turns, the desiccant passes first through the incoming air, where the desiccant adsorbs moisture from the air. Next, the desiccant passes through a regenerating zone, where the desiccant is dried by the hot regenerating air. The wheel continues to rotate and the adsorbent process is repeated. Typically, about three-fourths of the desiccant wheel is exposed to the incoming air throughout the process. The regeneration heat source can be a direct-fired gas burner, a hot water or steam coil, condenser air, or hot exhaust air. Figure 43-28 shows a typical desiccant wheel dehumidifier. Desiccant wheel dehumidifiers are generally designed for large commercial applications. One advantage desiccant drying has over refrigerated drying is that desiccant systems can operate at temperatures below freezing. There are a number of specialized systems that incorporate desiccant wheels along with refrigeration to meet the needs

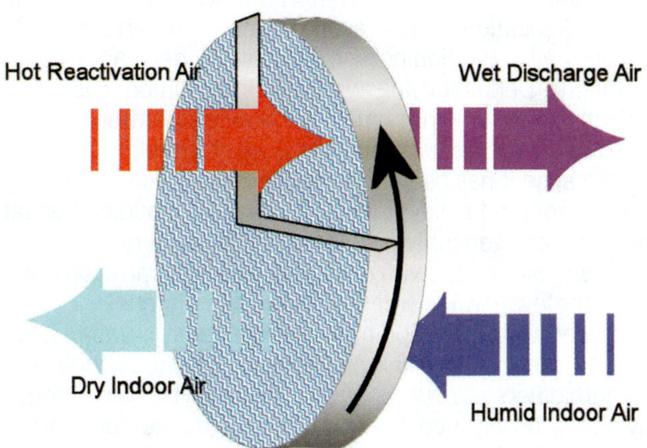

Figure 43-28 Illustration of how a desiccant wheel dehumidifier works.

Figure 43-29 Dehumidifier made for ice rink applications. *(Courtesy of BRR Technologies)*

of a particular application. Figure 43-29 shows a unit made specifically for conditioning ice-skating rinks.

UNIT 43—SUMMARY

Infiltration is the process of air leaking into a building; exfiltration is air leaking out of a building. Ventilation air blows into a building; exhaust air blows out. Building pressure becomes positive when ventilation air exceeds exhaust air. Building pressure becomes negative when exhaust air exceeds ventilation air. Ventilation systems can be grouped by building pressure as exhaust (negative pressure), supply (positive pressure), or balanced (neutral). Heat-recovery ventilators transfer heat between exhaust air and ventilation air to reduce the energy lost in the exhaust air. Energy-recovery ventilators use desiccant to transfer both temperature and humidity. The types of heat-exchange cores used in heat-recovery ventilators include flat plate, rotary wheel, and heat pipe.

Dehumidification can be as important as temperature reduction for maintaining comfort in humid areas. Dehumidification can be accomplished through air conditioning, air conditioning with reheat, or dedicated dehumidifiers. Dehumidifying with an air conditioner is more effective with long run times and slow evaporator air flow. The air leaving an air conditioner is at a high relative humidity because it has been cooled to the dew point. Reheating the air-conditioned air lowers the relative humidity. Reheat systems are generally not energy efficient. A dehumidifier is an air conditioner with a single blower that moves air first over the evaporator and then over the condenser. The air is cooled to the dew point by the evaporator, water is removed, and then the air is heated back up by the condenser. Dehumidifiers are available as console units, whole-house systems, or large commercial systems. Large commercial dehumidifiers typically use desiccant wheels. The desiccant in the wheel adsorbs moisture from one airstream and releases it to a heated, regenerating airstream.

WORK ORDERS

Service Ticket 4301

Customer Complaint: Exhaust Vent Not Working Properly

Equipment Type: Commercial Exhaust Hood

A technician is called to check the operation of a commercial exhaust hood. A new restaurant has just opened in an older part of downtown in an old retail store that has been converted into a restaurant. The technician notices that the door, which swings out, is hard to open and then slams shut behind him. The manager shows the technician the hood and complains that even though it is new, it does not seem to work. The kitchen is filled with smoke. The technician asks who installed it, and the manager admits that he did it himself to save money. He bought the hood over the Internet. The technician asks where the makeup-air duct is, and the manager looks puzzled. The technician explains that to exhaust air out, you have to bring air in to replace it. To illustrate, he props open the door and shows the manager that the fan operation is improved. He suggests that the manager make an appointment with one of the company engineers to determine if the hood is properly sized and suggest a solution, including installing makeup-air ducts.

Service Ticket 4302

Customer Complaint: Stale Air Inside House

Equipment Type: High-Efficiency Heat Pump Installed in a Super-Insulated House

A technician is called to take a look at a system in a new, super-insulated house. The system appears to be operating well—the temperature stays at the thermostat setpoint. However, the homeowners complain that the house just feels stuffy. They cooked fish last night and the technician can still smell it this afternoon. The technician notes that the house appears to be very tight. The homeowner proudly states that the people who did the blower door test said they had never seen a house with this low a leakage rate. The technician suggests bringing in outside ventilation air. The homeowner asks if that is not defeating the purpose of sealing up the house. The technician explains that they would use a heat-recovery ventilator that exchanges heat between the exhaust air and the incoming ventilation air. This allows controlled ventilation without incurring an energy cost to reheat the incoming ventilation air.

Service Ticket 4303

Customer Complaint: Musty Odor in Finished Basement

A technician is called about a musty odor in the basement. The owners have read about dirty sock syndrome and feel that they may need an epoxy coated coil. The technician asks if the smell is everywhere or just in the basement. They say mainly in the basement. The technician suggests that the problem may be high relative humidity in the basement. The technician measures the humidity with his electronic hygrometer and measures 70 percent relative humidity. He explains that anything over 60 percent will foster mold and mildew growth and suggests a dehumidifier. They say that they saw one at the drugstore for $300. He suggests that they may need a larger unit and offers to have a salesman come and provide a free estimate of the size and type of equipment that would best solve their problem.

Service Ticket 4304

Customer Complaint: Underground House Feels Sticky

A technician is called to look at an earth-bermed underground house. Even though the house has no air-conditioning system, the temperature inside is only 76°F compared to 90°F outside. However, the homeowners say they still feel warm. The technician measures the humidity and finds that the relative humidity is 65 percent. He explains that the high humidity makes the body feel warmer. He suggests a whole-house dehumidifier that would connect to the existing furnace ducts. The technician explains that dehumidification would make the house feel cooler.

UNIT 43—REVIEW QUESTIONS

1. What is the difference between infiltration and exfiltration?
2. What is the difference between exhaust air and ventilation air?
3. What benefit is there to supplying fresh air through ventilation versus simply allowing fresh air to come in through infiltration?
4. What is makeup air?
5. How does a building end up with a positive or negative e pressure?
6. Explain how a balanced ventilation system works.
7. Explain how a heat-recovery ventilation system works.
8. What is the difference between a heat-recovery ventilation system and an energy-recovery ventilation system?
9. How can a building's infiltration rate be accurately measured?
10. How can building pressure be accurately measured?
11. Why does the humidity in the air affect comfort?
12. How does the humidity in the air affect the load on an air-conditioning unit?
13. Explain how a reheat system operates.
14. Explain how a heat pipe works.
15. How can heat pipes be used to improve the dehumidification abilities of air-conditioning systems?
16. Describe how a basic compression-cycle dehumidifier works.
17. Describe the energy-efficiency ratings used for dehumidifiers.
18. Describe the system for rating dehumidifier capacity.
19. What effect do ambient temperature and relative humidity have on dehumidifier capacity?
20. How do console dehumidifiers keep from being blocked with frost when operating at temperatures below 65°F
21. Which type of dehumidifier is generally more efficient: small console units or larger whole-house units?
22. What advantages do whole-house dehumidifiers have over small console units?
23. Explain how a desiccant wheel dehumidifier works.
24. What types of dehumidifiers are used on large commercial applications?
25. What advantage do desiccant dehumidifiers have over refrigerated dehumidifiers?
26. What is the CFM requirement for a ventilation system to provide 1.5 air changes per hour in an 1,800 ft^2 house with a 10 ft ceiling?
27. What is an air change?
28. Why do commercial buildings require more air changes per hour than a residential house?

UNIT 44

Residential Air Conditioning

OBJECTIVES

After completing this unit, you will be able to:

1. explain the details of a unitary refrigeration system.
2. install a room or window air conditioner.
3. describe how a residential split air conditioning system is configured.
4. explain why split systems were developed.
5. discuss the proper arrangement and placement of equipment in a split system.
6. inspect and change air filters.

44.1 INTRODUCTION

Residential air conditioning generally includes single family, multifamily, and low-rise multifamily private household residences. The air-conditioning type is typically central forced air, central hydronic, or zoned. The capacities of residential installations typically range from less than 1 ton up to about 5 tons.

44.2 UNITARY SYSTEMS

Residential air-conditioning systems are often made up of unitary air-conditioning components. These are factory built and tested systems, complete as much as possible, with piping, controls, wiring, and refrigerant. Self-contained package units are usually simple to install, requiring only service connections and in some cases ductwork for field applications. The simplest of these is the window-mounted room air conditioner.

Unitary air-conditioning equipment consists of one or more factory-made assemblies, which normally include an evaporator or cooling coil, a compressor and condenser combination, and possibly a heating unit. When the air conditioner is connected to a remote condensing unit, such as in residential applications, the system is often referred to as a split system. A split-system installation will require more fieldwork for the technician as compared to the installation of a simple package unit. The sizes of unitary equipment range from small fractional-tonnage room coolers to large packaged rooftop units in the 100-ton category. Split-system equipment up to 5 tons in capacity may be classified as either commercial or residential. There is a wide range of applications in both markets using the same product; however, above 5 tons the application becomes distinctly commercial, and product designs use different components.

44.3 ROOM AIR CONDITIONERS

Room air conditioners were primarily developed to provide a simplified means of adding air conditioning to an existing room. These units are considered semiportable in that they can easily be moved from one room to another or from one building to another. They provide cooling, dehumidifying, filtering, and ventilation, and some units provide supplementary heating.

Figure 44-1 Window air-conditioning unit for cooling.

In numbers sold, room air conditioners such as the one shown in Figure 44-1 outsell all other types of unitary equipment. They are relatively low in cost, easy to install, and can be used in almost any type of structure, as shown in Figure 44-2.

Figure 44-2 Window air-conditioning unit on the Harry Elkins Widner building at Harvard University.

The disadvantage of room air conditioners is that they may either block part of the window area and prevent the window from being opened or require a special hole through the wall. Some people object to the operating noise that they produce close to the occupants. They are best used to condition a single room, but the spillover can supply some conditioning to adjacent areas.

TECH TIP

A common problem with window air-conditioning units is the nuisance trips of the residence circuit breakers. Window units are often used in older homes where the electrical service is not adequate for the air-conditioner load. It may be possible to rewire the residence for a single circuit to the air conditioner, or it may be best to downsize the system to one that will not trip the circuit breaker. Downsizing the unit may provide less than desired levels of cooling on extremely hot days. However, the constant shutting down of the oversized unit electrical service can result in the same lack of cooling. In addition, it may be possible to use more than one window unit in a room if the electrical outlets in that room are serviced by more than one circuit breaker.

Figure 44-3 Two basic parts of a window unit.

44.4 CONSTRUCTION AND INSTALLATION OF ROOM AIR CONDITIONERS

There are basically two parts to the unit, as shown in Figure 44-3. One section goes inside the room, where the evaporator fan draws in room air through the filter and cooling coil, delivering conditioned air back into the room. The other section extends outside the room, where the condenser fan forces outside air through the condenser, exhausting the heat absorbed by the evaporator. One motor operates both the indoor blower and outdoor fan (Figure 44-4). The motor shaft extends through the separating partition and drives both fans. Condensate from the evaporator coil flows into the drain pan, which extends below the condenser fan. The condenser fan tip dips into condensate, splashing it onto the hot condenser, where it evaporates and is blown into the outside air.

The window-mounted units are supplied with a kit of parts for installation. Sill brackets, window mounting strips, and sealing strips are set in place for installations in double-hung windows (Figure 44-5). Side curtains fold out to fill up the extra window space (Figure 44-6). A sponge rubber seal is provided for the opening where the sash overlaps, and a sash bracket is installed to lock the lower sash in place (Figure 44-7).

Room air conditioners are also available in a vertical configuration for mounting in sliding window openings (Figure 44-8). These vertical conditioners are held in place by the sides as opposed to the conventional-shaped room air conditioners that are secured in place on their tops and bottoms.

Installation Precautions

Be sure to plan the installation, even if you feel it will be simple! Very often there will be furniture in the way of the window. Ask for permission to move the furniture or have the customer clear the space. Drilling and mounting the unit will be somewhat messy. Warn the customer of this ahead of time. Place a drop cloth on the floor and then always remember to clean up any mess when you are finished with the job.

There are also a number of safety concerns that must be addressed. Some units are quite heavy and can easily be dropped from a high-story window during installation. Many units have sharp corners and are very slippery, so a good pair of gloves will help. Grabbing onto the tubing for a better hold will ruin the new unit before it is even installed. Don't drill through the tubing when installing the cabinet with the chassis inside. Make sure that the supporting bracket is installed properly and in place before aligning the unit. If the unit is too heavy, then get some additional help. When lifting the unit with another person, keep it level. The taller person will tend to tip the unit and the drain pan of water will spill all over you and the customer's floor.

44.5 EER AND SEER RATINGS

The typical size range for window air conditioning units is from 5,000 to 20,000 BTU/hr. These ratings are based on inside air at 80°F db temperature 67°F wb temperature, and outside air at 95°F db temperature. Smaller units normally operate on 115 V, while larger models operate on 230/208 V.

Equipment is generally rated for efficiency by the manufacturer to help determine which system is most suitable for a particular application. This energy efficiency ratio (EER)

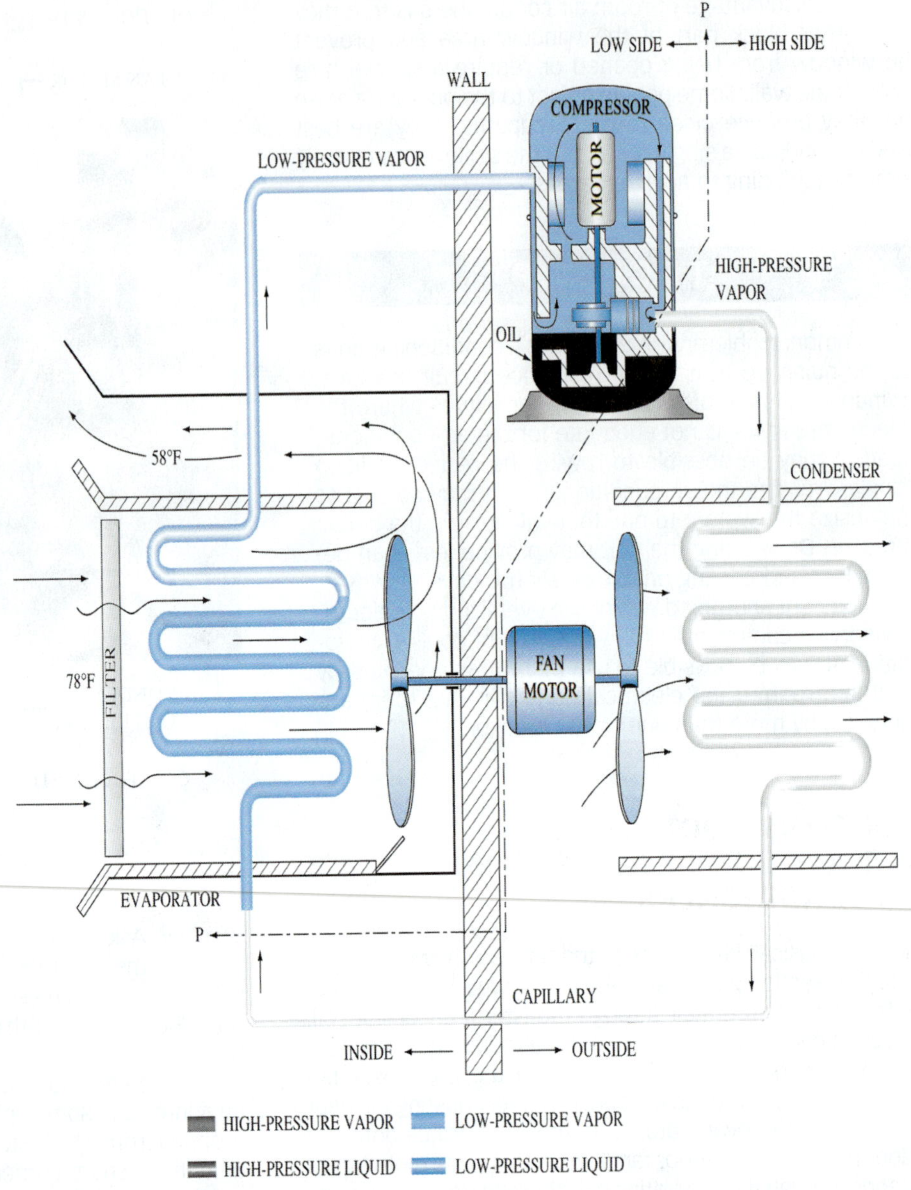

Figure 44-4 Refrigeration cycle of a room cooler.

is calculated by taking the output in BTU/hr and dividing it by the input in watts. The higher the EER rating, the better the efficiency.

The seasonal energy efficiency ratio (SEER) takes into account the cycling of the equipment on and off and is generally considered to be a better indicator of actual equipment efficiency. The Air Conditioning, Heating and Refrigeration Institute (AHRI) rates and publishes these ratings.

44.6 CONSOLE THROUGH-THE-WALL CONDITIONERS

A console through-the-wall conditioner is a type of room cooler that is designed for permanent installation. It was developed to provide individual room conditioning for hotels, motels, offices, and low-rise multifamily private household residences where it is impractical or uneconomical to install a central plant system. An opening needs to be made in the outside wall adjacent to the unit for condenser air and ventilation. These units are also known as packaged terminal air conditioners (PTAC), as shown in Figure 44-9.

TECH TIP

PTACs are most commonly used in hotels, nursing homes, apartment complexes, and other such areas where the area being cooled is limited. PTAC units may be straight air conditioning, heat pumps, heat pumps with electric resistance supplemental heating, or straight air conditioning with electric resistance heating. All of these units are similar in appearance, and most fit in a standard wall sleeve. These units are accessible from inside the occupied space, and in most cases the condensate is simply allowed to drain outside the building. Some cities and municipalities have ordinances controlling condensate from PTAC units, but many do not.

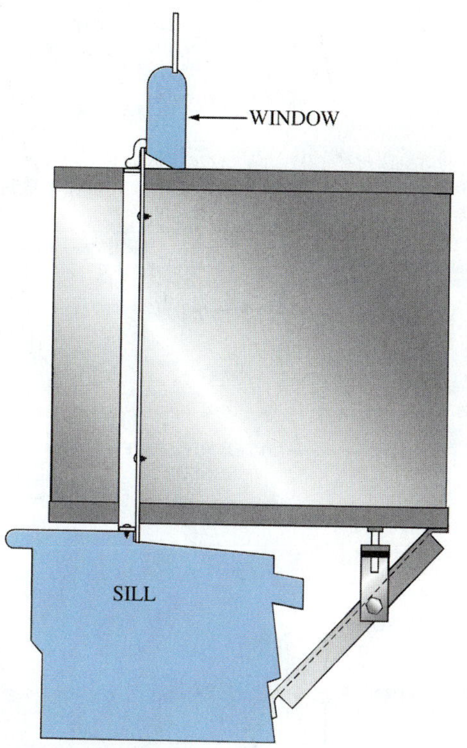

Figure 44-5 Installation diagram of a room cooler showing the supporting bracket from the outside.

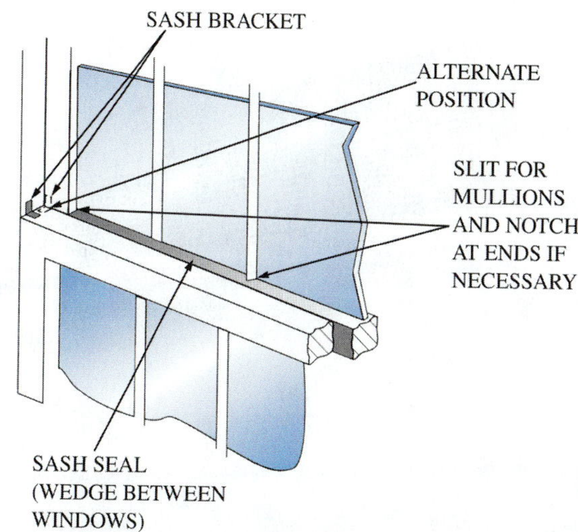

Figure 44-7 Installation diagram of a room cooler showing the sponge rubber seal between the top of the lower sash and the upper sash of a double window.

Performance

PTAC sizes range from 6,800 to 14,900 BTU/hr at standard rating conditions. The quantity of outside air that the units can admit ranges from 40 to 55 CFM depending on the size of the unit. Units are available for 115 and 208/230 V, single-phase, AC power. Electric heat capacity from 1.5 to 5.0 kW can be installed in any size unit. Power-receptacle configurations depend on the amperage drawn.

PTAC units are efficient, quiet, and easy to install. Temperature efficiency is usually stated in terms of EER. The EER is equal to the cooling output in BTU/hr divided by

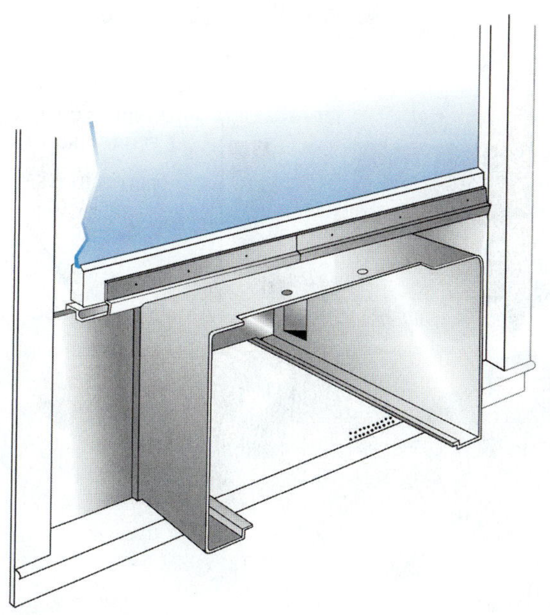

Figure 44-6 Installation of a diagram of a room cooler showing the window-sealing strips and filler boards.

Figure 44-8 Vertically constructed window unit for horizontal sliding window.

(a)

(b)

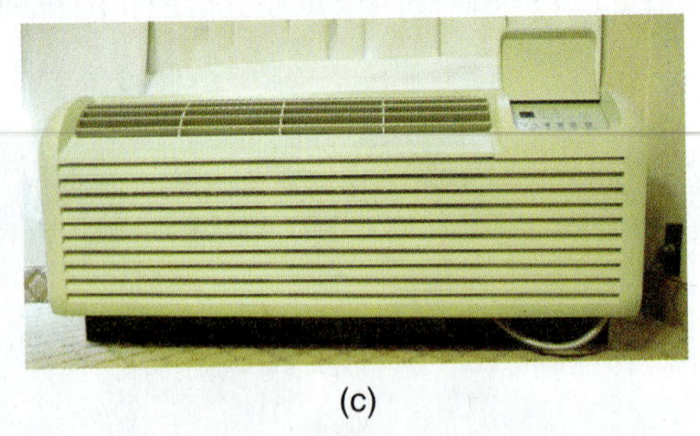

(c)

Figure 44-9 (a) Exterior view of a packaged terminal air conditioner (PTAC); (b) interior view; (c) console type unit.

the power input in watts under standard rating conditions. Standard rating conditions set up by AHRI are based on 80°F db, 67°F wb for the indoor entering air, and 95°F db for the outdoor ambient air. For example, a unit with an output of 6,600 BTU/hr and 660 W input, under standard conditions, would have an EER of 10 (6,600/660), which is considered a good rating.

44.7 PTAC COMPONENTS AND INSTALLATION

Figure 44-10 shows the location of the following essential parts: (1) indoor fan cover, (2) temperature control, (3) heat/cool/fan switch, (4) evaporator, (5) temperature sensor, (6) compressor, (7) condenser, (8) metering device, (9) condensate level float/drain, and (10) condensate pan. The air filter is located behind the front return-air panel and can be easily changed (Figure 44-11).

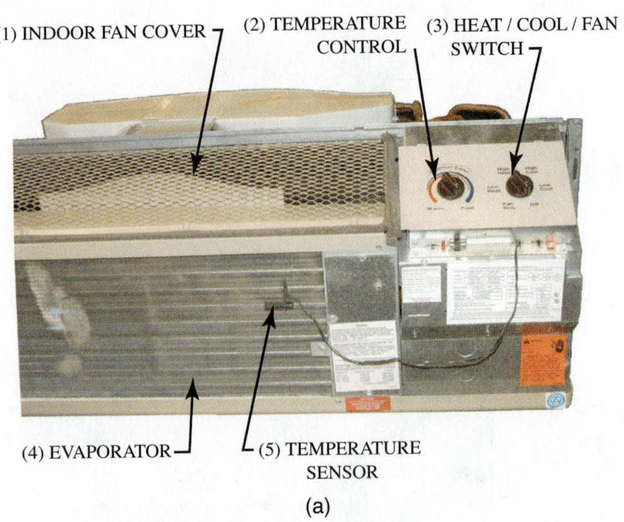

(a)

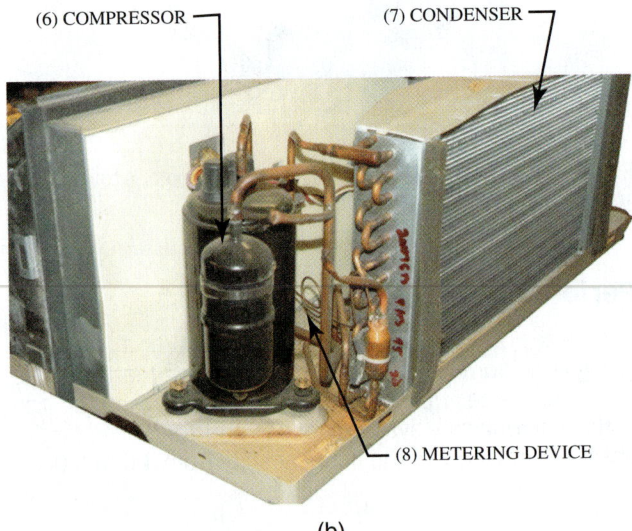

(b)

(c)

Figure 44-10 Expanded views of a PTAC unit listing major parts.

Figure 44-11 Air filters for PTAC.

SERVICE TIP

Most PTAC units use rotary compressors and capillary-tube metering devices. Rotary compressors have an excellent service life for this application, provided that the condenser is kept clean. A dirty condenser will raise the high-side pressure. High compression ratios will cause refrigerant to break down, and rotary compressors are unusually susceptible to damage from excessive heat because the motor is on the high side of the system. A dirty evaporator coil, as shown in Figure 44-12, should be cleaned and will also signal that the condenser needs to be checked. A small vacuum cleaner may be used for cleaning, but be very careful not to damage or bend any of the fins. These are very thin and are in place to increase surface area and heat transfer. If the condenser fins become bent, the air flow will be restricted, which will also create higher compressor discharge pressure.

44.8 PTAC CONTROLS

Most units are provided with the following choices (Figure 44-13):

OFF	turns unit off
FAN ONLY	indoor fan operates
COOL	provides cooling with indoor fan
HEAT	provides heating with indoor fan
HI	high fan speed
LOW	low fan speed

The following additional controls are provided:

- **Adjustable temperature limiting device** Limits the range of the room thermostat.

Figure 44-12 Dirty evaporator coil.

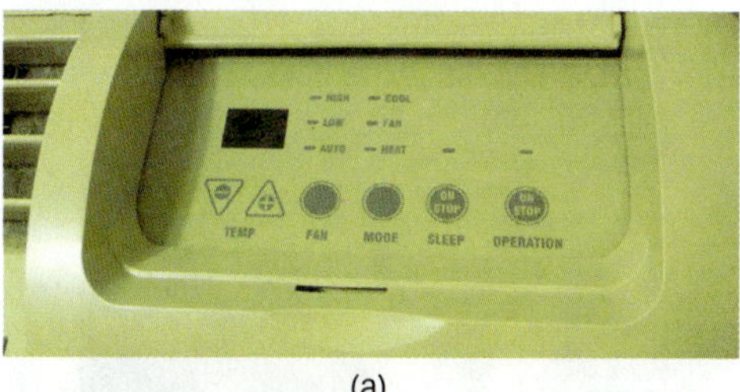

(a)

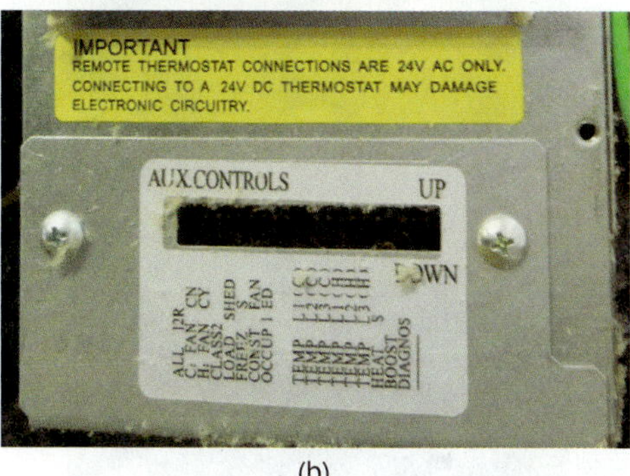

(b)

Figure 44-13 (a) Controls to operate unit; (b) DIP switches for controls.

- **Outside air damper** Control lever can be positioned to permit zero for fully open supply of outside air.
- **Fan cycle switch** Allows continuous or intermittent fan operation.
- **Remote thermostat (optional)** Unit can be wired to use a remote wall-mounted thermostat rather than the one normally supplied in the unit.
- **Front desk control interface (optional)** Units may be individually started and stopped with an energy-management panel from a central location.
- **Room freeze protection (optional)** Overrides OFF signal when room thermostat goes below 40°F.

44.9 STAND-ALONE AIR-CONDITIONING UNITS

A stand-alone air-conditioning unit for residential use requires an air handler outfitted with an evaporator coil, metering device, drain pan, transformer, blower assembly, and housing. The air handler is connected to its own duct system. There are three configurations of air handlers: vertical upflow, vertical downflow, and horizontal.

A vertical stand-alone is shown in Figure 44-14. The blower is drawing air across the refrigeration coil (Figure 44-15) and delivering it out through the top of the unit (Figure 44-16). The air passes across the coil, and refrigerant passes through the inside of the copper tubing (Figure 44-17). The refrigerant is supplied and returned from the coil through inlet and outlet distribution manifolds (Figure 44-18). The R-410A scroll compressor used for this unit is located below the blower and coil (Figure 44-19). The refrigerant lines pass upward through the air-handler casing to the refrigerant coil (Figure 44-20). A condensate drain line, often made up from PVC piping, will be connected to direct

Figure 44-14 Stand-alone AC unit.

Figure 44-15 Blower and refrigerant coil.

Figure 44-16 Blower discharge located at top of the unit.

Figure 44-17 Refrigerant flows through copper tubing.

Figure 44-18 Refrigerant is distributed to and from the coil.

water away from the drip pan located under the refrigerant coil (Figure 44-21).

The major disadvantages of a stand-alone air-conditioning unit are the type and location of the condenser. The stand-alone will be conveniently located inside the conditioned building. Therefore it is impractical to use an air-cooled condenser unless the heat can be rejected to the outdoors in some convenient fashion. Most stand-alone air-conditioning units will have a water-cooled condenser. This will require a continuous source of cooling water or some fashion of cooling water loop.

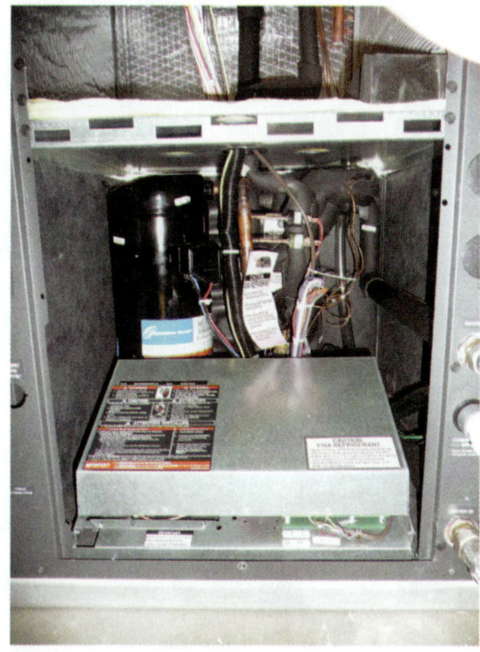

Figure 44-19 Compressor assembly.

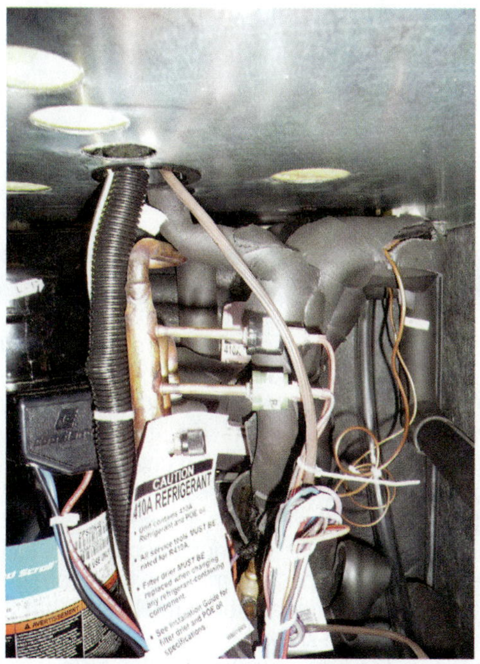

Figure 44-20 Refrigerant lines leading from compressor up through cabinet to the coil.

44.10 RESIDENTIAL SPLIT SYSTEMS

For many air-conditioning systems, it is not practical to place all components in a single package, particularly those that involve the use of air-cooled condensers. The air-cooled condenser must have access to outside air, so it is best placed outside. For this reason, split systems have been developed with the inside unit consisting of a fan coil unit, with or without heating; an outside mounted air-cooled condensing unit; and connecting refrigerant piping between the two, as shown in Figure 44-22.

Figure 44-21 PVC pipe condensate drain line.

TECH TIP

The final stage in manufacturing for split condensing systems is at the residence. Manufacturers have no control over the individuals who perform this last vital step in the production of their equipment. Manufacturers can't even insist that a matching system is installed. This final stage in manufacturing has a greater influence on the performance of a system than anything that the manufacturer can do. If the system is not installed with the correct size air duct systems, electrical wiring, and many other factors, the result can be the system providing poor performance or a reduction of its operational life. It is therefore very important that all of the manufacturer's specifications and guidelines be followed during installation so that your customer's system can be its most efficient and have the longest life possible.

Air-Cooled Condensing Unit

A typical split outdoor air-cooled condensing unit for residential settings is shown in Figure 44-23. This unit consists of a compressor, condenser coil, a condenser fan, and the necessary electrical control box assembly. On residential condensing units, a fully hermetic compressor is used.

The condenser coil is a fin-and-tube arrangement that varies in design from manufacturer to manufacturer. A large surface area is desirable, and many units offer almost a complete wraparound coil to gain maximum coil surface area. The coil tube depth is limited to reduce resistance to airflow.

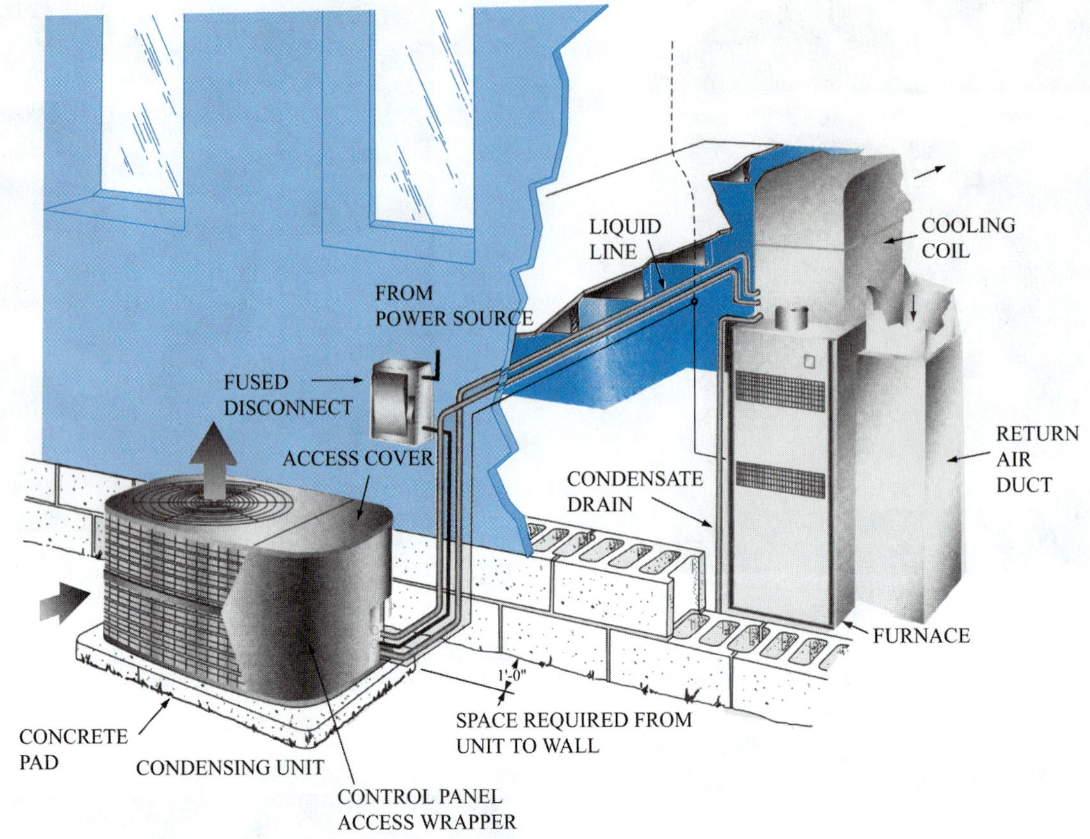

Figure 44-22 Residential add-on cooling to furnace installation.

Figure 44-23 Outdoor condensing unit for residential split system.

TECH TIP

Condenser fan motors turn at slower speeds than indoor blower motors. Not all condenser fan motors spin at the same rpm. Some operate around 1,000 rpm, while others may operate around 800 rpm. It is important when changing out a condenser fan that one of the exact same rpm be used. The fan blades are designed to move air at a certain speed. If the fan blade is spinning faster than it was designed for, it will not operate as efficiently. Increasing the rpm does not mean that you are going to increase the airflow. In addition, the height that the fan extends into the fan shroud is important. If the fan blade is too low or too high, it will affect the airflow from the condenser. Mark the height of the blade before removing it to be sure it is put back in at the same height.

The condenser fan varies in design but is usually a propeller-type fan, which can move large volumes of air through clean coils, which in turn offer low resistance (Figure 44-24). Airflow direction is a function of the cabinet and coil arrangement, and there is no one best arrangement. Most units, however, use draw-through operation over the condenser coil (Figure 44-25). Outlet air can have an effect on surrounding plant life. Top discharge is the most common arrangement. Fan motors

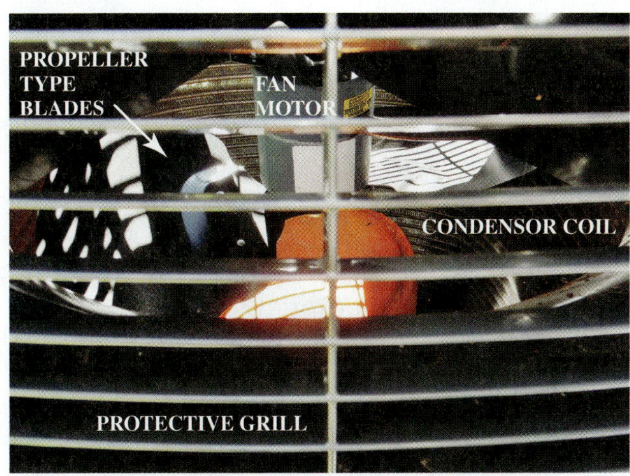

Figure 44-24 View of the interior of the condenser with propeller-type fan blades.

Figure 44-25 Air is drawn through the condenser from the outside and discharged out through the top.

are sealed or covered with rain shields. Fan blades are shielded with a grille for the protection of hands and fingers (Figure 44-26).

SERVICE TIP

Axial fans, like those on condensers, are high volume and low static. If the coils on a condenser are dirty, the fan will simply churn the air and throw it off the fan blade tips. You can actually take a very dirty condenser and place a sheet of paper in the center of the fan and it will suck it down while throwing the air off the fan tips. A clean condenser has a column of air that rises straight up from the fan. By feeling how the air is flowing, a technician can determine to some extent whether the coil is clean or dirty. An easy example of this is to place a box fan flat on a table, point it up, and turn it on. You will feel the air on the fan coming out the side and little or no air movement in the center. As the fan is lifted off the table, the correct airflow will be established, and the air will no longer come out the side of the blades.

Figure 44-26 Grille-type fan guard.

Figure 44-27 Typical evaporator used on a split AC.

Evaporator Coils

A typical evaporator used on a split air-conditioning system is shown in Figure 44-27. Coil cases or cabinets are insulated to prevent sweating, and all have pans for collecting condensate water runoff. Plastic pipe may be used to connect condensate drain water to the nearest drain. If no nearby drain is available, a small condensate pump may be installed to pump the water to a drain or to an outdoor disposal arrangement.

44.11 HEAT PUMPS

Heat pumps are commonly used for split systems. This is because heat can either be rejected or absorbed by the outside unit. On hot days in the summer when air conditioning is required, the unit will operate like a normal split air-conditioning system. The outside unit will operate as a

Figure 44-28 Outside unit for air-to-air heat pump.

condenser rejecting heat, and the inside unit will operate as an evaporator to cool the inside air. On colder days in the winter, the system will reverse and the outside unit will absorb heat from the outside air. This heat is transferred to the indoor unit, which now operates as the condenser and heats the inside air. This type of air-to-air heat pump is very common in climates where the outside temperatures do not fall very far below freezing in the winter. When outside temperatures do fall below the normal operating range for the unit, an electrical resistance heater will turn on. The cost for supplementing with this type of electric heat, however, is often far more expensive as compared to operating the normal heat pump cycle. Even so, heat pumps are used in colder climates, but they are arranged differently. For further information, all types of heat pump applications are explained in Section 8, Heat Pump Systems.

An outdoor unit for an air-to-air heat pump is shown in Figure 44-28. It can operate as either a condenser or an evaporator. The indoor unit is suspended from the roof in the attic by chain hangers to save space (Figure 44-29). The indoor unit has a drain pan located underneath it to catch any condensate to prevent leaks coming through the ceiling below (Figure 44-30). This indoor unit also has a water loop that is augmented by solar heating panels located outside, on the roof of the building.

44.12 MINI-SPLITS

Mini-split systems are similar to conventional split systems except they are smaller. They are typically used to cool or heat one space and not the whole building. The major advantage is that air ducting is not necessary. In some ways these are similar to window air-conditioning units and PTACs, but they do not require a window and they are more versatile. For example, mini-split indoor units can be placed anywhere in the home and multiple indoor units can be connected to a single outdoor unit. Mini-splits are available for cooling only

(a)

(b)

Figure 44-29 (a) Inside unit suspended by chain from the roof; (b) chain connection to inside unit.

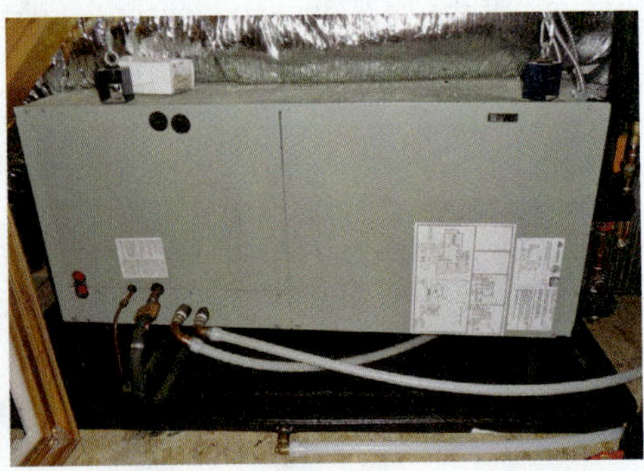

Figure 44-30 Inside unit for heat pump showing condensate drip pan located underneath.

or as heat pumps. This is possible because there is an outside unit and an inside unit (Figure 44-31). For installation, all that is required is a drilled hole through an outside wall for the piping and wiring connections.

44.13 AIR QUALITY

Air quality is sometimes neglected on air-conditioning systems, but an increasing emphasis is being placed on air quality both at home and the workplace. One simple task is to routinely change air filters (Figure 44-32). In some cases

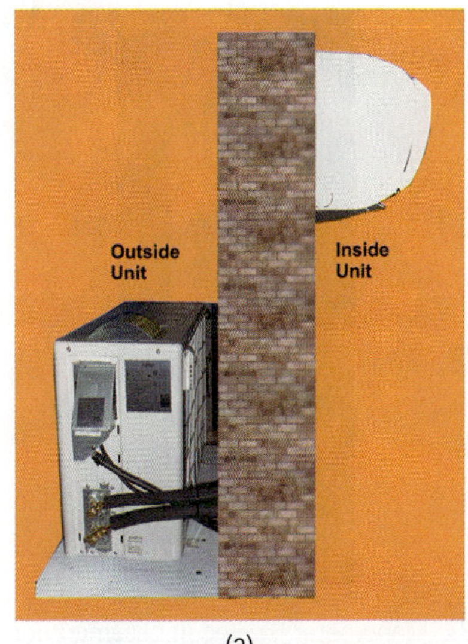

(a)

(b)

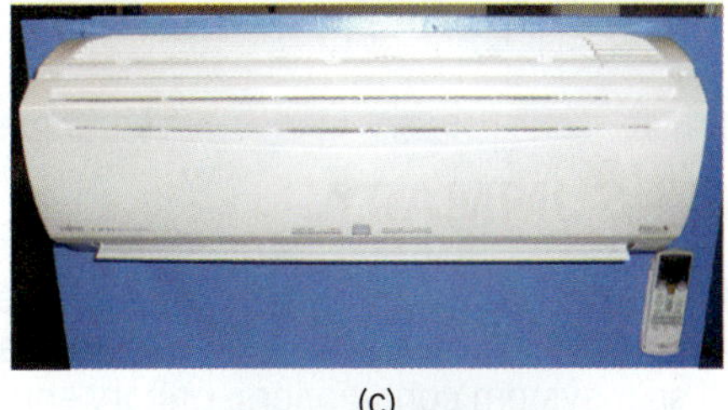

(c)

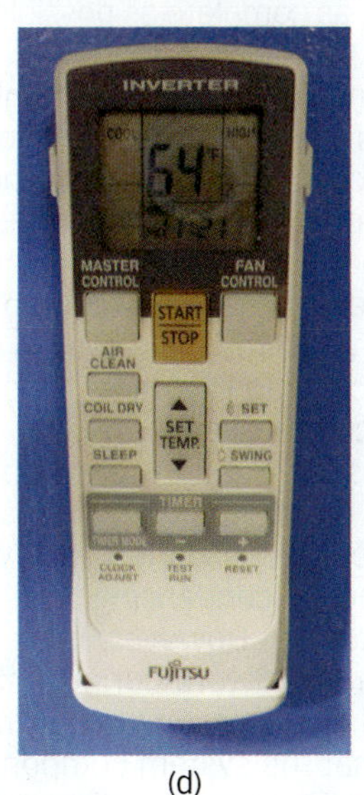

(d)

Figure 44-31 (a) Mini-split air-conditioning unit; (b) outside unit; (c) inside unit; (d) controls.

(a)

(b)

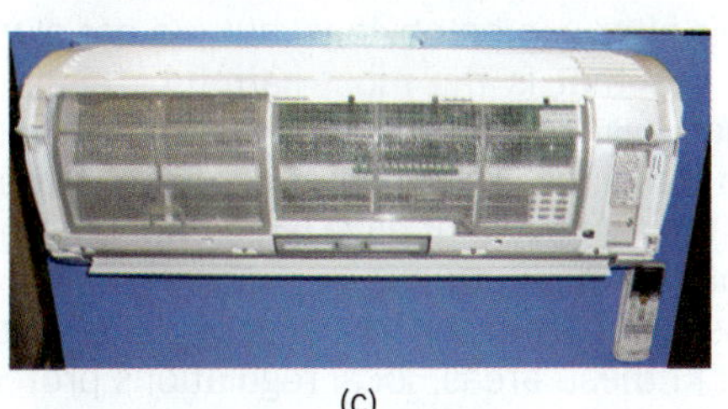

(c)

Figure 44-32 (a) Grille-type air filter cover on stand-alone air-conditioning unit; (b) air filters; (c) air filters for mini-split.

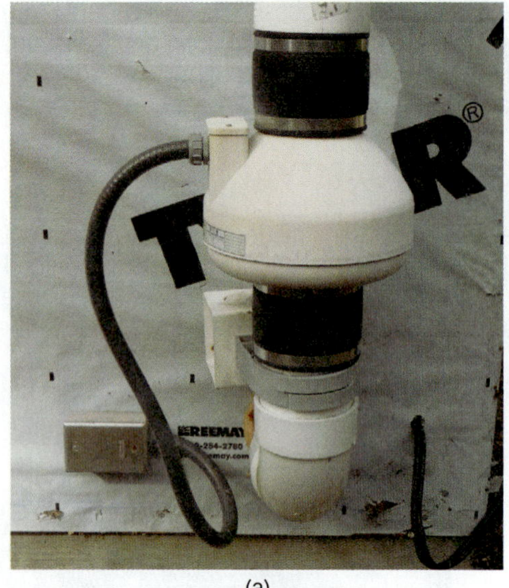

(a)

(b)

Figure 44-33 (a) Radon-mitigation blower; (b) radon gas is discharged above the roof line.

this can be done by the homeowner, but the homeowner might not be able to acquire the right type and style of filter required for the unit. A routine system check, and filter replacement, can often save the customer time and money in the long run.

Radon

A dirty air filter can be obvious, but it's not always what you can see that is the most harmful. This is the case with radon gas, which seeps up from the ground and enters into buildings. This type of seepage is prevalent in areas of the country that have large underground granite formations. Concentrations of radon gas over time can lead to serious health consequences for the building's occupants. In these areas, local regulations protect homebuyers. If a home is tested and found to be above the normal acceptable limits, then a radon-mitigation system must be installed before the home can be sold. There are different methods to accomplish this. One of the most common is to drill holes into the concrete basement floor or slab. A continuously operating blower draws the air and gas out of the building and terminates above the roof line (Figure 44-33). A manometer located in the basement provides a quick visual check that the blower is operating properly (Figure 44-34).

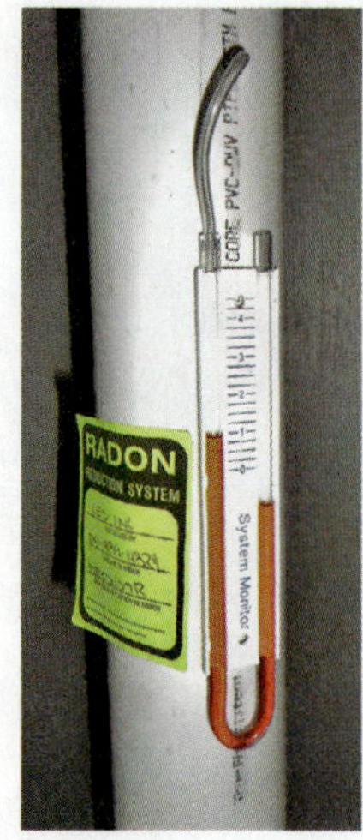

Figure 44-34 Manometer located in basement for radon-mitigation system.

UNIT 44—SUMMARY

In this unit you learned about different types of residential unitary air-conditioning systems, such as room air conditioners and split-system conditioners. Unitary equipment is factory built to be as complete as possible and fairly simple to install.

Room air conditioners are commonly found in many residential homes as well as small office buildings. They are relatively low in cost, easy to install, and can be used in almost any type of structure. A disadvantage of room air conditioners is that they may either block part of the window area and prevent the window from being opened or require a special hole through the wall. A common problem with window air-conditioning units is the nuisance trips of the residence circuit breakers.

Split systems have only the evaporator fan coil on the inside of the building. Split systems do not have the humming noise often associated with window units because the air-cooled condensing unit and compressor are mounted outdoors. However, these systems require considerably more fieldwork during installation because the refrigerant piping is assembled at the residence. It is very important that the system components be matched according to manufacturer specifications. Any change in design criteria in the field may adversely affect the system's performance.

WORK ORDERS

Service Ticket 4401

Customer Complaint: New Window Air-Conditioning Unit Keeps Tripping the Circuit Breaker

Equipment Type: 13,700 BTU Window Air Conditioner

After talking to the customer, the technician checks out the unit and finds nothing wrong. The unit is drawing 10 A and is on a 15-A breaker. There are numerous other household appliances on the same breaker. The technician checks the amperage draw at the breaker panel, and it measures 16.5 A. The technician recommends locating the unit to another room, where there is a lesser load on the wiring or to install a designated circuit for the new window AC unit. The customer is going to call an electrician to have a designated circuit installed.

Service Ticket 4402

Customer Complaint: Airflow Seems Low Since Adding Air Conditioning to the Existing Furnace

Equipment Type: 3-Ton 14-SEER Split System

The technician arrives at the job site, and after discussing the problem with the homeowner decides to check the indoor airflow. The technician discovers that there is not enough airflow for 3 tons of air conditioning. The furnace has a variable speed motor. The technician increases the motor speed to move more air. The technician rechecks the airflow and determines that the installers did not check to ensure there was sufficient airflow after installing the new coil. The airflow is now within industry standards, and the homeowner is satisfied.

Service Ticket 4403

Customer Complaint: Fan on Outside Unit Is Not Running

Equipment Type: 4-Ton 10-SEER Split System

The technician talks to the customer and then proceeds to analyze the system. The technician discovers that the fan motor is bad. The technician removes the fan motor and discovers the rated rpm is 875. The only motor on the service truck is rated for 1,050 rpm. Since the new motor must be within manufacturer's specifications, the technician goes to the supply house and buys a motor rated for 875 rpm. After installing the proper motor, the technician analyzes the system and determines that it is operating within the manufacturer's specifications.

UNIT 44—REVIEW QUESTIONS

1. List the primary advantages of a room air conditioner.
2. List the disadvantages of a room air conditioner.
3. What is a common problem with window air-conditioning units?
4. What are the two parts of a window air-conditioning unit?
5. How is the energy efficiency ratio (EER) calculated?
6. What is the primary difference between the EER and the SEER?
7. What is an advantage of a propeller-type fan?
8. Most PTAC units use what type of compressors and metering devices?
9. Why are there thin metal fins located on the tubes of evaporators and condensers on PTAC units?
10. What can happen if you tip the old window air-conditioning unit as you move it?
11. What are the three basic configurations of stand-alone air-conditioning air handlers?
12. What are the major disadvantages of a stand-alone air-conditioning unit?
13. Why are heat pumps often used on split systems?
14. What are some advantages of mini-split systems?
15. What is the concern about radon gas?
16. Explain why split systems were developed.
17. Why is a quality installation one of the most important phases of the manufacturing process?
18. Name the four major components of a condensing unit.
19. Describe why it is important to replace a fan motor with one of the exact same rpm.
20. In a condenser coil, why is the coil tube depth limited?

UNIT 45

Residential Split-System Air-Conditioning Installations

OBJECTIVES

After completing this unit, you will be able to:

1. list the various installation components of a residential split-system air conditioner.
2. explain the installation process of the outdoor condensing unit.
3. explain the installation process of the indoor evaporator and/or air handler.
4. explain the methods used to select and install the refrigerant line sets.
5. explain the electric wiring requirements of the system.
6. explain the various methods used to check the airflow across an evaporator.
7. explain the various methods used to check the refrigerant charge of the system.

45.1 INTRODUCTION

A popular style of air-conditioning system used in many homes is the split-system air conditioner. A typical split system consists of as few as two sections. There is an outdoor condensing unit and an indoor evaporator coil and air handler or an indoor evaporator coil and furnace. Many times these systems are connected to an existing forced-air furnace, but they can also be installed as a stand-alone system.

These systems must be assembled at the customer's home by a qualified technician. The installing technician must be proficient in the basic techniques used to run refrigerant lines, drain lines, and electric wiring. The technician must also possess the skills required to connect the system to the new or existing duct system.

45.2 PLANNING THE INSTALLATION

The success of any installation lies in its planning. Any equipment selection should be properly sized using the guidelines for load calculation as set forth in the Air Conditioning Contractors of America's (ACCA) Manual J. Proper planning allows for efficient and trouble-free installation. Without proper planning, costly mistakes can and probably will occur, increasing the costs for the installing contractor.

Part of the planning procedure requires a technician to reference the manufacturer's installation manual and determine the installation requirements of the system. This will allow the technician to determine the refrigerant line sizes to run, the electrical requirements of the system, and any other manufacturer's recommendations required to properly install the system. The installing contractor should also review the local code requirements and make sure any code-related issues are covered.

A technician should also develop a list of materials needed for the installation. This will ensure that all the correct materials arrive on the job. Leaving an installation to get more materials or the correct material will cost valuable time on the job.

Proper planning also allows a technician to choose the proper location of the equipment. This is especially important when deciding where to install the outdoor condensing unit. A technician should always discuss and get customer approval on the placement of the outdoor condensing unit. Installing it in the wrong location will cause problems for the contractor and the customer.

A technician should also develop a sequence for the installation to allow a more efficient installation process. A typical installation sequence is as follows:

- Determine the required BTU capacity of the system to be installed.
- Select the equipment and system components for the installation.
- If needed, configure the indoor section for an upflow, downflow, or horizontal installation.
- Set indoor coil in place.
- Install any needed ductwork.
- Set the outdoor pad in place, and level.
- Set condensing unit in place.
- Install refrigerant lines.
- Leak test and evacuate/dehydrate refrigerant lines.
- Run the required electrical wiring and connect up to the equipment.
- Install condensate drain line and pump (if needed).
- Weigh in the correct amount of refrigerant according to the manufacturer's specifications.
- Start up system, confirm airflow, and check/adjust refrigerant charge.
- Clean up work area.
- Explain operation, warranty, and any required maintenance of the unit to the customer.

Equipment Sizing

Part of the installation process is choosing the right system components. The installing contractor must first determine the proper capacity of the equipment needed to satisfy the cooling and heating requirements of the structure. To determine the capacity needed, a heat-gain calculation must be completed for the structure. For residential structures, the ACCA's Manual J is the industry standard. There are also many software programs available that will allow a contractor to perform this calculation.

It is important for a contractor to choose the right size system. Oversizing or undersizing a system can create problems for the contractor and the customer. An oversized system will not run long enough to properly dehumidify a structure. An undersized system will not be able to handle the load when the outdoor temperature is high.

Equipment Selection

Once the BTU requirements of a structure have been determined, the proper system components can then be selected. The outdoor condensing unit, such as the one shown in Figure 45-1, will be selected based on its BTU capacity for both sensible and latent cooling loads as outlined in ACCA Manual S, its seasonal energy efficiency ratio (SEER), and the type of condenser and voltage requirements. For residential systems, a 208/230 single-phase voltage, air-cooled condenser is the standard and is used on most installations. The BTU capacity of the outdoor condensing unit will be chosen to match the BTU requirements of the structure. The SEER rating will be chosen based on the customer's requirements. The SEER is the operating efficiency rating of the condensing unit. The higher the SEER rating, the more efficient the condensing unit will operate. The installing contractor will normally offer their customer the option of choosing the SEER rating. Currently the minimum SEER rating available for new systems is 13.

There are two types of installations for the indoor section of the system. One type is where the air-conditioning system is connected to an existing forced-air furnace. In this type of installation, an evaporator coil and drain pan are attached to the existing furnace, and the air-conditioning system will use the same blower assembly and duct system as the forced-air furnace. There are three different styles of evaporator coil and drain pan assemblies that can be used. One style is a cased-coil assembly, which consists of an evaporator coil, metering device, drain pan, and housing. The complete assembly is designed to fit directly on top of a forced-air furnace, and the system's plenum (duct section) is connected to the top of this assembly. An uncased coil can also be used. This assembly consists of just an evaporator coil, metering device, and drain pan. It does not have its own housing. The assembly is designed to fit directly in the plenum of the duct system. The third option is a half-cased coil, which is an evaporator coil, metering device, and drain pan contained in a housing covering only half of the coil.

The other type of indoor installation is where the air conditioner is a stand-alone system (Figure 45-2). An air handler is used for this type of installation, consisting of an evaporator coil, metering device, drain pan, transformer, blower assembly, and housing. The air handler is connected to its own duct system. There are three configurations of air handlers: vertical upflow, vertical downflow, and horizontal.

The BTU capacity of the evaporator coil assembly or air handler selected may be different from the capacity of the outdoor condensing unit. Many applications require a larger evaporator to obtain a particular SEER rating or sensible heat ratio.

The metering device used for a particular installation may also need to be selected. The indoor unit may come with a factory-installed metering device; however, it may

Figure 45-1 A typical outdoor condensing unit.

Figure 45-2 A stand-alone air-conditioning system. *(Courtesy of Goodman Manufacturing Company)*

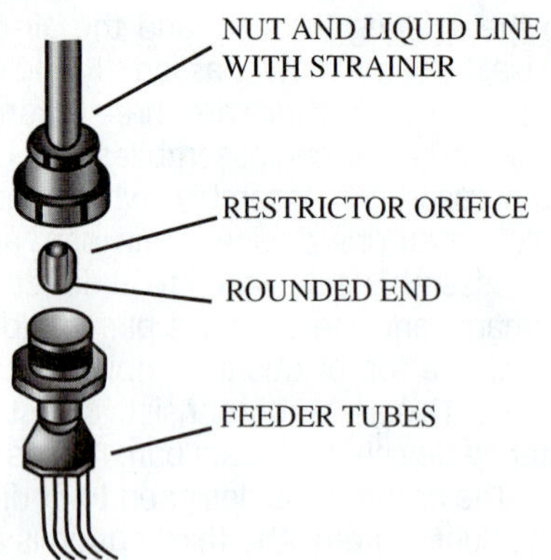

Figure 45-3 A fixed orifice used on a residential air-conditioning system.

need to be changed out if it does not match the requirements of the system. Some manufacturers ship coils today with no metering device installed, allowing the installer to use the one that matches both the unit capacity needs and the refrigerant type being used, such as R-22 or R-410A. The metering device must match the BTU capacity of the condensing unit and indoor evaporator. There are two general types of metering devices used: the fixed orifice (Figure 45-3), and the thermostatic expansion valve (TEV) (Figure 45-4).

Other Installation Components

There are several other installation components that must be chosen to complete an installation. The outdoor condensing unit will need an electrical disconnect switch (Figure 45-5). There are two types of disconnect boxes. One type contains only a switch, and it may or may not have a fuse. The other type uses a circuit breaker for both the switch and electrical overcurrent protection. An outdoor-rated watertight flexible electrical cable, generally referred to as a whip (Figure 45-6), connects the system's disconnect to the condensing unit.

An outdoor pad must also be selected to set the condensing unit. This pad can either be a poured concrete slab or a prefabricated reinforced plastic pad (Figure 45-7). Most

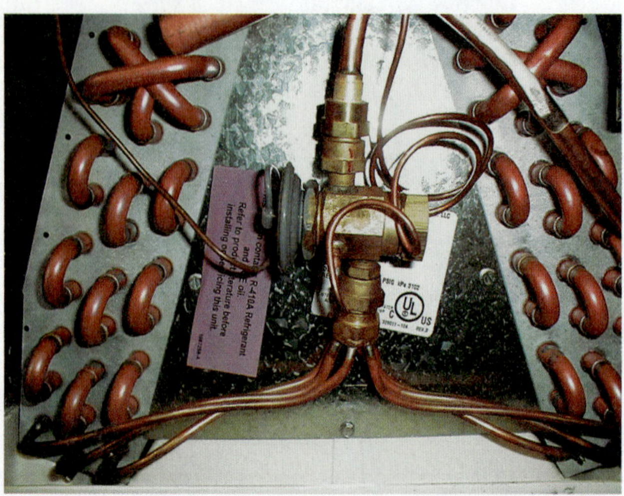

Figure 45-4 A TEV used on a residential air-conditioning system.

Figure 45-5 A disconnect switch for an outdoor condensing unit.

Figure 45-6 A whip used on an outdoor condensing unit.

Figure 45-7 Prefabricated reinforced plastic pad. *(Courtesy of DiversiTech Corp.)*

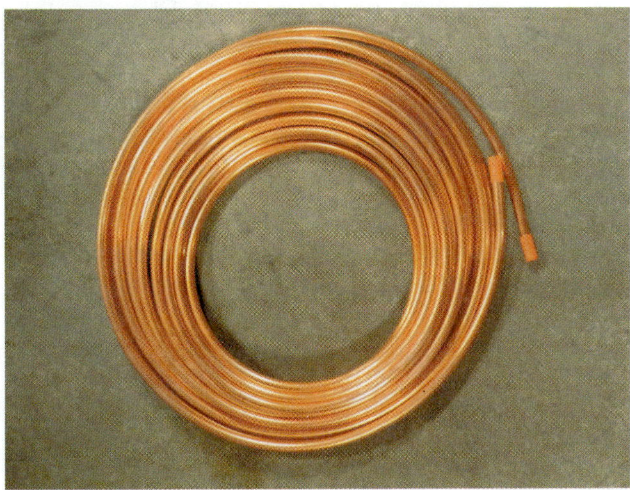

Figure 45-8 ACR tubing used for air-conditioning and refrigeration systems.

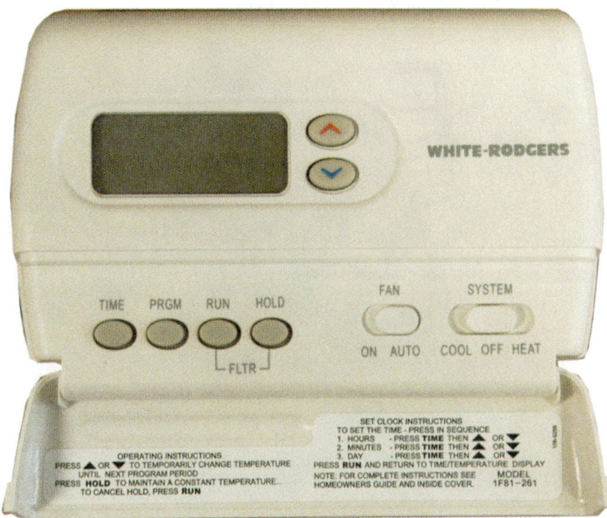

Figure 45-9 An air-conditioning thermostat.

equipment manufacturers recommend that the equipment pad be 2 to 3 in larger on all sides than the condenser to reduce dirt spattering into the coils.

Appropriately sized refrigerant lines will also need to be selected. Copper air-conditioning and refrigeration (ACR) tubing is used. ACR tubing is processed for use on air-conditioning and refrigeration systems. Its interior is processed to be clean and dehydrated and will not add any contaminants to a system. It is sized by its outside dimensions, which differs from pipe used in the plumbing industry. Copper pipe used in the plumbing industry is sized by its inside diameter.

Type L soft ACR tubing is sometimes used (Figure 45-8), and hard-drawn ACR can also be used. Soft ACR tubing can be purchased in rolls of 50-ft lengths and cut to size on the job. Hard-drawn ACR tubing can be purchased in 20-ft lengths and can also be cut to size on the job. To prevent condensation from forming on the suction line, it is typically insulated with closed cell foam (Armaflex® is often used). This insulation is sized based on the outside dimension tubing size and the wall thickness required. Insulation is available with wall thicknesses of ⅜ in, ½ in, ¾ in, and 1 in. Insulation with a ⅜-in wall thickness is normally used on residential split-system installations.

Wholesalers also sell ready-made line sets, consisting of an insulated suction line and a liquid line in various tubing diameters and lengths. These line sets are available in lengths of 15 ft, 25 ft, 35 ft, and 50 ft with tubing diameters of ⅝ in to 1⅛ in for the insulated suction lines and ⅜ in to ½ in for the liquid lines.

A thermostat (Figure 45-9) will also need to be selected. There are many different styles of thermostats. Thermostats are available as programmable and nonprogrammable as well as digital and mechanical. Different thermostats will have different features and capabilities depending on the system design. A contractor should consult with the customer as to the type of thermostat they prefer. A contractor will also need to select the wiring for the thermostat. Thermostat wiring is available in various gauge sizes and with different numbers of conductors. Residential installations will generally use a minimum of 18-gauge wire with either two, five, or eight conductors depending on the system and components to be installed. Some multistage heat pump applications may require as many as twelve wires.

Figure 45-10 A condensate pump used when no drain is close to the air conditioner's evaporator. *(Courtesy of DiversiTech Corp.)*

Condensate tubing and fittings will also need to be selected. Normally, the condensate drain lines used will be ¾ in Schedule 40 PVC. Be sure to check the local codes, as some may require 1¼-in drain lines after condensate leaves the unit. If there is not a convenient open drain close to the indoor assembly, a condensate pump (Figure 45-10) will also need to be selected to pump the condensate to an appropriate drain. The material used, as well as the location into which the condensate is drained, may be subject to local code.

45.3 INSTALLING THE INDOOR EQUIPMENT

Normally, the first part of the installation process is the installation of the indoor equipment. Depending on the type installed, the process will be slightly different. A cased coil, an uncased coil, and an air handler will all require different installation procedures.

Figure 45-11 An uncased evaporator coil.

Installing Cased Coils

When a cased coil is used on an installation it will generally sit on top of a forced-air furnace or below the furnace on a downflow system. If the coil is used on an existing system, the supply-air plenum will first need to be removed. Sometimes this also requires removing some other system components, such as the furnace's vent piping.

Once the supply-air plenum is removed, the new cased coil is set on top of the furnace and is leveled to ensure proper drainage of the condensate. A new supply air plenum is then fabricated and reinstalled between the top of the cased coil and the existing ductwork. All connections must be properly sealed to prevent air loss.

Installing an Uncased Coil

When an uncased coil (Figure 45-11) is used on an installation, it will generally be installed inside the supply-air plenum. One side of the supply-air plenum will need to be removed or cut to allow the uncased coil to be slipped into the plenum.

To install the coil, first shut off power to the furnace and remove any system components in front of the side of the plenum on which the coil is to be installed. Next, lay out a pattern on the supply-air plenum for the coil access opening and cut the opening. The access opening should allow the coil to sit close to the furnace but should be at least 4 in above the furnace's heat exchanger. Once an opening is created, install a set of brackets to hold the coil assembly, making sure the brackets installed are level to allow for proper drainage of the coil's condensate. Next, insert the coil into the plenum and place on the installed brackets. More than likely there will be open spacing between the coil and the inside walls of the plenum. These openings must be covered to prevent the supply air from bypassing the evaporator coil. Normally, a piece of sheet metal is installed between the coil and the inside walls of the plenum to cover this opening. Once the coil is set in place and leveled, a sheet metal patch will need to be fabricated to cover the opening on the side of the plenum. All joints must be sealed to prevent air leakage.

SERVICE TIP

Manufacturers provide extensive testing with their condensers and evaporator coils. Some manufacturers may have many different air-handler/evaporator-coil combinations for a single condenser. This research is carried out in scientific laboratories so that the most ideal combinations can be determined. Often, however, in the field, technicians simply purchase from a supply house a coil that is most convenient and not one specifically designed for the condenser. Any change in design criteria done in the field will adversely affect performance, not enhance performance. Manufacturers are under tremendous pressure from regulatory agencies to produce the highest possible system efficiency. If a change in coil size could enhance efficiency, it would have been done in the lab and there would be test criteria and procedures set up. It is not recommended in any way that technicians undertake design changes from the manufacturer's recommendations. These matched systems are available from the manufacturer and are available from the AHRI through publications and on their Web site.

Add-on Coils

Using a split system offers the opportunity to add cooling to an existing residential system heating unit where the necessary modifications are feasible. Add-on coils are available in a number of configurations to fit various types of heating units (Figure 45-12). This figure also shows the manner in which the position of the coils is related to the air-handling unit. Coils are supplied for upflow, horizontal, and downflow furnace applications.

SAFETY TIP

Codes prohibit the installation of an evaporator coil before a gas or oil furnace heat exchanger. If the coil is placed before the heat exchanger, then condensate will form in the heat exchanger, causing it to rust out quickly.

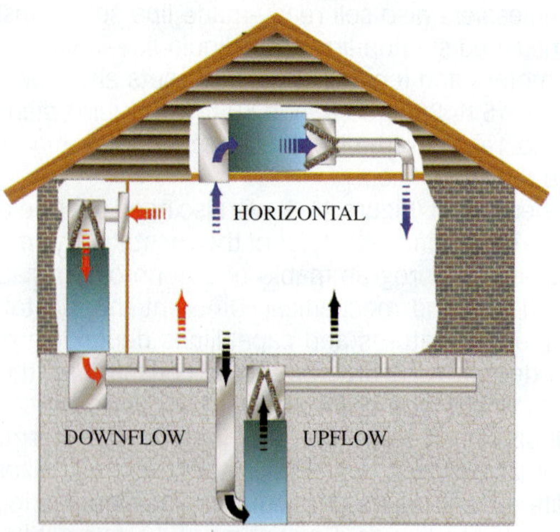

Figure 45-12 An uncased evaporator coil.

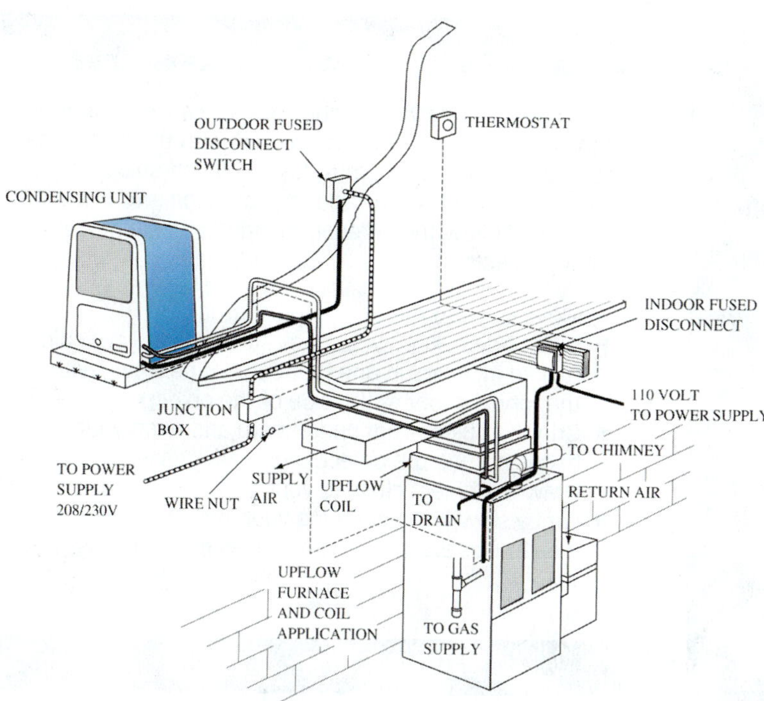

Figure 45-13 A single-circuit add-on split-system installation.

An important consideration in applying an add-on cooling coil to an existing furnace is the air resistance it adds to the furnace blower. During the heating cycle, the coil is inactive and dry. In summer, when the unit is cooling and dehumidifying, the coil is wet. The wet coil will add an average of 0.20 to 0.30 in of water column static pressure loss. This added resistance may require a change in furnace blower speed and an increase in motor horsepower. If the installation is a new system, a furnace blower should be selected that is capable of producing sufficient external static pressure to deliver the required amount of air. Wiring and piping a typical single-circuit add-on split system is shown in Figure 45-13.

Installing an Air Handler

Air handlers will often come as an assembly with a blower and refrigerant coil (Figure 45-14a). Since an air handler is a complete assembly, its installation consists of setting the air handler in place and connecting the ductwork, refrigeration lines, and condensate lines (Figure 45-14b). A supply plenum will be connected to the discharge of the air handler and a return boot connected to its inlet.

Most air handlers can be installed in the vertical upflow, the downflow, or the horizontal position. Normally the air handler will come ready for an upflow installation. If used in the downflow or horizontal position, some modifications will need to be made to the unit. Consult the installation manual to determine these modifications.

45.4 OUTDOOR CONDENSING UNIT INSTALLATION

The location of an outdoor condensing unit is an important part of the overall installation. An incorrectly placed condensing unit can cause problems for both the installing

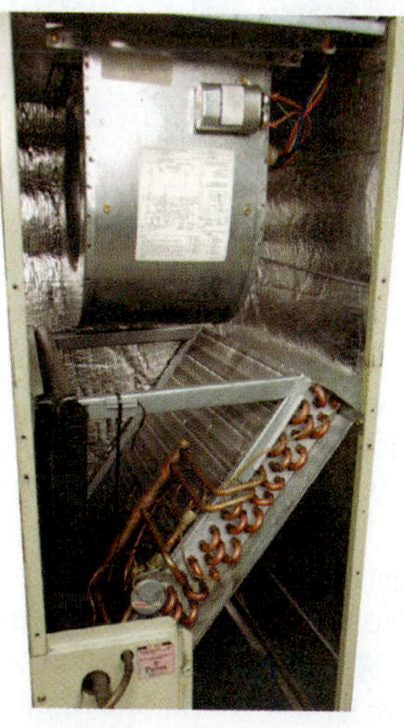

(a)

(b)

Figure 45-14 (a) Air handler with blower and refrigeration coil; (b) condensate drain for air handler.

Figure 45-15 A typical placement of an outdoor condensing unit.

contractor and the customer. The installer should always discuss the location of the condensing unit with the customer and verify the location chosen is not objectionable. Figure 45-15 shows the placement of an outdoor condensing unit. The following is a list of recommendations when deciding on the location of the outdoor condensing unit:

- Do not install a unit where the operating noise will be objectionable to the customer or their neighbors.
- Avoid locating the unit in direct sunlight.
- Do not locate where water, snow, or ice from the roof or eaves can fall directly onto the unit or where the roof overhang can cause recirculation of the air exhausted from the unit.
- Do not locate under decks or porches unless the clearance dimensions specified by the manufacturer are met.
- Locate the unit close to the building to minimize line length and to avoid installing underground piping.
- Locate the unit so that it can be readily serviced.
- Maintain required clearances around the unit for proper airflow.

Once the location has been chosen, the pad for the condensing unit can be set in place and leveled. The condensing unit can then be set in place on the pad. The condensing unit can be mounted and secured directly to the pad using a construction adhesive or tie-down bolts (some local codes may require this). If construction adhesive is used, make sure not to block the drain holes located on the bottom of the condensing unit.

SAFETY TIP

Improperly lifting heavy objects can cause serious injury. A technician should always lift with the legs rather than the back; the leg muscles are much stronger than the back muscles. For added protection, wear a back support. Follow these recommended steps when lifting heavy objects:

- Move close to the object to be lifted.
- Squat down. Keep your back straight and your chin tucked in. Position one foot behind the other, with the forward foot at the side of the object.
- Grip the object from underneath and wrap your hands around the object.
- Draw the object close to your body.
- Lift by slowly straightening your legs. Try to keep the weight centered over your legs as much as possible.

TECH TIP

Noise is recognized as an environmental pollutant. Outdoor air-cooled condensing units are sound-producing mechanical devices, which some cities and towns regulate. Early attempts to create local ordinances prompted action on the part of the ARI in 1971 to establish a sound rating standard that local communities could adopt.

AHRI Standard 270 applies to the outdoor sections of factory-made air-conditioning and heat pump equipment (unitary air conditioners). Under the program, all participating manufacturers are required to rate the sound power levels of their equipment in accordance with the technical specifications contained in this standard. Test results are submitted by the manufacturers to AHRI for review and evaluation. Units are sound rated with a single number: the sound rated number (SRN). Typical ratings are between 14 and 24.

45.5 INSTALLING THE REFRIGERANT LINES

To connect the indoor coil to the outdoor condensing unit, ACR tubing will need to be run and connected to each of these devices. One pipe run will be the vapor suction line and the other the liquid line of the system. These lines must be properly sized and installed to ensure that the system operates in accordance with the manufacturer's specifications. Incorrectly sized or installed refrigerant lines can cause the system to operate outside its rated capacity or cause damage to the system's compressor.

Before installing the refrigerant lines, make sure the metering device installed in the indoor unit is correctly sized for the application. If a fixed-orifice metering device is used, verify that it matches the rating for the outdoor condensing unit. Some new condensing units will come shipped with an appropriately sized fixed orifice, as shown in Figure 45-16. If the indoor unit's fixed orifice does not match the one that came

Figure 45-16 A fixed orifice shipped with a new condensing unit.

with the condensing unit, it should be replaced with the one supplied by the condensing unit manufacturer. Many coils now ship with no metering device, as one is added based upon capacity requirements and refrigerant type.

The overall length of the refrigerant line should be kept as short as possible. Do not exceed the maximum length as stated by the manufacturer. Generally pipe runs can be up to 175 ft. If a line set exceeds a distance specified by the manufacturer, special modifications will need to be done to the system. Most manufacturers will have an installation bulletin detailing their requirements for these long line applications.

The diameter of the refrigerant lines run must be properly sized; it is based on BTU capacity and the length of the refrigerant lines used on the system. Always follow the recommendations of the equipment manufacturer; the required line sizes will normally be given in the installation manual shipped with the equipment. When the condensing unit is installed above the evaporator unit, suction line size (if oversized) can affect the velocity of the refrigerant and the refrigerant's ability to carry oil.

TECH TIP

Do not base the line sizes on the stub connections at either the condensing unit or the indoor coil. These stub connections are based on an average installation and do not apply to every installation.

Suction lines should always be insulated to prevent water damage from condensation and to limit the refrigerant from becoming excessively superheated on the way back to the compressor. Liquid lines are normally not insulated

Figure 45-17 Soft ACR tubing is easier to bend and form.

except if there is a possibility that the liquid will gain excessive heat on its way to the metering device, such as if it is run through a hot attic.

Either soft or hard-drawn ACR tubing can be used. Soft tubing is normally used, as it is easier to bend and form (Figure 45-17), making the installation of the tubing easier for the technician. When bending this tubing, use a tube bender to prevent kinking. When elbows are required, use a long-radius elbow instead of short. Long-radius elbows offer less resistance to flow.

Many times the ACR tubing used will need to be cut to length. Make sure to use the appropriate tubing cutters (Figure 45-18); they allow for a clean and square cut. Avoid using a hacksaw, as it leaves too many burrs on the tube ends as well as filings that can contaminate the system. Always deburr the tubing ends after cutting them. Do not allow any of the burrs to fall into the copper tubing.

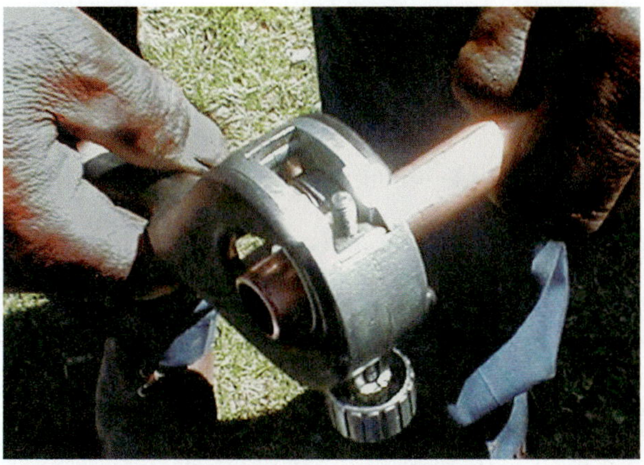

Figure 45-18 Use a tube cutter for a clean and square cut.

Figure 45-19 The isolation valves on an outdoor condensing unit.

Instead of using fittings to join sections of tubing, many technicians will use a swaging tool to increase the diameter of one tube end, allowing the opposite end to be inserted into the enlarged end. This method allows for less soldered or brazed joints, leading to less chance of refrigerant leaks.

The tubing connections are joined using the brazing process. Always flow an inert gas, such as nitrogen, through the lines when brazing. This prevents the copper from oxidizing due to the heat. Oxidation will tend to flake off during operation, which will plug up filter driers and metering devices.

The condensing unit will normally contain two isolation valves (Figure 45-19), each with a stub connection. One of these stub connections is for the suction line and the other is for the liquid-line connection. Each of these isolation valves will also have a Schrader port to allow a technician to attach the refrigeration manifolds. On new systems these isolation valves are initially front-seated to hold a refrigerant charge contained in the condensing unit. When brazing the tubing to these stub connections, the core inside the Schrader must be removed. The valve contains a rubber packing that if overheated will be damaged. A core remover can be used to remove this Schrader core. A wet rag or a heat-sink material must always be wrapped around the valves to prevent them from being overheated during brazing (Figure 45-20).

Figure 45-20 Wet rag wrapped around valves during brazing.

When running the refrigerant lines, there should be support lines every 6 to 10 ft and within 2 ft of any bends. Insulated supports should be used to avoid any noise transmission and to avoid having the tubing come in direct contact with any water pipes or ductwork.

Normally, the refrigerant lines will need to run through some type of wall structure. When cutting through any wall, always check and recheck before cutting access holes—mistakes can be costly. When running refrigerant lines through a masonry wall, the pipe must be sleeved with a section of PVC tubing to prevent the deterioration of the copper tubing. This reduces the chance of a leak developing over time from the tubing rubbing on the masonry due to vibration. Always seal pipe penetrations with an appropriate sealant.

A liquid-line filter drier should always be a component included in the system. The best location for the filter drier is indoors as close as possible to the metering device, although most installations have them outside near the condensing unit. The desiccant contained in the drier will normally hold more moisture in a cooler environment. Some manufacturers install a liquid-line drier at the factory. If this is the case, be sure not to add another one, as this would result in two driers in series, which could cause a restriction.

Leak Testing

Once all the refrigerant lines are run and connected, the system needs to be leak checked. The lines and coil should be leak tested with dry nitrogen. Nitrogen makes a good initial leak-test gas because it is inexpensive and legal to release to the air. Pressurize the lines and coil to between 100 psig and the system's low-side test pressure. Never pressurize a system more than 150 psi. Make a note of the time and pressure. Use soap bubbles or an ultrasonic leak detector to check the line connections at both the inside and outside units. Also check any other mechanical or brazed joints in between. Finally, if the unit uses a field-installed expansion valve, check all connections on the expansion valve. After making all these checks, recheck the pressure. The system has passed the initial leak test if the pressure is the same as when the system was first pressurized.

An alternative method for leak checking a system is to pressurize the system with a small amount of HCFC-22 and nitrogen. Add enough refrigerant to pressurize the system to 2 to 3 psig with the HCFC-22 and then add a nitrogen charge on top of the refrigerant. Pressurize the system again to a test pressure as recommended by the equipment manufacturer. Once the system is pressurized, an electronic leak detector can be used to locate the refrigerant leak. Only use this mixture with R-22 systems. Never mix R-22 with R-410A.

TECH TIP

Vacuum pumps are used in conjunction with a quality vacuum gauge to properly evacuate a refrigeration system before refrigerant is added to an empty system. Before opening a vacuum pump and gauge to a system, it is a good practice to perform a blank-off test. A blank-off test is when a vacuum pump only pulls a vacuum on the vacuum gauge and manifold.

Performing a blank-off test before opening a vacuum pump and gauge to a system will allow a technician to verify that the vacuum pump and gauge are operating normally. It is better to identify a problem with either a vacuum pump or gauge before running on a system. Running a defective vacuum pump on a system is a waste of time; a sufficient vacuum level will not be obtained. Using a defective vacuum gauge will not allow a technician to determine if the appropriate micron level for complete evacuation is reached.

To perform a blank-off test, turn off the valve between the system and the vacuum gauge, manifold, and vacuum pump. Then run the vacuum pump until a 500 micron level is achieved. If a 500 micron level is achieved, the technician can safely assume that the vacuum pump and vacuum gauge are operating normally.

System Evacuation

Once the system is leak tested and is determined to be leak free, it needs to be evacuated. Release the nitrogen to the air and connect a vacuum pump and micron vacuum gauge. The system should be evacuated to a 500 micron level or the level recommended by the equipment manufacturer. Do not rely on a compound gauge, as they are not designed for deep vacuum detection.

Once a 500 micron level is achieved, shut off the pump and watch the micron gauge. A rise of 100 to 200 microns is normal. If the micron level rises above this but then levels out, try running the vacuum pump again, because there may be some moisture left in the system. If the micron level rises and continues to rise, there is probably a leak in the system, which must be repaired. Once the line set and evaporator have been properly evacuated, the isolation valves on the condensing unit will need to be opened to allow the refrigerant to flow into the rest of the system if the new condensing unit has a precharge of refrigerant. Remove the valve stem caps from the isolation valves and turn the valve stem fully counterclockwise to open the valves.

If the new condensing unit does not have a precharge of refrigerant, weigh in the correct amount of refrigerant based on the system match and length of line set. Follow the manufacturer's recommendations, which is always the best way to charge a system.

45.6 FIELD WIRING

The installation of a residential split-system air conditioner requires the installation of two dedicated line-voltage and two low-voltage electrical circuits. One dedicated line-voltage circuit services the outdoor condensing unit through an (if required, fused) outside disconnect switch. The line-voltage service to the outdoor condensing unit is normally a 208/240 V single-phase service rated at a specific amperage. The amperage rating needed will depend upon the size of the condensing unit installed. A separate low-voltage circuit is also required from the furnace or air handler to the outdoor condensing unit. This is typically an 18-gauge two-conductor service used for the control circuit of the system of a cooling-only, single-stage system. The other dedicated line-voltage circuit is to the furnace or air handler. This is normally a 115 V or 230 V, 15 A or 20 A circuit. The second low-voltage circuit is for the system's thermostat connection. It is run from the furnace or air handler to the thermostat. Normally an 18-gauge, four-wire conductor is used. Some installations will require a five-conductor service if a 24 V digital thermostat is used. Newer multistage residential systems, which can have indoor air quality (IAQ) options, may require more wires. When connecting the low-voltage wiring of the system, always refer to the manufacturer's wiring diagram for the proper connections.

Choosing an Overcurrent Protective Device

When installing residential air-conditioning systems, the installing contractor must be sure that the appropriate type of overcurrent protection device is used with the outdoor condensing unit. There are three main types of overcurrent protection devices:

- Fuse
- Standard circuit breaker
- HACR circuit breaker

The type of overcurrent device to use will normally be stated in the installation instructions or on the rating plate of the condensing unit (Figure 45-21). Some manufacturers will use the following wording on the data plate to state the type of overcurrent device to use:

- If the rating plate states MAX FUSE _______ AMPS, then a fuse with the proper rating must be placed in the circuit. Normally, a fused disconnect will satisfy this requirement.
- If the rating plate states MAX FUSE or HACR CIRCUIT BREAKER ________ AMPS, then an HACR circuit breaker or fuse with the proper rating can be used.

Figure 45-21 A rating plate of an outdoor condensing unit.

- If the rating plate states MAX FUSE, HACR CIRCUIT BREAKER OR CIRCUIT BREAKER _____ AMPS or MAXIMUM OVERCURRENT PROTECTION DEVICE, then a fuse, HACR circuit breaker, or standard circuit breaker can be used.

The gauge of the electrical wire used also needs to be selected. The proper gauge must be selected to allow for the proper amperage draw without the possibility of overheating the wire. When wiring equipment, always use the type and gauge of wire recommended by the equipment manufacturer and only make adjustments based on local or national codes. The value marked MINIMUM CIRCUIT AMPS is used to determine the proper wiring size to use. The wire size rating is based on a maximum ambient temperature of 86°F. If this is exceeded, the wire size must be rechecked after being derated for its new value.

TECH TIP

Only use fuses or circuit breakers that are HACR rated. This type may be referred to as "slow-blow." These fuses and breakers can handle the momentary high current draw that occurs during compressor starting without blowing or tripping. The technician must be familiar with the appropriate National Electrical Codes, Local Codes and the Manufacturer Instructions.

45.7 INSTALLING THERMOSTATS

The placement of the thermostat is also important for the overall operation of the system. An incorrectly placed thermostat could cause the structure to be too cold or too warm. Following is a listing of some general requirements when mounting a thermostat:

- It should be installed on an inside wall and not on an outside wall or a wall exposed to any unconditioned space.
- It should be mounted 52–60 in above the floor in an area with good air circulation.
- It should be mounted level.
- It should not be located in dead-air spots.
- It should not be located near diffusers.
- It should not be located near any radiant heat sources, such as direct sunlight or in close proximity to floor and table lamps.
- It should not be located near any concealed pipes or ducts.
- It should not be mounted on a wall that separates conditioned and unconditioned space.

Normally, 18-gauge wire is run from the thermostat to the indoor coil assembly. If the thermostat is located over 100 ft from the cooling equipment, 16-gauge wire should be used. Remember, the smaller the number, the larger the wire size.

Wiring is normally color coded to simplify the connections between the thermostat and heating/cooling equipment. A typical color coding is shown below. Always check the manufacturer's instructions for the color coding of the low-voltage system, as not all manufacturers use this color coding.

- R Red: transformer power (24 V)
- G Green: fan control (the National Electrical Code allows the use of green as a current-carrying wire only on voltages less than 50 V)
- Y Yellow: compressor control
- W White: heating control

45.8 INSTALLING THE CONDENSATE DRAIN PIPING

Since the evaporator will often operate at a temperature below the dew-point temperature of the air drawn across it, water will be condensed from the air. The evaporator will have a drain pan located under its coil to collect this water. The water collected will need to be safely drained away from the evaporator. The installing technician needs to run a drain line from this pan to an open or vented drain. Laundry sink or basement floor drains are commonly used. Normally, ¾-in schedule 40 PVC is used as the drain line. However, other types of plastic pipe can be used, such as CPVC (chlorinated polyvinyl chloride), PP (polypropylene), and ABS (acrylonitrile-butadiene styrene).

The drain pan on the evaporator will normally have both a primary and secondary drain. Pipe connections are normally ¾-in FPT (female pipe thread) connection. The primary drain is the lower of the two, and a trap should be installed as close to the unit as possible if this location is under a negative pressure. The secondary drain is the upper connection and will allow water to drain from it, if the primary drain becomes clogged. Some applications do not require the use of a secondary drain; if it is unused it should be capped. These drain lines should be slightly pitched in the direction of flow and supported every 10 ft. A cleanout should be provided (Figure 45-22) to allow service and maintenance technicians to clean the drain line as needed.

The secondary drain connection serves a great purpose if the evaporator is installed in an area where water overflow can cause severe damage to a customer's property. If the primary drain becomes clogged, water will safely drain down through the secondary drain line. When used, the secondary drain should not be connected to the primary drain and should not be trapped. It should be piped to a location that is separate from the primary drain and visible to the owner. The line should be tagged to notify the customer to call for service if water is seen draining from this pipe.

For attic or ceiling installations, it is advisable to install a drain pan under the fan coil to also prevent water damage caused by an overflowing condensate line. If a drain pan is installed under the fan coil, the secondary drain from the evaporator pan can be directed to this pan. This pan must then be piped to a location that is both visible to the owner and will not cause any damage to the property. This line should then be tagged to notify the customer to call for service if water is seen draining from this pipe. Some units installed above finished ceilings have a safety switch located in the pan for shutdown in the event that the pan is

Figure 45-22 A clean-out for the primary drain line.

not draining properly. This limits expensive repairs due to water damage.

To cut and join PVC pipe as well as ABS and CPVC, the following procedure can be used:

- Use a plastic tubing cutter or a special cutter called a plastic tubing shear to cut the pipe. Avoid using a saw to cut this pipe. It will leave particles that could enter the pipe and cause a future clog.
- Deburr the pipe ends with a knife or file.
- Using an approved cleaning solvent, thoroughly clean the pipe ends and the inside of the fitting sockets.
- Apply a thin coat of primer to the tube end and the inside of the fitting sockets.
- Apply a thin coat of cement to the tube end and the inside of the fitting sockets.
- Immediately after applying the cement, push the pipe completely into the fitting sockets using a quarter-turn twisting motion until the pipe bottoms into the socket.

45.9 START-UP AND SYSTEM CHECKOUT

Once all of the system components have been installed, the system needs to be started up and checked out per the instructions supplied in the installation manual. If the compressor has a crankcase heater, many manufacturers recommend that the heater be energized for a minimum of 24 hours before starting the system. This prevents liquid refrigerant from being in the compressor on startup.

The system's fan should be energized and the airflow across the evaporator should be measured. There should be approximately 400 CFM (cubic feet per minute) per ton of airflow across the evaporator. For example, a 3-ton residential air-conditioning system should have approximately 1,200 CFM of air flowing across its evaporator coil. However, airflow can be as low as 350 CFM per ton or as high as 450 CFM per ton to allow for changes to the sensible heat ratio to match specific requirements. If the airflow is too low, increase the fan motor speed per the manufacturer's instructions. If the airflow is too high, decrease the airflow per the manufacturer's instructions.

TECH TIP

There is a great deal of information available to technicians on the manufacturer's technical data sheets. All of this information is important, and if you learn how to properly use it you will become more valuable to your company and your customers. Most manufacturers provide explanations as to what each column of data can be used for. If you have difficulty understanding any of the material, most manufacturers or their distributors will provide you with an in-depth explanation of the data and how they can be used. Also consult ACCA Manual S for detailed instruction regarding equipment selection.

Checking and Adjusting the Airflow

If too little or too much air is delivered across the evaporator coil, the system will not function properly. *Too much* air moving across an evaporator coil will result in poor humidity control. *Too little* airflow will result in an iced evaporator coil and possibly liquid flooding back to the compressor.

There are several ways a technician can measure the airflow across an evaporator. Two common methods are (1) measuring the temperature rise across the indoor system (this does require the air-conditioning system to be attached to a forced-air furnace or air-handling unit and (2) measuring the static pressure drop across the evaporator.

TECH TIP

Measuring various types of temperatures is a common task for technicians. The thermometers used must be accurate for a technician to be able to properly work on air-conditioning systems. Technicians should occasionally test the accuracy of their thermometers.

Testing a thermometer can easily be accomplished by placing it in a solution of crushed ice and water. Fill a bucket with crushed ice and add water so that three-fourths of the ice is immersed in the water. It is best to have a solution that has more ice than water. Let the temperature of the solution stabilize, which should happen close to 32°F.

Once the temperature of the solution has stabilized, place the thermometer in the ice/water solution. Let the temperature measured on the thermometer stabilize, and observe its reading. If the thermometer measures a temperature relatively close to 32°F, then it is relatively accurate. If not, the thermometer needs to be adjusted, repaired, or discarded.

Temperature-Rise Method

When an air-conditioning system is attached to a forced-air furnace, the furnace can be used to calculate the airflow across the evaporator. By measuring the temperature rise across the furnace in the heating mode and with the fan operating at the same speed as it would during the cooling mode, the airflow across the device can be calculated.

To use this method, a technician would need to know the BTU output of the furnace and the temperature rise across the furnace (temperature rise = supply air – return air). To obtain the BTU output of a furnace, a technician would need to know the BTU input and its actual efficiency or AFUE rating. The following steps can be used to estimate airflow through a furnace and its evaporator coil.

Step 1. Obtain the BTU output of the furnace.

Step 2. Drill a hole in the return-air duct near the furnace and insert a temperature probe into the hole. Drill another hole in the supply-air duct, out of the line of sight of the heat exchanger, and place a temperature probe in the hole (Figure 45-23). Make sure the instruments used to measure the temperature at these two areas are calibrated to each other, meaning when exposed to the same temperature they read the same temperature.

Step 3. Turn the system on, call for heat, and operate the fan at the same speed that it would run at in the cooling mode. On some systems, this is as easy as placing the fan selector switch to the ON position. On many systems, the fan speed will be the same as during the cooling mode. On some newer systems, when the fan selector is switched to the ON position, the fan will operate at a continuous speed, which may not be the same speed as in the cooling mode. On these systems, the technician may need to rewire the fan to allow it to operate at the same speed when in the cooling mode.

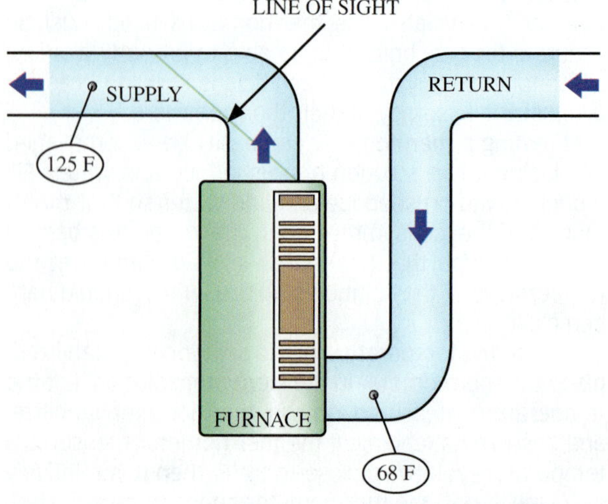

Figure 45-23 Measuring the temperature rise across a forced-air furnace.

Step 4. Let the system run for 10 to 15 minutes so the temperatures can stabilize.

Step 5. Determine the temperature rise across the system by subtracting the return-air temperature from the supply air.

Step 6. The airflow through the furnace as well as the air conditioning coil can then be calculated using the equation:

$$\text{CFM} = \frac{\text{Btu output}}{1.08 \times \Delta T}$$

where

CFM = airflow volume measured in cubic feet of air per minute

BTU output = the rated heat output for the furnace measured in BTU/hr

1.08 = a constant for air under normal conditions at sea level (sometimes 1.1 is used for this value)

ΔT = the temperature difference between entering and exiting air, sometimes expressed as ($T_1 - T_2$)

For example, suppose a technician is working on a 3-ton air-conditioning system with a forced-air furnace with a BTU output of 80,000 BTU. When operating in the heating mode with the fan selector switch at the ON position (the fan will operate at the cooling speed), a technician measures a supply-air temperature of 137°F and a return-air temperature of 75°F. The temperature rise across the system would be 62°F (137°F − 75°F = 62°F). Then the airflow through the furnace and the evaporator coil can be calculated as:

$$\text{CFM} = \frac{\text{Btu output}}{1.08 \times \Delta T}$$

$$\text{CFM} = \frac{80{,}000 \text{ Btu}}{1.08 \times 62^\circ \text{ F}}$$

$$\text{CFM} = 1{,}194.7$$

This will yield an estimated airflow value of 1,194.70 CFM across the evaporator coil. This is reasonably close to the desired value of 1,200 CFM, concluding that the airflow across the evaporator is sufficient.

Measuring the Static Pressure Drop Across an Evaporator

Another way a technician can measure the airflow across an evaporator is by measuring the static pressure drop across the evaporator coil. As air flows across an evaporator coil, it will lose some of its static pressure. As the volume of air flowing across the evaporator coil increases, so will its static pressure drop (loss). Equipment manufacturers publish specifications on the static pressure drop created by a volume (CFM) of air flowing across their evaporator coils. If the static pressure drop across an evaporator coil were measured, a technician could use the published specifications from a manufacturer to determine the actual airflow across the evaporator coil. Manufacturers will publish two different static pressure values: one for when the coil is dry and another for when the coil is wet. This is because there

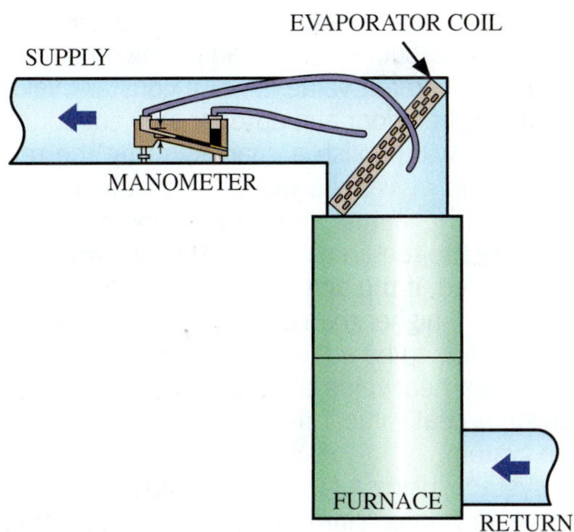

Figure 45-24 Measuring the static pressure drop across an evaporator coil.

is a difference in the resistance to airflow across a dry and a wetted coil.

To calculate the pressure drop across an evaporator coil, a technician will need to measure the static pressure at the inlet and the outlet of the coil. The pressures to be measured are relatively small, so a technician will need to use an inclined manometer, which has a range approximately of 0.00 in wc to 1.00 in wc.

Figure 45-24 shows a typical setup for measuring the static pressure drop across a coil. The steps for measuring the static pressure drop across an evaporator coil are:

Step 1. Drill a ¼-in hole on either side of the evaporator coil. Always refer to the manufacturer's instructions for the recommended location of these holes. *Be extremely careful not to drill into the evaporator coil*—this will take a rather simple task and turn it into a nightmare for a technician.

Step 2. Insert a ¼-in tube into each hole far enough so that it will measure the static pressure on each side of the coil. Seal the insertion point of each tube and connect to a manometer, as shown in Figure 45-24. Connect the inlet side of the evaporator coil to the lower end of the manometer, and connect the outlet side of the evaporator to the higher end of the manometer. Keep in mind that it is critical that the inserted tubes be held so they are perpendicular to the flow of air to avoid having the measurement tainted by velocity pressure. Use of a pitot tube or static pressure probe can prevent this possibility.

Step 3. Turn the indoor blower on and let it run for a few minutes so the airflow within the system can balance.

Step 4. Read the static pressure drop directly from the manometer.

Step 5. Using the information provided by the equipment manufacturer, read the amount of air flowing across the evaporator directly off the chart. Remember there will be a different value depending on whether the coil is dry or wet.

For example, a technician needs to determine the actual amount of air flowing across a dry 3 ton evaporator coil. The coil's model number is 036 and the measured static pressure drop is 0.18 in wc. Using the sample chart in Table 45-1, the airflow is determined to be 1,200 CFM. The conclusion is that the airflow is correct.

Table 45-1 Sample Performance Data: Coil Static Pressure Drop (in wc)

Unit Model	Air Quantity (CFM)													
	400	500	600	700	800	900	1,000	1,100	1,200	1,300	1,400	1,500	1,600	1,700
024	—	0.08	0.14	0.21	0.28	0.36	0.38	—	—	—	—	—	—	—
	—	0.07	0.11	0.15	0.20	0.25	0.30	—	—	—	—	—	—	—
030	—	—	—	0.12	0.17	0.22	0.28	0.34	0.42	0.50	—	—	—	—
	—	—	—	0.09	0.12	0.16	0.20	0.24	0.28	0.33	—	—	—	—
036	—	—	—	—	0.09	0.13	0.17	0.21	0.25	0.30	0.35	0.42	—	—
	—	—	—	—	0.07	0.10	0.12	0.15	0.18	0.21	0.25	0.28	—	—
042	—	—	—	—	—	—	0.10	0.13	0.17	0.20	0.24	0.28	0.32	0.36
	—	—	—	—	—	—	0.08	0.10	0.13	0.15	0.17	0.20	0.22	0.25
048	—	—	—	—	—	—	—	0.14	0.17	0.20	0.24	0.28	0.32	0.36
	—	—	—	—	—	—	—	0.11	0.13	0.15	0.17	0.20	0.22	0.25
060	—	—	—	—	—	—	—	0.12	0.16	0.18	0.22	0.26	0.30	0.34
	—	—	—	—	—	—	—	0.08	0.10	0.14	0.16	0.18	0.20	0.22

□ Wet coil

■ Dry coil

TECH TIP

Most manufacturers recommend that a system's airflow be set so that there is approximately 400 CFM per ton of air conditioning. In areas that have high relative humidity, a slightly slower fan speed can be used to aid in the dehumidification. In arid areas, the fan speed can be increased to provide more sensible cooling. As a rule of thumb, the fan speed should not be adjusted more than 10 percent above or below the manufacturer's recommendation. Excessively slow fan speeds can allow the evaporative coil to freeze up under light loads, and excessively high fan speeds can put too large a load on the condenser under heavy load conditions.

Checking and Adjusting the Refrigerant Charge

There are several methods a technician can use to determine the correct refrigerant charge of a residential split-system air conditioner. A technician should always first refer to the installation instructions and use the method recommended by the equipment manufacturer.

Many new residential condensing units are shipped from the factory with a refrigerant charge. The amount of refrigerant contained in the condensing unit will vary depending on its size and the manufacturer's engineering principles. Most manufacturers will ship their condensing units with enough refrigerant to properly operate a system with a 15-ft line set. This means that if there is 15-ft of suction line and liquid line between the condensing unit and the evaporator, no additional refrigerant will need to be added to the system. Always check the installation instructions of the condensing unit, as some manufacturers will add enough refrigerant for a different line length. If a system is installed with a line set greater than 15 ft (or the length stated in the installation instructions), the technician will need to add additional refrigerant to the system. The installation instructions will state the amount of refrigerant that needs to be added per foot of additional run of the liquid line used. Some manufacturers will also state the amount of additional refrigerant needed for the extended suction line, but this is normally a small amount and many times is omitted by the manufacturer.

For example, a new system is installed with a 25-ft line set, using a ⅜-in liquid line. According to the manufacturer's instructions, the factory charge is adequate for a 15-ft line set. For each additional foot of ⅜-in liquid run, there will need to be 0.6 oz of additional refrigerant added. As an example, if the length of the additional liquid line is 10 ft (25 ft − 15 ft = 10 ft), then 6 oz (10 × 0.6 = 6 oz) of additional refrigerant will need to be added to the system.

For a system using a fixed restriction metering device, such as a fixed orifice, there is another means of determining if a system has an adequate refrigerant charge. This can be accomplished by measuring the superheat value of the refrigerant at the suction line leading into the condensing unit. The superheat value is not a constant value and varies at different indoor and outdoor ambient conditions. Manufacturers will publish a chart showing the required superheat value at various indoor and outdoor ambient conditions. If the actual measured superheat value is within ±5°F of the required superheat value, the system is properly charged. If the actual measured superheat value is more than 5°F higher than the required superheat value, refrigerant needs to be added to the system. If the actual measured superheat value is more than 5°F lower than the required superheat value, refrigerant needs to be removed from the system.

For systems using a TEV metering device, there is another means of determining whether the system has an adequate refrigerant charge. This can be accomplished by measuring the subcooling value of the refrigerant at the liquid line leaving the condensing unit. Generally, a system is properly charged if the refrigerant is subcooled by 12 to 15°F at this location. Always check with the manufacturer of the system for their requirements. If the actual measured subcooling value is within ±3°F of the required subcooling value, the system is properly charged. If the actual measured subcooling value is more than 3°F lower than the required subcooling value, refrigerant needs to be added to the system. If the actual measured subcooling value is more than 3°F higher than the required subcooling value, refrigerant needs to be removed from the system.

When using either the subcooling or superheat methods of checking the refrigerant charge, allow the system to run approximately 15 minutes before measuring these values. This will allow the system's pressures and temperatures to stabilize so that an accurate reading can be obtained.

45.10 BEFORE YOU LEAVE THE JOB

Even after the installation is complete, the technician's job is not over. The work area should be completely cleaned. This includes wiping down the casing of the indoor and outdoor equipment as well as the thermostat. Any unused material needs to be removed from the property. If any of the customer's items were moved during the installation process, they would need to be returned to their original locations.

The technician should show the customer the complete system and explain its operation. This includes how to operate the system's thermostat and where the system's electrical disconnects are located. The technician should also explain any required maintenance, such as changing the filter and cleaning the outdoor condensing unit, and explain the system's warranty in detail.

After explaining the operation of the system and explaining the system's warranty, the technician should ask the customer if they have any questions, place a company sticker on the unit, and thank them for using their company. If a "use and care" manual is provided by the manufacturer, it should be given to the customer.

TECH TIP

The actual total BTU capacity of an air-conditioning system can be determined by measuring the airflow across the evaporator and the enthalpy difference of the air entering and leaving the evaporator. Then using the equation BTU capacity = 4.50 × CFM × enthalpy difference (ΔH), the actual total BTU capacity can be determined.

The enthalpy difference of the air entering and leaving the system's evaporator can be determined by first measuring the wet-bulb temperature of the air entering and leaving the evaporator. Then, using a psychometric chart, the enthalpy content of the air entering and leaving the evaporator can be determined.

For example, if the wet-bulb temperature of the air entering the evaporator is 66°F, its enthalpy content will be 30.8 BTU/lb. And if the wet-bulb temperature of the air leaving the evaporator is 57°F, its enthalpy content will be 24.5 BTU/lb. The enthalpy difference will then be 6.3 (30.8 − 24.5).

If the airflow across the evaporator were measured at 1,210 CFM, then using the equation BTU capacity = 4.50 × CFM × enthalpy difference, the BTU capacity of the system would equal approximately 34,304 BTU = (4.50 × 1,210 CFM × 6.3 BTU/lb).

UNIT 45—SUMMARY

The success of any installation lies in its planning. Part of the planning procedure requires a technician to reference the manufacturer's installation manual and determine the installation requirements of the system. There are two types of installations for the indoor section: cased/uncased evaporator assemblies and air handlers. The location of an outdoor condensing unit is an important part of the overall installation process.

Before installing the refrigerant lines, make sure the metering device installed in the indoor unit is correctly sized for the application. Do not exceed the maximum line set length as stated by the manufacturer; generally, pipe runs can be up to 175 ft. If the line set exceeds approximately 50 ft or a distance specified by the manufacturer, special modifications will need to be done to the system.

A standing pressure test is normally used to check for refrigerant leaks. Once the system is leak tested and is determined leak free, the system needs to be evacuated to 500 microns or less.

When installing residential air-conditioning systems, the installing contractor must be sure the appropriate type of overcurrent protection device is used with the outdoor condensing unit. The evaporator drain pan will normally have both a primary and a secondary drain. For attic or ceiling installations, it is advisable to install an additional drain pan under the fan coil to prevent water damage caused by an overflowing condensate line. There should be approximately 400 CFM per ton of airflow across the evaporator.

The correct refrigerant charge of a system can be determined by weighing the correct additional amount of refrigerant, measuring the superheat value of the refrigerant in the suction line at the inlet of the condensing unit, and measuring the subcooling value in the liquid line at the outlet of the condensing unit. For systems using a non-TEV metering device, the superheat method should be used to determine if a system has an adequate refrigerant charge. For systems using a TEV metering device, the subcooling method should be used to determine if a system has an adequate refrigerant charge.

After the installation is complete, the technician's job is not over. The work area should be completely cleaned and the system's operation and warranty explained to the customer.

WORK ORDERS

Service Ticket 4501

Customer Complaint: System Not Cooling Properly

Equipment Type: 3-Ton, 13-SEER R-22 Residential Split-System AC

Last week this new system was connected to an existing 80,000-BTU output forced-air furnace. This week the customer called complaining that the system is not cooling properly. A technician is called out to check the system.

When the technician arrives on the job, the outdoor condensing unit and indoor blower are running. After further investigation, the following conditions are observed:

- The outdoor ambient temperature is 95°F.
- Air filter is clean.
- Indoor blower is clean.
- Evaporator coil is clean.
- Condenser coil is clean.
- 208 V is measured at the common and run terminals of the compressor.
- Suction pressure is 76 psig.
- Discharge pressure is 275 psig.
- The wet-bulb temperature entering the evaporator is 70°F.
- A suction line temperature at the inlet of the compressor is 63°F.
- With the heating system turned on and with the fan running at the cooling speed, a temperature rise of 61°F is measured.

The technician then decides the system is operating normally and discusses the system's operation with the customer.

Service Ticket 4502

Customer Complaint: System Not Cooling Properly

Equipment Type: 3-Ton, 14-SEER R-410A Residential Split-System Air Conditioner

Last week this new system with a fixed orifice was connected to an existing 80,000-BTU output forced-air furnace. This week the customer called complaining that the system is not cooling properly. A technician is called out to check the system.

When the technician arrives on the job, the outdoor condensing unit and indoor blower are running. After further investigation, the following conditions are observed:

- Air filter is clean.
- Condenser coil is clean.
- Return air temperature is 79°F db and 64°F wb.
- System's supply air temperature is 69°F.
- Outdoor ambient temperature is 85°F.
- Suction pressure measured at the compressor is 105 psig.
- Suction line temperature at the inlet of the compressor is 65°F.
- Discharge pressure measured at the compressor is 330 psig.
- Discharge-line temperature at the compressor is 195°F.
- Liquid-line temperature at the outlet of the condenser is 99°F.
- Temperature rise across the furnace in the heating mode with the fan operating at the cooling speed is 65°F.

The technician then decides the system is operating with a low refrigerant charge and begins searching for the source of the leak.

Service Ticket 4503

Customer Complaint: System Not Cooling Properly When Outdoor Temperature Is High

Equipment Type: 2-Ton, 13-SEER R-22 Residential Split-System Air Conditioner

Last week this new system with a fixed orifice was connected to an existing 60,000-BTU output forced-air furnace. This week the customer called complaining that the system is not cooling properly when the outdoor temperature is high. A technician was called out to check the system.

When the technician arrives on the job, he discovers the outdoor condensing unit and indoor blower are running. After further investigation, the technician observes the following conditions:

- Outdoor ambient temperature is 95°F db.
- Return-air temperature is 70°F WB and 85°F db.
- Evaporator coil is clean and free of frost or ice.
- Filter is clean.
- Suction pressure is 80 psig.
- Discharge pressure is 305 psig.
- Suction-line temperature is 78°F.
- Liquid-line temperature is 121°F.
- Supply-air temperature at the outlet of the evaporator is 66°F db.

The technician then decides the system is operating with a high evaporator load, most likely due to an undersized air-conditioning system.

UNIT 45—REVIEW QUESTIONS

1. Why is it important to properly plan the installation of a residential split-system air conditioner?
2. What should a technician reference before installing a system?
3. What is the industry standard for determining the heat gain of a residential structure?
4. What happens if a system is oversized for its application?
5. What is the minimum SEER rating for new systems?
6. What is the name of the outdoor rated, watertight, flexible electrical cable connecting the system's disconnect to the condensing unit?
7. What gauge thermostat wire is typically used on residential systems?
8. Where should an outdoor condensing unit not be located?
9. Normally, what is the maximum length a line set can be run on many residential split-system air conditioners?
10. How should a technician determine the diameter of the line set to use?
11. Why are sections of tubing that are to be fitted together sometimes swaged?
12. What micron level should be achieved when evacuating a system?
13. What are the three main types of overcurrent protection devices that can be used on a condensing unit?
14. At what height should a thermostat be mounted above the floor?
15. What diameter condensate line is typically run from the drain pan of the evaporator?
16. What is the purpose of the secondary drain line?
17. What is the required airflow across the evaporator of a residential split-system air conditioner?
18. What two methods can be used to measure the airflow across an evaporator?
19. For a system using a non-TEV metering device, such as a fixed orifice, what method can be used to determine if the refrigerant charge is adequate?
20. When the installation is complete, what should a technician do before leaving the job?

UNIT 46

Duct Installation

OBJECTIVES

After completing this unit, you will be able to:

1. list the agencies that set standards for duct systems.
2. explain the difference between a flexible duct and a flexible connector.
3. list the materials used to construct duct systems.
4. explain how to seal duct systems.
5. describe how to join different types of duct.
6. discuss how to measure the amount of duct leakage.
7. describe the three methods of quantifying duct leakage rates.

46.1 INTRODUCTION

Duct systems are the distribution network for conditioned air to be moved throughout a building. Technicians need to understand duct systems and airflow to be able to troubleshoot and maintain HVACR systems. The most common materials used for duct construction include galvanized steel, fiberglass ductboard, and wire-helix flexible duct. Regardless of the materials used, a correctly installed duct system should last as long as the house, should not leak, and should be insulated well enough to prevent duct loss and gain to unconditioned space.

46.2 AIR-DISTRIBUTION SYSTEM COMPONENTS

Forced-air systems are used to distribute conditioned air in residential and small commercial buildings. Air is conditioned and distributed through the duct system throughout the building. The basic components of a forced-air system are the blower, the return-air ductwork carrying air to the blower, and the supply-air ductwork carrying air from the blower to the building.

Most systems have large boxes on both the return and supply ends of the blower called plenums (Figure 46-1). The plenum distributes the air to the ductwork attached to it. The duct design can be compared to a tree: the main ducts leaving the plenum are called trunk ducts, and the individual ducts running to each room are called branch ducts. The point where a branch duct comes off of a trunk duct is called a takeoff.

Figure 46-1 The supply and return plenums are connected on either end of the indoor blower. *(Courtesy of Stanfield Air Systems)*

The duct openings to the conditioned space are covered by registers, diffusers, or grilles. The return-air openings are covered by return-air grilles. The supply-air openings are covered by supply-air registers. They direct the airflow into the room. These are sometimes called diffusers because they spread, or diffuse, the air.

Between the blower and the diffuser, the airstream must change direction and shape. This is accomplished with duct fittings. Turns are made with elbows, often called ells (Figure 46-2). A wye fitting is used to split one large duct into two smaller ducts (Figure 46-3). A change in the size of a rectangular duct is accomplished with a transition (Figure 46-4). A change in the size of a round duct is done with a reducer (Figure 46-5). When the register at the end of a round branch duct is rectangular, a fitting called a boot is used to allow the round duct to connect to the rectangular register (Figure 46-6).

The blower provides the pressure difference to move the air through the ductwork. The amount of air the blower can move and the amount of energy needed to move the

Figure 46-2 Elbows are used for turns in the ductwork.

Figure 46-3 A wye is used to split one large duct into two smaller ducts.

Figure 46-4 A transition is used to change rectangular duct from one size to another.

Figure 46-5 A reducer is used to change round duct from one size to another.

Figure 46-6 A boot is used at the end of a round duct to allow connection of a rectangular register or grill.

air is controlled by the resistance to airflow from the ductwork and all the system components in the airstream. The duct offers resistance to airflow, creating a pressure drop as the air travels through it. Besides the ducts, every component that the air travels through adds pressure drop. This includes filters, humidifiers, heat exchangers, coils, registers, and grilles. The amount of pressure left for moving air through the ductwork is the difference between the amount of pressure the fan can produce and the amount of pressure drop from all these system components. Figure 46-7 shows the components of a forced-air system and the pressure drop created across each component.

46.3 DUCT LOCATION

The three most common areas to locate the duct system are in the attic, in the space between the joists in the ceiling or between floors, and under the building in either the basement or crawlspace . The structure and layout of the building determine the available location for system components. The building's design affects the air-handler location, which then controls the duct system layout. The

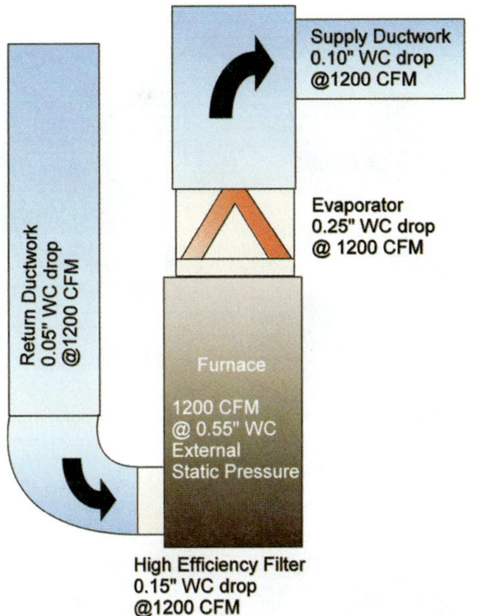

Figure 46-7 Basic components of a forced-air system.

duct location will determine the amount of duct insulation required. The building's designer may specify the layout or may leave it up to the air-conditioning contracting company.

For maximum system efficiency, the duct system should be located so that it is within the building's insulation envelope (Figure 46-8). In a residence with a full basement, the basement is an excellent location for the trunk ducts and air-conditioning equipment. Equipment can also be located in a closet space or utility room. All enclosures must meet local fire and safety codes.

Figure 46-8 Duct installed in joist space so it is inside the building envelope.

TECH TIP

A number of studies have shown that traditional duct-installation practices can result in systems that have excessive air leaks. New codes require the use of UL 181–approved duct tapes and UL 181–approved mastics to reduce air leaks and improve system efficiencies.

It is not always possible to locate the ducts within the building's envelope. For slab construction, ducts and even equipment can be placed in the attic. However, an attic duct system is a particularly poor choice for an air-conditioning system in a warm climate. Ducts located in unconditioned areas must be properly insulated, and an allowance must be made for heat loss included in the load calculation. Energy-efficiency guidelines require that ducts located outside of the building's envelope have R8 insulation value. Ducts for perimeter (heating only) systems can be located in the slab.

Before beginning the duct system layout and design for any building, check with the local building department, county, or state code and/or regulatory agencies to determine the specific insulation requirements for the building's type and location.

46.4 EQUIPMENT TYPES

The design of the duct system is affected by the type of equipment selected. Forced-air equipment is typically either packaged equipment or a split system. With packaged equipment, the entire system is located outside the structure and the ductwork is typically located in the basement or crawlspace (Figure 46-9).

Split systems consist of an inside blower and an outside condensing unit. Split systems offer more flexibility in duct location because of the many possible blower locations. The blowers can be upflow, downflow, or horizontal.

Figure 46-9 Packaged air-conditioning unit with ductwork in crawlspace.

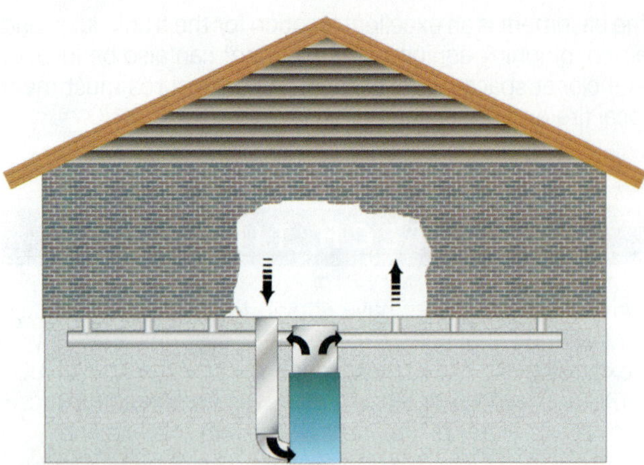

Figure 46-10 Upflow blower in the basement with all ducts in the basement.

The name describes the direction of airflow. Five common split-system duct configurations are as follows:

- Upflow blower in the basement with all ducts in the basement (Figure 46-10)
- Upflow blower located in the house with supply ducts located in the attic and return in the house (Figure 46-11)
- Downflow blower in the house with supply ducts in the crawlspace and return in the house (Figure 46-12)
- Horizontal blower in the crawlspace with all ducts located in the crawlspace (Figure 46-13)
- Horizontal blower located in the attic with all ducts located in the attic (Figure 46-14)

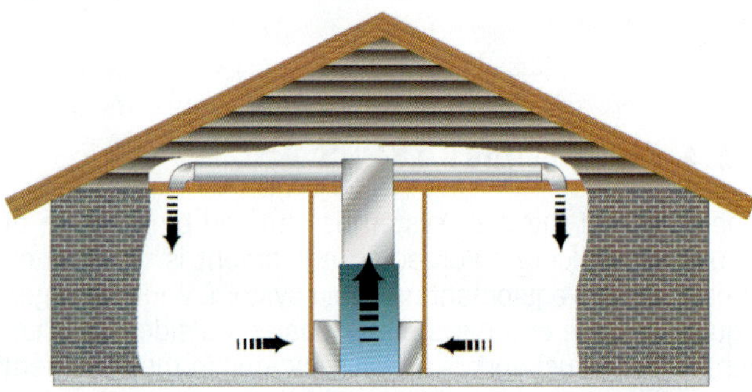

Figure 46-11 Upflow blower located in the house with supply ducts located in the attic and return in the house.

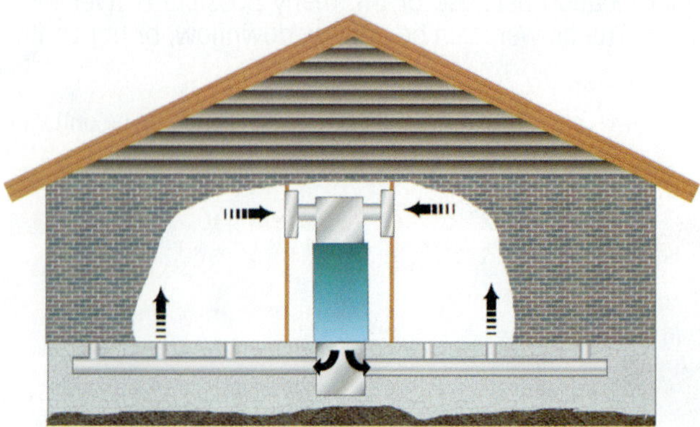

Figure 46-12 Downflow blower in the house with supply ducts in the crawlspace and return in the house.

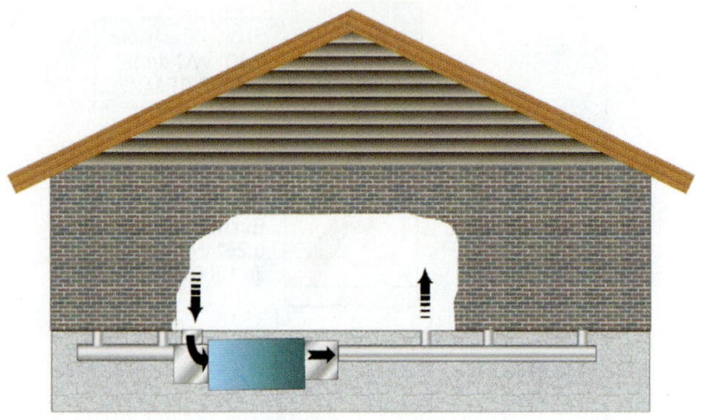

Figure 46-13 Horizontal blower in the crawlspace with all ducts located in the crawlspace.

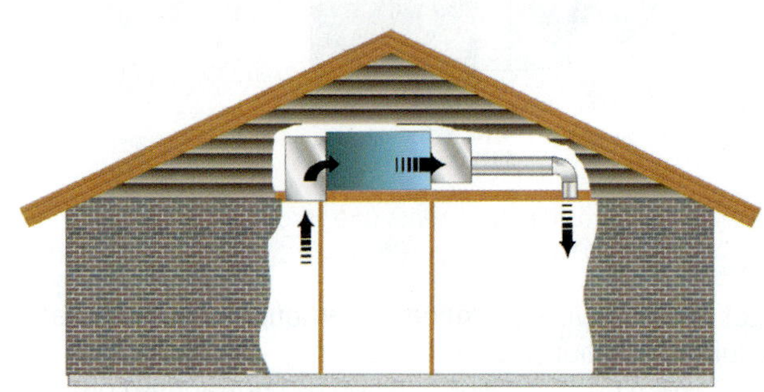

Figure 46-14 Horizontal blower located in the attic with all ducts located in the attic.

46.5 DUCT SYSTEM TYPES

The four most common duct configurations are radial, reducing radial, extended plenum, and reducing extended plenum. Two less common types are the perimeter loop system and the central plenum system.

Radial Duct Systems

Radial duct systems are designed so that all or almost all of the duct runs originate at the central plenum. In some cases, a few of the duct runs may have wyes or duct triangles as a means of joining additional ducts to an initial run (Figure 46-15). Radial systems are most frequently installed in attics, but they can also be installed in crawlspaces and basements. Radial systems are commonly used in small houses built on concrete slabs.

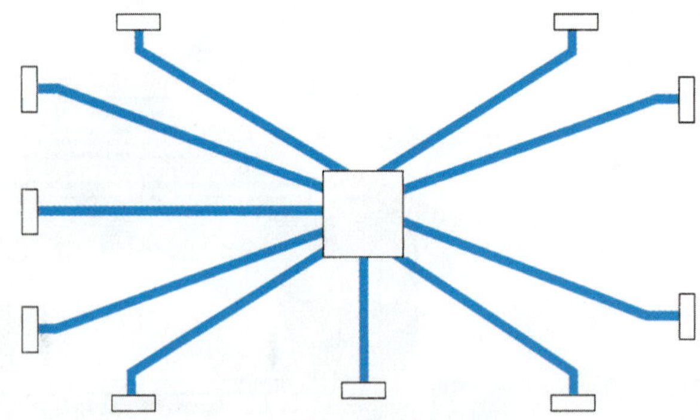

Figure 46-15 Radial duct system.

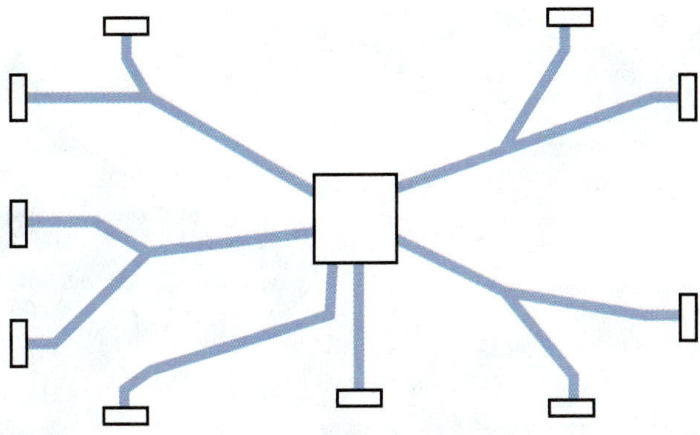

Figure 46-16 Reducing radial duct system.

Reducing Radial Duct Systems

A reducing radial system uses several larger ducts leaving the main plenum that branch into smaller ducts as they get closer to their destination (Figure 46-16). This reduces the connections at the plenum and reduces the overall amount of duct used.

Extended Plenum

The extended-plenum duct system uses a large trunk duct that travels the length of the building from the air handler. These systems are sometimes referred to as trunk duct systems (Figure 46-17). The trunk duct is considered an extension of the plenum. The trunk ducts in extended plenum systems do not reduce in size as they travel across the structure. Extended plenum systems can be located in the crawlspace, attic, or basement.

Reducing Extended Plenum

The trunk ducts in the reducing extended-plenum system reduce in size (Figure 46-18). Typical reducing-plenum trunk ducts will reduce after every three to four takeoffs. Not only does this save material cost, but it makes the duct system work better. The air velocity in the trunk drops as the volume of air traveling through the trunk drops. Reducing the size of the trunk duct restores the velocity of the air in the trunk duct. This helps keep the static pressure throughout the system more even, aiding in more even air distribution.

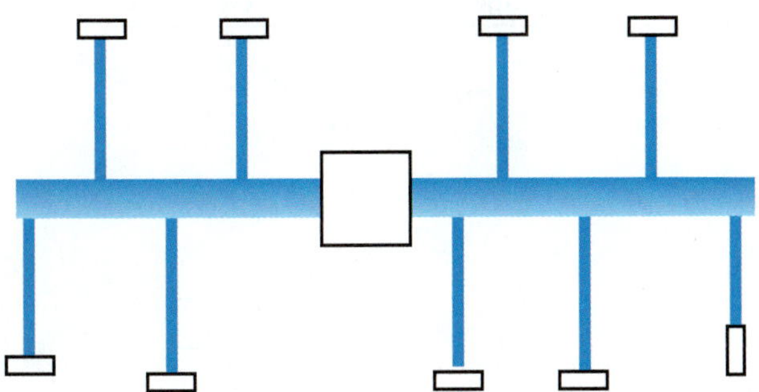

Figure 46-17 Extended plenum duct system.

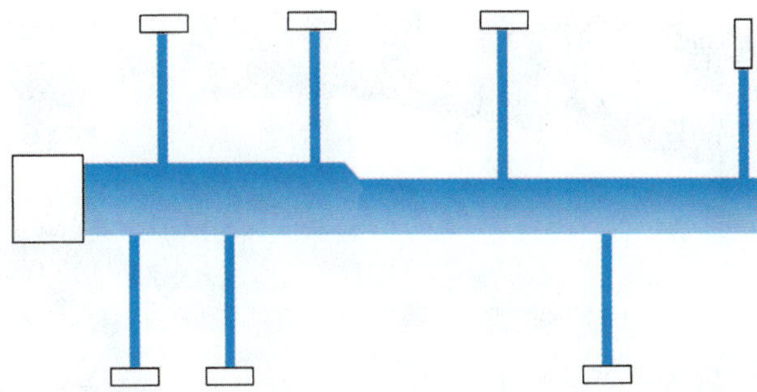

Figure 46-18 Reducing extended plenum duct system.

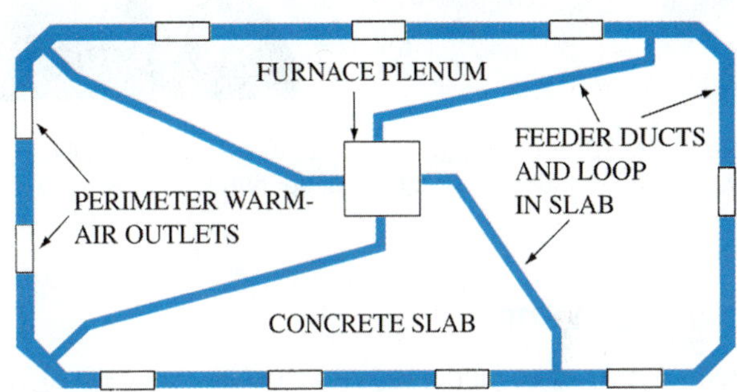

Figure 46-19 Perimeter loop duct system.

Perimeter Loop System

The perimeter-loop duct system uses radial feeder ducts from the blower that attach to one trunk that is installed around the outer edge of the foundation. Each supply is tapped off this trunk to provide equal airflow throughout the structure (Figure 46-19). Perimeter-loop systems are normally installed in concrete foundations and have the added efficiency of heating the slab. The duct for a perimeter-loop system must be placed before the slab is poured. These systems are generally used in single-story commercial office buildings built on slabs.

Central Plenum System

The central plenum duct system uses structural cavities as a pathway for supply- or return-air plenums. The basement, crawlspace, or space between floor joists can all serve as the plenum. Air is blown into the cavity, and holes are cut through the floor into the plenum cavity wherever a register is needed. To ensure proper operation of a plenum system, the cavity must be airtight and configured in a way that the airflow is not compromised.

This system is used because of its low cost. This system may not meet mechanical codes in some jurisdictions. Indoor air quality is a major concern due to the nature of these cavities. Moisture, mold, and odors are often hard to control.

46.6 DUCT MATERIALS

Air ducts can be made from many different materials. The most common types of ductwork are galvanized-steel sheet metal, spiral metal, fiberglass ductboard, flexible duct, and fabric duct.

Figure 46-20 Sheet metal is supplied in large sheets, which can be used to fabricate sheet-metal ductwork.

Galvanized Sheet Metal

Sheet metal can be fabricated into most any shape imaginable by a skilled sheet-metal worker. Galvanized sheet steel comes in large, flat sheets (Figure 46-20). Common sizes are 4 ft wide and 8–10 ft long, but other sizes are available depending upon the steel supplier. Sheet metal offers the least amount of resistance to airflow of any duct material because of its smooth surface.

Many localities adopt the metal duct standards developed by the Sheet Metal and Air Conditioning Contractor's National Association (SMACNA). The thickness of the metal is called its gauge. A guideline for selecting metal thickness is shown in Figure 46-21.

For many years, galvanized sheet steel was used exclusively for air-conditioning ductwork because of its workability and durability. However, the material is expensive and costly to install, so other types of duct material have become popular. These include spiral metal duct, fiberglass ductboard, and flexible duct.

Figure 46-22 Spiral metal duct system.

Spiral Metal Duct

Spiral duct is made from long strips of narrow metal and fabricated with spiral seams (Figure 46-22). Machines are available for making ducts on the job to fit required diameters and lengths. Spiral metal duct is used in commercial applications. It is frequently used where the ductwork will be exposed. It requires less support than other types of duct due to its inherent rigidity.

TECH TIP

The UL 181 standard covers factory-made ducts and air connectors. All fiberduct and flexible duct should be UL approved and meet the UL 181 standard.

Fiberglass Ductboard

Fiberglass ductboard is a rigid material made of compressed fiberglass with an outer vapor barrier (Figure 46-23). It comes

	Comfort Heating or Cooling			
	Galvanized Steel			Comfort Heating Only
	Nominal thickness (in inches)	Equivalent galvanized sheet gauge no.	Approximate aluminum B&S gauge	Minimum weight tin plate pounds per base box
Round ducts and enclosed rectangular ducts				
14 in or less	0.016	30	26	135
Over 14 in	0.019	28	24	—
Exposed rectangular ducts				
14 in or less	0.019	28	24	—
Over 14 in	0.022	26	23	—

Figure 46-21 Gauges recommended for sheet-metal ductwork.

Figure 46-23 Fiberglass fiberduct comes in large sheets and can easily be fabricated into duct.

in 1-in, 1½-in, and 2-in thicknesses. Fiberglass ductboard has the advantage of being an inherently good insulator for both heat and sound. This reduces the duct losses and provides sound-absorbing qualities. Fiberglass ductboard is less expensive than metal, and fabricating ductboard duct is generally easier to work with than fabricating sheet metal duct. Ductboard is less durable than sheet metal duct and has a higher resistance to airflow than sheet metal because of its rougher interior surface.

Flexible Duct

The most common duct material used today in residential duct systems is undoubtedly flexible duct. Flexible duct has a spiral metal wire for support, a smooth plastic inner liner, an outer cover that serves as a vapor barrier, and fiberglass insulation sandwiched in between the inner liner and outer cover (Figure 46-24). The outer cover is typically vinyl or Mylar. Flexible duct comes in 25-ft lengths compressed in a box that is about 3 ft long. When opened it expands lengthwise into ducts (Figure 46-25).

SERVICE TIP

It is very important to stretch flex duct before installing it. Leaving the duct slack can easily double the duct's resistance to airflow. Allowing extra length in the duct run is a particularly bad practice that may save time during installation but will significantly reduce the airflow through the duct.

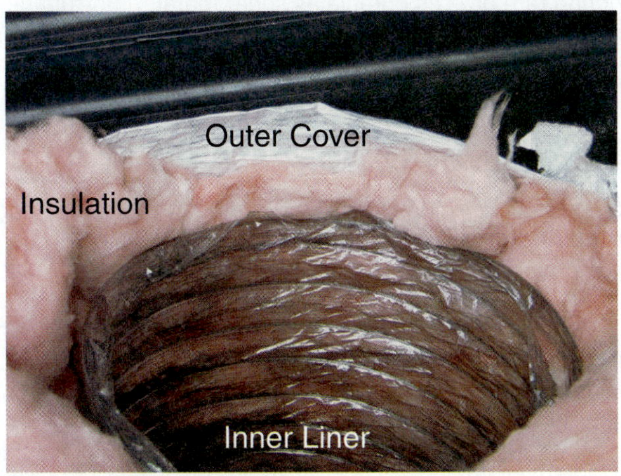

Figure 46-24 Flex duct is made of a helix wire for support, an inner liner, fiberglass insulation, and an outer vapor barrier.

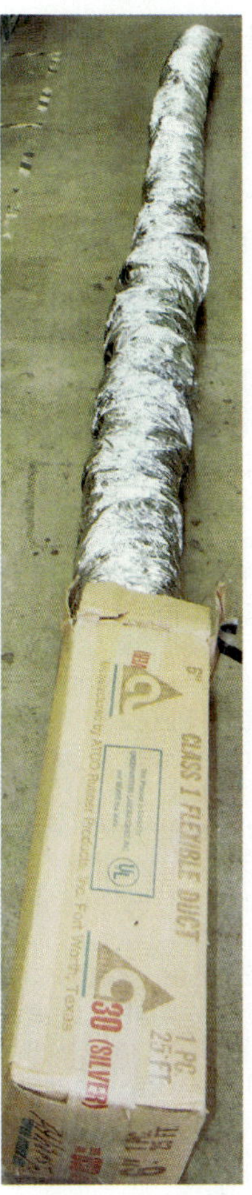

Figure 46-25 Flexible duct comes compressed in a box and expands when removed from the box.

Flex duct is very popular because it is the least expensive duct material and the easiest to install. Unfortunately, it is also the easiest to install incorrectly. Common problems with flex duct include tight radius turns and improper support. Many poorly installed flex duct jobs have given the material a bad name. However, it is possible to install a good duct system with flex duct by following the manufacturer's installation instructions. Flex duct is very quiet because its soft sides absorb sound. However, the soft, undulating sides also have the highest resistance to airflow of all of the most commonly used duct materials.

46.7 STANDARDS

Approved methods of manufacturing, installing, and sealing ductwork are published by several national agencies that publish codes, standards, and guides. These standards are often referenced in state and local codes. Organizations publishing duct standards include the following:

- Underwriters Laboratories
- National Fire Protection Association
- Sheet Metal and Air Conditioning Contractors National Association

- North American Insulation Manufacturers Association
- Air Diffusion Council
- Air Conditioning Contractors of America.

Underwriters Laboratories

Underwriters Laboratories Standards 181, 181A, and 181B are probably the most referenced duct standards. Any manufactured duct should comply with UL 181 standards. Standard 181 covers manufactured duct materials, such as fiberglass ductboard and wire-helix flexible duct. Two classes of manufactured duct are listed according to their flammability: Class 0 and Class 1. Class 0 ducts have a flame spread index of 0. Class 1 ducts have a flame spread index not over 25 and a smoke developed index of not over 50. The flame spread index is a number relating to ASTM E 84 flammability testing. A rate of 100 is assigned to untreated bare wood, while 0 is assigned to cement board. The smoke developed index compares the amount of smoke produced compared to concrete, at 0, and red oak, at 100.

Many mechanical codes require that mastic and tape be UL 723 rated for metal duct. UL 723 tests materials for flame spread and smoke development. Materials that are UL 723 rated can be used on and in air plenums. UL Standard 181A addresses closure and sealing systems for manufactured rigid ducts, such as fiberglass ductboard. UL Standard 181 B addresses closure and sealing systems for flexible ductwork. Note that a tape may be UL rated but not meet 181 standards. A tape that only displays the 723 rating and not the 181 rating is not intended for ductboard or flex. Figure 46-26 shows the marking on UL-rated duct tape and mastic.

National Fire Protection Association

The NFPA publishes two standards that are widely referenced: 90A and 90B. Standard 90A covers the installation of ventilation and air conditioning, while 90B covers the installation of warm-air heating and air conditioning. Both standards discuss the types of materials that may be used for ductwork and discuss the application difference between flexible air duct and flexible air connector. Air ducts must either be made of rigid, nonflammable materials, like sheet metal, or be tested and shown to meet the UL 181 standard. Flexible duct connectors are not certified to meet every provision of UL 181; therefore, their use is more limited.

Figure 46-26 Duct tape and mastic should be UL 181 approved.

Sheet Metal and Air Conditioning Contractors National Association

SMACNA publishes several duct and installation standards. The most commonly referenced is *HVAC Duct Construction Standards—Metal and Flexible*. It covers a lot of the nuts-and-bolts aspects of constructing and installing ductwork. Topics include duct joining, duct sealing, and duct hanging.

North American Insulation Manufacturers Association

NAIMA represents the fiberglass industry. They publish a manual called *Fibrous Glass Residential Duct Construction Standard*. This standard covers both the standard for fiberglass ductboard materials and how the materials should be installed.

Air Diffusion Council

The ADC represents flexible duct manufacturers and publishes standards relating to flexible duct systems. The most referenced is the *Flexible Duct Performance and Installation Standards*, which gives specific instructions for properly installing flexible wire-helix duct.

Air Conditioning Contractors of America

ACCA publishes many manuals for designing and installing air-conditioning systems. For duct installation, *HVAC Quality Installation Specification* is the most relevant. This manual covers the components of a quality installation, including the duct system.

46.8 SHEET-METAL DUCT

Galvanized sheet-metal duct has been the gold standard of duct systems for years. No other duct material has a lower friction rate or is as durable as sheet metal. Unfortunately, nothing else comes close to the cost of sheet metal either. Sheet metal is still used, but its prominence in residential duct systems is waning because of its high cost.

Round Galvanized Duct

Round duct is shipped in bundles with one piece of duct tucked inside another. Each piece needs to have its longitudinal seam snapped together before use. Figure 46-27 shows a typical snap-lock seam.

Snap-lock round duct is typically available in sections of 3, 4, and 5 ft depending upon the material, gauge, and duct diameter. Table 46-1 shows the lengths available from one supplier.

One end of the duct is crimped, making its outside diameter slightly smaller than the regular duct diameter. A bead behind the crimp serves as a stop when inserting one duct into another (Figure 46-28). The crimps should be

Table 46-1 Duct Section Length in Feet

Diameter (in)	30 Ga.	28 Ga.	26 Ga.	24 Ga.	Aluminum	Stainless
4–10	5		5	5	4	4
12	5		5	5	4	4
14		3	5	5	4	4
16			3	3	4	4
18			3	3	4	4
20			3	3	4	4
24			3	3	4	4

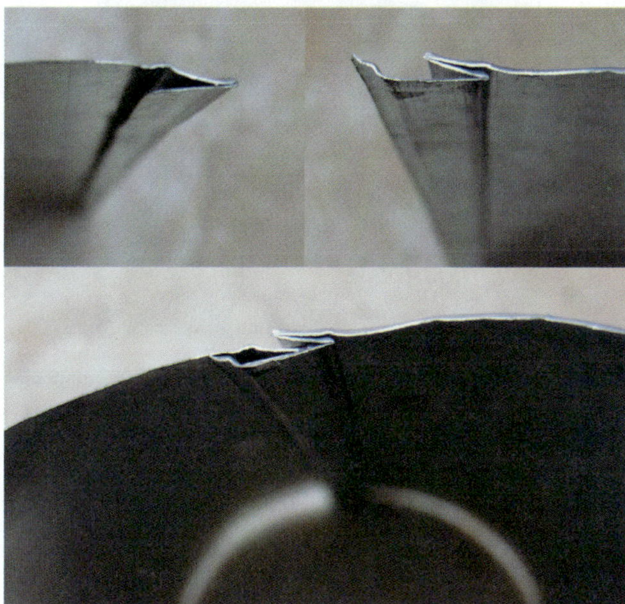

Figure 46-27 Snap-lock seam on round galvanized duct.

Figure 46-28 The crimped end on a round piece of galvanized duct.

Figure 46-29 Airflow should be toward the crimped end of the duct.

installed in the direction of airflow to reduce turbulence (Figure 46-29). All joints should be fastened with sheet-metal screws or rivets. Ducts 10 inches in diameter or smaller should have at least three equally spaced screws or rivets per joint. Larger ducts should have at least four equally spaced screws or rivets per joint. The quantity of screws may vary depending on the inspection authority in your area.

These joints are not airtight and need sealing. The outer insulation around the pipe does not help seal in the air. Take apart an older system with unsealed joints and there is a black line in the duct wrap where the joint was (Figure 46-30). This black ring is formed by tiny dirt particles that were caught in the fiberglass as air leaked out of the joint. Tape and/or mastic are required to make these joints airtight.

Figure 46-30 The stain on the duct wrap is caused by small dust particles escaping the leaky duct joint and getting caught in the duct insulation; it shows where the duct leaks air.

Figure 46-31 Wrinkles in duct tape can cause air leaks.

One method of sealing these joints is to use two wraps of UL 181A duct tape. The problem with this method is that little wrinkles are invariably left in the tape, and small amounts of air can escape through the wrinkles (Figure 46-31).

The preferred method is to use either mastic alone or tape and mastic. Mastic works fine by itself, provided that the cracks being sealed are not too large. Large cracks can be covered with UL 181A duct tape first and then covered with mastic.

Rectangular Galvanized Duct

Rectangular duct typically is made in two L-shaped pieces. Each piece has one duct width and one duct height (Figure 46-32). These pieces are normally joined by either a snap-lock joint or a Pittsburgh joint. The 0.25-in edge of the snap-lock snaps down into the seam. Typically it is hammered down using a sheet-metal hammer (Figure 46-33). The ¼-in edge in a Pittsburgh seam fits into the seam and the edge of the seam is hammered over (Figure 46-34). The Pittsburgh joint is more airtight but takes more material and time to produce.

Figure 46-32 Rectangular duct before assembly.

Figure 46-33 Hammering in a snap-lock seam on rectangular duct.

Rectangular ducts are joined by S-locks and drive cleats. An S-lock is a piece of sheet metal in a flattened S-shape. One duct slides into the S-lock in one direction and the other duct slides into the S-lock from the other direction (Figure 46-35). S-locks are used on the larger duct dimension, usually the top and bottom. The shorter dimensions are joined by drive cleats. The duct edges are turned back and the drive cleat slips over the edges and pulls the duct together (Figure 46-36). These are driven down with a mallet; thus the name "drive cleat." These joints are not airtight and need to be sealed, preferably with UL 181A mastic or UL 181A tape and mastic.

Hanging Metal Duct

One advantage of metal duct is that it has structural integrity, making it much easier to hang than flexible duct. Hangers should be spaced no farther than 12 ft apart. Many materials can be used as hangers, including wire, plumber's strap, or sheet-metal strips. Any screws that are run into the duct to fasten the hanger to the duct should be covered with mastic. When possible, long sections of metal duct are assembled on the ground and hung as a section. This makes it easier to assemble straight runs. Frequently these sections are also insulated before being hung. Attaching the duct one piece at a time while hanging the duct is much more difficult. Likewise, insulating while working around the hangers is more difficult than when the section of duct

Figure 46-34 Hammering a Pittsburgh seam on rectangular duct.

Figure 46-36 The short sides are joined with drive cleats.

Figure 46-37 Properly supported and insulated galvanized duct system. *(Courtesy of Stanfield Air Systems)*

is on the ground and can be freely moved around. Figure 46-37 shows properly supported duct.

Takeoffs

The first step in installing a takeoff is to cut a hole in the duct. The easiest, fastest, and cleanest method for cutting a round hole in the duct is to use an adjustable hole cutter (Figure 46-38). The hole cutter is set to the size of the hole needed. A drill is attached to the hole-cutter bit. First, drill a hole where the center of the hole will be. Next, place the pivot point in the hole and cut a full circle using the drill.

Figure 46-38 This hole was just cut using the adjustable hole cutter. *(Courtesy of Malco Products, Inc.)*

Figure 46-35 The long sides of the duct are joined using S-locks.

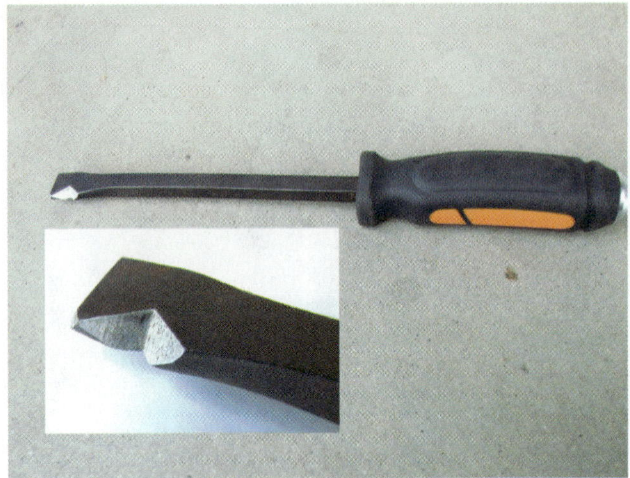

Figure 46-39 Duct-piercing tool.

Figure 46-41 Slicing the diameter of a circle using a duct-piercing tool.

Holes can also be cut using aviation snips. Hold a short piece of pipe the size of the takeoff up to the duct and draw its perimeter on the duct. Cutting this hole out is harder than it might appear at first. It is generally not possible to simply cut along the line in a circle. When sheet metal is cut with snips, one side must bend out of the way to let the snips pass by. Neither side can bend out of the way on a piece of duct. Two solutions are (1) making a spiral cut from the center and (2) using a duct-piercing tool to cut a line across the diameter of the hole and cutting out each semicircle.

For the spiral cut, a starter hole should be struck in the center using a duct-piercing tool (Figure 46-39). Cut in an increasing spiral using either left-hand or right-hand aviation snips (Figure 46-40). Continue to follow the line once the spiral reaches it.

For the semicircle cut, use a duct-piercing tool to cut across the diameter of the hole (Figure 46-41). Then use left- or right-hand aviation snips to cut out each semicircle (Figure 46-42).

Takeoffs are connected to metal duct using either dovetails or flanges. Dovetail fittings have tabs that are formed by a series of slits along the perimeter of the fitting (Figure 46-43). Every other tab is turned 90° to the fitting (Figure 46-44). The straight tabs are inserted into the hole, and the technician reaches through the hole and bends the straight tabs over. Sheet-metal screws are run through three or four

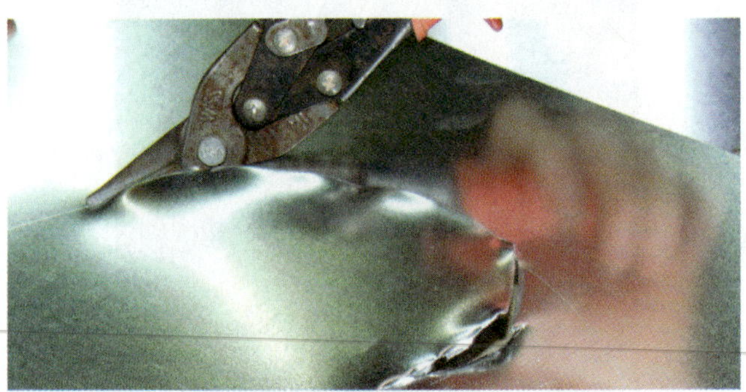

Figure 46-42 Once the diameter has been sliced, the circle can be cut in two semicircular sections.

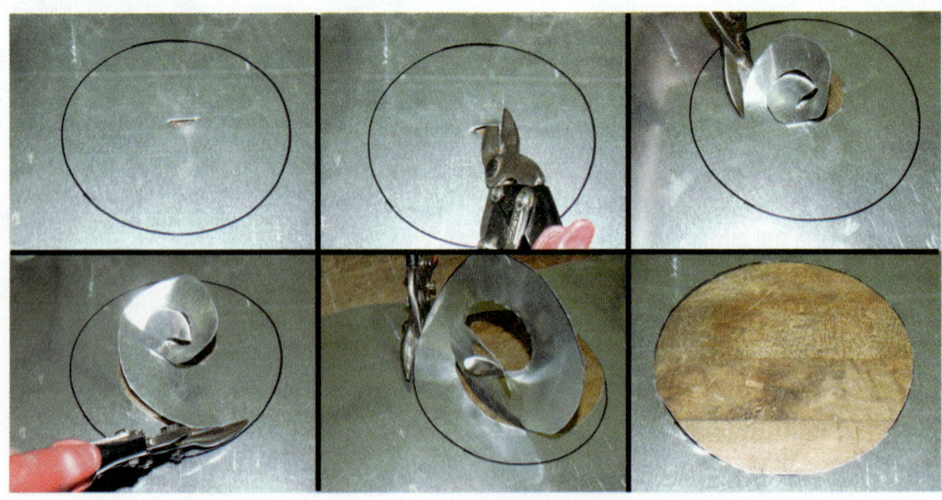

Figure 46-40 Round holes can be cut using a spiral pattern.

Figure 46-43 Traditional dovetail takeoff.

Figure 46-44 The tabs of the dovetail fitting folded over.

Figure 46-45 Flange take off with sticky pad.

evenly spaced tabs to securely hold the fitting in place. The intersection of the collar and ductwork is sealed with mastic. This is particularly important with dovetail fittings.

Flange fittings have a flange that fits on the outside of the duct. Some flange takeoffs have a self-adhesive pad that attaches the fitting to the duct and seals it (Figure 46-45). If the flange is not self-sealing, mastic should be applied to the flange before placing the fitting against the outside of the duct. The fitting is held in place by three to four equally spaced sheet-metal screws run through the flange.

46.9 DUCT INSULATION

One major disadvantage of metal duct compared to ductboard or flex is that metal duct must be insulated. Round duct is insulated with duct wrap. The most common type of duct wrap is made of fiberglass with an outer vapor barrier of vinyl, aluminum foil, or aluminum foil reinforced with kraft-back paper. Duct wrap comes in 4-ft rolls. One side has 2 in of vapor barrier that is uninsulated. This edge laps over the edge of the insulation already on the duct, providing a continuous vapor barrier. When insulating duct, the duct wrap is pulled around the duct and lapped over itself. The insulation should be pulled tightly around the duct to eliminate gaps and sags in the insulation. But compressing insulation reduces its R value, so it should not be pulled so tightly that the fiberglass is flattened. The most common method for fastening the duct wrap is stapling it to itself. A clinch staple goes through both layers of insulation and turns back (Figure 46-46). The insulation should be stapled every 6 in for the entire length. The exposed outer edge is taped with a UL 181A tape that matches the vapor barrier. When the job is done, no metal should be showing, no fiberglass should be

Figure 46-46 Clinch staples hold insulation together by turning outward in each direction.

Figure 46-47 Properly insulated ductwork. *(Courtesy of Stanfield Air Systems)*

showing, and the vapor barrier should cover the entire duct system (Figure 46-47).

The duct wrap should be cut in pieces that are large enough to wrap around the duct and have some overlap. For round duct, this is typically the duct circumference plus an adjustment for the thickness of the wrap plus some overlap. The adjustment is 9.5 in for 1.5-in-thick wrap, 12 in for 2-in-thick wrap, and 17 in for 3-in-thick wrap. Table 46-2 shows the size that duct wrap should be cut for different diameter ducts and different thickness of duct wrap. For example, 2-in-thick insulation for a 7-in-diameter duct should be 34-in wide.

The same basic concept applies to rectangular duct, but the adjustments are not quite as large. The adjustments are 7 in for 1.5-in-thick insulation, 8 in for 2-in-thick insulation, and 11.5 in for 3-in-thick insulation.

Duct Liner

Rectangular duct, plenums, and return-air boxes may be lined with insulation rather than wrapped (Figure 46-48). Duct liner is also fiberglass, but it does not have a vapor barrier. Duct liner is glued to the inside of the duct and connected with mechanical fasteners that compress the insulation. The fasteners should start within 4 in of the end of the duct and continue every 18 in for the length of the duct. A new row of fasteners should be installed every 12 in across the width the duct.

Lining the duct makes a system quieter and is usually easier than wrapping the duct. A disadvantage of lining duct is increased resistance to air flow. One of the advantages of metal duct, low air resistance, is negated by lining it with fiberglass. Liner also tends to collect dust. After a period of years, the liner will end up looking like a dirty filter because the fiberglass grabs and holds tiny dust particles. When a liner is used, the duct must be made larger to compensate for the space the liner takes up. When using 1-in liner, the duct must be 2 in larger in both dimensions to allow for the thickness of the duct liner.

Bubble Wrap

Bubble wrap is a new form of duct insulation. It is similar to the bubble packing material with outer layers of either metalized Mylar or aluminum (Figure 46-49). Bubble wrap may be used as liner or as duct wrap. Bubble wrap is typically not applied directly to the duct. Spacers are

Table 46-2 Duct Insulation Dimensions (in inches)

Duct Dimensions		Insulation Thickness		
Duct Diameter	Duct Circumference	1.5 in (9.5-in Adjustment)	2 in (12-in Adjustment)	3 in (17-in Adjustment)
5	16	25.5	28	33
6	19	28.5	31	36
7	22	31.5	34	39
8	25	34.5	37	42
9	28	37.5	40	45
10	31.5	41	43.5	48.5
11	35	44.5	47	52
12	38	47.5	50	55
14	44	53.5	56	61
16	50	59.5	62	67

Figure 46-48 Galvanized duct insulated with fiberglass liner.

Figure 46-49 Duct insulated with bubble wrap.

Figure 46-50 Fiberglass ductboard.

applied to the duct and the bubble wrap is applied to the spacers. Installers like bubble wrap because it does not irritate their skin and cause itching like fiberglass does. However, it is a new product without the long track record of success that fiberglass has.

46.10 DUCTBOARD

One alternative to sheet-metal duct is ductboard. Ductboard is a rigid fiberglass insulating board with an outer vapor barrier used to construct plenums and ductwork (Figure 46-50). The inner wall does not have a covering, but the fiberglass is matted and stiff, not fluffy like duct wrap. Quality ductboard should be UL 181 rated, and the tapes used to fabricate it should be UL 181A rated. Compared to metal duct, ductboard has several advantages. It is relatively low in both material and installation cost, it does not need insulating because it is inherently thermally efficient, it is generally easier to fabricate, and it produces quieter systems.

Despite these advantages, many contractors still consider a ductboard duct system inferior to a metal duct system. It is less durable than metal but holds up well if fabricated correctly. Unfortunately, the industry has seen a lot of ductboard that was not fabricated correctly and did not last long. Ductboard has a higher resistance to airflow because of its relatively rough interior surface. This surface also causes it to collect dirt. Small dust particles entrained in the air lodge in the surface of the ductboard; after several years of service, the interior surface of most ductboard systems is black.

Fabrication

Ductwork is constructed by cutting grooves or slots in the fiberglass where each corner will be. The tools are hand tools with razor-sharp knife blades that cut through the fiberglass to form the joints. Types of joints used are the V-groove, shiplap, and modified shiplap (Figure 46-51). Enough fiberglass is cut out to allow the ductboard to fold and form corners. The last edge is cut to form a flap. The ductboard is then folded at these corners. The two edges of the ductboard are mated and the flap is folded over and

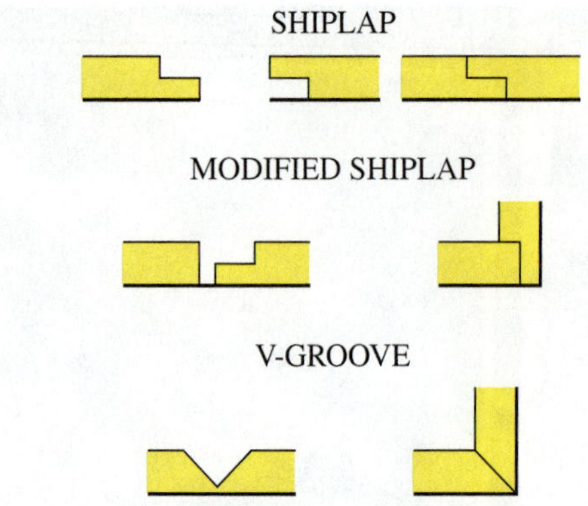

Figure 46-51 Types of ductboard joints.

secured using staples and UL 181A duct tape or just UL 181A duct tape. Pressure-sensitive and heat-activated tape are both available.

Pressure-Sensitive Tape

The edge of the flap should be positioned in the middle of the tape with a minimum overlap of 1 in on each side. Pressure-sensitive tape should be worked with a tool edge, not just not fingertip pressure. Rub tape firmly with a plastic sealing tool until the facing reinforcement shows through the tape. Avoid excessive pressure on the sealing tool that could cause the tape to be punctured at staple locations.

Pressure-sensitive tape should not be used below 50°F without heating. The duct must be preheated using an iron with the plate temperature set at approximately 400°F. Quickly position the tape on the preheated area and press in place. Pass the iron two or three times over the taped area using a rapid ironing motion. Complete the bond by rubbing the tape firmly with the plastic sealing tool until the facing reinforcement shows through the tape clearly.

Heat-Activated Tape

The edge of the flap should be positioned in the middle of the tape with a minimum overlap of 1 in on each side. The seaming iron should have a plate temperature of between 550°F and 600°F. Slowly pass the iron along the tape seam with sufficient pressure and dwell time to activate the adhesive. Heat-indicator dots on the tape will darken to indicate when a satisfactory bond has been achieved. Use a second pass of the iron to complete the bond. Avoid puncturing the tape at staple locations with excessive pressure from the iron. Allow all joints and seams to cool below a 150°F (66°C) surface temperature before any stress is applied.

SAFETY TIP

Always work with gloves and exercise caution to prevent burn injuries from contact with the iron or with heated surfaces. Even momentary contact with a 600° surface can cause a painful burn.

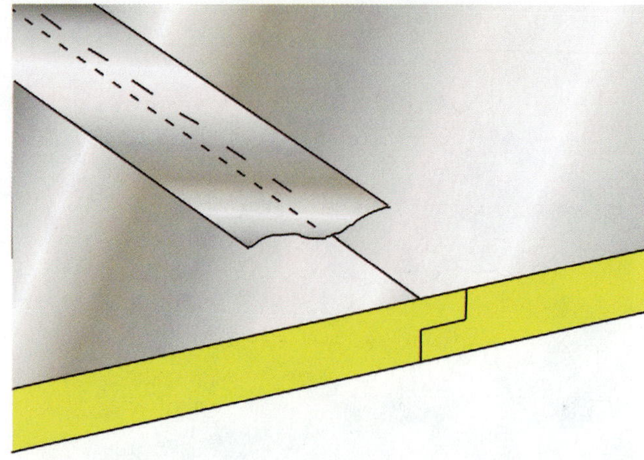

Figure 46-52 Joining ductboard sections.

Joining Ductboard Sections

Ductboard sections are joined to each other using a shiplap joint. The shiplap joint provides some structural rigidity to the joint. The sections are sealed using either pressure-sensitive or heat-activated tape (Figure 46-52).

Hanging Ductboard

Properly constructed, ductboard has good structural integrity. It is also lightweight, so hangers are not required to support a lot of weight. Duct hangers should be at least 1 in wide across the bottom. A common arrangement is to use an inverted 1-in metal channel that is suspended by wire, straps, or rods (Figure 46-53). Hangers should not be placed directly under joints between two pieces of duct to avoid stressing the joints. Hangers should be placed every 6 ft for duct that is less than 12-in high or every 8 ft for ducts that are 12-in high or larger. Duct should be suspended at least 1 in above ceiling insulation in attics. In crawlspace applications, it should be at least 4 in above the ground and 1 in below the floor insulation.

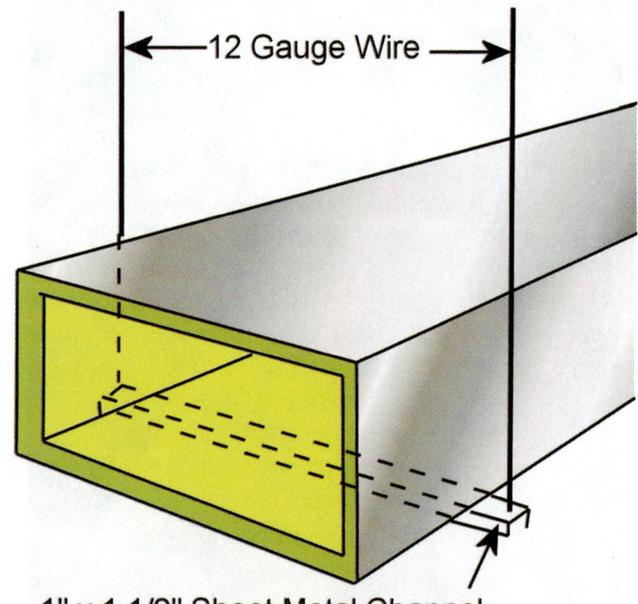

Figure 46-53 Properly supported ductboard duct.

Takeoffs

The most common use of ductboard is for plenums and trunk lines in systems that use flex duct for branch ducts. Takeoffs designed specifically for fiberduct should be used. Two types are available: spin-in collars and dovetail collars. The preferable method of cutting the hole is to use a hole cutter that is the same diameter as the collar. If using a spin-in, cut a 1-in slot on the outside of the hole. Cut a ring from a piece of ductboard to insulate the collar. Its inside diameter should be the same as the diameter of the sheet-metal collar. The width of the ring should correspond to the R value of the flexible duct insulation: 1 in wide for R-4, 1.5 in wide for R-6, and 2 in wide for R-8. Slide the ring over the collar and tape it to the collar. Apply mastic to the back of the collar flange and insert the collar into the hole until the flange seats against the face of the ductboard. For spin-ins, this is done by turning the collar clockwise and "threading" it into the hole. Dovetail collars will also go in more easily if they are turned while inserting them. Reach through the collar and bend the tabs on the dovetail over. The flex duct is connected to the takeoff, as described in Section 46.11.

46.11 FLEX DUCT

Both the UL 181 standard and the NFPA 90 standard differentiate between flexible air duct and flexible air connectors. A flexible air duct must pass all fifteen of the UL 181 tests, while air connectors are not required to pass the flame penetration, puncture, or impact tests in standard UL 181. Ducts meeting the more stringent air duct classification can be used in any length, while air connectors are limited to 14 ft. Air connectors are limited to temperature less than 250°F, are not allowed to pass through floors, and are not allowed to pass through walls that are required to have a fire rating.

The markings on the duct indicate whether it is rated as a flexible duct or a flexible connector. Flexible ducts have a square or rectangular label (Figure 46-54), while flexible connectors have a circular label (Figure 46-55). These markings should appear every 10 ft on the duct. Flexible duct is available in insulation R values of 4.2, 6.0, and 8.0. Many codes require a minimum R value of 6.

Figure 46-54 UL 181 flexible duct has a square or rectangle label and has no length restrictions.

Figure 46-55 UL 181 flexible air connector has a round label and cannot be used for lengths over 14 ft.

Flexible duct is the easiest ducting to install. Unfortunately, it is also the easiest to install incorrectly. Poor craftsmanship not only makes the job unprofessional, but it hurts the airflow as well. Flex duct that is kinked, flattened, or draped in lazy loops will not work correctly (Figure 46-56). The four basic requirements for correctly installing flex duct are as follows:

- Connecting flex duct correctly to takeoffs and fittings
- Stretching the duct to eliminate droops, sags, and crimps
- Trimming the duct to the correct length
- Supporting the duct every 5 ft with 2-in-wide material

Figure 46-56 Incorrectly installed flexible duct system.

Figure 46-57 The inner core is slid over the bead of the fitting and strapped.

Figure 46-58 The insulation and outer jacket are pulled over the inner core and strapped.

Connecting Flex Duct

The basic method of connecting flex duct is to slide the inner core at least 2 in over a metal beaded connector, roll back the outer insulation, fasten the inner core, and then roll the insulation back in place and fasten it. Two approved methods are used for fastening the core and insulation: using UL 181B duct tape or UL 181B mastic and straps.

When using duct tape, the inner core should be taped using at least two wraps of tape. Remember to leave enough metal exposed for the tape to grab. Slide the insulation back in place and tape the insulation in place with at least two wraps of tape.

When using mastic and clamps, UL 181B mastic must be applied to the metal connector before sliding the inner core on. Slide the inner core on past the bead and secure with a strap (Figure 46-57). Roll the insulation in place and secure it with a second strap (Figure 46-58). When the duct is installed vertically, three screws should be run into the metal behind the strap to avoid slipping.

Stretching Flex Duct

Flex duct by its very nature does not want to stay straight. It is necessary to pull it straight to eliminate sags. The maximum allowable sag between supports is 0.5 in/ft.

Trimming Flex Duct

Flex duct should be trimmed to the length needed and no more. Sloppy installers sometimes leave an extra foot or two on the run, knowing they cannot reuse that foot of duct. But leaving extra duct leads to sags, bends, and crimps that severely reduce airflow. Figure 46-59 shows the effect of leaving extra duct length. Figure 46-60 shows a duct that has been trimmed to the correct length.

Supporting Flex Duct

Flex duct is not heavy, but it needs more support than metal duct because it has no structural strength of its own. Flex

Figure 46-59 This run will have a high resistance to airflow.

Figure 46-60 This flexible duct is properly applied; it is tight, straight, and well supported by wide supports.

Figure 46-61 This flex duct is being pinched by the wire that is supporting it.

duct should be supported every 5 ft with straps or hooks that are at least 2 in wide. Galvanized wire and plumbing strap are definitely not appropriate for supporting flex duct. Figure 46-61 shows what happens when flex is supported by wire. Figure 46-62 shows a common abuse of flexible duct: squeezing and twisting it to fit through places that are simply not large enough for the duct. Figure 46-60 shows a properly supported flex run.

46.12 DUCT SEALING

The largest loss of efficiency in a ducted air system comes from air leaks. Unconditioned air leaking into the return-air side of the system can add a considerable heat load. Sealing duct leaks is the fastest way to improve the efficiency of a ducted heating and cooling system. Older homes that have poorly sealed duct systems typically lose 25 percent or more of their system capacity due to duct leaks. Sealing the duct system during installation is relatively simple and does not take a lot of extra time or money. The money spent sealing the system will be recovered during the first year of operation. Figure 46-63 illustrates the effect of sealing and insulating ductwork.

Figure 46-62 Not much air will make it through this duct.

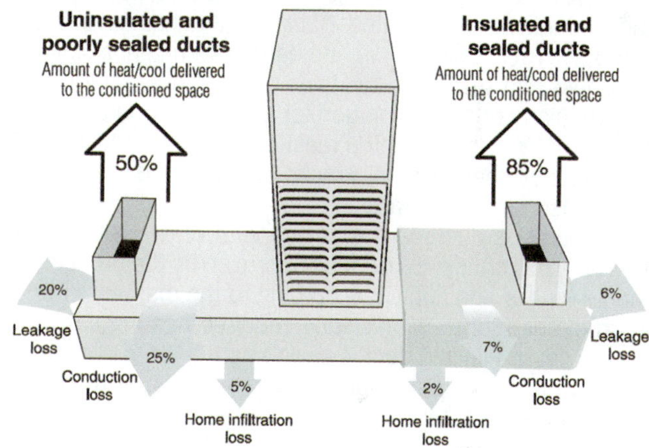

Figure 46-63 Improper duct sealing and poor duct insulation lose a significant portion of a system's capacity and increase utility bills. *(Courtesy of Office of Energy Efficiency and Renewable Energy, U.S. Department of Energy)*

Sheet-metal ductwork is most commonly sealed with mastic. Mastic has a consistency similar to sheetrock mud and is spread or "painted" on (Figure 46-64). It is stiff but still somewhat flexible after it dries. All duct sealants should be UL-181 approved. Duct tape was once the primary method

Figure 46-64 Applying mastic to sheet metal duct joints.

of sealing metal duct, but tape use is less common today. If duct tape is used, it should be UL-181 approved. There is an abundance of inexpensive duct tape available that is not UL 181 approved, but it should not be used for sealing ductwork.

The biggest leak source in many older systems is found when building cavities are used for return air. It is virtually impossible to seal these properly, resulting in large amounts of attic or crawlspace air being drawn into the system. The best way to correct this is to avoid using building cavities for duct.

Besides building cavities, the largest sources of leaks are around plenums and takeoffs. The holes are frequently not cut precisely, and many times the only attempted seal is duct tape. Close to half the duct leakage in many systems is at the supply and return plenums.

The use of garden-variety duct tape with no mastic is the next most often encountered problem. This type of tape is handy for many things, but sealing ductwork is not one of them. Any connecting and sealing products, including duct tape, should be UL 181 rated. Duct tape that has not passed the UL 181 tests will dry up and harden with age, losing its tackiness and sealing ability. Even the best duct tape is not a particularly effective solution for sealing around plenums and takeoffs; a UL 181 rated mastic is required. Mastic will work fine by itself on small cracks, but tape and mastic will be needed for larger cracks.

Fiberglass ductboard is sealed with UL-181 aluminum foil tape. It is critical to use tape approved by the ductboard manufacturer and apply it according to the manufacturer's instructions. Some of this tape requires heating with an iron to set it. Flexible duct is sealed with both UL-181 tape and mastic. Flexible duct only requires sealing at the ends where it is connected.

TECH TIP

The UL 181A standard covers closure systems for rigid ducts, while the UL 181B standard covers closure systems for flexible ducts and air connectors. Mastics, tapes, and sealants should meet either the UL 181A or the UL 181B standard, depending upon the type of duct.

46.13 TESTING THE DUCTWORK

CODE TIP

The 2009 International Energy Conservation Code (IECC) includes mandatory duct pressure testing coupled with maximum allowable duct leakage rates. These requirements are applicable when any portion of the ducts are outside the conditioned space.

Best-practice duct installation includes testing the duct system for leaks. The most common method is to use a duct blaster (Figure 46-65). All duct outlets are sealed, and a duct blaster is used to pressurize the entire duct system to 25–50 pascals (0.1–0.2 in wc). After the initial duct pressurization, the amount of airflow required to maintain the pressure is measured. For a completely tight duct system, the airflow would be 0 CFM. However, no system is that tight. The amount of air flowing is the amount leaking out of the ductwork. The amount of acceptable duct leakage depends upon several factors, including the size of the conditioned area the duct system serves and the total amount of airflow through the duct system.

Figure 46-65 A duct blaster is used to pressurize ducts to find leaks and verify duct system integrity. *(Photo provided by The Energy Conservatory)*

Three methods of quantifying duct leakage are used: percentage of design flow, airflow per 100 ft^2 of conditioned area, and leakage area. The percent of design flow states the measured leak rate as a percentage of the design flow rate. For example, a system designed to move 1,000 CFM that has a measured leak rate of 100 CFM has a 10 percent leak rate (100 CFM/1,000 CFM = 10 percent). Airflow per 100 ft^2 of conditioned area calculates the leak rate per 100 ft^2 of conditioned area and compares it to a maximum acceptable leak rate. For example, a 1,500 ft^2 house with a system that is leaking 90 CFM has a leak rate of 6 CFM per 100 ft^2 (1,500/100 = 15, 90/15 = 6). Duct leakage area specifies the size of the hole that would let out the measured CFM if all the combined leakage were combined into a single hole. This area is calculated using a formula that includes the duct pressure and the CFM airflow leak rate.

Energy codes and testing standards vary in what is considered acceptable. When using percent of airflow, less than 6 percent is generally considered good, while anything over 15 percent is generally considered poor. When using CFM per 100 ft^2, 6 CFM at rough-in or 8 CFM after construction is often considered the maximum acceptable leakage rate.

TECH TIP

Testing the duct leakage before insulating it will save time by avoiding the removal of insulation to get at the points that need sealing.

46.14 INSULATING DUCTWORK

Fiberglass ductboard does not need to be insulated because it is made of fiberglass and has inherent thermal insulating properties. Flexible duct does not need to be insulated either, because fiberglass insulation is part of its construction. Sheet-metal ducts do need to be insulated. Sheet-metal ductwork can be insulated on the inside with duct liner or on the outside with duct wrap.

Duct liner is applied to the inside of the duct. It is available in thicknesses from ½ in to 2 in. In general, duct liner is made of a denser fiberglass material than duct wrap. Duct liner reduces the interior size of the ductwork, so the ducts must be made with larger dimensions to accommodate the use of duct liner.

The most common duct wrap is fiberglass with an outer vapor barrier. It is commonly available in 1.5-in and 2-in thicknesses. Many energy-efficiency codes now require the 2-in-thick duct wrap. The vapor barrier prevents condensation on the outside of the duct during air conditioning when the duct temperature may be below the dew-point temperature of the area the duct is in. Condensation on the duct represents a significant loss of system efficiency and can cause property damage from dripping. The vapor barrier can be vinyl or paper and aluminum foil kraft back.

46.15 DUCT CLEANING

Duct-cleaning standards and duct-cleaning technician certifications are established by the National Air Duct Cleaners Association. A key point in discussing duct cleaning is that typically the entire system needs cleaning, not just the ducts. Blowers, coils, and filters should all be checked and cleaned whenever ductwork is cleaned. System components, including ducts, accumulate a fine layer of dust over time. In most systems there is more dust accumulation in the return than the supply. Mold can also sometimes grow in ductwork, but the presence of mold indicates a more serious problem than simply dirty ducts. The mold can be cleaned, but if the root cause is not addressed, the mold will return. Mold requires moisture to live; if mold is growing in the system, moisture is getting in. Finding and eliminating the source of moisture is necessary for long-term mold remediation.

A visual inspection is necessary to determine the need for cleaning. A visual inspection after cleaning is necessary to assess the results. Duct systems should have inspection ports at major turns and after any obstruction, such as coils or filters. Frequently these will have to be added the first time a system is cleaned. Tools are available that enable inspection of the duct system via remote video cameras

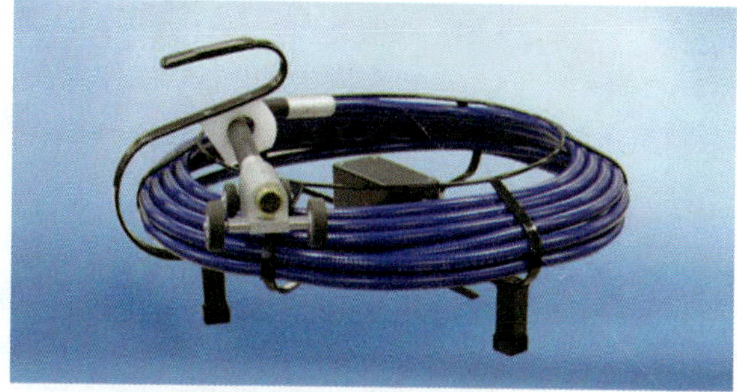

Figure 46-66 Video inspection tool for examining the inside of ductwork before and after cleaning. *(Courtesy of Abatement Technologies, Inc.)*

(Figure 46-66). This enables the technician to see what must be cleaned, even if the area is around a corner. Video inspection systems also allow the customer to see the results of the job.

Most duct-cleaning equipment is some form of specialized vacuum cleaner. Some systems are large units mounted on the back of a truck with large hoses connected to the ducts in the house. These systems move a large volume of air and rely on this air volume and velocity to remove debris from the duct. Other systems are much smaller and use a series of brushes along with inspection tools. Some systems use a percussive tool that flails around inside the duct to knock dirt loose. These systems will only work with metal ducts.

Metal ducts are the easiest to clean because they are not likely to be damaged by the cleaning equipment. Fiberduct and fiberglass-lined duct can be cleaned with soft-bristle brushes, but heavy agitation or stiff brushes can damage the fiberglass and do more harm than good. Flex duct can sometimes be cleaned with gentle brushes, but it may be more practical to replace it than clean it. Fabric duct systems (Figure 46-67) can literally be taken down and washed.

UNIT 46—SUMMARY

A properly designed and installed duct system:

- Allows for proper airflow through the system
- Delivers conditioned air throughout the structure according to the heat load
- Does not leak
- Is well insulated and thermally efficient
- Is quiet

This can be achieved using sheet metal, ductboard, flex duct, or any combination of these materials. The difference between a poor duct system and a good one is often just a matter of a few minutes and a little extra material. Sealing metal ducts, trimming flex ducts, or fabricating fiberboard ducts are not complicated tasks; they just take a little extra time and material.

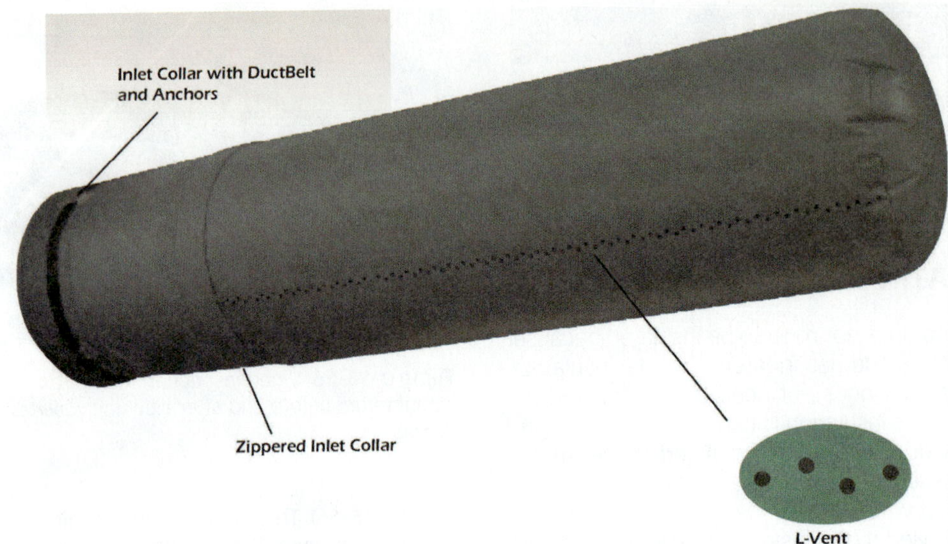

Figure 46-67 Fabric duct. *(Courtesy of DuctSox Corp.)*

WORK ORDERS

Service Ticket 4601

Job: Duct Tightening

Equipment Type: 20-Year-Old Split-System Air Conditioner with All Metal Duct

A customer contracted for a duct tightening job because the local electric utility has a rebate for duct tightening. The contractor is required to reduce the duct leakage by at least 50 percent. The technician starts by measuring the duct leakage with a duct blaster. Next, the old outer insulation around the plenum is removed. Many leaks are evident around both the supply and return plenums. The technician uses UL 181A tape and mastic to seal these leaks and retests the system and finds that simply sealing the leaks around the plenums reduced the duct leakage by 30 percent. The technician goes on to check the rest of the system and wraps the plenum with new higher-R-value insulation.

Service Ticket 4602

Complaint: Poor Airflow in Office

Equipment Type: Commercial System Using a Metal Trunk and Flex Duct Drops

The vice president of customer assurance at APEX Industries moves into a newly renovated office and finds it stuffy. There is very little air coming out of the ceiling register. The technician who is called to fix the problem looks above the ceiling and notices that the flex run is too long and is almost doubled over. The flex is trimmed to an appropriate length and stretched so that the air has a straight shot. The register now has adequate airflow.

Service Ticket 4603

Complaint: Musty Odor When Unit Runs

Equipment Type: Split System in Crawlspace with Ductboard Trunk Line and Flex Runouts

A technician is called to look at a system that is blowing musty air into the house and notices that the fiberduct plenums are only attached to the 0.5-in flanges on the unit with duct tape. They are both leaky, but the return plenum has sagged and there is a gap on one side between the plenum and the unit. The technician also notices that there is no vapor barrier on the ground. The plenums are removed, 2-in-wide sheet metal flanges are connected to the unit's 0.5-in flanges with sheet metal screws, and the plenums are reattached using three long sheet-metal screws with large washers on each side of the plenums to attach them to the new metal flanges. The plenums are then sealed to the unit with UL 181A duct tape. The technician recommends that the customer install a vapor barrier over the exposed dirt in the crawlspace.

Service Ticket 4604

Job: Duct Cleaning

Equipment Type: Metal Trunk Line and Runouts

A customer notices that the ductwork by his filter grille is very dirty on the inside and calls to have his ducts cleaned. The technician uses the video inspection tool to see the extent of the dirt, makes a tape of the inspection, and shows the customer. The ducts are cleaned using a portable duct-cleaning machine with soft bristles on the trunk lines. The blower wheel and evaporator coil are also cleaned. The technician makes a new tape showing the ducts after cleaning and suggests that the customer consider installing a high-efficiency air cleaner to keep the ducts clean.

Service Ticket 4605

Complaint: No Airflow in Bedroom

Equipment Type: Radial System with Lined Metal Plenum with Flex Runouts

A new system has no airflow in the master bedroom. The technician looks at the duct run to the bedroom and sees no obvious problems. After removing the flex duct from the plenum, the technician finds that the liner in the plenum was not entirely cut out. After cutting out the liner and reattaching the duct, the bedroom has good airflow.

UNIT 46—REVIEW QUESTIONS

1. What standard covers ductboard and methods of connecting and sealing ductboard?
2. What standard covers flexible duct and methods of connecting and sealing flexible duct?
3. What is the difference between Class 0 duct and Class 1 duct?
4. List the agencies that promote duct standards, and briefly describe what they cover.
5. Explain the difference between a flexible duct and a flexible air connector.
6. List the materials commonly used to construct duct systems.
7. Describe how two pieces of round sheet-metal pipe are joined.
8. Describe how sheet-metal ductwork is sealed.
9. Describe how to join fiberboard duct.
10. How far apart should hangers for metal duct be placed?
11. How far apart should hangers for fiber ductboard be placed?
12. Describe the proper method of hanging flexible duct.
13. List four methods of insulating metal ductwork.
14. Using Table 58-2, what width should 2-in fiberglass duct insulation be to wrap an 8-in-diameter round duct?
15. Describe how ductboard is formed into ducts.
16. List three types of takeoff fittings available for sheet-metal duct.
17. Describe how to install a dovetail takeoff fitting on a ductboard trunk.
18. What are some of the limitations on the use of flexible air connectors?
19. Describe how rectangular sheet-metal ducts are joined.
20. How do the staples that hold duct insulation together work?
21. What is the correct spacing for the mechanical fasteners used to hold duct liner in place?
22. Discuss some of the major problems that cause leaky duct systems in older homes.
23. How does a duct blaster measure the amount of duct leakage?
24. What are the two main types of duct-cleaning equipment being used?
25. Why is a visual inspection necessary when cleaning ductwork?
26. What are the three methods used to quantify duct leakage?

UNIT 47

Troubleshooting Air-Conditioning Systems

OBJECTIVES

After completing this unit, you will be able to:

1. troubleshoot air-conditioning system air-distribution problems.
2. adjust air-conditioning system airflow.
3. troubleshoot air-conditioning system electrical problems.
4. perform diagnostic tests on air-conditioning system electronic controls.
5. read and interpret troubleshooting charts.
6. troubleshoot air-conditioning system mechanical problems.

47.1 INTRODUCTION

Troubleshooting an air-conditioning system correctly will result in the proper repair for the unit in a timely fashion. It is discouraging for the customer and the technician if multiple return visits are required to correct a malfunctioning unit. The trouble described by the customer may initially help you to determine which course of action needs to be taken. It will save you time to take a few minutes to discuss the symptoms with them. They may be able to provide you with valuable information that may lead you to the problem.

Most manufacturers supply troubleshooting charts for their equipment. After speaking with the customer, a good next step is to review these to assist in troubleshooting the system. A quick simple fix may not always be the correct answer. For example, the unit may be tripped out due to a stuck high-water-level switch on the evaporator condensate drip pan, as shown in Figure 47-1.

Simply resetting the unit may bring it back on line, but if it trips out again you will be called to come back to the residence a second time. Before leaving the job, the drain pipe should be checked for an obstruction.

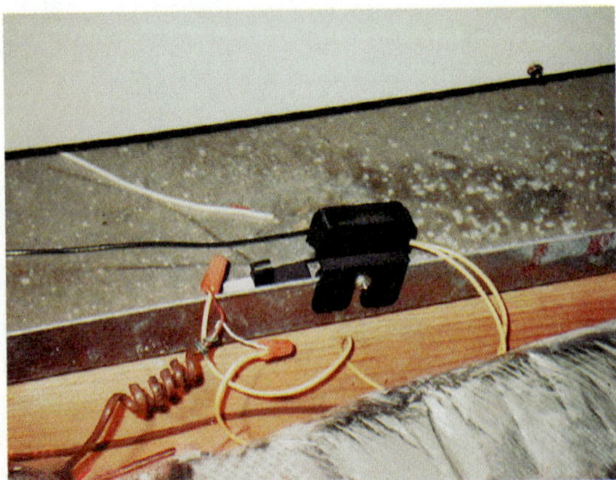

Figure 47-1 Float switch on condensate tray. *(Courtesy of InspectAPedia.com)*

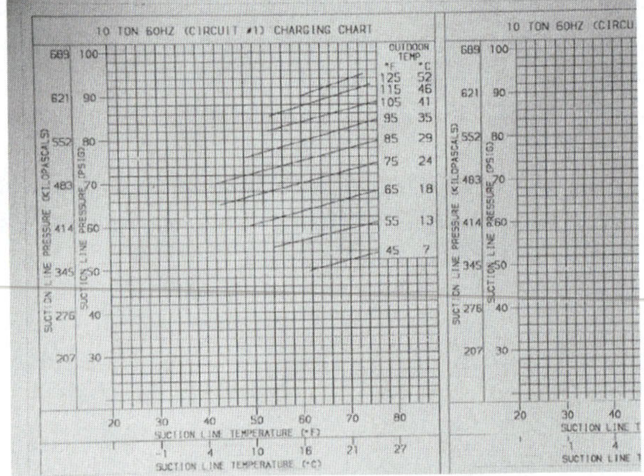

Figure 47-2 Refrigerant-charging service data tag. *(Courtesy of InspectAPedia.com)*

To avoid these types of situations, it is best to take the proper steps in evaluating the malfunction. Collect information about the problem. Read and calculate the system's vital signs, such as suction and discharge pressures. Compare the measured values to the expected manufacturer's recommendations (Figure 47-2). Consult the manufacturer's troubleshooting aids, if available, along with any company-specific troubleshooting guides.

SAFETY TIP

Electrical testing is generally a first step in any troubleshooting sequence. Make sure that you properly identify the power supply to the unit (Figure 47-3). Do not assume that the power is always shut off at the breaker. Sometimes breakers are mislabeled. Always use your meter to check for ground faults before energizing the unit to be serviced. At all times, use your meter to check for voltage before physically disconnecting wires or opening control boxes.

Figure 47-3 Service control switch circuit breaker.

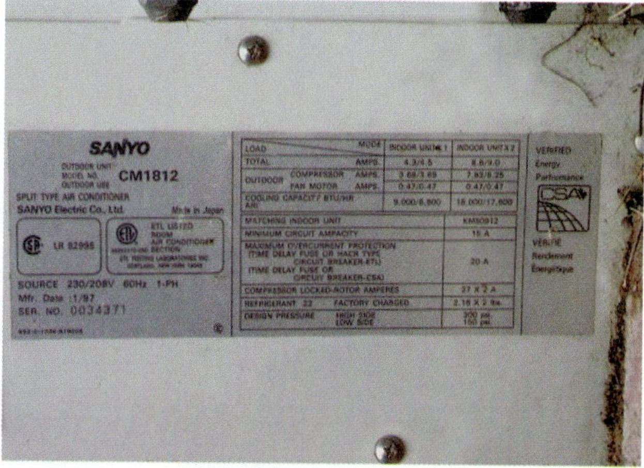

Figure 47-4 Split-system data tag. *(Courtesy of InspectAPedia.com)*

47.2 SYSTEM FAMILIARIZATION

The air-conditioning system design may affect the way you troubleshoot the system. It is important to familiarize yourself with the system components and their location. Is the evaporator coil located in an indoor air-handling unit or is it installed in conjunction with a furnace? Is the air handler a vertical upflow, vertical downflow, or horizontal unit, and where is it located? Where are the system controls and switches located? What type of safety controls does the unit have? Check the capacity ratings for the system (Figure 47-4). A split system has a fan coil unit with a refrigeration coil located on the inside of the building and an outside-mounted air-cooled condensing unit, connected together with refrigerant piping. A typical split system with an air-handling unit in the ceiling is shown in Figures 47-5 and 47-6.

Basically, there are three types of problems: trouble with the air system, the electrical system, or mechanical components. Within these there is much overlap, so whatever the nature of the problem, it is good practice to follow a logical, structured, systematic approach. In this manner, the correct solution is usually found in the shortest possible time.

47.3 AIR SYSTEM PROBLEMS

The primary problem that can occur in an air system is the reduction in airflow. Air-handling systems do not suddenly increase in capacity—that is, increase the amount of air across the coil. On the other hand, the refrigeration system does not suddenly increase in heat-transfer ability. First remove the panel to access the direct expansion (DX) coil in the air handler (Figure 47-7) so that you can inspect the coil for dirt and blockage. On many systems, removing a panel would change the airflow across the coil and change the return and supply reading. Take the temperatures of the return air as it enters and then the supply air as it leaves the coil. The difference between these two temperatures is referred to as the temperature drop or temperature difference of the air across the DX coil.

A sling psychrometer is used to measure the return-air dry-bulb and wet-bulb temperatures needed to determine the relative humidity. Many electronic psychrometers are able to display relative humidity directly. A chart as shown in Figure 47-8 or one supplied by the DX coil manufacturer is then used to determine the expected temperature drop. As an example from the chart, return air at a condition of 68°F and 30 percent relative humidity should have a 24°F air temperature drop across the coil.

This measurement will help to determine if the problem is a result of improper airflow or a refrigeration system error. If the actual air temperature drop is greater than the required temperature drop, then the air quantity has been reduced. In this case, look for problems in the air-handling system. This could be due to dirty air filters, a dirty evaporator coil, a problem with the blower, or an unusual restriction in the duct system.

Air Filters

Because this is the most common problem of air failure, check the filtering system first (Figure 47-9). Air filters of the throwaway type should be replaced at least once a month.

Blower Motor and Drive

Check the blower motor and drive in the case of belt-driven blowers to make sure that both the blower motor and blower bearing are properly lubricated and operating freely. The blower drive belt must be in good condition and properly adjusted. Cracked or heavily glazed belts must be replaced. Heavy glazing can be caused by too much tension on the belt, driving the belt down into the pulleys. Proper

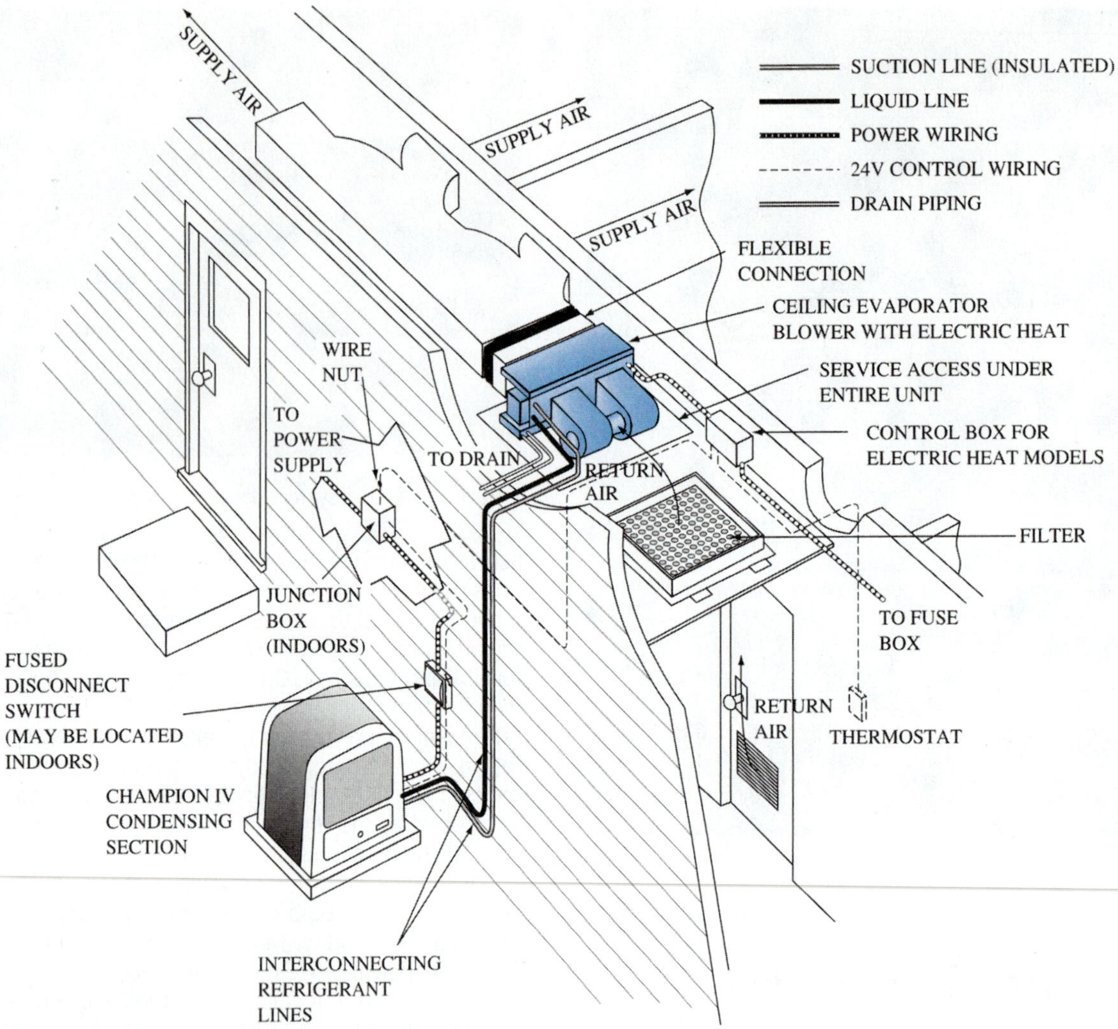

SUPPLY AIR
SUPPLY AIR
SUPPLY AIR
SUCTION LINE (INSULATED)
LIQUID LINE
POWER WIRING
24V CONTROL WIRING
DRAIN PIPING
FLEXIBLE CONNECTION
CEILING EVAPORATOR BLOWER WITH ELECTRIC HEAT
SERVICE ACCESS UNDER ENTIRE UNIT
CONTROL BOX FOR ELECTRIC HEAT MODELS
WIRE NUT
TO POWER SUPPLY
TO DRAIN
RETURN AIR
FILTER
JUNCTION BOX (INDOORS)
TO FUSE BOX
FUSED DISCONNECT SWITCH (MAY BE LOCATED INDOORS)
RETURN AIR
THERMOSTAT
CHAMPION IV CONDENSING SECTION
INTERCONNECTING REFRIGERANT LINES

Figure 47-5 Ceiling evaporator blower installation.

Figure 47-6 Attic-mounted air handler. *(Courtesy of InspectAPedia.com)*

Figure 47-7 Air-handler unit and coil. *(Courtesy of InspectAPedia.com)*

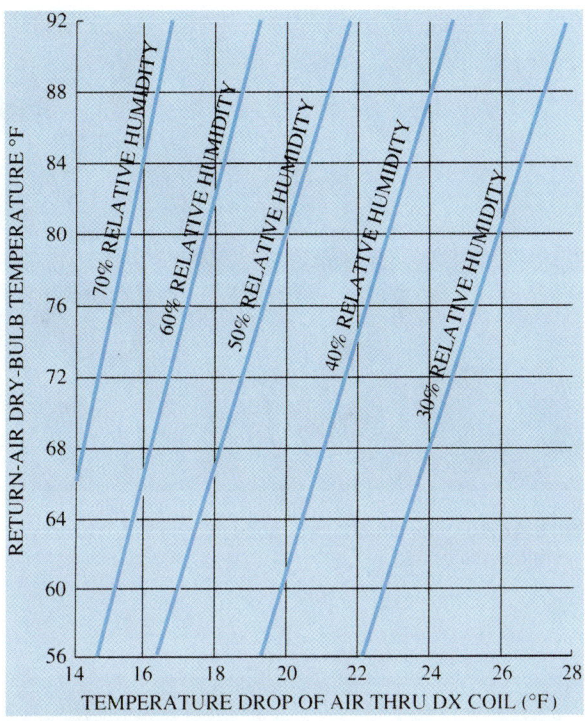

Figure 47-8 Air-temperature drop for various return-air conditions.

Figure 47-9 Dirty air filter *(Courtesy of InspectAPedia.com)*

Figure 47-10 Dirty blower fan. *(Courtesy of InspectAPedia.com)*

Figure 47-11 Mold growth on blower fan. *(Courtesy of InspectAPedia.com)*

adjustment requires the ability to depress the belt midway between the pulleys approximately 1 in for each 12 in between pulley centers.

The blower wheel should be clean. Dirt accumulation can sometimes fill in the area on a cupped blade, allowing it to spin freely but substantially reducing the airflow (Figure 47-10). If the wheel is dirty or has mold buildup (Figure 47-11), it must be removed and cleaned. Attempting to clean the wheel in place is never recommended. Do not try brushing only, because a poor cleaning job will cause an imbalance to occur on the wheel. Extreme vibration and noise will result. This could cause deterioration of the wheel, damage to the belt, and damage to the motor.

Unusual Restrictions in Duct Systems

Placing furniture or carpeting over return grilles reduces the air available for the blower to handle (Figure 47-12). Shutting off the air to unused areas will reduce the air over the coil. Covering a return-air grille to reduce the noise from the centrally located furnace or air handler may reduce the objectionable noise, but it also drastically affects the operation of the system by reducing air quantity.

The condition of the grille may also indicate potential problems. Water stains (Figure 47-13) can be caused by improper humidity conditions. Soot stains (Figure 47-14) may be a result of an improperly functioning furnace's exhaust leaking back into the air-supply line.

The collapse of the return-air duct system will affect the entire duct system performance (Figure 47-15). Air leaks in the return duct will raise the return-air temperature and reduce the temperature drop across the coil (Figure 47-16). Look for pinched ducts (Figure 47-17), sharp bends (Figure 47-18), and unnecessary duct length (Figure 47-19).

Figure 47-12 Blocked register. *(Courtesy of InspectAPedia.com)*

Figure 47-13 Supply register water stain. *(Courtesy of InspectAPedia.com)*

Figure 47-14 Supply register soot stain. *(Courtesy of InspectAPedia.com)*

Figure 47-15 Flex duct deterioration in hot attic. *(Courtesy of InspectAPedia.com)*

(a)

(b)

Figure 47-16 Leaky duct connections. *(Courtesy of InspectAPedia.com)*

Figure 47-17 Pinched duct. *(Courtesy of InspectAPedia.com)*

Figure 47-18 Duct sharp bend. *(Courtesy of InspectAPedia.com)*

Figure 47-19 Unnecessary duct length. *(Courtesy of InspectAPedia.com)*

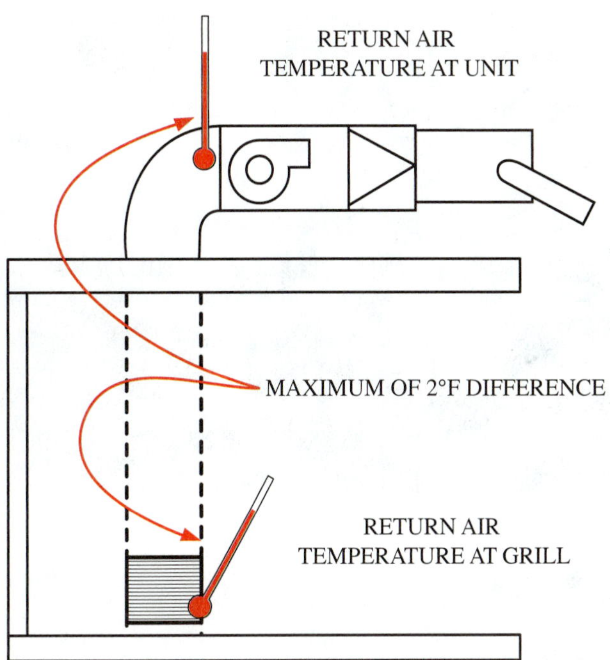

Figure 47-20 Checking return-air system for leaks.

TECH TIP

Air-distribution systems installed before the change in standards may have cloth or unapproved duct tape. These systems may need to be sealed properly for the central residential air-conditioning system to function properly. The technician should give particular attention to the integrity of the sealing of all joints and the insulation and tape if externally wrapped.

Measure the difference in temperature between the return air at the grille compared to the return-air temperature as it enters the unit. This difference should not exceed 2°F (Figure 47-20). If it does, the return duct needs to be insulated or there may be leak openings in the duct that need to be sealed.

SAFETY TIP

Some older fabric air-handler vibration dampers are made from asbestos (Figure 47-21). Always follow the proper procedures for handling asbestos materials. Never rip or tear the asbestos, and always wear an approved respirator.

Condensate Drain Pans

Condensate drain pans are located below the DX coil to collect any moisture that condenses from the air and drips off the coil. Depending on the humidity conditions, the amount

(a)

(b)

Figure 47-21 (a) Asbestos fabric in HVACR duct as vibration-damper material; (b) asbestos fabric air-handler vibration damper. *(Courtesy of InspectAPedia.com)*

of water drained away can be considerable. The condensate that collects will build up and overflow the pan if the drain becomes plugged or offers restricted flow. This is of particular concern with attic-mounted units because the condensate will spill over and damage the ceiling. Some condensate drip pans have condensate pumps that will stop and start automatically and are controlled by a float switch activated by the water level. Even if the condensate drain pan is empty, any signs of prior spillover should be investigated further.

Adjusting the Airflow

Most manufacturers supply operating data on their equipment (Table 47-1), indicating the total, sensible, and latent heat-removal rating at various outdoor dry-bulb and indoor wet-bulb temperatures at specific static pressures. For these same conditions, they supply the operating suction and discharge pressures of the equipment. This information is given so that the technician can match the actual conditions on the job with the performance conditions shown on the manufacturer's chart. Notice that in Table 47-2 a system with a low external static pressure of 0.30 would have an airflow of 625 CFM on LOW. But on a system with a higher external static pressure of 0.70, the blower would have to be set to HIGH to obtain the same airflow.

SERVICE TIP

Most manufacturers recommend that a system's airflow be set so there is approximately 400 CFM per ton of air conditioning. In areas that have high relative humidity, a slightly slower fan speed can be used to aid in the dehumidification. In arid areas, the fan speed can be increased to provide more sensible cooling. As a rule of thumb, the fan speed should not be adjusted more than 10 percent above or below the manufacturer's recommendations. Excessively slow fan speeds can allow the evaporator coil to freeze up under light loads, and excessively high fan speeds can put too large a load on the condenser under heavy load conditions.

47.4 ELECTRICAL PROBLEMS

Since the greatest numbers of malfunction problems are electrical, it is common practice to perform electrical troubleshooting (including controls) before mechanical troubleshooting. If the problem is mechanical, the electrical check will usually point the technician in that direction. If the system will not operate at all, it is probably an electrical problem that must be found and corrected.

TECH TIP

Before beginning any troubleshooting, you must locate the manufacturer's troubleshooting guide for the equipment you are working on (Table 47-3). Manufacturers have developed troubleshooting techniques that when followed will result in accurate and rapid location of the problem. These charts may seem complex and difficult to follow if you look at the entire chart, so to properly use one of these charts, start at the beginning and follow each and every step. Do not simply jump ahead assuming that you know what the answer is. Do the test and report the results, and then move to the next test as indicated by the manufacturer's guide.

Electrical Operating Sequence

The operating sequence is usually supplied by the manufacturer in the service instructions, or it can be determined by the technician by studying the schematic wiring diagram. The functions of the operating and nonoperating equipment are determined by examination and testing. Necessary test

Table 47-1 Example of Typical Air Conditioning Equipment Operational Specifications

95°F outside air temperature entering the outdoor coil

Entering Wet Bulb Temperature	Total Air Volume CFM	Total Cooling Capacity Btu/hr	Compressor Power Input kW	Sensible to Total Heat Ratio Coil Dry Bulb Temperatures		
				75°F	80°F	85°F
63 °F	450	15,900	1.39	0.74	0.86	0.98
63 °F	650	17,300	1.43	0.81	0.96	1.00
63 °F	850	18,300	1.45	0.87	1.00	1.00
67 °F	450	16,800	1.41	0.60	0.72	0.83
67 °F	650	18,200	1.45	0.64	0.79	0.93
67 °F	850	19,100	1.48	0.68	0.85	0.99
71 °F	450	17,500	1.43	0.47	0.58	0.69
71 °F	650	19,100	1.48	0.49	0.63	0.76
71 °F	850	20,000	1.50	0.51	0.67	0.83

105°F outside air temperature entering the outdoor coil

Entering Wet Bulb Temperature	Total Air Volume CFM	Total Cooling Capacity Btu/hr	Compressor Power Input kW	Sensible to Total Heat Ratio Coil Dry Bulb Temperatures		
				75°F	80°F	85°F
63 °F	450	15,000	1.47	0.76	0.89	1.00
63 °F	650	16,300	1.52	0.83	0.99	1.00
63 °F	850	17,400	1.56	0.91	1.00	1.00
67 °F	450	15,800	1.50	0.61	0.73	0.85
67 °F	650	17,200	1.55	0.66	0.81	0.96
67 °F	850	17,900	1.57	0.70	0.88	1.00
71 °F	450	16,600	1.53	0.48	0.59	0.71
71 °F	650	18,100	1.58	0.50	0.64	0.78
71 °F	850	18,900	1.61	0.52	0.69	0.86

115°F outside air temperature entering the outdoor coil

Entering Wet Bulb Temperature	Total Air Volume CFM	Total Cooling Capacity Btu/hr	Compressor Power Input kW	Sensible to Total Heat Ratio Coil Dry Bulb Temperatures		
				75°F	80°F	85°F
63 °F	450	14,100	1.55	0.78	0.92	1.00
63 °F	650	15,300	1.61	0.86	1.00	1.00
63 °F	850	16,400	1.65	0.94	1.00	1.00
67 °F	450	14,900	1.59	0.62	0.75	0.88
67 °F	650	16,100	1.64	0.67	0.84	0.99
67 °F	850	16,800	1.67	0.72	0.92	1.00
71 °F	450	15,700	1.62	0.48	0.61	0.73
71 °F	650	17,000	1.68	0.51	0.66	0.81
71 °F	850	17,800	1.71	0.53	0.71	0.89

Table 47-2 Typical Cooling Unit Air-Handler Performance Data

CB29M-21/26 Blower Performance (208/230 V)										
External Static Pressure		Air Volume and Motor Watts at Specific Blower Taps								
		Low			Medium			High		
In wg	Pa	CFM	L/s	Watts	CFM	L/s	Watts	CFM	L/s	Watts
.00	0	700	330	245	895	420	310	1,030	485	375
.05	10	690	325	240	875	415	305	1,010	475	370
.10	25	680	320	235	865	410	300	990	470	365
.15	35	665	315	230	850	400	290	970	480	355
.20	50	655	310	225	830	390	285	955	450	350
.25	60	640	300	220	810	385	280	925	440	345
.30	75	625	295	220	795	375	270	900	425	335
.40	100	595	280	210	750	355	255	850	400	320
.50	125	555	260	195	700	330	240	800	380	305
.60	150	510	240	185	640	300	225	725	340	290
.70	175	395	185	165	—	—	—	620	295	265
.75	185	—	—	—	—	—	—	570	270	255

Note: All air data are measured external to unit with air filter in place. Electric heaters have no appreciable air resistance.

instruments include the volt ohm meter, the clamp-on ammeter, the capacitor tester, and the temperature analyzer.

The power circuit is the first to be examined, because power must be available to operate the loads. For example, on a refrigeration system with an air-cooled condenser, the three principal loads that must be energized are the compressor motor, the condenser fan motor, and the evaporator fan motor. Before proceeding with anything else, the technician must be certain that the proper voltage is being supplied to the loads. The supply-line voltage should be tested with a voltmeter and then compared to the manufacturer's recommended values.

If the voltage is correct, then the circuit will need to be tested further. Before energizing the circuit, always use your meter to check for ground faults. Once energized, the voltage for each load in the circuit can be tested. A clamp-on ammeter can be used to check the current draw for each load one at a time. The measured current should be compared to the current rating listed on the motor nameplate. As an example, if the current draw is too high, this may indicate a mechanical problem such as a stuck fan or seized compressor.

In troubleshooting, when a load is not working, the technician must determine whether the problem is in the load itself or in the switches that control the load. If the proper supply voltage is available but the compressor or fan motor does not run, then there may be a fault in the control circuit and the control circuit voltage should be tested. Generally the control circuit voltage will be much lower than the line voltage. Each load, compressor, or fan is generally controlled by a relay switch operated by an electromagnetic coil energized by the control circuit.

Use your meter to check for voltage before physically disconnecting wires or opening control boxes. Always test to determine if the fan or motor is tripped out on a safety switch before replacing components. When checking components with an ohmmeter, it is important that all power is disconnected and the component part is electrically isolated. Short circuits are usually due to faulty loads. If a faulty component is located, the job is still not complete. The technician should make a concerted effort to determine why or how the component failure occurred.

Installation and Service Instructions

The installation and service instructions supply a wide variety of information that the manufacturer believes is necessary to properly install and service the unit. This bulletin includes the wiring diagram, the sequence of operation, and any notes or cautions that need to be observed in using them.

Wiring Diagrams

Wiring diagrams usually consist of connection diagrams and schematic diagrams. The connection diagram shows the wires to the various electrical component terminals in their approximate location on the unit. This is the diagram that the technician must use to locate the test points. The schematic diagram separates each circuit to clearly indicate the function of switches that control each load. This is the diagram the technician uses to determine the sequence of operation for the system.

TABLE 47-3 Troubleshooting Chart: Cooling Service

Problem	Cause	Remedy
Compressor and condenser fans will not start.	Power failure	Call power company.
	Fuse blown or circuit breaker tripped	Replace fuse or reset circuit breaker.
	Defective thermostat, contactor, transformer, or control relay	Replace component.
	Insufficient line voltage	Determine cause and correct.
	Incorrect or faulty wiring	Check wiring diagram and rewire correctly.
	Thermostat setting too high	Lower thermostat setting below room temperature.
Compressor will not start but condenser fans run.	Faulty wiring or loose connections in compressor circuit	Check wiring and repair or replace.
	Compressor motor burned out, seized, or internal overload open	Determine cause. Replace compressor.
	Defective run/start capacitor, overload, or start relay	Determine cause and replace.
	One leg of three-phase power dead	Replace fuse or reset circuit breaker.
Compressor cycles (other than normally satisfying thermostat).	Refrigerant overcharge or undercharge	Recover refrigerant, evacuate system, and recharge to nameplate.
	Defective compressor	Replace and determine cause.
	Insufficient line voltage	Determine cause and correct.
	Blocked condenser	Determine cause and correct.
	Defective run/start capacitor, overload, or start relay	Determine cause and replace.
	Defective thermostat	Replace thermostat.
	Faulty condenser-fan motor or capacitor	Replace.
	Restriction in refrigerant system	Locate restriction and remove.
Compressor (scroll only) makes excessive noise.	Compressor rotating in wrong direction	Reverse the three-phase power leads.
Compressor operates continuously.	Dirty air filter	Replace filter.
	Unit undersized for load	Decrease load or increase unit size.
	Thermostat set too low	Reset thermostat.
	Low refrigerant charge	Locate leak, repair, and recharge.
	Leaking valve in compressor	Replace compressor.
	Air in system	Recover refrigerant, evacuate system, and recharge.
	Condenser coil dirty or restricted	Clean coil, remove restriction.
Head pressure too high.	Dirty air filter	Replace filter.
	Dirty condenser coil	Clean coil.
	Refrigerant overcharged	Remove excess refrigerant.
	Air in system	Recover refrigerant, evacuate system, and recharge.
	Condenser air restricted or air short-cycling	Determine cause and correct.
Head pressure too low.	Low refrigerant charge	Check for leaks, repair, and recharge.
	Compressor valves leaking	Replace compressor.
	Restriction in liquid tube	Remove restriction.
Suction pressure too high.	High heat load	Check for source and eliminate.
	Compressor valves leaking	Replace compressor.
	Refrigerant overcharged	Recover excess refrigerant.
Suction pressure too low.	Dirty air filter	Replace filter.
	Low refrigerant charge	Check for leaks, repair, and recharge.
	Metering device or low side restricted	Remove restriction.
	Insufficient evaporator airflow	Increase air quantity. Check filter and replace as necessary.
	Temperature too low in conditioned area	Reset thermostat.
	Field-installed filter drier restricted	Replace.
Compressor no. 2 will not run.	Unit in economizer mode	Proper operation; no remedy necessary.

TECH TIP

Wiring diagrams are often included inside the equipment panels. If the diagrams are not there, then contact the manufacturer. They are available from local suppliers and are often available from the manufacturers themselves through their Web sites. If you do not have a diagram, troubleshooting electrical circuits can be lengthy, time-consuming, and inaccurate.

Troubleshooting Tables

Troubleshooting tables are helpful as a guide to corrective action. By a process of elimination, such tables offer a quick way to solve a service problem. The process of elimination permits the technician to examine each suggested remedy and disregard ones that do not apply or are impractical; leaving only the solution(s) that fits the problem.

Fault Isolation Diagrams

A fault isolation diagram (Figure 47-22) starts with a failure symptom and goes through a logical decision action process to isolate the failure.

Diagnostic Tests

Diagnostic tests can be conducted on electronic circuit boards, at points indicated by the manufacturer, to check voltages or other essential information critical to the operation of the unit.

Some electronically controlled systems have automatic testing features, which indicate by code number a malfunction in the operation of the equipment (Figure 47-23). Further tests are usually required to determine the action that is required.

47.5 MECHANICAL PROBLEMS

When the measured temperature drop across the DX coil is less than required, this means that the heat-removal capacity of the system has been reduced. This means that the amount of heat picked up in the coil plus the amount of motor heat added and the total rejected from the condenser is not the total heat quantity the unit is designed to handle. The problems associated with a system that starts and runs but does not produce satisfactory cooling can be simply divided into two categories: refrigerant quantity and refrigerant flow rate.

To determine the problem, all the information listed in Table 47-4 must be measured. These results compared to normal operating results will generally identify the problem. The use of the word *normal* does not imply a fixed set of pressures and temperatures. These will vary with each make and model of the system. A few temperatures are fairly consistent throughout the industry and can be used for comparison. These are the DX coil operating temperature, the condensing unit condensing temperature, and the refrigerant subcooling.

SERVICE TIP

These vital signs must also be modified according to the seasonal energy efficiency ratio (SEER) of the unit. The reason for this is that the amount of evaporation and condensing surface designed into the unit is directly related to the efficiency rating. A larger condensing surface results in a lower condensing temperature and a higher SEER. A larger evaporating surface results in a higher suction pressure and a higher SEER. The energy efficiency ratio is calculated by dividing the net capacity of the unit in BTU/hr by the watts input. Every central split cooling system manufactured in the United States today must have a SEER of at least 13.

47.6 DX COIL OPERATING TEMPERATURE

Normal coil operating temperatures can be found by subtracting the design DX coil split from the average air temperature going through the coil. The DX coil split will vary with the system design. The term *coil split* is also often referred to as the temperature split or the approach temperature. It is the difference in temperature between the coil (condenser or evaporator) and the air passing across it.

The energy efficiency ratio (EER) is the output in BTU/hr divided by the power input measured in watts. It is less meaningful than SEER because it is a steady state rating and does not account for the time the unit operates before reaching its peak efficiency. However, the EER is still useful in determining coil operating temperatures. Systems in the EER range of 7.0 to 8.0 will have DX coil design splits in the range of 25°F to 30°F. Systems in the EER range of 8.0 to 9.0 will have DX coil design splits in the range of 20°F to 25°F. Systems with 9.0+ EER ratings will have DX coil design splits in the range of 15°F to 20°F.

The normal DX coil operating temperature will be affected by the air passing across it. The air passing across the coil will have an inlet and an outlet temperature. The inlet air is the return air and can also be referred to as the entering air temperature (EAT). The outlet air is the supply air and can also be referred to as the leaving air temperature (LAT). The EAT − LAT is equal to the temperature difference (TD or ΔT) of the air across the DX coil. The average temperature of the air passing across the coil can be calculated as TD ÷ 2. This calculated average temperature of the air can be used to calculate the normal DX coil operating temperature. The formula used for determining coil operating temperature (COT) is

$$\text{COT} = \frac{(\text{EAT} + \text{LAT})}{2} - \text{DX coil split}$$

For example, a unit having an entering air condition of 80°F DB and leaving air temperature of 60°F will have an operating coil temperature determined as in the following example.

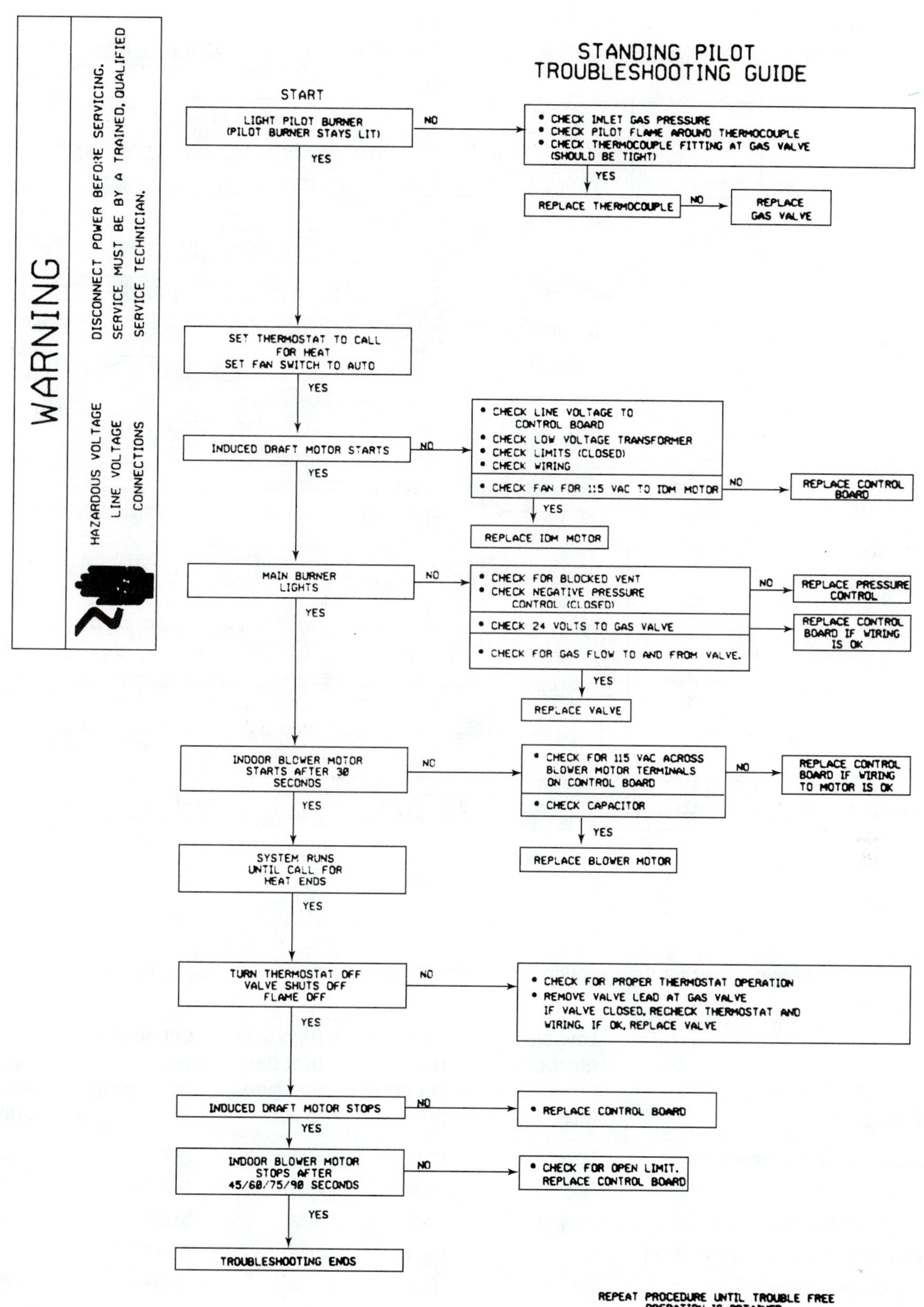

Figure 47-22 Typical fault-isolation diagram.

Figure 47-23 Fault codes.

Fault Codes

FAULT CODE	FAULT NAME IN PANEL	DESCRIPTION AND RECOMMENDED CORRECTIVE ACTION
1	OVERCURRENT	Output current is excessive. Check for excessive motor load, insufficient acceleration time (parameters 2202 ACCELER TIME 1, default 30 seconds), or faulty motor, motor cables or connections.
2	DC OVERVOLT	Intermediate circuit DC voltage is excessive. Check for static or transient over voltages in the input power supply, insufficient deceleration time (parameters 2203 DECELER TIME 1, default 30 seconds), or undersized brake chopper (if present).
3	DEV OVERTEMP	Drive heat sink is overheated. Temperature is at or above 115 C (239 F). Check for fan failure, obstructions in the airflow, dirt or dust coating on the heat sink, excessive ambient temperature, or excessive motor load.
4	SHORT CIRC	Fault current. Check for short-circuit in the motor cable(s) or motor or supply disturbances.
5	OVERLOAD	Inverter overload condition. The drive output current exceeds the ratings.
6	DC UNDERVOLT	Intermediate circuit DC voltage is not sufficient. Check for missing phase in the input power supply, blown fuse, or under voltage on main circuit.
7	AI1 LOSS	Analog input 1 loss. Analog input value is less than AI1 FLT LIMIT (3021). Check source and connection for analog input and parameter settings for AI1 FLT LIMIT (3021) and 3001 AI<MIN FUNCTION.
8	AI2 LOSS	Analog input 2 loss. Analog input value is less than AI2 FLT LIMIT (3022). Check source and connection for analog input and parameter settings for AI2 FLT LIMIT (3022) and 3001 AI<MIN FUNCTION.
9	MOT OVERTEMP	Motor is too hot, as estimated by the drive. Check for overloaded motor. Adjust the parameters used for the estimate (3005 through 3009). Check the temperature sensors and Group 35 parameters.
10	PANEL LOSS	Panel communication is lost and either drive is in local control mode (the control panel displays LOC), or drive is in remote control mode (REM) and is parameterized to accept start/stop, direction or reference from the control panel. To correct, check the communication lines and connections. Check parameter 3002 PANEL COMM ERROR, parameters in Group 10: Command Inputs and Group 11: Reference Select (if drive operation is REM).
11	ID RUN FAIL	The motor ID run was not completed successfully. Check motor connections.
12	MOTOR STALL	Motor or process stall. Motor is operating in the stall region. Check for excessive load or insufficient motor power. Check parameters 3010 through 3012.
13	RESERVED	Not used.
14	EXT FAULT 1	Digital input defined to report first external fault is active, See parameter 3003 EXTERNAL FAULT 1.
15	EXT FAULT 2	Digital input defined to report second external faults is active. See parameter 3004 EXTERNAL FAULT 1.
16	EARTH FAULT	The load on the input power system is out of balance. Check for faults in the motor or motor cable. Verify that motor cable does not exceed maximum specified length.
17	UNDERLOAD	Motor load is lower than expected. Check for disconnected load. Check parameters 3013 UNDERLOAD FUNCTION through 3015 UNDERLOAD CURVE.
18	THERM FAIL	Internal fault. The thermistor measuring the internal temperature of the drive is open or shorted. Contact your Carrier representative.
19	OPEX LINK	Internal fault. A communication-related problem has been detected between the OMIO and OINT boards. Contact your Carrier representative.
20	OPEX PWR	Internal fault. Low voltage condition detected on the OINT board. Contact your Carrier representative.
21	CURR MEAS	Internal fault. Current measurement is out of range. Contact your Carrier representative.
22	SUPPLY PHASE	Ripple voltage in the DC link is too high. Check for missing main phase or blown fuse.
23	RESERVED	Not used.
24	OVERSPEED	Motor speed is greater than 120% of the larger (in magnitude) of 2001 MINIMUM SPEED or 2002 MAXIMUM SPEED parameters. Check parameter settings for 2001 and 2002. Check adequacy of motor braking torque. Check applicability of torque control. Check brake chopper and resistor.
25	RESERVED	Not used.
26	DRIVE ID	Internal fault. Configuration block drive ID is not valid.
27	CONFIG FILE	Internal configuration file has an error. Contact your Carrier representative.

Table 47-4 Troubleshooting Chart for Refrigeration and Air-Conditioning Systems, Showing Symptoms and Probable Causes

Probable Cause	Low-Side (Suction) Pressure (psig)	D.X. Coil Superheat (°F)	High-Side (Hot Gas) Pressure (psig)	Condenser Liquid Subcooling (°F)	Cond. Unit Amperage Draw (A)
1. Insufficient or unbalanced load	Low	Low	Low	Normal	Low
2. Excessive load	High	High	High	Normal	High
3. Low ambient (cond. entering air °F)	Low	High	Low	Normal	Low
4. High ambient (cond. entering air °F)	High	High	High	Normal	High
5. Refrigerant undercharge	Low	High	Low	Low	Low
6. Refrigerant overcharge	High	Low	High	High	High
7. Liquid-line restriction	Low	High	Low	High	Low
8. Plugged capillary tube	Low	High	Low	High	Low
9. Suction-line restriction	Low	High	Low	Normal	Low
10. Hot-gas line restriction	High	High	High	Normal	High
11. Inefficient compressor	High	High	Low	Low	Low

EXAMPLE 47-1: SOLVING FOR DX COIL OPERATING TEMPERATURE

For an EER rating of 7.0 to 8.0, the DX coil split is typically 25–30°F.

$$\text{COT} = \left(\frac{80°\text{F} + 60°\text{F}}{2}\right) - (25 \text{ to } 30°\text{F}) = 40 \text{ to } 45°\text{F}$$

For an EER rating of 8.0 to 9.00, the DX coil split is typically 20–25°F.

$$\text{COT} = \left(\frac{80°\text{F} + 60°\text{F}}{2}\right) - (20 \text{ to } 25°\text{F}) = 45 \text{ to } 50°\text{F}$$

For an EER rating of 9.0+, the DX coil split is typically 15–20°F.

$$\text{COT} = \left(\frac{80°\text{F} + 60°\text{F}}{2}\right) - (15 \text{ to } 20°\text{F}) = 50 \text{ to } 55°\text{F}$$

This demonstrates that the operating coil temperature changes with the EER rating of the unit.

47.7 CONDENSING UNIT CONDENSING TEMPERATURE

The amount of surface in the condenser affects the condensing temperature the unit must develop to operate at rated capacity. The variation in the size of the condenser also affects the production cost and price of the unit. A smaller condenser will have a lower price, but it will also have a lower EER rating. In the same EER ratings used for the DX coil, at 95°F outside ambient, the 7.0 to 8.0 EER category will operate in the 25 to 30°F condenser split range, the 8.0 to 9.0 EER category in the 20 to 25°F condenser split range, and the 9.0+ EER category in the 15 to 20°F condenser split range.

The condenser split is the design temperature difference between the refrigerant condensing temperature (RCT) and the entering air temperature of the condenser (EAT).

$$\text{Condenser split} = \text{design RCT} - \text{design EAT}$$

To calculate the actual operating refrigerant condensing temperature in the condenser coil, the entering air temperature and the condenser split will be required.

$$\text{Operating RCT} = \text{EAT} + \text{Condenser split}$$

EXAMPLE 47-2: SOLVING FOR OPERATING RCT

Using the formula with 95°F EAT, the operating RCT for the various EER systems would be as follows:
For an EER rating of 7.0 to 8.0, the condenser split is typically 25 to 30°F

$$\text{Operating RCT} = 95°\text{F} + (25 \text{ to } 30°\text{F}) = 120 \text{ to } 125°\text{F}$$

For an EER rating of 8.0 to 9.0, the condenser split is typically 20 to 25°F

$$\text{Operating RCT} = 95°\text{F} + (20 \text{ to } 25°\text{F}) = 115 \text{ to } 120°\text{F}$$

For an EER rating of 9.0+, the condenser split is typically 15 to 20°F:

$$\text{Opcating RCT} = 95°\text{F} + (15 \text{ to } 20°\text{F}) = 110 \text{ to } 115°\text{F}$$

This demonstrates that the operating RCT varies not only from changes in outdoor temperatures but with the different EER ratings.

TECH TIP

As an air-conditioning system picks up heat, the temperature and pressure of the system increases on the suction side. If the air-conditioning system cannot pick up heat in the evaporator, the suction pressure and temperature will go down. If the condenser is unable to reject the heat, the condenser temperature and pressure will go up. If the condenser is not receiving heat, its temperature and pressure will go down. Heat and temperature and pressure are all interrelated. If a system has lower than normal pressure and temperature, then heat is not being picked up. This may be caused by low airflow or low refrigerant charge. If a system has higher temperatures and pressures than normal, it is either picking up more heat than can be rejected by the condenser or the condenser is unable to reject the heat normally. The bottom line is that if you follow the temperature and pressure, you will follow the heat, which will enable you to diagnose refrigerant circuit problems.

47.8 REFRIGERANT SUBCOOLING

The amount of subcooling produced in the condenser is affected by the quantity of refrigerant in the system. The temperature of the air entering the condenser and the load on the DX coil also has an effect on the amount of subcooling produced. Typically, it is desirable to have liquid subcooling of approximately 15–20°F.

If a refrigerant liquid line rises over 30 ft vertically, additional subcooling may be required to ensure that the metering device is receiving 100 percent liquid. The additional subcooling is required because of the pressure difference between the refrigerant at the condenser and pressure of the refrigerant at the metering device. This pressure drop is the result of the static head caused by the vertical lift of the refrigerant. Refer to the manufacturer's technical specifications to see what additional subcooling is required for unusually high vertical lifts.

47.9 INSUFFICIENT OR UNBALANCED LOAD

Insufficient air over the DX coil would be indicated by a greater than desired temperature drop through the coil. An unbalanced load on the DX coil would also give the opposite indication—some of the circuits of the DX coil would be overloaded, while others would be lightly loaded. This would result in a mixture of air off the coil that would cause a reduced temperature drop of the air mixture. The lightly loaded sections of the DX coil would allow liquid refrigerant to leave the coil and enter the suction manifold and suction line.

TECH TIP

The final step of installation that takes place at the residence is outside the control of the manufacturer. Many manufacturers have found that significant problems can exist in system installation, including the mismatching of equipment sizes. If you suspect that your refrigerant circuit problems are the result of mismatched equipment, call the equipment supplier or manufacturer and provide them the system component model numbers so they can inform you as to whether the system is mismatched in size and make recommendations to resolve the problems if a mismatched system exists.

In TEV systems, the liquid refrigerant passing the sensing bulb of the TEV would cause the valve to close down. This would reduce the operating temperature and capacity of the DX coil and lower the suction pressure. This reduction would be very pronounced. The DX coil operating superheat would be very low, probably zero, because of the liquid leaving some of the sections of the DX coil.

Discharge pressure (high side) would be low due to the reduced load on the compressor, reduced amount of refrigerant vapor pumped, and reduced heat load on the condenser. Condenser liquid subcooling would be higher than normal because of the reduction in refrigerant demand by the TEV. The condensing unit amperage draw would be down due to the reduced load.

In systems using fixed metering devices, the unbalanced load would produce a lower temperature drop of the air through the DX coil because the amount of refrigerant supplied by the fixed metering device would not be reduced; therefore, the system pressure (boiling point) would be approximately the same.

The DX coil superheat would drop to zero with liquid refrigerant flooding into the suction line. Under extreme cases of imbalance, liquid returning to the compressor could cause compressor damage. The reduction in heat gathered in the DX coil and the decrease of refrigerant vapor to the compressor will lower the load on the compressor. The compressor discharge pressure will be reduced.

The flow rate of the refrigerant will only be slightly reduced because of the lower discharge pressure. The subcooling of the refrigerant will be in the normal range. The amperage draw of the condensing unit will be slightly lower because of the reduced load on the compressor and reduction in head pressure.

47.10 EXCESSIVE LOAD

In the case of excessive load, the opposite effect exists. The temperature drop of the air through the coil will be less, because the unit cannot cool the air as much as it should. Air is moving through the coil at too high a velocity. There is also the possibility that the temperature of the air entering the coil is higher than the return air from the conditioned area. This could be from air leaks in the return duct system drawing hot air from unconditioned areas.

The excessive load raises the suction pressure. The refrigerant is evaporating at a rate faster than the pumping rate of the compressor. If the system uses a TEV, the superheat will be normal to slightly high. The valve will operate at a higher flow rate to attempt to maintain superheat settings. If the system uses fixed metering devices, the superheat will be high. The fixed metering devices cannot feed enough refrigerant to keep the DX coil fully active.

The discharge pressure will be high. The compressor will pump more vapor because of the increase in suction pressure. The condenser must handle more heat and will develop a higher condensing temperature. A higher condensing temperature means a greater discharge pressure. The quantity of liquid in the system has not changed, nor is the refrigerant flow restricted. The liquid subcooling will be in the normal range. The amperage draw of the unit will be high because of the additional load on the compressor.

47.11 LOW AMBIENT TEMPERATURE

In this case, the condenser heat transfer rate is excessive, producing an excessively low discharge pressure. As a result, the suction pressure will be low because the amount of refrigerant through the metering device will be reduced. This reduction will reduce the amount of liquid refrigerant supplied to the DX coil. The coil will produce less vapor and the suction pressure drops.

SERVICE TIP

Often air-conditioning and refrigerant equipment dies at night. This is frequently because of the lower ambient temperatures experienced in the evening. Low ambient operation of refrigeration and air-conditioning compressors can result in a liquid floodback or slugging of the compressor. If a system is to be operated on a regular basis during low ambient conditions, it should be equipped with a low ambient kit to protect the compressor.

The decrease in the refrigerant flow rate into the coil reduces the amount of active coil, and a higher superheat results. In addition, the reduced system capacity will decrease the amount of heat removed from the air. There will be a higher temperature and relative humidity in the conditioned area, and the discharge pressure will be low. This starts a reduction in system capacity. The amount of subcooling of the liquid will be in the normal range. The quantity of liquid in the condenser will be higher, but the heat transfer rate of the evaporator is less. The amperage draw of the condensing unit will be less because the compressor is doing less work.

The amount of drop in the condenser ambient air temperature that the air-conditioning system will tolerate depends on the type of pressure-reducing device in the system. Systems using fixed metering devices will have a gradual reduction in capacity as the outside ambient drops from 95°F. This gradual reduction occurs down to 65°F. Systems that use a TEV will maintain higher capacity down to an ambient temperature of 47°F. Below these temperatures the capacity loss is drastic, and some means of maintaining discharge pressure must be employed to prevent the evaporator temperature from dropping below freezing. The most reliable means is control of air through the condenser via dampers in the airstream, the condenser fan cycling on and off, a variable-speed condenser fan, or some combination of these components.

47.12 HIGH AMBIENT TEMPERATURE

When the outside air temperature rises on a hot day, the temperature of the air entering the condenser will be higher. This also increases the condensing temperature and pressure of the refrigerant vapor. The suction pressure will also be high because the pumping efficiency of the compressor is reduced. There will also be less liquid-line subcooling, which will increase the amount of flash gas across the metering device, further reducing the system efficiency. Due to the high ambient temperature, the discharge pressure will be high. There will be less liquid refrigerant in the condenser and reduced liquid subcooling. The system will run less efficiently and therefore will require more power, so the amperage draw of the condensing unit will be high.

The amount of superheat produced in the coil will be different in a TEV system as compared to a system using a fixed metering device. In the TEV system, the valve will maintain superheat close to the limits of its adjustment range even though the actual temperatures involved will be higher.

In a fixed metering device system, the amount of superheat produced in the coil is the reverse of the temperature of the air through the condenser. The flow rate through the fixed metering device is directly affected by discharge pressure. The higher air temperature will result in a higher discharge pressure and a higher flow rate. As a result of the higher flow rate, the amount of subcooling in the condenser is lower.

Table 47-5 shows the superheat that will be developed in a properly charged air-conditioning system using a fixed metering device. Do not attempt to charge a fixed metering device system below 65°F, as system operating characteristics become very erratic.

Table 47-5 The Effects of Outdoor (Ambient) Temperature on Superheat

Outdoor Air Temperature Entering Condenser Coil (°F)	Superheat (°F)
65	30
75	25
80	20
85	18
90	15
95	10
105 and above	05

47.13 REFRIGERANT UNDERCHARGE

With a shortage of refrigerant in the system, less liquid refrigerant enters the evaporator coil to pick up heat, creating a lower suction pressure. The smaller quantity of liquid supplied to the coil results in less active surface for the evaporator coil for vaporizing the liquid refrigerant and more surface to raise vapor temperature. The superheat will be high. There will be less vapor for the compressor to handle and less heat for the condenser to reject, leading to a lower condensing temperature and discharge pressure.

SERVICE TIP

The compressor in an air-conditioning system is cooled primarily by the cool returning suction gas. Compressors that are low on charge can have a much higher operating temperature. The temperature can be high enough so that the motor windings begin to break down. As this occurs, the motor can ultimately short out, resulting in a compressor change-out. If an air-conditioning system has a leak, it must be located so that a low refrigerant charge can be avoided.

The amount of subcooling will be below normal to zero, depending on the amount of undercharge. The system operation is usually not affected very seriously until the subcooling is zero and hot gas starts to leave the condenser together with the liquid refrigerant. The amperage draw of the condensing unit will be slightly less than normal.

47.14 REFRIGERANT-OVERCHARGE TEV SYSTEMS

In systems using a TEV, the valve will attempt to control the refrigerant flow into the coil to maintain the superheat setting of the valve. However, the extra refrigerant will back up into the condenser, occupying some of the heat-transfer area that would otherwise be available for condensing. As a result, the discharge pressure will be slightly higher than normal, the liquid subcooling will be high, and the unit amperage draw will be high. The suction pressure and DX coil superheat will be normal. Excessive overcharging will cause even higher discharge pressure and hunting of the TEV.

SERVICE TIP

It is unacceptable to add a little extra refrigerant to a system just in case it may have a leak. If you suspect that a refrigeration system has a leak, do not overcharge the system; instead, find the leak and fix it.

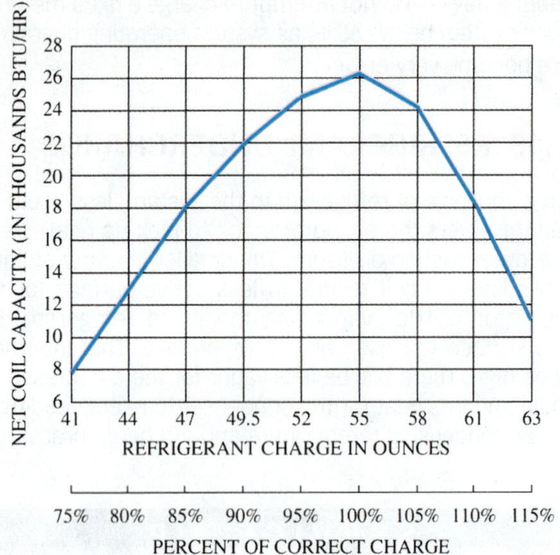

Figure 47-24 The effect of the refrigerant charge on the capacity of the unit.

47.15 REFRIGERANT OVERCHARGE IN FIXED METERING DEVICE SYSTEMS

The amount of refrigerant in the fixed metering system has a direct effect on system performance. An overcharge has a greater effect than an undercharge, but both affect system performance, EER, and operating cost.

As shown in Figure 47-24, at 100 percent of correct charge (55 oz), the unit developed a net capacity of 26,200 BTU/hr. When the amount of charge was varied 5 percent in either direction, the capacity dropped as the charge varied.

Figure 47-25 is a chart showing the amount of electrical energy the unit will demand because of pressure created by the amount of refrigerant in the system, with the only variable being the refrigerant charge. At 100 percent of charge (55 oz) the unit required 32 kW. As the charge was reduced, the wattage demand also dropped. When the unit was overcharged, the wattage required went up.

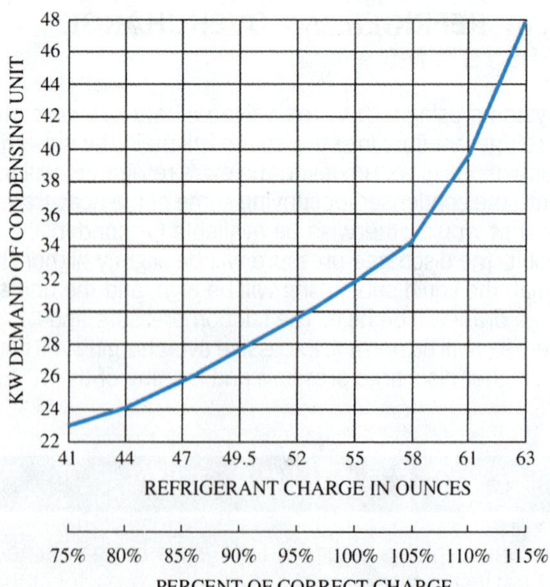

Figure 47-25 The effect of the refrigerant charge on the kW demand of the condensing unit.

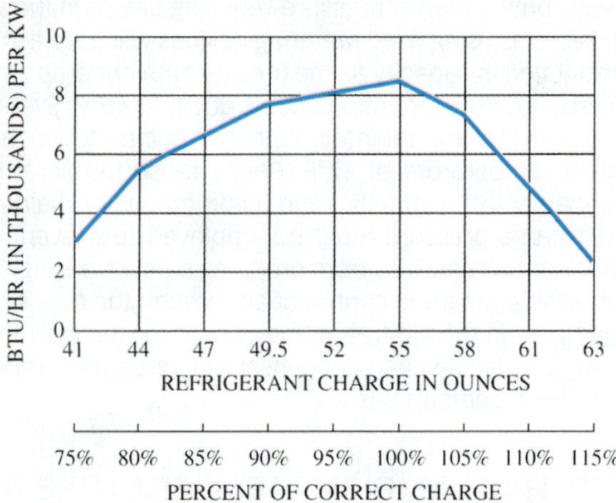

Figure 47-26 The effect of the refrigerant charge on the BTU/hr/kW ratio.

Figure 47-26 shows the EER of the unit based on the BTU/hr capacity of the system versus the wattage demand on the condensing unit. At the correct charge (55 oz), the EER was 8.49. As the refrigerant charge is either reduced or overcharged, the EER drops. From these charts, the only conclusion is that the capillary-tube systems must be charged to the correct charge with only a ±5% tolerance.

The effect of overcharge produces a high suction pressure because the refrigerant flow to the DX coil increases. Suction superheat will decrease because of the additional quantity to the DX coil. At approximately 8–10 percent of overcharge, the suction superheat becomes zero and liquid refrigerant will leave the DX coil. This will cause flooding of the compressor and greatly increases the chance of compressor failure. The discharge pressure will be high because of the extra refrigerant in the condenser. Liquid subcooling will also be high for the same reason. The amperage draw will increase due to the greater amount of vapor pumped as well as the higher compressor discharge pressure.

47.16 LIQUID-LINE RESTRICTION

A liquid-line restriction could be caused by a plugged liquid-line drier, a kink in the liquid line, or a solder joint filled with solder. A liquid-line restriction reduces the amount of refrigerant to the metering device. The suction pressure will be low because of the reduced amount of refrigerant to the DX coil. The suction superheat will be high because of the reduced active portion of the coil, allowing more coil surface for increasing the vapor temperature as well as reducing the refrigerant boiling point. The discharge pressure will be low because of the reduced load on the compressor. Liquid subcooling will be high. The liquid refrigerant will accumulate in the condenser. It cannot flow out at the proper rate because of the restriction. As a result, the liquid will cool more than desired. The amperage draw of the condensing unit will be low.

SERVICE TIP

If the symptoms indicate that there is a liquid-line restriction, it is sometimes easy to identify the liquid-line drier as the problem. As the refrigerant flow is reduced through the drier due to the restriction, the pressure of the refrigerant will drop. This reduction in pressure across the drier will produce some flash gas, similar to the situation that occurs across a capillary tube or TEV. The drier will be much colder on its outlet as compared to its inlet. A severely blocked drier may have frost accumulation on its outlet.

47.17 PLUGGED PRESSURE-REDUCING DEVICE

Either a plugged fixed metering device or plugged feeder tube between the TEV valve distributor and the coil will cause part of the coil to be inactive. The system will then be operating with an undersized coil.

Plugged Fixed-Bore or Capillary-Tube Metering Devices

The suction pressure will be low because the coil capacity has been reduced. The reduced amount of vapor produced in the coil and resultant reduction in suction pressure will reduce compressor capacity, discharge pressure, and the flow rate of the remaining active capillary tubes. The suction superheat will be high. The discharge pressure will be low. The liquid subcooling will be high because the liquid refrigerant will accumulate in the condenser. The unit amperage draw will be low. The possibility of moisture in the system that freezes and plugs the capillary orifice should be considered.

Plugged TEV Feeder Tube

A plugged feeder tube reduces the capacity of the coil. The coil cannot provide enough vapor to satisfy the pumping capacity of the compressor, and the suction pressure balances out at a low pressure. The remaining symptoms are similar to those of a plugged fixed metering device except that the superheat will be in the normal range. The TEV will adjust to the lower operating conditions and maintain the set superheat range.

47.18 SUCTION-LINE RESTRICTION

A suction-line restriction could be caused by a plugged suction-line strainer, a kink in the suction line, or a solder joint filled with solder. It results in a high-pressure drop between the DX coil and the compressor.

The suction pressure, as measured at the condensing unit end of the suction line, will be low. The superheat, as determined from the suction-line temperature at the DX coil and the suction pressure (boiling point) at the condensing unit, will be extremely high. The discharge pressure will be low because of the reduced load on the compressor. The amperage draw of the condensing unit will be low because of the light load on the compressor.

Be aware that the symptoms listed above usually indicate a refrigerant shortage. A major difference is that with a suction-line restriction, the subcooling will remain normal to slightly above normal. In an undercharged system, the amount of subcooling will be below normal to zero.

47.19 HOT-GAS-LINE RESTRICTION

For the condition of a gas-line restriction, the discharge pressure measured at the compressor outlet will be high and measured at the condenser outlet will be low. The suction pressure will be high due to the reduced pumping capacity of the compressor. The DX coil superheat is high because the suction pressure is high. Liquid subcooling is in the high end of the normal range. The compressor amperage draw will be above normal.

If the discharge pressure is only measured at the condenser outlet, the symptoms can be easily misinterpreted. High suction pressure and low discharge pressure will usually be interpreted as a low capacity compressor. However, the high amperage draw of the compressor indicates it is operating against a high discharge pressure. This will point toward a restriction between the outlet of the compressor and the pressure measuring point.

47.20 LOW-CAPACITY COMPRESSOR

The problem of a low-capacity compressor is last on the list because it is the least likely problem. Determining the age of the unit may be useful, as an older compressor is more likely to have a reduced efficiency. When the compressor will not pump the required amount of vapor, the suction pressure will balance out higher than normal. The DX coil superheat will be high. The discharge pressure measured at the compressor outlet will be low. Liquid subcooling will be low because not much heat will be in the condenser. The condensing will therefore be close to the entering air temperature. The amperage draw of the condensing unit will be extremely low, indicating that the compressor is doing very little work.

UNIT 47—SUMMARY

The referral to a troubleshooting chart will assist in locating the cause of the problem. As soon as the cause is verified, in most cases the remedy is self-evident. For example, if a wiring connection is loose, it needs to be tightened. When a motor is burned out, it needs to be replaced. Where the replacement or service of specialized parts is required, usually the manufacturer provides detailed instructions for performing the work. When maintenance is required, such as cleaning a dirty coil or replacing worn-out belts, these should be done following recommended standard procedures.

WORK ORDERS

Service Ticket 4701

Customer Complaint: Room Temperature Too High

Equipment Type: Residential Split Air-Conditioning System

A technician is called out to service a split air-conditioning system with an air handler and coil located in the attic. The technician arrives on the job, speaks to the customer, and discovers that the room temperature has been slowly rising during the past 3 to 4 weeks. The room temperature is higher than the thermostat setpoint, the air handler blower is running normally, and the outside condensing unit and compressor both appear to be running continuously.

The technician then measures the high- and low-side refrigerant system pressures, and they are both low. The compressor amperage draw is also low. The technician suspects that there may be a refrigerant leak, which has led to a shortage of refrigerant. Upon once again speaking with the customer, the technician discovers that some landscaping had been done about 3 weeks ago. Investigating the outside lines that lead into the home, the technician finds that the liquid line had been disturbed, and, using a leak detector, a leak is located in the area of a brazed fitting. The refrigerant will need to be recovered, the leak repaired, and the system evacuated and recharged.

Service Ticket 4702

Customer Complaint: Low Register Airflow

Equipment Type: Residential Split Air-Conditioning System

A technician is called out to service a split air-conditioning system with an air handler and coil located in the attic. The technician arrives on the job, speaks to the customer, and learns that the air flow through the registers seems very low. Even without taking a measurement, it is easy to feel along the register and determine that not much air is being delivered. The customer also tells the technician that the system has just been started for the summer season about a week ago and that it has not worked well from that time.

The technician decides to first check the air-handling unit in the attic, suspecting a clogged filter to be the problem. Upon investigating the unit, the technician discovers that squirrels had found an opening in the duct and built a nest in the path of the airflow during the winter months. The technician recommends that the customer contact an exterminator to remove the pests. Once this is done, the technician can come back and replace the filters and repair the duct.

Service Ticket 4703

Customer Complaint: Water Leaking from Ceiling

Equipment Type: Residential Split Air-Conditioning System

A technician is called out to service a split air-conditioning system with an air handler and coil located in the attic. The technician arrives on the job and speaks to the customer, who says that there is water leaking from the attic and it has stained his ceiling. The customer believes that the water is coming from the air-conditioning unit.

The technician is not sure whether this may be a humidity-related problem and first decides to check the air-handling unit in the attic. The technician quickly discovers that the coil drain pan is overflowing and has caused the stain. A section of the drain piping is blocked, and the technician cuts it out and replaces it. Once the blocked section is replaced, the water drains normally without backing up. The technician then recommends that the customer approve the installation of a float switch in the evaporator coil drain pan that will shut the unit down in the event of a similar occurrence in the future. This would reduce the chance of water ruining the ceiling again after the stain has been removed.

UNIT 47—REVIEW QUESTIONS

1. How is the temperature drop through a DX coil on an air-conditioning system determined?
2. What is the most common problem of air failure on an air-conditioning system?
3. If a split air-conditioning system does not operate at all, the problem is most likely what?
4. If an electrical load on an air-conditioning system is not working, what must the technician try to determine?
5. What does an electrical connection diagram show?
6. List the measurements that need to be taken to determine the extent of a mechanical problem in an air-conditioning system.
7. How are normal coil operating temperatures for an air-conditioning system determined?
8. How does condenser size affect the EER?
9. What has the predominant effect on the amount of subcooling produced in the condenser of an air-conditioning system?
10. What should be done if you suspect problems due to mismatched equipment sizes on a split air-conditioning system?

11. What type of temperature drop would you expect across the DX coil on an air-conditioning system with insufficient airflow?
12. What will happen to the suction pressure on an air-conditioning system that is running with an excessive load?
13. Why do split air-conditioning systems often fail during the night?
14. A split air-conditioning system that uses a fixed metering device should not be charged if the ambient temperature falls below what level?
15. Capillary-tube systems must be charged to within what tolerance?
16. What happens to the refrigerant superheat if an air-conditioning system using a capillary tube is overcharged?
17. What happens to the amperage draw of a condensing unit on an air-conditioning system if the liquid line is restricted?
18. How can one determine between a refrigerant shortage as compared to a suction-line restriction on a split air-conditioning system?
19. How can you determine between an inefficient compressor as compared to a hot-gas line restriction on a split air-conditioning system?
20. Is it acceptable to add a little extra refrigerant to a system just in case it may have a leak?

UNIT 48

Principles of Combustion and Safety

OBJECTIVES

After completing this unit, you will be able to:

1. list the most common types of fuel gases and compare their properties.
2. list the three essential components of complete combustion.
3. describe the two types of flame and list their characteristics.
4. list the products of complete combustion.
5. explain how CO is produced and how to keep CO out of combustion gases.
6. explain the combustion air requirements for gas furnaces.
7. define AFUE.
8. list the certifying agencies for gas-fired equipment.

48.1 INTRODUCTION

Nothing heats like fire. A large percentage of all home heating is provided by gas-fired heating systems. Gas flames provide large quantities of clean heat. A thorough understanding of the principles of combustion is necessary for technicians to safely work with gas and to keep gas-fired equipment operating safely and cleanly.

Safety should always be the primary concern for a technician when working with combustible gases. Safety starts with understanding the behavior of combustible gas and the principles of combustion. Technicians need to know the requirements for safe operation and recognize potential hazards. Customers depend on the technician to keep their gas-fired equipment operating safely and efficiently.

48.2 FUEL GASES

Fuel gases are used for many heating and air-conditioning processes and fall into three broad categories: natural, manufactured, and liquefied petroleum. Stories about the discovery and use of natural gas date back as far as 2000 BC. History notes that the Chinese piped gas from shallow wells through bamboo poles to boil seawater to obtain salt.

The great advantage of natural gas over other fuels is its relative simplicity of production, transportation, and use. When a sufficiently large quantity is discovered, it is pumped from drill wells to processing plants and on to refineries and industrial centers, where it has a large number of diverse uses in addition to heating (Figure 48-1).

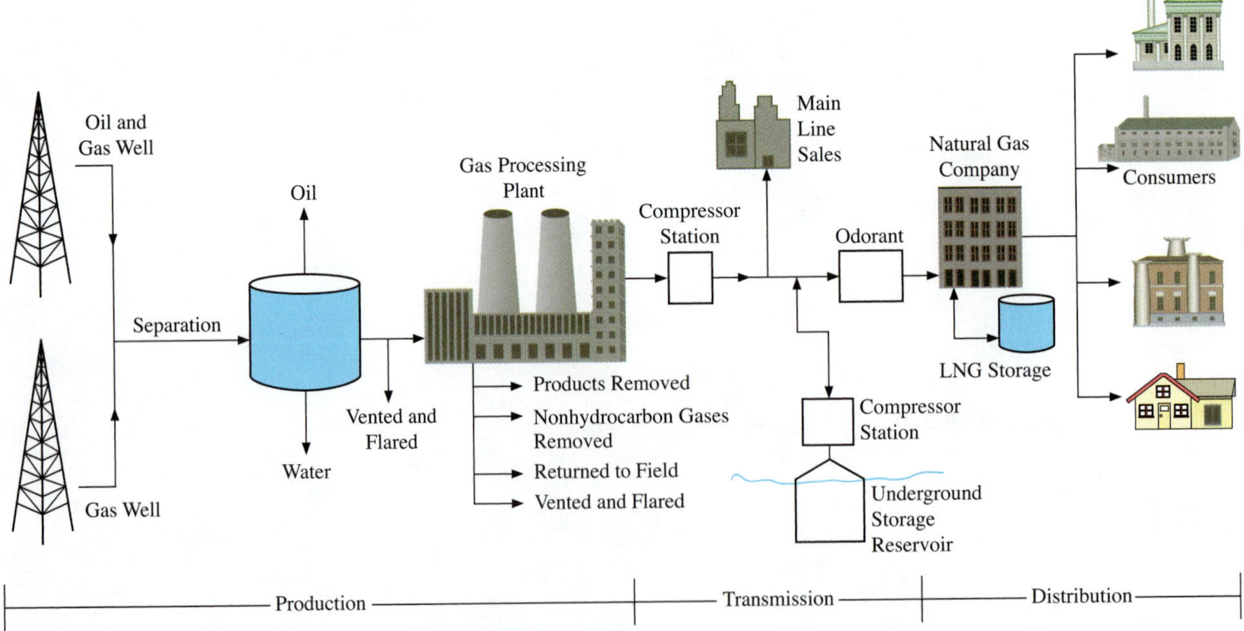

Figure 48-1 Natural gas distribution. *(Courtesy of the Energy Information Administration, Washington, DC)*

PETROLEUM AND NATURAL GAS FORMATION

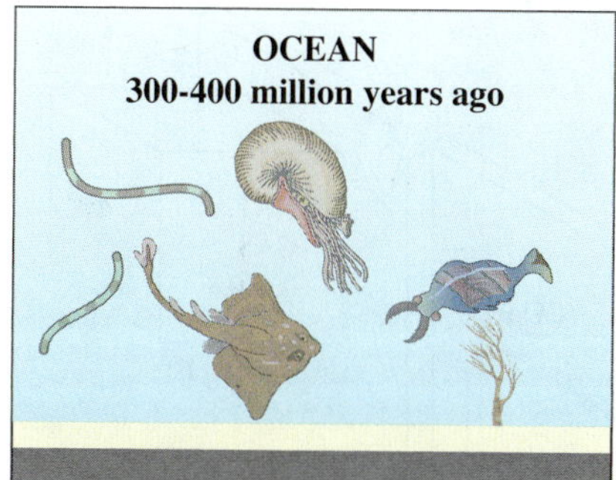

Tiny sea plants and animals died and were buried on the ocean floor. Over time, they were covered by layers of silt and sand.

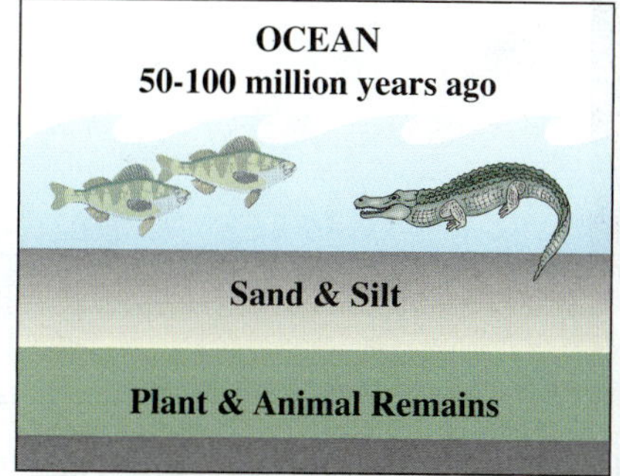

Over millions of years, the remains were buried deeper and deeper. The enormous heat and pressure turned them into oil and gas.

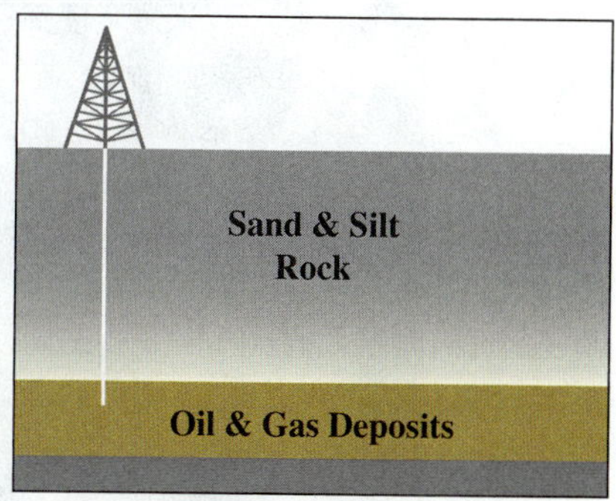

Today, we drill down through layers of sand, silt, and rock to reach the rock formations that contain oil and gas deposits.

Figure 48-2 Formation of natural gas. *(Courtesy of the Energy Information Administration, Washington, DC)*

Natural gas comes from the sedimentation of trillions of tiny organisms at the bottom of the sea that are buried and initially chemically converted into dense organic material. Over millions of years, pressure and heat gradually cracked this material into lighter hydrocarbon compounds. Liquids were liberated first, then the gases. Compaction squeezed these products from the source rock and they each migrated into the more porous reservoir rocks until stopped by impermeable barriers. Pools of gas and oil developed that are the targets of petroleum exploration (Figure 48-2).

Most fuel gases are hydrocarbons; they are made entirely of carbon and hydrogen (Figure 48-3).

The carbon atoms link to each other in a chain and are surrounded by hydrogen atoms. Table 48-1 shows properties of the most common hydrocarbons used in heating.

Natural gas comes out of the ground while liquid petroleum (LP) gases are the result of distillation. Natural gas is mostly methane, but it also contains other gases in small quantities, including ethane, propane, and butane (Figure 48-3).

LP gases are propane, butane, or a mix of the two. These fuel gases are obtained from natural gas or as a byproduct of refining oil. Propane is the LP gas mostly used for domestic heating. Butane has more agricultural and industrial applications. Figure 48-3 shows that propane and butane contain more carbon and hydrogen atoms than natural gases. LP gases are heavier and have more heating value per cubic foot than natural gas because of their increased number of carbon and hydrogen atoms. LP gas

```
     H               H   H
     |               |   |
 H - C - H       H - C - C - H
     |               |   |
     H               H   H
 METHANE CH4     ETHANE C2H6

   H   H   H         H   H   H   H
   |   |   |         |   |   |   |
H- C - C - C -H   H- C - C - C - C -H
   |   |   |         |   |   |   |
   H   H   H         H   H   H   H
 PROPANE C3H8       BUTANE C4H10
```

Figure 48-3 Chemical components of common fuels; natural gas is mostly methane and ethane, and LP is mostly propane and/or butane.

is stored and transported in liquid form. Unlike natural gas, LP is not transported through a network of pipelines. LP is shipped long distances on rail tankers and across shorter distances by tanker trucks. LP is delivered to the customer where it is pumped from the truck into the customer's storage tank (Figure 48-4). LP tends to be more expensive than natural gas because it must be refined and its delivery cost is higher.

Manufactured gas, as the name states, is manmade as a byproduct of other manufacturing operations. The use of manufactured fuel gases has declined greatly in the United States. Today over 99 percent of sales by gas-distribution and -transmission companies is natural gas. Manmade gas is still a popular fuel in Europe.

Table 48-1 Ultimate CO_2 Values, Flammable Limits of Fuels, and Heating Values of Fuels

Constituent	Chemical Formula	Ultimate CO_2 (%)	Lower Flammability Limit (%)	Upper Flammability Limit (%)	Ignition Temperature (°F)	Higher Heating Values (BTU/lb)	Lower Heating Values (BTU/lb)	Specific Volume (ft³/lb)
Methane	CH_4	11.73	5.0	15.0	1,301	23,875	21,495	23.6
Ethane	C_2H_6	13.18	3.0	12.5	968–1,166	22,323	20,418	12.5
Propane	C_3H_8	13.75	2.1	10.1	871	21,669	19,937	8.36
Butane	C_4H_{10}	14.05	1.86	8.41	761	21,321	19,678	6.32

(a)

(b)

Figure 48-4 (a) 100-gallon refillable propane tank for residential use; (b) tank level gauge for use with propane, butane, or 50/50 mix.

Mixed gas, as the name implies, is a manmade mixture of gases such as natural gas and manufactured gas.

It is important to know the density of a gas as expressed by its specific gravity. The specific gravity of a gas compares its density to the density of air. Standard air has a specific gravity of 1.0. The specific gravity of natural gas ranges from 0.4 to 0.8, making it lighter than air, as shown in Figure 48-5.

On the other hand, the specific gravity of propane is 1.5 and butane is 2.0. This means they are heavier than air (Figure 48-6). The specific gravity of the gases is important because it affects the flow of the gas through orifices (small holes) and then to the burner. Should a leak develop in a gas pipe, natural gas will rise and probably dissipate, while LP gases will drift to low spots and collect in pools, creating a greater hazard. Specific gravity also affects gas flow in supply pipes and the pressure needed to move the gas.

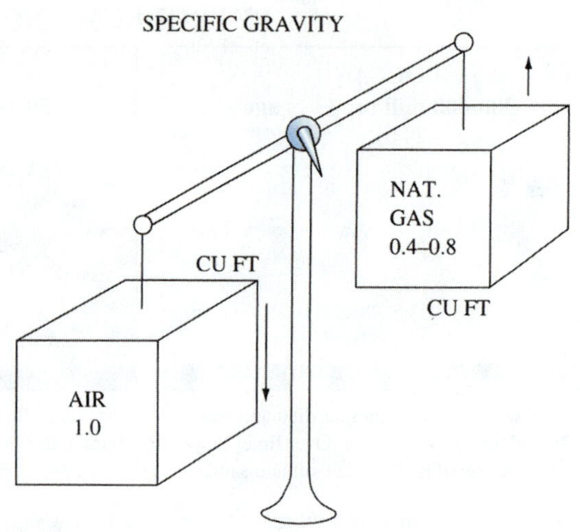

Figure 48-5 Weight of 1 ft^3 of air compared to 1 ft^3 of natural gas.

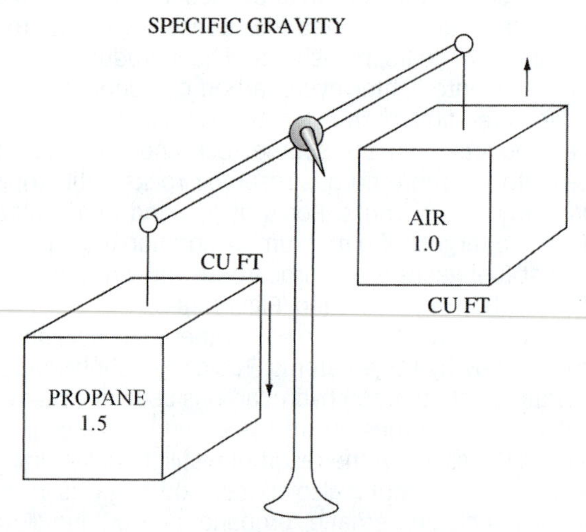

Figure 48-6 Weight of 1 ft^3 of air compared to 1 ft^3 of propane.

SAFETY TIP

Because LP gases are heavier than air, they collect in low areas. A leak in the gas line to an LP furnace in a crawlspace or basement will create a "puddle" of gas on the ground around the furnace. An unwary technician could be lying in this puddle of LP gas when it is ignited, surrounding the technician in flames.

The heat value of a gas is the amount of heat released when 1 ft^3 of the gas is completely burned (Figure 48-7).

Natural gas has a heating value range of 950 to 1,150 BTU/ft^3. The range of BTU per cubic foot for natural gas is because it is a mixture of gases, with methane making up the largest share. The heat value of 1,000 BTU/ft^3 is most often used for estimating performance for natural gas furnaces. Propane has a heat value of approximately 2,500 BTU/ft^3; butane's heat value is 3,200 BTU/ft^3. The exact heating value of gases in your local area can be obtained from the gas company or LP gas distributor.

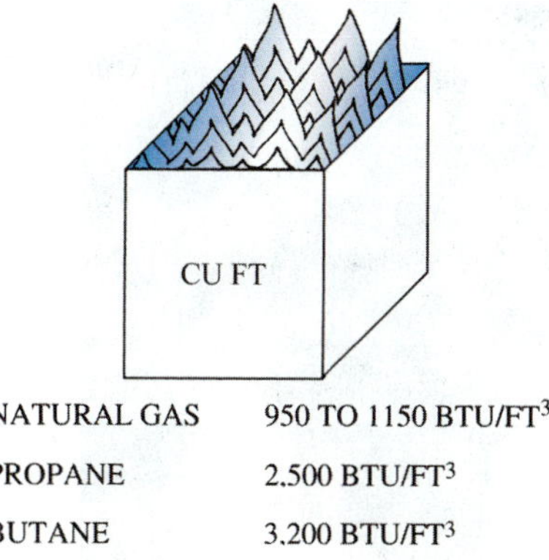

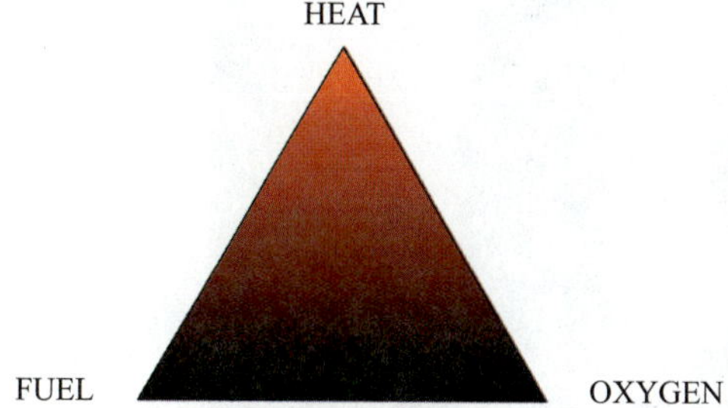

Figure 48-7 Heat delivered by gas fuels when burned.

Figure 48-8 The three things that must be present for combustion to occur are fuel, heat, and oxygen.

48.3 FUEL GAS COMBUSTION

Combustion is a term used to describe rapid oxidation. Oxidation is a chemical reaction between oxygen and another substance that produces a new compound and releases heat. Three ingredients are required for combustion: fuel, oxygen, and heat (Figure 48-8).

An example of combustion is the reaction between carbon and oxygen. They combine to form carbon dioxide. Great amounts of heat are generated within a short period of time because of the speed of the reaction.

TECH TIP

Large quantities of water vapor are produced as a product of combustion. The heating value of a fuel can be increased approximately 10 percent by condensing this water vapor from the flue gases. This is how manufacturers produce furnaces with efficiencies of 90–97 percent.

Most gaseous fuels consist of different mixtures of hydrocarbons. When oxygen and hydrocarbons are combined, water and carbon dioxide are produced. Both are extremely stable and safe compounds. The combustion of methane gas is illustrated in Figure 48-9. But if the reaction does not take place under the proper conditions (i.e., the correct temperature and gas-to-air ratio), then variable amounts of unstable and often toxic substances are produced due to incomplete combustion. Definite oxygen-to-hydrocarbon ratios are required for complete combustion; insufficient

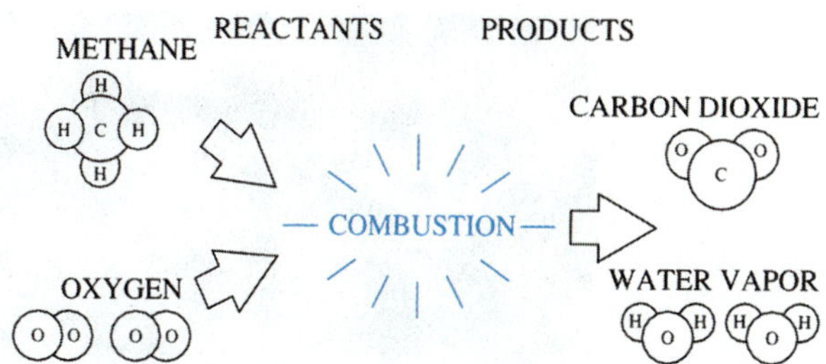

Figure 48-9 Combustion process when a fuel is burned.

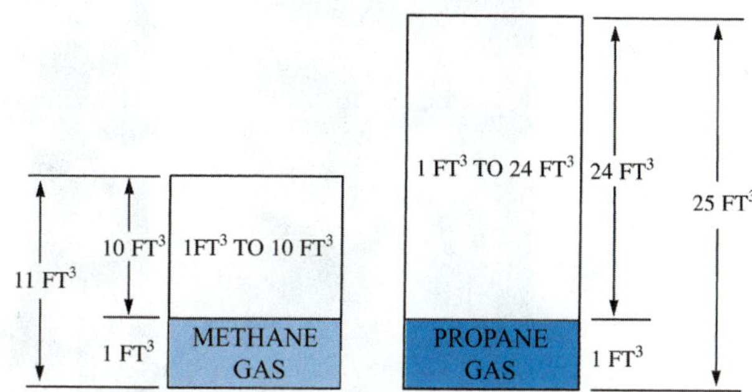

Figure 48-10 Amount of air required for combustion of methane compared to propane.

oxygen will produce incomplete combustion. For every 1 ft^3 of methane gas, 10 ft^3 of air is needed for complete combustion (Figure 48-10). Although natural gas requires a 10:1 ratio of air, LP fuels require much more, due to the concentration (greater density) of carbon and hydrogen atoms. Propane combustion must have more than 24 ft^3 of air per 1 ft^3 of gas to support proper combustion. When complete information about the fuel is not available, a frequently used value for estimating air requirements is that 0.9 ft^3 is required for 100 BTU of fuel.

In modern atmospheric burners, the combustion air enters the reaction at two separate points (Figure 48-11). These are called atmospheric burners because the air for the burning process is at atmospheric pressure. *Primary air* refers to air that is mixed with the fuel before the fuel is ignited. *Secondary air* refers to air that is drawn into the combustion process after the fuel has ignited. Air is introduced into the combustion process at two levels because of the difficulty in maintaining an exact mixture of oxygen and fuel before combustion. Primary air and secondary air

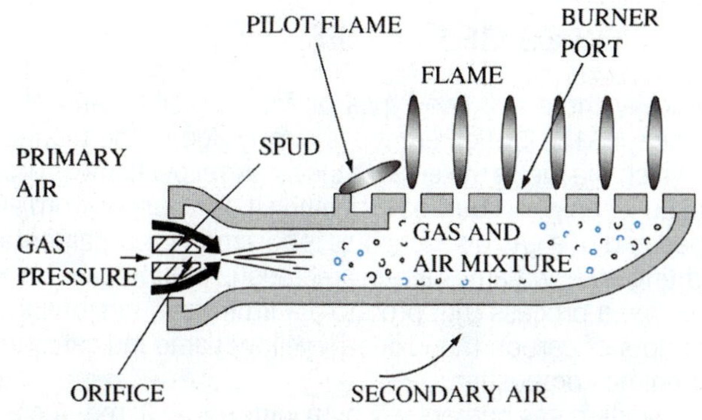

Figure 48-11 Cross-section of a gas burner.

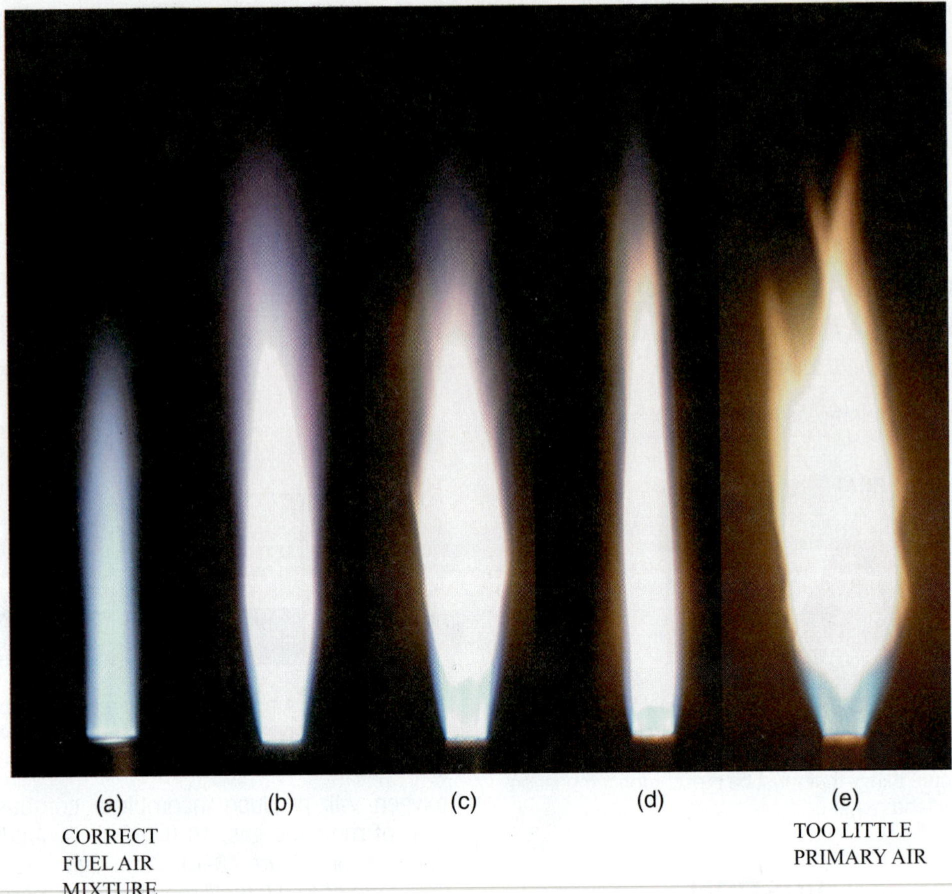

Figure 48-12 Natural gas flames: (a) correct mixture of fuel and air; irregular, (b) less primary air, cone reduced; (c) less primary air, flame wider; (d) less primary air, flame more irregular; (e) too little primary air, irregular, wide flame.

are generally referred to together as combustion air since they usually do come from the same place.

Heat is needed as a catalyst to initiate and maintain the combustion process. Without sufficient heat, the combustion process will not start; or, if started, it will result in incomplete combustion. Complete combustion occurs when the correct ratio of fuel and oxygen combine at a sufficiently high temperature to use up all the fuel, producing only water and carbon dioxide.

Incomplete combustion produces carbon monoxide (CO), an odorless, tasteless, and poisonous gas. Any condition that results in incomplete combustion should be corrected immediately.

48.4 TYPES OF FLAMES

Basically there are two types of flames: yellow and blue (Figure 48-12). The difference is mainly due to the manner in which the fuel is mixed with the air. A yellow flame is produced when gas is burned by igniting it as it gushes from an open end of a gas pipe, such as is common for ornamental lighting. Yellow flames are characteristic of luminous combustion; a process that produces a little heat, lots of light, and lots of carbon monoxide. A yellow flame indicates incomplete combustion.

Modern gas burners will burn with a blue flame. A blue flame is produced by a Bunsen burner such as those used in a laboratory, where 50 percent of the air requirement is mixed with the gas prior to ignition. Blue flames are characteristic of nonluminous combustion, a process that produces a little light, lots of heat, and no carbon monoxide. Complete combustion is indicated by an inner blue cone (Figure 48-13).

The manifold and gas spuds (Figure 48-14) are used on nearly 80 percent of gas furnaces.

A good example of the use of primary and secondary air is shown in Figure 48-11. This figure shows the primary air adjustment on a drilled burner port. The gas is supplied through a manifold and metered into the burner by a properly selected gas orifice. The air enters through adjustable openings around the gas orifice. The velocity of the gas jet creates a low-pressure area around it, which draws in

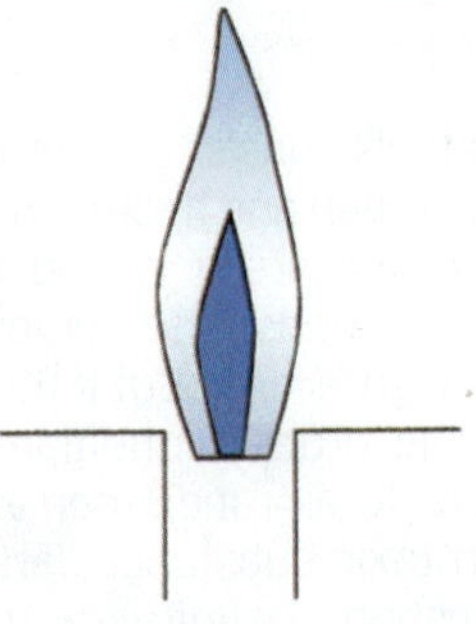

Figure 48-13 Inner blue cone indicates complete combustion.

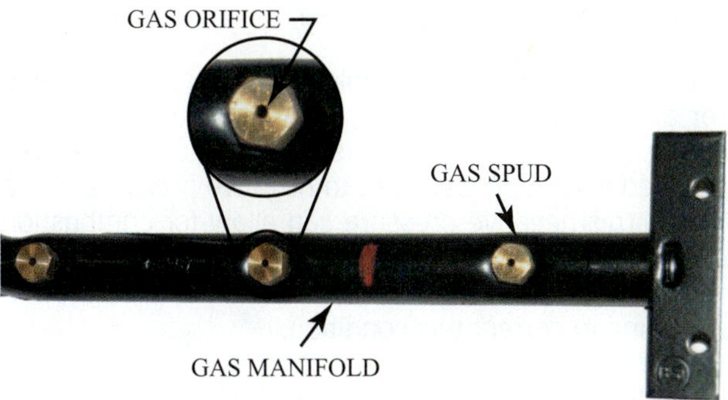

Figure 48-14 Gas burner manifold.

primary air. The gas and primary air pass through a toilet-bowl-shaped venturi. The venturi causes the gas and air to swirl as they pass through it, mixing them together. The secondary air is drawn in by the draft created from the flames. When the air supplies are properly adjusted for complete combustion, the burner operates with a blue flame.

48.5 PRACTICAL COMBUSTION CONSIDERATIONS

Perfect combustion will result in no leftover fuel, oxygen, or unstable compounds. Since absolute perfection in the gas-to-air ratio is not possible, most appliances introduce excess combustion air to ensure that no fuel or unstable compounds are left. Excess oxygen is left, which reduces combustion temperatures and lowers combustion efficiency. This is the tradeoff for safer operation. An absolute minimum of 10 percent excess air is required for safe operation. Practice has shown that 40–50 percent excess air provides safe operation in natural draft systems. This means that 14–15 ft^3 of combustion air per 1,000 BTU are needed for complete combustion and safe operation of a natural gas–fired appliance. The effect of excess air on combustion efficiency is shown in Table 48-2.

The heat catalyst is still required before complete combustion can take place. Methane, the primary component in natural gas, will ignite at temperatures as low as 1,076°F and can reach temperatures as high as 3,500°F. In general, a temperature of 1,200°F is the absolute minimum temperature to be maintained to keep the process going and achieve complete combustion. Commercial gas is almost always diluted with other gases, and its minimum combustion temperature is usually between 1,200°F and 1,220°F. But most appliances actually operate between 1,600°F and 1,800°F to compensate for the temperature drop in the combustion chamber due to the introduction of excess air. The heat carried away by the excess air represents an efficiency loss, but it is necessary in an atmospheric burner to ensure complete combustion.

Table 48-2 Effect of Excess Air on Combustion Efficiency

Excess %		Combustion Efficiency: Flue Gas Temperature Less Combustion Air Temp, °F				
Air	Oxygen	200	300	400	500	600
9.5	2.0	85.4	83.1	80.8	78.4	76.0
15.0	3.0	85.2	82.8	80.4	77.9	75.4
28.1	5.0	84.7	82.1	79.5	76.7	74.0
44.9	7.0	84.1	81.2	78.2	75.2	72.1
81.6	10.0	82.8	79.3	75.6	71.9	68.2

(Courtesy of the U.S. Department of Energy)

The byproducts of combustion in furnaces are called flue gases and are vented to the outside. Insufficient combustion air can produce hazardous conditions. If too little oxygen is supplied, part of the byproducts will be dangerous carbon monoxide gas (CO) rather than harmless carbon dioxide gas (CO_2). The lack of combustion air can also keep a natural draft vent from operating, stopping the flue gases from rising up the flue. Vent gases then begin to spill into the room. These flue gases in a living area cause serious health problems. Adequate combustion air is an essential element of the system's design and central to indoor air quality.

Air quality can be greatly affected by the formation of nitrous oxides. Nitrogen represents about 80 percent of the earth's atmosphere. Therefore, in conjunction with the excess air necessary for complete combustion, it is clear that nitrogen constitutes the major portion of the exhaust gases. Nitrogen is practically inert; however, at the temperatures generated in the combustion chamber, nitrogen is no longer inactive, and oxygen and nitrogen will react to form oxides of nitrogen (NOx). The two factors that most significantly influence the formation of NOx are temperature and oxygen concentration. A rule of thumb says that a temperature change of 212°F may change the NOx amount by a factor of 3.

Oxides of nitrogen in the atmosphere contribute to smog formation, which is considered to be a problem in many urban areas. Also, the conversion of NO to NO_2 will continue in the atmosphere and, as NO_2 is soluble in water, it will be washed out of the air by rain, creating undesirable acidic conditions. Oxides of nitrogen are also of concern because they are believed to be carcinogenic. Due to these factors, some cities and states require furnaces that have lower nitrous oxide values. They are referred to as low-NOx furnaces.

TECH TIP

Furnaces require a significant amount of combustion air. During new construction, the quantity of air required is carefully checked and inspected; however, over time in older residences insulation or other debris can inadvertently fall over the makeup-air vents for a furnace. When this happens it can adversely affect the furnace performance and create a potential safety problem. This problem is particularly a concern when the furnace is located inside of a dwelling space in an equipment closet. If the closet door is properly sealed, the lack of proper makeup air will be negated when the technician has the door to the furnace closet open for service. Therefore, technicians must visually inspect makeup-air vents to make sure they are free and open.

48.6 GAS SAFETY

The storage and distribution of gas should always follow all federal, state, and local code requirements. Even a small gas leak has the potential for massive destruction. In 2011, a natural gas blast in the United States demolished two houses, set fire to several more, and sent flames hundreds of feet into the air. By the time the fire was extinguished, six more homes were lost and over five hundred people were evacuated from the area.

Always inform your customer of the risks associated with gas. Alert them to some basic safety procedures if they smell gas:

- Do not try to light any appliance.
- Do not touch any electrical switch.
- Do not use any telephones in the building.
- Leave the building immediately.
- Call the gas supplier from a neighbor's phone or cell phone once outside and far away from the building.
- If the gas supplier cannot be reached, call the fire department.

Carbon Monoxide Poisoning

Never take any unnecessary risks when working with gas. Always read the equipment manual and follow all the manufacturer's safety recommendations. Devices that use gas as a fuel must be vented properly to the outdoors. Never allow the flue gases to enter the living spaces. These contain carbon monoxide, which is invisible, odorless, and toxic. Carbon monoxide detectors installed in the living spaces will alarm and alert the occupants before the carbon monoxide levels are allowed to reach toxic levels.

SAFETY TIP

Furnace vents are not allowed to be shared by any other appliances. Improper venting will allow combustion gases to collect in the building, which may lead to nausea and even death by asphyxiation.

Air Supply

Fuel-burning appliances require sufficient quantities of air for proper combustion and ventilation of the flue gases. Gas furnaces require air for combustion, and this should be taken from outside the living space. Make sure that combustion supply air is not blocked and adequate space is allowed around the exterior of the furnace. Common signals indicating insufficient supply air include the following:

- Complaints of headache, nausea, and dizziness
- Excessive humidity levels indicated by heavily frosted windows or a moist, "clammy" feeling in the living spaces
- For homes that have fireplaces, the smoke does not draw up the chimney properly and puffs back

Negative Building Pressure

The use of exhaust fans, fireplaces, clothes dryers, and other appliances will use air or vent air to the outdoors. This, along with new weather stripping, storm windows, and added insulation, can lead to negative pressure in the building. This negative pressure can allow for combustion gases and flue gases to more readily escape from leaks in the furnace. More outside air will need to be brought into the building to correct this condition.

Propane

Propane gas is heavier than air; therefore, leaking gas will settle into a low area. Propane is an odorless gas, but a distinct pungent aroma is added to it by the refinery to help detect leaks. Some furnaces come equipped to burn either natural gas or propane. Each gas requires a different size burner orifice. When converting from one gas to the other or when setting up a new system, make sure that the proper orifice kit is installed. Orifices are stamped to provide the size (twist drill number). Oversized orifices could result in hazardous conditions, especially if the ventilation is inadequate. Most new furnaces will have a label to designate the type of gas it has come equipped to handle (Figure 48-15).

SAFETY TIP

At high elevations, typically above 2,000 feet, the furnace input rating must be adjusted and the size of the burner orifices recalculated based on elevation and gas heating value. The burner orifices may (or may not) need to be changed.

(a)

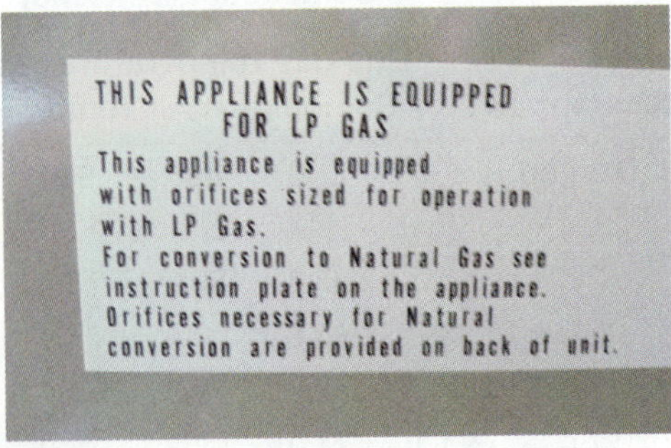

(b)

Figure 48-15 (a) Label for gas heater indicating propane gas use only; (b) label on gas heater providing orifice information.

Figure 48-16 Bank of gas meters and pressure regulator located outside an apartment complex.

Gas Pressure and Input Rate

The gas supply pressure should be checked with all the appliances attached to the gas supply line turned on. If possible, a separate gas supply line should be run directly from the meter to the furnace. The gas pressure required for natural gas is different than the pressure required for propane. Correct manifold gas pressure is required for proper ignition and burner operation. The input rate for natural gas furnaces can be measured by timing the gas meter (Figure 48-16). Propane gas installations do not have gas meters, and the input cannot normally be measured (Figure 48-17). Therefore, the proper gas pressure becomes even more critical (Figure 48-18).

Figure 48-17 Propane tanks have a pressure regulator located on the tank.

Figure 48-18 Gas valve on heater labeled with the correct input pressure for both natural gas and propane gas.

Gas Valves and Ignition

Make sure that flow is in the proper direction through the gas valve. There should be an arrow on the valve indicating the direction of gas flow. If flow is incorrect, the valve may not shut off, which can cause an uninterrupted gas flow and lead to a fire or explosion. Do not let fuel accumulate in the combustion chamber for longer than a few seconds without igniting. Fuel accumulation can lead to an explosion.

SAFETY TIP

Never attempt to manually light the burners with a match or other source of flame.

Checking for Gas Leaks

A gas leak check should be performed anytime work is done to the system. Isolate the furnace and gas control from the gas supply line during leak checks. This is because test pressures above ½ psi (14 in wc) could damage the gas control valve, causing it to leak and resulting in fire or explosion. Once pressure is applied to the piping, a commercial soap solution made to detect leaks can be applied around all piping connections. The formation of bubbles indicates gas leakage.

SAFETY TIP

NEVER CHECK FOR GAS LEAKS WITH AN OPEN FLAME!

Garage Installations

Some furnaces are installed in residential garages. If this is the case, then the burners and ignition source must be located no less than 18 in above the floor. This is to reduce the risk of igniting flammable vapors that may be present in a garage. Also, the furnace must be located so that it is protected from being damaged by a moving vehicle.

Water Damage

Do not operate a furnace if any part has been underwater. The ignition module can malfunction if it gets wet, leading to accumulation of explosive gas. Never install where water can flood, drip, or condense on the module.

48.7 CONDENSATION AND CORROSION

The heating technician must always be on the lookout for problems caused by condensation and corrosion produced by the flue gases. When the fuel-burning system cycles on and off to meet the demand, the flue passages cool down during the off cycle. When the system starts up again, condensate forms briefly on the surfaces until they are heated above the dew-point temperature. Low-temperature corrosion can occur in exhaust system components (heat exchangers, flues, vents, chimneys).

Figure 48-19 Rusty furnace vent.

The condensate includes such corrosive substances as sulfides, chlorides, and fluorides.

Corrosion increases as the condensate dwell time increases. One of the common signs of this is the corrosion that takes place in the flue pipe between the furnace and the chimney. Even though a galvanized flue pipe is used, this piping needs to be replaced periodically due to the corrosion effect (Figure 48-19). Currently, a large number of gas-fired furnaces are located in an unheated garage, and with the use of single wall pipe and cold ambient air in the garage, condensation occurs in the long run, where it goes through the roof in double-wall pipe. Evidence of this condensation in the form of rust can be seen at the vent pipe joints and elbow joints. This condensation usually runs to the bottom of the heat exchanger, resulting in corrosion.

Type B double-wall vent pipe helps prevent condensation in the vent by keeping the inner wall warm. The small space between the inner and outer wall acts like insulation. Oversizing of the vent also contributes to condensation. The flue gases move slowly through an oversized vent, giving them more opportunity to cool and condense.

Older, less-efficient natural draft furnaces have an opening in the draft hood called a draft diverter. If the wind conditions produce a downdraft, the air is diverted away, so as not to disturb the pilot and main burner flames. However, under normal operation, dilution air is sucked into the draft diverter opening and mixed with the flue gases. The addition of dilution air reduces the percentage of water vapor in the gases, which lowers the dew point of the flue gases.

Newer 80 percent efficient furnaces operate with much lower temperature flue gases than older, less-efficient furnaces. They also do not have an open draft diverter, so no dilution air is introduced into the vent gases. Instead they are equipped with draft inducers and reduced levels of excess air. Therefore, because of their design, the newer 80 percent efficient furnaces are particularly susceptible to vent condensation. Because of this, these furnaces are generally not approved for venting through masonry chimneys. The water condensation mixed with the flue gases will lead to the formation of sulfuric acid. The acid will eat away at the mortar over time, leading to masonry failure. In addition, the condensation absorbed by the masonry can freeze and

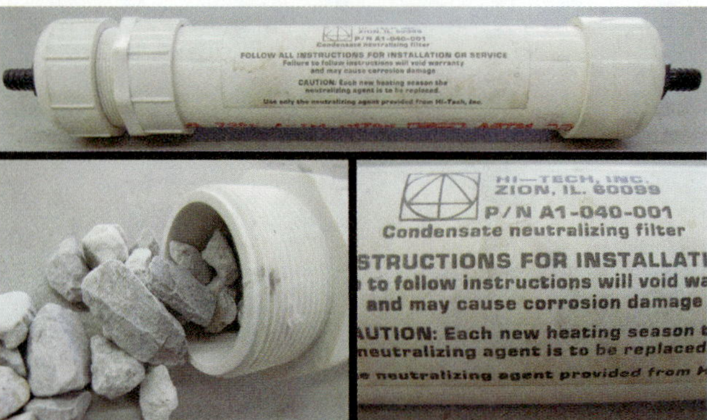

Figure 48-20 Acid-neutralizing drain trap.

lead to breakup of the chimney. Using type B double-wall vent pipe is critical for these furnaces.

In the high-efficiency furnace where condensation is allowed to occur inside the furnace on a continuous basis, the flue passages must be constructed of corrosion-resistant materials such as stainless steel and PVC. The material used for condensate drains for high-efficiency furnaces is usually corrosion- resistant PVC. The condensate from a condensing furnace is slightly acidic, about the same as a carbonated beverage. Some cities require neutralizing the condensate before it enters the public sewers. Running it through a filter (Figure 48-20), which is basically a PVC cylinder filled with limestone rocks, does this. The acid eats the rocks, neutralizing the acid.

48.8 SOOT BUILDUP AND PATTERNS

Soot is a form of carbon, the result of incomplete combustion. Complete combustion binds all the carbon into carbon dioxide, but incomplete combustion leaves some carbon to form soot. Yellow flames make soot; blue flames do not make soot. Soot can be deposited on flue surfaces and acts as an insulating layer, reducing the heat transfer and lowering the efficiency (Figure 48-21). Soot can also clog flues, reduce draft and available air, and prevent proper combustion. A large buildup of soot can create a safety hazard, as it is

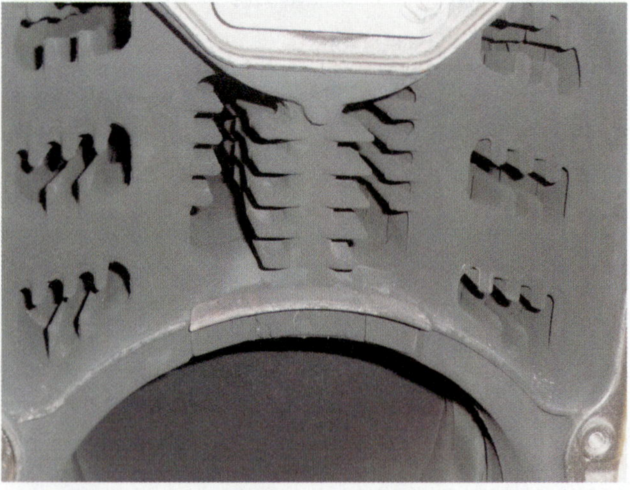

Figure 48-21 Soot deposits on heat exchanger.

very combustible! If the soot in a vent is ignited, the resulting fire in the vent can catch the house on fire.

A properly burning gas appliance will never make soot. Proper burner adjustment can prevent the formation of soot. Soot can be caused by too much fuel, too little air, or improper mixing. To prevent soot formation, the technician should do the following:

- Make sure the burner orifices are the correct size.
- Check the manifold gas pressure and adjust if necessary.
- Check the combustion air openings in the appliance room.
- Adjust the air shutters on the gas appliance to produce a blue flame.
- Perform a combustion analysis check.

48.9 AIR POLLUTION

The combustion processes constitute the largest single source of air pollution in a home. Some of the ways that the heating technician can help to reduce air pollution are as follows:

- Properly test installed furnaces and adjust the fuel-burning device for highest efficiency.
- Encourage the use of high-efficiency furnaces.
- Encourage the conservation of energy by recommending proper insulation and reduction of infiltration.
- Recommend the installation and operation of adequate outside air for ventilation and combustion.

Calculate the percentage of air required to determine if it meets guidelines:

$$\text{combustion air} = \frac{0.5\ \text{CFM}}{1{,}000\ \text{Btu}} \times \text{total Btu input}$$

To prevent air pollution, maintain clean plenum, air intakes, filters, ducts, and system components. And seal leaks, which is very important.

48.10 ANNUAL FUEL UTILIZATION EFFICIENCY (AFUE)

Furnaces are rated for their annual fuel utilization efficiency. This rating is obtained by applying an equation developed by the National Institute of Standards and Technology (NIST) for 100 percent efficiency and deducting losses for exhausted latent and sensible heat; the effects of cycling, infiltration; and the pilot burner effect. The AFUE is determined for residential fan-type furnaces by using the ANSI/ASHRAE Standard 103-1993 method of testing.

The federal law, effective January 1, 1992, requires that all new gas furnaces have a minimum AFUE efficiency of 78 percent. On June 27, 2011 the DOE published a direct final rule that splits the country into 3 regions: North, South, and Southwest. Beginning May 1, 2013, the minimum AFUE for furnaces in the South and Southwest regions will be 80% and the minimum AFUE for the North region will be 90%.

Most manufacturers offer two levels of furnace efficiency: 80 percent and 90 percent. There is no real difference in furnace cost between a 78 percent model and an 80 percent model, so very few 78 percent models are produced. Past experience has shown that furnaces operating at efficiencies higher than 80 percent AFUE are very likely to create condensation inside the heat exchanger or in the vent. Once a furnace passes the 80 percent threshold, the heat exchanger and vent must be designed to accommodate condensation, increasing the complexity of the furnace design. This limits low-end furnaces to 80 percent efficiency. The flue gas is too hot to use PVC vent material until the efficiency reaches 90 percent. So furnaces are basically available as 80 percent noncondensing furnaces and 90 percent condensing furnaces. Several manufacturers have tried to produce mid-efficiency furnaces with efficiencies between 80 and 90 percent, and the end result has always been unpleasant for contractors and customers. Some manufacturers also make super-efficiency models that exceed 90 percent, the highest being 98 percent.

The AFUE rating is only a basis of comparison. Obtaining that efficiency in any particular installation depends upon proper sizing and installation. Oversized furnaces will not deliver their rated efficiency due to the losses from heating up and cooling down the furnace. A smaller unit operating constantly will use less gas than a large unit cycling on and off. More operating efficiency can be gained by using a two-stage furnace. Operating on the lower firing rate allows for efficient operation. This is due in part to longer run cycles at a reduced capacity. This provides a closer track to the thermostat setpoint, ensuring a more comfortable environment. Table 48-3 shows the savings for every $100 spent on fuel when upgrading to a higher-efficiency furnace. As an example, changing to a 95 percent AFUE from a 50 percent AFUE would save $47 for every $100 spent for fuel.

TECH TIP

Gas furnaces are designed to last for many years. Many that were built before the change in efficiency laws have AFUE ratings of 60 percent. Over time, these furnaces have lost some of that original efficiency and can be operating at a level well below 60 percent efficiency. Therefore, it is often in your customer's best interest if you recommend that they trade up to a newer higher-efficiency model. Doing so will both save them money and reduce air pollution.

48.11 CERTIFICATION OF HEATING EQUIPMENT

All types of heating apparatus sold domestically have to be tested, certified, or listed by accredited testing agencies such as AHRI, CSA International or Underwriters Laboratories, Inc. (UL). Gas appliances must meet standards set by a large number of governmental and independent agencies. Typically, the logos of the approving agencies can be found

Table 48-3 Savings from Upgrading to a Higher-Efficiency Furnace

		AFUE of New System				
		75%	80%	85%	90%	95%
AFUE of Existing System	50%	$33	$37	$41	$44	$47
	55%	26	31	35	38	42
	60%	20	25	29	33	37
	65%	13	18	23	27	32
	70%	6	12	17	22	26
	75%		6	11	16	21
	80%			5	11	16
	85%				5	11

(Courtesy of the U.S. Department of Energy)

on the furnace nameplate. Some of these agencies just establish standards, while others perform testing to certify that equipment meets these standards. Three of these agencies, AHRI, CSA, and UL, are involved in both setting standards and certifying equipment.

Governmental Agencies

- **United States Department of Energy (DOE)** The primary governmental agency involved in gas furnace standards is the U.S. Department of Energy (DOE). One part of DOE's mission is to promote scientific and technological innovation in support of economic and energy security. DOE establishes the minimum AFUE that furnaces must meet. When developing their standards, DOE relies on research from two other governmental agencies: the National Institute of Standards and Technology (NIST) and the National Science Foundation (NSF).
- **National Institute of Standards and Technology (NIST)** Founded in 1901, NIST is a nonregulatory federal agency within the U.S. Commerce Department's Technology Administration. NIST's mission is to promote U.S. innovation and industrial competitiveness by advancing measurement science, standards, and technology in ways that enhance economic security and improve our quality of life. NIST establishes measurement standards of all types.
- **National Science Foundation (NSF)** The NSF is an independent federal agency created by Congress in 1950 "to promote the progress of science." NSF is the only federal agency whose mission includes support for all fields of fundamental science and engineering. NSF provides grants and oversight for basic science research.

Independent Agencies

- **Air-Conditioning Heating and Refrigeration Institute (AHRI)** AHRI is one of the largest trade organizations in the United States with over 300 member companies responsible for producing more than 90 percent of the residential and commercial air conditioning, heating, water heating, and commercial refrigeration equipment made in North America. AHRI writes the equipment standards that most manufacturers use to rate their equipment. AHRI also independently certifies the performance and efficiency of air conditioning, heating , and refrigeration equipment. AHRI's certification programs comply with ISO Guide 65, which sets the general international requirements for bodies operating product certification systems. Because of this recognition, AHRI's standards and certification programs are accepted and specified all over the world.
- **American Gas Association (AGA)** The AGA, founded in 1918, is the national trade association representing energy utilities that deliver natural gas. The AGA sets standards for gas appliance operation. By conforming to AGA guidelines, manufacturers can ensure that their equipment will operate properly on gas lines provided by AGA members.
- **Canadian Standards Association (CSA)** The Canadian Standards Association is a not-for-profit membership-based association serving business, industry, government, and consumers in Canada and the global marketplace. CSA America, Inc., a division of CSA Group, is a standards development body in the United States for appliances and accessories fueled by natural gas, liquefied petroleum, and hydrogen gas. The CSA Technical Committees establish minimum construction and performance standards for gas heating equipment.

CSA also maintains testing facilities to certify gas-fired heating equipment and electrical heating and air-conditioning equipment. CSA will also do testing at the manufacturer's facility. The testing certifies that equipment meets standards set by organizations like ANSI, UL, CSA America, or the NSF. CSA Group acquired the Certification and Testing business from the AGA in 1997.

Gas furnaces and other comfort heating gas appliances tested and found to be in compliance bear the CSA Blue Star mark and are listed in the CSA Certified Product Listings . Electrical products certified by CSA International to the applicable U.S. requirements are eligible to bear the CSA U.S. mark. This mark is accepted by regulatory authorities throughout the United States and is the equivalent of the UL mark.

- **Underwriters Laboratories (UL)** UL is an independent, not-for-profit product safety certification organization that has been testing products and writing standards for safety for over a century. UL is involved in setting standards as well as testing. There are many UL standards relating to gas heating equipment.

UL gets involved in the approval and listing of heating and air-conditioning equipment, including large centrifugal machines of 100 tons and above. Local city codes and inspectors are guided by UL standards, and failure to comply with them may be costly to the manufacturer and installer. Most people are familiar with the UL mark. UL maintains testing laboratories for certain types of products, but they often perform the necessary tests at the manufacturer's plant.

UNIT 48—SUMMARY

The three ingredients for combustion are fuel, oxygen, and heat. Complete combustion occurs when these are present in the correct quantities. The products of complete combustion of a fuel gas are carbon dioxide and water. Incomplete combustion occurs if any of the three ingredients are not present in the correct amount. Products of incomplete combustion include carbon monoxide and soot. Carbon monoxide is an extremely dangerous poisonous gas. It is odorless, tasteless, and lethal.

Gas-fired heating equipment standards are set by a number of separate agencies, including the Air Conditioning Heating and Refrigeration Institute, the American Gas Association, the American National Standards Institute, the Canadian Standards Association, the Department of Energy, the National Institute of Standards and Tech-nology, the National Science Foundation, and Underwriters Laboratories. Testing and certifying equipment is performed by the Air Conditioning Heating and Refrigeration Institute, the Canadian Standards Association and Underwriters Laboratories. Annual Fuel Utilization Efficiency is a measure of gas furnace efficiency for the entire heating season. The minimum AFUE is 78 percent, while the most efficient furnace available rates 98 percent.

WORK ORDERS

Service Ticket 4801

Customer Complaint: Furnace Does Not Heat the House

Equipment Type: Upflow Furnace Converted from Natural Gas to LP Gas

The furnace was recently installed. It was originally manufactured for natural gas and was converted to LP by the installer. The service technician notices soot buildup on the heat exchanger. The flames are lazy and yellow, indicating incomplete combustion. There appear to be adequate combustion air openings, and they are not obstructed. The manifold pressure reading is correct. The technician checks the gas burner orifices against the manufacturer's literature and finds that they are the original orifices sized for natural gas. Since a natural gas orifice is much larger than an LP gas orifice for the same capacity, the technician concludes that the oversized orifices are delivering too much gas, leading to incomplete combustion, soot formation, and poor heating. The orifices are changed to the correct size for LP gas and the soot buildup in the heat exchanger is removed.

Service Ticket 4802

Customer Complaint: Smoke Coming out of Fireplace

Equipment Type: Upflow Gas Furnace Installed in a Hall Equipment Closet

The customer complains that if the furnace comes on when there is a fire in the fireplace, smoke starts coming out of the fireplace into the room. The technician looks in the equipment closet and finds that the combustion air grilles have been covered in plastic and taped up. The homeowner says that he noticed cold air coming in through the grilles and covered them up to keep the cold air out. The technician explains that the grilles are there to provide combustion and ventilation air for the furnace. Without them, the furnace must draw in air from the rest of the house, creating a negative pressure in the house. This is why the fireplace smoke is coming out into the room. The customer is advised that this situation could be dangerous because it could lead to the furnace producing carbon monoxide and vent gas spilling into the house. The technician removes the plastic covering the grilles and operates the furnace with a fire in the fireplace to verify that there is enough combustion air for both.

Service Ticket 4803

Customer Complaint: Furnace Comes on with a Boom!

Equipment Type: Horizontal Furnace Installed in Attic

The customer complains that the furnace sometimes lights with a loud boom. It seems to happen more in colder weather. The technician observes that the burners are noisy and that the flames seem to be lifted away from the burner outlet. The air shutter adjustment is wide open. The technician reduces the primary air by closing the shutter until the flames start at the burner instead of away from it. The technician explains that the extra primary air was making the gas-air mixture hard to light. The delayed ignition caused gas to build up, resulting in a small explosion.

Service Ticket 4804

Customer Concern: Safety of Furnace

Equipment Type: Upflow Gas Furnace Installed in Equipment Closet

A customer has purchased a new house, which is heated by a gas furnace. She has always had electric heat and is concerned about the safety of gas appliances in general and fire safety in particular. The technician agrees to do a thorough safety inspection and leaves the customer a gas safety bulletin that the local gas utility produces. The technician explains that a key factor the customer can control is the combustion and ventilation air and shows the customer the combustion air grilles, explaining that they should not be covered up or blocked. The technician also warns against using the equipment room as a closet and recommends that the customer purchase a carbon monoxide alarm.

UNIT 48—REVIEW QUESTIONS

1. Discuss the origin of natural gas.
2. List the three broad categories of fuel gases.
3. Complete this table of fuel gas characteristics.

Gas	Heat Capacity	Specific Gravity
Methane		
Ethane		
Propane		
Butane		

4. What are the products of complete combustion of a hydrocarbon gas?
5. What are the products of incomplete combustion of a hydrocarbon gas?
6. Why are LP gases considered more hazardous than natural gas?
7. Describe the two types of flames and their characteristics.
8. Describe the gas combustion process as it applies to a gas furnace.
9. Explain the causes of flue gas condensation and how it can be stopped.
10. Explain the cause of soot formation on the heat exchanger and how it can be stopped.
11. Explain how carbon monoxide is formed.
12. Discuss furnace combustion air requirements.
13. List agencies that establish standards for gas furnaces.
14. List agencies that perform testing to verify equipment performance.
15. Explain AFUE.
16. What is the minimum AFUE rating for a new furnace in the United States?
17. Which agencies set equipment standards and test equipment?
18. Which government agencies are involved in setting equipment standards?
19. What can be done to minimize the pollutants produced by a gas furnace?
20. What is the purpose of excess air in furnace combustion?

UNIT 49

Gas Furnaces

OBJECTIVES

After completing this unit, you will be able to:

1. define the four categories of gas-fired furnaces.
2. list the five furnace cabinet configurations.
3. describe the operation of a standing-pilot, natural-draft furnace.
4. describe the operation of an 80 percent mid-efficiency furnace.
5. describe the operation of a 90 percent condensing furnace.
6. discuss the evolution of the heat exchanger.
7. discuss the operation of an atmospheric burner.
8. explain why furnace efficiencies jump from 80 to 90 percent.

49.1 INTRODUCTION

Gas furnaces have been one of the most common forms of heat for decades. The furnace has evolved to meet the challenge posed by rising energy costs. Furnaces today are smaller, lighter in weight, and more efficient. Understanding today's technologically advanced gas furnaces is crucial for the successful technician.

A wide variety of gas furnaces are available to meet many different applications. Factors to be considered when choosing a furnace include the fuel source, furnace location, efficiency, and venting. There are many furnaces to choose from because there are many options for each of these variables. In many cases, the contractor must choose between several possibilities because more than one type of furnace would work. Understanding all the choices will allow the installing contractor to make an informed, intelligent decision.

49.2 THE EARLY DEVELOPMENT OF HEATING SYSTEMS

The furnace of today has evolved from the old pot-bellied stove. Wood- and coal-burning stoves heated the room by a combination of convection and radiation. This type of heat was localized, and in very cold weather, people needed to huddle around the stove to stay warm. Over time, the basic principles of heating were improved to more efficiently heat the entire home.

Today's gas furnace heats by convection. Gas flames heat a heat exchanger that transfers heat to air circulating through the furnace. The heat is exchanged from the flames to the air while the combustion gasses are vented safely out through a vent. The furnace blower circulates warm air throughout the house to provide even temperatures.

Early Radiant and Convection Heat

All objects in a direct line of sight from the stove were heated by radiation. However, if something blocked the path between the object and the stove, the intermediate object absorbed all the heat. Another problem with radiant heat is that it is uneven by nature. The amount of radiant heat transferred diminishes by the square of the distance between the source of the radiant heat and the object absorbing the heat. In simpler terms, the closer you get to the stove, the hotter you get. This means that certain areas of the room will be very warm while others will be chilly. Convection is also taking place. Cool air near the floor is heated by the stove, rises to the ceiling, loses its heat along the way, cools off, and drops back down. This does help distribute the heat, but again, somewhat unevenly. The ceiling tends to stay about 20°F warmer than the floor. Unfortunately, most of the room's occupants are closer to the floor than the ceiling.

Beginning of Ductwork

The first central furnaces were essentially pot-bellied stoves inside a metal housing. Ductwork ran from the housing to all areas of the house, producing a more even distribution of heat. Since the stove was encased in a metal jacket, all radiant heat was converted to convection heat, eliminating the problem of hot and cold spots associated with radiant heat. These early furnaces did have a few problems of their own; for instance, the ductwork was enormous since the furnace was relying on natural convection to carry the heated air through the ducts to the house. If the ductwork was not gigantic, the restriction to flow did not allow the hot air to reach the house. Also, these furnaces lost more heat than they delivered. Since heat energy was used for venting the furnace as well as for air delivery, more heat went up the chimney than was delivered into the house.

Development of Forced Convection and Sectionalized Heat Exchangers

The next logical advance was the addition of a fan to force the air over the stove, which is now called a heat exchanger. This increased efficiency by wringing more heat out of the heat exchanger. It also reduced the size of the ductwork needed to deliver the air. Although this represented quite a step forward, the pot-belly shape of the heat exchanger

Figure 49-1 Sectionalized heat exchanger.

contributed to the problem of inefficiency. A large quantity of combustion gases never touched the sides of the heat exchanger due to its large round shape. If the combustion gas does not come in contact with the heat exchanger, its heat cannot be transferred. To increase heat transfer, the barrel-type heat exchanger was squeezed, or flattened, to a rectangular box that was not very wide. This made it more difficult for flue gas to leave the heat exchanger without touching the sides, thus increasing efficiency. Typically, several of these clamshell heat exchangers were combined together, each with its own burner. This became known as a sectionalized heat exchanger (Figure 49-1).

Development of Standard and Mid-Efficiency Gas Furnaces

Furnaces remained about the same for decades and performed reliably with 60–65 percent efficiency. There was not much incentive to change them until the energy crisis of the 1970s. Suddenly, the same old furnace, which had operated economically for years, seemed to turn into a fuel-hogging monster. Of course what changed was the cost of the fuel. Changes increased furnace efficiency by about 20 percent. Through experimentation and some unfortunate experiences, an efficiency of less than 80 percent was determined to be the highest efficiency possible without condensing water out of the flue gas. Units that exceeded this efficiency had the unfortunate experience of condensing water in both the heat exchanger and the flue. This water was very corrosive due to its mild acidity, and these units often died untimely deaths of flue or heat exchanger failure.

Development of High-Efficiency Gas Furnaces

The next major increase in efficiency took advantage of the early mistakes. By intentionally designing a furnace to condense water, efficiency could be increased. Now furnaces are available with efficiencies up to 98 percent. To do this, the point where condensation occurs has to be closely controlled, and the condensing section of the furnace is typically made of stainless steel.

49.3 GAS FURNACE TYPES

Gas heating has traditionally been mostly forced-warm-air furnaces. However, in colder climates, hydronic heating systems are often used rather than forced warm air. In these systems, hot water, rather than warm air, is circulated through the building. Hydronic systems have often been designed to utilize oil-fired boilers. This is now changing due to the increased efficiency and cleaner burn offered by gas-fired boilers. Many new installations and retrofits are using gas in lieu of oil. Gas is also a popular option for residential hot-water heaters, providing an almost continuous supply of hot water without the need for a hot-water holding tank.

Gas Forced Warm Air

Gas forced-warm-air furnaces are composed of two main sections: the heat exchanger compartment and the blower compartment. The arrangement of these two sections determines the general style of the furnace. Regardless of the style, the air always travels through the blower compartment first, before going to the heat exchanger. The reason is fairly simple: the fan motor would overheat otherwise. Furnaces can be categorized into five styles: upflow, lowboy, downflow or counterflow, horizontal, and multiposition.

For the most part, airflow through the furnace gives it its name. Air travels up in an upflow furnace, down in a downflow furnace, and horizontally in a horizontal furnace.

The upflow furnace (Figure 49-2) is most popular. Its narrow width and depth allow for location in first-floor closets and/or utility rooms. It can still be used in most basement applications for heating only or with cooling coils where headroom space permits. Blowers are usually directly driven by multispeed motors or variable speed electronically commutated motors (ECM). Return air can be either from the sides or from the bottom.

The counterflow or downflow furnace (Figure 49-3) is similar in design and style to the upflow, except that the air intake and fan are at the top and the air discharge is at the bottom. These are widely used where duct systems are set in concrete or in a crawlspace beneath the floor. A fireproof base is required when the furnace is installed on a combustible floor. An extra safety limit control in the fan compartment is required.

Horizontal furnaces are installed in low areas such as crawlspaces, attics, or partial basements (Figure 49-4). They require no floor space. Intake air enters at one end and is discharged out the other end. On older-style furnaces, burners are usually field changeable for left-hand or right-hand application.

Lowboy furnaces are built low in height to accommodate low ceilings. These furnaces are approximately 4 ft high, providing for easy installation in a low-ceiling-height

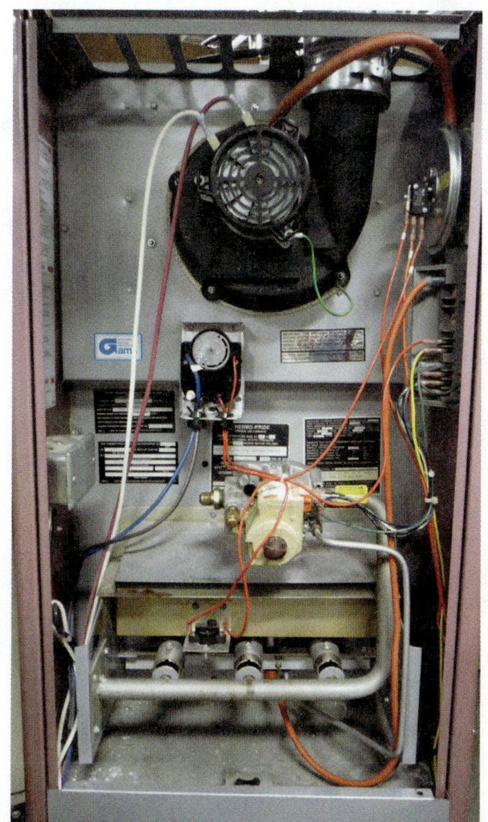

Figure 49-2 Upflow highboy furnace.

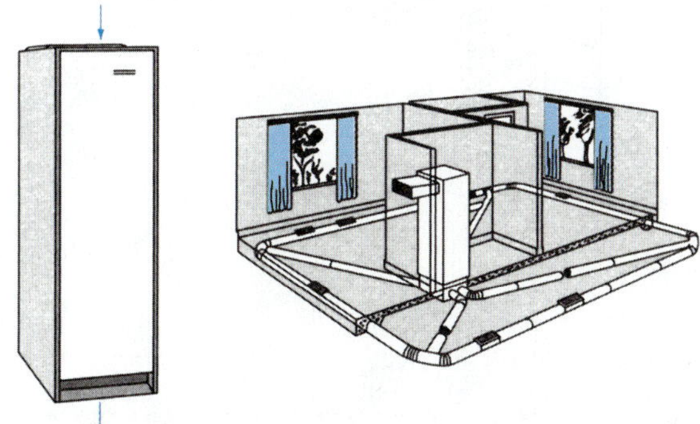

Figure 49-3 Counterflow furnace.

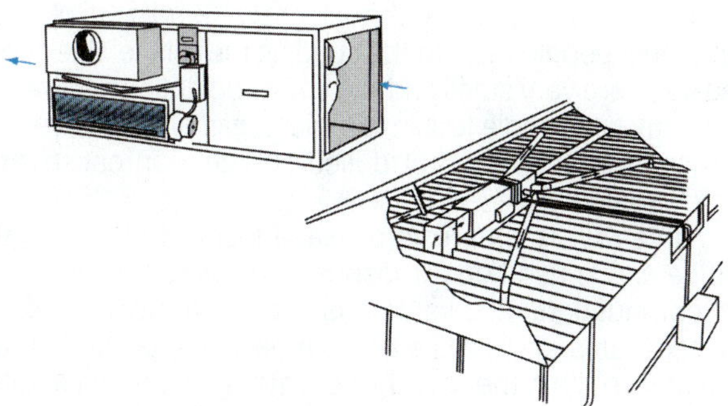

Figure 49-4 Horizontal furnace.

basement. Lowboy furnaces have two sections side by side, like horizontal furnaces. But the air goes in the top of the furnace, makes a 180° turn as it goes through the furnace, and then leaves out the top of the furnace. Blowers are commonly belt driven. Many of these furnaces are sold for retrofitting older homes.

The installation technician can configure the fifth style, the multiposition furnace, on the job site to be an upflow, horizontal, and/or downflow furnace. The introduction of the multiposition furnace has allowed manufacturers and contractors to reduce the number of furnaces they must keep in inventory to meet local demand. The vent typically needs to be configured for the installation. Figure 49-5 shows a vent configured for upflow, downflow, and horizontal installation. Typical installations of the horizontal, counterflow, and upflow furnaces are shown in Figure 49-6.

Gas-Fired Boilers

Gas-fired boilers circulate hot water through baseboard heaters or radiant floor piping. In addition to the boiler,

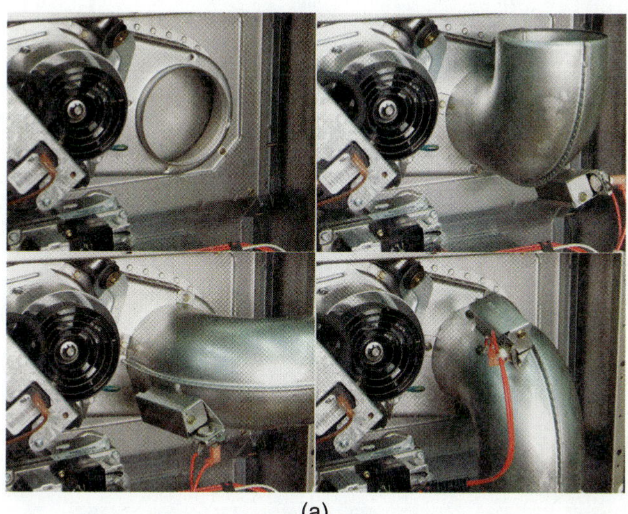

(a)

(b)

Figure 49-5 Multiposition furnace.

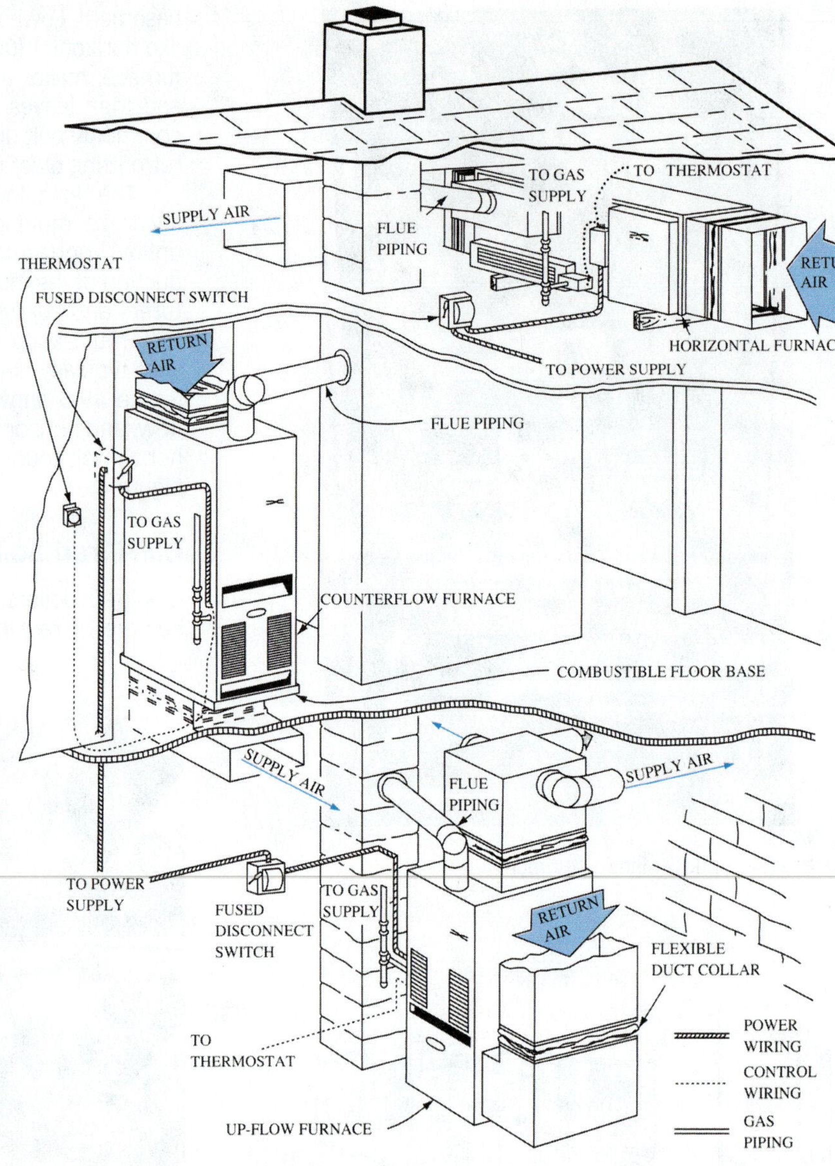

Figure 49-6 Composite diagram of three types of furnace installations: (1) attic installation of a horizontal furnace; (2) first-floor installation of a counterflow furnace; (3) basement installation of a lowboy furnace.

these systems include a number of additional components for the water loop. These include water-circulating pumps, an expansion tank, and water flow control valves. There will also be a heat exchanger to heat the water. A gas-fired hydronic heating system and its small plate-type heat exchanger are shown in Figure 49-7.

49.4 GAS FURNACE COMPONENTS: MANIFOLD

The purpose of the manifold is to supply gas to all the burners. The manifold is usually a length of pipe connected to the gas valve on one end and closed on the other end (Figure 49-8). There is a small plug, called a spud, along the manifold at each burner inlet (Figure 49-9) that threads into the manifold. Occasionally the manifold has a threaded plug that can be removed to check the manifold pressure.

49.5 GAS FURNACE COMPONENTS: ORIFICE

An orifice is a precisely drilled hole that meters the correct amount of gas into the burner. The orifice is drilled into a spud (Figure 49-10). Since the orifice size is what really matters, most people refer to the spud as the orifice. The specified sizes are very small and they must be drilled accurately. A slightly larger hole lets in too much gas and overfires the burner. A hole that is not drilled straight can cause even more problems.

A venturi is a narrow passage located just after the orifice and is used on atmospheric burners to draw in air for combustion. The venturi is the initial portion of the burner that is shaped like an hourglass. As gas and air are drawn through it they swirl, like water going down a toilet. The swirling mixes the gas and air. The space between the orifice and venturi has to be open so that primary air can

(a)

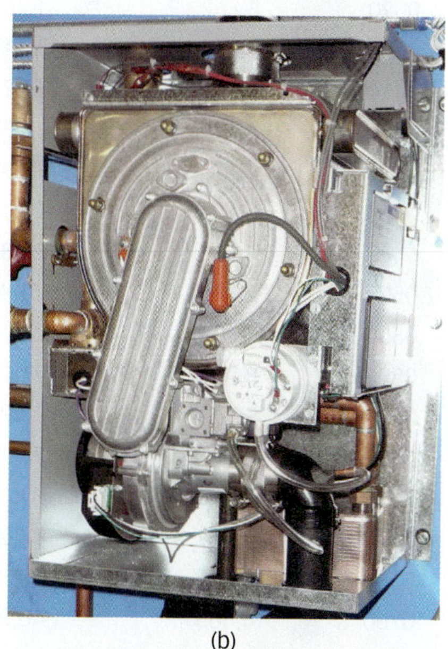

(b)

(c)

Figure 49-7 (a) Gas hydronic system; (b) gas hydronic furnace; (c) small plate-type heat exchanger is located inside the furnace.

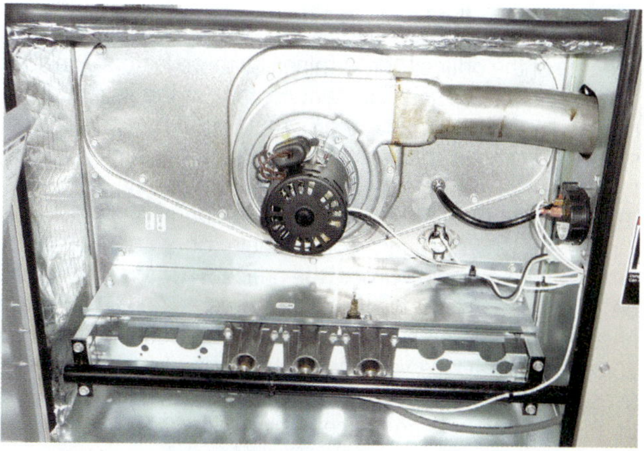

Figure 49-8 Gas manifold shown with burners located directly underneath the blower.

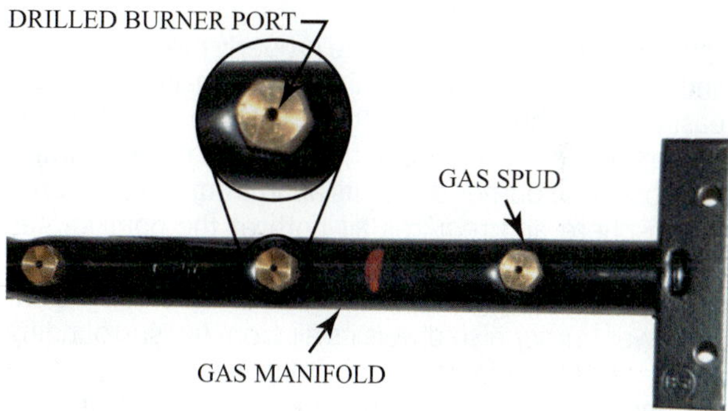

Figure 49-9 Gas manifold.

Figure 49-10 Spud with orifice number.

be injected by the jet action of the gas emitting from the orifice. If the gas stream does not aim straight inward, it may hit the side of the venturi, or worse, the gas stream may miss the venturi altogether. This causes poor ignition and even explosions. While it is not impossible for a factory orifice to be drilled incorrectly, dirt, dust, or other debris is more likely the cause of the gas stream being deflected.

Orifice Size Determination

The size of the orifice is determined by three factors: the specific gravity of the gas, the heat content of the gas, and the required BTU/hr rating of the burner. The first two factors, specific gravity and heat content, are generally set by the gas type, which is usually either natural gas or propane. The unit's BTU/hr rating divided by the number of burners determines the burner BTU/hr rating. For example: on a 40,000 BTU/hr furnace with a single burner, the orifice BTU/hr rating would be 40,000 BTU/hr. However, if the same 40,000 BTU/hr furnace had two burners, each burner orifice would be rated at 20,000 BTU/hr. Once you know what type of gas the burner uses and the BTU/hr required for each burner, you could look up the orifice size in an orifice chart (Table 49-1).

49.6 GAS FURNACE COMPONENTS: BURNERS

The purpose of the gas burner is to properly mix the primary air and gas and to deliver this mixture in a manner that can be easily ignited and burned. There are two general types of burners: the atmospheric burner and the power burner. The atmospheric burner draws its air in from the surrounding atmosphere and requires air both at the point of gas entry and at the point of combustion. Power burners, on the other hand, normally get all their air at the point of gas entry. A power burner also draws its air from the surrounding atmosphere, but a blower forces it in mechanically. Many power burners pressurize both the air and the gas and are able to burn large quantities of gas quickly. The basic residential burner is the atmospheric burner. Even induced- and forced-draft units normally use atmospheric burners, even though the "draft" through them is mechanically produced.

Atmospheric Burners

The burner generally slides onto the spud, with the spud located in the middle of the burner face. The primary air openings are located all around it. In older furnaces, these air openings were fitted with adjustable shutters to control primary air entering the burner (Figure 49-11). Air shutters are no longer used on modern burners.

Figure 49-11 Primary air shutter.

Table 49-1 Gas Orifice Capacity in BTU/hr

Drill	Size	Natural Gas 3.5 in wc BTU/hr	Propane 11 in wc BTU/hr
70	.0280	1,991	5,476
69	.0292	2,165	5,956
68	.0310	2,440	6,713
1/32	.0313	2,487	6,843
67	.0320	2,600	7,153
66	.0330	2,765	7,607
65	.0350	3,110	8,557
64	.0360	3,291	9,053
63	.0370	3,476	9,563
62	.0380	3,666	10,087
61	.0390	3,862	10,625
60	.0400	4,062	11,176
59	.0410	4,268	11,742
58	.0420	4,479	12,322
57	.0430	4,695	12,916
56	.0465	5,490	15,104
3/64	.0469	5,585	15,365
55	.0520	6,865	18,888
54	.0550	7,680	21,131
53	.0595	8,989	24,730
1/16	.0625	9,918	27,286
52	.0635	10,238	28,167
51	.0670	11,398	31,357
50	.0700	12,441	34,228
49	.0730	13,530	37,225
48	.0760	14,665	40,347
5/64	.0781	15,487	42,608
47	.0785	15,646	43,045
46	.0810	16,658	45,831
45	.0820	17,072	46,969
44	.0860	18,778	51,663
43	.0890	20,111	55,331
42	.0935	22,197	61,067
3/32	.0938	22,339	61,460
41	.0960	23,399	64,377
40	.0980	24,384	67,087
39	.0995	25,137	69,156
38	.1015	26,157	71,964
37	.1040	27,462	75,553
36	.1065	28,798	79,229
7/64	.1094	30,388	83,603
35	.1100	30,722	84,522
34	.1110	31,283	86,066

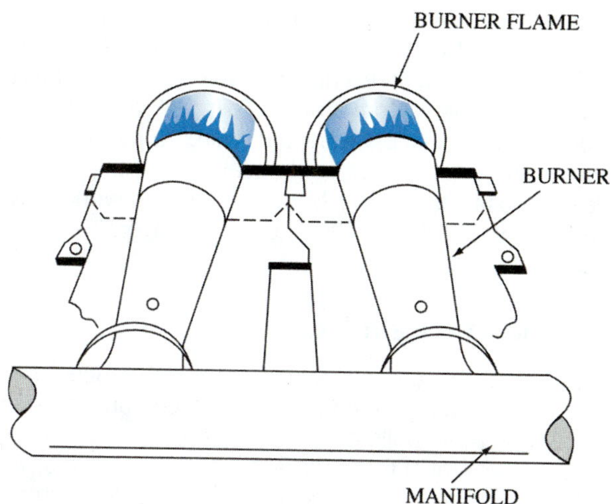

Figure 49-12 In-shot atmospheric gas burner.

Figure 49-14 Slotted atmospheric burner.

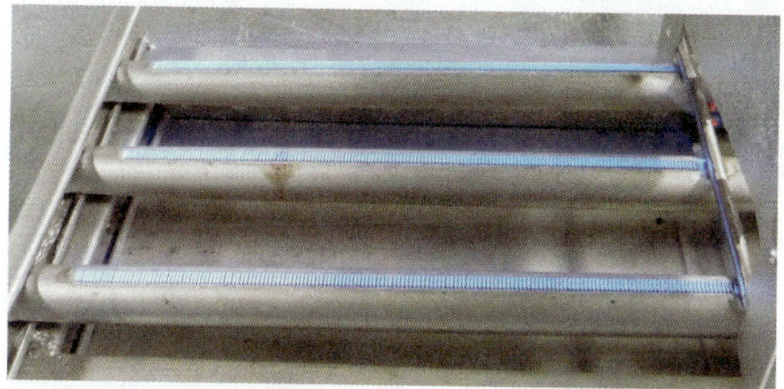

Figure 49-15 Slotted burners can produce long, thin, ribbon flames.

The gas shoots out of the orifice in a small, quick stream, which creates a negative pressure around it and draws in primary air through the primary air openings. The burner is shaped like an hourglass on the end. It has a large diameter at the face where the gas and air enter, a small diameter a short distance away, and returns to a large diameter at the burner's main body. This shape, called a venturi, causes the gas and air to swirl as they pass through the constricted opening, mixing them together (Figure 49-12). The orifice, primary air shutters, and venturi are crucially important because together they are responsible for mixing the gas and air in the correct proportions for complete combustion. Next, the gas travels through the venturi into the main burner body. Here, the gas-air mixture slows down and expands to fill the chamber. The mixture leaves through the burner ports.

There are several styles of ports, including in-shot (one large hole), drilled port (several small holes drilled in the burner), and slotted (ribbon). Drilled-port burners produce lots of small, distinct flames (Figure 49-13). These are usually quieter but less efficient than in-shot burners. Slotted-port burners are usually made of stamped sheet metal, and the slots are larger than the holes in a drilled-port burner (Figure 49-14). Ribbon burners consist of several slots, which run the length of the burner. These produce long, thin, ribbon flames (Figure 49-15).

Figure 49-13 Drilled-port burners produce small, distinct flames.

Virtually all furnaces today use in-shot burners because of their efficiency and because the flame from an in-shot burner fits the narrow heat exchangers used today (Figure 49-16). In-shot burners are generally the most efficient type of atmospheric burners, but they are also the noisiest. In-shot burners produce a flame like a jet or rocket tail.

Regardless of the burner style, a good flame should have a bright blue inner cone surrounded by a darker blue outer cone. There should be no yellow in the flame; yellow indicates either too little air or too much gas in the mixture. This should not be confused with intermittent orange flashes often caused by dust particles passing through the burner.

Power Burners

The other basic type of gas burner is the power burner. Power gas burners are seldom used on residential equipment, although a few units have used power burners. Power burners use a fan to force both the air and the gas into the combustion area. The air and gas can be mixed

(a)

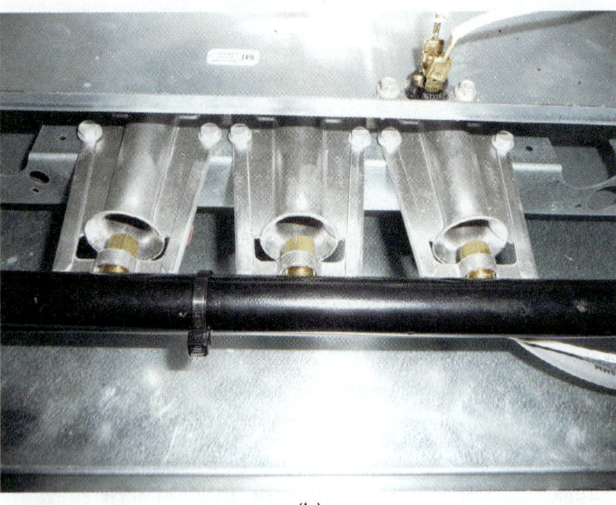

(b)

Figure 49-16 (a) In-shot atmospheric gas burner; (b) in-shot atmospheric gas burner, manifold, and spuds.

in the fan itself or after the fan inside a tube with angular vanes to create a vortex action that mixes the air and gas. Generally, power burners are ignited by spark igniters or glow coils. Using power burners can achieve higher burning efficiencies than atmospheric because they have better control over the fuel-air ratio. Power burners are used on commercial multi-fuel furnaces. Nongaseous fuels, such as oil, require power burners for operation. The use of a power burner can make changeover from gas to oil easier.

49.7 GAS FURNACE COMPONENTS: HEAT EXCHANGERS

The job of the heat exchanger is to separate the combustion process from the air being heated and, as its name implies, exchange the heat from the combustion process to the air being circulated over the heat exchanger. To perform the job of gas separation, a heat exchanger needs only to be an airtight vessel with a hole in the bottom for the burners and one at the top for the flue gas to escape. As long as there are no holes in between these two openings, there is no problem. The second job, efficient heat transfer, is not as simple.

The earliest heat exchangers were basically barrels with a hole in the bottom and a hole in the top. They were adequate for gas separation but were rather inefficient heat-transfer apparatuses. Large quantities of the relatively unrestricted flue gas found its way up the vent without ever contacting the surface of the heat exchanger. This translates into lost heat.

Sectionalized Heat Exchangers

By flattening the heat exchanger, more resistance to flow was created, which forced more of the combustion products to touch the walls of the heat exchanger, increasing efficiency. Several of these flattened heat exchangers were used together to create one unit. These are commonly called sectionalized heat exchangers because they are built in sections (Figure 49-17). The vent opening on sectionalized exchangers is on the side at the top. Flue baffles were added at the top of the sectionalized heat exchangers (Figure 49-18). The baffles restrict the flow of combustion

Figure 49-17 Sectionalized heat exchanger.

Figure 49-18 Heat exchanger baffles.

Figure 49-19 Serpentine heat exchanger section.

gases at a constriction near the top of the heat exchanger. This forces them to stay in the heat exchanger longer, thus increasing heat transfer.

Serpentine Heat Exchangers

This concept was taken further with serpentine heat exchangers. The serpentine path and the narrow width helped squeeze more heat out of the combustion gas (Figure 49-19). These features also required an induced-draft fan to draw the combustion gases through the heat exchanger. A fan pressure-breaking baffle was installed at the discharge of the induced-draft blower, making the vent pressure negative so that the furnace could be vented with standard type B vent. Furnace manufacturers began to use automotive exhaust tubing as the heat exchanger. By bending tailpipe material in a serpentine configuration, they managed to produce a heat exchanger that gave them 80 percent efficiency yet used considerably less material and energy to produce.

The most common type of heat exchanger is made from two formed or stamped steel sheets welded together (Figure 49-20). These sections are welded together and form a header at the top and bottom. The number of sections depends on the capacity of the furnace. The contours of the heat exchanger sections create turbulence to increase heat transfer.

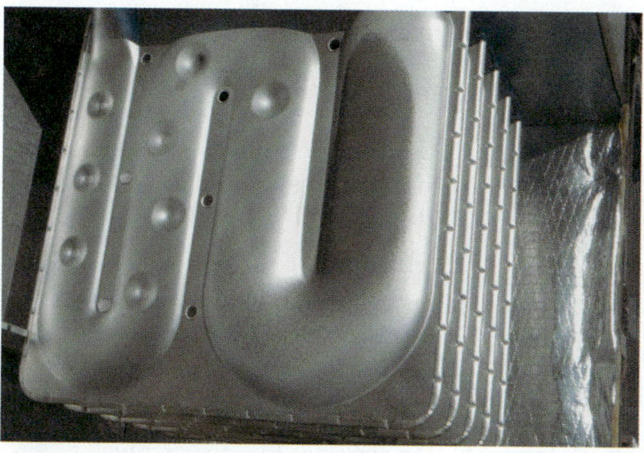

Figure 49-20 Typical furnace heat exchanger.

Recuperative Coils

To get into the 90+ percent range, you must condense. Most 90 percent furnaces are an 80 percent furnace with a stainless steel recuperative condensing coil. Most of these recuperative coils are built like a refrigeration coil, only from stainless steel. Some recuperative coils are just smaller stainless steel serpentine heat exchangers. Recuperative coils are installed ahead of the main heat exchanger in the airstream. Obviously, any unit employing such a coil must also provide a drain for the furnace.

So far, the experience with this approach has been very good. If you plan ahead for condensation to take place, then it ceases to be a problem. These units have another common characteristic: PVC positive pressure vents. The flue gas leaving a 90 percent efficient furnace feels warm and wet to the hand. PVC can handle the temperature easily. The positive pressure is no problem as long as PVC is used because it is relatively simple to ensure a leakproof PVC vent. The positive pressure also allows more flexibility in vent location and design since the flue gas is being pushed out, not floating out. Most major manufacturers now offer a 90+ percent condensing furnace as part of their product line, and installation practices are similar from one to another.

Burner Location

Traditionally, burners were located at the bottom of the heat exchanger and flue gases escaped out the top because the furnace relied on natural convection to move the flue gas through the heat exchanger. The addition of the induced-draft blower made it possible to have the flames at the top of the heat exchanger and the exhaust gases leaving the bottom of the heat exchanger. The counterflow process, where the combustion gas moves in the opposite direction from the air, is more efficient. Many manufacturers use the counterflow heat exchanger design in their 90 percent efficiency models.

Temperature Rise

A key operating characteristic of any furnace is its temperature rise. Manufacturers give the acceptable temperature rise range on the data plates of their units. For a standard furnace, the temperature rise range is usually 40 to 70°F. However, the exact range will change depending on the unit's capacity and airflow. The range will always be listed on the unit data plate. For example, a furnace with a temperature rise of 40 to 70°F and an entering air temperature of 75°F should have a leaving air temperature of 115 to 145°F. The higher the efficiency of the furnace, generally, the lower the temperature rise. To increase efficiency, increase the airflow through the unit compared to a standard furnace with the same BTU/hr rating. Thus, since there is more air to absorb the heat, more heat will be absorbed. However, that also means that less heat will go into each cubic foot of air, so the temperature rise of the air will be less. High-efficiency furnaces normally have a temperature rise of between 25°F and 65°F.

49.8 CATEGORY RATINGS AND EFFICIENCIES OF GAS FURNACES

Furnace manufacturers mostly offer two types of furnaces:

- Mid-efficiency category I fan assisted
- High-efficiency category IV condensing-type furnace

The mid-efficiency furnaces meet the minimum requirement of 78 percent AFUE (annual fuel utilization efficiency) and have a flue gas temperature at least 140°F above the dew point of the flue gas. The high-efficiency furnaces have flue gas temperatures below the dew point and permit condensation within the furnace to pick up the extra latent heat. In addition to these, there are standard low-efficiency furnaces, but they are no longer being manufactured. Even so, many of the earlier models of standard furnaces are still in operation and will require service. These utilize natural-draft venting, and most use standing pilots.

Standard Low-Efficiency Gas Furnaces

Lower-efficiency gas furnaces are the older units now in the field, all of which were manufactured prior to 1992. Since so many are in use and will require service for some time, this section describes them and indicates some differences from the newer designs.

A cross-sectional diagram for one of these units is shown in Figure 49-21. One distinctive characteristic of the older design is the convection flow of combustion gases from the burner, past the heat exchanger, and into the vent (Figure 49-22). This convection action of the flue gases requires a different type of venting as compared to newer, more efficient units that use fan assist.

Newer standard gas furnace burners like the one shown in Figure 49-23 use electronic igniters, while some of the older units used a standing, continuous-flame gas pilot. This pilot consumes a small amount of gas continuously. Along with the pilot is a runner attached to the main burner. The purpose of the runner is to carry the gas flame from the pilot to all of the burners. The pilot flame impinges on the thermocouple about ½ in, which provides a safety arrangement for turning off the gas if the flame is ever extinguished (Figure 49-24).

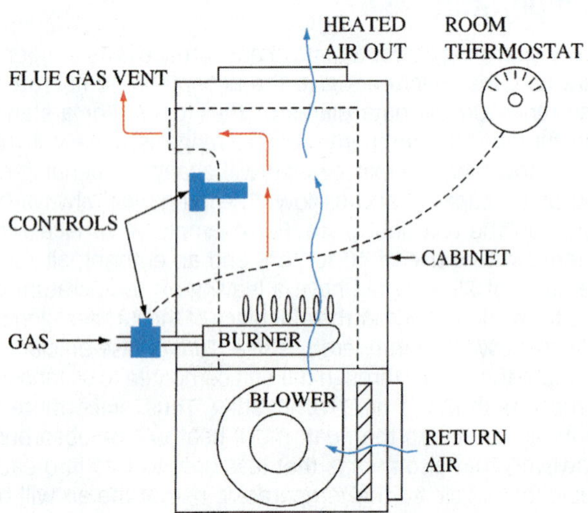

Figure 49-21 Cross-section of a gas upflow warm-air furnace.

Figure 49-22 Standard low-efficiency gas furnace.

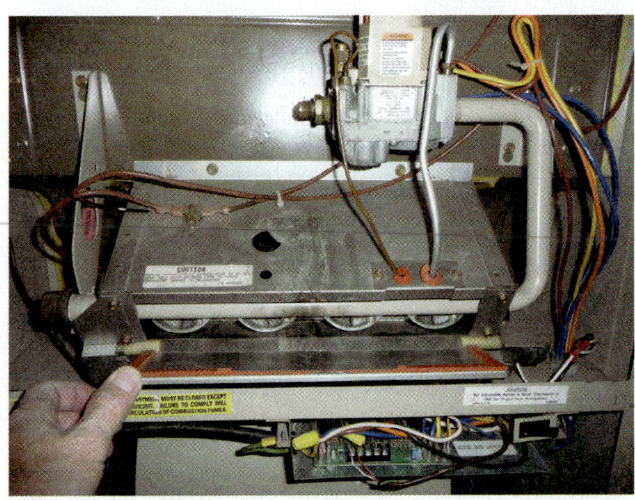

Figure 49-23 Standard low-efficiency gas furnace burners that use electronic ignition.

Category Ratings

A modern gas furnace's operational characteristics will place it into one of four categories based on its flue gas temperature and pressure.

- **Category I** An appliance that operates with a nonpositive vent static pressure and with a vent gas temperature greater than 140°F above the flue-gas dew point, which avoids excessive condensate production in the vent
- **Category II** An appliance that operates with a nonpositive vent static pressure and with a vent gas temperature less than 140°F above the flue-gas dew point, which may cause excessive condensate production in the vent
- **Category III** An appliance that operates with a positive vent static pressure and with a vent gas

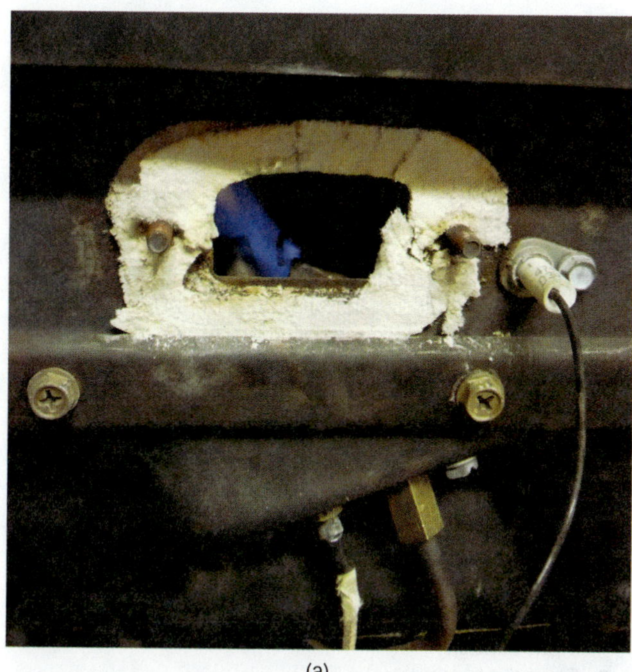

(a)

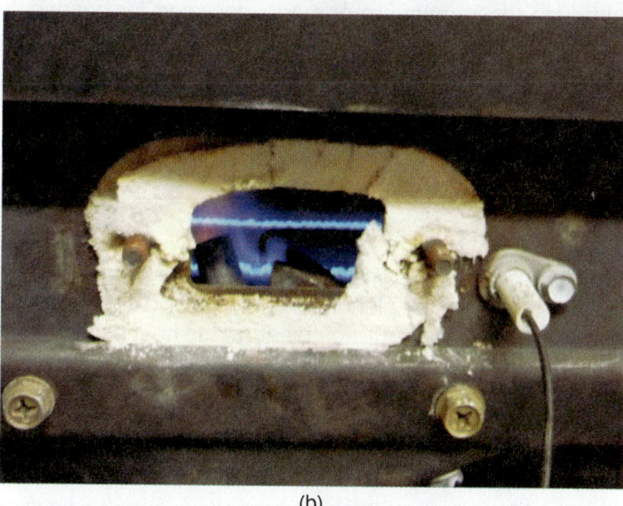

(b)

Figure 49-24 (a) Standing pilot and flame rod with burner flame off; (b) standing pilot with burner flame on.

temperature greater than 140°F above the flue gas dew point, which avoids excessive condensate production in the vent; however, the positive-pressure vent requires a sealed vent system

- **Category IV** An appliance that operates with a positive vent static pressure and with a vent gas temperature less than 140°F above the flue-gas dew point, which may cause excessive condensate production in the vent

Category I Furnaces

Category I furnaces can be divided into natural draft and fan assisted. Many furnaces sold today are Category I, fan assisted. The fan on most 80 percent furnaces does not produce a positive vent pressure because it is sized just large enough to overcome the resistance of the heat exchanger and because the temperature of the flue gas makes it lighter than air. A benefit to fan-assisted furnaces is that the heat exchanger operates in a negative pressure. This ensures added safety to the consumer because if a hole develops in the heat exchanger, the surrounding air rushes into the heat exchanger, preventing harmful flue gases from escaping into the occupied space.

Category II and III Furnaces

Category II and Category III furnaces are rare. Both of these require special vent materials, making their installation more costly. A special high-temperature plastic vent material was developed for use with Category II and III furnaces. This venting material is no longer manufactured because of high failure rates and the resulting legal action.

Category IV Furnaces

Category IV furnaces are also common; they are the 90 percent condensing furnaces. Since the flue gas is relatively cool, PVC can be used as a vent material. PVC is easy to seal airtight, and water will not bother it.

49.9 MID-EFFICIENCY FURNACES (80 PERCENT AFUE)

Furnace manufacturers achieve an efficiency of 80 percent by eliminating the pilot light, improving heat exchanger efficiency, and adding an induced-draft blower.

The internal components of a mid-efficiency furnace include the following:

- The inducer blower assembly, with an induced-draft blower to draw the flue gases through the heat exchanger and to the vent (Figure 49-25). The inducer motor draws the exact amount of air needed through the combustion exchanger. The assembly is resiliently mounted to provide quiet operation.
- The pressure switch, which proves that the inducer fan is operating before the fuel can be ignited.
- The gas control valve, which delivers the fuel gas into the burner. The gas valve opens slowly to provide a controlled ignition. It also provides for 100 percent shutoff to ensure safe operation.
- The burner assembly, which provides for proper mixing of fuel and air and ignites the fuel.
- The blower door safety switch, which disconnects the power supply to the unit whenever the front access panel is removed.
- The control box, which houses controls, including a microprocessor board that controls most furnace operations and functions. It provides a blower delay on startup and shutdown and monitors furnace performance. The technician can use this self-testing feature to identify a major component failure. The control board will check itself, then the inducer, silicon carbide ignition, low- and high-speed blower operation, and humidifier connections. Control boards often include a low-amperage fuse that protects the

Figure 49-25 An inducer blower used to force the combustion gases through the furnace.

transformer and control board. The boards may also include an LED status indicator light.

- The air filter retainer, which holds the air filter.
- The air filter, which is either the replaceable type or an optional electronic filter that can be installed in the return-air opening.
- The wraparound casing, which is a one-piece with seamless construction.
- The heat exchanger, which transfers the heat of combustion to the air-distribution system.
- The blower and blower motor, which force the air over the external surface of the heat exchanger to pick up the heat for delivery to the space. The motor is a direct-drive multiple-speed type.

TECH TIP

Some early Honeywell igniter controllers have an internal fuse. This fuse is not replaceable. If the leads are accidentally shorted, it will blow and the entire control board must be replaced. When working with any electronic device, be certain not to accidentally short any wires together for fear of damaging the component.

When the first 80 percent furnaces were being introduced, there were still many furnaces around with efficiencies in the 50–60 percent range. The 80 percent furnaces represented a midpoint in both efficiency and cost between the traditional standing-pilot, natural-draft furnace and the higher 90 percent condensing furnaces. Today, the 80 percent furnaces represent most manufacturers' bottom line. The minimum efficiency is 78 percent.

49.10 PRODUCT DATA (80 PERCENT AFUE FURNACE)

Following the information supplied by the manufacturer about the product during installation is important. The technician needs to check the installation, either at the time of startup or whenever a service problem arises to determine that the product has been properly installed. Some of this information appears on the nameplate, but not all of it may be listed. Data that is particularly important include:

- Minimum gas-line pressure
- Maximum gas-line pressure
- Manifold gas pressure
- Temperature rise
- Electrical characteristics
- Input capacity
- Output capacity

Although these units are primarily designed for heating, a number of accessories can be added, such as a cooling coil installed in the supply plenum, a humidifier installed on the supply duct, and an electronic air cleaner installed at the return-air entrance to the unit.

CAUTION

In most places, it is against code to install an air-conditioning coil in the return-air plenum of a furnace. If a coil were to be installed in the return-air plenum, the heat exchanger would be cooled below the dew point during the air-conditioning season. The resulting condensate would quickly rust out the furnace and heat exchanger.

The 80 percent AFUE units usually have output ranges of 35,000 to 124,000 BTU/hr. On larger jobs it may be necessary to install two of them linked together in a configuration called twinning. Twinning kits are available for field linking. A number of parts are required, and the manufacturer's instructions must be carefully followed to coordinate the two systems. When twinned, it is imperative that both blower motors energize simultaneously. Also, twinned furnaces typically result in at least a 10 percent loss of combined air volume (in CFM).

TECH TIP

When working on a customer's furnace equipment, avoid tracking grease and/or dirt into their home after the work is complete. If you are in a basement and must exit the house across a clean floor or rug, remove your shoes or cover them with paper shoe covers. You must never leave any tracks behind. A little time spent ensuring your footwear is clean will help ensure customer satisfaction.

49.11 GAS HIGH-EFFICIENCY FURNACES (90+ PERCENT AFUE)

High-efficiency or condensing-type gas furnaces are rated at 90 percent AFUE or better. They differ from the 80 percent AFUE mid-efficiency furnaces in that an extra heat exchanger (secondary heat exchanger) is added to extract more heat from the flue gases. This additional surface causes the moisture in the flue gas to condense. Additional heat is obtained by further lowering the temperature of the flue gas and picking up the heat rejected by the condensing moisture. Notice in Figure 49-26 that the burners are at the top, where the hot air is leaving.

This construction reduces the volume and temperature of the flue gas, making it possible to use smaller vent pipes and simplify venting arrangements. Since the temperature of the flue gas is lower, PVC vent piping can be used and vent outlets can be run to the side of a building, if more convenient.

A condensate line is required to dispose of the condensed water. This needs to be run to the suitable drain. The condensate usually contains some contaminants. Local codes need to be examined to determine the possible need for a condensate-neutralizer kit.

(a)

(b)

Figure 49-26 (a) High-efficiency furnace with burners located at the top where the hot air is leaving; (b) sealed combustion chamber with inspection port.

49.12 PRODUCT DATA (90+ PERCENT AFUE FURNACES)

90+ percent AFUE furnace product data cover information similar to that of the 80 percent AFUE models, but with a few additions. Since these are condensing-type furnaces, a drain for the condensate is needed.

Many high-efficiency furnaces have provisions for piping combustion air to the furnace from the outside (Figure 49-27). This keeps the furnace from drawing air out of the house. Piping in outside combustion air also keeps household chemicals that are in the air inside the house out of the furnace. Piping in combustion air means that two PVC lines are required: one for the combustion air and one for the vent gases (Figure 49-28).

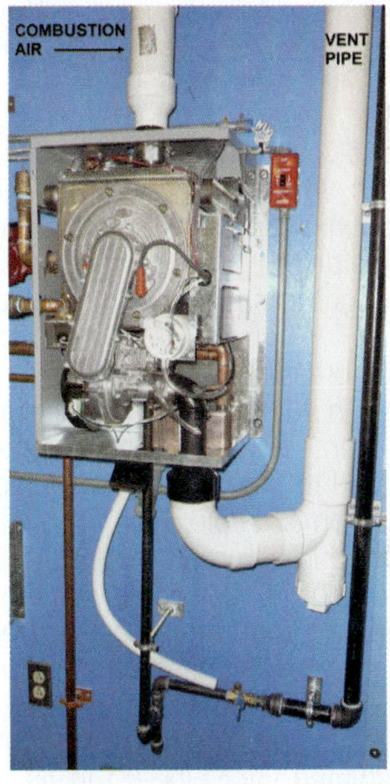

Figure 49-27 Gas furnace with piping for combustion air.

Figure 49-28 When possible, install the combustion air and vent pipes through the side wall of a building, as this method of side venting reduces the possibility of roof leaks.

Table 49-2 A Typical Table Giving Maximum Allowable Vent Pipe Length (ft)

	Number of 90° Elbows					
Pipe Diameter (in)	**1**	**2**	**3**	**4**	**5**	**6**
1½	70	70	65	60	60	55
2	70	70	70	70	70	70

Table 49-3 Limit on the Length of an Exposed Insulated Vent Pipe (ft) in an Unheated Area

		Insulation Thickness (in)				
Winter Temperature (°F)	**Pipe Diameter (in)**	**0**	**3/8**	**½**	**¾**	**1**
20	1½	31	56	63	70	70
0	1½	16	34	39	47	54
–20	1½	9	23	27	34	39

Some direct-vent furnaces use a concentric vent. This allows both vent pipe and combustion air pipe to terminate through a single exit in the roof or sidewall. One pipe runs inside the other, permitting venting through the inner pipe and combustion air to be drawn through the outer pipe.

The positive pressure in the vent pipe created by the draft-inducer fan in 90 percent furnaces is not very strong. Long runs or several elbows can reduce flow through the vent. This can lead to the furnace shutting off on the draft safety switch. As an example, Table 49-2 shows the maximum vent length and number of elbows allowable for one type of a 90 percent condensing furnace. When the vent pipe is exposed to temperatures below freezing, such as when it passes through an unheated space or when a chimney is used as a raceway, the pipe must be insulated with Armaflex-type insulation. There is a limit on the length of an exposed insulated vent pipe (ft) in an unheated area, as shown in Table 49-3. Vent piping should not have any traps or low places where water can collect and form restrictions. In general, the vent should slope back toward the furnace, where provisions are made for collecting and disposing of the condensate. It is possible to install the vent horizontally if the manufacturer's instructions are followed. Typical installations of units and venting arrangements are shown in Figures 49-29 and 49-30.

These values are for altitudes of 0 to 2,000 ft above sea level, using either two separate pipes or a 2-in concentric vent.

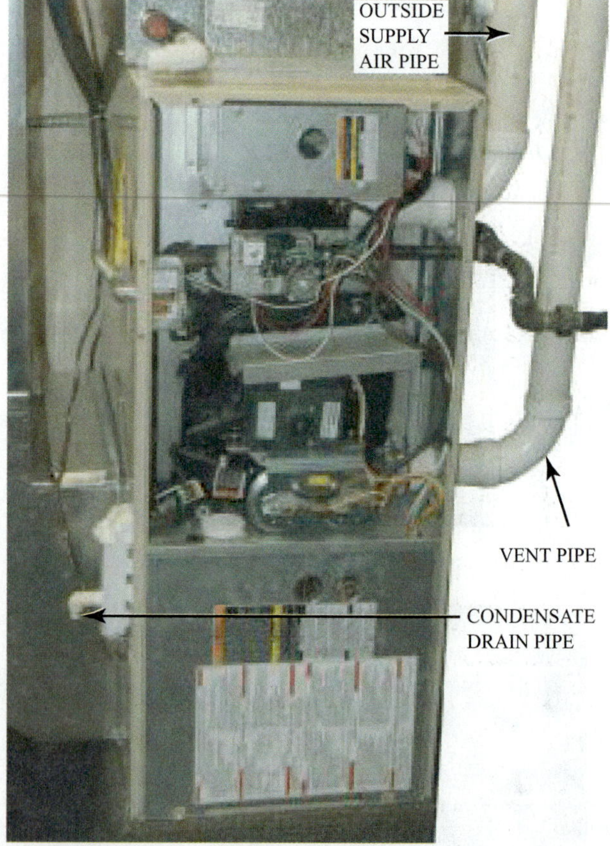

Figure 49-29 Side-venting condensing gas furnace.

SAFETY TIP

Because vent sizing is so important, you must check with the equipment manufacturer and with local and state regulations before designing and installing furnace vent piping.

49.13 COMBUSTION CONDENSATE PIPING

The condensate drain line should be trapped. However, many furnaces have a condensate trap built into the furnace, eliminating the need for another trap. Some condensing

Figure 49-30 Top-venting condensing gas furnace.

furnaces require an external trap in the condensate drain line. The field drain connection is typically sized for ½-in CPVC or PVC. The condensate drain should be run with schedule 40 PVC or CPVC.

Not all condensate pumps are approved for installation on condensing furnaces because furnace condensate is mildly acidic, typically in the pH range of 3.2 to 4.5. When a condensate pump is required, select a pump that is approved for condensing furnace applications. Ideally, the pump should also have an overflow switch to avoid condensate spillage in the case of pump failure or a stopped drain.

Due to corrosive nature of this condensate, a condensate pH-neutralizing filter may be desired. These filters are typically PVC canisters filled with limestone. Check with local authorities to determine if a pH neutralizer is required.

Do not normally combine a furnace condensate drain and evaporator condensate drain. Pressure produced in an upflow coil plenum can pressurize the drain and the secondary heat exchanger in the furnace. This stops the furnace from draining, pressurizes the heat exchanger, and causes the furnace to lock out on its pressure-switch safety.

If the furnace, air-conditioner, and humidifier drains are combined and drained together, then the air-conditioner drain must have an external, field-supplied trap prior to the furnace drain connection. The drain should be ¾-in PVC or CPVC when combining with the air-conditioner drain. When terminating the drain inside, it should be run into an open or vented drain as close to the respective equipment as possible. Outdoor draining of the furnace is permissible if allowed by local codes. Caution should be taken in cold climates because low temperatures may freeze the drainpipe and prohibit draining.

When condensing furnaces are used in an attic application or over a finished ceiling, local codes may require a drain pan under the entire furnace to prevent damage to the ceiling in the event of a plugged drain. The auxiliary drain should terminate in an obvious location so that any condensate from it will be noticed.

49.14 VARIABLE-CAPACITY FURNACES

Two ways manufacturers have to offer a premium furnace are by offering two-stage combustion and ECM blower motors. One of the most wasteful periods of a furnace operation cycle is the warm-up time when the burners are firing and the fan is not on. Increased cycling leads to decreased efficiency. The furnace run time can be extended and the cycling reduced by using a two-stage gas valve. On low fire, the furnace capacity is approximately half of the furnace's full capacity. This allows the furnace to operate for longer periods of time in milder weather. When the full capacity is needed, high fire can still deliver the full furnace capacity. On low fire, the gas valve operates at a lower manifold pressure so not as much gas goes into the burners. On high fire, the manifold pressure is normal, providing full capacity. Two-stage furnaces are available in both 80 percent and 90 percent AFUE ratings.

Customers are looking for more than just efficient gas usage; they want electrical savings also. ECM blower motors are more efficient than the standard PSC blowers and offer many operational advantages. ECM blowers are a natural pairing for furnaces with two-stage operation. ECM blower motors can deliver the correct amount of airflow for both low-fire and high-fire operation and consume less electricity as well. Some manufacturers now produce full modulating furnaces that effectively provide for upwards of one hundred stages of capacity.

49.15 PRODUCT DATA (VARIABLE-CAPACITY FURNACES)

The distinguishing features of variable capacity furnaces are the components that allow for two firing rates, including two-stage gas valves, two-speed induced-draft fan, two-draft proving switches, and the use of a two-stage thermostat.

The two-stage gas valve is the primary component that makes these furnaces possible. It has adjustments for both the low-fire and high-fire manifold pressures. Common settings for manifold pressures are 3.2 in wc to 3.8 in wc for high fire, and 1.4 in wc to 1.7 in wc for low fire. The BTU value of the gas and the altitude of the installation will determine the correct manifold pressures for any particular installation.

Figure 49-31 Two-stage thermostat sub-base.

SAFETY TIP

Always follow the manufacturer's installation instructions when adjusting a furnace firing rate for high altitude. Manufacturers specify he burner orifices and manifold pressure for different altitude applications.

The induced-draft fan motor must also move less air for low fire than it does for high fire. This can be done using a multispeed PSC motor or by using an ECM motor. Two draft-proving switches are also necessary—one for low fire and one for high fire.

Finally, a two-stage thermostat is needed to control the furnace. The thermostat should have both W1 and W2 terminals (Figure 49-31). A heat pump thermostat will not work. Some systems also have provisions for using an outdoor thermostat to control when the furnace runs in low fire and when it runs in high fire.

UNIT 49—SUMMARY

Furnaces can be grouped many different ways: by their fuel source, venting characteristics, cabinet configuration, and efficiency. Most manufacturers offer two types of gas furnaces: 80 percent efficient Category I fan assisted, and 90 percent efficient Category IV condensing furnaces. All furnaces made today use an induced-draft combustion blower and an improved efficiency heat exchanger. High-efficiency furnaces achieve 90+ percent efficiencies by condensing water from the flue gas using a recuperative heat exchanger made of stainless steel. Medium-efficiency furnaces are vented with type B vent, while high-efficiency furnaces are vented with PVC. Some high-efficiency furnaces require the combustion air to be piped in from outside.

WORK ORDERS

Service Ticket 4901

Customer Request: Old Furnace Needs Replacement

Equipment Type: Natural-Draft Horizontal Gas Furnace

A sales consultant is contacted by a customer and is asked for recommendations to replace an old natural-draft horizontal gas furnace installed in a crawlspace. The first step in this process is for the sales consultant to visit the residence and determine what will work best. The structure of the building and the heating load requirements need to be taken into account.

The consultant observes that the old furnace is vented into a masonry chimney. This will lead to additional costs because an 80 percent furnace cannot be vented into the masonry chimney without installing a liner. The customer asks if there are any other alternatives to consider. The consultant does have an option. The money that would be spent on the liner may be invested in a 90 percent furnace, which can be vented with PVC out the side of the crawlspace. This will not only save on fuel bills, but it will also eliminate the need for lining the existing chimney.

Service Ticket 4902

Customer Complaint: Old Furnace Needs Replacement

Equipment Type: Natural-Draft Upflow Gas Furnace

A customer would like to replace an old natural-draft upflow gas furnace installed in a basement. After visiting the residence, the sales representative explains to the customer that an 80 percent Category I furnace can use the old venting system.

However, after careful consideration, it is noted that the old furnace used a 7-in vent and the new 80 percent model only needs a 4-in vent. The increased size of the vent could lead to condensation and premature failure of the furnace. The consultant explains to the customer that although the new 80 percent model will work just fine, the vent system must also be replaced with the furnace.

Service Ticket 4903

Customer Request: New Home Furnace Installation

Equipment Type: 80 Percent Two-Stage Furnace

A customer wants to install a 90 percent condensing furnace in his new home, but the only place available is in the attic. The house is located in a very cold climate and the attic frequently drops below freezing in the winter. In this type of installation, the furnace condensate could freeze, especially if the system is operated and then shut off for several hours.

The sales consultant visits the residence and spends some time discussing the possibilities with the customer. Although a 90 percent condensing furnace may save in heating costs, the potential for freeze-up is problematic. To avoid this, the furnace would need to run more frequently, and there would still be no guarantee that the condensate would not freeze in the cold attic. The sales consultant instead recommends an 80 percent two-stage furnace with an ECM fan motor. In the long run, this would be the best alternative for the customer.

Service Ticket 4904

Customer Request: New Installation in Closet

Equipment: 90 Percent Counterflow Furnace

A sales consultant is asked for recommendations for a counterflow furnace to be installed in a closet in a new home. The design calls for the ductwork to be located within the house's concrete slab. The homeowner asks the sales consultant if this type of design is acceptable.

The sales consultant explains that with this type of installation there could be concerns regarding the proper supply of combustion air. This is mainly because the furnace will be located in the living space. Provided the design criteria and after consulting with the customer, the sales consultant recommends a 90 percent furnace with combustion air piped in from the outside. This will provide sufficient air for operation of the furnace and will not take away from any air in the living space.

Service Ticket 4905

Customer Request: New Furnace Relocation

Equipment Type: Upflow/Horizontal Flow Furnace

A homeowner has decided to install a furnace in the basement of his house. He purchases a furnace on sale from Furnace Mart and begins to install the furnace in the basement. After the job gets started, his wife inspects the furnace in the basement and does not like its location. She would prefer that the furnace be installed in the attic like the neighbor's.

The homeowner explains to his wife that he has already purchased the furnace and it is designed to be installed in the upflow position. His wife tells him to call a professional and find out if anything else can be done. At first the homeowner is reluctant but then after looking over the system he decides that maybe his wife is right. He is relieved after talking to a technician. It appears that the attic installation will work out just fine and the furnace he purchased can be adapted for this configuration. Although it costs the homeowner more money than if he installed the furnace himself, he feels reassured that the job was done professionally and with every safety requirement in mind.

UNIT 49—REVIEW QUESTIONS

1. List the four categories of gas-fired furnaces based on flue-gas temperature and pressure.
2. Explain the difference between a Category I furnace and a Category IV furnace.
3. List the five furnace cabinet configurations.
4. Why are standing-pilot natural-draft furnaces no longer manufactured?
5. What changes were made in furnace design to achieve 80 percent efficiency?
6. What is the difference between an 80 percent efficient furnace and a 90 percent efficient furnace?
7. Discuss the evolution of the heat exchanger.
8. Discuss the operation of an atmospheric burner.
9. Explain why furnace efficiencies jump from 80 percent to 90 percent.
10. What is the purpose of the orifice?
11. Why is it not a good idea to drill out orifices in the field?
12. What components are needed for a furnace to operate as a two-stage furnace?
13. Why are some condensate pumps not approved for furnace condensate?
14. Can the furnace drain and the air-conditioning drain be run with a common drain?
15. List the different types of gas burners.
16. Why do most furnaces today use burners?
17. What is the temperature rise of a furnace?
18. What part of the furnace does the air travel through first in all furnace cabinet configurations?
19. Why should technicians be familiar with standing-pilot, natural-draft furnaces, since they are no longer manufactured?
20. Why are draft-inducer fans necessary in today's furnaces?

UNIT 50

Gas Furnace Controls

OBJECTIVES

After completing this unit, you will be able to:

1. identify and name the different types of control components.
2. describe the operation of flame safety devices.
3. explain the function of gas pilots.
4. explain the operation of a direct-spark ignition system.
5. describe how gas valves operate with thermostat and limit controls.
6. determine the maximum length of wire run to limit voltage drop.
7. explain the operation of a control circuit.
8. wire a control circuit.

50.1 INTRODUCTION

Controls can be used to accomplish a number of different tasks in a system. Most commonly they will function to operate a system within prescribed limits, such as a thermostat control that starts a furnace when the temperature falls below the set point. Controls can also serve as safety devices—for example, to shut down a furnace if there is a flame failure. They can be designed with remote operation for convenience purposes, making it easier for the customer to change the operational settings. In addition, controls can be designed to cycle the furnace on and off in such a manner as to operate the most efficiently.

50.2 TYPES OF CONTROL SYSTEMS

Control systems will vary depending on the furnace type and design. Gas warm-air furnaces will have slightly different controls than gas hydronic furnaces. There are also differences between single-stage, multistage, and variable-speed furnaces.

Single-Stage Gas Furnaces

A single-stage gas warm-air furnace will have a very simple type of control system. A thermostat located in the space to be heated is one of the major components. When the thermostat calls for heat, the furnace will start. The fan switch will turn on the blower at a set temperature or time-delay setting. There are also safety controls such as the limit switch. This prevents the furnace from overheating. Gas valves are also designed for safety and will close to stop the gas flow upon an unexpected flame out.

The control components for a single-stage furnace can be very basic, such as a simple snap-action thermostat, some basic relays, a fan switch, a limit switch, and a combination gas valve. The major drawback to this type of furnace is that the heat output is constant. This means that the furnace is either on or off. To keep the building at a constant temperature, the furnace will continually cycle.

Multistage Gas Furnaces

Multistage gas furnaces are more efficient. They will produce heat at two or more different rates. Two-stage gas furnaces are a very common type of multistage furnace. As an example, there will be a high and a low setting. When less heat is required, the furnace will operate on its low setting rather than cycling repeatedly from high to off. Multispeed blowers and two-stage gas valves are required. The control system will be much more sophisticated, often using electronic thermostats and controls.

Not only are multistage gas furnaces more efficient than the single-stage types, but because of the steadier operation, there is less fluctuation in room temperature. The slower speed of the blower at low firing reduces drafts and is quieter. Due to these reasons, multistage gas furnaces have been replacing the single-stage types.

Variable-Speed Furnaces

Variable-speed gas furnaces will use a variable-speed ECM (electronically commutated motor)-driven blower in combination with a computer-controlled thermostat. The furnace control system will adjust automatically to maintain a constant temperature. They are the most efficient and quietest type of gas furnace. Variable-speed blowers use significantly less electricity than traditional single-stage furnace blowers. Although variable-speed furnaces will cost more initially, they will save the customer money over time.

Warm-Air Gas Furnaces Compared to Hydronic Gas Furnaces

Hydronic gas furnaces heat water that is circulated through the building rather than circulating forced warm air. This is very common for oil heat. Hydronic systems are explained in detail in Unit 80, Hydronic Heating Systems. Ductwork is not required for a hydronic system. This can be seen as an installation advantage for areas of the country were central cooling is not always necessary.

Hot water is circulated through baseboard heaters or through radiant-heat piping installed under the floor joists or

Figure 50-1 Low-voltage thermostat used for heating only.

in the slab. Radiant heat piping has become increasingly popular and in some designs will also include a solar hot-water loop. A hydronic gas furnace can also be used in combination with water-to-water heat pumps and geothermal systems.

In general, however, traditional hydronic systems are difficult to adapt for cooling because additional ductwork just for cooling would need to be installed. Therefore, many of the gas hydronic gas furnaces are heat only (Figure 50-1).

In comparison, warm-air gas furnaces will have an installed ductwork system that can be used for both cooling and heating. This is ideal for the many areas of the country that require both. Therefore, the control thermostat for warm-air gas furnaces often includes both heating and cooling modes (Figure 50-2).

50.3 THERMOSTAT OPERATION

Thermostats automatically measure and maintain a desired room temperature. The key components of a thermostat are the temperature sensor and the temperature switch. The most basic type of thermostat configuration will send an electrical signal to turn the heating equipment on or off. Electromechanical thermostats utilize a snap-action effect to open or close contacts. They are designed to sense the temperature change of the room by using a bimetallic element. Electronic thermostats utilize thermistors or other integrated circuit sensors to sense temperature changes.

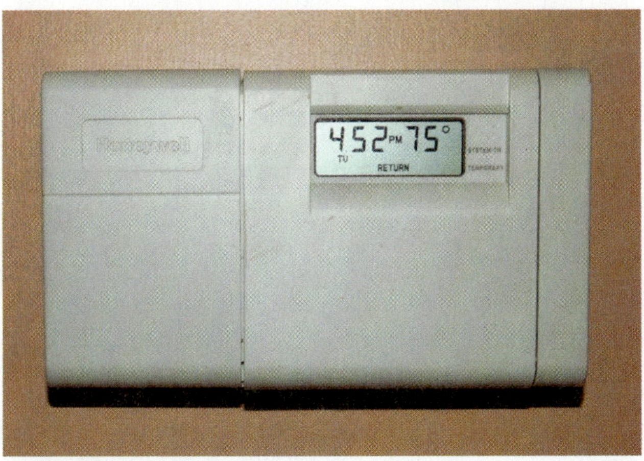

Figure 50-2 Low-voltage thermostat used for both heating and cooling.

The cut-in point is the room temperature at which the thermostat will turn the furnace on, and the cut-out point is when the furnace is turned off. The difference between the cut-in and the cut-out is called the differential. A large differential will lead to a greater temperature swing in the room. A low differential will cause the furnace to cycle on and off more frequently.

Even though the cut-in temperature turns the furnace on, it will take some time before the room begins to warm up. This is known as system lag. Likewise, there may be some residual heat from the furnace even after it has shut off on cut-out, and this is called system overshoot.

Temperature swing is the range between the highest and the lowest room temperature controlled by the thermostat. This is affected by the differential, the system lag, and the system overshoot. Thermostats should be installed in a proper location and adjusted correctly to provide for comfortable room conditions with minimal temperature swings.

50.4 MECHANICAL THERMOSTATS

Mechanical snap-action thermostats (Figure 50-3) have been widely used. The use of spiral-shaped, lightweight bimetallic elements increases the effective length and thus the sensitivity to temperature change. Sealed contacts eliminate the problem of dirt and dust. Although there are minor variations among different manufacturers, the contacts are always sealed in a glass tube.

A single-action mercury bulb design is shown in Figure 50-4. As the bimetallic strip expands and curves,

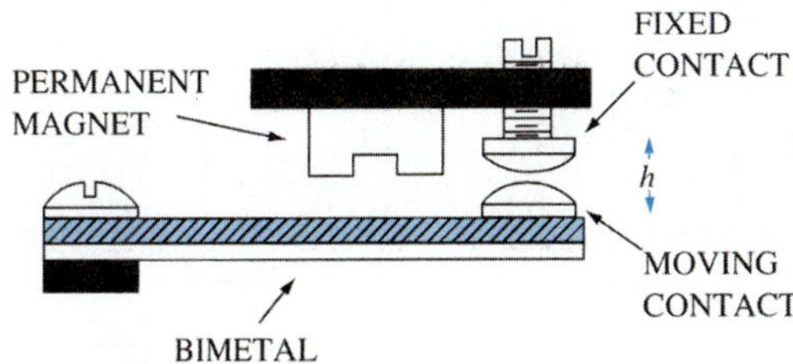

Figure 50-3 Snap action thermostat.

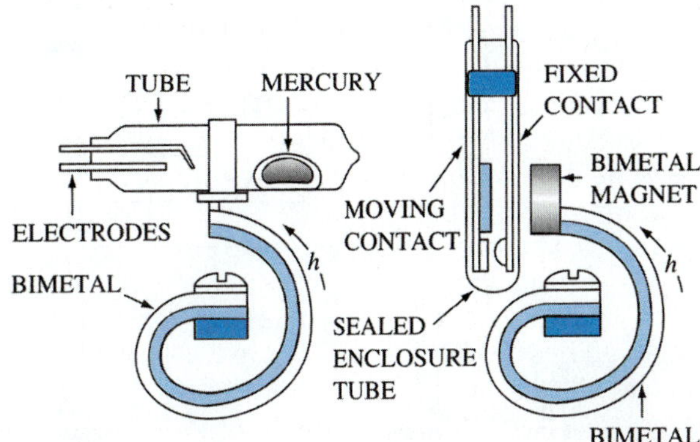

Figure 50-4 A single-action mercury-bulb thermometer on the left; a sealed tube with metal-to-metal contact on the right.

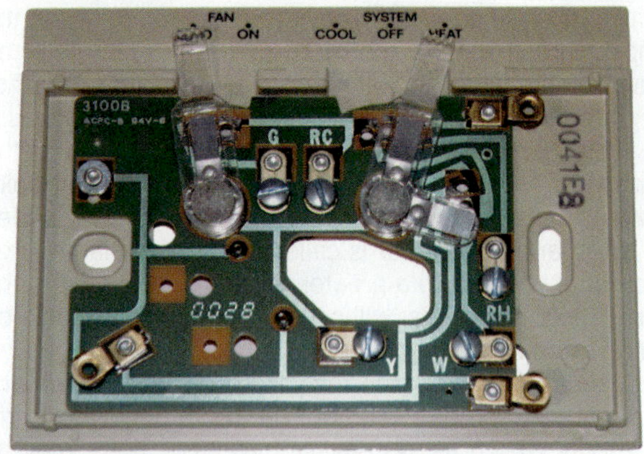

Figure 50-5 Sub-base portion provides a mounting base for the thermostat and for system controls.

the mercury fluid moves to the left, completing an electric circuit between the two electrodes, which carry only 24 V. The differential gap between OFF and ON is very small: 0.75°F to 1°F from the setpoint. On the right is a sealed tube that has a metal-to -metal contact. The magnet provides the force that closes the contacts.

The sub-base portion of the thermostat assembly (Figure 50-5) not only provides a mounting base for the thermostat but is also used to control the system operation through a series of electrical switches. The system switch selects COOL, OFF, or HEAT. The blower operation is controlled by the fan switch, which is a simple two-position switch. When set on automatic, the fan will cycle on and off with the system whenever there is a call for heating or cooling. In the ON position, the fan will run continuously.

More complex thermostats contain two adjustments that allow different control points for heating and cooling to be set. These thermostats may have an automatic changeover from heating to cooling. Figure 50-6 represents a two-stage cooling schematic wiring diagram. Two-stage thermostats are common in heat pumps or commercial rooftop equipment, which use multiple compressors for cooling and two or more stages of heating. Note that seven electrical connections are required; however, RH and RC terminals are the same power source, so six wires are needed. With color-coded low-voltage wire, however, it is no problem to connect the thermostat to the mechanical equipment.

GREEN TIP

The mercury found in mercury bulb thermostats is a deadly toxin. Mercury bulb thermostats should not be thrown away. They should be recycled so that the mercury can be safely recovered.

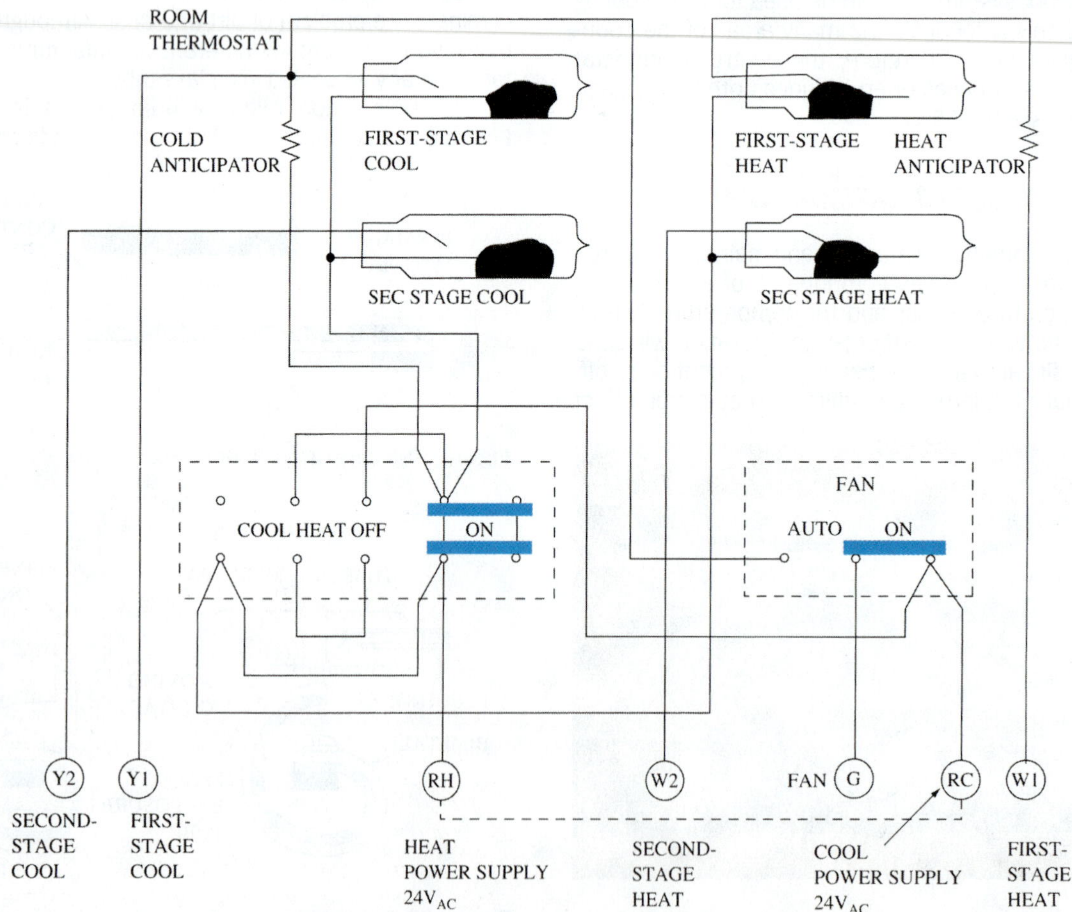

Figure 50-6 A two-stage heating and two-stage cooling schematic wiring diagram.

50.5 ELECTRONIC THERMOSTATS

Most thermostats today are electronic because they can maintain room temperature within less than a degree, do not use mercury, and can perform more functions than traditional mechanical thermostats. Electronic thermostats can be programmable or nonprogrammable. These thermostats often use thermistors or other integrated circuit sensors to sense temperature changes. Thermistor operation is based on the fact that the electrical resistance of a ceramic semiconductor changes as its temperature changes. Generally these thermostats have an LED or LCD display for enhanced readability. Electronic thermostats use relays or electronic switches to control system operation.

Programmable electronic thermostats can be programmed to set back the temperature at predetermined times and days (Figure 50-7). This allows for additional energy savings by automatically reducing the demand for heating or cooling—for example, at night or during the day when no one may be home. These can also provide additional information such as filter change reminders and system runtime data. Many of these thermostats provide for adjusting burner on-time and fan operation.

Single-Stage, Two-Stage, and Modulating Thermostats

Some furnaces can be configured with different thermostat arrangements. As an example, assume a furnace that modulates between 40 percent and 100 percent capacity is to be wired with a single-stage thermostat. The thermostat would need to be connected according to the manufacturer's wiring diagram, and the correct dual in-line package (DIP) switches would need to be turned off. With the switches off, the lack of the modulating signal will automatically be sensed as a single-stage thermostat. Instead of operating as a fully modulating furnace, this furnace will operate at the three different firing rates of 40 percent, 65 percent, and 100 percent. For the first 5 minutes of operation, the furnace will operate at 40 percent capacity. For the next 7 minutes, the furnace will operate at 65 percent furnace capacity. After 12 minutes, the furnace will operate at 100 percent capacity. If the call for heat ends, the furnace will shut off regardless of the firing rate at that time. When restarting, the furnace will once again start at 40 percent capacity.

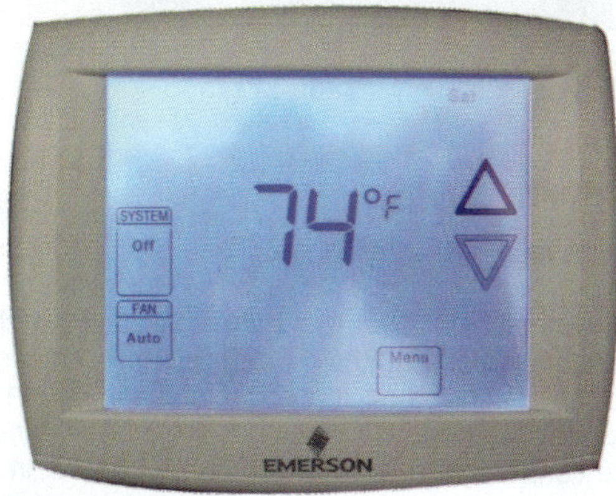

Figure 50-7 Programmable electronic thermostat with LED display.

In comparison, a two-stage thermostat wired to the same furnace would provide different operating conditions. The first-stage heat would allow for continuous operation at 40 percent capacity. The second-stage heat would allow 65 percent capacity for the first 5 minutes and then 100 percent capacity continuously after that.

A modulating thermostat connected to this same furnace would allow for full modulation between 40 percent and 100 percent capacity. The firing rate is first determined by the thermostat and then sent to the furnace. This type of control would provide the best temperature control and furnace efficiency.

50.6 HEAT ANTICIPATION

Heat anticipators are used in mechanical low-voltage thermostats to reduce the temperature swing caused by the systems. Heating anticipators are small electrical resistors that generate a little heat inside the thermostat when the heat is on. This "false heat" on the bimetal tricks the thermostat into thinking the room has reached the set temperature. The furnace burner goes off when the thermostat thinks the room is warm enough, but the fan keeps running until the heat exchanger cools down. Without a heating anticipator, the room temperature would overshoot the set temperature. Temperature swing is the difference between the temperature when the furnace comes on and goes off.

Although heating anticipators result in closer room temperature control and less overshooting of heating, they can shorten the heating cycle time. If the heating anticipator is set too high, the furnace will short cycle.

Some heating anticipators are fixed, while others are wire-wound variable resistors wired in series with the load. They are rated in fractions of amperes. On the adjustable type, the installer will position the sliding arm to the proper load rating. A heating anticipator is shown in Figure 50-8. The initial setting should match the amp draw of the gas valve. But if on/off cycles are too long or too short, the system operation can be changed by adjusting the anticipator control lever to give a faster or slower response.

Electronic thermostats do not use false heat to anticipate when to cycle heating and cooling equipment. They can measure temperature changes of less than 1°F and are able to anticipate based on temperature changes and logic.

SERVICE TIP

Some customers are more sensitive to temperature change than others. These customers might have a complaint that the room gets too cold before the heat comes on. They might also complain that this occurs even when the temperature has been turned up slightly. What they are feeling is the temperature swing resulting from the heating anticipator being set too low. By raising the current setting on the heating anticipator, the heating cycle will shorten, and the temperature swing will decrease.

Figure 50-8 An adjustable heating anticipator.

Figure 50-9 Blower off delay adjustment on furnace board.

TECH TIP

The longer a furnace runs during its heating cycle, the more efficiently it operates. Setting the heating anticipator so that a larger temperature swing occurs will result in energy savings. Conversely, a short-cycling heating system is less efficient to operate.

SERVICE TIP

Adding humidity to a residence during the heating season will allow your customer to feel more comfortable at a lower thermostat setting. A large part of a person's comfort is the rate at which perspiration evaporates from the skin. As the relative humidity goes up, that rate of evaporation decreases; thus people feel warmer even at a lower actual room temperature. A humidistat can provide that control, which will increase your customer's comfort and save heating costs.

50.7 FAN AND LIMIT SWITCHES

Fan Switch

The fan switch controls the starting and stopping of the blower motor. The fan does not start when the burner starts. Instead, the fan switch closes and starts the blower when the air temperature inside the furnace plenum chamber has warmed up to the FAN ON setting. The fan switch remains closed when the burner stops, keeping the blower in operation until the air temperature inside the furnace cools down to the FAN OFF setting. Modern furnaces cycle the fan based on time rather than temperature. A relay built into the control board energizes the fan 1-2 minutes after the burner lights. The relay turns the fan off 1-2 minutes after the burners shut off. The off delay is usually adjustable, Figure 50-9.

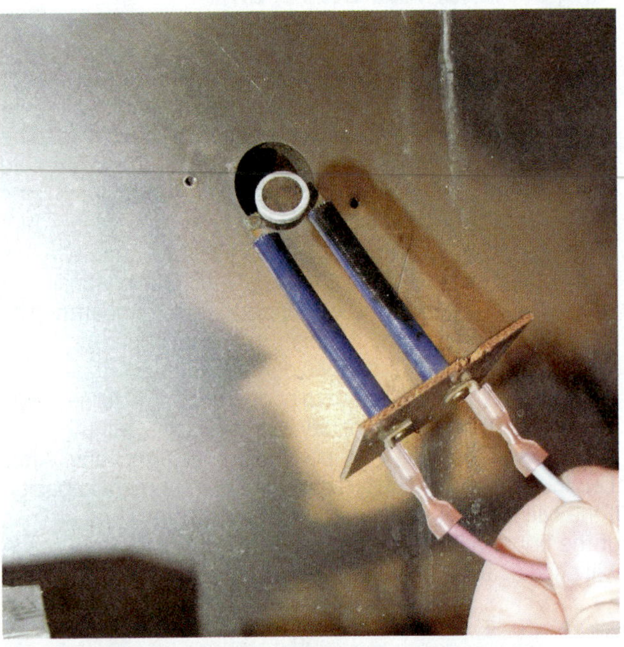

Figure 50-10 Limit switch.

Limit Switch

The limit switch is a safety control that prevents the furnace from overheating Figure 50-10. As long as the plenum chamber temperature is below the setting of the limit switch, the switch remains closed. If the plenum chamber temperature rises to the switch setting, it opens to de-energize the heating control circuit and close the gas valve. Most limits today are connected to the furnace control board. In older furnaces the limit was either wired directly in the 24 volt circuit to the gas valve or in the line voltage circuit to the transformer primary.

Figure 50-11 Auxiliary limit switch.

Auxiliary Limit Switch

Downflow, horizontal, and multi-poise furnaces also have auxiliary limit switches located in the fan compartment because the blower compartment can become overheated by natural convection, Figure 50-11.

Combination Fan and Limit Switch

The combination fan and limit switch (Figure 50-12) is actually two switches in one. As a safety limit, some have fixed limit temperature settings; others are adjustable (approximately 180–200°F is the usual range). This allows a 50 to 60°F rise above normal operation before it opens. The fan control switch is also a temperature-sensing device that is set to turn on the fan after the furnace has warmed up at least 15 to 20°F above room conditions, so that cold drafts are not experienced. It also stops the blower after the burner cuts off, so again there are no uncomfortable drafts. It is important to note that some systems employ constant fan operation and thus override this switch.

Other models may use spiral, flat bimetallic, or even liquid-filled elements. Some forms of duct-mounted limit controls use a rod and tube or a liquid-filled bulb to sense the air conditions.

Figure 50-12 Combination fan and limit switch acts as a safety limit and temperature-sensing device.

Figure 50-13 Draft pressure switch.

Figure 50-14 Dual draft pressure switches for two stage furnace.

Air-Pressure Switch

Furnaces that have a draft inducer may have an air-pressure switch that verifies air flow by monitoring the pressure differential between the draft inducer and the atmosphere Figure 50-13. Insufficient negative pressure indicates a lack of air flow and the ignition control module will shut off the gas supply. The draft switch is the first safety in the ignition sequence of most furnaces. The board checks to see that the draft switch is open before starting the draft inducer. After starting the draft inducer, the board checks to see that the draft switch is closed. If the draft switch does not close, the ignition process will go no further. Two-speed draft inducers require a dual air-pressure switch for both high- and low-speed air pressures Figure 50-14.

Flame Rollout Switch

Flame rollout can be caused by low air flow for a given burner firing rate or blockage in the vent system or heat exchanger. Flame rollout switches are located around the furnace cabinet near the burners, Figure 50-15 If flame rollout occurs and the temperature at the switch rises, the rollout switch breaks the circuit to the control module and the control module shuts off the gas valve. Normally the flame rollout switch must be reset manually before the furnace will restart.

Figure 50-15 Roll-out switch mounted above burners.

50.8 IGNITION SYSTEMS

Standing Pilots

Older furnaces use standing pilot lights for ignition, but standing pilot lights are no longer used in modern gas furnaces. A standing pilot is a small burner that is lit continuously. If the pilot flame goes out, the LP gas, which is heavier than air, could collect over a prolonged period to be a hazard. There are three types of flame-proving devices that will shut off the gas flow if the pilot flame goes out: thermocouples (thermopiles), bimetallic safety devices, and liquid-filled remote bulbs. As long as the pilot flame heats the flame-proving device, the gas safety valve remains open and the gas will flow when called for. If the pilot flame goes out and the flame-proving device cools, the gas safety valve will close. In most cases, the pilot will need to be manually reset.

Intermittent Pilot Ignition

An electronic intermittent pilot-ignition system is shown in Figure 50-16. Pilot gas is fed to the assembly and burner orifice. The spark electrode is positioned to ignite the gas on a signal from the room thermostat. With the pilot ignited and burning, a sensing probe establishes a small DC current of 2 to 50 microamps by using flame rectification. The control board energizes the main gas valve and de-energizes the spark-ignition circuit. As long as the sensing probe recognizes the pilot flame, the gas valve will be energized and the spark-ignition system will remain de-energized.

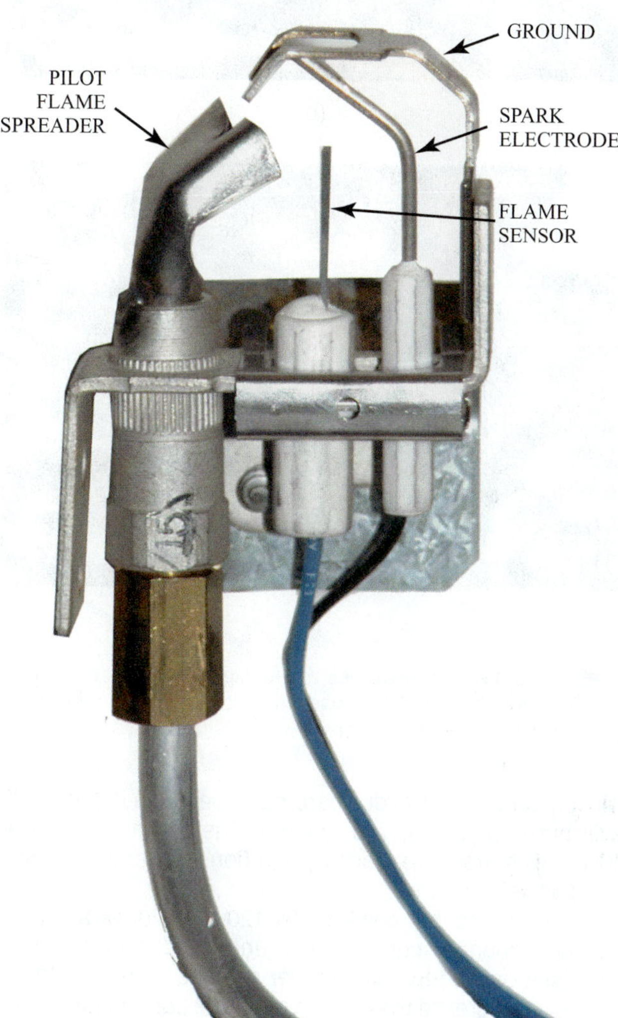

Figure 50-16 Gas furnaces that do not have a standing pilot are often referred to as being "pilotless."

SAFETY TIP

Approximately 10,000 V are generated in the spark-ignition circuit. Although the spark has very low amperage and is not considered a major safety hazard, it can cause a great deal of discomfort if it is energized while you are working in the area of the spark igniter or spark-igniter wire.

Direct Spark Ignition

The direct spark-ignition system (Figure 50-17) uses an electrical flame sensor rod mounted so that it has direct contact with the main burner flame. Because the gas flame will carry electrical energy by means of electrically charging the carbon atoms in the gas before combustion takes place, a current flow can be passed from the burner to a positively charged flame rod. When current is allowed to flow in only one direction, that current is said to be rectified. The flame rectifies the AC current so that it flows out of the flame as pulsed DC. This current flow controls a circuit in the solid-state control module and keeps the gas

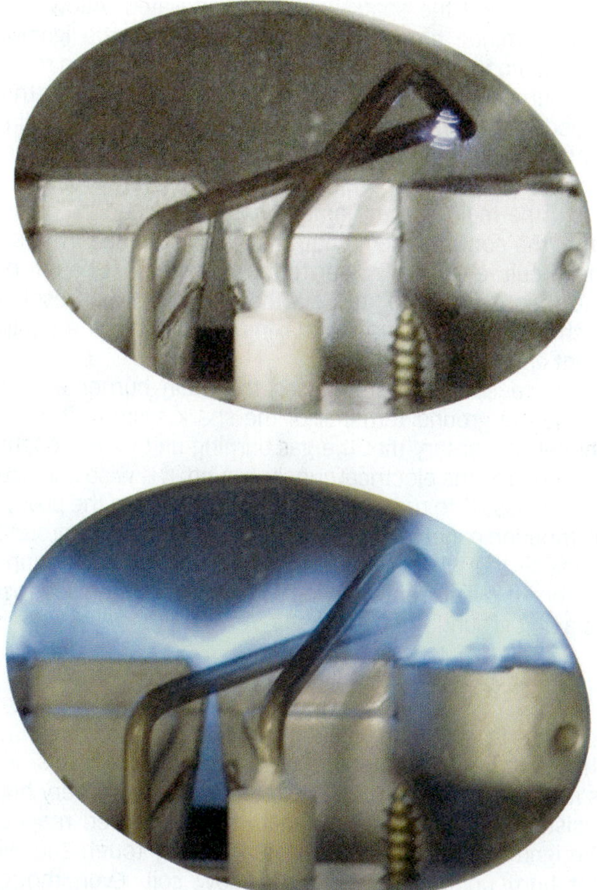

Figure 50-17 Spark-ignition pilot assembly.

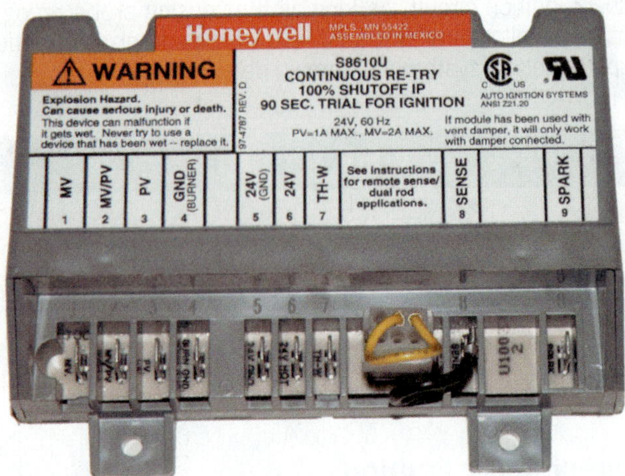

Figure 50-18 Pilot and gas valve controller.

valve energized (Figure 50-18). Flame rectification provides several unique features for a flame sensor. For example, if the flame rod has shorted against the furnace, it will pass the AC current through the furnace and the controller will not open the gas valve because pulsed DC is not present. Also, if the flame does not light, no current will flow, and again the controller will not open the gas valve. If the flame goes out, the controller will shut off the gas valve once the pulsed DC signal is lost.

Upon a call from the room thermostat, both the main gas valve and the spark igniter are activated. Allowing a predetermined time for main flame ignition, the ignition control module will shut down the lockout circuit and maintain burner operation if main flame ignition occurs in that period. The period may be from 4 to 21 seconds, depending on the model of control module used. Generally, a higher input to the gas unit results in a shorter proving time.

If the main burner flame is not established in the set time, the control module automatically locks out. To reset the circuit, electrical power to the system must be cut off and then back on to start another cycle. Manual reset of such a system is used for maximum safety to the equipment and building.

Because this system uses the main burner assembly as the ground terminal of the spark system, it is absolutely necessary that the gas burning unit be thoroughly grounded to the electrical supply ground. It is wise and usually necessary to run a ground (green) wire from the power-distribution panel to the unit to provide this positive ground. Because the white or neutral wire is a 120/240 V supply system, it is a current-carrying wire. It is not suitable for use as a unit ground.

Hot-Surface Ignition

A hot-surface ignition unit (Figure 50-19) is used for igniting the gas burners on many of the latest furnaces. This unit is made of silicon carbide, a material that has a very high resistance to current flow and when energized reaches the ignition temperature of gas. It is very tough and will not burn up, something like a glow coil. Even though it will last a long time, it is also brittle and will break easily with rough treatment. Make sure you do not touch the silicon carbide with your bare hand because it eventually will burn out where you touch it. The control system allows this material to reach the ignition temperature before the gas valve opens.

These units are powered by 120 V and draw a considerable amount of current when energized; however, they are used only a few minutes per day and therefore do not materially increase the electric bill. If the burner fails to light or the flame sensor fails to detect a flame after the gas is turned on for a few seconds, a safety lockout will occur to stop the flow of gas.

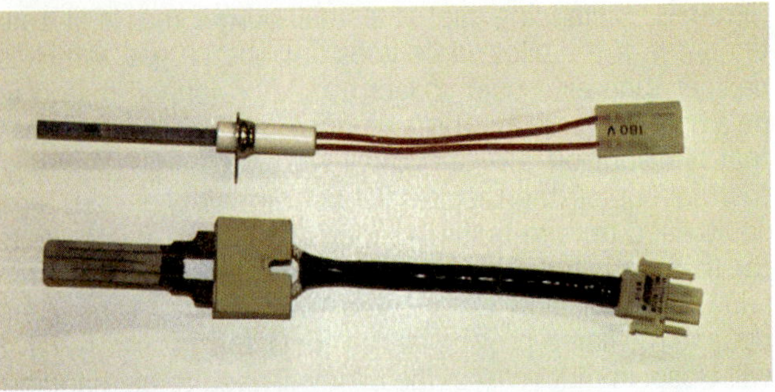

(a)

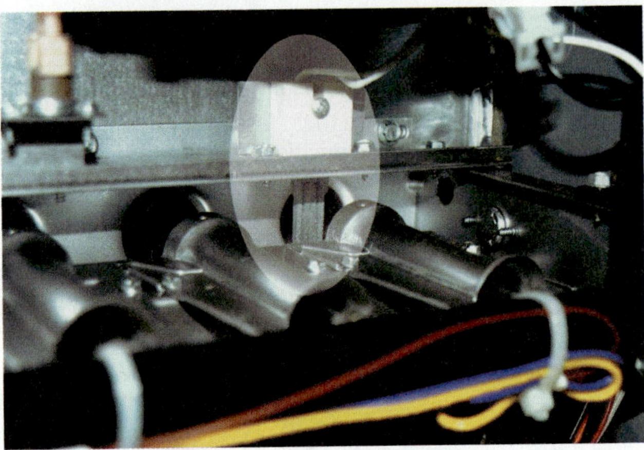

(b)

(c)

Figure 50-19 (a) Hot surface gas fuel igniter; (b) hot surface gas fuel igniter located in furnace; (c) hot surface gas fuel igniter glows red before gas valve opens.

Figure 50-20 Combination gas valve.

50.9 COMBINATION GAS VALVES

At one time, supplying fuel gas to a heating unit was done by a combination of controls consisting of a gas pressure regulator, a solenoid valve, and a pilot safety valve. These controls were combined into a single valve assembly (Figure 50-20) to meet the standards for proper ignition, input control, and quiet cutoff of gas unit operation.

Combination gas valves have the following functions:

- Manual control for ignition and normal operation
- Pilot supply, adjustment, and safety shutoff
- Pressure regulation of burner gas feed
- On/off main gas valve controlled by the room thermostat

CAUTION

Unless you have received specific manufacturer's training on servicing gas valves, do not try to repair one that is not working. It is not safe for technicians to disassemble and reassemble gas valves and gas regulators. If it is determined that the problem is in the gas valve, it should be replaced and not repaired.

50.10 MAIN GAS VALVES

Main gas valves are used to control the flow of gas to the burner. The valve breaks down into four main parts: the valve body, seat, disk, and stem. Anytime the disk is lifted off the valve seat, gas will flow through the valve. Gas valves can be classified as either direct acting or pilot operated. The solenoid, bimetal, and heat motor vales are all direct acting because the mechanism that is energized electrically to open the valve is also the device that mechanically opens the valve. Diaphragm valves are pilot acting; the diaphragm is what actually opens and closes the valve, but it is opened and closed by pressure difference. A small valve called an operator is energized electrically to control the pressure that opens and closes the diaphragm valve.

Direct-Acting Valves

The solenoid valve is the most familiar type of direct-acting valve. Solenoids may be normally closed or normally open, but most are normally closed. When power is applied to the coil, the magnetism created pulls the stem up, lifting the valve disk off of its seat. When power is turned off to the coil, the weight of the stem and/or spring pressure forces the stems back down, reseating the disk. Solenoids open and close quickly. This presents problems for some types of burners that work better if the gas pressure rises gradually, rather than instantly. To help solve this problem and reduce noise, solenoids have been made with springs and fluid to slow down the stem. Rubber and plastic seats and/or disks also help reduce noise and provide a more positive seal.

Another type of direct-acting valve is called bimetal. Bimetal valves are a slow opening by nature. They use a heater wrapped around a bimetal strip to open and close the valve. Voltage to the heater warms the bimetal, which causes it to warp, moving a soft disk or plug and opening the valve. Bimetal valves also close slowly since the bimetal must cool of before closing the valve.

Heat motor valves also use heat from an electrical heater to open the valve, but they use an expansible rod instead of a bimetal. An expansible rod is a rod inside a sealed cylinder that is filled with fluid. Heating the cylinder causes the fluid to expand and push the rod out. A disk is connected to the end of the rod and rests on the underside of the valve seat. Unlike the other direct-acting valves discussed, heat motor valves push the disk down to open the valve and spring pressure pushes the disk up to close the valve.

Diaphragm Valves

Most furnaces today use pilot-acting diaphragm valves. Operators control small pilot valves that control pressure to the underside of a diaphragm, causing the diaphragm to open or close. Operators can be a small solenoid, bimetal, or heat motor. The pilot valve controls a bleed port, which controls the pressure difference across the diaphragm to open and close the diaphragm. Underneath the diaphragm, the incoming gas pressure pushes up, tending to open the valve. Above the diaphragm is a bleed chamber. When it is open to the incoming pressure, the gas pressure also tends to push down, closing the valve. In other words, the same gas pressure is exerted on both sides of the diaphragm. The weight of the diaphragm and a slight spring pressure ensure that the valve remains closed when there is no pressure difference across the diaphragm. When the bleed valve is activated, the pressure in the bleed chamber is opened to the outlet side of the valve. Since this is usually atmospheric pressure at the beginning, it is lower than the incoming gas pressure, which is pushing up on the diaphragm. The diaphragm lifts due to the higher pressure beneath it and lets gas flow through the valve. Even after gas enters the manifold, the pressure after the gas valve will be slightly lower than the pressure entering the gas valve, and the diaphragm will remain open as long as the bleed chamber is open to the outlet of the gas valve. When the operator is de-energized, the bleed chamber is closed to the valve outlet and opened to the gas inlet. The chamber pressure climbs to the incoming gas pressure and forces the diaphragm back down, closing the valve. Figure 50-21 shows a typical combination gas valve with a diaphragm-operated main valve.

Redundant Gas Valves

On standing-pilot gas valves, gas must pass through two shutoffs to exit the valve: the millivolt pilot safety solenoid and the main gas control. Because standing pilots are rarely used today, this has eliminated the need for the millivolt pilot safety shutoff. Most furnaces now use intermittent pilot ignition or direct ignition controls. They will use two automatic shutoffs that are mechanically in series to provide the same level of safety. These valves are called redundant gas valves because they have two valves to shut off the gas flow. Figure 50-22 shows a typical redundant gas valve.

Redundant Valves for Intermittent Pilot Ignition Systems

These systems replace the millivolt solenoid with a 24 V solenoid in line to the pilot burner and the main burner. There is still a diaphragm valve for the main burner. The gas to the main burner has to go through both the solenoid and the diaphragm valve, providing redundant safety. The first solenoid is energized along with a spark igniter to light the pilot burner. Flame rectification is most often used for pilot proving. After the pilot flame is proved, the spark igniter is de-energized and the main burner valve is energized to light the main burner. If the pilot is not proved, the main burner valve will not be energized. On some systems, the spark igniter will keep trying to light the pilot while others will "lock out" after several failed attempts at lighting. "Locking out"

Figure 50-22 Redundant gas valve.

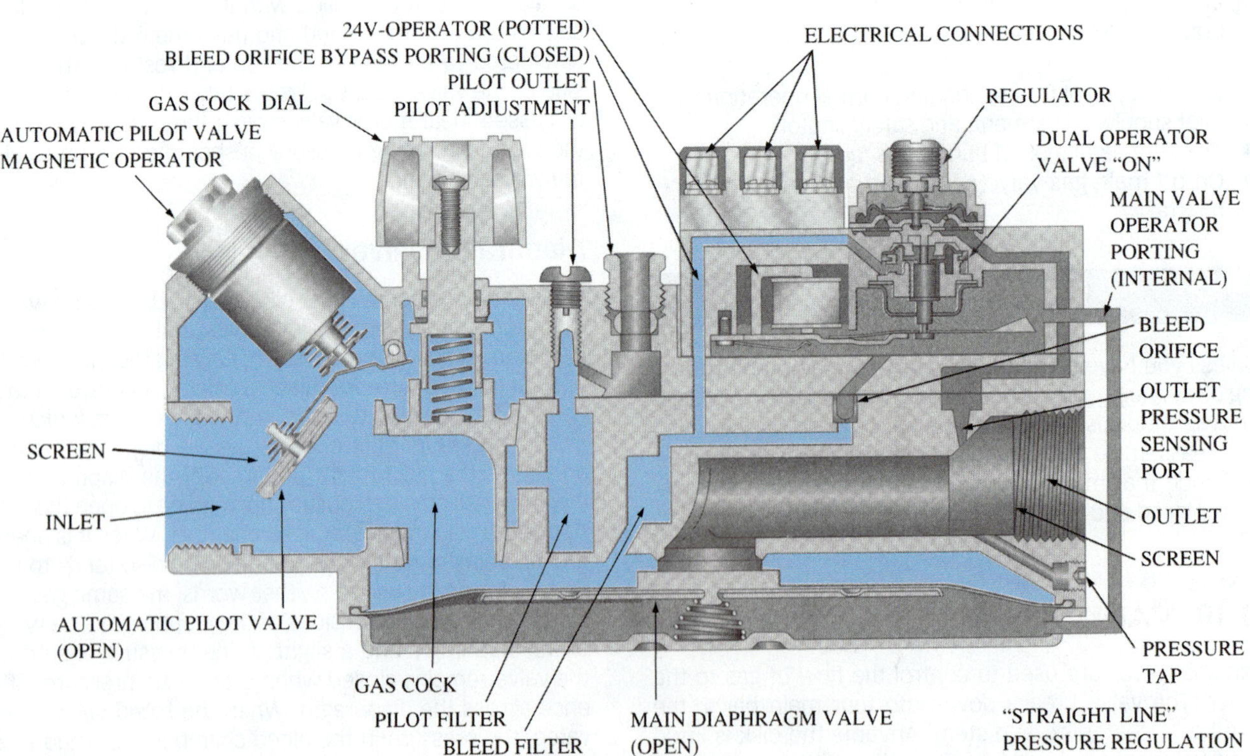

Figure 50-21 Cutaway of a combination gas valve in open position.

simply means that the spark igniter and first solenoid are de-energized and remain de-energized until the call for heat is killed and reinitiated.

Redundant Valves for Direct Ignition Systems

These systems do not use pilot lights, so the only gas flow is to the main burner. The two operators on these valves are energized at the same time. The gas must pass through both valves, providing two methods of shutting down the gas flow.

Two-Stage Valves

Many furnaces today achieve increased efficiency using two-stage gas valves (Figure 50-23). These gas valves have two firing rates: the first stage maintains a lower manifold gas pressure, and the second stage provides a higher manifold pressure (Figure 50-24). The lower firing rate of the first stage allows the furnace to operate longer in mild weather. This increases system efficiency by reducing startup and shutdown losses. Two-stage gas valves can be designed to operate with standing pilots, intermittent pilots, direct-spark ignition, and hot-surface ignition systems.

These valves use two pressure regulators (low fire and high fire) and two solenoids to provide two distinct stages of pressure regulation. Unlike the operation of the single-stage diaphragm valve described previously, the bleed gas flows through a first stage (low fire) regulator first rather than directly to the outlet side of the valve. This maintains the outlet at a pressure controlled by the low-fire regulator. When the second-stage solenoid is energized, the pressure above the main diaphragm is reduced. This causes the main valve to open more. The resulting increased supply of gas forces the low-pressure regulator to close and now allows the gas pressure to be controlled by the high-fire regulator.

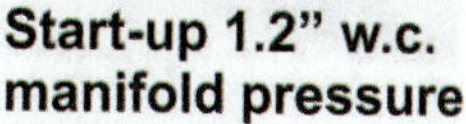

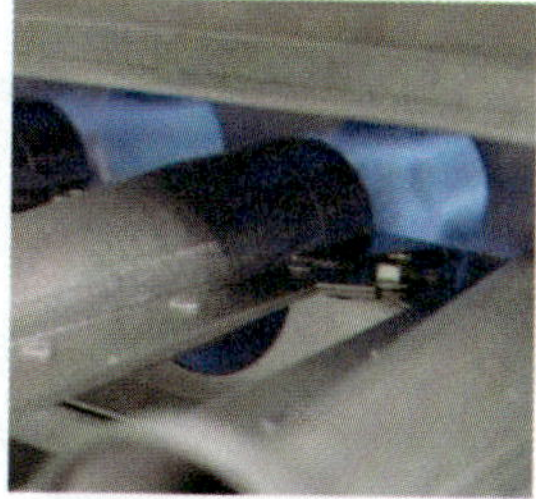

Figure 50-24 Start-up and high-fire burner flames.

The first-stage pressure setting is determined as a percentage of the full output of the valve, and often the pressure regulator is factory set. On some valves, both the low-fire and high-fire regulators can be adjusted separately (Figure 50-25). The highest low-fire pressure regulator adjustment will always be less than the lowest high-fire pressure regulator adjustment. The first-stage low-fire solenoid must always be energized first before the second-stage high-fire solenoid can energize.

Modulating Gas Valves

This type of gas valve will be used along with an integrated furnace control system. A modulating gas valve is shown

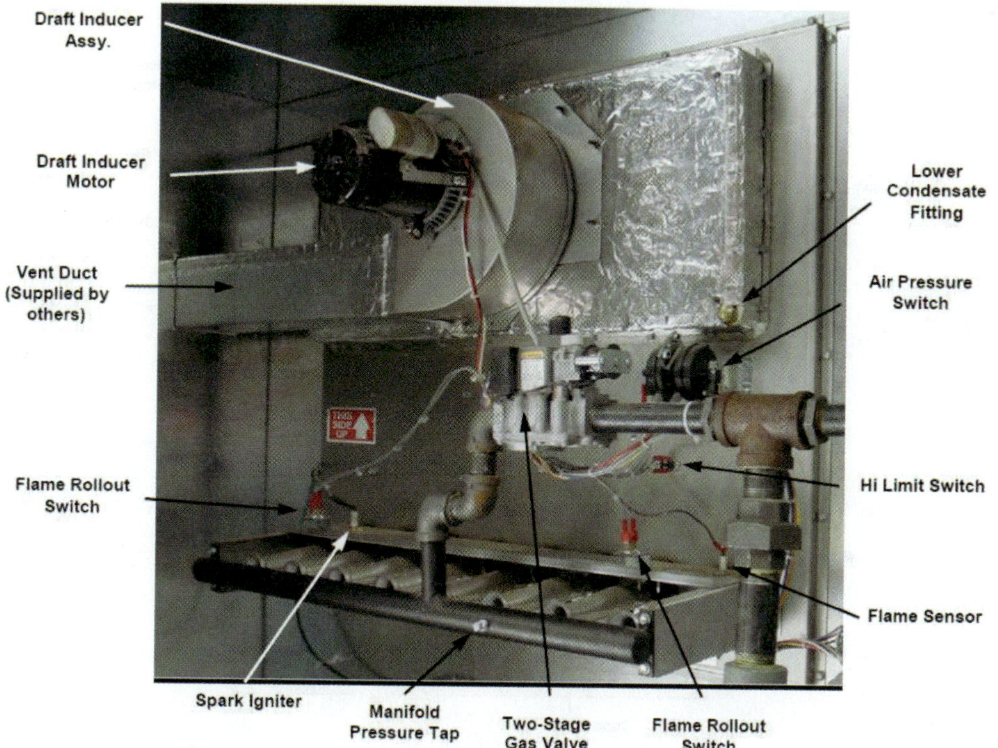

Figure 50-23 Two-stage gas valve location in furnace.

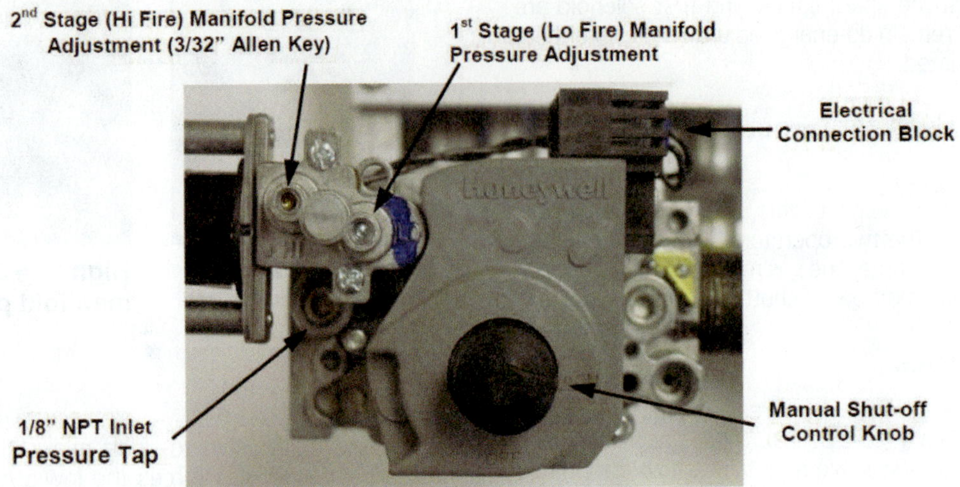

Figure 50-25 Two-stage valve showing adjustment for high- and low-fire manifold pressure.

in series with a two-stage gas valve in Figure 50-26. Both the gas flow and air flow are matched to correctly provide the proper firing rate. The furnace will operate over a heating output range of from 40 to 100 percent. The controller will sense both the supply- and return-air temperatures to determine the temperature rise. Based upon this, along with the thermostat input, the modulating gas valve and variable-speed ECM blower will be proportionally adjusted to maintain the proper firing rate and air-discharge temperature. Some advanced systems can also take into account other variables, such as gas heating values and air density.

50.11 GAS FURNACE WIRING DESIGNATIONS

Different manufacturers will have variations in the physical wiring, but the function remains the same in all systems. With a firm understanding of the construction and use of schematic wiring diagrams and sequence of operation, the student should be able to wire and understand any manufacturer's product. The industry uses standardized electrical symbols that help the technician identify the common controls. Also, alphabetic legend designations are used to help

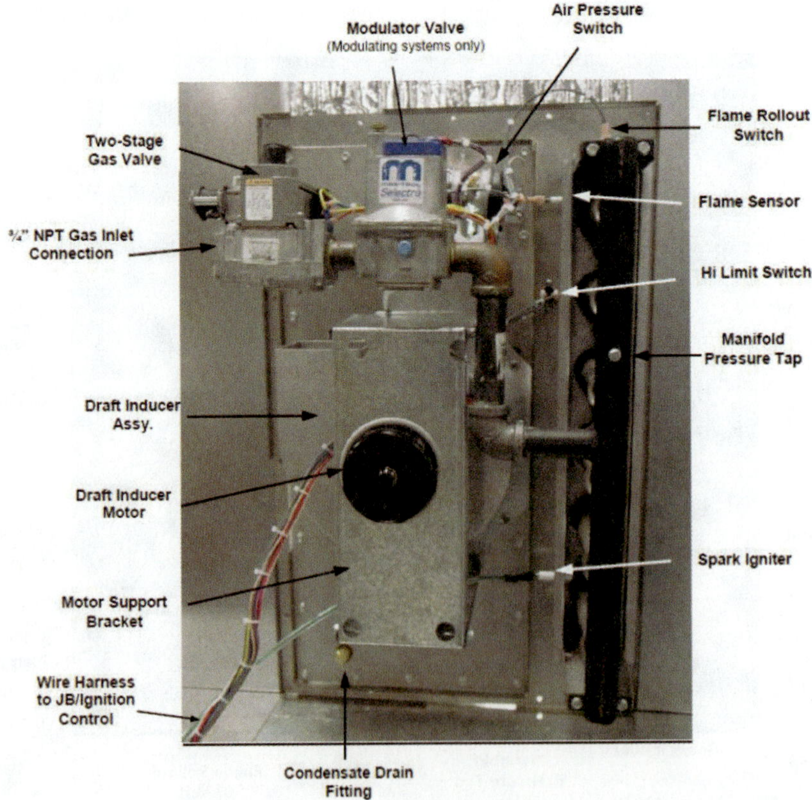

Figure 50-26 Modulating gas valve in series with two-stage gas valve.

Table 50-1 Maximum Length of Two-Wire Run Based on Length and Current Capacity*

	Amperes												
Wire Size	**5**	**10**	**15**	**20**	**25**	**30**	**35**	**40**	**45**	**50**	**55**	**70**	**80**
14	274	137	91										
12		218	145	109									
10			230	173	138	115							
8					220	182	156	138					
6								219	193	175	159		
4									309	278	253	199	
3										350	319	250	219
2											402	310	276
1												399	349
0												502	439
00													560

*To limit voltage drop to 3 percent at 240 V. For other voltages, use the following multipliers:

110 V	0.458	220 V	0.917
115 V	0.479	230 V	0.966
120 V	0.50	250 V	1.042
125 V	0.521		

For example, the maximum run for #10 wire carrying 30 amps at 120 V is 115 × 0.5 = 57 ft.
Note that if the required length of run is nearly equal to, or even somewhat more than, the maximum run shown for a given wire size, select the next larger wire. This will provide a margin of safety. The recommended limit on voltage drop is 3 percent. Something less than the maximum is preferable.

identify the devices they represent. For example, R is a general relay, while CR is a cooling relay, HR is a heating relay, and DR is a defrost relay. Legends may vary with the particular wiring diagram or the manufacturer's method of labeling.

SERVICE TIP

Manufacturers include wiring diagrams inside of the furnace cabinet. Diagrams are also available from the manufacturer and supply house for each piece of equipment. Be sure to study these diagrams carefully because manufacturers may make minor changes in the circuits within the furnace from year to year. Do not assume that the diagram that you are familiar with from previously working on a similar unit is exactly the same. Check it out carefully.

50.12 WIRE SIZE

Wire sizes and fusing for independent appliances will be determined by the manufacturers and specified on their wiring diagrams. Local codes should be examined to determine whether conduit is required in bringing the wire from the main circuit breaker to the furnace or air conditioner. Some codes even require conduit for parts of the circuit operating on 120 V. Specification sheets will give the minimum wire and fuse sizes that will meet the requirements of the National Electric Code, and this wire size is based on 125 percent of the full load current rating.

The length of the wire run will also determine the wire size, and Table 50-1 shows the maximum length of two-wire runs for various wire sizes. This table is based on holding the voltage drop to 3 percent at 240 V. For other voltages, the multipliers at the bottom should be used. If the required run is nearly equal to or somewhat more than the maximum run shown by the calculation in the table, the next larger wire size should be used as a safety factor. It is always possible to use larger wire than called for on the specifications, but a smaller wire size should never be used, as this will cause nuisance trips, affect the efficiency of operation, and create a safety hazard. All replacement wire should be the same size and type as the original wire and should be rated at 90–105°C (194–221°F).

50.13 CONTROL SYSTEM TRANSFORMERS

The step-down or low-voltage transformer that comes inside the furnace (Figure 50-27) is used in heating and air-conditioning control systems. This transformer is used to reduce line voltage to operate the control components. Inside a simple step-down transformer are two unconnected coils of insulated wire wound around a common iron core, as shown in Figure 50-28. To go from a 120 V (primary) to a 24 V (secondary), there are five primary turns to one secondary turn. For a 240 V, primary the ratio would

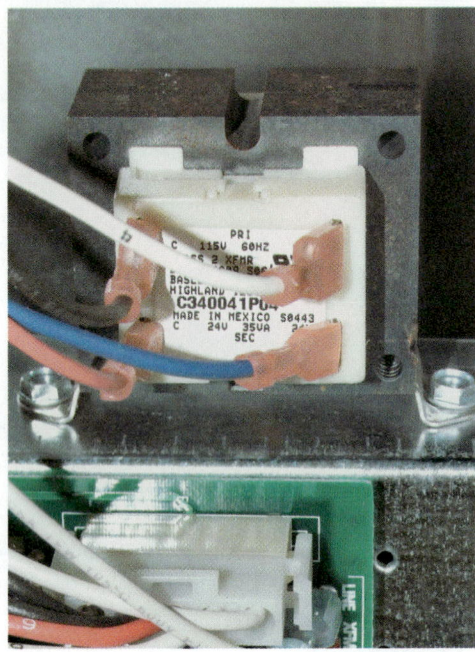

Figure 50-27 AC voltage transformer for low-voltage controls.

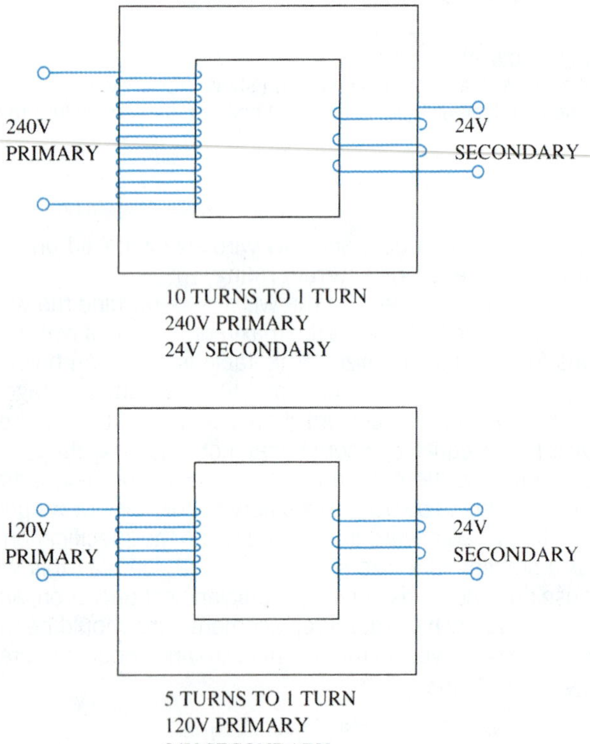

Figure 50-28 Construction of a step-down transformer.

be 10:1, and so on. This, the induction ratio, is a direct proportion.

Transformers used in the HVACR field are rated by their secondary power output. The most common size transformer found in residential gas furnaces is 40 VA. The term VA (volt-amp) is used to express the watts of secondary power output. To convert VA to amps of output, divide the VA rating by the secondary voltage rating.

EXAMPLE 50-1: SOLVING FOR TRANSFORMER SECONDARY CURRENT

A given transformer has a primary voltage of 120 V and a secondary voltage of 24 V. Assuming that the transformer is rated at 40 VA, what is the secondary current?

$$\text{Secondary (A)} = 40\ \text{VA}/24\ \text{V} = 1.67\ \text{A}$$

SERVICE TIP

A standard 40 VA furnace control transformer has a very limited amperage capacity. If additional components such as humidifiers, duct boosters, and dampers are added to a system's control circuit, it may be necessary to increase the size of the transformer used.

Transformers are available in a variety of voltages and capacities. The capacity refers to the amount of electrical current expressed in volt-amperes. A transformer for a control circuit must have a capacity rating sufficient to handle the current (amperage) requirements of the loads connected to the secondary. Ratings of 40 VA are needed for air conditioning because electrical devices containing a coil and iron, such as solenoid valves and relays, have a power factor of approximately 50 percent. Thus, for secondary circuits with such controls, the capacity of a transformer must be equal to or greater than twice the total nameplate wattages of the connected loads.

The proper transformer rating for the electric control circuits will have been selected by the equipment manufacturer. If accessory equipment is added, however, the additional power draw must be considered. You may need to install a new larger VA transformer. This situation is common, for example, when cooling is added to an existing furnace and where the original transformer is too small. When replacing a defective transformer, make sure that the rating is equal to or greater than the original equipment.

50.14 SERVICING ELECTRONIC CONTROL BOARDS

Much of the equipment comes from the factory with electronic boards to control specific functions. When working on these units, it is necessary to get the service manual or technical sheet from the manufacturer's supply house. To help simplify the troubleshooting at the board level, identify four areas on the board: the input, output, board ground, and power supply Figure 50-29. The input is probably an on/off signal from a thermostat or variable signal from a thermistor that indicates a change has occurred. The output is the signal coming out of the board to activate a relay or controlled device. Most solid-state circuit boards rely on direct current to operate the electronic components, so they must be properly grounded. The power supply required could be high or low voltage, either 24 V AC or 5 V DC, or line voltage. If all four of these areas on the board are being provided as specified from the factory, there is a good chance the board has failed. It is not normally repaired in the field and

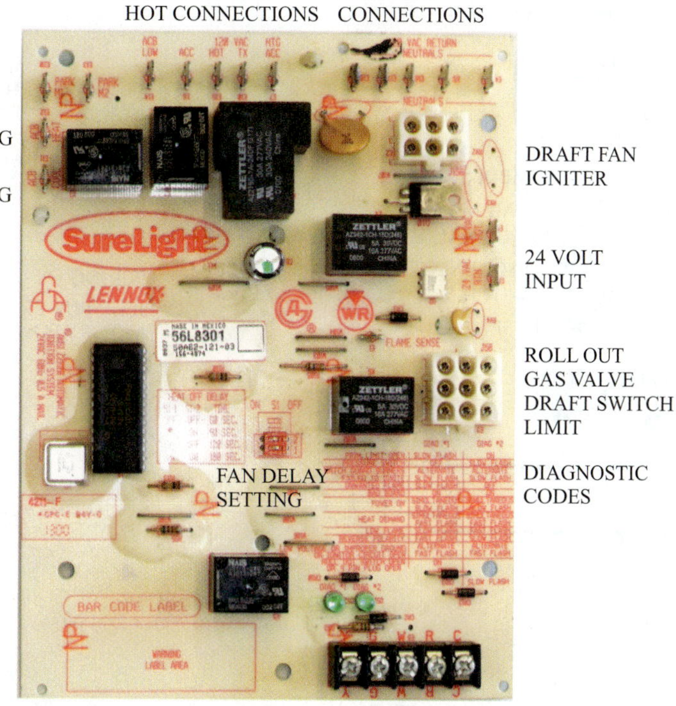

Figure 50-29 Input and output connections on furnace control board.

should be replaced. Prior to replacement, however, verify there are no electrical shorts or wiring problems that may have caused premature failure.

50.15 HEATING CIRCUIT

A simple schematic diagram for a typical gas-fired upflow air furnace is shown in Figure 50-30. The wire conductors are represented by lines and the other components by symbols and letter designations. Note the very important legend that identifies these. Typically, the schematic is the type of diagram used by service personnel to analyze the system.

Only five major electrical devices are in the diagram:

- The fan motor, which circulates the heated air
- The 24 V automatic gas solenoid valve, to control the flow of gas to the burner
- The combination blower and limit control, which controls fan operation and governs the flow of current to the low-voltage control circuit
- The step-down transformer, which provides 24 V current to operate the automatic gas valve
- The room thermostat, which directly controls opening and closing the automatic gas valve

To build this same schematic diagram, begin with the power supply. The wires for the power supply are usually run by the installing technician. They must carry line voltage (single-phase, 60 Hz, 120 V) to operate the blower motor and are represented by heavy broken lines to indicate field wiring (Figure 50-31). As a protection to the circuit, the power supply must be run through a fused disconnect switch. This is the main switch to the entire system.

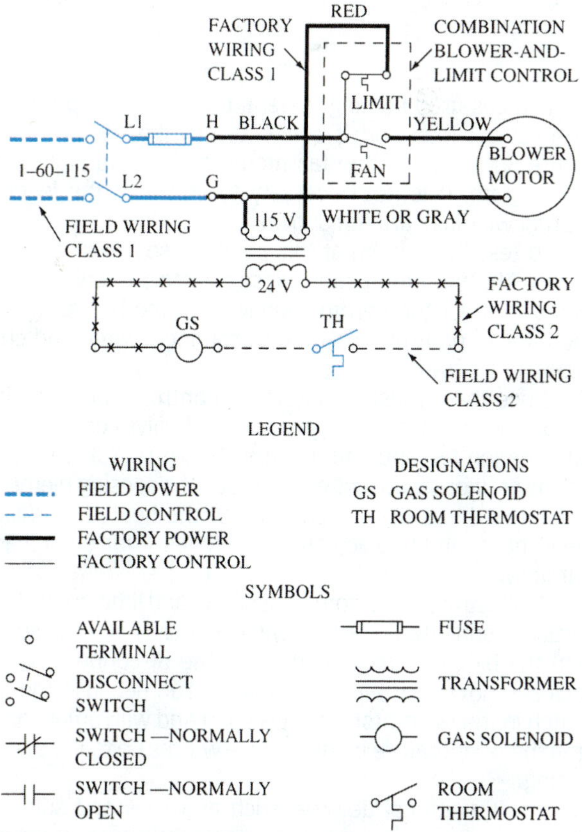

Figure 50-30 Heating circuit diagram for gas-fired upflow furnace.

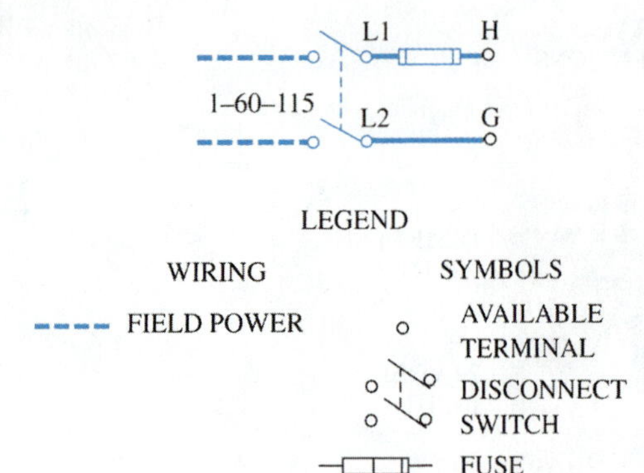

Figure 50-31 Diagram for a fused disconnect.

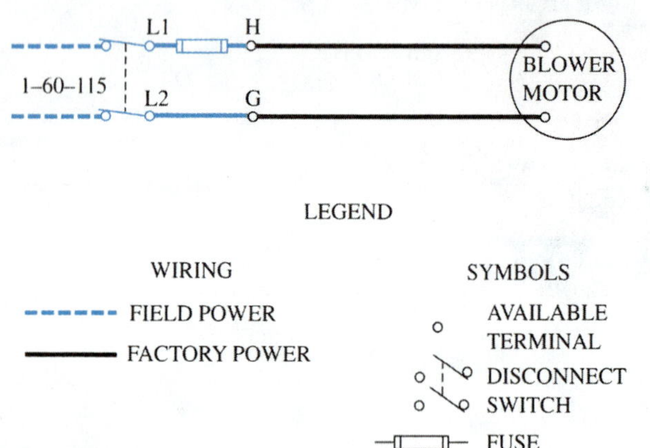

Figure 50-32 Diagram showing the fused disconnect wired to the blower motor.

In a gas-fired heating system, the fan motor produces the heaviest load and is connected to the power supply (Figure 50-32). Since the fan motor is clearly identified on the diagram, it is not necessary to add it to the legend. Internal windings are not shown.

To test the system at this point, close the disconnect switch. The fan motor runs, since a simple circuit is completed from L1 through the windings of the blower motor and back to neutral. Open the disconnect switch and continue the diagramming.

Since there must be automatic control of both the fan motor and the automatic gas solenoid valve, connect the combination fan and limit control (Figure 50-33). This control, as previously described, consists of a sensing element, which measures the temperature of the heated air inside the furnace, and two adjustable switches: a fan switch and a limit switch.

As illustrated, the combination fan and limit control are partially connected. The fan switch is wired into L1 in series with the blower motor windings. If the disconnect switch is closed now, the fan motor will not run because the fan switch in the fan motor circuit is open and will remain open until the temperature in the furnace warms up to the switch setting.

Small electrical devices, such as relays and solenoid valves, do very little work and require little current for operation. A step-down transformer to reduce line voltage to 24 V is required and properly installed in the wiring diagram

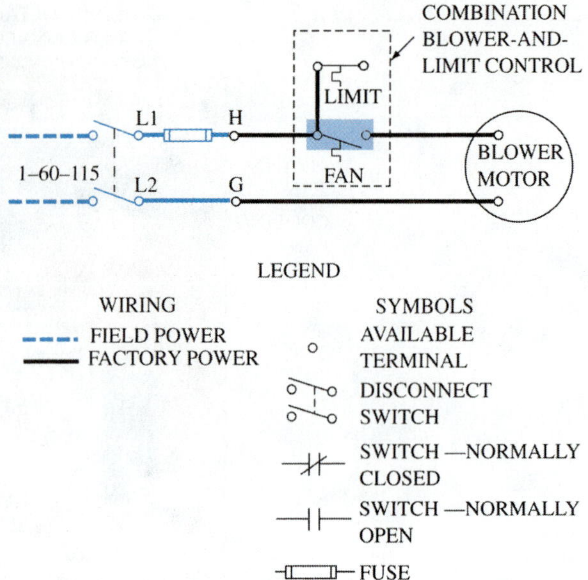

Figure 50-33 Wiring diagram showing the use of a fan limit control wired to a blower motor.

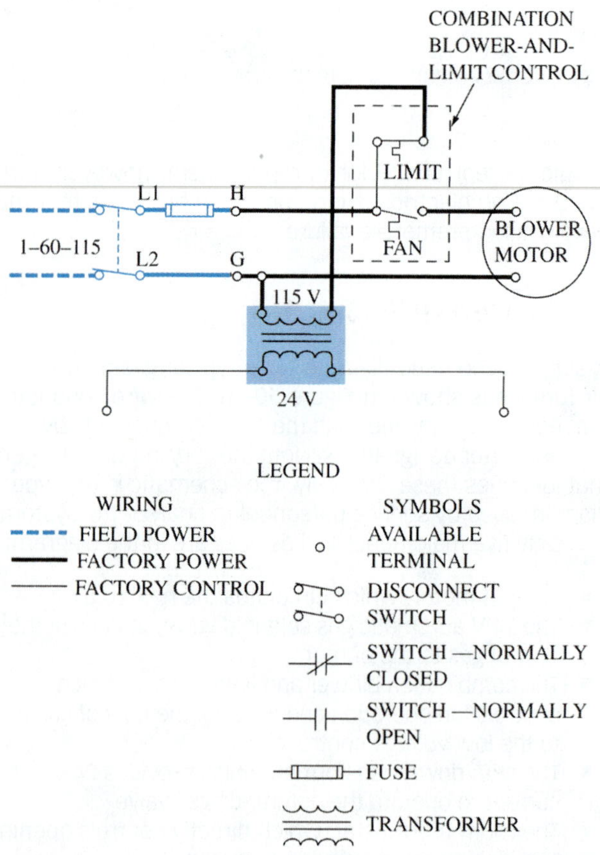

Figure 50-34 Schematic diagram showing the addition of a control circuit transformer.

(Figure 50-34). One side is connected to neutral and the other side is connected to L1 through the limit switch. This places the limit switch in control of the entire 24 V circuit.

If the disconnect switch is closed, the blower motor still cannot run since the fan switch must remain open until the air temperature warms up. The transformer and the

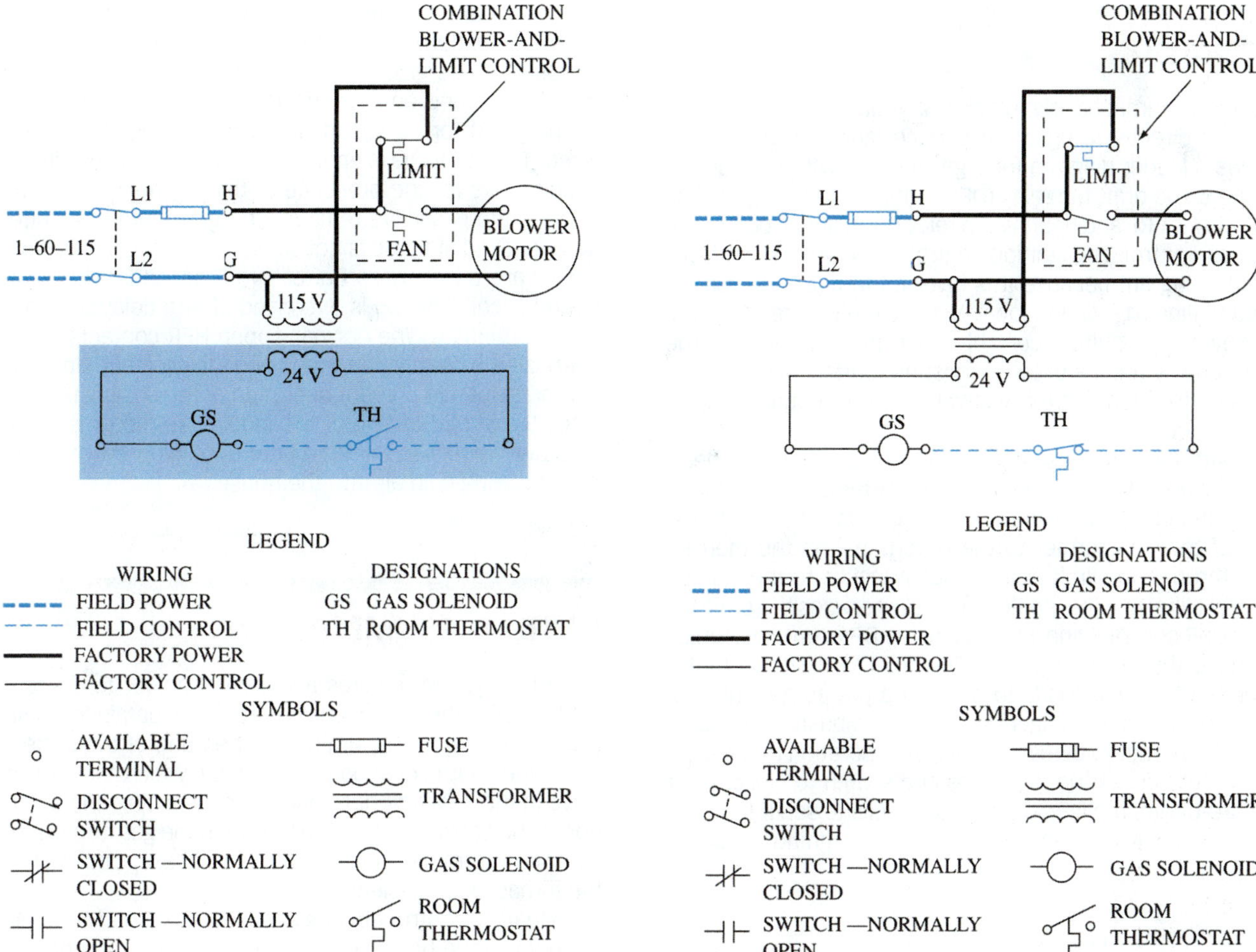

Figure 50-35 Schematic diagram of the low-voltage circuit showing the thermostat operation of the gas valve.

Figure 50-36 Schematic diagram showing the wiring of the limit control in series with the line-voltage side of the transformer.

24 V power source, however, are energized immediately from L1 through the closed limit switch and back to L2. Although there is now a source of 24 V current, nothing can be accomplished until there is a circuit across the 24 V supply.

The automatic gas solenoid valve and room thermostat are wired in series across the 24 V circuit (Figure 50-35), since the opening and closing of the gas solenoid valve is controlled by the thermostat. The gas valve is factory wired as indicated on the diagram by a lightly shaded, unbroken line between terminals. Normally, the thermostat is located in a room away from the furnace and must be field wired, as indicated by the lightly shaded, broken lines used to represent control circuit field wiring.

The circuit is now completely wired. Close the disconnect and limit switches and open the thermostat. The transformer and low-voltage circuit are immediately energized from L1 through the closed limit switch back to neutral. With the thermostat open, the automatic gas valve remains closed. The fan motor cannot operate because the fan switch is open, and it will remain open until the air temperature in the furnace warms up to the FAN ON setting.

When the thermostat calls for heat, the automatic gas valve is energized (Figure 50-36) and the valve opens to admit gas to the burner. The fan switch would still be open so that the fan motor cannot run until the air temperature in the furnace plenum chamber warms up to the FAN ON setting.

Thermostat and gas valve circuits are reopened when the room-temperature setting is reached. Fan and blower contacts remain closed until the temperature of the heated air cools to the FAN OFF setting.

If at any time during furnace operation the plenum chamber overheats, the air temperature soon reaches the setting of the limit switch. This is an important safety switch, which opens at the overheating setting to de-energize the entire 24 V circuit. When this happens, the automatic gas valve closes, the burner is extinguished, and the thermostat is overridden. The fan switch remains closed and the blower continues to run as long as the heated air remains above the FAN OFF setting. The limit switch is manually reset by the service person after correcting the problem.

The operating sequence just completed proves that the wiring diagram is correct. The disconnect switch controls the entire electrical system; the blower motor operates independently of the burner, under the control of the fan switch. The automatic gas valve is cycled on and off by the room thermostat, under the control of the limit switch.

50.16 HOT SURFACE IGNITION OPERATION

Most gas furnaces manufactured today ignite the main burners directly using a hot surface igniter. Figure 50-37 shows a ladder diagram for a gas furnace with hot surface ignition. The draft fan relay (DFR), ignition relay (IR), cooling fan relay (CFR), and heating fan relay (HFR) are all part of the ignition control. The ignition control is shown in two places on the diagram because it is involved in both line voltage circuits (lines 1—6) and low-voltage circuits (lines 10—17). However, the ignition control is one physical device. The operating sequence used by most hot surface ignition systems is draft fan, igniter, gas valve, flame rectification, indoor blower.

With the system switch in the heat position, the heating thermostat closes when the room temperature drops. This initiates a call for heat, creating a path from the Red side of the transformer to the W terminal on the thermostat, through the limit and roll-out switches to the ignition control (line 14). The ignition control energizes the draft fan relay DFR coil (DRF line 14), closing the DFR contacts, which energize the draft fan motor (line 3). The draft fan motor operates for 60 seconds to perform a pre-purge. The pre-purge removes any lingering gas or combustion products from the burner and heat exchanger. Operation of the draft fan motor causes the draft pressure switch to close. After the pre-purge, the ignition control checks to see if the draft pressure switch is closed (line 13). The ignition process does not continue until the draft pressure switch closes. After the ignition control sees that the draft pressure switch is closed, it energizes the ignition relay coil line 16), closing its contacts and energizing the igniter (line 1). The igniter is energized for 60 seconds before the gas valve is energized. The gas valve is energized and the flame rectification circuit looks for the presence of flame (line 12). If flame is sensed, the gas valve remains energized and the igniter is de-energized. If flame is not sensed, both the gas valve and igniter are de-energized after five 5 seconds. The heating fan relay coil (line 17) is energized after a delay of two 2 to three 3 minutes. The normally open HFR contacts close to energize the heating fan speed and the normally closed HFR contacts open to prevent operation of the cooling fan speed (lines 5 and 6). Starting of the indoor fan motor is delayed to allow the furnace time to heat up. This prevents the fan from blowing cold air into the house.

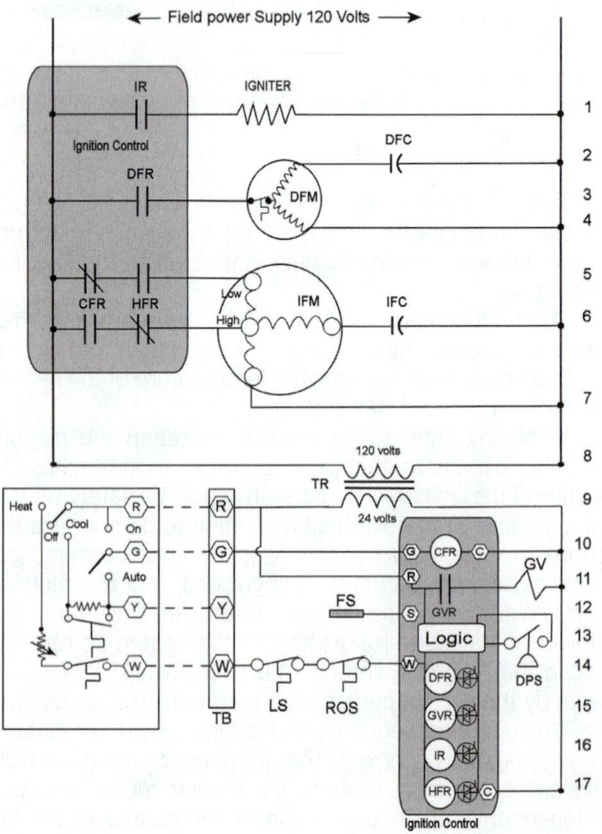

Figure 50-37 Schematic diagram for hot surface ignition furnace.

UNIT 50—SUMMARY

A control system requires a source of power to operate, a load to utilize the power, and controllers to obtain the desired level of operation. Generally the operation of the system is under the control of a room thermostat, which controls a gas valve in the heating phase and the condensing unit's operation in the cooling phase. The blower in the gas furnace must operate intermittently in the heating phase, depending on the furnace supply-plenum temperature.

Heating control circuits are often designed to operate on low voltage because the wiring is simplified and safer. Furthermore, low-voltage thermostats provide closer temperature control than do line-voltage thermostats. A step-down transformer is used to reduce line voltage to operate control components. Heat anticipators are used in low-voltage thermostats to reduce the temperature swing caused by the systems.

Burner ignition methods are varied. A pilot is a small burner used to light the main burner when the gas valve opens. Standing pilots operate continuously while an electronic ignition system utilizes a spark electrode to ignite the gas pilot on a signal from the room thermostat. The direct-spark ignition system uses an electrical flame sensor rod rather than a pilot. A hot-surface ignition unit, rather than using a pilot, uses a silicon carbide material that when energized heats up to the gas ignition temperature.

WORK ORDERS

Service Ticket 5001

Customer Complaint: Furnace Fan Keeps Short-Cycling

Equipment Type: 80,000 BTU Upflow Natural Gas Furnace

The technician analyzes the unit and discovers that the fan is short cycling with the burners off. The technician discovers the fan switch differential is set too narrow

(20 percent). The technician sets the fan to come on at 140°F and cycle off at 90°F, which is a 50°F differential. The technician runs the system through several cycles to ensure that it is operating properly. The system is now operating as designed.

Service Ticket 5002

Customer Complaint: Pilot Light Comes on Briefly, but the Burners Will Not Light

Equipment Type: 100,000 BTU Horizontal Gas Furnace

After talking to the customer, the technician decides to try running the system. After setting the thermostat to heat, the technician observes the spark igniter come on and light the pilot. After about 21 seconds, the pilot turns off without lighting the furnace. The technician determines that the flame sensor is bad. After disconnecting the main power, the technician replaces the flame sensor. The technician turns the power back on and sets the thermostat to heat. The spark igniter lights the pilot, and in just a few seconds the burner lights and the system operates normally.

Service Ticket 5003

Customer Complaint: The System Runs Fine on Cooling but Not on Heat

Equipment Type: 125,000 BTU Upflow NG Furnace

The technician checks the system and finds that it is not getting voltage between the common or C and W terminals. When the furnace is getting a heating signal, there will be no voltage read between R and W. It is probably less confusing to compare voltages to C common terminals in the furnace. The technician de-energizes the system and performs a continuity check back to the thermostat and finds that there was no continuity. The technician then checks for continuity between the R and W terminals on the thermostat with the thermostat set to call for heat. With no continuity between the thermostat terminals R and W, the technician replaces the thermostat. The technician returns power to the system and sets the new thermostat to call for heat. The system operates normally in the heat settings. The technician cycles the thermostat a few times to ensure that the system is operating as designed.

UNIT 50—REVIEW QUESTIONS

1. List the four areas of an electronic control board that should be identified when troubleshooting.
2. Identify the reasons a low-voltage control circuit is more desirable than a line-voltage control circuit.
3. A given transformer has a primary voltage of 240 V and a secondary voltage of 24 V. Assuming the transformer is rated at 40 VA, what is the secondary current?
4. What is the standard transformer VA rating for most HVACR systems?
5. Why are heat anticipators used in low-voltage thermostats?
6. State the problem associated with setting a thermostat heat anticipator too high.
7. How is a humidistat similar to a low-voltage switch?
8. Explain the reason why adding humidity to a residence increases customer comfort.
9. On a fan/limit switch, what is the usual setting for the limit?
10. List the four functions of a furnace gas valve.
11. Describe how a main gas valve gets electrical power to open.
12. How many volts are generated in the spark-ignition circuit?
13. Why is it necessary that the gas-burning unit be grounded on a DSI system?
14. If a burner fails to light or the flame sensor fails to detect a flame, what happens to ensure the system is safe?
15. Identify the standardized electrical symbols for the following common controls: indoor fan relay, humidistat, manual reset, and crankcase heater.
16. If there is a 240 V circuit that will be carrying 30 A using #10 wire size, what is the maximum length of the two-wire run that would limit the voltage drop to 3 percent?
17. In question 16, what would be the answer if the wire size is #12, the voltage is 120 V, and the amperage is 15 A?
18. List the five major electrical devices in the heating circuit diagram in Figure 50-30.
19. Explain what a limit switch is, its function, and how it affects the control circuit.
20. State why the legend and the notes are important parts of an electrical diagram.

UNIT 51

Gas Furnace Installation

OBJECTIVES

After completing this unit, you will be able to:

1. explain the difference between natural gas and LP gas.
2. use a gas meter to check the gas input of a gas furnace.
3. use gas piping tables to size gas piping for a furnace.
4. use GAMA vent-sizing tables to determine the correct vent size for a furnace.
5. determine the correct size openings to provide combustion air for a furnace.
6. determine furnace temperature rise.
7. list acceptable ranges for flue gas CO percent, O_2 percent, excess air percent, and net stack temperature.
8. list safety precautions that should be practiced when starting a gas furnace.

51.1 INTRODUCTION

Modern gas furnaces are an efficient and safe way to provide whole house comfort. Increasingly, customers are choosing higher-efficiency models to control heating costs. However, the full energy savings will only be realized if the furnace is installed correctly.

Design innovations in modern furnaces have produced more efficient, safer systems than those of just a few years ago. Some of these innovations have changed the way furnaces are installed. Induced-draft furnaces, two-stage furnaces, condensing furnaces, and direct-vent furnaces all require special installation techniques. These innovative furnaces will not perform as designed if installed incorrectly. Proper installation is also crucial for safe operation. Despite the many safety features incorporated into today's furnaces, they cannot always guarantee safe operation if the installation is unsafe.

51.2 NATURAL GAS

To check, test, and adjust a gas-burning unit for the highest operating efficiency, the unit must have the proper gas input, the proper adjustment of the burners, the correct amount of combustion air, proper venting, and the correct amount of air for heat distribution.

To arrive at the correct gas input, two important factors must be known about the gas:

- Heat content in BTU/ft^3
- Specific gravity

Natural gas is a mixture of methane and ethane. Various sources have different heat contents in the range of 950 to 1,150 BTU/ft^3. The specific gravity also varies between 0.56 and 0.72. These characteristics relate to sizing the piping and determining the amount of gas to be supplied to each burner.

Heating units are constructed to use gas-pressure regulators set at an output of 3½ in wc (water column, a pressure measurement). Burner designs will allow operation between 3 and 4 in wc manifold pressure. These pressures must not be exceeded. The actual operating pressure for any particular installation is determined by the gas heating value and the altitude. Table 51-1 shows a typical manifold pressure chart supplied with the installation instructions.

CAUTION

The diaphragm in a gas regulator is easily damaged with excessive pressure. Before pressure testing a gas piping system, the gas line should be removed from the gas valve and capped. This should be done even though there is a gas-line shutoff so that the pipe joints from the shutoff to the gas valve can also be leak checked.

51.3 CHECKING THE GAS INPUT

The quantity of gas a customer uses is recorded by a gas meter (Figure 51-1). Gas meters use dials to indicate the gas quantity (Figure 51-2). The two smaller dials, one for ½ ft per revolution and the other for 2 ft per revolution, can be used to measure a furnace's gas consumption. To determine if the correct amount of gas is being fed to the heating unit, it is necessary to find the feed rate through the meter. For accuracy, all other appliances must be turned off. If the pilot lights of the other appliances, including water heaters, are turned off, be sure to relight them before leaving. Usually, the requirements of pilot lights are so small that they are ignored.

Gas flow through the meter is determined by the time it takes the test dials to turn one revolution. To determine this time, the following formula can be used:

$$\text{Time}(\text{sec}/\text{ft}^3) = \frac{\text{seconds per hour}}{\text{ft}^3/\text{hr of gas}}$$

Table 51-1 Manifold Pressure Chart

	Altitude Range (ft)	Avg. Gas Heat Value at Altitude (BTU/cu ft)	Specific Gravity of Natural Gas 0.58 Orifice No.	0.58 Mnfld Press High/Low	0.60 Orifice No.	0.60 Mnfld Press High/Low	0.62 Orifice No.	0.62 Mnfld Press High/Low	0.64 Orifice No.	0.64 Mnfld Press High/Low
USA and Canada		900	43	3.5/1.5	43	3.6/1.5	43	3.8/1.6	42	3.2/1.3
		925	44	3.8/1.6	43	3.5/1.5	43	3.6/1.5	43	3.7/1.6
	0	950	44	3.6/1.5	44	3.8/1.6	43	3.4/1.4	43	3.5/1.5
		975	44	3.4/1.5	44	3.6/1.5	44	3.7/1.6	44	3.8/1.6
	to	1,000	44	3.3/1.4	44	3.4/1.4	44	3.5/1.5	44	3.6/1.5
		1,025	**45**	**3.8/1.6**	44	3.2/1.4	44	3.3/1.4	44	3.4/1.5
	2,000	1,050	**45**	**3.6/1.5**					44	3.3/1.4
		1,075	**45**	**3.4/1.4**	**45**	**3.7/1.6**	**45**	**3.8/1.6**		
		1,100	**45**	**3.3/1.4**	**45**	**3.5/1.5**	**45**	**3.7/1.5**		
					45	**3.4/1.4**	**45**	**3.5/1.5**		
									45	**3.8/1.6**
									45	**3.6/1.5**
USA and Canada	USA	800	43	3.8/1.6	42	3.2/1.4	42	3.3/1.4	42	3.5/1.5
	Altitudes	825	43	3.6/1.5	43	3.7/1.6	43	3.8/1.6	42	3.2/1.4
	2,001	850	43	3.4/1.4	43	3.5/1.5	43	3.6/1.5	43	3.7/1.6
	to 3,000	875	44	3.7/1.5	44	3.8/1.6	43	3.4/1.4	43	3.5/1.5
	or	900	44	3.5/1.5	44	3.6/1.5	44	3.7/1.6	44	3.8/1.6
	Canada	925	44	3.3/1.4	44	3.4/1.4	44	3.5/1.5	44	3.6/1.5
	Altitudes	950	**45**	**3.7/1.6**	44	3.2/1.4	44	3.3/1.4	44	3.4/1.4
	2,001	975	**45**	**3.6/1.5**					44	3.2/1.4
	to 4,500	1000	**45**	**3.4/1.4**	**45**	**3.7/1.6**	**45**	**3.8/1.6**		
					45	**3.5/1.5**	**45**	**3.6/1.5**	**45**	**3.7/1.6**
USA Only		775	43	3.7/1.6	42	3.2/1.3	42	3.3/1.4	42	3.4/1.4
	3,001	800	43	3.5/1.5	43	3.6/1.5	43	3.8/1.6	42	3.2/1.3
		825	44	3.8/1.6	43	3.4/1.4	43	3.5/1.5	43	3.7/1.5
	to	850	44	3.6/1.5	44	3.7/1.6	44	3.8/1.6	43	3.4/1.5
		875	44	3.4/1.4	44	3.5/1.5	44	3.6/1.5	44	3.7/1.6
	4,000	900	44	3.2/1.3	44	3.3/1.4	44	3.4/1.4	44	3.5/1.5
		925	**45**	**3.7/1.5**	**45**	**3.8/1.6**	44	3.2/1.4	44	3.3/1.4
		950	**45**	**3.5/1.5**	**45**	**3.6/1.5**				
							45	**3.7/1.6**	**45**	**3.8/1.6**

Note: Orifice numbers shown in **bold** are factory installed.

To determine how long it would take for 142.8 ft^3/hr of gas to flow, the formula would be set up as follows:

$$\text{sec}/\text{ft}^3 = \frac{60 \times 60}{142.8} = \frac{3{,}600}{142.8}$$

It will require 25.2 seconds for 1 ft^3 of gas to go through the meter. Thus, the ½ ft^3 dial would require 12.6 seconds per revolution, and the 2 ft^3 dial, 50.4 seconds per revolution. A stopwatch or a digital watch timing function is recommended.

The firing rate of the furnace will vary with altitude. Furnaces installed in higher altitudes have a lower firing rate because the lower air pressure delivers less oxygen to burn the gas. The gas flow is reduced to match the available

Figure 51-1 Residential gas meter.

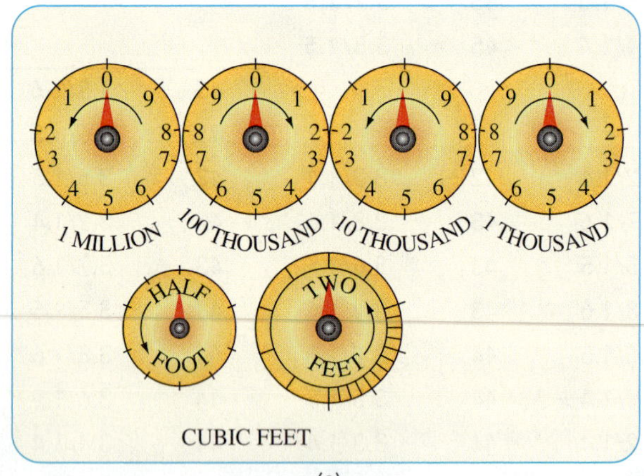

(a)

(b)

Figure 51-2 Typical domestic gas meter dials in ft^3.

oxygen. The gas flow is also adjusted to account for differences in gas heat capacity.

If adjustments in the gas flow need to be made to produce the required input, it can be done by adjusting the gas-pressure regulator, as long as the input pressure stays within the 3–4 in wc range. If the adjustment cannot be made within this range, the gas orifice size for the burners needs to be changed. The manufacturer's installation data supply information for selecting the orifice size and adjusting the manifold pressure to achieve the correct capacity (see Table 51-1).

Table 51-2 Schedule 40 Iron Gas Pipe Support

Pipe Size	Support Interval
1/2 in	6 ft
3/4 in	8 ft
1 in	8 ft
1 1/4 in	10 ft

51.4 GAS PIPING

Fuel gas piping is specified in national codes such as the National Fuel Gas Code or the International Residential Code, Section VI, Fuel Gas. Natural gas piping is typically schedule 40 iron pipe or corrugated stainless steel (CSST) designed specifically for fuel gas piping. The iron pipe is joined with threaded fittings using thread sealing compound that is rated for use with gas. CSST is joined using special fittings designed specifically for CSST. All gas piping should be supported at regular intervals. Typical support intervals are listed in Table 51-2. Pipe should be sloped ¼ in per 15 ft toward a drip leg. The gas piping at the furnace should include a shutoff valve, a drip leg, and a union (Figure 51-3). A furnace connected with CSST or a flexible gas connector does not need a union.

Before the gas is turned on, the gas piping should be pressure tested to at least 1.5 times the maximum working pressure or 3 psig, whichever is higher. Air, nitrogen, or carbon dioxide can all be used for pressure testing. Oxygen should never be used! To pressure test the gas line: close the manual shutoff valve to the unit, connect the pressure gauge to the gas piping, pressurize the piping, and then close the pressure source. The gas piping must hold the test pressure for at least 10 minutes. In many areas, the local inspector must see both the initial pressurization and the final pressure. After the inspector has signed it off, the air or nitrogen in the line is released outside. Leaving the pressure in the line can damage the gas controls in the furnace. It is important to use quality shutoff valves that will hold back the test pressure. In some areas, the line is not connected to the furnace until after this initial pressure test to avoid accidentally damaging the gas controls in the unit.

51.5 SIZING GAS LINES

Factors used to determine gas-piping size are as follows:

1. Gas line pressure
2. Specific gravity of the gas

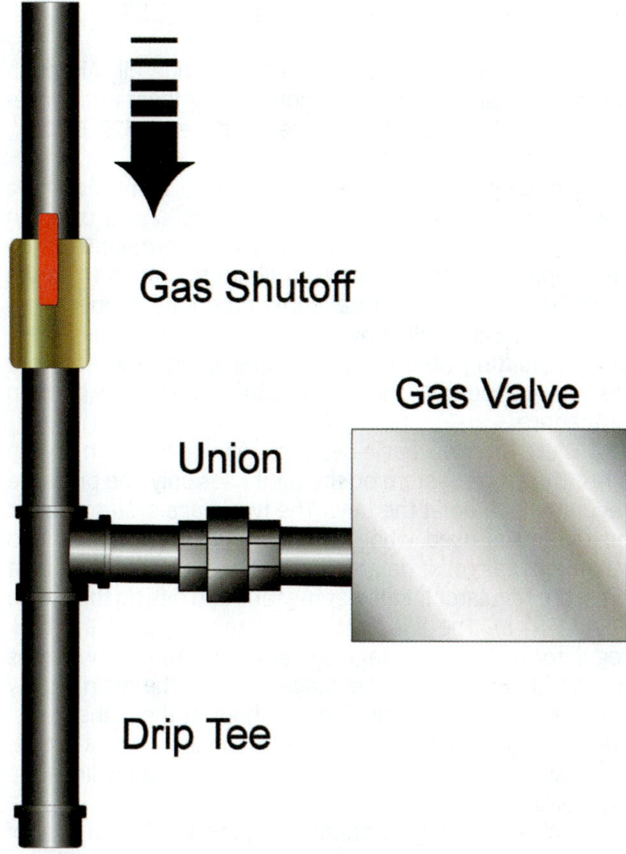

Figure 51-3 The gas line to a furnace should have a manual shutoff valve, a drip leg, and a union.

3. Allowable pressure loss from the gas source to the unit
4. Maximum gas consumption
5. Piping length plus the equivalent length of all fittings and valves

Line Pressure

Gas lines can be operated at different pressures. Higher-pressure lines can carry more gas through smaller lines. Gas piping tables are built for a specific delivery pressure. Most residential systems operate at less than 2 psi, often as low as 6 in wc (1/4 psi). Commercial systems often operate at higher pressures, from 1 to 10 psi.

Specific Gravity

The pipe sizing tables are for a particular specific gravity gas. The specific gravity of natural gas is 0.6; the specific gravity of propane is 1.5.

Maximum Allowable Pressure Drop

The maximum allowable pressure drop is the difference in the gas pressure at the source and the appliance. Generally, residential gas piping is sized for a maximum pressure loss of between 0.35 in wc and 0.5 in wc.

Maximum Consumption

The maximum consumption of the unit is the most cubic feet per hour that the connected appliance will use. This is easily found by dividing the firing rate in BTUs per hour for the appliance by the BTU content of the gas. Natural gas has a heat content of approximately 1,000 BTUs per cubic foot. Propane's heat content is approximately 2,500 BTUs per cubic foot. These figures are close enough for most calculations, but the heat content does vary from one location to another. To make accurate calculations, contact the local gas distributor for exact heat content figures.

Piping Length

The piping length is the actual length in feet between the gas source and the gas appliance plus the equivalent length in feet of all fittings, valves, or obstructions. Most gas pipe sizing tables include allowance for a "normal" amount of fittings, eliminating the need to calculate equivalent length separately. For our purposes, we'll assume that all systems in this section are "normal" systems.

Using Piping Tables

To use the pipe sizing Table 51-3:

1. Determine the greatest length of piping from the source to the appliance.
2. Find a row with a length equal to or greater than that length.
3. This row is used for ALL appliances and ALL sections of pipe.
4. Use this row to find a cubic foot capacity equal to or greater than the capacity needed for each section of gas pipe.

Table 51-3 Capacity in Cubic Feet of Gas for 0.6 SG with 0.5" wc Pressure Drop

	Nominal Pipe Size Rigid Schedule 40 Iron Pipe				
Length	**1/2**	**3/4**	**1**	**1 1/4**	**1 1/2**
10	172	360	678	1390	2090
20	118	247	466	957	1430
30	95	199	374	768	1150
40	81	170	320	657	985
50	72	151	284	583	873
60	65	137	257	528	791
70	60	126	220	486	728
80	56	117	207	452	677
90	52	110	195	424	635
100	50	104	173	400	600
125	44	92	157	355	532
150	40	83	144	322	482

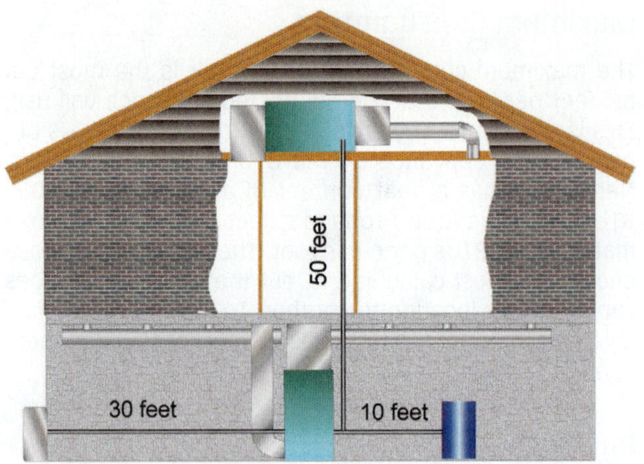

Figure 51-4 Gas piping example.

51.6 PIPE SIZING EXAMPLE

Figure 51-4 shows our example: a house with two furnaces and one water heater operating on natural gas with a heat content of 900 BTUs/ft^3. The system is piped with schedule 40 iron pipe. The basement furnace firing rate is 100,000 BTUs/hr, the water heater firing rate is 50,000 BTUs/hr, and the upstairs furnace firing rate is 80,000 BTUs/hr. The gas pipe runs 30 ft from the meter to the basement furnace, 10 ft from the basement furnace to the water heater, and 50 ft from the basement to the upstairs furnace. The basement furnace is 30 ft from the meter, the water heater is 40 ft from the meter, and the upstairs furnace is 80 ft from the meter. Therefore, the farthest distance is 80 ft. The 80-ft row in Table 51-3 will be used to look up all the gas pipe sizes, even the shorter runs. The CFH consumption of the appliances is 100,000 /900 = 111 CFH, 50,000/900 = 56 CFH, and 80,000/900 = 89 CFH. The main line must carry the total CFH capacity of 111 CFH + 56 CFH + 89 CFH = 256 CFH. Looking across the 80-ft row, the first column that has a capacity equal to or greater than 256 is 1 1/4 in pipe at 452 CFH. The main line will be 1 1/4 in. The 3/4 in pipe size, can carry 117 CFH, which is more than enough for each of the furnaces. Each furnace will be piped with ¾ in. The water heater can be piped with 1/2 in since ½-in pipe will carry 56 CFH, which is the exact consumption of the water heater.

51.5 LIQUID PETROLEUM (LP) GAS

Propane and butane are refined from crude oil. Although mixed, they are separable by condensing them at their respective boiling points. Propane produces 2,522 BTU/ft^3, and butane yields 3,261 BTU/ft^3.

The LP gas industry has established a set incoming line pressure for all LP gas-burning appliances of 11 in wc. Therefore, it is only necessary to determine that the LP supply system is large enough to maintain 11 in wc at the units when the total connected load is operating. The LP supplier will provide the tank installation, pressure-regulating devices, and piping to the furnace, but the service person should be familiar with the hookup procedure.

There are two basic systems, as illustrated in Figure 51-5. The single system on the right uses only one pressure regulator located at the tank. The two-stage system shown on the left is used where the number of appliances and volume of gas must be greater. In this case, there will be a pressure regulator located at the tank and one on the house (Figure 51-6). The line pressure at the outlet of the tank regulator on the two-stage system is 10–15 psig, whereas it is 11 in wc on the single-stage system. The main supply line will carry more cubic feet per hour at the higher pressure. Sometimes it is necessary to connect from a single-stage system to a two-stage system if the supply lines are undersized.

With single-stage systems, the pressure on the inlet of the regulator will be direct tank pressure, and this will vary with fuel and temperature.

To ensure that the gas will flow from the tank into the supply system to the heating unit, the tank pressure must at all times be higher than the line pressure required to supply the system. To maintain a pressure of 11 in wc at the heating unit, allowing for a pressure loss in the line of ½ in wc, the minimum tank pressure would be 2 psig. This means that the minimum outside temperature must be considered when selecting the mixture of LP fuel.

Table 51-4 shows the tank pressure that will occur at temperatures of −30°F to +110°F for mixtures of propane and butane. From this table it can be seen that butane is not usable below +40°F and even propane will not develop sufficient pressures below −30°F. In extremely cold climates, tank heaters are used to ensure adequate fuel supply.

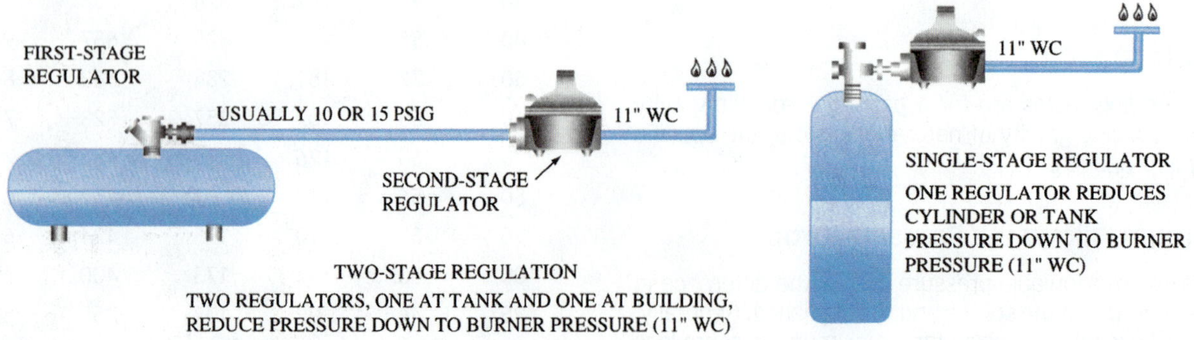

Figure 51-5 LP gas supply systems.

(a)

(b)

Figure 51-6 (a) First pressure regulator located at the tank; (b) second pressure regulator located at the house.

TECH TIP

Some local and state governments require specific licenses to work on LP systems.

51.6 COMBUSTION GAS VENTING

Gas manufacturers now recognize four distinct categories of vented appliances. Only Category I appliances may be connected to a traditional venting system. Building codes usually just state that Categories II, III, and IV should be vented per the manufacturer's instructions. Category IV furnaces are vented with PVC or other noncorrosive materials, such as stainless steel. The required material for any furnace is identified in the installation instructions. Materials other than those listed in the manufacturer's instructions are not permitted. Categories II and III require special venting systems, and these are normally expensive. For this reason, most furnace manufacturers no longer offer any Category II and III appliances. The rest of this section deals only with Category I and Category IV appliances.

Category I Appliance Venting

The development of 80 percent furnaces using draft-inducer fans created problems with the old vent tables that had been used for years. Vent sizes for fan-assisted appliances are smaller because of the fan assist and because vent gases with these furnaces are cooler, so condensation is more of a concern. This is why the vent tables now have separate listings for natural-draft appliances and fan-assisted appliances. The fan-assisted tables list two columns, a minimum rating and a maximum rating. If a vent that is too big is used with a fan-assisted furnace, condensation will occur in the vent.

To understand the GAMA Category I venting tables, you must first understand the terminology used in the tables, which is listed below.

Vent Table Terminology

- **Fan-assisted combustion systems** Fan-assisted combustion systems are appliances equipped with an integral mechanical means to either draw or force products of combustion through the combustion chamber and/or heat exchanger.

Table 51-4 LP Gas Tank Pressures (psig)

	Outside Temperature (°Fahrenheit)														
	–30	–20	–10	0	10	20	30	40	50	60	70	80	90	100	110
100% Propane	6.8	11.5	17.5	24.5	34	42	53	65	78	93	110	128	150	177	204
70% Propane	—	4.7	9	15	20.5	28	36.5	46	56	68	82	96	114	134	158
30% Butane 50% Propane	—	—	3.5	7.6	12.3	17.8	24.5	32.4	41	50	61	74	88	104	122
50% Butane 70% Butane	—	—	—	2.3	5.9	10.2	15.4	21.5	28.5	36.5	45	54	66	79	93
30% Propane 100% Butane	—	—	—	—	—	—	—	3.1	6.9	11.5	17	23	30	38	47

- **Fan min** Fan min refers to the minimum appliance input rating of a Category I appliance with a fan-assisted combustion system that could be attached to the vent.
- **Fan max** Fan max refers to the maximum appliance input rating of a Category I appliance with a fan-assisted combustion system that could be attached to the vent.
- **Nat max** Nat max refers to the maximum appliance input rating of a Category I appliance equipped with a draft hood that could be attached to the vent. There are no minimum appliance input ratings for draft hood equipped appliances.
- **Fan+fan** Fan+fan refers to the maximum combined input rating of two or more fan-assisted appliances attached to the common vent.
- **Fan+nat** Fan+nat refers to the maximum combined input rating of one or more fan-assisted appliance and one or more draft hood–equipped appliances attached to the common vent.
- **Nat+nat** Nat+nat refers to the maximum combined input rating of two or more draft hood–equipped appliances attached to the common vent.
- **NR** NR means not recommended due to potential for condensate formation and/or pressurization of the venting system.
- **NA** NA means not applicable due to physical or geometric constraints.
- **Draft hood** A draft hood is a device built into an appliance, or made a part of the vent connector from an appliance, which is designed to
  - **a.** Provide for the ready escape of the flue gases from the appliance in the event of no draft, backdraft, or stoppage beyond the draft hood
  - **b.** Prevent a backdraft from entering the appliance
  - **c.** Neutralize the effect of stack action of the chimney or gas vent upon the operation of the appliance
- **Vent** The vent is a passageway used to convey flue gases from gas-utilization equipment, or their vent connectors, to the outside atmosphere.
- **Vent connector** The vent connector is the pipe or duct that connects a fuel-gas-burning appliance to a vent or chimney.
- **Flue collar** The flue collar is that portion of an appliance designed for the attachment of a draft hood, vent connector, or venting system.
- **Categorized vent diameter** The categorized vent diameter is the minimum vent diameter permissible for Category I appliances to maintain a negative vent static pressure when tested in accordance with nationally recognized standards.

Venting Tables

The most authoritative vent-sizing tables are published by GAMA, the Gas Appliance Manufacturers Association (GAMA merged with ARI in 2007/2008 to form AHRI). Because there are many vent materials used and many types of equipment connected to vents, there are a large number of tables. Figure 51-7 illustrates the dimensions used in the GAMA venting table. This text covers only tables for use with single appliances vented with type B double-wall vents and vent connectors. Examples of type B venting are shown in Figure 51-8.

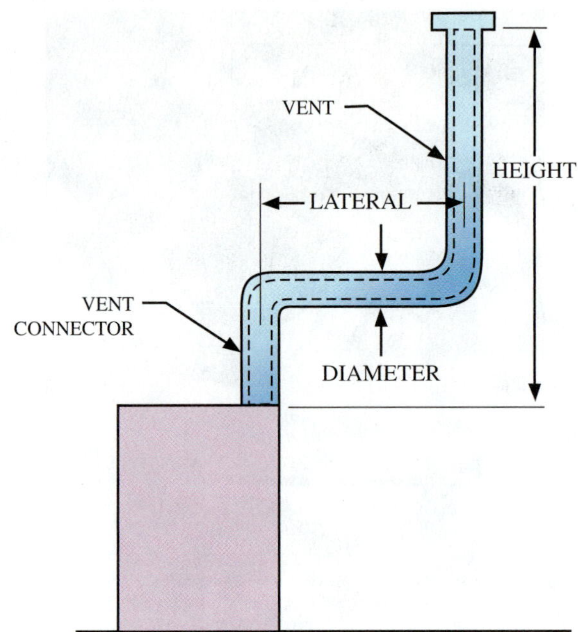

Figure 51-7 Illustration of terminology used in venting tables.

Vent Sizing for a Single Appliance

Table 51-5 shows a vent-sizing table for a single Category I gas-fired appliance using a type B vent and vent connector. First, determine the total height of the vent from the top of the draft diverter to the cap. Next, determine the lateral distance the vent will be traveling. The lateral distance should never be more than the vertical height of the vent. Find the vent height on the left. If the actual distance is in between sizes, round down because taller vents work better than shorter ones. That way you are selecting the worst-case scenario. Find the lateral length in the column immediately to the right of the height column. If the actual length is in between sizes, round up, because longer lateral lengths reduce the vent capacity.

For natural-draft systems, read across under the Nat columns until you find the first BTU input rating equal to or greater than the input rating of the appliance. Read up to find the vent size. In no case should you use a vent size smaller than the connection on the draft hood.

For fan-assisted appliances, read across under the Fan min columns until you find the first BTU input rating equal to or less than the input rating of the appliance. Next, check the Fan max column, which is located next to the Fan min column. The furnace BTU capacity must be equal to or less than the Fan max rating and equal to or greater than the Fan min rating.

Category IV Furnace Venting

Category IV furnaces are generally vented with PVC because they operate with a positive vent pressure and the vent gases tend to condense in the vent. PVC can easily withstand the low vent-gas temperatures of a Category IV

(a)

(b)

(c)

Figure 51-8 (a) Type B combined venting for two appliances; (b) type B vent passing through roof; (c) examples of vent pipe flashing and round-top arrangements.

furnace, and PVC is easily sealed to contain the positive vent pressure. Other material can also be used, such as stainless steel, if approved by the manufacturer. Many Category IV furnaces are also direct-vent furnaces; their combustion air is piped in through another PVC pipe.

TECH TIP

On many direct-vent Category IV furnaces the air intake enters near the top of the furnace and the vent exits near the middle of the furnace (Figure 51-9). This is because they are using counterflow heat exchange to increase their efficiency. It is easy to get these backward because the air intake is often located where the vent is on a regular furnace.

Care must be taken when running the combustion-air line and the vent so that the vent products do not blow right back in the combustion-air line. One way to accomplish this is to turn the vent up and the air intake down (Figure 51-10). The vent and air intake should also terminate at least 12 in above the normal season snow line. Always follow the manufacturer's instructions when terminating a vent system. Failure to do so can have disastrous results, and in most cases the system will not pass inspection.

Category IV furnaces condense water as part of their operation. Additional water forms in the vent and drains back to the furnace. Part of running the vent system to one of these furnaces is running the drain. If the furnace is installed where the drain will not operate by gravity flow, a condensate pump will be required. Make sure the condensate pump is rated for furnace duty; the condensate from condensing furnaces can eat up the materials on some condensate pumps.

51.7 COMBUSTION AIR

Gas furnaces cannot operate without an adequate supply of combustion and ventilation air. As the furnace operates, it draws air out of the room it is in. Air is needed from outside to keep the furnace operating. In many older installations, the assumption was that enough air would infiltrate through leaks in the house to supply the combustion-air needs. This is not a safe assumption, especially as home construction actively tries to eliminate leaks. It is now common to wrap the entire house in a house wrap to seal out air leaks. Air must be supplied from outside for safe operation of gas appliances.

This can be accomplished with two rectangular openings, one within 12 in of the ceiling and one 12 in from the floor. These openings should be open to the outside. Each opening should have a minimum free area of 1 in^2 per 4,000 BTU of the input rating of all gas appliances. Ventilated attics and ventilated crawlspaces count as outdoors. Figures 51-11 and 51-12 show two ways to provide adequate combustion air.

Sometimes ducting will be needed to reach the outside. If vertical ducting is used, the free area stays the same. However, if horizontal ducting is used, the free area must be based on 2,000 BTU/hr per in^2 of the total input rating

Table 51-5 Vent Table

Capacity of Type B Double-Wall Vents with Type B Double-Wall Connectors Serving a Single Category I Appliance

		Vent and Connector Diameter – D (inches)																				
		3			**4**			**5**			**6**			**7**			**8**			**9**		
Height	Lateral	**Appliance Input Rating in Thousands of BTU per Hour**																				
H	L	Fan		Nat	Fan		Nat	Fan		Nat	Fan		Nat	Fan		Nat	Fan		Nat	Fan		Nat
(ft)	(ft)	min	max	max	min	max	Max	min	max	max	min	max	max	min	max	max	min	max	max	min	max	max
6	0	0	78	46	0	152	86	0	251	141	0	375	205	0	524	285	0	698	370	0	897	470
	2	13	51	36	18	97	67	27	157	105	32	232	157	44	321	217	53	425	285	63	543	370
	4	21	49	34	30	94	64	39	153	103	50	227	153	66	316	211	79	419	279	93	536	362
	6	25	46	32	36	91	61	47	149	100	59	223	149	78	310	205	93	413	273	110	530	354
8	0	0	84	50	0	165	94	0	276	155	0	415	235	0	583	320	0	780	415	0	1,006	537
	2	12	57	40	16	109	75	25	178	120	28	263	180	42	365	247	50	483	322	60	619	418
	5	23	53	38	32	103	71	42	171	115	53	255	173	70	356	237	83	473	313	99	607	407
	8	28	49	35	39	98	66	51	164	109	64	247	165	84	347	227	99	463	303	117	596	396
10	0	0	88	53	0	175	100	0	295	166	0	447	255	0	631	345	0	847	450	0	1,096	585
	2	12	61	42	17	118	81	23	194	129	26	289	195	40	402	273	48	533	355	57	684	457
	5	23	57	40	32	113	77	41	187	124	52	280	188	68	392	263	81	522	346	95	671	446
	10	30	51	36	41	104	70	54	176	115	67	267	175	88	376	245	104	504	330	122	651	427
15	0	0	94	58	0	191	112	0	327	187	0	502	285	0	716	390	0	970	525	0	1,263	682
	2	11	69	48	15	136	93	20	226	150	22	339	225	38	475	316	45	633	414	53	815	544
	5	22	65	45	30	130	87	39	219	142	49	330	217	64	463	300	76	620	403	90	800	529
	10	29	59	41	40	121	82	51	206	135	64	315	208	84	445	288	99	600	386	166	777	507
	15	35	53	37	48	112	76	61	195	128	76	301	198	98	429	275	115	580	373	134	755	491
20	0	0	97	61	0	202	119	0	349	202	0	540	307	0	776	430	0	1,057	575	0	1,384	752
	2	10	75	51	14	149	100	18	250	166	20	377	249	33	531	346	41	711	470	50	917	612
	5	21	71	48	29	143	96	38	242	160	47	367	241	62	519	337	73	697	460	86	902	599
	10	28	64	44	38	133	89	50	229	150	62	351	228	81	499	321	95	675	443	112	877	576
	15	34	58	40	46	124	84	59	217	142	73	337	217	94	481	308	111	654	427	129	853	557
	20	48	52	35	55	116	78	69	206	134	84	322	206	107	464	295	125	634	410	145	830	537
30	0	0	100	64	0	213	128	0	374	220	0	587	336	0	853	475	0	1,173	650	0	1,548	855
	2	9	81	56	13	166	112	14	283	185	18	432	280	27	613	394	33	826	535	42	1,072	700
	5	21	77	54	28	160	108	36	275	176	45	421	273	58	600	385	69	811	524	82	1,055	688
	10	27	70	50	37	150	102	48	262	171	59	405	261	77	580	371	91	788	507	107	1,028	668
	15	33	64	NR	44	141	96	57	249	163	70	389	249	90	560	357	105	765	490	124	1,002	648
	20	56	58	NR	53	132	90	66	237	154	80	374	237	102	542	343	119	743	473	139	977	628
	30	NR	NR	NR	73	113	NR	88	214	NR	104	346	219	131	507	321	149	702	444	171	929	594
50	0	0	101	67	0	216	134	0	397	232	0	633	363	0	932	518	0	1,297	708	0	1,730	952
	2	8	86	61	11	183	122	14	320	206	15	497	314	22	715	445	26	975	615	33	1,276	813
	5	20	82	NR	27	177	119	35	312	200	43	487	308	55	702	438	65	960	605	77	1,259	798
	10	26	76	NR	35	168	114	45	299	190	56	471	298	73	681	426	86	935	589	101	1,230	773
	15	59	70	NR	42	158	NR	54	387	180	66	455	288	85	662	413	100	911	572	117	1,203	747
	20	NR	NR	NR	50	149	NR	63	275	169	76	440	278	97	642	401	113	888	556	131	1,176	722
	30	NR	NR	NR	69	131	NR	84	250	NR	99	410	259	123	605	376	141	844	522	161	1,125	670
100	0	NR	NR	NR	0	218	NR	0	407	NR	0	665	400	0	997	560	0	1,411	770	0	1,908	1040
	2	NR	NR	NR	10	194	NR	12	354	NR	13	566	375	18	831	510	21	1,155	700	25	1,536	935
	5	NR	NR	NR	26	189	NR	33	347	NR	40	557	369	52	820	504	60	1,141	692	71	1,519	926
	10	NR	NR	NR	33	182	NR	43	335	NR	53	542	361	68	801	493	80	1,118	679	94	1,492	910
	15	NR	NR	NR	40	174	NR	50	321	NR	62	528	353	80	782	482	93	1,095	666	109	1,465	895
	20	NR	NR	NR	47	166	NR	59	311	NR	71	513	344	90	763	471	105	1,073	653	122	1,438	880
	30	NR	NR	NR	NR	NR	NR	78	290	NR	92	483	NR	115	726	449	131	1,029	627	149	1,387	849
	50	NR	NR	NR	NR	NR	NR	NR	NR	NR	147	428	NR	180	651	405	197	944	575	217	1,288	787

Figure 51-9 The air intake is near the top and the vent exits near the middle on this direct-vent high-efficiency furnace.

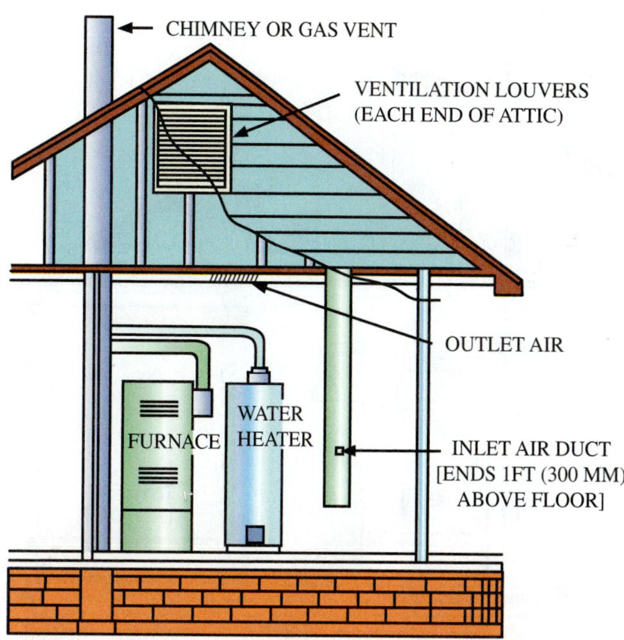

Figure 51-11 Installation with all outdoor inlet and outlet air openings in the attic; Openings are 1 sq in per 4000 BTU.

of all gas appliances. The minimum dimension of any duct used is 4 in. Figure 51-13 shows an installation using horizontal ducts for outside combustion air.

Note that these areas are free areas. If a 6 in × 10 in grill is placed over a 6 in × 10 in hole, the hole no longer has a free area of 60 in^2 because the grill free area is only about 40 in^2. The free area of the grill must be taken into account.

Direct-vent furnaces do not need a combustion air opening to the room because they draw their combustion air in through a PVC pipe. Their combustion process is sealed; they pipe air in and vent combustion products out. The concentric vent (Figure 51-14) allows both vent pipe and combustion-air pipe to pass through a single opening in the roof or sidewall.

Many burners have an adjustment for the primary air. Primary air is the air that is mixed with the fuel in the burner before combustion. The air adjustment should be set to produce a soft blue flame. Too much primary air will produce a hard blue flame that will waste fuel. Insufficient

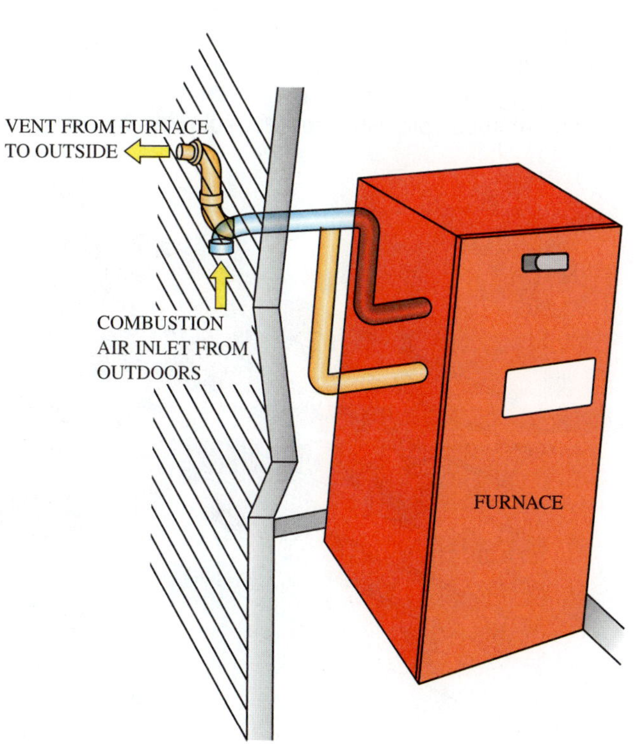

Figure 51-10 Category IV furnace venting.

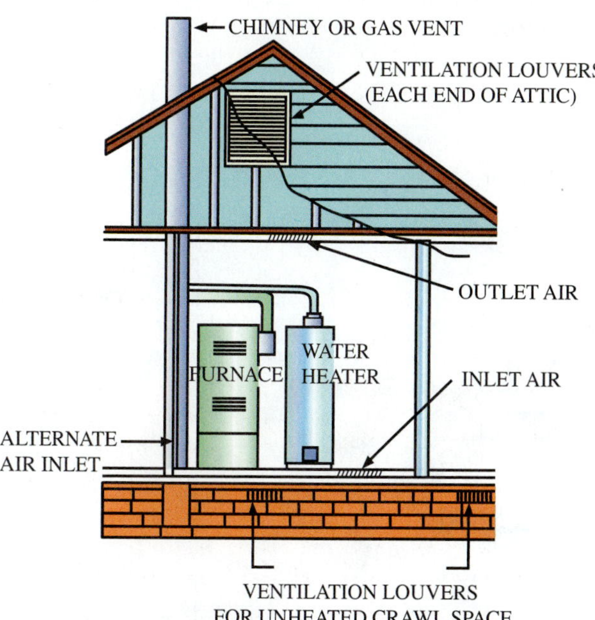

Figure 51-12 Installation with inlet air opening in the crawlspace and outlet air opening in the attic. Openings are 1 sq in per 4000 BTU.

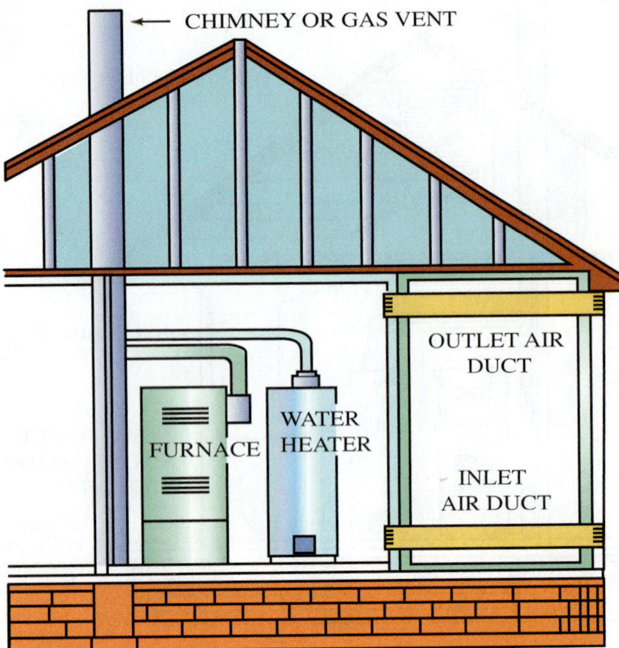

Figure 51-13 Installation with inlet and outlet air openings through horizontal ducts. Openings are 1 sq in per 2000 BTU.

primary air will produce a flame with yellow tips. Insufficient secondary air causes incomplete combustion, which wastes fuel and forms carbon monoxide (CO) and soot.

Most high-efficiency furnaces use an in-shot gas burner and an induced-draft fan. Normally the only adjustment that is made on these burners is the gas pressure. Typical settings are between 3.2 and 3.8 in wc for natural gas and 11 in wc for propane. These burners do not have a primary air adjustment. Regulation of the gas flow is by the gas pressure and the size of the orifice. The flame characteristic is clear blue.

After checking and setting the BTU/hr input and adjusting the primary air shutters, the vent draft condition should be checked.

51.8 AIR TEMPERATURE RISE TEST

The temperature rise of the air through the heating unit should be set as specified by the furnace manufacturer. Figure 51-15 shows the proper insertion of dial-type thermometers into the supply and return plenums of the heating unit. Use a drill to make a hole large enough to take the ⅛-in-diameter stem of the dial thermometer. Close the hole with a sheet-metal screw after the test is completed.

The supply-air thermometer should be located far enough away from the surface of the heat exchanger to prevent the effect of radiant heat on the thermometer from the heat exchanger. This is usually a minimum distance of 12 in. If the thermometer can be located in the main duct off the supply-air plenum, the radiant effect will be eliminated.

TECH TIP

It is not always possible to get an ideal temperature location on furnaces that are located in equipment closets. Often the furnace plenum is not accessible. For these installations, a temperature taken at the closest register to the furnace can be used.

After the unit has operated long enough for the supply-air thermometer to hold a steady reading (a stabilized reading), the return-air temperature should be subtracted from the supply-air temperature and the temperature rise recorded. If the temperature rise is below the specified temperature, the heating unit blower is moving too much air, and the speed should be reduced. If the temperature rise is above the specified temperature, the blower speed should be increased.

The procedure for this would depend on the type of drive used on the blower. Belt-driven blowers are adjusted by changing the size of the motor or driving pulley, direct-drive permanent split capacitor (PSC) blowers by changing

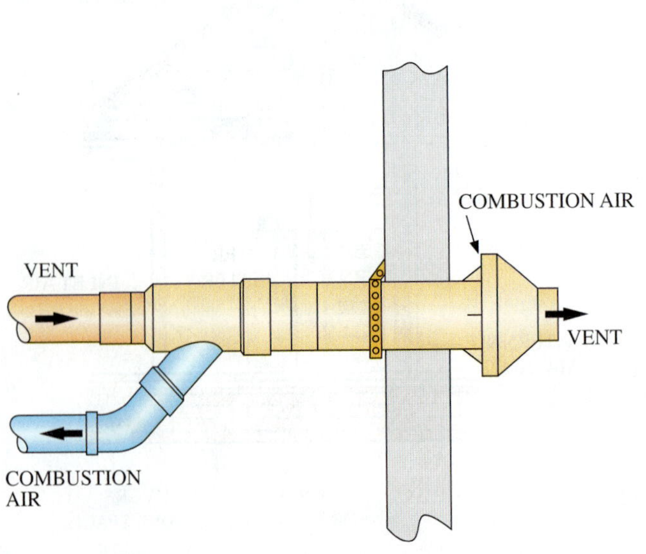

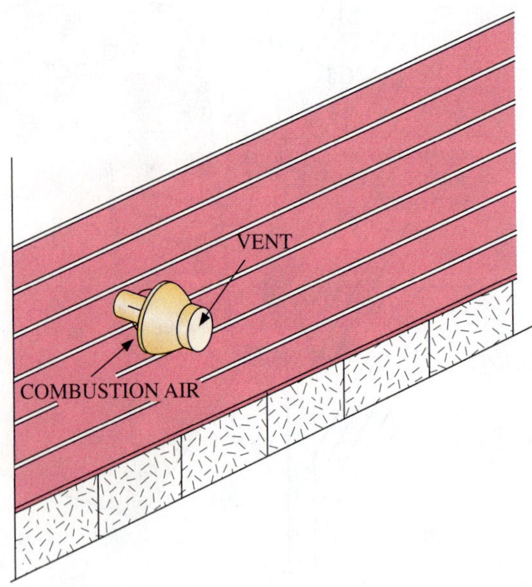

Figure 51-14 Concentric vent passing through the sidewall should exit at least 12 in above the normal season snow line.

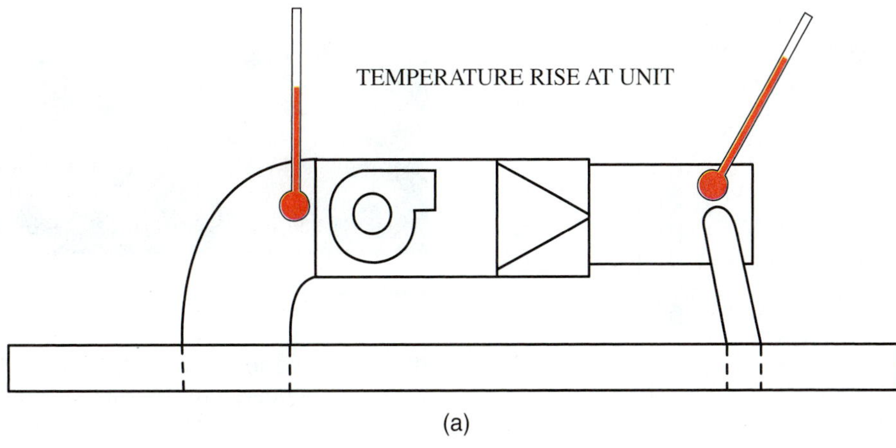

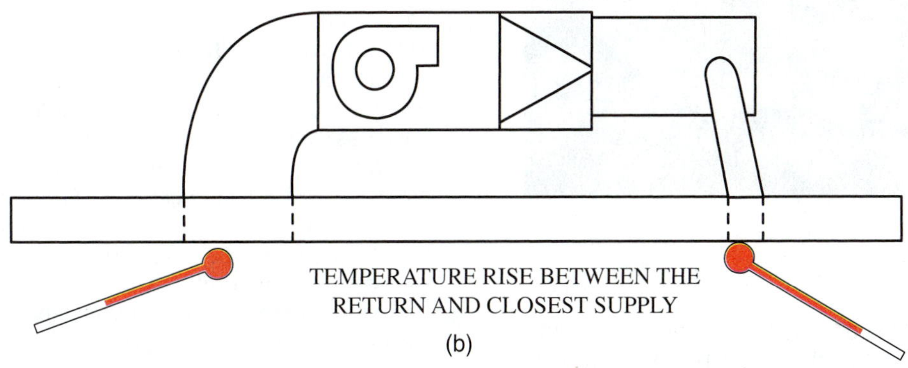

Figure 51-15 Measuring the air temperature rise though a furnace: (a) at the furnace; (b) at the return-air and closest supply.

the electrical connections to the motor, and direct-drive electronically commutated motors (ECMs) by changing the jumper on the blower board.

Belt-driven blowers use a combination of adjustable motor pulley (the driving pulley), blower pulley (the driven pulley), and drive belt (Figure 51-16). The speed of the blower is adjusted by changing the spread of the flanges of the driving pulley. Opening the pulley by spreading the flanges allows the belt to ride lower in the pulley, thus reducing the effective diameter of the pulley. This in turn reduces the pulley diameter rotation between the two flanges and reduces the blower speed.

Closing the pulley spread increases the drive pulley diameter, resulting in an increase in blower speed. The pulley usually has two set-screw flats to allow adjustment of the pulley in half-turn increments. Use these flats! Do not drive the set screw into the pulley adjustment threads, as this ruins the chance for future pulley adjustments!

After the blower speed has been set, the belt tension and alignment should be checked (Figure 51-17). The alignment is adjusted by changing the position of the pulley on the shaft. In Figure 51-18, the driven pulley on the left has a set screw that holds the pulley in position. If the set screw is loosened, the pulley can be moved up or down the shaft until

Figure 51-16 Motor pulley, blower pulley, and drive belt.

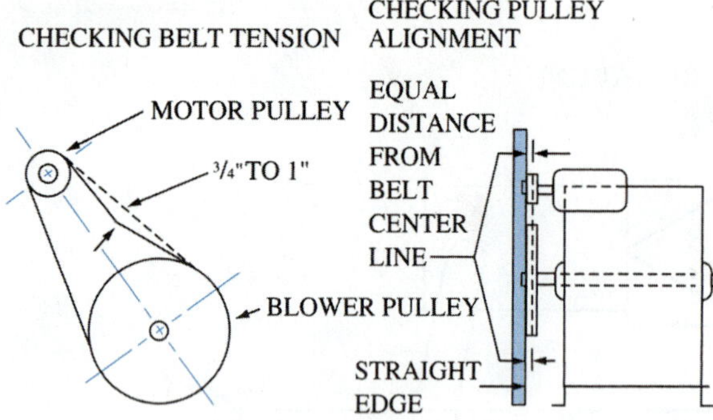

Figure 51-17 Aligning pulleys and tightening belts.

Figure 51-18 The driven pulley on the left can be adjusted for alignment by loosening the set screw.

the proper alignment is obtained. The pulley alignment can be checked by using a straight edge. Once properly aligned, the set screw can be tightened to hold the pulley in place. The belt tension must also be adjusted. It should be tight enough to avoid slippage but not so tight to cause excessive wear. Approximately ¾ to 1 in of play should be allowed for each 12-in distance between the motor and blower shafts.

TECH TIP

After the pulley has been adjusted and tightened, close the blower door and check the motor amperage. If the motor amperage is at or near the rated load amperage for the motor, it may be necessary to slow the blower down slightly to ensure that the motor does not overheat.

On direct-drive blowers, the choice of blower speeds is limited to the number of speeds built into the blower motor. On some heating-only units, the blower has only one speed, and no choice of temperature rise is available. If air conditioning is added to the unit, a change in blower motor or blower assembly is required to accommodate the additional pressure loss through the coil.

If the heating unit is equipped with a multispeed blower motor, observe which fan speed lead is connected: high, medium, or low. For the initial startup, it is recommended

Figure 51-19 Fan settings on furnace control board.

that the fan be set for medium speed, subject to verification when the temperature rise is obtained.

Most furnaces built today use control boards that have fan connections marked Heat, Cool, Fan, and Park. The blower speed used for heating goes on the Heat terminal. The terminals marked Park are dummy terminals that just give the unused speed wires a place to sit safely while not in use. Figure 51-19 shows the fan settings on a typical furnace control board.

On furnaces using ECM blower motors, the change in speed is accomplished by moving a programming jumper. Many boards actually have an airflow written on the board for each available programming option. Figure 51-20 shows a typical ECM blower control with dual inline processor (DIP) switches to adjust blower settings.

No matter how it is accomplished, increasing the blower speed will increase the CFM of air through the unit and lower the temperature rise. Conversely, lowering the blower speed will decrease the CFM through the unit and increase the temperature rise.

When increasing the blower speed, which increases the load in the motor, the amperage draw of the motor will also increase. A clamp-type ammeter should be used to check the motor operating amperage. If the required CFM causes the motor to draw more than its amperage rating, a motor or blower assembly of larger capacity will have to be substituted.

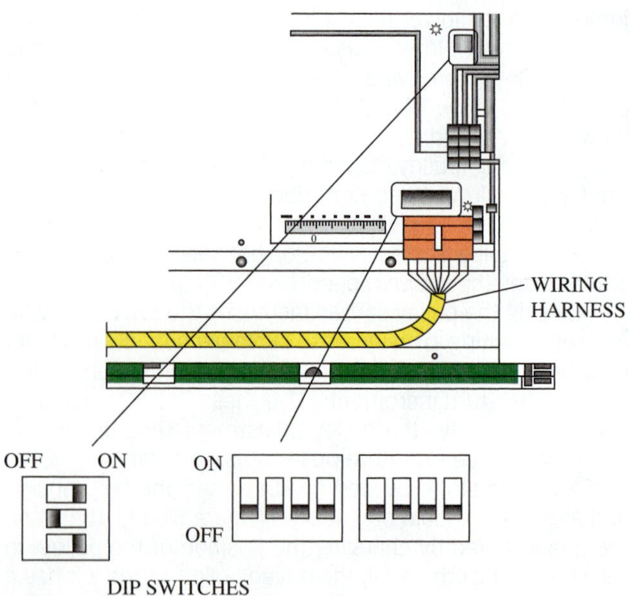

Figure 51-20 ECM blower control jumper setting.

51.9 EFFICIENCY TESTING

With the heating unit input set, the burners properly adjusted, and the unit operating at as close to desirable temperature rise as possible, the unit can be tested for operating efficiency. The objective of efficiency testing is to obtain as high an efficiency rating of the heating unit as possible, taking into account the operating cost, the equipment operating life, and the comfort obtained in the conditioned area.

It is necessary to reach adjustments that achieve a balance between operating cost (efficiency) and comfort. Standards that are used for efficiency testing are as follows:

- **Input** The unit must be supplied with the correct amount of fuel at 90–100 percent of its rated capacity.
- **Burner primary air adjustment** Burner flames should be soft blue without yellow color or flame lift.
- **Air temperature rise** Adjust the fan to obtain the manufacturer's specified temperature rise.
- **Fan control settings** The fan control should be set according to the manufacturer's specifications.
- **O_2** The percentage O_2 in the flue gas is an indicator of the amount of excess combustion air and overall combustion efficiency. It may be measured with an O_2 analyzer (Figure 51-21).

No O_2 in the flue gas indicates incomplete combustion and possible production of carbon monoxide. Acceptable readings for safe operation range from 7 to 9 percent. Readings above 9 percent indicate too much combustion air, which is detrimental to efficiency.

- **CO_2** The percentage of CO_2 in the flue gas is also an indicator of overall combustion efficiency. Very often the flue gas is tested for CO_2 rather than O_2. If possible, it is recommended to test for both. The operation range for CO_2 is the same as for O_2, which is 7–9 percent.

TECH TIP

Checking a gas furnace's fuel consumption as compared to heat output is a way of determining the actual operating efficiency of a particular system. This performance check is much like a driver confirming the fuel economy (miles per gallon) that his car is getting.

A means of recording such information is necessary so that it will become a permanent part of the unit operating and service history. This efficiency check sheet should include the information shown below.

Input

1. Type of gas: Nat. _____ Mixed _____ Mfg. _____ Prop. _____ Bu. _____
2. Heat content (BTU/ft^3) ___________
3. Specific gravity of the gas: ___________
4. Main burner orifice drill size: Found _____ Left _____
5. Manifold pressure (in wc): Found _____ Left _____

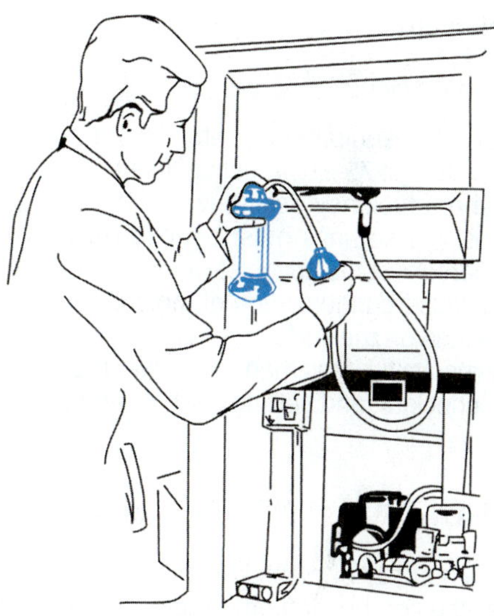

Figure 51-21 Testing the percent of O_2 in the flue gas.

6. Meter test dial size: _____ ft^3 per rev.
7. Seconds required per rev. of test dial: Found _____ Left _____

Primary Air Adjustment

1. Flame before adjustment: Sharp blue _____ Soft blue _____ Yellow tips _____
2. Flame after adjustment: Soft blue _____

Air Temperature Rise

1. Supply-air temperature: First test _____ Second test _____ Left test _____
2. Return-air temperature: First test _____ Second test _____ Left test _____
3. Air temperature rise: First test _____ Second test _____ Left test _____

O_2 Percent

1. First test _____%
2. Second test _____%
3. Left test _____%

CO_2 Percent

1. First test _____%
2. Second test _____%
3. Left test _____%

Stack Temperature Rise

1. Stack temperature: First test _____ Second test _____ Left test _____
2. Combustion-air temperature: First test _____ Second test _____ Left test _____
3. Stack temperature rise: First test _____ Second test _____ Left test _____

Combustion Efficiency

1. Percent efficiency _____ _____ _____

Gas-burning equipment of standard design should always be capable of 75–80 percent efficiency. Unless the unit is of a higher-efficiency design (when the manufacturer's instructions and settings must be followed), an efficiency above 80 percent could adversely affect the draft of the unit as well as cause condensation of moisture in the chimney or flue pipe and on the surfaces of the heat exchanger. If the efficiency results are less than this range, the test should be repeated, checking and setting the proper input.

51.10 COMBUSTION EFFICIENCY

There are a number of measurements that can be taken to determine the combustion efficiency of a gas furnace:

- Ambient temperature (temperature around the furnace)
- Stack temperature (vent temperature)
- Oxygen percent in the flue gas
- Carbon dioxide percent in the flue gas

The first two measurements are used to calculate the net stack temperature, which is found by subtracting the ambient temperature from the stack temperature. Take the temperature before the draft diverter when taking it on a natural-draft furnace. The oxygen and carbon dioxide readings are taken by drilling a small hole in the vent and taking a sample of the gas. This can be done with an aspirator connected to an hourglass shaped tester (Figure 51-22). The combustion efficiency is found using a slide rule provided with the combustion efficiency test kit.

All of these measurements can also be taken using an electronic combustion analyzer. The analyzer takes the sample, measures the oxygen percentage in the flue gas, measures the net stack temperature, and calculates the combustion efficiency. The analyzer can also calculate the CO_2 and excess air percentages. Figure 51-23 shows a typical electronic combustion efficiency analyzer.

Acceptable readings for atmospheric gas burners are the following:

Oxygen percent	7–9%
Carbon dioxide percent	7–9%
Excess air percent	50%
Net stack temperature	325°F–500°F

Figure 51-22 Carbon dioxide (CO_2) analyzer.

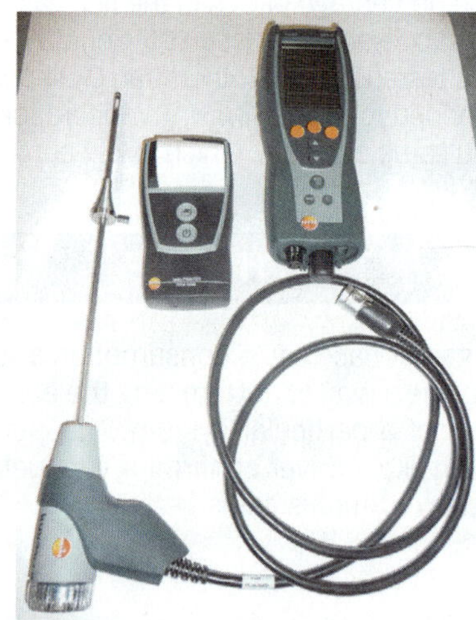

Figure 51-23 Electronic combustion analyzer.

51.11 STARTUP SAFETY

Technicians should always be conscious of the unique hazards posed by combustible gases when working with gas furnaces. Gas appliances are safe when installed and operated according to the manufacturer's instructions, but technicians should not assume that everything is per the manufacturer's instructions. In fact, they should assume the opposite. It is reasonable to assume that a system that is not operating properly is either installed incorrectly or has a component that is not performing as designed. Either situation could create a hazard for the technician. There are a few precautions a technician can take that will reduce the chance of injury even if the furnace is potentially dangerous.

- Do not turn anything on until you have inspected the furnace! Check for the key safety items listed below. For example, turning on a system with a leak that has caused a buildup of gas could create an explosion.

- Check for the presence of gas in the area around the furnace. Use your nose or an electronic combustible gas sniffer to check for the presence of gas around the furnace. If gas is detected in the area of the furnace, the cause should be found and corrected before proceeding.
- Make sure there are no combustible items stored near the furnace. This is especially true of any volatile liquids that create vapors that can be ignited. These should not be in the same room!
- Check the vent. Make sure there is a properly connected vent and that the vent is in good shape. Vent gases spilling out into the area the technician is working in can overcome the technician.
- Check for combustion air. Do a quick visual check to determine where the combustion air will come from. A furnace operating without enough combustion and ventilation air will start producing carbon monoxide after a few minutes of operation.
- Do not position yourself directly in front of the furnace while it is lighting. If the problem with the system causes delayed ignition, a flashback or mini-explosion can occur. The flames normally will shoot straight out of the furnace; you do not want to be in their path.
- Once the furnace lights, check the vent to make sure it is working. This can be done by checking the draft pressure in the vent.

These safety precautions will safeguard the customer as well. There are a few additional items that may not present an immediate hazard to the technician but can present a hazard to the customer.

- Check the clearance to combustibles. Unit installation instructions will give specific clearances. A single-wall vent should have a clearance of at least 6 in; a double-wall vent should have a clearance of at least 1 in. Downflow furnaces installed on a wood floor generally require a noncombustible base to shield the combustible flooring under them.
- Make sure the door safety switch works. The door switch usually needs to be disabled to work on the furnace. Make sure to restore it to operating condition before leaving. The purpose of this switch is to prevent the furnace from operating with the door removed. Operation with the door removed would allow the combustion gases to be sucked out of the vent by the indoor blower, and they would begin circulating throughout the house.

51.12 OPERATIONAL CHECK

For Category I and II furnaces, place a draft gauge in the vent and check the draft with the burners operating. You should have a draft of at least −0.02 in wc. Watch the draft reading when the indoor blower comes on. A reduction in draft or an increase to a positive pressure indicates a leak in the heat exchanger. This is a serious safety issue that generally warrants replacing the furnace.

Check the manifold gas pressure after the burners light. The correct manifold gas pressure at sea level is indicated on the furnace nameplate. The manifold pressure is usually slightly less at higher altitudes. The manufacturer's installation instructions are the best source of information for de-rating gas appliances.

Look at the color of the flames: they should be blue. Yellow flames indicate incomplete combustion and the presence of carbon monoxide.

Yellow flames are the result of too much gas, too little air, or faulty burners. Older gas furnaces have adjustable primary air intakes. The primary air can be adjusted until the yellow disappears. However, opening the shutters too far can cause delayed ignition, especially on cold days. Too much primary air also causes the flames to lift away from the burner face. In-shot burners used in most modern furnaces do not have any primary air adjustment. Yellow flames from an in-shot burner indicate a lack of combustion air in the room or too much gas. Check the orifice size against the manufacturer's specifications or an orifice chart to make certain the furnace has the correct size gas orifices.

The combustion gases in the vent should be checked for carbon monoxide. Ideally, the carbon monoxide level should be 0 parts per million for complete combustion. Realistically, it is common for some low level of CO to be present in the combustion gases. Levels of CO in the vent gases exceeding 50 parts per million will warrant a closer look at the combustion process (Figure 51-24).

After the blower has been operating for a few minutes, check the temperature of the air entering and leaving the furnace. The difference is the temperature rise. Be sure to check the leaving air temperature in a location that is not directly in the line of sight of the heat exchanger. If the thermometer can "see" the heat exchanger, it may read higher due to the heat exchanger's radiant heat. If the temperature rise is too high, the airflow needs to be increased. If the temperature rise is too low, the airflow needs to be reduced.

Figure 51-24 Portable CO (carbon monoxide) analyzer.

51.13 PRACTICING PROFESSIONAL SERVICES

Service technicians should always remember that their job is not simply to push reset buttons and sell parts but to provide a service to the customer. Generally, the customer is paying more for the services being rendered than for any parts being used. They have a right to expect professionalism. Technicians should maintain an active interest in training and education and pursue professional NATE certification.

TECH TIP

Part of being a professional is practicing safe work habits. Keeping yourself and your customers safe is a professional responsibility all technicians should take very seriously.

UNIT 51—SUMMARY

Furnaces must be installed according to the manufacturer's instructions if they are to perform safely and efficiently. When installing a gas furnace, particular attention should be paid to clearances, combustion air, and venting. When performing the operational check, the manifold pressure should be set per the manufacturer's instructions and the primary air to the burners adjusted if necessary. The vent should be checked for proper operation, including a check of the drain for 90 percent annual fuel utilization efficiency (AFUE) furnaces. The furnace temperature rise should be checked and the airflow adjusted to keep the temperature rise within the manufacturer's specifications.

WORK ORDERS

Service Ticket 5101

Customer Complaint: Furnace Will Not Operate

Equipment Type: 90 Percent Direct-Vent Condensing Furnace Installed in Basement

A new 90 percent direct-vent furnace was installed in the fall and worked without any problems until recently. The technician observes that the induced-draft blower starts, but nothing else happens. The technician checks the pressure in the combustion air pipe and finds it in a vacuum of 0.5 in wc. The combustion air and vent pipe exit high on the basement wall, just above the grade level. Walking outside, the technician sees that the combustion air pipe is buried in snow. After digging the snow away from the air intake, the furnace operates. The technician re-pipes the combustion air and vent pipes to be well above the snow line.

Service Ticket 5102

Customer Complaint: Water All over the Floor

Equipment Type: 90 Percent Direct-Vent Condensing Furnace Installed in Basement

A customer has called about a furnace that was installed 6 months ago. When she got up this morning, there was water all over the floor, mostly by the furnace. The technician observes that the condensate pump can be heard running, but the water level in the sump is not going down. The technician checks the condensate water discharge line for kinks, but the line appears to be fine. The end of the line appears to be open as well. The technician lifts the pump out of the sump and opens it to find that the pump impeller has disintegrated. While inspecting the pump, the technician notices a sticker that reads, "Not for use with condensate from condensing furnaces." The technician explains that the pump was not designed for furnace use and recommends replacing the pump with one that is approved for furnace use.

Service Ticket 5103

Customer Complaint: CO Alarm Sounding

Equipment Type: Upflow 80 Percent Induced-Draft Furnace Installed in Hall Closet

A customer complains that since he moved in to his newly constructed house, the CO alarm in the hall periodically goes off. It seems to happen more often when there is a fire in the fireplace. The technician looks at the furnace closet and notices several boxes stacked up beside it from floor to ceiling, but no combustion air vents are visible. The technician finds the combustion air vents after moving the boxes. The technician asks the homeowner to start a big fire in the fireplace and then proceeds to check the furnace operation. The draft in the vent reads –0.03 in wc, and no CO is detected around the furnace. The technician explains that the boxes blocked the combustion air vents, creating a negative pressure in the furnace closet and resulting in the production of carbon monoxide. The technician advises the customer that the furnace closet should not be used for storage.

Service Ticket 5104

Customer Complaint: Supply Air Not Hot Enough

Equipment Type: 80 Percent Induced-Draft Furnace Installed in Crawlspace

A customer complains that the air coming out of the registers is not hot enough. The technician checks the manifold pressure and temperature rise. The manifold pressure is 3.5 in wc and the temperature rise is 20°F. The furnace data plate calls for a temperature rise of 25°–50° F. The technician checks the fan taps on the control board and notes that the blower's highest speed is on the Heat position and the blower's lowest speed is on Park. The low speed is moved to Heat and the high speed to Park. The technician rechecks the temperature rise and finds that it is 35°F.

Service Ticket 5105

Customer Complaint: Furnace Only Runs for 10 Minutes

Equipment Type: 80 Percent Multiposition Induced-Draft Furnace Installed in Crawlspace

A technician is called to check a furnace that was recently installed to replace an old horizontal furnace in a crawlspace. The homeowner explains that a friend of his did the change-out to save him money, but he is too busy at work to come by and take a look. The furnace will start and operate correctly for a few minutes, then the vent gas safety switch trips. The technician notices that the furnace was never reconfigured for horizontal installation and that it is connected to the old single-wall vent connector, which runs horizontally for 30 ft before entering a masonry chimney. The technician explains to the customer that the vent is not operating properly and that is causing the safety switch to trip. The technician recommends orienting the furnace for horizontal operation, replacing the single-wall vent with double wall, and lining the masonry chimney with a metal chimney liner so the furnace can vent properly.

UNIT 51—REVIEW QUESTIONS

1. List the safety precautions that should be observed when starting a gas furnace.
2. Explain the difference between natural gas and propane.
3. What is the most common manifold pressure for natural gas furnaces?
4. What is the most common manifold pressure for propane furnaces?
5. A technician is clocking the gas usage of a gas furnace. It takes 30 seconds for the ½-ft dial to make one revolution. What is the rate of gas flow in cubic feet per hour?
6. Use the gas flow rate calculated from question 5 to determine the input in BTU/hr if the heating capacity of the gas is 950 BTU/ft^3.
7. What is the correct vent size for a 50,000 BTU/hr input induced-draft furnace with a 15-ft lateral run and a 30-ft vent height?
8. What is the smallest fan-assisted furnace that can be connected to an existing 6-in type B vent that is 30 ft high with a lateral run of 15 ft?
9. What size should the combustion air openings be for a room with an 80,000 BTU/hr input furnace and a 35,000 BTU/hr water heater if the openings are opening directly into a ventilated attic and a ventilated crawlspace?
10. What size should the openings be in question 9 if the air is traveling through horizontal ducts to the outside?
11. Where do you measure the temperature when checking furnace temperature rise?
12. What is the temperature rise of a furnace with a supply-air temperature of 130°F and a return-air temperature of 75°F?
13. How can the temperature rise of a furnace be reduced if it is too high?
14. What are the normal oxygen and carbon dioxide percentages in the flue gas in a properly operating gas furnace?
15. How is net stack temperature determined?
16. What is the lowest temperature at which butane can be used as a heating fuel?
17. What is the difference between vent sizing for natural-draft appliances and fan-assisted appliances?
18. Why are most furnaces manufactured today either Category I or Category IV furnaces?
19. What is the difference between a vent connector and the vent?
20. How is the correct manifold pressure for any particular installation determined?

UNIT 52

Troubleshooting Gas Furnaces

OBJECTIVES

After completing this unit, you will be able to:

1. list safety precautions to follow when working on gas furnaces.
2. list the general sequence of operation for an induced-draft furnace.
3. interpret an LED fault code on a furnace board.
4. diagnose a system problem using a troubleshooting flowchart.
5. use a wiring diagram to identify system problems.

52.1 INTRODUCTION

Any heating service organization is aware of the accumulation of service calls during the fall of the year. When homeowners first try to start their systems, the problems reveal themselves. The most serious of these is "no heat." Whenever a no-heat call is received, it receives top priority. It means that for some reason the furnace will not run. The technician must solve the problem as quickly as possible.

Preliminary information is essential. What type of furnace is being used? What type of fuel is being supplied? Has the unit operated properly in the past, or is it a new installation? What has the customer done to get the unit started? Does any part of the unit run, or is it completely "dead"?

52.2 TROUBLESHOOTING GAS FURNACES

Gas furnace problems generally fall into two broad areas: electrical problems and combustion problems.

Troubleshooting the electrical problems is similar to isolating electrical problems with any appliance. If nothing is working, you start by checking power and control voltage to the unit. If the unit operates incorrectly, you determine what is not operating, determine if it has the correct operating voltage, consult the diagram, and check the circuits systematically to isolate the cause.

SAFETY TIP

Technicians should always be conscious of the unique hazards posed by combustible gases when troubleshooting gas furnaces. Even problems that are electrical or mechanical in nature can create explosion hazards. A safety control that fails to shut off a gas valve after a failed ignition attempt can allow a buildup of gas that can explode on the next ignition trial.

Combustion problems can be tricky, but understanding the effects of the three necessary components of combustion will help. Combustion requires three things: fuel, oxygen, and heat. If any one of these is not present, there will be no combustion. Further, if the gas and air are not mixed in the proper ratio, combustion will either not occur or will be incomplete. A visual inspection of the burner can often identify problems that should be followed up with a thorough combustion analysis. Figure 52-1 shows a correctly operating gas burner.

Figure 52-1 Correct burner flame for gas heater with standing pilot and flame rod.

Too Little Gas at the Burners

The gas is normally delivered through an orifice at a pressure established by the manufacturer. Common manifold pressures are 3.5 in wc for natural gas and 11 in wc for propane. Low gas pressure can cause smaller than normal flames or delayed ignition. Figure 52-2 shows a gas burner operating with a low gas pressure. Check to see that the pressure at the inlet of the gas valve is at least the minimum supply pressure listed on the unit, usually 4.5 in wc for natural gas and 11 in wc for propane. Low gas supply pressure can be caused from undersized gas piping, restrictions in the gas line, or supplier problems. Next, check the manifold pressure leaving the gas valve; it should be close to the

Figure 52-2 Burner with too little gas.

manifold pressure rating listed on the unit. Try adjusting the outlet pressure if it is too low. If the inlet pressure is all right but the outlet pressure cannot be adjusted to the correct pressure, then the gas valve is defective.

Too Much Gas at the Burners

Too much gas will cause large, lazy, yellow flames that produce soot. Too much gas is usually caused by oversized orifices. Check the orifice number against the manufacturer's specifications. This can occur when changing a furnace from natural gas to propane because natural gas orifices are much larger than propane orifices for the same firing rate. A defective gas valve can also cause this. Check the manifold pressure and compare it to the manifold pressure listed on the unit. Try adjusting the outlet pressure if it is too high. If the inlet pressure is all right but the outlet pressure cannot be adjusted to the correct pressure, then the gas valve is defective. Figure 52-3 shows a gas burner operating with too much gas pressure.

Figure 52-3 Burner with too much gas.

Figure 52-4 Burner with too little primary air.

Too Much Primary Air

Too much primary air will cause delayed ignition and noisy flames that lift off the burner. This problem gets worse when the air is colder. The solution is to close off the primary air shutter until the flames have yellow tips, then gradually open the shutter until the yellow tips disappear.

Too Little Primary Air

If the primary air is closed completely, the flames will be lazy and yellow, similar to having too much gas. If the primary air is just not open enough, the flames will be blue at the bottom and yellow at the tips (Figure 52-4). The solution is to gradually open the shutter until the yellow tips disappear.

Improper Venting

If the vent does not work properly, the combustion products will build up in the vent and the flames will start to come out of the combustion chamber. Usually the furnace lights and burns correctly for a minute or two before this starts. The flames may also turn from blue to yellow. This can be caused by a blocked vent or by a severe lack of combustion air to the room. This condition is extremely dangerous because it releases carbon monoxide!

52.3 A PROGRESSION OF ANALYSIS FOR TROUBLESHOOTING

One way of increasing speed and accuracy when troubleshooting is to try to eliminate possible problems with each test made. For example, on a no-heat service call, many technicians will turn the fan switch to ON to see if the indoor fan comes on. If it does come on, two possible problems have been eliminated: line voltage to the furnace and control voltage from the transformer. This eliminates them as possible problems. There should be a logical reason for each test that is performed. As successive tests are made, the possibilities should narrow, eventually leading to the cause of the problem. Understanding the normal operational sequence of the furnace is crucial. The following operational sequence works for most furnaces made today.

1. The thermostat calls for heat.
2. The induced-draft fan motor energizes.
3. The draft switch closes to prove the draft.
4. The igniter is energized for approximately a minute.
5. The gas valve is energized.
6. The burners light.
7. The indoor blower comes on after a delay of a few minutes.

The technician can learn where to start troubleshooting by observing the system and understanding the checks the system makes along the way. For example, if the igniter is coming on, there is no need to check the draft switch because the draft switch must close before the igniter can come on.

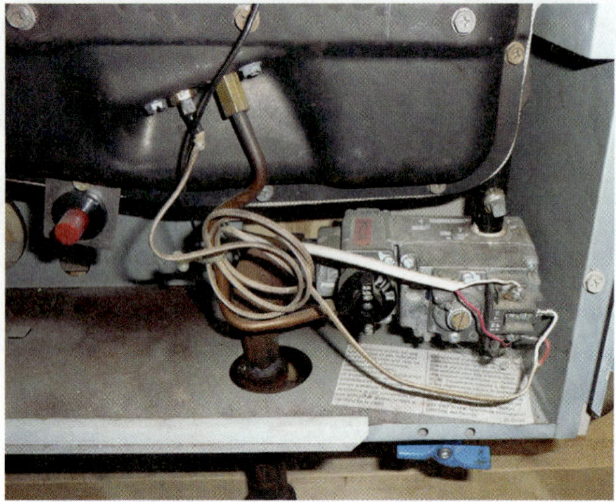

(a)

(b)

(c)

Figure 52-5 (a) The igniter is manually engaged by pushing the red button; (b) a spark is generated to light the pilot; (c) the pilot is lit.

52.4 STANDARD FURNACES

Since many thousands of standard furnaces are in residential use that were manufactured before the 78 percent AFUE minimum requirements, the service technician must be prepared to troubleshoot these units. Some important differences from the newer designs are natural-draft venting and standing gas pilots.

The sequence of operation for a standing pilot, natural-draft furnace is not complicated, but it is different from furnaces manufactured today.

The standing pilot light must be lit before the furnace can operate. The gas valve selection lever is turned to the pilot position and manually depressed so that gas will come out of the pilot burner. The pilot burner is lit with a match or igniter (Figure 52-5). The valve selection lever must be held in for a minute to allow the thermocouple to heat up enough to hold in the safety solenoid. After letting up on the valve selection lever, rotate it to the ON position (Figure 52-6).

The thermostat calls for heat and energizes the gas valve. The 24 V circuit to the gas valve passes through all the system safeties, like the limit switch. If the furnace overheats, the limit switch will de-energize the gas valve. After the burners light, the furnace heat exchanger heats up. When the fan switch senses the fan-on temperature, the fan motor is energized by the closing of the fan switch. When the thermostat is satisfied, the circuit to the gas valve is opened and the gas valve shuts down. The indoor blower continues to run, cooling down the heat exchanger. When the fan switch reaches the fan-off temperature, the fan shuts off.

TECH TIP

Standing pilots have been mostly replaced by intermittent pilots. However, the distinct advantage to a standing pilot gas space heater is that no outside source of electricity is required to operate it. Gas-fired space heaters do not have blowers. The only requirement is for the pilot to be initially lit. Depending on the BTU output, they can heat a small room. The thermostat requires no external power and will cycle the main burner valve on and off dependent on the temperature setting. This type of space heater is ideally suited for a remote camp and is often used for gas-fired fireplaces and gas-fired wood stoves. These can provide heat during a power outage when no other source of heat is available.

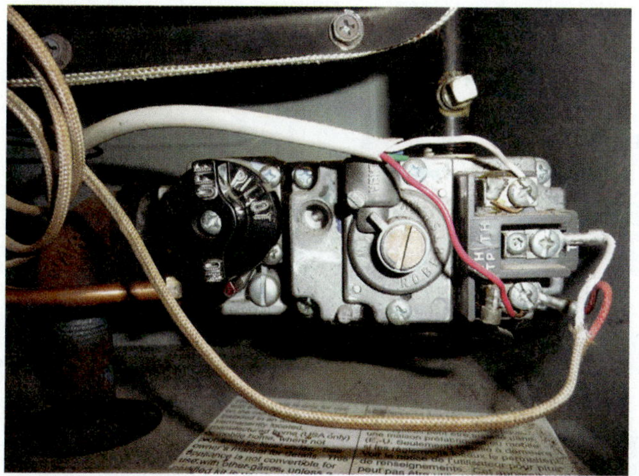

Figure 52-6 The gas valve for the standing pilot has a manual knob that must be pushed in and turned.

52.5 INTERMITTENT PILOT SYSTEMS

A typical ignition control module for an intermittent pilot system installed on a furnace with atmospheric burners is shown in Figure 52-7. This type of module can be used for different furnace configurations. As an example, if the pilot has a separate flame sensor, it would need to be connected slightly different, as shown in Figure 52-8. This type of module should not be allowed to get wet or be exposed to high humidity, corrosive chemicals, dust, grease, or excessive temperatures. The starting sequence begins with a trial for ignition. An electronic spark generator in the module produces a pulse output of over 10,000 V, and at the same time, the first main valve operator opens, allowing gas to flow to the pilot burner. If the pilot does not light or if the pilot flame current is too low and not steady, the module will not energize the main valve and the main burner will not light.

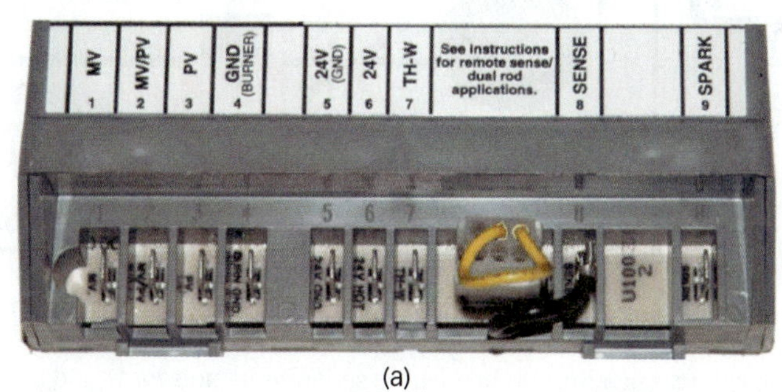

(a)

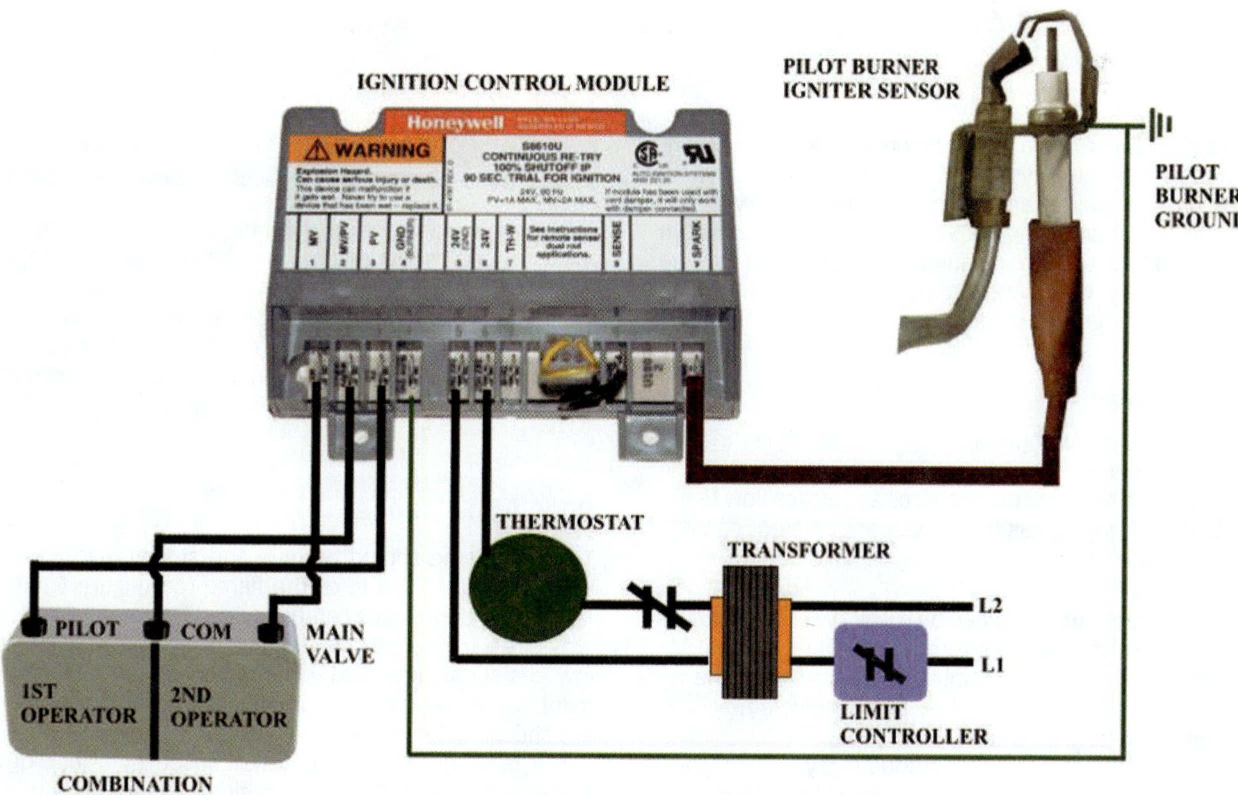

(b)

Figure 52-7 (a) Ignition control module connections; (b) ignition control module configuration for a furnace with atmospheric burners.

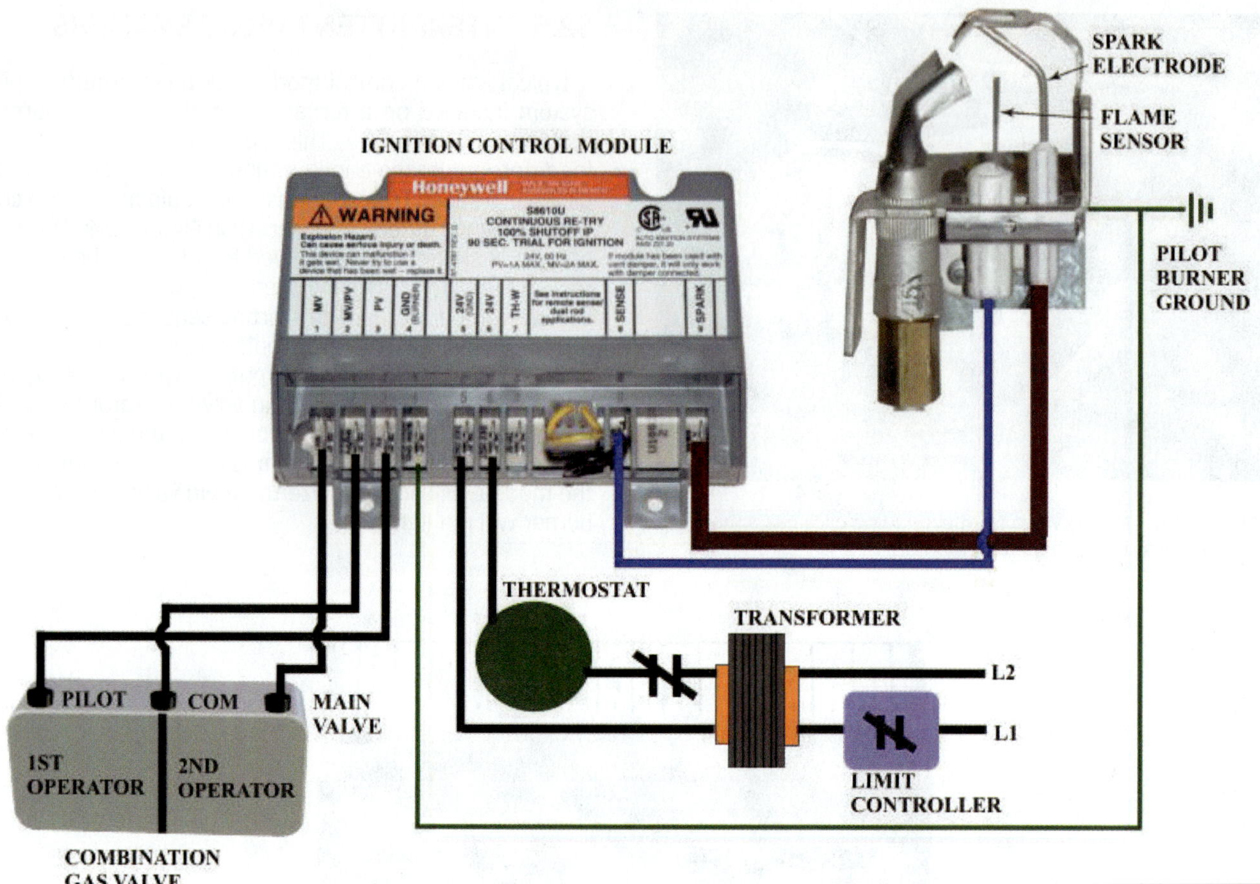

Figure 52-8 Ignition control module configuration for furnace with a separate flame sensor and atmospheric burners.

The sequence for a trial ignition will differ for a fan-assisted combustion system. In this case, there will be an air proving switch and a combustion air blower motor relay, as shown in Figure 52-9. On a call for heat from the thermostat, the blower initially starts. There will be approximately a 45-second pre-purge before the module begins the pilot ignition sequence.

Safety Lockout

The ignition spark will only continue until the timed trial for the ignition period ends. If the trial for ignition fails, then the first main valve operator de-energizes and closes flow to the pilot burner, stopping gas flow. There are different configurations for lockout.

- **Safety lockout** This can be reset by setting the thermostat below room temperature for a specific time (approximately 1 minute) or by turning off the power to the ignition module for a specific time (approximately 1 minute).
- **Safety shutoff with continuous retry** If the ignition trial initially fails, the module will restart the ignition sequence over again after a time delay period (approximately 5 minutes).

TECH TIP

The ignition process may repeat several times before lockout occurs. Depending on the manufacturer's setup, lockout may automatically reset after an hour or more, or it may have to be reset manually. Both the automatic and manual ignition lockouts can be reset manually by turning the power off for a few seconds, then powering up again.

Pilot Appearance

The pilot flame should be blue and steady, and it should envelope 3/8 to 1/2 in of the flame rod (Figure 52-10). A flame that is too small can be caused by a lack of gas. This would result from a clogged orifice filter, clogged pilot filter, low gas supply pressure, or an incorrect pilot adjustment (pilot set at minimum). Examples of incorrect pilot flames are shown in Figure 52-11.

A lazy yellow flame can be caused by a lack of air. This could result from a large orifice, dirty lint screen, dirty primary air opening, or an incorrect pilot adjustment. A waving blue flame is caused by an excessive draft at the

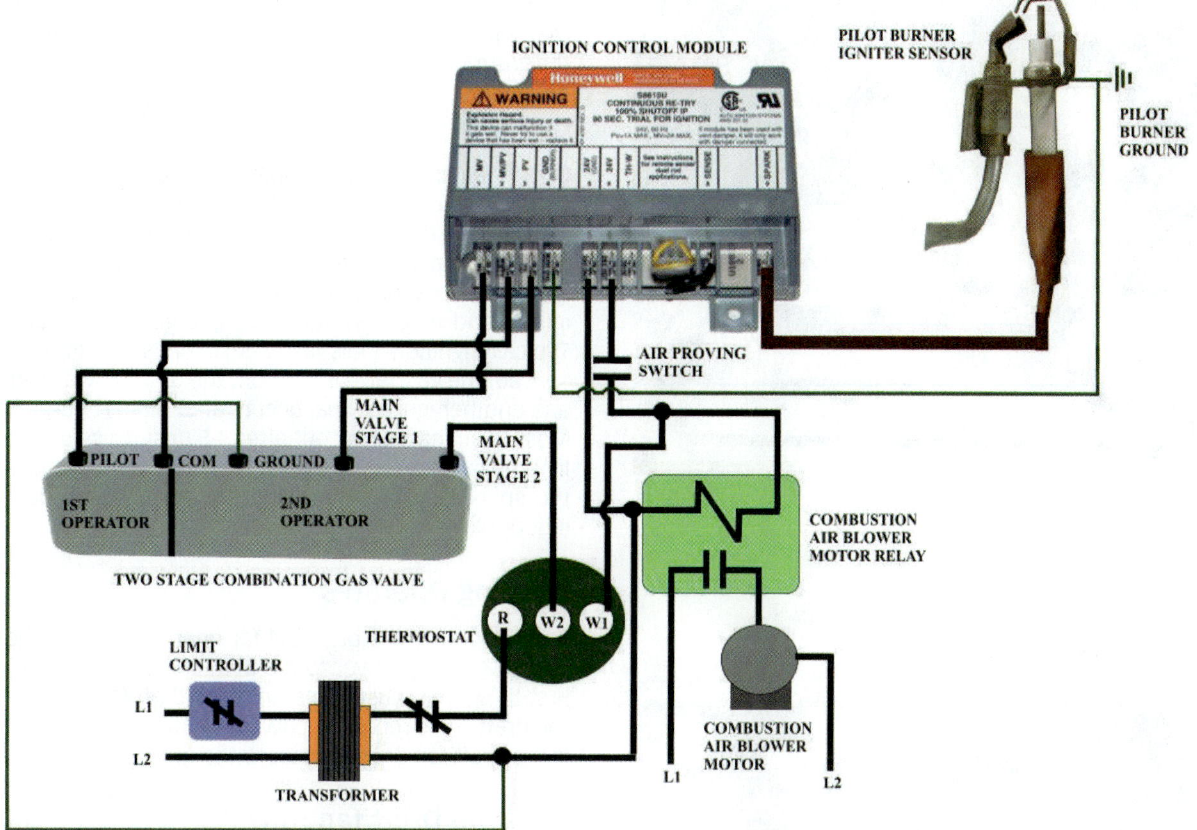

Figure 52-9 Ignition control module configuration for a fan-assisted furnace.

pilot location or recirculating products of combustion. A noisy, lifting, blowing flame is caused by high gas pressure. A hard, sharp flame is caused by high gas pressure or an orifice that is too small.

Ignition Trial Successful

If the pilot flame is established, then the flame rectification circuit is completed between the sensor and the burner ground. The spark is shut off and the second main valve operator is energized and opens the main valve to allow gas to flow to the burner, where it is ignited by the pilot.

Figure 52-10 Pilot flame envelopes 3/8 to 1/2 in of the flame rod.

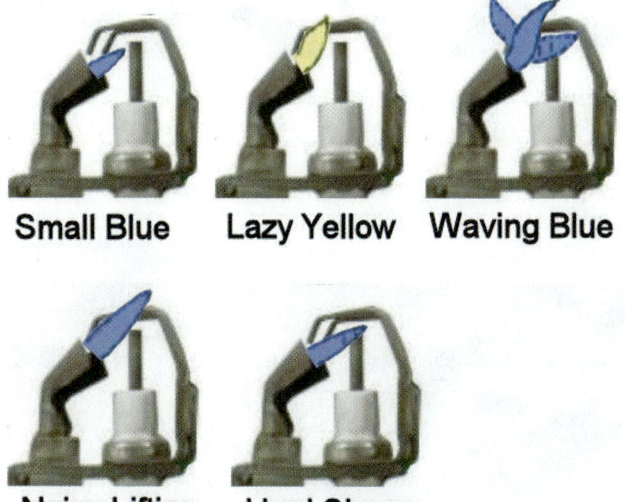

Figure 52-11 Incorrect pilot flames.

52.6 DIRECT SPARK IGNITION SYSTEMS

Outdoor furnaces and packaged units with gas heat often use direct spark ignition (DSI) systems. A direct spark ignition system does not have a pilot light. Instead, DSI systems use a spark to ignite the main burners. They consist of a DSI control board (Figure 52-12), a sparker (Figure 52-13), and a flame sensor (Figure 52-14). Flame rectification is used to

Figure 52-12 DSI Control.

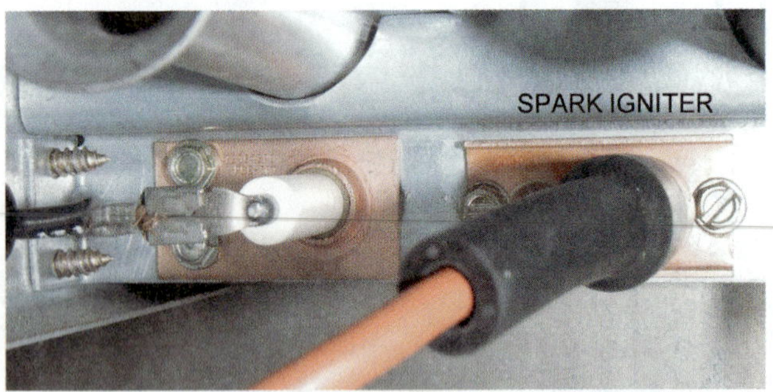

Figure 52-13 Sparker on DSI system.

Figure 52-14 Flame sensor on DSI system.

sense the presence of flames because it is the only method that is fast enough. Since gas is flowing to the main burners, the gas must be shut down quickly if flames are not sensed.

Operating Sequence

Understanding the operating sequence of a Direct Spark Ignition System (DSI) will make troubleshooting it much easier. First the draft inducer fan is energized. It runs for 30-45 seconds to perform a pre-purge to vent any accumulated gas or combustion products. After the pre-purge, the spark igniter and main gas valve are energized simultaneously. Typically ignition trials last 3-5 seconds. If flames are not sensed, the gas valve is shut off, the draft blower continues and another ignition trial occurs after a short purge period. Most DSI controls lock out after 3-5 unsuccessful attempts. If flame is established, the gas valve remains energized and the spark igniter is de-energized. After a time delay, the indoor blower is energized.

Nothing Operates

If nothing happens on a call for heat, first check the thermostat setting, power to the unit, and control voltage. The first thing that must come on is the draft fan. Test to see if the draft fan is receiving power to determine if the problem is the draft fan or the circuit to the draft fan.

Only the Draft fan runs

Draft fan operation is sensed by a pressure switch (Figure 52-15), a centrifugal switch, or a hall effect switch. If the fan is operating and the switch is not closing, make sure the draft fan is actually creating a draft. Sometimes the inducer fan rusts out or disintegrates. The motor operates but there is no fan to create a draft. The switch may be bad, or the hose to the draft switch may be loose or cracked.

Draft Switch Closing but No Ignition

When troubleshooting a DSI system that is not igniting, try to determine if the igniter is sparking. For safety, turn the gas off, call for heat, and listen for the "clicking" sound of the

Figure 52-15 Draft sensing switch.

spark. If you do not have spark, check the power to the DSI control, the sparker wire and connections at both the board and at the igniter. A break in the insulation or the wire can prevent sparking. check the grounding of the control and the unit. Check all safeties such as limits and roll out switches. An open safety can prevent the board from trying to spark.

Sparking but No Ignition

If the igniter is sparking, try to determine if the gas valve is energized. For safety, this can be done with the main gas valve closed. You need to check the voltage to the gas valve during an ignition trial. You need to have your meter ready to read voltage because you only have a few seconds to take the reading. Leads with Alligator clips are useful. If the valve is not receiving voltage but the igniter is sparking, either the ignition board is bad or there is a bad connection between the board and the valve.

If the valve is receiving voltage, set two manometers up to read the pressure entering and leaving the valve, turn the gas on, and call for heat. If the valve has the correct gas inlet pressure, there should be outlet pressure when the valve is energized. If not, the valve is bad.

Flames Ignite But Go Out

If the flames light but extinguish, the flame rectification circuit is not sensing the flames. This can be due to poor flame sensor positioning, poor ground, a defective flame sensor, or a defective ignition board. The flame sensing circuit can be checked using a multi-meter with a DC microamp setting. The meter must be wired in series with the circuit. The wire to the flame sensor is removed, the red meter lead connects to the wire and the black meter lead connects to the flame sensor, putting the meter in series between the ignition board and the flame sensor, Figure 52-16. When the flame touches the flame sensor, a DC microamp signal of 2-50 microamps indicates that the flame sensing circuit is working.

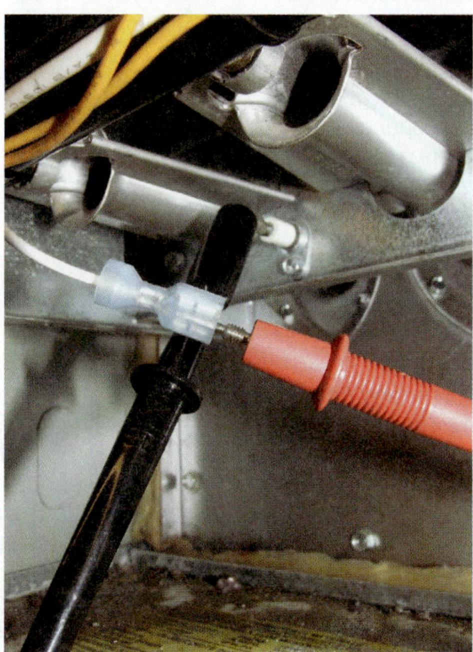

Figure 52-16 Multimeter is placed in series to check the microamp signal.

52.7 HOT SURFACE IGNITION SYSTEMS

Hot surface ignition systems are the most common among furnaces today. Like the DSI systems, hot surface systems do not have a pilot light; they light the main burners. The primary difference is that hot surface ignition systems use igniters that heat up instead of a spark. The spark on most DSI systems is very small, and its location is critical. Hot surface igniters are large in comparison, so their placement is not as critical. It is easy to tell when an igniter is working because it becomes so bright that it is hard to look at, Figure 52-17.

Operating Sequence

Understanding the operating sequence of a Hot Surface Ignition System (HSI) will make troubleshooting it much easier. The operating sequence is similar to a DSI system with the addition of a time delay to wait for the hot surface igniter to heat up. First the draft inducer fan is energized. It runs for 30-45 seconds to preform a pre-purge to vent any accumulated gas or combustion products. After the prepurge, the draft switch looks to see if there is a draft. After the draft switch closes, the hot surface igniter is energized for 30-60 seconds to allow it to heat up. The gas valve is then energized and the flame sensor looks to see if flame has been established. Typically ignition trials last 3-5 seconds. If flames are not sensed, the gas valve is shut off, the draft blower continues and another ignition trial occurs after a short purge period. Most HSI controls lock out after 3-5 unsuccessful attempts. If flame is established, the gas valve remains energized and the blower is energized after a time delay.

Nothing Operates

If nothing happens on a call for heat, first check the thermostat setting, power to the unit, and control voltage. The first thing that must come on is the draft fan. Test to see if the draft fan is receiving power to determine if the problem is the draft fan or the circuit to the draft fan.

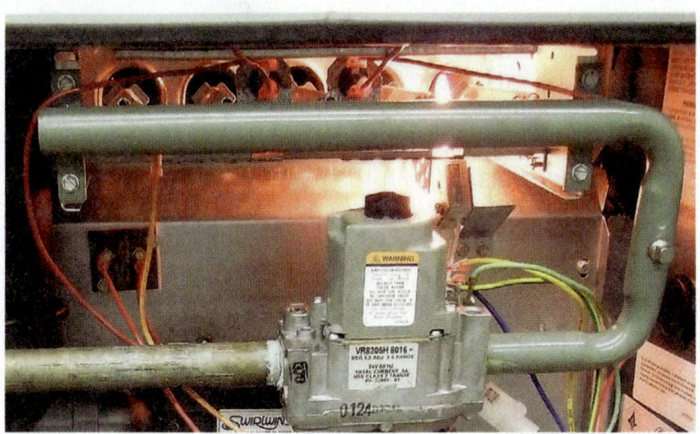

Figure 52-17 Hot surface igniter in operation.

Only the Draft fan runs

Draft fan operation is sensed by a pressure switch. If the fan is operating and the switch is not closing, make sure the draft fan is actually creating a draft. Sometimes the inducer fan rusts out or disintegrates. The motor operates but there is no fan to create a draft. The switch may be bad, or the hose to the draft switch may be loose or cracked.

Draft Switch Closing but Igniter Not Glowing

It is easy to tell if the igniter is working because it gets red hot and then glows. If you cannot see the igniter, an amp reading will tell if it is operating. Igniters can be checked using an ohm meter. A bad igniter will ohm out open—an infinite resistance. The voltage to the igniter should normally be 70-120 volts depending on the control.

Igniter Glowing but No Ignition

If the igniter is glowing, try to determine if the gas valve is energized. Use the same procedure as outlined for DSI controls.

Flames Ignite But Go Out

If the flames light but extinguish, the flame rectification circuit is not sensing the flames. Use the same procedure as outlined for DSI controls.

Induced Draft Blower and Fan Operating Constantly

Many HSI controls will operate both the draft fan and the indoor blower if one of the limits or roll out switches opens. Typically all the safeties are wired in a big series loop that wires to the control board. If any safety opens, the board shuts down the gas and the igniter and operates both fans to cool the furnace down.

52.8 MICROPROCESSOR FURNACE CONTROLS

Many of the newer furnaces use microprocessor controls and circuit boards. Keep in mind that while many controls may appear identical, timing sequences and programmed delays can vary between furnace manufacturers. Never substitute controls without the manufacturer's consent, as it can result in very dangerous conditions.

One helpful feature of the solid-state control system is its ability to diagnose its own service problems. For example, a component test is available that allows all components, except the gas valve, to run for a short period of time to reveal any service problems or indicate a component failure. Some advanced boards can store a history of the last several fault codes. This is especially helpful when diagnosing intermittent problems.

SERVICE TIP

Many of the components found in medium- and high-efficiency furnaces contain computer chips. These chips are protected from accidental damage as a result of a minor static discharge. However, they may not withstand the current from an improperly installed jumper. Before jumping terminals as part of board troubleshooting, make certain that you are following the correct procedure.

52.9 GENERAL PROCEDURES

A suggested progression of analysis for troubleshooting starts with identifying the common categories of complaints or problems, then the possible problems in the system, followed by the symptoms and causes of specific problems. Causes can be either electrical or mechanical and related to the use of gas equipment.

TECH TIP

Before beginning the actual testing or troubleshooting of a unit, take a few moments to check the system visually. Look for indications of arcing, burned or overheated wires, loose wires, or other easily visible mechanical signs of damage. Often the few minutes spent checking the system over can result in identifying the problem more quickly.

The categories of complaints or problems fall under the following headings:

1. Entire system operation
2. Unit operation
3. Burner operation
4. Blower operation
5. Heat exchanger complaints
6. Cost of operation
7. Noise

52.10 DIAGNOSTIC FLOW CHARTS

Many technicians find it helpful to use a troubleshooting flowchart. The first thing to check is always the power supply. Is proper power being supplied to the unit? A voltmeter is a handy tool to use in checking the power supply. The second item to check is the thermostat. Is the thermostat calling for heat? If both of these conditions are satisfactory, the wiring diagram should be referenced. Prior to cycling power or removing a door that would de-energize the unit, inspect the furnace for the presence of a fault light designed to give the technician insight into the cause of a shutdown or fault. Many times, cycling power will cause these valuable data to be lost.

TECH TIP

Manufacturers have done extensive research into the diagnostic testing of equipment. They provide this material in the form of flowcharts or troubleshooting tables. Using the manufacturer's troubleshooting information will make it much easier to locate and repair problems.

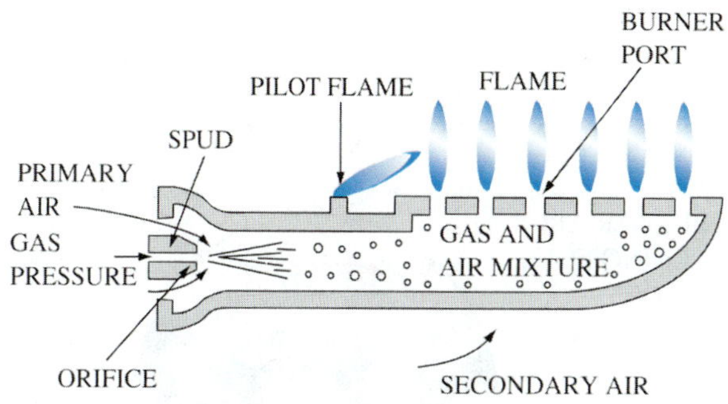

Figure 52-18 Primary and secondary air supply to the gas burner.

52.11 GAS SYSTEM PROBLEMS

Many of the solutions to problems are not always evident when the cause is determined. The following list of problem categories includes additional information regarding some of the remedies that can be tried.

- Control transformer burned out
- Pilot outage
- Gas valve stuck open or closed
- Improper heat anticipator setting
- Cycling on the limit control
- Fan control settings
- Improper burner adjustment
- Improper combustion air supply
- Improper venting
- Blower drive
- Expansion noise in heat exchanger

1. **Control transformer burned out** An ohmmeter should be used to determine whether the primary or the secondary of the transformer is burned out. The cause is probably an overload in the secondary, which needs to be corrected before the transformer is replaced. If the overload cannot be reduced, a larger transformer must be used for the replacement.
2. **Pilot outage** One of the most common causes of pilot failure is improper impingement of the flame on the thermocouple. Sometimes the pilot is extinguished by the gas burner during lighting. It is actually blown out by the burner flame. To correct this, it is often necessary to reposition the pilot to a more favorable location.
3. **Gas valve stuck open or closed** A malfunctioning gas valve should be replaced.
4. **Improper heat anticipator setting** The anticipator should be set at the amount of current traveling in the control circuit when the unit is operating. The current can be measured using a multiplier coil and a clamp-on ammeter.
5. **Cycling on the limit control** Occasionally a limit control will weaken and lower the operating range of the control. The normal range is to cut off between 140°F for a counterflow unit to 160–220°F on upright and horizontal units. If the control is cycling at a lower range, replace it.
6. **Fan control settings** Almost all gas-fired units operate best with fan control settings of 125–130°F fan on and 100–105°F fan off. If the unit blows cold air on start-up, the fan-on temperature can be changed to 145–150°F.
7. **Improper burner adjustment** A properly adjusted burner will have approximately 40 percent of the combustion air mixing with 100 percent of the gas in the burner and 60 percent of the air mixing with the flame above the burner to complete the combustion process. Figure 52-18 shows a typical burner arrangement.

The proper setting of the primary air quantity is the beginning step in producing high unit operating efficiency. Improper setting of the primary air shutter can contribute to pilot outage by producing extinction flashback, called *extinction pop*.

As the gas is emitted from the orifice, it expands and hits the proper place in the throat of the burner venturi. This produces maximum pull of primary air into the burner. Setting the burner for the correct flame condition will mean a minimum opening in the primary air control.

Figure 52-19a shows the size of the opening in an ordinary butterfly air control. When the burner and orifice are aligned properly and the burner is working correctly, a small opening in the primary air control will produce the soft blue flame that gives best overall unit performance (Figure 52-19b).

An all-blue sharp flame is one that is receiving too much primary air. This means that there is less radiant heat to heat the lower portion of the heat exchanger. Also, the excess air drives the flame products from the heat exchanger before good transfer of heat occurs from the flue product to the heat exchanger. Flue product temperatures rise and the unit efficiency drops.

On the contrary, if the primary air is reduced too much, heavy yellow tips of improperly burning carbon are produced. These result in much lower temperatures, and less heat is produced. Therefore, the unit efficiency will drop. In addition, the carbon will release from the flame and collect in the heat exchanger, to causing sooting and plugging of the flue passages.

TECH TIP

Soot is most often caused by improper fuel/air adjustments. However, if the flame impinges on internal metal parts of the furnace, then heat can be pulled out of the flame rapidly enough to retard proper combustion. Flame impingement can result in carbon forming. An example of carbon formed as the result of flame impingement can be seen if a copper wire is held in a match flame. Within moments, black carbon will form around the copper wire as it withdraws the heat from the flame and retards the complete combustion process.

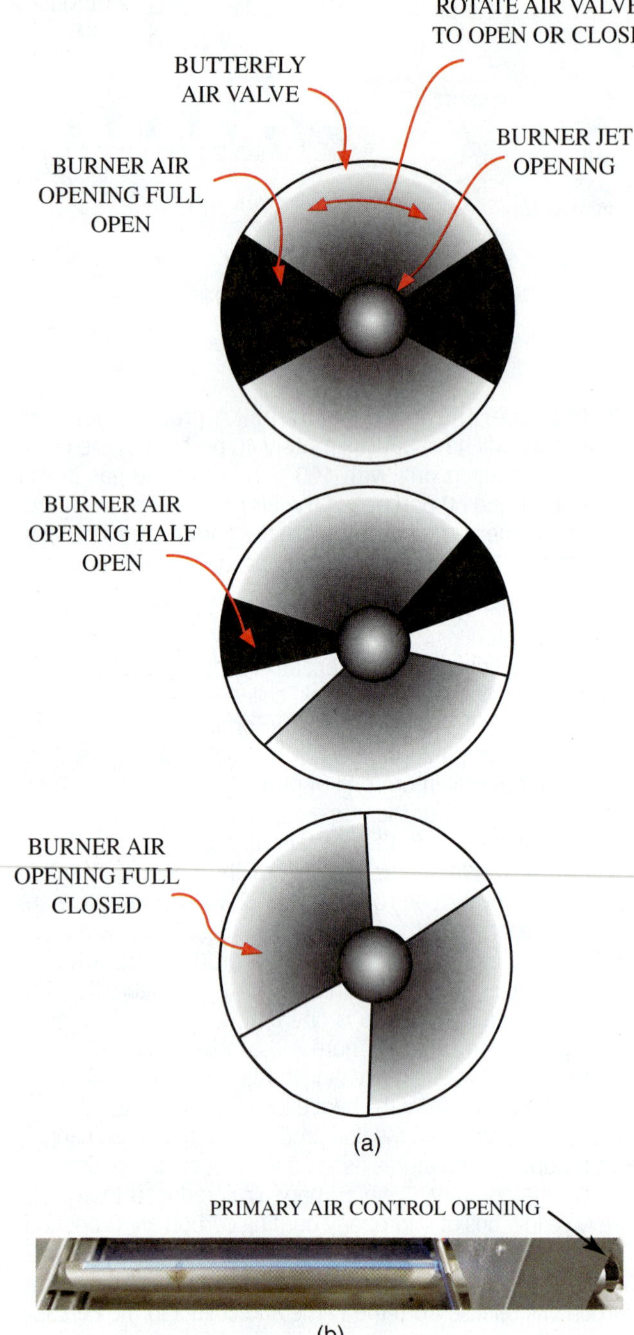

Figure 52-19 (a) Primary air control opening size; (b) soft blue flame with proper air control.

When the burner is operating, the gas/air mixture is blowing upward through the burner port at a given speed or velocity (determined by the burner design and type of gas). There is also a downward force or burning velocity, which is equalized by the gas/air outward velocity when burning is taking place. If, however, the burning velocity were to increase due to shutoff of the gas supply, the flame could approach the burner. Either the burner would absorb the heat below the combustion point and extinguish the flame or the flame would burn down through the burner port and ignite the mixture in the burner. This ignition produces the extinction pop.

Figure 52-20 Rollout, due to insufficient secondary air.

8. **Improper combustion air supply** Excess air is needed for proper combustion, even though changes in gas pressure, heat content of the gas, and barometric pressure may occur. Draft conditions also change with barometric pressure and wind conditions.

When a unit encounters insufficient combustion air, the flame tends to become lazy and erratic and may even roll over the edge of the burner and out the burner pouch opening. The flame will seek air. Figure 52-20 shows the effect of insufficient secondary air causing a floating flame.

9. **Improper venting** The mixture of gas and air produces a mixture of water, carbon dioxide, nitrogen, and excess air. All of this has to be removed from the heat exchanger. This removal process is called venting.

TECH TIP

The inside lining of a double-walled vent pipe is made of aluminum. Aluminum is used because it heats up very quickly once the furnace starts. If heavier metal were used that warmed up slower, the water vapor in the vent gases would condense inside of the flue pipe until it heated up. This condensed water would then run down and corrode the furnace and/or heat exchanger. Even with an aluminum-lined vent pipe, some condensation occurs, and in extremely cold climates enough water can be condensed to still run down to the furnace. Fortunately, in cold climates the furnace will run long enough to evaporate most of that moisture before it has a chance to cause damage. Traces of the moisture can be found around the furnace vent. Technicians may assume that water is from a leaking roof stack, but often it is just condensation.

There are two types of venting: active (power) venting and atmospheric, or passive (gravity), venting. Active venting uses a mechanical device such as a motor-driven blower to either draw flue products from the heat exchanger or to force combustion air into the heat exchanger. The most popular type is the draw type, where the blower is mounted on the flue outlet and creates negative pressure in the outlet of the heat exchanger to get the desired combustion efficiency. Because the pressure difference is caused by mechanical power, wind and/or atmospheric conditions have little effect on the venting performance.

In atmospheric, or passive, venting, hot flue gases pass from the heat exchanger into a flue pipe, chimney, or vent stack. The driving force for a passive vent is obtained from the hot gases rising in the surrounding cooler air. The amount of force depends on the temperature of the hot gases and the height of the gravity vent. The hotter the gases and/or the higher the vent, the greater the amount of driving force or pull that is produced. Also, the greater the pull, the more secondary air is drawn through the heat exchanger.

Enough air must be drawn through to provide complete combustion as well as complete venting of the flue products. If too much air is drawn out of the heat exchanger before the correct amount of heat extraction is done, this results in higher flue temperatures and reduced unit efficiency.

If the passive vent pipe were connected directly to the flue outlet, the amount of air drawn through the heat exchanger would vary with factors like the pull of the vent stack, the wind effect on the vent stack, and outside temperature. Control of the venting rate on the heat exchanger would be impossible. Further, under some atmospheric conditions it may be possible to have a higher pressure at the outlet of the vent than the combustion process can overcome. This can produce poor combustion, with the production of CO as well as the usual products of CO_2 and H_2O.

To overcome the effect of atmospheric conditions, all units use an opening in the venting system called a draft diverter. Figure 52-21 shows four typical heating unit draft diverters. These all consist of an opening from the flue outlet of the heating unit, an opening into the vent pipe, and a relief opening to the surrounding atmosphere.

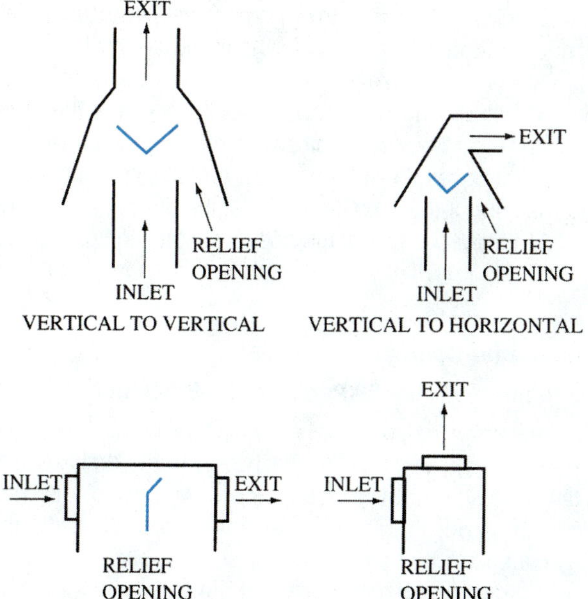

Figure 52-21 Typical gas appliance draft diverters.

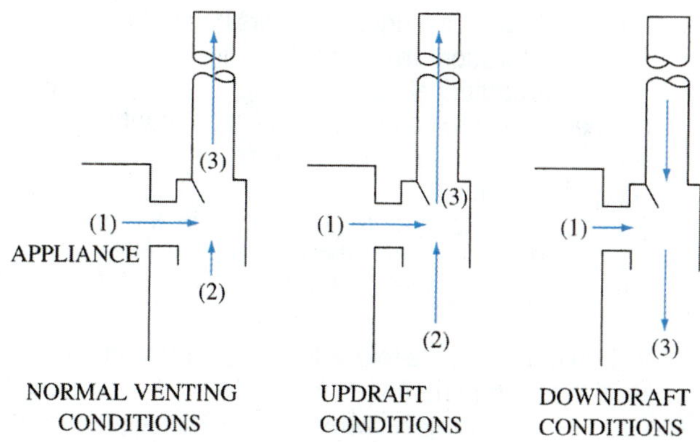

Figure 52-22 Operation of a draft diverter under various wind conditions.

Figure 52-22 shows the operation of a typical draft diverter under no-wind conditions and with updraft and downdraft conditions. The amount of flue products, the dilution air entering the relief opening, and the amount of vent gases are indicated by the length of the arrows. With normal venting, some air is pulled into the draft diverter by the pull of the passive vent. The mixture of flue products and surrounding air (called dilution air) that blows up the vent is called vent gas. The action of the draft diverter is to break the effect of the vent by introducing surrounding air and neutralizing the pull at the flue outlet. The heat exchanger then operates at approximately equal pressure from burner opening to flue outlet. The amount of air for combustion is then controlled by the flue restrictors.

If the conditions surrounding the vent stack increase the stack pull, additional air is drawn into the draft diverter to compensate for the increased pull. There is little effect on the heat exchanger performance.

Under conditions where the vent stack pull is reduced or even reversed, creating a downdraft, all combustion flue products are forced into the surrounding area. In addition, the increased pressure in the flue outlet will reduce the flow through the heat exchanger. This can cause incomplete combustion and produce odors carried by the gases and moisture produced by the combustion process. Even though no odors may result, the large amount of moisture produced in the combustion process can accumulate in the occupied area and create adverse living conditions or possible structural damage.

To check for proper operation of the vent system, use a candle placed below the bottom edge of the diverter opening. With the unit operating and up to temperature, the candle flame should bend in the direction of the opening in the diverter. If the flame is neutral, the draft is on the weak side. Possibly the vent stack is not high enough or large enough. If the candle flame bends outward, a draft problem definitely exists that must be corrected. If the vent stack cannot be lengthened or enlarged, a forced-draft unit must be installed to overcome the problem.

10. **Blower drive** Belt driven blowers have a higher probability of vibration problems than direct drive blowers due to the additional parts involved. The most common problem is due to belt tension. It is commonly believed that the tighter the belt, the better the performance, but the opposite is true. The tighter the belt, the harder the motor has to work to get the belt in and out of the pulleys. The belt should therefore be as loose as possible without slipping on startup.

It should be possible to easily depress the belt midway between the motor shaft and blower shafts ¾ in to 1 in for each 12 in of distance between the shafts. Alignment of the motor and blower pulleys is important to keep vibration to a minimum as well as to reduce wear on the sides of the belt.

Both motor pulley and blower pulley should be checked. They should be parallel to each other and perpendicular to the shaft. A warped pulley has a visible wobble that will create vibrations. Warped pulleys should be replaced.

11. **Expansion noise in heat exchanger** Figure 52-23 shows a four-section unit, each section composed of a right-hand and left-hand drawn steel "clamshell" welded together. The sections are then joined and welded into an assembly by fastening them to the front mounting plate and rear retainer strap. Sometimes in the welding process stresses will be set up if the two metals are at different temperatures when the bond is made. This results in expansion noises, ticking, and popping as the heat exchanger heats and cools. Most of the time, these noises are muffled by the unit casing and duct system to a level where they are not objectionable.

In extreme cases, it is possible to reduce the noises by operating the unit with the blower disconnected, allowing the limit control to turn the unit off and on. This should be done through several cycles of the limit control. Cycling on this extreme heat will cause metal to stretch beyond the normal operation range and eliminate the sound. If this does not produce satisfactory results, the only cure is to change the heat exchanger.

12. **Oil-can effect** This effect is caused by the sudden movement of a flat metal surface where a forming stress has been left in the surface. This stress causes the metal to have a slightly concave or convex position rather than a flat plane surface.

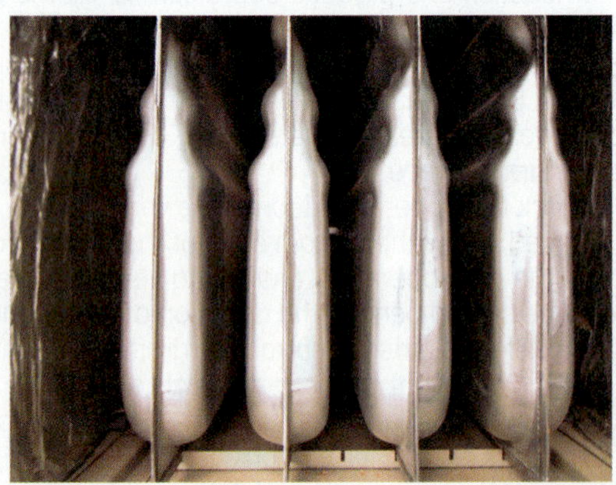

Figure 52-23 Clamshell gas heat exchanger.

Temperature change will cause a stress increase in the material until the metal rapidly changes position to the opposite of its original position. This change will produce a loud bang. Ductwork is very prone to this action and must be cross-broken over any large panel areas. Unless this surface is cross-broken, it is subject to the oil-can effect. The best correction is to remove the panel and cross-break it to relieve temperature stresses.

UNIT 52—SUMMARY

When troubleshooting a gas furnace, the technician's first responsibility is to perform a safety check. Look for visual signs of problems. After verifying that the system is safe to operate, proceed in a systematic manner. Use available resources such as troubleshooting flowcharts and diagnostic codes. Perform tests that can eliminate possible problem causes and narrow your search. Try to look for underlying causes, not just symptoms. For example: if a furnace is tripping out on the high limit, check for airflow problems. After isolating and repairing the problem, retest the system to ensure safe and proper operation.

WORK ORDERS

Service Ticket 5201

Customer Complaint: No Heat

Equipment Type: 80 Percent Mid-Efficiency

A technician responds to a no-heat call. The technician sets the thermostat to call for heat and the draft-inducer fan starts, but nothing else happens. A check of the draft switch shows that it has not closed. A check of the pressure on the hose to the draft switch reads 0 in wc. The technician removes the tube to see if it is plugged and water comes out of the tube. Rechecking the draft, it now reads 0.25 in wc vacuum. The furnace now lights and operates.

Service Ticket 5202

Customer Complaint: No Heat

Equipment Type: 80 Percent Mid-Efficiency

A technician responds to a no-heat call. The technician sets the thermostat to call for heat and the draft-inducer fan starts, but nothing else happens. A check of the draft switch shows that it has closed. A voltage check at the igniter shows that the igniter is receiving voltage but is not operating. The technician checks the resistance of the igniter with an ohmmeter, and it reads infinity. The technician replaces the igniter. The furnace now operates correctly.

Service Ticket 5203

Customer Complaint: Insufficient Heat

Equipment Type: 80 Percent Two-Stage Mid-Efficiency

A technician responds to an insufficient heat call. The technician sets the thermostat to call for high fire. The furnace lights and operates. The technician checks the voltage at the high-fire solenoid on the gas valve and reads 24 V. Next, the gas pressure is checked. It reads 1.5 in wc at the manifold and 7 in wc at the gas valve inlet. The furnace data plate reads 1.5 in low-fire manifold pressure and 3.5 in high-fire manifold pressure. The technician unsuccessfully tries to adjust the high-fire manifold pressure. The gas valve is determined to be defective and is replaced. When the system is rechecked, the high-fire manifold pressure reads 3.5 in wc.

Service Ticket 5204

Customer Complaint: Furnace Sometimes Does Not Heat

Equipment Type: 90 Percent Condensing Furnace Installed in Crawlspace

A technician is called to look at a 90 percent condensing furnace that works sometimes and sometimes does not. The furnace is operating when the technician arrives. The customer says that he turns the heat off when he leaves for work and back on when he gets home. Occasionally on really cold days, the furnace will not come on. Yet it comes on later, usually after the outside temperature has warmed up a little. The furnace has a microprocessor control that stores fault history. The technician checks the fault history and sees that the furnace has shut down on draft safety several times. The technician notices that the drain from the condensate pump runs a long way through the crawlspace and that the pump is equipped with an overflow kill switch. The technician concludes that water is freezing in the condensate pump discharge line, causing the sump to fill and the kill switch to shut down the furnace. The technician replaces the soft plastic tubing used for the condensate pump discharge line with PVC, wraps the PVC with heat tape, and places pipe insulation over the new PVC condensate discharge line. Before leaving, the technician clears the fault codes so any faults that are stored will be after the drain fix. The mysterious cold weather shutdowns disappear. The technician checks the fault codes during the subsequent spring checkup and finds none.

Service Ticket 5205

Customer Complaint: No Heat

Equipment Type: Standing Pilot, Natural-Draft Furnace

A technician responds to a no-heat call on an old standard furnace. The technician notices the pilot light is not burning and attempts to light it. The pilot light will light, but it will not stay lit. The technician checks the millivoltage output of the thermocouple, reads 0 mV, and determines that the thermocouple is defective. The technician changes the thermocouple, lights the pilot light, and the furnace operates correctly.

Service Ticket 5206

Customer Complaint: No Heat

Equipment Type: Standing Pilot, Natural-Draft Furnace

A technician responds to a no-heat call on an old standard furnace. On a call for heat, the main burners light. After burning for several minutes, the burners shut down. The indoor blower never comes on. The technician checks voltage to the blower motor and finds 0 V. The technician observes that the fan switch has rotated past the fan on the cut-in point and decides to check the fan switch next. Entering the fan switch is 120 V, but leaving is 0 volts. The technician concludes that the fan switch is defective. The flames had shut down on limit since the fan never came on. The fan switch is changed and the system rechecked. This time the fan comes on about 3 minutes after the burners light and the burners stay lit.

Service Ticket 5207

Customer Complaint: Carbon Monoxide Alarm

Equipment Type: 80 Percent Mid-Efficiency Upflow Replacement Furnace

A technician is sent to look at a 1-year-old mid-efficiency replacement furnace. The original installing contractor is no longer in business. The furnace is installed in the basement and connected to the old vent system. The technician observes that the old vent is a 7-in single-wall vent and the flue collar on the new furnace is 4 in. Streaky rust stains appear on the bottom of every vent joint. The flames in one tube are visibly different from the others. The flames in that tube blow all around when the indoor blower comes on. The technician shuts down the furnace, removes the burner from that tube, and takes a closer look at the tube. There is a lot of rust and a hole at the back of the tube. The technician concludes that the heat exchanger has rusted out as a result of condensation in the oversized single-wall vent. The technician recommends that the heat exchanger be replaced and a new correctly sized double-wall vent run.

Service Ticket 5208

Customer Complaint: No Heat

Equipment Type: 90 Percent Condensing Furnace with Combustion Air Piped In

A technician responds to a no-heat call. The induced-draft blower starts on a call for heat, but nothing else happens. The technician checks the draft switch and finds that it has not closed. Next, the vent pressure is checked and found to be −0.3 in wc in the combustion air pipe and

+0.05 in wc in the vent pipe. The technician checks the ends of the air intake and vent pipes and notices leaves and grass in the end of the air intake pipe. Closer inspection reveals a bird's nest in the air intake. The technician removes the bird's nest and checks the vent screen at the furnace, where a dead bird is found. The furnace operates correctly after removing the dead bird and bird's nest.

Service Ticket 5209

Customer Complaint: No Heat, Fan Runs Continuously

Equipment Type: 90 Percent Condensing Furnace

A technician is sent to a no-heat call on a high-efficiency furnace. Upon arriving, the furnace fan and combustion blower motor are running even though the thermostat is turned off. The technician checks the voltage at the W terminal and finds 0 V. The diagnostic light is flashing a code that indicates the system shut down because of an open limit switch. The technician checks the resistance of the limit switch and it is indeed open, even though it is not hot. The technician replaces the limit switch and checks the air filter. The air filter is completely blocked. Because the filter is so dirty, the technician decides to check the underside of the recuperative coil. The condensing recuperative coil is checked after removing the blower; it is matted with dirt and animal hair. The coil is cleaned and the blower replaced. The technician explains that blocked airflow from the dirty filter and coil made the furnace cycle on the limit and eventually killed the limit switch. The customer is advised that the air filter needs changing regularly, at least once a month, or the problems will repeat.

UNIT 52—REVIEW QUESTIONS

1. What is the one advantage of a standing pilot for a gas-fired space heater?
2. List the general sequence of operation for an induced-draft furnace.
3. What happens if the pilot fails to light during a trial ignition?
4. List the seven categories of complaints with gas furnaces.
5. What is the purpose of the draft diverter on natural-draft units?
6. What is an indication of too much gas at the burners?
7. What is an indication of too much primary air?
8. What is an indication of too little primary air?
9. What happens if there is improper venting of the furnace?
10. What is a safety lockout?
11. A waving blue pilot flame can be caused by what?
12. How does the trial ignition period differ between a fan-assisted furnace as compared to a furnace with atmospheric burners?

Use Table 52-1, Gas Systems Problems, to answer the following questions.

13. What are some probable causes of noise and vibration?
14. What causes rollout of burner flames?
15. What are some possible causes of burner flashback?
16. What can cause the blower to come back on for a short time after cycling off?
17. What are some possible causes of delayed ignition?
18. What can cause soot accumulation on the heat exchanger?
19. On some furnaces, little pops and pings can be heard before the blower comes on. What causes this noise?
20. A gas burner is noisy and the flames are lifting off the burner. What is the most likely cause?

UNIT 53

Oil-Fired Heating Systems

OBJECTIVES

After completing this unit, you will be able to:

1. list the grades of fuel oil and compare their properties.
2. discuss the types of nozzles available.
3. explain the operation of a fuel oil burner.
4. list the types of oil furnaces and boilers available.
5. compare the efficiency ratings of different types of oil furnaces.
6. explain the operation of oil furnace primary controls.

53.1 INTRODUCTION

Oil is still a popular form of heat in colder areas of the country such as the Northeast, which accounts for approximately 82 percent of the total usage. Today, manufacturers offer a wide array of choices in oil heat. These designs include condensing furnaces with efficiencies reaching 95 percent, furnaces with firing rates as low as 50,000 BTU/hr, two-stage residential furnaces, mid-efficiency noncondensing furnaces with efficiency ratings reaching 89 percent, and sealed combustion furnaces.

53.2 FUEL OILS

Fuel oils are mixtures of hydrocarbons made of long hydrocarbon chains derived from crude petroleum by distillation. Several different grades of fuel oil are available. Oil is divided into grades according to the oil characteristics, including viscosity, flash point, pour point, water and sediment content, carbon residue, and ash. The oil grades are classified according to their varying combinations of distillates and residuals. A distillate evaporates at relatively low temperatures. Light distillates are clear and will pour like water at normal room temperature. Residuals are what remain at the end of the distillation process. Residuals are black and tarry in appearance with the consistency of toothpaste at room temperature. Distillates are made of shorter carbon chains than residuals and have a lower heating capacity, but they are easier to store, pump, atomize, and burn. Oil grades range from 1 to 6, with 1 being 100 percent light distillates and 6 being nearly 100 percent residuals.

Oil Viscosity

Viscosity is a measure of how fast oil flows. A high viscosity number means that the oil is thicker and flows slowly, while a low viscosity number means that the oil is thinner and flows more readily. Oil's viscosity determines whether the fuel oil must be heated for it to flow. Table 53-1 shows the viscosity of several grades of fuel oil over a range of temperatures.

Table 53-1 Properties of Fuel Oils.

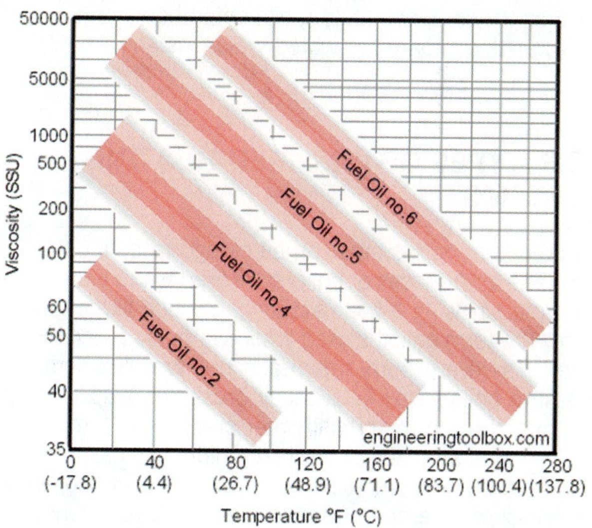

SERVICE TIP

All heating oil has a very distinctive odor. You must be very careful when working in residences not to contaminate the residence property with oil. Careless work habits can result in the home smelling of oil for days after a technician leaves. Often oil-fired furnaces are located in basements, some of which do not have outside access. If it is necessary for you to access the system through the residence, you should wear paper shoe covers as you enter and leave the residence and each time you enter and leave the basement. In addition, you must clean your hands thoroughly before entering the residence to adjust the thermostat, set dampers or vents, and any other such service.

Oil Grades

Grades 1 and 2 are the grades of fuel oil primarily used for comfort heating applications. Both contain high quantities of carbon and hydrogen with traces of sulfur. The carbon content for these fuel oils ranges from 84 percent to 86 percent, with a maximum of 1 percent sulfur. The sulfur content of fuel oils is kept as low as possible to reduce air pollution.

Grade 1 fuel oil is considered premium quality. It is used in room space heaters, which do not use high pressure burners and depend on gravity flow, thus the need for the lower viscosity. Grade 2 oil is the standard grade heating oil sold. Grade 2 oil is used in equipment that has pressure atomizing, which includes most forced-air furnaces and boilers. The heating value for grades 1 and 2 oil falls in the range of 135,000 to 142,000 BTU/gal.

TECH TIP

It is often assumed that because heating oils are thicker than water they are heavier than water. This is not true. Oil floats on water. Although it is thicker than water, it is less dense. Most heating oils range between 6.8 and 7.2 lb/gal, compared to water, which weighs 8.3 lb/gal.

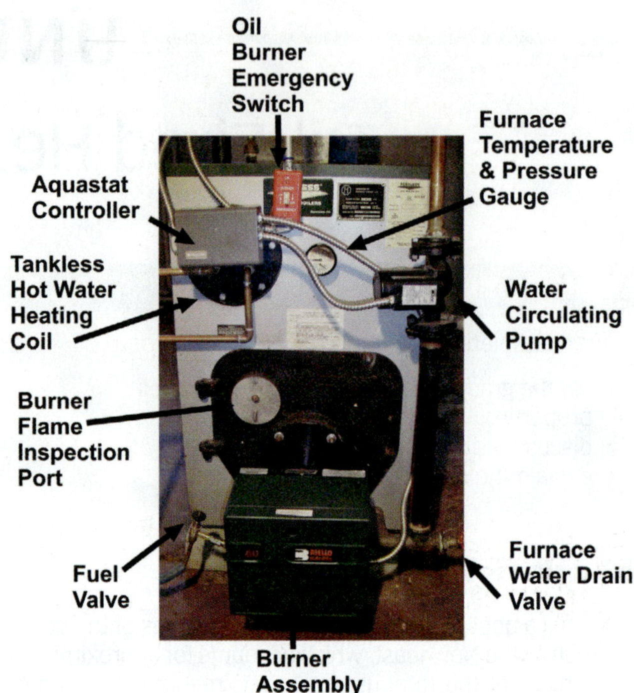

Figure 53-1 Hot-water oil-fired boiler.

53.3 TYPES OF OIL-FIRED HEATING SYSTEMS

Oil fired heating systems can be categorized as either being warm air furnaces or boilers. Oil fired warm air furnaces heat the air using a heat exchanger, much like a gas furnace. The term boiler is used when water is heated instead of air. There are both steam and hot water boilers. Low pressure boilers have a working pressure of up to 15 psi steam and/or up to 160 psi water pressure at a maximum of 250°F operating temperature. Medium and high pressure boilers operate above 15 psi steam and/or over 160 psi water pressure at temperatures above 250°F. Today most oil-fired boilers used for residential heating are low pressure and do not convert the water to steam but instead are specialized hot water heaters.

Oil-Fired Boilers

Many older oil-fired heating systems were used for making low-pressure steam, thus called boilers. This steam was sent to radiators located in the living spaces. The radiators gave off heat and the steam was condensed back to water by a steam trap at the outlet of the radiator before being returned to the boiler. In this way the latent heat of the steam was transferred to the living space. Today most oil-fired boilers used for residential heating do not convert the water to steam but instead are specialized hot-water heaters. Figure 53-1 shows a typical oil-fired hot-water boiler. The hot water produced by the boiler is circulated through radiant baseboard heaters and radiant heating floor systems located in the living spaces.

Waste Oil Furnace

One product that is unique to oil-fired systems is the waste oil furnace. These are warm-air furnaces that are specifically designed to burn waste oil. Waste oil burners are particularly attractive to companies that generate waste oil. Normally, the company would have to pay to dispose of the waste oil. By using waste oil as fuel, the company not only saves on disposal costs but gets free heat in the bargain. Most modern recently manufactured oil burners are approved for use for B5 or lower grades of biofuels. However, older oil burners may need to be replaced or upgraded with all involved components to be compatible.B5 is a blend of biomass and petroleum diesel. Biomass consists of organic materials such as vegetable oils and animal fats and is a source of renewable energy. Figure 53-2 shows a waste oil furnace.

Oil-Fired Warm-Air Furnaces

Oil-fired warm-air furnaces heat the air using a heat exchanger, much like a gas furnace. Oil-fired forced-air furnaces are designed in five basic configurations:

- Upflow
- Downflow (counterflow)
- Horizontal
- Lowboy
- Multiposition models

These configurations describe the airflow through the furnace. Air travels up in an upflow furnace, down in a downflow furnace, and horizontally (left or right) in a horizontal furnace. Lowboy furnaces have two sections side by side, as do horizontal furnaces; but the air goes in the top of the furnace, makes a 180° turn as it goes through the furnace, and then leaves out the top of the furnace. Lowboy furnaces are built low in height to accommodate areas where headroom is minimal. These furnaces are approximately

Figure 53-2 Waste-oil furnace. *(Photo courtesy of Clean Burn)*

4 ft high, providing for easy installation in a low-ceiling-height basement. Blowers for lowboy furnaces are typically belt driven. A multiposition furnace can be used in more than one configuration.

53.4 OIL FURNACE EFFICIENCIES

Oil furnaces are available in standard efficiency, mid-efficiency, and high-efficiency condensing models. Older-design oil furnaces had combustion efficiencies of around 60 percent. Added to this is the heat lost through the large amount of air required for combustion air and dilution air. Taking this additional heat loss into account, they perform no better than 50 percent efficiency overall. Most new "standard" oil furnaces now use high-pressure burners that use a fuel pump to pressurize the oil to 100 psig or more. These operate with efficiencies closer to 80 percent, but the losses from combustion and dilution air are still present.

Many mid-efficiency furnaces, depending on their design and application, use high-pressure burners along with an increased combustion supply-air pressure. With this design, the burner flame is held very close to the face of the combustion head and the flame is smaller and more compact than with older burners. This type of burner design is referred to as flame retention and is the standard for most burners today.

With flame-retention burners, the increased combustion supply-air static pressure will push the exhaust gases up and out the chimney so a barometric damper might not be required. An advantage to this is the loss from dilution air is eliminated. The main determination for a barometric damper is the chimney. If the draft is excessive, the system may still require a barometric damper even with a high-pressure burner. Some furnaces add a sealed combustion chamber, so all the combustion air comes from outside. These furnaces operate with efficiencies of 83 to 89 percent.

Both mid-efficiency and high-efficiency furnaces are available in two-stage models that have a low and high firing rate. This improves the overall furnace efficiency by reducing the furnace cycling and extending the run times.

In the past, the number of sizes available for oil-burning furnaces was limited. Most furnace lines started big and got bigger. Offerings are more varied today, both in terms of size and efficiency. Table 53-2 lists some examples of models available from one manufacturer.

Table 53-2 Oil Furnace Models.

Firing Rate	Efficiencies	Configurations
50,000	98%	Horizontal, unit heater
70,000	83%	Upflow
75,000	98%	Horizontal, unit heater
90,000	84%	Horizontal, upflow, downflow, lowboy
100,000	94%	Horizontal, unit heater
105,000	83%	Horizontal, upflow, downflow, lowboy
120,000	83%	Horizontal, upflow, downflow, lowboy
125,000	92%	Horizontal, unit heater
140,000	84%	Horizontal, upflow, downflow, lowboy
150,000	98%	Horizontal
200,000	94%	Horizontal
250,000	92%	Horizontal

53.5 OIL-FIRED CONDENSING FURNACES

The combustion products from an oil furnace contain only half as much water as the combustion products from a gas furnace. Since this lowers the dew point, the combustion gases from an oil furnace can be cooled further before condensation starts to take place. Therefore, mid-efficiency oil furnaces are less likely to condense water in this efficiency range as compared to gas furnaces, although this is subject to the furnace application and the operating environment.

The high-efficiency furnaces rated in the 90–95 percent efficiency range condense water out of the vent gases using a stainless steel recuperative secondary coil. The combustion gases pass through this stainless-steel recuperative

Figure 53-3 High-pressure atomizing gun-type oil burner.

secondary coil before traveling out the vent. Building a reliable condensing oil furnace can be somewhat of a challenge. This is because there is less latent heat to capture with only half as much water in the combustion gas of an oil furnace as compared to a natural gas furnace. Also, oil has considerably higher sulfur content, so the condensate is more acidic than the condensate from condensing gas furnaces. This means that the condensing recuperative secondary coil must be more resistant to corrosion. Despite these obstacles, there are over 300 furnaces listed by the Department of Energy as Energy Star furnaces. These furnaces have efficiencies in the range of 90 to 95 percent.

53.6 OIL BURNERS

The high-pressure atomizing gun burner shown in Figure 53-3 is the type used on most residential and small commercial oil-fired heating systems. Traditional high-pressure oil burners will burn with a yellow flame, as shown in Figure 53-4. High-pressure oil burners have the following components:

- Oil pump
- Air blower
- Electric motor
- Ignition transformer
- Blast tube, with nozzles and ignition system
- Primary control

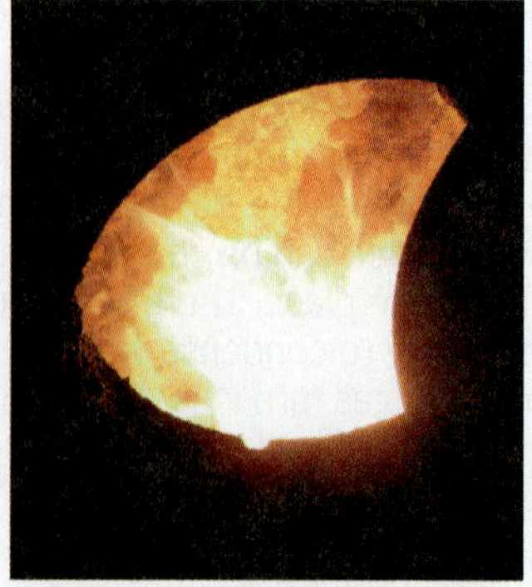

Figure 53-4 Oil burner flame.

The burner illustrated in Figure 53-5 shows a cutaway view of the internal construction of a burner assembly. The pump and blower are driven from a common motor shaft. Oil is pumped through the oil burner nozzle and ignites as it passes from the burner head into the combustion chamber.

Flame Retention Burners

In a flame-retention burner, the air flow is a swirling pattern that recirculates combustion products for a more complete mixing of the fuel and air. Air from the combustion chamber

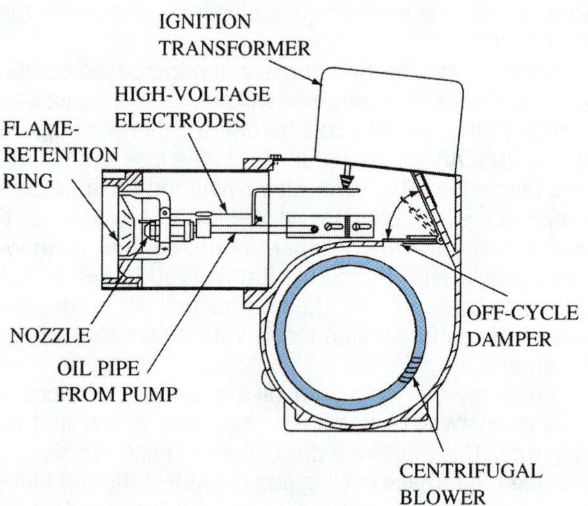

Figure 53-5 Cutaway view of the internal construction of an oil burner.

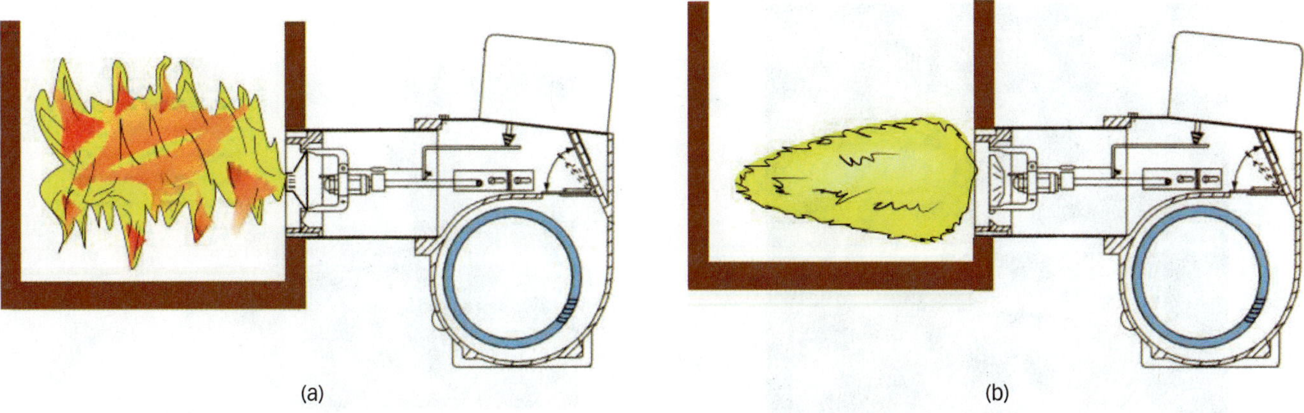

Figure 53-6 (a) Ragged flame pattern of standard burner; (b) uniform flame pattern of flame-retention burner.

is sucked back toward the burner and pulls the flame back with it. The flame stabilizes closer to the burner head and produces a more uniform flame pattern (Figure 53-6). Flame-retention burners have lower smoke levels and higher efficiencies than standard burners.

Primary air passes through the center opening, where the oil from the nozzle is sprayed (Figure 53-7). The secondary air passes through slots in the flame-retention ring (Figure 53-8). This secondary air helps to keep the burner head clean and free of carbon and by its spinning action also helps to pull the flame back toward the flame-retention ring. Tertiary air ensures that all of the fuel oil, even in the surrounding outer edges of the spray pattern, will be burned. A variable burner head allows for adjustment either back or forth to change the tertiary opening to allow fine-tuning of the burner.

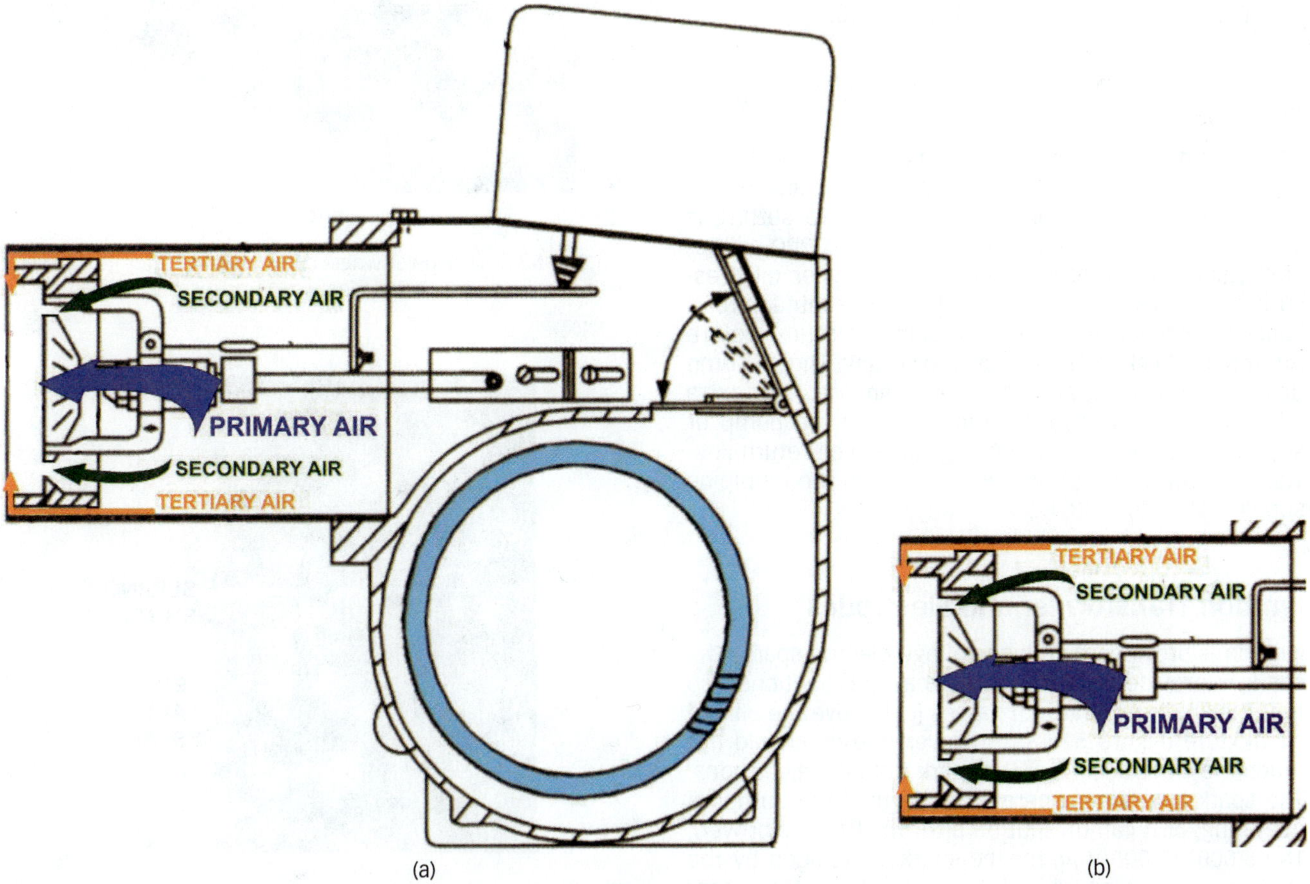

Figure 53-7 (a) Cutaway view of an oil burner showing primary, secondary, and tertiary air flow; (b) air flow through the burner head.

Figure 53-8 Slots cut into flame-retention ring to provide a path for the secondary air.

Burner Blower

The burner blower includes a centrifugal wheel mounted on the motor shaft (Figure 53-9). It furnishes air through the blast tube, which produces air/fuel turbulence for proper mixing and combustion. The amount of air is controlled by sliding an air shutter or air band on the blower housing section, as shown in Figure 53-10.

Oil Pump

The oil pump is a positive-displacement gear type (Figure 53-11) and is driven by the same motor that also drives the blower. The normal discharge pressure setting is 100 psi. This can be adjusted to as high as 200 psi because some burners are designed for higher oil pressures. However, always refer to the manufacturer's specifications for recommended pump pressures before attempting to change the pressure setting. The oil pump delivers more oil than the burner can use. The extra oil must either be bypassed internally in the pump or sent back to the tank through a separate oil return line. Thermal safety shutoff valves are used in the oil piping supply line (Figure 53-12).

Ignition Transformer and Electrodes

Ignition is provided by a high-voltage electric spark. The spark is provided by electrodes that are positioned so that the arc between them just above the oil and air mixture (Figure 53-13). However, the oil should not touch the electrodes, just the spark from the electrodes. The spark can either operate all the time the burner is operating, or it can be shut off after the flame is proved. The 8,000–15,000 V AC for the spark is provided by the ignition transformer, which is located on top of the burner housing (Figure 53-14).

(a)

(b)

Figure 53-9 (a) Blower wheel connected to motor; (b) blower wheel showing beveled blades.

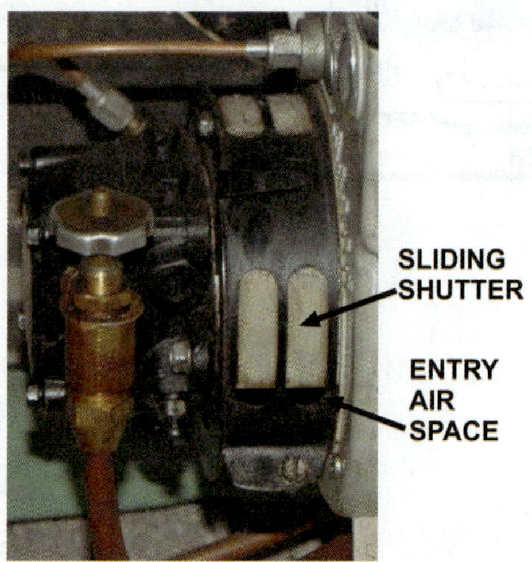

Figure 53-10 The sliding air shutter can be positioned to allow for more or less air flow to the blower wheel.

Figure 53-11 Typical oil pump.

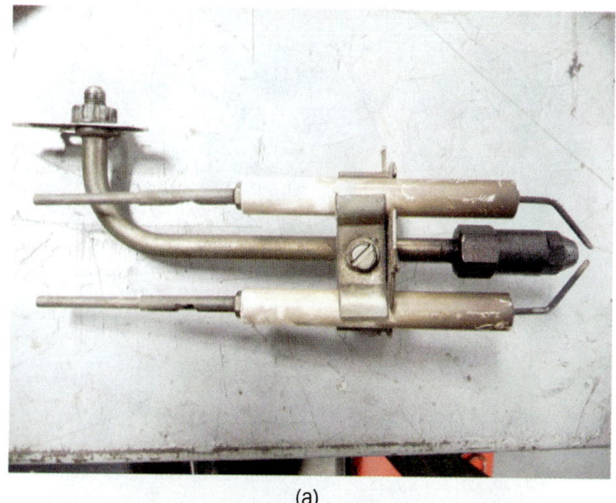

(a)

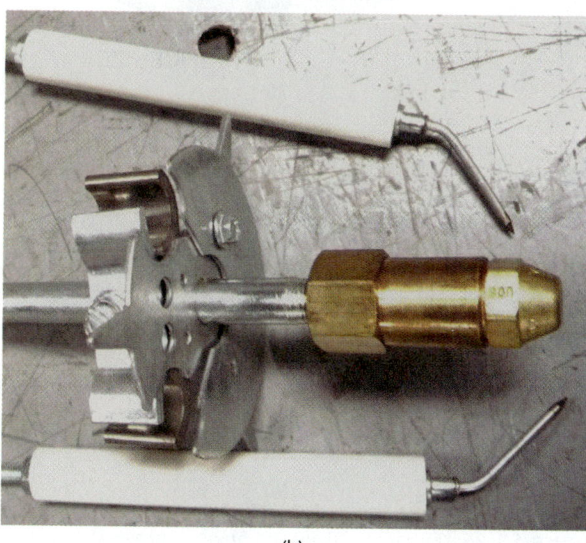

(b)

Figure 53-13 (a) Nozzle assembly being removed for replacement; (b) new nozzle and electrodes must be positioned properly before installation.

Figure 53-12 Thermal safety shutoff valve.

53.7 OIL BURNER NOZZLE SPRAY PATTERNS

Nozzles are described by their individual physical characteristics. These characteristics are the spray angle, the spray pattern, and the gallons per hour that pass through the nozzle at a given pressure (Figure 53-15). The nozzle is responsible for atomizing the fuel oil. The nozzle has an orifice that is factory bored to produce the correct firing rate and must not be altered or changed in any way. Under high pressure, the oil is atomized into fine droplets and mixes with the primary air. The mist from a nozzle makes a cone shape, with the point of the cone at the tip of the nozzle.

The spray angle describes the included angle of the sides of this cone. The spray patterns created by the nozzles tend to be listed as solid, semisolid, or hollow cone (Figure 53-16). A solid-cone spray pattern has a more or less uniform density of spray droplets throughout the cone shape. A hollow cone concentrates most of the droplets

Figure 53-14 Burner assembly showing position of ignition transformer.

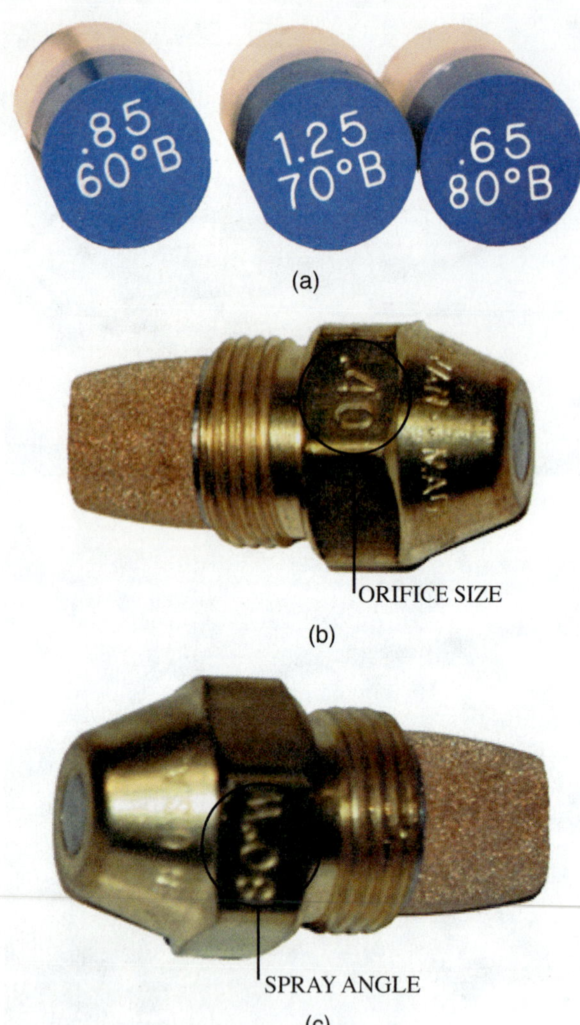

Figure 53-15 (a) Various sizes and types of nozzles; (b) nozzle size is stamped on the nozzle; (c) nozzle angle is stamped on the nozzle.

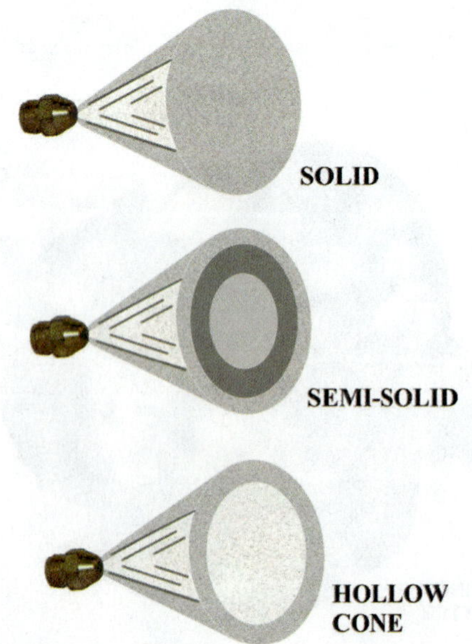

Figure 53-16 Comparison among solid, semisolid, and hollow nozzle spray patterns.

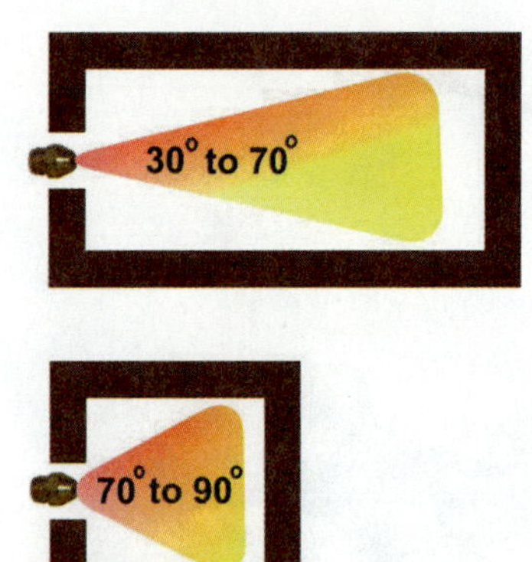

Figure 53-17 Nozzle spray angles will vary for different size combustion chambers.

toward the outside of the cone, creating a hollow center in the cone. The semi-hollow core would be someplace between these two patterns.

The nozzle must be matched to the burner and combustion chamber. For example, a 30° spray pattern would be used for heat exchangers that are long and narrow, while a 90° spray pattern would be used for a more square shape (Figure 53-17). The mix of oil and air is controlled by the spray pattern of the nozzle and the air pattern of the burner. The burner air pattern and the nozzle spray pattern must be compatible for proper mixing. The oil spray should not touch the sides of the combustion chamber or the ignition electrodes.

TECH TIP

Although a nozzle appears to be a simple device, there are many different designs of nozzles. Having a nozzle that matches the burner design is crucial. Using a nozzle that does not match the burner could result in incomplete combustion and inefficient operation.

53.8 HEAT EXCHANGER

The typical oil-fired heat exchanger (Figure 53-18) is a cylindrical shell of heavy-gauge steel or cast iron in which combustion takes place. It offers additional surfaces for heat transfer from the products of combustion inside to the air around the outside of the chamber. This type of heat exchanger is called a drum and radiator. The inside, containing the flame, is called the primary surface, and the outside is called the secondary surface.

Some manufacturers add baffles, flanges, fins, or ribs to the surfaces to provide increased surface area to allow for faster heat transfer to the air passing over the surfaces.

The burner assembly is bolted to the front of the heat exchanger (Figure 53-19). The firing assembly and blast tube extend into the primary surface in correct relationship to the combustion chamber or refractory. A flame-inspection

(a)

(b)

Figure 53-18 (a) Oil furnace heat exchanger opened; (b) heat exchanger closed, top view.

port is provided just above the upper edge of the refractory. This is used to observe proper ignition, flame shape, and flame size and to measure over-fire draft for startup and service operations (Figure 53-20).

SAFETY TIP

Never open the inspection port unless the power to the furnace is off or there is a flame.

53.9 REFRACTORY

High temperatures are required in the flame area of the heat exchanger to produce maximum burning efficiency of the oil/air mixture. To obtain this high temperature, a reflective material, called the refractory, is installed around

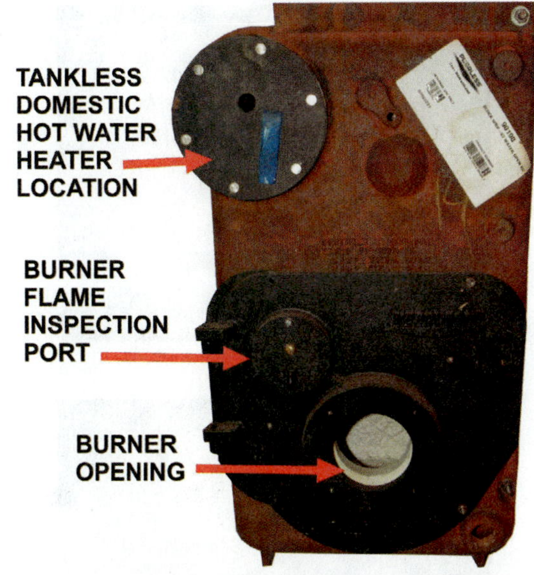

Figure 53-19 Heat exchanger showing burner opening, inspection port, and tankless domestic hot-water heater location.

Figure 53-20 Burner flame inspection port.

the combustion area. It is an insulation type of material designed to reach white-hot surface temperatures quickly, with minimum deterioration. A refractory fire box is shown in Figure 53-21. It should be noted that caution should be taken not to damage the refractory if a vacuum cleaner is to be used for cleaning during routine service.

SAFETY TIP

Occasionally the refractory firebox on an oil furnace can be flooded with oil. When this happens, the excess oil must be removed before the furnace can be restarted. With the oil burner assembly removed, use dry paper towels or rags to absorb the remaining oil before attempting to relight the furnace. Never attempt to relight a furnace with liquid oil in the bottom of the refractory firebox. If the oil in the refractory firebox is ignited, it is very difficult to extinguish.

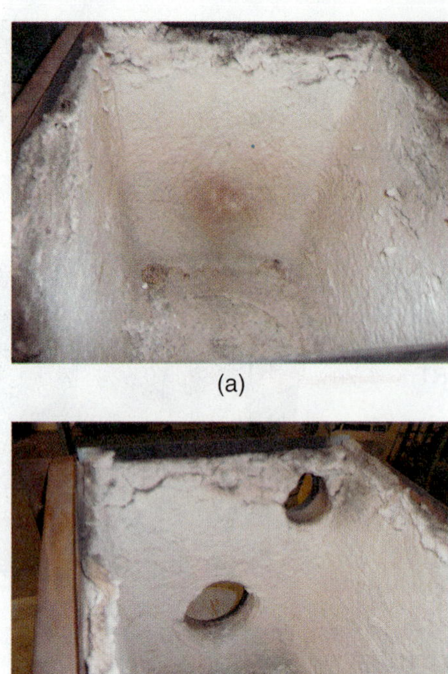

(a)

(b)

Figure 53-21 Oil furnace refractory.

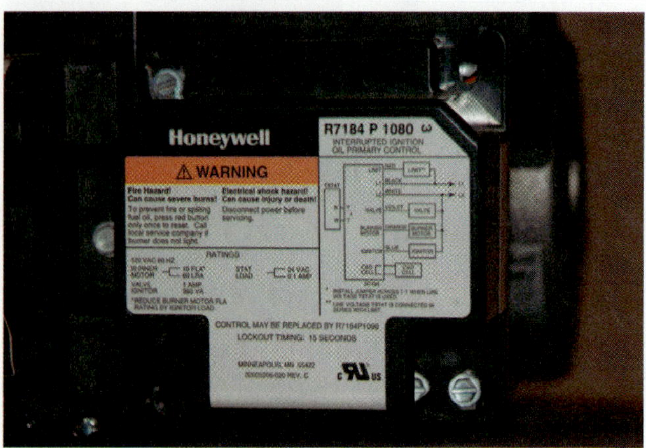

Figure 53-22 Oil burner primary control.

SAFETY TIP

Pushing the reset button multiple times can set up a potentially dangerous situation by soaking the combustion chamber with oil. Find out if the customer has already pushed the reset several times before trying to operate the furnace. Oil does not evaporate, so the oil will still be sitting in the combustion chamber even if the reset button was pushed several times the day before. If the furnace does light with a combustion chamber full of oil, the combustion chamber may not be able to contain the flames.

CAUTION

Only CO_2-type extinguishers should be used on oil burner fires because they leave no residue. Dry chemical fire extinguishers will put out an oil fire, but the residue can cause an extensive and lengthy cleanup.

53.10 PRIMARY OIL BURNER CONTROLS

The primary control governs the operation of the oil burner (Figure 53-22). It responds to the signal from the thermostat and will control the start-up and run cycle. The primary control will also shut down the burner when the thermostat is satisfied or if the limit control opens. At one time, stack-type relays were the standard, but today mostly cad cell primary controls are used.

Cadmium Sulfide Cell (Cad Cell)

Flame safety is provided by a light-sensitive cadmium sulfide cell (Figure 53-23). The "cad cell" has a very high electrical resistance in the dark but a low electrical resistance when exposed to light. The cad cell senses when there is no flame because it cannot see the light from the flame. When the cad cell does not see the light of the flame, the primary control will stop the burner motor and ignition transformer, thus preventing oil from flowing into the heat exchanger. Most furnaces have a lockout system requiring the control to be reset manually before the burner can run again.

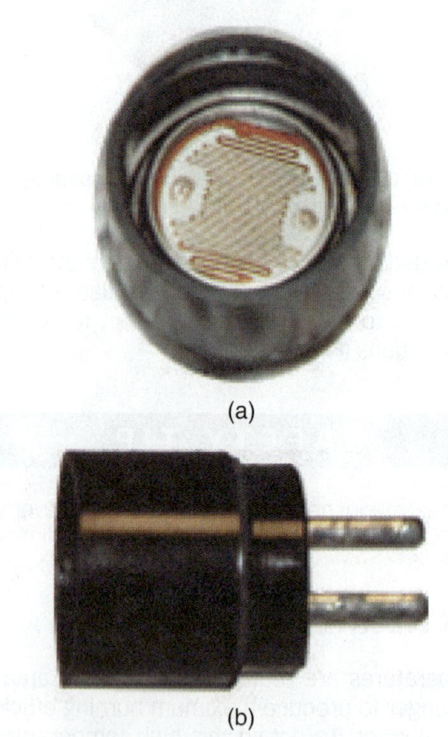

(a)

(b)

Figure 53-23 Cadmium sulfide (cad) cell.

Figure 53-24 Burner emergency switch.

Oil Burner Emergency Switch

The oil burner emergency switch has a red cover plate and is labeled for emergency (Figure 53-24). This is usually located at a convenient shutoff point for the customer to access. As an example, if the furnace is located in the basement, then it would be located at the top of the basement stairway. This allows the customer to shut down the furnace upon a malfunction, such as excessive smoking, without the need to enter the basement. There will also be a second disconnect switch located on or near the furnace for the technician to shut off whenever servicing the furnace.

Warm-Air Oil Furnace Fan and Limit Switch

One type of combination fan and limit control for a typical gas or oil warm-air furnace (Figure 53-25a-d) uses the power created by the rotary movement of a helix bimetal. The rotary motion comes about from the expansion and contraction of this bimetallic element. Since it is helical in shape, it turns in a circular motion, either clockwise or counterclockwise, depending on whether the furnace air is becoming hotter or cooler. The rotating cam makes or breaks the separate fan and limit electrical contacts. The control board in many newer oil furnaces operates the indoor blower on a timing circuit. These furnaces do not have a fan switch, only a limit switch.

Direct- and Reverse-Acting Aquastats

Aquastat controls are used for hot-water boilers. They maintain the boiler water at the proper temperature. There is a high-limit setting and a low-limit setting (Figure 53-26). Since these hot-water boilers are not designed to produce steam, the high limit will regularly shut off the burner at a preset temperature. This will prevent the water from becoming hot enough to produce steam. If the thermostat is calling for heat, the circulating pump (called a circulator) will continue to operate and deliver hot water through the closed heating loop even though the oil burner is off. When the water temperature reaches the low limit, the burner will start up again so that the water temperature does not drop too low.

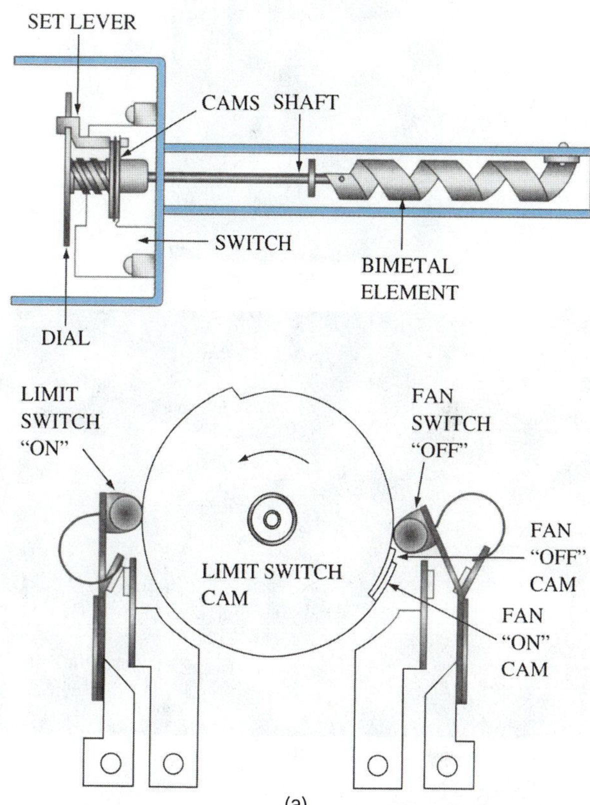

(a)

(b)

Figure 53-25 (a) Construction of combination fan and limit control for gas- or oil-fired warm-air furnaces; shows how twisting motion of heated helix activates switches; (b) control is located above and to the left of the burner; (see next page) (c) close-up of control; (d) close-up of helix

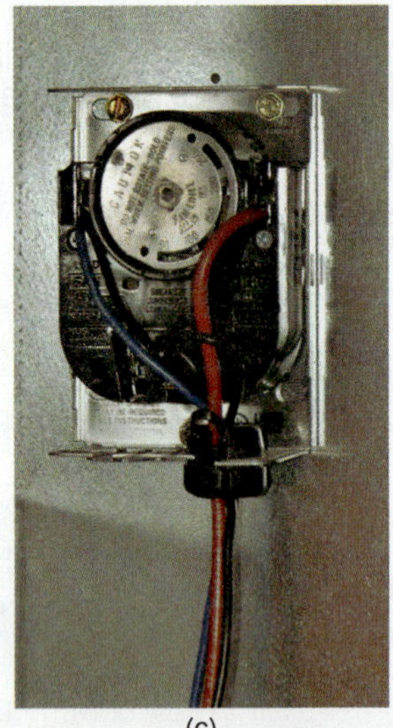

(c)

(d)

Figure 53-25 *(Continued)*

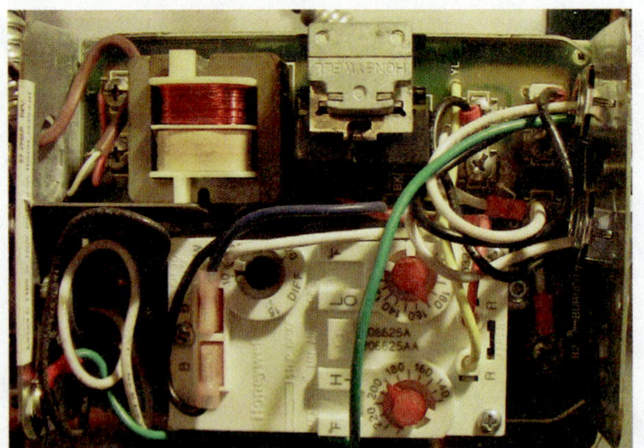

Figure 53-26 Aquastat with high-limit, low-limit, and differential adjustment.

The aquastat will also control the hot-water circulator (Figure 53-27). A direct-acting aquastat will start the circulator upon a drop in temperature. As an example, the thermostat calls for heat, the burner turns on, and the circulator starts. When this happens, the hot water is delivered from the boiler to the baseboard heaters and cold-water returns. Since the oil burner has just started, it will take a few minutes to warm up the heat exchanger. Because of this delay, the flow produced by the circulator to push the cold water back to the boiler will cause the temperature to drop too low.

Figure 53-27 Hot-water circulating pump.

A reverse-acting aquastat corrects this. It will start the circulator upon a rise in temperature instead of a drop. This allows the boiler enough time to heat up before the circulator begins pushing cold water back to it. This is important because many boilers also provide domestic hot water. With a reverse-acting aquastat, a person taking a shower would not be affected by a thermostat call for heat. This is because the circulator would not come on until the water temperature was high enough. If the circulator came on earlier, it would rob heat from the shower because of the resulting drop in the boiler temperature.

This drop in boiler temperature is also detrimental to high-efficiency boilers. The already low-temperature combustion gases would be even cooler. This causes even more water vapor to condense in the combustion chamber and exhaust vent than normal. The sulfur contained in the fuel oil will carry over in the combustion process, forming sulfuric oxides and mix with the water to create sulfuric acid. Not only is sulfuric acid highly corrosive, but it will also lead to a buildup of scale on the heat exchanger. The reverse-acting aquastat does not start the circulator until the oil burner raises the boiler temperature. This helps to keep the combustion gases above the dew-point temperature.

Hot-Water Boiler Auxiliary Components

An oil-fired hot-water boiler will have a low-water cutoff to stop the oil burner if the water level becomes too low (Figure 53-28). If more water is required, a pressure-reducing valve will automatically add makeup water from the building's water supply piping (Figure 53-29). This

(a)

(b)

(c)

Figure 53-28 (a) Low-water cutoff; (b) with cover off; (c) connection to the boiler.

Figure 53-29 Pressure-reducing valve adds makeup water to the boiler.

Figure 53-30 Boiler pressure-relief valve.

reducing valve for residential systems is set at 12 psi. In addition to this there will be a system-relief valve set for 30 psi (Figure 53-30).

The boiler will have a temperature/pressure gauge, as shown in Figure 53-31. Notice the boiler is shut down as the temperature reads low. The pressure on this gauge can be read in feet of water (outer scale) or psi (inner scale). The black needle pressure reading indicates approximately 12 to 15 psi. The red indicator arrow is pointing toward the set point for the pressure-relief valve at approximately 30 psi.

Figure 53-31 Boiler temperature and pressure gauge.

UNIT 53—SUMMARY

Today's oil heating systems are more efficient and adaptable than the large, inefficient furnaces from previous decades. Now oil heating systems are available in smaller, more efficient furnaces and boilers with efficiencies up to 95 percent. Two-stage furnaces, condensing furnaces, sealed combustion furnaces, oil-fired unit heaters, and furnaces designed to burn waste oil are some of the innovations in oil heat.

Fuel oil must be atomized to be burned. Because of this, oil furnaces have more parts and are somewhat more complex than gas furnaces. Oil burner components include an oil pump, a combustion air blower, a burner motor, a nozzle, an ignition transformer, electrodes, a primary control, and a cad cell. The oil is forced through a nozzle at high pressure to atomize it. At the same time, air is forced into the combustion chamber, where it mixes with the oil spray. The transformer delivers 10,000 V AC to the electrodes, which create a large electric arc that ignites the oil. The cad cell senses the light from the flames, and resistance drops and allows the primary control to continue operation. Should the flames extinguish, the cad cell resistance increases, and the primary control shuts off the burner motor.

WORK ORDERS

Service Ticket 5301

Customer Complaint: Expensive Heating Bills

Equipment Type: Propane Unit Heater

Precision Auto, located in rural Wisconsin, has asked their local heating contractor if there is a less expensive way for them to heat their garage. They are using propane-fired unit heaters, which run a lot in the winter. The sales consultant suggests a waste-oil-fired furnace. The waste-oil furnace can use old motor oil, making their fuel free. In addition to saving on heating bills, they will save on used oil disposal fees.

Service Ticket 5302

Customer Request: New Construction

Equipment Type: Central Hot Water Boiler

The architect for a new high-rise being planned has asked for ideas on keeping heating costs controlled. She wants to use a large central boiler with hot water distributed throughout the building. She was considering a natural gas boiler, but the local gas company wants to put the complex on an interruptible service in exchange for favorable prices. The sales consultant suggests using a boiler fired with grade 6 fuel oil. The equipment will require a heated storage tank and the oil will have to be preheated before combustion, but the price per BTU is much lower for grade 6 fuel oil.

Service Ticket 5303

Customer Request: Furnace Replacement

Equipment Type: Oil-Fired Furnace

Dr. Greene has just purchased a house in the historic district that has an old oil-fired furnace. He wants to replace the aging, inefficient furnace with a modern, energy-efficient gas furnace. The sales consultant mentions that he does not have to switch to gas to get an efficient furnace and suggests a 95 percent efficient condensing oil furnace.

Service Ticket 5304

Customer Request: New Construction

Equipment Type: Mid-Efficiency Oil-Fired Furnace

A developer has asked a contractor for ideas for efficient heating for the homes in her subdivision in rural Canada. The developer stresses that operating efficiency is important, but so is keeping costs controlled. The sales consultant suggests a mid-efficiency oil-fired furnace as a good compromise between efficiency and cost. The consultant points out that a mid-efficiency oil-fired furnace operates about 5 percent more efficiently than a mid-efficiency gas-fired furnace.

UNIT 53—REVIEW QUESTIONS

1. List the grades of fuel oil, and give a brief description of each type.
2. What is the difference between distillate oil and a residual oil?
3. How are nozzles rated?

4. What does oil viscosity mean?
5. Which grade of fuel oil has the highest viscosity?
6. Why can oil furnaces operate at efficiencies between 80 percent and 90 percent without condensing water in the heat exchanger or vent?
7. List the available warm-air oil-fired furnace configurations.
8. Compare the efficiency of a mid-efficiency oil furnace to an older, traditional oil furnace.
9. What is the advantage of a high-pressure oil burner?
10. What are some of the challenges in building a condensing oil-fired furnace?
11. How is an oil furnace heat exchanger shaped?
12. What is the purpose of the refractory?
13. What is the typical output pressure of an oil-fired furnace oil pump?
14. What are the typical pressure settings for a pressure-reducing valve and system relief valve in a hot-water oil-fired boiler?
15. What is the purpose of an aquastat?
16. What is the difference between a direct-acting and a reverse-acting aquastat?
17. What voltage does the ignition transformer provide to the electrodes?
18. How does a cad cell sense the flames?
19. What does the primary control do?
20. What is the heating value of grade 2 fuel oil?

UNIT 54

Oil Furnace and Boiler Service

OBJECTIVES

After completing this unit, you will be able to:

1. list the operating sequence of a typical oil-fired furnace.
2. describe the procedure for purging the oil lines on an oil burner.
3. list the operating characteristics required to determine combustion efficiency.
4. explain the relationship between CO_2 flue gas percentage and combustion efficiency.
5. list the items that should be checked as part of planned maintenance on an oil burner.

54.1 INTRODUCTION

Annual service is a must for oil burners. Oil must be atomized before burning because oil is a liquid fuel. The process of atomizing the oil, mixing the combustion air with the oil, and delivering the correct amount of both oil and air has many variables that all must be correct for the process to work. Some components, such as the nozzle and oil filter, need to be changed regularly. Operating characteristics such as the oil pressure, smoke spot readings, stack temperature, and CO_2 percentage need to be checked. Failure to perform the necessary annual maintenance will result in inefficient operation and dirty combustion.

54.2 PLANNED MAINTENANCE

Good routine maintenance includes annual start-up and calibration on oil heat equipment prior to the heating season. Prior to start-up and calibration, a number of maintenance tasks should be completed, such as checking air filters, oiling motors, replacing oil filters, cleaning strainers, checking electrodes, and cleaning the cad cell. The furnace combustion chamber should be cleaned as well as the vent pipe. After the routine cleaning and replacement of components is completed, the furnace should be tested and verified during normal operation to determine that the operating and safety controls are working as designed.

Initial Preparation for Annual Tune-up

Speak with the customer to find out where the furnace is located and the best way to gain access. If the furnace is located in the basement, there may be an outside entry so you do not have to keep going back and forth through the living spaces. This way you are less likely to track dirt and soot through the house. Ask if there have been any operational problems with the furnace. Also check to see if a service record has been attached to the furnace listing the dates of previous service visits and what was done

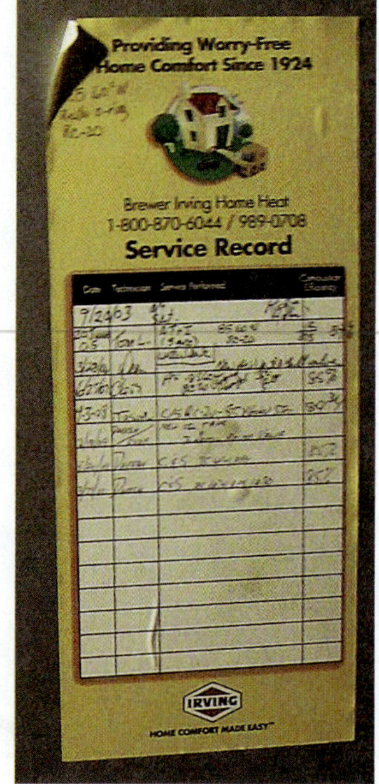

Figure 54-1 Furnace service record attached to side of the furnace.

(Figure 54-1). If it has been more than a year since the last service, the furnace will require a more thorough cleaning. Check the furnace make and model to determine if you have the proper spare parts in the service van. Place clean drop cloths around the furnace and oil tank, and remember to clean up when you are finished.

Check the Furnace Operation

Check the high- and low-limit control settings to make sure they are correct. Use the thermostat to turn the furnace on and check the operation of the circulator. In addition to checking the operation of the burner and controls, this also provides time for the heat exchanger surfaces to dry out. This is important if the furnace has been out of operation for a period of time. A draft drop test can be performed at this time by testing at the breech and over the fire. If the drop is greater than −.04 in, there is probably a buildup of soot and scale on the heat exchanger surfaces. After the furnace has run for about 5 minutes and your draft drop test is completed, it can be shut off at the emergency switch. This will test the emergency switch and also secures the power to the furnace so that you can continue with the service.

Oil Tank Filter

This is a good time to inspect the oil tank and change oil filters, as the furnace is cooling down. Many residential tanks are made of steel and will rust out over time due to the water that tends to collect in the bottom of the tank. The tank may appear to be in good condition on the outside but may be rusting from the inside out. The tank filter is connected at the bottom of the tank by a pipe nipple. Look for any signs of oil leakage in this area.

Before replacing the filter, make sure the thermal safety shutoff valve is completely closed (Figure 54-2a-c). A drip pan should be placed under the filter housing to collect any oil that spills when the dirty cartridge is removed (Figure 54-3a). Loosen the vent located on top of the filter housing to make sure that there is no oil pressure, and catch any oil that comes out in the drip pan. The main bolt that holds the filter housing in place can now be removed to access the dirty cartridge. Check for excessive water in the filter housing (Figure 54-3b).

Place the dirty cartridge into a plastic bag. Do not drain any remaining oil back into the tank, as it will be very dirty. Wipe out the filter housing with a rag and then install the new cartridge and all new gaskets and tighten the filter housing (Figure 54-4). Open the shutoff valve at the inlet of the filter housing and check for any leaks.

SAFETY TIP

Thermal safety shutoff valves have a fusible handle that melts at 165°F. In the event of a fire, the valve will automatically close once this temperature is reached.

Oil Pump Inlet Strainer

Even though there is a filter located on the outlet of the oil tank, the oil pump inlet strainer should also be replaced. This can affect the suction produced by the fuel pump which is measured in inches of mercury vacuum (in Hg). Checking the oil pump operation is further described in Unit 55.

The inlet strainer is located in the oil pump housing (Figure 54-5). Place the drip pan under the oil pump as you remove the strainer to catch any oil that will drain from it. Close the thermal safety shutoff valve located at the pump inlet. Remove the oil pump cover and its

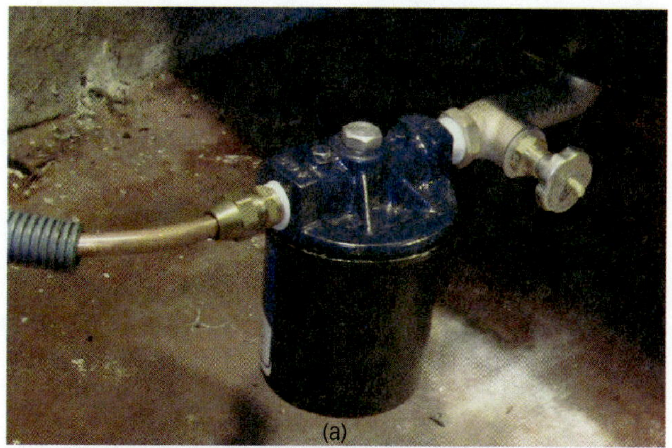

Figure 54-2 (a) The oil tank filter is connected by a pipe nipple to the bottom of the tank; (b) Thermal shutoff valve is open; (c) thermal shutoff valve is closed.

(a)

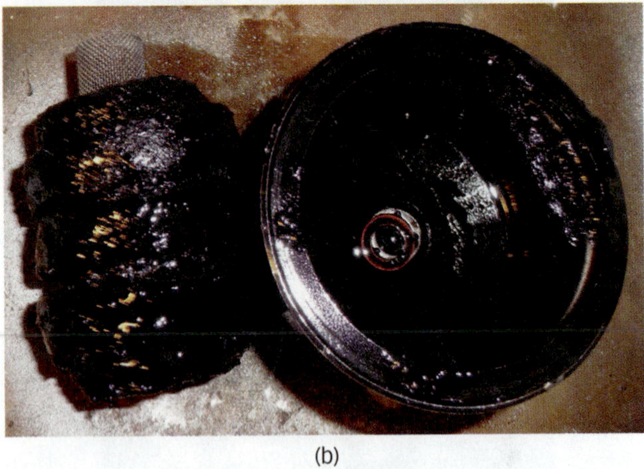

(b)

Figure 54-3 (a) Place a drip pan under the filter housing; (b) dirty filter cartridge and housing.

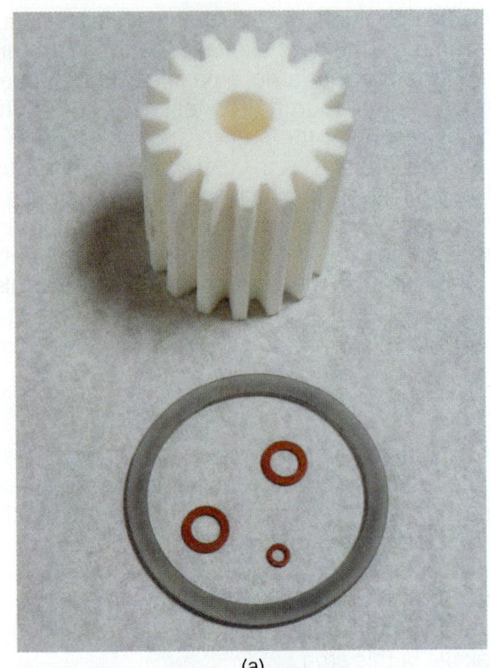

(a)

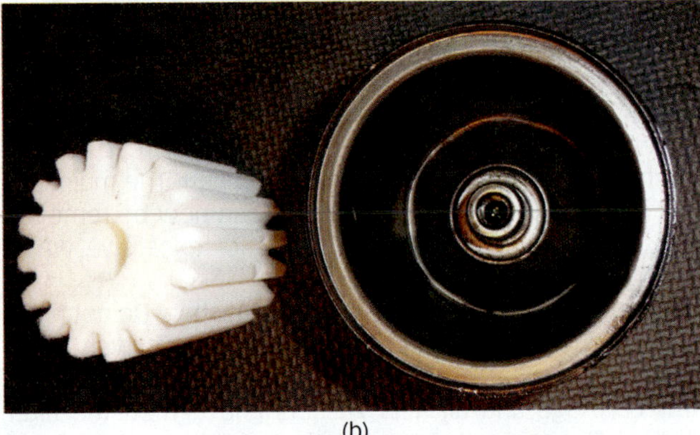

(b)

Figure 54-4 (a) Replacement cartridge and gaskets; (b) filter housing after cleaning.

gasket to access the strainer. After the inlet strainer and gasket have been replaced and the cover fastened securely, open the safety shutoff valve and check for leaks.

Purging Air from the Oil Line

On the single-pipe system, it is necessary to purge the air from the oil filter, fuel line, and fuel pump. Connect a pressure gauge on the oil line leading from the pump to the nozzle (Figure 54-6). Open the bleed connection located on the bottom of the pump housing using a small wrench (Figure 54-7). It is only necessary to open this valve until the air in the system starts to flow out. Usually, this is one half to one full counterclockwise turn of the valve. Make sure the drip pan is in place to catch the oil. When the burner is turned on, oil will flow from the bleed connection. Allow the oil pump to run until all the air is bled from the suction line.

SERVICE TIP

Use this tip during purging. Place a clear plastic hose on the air purge valve and put the other end of the hose in a container. Run the pump until the hose is full of oil with no air bubbles.

A two-pipe system has both a supply line to the burner and a return line back to the fuel oil tank. Because of this, a two-pipe system is self-purging. The pump will force the air down the oil-return line and back into the tank. Even so, it is good practice when bleeding a two-pipe system to have your pump gauge connected. This will ensure that no uncontrolled oil will spray into the combustion chamber.

After the line is bled, operate the burner until the pump pressure holds steady on the gauge and then turn the pump

(a)

(b)

Figure 54-5 (a) Removing oil pump inlet strainer; (b) oil pump inlet strainer.

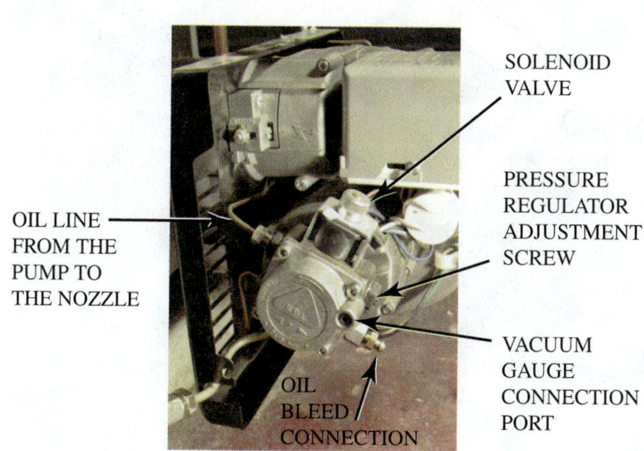

Figure 54-6 Oil pump connections and parts identified.

Figure 54-7 Use a small wrench to open the oil pump bleed valve.

adjustment screw. The pressure readings should change smoothly and not fluctuate up and down.

Nozzle and Electrodes

The nozzle assembly should be removed and inspected and the nozzle replaced. The electrodes' porcelain insulators should be checked for cracks, and any oil or carbon should be cleaned off (Figure 54-8). Check the setting for the electrodes. Generally speaking, they should be ⅜ in apart, slightly ahead of the nozzle, and set at an angle specified by the burner manufacturer (Figure 54-9). Many

(a)

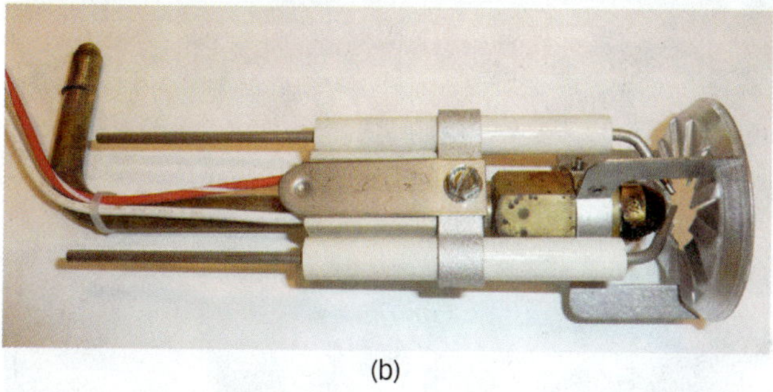

(b)

Figure 54-8 (a) The nozzle needs to be replaced and the igniters cleaned and readjusted; (b) replacement nozzle assembly.

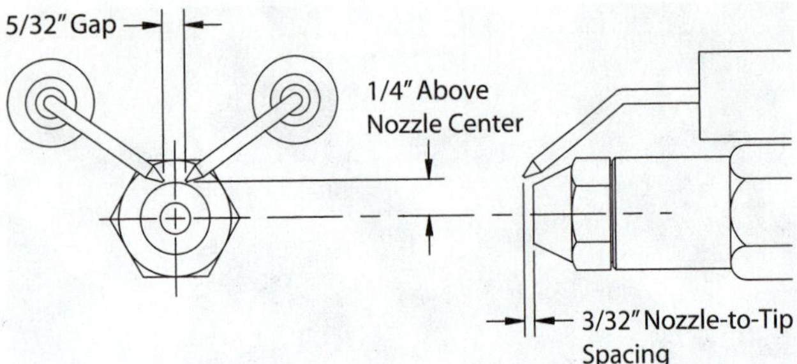

Figure 54-9 Oil burner electrode spacing dimensions. *(Courtesy of the R.W. Beckett Corporation)*

manufacturers have plastic gauges that fit over the nozzle to assist in setting the electrodes.

SERVICE TIP

Replacing the nozzle as part of your routine maintenance is important because over a season the nozzle bore will be worn slightly larger as the result of oil flow. As this wear occurs, the burner efficiency is affected. Replacing the nozzle each season ensures that the burner will operate properly.

Ignition Transformer and Cad Cell

Check the springs and contacts that transfer voltage from the transformer to the electrodes and clean the cad cell eye and wires (Figure 54-10). Test the ignition transformer (Figure 54-11) for the correct voltage output to the electrodes. This is done with a special high-voltage probe (Figure 54-12). The probe reduces the voltage by a factor of 1,000; 10,000 V will show 10 V on the meter. A Donogan transformer tester or a Mitco ignitor tester (Figure 54-13) is also often used. Alternatively, a resistance test with a multimeter can be performed to measure the transformer's resistance.

Figure 54-10 Ignition transformer with cad cell located directly between the springs.

Figure 54-11 Ignition transformer.

Figure 54-12 High-voltage probe.

Figure 54-13 Mitco ignition tester. *(Courtesy of Sid Harvey's Inc.)*

Figure 54-14 Flue pipe should be removed and cleaned as necessary.

Figure 54-15 Make sure barometric damper swings freely.

SAFETY TIP

The output of the ignition transformer cannot be tested with a standard set of leads. A high-voltage probe is required. Attempting to test the ignition transformer output with regular multimeter leads will destroy your meter and possibly injure you.

Blower Wheel and Motor

Clean the blower wheel, making sure it is tight on the motor shaft. Pay attention to the air-intake slots on the air-adjustment shutter and clean if required. Check the motor shaft endplay. While in the blower compartment, check the oil pump coupling for alignment and tightness on the motor shaft to make sure it is not worn or stripped.

Flue Pipe and Draft Regulator

The flue pipe should be removed and cleaned (Figure 54-14). Check the barometric damper that is usually installed in the horizontal vent pipe between the furnace and the chimney. Make sure it swings freely (Figure 54-15). The damper has a movable weight so that it can be set to counterbalance the suction and to maintain reasonably constant flue operation (Figure 54-16). It is adjusted while the furnace is in operation and the chimney is hot.

The clean-out at the base of the chimney should not be neglected, because this can fill up with deposits over time (Figure 54-17). Once the clean-out is cleared, the chimney lining can be inspected. Place a mirror in the opening and there should be a clearly visible passage extending all the way to the top with no broken sections of lining or brick extending into the interior of the chimney.

For furnaces with power venting, the fan blower wheel and motor should be checked (Figure 54-18) and the outside hood should be checked and cleaned.

Figure 54-16 Damper adjustment counterweight.

Figure 54-17 Clean-out at base of chimney.

(a)

Figure 54-18 Power vent.

(b)

Figure 54-19 (a) Hinges on furnace front allow it to be opened; (b) clean out door swings open.

Combustion Area and Heat Exchanger

The furnace front is mounted on hinges that allow it to swing open (Figure 54-19a) or there may be a clean out door (Figure 54-19b). Electrical connections may need to be disconnected to swing the front in the fully open position (Figure 54-20). There is a braided gasket around the entire perimeter of the furnace front where it seals against the furnace (Figure 54-21). If the gasket is torn or damaged, it must be replaced. To replace the gasket, the groove that it sits in must be cleaned. A high-temperature furnace sealant can be applied to the groove to help hold the new gasket in place.

Figure 54-20 Disconnect wires to swing furnace front open fully.

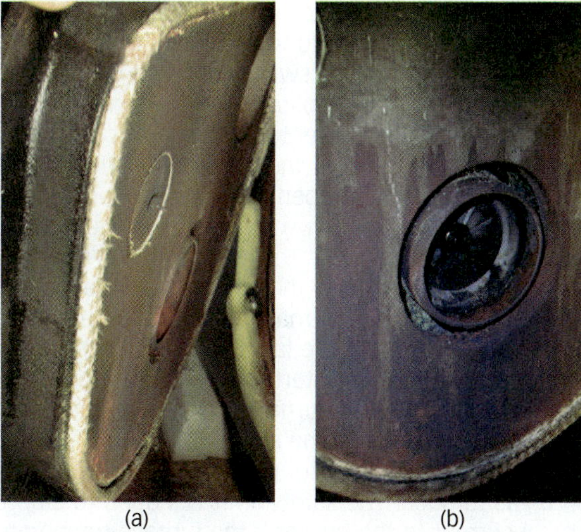

Figure 54-21 (a) Braided gasket seals around the furnace front; (b) inspect the burner opening on the furnace front.

Once the furnace front is opened, the combustion chamber can be inspected and cleaned (Figure 54-22). A small vacuum cleaner may be used, but be very careful not to damage the refractory. Some of the metal surfaces will need to be lightly scraped to remove scale buildup. The combustion chamber will never look like new, but it should be cleaned enough so that no loose scale or soot remains (Figure 54-23).

TECH TIP

Carbon or soot sublimes at a temperature slightly above 800°F. Subliming is the process of changing from a solid to a vapor without becoming a liquid. This means that the area of refractory material that is carbon free was at a temperature slightly above 800°F. The line where the carbon first begins to show is the beginning of the area that was below 800°F during system operation. If you observe a furnace firebox that has far less carbon-free refractory material, this may be an indication of a low flame temperature.

The heat exchanger will need to be accessed through a cover in the top of the furnace. This cover will also have a gasket. A long-handled wire brush is used to clean the surfaces of the heat exchanger, and the soot and scale removed can be picked up with a vacuum cleaner. Figure 54-24a shows a new heat exchanger surface and Figure 54-24b,Figure c shows a heat exchanger before and after cleaning.

TECH TIP

Soot deposited on flue surfaces acts as an insulating layer, reducing the heat transfer and lowering the efficiency. Soot can also clog flues, reduce draft and available air, and prevent proper combustion. Proper burner adjustment can minimize soot accumulation. An accumulation of 0.25 in of soot is equal to 1 in of insulation, and ⅛ in of soot decreases the efficiency by around 9 percent.

Figure 54-22 Soot and scale buildup in combustion chamber.

54.3 START-UP

The initial system start-up requires more than just turning on the burner. It is important that the entire operating sequence of the furnace is understood. This allows for checks of the safety system during the initial start-up.

As an example, the furnace may not light properly if there is still air remaining in the fuel lines. The result of air in the fuel line will be different for a single-pipe system as compared to a two-pipe system. In a two-pipe system, it may take longer than the cycle time of the safety switch to purge the system, and the burner may shut down before a

Figure 54-23 Combustion chamber after cleaning.

(a)

(b)

(c)

Figure 54-24 (a) New heat exchanger surface; (b) heat exchanger that requires cleaning; (c) heat exchanger after cleaning.

flame is established. It is then necessary to allow the safety switch to cool and reset before the burner can be started again. This must be repeated until the oil lines are completely purged of air and the fire is established.

However, repeated start-up and shutdown of the burner can lead to a buildup of oil in the firebox. This oil can ignite all at once and cause a flareback, which can damage the furnace and cause personal injury to someone standing close by. If you are trying to purge air from a two-pipe system, make sure that excessive amounts of oil are not collecting in the firebox. If the air is not purged after more than a few attempts, check to see if there are other operational problems.

This is why it is important to understand the operating sequence and the wiring diagrams for the equipment you are servicing.

54.4 OIL FURNACE WIRING DIAGRAMS

The primary control is the main operating control of the furnace. It controls the oil burner motor and the ignition transformer and provides flame safety. Flames are proved using a cad cell flame sensor that senses the light of the flame. Safety against the blower circulating vent gases is provided by a blower door switch in series with the low-voltage circuit to the thermostat. Over-temperature safety is provided by the limit. Power to the primary control passes through the limit control. If the furnace reaches the limit temperature, the limit will open to shut off the power to the primary control. This in turn will shut down the burner to prevent overheating the unit. The fan switch is in series with the blower motor to control the operation of the fan. After the burner lights and the furnace has warmed up, the fan control closes to operate the fan. Properly set, this control will close at 125–130°F to start the blower motor. The fan continues to run after the burner stops, and the fan switch opens at 95–100°F to stop the blower motor.

The fan and limit control perform the same function as on a gas furnace. The safety limit will open to stop the electrical current to the burner if the furnace temperature exceeds the limit setpoint. The fan switch is set to cycle the furnace blower on when the heat exchanger heats up to the fan switch ON setting and keeps the blower operating until the heat exchanger cools to the fan switch OFF setting. The space thermostat (24 V) feeds directly to the burner low-voltage secondary control circuit. This actuates the relay and flame-detector circuit, feeding the primary line voltage to the burner motor and ignition transformer.

SAFETY TIP

The ignition circuit on an oil-fired furnace uses approximately an 8,000–15,000 V arc to continuously ignite the oil. Accidentally touching the "live" ignition transformer contacts can result in a severe electrical shock.

Cad Cell Primary Control

The original primary oil burner control, known as a stack relay, was mounted on the vent connector pipe. The bimetallic sensing element in the sensing tube would react to a rise in flue gas temperature and cause the relay to keep the oil burner operating. The stack mounted protector relay required extra wiring, as well as time and cost for manufacturing assembly or installation. To overcome this, a flame-detection relay was developed.

The cad cell is mounted directly in the oil burner housing and sees the light of the burner from the instant the flame is established. Two relays are used: 1K for control of the oil burner and ignition transformer and 2K, the sensitive relay, controlled by the cad cell. Following the circuit action in the schematic in Figure 54-25, when the thermostat calls for heat and closes the contact between T and T, current flows from the transformer through the thermostat through relay coil 1K, the safety switch (SS1), the timer contact (T2), the safety switch heater, and contact 2K1 and to the other side of the transformer. As relay coil 1K energizes, it pulls in to close contacts 1K1, 1K2, and 1K3. When 1K1 closes, it energizes the oil burner motor, oil valve (if used), and the ignition transformer. At the same time, contact 1K2 closes, which reduces the current through the safety switch heater. 1K3 is in series with the timer heater.

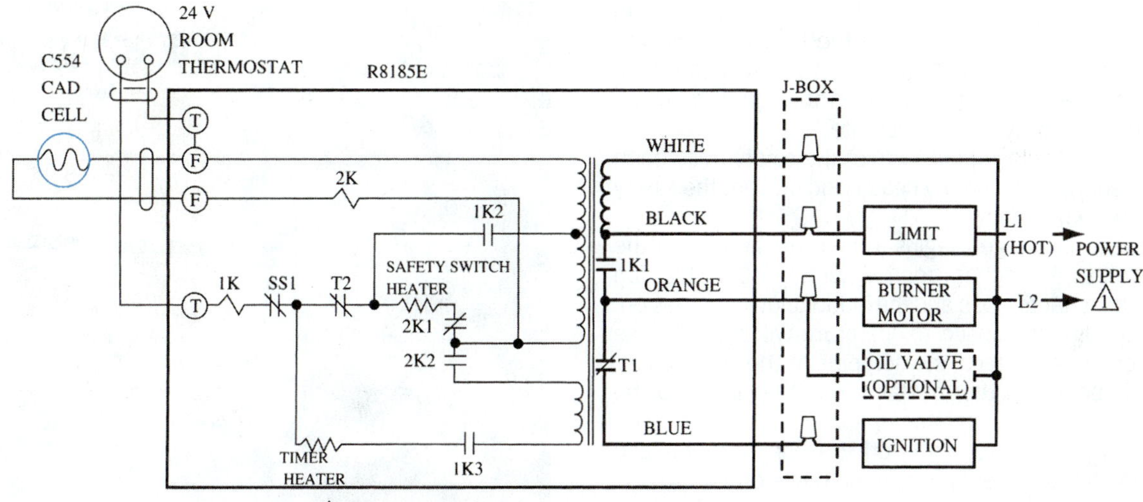

Figure 54-25 Oil burner wiring diagram, using a cad cell relay.

The safety switch heater is a type of time delay. As electrical current flows through the heater, its temperature increases to the point where it acts to open the safety switch, which de-energizes the entire circuit and shuts the burner down. This will happen if no flame is established. The delay allows enough time for the burner to light. If the burner lights, then the safety heater will be de-energized.

When a flame is established, the light strikes the cad cell and immediately reduces the resistance of the cell from over 10,000 Ω to about 1,500 Ω or less. This increases the current flow through the sensitive relay coil 2K, allowing it to pull in and close contact 2K2 and open contact 2K1, which de-energizes the safety switch heater.

The timer heater is another time delay similar to the safety switch heater in operation but not function. It thermally opens contacts T1 and T2 after a set period of time. When T1 opens, it de-energizes the ignition transformer. This allows enough time for the ignition transformer to create a spark across the burner electrodes to ignite the fuel. In this way, once enough time has passed for the burner to light, the ignition transformer is de-energized. It is only required to create an initial spark for start-up.

If flame failure should occur, the cad cell resistance will increase and current flow through the sensitive relay 2K will decrease, and the relay will drop out. This opens contact 2K2, which opens the circuit to relay 1K, and the burner shuts down.

The burner cannot come on until the timer heater has cooled sufficiently to close ignition contact T1 and relay 1K circuit contact T2. This ensures that the unit will have sufficient time to vent the furnace heat exchanger of unburned vapors as well as ensuring ignition at the next startup.

Most furnaces today still use the cad cell to sense the flame, but they now use electronic switching and circuitry rather than heaters and sensitive relays. The operating principle remains the same in that the resistance of the cad cell decreases when it detects light. These controls typically lock out after an ignition failure and must be manually reset.

54.5 OPERATING SEQUENCE

The step-by-step operating sequence is as follows:

1. The room thermostat closes to initiate a call for heat in the conditioned area.
2. If the blower door switch is closed, this causes current flow from the 40 VA transformer through the control circuits to the burner motor relay coil.
3. The relay contacts close, allowing current flow from the H, or hot, side of the supply line through the limit switch to the burner motor and ignition transformer.
4. At the same time, two other 24 V control circuits exist. One circuit is through the cad cell and the safety relay coil. The second circuit is through the safety relay normally closed contacts and the safety heater.
5. The lack of a flame causes a high resistance in the cad cell, which drops nearly all the voltage in its circuit, so the safety relay does not operate. The relay's normally closed contacts remain closed and the safety heater starts producing heat. Note that the safety heater is simply a time delay. The safety heater acts upon the safety switch contacts, which are opened by the expansion of a bimetallic element. If the burner does not light, the heater will remain on and continue to produce heat until the bimetallic element expands to the point where the safety switch contacts open and the burner motor shuts down.
6. The ignition transformer establishes a spark across the electrodes located above the nozzle. The spark at this time is only ⅛ in long, the distance between the electrode tips.
7. The burner motor, which is energized at the same time as the ignition transformer, also operates the combustion air blower. It reaches full load speed within 1 second. When the full amount of air is delivered by the blower, the ignition spark is blown forward into the oil spray and the oil is ignited. This occurs directly in front of the nozzle and quickly expands into a full burst of fire in the combustion chamber.

8. The electrical resistance of the cad cell located in the fire tube decreases when the light from the fire reaches the cad cell. This decrease in resistance increases the current flow through the safety relay, allowing the safety relay to operate.
9. The normally closed contacts of the safety relay open to de-energize the safety heater and prevent the safety switch from opening.
10. The heat exchanger begins to heat up, and when the temperature reaches approximately 130°F, the fan switch contacts close and the indoor blower comes on.
11. The conditioned space temperature will begin to rise and eventually reach the setpoint of the thermostat. At this point the thermostat opens the circuit to the primary control.
12. The primary control shuts off the burner motor and ignition transformer, shutting the flame down.
13. The indoor blower continues to run until the heat exchanger cools to approximately 100°F. At this point, the fan switch contacts open and the indoor blower turns off.

SERVICE TIP

One way to test a cad cell primary control is to cover the cad cell with black electrical tape to simulate a flame failure. The control should shut down the burner the same as it would during a no-flame condition.

54.6 DELAYED-ACTION SOLENOID OIL VALVES

For oil to burn cleanly, it needs to be atomized and mixed with combustion air. Incomplete combustion will result if the oil is not delivered at the correct pressure or air is not available in sufficient quantity. Ideally, when the oil burner starts, it should reach operating speed and have the correct amount of combustion air volume before oil is supplied and combustion is established. When the burner shuts off, the oil flow and combustion should shut off instantly. Both of these are accomplished by the oil valve.

A delayed-action solenoid oil valve may be installed between the oil pump and the burner firing assembly. This type of valve, although wired to be energized at the same time as the oil burner motor (wired in parallel), has a delayed opening. This allows full operation of the burner blower and pump before oil flow is allowed. A thermistor is wired in series with the oil valve coil circuit. The thermistor has a high resistance when it is cold, limiting the current to the coil on start-up. The voltage across the thermistor causes it to heat up, and its resistance drops as its temperature rises. The thermistor's reduced resistance causes an increase in the valve coil circuit. After 8–10 seconds, the current increases sufficiently to cause the magnetic coil pull to open the valve. The valve acts the same as any other type of solenoid valve on cutoff: it closes immediately upon de-energizing the valve and motor. Thus, instant cutoff of combustion occurs.

Figure 54-26 Energy efficiency sticker.

54.7 EFFICIENCY TESTING

Checking the combustion efficiency is an important part of servicing oil-fired furnaces. Figure 54-26 shows the label on all new models listing the expected efficiency. Efficiency testing is crucial for oil furnaces because there are a number of variables that may need to be adjusted in the field. As an example, the flame efficiency increases as the refractory temperature rises. The burner is operating at peak efficiency when the surface temperature of the refractory is white hot. At this point, efficiency tests can be taken. To reach these conditions, the burner should be allowed to operate for at least 5 minutes.

The general order for taking measurements is as follows:

- Stack temperature
- Draft
- Smoke
- CO_2 or O_2
- Calculate efficiency

The make, model number, and serial number of the burner are required when referring to the manufacturer's recommendations. Check the nozzle size, type, and spray angle to ensure that the unit has been checked for proper input as well as spray angle for the shape of the refractory. Figure 54-27 shows a typical completed efficiency test sheet.

To use for future reference, a test sheet should be filled out recording the conditions to which the burner has been adjusted. The following information should be included on the test sheet.

ACCURATE PRINTING CO • 104 PERIMETER ROAD • NASHUA, NH 03063

COMBUSTION EFFICIENCY TEST RESULTS

DATE [illegible]

NAME

ADDRESS 58 Woodward Rd

TOWN Milford PHONE:

	BEFORE	AFTER
SMOKE		0
CO_2		8
DRAFT		.04
NET TEMP		450
EFFICIENCY		77½%
DRAFT O.F.		.02
CO_2 O.F.		—
NOZZLE SIZE		1.00
NOZ. ANGLE		80
NOZ. PATTERN		[illegible]

OIL COMPANY, INC.

SER. TECH. Rob

DEALER COPY

Figure 54-27 Efficiency inspection tag.

Data

Make of oil burner: ______________________

Model no.: ____________ Serial no.: ____________

Nozzle: [size] (gal/hr) __________ Type: __________

Angle: ______________________

Refractory: [shape] ______________________

Operation

Over-fire draft: ______________________

Stack draft: ______________________

Heat exchanger flow resistance: ______________

CO_2%: ______________________

Net stack temperature: ______________________

Efficiency: ______________________

Smoke number: ______________________

Air temperature (supply plenum): ______________

Air temperature (return plenum): ______________

Air temperature (rise): ______________________

54.8 CARBON MONOXIDE (CO)

Carbon monoxide is a colorless, odorless, tasteless poison. It is the product of incomplete combustion. Usually it is produced by too little combustion air and a smoky fire, which is an indication of unburned fuel. Sometimes too much air can also lead to increased levels of CO. Therefore, always operate a furnace at peak efficiency with good combustion. This will result in very low levels of CO. Normal levels within a house would be about 1 to 9 ppm. Levels above 10 ppm necessitate action being taken to bring the CO levels down. At levels above 36 ppm, the occupants should go outside to fresh air and the space ventilated. Levels above 400 ppm can lead to death.

Figure 54-28 High-temperature stack thermometer.

54.9 STACK TEMPERATURE

A high-temperature thermometer, such as that shown in Figure 54-28, is inserted through a small 3/8-in hole that has been drilled through the flue pipe at the furnace outlet (Figure 54-29). The temperature of the flue products will also help to determine the efficiency of the heating unit.

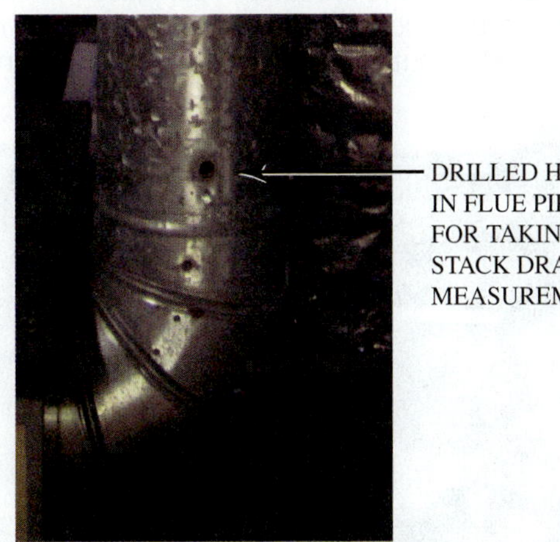

Figure 54-29 Drilled hole in the flue pipe to measure stack temperature and stack draft.

The cleaner the heat exchanger, the more efficiently it removes heat from the flue products as they pass through. If the CO_2 content is correct but the stack temperature is too high, this is usually a sign of excessive carbon deposit on the heat exchanger surfaces. To determine the net stack temperature, subtract the temperature of the air entering the burner from the stack temperature.

Stack temperatures for 60 to 79 percent efficient furnaces range from 400°F to 600°F. Furnaces with 80+ efficiencies have lower stack temperatures that range from 330°F to 450°F. High efficiency 90+ furnaces have the lowest stack temperatures—less than 125°F.

54.10 FURNACE DRAFT MEASUREMENTS

Draft is created by the pressure difference between the top and bottom of the chimney. This is affected by the temperature of the exhaust gases, the temperature of the outside air, barometric pressure, and humidity. A negative pressure creates more suction to pull in air, and therefore the lower the pressure, the higher the draft. If the draft is too low, then less air is drawn into the burner and the fire can become smoky. If the draft is too high, then too much air is drawn into the burner and the furnace efficiency will drop.

Draft is measured in inches of water column (in wc) because the pressure is too small for a regular pressure gauge to measure. When reading draft measurements, it must be understood that a more negative reading indicates a higher draft. As an example, a draft of −0.5 in wc would be a higher draft than −.02 in wc.

Over-Fire Draft

For this measurement, a draft gauge is required, such as those pictured in Figure 54-30 and Figure 54-31. Insert the test pipe for the gauge through the ¼-in hole in the observation port over the burner until the end of the pipe is beyond the inner edge of the combustion chamber. The inspection port often has a threaded bolt that can be removed for insertion of the probe pipe (Figure 54-32). To accomplish this, it is sometimes necessary to substitute a longer piece of ¼-in copper tubing for the gauge probe pipe.

A negative pressure must be maintained on the heat exchanger to prevent the products of combustion from being forced into the occupied area. Not only do they carry free carbon or soot, but they also contain a high percentage of CO. This is especially true at start-up. An over-fire draft of −.01 to −.02 in wc is usually required for proper operation.

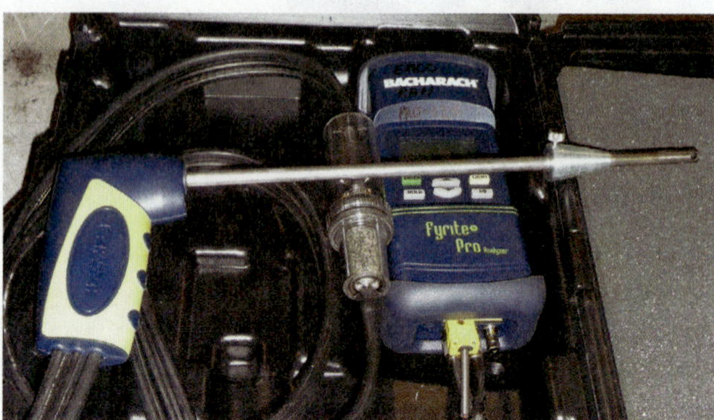

Figure 54-30 Electronic combustion analyzer.

Figure 54-31 Stack draft gauge.

CAUTION

A carbon monoxide (CO) detector is a good recommendation for the homeowner to purchase and install near any combustion equipment, whether it is gas or oil. Carbon monoxide is often referred to as the silent killer because it can go unnoticed in a home for a long period of time if the home is not protected with a carbon monoxide monitor.

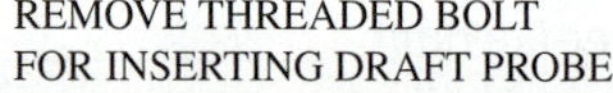

Figure 54-32 Remove threaded bolt in center of inspection port cover to insert probe.

If the over-fire draft is higher than the manufacturer's specifications (generally above –.02 in wc), the draft can be adjusted using the weight on the barometric damper. The draft will be decreased when the barometric damper opens. This is because the cool air mixes with the stack gases to reduce their temperature. Since cooler air rises slower than hotter air, the draft is reduced. When the draft drops below the damper setting, the counterweight closes the damper.

Some units can operate with a positive pressure of over fire, and it is important to be able to identify this type and make adjustments according to the manufacturer's specifications.

Stack Draft

The draft in the furnace stack will be higher than the over-fire draft. This measurement should be taken through the same opening in the flue pipe where the stack temperature was taken. The reading should be in the range of –.03 to –.06 in wc when the over-fire draft is –.01 to –.02 in wc.

Some manufacturers specify stack draft for a setting. This means that the flow resistance of the heat exchanger must be taken into account for burner operation. By measuring both the over-fire draft and stack draft on the new unit, the design draft resistance can be determined. To produce a negative pressure of a given amount over the fire, the stack draft must be a greater negative amount. The difference between the over-fire draft and the stack draft will be the heat exchanger draft flow resistance. For example, if it is necessary to have a stack draft of –.07 in wc to produce an over-fire draft of –.02 in wc, the heat exchanger resistance would be 0.05 in wc. When checking the unit performance after a period of operation, if the heat exchanger flow resistance has increased, then it is necessary to clean the soot from the heat exchanger flue passages.

54.11 SMOKE TESTING

A smoke tester, such as the one shown in Figure 54-33, uses a pump piston to draw flue gas products through a filter inserted in the head of the pump between the sample tube connection and the piston body. The sample tube can be inserted into the same hole in the flue pipe where the draft measurement was taken. This particular instrument requires ten slow strokes of the piston to draw the required amount of flue gas products through the filter. Some technicians want to make one more stroke on the piston to be sure they get a good test. This does not make a good test; it invalidates the test. Make only the exact number of strokes with the piston pump as specified by the manufacturer.

The filtered sample is compared to numbered rings on the test card (Figure 54-34). The smoke spot should never be higher than the No. 1 ring. If higher, the burner is receiving insufficient air and is not burning the carbon sufficiently. This will quickly produce carbon deposits and plug heat exchanger passages. On older units, if the sample is less than the No.1 ring, too much air is being allowed to the burner. The CO_2 content is too low, and too much heat is being forced out of the heat exchanger. This results in considerably lower efficiency and a much higher operating cost. On newer units, the smoke spot reading should be as low as possible.

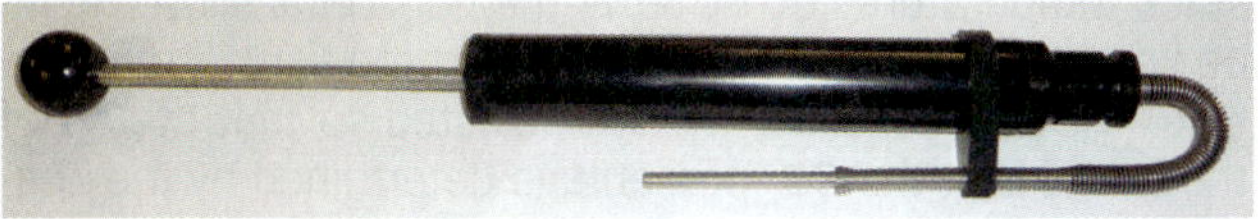

Figure 54-33 Smoke tester.

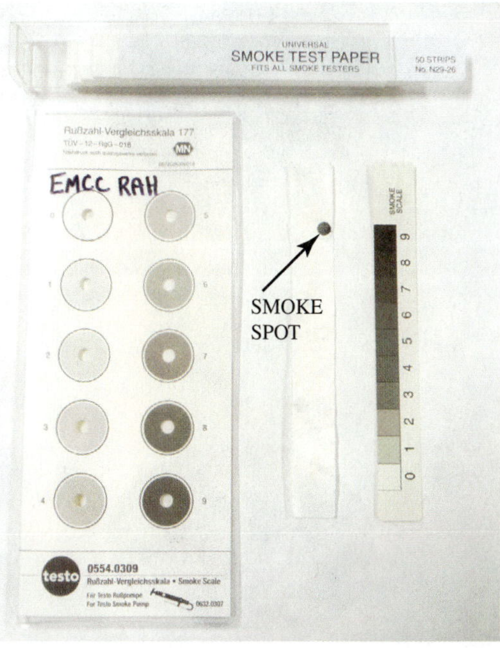

Figure 54-34 Comparing smoke spot to the chart.

SERVICE TIP

Do not let the tip of the tube touch the far side of the flue pipe, as you may pull soot off the pipe. Be sure to clean and inspect your smoke pump regularly.

SERVICE TIP

Some furnaces are located in poorly lit basement areas. It may be necessary for you to use a drop light or flashlight to determine the difference between the shades of gray of the smoke spot test rings.

54.12 CO_2 RANGES

The theoretically perfect mix of fuel and air to burn completely with no fuel or air remaining is called stoichiometric combustion. The CO_2 percentage in the flue gas will be 15.5 percent at this point. The CO_2 percentage will drop from 15.5 percent if there is not enough air or if there is air remaining. For safety reasons, the correct air adjustment

Figure 54-35 Carbon dioxide (CO_2) analyzer.

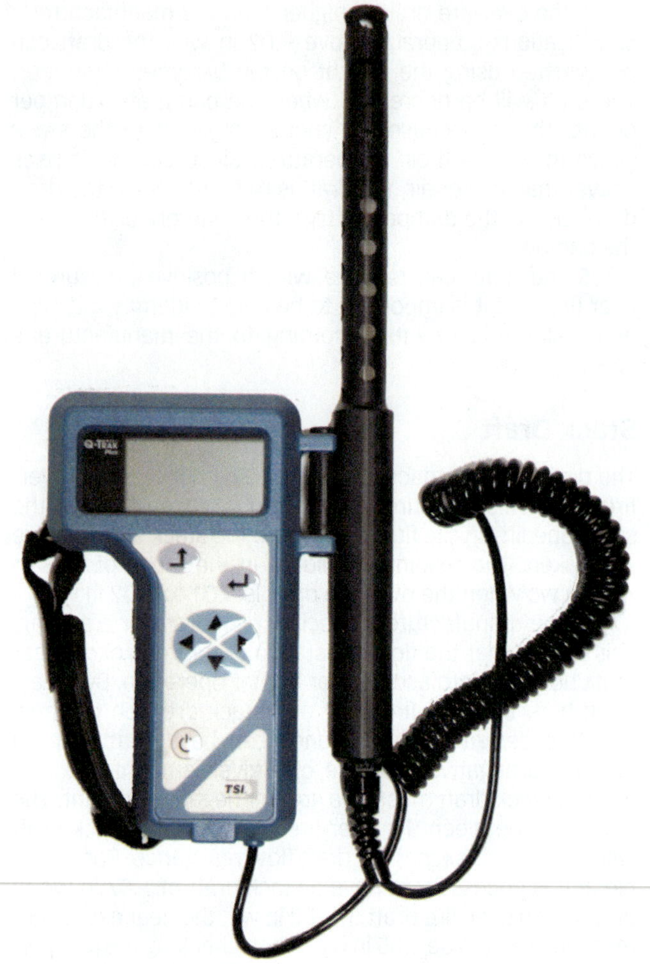

Figure 54-36 Electronic CO_2 tester.

will always result in some amount of excess air. Therefore, the CO_2 percentage should be less than 15.5 percent. How much lower depends upon the burner design. In general, more efficient burners should be adjusted to produce higher CO_2 percentages than less efficient burners.

To obtain a good sample of flue gas, the CO_2 sample is taken at least 6 in upstream of the barometric draft damper and a distance twice the diameter of the flue pipe from an elbow. This will help to avoid the possibility of outside air mixing with the flue gas sample. Again, this may mean a longer sampling tube on the analyzer. Use a CO_2 analyzer such as the ones pictured in Figures 54-35 and 54-36, following the manufacturer's instructions.

Common operating CO_2 percentages for different types of burners are as follows:

- **Old-style gun burners** These burners have no special air-handling parts other than an end cone and a stabilizer. A CO_2 reading of 7 to 9 percent should be obtained unless a CO_2 reading in this range results in more than a No. 2 smoke. If so, the CO_2 reading should be reduced until the smoke test results in below a No. 2 smoke.
- **Newer-style gun burners** Burners with special air-handling parts should be set in the range of 9 to 11 percent. This should result in less than a No. 2 smoke (closer to a No. 0 smoke).
- **Flame-retention gun burners** These burners should be set in the range of 10 to 12.5 percent with a No. 0 on the smoke test results.
- **Rotary burners** These burners should be set the same as old-style gun burners.
- **Rotary wall flame burners** Higher CO_2 settings up to 13.5 percent are available with these older burners, but the maximum of a No. 2 on the smoke test is required.

54.13 CALCULATION OF EFFICIENCY

Three pieces of information must be known to calculate the combustion efficiency:

- The ambient temperature around the furnace
- The stack temperature
- The CO_2 or O_2 percentage of the flue gas

The ambient temperature is subtracted from the stack temperature to get the net stack temperature. This is the temperature that is added by the combustion process. Either CO_2 or O_2 percentage may be used; it is just necessary to have the correct slide rule. It is also important that the card in the slide rule be for the type of fuel being used, in this case, oil. Figure 54-37 shows a typical combustion efficiency calculator. The horizontal slider is set to the net stack temperature. Since the calculator only displays net stack temperatures every 50°F, the horizontal slide should be lined up to the temperature nearest the actual temperature. The vertical slider is then adjusted so that the tip of the arrow is at the CO_2 percentage determined by the analyzer test. The efficiency of the unit is then shown in the arrow in the vertical slider.

Figure 54-37 Efficiency calculator.

An example will illustrate how this works. An oil furnace is operating with the following characteristics:

- Stack temperature of 500°F
- Ambient temperature of 55°F
- CO_2 of 10 percent

First calculate the net stack temperature: 500°F – 55°F = 445°F

Line up the vertical window in the horizontal slide under 450°F (Figure 54-38) (this is the calculator's closest value to 445°F).

Line up the vertical slider so that the arrow points to 10 percent (Figure 54-39).

Read the efficiency in the window next to the arrow as 80.5 percent.

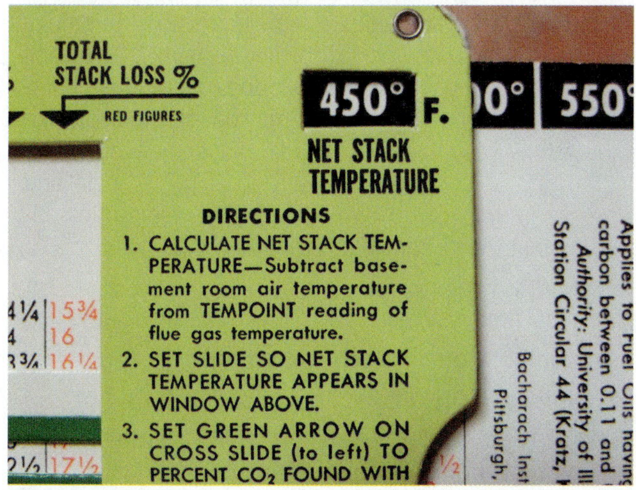

Figure 54-38 Efficiency calculator lined up on 450°F.

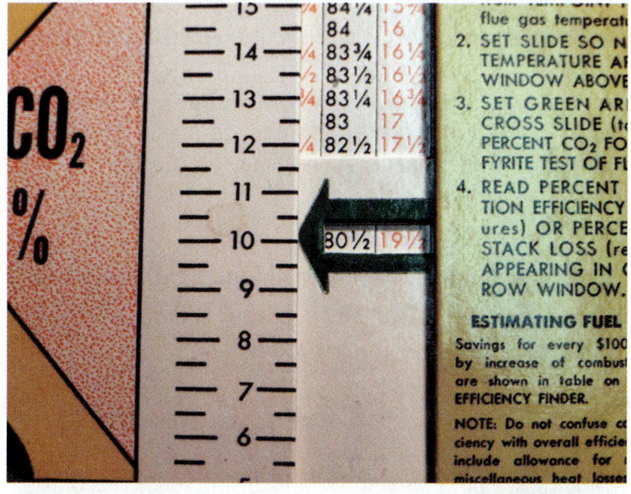

Figure 54-39 Efficiency calculator lined up on 10 percent.

The efficiency results should be within the smoke test ratings as described above or the manufacturer's specifications. If not, the unit should be examined for the cause of the difference. Usually, a cleaning of the heat exchanger is required.

UNIT 54—SUMMARY

Oil systems can burn cleanly if properly maintained. The fuel-line filter and nozzle should be replaced annually. Operational characteristics that should be tested include the oil pump inlet vacuum, the oil pump outlet pressure, the stack temperature, and the CO_2 flue gas percentage. The combustion airflow should be adjusted to obtain a smoke spot number of less than 2 while still achieving the correct CO_2 range in the flue gas. The money spent on keeping the system operating correctly will be repaid in lower operating costs and longer equipment life.

WORK ORDERS

Service Ticket 5401

Customer Request: New Furnace Start-up

Equipment Type: Oil Furnace Using a Two-Pipe Fuel Delivery System

Furnace Is Above the Oil Tank

A technician attempts to purge a new two-pipe system by operating the system. After pushing the reset button twice, the technician inspects the combustion chamber for oil and finds none. Next, the technician installs a vacuum gauge on the suction side of the oil pump. With the pump running, the vacuum is only about 1 in Hg. The technician inspects the line for leaks and finds that the suction line connection is loose at the tank. The connection is tightened and the vacuum rechecked. This time the vacuum reads 10 in Hg. After pushing the reset button once, the furnace operates and fires briefly, but it goes out again. Before pushing the reset button a second time, the technician must check the combustion chamber for a buildup of oil. The inspection port is opened and the combustion chamber is inspected. Since there is no buildup of oil, the technician pushes the reset button a second time and allows the furnace to start one more time. This time, after the furnace fires, it stays going.

Service Ticket 5402

Customer Request: Annual Maintenance of Equipment

Equipment Type: Oil-Fired Furnace with a Flame-Retention Burner

A technician is called to perform an annual maintenance service on a flame-retention burner. During the efficiency testing portion of the maintenance, the CO_2 reading is 9 percent and the smoke spot reading is 0. The technician remembers that while 9 percent is fine for an older-style burner, it is too low for a flame-retention burner. Leaving it this way would result in inefficient operation. After reducing the combustion air, the CO_2 increases to 12 percent and the smoke spot remains 0.

Service Ticket 5403

Customer Complaint: High Operating Cost

Equipment Type: Oil Furnace with Older-Style Cast Iron Burner

A technician is asked to look at an older furnace whose owner is complaining about the operating cost. The technician operates the furnace and takes CO_2 percentage, stack temperature, over-fire draft, and stack draft measurements. The combustion efficiency seems appropriate for a furnace its age, but the over-fire draft is low, only −.01 in wc, while the stack draft is fine at −.04 in wc. The technician pulls the burner out and inspects the heat exchanger. The heat exchanger is covered with a thin layer of fine soot. The technician explains to the homeowner that the soot is acting like insulation, reducing the efficiency of the furnace by keeping the heat from transferring to the air flowing over the heat exchanger. After cleaning the heat exchanger, the over-fire draft improves to −.02 in wc.

Service Ticket 5404

Customer Complaint: High Operating Cost

Equipment Type: Oil Furnace with Newer-Style Gun Burner

A technician is called to check a furnace that is operating inefficiently. The customer says that a previous technician had done an annual maintenance service on the burner but left no paperwork. The furnace is operating with a low smoke spot of 0 to 1, but the CO_2 percentage is only 7 percent, indicating too much combustion air. The owner asks the technician about the CO_2 analyzer and remarks that the other technician did not have one. After reducing the combustion air, the CO_2 percentage increases to 10 percent and the smoke spot still reads between 0 and 1. The technician explains that it is possible to have too much combustion air, reducing efficiency.

Service Ticket 5405

Customer Complaint: Poor Heat, Smoky Vent Gases

Equipment Type: Newer Gun-Style-Burner Oil Furnace

A technician is called to do an annual maintenance on an oil-fired furnace. The customer mentions that the furnace is not heating as well as in previous years and that the vent gases are smoky. The technician inspects the heat exchanger and finds that it needs cleaning. The electrodes appear as if oil has been hitting them. The technician cleans the heat exchanger, replaces the nozzle, and cleans and adjusts the electrodes. The smoky flue gas clears up. The technician explains that the oil mist should not hit anything; if it does, incomplete combustion and smoke will result. The soot created acts like insulation, reducing the furnace efficiency even more. Changing the nozzle ensures a good spray angle and proper atomization, while adjusting the electrodes ensures that the oil will not hit them.

Service Ticket 5406

Customer Request: Replacement Unit Start-up

Equipment Type: Flame-Retention Burner

A new technician was sent to perform a system start-up on a new oil furnace that has just been installed to replace an older furnace. While taking over-fire draft readings, the technician notices that the readings are lower than the manufacturer's recommendations and that the barometric damper is staying wide open most of the time. After adjusting the weight on the barometric damper so that it stays shut more, the over-fire draft increases to the manufacturer's recommended −.02 in wc draft.

UNIT 54—REVIEW QUESTIONS

1. How does the refractory temperature affect the combustion efficiency?
2. What are normal on and off temperature ranges for the settings on the fan control?
3. Describe the procedure for purging a single-pipe gravity feed system.
4. Describe the procedure for purging a two-pipe system.
5. What pieces of information are necessary to determine combustion efficiency?
6. What can cause the heat exchanger flow resistance to increase after several years of operation?
7. Why must a negative pressure be maintained on the heat exchanger?
8. How is the door switch wired into the system in Figure 54-25?
9. What opens the contact on the safety switch in Figure 54-25?
10. Why does soot accumulation on the inside of the heat exchanger reduce system efficiency?
11. How does the cad cell operate to secure the burner in the event of a flameout?
12. What smoke spot number represents correct combustion?
13. How is net stack temperature calculated?
14. Discuss the relationship between the over-fire draft and the stack draft.
15. What is the CO_2 percentage for a perfect stoichiometric mix of oil and air?
16. Why do we not try to adjust the CO_2 percentage to the perfect stoichiometric percentage?
17. Why does excess combustion air reduce combustion efficiency?
18. How should the output of an oil furnace ignition transformer be tested?
19. What components on an oil furnace should be changed annually?
20. What angle nozzle is usually used with a round combustion chamber?

UNIT 55

Residential Oil Heating Installation

OBJECTIVES

After completing this unit, you will be able to:

1. calculate the minimum required combustion air openings.
2. discuss common applications for different oil furnace configurations.
3. discuss oil tank installation.
4. discuss factors in oil piping.
5. calculate the equivalent length of a chimney connector.

55.1 INTRODUCTION

Oil-fired furnaces provide clean, reliable heat when properly sized and configured for the application. However, they will not be clean and efficient unless they are installed correctly and maintained regularly. Proper planning for the installation is very important. The furnace should meet the heating requirements of the space. It should be located in a convenient area for the customer and for the technician who will be servicing it. As an example, furnaces installed in cramped crawlspaces are difficult to access and work on. Oil tank location also has to be considered when planning where to locate the furnace as well as flue pipe placement. Oil-fired furnaces can be very heavy, so help may be required in moving the furnace on the initial day of delivery.

55.2 EQUIPMENT SELECTION

Factors to be considered when selecting an oil furnace include the following:

- Required heating capacity
- Desired efficiency
- Systems designed for both heating and cooling
- Furnace location

Heating Capacity

The required heating capacity should be determined by using a recognized load calculation method, such as outlined in the ACCA Manual J. The furnace heating capacity should be the lowest standard capacity that is at least equal to the calculated required capacity. Since furnaces are only available in a few sizes, some oversizing is acceptable. Gross oversizing should be avoided because furnaces operate most efficiently with longer running cycles than with several short on and off cycles. Note that the heating capacity is listed as the output or bonnet capacity, not the firing rate.

A range of capacities can be achieved with the same burner by changing the nozzle. Table 55-1 shows the capacities available with one model furnace using the same burner and changing the nozzle size. Always refer to the manufacturer's specifications when selecting different nozzle sizes.

Table 55-1 Firing Capacity with Beckett AFG-F3 Burner

Firing Rate (BTU/hr)	Firing Rate (gph)	Nozzle
70,000	0.5	0.50 gph 70-W
91,000	0.65	0.65 gph 70-W
105,000	0.75	0.75 gph 70-B
119,000	0.85	0.85 gph 70-B
140,000	1.00	1.00 gph 70-W

Efficiency

Oil furnaces are now available in efficiencies as high as 95 percent AFUE. The efficiencies and capacities available may be different from one configuration to another. The efficiency can affect the installation. If a condensing-type furnace is chosen, the venting is different than for noncondensing models. Also, condensing furnaces require drains; noncondensing furnaces do not.

Systems Designed for Both Heating and Cooling

If an air conditioner is to be installed with the oil furnace, the airflow capacity of the furnace must be adequate for the size of the air conditioner. Most manufacturers have furnaces that are considered heating-only furnaces. These furnaces do not have blower motors capable of moving the higher airflow required for air conditioning. If the furnace will be used to supply the airflow for an air-conditioning system, make certain that the furnace airflow capacity meets the requirements of the air-conditioning system.

Table 55-2 Furnace Application

Configuration	Typical Application
Lowboy	Basement, particularly basements with low headroom
Upflow	Basement with good clearance; utility room or closet with duct in attic
Downflow	Utility room or closet with duct in crawlspace
Horizontal	Attic or crawlspace with duct in same space as furnace
Multiposition	Usually any of the above applications

Location

The proposed furnace location will determine the furnace configuration. If the furnace is to be installed in an attic, a lowboy would not work. Table 55-2 summarizes the available oil furnace configurations and their typical applications. Furnaces should not be installed under stairways. Oil furnaces can be heavy and are often mounted on cement blocks rather than directly on the basement floor (Figure 55-1). When installing furnaces in the attic, make certain that the structural components of the house can support the furnace. Generally speaking, the furnace return air must not be taken from the same room that the furnace is located in.

CAUTION

Many oil-fired furnaces are located in open basement areas where they are subject to having items stored on, against, or near them. Such items constitute a major fire hazard and must be removed before you light the furnace. Caution the homeowner or resident against such storage practices.

Figure 55-1 Furnace installed in basement on top of flat cement blocks.

55.3 CLEARANCE

Furnace clearances should be maintained according to the manufacturer's specifications. Cabinet temperatures and flue pipe temperatures run warmer for oil-burning equipment, so clearances from combustible material should be adjusted accordingly. One-inch clearance is common for the sides and rear of the cabinet, compared to zero clearance for many gas units. Flue pipe clearance of 9 in or more is needed for oil-fired furnaces, whereas only 6 in of clearance is needed for natural gas.

Front clearance is extremely important, as space is needed to remove the burner assembly. Remember to leave sufficient clearance around any panel that must be removed for service. This is usually at the front of the furnace. Although many codes specify a 24-in service clearance, do yourself a favor and leave at least 36 in.

Floor bases over combustible surfaces are generally increased for oil as compared to gas. For example, in horizontal oil furnaces installed in attics, the installer must pay particular attention to recommendations related to adjoining surfaces.

CODES AND STANDARDS

Clearance is an important part of the compliance and inspection procedures required by Underwriters Laboratories (UL) for securing their seal of approval. Oil furnace installations should follow the manufacturer's instructions and all state and local codes. NFPA 31, Standard for the Installation of Oil Burning Equipment, is an excellent resource and is referenced in many state codes.

55.4 COMBUSTION AIR

Oil-fired furnaces must have an ample supply of makeup air for combustion, and the methods of introducing makeup air for a gas furnace will also be adequate for oil equipment. Combustion air should be provided from the outside near the ceiling and floor. If the openings go directly outside, each opening should be at least 1 in^2 for every 5,000 BTU input. If the openings pass through vertical ducts, each opening should be at least 1 in^2 for every 4,000 BTU input. If the openings pass through horizontal ducts, each opening should be at least 1 in^2 for every 2,000 BTU input.

55.5 FLUE VENTING

The flue venting system will often be dictated by the building design and structure. Many homes are built with masonry chimneys that because of their weight will extend from the basement or slab, up through the roof. Lightweight prefabricated metal chimneys are often installed as acceptable alternatives. In structures where no chimney is desired, power venting and direct venting systems can be installed.

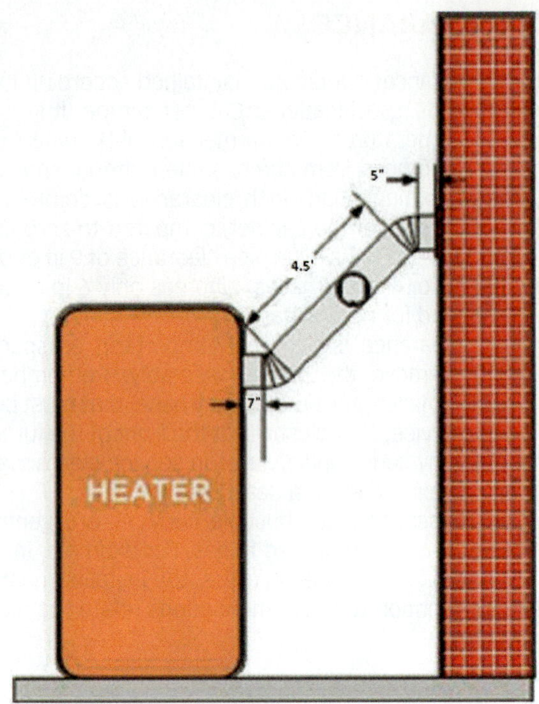

Figure 55-2 Oil furnace connected to masonry chimney approximate dimensions. *(From* Advanced Residential Oilburners *by George Lanthier, © 2007, by permission of George Lanthier.)*

Masonry Chimneys

The most common venting practice for oil furnaces is to use a metal chimney connector and a masonry chimney (Figure 55-2). The connector is the pipe running from the furnace to the chimney. It should be kept as short as possible, with as few turns as possible, and it should have a rise of at least 0.25 in per foot of run.

In basements and equipment rooms, the connector may be either single-wall metal or type L manufactured venting material. In attics and crawlspaces, it must be type L. The connector should not pass through any floor, ceiling, or combustible walls, and the entire length of the connector should be accessible for inspection and replacement. Masonry chimneys used for oil-fired furnaces should be constructed as specified in the National Building Code of the National Board of Fire Underwriters.

The chimney should be sized according to NFPA 211 and all local codes. The cross-sectional area of the chimney should not be less than the cross-sectional area of the connector. If a metal chimney liner is used, it should comply with UL 64. Oil furnaces should not be common vented with wood-burning appliances, fireplaces, or gas-burning appliances. They may be common vented with other oil-fired appliances. If more than one oil-fired appliance is vented into the same chimney, the cross-sectional area of the chimney must be at least equal to the combined cross-sectional areas of the connectors.

Table 55-3 Minimum Gauge for Single-Wall Chimney Connector

Connector Diameter	Minimum Gauge
Less than 6 in	26
6–10 in	24

Masonry chimney connector size If a single-wall connector is used, the metal needs to be a much heavier gauge than the galvanized metal normally used for ductwork. Table 55-3 shows the minimum gauge for different connector sizes.

The connector must not be smaller in cross-sectional area than the flue collar on the furnace. The chimney connector should have a total equivalent length of no more than 75 percent of the chimney height. The total equivalent length is based on the actual length of the connector plus the equivalent length of any fittings. Table 55-4 shows the equivalent lengths for several common fittings.

The chimney height is based on the height measured from where the connector enters the chimney. For example, a 6-in connector that is 8 ft in actual length with two 90° ells has a total equivalent length of 8 + 11 + 11 = 30 ft. This must be no more than 75 percent of the chimney height above where the connector enters. Thus the chimney above the connector entrance must be 30 ÷ 0.75 = 40 ft.

Unfortunately, this is higher than many residential chimneys. Replacing the 90° ells with 45° ells reduces the total equivalent length to 8 + 5 + 5 = 18 ft. By using the 45° ells, the chimney height above the connector entrance need only be 18 ÷ 0.75 = 24 ft.

Prefabricated Metal Chimneys

Prefabricated lightweight metal chimneys are rated for all flues, Class A and Class B. Class A flues are used for solid and liquid fuels, while Class B flues are made specifically for gas-fired equipment. Oil furnaces may also be vented using Type L vent, which is specifically made for oil furnaces.

The entire vent system can be constructed of type L venting material, providing the furnace manufacturer approves

Table 55-4 Connector Fitting Equivalent Length in Feet

Fitting	3 in	4 in	5 in	6 in	7 in	8 in	9 in	10 in
Tee	19 ft	25 ft	31 ft	38 ft	44 ft	50 ft	56 ft	63 ft
Wye	10 ft	13 ft	16 ft	20 ft	23 ft	26 ft	29 ft	32 ft
90° ell	5 ft	7 ft	9 ft	11 ft	12 ft	14 ft	16 ft	18 ft
45° ell	3 ft	4 ft	4 ft	5 ft	6 ft	7 ft	8 ft	9 ft

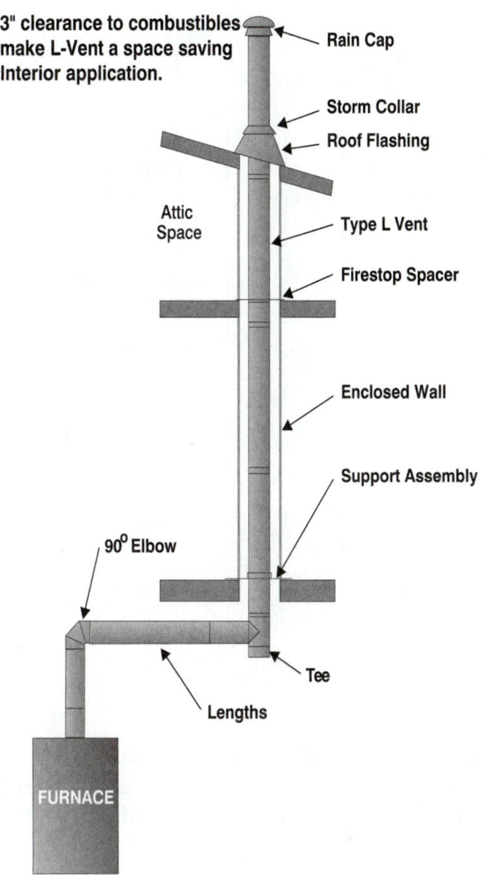

Figure 55-3 Type L vent *(Courtesy of Selkirk Corp.)*

the furnace for use with type L venting systems. Type L venting material should not be used with appliances with vent gas temperatures exceeding 550°F. Figure 55-3 shows a type L venting system for connection to an oil-fired furnace.

The roof dimension of the vent or chimney is the same for both gas and oil. Since oil-fired furnaces operate on positive pressure from the burner blower, it is important to have a chimney that will develop a minimum draft of −.01 to −.02 in wc as measured at the burner flame-inspection port. However, there are some newer residential units that can be designed to operate with a positive over-fire draft.

Power Venting

The advantage of power venting is that no chimney is required, but it can cause noise and vibration in the building. Power venting systems are designed with a blower located in the flue pipe either outdoors or indoors just before the outside wall (Figure 55-4). Power venting systems require that the furnace primary control allows for a delayed burner firing at start-up. This allows the vent blower to come on and produce a draft before the burner firing sequence begins. The blower must also continue to run for a period of time after the burner shuts down to ensure venting of all the combustion gases.

Figure 55-4 Power venting.

Direct Venting

Similar to power venting, direct venting systems require no chimney. However, direct vent systems do not use a blower but instead rely on the power of the burner fan to push the combustion gases out of the building. They also use a sealed combustion chamber with air brought in from the outside. A disadvantage to this type of system is that the pressure inside the furnace vent is higher than the pressure in the building. Because of this, the vent system needs to be properly sealed to prevent leakage of combustion gases into the building.

Barometric Damper

Standard efficiency oil furnaces require a barometric damper, in the chimney connector. Some sealed combustion furnaces do not use a barometric damper. Condensing furnaces also do not use a barometric damper.

The barometric damper should be installed in the chimney connector as close to the furnace as possible and at least 18 in away from any combustible wall or ceiling. Older units may have a stack safety switch, and if that is the case, the barometric damper should be at least 18 in downstream from the stack switch, as in Figure 55-5.

The barometric damper should not be installed on the end of a bull-headed tee but on the branch of a tee that is in line with the connector. Make sure the damper is oriented the way it was designed. A damper that is designed to be installed with the pivots horizontal to the ground will not work if it is installed on its side (Figure 55-6). The draft in the connector will need to be measured and the barometric damper adjusted during the initial start-up.

55.6 SYSTEM WIRING

The vast majority of oil furnaces and boilers operate on 115 V circuits. The wire conductors must be sized to carry the maximum circuit amperage of the furnace or boiler.

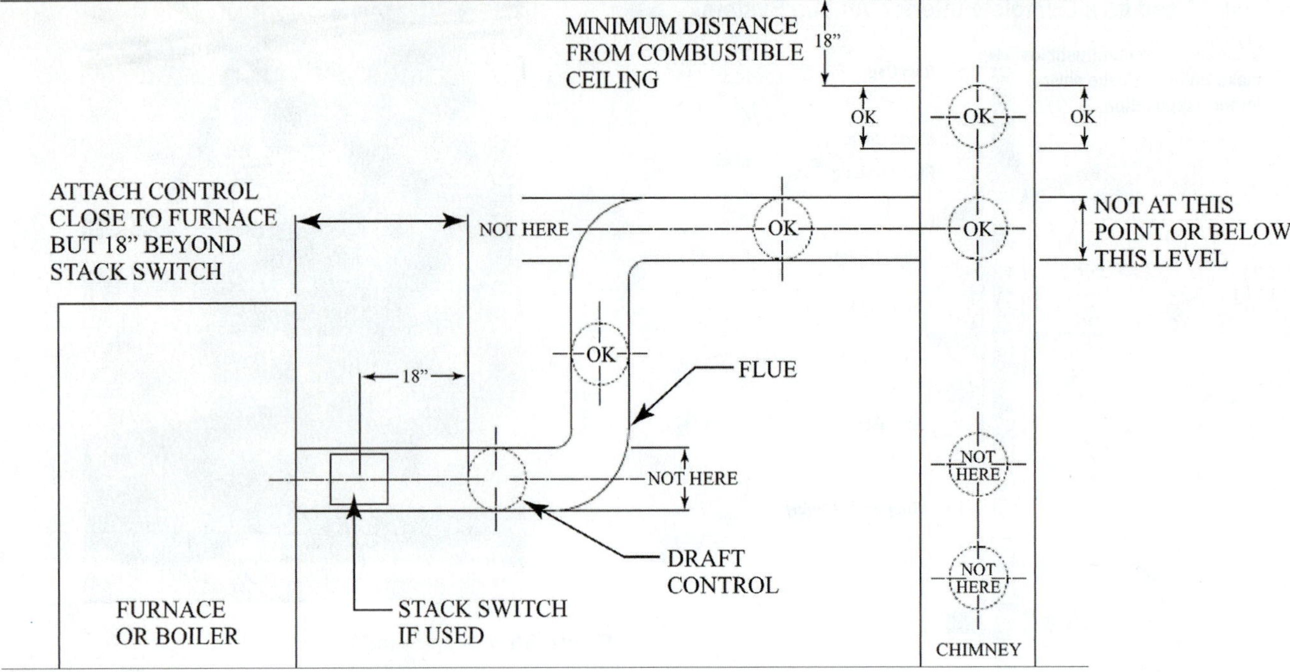

Figure 55-5 Barometric damper location. *(Courtesy of Field Controls)*

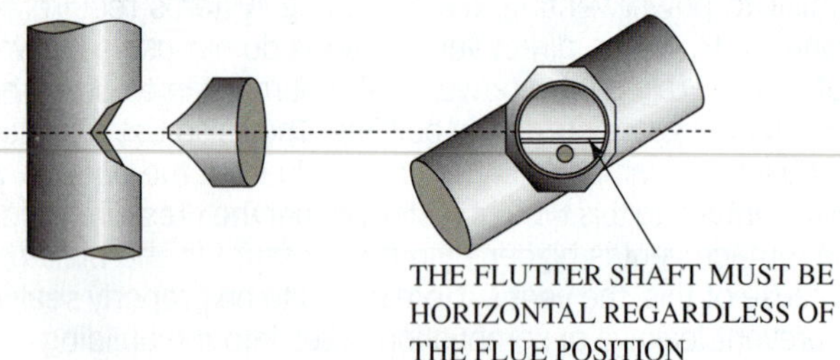

Figure 55-6 Barometric damper.

The manufacturer's instructions typically specify the correct wire size (Table 55-5).

Three wires are typically connected for the power wiring: a hot wire, a neutral wire, and a ground wire. The ground wire is connected to the ground lug on the equipment cabinet. It is important that the polarity of the power wiring is connected correctly. The ground is important for safety. The connections to the hot wire and neutral must not be swapped. Some units with electronic control boards will not operate if the hot and neutral are swapped. Typically, the hot connection is black and the neutral is white.

SAFETY TIP

Many older furnaces were originally wired without a bonding safety ground to the cabinet. When replacing these furnaces, you should always run a new power wire with a hot, neutral, and ground so that the furnace can be properly grounded. Leaving the furnace cabinet ungrounded is dangerous.

Control Wiring

The control wiring for a heat-only application consists of two wires. Typically the control wires connect to the two T

Table 55-5 Wire Sizing Guide

Unit Size	Volts–Hertz–Phase	Operating Voltage Range Max*	Min*	Max Unit Amps	Min Wire Gauge	Max Wire Length (ft)†	Max Fuse or Circuit Breaker Amps‡
105-12	115–60–1	132	104	12.2	14	26	15
120-20	115–60–1	132	104	15.7	14	26	20

*Permissible limits of voltage range at which unit will operate satisfactorily

†Length shown is as measured one way along with path between unit and service panel for maximum 2 percent voltage drop.

‡Time-delay fuse is recommended.

terminals on the primary control and to the R and W terminals on the thermostat.

The control wiring for both a heating and cooling application consists of five wires. Two wires are used for heat and three more for the cooling and fan operation. Typically, separate transformers are used for heating and cooling because the primary control has its own transformer.

SERVICE TIP

Most heating and cooling thermostats have connections for two 24 V inputs. These are normally labeled RH (red for heating) and RC (red for cooling). Frequently, the thermostats come with a jumper already installed between the two terminals, because most systems only have one transformer. Oil heating systems with air conditioning typically have a separate cooling transformer, so the jumper should be removed.

Wireless Thermostats

The thermostat should be located about 5 ft from the floor on an inside wall where there is good natural circulation. Running thermostat wires for a newly constructed building is generally fairly simple before the drywall is hung. It will be more difficult running thermostat wires once the interior finish work of the building is completed. Some heating systems also have multiple zones that require multiple thermostats located in different rooms or different stories of the building. This involves running additional wires.

Wireless thermostats are now available that eliminate the need for running wires to individual zones. The control unit is connected to the furnace control wiring and to the wireless transmitter. The transmitter responds to wireless signals from the individual zone thermostats, which are then fed back to the control unit. An example of a system using two zone thermostats to control damper position is shown in Figure 55-7.

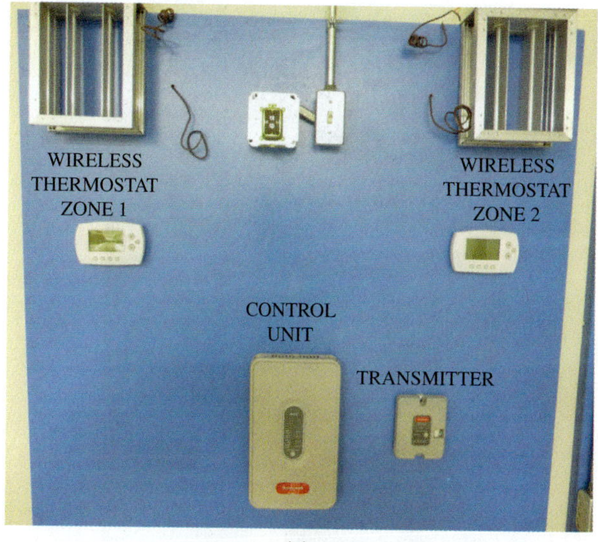

(a)

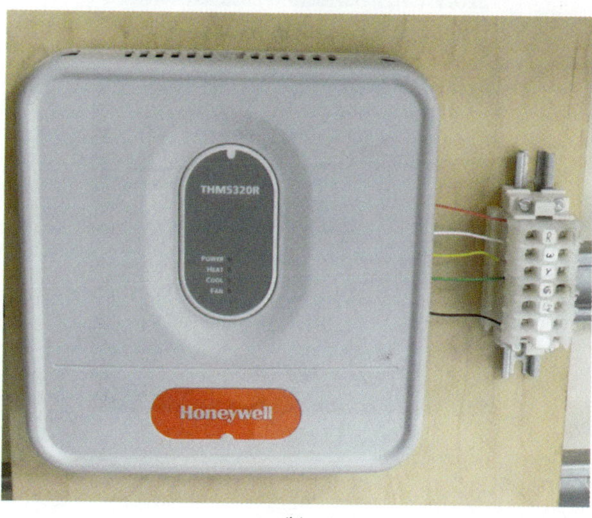

(b)

Figure 55-7 (a) Two-zone wireless thermostat system; (b) wireless transmitter.

TECH TIP

When the need arises to rewire a thermostat, use the old thermostat wiring to help pull the new wire through the wall. Attach the new wire securely to the old wire on one free end. Then pull slowly from the other end until the new wire comes through.

55.7 OIL STORAGE TANK TYPES

Oil storage tanks for residential furnaces typically hold anywhere from 100 to 330 gallons. The most common storage tank is typically made of steel and holds 275 gallons of fuel oil (Figure 55-8). Smaller tanks are used for mobile homes or in locations where a larger tank cannot be easily installed. Some tanks are coated on the outside with polyethylene to reduce external corrosion. Double-wall tanks are designed

Figure 55-8 Metal 275-gallon oil storage tank.

(a)

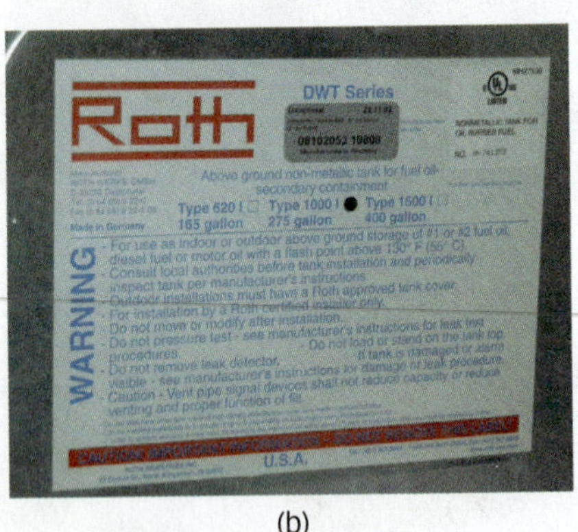

(b)

Figure 55-9 Double-wall oil storage tank.

(a)

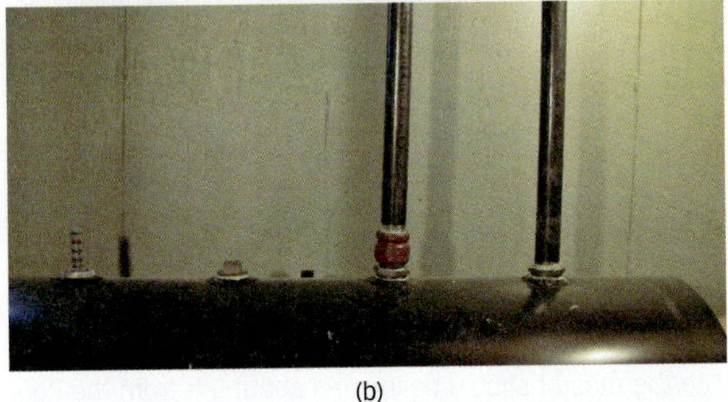

(b)

Figure 55-10 (a) Outside oil fill and vent pipes; (b) inside oil fill and vent pipes connected to the oil storage tank.

to prevent spillage if the inner tank leaks. These can be made of fiberglass, while others have an inner tank made of polyethylene with an outer steel tank surrounding it. A double-walled 275-gallon capacity with an inner nonmetallic tank and a galvanized steel exterior is shown in Figure 55-9.

The tank should be sized properly for the customer. Undersized tanks will need to be filled too frequently. Oversized tanks are subject to excessive condensation and poor fuel quality due to long storage periods. A properly sized tank should supply about one-third of the expected annual fuel consumption. As an example, a 275-gallon tank filled three times a year would be an appropriate size for an annual consumption of 825 gallons.

Residential fuel oil storage tanks are nonpressurized and vented to the air. There are two pipes connected to the top of the oil storage tank. One pipe is used for filling the tank with oil, and the other pipe is a vent. Figure 55-10 shows the typical arrangement of these pipes outside the residence. As the oil fills the tank, air leaves through the vent.

The fill pipe has a cap that is clearly marked, and it should contain an insect screen. The vent pipe cap is designed to whistle while the tank is being filled and air is rushing out through it (Figure 55-11). This whistling will stop

(a)

(b)

Figure 55-11 (a) Top of vent cap; (b) underside of vent cap with holes arranged for whistling.

Figure 55-12 Oil storage tank level indicator.

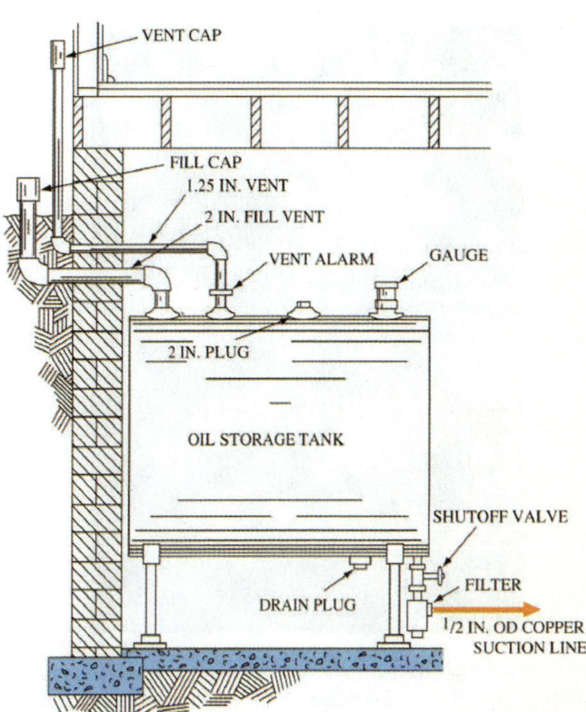

Figure 55-13 Oil storage tank located inside the basement.

sounding when the level of oil in the tank rises up to the full level in the tank. The tank will also have a level indicator mounted on top (Figure 55-12).

TECH TIP

When planning an oil-fired furnace installation, the placement of the oil fill line must be considered. It should be located conveniently for the oil delivery truck to access. However, do not allow the piping to pass under stairs or entries.

55.8 OIL STORAGE TANK LOCATIONS

Because of environmental concerns, many local and state governments have outlawed the installation of oil storage tanks below ground. Check the local building codes to see whether it is possible to install a tank underground. In some localities, existing underground tanks are not allowed to be replaced. Regardless of location, the installation of the fuel oil tank and connecting piping must conform to National Fire Protection Association (NFPA) standard 31 and to local code requirements.

Inside Tanks

Typically, most residential storage tanks are located inside, often in the basement (Figure 55-13). This is preferred because the oil is warmer and the tank is subject to less external corrosion. The inside temperature will be steady, leading to less condensation in the tank. Leaks will be identified quickly due to the strong odor of the fuel oil. When installed inside, the tank is required to be at least 5 ft from any flames or heat-producing equipment.

Outside Tanks

These are used if the space inside is limited—for example, if there is no basement to hold the tank. This type of arrangement does not do well in very cold climates. The fuel will gel due to the low outside temperatures and plug filters and lines. Some fuel suppliers will mix a grade 1 fuel or kerosene with the normal grade 2 oil. This lighter mix of fuel will flow better during conditions of low outside temperatures. Outside tanks will have the fuel line coming from the top of the tank, and it should be insulated up to the point where it enters the building. If the oil tank is placed outdoors aboveground, firm footings must be provided (Figure 55-14).

55.9 OIL LINES

Residential furnaces normally use ½-in outside diameter (OD) copper oil lines connected with flared fittings. The total number of fittings should be kept at a minimum, and compression fittings are not to be used. The connection from the oil storage tank to the burner oil pump can be arranged as a one-pipe or two-pipe system. For either type, there will be a thermal safety shutoff valve and filter assembly at the tank and a thermal shutoff valve located at the inlet to the oil pump.

For an inside installation, it has been common practice to run the oil line through the concrete in the basement floor. However, the line should not be allowed to touch concrete or soil and should be shielded by flexible plastic conduit or a coated copper line. Running the oil line through the floor is done to limit the amount of suction required by the pump. This arrangement allows for the oil to flow to the pump by gravity and reduces the total vacuum required at the suction of the pump.

Figure 55-14 The outside oil storage tank is mounted on a cement slab, and the oil line is connected at the top of the tank.

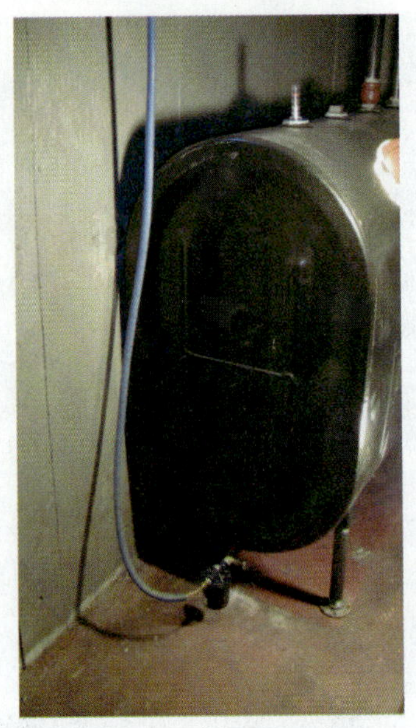

(a)

(b)

Figure 55-15 (a) The oil line shown runs from the tank to the floor joists and then down again to the furnace; (b) oil lines should be shielded with flexible plastic conduit.

It is also common practice to run the oil pipe up and across the basement floor joists and back down to the oil pump (Figure 55-15). This eliminates the trouble involved with running the oil line through the concrete floor. It is also preferred to running the oil line on top of the basement floor, where it may be easily subject to damage. In either case, the oil line must be shielded with flexible plastic conduit.

Placement of the oil line depends on the vertical lift and horizontal run of the oil line, which will affect the suction required by the oil pump. The suction produced by the fuel pump is measured in inches of mercury vacuum (in Hg). About .75 to 1 in of Hg vacuum is required for each foot of vertical lift and for every 10 ft of horizontal supply piping. The vacuum required for lifting the oil due to the combined resistance of vertical lift and piping length should never be allowed to exceed 6 in Hg with a single-stage single-pipe, 12 in Hg with a single-stage two-pipe, or 17 in Hg with a two-stage two-pipe system. If the vacuum at the pump exceeds these limits, entrained air in the oil will expand and cause capacity and noise problems. This will reduce the life of the oil pump and can also lead to unstable flame conditions.

Single-Pipe System

A single-pipe system uses one oil line from the tank to the oil pump. This is the preferred method of installation. If the tank and the oil lines are installed such that the total vertical lift and horizontal run do not exceed 6 in Hg vacuum, then a single-pipe system should be installed.

Two-Pipe System

Two-pipe systems are required when the oil tank is located below the fuel unit and a high lift and/or long run of fuel line exists. If the total vacuum required at the suction of the oil pump is greater than 6 in Hg, then a two-pipe system will need to be installed. A return line from the pump delivers oil back to the fuel tank (Figure 55-16) or to a fuel de-aerator to allow for the entrained air to be removed from the oil.

Many oil pumps are designed for both single-pipe and two-pipe systems. By placing a bypass plug in the housing, the pump can operate as a single-pipe or a two-pipe system. The plug is removed for single-pipe operation. Generally the bypass plug must be installed for two-pipe systems, but some newer pumps today do not require a bypass plug. The major disadvantage of the two-pipe system is the potential for a leak to develop in the fuel return line, which is under positive pressure. A small leak in this line could result in an extensive oil spill over the course of several days of operation.

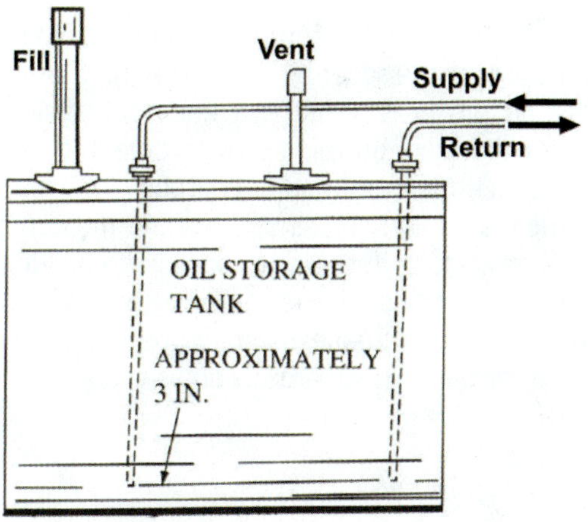

Figure 55-16 An oil storage tank for a two-pipe system showing the supply and return lines.

55.10 OIL SAFETY VALVES

Some fuel oil systems are installed so that the oil tank is at a higher elevation than the furnace. This makes getting the fuel to the burner easier. Unfortunately, it also makes it possible for any leaks in between the tank and the pump to leak oil continuously, especially when the furnace is off. In a worst-case scenario, an entire tank of oil can leak out. The purpose of the oil safety valve is to prevent line leaks from draining the tank and causing environmental damage. Many state and local codes require oil safety valves. Refer to the latest edition of the National Fire Protection Association Standard (NFPA 31) for the installation of oil burner equipment.

If the top of the oil supply source is more than 8 ft above the fuel unit, an oil safety valve will need to be installed. The valve is normally closed and will only open if there is a vacuum on the outlet side of the valve. When the burner is off, if there is no vacuum, then the valve remains closed and oil flow is stopped. If the fuel lines are tight, a vacuum is created when the burner starts, allowing the oil to flow. However, air will be sucked into any leaks, reducing the vacuum on the line and keeping the valve closed. Since the valve remains closed, oil will not leak out. This not only prevents leaks but also encourages customers to get the leaky lines repaired because the furnace will not fire. Without the oil safety valve, small leaks can be ignored because the unit will still operate and heat.

55.11 START-UP PROCEDURES

Some preliminary checks and steps are necessary before attempting to start the furnace. With the main power supply turned off, check to see that the following are in place:

- The correct nozzle size has been selected for the desired input rate.
- All shipping supports are removed.
- Power wiring is properly connected.
- Control wiring is properly connected.
- Blower access door is secured in place.
- Valve on the oil supply line is open.
- Reset button on the primary control is pushed down.
- Flame observation door and all clean-out access covers are closed.
- The thermostat is set to HEAT at a temperature higher than the room.

After all of the above items have been checked, set the disconnect switch to OFF and energize the circuit to the furnace. Check the voltage at the disconnect switch and verify that it is the correct voltage, usually 115 V.

Operational Check

To correctly adjust the burner, it will be necessary to measure the oil pump pressure, stack temperature, and carbon dioxide flue gas percentage.

All air should be purged from the fuel lines and a pressure gauge installed on the pressure port of the oil pump. The manufacturer's literature or a data plate on the oil burner can be used to identify the correct pump pressure. The most common pressure has been 100 psig for many years, but more recent oil burners may operate at a higher pressure. Start the burner and compare the oil pump pressure to the recommended pressure and adjust it as necessary.

The operation of the barometric damper should be checked. The manufacturer's instructions should list the recommended draft pressure between the damper and the furnace; –0.02 in wc is common. Adjust the weight on the barometric damper to achieve the correct draft.

The correct flame should have no smoke, as measured by a smoke spot test. The CO_2 level should be 10–12.5 percent. If there is smoke in the flame or if the CO_2 level is outside of the 10–12.5 percent range, you need to adjust the air. To do this:

- Adjust the air-inlet vane until you see a trace of smoke in the flame.
- Measure the CO_2 content of the flue gas.
- Open the air until you have lowered the CO_2 by 1 to 2 percent.
- The correct flame should have no smoke and the CO_2 level should be 10 to 12.5 percent.

The flue gas should be checked to determine the combustion efficiency. To do this, measure the actual stack temperature and then subtract the equipment ambient temperature. The result will be the net stack temperature. The combustion efficiency can then be calculated using the CO_2 percentage, the net stack temperature, and a combustion efficiency slide rule.

Other items to check are the carbon monoxide level of the flue gas, the temperature rise, and blower current draw once the indoor blower comes on. Ideally, the flue gas should have no carbon monoxide, but trace amounts of up to 50 ppm are acceptable.

SERVICE TIP

Pushing the safety reset several times on an existing furnace is not a safe practice because oil may already be sitting in the combustion chamber and every push delivers more oil. On a new installation, the lines are full of air. A two-pipe system will purge itself, but it may take several pushes. Always check the combustion chamber for excess oil accumulation before pushing the safety reset.

UNIT 55—SUMMARY

Proper installation and start-up are required for oil-fired appliances to burn cleanly and efficiently. There are several key components to ensure a successful installation. The furnace capacity needs to match the heating load. The proper oil tank must be installed along with the correct fuel line arrangement. There must be adequate combustion air supplied to the furnace as necessary. Always refer to both the NFPA and local codes for guidance. The correct exhaust vent and vent connector need to be installed at the proper locations and at the proper heights. Power and control wiring need to comply with the manufacturer's recommendations for voltage and current demand along with wire sizing requirements. Final installation will include combustion measurements and adjustment.

WORK ORDERS

Service Ticket 5501

Customer Complaint: Unit Will Not Operate

Equipment Type: Oil-Fired Furnace with Electronic Controls

A technician answers a service call to start a newly installed oil furnace. After making the necessary prestart checks, the technician tries to start the furnace and nothing happens. The technician investigates further and finds that there is a fault code flashing on the control board LED.

The technician refers to the installation manual that came with the furnace and finds that the flash pattern is listed as "incorrect power polarity." This leads the technician to check the power wire connections. The black power wire is connected to the white on the furnace, and the white power wire is connected to the black on the furnace. The technician swaps the wires and the furnace operates properly.

Service Ticket 5502

Customer Complaint: Unit Does Not Fire When Started

Equipment Type: Oil-Fired Furnace with Two-Pipe Fuel Supply, Tank Below Furnace

A technician is sent to start a newly installed oil-fired furnace. Initially the furnace appears to start when it receives a call for heat, but it does not fire. After 2 minutes, the primary control shuts down on safety. Since this is a new installation and the furnace has not fired, the technician assumes that the lines have not been completely purged of air.

Sometimes with a two-pipe fuel supply such as this, purging can be accomplished by restarting the furnace, thereby allowing the air to be pushed through the supply line into the return line and back to the tank. After pushing the reset button once, the furnace operates and the fires briefly but goes out again. Before pushing the reset button a second time, the technician must check the combustion chamber for a buildup of oil. The inspection port is opened and the combustion chamber is inspected. Since there is no buildup of oil, the technician pushes the reset button a second time and allows the furnace to start one more time. This time, after the furnace fires, it stays going.

Service Ticket 5503

Customer Request: Furnace Change-out: Reconnect Vent

Equipment Type: Oil-Fired Furnace Vented into Masonry Chimney

A technician is sent to install the chimney connector to a new oil-fired furnace that has just replaced an older unit. The customer complained that the old furnace would sometimes admit smoke into the basement. The technician realized that this may have resulted from an improper vent connection.

The old furnace metal connecter was single wall and the draft regulator was connected to the end of a bull-headed tee. The technician decides to run a type L chimney connector and places the draft regulator on the branch end of an in-line tee. Also, the 90° ell and tee are replaced by two 45° ells. The draft regulator is adjusted and the over-fire draft is measured to be –.02 in wc. The draft in the chimney connector is measured to be –.04 in wc. The system operates properly and there is no oil smell or smoke in the basement.

Service Ticket 5504

Customer Complaint: Unit Does Not Operate

Equipment Type: Oil-Fired Furnace with Electronic Controls

A technician has installed a new electronically controlled oil furnace to replace an aging furnace and used the same power wire configuration on the new furnace. The technician had not paid attention to the fact that the power line to the old furnace did not have an equipment ground.

After conducting the routine prestart checks, the furnace was started but nothing happened. The technician investigated further and found that the control board LED was flashing. The technician refers to the installation manual that came with the furnace and finds that the flash pattern is listed as "furnace not grounded." The technician quickly realizes what the mistake is and runs a new grounded power supply. With the unit grounded, the furnace starts and operates correctly.

Service Ticket 5505

Customer Complaint: New Hydronic System Has Cold Zone

Equipment Type: Oil-Fired Boiler with Zoned Hydronic Radiant Floor Heat

A technician answers a service call for a recently installed hydronic heating system. The customer says that the problem is that one zone is colder than the others. This is especially true on cold days when all zones are calling for heat.

The first thing the technician checks is the zone valve, and it is opening and hot water is passing through it. However, the return pipe from that zone is 15°F cooler than the other two zones. This low return temperature indicates poor water flow to that zone. The technician fully opens the balancing valve for the cold zone and partially closes the balancing valves for the other two zones. After 30 minutes of operation, the return temperature of all zones is within 5°F. A check of the floor temperature in all three zones with an infrared thermometer shows that the floor temperatures are now around 90°F.

UNIT 55—REVIEW QUESTIONS

1. What factors should be considered when selecting an oil-fired furnace?
2. What is the proper service clearance that should be allowed when installing a furnace?
3. What is the firing rate of a furnace with a Beckett AFG-FG burner and a 0.75 gph 70-B nozzle? (Refer to Table 55-1.)
4. What is the heating capacity of an 80 percent efficient furnace with a Beckett AFG-FG burner and a 0.75 gph 70-B nozzle? (Refer to Table 55-1.)
5. What would be the required vacuum for an oil furnace located 5 ft above the oil tank with a horizontal run of 20 ft?
6. What national standard is a good reference when installing oil furnaces?
7. What free air opening would be required for an oil furnace installed in a confined space if the air was coming through horizontal ducts and the furnace had a Beckett AFG-FG burner and a 0.65 gph 70-W nozzle? (Refer to Table 55-1.)
8. Calculate the equivalent length of a 6-in-diameter chimney connector with two 45° ells and 8 ft of pipe.
9. How high must the chimney be above the connector entrance for the connector in question 8 to operate properly?
10. What type of material must be used for a chimney connector in a crawlspace?
11. Describe where the barometric damper should be installed in the chimney connector.
12. How would the primary control need to differ to accommodate power venting?
13. What is a disadvantage to direct venting an oil-fired furnace?
14. What problems can arise from an oversized oil storage tank?
15. How much fuel should a properly sized oil storage tank hold?
16. What is the function of an oil safety valve?
17. Give a prestart checklist when checking a newly installed oil furnace.
18. What measurements should be taken when setting up a new oil furnace?
19. What national publication is a good reference for chimney construction?
20. What types of appliances may oil furnaces be common vented with?

UNIT 56

Troubleshooting Oil Heating Systems

OBJECTIVES

After completing this unit, you will be able to:

1. discuss common oil delivery problems.
2. discuss common ignition system problems.
3. discuss common primary control problems.
4. explain how soot forms and what problems soot buildup can lead to.
5. describe safety precautions that should be followed when checking a system with a tripped primary control.
6. determine likely causes of oil heating system failures using a troubleshooting chart.

56.1 INTRODUCTION

Oil furnaces use a liquid fuel that must be atomized before burning, which provides for a unique set of problems when compared to gas furnaces. Operational troubles associated with oil furnaces can be related to the process of atomizing the fuel, mixing it with air, and igniting it with a high-voltage spark. Though these are typical, in some cases the problems are much simpler. A thorough understanding of how all the furnace components operate will help you to identify and troubleshoot system malfunctions.

56.2 OIL BURNER PROBLEMS

Oil burner problems can be grouped by the main components of the oil burner:

- Oil burner motor problems
- Oil pump problems
- Nozzle problems
- Ignition system problems
- Primary control problems

Oil Burner Motor Problems

The burner motor is usually a split-phase motor with a centrifugal switch and a manual-reset-type overload (Figure 56-1). Motor problems can include open, shorted, or grounded windings, a defective centrifugal switch, or even on rare occasions a defective overload. Regardless of the specific motor malady, the solution is generally motor replacement. It is usually not practical to repair motors in the field.

Symptoms that indicate problems with the burner motor include the following:

- Burner motor does not start
- Burner motor runs but trips out on overload
- Burner motor runs but does not produce enough oil pressure

Burner motor does not start If the motor will not start, begin by checking to see if the motor is receiving the correct

Figure 56-1 Oil burner motor reset button. *(Courtesy of InspectAPedia.com)*

voltage. If the motor has no measured voltage, then the problem is not with the motor. If the motor is receiving voltage and not operating, try pushing the manual reset for the motor overload. If the overload has tripped, the red button will be sticking out. Try pushing it to reset it. If the motor tries to start but cannot, either the motor is stuck or it is defective. Turn the power off and check to see if the motor is binding by spinning the combustion blower by hand. If it will spin easily, the motor is not bound. If the motor overload is not tripped, it is receiving voltage; if it is not bound but still will not run, then the motor is damaged. This may be traced to the windings or the centrifugal switch; however, rather than trying to repair it, the recommended remedy is to replace the motor.

If the fan will not spin, or is very difficult to turn, then the fuel pump is binding, the fan is hitting the housing, or the motor bearings are seizing. Disconnect the motor from the coupling that drives the oil pump (Figure 56-2). If the oil pump is the problem, the motor will now turn freely. If it does not turn freely, then the motor bearings are most likely the problem and the motor should be replaced. Once

(a)

(b)

Figure 56-2 (a) Oil pump drive coupling location; (b) oil pump and drive shaft.

the bearings become so dry that they seize, they cannot be fixed by lubricating them.

Burner motor runs but trips out on overload If the motor runs but trips out on overload, check the motor's operating current and compare it to the full load amperage (FLA) marked on the motor. If the overload trips without the current exceeding the rated FLA, the overload is weak and the motor needs to be replaced. If the operating current exceeds the FLA, then look for items that might be binding the motor. Disconnect the coupling to the oil pump to see if the motor will operate with the correct operating current. If the oil pump is the problem, the motor will now turn freely. If the motor bearings are the problem, the motor should be replaced.

Burner motor runs but does not produce enough oil pressure If the motor runs but the oil pressure is low, check the coupling that drives the oil pump. If the coupling is cracked or stripped out, the motor will run but the oil pump will turn slowly, or not at all. Replacing the coupling will correct this.

Oil Pump Problems

Oil pump problems include the following:

- Seized oil pump
- Oil pump pressure not correct
- Bypass plug incorrectly applied
- Leaky shutoff

Seized oil pump If the oil pump will not turn, it is seized and should be replaced. If this is the case, then the quality of the oil should be checked. Water and sludge in the pump can cause it to seize. Cleaning the oil tank to remove accumulated debris and sludge is a good precautionary measure whenever an oil pump is replaced.

Oil pump pressure not correct Check the inlet strainer screen if the pump will not maintain the correct pressure (Figure 56-3). If the system is a two-pipe system, check to see that the bypass plug is installed. If the system is a one-pipe system, check to see that the bypass plug has been removed. If the screen is clear and the pressure is still too low, try adjusting the pump's output pressure. Turn the adjustment screw clockwise to increase the pressure.

A leak in the suction line can also cause poor oil pressure by letting air in. Disconnect the nozzle line connector and direct its output to a container. Run the burner and observe the oil stream. The stream should be clear and steady. If air is being drawn in through a leak in the suction, the oil stream will be frothy or bubbly. Finding and repairing the leak is necessary before proceeding. If none of the troubleshooting methods result in proper operation, then the pump is damaged and should be replaced.

Bypass plug incorrectly applied The oil pump delivers more oil than the burner can use. The extra oil must either

Figure 56-3 Oil pump inlet screen.

be bypassed internally in the pump or sent back to the tank through a separate oil-return line. By placing a bypass plug in the housing, the pump can operate as a single-pipe or a two-pipe system. The plug is removed for single-pipe operation. Single-pipe operation is used where the oil tank is above the burner, and gravity oil feed to the burner is permitted. Generally the bypass plug must be installed for two-pipe operation. However, some newer pumps today do not require a bypass plug. Two-pipe operation is needed when the oil tank is located below the fuel unit and a high lift and/or long run of fuel line exists.

Leaky shutoff When the burner motor stops, the oil pump should shut off tightly to prevent continued oil delivery into the combustion chamber. If the pump cutoff is working properly, the output pressure should not drop past 80 psig or not more than 20 percent below the operating pressure. If the pressure drops below these levels, the cutoff is leaking and the pump should be replaced.

Nozzle Problems

The oil mist from the atomized spray cone should not come into contact with any part of the combustion chamber. If it does, the oil will not burn completely, and sooty, inefficient combustion will result. Nozzles become worn over time, and the atomization becomes less complete, resulting in incomplete combustion. The nozzle filter can become clogged over time and produce a poor spray pattern. Poor combustion that persists and cannot be adjusted can be a sign of a worn or dirty nozzle. Typically, the nozzle should be changed every year to ensure correct operation.

Ignition System Problems

Ignition problems can result in delayed ignition or no ignition. Ignition components include the ignition transformer, the electrodes, and the contacts that transfer electricity from the transformer to the electrodes. The transformer delivers 10,000 V AC to the electrodes, producing an arc between them.

This high voltage arc can be diminished by the following:

- Low transformer primary voltage
- Improper electrode gap
- Cracked or dirty porcelain insulators on the electrodes
- Poor contact between the electrodes and the transformer
- A weak transformer

If the oil burner has been delivering oil and it has not been igniting, the first thing that needs to be done is to clean the excess oil out of the combustion chamber. Turn off the power and disconnect the control wires so that the burner will not try to light. Some furnaces have a clean-out port that can be used to gain access to the combustion chamber. If the furnace does not, it will be necessary to remove the burner assembly to allow for access. Carefully clean out any accumulated oil, but be careful not to damage the refractory.

When the furnace starts, the ignition spark should make a buzzing sound. If you do not hear the arc, check the transformer primary voltage. If the transformer is not getting the correct voltage, usually 115 V AC, then the primary control and wiring should be checked.

If the transformer is receiving the correct voltage, check the output of the transformer with a high-voltage probe. Turn the power off, swing the transformer up, turn the system on, and check the output terminals with a special probe made for the 10,000 V secondary winding. If the voltage is less than the transformer rating, usually 10,000 V, the transformer must be replaced.

Primary Control Problems

The burner should be operating if the primary control is receiving 120 V, the circuit is closed across the two T terminals, and the primary control has not tripped. An easy way to test the primary control is to jumper the T terminals (Figure 56-4). The burner should start, and after lighting it should stay on. If the primary control trips out even though a flame is present, the problem could be in either the primary control or the cad cell.

After starting the burner, connect the leads to jumper the F terminals to test the cad cell. This will act to bypass the cad cell. If the control still trips out, it cannot be the fault of the cad cell so the primary control is bad. If the burner continues to fire, disconnect the cad cell and check its resistance. When a flame is established, the light strikes the cad cell and immediately reduces the resistance of the cell from over 10,000 Ω to about 1,500 Ω or less. A good reading would be between 300 and 1,000 Ω. If the resistance is over 1,600 Ω, the cad cell is misaligned, dirty, or defective.

If the reading is high, turn off the burner and recheck the cad cell resistance with no flame present. It should rise to 100,000 Ω or more. While the burner is off, clean the cad

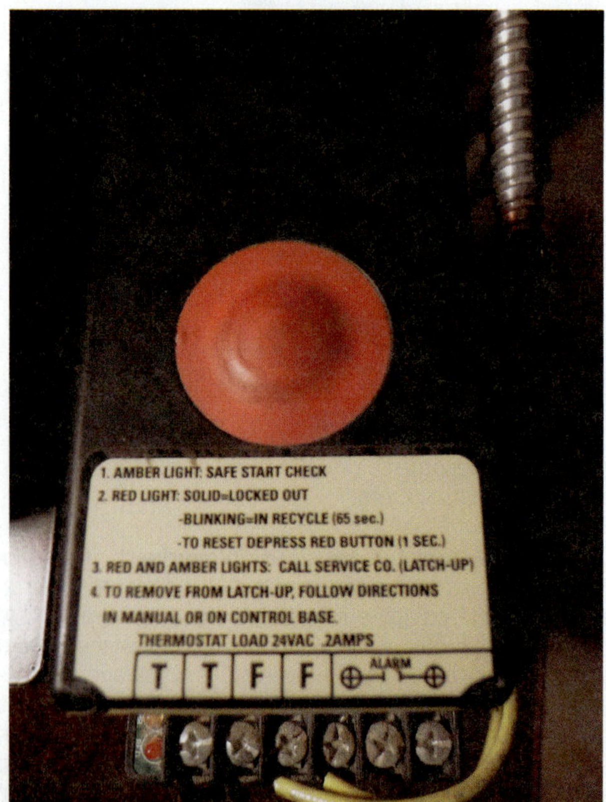

Figure 56-4 Primary control T terminals.

cell and check its alignment. Start the burner again and recheck the cad cell resistance with the flame established. If the cad cell still has too high a resistance, replace it.

56.3 OIL SYSTEM PROBLEMS

Not all problems are caused by the oil burner. The rest of the system, from the tank to the vent, can be the cause of system malfunctions.

Oil Tank–Related Problems

A common oil furnace problem is the lack of fuel. Check to see that there is fuel in the tank before spending a lot of time chasing down nonexistent problems. If the fuel tank is empty, the lines will need to be purged of air after the tank is filled. If the oil tank was emptied, it is a good idea to change the fuel-line filter, pump strainer, and nozzle. The last few gallons of oil are always the dirtiest, and filling the tank stirs up the sediment on the bottom. A new filter will help protect against future problems.

Suction-line leaks can cause two types of problems. On systems with oil safety valves, a suction leak may prevent the oil safety valve from opening, and this will shut off all oil flow. Even without the safety valve, suction leaks can introduce air into the oil. A frothy oil/air mixture is delivered to the pump, and uneven, dirty burning results.

If the vent on the oil tank is obstructed or plugged, the tank pressure can drop to the point that the fuel will not move.

Electrical Problems

Although oil furnaces are fairly simple to wire, they can still be wired incorrectly. If the furnace does not receive the proper voltage, nothing will operate. Modern furnaces with electronic control boards must be properly grounded and connected with the correct polarity. If the furnace is not grounded or the hot and neutral power wires are swapped, the furnace will not operate. This is usually indicated by a fault code from the diagnostic LED. On the control side, a bad thermostat or broken thermostat wire can cause the furnace to not operate.

Venting and Combustion Air Problems

Like any combustion appliance, oil furnaces need combustion air. In older leaky houses, combustion air is usually assumed to enter through infiltration, provided the furnace is located in an unconfined space. Newer houses should always have combustion air provided through grilles or ducts. It is never wrong to bring in combustion air, even with older houses. A lack of combustion air will cause poor combustion and can lead to carbon monoxide poisoning.

The draft regulator should be adjusted to maintain the correct over-fire draft, usually at least –.02 in wc draft. The vent connector should be as short and straight as possible. Long chimney connectors with lots of sharp turns will lead to poor venting and poor combustion. If the draft regulator cannot maintain the draft at the proper setting, the chimney connector may be the problem.

If the over-fire draft rises significantly after the indoor fan comes on, the heat exchanger is cracked. This is a dangerous situation that should be corrected immediately. The furnace must usually be replaced if the heat exchanger is cracked. Heat exchangers are normally only changed under warranty.

Soot

An oil furnace that is not properly installed and maintained will produce soot. Soot is unburned carbon and oil that forms because of incomplete combustion. Soot can build to the point that heat exchangers and flues are clogged with it. When this happens, an extensive cleaning of the heat exchanger, chimney connector, and chimney is required. Soot buildup can lead to a condition called puff-back. This occurs when the combustion products cannot leave the furnace. Usually the incomplete combustion of the oil/air mixture is due to a blocked combustion chamber, leading to delayed ignition. When the oil does ignite, the small explosion produces a puff of soot and unburned oil forces its way out into the area around the furnace. This greasy black cloud leaves stains on surrounding areas. The way to prevent puff-back is to keep the furnace properly maintained so that soot buildup does not occur.

Fan Problems

If the burner runs but the indoor fan never comes on, the problem could be with the fan motor, the fan switch, or the primary control. If the fan motor is not operating, check the voltage to the motor. If the motor is not operating and it is receiving the correct voltage, it may need to be replaced. However, before replacing the motor, check the run capacitor. If the capacitor is good and the motor is still not running, replace the motor.

If the motor is not receiving voltage and the furnace has heated up, then the fan switch or primary control is bad. Some furnace fans are controlled by a temperature sensing bimetal fan switch (Figure 56-5). When the furnace heats up to the fan setpoint, the switch closes to start the fan. When the furnace temperature drops below the fan setpoint, the switch opens to stop the fan. Other furnaces use primary controls that bring the fan on using a time delay relay.

56.4 OIL FURNACE TROUBLESHOOTING PROCEDURES

A good practice when troubleshooting oil furnace problems is to check the simple things first. This takes very little time and often saves lots of time and frustration.

- Check the fuel. Start by checking to see if there is oil in the tank and if the oil is clean.
- Make sure all valves in the fuel lines are open.
- Check the combustion chamber for unburned oil—better safe than sorry!
- Check the voltage to the furnace, including the grounding and polarity.
- Check to see that the thermostat is calling for heat. The furnace will not operate unless it is told to.
- Check to see if the primary control has tripped.

(a)

(b)

(c)

Figure 56-5 (a,b) Combination fan and limit control switch located above burner; (c) bimetal element connected to fan switch for sensing the furnace warm-air temperature.

Before starting the burner, check for liquid oil in the bottom of the combustion chamber. The customer knows that the burner should start if the reset button is pressed. If the reset button has been pressed a number of times, it is possible to have a considerable quantity of oil sprayed into the combustion chamber before the owner calls for help.

If any oil is pooling in the bottom of the combustion chamber, it must be removed by soaking it up with sponges or rags. When reaching into the combustion chamber, make sure that the power to the burner is off and your arm is covered. A fire extinguisher must be within reach and the observation door must be secured in the open position. When the flame is established, there will be a considerable fire developed until all of the oil has burned out of the bottom where it has soaked into the refractory. If the furnace is a large unit of 2.5 gph input or more, it is advisable to call the fire department to stand by before lighting the burner.

A manufacturer's troubleshooting logic chart can be useful for more difficult problems. Table 56-1 provides common causes of oil fired system problems.

Many of the solutions to the problems in Table 56-1 are evident when the cause is determined. The following list of remedies from 1 through 7 provides further information. The numbers refer to the references in Table 56-1.

TECH TIP

Listen to the customer's comments carefully when talking with them. Their description may shed light on the problem. But more important, being interested in the customer and their concerns is a big part of a technician's job. Displaying a lack of interest in the customer's concerns will alienate the customer. Alienating the customer can result in their loss of confidence in you and your company. Ultimately, this results in reduced income for you.

1. Protector Relay Defective

There are two types of protector relays: a cad cell that operates the relay from the light of the fire and a bimetallic control in the hot flue gases of a smoke pipe mount.

Cad cell This type of protector relay uses a light-sensitive cadmium sulfide flame detector, as pictured in Figure 56-6, mounted to the firing tube of the oil burner. When the cell is exposed to light, its resistance is very low, which allows current to flow through it. This current is sufficient to pull in the sensitive relay in the protector relay.

When the relay pulls in, it opens the circuit to the safety switch heater and prevents cutout of the burner. If the cell is not exposed to the light of the fire or if the cell becomes dirty or covered with soot, then it will not allow the current to pull in the sensitive relay; the safety switch heater remains in the circuit until the safety switch opens. This breaks the thermostat circuit and the burner stops, as shown in the circuit in Figure 56-7.

TABLE 56-1 Oil System Problems

Problems	Possible causes (see the Following Number References for Further Information)
Will not start	Season switch open
	Room thermostat improperly set
	Safety switch open on protector relay
	Fuse blown
	Limit control open
	Protector relay transformer defective (1)
	Protector relay defective (2)
Starts, but will not continue to run	
A. No fire established	No fuel (6)
	No ignition
	Protector relay not functioning
	Nozzle defective
	Fuel pump defective
B. Fire established	Defective oil burner component
	Defective protector relay
Runs, but short-cycles	Improper heat anticipator setting
	Cycling on limit control
Runs, but room temperature too high	High setting of thermostat
	Improper heat anticipator setting
Runs continuously	Short in thermostat circuit
	Stuck contacts in protector relay
Blower cycles after thermostat is satisfied	Incorrect blower CFM
	Incorrect fan control setting
	Heat exchange heavy with soot
Blows cold air on start	Incorrect blower CFM
	Incorrect fan control setting
Start-up/cool-down noise	Expansion noise in heat exchanger
	Duct expansion
Noise/vibration	Burner pulsation
	Blower wheel unbalanced
	Blower drive problems
Odor	Oil odor on unit start-up
	Improper venting
	Cracked heat exchanger
High fuel cost	Improper input
	Improper burner adjustment
High electric cost	Improper load on blower motor
Smoke or odor from observation door	Improper draft or draft control setting
	Improper venting
	Improper input
	Delayed ignition
Burner pulses	Improper draft
	Improper venting
	Burner cuts off on safety switch
A. No fire established	No fuel
	No ignition
	Protector relay defective
	Components of burner defective
B. Fire established	Components of burner defective
Delayed ignition	Carbon deposit on firing head
	Improper electrode adjustment (3)
	Cracked electrode insulator (4)
	Ignition leads burned out
	Ignition transformer burned out (5)
Burns inside burner after cutoff	Fuel unit defective
Carbon deposits on refractory	Defective nozzle
Noisy flame	Improper air adjustment
	Improper nozzle
Fuel pump sings	Fuel unit defective (7)
Heat exchanger burnout	Input too high
	Defective refractory
	Defective nozzle
Blower short-cycles	Input too low
	Blower CFM too high
	Fan control defective

Figure 56-6 The ignition transformer cover is swung open to access the light-sensitive cadmium sulfide cell for one style of burner.

The safety switch is manually reset. When checking the protector relay for repeated burner cutoff even though the flame is established, make sure that the cad cell is clean. The cad cell can also be checked using an ohmmeter. Connect the ohmmeter leads across the leads of the cad cell. For a properly adjusted burner during operation, the cad cell resistance should be approximately 300–1,000 Ω but not more than 1,600 Ω. Understand that this is the expected resistance when the cad cell is exposed to a bright flame. The reading measured from exposure to room lighting or a flashlight will be different. Placing a finger over the cell, cutting off the light, will raise the resistance to 100,000 Ω or higher.

SERVICE TIP

When replacing a cad cell, be sure that it is completely and squarely seated. If the cad cell is not seated properly, it might not be "looking" at the flame, so it will not operate properly. An improperly seated cad cell can be the cause of an intermittent problem.

If the cad cell checks out, check the protector relay by placing a jumper wire across the F-F terminals of the protector relay. The burner should start and continue to operate. If it cuts off, the timer contacts and heater circuit are defective, and the relay should be replaced.

Bimetallic controls This type of control is clutch-operated to move the hot and cold controls through their proper sequence. Any sharp blow to the control can release this clutch and throw the control out of sequence. The clutch also wears and can loosen the hold on the contact fingers.

If the control refuses to operate, pull the drive shaft lever forward until the stop is reached. Slowly release the drive shaft lever to the cold position. This should close the contacts and the unit should operate. If not, the bimetallic element could be jammed or the contacts could be defective. To check the bimetallic element, remove the protector relay from the vent stack and check it. Usually carbon (soot) buildup through the bimetallic helix will be the cause of jamming the drive shaft lever. When cleaning the helix, be careful not to bend or break it. With a clean helix, if the hot and cold contacts still do not hold, then the clutch and contact leaves are worn and the entire control should be replaced.

SERVICE TIP

If a bimetal helix is not operating smoothly after cleaning, it must be replaced. Any slight roughness in operation will only get worse and will result in a recall to the residence to fix the problem later.

2. Protector Relay Transformer Burned Out

It is usually not possible to open the protector relay to reach the transformer to test it. It is not necessary since connections to the transformer are done through the T-T terminals on the outside terminal board.

Figure 56-7 shows the inside wiring diagram of a typical protector relay circuit. The control circuit is completed from the top terminal T through the transformer, the normally

Figure 56-7 Oil burner wiring schematic diagram using cad cell protector relay.

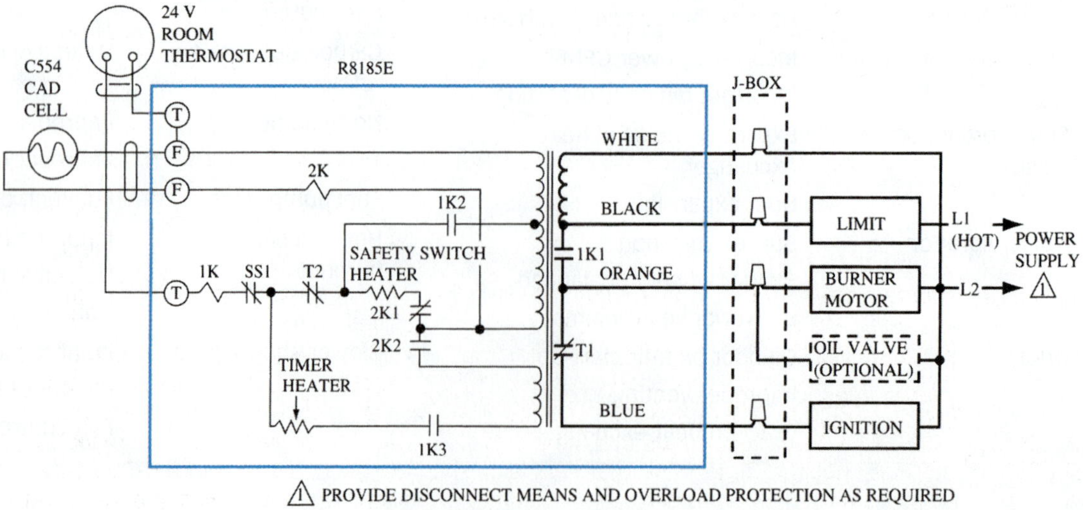

⚠ PROVIDE DISCONNECT MEANS AND OVERLOAD PROTECTION AS REQUIRED

closed contact (2K1), the safety switch heater, the normally closed contact (T2), the safety switch (SS1), and the relay coil (1K) to the bottom terminal T. With the control wires removed from terminals T-T, 24 V should be measured across them when there is 120 V across the white and black leads to the relay. If there is no voltage measured at T-T, replace the protector relay. If there is voltage at T-T and the thermostat will not operate the relay, check the thermostat and sub-base on the control cable.

3. Improper Electrode Adjustment

To establish the spark necessary to ignite the oil, the electrodes have to be close enough together to present a minimum gap resistance to the 10,000–15,000 V supplied by the ignition transformer. The size of the gap between the electrodes is normally set at a distance of ⅛ in. The electrodes must be positioned out of the oil spray but still have the arc flame close enough for ignition. The electrodes must be positioned at the proper height above the hole in the nozzle tip and at the proper distance ahead of it. These dimensions will vary between different burner types.

Another measurement, called the Z dimension, is the distance from the nozzle tip to the end of the firing-tube air turbulator. If the distance is too close, then oil spray can impinge on the turbulator and lead to carbon buildup. If the distance is too far, then the nozzle will be exposed to more heat from the furnace. The nozzle should be positioned back far enough to keep the effect of the heat from the fire to a minimum.

4. Cracked Electrode Insulators

The electrodes are held in place by clamps to ensure stability in the proper position. Where the clamps are fastened, a ceramic insulator surrounds each electrode to insulate the spark voltage from the grounded assembly. The ceramic must insulate against 5,000–7,500 V (one-half the spark voltage) and still stand up against the heat of the burner.

Never overtighten the electrode clamps, because the ceramic insulators are hard and brittle and crack easily. Loosen the electrode clamps before applying any twisting or bending pressure to change the position of the electrodes. Do not attempt to bend the electrode wires because they are harder than the ceramic.

SERVICE TIP

If any fine cracks (often referred to as crazing) are noted on the surface of the insulators, replace them. Do not take a chance on the old ones.

5. Ignition Transformer Defective

Ignition transformers are 10,000–15,000 V with a grounded center tap on the high-voltage side. With 12,000 V between the terminals and 6,000 V from either terminal to ground, attempting to check the transformer by producing a spark with a wire, screwdriver, or other shorting means can be dangerous.

In time, the heat of operation dries the transformer, and cracks develop. Moisture enters the assembly and is absorbed into the windings. This leads to shorts between the windings, which lower the output voltage to the point of failure of the spark across the ignition gap, and faulty ignition results.

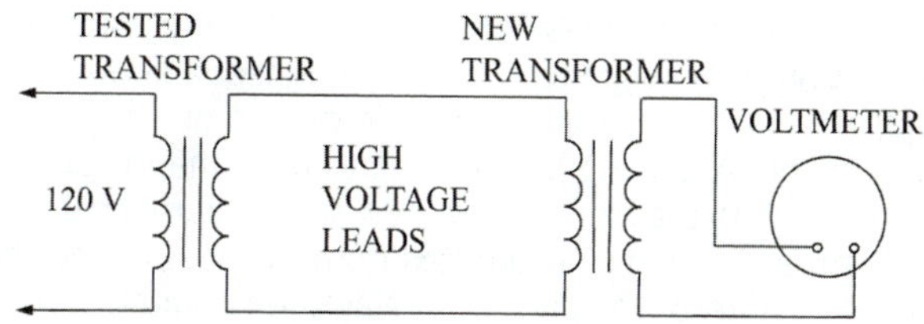

Figure 56-8 Transformer test wiring diagram.

The best way to check a transformer is with a high-voltage meter in the range of 10,000 to 15,000 V. If this meter is not available, two 120 V voltmeters and a new ignition transformer can be substituted. This transformer needs to have the same input and output voltage rating as the transformer being tested. Figure 56-8 shows the wiring diagram for this test.

The secondary of each of the transformers (high-voltage terminals) are connected together. The new transformer will step down the voltage output of the transformer being tested. With 120 V applied to the test transformer (measured with one of the voltmeters), the output of the new or testing transformer should be within 10 percent of the applied voltage. In addition, the output voltage should hold steady. If there is more than a 10 percent difference or if the output voltage varies, the original transformer has internal shorts and should be replaced.

6. No Fuel

Before taking the burner apart, first check the quantity of oil in the supply tank (the oil gauge on the inside tank or the dip rod for the outside buried tank). Second, close the tank outlet valve, open the filter cartridge case, and put in a new filter cartridge. Bleed the air from the filter cartridge after assembly. Third, after replacing the filter cartridge, make sure that the tank outlet valve is open.

If this procedure still does not supply oil to the burner and produce fire, remove the oil burner nozzle and check the nozzle filter. If this is plugged, replace the entire nozzle assembly with one of like capacity, spray angle, and cone type.

SERVICE TIP

If the oil filter or oil strainers are plugged with rust, the oil storage tank must be cleaned or replaced. A dirty oil storage tank will continually cause the filter and strainers to be plugged, resulting in numerous callbacks.

If these steps do not produce oil flow when the unit runs, check the inlet screen of the fuel pump. Inside the unit is located a fine mesh screen filter at the inlet to the pump assembly. This screen can be removed and replaced. Do not leave this screen out of the unit.

After any part of the fuel supply system has been opened, the system must be purged of air. A two-pipe system will automatically purge itself of air. The single-pipe gravity feed system does not have an automatic purge feature. To purge a single-pipe system, connect a short piece of plastic hose from the bleed valve located on the side of the fuel unit. Allow the plastic hose to drain into a suitable container. With the bleed valve open, run the furnace until a clean stream of fuel oil is emitted from the hose. It may be necessary to reset the protector relay safety switch several times before the supply line is completely purged.

If it is not possible to obtain a flow of clear fuel oil, it is possible that the fuel unit is drawing in air through a line or fitting leak. If this is the case, then the leak must be located and corrected.

7. Fuel Unit Poor Cutoff

The most common problem in fuel units is poor cutoff of the oil supply when the unit cycles off. Correct cutoff will provide instantaneous cutoff of the oil to the fuel pipe and nozzle when the pump pressure drops to 80 percent of the operating pressure. At 100 psig operating pressure, the cutoff pressure is 80 psig. This is accomplished as the spring forces the control piston in the fuel unit against the fuel outlet seat and cuts off the flow of oil. The nozzle pressure drops immediately.

If particle buildup occurs on the face of the neoprene seat on the end of the piston, this prevents full-circle contact of the neoprene disk against the seat. Instead of positive cutoff, leakage occurs and the pressure gradually decreases in the nozzle. This gradual pressure reduction causes oil flow from the nozzle after the unit stops and the air supply disappears. The oil now burns with very little combustion air, producing a very smoky flame.

Carbon builds up on the turbulator end of the firing tube as well as the firing assembly. If the burner is not slanted at least 2° downward toward the combustion chamber, burning oil can flow back toward the blower. This can burn or smoke up the cad cell, with resulting cutoff of the burner on the safety switch. The correction for this is to clean the piston chamber of the fuel unit as well as the intake screen.

A persistent cause of this problem due to the quality of fuel oil supply can be reduced by double filtering the oil before it reaches the unit and using a delayed-action oil valve, as shown in Figure 56-9.

Figure 56-9 Delayed-action oil valve used with fuel unit.

Located in the fuel line between the outlet of the fuel unit and the firing head, a delayed-action oil valve provides a time delay between the blower starting and the start of oil spray. This ensures that air for combustion as well as airflow through the heat exchanger are established before combustion starts. The valve also provides instant cutoff of oil pressure to the nozzle, regardless of the action of the fuel unit. This is a highly recommended accessory for any oil-fired unit.

UNIT 56—SUMMARY

Remember when troubleshooting oil heating furnaces that the entire system is important, not just the burner. Problems in the oil supply or vent can shut the system down just as readily as problems with the burner. Try not to overlook the obvious: many oil furnaces quit heating because they run out of oil. Work safely. When the problem is related to the burner, remember to always check the combustion chamber for oil anytime the primary control is tripped when you arrive. Also remember that the output of the ignition transformer is 10,000 V and is more than capable of causing you serious harm.

WORK ORDERS

Service Ticket 5601

Customer Complaint: No Heat

Equipment Type: 80 Percent Oil Furnace with Standard Burner

A technician answers a service call for an oil furnace that is not heating. Upon inspection, the first thing that the technician finds is that the primary control safety has tripped. The first step taken is to shut off the power supply to the furnace and check the combustion chamber for oil. Fortunately, the combustion chamber is dry.

After completing this first safety check, the technician resets the primary control and starts the furnace. There is the normal buzzing sound coming from the ignition system but an unfamiliar fluttering noise coming from the burner. The furnace does not fire and the primary control trips out. The technician once again checks the combustion chamber and finds that it still dry, indicating that most likely no fuel had passed through the nozzle.

Given this, along with the fluttering noise, the technician suspects the problem to be in the fuel unit. Inspecting the oil pump, it is found that the coupling is stripped so that the pump is not turning as it should. The technician replaces the coupling and installs a pressure gauge to monitor the pump pressure. Pushing the reset, the burner fires and the gauge shows that the operating pressure is 100 psig.

Service Ticket 5602

Customer Complaint: Poor Heat

Equipment Type: 84 Percent Efficient Lowboy Oil Furnace

The furnace is running when the technician arrives to answer a call for an oil furnace that is operating but with poor heat output. It is immediately apparent that the furnace flame is unsteady. The technician conducts a smoke spot test and finds that there is enough soot to produce a No. 3 smoke spot.

A pressure gauge is installed to monitor the pump pressure, and it is fluctuating between 85 psig and 95 psig. A check of the pressure on the suction line reveals a vacuum of around 18 in Hg. This vacuum is much too high given that the oil tank is located in the basement with the furnace. The furnace is shut off to allow the filter on the fuel unit to be checked, and it is found to be extremely clogged. The technician not only replaces the filter but also the nozzle, because it will also prove to be partially blocked. The furnace is started, and now the suction vacuum is around 2 in Hg and the output pressure is 100 psig. The flames are steady and the smoke spot reading drops to No. 1.

Service Ticket 5603

Customer Complaint: Poor Heat

Equipment Type: 86 Percent Efficient Oil Furnace with Flame-Retention Burner

Arriving at a house for a poor-heat call, the technician immediately notices smoke coming from the chimney and the distinctive odor of unburned fuel oil. Upon inspection, the technician finds that the furnace is operating but putting out very little heat. The flame pattern is ragged and the color is not uniform but streaked.

The furnace is shut off and the combustion chamber is inspected. It is found to be full of soot. Another thing the technician finds is that the retention ring is missing from the burner. It is in the bottom of the combustion chamber covered in soot. The technician explains to the customer that without the retention ring in place, the burner cannot properly mix the fuel and air, so a low-temperature, sooty flame is the result. The customer is also informed that besides replacing the retention ring, the heat exchanger, chimney connector, and chimney will all need cleaning before the furnace can operate safely again.

Service Ticket 5604

Customer Complaint: No Heat

Equipment Type: Standard 80 Percent Efficient Oil Furnace in Auto Garage

A technician is called to a no-heat call on an oil furnace and finds that both the primary control and the burner motor overload have tripped. The technician checks for oil in the combustion chamber and finds it dry. The burner motor overload and primary control are reset and the furnace begins to start. The burner motor makes a loud hum and trips out again along with the primary control.

The technician turns off the power and then tries to turn the combustion blower by hand. It will not turn freely. The drive coupling is removed from the pump and the blower is turned once again. Now the motor and fan turn freely, so evidently the oil pump is locked up. The technician also finds that the fuel filter is very dirty and it has an unusual black, tarry look to it. This leads the technician to talk to the customer about the fuel supplier.

After a brief discussion, the customer confesses that he has been pouring used motor oil into the oil tank. The technician politely explains that the furnace is not designed to burn used motor oil and now the furnace needs a new oil pump, a new nozzle, and a new filter. The heat exchanger and chimney will need to be cleaned. In addition, the oil tank should be drained and the lines flushed. The customer asks if there is any alternative, and the technician explains that furnaces are available that are designed to operate on used oil. The technician sets up an appointment for the customer with a company sales representative.

Service Ticket 5605

Customer Complaint: No Heat

Equipment Type: Oil-Fired Hydronic Boiler

Upon arriving, the technician finds that the primary control has tripped. So first the combustion chamber is inspected for oil. As to be expected, a puddle of oil is found on the bottom of the combustion chamber and is cleaned out. After this, the technician disconnects the nozzle line connector and directs it toward a container. When the primary control is reset, the burner comes on. A clear steady stream of oil shoots into the container, but no characteristic buzz is heard from the electrodes.

The nozzle line connector is replaced and the wires to the burner motor are disconnected. The technician starts the furnace again and this time tests the transformer output with a special high-voltage probe. The transformer is not producing any secondary voltage. The technician turns off the power and measures the resistance of the secondary winding. The reading shows an infinite resistance, indicating it is open. After changing the ignition transformer and checking the electrode gap, the boiler is started and found to run correctly.

Service Ticket 5606

Customer Complaint: No Heat

Equipment Type: 84 Percent Efficient Oil Furnace with Flame-Retention Burner

When the technician arrives, the furnace is not operating and the primary control has tripped. A check of the combustion chamber shows no accumulated oil.

The technician starts the furnace, it lights, and a short while later the primary control trips and the furnace shuts down.

The technician disconnects the cad cell, resets the primary control, and jumpers the F terminals once the burner has lit. This time the furnace remains firing and the cad cell resistance is measured to be 20,000 Ω. The resistance is too high so the cad cell is cleaned and then checked to make sure it is positioned correctly. Even so, the resistance reading remains high, and this indicates that the cad cell needs to be replaced. The technician replaces the cad cell and the furnace lights and stays running.

UNIT 56—REVIEW QUESTIONS

1. What is indicated by a frothy stream of oil in the oil supply line?
2. List the general types of oil furnace problems.
3. How can the cause of a locked burner motor be determined?
4. How can a defective ignition transformer be diagnosed?
5. What are some common types of ignition system problems other than a defective ignition transformer?
6. What are the symptoms of a defective oil pump drive coupling?
7. What are the symptoms of an oil pump with a leaky shutoff?
8. What resistance readings indicate a good cad cell?
9. What causes soot buildup?
10. What is puff-back?
11. What safety precaution should be followed when working on a furnace with a tripped primary control?
12. According to Table 56-1, what can cause carbon deposits on the refractory?
13. List some types of oil pump problems.
14. When should an oil pump internal bypass plug be removed?
15. How can a leak in the fuel suction line cause no oil to be delivered to the pump?
16. What two methods are used to control the indoor fan?
17. According to Table 56-1, what are some causes of a short-cycling indoor blower?
18. Provide a list of preliminary checks when troubleshooting oil furnaces.
19. What unique problems can arise in oil furnaces because oil is a liquid fuel?
20. What two manual resets do all oil burners have?

UNIT 57

Space Heaters

OBJECTIVES

After completing this unit, you will be able to:

1. list the different types of gas space heaters.
2. list the different types of electric space heaters.
3. explain which type of space heater is best suited for a specific application.
4. determine the best location for an infrared-type space heater.

57.1 INTRODUCTION

A space heater is used to heat a dedicated space rather than to supply heat for the entire building. There are a wide variety of types, with some being portable while others are permanently installed in the space. Space heaters employ a combination of radiation, natural convection, and forced convection to transfer the heat produced. The energy source may be liquid, solid, gaseous, or electric. Space heaters are often used for garages, workshops, and utility rooms that may not be a part of the regular house heating circuit. They are also popular in applications where heat is to be used only on a periodic basis, such as for a vacation home. Industrial space heaters can be used to heat a warehouse by staggering them at different locations throughout the building with each unit heating a specific area. Space heaters are designed to be easily installed and operated, and due to the variation in their type and flexibility, they are commonly used for many different applications.

57.2 GAS ROOM HEATERS

Gas room heaters are self-contained, freestanding units that transfer the heat produced from fuel combustion to room air through a heat exchanger system. The heat is transferred by radiation and convection without mixing the flue gas with the circulating room air. Since air for combustion is supplied from the room air, these units must be installed in a space that has sufficient infiltration to supply the necessary combustion air. Room heaters are made in a number of different constructions: vented, unvented, and catalytic.

TECH TIP

Many local codes do not allow unvented space heaters to be used in many occupied areas. This may mean that when replacing an unvented space heater with new equipment, a vented space heater may need to be used. The primary concern for unvented space heaters is the oxygen depletion and possible carbon monoxide buildup. For that reason, it is strongly recommended and may be mandated by ordinance that carbon monoxide detectors be located in any area where unvented space heaters are being used.

Vented Room Gas Heaters

Vented room gas heaters have an opening that is permanently connected to the chimney to convey the flue gas to the outdoors. The combustion gases pass through a heat exchanger that transfers the heat to the room air by radiation and convection. Cool room air enters the grille at the bottom of the unit and the heated air is distributed from the top. These heaters often have a glass panel on the front to supply radiant heat. Room air is completely separated from the combustion gases.

Some room heaters are operated entirely by gravity. Others use a fan to circulate the room air through the unit. The fan increases the efficiency of the heater and provides better distribution of the heated air. Since the air for combustion is taken from the room, applications are limited to spaces where there is sufficient infiltrated air to supply the required combustion air. These units are available in sizes of 10,000 to 75,000 BTU/hr.

Unvented Room Gas Heaters

Unvented room gas heaters discharge the products of combustion into the room. They are limited in application to commercial projects where the area is relatively open. They are often used during building construction to supply temporary heat. One type commonly used is called a "salamander." It is portable and can be located where needed.

Catalytic Room Gas Heater

The catalytic heater transfers heat from a glowing heat exchanger. It has no flame. The heat exchanger is constructed of fibrous material impregnated with a catalytic substance that accelerates the oxidation of a gaseous fuel. Catalytic heaters transfer heat by radiation and convection. The surface temperature is below red heat, usually about 1,200°F.

57.3 GAS WALL FURNACES

Gas wall furnaces are designed in a vertical configuration to fit into the stud space. They heat a single room or have a rear boot to also supply heat to an adjacent room. The units are usually 6 or 8 in deep, so that part of the cabinet protrudes into the room. The units are completely self-contained and

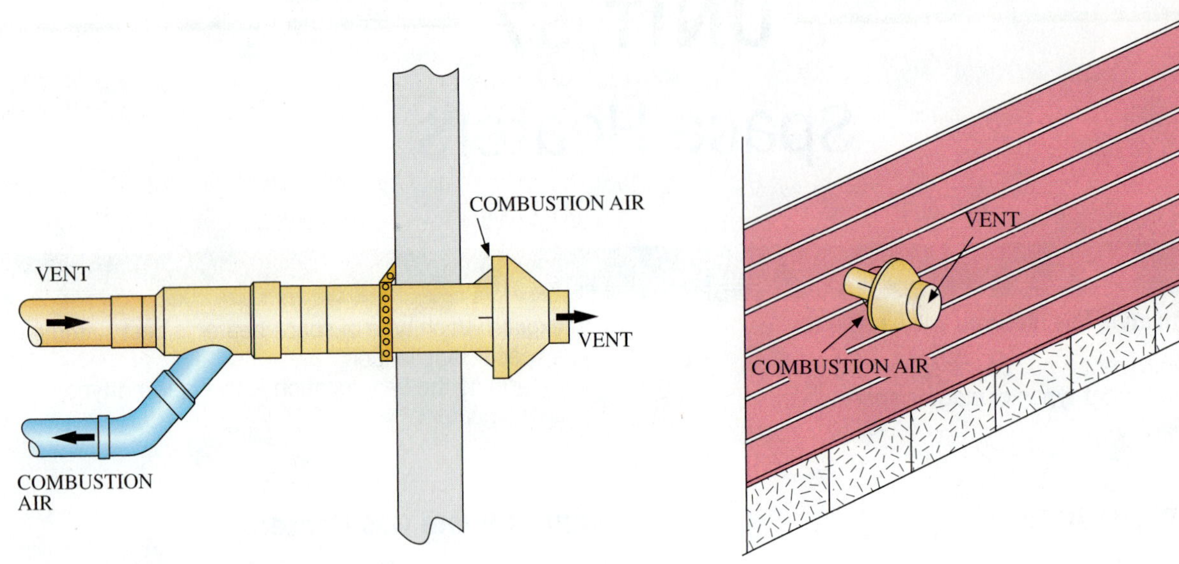

Figure 57-1 Direct vent for gas heater.

operate from a room thermostat. Some units have gravity air circulation, while others use a small blower. Some units have conventional venting arrangements with a flue extending above the roof. Others, located on the outside wall, have direct vents.

The installation of a direct-vent unit requires one penetration to be made through the exterior wall. The exhaust and intake are a double-pipe configuration (Figure 57-1). For small gas space heaters, the direct-vent connection is often supplied with the unit (Figure 57-2a-d). The gas connection and operation are similar to regular gas furnace types. A typical sealed combustion chamber and igniter are shown in Figure 57-3. These units pull the air for combustion from the outside and therefore can be placed in a tight room. Heating capacities are available from 6,000 to 65,000 BTU/hr.

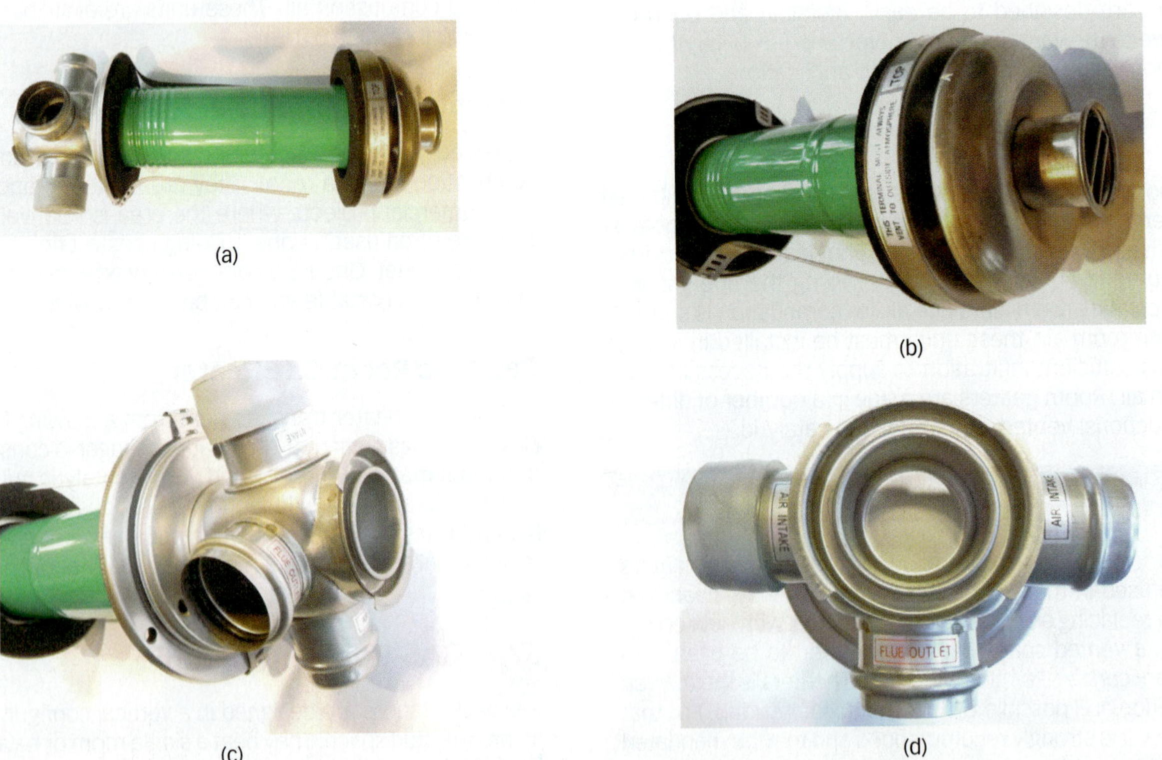

Figure 57-2 (a) Direct-vent piping for small gas space heater; (b) outside vent and air intake; (c) connections for flue gases and combustion air; (d) connection to gas space heater.

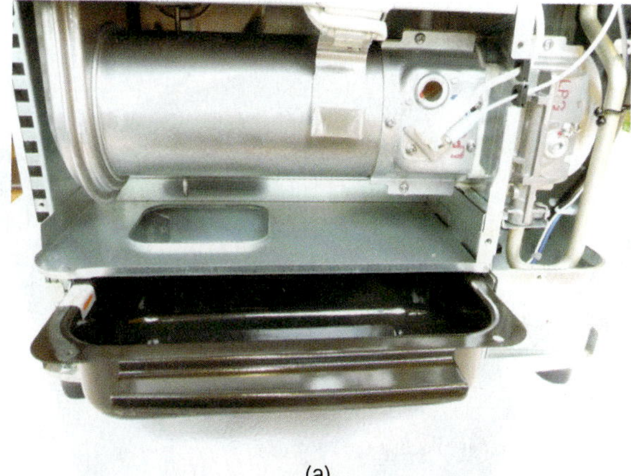

(a)

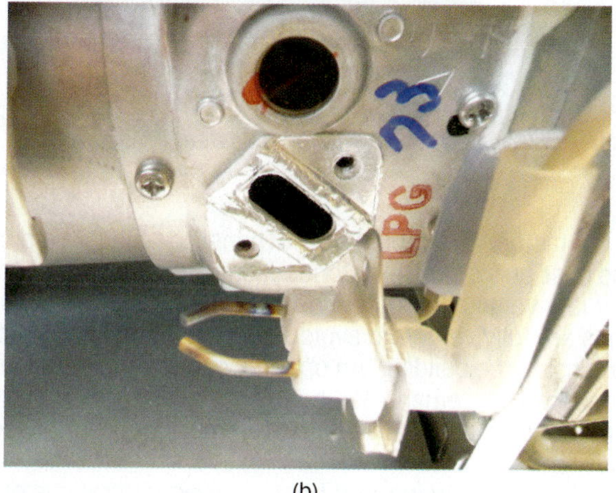

(b)

Figure 57-3 (a) Sealed combustion chamber for small gas space heater; (b) igniter for gas space heater.

57.4 GAS FLOOR FURNACES

Gas floor furnaces are constructed for suspension from a floor. The unit is constructed for supplying heated air through the center of the grille, with the return air entering through the outside corners. Combustion air is drawn from the outside of the building. The common application of these units is in a central room of a small house, where often a single unit is used to circulate heated air through the entire house.

SERVICE TIP

Some wall furnaces and floor furnaces use micro-voltage thermostat and gas valves. These units are typically ones that do not have circulating blowers, so there is no need to supply line voltage to the furnaces. The voltage to operate the system comes from a thermopile. A thermopile is much like a thermocouple except that it generates between 600 and 800 mV, which is significantly more than the 20 to 30 mV produced by a thermocouple. The thermopile's voltage is enough to operate the gas valve. Because the operative voltage is so slight, any loose connections in the control wiring can cause the system to fail to operate.

Table 57-1 Gas-Fired Direct Heating Equipment Efficiency Requirement

Heater Type	Input (BTU/hr)	Standard Level
Wall fan heaters	up to 42,000	AFUE = 75%
Wall fan heaters	over 42,000	AFUE = 76%
Wall gravity heaters	up to 27,000	AFUE = 65%
Wall gravity heaters	27,000 to 46,000	AFUE = 66%
Wall gravity heaters	over 46,000	AFUE = 67%
Floor heaters	up to 37,000	AFUE = 57%
Floor heaters	over 37,000	AFUE = 58%
Room heaters	up to 20,000	AFUE = 61%
Room heaters	20,000 to 27,000	AFUE = 66%
Room heaters	27,000 to 46,000	AFUE = 67%
Room heaters	over 46,000	AFUE = 68%
Hearth heaters	up to 20,000	AFUE = 61%
Hearth heaters	20,000 to 27,000	AFUE = 66%
Hearth heaters	27,000 to 46,000	AFUE = 67%
Hearth heaters	over 46,000	AFUE = 68%

57.5 GAS HEATER MINIMUM EFFICIENCY REQUIREMENTS

The National Appliance Energy Conservation Act of 1987 established minimum efficiency requirements for all gas-fired direct heating equipment (Table 57-1). The AFUE values are obtained by the test methods set up by the Department of Energy.

57.6 GAS HEATER CONTROLS

The requirements for gas space heater controls are different from conventional furnaces. Since a space heater is located in the room, a wall-mounted thermostat is more often an option as compared to a thermostat that is built directly into the heater.

Wall Thermostats

These thermostats can be either 24 V or use millivolt power for thermopile-type gas heaters. They are selected to operate whatever gas valve is being used. A suitable source of power must be supplied.

Built-in Hydraulic Thermostats

These are made in two types: a snap-action two-position thermostat with a liquid-filled capillary-tube temperature-sensing

Figure 57-4 Gas fired infrared heater for large building or warehouse.

element, and a modulating-type thermostat, similar to the first type, except the temperature alters the position of the gas valve between off and fully open.

57.7 GAS-FIRED INFRARED HEATERS

A typical gas-fired infrared heater installation for a warehouse or large open area is shown in Figure 57-4. Heaters are located overhead and radiate heat to the area below. By using radiation, heat is directed to the people and objects in the spaces that require warming, rather than wasting it to the air or other nonessential surfaces. This makes possible maintaining lower room temperatures, yet creating comfortable conditions and saving energy.

An operating heater is shown in Figure 57-5. Individual burners of this type can the range from 20,000 to 120,000 BTU/hr, fired in series, with branches connected to one induction blower. These systems are custom engineered to meet specific floor plans as well as heating requirements. They can be operated in a condensing mode for the ultimate in heating efficiency. Direct-spark ignition is used. Combustion chambers can be either fabricated steel or optional cast iron. The tailpipe is porcelain lined.

Figure 57-5 Gas fired infrared heater in operation..

Figure 57-6 High-intensity infrared heater.

A factory-assembled control panel combines pre-purge and post-purge controls. Connections are provided for up to four low-voltage thermostats for zone temperature control. Units are fully vented to avoid release of combustion moisture inside the building. An optional controller continuously monitors the demand for heat and adjusts firing cycles. Automatic setback can be programmed for nights, weekends, and holidays.

High-intensity infrared heaters using natural gas or LP are also available for direct radiant heating, Figure 57-6.

57.8 VAPORIZING OIL POT (KEROSENE) HEATERS

Vaporizing-type oil-burning units use grade 1 fuel oil (kerosene). The oil is burned in a bowl or pot. A constant level metering device, with an adjustable needle, controls the flow of oil into the combustion changer and pilot burner through a burner oil supply pipe. The adjustable needle can be manually placed in an OFF, PILOT, or VARIABLE FLOW setting. The oil flows by gravity from a 2- or 3-gallon tank attached to the unit or a larger tank located outside.

Oil is vaporized by the heat of combustion and by contact with the hot metal surfaces of the combustion chamber. The combustion air enters from the room and mixes with the vaporized oil. Units can be supplied for gravity air circulation or with an air circulator. A port opening for cleaning the unit and for access for lighting the pilot is provided on the front of the drum-type heat exchanger. An automatic draft controller is used on the flue pipe to maintain the required draft. Flue gases are vented to the outside. A perforated metal grille is located on the lower front of the cabinet to admit return air for heating. The supply of heated air enters the room from the top of the unit. The steel heat exchanger furnishes radiant heat to the room by heating the side walls of the cabinet. A thermostat can be used to operate the unit by turning on and off the unit at a selected firing rate.

57.9 ELECTRIC IN-SPACE HEATERS

There are various types of electrical in-space heaters. Because there are no products of combustion and no flue, their location is very flexible.

TECH TIP

Although electric heat is the most expensive type for heating, it is the least expensive as far as equipment and maintenance are concerned. Electric space heaters are ideal for applications such as sunrooms, vacation homes, and other areas where heating would only be used on a limited basis. Electric space heaters can sit for long periods of time unused and still function reliably.

Forced-Air Electric Heaters

Forced-air electric heaters add a measure of comfort to electric space heating. In homes and offices, the recessed models combine style with the forced-air circulation, resulting in more uniform room temperature control. Most are wall mounted, but models are also available for recessing into the ceiling and even under kitchen counters. Wall-mounted units provide up to 3,000 W (10,240 BTU/hr), which can warm larger rooms or offices.

Suspended-Force Fan-Unit Electric Heaters

Suspended-force fan-unit electrical heaters are most functional for garages, workrooms, and playrooms where large capacity and positive air circulation are important. These are also popular in commercial and industrial establishments of all kinds. They may be suspended in vertical or horizontal fashion, as shown in Figure 57-7, with a number of control options, such as wall switch only, thermostats, timers, and night setback operation. Capacities generally range from 3 to 12 kW and above. Discharge louvers may be set to regulate air motion patterns. The heavy-duty construction of this heater category is geared for commercial application.

Cabinet-Type Electrical Space Heaters

Cabinet-type electrical space heaters are found in classrooms, corridors, foyers, and similar areas of commercial and institutional buildings where cooling is not a consideration. Heating elements go up to 24 kW, and the fans are usually the centrifugal type for quiet but high volume airflow. Cabinets have provision for introducing outside air where use and/or local codes require minimum ventilation air.

Outdoor air intake damper operation can be manual or automatic. Control systems can be simple one-unit operation or multizone control from a central control station.

Figure 57-7 Suspended space heater.

Convection Baseboard Electrical Heaters

This is the most frequently used electric space heater for residential and commercial buildings. The heater provides for convective airflow across the finned-tube heating element. The contour of the casing provides warm air motion away from the walls, thus keeping them cooler and cleaner, as shown in Figure 57-8.

It is important that drapes do not block the airflow and cause overheating or that the carpeting block the inlet air. Most baseboard units have a built-in linear-type of thermal protection that prevents overheating. If blockage occurs, the safety limit stops the flow of electrical current. It cycles off and on until the blockage is removed. Baseboard heaters come in lengths of 2–10 ft in the nominal 120, 208, 240, and 277 V rating. Standard wattage per foot is 250 at the rated voltage. But lower heat output can be obtained by applying heaters on lower voltage.

For example, a standard 4-ft heater rated at 250 W/ft at 240 V will produce 1,000 W of heat. Most manufacturers offer these reduced-output (low density) models by changing

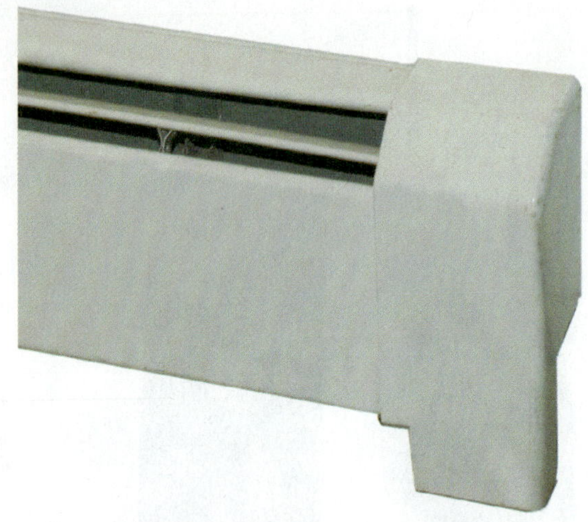

Figure 57-8 Baseboard space heater.

the heating element to 187 W/ft. Low density is generally preferred by engineers and utilities and is recommended for greatest comfort. Controls may be incorporated in the baseboard or through the use of wall-mounted thermostats. Accessories such as corner boxes, electrical outlets, and plug-in outlets for room air conditioners are available.

Electric Infrared Heaters

A multimount electric infrared heater has many applications, including total area heat, spot heat, and snow and ice control. They can be mounted horizontally, or at a 30°, 60°, or 90° angle. The double reflectors can be placed to provide a wide variety of patterns. Two elements are used per fixture. Three different lengths are available, and units can be mounted end to end for additional lengths. Units are available in sizes of 3,200, 5,000, and 7,300 W. Any surface or object should be 24 in away from direct radiation from the unit. For mounting, UL requires fixtures to be at least 3 in from the ceiling and 24 in from a vertical surface. Rows of fixtures should be separated by a minimum of 36 in.

Electric Radiant Heating Cable

Electric radiant heating cable was one of the most popular methods of heating early in the development of radiant heating. It is an invisible source of heat and does not interfere with the placement of furnishings. The source of warmth is spread evenly over the area of the room. It can be installed within either plaster or drywall ceilings, as shown in Figure 57-9. Staples hold the wire in place until the finish layer of plaster or drywall is applied. Wall-mounted thermostats control space temperature.

Electric Heat Floor Panels

Electric heat floor panels can be constructed using perforated mats consisting of PVC-insulated heating cable woven in or attached to metal or glass fiber mesh. Such assemblies are available in sizes from 2 to 100 ft^2, with various watt densities ranging from 15 to 25 W/ft^2. Another effective method of slab heating uses mineral insulated (MI) heating cable. MI cable is small-diameter, highly durable, solid, electrical-resistance heating wire surrounded by compressed magnesium oxide electrical insulation and enclosed in a metal sheath. Several MI cable constructions are available, such as single conductor, double conductor, and double cable, as well as custom designed cables. Any of these constructions can be embedded in concrete.

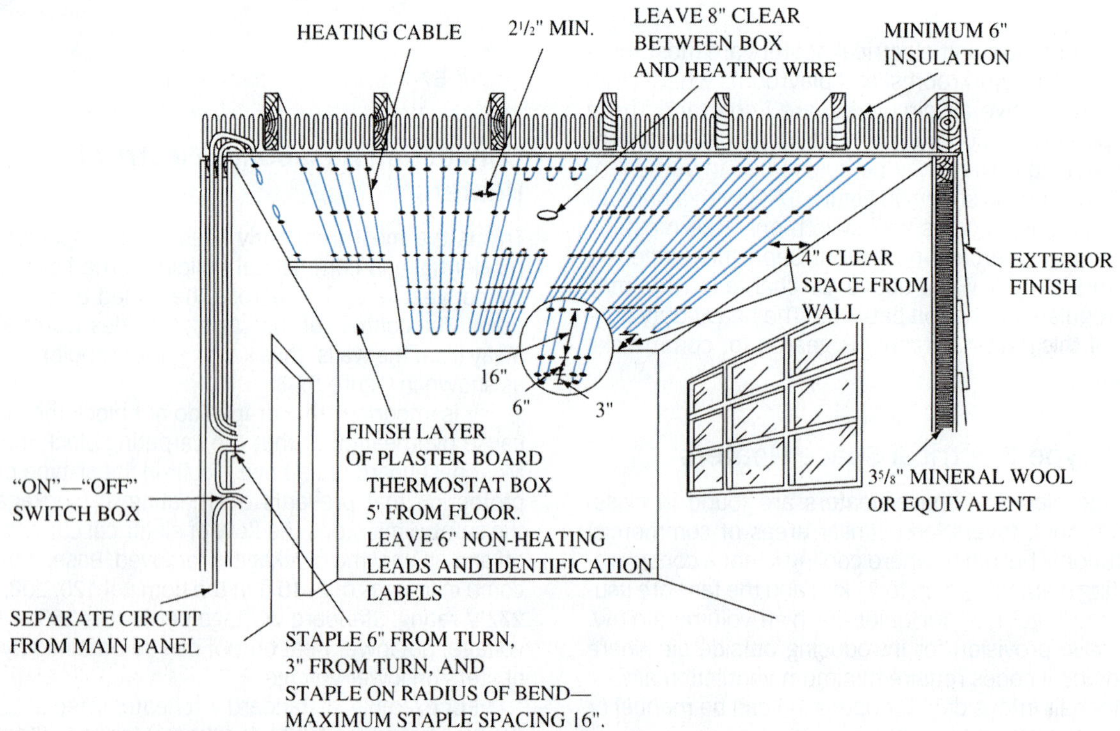

Figure 57-9 Diagram showing installation of ceiling cable for radiant heating.

Figure 57-10 Room steam space heater.

57.10 STEAM SPACE HEATERS

Steam space heaters are used in applications where a ready supply of low-pressure steam is available. This type of heater is often suspended from the ceiling, as shown in Figure 57-10. The steam control valve is generally operated by a thermostat. The return line for the condensate will include a strainer and a steam trap. When this type of heater is used in a space that may fall below the freezing temperature, it should be turned on and operating. If the space is to be unheated for a period of time, a provision must be made for draining the heating coil. If the coil is not drained and freezes, the water inside will expand and split the coil. This type of damage is difficult to repair, and most likely the coil will need to be replaced.

UNIT 57—SUMMARY

There are many different types of space heaters available for different applications. Common types are gas or electric, but steam and kerosene-type heaters are also used. Direct-vent gas space heaters are fairly simple to install as there is no requirement for a chimney. Electric space heaters do not require any fuel and can be left for long periods without the need for layup or start-up maintenance. When considering the type of heater to install, consider the venting requirements, the availability of a fuel source, and the local code requirements. Larger space heaters may have repairable components, while smaller units may be designed for total replacement rather than repair. Space heaters are available in many different heating capacities, which allows them to be closely matched to the heating requirements for the space.

WORK ORDERS

Service Ticket 5701

Customer Complaint: Steam Leaking from Heater

Equipment Type: Steam Space Heater

The technician is told that steam began leaking from the heater when it was turned back on after the 3-week winter shutdown at the facility. The technician inspects the heater and then carefully opens the steam valve a small amount. Steam begins leaking out through a number of splits in the heating coil. The technician informs the facilities manager that the heater is beyond repair and needs to be replaced. It is also recommended that the coil be drained the next time a winter shutdown takes place.

Service Ticket 5702

Customer Complaint: Pilot Will Not Light

Equipment Type: Direct-Vent Gas Space Heater

The technician is told that the pilot will not light even though there seems to be gas available. The technician inspects the heater and asks if it has ever been cleaned before and is told it has not. The combustion chamber is removed and the rust that has collected inside the unit is vacuumed out and the thermopile and pilot are cleaned. Now the pilot lights normally and the unit works and provides the proper flame pattern.

Service Ticket 5703

Customer Complaint: Replacement Installation

Equipment Type: Unvented Gas Space Heater

The customer explains that the old gas space heater in the workshop needs to be replaced. It has worked well for many years, and an exact duplicate is preferred. The technician inspects the unit and explains that the local code no longer allows an unvented space heater to be used in an occupied area. This is because of the possibility of oxygen deprivation and possible carbon monoxide buildup. The customer complains that this will be more expensive if a chimney needs to be added to the workshop. The technician explains that a direct-vent-type gas space heater can easily be installed, and that would not require a chimney. It would only cost slightly more than the unvented gas heater to install. This would meet the code requirements and be much safer overall.

UNIT 57—REVIEW QUESTIONS

1. What is a restriction to be considered when installing a gas room heater?
2. What safety device should be installed where unvented space heaters are being used?

3. How does a vented room gas heater differ from an unvented room gas heater?
4. What is a "salamander"?
5. What type of gas space heater has no flame?
6. What is meant by a "direct-vent" heater?
7. Why do some gas furnaces not require an electrical power supply?
8. How do infrared heaters transfer heat?
9. What type of fuel does a vaporizing oil pot heater use?
10. How is the oil vaporized in a vaporizing oil pot heater?
11. What happens to an electrical space heater if it sits for long periods of time unused?
12. What is an advantage of a forced-air electric heater as compared to natural convection?
13. What type of electric space heater is used when large capacity and positive air circulation are important?
14. Where are cabinet-type electrical space heaters typically installed?
15. What is the most frequently used type of electric space heater for residential and commercial buildings?
16. How does a standard convection baseboard electrical heater compare to a low-density type?
17. How can electric infrared heaters be mounted?
18. What distance should an electric infrared heater be from any other surface?
19. How are mats for electric heat floor panels constructed?
20. What is mineral insulated (MI) heating cable?

UNIT 58

Humidifiers

OBJECTIVES

After completing this unit, you will be able to:

1. list the types of humidifiers.
2. recommend humidifiers that are appropriate for heat pump systems.
3. explain why humidification is needed in the winter.
4. describe humidifier maintenance procedures.
5. discuss humidifier control systems.

58.1 INTRODUCTION

Total comfort involves more than heating and cooling the air. In the heating season, the air is not truly conditioned until it has been humidified. Heating the air causes a natural decrease in relative humidity. Without humidification, the relative humidity in a heated house is similar to the Sahara Desert. Humidifiers that are not properly maintained will lose their effectiveness and can become a source of air contamination. Installing and servicing humidifiers is part of an air-conditioning technician's job description.

58.2 WHY HUMIDIFICATION IS NEEDED

The heated air in homes and buildings becomes dry during the heating season. The outdoor air cannot hold as much moisture at the lower temperatures. When the cold winter air is heated, the increased volume increases the air's ability to hold moisture, but the amount of moisture actually in the air remains the same. As a result, the relative humidity drops drastically.

This super-dry air sucks water out of everything it contacts, including people. This impacts comfort by making us feel colder and by drying out our mucous membranes. Other undesirable conditions that result from low relative humidity include increased static electricity and shrinking and cracking of wood furniture. Adding humidity to the air can also help save energy in the winter by promoting comfort at lower temperatures. It is desirable to maintain 30–60 percent relative humidity in the house. Figure 58-1 shows that many undesirable conditions that are detrimental to health are decreased between 30 and 60 percent relative humidity.

Air at 35°F and 100 percent relative humidity contains less water than air at 70°F and 30 percent relative humidity. Air at 35°F and 100 percent relative humidity contains 30 grains of moisture per pound of air. When air is heated to 70°F, it can hold 110 grains of moisture per pound. Thus, to maintain 50 percent relative humidity in a 70°F house, the grains of moisture per pound must be increased to 55 ($110 \times 0.5 = 55$). One pound of air entering a house from the outside will require the addition of 25 grains of moisture ($55 - 30 = 25$). The amount of infiltration depends on the tightness of the windows and doors and other parts of the construction.

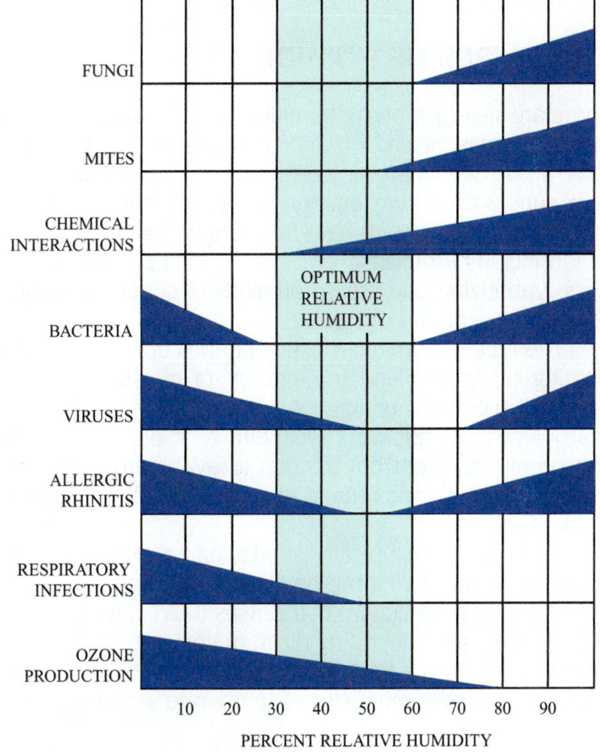

Figure 58-1 Maintaining the relative humidity between 30 and 60 percent has numerous health benefits.

It is impractical in most buildings to maintain high relative humidity when the outside temperature is very low. Condensation forms on the inside of cold building surfaces when the surface temperature drops below the dew point of the air. This occurs first on the windows, but it can also occur inside the walls, contributing to rotting of walls from the inside out. For this reason, the recommended humidity level is lower for low outside temperatures than for higher outdoor temperatures. The amount of water that can be safely added to the air depends on the following:

1. Outside temperatures
2. House construction
3. Amount of relative humidity that the interior of the house will withstand without condensation problems

OUTDOOR TEMPERATURE	RECOMMENDED SETTING
-20°F	15%
-10°F	20%
0°F	25%
10°F	30%
20°F	40%
30°F	45%
40°F	50%

Figure 58-2 The recommended humidity setting is 50 percent at 40°F; the recommended setting decreases 5 percent for every 10°F decrease in outdoor temperature.

Figure 58-2 shows the recommended humidity levels for different outdoor temperatures.

58.3 TYPES OF HUMIDIFIERS

There are several types of humidifiers. Generally, the different types may be divided into two general classes: those that rely on heat in the air to evaporate the water and those that supply their own energy for evaporating the water. Humidifiers that require heat from the air for water evaporation include atomizing, evaporative, and bypass humidifiers. Vaporizing and steam humidifiers supply their own energy to evaporate the water.

The type of humidifier used depends upon the type of heating equipment and the amount of moisture needed. While all types of humidifiers will operate with furnaces, many types will not work well with heat pumps because the supply air is not hot enough to evaporate the water. Typically heat pumps require vaporizing, steam, or bypass humidifiers.

Humidifiers are typically wired to operate whenever the fan is operating during the heating cycle. They are usually controlled by a humidistat that senses the relative humidity of the return air. The wiring diagram shown in Figure 58-3 shows wiring with a single-speed fan motor. Vaporizing and steam humidifiers can be wired to operate whenever there

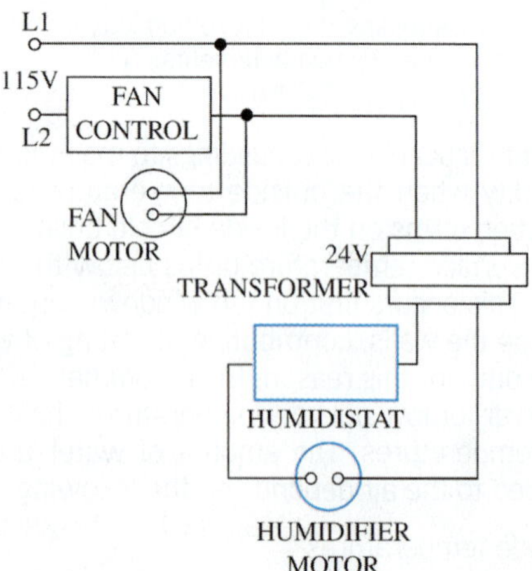

Figure 58-3 Wiring diagram for a humidifier with a single-speed motor.

Figure 58-4 Motorized atomized humidifiers can deliver 6–11 gal of water per day.

is a demand for more humidification. They can actually turn on the indoor blower and operate even if the heat is not operating.

58.4 ATOMIZING HUMIDIFIERS

Atomizing humidifiers work by spraying an atomized water mist into the airstream. Heat in the air then evaporates these small atomized droplets. Atomizing humidifiers typically require a 100°F minimum air temperature to operate. The water is atomized by either a motor that slings the water around or a spray nozzle that sprays a water mist directly into the duct. A motorized humidifier will not put more water into the air than it can evaporate, but these humidifiers can have a low capacity. Figure 58-4 shows a motorized atomizing humidifier.

The capacity of mister-style humidifiers is only limited by the nozzle size. However, it is easy to spray more water into the duct than the air can evaporate, making a mess. Mister humidifiers typically use a temperature sensor wired in series with the water solenoid to prevent spraying water into the duct when the air is not hot enough to evaporate the water (Figure 58-5). The nozzle has a built-in filter to prevent the nozzle from clogging (Figure 58-6). The nozzle should be changed annually to ensure correct operation.

58.5 EVAPORATIVE HUMIDIFIERS

Evaporative humidifiers provide a wetted surface that adds moisture to the heated air. There are three types of evaporative humidifiers:

1. Plate
2. Rotating drum
3. Fan powered

Plate Humidifiers

The plate evaporative humidifier has a series of porous plates mounted in a rack. The lower section of the plates

Figure 58-5 The mister will not energize the water solenoid until its thermostat senses that the air is warm.

Figure 58-6 Typical humidifier nozzle manufactured by Delavan. *(Courtesy of Delavan)*

extends down into water contained in the pan. A float valve regulates the supply of water to maintain a constant level in the pan. The pan and plates are mounted in the warm-air plenum. Plate humidifiers have a relatively low capacity compared to other humidifiers. The plates become brittle as mineral deposits from the evaporated water build in the plates. Annual maintenance includes changing the plates and cleaning the water sump.

Rotating-Drum Humidifiers

The rotating-drum evaporative humidifier has a slowly revolving drum covered with a polyurethane pad partially submerged in water (Figure 58-7). As the drum rotates, it absorbs water. The water level in the pan is maintained by a float valve. The humidifier is mounted on the side of the return-air plenum. Air from the supply plenum is ducted into the side of the humidifier. The air passes over the wetted surface, absorbs moisture, and then goes into the return-air plenum. Annual maintenance includes changing the drum pad and cleaning the water sump.

Figure 58-7 Rotating-drum humidifier.

Rotating Plate Humidifiers

The rotating plate evaporative humidifier is similar to the drum type in that the water-absorbing material revolves; however, this type is normally mounted on the underside of the main warm-air supply duct (Figure 58-8). Annual maintenance includes changing the plates and cleaning the water sump.

Bypass Evaporative Media Humidifier

The bypass-style humidifier is mounted on either the supply- or return-air plenum. A duct is run from the humidifier to the other plenum (Figure 58-9). When the fan runs, air

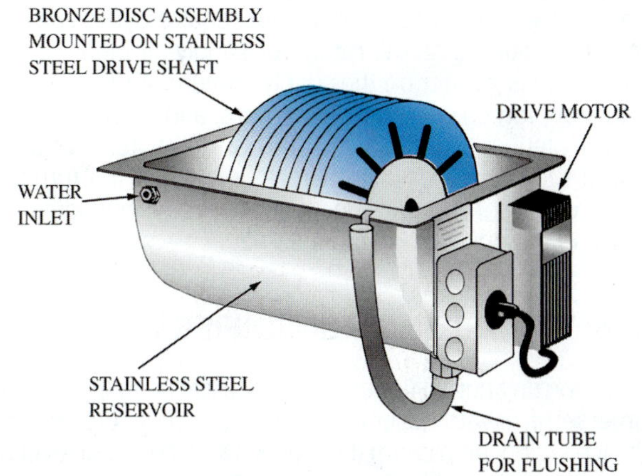

Figure 58-8 Rotating-plate evaporative humidifier.

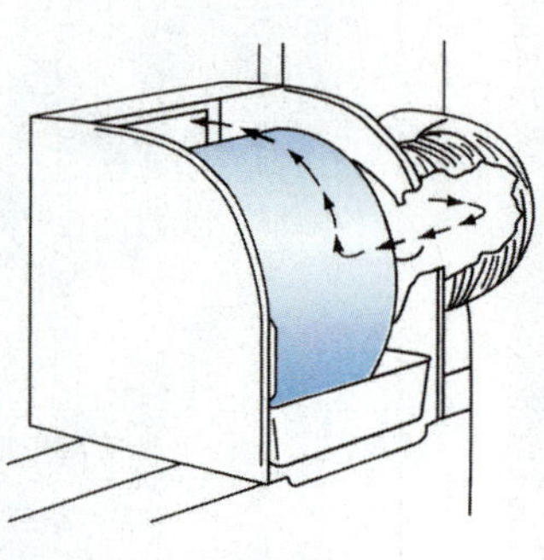

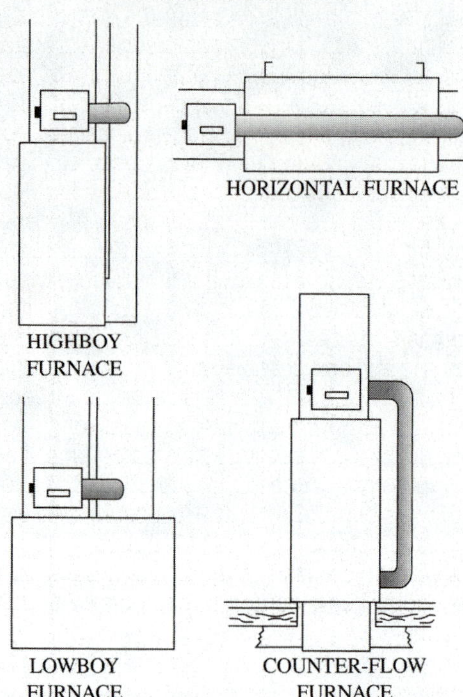

Figure 58-9 In a bypass humidifier installation, air flows from the supply plenum, through the humidifier, and into the return plenum.

passes from the supply plenum, through the humidifier, into the return-air plenum. A manual damper is installed to shut off the bypass air when the system is operated in cooling.

Air passes over a media pad inside the humidifier. Typically, the media is expanded metal (Figure 58-10). A humidistat is used to control a water solenoid valve that controls the flow of water over the media. The control system is set up so that the humidifier can operate only when the furnace fan is running. This type of humidifier usually does not require much maintenance because the water is not recirculated and does not sit in a sump. The minerals left behind from evaporation are washed away with the drain water. However, it does use more water than other types with similar capacities because the water that is not evaporated washes down the drain.

Figure 58-10 Expanded metal pad used in a bypass humidifier.

Fan-Powered Evaporative Humidifier

The fan-powered evaporative humidifier is similar to the bypass type, except that it does not require bypassing air. It can be mounted on the return- or supply-air plenum, but its capacity is greater on the supply plenum. Air is drawn in by the fan, forced over the wetted core, and delivered back into the plenum (Figure 58-11). The control system is set up so that the humidifier can operate only when the furnace fan is running. The wiring for a fan-operated humidifier is shown in the diagram in Figure 58-12.

58.6 VAPORIZING HUMIDIFIERS

The vaporizing humidifier uses an electrical heating element immersed in a water reservoir to evaporate moisture into the furnace supply-air plenum (Figure 58-13). A constant level of water is maintained in the reservoir. These humidifiers can operate even though the furnace is not supplying heat. The humidistat not only starts the water heater but also turns the furnace fan on if it is not running. These humidifiers have among the highest capacities of any type humidifier. They require more electricity than other types because the electrical element is heating water to 120°–160°F to increase the

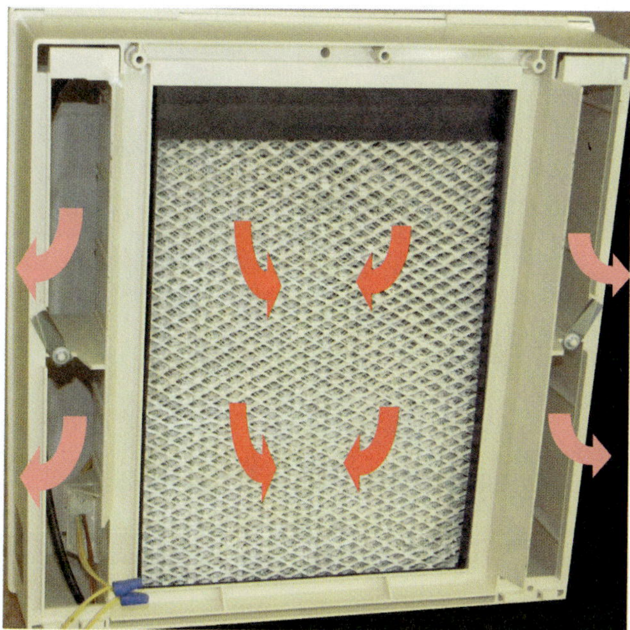

Figure 58-11 Air is drawn in by the fan, forced over the wetted core, and delivered back into the plenum.

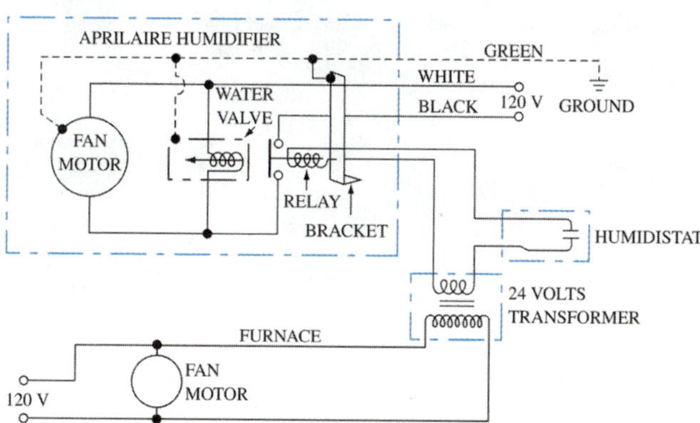

Figure 58-12 Wiring diagram for a fan-powered evaporative humidifier.

water evaporation. The power wiring required for vaporizing and steam humidifiers is much heavier than other types of humidifiers because of the increased electrical power needed to heat the water. Minerals from the water deposit on the element and in the sump, necessitating frequent cleaning. This type of humidifier often needs cleaning more than once a year. The heating element will burn out quickly if it is allowed to operate without water in the humidifier. These are both effective and simple, but frequent maintenance is absolutely essential.

Steam Humidifiers

The steam humidifier is a type of vaporizing humidifier (Figure 58-14). Steam humidifiers are ideal for use with systems like heat pumps, electric furnaces, high-efficiency furnaces, and furnaces using night setback thermostats that do not generate the high air temperatures necessary for evaporative humidifiers. The internal water temperature-sensing device

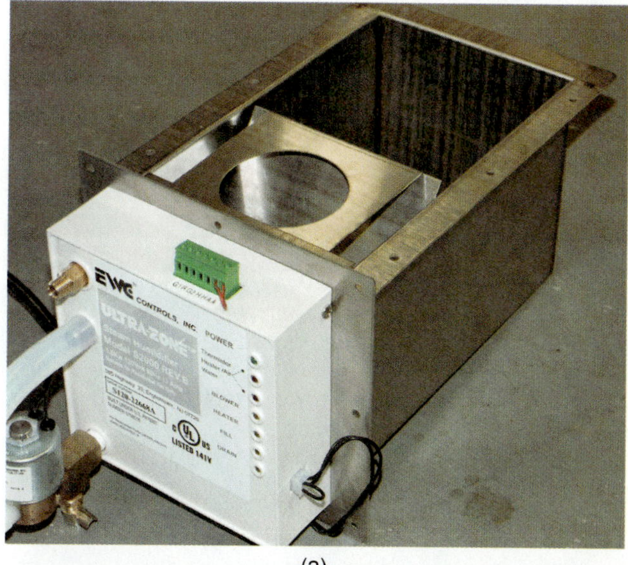

(a)

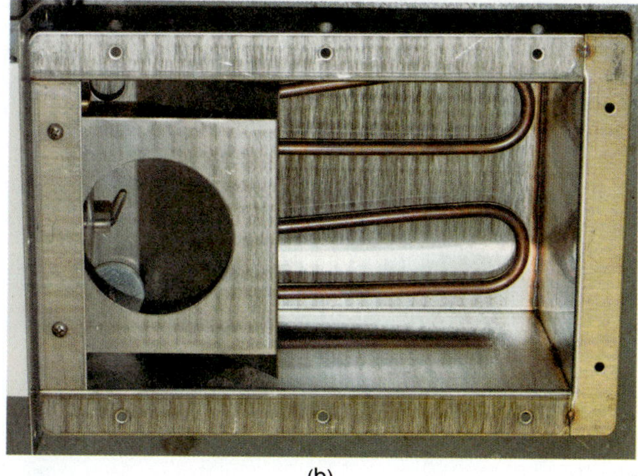

(b)

Figure 58-13 Vaporizing humidifier.

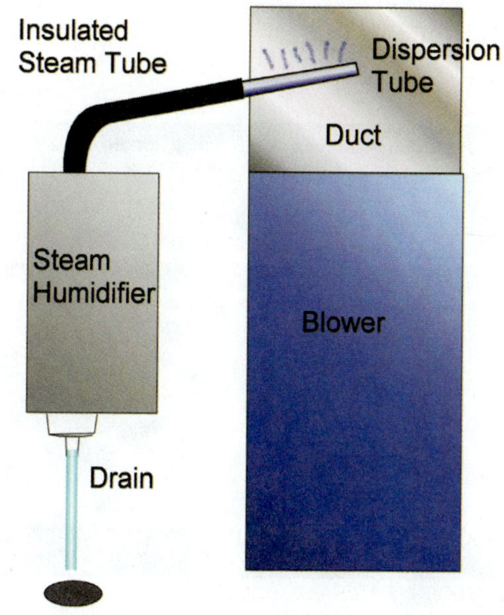

Figure 58-14 Steam humidifier.

operates the fan independently of the heating system. To reduce scaling and mineral deposits, some steam humidifiers have a flushing timer and chlorine-removal filters. The water in the unit is flushed out every 12 hours, removing accumulated solid materials.

SERVICE TIP

Mineral deposits are a problem with all types of humidifiers. When water evaporates, the minerals are left behind. This can be reduced by using filters in the water feeding the humidifier. Carbon filters can be used to remove chlorine from the water supply, thus eliminating the corrosive effects of chlorine on the unit. These filters also reduce mineral buildup.

SERVICE TIP

It is very important that humidifiers be regularly cleaned and checked for proper operation. The water sump in some types of humidifiers can grow algae and mold if not properly maintained. It is especially important to ensure that excessive moisture is not released into the house or duct system. Excessive moisture has been associated with mold growth, and mold growth is associated with poor indoor air quality.

58.7 HUMIDISTATS

A humidistat is used to control when the humidifier operates (Figure 58-15). Humidistats used with humidifiers open on an increase in humidity and close on a decrease in humidity. The setting of a humidistat can be changed to comply with changing outside air temperatures. If the setting on the humidistat is too high, it may cause sweating on interior walls and windows (Figure 58-16). As it gets colder outside, the humidistat setting may need to be lowered.

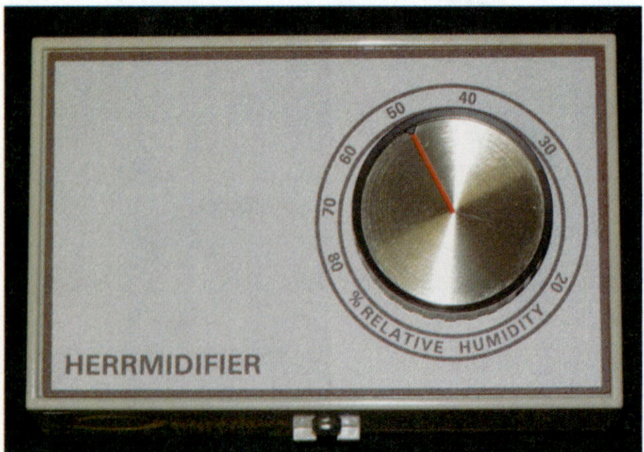

Figure 58-15 Typical humidistat for controlling a humidifier.

Figure 58-16 When too much moisture is added, the increased dew point causes water to condense on inside surfaces.

Figure 58-17 The clear nylon ribbon is the controlling element in this humidistat.

The least expensive and most common humidistats are mounted on the return-air plenum. They typically use a nylon ribbon for sensing the humidity in the air (Figure 58-17). Better-quality humidistats use hair to sense humidity change (Figure 58-18). Hair shrinks when it gets wet and stretches when it dries. Several hundred hairs are stretched across a frame that controls the opening and closing of a set of points.

Some electronic humidity controls adjust themselves automatically to changing outdoor temperature. They sense the outdoor temperature and reset the humidity to a safe level for the outdoor temperature (Figure 58-19).

A few advanced electronic thermostats also incorporate humidity control. These controls can be used to control the total system operation, including humidification (Figure 58-20).

Figure 58-18 Hair is the controlling element of this humidistat.

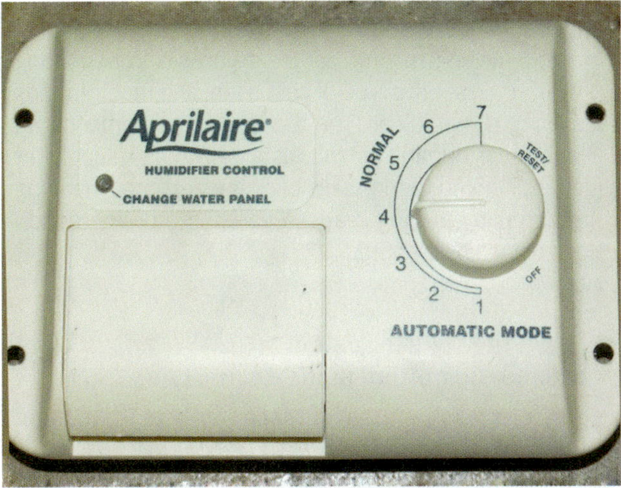

Figure 58-19 Advanced electronic humidity controller.

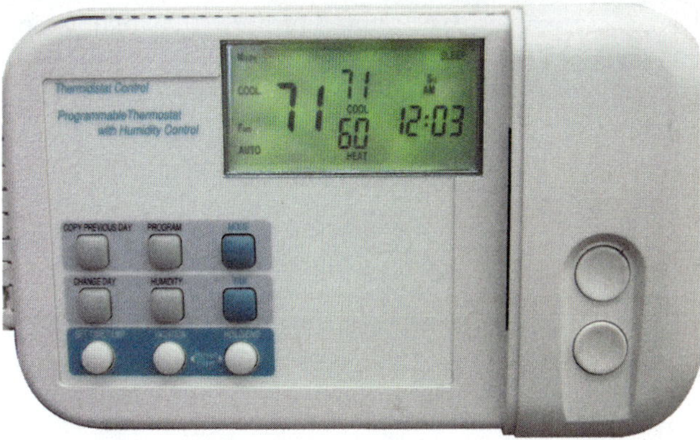

Figure 58-20 This thermostat incorporates humidity control in addition to temperature control.

UNIT 58—SUMMARY

Humidification is needed in the winter because heating the air reduces its relative humidity to desert levels. Humidifiers may be divided into two groups: those that rely on heat in the air to evaporate the water and those that supply their own energy for evaporating the water. Humidifiers that require heat from the air for water evaporation include atomizing, evaporative, and bypass humidifiers. Vaporizing and steam humidifiers supply their own energy to evaporate the water. Humidistats are used to control the operation of humidifiers. Mineral deposits are a common problem with many types of humidifiers. Humidifier service includes changing media and cleaning mineral deposits.

WORK ORDERS

Service Ticket 5801

Customer Complaint: Morning Dry Mouth and Irritated Sinuses

Equipment Type: Gas Furnace

A customer complains that the family wakes up in the morning with dry, cracked lips and irritated throat and sinus passages. They feel that the gas heat is to blame and ask if another form of heat would cause less drying. The technician explains that the dryness is from heating the air, not from the type of heat used, and recommends installing a humidifier. The technician recommends an evaporative humidifier and installs it in the supply-air plenum. The customer calls back a week later thanking the company, saying that the parched morning conditions have disappeared.

Service Ticket 5802

Customer Complaint: Static Electricity

Equipment Type: Gas Furnace with a Plate-Type Humidifier

A customer has noticed that the amount of static electricity in her home seems high. She suspects that their humidifier is not working. The technician arrives and notices that the plates in the humidifier are hard and brittle, instead of soft. The technician explains to the customer that the plates are not very effective when they are calcified with mineral deposits. The humidifier is cleaned and the plates are replaced. The next day the static electricity in the house goes away.

UNIT 58—REVIEW QUESTIONS

1. What advantages are there to adding humidity to heated air?
2. What is the desirable amount of relative humidity that should be maintained?
3. What are the two general categories of humidifiers?
4. What types of humidifiers work well with heat pump systems?
5. Which types of humidifier require the least cleaning?
6. Why do most humidifiers require annual cleaning and media replacement?
7. Why should the humidity level in the house be reduced as the outdoor temperature drops?
8. Explain how a bypass humidifier works.
9. What are the two most common controlling elements in a mechanical humidistat?
10. How is steam humidifier operation different from evaporative humidifier operation?
11. Why should a steam humidifier be filled with water before it is energized?
12. How can an improperly maintained humidifier contribute to bad indoor air-quality issues?
13. Why do steam humidifiers require heavier electrical wiring than bypass-type humidifiers?
14. What is the purpose of the flush cycle used in some steam humidifiers?
15. On what types of heating systems can atomizing humidifiers be used?

UNIT 59

Electric Heat

OBJECTIVES

After completing this unit, you will be able to:

1. describe the different applications for electric heating.
2. explain the purpose of sequencers used for electric heating elements.
3. describe how the safety devices of electric furnaces operate.
4. determine the best application for supplemental electric heating.
5. explain the operation of a multistage thermostat for electric heating.

59.1 INTRODUCTION

The growth of electric heating began in the 1950s and first became popular in areas not served by natural gas pipelines. Throughout this same period, the electrical demand for air conditioning during the summer months was increasing. To help boost the winter demand, electric utilities and heating manufacturers began to encourage the use of electric heating. Special rates and promotional programs like "Live Better Electrically" and "Total Electric Home" were introduced in the 1960s, and the market developed rapidly.

Today, many systems provide both heating and cooling with a single air-distribution system. Therefore, most air handlers are configured to accept both cooling coils and heating elements. This has reduced the number of installations for straight stand-alone electric furnaces, and instead electric heat is often used in combination with a cooling system or heat pump.

59.2 WARM AIR ELECTRIC FURNACES

The typical residential electric furnace (Figure 59-1) consists of a cabinet, blower compartment, filter, and resistance heating section. Most manufacturers provide a place for a cooling coil. An electrical furnace has many advantages:

- It is more compact than the equivalent gas or oil furnace.
- Due to cooler surface temperature, it has "zero" clearance requirements on all sides. It may, therefore, be located in small spaces, such as closets.
- Since there is no combustion process, there are no requirements for venting pipes, chimney, or makeup air. This reduces building costs and simplifies installation.
- Units may be mounted for upflow, downflow, or horizontal airflow applications (Figure 59-2).

With electric heat there are no losses such as those experienced with oil and gas combustion processes. Therefore electric heat is considered to be close to 100 percent efficient.

Figure 59-1 External view of an electric furnace.

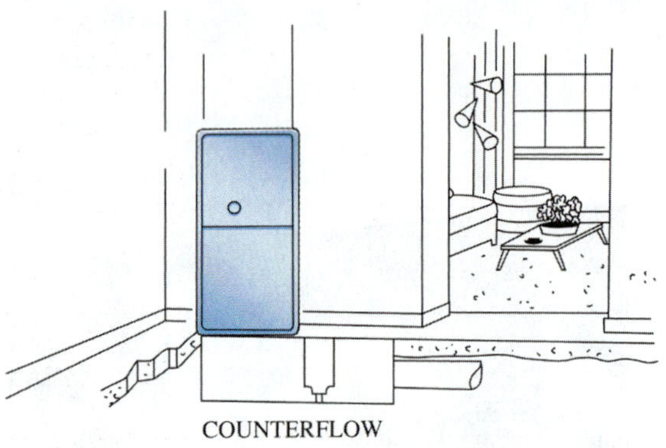

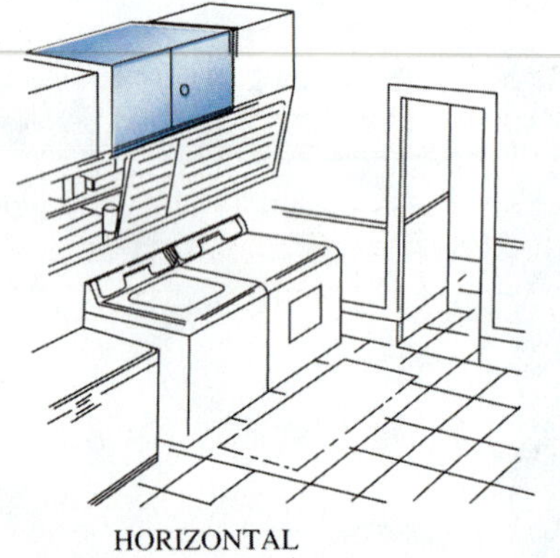

Figure 59-2 Various types of electric furnaces.

TECH TIP

Always try to help your customer understand the different cost alternatives available. Electric heat is almost always more expensive than other types. Think of electrical power to be a finished product as compared to a raw material. As an example, a sheet of plywood will cost more than rough sawn boards. In the same way, electrical power will cost more than gas or oil. Sometimes there are discounts offered by power companies. Often large users of electrical power can take advantage of lower rates and off-peak reductions.

The blower compartment is a housing that includes the filters and the blower fan for circulating the air through the system. Smaller units will use a centrifugal multispeed direct-drive fan while larger units use a belt-driven fan. An electric furnace has less resistance to airflow as compared to an oil or gas furnace, and fan performance is more efficient.

The heating section consists of banks of resistance heater coils of nickel chrome wire held in place by ceramic insulators (Figure 59-3). The heater normally operates on a 208 V to 240 V power source. The amount of heat produced by each heating element is affected by the supply voltage and the current flow, which together determine the total wattage.

One watt of power produces an equivalent 3.412 BTU/hr of resistance heat. Electric power is measured in kilowatts (kW), with 1 kW being equal to 1,000 W. The term kilowatt-hour (kWh) is used to express the amount of energy used in 1 hour. The input for an electric furnace is given in kWh and the output is given in BTU/hr (also referred to as BTUH, but either way, it is BTU per hour). To find the heat output in BTU/hr, multiply the kW input by 3,412. Conversely, the input kW can be calculated by dividing the furnace BTU/hr output by 3,412.

The amount of heat produced by an electric furnace is a function of the supply voltage and the current flow.

$$\text{Watts} = \text{Volts} \times \text{Amps}$$

A furnace rated at 240 V with a current draw of 125 A would have a kW input of 240 V × 125 A = 30,000 W, which

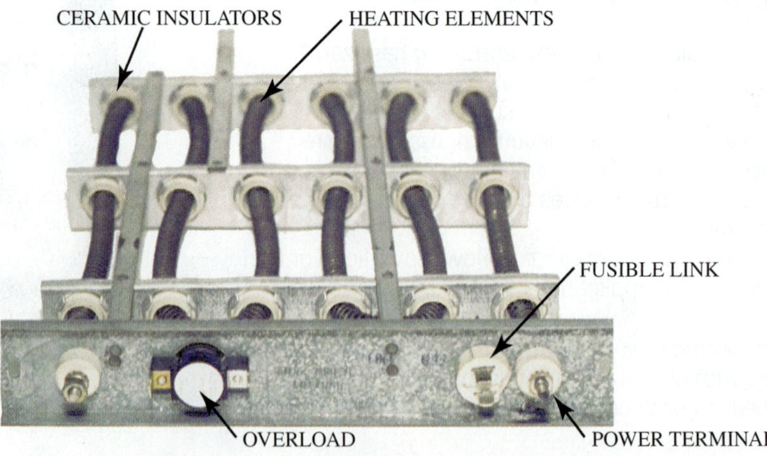

Figure 59-3 Electric heating element rack.

is 30 kW. Residential furnace output capacities normally range from 5 kW (17,060 BTU/hr) to 30 kW (102,360 BTU/hr).

A 30 kW, 240 V electric furnace has a current draw of 125 A, which for a 200 A entrance panel would leave only 75 A for other electrical uses. If the total heating demand is greater than this, then two electric furnaces would be used, with sequencing and zoning.

SAFETY TIP

All electrical work MUST conform to the requirements of local codes and ordinances and the National Electrical Code NFPA 70, current edition.

59.3 SEQUENCERS

The sequencer control is an important operation of the electrical furnace. When there is a call for heat, the electric sequencer contacts close and the circuit is energized. As the sequencer begins to heat up, the blower and first heater come on. To stage the elements, there is a time delay (in seconds) before each additional heater stage is energized.

Electric furnaces that have multiple heat strips are designed with a sequencer that allows each heat strip to come on individually or in small groups. Most systems use sequencers to sequence or stage on elements. A sequencer uses a small heating element wrapped around a bimetal strip or a PTC control. When the bimetal strip warms up enough, it snaps or warps to close a set of contacts to turn a heat strip on. Some sequencers have multiple contacts that are staged on. Staging can be accomplished by using a center pin that pushes upward on each successive set of sequencer contacts (Figure 59-4). In some cases where several heating elements are sequenced to turn on, multiple sequencers are used (Figure 59-5).

To prevent the heating elements from overheating, the unit blower must come on with the first heating element and

Figure 59-4 White center pin shown pushes upward on contact.

Figure 59-5 Multiple sequencers.

it must remain on throughout the heating cycle until the last-stage heating element has cycled off. Some sequencers have specific terminals that are designated for the blower. This set of contacts is the first on and the last off in the sequence.

SERVICE TIP

UL requires that the blower come on with or before the first heat element and must remain on until the last heating element has been de-energized.

The time interval between when a sequencer is energized and when its contacts close is predetermined, within a range, by the sequencer's manufacturer. Some manufacturers offer sequencers with different time cycles. Likewise, the time interval between when a sequencer is de-energized and when each of the contacts drops out is preset by the manufacturer. In some cases, although the sequencers may come on one at a time sequentially, they may all go off at the same time together. Check with your supply house or the manufacturer to see what time cycle is required for the unit you are working on.

59.4 HEATING ELEMENTS

The heating elements used in electric furnaces are made out of a nickel chrome alloy. The nickel chrome alloy is used because it has both a high electrical resistance and slows down oxidation. Many metal alloys have an electrical resistance too low to be effective heating elements. Also, many alloys are susceptible to extreme oxidation at elevated temperatures. Heating elements appear to be long coil springs and are bright in color until the coil is heated. As the elements are used, they will begin to discolor. Eventually, they will become dull black in color. The length of the heating element determines its kW capacity. Heating elements are available in 5 kW through 30 kW capacities for residential electric furnaces and may be as large as 60 kW or more for commercial electric furnaces. The typical kW options of a heating unit are shown in Figure 59-6. Figure 59-7 shows the heating elements held in place by ceramic spacers.

Electric heating elements are rated by their kW capacity and come in a prewired rack. Each electric furnace is designed to hold a specific type and size of heating element rack. These racks are not interchangeable from manufacturer to manufacturer, or in many cases from one model furnace to

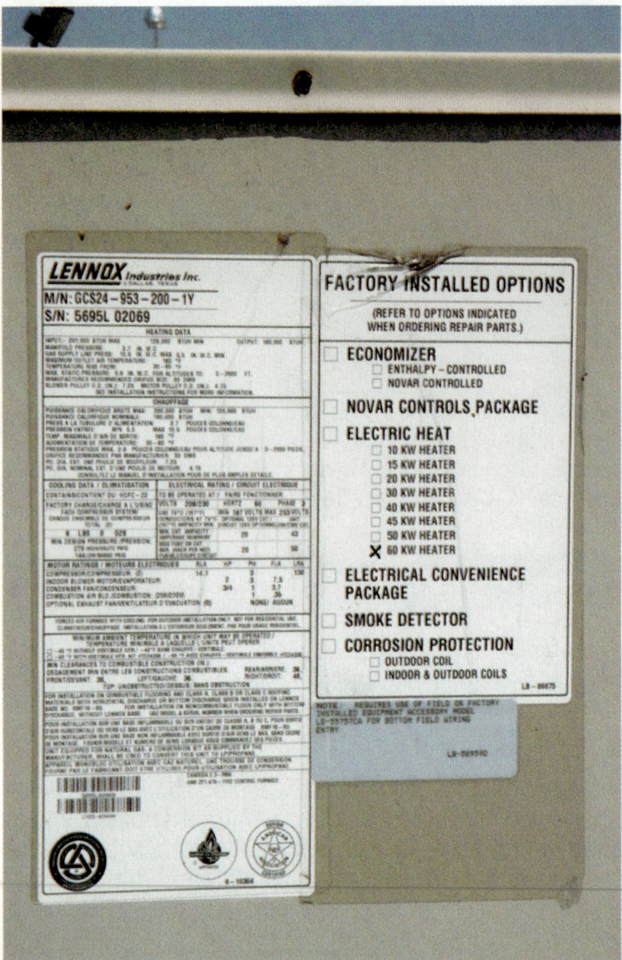

Figure 59-6 Nameplate for 60 kW commercial heater.

Figure 59-7 Heating element for 60 kW commercial heater.

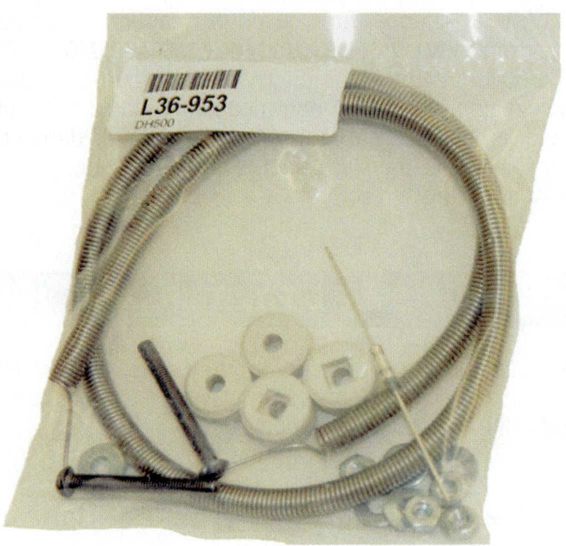

Figure 59-8 Electric heating element restringing kit.

another within the same manufacturer's product line. When replacing a heating element rack, be sure that you specify the model and serial number of the furnace the rack is to fit into.

Some heating element racks can be restrung with a replacement heating element (Figure 59-8). An advantage to restringing heating elements is that only a limited number of heating element sizes need to be stocked in your service van. However, replacing the heating element rack results in a significant time savings over restringing, and you are providing the customer with an entirely new part. Sometimes the ceramic insulators can be damaged during restringing, such as the one shown in Figure 59-9. A damaged insulator may fail sometime after the installation job is complete.

59.5 APPLICATIONS

One common application of an electric furnace is to combine it with a split-system heat pump. Whenever the outdoor temperature drops below the thermal balance point of the structure, the heat pump alone may not be able to maintain a comfortable indoor temperature. This is when the electric furnace will supply the additional heat necessary to maintain the thermostat setpoint.

Figure 59-9 Damaged ceramic insulator.

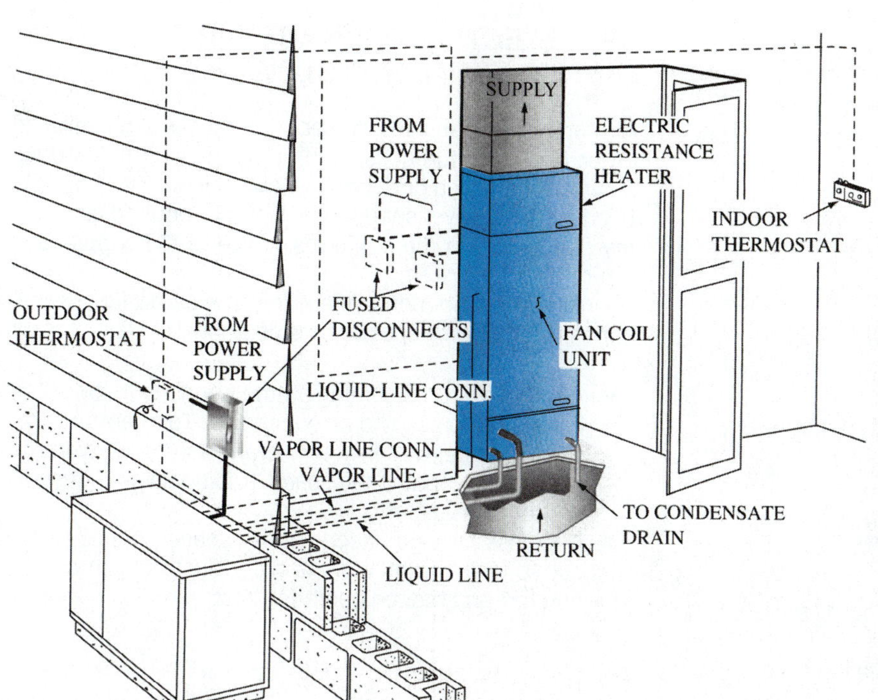

Figure 59-10 Split-system heat pump.

An example of an electric furnace and heat pump combination installation is shown in Figure 59-10. For these installations, the electric furnace acts as a fan coil unit. The refrigerated coil is installed in the furnace return-air stream before the blower. The coil is connected by refrigerant tubing to the outdoor heat pump located on a slab outside the building. The thermostat is located in the conditioned space and controls the operation of the complete system to produce the heating or cooling as required.

An advantage of using an electric furnace as a supplemental heat source in a heat pump application is that this creates a draw-through coil for the heat pump. This is the most efficient application for an indoor heat pump coil. This is in contrast to gas or oil supplemental heating systems, where the coil must be located after the firebox to prevent condensation from forming on the heat exchanger.

Because of energy conservation concerns, some cities, counties, and states have ordinances or laws that prevent straight stand-alone electric furnaces from being installed in new construction. This is because heat pumps are more efficient and under normal operating conditions will use less electricity. In those localities, heat pumps are used for primary heating in conjunction with electric furnaces that supply only supplemental heat.

59.6 ELECTRIC FURNACE CONTROLS

There are single-stage and two-stage thermostat systems, depending on the size of the unit. Figure 59-11 shows a thermostat that has a single stage for heating and a single stage for cooling. A fan control switch lever can be set to operate at low, medium, or high speed whenever the heating or cooling equipment cycles on.

Larger furnaces and some heat pumps use multistage thermostats. For electric furnaces, the first stage would bring on at least 50 percent of the total capacity. The second stage of the thermostat would respond only when full heating capacity is needed. With this added control, wide variations in indoor temperature are avoided. For heat pumps, the first stage brings on the heat pump and the next stage brings on the electric strips as needed.

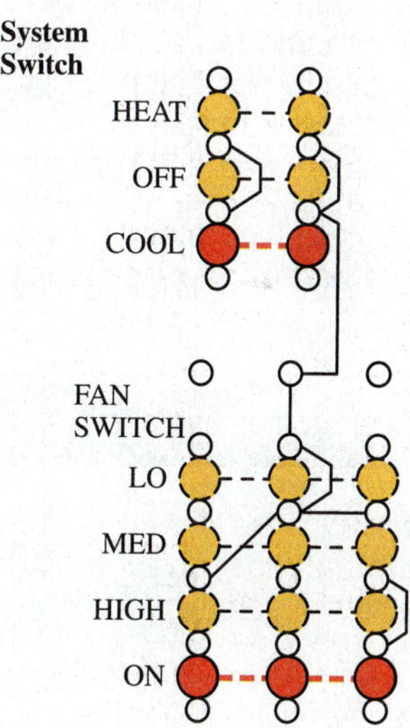

Figure 59-11 Single-stage heat/cool thermostat.

Figure 59-12 shows a thermostat with two-stage heating and cooling, and the system switch settings to control the thermostat operation are as follows:

- **OFF** Heating and cooling systems are off. If the fan switch is in the AUTO position, the cooling fan is also off.
- **COOL** Thermostat controls the cooling system. Heating system is off.
- **AUX. HEAT** Auxiliary heat is on. Cooling system is off.

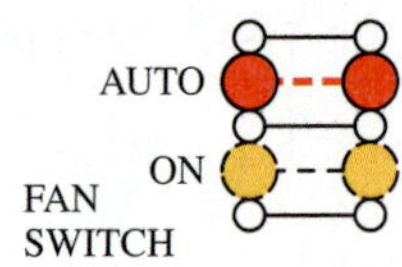

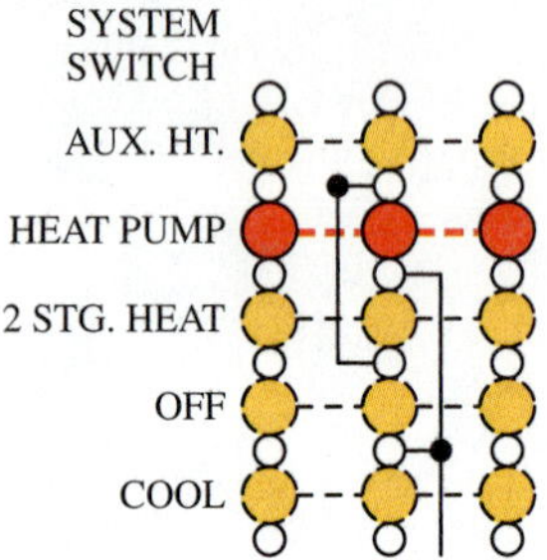

Figure 59-12 Two-stage heat/cool thermostat.

- **HEAT PUMP** First-stage heat is on. Cooling system is off.
- **2 STG. HEAT** First- and second-stage heat is on. Cooling system is off.

Electronic or digital thermostats use thermistors or other integrated circuit devices to sense temperature changes. The system and fan selections can be operated by a keyboard (Figure 59-13). The installer setup is used to customize the thermostat to specific systems. The fan can be configured to be operated in the heat mode either by the equipment or by the thermostat. The heating cycle rate can be adjusted for stage-two heating. Fan operation can be extended to continue after the call for heating or cooling ends.

SERVICE TIP

While sequentially staged design features are feasible for modifying the load, code assumes all installations are used continuously, for safety purposes.

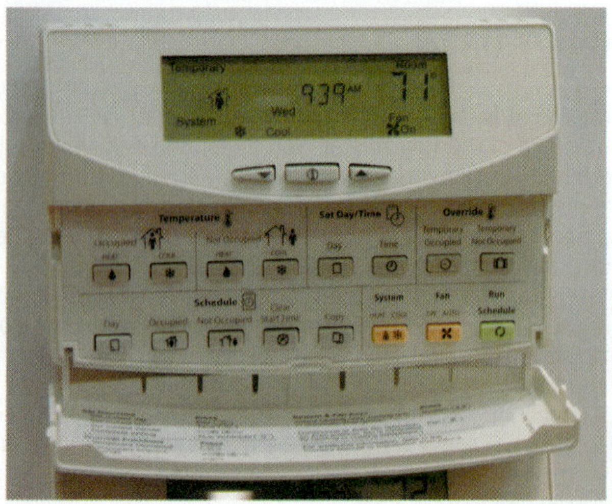

Figure 59-13 Programmable thermostat keyboard.

59.7 ELECTRIC FURNACE AIR DISTRIBUTION AND BLOWERS

Air-handling capabilities of electric furnaces are similar to other forced-air furnaces, using a multispeed direct-drive blower. Multispeed direct-drive blowers have the capability to adjust the blower speed to deliver the required air quantity depending on the static resistance of the supply- and return-air duct system.

The air-distribution system for an electric furnace requires the following special considerations:

- Air temperatures coming off the furnace are lower, as compared to gas and oil equipment. Temperatures of 120°F and below can create drafts if improperly introduced into conditioned space. Additional air diffusers are recommended.
- Duct loss through unconditioned spaces can be critical, so well-insulated ducts are a must to maintain comfort and reduce operating cost.

59.8 SAFETY DEVICES

All electric resistance heating circuits have one or more safety devices to protect the system from overheating and overcurrent. Electric heating elements are protected from any overheating that may be caused by fan failure or a blocked filter. High limit switches sense air temperature and open the electrical circuit when overheating occurs (Figure 59-14).

As a backup, some furnaces have fusible links that melt when the heating coil temperature goes too high due to overcurrent conditions. This opens and protects the circuit. A fusible link (Figure 59-15) is a small one-time-use device that is

Figure 59-14 Limit switch for electric heat elements.

Figure 59-15 Fusible link.

Figure 59-16 Overload protector.

wired in series with the heating element. Some manufacturers use fusible links that are mounted in ceramic. Others use fusible links that are spot welded to the heating element. Fusible links are made of an alloy that melts at a specific temperature, generally at about 300°F. The advantage of a fusible link is that the melting temperature of the alloy remains constant over time. This provides the maximum level of safety because the fusible link will open instantly anytime an overcurrent situation occurs. Because the opening function of fusible links is based on the melting temperature of an alloy, they are highly reliable overload protectors.

A high limit is another heat protector for electric heating elements (Figure 59-16). High limits sense temperature so that in a high-temperature situation the limit will open. High limits will automatically reset once the bimetal strip in the limit has cooled sufficiently.

Inline cartridge (BUSS) fuses (Figure 59-17) are used to protect individual heating elements, groups of heating elements, or the entire furnace. Fast-acting fuses should be used instead of dual-element time-delay types. The metal clips that hold cartridge fuses in place have sufficient spring tension when they are new to prevent heating of the connection between the fuse and the fuse holder. However, over time these connections can become loose, resulting in resistance that will cause them to heat up and lose their spring tension. If these clips have become discolored—even the slightest amount—they should be replaced before the resistance gets high enough that an arc occurs between the fuse and fuse clip. To reduce damage to the fuse spring clips and work more safely, you should remove the fuses with a fuse-puller tool. Do not pry them out with a screwdriver.

SERVICE TIP

Some inexpensive cartridge fuses will deteriorate over time when used in an electric furnace. These fuses can actually burn through the outer casing when they fail. This can result in damage to nearby wiring. Use only quality fuses such as those purchased from air-conditioning and refrigeration supply houses. If a customer replaced fuses on their unit with lower-grade fuses, you should change them (or at least note it on your service ticket).

Circuit breakers (Figure 59-18) may be used as overcurrent protection. Circuit breakers used on the 208 V to 240 V systems have two breakers that are interlocked so that overcurrent in one leg causes both breakers to open, removing power from the load. Some furnaces use more than one set of breakers so that individual parts of the heating circuit can be protected separately. An advantage of circuit breakers is that they can be reset.

(a)

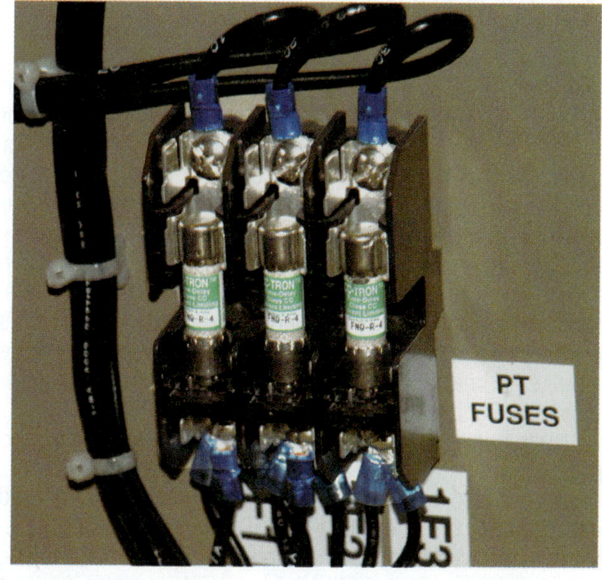

(b)

Figure 59-17 (a) Cartridge-type fuse; (b) fuses are held by metal clips and should be removed with a fuse-puller tool and not by a screwdriver to preserve the spring tension of the clips.

Figure 59-18 Dual circuit breaker.

SERVICE TIP

Built-in fusing or circuit breakers are provided by some manufacturers in compliance with the National Electric Code. This eliminates the need for external fuse boxes.

CAUTION

All replacement wiring used in an electric furnace must have a rating of 105°C (221°F) with an AWM-type appliance wire rating. This wire will stand a much higher environmental temperature, which can be experienced within an electric furnace cabinet.

59.9 AIRHANDLING UNITS AND DUCT HEATERS

A variation of the electric furnace is an air-handling unit with duct-type electric heaters. The air handler consists of a blower housed in an insulated cabinet with openings for connections to supply and return ducts. Electric resistance heaters are installed either in the primary supply trunk or in branch ducts leading from the main trunk to the rooms in the dwelling. When heaters are installed in branch runs, there is comfort flexibility by zone control; for example, room temperature can be individually controlled. This system has not been as popular as the complete furnace package, primarily because it complicates installation requirements and adds costs.

Duct heaters are made to fit standard-size ducts and are equipped with overheat protection devices. Electric duct heaters can also be used to supplement other types of ducted heating systems, to add heat in remote duct runs, or to beef up the system if the house has been enlarged. They may be connected to come on with the furnace blower and are thus controlled by a room thermostat. Additional safety may be achieved by installing a sail switch in the duct that must sense air movement before the duct heater will come on.

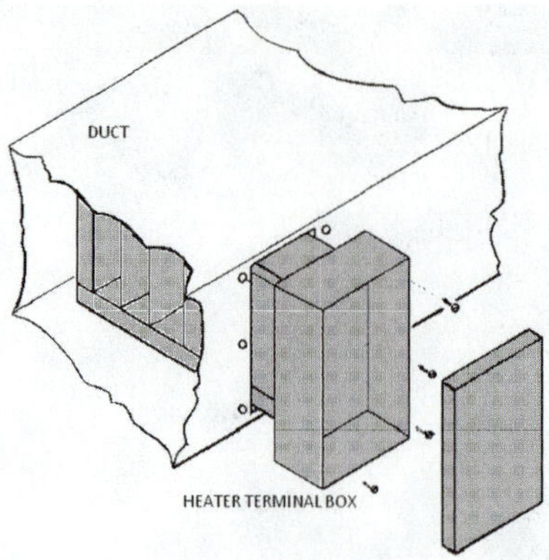

Figure 59-19 Installation of a slip-in-duct heater. *(Brasch Mfg. Co., Inc.)*

Slip-in-duct–type heaters are the most widely used because they are the easiest to install, as shown in Figure 59-19. A hole is cut into the duct and then the heater is secured with sheet-metal screws.

Another type of duct heater is the flanged type. The flanges on the heater match those of the duct and are screwed into place with sheet-metal screws, as shown in Figure 59-20. There are also heaters that can be installed into round ducts (Figure 59-21).

Multistep heaters use electronic modulation with solid-state electronic step controllers. Some of these can switch up to ten contactor-holding coils apiece and may be wired in series for a maximum of thirty steps. They are available with a single input from all commonly used thermostat input signals.

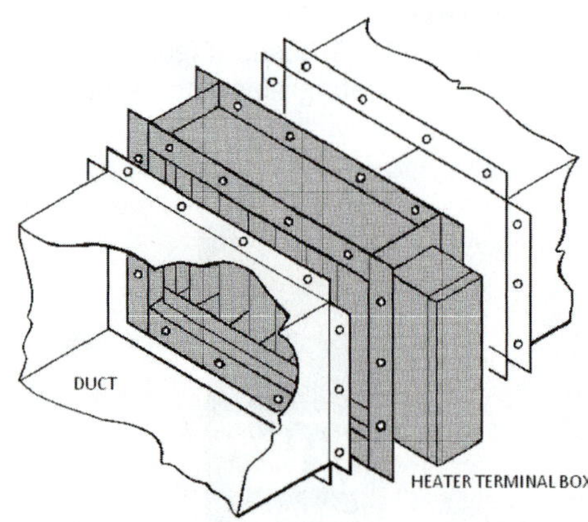

Figure 59-20 Installation of a flanged duct heater. *(Brasch Mfg. Co., Inc.)*

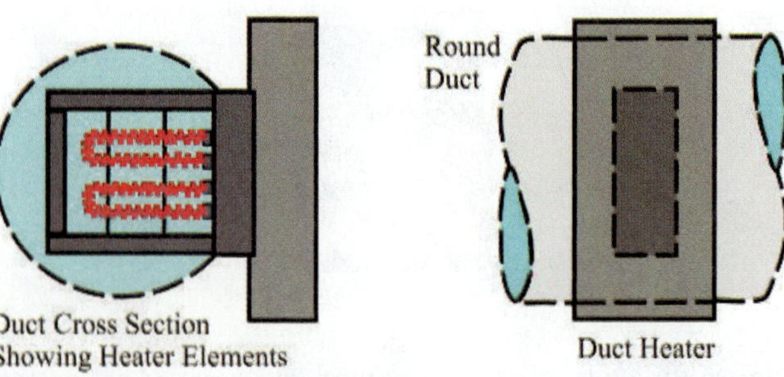

Figure 59-21 Round duct heater.

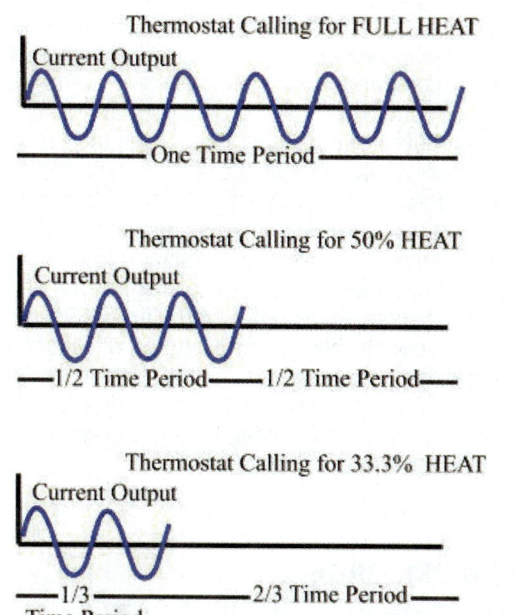

Figure 59-22 Modulating thermostat signal.

Silicone-controlled rectifiers can be used to provide very close heat control and/or silent operation for critical areas such as laboratories, computer rooms, and executive offices. Power and heat output are precisely controlled from 0 to 100 percent in direct response to the modulating thermostat signal (Figure 59-22). All the heating elements are simultaneously controlled.

59.10 ELECTRIC HYDRONIC BOILERS

Electric hydronic boilers heat water that is then circulated to the spaces to be heated rather than by blowing hot air through a duct system. This type of boiler can be used for standard baseboard radiant heaters or for under-floor radiant heat. Other applications include supplemental heat for oil, gas, and wood-fired systems.

Hydronic systems are popular in geographical locations where there is little need for air conditioning. Most heating/cooling systems will utilize the same air-distribution system that requires supply and return ducting throughout the residence. Without the need for air conditioning, this ducting is not necessary, and instead water circuits can be installed. Each circuit (zone) can be configured with its own circulation pump and temperature control. In-the-floor radiant heating has grown in popularity and provides for an even, prolonged heat once the entire floor has come up to temperature.

59.11 BASEBOARD HEATERS

Electric baseboard heaters can easily be installed in spaces where additional heating is required (Figure 59-23). Typically they can be wired for 120 V, 208 V, or 240 V applications, but 240 V is recommended. This type of heater operates on the principle of natural convection, and there is no fan or blower attached to it. The best location for an electric baseboard heater is along an outside wall, as close as possible to an outside door, or directly under a window, which takes advantage of the natural air currents: the cold air falls from the window area and the warm air rises.

Figure 59-23 Electric baseboard heater.

SAFETY TIP

Do not block an electric baseboard heater. It must be kept clear of all obstructions with at least 12 in minimum clearance in front and above the heater with a 6 in minimum on each side. Do not place the heater against paperboard or low-density fiberboard surfaces. Do not install the heater behind doors or place the heater below an electrical receptacle.

Electric baseboard heaters come in many sizes, from as short as 2 ft in length up to 8 ft or more. They can range from as low as 250 W up to 2,500 W. Being quiet, they are ideally suited for bedrooms. A thermostat is required, and this may be of the type that is mounted on the wall or built in to the heater. When servicing a unit, check for any accumulation of lint or dust. This can be vacuumed away carefully, while taking care not to damage any of the aluminum fins on the heat exchanger.

A variation of an externally mounted electrical baseboard heater is a register type of electrical heater. This type of heater is more difficult to install but has the advantage of being mounted flush into the wall. They also come equipped with a fan to provide forced-air circulation, resulting in better heat transfer. However, due to the blower assembly, this type of heater is not as quiet as a standard baseboard type.

Register-type electrical heaters can also be configured for 120 V, 208 V, or 240 V, with 240 V normally recommended. They can range in output from 500 W up to 2,000 W. A hole will need to be cut into an existing wall for installation. After the wall can is secured, the heater assembly with the blower wheel is screwed into position (Figure 59-24). During installation, the manufacturer's recommendations for top, front, bottom, and side clearances should be followed. Thermostat and periodic cleaning requirements are similar to those for electric baseboard heaters.

Line-Voltage Thermostats

Most electrical baseboard-type heaters will utilize line-voltage thermostats. Typically these heating units do not have step-down transformers to supply 24 V to the thermostat. Instead,

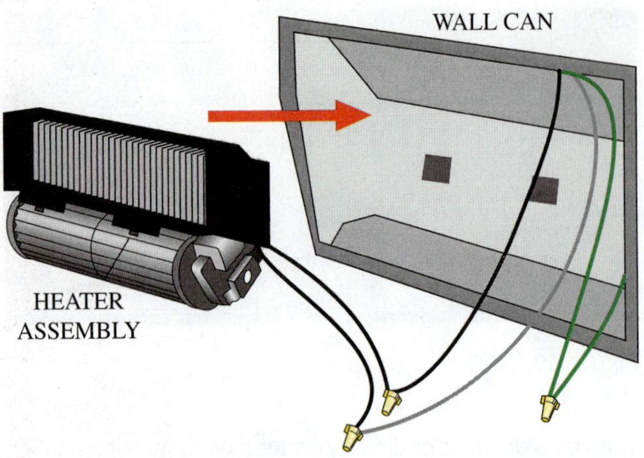

Figure 59-24 Register electric heater blower assembly mounting.

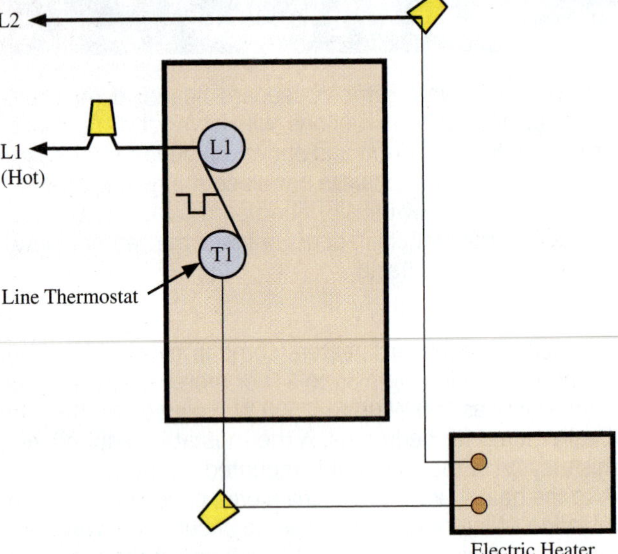

Figure 59-25 Typical line-voltage thermostat.

the thermostats will be wired directly into the operating circuit and therefore at line voltage (Figure 59-25). All wiring must comply with local codes and be of sufficient size for the circuit requirements.

CAUTION

Line-voltage thermostats control 120–240 V circuits. If improperly handled, there can be risk of 120–240 V electric shock hazard that can cause serious injury or death.

UNIT 59—SUMMARY

Electric furnaces are somewhat compact, need only minimal clearance, and require no makeup air or exhaust vents. Although there are applications for electric heat stand-alone systems, very commonly it is used as a supplemental form of heat. Although very efficient, electric heat can often cost more than comparable oil, gas, and heat pump systems. However, electric heat can be added to an existing room with only moderate modifications through the use of baseboard heaters, in-the-wall duct registers, and duct heaters.

When used with a heat pump, the electric heat turns on as outside temperatures drop below the normal heat pump operating range. This type of two-stage heating is common, and many thermostats are configured for these applications. Staging can also be controlled between heating elements, allowing them to turn on and off depending on the thermostat setting and the heating demand. It is always important to note that while sequentially staged design features are feasible for modifying the load, code assumes all installations are used continuously.

WORK ORDERS

Service Ticket 5901

Customer Complaint: Drafty Alcove

Equipment Type: 4-Foot Electric Baseboard Heater

The technician met with the customer in regard to upgrading an existing heating system. The complaint was that an alcove near the front entrance to the home was drafty. Although this area was used mostly as an entrance, the customer would sometimes work at a small desk located there and feel cold.

The technician inspected the existing system, which was gas-fired forced hot air. He explained to the customer that to supply heat to this space, new ductwork would need to be added and a hole for a register cut into the floor. Some additional calculations would also need to be considered to determine if the existing airflow and furnace size could accommodate the additional length of duct. In any event, this modification was not simple and could be expensive. The technician recommended an electric space heater that could be turned on whenever the customer worked at the desk. The customer did not like this idea because the space heater would always be in the way.

Agreeing with this, the technician recommended two other alternatives. One was an electric baseboard heater and the other was a register-type forced-air electric heater mounted into the wall. The customer decided on the register heater with a wall-mounted thermostat. The existing furnace would not need to be modified. The electric heater could be set to turn on through the thermostat setting whenever the customer desired heat.

Service Ticket 5902

Customer Complaint: Supplemental Heating

Equipment Type: Wood-Fired Furnace with Oil Backup

The customer set up an appointment to meet with a sales representative to discuss the alternatives available for supplemental heat. Originally the residence was heated by hot water baseboard heat supplied from an oil furnace, but a new wood-fired boiler was installed 2 years ago. This was done to offset the high cost of heating as the price of firewood was very reasonable in this geographical area.

This new system worked very well, but the wood furnace needed to be tended on a regular basis. If the customer was home late from work or away for a few days, the oil furnace would come on and provide heat to prevent any freeze-up in the house. However, the existing oil furnace was now in need of replacement. The oil tank was also starting to develop leaks. A new oil furnace and tank would be a costly proposition. The sales representative discussed the possibilities for supplemental oil, gas, or electric heat. The customer decided on electrical heat because there was no convenient gas supply, and an added convenience of electric heat is that no fuel oil deliveries would need to be regularly scheduled.

Service Ticket 5903

Customer Complaint: Residence Too Cold

Equipment Type: Forced-Air Heat Pump

The technician first spoke with the customer and then checked the system and found everything to be operating correctly. Apparently from what the customer had explained, in the evening as the temperature dropped, the house would feel chilly. The customer also noticed that the room temperature would fall well below the thermostat setting.

The house was new and this was the first winter that the customer had spent there. During the summer months, the house temperature was very comfortable no matter how hot it became outside. The technician checked the owner's manual for the heat pump and found that its efficiency dropped when the outside air temperature fell below 40°F. There was no supplemental heating element installed in the existing system.

The technician recommended that supplemental electrical heating elements be installed into the air handler. The unit was adaptable for this add-on, so the existing system could be used. The single-stage thermostat would need to be changed out and a programmable two-stage was recommended. Once the heating elements and the new thermostat were in place, the room temperature remained at the thermostat setpoint level even during the cool hours late into the evening.

UNIT 59—REVIEW QUESTIONS

1. List three advantages of a warm-air electric furnace as compared to an oil or gas furnace.
2. What material are electric heater coils often made from?
3. An electric furnace rated at 240 V and 30 A would have an output capacity of how many kilowatts?
4. What happens to the current draw as an electric heating element warms up?
5. When would the blower come on in an electric furnace?
6. Why is nickel chrome alloy used for electric furnace heating elements?
7. How are electric heating elements rated?
8. How would a multistage thermostat operate for a heat pump system with supplemental electric heat?
9. How do air temperatures compare between electric furnaces and those that are gas or oil fired?
10. What is the function of a high-limit switch on an electric heater?
11. What is a fusible link on an electric heater?
12. How does an overload protection device on an electric heater differ from a fusible link?
13. What happens when the spring clips holding an inline cartridge fuse become loose?
14. What are the requirements for replacement wiring in an electric furnace?
15. What is a term used for a furnace that heats hot water that is then circulated for space heating by a circulation pump?
16. What is an advantage of an electric baseboard heater?
17. In what locations should electric baseboard heaters be installed?
18. What precaution should be taken when locating an electric baseboard heater near an electrical outlet?
19. How can an electric baseboard heater be cleaned?
20. Why are hydronic furnaces popular in geographical areas that do not require air conditioning?

UNIT 60

Electric Heat Installation

OBJECTIVES

After completing this unit, you will be able to:

1. list the installation methods used to support an electric furnace.
2. discuss sizing of the power wire, overcurrent protection, and disconnect.
3. discuss the effect of reduced voltage on furnace capacity.
4. list the operational checks for a new electric furnace installation.

60.1 INTRODUCTION

Electric heat installation is less complicated to install than most other types of heating. The only connections required are the electrical wiring, the thermostat controls, and the air ducting. This makes electric heating very desirable for many multifamily residential, light commercial, and commercial applications, especially when only limited heating is required.

This limited heating is typical in areas where the cooling load is predominant. Therefore it is very common to find electric heat used in tandem with cooling units. Unlike a stand-alone electric furnace, most air handlers can be adapted for both electric heating elements and cooling coils. In this way, the same air-distribution system can be used for both the heating and cooling modes. The installation of a heating/cooling air handler is more involved than a stand-alone electric furnace partly due to the cooling coil refrigerant connections. However, the basic location, support, and duct connections for an air handler as compared to a stand-alone electric furnace can be very similar.

60.2 WARM-AIR ELECTRIC FURNACE INSTALLATION

The installation of a warm-air electric furnace requires a few preliminary considerations. The furnace should be located in an area that will provide for proper clearance from combustible materials and also allow for good support. The plenum connection should be adjusted to fit properly and sealed to prevent air leakage.

Furnace Location

An electric furnace's installation is flexible; it may be installed almost anywhere in a building. Electric furnaces do not need a vent, so they can be located in places where fossil fuel furnaces, such as those burning gas or heating oil, cannot be located. Electric furnaces also weigh much less than fossil fuel furnaces, so they are easier to support or hang. Most electric furnaces are multipositional, so the same furnace may be installed in the upflow, downflow, or horizontal position without modifying the unit. While multipositional fossil fuel furnaces are available, they typically require the removal and relocation of several parts to change from one position to another.

Furnace Support

In slab on-grade and basement installations, the furnace would typically rest on a return-air box called a plenum. Return-air boxes may be made out of sheet metal or duct board or framed in with standard building materials. A sheet-metal return-air box is shown in Figure 60-1. The

Figure 60-1 Sheet-metal return-air box.

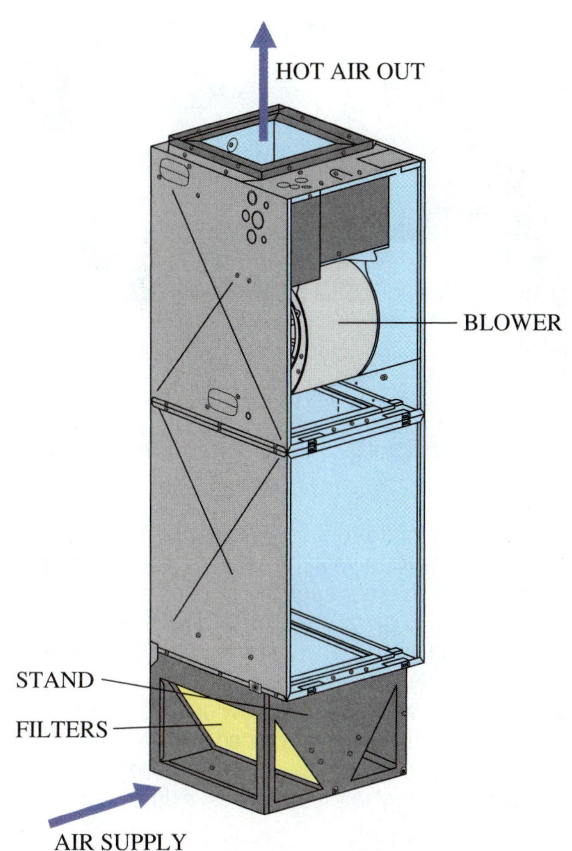

Figure 60-2 Upflow electric furnace installation.

electric furnace would be positioned to sit on top of the return-air box, which houses the filters (Figure 60-2).

In crawlspace installations, the furnace is suspended from the floor joists; in attic installations, it can be suspended from the roof rafters.

Because the fan noise and vibration will be transmitted to the ceiling through the joists in attic installations, the furnace should not rest directly on the ceiling joists or on plywood on top of the ceiling joists. The ceiling acts like a soundboard and amplifies the sound of the furnace. An air handler located in an attic and resting on vibration isolation pads is shown in Figure 60-3.

Furnace Clearance

Always refer to the manufacturer's literature and appropriate local codes to determine the furnace clearances. The *furnace clearance minimums* are the minimum distances from each of the furnace surfaces to any combustible materials. The manufacturer's literature and/or local codes will provide these distances. An example of a furnace installed in a closet is shown in Figure 60-4. Not all of the furnaces' surfaces will require the same minimum clearances. The bottom of the furnace may rest directly on a plywood base with 0 in clearance, but the front may require 24 in or more for service. However, ideally, the service clearance should be at least 36 in.

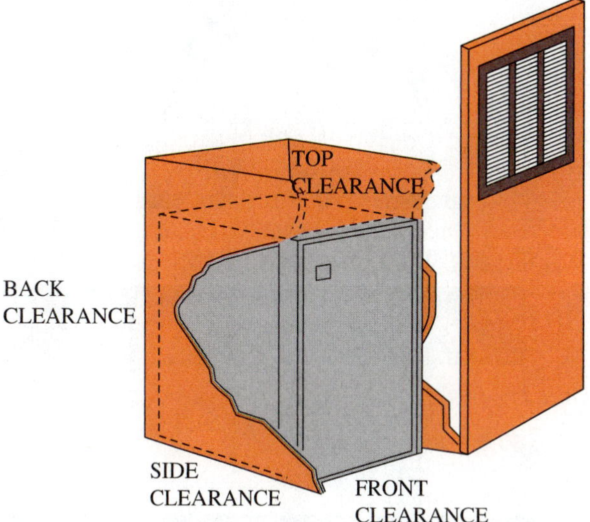

Figure 60-4 Minimum clearances for a closet installation.

Plenum Connections

There is a ¾-in flange on the furnace supply-air and return-air outlets. To protect the ¾-in furnace supply-air plenum flange from damage during shipping and handling, some manufacturers leave the flange flat. On these furnaces, the technician will have to bend the flange 90° outward so it can be used to attach the supply-air plenum.

Care should be taken when bending the flange outward not to distort it excessively. To avoid excessive distortion, do not bend the flange 90° all at once. It is important that you bend the tab up as uniformly as possible so that the top edge does not stretch and warp excessively. To minimize distortion, make a series of smaller angled bends along the length of the flange. Start at one end, and bend it slightly upward. Move along the flange, and repeat that slight bend. Start over again, and make another slight bend. Repeat this process until the desired 90° bend is completed.

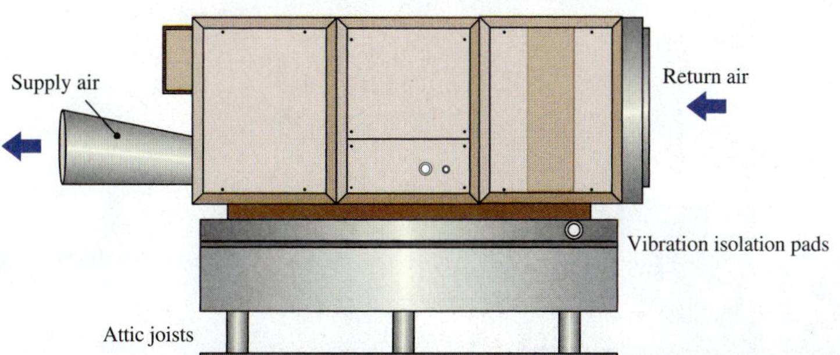

Figure 60-3 Air handler resting on vibration isolation pads.

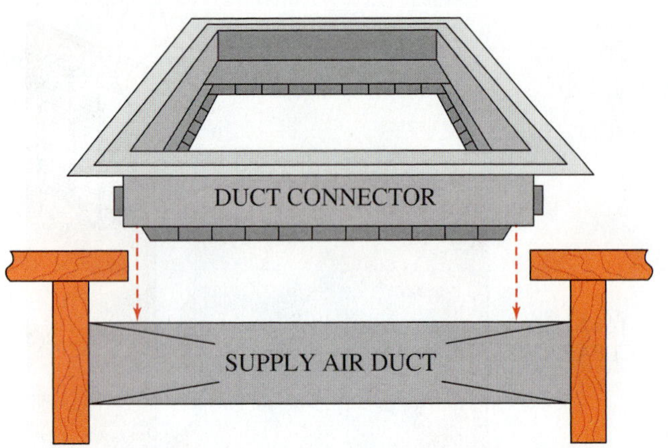

Figure 60-5 Duct connector.

TECH TIP

The best tool to use for bending the supply-air flange are hand seamers, sometimes called hand brakes.

If the furnace is being installed in the vertical up position, the furnace flange does not need to be bent up. The furnace can be secured directly to the plywood base with sheet-metal screws. Be sure to seal the connection between the base and furnace with duct mastic and fiberduct tape.

The duct connections will vary between furnace types as to whether they may be upflow, downflow, or horizontal configurations. Figure 60-5 shows a reducer for connecting the furnace to the duct in an under-the-floor supply duct system for a downflow furnace. A duct opening would need to be cut out in the floor as shown in Figure 60-6. The connector tabs will need to be bent over to fasten the connector to the duct, as shown in Figure 60-7. Depending on the air-supply and return system, a rectangular-to-round connector may be required for flex duct, as shown in Figure 60-8. If flex duct is used, it must have a minimum temperature rating of 200°F and meet all other applicable codes and standards.

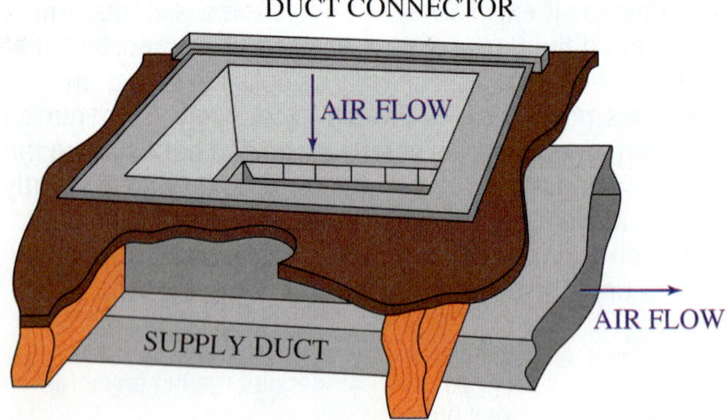

Figure 60-6 Duct cutout.

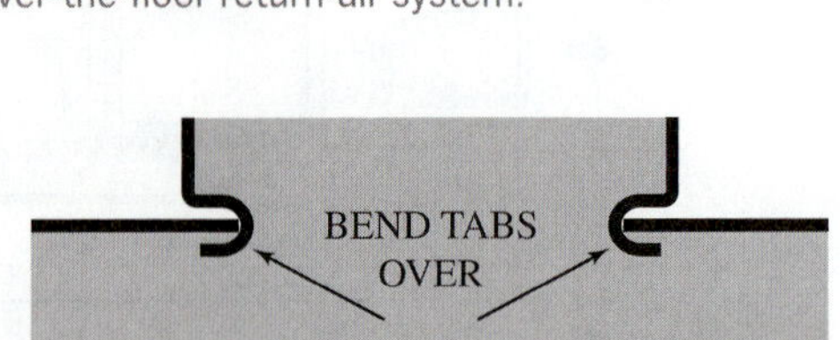

Figure 60-7 Bending tabs on connector.

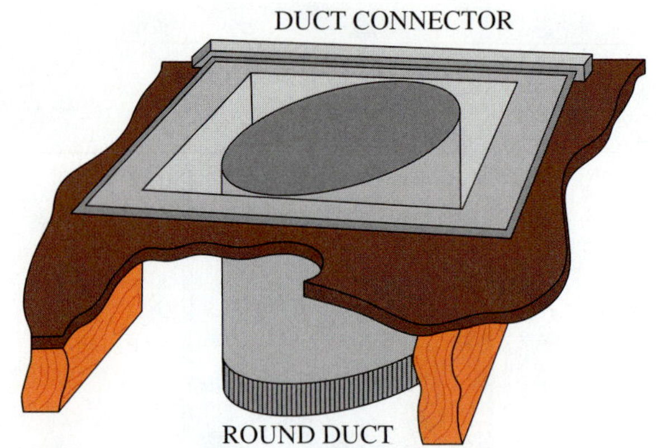

Figure 60-8 Rectangular-to-round duct connection.

An over-the-floor return-air system for an upflow furnace is shown in Figure 60-9. In this installation, the furnace is mounted slightly above the floor and the return-air register is mounted in the wall. For an upflow or horizontal furnace, a sheet-metal plenum can be attached directly to the furnace flange with sheet-metal screws. Fiberduct board should not be connected directly to these flanges. If duct board or other

Figure 60-9 Over-the-floor return-air system.

similar materials are to be used, a 3-in-wide sheet-metal band must be attached to the furnace flange first. The metal collar is attached to the flanges with sheet-metal screws, and the fiberduct is connected to the collar using long sheet-metal screws with 2½-in-square large washers under the heads.

Plenum Sealing

To prevent air leaks, the supply and return plenums must be sealed with mastic. Mastic may be applied directly to sheet-metal plenums. Be careful when applying mastic. The sheet-metal edges may be sharp or have burrs. Fiber tape must be used when duct board or other such materials are used as plenums. Air leakage can also occur in the space in which the furnace is installed if it is not mounted and sealed properly.

60.3 SYSTEM SELECTION

A furnace should be chosen with a heating capacity at least equal to the heat loss of the room, area, or building being heated. The heat strips used in most electric furnaces are rated at approximately 5 kw per strip when operated on 240 V. Since each kilowatt produces 3,410 BTU/hr, the heat output of each strip is 17,050 BTU/hr. Therefore, electric furnaces typically are available in multiples of 17,000 BTU/hr.

The electric service capacity must be considered when selecting electric heat, because all forms of electric heat use large amounts of current. A 20 kw electric furnace will use approximately 84 A. A typical modern residential electric service connection is 150–200 A, while older houses have service connections of 60–100 A. In some cases, the home's electric service connection may have to be upgraded before an electric furnace can be installed.

The actual voltage the furnace will be operating on is also important. Although most electric furnaces will operate at voltages less than 240 V, there is a significant loss of heating capacity when the furnaces are installed on lower-voltage supplies. Table 60-1 compares the output and operating characteristics of furnaces operating on 240 V and 208 V.

60.4 ELECTRICAL WIRING

Power Wiring

The most important part of an electric furnace installation is sizing the power wiring. Due to the fact that electric furnaces use large amounts of current, the power wiring is critical. The conductor should be sized using the wire-sizing tables of the National Electrical Code to handle the minimum circuit amperage rating listed on the unit data plate. In some cases the data plate lists data for several units. As an example, a furnace model identification number could vary from 10 kW to 23 kW. Generally the only difference is the number of strip heaters in the furnace. The installer must determine which data matches the actual furnace being installed and mark it on the data plate.

Sometimes aluminum conductors are used to save on installation cost, because large-gauge copper wire can be expensive. Before aluminum wiring can be considered for use, check the furnace manufacturer's warning label. Figure 60-10 shows a manufacturer's installation guide that prohibits the use of aluminum wiring. If no such guidance is provided, the installer must check to see if the connections on the furnace are designed for aluminum conductors. Typically, they will be marked "Cu/Al" if both copper and aluminum wire types are acceptable. If only copper conductors can be used, the lugs will be marked "Cu only."

Aluminum conductors require application of antioxidation paste to prevent a poor connection, which will result in oxidation.

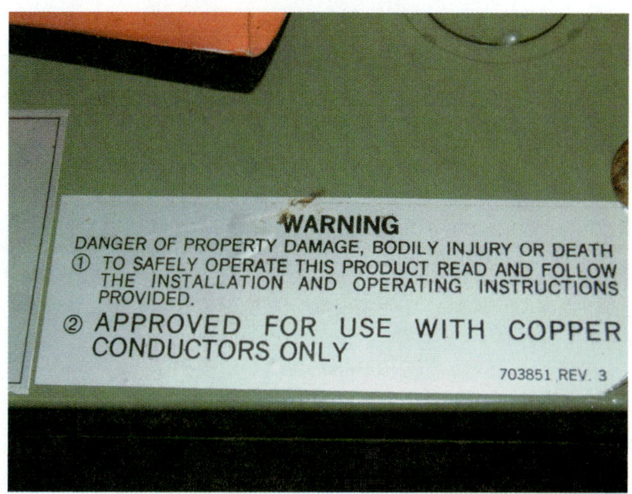

Figure 60-10 Warning label requiring use of copper conductors.

TABLE 60-1 Electric Furnace Specifications

Strips	Kilowatts		BTU		Amps	
	240 V	208 V	240 V	208 V	240 V	208 V
1	5	3.7	17,050	12,786	20.8	18
2	10	7.5	34,100	25,534	41.7	36
3	15	11.2	51,150	38,301	62.5	54
4	20	15	68,200	51,068	83.3	72

CAUTION

It is dangerous to use improperly sized wire or the wrong type of wire on an electric furnace. This is true with any equipment, but it is even more of a concern with equipment that has a high current draw. The wire and wire connections may become hot if the wire is undersized or not compatible with the furnace connectors. This can cause fires in extreme cases.

The Overcurrent Protection and Disconnect

The furnace should be protected with a circuit breaker no larger than the maximum overcurrent rating on the data plate. In some cases the technician will have to determine which data to use. There should be a disconnect switch located within sight of the furnace sized at 115 percent of the unit's minimum circuit ampacity.

Control Wiring

The control wiring should be done with 18-gauge solid copper wire. Typical mechanical-heating-only thermostats will have just two connections: one connection for the 24 V furnace transformer and another connection for the furnace controls (Figure 60-11).

With only two wires, it really does not matter which of the control terminals the wires are connected to. Some systems may have a third wire to control the fan. With three wires, the connections do matter. Some digital thermostats also require a common wire from the transformer to operate the thermostat. Table 60-2 shows typical control wiring connections between the thermostat and the electric furnace.

Some electric furnaces require a thermostat that is designed for electric heat because the thermostat controls the fan operation in the heating mode. These systems would require three control wires.

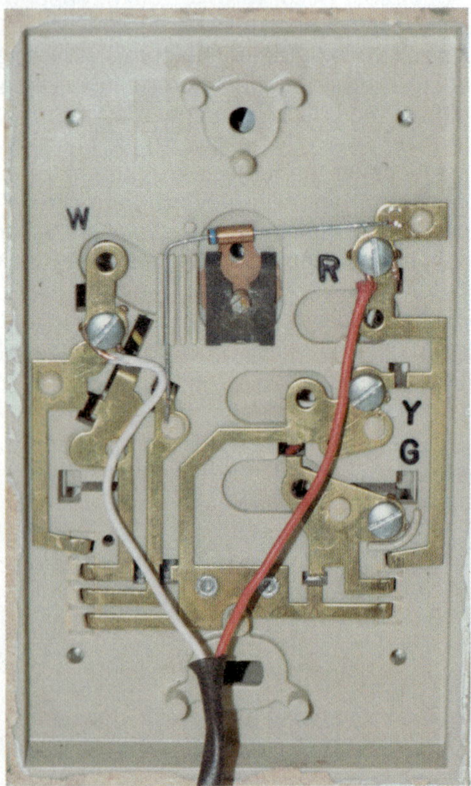

Figure 60-11 Two-wire thermostat connection.

TABLE 60-2 Electric Furnace Control Wiring

Furnace Terminal	Wire Color	Thermostat Terminal	Purpose of Wire
R	Red	R	24 V power
W	White	W	Heat signal
G	Green	G	Fan signal
C	Any color	C Only used for digital thermostats	Common side of transformer

TECH TIP

Some thermostats have a jumper that configures the thermostat for either gas or electric furnaces. On the gas position, the thermostat does not send 24 V to the G terminal on a call for heat, so the furnace controls must energize the fan. On the electric position, the thermostat does send 24 V to the G terminal on a call for heat, energizing the fan.

60.5 START-UP PROCEDURES

Newly installed equipment should always be started to verify that everything is operating properly. This begins with a prestart inspection followed by an operational check once the equipment is running.

Prestart Inspection

The incoming voltage should be checked and compared to the data plate rating. Perform a visual inspection on all wires and wire connections. Loose connections will cause overheating and degradation of wires, wire lugs, and fuse blocks. If aluminum power wire is used, make sure the furnace connections are rated for aluminum conductors and antioxidation paste is used on the connections.

Operational Check

Checking the power source and all electrical connections is essential when starting an electric furnace for the first time. Because electric furnaces use large amounts of power, any resistance in any of the power-handling components causes a voltage drop and leads to heat and failure of that component.

The voltage to the furnace should be checked before and after starting the furnace. A voltage drop indicates a problem with the power wiring to the furnace. The voltage at the heating elements should be checked and compared to the voltage entering the furnace; they should be the same. If they are not, then there is a problem in the furnace.

A furnace problem can be located by measuring the voltage across all the power-handling components. An infrared thermometer can also help. The point of voltage drop will be hotter than the rest of the wires and components. The current draw of each heating element should be checked and compared to the data plate rating. After all heating elements are operating, measure the temperature of the air entering and leaving the furnace. The difference between these two temperatures is the temperature rise. The temperature rise should fall within the range listed in the furnace installation instructions. If the temperature rise is too high, the fan speed should be increased. If the temperature rise is too low, the fan speed should be decreased.

UNIT 60—SUMMARY

A successful installation starts with proper selection, careful planning, and attention to detail. Equipment should be sized to meet the building's heat loss, the building's electrical service must be adequate, and the equipment should be located where it may be easily serviced. Sizing and selection of the power wire, the overcurrent protection, and the disconnect are all vital for the safe and efficient operation of the furnace. Finally, the furnace operation should be checked to verify that it is operating within manufacturer's specifications.

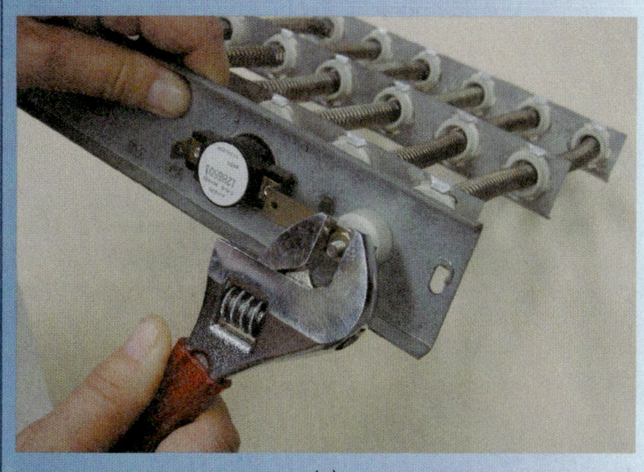

(a)

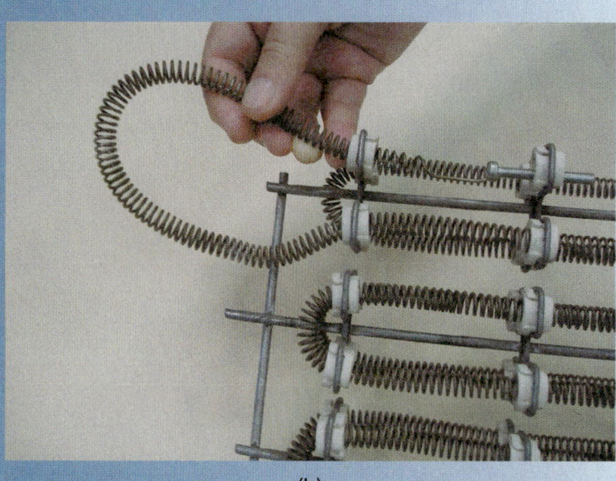

(b)

Figure 60-12 Restringing a broken heating element.

WORK ORDERS

Service Ticket 6001

Customer Complaint: Poor Heating on Cold Days

Equipment Type: 15 kw Electric Furnace

A technician is called to check an electric furnace that was installed several years ago. The customer complains that the furnace will no longer keep the house warm on cold days.

The technician visually checks the furnace and sees no obvious problems. The furnace is started and the voltage and amperage are checked to each of the heating elements. A reading of 240 V is found to be supplied to each of the heating elements except one that shows no current draw. After turning the furnace off, the technician shuts off the main power supply and locks out the circuit as a safety precaution. The circuit is also checked to verify that there is no voltage before the technician begins disconnecting the power leads from the troublesome heating element. An ohmmeter measurement of infinity indicates that the element is open, and a visible break can be seen in the heating element.

The broken element is removed and a new element is restrung (Figure 60-12). After the heating element is replaced and reconnected, the power is turned on and all of the elements are rechecked. A heat rise test is performed, and the furnace is found to be operating correctly according to the manufacturer's data plate.

Service Ticket 6002

Customer Complaint: Poor Heating on Very Cold Mornings

Equipment Type: 10 kW Electric Furnace

A technician is called to check an electric furnace that was recently installed. The customer complains that the heating capacity of the furnace seems inadequate on very cold mornings. The technician is also informed that a load study was performed by the local utility company just a few months back. This energy audit was part of a campaign to help homeowners find potential energy losses and hopefully lower their electricity costs.

In this instance the load study resulted in a design heat loss of 34,000 BTU/hr. The technician checked the furnace rating and found that it matched the design load with an operating voltage of 240 V. Checking the voltage to the furnace, the technician measured only 210 V before the heating elements were energized and only 208 V when the elements were on. Other than the low voltage supply, there seemed to be no other problems with the operation of the furnace. The technician explained to the customer that the furnace is rated for 240 V and that the heating capacity will be reduced when operating at a lower voltage.

Service Ticket 6003

Customer Complaint: No Fan on Auto

Equipment Type: 15 kW Electric Furnace

A technician is called to look at a newly installed electric furnace whose fan does not seem to be operating properly. The customer explains that whenever the fan switch is set to auto, the furnace does not seem to work. The technician then tries to start the furnace with the thermostat set in this position and finds that the heating elements are energizing and then cycling on overload because the indoor fan does not come on.

The technician then changes the thermostat fan switch setting to ON. In this instance, the fan operates normally and the furnace heating elements remain energized. The technician notes that the thermostat is the type that can be configured for both electric and gas furnaces, and it is found that the fan jumper in the thermostat is set to "gas." The technician moves the jumper to the electric position and now the fan comes on whenever there is a call for heat.

Service Ticket 6004

Customer Complaint: Air Temperature Too Low

Equipment Type: 10 kW Electric Furnace

A technician is called to check a newly installed electric furnace. The customer complains that the air being supplied to the room is not hot enough. The first check that the technician makes is to measure the furnace temperature rise. This is found to be only 30°F. The technician reviews the furnace operating instructions and finds that the recommended temperature rise is listed at 40°F.

Next the technician measures the furnace and heating element voltages. This seems to check out as it should, with 240 V coming into the unit and 240 V measured at the heat elements. A current measurement provides an amperage reading of 20 A on each of the two heat elements. The furnace seems to be operating properly, so the technician checks the blower and finds that it is set to the highest speed tap. The technician moves the blower speed to low, starts the furnace, and then performs another check on the temperature rise. Now the temperature rise is 40°F and the supply air feels noticeably warmer to the customer.

UNIT 60—REVIEW QUESTIONS

1. Why should furnaces not rest directly on the ceiling joists?
2. How is fiberduct connected to the furnace?
3. How is the power wire sized for an electric furnace?
4. What is necessary to make a safe connection with aluminum wire?
5. An electric furnace has a minimum circuit ampacity rating of 50 A. What is the minimum current rating for the disconnect switch?
6. Where should the disconnect switch for the furnace be located?
7. How is the overcurrent protection sized for an electric furnace?
8. Discuss the effect of reduced voltage on electric furnace capacity.
9. Discuss the operational checks that should be made on a new electric furnace installation.
10. Draw a connection diagram for the 24 V controls on an electric furnace.
11. Why are all aspects of the power wiring circuit so important when installing electric furnaces?
12. Why should the electrical service to the house be checked before installing an electric furnace in an older house?
13. What can be done to remedy a furnace whose temperature rise is too high?
14. What is the most common size for heat strips in an electric furnace?
15. What is the current draw of the heat strips on a 15 kw electric furnace operating on 240 V?
16. List the installation methods used to support an electric furnace.
17. Is an electric furnace installation simpler than that for a gas furnace?
18. How are furnace clearances determined?
19. What is the best tool to use for bending a furnace air-supply flange?
20. Can mastic be used for duct board that will be used as plenums?

UNIT 61

Troubleshooting Electric Heat

OBJECTIVES

After completing this unit, you will be able to:

1. check electric furnace heating elements.
2. check electric furnace sequencers.
3. check electric furnace safety controls.
4. check the control circuit on an electric furnace.
5. measure the airflow of an electric furnace.

61.1 INTRODUCTION

Electric furnaces are the simplest form of warm-air furnace being sold today. However, they do still require service. Understanding how to properly check all the furnace components will save the technician time and frustration.

Service technicians must develop an understanding of the high-current circuit behavior to service electric furnaces. Due to the high current required for electric furnaces, even a minor connection resistance can lead to problems. Patience is also required because the controls are normally time delayed, and multiple time delay relays are common.

61.2 HEATING ELEMENT SERVICE

From time to time the heating element in an electric furnace must be replaced because it breaks. The most common cause of heating element failure is poor airflow. Reduced airflow causes increased temperature in the heating element, leading to failure. When a heating element breaks (Figure 61-1), it cannot be repaired and must be replaced.

There are several ways to identify a broken heating element: by amp draw, ohm readings, or by visual inspection. Visual inspection is usually not practical in the field because the strips are not visible on an installed unit. The strips have to be removed to see them. The most commonly used method is to check the amperage draw of the heater elements.

Figure 61-1 Broken heating element.

Checking Heater Element Amperage Draw

When identifying a broken heating element by measuring the amperage draw, first turn on the system and allow time for all of the sequencers to energize. Next, use a clamp-on ammeter to carefully check each of the circuit's power wires for amperage and voltage (Figure 61-2). If an

Figure 61-2 Clamp-on ammeter used to measure amp draw.

element has a measured voltage but no amperage draw, then it is open.

CAUTION

Some newer digital multimeters have internal protection to prevent damage if a live circuit is touched when set to measure the resistance in ohms. However, not all meters have this protection, and accidentally touching your ohmmeter to a live circuit may blow an internal fuse or destroy the meter. If your meter does not have automatic protection as described in the owner's manual, you should check the circuit for voltage before switching the meter to measure resistance in ohms. This will help you prevent meter damage.

Checking Heater Element Resistance

The power to the furnace must be turned off before performing a resistance check of the heating elements. When using an ohmmeter to check for broken heating elements, first isolate the heating element to be measured. This is done by removing one of the power leads to the element. It is not necessary to remove both to isolate the circuit. Using your ohmmeter probes, check for continuity through each of the heating elements one at a time. A good element will have a measurable resistance reading that is fairly low, as shown in Figure 61-3a. A broken heating element will have an infinite resistance, shown as OL (overload) in Figure 61-3b.

Visual Heater Element Inspection

The heating elements must be removed from the furnace to conduct a visual inspection. Most furnace heating element racks can be removed once the power wires have been disconnected. Before disconnecting the wiring, make a sketch of the wire location so that you can easily reconnect them once the heating element racks have been replaced. There are typically two, four, or six sheet-metal screws that hold the rack in place. Once the rack is out, it can be inspected for broken insulators, broken heating elements, or any other visible damage.

When reinstalling the heating element rack, you must make certain that the alignment pin or bracket on the heating element rack fits into the slot at the back of the furnace. In most cases, the opening at the back of the furnace is slightly indented so that the alignment pin will track into the center opening. If the pin is not seated properly, the heating element rack will not slide completely into place or it may align at a slight angle. If this happens, do not force the rack into place with the screws. You must reposition it until it fits properly into place. Figure 61-4 shows an electric heating element rack.

Use both the manufacturer's wiring diagram and your own disassembly sketch as guides when reconnecting the wiring. In some cases, components may have been changed in older furnaces so that the manufacturer's diagram is no longer exactly correct. Therefore, you may need to rely mainly on your sketch when reconnecting the furnace wiring.

(a)

(b)

Figure 61-3 Using an ohmmeter to check heating element continuity: (a) a reasonable value of ohms indicates a working element; (b) a reading of ∞ or OL for overload indicates a broken element that must be replaced.

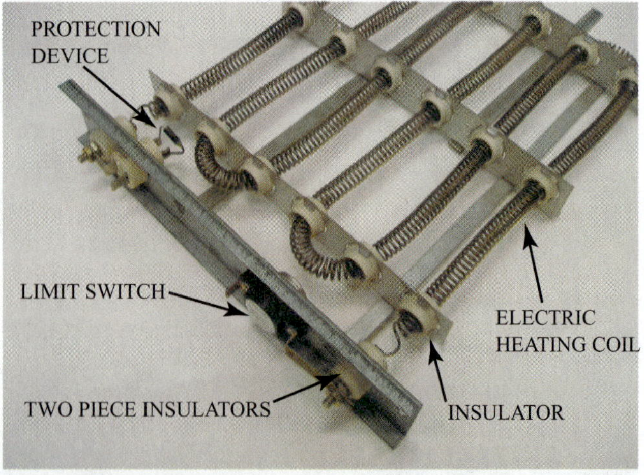

Figure 61-4 Electric heating element rack. *(Courtesy of HVAC Department, Terra Community College, Fremont, Ohio)*

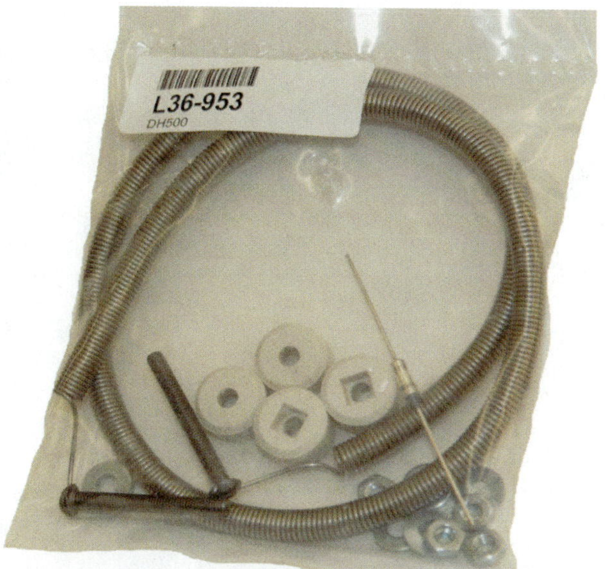

Figure 61-5 Electric heating element restringing kit.

Replacement heating elements are available either prestrung in a replacement rack or as individual elements that must be restrung by the technician (Figure 61-5). Make certain that the replacement heating element has the exact same kilowatt capacity as the one you are replacing. If a larger element is used, you might exceed the amperage capacity of the main power circuit. If a lower-capacity element is used, the furnace may not produce sufficient heat to keep the residence warm.

61.3 FUSIBLE LINKS

Fusible links are intended to protect the heating element from overheating. They are designed to open before the heating element burns apart. The most common cause for fusible links to open is poor airflow. If a furnace has one or more fusible links open, the airflow needs to be checked.

Fusible links are replaceable only if they are a separate component in the furnace heating circuit. Some fusible links are welded to the end of the heating element and are therefore not replaceable. Most replaceable fusible links are mounted on ceramic bases that attach in the sheet-metal heating-element rack plate (Figure 61-6). Make certain that you order a replacement fusible link that is specifically designed for your particular furnace. There are a number of different designs, and all are not interchangeable.

TECH TIP

Ceramic insulators are brittle, so be sure that you do not force them as you are inserting them into their fitted location on the furnace heating rack. Do not attempt to hold a ceramic fitting down into place with the fastening screws because this will most often result in a cracked insulator. If the insulator is cracked during installation, the part must be replaced. Figure 61-7 shows a cracked ceramic insulator. It is not safe or acceptable to use damaged ceramic insulators.

(a)

(b)

Figure 61-6 Fusible links: (a) installed in ceramic bases; (b) replacement for fusible link.

61.4 SEQUENCERS

All of the heating elements are not energized at the same time to avoid a large initial current draw at startup. The job of the sequencer is to turn on the heating elements in a specific order, or sequence. There is a wide variety

Figure 61-7 Broken ceramic insulator.

of sequencers used in electric heating systems. It is important to use a matched sequencer when making a repair or replacement. Two general types of sequencers are commonly used: individual time-delay relays and all-in-one controls.

Individual time-delay relays usually have one or two sets of contacts. A common application is for each control to energize a single heating element and also the next sequencer in line. That energized sequencer will likewise energize one element and the next sequencer in line. This pattern continues until all the elements are energized. Furnaces using this type of control system stage the strips on and off in that same order. The first set energized is also the first set shut off, and the last set energized is also the last set shut off. Figure 61-8 shows a typical time-delay relay type of sequencer.

Figure 61-8 Sequencer.

Figure 61-9 M terminal on a sequencer.

The all-in-one sequencers have a sufficient number of contacts to control all the heating elements and the fan. These sequencers have terminals specifically designed for the blower (Figure 61-9). The fan terminals should not be used for electric elements because their current rating is much lower than the rating designed for the heating elements. Unlike the individual time-delay relays, the all-in-one sequencers de-energize the strips in the reverse order that they are turned on. The first strip on is also the last strip off.

A sequencer can be checked before it is installed in the furnace by energizing the relay's sequencer heater terminals with the correct voltage and testing each set of relay contacts for continuity as they snap on. The relay sequencer heater voltage can be 24 V, 120 V, or 240 V, which can be very different from the voltage requirement of the heating element at the relay contact. Check the relay sequencer heater voltage carefully; applying an incorrect voltage to the relay sequencer heater can destroy it. You should also test the sequencer as it is de-energized to see that each of the contacts opens as it should.

Testing a sequencer is similar to testing a relay or contactor, except with a time delay. First check to see that the relay sequencer heater has the correct voltage. Once the relay sequencer heater is energized, a timing period will start. After a time delay of 30 seconds to 2 minutes, the relay contacts should close. The closing of the contacts can be detected by measuring the amperage draw to the heater. The measured voltage drop across the contacts after they close should be zero. A measurable voltage drop across the closed contacts indicates pitting, and the sequencer should be replaced.

When replacing a sequencer, cycle the furnace on and off several times to make sure that all the heating elements and the blower are switching on and off in the correct time cycle. The blower should be operating anytime one of the heating elements is energized.

TECH TIP

The contacts inside a sequencer cannot be visually inspected for wear. However, with an ohmmeter set to the R × 1 scale and the relay energized so that the contacts are made, you can test for resistance between the contact terminals. The resistance of a good set of closed contacts should be between 0 and 0.1 Ω. A slightly higher resistance means that the contacts are worn, while a high ohm reading indicates that the contacts are extremely pitted. In either case, the contacts must be replaced. An infinite reading means they are still open and should be replaced.

61.5 OVERLOAD

The limits in an electric furnace sense temperature. They are wired in series with the strips and open when the strips overheat. Limit problems can usually be traced to airflow problems. Low airflow will cause the strips to overheat, tripping the limits. If limits are tripping, the air filter and airflow need to be checked.

The high-limit switch (Figure 61-10) is not easily tested because it is auto-resetting. The service technician has to catch it when it opens. A voltage drop will be read across the overload if it has opened.

Another way of testing limits in an electric furnace is to use your clamp-on ammeter at the main power feed to the unit. Allow the unit to run for 15 to 20 minutes as you perform other checks. If any of the limits are opening, the total current draw to the unit will drop. Some digital ammeters have data-recording capacity so that the high and low amp readings remain in the instrument. These readings can be reviewed at the end of the test. This type of meter makes a limit test much easier.

Figure 61-10 Overload protector.

Figure 61-11 Tattle Tale for testing overload protectors.

A way of checking limits without being present is to place an electronic Tattle Tale (Figure 61-11) across the limit. If the limit opens while the heating element is energized, the Tattle Tale will indicate that the limit opened. After the heating element cools, the limit will close; however, the Tattle Tale will continue to indicate that the limit has opened.

Since the limits are auto-resetting, a problem that causes a limit to trip will often cause the limit to open and close repeatedly before it is noticed. This repeated opening and closing with a high current draw on the circuit will eventually take its toll on the limit and it will stay open permanently. This is easily tested by checking the voltage drop across the limit or by turning off the power and measuring the resistance of the limit. If it is open, the reading will be infinite.

Any limit problem should prompt the technician to check for airflow problems because airflow issues are usually the root cause of limit problems on electric furnaces.

61.6 WIRES AND TERMINALS

Electric furnace terminals carry large amperage loads. Any unusual resistance, such as that resulting from a loose slip-on connector at the terminals, will cause the terminal to discolor. A set of contacts in a relay or a sequencer that are pitted will heat up, and that heat will be transmitted to the slip-on connectors and connecting wire. If the resistance

Figure 61-12 Overheating causes the terminal to discolor and could cause wire insulation to melt.

is significant enough, heat can be generated to both discolor the terminal and melt the wire insulation, as shown in Figure 61-12.

SAFETY TIP

Any wiring used in an area subject to high temperatures must have an appliance rating. These wires will have labels stating that they are suitable for internal wiring of appliances where the wire may be exposed to temperatures not to exceed 105°F, as shown in Figure 61-13. Any other nonapproved wiring insulation may melt. This nonapproved wire can cause a short if used inside an electric furnace.

It is not recommended that you simply replace the terminal end if a wire and terminal have become overheated. The overheated copper conductor will have oxidized some distance back inside of the insulation. Merely cutting off the end, even if you remove several inches, will not prevent this wire from overheating again. You must replace the entire wire with another wire of the same or larger gauge that has the proper AMW appliance rating. Any wires that are not properly rated will present a safety hazard.

If the slip-on terminal on a relay, sequencer, fuse block, or other component has become discolored as a result of overheating, that component should be replaced. It is not recommended to simply sand or buff off the terminal's oxidization. Terminals have a thin coating of highly conductive material that resists oxidation layered over a base of less resistant copper, steel, or brass. When the terminal is sanded, the outer coating is abraded away, exposing the sub-base material. These base metals are not as resistant to oxidation or as efficient as conductors when compared to the highly conductive surface coating material. Cleaning these terminals will only serve as a temporary fix and not a permanent repair because oxidation will occur more rapidly once the sub-base material is exposed.

TECH TIP

It is important that you use a quality crimping tool anytime you are putting a new end on a wire that is going to be used in an electric furnace. Furnaces have fairly large amperage draws, and if the terminal is not extremely tight, the slightest resistance can cause the terminal and wire to overheat.

61.7 BLOWER MOTOR

The blower motor used on an electric furnace is typically a multispeed PSC electric motor. The variable-speed operation allows the installing technician to match the system blower to the duct system the unit is to be installed on.

Some new blower motors do not have oil ports. These motors are supplied with a "lifetime" of lubrication sealed in the motor bushings. If one of these sealed motors has developed a bearing problem, it cannot be lubricated and must be replaced. Motors that have oil ports should be

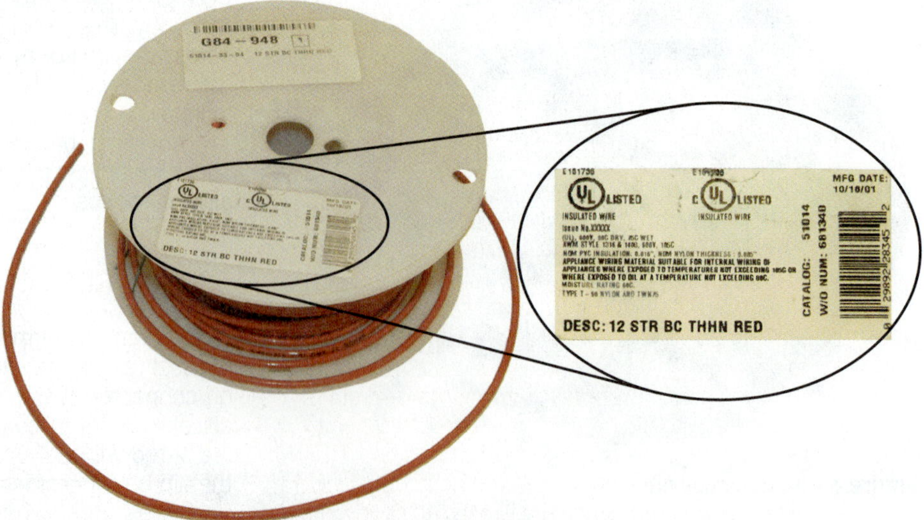

Figure 61-13 High-temperature wire for appliance use.

Figure 61-14 (a) Zip oiler; (b) telescoping spout.

lubricated no more than once annually for heating furnaces and no more than twice annually for heating/cooling furnaces. The oil used to lubricate a blower motor must be specifically labeled as electric motor lubrication oil. Figure 61-14 shows oil that is commonly used to lubricate electric motor bearings. General purpose or penetrating lubricants like Three in One and WD40 are not acceptable. Some of these oils have detergents or solvents that will actually damage the motor's bearings over time.

Do not overlubricate blower motors, and follow any manufacturer's instructions (Figure 61-15). Most motors require just a few drops of lubricant in the oil fill ports located on each end. Excessive lubrication will result in oil running out of the bearing seals, where it will deposit on the inside of the motor housing and windings as a thin film. The dust present in an air handler will collect as a layer on this oil film, which then acts as an insulator and restricts airflow. In both cases, the motor will run hotter and the motor winding insulation will begin to break down. Over a relatively short period of time, this can cause the motor windings to short out.

If the motor has to be removed to lubricate the inner bearing, make certain that it is reinstalled with the oil fill ports facing up. Always make certain that the oil fill port plugs have been replaced before the motor is put back in service. Failing to cap these ports will allow dirt and dust

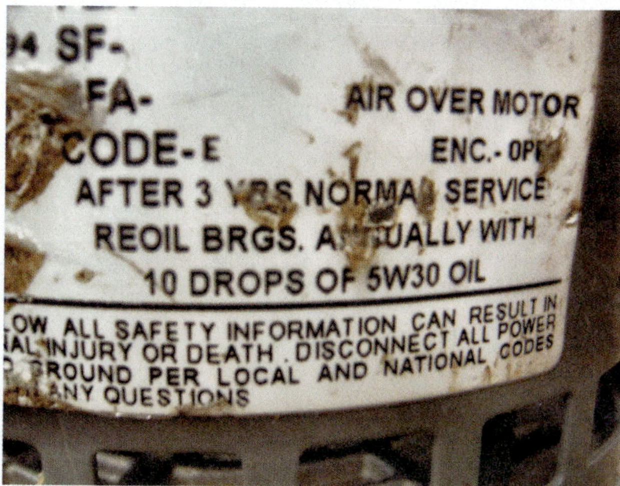

Figure 61-15 Oiling instructions on motor cover.

to collect, which will be carried into the bearing the next time the motor is lubricated. This can cause damage to the motor bearings.

61.8 FURNACE EFFICIENCY AND CFM

Compared to most furnaces, it is easier to accurately determine the exact airflow through the air handler of an electric furnace. It is fairly simple to set the fan speed properly to provide the 400 CFM of airflow per ton that is recommended for most air-conditioner air handlers. To calculate the actual CFM of an air handler, you must take the following readings:

- Total amperage draw of the heating elements. This amperage must be of the heating elements only; it should not include the blower.
- Voltage at the heating elements as the unit is operating.
- The return-air temperature.
- The supply-air temperature. Make certain that this temperature is not taken with a probe inserted in the supply plenum so that it would be in a direct line of sight with the heating elements. The radiant heat that the elements produce would give a false high reading.

Use the following formula:

$$\text{CFM} = \frac{V \times A \times 3.412}{\Delta T \times 1.08}$$

EXAMPLE 61-1: SOLVING FOR ACTUAL AIR FLOW IN CFM

A unit with the following operating characteristics:

Voltage at strips	220 V
Amp draw of strips	27.8 A
Return-air temperature	70°F
Supply-air temperature	117°F

Temperature difference = 117°F − 70°F = 47°F

$$\text{CFM} = \frac{220 \times 27.8 \times 3.412}{47 \times 1.08} = \frac{20{,}868}{50.76} = 411\ \text{CFM}$$

Figure 61-16 Control voltage transformer.

61.9 TRANSFORMER

The transformer (Figure 61-16) in an electric furnace is subject to a greater load than most HVAC transformers because the system may have a number of sequencers, all of which have heat motors that use 24 V. It is important to check the secondary voltage supplied by a control voltage transformer while the system is operating to determine if the transformer can supply 24 V under load. When a transformer is overloaded or weak, the secondary voltage drops. A control voltage of 18 V or lower indicates a problem in the 24 V control system.

Transformer problems are usually caused by control problems. A shorted relay sequencer heater will overload the transformer, causing it to fail. Simply replacing the transformer on a system with a shorted relay sequencer heater will only result in another failed transformer. The technician should check the resistance of all 24 V controls whenever replacing a transformer. Any control with a very low resistance across a relay coil should be replaced.

TECH TIP

Electric furnaces use 208–240 V as their primary power. Most replacement transformers have multiple power taps. Refer to the transformer's specifications to make certain that you are connecting the proper leads for the voltage your unit operates on. It is not an acceptable practice to simply assume that you are on a 240 V tap. If you do this on a 208 V system, the secondary voltage will be lower than 24 V. The voltage will be low under some circumstances where there are a number of current-drawing devices on the secondary, such as humidifiers, multiple sequencers, and dampener motors. For example, under these heavy loads, there may not be sufficient voltage and current to operate all of these devices.

61.10 BLOWER RELAY

The blower should come on when the first heating strip comes on and go off when the last heating strip goes off. How this is done depends on the furnace's control system. Electric furnaces that use separate sequencers for each heating strip typically use a second set of contacts on each sequencer to control the fan. All of these contacts are wired in parallel so that if any sequencer is closed, the fan operates. The blower relay on these furnaces is only used for continuous fan operation or for air conditioning.

Other furnaces use the blower relay to control the fan during heating. These usually require an electric heating thermostat. An electric heating thermostat energizes the G terminal during a call for heat to start the fan. The last sequencer to shut off has a set of contacts wired to control the blower on low speed. This keeps the blower operating when the thermostat is satisfied and the G terminal is no longer energizing the blower relay.

The all-in-one sequencers use a separate set of contacts to control the blower. These contacts will open and close at the same time as those controlling the first heating strip. On this type of system the first heating strip on is also the last one off, so the blower is operating anytime any one of the heating strips is on.

A blower that runs constantly on an electric furnace can be a sign of a sequencer with stuck contacts. This can be checked by measuring the furnace amperage draw. A high amperage draw with no call for heat is an indication of a stuck sequencer. The sequencer can be identified by checking the amperage draw leaving each sequencer until you find the one that is operating all the time. Check the relay sequencer heater voltage to be sure the problem is not in the control circuit.

TECH TIP

Occasionally during new construction metal debris can accidentally be left in a duct system. This debris can fall onto the heating elements because they are located at the top of the unit and are not protected by an evaporative coil the way that gas- and oil-burning systems are. Metal debris falling onto the heating elements will short a portion of the element to ground, turning it on.

UNIT 61—SUMMARY

The voltmeter and clamp-on ammeter are the service technician's most effective tools for tracking down problems in electric furnaces. They can be used to locate defective heating elements, open safety devices, and to check the airflow through the furnace.

Small resistances from poor connections can turn into large problems. Insufficient airflow can be at the root of many component failures in electric furnaces. Broken heating strips, open fusible links, and open thermal overloads are all caused by inadequate airflow.

WORK ORDERS

Service Ticket 6101

Customer Complaint: Poor Heating

Equipment Type: Electric Furnace Installed at the Local Animal Shelter

A service technician is called to inspect an electric furnace that is heating poorly. The furnace has three heating strips, each rated at 5 kW. The technician finds that only one heating strip is pulling current, while the other two have no amperage draw. The technician checks the voltage at these two heating strips and measures a reading of 0 V. He then checks the fusible links and measures 240 V across one of them. Likewise, there is a reading of 240 V across the limit on the other strip.

The technician also notices that the airflow coming out of the registers is very low. Inspecting the air filter, the technician finds that it is packed with animal hair. The technician informs the manager of the animal shelter that the dirty filter reduced the airflow to the point that the heating elements overheated. This in turn caused the fusible link and limit to fail. The technician replaces the fusible link, limit, and air filter and recommends to the manager that in the future the filter be changed once a week.

Service Ticket 6102

Customer Complaint: No Heat

Equipment Type: 15 kW Electric Furnace Installed in a Basement

A service technician answers a no-heat call. Upon arrival, the technician finds that nothing is operating. An initial reading shows that the furnace has 240 V at the fuse block. The technician then measures the voltage at the R and C terminals on the furnace and reads 0 V.

In a hurry to get home, the technician diagnoses a bad transformer and quickly replaces it and then turns the furnace back on. The unit immediately starts, but no sooner has the technician packed up tools but a puff of smoke comes out of the transformer and the furnace stops. Unfortunately, upon checking the new transformer, the technician finds that it is now failed, too.

The technician realizes the unwise mistake that was made in jumping to an immediate conclusion. This time, skipping no steps, the technician begins by checking the 24 V circuits. The second relay sequencer heater is shorted, which led to the transformer failure. The technician replaces the sequencer and transformer and retests the system. This time the system runs and operates correctly.

Service Ticket 6103

Customer Complaint: High Bills

Equipment Type: 15 kW Electric Furnace Installed in Crawlspace

A customer has called complaining of higher than normal electric bills. The technician arrives and the first obvious thing is that the furnace fan is running continuously. This is the case even with the fan switch set to AUTO and the system switch set to OFF.

The technician measures the furnace amperage draw at 23 A, which indicates that one of the heating strips is operating continuously. Using an ammeter, the technician identifies which set of heating strips it is. The technician then verifies which sequencer is controlling that set of heating strips.

The technician finds that there is no voltage to the relay sequencer heater. This is an indication that the sequencer contacts are not being held closed by the relay but instead must be stuck together. In this case, the sequencer must be replaced. The technician makes the replacement, tests the furnace, and finds that it no longer runs continuously.

Service Ticket 6104

Customer Complaint: Furnace Will Not Run

Equipment Type: 20 kW Electric Furnace with an Electronic Setback Thermostat

A customer calls for help with an electric furnace. When the technician arrives at the residence, the customer explains the problem and provides some detail on the past history of the unit. This past spring, the customer purchased and installed a new electronic setback thermostat from the home supply store down the street. The air conditioner worked fine through the spring and summer months. However, turning the furnace on this fall, nothing happened.

The technician inspects the thermostat wiring, and everything appears to be connected tightly with no loose wires. While inspecting the furnace wiring diagram, the technician notices that the blower motor is controlled by the blower relay in the heat mode and the blower relay is energized by the G terminal. Realizing that this could be the problem, the technician asks the customer for the instructions that came with the thermostat. A section on configuring the thermostat for different systems is shown. The technician modifies the thermostat to bring the fan on in the heat mode, and now the furnace operates correctly.

UNIT 61—REVIEW QUESTIONS

1. List three ways of checking the heating elements of an electric furnace.
2. Describe how a clamp-on ammeter can be used to check an electric furnace.
3. What would be the resistance reading of an open heating element?
4. A technician reads 240 V across the thermal overload to the first strip heater on an electric furnace. What does this mean?
5. List three ways of checking an overload in an electric furnace.

6. What should always be checked when replacing an overload?
7. What should always be checked when replacing a transformer?
8. What is the airflow through a furnace with the following readings:
   Voltage at the strips of 235 V
   Current draw of 39.5 A
   Return-air temperature 72°F
   Supply-air temperature 120°F
9. What type of oil should be used when lubricating the motor bearings?
10. Explain how a stuck heating sequencer can keep the fan operating continuously.
11. What is the correct remedy for a wire with one end that has overheated?
12. Why should relay slide-on posts not be sanded to clean off oxidation?
13. How can overlubricating a blower motor be detrimental?
14. What are the two general types of sequencers used on electric furnaces?
15. Describe two ways that electric furnaces control the blower motor.
16. What is the purpose of the white ceramic pieces on the strip heaters?
17. Why is a visual check of the heating strips on an electric furnace difficult to accomplish?
18. Why are terminal and wire resistance values critical in electric furnaces?
19. What are two types of safety that most electric furnaces utilize?
20. Why are the strips on an electric furnace staged to turn on in sequence instead of being energized all at one time?

UNIT 62

Heat Pump System Fundamentals

OBJECTIVES

After completing this unit, you will be able to:

1. describe what a heat pump is.
2. describe the cooling, heating, and defrost cycles.
3. explain the operation of the reversing valve.
4. define *heat source* and *heat sink*.
5. list the types of heat pump systems and configurations.
6. discuss the operation of metering devices and check valves used in heat pumps.
7. explain the purpose of suction accumulators and heating-cycle charge compensators.
8. explain the different types of efficiency ratings used for heat pumps.

62.1 INTRODUCTION

Heat pump systems lead all other forms of heat in terms of efficiency and low operating cost. The demand for heat pumps has increased because of the dramatic increase in the cost for all forms of energy. They must be installed correctly to take advantage of their energy-saving technology. The least efficient system installed correctly will outperform a more efficient system that is poorly installed.

Heat pumps derive their name from the manner in which heat is delivered into the home: they are not "producing" heat, as in the case of fuel-fired or electrical furnaces, but actually "pumping" heat from one area to another. Their efficiency is rated not by how efficiently they consume energy but rather by how efficiently they move it.

62.2 HISTORY OF HEAT PUMPS

The popularity of the heat pump rose with the increasing cost of electrical energy used for heating in residential and small commercial buildings. Heat pumps generally cost less to operate than fossil fuel furnaces and cost only about a third of what it costs to operate electric strip heaters. The heat pump was actually developed by Lord Kelvin in 1852. The first practically applied heat pump was installed in Scotland in 1927. Between 1927 and 1950, heat pumps were installed in hundreds of residential and small commercial settings throughout Europe and the southern part of the United States. Many of these installations were merely air-conditioning units converted to heat pumps by the addition of reversing valves and applicable controls. These early heat pumps had high failure rates because they did not have defrost controls, crankcase heaters, or suction accumulators. Heat pumps acquired a bad reputation as a result of this high failure rate. Although improvements in design made heat pumps more reliable, the early high failure rates caused problems in the market for years. Further, the relatively low cost of energy during this same time period almost destroyed this market.

The demand for heat pumps has grown over the past several decades. The reason for this new interest in heat pumps is both economic and environmental in nature. Interest in heat pumps has increased in recent years because of growing interest in energy conservation, the rising cost of heating, and mounting concern over the pollution produced by power plants.

62.3 HEAT PUMP CONCEPTS

Like all refrigeration systems and air conditioners, heat pumps transfer heat from one place to another by changing the state of refrigerant. Basically all refrigeration systems are heat pumps; they transfer heat from a heat source at a low temperature to a heat sink at a higher temperature. What makes the heat pump unique is the ability to reverse the direction of refrigerant flow. In the summer it removes heat from the inside air and moves the heat to the outside air. In the winter it will remove heat from the outside air and move it to the inside air. Heat pumps are also called reverse-cycle refrigeration systems.

A review of some basic physics principles will aid in your understanding of heat pumps. Five basic concepts will be discussed:

- Heat exists down to absolute zero, which is –460°F.
- Heat will only flow from a higher temperature region to a lower temperature region.
- Gases become warmer when compressed.
- Most matter can be in a solid (ice), liquid (water), or gas (steam) state.
- The temperature at which a material changes from a liquid to a gas or from a gas to a liquid depends on the pressure at which it is contained.

Concept 1: Heat in Cold Air

For many people, one of the biggest obstacles to overcome in understanding heat pump operation is accepting that heat exists even at very cold temperatures. Since heat is the form of energy that gives molecules their motion and all molecules are in motion, all molecules contain heat. The single exception to this is absolute zero. At absolute zero, molecules do not move, and there is no heat.

Concept 2: Heat Flows from High to Low

Heat will only flow from a higher temperature to a lower temperature, period! The job of the heat pump is to make heat flow from a cooler area to a warmer area. To do this, the heat pump has a low side, which is colder than the cool area, and a high side, which is warmer than the warm area. Moving the heat from the low side to the high side is where the name "pump" comes from.

Concept 3: Compression Raises Temperature, Expansion Lowers Temperature

The high side is created by compressing, or squeezing, the gas. When a gas is compressed, the heat in the gas is concentrated and the temperature rises. This is done by a component called a compressor. Conversely, expanding a gas or liquid causes its pressure and temperature to drop. This is done by the expansion valve. The high side and low side of the system are created by the compressor and expansion valve. Note that the name of the component tells you what it does.

Concept 4: Change of State

Although it is possible to build a heat pump just by compressing and expanding gas, this type of system is not particularly efficient. The change in state of a refrigerant will more dramatically increase the efficiency of the heat pump. Liquids absorb hundreds of times more heat when they change state than when they change temperature. With this in mind, the refrigerant evaporates from a liquid to a gas in the evaporator and condenses from a gas to a liquid in the condenser. Again, note that the name indicates the component's purpose.

Concept 5: Saturation Temperature and Pressure

Fortunately, the temperature at which evaporation and condensation will take place is controlled by pressure. The increased pressure on the high side raises the condensing temperature, allowing the condenser to reject large amounts of heat while condensing the refrigerant. The decreased pressure on the low side lowers the evaporation temperature, allowing the evaporator to absorb large amounts of heat while the refrigerant evaporates.

Low Supply-Air Temperatures

One obstacle to accepting heat pumps has been the relatively cool supply-air temperatures that heat pumps produce. Customers who have been accustomed to traditional furnaces often object to the cooler supply air from the heat pump. Typical heat pump supply-air temperatures range from 95°F to 110°F. Electric heat has a discharge temperature of 110–125°F, and gas and oil heat supply air in the range of 120–140°F (Table 62-1).

Air at 95° blowing across your body in the winter does feel drafty. The solution to this problem is proper supply register placement. The air from the supply registers should not blow directly on the occupants. The heat pump is designed to heat your house, not you!

Table 62-1 Supply-Air Temperature Ranges for Various Heating Systems

Heating Type	Discharge Temperature
Electric strip heat	110–125°F
Gas heating	120–140°F
Heat pumps	95–115°F

62.4 HEAT SOURCE AND SINK

Heat pumps are often identified by their heat source and their heat sink. The heat source is where the heat comes from when the heat pump is operating in the heating cycle. Typically, the source can be either air or water. Air-source heat pumps pull heat out of the surrounding air. Water-source heat pumps pull heat out of water circulated through them. The sink is where the heat pump puts the heat. The sink is most often assumed to be air since most heat pumps are designed to condition the air in a building. For this reason, the sink is usually not specified unless it is something other than air. For example, water-source heat pumps only mention the source because the sink is assumed to be air. On the other hand, water-to-water heat pumps use water as both the source and the sink. Table 62-2 lists different heat pump applications by heat source and sink.

62.5 PHYSICAL CONFIGURATION

Heat pumps are available in different physical configurations to match many different applications, including packaged units, split systems, and ductless mini-split systems. Packaged units have all the system components assembled together in one package. They are available to fit different physical space requirements, including units made

Table 62-2 Types of Heat Pumps Listed by Source and Sink

Type of Unit	Source	Sink	Application
Air source	Outdoor air	Indoor air	Heating homes with forced air
Water source/ geothermal	Ground water	Indoor air	Heating homes with forced air
Air to water	Indoor air	Domestic hot water	Heating domestic hot water
Water to water	Ground water	Water loop	Heating homes with hydronic radiant heat

for installation outdoors, over drop ceilings, and through walls or mounted on a wall. Traditional air-source packaged heat pumps are designed for the entire unit to sit outside. Ductwork from inside the building is connected to them. Horizontal water-source packaged units are designed to be installed above a drop ceiling. They are very compact and can easily fit in the small space above a drop ceiling. Packaged terminal heat pumps fit through a sleeve in the wall in motel rooms. Wall-mount packaged units are vertical air-source units that bolt directly to the wall. Ductwork or sleeves pass through the wall into the inside. They are often used for large open meeting rooms.

Split systems break the unit into two separate components that are connected through refrigeration lines. Split systems usually consist of an indoor unit and an outdoor unit connected together through a large gas line and a small liquid line. The large gas line is a low-pressure suction line when operating in cooling and a high-pressure discharge line when operating in heating. Typically, the outdoor section contains the compressor, a coil, a metering device, the reversing valve, and most of the system controls. The indoor unit contains a blower, a coil, and a metering device.

Ductless mini-split systems are similar to split systems, except they are not designed to use ductwork. Instead, the indoor section is designed to be installed in the space being conditioned. Mini-split systems are available that connect up to eight indoor sections to a single outdoor section, providing zoned comfort without ductwork. Another significant difference between traditional split systems and mini-split systems is that the only metering device in a mini-split system is located in the outdoor section. In cooling, the line going inside is cool, low-pressure, saturated liquid because the refrigerant has already passed through the metering device. As a result, both lines are insulated.

62.6 HEATING AND COOLING CYCLES

Heat pumps move heat against a temperature gradient. In the summer they move heat from the inside to the outside, and in the winter they move heat from the outside to the inside. The direction of heat flow is controlled by the direction of the refrigerant flow.

Basic Reverse-Cycle Concept

In the conventional refrigeration cooling cycle (Figure 62-1), heat is absorbed by the indoor coil, the evaporator, and discharged to the outside by the outdoor coil, the condenser. To change from cooling to heating, the indoor and outdoor coils trade functions. The outdoor coil becomes the evaporator, and the indoor coil becomes the condenser. In the heating cycle, heat is absorbed by the outdoor coil, which is now the evaporator, and discharged to the inside by the indoor coil, which is now the condenser.

The heat pump cycle is shown in Figures 62-2 and 62-3. The coils are relabeled as indoor and outdoor because they are now dual purpose. This reduces confusion because the outdoor coil is the condenser in the cooling cycle and the evaporator in the heating cycle. The indoor coil is the evaporator in the cooling mode and the condenser in the heating cycle. The process is reversed by means of a four-way valve called a reversing valve (Figure 62-4).

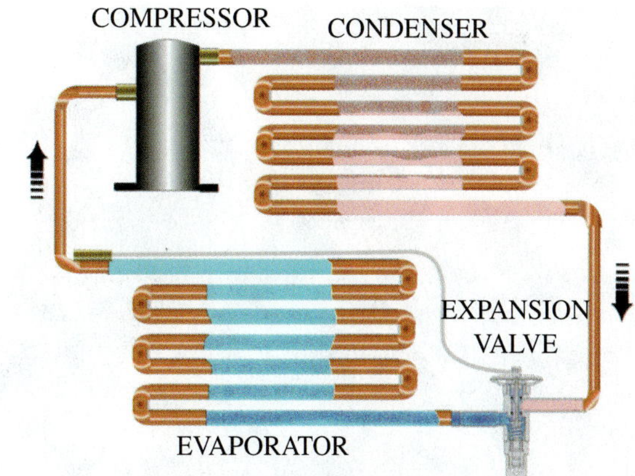

Figure 62-1 A conventional refrigeration cycle.

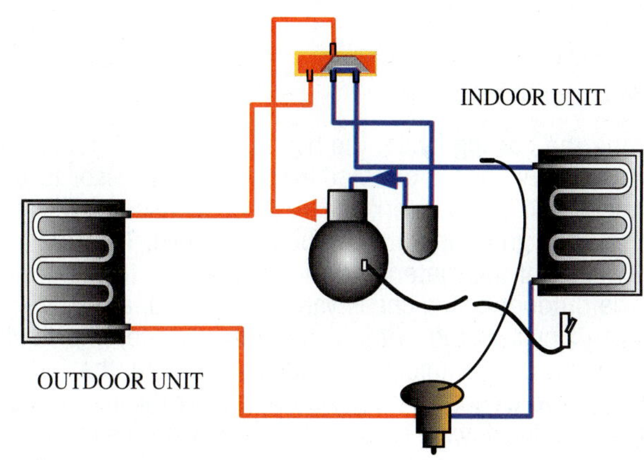

Figure 62-2 Heat pump system in cooling mode.

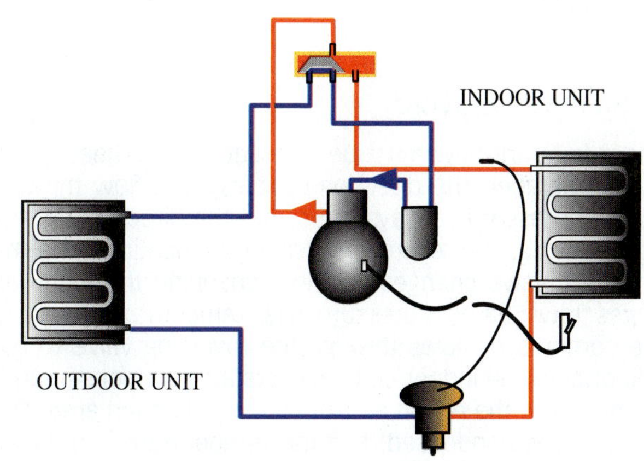

Figure 62-3 Heat pump system in heating mode.

TECH TIP

Remember to tell the owner that the outdoor section will run during the winter heating mode. Some customers may make unnecessary service calls because they believe that the outdoor section only operates in the cooling mode.

Figure 62-4 Heat pump reversing valve.

The Cooling Cycle

During the cooling cycle, the high-pressure/high-temperature discharge gas is directed from the compressor to the outdoor coil (condenser), where it condenses to a high-pressure, high-temperature subcooled liquid. To avoid the restriction of the metering device connected to the outlet of the outdoor coil, a check valve is installed, allowing the liquid to bypass the outdoor metering device. The liquid travels inside through the liquid line, where the indoor check valve closes, forcing the refrigerant through the indoor metering device. The indoor metering device drops the refrigerant pressure and temperature so that heat can be absorbed in the indoor coil. The refrigerant vapor produced in the indoor coil now travels through the reversing valve to the suction accumulator and on to the compressor. The cooling cycle is complete.

The Heating Cycle

The refrigerant flow must be reversed for the heating cycle. Remember, the direction of refrigerant flow through the compressor is always the same; the direction of flow is changed by the reversing valve. In the heating cycle the reversing valve changes position, changing the direction of gas flow. The high-pressure, high-temperature gas from the compressor flows through the reversing valve to the indoor coil. The indoor coil now acts as a condenser, adding heat into the return air from the conditioned area. The hot vapor is condensed, and the temperature is reduced to produce a high-pressure, high-temperature subcooled liquid. The indoor check valve opens and allows the liquid refrigerant to flow around the indoor metering device. Continuing through the liquid line, the liquid refrigerant is forced through the outdoor coil metering device by the closing of the check valve connected in parallel with it. The low-pressure, low-temperature mixture of liquid and vapor refrigerant flows into the outdoor coil, which is now the evaporator. The low refrigerant pressure keeps the refrigerant temperature lower than the outdoor temperature. Since the outdoor air is warmer than the refrigerant, heat is picked up from the outdoor air, evaporating the liquid refrigerant. This refrigerant vapor then flows through the vapor line, reversing valve, and accumulator to the compressor. The heating cycle is now complete.

62.7 THE REVERSING VALVE

The reversing valve is composed of three main parts: the pilot valve, the main valve, and the solenoid coil. The solenoid coil controls the pilot valve, which in turn controls the suction bleed to the main valve. The purpose of the main valve is to reverse the refrigerant route through the indoor and outdoor coils, thereby exchanging the functions of the condenser and evaporator coils.

The basic operation of a reversing valve can be compared to a solid cylinder sliding inside a hollow tube. Blowing in one end of the tube will shift the cylinder to the other end of the tube. This is how the cylinder inside the reversing valve is shifted, by pressure difference. By imposing the discharge pressure on one end of the reversing valve and the suction pressure on the other end of the valve, the pressure difference will shift the internal cylinder toward the side that is connected to the suction pressure.

Figure 62-5 shows a typical reversing valve. The main valve body is constructed with the permanent suction in the middle of the valve on one side and the permanent discharge in the middle of the valve directly opposite the permanent suction port. The ports leading to the indoor and outdoor coils are located on either side of the permanent suction port. When at rest, the sliding cylinder always straddles two tube openings. One of these openings is always the permanent suction port, which is connected to the low side of the system. The other opening will be either the indoor coil port or the outdoor coil port. This makes whichever coil is connected to the permanent suction via the sliding valve the evaporator. The port that is not covered by the sliding valve is open to the permanent discharge port located opposite the permanent suction port. This makes whichever coil is connected to the uncovered port the condenser. In Figure 62-6, the tube connected to the blue arrow becomes the low side and the tube connected to the red arrow becomes the high side. Note that the two outer tubes swap between the two valve positions.

Heating Cycle: Valve De-energized

Figure 62-7 shows a reversing valve that is de-energized for heating. The solenoid valve is de-energized and the control plunger is down, closing the bottom port vent line connected to the right end of the piston. The top port vent line connected to the left end of the piston is open to the equalizing line. This allows the pressure to bleed off from the left end of the piston to the permanent suction line. This creates a pressure difference across the left-side piston (usually 75–100 psig or greater), which causes it to travel to the extreme left. This movement repositions the slide valve to align the control port from the outdoor coil with the control port leading to the compressor suction, making the outdoor coil the evaporator. This same movement also aligns the compressor discharge with the indoor coil, making the indoor coil the condenser.

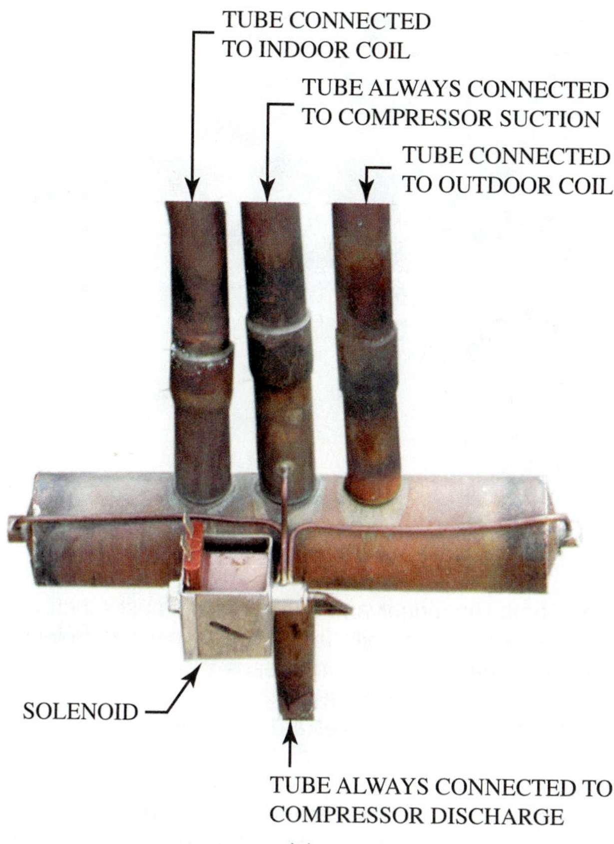

(a)

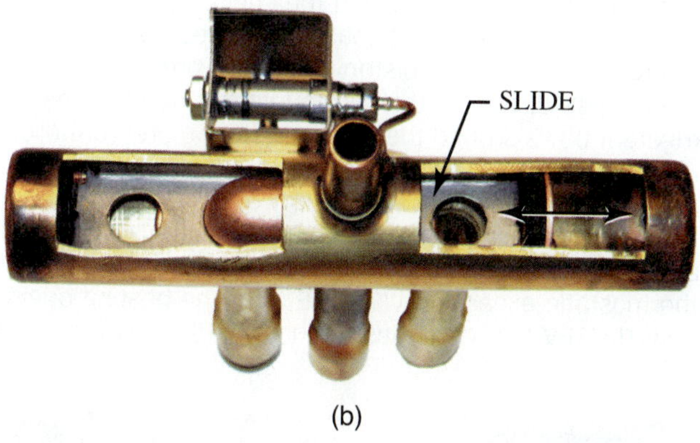

(b)

Figure 62-5 Connections for a typical heat pump reversing valve.

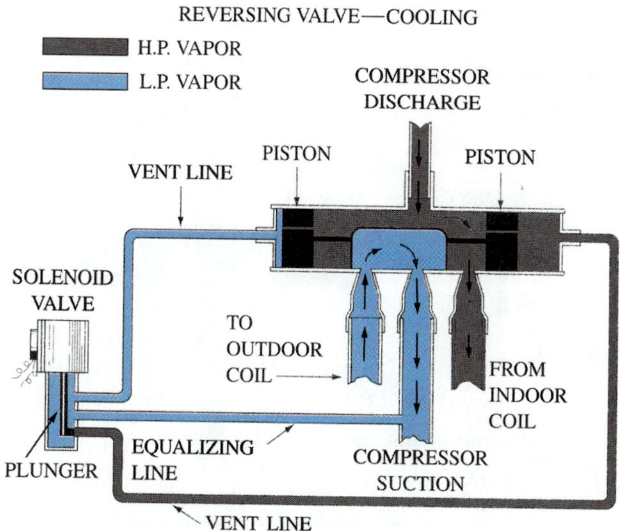

Figure 62-7 Reversing valve in heating position.

TECH TIP

When a reversing valve repositions, refrigerant from the compressor discharge will pass through the piston bleed hole to the vent line, then into the equalizing line and back to the compressor suction. Once the pistons are positioned either to the extreme right or left, the vent line is closed off. On the end of each piston at the vent passage is a tip that seals into a seat to prevent continuous bypass of the hot vapor from the cylinder through the piston bleed hole into the open vent line.

Cooling Cycle: Valve Energized

In Figure 62-8, the solenoid valve is energized and the control plunger lifts, closing the top port vent line connected to the left end of the piston. The bottom port vent line connected to the right end of the piston is open to the equalizing line. This allows the pressure to bleed off from the right end of the piston, which causes it to travel to the extreme right, repositioning the slide valve to align the control port from the indoor coil with the control port leading to the compressor suction, making the indoor coil the evaporator. This same movement also aligns the compressor discharge with the outdoor coil, making the outdoor coil the condenser.

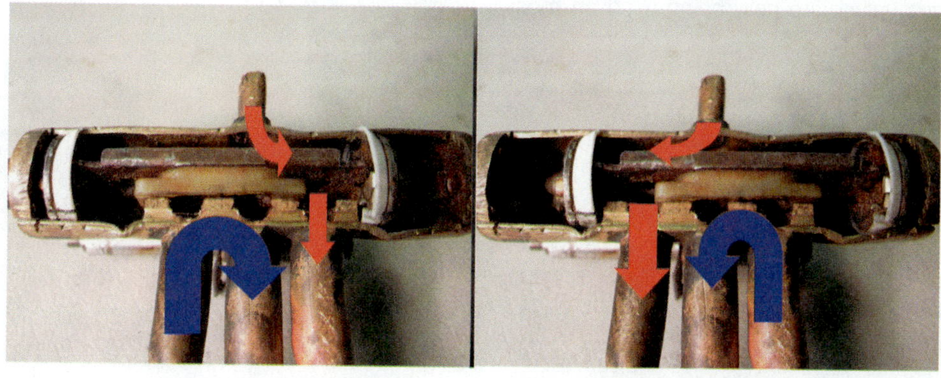

Figure 62-6 Internal view of heat pump reversing valve.

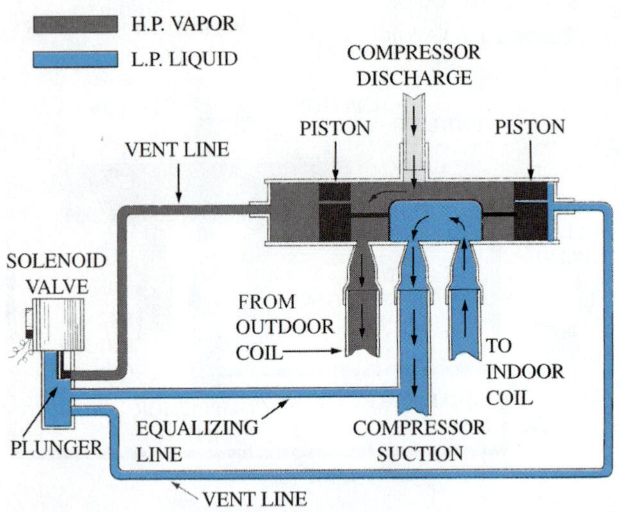

Figure 62-8 Reversing valve in cooling position.

SERVICE TIP

On each end of the sliding piston is a tip that seals into a seat to prevent continuous bypass of the hot vapor from the cylinder through the piston bleed hole into the open vent line. Once the piston is positioned on either side, the tip should seal off the vent line. Leaks at the tip can be detected by a very hot capillary line on the end of the valve.

High efficiency units have less pressure difference between the high and low sides. This reduced pressure difference sometimes causes reversing valves to get stuck halfway through a shift. When this happens, the high- and low-side pressures will equalize through the valve. Valves with four pilot tubes, instead of the usual three, were developed to attack this problem (Figure 62-9). As before, the pilot valve shifts the suction gas to either end of the valve. The extra fourth line on the pilot valve goes to the discharge line and shifts discharge gas to the opposite end. Instead of relying on a bleed port for the high-pressure gas, it is carried via a pilot line directly from the discharge line coming into the reversing valve. No bleed is designed into the valve at all, reducing the chance of unwanted bleed-through during operation and creating a more positive shift. Four-line pilot valve–reversing valves are now being used on most heat pumps.

SERVICE TIP

A reversing valve has a relatively thin copper or brass outside skin. It can be easily damaged if struck by a hammer or wrench. Some technicians feel that reversing-valve problems may be caused by a stuck reversing-valve slide and strike the reversing-valve tube in an attempt to dislodge the stuck slide. Striking a reversing valve with anything can only result in damage to the reversing valve and can never resolve a stuck slide. Never strike a reversing valve.

Figure 62-9 Four-pilot-tube reversing valve.

62.8 THE METERING DEVICES

Since both the indoor and outdoor coils must function as the evaporator, both coils need metering devices. However, only the metering device that feeds the coil acting as the evaporator should be in the refrigerant circuit. In some heat pump designs, check valves are connected in parallel with the metering devices to permit bypassing the metering devices when they are not needed in the circuit (Figure 62-10).

Recently, most manufacturers have been using metering devices that act as both a metering device and a check valve. Flow restrictors with pistons in them are popular in lower-end models. These are like a check valve with a hole drilled through the ball in the check valve. In the forward direction, the piston seats and forces the refrigerant through the hole in the piston, creating a pressure drop. In the reverse direction, the piston lifts off the seat and refrigerant flows around the piston without a pressure drop. Figure 62-11 shows a flow check device in a heat pump and a close-up of the piston from a flow check device.

Some manufacturers design a heat pump system with a fixed metering device for the cooling operation and the thermostatic expansion valve (TEV) for the heating operation. The TEV is a better design choice in the heating mode

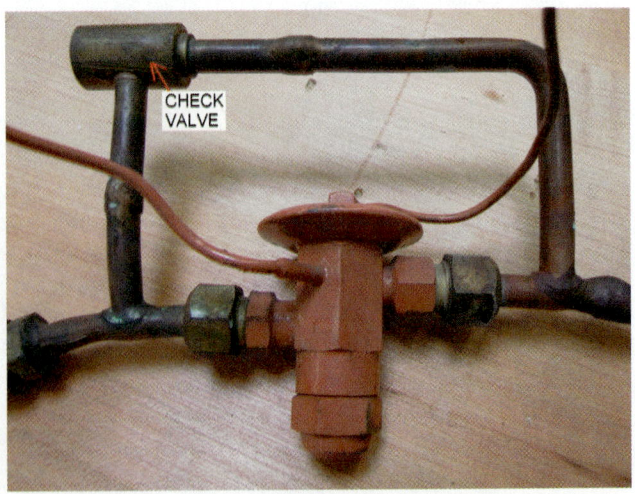

Figure 62-10 Check valve in parallel to TEV.

Figure 62-11 Flow check orifice.

compared to the fixed metering device. The TEV can modulate the refrigerant flow over the wide range of temperatures experienced in the heating mode.

Bi-flow TEVs with a built-in check valve are now popular (Figure 62-12). In the forward direction, the check valve seats and forces the refrigerant through the orifice in the expansion valve, creating a pressure drop. In the reverse direction, the ball lifts off the seat and refrigerant flows around the valve seat without a pressure drop.

SERVICE TIP

A bi-flow TEV still has a specific inlet and outlet. If a bi-flow TEV with a check valve is installed backwards, it will not properly meter refrigerant, flooding the evaporator coil.

Figure 62-12 Bi-flow TEV on heat pump split system.

Figure 62-13 Bi-flow TEV on water-source heat pump.

Another form of bi-flow TEVs is used on water-source heat pumps. These valves meter in both directions (Figure 62-13). It is possible to have a single expansion valve on packaged water-source units because the two coils are so close together. Using a bi-flow valve that meters in both directions allows manufacturers to build a unit with only one metering device, eliminating check valves completely. The sensing bulb is not located on the outlet of either coil. Instead, the sensing bulb is located after the reversing valve and before the suction accumulator. In this way the bulb will respond to the refrigerant condition leaving either coil, regardless of mode.

62.9 CHECK VALVES

Checks valves are used in some heat pump designs to direct the flow of refrigerant around the metering device not being used. Two check valves are used: one connected in parallel with the inside metering device and one in parallel with the outdoor metering device. The check valves in the heat pump circuit are vital to ensure that proper pressures are maintained. Today, most heat pump systems use metering devices with built-in check valves, rather than separate check valves. The check valves are still there but are incorporated into the metering devices. (Figure 62-14) shows the flow through the inside and outside flow-check metering devices in both the cooling and heating cycles.

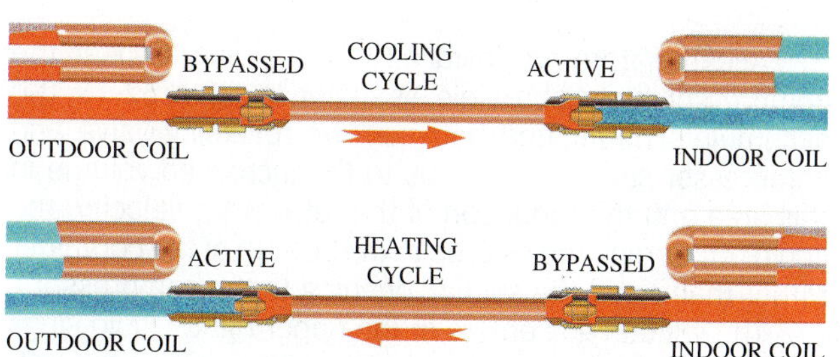

Figure 62-14 The active metering device is shown in the cooling and heating cycle.

The first check valves used in residential and small commercial heat pumps were of the disk type. This valve used a flat disk over a circular flat inset with a spring that had just a light spring constant to aid in closing and still cause very little resistance to refrigerant flow. Where pressure differences occurred rapidly, the disk would be sometimes forced to one side and hang up. This prevented the shutoff of reverse refrigerant flow. An improved version using a steel ball instead of a flat disk was developed. This type of check valve can withstand a heavy reversal of refrigerant flow pressure, especially when the unit completes the defrost cycle.

62.10 REFRIGERANT STORAGE

Since the heat pump operates year round, it must be adaptable to a wide range of operating conditions. These varying conditions translate into significant difference in unit capacity. In simple terms, the heat pump is not always circulating the same amount of refrigerant, so it needs a place for the unused refrigerant to collect. Liquid receivers do not serve this purpose for heat pumps because they are located in the liquid line, where the refrigerant reverses between cycles. Receivers have a definitive inlet and outlet and cannot be reversed. The two storage vessels used in heat pumps are the suction line accumulator and the heating-cycle charge compensator.

62.11 SUCTION-LINE ACCUMULATOR

Accumulators protect the compressor from liquid refrigerant that will leave the evaporator coil during light loads or if the airflow across the coil is reduced for any reason. A suction-line accumulator (Figure 62-15) is a cylinder placed in the suction line, ahead of the compressor, to catch any liquid refrigerant that has not evaporated before it reaches the compressor. In the heat pump, there are three situations where an accumulator can provide compressor protection is needed against liquid slugs:

- **Floodback on the cooling cycle** An air restriction through the indoor coil will lead to a light load, low superheat, and liquid floodback.
- **Floodback on the heating cycle** Excessive frost buildup on the outdoor coil as well as low airflow will lead to a low superheat and liquid floodback.
- **Termination of the defrost cycle** Liquid floodback will always occur when the defrost cycle is terminated.

Figure 62-15 Suction-line accumulator.

Accumulators are installed in the piping as near the compressor inlet as possible, as shown in Figure 62-16. The accumulator is located between the reversing valve and compressor suction inlet. Due to the increased volume in this area and the reduction of the refrigerant velocity, the liquid refrigerant and oil drop to the bottom of the container rather than enter the suction opening to the compressor.

The suction gas enters at the upper right. Liquid slugs are directed toward the right-hand wall and run to the bottom. The shape of the deflector directs the inlet flow in a slightly downward tangential direction. There is a metering orifice at the bottom of the U-tube return pipe that slowly returns oil to the compressor through the suction line, along with a limited amount of liquid refrigerant. The liquid refrigerant in the accumulator gradually evaporates, entering the compressor as vapor.

Termination of the defrost cycle can cause liquid to flood back to the compressor. The hot gas used to defrost the outdoor coil will condense and turn to liquid refrigerant as it travels through the coil. When the defrost cycle terminates and the reversing valve switches over, the outdoor coil once again becomes the evaporator, and this liquid refrigerant will enter the compressor suction. Without

Figure 62-16 The accumulator goes in the suction line just after the reversing valve.

Figure 62-17 Heating-cycle charge compensator tank.

the accumulator to catch and hold the liquid refrigerant, a liquid surge into the compressor can ruin valves. The accumulator is often sized to hold the entire refrigerant charge of the system to prevent the possibility of liquid carryover into the compressor. Therefore, the refrigerant charge of a heat pump system is very critical.

When a compressor is changed because of a burnout, the accumulator must also be changed because the contamination trapped in the accumulator cannot be removed. Purging or flushing will not clean the accumulator or clear the oil return orifice. Even if the accumulator does not become clogged after a compressor burnout, it can hold a lot of contaminants that the system would be better off without.

SERVICE TIP

During refrigerant recovery, it is sometimes difficult to remove all refrigerant from an accumulator. Some refrigerant can be trapped in the oil at the bottom of the accumulator. As refrigerant is recovered, the accumulator may ice up around the base. If this occurs, recovery can be helped by warming the accumulator. Never use a torch to warm any part of a system during a recovery. Warm water or a heat lamp can be used to warm the accumulator to speed refrigerant recovery.

SERVICE TIP

Pay attention to the inlet and outlet of the accumulator. They are marked and must not be reversed!

62.12 HEATING-CYCLE CHARGE COMPENSATOR

A few manufacturers now use heating-cycle charge compensator tanks (Figure 62-17). The compensator tank is basically an empty can with the gas line running straight through it. The gas line is not open to the tank; it just passes through the tank. Liquid refrigerant enters and leaves the tank through a single small line. This line is connected to the liquid line.

In the heating cycle, the gas line is cold, making the refrigerant in the compensator tank cold. Refrigerant in the liquid line migrates to the charge compensator tank because the refrigerant in the liquid line is at a higher temperature and pressure than the liquid sitting in the compensator tank. This stores the extra refrigerant that is not circulated during the heating cycle (Figure 62-18).

During the cooling cycle, the gas line is hot, making the refrigerant in the compensator tank hot. The liquid refrigerant in the compensator tank leaves the tank and goes into the liquid line because the temperature and pressure of the refrigerant in the tank are now higher than the temperature

Figure 62-18 Compensator tank full during the heat cycle.

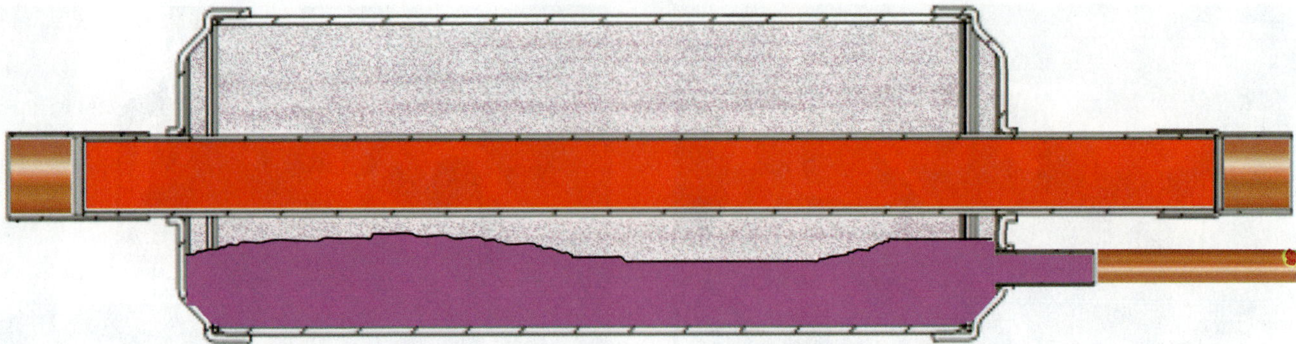

Figure 62-19 Compensator tank empty during the cooling cycle.

and pressure of the liquid line. This returns refrigerant to the cycle to be circulated during the cooling cycle (Figure 62-19).

62.13 BI-FLOW FILTER DRIER

The purpose of the filter drier is to remove suspended particles and moisture from the refrigeration system. The bi-flow filter drier (Figure 62-20) is used in the liquid line to prevent contamination of the refrigerant metering devices. Refrigerant flows through the liquid line in one direction in cooling and in the other in heating. If a conventional filter drier were installed in a flow-reversing refrigerant line, most of the contaminants collected in one direction of flow would be flushed back into the system when the reverse flow starts. The design of the bi-flow drier prevents this from happening. The bi-flow drier is similar to having two parallel filter driers with internal check valves that prevent the loss of trapped contaminants. Figure 62-21 shows the internal arrangement of a bi-flow drier. A double-sided arrow is used to indicate the bidirectional flow of the refrigerant through the drier. These driers should be changed every time the system is exposed to the atmosphere.

SERVICE TIP

Filter driers can be used to clean up a system after a compressor burnout. Once the new compressor or condenser is in place, a less expensive model of filter drier can be temporarily installed to trap the major contaminations in the system. A severe burnout may require more than one filter drier change-out. Once the system has been cleaned up, a bi-flow filter drier can be installed permanently.

Figure 62-20 Heat pump bi-flow filter drier.

62.14 HEAT PUMP EFFICIENCY RATINGS

Several different efficiency ratings are used to measure heat pump efficiency. The next two sections describe the different heat pump efficiency ratings and how they are used to compare different units. The four common measures of heat pump efficiency are as follows:

- Energy efficiency ratio (EER)
- Seasonal energy efficiency ratio (SEER)
- Coefficient of performance (COP)
- Heating seasonal performance factor (HSPF)

All of these ratings are used to compare the efficiency of like capacity units. The higher the rating, the higher the efficiency. Higher efficiencies mean lower operating costs. No single rating covers both heating and cooling. EER and SEER rate the cooling efficiency, while COP and HSPF rate the heating efficiency. The temperature and humidity conditions the unit operates in affect the ratings. When comparing ratings, it is important to know the conditions used to produce the rating.

62.15 COOLING EFFICIENCY RATINGS: EER AND SEER

A system's EER and SEER are crucial specifications when choosing equipment. The Federal Department of Energy (DOE) mandates minimum EER and SEER ratings for different types of equipment. The EPA Energy Star program also specifies EER and SEER ratings for units to receive the Energy Star label.

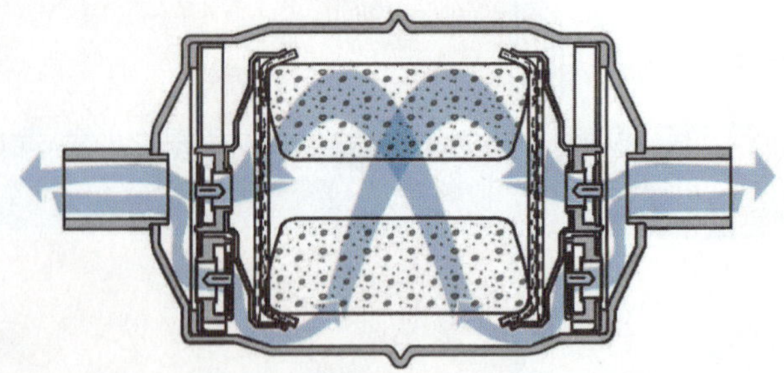

Figure 62-21 Internal operation of a bi-flow heat pump filter drier.

EER: Energy Efficiency Ratio

The energy efficiency ratio (EER) is the number of cooling BTUs accomplished by each watt of electricity used at a given condition. EER is calculated by dividing the cooling capacity in BTUs per hour by the electrical consumption in watt-hours. EER is calculated for a specific condition, normally AHRI rating condition A: 95° outdoor ambient, 80°F dry-bulb, and 67°F wet-bulb indoor air. EER is a steady state measurement. The system is operated long enough to ensure that any startup fluctuations in energy use and cooling capacity have passed. EER calculates the efficiency at one specific condition, but air-conditioning systems operate across a range of conditions. Therefore, the SEER test standard was developed to check the equipment operation across the normal operating outdoor conditions.

SEER: Seasonal Energy Efficiency Ratio

The seasonal energy efficiency ratio (SEER) is the number of cooling BTUs accomplished by each watt of electricity used over a season. Readings are not actually taken over an entire season. To simulate the total cooling capacity and watt-hour use over a season, units are operated at three different specific steady state test conditions (Table 62-3). A fourth test, which cycles the unit on and off during the test, is used to determine the effect of cycling. Additional tests are required for two-stage and variable-capacity equipment. The results of these tests are plugged into a complex formula that calculates the SEER rating. For most equipment, this is very close to the efficiency of the equipment at 82°F outdoor ambient with 80°F dry-bulb and 67°F wet-bulb indoor return air. Since most air conditioners are more efficient at 82°F outside than at 95°F outside, their SEER rating is normally higher than their EER rating.

TECH TIP

It is important to understand that EER and SEER ratings for split systems are for matched systems, not a condenser by itself. The equipment must be installed with an AHRI matched coil for the rating to be accurate.

Table 62-3 Seasonal Energy Efficiency Testing Conditions

Condition	Outdoor Ambient	Indoor Temperature
A	95°F	80°F dry bulb 67°F wet bulb
B	82°F	80°F dry bulb 67°F wet bulb
C	82°F	80°F dry bulb 57°F wet bulb

62.16 HEATING EFFICIENCY RATINGS: COP AND HSPF

A system's COP and HSPF are crucial specifications when choosing equipment. The DOE mandates minimum HSPF ratings for different types of equipment. The EPA Energy Star program also specifies HSPF ratings for units to receive the Energy Star label.

COP: Coefficient of Performance

The coefficient of performance (COP) is the ratio of total heat output divided by heat input. COP excludes any supplementary resistance (electric) heat. Since the fan motor adds heat to the air, it is included in this rating. COP is represented by the following formula:

$$\text{COP} = \frac{\text{BTU heat output}}{\text{BTU heat input}}$$

The COP for an electric resistance heater is 1. In other words, the electric heat output in BTUH is equal to the heat equivalent of the input wattage. Watt-hours are converted into BTUH by multiplying the watt-hours by 3.41.

The COP rating for a heat pump is variable. The heat output for an air-source heat pump will depend on the outdoor temperature. The cooler the outdoor temperature, the less heat in the air and the lower the heat capacity of the heat pump. As the temperature drops, the COP drops. The COP of a heat pump is always comparably higher than electric heat except at very low temperatures. The COP of an air-source heat pump can be as low as one at very low temperatures and may approach four at ideal conditions. The higher the COP, the more efficient the operation and the lower the operating cost. Most air-source heat pumps operate in a COP range of 2.0 to 4.0.

Generally, the cost of operating an air-source heat pump is one half to one fourth the cost of operating straight electric heat. The COP rating is not the best way to compare heat pump efficiency because the COP varies with the outdoor temperature. COP also does not take defrosting into account. To truly compare two units on the basis of COP, you would need to know the COP of both units over the range of operating conditions, not just the system design condition.

HSPF: Heating Seasonal Performance Factor

The heating seasonal performance factor (HSPF) is like a SEER for the heating season. Unlike the COP rating, the HSPF is designed to measure performance over an entire season, not just one specific operating point. HSPF is the total heat output of a heat pump, including energy used in defrost, divided by the total electric power in watt-hours. It could be stated as the following formula:

$$\text{HSPF} = \frac{\text{Total Btu heating season output}}{\text{Total heating season watts}}$$

Testing for the HSPF is designed to simulate operation over a season using data collected from three specific operating conditions (Table 62-4). A fourth test is used to

Table 62-4 Heating Seasonal Performance Factor Conditions

Condition	Outdoor Ambient	Indoor Temperature
H1	47°F dry bulb 43°F wet bulb	70° dry bulb
H2	35°F dry bulb 33°F wet bulb	70° dry bulb
H3	17°F dry bulb 15° wet bulb	70° dry bulb

determine the effect of frost accumulation and defrosting. Data from these four tests is fed into a complex formula to calculate the unit's HSPF. The HSPF of a heat pump is always lower than its SEER.

UNIT 62—SUMMARY

Heat pumps are available as air-to-air, water-to-air, air-to-water, and water-to-water units. Heat pump systems are the most efficient and economical heating systems available because they operate by moving heat instead of directly converting another form of energy into heat. The two coils in the heat pump act as either evaporators or condensers, depending upon the direction of the refrigerant flow. Heat pumps change the direction of refrigerant flow using a reversing valve. A metering device is required for both coils. Check valves built into the metering devices control which metering device is active. Refrigerant is stored in either a suction accumulator or a heating-cycle charge compensator.

Efficiency ratings for heat pumps include energy efficiency ratio (EER), seasonal energy efficiency ratio (SEER), coefficient of performance (COP), and heating seasonal performance factor (HSPF). The EER and SEER are for cooling; COP and HSPF are for heating. EER and COP are steady state ratings, while SEER and HSPF are calculated to simulate seasonal efficiency.

UNIT 62—REVIEW QUESTIONS

1. List the types of heat pumps available by combinations of heat source and heat sink.
2. Why is a heat pump called a heat pump?
3. How is the refrigeration cycle reversed in a heat pump?
4. Why are accumulators and/or charge compensator tanks necessary in air-source heat pumps?
5. Describe the operation of the reversing valve.
6. What is the difference between an EER and a SEER?
7. What is the difference between a COP and an HSPF?
8. Why are the coils in an air-source heat pump referred to as the outdoor coil and indoor coil instead of the condenser and evaporator?
9. Explain why there is heat in the air even at cold temperatures.
10. List the physics principles that make heat pump operation possible.
11. Why do most heat pumps have two metering devices?
12. What keeps both metering devices from being in the refrigeration circuit at the same time?
13. What is a bi-flow thermostatic expansion valve?
14. What is the difference between the defrost cycle and the cooling cycle?
15. Show the function of the coils by writing "Evaporator" or "Condenser" in the appropriate box for the following table.

	Cooling	Heating
Outdoor coil		
Indoor coil		

16. Show the function of the metering devices by writing "Active" or "Bypassed" in the appropriate box for the following table.

	Cooling	Heating
Outdoor metering device		
Indoor metering device		

17. Where is the permanent suction line always located on a reversing valve?
18. Where is the permanent discharge line always located?
19. Why are bi-flow filter driers used on heat pumps instead of standard filter driers?
20. How does a flow check piston-type metering device work?

UNIT 63

Air-Source Heat Pump Applications

OBJECTIVES

After completing this unit, you will be able to:

1. describe the operation of a typical heat pump thermostat.
2. explain the effect that outdoor temperature has on heating capacity.
3. explain the purpose of supplemental heat.
4. describe the operation of heat pump defrost controls.
5. list the advantages of air-to-water heat pump systems.
6. list the advantages of mini-split heat pump systems.
7. describe the operation of dual-fuel heat pump systems.

63.1 INTRODUCTION

Air-source heat pumps provide an efficient and economical alternative to fossil fuel furnaces. They are popular because of their ability to provide heating and cooling efficiently from a single system. Air-source heat pumps are the dominant heating and cooling system in the southern portion of the country, surpassing fossil fuel furnaces. Air-to-air heat pumps are the most common type of heat pump. They can be readily applied for residential use. They are less expensive than water-source heat pumps and easier to install than a furnace and air conditioner. Air-to-air systems use air as both the heat source and the heat sink. In heating, the outdoor air is the heat source and the indoor air is the heat sink. They are available as packaged units, with all the components in one cabinet, or as split systems, with the compressor and condenser in one cabinet and the evaporator in another.

63.2 AIR-TO-AIR HEAT PUMPS

Air-to-air heat pumps are the most common heat pumps in use today because they are generally less expensive and easier to install than other types of heat pumps. The basic cycle for an air-to-air heat pump consists of absorbing heat from the air in one space and then rejecting this heat to the air at a different location. In the cooling cycle, the indoor coil is the evaporator and the outdoor coil the condenser. The indoor coil absorbs heat from the inside air and the outdoor coil rejects heat to the outside air. In the heating cycle, this transfer of heat is reversed. The outdoor coil is the evaporator and the indoor coil is the condenser. The outdoor coil absorbs heat from the outside air and the indoor coil rejects heat to the inside air.

TECH TIP

Heat pump indoor coils must have a higher design pressure than regular air-conditioning evaporator coils. Since the indoor coil of a heat pump serves as the condenser during heating operation, it must withstand system high-side pressures.

The indoor coil on a heat pump must be able to withstand the high-side system pressure because it is the condenser in the heating cycle. Typically, air movement across the indoor coil is provided by a blower assembly in a fan coil unit that circulates the air through the supply and return duct system. The coil in a heat pump is normally located on the return side of the blower: the blower draws the air across the coil (Figure 63-1). A propeller fan assembly in the outdoor section provides the airflow through the outdoor coil. Most indoor coils are tube-and-fin coils. The outdoor coils can be tube and fin, spiny fin, or microchannel coils (Figure 63-2).

Figure 63-1 The blower on a heat pump draws the air across the coil.

Figure 63-2 The outdoor coil on a heat pump may be finned tube, spiny fin, or microchannel.

SERVICE TIP

Because the indoor blower on a heat pump draws air across the indoor coil, heat pumps will not operate properly with the panels removed from the indoor blower section. With a panel off, most of the air just enters through the open panel, bypassing the coil altogether. Operating a heat pump with the indoor panel off will cause the indoor coil to freeze up in the cooling mode. In heating, operating with the panel off will cause the unit to shut off on high head pressure.

63.3 HEAT PUMP THERMOSTATS

Most air-to-air heat pump systems use a single-stage cooling and two-stage heating thermostat. Figure 63-3 shows a common heat pump thermostat. A two-stage thermostat is required for the heating operation. The first stage of heat energizes the refrigeration circuit of the heat pump: the compressor and both indoor and outdoor fan motors. The second stage energizes the auxiliary heat when the heat pump cannot maintain the temperature on its own. Because the compressor operates in both heating and cooling, the Y terminal that controls the compressor contactor is energized in both heating and cooling on a heat pump thermostat. The G terminal that controls the fan relay is also energized in both heating and cooling. The thermostat controls system changeover from heating to cooling by controlling the reversing valve. Most heat pumps today energize the reversing valve in cooling. The O terminal on the thermostat is energized whenever the system switch is placed in the cooling position. The O terminal is used to control the reversing valve on heat pumps that energize the reversing valve in cooling. However, not all heat pumps energize the reversing valve in cooling: some energize it in heating. These systems use the B terminal to control the reversing valve. The B terminal is energized whenever the system switch is moved to the HEAT position. Since few systems use both the O and B terminals, many thermostats now have an O/B terminal that is selectable between the two (Figure 63-4).

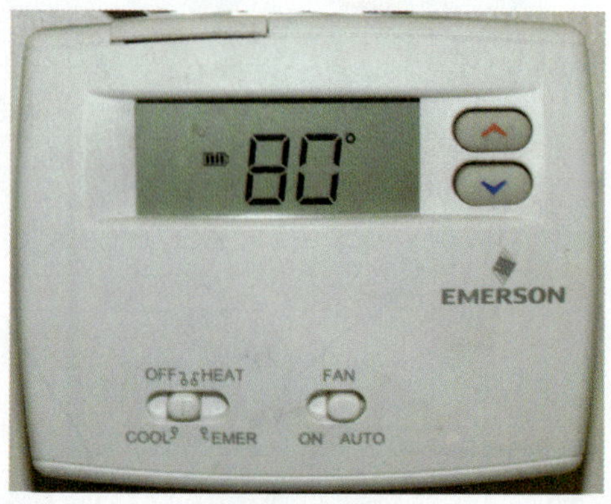

Figure 63-3 Typical heat pump thermostat.

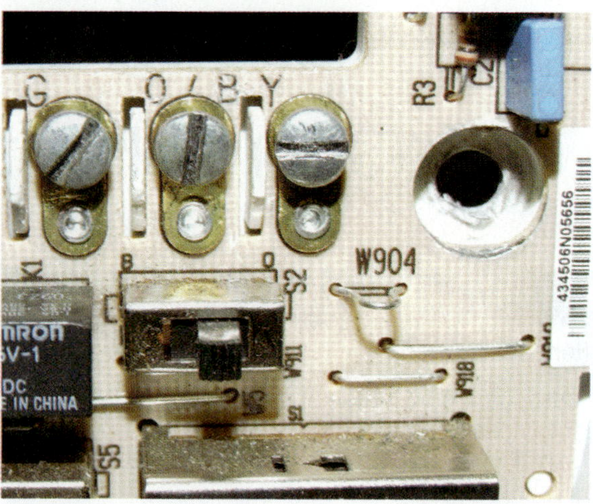

Figure 63-4 Selectable O/B terminal on heat pump thermostat.

SERVICE TIP

Trane and some other manufacturers use B to designate the common side of the 24 V control system. You should never connect a B terminal on an over-the-counter thermostat to the B terminal on one of these units. The thermostat will create a direct short across the secondary side of the transformer when it is placed in the HEAT position. A careful inspection of the unit wiring diagram should indicate if B is being used for common.

If the heat pump is not able to maintain the space temperature within a degree or two of the setpoint, the second stage of heat will be energized. When the heat pump and second-stage heat are operating together, the second stage of heat is known as supplemental heat. Second-stage heat is normally controlled by the W or W_2 terminal. Supplemental, or auxiliary heat, is usually electric strip heat. Less commonly, gas or oil heat can also be used as the auxiliary or second-stage heat.

Besides the familiar HEAT, COOL, and OFF system settings, heat pump thermostats also have an emergency heat setting (Figure 63-5). Emergency heat energizes the auxiliary heat on first stage instead of second stage and does not energize the compressor and outdoor fan at all. This setting is used to provide the customer with heat when a problem exists with the refrigeration system. Emergency heat should be reserved for emergency use since the auxiliary heat is typically more expensive to operate than the heat pump. Heat pump thermostats often have lights to indicate when emergency heat is energized. On some models, a light will indicate that the

Figure 63-5 Thermostat showing emergency heat setting.

second stage of heat is energized and a different light will indicate that the emergency heat is operating. Digital thermostats will also provide operating information within the display screen (Figure 63-6). Some messages may flash on and off, such as built-in compressor protection, which does not allow the compressor to restart too soon after a shutdown.

63.4 AIR-TO-AIR HEAT PUMP CAPACITY

Air-source heat pumps are rated for both heating and cooling capacity. Heat pump cooling capacity is rated like air-conditioner capacity, using the AHRI rating condition of 95°F outdoor ambient and 80°F db 67°F wb return air. Heat pump cooling efficiency is also rated like an air-conditioning system using either EER or SEER. The energy efficiency ratio (EER) gives the cooling BTUH produced per watt-hour of electricity used at a given operating condition. The seasonal energy efficiency ratio (SEER) approximates the cooling BTUH per watt-hour produced over a season. See Unit 62, Heat Pump Fundamentals, for more information on EER and SEER. Currently, the minimum SEER rating is 13. The highest EER in a heat pump today is 16, and the highest SEER in a heat pump is 22.

Air-source heating capacity is rated at two conditions: 47°F outdoor ambient and 17°F outdoor ambient. The refrigerant system heating capacity is the combination of the heat absorbed from the outside air along with the heat added by the compressor. The more heat absorbed in the outdoor coil, the higher the heat pump heating capacity is. As the outdoor temperature drops, the heating capacity also drops because the heat pump cannot remove as much heat from the air. Table 63-1 shows the capacity of a typical air-source heat pump. At 47°F, the heating capacity of an air-source heat pump is similar to its cooling capacity. However, the heating capacity at 17°F is typically only about half of the heating capacity at 47°F. The heating seasonal performance factor (HSPF) is used to specify the heating efficiency. It approximates the heating BTUH per watt-hour for a season. The minimum HSPF is currently 7.7; the highest HSPF currently available is 13.

Energy Star is a voluntary awareness, communications, and marketing campaign developed by the U.S. Department of Energy, Environmental Protection Agency (EPA), in 1992 to promote energy efficiency in a multitude of products, including heating/cooling equipment. There are often federal programs that reward consumers financially for choosing Energy Star appliances, including tax credits. The Energy Star label is awarded to those air-to-air packaged units with an EER of 11 or greater, a SEER of 14 or greater, and an HSPF of 8 or greater (Figure 63-7). For split systems, the requirements are an EER of 12 or greater, an SEER of 14.5 or better, and an HSPF of 8.2 or better.

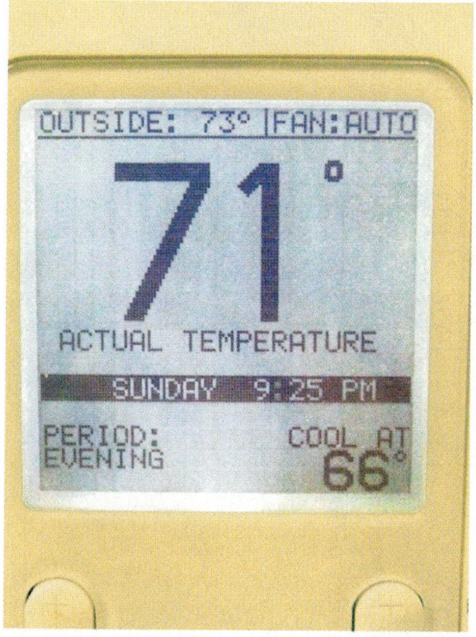

Figure 63-6 Digital thermostat display status.

Figure 63-7 Energy Star label. *(Courtesy of U.S. Environmental Protection Agency, Energy Star Program)*

Table 63-1 Heating Capacity at Different Outside Temperatures

	Outdoor Ambient Temperature*																	
	65	**60**	**55**	**50**	**47**	**45**	**40**	**35**	**30**	**25**	**20**	**17**	**15**	**10**	**5**	**0**	**−5**	**−10**
MBh	29.4	27.8	26.2	24.5	23.4	22.7	21.1	19.4	15.6	14.4	13.2	12.5	12.0	10.8	9.6	8.4	7.1	5.8
ΔT	31.7	30.0	28.2	26.4	25.2	24.4	22.7	20.9	16.8	15.5	14.3	13.5	13.0	11.6	10.3	9.0	7.7	6.3
KW	2.08	2.04	2.00	1.96	1.94	1.92	1.88	1.84	1.81	1.77	1.73	1.71	1.69	1.65	1.61	1.57	1.53	1.49
Amps	10.8	10.2	9.6	9.2	8.9	8.8	8.4	8.0	7.8	7.5	7.2	7.1	7.0	6.8	6.4	6.2	5.8	5.4
COP	4.14	4.00	3.84	3.66	3.54	3.46	3.28	3.09	2.52	2.38	2.24	2.15	2.09	1.92	1.74	1.56	1.36	1.15
EER	14.2	13.7	13.1	12.5	12.1	11.8	11.2	10.6	8.6	8.1	7.7	7.3	7.1	6.6	6.0	5.3	4.7	3.9
Hi PR	388	372	358	342	334	328	315	302	290	277	266	259	255	245	235	226	218	210
Lo PR	45	134	126	115	109	105	96	86	77	69	61	57	55	46	40	34	29	23

*For nominal CFM and 70°F indoor dry bulb. Instantaneous capacity listed.

(Courtesy of Goodman Global Group, Inc.)

63.5 AUXILIARY HEAT

Air-source heat pumps heat by removing heat from the outside air and delivering it to the inside. As the outside air temperature drops, the evaporator temperature and pressure must also drop in order to continue to absorb heat. This increases the compression ratio of the compressor pump and decreases its volumetric efficiency. In short, the system loses capacity. At the same time, the system load actually increases due to the increased demand for heat at lower temperatures. At some point, the heat pump's decreasing capacity will not be able to keep up with the structure's increasing load. The lowest temperature at which the heat pump can heat the structure is called the balance point. Below this temperature, auxiliary heat will be required to maintain comfort conditions.

Normally the auxiliary heat is not energized until the system reaches its balance point because auxiliary heat is more expensive to operate. In practice, the auxiliary heat is energized 5°F above the balance point to ensure that the system can handle the demand. The auxiliary heat is also used to reheat the cool air the heat pump circulates in the defrost cycle.

The auxiliary heat is controlled by the second stage of a two-stage thermostat. The second stage closes approximately 1.5°F below the thermostat setpoint. If the heat pump cannot meet the heating demand, the second stage energizes to bring on the auxiliary heat. Some units also use outdoor thermostats set to 5°F above the balance point. These outdoor thermostats will not let the auxiliary heat come on until the temperature outside drops to their setpoint.

Electric strip heaters are the most common auxiliary heat (Figure 63-8). Since the refrigeration circuit can move heat more inexpensively than the electric auxiliary heaters can produce heat, the auxiliary electric is the second stage of heat. It cycles on and off while the refrigeration system continues to operate. The electric strip heaters are normally installed after the blower. Most units have access panels where the strip heaters are installed when the heat pump is installed (Figure 63-9). Electric strip heaters normally come in 5-kW increments. A 5-kW strip heater produces 17,000 BTUH when operated on 240 V. The same heater operated on 208 V produces only 3.6 kW and 12,300 BTUH.

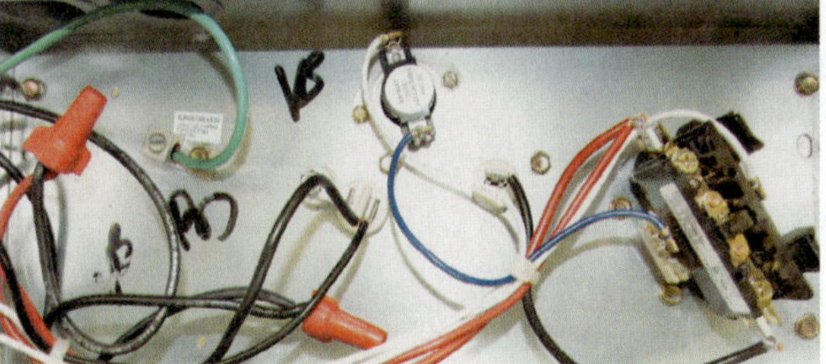

Figure 63-8 Electric strip heater.

Figure 63-9 Strip heaters can be installed in space provided in heat pump.

The load of the building being conditioned and the capacity of the equipment being used to condition it determine a system's balance point. This means that the balance point of a heat pump installed in your house would be different than the balance point of the same system installed in your neighbor's house. The amount of auxiliary heat required increases substantially as the outside air temperature decreases. This can be seen in Table 63-2, which is based on an inside temperature of 70°F. At an outdoor temperature of 40°F and 5 kW of auxiliary electric heat, the heating capacity for the unit would be 37,440 BTU/hr. If the outside temperature dropped to 10°F, an 8-kW heater would be required to maintain roughly the same heating capacity, which would now be 38,110 BTU/hr. If the temperature dropped as low as −10°F, a 10-kW heater would be required.

TECH TIP

An air-to-air heat pump will operate continuously when the outdoor temperature is below the balance point. The auxiliary heat is turned off before the first stage of the thermostat is satisfied, so the heat pump continues to run. Since the refrigeration system cannot provide enough heat below the balance point to meet the load, the first-stage heating thermostat will not be satisfied. The auxiliary heat will cycle on and off to maintain a temperature just below the thermostat setpoint.

Table 63-2 Heating Capacity at Different Ambient Temperatures

Cooling and Heating Capacity				
Cooling capacity				
Total BTU/hr	24,000	35,400	46,600	55,600
Sensible BTU/hr	18,000	26,700	36,600	39,300
SEER/EER	15/11.3	15/12.0	15/11.3	14/10.3
Decibels	76	75	78	78
Heating capacity				
BTU/hr (47°F)	23,400	35,500	45,600	56,000
COP (47°F)	3.5	3.5	3.4	3.3
BTU/hr (17°F)	12,500	18,600	24,600	31,200
COP (17°F)	2.2	2.4	2.1	2.1
HSPF	8.0	8.0	8.0	8.0

(Courtesy of Goodman Global Group, Inc.)

63.6 OUTDOOR AMBIENT THERMOSTAT

The outdoor ambient thermostat (OAT) controls the operations of the supplemental heat. The remote sensing bulb is mounted outside the unit in a location where it will not be influenced by airflow from the unit (Figure 63-10). The outdoor thermostat is set low enough to prevent the operation of all the electric heating elements when the outdoor temperature is mild (Figure 63-11). For example, a common setting is 40°F. The purpose of this control is to prevent unnecessary operation of electric heat should the thermostat operator turn the temperature setting high enough to operate the heat pump and the supplemental heat. The outdoor thermostat is bypassed when the heat pump thermostat is switched to the emergency heat mode.

Figure 63-10 Outdoor ambient thermostat sensing bulb.

Figure 63-11 Outdoor ambient thermostat.

> **TECH TIP**
>
> New energy codes that have been adopted in many cities and states require an outdoor thermostat on all installations. The outdoor thermostat must lock out the supplemental heat when the outdoor temperature is above 50°F. This allows the compressor to provide heat for the resident and keeps the more expensive energy-consuming supplemental heat strip off.

63.7 DEFROST CYCLE

Frost forms on the outdoor coil at low ambient temperatures because the outdoor coil must operate below the outdoor temperature. It is important to keep the outdoor

Table 63-3 Heating Capacity for Air-to-Air Heat Pump with Auxiliary Heat

Outdoor Ambient °F	Basic Unit Without Auxiliary Heat		Capacity of Unit with kW of Auxiliary Heat				
	Capacity BTUH	COP	5k	8k	10k	15k	20k
65	29.41	4.14	45.80	56.72	62.18	—	—
60	27.85	4.00	44.23	55.15	60.61	—	—
55	26.21	3.84	42.59	53.51	58.97	—	—
50	24.50	3.66	40.88	51.80	57.26	—	—
45	22.67	3.46	39.06	49.98	55.44	—	—
40	21.06	3.28	37.44	48.36	53.82	—	—
35	19.42	3.09	35.80	46.73	52.19	—	—
30	15.59	2.52	31.97	42.89	48.35	—	—
25	14.39	2.38	30.77	41.69	47.15	—	—
20	13.25	2.24	29.63	40.55	46.01	—	—
15	12.05	2.09	28.43	39.35	44.81	—	—
10	10.81	1.92	27.19	38.11	43.57	—	—
5	9.58	1.74	25.97	36.89	42.35	—	—
0	8.36	1.56	24.74	35.66	41.12	—	—
–5	7.13	1.36	23.51	34.43	39.90	—	—
–10	5.84	1.15	22.22	33.15	38.61	—	—

Conditions: 860 CFM; indoor air at 70°F DB

(Courtesy of Goodman Global Group, Inc.)

coil cleared of frost because the system efficiency drops when frost builds to the point that it blocks airflow. The defrost cycle is required to melt the frost before it interferes with system operation. The defrost cycle is similar to the cooling cycle, except that the outdoor fan is de-energized and the strip heaters are energized. The outdoor coil switches from being an evaporator to now being the condenser. This allows for hot vapor to pass through the outdoor coil to help defrost any frost buildup. However, an adverse result from operating in this mode will be that the unit is air-conditioning the house in the middle of winter! To offset this effect, the outdoor fan is de-energized and the auxiliary heat is energized during the defrost cycle. The auxiliary heat is energized to prevent 60°F air from blowing in the house.

When the defrost cycle is initiated, the reversing valve reverses to send hot gas to the outdoor coil and the outdoor fan is de-energized. In defrost mode, the indoor coil becomes an evaporator, removing heat from the indoor air. Without energizing the auxiliary heat, the air delivered to the conditioned space would be cold air rather than hot air. To reduce this effect, the auxiliary heat comes on for the duration of the defrost cycle to reheat the air. The result is that the air is a slightly warm 80°F instead of a chilly 50°F.

TECH TIP

Tell your customer about the steam that will come off the outdoor unit during winter defrost cycles. Some owners may be startled when they see the steam coming from the outdoor coil. The owner may think this steam is smoke and may believe the outdoor section is on fire. Prepare owners so that when they see this, they will be assured the system is running properly.

The defrost control must determine when to start a defrost cycle and when to stop it. Starting a defrost cycle is called initiation; stopping a defrost cycle is called termination. If the defrost cycle is not initiated frequently enough, the outdoor coil will become frozen over with a sheet of ice. If the defrost cycle is initiated too often, the efficiency of the system is decreased and energy is wasted, increasing the cost of operation. Some method in addition to temperature must be used for initiation. If temperature alone were used for initiation, the system would not be able to operate normally below freezing. Most defrost controls use temperature to sense when they should terminate the defrost

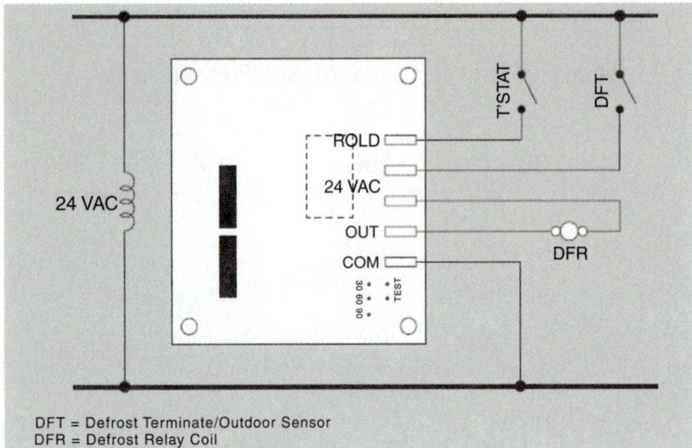

Figure 63-12 Simple time temperature defrost timer.

cycle. The two most common types of defrost controls are time temperature and temperature differential.

63.8 TIME TEMPERATURE DEFROST CONTROLS

The time-initiated temperature-terminated defrost control board uses a timing circuit and an open-on-rise defrost thermostat to control the defrost cycle (Figure 63-12). The timing circuit times the amount of compressor operating time during the heating cycle. It will not allow a defrost cycle until after a specific amount of compressor operating time. The most common defrost periods are 30, 45, 60, and 90 minutes. This time period is adjustable on the defrost board by moving a jumper or a selector switch (Figure 63-13). When the defrost time is reached, the timer will try to initiate the defrost cycle. However, if the temperature is not low enough, generally below 26°F, the defrost thermostat (Figure 63-14)

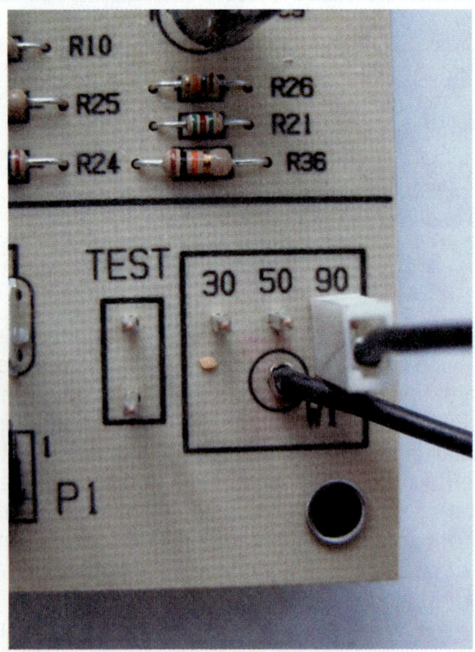

Figure 63-13 The defrost period is selectable with a jumper.

Figure 63-14 Defrost thermostat.

will not close and the defrost cycle will not begin. The timer will reset, and a new countdown starts until the next defrost begins. Once a defrost cycle has been started, it will continue until the defrost thermostat opens, usually somewhere around 50°F. The placement of the defrost thermostat is critical for correct operation. Its location can vary a great deal from one model unit to another. The timer has a failsafe defrost period, usually 15 to 30 minutes. Even if the defrost thermostat fails to open after the failsafe period of defrost operation, the defrost control terminates the defrost cycle and puts the heat pump back into heating mode.

63.9 DEMAND DEFROST CONTROLS

The major advantage of time temperature defrost controls is reliability. The main disadvantage is that forcing a defrost every hour of operation in cold weather can cause many unnecessary defrosts that waste energy. Just because the coil temperature is below freezing does not necessarily mean there is frost on the coil. A demand defrost control senses some aspect of unit operation to determine when there is actually frost on the outdoor coil. Methods used for demand defrost controls have included the following:

- Air pressure drop across the outdoor coil
- Compressor amp draw
- Temperature differential

Temperature differential has proven to be the most reliable. Nearly all demand defrost controls today use the temperature differential method. A temperature differential defrost control uses thermistors to sense the outdoor coil temperature and the outdoor air temperature. Under normal operating conditions, the temperature difference between the coil and the outdoor air remains fairly constant. When the frost on the coil builds to the point that it is blocking airflow, the coil temperature starts to drop. Since only the coil temperature is dropping, the temperature difference between the coil and the air is increased. The temperature differential control initiates a defrost whenever the difference between the coil and the air increases for a prolonged period of time. Most demand defrost controls have some failsafe period built in to force a defrost after several hours of operation below freezing. Like time temperature controls, temperature differential controls terminate the defrost when the coil temperature has reached around 50°F. They also have a failsafe defrost operation time that will terminate a defrost cycle after 15 to 30 minutes.

TECH TIP

The defrost thermostat on a time temperature defrost control is usually an open-on-rise thermostat. It is a mechanical switch that is either opened or closed. Its resistance should be 0 when it is closed, calling for a defrost period, or infinite when it is open, terminating the defrost period. The thermistor sensors on a demand defrost control are temperature sensors with a measureable resistance at all times. Their resistance changes with temperature. The two are not interchangeable.

63.10 CRANKCASE HEATER

The crankcase heater (Figure 63-15) is an electrical heating device that is inserted in the compressor crankcase or around the lower part of a hermetically sealed compressor shell. It provides heat to evaporate any liquid refrigerant that reaches the crankcase during the OFF cycle. At shutdown, the compressor will often become the coldest part of the system. The cold compressor creates a low-pressure condition and condenses the vapor refrigerant into a liquid. This process of attracting and condensing refrigerant is known as migration.

The crankcase heater protects the compressor when starting by vaporizing any liquid refrigerant that may enter the compressor during the OFF cycle. On many compressors, the crankcase heater only operates during the OFF cycle. On others it may be energized all the time. The crankcase heater may be fastened to the bottom of the compressor crankcase or it may be inserted in a tube located in the crankcase. When it is wrapped around the compressor shell it is referred to as a bellyband crankcase heater.

Liquid refrigerant in the crankcase can be very damaging to the compressor. If refrigerant collects in the crankcase, it mixes with the oil. This creates a condition known as a flooded start. When the compressor cycles on, the crankcase pressure quickly falls to the level of the suction pressure. This causes the liquid refrigerant mixed with oil to boil violently, creating foam, which can result in compressor slugging and the oil being pumped from the compressor crankcase. At a minimum, the liquid refrigerant in the compressor crankcase dilutes the oil, causing improper lubrication, which shortens the life of the compressor mechanical parts.

62.11 PRESSURE SWITCHES

Most heat pumps have a high-pressure switch that detects the pressure on the discharge side of the compressor (Figure 63-16). The high-pressure switch has a set of open-on-rise

(a)

(b)

Figure 63-15 (a) Immersion-type crankcase heaters must be installed with thermal mastic; (b) external-type crankcase heater (sometimes referred to as "bellyband").

Figure 63-16 High-pressure switch.

Figure 63-17 High-pressure switch with manual reset.

contacts that open when the system high-side pressure exceeds a safe level. Many high-pressure switches are the automatic reset type. Once the high-pressure condition drops, the switch will close, allowing the heat pump to operate again. One problem with this type of control is that the condition causing the pressure switch to trip may go undetected for a long time. A manual-reset high-pressure switch does not automatically close its contacts on a drop. A technician will need to push a button on the pressure switch to complete the electrical circuit and allow the heat pump to operate again (Figure 63-17). You should always determine the reason for the high-pressure condition after resetting it. The tripped switch is not the problem but a symptom.

Air-source units typically do not use a traditional low-pressure switch because the suction pressure can be very low in normal operation during cold weather heating. Instead, they use a loss-of-charge low-pressure switch. A loss-of-charge switch is installed in the high side of the system but has a close-on-rise switching action similar to a low-pressure switch. The switch is designed to open when the high-side pressure drops abnormally low due to lack of refrigerant.

Figure 63-18 Discharge line thermostat.

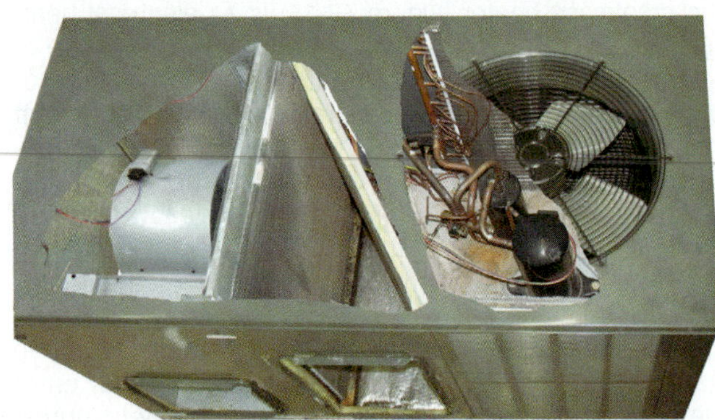

Figure 63-19 Packaged heat pump components

63.12 DISCHARGE-LINE THERMOSTAT

A discharge-line thermostat is an open-on-rise thermostat that opens to keep the compressor from overheating. It can be attached to the discharge line (Figure 63-18) or to the head of compressor. A discharge line temperature in excess of 250°F indicates overheating of the compressor. If the discharge line temperature is 250°F, the internal temperature of the compressor is around 300°F. At this temperature, the oil will begin to break down. Discharge-line thermostats typically open between 200°F and 250°F. Excessive discharge temperatures are caused by low charge, high compression ratios, and high discharge pressures.

63.13 PACKAGED UNITS

Packaged systems have all of the components in one preassembled unit, as shown in Figure 63-19. Typically included are electric heat strips located in the blower discharge area. Some type of auxiliary heat is required, except in semitropical climates such as that in Southern Florida. The easiest type of auxiliary heat system to incorporate into the heat pump assembly is the electric strip heater. When a gas or oil furnace is used for auxiliary heat, the combined system is referred to as being duel fuel.

Air-source heat pump package systems require no refrigerant line connections to be assembled. This advantage reduces installation costs. However, a disadvantage is that packaged units are somewhat limited in configuration types and to where they must be located. They are available as side discharge, bottom discharge, or wall mount. Bottom-discharge units are typically mounted on flat roofs on a curb. They are sometimes referred to as curb-mount units. Side-discharge duct connections can be arranged in two ways: over-under or side-by-side. Over-under units have the supply duct over the return duct (Figure 63-20). Side-by-side units have the return and supply ducts beside each other (Figure 63-21).

Some units are convertible from side discharge to bottom discharge. They accomplish this with removable panels

Figure 63-20 Over-under package unit with panels removed.

Figure 63-21 Side-by-side packaged unit.

on the bottom and sides of the unit where duct can connect. Typically these units arrive with both sets of panels in place. For bottom discharge, remove the bottom panels; for side discharge, remove the side panels (Figure 63-22).

Wall-mount packaged units mount to the exterior wall of a building. The air moves through two large holes in the wall. The return air is drawn in at the bottom of the unit, and the supply air leaves at the top. Many wall-mount units are installed like giant room air conditioners, with no ductwork. The unit is bolted to the wall and two large grills cover the return- and supply-air openings. This is typical in trailers that serve as temporary offices or classrooms (Figure 63-23).

Figure 63-22 Convertible packaged unit.

Figure 63-23 Wall-mount unit.

63.14 SPLIT SYSTEMS

Split systems have two components: a condensing unit that is installed outside and a blower coil that is installed inside (Figure 63-24). The condensing unit consists of a compressor, reversing valve, outside air coil, outdoor fan, and outdoor metering device (Figure 63-25). The blower coil contains the indoor fan, indoor air coil, and indoor metering device (Figure 63-26). The indoor coil is located in front of the blower so that air is drawn through the coil. By drawing the air through the coil, air travels more evenly over the coil and all of the coil area is used. With electric resistance auxiliary heat, the electric heat strip is located in the upper section of the blower discharge area, behind the control assembly and enclosing panel.

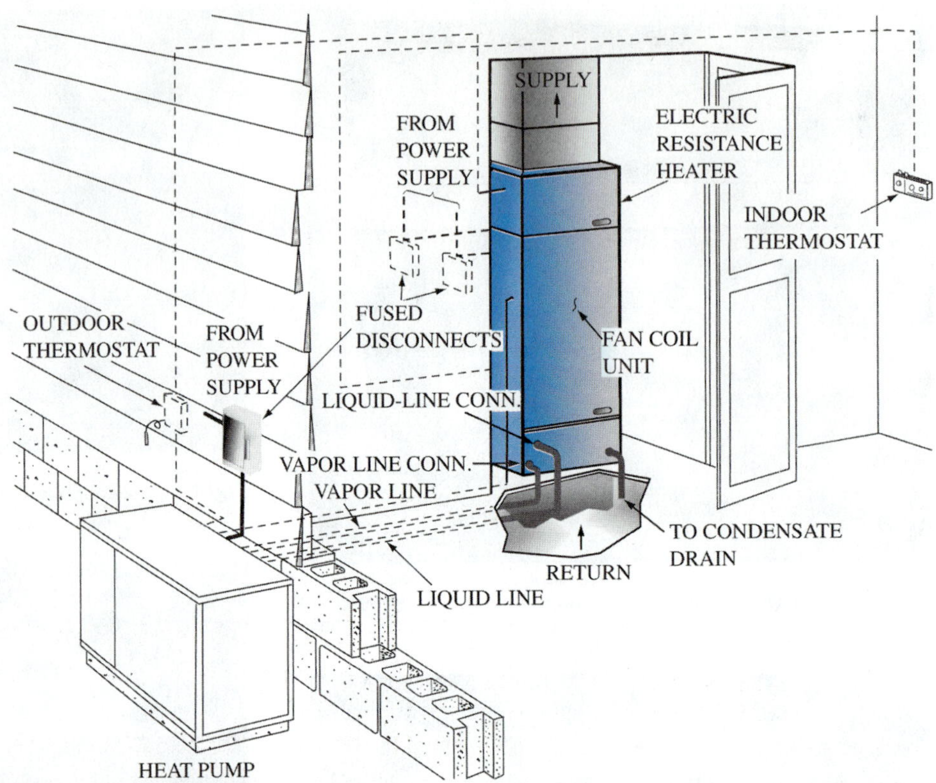

Figure 63-24 Split-system air-to-air heat pump.

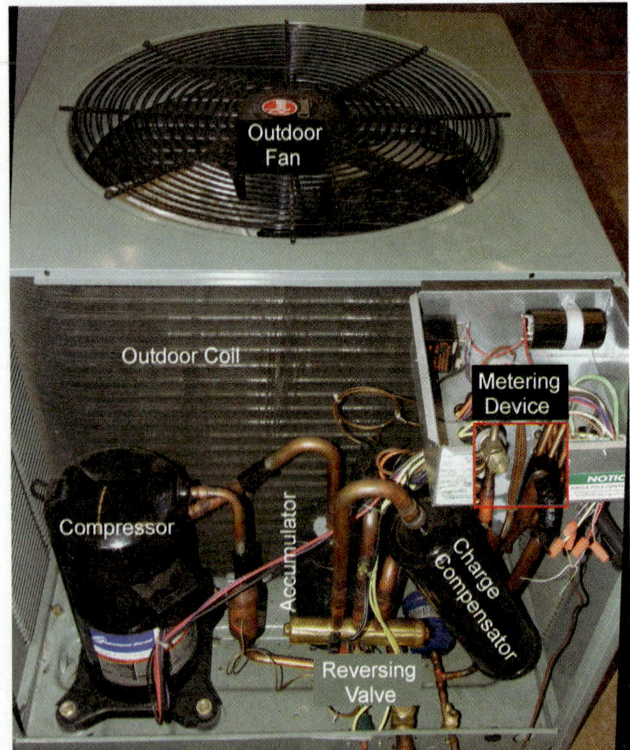

Figure 63-25 Split-system heat pump outdoor unit components.

The condensing unit and blower coil are connected by two refrigerant lines: a gas line and a liquid line. The liquid line always contains high-pressure liquid, regardless of the cycle; however, the direction of the liquid changes. Only one metering device is active at a time. This is determined by the internal check valve inside the metering device. When liquid flows from the liquid line into the metering device, the check valve closes, and the metering device becomes active. When liquid flows from the coil out of the metering device, the check valve opens, allowing full, unrestricted flow. In cooling, the liquid is traveling from outside to inside and the indoor metering device is active. In heating, the liquid is traveling from inside to outside and the outside metering device is active. The gas line changes from a low-pressure suction line in cooling to a high-pressure discharge line in heating. That is why it is called the gas line: the refrigerant in it is always gas.

SERVICE TIP

There is a gauge access port on the gas line at the condensing unit. However, it cannot be used to measure the system suction pressure in heating because the gas line is a discharge line during the heating cycle. Heat pumps all have a third port that is always low pressure. You must find it and install your gauges there.

An advantage of the split system is that it can be used in a wide range of vertical, horizontal, or downflow applications. However, the installation cost is higher than for a comparable packaged system. Split systems are brought to the job site in several individual parts, including an air handler, indoor coil, auxiliary heat source, outdoor unit, and interconnecting wiring and copper tubing (Figure 63-27). All of these components need to be connected together. Installing a split system requires brazing, cutting pipes to length, evacuating, and charging the system.

Figure 63-26 Heat pump blower and cooling coil indoor unit.

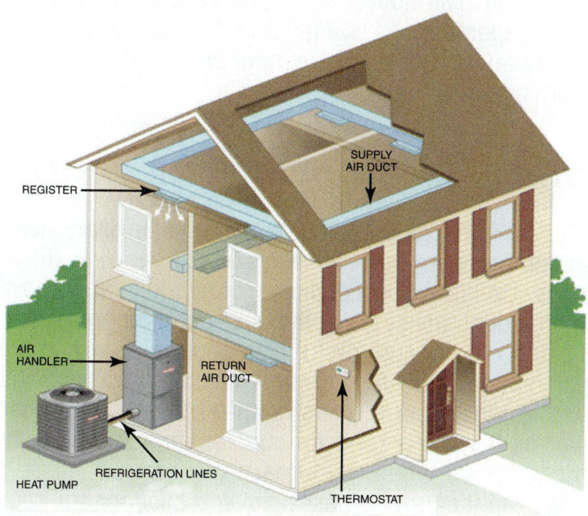

Figure 63-27 Split air-to-air heat pump system. *(Courtesy of Goodman Global Group, Inc.)*

Figure 63-28 Split-system heat pump mounted above the ground for snow.

TECH TIP

The outdoor unit on a heat pump has to be mounted above the ground in areas where snowfall accumulates, at a height above the average annual snowfall for the region (Figure 63-28).

63.15 TWO-STAGE HEAT PUMPS

Many manufacturers now offer heat pumps with two-stage compressors (Figure 63-29). These systems have two stages of cooling and three stages of heat: two stages of compressor heat and one stage of supplemental heat. Staged systems operate at the low compressor capacity most of

Figure 63-29 Two-stage compressor.

Figure 63-30 Board to control two-stage compressor.

the time. This allows them to use less energy when the full system capacity is not needed. Operating on low capacity also produces longer run times, reducing the energy loss from cycling on and off. These systems require thermostats capable of two stages of cooling and three stages of heating. The two-stage compressors are controlled by a board that controls the operation of an unloading solenoid in the compressor (Figure 63-30). Many heat pumps with two-stage compressors have built-in communicating controls. They are designed to work with indoor blowers that use variable-speed brushless DC motors (ECM) and communicating controls. When the heat pump is operating at low capacity, the indoor blower speed reduces to match the system capacity. When the thermostat calls for high speed, the indoor blower speed ramps up to match the increased compressor capacity. Heat pump systems using two-capacity compressors and ECM motors typically achieve efficiencies of 20 SEER and 10 HSPF.

63.16 MINI-SPLIT HEAT PUMP

Most mini-split heat pumps do not require ducts. This makes them easier to install. This type of unit is generally sized to heat and cool one room and have its own thermostat for that space. As an example, a building with four major rooms would have a mini-split located in each one. The indoor unit is relatively small, as little as 7 in deep, and can be mounted to the ceiling or hung from a wall (Figure 63-31). The thermostat will generally have a remote control for ease of operation. A significant difference between most mini-split systems and traditional split systems is the location of the metering devices. Most mini-split heat pumps have a single electronically controlled metering device located outside in the condensing unit. There are no metering devices in the indoor coils. As a result, both refrigerant lines are insulated, because in cooling, both lines are cold.

Figure 63-31 Mini-split heat pump system.

Some mini-split units, called multi units, have multiple indoor coils attached to a single outdoor unit. Multi units are available that allow up to eight indoor units to connect to a single outdoor unit. These usually have a variable-speed compressor, allowing the compressor capacity to match the load. The unit uses less energy at reduced capacity, saving energy. This type of system will operate more efficiently than several smaller systems cycling on and off.

There are some distinct advantages for this type of system. Since a duct network is not required for the air to travel through, these units have no duct losses. In poorly insulated duct systems, these losses can be as high as 30 percent or more, especially for ducts running through an unconditioned space, such as an attic. The connection between the outdoor and indoor units is also usually fairly simple, with the only requirement consisting of a 3-in-diameter hole through the exterior wall. The indoor section is sleek and attractive and will not block incoming light, as a standard window-mounted air conditioner will. Each space has its own thermostat, which allows for reducing the cooling or heating supply when no one is there, which will save on energy costs.

Some disadvantages of the mini-split include its high initial cost and more refrigerant lines, wiring, and drains to run. Installing multiple mini-split systems in a house will generally have a higher initial cost than a single central system with air distribution. Every indoor unit needs its own set of refrigerant lines, wiring, and drain. So while there is no ductwork to run, there are more wires, refrigerant lines, and drain lines to run.

63.17 DUAL-FUEL HEAT PUMP SYSTEMS

Gas or oil furnaces can also be used for auxiliary heat. A heat pump system that uses a gas or oil furnace as the auxiliary heat is known as a dual fuel, or hybrid heat pump system. In a dual-fuel system, the heat pump operates above the balance point and the furnace operates below the balance point. In a dual-fuel system, the indoor coil is installed after the furnace in the airstream (Figure 63-32). When a gas or oil furnace is used for supplemental heat, the heat pump shuts off when the furnace comes on because the hot air from the furnace passing over the indoor coil causes extremely high head pressures in the heat pump. This type of system requires the coordination of the heat pump and the furnace operation. This is normally accomplished using

Figure 63-32 Heat pump coil mounted on furnace to form dual-fuel heat pump system.

either a separate dual-fuel add-on control board or a thermostat designed for a dual-fuel system. A few heat pumps and furnaces now have dual-fuel control capability built into their control systems.

63.18 DUAL-FUEL HEAT PUMP PACKAGED SYSTEM

A packaged dual-fuel heat pump system combines both an air-to-air heat pump and a gas furnace into one packaged system (Figure 63-33). The indoor coil on these systems is located in the return air, before the gas heat exchanger. This type of unit often uses a two-stage compressor, a variable-speed blower motor, and a dual-fuel thermostat.

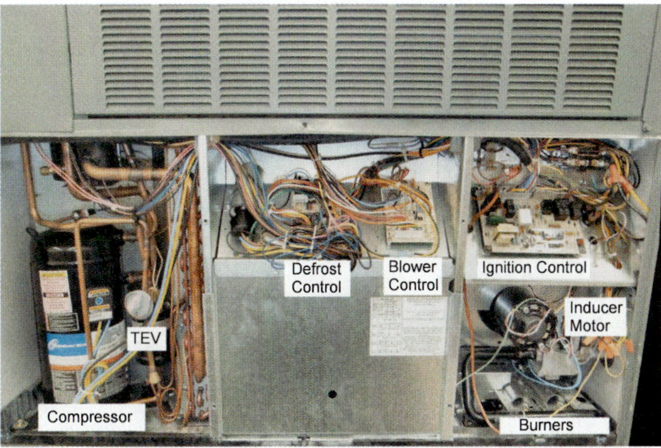

Figure 63-33 Dual-fuel heat packaged heat pump.

The heat pump will provide cooling in the summer season and heating during moderate winter conditions. When the temperature drops below the balance point, the system automatically switches to gas furnace heating. At this lower outside temperature, gas becomes the more economical heating source and also provides for greater heating capacity than the heat pump.

63.19 SOLAR-ASSISTED HEAT PUMPS

Two types of solar-assisted heat pump systems are currently available: thermal and photovoltaic. Thermal solar panels heat water circulated through them. Photovoltaic solar cells generate electricity. Thermal solar-assisted systems use thermal solar collectors to heat water. This hot water is used for both domestic hot water and additional space heating in the winter. Heat is added by circulating the solar-heated water through a water coil in the heat pump indoor blower section (Figure 63-34). Some heat pump systems are now available with a hot-water coil and controls preinstalled for adding thermal solar hot water. This system

Figure 63-34 This air-to-water heat pump system also can use solar energy from thermal solar collectors for heat.

Figure 63-35 Installing a 5-kW solar array for use with two solar-assisted air-source heat pumps.

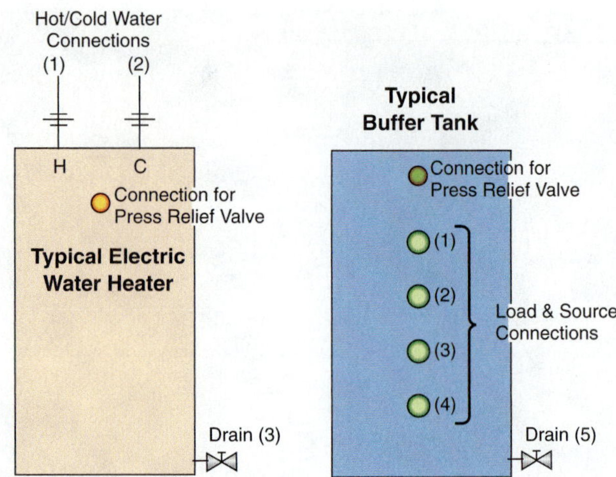

Figure 63-36 A buffer tank as compared to an electric hot-water heater.

required the addition of solar thermal collectors, hot-water storage, circulating pumps, and solar controls.

Photovoltaic-assisted heat pumps have two power supplies: one from the power grid and another from the solar cells (Figure 63-35). The heat pump uses the solar-generated electricity to reduce its power consumption from the power company when the unit is operating and solar power is available. When solar power is available and the unit is not operating, the power is sold back to the power company. If the solar collectors are generating more electricity than the system requires, it is possible to operate the system entirely from solar power and sell the surplus back to the power company. These systems require photovoltaic solar cells and an extra disconnect. Other electrical components normally required to use photovoltaic cells with alternating current devices include a voltage regulator and power inverter. The solar cells used with solar-assisted heat pumps greatly simplify the installation by having internal voltage regulators, power inverters, and safety shutoff controls built in. This eliminates the need for these items and the electrical connections between them. Installation is very similar to a regular air-to-air heat pump, except that there are two power supplies feeding the unit. As a safety feature, these systems are specifically designed to operate only when they are connected to the power grid. If the solar cells do not see the power from the power company, they automatically shut down their output.

63.20 AIR-TO-WATER HEAT PUMP SYSTEM

The air-to-water system is a relatively new application and involves more equipment as compared to a basic air-to-air system. These are often referred to as reverse-cycle chillers (RCCs). RCCs have an outdoor air coil and a water coil connected to hydronic systems inside the house. In heating, a reverse cycle chiller moves heat from the outdoor air into the water storage tank inside. The outdoor air coil is the evaporator and the water coil is the condenser. This hot water is circulated to a large insulated tank of water called a buffer tank. It is called this because it balances out any fluctuations in load to provide more constant and steady heating and cooling. The water from the buffer tank can be circulated to a coil in an air handler for forced-hot-air systems. This application can also be arranged to accommodate forced-air systems with multiple zone radiators, as well as radiant floor heating systems, blower coils, or even to domestic hot-water heaters. Figure 63-36 shows a small buffer tank that is comparable in size to a hot-water heater but with additional connections for the refrigerant coil and a connection for the air-handler coil.

In cooling, the water circulating through the water coil does not circulate through the buffer tank. Instead, it circulates only between the heat pump and a blower coil. Heat is absorbed from the water and rejected to the outdoor air at the air coil.

There are several advantages of an air-to-water system. There is little to no effect on the warm water air temperature being supplied to the building when the outdoor coil enters the defrost mode because the heat for defrost comes from the buffer tank, not the building. Cold air during defrosting is a common complaint for air-to-air systems. The electric heat that is required to come on during the defrost mode for an air-to-air system is not required because the heat is coming from the buffer tank , therefore saving energy. The air-to-water system will generally operate with higher efficiencies at lower outdoor temperatures as compared to air-to-air systems. The large heat sink of water will maintain a more constant temperature when the outside temperature drops, reducing the amount of supplemental electric heat required.

63.21 HEAT PUMP HOT WATER HEATERS

Heat pump hot water heaters are air-to-water systems that use the air surrounding the unit as the heat source and the water in the hot water system as the heat sink. They move heat from the air surrounding them into the domestic hot water tank. These heat pump systems are marketed in small unit sizes for water heating. This system is used to replace traditional electric water heaters. The initial cost

Figure 63-37 Heat pump hot-water heater.

of a heat pump water heater is 4 to 5 times as much as a traditional electric water heater, but they operate for half the cost. Payback on the additional cost is typically less than 5 years. Residential units are available in the capacity range of 6,000 to 12,000 BTUH. They may be incorporated into a package unit along with a hot-water storage tank or as an add-on unit for existing storage tanks (Figure 63-37).

Larger units are available for commercial application to provide hot water and spot cooling. The unit becomes the primary heat source for domestic or small commercial hot-water use, with the heat source of the water heater as backup. The higher efficiency rating or COP (coefficient of performance) of the heat pump water heater will supply hot water at approximately half of the cost of a traditional electric water heater.

UNIT 63—SUMMARY

Air-to-air heat pumps are the most common type of heat pump. They use air as both the heat source and the heat sink. In heating, the outdoor air is the heat source and the indoor air is the heat sink. They are available as packaged units, with all the components in one cabinet, or as split systems, with the compressor and condenser in one cabinet and the evaporator in another. Air-to-air heat pumps require supplemental heat, which is normally provided by electric heat strips. A dual-fuel heat pump uses a gas or oil furnace for supplemental heat. Air-source heat pumps must have a defrost cycle to clear them of frost during operation at temperatures below freezing. The two common types of defrost controls are time temperature and demand defrost. The best system is the one that is properly designed, sized, and installed while meeting the needs of the specific application. Properly designed and installed heat pump systems can save the customer money and provide for four-season heating and cooling needs.

WORK ORDERS

Service Ticket 6301
Customer Request: Split-System Retrofit

Equipment Type: Existing Split System to Be Replaced by Air-Source Heat Pump

The sales representative meets with the customer to discuss the options for upgrading her current heating and cooling system. During a recent service call, the technician recommended that the outside air-conditioning system be replaced, as it was old and becoming expensive to maintain. However, the small gas furnace and cooling coil located in the basement are still in good operating condition. When the homeowner told the technician that the price of heating with gas was becoming increasingly expensive, he recommended a heat pump system. The technician had recently been installing heat pumps throughout the community and their performance was receiving favorable reviews.

The sales representative agrees with the technician's advice. The latest 15-SEER air-to-air heat pump systems the company has been carrying as a product line are much improved over earlier versions. The energy efficiency is very good, and in this moderate climate the air-to-air system with on-demand defrost works exceptionally well. Rather than replace the gas furnace, it can remain in place and be used for auxiliary heat if necessary. The customer agrees to the retrofit, and the existing outdoor unit is replaced with an air-to-air heat pump. The existing coil is replaced with one that is designed for heat pumps and is an AHRI match for the new heat pump condensing unit.

Service Ticket 6302

Customer Complaint: Frequent Cycling and High Operating Cost

Equipment Type: Mini-Split Heat Pump

The customer complains to the technician that the new mini-split system that was recently installed by a local discount supply house is frequently cycling, using more electricity than originally anticipated. Almost immediately, the technician realizes that the system is much larger than what would normally be required for a space of this size. The technician asks the customer who had determined the sizing requirements for the mini-split. The customer explains that the unit was a closeout model and on sale for a very reasonable price. As a further incentive, the discount supply house where the mini-split was purchased offered free installation.

The technician asks if the mini-split could be returned, because it is oversized. The customer shakes his head no, because this was a sale item and no returns were allowed. The technician says that someone from the supply house should have offered to help in sizing the unit to make sure it would meet the application. The technician recommends that the customer talk to someone at the discount supply house about this situation but cannot offer any solution other than replacing the unit with an appropriately sized mini-split.

Service Ticket 6303
Customer Request: New Installation

Equipment Type: Wall-Mounted Packaged Heat Pump

A sales consultant is asked to recommend a heating and cooling system for an add-on classroom trailer that is to be located outside the main building at the local high school. The building maintenance supervisor has requested that the installation be completed quickly and within a limited budget. There are a number of options that can be considered. One recommendation is for a split-system air-conditioning unit with a small gas-fired furnace.

The building maintenance supervisor asks the sales consultant about the possibility of installing a heat pump system. The sales consultant acknowledges that these are being used more often, but auxiliary electric heat will need to be included for the colder months. This will add to heating costs come winter. However, a wall-mount packaged heat pump is the least costly alternative for installation. No ductwork will be needed because the building is one large room. The installation can be done in 1 or 2 days. Since the classroom trailer is not a permanent structure, this is considered to be the best solution for the building maintenance supervisor.

Service Ticket 6304
Customer Request: New Installation

Equipment Type: Air-Source Split-System Heat Pump

Dr. Studdart is the dean of the Technical Division of the local Technical College and has learned about ground loop systems from the air-conditioning faculty. He is in the process of building a new home, and this type of system for heating and cooling is intriguing to him. Dr. Studdart contacts a local company requesting a quote for the installation of a ground loop system. He is surprised when he finds how expensive it will be. The sales consultant explains that a ground coil requires considerable excavation and long lengths of piping.

Instead, the consultant suggests a very efficient air-source split-system heat pump with a SEER of 18 and an HSPF of 10. This is a much less costly alternative that still delivers impressive energy savings by using a two-capacity compressor and ECM fan motors. Although it is not exactly what Dr. Studdart originally had in mind, it is certainly the most practical solution for his heating and cooling needs.

UNIT 63—REVIEW QUESTIONS

1. What type of heat pump is the most common, and why?
2. What is the relationship between air-to-air heat pump heating capacity and outside air temperature?
3. What is a dual-fuel heat pump?
4. What is the HSPF of the most efficient air-to-air heat pumps?
5. What are the efficiency requirements for the Energy Star label to be awarded for an air-to-air heat pump?
6. What are some advantages of the mini-split heat pump?
7. How should heat pump outdoor units be mounted in areas where snowfall accumulates?
8. What does the Y terminal on most heat pump thermostats control, and when is it energized?
9. Why do air-source heat pump thermostats have two stages of heat?
10. What is usually used for auxiliary heat?
11. What is the difference between auxiliary heat and emergency heat?
12. What is the purpose of the crankcase heater?
13. What two types of solar-assisted heat pump systems are available?
14. What are the two main types of defrost controls used on heat pumps today?
15. Describe what happens during the defrost cycle.
16. When describing a defrost control, what do the terms *initiation* and *termination* mean?
17. Why must a heat pump indoor coil be able to withstand higher pressures than a regular air-conditioning evaporator coil?
18. Where is the indoor coil in relation to the blower on a heat pump blower coil?
19. What is the system balance point?
20. What is the purpose of the outdoor air thermostat?
21. What determines which metering device is active?
22. Where in the airstream are electric strip heaters installed?
23. Describe the operation of a heat pump with electric strip heat operating below the balance point.
24. Describe the operation of a dual-fuel heat pump operating below the balance point.
25. What is the difference between demand defrost control and time temperature defrost control?
26. What are some of the advantages of an air-to-water heat pump system?
27. How does a heat pump hot-water heater work?
28. What is the difference between a mini-split system and a traditional split system?
29. What controls the operation of the furnace and heat pump in a dual-fuel system?

UNIT 64

Geothermal Heat Pumps

OBJECTIVES

After completing this unit, you will be able to:

1. describe the applications best suited for air-source and water-source heat pumps.
2. recommend the most efficient heat pump system depending on the application.
3. describe the different configurations for all of the major heat pump types.
4. explain why some heat pump systems are better suited than others depending on the geographical location.
5. list the advantages of air-to-water systems, mini-split systems, and water-source systems.

64.1 INTRODUCTION

Geothermal heat pump systems move heat from the earth into your house. They take advantage of the moderating effect of the earth on seasonal temperatures. Unlike the air temperature, which can change 30°F from day to night, earth temperatures take much longer to change. At depths below 28 feet, the earth temperature is nearly constant. Geothermal systems use buried loops in which water is circulated. Water-source heat pumps are used to extract heat from the water. Geothermal water-source heat pump systems are more efficient than air-source heat pumps because water is a more efficient heat-transfer medium, the temperatures they operate at are more moderate due to the moderating effect of the earth, they do not need a defrost cycle, and their capacity does not drop off as the temperature drops outside.

64.2 WATER-SOURCE HEAT PUMP SYSTEMS

Figure 64-1 Horizontal-style water-source heat pump with panels removed.

The operating concept of a water-source heat pump is similar to an air source system: heat is transferred from one coil to another. Instead of two air coils, water-to-air heat pumps use one water coil and one air coil (Figure 64-1). In heating, water-source systems use water as their heat source and the indoor air as the heat sink. The water coil transfers heat from the water to the refrigerant, and the air coil transfers heat from the refrigerant to the air. In cooling, water-source systems use the indoor air as their heat source and the water as the heat sink. The air coil transfers heat from the air to the refrigerant, and the water coil transfers heat from the refrigerant to the water. The refrigeration cycle for a water-source heat pump is a little simpler because water-source units have a much narrower operating temperature range than air source units. The water-source unit's capacity is less variable because the water being supplied to it must be moderate in temperature. This means that water-source units do not require suction line accumulators or charge compensators. Also, most water-source packaged units use a single bi-flow thermostatic expansion valve, so there are no check valves or extra metering devices. Since the fluid circulating through the water-source unit cannot be allowed to freeze, this type of heat pump does not need a defrost cycle.

The water coil in a water-source unit is a coaxial tube in tube coil. These heat exchangers are made of copper, cupronickel, or stainless steel. Figure 64-2 shows a water-to-refrigerant coil and a cutaway of the coil. Water travels through the inner tube, and refrigerant flows through the outer tube. The water and refrigerant always travel in opposite directions. This counterflow design improves efficiency by placing the coldest water against the coldest refrigerant and the warmest water against the warmest refrigerant. The water coil should be located in the heated area to prevent possible freeze-up if the outdoor ambient temperature should drop below 32°F. If the

Figure 64-2 Cutaway view of a water coil.

water-to-refrigerant heat exchanger is part of a packaged unit, it would be located in the conditioned area.

Water-source systems can be connected to an open or closed water loop. Open loops are open to the air and are supplied with water from a well. The water for an open loop is not recirculated. Figure 64-3a shows a typical open-loop system. Closed loops are pressurized loops of water buried in the ground that recirculate the same water. The water transfers heat between the ground and the refrigerant in the unit. Figure 64-3b shows a typical closed-loop system.

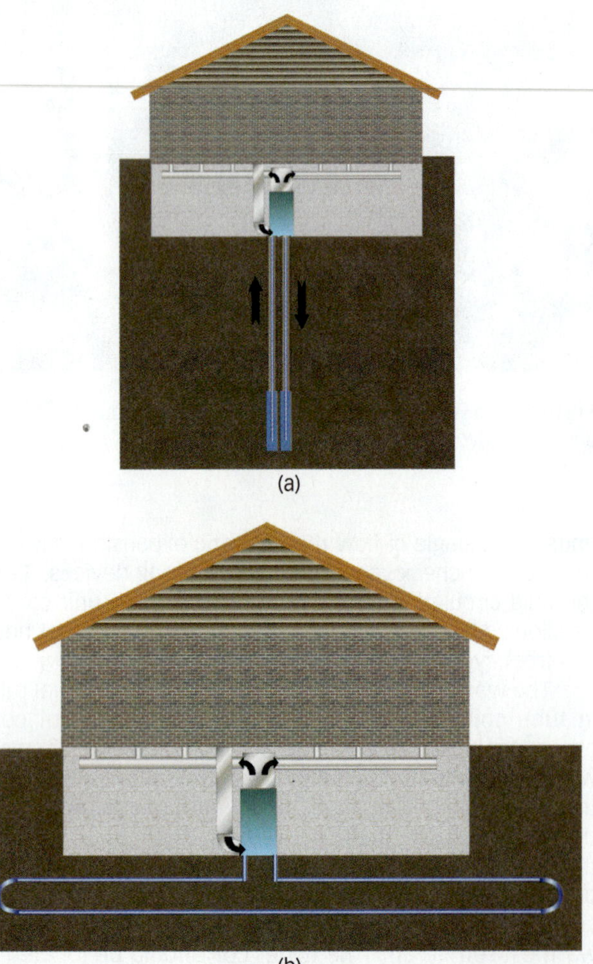

Figure 64-3 (a) Open-loop system; (b) closed-loop system.

Closed-loop systems use much smaller water pumps than open-loop systems. A closed-loop pump only has to overcome the resistance of the loop; it does not have to work against gravity. Even in a vertical loop, the weight of the water flowing down balances the weight of the water being lifted, so gravity is not a factor. In an open loop, the pump must lift the water against gravity, so a much stronger pump is required.

Water-source packaged units are available as either vertical or horizontal discharge. Many water-source systems also have supplemental heat, but it is not as necessary for them because they do not lose capacity as the outdoor temperature drops and they do not have a defrost cycle. However, a unit sized correctly for cooling may still be undersized for heating and require additional heat.

64.3 WATER-TO-AIR PACKAGED SYSTEM

Packaged water-to-air heat pumps are the most popular type of water-source heat pumps. The building is heated by a forced-air duct-distribution system similar to that of an air-to-air system. However, there are some significant differences. The most notable is that the outdoor coil absorbs heat from a water source rather than the outside air (Figure 64-4; notice that domestic hot water heat can also be added to the system).

Whether using a shallow pond, well, or groundwater, the temperature remains fairly consistent and above freezing even as outside temperatures drop. Depending on latitude, ground temperatures range from 45°F to 75°F, so that any water below the ground-level frost line should be somewhere within this range. Because the outdoor coil never drops below freezing, it never needs to be defrosted.

With all water-source heat pumps, the compressor and refrigerant components can all be located inside the building (Figure 64-5). The packaged-unit water-to-air configurations can come as upflow, downflow, and horizontal

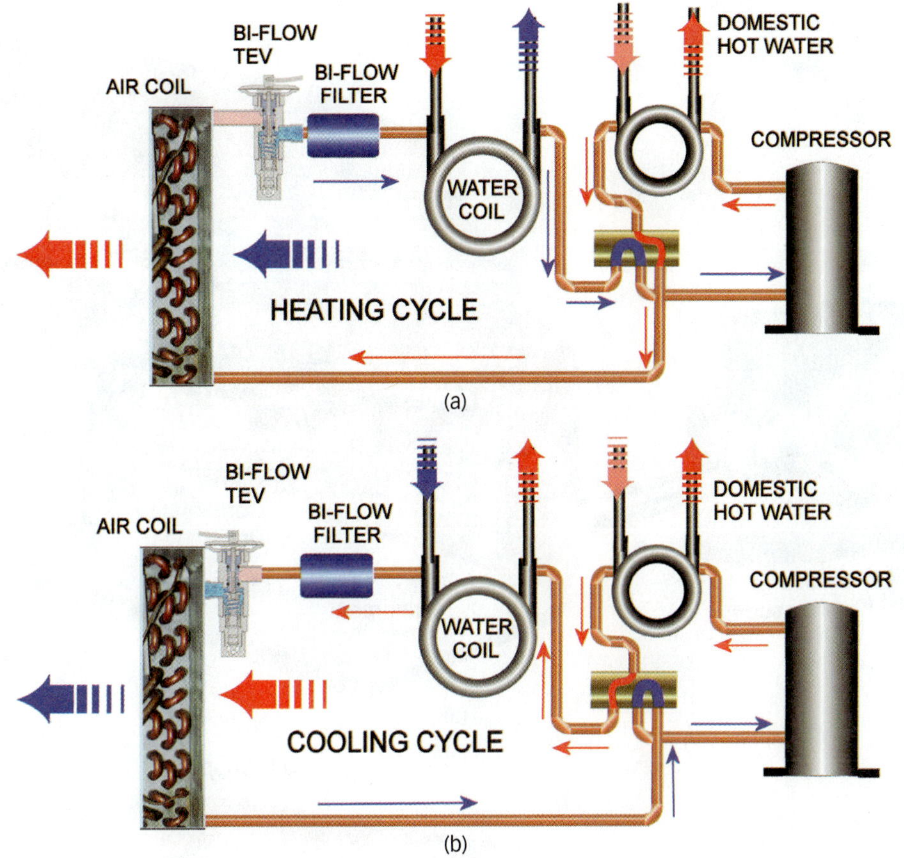

Figure 64-4 Water-to-air cycle.

supply-air options with ECM variable-speed fan motors. Figure 64-6a shows a vertical water source unit and Figure 64-6b shows a horizontal water source unit.

64.4 WATER-TO-AIR SPLIT SYSTEM

Water-to-air heat pumps can also come as split systems. These are more common for retrofit applications where the outdoor coil can be replaced with the heat pump but

Figure 64-5 Water-source heat pump compressor can be located indoors.

(a)

(b)

Figure 64-6 (a) Vertical water-to-air packaged heat pump; (b) horizontal water-to-air packaged heat pump.

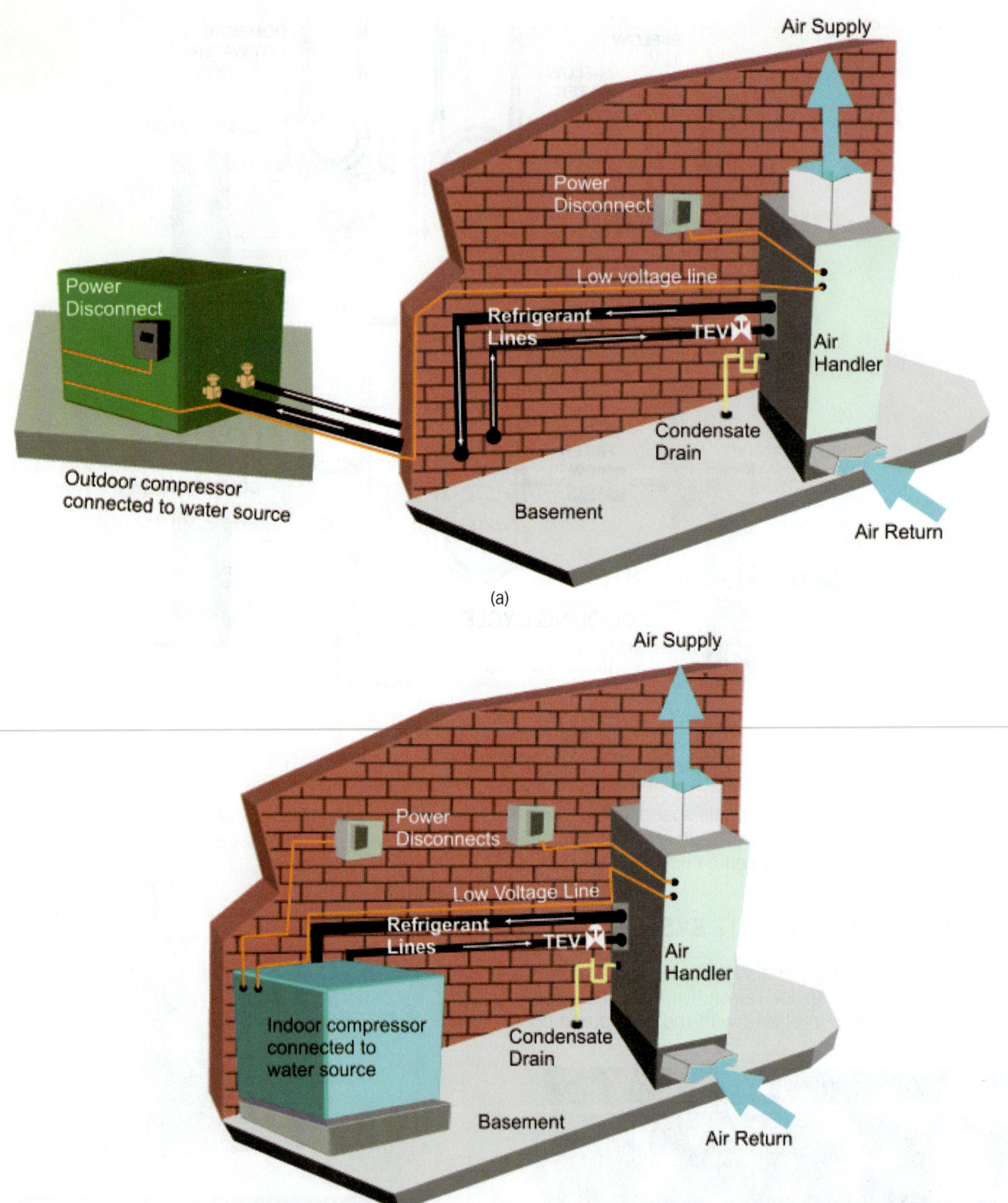

Figure 64-7 (a) Split water-to-air heat pump air-handler installation with outdoor compressor; (b) split water-to-air heat pump air-handler installation with indoor compressor.

the indoor air-handler wiring and configuration remain the same (Figure 64-7a). This type of system is popular if indoor space is limited and the air handler can be located in the most convenient location. A narrow closet may not be large enough for a packaged unit but may be perfectly suitable for an air handler. If space allows, the compressor section can also be located indoors (Figure 64-7b).

64.5 GEOTHERMAL HEAT PUMP SYSTEM

Geothermal systems are water-source heat pump systems whose heat source comes from the ground. This can either be water supplied directly from a well or water circulated in a loop buried in the ground. The most common geothermal systems are closed-loop systems with water circulated through ground loops. The water acts as a secondary refrigerant, absorbing heat from the soil and transferring it to the refrigerant in the heat pump. This is less efficient than absorbing heat from groundwater, but it avoids many problems associated with open loops, such as mineral deposits in the water coil. The difference in efficiency can be offset by increasing the loop size, effectively making the heat exchanger larger.

Ground-source heat pumps are well suited for cold climates because the temperature of the earth below the frost line remains above freezing and relatively constant all year long. However, there are a number of factors to consider before recommending the installation of this type of system. The site must be evaluated with regard to geology, hydrology, and land availability. Extremely rocky soil or land on top of a rock outcropping may not be practical due to the expense in drilling or trenching. A large area of land is required for a horizontal installation—a half acre minimum for a small system. All soil does not transfer heat readily. Sandy, dry soil is a poor heat conductor, while moist clay is ideal. Although geothermal systems use less energy than most other forms of heating, they are also the most expensive to install.

A closed loop located in a body of water has the advantage of the natural circulation of water where the warm water rises and the cold water falls, allowing for heat transfer through convection. In contrast, a ground loop has a limited ability to remove heat from the ground. As the ground surrounding the loop grows colder, it will take time for additional heat to travel through the soil by conduction. It is therefore essential that a large enough area is excavated for the ground loop to be effective as a heat sink. Horizontal ground loops are the most economical but can take up a lot of space (Figure 64-8). Vertical ground loops take up less space but are more difficult to install, as they need to be drilled rather than dug with a backhoe or trenched with an excavator (Figure 64-9).

The soil and rock that make up the area to be used for the ground loop need to be considered. Difficulties can be encountered in areas with an abundance of hard rock with little soil available to trench. Ground surface water can sometimes be desirable for the ground loop as long as it does not interfere with the installation process. A trench that fills with water after it is dug and the piping has been installed will provide for better heat transfer. There will be some natural circulation of the ground water that will improve the ground-loop efficiency as long as the operating conditions are set so as not to freeze the water on the outer surface of the piping, which would then only reduce rather than increase heat transfer.

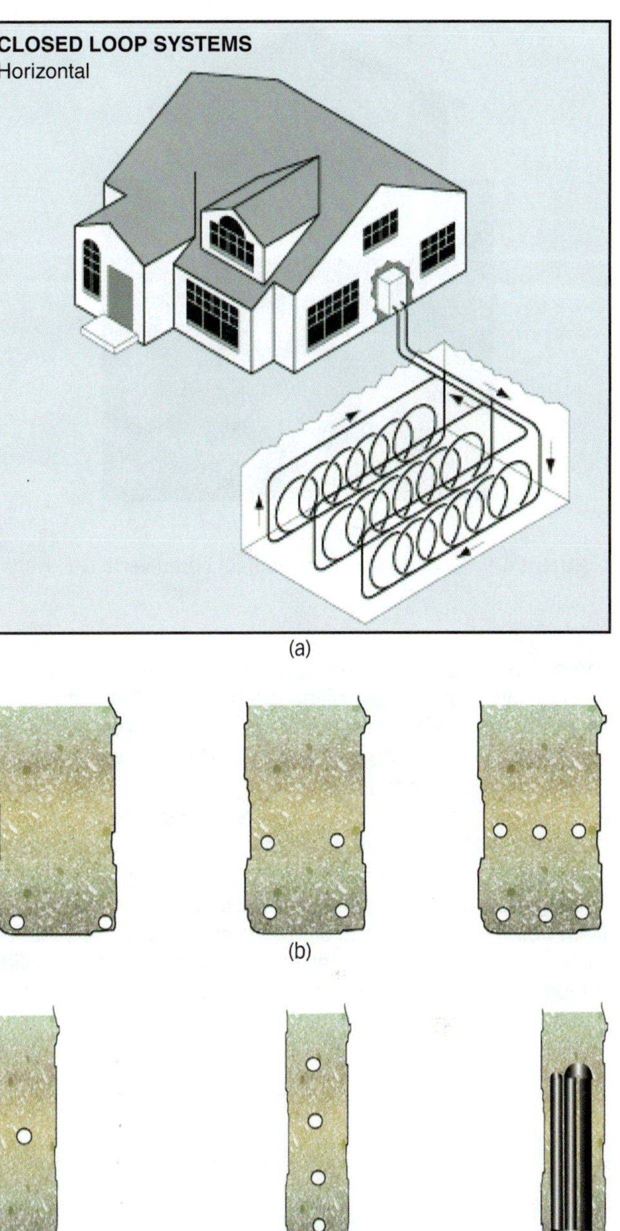

Figure 64-8 (a) Horizontal closed loop; (b) backhoe horizontal closed loop; (c) trenched horizontal closed loop. *(Courtesy of Office Energy Efficiency and Renewable Energy, U.S. Department of Energy)*

TECH TIP

A closed-loop ground system is practical only when the geology, hydrology, and land availability meet the requirements for the system to be installed. This type of system will have an extensive ground loop through which a secondary refrigerant circulates. Always check with all local codes and environmental regulations prior to installing this type of system. In some areas, any leakage of the secondary refrigerant from the ground loop into the soil could be deemed unacceptable and subject to a possible fine.

64.6 GROUND-SOURCE HEAT PUMP CAPACITY

Although the capacity of a ground-source system does not fall off in cold weather like the capacity of an air-source system, it may not always be practical to size the unit to meet the entire heating demand in cold climates. Sizing to the heating load in a cold climate could oversize the system for cooling. Instead, the unit can be sized for the cooling load and supplemental heat added to make up the difference. The ground-source heat pump can be used in combination with gas, oil, or electric heat. Figure 64-10a shows an outdoor ground-source heat pump used in combination with a

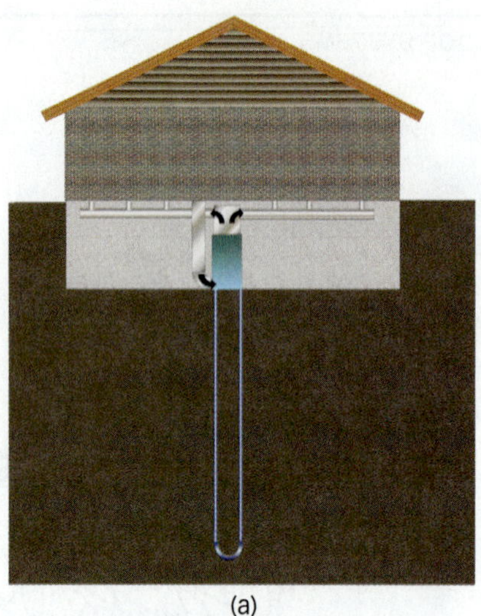
(a)

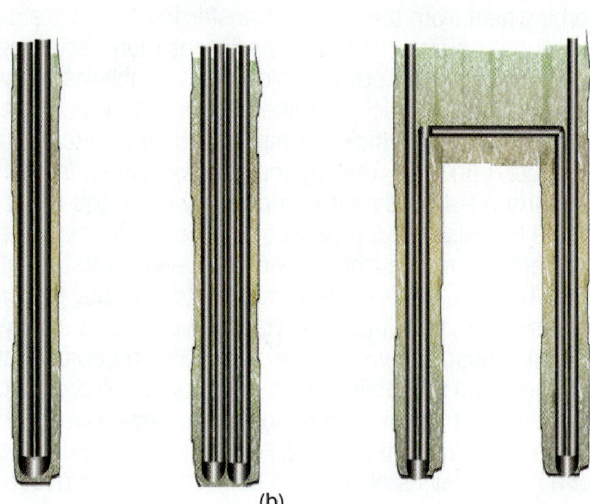
(b)

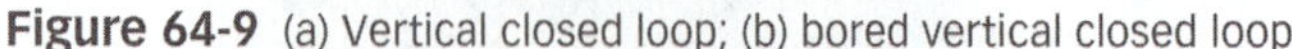

Figure 64-9 (a) Vertical closed loop; (b) bored vertical closed loop.

furnace. Figure 64-10b shows a similar application but with the compressor located inside the building. The installation of a coil on an existing furnace may help to reduce energy costs. The ground-source heat pump will supplement the heat required for the building.

Another solution is to use a water-source heat pump that has a two-stage compressor. This way, the unit can be sized to meet the heating load when operated at high capacity. Then in cooling, operating at the lower capacity can keep the system from being grossly oversized. In a warm climate, the situation may be reversed. The higher capacity will meet the cooling needs and the lower capacity can allow extended run times and more efficient operation in mild winter weather.

64.7 WATER-TO-AIR OPEN-LOOP SYSTEM

An open-loop system is the simplest type of application for water-source heat pumps (Figure 64-11). Water is drawn from a well (Figure 64-12) and pumped through a refrigerant coaxial heat exchanger (Figure 64-13). In the heating cycle, heat will be absorbed from the water by the refrigerant. In the cooling cycle, heat will be given off to the water. Water-to-air heat pumps are more efficient than air-to-air heat pumps for two reasons: water is a better heat transfer medium than air, and the water temperature will remain fairly constant and not drop below freezing.

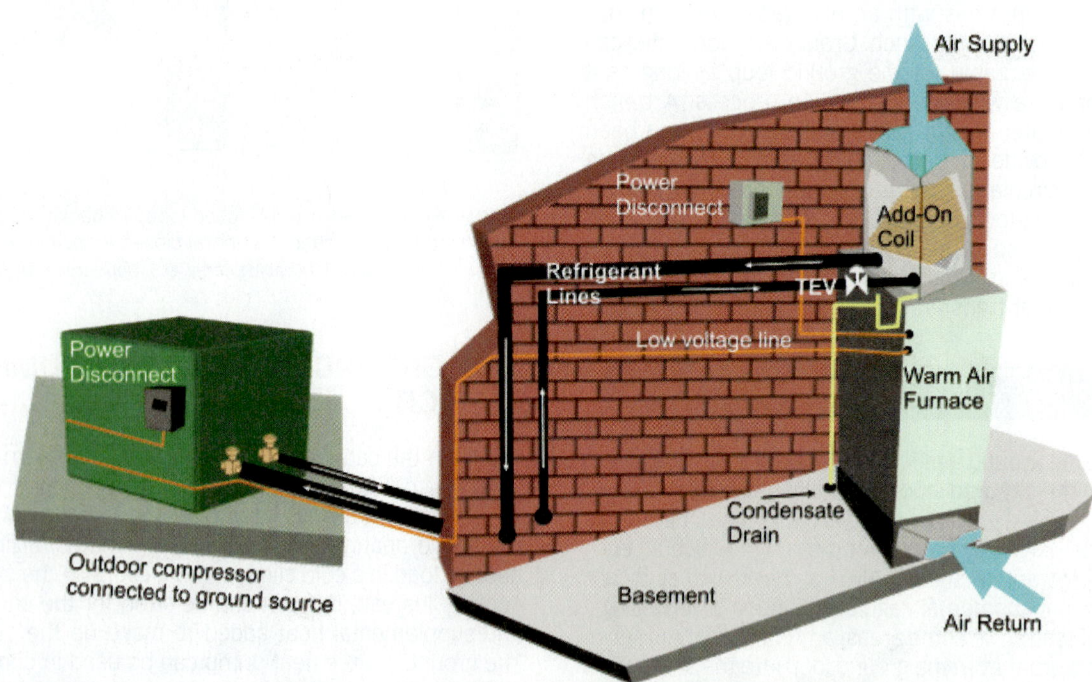

Figure 64-10a (a) Split ground-source heat pump add-on coil for fossil fuel furnace with outdoor compressor

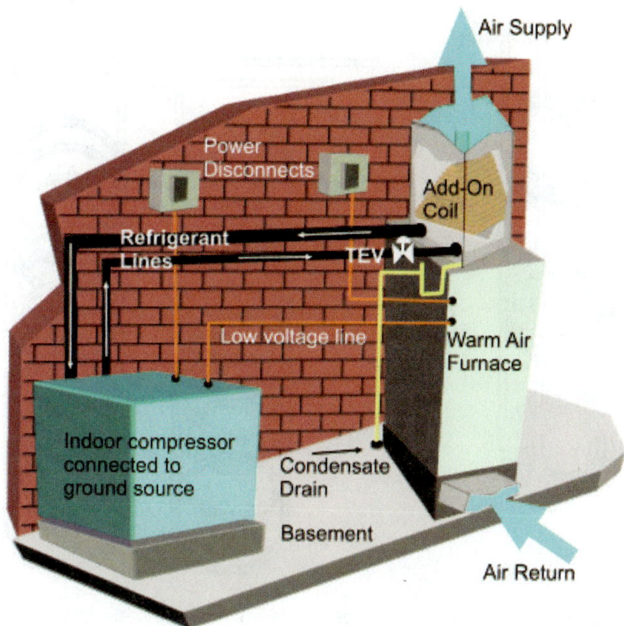

Figure 64-10b (b) split ground-source heat pump add-on coil for fossil fuel furnace with indoor compressor.

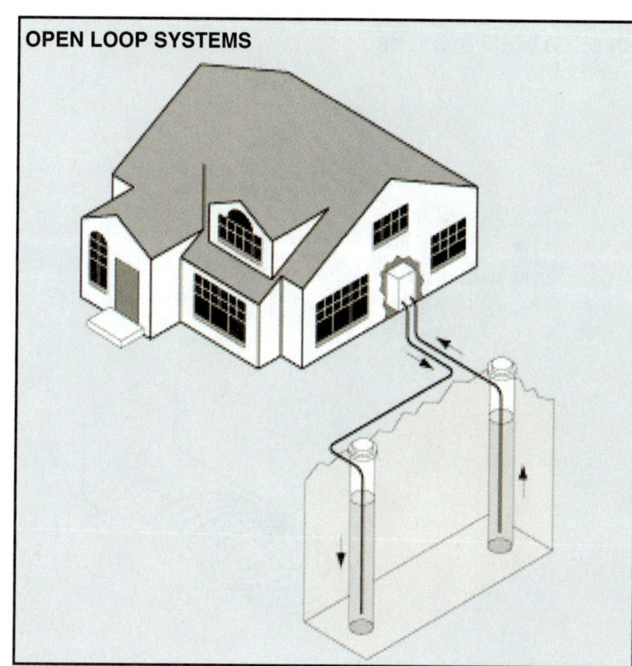

Figure 64-12 Open-loop system. *(Courtesy of Office of Energy Efficiency and Renewable Energy, U.S. Department of Energy)*

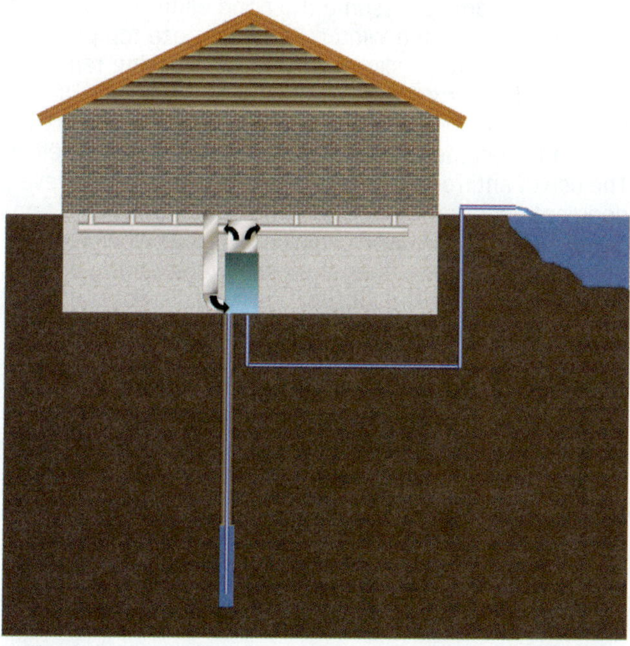

Figure 64-11 Water-to-air packaged heat pump open-loop application.

Figure 64-13 Coaxial coil. *(Climate Master, Inc.)*

TECH TIP

An open-loop system is practical only where there is an adequate supply of water and it is of good quality. Although the discharge water from the unit should not be contaminated, make sure that you check with all local codes and regulations regarding groundwater disposal.

SERVICE TIP

The water coil in an open-loop system can lose efficiency over time because of scale deposits inside the water coil. The constant supply of new groundwater also delivers a constant supply of minerals that can settle in the heat exchanger. Keeping a clean water-inlet filter on machines in hard-water areas will prevent this. Once a coil is scaled, the unit will not operate efficiently, and it is very difficult to remove the scale.

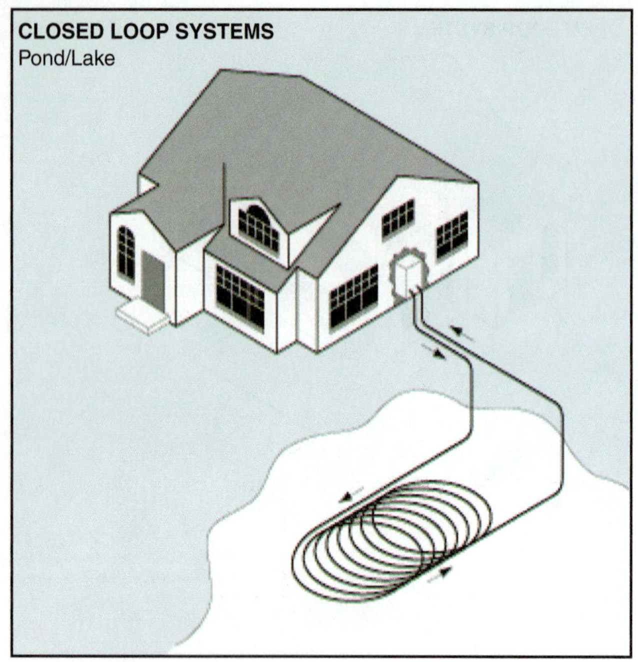

Figure 64-14 Pond closed-loop system. *(Courtesy of Office of Energy Efficiency and Renewable Energy, U.S. Department of Energy)*

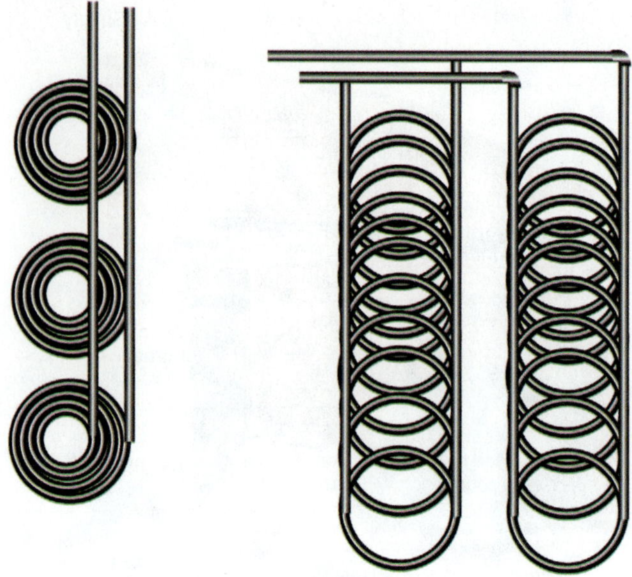

Figure 64-15 Piping arrangements for pond closed loop.

64.8 WATER-TO-AIR CLOSED-LOOP SYSTEM

A closed-loop system involves indirect heat exchange, unlike an open-loop system. This type of system is suitable where a pond or lake is located fairly close to the building (Figure 64-14). Most pond and lake water would not be suitable for an open-loop system because it contains too many minerals and contaminants that would lead to scale buildup inside the direct expansion evaporator coil. This type of system requires a secondary refrigerant, normally water or a water and antifreeze mixture. The closed loop is buried in the ground or sunk in a pond, and the water is circulated through this loop to the heat pump. Heat is transferred from the ground or pond to the secondary refrigerant in the loop, and the secondary refrigerant transfers the heat to the heat pump through its water-to-air coil.

Minimum pond sizes are roughly ½ acre with at least an 8–10 ft depth. Coiled piping, typically polyethylene, is placed below the expected ice cap level in a series of loops or coils to increase the total surface area exposed to the water (Figure 64-15). A secondary refrigerant must be circulated through the piping to absorb heat from the pond water. Generally, an antifreeze solution with varying percentages of propylene glycol, methanol, or ethanol is used.

The reason for the use of antifreeze is primarily one of heat transfer. The secondary refrigerant must be at a lower temperature than the pond water for heat transfer to occur. The greater the temperature difference, the better the heat transfer. In a direct expansion system, the primary refrigerant evaporates and expands, absorbing both sensible and latent heat. In contrast, a secondary refrigerant does not evaporate and therefore only absorbs sensible heat, which makes the temperature difference and surface area requirements greater than those needed for a direct expansion system.

As an example, assume the pond water temperature to be 40°F during the winter. The antifreeze temperature must be close to or slightly below the freezing temperature of water for good heat transfer to take place. If only pure, clean, distilled water were used as a secondary refrigerant, then there is the likely possibility for it to freeze. The use of antifreeze as a secondary refrigerant, however, does not necessarily suggest that the piping surface temperature of the pond loop could be lowered below freezing. If this were the case, then the pond water on the outside surface of the closed-loop piping would freeze and act as insulation, reducing the efficiency of the system dramatically. Due to the heat transfer through the piping and the flow velocity of the secondary refrigerant, its temperature might be at or slightly below freezing, while the outside surface temperature of the piping would still remain just above freezing. This type of system requires close control of the secondary refrigerant loop temperature to operate properly and efficiently.

In a water-to-air open-loop system, the heat from the water source is directly absorbed by the primary refrigerant and transferred to the air supply for the building. With a water-to-air closed-loop system, the heat from the water source must first be absorbed by the secondary refrigerant, which must then be absorbed by the primary refrigerant before being transferred to the air supply for the building. Therefore, in general, water-to-air open-loop systems will be less complex and more efficient than water-to-air closed-loop systems. However, the open loop will require more pumping horsepower than the closed loop. Both of these systems have efficiency and operating advantages when compared with air-to-air systems that operate in cold climates. An air-source heat pump must operate at higher compression ratios than water-source units because the air temperature is much colder than the groundwater temperature.

TECH TIP

Local codes and regulations must be checked carefully before installing a closed-loop pond system. There may be restrictions with regard to the type of secondary refrigerants that can be used because there is always a potential for leakage from a damaged pipe into the pond water. Another consideration often overlooked is the thermal shock to the natural pond environment. Changing the water temperature can change the type of plants and animals that can live in and around the water. There are many regulations today dealing with acceptable levels and restrictions on thermal pollution.

64.9 WATER-TO-WATER HEAT PUMP SYSTEM

Water-to-water systems use water as both the heat source and heat sink. They use two water coils and no air coils and are very compact units. Water-to-water heat pumps take heat out of the geothermal ground loop and transfer it to the water in a large insulated tank called a buffer tank. The buffer tank smooths out, or buffers, the effect of changing heat loads. The buffer tank allows the system to store heat, so that the amount of heat going into and out of the tank does not have to be exactly the same at all times. Water-to-water heat pumps are used with radiant floor-heating systems. Radiant floor-heating systems circulate hot water through tubing in the floor. Because radiant systems heat objects directly, they can provide comfort at lower room temperatures. Water-to-water heat pumps take heat out of the geothermal ground loop and transfer it to the buffer tank, and from the buffer tank to the water circulating in the floor (Figure 64-16).

If the heat pump capacity is insufficient for all of the heating needs, this type of system can be used in combination with an oil, gas, or electric furnace. If hot water for heating and chilled water for cooling are required, two separate buffer tanks may be used.

There are several advantages of radiant floor systems. Water transfers heat better than air. The amount of heat that can be carried by water in a small-diameter length of piping will be equal to or greater than the same amount that can be carried with the air in a large duct, thereby saving space. More power would be used to circulate an equivalent amount of air required for heating as compared to water. Radiant floor systems provide heat at the floor level, heating objects in the space and not the air directly. Less of the heat will be wasted by rising as hot air to collect at the top of the ceiling, which reduces efficiency.

A reversible water-to-water heat pump can provide chilled water to cool the building and hot water for the heating system. This type of system is in some ways similar to the air-to-water heat pump configuration. The basic cycle for a water-to-water heat pump consists of absorbing heat from a water source through an open- or closed-loop system and rejecting this heat to a buffer tank. The water from the buffer tank can be used with radiant flooring or hot-water baseboard heaters and in applications for melting ice.

(a)

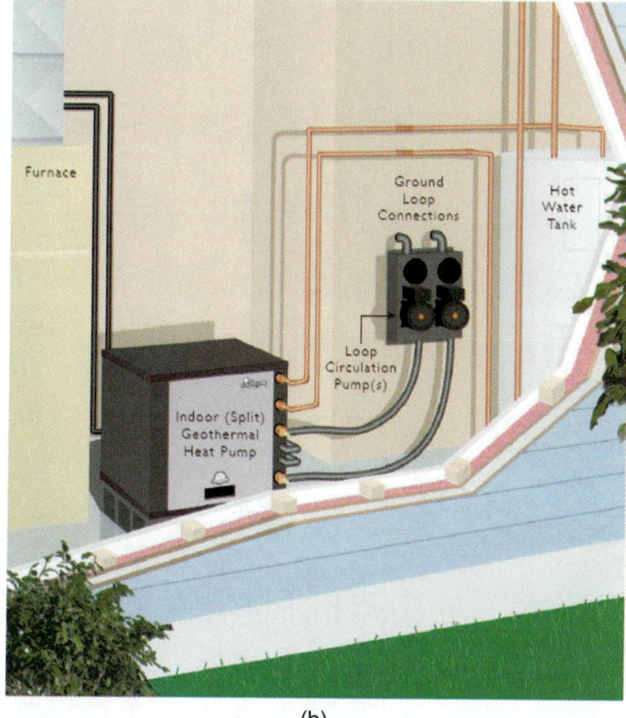

(b)

Figure 64-16 (a) Heat pump groundwater installation; (b) geothermal heat pump. *(Climate Master, Inc.)*

64.10 WATER-TO-WATER HEAT PUMP EFFICIENCIES

Water-source heat pump efficiencies are often best determined by the coefficient of performance (COP). This can be seen from the water-to-water performance data from Table 64-1. The source EWT (entering water temperature) will have the most effect on the system performance. This is similar to the effect the outdoor air temperature has on air-to-air systems. However, as noted earlier, the water source will generally have a fairly constant temperature so that performance is more consistent over the entire heating season.

Table 64-1 Water to Water Heat Pump Performance Data

SOURCE WATER		LOAD WATER TEMP	LOAD WATER FLOW = 15 GPM					LOAD WATER FLOW = 23 GPM					LOAD WATER FLOW = 30 GPM				
Temp	Flow		Heat Output	Energy Use	Heat Extract	Leaving Temp		Heat Output	Energy Use	Heat Extract	Leaving Temp		Heat Output	Energy Use	Heat Extract	Leaving Temp	
°F	GPM	°F	MBtuh	KW	MBtuh	°F	COP	MBTUH	KW	MBtuh	°F	COP	MBTUH	KW	MBtuh	°F	COP
20	30	60	74	5.1	56	70	4.3	74	4.9	58	67	4.5	75	4.8	59	65	4.6
		80	70	6.3	48	89	3.2	71	6.1	50	86	3.4	71	6.0	50	85	3.5
		100	68	8.0	40	109	2.5	68	7.7	42	106	2.6	68	7.6	42	105	2.6
		120	67	10.2	33	129	1.9	67	9.8	34	126	2.0	67	9.6	35	125	2.1
	15	60	77	5.1	59	70	4.4	77	4.9	61	67	4.6	78	4.8	61	65	4.7
		80	73	6.4	51	90	3.3	73	6.1	52	87	3.5	74	6.0	53	85	3.6
		100	70	8.1	43	109	2.6	71	7.8	44	106	2.7	71	7.6	45	105	2.7
		120	70	10.2	35	129	2.0	70	9.8	36	126	2.1	70	9.7	37	125	2.1
30	30	60	81	5.2	63	71	4.6	82	5.0	65	67	4.8	82	4.9	66	66	5.0
		80	77	6.4	55	90	3.5	78	6.2	56	87	3.7	78	6.0	57	85	3.8
		100	75	8.2	47	110	2.7	75	7.8	48	107	2.8	75	7.7	49	105	2.9
		120	74	10.3	39	130	2.1	74	9.9	40	127	2.2	74	9.7	41	125	2.2
	23	60	85	5.2	67	71	4.8	86	5.0	69	68	5.0	86	4.9	69	66	5.2
		80	80	6.5	58	91	3.7	81	6.2	60	87	3.8	81	6.1	61	85	3.9
		100	78	8.2	50	110	2.8	78	7.9	51	107	2.9	78	7.7	52	105	3.0
		120	77	10.4	42	130	2.2	77	10.0	43	127	2.3	77	9.8	44	125	2.3
	15	60	88	5.2	70	72	4.9	89	5.0	72	68	5.2	89	4.8	72	66	5.4
		80	86	6.6	64	92	3.8	87	6.3	65	88	4.0	87	6.1	66	86	4.2
		100	84	8.4	55	111	2.9	84	8.0	57	108	3.1	85	7.8	58	106	3.2
		120	81	10.6	45	131	2.2	81	10.0	47	127	2.4	82	9.8	48	125	2.4

SOURCE WATER		LOAD WATER TEMP	LOAD WATER FLOW = 15 GPM					LOAD WATER FLOW = 23 GPM					LOAD WATER FLOW = 30 GPM				
Temp	Flow		Heat Output	Energy Use	Heat Extract	Leaving Temp		Heat Output	Energy Use	Heat Extract	Leaving Temp		Heat Output	Energy Use	Heat Extract	Leaving Temp	
°F	GPM	°F	MBTUH	KW	MBTUH	°F	CoP	MBTUH	KW	MBTUH	°F	CoP	MBTUH	KW	MBTUH	°F	CoP
40	30	60	93	5.3	75	72	5.2	94	5.1	76.7	68	5.5	94	4.9	77	66	5.6
		80	91	6.7	68	92	4.0	92	6.3	70.2	88	4.3	92	6.2	71	86	4.4
		100	89	8.5	60	112	3.1	89	8.1	61.8	108	3.3	90	7.8	63	106	3.4
		120	85	10.7	49	131	2.4	86	10.2	51.4	128	2.5	86	9.9	53	126	2.6
	23	60	98	5.3	79	73	5.4	98	5.1	80.8	69	5.7	98	4.9	81	67	5.9
		80	95	6.7	72	93	4.2	96	6.	54.2	89	4.4	96	6.2	75	86	4.6
		100	93	8.5	64	112	3.2	93	8.2	65.7	108	3.4	94	7.9	67	106	3.5
		120	89	10.7	53	132	2.4	90	10.3	55.1	128	2.6	90	9.9	56	126	2.7
	15	60	100	5.3	82	73	5.5	101	5.2	83.3	69	5.8	101	4.9	84	67	6.0
		80	98	6.8	75	93	4.2	98	6.5	76.6	89	4.5	99	6.2	78	87	4.7
		100	95	8.6	66	113	3.3	96	8.2	68.0	109	3.5	96	7.9	69	106	3.6
		120	91	10.8	55	132	2.5	92	10.3	57.3	128	2.6	93	10.0	59	126	2.7
50	30	60	106	5.4	88	74	5.8	106	5.2	89.0	69	6.1	107	5.0	90	67	6.3
		80	104	6.8	80	94	4.5	104	6.5	82.3	89	4.7	105	6.3	83	87	4.9
		100	101	8.6	71	113	3.4	101	8.2	73.4	109	3.6	102	8.0	75	107	3.7
		120	97	10.8	60	133	2.6	98	10.3	62.5	129	2.8	98	10.1	64	127	3.0
	23	60	111	5.4	92	75	6.0	111	5.1	93.7	70	6.4	111	5.0	94	67	6.6
		80	108	6.9	85	94	4.6	109	6.5	86.8	90	4.3	109	6.3	88	87	5.1
		100	105	8.7	76	114	3.6	106	8.2	77.8	109	3.8	106	8.0	79	107	3.9
		120	101	10.9	64	134	2.7	102	10.4	66.7	129	2.9	103	10.1	68	127	3.0

The load EWT and LWT are the temperatures of the water entering and leaving the heat pump to be circulated through the heating loop (radiant floor, baseboard, etc.). The water leaving the heat pump is at a higher temperature (LWT) as it enters the heating loop and comes back to the heat pump at a lower temperature after giving off heat to the space. Notice that the COP decreases as the required hot-water temperature for the heating loop rises. It would be more efficient to maintain a radiant floor temperature from between 80°F to 100°F, which generally should be sufficient. However, many hot-water baseboard heating systems require water temperatures of at least 120°F.

Also, notice that the COP increases at higher flow rates. Therefore, careful consideration should be given to the piping system layout and circulating pump requirements when sizing an appropriate system.

64.11 DOMESTIC HOT WATER

The hot discharge gas leaving the compressor is hot enough to heat domestic hot water. The condenser saturation temperature is typically too low to provide hot water, so all the hot-water heating is done from the discharge gas superheat. Many water-source heat pump systems can be configured to supply hot domestic water using de-superheating coils and water-circulating pumps built into the system (Figure 64-17). During cooling, the heat removed from the house is being wasted. Putting it in the hot water not only makes the heat pump more efficient by keeping the loop cooler, but it reduces the energy required to heat the hot water by using waste heat. In heating, the heat is not free, but it is still far more efficient than operating an electric hot-water heater. Water temperatures leaving the water-heating loop are generally no higher than 145°F.

Domestic hot-water heating can be configured with outdoor split systems (Figure 64-18) and indoor package units. Water-source heat pumps typically use an indirect water-to-water heat exchanger. This coil located inside the tank isolates the domestic (potable) water from the heating water. Some units have electric resistance heating elements for use as a backup.

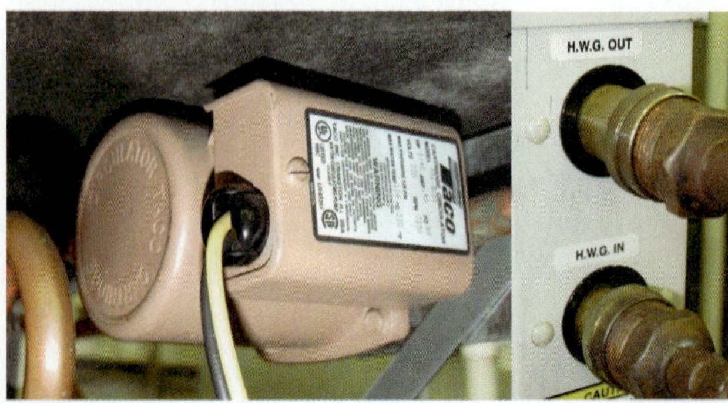

Figure 64-17 Circulating pump and connections for domestic hot-water heating.

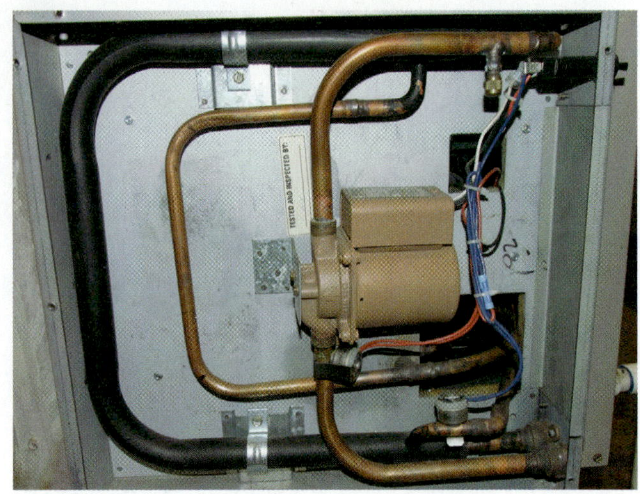
Figure 64-18 Domestic hot-water split system with outside compressor.

UNIT 64—SUMMARY

A geothermal heat pump system is the least expensive way to heat. Geothermal systems circulate water through closed loops buried in the ground to transfer heat from the ground into the house using a water-source heat pump. Geothermal systems do not lose capacity as the air temperature drops because they extract heat from the ground, which is a stable temperature. Water-source systems are more efficient than air-source systems because water is a more efficient heat-transfer medium, the temperatures they operate at are more moderate due to the moderating effect of the earth, they do not need a defrost cycle, and their capacity does not drop off as the temperature drops outside. Water-source units can be connected to either open or closed loops. An open loop does not recirculate the water. Closed loops recirculate water in closed loops buried in the ground or in the bottom of a large body of water. Loops can be horizontal or vertical. Water-to-water systems can be used to provide warm water for radiant heating systems. Some water-source systems have de-superheating coils and circulating pumps to provide domestic hot water as well as space heating and cooling.

WORK ORDERS

Service Ticket 6401

Customer Complaint: Low Water Temperature

Equipment Type: Water-Source Heat Pump with Open Loop

The customer complains to the technician that the heat from the radiant floor is not sufficient. Dial thermometers were installed in the lines leaving the heat pump and the customer notes that the water temperature has been gradually decreasing since the system was installed more than a year ago.

Having never been to this residence before, the technician begins to review the specifications for the system. The technician is surprised to find that the system is operating with an open loop from a small man-made pond located in the backyard. The water quality is obviously poor and no doubt has led to scale and deposits on the water side of the refrigeration coil. This would reduce the heat transfer and lead to a gradual decrease in amount of heat absorbed from the pond water. The technician also finds that the auxiliary electric heat for the buffer tank is disconnected.

The customer explains that during the initial installation, the contractor recommended a closed loop, but it was going to be much more expensive to install. It was the customer's decision to go with the open loop. A few months ago, when the auxiliary electric heat began running a majority of the time, the customer simply disconnected it.

The technician explains that it is not worthwhile to try to cut costs by trying to change the parameters for the application involved. It does no good to try to save a few dollars on the installation to only have a system that does not work properly. The technician politely tells the customer that the technician is only willing to work on the system if the customer follows the suggestions for improvement. At the very least, a closed-loop system needs to be put into place if the existing system is to be used.

Service Ticket 6402
Customer Request: New Installation

Equipment: Water-Source Heat Pump

Big Lake Power and Light is building a new office building that they would like to make a showcase for innovative energy-conservation measures. A sales consultant is called in to help them decide which type of system would work best. The building is fairly large and there are a large number of different spaces. The best way to configure the space would be to set up a number of different heating zones.

The building is located in a rather cool climate, which would certainly have an adverse effect on the efficiency of any heat pump system. However, the site is adjacent to a large lake, which would allow for water-source heat pumps connected together on a closed-loop system. The lake water temperature at the location of the coils would remain fairly constant and provide enough heat for the building. In terms of energy conservation, this is exactly what the customer had in mind.

Service Ticket 6403
Customer Request: New Installation

Equipment Type: Water-to-Water Heat Pump

Tiles R Us is building a showroom to show off some of their tile floor products. They want to use radiant floor heating because they do not want the customers to perceive the flooring as cold. One method suggested by the sales consultant would be to use an LP gas-fired hydronic system. This would be the least complicated system for them to install. However, the owner of the showroom is concerned about rising energy costs and would like to take advantage of heat pump energy savings if possible.

The sales consultant notices that space for a geothermal application is limited. However, a vertical loop, which would reduce the amount of area that is disturbed when the loop is installed, would probably work. The consultant would need to perform some additional calculations to make sure. The sales consultant recommends a water-to-water geothermal heat pump application but cautions that he needs to make sure the loop sizing would work for a vertical system at the site. The heat pump system would still provide warm water for a warm floor, and the heat pump would provide operating savings over the gas boiler. The initial installation cost will be greater, so this would need to be calculated against the annual fuel savings to see if the decision would make economic sense.

UNIT 64—REVIEW QUESTIONS

1. What is the difference between a water-source heat pump and an air-source heat pump?
2. What type of metering device do most packaged water-source heat pumps use?
3. Why do the refrigerant and water travel in opposite directions through the water coil?
4. Why don't water-source systems have a defrost cycle?
5. Explain the difference between an open loop and a closed loop.
6. What is a geothermal heat pump system?
7. What site conditions should be taken into account when considering the installation of a ground-source heat pump?
8. What is a water-to-water heat pump?
9. What are some of the advantages of radiant floor heating?
10. Why are the insulated water tanks for heat pump systems referred to as "buffer" tanks?
11. Where is the compressor located in most water-source systems?
12. Why do geothermal heat pump systems work better in cold climates than air-source systems?

13. How can heat be considered a form of pollution?
14. How is heat transferred in a heat pump ground loop?
15. What type of system requires a secondary refrigerant?
16. What happens if water freezes on the outside of a closed-loop pond heat pump system?
17. What is the difference between the way a secondary refrigerant absorbs heat and the way a primary, direct expansion refrigerant absorbs heat?
18. Why would a water source heat pump be more efficient than an air-to-air heat pump?
19. What is the problem with sizing a geothermal heat pump to the full heat loss of a home in a cold climate?
20. What is the advantage of a dual-capacity compressor in a water-source heat pump?
21. How can water-source heat pumps help reduce domestic hot-water cost?
22. Which type of system requires more pumping horsepower for the water: open loop or closed loop?
23. What is most often used as a secondary refrigerant in geothermal systems?

UNIT 65

Heat Pump Installation

OBJECTIVES

After completing this unit, you will be able to:

1. discuss the decisions involved in selecting an air-source heat pump.
2. discuss the decisions involved in selecting a water-source heat pump.
3. plot a system balance point.
4. discuss the correct way to run the condensate drain lines in an attic installation.
5. calculate the correct refrigerant charge for a split-system heat pump.
6. list the types of water loops for water-source heat pumps.
7. list the steps for filling a ground-loop system for the first time.
8. discuss the process of checking water-source system performance.

65.1 INTRODUCTION

Heat pump systems lead all other forms of heat in terms of efficiency and low operating cost. The demand for heat pumps has increased because of the dramatic increase in the cost of all forms of energy. They must be installed correctly to take advantage of their energy-saving technology. The least efficient system installed correctly will outperform a more efficient system that is poorly installed.

Correct installation starts with selecting a system that meets the job's application requirements. This includes choosing a configuration, sizing the equipment, and sizing the auxiliary heat. Reading and following the manufacturer's installation instructions is the single most important step any installer can take. Attention to detail in locating the unit, refrigerant piping, wiring, duct connection, airflow, and water flow will pay dividends for the customer in the form of efficient operation and long equipment life.

65.2 EQUIPMENT SELECTION

Four decisions are involved in choosing a heat pump system:

- Air source versus water source
- Equipment configuration
- Equipment size
- Supplemental heat source

Air Versus Water

The first decision to make when selecting a heat pump system is to use either an air-source unit or a water-source unit. Air-source systems have a lower installed cost, while water-source systems usually have a lower operating cost. The decision is primarily an economic one: how much money is available to invest versus how long it will take to recover the investment. A water-source system will require either an abundant supply of water or installation of a ground loop for recirculation of the water.

Configuration

The type of building construction and the location of the unit helps determine the best configuration for any particular job. Buildings on slabs almost always have the ductwork in the attic. Horizontal split systems located in the attic are frequently used on this type of construction. Upflow split systems can be used if the air handler is located in a closet. For houses on a foundation with a crawlspace, side-discharge packaged units are popular. The unit sits on the end of the house and the duct runs in the crawlspace. Horizontal split systems are also popular for crawlspace construction. A downflow unit in a closet can be used with crawlspace construction, but this is less common. Buildings with basements usually have upflow systems with both the system and the ductwork located in the basement. Table 65-1 shows typical applications for several types of units.

Table 65-1 Typical Heat Pump Applications

Source	Type of Unit	Application
Air	Bottom-discharge packaged	Rooftop mounted on curb
Air	Wall-mount packaged	Schools mounted on outside wall
Air	Split system	House with blower in attic, crawlspace, or basement
Water	Vertical-discharge packaged	Unit located in basement or closet
Water	Horizontal discharge	Unit located above dropped ceiling

Equipment Size

The challenge in sizing a heat pump is matching both the cooling and heating load of the structure, which are usually not the same. Meeting the building heat loss with the heat pump alone is difficult because the heating capacity of the heat pump declines as the heat loss of the building increases. In most cases, a heat pump sized to meet the heat loss at the winter design temperature listed in the ACCA Manual J would be grossly oversized in cooling. Systems that are grossly oversized in cooling are very energy inefficient, increasing the cost of operation. They also do a poor job of controlling humidity because they do not operate long enough to remove much moisture. In severe cases, this can contribute to mold and mildew formation in the structure.

The most common practice in sizing heat pumps is to size the system for the cooling load and add supplemental heat to make up the difference in heating. Some engineers will select a unit with a cooling capacity of a half ton larger than necessary to help increase the heating capacity without causing significant problems in cooling operation.

Two-stage heat pumps are now being offered that help address the issue of sizing. These systems use compressors with two pumping capacities (Figure 65-1).

These systems can be sized by matching the cooling load to the low-capacity cooling. The compressor runs at low capacity most of the time, allowing efficient operation and dehumidification. The system can operate at full capacity when more capacity is needed, such as at lower ambient temperatures in heating.

DIGITAL SOLENOID VALVE

Figure 65-1 Two-capacity compressor.

Determining the Balance Point

The heat loss of the building and the capacity of the equipment determine a system's balance point. This means that the balance point for a 2-ton brand X heat pump would be different for your house than your neighbor's house. It also means that the balance point for a 2-ton brand Y heat pump could be different from a 2-ton brand X heat pump, even on the same house! When selecting equipment, it is useful to plot the balance point of the systems being considered. To calculate a balance point correctly, you need the following:

- A heat load calculation for the specific house
- The capacity ratings for the specific model unit you propose to use—two rating points are necessary; usually this will be at 47°F and 17°F
- A piece of graph paper and a straightedge or a balance point graph

First, plot the structure heat loss. Mark a point on the graph that represents the heat loss at the Manual J winter design condition. Mark another point along the bottom of the graph to represent 0 BTU heat loss at 65°F. Draw a straight line between these two points. This line represents the building load.

To plot the equipment capacity line, mark a point that represents the equipment capacity at 47°F and another that represents the equipment capacity at 17°F and draw a straight line through them. This is the equipment capacity line. The point at which the equipment capacity line intersects the structure load line is the balance point. To find this temperature, read straight down from the intersection of the two lines. Figure 65-2 shows a balance point graph for a house with a heat loss of 45,000 BTU/hr at 18°F and a heat pump with heating capacities of 35,000 BTU/hr at 47°F and 20,000 BTU/hr at 17°F. At the balance point of 32°F, the heat loss is 29,000 BTU/hr.

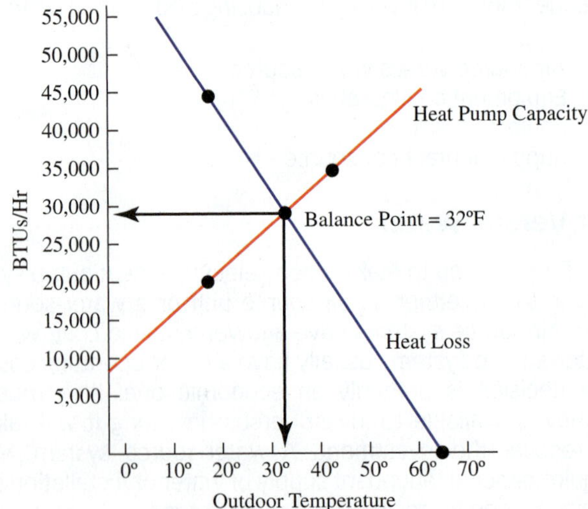

Figure 65-2 Balance point graph.

Supplemental Heat

Auxiliary heat will be required for temperatures below the balance point. The most common form of auxiliary heat is electrical strip heat. This has the advantage of being relatively inexpensive, easy to control, and easy to install. The overall efficiency of the heat pump is reduced anytime the strips are operated. Outdoor thermostats are often used to avoid operating this form of heat whenever the ambient temperature is above the balance point. This is why establishing the system balance point is important. The goal is to deliver comfort at the most reasonable operating cost possible. In practice, the outdoor thermostat is set 5°F above the actual system balance point to ensure that the equipment can keep up with the load.

The next decision is to determine how much auxiliary heat is needed. One possible solution is to add enough auxiliary heat to make up the difference between the building load and the equipment capacity. This would be considered the absolute minimum amount of auxiliary heat, since anything less would leave the customer without sufficient capacity at cold temperatures. The only problem with this approach is that it assumes the heat pump capacity will always be available. If any problems occurred in the refrigeration system, the customer would be left without adequate heating capacity. If this is a concern, the auxiliary heat can be sized to meet the entire load at the design condition.

Auxiliary heat does not have to be electric strip heat; it can also be a fossil fuel furnace. Heat pump systems installed with fossil fuel auxiliary heat are called dual-fuel systems. Typically, these systems take advantage of the heat pump's capacity to deliver economical heat at mild temperatures and the furnace's capacity to deliver large amounts of heat with no capacity loss at low temperatures.

65.3 AIR-SOURCE SYSTEMS

Steps involved in installing an air-source heat pump are as follows:

- Locate the equipment.
- Connect the ductwork.
- Size and connect the power wiring.
- Wire the controls.
- Connect the refrigerant lines.
- Evacuate and charge the system.

Outdoor Unit Location

Ideally, heat pumps should keep the house comfortable without being noticed. All equipment should be located where sound from the equipment will not be a nuisance, away from bedroom windows and outdoor patios. It is also important that the equipment is not unsightly to the homeowner. The front of most houses is usually off limits. In many places the outdoor equipment is located on the side or back of the house. However, in areas with little or no yard, equipment is often located on the roof. In Georgia, locating a unit on someone's roof is considered abominable. However, in Arizona, where swamp coolers are common, locating equipment on the roof is normal.

Figure 65-3 Condensing unit elevated above pad.

Packaged units and split-system condensing units should be located above the average winter snow line. In Florida, the unit can rest directly on the pad, while in Michigan, the unit should be elevated several feet above the ground (Figure 65-3). Systems buried in snow will not work well.

Airflow should be considered when locating the equipment. Vertical discharge units should not be located under a deck or overhang. Horizontal discharge units should not be located with their discharge air blowing directly into a wall. The amount of space required around a unit to ensure proper airflow is listed in the installation instructions.

Indoor Unit Location

Both air- and water-source heat pumps can be split systems. The indoor blower coils for a split system may be located in a basement, closet, crawlspace, or attic. When installed in the attic, the unit should not rest directly on the ceiling joists or on plywood on top of the ceiling joists because fan noise and vibration will be transmitted into the house through the ceiling. To avoid unwanted sound transmission, the blower coil may sit on vibration isolation pads or be suspended from the roof rafters as long as the rafters will support the weight. It should be noted that this is generally not allowed on most trussed roof systems.

Indoor Unit Configuration

Many heat pump air handlers can be installed in the upflow, downflow, or horizontal position. However, this is not simply a matter of flipping the unit on its side. The installer must read and follow the installation instructions carefully. Frequently, parts of the unit will need to be repositioned to change the orientation of the air handler. The installer must make sure of the following:

- The coil orientation in the cabinet is correct for the position of the unit.
- The drain pan is positioned under the coil.
- The correct drain opening is being used.
- The service access panel has enough clearance to service the unit.

Drain Connections

When the unit operates in cooling, the indoor coil creates condensate water that needs to be drained. The drain should be at least ¾-in PVC and should slope downward at least 1/8 inch per foot. The air velocity rushing past the drain hole creates a vacuum that can hold the water in the drain pan. Figure 65-4 shows water standing in the drain of an operating unit with the drain line removed.

The drain is normally trapped to keep this from happening. The water in the trap stops air from coming up the drain line and allows the water in the drain pan to drain out. Some units have traps built into the unit; these should not be trapped. The installation instructions will specify whether to trap the drain (Figure 65-5).

Units installed in the attic should have an auxiliary drain pan under them that is larger than the overall dimensions of the unit (Figure 65-6).

The purpose of the auxiliary drain is to prevent damage to the ceiling when the main drain clogs. The auxiliary drain opening in the unit should be run to the auxiliary drain pan. The drain line from the auxiliary pan should be run out separately from the main drain; the two drain lines should not be tied together. The auxiliary drain should exit in a location

Figure 65-4 Standing water in a drain outlet.

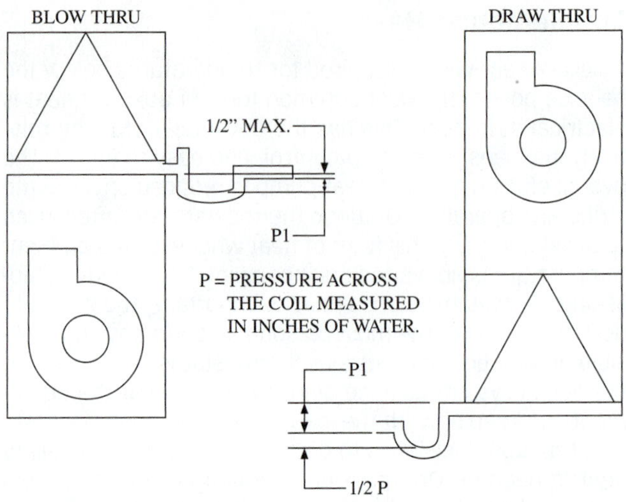

Figure 65-5 Drain trap.

Figure 65-6 Drain pan under an attic unit.

where the water will be obvious to encourage investigation of the source of the water (Figure 65-7). Float switches can be installed to shut the unit down in the event the drain becomes plugged. The increasing water level will cause the float to rise and activate the shutoff switch.

Units installed in basements and crawlspaces usually require a condensate pump (Figure 65-8).

The water drains into the condensate pump sump and the condensate pump pumps the water out when the sump fills. Small ⅜-in inside diameter plastic tubing runs from the condensate pump outside. This line should have a check valve to prevent water from flowing back into the sump when the pump shuts down. Figure 65-9 shows a system connected to a condensate pump. A float activated shutoff switch can also be installed on this type of system.

Connecting the Ductwork

Metal ductwork may be fastened directly to the ¾-in flanges of the blower coil (Figure 65-10).

Sheet-metal screws are used to fasten the ductwork to the flanges. Fiber ductboard should not be connected

Figure 65-7 Drain outlet.

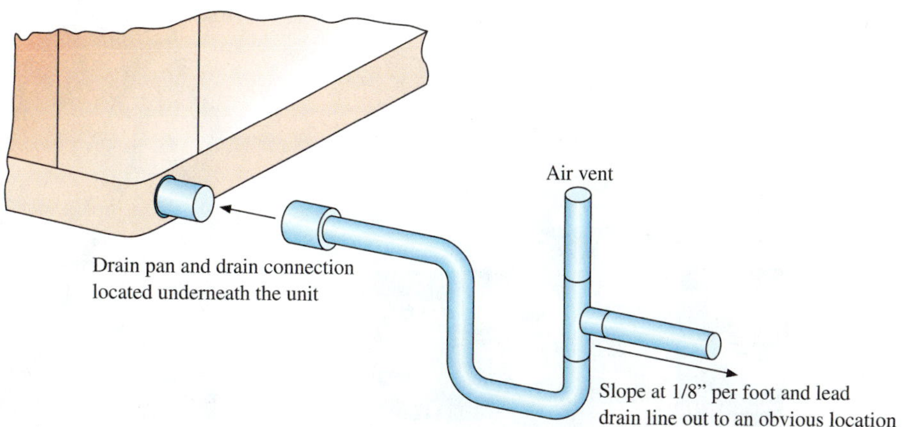

Figure 65-8 Condensate pump.

Figure 65-9 Installed condensate pump. *(Courtesy of InspectAPedia.com)*

directly to these flanges. A metal collar at least 3 in wide should be attached to the flanges first before attaching fiber ductboard. The metal collar is attached to the flanges with sheet-metal screws and the fiber duct is connected to the collar using long sheet-metal screws with large washers under the heads.

General Power Wiring Requirements

The type of wire that may be used is listed in the installation instructions and frequently also on the unit. If aluminum wiring is used, the installer must check to see if the connections on the unit are designed for aluminum conductors. Using aluminum conductors on units that do not have aluminum rated connections will void the system warranty! Aluminum conductors also require application of antioxidation paste on the connection to prevent poor connection as a result of oxidation (Figure 65-11).

The conductor should be sized using the wire sizing tables of the National Electric Code to handle the minimum circuit ampacity listed on the unit data plate or installation instructions. The disconnect should be located within sight of the equipment and rated no smaller than 115 percent of this minimum circuit ampacity. The overcurrent protection

Figure 65-10 Metal duct connected to flanges with sheet-metal screws.

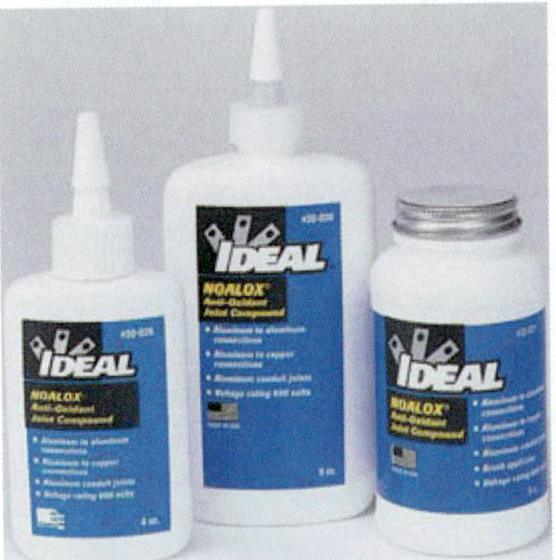

Figure 65-11 Antioxidation paste. *(Courtesy of IDEAL Industries, Inc.)*

should be no larger than the maximum overcurrent protection listed on the data plate or installation instructions. All wiring should pass through strain-relief connectors whenever it passes through the cabinet or disconnect.

Power Wiring for Packaged Units

Supplemental electric heat strips are sized for each job and are added in the field. Because the electric specifications will vary depending upon which heat kit is being installed, the installation instructions list specifications for many different combinations of units and heat kits. The installer must identify the data for the particular unit and heat kit being installed.

Power Wiring for Split Systems

Split systems require two power circuits: one for the indoor blower and one for the outdoor condensing unit. Each should have its own disconnect and overcurrent protection. The electrical data for the circuit to the indoor blower must be selected from a list of data because the supplemental heat is sized for each job and installed in the field.

Control Wiring

All control wiring should be solid copper and at least 18 gauge. Control wires are typically bundled together in groups of wires with different colors (Figure 65-12). Heat pump controls require anywhere from four wires per bundle for digital controls like the Carrier Infinity system, to 10-wire bundles for some multistage systems using traditional controls.

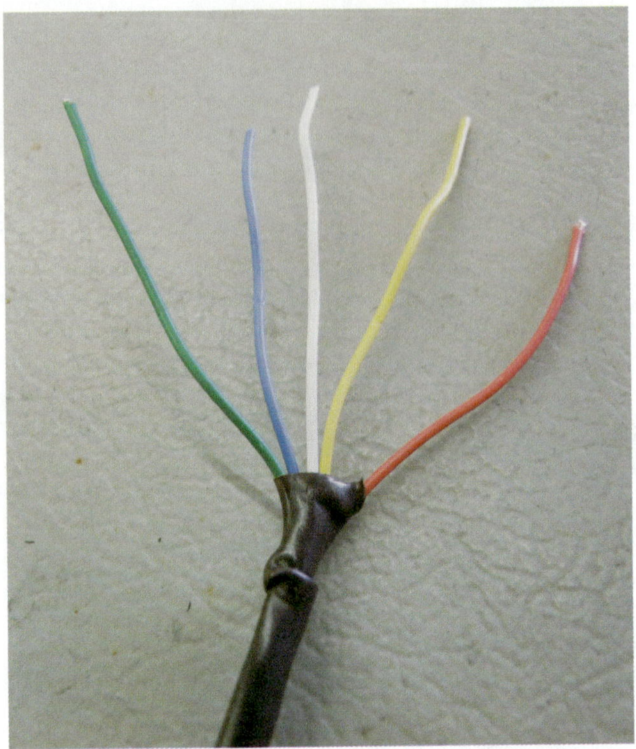

Figure 65-12 Thermostat wire bundle.

Packaged Unit Control Wiring

The control wiring for packaged units is straightforward: there is one bundle of control wires from the thermostat to the unit. The control wiring connects to the thermostat subbase terminals on one end and to the unit terminal board on the other end. The ends of the control wires should only be stripped far enough to fit completely in the terminal; bare wire should not be hanging out beyond the connection points (Figure 65-13).

The installation instructions will identify which terminals on the thermostat should be connected to which

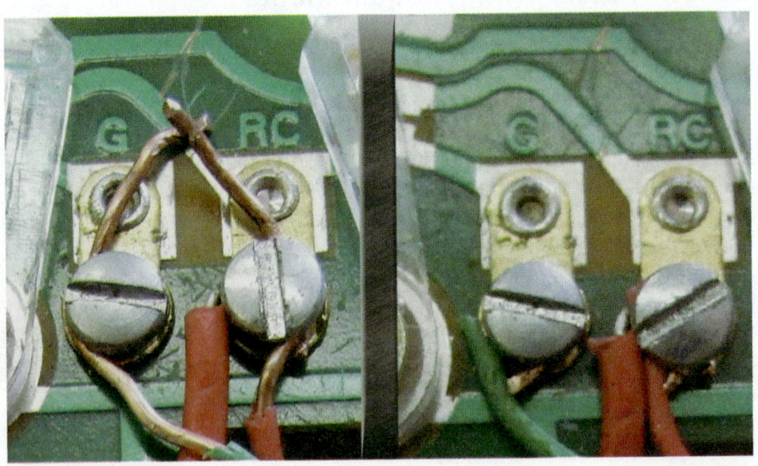

Figure 65-13 Right and wrong ways to strip and connect thermostat wire.

Figure 65-14 Control wire pigtails.

terminals on the unit. The majority of manufacturers keep it simple by using the same letters on the thermostat and the unit. Some manufacturers do not use a terminal board on the unit; they have color-coded control wire pigtails instead. The control wires from the thermostat are connected to these using wire nuts (Figure 65-14).

Split-System Control Wiring

Split systems require one wire bundle running from the thermostat to the indoor unit and another running from the indoor unit to the outdoor unit. Some of the wires in these bundles perform the same function in each bundle, but other wires in the bundles have functions that are only necessary for the indoor or outdoor unit. Manufacturers try to simplify this by using common terminal designations, but it is not possible to wire most split-system heat pumps by just connecting all the same color wires together. Table 65-2 shows common terminal designations used in the industry. While it is useful to know the most common control wire letter designations, installers should always refer to the installation instructions when wiring a system.

CAUTION

Note that the green wire in the indoor bundle is used to energize the fan and is not needed in the outside bundle. Many installers use the green wire in the outside bundle for the common connection. If green is used for common in the outside bundle, connecting the green wires of the two bundles together will short out the transformer when the fan is energized.

CAUTION

Note that there are two definitions of the B terminal. On most over-the-counter thermostats, B is connected to the red power terminal whenever the thermostat is set to heat. Heat pumps that energize the reversing valve in heat use this terminal. Some manufacturers use the B terminal for the common side of 24 V. If the B terminal of an over-the-counter thermostat is connected to the B terminal on a unit that uses B for common, the transformer will short out when the system is placed in heat.

Table 65-2 Common Industry Terminal Designations

Thermostat			Indoor Unit			Outdoor Unit	
Terminal	**Function**	**Wire**	**Terminal**	**Function**	**Wire**	**Terminal**	**Function**
R	24 V power in	Red	R	24 V power out	Red	R	24 V power in
Y	Energize contactor	Yellow	Y	None	Yellow	Y	Energize contactor
O	Energize reversing valve in cooling	orange	O	None	Orange	O	Energize reversing valve in cooling
B	Energize reversing valve in heating	Blue	B	None	Blue	B	Energize reversing valve in heating
B	Common side of 24 V	Blue	B	Common 24 V out	Blue	B	Common 24 V in
C	Common side of 24 V	Installer's choice	C	Common 24 V out	Installer's choice	C	Common 24 V in
G	Fan	Green	G	Energize fan	None	None	None
W	Energize aux heat	White	W	Energize aux heat	white	W	Energize aux heat
W2	Energize aux heat	White	W2	Energize aux heat	white	W2	Energize aux heat

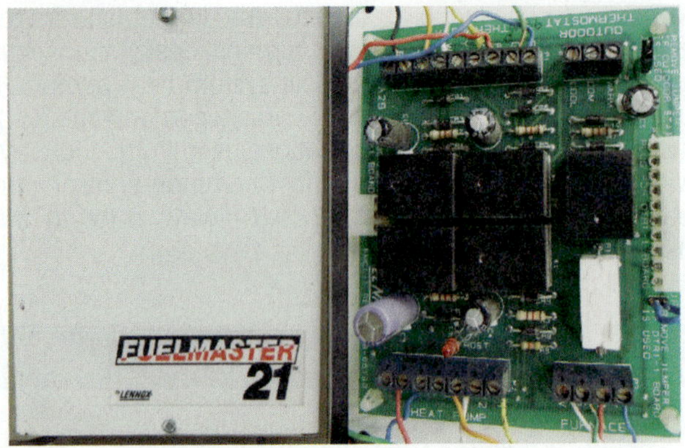

Figure 65-15 Dual-fuel board.

Control Wiring for Dual-Fuel Heat Pumps

When a fossil fuel furnace is used for the auxiliary heat, a separate control board is added to coordinate the operation of the furnace and heat pump (Figure 65-15).

These systems use three wire bundles:

1. Thermostat to the dual-fuel board
2. Dual-fuel board to the furnace
3. Dual-fuel board to the heat pump

Because the wiring for these is considerably more complex than traditional split systems, it is imperative that the installer read all the instructions and follow the wiring diagram.

Infinity Control Systems

Carrier products offer systems with a digital control system. All components wire with exactly four wires. Two wires provide the control power and two more are the communication wires. These systems use serial communication to control all pieces of equipment in the system. Installation is greatly simplified because everything is wired with the same four wires (Figure 65-16).

(a)

Indoor Blower

(b)

Outdoor Unit

(c)

Figure 65-16 Infinity control system wiring.

Figure 65-17 Split-system refrigerant line set.

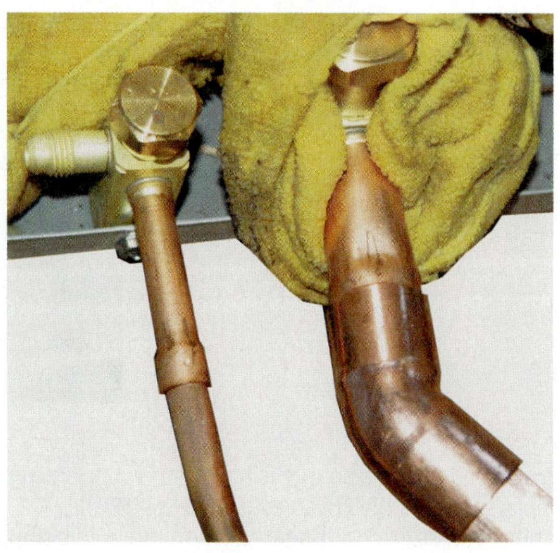

Figure 65-18 Cooling valves while brazing.

Running Refrigerant Lines

Split-system heat pumps require two refrigerant lines running between the indoor and outdoor units: a large gas line and a smaller liquid line. The lines can be connected to the units by brazed joints, flare fittings, or compression fittings. The connection method is determined by the stub-outs on the equipment. The lines are typically sold in line sets with insulation already on the gas line (Figure 65-17).

Line sets are made of soft ACR copper tubing. ACR tubing is type L copper tubing that has been dehydrated. The insulation used on the gas line of a heat pump must be able to withstand high temperatures. Before using any insulation on a heat pump gas line, check to be sure it is rated for heat pump duty.

SAFETY TIP

Never use plumbing copper to run refrigeration lines! Standard plumbing copper is type M, which has a thinner wall than type L copper used in ACR tubing. Plumbing tubing is also not dehydrated, so the risk of introducing moisture into the system is increased.

Most manufacturers limit the line length on residential systems to 150 ft. Long lines have increased pressure drop, lowering system efficiency. This can be offset by increasing the size of the lines. However, manufacturers differ in philosophy on line sizing for residential systems. At least one major manufacturer does not want the line size increased from the stub-outs on the equipment. Other manufacturers recommend increasing the line size to offset length and provide charts to determine the correct line size. Either way, the best practice is to keep lines as short as possible and follow the manufacturer's instructions.

TECH TIP

One change most manufacturers made to increase the efficiency of their systems was increasing the size of the suction line on most residential systems. Suction lines as small as $\frac{5}{8}$ in outside diameter (OD) were common at one time on 2-ton systems. Today most 2-ton units use $\frac{3}{4}$-in or even $\frac{7}{8}$-in suction lines. This increases the efficiency by reducing the pressure drop in the suction line.

Connecting Brazed Lines

For units with brazed connections, pull back the insulation on the gas line before brazing to avoid burning the insulation. To protect the rubber O-rings in the valves, cool the installation valves with thermal absorbing paste or wet rags during brazing (Figure 65-18).

Connecting Compression Lines

Manufacturers used to offer equipment with compression fittings for the stub outs. The copper tubing should be straight and clean where the ferrule slides over it and goes into the compression fitting. Any dents in the tubing or grit under the ferrule will cause a leak. Make sure the tubing is seated fully into the fitting before tightening. Once the ferrule starts to compress, it cannot be moved. Mistakes must be cut off and a new attempt made with a new ferrule. Only use compression fittings supplied by the air conditioning manufacturer. Never try yo use plumbing fittings on refrigeration systems.

Evacuation and Charging

Packaged units do not need to be evacuated and charged; they arrive from the factory with the correct charge. Split systems also are shipped with a refrigerant charge, but they usually do require some adjustment to the refrigerant charge to account for their refrigerant lines. Most manufacturers ship their condensing units with enough charge for the outdoor unit, the indoor coil, and some assumed length

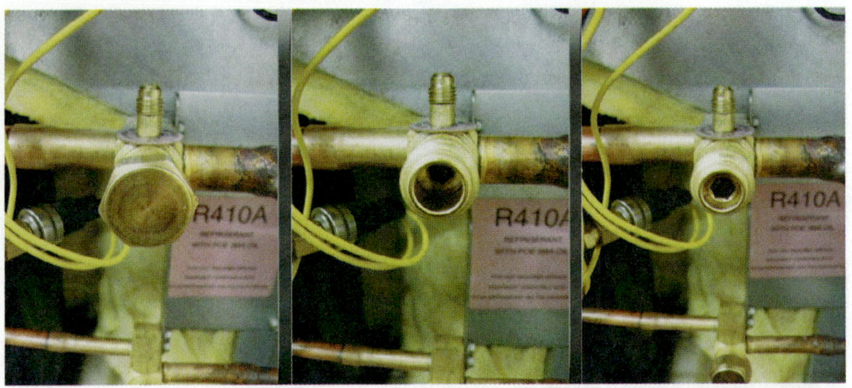

Figure 65-19 Split-system shutoff valves.

of refrigerant line. This charge is held in the condensing unit with split system shutoff valves (Figure 65-19).

These valves are front-seated when the unit is shipped. The Schrader access fittings on these valves are always open to the refrigerant line regardless of the valve's position. The shutoff valves are opened only after the following steps are complete:

- Installing the lines and coil
- Leak testing the lines and coil
- Evacuating the lines and coil

Leak Testing

The lines and coil should be leak tested with dry nitrogen. Nitrogen makes a good initial leak test gas because it is inexpensive and legal to release to the air. Pressurize the lines and coil to between 100 psig and the system's low-side test pressure. Make a note of the time and pressure. Use soap bubbles or an ultrasonic leak detector to check the line connections at both the inside and outside units. Also, check any other mechanical or brazed joints in between. Finally, if the unit uses a field-installed expansion valve, check all connections on the expansion valve. After making all these checks, recheck the pressure. The system has passed the initial leak test if the pressure is the same as when the system was first pressurized.

Evacuation

Release the nitrogen to the air and connect a vacuum pump and vacuum gauge. Evacuate the lines and coil until you reach a vacuum level below 500 microns. Shut the valve off at the vacuum pump and wait to see if the vacuum holds. A rise of 100–200 microns is normal. If the vacuum holds, the system has passed its second leak test.

CAUTION

Do not overtighten the installation valves when opening them. Stop turning the valve as soon as you feel increased resistance or see the top of the plug. A lock ring is all that keeps the valve plug in. Overtightening it in the counterclockwise direction will result in the valve plug shooting out at a pressure exceeding 100 psig, and all of the refrigerant will come with it! This would be a physical danger to the technician. There could also be possible damage to the unit.

Adjusting the Charge

It is now time to open the installation valves. Opening the valves allows the refrigerant that is trapped in the condenser to travel throughout the system. If the actual line length is exactly the same as the manufacturer's assumed line length, you are done. If the lines are longer than the assumed length, you will need to add refrigerant. If the lines are shorter, you will need to remove refrigerant. The amount of assumed line length varies from 15 to 30 ft depending upon the manufacturer. Many systems now include information on the unit that lists the assumed line length and gives the required per-foot charge adjustment. Table 65-3 shows the allowance per foot for different liquid line sizes. No allowance is made for the vapor line.

Digital scales are often used for measuring the refrigerant charge. Most are battery operated and can be zeroed in to count up or down. To do this, place the refrigerant cylinder securely on the scale, making sure that it is fairly level and not rocking. This type of scale is very sensitive, and just a little motion can lead to a false reading. Once the cylinder is in place and the system is ready to charge, zero the scale with the scale control panel. As the refrigerant leaves the cylinder, the amount of weight change will be indicated as a negative number.

65.4 WATER-SOURCE SYSTEMS

The first decision to make when planning a water-source installation is where the water will come from. Water-source heat pumps use lots of water, up to 4 gallons per minute per ton. Typically, the water is either supplied from a deep well or is recirculated in a ground loop.

Table 65-3 Charge Allowance Adjustment per Foot for Different Liquid Line Sizes

Liquid Line Size	Ounces per Foot
¼ in OD	0.3 oz
5/15 in OD	0.4 oz
3/8 in OD	0.6 oz
½ in OD	1.2 oz

Types of Water Loops

There are many variations on ground loops, including the following:

- Open
- Horizontal loops
- Vertical loops
- Slinky loops
- Lake loops

Open Loops

The open loop draws water from a well and deposits the water in another well or in nearby surface water. Wells for open loops must be deep and produce an abundance of water. The water quality is also a concern when planning an open-loop system. The amount of minerals in the water, the pH of the water, and the amount of biological growth all play a part in determining how suitable the water is for use in an open-loop system. Table 65-4 shows the acceptable levels of contaminants for open-loop use.

Open loops can be restricted by local government regulations. Regulations in some states do not allow water to be deposited back in the groundwater. In these states, open loops must be disposed of on top of the ground. Methods for handling the water include using holding ponds, draining to nearby running surface water, watering, and irrigation. Some states have regulations that require all discharged water to stay on the property. This makes running the water off on the surface impractical in most instances.

Open-Loop Well Systems

Water-source units are available with either copper or cupronickel water coils. Copper is fine for closed-loop systems, but open systems should use the more corrosion-resistant cupronickel. All piping should be either copper or schedule 80 PVC. Schedule 40 PVC is not acceptable due to the pressure and temperature extremes to which the piping will be exposed.

Shutoff valves should be included for ease of servicing. Boiler drains or other valves should be "teed" into the lines to allow acid flushing of the heat exchanger. Shutoff valves should be positioned to allow flow through the unit from the boiler drains without allowing flow into the piping system. P/T plugs should be used so that pressure drop across the unit and water temperature can be measured.

Ground-Loop Systems

All piping materials should be polyethylene for in-ground sections of the loop. All in-ground joints should be heat-fusion joined.

Galvanized or steel fittings should not be used at any time due to their tendency to corrode. Plastic to metal threaded fittings should also be avoided; due to the potential to leak in earth-coupled applications, a flanged fitting should be used instead. P/T plugs should be used going in and out of the unit so that the pressure drop across the unit can be measured to determine the water flow. Manual shutoff valves should be located in the piping entering and leaving the unit. They should be able to isolate everything else from the loop.

65.5 SOLAR-ASSISTED SYSTEMS

Heat pumps and solar systems may be combined to improve the efficiency of both. Solar collectors lose efficiency as the temperature of the fluid in them increases. The fluid temperature in the solar collectors used for direct heat needs to be over 100°F, making the solar collectors inefficient. However, water-source heat pumps only need moderate-temperature water to operate. To a heat pump, 60°F entering water is hot. Solar systems can be used to raise the water temperature entering a water-source heat pump and increase the efficiency of both the solar collectors and the heat pump. The solar system transfers the heat to the water, and the heat pump transfers the heat from the water to the air.

65.6 START-UP PROCEDURES

A quality installation includes a careful system start-up procedure. It is important to follow a systematic start-up procedure to insure that the heat pump is performing as the manufacturer intended. Start-up should include a visual inspection, a prestart check, and system operational checks.

Visual Inspection

Make a thorough visual inspection of all equipment before turning anything on! Check to see that all packing materials and shipping bolts have been removed. Make sure all fans turn freely. Check to see that all wires have mechanical strain relief, that all equipment is properly grounded, and that all electrical connections are tight and properly made.

Prestart Inspection

Turn the thermostat system switch to OFF and the fan switch to AUTOMATIC so that the unit does not try to start. Turn the disconnect switch ON and check the incoming voltage. Compare the incoming voltage to the minimum and maximum voltage ratings on the data plate. Do not turn the unit on if the voltage is above or below these limits. Some units require that power be applied to the equipment for a period of several hours before turning it on. This is to allow any crankcase heat to drive refrigerant out of the oil in the compressor. If the equipment requires a long power-up period before operating:

1. Turn the power to the unit off.
2. Disconnect and tape the control wire that feeds the thermostat. (This is usually a red wire on the R terminal.)
3. Turn the power back on.
4. Return to complete the job the following day.

Airflow Check

Leave the system switch off and turn the fan switch to ON. The indoor fan should start. Some units will have a delay of several minutes before anything will happen after being

Table 65-4 Water Quality Standards

WATER QUALITY PARAMETER	HEAT EXCHANGER MATERIAL	RECOMMENDED SAFE LIMITS		
Scaling Potential—Primary Measurement				
Above the given limits, scaling is likely to occur. Scaling indexes should be calculated using the limits below.				
pH/Calcium Hardness Method	**All**	**pH < 7.5 and Ca Hardness <100 ppm**		
Index Limits For Probable Scaling Situations (operation outside these limits is not recommended)				
Scaling indexes should be calculated at 150°F for direct use and HWG applications and at 90°F for indirect HX use. A monitoring plan should be implemented.				
Ryznar stability index	All	**6.0–7.5** If >7.5, minimize steel pipe use.		
Langelier saturation index	All	**−0.5 to +0.5** If <−0.5, minimize steel pipe use. Based upon 150 °F HWG and direct well, 85°F indirect well HX.		
Iron Fouling				
Iron Fe^{2+} (ferrous)(bacterial iron potential)	All	**<0.2 ppm (ferrous)** If Fe^{2+} (ferrous) > 0.2 ppm with pH 6–8, O2 <5 ppm check for iron bacteria.		
Iron fouling	All	**<0.5 ppm of oxygen** Above this level deposition will occur.		
Corrosion Prevention				
pH	All	**6–8.5** Minimize steel pipe below 7 and no open tanks with pH <8.		
Hydrogen sulfide (H_2S)	All	**<0.5 ppm:** At H_2S > 0.2 ppm, avoid use of copper and copper nickel piping or HXs. Rotten egg smell appears at 0.5 ppm level. Copper alloy (bronze or brass) cast components are OK to <0.5 ppm.		
Ammonia ion as hydroxide, chloride, nitrate, and sulfate compounds	All	**<0.5 ppm**		
Maximum chloride levels		Maximum allowable Chloride at maximum water temperature.		
		50°F (10°C)	75°F (24°C)	100°F (38°C)
	Copper	<20 ppm	NR	NR
	Cupronickel	<150 ppm	NR	NR
	304 SS	<400 ppm	<250 ppm	<150 ppm
	316 SS	<1,000 ppm	<550 ppm	<375 ppm
	Titanium	>1,000 ppm	>550 ppm	>375 ppm
Erosion and Clogging				
Particulate size and erosion	All	<10 ppm (<1 ppm "sandfree" for reinjection) of particles and a maximum velocity of 6 fps. Filtered for maximum 800 micron size. Any particulate that is not removed can potentially clog components.		

powered up. This delay only happens when power is first applied to the unit. Some units do have a slight delay after every call for fan operation. This delay is usually no longer than 30 seconds. After the fan starts, check the system airflow and compare it to the manufacturer's specifications. Any duct or airflow problems should be corrected before continuing. Note that the airflow in fan-only operation may not be the same as in cooling or heat.

TECH TIP

Units with ECM blowers usually circulate less air in fan-only operation than they do in cooling or heating. Make sure to compare the airflow to the fan-only specifications when checking airflow in the fan-only mode.

Air-Source Operational Check

Move the fan switch to the AUTO position. If the outdoor temperature is above 60°F, set the thermostat to call for cooling; otherwise, set it to call for heat. Check the voltage to the unit after the compressor has started and compare it to the voltage with the unit set to OFF. The operating voltage should be within 2 percent of the original voltage and should still be within the system parameters. Check the amp draw of the compressor and fans after the unit has operated for several minutes and the pressures have stabilized.

Compare the compressor reading to the rated load amps (RLA) on the data plate. Compare the blower readings to the full load amps (FLA) listed for each motor. The readings will usually not be exactly the same as the listing on the data plate, but they should be reasonably close. The actual current draw is affected by the load on the compressor and fan motors, which changes as the operating conditions change.

Check the system operating pressures and charge using whatever method is recommended by the manufacturer. Do not be too quick to adjust the charge. Look for other causes first if the pressures are incorrect! This is especially true with packaged units.

Filling the Water Loop

Newly installed loops normally have some debris in them, such as pipe shavings and dirt. The ground loop should be flushed with water to remove any debris before operating the system. Leaving the pipe shavings and dirt in the loop can damage the pumps and restrict the water flow. Flushing is done before the loop is connected to the system. After flushing, the loop is connected to the system.

The ground loop and heat pump must also be purged of any trapped air. Trapped air can block water flow, corrode metallic components, and cause pump failure. A purge unit is temporarily connected to the system to purge the system and to add antifreeze if necessary. If antifreeze is used, it is added to the water in the purge unit before purging starts. Only low-toxicity antifreeze solutions should be considered. In the event of a loop leak, the system could release toxic chemicals into the ground water if toxic antifreeze is used.

A water flow of 2 ft/sec is required to completely remove any trapped air. After purging the loop, pressurize it to at least 40 psig.

Water-Source Operational Check

The water flow on water-source systems needs to be checked immediately after calling for the unit to operate. The water flow rate can be checked by checking the pressure drop through the unit and comparing it to the manufacturer's specifications (Table 65-5).

Adjust the balancing valve to achieve a water flow within the manufacturer's specifications. The water temperature should be checked to be certain it falls within the allowable operating range (Table 65-6).

All other checks on a water-source system are useless if the water flow and temperature are not right! Once the water flow is correct, you can proceed to check the other system operating characteristics. The installation instructions for water-source heat pumps usually provide very detailed operational characteristics that can be used to check the system operation (Table 65-7).

Table 65-5 Example of Water-Source Heat Pump Pressure Drop and Flow Rate

	Pressure Drop as Compared to Flow Rate and Entering Water Temperature*			
Flow Rate Gallons Per Minute (GPM)	Entering Water Temperature 30 °F	Entering Water Temperature 50 °F	Entering Water Temperature 70 °F	Entering Water Temperature 90 °F
4 GPM	1.5 psi	1.3 psi	1.1 psi	1.0 psi
6 GPM	3.1 psi	2.6 psi	2.3 psi	2.1 psi
7 GPM	4.1 psi	3.4 psi	3.0 psi	2.7 psi
8 GPM	5.1 psi	4.3 psi	3.8 psi	3.4 psi
9 GPM	5.7 psi	5.2 psi	4.8 psi	0.6 psi
Notice that as the flow rate increases, so does the pressure drop	Notice that as the entering water temperature increases the pressure drop decreases			

**Entering Water Temperature When in the Heating Mode.*

Table 65-6 Example of Water-Source Heat Pump Allowable Operating Range

	Ambient Air Limits*			Entering Air Limits*			Entering Water Limits*		
	Rated	**Min**	**Max**	**Rated**	**Min**	**Max**	**Rated**	**Min**	**Max**
HEATING MODE	68 °F	39 °F	85 °F	68 °F	40 °F	80 °F	30 °F to 70 °F	20 °F	90 °F
COOLING MODE	80.6 °F	45 °F	110 °F	DB 80.6 °F and WB 66.2 °F	50 °F	DB 110 °F and WB 83 °F	50 °F to 110 °F	30 °F	120 °F

Normal Water Flow of 1.5 to 3.0 gallons per minute per ton of capacity.
**Temperatures are dry bulb unless noted.*

Table 65-7 Example of Water-Source heat Pump Operating Parameters

		Entering Water Temperature				
		30 °F	**50 °F**	**70 °F**	**90 °F**	**110 °F**
COOLING MODE	**Suction Pressure (psig)**	118–128	128–138	136–146	139–149	143–153
	Discharge Pressure (psig)	146–166	172–192	267–287	354–374	450–470
	Superheat (°F)	25–30	18–23	7–12	6–11	6–11
	Subcooling (°F)	7–12	6–11	5–10	5–10	5–10
	Water Temperature Rise (°F)	12–14	12–14	12–14	11–13	10–12
	Air Temperature Drop (°F)	20–26	20–26	19–25	18–24	17–23
HEATING MODE	**Suction Pressure (psig)**	75–85	106–116	134–144	166–176	
	Discharge Pressure (psig)	275–295	303–323	332–352	372–392	
	Superheat (°F)	6–11	8–12	10–15	15–20	
	Subcooling (°F)	3–8	6–11	8–13	10–15	
	Water Temperature Drop (°F)	4–6	7–9	9–11	11–13	
	Air Temperature Rise (°F)	17–23	23–29	28–35	34–42	

**Entering Water temperature for a Flow Rate of 2.25 Gallons per Minute per Ton of Capacity.*

UNIT 65—SUMMARY

Installing a heat pump is not easy. There are many decisions to be made and many specific details to follow. The complexity is increased by the many types of heat pump systems available. There is simply not one way to install a heat pump, but many depending upon the specific application and system. One thing remains constant: reading and following all the manufacturer's installation instructions is the key to installing any type of heat pump system.

WORK ORDERS

Service Ticket 6501

Customer Complaint: High Heating Bills

Equipment Type: Split-System Heat Pump, Air Handler in Basement

A technician is called to look at a system that was installed 3 months ago. The customer complains that it keeps the house about 2° colder than the thermostat setpoint and costs almost as much to operate as the old electric furnace. The technician notes that even though the indoor unit is operating in the heating mode, the outdoor unit is not operating. The high-pressure switch is tripped. The unit starts after resetting the high-pressure switch but only stays on for about 10 minutes before it trips the pressure switch again. Reasoning that poor airflow can cause high pressures in heating, the technician measures the airflow and finds that it is much lower than the manufacturer's specification. In the installation manual, a drawing shows a minimum distance of 24 in before any takeoffs are to come off the plenum, but the installation has many takeoffs within the first foot. The technician asks the installation crew to return and install the duct according to the instructions.

Service Ticket 6502

Customer Complaint: No Heating

Equipment Type: Air-Source Packaged Heat Pump

A technician is called to check a heat pump that was installed 6 months ago. The customer complains that the unit operated fine all summer, but when she turned the thermostat to heat this fall, nothing happened. The technician notes that when the fan switch is switched to ON, the blower does not come on. The unit has 230 V at the contactor between L1 and L2, but there is no hum and the transformer is cold. The voltage between R and B, the common terminal, is 0. The transformer secondary winding resistance reading indicates an open circuit. The technician carefully inspects all the control wiring and finds that the blue wire in the thermostat bundle is connected to B both at the unit and at the thermostat. However, the common terminal on the thermostat that was installed is terminal C. The B terminal is energized in heating. The technician moves the blue wire to the C terminal on the thermostat and replaces the transformer.

Service Ticket 6503

Customer Complaint: Warm Air in Cooling

Equipment Type: Water-Source Heat Pump Connected to Ground Loop

A technician is called to check a geothermal system that was just installed. The customer complains that the air leaving the registers is not very cold. The technician observes that the blower is running, but the compressor is not running. Further inspection reveals that the high-pressure switch is tripped. The water pump is drawing amps and is very hot, but there is very little pressure drop across the unit, indicating little or no water flow. When one of the purge valves is opened, air and water sputter out. The technician concludes that the loop still has air in it and needs to be purged and refilled.

Service Ticket 6504

Customer Complaint: Smoke Coming out of Heat Pump

Equipment Type: Split-System Air-Source Heat Pump

The customer has called in a panic. He looked out the window and saw billows of smoke coming out of the unit. A few minutes later the unit outside made a loud whooshing sound and got noticeably louder. The customer shut off the unit and called the shop immediately for help. The technician explains that it sounds like the customer just happened to see the unit at the end of the defrost cycle, which causes steam in cold weather. The whooshing noise was the system returning to heating mode. The technician completes a thorough system check to assure the customer that the system is operating correctly.

Service Ticket 6505

Customer Complaint: Cold Air in Heating Cycle

Equipment Type: Water-Source Heat Pump Connected to a Well

A technician is called out to check an open-loop ground-source system that has just been installed. The customer complains that the air in the heating cycle is cold. The technician notices that the isolation valves on the water lines to the unit are closed and the compressor is not running. The high-pressure switch has tripped because of the lack of water flow. The compressor comes back on after opening the water valves and resetting the high-pressure switch.

UNIT 65—REVIEW QUESTIONS

1. List the different configurations of air-source packaged heat pumps.
2. Give a common application for horizontal water-source heat pumps.
3. Explain the difference between an open loop and a closed loop.
4. List the types of loops geothermal heat pumps can use.
5. What are the main considerations when sizing an air-source heat pump?
6. How does oversizing a heat pump affect its efficiency?
7. What are some of the factors that should be considered when choosing a location for outdoor equipment?
8. Explain the following data plate ratings:

   RLA
   FLA
   Minimum voltage
   Maximum voltage
   Minimum circuit ampacity
   Maximum overcurrent protection

9. Plot the balance point for the following system.

   Heating capacity at 47°F = 30,000 BTU/hr
   Heating capacity at 17°F = 18,000 BTU/hr
   House design heat loss at 20°F = 38,000 BTU/hr

10. How does the staging and sequence of operation for a dual-fuel heat pump system differ from a traditional heat pump with electric strip heat?
11. How should fiberboard duct be connected to a heat pump?
12. When are condensate pumps necessary?
13. How is the tubing used in ground-source systems joined?
14. Describe how a solar-assisted heat pump works.
15. List the steps required to fill a ground-source water loop for the first time.
16. Use Table 65-5 to determine the water flow through a model 038 water-source heat pump with a pressure drop across the unit of 2.3 psi if the water is at 70°F.
17. Use Table 65-6 to determine the maximum inlet water temperature for a GC model operating in heating.
18. How is the water flow through a water-source heat pump measured?
19. What steps must be taken when installing split-system units before opening the unit installation valves?
20. What items should be checked when converting an indoor air handler from upflow configuration to horizontal configuration?
21. How does the Carrier Infinity control system differ from traditional low-voltage control systems?
22. What type of antifreeze solutions may be used in ground loops?
23. Explain the difference between the two uses of the B terminal in heat pump controls.
24. What type of heat exchanger coil is best suited to open-loop well systems?

UNIT 66

Troubleshooting Heat Pump Systems

OBJECTIVES

After completing this unit, you will be able to:

1. list the common heat pump operating problems.
2. diagnose the symptoms and the corrective actions for heat pump problems.
3. describe the different types of heat pump defrost systems.
4. list the common problems associated with heat pump check valves.
5. discuss how to troubleshoot a heat pump reversing valve.
6. discuss how to troubleshoot heat pump defrost problems.
7. describe the airflow problems associated with heat pump operation.

66.1 INTRODUCTION

The heat pump is unique in design compared to other air-conditioning apparatus in that it operates year-round, both in summer and winter, cooling in one season and heating in the other. Because of this, the total operating time for a heat pump will be far greater, which invariably means more service. Even so, a properly maintained heat pump will last for many years.

Since heat pumps are a form of air conditioning, the technician's basic understanding of the fundamentals of air conditioning applies. Even with a water-source heat pump, air is still the principal medium used for transferring heat to the space to be conditioned. The correct quantity must be delivered at the correct temperature over the variations of winter and summer to provide comfortable conditions.

66.2 TROUBLESHOOTING GUIDELINES

In troubleshooting heat pumps, the technician must become thoroughly familiar with the controls and operating sequence of the particular design being serviced. Each design has its own selection of components and method of operation. To assist in finding a problem, many manufacturers describe helpful procedures in their installation and service bulletins.

It is important in troubleshooting heat pumps to find out as much as possible about the problem before starting a test procedure. Probably the most important question to ask first is, "Does the unit run or not run?" If the unit does not run, the first step is to check the power supply. Many heat pump failures are caused by thermostat malfunctions, a tripped circuit breaker, or a blown fuse. If the unit does run, find out what components run and what components do not run. If all the components operate, then investigate whether the problem occurs during the heating mode or the cooling mode.

Basically, the service problems with heat pumps fall into three main categories: air circulation problems, refrigeration problems, and defrost problems, each of which is discussed in the following sections.

SERVICE TIP

Manufacturers have produced troubleshooting charts for their equipment. These charts are extremely beneficial in resolving problems, but they do not work as well when a system is mismatched. A mismatched system is one that has one manufacturer's outdoor unit and another manufacturer's indoor unit. In these cases, you will need to follow the troubleshooting skills that you will develop from the following unit material.

66.3 AIR SYSTEM PROBLEMS

The first check is the temperature drop through the evaporator. Assuming that the system airflow is correct, if the temperature drop is higher than specified by the manufacturer, the amount of air through the coil has decreased. Using the same refrigeration capacity on less air produces colder air. When in the cooling cycle, the inside coil must be able to remove the correct amount of sensible and latent heat from the air to produce the desired room conditions. Manufacturers publish the expected temperature drop through the coil (Table 66-1). The higher the indoor wet-bulb (wb) temperature, the lower the expected temperature drop. At 72°F, most systems have a temperature drop of only 10°F. This is because more of the unit's capacity is being used to remove water, so less is available to reduce the air temperature. At lower wet-bulb temperatures, the system is removing very little water and most of the capacity is going toward reducing the air temperature. At a 62°F wb, a temperature drop of 20°F is common. At AHRI design conditions of 67°F wb, systems typically have a temperature drop of approximately 15°F.

Table 66-1 Temperature Drop for Different Operation Conditions

		Temperature Drop in Cooling			
Airflow	**Wet Bulb**	**Indoor Dry Bulb**			
		70°F	75°F	80°F	85°F
956 CFM	59°F	17°F	20°F	22°F	23°F
	63°F	15°F	18°F	21°F	23°F
	67°F	11°F	15°F	19°F	22°F
	71°F	NR	10°F	15°F	19°F
850 CFM	59°F	18°F	21°F	23°F	25°F
	63°F	16°F	19°F	22°F	24°F
	67°F	12°F	16°F	19°F	23°F
	71°F	NR	11°F	15°F	20°F
744 CFM	59°F	18°F	21°F	24°F	25°F
	63°F	16°F	19°F	23°F	25°F
	67°F	12°F	16°F	20°F	23°F
	71°F	NR	11°F	16°F	20°F

If the unit is in the heating cycle, the inside coil now acts as the condenser. The air will increase in temperature as it goes through the coil. The temperature rise across the coil decreases as the outdoor temperature drops. Table 66-2 shows common temperature rise information for different outdoor ambient temperatures based on nominal airflow through the coil of 400 CFM per ton and an indoor temperature of 70°F. A decrease in airflow will cause an increase in the temperature rise of the air across the inside coil. If the airflow reduction is severe enough, the unit will cut out on high head pressure. A repetitive cutting out of the compressor should be investigated and corrected before damage to the compressor results.

The heat pump operates on a year-round basis, and the amount of air through the inside coil is more critical than in a heating or air-conditioning system. Throwaway air filters should be checked every 30 days (Figure 66-1). For high-efficiency air cleaning, it is advisable to use electronic air cleaners with heat pumps because these filters have a low static resistance even when dirty.

Table 66-2 Temperature Rise in Heating

(Airflow = 400 CFM per ton, 70°F return air)			
Outdoor Ambient	**Temperature Rise**	**Outdoor Ambient**	**Temperature Rise**
−10°F	7°F	25°F	18°F
−5°F	9°F	30°F	20°F
0°F	11°F	35°F	23°F
5°	12°F	40°F	25°F
10°F	14°F	45°F	27°F
15°F	15°F	50°F	29°F
17°F	16°F	55°F	31°F
20°F	17°F	60°F	33°F

Figure 66-1 Disposable panel filters should be checked monthly.

SERVICE TIP

Many customers interested in better filtration and cleaner air buy 1-in high-efficiency filters at home improvement stores. Unfortunately, many of these have a very high static pressure drop even when the filter is new. If a customer wants a better filter, suggest installing a 4- or 5-in media filter. These offer the filtration they want without choking off the airflow.

66.4 BLOWER MOTOR PROBLEMS

Inspection of the blower and drive should be done at least once a year. Most heat pump blower motors today use permanently lubricated bearings that do not require lubrication. You should lubricate the blower motor only if it has oiler holes. The motor should be lubricated with no more than ten drops of No. 20 electric motor oil. Excessive lubrication can lead to dirt sticking to the oiling ports, which can contaminate the bearings. Always use detergent-free oil. Automobile oil is discouraged because it has detergent (soap), which coats the outer surface of the sintered bronze bearing and prevents oil passage through the bearing.

SERVICE TIP

Never lubricate blower motors anywhere except at the oiler holes. Trying to lubricate motor parts by just squirting oil on them will likely do more harm than good. If a motor does not have oiler ports, do not try to oil it.

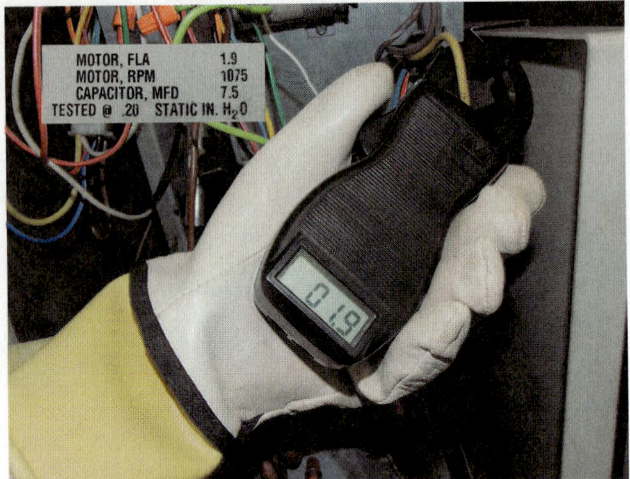

Figure 66-2 Check the blower motor amp draw and compare it to the motor full-load amp (FLA) rating.

Check the amp draw of the blower motor while it is operating and compare it to the full-load amp rating (Figure 66-2). An amp draw significantly higher than full-load amps could indicate a problem. Keep in mind that the amp draw will vary based on the load on the motor. If the motor is suspected to be faulty, a resistance check of the motor windings should be performed. The motor windings should show a measureable resistance: not infinite (OL) and not 0 (Figure 66-3). An infinite reading indicates an open winding; a reading of less than1 ohm indicates a shorted winding. However, if the motor is hot, an infinite reading ,may just indicate that the internal thermal overload is open. Wait for a hot motor to cool before condemning it. If the windings are either open or shorted, the motor will need to be replaced. If the motor needs replacing, it is generally recommended to replace the capacitor as well.

Figure 66-3 The motor windings should have a measurable resistance.

GREEN TIP

Brushless DC motors are available to replace standard PSC motors used in many heat pump blowers. The brushless DC motors are considerably more energy efficient and will save on the system operating cost and energy use.

66.5 MEASURING SYSTEM AIRFLOW

The two most common methods of measuring the airflow is to either read the static pressure across the unit and compare it to the unit specifications or to operate the unit in emergency heat and measure the temperature rise.

Measuring External Static Pressure

Manufacturers specify the unit airflow at different external static pressures. External static pressures are created by the resistance of the ductwork and other components mounted outside of the unit. This can be measured using a magnehelic gauge, an inclined manometer, or a digital manometer. To measure the external static pressure, drill a ¼-in hole in both the return and supply plenum. Place a static pressure probe in each hole. Connect the return probe to the low side of the manometer or magnehelic and the supply probe to the high side of the manometer or magnehelic (Figure 66-4).

SERVICE TIP

External static pressure readings are taken using the filter that ships with the unit. If the low-pressure drop filter that comes with the unit has been replaced with a filter with a higher pressure drop, readings taken at the return plenum will be inaccurate. In this case, drill a hole in the return side of the blower cabinet. Be careful not to drill into components located in the lower half of the cabinet. Take off the unit covers to make sure you know where you are drilling.

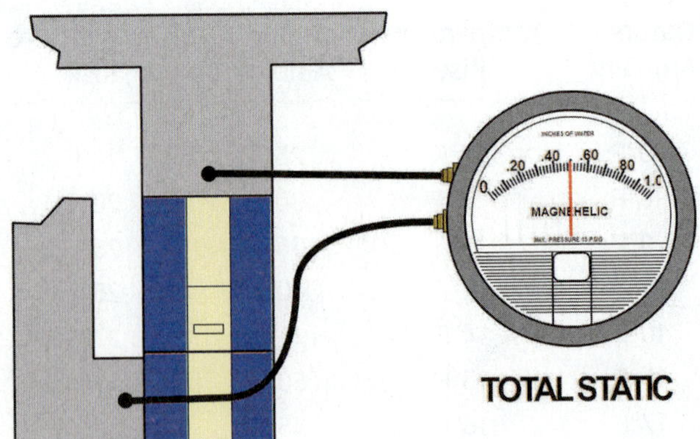

Figure 66-4 Measuring the total external static pressure.

CFM Delivered Against Total External Static Pressure					
Speed	0.1 in wc	0.2 in wc	0.3 in wc	0.4 in wc	0.5 in wc
High	1,155	1,090	1,025	950	895
Medium	940	890	860	815	755
Low	695	665	650	610	550

Figure 66-5 Blower performance table showing airflow against external static pressure.

Operate the blower and read the external static pressure. Compare your reading to the manufacturers chart. For example, in Figure 66-5, a blower operating on low speed with an external static pressure reading of 0.4 in external static would indicate an airflow of 610 CFM. You should have approximately 400 CFM per ton. An airflow of 610 CFM would not be enough for a 2-ton unit. On the other hand, the same reading with the blower operating on medium speed would indicate an airflow of 815 CFM, which would be fine for a 2-ton unit. To make much sense of external static pressure readings, you must have the manufacturer's data for that particular model unit.

Measuring Temperature Rise

Many manufacturers recommend using the temperature-rise method with heat pumps because it only requires measuring voltage, amperage, and air temperature. Operate the unit in emergency heat mode and measure the voltage and amperage of the electric strip heaters (Figure 66-6). Measure the return-air and supply-air temperatures near the unit. Calculate the BTUH using this formula:

$$\text{CFM} = (\text{volts} \times \text{amps} \times 3.41) / (1.1 \times (\text{supply-air temperature} - \text{return-air temperature}))$$

Figure 66-6 Measuring the voltage and amp draw of the strip heaters.

For example, strip heaters pulling 40 A on 230 V with a return-air temperature of 70°F and a supply-air temperature of 100°F would be

$$\text{CFM} = (230\text{ V} \times 40\text{ A} \times 3.41\text{ BTUH/watt}) / (1.1 \times (100°\text{F}–70°\text{F}))$$

$$\text{CFM} = (9{,}200\text{ watts} \times 3.41\text{ BTUH/watt}) / (1.1 \times 30°\text{F})$$

$$\text{CFM} = 31{,}372 / 33 = 950\text{ CFM}$$

SERVICE TIP

When taking the supply-air temperature, make sure that there is not a direct line of sight between the thermometer and the heat strips. If the thermometer can "see" the strips, it will read slightly higher temperature.

Duct System Problems

High external static pressures can be caused by poor ductwork. Unfortunately, poor duct systems are all too common. Unusual restrictions in the duct system, such as closing off unused rooms and placing furniture or carpeting over supply and/or return grilles, will cause coil frosting in the cooling cycle and unit cutoff in the heating cycle. This practice has a greater effect on the heating cycle than on the cooling cycle.

Failure of the duct system also has a greater effect in the heating cycle than in the cooling cycle. With 70°F in the occupied area, the supply-air temperature will only be in the range of 100°F to 105°F when the outdoor temperature is 60°F to 65°F and down to the 80°F range when it is −10°F outside. Any leakage in the duct system will therefore seriously reduce the capacity of the unit to handle the heating load. When this happens, more auxiliary heat will be required and as a result, the operating cost will be higher than normal.

SERVICE TIP

Heat pumps have higher airflows for the same heating capacity than other heating systems, and their air supply temperature is lower than other heating systems. This often results in customers complaining that heat pumps are cold and drafty. To prevent this problem, the supply registers should have deep curved louvers so that the air is directed across the ceiling and not down into the room. By directing the air away from room occupants, they are not as likely to feel a cold draft. Most register manufacturers have a specific designed series of registers for heat pumps.

SERVICE TIP

The first step in checking a suspected refrigeration system problem is to check the airflow. Many systems with airflow problems are misdiagnosed as having refrigeration problems. If the airflow is not correct, you are wasting your time looking for a refrigeration problem.

66.6 REFRIGERATION SYSTEM PROBLEMS

When the temperature change across the inside coil is less than it should be on the cooling or heating cycle, the refrigeration system should be suspected. As in an air-conditioning system, this is generally classified into two categories:

- Refrigerant quantity
- Refrigerant flow rate

If the system has the proper amount of refrigerant and it is flowing at the desired rate, the system will work properly and deliver the rated capacity. Any problem in either category will affect the temperatures and pressures developed in the unit when the correct amount of air is supplied over the inside coil.

66.7 REFRIGERANT QUANTITY PROBLEMS

If the system is low on refrigerant, the measured operating pressures will be below normal, the superheat will be high, and the subcooling will be low. In addition, the compressor will run for longer periods even at times when the cooling demand is not extremely high. The compressor may be running hotter than normal due to a lack of refrigerant flow, which helps to cool it. On a newly installed system, a low refrigerant charge may be the result of improper installation. Although many new units come precharged, additional refrigerant may be required depending on the length and size of refrigerant piping. If this was not taken into account on installation, the system will require additional refrigerant for proper operation.

SERVICE TIP

Some systems have a hot gas sensor control that is wired in series with the compressor contactor and shuts down the unit if excessively high discharge temperatures are reached. If the hot gas sensor switch opens, this could be the result of a low level of refrigerant in the system. Due to the shortage of refrigerant, the compressor motor normally cooled by the suction vapor will heat up. The gas passing through the compressor will therefore be hotter than normal, indicated by a high compressor discharge temperature.

Whenever adjusting the charge for any refrigeration system, the manufacturer's recommendations must be followed (Figure 66-7). Too much refrigerant will increase the system's operating pressures and temperatures, making the compressor work harder. This leads to an increase in the power required, and the compressor will run hotter than normal.

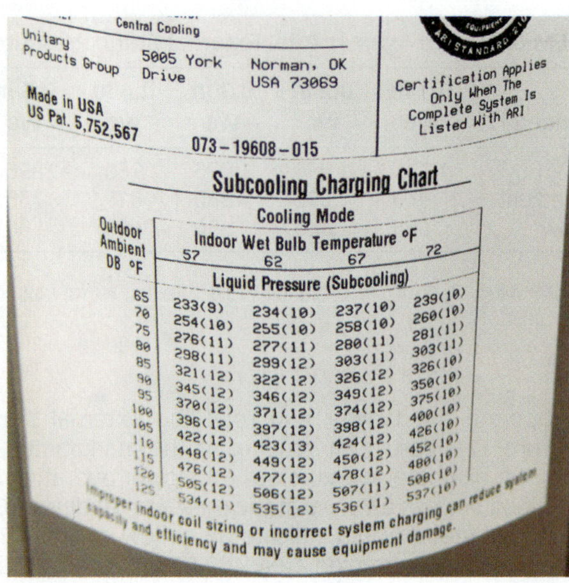

Figure 66-7 Subcooling charging chart.

SERVICE TIP

The high-pressure switch is wired in series with the compressor contactor and shuts down the unit if excessively high discharge pressures are reached. One possible cause of this switch opening is a refrigerant overcharge, which will lead to an increase in the compressor discharge pressure.

Too much refrigerant could also lead to an increased amount of liquid in the evaporator coil and possible floodback to the compressor. If a system is experiencing floodback, the suction line superheat will be low. Many systems have suction-line accumulators located after the evaporator that are sized to hold the full system charge. If the accumulator capacity is reached, the liquid refrigerant will flow from the accumulator into the compressor suction line, possibly causing damage. This is most likely to happen at the completion of the defrost cycle when the system is switching back into the heating mode.

Frost buildup on the indoor coil, as shown in Figure 66-8, is normally caused by either low airflow or too little refrigerant. On an existing unit, a low refrigerant level is generally an indication of a leak. If this is the case, then simply adding additional refrigerant is never recommended. Any remaining refrigerant must be recovered from the system, and the leak must be found and repaired. The system must then be evacuated thoroughly and recharged with the correct amount of refrigerant.

CAUTION

During the heating mode, the vapor line on a heat pump is at the higher discharge pressure of the system. Make sure not to automatically attach the low-pressure gauge to the vapor line. To do so can damage your low-pressure gauge. There is a low-pressure port that is between the reversing valve and the compressor.

Figure 66-8 Ice buildup on coil. *(Courtesy of InspectAPedia.com)*

66.8 COLD WEATHER REFRIGERANT CHECK

As the outdoor temperature drops, the capacity of an air source heat pump drops because the compressor must work against increased pressure difference due to the lower evaporator temperature and pressure. Less refrigerant is circulated because of the increased compression ratio and diminished compressor capacity. The refrigerant that is not being circulated must sit somewhere, usually in the accumulator. For this reason, a charge which is adequate to maintain correct pressures and temperatures at 25°F may not be adequate at 35°F. Just adding a little extra refrigerant is not a good idea either because there is no way to determine how much extra to add. This is why manufacturers say that a heating performance check chart should not be used to charge the system, only to check its operation. If the system is undercharged, the recommendation from many manufacturers is to recover the refrigerant in the unit, evacuate the system, and weigh in a total system charge according to the manufacturer's instructions.

Methods used to check system performance in the heating cycle include

- System pressure charts that correspond to indoor and outdoor temperatures
- Indoor air temperature rise that corresponds to outdoor temperature

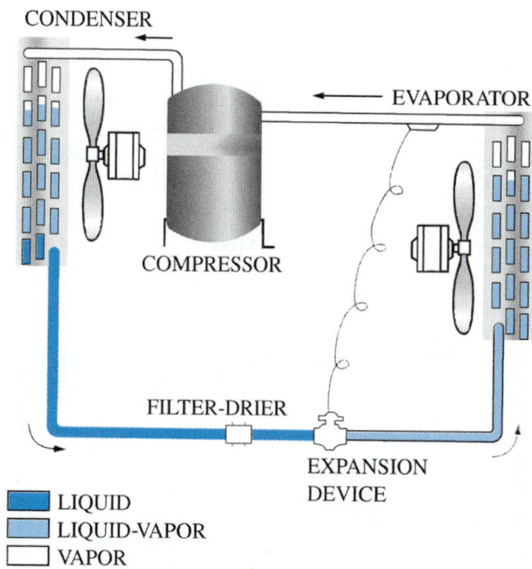

Figure 66-9 Schematic diagram of a conventional air-conditioning/refrigeration system.

- Discharge line superheat (usually 50°F to 60°F)
- Discharge line approach (usually 100°F to 110°F above ambient)

66.9 REFRIGERANT FLOW PROBLEMS

To troubleshoot refrigerant flow problems, a detailed understanding of the system configuration is essential. Figure 66-9 shows a conventional air-conditioning system. This system shows the refrigerant flow from the discharge of the compressor to the outside coil, in this case the condenser. The refrigerant condenses and flows from the outside coil through the liquid line, through the filter drier and expansion device, and into the inside coil, in this case the evaporator. The refrigerant expands, picking up heat in the evaporator, and then flows as a vapor to the compressor.

In Figure 66-10, the same action takes place, except that additional devices to accommodate the reverse flow have been added. A reversing valve is located between the suction and discharge lines to enable the system to reverse the flow of refrigerant. In the cooling mode, the position of the reversing valve directs the flow the same way as in the conventional air-conditioning system. However, unlike the conventional system, an expansion device has been added to the outside coil, allowing it to operate as an evaporator during the heating cycle. In addition, two check valves are added to allow for bypassed flow around each expansion device dependent on the operating mode.

Figure 66-11 shows the flow of refrigerant through the system in the heating mode. The reversing valve directs the hot refrigerant vapor to the inside coil, in this case the condenser. Heat is transferred to the air supplying the conditioned space. The hot refrigerant vapor cools, condenses, and flows out of the bottom of the condenser. With the flow reversed, check valve #2 opens and allows the liquid refrigerant to flow around the cooling-cycle expansion device to eliminate any pressure loss. Since check valve #1 will not allow for flow in this direction, the refrigerant passes

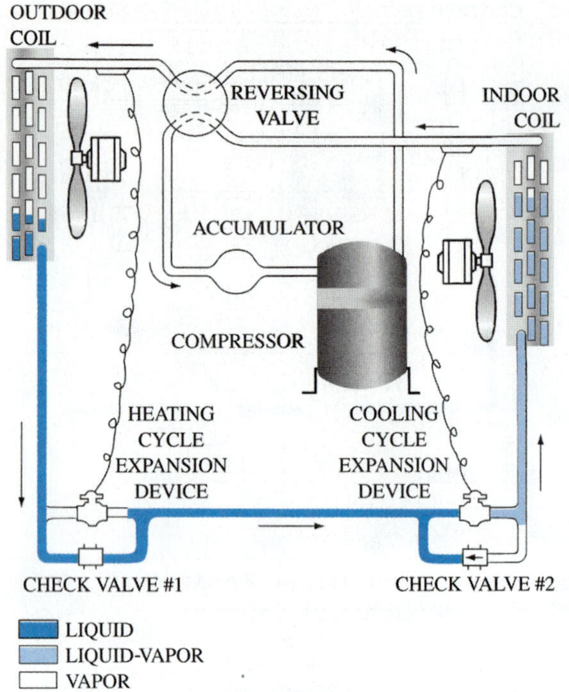

Figure 66-10 Schematic diagram of a heat pump system used for cooling, showing the position of the reversing valve.

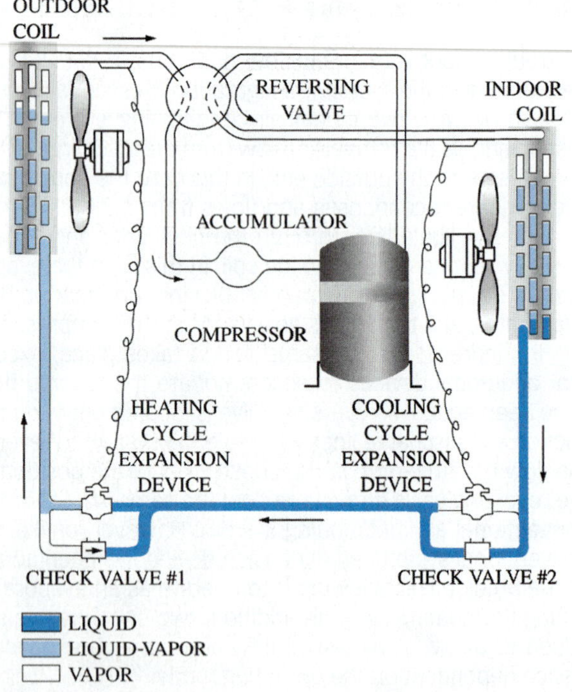

Figure 66-11 Schematic diagram of a heat pump system used for heating, showing the position of the reversing valve.

through the heating-cycle expansion device. Entering the outdoor coil, in this case the evaporator, the liquid refrigerant, at a lower pressure and boiling point, vaporizes as it picks up heat from the outside air. The vaporized refrigerant flows through the reversing valve and the accumulator to the compressor.

TECH TIP

The only reversing of the refrigerant flow is from the reversing valve through the coils, pressure-reducing devices, and check valves. The refrigerant vapor always flows from the reversing valve, accumulator, and compressor back to the reversing valve. Although the flow of refrigerant through the system reverses, the flow through the compressor never changes.

66.10 CHECK VALVES

Check valves allow for flow in one direction only. An operational problem encountered with these valves is sticking in either the open or closed position. With this type of malfunction, the flow in only one direction would be affected. The system will therefore work correctly in one cycle but not the other, dependent on the valve's location in the system. This type of problem is somewhat easy to identify, because if the valve is doing what it is supposed to, the system will operate properly. If not, the problem will show up.

Common problems include the following:

- *The check valve installed to bypass the cooling-cycle expansion device on the indoor coil sticks in the open position.* The system will operate properly in the heating mode, as the check valve is normally open in this cycle. In the cooling mode, however, the cooling-cycle expansion device is bypassed, allowing for uncontrolled refrigerant flow into the indoor coil. The refrigerant will flood through the coil, the suction pressure will be high, the discharge pressure will be low, and the suction-line accumulator will fill with liquid refrigerant. In this case, the possibility of liquid flooding back to the compressor exists.
- *The check valve installed to bypass the cooling-cycle expansion device on the indoor coil sticks in the closed position.* The system will operate properly in the cooling mode, as the check valve is normally closed in this cycle. In the heating mode, however, the suction pressure will be much lower than normal. This is because the normal flow, rather than being bypassed, is restricted by the expansion device. This restriction to flow allows for a reduced amount of refrigerant to enter the outdoor coil and return in the suction line to the compressor. At the same time, the discharge pressure will be low due to the reduced amount of vapor available for the compressor to pump. The symptoms of this condition are similar to those of an undercharge, but the difference is that the system will operate normally in the cooling mode. This can be easily checked by switching to the opposite cycle.
- *The check valve installed to bypass the heating-cycle expansion device on the outdoor coil sticks in the open position.* The system will operate properly in the cooling mode, as the check valve is normally open in this cycle. In the heating mode, however, the heating-cycle expansion device is bypassed, allowing for uncontrolled refrigerant flow into the outdoor coil.

The result is similar to what happens when the check valve installed to bypass the cooling-cycle expansion device sticks open.

- *The check valve installed to bypass the heating-cycle expansion device on the outdoor coil sticks in the closed position.* The system will operate properly in the heating mode, as the check valve is normally closed in this cycle. In the cooling mode, however, the suction pressure will be much lower than normal. This is because the normal flow, rather than being bypassed, is restricted by the expansion device. The result is similar to what happens when the check valve installed to bypass the cooling-cycle expansion device sticks closed.

Usually, a check valve can be released from the stuck-open position by means of a magnet placed against the outlet end of the valve. Moving the magnet toward the center will force the ball or flapper to move to the seat. If in the stuck-closed position, place the magnet at the middle of the valve and move it to the outlet end. If this does not work, replace the valve. To reduce the possibility of future problems, use a ball-type check.

66.11 REVERSING VALVES

Figure 66-12 shows the exterior view of a reversing valve. The single tube connection is always connected to the compressor discharge. The bottom middle connection is always connected to the compressor suction connection. When an accumulator is in the circuit, this connection is to the accumulator inlet. The accumulator outlet is then connected to the compressor suction. This puts the accumulator upstream from the compressor and provides protection in either the heating or cooling mode.

The right and left connections are to the vapor line connection of the indoor and outdoor coils. Which connection goes where depends on whether the operating solenoid coil is energized for heating or energized for cooling. If the solenoid coil is energized in cooling, a bad solenoid coil will cause the unit to operate in the heating mode anytime the compressor runs. If the solenoid coil is energized in heating, a bad solenoid coil will cause the unit to operate in the cooling mode anytime the compressor runs.

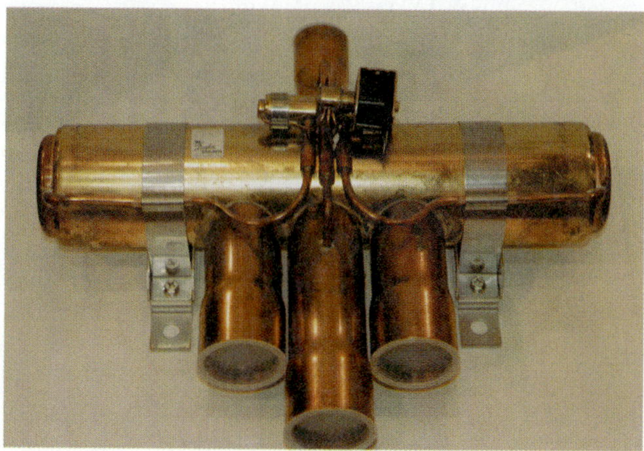

Figure 66-12 Reversing valve.

Before checking the position of the valve operation, the electrical control system should be checked to determine the operating requirements of the valve. Valves that operate in cooling are usually energized by the O terminal on the thermostat, while valves that are energized in heating are usually energized by the B terminal on the thermostat.

CAUTION

Do not energize the reversing-valve solenoid coil when it is not installed on the reversing valve. The solenoid coil can quickly overheat and burn out.

Problems in reversing valves are either electrical or mechanical. Electrical problems are confined to the solenoid coil on the pilot valve. When the solenoid coil is energized and nothing happens, test the coil as follows:

- Feel the solenoid coil. If it feels hot, the coil is energized.
- Make sure that voltage is applied to the solenoid coil. Some units have the coil in the 240 V portion of the system, while others use it in the 24 V portion. Check the wiring diagram for coil voltage before applying the leads of the voltmeter. Also, start the test with the voltmeter set to the higher range. This helps reduce meter damage.
- With voltage applied to the solenoid coil, remove the coil holding nut and attempt to pull the coil off the pilot valve plunger casing. If you feel resistance to removing the coil, the coil is active (Figure 66-13). If no pull is felt, shut off the power, remove the coil heads, and check for continuity with the ohmmeter (Figure 66-14). If open, replace the coil. If a circuit exists, check the leads and the connections for continuity.

Figure 66-13 Briefly pull the solenoid coil part way off while energized to check the pilot valve operation.

Figure 66-14 Checking the resistance of the reversing-valve solenoid coil.

- If the solenoid coil is active, when removing the coil, a click should be heard when the pilot plunger returns to the normally closed position. If no clicks are heard, the pilot valve is stuck. The only repair is to replace the reversing valve.

When the pilot valve checks out satisfactorily and the main valve does not shift, make sure that suction and discharge show more than a 100 psig difference. If the valve is in the cooling position, block off the air to the condenser with plastic on the inlet face of the coil. Allow the unit to operate until the condensing saturation temperature rises to 130°F. With the unit operating, cycle the valve on and off several times. This will usually free the main valve to operate again. If no results are achieved, change the valve.

SERVICE TIP

If the heat pump is operated for any period of time in the heating mode in warm weather, the compressor can become overloaded. If it is necessary to run the heat pump in the heating mode for testing purposes when the weather is warmer than 60°F, you should cover the coils to restrict the airflow. This will prevent the heat pump from picking up excessive amounts of heat and overloading the compressor while the system is being tested.

When changing a reversing valve, be sure to read the installation instructions supplied with the valve. The first step is always to recover the entire system charge. Replace the valve with one of a comparable size. Never use a valve with a smaller body or line ports. Always position the valve so that the main piston is in a horizontal position and the pilot valve is higher than the main valve. This is to keep oil from gathering in the pilot valve and affecting its operation.

The valve body must always be protected from heat when installing by wrapping the body with some type of thermoplastic material or wrapping with a water-saturated cloth. The maximum temperature the valve body will tolerate is 250°F.

SERVICE TIP

If it is necessary to replace the heat pump reversing valve, the new valve must be protected from overheating during brazing. The valve contains plastic gaskets and seals that are easily damaged from excessive heat. Wrapping a wet rag around the valve body will help prevent overheating. In addition, an oxyacetylene torch with a large enough flame to heat the fitting quickly should be used.

66.12 DEFROST SYSTEM PROBLEMS

Defrosting is required when the unit is running in the heating mode. In this case, the outdoor coil is the evaporator and is absorbing heat from the outside air. Depending on the outside air temperature and humidity conditions, the cold outdoor coil will develop frost. When the defrost cycle is initiated, the system temporarily reverses to send hot gas to the outdoor coil, which melts any frost that has formed. However, in this defrost mode, cold rather than hot air is delivered to the conditioned space. To reduce this effect, the fan for the outdoor coil shuts off and the auxiliary heat comes on for the duration of the defrost cycle.

If the defrost cycle is not initiated frequently enough, the outdoor coil will develop more and more frost until the frost turns to sheets of ice on the coil (Figure 66-15). If the defrost cycle is initiated too often, the efficiency of the system is decreased and energy is wasted, increasing the cost of operation. To diagnose problems in the defrost system, the type of defrost must first be determined. The four most predominant defrost systems are covered here. On those systems used by individual manufacturers, the manufacturer should be contacted for service information.

Figure 66-15 Heavy frost buildup starting to form into an ice sheet.

66.13 TEMPERATURE-DIFFERENTIAL DEFROST SYSTEM

This defrost system measures the temperature difference between the outdoor ambient temperature and the coil temperature. Although the coil temperature will drop as the outdoor temperature drops, the temperature difference between the outdoor air and the coil remains about the same under normal operation. However, when enough frost forms on the coil to restrict airflow across it, the coil temperature will drop even if the outdoor temperature remains the same, increasing the temperature differential between the outdoor air and the outdoor coil. Temperature differential defrost controls initiate a defrost cycle whenever the temperature differential between the outdoor temperature and the coil temperature increases beyond the normal operating differential.

The Unit Will Not Go into Defrost

This is the most common complaint on this type of system. The control is operated by the difference between the temperature of the air entering the outdoor coil (the evaporator) and the coil temperature itself. The predominant reason that the unit will not go into defrost is that the coil will not reach a low enough temperature to provide an increase in the control differential to activate the defrost cycle. The major reason for this is that the discharge pressure is too high, which forces the suction pressure and coil boiling point up. Only a 2°F rise in boiling point can result in the system needing complete coverage of the coil with frost and ice before the system initiates the defrost cycle.

The technician should never attempt to adjust the control, which is not possible in the field. Before replacing the control, the technician should determine the cause of the problem. If the discharge pressure is too high, this can very often be attributed to an airflow problem. The duct system may be inadequate, the indoor coil may be dirty or plugged, or the blower may not be operating properly. With a reduced airflow through the indoor coil, the discharge pressure will increase. The correct solution to this is to restore the correct flow of air over the indoor coil to design conditions. This will lower the head pressure and the suction pressure.

The Unit Goes into Defrost with Very Little Frost on the Bottom of the Outdoor Coil

The location of the outdoor coil temperature-sensing thermistor is critical. The thermistor should be located in the original location selected by the manufacturer. This can vary quite a bit from one manufacturer to another. If the coil temperature sensing thermistor has come off, try to determine where it was originally located.

The Unit Does Not Defrost the Entire Coil

This can result in an ice ring building up around the bottom of the coil. The ice buildup indicates that the coil temperature bulb is located too high on the coil. The coil has to reach a preset temperature to terminate the defrost cycle. Usually, this temperature is 50–60°F. Most demand boards have a programmed maximum defrost time. They will terminate the defrost cycle even if the defrost is not complete. Failure to complete the defrost can be caused by not defrosting frequently enough, a slight refrigerant undercharge, the outdoor fan continuing to run in defrost, or incorrect coil thermistor sensor placement. A slight undercharge can cause more frequent icing. It also reduces the amount of heat available during defrost. If the fan continues to operate and blow cold air across the coil, there will not be enough heat to defrost it.

66.14 PRESSURE-TEMPERATURE DEFROST SYSTEM

In this system, the airflow pressure differential across the outdoor coil initiates the defrost cycle and the temperature of the liquid leaving the coil terminates it. When the frost buildup causes enough airflow resistance to close the differential pressure switch, the defrost cycle is initiated. The cycle is reversed and hot gas flows into the outdoor coil, which is now acting as a condenser. The defrost cycle will end as the outdoor coil begins heating up and the liquid leaving reaches 55°F.

The defrost thermostat also controls the possibility of a defrost cycle. A thermostat fastened to the coil is exposed to the expanding liquid refrigerant when the unit is operating in the heating cycle. If the evaporator coil is operating with an entering liquid of 26°F or higher, there is little chance that frost will form on the coil. If this temperature is below 26°F, the thermostat closes and completes the circuit through the defrost relay to the differential pressure switch.

The Unit Will Not Go into Defrost

Two possible causes can prevent the unit from going into defrost. The most probable cause is that the termination thermostat has come loose from the coil outlet pipe. The thermostat cannot reach 26°F or below to close and allow the defrost circuit to initiate. The other probable cause is plugged tubes to the differential pressure switch. Insects sometimes plug these pipes. Check to see that the tubes are located correctly, connected to the differential air switch, clear of insects and debris, and not cracked or broken.

The Unit Goes into Defrost with Very Little Frost on the Bottom of the Coil

Because the pressure drop of the air through the coil is the determining factor for initiating the defrost cycle, both the cleanliness of the coil and the frost buildup make up this pressure drop. Therefore, the more dirt buildup on the coil, the less frost has to form to reach the required pressure drop to initiate the defrost cycle. When the unit starts defrosting with little frost buildup, clean the coil.

The Unit Does Not Defrost the Entire Coil

This can result in an ice ring forming at the bottom of the coil. This is an indication that the defrost cycle is interrupted before the defrost cycle is completed. Many systems have an automatic defrost cycle termination anywhere from 5 to 12 minutes, which can be set dependent on the outside weather conditions and humidity. This ensures that no matter what may affect any of the other controls, the defrost cycle will not continue indefinitely. If the defrost cycle is ending too quickly, then the automatic termination needs to be adjusted to extend the period of the defrost cycle.

Another reason for incomplete defrost can be from a combination of the indoor thermostat setting and the auxiliary heat. In the defrost cycle, auxiliary heat comes on to offset the effect of supplying cold air to the conditioned space. If the auxiliary heat is great enough to warm the occupied area, the room thermostat may open before the defrost cycle is completed and the unit will shut down. The incomplete defrost cycle will leave an ice buildup at the bottom of the coil until the unit restarts on the next heating demand. In most cases, the unit will usually restart in the defrost cycle and then continue until defrosting is completed.

The alternate thawing and freezing of the water held on the coil by the ice coating can lead to a collapse of the coil tubes and loss of refrigerant. To correct this problem, reduce the amount of auxiliary heat used in the defrost cycle to an amount not more than the sensible capacity of the unit on the cooling cycle.

66.15 TIME-TEMPERATURE DEFROST SYSTEM

This system uses a board with timing circuit in combination with a temperature-sensing device called a defrost thermostat (Figure 66-16). The timer counts down the defrost starting time whenever the unit is running in heat. When the defrost time is reached, the timer will always try to initiate the defrost cycle. However, if the defrost temperature is not low enough, generally below 26°F, the defrost will not begin. The timer will reset, and a new countdown starts until the next defrost begins. The time periods are adjustable by moving a jumper or switch on the board.

The Unit Will Not Go into Defrost

The most common cause of this problem is a loose defrost thermostat. A poor contact between the thermostat and the coil tube prevents the thermostat temperature from dropping below 26°F. If poor contact is not the problem, then the thermostat should be checked to determine if it closes at a temperature below 26°F. Immersion in an ice and saltwater bath using an immersion thermometer will accomplish this.

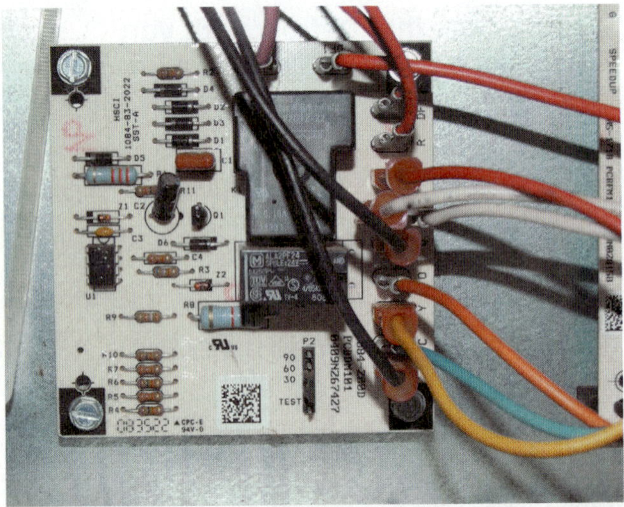

Figure 66-16 Standard time-temperature defrost control.

The Unit Goes into Defrost with Very Little Frost on the Bottom of the Coil

The timer may be operating on too short a cycle time. Most defrost boards have several possible time settings between 30 and 90 minutes. Change the timer to a higher setting if the unit defrosts too frequently.

The Unit Does Not Defrost the Entire Coil

This can result in an ice ring forming at the bottom of the coil. This is an indication that the defrost cycle is interrupted before the defrost cycle is completed. As noted earlier in Section 66-13, one correction to this problem is to adjust the automatic termination to extend the period of the defrost cycle, and another is to reduce the amount of auxiliary heat used in the defrost cycle to an amount not more than the sensible capacity of the unit on the cooling cycle. Also, check to see that the outdoor fan shuts off during defrost. Finally, a slight undercharge can cause this problem.

SERVICE TIP

Time defrost systems can be set for various defrost intervals. These are typically 30-, 60-, and 90-minute intervals. The higher the relative humidity, the more frequently the system should defrost to reduce the chance of excessive frost buildup. However, the more frequent the defrost time, the less efficiently the system will operate. Therefore, the defrost cycle should be adjusted to go on and off no more than what is required to eliminate any frost buildup. It is recommended that you check with your local supplier for their recommendation on defrost time settings for your geographical area.

66.16 PRESSURE-TIME-TEMPERATURE DEFROST SYSTEM

This system uses a differential airflow pressure switch to measure the amount of frost on the outdoor coil. When the pressure drop through the coil reaches a preset amount, the switch closes the timer circuit. If the defrost thermostat is below 26°F, the timer will operate. This timer usually requires 5 minutes of pressure switch closure time to initiate the defrost cycle by energizing the defrost relay. When the termination thermostat reaches 55°F, the power to the defrost relay is interrupted and the unit changes over to the heating cycle.

The Unit Will Not Go into Defrost

The most common cause of this is a loose or poorly connected termination bulb of the defrost control. It is very important that the bulb be securely fastened with heat-transfer compound on the joint, insulated, and sealed from moisture. Ice can build up at this location and loosen the bulb fastening. The second cause can be failure of the pressure switch due to dirt and/or insects in the pressure switch tube connections. There is always the possibility of failure of the timer control, but this is remote. Check the previous two items before replacing any parts.

The Unit Goes into Defrost with Very Little Frost on the Bottom of the Coil

The major cause of this action is a dirty outdoor coil. Its frost-free resistance is so high that it takes very little increase in pressure drop to start the defrost cycle. A thorough coil cleaning is in order.

The Unit Does Not Defrost the Entire Coil

This is an indication that the defrost cycle is interrupted before the cycle is completed. As noted earlier in Section 66-13, one way to correct this problem is to adjust the automatic termination to extend the period of the defrost cycle, and another is to reduce the amount of auxiliary heat used in the defrost cycle to an amount not more than the sensible capacity of the unit on the cooling cycle.

66.17 FORCING A DEFROST CYCLE

Most defrost boards have test jumpers to allow technicians to check the defrost board operation. The first step is to make sure the defrost thermostat is cold enough to allow a defrost. If the coil is not frosted, disconnect the outdoor fan and operate the unit in heating until the area where the defrost thermostat is located is frosted. Next, find the test pins and short them together (Figure 66-17). This speeds up time. Normally, minutes become seconds, so a 60-minute timing period becomes a 60-second timing period. The unit should go into defrost after a short delay. Typically, the unit will remain in defrost until the defrost thermostat senses that the coil is 50–70°F. If the system will not go into defrost, you need to try fooling the board into thinking the coil is cold. With time-temperature boards that use defrost thermostats, the defrost thermostat connections can be jumped out on the board. For boards that use thermistors, disconnecting the coil thermistor simulates a very cold temperature because most of the thermistors used with defrost boards are negative-temperature coefficient. A high resistance means a low temperature. If the board still will not go into defrost, the board needs to be replaced. It is important to know which type you have. Shorting out the thermistor could damage the board. Defrost thermostats are larger and have larger-gauge wires. Thermistors are small with much smaller wires. Figure 66-18 shows a defrost thermostat on the top and a defrost thermistor on the bottom.

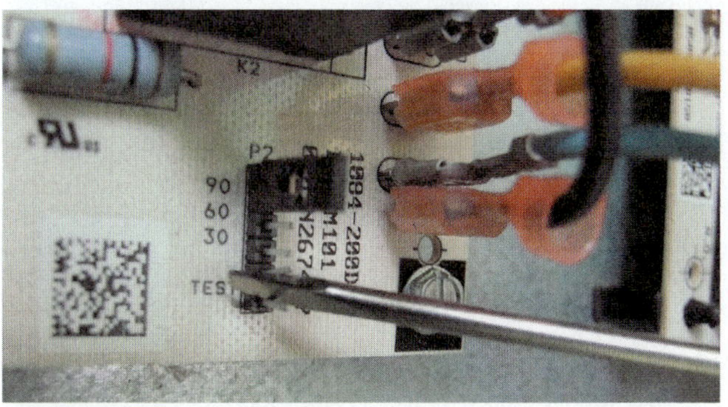

Figure 66-17 Shorting across the test pins speeds up the timing functions of the board.

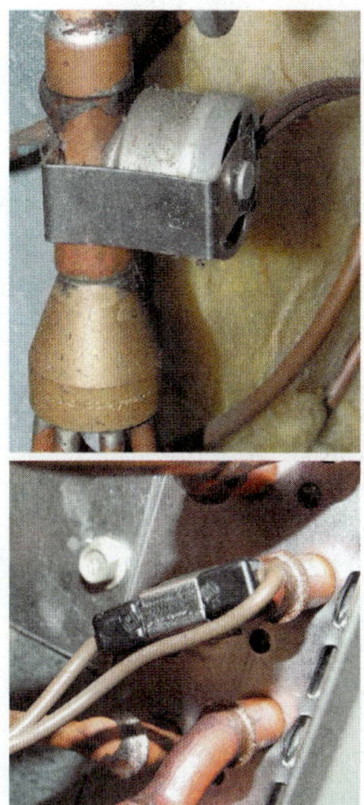

Figure 66-18 Defrost thermostat and defrost thermistor.

UNIT 66—SUMMARY

Service problems with heat pumps fall into three main categories: air circulation problems, refrigeration problems, and defrost problems. Air circulation problems for heat pumps are similar to those encountered with general air-conditioning systems. However, one significant difference is that heat pumps have higher airflow per ton than other heating systems and their air supply temperature is lower than other heating systems.

Refrigeration problems are generally categorized as refrigerant quantity or refrigerant flow problems. Heat pumps have a number of different components compared to conventional refrigeration systems, such as reversing valves, check valves, and bidirectional expansion devices. It is important for the technician to understand why these components are required and how they are installed and operated.

The heat pump defrost system is essential to maintain proper operation during the heating mode. During this cycle, the outdoor coil is subject to icing dependent on the outside air temperature and humidity. The defrost cycle needs to be initiated early enough to eliminate the formation of ice but not so often as to reduce the unit's efficiency.

WORK ORDERS

Service Ticket 6601

Customer Complaint: Room Too Drafty

Equipment Type: Air-Source Packaged Heat Pump

The technician answers a service call for a customer complaining that the home is too drafty and that cold air is blowing from the registers all of the time. The first thing that the technician checks is the thermostat setting. There are three settings available: ON, OFF, and AUTO. The blower is set to the ON position, which allows it to run continuously whether the heat pump is operating or not.

The technician asks the customer about this, and the customer explains that the thermostat was originally set for the AUTO position, but because it always seemed so drafty she changed it to the ON position hoping to blow more air into the room.

The technician tells her that this would not increase the heat and that the unit is designed to work best with the blower in the AUTO position. If it is not supplying enough heat, then there could be other reasons. The technician reviews the original specifications for the unit, and it seems to be sized properly for the heating load. The air supply registers are also properly located so as not to induce unnecessary drafts.

However, it has been unusually cold in the area during the past month, and this type of a system would require additional auxiliary heat as the temperature has been dropping below freezing. The technician finds that one of the electrical strips for the auxiliary heat has failed and so that is replaced. With the blower on AUTO and the electric auxiliary heat operational, the house is no longer drafty for the customer.

Service Ticket 6602

Customer Complaint: High Energy Cost

Equipment Type: Split-System Air-Source Heat Pump, Air Handler in Basement

The customer tells the technician that his electricity costs have been fairly stable since installing the heat pump 5 years ago. However, most recently, the electricity cost has increased significantly and he is wondering if this is because the system is getting older and worn out.

The technician tells him that that should not be the case as long as regular maintenance is conducted, which the records indicate. He does mention, however, that some landscaping was done just a month or so back in the vicinity of the outdoor coil. Inspecting the manual for the unit, the technician finds it has a pressure-time-temperature defrost system. This is a fairly elaborate system, but it is also generally very efficient in regulating the timing for the defrost cycle.

Removing the outside protective grille from the outdoor coil, the technician finds it to be extremely dirty. It is covered with dust, leaves, and grass. Evidently, the landscaping work had stirred up this debris, which was drawn into the outdoor coil. The pressure drop of the air through the coil is the determining factor for initiating the defrost cycle for this type of unit. Due to the dirt buildup on the coil, less frost has to form to reach the required pressure drop to initiate the defrost cycle. The unit is defrosting far too often and wasting energy. The technician cleans the outdoor coil and tells the customer that the electrical bill should return to normal.

Service Ticket 6603

Customer Complaint: Not Enough Heat

Equipment Type: Split System Air Source Heat Pump

The technician answers a service call for a customer complaining that the home is too cold and not getting enough heat. This has only been happening within the past few days, and before that time the house was warm enough. The technician checks the thermostat settings, and they seem to be OK, but the room temperature of 65°F is considerably lower than the setpoint of 70°F. The blower for the air handler in the basement appears to be operating normally, but the temperature of the indoor coil is lower than expected and the auxiliary electric strip heaters are on.

The technician goes outside and removes the protective grille from the outdoor coil and finds it to be covered in ice. This most likely indicates a problem with the defrost cycle. The defrost system for this unit is the time-temperature type. The problem could be a loose or faulty defrost thermostat. A poor contact between the thermostat and the coil tube prevents the thermostat temperature from dropping below 26°F. However, the technician cannot check this because the entire coil is covered in ice, including the thermostat.

The technician decides to initiate the defrost cycle manually while at the same time testing the defrost timer control board. After turning off the power, a jumper is placed across the outdoor thermostat connections DFT and R (see Figure 66-17). Then turning the power back on, the heat pump immediately shifts into the defrost mode. This indicates that the defrost timer is working properly. After enough of the ice is clear, a visual inspection seems to indicate that the outdoor thermostat sensor is mounted correctly with good contact. Therefore, suspecting a faulty thermostat, the technician replaces it.

UNIT 66—REVIEW QUESTIONS

1. When a heat pump system is in the heating mode and there is an increase in the temperature rise, what does this indicate?
2. How often should throwaway filters be checked?
3. Why should automotive oil never be used to lubricate blower motors?

4. What are the three major differences between an air-conditioning and a heat pump system?
5. When changing a reversing valve, in what position should the main piston and also the pilot valve sit?
6. Why must the pilot valve be positioned as described in question 5?
7. When installing a new valve, it should be protected from getting too hot. What is the maximum temperature the valve body will tolerate?
8. What is the primary refrigeration problem in a heat pump system?
9. List four common problems with check valves on a heat pump system.
10. What are the indications a check valve has stuck closed on an outside coil?
11. Name the four most predominant defrost systems.
12. A temperature-differential defrost system will not go into defrost. This is the most common complaint on this control. Describe the correct solution to this problem.
13. A pressure-time-temperature defrost system goes into defrost with very little frost on the bottom of the coil. Describe the major cause of and solution to this problem.
14. If the unit does not run, what is the first thing to check?
15. Basically, the service problems with heat pumps fall into three main categories. What are they?
16. Why might customers complain about their heat pump systems being cold and drafty?
17. If the hot gas sensor located in the compressor discharge line shuts down the compressor, this may indicate what?
18. If the high-pressure switch located in the compressor discharge line shuts down the compressor, this may indicate what?
19. How should the accumulator for a heat pump system be sized, and why?
20. Why is it not recommended to energize the solenoid coil when it is not installed on the reversing valve?

UNIT 67

Basic Building Construction

OBJECTIVES

After completing this unit, you will be able to:

1. recognize the different materials used in basic construction and select the proper procedures and methods of construction for each.
2. describe the different insulation R values of common building materials.
3. relate the importance of fire dampers in construction.
4. incorporate the different methods of construction necessary for a successful residential or commercial project.
5. forecast what part of an HVACR system needs to be installed during each phase of construction.
6. explain the importance of LEED certification.

67.1 INTRODUCTION

The heating and air-conditioning system becomes an integral part of the building. System components, ductwork, and piping must all must be supported by the structural members of the building. Wiring, ductwork, and piping must pass through floors, walls, and ceilings. An understanding of common building construction is necessary to ensure safe installations. Improperly supported equipment can pose a safety hazard. Cutting the wrong structural components of a building can compromise the integrity of the entire structure. Air-conditioning technicians must have an understanding of basic building construction so that they can intelligently discuss aspects of the building construction that will affect the installation of the HVACR system with the general contractor.

67.2 BASIC RESIDENTIAL CONSTRUCTION

There are four major parts of a residential building:

- Foundation
- Floor
- Walls
- Roof

Foundation Types

The foundation supports the rest of the building. As seen in Figure 67-1, the foundation of a residential building is generally one of three construction types:

- Slab on grade
- Crawlspace
- Basement

All foundations, regardless of type, rest on a footing. The footing supports the weight of the building, spreading it over the entire area of the footing. The footing dimensions vary depending upon the type of soil and the local frost line. Figure 67-2 shows the cross-section of a typical footing. Most footings are poured concrete reinforced with steel reinforcing rod.

67.3 SLAB ON GRADE

Concrete slab-on-grade floors are poured on the ground with reinforcing. The reinforcing may be post tension cables, reinforcement rods, or wire mesh. These materials are used to help prevent cracking of the slab over time. Typical residential slabs are from 4 to 6 in thick and are poured continuously, incorporating the footing and crossbeams where they are used (Figure 67-3). The concrete is deeper at the footing area, and the footing reinforcing material is heavier than the reinforcing material used in the rest of the slab. Figure 67-4 shows a slab that is ready to pour.

Running Lines Under a Slab

Refrigerant lines, plumbing, and wiring must sometimes be located under a concrete slab. If refrigerant lines or condensate plumbing are to be placed below grade, they must be placed before the slab is poured. They should be installed inside a 3–4-in chase or conduit pipe. The conduit is buried in sand to protect it from sharp objects such as stones (Figure 67-5).

CAUTION

Copper piping, including refrigerant lines, reacts to concrete. Therefore, copper tubing should not be buried directly in the concrete.

When installing refrigerant lines and condensate drain lines before a slab is poured, make certain that your dimensions are taken from the outside of the building to locate these pipes. It is often the case that dimensions are given

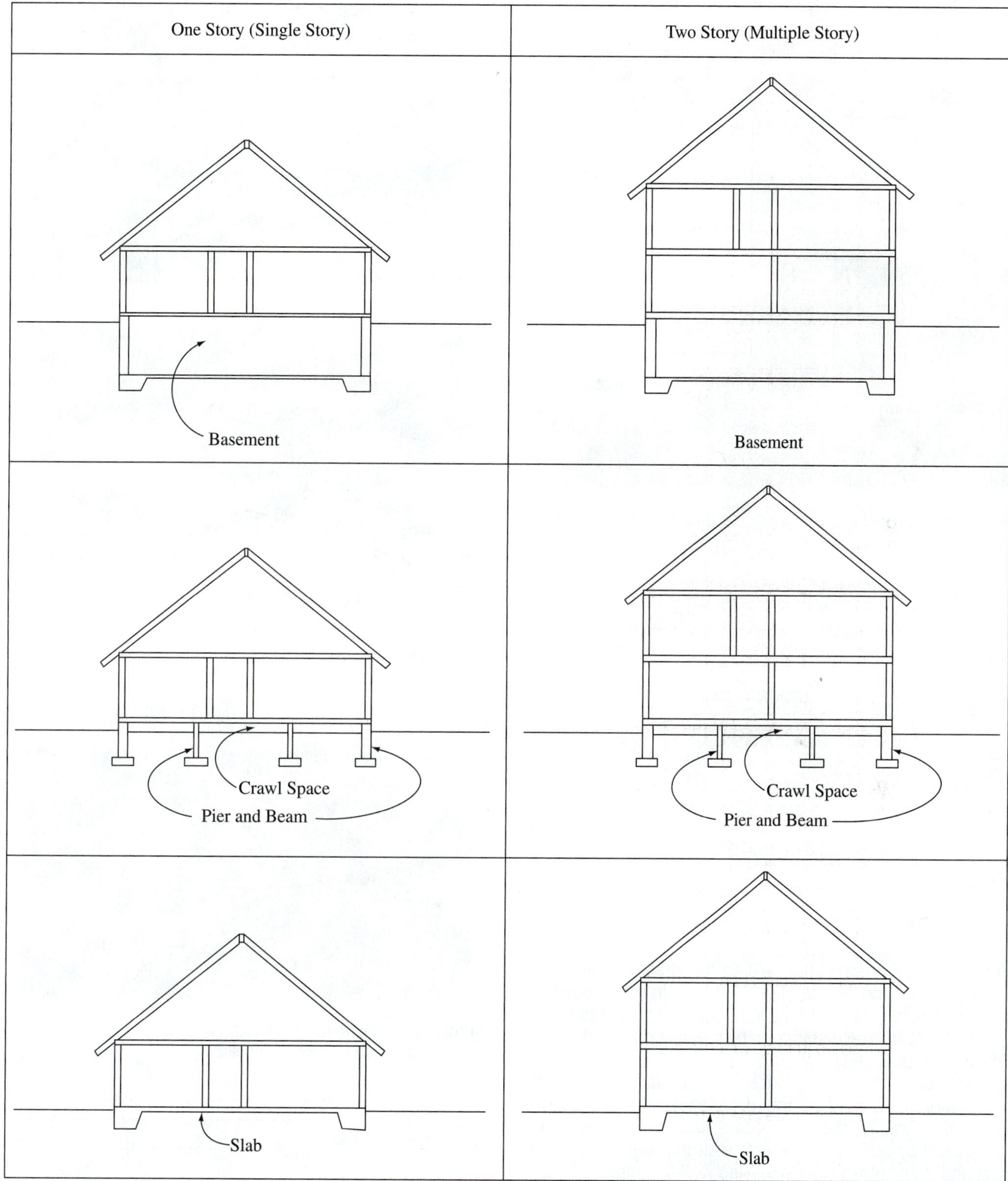

Figure 67-1 Basic house design and types of foundation used in residential construction.

from the inside wall. If the outside building dimensions are not given, you must look at the wall thickness as provided on the detailed wall section and add that dimension to your interior dimensions so that your pipes will come up in the appropriate location.

For example, in the condensate drain in the center of Figure 67-6, the center of the drain is approximately 2 in inside the air-conditioning equipment room. The distance from the outside kitchen wall to the condensate drain is calculated by adding the wall thickness and the distance

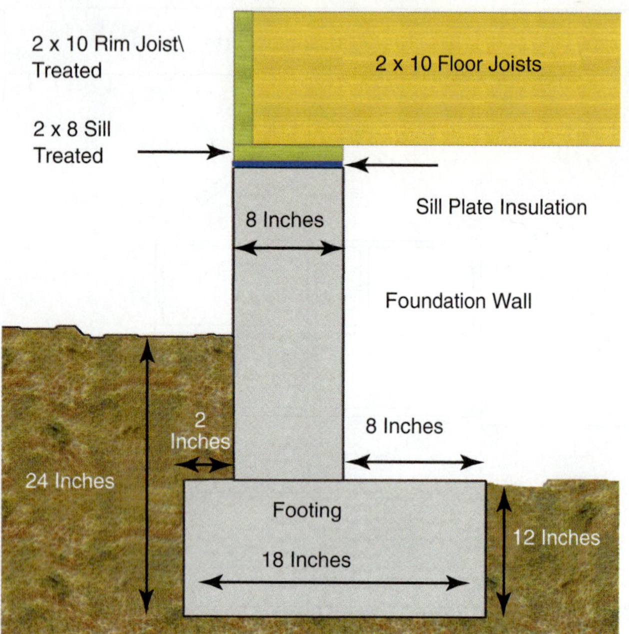

Figure 67-2 Detail of typical footing and foundation wall.

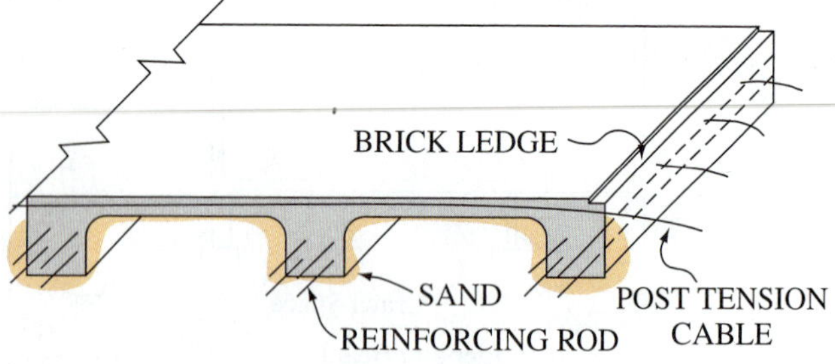

Figure 67-3 The footing and support beams are formed along with the slab.

from the equipment room wall to the drain to the 13 ft by 4 in dimension on the plan. 13 ft by 4 in + 4 in outside wall thickness + 2 in distance from equipment wall to drain = 13 ft by 10 in. The distance to be measured is 13 ft by 10 in from the outside kitchen wall.

However, this only locates one dimension. It is also necessary to determine the distance from the outside bathroom wall to the condensate line. This distance is calculated by adding the outside wall thickness to 9 ft by 0 in and then subtracting the inside wall and pipe distance inside the room.

9 ft by 0 in + 4 in outside wall − 4 in inside wall
− 2 indistance to drain = 8 ft by 10 in

The distance to the condensate pipe from the outside bathroom wall is 8 ft by 10 in. Always double-check your dimensions before leaving the job site after roughing in these lines. Figure 67-7 shows lines being installed in a concrete slab.

(a)

(b)

Figure 67-4 Site preparation for pouring a slab.

Figure 67-5 Plumbing, including refrigerant, lines are buried in sand-filled trenches beneath the slab.

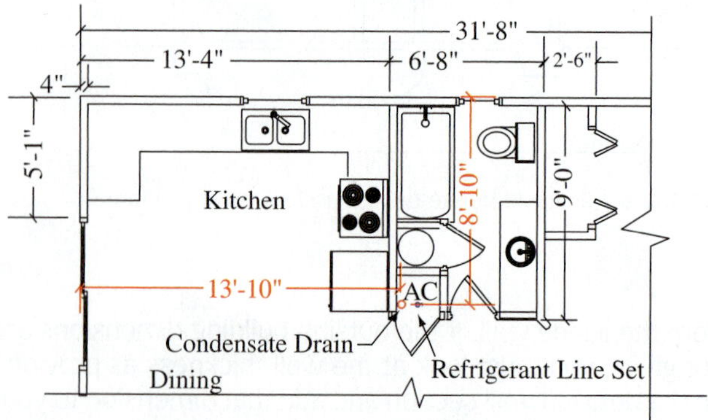

Figure 67-6 Locating refrigerant and condensate drain lines.

(a)

(b)

Figure 67-7 All copper lines are protected with foam insulation so that they do not come in direct contact with the slab concrete.

67.4 CRAWLSPACE FOUNDATIONS

The foundation wall rests on the footing. The floor will be constructed and supported by the foundation wall along the perimeter of the house and on beams supported by piers in the interior portion of the house (Figure 67-8). A crawlspace is an unfinished space beneath the floor that is enclosed by the foundation wall. Crawlspaces get their name from their height. Typically, most crawlspaces are too low to walk in, necessitating crawling to move from one area to another. There is no floor, just dirt. The dirt should be covered with a vapor barrier (Figure 67-9).

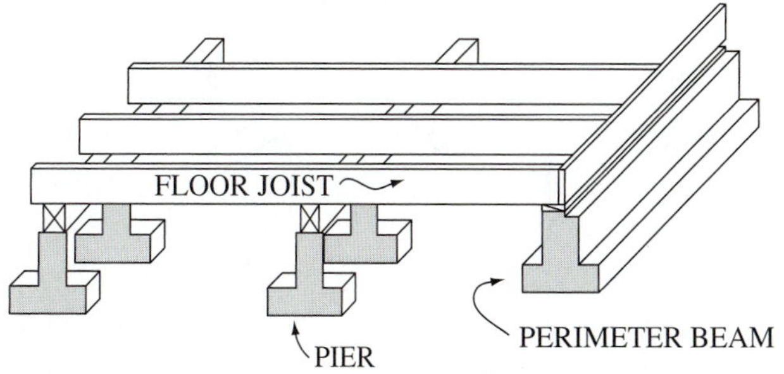

Figure 67-8 Pier-and-beam foundation for supporting floor joists.

The wiring, plumbing, air-conditioning equipment, and ductwork are usually located in the crawlspace of houses with crawlspace construction.

TECH TIP

The vapor barrier is extremely important. Moisture accumulation under the house can create unhealthy conditions that promote mold and attack the wood members of the floor. The vapor barrier is also important for comfort. Houses with crawlspaces that lack vapor barriers can feel clammy in the summer, even at relatively cool temperatures, because of the increased humidity entering the house through the crawlspace.

67.5 BASEMENTS

Residential construction sometimes includes a basement, as shown in the drawings in Figure 67-1. Basements have a floor and are high enough to walk in. While the standard ceiling height of a normal room in a residence is 8 ft, basement ceilings may be lower. When the basement ceilings are less than 7 ft high, laying out and installing ductwork so it does not interfere with the use of the basement can be a problem. Basements are constructed in houses in many states and regions but may not be very common in other areas. Area soil conditions and the level of groundwater both affect the construction of residential basements.

Basement walls are typically constructed with poured concrete or cinderblock walls (Figure 67-10, top, middle). They usually have concrete floors. However, in some arid (dry)

Figure 67-9 Crawlspace with a vapor barrier. *(Courtesy of Basement Systems, Inc.)*

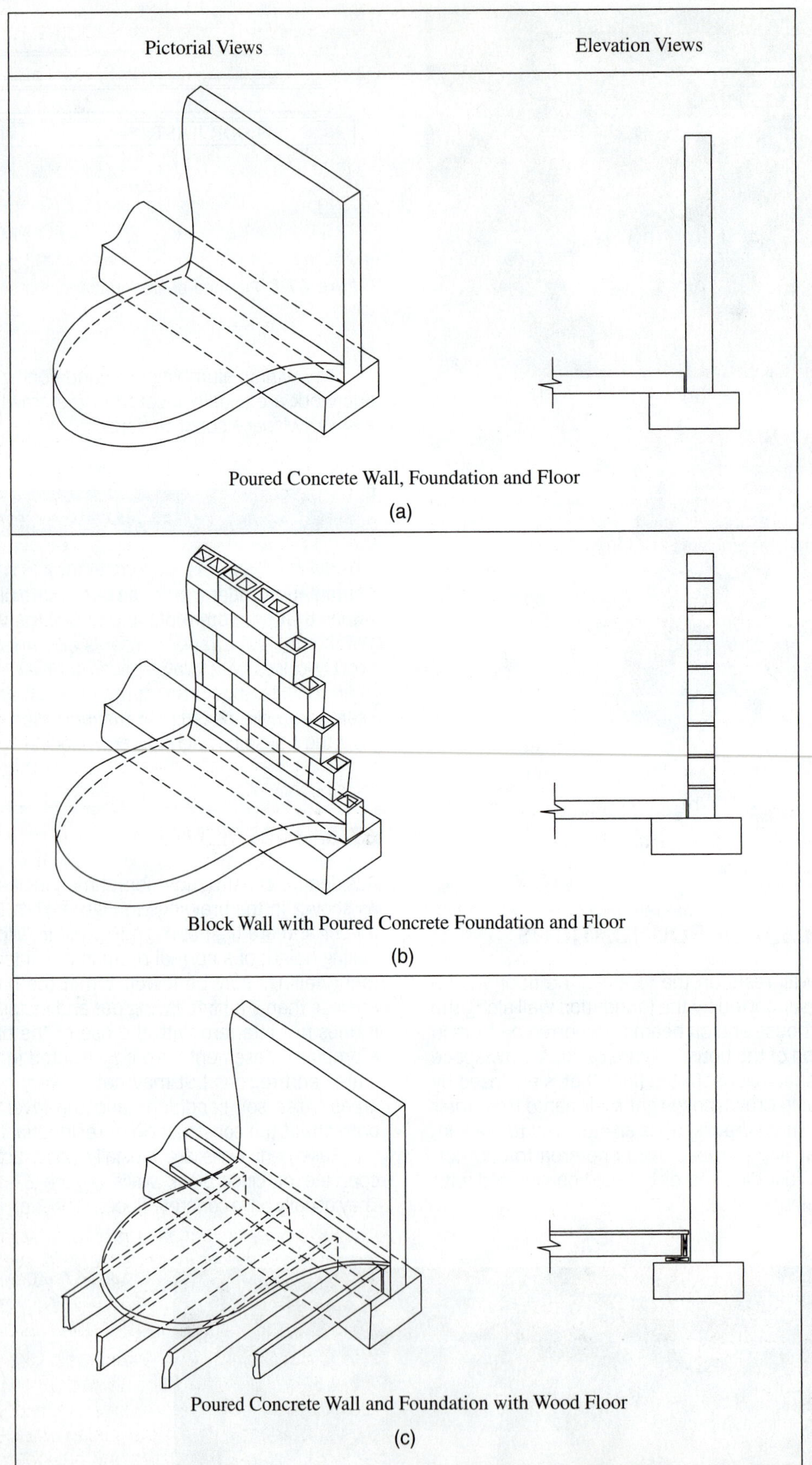

Figure 67-10 Types of basement wall and floor construction techniques used for residential homes.

Figure 67-11 Typical framing is installed on 16-in centers; blocking is used at wall intersection to provide easier sheetrock installation.

climates, such as Denver, Colorado, basements may have a wooden joist floor structure (Figure 67-10, bottom). Properly designed and constructed basements provide a relatively low cost per square foot additional space to residences.

Moisture and poor ventilation are the two most common problems associated with basement areas. Without adequate dehumidification, high levels of moisture can contribute to the growth of mold and mildew. Proper ventilation can help control mold and mildew growth as well as provide better indoor air quality.

67.6 HOUSE FRAMING

Most residential construction uses wood as the primary structural material for walls, floors, ceilings, rafters, and decking material (Figure 67-11). Steel studs and beams can be used in some residential construction but are not in widespread use. Some residential structures are made of masonry materials such as concrete blocks or poured concrete walls. Typical "brick" houses are actually brick veneer; the walls are still framed. The brick wall is not structural but built in front of the frame wall and tied to it to hold the brick wall in place (Figure 67-12).

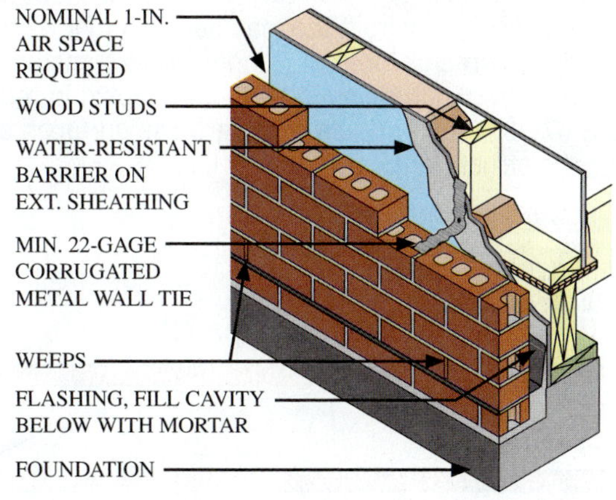

Figure 67-12 The brick veneer wall is tied to the frame structural wall.

Nominal Size	Actual Dimensions
2 × 4	1½ × 3½
2 × 6	1½ × 5½
2 × 8	1½ × 7¼
2 × 10	1½ × 9¼
2 × 12	1½ × 11¼

Figure 67-13 Lumber dimensions in inches.

Framing Members

Nearly all framing members in a traditional wood frame house are "2 by" lumber with a nominal 2-in thickness. It is rough cut at 2 in, but the actual thickness is closer to 11/2 in by the time the lumber is planed. The nominal dimensions and the actual dimensions for most common framing lumber are listed in Figure 67-13.

Floor joists support the floor. They rest upon the foundation wall and on beams supported by piers. Typically, floor joists are placed 16 in on center. This means that the measurement from the center of one framing member to the center of the next one is 16 in. Floor joists vary in size from 2 × 6 up to 2 × 12 depending upon the span, the type of wood used, and the floor load. The walls are constructed of studs. Wall studs are usually 2 × 4 on 16 in centers. Ceiling joists lie on top of the walls and support the weight of the ceiling. Ceiling joists are usually 2 × 6 and are usually placed 16 in on center. Rafters rest on the exterior walls and support the roof. Rafters can be as small as 2 × 6 or as large as 2 × 12 depending upon the span, roof pitch, type of wood, and snow load.

GREEN TIP

Many contractors are using 2 × 6 wall studs on 24-in centers because the larger space formed between the inner and outer walls allows for more insulation.

67.7 WOOD FLOOR CONSTRUCTION

Floor construction for crawlspace and basement foundations is similar. Crawlspace foundations use piers and beams to support the floor joists; basement foundations can use either piers and beams or load-bearing walls to support the floor joists. Floor joists are spaced either 16 or 24 in apart. These joists can be solid wood, wood trusses, or engineered wood I-beams. Figure 67-14 shows a floor system framed with dimensional lumber. Figure 67-15 shows floor joist framing using engineered wood I-beams, and Figure 67-16 shows a floor framed with open-web wood trusses. When solid wood joists are used, any ductwork running perpendicular to the joists must be located below the joists. Open-web trusses and engineered I-beams may allow the ductwork to pass through precut or fabricated openings. With permission from the construction manager and the manufacturer of the laminated beams being used,

Figure 67-14 Dimensional lumber floor joists.

Figure 67-15 Wood I-beam floor joists.

Figure 67-16 Open-web floor joists.

(a)

(b)

Figure 67-17 Vents going through two types of floor joists: (a) truss and (b) beam.

you may be allowed to cut holes on site that will allow the ducts to pass through (Figure 67-17).

SAFETY TIP

Do not cut a hole in any structural member without first obtaining permission from the manufacturer of the product and the job site foreman. Under no circumstances can a beam's structural surface be cut, shown in red in Figure 67-18. Cuts or gouges along these surfaces can cause the structure to prematurely fail.

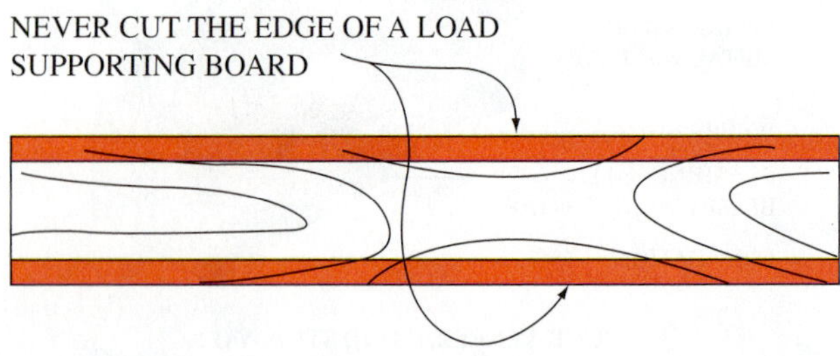

Figure 67-18 Cutting the edge of a horizontal supporting beam will significantly reduce the beam's strength.

Figure 67-19 Tongue-and-groove flooring material.

Flooring Material

Wood floors consist of multiple layers: a subfloor, the underlayment, and the actual flooring surface. The subfloor is commonly plywood or OSB (oriented strand board) with tongue-and-groove edges along its length (Figure 67-19). The subfloor ranges in thickness from ⅝ in to 1⅛ in. When ⅝-in material is used for subflooring, another layer of plywood or OSB is applied at right angles to the subfloor. This layer is called the underlayment. It normally has a smooth finish to facilitate applying the finished flooring materials. The total thickness typically must be at least 1 in. The 1⅛-in flooring material does not need an extra layer of underlayment.

Subflooring and underlayment are often glued and nailed down. Sometimes screws are used in place of nails. Using mastic increases the overall floor strength and reduces future floor squeaks. Figure 67-20 shows how the plywood decking is offset—that is, each new length starts at a different place from the one next to it so that the seams between pieces of plywood do not occur next to each other. This increases the strength and durability of the floor.

67.8 WALL CONSTRUCTION

Figure 67-21 shows the typical wall sections for the four most common types of residential wood frame construction. You will note from the first-floor construction level that each of the typical wall sections is similar.

Most residential walls are constructed with 2 × 4s located 16 in on center (see Figure 67-11). In some cases, 2 × 6s are used on the exterior walls. This allows for more insulation to both reduce the heat load and the sound level in the dwelling. In some areas, when 2 × 6s are used in the perimeter wall, they may be spaced 24 in on center. The 2 × 4s or 2 × 6s used for walls are located on a horizontal board called a base plate. If this base plate is located on a concrete floor, it must be pressure-treated lumber. This is to reduce both wood rot and insect attacks. Two horizontal boards are placed on top of the stud wall. These are called the top plate. Typical wall heights for residential construction are 8 ft, 9 ft, and 10 ft, with 8 ft being the most common. Vapor barriers are frequently used in wall construction to control the flow of water vapor through the wall. The location and type of vapor barrier is in part determined by the climate. A vapor barrier is placed on the outside of the 2 × 4 wall in warm climates and on the inside in cold climates before the finished siding or masonry work is performed, as shown in Figure 67-22. House wraps have become popular for controlling infiltration through the wall. A house wrap is a large roll of material that can literally be wrapped around the exterior of the house before the exterior siding is installed (Figure 67-23). House wraps differ from vapor barriers in that they are specifically designed not to be a vapor barrier. They prevent air leaks through the wall without impeding the flow of water vapor.

67.9 CEILING AND ROOF CONSTRUCTION

Some builders use prefabricated roof trusses, as shown in Figure 67-24. These are almost always fabricated out of 2 × 4 materials with metal cleats connecting the joints. Contractors often favor such trussed roof systems

(a)

(b)

Figure 67-20 (a) Plywood decking is glued and nailed to the floor joists; (b) plywood joints.

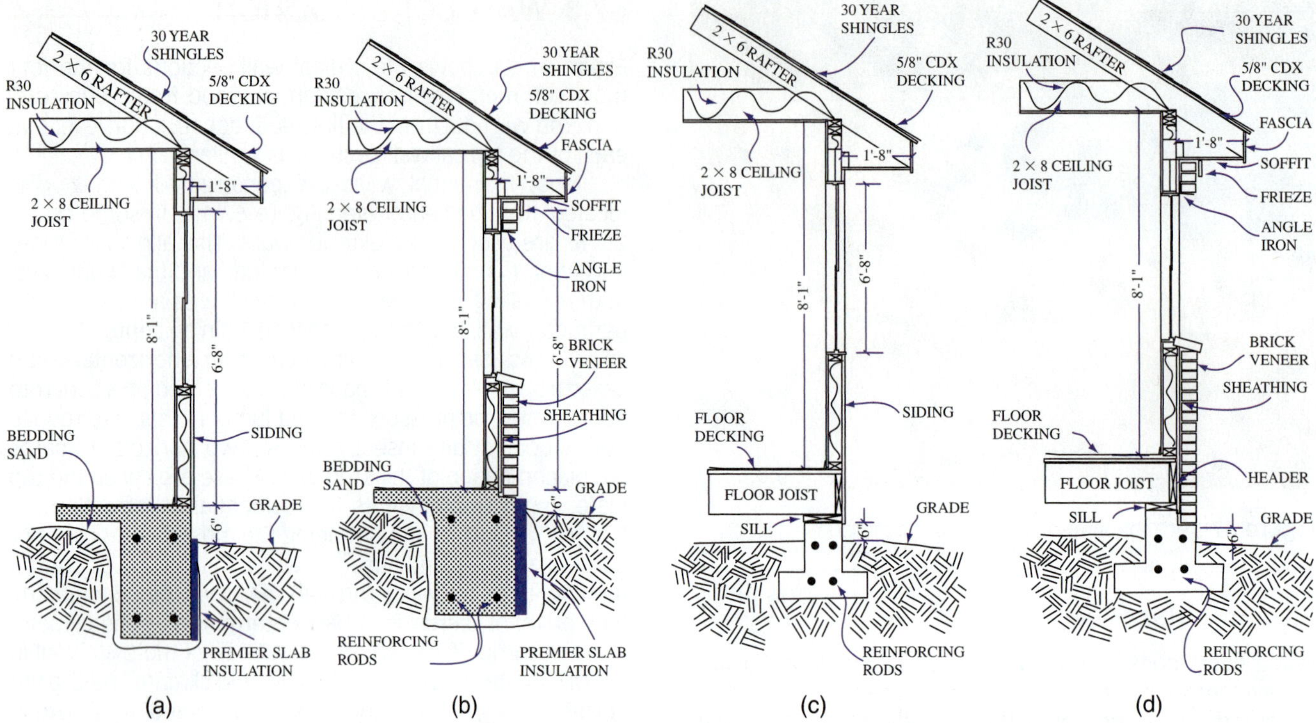

Figure 67-21 Typical wall sections: (a) slab with siding; (b) slab with brick veneer; (c) pier-and-beam with siding; (d) pier-and-beam with brick veneer.

Figure 67-22 Vapor barrier is installed before brick veneer is applied.

Figure 67-23 House covered with house wrap to reduce infiltration.

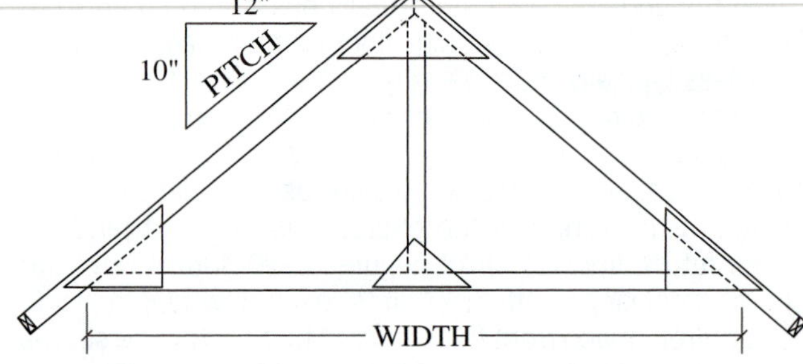

Figure 67-24 Prefabricated trusses are available in a variety of widths.

because they are strong, durable, and easily assembled on the job site. They make it slightly more difficult for the air-conditioning technician to locate a unit in an attic and make straight duct runs.

If trusses are not used, the ceiling rafters are typically 2 × 6 to 2 × 12 dimensional lumber depending on the room span, as shown in Figure 67-25. Roof rafters are typically 2 × 6s, but, depending on the height of the roof, width of the span, and possible snow load or northern climates, 2 × 8, 2 × 10, and even 2 × 12 may be used for rafters. The entire roof area will be covered by thick plywood roof decking material for composite roofs, asphalt shingles, and sheet-metal roofs. When wood shingles are used, a lapped system will be installed across the top of the rafters as opposed to roof decking. An advantage of shingle construction is that with this lapped substructure the attic area will vent freely, which reduces the heat load on the residence during summer.

Figure 67-25 Ceiling joists can be made from a variety of sizes of conventional lumber depending on the roof slope, load, and span; the flatter roof section here uses 2 × 6s, but the steeper section uses 2 × 12s.

Figure 67-26 Wires that go through the top plate of a wall must be either caulked or filled with expanding foam.

SAFETY TIP

A safety harness and lanyard are required under OSHA regulation for construction workers anytime they will be working more than 6 ft above the floor.

67.10 INSULATION

There are four common types of insulation used in residential construction:

- **Rock wool** Rock wool is a synthetic material fabricated from a mineral.
- **Fiberglass** Fiberglass insulation is a spun glass. Fiberglass is available in rolls and bats with or without an attached vapor barrier. Most fiberglass bat insulation is pink or yellow. Blown-in fiberglass is usually white. The color of the insulation has no effect on the capacity of the insulation to resist heat.
- **Cellulose insulation** Cellulose insulation is a synthetically produced insulation product that uses recycled newspaper as its basic material. Cellulose insulation can either be blown loose or else a binder can be added so that it can be blown in wall cavities as a solid fill material.
- **Foam insulation** There are a number of foam insulations that can be used in residential construction as either a sheet or a product that is blown in on site.

SAFETY TIP

When working with insulation, wear a respirator to avoid breathing the fine insulation dust. Avoid touching insulation with unprotected skin. Wearing a long-sleeved shirt, gloves, and a cap will reduce your exposure to insulation that can cause irritation to your skin.

For all insulation, the ability to resist the transfer of heat is directly related to its thickness. Some materials may have greater resistance per inch of material than others. However, compressing the insulation material significantly reduces its heat resistance. Squeezing a 6-in fiberglass batt into a 3½-in wall space will not create a superinsulated wall. Fiberglass batts designed for insulating standard 3½-in walls are available in R-11 and R-13 densities.

Often the 2 × 4 top plates and bottom plates on a wall must have holes drilled in them for wires or plumbing to pass through. These holes must be sealed to prevent air infiltration. The holes are typically filled using expandable foam, as shown in white in Figure 67-26. Any holes drilled horizontally in 2 × 4 wall studs are typically not sealed.

The ceiling is the area in a residence that loses and gains most of the heat. Since ceilings are so crucial for heat gain and loss, code requires that a much thicker layer of insulation be installed in the attic. Attic insulation can be blown, loose insulation, or batt insulation. Table 67-1 lists common insulating materials and their R values based on thickness.

67.11 WALLS AND CEILINGS

The most common material found on interior walls for residences is sheetrock. Sheetrock is a product that has paper on both sides of a powdered gypsum. Figure 67-27 shows a room before (a) and after (b) sheetrock installation. Sheetrock is not a structural material. If it is struck very hard, the paper will become damaged and the gypsum can be pulverized. Once sheetrock has been damaged, the damaged portion must be removed and replaced with a new section of sheetrock. Sheetrock is relatively soft and can be easily pierced with a screwdriver.

Sheetrock is attached to the stud using either sheetrock nails or screws. The sheetrock joints are covered with a sheetrock compound and paper tape. The paper tape does not have adhesive. It is held in place by the sheetrock compound. Sheetrock is finished by sanding the seams and joints and most often has a texture sprayed on as a finishing step.

Table 67-1 R Values of Common Building Materials

Material	R per Inch
Batt and blanket rock wool	R3.7
Blown cellulose	R3.1–R3.7
Blown mineral wool—horizontal fill	R2.9–R2.2
Blown mineral fiber	R2.2–R2.9
Expanded polystyrene board (white bead board)	R4.0
Fiberglass batt, high density	R3.5–R4.3
Fiberglass batt, standard density	R2.9–R3.7
High-density batt and blanket fiberglass	R4.3
Loose-fill cellulose	R3.7
Loose-fill fiberglass	R2.2
Loose-fill rockwool	R2.9
Low-density fiberglass batt and blanket	R3.1
Medium-density fiberglass batt and blanket	R3.7
Mineral wool batt	R3.14–R3.80
Polystyrene board (blue or pink board)	R6.3–R5.6
Polyurethane board (open cell)	R5.50–R6.25
Rigid foam board	R3.7–R8
Spray-on polyurethane foam	R6.3–R5.6
Spray-on cellulosic fiber	R3.4–R2.9

(a)

(b)

Figure 67-27 Sheetrock installation: (a) sheetrock installed on ceiling first; (b) sheetrock installed on walls with joints taped and bedded.

SAFETY TIP

Sheetrock will not support your weight. If you intentionally or accidentally step on a sheetrock ceiling, you will fall through. Pay very close attention to where you are stepping when working in an attic to avoid a fall.

67.12 CONSTRUCTION PHASES

Construction projects typically have four major phases of construction:

- Site preparation and foundation work
- Framing
- Dried in
- Exterior and interior finish work

Figure 67-28 shows these four phases in a house under construction. The site preparation phase prepares the building site, including any required excavation or grading. No air-conditioning work is done during the site preparation of houses built on crawlspace and basement foundations. Placement of refrigerant lines, wiring, drain lines, and sometimes even ductwork is done during site preparation for houses built on a slab.

The framing phase is when the basic structure of the house is erected. Roughing in the duct system starts at the end of the framing phase. Roughing in a system includes making any cuts in the structure for ducts to pass through, locating return-air boxes, and locating boots for supply registers. Rough-in typically continues into the dried-in phase. It is important to get the boots roughed in before the sheetrock is installed.

A house is dried in when it is weatherproof. At this point it has a roof, exterior walls, windows, and doors. Typically most of the air-conditioning installation is performed after the house is dried in. This includes setting the equipment, pulling the control and power wires for the equipment, running the refrigerant lines, and running the ductwork.

Most of the air-conditioning work should be completed before the interior finish work is done. Typically, the supply-air registers and return-air grilles are installed after the walls have been painted. The thermostat is also usually not installed until after the walls are painted. The final wiring and commissioning of the system takes place toward the end of the finishing phase.

(a)

(b)

(c)

(d)

Figure 67-28 Major steps in house construction: (a) site preparation; (b) framing; (c) dried in; (d) exterior flat work.

67.13 LIGHT COMMERCIAL CONSTRUCTION

Buildings constructed for light commercial use may either be built in a fashion similar to residential, with all wood construction, or they may be constructed using metal studs and steel structures. When wooden structures are used, the construction of commercial buildings is very similar to residential construction.

Fire codes often require light commercial buildings to be constructed using metal studs (Figure 67-29). Metal studs are preformed out of 16-, 18-, or 20-gauge galvanized sheet metal and are assembled using sheet-metal screws.

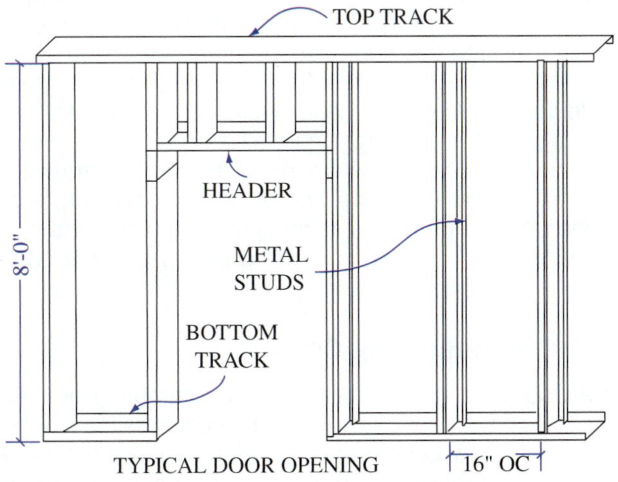

Figure 67-29 Typical metal stud wall layout with door opening.

67.14 FLAT ROOFS

Commercial buildings often have flat roofs. A flat roof is constructed with a roof decking that may be either sheet metal or wood. Sheet-metal decking is often covered with a thin layer of lightweight concrete. Insulating board is placed on this deck and the top surface is then sealed using hot tar, thin layers of roofing paper, or rubber roofing material layered in sheets and glued together. Usually a thin layer of gravel is applied to the top surface. A roof with a layer of gravel will stay cooler because the gravel reflects a lot of the heat.

When rooftop air-conditioning units are used, a curb must be first put in place on the roof and sealed completely. A roof curb is a raised metal frame that rooftop air conditioners rest on. The curb is sealed to the roof along the outside and has holes in the inside to allow ductwork to pass through. Curbs are typically installed by a roofing contractor. Water leaks are a constant potential problem for flat roofs, and it is essential that these air conditioning curbs be sealed completely and properly. Water leaks are such an ongoing issue for flat roofs that many, such as the one pictured in Figure 67-30, are bonded. Bonding is insurance that the roofing contractor provides the building owner to cover water damage if the roof were to leak. Because of this insurance bond provided by the roofing contractor, it is essential that anytime air-conditioning work is being performed that would require a new opening to be cut through the roof, the roofing contractor be contacted before the work begins. This must be done whether the hole is for electrical, refrigerant lines, or the installation of a new rooftop unit. Failure to contact the roofing contractor may result in your air-conditioning company being held liable for any water leaks that occur in the building.

Figure 67-30 Rooftop unit on a composite flat roof.

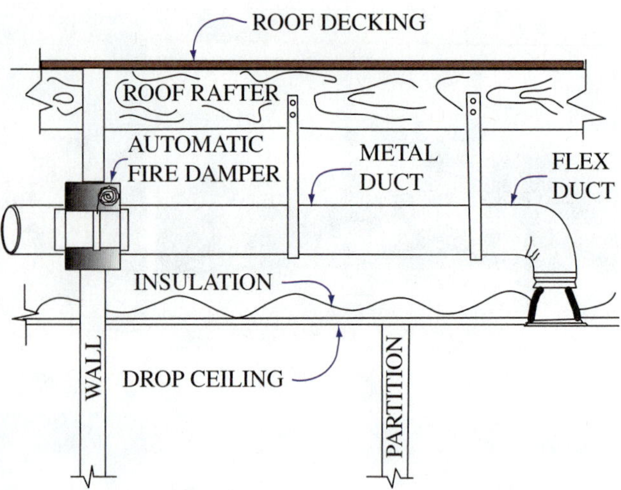

Figure 67-31 Partitions in commercial construction are often used as room dividers and do not extend beyond the dropped ceiling; because partitions are not structural, automatic fire dampers are not required when a duct passes over or through partitions; automatic fire dampers are required anytime a duct passes through an interior wall.

SAFETY TIP

A roof's slope is given as its height in inches as compared to each 12 in of horizontal run. Roofs that have slopes less than 8 in for each 12 in can be worked on relatively safely. Roofs that have slopes greater than 8 in are very hard to impossible to stand on without slipping. If you must work on a sloped roof, you must wear a safety harness.

67.15 DROPPED CEILINGS

Dropped ceilings are commonly used in light commercial buildings. They allow for easy reconfiguring of the interior space as the business changes. Dropped ceilings are constructed by suspending a grid of metal rails. Ceiling tiles are placed in the grid from above. These tiles are easily damaged, so be very careful if you must move them. Do so with clean hands, and be very careful not to scar or scratch the surface of the tile. If the tile becomes damaged, you may have difficulty locating an exact replacement pattern, which can become time-consuming and expensive. Often the space above a building's dropped ceiling is used as the return-air plenum for light commercial buildings.

SAFETY TIP

Dropped-ceiling grids will only support the ceiling! They will not support equipment, refrigeration lines, ductwork, wiring, or you.

67.16 WALLS AND PARTITIONS

Walls and partitions are similar in commercial construction. Walls, however, extend all the way up through the ceiling to the roof of the building. Partitions only extend up to the ceiling. That makes partitions very easy to move as the business reconfigures as it grows. Some walls extend all the way up to the ceiling and are classified as firewalls. These walls are constructed with fire-resistant materials. Building and fire codes regulate the construction of firewalls. Anytime supply or return ducts must pass through a firewall, the duct must have a fire damper (Figure 67-31). Fire dampers will automatically close, sealing off the duct to prevent the spread of smoke and fire throughout the building.

67.17 GREEN BUILDING

Green building is a general term used to indicate construction techniques and designs that minimize energy use and reduce water consumption, waste, and a building's impact on the environment. The increased demand for energy, water, and land resources has moved a number of professionals in the construction industry to look for ways of building that use these resources more than intelligently. Rather than simply trying to find more energy and water, they seek ways to use less of these precious resources through design that looks at the entire cost of a building, including its operation. The cost of owning a green building is no more than, and often less than, a traditionally constructed building. Figure 67-32 compares the construction, finance, and operating costs of two comparable homes, one built to code, the other built to incorporate green building principles. Although the green home cost more to build, it costs no more to own because it costs much less to operate.

U.S. Green Building Council

The USGBC is a 501(c)(3) nonprofit community of leaders working to make green buildings accessible to everyone within a generation. A major part of their work involves establishing standards to encourage green building. They have developed the Leadership in Energy

Comparison	Standard Construction ($)	LEED Construction ($)	Cost Difference ($)
Initial price	300,000	308,500	+ 8,500
Mortgage	1,890	1,945	+ 55
Energy bill	150	105	−45
Water bill	30	20	−10
Net ownership cost	2,070	2,070	Same

Figure 67-32 Cost comparison between standard and green construction.

Green Facts

Project Title	1125 Main Street
Building Use	Multi-Use
Location	Bogart, GA, USA
Size	135,000 SF
Cost	$15,100,000
LEED for core Shell rating out of	61
Total Score	31
Sustainable Sites	7
Water Efficiency	2
Energy & Atmosphere	7
Materials & Resources	6
Indoor Environmental Quality	6
Innovation & Design Process	3
Certification Level	Silver
Energy Savings	314,551 kWh/yr
Carbon Emissions Avoided	NA
Water Savings in gallons/year	191,425 gallons
Waste Diverted (tons)	84 tons
Project Team Profile	
Owner/ Developer	Mover & Shakers Corp
Architect	Environmental Designs
Engineers	Clean Design Inc
Contractor	Green Build
Landscape Architect	Green Scapes
Commissioning Agent	EPT

Figure 67-33 Green fact sheet for a LEED-certified building.

Efficiency and Environmental Design (LEED) program to certify green building standards. Figure 67-33 shows the "green facts" of LEED-certified buildings. Studies have shown that LEED certified buildings cost less to own and operate over their lifetime. This has gotten the attention of people whose only interest in LEED construction is financial; they have discovered that building green saves money.

Air Conditioning and LEED Certification

Heating and air conditioning account for approximately half of a building's energy use. A major part of LEED certification is reducing a building's overall energy use. The building's air-conditioning system can have a major impact on LEED certification because it represents such a large portion of the building's energy use. Energy-efficient equipment is only part of the answer. The equipment must be properly applied and installed. Air-conditioning technicians can be part of the green wave by ensuring that systems are properly applied, installed, and serviced. Bad installation and poor service will keep even the most energy-efficient equipment from being green.

UNIT 67—SUMMARY

The construction techniques used in different regions of the country vary. In some areas of the country, basements are commonly found in residential construction, while in other parts of the country, most residential construction is built on concrete slabs. However, the basics of construction remain the same throughout the country.

Construction usually progresses through four phases: site and foundation preparation, framing, drying in, and interior and exterior finish work. Concrete slabs, crawlspaces, and basements are three common foundation systems. Regardless of the foundation system, all foundations rest on a footing.

Wood framing is the most common structural system used in residential construction. Floor joists can be made of lumber, open-web wood trusses, or engineered wood I-beams. Walls are typically constructed of 2 × 4 studs placed 16 in on center. Ceilings and roofs can either be framed with joists and rafters or with roof trusses.

Insulation is especially important in reducing the heating and cooling load of the structure.

Light commercial and residential construction techniques are similar. Commercial construction often uses large steel and concrete beams, but much of the interior finish and detail work is similar to light commercial. Therefore, understanding the basic construction techniques for light commercial can help with commercial jobs.

WORK ORDERS

Service Ticket 6701

Equipment Type: New Split-System Installation in House on Concrete Slab

In the site preparation phase, the general contractor is preparing a slab to be poured and needs any work under the slab to be completed. Refrigerant lines and the condensate drain need to be installed.

The technician arrives at the job site, lays out the location of furnace/AHU and also the location of condensing unit, and runs a 4-in PVC conduit under the future concrete slab. The technician then rolls out both the suction and liquid lines, insulates the suction line, and pulls them along with thermostat wire inside the conduit, being careful not to kink the copper tube or tear the insulation. The technician checks to make sure there is no place where concrete from the poured slab will contact any part of either line and makes sure there is plenty of tubing length on each end to connect the furnace/AHU and the condensing unit. The technician makes sure that the tubing caps remain in place on each end of the tubing so the system will be free of moisture and foreign matter. Before leaving, the technician takes location dimensions of the line location and records this on the blueprints. The job is ready for slab inspection.

Service Ticket 6702

Equipment Type: New Split-System Installation in House on Concrete Slab

In the framing phase, the general contractor calls and the framing is near completion. The ductwork and wiring need to be roughed in. The refrigerant lines were pulled during the slab preparation.

The technician roughs in the return-air box in the hall and the supply-air drops in the ceiling. The control wiring is run from the thermostat location in the hall to the air handler in the equipment closet. The control wiring between the air handler and the condenser was run with the refrigerant lines during slab preparation. The electrician pulls the power wire for the air handler inside and the condensing unit outside. The technician installs nail plates where appropriate. The technician then calls for a rough-in inspection.

Service Ticket 6703

Equipment Type: New Split-System Installation in House on Concrete Slab

The general contractor notifies the technician that the house is dried in. It is time to set the equipment, run the ductwork, connect the refrigerant lines, and complete the wiring.

The technician sets the air-handling unit in the equipment closet and the condensing unit on the pad outside. The refrigerant lines are connected to the air-handling unit. The condensate line is run from the air-handling unit to the condensate stub provided during slab preparation. The electrician wires the power wire to the air handler and the condensing unit. The technician wires the low-voltage thermostat wiring to the air handler and condensing unit.

The return- and supply-air plenums are installed, and the ductwork is run. The supply duct is run in the attic and the return runs to a return box in the hall. Since the supply duct is in the attic, duct with an insulation value of R-8 is used. The return duct is insulated with R-6 because it is not exposed to attic conditions. All ductwork is properly connected, sealed, and supported to prevent air leaks and restrictions.

Service Ticket 6704

Equipment Type: New Split-System Installation with Lines Under Concrete Slab

The home is complete and the new owner is ready to move in. The technician is called to do the initial system start-up. The technician leak checks the lines and coil using nitrogen. Finding no leaks, the lines and coil are evacuated. While the system is being evacuated, the technician checks all electrical connections. With the disconnect remaining off, the voltage to the lugs at the top of the disconnect is checked and compared to the voltage on the system data plate. The thermostat is mounted and wired. The system is valved off when the vacuum reaches 400 microns. The vacuum holds for 10 minutes, rising slightly to 450 microns and leveling off. After determining the system is tight, clean, and dry, the technician releases the refrigerant from the condenser into the system. The technician checks the voltage before starting the unit, starts the unit, and then checks the voltage again after starting the unit to check for voltage drop in the power wire. The information in the manufacturer's installation instructions is used to check the system charge. The technician records the system operating data on the start-up form used by the company. The technician calls for final inspection when the system is performing per the manufacturer's specifications.

UNIT 67—REVIEW QUESTIONS

1. What is the standard ceiling height in a residence?
2. Name the two most common problems with basements.
3. Name the reinforcement materials used in concrete slabs.
4. Why should refrigerant lines or condensate plumbing be installed in a chase or conduit if they are to be under a poured concrete slab?
5. How is ductwork location affected by solid wood floor joists?
6. Why is plywood decking offset, as seen in Figure 67-20b?
7. Why do contractors often favor a trussed roof system?
8. Name four common types of insulation used in residential construction.
9. Using Table 67-1, what would be the insulation R-factor of 4 in of medium-density fiberglass batt and blanket material?
10. What is the most common material used for interior walls of residences?
11. What material is sometimes required by the fire code for studs?
12. What is a roof curb?
13. What consequences can result from failing to contact the roofing company before making a roof penetration on a bonded roof?
14. What is the space above the drop-in ceiling in a commercial building used for in the HVACR industry?
15. What must be included in any duct that passes through a firewall?
16. What is the purpose of a fire damper?
17. Why is the cost of owning a "green" building often less than the cost of owning a conventionally constructed building?
18. What is the mission of the USGBC?
19. What does LEED certification mean?

UNIT 68

Green Buildings and Systems

OBJECTIVES

After completing this unit, you will be able to:

1. describe how the term *sustainability* applies to green building design.
2. recognize the difference between green building standards, codes, and rating systems.
3. explain the basic concept behind the LEED rating system and understand the process required to earn LEED credentials.
4. determine which codes or standards may apply for a given project.
5. describe what type of site-selection, material-selection, and system-design components satisfy the requirements for commercial and residential green building certification.

68.1 INTRODUCTION

The U.S. Department of Energy (DOE) reports that buildings use more energy than any other sector of the U.S. economy. The combined impact of all buildings, both commercial and residential, accounts for more than 70 percent of all electricity usage and 50 percent of all natural gas usage. According to the U.S. Green Building Council (USGBC), buildings in the United States are responsible for 39 percent of CO_2 emissions, 40 percent of the total energy consumption, and 13 percent of all water consumption.

"Green" construction practices are being used today that efficiently use energy, reduce waste and pollution, and create a healthy environment for the occupants. As an example the main goal of the DOE's Building Technologies Program (BTP) is to develop affordable, energy-efficient homes by 2020 and high-performance commercial buildings by 2025. These high-performance buildings will be 60 to 70 percent more energy efficient than today's typical buildings.

The HVACR industry must also adjust to meet the demands of government programs and regulations as they slowly develop and become mandatory. There will be increasingly stringent regulations regarding energy efficiency, alternative energy use, water conservation, and environmentally friendly building materials. As of January 1, 2011, California became the first state to develop a mandatory green building standards code (CALGreen Code) that applies to most all-new building design and construction.

CODE

The 2010 California Green Building Standards Code, which took effect on January 1, 2011, is a code with mandatory requirements for new residential and nonresidential buildings (including buildings for retail, office, public schools, and hospitals) throughout California. The code is Part 11 of the California Building Standards Code in Title 24 of the California Code of Regulations and is also known as the CALGreen Code. In short, the code is established to reduce construction waste, make buildings more efficient in the use of materials and energy, and reduce environmental impact during and after construction.

68.2 SUSTAINABILITY

Sustainability is a term often used when describing green technology and innovative building design. For a building to be sustainable, it must be designed and built to meet the needs of society in ways that can continue indefinitely into the future without damaging or depleting natural resources. These buildings are designed to be energy smart and minimize electrical power consumption. High-performance buildings are able to generate their own power, and some can also supply power back to the utility grid. Not only do green buildings use fewer resources through improved energy and water efficiency, but studies have shown that they also offer a better place to live and work due to better air quality and access to natural daylight. People care about these health impacts, and so from a financial perspective, owners of sustainably managed buildings expect a higher return on investment because of the greater productivity, satisfaction, health, and well-being of the buildings occupants.

68.3 BUILDING ENERGY CODES

Energy efficiency for a home, factory, or office building is often determined by decisions made at the time of initial construction. Choosing less energy-efficient methods and materials may save money at the outset, but this will most likely result in increased energy costs for many years to come. Energy codes and standards set minimum requirements for energy-efficient design and construction for new and renovated buildings that impact energy use and emissions for the life of the building. Most often these standards set mandatory requirements that can be enforced. Because it is difficult to impossible to incorporate some forms of building efficiency after construction, incorporating sustainable features at the outset makes sense. Economically, it is less expensive to incorporate sustainable features during new construction than to try and add them later. The benefits of more efficient construction today are enjoyed for thirty to fifty years into the future.

68.4 GREEN BUILDING RATING SYSTEMS

Green building rating systems typically accumulate overall point totals based on factors such as energy savings, accessibility to mass transit, reductions in storm-water runoff, and a host of prevailing factors that result in maximum sustainability. The higher the overall point total, the more sustainable the building rating. These rating systems have been mostly voluntary and until recently have not been enforced as mandatory requirements. They often provide a limited number of prerequisites with many optional credits to allow more focus on the green building aspects deemed important to the user of the system. One recognized rating system is LEED (Leadership in Energy and Environmental Design). This system was created by the U.S. Green Building Council (USGBC).

CODE

Standards and codes are typically mandatory requirements that must be followed. These requirements can be addressed at federal, state, and local levels (city codes). Local codes are often referred to as jurisdictional codes. Rating systems are different in that they are not necessarily mandatory but can often be voluntary. However, the rating systems are often based upon codes and standards. Green building codes and green rating systems are somewhat unique in that they have been developed with mutual cooperation between standards organizations and rating agencies.

USGBC

The U.S. Green Building Council (USGBC) was founded in 1993, and its membership includes corporations, builders, universities, government agencies, and other nonprofit organizations. USGBC is committed to a prosperous and sustainable future through cost-efficient and energy-saving green buildings. The USGBC introduced the LEED rating system into the marketplace in 2000 and by 2010 had a total footprint of certified commercial projects that surpassed 1 billion ft^2. Over 36,000 commercial projects and 38,000 single-family homes have participated in LEED. USGBC is also a cooperating sponsor in the development of the two primary baseline building energy codes: ANSI/ASHRAE/USGBC/IES Standard 189.1-2009 and the International Green Construction Code (IGCC).

CODE

The IGCC integrates with the existing family of I-Codes to ensure that building designers, contractors, owners, and inspectors all work from a common platform. The IGCC acts as an overlay on other existing codes, including the provisions of the International Energy Conservation Code (IECC) as a baseline.

68.5 INTERNATIONAL GREEN CONSTRUCTION CODE (IGCC)

The International Green Construction Code (IGCC) was introduced by the International Code Council (ICC) in 2010. This is a first-of-its-kind collaboration with cooperating sponsors that include the American Institute of Architects (AIA), the American Society for Testing and Materials (ASTM International), ASHRAE, USGBC, and the Illuminating Engineering Society (IES). The IGCC provides a comprehensive set of requirements intended to reduce the negative impact of buildings on the natural environment.

The IGCC is intended to be used by manufacturers, design professionals, and contractors, but unlike voluntary ratings systems, it is written in mandatory language. The model code language becomes law when it is adopted by the appropriate state or local authority. To this point, voluntary programs have been the driving force for green construction. The IGCC has been designed to complement these programs as well as to be applied to construction projects that are unlikely to adopt the voluntary programs.

Many building rating systems offer many choices to the owner and designer but do not require increased performance in specific areas such as energy, water, natural resources, or material conservation, which may be of importance to the local jurisdiction. The IGCC is composed primarily of mandatory requirements but also allows local jurisdictions to regulate owner/designer choices for a minimum number of project electives. This allows the IGCC to produce more predictable results that are closely aligned with the local jurisdiction's requirements.

International Code Council (ICC)

The ICC is a U.S.-based nonprofit, nongovernmental membership association dedicated to building safety, fire prevention, and energy efficiency. The International Codes, or I-Codes, published by the ICC, provide minimum safeguards for people at home, at school, and in the workplace. Most U.S. cities, counties, and states adopt the building safety codes developed by the ICC.

American Institute of Architects (AIA)

AIA has been the leading professional membership association for licensed architects, emerging professionals, and allied partners since 1857.

ASTM International

ASTM International is one of the largest voluntary standards-development organizations in the world. Today, some twelve thousand ASTM standards are used around the world to improve product quality, enhance safety, facilitate market access and trade, and build consumer confidence.

Illuminating Engineering Society of North America (IES)

The IES is the recognized technical authority on illumination. For over one hundred years, its objective has been to communicate information on all aspects of good lighting practices to its members, to the lighting community, and to consumers, through a variety of programs, publications, and services.

ASHRAE

American Society of Heating, Refrigerating and Air-Conditioning Engineers (ASHRAE), founded in 1894, is an international organization of fifty-one thousand persons. ASHRAE fulfills its mission of advancing heating, ventilation, air conditioning, and refrigeration to serve humanity and promote a sustainable world through research, standards writing, publishing, and continuing education.

CODE

ANSI/ASHRAE/USGBC/IES Standard 189.1-2009 addresses the same building spaces as ANSI/ASHRAE/IESNA Standard 90.1-2007 except with a calculated 27 percent increase in energy savings.

68.6 ANSI/ASHRAE/USGBC/IES STANDARD 189.1-2009

Standard 189.1-2009 (Standard for the Design of High-Performance Green Buildings Except Low-Rise Residential Buildings) covers site sustainability, water-use efficiency, energy efficiency, indoor environmental quality, and the building's impact on the atmosphere, materials and resources. The standard is not intended to be a design guide or a rating system, but it is expected that voluntary building rating systems will integrate this standard into their programs. The standard is based on the mandatory requirements that establish baseline criteria for a high-performance green building, and as such, these standards must be met to guarantee compliance.

American National Standards Institute (ANSI)

ANSI is a private, nonprofit membership organization founded in 1918. Its primary goal is to promote and facilitate voluntary consensus standards and conformity assessment systems. ANSI serves as an administrator and coordinator of the U.S. private-sector voluntary standardization system. Accreditation by ANSI signifies that the procedures used to create new standards meet the Institute's essential requirements for openness, balance, consensus, and due process.

CODE

ANSI/ASHRAE/USGBC/IES Standard 189.1-2009 is a jurisdictional compliance option to the International Green Construction Code (IGCC). IGCC Section 302.1 and Table 302.1 provide the framework that allows the jurisdiction to select Standard 189.1 as a compliance option.

68.7 NATIONAL GREEN BUILDING STANDARD

In 2007, the National Association of Home Builders (NAHB) and the ICC partnered to establish the National Green Building Standard. This standard applies to all residential construction work—including single-family homes, apartments and condos, land development, and remodeling renovations—and is approved by the ANSI. This standard provides homebuilders and remodelers with a third-party rating system that they can use to achieve green certification under NAHB Green and the National Green Building Certification Program.

The standard offers green building practices in six categories: lot design, resource efficiency, energy efficiency, water efficiency, indoor environmental quality, and operation maintenance. To achieve green certification, a project must have minimum point total score. Certification levels are based on progressively higher point totals and can be awarded as bronze, silver, gold, or emerald.

National Association of Home Builders (NAHB)

The NAHB is a trade association that helps to promote policies that make housing a national priority. Its primary goal is to provide and expand opportunities for all consumers to have safe, decent, and affordable housing. The NAHB has been serving its members, the housing industry, and the public at large since 1942.

68.8 ANSI/GBI 01-2010 STANDARD

The Green Building Assessment Protocol for Commercial Buildings is an assessment rating system for new construction formally approved in 2010. This standard was derived from the Canadian Green Globes environmental assessment and rating system. There are seven areas of assessment that are awarded points: project management, site, water, energy, emissions, indoor environment, and resources. There are minimum point requirements in each of the seven areas, and the level of achievement for certification range from 1 to 4.

Green Building Initiative

The Green Building Initiative (GBI) is a not-for-profit organization whose mission is to accelerate the adoption of building practices that result in energy-efficient, healthier, and environmentally sustainable buildings by promoting credible and practical green building approaches for residential and commercial construction. The GBI is governed by a multi-stakeholder board of fifteen directors featuring representatives from industry.

68.9 DOE BUILDING TECHNOLOGIES PROGRAM (BTP)

The Department of Energy's Building Technology Program (BTP) is a partnership with the building industry to research, develop, and deploy energy-efficient building technologies and practices. BTP is working with organizations such as ASHRAE and ICC to achieve a 30 percent energy improvement in commercial and residential building codes in the United States. Another BTP initiative is to promote and support the ENERGY STAR program to ensure that appliance and equipment standards are met in terms of energy efficiency. BTP is also actively educating homeowners, builders, and developers about the benefits of embracing energy-efficient technologies and practices.

ENERGY STAR

ENERGY STAR is a joint program of the U.S. Environmental Protection Agency and the U.S. Department of Energy designed to save consumers money and protect the environment through energy-efficient products and practices. This program helped to save enough energy in 2009 to avoid greenhouse gas emissions equivalent to those from 30 million cars, while saving American consumers nearly $17 billion on their utility bills.

68.10 LEADERSHIP IN ENERGY AND ENVIRONMENTAL DESIGN (LEED)

The LEED standards are developed by the USBGC, while the Green Building Certification Institute (GBCI) runs the LEED Building Certification. LEED recognizes five areas of performance: site, water, energy, materials, and indoor environmental quality. The LEED rating systems are grouped into five main categories: Building Design & Construction, Interior Design and Construction, Operations & Maintenance, Homes, and Neighborhood Development. To achieve LEED certification, a project must have minimum point total score. Certification levels are based on progressively higher point totals and can be awarded as certified, silver, gold, or platinum.

First Category: Green Building Design & Construction

LEED for New Construction
LEED for Core & Shell
LEED for Schools
LEED for Healthcare
LEED for Retail

Second Category: Green Interior Design & Construction

LEED for Commercial Interiors
LEED for Retail Interiors

Third Category: Green Building Operations & Maintenance

LEED for Existing Buildings
LEED for Existing Schools

Fourth Category: Single- & Multi-family Residential Homes

LEED for Homes Three Stories or Fewer

Fifth Category: Neighborhood Design

LEED for Neighborhood Development

LEED Credentialing

The LEED professional credentials and exams are administered by the GBCI. As of May 2009, the new credentialing system is broken into three tiers. The only Tier I credential available is the LEED Green Associate. This can be earned by candidates who demonstrate that they possess the knowledge and skill to support green design, construction, and operations. The candidates must also successfully pass the two-hour LEED Green Associate exam.

There are five different credentials for Tier II, LEED AP with Specialty. The LEED AP with Specialty credential signifies an advanced depth of knowledge in green building practices. Two-hour specialty exams are available for each of the five major categories of LEED. Only one of the five specialty exams must be passed for the LEED AP credential. Candidates for the Tier III LEED AP Fellow must be nominated by their peers and have at least ten years of experience in the green building field with a minimum of eight cumulative years as LEED AP.

68.11 PLATINUM LEED COMMERCIAL APPLICATION: HANNAFORD SUPERMARKET & PHARMACY

Hannaford Supermarket & Pharmacy, located in Augusta, Maine, was recognized by the USGBC in 2009 as the first supermarket in the world to receive a LEED Platinum certification (Figure 68-1). The supermarket's 50,000-square-foot building is designed to use about 50 percent less energy than a conventional building of the same size.

Site Development

This was not a new site, as the building was constructed on the grounds of a former high school. The existing site was demolished and 96 percent of the existing building debris and contents were either recycled or reused (Figure 68-2). Parking spaces close to the store entrance are reserved for van pools and low-emission hybrid vehicles. There are

Figure 68-1 Exterior of a "green" supermarket.

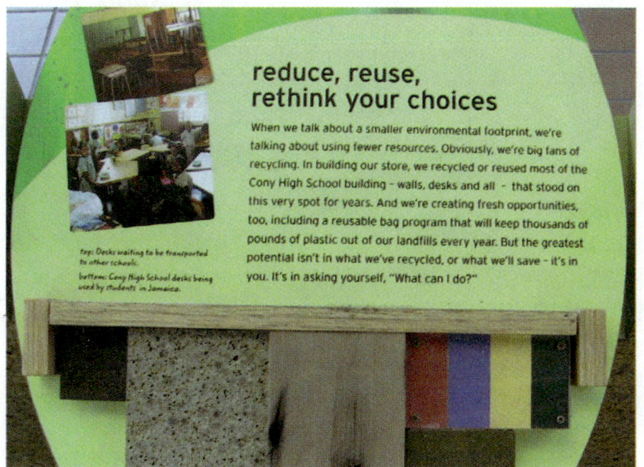

Figure 68-2 Learning center that describes the benefits of recycling.

carpool spaces for employees and conveniently located bicycle racks. Reflective material on the surface of the parking lot has been designed to reduce heat-island effects.

Water

The building has a 7,000 square feet of vegetated roof comprising a mixture of green and red sedums (Figure 68-3). This "green roof" reduces water runoff, helps to insulate the store, and reclaims green space.

(a)

(b)

Figure 68-3 (a) Vegetated roof; (b) Vegetated roof and skylights.

Figure 68-4 41-kW rooftop solar panel array.

To conserve on water usage, the restrooms utilize low-flow faucets and low-flow, dual-flush toilets. The urinals are the waterless type. The seafood department uses ice-free cases. It is estimated that this combined conservation effort reduces water usage by 38 percent, equating to a savings of roughly a half million gallons of water annually.

Energy

The supermarket has been designed to use 50 percent less energy than a typical grocery store. With conventional supermarket design, the refrigeration system consumes the most electricity, followed next by lighting and then HVAC. To supplement the store's energy supply, there is a 41-kW solar photovoltaic system located on the roof, which provides on-site power (Figure 68-4).

To reduce energy consumption, a high-performance Green Chill refrigeration system is employed at the store. The Green Chill heat-reclaim system meets almost 100 percent of the store's heating needs by capturing the heat from the coolers, which is then used to warm the store (Figure 68-5). In addition, there are two geothermal wells located 750 feet below the ground that help to regulate the store's interior temperature year round (Figure 68-6). Radiant heat is used to warm the concrete floors in the checkout areas and to melt snow at the outside sidewalk entrance to the store.

The building was designed to take advantage of natural lighting with an array of fifty skylights, a dozen solar tubes, large windows, and transom glass on three walls (Figure 68-7). Located in the ceiling near the center of the building is a panorama of windows designed to direct the incoming

Figure 68-5 Heat-reclaim piping.

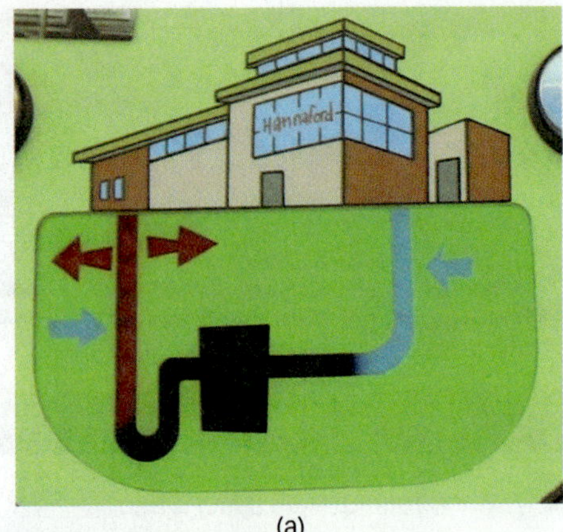

(a)

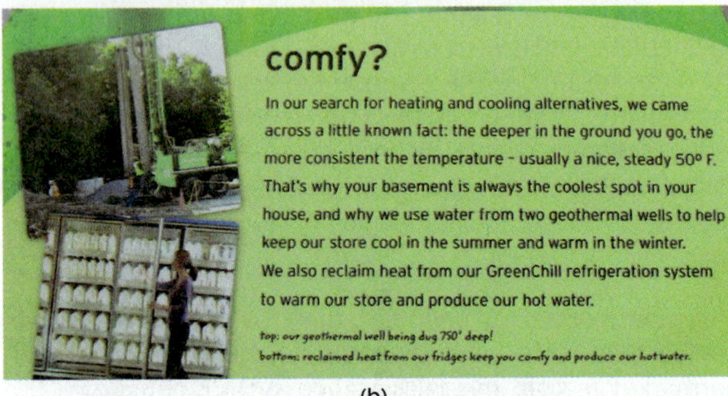

(b)

Figure 68-6 (a) Learning center geothermal informational display; (b) Learning center description of geothermal benefits.

(a)

(b)

Figure 68-7 (a) Rooftop view of skylights; outside view of glass pavilion and solar panel array. (b) Close-up view of rooftop skylight.

Figure 68-8 Inside view of illumination from fabric light shelves and glass pavilion located in the center of the store.

sunlight onto fabric-covered shelves that reflect the light back toward the center of the ceiling for improved illumination (Figure 68-8). High-efficiency fixtures that come on and off automatically depending on the amount of natural sunlight entering the building are used for the interior lighting (Figure 68-9). In this way, when the sunlight is at its maximum, much of the electric lighting in the store is automatically turned off to save energy.

Conventional supermarkets often have an assortment of open freezers and cold cases to allow easy access to the product. However, new energy-efficient designs restrict the use of open cases and use mostly closed cases with glass doors (Figure 68-10). Many are equipped with motion sensors that allow LED lighting to turn on when someone approaches (Figure 68-11). The lighting then turns off again automatically when motion is no longer detected as another means to conserve power.

Materials and Resources

Over 70 percent of the wood used in the store is Forest Stewardship Council (FSC) certified. The FSC sets forth principles, criteria, and standards that span economic, social, and environmental concerns for guiding forest management toward sustainable outcomes. The interior finishes are low in volatile organic compounds (VOC). VOCs are both manmade and naturally occurring organic chemical compounds that can affect the environment and human health. Recycled

Figure 68-9 Energy-efficient lights turn on and off automatically dependent on the amount of outside light entering the building.

(a)

(b)

Figure 68-10 (a) Sign indicating that the freezer glass door slides open; (b) Freezer with sliding glass doors.

(a)

(b)

Figure 68-11 (a) The freezer interior lighting only comes on when motion is detected; (b) The lighting motion sensors are located above freezer doors.

Figure 68-12 Increased window area and outside lighting provide for a cheery atmosphere leading to improved customer satisfaction.

materials have been used for the store's interior surfaces, such as the pharmacy's counters. There is ongoing education awareness for recycling materials, and a storewide program is in place for recycling cardboard, plastics, paper, lightbulbs, and batteries that is also made available for customers.

Indoor Environmental Quality

The maximized natural lighting provides for a bright and cheery atmosphere that promotes customer satisfaction (Figure 68-12). The glass doors for the freezers and cold cases not only save energy but also maintain a more comfortable indoor temperature for the shoppers. The combination of a low-VOC interior and increased airflow reduces airborne contaminants (Figure 68-13). At the entrance to the store, a large informational display was built to help

take a deep breath

Such a simple thing, fresh air – but we've thought a lot about it... We ditched materials made with a higher percentage of potentially airborne chemicals in favor of safer, friendlier alternatives, at the same time maximizing airflow and ventilation in our store... no stale air here.

(a)

(b)

Figure 68-13 (a) Increased airflow and a reduction in the use of VOCs reduces airborne chemicals; (b) This design provides for cleaner air.

(a)

(b)

Figure 68-14 (a) Public information center located at the entrance to the supermarket; (b) Individual learning stations provide information about sustainable technologies and applications.

educate customers and employees about "green" building practices to promote a public awareness (Figure 68-14).

68.12 PLATINUM LEED RESIDENTIAL APPLICATION: BRIGHTBUILT BARN

The BrightBuilt Barn is a Platinum LEED-certified, 700-square-foot, single-story residence and loft that was built in 2008 (Figure 68-15). It is one of the few buildings in the world

Figure 68-15 Exterior entrance view of the BrightBuilt Barn.

Figure 68-16 View of the 6-kW solar panel array located on the roof.

designed to be truly carbon neutral. This is because on a typical sunny day, the Barn produces more renewable energy in the form of electricity than it uses. This additional electricity is fed back to the utility grid and acts as an offset to the carbon footprint that resulted from its initial building phase. Over time, this surplus energy will balance the carbon dioxide released directly into the atmosphere during construction and the carbon indirectly attributed to the building materials.

Guiding Principles

The five guiding principles integrated into the project from its conception are livability, sustainability, replicability/affordability, disentanglement, and education. In terms of livability, the building needed to be comfortable, functional, and attractive. Long-term sustainability for the future is achieved by reducing the carbon footprint used to create the structure. Replicability and affordability result from the Barn being built as a prototype so that others can duplicate the design to the extent that is economical for them. Disentanglement helps to provide for adaptability, allowing the building to be changed over time. Education is provided to the greater community through regularly scheduled open houses that allow others to study and learn from the project.

Design Features

The Barn features a 6-kW photovoltaic panel array located on its roof (Figure 68-16). This array comprises thirty individual 32×62–inch solar panels able to produce approximately 20 kWh per day (Figure 68-17). The electrical use

Figure 68-17 Close-up view of 32×62–inch solar panels.

Figure 68-18 A solar panel array's electricity-tracking monitor.

is continuously monitored by a SunPower Monitor (Figure 68-18) and can be tracked on the BrightBuilt Barn Web site. The Barn generally produces more electricity than it uses. High-efficiency LED lighting is strategically positioned throughout the building for the best illuminating effect (Figure 68-19).

The roof and walls are made from structural insulated panels (SIP). These panels are designed to form a complete

(a)

(b)

Figure 68-19 (a) Energy-efficient LED lighting; (b) The lights are positioned to provide the appearance of outside sunlight.

(a)

(b)

Figure 68-20 (a) Polycarbonate light panels above and below the window as viewed from the inside of the building; (b) Polycarbonate light panels above and below the window as viewed from the outside of the building.

wall system that eliminates air infiltration. This superinsulated building envelope has a continuous R-40 value. In addition to the triple-glazed windows, there are light-transmitting, nanogel-filled polycarbonate panels that impart an effect of a semi-transparent wall. These panels offer an abundance of light but provide twice the insulation value (R-12) as compared to the triple-glazed windows alone (Figure 68-20).

To provide for disentanglement, the Barn is designed so that the systems are separated physically from the structure. Electrical and mechanical systems are run through integral chassis hidden in the baseboard rather than being located in the walls themselves (Figure 68-21). Interior walls can be easily moved as future use of the building changes (Figure 68-22).

The mechanical/storage loft holds an insulated solar storage tank for domestic hot water, the heat pump air handler, and a heat-recovery ventilator. The hot-water storage tank is heated by propylene glycol circulating through a heat-exchange coil (Figure 68-23). The glycol is heated by the sun's energy in two thirty-tube evacuated solar hot-water collectors located on the roof (Figure 68-24). Whenever the glycol temperature becomes warm enough, a differential

(a)

(b)

Figure 68-21 (a) Separate chassis for wiring located in the floor baseboard showing a receptacle and a light switch; (b) Corner view of electrical chassis.

Figure 68-22 Room dividing wall (partition) that can be easily moved to alter the dimensions of the space.

(a)

(b)

Figure 68-23 (a) Hot-water storage tank located in the loft; (b) Glycol is circulated between a heat exchanger located in the hot-water storage tank and the solar hot-water collectors located on the roof.

(a)

(b)

Figure 68-24 (a) View of the two 30-tube evacuated solar hot-water collectors located just above the electrical solar panels; (b) Close-up of the solar hot-water collector.

(a)

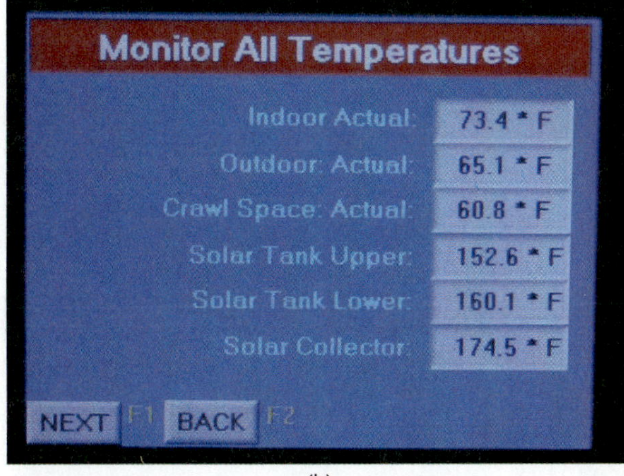

(b)

Figure 68-25 (a) System temperature monitor; (b) System temperature display.

temperature controller automatically activates a circulating pump attached to the tank to circulate the glycol through the heat exchanger. In winter, the hot water is also used to heat the building. No furnace is used for heating. An outside air-to-air heat pump and electric heating coils within the storage tank supplement the solar collectors on very cold days. The operating temperatures are continuously monitored and can be tracked on the BrightBuilt Barn Web site (Figure 68-25).

UNIT 68—SUMMARY

There has been a substantial movement toward sustainable green building design. Both government agencies and private organizations are expending considerable resources to provide education, establish standards, and certify compliance to ensure a sustainable future. The HVACR industry will be tasked to provide energy-efficient equipment that is installed properly and maintained to guarantee energy savings. While the initial rating systems for green design were most often voluntary, the trend is moving toward an increasing number of mandatory standards and code requirements. You will need to rely on professional journals and regular attendance at continuing education seminars and workshops to keep abreast of the most current regulations.

UNIT 68—REVIEW QUESTIONS

1. According to the Department of Energy, the combined impact of all buildings, both commercial and residential, accounts for more than _________ percent of all electricity usage and ______ percent of all natural gas usage.
2. The DOE's Building Technologies Program has a goal to increase the number of high-performance buildings, which will be ______ percent more efficient than today's typical buildings.
3. What was the first state to develop a mandatory green building standards code?
4. The CALGreen code is established to do what?
5. What makes a building sustainable?
6. From a financial perspective, should owners of a sustainably managed building expect a higher return on investment?
7. Why is it important to meet the building standards during new construction rather than waiting until some later time to incorporate them?
8. What does the acronym LEED stand for?
9. Who developed the LEED rating system?
10. What is a jurisdictional code?
11. What is the IGCC, and what does it serve to accomplish?
12. What are the I-Codes?
13. What is the purpose of the ASTM standards?
14. What is Standard 189.1-2009?
15. What is the National Green Building Standard?
16. What is the ANSI/GBI 01-2010 standard?
17. What are the five main rating system categories for LEED certification?
18. What are the three different tier credentials that can be earned through the LEED program?
19. What are VOCs, and how does it apply to green building programs?
20. How are freezer and cold cases different in a "green" supermarket as compared to a conventional supermarket?

UNIT 69

Indoor Air Quality (IAQ)

OBJECTIVES

After completing this unit, you will be able to:

1. define the meaning of indoor air quality (IAQ).
2. state the HVACR technician's role as it relates to indoor air quality.
3. identify various pollutants and pollutant pathways that affect indoor air quality.
4. list several tools and instruments to help measure and evaluate indoor air quality.
5. explain appropriate strategies to prevent, control, and resolve certain issues related to indoor air quality.
6. set limits that guide when other professionals need to be involved in addressing indoor-air-quality issues.

69.1 INTRODUCTION

A properly trained and certified HVACR technician should be able to communicate how the interacting systems of a building affect indoor air quality with the building owner and the occupants. An HVACR technician who takes the time to expand his knowledge by learning and applying building science will be an extremely valuable resource in the marketplace. These technicians will be far better prepared to positively affect the building environment from an energy and environmental perspective in addition to improving indoor air quality. Providing the occupants with a healthy, safe, comfortable, and sustainable indoor environment often also saves energy and reduces the building's impact on the environment. HVACR technicians have a responsibility to the customer to ensure that the HVACR systems do not negatively affect the indoor air quality or cause other unintended consequences with building systems. Learning the concepts of building science and being able to apply them are parts of an HVACR technician's role in today's marketplace. Properly trained technicians can use their knowledge of building science to provide a building whose systems and occupants are able to work together.

69.2 INDOOR AIR QUALITY DEFINED

Simply defined, indoor air quality (IAQ) is the state of the air contained within an enclosed building. The air may be conditioned or unconditioned, and it may be acceptable or objectionable to the occupants. To achieve acceptable IAQ, the air's temperature, humidity, and velocity may need to be conditioned and pollutants must be removed. Often, when people talk about IAQ they are talking about identifying any source of pollutants or contaminants and taking appropriate actions to remove, control, or prevent their amplification. Amplification in IAQ terms means the contaminant's ability to grow and spread.

"Indoor air quality" has rapidly advanced from a cutting-edge buzzword to an important consideration in building construction and operation. IAQ is now widely acknowledged and accepted as being important in the marketplace. However, the perception of what constitutes good IAQ and the associated improvements needed to achieve it remain an interesting topic of discussion and debate by those in the industry. For some, improving IAQ has meant simply upgrading air filters or the addition of an advanced air cleaner. For others, it has involved installing humidification or dehumidification systems, coil and drain pan cleanings, or duct cleanings, or even installing heat-recovery ventilation systems. Still to others it means looking at the entire building as a system and looking beyond just the HVACR systems to truly comprehend what is occurring in a building and why it is happening. It involves getting into the science that is affecting the air from a chemical and biological perspective as well as the physics of the air. To truly make a difference and solve IAQ issues properly, a technician needs to understand how a building and its systems interact. HVACR technicians need to understand that the most effective way to address IAQ is by using techniques that eliminate or reduce the source of the pollutants. It will sometimes take multiple strategies working together to fully address complex IAQ issues.

69.3 POLLUTANTS

There are a variety of potentially harmful or damaging pollutants that must be considered when addressing indoor air quality in buildings. These pollutants can affect human occupants and building systems in a variety of ways. There are two major categories of pollutants that must be controlled: chemical and biological pollutants. Within each of these broad categories are many subcategories that an HVACR technician should have a working knowledge of. This is for the safety of both the technician and the occupants. Each pollutant has its own unique characteristics. Where the various pollutants may be encountered in the building varies. There may also be many ways to prevent, control, or resolve issues that might arise from the pollutant being encountered in the indoor environment.

Pollutants can be produced naturally by the earth or may be dependent on some level of human intervention. Radon is a radioactive gas that is produced naturally by the earth. There are also certain elements or compounds that may mix and interact to produce a potentially harmful contaminant. Some household cleaners can oxidize rapidly and emit a strong, gaseous odor when contacting materials that are commonly found in buildings. Chlorine bleach and ammonia cleaner can react and form chlorine, a poisonous gas. Still other pollutants may result from certain processes taking place. For example, a copier may be producing ozone at a level that it is harmful to the operator and perhaps even coworkers because it is not properly ventilated. Another example is the buildup of carbon dioxide and moisture from the building occupants' breathing. Both water and carbon dioxide can cause IAQ problems in high enough concentrations. Improper operation of a piece of heating equipment can be the source of deadly carbon monoxide from incomplete combustion. Many building materials contain formaldehyde and will produce formaldehyde gas when exposed to higher temperatures and humidity. This pollutant can be harmful even at relatively low levels. These are all examples of situations commonly encountered in the marketplace that an HVACR technician might encounter and have to resolve.

CAUTION

It should be emphasized that this is a basic introduction to the pollutants. It is not intended to be an all-inclusive or exhaustive set of information. What is contained in this unit should help HVACR technicians become familiar with the concepts and to gain a basic understanding of what they might encounter. However, it should not be the only source of information, and it is not intended to instruct technicians how to resolve every IAQ issue they will encounter.

69.4 CHEMICAL POLLUTANTS

Asbestos

Asbestos is a naturally occurring material mined from rocks that occurs in fiber bundles having unusual tensile strength and fire resistance. Asbestos has been banned for use in the United States in certain products since the late 1980s, but it is frequently encountered in older buildings. It can be found in the older buildings in joint compounds, vinyl floor tiles, pipe and boiler insulation, fireproofing, caulks, shingles, and a variety of coatings. Asbestos becomes a potential IAQ problem when it is in a "friable" condition. Friable means that the fibers can easily be released into the air with minimal disturbance. Asbestos also becomes a serious issue when it might be disturbed for any cutting or sanding operations or any demolition. The HVACR technician should not disturb or remove asbestos-containing materials. In many cases, it is best to simply leave material alone that it is believed to be asbestos if it is in good condition and intact. Outside assistance will almost certainly be required if the project requires disturbing asbestos material. Asbestos cannot be positively identified without the services of an outside laboratory to do sampling.

SAFETY TIP

In all cases where asbestos-like material is encountered, the HVACR technician needs to take extra care and caution, as the fibers can cause diseases of the lung, including asbestosis, mesothelioma, and lung cancer. HVACR technicians should never enter a posted area for any reason when working in an environment where asbestos remediation is underway in a part of a building.

SERVICE TIP

When a material suspected of containing asbestos is believed to be the cause of an IAQ problem, the services of another professional should be retained. It is good to have a relationship with a local laboratory and properly qualified asbestos remediation contractor who can address asbestos if it is encountered in the area of operations where an HVACR technician works.

CAUTION

Under most circumstances it is illegal for technicians to remove asbestos from the job site. Asbestos removal must be carried out by licensed asbestos abatement companies. Removing asbestos without the proper equipment can be hazardous to your health and could result in fines for you and your company. If you suspect that an older installation has asbestos, you must request that an approved laboratory test the material to see whether asbestos is present.

Lead

Lead is another pollutant that HVACR technicians will encounter as they work in older buildings. The HVACR technician will need to know how to address lead when it is encountered. While paint that contains lead is often a primary source of airborne contamination, lead can also be found in water and contaminated soils. Lead paint that is in good condition should remain intact. If the paint suspected of containing lead is found to be in poor condition, then it will need to be tested by a laboratory to confirm that it does indeed contain the pollutant, and then it must be properly removed. The paint should not be scraped, sanded, or burnt off by untrained individuals. Lead poisoning can cause damage to many of the body's organs, and lead dust can be easily brought home to spouses and children on clothes. Decontamination is part of the requirement for performing lead remediation work.

CODE TIP

The Environmental Protection Agency (EPA) requires contractors to be certified and follow specific work practices to prevent lead contamination when performing renovation, repair, and painting (RRP) projects that disturb lead-based paint in homes, child-care facilities, and schools built before 1978. Contractors should test for lead paint in dwellings built before 1978 before beginning any remodeling or reconstruction work. If lead is discovered, all contractors who work on the project must be EPA RRP certified and must use Lead Safe Work Practices. Any contractor who neglects to comply could be penalized up to $37,500 each day.

Radon

Radon is a gas that is produced from radium, which is found in rocks and soil in the ground. A building that is constructed in proximity to a source of this gas can become contaminated easily if the pollutant is not controlled. It should also be noted that radon has been found in water from wells, and this can be a source for the pollutant to enter a building. Radon gas is colorless and odorless. Radon testing should be performed using methods that meet or exceed established industry testing standards. The standard remediation for radon gas is tightening the building envelope and capturing and removing the gases. This is usually done through ventilation. Radon gas that is not addressed and allowed to enter the living space can lead to lung cancer for building occupants.

Combustion

Combustion is a chemical reaction that occurs most commonly inside equipment that is installed in buildings. The byproducts of combustion have the potential to cause significant IAQ issues when they do not vent properly or something causes the combustion process to go wrong. HVACR technicians may need to perform combustion analysis, draft testing, spillage testing, or building depressurization testing when investigating IAQ issues relating to combustion.

CO

Carbon monoxide is a colorless, odorless, and tasteless gas. CO can be produced in substantially harmful quantities when the combustion process becomes incomplete from lack of oxygen or low combustion temperatures. Incomplete combustion and CO may occur from a flame impingement, poor fuel-air mixture, incorrect flame pattern, or combustion air starvation.

Carbon monoxide inhibits the blood's ability to carry oxygen. Extremely low concentrations of CO have been associated with long-term health effects, especially in children, the elderly, and those who have their health compromised by other diseases or ailments. CO levels as low as 9 ppm (parts per million) are deemed as actionable by some credible industry standards. The technician should always exercise extra caution when working in or around combustion appliances to ensure that the levels do not exceed the OSHA-permissible exposure limits of 50 ppm for an 8-hour time weighted average.

The HVACR technician needs to understand that proper combustion is not only essential to aid them in preventing an opportunity for an IAQ problem to occur, but it is also essential to enable the equipment to operate efficiently, meet manufacturer specifications, and remain clean and serviceable.

SERVICE TIP

The source of CO contamination may not be the appliances but the garage. Cars produce large volumes of CO. CO produced by a car operating in an attached garage can be pulled into the house and distributed by the air conditioner or furnace. This is especially likely if the air-conditioning equipment is located in a mechanical room adjoining the garage.

CO_2

Carbon dioxide is a normally occurring trace gas in the earth's atmosphere. It has been used as one means for measuring IAQ relative to outdoor air quality. When measuring carbon dioxide, the HVACR technician needs to recognize that outdoor air will provide a baseline against which the indoor air can be evaluated. A technician addressing IAQ should recognize when the indoor air concentrations of CO_2 are higher than the outdoor concentrations. Demand-controlled ventilation systems are available that operate when carbon dioxide levels exceed a certain operating set-point in a building. Keeping inside levels close to the outdoor baseline is the goal. Since CO_2 concentration is addressed by outside air ventilation, it is not possible to reduce the CO_2 level inside any lower than the CO_2 level outside.

NO_2

Nitrogen dioxide is a gas that is poisonous when inhaled. It is a reddish-brown color and has an odor. It can cause serious injury or death even at extremely low levels of exposure, especially to those individuals at risk, such as children, the elderly, and those with respiratory diseases and ailments. Another closely related pollutant is nitric oxide. Together these two pollutants are commonly referred to as NO_X. These are both byproducts of combustion and contribute to acid rain.

Environmental Tobacco Smoke

Environmental tobacco smoke is becoming less of an IAQ problem as more and more authorities having jurisdiction are restricting the use of tobacco products in public places. Where smoking does still occur, it can become an IAQ issue that an HVAC R technician may be asked to address. Since environmental tobacco smoke has an estimated 4,000 compounds in it, it is typically a challenge to address using a single method of remediation. Most often the best solution is a combination filtration and exhaust system.

Mercury

Mercury is a heavy metal. Mercury is also found in the air in particulate form around industrial plants that consume fuels with a high sulfur content. It can be found in buildings in a number of other forms, such as in fluorescent lightbulbs, batteries, thermostats, and thermometers. It is important that HVACR technicians who are addressing IAQ where mercury is found or suspected take special precautions to protect themselves from unnecessary exposure. Mercury must be disposed of properly to avoid cross-contamination.

Volatile Organic Compounds

VOCs are a class of chemical compounds that give off gas due to their high vapor pressure under standard conditions. IAQ is quite often compromised by these compounds, which are found in a number of building materials. They come from paints, carpeting, manufactured woods, and laminates. There are several types of VOCs encountered, include aldehydes, ketones, mercaptans, and hydrocarbons.

- **Aldehydes** Aldehydes are a class of organic chemical compounds. Formaldehyde is the most commonly encountered aldehyde. Varying forms of this chemical are used in pressboard materials like medium-density fiberboard, plywood, particleboard, or oriented strand board (OSB). It is also used in the production of insulation, paints, coatings, glues, and certain fabric materials and is a byproduct produced during combustion, especially from unvented fossil fuel burning appliances.

People are typically exposed to formaldehyde when they breathe air that contains formaldehyde fumes. Formaldehyde can cause burning and watery eyes and irritate nasal cavities and mucous membranes. Other reactions include chest tightness, difficulty breathing, and allergic reactions. The first line of defense against formaldehyde exposure is to use formaldehyde-free products or low-formaldehyde-containing products. Ventilation is a common approach to reducing formaldehyde concentrations when the formaldehyde is coming from products inside the building.

- **Ketones** Ketones are often used in paints and perfumes. The most commonly encountered ketone encountered in IAQ will likely be in paints. When an IAQ issue is raised due to odor from painting, it is often related to the ketones in the paint. Selecting paint products that are low in VOCs will typically lower the amounts of ketones.
- **Mercaptans** Mercaptans are often used as an odorant in natural gas for safety, since the gas is odorless. Those who are chemically sensitive can react to even trace amounts in the parts-per-billion range of these odorants. It does not take a large dose to affect IAQ for these people. Mercaptans can enter the building through gas leaks and combustion products entering the building. To reduce mercaptan levels, eliminate any unvented combustion appliances and make sure the vents on vented appliances are operating correctly.
- **Hydrocarbons** Hydrocarbons are substances that contain only the elements hydrogen and carbon. They are often used as fuels. Many of our rubber products, asphalts, plastics, and fiber materials contain hydrocarbons in aromatic form. These hydrocarbons can cause IAQ issues in buildings.

Particulates

Particulates can arise from natural processes such as wind erosion, sea spray evaporation, volcanic eruption, and biological processes. However, natural sources usually create far less contamination than manmade activities. Some of the manmade activities that cause air contamination are power plant operation, industrial processing, various types of transportation, and agricultural activities.

A micron is a millionth of a meter, or one thousandth of a millimeter. Figure 69-1 shows the relative size of some particulates in microns. By comparison, human hairs range

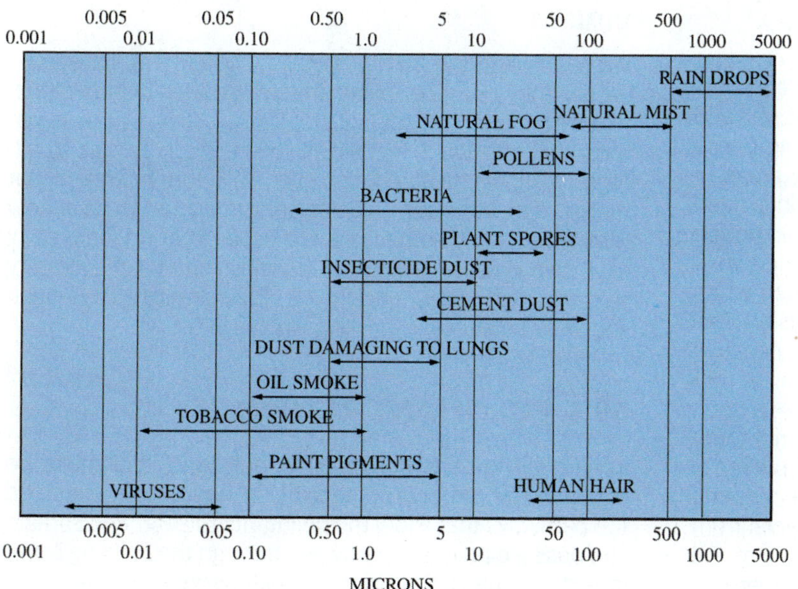

Figure 69-1 Sizes and characteristics of airborne solids and liquids.

in size from 25 to 300 microns. Particles smaller than 10 microns are not visible to the naked eye. Smaller particles, such as smoke, are visible only in high concentrations. Health authorities are concerned with particles that are 2 microns or less because they can be retained in the lungs. Note the size range of a few common pollutants: pollen, 10–100 microns; bacteria, 0.4–5.0 microns; tobacco smoke is a solid ranging in size from 0.1 to 0.3 microns; and viruses, 0.003–0.06 microns.

Pesticides

Pesticides as well as insecticides, rodenticides, herbicides, and fungicides are chemicals used for controlling certain plants, animals, and parasitic organisms. They are designed to kill plants and animals and can be extremely harmful or even fatal to humans. They are important to the technician addressing IAQ because the compounds can be carried in on the feet of occupants of the building and become airborne, causing reactions in some individuals. These items may be in high enough concentration on surfaces of or around a building that they can actually be drawn in by natural infiltration entering the building through the cracks and crevices.

Ozone

Ozone (O_3) is an oxidizing agent with a very pungent odor. It is harmful to humans and animals and will cause severe irritation of the mucous membranes in the airway. Ozone is produced by electrical discharge and certain types of ultraviolet radiation. Ozone is an excellent agent for breaking down certain hydrocarbon substances as well as deodorizing air and purifying water, but it must be used carefully. The levels of ozone present in a space must be checked throughout the process of sanitizing with proper testing equipment because ozone can cause damage to or harm items it comes into contact with through rapid oxidation. Ozone should never be introduced into occupied living spaces, as it will cause harm to humans and animals.

69.5 BIOLOGICAL POLLUTANTS

Pests

Pests have the potential to introduce biological contaminants into an indoor air environment. In the case of bats and mice, their fecal matter can be hazardous to humans. These droppings are often encountered when entering areas of buildings like ceilings, soffits, or attics. The droppings need to be carefully dealt with. The fecal material should not be inhaled or allowed to be dispersed, as it could cause new or additional IAQ issues. Care should also be taken when performing pressure diagnostics on buildings using fan-assisted pressurization devices. Creating a large negative pressure in the building can pull these contaminants into occupied space. Mice can carry foodborne bacteria such as salmonella, and deer mice in some parts of the country carry the hantavirus. The disease can spread when a person breathes in air or comes into contact with contaminated surfaces exposed to the saliva, urine, or feces from an infected mouse. From an IAQ perspective, one of the best ways to prevent infection in a building is to keep the building tight and not let anything in. If there is evidence of infestation from pests, then this needs to be properly addressed by trained professionals.

Bacteria

Bacteria are common in the indoor and outdoor environment and can be the cause of diseases in plants, humans, and animals. All bacteria are not harmful. For example, bacteria are used in pickling and fermenting processes. Bacteria are classified by their shape primarily as well as their need for oxygen and whether they are affected by heat. The quantity of bacteria growing is largely impacted by the host it is living on or in as well as temperature, air movement, light, and moisture. Bacteria may grow in soil, water, food, and in humans and animals, to name a few potential hosts. Bacteria may often be found as a result of a water or moisture problem. These bacteria tend to live off of dead organic matter and will thrive in many of the materials commonly found in and around our buildings.

Fungi

Fungi are organisms that absorb their food from an external source. This group includes molds, mildew, yeasts, and mushrooms. An HVACR technician will have numerous encounters with items that fall into this category. Fungi generally thrive on organic materials, although there are some that require a living host organism and are parasitic in nature. Much like bacteria, there are fungi that are good and bad. Fungi can cause disease in humans. From an IAQ perspective, it is often the airborne mold spores containing mycotoxins that need to be addressed.

It is important to recognize that most of these items occur everywhere. However, an HVACR technician should learn to recognize that the presence of fungi may be an indication of other problems. For example, *Stachybotrys chartarum* is a mold that thrives in materials containing cellulose that have been water damaged. If it is found, then there is a need to investigate where the water is coming from. The likelihood of finding water damage along with the source of the contaminant is extremely high.

Viruses

Viruses invade the human immune system and cause diseases. They produce toxins and are smaller than bacteria. Viruses are often very difficult to destroy. They are not something that an HVACR technician will normally be able to address. It should be noted, however, for future reference that there is ongoing research being performed that looks at how HVAC and IAQ equipment may be used to help in the control of certain viruses, particularly in highly controlled laboratory settings. High concentrations of ultraviolet light have been shown to be effective in killing viruses in the air.

69.6 POLLUTANT PATHWAYS

In addition to understanding the pollutants that might be encountered, an HVACR technician needs to understand the pathways that allow the pollutants to enter and spread. The possible pollutant pathways through the building envelope or mechanical systems are extensive.

Figure 69-2 Leaks in return ductwork located in a crawlspace can pull in contaminants from the crawlspace.

Outside to Inside

One of the more common pollutant paths occurs from outside the building to inside the building. For example, window or door openings, plumbing stacks, and chimney chases are obvious examples. Some common pathways would include a duct system (Figure 69-2), wall chase (Figure 69-3), plumbing chase (Figure 69-4), chimney stack (Figure 69-5), or an electrical penetration (Figure 69-6).

CAUTION

It is imperative when looking to resolve IAQ problems that one identify the pollutant pathways to determine if there is any cause-and-effect relationship between them. Failure to identify these pathways may result in providing a solution that is ineffective or causes other unintended consequences.

Figure 69-3 Panned floor joists can suck air from both the crawlspace and the attic.

Figure 69-4 Leaks around plumbing can provide a path for pollutants.

There are also less obvious examples, such as the space between the sill plate and foundation (Figure 69-7), foundation wall cracks (Figure 69-8), and voids in air barriers, vapor barriers, and thermal materials (Figure 69-9), attic hatches (Figure 69-10), and top plates (Figure 69-11). There are also substantial pathways through vent pipes and exhausts, fresh air intakes, and other openings made for mechanical devices such as attic fans. These can be pathways under some conditions but not under other conditions. Each of these is an

Figure 69-5 Pollutants can be pulled in through a chimney.

Figure 69-6 The holes in framing for wires can provide a pathway for pollutants.

Figure 69-7 The crack between the sill and the foundation can allow infiltration to carry pollutants inside.

Figure 69-8 Cracks in the foundation wall can provide a pathway for pollutants to enter.

Figure 69-9 Gaps in crawlspace vapor barriers can increase the pollutants inside the crawlspace.

Figure 69-10 Air gaps around attic hatches can provide a way for infiltration to carry pollutants into the house from the attic.

Figure 69-11 The outer wall top plate is exposed to unconditioned air; cracks and penetrations in the top plate can create a path for infiltration and pollutants.

Figure 69-12 The holes in the framing above this power panel create a conduit for air and pollutants from one part of the house to the other.

outdoor-to-indoor connection. HVACR technicians often look first at the mechanical system as the source for IAQ problems since they do not typically focus much attention on the integrity of the building envelope. Technicians should know where and how to identify issues in the building envelope. They should also be able to measure the amount of building leakage to the outside and understand how this interacts with the building mechanical systems. Understanding how a building's construction and its systems interact is called building science.

One Location to Another Inside the Building

Another common pollutant path is from one location inside the building to another location inside the building. Walls, floors, ceiling cavities, soffits, chases, and ductwork are all typical paths. Holes through framing can create an unintended pathway between adjacent areas of a building (Figure 69-12). One common situation that an HVACR technician is often asked to address is carbon monoxide presence in a building. It is important that an HVACR technician understand the sources of CO because some may be more effectively controlled at the source, while other sources, such as an attached garage, may require addressing the pressure boundary between the garage and the occupied space. Still other situations, like CO generated from a kitchen area where cooking appliances are located, may require some spot ventilation. Each of these situations might be encountered by technicians. Knowing how to address the suspected problems is important.

Benefits of Locating and Sealing Pollutant Pathways

Another reason that a technician needs to address pollutant pathways has to do with energy effects. This is an IAQ issue because as energy costs increase, HVACR systems may be operated less or at the extremes of the typical comfort ranges to try to save money. This might involve higher or lower temperatures and more substantial variations in relative humidity levels; and outdoor air ventilation rates may also be affected. A building that has high rates of infiltration will add additional load to the HVAC systems. This may cause new or replacement systems to be sized larger than is truly needed. Oversized systems also use more materials, which increases the building's impact on the environment. The same pathways that can let in pollutants are also a major area for energy waste. Defects in the building envelope affect HVAC system performance, energy use, and the IAQ of the building.

69.7 BUILDING SYSTEMS

The air pressure inside a building can be about the same, lower than, or higher than the air pressure outside the building. Residential buildings seldom seek to control the pressure inside the building, but commercial structures typically try to maintain a positive pressure. A building with a negative pressure will experience infiltration (air leaking in) through all the cracks and holes between the outside and inside. Drafts by windows and doors in the winter are a good example of infiltration. Buildings with a positive pressure will experience exfiltration (air leaking out) through all the cracks and holes in the building. Outside air must be introduced into the building to maintain a positive building pressure.

Ventilation Versus Infiltration

Infiltration is air that leaks in on its own because the pressure in the building is lower than the pressure outside. Infiltration provides a path for pollutants to enter the building. Ventilation air is intentionally brought in to maintain a positive building pressure and to provide air for vent fans and draft hoods. Air introduced through ventilation can be controlled, filtered, and conditioned before being introduced into the building. Diagnostic testing can help quantify the amount of air infiltration/exfiltration and locate the pathways it is capable of taking. These pathways can be critical pieces of information when evaluating the IAQ in a building. A building's air change information will also have an important impact on load calculations and may affect the size of the heating or cooling equipment.

69.8 TOOLS AND INSTRUMENTS

There are many tools and instruments that can be used by a technician when addressing IAQ in a building. Each of these tools and instruments will require that technicians take the time to learn how to use and master them. A technician learning how to use these tools and instruments should plan to work with another technician who has extensive experience using the devices to ensure that they are being used and understood correctly. Failure to use the tools and instruments properly might lead a technician to make decisions that could expose the occupants to unnecessary risks, jeopardize their own safety, and increase the liability exposure of their employer should their actions be determined to be the cause of death, injury, or destruction.

INDOOR AIR QUALITY CHECKLIST

Collect and Review Existing Records

- HVAC design data, operating instructions, and manuals.
- HVAC maintenance and calibration records, testing and balancing report.
- Inventory of locations where occupancy, equipment, or building use has changed.

Conduct a Walkthrough Inspection of the Building

- List of responsible staff and/or contractors, evidence of training, and job descriptions.
- Identification of areas where positive or negative pressure should be maintained.
- Record of locations that need monitoring or correction.

Collect Detailed Information

- Inventory of HVAC system components needing repair, adjustment, or replacement.
- Record of control settings and operating schedules.
- Plan showing airflow directions or pressure differentials in significant areas.
- Inventory of significant pollutant sources and their locations.
- MSDSs for supplies and hazardous substances that are stored or used in the building.
- Zone/room record.

IAQ MANAGEMENT PLAN

Select IAQ Manager

Assign Staff Responsibilities/Train Staff

- Facilities operation and maintenance.
- Confirm that equipment operating schedules are appropriate.
- Confirm appropriate pressure relationships between building usage areas.
- Compare ventilation quantities to design, codes, and ASHRAE 62-1989.
- Schedule equipment inspections per preventive mainte-
nance plan or recommended maintenance schedule.
- Modify and use HVAC checklist(s); update as equipment is added, removed, or replaced.
- Schedule maintenance activities to avoid creating IAQ problems.
- Review MSDSs for supplies; request additional information as needed.
- Consider using alarms or other devices to signal need for HVAC maintenance (e.g., clogged filters).

Housekeeping

- Evaluate cleaning schedules and procedures; modify if necessary.
- Review MSDSs for products in use; buy different products if necessary.
- Confirm proper use and storage of materials.
- Review trash disposal procedures; modify if necessary.

Shipping and Receiving

- Review loading dock procedures. (NOTE: If air intake is located nearby, take precautions to prevent intake of exhaust fumes.)
- Check pressure relationships around loading dock.

Pest Control

- Obtain and review MSDSs; review handling and storage.
- Review pest control schedules and procedures.
- Review ventilation used during pesticide application.

Occupant Relations

- Establish health and safety committee or joint tenant/management IAQ task force.
- Review procedures for responding to complaints; modify if necessary.
- Review lease provisions; modify if necessary.

Renovation, Redecorating, Remodeling

- Discuss IAQ concerns with architects, engineers, contractors, and other professionals.
- Obtain MSDSs; use materials and procedures that minimize IAQ problems.
- Schedule work to minimize IAQ problems.
- Arrange ventilation to isolate work areas.
- Use installation procedures that minimize emissions from new furnishings.

Smoking

- Eliminate smoking in the building.
- If smoking areas are designated, provide adequate ventilation and maintain under negative pressure.
- Work with occupants to develop appropriate non-smoking policies, including implementation of smoking cessation programs.

Figure 69-13 Example of an IAQ checklist.

Checklists

Checklists ensure that critical steps in a process are not missed (Figure 69-13). Establishing and using methodical procedures will improve accuracy and consistency. This will also ensure that a technician has all information needed to make a proper recommendation or convey correct information to others who will rely on it. These checklists can be organized into a book format or used individually for various test procedures, evaluation procedures, and even common daily routines.

Surveys

Surveys are an essential instrument for every technician to have and use when addressing IAQ issues. They will guide the technician through a series of questions that should be asked before, during, and after the assessment is performed. This process will systematically allow the technician to identify building areas for IAQ improvement and to understand the wants and needs of the occupants.

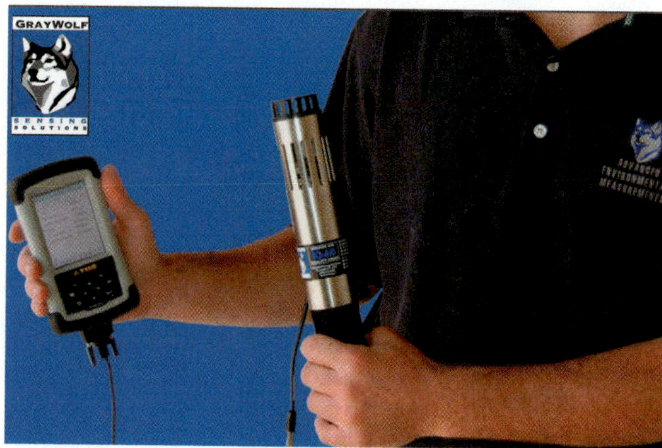

Figure 69-14 This handheld IAQ meter can monitor VOCs, CO_2, CO, %RH, and °C/°F in one place. *(Courtesy of Gray Wolf Sensing Solutions)*

Detectors

Detectors may be used for identifying a variety of substances, such as gases, refrigerants, and other indoor air contaminants such as volatile organic compounds (Figure 69-14). Detectors are typically not capable of quantifying the amount of a pollutant found; they will only indicate that it is present. Where quantifiable data are needed, such as in CO levels, a monitor or instrument with an ability to chart, record, or in some way quantify the amount of pollutant being collected is necessary (Figure 69-15).

Fan-Assisted Pressurization Devices

Blower doors and duct blasters are excellent tools that an HVACR technician can use for the purposes of identifying and quantifying the air leakage rates, locations in buildings, and duct systems. These tools allow measurement of leakage rates. Blower doors are used to pressurize an entire building (Figure 69-16). Duct blasters are used to pressurize just the duct system (Figure 69-17). The information gained from this diagnostic testing will serve to provide accurate input into load calculations being performed related to infiltration. An IAQ investigation may reveal pressure differences from one part of a building to another, which would cause pollutants to flow through pathways that were not detectable without instruments. In ducted systems, the identification of the leak points and amount of air leakage can be very useful in determining why heating or cooling is not working as expected.

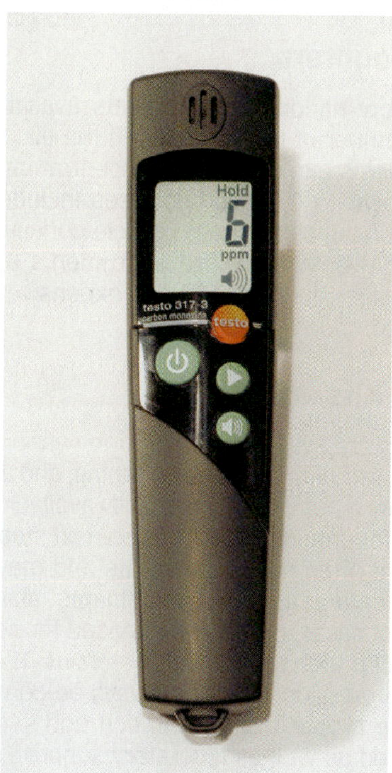

Figure 69-15 Carbon monoxide detector. *(Courtesy Testo, Inc.)*

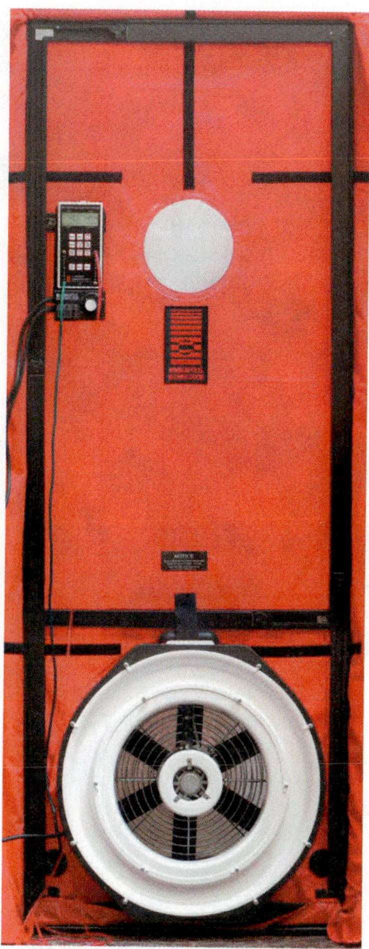

Figure 69-16 Blower doors are used for pressurizing entire buildings for testing. *(Courtesy of the Energy Conservatory)*

Manometers

Manometers should be a familiar tool to an HVACR technician. These devices are used in measuring the pressure of liquids and gases in relation to atmospheric pressure. There are a number of different types of manometers used for the work of an HVACR professional. Common manometer types include the U-tube manometer, commonly used in measuring gas pressure, and the inclined manometer, used in measuring static pressure and velocity pressure in ductwork. Digital manometers are commonly used for IAQ work (Figure 69-18).

Figure 69-17 Duct blasters are used for pressurizing the duct system for testing. *(Courtesy of the Energy Conservatory)*

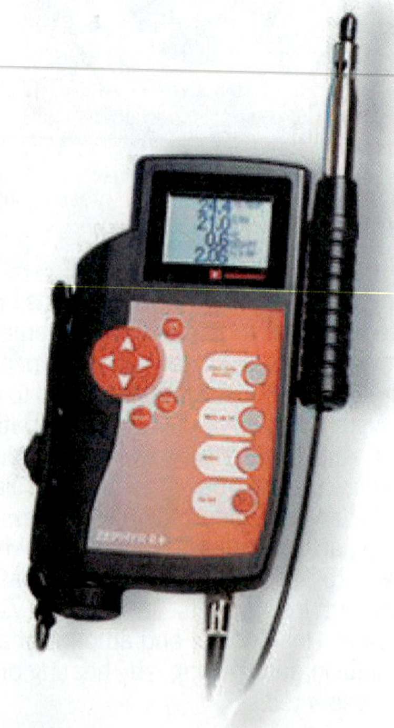

Figure 69-18 Digital manometers are frequently used in IAQ work. *(Courtesy of Gray Wolf Sensing Solutions)*

Depending on the instrument and its precision, these devices may enable accurate measurement of pressure down to 1 Pa or .004 in wc. This is important when evaluating the small pressure differences between different parts of a building or draft pressures where relatively small variations in pressure can mean the difference between something working properly and something having serious problems.

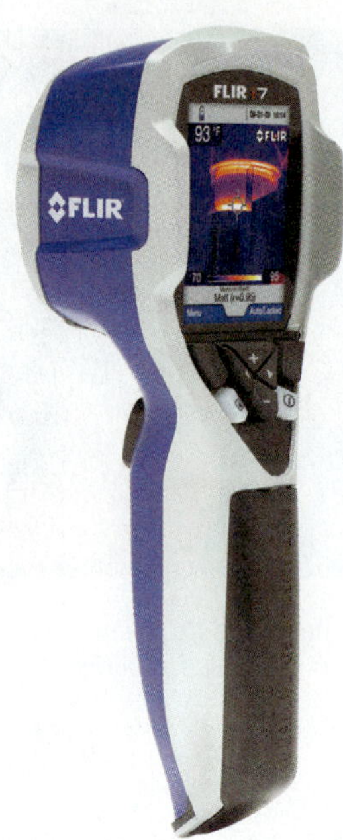

Figure 69-19 Infrared cameras can help spot areas that have thermal leakage. *(Reproduced with permission of FLIR Systems)*

Infrared Cameras and Thermal Imagers

Infrared cameras are a tool that can provide the HVACR technician addressing IAQ with a visual picture to determine what is happening inside or outside of a building (Figure 69-19). Thermal imagers are an excellent tool for use in explaining to owners and occupants of a building why something is or is not happening. The best feature of an IR camera is that it gives the technician the ability to provide a graphic record of the before and after conditions.

Particle Counters

There are now handheld instruments available that can count the number of fine particles in the air. These handheld instruments can count the concentration of particles in the air down to 0.3 microns. Uses include testing air filtration, verifying clean room particle concentration, and investigating IAQ. While these instruments are still relatively expensive, they are far less expensive than just a few years ago.

Sampling Kits and Devices

Sampling kits are available for capturing various pollutants through bulk sampling, surface swabbing, and airborne capture. There are a number of kits readily available that can be used simply for providing guidance on next steps. However, these kits are often not too accurate and may provide no quantifiable data as to the contaminants' ability to cause harm. In most cases, precision is needed for decision making and in fact preferred for liability reasons. The services of an outside professional should always be considered who will have the proper instrumentation and sampling techniques needed as well as laboratory support. An example where testing needs to be precise is evaluating suspected mold, lead, asbestos, and volatile organic compounds.

69.9 PREVENTION, CONTROL, AND REMEDIATION STRATEGIES

There are many different strategies available for preventing, controlling, and eliminating IAQ issues. The three most common strategies for reducing indoor pollutants are eliminating the source of the pollutants, ventilating to dilute the pollutants and reduce their effect, and using filters and cleaners to remove the pollutants from the air. Another strategy that will be discussed is called prevention through education.

Source Control

Source control may also be described as elimination of the pollutant. It is the best choice when it comes to managing IAQ. Source control typically involves the physical removal of a known or suspected pollutant from the building. For example, removal of mold-contaminated materials would be a form of source control. Another example of source control would be removing an element contributing to the problem—for example, for mold to flourish it needs a moisture source; if we remove the source of moisture, then we can remove the ability for mold to grow.

Substitution

Substitution is a variation of source control that is best described as care in product selection. This strategy is becoming increasingly important not only for IAQ purposes but also as the push for a greener planet becomes the focus of so many. Green products are typically evaluated based on their use of recycled content, reusability, renewability, embodied energy, durability, and overall impact on the environment. It is this last attribute about overall impact on the environment that is the most important for IAQ. The HVACR technician who is addressing IAQ can rely on the strategy of substitution to find the right product or material that can be used to improve or prevent an IAQ issue from occurring. Some good examples of alternative products that might be selected include alternative duct lining and insulating products that do not negatively impact IAQ. Duct sealing and other sealants that are low in or contain no VOCs might be another product decision that an HVACR technician addressing indoor quality can rely upon. One of the keys to IAQ-friendly product selection might lie in a careful review of the product's material safety data sheet (MSDS). An MSDS contains a wealth of information on the material and its properties that can guide HVACR technicians in understanding if they are contributing to an IAQ issue.

Isolation

Isolation is also a variation of source control. Isolation may also be described as separation or encapsulation of the pollutant or contaminant. This strategy does not remove the pollutant or contaminant from the building but will serve to keep it from entering the occupied spaces. Isolation may involve installing an encapsulating-type product over the top of an existing product, or it may involve the sealing of one building space from another to eliminate the pollutant pathways. Oftentimes the latter strategy is coupled with some other strategy, such as dilution, to reduce the potential for harm when someone enters the isolated space. A good use of the encapsulating strategy might when encountering existing asbestos-covered pipe that has been tested and is not considered friable. While airborne asbestos fibers would become harmful if the material is disturbed, simply encapsulating with another product may be a viable option to prevent that from occurring and allow it to be left in place without the need to be concerned about its presence. This would be an example of isolation without removal.

Another strategy would be where a particular commercial process is taking place—for example, a printing and copying room in an office facility. The process is not something that can be stopped, as it is critical to a successful businesses operation. In this example, it would be necessary for the room where the process is taking place to be isolated from another room, and then for the safety of the users there would need to be a system of supply and exhaust ventilation. We cover this concept in the next section.

Dilution

Dilution may also be described as a reduction strategy, as it involves a form of exhaust or ventilation but may not completely remove the contaminant. This strategy may involve bringing in air from outside to dilute the harmful pollutant or contaminant being generated in that space and exhausting air from the space to reduce the impact of the pollutant or contaminant. Care needs to be taken when using dilution to ensure that the space remains properly balanced in terms of its pressure relationship with other rooms in the building. Care also needs to be taken to ensure that vents on furnaces, water heaters, and fireplaces work properly. Bringing in more air than is capable of being exhausted can force the pollutants into other connected spaces. Exhausting more air than is being brought in can draw other contaminants in from adjoining connected spaces. To keep the pressure in an area of the building the same, as much air must be brought in as is exhausted out (Figure 69-20). Air that is brought in to allow proper ventilation is called makeup air. Using an energy-recovery ventilation system is one way to maintain a proper balance between makeup air and exhaust air without losing large amounts of heating or cooling in the exhaust air (Figure 69-21).

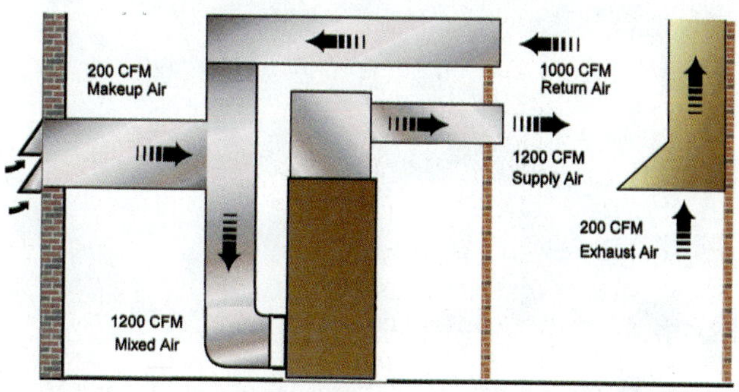

Figure 69-20 It is necessary to provide makeup air from outside for ventilation to work properly.

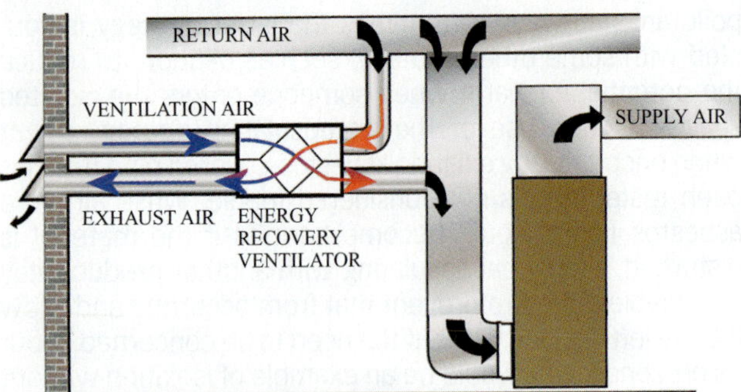

Figure 69-21 An energy-recovery ventilation system can reduce the energy cost of ventilation.

Filtration

Filtration is a reduction strategy and involves air cleaning that typically involves moving air across a surface that is made of various types of filter media material. Filtration is covered in a separate unit and will not be discussed in detail in this section. Care should be exercised to ensure that fan curves and motor capabilities are not exceeded when advanced filtration systems are used on the equipment. In some cases, the attempt to filter pollutants may actually be negatively affected by the installation of advanced filters. This can also cause other problems, such as premature motor failures and excessive use of energy. If filters are not replaced or cleaned regularly, the filter itself can become a pollutant source (Figure 69-22).

Prevention by Education

This is a strategy that is instrumental in achieving long-term and sustainable goals of an IAQ program. It should be coupled with any number of the existing strategies.

Figure 69-22 This filter has collected enough dust that it is now a source of pollutants.

When properly executed, this strategy involves the building owner, occupants, and mechanical systems contractor to address activities that affect the building's indoor air environment. For example, maintenance staff may receive training on cleaning drain pans and condensate lines as well as on the use of chemical-free cleaning methods. Occupants might receive training on the use of equipment or disposal of certain products or waste materials that can cause potential IAQ issues, such as lightbulbs containing mercury, cooking exhaust fans, or bathroom ventilation, to reduce potential moisture problems. The owners of buildings might receive training on systems that can reduce their risk for an IAQ problem becoming a crisis. For example, the owner might be trained on how combustion appliances that are not functioning properly can cause the potential for a CO problem or why oversizing air-conditioning equipment can lead to moisture problems. The key is to condition the behavior of the individuals to prevent problems instead of waiting until the problems have occurred.

69.10 KNOWING YOUR LIMITS

There are likely to be certain aspects of IAQ that are going to be beyond the HVACR technician's capabilities without additional specialized training and professional credentialing. That is not to say that the individual cannot pursue the opportunity and achieve the requirements. However, HVACR technicians need to be careful that in the course of their work they do not overstep their own limits, violate laws and regulations, and, most important, jeopardize life, limb, or property through actions that cause death, injury, or destruction. Clear boundaries need to be established by an HVACR technician addressing IAQ that will ensure that the most appropriate action is taken when a certain situation is encountered. For example, when asbestos, lead, radon, or mold is suspected, the HVACR technician needs to do certain things differently to address the issue than perhaps if a combustion safety problem were encountered. It is far too easy for a human being to get caught up in the moment when attempting to resolve an issue and to overstep one's abilities. Having clear, preestablished limits as well as systems of checks and balances will make it much easier for HVACR technicians to perform their jobs in an organized and methodical manner. It is also valuable for an HVACR technician to set up checklists to guide actions to be taken. This is true for any work being done, not just that related to IAQ. Checklists are a tool used to protect the interests of both the customer and the technician. The customer is going to be much more comfortable knowing that a methodical system has been followed instead of someone simply winging it. It is also a well-known fact that our airline industry uses checklists to ensure that critical steps are not missed in the process of flying airplanes. While we recognize this is not a foolproof method, it helps to manage risk, as you do not get a second chance when you are in the air and mistakes are made. An HVACR technician who is addressing IAQ should give serious consideration to using this approach.

69.11 SAFETY IN IAQ

While this unit is not specifically focused on safety, safety is paramount in all that we do. It is imperative that we do not forget the relevance of being safe when it comes to IAQ work that an HVACR technician may become involved in. There are countless potential hazards that might be encountered, including water hazards, confined spaces, poisoning from ingestion or exposure to harmful or potentially toxic chemicals or biological contaminants, climbing on ladders or scaffolds or surfaces where slips and falls might occur and working in and around systems that might cause electrical, fire, or explosion hazards. Always use personal protection items such as hard hats, gloves, eye protection, hearing protection, protective suits, boots, respirators, and fall protection. Safety procedures should also be practiced, including lockout, tag-out, confined-space safety practices, electrical safety, and fire safety. The importance of safety in and around the job site should never be secondary. An HVACR technician is a first line of defense in the safety effort, and all work must be performed safely at all times. Your life and that of others will depend on it.

69.12 MAKING IAQ HAPPEN

To make IAQ happen, the technician needs to understand what the customer wants and provide solutions that address the customer's concerns. The technician needs to provide guidance because the customer will be relying on the expertise of the HVACR technician to recommend appropriate strategies. If the customer does not know what it takes to make something work right and the technician does not tell them, then IAQ will not happen. It is better to discuss IAQ strategies before there is a problem. The cost of remediation and control after a problem has occurred is usually higher than taking preventative measures ahead of time. So the best course of action when it comes to IAQ is to make it happen right, the first time.

UNIT 69—SUMMARY

After completing this unit, the HVACR technician should be able to define the meaning of IAQ; state the HVACR technician's role as it relates to IAQ; identify various pollutants and pollutant pathways that affect IAQ; list several tools and instruments to help measure and evaluate IAQ; explain appropriate strategies to prevent, control, and resolve certain IAQ issues; and to set limits that guide when other professionals need to be involved in addressing IAQ issues. The HVACR technician who begins to understand and articulate these concepts will be well on the way to becoming a valuable asset to consumers and contractors in the marketplace.

WORK ORDERS

Service Ticket 6901

Customer Complaint: Visible Airborne Contaminant

Equipment Type: Commercial Building with Drop Ceiling Return

An HVACR technician is brought in to address an IAQ problem where the air conditioning is suspected of spreading an airborne contaminant, most likely pollen, throughout the building. The technician encounters some slightly open windows in an office and recognizes that allergens like pollen from some nearby flowers and trees is being allowed to get inside an office space, triggering the suspected IAQ problems for a number of sensitive occupants. The technician decides to close the windows and instructs the occupants to use the air conditioning, believing the problem would simply go away.

Unfortunately, the problem proceeded to get worse and become more widespread after this action was taken. The allergens already in the building were further spread through the building by the air-conditioning ductwork. This caused even more of the sensitive occupants in other parts of the building to have allergic reactions. In addition, the presence of the contaminants seemed to increase exponentially, which was puzzling. The technician decided to recommend using some advanced air-cleaning devices inside the building and in the mechanical system as a solution to the problem. The devices indeed lowered the impact of the allergens spreading throughout the system, but this approach failed to address the source of the problem, which is still unidentified.

In this case, the pollutant pathway that is allowing the contaminant to enter the building is from outside. The technician has failed to understand that pollutants enter a building through not only mechanical systems but also the building envelope and that these two interact. In this situation, the technician should have taken the time to determine where the pollutant was coming in from, not to simply assume it was from open windows or from contaminants already inside the building. It was actually being caused by pressure imbalances in the building space. This was allowing the allergen to be conveyed throughout not only in the ductwork but also a number of open building cavities, including the open return-air plenum above the dropped ceiling, which was in fact drawing air in from outside through the building envelope. The proper solution to this problem would have been to air-seal the outdoors from the indoors to prevent the contaminant from being drawn in and ideally ducting and sealing the return-air system to eliminate the chance for this to happen. Filtering this air certainly helped reduce the impact of the problem, but it did not fix the problem because it was more complicated and the technician did not understand this interaction.

Service Ticket 6902

Customer Complaint: CO Alarm Going Off

Equipment Type: Residence with Attached Garage and Mechanical Room off of Garage

A customer complains that his CO alarm sounds occasionally, and he has called a technician to check the combustion of the furnace and water heater. The technician finds that the furnace and water heater are both operating correctly and that their vents are operating properly. The technician notices that the combustion air grille in the door leads to the garage and asks the customer when the alarm sounds most often. The customer says that the alarm goes off typically after he has left for work in the morning, annoying his wife. The technician asks the customer if he runs his car in the garage. The customer says that he warms his car up in the mornings with the garage door open. The technician shows the customer the CO level in the garage with the car operating and the door up to demonstrate the high levels of CO that build up even with the door open. The technician suggests that the CO is coming from the car, traveling into the mechanical room through the combustion air grille, and entering the ducts through small leaks in the ductwork and furnace cabinet. The technician recommends moving the combustion air grille to an outside wall, replacing the door with a tight-sealing door, and sealing the ductwork to prevent leakage into the return ducts. However, the primary recommendation is to quit operating the car while parked in the garage.

Service Ticket 6903

Customer Complaint: Musty Smell from Air Conditioner

Equipment Type: Split-System Air Conditioner with Blower and Coil in Crawlspace and a Filter Grille in the Hall

A customer complains that the air leaving her vents stinks with a musty odor. The customer says she purchased a high-efficiency panel filter at the local home improvement warehouse and installed it in the filter grille in the house but that it has not helped solve the problem. The technician finds that the dirt in the crawlspace is not covered by a vapor barrier and that the floor joist space is used as a chase for the return air. The technician explains that the seal between the floor joists and the sheet metal is not airtight, so the return air is constantly sucking in crawlspace air. Since the dirt in the crawlspace is exposed, dirt, mildew, and mold spores are also drawn into the system. To stop the entry of contaminants, the technician recommends covering the crawlspace with a vapor barrier and replacing the panned returns with ducted and sealed returns. The technician recommends a duct cleaning to help remove the contaminants already introduced into the system. Finally, the technician recommends removing the filter in the filter grille and installing a high-efficiency air cleaner at the unit, explaining that the high-efficiency 1-in filter made the problem worse by creating a lower pressure in the return ducts due to the pressure drop across the filter.

UNIT 69—REVIEW QUESTIONS

1. Define IAQ.
2. What needs to be addressed to achieve acceptable IAQ?
3. What is amplification?
4. Explain the technician's role in IAQ.
5. Name five pollutants that would be categorized as chemical in nature.
6. Name three pollutant areas that would be categorized as biological in nature.
7. What odorant is commonly used in natural gas?
8. What are the two most common pollutant pathways?
9. What is the best method to use for controlling IAQ?
10. What are other methods used for controlling IAQ?
11. What tool or instrument can be used for ensuring that critical steps are not missed in a test procedure?
12. What tool or instrument can be used for ensuring that critical steps are not missed when interviewing occupants?
13. State the primary difference between a detector and a meter.
14. What tool or instrument can be used for measuring infiltration or exfiltration in a building?
15. What tool or instrument can be used for measuring leakage in ductwork?
16. Why is environmental tobacco smoke difficult to remove from the air?
17. Explain why safety is so important when addressing IAQ.
18. Name some pollutants that require lab testing for positive identification.
19. What is the viral disease that certain mice can carry?
20. Why is it necessary for an HVACR technician to understand the building systems?

UNIT 70

Residential Load Calculations

OBJECTIVES

After completing this unit, you will be able to:

1. discuss the importance of a heat load calculation.
2. explain the difference between heat loss, sensible heat gain, and latent heat gain.
3. construct a U factor for a building panel given the R values of the components.
4. construct an HTM given a U factor and a design temperature difference.
5. discuss the effect that daily temperature swings have on heat gain.
6. explain the difference that house orientation has on solar heat gain.

70.1 INTRODUCTION

Heating and air-conditioning equipment provides the most comfort at the least cost when it is properly sized. The calculations done to determine the amount of heating and cooling required for a particular building are called heat load calculations or a heat load study. In the winter, a load calculation is done to determine how fast the building loses heat. In the summer, the load calculation is done to determine how fast the building gains heat. The calculations determine the size of the equipment that will be used to condition the building. The load calculation is the first step in choosing the equipment and sizing the ductwork. The purpose of this unit is to raise the student's awareness of the factors involved in performing accurate heat load calculations; it is not intended as a complete calculation procedure.

70.2 WHY LOAD CALCULATIONS ARE NECESSARY

Heat load calculations are necessary to properly match the equipment capacity to the building requirement. The two primary reasons for doing a heat load study are to properly size the heating and cooling equipment and to design the duct system to deliver the correct amount of air to each room or zone. A shell load is a calculation treating the entire house like one big room. This is fine for sizing the equipment, but is no help for sizing the ductwork. A room-by-room study is necessary to properly size the duct system because the duct design is based on the required heating and cooling load for each room or zone.

There are many variables to take into account when performing load calculations. The sheer number of factors to consider and the amount of number crunching involved can discourage contractors from performing accurate load calculations. One of the most common quick-and-dirty methods involves dividing the building's square footage by a factor to arrive at an equipment size. There are two obvious problems with this method: all buildings are not the same, and all rooms within a single building are not the same. The time spent doing a careful survey of the building and performing accurate load calculations will save time and money. It is time-consuming and expensive to attempt to make an incorrectly sized and applied system work. This unit concentrates on residential applications.

70.3 LOAD CALCULATION METHODS

Residential and commercial load calculations have unique characteristics that are best served by using a procedure especially designed for each specific system type. A few of the differences between residential and commercial load calculations are as follows:

- **Ventilation** Commercial buildings typically have much higher ventilation rates.
- **Lights** The heat gain from lights in most commercial buildings is a major heat load.
- **People** The people in many commercial buildings are a major source of heat gain.
- **Machinery** The heat from equipment and machinery is often a major source of heat gain in commercial buildings.

The Air Conditioning Contractors of America (ACCA) has developed load calculation procedures that help organize the required information and simplify the process of performing an accurate load calculation. Current manuals include Manual J, 8th edition, and Manual J, 8th edition, Abridged, for residential calculations. Manual N, 5th edition, applies to commercial calculations. These calculation procedures provide tables with factors and data necessary to produce a handwritten spreadsheet. Many computer-based solutions are now available to automate the process.

Manual J, 8th edition, provides tables and procedures for two types of load calculations: peak load procedure and average load procedure. The peak load procedure determines the peak load for each room or zone. Loads are calculated for each room or zone at the time of day in midsummer that represents the highest load for that exposure. Peak load procedure can be used for zoned systems or for rooms with large exposures of glass. The average load procedure simplifies the calculations by averaging the factors over different times of the day to arrive at a single factor for each exposure, regardless of the time of day. The majority of residential load calculations can be done using the average load procedure. Manual J, 8th edition, Abridged, only includes the average load procedure. The abridged edition also only has data on common forms of residential construction.

TECH TIP

There are many computer programs available that automate load calculation procedure. However, it is still necessary to collect all the relevant data from the plans or a building survey. The programs primarily save you the time required to look up factors in the tables and do the calculations. Probably the biggest advantage of computer-aided load design is the ability to make changes and quickly see their effect. A customer might be better off spending more money on sealing and insulation than on purchasing a larger unit. This can be easily demonstrated using computer-aided design.

70.4 RESIDENTIAL HEAT-LOSS FACTORS

Heat loss occurs in the heating season when the building is warmer than the outside. Figure 70-1 illustrates common types of heat losses. The primary types of heat loss in a residential building are the following:

- Transmission losses
- Infiltration losses
- Ventilation losses
- Duct losses

Transmission Losses

Transmission losses occur because of a temperature difference between the inside and outside. Heat is transmitted through walls, windows, doors, floors, and ceilings from the inside to the outside.

The heat loss through these surfaces varies with each of the following factors:

- The type of material
- The thickness and thermal conductivity of the material
- The area that is exposed to different temperatures
- The temperature difference between the inside and the outside

Infiltration Losses

Infiltration is air that leaks into the house through cracks and openings. Infiltration can be a significant heat loss because the warm, conditioned air is replaced with cold air from outside. The most obvious places where infiltration occurs are windows and doors. However, infiltration also occurs in walls between the sill and the foundation and in other openings, such as fireplaces.

Ventilation Losses

Ventilation air is air that is brought in on purpose. Like infiltration, ventilation is a heat loss because warm, conditioned air is replaced with cold, outdoor air. However, a big advantage of ventilation is that it can be controlled.

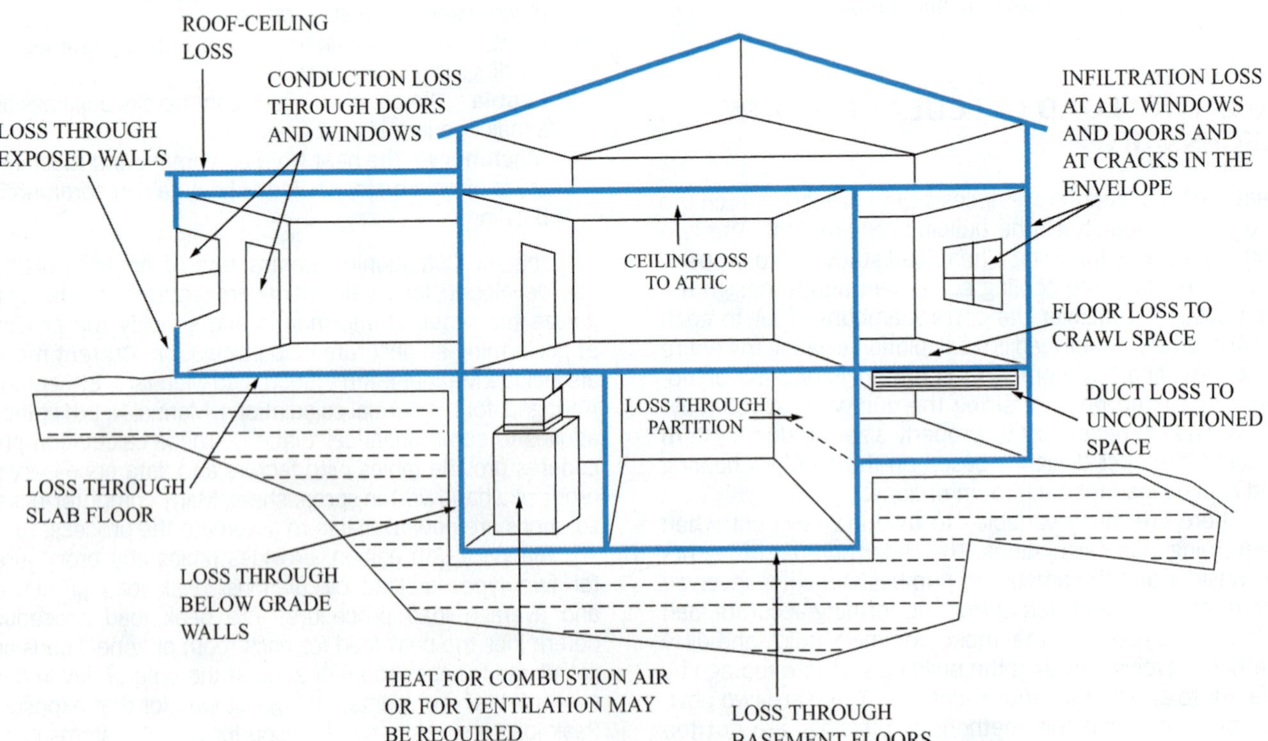

Figure 70-1 Winter heating loads. *(Courtesy of Air Conditioning Contractors of America (ACCA). Copyright ACCA.)*

Ventilation air can be filtered, and the amount entering can be controlled. Also, ventilation losses can be reduced by using an energy-recovery ventilator. Ventilation losses are significant in commercial structures but play only a minor role in most residential buildings.

Duct Losses

There are really two types of duct losses: leaks and transmission losses through ducts located outside the conditioned area of the building. Duct leaks add to the overall infiltration in the building. The infiltration due to duct leaks can easily exceed the infiltration in the rest of the building. This can be addressed through proper duct installation. Duct transmission losses occur when the duct is located in an area exposed to outdoor temperatures, such as an attic or crawlspace. These can be reduced by insulating the duct, but they cannot be eliminated. The only ways to totally eliminate duct losses is to locate the duct inside the conditioned space or use ductless systems.

70.5 RESIDENTIAL HEAT-GAIN FACTORS

Heat gain occurs in the cooling season when the building is cooler than the outside. Figure 70-2 illustrates common types of heat gain. The same factors that affect heat loss also affect heat gain:

- Transmission gains
- Infiltration gains
- Ventilation gains
- Duct gains

However, there are additional factors that play an important role in heat gain:

- Solar gain through glass
- Internal heat gain from lights, machines, and appliances
- Internal gain from people
- Latent heat gain required for dehumidification

Solar Gain

Solar gain is radiant heat gained from sun shining in through glass. Solar gain varies with the direction of the glass, the latitude of the building, the amount of external shade, and even the time of day. All directions have some solar gain. Northern-facing glass has very little solar gain because the sun never shines directly into northern-facing glass. In commercial load calculations, solar gain is determined separately from the transmissive gains through the windows. Residential calculations, like Manual J, combine the transmissive and solar gains into a single factor.

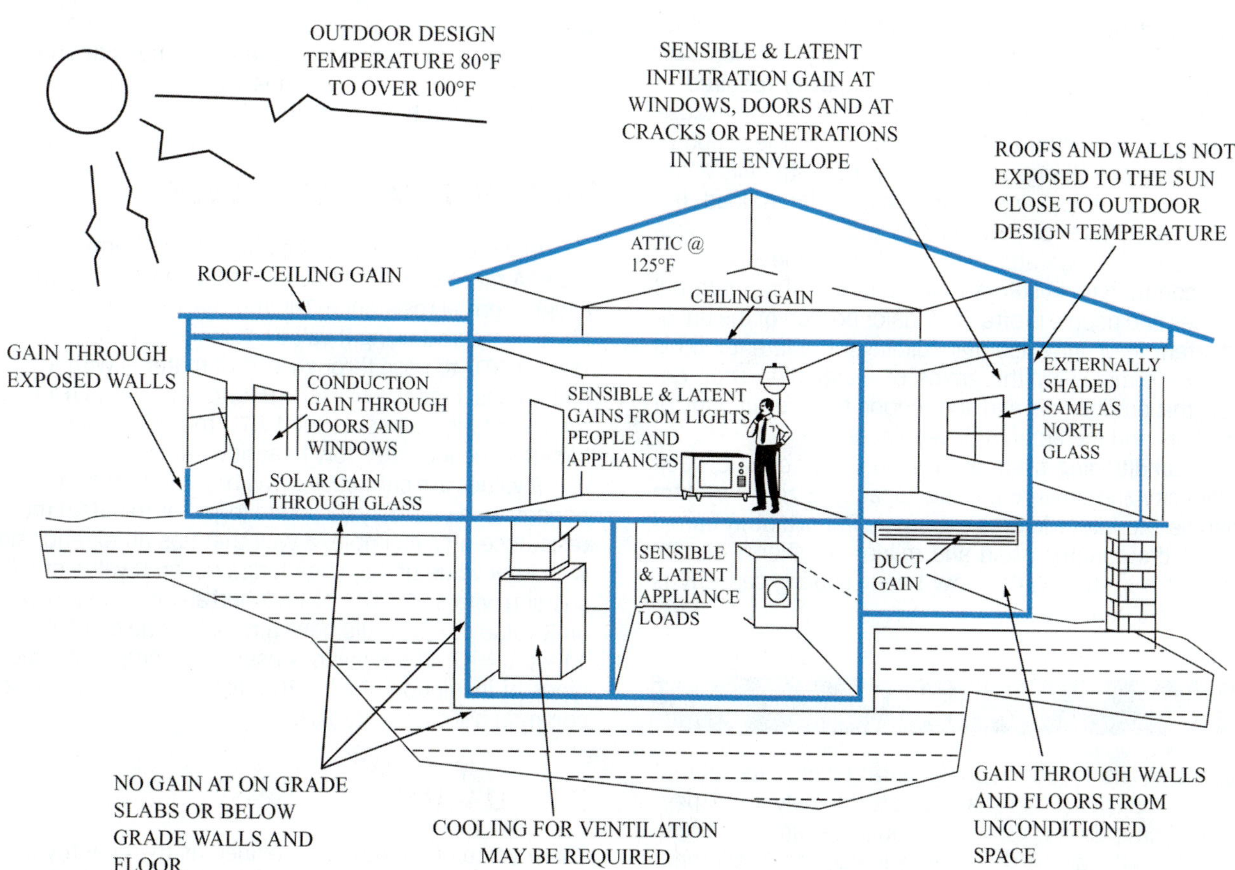

Figure 70-2 Summer cooling loads. *(Courtesy of Air Conditioning Contractors of America (ACCA). Copyright ACCA.)*

Internal Gain from Lights, Machines, and Appliances

Lights, machines, and appliances all create heat. In many commercial applications, the internal gain exceeds the gain from the walls, ceilings, and floors. Because of this, many commercial buildings need air conditioning even when it is cool outside. The internal light and appliance gains in a typical residence are much smaller. For years, these gains were estimated at 1,200 BTU/hr for a residence. Typically, this amount was added to the kitchen gain. Manual J, 8th edition, has a much more complete listing of appliances and their associated heat gain. The internal gain is still much smaller in residential buildings than in commercial buildings, but it is large enough to warrant a careful assessment.

Internal Gain from People

People produce heat. The gain from people in a theater is usually the major gain for that building. The gain from people in residential buildings is much smaller. The amount of heat produced by each person varies with activity level. Sedentary people add approximately 300 BTU/hr in total heat gain, while people engaged in heavy work can produce approximately 1,500 BTU/hr in total heat gain.

Latent Heat Gain Required for Dehumidification

A big difference between heating loads and air-conditioning loads is that air conditioning also involves dehumidification in most parts of the country. The amount of system capacity devoted to changing the air temperature is called sensible load. The amount of equipment capacity required to remove water from the air is called a latent load because the energy required for condensing water on the evaporator coil does not change the temperature of the air. When an air conditioner is operating with a large latent load, the temperature difference between the return and supply air is less than when it operates with a large sensible load.

In commercial buildings, the latent load from internal machines and people is often the major portion of the building's latent load. In residential buildings, the latent load is largely determined by the grains of water difference between the outdoor air and the indoor air, the amount of infiltration, and the amount of ventilation.

Air-conditioning systems have both a sensible cooling capacity and a latent cooling capacity. Sizing a system to handle the latent load is extremely important in humid areas of the country. Mold and mildew problems can be created if a system with a low latent cooling capacity is installed in a humid area.

TECH TIP

Oversizing an air-conditioning unit reduces its latent cooling capacity. Units that operate for only short periods of time do not run long enough to remove much water from the air, so they have a lower effective latent cooling capacity.

70.6 HOUSE ORIENTATION AND SOLAR GAIN

East-facing windows have a high solar heat gain in the morning, west-facing windows have a high solar gain in the afternoon, and south-facing windows receive solar radiation all day. Northern windows receive no direct solar gain. Most houses have at least one side that has more windows and glass than the other sides. The direction the house is facing will significantly affect the heat gain for cooling load calculations because of the difference in solar heat gain. The amount of heat gained from the floor and ceiling will not change as the house is rotated to face north, south, east, and west.

The effect of house orientation on heat gain will be demonstrated using the house floor plan shown in Figure 70-3. This house has 40 ft^2 of glass on the front, 75 ft^2 on the back, 13 ft^2 on the right side, and 24 ft^2 on the left side. The table in Figure 70-4 compares the solar load when the house faces south and when the house faces east. The solar gain nearly doubles from 13,400 BTU/hr when facing south to 25,920 BTU/hr when facing east.

70.7 THERMAL CONDUCTIVITY

All materials have different levels of thermal conductivity. The thermal conductivity of different materials is compared using *k* values. A *k* value states the amount of heat that will transfer through each square foot of material for each inch of thickness for every degree in temperature difference. It is expressed in BTU/hr, per square foot of area, per inch thickness, per °F of temperature difference (TD). Materials with high *k* values transmit heat more easily than materials with low *k* values. Figure 70-5 compares the thermal conductivities of common building materials.

70.8 THERMAL RESISTANCE

Thermal resistance is just the opposite of thermal conductance. A material with a low thermal conductance will have a high thermal resistance. The thermal resistance of materials is compared using R values. R is the number of hours it takes 1 BTU to pass through 1 ft^2 of material with a 1°F TD. R11 means it takes 11 hr for 1 BTU to pass through 1 ft^2. R32 means it takes 32 hours for 1 BTU to pass through 1 ft^2. A high R value indicates low heat flow rates.

R values are cumulative. The resistance of several components of a wall may be added together to obtain the total resistance. For example, a wall that has an exterior siding with an R value of 0.5, sheathing with an R value of 5.0, insulation with an R value of 11.0, and an interior surface with an R value of 0.5 would have a total R value of 17: $0.5 + 5 + 11 + 0.5 = 17$. The R value is useful for comparing different types of insulation. Figure 70-6 lists the R values of some common building materials.

70.9 U VALUES

The U value of a material is defined as the quantity of heat in BTU that will flow through 1 ft^2 of material in 1 hr with a 1°F TD. The U value of any part of a structure is the total

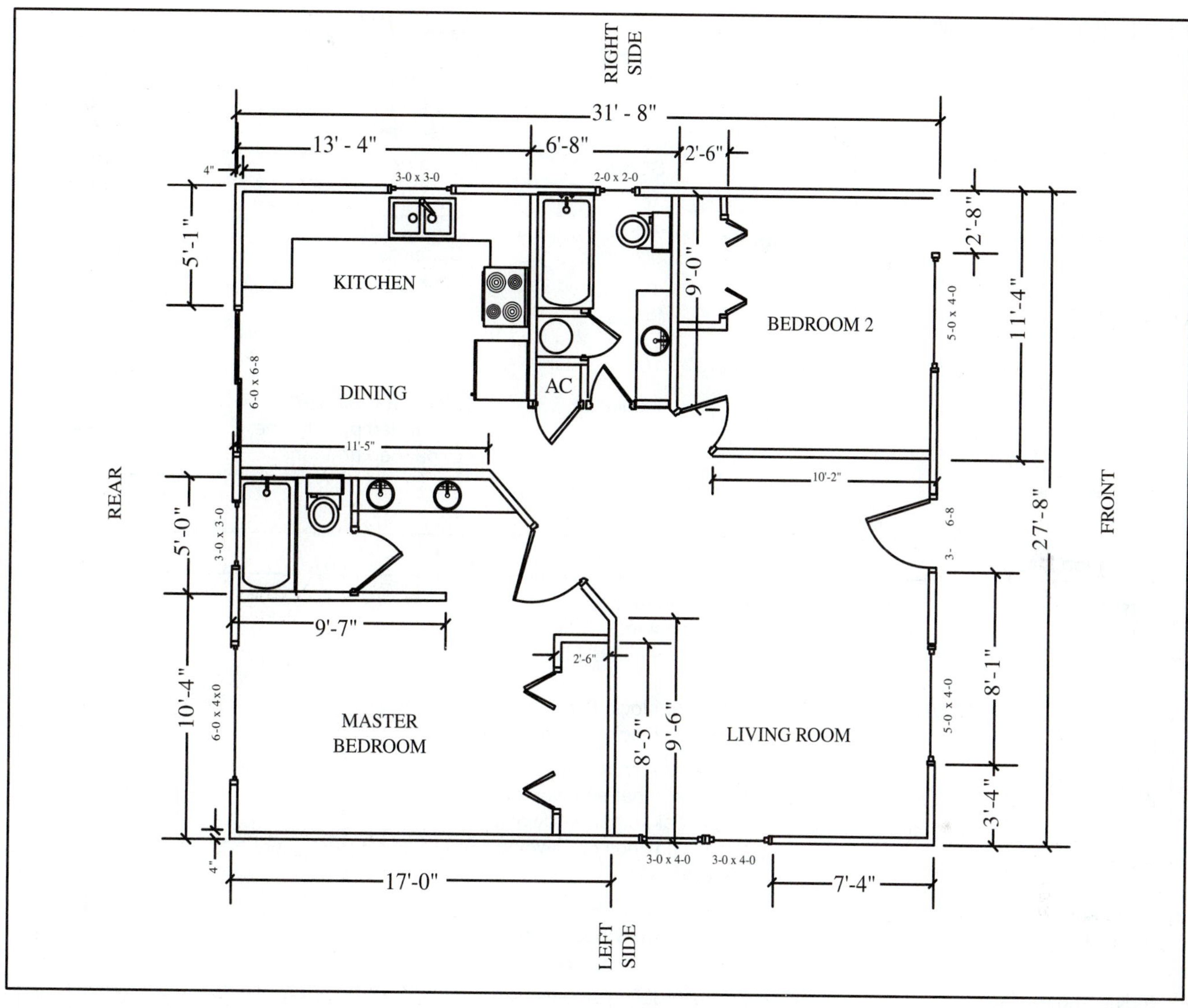

Figure 70-3 Basic residential floor plan.

Facing Direction		South Factor	South Area	South BTU Gain	East Direction	East Factor	East Area	East BTU Gain
Front	South	100	40	4,000	East	200	40	8,000
Right	East	200	13	2,600	North	40	13	520
Back	North	40	75	3,000	West	200	75	15,000
Left	West	200	24	4,800	South	100	24	2,400
		Total Solar Gain		**13,400**		**Total Solar Gain**		**25,920**

Figure 70-4 The solar load for this house nearly doubles when it faces east instead of south.

heat transfer coefficient for that component. The U value is the reciprocal of the component's R value. For example, if a material had a resistance of R-5, its U value would be 0.2: $\frac{1}{5} = 0.2$. In the example of the wall with a total R value of 17, its U value is 0.06: $\frac{1}{17} = 0.06$ (rounded). The U value gives the BTU/hr per 1 ft^2 of that component per TD. In this case, 0.06 BTU/hr will transfer through each ft^2 of wall for every 1°F TD.

70.10 HEAT TRANSFER MULTIPLIERS—HTM

U values give the heat transfer for each degree of temperature difference between the inside and outside temperatures. Heat transfer multipliers (HTMs) give the heat transfer for a specific temperature difference. As the temperature

Figure 70-5 Thermal conductivity of common building materials.

	Material	*K* value
Metal	Copper	223.00
	Aluminum	118.00
	Cast iron	40.00
Stone	Granite	24.29
	Marble	17.35
	Sandstone	11.10
Masonry	Concrete	12.00
	Brick	5.00
Wood	Kiln-dried spruce	1.60
	Red oak	1.20
	Plywood	0.89
Insulation	Fiberglass blanket	0.32
	Expanded polystyrene	0.24
	Expanded urethane	0.17

Insulation Material	Kind	R value
Masonry	Brick, 4-in face	0.44
	Common 4 in	0.80
	Stone, lime/sand 1 in	0.08
	Stucco 1 in	0.20
	Concrete block, 8 in	1.04
	Concrete block, 8 in, cores filled	1.93
Concrete	Poured concrete 1 in	0.08
Wood	Fir, pine, other soft woods 1 in	1.25
	Maple oak, other hardwood 1 in	0.91
	Wood shavings or sawdust 1 in	2.20
	Plywood, ½ in	0.63
Manufactured wood Products	Gypsum board ½ in	0.45
	Plywood, softwood, ¼ in	0.31
	Hardboard, underlayment ¼ in	0.31
Siding and roofing	Asphalt roofing shingles	0.44
	Wood bevel siding ½ in	0.81
	Building paper, felt 15	0.06
Insulation	Glass fiber, batt 1 in	3.13
	Blown cellulose 1 in	3.4
	Expanded polystyrene 1 in	3.85
	Expanded polyurethane 1 in	6.64
	Extruded polystyrene 1 in (styrofoam blue board)	4.92

Various Types of Fiberglass Insulation

R value	Applications(s)	Width (in)	Thickness (in)
R-11	Interior walls/noise control/basement walls	15, 23	3.5
R-13	Exterior walls/basement walls	15, 23	3.5
R-15	Exterior walls	15	3.5
R-19	Attics/exterior walls/crawlspaces	15, 23	6.25
R-21	Attics/exterior walls	15, 23	5.5
R-30	Attics	16, 24	9.5
R-38	Attics	16, 24	12

Note: Insulation comes both faced and unfaced, in rolls and batts.

Figure 70-6 Value of common building materials.

difference increases, so does the rate of heat transfer. HTMs are calculated by multiplying the U value times the temperature difference. In the example of the R-17 wall with the U value of 0.06 per ft^2, the HTM for heating with a 50°F TD between outside and inside would be 3 BTU/hr per ft^2: 0.06 BTU/hr/F° per $ft^2 \times 50°F = 3$ BTU/hr per ft^2. Removing the terms can simplify the math: $0.06 \times 50 = 3$.

70.11 STORAGE AND DAILY RANGE

The temperature difference used for determining most heat loss HTMs is simply the difference between the inside and outside temperature. The temperature difference used for determining heat gain HTMs is called the effective temperature difference. The effective temperature difference takes into account the temperature swings throughout the day and thermal storage in some parts of a building, such as attics.

Storage

When the sun shines in a window, its radiant heat is stored when it reaches the interior surfaces of the house and later gradually heats room air. The same kind of action occurs on the exterior of structures. Radiant heat from the sun warms the surfaces and is stored. Gradually, at a later time, the heat reaches the interior, where it warms the interior air. The net effect of storage is to delay and smooth out solar loads. The delayed effect of the sun's load is incorporated in the procedures used for calculating heat gain.

Daily Range

The difference between the average high and low temperatures is the daily range. The high usually occurs late in the afternoon and the low at about daybreak. The daily range affects the cooling load since a low night temperature can reduce the daytime load due to the storage factor. The daily range factors are given for three ranges: low, medium, and high. The low daily range applies to climates where the difference between the high and low temperatures for the day is 15°F or less. This means there is less natural cooling effect to take advantage of. Areas with a low daily range have the highest effective temperature differences and therefore the highest heat-gain HTMs. The medium daily range applies to climates where the difference between the high and low temperatures for the day is between 16°F and 25°F. These areas can take advantage of some natural cooling. The high daily range applies to climates where the difference between the high and low temperatures for the day is greater than 25°F. This means there is a large natural cooling effect to take advantage of. Areas with a high daily range have the lowest effective temperature differences and therefore the lowest heat-gain HTMs.

Cooling Temperature Difference Versus Design Temperature Difference

The design temperature difference is the difference between the outside and inside design temperatures. The outside design temperature is taken from weather data tables; the inside design temperature is typically 75°F for cooling and 70°F for heating. The cooling temperature difference is the temperature difference across a building component after taking into consideration factors like thermal storage and daily temperature range. The temperature of the surface of most roofs in the summer is considerably higher than the outdoor temperature. This makes the cooling temperature difference for most ceilings greater than the temperature difference between inside and outside. The cooling temperature difference across walls is often less than the difference between inside and outside because of the natural cooling effect of the daily range.

70.12 OUTSIDE DESIGN CONDITIONS

The outdoor temperature used for determining the temperature difference is not the coldest possible day in the winter or the hottest possible day in summer. The outside design conditions are selected from ASHRAE weather data. The data are assigned a percentage that represents the percentage of time during the year that the temperature exceeds that temperature. Manual J uses 99 percent for heating. This means that 99 percent of the time, the temperature will be above the heating design temperature. The temperature should dip below the heating design temperature for only 88 hr a year. Manual J uses 1 percent for cooling. This means that only 1 percent of the time will the temperature be above the cooling design temperature. The temperature should exceed the heating design temperature for only 88 hr a year. Using the 99 percent and 1 percent conditions prevents excessive oversizing of the equipment to handle a temporary high load. The recommended design temperatures are based on the average temperature of an entire day. Although people tend to think of the highs and lows they hear on the news, these temperatures typically only last for a few hours and do not represent the entire day.

TECH TIP

The design temperature for any locale is based on a 10-year average based on data provided by the U.S. Weather Service. Over time, a design temperature for a locale may vary to match the data collected.

70.13 MANUAL J, 8TH EDITION, WORKSHEET

The Manual J worksheet is like a computer spreadsheet, only in handwritten form (Figure 70-7). It helps organize the data and values for the many calculations required to perform an accurate heat load study. Rows 1–5 across the top contain information about the size of the house or room and the amount of exposed wall. This is shown in Figure 70-8 in yellow. The large block of columns on the left-hand side of the form are for listing the types of materials used and their corresponding heat-transfer multipliers. This is shown in Figure 70-8 in blue. The materials are assigned construction numbers in Manual J to make referencing them easier. The next block over is for calculating the heat loss and heat gain for the entire house. This is shown in Figure 70-8 in green.

Figure 70-7 Worksheet for Manual J, 8th edition. *(Courtesy of Air Conditioning Contractors of America (ACCA). From Manual J 8th Edition. Copyright ACCA.)*

#	Item		Const.. Number	Panel Faces	HTM Htg.	HTM Clg.	Area or Length	Btuh Heating	Btuh S-Clg.	Btuh L-Clg.	Area or Length	Btuh Heating	Btuh S-Clg.
1	Name of Room						Entire House						
2	Running Feet of Exposed Wall												
3	Ceiling Height At Walls (Ft) and Gross Wall Area (SqFt)												
4	Room Dimensions (Feet) and Floor Plan Area (SqFT)												
5	Ceiling Slope (Degrees) and Gross Ceiling Area												
	Type of Exposure												
6a	Windows and Glass Doors	a											
		b											
		c											
6b	Skylights	a											
		b											
7	Metal Door	a											
		b											
8	Above Grade Walls	a											
		b											
		c											
9	Below Grade Walls	a											
		b											
		c											
10	Ceiling	a											
		b											
11b	Passive Floors	a											
		b											
12	Infiltration	Heat Loss				Btuh	WAR				WAR		
		Sensible Gain				Btuh							
		Latent Gain				Btuh							
13	Internal	a Occupants at 230 and 200 Btuh					#				#		
		b Scenario Number											
		c Default Adjustments											
		d Individual Appliances											
		e Plants											
14	Subtotals	Sum lines 6 through 11a + line 12 + line 13											
15	Duct Loads	ELF-Loss and ELF-Gain											
		Latent Gain											
16	Ventilation Loads	Vent CFM			Exh								
17	Winter Humidification load			Gal / Day									
18	Piping Load												
19	Blower Heat												
20	Total Load	Sum lines 11b + lines 14 through 19											

The rest of the blocks are for calculating the heat loss and gain for each individual room. This is shown in Figure 70-8 in purple. Each of these blocks has columns for area, heating, and sensible cooling. The area of each material listed in the area column is multiplied by the HTMs for heating and cooling found on the left-hand side of the form. The answers are written in the heating and cooling columns. Totals are tabulated at the bottom of each column.

TECH TIP

There are many multipliers in Manual J that are decimal numbers with two digits to the right of the decimal. There is no reason to keep values in the BTU loss and gain columns in decimal form. A tenth of a BTU has no practical effect on the size of the equipment. For simplicity's sake, most designers round off the BTU loss and gain to whole numbers. Many round to the nearest hundred BTUs because equipment is sized in thousands of BTUs. For example, a BTU/hr loss of 2,517.6 would be rounded to either 2,518 for whole numbers or simply 2,500 for even hundreds.

70.14 MANUAL J TABLES

At this point, the reader may be thinking that using a quick-and-dirty method is beginning to look pretty good. Keeping track of effective temperature differences, R values, U values, and heat-transfer multipliers can appear to be a difficult task. The purpose of Manual J is to do most of the heavy lifting for you. The tables in Manual J, 8th edition, list heat-transfer multipliers for many different construction types. Factors like the direction that windows face, the daily range, and equivalent temperature differences have all been factored in for you. All you have to do is learn to use the tables. We are going to go through an example whole-house calculation using the plan in Figure 70-3. An abbreviated worksheet will be used that only shows the items being used. The necessary tables will be described along the way.

70.15 WEATHER DATA TABLES 1A AND 1B

Table 1A contains weather data for cities throughout the United States and Canada. The table is organized with cities in the United States in the first part of the table and cities

1	Name of Room						Entire House						
2	Running Feet of Exposed Wall												
3	Ceiling Height At Walls (Ft) and Gross Wall Area (SqFt)												
4	Room Dimensions (Feet) and Floor Plan Area (SqFT)												
5	Ceiling Slope (Degrees) and Gross Ceiling Area												
	Type of Exposure		Const.. Number	Panel Faces	HTM		Area or Length	Btuh			Area or Length	Btuh	
					Htg.	Clg.		Heating	S-Clg.	L-Clg.		Heating	S-Clg.
6a	Windows and Glass Doors	a											
		b											
		c											
6b	Skylights	a											
		b											
7	Metal Door	a											
		b											
8	Above Grade Walls	a											
		b											
		c											
9	Below Grade Walls	a											
		b											
		c											
10	Ceiling	a											
		b											
11b	Passive Floors	a											
		b											
12	Infiltration	Heat Loss				Btuh	WAR				WAR		
		Sensible Gain				Btuh							
		Latent Gain				Btuh							
13	Internal	a	Occupants at 230 and 200 Btuh				#				#		
		b	Scenario Number										
		c	Default Adjustments										
		d	Individual Appliances										
		e	Plants										
14	Subtotals	Sum lines 6 through 11a + line 12 + line 13											
15	Duct Loads	ELF-Loss and ELF-Gain											
		Latent Gain											
16	Ventilation Loads	Vent CFM			Exh								
17	Winter Humidification load			Gal / Day									
18	Piping Load												
19	Blower Heat												
20	Total Load	Sum lines 11b + lines 14 through 19											

Figure 70-8 Major sections of the Manual J, 8th edition, worksheet. *(Courtesy of Air Conditioning Contractors of America (ACCA). From Manual J 8th Edition. Copyright ACCA.)*

in Canada in the second part. The states and provinces are listed in alphabetical order, and the cities are listed under their state in alphabetical order. The most important data found in this table are the following:

- **Latitude** Latitude is used to calculate shade from overhangs above windows.
- **Heating 99% dry bulb** Heating 99% dry bulb is used for the heating outdoor design temperature.
- **Cooling 1% dry bulb** Cooling 1% dry bulb is used for the cooling outdoor design temperature.
- **Design grains RH** Design grains RH lists the grains of difference between inside and outside in the summer. This is used for calculating the latent cooling load. Manual J has three columns, one each for 55%, 50%, and 45% indoor relative humidity.

Table 1B lists the same information as Table 1A but has more locations in the states of Arizona, California, Hawaii, and Nevada.

For our example, the house will be located in Selma, Alabama. The winter design for Selma is 26°F and the summer design temperature is 95°F (Figure 70-9). The design grains of difference will be 49 grains for 50 percent relative humidity. The winter design temperature difference will be 44°F: 70°F − 26°F = 44°F. The summer design temperature difference will be 20°F: 95°F − 75° = 20°F. The house will be facing east and the daily range is medium.

70.16 ROOM DIMENSIONS

The first line is for the room name—in this case it is the entire house. The running feet of exposed wall is the length of wall that is exposed to the outside. Only walls that are exposed to the outside are listed. For example, the master bathroom has only 5 ft of exposed wall. The house as a whole has two walls of 32 ft and two walls of 28 ft, for a total of 120 ft. The ceiling in our example will be a standard ceiling: 8 ft high with no slope. Figure 70-10 shows this part of the form completed.

70.17 FENESTRATION HEAT LOSS TABLES 2A AND 2B

In Manual J, fenestration refers to anything that light can pass through, including windows, skylights, and glass doors. Table 2A lists U values for generic windows, skylights, and glass doors. These U values are used to construct heat-loss HTMs.

Table 1A

Outdoor Design Conditions For the United States and Canada

Location	Elevation	Latitude	Winter	Summer						
	Feet	Degrees North	Heating 99% Dry Bulb	Cooling 1% Dry Bulb	Coincident Wet Bulb	Design Grains 55% RH	Design Grains 50% RH	Design Grains 45% RH	Daily Range (DR)
Alabama									
Alexander City	686	33	22	93	76	39	46	52	M
Anniston AP	612	33	24	93	76	39	46	52	M
Auburn	776	32	22	93	76	39	46	52	M
Birmingham AP	644	33	23	92	75	34	41	47	M
Decatur	592	34	16	93	74	27	34	40	M
Dothan AP	401	31	32	93	76	39	46	52	M
Florence AP	581	34	21	94	75	31	38	44	M
Gadsden	569	34	20	94	75	31	38	44	M
Huntsville AP	629	34	20	92	74	28	35	41	M
Mobile AP	218	30	30	92	76	41	48	54	M
Mobile CO	26	30	29	93	77	46	53	59	M
Montgomery AP	221	32	27	93	76	39	46	52	M
Ozark, Fort Rucker	356	31	31	94	77	44	51	57	M
Selma-Craig AFB	166	32	26	95	77	42	49	55	M
Talladega	528	33	22	94	76	37	44	50	M
Tuscaloosa AP	170	33	24	94	77	44	51	57	M

Figure 70-9 Weather data selection from Table IA of Manual J, 8th edition. *(Courtesy of Air Conditioning Contractors of America (ACCA). From Manual J 8th Edition. Copyright ACCA.)*

1	**Name of Room**						**Entire House**			
2	**Running Feet of Exposed Wall**						**120**			
3	**Ceiling Height and Gross Wall Area**						**8**	**120 × 8 = 960**		
4	**Room Dimensions and Floor Plan Area**						**32 × 28**	**32 × 28 = 896**		
5	**Ceiling Slope**						**0**	**896**		
	Type of Exposure		Const #	Facing	HTM		Area or Length	BTU per Hour		
					Heating	Cooling		Heating	Sensible	Latent
6a	Windows and Glass Doors	a								
		b								
		c								
		d								
7	Wood and Metal Doors	a								
8	Above Grade Walls	a								
10	Ceilings	a								
11a	Passive Floors	a								
12	Infiltration	Heat Loss					WAR 1.0			
		Sensible Gain								
		Latent Gain								
13	Internal	a								
		b								
14	Subtotals	Sum lines 6–11a + lines 12 and 13								
15	Duct Loads	ELF Loss and ELF Gain								
		Latent Gain								
20	Total Load	Sum lines 11b + lines 14–19								

Figure 70-10 Lines 1–5 completed on Manual J worksheet.

Construction Numbers 1 Though 7 Window and Glass Door Assemblies Reference Area = Area of Rough Opening (SqFt). Bold Face denotes U-Values used for Tables 3A, 3B, 3F and 3G.	Type of Frame Construction			
	Metal No Break	Metal With Break	Wood, Wood With Metal Clad or Vinyl	Insulated Fiberglass
Number 1 — Clear, Heat Absorbing or Reflective Glass	U-Value			
A. Single pane operable window or sliding glass door	1.27	1.08	0.90	0.81
B. Single pane window, fixed sash	1.13	1.07	0.98	0.94
C. Single pane operable window with storm	0.87	0.65	0.57	0.49
D. Double pane operable window or sliding glass door	0.87	0.65	0.57	0.49
E. Double pane window, fixed sash	0.69	0.63	0.56	0.53
F. Triple pane window, fixed sash	0.55	0.48	0.42	0.40
G. Products rated and labeled by the NFRC (see Table 2B-1)	Refer to label, NFRC Directory or manufacturer's engineering data.			
Number 2 — Double Pane Low-e Glass Designed For Cold Climate (Emissivity of Low-e coating = 0.60)	U-Value			
A. Operable window or sliding glass door	0.84	0.63	0.55	0.47
B. Window with fixed sash	0.67	0.60	0.54	0.51
C. Products rated and labeled by the NFRC (see Table 2B-1)	Refer to label, NFRC Directory or manufacturer's engineering data.			
Number 3 — Double Pane Low-e Glass Designed For Cold Climate (Emissivity of Low-e coating = 0.40)	U-Value			
A. Operable window or sliding glass door	0.82	0.61	0.53	0.45
B. Window with fixed sash	0.64	0.58	0.51	0.49
C. Products rated and labeled by the NFRC (see Table 2B-1)	Refer to label, NFRC Directory or manufacturer's engineering data.			
Number 4 — Double Pane Low-e Glass Designed For Cold Climate (Emissivity of Low-e coating = 0.20 or less)	U-Value			
A. Operable window or sliding glass door	0.77	0.56	0.47	0.41
B. Window with fixed sash	0.59	0.52	0.45	0.43
C. Products rated and labeled by the NFRC (see Table 2B-1)	Refer to label, NFRC Directory or manufacturer's engineering data.			
Number 5 — Jalousie Window	U-Value			
A. Jalousie window only	1.27	~	~	~
B. Jalousie window with storm	0.69	~	~	~
Number 6 — Projected Window Assemblies	U-Value			
A. Bay window with any fixed sash window listed above	Multiply flat panel, fixed sash U-Value by 1.15.			
B. Garden window with any fixed sash window listed above	Multiply flat panel, fixed sash U-Value by 2.75.			
C. Products rated and labeled by the NFRC (see Table 2B-1)	Refer to label, NFRC Directory or manufacturer's engineering data.			
Number 7 — Glass or Plastic Block	U-Value			
A. Glass block with mortar joints	Use 0.60 U-Value for all products.			
B. Products rated and labeled by the NFRC (see Table 2B-5)	Refer to label, NFRC Directory or manufacturer's engineering data.			

T2A-1

Figure 70-11 Window construction number and U value from Table 2A of Manual J, 8th edition. *(Courtesy of Air Conditioning Contractors of America (ACCA). From Manual J 8th Edition. Copyright ACCA.)*

In Table 2A, the column on the left describes the type of glass, the number of panes, and the general construction. The column on the right lists the type of frame. To use the table, locate the row in the left column that describes the type of glass and window construction. The U value is listed over on the right under the column that describes the frame style. The windows in our example house are fixed sash, wood frame, double-pane windows with a low e coating of 0.4 emissivity. Figure 70-11 shows the construction number is 3B and the U value is 0.51.

Once the U value has been determined, the HTM is calculated by multiplying the winter temperature difference times the U value. In this case, the HTM would be 22.4 BTU/hr for a winter temperature difference of 44°F: 0.51 BTU(hr°F × 44°F = 22.4 BTU/hr. The heat loss for the windows is calculated by multiplying this HTM by the area of all the windows (Figure 70-12).

Table 2B is for windows, glass doors, and skylights that are rated by the National Fenestration Rating Council (NFRC). NFRC-rated items have a label that lists important energy performance data, including the U value, solar heat-gain coefficient, and air leakage. For typical windows and glass doors, the HTM can be calculated by multiplying the U value listed on the window by the winter temperature difference. Table 2B lists formulas to help determine the HTM values for more complicated fenestrations, such as bay windows and skylights.

70.18 FENESTRATION HEAT-GAIN TABLES 3A–3I

Tables 3A–3I provide HTM values for heat gain, including solar gain. The large number of tables is for specifying the amount of external shade, the type of fenestration, and the load procedure used. Tables 3A–3D are used for the average load procedure. Tables 3F–3I are used for the peak load procedure. Table 3E is used to adjust the HTM for factors such as overhang, foreground reflectance, and latitude. We will only be discussing Tables 3A–3D. Briefly, the tables are as follows:

- **3A** Generic windows and glass doors with no external sun screen
- **3B** Generic windows and glass doors behind an external sun screen
- **3C** Generic skylights
- **3D** NFRC-rated windows and doors

1	Name of Room						Entire House			
2	Running Feet of Exposed Wall						120			
3	Ceiling Height and Gross Wall Area						8	120 × 8 = 960		
4	Room Dimensions and Floor Plan Area						32 × 28	32 × 28 = 896		
5	Ceiling Slope						0	896		
	Type of Exposure		Const #	Facing	HTM		Area or Length	BTU per Hour		
					Heating	Cooling		Heating	Sensible	Latent
6a	Windows and Glass Doors	a	**3B**	**North**	22.4		13	291		
		b	**3B**	**East**	22.4		**40**	896		
		c	**3B**	**South**	22.4		**24**	538		
		d	**3B**	**West**	22.4		**75**	1,680		
7	Wood and Metal Doors	a								
8	Above Grade Walls	a								
10	Ceilings	a								
11a	Passive Floors	a								
12	Infiltration	Heat Loss					WAR 1.0			
		Sensible Gain								
		Latent Gain								
13	Internal	a								
		b								
14	Subtotals	Sum lines 6–11a + lines 12 and 13								
15	Duct Loads	ELF Loss and ELF Gain								
		Latent Gain								
20	Total Load	Sum lines 11b + lines 14–19								

Figure 70-12 The heat loss portion of line 6a of the Manual J worksheet is completed.

Tables 3A and 3B both have multiple pages. In Table 3A, each page covers a different type of glass: clear, heat absorbing, reflective, and glass block. In Table 3B, each page covers a different amount of external sun screen: 15, 25, 35, and 45 percent. Each page in both tables is organized into three columns: single pane, double pane, and triple pane. Each column is further divided into columns for temperature difference and rows for the direction the window faces. In our example, the HTM for a double-pane window with no external sun screen, heat-absorbing glass, and no internal shade would be found in Table 3A, page 2. To find the HTM for the east- and west-facing windows in the house, first look at the large row at the top of Table 3A, page 2, described as *No internal shade*. Find the column in the middle listed as *Double pane*. The HTM is at the intersection of the row for *East or west* and the *20* column (the summer design temperature difference). In this case, the HTM is 52 (Figure 70-13). The HTM values for the north-facing and south-facing windows are found in a similar manner. The HTM for the north-facing windows is 20 and the HTM for the south-facing windows is 29. The heat gain through the windows is then calculated by multiplying these HTM values by the area of the windows. (The heat gain for each window is recorded in the sensible column as shown in Figure 70-14).

70.19 HEAT LOSS AND GAIN FOR OPAQUE SURFACES, TABLES 4A–4C

Opaque surfaces are simply surfaces that light will not pass through, including doors, walls, floors, and ceilings. Tables 4A and 4B are used for the average load procedure; Tables 4C and 4D are used for the peak load procedure. We will only be discussing Tables 4A and 4B. Table 4A covers several pages and is organized into construction numbers:

11 Wood and metal doors
12 Frame walls and partitions
13 Block walls and partitions
14 Alternative constructions of walls and partitions
15 Basement walls
16 Ceiling under attic and knee walls
17 Ceiling on exposed beams
18 Ceiling below roof joists
19 Floor over enclosed crawlspace or unconditioned basement

Table 3A — Continued

Fenestration Heat Gain — No External Sun Screen
Generic HTM Values For The Average Load Procedure

Table 3A-2 — Heat Absorbing Glass

No Internal Shade

Default Assembly Performance	Single Pane						Double Pane						Triple Pane					
	U-Value		SC		SHGC		U-Value		SC		SHGC		U-Value		SC		SHGC	
	0.98		0.60		0.52		0.56		0.50		0.44		0.42		0.35		0.30	
Design CTD	10	15	20	25	30	35	10	15	20	25	30	35	10	15	20	25	30	35
Exposure	HTM For Rough Opening						HTM For Rough Opening						HTM For Rough Opening					
North	20	25	30	35	39	44	14	17	20	22	25	28	10	12	14	16	18	21
NE or NW	42	47	52	57	62	67	33	35	38	41	44	47	23	26	27	29	32	34
East or West	59	64	69	74	79	84	47	49	52	55	58	61	33	35	37	39	41	43
SE or SW	51	56	61	66	71	75	40	43	46	48	51	54	28	30	32	35	37	39
South	31	36	41	46	51	56	23	26	29	32	35	37	17	19	21	23	25	27

Vertical or Horizontal Blinds With Slats At 45 Degrees

Default Assembly Performance	Single Pane						Double Pane						Triple Pane					
	U-Value		SC		SHGC		U-Value		SC		SHGC		U-Value		SC		SHGC	
	0.98		0.50		0.44		0.56		0.35		0.30		0.42		0.30		0.26	
Design CTD	10	15	20	25	30	35	10	15	20	25	30	35	10	15	20	25	30	35
Exposure	HTM For Rough Opening						HTM For Rough Opening						HTM For Rough Opening					
North	15	20	25	30	34	39	9	12	15	18	20	2	7	9	11	14	16	18
NE or NW	31	36	41	46	51	56	21	24	26	29	32	35	17	19	21	23	26	28
East or West	44	49	54	59	64	69	30	33	35	38	41	44	25	27	29	31	33	35
SE or SW	38	43	48	53	58	63	25	28	31	34	37	39	21	23	25	28	30	32
South	22	27	32	37	42	47	14	17	20	23	26	28	12	14	16	18	20	22

Drape or Roller Shade Half Drawn

Default Assembly Performance	Single Pane						Double Pane						Triple Pane					
	U-Value		SC		SHGC		U-Value		SC		SHGC		U-Value		SC		SHGC	
	0.98		0.50		0.44		0.56		0.40		0.35		0.42		0.30		0.26	
Design CTD	10	15	20	25	30	35	10	15	20	25	30	35	10	15	20	25	30	35
Exposure	HTM For Rough Opening						HTM For Rough Opening						HTM For Rough Opening					
North	15	20	25	30	34	39	10	12	15	18	21	24	7	9	11	14	16	18
NE or NW	31	36	41	46	51	56	23	26	28	31	34	37	17	19	21	23	26	28
East or West	44	49	54	59	64	69	33	36	39	42	44	47	25	27	29	31	33	35
SE or SW	38	43	48	53	58	63	28	31	34	37	40	42	21	23	25	28	30	32
South	22	27	32	37	42	47	16	19	21	24	27	30	12	14	16	18	20	22

Drape or Roller Shade Fully Drawn

Default Assembly Performance	Single Pane						Double Pane						Triple Pane					
	U-Value		SC		SHGC		U-Value		SC		SHGC		U-Value		SC		SHGC	
	0.98		0.40		0.35		0.56		0.30		0.26		0.42		0.25		0.22	
Design CTD	10	15	20	25	30	35	10	15	20	25	30	35	10	15	20	25	30	35
Exposure	HTM For Rough Opening						HTM For Rough Opening						HTM For Rough Opening					
North	14	19	24	29	33	38	9	11	14	17	20	23	7	9	11	13	15	17
NE or NW	27	32	37	42	47	52	19	21	24	27	30	33	15	17	19	21	23	26
East or West	37	42	47	52	57	62	26	29	32	35	38	40	21	24	26	28	30	32
SE or SW	33	37	42	47	52	57	23	25	28	31	34	37	18	21	23	25	27	29
South	20	25	30	35	40	44	13	16	19	22	24	27	11	13	15	17	19	21

Figure 70-13 Window construction number and U value from Table 3A of Manual J, 8th edition. *(Courtesy of Air Conditioning Contractors of America (ACCA). From Manual J 8th Edition. Copyright ACCA.)*

20 Floor over open crawlspace or garage
21 Basement floor
22 Slab on grade

Heat Loss

The tables in 4A all include U values. The heat-loss HTM is calculated by multiplying the U valve by the heating temperature difference.

Heat Gain

The tables for construction numbers 11, 16, 17, 18, 19, and 20 list a cooling temperature difference. The cooling temperature difference and the U value are used to construct a heat-gain HTM. The tables for construction numbers 12, 13, 14, and 15 do not give a cooling temperature difference but list a group number instead. The group number is used to look up the cooling temperature difference using Table 4B. The cooling temperature difference and the U value are then used to construct a heat-gain HTM.

1	Name of Room						Entire House			
2	Running Feet of Exposed Wall						120			
3	Ceiling Height and Gross Wall Area						8	120 × 8 = 960		
4	Room Dimensions and Floor Plan Area						32 × 28	32 × 28 = 896		
5	Ceiling Slope						0	896		
	Type of Exposure		Const #	Facing	HTM		Area or Length	BTU per Hour		
					Heating	Cooling		Heating	Sensible	Latent
6a	Windows and Glass Doors	a	**3B**	**North**	22.4	**20**	**13**	291	**260**	
		b	**3B**	**East**	22.4	**52**	**40**	896	**2,080**	
		c	**3B**	**South**	22.4	**29**	**24**	538	**696**	
		d	**3B**	**West**	22.4	**52**	**75**	1,680	**3,900**	
7	Wood and Metal Doors	a								
8	Above Grade Walls	a								
10	Ceilings	a								
11a	Passive Floors	a								
12	Infiltration	Heat Loss					WAR 1.0			
		Sensible Gain								
		Latent Gain								
13	Internal	a								
		b								
14	Subtotals	Sum lines 6–11a + lines 12 and 13								
15	Duct Loads	ELF Loss and ELF Gain								
		Latent Gain								
20	Total Load	Sum lines 11b + lines 14–19								

Figure 70-14 The heat gain portion of line 6a of the Manual J worksheet is completed. *(Courtesy of Air Conditioning Contractors of America (ACCA). From Manual J 8th Edition. Copyright ACCA.)*

Door

In our example, the U value for a polyurethane core metal door would be 0.29 (Figure 70-15). The heat loss HTM for this door in our house with a 44°F winter temperature difference would be 12.8: 0.29 × 44 = 12.8. The heat loss through the door would be calculated by multiplying this HTM by the area of the door, 20 ft^2. For a 20 degree summer temperature difference and a medium daily range the table shows the cooling HTM as 31. The heat gain is calculated by multiplying this HTM by the area of the door. Figure 70-16 shows the calculations for both heating and cooling.

Figure 70-15 Table 4A, Construction Number 11 Selection. *(Courtesy of Air Conditioning Contractors of America (ACCA). From Manual J 8th Edition. Copyright ACCA.)*

Construction Number 11
Wood and Metal Doors
Referance Area = Area of Rough Opening (SqFt)

Wood Door	U-Value
A. Hollow Core	0.47
B. Hollow Core With Wood Storm	0.30
C. Hollow Core With Metal Storm	0.32
D. Solid Core	0.39
E. Solid Core With Wood Storm	0.26
F. Solid Core With Metal Storm	0.28
G. Panel	0.54
H. Panel With Wood Storm	0.32
I. Panel With Metal Storm	0.36
Metal Door	**U-Value**
J. Fiberglass Core	0.60
K. Fiberglass Core With Storm	0.36
L. Paper Honeycomb Core	0.56
M. Paper Honeycomb Core, With Storm	0.34
N. Polystyrene Core	0.35
O. Polystyrene Core With Storm	0.21
P. Polyurethane Core	0.29
Q. Polyurethane Core With Storm	0.17

CLTD Values For The Average and Peak Load Procedures
Medium Color Wood or Metal Doors

10		15			20			25		30	35
L	M	L	M	H	L	M	H	M	H	H	H
25.0	21.0	30.0	26.0	21.0	35.0	31.0	26.0	36.0	31.0	36.0	41.0

Wood and metal doors do not have a group number.

1	Name of Room						Entire House			
2	Running Feet of Exposed Wall						120			
3	Ceiling Height and Gross Wall Area						8	$120 \times 8 = 960$		
4	Room Dimensions and Floor Plan Area						32×28	$32 \times 28 = 896$		
5	Ceiling Slope						0	896		
	Type of Exposure		Const #	Facing	HTM		Area or Length	BTU per Hour		
					Heating	Cooling		Heating	Sensible	Latent
6a	Windows and Glass Doors	a	3B	North	22.4	20	13	291	260	
		b	3B	East	22.4	52	40	896	2080	
		c	3B	South	22.4	29	24	538	696	
		d	3B	West	22.4	52	75	1,680	3,900	
7	**Wood and Metal Doors**	**a**	**11 N**		**12.8**	**31**	**20**	**256**	**180**	
8	Above Grade Walls	a								
10	Ceilings	a								
11a	Passive Floors	a								
12	Infiltration	Heat Loss					WAR 1.0			
		Sensible Gain								
		Latent Gain								
13	Internal	a								
		b								
14	Subtotals	Sum lines 6–11a + lines 12 and 13								
15	Duct Loads	ELF Loss and ELF Gain								
		Latent Gain								
20	Total Load	Sum lines 11b + lines 14–19								

Figure 70-16 Door calculation on line 7 completed on Manual J worksheet.

Walls

The frame walls in our house are construction number 12B-4b: wood frame wall with R-11 insulation, R-4 insulating sheathing, and a brick veneer. They have a U value of 0.073 and are listed in Group J (Figure 70-17). The heating HTM is 3.2: $0.073 \times 44 = 3.2$.

Our house is in a climate with a medium daily range, and the summer design temperature difference is 20°F. Table 4B shows that group J would have a cooling temperature difference of 16.4 (Figure 70-18). The heat-gain HTM is calculated by multiplying the U value from Table 4A (Figure 70-17) by the cooling temperature difference from Table 4B (Figure 70-18): HTM $= 0.073 \times 16.4 = 1.2$.

The heat loss and gain for the walls are calculated by multiplying the HTM by the net wall area. The net wall area is the wall area left after subtracting the window and door areas from the gross wall area. In our example, there is a total of 172 ft^2 of doors and windows. The gross wall area is 960 ft^2. The net wall area is $960\ ft^2 - 172\ ft^2 = 788\ ft^2$ net wall area. Figure 70-19 shows the wall calculations.

Ceiling

The ceiling in our example has a construction number of 16B-30: ceiling under a vented attic with R-30 insulation. The U value is 0.032. The heat-loss HTM is 1.4: $44 \times 0.032 = 1.4$. Our house is located in an area with a medium daily range and has a design temperature difference of 20°F in the summer. The table shows a cooling temperature difference of 55°F (Figure 70-20). The cooling HTM would then be 1.8: $0.032 \times 55 = 1.8$. Figure 70-21 shows the ceiling calculations.

Floor

The slab-on-grade construction number 22 is unique in that it lists F values rather than U values. The HTM is calculated the same: F value times the heating temperature difference equals the HTM. The concrete slab floor in our example has a construction number of 22A-ph: no insulation on heavy, moist soil. It has an F value of 1.358 (Figure 70-22). The heat-loss HTM for a house with a heating temperature difference of 44°F would be 60: $44 \times 1.358 = 60$ (approximately). However, the resulting HTM is not for square feet but for linear feet of exposed edge. So the heat loss of the floor is $60 \times 120 = 7{,}200$ BTU/hr. Figure 70-23 shows the floor calculations.

Tables for construction numbers 21 and 22 do not include any heat-gain information because there is no heat gain from a concrete floor on the ground.

Table 4A — Continued

Heat Loss and Heat Gain For Opaque Surfaces
U-Values and Group Numbers or CLTD Values

Construction Number 12
Frame Walls and Partitions

Wall or partition with brick veneer, plus interior finish (40 to 50 Lb / SqFt)
Wall with siding or stucco, or light partition, plus interior finish (7 to 20 Lb / SqFt)
Exterior finish code: b = brick veneer; s = stucco or siding
Framing code: w = wood, m = metal (studs 16 inches on center, 75% cavity, 25% framing)
Reference Area = Gross Wall Area - Area of Window and Door Openings

Construction Number	Insulation R-Values	Description of Construction	Exterior Finish	U-Value With Wood Studs	U-Value With Metal Studs	Group Number
12A — No Insulation In Stud Cavity						
12A-0b w/m 12A-0s w/m	Cavity: None Board: None	Frame construction, no cavity insulation, no board insulation, wood sheathing	Brick Siding	0.253 0.240	0.315 0.295	E A
12A-2b w/m 12A-2s w/m	Cavity: None Board: R-2	Frame construction, no cavity insulation, R-2 board insulation	Brick Siding	0.194 0.186	0.230 0.219	E A
12A-3b w/m 12A-3s w/m	Cavity: None Board: R-3	Frame construction, no cavity insulation, R-3 board insulation	Brick Siding	0.162 0.157	0.187 0.180	F B
12A-4b w/m 12A-4s w/m	Cavity: None Board: R-4	Frame construction, no cavity insulation, R-4 board insulation	Brick Siding	0.139 0.135	0.157 0.152	F B
12A-5b w/m 12A-5s w/m	Cavity: None Board: R-5	Frame construction, no cavity insulation, R-5 board insulation	Brick Siding	0.122 0.119	0.136 0.132	G C
12A-6b w/m 12A-6s w/m	Cavity: None Board: R-6	Frame construction, no cavity insulation, R-6 board insulation	Brick Siding	0.109 0.106	0.120 0.117	G C
12B — R-11 Insulation In 2 x 4 Stud Cavity						
12B-0b w/m 12B-0s w/m	Cavity: R-11 Board: None	Frame construction, R-11 cavity insulation, no board insulation, wood sheathing	Brick Siding	0.097	0.122	H B
12B-2b w/m 12B-2s w/m	Cavity: R-11 Board: R-2	Frame construction, R-11 cavity insulation, R-2 board insulation	Brick Siding	0.086	0.106	I C
12B-3b w/m 12B-3s w/m	Cavity: R-11 Board: R-3	Frame construction, R-11 cavity insulation, R-3 board insulation	Brick Siding	0.079	0.096	J D
12B-4b w/m 12B-4b w/m	Cavity: R-11 Board: R-4	Frame construction, R-11 cavity insulation, R-4 board insulation	Brick Siding	0.073	0.088	J D
12B-5b w/m 12B-5s w/m	Cavity: R-11 Board: R-5	Frame construction, R-11 cavity insulation, R-5 board insulation	Brick Siding	0.068	0.081	K E
12B-6b w/m 12B-6s w/m	Cavity: R-11 Board: R-6	Frame construction, R-11 cavity insulation, R-6 board insulation	Brick Siding	0.064	0.075	K F
12C — R-13 Insulation In 2 x 4 Stud Cavity						
12C-0b w/m 12C-0s w/m	Cavity: R-13 Board: None	Frame construction, R-13 cavity insulation, no board insulation, wood sheathing	Brick Siding	0.091	0.115	I C
12C-2b w/m 12C-2s w/m	Cavity: R-13 Board: R-2	Frame construction, R-13 cavity insulation, R-2 board insulation	Brick Siding	0.081	0.101	J D
12C-3b w/m 12C-3s w/m	Cavity: R-13 Board: R-3	Frame construction, R-13 cavity insulation, R-3 board insulation	Brick Siding	0.075	0.092	K E
12C-4b w/m 12C-4s w/m	Cavity: R-13 Board: R-4	Frame construction, R-13 cavity insulation, R-4 board insulation	Brick Siding	0.069	0.084	K E
12C-5b w/m 12C-5s w/m	Cavity: R-13 Board: R-5	Frame construction, R-13 cavity insulation, R-5 board insulation	Brick Siding	0.064	0.078	K F
12C-6b w/m 12C-6s w/m	Cavity: R-13 Board: R-6	Frame construction, R-13 cavity insulation, R-6 board insulation	Brick Siding	0.060	0.072	K G

T4A-2

Figure 70-17 Wall U factor and group number selection from Table 4A, construction number 12. *(Courtesy of Air Conditioning Contractors of America (ACCA). From Manual J 8th Edition. Copyright ACCA.)*

70.20 INFILTRATION TABLES 5A–5E

Tables 5A and 5B show a simplified method of estimating the amount of infiltration in a house. Tables 5C–5E offer a more detailed method of calculating the amount of infiltration air based on specific construction details. We will only be discussing Tables 5A and 5B. Table 5A is for homes with three or four exposures. Most single-family houses fall into this category. Table 5B is for homes with only one or two exposures. This would apply to many apartments, condominiums, or townhomes.

Tables 5A and 5B are used in the same manner because they are organized identically. These tables estimate the number of air changes per hour based on general construction criteria. An air change occurs when all the air in a building has been replaced with new, outside air. The phrase "air changes per hour" describes how often this occurs. The amount of infiltration can be calculated by multiplying the total house volume by the number of air changes per hour. For example, in a 1,000-ft^2 house with an 8-ft ceiling, the volume is 8,000 ft^3: $1{,}000 \times 8 = 8{,}000$. If this house had an air change rate of 1.5 air changes per hour, the amount of infiltration would be $1.5 \times 8{,}000 = 12{,}000$ ft^3/hr.

Each table has two sections: one for heating air changes and one for cooling air changes. Heating air changes tend to be higher than cooling air changes, primarily because the average wind velocity in the winter is assumed to be higher than in the summer. Using the general guides at the bottom

Table 4B
CLTD Values For The Average Load Procedure
Construction Number 12 (Frame Walls), 13 (Block Walls) and 14 (Alternative Construction)
Medium Color

Group Number	Construction	Design Cooling Temperature Difference (CTD) and Daily Range											
		10		15			20			25		30	35
		L	M	L	M	H	L	M	H	M	H	H	H
A	Wall	25.5	21.5	30.5	26.5	21.5	35.5	31.5	26.5	36.5	31.5	36.5	41.5
	Partition	14.8	10.8	19.8	15.8	10.8	24.8	20.8	15.8	25.8	20.8	25.8	30.8
B	Wall	23.1	19.1	28.1	24.1	19.1	33.1	29.1	24.1	34.1	29.1	34.1	39.1
	Partition	12.7	8.7	17.7	13.7	8.7	22.7	18.7	13.7	23.7	18.7	23.7	28.7
C	Wall	20.5	16.5	25.5	21.5	16.5	30.5	26.5	21.5	31.5	26.5	31.5	36.5
	Partition	10.6	6.6	15.6	11.6	6.6	20.6	16.6	11.6	21.6	16.6	21.6	26.6
D	Wall	18.5	14.5	23.5	19.4	14.5	28.5	24.5	19.5	29.5	24.5	29.5	34.5
	Partition	9.4	5.4	14.4	10.4	5.4	19.4	15.4	10.4	20.4	15.4	20.4	25.4
E	Wall	16.4	12.4	21.4	17.4	12.4	26.4	22.4	17.4	27.4	22.4	27.4	32.4
	Partition	8.0	4.0	13.0	9.0	4.0	18.0	14.0	9.0	19.0	14.0	19.0	24.0
F	Wall	14.3	10.3	19.3	15.3	10.3	24.3	20.3	15.3	25.3	20.3	25.3	30.3
	Partition	6.9	2.9	11.9	7.9	2.9	16.9	12.9	7.9	17.9	12.9	17.9	22.9
G	Wall	12.3	8.3	17.3	13.3	8.3	22.3	18.3	13.3	23.3	18.3	23.3	28.3
	Partition	5.6	1.6	10.6	6.6	1.6	15.6	11.6	6.6	16.6	11.6	16.6	21.6
H	Wall	11.7	7.7	16.7	12.7	7.7	21.7	17.7	12.7	22.7	17.7	22.7	27.7
	Partition	5.3	1.3	10.3	6.3	1.3	15.3	11.3	6.3	16.3	11.3	16.3	21.3
I	Wall	11.0	7.0	16.0	12.0	7.0	21.0	17.0	12.0	22.0	17.0	22.0	27.0
	Partition	4.9	0.9	9.9	5.9	0.9	14.9	10.9	5.9	15.9	10.9	15.9	20.9
J	Wall	10.4	6.4	15.4	11.4	6.4	20.4	16.4	11.4	21.4	16.4	21.4	26.4
	Partition	4.7	0.7	9.7	5.7	0.7	14.7	10.7	5.7	15.7	10.7	15.7	20.7
K	Wall	9.6	5.6	14.6	10.6	5.6	19.6	15.6	10.6	20.6	15.6	20.6	25.6
	Partition	4.5	0.5	9.5	5.5	0.5	14.5	10.5	5.5	15.5	10.5	15.5	20.5

Figure 70-18 Cooling temperature difference from Table 4B. *(Courtesy of Air Conditioning Contractors of America (ACCA). From Manual J 8th Edition. Copyright ACCA.)*

of the page, the house is analyzed as being tight, average, or leaky. Select the description that best fits the house. The floor area of the house is listed in columns. The value for the average number of air changes is at the intersection. For our example house with 876 ft^2, the table shows a winter air change rate of 1.05 and a summer air change rate of 0.55 (Figure 70-24). Converting these air changes to airflow:

Winter 896 × 8 × 1.05 = 7,526 ft^3/hr (7,526 ft^3/hr)/60
= 125 ft3/min

Summer 896 × 8 × 0.55 = 3,942 ft^3/hr (3,942 ft^3/hr)/60
= 66 ft^3/rom

The infiltration loss and the sensible infiltration gain are calculated by multiplying the design temperature difference by 1.1 and by CFM of infiltration air because it takes approximately 1.1 BTU to raise 1 CFM 1°F. The infiltration loss for our example house is:

Infiltration loss 44°F × 1.1 BTU/hr/CFM × 125 CFM
= 6,050 BTU/hr

Sensible infiltration gain 20°F × 1.1 BTU/hr/CFM° × 66 CFM
= 1,452 BTU/hr

The latent infiltration is calculated by multiplying 0.68 × grains difference × CFM. In our example for Selma, Alabama there is a 49 grain difference at the summer design condition of 50% relative humidity. The result is 0.68 × 49 grains × 66 CFM = 2200 Btu/hr.

.

The infiltration loss and gain is distributed throughout the house based on the ratio of the outside wall area for that room compared to the entire outside wall area. This is called the wall area ratio, or WAR. The WAR in our example is 1.0 because we are calculating the load for the entire house. Figure 70-25 shows the infiltration calculations.

1	Name of Room						Entire House			
2	Running Feet of Exposed Wall						120			
3	Ceiling Height and Gross Wall Area						8	120 × 8 = 960		
4	Room Dimensions and Floor Plan Area						32 × 28	32 × 28 = 896		
5	Ceiling Slope						0	896		
					HTM		Area or Length	BTU per Hour		
	Type of Exposure		Const #	Facing	Heating	Cooling		Heating	Sensible	Latent
6a	Windows and Glass Doors	a	3B	North	22.4	20	13	291	260	
		b	3B	East	22.4	52	40	896	2,080	
		c	3B	South	22.4	29	24	538	696	
		d	3B	West	22.4	52	75	1,680	3,900	
7	Wood and Metal Doors	a	11 N		12.8	31	20	256	180	
8	**Above Grade Walls**	**a**	**12B-4b**		**3.2**	**1.2**	**788**	**2,521**	**946**	
10	Ceilings	a								
11a	Passive Floors	a								
12	Infiltration	Heat Loss					WAR 1.0			
		Sensible Gain								
		Latent Gain								
13	Internal	a								
		b								
14	Subtotals	Sum lines 6–11a + lines 12 and 13								
15	Duct Loads	ELF Loss and ELF Gain								
		Latent Gain								
20	Total Load	Sum lines 11b + lines 14–19								

Figure 70-19 Wall calculations on line 8 completed on Manual J worksheet.

We will calculate the WAR for the kitchen to show how the WAR is calculated for a room. The total outside wall area for our example house is 960 ft^2. The outside wall area for the kitchen is 216 ft^2: 27 ft × 8 ft = 216 ft^2. The WAR for the kitchen is 216 ft^2/960 ft^2 = 0.225. This factor would be multiplied by the total infiltration loss and gain for the house to determine the infiltration loss and gain for the kitchen.

70.21 INTERNAL LOADS, TABLES 6A AND 6B

Table 6A contains general information on sensible and latent loads added by appliances and people. It starts with the same 1,200 BTU/hr for appliances that has been used for 30 years but then offers other scenarios. A brief description is provided with each scenario to help determine which is the most appropriate. Extra information is also provided for adjusting the appliance allowance. For example, most families today would require the 1,400 additional BTU/hr for using electronic entertainment equipment.

Notice that some loads add to both the sensible and the latent load. People are an example of a load that is both sensible and latent. Plants add only to the latent load. People are estimated at one person per bedroom plus one more person. Our example house has two bedrooms, so we will use three people. For the appliances we will use Scenario 1 with 2,400 BTU/hr (Figure 70-26). The internal load calculations are shown in Figure 70-27.

Table 6B has information on specific appliances. This table can be used to produce a more detailed estimate of internal appliance load when the specific types of appliances being used are known.

70.22 SUBTOTAL

The subtotal sums all the loads for heat loss, sensible heat gain, and latent heat gain. The subtotal does not include duct load, ventilation load, winter humidification load, piping load, or blower heat. The subtotal amounts will be used to calculate the duct loss and gain. The subtotal calculations for the example are shown in Figure 70-28.

70.23 DUCT LOADS, TABLES 7A–7L

The duct loss in heating and duct gain in cooling take two forms: heat transfer between the duct and the surrounding air and air leakage in the duct. The heat gain that can occur from putting an air-conditioning duct in a hot attic is

Construction Number 16
Ceiling Under Attic or Attic Knee Wall

Ventilation options: Unvented or vented to FHA specifications.
Roofing material options: Asphalt shingles, wood shakes, tile, slate, metal, concrete, tar and gravel or membrane.
Roof color options: Dark, red or solid bold color; light color, light gray, silver or unpainted metal and white (see absorptivity notes).
Reference Area = Gross Area - Skylight Area (SqFt)

CLTD Values For The Average and Peak Load Procedures
Ceilings Under An Attic or Attic Knee Wall

16A Attic Temperature = 150 °F
Unvented Attic, No Radiant Barrier, Any Roofing Material, Any Roof Color

Number	Construction Notes	Insulation R-Value	U-Value
16A-0	16A Unvented attic over ceiling or same type of air space behind an attic knee wall.	None	0.408
16A-7		R-7	0.112
16A-11		R-11	0.081
16A-13		R-13	0.070
16A-15		R-15	0.061
16A-19		R-19	0.049
16A-21		R-21	0.044
16A-25		R-25	0.038
16A-28		R-28	0.034
16A-30		R-30	0.032
16A-38		R-38	0.026
16A-44		R-44	0.022
16A-56		R-56	0.018

Design Temperature Difference and Daily Range

10		15			20			25		30	35
L	M	L	M	H	L	M	H	M	H	H	H
69	65	74	70	65	79	75	70	80	75	80	85

Roofs and ceilings do not have a group number.

16A
Roofing material code: None required
Roof color code: None required

16B and 16BR Attic Temperature = 130 °F
16B = Vented Attic, No Radiant Barrier, Dark Asphalt Shingles or Dark Metal, Tar and Gravel or Membrane
16BR = Unvented Attic With Radiant Barrier, Any Roofing Material, Any Roof Color

Number	Construction Notes	Insulation R-Value	U-Value
16B-0	16B FHA vented attic with no radiant barrier over ceiling or same type of air space behind an attic knee wall.	None	0.408
16B-7		R-7	0.112
16B-11		R-11	0.081
16B-13		R-13	0.070
16B-15		R-15	0.061
16B-19	16BR Unvented attic with radiant barrier over ceiling or same type of air space behind an attic knee wall.	R-19	0.049
16B-21		R-21	0.044
16B-25		R-25	0.038
16B-28		R-28	0.034
16B-30		R-30	0.032
16B-38		R-38	0.026
16B-44		R-44	0.022
16B-50		R-50	0.020
16B-56		R-56	0.018

Design Temperature Difference and Daily Range

10		15			20			25		30	35
L	M	L	M	H	L	M	H	M	H	H	H
49	45	54	50	45	59	55	50	60	55	60	65

Roofs and ceilings do not have a group number.

16B
Roofing code: a = asphalt shingles, m = metal, x = tar/gravel, z = membrane
Roof color code: d = dark (absorptivity of roofing material exceeds 0.75)
Red or solid bold color shingle = dark color

16BR
Roof material code: None required
Roof color code: None required

T4A-18

Figure 70-20 U value and cooling temperature difference for ceiling from Table 4A, construction number 16. *(Courtesy of Air Conditioning Contractors of America (ACCA). From Manual J 8th Edition. Copyright ACCA.)*

obvious. Less obvious is the large heat gain that can occur from a leaky return in the same hot attic. In fact, the air leak in the return is probably a more significant heat gain. The duct loss and gain tables take the type of duct system, duct sealing, duct location, the temperature of the air in the duct, and the temperature of the air around the duct into account. As a result, the duct heat-loss and heat-gain tables cover over 40 pages. However, you will normally only be using two of these pages for any particular job.

The first page is like a table of contents for the other pages. Descriptions of the duct location, the type of return, and the type of supply are used to select the appropriate table. In our example, a sealed radial duct system in a 120°F vented attic would use Table 7C-RS (Figure 70-29). The following pages all work similarly. The box in the upper left-hand corner describes the duct system, outlined in green on Figure 70-30. The long vertical box on the right-hand side of the page is used for determining the heat loss, outlined in red on Figure 70-30. The box in the middle of the left-hand column is for determining sensible heat gain, outlined in blue on Figure 70-30. The box in the lower left-hand corner is for determining latent heat gain, outlined in purple on Figure 70-30.

Duct Loss

The outdoor air temperature, the heated supply-air temperature, and the floor area are used to determine the duct loss. In our example, a system with a supply air temperature of 100°F and an outdoor air temperature of 30°F in a 1,000-ft^3 house would have a duct loss factor of 0.08 (Figure 70-31). This factor is multiplied by the heat loss for the entire house to

1	Name of Room						Entire House			
2	Running Feet of Exposed Wall						120			
3	Ceiling Height and Gross Wall Area						8	120 × 8 = 960		
4	Room Dimensions and Floor Plan Area						32 × 28	32 × 28 = 896		
5	Ceiling Slope						0	896		
Type of Exposure			Const #	Facing	HTM		Area or Length	BTU per Hour		
					Heating	Cooling		Heating	Sensible	Latent
6a	Windows and Glass Doors	a	3B	North	22.4	20	13	291	260	
		b	3B	East	22.4	52	40	896	2,080	
		c	3B	South	22.4	29	24	538	696	
		d	3B	West	22.4	52	75	1,680	3,900	
7	Wood and Metal Doors	a	11 N		12.8	31	20	256	180	
8	Above Grade Walls	a	12B-4b		3.2	1.2	788	2,521	946	
10	**Ceilings**	**a**	**16A-30**		**1.4**	**1.8**	**896**	**1,254**	**1,613**	
11a	Passive Floors	a								
12	Infiltration	Heat Loss					WAR 1.0			
		Sensible Gain								
		Latent Gain								
13	Internal	a								
		b								
14	Subtotals	Sum lines 6–11a + lines 12 and 13								
15	Duct Loads	ELF Loss and ELF Gain								
		Latent Gain								
20	Total Load	Sum lines 11b + lines 14–19								

Figure 70-21 Ceiling calculations for Manual J, line 10.

Construction Number 22
Concrete Slab-On-Grade Floor

For passive floors: ***Heat Loss HTM = F-Value x HTD***
For radiant floors: ***Heat Loss HTM = F-Value x (HTD + 25)***
For this construction only: ***Heat Loss = HTM x Running Feet of Exposed Edge***
Use Construction 22 for a floor that is less than 2 feet below grade. See Construction 21 for a floor that is more than 2 feet below grade.
Select passive floor when radial duct system is installed below floor.
Floor construction code: p = passive (no radiant heat), r = floor over radiant heating coils
Soil condition code: h = heavy moist soil, m = heavy dry or light moist soil, l = light dry soil
Reference Area does not apply, use running feet of exposed edge.

Number Passive Floor	Number Radiant Floor	Insulation Arrangement	Insulation R-Value	Soil Condition Below Slab	F-Value
22A — No Edge Insulation, No Insulation Below Floor, Any Floor Cover					
22A-ph	22A-rh	None	R-0	Heavy Moist Soil	1.358
22A-pm	22A-rm			Heavy Dry or Light Wet Soil	1.180
22A-pl	22A-rl			Light Dry Soil	0.989

Figure 70-22 Floor F value from Table 4A, construction number 22. *(Courtesy of Air Conditioning Contractors of America (ACCA). From Manual J 8th Edition. Copyright ACCA.)*

1	Name of Room							Entire House			
2	Running Feet of Exposed Wall							120			
3	Ceiling Height and Gross Wall Area							8	120 × 8 = 960		
4	Room Dimensions and Floor Plan Area							32 × 28	32 × 28 = 896		
5	Ceiling Slope							0	896		
	Type of Exposure		Const #	Facing	HTM			Area or Length	BTU per Hour		
					Heating	Cooling			Heating	Sensible	Latent
6a	Windows and Glass Doors	a	3B	North	22.4	20		13	291	260	
		b	3B	East	22.4	52		40	896	2,080	
		c	3B	South	22.4	29		24	538	696	
		d	3B	West	22.4	52		75	1,680	3,900	
7	Wood and Metal Doors	a	11 N		12.8	31		20	256	180	
8	Above Grade Walls	a	12B-4b		3.2	1.2		788	2,521	946	
10	Ceilings	a	16A-30		1.4	1.8		896	1,254	1,613	
11a	**Passive Floors**	**a**	**22A-ph**		**60**	**0**		**120**	**7,200**	**0**	
12	Infiltration	Heat Loss						WAR 1.0			
		Sensible Gain									
		Latent Gain									
13	Internal	a									
		b									
14	Subtotals	Sum lines 6–11a + lines 12 and 13									
15	Duct Loads	ELF Loss and ELF Gain									
		Latent Gain									
20	Total Load	Sum lines 11b + lines 14–19									

Figure 70-23 Floor calculations for Manual J, line 11a.

Table 5A
Default Air Change Values For Three or Four Exposures

Air Changes Per Hour — Heating						
Construction	Floor Area In SqFt					Each Fireplace
	900 Or Less	901 to 1500	1501 to 2000	2001 to 3000	More than 3000	
Tight	0.40	0.30	0.25	0.20	0.20	0.10
Average	1.05	0.75	0.60	0.50	0.45	0.20
Leaky	2.40	1.70	1.45	1.15	1.00	0.60
Air Changes Per Hour — Cooling						
Construction	Floor Area In SqFt					Each Fireplace
	900 Or Less	901 to 1500	1501 to 2000	2001 to 3000	More than 3000	
Tight	0.20	0.15	0.10	0.10	0.10	0.05
Average	0.55	0.35	0.30	0.25	0.20	0.10
Leaky	1.20	0.90	0.75	0.60	0.50	0.20

Figure 70-24 Default infiltration values, Table 5A. *(Courtesy of Air Conditioning Contractors of America (ACCA). From Manual J 8th Edition. Copyright ACCA.)*

1	Name of Room						Entire House			
2	Running Feet of Exposed Wall						120			
3	Ceiling Height and Gross Wall Area						8	120 × 8 = 960		
4	Room Dimensions and Floor Plan Area						32 × 28	32 × 28 = 896		
5	Ceiling Slope						0	896		
	Type of Exposure		Const #	Facing	HTM		Area or Length	BTU per Hour		
					Heating	Cooling		Heating	Sensible	Latent
6a	Windows and Glass Doors	a	3B	North	22.4	20	13	291	260	
		b	3B	East	22.4	52	40	896	2,080	
		c	3B	South	22.4	29	24	538	696	
		d	3B	West	22.4	52	75	1,680	3,900	
7	Wood and Metal Doors	a	11 N		12.8	31	20	256	180	
8	Above Grade Walls	a	12B-4b		3.2	1.2	788	2,521	946	
10	Ceilings	a	16A-30		1.4	1.8	896	1,254	1,613	
11a	Passive Floors	a	22A-ph		60	0	120	7,200	0	
12	**Infiltration**	**Heat Loss**			**6,050 BTU/hr**		**WAR 1.0**	**6,050**		
		Sensible Gain			**1,452 BTU/hr**				**1,452**	
		Latent Gain			**2,200 BTU/hr**					**2,200**
13	Internal	a								
		b								
14	Subtotals	Sum lines 6–11a + lines 12 and 13								
15	Duct Loads	ELF Loss and ELF Gain								
		Latent Gain								
20	Total Load	Sum lines 11b + lines 14–19								

Figure 70-25 Infiltration calculations for Manual J, line 12.

determine the BTU/hr duct loss. In our example, the duct loss for the house would be 20,686 BTU/hr × 0.08 = 1,655 BTU/hr duct loss. The duct loss tables assume a duct insulation value of R-6. Factors for other insulation values are given at the bottom of the duct loss column (Figure 70-32).

Sensible Duct Gain

There are two types of duct gain: sensible and latent. The outdoor air temperature and house size are used to determine the sensible duct gain. In our example, a system in a 1,000 ft^3 house with an outdoor air temperature of 95°F would have a sensible duct gain factor of 0.12 (Figure 70-33). The sensible duct gain for our example house would be 14,217 BTU/hr × 0.12 = 1,706 BTU/hr sensible duct gain. The sensible duct gain tables assume a duct insulation value of R-6. Factors for other insulation values are given at the bottom of the sensible duct gain box (Figure 70-32).

Latent Duct Gain

Unlike the duct loss or the sensible duct gain, the latent duct gain tables give a BTU/hr amount instead of a factor. The grains of water difference and house size are used to determine the latent duct gain. For our example, a system in a 1,000-ft^3 house with a grains of water difference of 50 grains would have a latent duct gain of 283 BTU/hr (Figure 70-30). Figure 70-35 shows all the duct load calculations.

70.24 TOTAL LOAD

The last line on the form is for tallying the total load. There will be three totals: heat loss, sensible heat gain, and latent gain. These are the design loads that will be used to select the equipment. Figure 70-36 shows the total calculations for our example house. Ideally, the equipment should be sized as close to the total design load as possible without oversizing. Remember that most of the time the equipment will be operating at loads that are less than the design condition. In effect, the equipment is already oversized for most of its operating life and does not need to be oversized any further. Oversized equipment operates less efficiently than undersized equipment. Furnaces, air conditioners, and heat pumps all must operate for several minutes before they are operating at their peak efficiency. A unit that only operates for 2 minutes at a time never operates at its peak efficiency.

Appliance, Equipment and Lighting Loads			
Default Appliance Load	**Sensible Btuh**	**Latent Btuh**	**Notes**
Refrigerator and range with vented hood.	1,200	~	1,200 Btuh applied to the kitchen.
Scenario Options			
1) Refrigerator, range with vented hood, dish washer, clothes washer and vented clothes dryer, electronic equipment and lighting allowance.	2,400	~	1,000 Btuh for the kitchen, 500 Btuh for the utility room, 900 Btuh allowance for a TV or computer and a few lighting fixtures.
2) Two refrigerators or one refrigerator and one freezer, dish washer, range with vented hood, clothes washer and vented clothes dryer, electronic equipment and lighting allowance.	3,400	~	2,000 Btuh for the room or rooms equipped with a refrigerator, 500 Btuh for the utility room, 900 Btuh allowance for a TV or computer and a few lighting fixtures.
Adjustment Options			
A) Cooking range not equipped with a hood that is vented to outdoors, or an unvented dishwasher operating during the late afternoon in mid summer, or simultaneous use of unvented range and dishwasher.	+ 850	+ 600	Light duty cooking, 25 percent of the available range capacity used for 15 to 20 minutes. One dishwasher-cycle load spread over a two hour recovery period.
B) Water bed heater (400 Watts, 33 percent duty cycle).	+ 450	~	Apply to each bedroom equipped with a water bed .
C) Ceiling fan (75 Watts).	+ 250	~	Apply to the room where the fan is located.
D) Large family using TV's stereos, computers and laundry room during the late afternoon in mid summer.	+ 1,400	~	This is an additional 410 Watt allowance for A/V or computer equipment.
E) Allowance for above average lighting load.	+ 1,705	~	Five 100 watt lights.
F) Unvented clothes dryer.	Clothes dryers must be vented for air quality and efficiency.		
Assignment			
For room heat gain estimates, correlate the equipment loads with the rooms that contain the equipment (existing construction) or which will logically contain the equipment (working from drawings).			

Internal Loads for Full Time Occupants, Guests and Plants			
Item	**Sensible Btuh**	**Latent Btuh**	**Notes**
Full Time Occupant	230	200	Occupancy = Number Bedrooms + 1
Guest	230	180	Number of guests specified by owner
Each Small Plant	~	10	Less than 12 inches high
Each Medium Plant	~	20	12 inches to 24 inches high
Each Large Plant	~	30	More than 24 inches high

Figure 70-26 Internal heat load selection from Table 6A. *(Courtesy of Air Conditioning Contractors of America (ACCA). From Manual J 8th Edition. Copyright ACCA.)*

However, equipment is only available in certain sizes. In our example house, almost any gas or oil furnace will be oversized. Typically, the smallest gas furnaces available have a heat output capacity of approximately 32,000 BTU/hr, and the house only requires 22,341 BTU/hr. A two-stage furnace could solve this problem. The first stage would not be oversized and would allow the equipment to operate efficiently.

The cooling load has two components: sensible and latent. The capacity of air-conditioning equipment varies based on the coil match and airflow. It is important to look at both the sensible and latent capacity when sizing air-conditioning equipment. It would be possible for a piece of equipment to match the total of the sensible and latent loads but not provide enough latent capacity. This house falls in between a 1.5-ton and a 2-ton air conditioner. In general, using the smaller size will give better results for latent cooling, comfort, and operating efficiency because it will run longer. Not all companies make a 1.5-ton system. A 2-ton system would also be acceptable if a system with a variable-speed blower were selected that can operate in dehumidification mode. This particular house would be a good candidate for a heat pump because the heat loss and heat gain are close to each other in capacity.

TECH TIP

It is important to note that properly sized heating and cooling systems will provide your customers with the best system for their homes. Oversizing causes the system to start and stop too frequently, just like driving your car in stop-and-go traffic. When you drive a car in stop-and-go traffic it does not get as good gas mileage as it does when on the highway. A car driven mostly on the highway will last longer than one that is driven mostly in stop-and-go traffic. That is the same with all equipment, including heating and air-conditioning systems. Therefore, longer periods of operation are better for both the efficiency and life expectancy of all HVAC equipment.

Line	Type of Exposure		Const #	Facing	HTM Heating	HTM Cooling	Area or Length	BTU per Hour Heating	BTU per Hour Sensible	BTU per Hour Latent
1	Name of Room						Entire House			
2	Running Feet of Exposed Wall						120			
3	Ceiling Height and Gross Wall Area						8	120 × 8 = 960		
4	Room Dimensions and Floor Plan Area						32 × 28	32 × 28 = 896		
5	Ceiling Slope						0	896		
6a	Windows and Glass Doors	a	3B	North	22.4	20	13	291	260	
		b	3B	East	22.4	52	40	896	2,080	
		c	3B	South	22.4	29	24	538	696	
		d	3B	West	22.4	52	75	1,680	3,900	
7	Wood and Metal Doors	a	11 N		12.8	31	20	256	180	
8	Above Grade Walls	a	12B-4b		3.2	1.2	788	2,521	946	
10	Ceilings	a	16A-30		1.4	1.8	896	1,254	1,613	
11a	Passive Floors	a	22A-ph		60	0	120	7,200	0	
12	Infiltration	Heat Loss			6,050 BTU/hr		WAR 1.0	6,050		
		Sensible Gain			1,452 BTU/hr				1,452	
		Latent Gain			2,200 BTU/hr					2,200
13	**Internal**	**a**	**People 230 sensible 200 latent**				**3**		**690**	**600**
		b	**Scenario # 1**						**2,400**	
14	Subtotals	Sum lines 6–11a + lines 12 and 13								
15	Duct Loads	ELF Loss and ELF Gain								
		Latent Gain								
20	Total Load	Sum lines 11b + lines 14–19								

Figure 70-27 Internal heat load calculations for Manual J, line 13.

Line	Type of Exposure		Const #	Facing	HTM Heating	HTM Cooling	Area or Length	BTU per Hour Heating	BTU per Hour Sensible	BTU per Hour Latent
1	Name of Room						Entire House			
2	Running Feet of Exposed Wall						120			
3	Ceiling Height and Gross Wall Area						8	120 × 8 = 960		
4	Room Dimensions and Floor Plan Area						32 × 28	32 × 28 = 896		
5	Ceiling Slope						0	896		
6a	Windows and Glass Doors	a	3B	North	22.4	20	13	291	260	
		b	3B	East	22.4	52	40	896	2,080	
		c	3B	South	22.4	29	24	538	696	
		d	3B	West	22.4	52	75	1,680	3,900	
7	Wood and Metal Doors	a	11 N		12.8	31	20	256	180	
8	Above Grade Walls	a	12B-4b		3.2	1.2	788	2,521	946	
10	Ceilings	a	16A-30		1.4	1.8	896	1,254	1,613	
11a	Passive Floors	a	22A-ph		60	0	120	7,200	0	
12	Infiltration	Heat Loss			6,050 BTU/hr		WAR 1.0	6,050		
		Sensible Gain			1,452 BTU/hr				1,452	
		Latent Gain			2,200 BTU/hr					2,200
13	Internal	a	People 230 sensible 200 latent				3		690	600
		b	Scenario # 1						2,400	
14	**Subtotals**	**Sum lines 6–11a + lines 12 and 13**						**20,686**	**14,217**	**2,800**
15	Duct Loads	ELF Loss and ELF Gain								
		Latent Gain								
20	Total Load	Sum lines 11b + lines 14–19								

Figure 70-28 Subtotal calculations for Manual J, line 14.

Summary of Default Duct Load Factor Tables			
Location	**Supply System**	**Return System**	**Table Number**
In unvented attic or attic knee wall space above ceiling 16A (150 °F attic). Default for duct in unvented ceiling cavity.	Radial with outlets in center of rooms.	Radial, 400 CFM per return, returns close to air handler.	7A-RN Not sealed 7A-RS Sealed
	Trunk and branch with outlets in center of rooms.	Trunk and branch, 400 CFM per return, returns close to air handler.	7A-TN Not sealed 7A-TS Sealed
In vented attic or attic knee wall space above ceiling 16B (130 °F attic).	Radial with outlets in center of rooms.	Radial, 400 CFM per return, returns close to air handler.	7B-RN Not sealed 7B-RS Sealed
	Trunk and branch with outlets in center of rooms.	Trunk and branch, 400 CFM per return, returns close to air handler.	7B-TN Not sealed 7B-TS Sealed
In vented attic or attic knee wall space above ceiling 16C (120 °F attic).	Radial with outlets in center of rooms.	Radial, 400 CFM per return, returns close to air handler.	7C-RN Not sealed 7C-RS Sealed
	Trunk and branch with outlets in center of rooms.	Trunk and branch, 400 CFM per return, returns close to air handler.	7C-TN Not sealed 7C-TS Sealed
In vented attic or attic knee wall space above ceiling 16D (110 °F attic).	Radial with outlets in center of rooms.	Radial, 400 CFM per return, returns close to air handler.	7D-RN Not sealed 7D-RS Sealed
	Trunk and branch with outlets in center of rooms.	Trunk and branch, 400 CFM per return, returns close to air handler.	7D-TN Not sealed 7D-TS Sealed
In vented attic or attic knee wall space above ceiling 16E (105 °F attic).	Radial with outlets in center of rooms.	Radial, 400 CFM per return, returns close to air handler.	7E-RN Not sealed 7E-RS Sealed
	Trunk and branch with outlets in center of rooms.	Trunk and branch, 400 CFM per return, returns close to air handler.	7E-TN Not sealed 7E-TS Sealed
In vented attic or attic knee wall space above ceiling 16F (95 °F attic).	Radial with outlets in center of rooms.	Radial, 400 CFM per return, returns close to air handler.	7F-RN Not sealed 7F-RS Sealed
	Trunk and branch with outlets in center of rooms.	Trunk and branch, 400 CFM per return, returns close to air handler.	7F-TN Not sealed 7F-TS Sealed
In unvented cavity below roof joists or in cavity between roof joists (roof construction 18).	Radial with outlets in center of rooms.	Radial, 400 CFM per return, returns close to air handler.	Use Tables 7A to 7F. Use estimated cavity temperature to select a table.
	Trunk and branch with outlets in center of rooms.	Trunk and branch, 400 CFM per return, returns close to air handler.	
In open crawl space or garage.	Radial with outlets at perimeter of rooms.	Radial, 400 CFM per return, returns close to air handler	7G-RN Not sealed 7G-RS Sealed
	Trunk and branch with outlets in center of rooms.	Trunk and branch, 400 CFM per return, returns close to air handler.	7G-TN Not sealed 7G-TS Sealed
In closed crawl space below insulated floor, no wall insulation.	Radial or trunk and branch with outlets at perimeter of rooms.	Radial or trunk and branch, 400 CFM per return, returns close to air handler	7H-N Not sealed 7H-S Sealed
In unconditioned basement or closed crawl space with: o *No wall or ceiling insulation* o *Wall insulation only* o *Wall and ceiling insulation*	Radial or trunk and branch with outlets at perimeter of rooms.	Radial or trunk and branch, 400 CFM per return, returns close to air handler.	7I-N Not sealed 7I-S Sealed
Supply runs below slab. Return runs in conditioned space or in attic.	Radial with outlets at room perimeter. No supply leakage.Sealing options apply to the return runs.	Duct runs in conditioned space.	7J-1 No Leakage
		Radial system in attic, 400 CFM per return, returns close to air handler.	7J-2N Not Sealed 7J-2S Sealed
Riser or drop in exterior wall.	Rectangular or round airway.		7K-N Not sealed 7K-S Sealed
		Rectangular or round airway.	7L-N Not Sealed 7L-S Sealed

Figure 70-29 This page acts like a table of contents for the duct load section. *(Courtesy of Air Conditioning Contractors of America (ACCA). From Manual J 8th Edition. Copyright ACCA)*

Table 7C-RS FHA Vented Attic or Vented Attic Knee Wall Space Over a 16C Ceiling		
	Supply-Side	Return-Side
Configuration	Radial	Radial
System Tightness	Sealed	Sealed
Grille Location	Center of Room	Close to Unit
Risers or Drops	None	None
Heating Ambient	Outdoor Temperature + 11 °F	
Cooling Ambient	Outdoor Temperature + 25 °F	

Sealed: Methods and materials used for fabrication and sealing work conform to industry standards.

Not Sealed: No sealing work, or fabrication and /or sealing work does not comply with industry standards.

Sensible Gain Factor As a Fraction of The Sensible Envelope Load					
OAT °F Table 1	Floor Area of Conditioned Space (SqFt)				
	1000	1500	2000	2500	3000
85	0.10	0.12	0.14	0.15	0.17
90	0.11	0.13	0.14	0.16	0.18
95	0.12	0.14	0.16	0.18	0.21
100	0.12	0.15	0.17	0.18	0.20
105	0.13	0.16	0.18	0.20	0.22
Wall Insulation Correction					
Wall R	R-0	R-2	R-4	R-6	R-8
Scaler	NA	2.15	1.33	1.00	0.82

Latent Gain Value (Btuh) Add This Gain (or Loss) to The Latent Envelope Load					
Grains Table 1	Floor Area of Conditioned Space (SqFt)				
	1000	1500	2000	2500	3000
±10	90	139	186	227	266
±20	136	210	280	342	402
±30	183	283	378	462	542
±40	232	359	479	586	688
±50	283	438	584	714	839
±60	336	519	693	847	995
±70	390	603	805	984	1,155

Interpolate to find latent gain for intermediate grains difference values. Use zero or negative latent gain when the Table 1 grains difference value is zero or negative.

Heat Loss Factor As a Fraction of The Envelope Heating Load					
OAT °F Table 1	Floor Area of Conditioned Space (SqFt)				
	1000	1500	2000	2500	3000
Discharge Air @ 90 °F					
-10	0.17	0.20	0.24	0.26	0.29
0	0.15	0.18	0.21	0.23	0.26
10	0.14	0.15	0.18	0.20	0.23
20	0.11	0.13	0.15	0.17	0.19
30	0.09	0.12	0.13	0.14	0.16
40	0.07	0.08	0.10	0.11	0.13
Discharge Air @ 100 °F Heating-Cooling Default					
-10	0.13	0.16	0.18	0.19	0.21
0	0.11	0.14	0.16	0.17	0.19
10	0.10	0.12	0.14	0.15	0.17
20	0.09	0.11	0.13	0.14	0.15
30	0.08	0.09	0.10	0.11	0.13
40	0.06	0.07	0.09	0.09	0.10
Discharge Air @ 110 °F					
-10	0.10	0.13	0.14	0.16	0.18
0	0.10	0.11	0.13	0.15	0.16
10	0.09	0.10	0.12	0.14	0.14
20	0.08	0.09	0.10	0.12	0.13
30	0.07	0.08	0.09	0.10	0.11
40	0.06	0.07	0.08	0.09	0.09
Discharge Air @ 120 °F					
-10	0.09	0.11	0.13	0.14	0.16
0	0.08	0.10	0.12	0.13	0.14
10	0.08	0.09	0.10	0.11	0.13
20	0.07	0.08	0.09	0.10	0.11
30	0.06	0.07	0.08	0.09	0.10
40	0.05	0.06	0.07	0.08	0.09
Discharge Air @ 130 °F or 140 °F					
-10	0.08	0.10	0.12	0.12	0.14
0	0.08	0.09	0.11	0.11	0.12
10	0.07	0.08	0.10	0.11	0.12
20	0.06	0.07	0.09	0.10	0.11
30	0.06	0.07	0.08	0.09	0.09
40	0.05	0.06	0.07	0.08	0.08
Wall Insulation Correction					
Wall R	R-0	R-2	R-4	R-6	R-8
Scaler	NA	1.99	1.37	1.00	0.87

Figure 70-30 Major sections of a duct load table. *(Courtesy of Air Conditioning Contractors of America (ACCA). From Manual J 8th Edition. Copyright ACCA.)*

40	0.05	0.06	0.07	0.08	0.08
Wall Insulation Correction					
Wall R	R-0	R-2	R-4	R-6	R-8
Scaler	NA	1.99	1.37	1.00	0.87

Figure 70-32 Duct insulation correction factors. *(Courtesy of Air Conditioning Contractors of America (ACCA). From Manual J 8th Edition. Copyright ACCA.)*

Sensible Gain Factor As a Fraction of The Sensible Envelope Load					
OAT °F Table 1	Floor Area of Conditioned Space (SqFt)				
	1000	1500	2000	2500	3000
85	0.10	0.12	0.14	0.15	0.17
90	0.11	0.13	0.14	0.16	0.18
95	0.12	0.14	0.16	0.18	0.21
100	0.12	0.15	0.17	0.18	0.20
105	0.13	0.16	0.18	0.20	0.22
Wall Insulation Correction					
Wall R	R-0	R-2	R-4	R-6	R-8
Scaler	NA	2.15	1.33	1.00	0.82

Figure 70-33 Sensible duct gain factor selection. *(Courtesy of Air Conditioning Contractors of America (ACCA). From Manual J 8th Edition. Copyright ACCA.)*

Heat Loss Factor As a Fraction of The Envelope Heating Load					
OAT °F Table 1	Floor Area of Conditioned Space (SqFt)				
	1000	1500	2000	2500	3000
Discharge Air @ 90 °F					
-10	0.17	0.20	0.24	0.26	0.29
0	0.15	0.18	0.21	0.23	0.26
10	0.14	0.15	0.18	0.20	0.23
20	0.11	0.13	0.15	0.17	0.19
30	0.09	0.12	0.13	0.14	0.16
40	0.07	0.08	0.10	0.11	0.13
Discharge Air @ 100 °F Heating-Cooling Default					
-10	0.13	0.16	0.18	0.19	0.21
0	0.11	0.14	0.16	0.17	0.19
10	0.10	0.12	0.14	0.15	0.17
20	0.09	0.11	0.13	0.14	0.15
30	0.08	0.09	0.10	0.11	0.13
40	0.06	0.07	0.09	0.09	0.10

Figure 70-31 Example duct loss factor selection. *(Courtesy of Air Conditioning Contractors of America (ACCA). From Manual J 8th Edition. Copyright ACCA.)*

Table 7C-RS FHA Vented Attic or Vented Attic Knee Wall Space Over a 16C Ceiling		
	Supply-Side	Return-Side
Configuration	Radial	Radial
System Tightness	Sealed	Sealed
Grille Location	Center of Room	Close to Unit
Risers or Drops	None	None
Heating Ambient	Outdoor Temperature + 11 °F	
Cooling Ambient	Outdoor Temperature + 25 °F	
Sealed: Methods and materials used for fabrication and sealing work conform to industry standards. Not Sealed: No sealing work, or fabrication and /or sealing work does not comply with industry standards.		

Sensible Gain Factor As a Fraction of The Sensible Envelope Load					
OAT °F Table 1	Floor Area of Conditioned Space (SqFt)				
	1000	1500	2000	2500	3000
85	0.10	0.12	0.14	0.15	0.17
90	0.11	0.13	0.14	0.16	0.18
95	0.12	0.14	0.16	0.18	0.21
100	0.12	0.15	0.17	0.18	0.20
105	0.13	0.16	0.18	0.20	0.22
Wall Insulation Correction					
Wall R	R-0	R-2	R-4	R-6	R-8
Scaler	NA	2.15	1.33	1.00	0.82

Latent Gain Value (Btuh) Add This Gain (or Loss) to The Latent Envelope Load					
Grains Table 1	Floor Area of Conditioned Space (SqFt)				
	1000	1500	2000	2500	3000
±10	90	139	186	227	266
±20	136	210	280	342	402
±30	183	283	378	462	542
±40	232	359	479	586	688
±50	283	438	584	714	839
±60	336	519	693	847	995
±70	390	603	805	984	1,155
Interpolate to find latent gain for intermediate grains difference values. Use zero or negative latent gain when the Table 1 grains difference value is zero or negative.					

Heat Loss Factor As a Fraction of The Envelope Heating Load					
OAT °F Table 1	Floor Area of Conditioned Space (SqFt)				
	1000	1500	2000	2500	3000
Discharge Air @ 90 °F					
-10	0.17	0.20	0.24	0.26	0.29
0	0.15	0.18	0.21	0.23	0.26
10	0.14	0.15	0.18	0.20	0.23
20	0.11	0.13	0.15	0.17	0.19
30	0.09	0.12	0.13	0.14	0.16
40	0.07	0.08	0.10	0.11	0.13
Discharge Air @ 100 °F Heating-Cooling Default					
-10	0.13	0.16	0.18	0.19	0.21
0	0.11	0.14	0.16	0.17	0.19
10	0.10	0.12	0.14	0.15	0.17
20	0.09	0.11	0.13	0.14	0.15
30	0.08	0.09	0.10	0.11	0.13
40	0.06	0.07	0.09	0.09	0.10
Discharge Air @ 110 °F					
-10	0.10	0.13	0.14	0.16	0.18
0	0.10	0.11	0.13	0.15	0.16
10	0.09	0.10	0.12	0.14	0.14
20	0.08	0.09	0.10	0.12	0.13
30	0.07	0.08	0.09	0.10	0.11
40	0.06	0.07	0.08	0.09	0.09
Discharge Air @ 120 °F					
-10	0.09	0.11	0.13	0.14	0.16
0	0.08	0.10	0.12	0.13	0.14
10	0.08	0.09	0.10	0.11	0.13
20	0.07	0.08	0.09	0.10	0.11
30	0.06	0.07	0.08	0.09	0.10
40	0.05	0.06	0.07	0.08	0.09
Discharge Air @ 130 °F or 140 °F					
-10	0.08	0.10	0.12	0.12	0.14
0	0.08	0.09	0.11	0.11	0.12
10	0.07	0.08	0.10	0.11	0.12
20	0.06	0.07	0.09	0.10	0.11
30	0.06	0.07	0.08	0.09	0.09
40	0.05	0.06	0.07	0.08	0.08
Wall Insulation Correction					
Wall R	R-0	R-2	R-4	R-6	R-8
Scaler	NA	1.99	1.37	1.00	0.87

Figure 70-34 Latent duct gain selection. *(Courtesy of Air Conditioning Contractors of America (ACCA). From Manual J 8th Edition. Copyright ACCA.)*

1	Name of Room						Entire House			
2	Running Feet of Exposed Wall						120			
3	Ceiling Height and Gross Wall Area						8	120 × 8 = 960		
4	Room Dimensions and Floor Plan Area						32 × 28	32 × 28 = 896		
5	Ceiling Slope						0	896		
	Type of Exposure		Const #	Facing	HTM		Area or Length	BTU per Hour		
					Heating	Cooling		Heating	Sensible	Latent
6a	Windows and Glass Doors	a	3B	North	22.4	20	13	291	260	
		b	3B	East	22.4	52	40	896	2,080	
		c	3B	South	22.4	29	24	538	696	
		d	3B	West	22.4	52	75	1,680	3,900	
7	Wood and Metal Doors	a	11 N		12.8	31	20	256	180	
8	Above Grade Walls	a	12B-4b		3.2	1.2	788	2,521	946	
10	Ceilings	a	16A-30		1.4	1.8	896	1,254	2,150	
11a	Passive Floors	a	22A-ph		60	0	120	7,200	0	
12	Infiltration	Heat Loss			6,050 BTU/hr		WAR 1.0	6,050		
		Sensible Gain			1,452 BTU/hr				1,452	
		Latent Gain			2,200 BTU/hr					2,200
13	Internal	a	People 230 sensible 200 latent				3		690	600
		b	Scenario # 1						2,400	
14	Subtotals	Sum lines 6–11a + lines 12 and 13						20,686	14,217	2,800
15	**Duct Loads**	**ELF Loss and ELF Gain**			**0.08**	**0.12**		**1,655**	**1,706**	
		Latent Gain								**283**
20	Total Load	Sum lines 11b + lines 14–19								

Figure 70-35 Duct load calculations on line 15 of Manual J worksheet.

1	Name of Room						Entire House			
2	Running Feet of Exposed Wall						120			
3	Ceiling Height and Gross Wall Area						8	120 * 8 = 960		
4	Room Dimensions and Floor Plan Area						32 * 28	32 * 28 = 896		
5	Ceiling Slope						0	896		
	Type of Exposure		Const #	Facing	HTM		Area or Length	BTU per Hour		
					Heating	Cooling		Heating	Sensible	Latent
6a	Windows and Glass Doors	a	3B	North	22.4	20	13	291	260	
		b	3B	East	22.4	52	40	896	2,080	
		c	3B	South	22.4	29	24	538	696	
		d	3B	West	22.4	52	75	1,680	3,900	
7	Wood and Metal Doors	a	11 N		12.8	31	20	256	180	
8	Above Grade Walls	a	12B-4b		3.2	1.2	788	2,521	946	
10	Ceilings	a	16A-30		1.4	1.8	896	1,254	2,150	
11a	Passive Floors	a	22A-ph		60	0	120	7,200	0	
12	Infiltration	Heat Loss			6,050 BTU/hr		WAR 1.0	6,050		
		Sensible Gain			1,452 BTU/hr				1,452	
		Latent Gain			2,200 BTU/hr					2,200
13	Internal	a	People 230 sensible 200 latent				3		690	600
		b	Scenario # 1						2,400	
14	Subtotals	Sum lines 6–11a + lines 12 and 13						20,686	14,217	2,800
15	Duct Loads	ELF Loss and ELF Gain			0.08	0.12		1,655	1,706	
		Latent Gain								283
20	**Total Load**	**Sum lines 11b + lines 14–19**						**22,341**	**15,923**	**3,083**

Figure 70-36 Total load calculations for Manual J worksheet.

UNIT 70—SUMMARY

Heating and air-conditioning equipment operates most efficiently when it is properly sized. The calculations done to determine the amount of heating and cooling required for a particular building are called heat load calculations or a heat load study. Winter calculations are called heat loss, and summer calculations are called heat gain. There are two types of heat gain: sensible and latent. Calculations are necessary to take into account the many variables that affect heat loss and heat gain. The resistance of a material to heat flow is measured in R value. Heat travels more slowly through materials with higher R values. U factors are used to calculate the amount of heat traveling through a building panel. U factors are built by adding up all the R values of the components in a building panel and finding the reciprocal. A panel with a total resistance of R-10 has a U factor of 1⁄10, or 0.10. Heat-transfer multipliers, or HTMs, are built by multiplying a panel's U factor by the temperature difference across the panel. A panel with a U factor of 0.10 and a temperature difference of 50°F will have an HTM of 5. The heat transferred through a section of a building is determined by multiplying that panel's HTM by the square-foot area of the panel. The most accepted method for organizing and simplifying heat load calculations is Manual J. Manual J worksheets are like a handwritten spreadsheet. The information is organized in tables that make calculations easier.

UNIT 70—REVIEW QUESTIONS

1. Why are heat load calculations necessary?
2. What is the difference between a shell load study and a room-by-room load study?
3. What advantage is there to doing a room-by-room load study over a shell load study?
4. Why does oversized equipment operate inefficiently?
5. What is the difference between heat loss and heat gain?
6. What is the difference between sensible heat gain and latent heat gain?
7. What is the R value of a material with a thermal conductivity of 2.0 that is 4 in thick?
8. How is a U factor calculated for a building panel built of several different materials?
9. Determine the U value for a wall with the following construction:

Exterior siding	R-0.75
Cavity insulation	R-13
Interior wall	R-0.5

10. How is an HTM calculated?
11. What would the HTM be for a building panel with a U value of 0.8 and a temperature difference of 40°F?
12. What is meant by the temperature daily range?
13. What effect does the daily range have on heat gain?
14. Why is the cooling temperature difference often different from the difference between the indoor and outdoor temperatures?
15. How are the outside design conditions chosen?
16. List some of the differences between residential and commercial load calculations.
17. What is Manual J?
18. Using the table in Figure 70-20, determine the heat-gain HTM for a ceiling with the following specifications:
    150°F unvented attic
    R-19 insulation
    Outside design temperature 90°F
    Inside design temperature 75°F
    Low daily temperature range
19. What is the heat-gain HTM for a concrete slab on a grade floor?
20. Calculate the heat loss for a concrete slab on a grade floor for a room with the following specifications
    10 ft by 15 ft bedroom on corner of house
    Two sides exposed
    HTM of 40

UNIT 71

Duct Design

OBJECTIVES

Upon completion of this unit, you will be able to:

1. list the three overall goals in duct design.
2. list the different duct design methods.
3. discuss how to determine the duct system design static pressure.
4. explain the difference between design friction loss rate and the actual static pressure loss through a duct.
5. explain how a duct's equivalent length is determined.
6. use a duct calculator to look up the duct size, air velocity, and friction rate.
7. list the criteria for selecting grilles and registers.

71.1 INTRODUCTION

Duct systems are the distribution network for conditioned air to be moved throughout a building. Duct systems are designed to provide conditioned air that matches the needs of the structure. Proper duct system design is critical to the energy efficiency of a building. Duct designers must use information such as heat load calculations, cooling load calculations, equipment selection, and architectural design to produce specialized duct plans for each structure. Technicians need to understand duct systems and airflow to be able to troubleshoot and maintain an HVAC system. Factors that adversely affect system airflow will also have a negative impact on system efficiency and reliability.

71.2 DUCT DESIGN METHODS

The two most common methods of duct design are static regain and equal friction. In the static regain method, the air velocity is reduced at the end of long trunk-duct runs to increase the static pressure. The static pressure increase is calculated to offset the static pressure lost through the run, so the static is regained. This method is primarily applied to large commercial installations and is not used in residential or light commercial applications. In the equal-friction method, the duct is sized to offset differences in duct length. When selecting the duct size, long runs use a lower friction rate than short runs. When applied properly, the equal-friction method produces a well-balanced system, delivering air where it is needed. A practical limitation of equal-friction design is that duct is only available in certain physical sizes. This often forces the designer to choose a duct that is a little larger or smaller than the ideal duct size. Balancing dampers can be added so that final adjustments can be made to produce a well-balanced system.

Two commonly used guides for duct design are the ASHRAE duct design method and the ACCA Manual D. The current Manual D method is a modified equal-friction method because the friction rate is not adjusted for each run. Rather, all runs are designed at a friction rate that will work for the run with the highest equivalent length. This ensures that the pressure drop through the duct system will not be excessive while keeping the calculations simpler. However, it produces an inherently unbalanced system because all runs are treated the same, even though they have different equivalent lengths. Balancing dampers are a necessity with this design system. To have balanced airflow, the system must be commissioned by taking airflow measurements and making adjustments to the dampers to adjust the airflow through each run. The disadvantage of this system is simply that the system must be operated and balanced after installation to perform well. While this is certainly a step that should be done, many residential systems never undergo a commissioning process. For the remainder of this unit, we discuss the equal-friction method.

71.3 EQUIVALENT LENGTH

The amount of pressure lost through a duct is directly related to its equivalent length. Many runs in a normal residential house will be longer than 100 ft in equivalent length. Every time the air changes direction it loses some of its pressure. This loss in pressure is most commonly stated as equivalent to the pressure loss through a length of duct. An elbow with a large radius will have an equivalent length of approximately 10 ft. This means that the pressure lost when the air moves around this elbow is equivalent to the pressure lost when traveling through 10 ft of pipe. Figures through 71-4 show the equivalent length for several common fittings when duct static and velocity are within the normal range of residential systems.

To find a duct's total equivalent length, add up the trunk duct length to the takeoff, the branch duct length, and the equivalent lengths of all the fittings used from the blower to the register. For example, take a branch duct run on an extended plenum system (Figure 71-5). The trunk duct leaving the plenum is equivalent to 35 ft; the branch takeoff from the trunk duct is another 35 ft. The torpedo boot on the end of the run adds 50 ft. The run has 120 ft of equivalent length in fittings alone. After adding 10 ft for each downstream

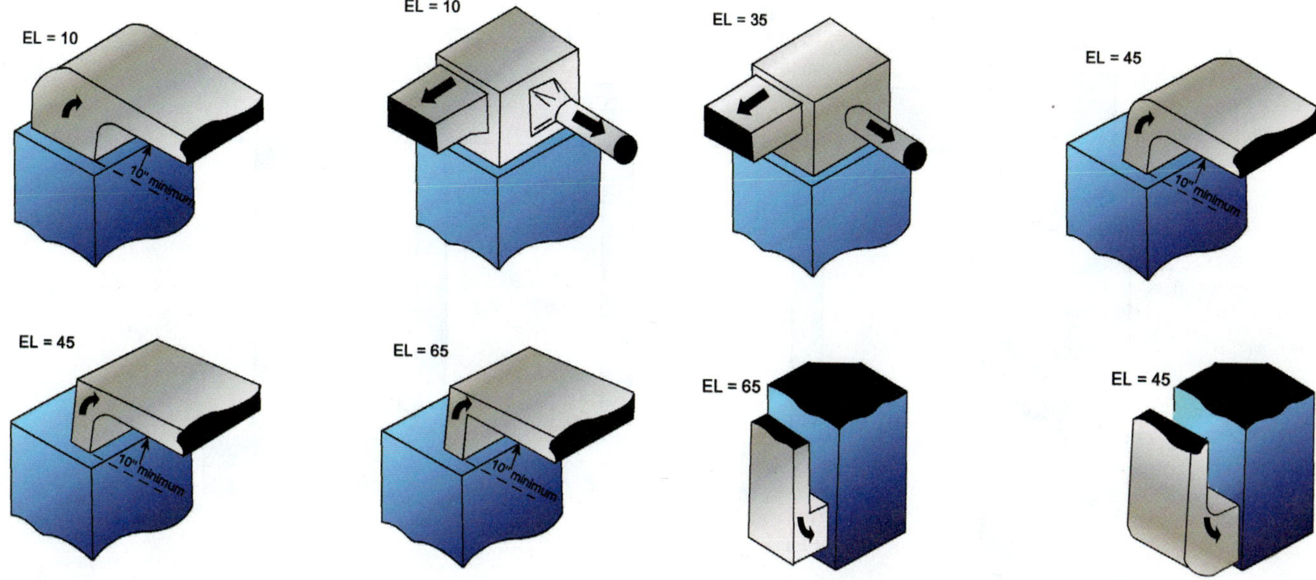

Figure 71-1 Equivalent lengths: supply- and return-air plenum fittings.

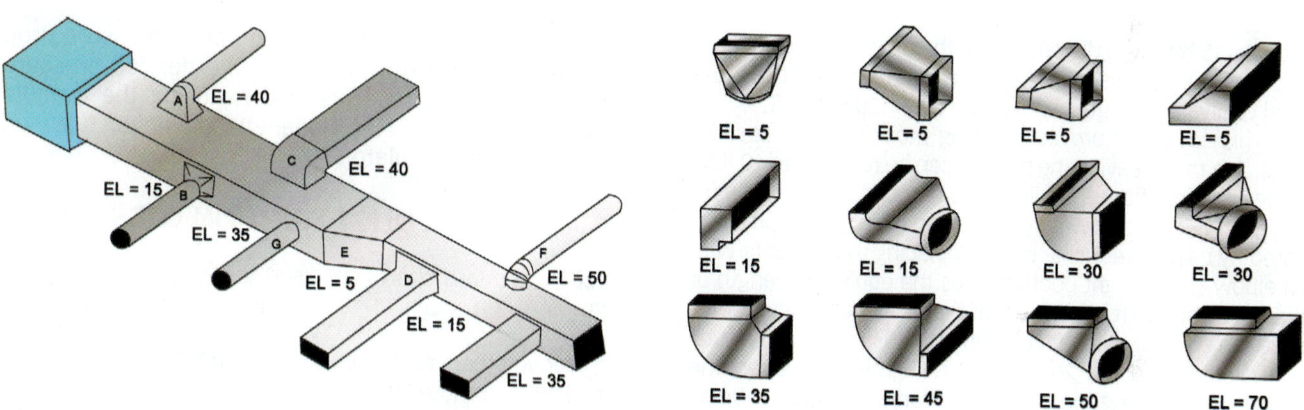

Figure 71-2 Equivalent lengths: extended-plenum fittings.

Figure 71-4 Equivalent lengths of boot fittings.

Figure 71-3 Equivalent length of round supply system fittings.

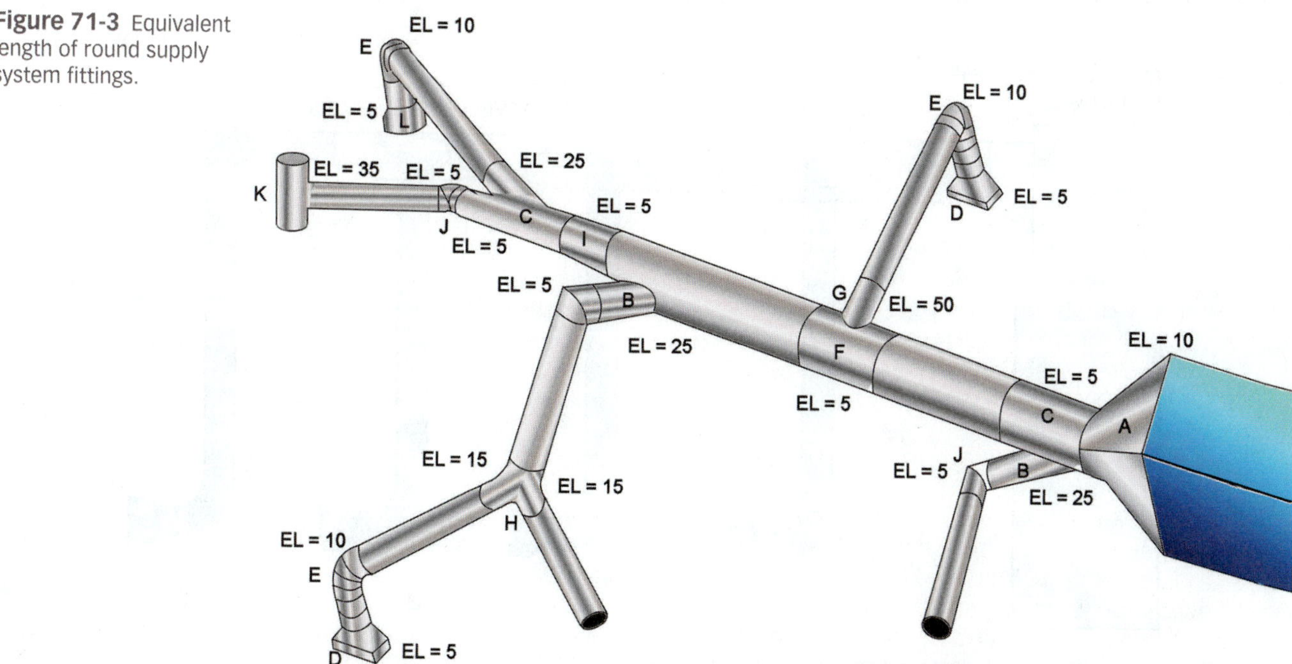

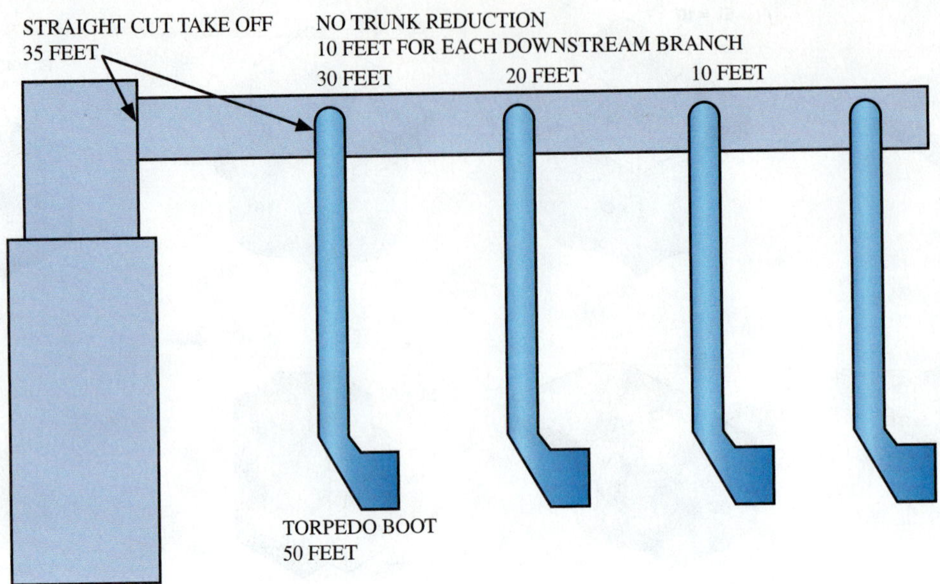

Figure 71-5 Example of branch duct total equivalent length using inefficient fittings.

branch, 15 ft for the distance from the blower to the takeoff, and 20 ft for the distance from the takeoff to the boot, the total equivalent length is 30 + 15 + 20 + 120 = 185 ft.

This can be improved by using more efficient duct fittings. The trunk leaving the plenum can be reduced to 10 ft by using a tapered fitting. Similarly, the branch takeoff can be reduced to 15 ft. The downstream branch factor has been eliminated by using a reducing trunk. Replacing the 90° boot with an elbow and straight boot changes the equivalent length of the boot from 50 ft to 15 ft. All of these changes together reduce the equivalent length of fittings to 40 ft, making the total equivalent length 15 + 20 + 40 = 75 ft (Figure 71-6). Often, the added cost of using more efficient fittings is offset by being able to use a smaller duct at a higher friction rate.

The bottom line is that many duct runs exceed 100 equivalent feet. The design friction rate will have to be adjusted lower to keep the actual pressure drop through the run from exceeding the desired pressure loss through the run. This is done using the formula shown in section 71.4 or the length conversion scale on the ACCA duct calculator.

71.4 USING A DUCT FRICTION LOSS CHART

Duct friction loss charts are used to design duct systems. Figures 71-7 through 71-10 show the standard scales:

- Air volume, vertical lines shown in red (Figure 71-7)
- Static pressure loss (friction loss), horizontal lines shown in blue (Figure 71-8)
- Duct size, diagonal lines shown in green (Figure 71-9)
- Air velocity, diagonal lines shown in yellow (Figure 71-10)

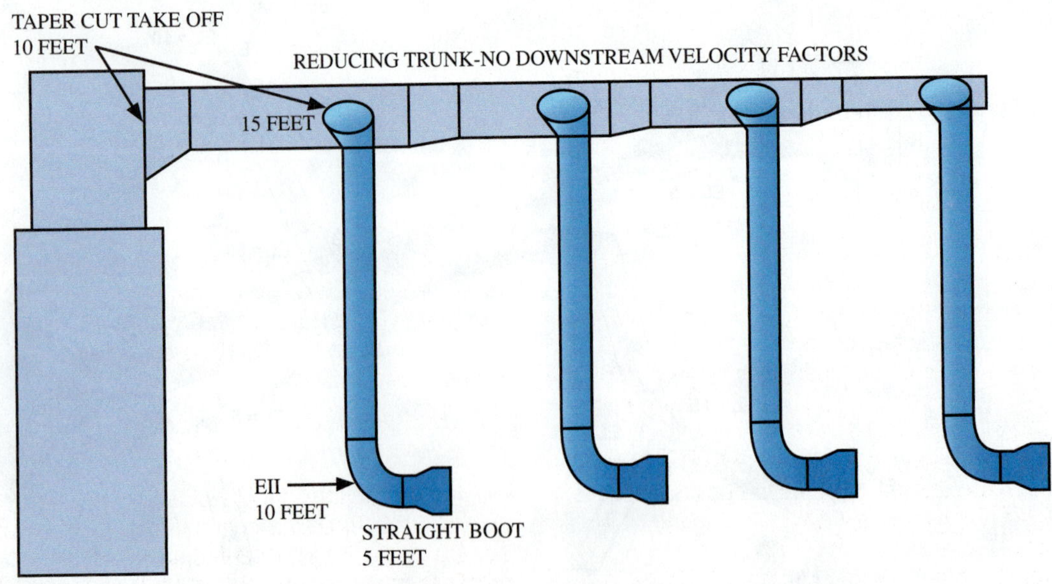

Figure 71-6 Example of branch duct total equivalent length using efficient fittings.

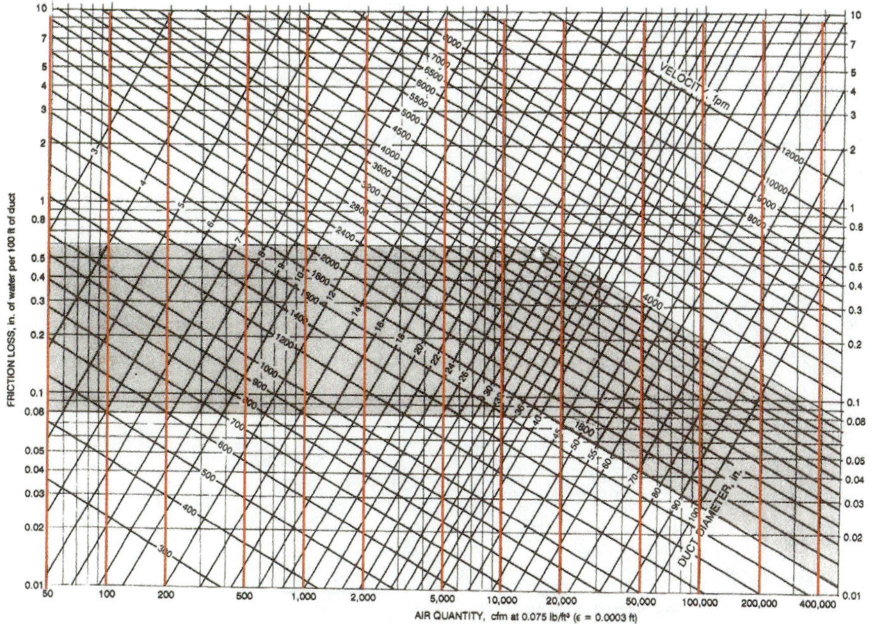

Figure 71-7 The air volume lines are highlighted in red on this duct friction chart. *(2009 ASHRAE Handbook—Fundamentals. © ASHRAE., www.ashrae.org.)*

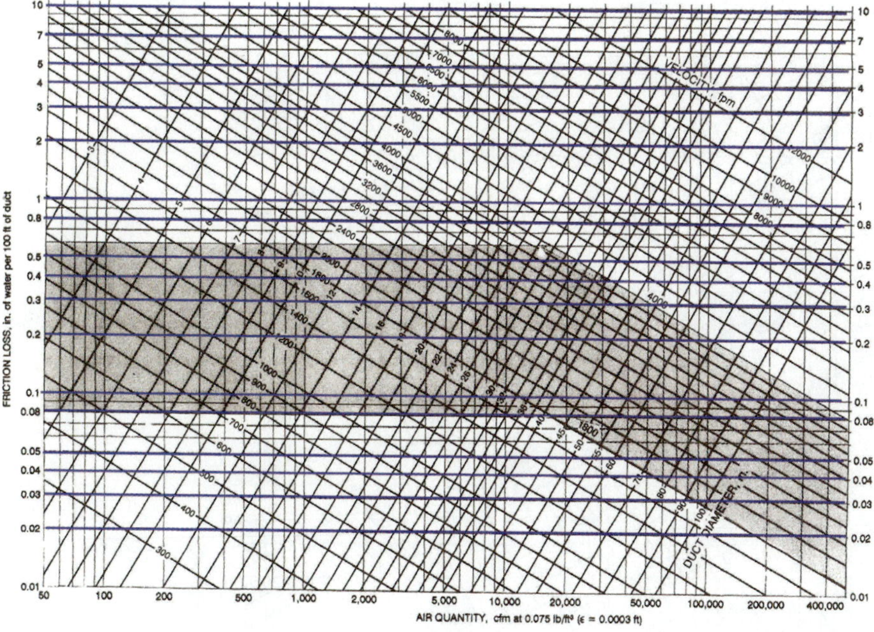

Figure 71-8 The friction loss lines are highlighted in blue on this duct friction chart. *(2009 ASHRAE Handbook—Fundamentals. © ASHRAE., www.ashrae.org.)*

If any two values are known, all the other values can be determined from the chart. For example, 1,000 CFM moving through a 12-in duct will have a velocity of approximately 1,300 fpm and the rate of static loss will be just under 0.2 in wc per 100 equivalent feet of duct (point A on Figure 71-11). Most commonly, the design friction rate and airflow volume are the two known quantities. To select a duct size for a specific friction rate, find the intersection of the air volume and friction rate lines. Moving 1,000 CFM at a friction rate of 0.1 in wc per 100 ft of duct will require a 14-in duct (point B on Figure 71-11).

It is important to recognize the difference between rate and absolute value. The friction rate chart does not tell the designer what the actual loss through a duct will be because that depends upon the duct length. Rather, it tells the designer what the loss will be for every 100 ft of duct length. To determine the actual static pressure loss for a given CFM with any length run

$$\text{Actual static loss} = (\text{total equivalent length}/100) \times \text{friction rate per 100 ft}$$

A 50-ft section of duct would be half as long as a 100-ft duct and would create only half the static pressure drop. Since a 100-ft length of 6-in duct loses 0.085 in wc static pressure while moving 100 CFM, it will lose half of that, 0.0425 in wc, to move 100 CFM through a 50-ft section.

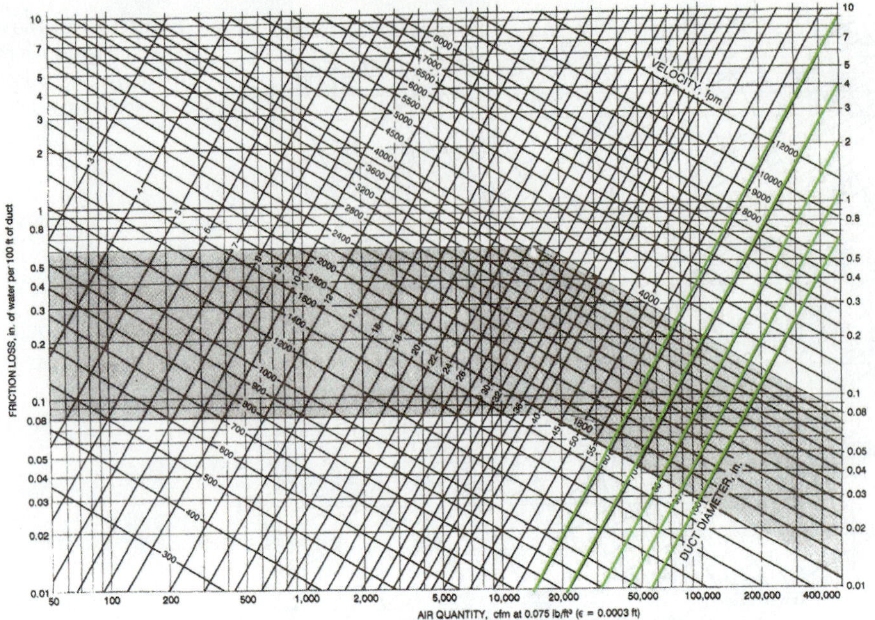

Figure 71-9 The duct size lines are highlighted in green on this duct friction chart. *(2009 ASHRAE Handbook—Fundamentals. © ASHRAE., www.ashrae.org.)*

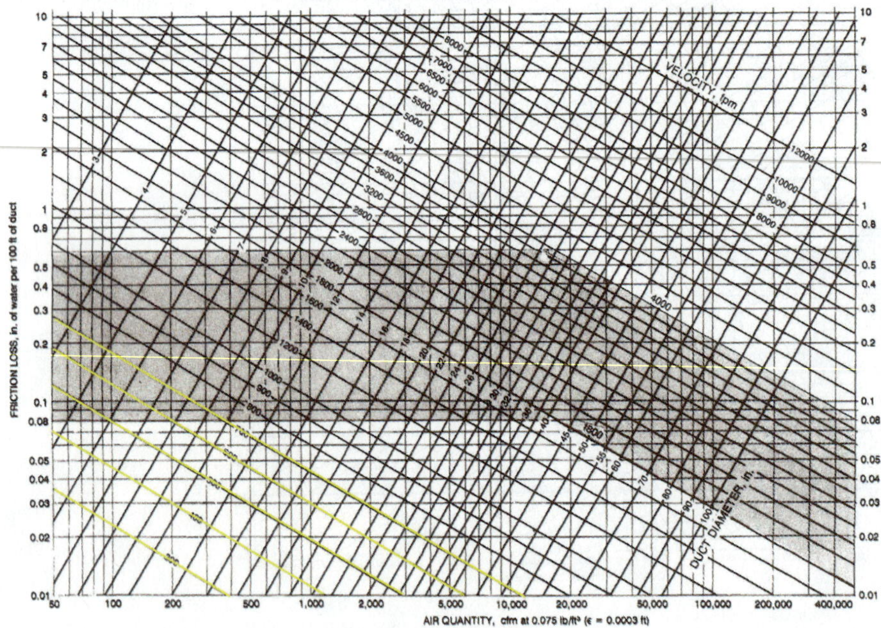

Figure 71-10 The air velocity lines are highlighted in yellow on this duct friction chart. *(2009 ASHRAE Handbook—Fundamentals. © ASHRAE., www.ashrae.org.)*

A 200-ft section of 6-in duct would lose twice the static pressure, 0.17 in wc, while moving 100 CFM. Since all duct runs are not the same length, they will not all deliver the same amount of air, even if they are all the same size duct. This is why it is inaccurate to state that a 6-in duct will carry 100 CFM at a static pressure of 0.085 in without any reference to its length.

Most of the time the available static pressure is known and what the designer is looking for is what size duct will deliver the correct amount of air. In this case, the friction rate is adjusted so that the end result is equal to the design static pressure. The friction rate can be adjusted for different lengths of duct using the following formula:

Adjusted friction rate = design friction rate × (100/equivalent length)

For example, the adjusted friction rate using a system design friction rate of 0.10 in wc and a duct length of 50 ft would be 0.1 in wc × (100/50) = 0.2 in wc. At 0.2 in wc, the air will lose exactly 0.1 in wc static pressure traveling 50 ft. The adjusted friction rate for a 200-ft run would be 0.1 in wc × (100/200) = 0.05 in wc. At 0.05 in wc, the air will lose exactly 0.1 in wc by the time it travels 200 ft.

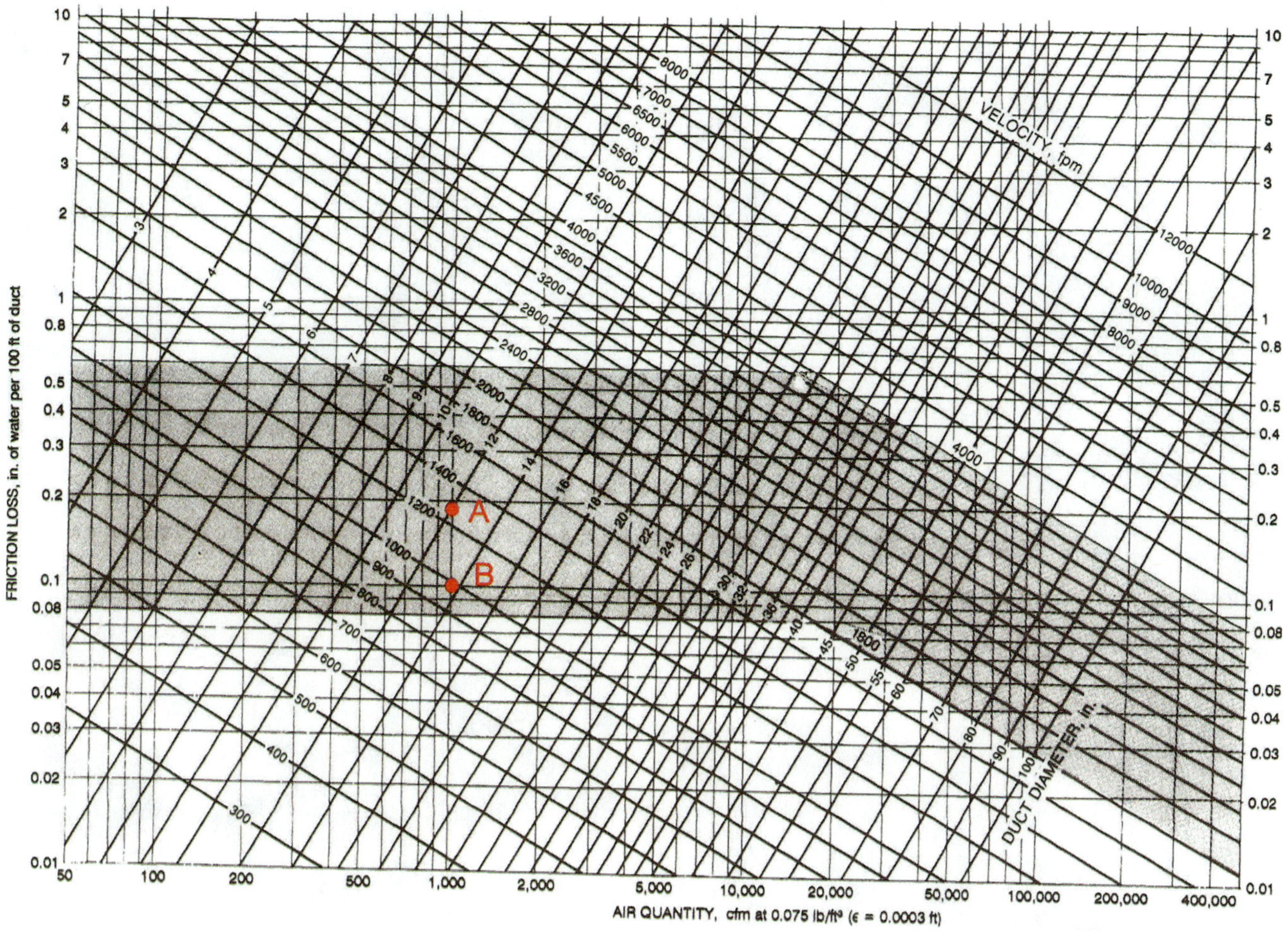

Figure 71-11 Friction chart examples. *(2009 ASHRAE Handbook—Fundamentals. © ASHRAE., www.ashrae.org.)*

71.5 USING A DUCT CALCULATOR

Duct calculators are the most commonly used duct design tools. They take the place of friction charts and moderately complex calculations. All duct calculators have a few standard scales: round duct size, square-to-round conversion, friction rate per 100 ft, velocity, and volume. Lining up any two of these variables will allow you to look up the others. Most duct calculators are designed for smooth galvanized metal duct. Some of these will give conversion factors to accurately calculate other types of duct. The ACCA duct calculator offers a number of advantages over most other duct calculators. It gives friction rates for many different types of duct, converts velocity pressures to velocity, and has a convenient scale for determining the actual static pressure loss of a duct whose equivalent length is not 100 ft (Figure 71-12).

Sample Calculations

Place the arrow on "6" in the round duct diameter window. Find 100 in the volume window. Directly across from 100 CFM in the galvanized metal friction rate window you should see approximately 0.085 in wc (Figure 71-13). In the flexible duct friction rate window you should see approximately 0.15 in wc (Figure 71-14). These numbers represent

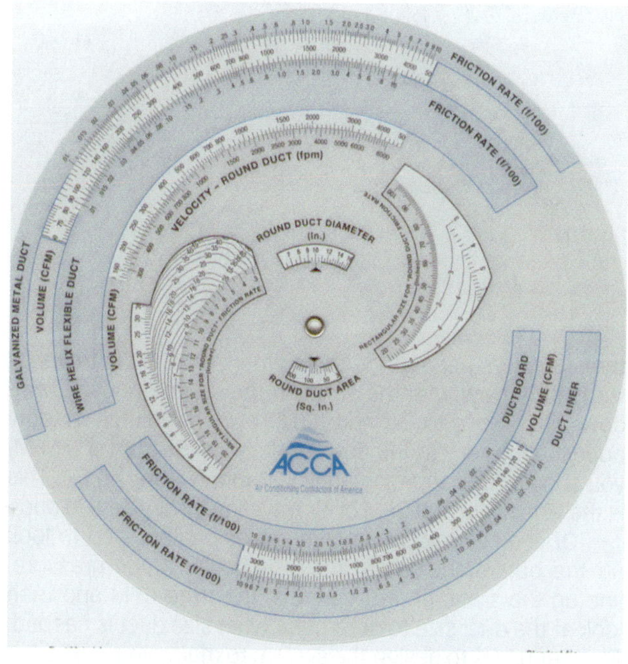

Figure 71-12 ACCA duct calculator. *(© Air Conditioning Contractors of America, Arlington, VA ACCA.org)*

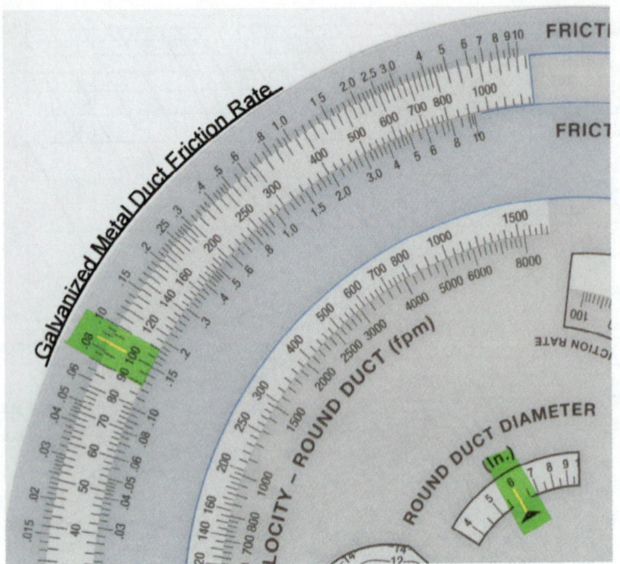

Figure 71-13 A 6-in metal duct carrying 100 CFM has a static pressure loss of 0.085 in wc per 100 ft. *(© Air Conditioning Contractors of America, Arlington, VA ACCA.org)*

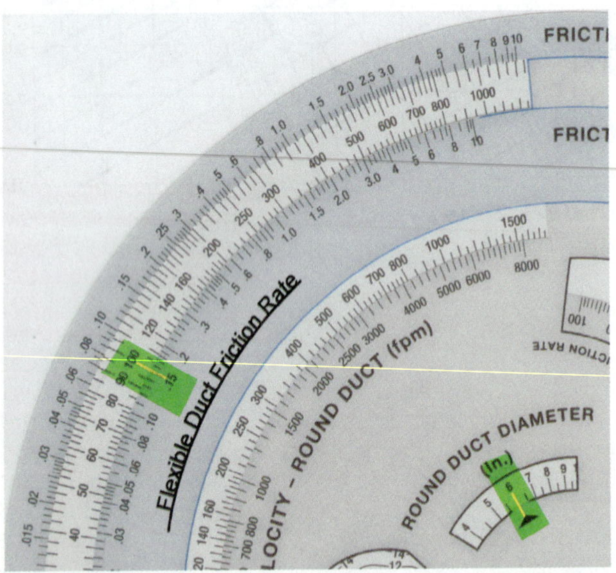

Figure 71-14 A 6-in flex duct carrying 100 CFM has a static pressure loss of 0.15 in wc per 100 ft. *(© Air Conditioning Contractors of America, Arlington, VA ACCA.org)*

the static loss per 100 ft of duct when 100 CFM is traveling through it. Now with the arrow in the round duct size window still pointing to "6," find 100 CFM in the volume window opposite the velocity window. Directly across from 100 CFM, you should read approximately 520 fpm (Figure 71-15). This is the velocity of 100 CFM of air traveling through a 6-in duct.

Of course, any two known values can be used to look up the others. To design for a particular static pressure, line up the static pressure across from the CFM and then look at the duct size to determine what size duct is needed. Another use is to design the system to maintain a particular duct velocity. The required duct size can be determined by lining up the design velocity across from the needed CFM.

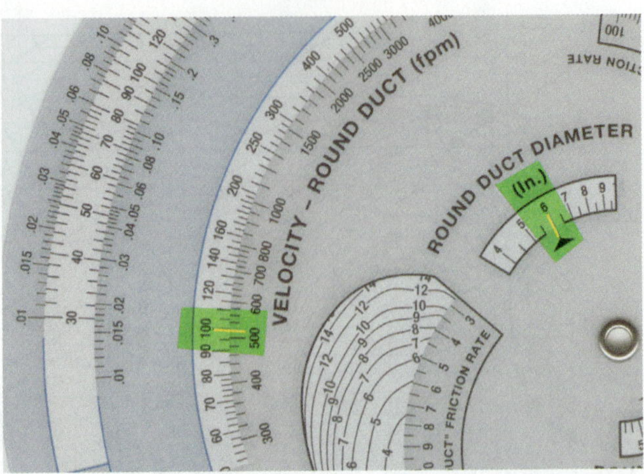

Figure 71-15 100 CFM of air traveling through a 6-in duct has a velocity of approximately 520 fpm. *(© Air Conditioning Contractors of America, Arlington, VA ACCA.org)*

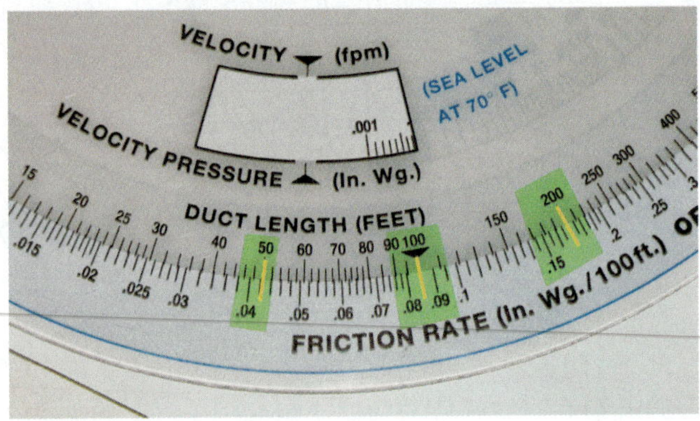

Figure 71-16 At a friction *rate* of 0.85 in wc, a 50-ft length of duct will lose 0.042 in wc; a 200-ft length will lose 0.17 in wc. *(© Air Conditioning Contractors of America, Arlington, VA ACCA.org)*

The ACCA duct calculator has a handy conversion chart at the bottom on the side that says "Auxiliary Calculations." Using the earlier example of 100 CFM at 0.085 in wc, put the 100-ft arrow across from 0.085 in wc. Now look across from the 50-ft mark and see an adjusted static of 0.042 in wc. Without moving the duct calculator, look across from 200 ft and you should see 0.17 in wc (Figure 71-16). This scale can also be used to determine what design friction rate is needed to achieve a desired design static pressure drop for a duct run that is not 100 ft long.

Suppose a system is to be designed with a supply duct static loss of 0.085 in wc. What friction rate would achieve 0.085 in wc pressure drop in a 50-ft duct? Note that in this case the end result is 0.085 in wc, not the rate per 100 ft. Line up the 0.085 in wc with 50 ft. Now look across from the arrow at 100 ft and you should see approximately 0.17 in wc (Figure 71-17). A 50-ft duct with a friction rate of 0.17 in wc will give exactly 0.085 in wc drop. Looking in the round duct diameter window you should see that a 5-in duct will handle 100 CFM at this friction rate (Figure 71-18). Now line up 0.085 in wc with the 200-ft mark. The arrow at the 100-ft mark indicates that we should use 0.042 in wc to design a 200-ft duct

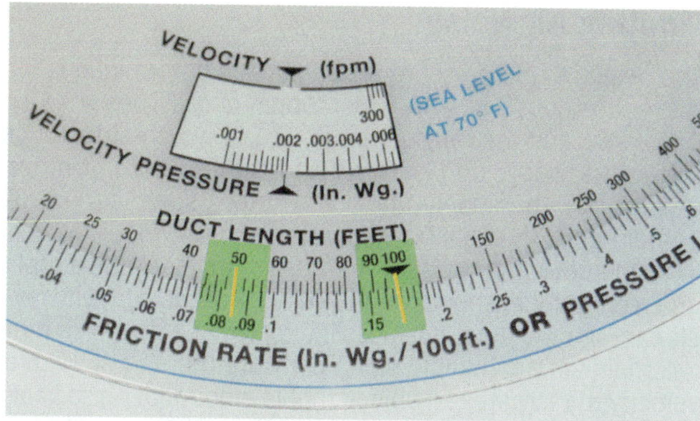

Figure 71-17 For a 50-ft section of duct to lose 0.085 in wc, the friction *rate* should be 0.17 in wc. *(© Air Conditioning Contractors of America, Arlington, VA ACCA.org)*

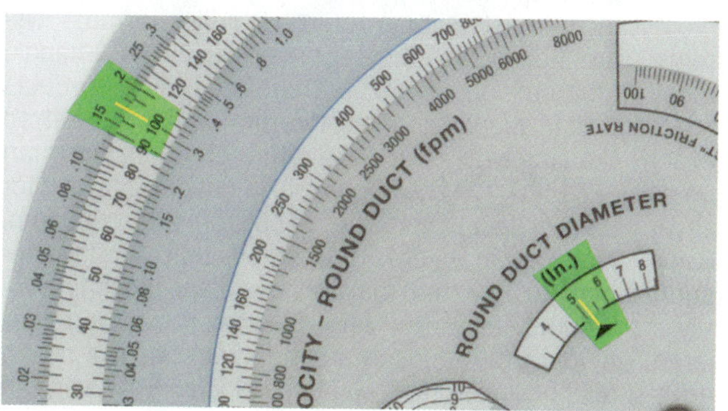

Figure 71-18 A 5-in duct can handle 100 CFM at a friction rate of 0.17 in wc. *(© Air Conditioning Contractors of America, Arlington, VA ACCA.org)*

(Figure 71-19). Line up 0.042 in wc with 100 CFM and look in the round duct size window. You should see that a 7-in duct is required for this 200-ft duct (Figure 71-20). Notice that the 50-ft 5-in duct, the 100-ft 6-in duct, and the 200-ft 7-in duct will all deliver the same amount of air with a static pressure loss of 0.085 in.

To design a truly balanced system, the design friction rate for each run should be adjusted. This scale comes in *very* handy when designing duct systems.

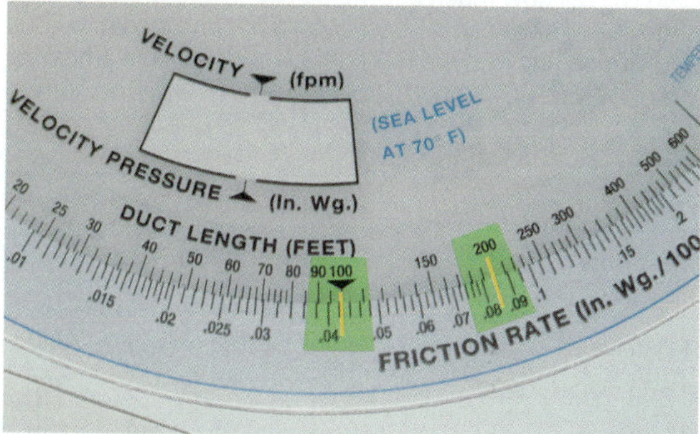

Figure 71-19 For a 200-ft section of duct to lose 0.085 in wc, the friction *rate* should be 0.042 in wc. *(© Air Conditioning Contractors of America, Arlington, VA ACCA.org)*

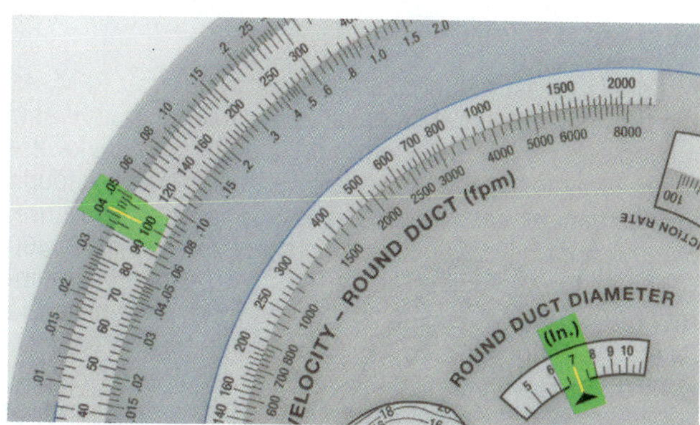

Figure 71-20 A 7-in duct can handle 100 CFM at a friction rate of 0.042 in wc. *(© Air Conditioning Contractors of America, Arlington, VA ACCA.org)*

71.6 DUCT DESIGN GOALS

There are three basic goals when designing a duct system. In order of importance, they are as follows:

1. Allow the system to circulate enough air to operate properly.
2. Distribute the air proportionally to where the heat load is.
3. Keep airflow noise below objectionable levels.

Minimum Duct Restriction

The first goal is the most important. All heating and cooling equipment that circulates air has some minimum requirement for airflow. A duct system that will not allow the unit to deliver this minimum airflow will cause inefficient operation at best and equipment problems at worst. A fan's capacity is stated as the amount of air it can move against a particular pressure. This pressure is usually stated in inches of water column (in wc). Friction charts and/or duct calculators help determine the amount of static pressure drop a particular type and size of duct will offer to any given airflow. In general, as more air is pushed through the same size duct, the pressure drop through the duct increases. Duct calculators or friction charts show the amount of resistance a duct has to airflow. This is stated as static pressure loss per 100 ft of duct, measured in inches of water column. It is important to keep in mind that this is a friction *rate*, not an absolute quantity. In other words, a duct length is necessary to accurately determine what the static loss will be for a particular duct.

Required Information

To design a duct system that will allow the fan to move an adequate amount of air, the designer needs to know the following:

- **System CFM** The amount of air the system must move
- **Maximum available system static** The maximum static pressure available for moving that quantity of air
- **Duct friction rate** The airflow resistance of the duct material has to airflow

System Airflow

Airflow is measured in cubic feet per minute, or CFM. As the term suggests, CFM measures the volume of air that is moved every minute. Systems can operate correctly over a range of airflows. The proper airflow requirement is found in the manufacturer's design and specification literature. If you do not have a specification sheet to work from, 400 CFM per ton of cooling works for most residential equipment. The ductwork should be designed to handle the higher of the heating or cooling system airflow. The cooling airflow is usually used because the required airflow for cooling exceeds the required airflow for heating in most situations. Cool air is heavier and therefore harder to move.

In air conditioning, the airflow across the evaporator helps determine the amount of sensible cooling versus the amount of latent cooling accomplished. Reducing the airflow will increase the amount of moisture removed from the air but decrease the amount of sensible cooling accomplished. Conversely, increasing the airflow reduces the latent cooling but increases the sensible cooling. The amount of sensible and latent cooling required can be determined by using ACCA Manual J. It is important to recognize that duct sizing is done on the basis of sensible cooling capacity, not total cooling capacity. Sensible capacity is what is going to actually cool the room down, so the sensible heat load for each room is used to determine how much air that room requires.

Maximum Available System Static

The maximum static pressure the equipment can operate against is usually specified by the equipment manufacturer. Keep in mind that anything the air must travel over or through will create some static pressure drop. Examples of devices that produce a static pressure drop include evaporator coils, humidifiers, air filters, grilles, and registers. The amount of static pressure loss for any of these devices depends upon the airflow rate. More airflow creates a higher static pressure loss. The maximum available static pressure for the duct system can be determined by subtracting all losses except ductwork from the maximum operating static pressure of the equipment at the design airflow.

For example, take a 2-ton system moving 800 CFM of air with a total available static pressure of 0.55 in wc, an evaporator coil with a static pressure drop of 0.25 in wc, a filter with a static pressure drop of 0.10 in wc, and supply registers and return grilles with a static pressure drop of 0.05 in wc. All the fixed losses add up to 0.4 in wc, leaving 0.15 in wc of pressure for ductwork.

Limiting Air Noise

The design static pressure that has been determined is for the entire duct system. If the supply duct is designed at 0.15 in wc and the return duct at 0.15 in wc, the total duct static loss would be 0.3 in wc and the fan will not move enough air. The supply static and return static added together needs to equal 0.15 in wc.

One determining factor for design static for the supply and return ducts should be the velocity of the air as it travels through the duct. It is possible to select a duct size that will maintain a workable static pressure but end up with a velocity that creates an objectionable amount of noise. If air moves faster than 600 fpm it begins to make noise. Above 1,000 fpm, the customer will have to turn up the television when the air conditioning starts.

In general, as duct sizes increase, design static must decrease to maintain the same velocity. There are normally more supply registers than return-air grilles in most systems. This means that the return-air ducts must handle more air than the supply ducts. Therefore, return-air ducts are larger than supply-air ducts. If return-air ducts are sized exactly like supply-air ducts, the return-air ducts will often be noisy. Figure 71-21 shows the maximum design static for several sizes of duct at velocities of 600 fpm and 1,000 fpm.

A quick look at Figure 71-21 shows that if a designer sticks strictly to a 600 fpm maximum velocity, the main supply trunk ducts and all the return ductwork will be rather large. Any duct that ends in a supply register or a return grille should be limited to a 600 fpm velocity. However, the trunk ducts that supply the branch ducts can operate at higher velocity and noise levels because they do not open up into the living area. Still, 1,000 fpm is considered the outer limit for velocity in trunk ducts in residential systems.

71.7 DETERMINING BRANCH DUCT AIRFLOW

To have an even temperature throughout the house, the amount of air each room receives should be based on the heat load for each room. Typically, the airflow required for both heating and cooling is calculated and the higher value chosen.

Dividing the entire building heat load by the system CFM will tell how much BTU/hr is delivered for every CFM of airflow. For example, a house with a heating load of 30,000 BTU/hr and an airflow of 600 CFM would be have a heating airflow divisor of 30,000/600 = 50 BTU/CFM. If a room in

Figure 71-21 Airflow through metal duct at 600 fpm and 1,000 fpm velocities.

@600 fpm Duct Velocity			@1000 fpm Duct Velocity		
Duct Size	Friction Rate (in wc)	CFM	Duct Size	Friction Rate (in wc)	CFM
6	0.11 (in wc)	120	6	0.3 (in wc)	290
8	0.085 (in wc)	220	8	0.2 (in wc)	350
10	0.07 (in wc)	330	10	0.15 (in wc)	550
12	0.055 (in wc)	480	12	0.12 (in wc)	780
14	0.045 (in wc)	650	14	0.10 (in wc)	1,050

this house had a heating load of 4,000 BTU/hr, the required airflow for heating that room would be 4,000/50 = 80 CFM.

To determine the cooling airflow divisor, divide the total calculated sensible cooling load by the amount of airflow the system will move to arrive at a capacity divisor. For example, a house with a sensible cooling load of 16,000 BTU/hr and an airflow of 800 CFM would have a capacity divisor of 16,000/800 = 20 BTU/CFM. If a room in this house had a sensible cooling load of 2,400 BTU/hr, the required airflow for cooling that room would be 2,400/20 = 120 CFM.

TECH TIP

Airflow divisors are simply reciprocals of airflow factors, as used in ACCA Manual J. Using the example of 30,000 BTU and 600 CFM, the airflow factor would be calculated by dividing 600 CFM by 30,000 BTU/hr to get 0.02 CFM per BTU/hr. This number is then multiplied by the room load to get the required airflow. For example, 0.02 × 4,000 BTU/hr = 80 CFM. By contrast, the airflow divisor has the advantage of being a "normal" number, rather than a tiny decimal number. This reduces rounding error and keeps the process simpler. The air flow divisor is 50 for the same system. So the same calculation is 4,000 BTU/hr /50 = 80 CFM.

71.8 BRANCH DUCT SIZING

Three pieces of information are required to size each branch duct: the system design static pressure, the total equivalent length of each run, and the required airflow for each run. A friction chart or duct calculator can be used to determine what size duct will be required to deliver the correct amount of air at the design static. Line up the duct's total equivalent length, including fittings, with the adjusted design static on the scale for duct length adjustments. Now look at the static pressure indicated by the arrow at the 100-ft mark. This is the value that will be used to look up the duct size on the other side of the duct calculator. On the other side of the duct calculator, line up this adjusted static with the required CFM for the duct run. The arrow in the round duct window will indicate the size of round duct that will deliver the required amount of air at the design pressure loss for that duct run.

You have many of the same issues on the return side. Ideally, each return should handle the air from the supply registers in the portion of the house where the return is located. It is critical that the total amount of air handled by all the returns adds up to the total system CFM. It is also important that air be able to move from one room to another easily. Air will not enter a room if the air that is already in the room cannot leave.

SERVICE TIP

Undersized returns are a common problem found in duct systems today. A marginal duct system can often be dramatically improved by adding additional return-air runs.

Airflow (CFM)	Adjusted Design Static (in wc)
120	0.12
80	0.08
100	0.10
200	0.11

Figure 71-22 Trunk duct sizing example data.

71.9 TRUNK DUCT SIZING

Each trunk duct should be able to carry the total volume of air for all its connected branches at the static pressure loss of the branch with the lowest adjusted static. For example, a trunk duct to supply the four branch ducts listed in Figure 71-22 would be sized for 400 CFM at a design static of 0.08 in wc.

The trunk duct is sized using a friction chart or duct calculator. In this example, a 10-in-diameter duct would be the closest actual duct size (Figure 71-23).

71.10 GRILLE AND REGISTER SELECTION

Supply-air registers or diffusers are responsible for distributing the air into the room. Without the correct diffuser, the room can still be uncomfortable even if the system and ductwork are sized correctly. The overall goal is to ensure even temperature throughout the room without creating noise and drafts. In general, the supply air should not blow directly on the people in the room. The primary factors to be considered when selecting supply-air registers are the following:

- Required CFM
- Throw

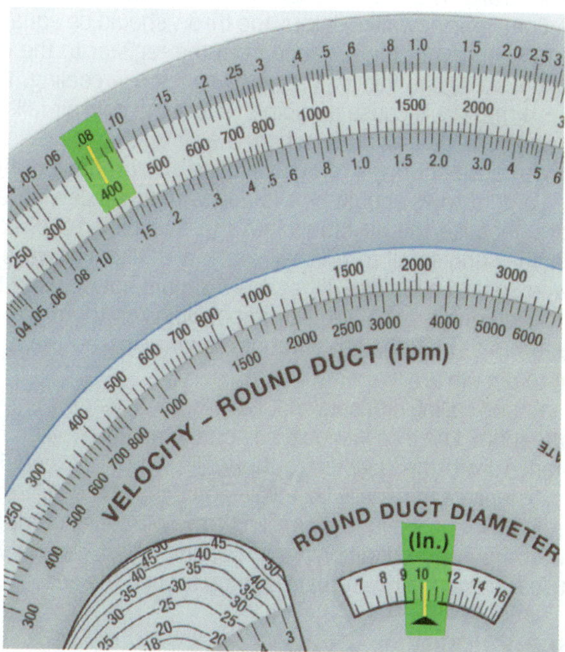

Figure 71-23 Duct selection for 400 CFM at 0.085 in wc. *(© Air Conditioning Contractors of America, Arlington, VA ACCA.org)*

Table 71-1 Terminal Velocity for Common Residential Diffusers

Register Type and Location	Terminal Velocity (fpm)
Floor register	50
Sidewall register	75
Round ceiling diffuser	50
Square ceiling register	75–150

- Face velocity
- Static pressure drop

Required CFM

The required air volume is measured in cubic feet per minute, or CFM. The CFM for the branch duct is determined based on the heat load for the area the register serves. The register must be designed to handle this quantity of air.

Throw

The throw is important because it determines how far into the room the conditioned air travels. The register throw is the distance from the register that the air will travel before it slows down to the terminal velocity listed in the engineering data. Table 71-1 shows the terminal velocity for common residential diffusers.

The desired throw is also determined by register location. Floor registers should have a throw of at least 6 ft to get the air up to head height. This is especially true for cooling applications with floor registers. If the throw is too little, the cold air will stay near the floor. However, the throw should be less than the ceiling height. Using a floor register with a 10-ft throw in a room with an 8-ft-high ceiling will create drafts as the air bounces off the ceiling.

For high-sidewall registers, the throw should be equal to 75–100 percent of the distance from the register to the opposite wall. Again, this is primarily a concern in cooling. Too small a throw will cause dumping, where the cold air falls to the floor. This causes drafts where the air is falling. Excessive throws cause bouncing off the wall and also produce drafts.

Large throws should be avoided with low-sidewall registers because the air blows into the occupied zone. Low velocities and small throws will work with low sidewalls in heating only. Low-sidewall registers should not be used for air conditioning because it is impossible for them to get the air up at head height without blowing air into the area where people are.

Round ceiling diffusers should be located near the center of the room. The throw should be approximately equal to the distance from the register to the walls. Throws exceeding this distance can usually be tolerated because the air tends to roll down the walls. Too small a throw will cause dumping and drafts in air conditioning. In heating, short throws will cause the warm air to remain high up around the ceiling.

Face Velocity

The velocity of the air at the face of the register determines the throw and the noise. Higher velocities give larger throws and also more noise. Lower velocities yield less noise and shorter throws. The main limitation of register face velocity is the noise created by the moving air. Table 71-2 shows recommended maximum face velocities for several different applications. Note that the quieter applications have lower recommended face velocities.

Table 71-2 Recommended Register Face Velocities

Type of Location	Recommended Face Velocity (fpm)
Broadcasting studios	500
Residences	500–750
Apartments	500–750
Churches	500–750
Hotel bedrooms	500–750
Theaters (live)	500–1,000
Private offices, acoustically treated	500–1,000
Motion picture theaters	1,000–1,250
Private offices, not treated	1,000–1,250
General offices	1,250–1,500
Retail stores	1,500
Industrial buildings	1,500–2,000

Static Pressure Drop

Since supply registers and grilles add restriction to the free flow of air out of the ductwork, air pressure is required to force the air through them. This is referred to as the static pressure drop across the register. Ideally, it should be kept as low as is possible, but it must be accounted for when designing a duct system. The registers selected should have a static pressure drop at the design airflow that is equal to or less than the assumed register pressure drop used to design the duct system. Like the noise, the pressure drop increases as the velocity increases. Most residential registers have a static pressure drop between 0.03 and 0.05 in wc when the face velocity is between 500 and 750 fpm.

Noise Criteria

The industry uses a noise criteria (NC) system to designate the level of noise in the area where the registers will be used. Each NC curve is assigned a number, in 5-dB increments. The number corresponds to the decibel reading at approximately 1,500 Hz. NC 25 is characteristic of areas with very little background noise, and NC 50 is characteristic of areas with so much background noise that people must be within a few feet of each other to talk. Most register manufacturers now include NC recommendations in their commercial register selection guides (Figure 71-24).

Engineering Data

The tables with the information on register performance are in the engineering data section of the manufacturer's

NC Rating	Application
NC-20	Concert halls, recording studios, large auditoriums, recital halls
NC 20 to NC 30	Small auditoriums, theaters, music practice rooms, court rooms
NC 25 to NC 35	Bedrooms, sleeping quarters, hospitals, apartments, hotels and motels
NC 30 to NC 35	Private offices, small conference rooms, classrooms, libraries
NC 35 to NC 40	Large offices, reception areas, retail shops, restaurants, gymnasiums
NC 40 to NC 45	Lobbies, maintenance shops
NC 45 to NC 55	Kitchens, laundry facilities and computer equipment rooms

Figure 71-24 Register selection table with integrated noise criteria information.

literature. Figure 71-25 shows the engineering data for a series of residential floor registers. The register face velocities are listed across the top with the static pressure loss across the registers listed immediately underneath the velocities. The register sizes are on the left-hand side with the effective face area of the register listed under each size. In each box in the table are listed the CFM and throw for that size and face velocity. The easiest way to locate an acceptable register is to start with one criterion and look for the rest. This can be demonstrated by an example.

Suppose a register is needed for an air volume of 100 CFM, a throw of 6 to 8 ft, a pressure drop of no more than 0.03 in wc, and a face velocity below 700 fpm. Under the 700 fpm velocity column, note that a 4 x 10 will handle 120 CFM at the maximum acceptable velocity. Move to the left to find 100 CFM listed under the 600 fpm velocity column. This selection matches our volume, velocity, pressure drop, and throw requirements. There are many register sizes that can handle 100 CFM, but only a few that will meet all our requirements. For example, a 6 × 12 will handle 100 CFM but with a throw of only 4 ft.

Return-Air Grilles

Return-air grilles primarily serve to hide what would otherwise be a large unsightly hole. They do not direct the air like supply registers do. A generally accepted rule for estimating the required return-air grille size is that it should have 1 in^2 of gross area for every 2 CFM of air it handles. However, manufacturers' selection charts are the most accurate way to size return grilles. The selection criteria for return-air grilles are as follows:

- Air volume
- Pressure drop
- Face velocity

Typical Floor Register Performance Data

Face Velocity FPM		300	400	500	600	700	800	900	1000
Pressure Loss in wc		.006	.010	.016	.022	.031	.040	.050	.060
4 × 10	CFM	50	70	85	100	120	135	155	170
	SPREAD ft	4	6	8	9	11	12	13	15
Ak 0.17	THROW ft	3	4	5	6	7	8	9	10
4 × 12	CFM	65	85	105	125	145	170	190	210
	SPREAD ft	5	7	9	10	12	14	16	17
Ak 0.21	THROW ft	3	4	5	6	7	8	9	10
4 × 14	CFM	75	100	125	150	175	200	225	250
	SPREAD ft	6	8	10	11	13	15	17	19
Ak 0.25	THROW ft	4	5	6	7	8	9	10	11
6 × 10	CFM	80	110	135	160	190	215	245	270
	SPREAD ft	6	8	10	12	14	16	18	20
Ak 0.27	THROW ft	4	5	6	7	8	9	10	11
6 × 12	CFM	100	130	165	200	230	265	295	330
	SPREAD ft	7	9	11	13	15	17	19	21
Ak 0.33	THROW ft	4	5	6	8	9	10	11	12
6 × 14	CFM	115	155	195	235	275	310	350	390
	SPREAD ft	8	10	12	14	17	19	21	23
Ak 0.39	THROW ft	5	6	7	9	10	11	12	14

Figure 71-25 Floor register selection data.

Typical Return Grille Performance Data								
Face Velocity FPM		400	500	600	700	800	900	1000
8 × 10	CFM	195	245	295	345	390	440	490
Ak 0.49	ΔP in wc	0.010	0.015	0.021	0.029	0.038	0.047	0.059
8 × 12	CFM	235	295	350	410	470	525	585
Ak 0.57	ΔP in wc	0.009	0.014	0.020	0.028	0.036	0.046	0.057
8 × 14	CFM	270	340	410	475	545	610	680
Ak 0.68	ΔP in wc	0.009	0.014	0.019	0.027	0.035	0.044	0.055
10 × 12	CFM	290	360	430	505	575	650	720
Ak 0.72	ΔP in wc	0.009	0.014	0.019	0.027	0.035	0.043	0.054
12 × 12	CFM	340	425	510	595	680	765	850
Ak 0.85	ΔP in wc	0.009	0.014	0.019	0.026	0.033	0.042	0.053
6 × 30	CFM	410	515	620	720	825	925	1030
Ak 1.03	ΔP in wc	0.008	0.012	0.018	0.024	0.032	0.040	0.049
14 × 20	CFM	630	790	950	1105	1265	1420	1580
Ak 1.58	ΔP in wc	0.007	0.011	0.016	0.022	0.029	0.036	0.045
12 × 30	CFM	800	1000	1200	1400	1600	1800	2000
Ak 2.00	ΔP in wc	0.007	0.011	0.015	0.021	0.028	0.035	0.043

Figure 71-26 Return-air grille selection chart.

Ideally, the pressure drop should be kept as low as is possible, but it must be accounted for when designing a duct system. The grille's static pressure drop at the design airflow should be equal to or less than the assumed pressure drop used to design the duct system. Just as in sizing supply registers, the pressure drop and noise both increase as the velocity increases. The face velocity should be kept under 600 fpm; faster velocities lead to noticeably increased air noise.

Figure 71-26 shows a typical return-air grille selection chart. The face velocities are listed across the top. The grille sizes are on the left-hand side with the effective face area of the grille listed under each size. In each box in the table are listed the CFM and static pressure drop for that size and face velocity. Suppose a return grille was needed to handle 400 CFM with a face velocity of less than 600 fpm and a static pressure drop of 0.03 in wc or less. Under the 600 fpm column, a 8 × 14 will handle 410 CFM, with a pressure drop of 0.019. Other sizes also meet our criteria: a 10 × 12 has a pressure drop of 0.019 in wc with a volume of 430 CFM.

Filter Grilles

Filter grilles should be sized larger than standard return-air grilles because the filter reduces the effective area of the grille and because a lower velocity across the filter is needed for it to be effective. Maximum face velocities of 300–400 fpm are recommended.

UNIT 71—SUMMARY

Duct systems distribute conditioned air through a building. The basic components are a blower, supply-air ducts, and return-air ducts. Duct systems are designed using load calculations, equipment performance specifications, and building plans.

All components in the airstream create static pressure loss, including the air filter, evaporator coil, electric heaters, humidifiers, supply registers, and return-air grilles. The static pressure available for moving air through the duct system is found by subtracting all losses other than duct losses from the blower total external static pressure at design airflow. The pressure lost as air travels through duct fittings is measured in equivalent feet of pipe. The total equivalent length of a duct run is found by adding the length of the trunk duct, branch duct, and the equivalent length of all the fittings. Duct is sized using a friction chart or duct calculator. Friction charts and duct calculators give the friction rate per 100 ft of duct. The actual pressure drop through any duct will also depend upon its total equivalent length.

UNIT 71—REVIEW QUESTIONS

1. What friction rate should be used to size a duct for a static pressure drop of 0.1 in wc if the duct has a total equivalent length of 150 ft?

2. What size metal duct should be used to deliver 170 CFM with a pressure drop of 0.15 in wc if the total equivalent length is 130 ft?
3. What friction rate should be used to size a duct for a static pressure drop of 0.1 in wc if the duct has a total equivalent length of 50 ft?
4. What size metal duct should be used to deliver 290 CFM with a pressure drop of 0.15 in wc if the total equivalent length is 80 ft?
5. What is the velocity of 500 CFM of air moving through a 10-in duct?
6. Why should all ductwork designs start with an accurate heat load study?
7. What is the total available static pressure for the ductwork for a system with the following specifications?
Blower: 0.55 in wc total external static at design airflow
Evaporator coil: 0.25 in wc static pressure drop at design airflow
Filter: 0.1 in wc static pressure drop at design airflow
Registers and grilles: 0.05 in wc static pressure drop at design airflow
8. Why should air velocity in branch ducts be limited to 600 fpm?
9. Why are duct friction charts designed for sheet-metal ducts not accurate for fiberglass ductboard or flexible ducts?
10. Why do fiberglass ductboard and flex duct have higher friction loss rates than sheet metal?
11. What is the purpose of reducing the trunk duct size in a reducing extended-plenum duct system?
12. What is the maximum static pressure difference that most residential blowers are designed for?
13. What is the recommended face velocity for a return-air filter grille?
14. Explain what manufacturers mean by a register's throw.
15. List the factors to consider when choosing a supply register.
16. What are the noise criteria curves?
17. Using Figure 71-25, select a floor register that will provide an 8–10 ft throw for 150 CFM with a face velocity less than 700 fpm.

UNIT 72

Zone Control Systems

OBJECTIVES

After completing this unit, you will be able to:

1. explain the purpose of zone control systems.
2. list the methods used to control excess airflow in a zone system.
3. explain the purpose of variable air volume control systems.
4. describe the operation of a basic zone control system.
5. list the components commonly used in a zone control system.
6. explain the difference between a basic zone control system and a communicating zone control system.

72.1 INTRODUCTION

Zone control gives customers increased control over their comfort systems. Customers are no longer satisfied with systems that leave one part of their house hot while another part is cold. Zone systems allow control over different portions of the house without installing multiple units.

Many heating and air-conditioning installations are installed in buildings where the floor plan is spread out, such as ranch-style and split-level houses. On these jobs it is not practical to control comfort throughout the entire area using one thermostat. These layouts require zoning, because the loads differ from one area to another. For example, a basement area typically needs a significant amount of heat to stay warm in the winter but needs virtually no cooling in the summer. The amount of air delivered to the basement needs to change between seasons. Zoning solves this problem.

72.2 PRINCIPLES OF ZONING

Zoning is a method of controlling the supply of conditioned air to each unique area to match the load requirements. This is done by separating the air supply to each unique area by the use of a zone damper in the ductwork and controlling each zone damper with an individual thermostat.

The first consideration when selecting a zoning system is to determine what areas of the building are to be zoned. This can range from a two-zone system for a two-level house to a system that handles each room as a separate zone. Areas to be zoned are selected by building use, building load, or building orientation.

Building-use zoning is based on separating the areas depending on how they are used. For example, bedrooms can be on one zone and living areas on another. This is common for residential zoning.

TECH TIP

Occupancy sensors, devices that detect movement of people in an area, can be used to enhance the efficient operation of zoned systems. Occupancy sensors will allow a zone to be activated only when it is occupied. The popularity of occupancy sensors has resulted in a decrease in their cost so that they now can be used for many more applications. In fact, occupancy sensors have already found their way into residences for light switches.

Building load and orientation zoning considers the areas of a building with unique heat loads. For example, the south side of a building would be on a separate zone from the north side. This type of zoning is common for commercial structures.

The following are some examples of zone controlled systems:

1. A three-zone split-level system is shown in Figure 72-1. Each level of this residence has a separate trunk duct from the supply-air plenum with a zone damper installed. Each of these dampers is controlled from an individual thermostat located in that zone.
2. Zone control for a bilevel house is shown in Figure 72-2. This system has three zones: living area, bedrooms, and recreation room. The system uses a zone control panel, a 40 VA transformer, three zone dampers, and three thermostats.
3. A room-by-room temperature control is shown in Figure 72-3 for a five-zone radial system. These ducts can be located overhead or in the slab floor.

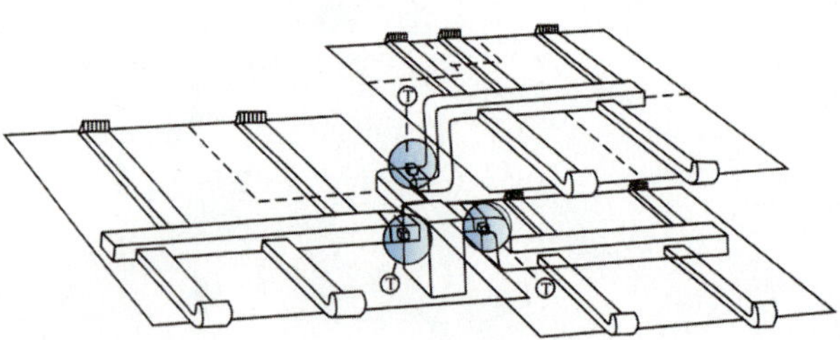

Figure 72-1 Three-zone split-level system.

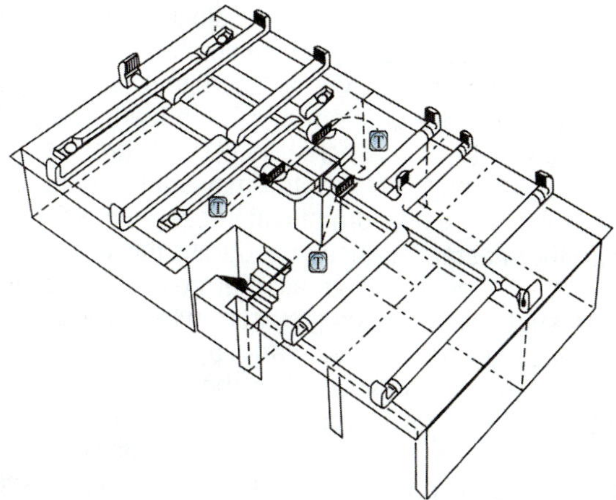

Figure 72-2 Zone control for a bilevel house.

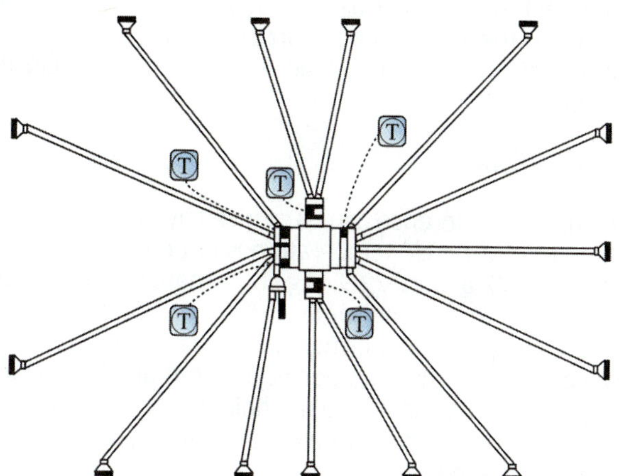

Figure 72-3 Five-zone radial system.

4. A two-zone ranch house is shown in Figure 72-4. Each zone has a thermostat controlling a zone damper.
5. A four-zone office building system is shown in Figure 72-5 using a zone control panel, four dampers and thermostats, and a 40 VA transformer.
6. A room-by-room comfort control layout is shown in Figure 72-6. In this illustration each outlet has an automatic square-to-round transition damper.

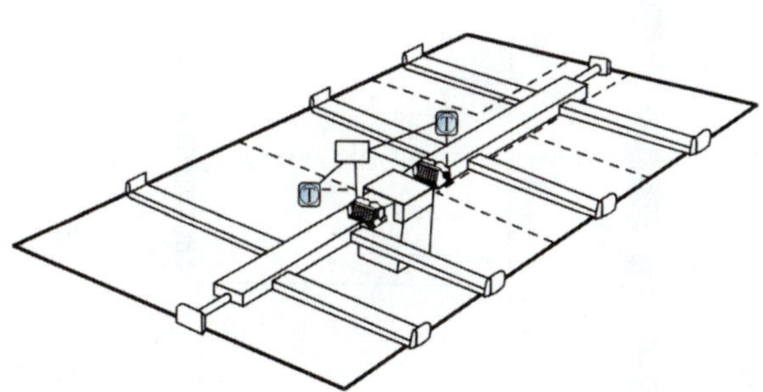

Figure 72-4 Two-zone ranch-style house.

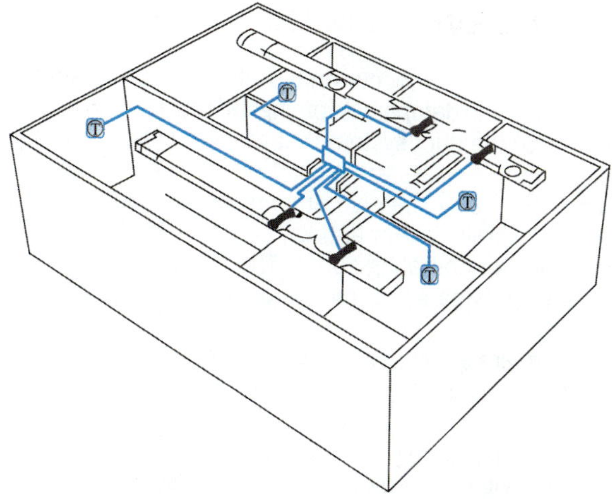

Figure 72-5 Four-zone professional office system.

SERVICE TIP

Some homeowners will close off the vent to rooms that are not in use or commonly used, such as spare bedrooms. If enough of the system's capacity is shut down as a result of this attempt to economize, it can actually cause major problems with the system itself. For example, reducing the airflow below the manufacturer's specifications can reduce the heat load on the evaporator. This can cause refrigerant floodback to the compressor, shortening its life. Another example is raising the discharge-air temperature of a gas furnace by restricting the airflow through it. This causes the high limit to cycle the furnace off and on, reducing its efficiency. An attempt to economize in this manner may ultimately result in higher utility bills and shorter equipment life for the owner.

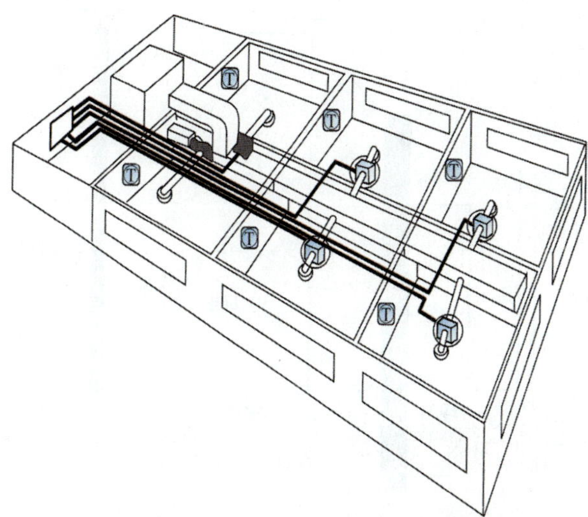

Figure 72-6 Individual room zone control system.

72.3 RELIEVING EXCESS AIR PRESSURE

A problem that can occur on a system of thermostatically controlled zone dampers is forcing too much air through the portions of the duct system that remain open. When some major zone dampers are closed off, the remaining zones must handle an excessive amount of air. This can cause reduced overall system airflow, air noise, drafts, and poor control in the operating zones. The solution to this condition is to provide some means of relieving this excessive air pressure.

There are a number of ways to accomplish this, including the following:

- Bypassing supply air to the return
- Dumping excess air into an area whose temperature is not critical
- Using dampers that allow some air to pass even when they are closed
- Oversizing the ductwork so that a single zone can handle the system airflow
- Variable air volume

Air Bypass Method

The most commonly used method is to bypass supply air back into the return-air duct as shown in Figure 72-7. This is usually accomplished by adding a barometric damper between the supply and return. When the pressure differential exceeds the setpoint of the damper, it opens to relieve the pressure. This ensures that the static pressure across the blower never exceeds the setpoint of the bypass damper.

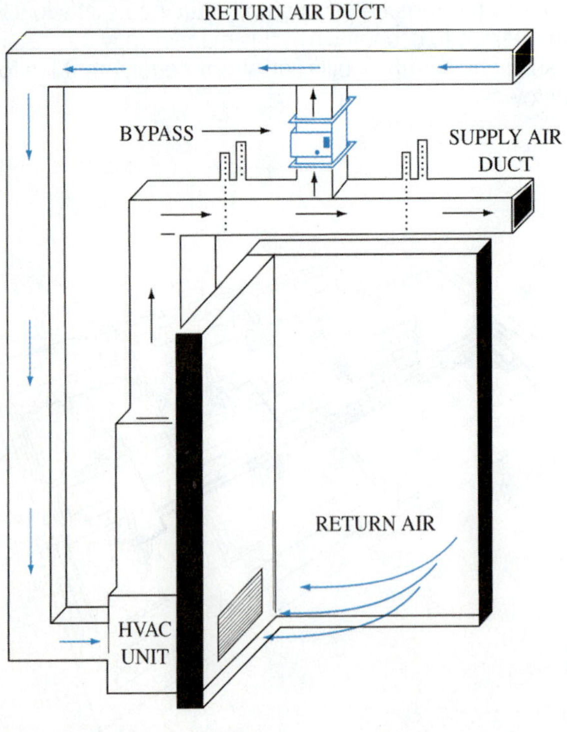

Figure 72-7 Relieving air pressure through bypassing air from the supply plenum back to the return.

Keeping the system pressures and velocities under control eliminates air noise and objectionable drafts.

The disadvantage of this arrangement is that it does nothing to solve the problem of system load. The system is operating with only a partial load. With prolonged use on heating, the high limit control may shut down the system. In cooling, the evaporator can freeze from lack of load. Plenum thermostats are used to address this problem.

A plenum, or discharge air thermostat, is frequently used to shut off the heating system when the discharge air starts to get too hot. Similarly, a low limit is used to shut down the compressor when the discharge air starts to get too cold. On systems that use thermistor sensors, a single discharge air sensor is used. The discharge thermostat solves the problem of equipment capacity exceeding the load, but it creates a new problem: short-cycling. The discharge-air temperature can change rapidly when only a single zone is open, causing the system to shut off before the demand in that zone is met. The system will come back on as soon as the required off time is satisfied. Thus, the system can short-cycle trying to meet the demand of a zone whose load is much smaller than the system capacity.

Dump Zone

The pressure difference can be relieved by dumping excess air into a zone whose temperature is not critical, as shown in Figures 72-8 and 72-9. Again, a barometric damper can be used to accomplish this. If the dump zone is large enough, discharge air thermostats are usually not required. This solves both the air velocity problem and the system load problem but can create a problem with the comfort in the dump zone. Using a dump zone can also have an impact on energy cost because the system is over-conditioning an area that might not be conditioned at all if it were not a dump zone.

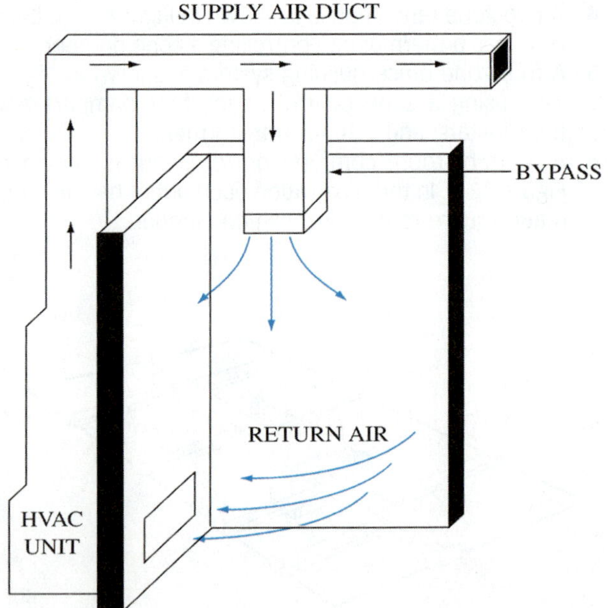

Figure 72-8 Duct arrangement for a "dump" zone.

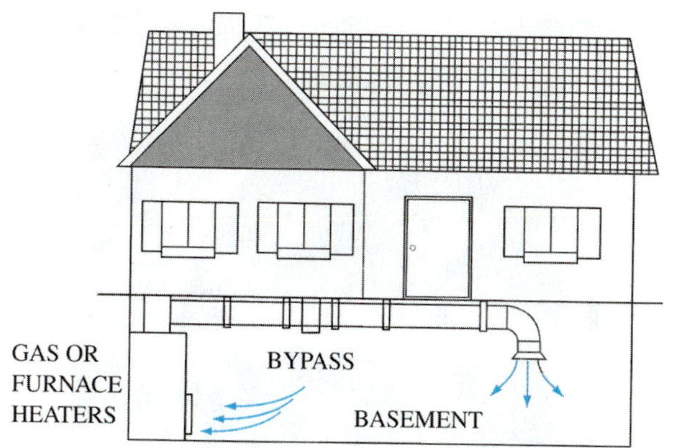

Figure 72-9 Relieving air pressure by dumping it into the basement.

Damper Leakage Method

The damper leakage method essentially uses the entire building as a dump zone. Some air passes by all the dampers, but more is directed through the open damper. This system is simple; it does not involve any bypass dampers or discharge air thermostats.

This can be accomplished by adjusting standard dampers to still be slightly open in the closed position or using dampers that are slightly smaller than the duct they are in. Using this method, a small amount of air will be supplied all the time; even to the zones that are not calling.

The system performance of a damper leaking system is difficult to predict. If the loads in the zones are not close to the same, the system will not maintain the correct temperature in any zone except the zone with the highest load.

Oversized Duct

Another approach is to size each zone duct to handle 60–70 percent of the total system air volume. This keeps the equipment operating correctly and puts the air where the load is. It can create problems with the air distribution in the rooms. As the volume of air leaving the registers changes, so does the throw. If the registers are sized for the largest volume of air that will be moving through them, the throw will be too little when all the zones are open. On the other hand, if the registers are sized for a smaller amount of air, the registers will be noisy and drafty when the full system air volume is going to just one zone. This arrangement is really only practical on small jobs with two or three zones.

Variable Air Volume

A variable-air-volume system changes the amount of air the fan delivers to match the zone requirements, eliminating the need for bypassing or dumping air. When applying variable-air-volume zone systems, heating and cooling systems with staging or modulating capacity are used so that the system capacity can match the load on the system. In the past, this type of control was only available for commercial systems. With the advent of ECM variable-speed motors, variable-capacity residential equipment, and communicating controls, these systems are now available for residential systems.

One example of a variable-air-volume residential system is the Carrier Infinity Zone controller used on a variable-capacity heat pump. The control system uses only four wires: two for power and two for communications. The blower communicates zone airflow characteristics back to the system controller. Thus, the controller knows what the airflow and resistance of each zone is.

72.4 AUTOMATIC RELIEF DAMPERS

During relief operations, it is important to maintain at least the minimum airflow through the air-handling unit; otherwise the coil performance will be unsatisfactory. The following formula can be used to determine the quantity of bypass air:

$$\begin{aligned} &\text{air handler (CFM)} - \text{smallest zone peak (CFM)} \\ &\quad - \text{leakage of all closed dampers (CFM)} \\ &\quad = \text{bypass airflow (CFM)} \end{aligned}$$

Automatic dampers are used when either bypassing or dumping is required to relieve excess air pressure. Two types of dampers are used for this: barometric dampers and motorized dampers.

Barometric Dampers

Barometric static-pressure-relief dampers (Figure 72-10) require no electrical connections. They are used on low-pressure systems with static pressures usually below 0.5 in wc. They are held closed with either an adjustable weight or by spring pressure. When the pressure difference across the damper exceeds the weight or spring pressure, the damper opens. The amount of opening is regulated by the amount of air to be relieved. The advantage of these dampers is that they are self-regulating and require no power or control signals. The disadvantages are that they must be adjusted for each installation to work correctly and they tend to leak a little all the time.

Motorized Dampers

Motorized static-pressure-relief dampers are electrically or electronically operated by a zone control in response to

Figure 72-10 Static-pressure-relief damper.

a signal from a remotely located duct pressurestat. These dampers have the advantage of closing tighter when bypass is not needed, and they generally do not require adjustment. They do require a more sophisticated control system and a static pressure sensor.

72.5 TYPES OF DAMPERS

There are many shapes, sizes, and styles of zone dampers designed to fit different requirements and applications. The flat piece of metal that opens or closes the airstream in a damper is called a blade. Dampers are often listed by the blade style used in the damper. Some different damper styles available are the following:

- Opposed-blade damper for rectangular ducts
- Parallel-blade damper for rectangular ducts
- Single-blade dampers for round duct
- Retrofit dampers

Opposed-Blade Dampers

Opposed-blade dampers use flat, rectangular blades that pivot in opposite directions to each other. They form a V shape when they are closed. Opposed-blade dampers are most often used in commercial systems. They are well suited for variable volume applications where the damper will be operated in a partially open position.

Parallel-Blade Dampers

Parallel-blade dampers use flat, rectangular blades that pivot together in parallel with each other (Figure 72-11). Parallel-blade dampers are often used in commercial systems for controlling the return air and outdoor air. They are also commonly used in residential applications for rectangular duct dampers.

Figure 72-11 Parallel-blade damper. *(Courtesy of Greenheck Fan)*

Figure 72-12 Single-blade damper.

Single-Blade Dampers

Single-blade dampers are common in residential applications with round duct. They consist of a single piece of metal that pivots in the middle (Figure 72-12). Single-blade dampers are mainly used in residential applications

Retrofit Dampers

Adding regular dampers to an existing system requires duct modification. The labor and cost involved can be a deterrent to adding a zone controls to an existing system. Retrofit dampers are single-blade dampers designed specifically to be inserted into an existing duct. They are added by making a slot in the side of the duct, inserting the damper blade into the duct, and mounting the damper to the outside of it. Figure 72-13 shows a retrofit damper.

72.6 DAMPER LOCATION

The type of dampers used is partly determined by the general design of the duct system. In new systems, the duct can be designed with a separate trunk duct for each major zone. These zones are then controlled by large dampers that control flow through the entire trunk duct.

Figure 72-13 Retrofit damper. *(Courtesy of Jackson Systems, LLC)*

Figure 72-14 Zone damper for rectangular duct.

Trunk Duct Dampers

Most zone systems control airflow in the trunk ducts. This has the advantage of minimizing damper cost since a single trunk damper controls the airflow to several branch ducts. This method also keeps air noise to a minimum by placing the dampers away from the point where the air enters the house. The disadvantage of trunk duct dampers is that multiple trunk lines are required. This is more of a concern for retrofit applications than for new construction. Figure 72-14 shows a damper used for a rectangular trunk duct, while Figure 72-15 shows a damper used for a round trunk duct.

Branch Duct Dampers

Another strategy is to put a damper in every branch duct. The advantage of this strategy is that multiple trunk ducts are not required. However, the system expense goes up rapidly because of the cost of the extra dampers. The system does not need to have a zone for every room when using branch duct dampers; several dampers can be slaved together and controlled from one zone. This is usually preferable to having a zone for every room. It is hard to control the system operation correctly if the zone is too small compared to the overall system capacity. The dampers used for branch ducts are smaller versions of dampers used for trunk ducts as shown in Figure 72-14 and 72-15.

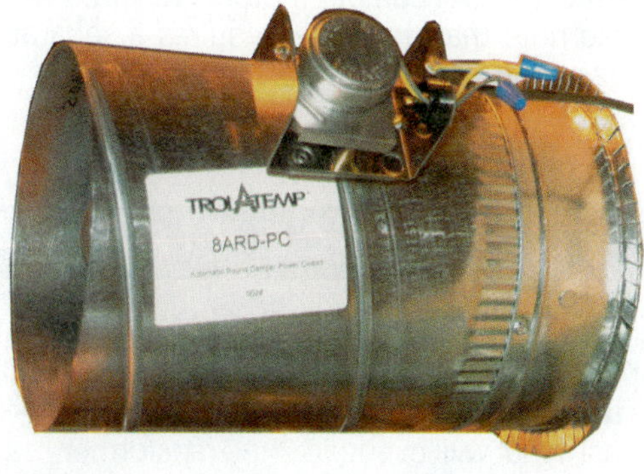

Figure 72-15 Zone damper for a round duct.

Register and Diffuser Dampers

The damper can be built into the supply registers and diffusers. The advantage of this method is that changing the registers is usually easier than changing the ductwork. The disadvantages are the cost of buying a diffuser damper for every diffuser in the house and the noise created by throttling the air right where it enters the room.

72.7 DAMPER CONTROL

The two methods of controlling zone dampers are two-position control or modulating control. Some dampers may be used with either type of system, while others are designed exclusively for either two-position control or modulating control.

The dampers in a two-position control system are either open or closed; there is no partial opening or closing. Two-position control is simpler but generally does not provide as even a temperature control as modulating control. Dampers designed for two-position control can be either motor driven in both directions or motor driven in one direction and returned in the other direction by a spring. The control system for a spring-return damper can be simpler because it is energized in only one direction. Dampers that are motor driven in both directions can usually be applied to both two-position and modulating systems, while spring-return dampers are only applicable to two-position systems. The motor on a spring-return damper is usually stronger because it must overcome the spring as well as operate the damper.

Modulating controls can open and close the dampers in increments, rather than completely opening or closing the zone dampers. This type of control requires dampers that are powered in both directions. The zone control system for a modulating damper system is usually more complex. Modulating zone damper control can be used in conjunction with air bypass to produce variable-air-volume control to a system without changing the fan speed.

SERVICE TIP

Zone dampers must be located as close as possible to the plenum or supply trunk. If they are too close to the register, there is a possibility that air passing a partially opened damper will cause a whistling sound that can be heard in the room. If a sound can be heard and it is possible to move the damper farther away, one method of eliminating the sound is to put one or more 90° turns in the duct after the damper and before the room register.

72.8 TYPES OF ZONING SYSTEMS

A number of manufacturers make control systems for installations requiring zoning. These systems differ in number of zones handled, types of applications, methods of control, and types of adjustments provided. The types of systems available generally fall into two categories: basic systems that stress simplicity and advanced systems that stress performance.

Basic Zone Control Systems

The most basic zone control systems are typically limited to a maximum of four zones, single-stage heating and cooling, and two-position damper control and require either bypassing or dumping to control the air pressure. These systems typically are designed to work with standard off-the-shelf thermostats. Wiring for these systems is usually letter for letter: R to R, G to G, and so on. The emphasis of these systems is simplicity. On many of these, one of the thermostats is considered the master, and the system selection is determined by its system setting. The system will cycle on by any of the thermostats, but only the master determines whether the system is in heating or cooling.

Added Features

Companies offering these basic types of zoning panels also offer systems with more features. Items like two-stage heating or cooling or dual-fuel heat pump integration can be included. These systems still use traditional 24 V thermostats, two-position dampers, and bypass air, but they do have the ability to work with more efficient staged heating and cooling equipment.

Advanced Zone Systems

Other zoning systems stress a high level of technology to achieve enhanced performance and comfort. Typically these systems take advantage of equipment with variable-speed ECM blower motors, modulating damper control, and staged-capacity heating and cooling. Normally, the controls are proprietary, not common over-the-counter controls. Temperature is typically measured using thermistor sensors.

72.9 BASIC ZONE CONTROL SYSTEM

Ultrazone EWC-ST-3E Zone Control Panel

Ultrazone EWC-ST-3E zone control (Figure 72-16) is an example of a basic zone control system. The system is designed to work with standard heating and air-conditioning equipment. Either mechanical or non-power-robbing electronic thermostats may be used. A single control is capable of controlling a two-zone or three-zone system. If a greater number of zones are required, controls may be interconnected for additional zones.

A thermostat and a damper are required for each zone. The zone dampers are controlled by the zone control panel and operate in two positions; dampers are either fully open or fully closed. The zone thermostats send a signal to the zone control and the control operates their respective dampers.

Figure 72-16 Basic three-zone control panel.

The zone 1 thermostat determines whether the system is operating in heating, cooling, or is off. The zone 1 thermostat must have an O terminal that is energized in cooling and a B terminal that is energized in heating. These signals tell the zone control whether the system will operate in heating or cooling. There will be either six wires connecting the panel to the zone 1 thermostat, or seven wires if the thermostat requires a common terminal connection. All other zone thermostats can be simple three- or four-wire thermostats.

Since the EWC-ST-3E zone control panel is designed for limited applications, there is no programming and no mode switches. The panel has a single J1 jumper that determines fan operation in heat. With the jumper in place, the zone panel sends 24 V to the system's G terminal during heat. With the jumper cut, the panel does not send 24 V to the unit in heat, as in a gas furnace.

Control Panel Wiring

A wiring diagram for a typical zone control panel is shown in Figure 72-17. The system shown is for three zones. For more than three zones, a slave panel would be used. Note that there are seven wires between the panel and the zone 1 thermostat and only five wires to the other two-zone thermostats. All damper motors are wired with three wires. Also note that the panel requires a separate field-supplied 24 V, 40 VA transformer.

SERVICE TIP

A separate 24 V, 40 V A transformer is required for most zone panels. The transformer on the indoor unit should not be used to supply 24 V to the zone control panel. The additional load will overload the transformer, causing transformer failure and erratic system operation.

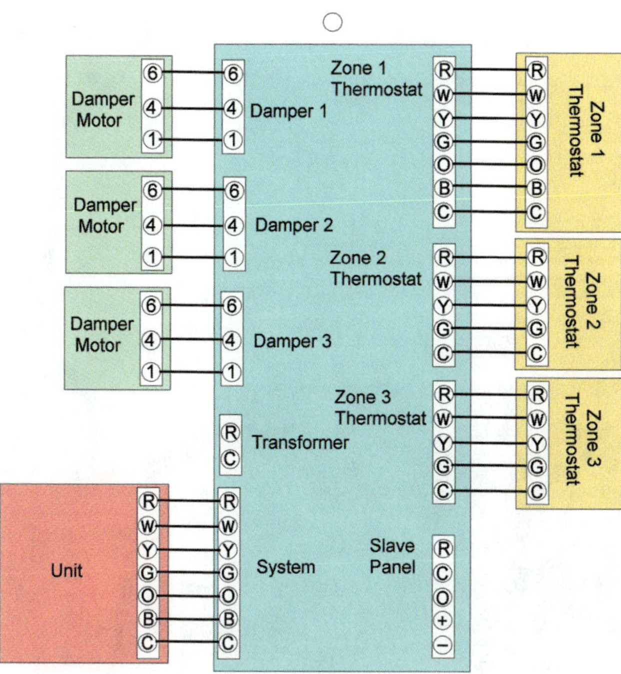

Figure 72-17 Typical wiring for basic relay based zone panel.

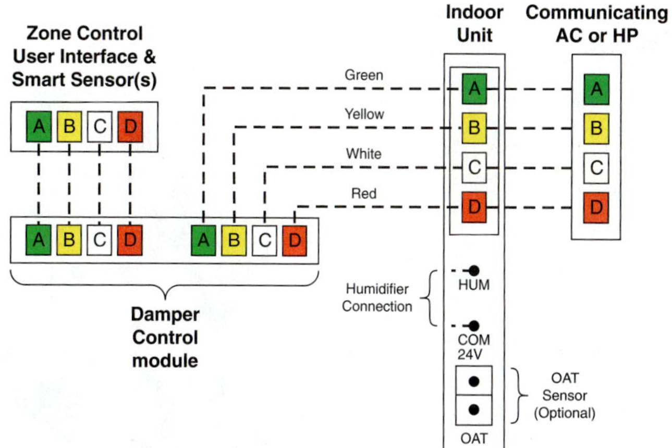

Figure 72-18 Four-wire connections for the Infinity zone control system.

72.10 BASIC ZONE CONTROL SYSTEM OPERATION

System switching from heating to cooling is provided through the zone 1 thermostat. This thermostat is normally placed in the zone that has the greatest usage. In a residential installation, the best location would probably be the living room.

The zone panel will respond to a call from any thermostat, provided the call is for the same mode that the zone 1 thermostat is in. If zone 1 is on cooling, a call for heat from another thermostat will have no effect. When any zone calls, its zone damper opens and zones not called are closed. The system is started if it is not already on. When all the zones are satisfied, all dampers go to a fully open position and the system cycles off. This permits the fan to dissipate any residual heat that may be left in the furnace and allows fan-only operation.

72.11 ADVANCED COMMUNICATING ZONE SYSTEMS

Carrier Infinity Zone System

The Infinity zone control is designed to work with Carrier's Infinity line of products. The Infinity zone system consists of several intelligent communicating components:

- Infinity zone control (or user interface)
- Smart sensors
- Damper control module
- A variable-speed furnace or fan coil
- Two-stage air conditioner or heat pump
- Infinity packaged products

The components continually communicate with each other via a four-wire connection called the ABCD bus. Commands, operating conditions, and other data are passed continually between components over the ABCD bus. Figure 72-18 shows the ABCD bus connections between communicating components. The communications allow for more intelligent system control. For example, the Infinity zone control knows the real-time airflow through each zone because the blowers communicate this information back to the control.

Unlike the other zone control systems discussed, the Infinity zone control is the component that would normally be considered the thermostat. The user interface that mounts on the wall is actually the system zone controller. The Infinity system does have a damper controller, but its job is simply to operate the dampers based on commands from the Infinity controller.

The Infinity zone control is also the user interface. It functions as a programmable thermostat as well as a system controller and zone controller. In a standard setup, the Infinity controller is used to program all the zones and enter the temperature settings for all zones. Temperature sensors that have no user interface are used in the other zones (Figure 72-19). Optional smart sensors can be used in place of the standard zone sensors. These allow users to change

Figure 72-19 Infinity remote room sensor.

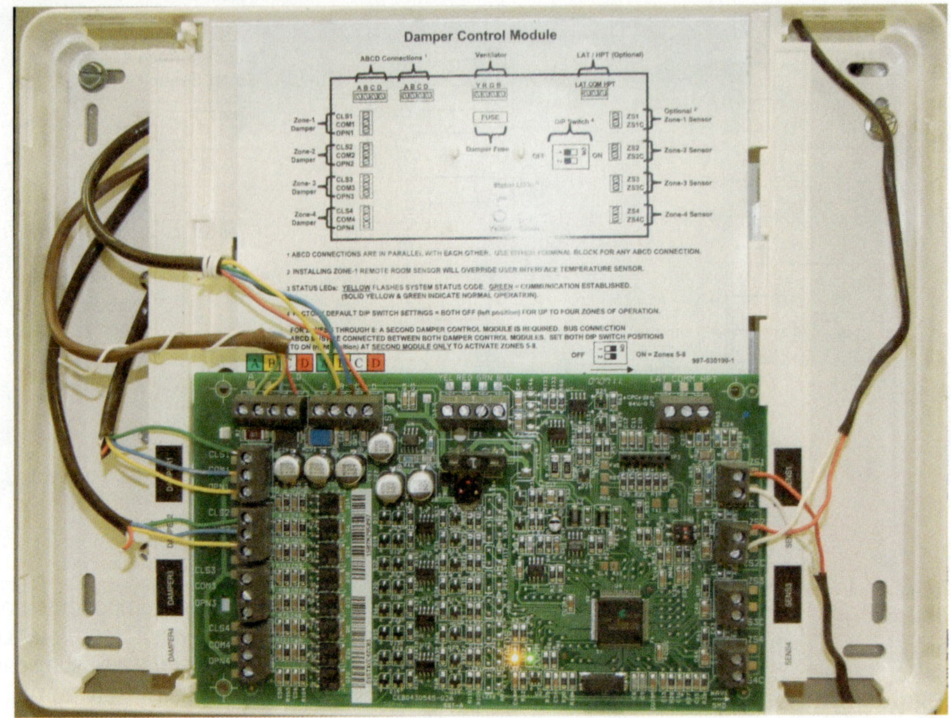

Figure 72-20 Wiring of complete Infinity zone control system.

the temperature setting for a zone at the zone sensor rather than at the Infinity controller.

The wiring and connections for an Infinity zone control system are shown in Figure 72-20. The indoor unit, outdoor unit, zone control, and damper control are all connected with the four wires of the ABCD buss. The sensors are connected with two wires and the damper motors with three wires.

All Infinity furnaces or fan coils are variable speed and multistage. The infinity system controls the fan speed and system capacity to match the demands of the system, so bypassing air is not required. The control can be used with furnaces, air conditioners, heat pumps, or dual-fuel heat pumps.

New installations do not require jumpers or programming; the control recognizes the communicating components connected to the system and configures itself on initial power-up. First the control will locate all communicating components. Next, the system will open all dampers and perform a static pressure check. This process will take about 1½ minutes to complete. When completed, a screen will appear displaying the static pressure (in inches) across the equipment at the expected highest delivered airflow. If the static pressure is over 1 in, a warning will appear, but equipment operation and the True Sense dirty-filter detection operation will not be affected. After the static test, the control does a duct assessment. The duct assessment will measure the relative size of the ductwork, up to and through the dampers. These measurements are used to control the correct amount of airflow in the zoned system. Status messages will appear on the screen to indicate what the system is doing. The process will take approximately 1 minute per zone. The duct assessment will override a call for heat or cool. A duct assessment will automatically occur each day at a user-selectable time. The factory default time is 1 PM but may be changed by entering the Zoning Setup menu.

During the duct assessment, the system will first open all zones and drive the blower to 175 CFM/ton of cooling (or the minimum indoor unit's airflow, whichever is greater). It will then take a static pressure measurement. The system will then close all zones and open one zone at a time, taking a static pressure measurement for each zone. The system will then close all zones and take a pressure measurement, getting a value for the duct leakage up to and through the dampers. With these static pressure measurements, the system will calculate the relative size of each zone as well as the percent leakage through the dampers. At the end of the process, the display will show the relative size of each zone duct.

72.12 COMMUNICATING ZONE CONTROL SYSTEM OPERATION

The primary difference in operation between most other zone control systems and the Infinity zone control system is that the Infinity system is an intelligent communicating system. The zone controller is collecting real-time operating data and making decisions about system operation based on this data. Other systems rely on mode switches, jumpers, or programs that are configured at the time of installation. These systems can only respond to on or off signals. The Infinity can determine when the air filter is stopped up based on changes in the fan. Staging of multiple-stage equipment is done based on demand, rather than any predetermined temperature or zone arrangement. Air bypass is not required because the fan speed is varied based on the demand and the number of operating zones. The Infinity also responds to demands for dehumidification by reducing the fan speed and increasing the latent heat capacity of the evaporator coil. The Infinity controller has a service mode

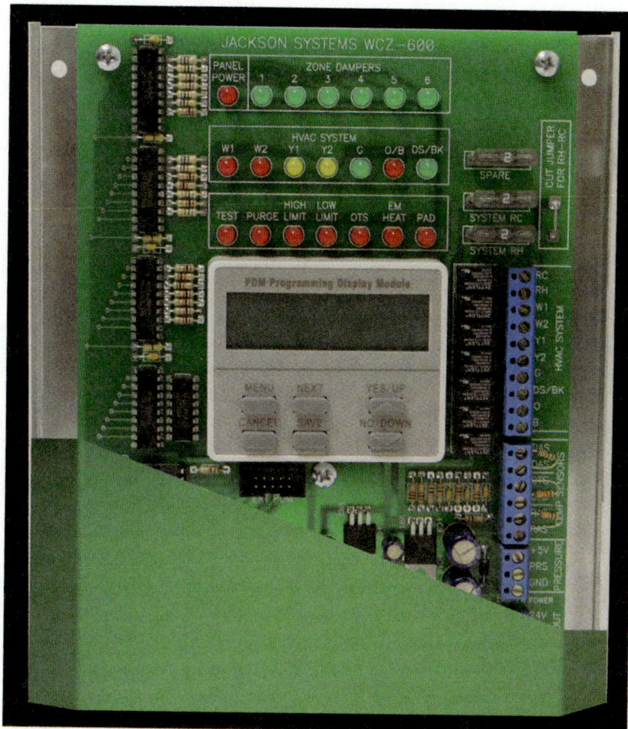

Figure 72-21 This zone control panel uses wireless communication. *(Courtesy of Jackson Systems, LLC)*

that displays the system history and any system faults, allowing technicians to see faults when diagnosing system problems. All of this is done by collecting real-time operating data and making decisions.

72.13 WIRELESS ZONE CONTROL SYSTEM

One of the newest types of zone control systems uses wireless communication instead of wires. Figure 72-21 shows a wireless zone panel, while Figure 72-22 shows a wireless thermostat used with the wireless zone panel. No control wires are required between the thermostats, zone panel, and dampers. The zone panel does require power to operate and is wired to the control system of the air-conditioning equipment. Duct supply, return, and outdoor air tempera-

Figure 72-22 This wireless thermostat is used in conjunction with the wireless zone panel. *(Courtesy of Jackson Systems, LLC)*

Figure 72-23 Wireless Zone One single-zone damper. *(Courtesy of Jackson Systems, LLC)*

ture sensors are still wired as well. Since no control wires are required between the zone panel and the thermostats, thermostat placement is much simpler. The zone dampers require wired power to operate, but no control wires are necessary. Wireless zone systems are ideal for adding zone control to an existing system where pulling control wires would be difficult.

72.14 SINGLE-ZONE DAMPER SYSTEMS

A complete zone control system might be overkill if the problem is a single room or area that is overconditioned compared to the rest of the house. This can be solved using a system with a single zone damper that does not turn the heating or cooling system on but closes off the damper to the room when the temperature is satisfied to prevent overheating or overcooling. Figure 72-23 shows a wirelessly controlled Zone One damper for a single-zone system.

It uses a duct sensor located on the damper to sense if the air in the duct is warm air or cool. The control then operates the damper based on the signal from the wireless thermostat and the air temperature in the duct. If the Zone One thermostat is calling for cooling and the duct sensor determines there is cool air in the duct, the damper will power open; otherwise, the damper will remain closed. If the Zone One thermostat is calling for heating and the duct sensor determines there is warm air in the duct, the damper will power open; otherwise, the damper will remain closed. The wireless relay control module has LED indication to confirm the operating mode and damper position.

UNIT 72—SUMMARY

Residential zone systems have moved from a luxury reserved for only a few high-end homes to a viable strategy for comfort and economy in a much broader range of applications. Many zoning products are available, from simple relay logic, two-position systems with air bypass to sophisticated communicating systems with variable air volume and capacity control. A zone system may be as simple as a two-zone system for a two-story house or as complex as a zone for every area in the house with different temperature and setback schedules for each area. A decade ago, technicians might have seldom been required to work on a zone system; today a technician is almost certainly going to need to work on residential zone systems.

WORK ORDERS

Service Ticket 7201

Customer Complaint: Basement Too Cold in Summer

Equipment Type: Basic Zone Control Using Two-Position Dampers, Single-Stage Heating and Cooling

A technician is called to look at a zoned system with three zones: one in the finished basement, one in the living areas, and one in the bedrooms. The basement is staying below 70°F even though the thermostat for that zone is set to 78°F. The technician notices that the system is operating and cold air is coming out of the basement diffusers. A visual check of the damper shows that it is open. However, a voltage check at the damper indicates that it is receiving a signal to close. The technician determines that the damper is stuck open and recommends replacing the damper.

Service Ticket 7202

Customer Complaint: No Heat

Equipment Type: Heat Pump with ECM Motor and an Advanced Zone Control

A technician is called to look at a zoned heat pump system that is not heating. The technician notices that all zones are calling for heat, but no air is coming out of the registers in one zone. Air is coming out of the other zones, but it is not warm. The temperature outside is above the heat pump balance point. A check of the outside unit reveals that it is not operating and the high-pressure switch has tripped. The technician resets the pressure switch and the compressor comes on. Heat is now coming out of two zones, but one zone still has no airflow. The technician checks the damper and finds it closed even though it is receiving a signal to open. The technician concludes that the stuck damper caused an airflow problem when only that zone was calling, which led to the high-pressure switch tripping. The tripped pressure switch then prevented the system from operating and heating the remaining zones. The technician recommends replacing the stuck damper.

Service Ticket 7203

Customer Complaint: Delay in System Starting

Equipment Type: Advanced Zone Control

A customer is concerned because the system waited for several minutes before it came on after she adjusted the temperature. The customer had just returned home from running and heard the system cycle off. Because she was hot, she adjusted the temperature down several degrees to bring on the air conditioner. However, nothing happened. While the customer is on the phone with the service company, the system comes on. The technician explains that the system has a built-in off-cycle timer that makes the system stay off for 5 minutes anytime the system cycles off. This is for the protection of the compressor.

Service Ticket 7204

Customer Complaint: Hot Air in Cooling Mode

Equipment Type: Basic Zone Control on Heat Pump

A technician is called to look at a newly installed system that is delivering hot air in the cooling mode. The technician notices that the heat pump is operating in the heating mode even though the master thermostat is set for cooling. A look at the diagram for the outdoor unit shows the reversing valve should energize in cooling with the O terminal. Checking the zone panel jumpers, the technician discovers that the reversing valve switchover jumper has been set to reverse in heat instead of cooling. The technician moves the jumper to the cooling position and the heat pump operates correctly.

Service Ticket 7205

Customer Complaint: No Airflow in the Bedroom Zone

Equipment Type: Advanced Zone Control System, Single-Stage Heating and Cooling

A technician is called to look at a newly installed system. The zone serving the bedroom area is not delivering any air. The technician notices that there is no air delivery in the zone, even with the system switch set to off and the fan switch set to on. A check of the zone damper shows it is closed. A voltage check shows voltage between the common and close terminals on the damper. At the zone control, a voltage check shows voltage between the common and open terminals. The technician looks at the wire colors and determines that the open and close signal wires to the zone damper have been reversed at the damper. The wires are swapped at the damper and the system operates correctly.

UNIT 72—REVIEW QUESTIONS

1. Why are zone control systems used?
2. List the components found in a basic zone control system.
3. What two general criteria are used to select zones for a building?
4. Explain the reason zone control systems need a method for handling excess air.
5. List the methods of handling the excess air in a zone system.
6. Discuss the disadvantages of air bypass.
7. Explain the concept of variable air volume in a zoning system.
8. Explain the difference between two-position and modulating damper control.
9. Explain how a barometric damper operates.
10. List several types of zone dampers.
11. Explain the difference between a basic zone control system and an advanced zone control system.
12. The PIAB jumper on zone 1 of a Lennox Harmony III control is set at 40 percent. The minimum airflow is 900 CFM and the maximum is 1,800 CFM. How much air will this zone receive? Show all work.
13. Explain the purpose of using a dump zone in a zone control system, and give an example.
14. What is the difference between a communicating zone control system and a basic zone control system?
15. What is the purpose of the master thermostat in the basic zone control system?
16. What is the purpose of discharge air sensors?
17. What is the purpose of the outdoor air sensor on a zone control system?
18. How are motorized dampers controlled when they are used as bypass dampers to relieve a pressure differential across the blower?
19. Describe the added features available on some basic zone control systems.
20. How is an Infinity zone control system configured?

UNIT 73

Testing and Balancing Air Systems

OBJECTIVES

After completing this unit, you will be able to:

1. explain why TAB (test and balance) is critical to the operation of an air-conditioning system.
2. explain why TAB is critical to energy consumption of an air-conditioning system.
3. list the organizations involved in the TAB industry.
4. list the instruments used to perform TAB procedures.
5. explain proper use of instruments used in TAB procedures.
6. describe TAB procedures.

73.1 INTRODUCTION

Test and balance, sometimes referred to as test, adjust, and balance or just TAB, is the adjustment of the air side and water side of an HVACR system to the operational specifications of the designer's original intent. In some cases these adjustments must be redone due to seasonal changes or changes in the structure. This unit will discuss the air side of an HVACR system.

Commercial projects recognized the need for TAB as early as the 1960s. Originally the installation contractor completed any test and balance that was needed at start-up along with other commissioning activities. As commercial projects grew and indoor air quality became a more clearly defined issue, independent TAB contractors began to operate as a quality-control check on the system and the project. The TAB industry has grown into a large part of the commercial HVACR industry. There are many measurements and adjustments that are specific to testing and balancing that are not normally performed in regular HVACR work. Technicians working for TAB contractors go through extensive training and certification, which can take up to 5 years to complete.

The residential TAB market has seen large growth recently due to energy costs and increased health concerns related to residential indoor air quality. The largest energy consuming appliance in the home is the HVACR system, and most of these systems are estimated to have a 30 percent loss of efficiency due to improper air-side design and installation. Some states and the federal Department of Energy have adopted standards for the testing of HVACR residential systems and have offered incentives to builders who will have their homes certified under the programs. At the local level, many power companies offer TAB analysis for little or no cost to the consumer, in hopes of lowering energy consumption.

Health concerns have increased in the residential environment as home construction techniques and materials have improved. As homes have become increasingly tighter due to energy concerns, the infiltration of outside air has been reduced, causing indoor pollutants and stagnant air to affect those more susceptible to such pollutants, such as infants and the elderly. Many respiratory afflictions, such as asthma, are on the rise, while at the same time studies show indoor pollution rising as well. A combination of ventilation and filtration can be used to introduce filtered, fresh air into the house. TAB is seen as part of the solution because the airflow in a properly balanced system is more predictable and controllable. A home that is properly balanced will prevent air from stagnating and help to reduce pollutants by the conditioning and filtering of the indoor air.

73.2 TAB ORGANIZATIONS

In the TAB industry there are several organizations committed to developing standards, procedures, training courses, and to certification of technicians and contractors. As in the rest of the HVAC industry, these organizations overlap on missions and compete in several ways. Associated Air Balance Council (AABC), National Environmental Balancing Bureau (NABB), and Testing Adjusting and Balancing Bureau (TABB) are the major organizations that specialize in test and balance of HVAC systems. Other groups are also involved in TAB but not exclusively. ASHRAE, the American Society of Heating and Air-Conditioning Engineers, is a good example of one of these organizations.

The Associated Air Balance Council is the oldest of the three and started in 1960s. The AABC is an independent association, and its member's firms can have no affiliation with manufacturers or HVAC contractors. Member firms must also enroll their technicians in AABC training courses. These courses, combined with 4 years of industry experience, lead to technician certification. AABC publishes standards for the industry and a newsletter. Member firms are required to participate in a quality-assurance program where the association receives feedback from the client on the work performed.

The National Environmental Balancing Bureau was started in the 1970s. NEBB certifies firms and provides training at their Arizona facility. The training includes labs and practical exams. NEBB also publishes standards and manuals that cover all aspects of the TAB industry.

The Testing Adjusting and Balancing Bureau is the youngest of the three. TABB is an organization started in 2001 to work with Sheet Metal and Air Conditioning

Contractors Association (SMACNA). TABB publishes standards, manuals, and a newsletter, and all these publications are produced in partnership with SMACNA. TABB certifies technicians, supervisors, and contractors because TABB's approach is that everyone involved in the project be TABB trained from designer to installer. TABB produces industry training for designers, supervisors, and tradesmen. TABB training is offered to tradesman at the union hall during apprenticeship and as continuing education throughout their career. TABB takes a "total quality" approach instead of a "quality by inspection" approach. This is possible because of the relationship TABB has with all of the participants in the project.

Although each organization takes a different approach, the outcome of having a well-installed and energy-efficient system that provides conditioned air and comfort to the people using the space is met by using standards and methods developed by these associations. Although many smaller projects may not use independent firms or union contractors, TAB is a critical part of the commissioning of each installation and modification of all HVAC systems. Not testing and balancing every system adds to the energy cost for the owner and society, as well as possible medical costs for the occupants.

73.3 TAB TEST INSTRUMENTS

In smaller companies, contractors may have the start-up technician perform the required TAB tests. Measurements taken for TAB applications are temperature, pressure, velocity, rotating speed (rpm), and electrical. Many instruments used by service technicians are also used by TAB technicians, such as digital thermometers, psychrometers, anemometers, manometers with a pitot tube, and digital multimeters. Other instruments used in TAB tests that would not be commonly found on a typical service van are chronometric tachometers, digital or analog capture hoods, and air-differential pressure gauges. These are specialized tools that TAB technicians may choose to use depending on the application.

TECH TIP

As with most test instruments on the market today, the accuracy of digital test equipment is better and the ease of use is often greatly improved over analog models. The sling psychrometer is the clearest example. The drawback to digital instruments is price, and in some cases digital versions can run 10 times more than analog versions. This is why sling psychrometers are still in use today.

73.4 TEMPERATURE MEASUREMENTS

Temperature measurements are taken to compare return- and supply-air temperature in the conditioned space and entering- and leaving-air temperature at the furnace or evaporator coil. In some cases, suction and discharge temperature are also measured by TAB technicians to confirm proper system refrigerant charge. Instruments used for these measurements are as follows:

Glass-tube thermometers (Figure 73-1) are available in a number of ranges, scale graduations, and lengths. They have a useful range from −40° to over 220°F.

Dial thermometers are available in two general types: bimetal stem (Figure 73-2) and remote bulb with flexible capillary tube (Figure 73-3). Bimetal stem thermometers use a spiral formed bimetal inside the tube. As the bimetal expands and contracts, it produces a

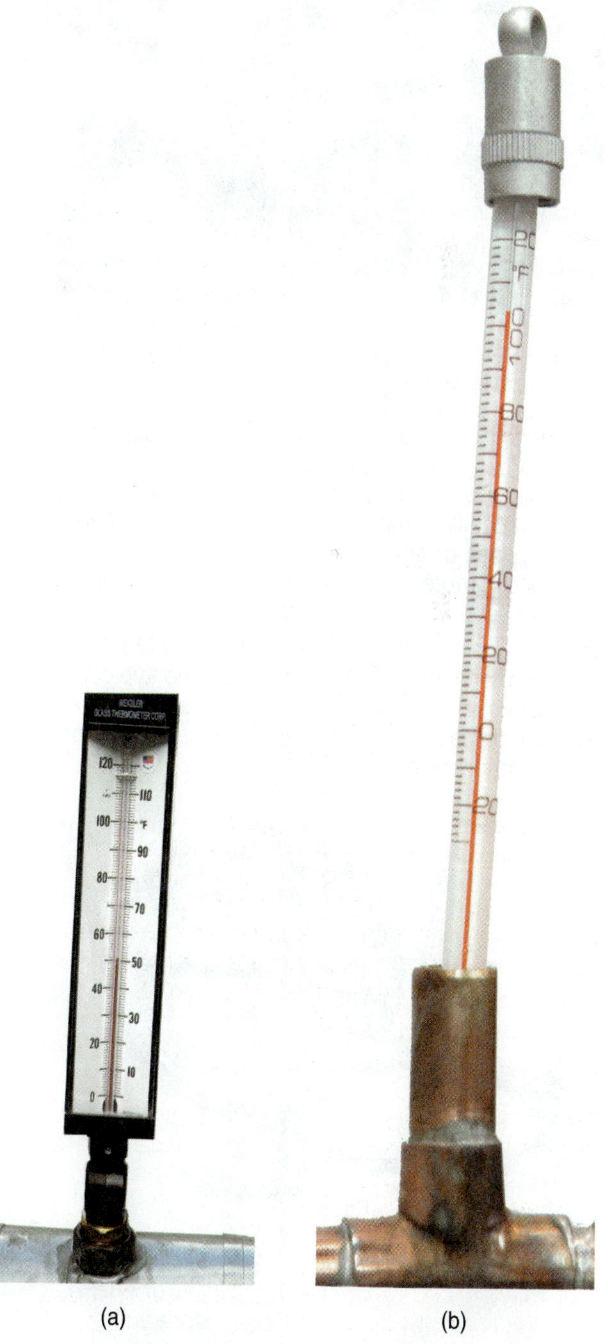

Figure 73-1 Glass-tube thermometers: (a) in protective case; (b) with glass tube exposed.

Figure 73-2 Bimetal dial thermometer reading air temperature leaving a supply-air register.

twisting motion that moves the needle on the thermometer. The remote bulb model permits temperature measurements from a remote location. The bulb and capillary tube are liquid or gas filled, and the fluid expands or contracts to operate a Bourdon tube. Dial thermometers usually have a longer reading time lag than glass-tube thermometers.

Digital pocket thermometers (Figure 73-4) are electronic, digital replacements for the typical dial-type pocket thermometer. They are generally more durable and more accurate than mechanical stem-type pocket thermometers.

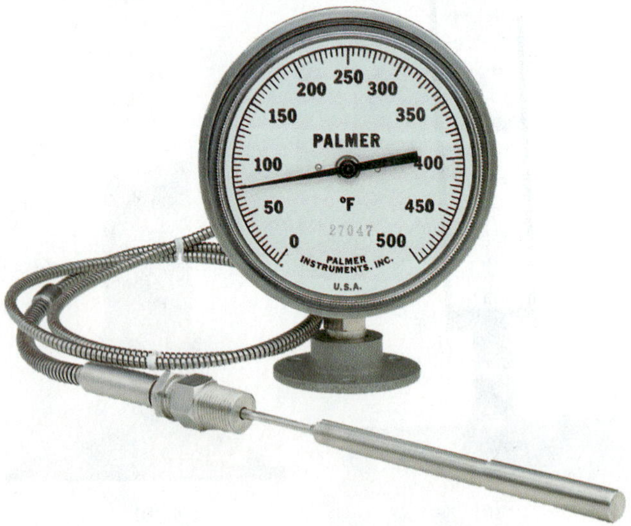

Figure 73-3 Remote bulb and capillary-tube dial-type thermometer. *(Palmer Wahl Instrumentation Group)*

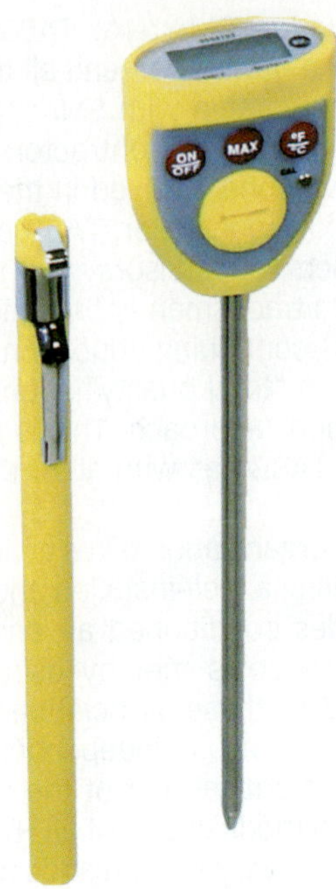

Figure 73-4 Digital pocket thermometer.

Electronic thermometers have interchangeable probes to permit more accurate reading in selected ranges. There are three basic types of temperature-sensing elements used: resistance temperature detectors (RTDs), thermistors, and thermocouples. The meters are built for use with a particular type of sensor. Thermometers with interchangeable probes, and multiple connections are useful for reading temperatures at a number of locations. They can also be used to read the temperature difference between two probes for comparative temperature readings like superheat, subcooling, or delta T measurements (Figure 73-5).

Figure 73-5 This dual-input electronic thermometer can be used for comparative temperature measurements like superheat, subcooling, and delta T.

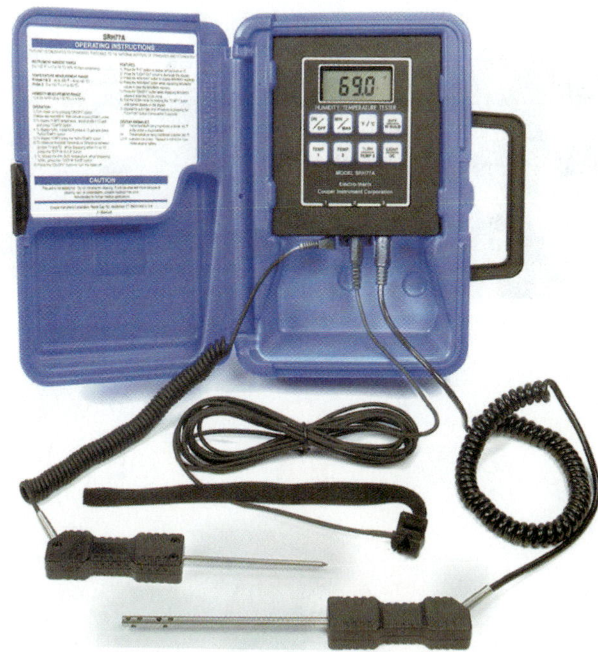

Figure 73-6 This electronic thermometer uses thermistor sensors. *(Courtesy of Cooper-Atkins Corporation)*

Resistance temperature sensors are wound wire with known resistance values. The resistance of the wire increases as its temperature increases, producing a very accurate reading. RTD sensors typically have a very narrow operating range. They also have a longer time lag than the thermocouple type due to the time required for the wire's temperature to change.

Thermistor sensors also change resistance with temperature, except they use a semiconductor instead of wire. Most thermistors used for electronic thermometers decrease in resistance as their temperature increases. Thermistor sensors typically read faster than RTD sensors. Figure 73-6 shows an electronic thermometer that uses thermistor-type sensors.

Thermocouple sensors produce a small DC voltage that is proportional to the temperature difference between the two ends of the thermocouple. Thermometers that use thermocouple sensing devices are calibrated to read this voltage as temperature and read temperature directly. There are two common types of thermocouple sensors: type J and type K. Figure 73-7 shows an electronic thermometer that uses thermocouple sensors.

SERVICE TIP

Electronic thermometers with dual inputs make reading differential temperature measurements such as superheat, subcooling, and delta T much easier. Most dual-input electronic thermometers have a function that automatically displays the difference between the two temperature probes.

Figure 73-7 Thermocouple thermometers.

Figure 73-8 Sling psychrometer.

73.5 ENTHALPY MEASUREMENTS

Enthalpy measurements are calculated from wet-bulb and dry-bulb temperatures. Measurements are then plotted on a psychrometric chart to find the enthalpy of air in the measured space or to compare entering-air enthalpy to leaving-air enthalpy across an evaporator coil. Instruments used are, the sling psychrometer and the electronic hygrometer.

The sling psychrometer consists of two glass thermometers, one of which has a wick wetted by water surrounding the bulb. The frame that supports the two thermometers is hinged to permit revolving the wetted instrument in the air. The unit should be whirled at a rate of two revolutions per second for most accurate results. Readings are taken on both thermometers after the temperatures have stabilized. Readings indicate dry-bulb and wet-bulb temperatures, which can be plotted on a psychrometric chart to determine numerous properties of the air sampled. Figure 73-8 shows a sling psychrometer.

The electronic hygrometer is usually constructed with a thin-film capacitance sensor. As the moisture content and temperature change, the resistance of the sensor changes proportionately. These instruments usually read directly in relative humidity. No wetted wick is necessary, and the instrument can remain stationary when readings are taken. Figure 73-9 shows an electronic hygrometer.

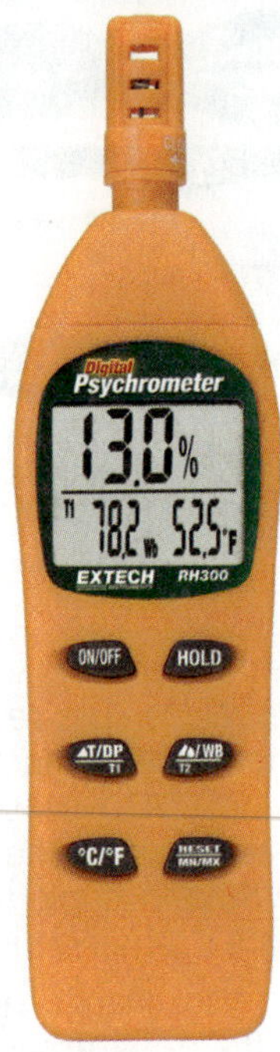

Figure 73-9 Digital psychrometer. *(Courtesy of Extech Instruments, A FLIR Company)*

73.6 PRESSURE MEASUREMENT

The pressure measurements that technicians take while testing and balancing systems range from the air pressure inside ductwork to the pressures in a refrigeration system. The instrument used to take the pressure reading depends largely upon the level of the pressure. Inclined manometers, magnehelic gauges, and digital manometers are used for small pressures of less than 1 in wc, such as the air pressure drop across a coil. Bourdon tube gauges are used for higher pressures measured in psig, such as the refrigerant pressure drop across a refrigeration system component.

Bourdon Tube Gauges

Calibrated pressure gauges typically use a Bourdon tube assembly for sensing pressure (Figure 73-10. The Bourdon tube may be constructed of stainless steel, copper, or bronze. The dials are available with pressure, vacuum, or compound gauges showing both pressure and vacuum. Gauges are used for checking pump pressures; coil, chiller, and condenser pressure drops; and pressure drops across orifice plates, valves, and other flow-calibrated devices.

Most flow measurements require determining the pressure drop across a component. Pressure-differential readings can be done with two separate gauges or a single gauge with a rotating dial or a special differential pressure gauge. When two gauges are used, it is essential that the gauges be calibrated to each other. This can be checked by comparing the two gauges when there is no flow and they are reading only static pressure. They should register the same pressure.

A single gauge with a rotating dial can be used to measure differential pressure. The gauge is installed with two valve connections: one to the lower of the two pressures and one to the higher pressure. The low-pressure line is opened while the high-pressure line valve remains closed. The gauge dial is then rotated to zero. The low-pressure valve is closed and the high-pressure line valve opened. The pressure reading will be the difference between the two

Figure 73-10 The most common type of gauge used for higher pressure is the Bourdon tube gauge.

pressures. For example, if the lower pressure is 15 psi and the higher pressure is 20 psi, the gauge will read 5 psi.

A differential pressure gauge uses two inputs and measures the difference between the two pressures.

Manometers

U tube manometers (Figure 73-11) are used for measuring gas pressure. When not connected, both ends of the U tube are open to the atmosphere. The tube is partly filled with liquid. Both tubes register zero when they are both open to the atmosphere. When one tube is attached to gas pressure, the side connected to gas pressure drops and the open side rises. The pressure is indicated by the difference in the height of the two columns. U tube manometers are recommended for measuring pressures above 1.0 in wc up to 15 in wc.

Inclined Manometers

The inclined manometer (Figure 73-12) is useful for measuring static pressure, velocity pressure, and total pressure in a duct when connected to a pitot tube. The pitot tube is constructed with a tube-in-a-tube design (Figure 73-13).

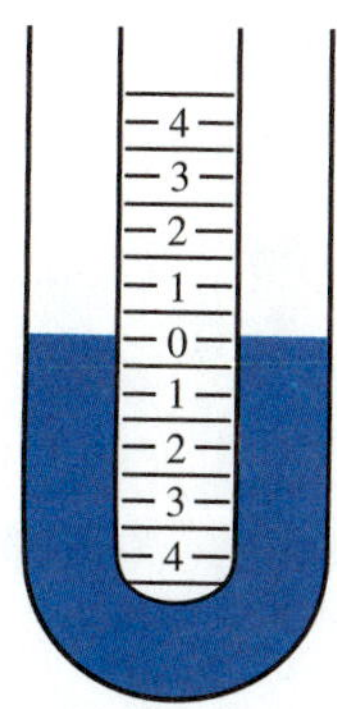

Figure 73-11 U tube manometer.

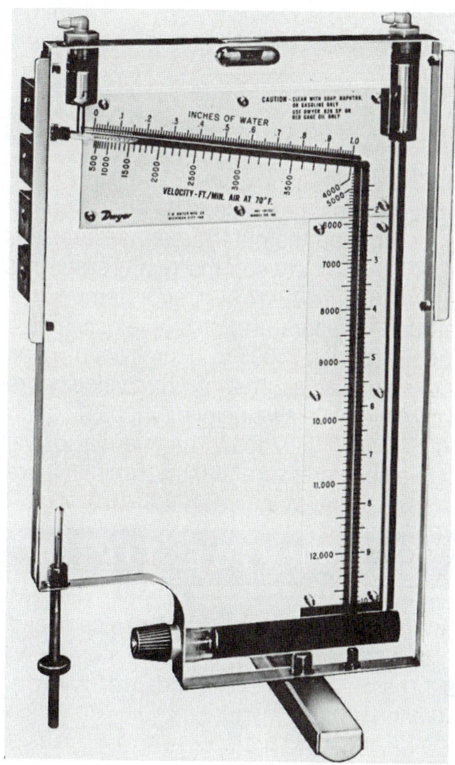

Figure 73-12 Inclined/vertical manometer. *(Courtesy of Dwyer Instruments, Inc.)*

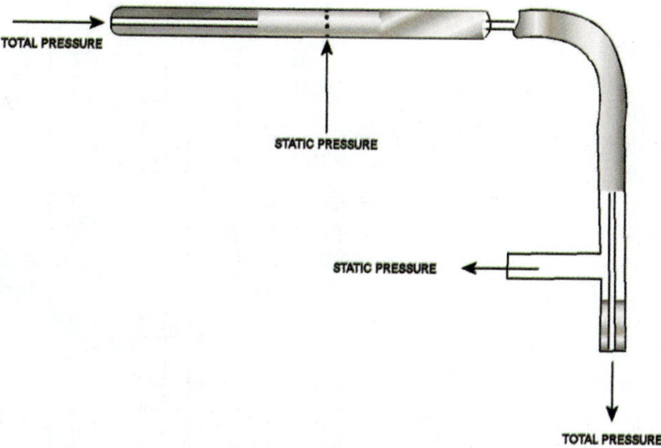

Figure 73-13 Detailed construction of a pitot tube.

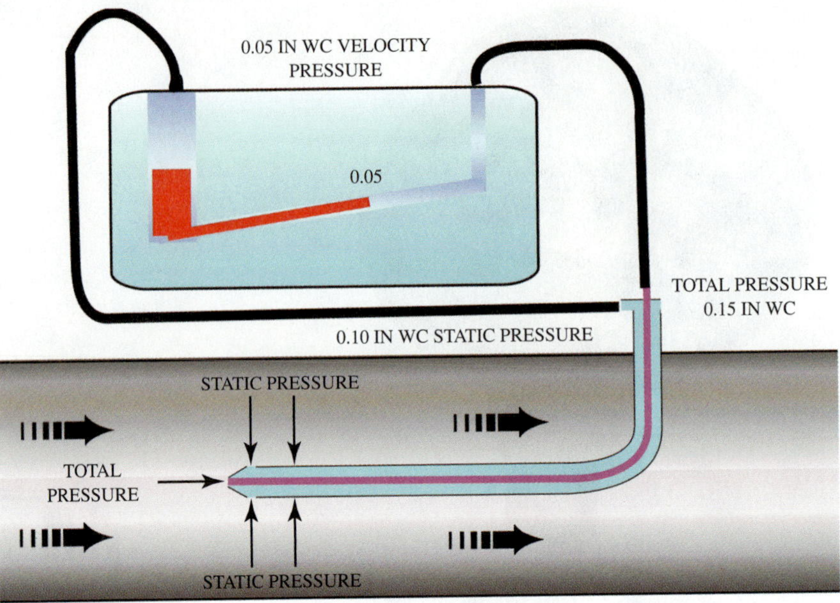

Figure 73-14 Standard manometer connection to a pitot tube.

When pointed against the airflow, the inner tube measures total air pressure and the outer tube measures static pressure. The difference between the two pressures is the velocity pressure (Figure 73-14). For air at temperatures close to normal room temperature, the air velocity can be calculated using the formula: 4005 × square root of velocity pressure. For example, an airstream with a velocity pressure of 0.16 in wc would have a velocity of: 4,005 × square root of 0.16 = 4,005 × 0.4 = 1,602 fpm. Charts are also available for converting velocity pressure to fpm (ft/min) (Figure 73-15).

SERVICE TIP

Digital manometers are available that can read in the same pressure ranges as inclined manometers. These are much easier to transport and use than traditional inclined manometers.

73.7 AIRFLOW

Airflow measurements measure the rate of airflow. The two most common airflow measurements are air velocity, in feet per minute (fpm), and air volume, in cubic feet per minute (CFM). Feet per minute measures how fast the air is traveling. CFM measures the volume of airflow. Having the correct air volume, CFM, is crucial to any air-conditioning application.

Air Velocity

Air velocity is not the same through all parts of the duct. It tends to move faster toward the middle and slower toward the outside of the duct. A procedure called traversing the duct is used to get an accurate velocity reading. Readings are taken at several positions in the duct to get the average air speed through the duct. The combination of the inclined manometer and pitot tube is used in measuring air volume in a duct by making a traverse (Figure 73-16).

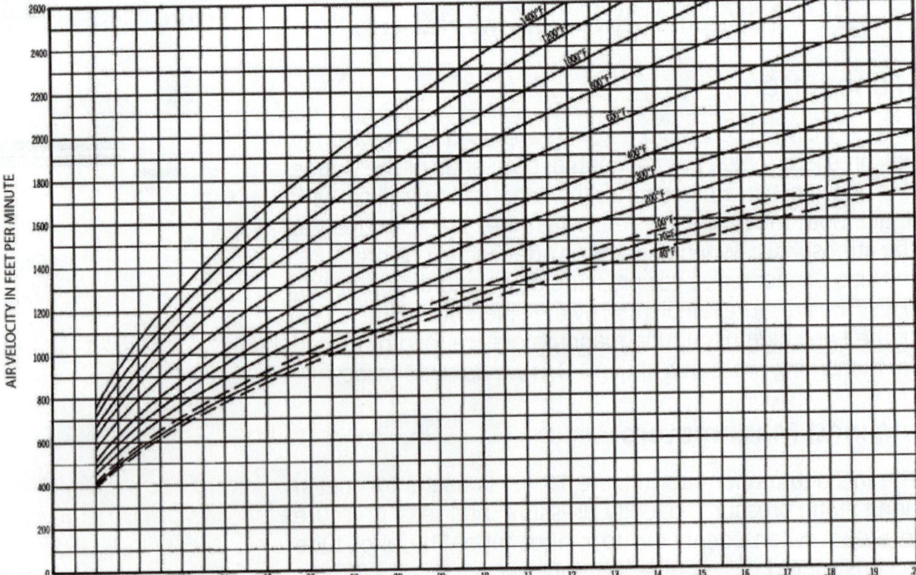

Figure 73-15 Chart for determining air velocity pressure. *(Courtesy of Dwyer Instruments, Inc.)*

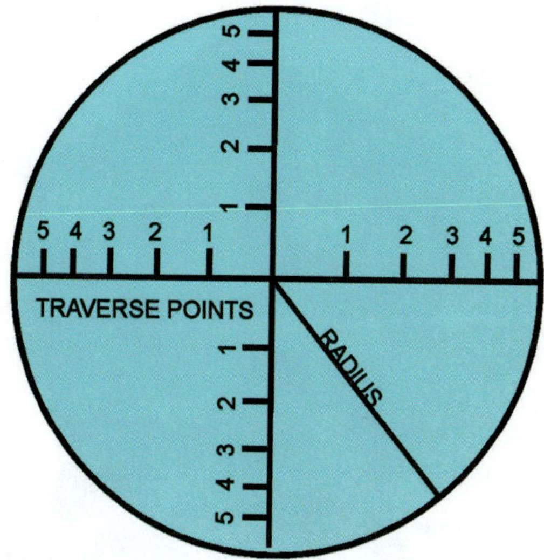

Figure 73-16 Round duct traverse.

Air Volume

The air volume can be calculated in cubic feet per minute using the velocity in feet per minute and the cross-sectional area of the duct in square feet (ft^2). The formula is:

$$\text{CFM} = \text{area} \times \text{velocity}$$

For example, a 12 in × 24 in duct has a cross-sectional area of 2 ft^2:

$$12 \times 24 = 288\ \text{in}^2$$
$$288\ \text{in}^2/144\ \text{in}^2/\text{ft}^2 = 2\ \text{ft}^2$$

If the velocity through this duct is 350 fpm, the air volume is 2 ft^2 × 350 fpm = 700 CFM.

Air Velocity/Volume Measuring Instruments

Inclined manometers and pitot tubes read airflow by reading the pressure in a duct. This will not work once the air has left the duct and is entering the room. Instruments are available that read fpm, CFM, or both directly without having to read a pressure in a duct. Some direct-reading airflow instruments used are the rotating vane anemometer, the thermal anemometer, the velometer, and the flow hood.

The rotating-vane anemometer consists of a lightweight air-propelled wheel, geared to dials that record the linear feet of airflow passing through the instrument. Anemometers read feet of air. The instrument is placed in the airstream and readings are then taken for a measured amount of time. The anemometer reading is divided by the time to arrive at a velocity reading in fpm. Anemometers can be either analog or digital. The analog instrument is a direct reading with the choice of a number of different velocity scales. The digital instrument will automatically average the readings and display the velocity readout in fpm (Figure 73-17).

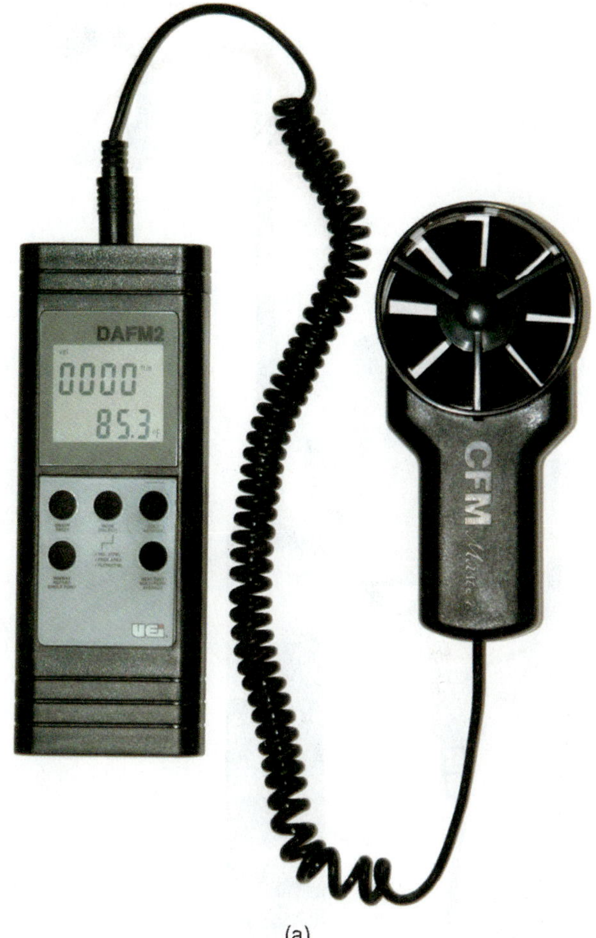

(a)

(b)

Figure 73-17 Two examples of a rotating vane anemometer.

The thermal anemometer operates on the principle that the resistance of a heated wire will change with its temperature (Figure 73-18). The probe is placed in the airstream and the velocity is indicated on the scale of the instrument. Thermal anemometers can be used for measuring very low velocities of air, such as that found in a room. They can also be used in a duct to determine air velocity.

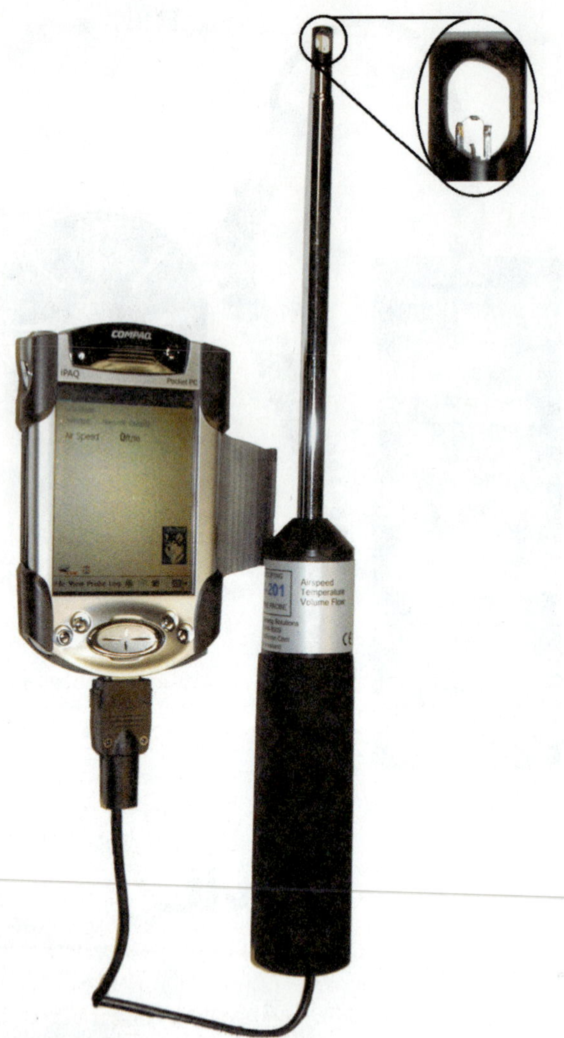

Figure 73-18 Thermal anemometer.

Figure 73-19 Flow-measuring hood.

A velometer reads the instantaneous speed of the air, not the averaged speed. A velometer is similar to the speedometer on a car, which tells how fast the car is going at that instant. Velometers are also available in both analog and digital form. Many digital velometers can function as either a velometer or an anemometer.

The flow hood is useful in accurately determining the airflow volume in CFM from an outlet (Figure 73-19). It completely collects the air supply and directs it through a 1-ft^2 opening where a calibrated manometer provides the CFM readout. Hoods should be selected to as nearly as possible fit the diffuser being measured and have a tight seal around the opening. Flow hoods are considered the most accurate measurement of airflow through registers and grilles.

73.8 ELECTRICAL MEASUREMENTS

Electrical measurements are taken to determine voltages, current, and wattage used by electrical loads such as electric motors or electric furnaces. The wattage can be used to find variables such as BTU or rpm (revolution/minute), which are used in duct system performance equations. The instrument used may display units of measure such as volts, amps, or watts, and these units are converted by technicians as needed. Common electrical meters include the digital multimeter and the clamp-on ammeter.

The digital multimeter is useful for making electrical measurements in the field. Most meters have several scales, including volts, ohms, and amperes. Multimeters typically read small currents inline, such as milliamps.

The clamp-on ammeter is used to read large currents. The clamp-on transformer jaws permit reading current flow without disconnecting the circuit. Care must be exercised when reading ampere flow to only clamp onto one wire at a time. Enclosing two wires may result in a zero reading. The amp draw of a blower motor can be used as a general indicator to show the amount of air that a blower is moving. The amp draw increases as the fan moves more air and decreases as the blower moves less air.

A wattmeter is needed to take accurate wattage readings in most AC circuits. Apparent power can be determined by multiplying the voltage and amperage, but the apparent power in an AC circuit is usually higher than the actual wattage because of the effects of inductive and capacitive reactance. Purely resistive AC circuits are an exception because they are not subject to inductive or capacitive reactance.

73.9 ROTATION MEASUREMENT

Rotation measurement of fans by TAB technicians may be required to determine fan rpm. The airflow rate produced by a fan is directly related to fan speed. Measuring

(a)

(b)

Figure 73-20 Tachometer: (a) front view; (b) side view.

the fan speed before and after any adjustments ensures the proper adjustment is made. Instruments used to measure rotation are the mechanical tachometer, digital photo-tachometer, the chronometric tachometer, and the stroboscope.

A mechanical tachometer requires contact to the center of the rotating shaft that is being measured (Figure 73-20). A rubber tip transfers the rotational motion to the tachometer, which reads in revolutions per minute.

The chronometric tachometer combines a revolution counter and a stopwatch in one instrument. The spindle is placed in contact with the rotating shaft. This sets the meter hand to zero and starts the stopwatch. After a fixed amount of time, usually 6 seconds, the counting mechanism is automatically uncoupled, and the instrument can be removed from the shaft and read.

A digital photo-tachometer uses reflective tape on the outside on the rotating device. It calculates the speed by measuring the frequency of the reflective tape passing by the stationary position of the photo tachometer. The instrument is battery operated and uses a digital display to show the speed in rpm. Digital photo-tachometers can take measurements that contact tachometers cannot take because they do not require physical contact. Some models

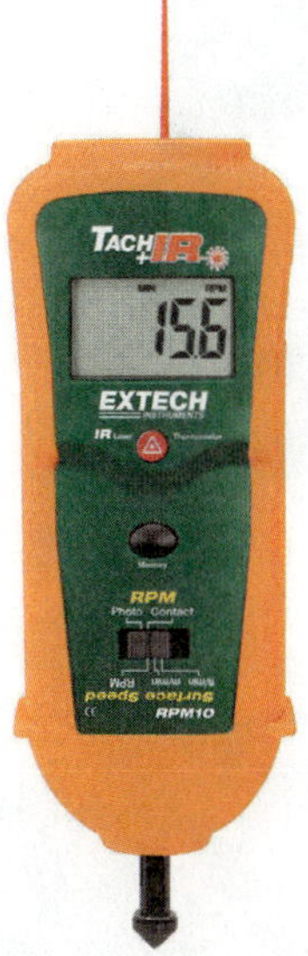

Figure 73-21 This tachometer can be used for either contact or non-contact RPM measurement. *(Courtesy of Extech Instruments, A FLIR Company)*

Figure 73-22 Digital photo-tachometer. *(Courtesy of Extech Instruments, A FLIR Company)*

can save the readings in memory. Figure 73-21 shows a tachometer that combines contact measurement, photo tach measurement, and infrared temperature measurement. Figure 73-22 shows a digital photo-tachometer.

The stroboscope is an electronic tachometer that uses an electronically flashing light. The frequency of the light flashes can be adjusted to equal the frequency of the rotating object. When the two are synchronized, the rotating object appears to be standing still. The rpm value can then be read on the stroboscope's dial Figure 73-23.

73.10 USING TEMPERATURE RISE TO MEASURE SYSTEM AIRFLOW

One of the simplest ways to measure airflow is the temperature-rise method. One advantage to this method is that the equipment used in the measurement is standard

Figure 73-23 The gear on the back of the pump motor appears "frozen" by the stroboscope, which shows the flashes per minute.

on all service trucks. Using the temperature-rise method on a heating system is straightforward because it is entirely a sensible heat process. All the heat that goes into the air changes the air temperature. Dry-bulb temperature measurements are taken both of entering and leaving air, and the difference of the two numbers gives the technician the TD, or temperature difference.

$$\text{Entering air temperature} - \text{leaving air temperature} = \text{temperature difference}$$

Temperature-Rise Measurements

Find the furnace output ratings from the equipment information sheets. These should be listed in BTU/hr. The CFM can be calculated using the heat output in BTU/hr, the temperature difference, and 1.08, which is the constant for standard air. The factor 1.08 is the mathematical combination of all the other factors in the formula: 1.08 = 60 min/hr × 0.075 lb/ft^3 × 0.24 BTU/lb. To calculate CFM:

$$\text{CFM} = \frac{\text{furnace output in BTU/hr}}{(\text{TD} \times 1.08)}$$

Air-Conditioning System Temperature Drop

The temperature difference across an air-conditioning coil cannot be used to calculate airflow because all the system capacity does not go into changing the air temperature; some capacity is used to condense water on the evaporator coil. The system capacity and the amount of latent cooling versus sensible cooling varies depending upon the system operating conditions, making temperature-drop calculations inaccurate.

Wet-bulb temperatures can be used to find the enthalpy of the air entering and leaving the evaporator. These measurements can be used to calculate the total system capacity. However, since the system capacity is not fixed, but varies depending upon its operating conditions, the enthalpy change through the evaporator coil is not an accurate way to determine system airflow.

TECH TIP

TAB technicians are keenly aware of the impact that improper unit sizing has on occupants and society as a whole because their training is focused on efficiency and indoor air quality. Many times "rule of thumb" sizing is used by untrained workers to speed a project along. Problems with these systems often stick out like a *sore* thumb once TAB technicians take their measurements and apply formulas on those units that have been sized using a rule of thumb. Problems resulting from improper sizing can affect the reputation of the worker and the employer.

73.11 USING TOTAL EXTERNAL STATIC PRESSURE TO MEASURE SYSTEM AIRFLOW

The amount of air moved by any blower or air-conditioning component containing a blower is dependent on the pressure that the blower is working against. The total difference in static pressure between the supply side of the blower and the return side of the blower is called the total external static pressure. In general, the higher the external static pressure on the blower, the lower its CFM output.

The airflow a blower is moving in a system can be accurately determined by measuring the pressure drop across the blower and comparing that pressure to the manufacturer's specifications. To determine the amount of air a blower is moving, do the following:

1. Measure the total external static pressure difference across the blower.
2. Compare the external static pressure difference across the blower to the manufacturer's specifications to determine the system airflow.

Figure 73-24 shows how to measure the pressure drop across a blower. For example, using Table 73-1 to check the blower performance of a blower operating on high speed against a total external static pressure of 0.3 in wc, we see that it can move 1,025 CFM at this pressure. The same blower operating against 0.5 in wc external static pressure can move only 915 CFM.

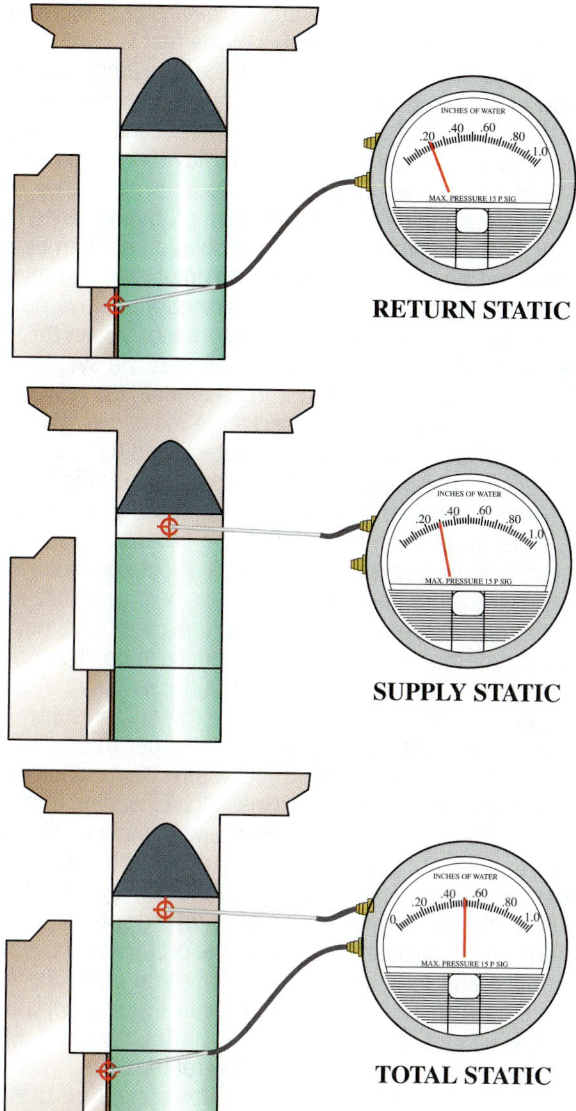

Figure 73-24 The static pressure difference across a blower can be used to determine the airflow through the blower.

Table 73-1 Blower Performance Data

CFM	High	Medium	Low
0.1 in wc	1,135 CFM	860 CFM	465 CFM
0.2 in wc	1,085 CFM	825 CFM	500 CFM
0.3 in wc	1,025 CFM	780 CFM	545 CFM
0.4 in wc	965 CFM	750 CFM	570 CFM
0.5 in wc	915 CFM	680 CFM	600 CFM

73.12 USING STATIC PRESSURE DROP TO MEASURE AIRFLOW

Any component in the airstream causes a pressure drop across it. This pressure drop can be used to determine the CFM traveling across the component if you have the manufacturer's specifications for that component. Measuring the pressure drop across a component like a coil or filter works backward from the blower. In general, higher pressure drops across a device indicate that more air is moving through it. For example, measuring the static pressure drop across a coil and comparing it to the manufacturer's literature can show the amount of air moving across it. Table 73-2 shows the airflow characteristics for an evaporator coil. A measured dry-coil pressure drop of 0.15 in wc would mean that 800 CFM are moving across the coil, while a measured pressure drop of 0.25 in wc would indicate that 1,100 CFM are moving through the coil.

Table 73-2 Coil Airflow Data

CFM	Dry Coil Δ p	Wet Coil Δ p
600	0.10 in wc	0.11 in wc
700	0.12 in wc	0.12 in wc
800	0.15 in wc	0.15 in wc
900	0.17 in wc	0.18 in wc
1000	0.20 in wc	0.22 in wc
1100	0.25 in wc	0.26 in wc
1200	0.29 in wc	0.30 in wc
1300	0.33 in wc	0.35 in wc
1400	0.38 in wc	0.40 in wc
1500	0.43 in wc	0.46 in wc
1600	0.48 in wc	0.51 in wc

SERVICE TIP

The pressure drop stated in manufacturers' literature assumes clean components. A centrifugal fan with the cups full of dirt will not move as much air, causing a lower than normal pressure drop for that blower. The blower data will indicate that the fan is moving more air than it really is. Other components such as coils and filters will have an increased pressure drop when they are dirty.

73.13 USING VELOCITY PRESSURE TO MEASURE AIRFLOW

Sometimes it is necessary to measure the airflow moving through a section of duct. This is done using a manometer or magnehelic and a pitot tube to measure the velocity pressure. The pitot tube picks up both the total pressure and the static pressure. Velocity pressure is the difference between the two. Place the total-pressure tube on the high-pressure input of the manometer and the static pressure tube on the low-pressure input of the manometer to read the velocity pressure. Since air velocity is not steady, but is different in different parts of the duct, an average, or traverse, will need to be taken. This can be easily done using a digital manometer that offers real-time averaging. For normal room temperatures, the air velocity can be determined using the formula:

$$\text{Velocity in feet per minute} = 4{,}005 \times \sqrt{\text{velocity pressure}}.$$

Once you have the velocity, multiply the velocity times the cross-sectional area of the duct. For example, a 12 x 24 in duct with an air velocity pressure of 0.04 in wc:

$$4{,}005 \times \sqrt{0.04} = 4{,}005 \times 0.2 = 801 \text{ FPM velocity}$$
$$12 \times 24 \text{ in} = 288 \text{ in}^2 = 2 \text{ ft}^2 \text{ area}$$
$$801 \text{ FPM} \times 2 \text{ ft}^2 = 1{,}602 \text{ CFM}.$$

73.14 MEASURING AIRFLOW AT REGISTERS AND GRILLES

The airflow from a register or grille can be measured using an anemometer. The velocity needs to be read at several points and averaged. A digital rotating-vane anemometer can keep a real-time average, making this much easier. The air velocity is multiplied by the grille-free area to arrive at the CFM. The free area is the area of the grille that is not taken up by the louvers. The free area can be as little as 50 percent of the overall area. Drop-ceiling perforated grilles typically have a free area of on 50 percent. The difficulty in using this method is in knowing what the free area is. The only way to know accurately is to have the manufacturer's specifications.

The easiest and most accurate method for measuring the airflow at registers and grilles is with a flow hood. The flow hood is placed over the register or grille and all the air moving through the register or grille must pass through the flow hood. The flow hood can easily calculate the CFM since its free area is known, The CFM is read directly on the flow hood.

SERVICE TIP

A cone or pyramid with a known area can be fabricated and placed over the register. This eliminates the need to determine the free area. The airflow through the opening of the measurement cone is generally more even than at the face of the register, making it easier to take an average velocity reading.

73.15 DUCT LEAK TESTING

The *T* in TAB is for testing, and this starts with a general inspection. If the testing is part of the commissioning of a new or modified system, then comparisons should be made between the designer's plans and the physical layout of the project. This comparison will point to possible problems before measurements are actually taken and could save the technician considerable time. If the testing is taking place on an existing system, then a review of designer plans may not be an option. In this case, gathering manufacturer's data is the first step prior to inspection. System tonnage, blower CFM, and heat exchanger data should be noted by the technician on the paperwork for easy reference. This information will be included later in the report submitted to the client along with the findings of the test.

The system should be started prior to testing to confirm airflow at the return and that the system is supplying airflow at the supply terminals. At this point the technician is simply concerned about the presence of airflow. In the balancing phase of TAB we will measure for quantity of airflow. Low or no flow can indicate duct or system problems. Once flow is confirmed, an amp draw reading should be taken before shutting the system down. If the amp draw is not in range of what the manufacturer requires, restrictions above design specifications or duct leaks may be present. Again this will give the TAB technician an indication that the system has apparent problems before starting the duct leak test.

Once these preliminary checks are made, the duct leak test can be done. Duct leakage is determined by blocking all duct outlets, pressurizing the duct, and measuring the amount of air that is required to maintain that pressure. As air leaks out, more air is let in to maintain the pressure in the duct system. In theory, no additional airflow should be required to maintain the pressure once the duct has been pressurized.

There are several ways to complete a leak test, but the standard method is the use of an orifice tube. The orifice is certified and designed for a specified flow curve. A pressure source is installed to pressurize the duct system. This is normally an axial fan with an inlet damper to control airflow and limit pressure to the duct system. Two manometers are used during these checks. One will be installed to measure pressure drop across the orifice, and the other will measure static pressure in the duct system at the farthest distance from the orifice. Pressure supplied to the duct system should not exceed maximum design ratings that are specific to each system. The technician should read the pressure drop across the orifice and then compute the flow rate using the flow curve data. The actual leakage rate can then be compared with the allowable leakage rate of the duct system. Most systems should have a loss rate no greater than 2 percent. A review of standards and local codes should be used to determine each project's maximum loss rate.

Once the inspection and leak tests are completed, a report should be generated for the client. If the duct system is out of design tolerance, the system should be corrected to meet design conditions before attempting to balance it. Most duct systems are installed in a unconditioned space, so ducts that leak will increase energy costs and prevent the system from controlling humidity, temperature, and indoor air quality to the efficiency level that the system was designed to meet.

73.16 SYSTEM BALANCING

Balancing in its broadest meaning includes testing and adjusting a system to produce its design specifications. It is the final step in the HVAC project prior to occupancy and use by the owner. It is an important phase of the work, since many components from many sources have been assembled to perform a valuable service. Usually equipment adjustments and control settings are necessary to prepare for the desired operation.

General concepts and measurements should be understood before focusing on particular systems. It is easier to balance a heating system during cold weather and a cooling system during warm weather, but this is not always possible. It is recommended that the first balance be made when the job is completed or before occupancy and the final balance made during a period as near design temperature conditions as possible.

All duct systems should have balancing dampers installed at the trunk of the duct system. All branches should have the airflow adjusted at these dampers. Supply registers should be used as balance adjustment points only if a branch feeds more than one supply register. If supply registers are used as balancing dampers, they will produce excessive noise in the space. If a duct system has no balancing dampers installed at the trunk, the owner should be notified in writing that the system cannot be properly balanced without their installation.

The blower must be sized and adjusted to supply the needed output for the whole duct system. The blower is sized by the design engineer, but the blower rpm may need to be adjusted by the technician to meet actual demand. The TAB technician is expected to measure the CFM produced by blower and complete the adjustments necessary for the actual performance to match the design performance.

73.17 GENERAL AIR-BALANCING PROCEDURE

First, collect complete information about the installation. This should include unit specifications, control systems details, job layout, and airflow quantities for each register and grille. Examine the entire installation. Look for problems that need to be corrected, such as leaky ductwork, missing parts, and incomplete wiring.

After gathering system data, the overall system airflow should be measured and compared to design specification. There is no point in adjusting the airflow to individual areas until the overall system airflow is correct. Open all dampers, operate the system, and measure the overall system airflow in CFM. Compare the actual system airflow to the design airflow. If the actual delivered airflow is more than 5 percent out of specification, adjust the blower speed to achieve design airflow.

The next step is to compare the airflow delivered in the space to the airflow at the blower. Use a flow hood, velometer, or anemometer to measure the airflow at each supply register and return grille. The sum of all the CFM measurements at the individual return grilles should be within 10 percent of the system CFM at the blower. Leaks in the return-air ducts will cause the air entering the return grilles to be less than the total amount of air the blower is moving. If the difference is greater than 10 percent, the return ducts should be checked for leaks. Similarly, the sum of all the CFM measurements at the individual supply registers should be within 10 percent of the system CFM at the blower. Leaks in the supply-air ducts will cause the air leaving the supply-air registers to be less than the total amount of air the blower is moving. If the difference is greater than 10 percent, the supply-air ducts should be checked for leaks.

Once the total delivered airflow is within specification, the supply register measurements should be compared to design specifications for each area. Dampers can now be adjusted to achieve the correct airflow at each register. Since the adjustments are being made starting with all dampers open, start by partially closing the dampers to registers that are delivering too much air. Begin with the register that exceeds design specifications the most.

Physical adjustments to dampers change the overall system resistance, affecting the CFM throughout the duct system.

Ideally, measurements should be taken after each adjustment to see the effect of the adjustment on the entire system. However, this can be very time-consuming. An alternative is to measure only the output of the zone you are adjusting until it is correct and then retake the overall system CFM to ensure that the overall system airflow has not dropped as a result of the adjustment. Recheck all measurements periodically to see the overall effect of your adjustments. Using reliable equipment and taking accurate measurements can save large amounts of time.

Last, reset the thermostat to the proper room temperature. Instruct the owner on proper operation of the system. Record the information used in balancing the system, including the weather conditions on the day the job was balanced.

73.18 BALANCING A ZONED AIR SYSTEM INSTALLATION

The same procedure is used for balancing a zoned air system, except that zone dampers are adjusted to produce the correct airflow to each zone. The overall system airflow is established, the zone dampers are adjusted, and finally the dampers to individual runs are adjusted.

UNIT 73—SUMMARY

Testing and balance (TAB) procedures are the final check to confirm that the operation of a heating and air-conditioning system is within design specifications. A system that has not been tested and balanced can allow unhealthy conditions to develop in the conditioned space and could affect the long-term health of the occupants.

TAB is considered the first step in improving energy efficiency of an air-conditioning system.

TAB procedures are designed to balance airflow evenly in a structure, and this includes comparing design plans to actual measurements of duct system CFM along with all factors contributing to duct system CFM.

In addition to standard tools used by service technicians, TAB technicians may also use a chronometric tachometer, digital or analog capture hood, air-differential pressure gauges, and recording thermometers to perform their work.

The TAB industry is supported by many organizations and associations that produce standards and certifications for contractors and technicians as well as publications for consumer information.

WORK ORDERS

The service tickets are all on a small, single-zone commercial air-conditioning system with ducted return and supply air having the following specifications:

Design airflow: 4,000 CFM

Blower design pressure drop: 0.7 in wc

Design supply-air static pressure: 0.5 in wc

Design return-air static pressure: –0.2 in wc

Blower full load amp rating: 5.5 A

Service Ticket 7301

The technician measures the following system performance data:

Actual pressure drop across blower: 0.68 in wc

Measured airflow at blower: 4,050 CFM

Actual measured blower amp draw: 5.5 A

Total airflow at supply registers: 3,900 CFM

Total airflow at return grilles: 3,000 CFM

From this data, the technician concludes that the total airflow is within specifications but that the return-air duct has one or more leaks.

Service Ticket 7302

The technician measures the following system performance data:

Actual pressure drop across blower: 1.2 in wc

Actual measured supply-air static: 0.3 in wc

Actual measured return-air static: –0.9 in wc

Measured airflow at blower: 3,100 CFM

Actual measured blower amp draw: 2.5 A

From this data the technician determines that the overall airflow is low due to airflow restrictions that are causing an excessive pressure drop across the blower and reducing the airflow. The technician examines the supply- and return-air static pressure readings and notices that the return-air duct has a lower than normal static pressure reading. Static pressure readings across the HEPA filters show that the majority of the pressure drop is occurring at the filter. The filters appear clean. The technician looks up the application data for the filters. They are designed to have a static pressure drop of 1 in wc at their rated airflow. The technician informs the company that the filters will not work with the system.

Service Ticket 7303

The technician measures the following system performance data:

Actual pressure drop across blower: 0.5 in wc

Actual measured supply-air static: 0.3 in wc

Actual measured return-air static: –0.2 in wc

Measured airflow at blower: 3,000 CFM

Actual measured blower amp draw: 2.4 A

The technician notices that the pressure drop across the blower is lower than the design specification and that both the supply and return static pressures are lower than design. From this data the technician concludes that the blower speed should be increased to increase the system airflow.

UNIT 73—REVIEW QUESTIONS

1. List six types of measurements used to perform TAB procedures.
2. List five instruments used in TAB procedures.
3. Outline the general TAB procedure for system airflow.
4. Explain how to measure temperature rise.
5. Why is the temperature difference across an air-conditioning coil not used to calculate airflow?
6. Explain how to measure airflow using temperature rise.
7. How do the TAB procedures for a zoned system differ from a nonzoned system?
8. How is the CFM at a supply register determined using a velometer?
9. What does amp draw have to do with TAB procedures?
10. Why are the dampers on registers normally not used for balancing the system airflow?
11. Can a proper balance be performed on a duct system that does not have balancing dampers for each branch?
12. What can cause the total airflow leaving the supply registers to be much lower than the amount of air the blower is moving?
13. Why does a technician working for a TAB firm require specialized training?
14. How can TAB improve the health of the occupants?
15. Why is interest in TAB picking up in the residential market?
16. Can energy costs be affected by TAB of a system?
17. How are feet per minute measurements (fpm) different from cubic feet per minute measurements (CFM)?
18. Name two organizations that support the TAB industry by developing standards and offering certification.
19. How is a single pressure gauge with a rotating dial used to measure the pressure difference in a circulating water system?
20. Why would having a system tested and balanced be a good first step toward reducing energy cost?

UNIT 74

Commercial Air-Conditioning Systems

OBJECTIVES

After completing this unit, you will be able to:

1. explain the differences between residential, commercial, and industrial air-conditioning applications.
2. list the different types of commercial air-conditioning applications.
2. identify which type of equipment configuration best suits each commercial application.
4. list the basic chilled-water system compressor and condenser types.
5. describe the differences between single packaged conditioners and rooftop units.
6. describe the general arrangement of a direct expansion coil in an air-handling unit.
7. list the different types of heating and cooling coils.
8. select the proper coil for a given application.

74.1 INTRODUCTION

Air-conditioning systems may vary widely depending on the size of the space to be conditioned, the geographical location, and specific applications required for the space. For example, a hospital operating room demands reduced room temperatures along with highly filtered dry air. The requirements for such a system are far different from a hotel room that requires only minor levels of filtration along with the ability to supply both heating and cooling through simple operation.

Most systems utilize the standard mechanical refrigeration cycle components of compressors, condenser, flow control, and evaporator, but the physical size and location of the components will differ. In addition, the air-distribution systems for smaller units can be very simple by means of only a self-contained fan, while larger systems could have a complex layout of ducting that may include an economizer or dehumidifier function.

A large building in a dry, hot climate such as Arizona may be designed with an economizer that will use some outside air for cooling, particularly when the temperature begins to drop in the early evening. A supermarket in the Northeast may use a heat-reclaiming system to recirculate some of the heat removed by the condenser back into the store during the cold winter months. The air-conditioning system for the control room of a power plant is designed to keep the computer systems dry and cool, and therefore human comfort is only secondary to the needs of the equipment.

74.2 AIR-CONDITIONING APPLICATIONS

The type of air-conditioning system installed will depend on the geographical location, the size of the building, and the occupant use. There are many types of systems and designs, but they can generally be classified as being residential, commercial, or industrial.

Residential Air-Conditioning Systems

This category generally includes single-family, multifamily, and low-rise multifamily private household residences. The air-conditioning type is typically central forced air, central hydronic, or zoned. The capacities of residential installations typically range from less than 1 ton up to about 5 tons.

Commercial Air-Conditioning Systems

The commercial sector is generally defined as nonmanufacturing business establishments, including hotels, motels, restaurants, wholesale businesses, retail stores, and health, social, and educational institutions. The air-conditioning type varies considerably depending on the application. Systems can be as simple as a standard console through-the-wall conditioner for a hotel room. For conditioning multiple rooms or spaces in one building, single packaged conditioners, split-system conditioners, and rooftop conditioners are most commonly used. Many of the smaller systems ranging from fractional-tonnage room coolers up to about 20 to 30 tons are often referred to as "light commercial" systems (Figure 74-1). For substantial cooling loads,

Figure 74-1 Commercial rooftop single packaged unit.

large central forced air systems and chiller systems with ratings of 100 tons and above may be required.

Industrial Air-Conditioning Systems

Both residential and commercial air-conditioning systems are designed primarily for comfort. Industrial air-conditioning systems are designed for processes and environmental conditions that may require greater levels of filtration, more stringent humidity control, and special requirements for treatment of airborne contaminants that may result from a manufacturing process. Some examples of industrial facilities include laboratories, printing plants, textile processing plants, environmental control for animals and plants, wood and paper product facilities, nuclear facilities, and mine air-conditioning and ventilation. Many of the system types are similar in arrangement and equipment requirements as compared to large commercial systems.

74.3 COMMERCIAL AIR-CONDITIONING APPLICATIONS

There are many different types of commercial air-conditioning systems because there are such a wide range of establishments. For example, a system designed for a hotel or motel may be very different from one that would be found in a movie theater. There are many different commercial applications, including restaurants, shopping malls, hospitals, and train stations, just to name a few.

Hotels, Motels, and Dormitories

Very often each room will have its own unitary system, such as a through-the-wall packaged terminal air conditioner (PTAC) for heating and cooling. This type of unit is generally simple to operate and can be turned down or shut off since the room is not occupied at all times. The typical system consists of an indoor and outdoor coil, compressor, and refrigerant flow control. Electric heat is used either as a primary source or as an auxiliary source in the case of a heat pump application.

For larger areas where a single PTAC cannot meet the load requirements, single packaged vertical and horizontal units are typically installed. A horizontal unit can come as one complete assembly, however the air cooled condenser section must be placed outside. If that is a problem, then a mini-split system may be preferable. In this case, the compressor and condenser are located outdoors, thereby saving some inside space and reducing noise inside the building. These are also popular in residential applications and are described in Unit 44.

Restaurants, Nightclubs, and Casinos

Small cafeterias and bars with loads up to 10 tons will typically use direct expansion, self-contained packaged units located within the space itself. Large establishments will use some form of central plant with either forced air or a chilled-water system. These spaces require high ventilation rates to exhaust odors and fumes, which will require increased amounts of outside air. The increased air supply, however, should be admitted to the space without causing drafts for people who are dining. The considerable latent heat load from cooking, food, and people who may be active or dancing requires a system with an increased cooling coil size to remove the excess moisture from the air. Special consideration needs to be given to odors. As an example, air should circulate from the dining area into the kitchen. This will keep odors out of the dining areas and serve to help the kitchen cool. Ideally, larger commercial kitchens should have a HVAC system separate from the dining area.

Supermarkets, Department Stores, and Malls

Supermarket systems are covered in detail in Unit 86. Department stores today are offering a number of conveniences, including supermarket areas, lunch counters, auto service areas, and garden shops. This variety of merchandise will affect the air-conditioning system type and requirements. Special ventilation requirements may be necessary if forklift trucks are entering the space. Major considerations to be taken into account for department store air-conditioning systems are the installation cost, floor space requirements, and simplicity of control. For these reasons, rooftop-mounted units are most commonly used. Usually the controls are operated by personnel who have little knowledge of air-conditioning systems, so they should be simple and fully automatic.

Malls often utilize a central plant to distribute chilled air for the enclosed mall and then to the individual tenant stores. Variable volume control of the air supply and electric reheat is used for the temperature control of each individual space or store. Generally the enclosed mall will be at a slight positive air pressure relative to atmospheric. Individual stores will maintain an air pressure that is best suited for their use. As an example, a store may operate under a negative pressure for better odor control.

Office Buildings

The variety of office building configurations allows for the use of almost every type of available air-conditioning system. A large office building will have some offices located near windows around the outer walls of the building, while others will be located on the interior of the building. The outer office's cooling load may vary considerably during the day as the sun rises and as the seasons change. The interior office's load will remain more uniform. Another consideration is the occupancy. Most office buildings are occupied from 8 AM until 6 PM. Lighting is also a consideration, as it is a significant part of the cooling load.

Small one-story office buildings may have individual unitary packaged systems for individual spaces. Multiple-story buildings may use forced air with a rooftop central unit. This could be a rooftop forced-air system or an air-cooled chiller as shown in Figure 74-2. Very large multistory office buildings often use a large central chilled-water system located in a mechanical equipment room (Figure 74-3). Each floor will have its own fan room where the chilled water supplied from the central system is used to indirectly cool the air. One mechanical equipment room usually services eight to twenty floors. The fewer the floors served by an equipment room, the greater the flexibility in controlling

Figure 74-2 Air-cooled chiller.

the space temperatures. One equipment room per floor will allow for better control, but there will be more units that require maintenance and increased installation costs. There may be some savings in efficiency because equipment can be shut off in unoccupied areas.

Hospitals and Nursing Homes

Of all the commercial areas, hospital air conditioning is the most demanding in terms of the overall requirements necessary. For these systems, the air-handling system can be very complex and sophisticated. Air circulation between spaces is often restricted. This is because many bacterial and viral infections, such as Legionnaires' disease or chicken pox, can be transported within the air. A good source of outside air will help to dilute viral and bacterial contaminants within a hospital. Special requirements for high-efficiency filters must be followed to make sure that the correct filters are installed and then replaced at regular intervals.

Temperature and humidity are of extreme concern for patient healing and recovery. Dry conditions can contribute to secondary infection and should be avoided. However, an increased relative humidity is not necessarily desired, as some bacteria survive more readily in a humid environment. Therefore, general clinic areas of the entire hospital

Figure 74-3 Semihermetic centrifugal water-cooled chiller.

are typically maintained at 30–60 percent relative humidity. However, individual operating rooms, delivery rooms, recovery rooms, intensive care units, and all other special treatment spaces will all have varying requirements that must be maintained.

Most hospitals will have a fairly large hot-water central heating system utilizing steam boilers. Steam is also used for the sterilization of utensils, instruments, and equipment. Large centrifugal or screw-type compressors are often used for the chilled-water cooling plant. Most hospitals have backup emergency power in the event that the local power is disrupted. This backup power may be supplied by gasoline or diesel-powered generators. Some of the newest systems use the combined cycle of gas and steam turbine-driven generators. The gas turbine exhaust passes through a heat exchanger to boil water and produce steam. This steam, which is generated from heat that would normally be lost up the stack, is supplied to the steam turbine-driven generator. Because there is an abundance of heat in this type of system, absorption-type chillers can be used to reduce total operating costs. An example would be a gas turbine-driven refrigeration compressor with the waste heat being used for an air-conditioning absorption system.

Nursing homes have many of the same requirements as hospitals in regard to air quality, temperature, and humidity. However, since normally there are few special treatment spaces, such as operating rooms, the overall demands for the system are far less complex. Individual room temperature control should be allowed for, and the registers should be located to reduce drafts. It is not uncommon for these systems to have large volumes of outside air admitted for odor control. To reduce operating costs, odor may be controlled with activated carbon or potassium permanganate–impregnated activated alumina filters.

Schools, Colleges, and Universities

The type of system normally found will vary depending whether the facility is new or existing and whether it is to be totally or partially renovated at some point. It should be a system that can be easily maintained by the institution's maintenance staff. Educational facilities typically have very limited budgets, so energy savings are often extremely important, and this can affect the type of system that is installed. Schools that need to grow with new additions or wings may not necessarily add new heating and cooling systems but instead may just upgrade and add on to the existing system.

There are a number of methods to reduce energy consumption that can be somewhat straightforward to incorporate. Outdoor economizer cycles use the cooler outdoor air whenever the outside air temperature begins to drop. This helps to offset the high inside load, which has increased during the peak load part of the day. Night setback conserves energy by resetting the heating and cooling space temperatures during the evening hours when no one is in the building. Morning warmup and cool-down will reset the temperature back to normal with enough time allowed for the space to be comfortable when the students arrive at school.

The use of the facility is normally taken into account. As an example, a preschool should be configured to minimize drafts near the floor level. This is because the children are small and will most likely be playing on the floor a majority of the time. Large universities will have multiple buildings that may each utilize its own central cooling system. Many schools in northern climates will not have any air-conditioning system at all. This is because during the summer months when it would be required, school is normally out of session. If summer sessions are held, windows are opened to provide for air circulation.

Airports, Bus Stations, Ship Docks, and Train Stations

These are generally large open areas one or more stories high. Heating and cooling is typically supplied from a central system through an air-distribution system that has multiple zones. Some problems encountered relate to the air balancing of the space due to the many outside openings, high ceilings, and long, low passageways. The requirements for these spaces are different from office buildings in that often transportation centers are occupied on a 24-hour-a-day basis.

Auditoriums, Movie Theaters, Arenas, and Stadiums

These spaces are subject to high peak loads during an event and then very reduced loads at most other times. Many events are held in the evening rather than at the peak load portion of the day, which helps to minimize the cooling demand. The major cooling load results from the people in the space, and relative humidity is a concern. Large systems will be designed to cool the air down below its dew point for the removal of moisture and then utilize some form of reheat. Often large central chilled-water systems are used for these facilities. The air supply would be too noisy if located near the seating area, particularly for a theater where a performance is being held. In most cases, air is supplied from overhead at a high enough velocity to reach the seating area without being too noisy. The return-air registers are located near the seats to help circulate the air through the space. Many auditoriums and arenas are designed to operate on all outside air, because during setup time trucks may be used to bring in equipment and supplies.

74.4 INDUSTRIAL AIR-CONDITIONING APPLICATIONS

Residential and commercial air-conditioning systems are generally designed for human comfort. Industrial systems are different, as more often the main purpose is to ensure product quality at a manufacturing facility. Air quality, temperature, and humidity can be critical in many manufacturing environments.

Manufacturing Plants

There are many different types of manufacturing facilities, and each has its own needs and requirements for air conditioning. Some typical applications include the manufacture or processing of ceramics, plastics, leather, plywood, textiles, rubber dipped goods, floor coverings, and electrical products. Then there are consumer products such as tea, tobacco, and distilled liquors. Many applications require some form of painting process that requires special air-filtration systems.

These facilities often require the use of large amounts of energy for the manufacturing processes involved and also for the air-conditioning systems. Many processes require specific temperature and humidity levels. As an example, in the manufacture of rayon, temperature controls the rate of reaction and humidity maintains the solution at a constant strength. For varnish drying, the temperature is critical, and proper humidity levels reduce the formation of bubbles on the surface.

Moisture regain is important for the processing of hygroscopic materials such as textiles, paper, wood, leather, and tobacco. A hygroscopic material readily absorbs moisture from the air. Moisture regain is the percentage of absorbed moisture in a material as compared to its bone-dry mass. Conditioning and drying are the typical processes utilized for removing or adding hygroscopic and free moisture. These two are combined to accurately regulate the final moisture content in products such as textiles and tobacco. Large dryer drums are used for the papermaking process.

In the manufacturing process, many airborne particles may be released. Some of these include dust, bacteria, smoke, spores, pollen, and radioactive particles. If left unfiltered, these may spoil perishable goods and clog small openings in precision machinery. Static electricity can be dangerous in explosive atmospheres, and generally its effect is minimized at humidity levels above 35 percent.

The most important factor to be considered given all of these requirements is to also provide an environment that is suitable for employees working at the facility. High temperature and humidity levels can lead to worker fatigue. An employee who is sweating can leave fingerprints on finished metal products, and the salt and acid from perspiration can cause corrosion and rust on metal surfaces. Most of the workplace standards are set and enforced by the Occupational Safety and Health Administration (OSHA). Industrial manufacturing plants are usually designed for an internal temperature of 60°F to 90°F and a maximum of 60 percent relative humidity.

With the many different types of requirements and the many diverse industrial manufacturing processes, there is no one common air-conditioning system that is typically used. For large facilities, central systems generally are the most cost effective. Very often there will also be refrigeration systems and heating systems located within the plant that are required for the process. If there is a need for process heating with hot water or steam, then there may be waste heat available or conditions suited for a combined cycle application. Whenever this is the case, an absorption chiller system may prove to be the most cost effective. As an example, steam may be used for a drying process. Once the steam has been used for heating, it may need to be condensed back into water and then sent back to the boiler to be heated again. However, if there is still enough heat available in the steam to be used in an absorption chiller, this may be a better route. The steam condenses in the absorption chiller, giving off its latent heat before traveling back to the boiler.

74.5 SINGLE PACKAGED CONDITIONERS

A single packaged conditioner, often called a Unitaire, is a complete self-contained factory-built unit, for permanent installation, to condition larger spaces than would be practical using single room conditioners. In this category there are two variations of available equipment:

- **Horizontal conditioner** Horizontal packaged heating and cooling units, with integral air cooled condensers, in the size range of 1½ to 5 tons
- **Vertical conditioner** Vertical self-contained air conditioners, with water-cooled or remote air-cooled condensers, in the size range of 3 to 15 tons

74.6 HORIZONTAL PACKAGE CONDITIONER

The horizontal unit usually uses ductwork for air distribution. The horizontal unit is completely self-contained, including the air-cooled condenser. It therefore must be placed either entirely outside or at least with the condenser section outside. A schematic diagram for the arrangement of parts is shown in Figure 74-4.

The unit is primarily a cooling unit, although electric heaters can be installed in the unit as shown in Figure 74-5. The unit can be equipped with supply- and return-air duct connections. Ducts can be entered from the side or bottom, depending on the application. The unit can be installed on the roof, as shown in Figure 74-6. This type of application is used for shopping malls, factories, and other commercial buildings.

Units range in capacity from 18,000 to 60,000 BTU/hr and range in air quantity from 600 to 2,000 CFM at standard rating conditions. All units are available for 208/230 V, single-phase, 60 Hz power. Efficiencies range from 10.0 EER and higher depending on the size. Optional electric heaters range in size from 3.74 to 29.80 kW.

The horizontal unit comes equipped with the following features:

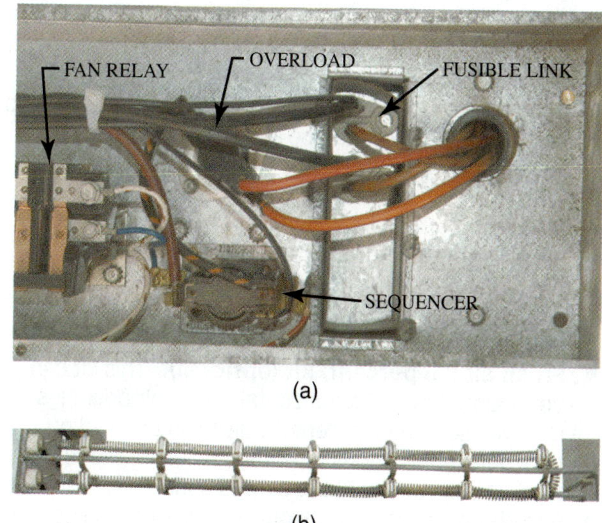

Figure 74-5 (a) Electric heaters for a packaged air conditioner; (b) Nichrome resistance heating wire.

Figure 74-6 Roof installation of a single packaged air conditioner.

- **Water protection** A weather-resistant cabinet along with a water-shedding base pan with elevated downflow openings and a perimeter channel prevent water from draining into the ductwork.

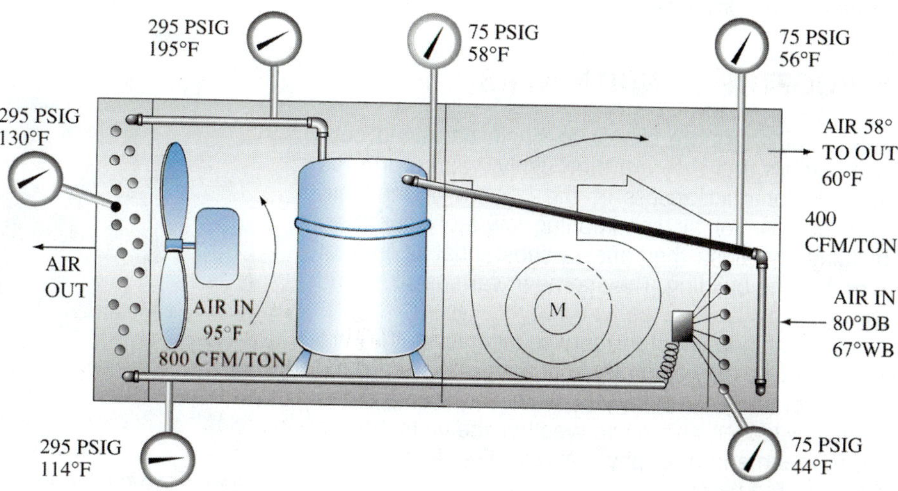

Figure 74-4 Component arrangement for a single packaged conditioner.

- **Low ambient control (optional)** Kits are available that control the condenser head pressure to permit the unit to cool even at low ambient temperatures.
- **Economizer (optional)** An economizer and dry-bulb temperature sensor can be supplied for downflow installations. This makes it possible to use outside air for cooling when outdoor temperatures and humidity permit.
- **Enthalpy control kit (optional)** This can be supplied in place of the dry-bulb sensor, or two enthalpy controls can be paired to provide differential enthalpy control.
- **Fresh air (25 percent) kit (optional)** This kit can be mounted over the horizontal return-air openings for downflow requirements. It also can be used on horizontal applications by cutting a hole in the return-air duct or in the unit filter access panel.
- **Fan-delay relay kit (optional)** This control keeps the indoor blower on for about 90 seconds after the unit stops to improve the EER.
- **Anti-short cycle timer** A time-off device ensures a minimum of 5 minutes off between compressor cycles.

74.7 VERTICAL PACKAGED CONDITIONER

This is a packaged unit for installation, usually inside the space being conditioned. The condenser is water cooled. Water from a remotely located cooling tower needs to be piped in and out of a water-cooled condenser.

Discharge air can either be free throw or ducted horizontally or vertically. An accessory plenum and grille that fits on top of the unit is supplied for the free-throw arrangement. When ductwork is used, a number of discharge configurations can be supplied.

Application features that are incorporated into the unit are as follows:

- The evaporator fan speed is adjustable, affecting the air-delivery CFM and the available ductwork static pressure.
- Thermostats can be supplied as an integral part of the unit, or provision can be made for remote mounting.
- An anti-short-cycle timer is provided to protect the compressor from excess cycling.

74.8 ROOFTOP CONDITIONERS

Rooftop conditioners are similar to single packaged conditioners except that they are thoroughly weatherproofed and provide for duct access at the bottom of the unit. They are popular for air conditioning low-story commercial buildings because they offer a substantial savings of space within the building. They come in various sizes and configurations.

From the standpoint of the service technician, they are desirable because they offer plenty of access space around the unit for servicing. On many jobs, however, access to the roof is only by ladder, and in bad weather the units may be difficult to reach and offer physical restraints in supplying needed tools and parts.

SERVICE TIP

One of the prime concerns and problems with all rooftop installations is water leakage. Most commercial buildings use flat roofs, and water in a rainstorm can build up to several inches in depth. If the equipment panels are not properly reinstalled to provide an adequate water seal, rainwater can easily enter the building. In addition, if the proper curb is not provided for the unit, then the system will leak. Many commercial roofs are "bonded," which means that there is a specific insurance policy taken out on the roof by the roofing company at the time of installation. This provides the building owner with insurance that the roof will not leak. That roofing contractor is the only one who can do any work on the roof that would affect the integrity of that roof without violating that bond. Violating the bond on the roof by putting in a vent pipe could ultimately make you liable for any water damage resulting from leaks. Check with the building owner before any work is done on bonded roofs.

Rooftop self-contained air conditioning units are commonly used on commercial installations (Figure 74-7). The sizes range from 3 tons to 130 tons of cooling capacity under standard rating conditions. Besides the difference in size of the components, individual units differ in the type of heating supplied with the package.

For the units in the 3–25-ton range, gas-fired or electric heat can be supplied. Larger units can also be equipped with hot-water or steam coils. When the units are located outside, adequate freeze protection needs to be provided where the unit contains water. Condensate drain pans must be free draining.

Reciprocating or scroll-type, direct-drive, semihermetic, and hermetic compressors are used. Compressor motors are suction-gas cooled and are protected with temperature- and current-sensitive overloads. Crankcase heaters are standard equipment. Dual-compressor models are usually available starting at 7½ tons and higher. When dual compressors are furnished, dual-refrigeration circuits are also supplied. This arrangement makes possible better performance ratings and increased energy savings at partial loads.

Figure 74-7 Rooftop packaged system.

Cabinets are constructed of zinc-coated, heavy-gauge steel and are weather tight. All services can be performed through access panels on one side. Supply and return ductwork connections can be made at the bottom or side of the unit. Roof curb frames are available for roof mounting.

74.9 CENTRAL SYSTEM CHILLERS

For large air-conditioning loads, central systems utilizing chilled water are often used. There are many different configurations that can range from capacities of more than 700 tons for large absorption systems to as little as 15 tons for small reciprocating chillers.

Chilled-water systems can utilize reciprocating, centrifugal, and screw-type compressors. These can be configured as open or semihermetic systems. Condensers are often water cooled, and if no ready source of water is available, then cooling water towers are used. Chillers can also come as air-cooled units, but the compressor location is then limited. The compressor needs to be located relatively close to the condenser. With a water-cooled unit, the compressor can be located most anywhere with the condenser water pumped to its location.

Chillers can also be configured as absorption systems. A major advantage is that no compressor is required. Another advantage of this type is that waste heat can be utilized as the driving heat source, which provides for economical operation.

74.10 AIR-HANDLING UNITS

A small residential horizontal air handler would have a direct expansion (DX) coil and a blower as shown in Figure 74-8. This would be a fairly simple type of air handler to install and service. Larger air handlers are more complex. In large air-handling systems, the chiller supplies chilled water to the cooling coil located in the air-handler assembly. This cooling circuit consists of a chiller, a pump, the cooling coil, and all necessary piping. Separate from the air handler, the compressor for the chiller system utilizes a water-cooled condenser.

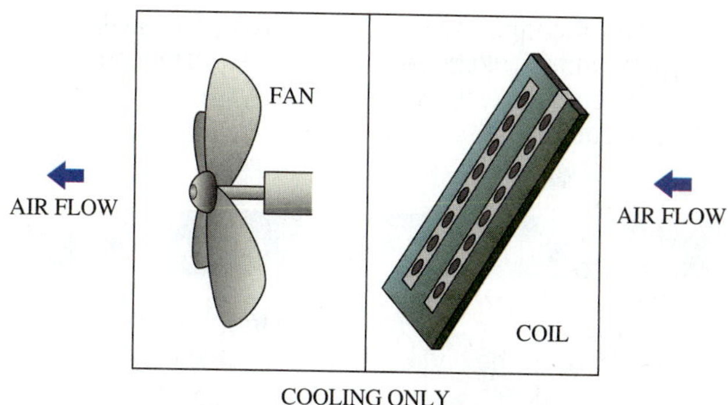

Figure 74-8 Cooling-only horizontal flow air-handler with DX coil.

The condenser-cooling water circuit consists of a cooling tower, a pump, and a water-cooled condenser (part of the chiller). This is actually a subcircuit to the cooling circuit. This type of configuration will also have a heating circuit that generally consists of the boiler, a pump, and the heating coil.

SERVICE TIP

A common maintenance task is the replacement of the filters at the inlet of the air handler chilled/heat coil. A building may have large and small air-handler units (all usually located in relatively inaccessible locations and environments) with different filter demands. It is important to make sure that filters are changed when needed and not overlooked.

74.11 DIRECT EXPANSION (DX) COILS

Many residential and light commercial air handlers will have DX coils (Figure 74-9). These may come from the factory without refrigerant and pressurized with nitrogen. If this is the case, they should be checked for pressure with the installed Schrader valve prior to installation. If no pressure

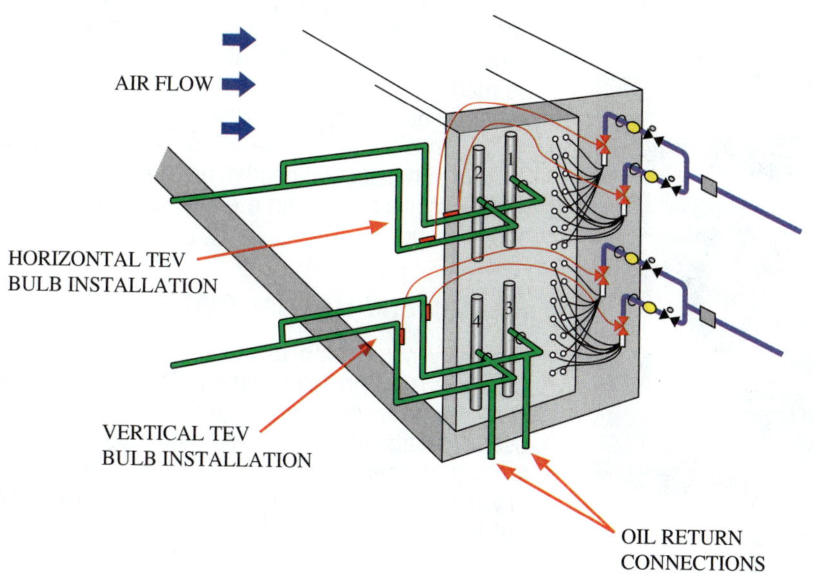

Figure 74-9 DX coils for air-handler applications.

is present, then the coil may have developed a leak during shipment and should be returned. Before installation, all of the nitrogen should be released and the ends of the connections cut off.

SAFETY TIP

Make sure to bleed all the pressure from the DX coil before cutting off the end caps. Also make sure to protect your eyes and face while doing so.

Once the connections have been cut off (Figure 74-10), they should be prepared for flaring or brazing. A liquid-line filter drier should be installed as close to the coil as possible. When brazing a thermostatic expansion valve (TEV), always direct the flame away from the valve body. As an added precaution, a wet cloth should be wrapped around the body and element during the brazing operation (Figure 74-11). After the proper fittings have been soldered, the TEV can be installed. If the TEV is externally equalized, run both the temperature-sensing and pressure-sensing lines to the coil outlet (Figure 74-12).

Figure 74-10 Connection after the end has been cut off.

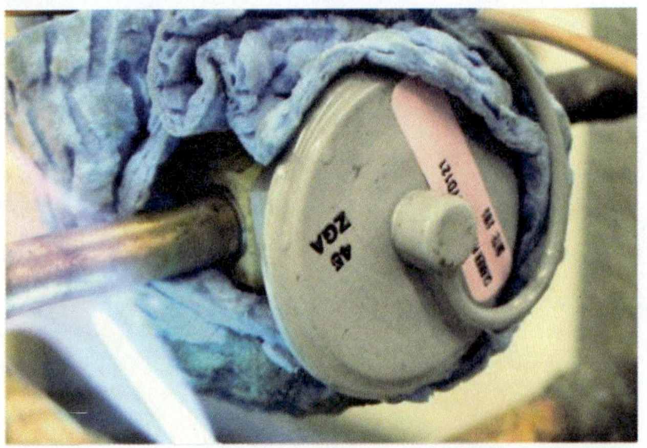

Figure 74-11 Wrap a wet cloth around the valve body when brazing.

Figure 74-12 Properly locate and install both the temperature-sensing and pressure-sensing lines.

Make sure that the TEV temperature-sensing bulb is attached to the coil and properly insulated. The sensing bulb can be installed in a vertical or horizontal position. as shown previously in Figure 74-9. The external equalizing connection is typically used on large evaporator sizes and should be located close to and downstream of the sensing bulb. On horizontal runs, the connection should be made to the top of the tube. On vertical runs, the connection should be made well above the beginning of the riser. Some DX coils have an oil-return connection at the bottom of the suction header, as shown previously in Figure 74-9. This is a small-diameter line that will return oil to the compressor.

Staging and Hot Gas Bypass

Generally some type of staging control system is used. With a change in load demand, the compressor may be staged to start or stop. A liquid-line solenoid valve can be used for capacity control. Another type of staging is unloading. Rather than allowing the compressor to continually cycle with demand, unloaders are often used to cut out individual compressor cylinders or banks of cylinders (Figure 74-13).

Figure 74-13 Compressor unloader solenoid.

Often two compressors are staged together. As an example, at maximum demand, both compressors operate at full load. As the demand drops, the lag compressor begins unloading cylinders. As demand drops further, the lag compressor shuts down and only the lead compressor remains in operation. If the demand drops even further, then the lead compressor begins to unload cylinders.

Larger light commercial systems may require hot gas bypass for low load operation. The hot gas is introduced between the expansion valve and the distributor.

74.12 CHILLED-WATER COOLING COILS

Chilled-water systems are referred to as indirect expansion systems. A primary refrigerant will be used to remove heat from a secondary refrigerant in a chiller unit. The primary refrigerant remains in the chiller unit and operates on the normal refrigeration cycle. The secondary refrigerant (chilled water) will be pumped to the air-handler cooling coils and then return to the chiller. An example of a chiller is shown in Figure 74-14.

The secondary refrigerant does not change state and expand when it removes heat from the air, and so this type of system is known as indirect expansion. The secondary refrigerant is often a glycol and water mixture. The water should be of high quality, preferably distilled water, to avoid the buildup of scale on the inside tubing surfaces due to mineral deposits and sediments contained within the water. The glycol (antifreeze) will help to prevent freezing of the cooling coil if the outside air temperature drops too low. Coil-freeze-up thermostats can be installed to shut down the system if the air temperature drops below 36°F. Thermostats can also measure water coil temperature to shut the unit down if it gets to or near the freezing point. For cold weather winter conditions, it is recommended that the coil be drained of water and flushed with antifreeze.

TECH TIP

In addition to protecting the cooling coil from freezing, glycol solutions generally have corrosion inhibitors added. These will protect the interior of the cooling coils and associated piping from prematurely rusting away. One drawback of glycol is that its specific heat is lower than that of water so it does not transfer heat as well.

Chilled-water cooling coils are generally installed for counterflow. They should be piped so that the coldest water meets the coldest air. After installation, chilled-water coils need to be purged of air through the bleed valve (vent).

74.13 HOT-WATER COILS

Air handlers can be configured with both cooling and heating coils (Figure 74-15). The heating coil is located in the airstream before the cooling coil. This arrangement will help to reduce the possibility of the cooling coil freezing during cold weather conditions. Steam heating coils are more efficient than hot-water heating coils, but boilers that generate steam fall under stricter regulation and require more safety controls. Therefore, many heating systems for residential and light commercial applications use circulated hot water rather than steam.

The hot water for the heating coil is normally supplied from a boiler and will circulate from the boiler to the air handler and back again. This type of heating coil will have a bleed valve (Figure 74-16). Hot water coils should be piped for counterflow so that the warmest water meets the warmest air. The hot-water loop will include an expansion tank and a fill connection. Similar to a cooling coil, high-quality distilled water should be used for the circulated hot water to prevent scale formation on the inside of the coil surfaces.

74.14 STEAM HEATING COILS

Steam heating coils will be smaller in physical size than a comparable hot-water heating coil. This is because the coil not only transfers the sensible heat of the steam to the air but also the latent heat as the steam changes back to water within the heating coil. A steam trap is located at the outlet of the heating coil. To gain the most efficiency, it is

Figure 74-14 Large commercial chiller unit.

important that the steam turn back to water, thereby giving up as much heat as possible before leaving the heating coil. The water formed from the steam condensing inside the heating coil is commonly referred to as condensate. The condensate will return to the boiler and be heated again to turn back into steam.

The water quality for this type of system is even more important than a chilled-water or hot-water loop. Since the water is boiling in the furnace, there is an even greater tendency for minerals and deposits to come out of the water

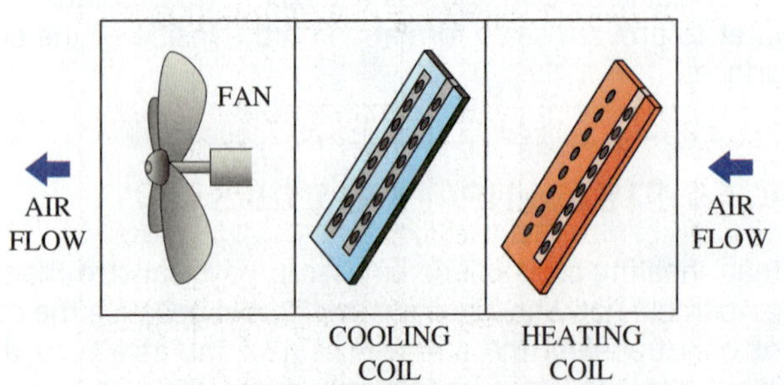

Figure 74-15 Horizontal flow configuration with heating and cooling coils.

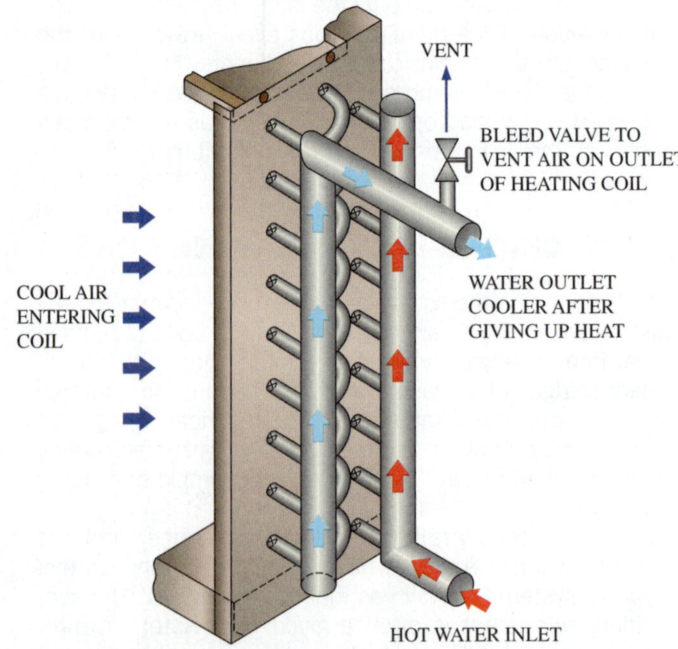

Figure 74-16 Hot-water coil with bleed vent.

and collect on the boiler heat-transfer surfaces. It is essential that only high-quality distilled water be used for this type of system.

SAFETY TIP

When working on a steam coil or steam trap, always make sure that the steam supply is shut off. Wear heavy gloves and a long-sleeve shirt whenever disconnecting steam line components. Always bleed off any pressure in the line before opening or disconnecting components.

An advantage to a steam system is that circulating pumps are not required. In a hot-water system, pumps must run to circulate the hot water to the heating coil and then back to the furnace. For steam systems, supply pumps are not required. The steam under pressure will travel through the piping and into the heating coil. The condensate leaves the heating coil through a steam trap. This is generally piped as a gravity or vacuum return to a condensate return tank (Figure 74-17). The condensate from the drain tank will be drawn off as feed water for the boiler by a feed water pump.

TECH TIP

Steam traps will generally have a bypass valve around them. If a steam trap happens to freeze in a stuck-closed position, the steam flow through the coil will stop. The bypass valve can be opened to bring the coil temperature back up to normal and the trap can be isolated and replaced.

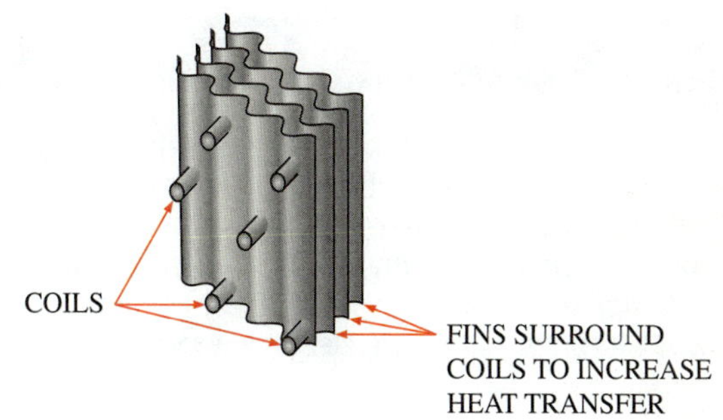

Figure 74-18 Standard finned-coil construction.

74.15 COIL CLEANING AND CORROSION

Chilled water, hot water, steam, and DX coil exterior surfaces must be cleaned regularly. The exterior surfaces of the coils are often finned to help increase the heat-transfer surface area (Figure 74-18). The finned coils provide an opportunity for dirt and grime to collect. Dirty coils reduce heat transfer and lead to increased operating costs due to inefficiency. On direct expansion systems, a dirty coil can lead to liquid floodback and compressor slugging.

Water that forms and collects on the coil surface will drip off and fall into the drain pan located beneath the coil. Dirt and grease collecting on the fins and coil reduce the wettability of the coil surface. Rather than collecting on the coil and dripping into the drain pan, the water will be moved through the system with the air to potentially create moisture problems in other areas of the air-distribution system. Surface grime and wet conditions can also lead to microbial growth (mold), which leads to foul odors and health-related indoor air-quality problems. Coils can be pressure cleaned with hot water and detergent or a commercial coil cleaner.

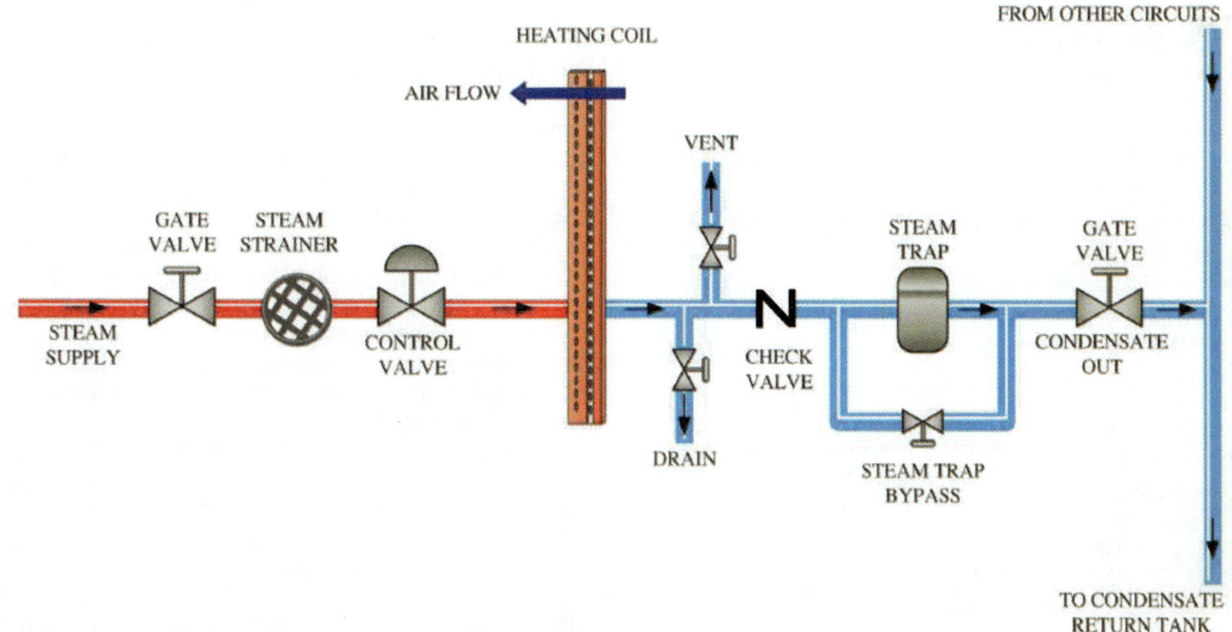

Figure 74-17 Steam coil supply and return.

TECH TIP

Some chemical cleaning solutions are self-rinsing, and over time the solution will be rinsed by the condensate and the coil will not be harmed. However, if a standard chemical cleaning solution is used when cleaning the coil exterior surfaces, always make sure to rinse the coil surface thoroughly after the cleaning is complete

Coil Corrosion

Corrosion can be caused by chemical, physical, or electrochemical reaction due to exposure from elements in the environment. Coils located in coastal and marine areas are subject to corrosion from sodium chloride (salt) that is carried by sea spray, mist, or fog. Units located near industrial areas may be exposed to corrosive emissions such as sulfur oxides (SO_2 and SO_3) and nitrogen oxides (NO_2), which lead to the formation of acid rain. In addition, dust particles containing harmful metal oxides, chlorides, sulfates, sulfuric acid, carbon, and carbon compounds may also be prevalent. Coils located in urban areas are susceptible to high sulfur oxide and nitrogen oxide levels generated from automobile and heating emissions.

The two common types of coil corrosion are galvanic and general corrosion. Galvanic corrosion occurs when two dissimilar metals are in contact with one another and an electrolyte is present. A coil with copper tubes and aluminum fins will be subject to galvanic corrosion in the presence of sodium or calcium chloride compounds. These are most commonly found in sea spray near coastal areas. Aluminum is a less noble metal than copper and so the fins will begin to deteriorate first.

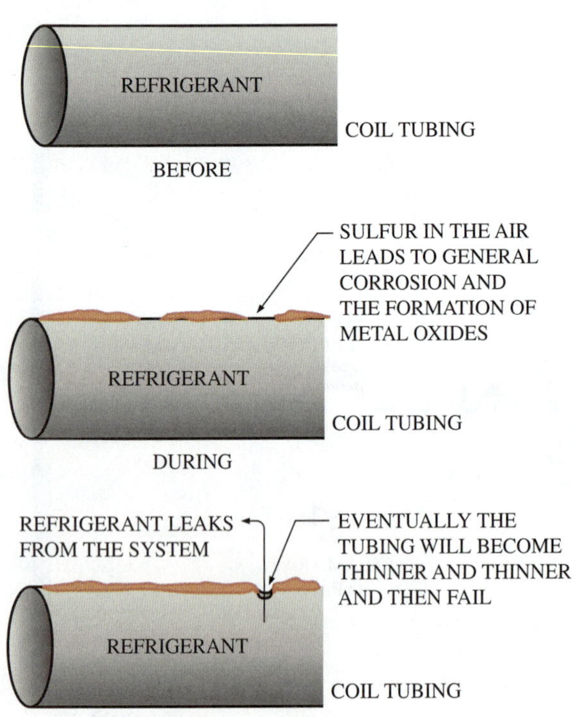

Figure 74-19 Coil tubing failure due to general corrosion.

General corrosion is caused by oxidation and chemical attack on the coil tubing surfaces. As an example, copper is susceptible to attack from gases containing sulfur. This leads to general corrosion that consumes some of the copper to form metal oxides on the tube surface. Eventually the copper tubing becomes thinner and thinner and turns brittle, resulting in tube failure (Figure 74-19).

There are a number of processes that can be applied to coils to reduce the potential for corrosion. Aluminum fin coils can be precoated with a durable epoxy coating for mildly corrosive environments. Aluminum fins may also be electrocoated through a multistep process that ensures the ultraclean coils are properly coated and cured. To reduce the chance of galvanic corrosion resulting from the use of dissimilar metals, copper fins rather than aluminum are mechanically bonded to copper tubes. A disadvantage, however, is that this type of coil can be rather expensive in comparison to standard aluminum fins. Refer to manufacturers' supply information to help determine which type of coil is best suited for any particular application.

UNIT 74—SUMMARY

In this unit you have learned about different types of air-conditioning systems, such as console-type through-the-wall conditioners, single packaged conditioners, rooftop conditioners, and chillers. Unitary equipment is factory built to be as complete as possible and fairly simple to install.

Packaged terminal air conditioners (PTAC) are most commonly used in hotels, nursing homes, and apartment complexes. Single packaged conditioners can be of the horizontal or vertical type and are permanently located in a separate space away from the room being cooled. Ductwork is often used for air distribution, and horizontal units are generally sized from 1½ to 5 tons, while vertical units range from 3 to 15 tons.

Rooftop conditioners are similar to single packaged conditioners except that they are thoroughly waterproofed. In this way they can be conveniently located on the building's roof, thus conserving valuable inside building space. Large air-conditioning load capacities often require chillers, which can operate with mechanical compressors or as absorption-type systems.

Air handlers come in many varied configurations and sizes. Many commercial and light commercial handlers are available in sectional units that can be put together like building blocks to meet the requirements for a specific application. All air handlers will have a fan and cooling/heating coils. However, the fan size, speed, and type of control will vary depending on the space requirements. There are also different types of coils. These can be direct expansion, chilled water, hot water, or steam type.

WORK ORDERS

Service Ticket 7401

Building Conversion: New Installation

Equipment Type: Rooftop Packaged Unit/PTAC

The company sales representative visits with the owner of a small business who is in the process of converting an older one-story building into office spaces. The old air-distribution system is to be entirely replaced. The owner is wondering if through-the-wall individual packaged terminal units for each space will be sufficient.

The sales representative uses information about the building dimensions, construction materials, and outside/inside design temperatures to determine the total heating and cooling load for the building. He then explains to the owner that a PTAC for each space would be sufficient to handle the load. Some advantages to the PTACs are that they can be turned down anytime someone is not in that particular office space, and no new duct system needs to be installed during the building conversion. However, the inner offices would require mini-split systems rather than using through-the-wall units.

Although a packaged rooftop unit would involve the installation of an air-distribution system, this would be the best time to do that, while the building interior is being torn apart. The salesman recommends a cooling/heating rooftop unit with a gas furnace. This would prove much more economical over time for heating as compared to the electric resistance heat of a PTAC. The initial cost would be higher and the control system somewhat more complex, but the overall comfort for both the heating and cooling season would be better. The PTACs also could be noisy for the office space, so the owner decides on installing the rooftop packaged unit.

Service Tic ket 7402

Customer Complaint: Cold Conference Room

Equipment Type: Central Chilled-Water System

The technician answers a call at a large office building downtown. There is a problem with the cooling system in the conference room on the sixth floor. The 120-ton semihermetic centrifugal chiller located in a machinery room in the basement is operating normally. Everything also appears normal in the sixth floor fan room, with the chilled-water supply temperature at 45°F and the return at 55°F.

When the service technician enters the conference room, it immediately feels cold. The supply register has cardboard taped over it, and air is whistling through the cracks. The technician uses a small stepladder and removes the ceiling panels from around the supply register. This system has steam reheat, and the steam coil is now exposed. Without any reheat, cold air is being supplied regardless of the setting on the space thermostat. The supply damper is fixed in place and reheat is the only adjustment for space temperature. The technician finds that the steam-regulating supply valve to the reheater is frozen shut. The steam supply to the line is secured and the valve replaced. Soon after, the space temperature starts returning to a more comfortable level.

Service Ticket 7403

New Installation: System Replacement

Equipment Type: Packaged Rooftop Unit

The service technician arrives at the job site and finds that the new packaged rooftop unit has already been lifted on to the roof of the building. Climbing up the fire escape to where the unit is located, the technician brings along a tape measure and the installation specifications for the new unit. Before even beginning to make measurements, it is obvious that the footprint for the new unit is going to be quite a bit larger than the old unit. There are also structural members that will need to be reinforced for the rooftop foundation.

The service technician goees back down and locates the building superintendent. The roof is "bonded," and any new penetrations need to be approved by the roofing contractor. The technician arranges for the installation to begin in coordination with a crew from the roofing company. They need to work together to make sure that the old unit is properly removed and the new unit installed with adequate support.

UNIT 74—REVIEW QUESTIONS

1. Define *residential air conditioning*.
2. List at least five types commercial air-conditioning applications.
3. If a direct expansion coil delivered from the factory has lost its nitrogen pressure charge, what should be done?
4. How should an external equalizing line for a TEV be connected to a horizontal run?
5. List the different types of staging that can be used for air-handler compressors using direct expansion coils.
6. What is the difference between a primary and a secondary refrigerant?
7. What is a special ventilation consideration for restaurants, nightclubs, and casinos?
8. What type of water should be used in a chilled-water system?
9. What consideration should be given to office spaces when determining space heating and cooling loads?
10. State the two variations of single packaged conditioners and the tonnage capacity ranges of each.
11. The single packaged conditioner in the horizontal variation may come equipped with seven features. What are these?
12. What are the differences between a rooftop air conditioner and a single packaged air conditioner?

13. Why are rooftop units popular on small commercial buildings?
14. Explain what would happen if an HVACR contractor violated a bonded roof by putting in a vent pipe.
15. Could a packaged rooftop unit be adapted for both heating and cooling?
16. How are hermetic compressors cooled?
17. What types of compressors are commonly used for chillers?
18. Why do many central system chillers use water-cooled condensers?
19. What is moisture regain?
20. List some of the airborne particles that may be released in an industrial manufacturing process.
22. How should hot-water heating coils be piped?
23. What happens if the steam trap at the outlet of a heating coil is stuck in the closed position?
24. Should heating and cooling coil exterior surfaces ever be cleaned?
25. What are the two common types of coil corrosion?
26. What happens when metal oxides are formed on the coil tube surfaces?
27. What is galvanic corrosion?

UNIT 75

Fans and Air-Handling Units

OBJECTIVES

After completing this unit, you will be able to:

1. describe the different types of air handlers.
2. explain the operation of a fan coil unit.
3. list the different types of coils used for heating and cooling.
4. identify types of fans and describe their uses.
5. explain the differences between centrifugal and axial fan arrangements.
6. use the fan laws to predict the end result of changing fan speed, air volume, static pressure difference, or fan horsepower.
7. use a fan curve to predict fan performance.
8. list the different types of air filters used in air handlers.
9. explain how humidifiers operate.
10. describe the operation of economizers and energy-recovery ventilation sections.

75.1 INTRODUCTION

Air conditioning involves more than just cooling or heating air. The air also needs to be filtered, humidified or dehumidified, and distributed to the proper location in the appropriate amount at the designed velocity. Too little outside air will lead to an increase in the level of indoor pollutants. Too much outside air increases the load on the system and leads to lower efficiencies and higher operating costs. Air balance is important so that doors inside a building open easily and close without slamming shut. Too little exhaust air can lead to overpressurization of a space.

The air-handler unit (AHU) should be selected and installed with all of the major requirements for the space taken into account. These include the air handler's location, type of coil design, and filtration ability. A central station AHU may offer economy of operation as compared to individual space fan coil units. However, individual units may provide for better temperature control and eliminate the mixing of airstreams between rooms, such as may be required in a laboratory or hospital. Individual fan coil units can also be shut down when the space is unoccupied, such as in a hotel room. The decision to install one large central air handler, multiple air handlers, or individual fan coil units requires careful planning and consideration of the building's main functions and requirements.

75.2 TYPES OF AIR-HANDLING UNITS

AHUs come in various sizes and configurations. These can be arranged as complete packaged systems for residential, light, and moderate commercial applications. Large central-air-conditioning systems typically have built-up units that are assembled component by component. Sometimes these come as modular building blocks and can be arranged in a variety of configurations (Figure 75-1). Depending on the application, the air handler may be located in the conditioned area. However, installation in the basement, on a rooftop, or in some other isolated area saves on valuable inside space and reduces the fan noise heard by the people indoors. Figure 75-2 shows the proper installation for ceiling-mounted units that may be found in a large retail store or in warehouse applications. Many of these units have double-wall insulation, which reduces sweating on the outside surfaces and noise from the fan compartment.

The location for the air handler is usually flexible because it does not need to be located alongside the heating and cooling equipment. The handler can be as simple as a fan, a filter, and a coil. For example, a residential split system has refrigerant supplied from the outdoor unit to an air handler that may be located in the basement. This type of system would require just the basic components and would be considered a direct expansion system. However, for larger buildings that require more than one air handler, it would be impractical to deliver refrigerant to each one. In this case, a central plant will cool a secondary refrigerant (chilled water) that is then delivered to each air-handler coil. This type of air handler may also have additional hot-water or steam coils for heating and mixing and dampers for energy-recovery applications. Figure 75-3 shows a large indoor air handler.

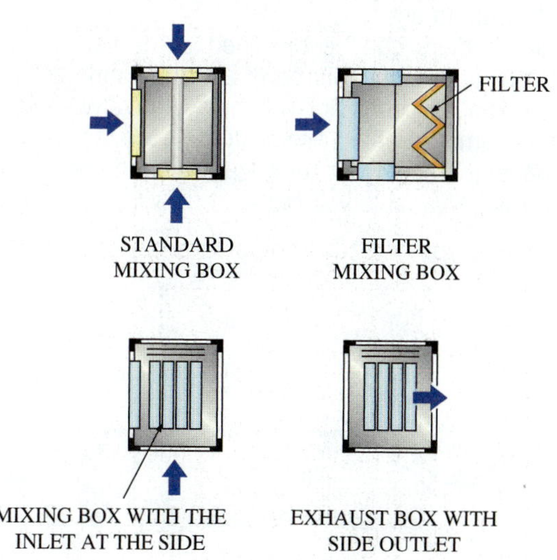

Figure 75-1 Modular air-distribution components.

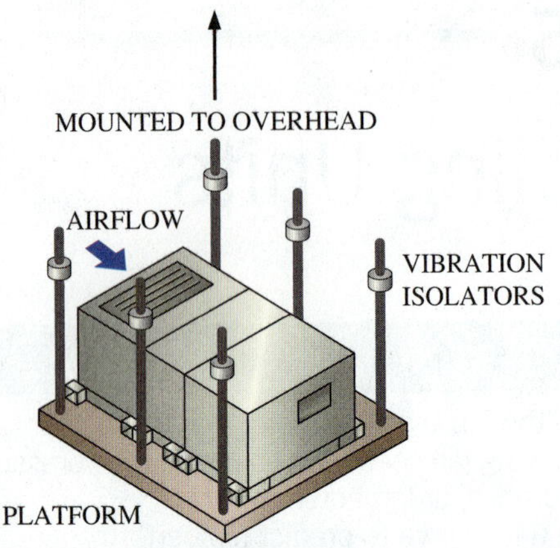

Figure 75-2 Ceiling-mounted air handler.

Figure 75-3 Indoor air handlers.

Most air handlers use a fan (or blower) to force the air at low, medium, or high velocity through the air-distribution system. Draw-through units are designed so that the fan is at the outlet of the air handler to draw the air across the coils (Figure 75-4). This is more common than a blow-through type where the fan is located at the inlet to the coils. Sometimes blow-through systems can be used if sound levels are a major consideration. Noise levels can be lowered because sound is absorbed by the coil section downstream from the fan. Less common are induction units that are designed to operate without any fan.

Air handlers can be designed to deliver a constant volume or a variable volume of air to a single zone or to multiple zones. A constant-volume system always supplies the same amount of air, and to control the temperature of the space, the supply-air temperature has to be regulated. In contrast, a variable-volume system will adjust to supply more or less air depending on the space temperature. In this case, the supply-air temperature does not need to change.

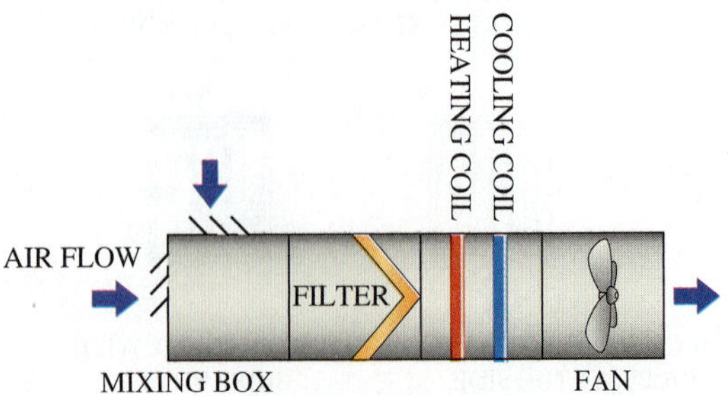

Figure 75-4 Draw-through horizontal air handler.

75.3 FAN COIL UNITS

Fan coil units (FCUs) are supplied with chilled water directly from a central plant and can also include heating coils or electric heat. They are primarily designed to supply conditioned air to individual rooms, but larger FCUs are capable of conditioning multiple spaces. Each one of these units has fans, cooling coils, filters, a fresh air supply, and a control system.

FCUs for combination heating and cooling come in a variety of designs, to fit under windowsills, above furred ceilings, and in vertical supports built into walls. FCUs have many of the same advantages of packaged terminal air conditioners (PTACs). These include individual space temperature controls and no requirements for a large duct system. However, unlike packaged air conditioners, the efficiency of a central plant can be realized. An indoor water-cooled condensing unit or split-system outdoor condenser is not needed. All that is required is an insulated chilled-water loop piped from the central plant chiller.

FCUs have a wide variety of applications. One advantage is the ability to provide individual room conditioning. This is desirable in applications for hospital rooms, motel rooms, and individual private offices. An accessory fresh-air duct can be provided in the bottom rear of the unit to permit the entrance of outside ventilation air. Each room unit can be individually controlled without mixing air from an adjacent room.

A common type of FCU, the individual room conditioner, is shown in Figure 75-5. This type of unit consists of a filter, direct-driven centrifugal fan(s), and a coil suitable for handling chilled or hot water. The size of the unit is based on its cooling ability; it is usually more than adequate for heating. A variety of water-flow control packages are available for manual, semiautomatic, or fully automatic motorized or solenoid valve operation.

TECH TIP

Often around fluorescent lights in a dropped ceiling there is a narrow slit. That opening is the return air. In this location the heat generated by the light is drawn into the return airstream, allowing for less heat to be radiated into the occupied space. The high artificial light levels in most commercial buildings represent a significant cooling load.

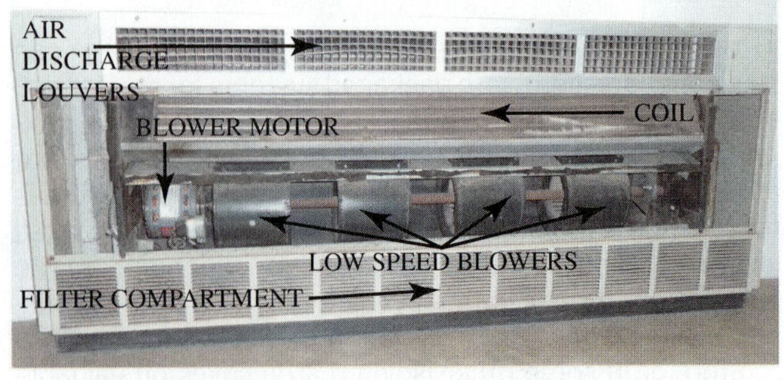

Figure 75-5 Typical fan coil unit, showing component parts.

Unit capacity is controlled by coil water flow, fan speed, or a combination of the two. A wall thermostat or the return-air temperature can be used to control the water flow. Airflow can be controlled by manual or automatic fan speed adjustment. Some units utilize variable-speed motors for modulated speed control. Outside air is introduced through a damper-controlled opening in the outside wall. The size of the cabinet is designated in CFM of airflow and ranges from 200 to 1,200 CFM. These units may be installed on two-, three-, or four-pipe system designs, as illustrated in Figure 75-6.

With modifications to the cabinet, the same components are assembled in a horizontal ceiling-mounted version (Figure 75-7). Where hot water is not available for heating,

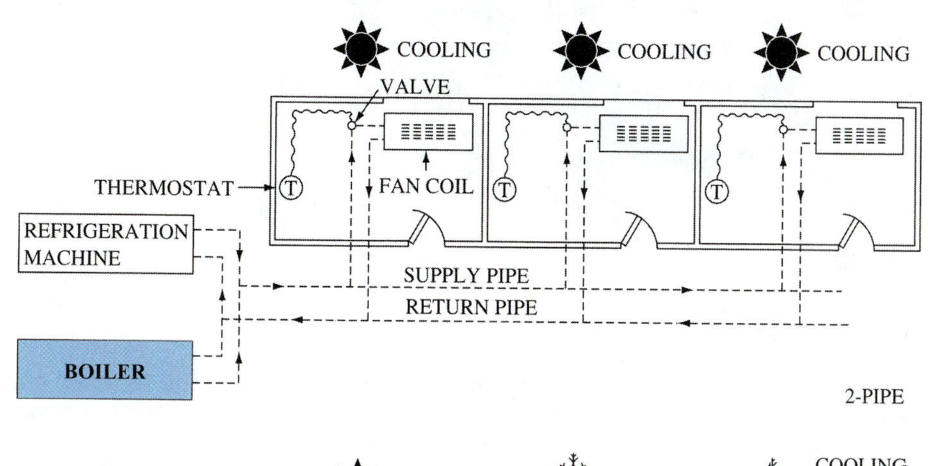

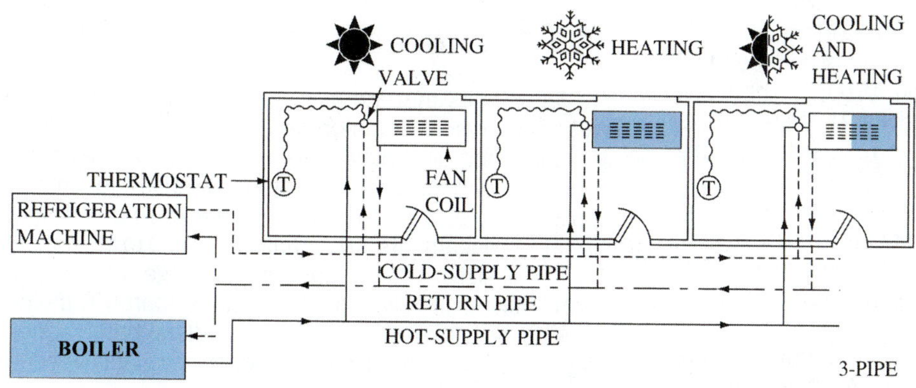

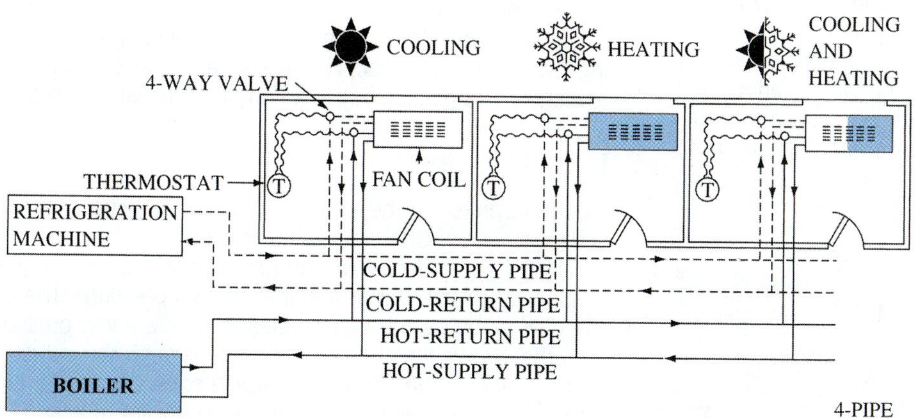

2-PIPE SYSTEM
EITHER HOT OR CHILLED WATER IS PIPED THROUGHOUT THE BUILDING TO A NUMBER OF FAN-COIL UNITS. ONE PIPE SUPPLIES WATER AND THE OTHER RETURNS IT. COOLING OPERATION IS ILLUSTRATED HERE.

3-PIPE SYSTEM
TWO SUPPLY PIPES, ONE CARRYING HOT AND THE OTHER CHILLED WATER, MAKE BOTH HEATING AND COOLING AVAILABLE AT ANY TIME NEEDED. ONE COMMON RETURN PIPE SERVES ALL FAN-COIL UNITS.

4-PIPE SYSTEM
TWO SEPARATE PIPING CIRCUITS – ONE FOR HOT AND ONE FOR CHILLED WATER. MODIFIED FAN COIL UNIT HAS A DOUBLE OR SPLIT COIL. PART OF THIS HEATS ONLY, PART COOLS ONLY.

Figure 75-6 Three types of piping arrangements for fan coil units.

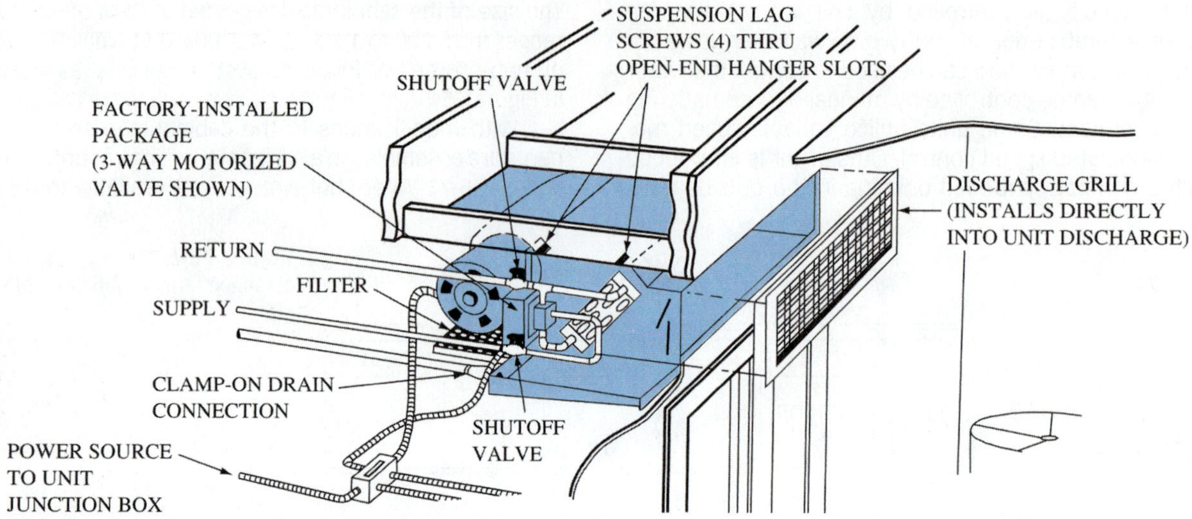

Figure 75-7 Ceiling-mounted fan coil unit.

electric resistance heaters are installed in the cabinet after the cooling coil in the path of the air leaving the unit.

Another version of the room FCU is a vertical column design, which may be installed exposed or concealed in the wall. These are placed in common walls between two apartments, motel rooms, and so on. Water piping risers are also included in the same wall cavity. They are not designed for ducting but can serve two rooms through the addition of air supply grilles.

Small ducted FCUs with water coils may be installed in a dropped ceiling or in a closet, with ducts running to individual rooms in an apartment. These range in size from 800 to 2,000 CFM. The unit's cooling capacity is selected based on the desired airflow in CFM.

Larger FCUs, sometimes called unit ventilators, are used for school classroom conditioning. A single unit ventilator is installed below window level in each classroom with extension supply-air ducts and grilles running along the sill for the full length of the room. An opening in the back supplies outside air from a through-the-wall opening. The units include air filters to clean the air. Units are individually controlled, and this allows for differences in sun load, depending on the orientation of the room. They also meet the ventilation requirements of local codes for classrooms.

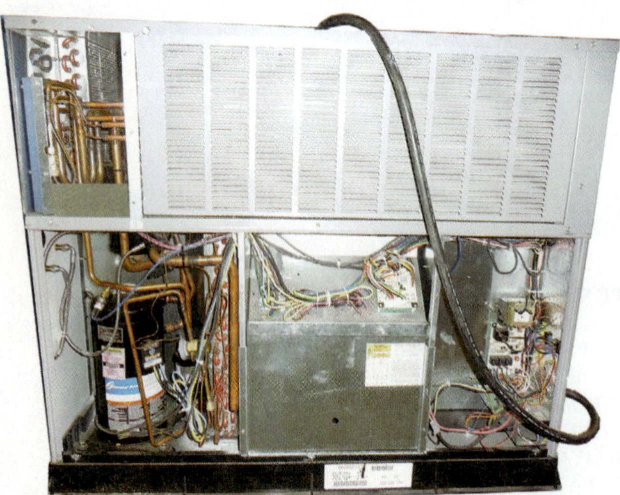

Figure 75-8 Packaged Unit.

75.4 PACKAGED UNITS

Packaged units that are used to condition offices, stores, and so on, where common air distribution is feasible, are self-contained air conditioners (Figure 75-8). They can be equipped with supply- and return-air grilles for in-space applications, or they may be remotely located with minor ducting arrangements. Sizes range from 800 to 15,000 CFM. For larger sizes, cabinet and fan-discharge arrangements permit flexible installations. These units approach the next category of equipment, called air handlers, but in general, they do not have the size and functions available in central station air handlers.

For all installations that include cooling, it is important to provide a drain for any water vapor condensing on the cooling coil. This water collects in a pan located just below the coil and should be piped away to an appropriate sanitary drain.

Induction Units

Induction units can be located for individual space cooling similar to FCUs. One significant difference is that induction units do not require fans. Centrally conditioned primary air is supplied to the unit at medium to high pressure. The primary air will pass through nozzles to create a low-pressure area that draws the air from the room through the induction unit and across the cooling or heating coils. Due to the high airflow required, this type of unit can be noisier than an FCU. Induction units are not designed to handle condensation, therefore the temperature and humidity of the primary air supplied from the central system must be properly controlled.

75.5 CENTRAL STATION AIR-HANDLING UNITS

Central station systems use one or more large AHUs equipped with fans, heating coils, cooling coils, and other accessories. They are centrally located in the building and

Figure 75-9 Central multizone AHU, modular construction.

connected to ductwork that distributes the conditioned air to the desired areas (Figure 75-9).

Often these units are made in modules or sections that can be joined together on the job. This arrangement offers flexibility in the selection of components required for a specific installation.

The typical components found in a central station AHU with a single fan system are the filter, preheater, and cooling coil. These may all be enclosed in one cabinet, in a series of cabinets, or installed as separate components. For example, the preheating coil may be installed away from the cabinet in the ductwork near the entrance of the outside air into the building.

With a single fan, the suction pressure in the return system can create a negative pressure in many parts of the building. This can cause undesirable infiltration and put an extra resistance on the operation of the fan. The air imbalance may also affect the operation of doors within the conditioned space, making them hard to open or causing them to slam shut. For these reasons, it is generally desirable to maintain a slightly positive building pressure. On a large system, the return-air fan is used to control the building pressurization. A schematic diagram of a two-fan central station air-handling system is shown in Figure 75-10. The return-air fan may be installed in the ductwork and not as part of the AHU.

TECH TIP

Central station air handlers can be very large. Some blower wheels are more than 20 ft in height and are powered by very large variable-speed motors. These blowers handle tens of thousands of cubic feet of air per minute. The blower area is actually a large room that may be located in a subbasement or on a separate floor in a high-rise building. When you enter this area you are actually walking into the blower chamber. Be careful when opening any door in this area or accessing the area because there may be a substantial pressure difference and the door can lunge open or closed.

75.6 AIR-HANDLER CABINET

Many air handlers are factory assembled as complete units, while others are factory assembled in sections and then put together at the job site. These units are enclosed in a cabinet typically made of galvanized steel that is coated on the outside with a polyurethane painted finish

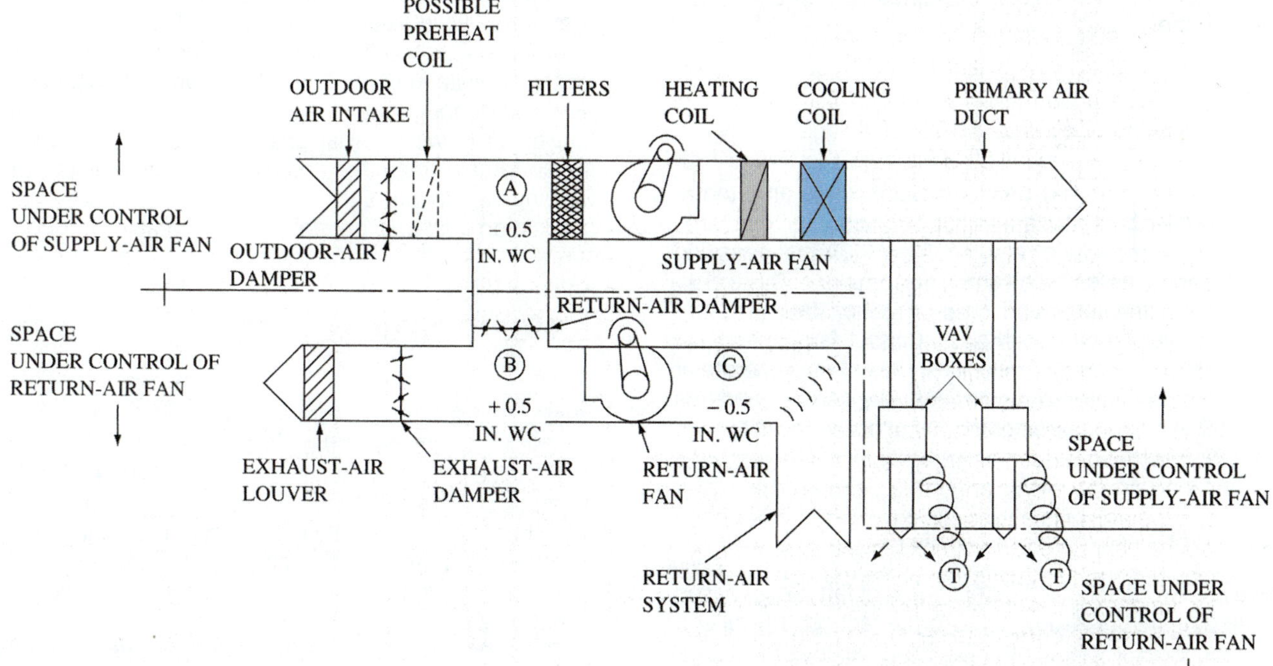

Figure 75-10 Schematic view of a two-fan central station AHU.

Figure 75-11 Air-handler door with painted outside coating.

(Figure 75-11). The inside can be coated with an antimicrobial finish. The cabinet is insulated to reduce sweating on the outside, which will occur during periods of high humidity. Some cabinets utilize double-wall construction with a layer of insulation sandwiched between an inner and outer wall. This provides for some rigidity that reduces the vibration of the cabinet. This also helps to provide a sound barrier, thereby reducing noise levels in the space created by the air handler. Access panels are provided to gain entry into the air handler.

TECH TIP

Although air-handler cabinets may be made of galvanized steel, there are still areas that may be subject to corrosion. The sheet-metal edge and any scratches deeper than the galvanized surface will expose the bare steel. This will eventually lead to rust, so be careful when installing or performing maintenance on a unit not to scratch the surface.

75.7 FAN CLASSES

The Air Movement and Control Association is an international standards organization that certifies fans, dampers, and other air-movement devices. The AMCA has classified fans into four classes based on fan performance. These classifications define upper and lower ranges of static pressure and outlet velocity for each class. Class 1 fans operate at the lowest pressures and velocities; Class 4 fans operate at the highest pressures and velocities. The AMCA standards require that the fan operate safely on or below the minimum rating for its class.

75.8 TYPES OF FAN DESIGN

There are two distinctly different types of fan design based on the direction of airflow through the fan: centrifugal and axial. Air flows into a centrifugal fan parallel to the motor shaft and is then spun outward, through the fan blades, by the centrifugal motion of the fan. The fan housing then channels the airflow toward the fan outlet. The air movement

Figure 75-12 Many units use propeller fans for condenser fans.

in an axial fan is parallel to the shaft. A propeller fan is an example of an axial fan. Propeller fans are often used in residential condensing units (Figure 75-12).

TECH TIP

Centrifugal fans are designed to produce high static pressures, while axial fans provide high volumes of air. Centrifugal fans work very well on duct air-distribution systems, while in contrast axial fans are not capable of producing enough static pressure to move air through long ducts. However, axial fans are very effective in moving air through the large openings typically required for building ventilation. Each fan type has its purpose and application, and they are not interchangeable.

75.9 CENTRIFUGAL FANS

Centrifugal fans are the workhorse of the air-conditioning industry. These fans are sometimes called "squirrel cage" fans due to the shape of the impeller wheel. They are the fan of choice when "pumping" air in high volumes, against high static pressure. A ductless system may use a centrifugal fan to deliver air directly to a room or use a blower in a split system to pump air through an elaborate duct system to condition an entire residence. These fans are also used in many ventilation applications.

There are several different blade designs used with centrifugal fans. The two main types of centrifugal fans used in air-conditioning work are the forward-curved blade, which is most common, and the backward-curved blade.

Figure 75-13 Centrifugal fan with a scroll housing.

The advantage of the backward-curved blade is that it is nonoverloading. The disadvantage is that it is noisier.

Centrifugal Fan Housings

Centrifugal fans can be divided into three housing types: scroll, tubular, and plug, the last of which has no housing. The scroll housing is the most widely used for air conditioning and heating. This housing design directs the airflow from the fan blades down a channel at a 90-degree angle from the drive shaft. Figure 75-13 shows a centrifugal fan with a scroll housing.

Tubular centrifugal fans have a housing that looks like a tube or pipe. The air is drawn in parallel to the drive shaft and spun out of the blades. It is then redirected out the other end of the "tube," giving it parallel airflow to the drive shaft when exiting the fan housing. This centrifugal fan is used to ventilate heat, fumes, or smoke in commercial or industrial applications.

The third and least used centrifugal fan is called a plug fan. It is a design that has no housing. These fans are installed directly into plenums of AHUs to pressurize the complete unit. This fan design is less efficient but offers flexibility in unit design and can reduce unit size.

SERVICE TIP

The amp draw of a centrifugal blower will actually decrease if the air to the blower is blocked. A centrifugal fan will turn faster if it is not actually moving air, and the motor amp draw decreases because of the reduced load on the blower.

75.10 BLADE DESIGN

There are six types of centrifugal fans based on the design and shape of the blades used to move the air: forward curved, backward curved, airfoil, radial, radial tip, and backward inclined.

Forward-Curved Centrifugal Fans

Forward-curved centrifugal fans are the most popular blade design for residential systems. This blade design curves the blade's leading edge toward the direction of airflow. This adds resistance to the forward motion of the fan wheel, reducing fan speed, noise, and efficiency. This blade design can cause motor overloading if the static pressure is decreased past design limits of the fan. One advantage to this design is that bearing size is reduced and bearing life extended because of the fan's lower rpm.

SERVICE TIP

The indoor blower in most furnaces and heat pumps is a forward-curved centrifugal fan. These blowers are not designed to operate in free air; they need to have some resistance to airflow. Technicians often try and use blowers removed from old equipment as fans to cool the work area. Some restriction must be added to the opening of either the fan inlet or outlet to keep the fan operating without overloading the motor.

Backward-Curved Centrifugal Fans

Backward-curved fan blades have a leading edge that trails the airflow, producing little resistance to the fan wheel's forward motion. The speed of one of these blowers is nearly twice the speed of a forward-curved blower. This low-resistance design increases both efficiency and noise due to its higher rpm. These fans are larger in design and require larger drive shafts and bearings than do forward-curved blowers. Figure 75-14 shows a centrifugal fan with backward-curved blades. A comparison of forward- and backward-curved blades is shown in Figure 75-15.

Airfoil Centrifugal Fans

Airfoil blade blowers are a variant of the backward-blade design. This is the most efficient centrifugal blade design

Figure 75-14 Backward-curved centrifugal fan.

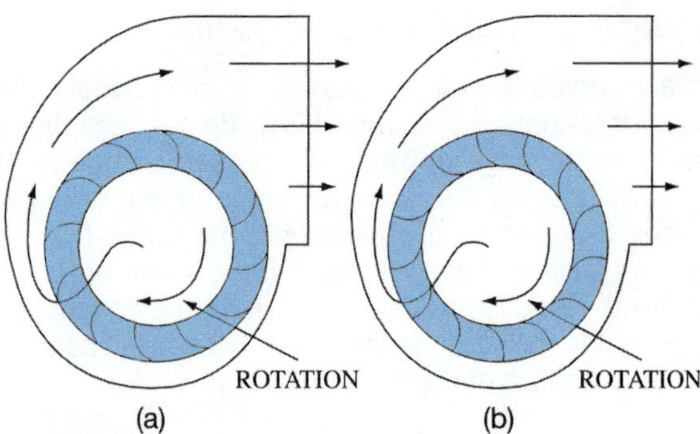

Figure 75-15 Schematic view of the construction of (a) backward-curved fan and (b) forward-curved fan.

due to its very low resistance to airflow. However, this resulting higher efficiency covers only a small area of its performance curve in relation to other backward-curved fan blades; and has a higher cost associated with it. This often reduces its use to only high-static-pressure applications where additional cost can be justified. In most other ways, the airfoil blade design reflects the backward-blade design's operational advantages.

75.11 AXIAL (IN-LINE) FAN

Axial fans are divided into three types based on housing design: the propeller fan, the tube axial, and the vane axial. The propeller type will handle large volumes of air for low-pressure applications. It has high usage for exhaust fans and condenser fans. Vane-type fans are highly efficient but noisy. The blade pitch can be adjusted to control the amount of air the fan handles.

SERVICE TIP

The amp draw of a propeller fan will increase if the air leaving the blower is blocked. Propeller fans do not work well against the high-pressure difference created when the fan outlet is blocked. They slow down, putting more load on the fan motor and increasing its amp draw.

Propeller Fans

The propeller fan shown Figure 75-16 is the most familiar axial fan design. It consists of a fan shaft with two or more blades attached at a 90-degree angle to the shaft. Some of these fans will have a hub assembly that attaches the blades to the shaft. This hub assembly allows for maintenance and modification. Air is pulled through the fan blades parallel to the shaft and is blown outward. These fans are high speed and as a result produce high noise levels. Axial fans operate well in low-static environments and can move high volumes of air.

The propeller fan is generally is used as a condenser fan on air-conditioning systems. These fans are generally shrouded to improve performance but do not use a true housing. Propeller fans typically have two to six blades. This fan works well in a low static pressure and relies on blade pitch to determine efficiency.

Figure 75-16 Propeller fan.

SERVICE TIP

The propeller fans used with most air-conditioning applications have a cup shape. The cup pushes the air forward. The rotational direction of a propeller fan can be determined by observing the direction of the blade cup. When looking at the blade so that the cups are visible, the edge of the blade cup that is farthest away from the observer is the leading edge. The blade should turn toward the leading edge. Figure 75-17 illustrates the rotational direction and airflow direction of a typical propeller fan.

Tube Axial Fan

The tube axial fan has a housing that resembles a pipe or tube. The blades are in close tolerance to the housing. Flat blade or foil design blades are most common. These fans are widely used in ventilation, such as warehouses or in structures where fume venting is required.

Vane Axial Fan

The vane axial fan is the most efficient axial design and can operate at higher static pressures than other axial designs. The efficiency of this housing comes from vanes added in the path of the airflow to reduce "swirling" of the air. These fans move high volumes of air and produce high noise levels

Figure 75-17 A propeller fan turns toward the leading edge of the fan; the leading edge is farthest from the viewer.

and may require noise dampers in some applications. This type of design can offer an adjustable blade pitch that can be used to vary the amount of air moved by the fan. These fans are often applied to variable-air-volume systems because of their ability to easily vary the airflow.

75.12 TYPES OF FAN DRIVES

The fan drive type refers to how the motor motion is transferred to the blower. There are two types of drive arrangements for air-conditioning fans: direct drives (Figure 75-18) and belt drives (Figure 75-19). With direct drive, the fan driveshaft or fan wheel is attached to the motor's driveshaft. This method is normally applied to lower-torque applications and lower-horsepower motors. This is the most efficient design because there is no loss due to slip. Belt-driven motors use belts, such as v-belts, to attach the motor driveshaft to the fan driveshaft. Normally, large-torque applications use belt drives. Torque is improved because the motor turns over several revolutions for each revolution of the blower. Other design requirements, such as size or

Figure 75-18 On the left is a direct drive centrifugal fan; on the right is a direct drive propeller fan.

(a)

(b)

(c)

Figure 75-19 (a) Belt drive blower; (b) fan, belt, and motor arrangement for large air-handling unit; (c) belt-driven blower for a stand-alone computer room HVAC unit.

maintenance requirements, may favor a belt-drive design. The fan speed can be changed on a belt-drive blower by changing the pulley size or in some cases by adjusting the pulley sheaves without changing the entire pulley out.

SERVICE TIP

The motors used in most fans and blowers are called air-over motors. They depend on airflow over the motor to cool the motor. These motors have openings in the end bells of the motor for airflow. If these openings are plugged with lint, dust, pet hair, or other particulate matter, the motor will overheat. Dirty air filters or dirty evaporator coils can severely reduce system airflow and can also cause the blower motors to overheat from lack of cooling. Most of these motors have internal thermal overloads that will shut down the motor until it cools off. This can take over an hour. (Note: Some totally enclosed motors are also listed as air-over motors and may be designated as totally enclosed air-over or TEAO motors.)

Belt-Drive Blowers

For belt-driven blowers, the speed of the fan is determined by the ratio of the motor pulley to the fan pulley. Increasing the size of the motor pulley will increase the rpm of the blower. Increasing the size of the blower pulley will decrease the rpm of the blower. For example, to calculate the fan rpm with a motor rpm of 1,750, a 3-in-diameter motor pulley, and a 6-in-diameter fan pulley, use the following formula:

$$\text{Fan rpm} = \frac{\text{Diameter of motor pulley} \times \text{motor rpm}}{\text{diameter of fan pulley}}$$

$$\text{Fan rpm} = \frac{3}{6} \times 1750$$

$$= 875 \text{ rpm}$$

SERVICE TIP

A relatively small change in pulley size can make a big difference in motor load. When adjusting the pulleys or belt on a belt-drive blower, it is important to check the amp draw once the motor pulley and belt have been adjusted to make certain that the motor is not drawing excessive current, which would result in its overheating and failing.

SAFETY TIP

Belts and pulleys are major pinch points and should be regarded with caution when changing belts or making adjustments. The power source to the equipment should be locked out when changing or adjusting drive components. Any belt or fan guards that are removed to work on the equipment should be replaced before putting the equipment back into service.

Direct-Drive Blowers

Direct-drive motors are available with multiple speed windings. The speed can be changed by switching wires in the motor terminal box. This can be done by the control system so that a different speed can be used for heating and cooling, or it can be a manual setting used to match the blower to the resistance of the duct system. Direct-drive fan motors are usually the PSC (permanent split capacitor) type. Higher-efficiency systems typically use ECM brushless DC motors. These motors can be programmed for a wide range of operating conditions. They can vary the airflow based on the equipment operating cycle or based on the resistance to airflow.

Variable-frequency drives are also becoming popular to control the speed. This type of system uses an inverter to modify the drive power output into many steps rather than into a smooth sine wave. Instead of a fixed voltage and frequency, the motor input is variable, which allows for changeable motor speed without losing torque. However, this requires motors rated for inverter duty, as standard motors will overheat. Although the variable-frequency drive motor efficiency is lower than a standard motor drive, the overall power required should be less. This is because the motor power requirement can be more evenly matched to the load. Systems can be designed to allow for the control of fan speed down to as low as 10 percent of the normal design speed.

SERVICE TIP

When adjusting the speed of a blower, check the motor amperage with the blower door closed. With the blower door open, more free air is allowed to enter the blower and the amp draw will be higher than it would be with the door closed.

75.13 FAN LAWS

A fan's performance varies depending upon the conditions it operates under. There is no such thing as a 400 CFM fan. A fan that moves 400 CFM of air at one condition might only move 200 CFM at another condition. The amount of air a fan moves, the speed of the fan, the amount of pressure difference produced across the fan, and the motor horsepower required are all interrelated.

Fan speed is directly related to the amount of air a fan can move: doubling the fan speed doubles the airflow. Take, for example, a fan moving 400 CFM against a static pressure of 0.1 in wc turning 500 rpm and using a ¼-hp motor. Doubling the speed from 500 rpm to 1,000 rpm will double the CFM from 400 CFM to 800 CFM. The static pressure difference across the fan increases more rapidly. Doubling the fan speed increases the static pressure by a factor of 4. Doubling the speed of the same fan from 500 rpm to 1,000 rpm will increase the static pressure across the fan from 0.1 in wc to 0.4 in wc. The horsepower requirement to accomplish this increases even more dramatically. Doubling the fan rpm will require 8 times the horsepower! So increasing the fan speed from 500 rpm to 1,000 rpm will increase the horsepower requirement from ¼ hp to 2 hp.

The relationship between changes in speed (rpm), air volume (CFM), static pressure (SP), and power in brake horsepower (Bhp) are described by a set of formulas called fan laws. The fan laws in classic form are:

$$\frac{CFM_2}{CFM_1} = \frac{rpm_2}{rpm_1}$$

$$\frac{SP_2}{SP_1} = \left(\frac{rpm_2}{rpm_1}\right)^2$$

$$\frac{Bhp_2}{Bhp_1} = \left(\frac{rpm_2}{rpm_1}\right)^3$$

These can be algebraically rearranged to find the new airflow, speed, static pressure, or horsepower.

$$\text{New CFM} = \text{Old CFM} \times \frac{\text{New RPM}}{\text{Old RPM}}$$

$$\text{New RPM} = \text{Old RPM} \times \frac{\text{New CFM}}{\text{Old CFM}}$$

$$\text{New SP} = \text{Old SP} \times \left[\frac{\text{New RPM}}{\text{Old RPM}}\right]^2$$

$$\text{New BHP} = \text{Old BHP} \times \left[\frac{\text{New RPM}}{\text{Old RPM}}\right]^3$$

EXAMPLE 75-1: RESULT FROM INCREASING FAN CFM

What would be the resulting operating condition if the airflow moved by a fan operating at 300 rpm, 0.15 in wc static, and 0.2 Bhp was increased from 600 CFM to 800 CFM?

First find the new speed.

$$\text{New rpm} = 300 \text{ rpm} \times (800 \text{ CFM}/600 \text{ CFM}) = 400 \text{ rpm}$$

Next, use the new rpm to find the new static pressure.

$$\text{New SP} = 0.15 \text{ in wc} \times (400 \text{ rpm}/300 \text{ rpm})^2$$
$$= 0.15 \text{ in wc} \times 1.78 = 0.27 \text{ in wc static}$$

Finally, use the new rpm to find the new horsepower.

$$\text{New Bhp} = 0.2 \text{ Bhp} \times (400 \text{ rpm}/300 \text{ rpm})^3 =$$
$$0.2 \text{ Bhp} \times 2.37 = 0.47 \text{ bhp}.$$

The new operating conditions are at 400 rpm, 0.27 in wc static, and 0.47 Bhp to increase the airflow from 600 CFM to 800 CFM.

Fan Performance Tables

Manufacturers publish performance tables for each specific fan. Performance tables show the fan performance at specific operating conditions. Performance tables show the fan CFM compared to the static pressure. Other information included in fan performance tables are the outlet velocity, fan rpm, and motor horsepower requirement for the selected condition.

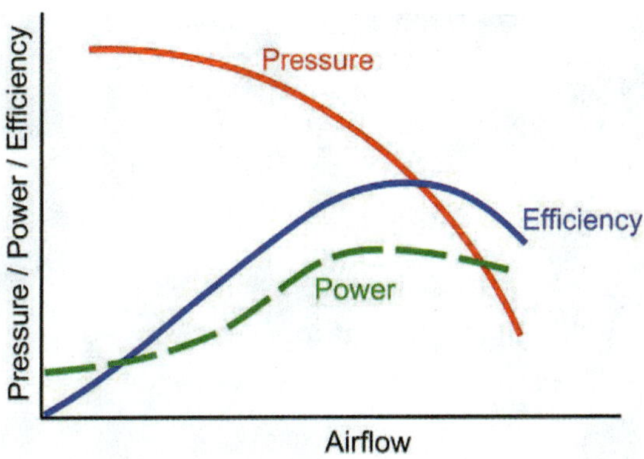

Figure 75-20 Typical fan curve with CFM along the bottom and all other data on the side.

SERVICE TIP

Very often the fans, motors, and drives are all located in the airstream. If this is the case, the heat from the motor, the frictional heat from the belt drives, and the heat from the bearings need to be added to the total cooling load capacity.

Fan Performance Curves

The fan curve is a graphical representation of fan performance. Airflow is typically plotted along the *x*-axis, and power, pressure, and efficiency are plotted along the *y*-axis (Figure 75-20). Fan curves are used in system design to select a fan that is capable of moving the correct amount of air at the desired system operating condition. There will be a best match for air flow, efficiency, and power for any given fan. Fan curves can also be used to predict the effect any change in operating conditions will have on fan performance. A fan or a family of fans may be compared on one chart.

75.14 MECHANICAL AIR FILTERS

Mechanical air filters can be of the throwaway or the cleanable type. The simplest and cheapest construction is the dry disposable type. The frame is generally made of cardboard and the filtering medium is often bonded glass fiber, wool, felt, or synthetics. Extended surface filters are pleated to create a larger filter surface. If they are cleanable, they are more rugged than the disposable type with sturdier metal frames. Cleaning can be accomplished using steam, water, or detergents.

Viscous impingement-type filters will be coated with a sticky substance such as oil and come in both the throwaway and cleanable types. The sticky surface will help to catch the contaminant particles as they pass through the screen. Because of this, the filter does not need to be as dense and can be made with a higher porosity. The advantage to higher porosity is that there is less restriction to airflow. However, impingement-type filters will not work well if the adhesive surface is full of dirt. The air velocity across the filter is also important. If the air travels too quickly,

Figure 75-21 Roller-type air filter.

some contaminants will slip through the filter. If the velocity is too low, the contaminants will not be forced into the adhesive and may simply pass through. If a viscous impingement filter is cleaned, the adhesive needs to be reapplied. A roller-type cleanable air filter used for a large industrial air filtration system is shown in Figure 75-21. The filter media is periodically rolled as it becomes loaded with dirt.

SERVICE TIP

MERV stands for Minimum Efficiency Reporting Value. Higher MERV ratings are provided for more efficient air filters.

High-efficiency particulate air (HEPA) filters have very high filtration efficiencies. Typical cartridge filters range from 65 to 95 percent, while HEPA filters can be as high as 99.9 percent. This type of filter is often located in the air supply for surgical suites, intensive care units, and electronic or microprocessing clean rooms. Due to its high efficiency and dense structure made of submicron glass asbestos fiber, the filter will plug up quickly if no provision is made for a prefilter system.

Antimicrobial Treatment

Some air handlers now utilize antimicrobial-coated steel on the inner sections of the unit. Silver ions are blended into the paint that is used for the inner surfaces. The silver is slowly released to react with any bacteria or other microbes that try to accumulate and grow. This same application is also used to treat filter media. This type of application is particularly well suited for hospitals, schools, and restaurants.

75.15 ELECTRONIC AIR FILTERS

Like applications for the HEPA filter where high filtering efficiency is required, electronic air filters are often used. An advantage to electronic air filters is that they can be cleaned, while most HEPA filters are replaced once they become dirty. One type of electronic air filter applies a charge to the dust particles by passing them through an ionizing zone of vertical wires that have a positive charge. The dust passing through the filter will be positively charged and attracted to negatively charged plates also located in the airstream. The plates are cleaned by turning off the power and washing the plates with a nontoxic detergent solution. Prefilters or screens are commonly used with this type of filtering system.

CAUTION

Electronic air filters can have charges as high as 10,000 to 12,000 V. All electrical power must be disconnected from this type of filter prior to service.

Another type of electronic filter uses a charged filter medium of glass fiber held in place by nonconductive frames. A charge of 10,000 to 12,000 V (DC) is common. Unlike the electronic air cleaners that use metal plates, this type of cleaner also acts as an impingement filter due to the construction and arrangement of the glass fiber filter media.

75.16 ULTRAVIOLET GERMICIDAL LAMPS

Odor and fungus may develop in the wet sections of an air handler. Ultraviolet lamps can be located directly after the cooling coil section and directed toward the coil and the drain pan (Figure 75-22). The lamps should be powered by a separate power source so as not to cycle off and on with the fans. Ultraviolet lamps should not be used with polyester or cotton-type filter elements, as the ultraviolet light will break down the filter media over time. Fiberglass filter media is recommended. The normal color for the lamps is blue, and as the color begins to change to red, the bulbs should be replaced.

CAUTION

Ultraviolet light can be harmful to your eyes and skin. Never open up an air handler that has ultraviolet lamps until the lamps have been turned off.

Figure 75-22 Ultraviolet germicidal lamp.

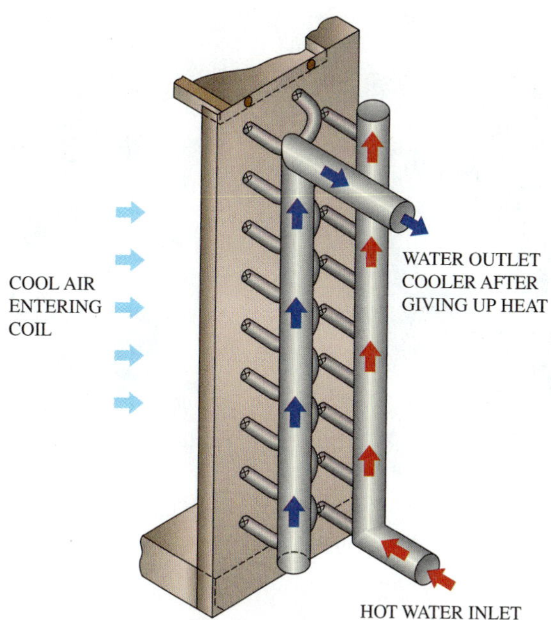

Figure 75-23 Counterflow water coil.

SERVICE TIP

The surface of an ultraviolet lamp is very sensitive and just the oil from your fingerprints can lead to premature failure of the lamp. Always use gloves when handling the lamp and wipe away any film to clean the lamp after handling.

SERVICE TIP

Central AHUs are generally arranged so that it is not possible to operate both the heating and cooling coils at the same time. This is because the energy-conservation codes do not permit it and there would be no advantage since a single-zone unit can provide only heating or cooling at one time.

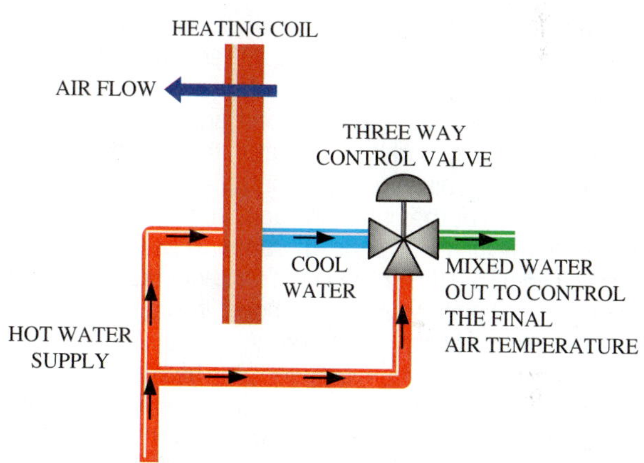

Figure 75-24 Flow through a water heating coil.

75.17 WATER HEATING COILS

The heating coil is often selected for hot water heating because most furnaces today are either forced hot air or hydronic systems rather than steam. A hot-water system is also simpler in its design. Water coils are available in depths of one, two, three, four, five, six, and eight rows deep with various types of circuiting. They are free draining and are provided with ¼-in NPT fittings in the header for draining and venting. Tubes are staggered to improve thermal efficiency.

The water heating coil is most always a counterflow design for the best thermal efficiency. The incoming water will be piped to enter at the air outlet side of the coil (Figure 75-23). In this way there is more even heat transfer across the entire coil from the air inlet to the outlet. The water flow through the coil is generally regulated by a valve on the outlet side. A three-way mixing valve is shown in Figure 75-24. Some air handlers utilize heat-recovery coils, which circulate water or glycol that has absorbed waste heat (Figure 75-25).

Figure 75-25 Piping for a hot-water heat-recovery coil.

Figure 75-26 Steam traps are located in the condensate return line from the steam heating coil.

75.18 STEAM HEATING COILS

If the building has a ready supply of steam available, then steam heating coils may be used rather than water. Low-pressure steam applications are rated at 15 psig or less. Medium-pressure steam applications are rated at greater than 15 psig but generally no greater than 35 psig. Steam flow through the coil is typically controlled by a valve at the inlet rather than at the outlet. In addition, a steam strainer is required at the inlet, and a steam trap must be located at the outlet. A typical piping arrangement for a low-pressure steam heating condensate return line is shown in Figure 75-26.

More heat will be transferred in a steam coil as compared to a water coil where only sensible heat is transferred. A steam coil transfers both sensible and latent heat. Due to this additional heat transfer, a steam coil will be smaller in overall size as compared to a water coil. This allows for a much larger water coil to be replaced by a smaller steam coil if space limitations are a concern.

It is important for all of the steam to condense in the coil, and that is the function of the steam trap. If the trap is hung open and steam is allowed to leave the coil, the efficiency will decrease because latent heat is not being transferred. If the trap is frozen shut, the steam flow will back up and the air passing across the coil will no longer be heated. Opening the bypass valve around the trap will bring the air temperature back up, but this should only be a temporary correction and the trap should be isolated and then repaired or replaced.

CAUTION

Most steam valves for heating systems are designed to operate with maximum steam pressures of up to 35 psig and no more. Be sure to check the rating of any steam valve that you install to make sure it is correct for that system.

Figure 75-27 Water drain located at low point of air duct.

75.19 WATER-COOLING COIL

The chilled-water cooling coil is similar in construction to the hot-water heating coil. The cooling coil, however, will probably be deeper (more rows deep), since more surface is necessary for dehumidification. Also, due to the air dehumidification, it is necessary to allow for condensate to be drained away from the unit. This will require a condensate pan and drain and possibly a condensate pump depending on the location of the air handler. Some systems have drain lines located in the air duct to remove any additional water that may condense in the ductwork (Figure 75-27).

Water-cooling coils should also be piped as counterflow heat exchangers. The chilled water entering should be supplied to the coil at the air outlet side of the coil. In this way the chilled water will first absorb heat from the coldest air leaving the coil. A typical chilled-water piping system for a large air handler is shown in Figure 75-28.

Figure 75-28 Typical chilled-water piping system for a large air handler.

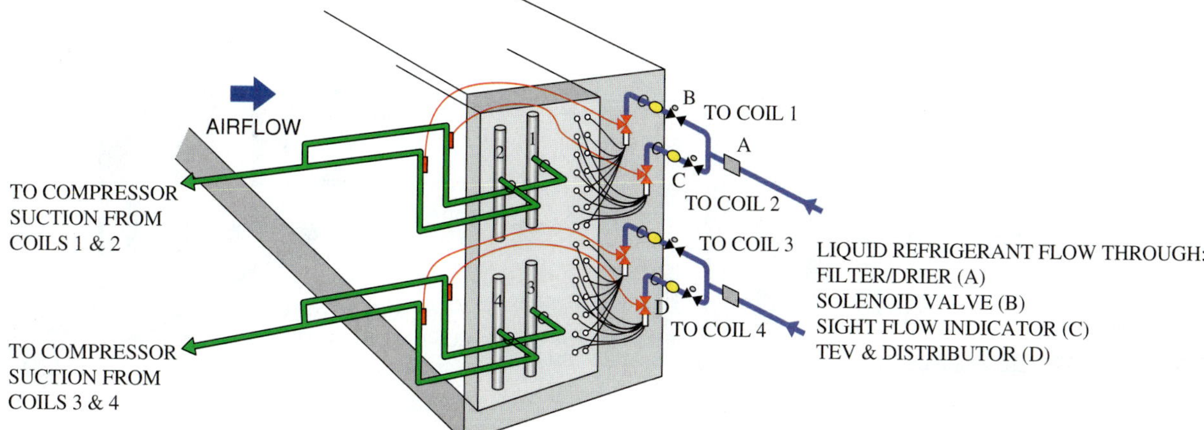

Figure 75-29 Direct expansion four-row face-split cooling coil.

75.20 DIRECT EXPANSION COOLING COILS

Direct expansion cooling coils can be configured for multiple stages of capacity control. Multiple expansion valves along with unloader solenoid valves, hot-gas bypass valves, and liquid-line solenoid valves are used. The direct expansion coil shown in Figure 75-29 is referred to as a four-row face-split coil. This coil uses four expansion valves with two supply lines. The refrigerant flow can be sequenced through the different coils—one, two, three, or four—depending on the load demand. The flow operating sequence is generally that the first coil to come on is the last one to come off.

SERVICE TIP

Many direct expansion coils come sealed with pressurized dry air inside the coil. If the protective caps are removed from the coil, air and moisture will be allowed to enter. If the coil is being installed and the connections are not complete, the coil should be charged with nitrogen and recapped until the final installation can be completed.

The suction connections on a direct expansion cooling coil are typically on the air inlet side of the coil. Expansion valves should be mounted as close to the coil as possible. The location of the expansion valve thermal bulb is very important. These are commonly mounted on the vertical line located after the refrigerant outlet header. In addition to the expansion valve for the control of refrigerant flow, distributors are often used to channel the refrigerant evenly to all sections of the coil (Figure 75-30).

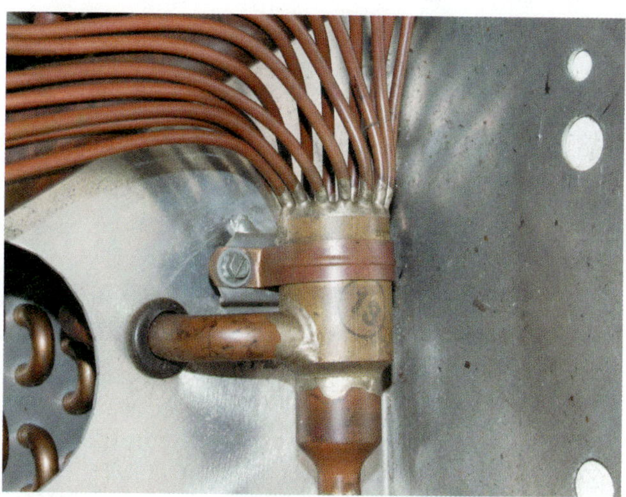

Figure 75-30 Refrigerant distributor.

75.21 HUMIDIFIERS

Comfort is greatly affected by the relative humidity of the air. At a temperature of 72°F and a high relative humidity level of 75 percent, you would feel very uncomfortable. At the same temperature of 72°F with a lower relative humidity of 35 percent, you would feel cozy. At 72°F and a 10 percent relative humidity, you would probably feel cool. Too much or too little humidity is unpleasant, so it is important to maintain the proper humidity.

Cooling coils often are used for dehumidification purposes, particularly for moist air. The air temperature falls below the dew point as it passes across the coil and the water is removed. The opposite is true during the heating season. Very often in the heating months, the air may be too dry. Hygroscopic materials such as wood, leather, paper, and cloth require a fixed amount of moisture to preserve their proper condition. These materials shrink when they dry out. If moisture loss is rapid, warping and cracking can take place. To retain their quality, proper humidity must be maintained. A low humidity can cause shrinkage of the framing around doors and windows. This increases the infiltration of outside air into the building, which leads to greater energy usage.

From a health standpoint, medical science reveals that the nasal mucus contains some 96 percent water. Doctors indicate that the drying out of the nasal tissues in winter

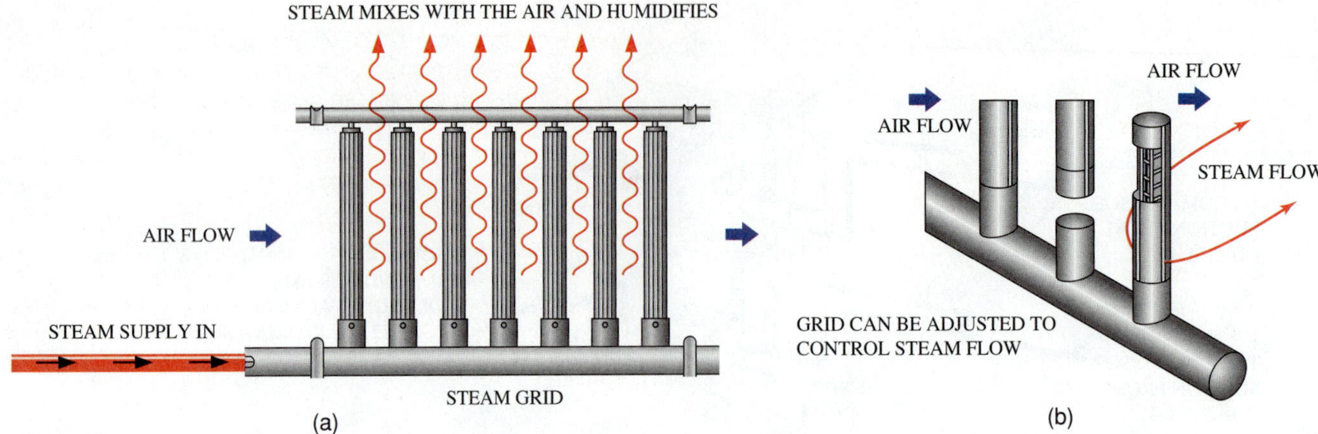

Figure 75-31a (a) Steam humidifier; (b) steam humidifier vertical manifold.

helps to initiate the common cold. Maintaining proper humidity can greatly affect one's susceptibility to colds. Also, proper humidity levels permit lower comfort temperatures, allowing thermostats to be set lower, which results in energy conservation.

TECH TIP

One of the limiting factors in maintaining proper humidity is the condensation that occurs on windows in winter. Table 75-1 shows the outside temperature at which condensation will occur for various types of glass.

Humidifiers are used to add moisture to the air. The steam-grid type is highly recommended because it offers simplicity in construction and operation, and humidification can be closely controlled (Figure 75-31a). The live steam is directly admitted through a control valve into the airstream (Figure 75-31b). The steam will heat the air and add water vapor to it. Another style of humidifier, referred to as the steam-pan type, is used when the introduction of steam directly into the airstream is undesirable. The vaporization of water from a pan provides moisture to the conditioned air. In applications where steam is unavailable, hot water sprayed directly into the airstream can be used. Water spay will provide optimum performance in applications where the humidity level is fairly low and precise control is not required.

Table 75-1 Outside Air Temperature Effect on Window Condensation

Window Type (@ 70 °F Inside Temperature)			Inside Relative Humidity Resulting In Condensation
Single Pane	**Double Pane**	**Triple Pane**	
43 °F	18 °F	10 °F	50%
38 °F	11 °F	0 °F	45%
34 °F	2 °F	–10 °F	40%
28 °F	–8 °F	–25 °F	35%
22 °F	–20 °F	–35 °F	30%
15 °F	–30 °F		25%
8 °F			20%
0 °F			15%
–11 °F			10%

TECH TIP

Humidifiers that are not properly maintained are subject to mold and mildew growth. Because of concerns with indoor air quality, it is very important that any humidifier be properly maintained to prevent and remove mold and mildew formation. When mold and mildew dry out, spores are released that become primary irritants in a building. Before a system is restarted, you must remove all mold and mildew. Follow the equipment manufacturer's procedures and all local health code requirements when removing mold and mildew from humidifiers or any other part of a building.

UNIT 75—SUMMARY

Air-handling units come in many sizes and configurations depending on the application. For individual space heating and cooling, fan coil units may be used supplied with hot or cold water from a central plant. Packaged air handlers can be factory assembled or put together as sections on the job site. Cooling coils can be of the indirect expansion type using chilled water or the direct expansion type with multiple expansion valves for capacity control.

There are two distinctly different types of fans: centrifugal and axial, based on the direction of airflow through the impeller. The centrifugal fan is used the most for moving air through duct systems. Propeller fans are used for moving air at relatively low-pressure differences, as in condenser fans. The fan laws are a group of formulas that describe the relationship between the variables affecting airflow, including fan speed, air volume, CFM, static pressure, and horsepower. The fan curve is a graphical representation of fan performance. Fan curves generally show the volume of air moved in CFM compared to the power required in Bhp (brake horsepower) and the resistance in inches of water column. The performance of any particular fan can be accurately predicted using fan performance curves.

Filters in the airstream can be disposable or cleanable depending on their design. HEPA and electronic filters are used where high-efficiency filtration is required. Humidification of the airstream allows for increased human comfort and reduces the drying out and shrinking of windowsills and doors. Microbial treatments and ultraviolet lamps are used to reduce the buildup of mold and mildew. Depending on the application, an air handler may include any arrangement of these accessories to provide for the proper distribution and treatment of air supplied to a building.

WORK ORDERS

Service Ticket 7501

Customer Complaint: AHU Supply Fan Motor Is Tripping

Equipment Type: 75-Ton AHU with Centrifugal Fan

The technician arrives at the site and speaks to the maintenance department supervisor. They recently replaced belts on the air-handler supply fan and replaced a worn adjustable pulley sheave and now the AHU will not stay online. First the technician starts the unit and measures the amperage drawn by the supply fan motor and finds it to be too high. Next the technician visually checks the supply fan belts and finds that one of the new belts is already wearing on one side.

It is now obvious that the sheave-shaft alignment and overhang are not correct (Figure 75-32). The belt tension was also too tight (Figure 75-33). When maintenance replaced the motor sheave, they did not align it

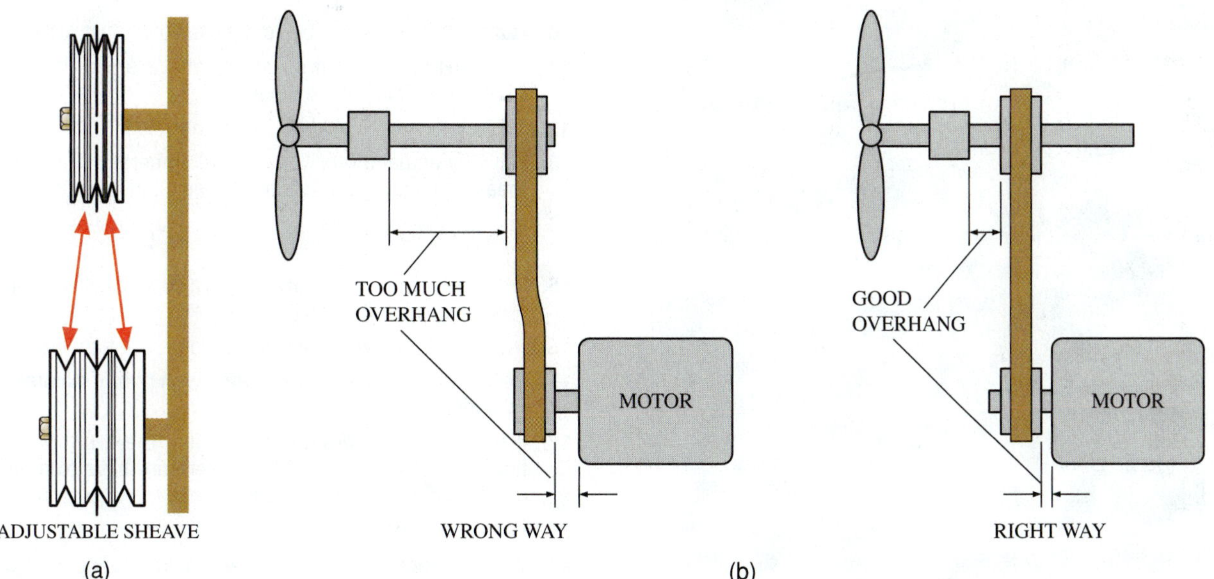

Figure 75-32 (a) Sheave-shaft alignment; (b) sheave-shaft overhang.

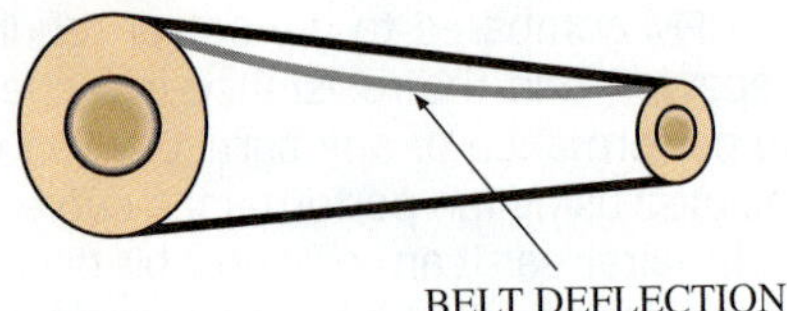

Figure 75-33 Supply fan belt deflection.

properly, causing the motor to overload. The technician removes the belts and adjusts the sheave alignment correctly. The technician reinstalls the belts, aligns the belts, and restarts the AHU. The unit is now operating within design specifications.

Service Ticket 7502

Customer Complaint: Humidifier Will Not Work

Equipment Type: Electric-Powered Steam-Pan Humidifier

The technician answers a complaint about very dry air in a dental office. The office has a small central station air handler located in a utility room in the back of the building. Opening the access panels to the unit, the technician finds an electric-powered steam-pan humidifier. Measuring a zero voltage drop across the electric heater element indicates that it needs replacement.

The technician removes the old element and cleans the pan of scale and mildew. The scale and mildew are the result of normal use, and the minerals normally found in the makeup water tend to collect in the pan. The technician replaces the element with the exact factory replacement part and restarts the humidifier. The humidifier is now operating normally. The technician recommends regular maintenance and a water softener installed on the makeup water to the pan.

Service Ticket 7503

Customer Complaint: No Heat

Equipment Type: Weather-Tight Central Station Air Handler

The technician answers an emergency no-heat call at a large dry-cleaning establishment. The central station air handler is a large weather-tight unit located on the building's roof. The technician pulls hard to get the door to open as it is under a slight negative pressure inside the air handler. There are multiple fan and coil arrangements located inside supplying air individually to multiple zones. Each fan arrangement has its control mounted beside it. One of these is flashing an alarm signal. The alarm code manual is hanging by a small piece of chain next to the control panel, and the technician looks up the fault code to find that it is for a low air-supply temperature.

The heating coil is easy to distinguish from the cooling coil because of the steam supply and return piping attached to it. The steam control valve is in the fully stroked wide-open position as it should be calling for heat. The technician uses a heat gun to check the surface temperature of the steam coil and finds that it is no hotter than the airstream passing over it. The smaller steam preheat coil at the entrance to the air stream registers a temperature of over 200°F and is the only source of heat. The technician locates the steam trap at the outlet of the heating coil and opens the bypass valve around it slowly to minimize any water hammer. Almost immediately the heating temperature begins to increase. Next the technician closes the valves on either side of the steam trap and slowly breaks the connection to make sure that there is no steam pressure remaining. The steam trap is replaced and put back into service, and now the supply air temperature is normal.

UNIT 75—REVIEW QUESTIONS

1. Although air-handler cabinets may be made of galvanized steel, there are still areas that may be subject to corrosion. Why?
2. Describe the two general types of AHUs.
3. List the components of an FCU.
4. Often around fluorescent lights in a dropped ceiling there is a narrow slit. What is the purpose and advantages of this application?
5. Where are larger FCUs most commonly used?
6. In FCUs that have cooling coils, how is the dehumidification condensate disposed of?
7. Explain the advantages that unit-ventilator-type FCUs have in schools.
8. List the major components in most AHUs.
9. State the reason a properly sized AHU fan and motor usually requires field adjustment during the final start-up/commissioning phase.
10. Name the two basic types of fans used in most AHUs.
11. On an AHU with a centrifugal fan, what are two advantages of backward-curved airfoil blades?
12. State where axial fans are commonly used.
13. Tell why actual motor voltage and amperage should be field measured and logged under full load conditions.
14. In a heating coil, what advantage does heating water have over steam?
15. When referring to humidity, why is a cooling coil deeper (more rows) than a heating coil?
16. How should ultraviolet germicidal lamps be handled?
17. Describe the problems associated with poorly maintained humidifiers.
18. If the inside temperature is 70°F and the RH is 50 percent, how low would the outside temperature have to be to form condensation on a double-pane window? Refer to Table 75-1.
19. Why is the steam-grid humidifier recommended?
20. What horsepower motor will be required to increase the airflow of a fan to 1,000 CFM if a ¼-hp motor is required to move 800 CFM?

21. What are the characteristics that AMCA fan classes are based on?
22. What are the two broad types of fan designs?
23. What is the difference in operating characteristics between centrifugal fans with forward-curved blades and centrifugal fans with backward-curved blades?
24. Explain how a centrifugal fan moves air.
25. List the two main types of fan drives.
26. What speed does a fan need to turn to increase its airflow from 800 CFM to 1,000 CFM if the original speed is 230 rpm?
27. What happens to the amp draw of a centrifugal blower motor if the airflow to the blower is restricted?
28. What is an "air-over" motor?
29. What is the most common fan used in residential air conditioning?
30. Why should a current reading be taken anytime a belt or pulley adjustment is made to a belt-drive fan?

UNIT 76

Single-Zone Rooftop Unit Installation

OBJECTIVES

After completing this unit, you will be able to:

1. apply the manufacturer's installation instructions to the proper locating of a rooftop/package unit.
2. select the proper clearances around and above the rooftop unit.
3. evaluate the roof structure for proper weight-bearing capabilities.
4. follow instructions on how to install a roof curb adaptor for a rooftop unit.
5. perform the correct hookup of the utilities through the curb.
6. design and install the condensate drain.
7. follow directions on installing the vent hood.

76.1 ROOFTOP UNIT INSTALLATION

The rooftop unit is only as good as its installation. The majority of rooftop units are packaged systems. The longevity, operation, and servicing of the rooftop unit depend on how well the unit is installed relative to the manufacturer's installation recommendations and common sense. If the rooftop unit is installed improperly, the system will not work according to its design specifications or only marginally at best. Improper operating conditions will lead to a shorter lifespan for the unit. Poor installation practices also make it more difficult to service the unit, can create unsafe work conditions, and can possibly lead to property damage.

SAFETY TIP

Safety precautions should always be taken whenever accessing a rooftop unit. Ladders are normally used to gain access to the roof. Only occasionally will there be stairs in a stairwell that will grant roof access. The base of the ladder should be placed the correct distance from the roof. A general rule is for every 4 ft the ladder is extended, the ladder's base should be 1 ft away from the wall it is leaning on. Another technique is to set up the ladder placing your toes at its base. Standing straight up with your arms extended, you should be able to touch the ladder rails with the palms of your hands.

The ladder should extend at least two rungs above the roof, which should be about 3 ft. The ladder should be secured at the top to keep it from blowing down in the wind. This would leave you stranded on the roof, and the falling ladder could cause damage to objects or people below.

The ladder should also be inspected periodically to make sure it is sound. As an example, ladder rails can become worn from constantly rubbing against the ladder rack on the service van. This might not be noticed without careful inspection. A ladder in this condition may be set up correctly but collapse under your weight as you begin to climb it, leading to serious injury.

Pay special attention to the following rooftop installation considerations:

- Unit location
- Clearances
- Roof support: whether the roof is strong enough for the weight of the unit
- Rooftop curb choices and installation
- Electrical specifications and hook-up
- Gas piping, if gas heat
- Condensate drainage
- Venting if a gas heat unit

76.2 UNIT LOCATION

Installation of a rooftop air-conditioning/heating unit should conform with the manufacturer's recommendations and all local and national codes, such as the National Electric Code, the National Fuel Gas Code, the Uniform Mechanical Code or the International Mechanical Code.

Considerations must be given for rooftop units so they are not subjected to unduly harsh environmental conditions that could cause corrosion of the metallic parts of the system. Examples leading to corrosion would be salt spray near seacoast areas and fumes or mists from nearby manufacturing facilities. If the unit is to be located in a seacoast area, it should be sheltered from the predominant ocean wind direction. This may be difficult, so one option may be to install the unit on the ground if that is a better choice than the roof (Figure 76-1).

If fumes or mists are present, locate the unit in the best possible place relative to their source. Offending smells can also create a problem. Some outside air is always required, and the amount will depend on the application. Location of the unit is important so objectionable smells are not introduced to the conditioned space.

(a)

(b)

Figure 76-1 (a) Rooftop unit installed on the ground; (b) rooftop unit installed on the roof.

TECH TIP

The location and orientation of a rooftop unit on a shopping mall installation should include an assessment of the exhaust vents and makeup-air vents for other units located within the same general vicinity. It is important to avoid installations that can lead to such things as hot-dog-flavored dresses in a clothing store or loading dock truck fumes in a sporting goods shop.

If the rooftop unit cannot be installed in a favorable location away from contaminants, then frequently wash the coils, cabinets, and fan blades. This washing will slow down any corrosive effects. The condenser coil is the component most likely to be subject to the damaging effects of corrosion eventually affecting the operation of the unit. There are aftermarket coatings that can be applied to reduce corrosion. However, always check with the manufacturer before making any modifications to the equipment so as not to void any warranties.

Another factor to consider in locating the rooftop unit is water runoff. A poor location increases the risk of water leaking into the building. Choose a location least affected by rain and water runoff. Visual inspection of the roof after a rainfall would be one way of determining a good location. Check for any standing water puddles on the roof. For a new construction project, the building plans should be consulted. There will always be limitations to where the unit can be installed as determined by access to ductwork and utilities.

The unit should be installed high enough off the roof to prevent water from collecting in the unit itself. If it is ducted from the side, place it on 4 × 4 treated wood posts cut to fit the length or width of the unit. Generally two or three posts will be necessary to support the unit properly, depending on the size of the system. A transition of some kind, such as a piece of felt or tar paper, should be placed between the wood and the roof. The rooftop unit should be set on a roof curb if the system utilizes a duct return and supply.

TECH TIP

When deciding on the location for the rooftop unit, keep in mind that a crane will be picking the unit off the ground and setting it on the roof. Cranes will have a maximum distance they can reach out over the edge of the roof. If the designated location is too far from the edge of the roof, the crane cannot be used to move it all the way. Some other means will have to be considered for the final positioning of the unit. In this case, use the crane initially for setting the unit down on a dolly or on round posts. If posts are to be used, then simply push the unit and the posts will roll with it. As the unit rolls off the last post in line, move that post to the front to keep the process going.

76.3 LEVELING THE UNIT

The rooftop unit has to be installed in a level position. This is so condensate collecting from the evaporator cooling coil surface drains properly. On a flat roof, leveling the unit is typically not much of a problem. A level placed on the unit can be used to check that it is setting properly. When a curb is utilized, it must be leveled during its installation. If treated wood posts are used, make sure they create a level platform before setting the packaged unit on them. When the unit is to be set on a sloped roof, other considerations will have to be made. Build a platform, usually made from steel, to keep the unit level on a sloping roof.

76.4 CLEARANCES

Maintain proper clearances around the rooftop unit. This allows for proper condenser airflow, room for service, room for a possible economizer, room for the proper exhaust venting from a gas heat unit, room for side duct connections, and room for electrical and gas connections.

Clearances will vary by manufacturer and style of unit. Typical manufacturer's recommendations call for at least 60 in of clearance above the condenser fan to prevent recirculation of air back into the condenser. A clearance of 60 in is common for most air-conditioning units if the unit has top discharge. Maintain an 18-in clearance around the condenser. Maintain a 48-in clearance in the front to access controls.

Another issue related to clearance is snowfall. Make sure the unit is set above the highest anticipated snowfall for the geographical location. This is important for proper venting of flue gases and flow of combustion air when a gas heater is utilized. If a heat pump rooftop unit is used, then the outdoor coil has to be above the highest anticipated snowfall for proper airflow through the coil.

TECH TIP

While servicing a rooftop unit, care must be taken not to damage the roof. When panels are removed to access electrical or refrigerant connections, make sure the panels are secured. Unsecured panels can become dangerous flying objects when blown by the wind. Also be careful with the corners of panels when setting them down, as these can damage the roof. A worst-case scenario is for inside water damage from a heavy rain shortly after the unit has been serviced. This most likely indicates that the roof was damaged due to poor maintenance procedures.

76.5 ROOF SUPPORT

Roof strength has to be considered when installing a rooftop unit. First determine if the roof is strong enough to support the weight of a rooftop unit. The weight of the unit can be found in the manufacturer's installation instructions. The total weight will depend on the size of the unit. Some examples are as follows:

- Rheem 180,000 BTU of cooling—1,702 lb
- Rheem 240,000 BTU of cooling—1,796 lb
- Lennox 60,000 BTU of cooling—535 lb
- Rheem 36,000 BTU of cooling —520 lb
- Lennox 36,000 BTU of cooling—355 lb

Once the weight of the unit is known, consulting with the architect or an engineer may be required to determine if the roof is strong enough for a particular unit.

SAFETY TIP

The weight of the rooftop unit will have to be known to make sure the proper size crane is used to set the unit on the roof.

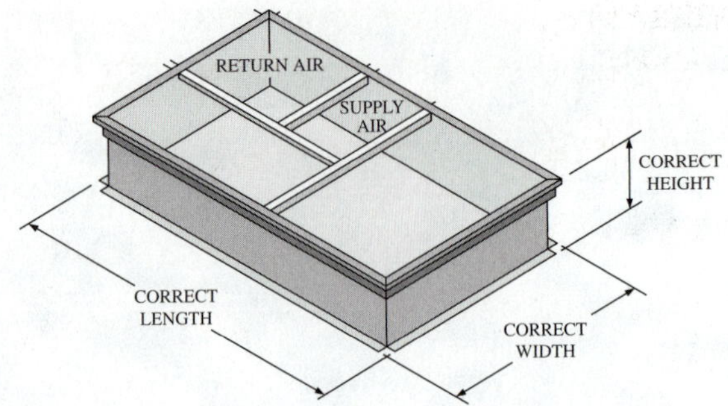

Figure 76-2 Typical roof curb dimensions and openings.

76.6 ROOF CURB

When the installation requires a roof curb, follow the installation instructions to install the curb properly and safely. The correct curb has to be matched to the rooftop unit (Figure 76-2). Normally the manufacturer supplies the correct curb for a particular rooftop unit, but when replacing an existing rooftop unit with another brand and/or size of system, a roof curb adaptor may have to be used. Replacing an existing roof curb should be avoided to preserve the integrity of the roof.

The roof curb from the manufacturer may come unassembled, but even so, field assembly is still relatively easy. Matching up the correct sides and inserting corner brackets will typically connect them together. In some cases, the curb will be a generic one that will fit different sizes of packaged units. With these, the curb package will come with rails to attach to the inside of the curb frame, relative to return-air and supply-air openings. Consulting the installation instructions makes putting things together simpler than guesswork.

For a new construction project, the roof curb should be delivered and installed (per installation instructions) in the chosen location when the rafters are up and before the roof is decked. This allows for the curb to be flashed and installed as an integral part of the roof.

The typical design for roof curbs is as follows:

- The heights of the curbs may vary with manufacturers. An example would be a 14-in or 24-in curb, depending on how high the unit needs to be off the roof.
- The curb, when put together, is normally in a rectangular shape to reflect the shape of the rooftop unit.
- The top of the curb will be flat all the way around. The rooftop unit will set on top of this flat spot and have a lip that extends down and on the outside of the curb (Figure 76-3).
- Use a gasket material between the rooftop unit and the curb to prevent water leakage, which would flow inside of the structure.
- The rooftop unit is secured to the curb by some means, such as screws.
- The roof curb will generally have a nailer strip around its top edge for the roof flashing to attach to.
- The bottom of the curb will have a 90-degree turnout that would set on the frame of the roof and that the roofer would set the decking on top of.
- The curb will have rails that are positioned inside the curb relative to how the ductwork will be installed. If

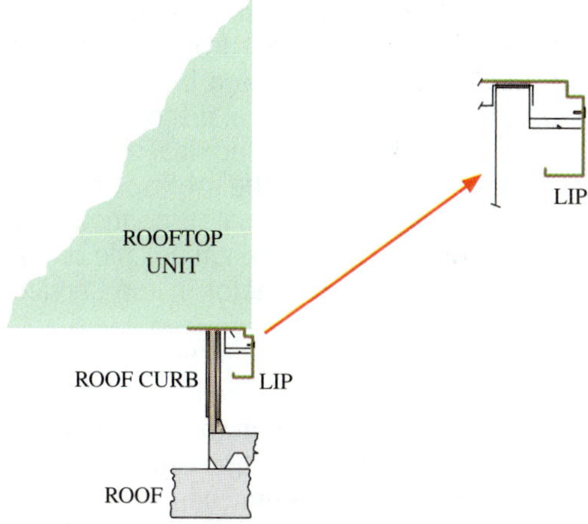

Figure 76-3 Cutaway view of a roof curb.

the duct comes out of the sides of the rooftop unit, then this position is not critical. If the ductwork is to come out of the bottom of the unit, then the rails will have to be set accordingly.

- The curb may have insulated panels that will be set between the rooftop unit and the portion of the curb under the condenser.
- The roofer provides counterflashing, insulation, and cant strip, which all go on the outside of the roof curb.

The roof curb has to be installed square to the roof relative to the roof joists. The purpose is so that the plenums, when dropped into their correct positions, will line up with the roof joists. Once finished and the curb is set, the building will most likely still be in the rough-in phase of construction. Therefore, the rooftop unit will not need to be set in place until later in the construction process. If that is the case, close off the top of the curb with plywood or sheet metal. This prevents water from leaking into the structure.

The advantage of installing the curb on a roof during the construction phase should now be apparent. No holes for the return- and supply-air plenums must be cut into the roof. The roofer simply decks to the curb, insulates around the outside, applies flashing, and installs the cant strips along the bottom edge of the curb. The roof is finished with whatever material is used as a sealant, leaving the portion of the roof inside the curb open.

For installation on an existing roof, holes have to be cut through the roof for the return-air and supply-air plenums. This can be a complicated procedure. Concrete roofs require special concrete saws. If the decking is wood, then the procedure is slightly easier. However, no matter the roofing structure, consideration should be given for subcontracting this portion of the installation to a roofing company. They normally have the proper tools to penetrate a roof, whatever the material. Once the holes have been cut through the roof for the plenums, the curb is placed into position, and the roofer takes care of the flashing and sealing.

When preparing to set the rooftop unit, first the plenums are installed through the roof curb. The plenum's open end will have a lip around the edge, and this lip will rest on rails laid in position. Gasket material is placed on the top edge of the curb, and the rooftop unit is lowered onto it (Figure 76-4). Generally a crane company will be contracted

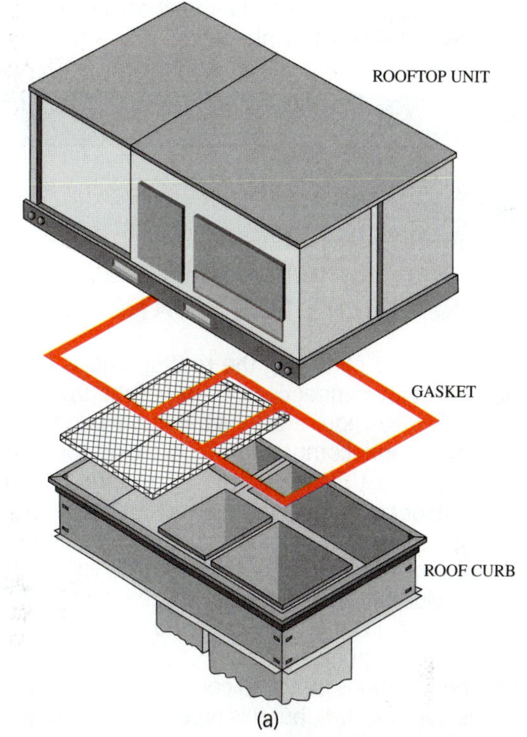

(a)

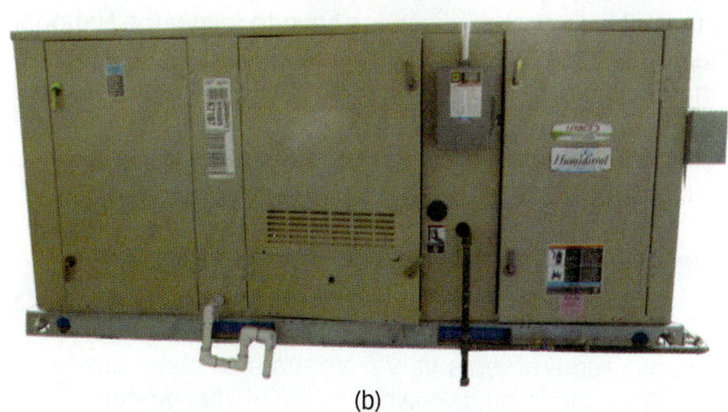

(b)

(c)

Figure 76-4 (a) Rooftop unit, curb, and gasket; (b) exterior view of rooftop unit; (c) air flow through the filter and down through the opening in the rooftop unit to the building's duct system.

to pick the unit up, move it over the roof and above the curb, and lower it down and into place.

SERVICE TIP

Before the rooftop unit is set, always check the level of the curb.

76.7 CURB UTILITY PENETRATIONS

Another key aspect of setting the rooftop unit is the consideration for utility connections. Some of the utilities that a rooftop unit may require are high-voltage wiring to run the fan motors and compressor or compressors, wiring for electric strip heat if the unit utilizes electric heat, low-voltage wiring for connecting the thermostat to the package unit, and a gas line if the system has gas heat. Rooftop units normally have openings in their bottom pans to match the openings provided in the roof curb. Electrical connections can be brought in from the building and connected to the unit through these openings.

Utility penetrations may also come directly through the roof outside the roof curb, but this type of setup offers more chances for water leaks. One important point to remember is that all electrical connections have to follow the National Electrical Code. In addition, local and state laws will dictate what utilities the HVACR installer/service technician can run and hook up. Also, what licenses the HVAC company holds will determine what work the HVACR installer/service technician can do.

TECH TIP

License requirements vary from state to state. License law and codes dictate what parts of the heating and cooling system the HVACR installer/service technician can hook up. Certain portions of the electrical and gas installation can be performed by the HVACR installer/service technician, while some parts of the installation cannot. If in doubt, contact the local code officials or state license law officials.

On a new installation, the voltage requirements of the unit must be determined so the electrician can connect the proper supply voltage for the unit. Voltages available vary and may be 208–240 V single-phase, 208–240 V three-phase, or 460–480 V three-phase all at the same location. If an existing rooftop unit is to be replaced, then the new unit will need to match the voltage currently used, unless another voltage is available. Installing a new unit with a different voltage or phase would increase the cost of replacement. It is also important to note that a balanced three-phase circuit means more than just a voltage balance. It also includes a balance of frequency and sine wave, which can be tested with some higher-end meters.

The high-voltage circuit from the main electrical panel should be wired to the disconnect switch mounted on the rooftop unit. This gives the service tech an easy and quick way to disconnect the unit from its power source. Some manufacturers will provide a bracket to attach to the rooftop unit, and the electrical disconnect will attach to it. The main thing is that the electrical disconnect must not block access to any part of the rooftop unit that might need servicing. From the electrical disconnect, the high-voltage wiring normally goes through flexible or rigid conduit and connects to the contactor in the service panel inside of the unit.

Low-voltage wiring will have to be run from the unit down to the system thermostat. The thermostat should be positioned at a central location in the conditioned space, close to the return air if possible. The return air is what the thermostat should be monitoring. Thermostat locations to be avoided are those that may be influenced by blowing supply air, drafts from doors to the outside, or direct sunlight. The thermostat should be mounted approximately 5 ft high.

If the unit utilizes resistance heat, then a separate electrical circuit may have to be wired into the heating package. (The manufacturer's installation data should be consulted for proper field wiring of the low- and high-voltage circuits.) Proper wire size is a must. The size of wire is dependent on the load demand. (Manufacturer's data can be consulted for minimum and maximum ampacities of their systems, which can be used to determine proper wire sizes.) Voltages should not vary more than 10 percent from nameplate data. If the rooftop unit is three phase, then the phases must be balanced within 3 percent. This means that the voltage between any two power leads has to be within 3 percent of one another.

76.8 GAS PIPING

If the unit utilizes gas heat, then a gas line will need to be connected to the rooftop unit's gas valve. A gas stop valve must be installed in the line ahead of the rooftop unit's gas valve. In addition, a regulator to reduce the pressure will have to be installed, unless there is a regulator installed on the ground at a different location.

SERVICE TIP

Before opening the gas stop valve, make sure the gas pressure has been reduced from the main line pressure. Gas pressure should not exceed 10.5 in wc at the gas valve for natural gas. If it is higher than this, damage will likely occur to the gas valve.

The gas line should be sized to prevent any undue pressure drop. A drip leg has to be installed in the gas line as close to the unit as possible. This provides a place for sediment and moisture to collect and is easy to clean out. It also can be used for measuring the gas supply pressure (Figure 76-5). The manufacturer's installation instructions should be consulted to determine the proper pipe size gas line for the capacity of the unit.

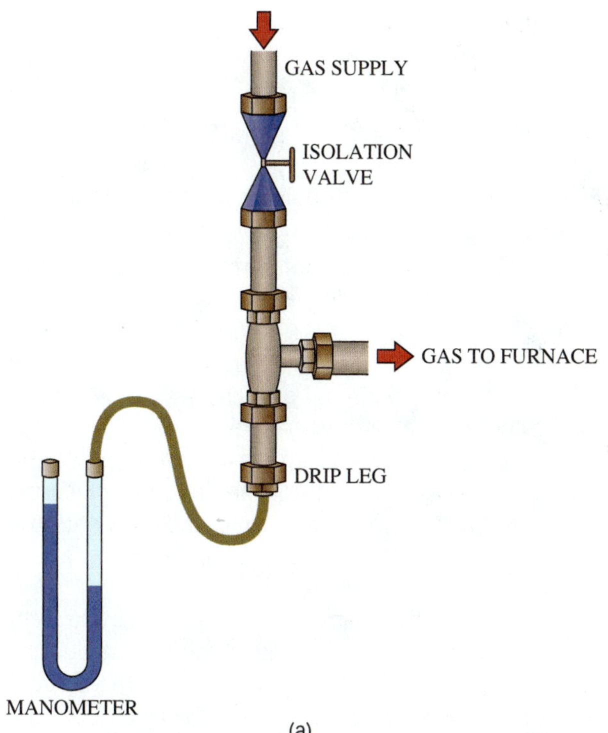

(a)

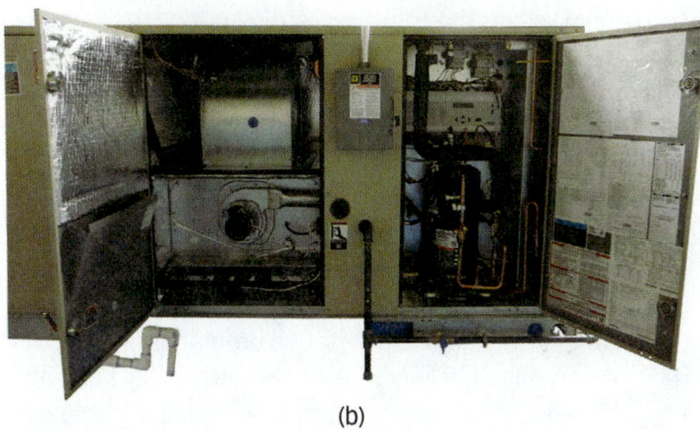

(b)

Figure 76-5 (a) Measuring gas supply pressure at the clean-out connection with a manometer; (b) rooftop unit with both a gas line and a condensate line.

76.9 CONDENSATE DRAIN

Part of the process of air conditioning is moisture removal from the air for dehumidification. The evaporator coil is below the dew point of the air being conditioned. Water in the air condenses on the cold coil and follows the fins down into the drain pan located beneath it. From there, the condensate drains out into the drain line through the drain connection.

For this drain process to work properly, the condensate drain line should have a trap installed in it. The trap serves a couple of purposes. One is to prevent insects from finding their way into the rooftop unit, where they could then enter the structure. The other purpose is so the condensate will drain properly.

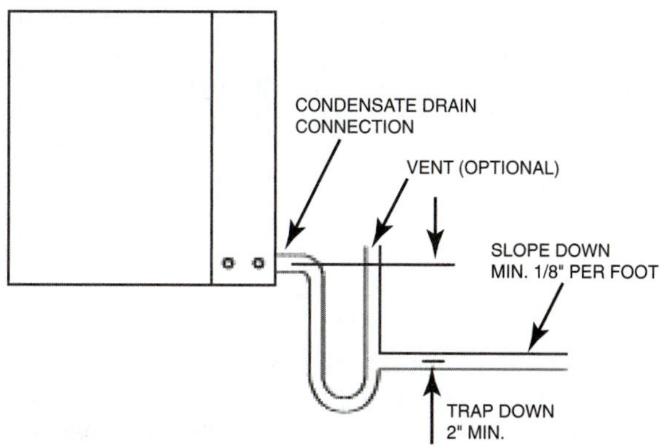

Figure 76-6 Condensate trap. *(Copyright Climate Master, Inc.)*

Many packaged/rooftop units have draw-through evaporator coils. The blower is located downstream of the evaporator coil and creates a negative pressure across the coil. Therefore, the drain will also be at a negative pressure whenever the blower is running. This can prevent the condensate from draining properly. A water-filled trap installed in the drain line will provide an air break. Because of the trap, water will not be pulled back up through the drain while the blower is running. There are also many examples of older units where the evaporator coil is located in the discharge airstream. In this case, the primary function of traps is to prevent the cooled air from escaping the unit and being wasted.

When trying to decide the best way to design a trap, you should simply follow the recommendations of the manufacturer. Normally a schematic with the proper drain trap dimensions is provided with the unit (Figure 76-6).

Although a condensate trap is a necessity, it can also create problems. One of these is the possibility of freeze-up. There will be instances when the rooftop unit may be cooling in the afternoon but by morning the outside temperature has dropped below freezing. The condensate trap full of water from the cooling cycle will freeze and split. Now there will be a problem during the next cooling cycle. If this is a common occurrence, corrective steps can be taken. The trap can be insulated, and if necessary heat tape can be installed around the outside of it.

Another problem with drain lines on rooftops is that they are exposed to ultraviolet radiation from the sun. UV light will break down PVC pipe. One way to correct this condition is to insulate the pipe. Another option is to use an alternative piping material, such as copper tubing. However, this is generally a more expensive remedy.

SERVICE TIP

When installing the drain line for a rooftop unit, attention to detail will guarantee proper condensate removal. If the drain line is going to be run a considerable distance, an air vent should be installed. The location for the vent is important. It should be placed right after the trap at the point where the condensate exits. The vent will help prevent vapor locking of the water in the drain. It is also important to correctly pitch the drain line away from the unit, typically a 1-in drop for every 10 ft of distance. There should also be no dips in the drain line.

76.10 VENTING

Rooftop units can be manufactured with cooling only or with cooling/heating. If the unit has electric heat or is a heat pump, there are no considerations required for proper venting or combustion air supply. If the unit has gas heat, there needs to be accommodation for venting and combustion air. Before the introduction of forced-draft or induced-draft rooftop units, the heating part of the unit was vented from the top through a vent pipe and cap. These older-model rooftops used standing-pilot ignition systems, and nuisance blowouts of the pilot occurred frequently. Most of the units used today, which are forced- or induced-draft systems with automatic ignitions, use only an exhaust/combustion air inlet hood. The exhaust/combustion hood normally comes with the unit uninstalled. After the initial setting of the rooftop unit, the hood is screwed into its proper place.

UNIT 76—SUMMARY

There is nothing that can replace the experience gained from working in the field, but even an experienced technician must still keep abreast of changes in installation procedures and equipment design. Reading current literature and trade journals, attending seminars and training sessions, and referring to the manufacturer's installation instructions are the keys for the HVACR installer/service technician to stay current with installation procedures. HVAC equipment manufacturers want their equipment installed correctly and operated properly and safely.

There are a number of important considerations to be taken into account when installing a rooftop unit. The strength of the roof to support the HVAC unit needs to be determined. The proper location of the unit has to be selected to minimize the chance of water leakage into the building and to avoid damage from corrosive, poor-quality outside air conditions. Proper voltage and phasing must be applied to the unit. Drain lines must be installed correctly for proper condensate removal. Clearances on top and around all sides of the unit must be adequate for correct operation, safety, and servicing. The correctly sized roof curb must be installed correctly.

Installing a rooftop unit is a multistep process. Any one improperly completed step can lead to poor performance from the system and premature failure of the unit. The manufacturer's installation instructions the relevant codes, and the use of common sense, will go a long way in making sure a rooftop HVAC system runs like it was designed and in providing the customer with what has been paid for.

WORK ORDERS

Service Ticket 7601

Customer Complaint: Water Dripping Through Ceiling Tile

Equipment Type: Rooftop Package Air-Conditioning Unit

A business customer calls in a problem complaining of water dripping through the ceiling tile. There has not been a rain in several weeks, so the rooftop air conditioner is suspected. The service technician gets the call and arrives, goes inside the establishment, and locates the ceiling tile with the water damage and drip. Now the service technician has a good idea of where on the roof to start looking for the leak. Setting up a ladder and buckling on a tool belt, the technician climbs onto the roof. Just as suspected, a rooftop unit is directly above the spot in the ceiling where the leak is.

It is a spring day, warm and humid. The service technician knows the air-conditioning unit will be condensing a lot of water from the air being conditioned. The condensate trap is found lying on the roof, detached from the main drain leaving the unit. It looks as though the trap broke at about the middle of an elbow, probably due to water being still in the trap and freezing during the winter.

The air conditioner is running, and the service technician hears the condensate gurgling as it tries to drain out, but the negative pressure will not allow it. The service technician takes the panel off of the evaporator section and secures it. The condensate is flowing over the edges of the drain pan and through a gap between the return duct and the roof curb. Then the condensate is spilling down through the attic and on top of the ceiling tile.

The service technician installs a prefabricated trap to replace the damaged one. To help prevent the freezing problem from occurring again, the technician insulates the trap. Following the condensate drain line down along to the roof drain, the technician sees a steady stream of water running into the roof drain. The inside of the unit is checked one more time to make sure the condensate drain pan for the coil is not overflowing. The service technician reinstalls all panels securely and cleans up.

Service Ticket 7602

Customer Complaint: Compressor Noise on Start-up

Equipment Type: 3-Ton Rooftop Packaged Air Conditioner

The 3-ton rooftop unit's installation is complete. It is now time to start the unit up. With the correct panels removed, the service technician has the proper refrigerant gauges installed and a voltmeter ready to check voltages. A clamp-on ammeter is placed around one of the power leads to the contactor. The system is a three-phase, 240 V unit.

The service technician does one more check to make sure all the wiring has been done correctly. Taking out a set of jumper cords, the technician carefully jumps R to Y and R to G and then turns on the disconnect switch. The compressor contactor and fan relay energize. The blower comes on correctly, the condenser fan begins running, and the compressor starts up, but it makes a terrible noise. The service technician looks at the gauges and notices that the suction pressure does not drop like it should if the compressor were pumping.

The unit is equipped with a scroll-type compressor. Quickly turning off the electrical disconnect, the service technician notes that the compressor scroll is orbiting backward. The scroll compressor will only compress refrigerant if the rotor is running in the correct direction. The service technician disconnects two of the high-voltage leads from the contactor and reverses them, tightening them securely. The disconnect switch is then turned back on and the unit starts up and begins running normally. The compressor runs smoothly, and the suction pressure starts to pull down like it should.

Service Ticket 7603

Customer Complaint: Several Offices Warmer than Usual, Unit Not Cooling

Equipment Type: Rooftop Air-Conditioning Unit

The building manager calls in and explains that several offices are running warmer than usual. Contacted by the dispatcher, the service technician heads out to the location. This is a new account, so there is no record of what kind of equipment will be involved. The only information received from the dispatcher was a no-cool call, an address, and who to talk to.

Arriving on the job, and after finally locating the right contact person, the service technician finds out which offices are having problems. The HVAC unit is located on the roof. The service technician correctly sets up a ladder, buckles on a tool belt, and climbs up the ladder to the rooftop unit. The unit is running, so the disconnect switch is shut off to remove power to the unit.

The service technician pulls the panel to access the refrigerant service ports and secures the panel. The service port caps are unscrewed and a gauge manifold with low-loss fittings is installed. The panel providing access to the evaporator coil is next removed and secured. One temperature thermister is positioned in the return air and another in the supply airstream. The service technician puts this panel back in place, so the airflow is correct through the coil. A temperature probe is positioned on the suction line about 10 in before the line enters the compressor.

The service technician then powers the unit back up. All components seem to be running, so the panel removed to access the service ports is put back in place. If this panel is not in place, then the airflow would be around the condenser coil and not through it.

After the unit has run about 10 minutes, the service technician takes some readings. The discharge pressure appears to be reasonable for R-22 at 275 psig, as the outdoor ambient temperature is 94°F. The suction pressure of 80 psig is high. The service technician then checks the temperature difference across the evaporator coil, and it is 10°F, which is too little. Upon further investigation, the service technician can feel hot air being pulled in from a gap between the rooftop unit and the curb. The evaporator coil cannot overcome the 94°F air that is being pulled in from the roof.

This is a job for the installers to fix. The service technician calls in to the dispatcher and fills in the details. The dispatcher says an installer will be over right away.

UNIT 76—REVIEW QUESTIONS

1. Describe one method to determine if a ladder is correctly set up.
2. Besides the manufacturer, where can the rooftop unit installer consult for proper unit installation?
3. If the rooftop unit is installed in a seacoast area, what is one consideration as to its location?
4. If the rooftop unit is subjected to a corrosive environment, and cannot be installed in an area away from the offending corrosion, what would be a possible alternative?
5. What is one of the main reasons the rooftop unit has to be level?
6. Proper clearances are required for what purpose concerning the condenser?
7. What are the two possible duct configurations for a rooftop unit?
8. If using a heat pump rooftop unit, why is proper height off of the roof important?
9. Why must the technician secure the rooftop unit's panels when servicing the unit?
10. Where can the weight for a rooftop unit be found?
11. The gasket material used between the rooftop unit and the roof curb is used for what purpose?
12. Why is it important for the roof curb to be square with the roof joists or rafters?
13. Utilities can be run to the rooftop unit in what two ways?
14. What code specifies the correct way for electrical connections to be made?
15. When installing units requiring three-phase power, what is one of the concerns with the three-phase installation?
16. What process allows for condensation of moisture on the evaporator coil?
17. What must be installed in the drain line to ensure proper condensate removal?
18. If PVC pipe is used for the construction of the drain line, what must be done to prevent deterioration of the pipe, and what causes this deterioration?
19. If the rooftop unit utilizes gas heat, what two processes have to be considered in the installation of the unit?
20. What is the customer's number-one concern relative to the rooftop unit?

UNIT 77

Zoned Systems

OBJECTIVES

After completing this unit, you will be able to:

1. explain how the outside air supply to a conditioned space is normally controlled.
2. identify the differences between air-distribution systems for core cooling as compared to perimeter cooling applications.
3. describe how a constant-air-volume system is normally configured.
4. explain how a variable-air-volume multiple-zone air system is normally configured.
5. explain how an energy-recovery wheel operates.

77.1 INTRODUCTION

Air-distribution control systems vary depending on the type of application, building design, and zone requirements. A single space maintained at one constant temperature would require the simplest form of control system. The amount of air supplied to the space can be allowed to remain constant with the control system configured to control the air temperature through the air handler. The duct static air pressure and the load on the supply fan remains relatively constant.

A large multistory building will have conditioned spaces of all different sizes. Some of these will be in the center core of the building, while others will be around the outside located in the outer perimeter. The heating and cooling demands for one space will vary from another. For these reasons, a building of this type will require more complex control systems. Each space may require its own terminal unit with dampers that are adjusted to meet the individual space temperature requirements. As the position of air dampers changes throughout the system, the duct static pressure and loading of the supply fan also change. Variable-frequency drive supply fans can be used that allow for speed variations to offset the changing duct static pressure.

77.2 AIR HANDLING

A typical air-distribution system is shown in Figure 77-1. This circuit has many different components. In tracing this circuit, we will start at the entrance to the air-handling unit (AHU). Air that enters the unit is made up of both return air from the conditioned space and outside ventilation air. In the AHU, air passes through the filters, preheater, cooling coil, heating coil, reheater, and finally the fan. This type of system is referred to as a draw-through system because the fan is located after the coils and draws the air across them. A blow-through system would locate the fan before the coils.

Outside Air Supply

The air entering the air handler will be a combination of outside air mixed with return air from the space being conditioned. If only outside air were used, the load on the heating or cooling system would be excessive. To conserve energy, the outside air supply can be regulated to a minimum amount. Normally, the outside air supply is never completely shut off because some fresh air should always be supplied to the space. The amount required typically depends on the number of people occupying the space and the usage of the space. As the amount of outside air is reduced, the carbon dioxide (CO_2) levels in the space will increase (Figure 77-2). An air-quality sensor can be installed in the conditioned space or the return-air duct to detect levels of CO_2 in the air. The sensor can be used as an input for control of the outside air damper. This allows only enough outside air to maintain the proper air quality while not oversupplying outside air, which would increase the total energy demand.

TECH TIP

Generally some outside air is always supplied to a space, but there are exceptions for fire containment. Fire dampers close off the supply of outside air completely so that the fire is not fed any additional oxygen. There are also exclusive systems for special applications that totally shut off the supply of outside air in the event of chemical or biological attack. These dampers should not interfere with normal operation.

Filters

Filters may vary greatly depending on the building type. On packaged air handlers, disposable filters are often used. As the filters become dirty, the airflow will become restricted. This can lead to increased energy use due to inefficient system operation. A filter status differential pressure switch can be used to trigger an alarm when the filters become dirty. Filter manufacturer literature should supply information on the differential pressure limits for dirty filters. Generally, the differential pressure is set from 0.05 to 2.0 in wc, depending on the filter and system type.

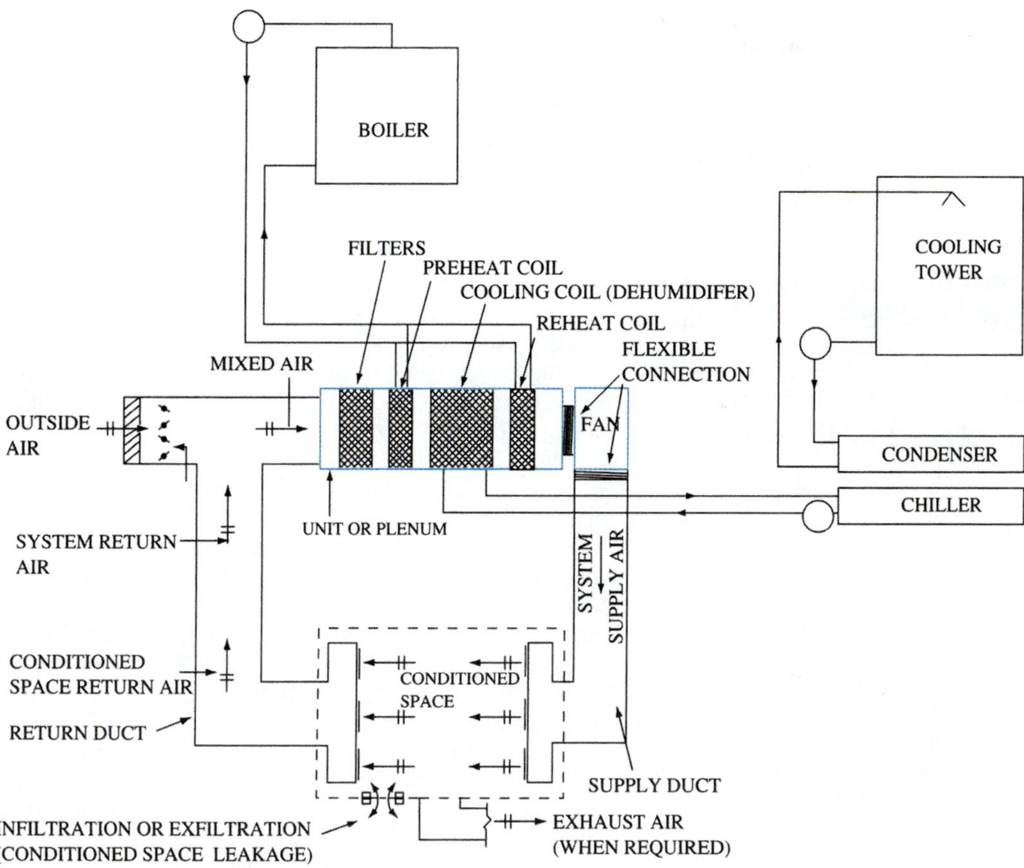

Figure 77-1 Typical central HVAC system showing air handling unit (AHU).

Cooling Coil

Cooling coils are typically direct or indirect expansion. The evaporator for a direct expansion system is directly in the air stream. The evaporator for an indirect expansion system is in a chiller, which cools a secondary refrigerant. The secondary refrigerant is circulated to water coils located in the air stream. The heat from the air gets to the refrigeration system indirectly, through a secondary refrigerant. The control of air temperature when using direct expansion coils is accomplished by regulating the refrigerant flow through the coil at the inlet. Coils may have multiple expansion valves and be staged to supply more or less cooling depending on the final air temperature. Indirect expansion coils use chilled water and typically the air temperature is controlled by regulating the water flow through the coil at the outlet. Face and bypass dampers can also be used to vary the airflow across the coil as an added control of the air temperature.

Preheater

The preheater is located in the airstream before the cooling coil. With indirect expansion chilled-water coils, there is always a possibility of the coil freezing during cold weather conditions. This is because no matter how cold it may become outside, some outside air needs to enter the space for proper air quality. The low air temperature could freeze the water inside the coil, causing it to split and then leak. The coil would need to be repaired or replaced. To avoid this situation, one option is to drain the chilled water loop every time the outside air temperature drops, but this would be impractical. More commonly, preheaters are utilized. These

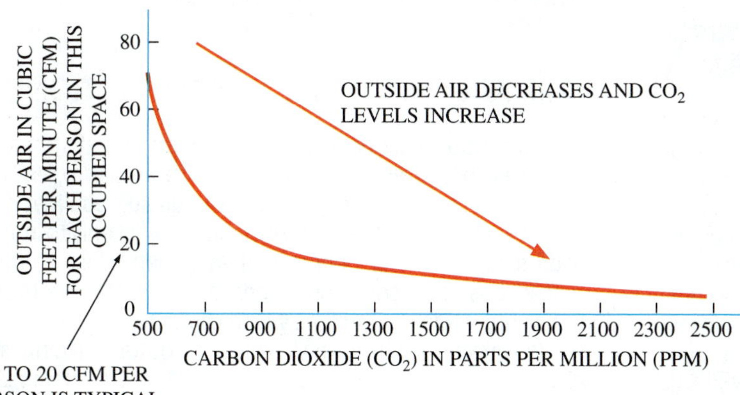

Figure 77-2 Ventilation air based on CO_2 setpoint.

may be electric, steam, or hot water. The preheater will turn on when the incoming air drops below a preset temperature. This should be well above the freezing temperature of the coil, which may vary from system to system depending on the amount of glycol added to the chilled-water loop. In addition to protecting the cooling coil from freezing, the preheater also provides some heat to the conditioned space.

TECH TIP

Preheaters are used to prevent the cooling coil from freezing. Many systems also include a low-temperature switch as a backup that will turn off the fan if the air temperature becomes too low.

Heating Coil

Heating coils may be electric, steam, or hot water. Electric heat is the most expensive but also the simplest. Hot water is also very common on small commercial systems. Many furnaces in this size category are forced hot air or hot water rather than steam. Hot-water heating coils are also simpler and easier to control than steam coils. However, steam coils provide for increased heat transfer for large-capacity systems and are preferable if an adequate steam supply is available.

Reheater

Reheaters can be electric, steam, or hot water. These are used on systems that require increased dehumidification. In geographical locations of high humidity, the air drops down below its dew point temperature as it passes across the cooling coil. The water vapor in the air will condense and be collected and drained away as condensate from a pan beneath the cooling coil. Although moisture has been removed from the air, it is now cold with a high relative humidity and will create drafts that could be uncomfortable for the people in the space. The reheater will increase the temperature of the air and reduce its relative humidity to more comfortable levels. This type of system is inefficient because the air is being cooled, which requires energy, and then it is heated back up again, which requires additional energy. Instead of this form of reheat, it is preferable to use the warmer return air or outside air mixed with the cold air supply rather than a separate reheater whenever practical.

TECH TIP

In older building design, zone reheaters were used to control the final temperature of individual conditioned spaces. As an example, in a central system, the air supplied to multiple spaces will be at the same temperature. However, individual thermostats can be used for each space. If a person is too cold in one space then he or she can adjust the thermostat setting. This does not affect the central system but only the reheater, turning it on for that particular space. Due to current energy codes, these types of systems are no longer used.

77.3 AIR-DISTRIBUTION SYSTEMS

For simplicity, types of commercial air-distribution may be classified as low-, medium-, or high-velocity systems. The required velocity of the air is dependent on the length and configuration of the duct system along with the flow rate required for the space. Cool air is admitted to a space to absorb heat, thereby maintaining a comfortable temperature. The amount of cool air (i.e., the flow rate in CFM) required will depend on the total cooling load for the space (how much heat will need to be absorbed). Velocity should not be confused with flow rate. A large axial fan used for ventilation can deliver high flow rates of air, but this would be at a relatively low velocity. For a given flow rate (dependent on cooling load), if duct size decreases then the static pressure and the power required to deliver the air will increase.

Low-velocity systems are often associated with the application of smaller unitary packaged and split-system units. The use of limited ductwork or none at all (free blow) is typical of the classification. Where ductwork is used, the external static pressure is held down to the range of 0.25 to 0.50 in wc. The use of concentric supply and return ducts is a common application. The type of duct design is the same as that of the equal friction method, which permits the prediction or control of total static.

Medium- to high-velocity systems are often used in large central air-distribution systems. Smaller ducts use up less space, and this can lead to installation cost savings. With smaller ducts, the ceiling height can be lowered by as much as 1 ft. For a large twelve-floor building with 10-ft ceilings, this will mean that one additional floor can be added that would have been occupied by the total sum of each floor's duct system. Smaller duct systems also provide for added installation choices. Older buildings with narrow areas and space limitations due to the structural design of the building can be better accommodated. Despite these space and installation savings, it should be realized that high-velocity systems will have higher operating costs and higher noise levels.

TECH TIP

The higher the air-distribution system's operating static pressure, the lower the system efficiency. This is because the blower has to work harder to move enough air. To reduce the static pressure, larger ducts are required. The savings from having a lower static pressure must be balanced against the higher cost of the larger duct system.

77.4 AIR-DISTRIBUTION APPLICATIONS

Air-distribution applications in central station systems fall into three broad areas: interior or core areas, exterior or perimeter areas or zones, and entire building applications.

Core areas have conditioning loads subject only to interior loads, such as lighting, people, and equipment. Consequently, they are basically cooling-only loads, except for top floors and/or warmup cycles in extremely cold climates.

Perimeter zones are exposed to outer building skin variables, such as wall and window loads, wind effects, and exposure effects, as well as the items in the core areas. The

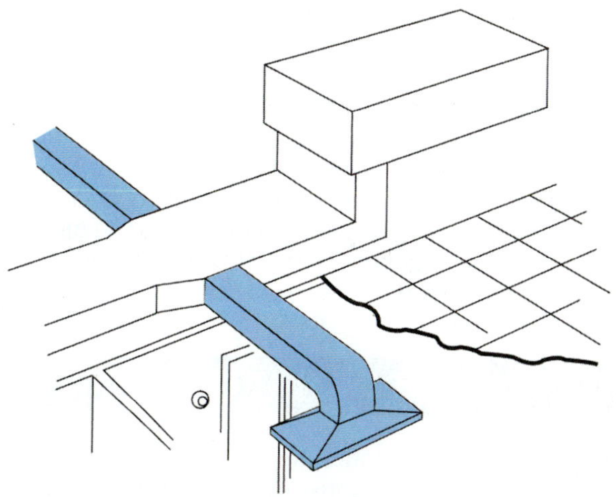

Figure 77-3 Single-zone system using constant volume.

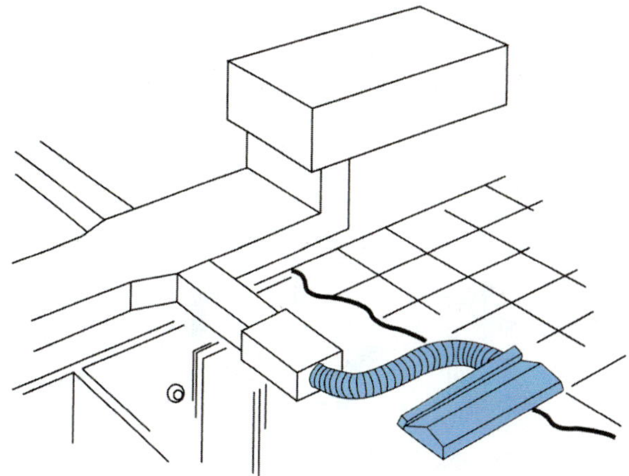

Figure 77-4 Single-zone system using variable air volume (VAV).

perimeter zones have to handle a wide range of conditions, including such variances as high solar gain on a winter day requiring cooling, while shaded parts of the building require heating.

Entire building applications are a combination of these systems, the selection of which depends on building size and economics. Small buildings frequently cannot justify one system for the core and another for the perimeter. In these instances, any one of the perimeter systems can be used for the entire building.

The systems are all of the type that use primary (supply) air from the air conditioner, and the maximum quantity needed is based on the maximum load conditions of the area.

77.5 CORE AREA APPLICATIONS

If the interior core of a building is a large open area, a single-zone constant-volume system (Figure 77-3) can be used at a reasonable initial cost. The amount of air supplied is always the same. The desired supply-air temperature is maintained by the heating and cooling coils only. A single HEAT/COOL thermostat with automatic changeover provides year-round control of the coil temperatures. Top-floor and ground-floor systems will have heating for morning warmup. Intermediate floors usually only have provisions for cooling. The duct design follows conventional practice, with room air distribution from the ceiling.

For core areas divided into smaller spaces, with variations in lighting and people loads, a variable air volume (VAV) single-zone system has typically been used. Rather than always supplying the same amount of air to the space, the amount of air can be adjusted to better meet the load demands of the space. There are several approaches to this application.

Constant-volume fans may be used with individual VAV terminals located for each space that increase or decrease the air supply with dampers. Terminal units can be designed to control air velocity, pressure, temperature, and flow rate. The final temperature is a result of mixing airstreams of different temperatures or humidity levels. Figure 77-4 shows a constant-volume air handler connected to individual VAV terminals. The terminal may be an individual control box, a system air-powered VAV slot diffuser, or a self-contained temperature-actuated VAV diffuser. In all these cases, the variable air volume is accomplished by throttling the airflow at the individual duct run, causing the constant volume air handler to "back up" the fan curve to a new balance point of lower CFM at a higher static pressure. The controls for this type of system are self-contained within the air-distribution system and do not require any more control than the thermostat noted above.

TECH TIP

Constant-volume supply fans can sometimes be used on variable-air-volume systems. However, as the demand for air decreases and dampers begin to close, the duct static pressure will increase as well as the load on the fan. This type of configuration should only be used in locations where heating and cooling loads are relatively steady.

Another approach for the core area is to use VAV terminals, with the capability of handling several duct runs from each terminal. In this case, the air handler is also constant volume, but the VAV terminal throttles the air supply to the ducts and dumps the remainder back into the ceiling-space return-air plenum. The space is controlled with variable air volume, but the air handler is operating at constant volume. The control for this subsystem requires that the space thermostat regulates an actuator on the VAV terminal, as well as the cooling and heating demand.

In either of these cases, a single air handler is sufficient in the core zone and can provide proper comfort conditions. The air handler is typically constant volume because the loads are relatively steady, regardless of the season or ambient air temperature.

An approach often seen on older systems, but prohibited by current energy codes, is commonly called zone reheat. Reheat terminals differ from this basic mixing design by controlling the final temperature of the air with hot-water coils, steam coils, or electric resistance heaters. As seen in Figure 77-5, an air handler delivers constant-volume, constant-temperature air to the distribution system. Individual zones or branches have heaters in them,

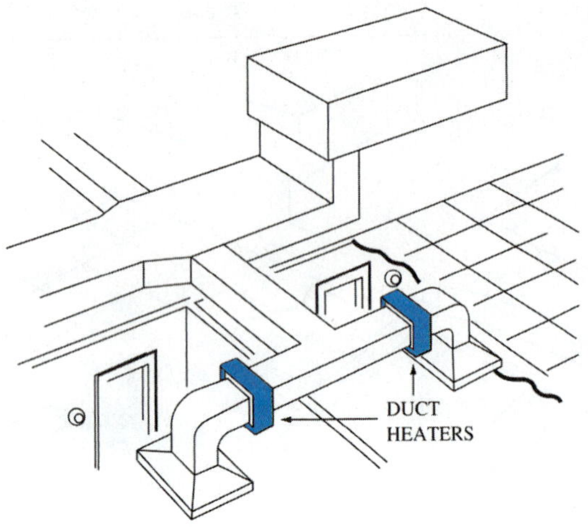

Figure 77-5 Constant-volume system using zone duct heaters.

which reheat the cool air to a comfort level for each space. These heaters may be individually controlled electric heaters, hot-water coils, or steam coils. In many cases, these heaters are applied to a problem area in a system, to provide adequate heat where the original design was insufficient.

77.6 PERIMETER ZONE APPLICATIONS

Perimeter zones require terminal systems that can handle the wide range of conditions, from the coldest morning warmup to the hottest solar gain cooling load. In many systems, this range was handled by providing perimeter-radiant or forced-air heat systems on the outer wall and under windows to furnish the necessary heat for the cold loads. A separate air system provided the necessary air movement, outside air requirements, and the cooling load responsibility. The air systems could be either constant or variable volume, but it has been found that only VAV really can provide the necessary control for comfort. In milder climates, the perimeter heating system is eliminated, without much loss of comfort.

TECH TIP

In some cases, the entire perimeter heating and cooling load will not be controlled by the central system. Instead it may be handled by console units on the outside wall, with individual outside air sources and temperature controls.

The VAV systems for perimeter applications are generally provided with primary air from either a central air handler for the entire building or an entire floor, depending on the size of the system (Figure 77-6). A single air handler may handle up to 40,000 ft^2 of occupied space, but in many cases the ductwork becomes unmanageable. Variable volume systems may be low velocity or high velocity, depending on the type of controller and the terminal units used. The volume of primary air is automatically adjusted to the

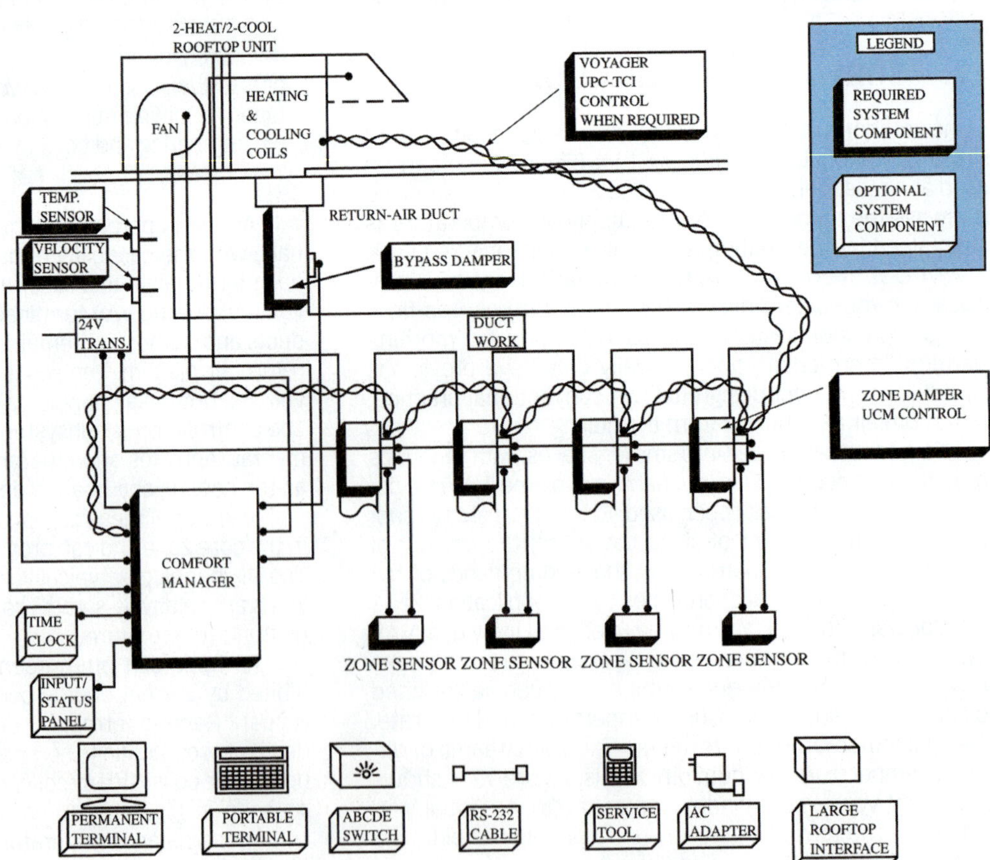

Figure 77-6 Schematic view of a typical variable air volume (VAV) system.

total cooling demand by duct sensors controlling fan volume controls.

The fan output may be adjusted by using variable inlet guide vanes. Throttling the air inlet to the fan will reduce the air flow during lower demand operation. This method is often used because it is more efficient as compared to reducing the airflow by placing variable vanes at the fan outlet. Reducing the air flow at the fan outlet leads to an increased static pressure. The increased pressure on the outlet side of the fan will make it work harder, and this can lead to an increase in power consumption.

Another option is for stepped speed control settings such as slow, medium, and fast. Stepped speed control and inlet guide vanes may also be used in combination with one another. However, the most efficient method is for the use of a variable-frequency drive (VFD) fan motor, which allows for the full range of speed control. VFD's are considerably more expensive than most other alternatives, but as the fan slows down under light loads, less energy is used, reducing long-term operating costs.

An older type of VAV perimeter zone system is the induction type. High-velocity primary air is delivered to induction room terminals. This saves space, since the ductwork is much smaller than conventional ducting. The terminal unit takes the high-velocity air through a nozzle arrangement, as shown in Figure 77-7. This induces room air into the unit through a heating coil. The primary air may be unconditioned, as there can be both heating and cooling coils located in the induction terminal. Figure 77-8 shows an induction room terminal that is constant volume, with temperature control achieved by reheat. Note that these terminal units do not rely on fans; the induced airflow provides the necessary secondary air.

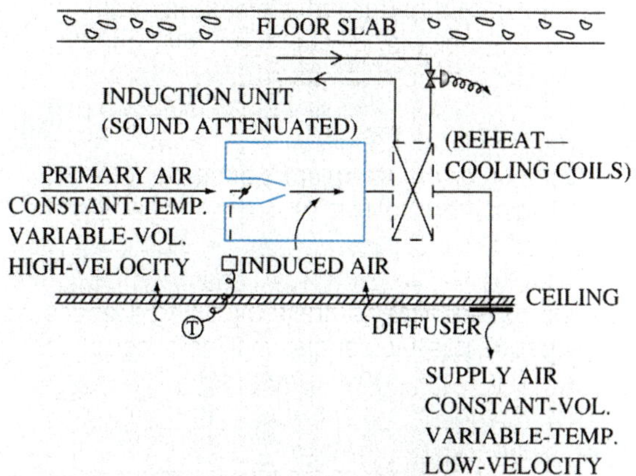

Figure 77-7 Induction terminal.

SERVICE TIP

Induction air systems have many drawbacks. High-pressure air requires very tight and sealed ductwork, and air leaks are very noisy. These systems require considerable energy, since high duct static pressure requires increased fan power. VAV aspects are also limited because the induction effect falls off quickly with a reduction of airflow.

A type of VAV perimeter zone system operating at low pressure uses system-powered boxes or diffusers. System air-pressure controls the inflation of bladders that in turn control the air volume in the unit. These are simple in concept and effective at a reasonable cost. They provide VAV

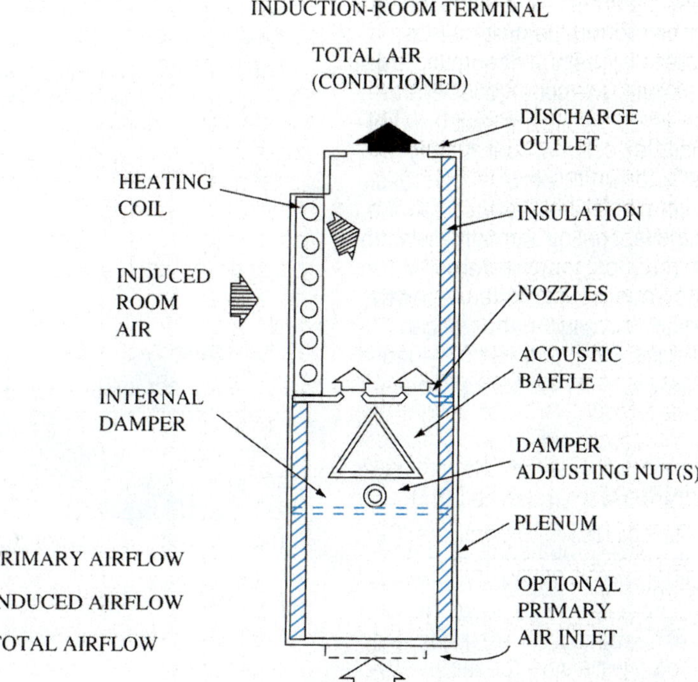

Figure 77-8 Sectional view of an induction-type room terminal unit.

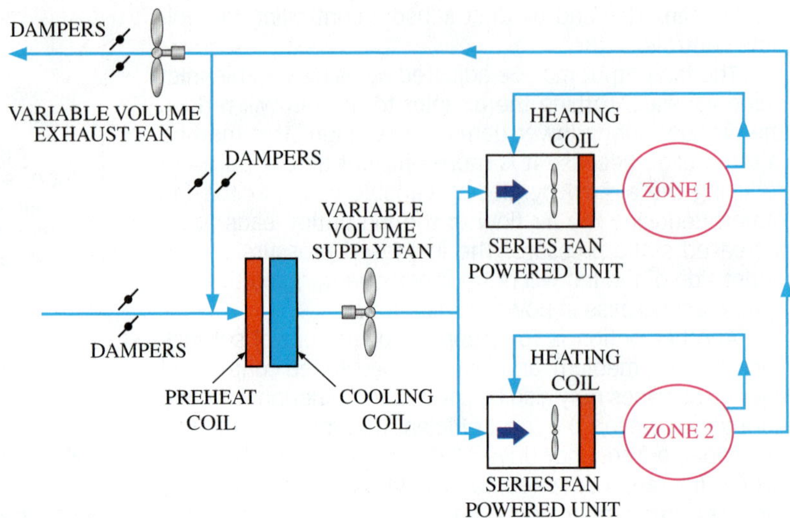

Figure 77-9 Series fan-powered variable-volume system.

space conditioning, but usually with constant-volume supply fans. Some of these diffuser units use self-contained, temperature-powered actuators to modulate the diffuser openings. These can provide fairly simple, reasonably priced VAV to smaller systems.

Other systems provide cost-effective zone control with a combination of VAV duct dampers and a sophisticated temperature control system. These allow the system to provide only heating or cooling at any given time. The demands of a number of zones are programmed to switch the system to that mode—for example, cooling. Any zones requiring heat are temporarily closed off. With a heating demand in place, the system controls will then switch the system back to heating, closing off the zones requiring cooling. The individual zone dampers are modulating, so there are no abrupt changes. A fan-speed controller is part of the system, reducing airflow to a practical minimum, as allowable. These systems are generally designed for unitary equipment, but they can be used as part of a central system.

VAV systems can be used to provide minimum airflows to meet ventilation requirements. In this type of system, the terminal units are fan powered, providing constant volume in the conditioned zone. The primary air supply will vary from 100 percent down to about 20 percent of total air requirements. The primary air handler is variable volume, constant temperature, with the cooling system demand proportional to the airflow. The low-powered terminal fan provides terminal reheat at reduced air conditions for the zone (Figure 77-9). The individual zone controls can be stand-alone systems but usually are integrated through a direct digital control (DDC) system that manages the overall operation of all terminal units.

77.7 ENTIRE BUILDING APPLICATIONS

Some older building designs utilized multizone or double-duct applications. The major limitation for this type of system is the cost of energy to provide simultaneous heating and cooling. Different types of heat-recovery methods were tried in an attempt to gain efficiency. As an example, the heat rejected from the chiller was used to provide warm water for heating. Other heat-recovery approaches have been tried, but the fully energy-efficient approaches were not always the simplest or most economical to install.

TECH TIP

Multizone and double-duct applications were popular before energy codes were enacted; these were the ultimate in zoned comfort systems. By providing a blend of always available heated and cooled air to any zone, with the addition of good pneumatic control systems, these systems supplied excellent building comfort.

Multizone systems provide one trunk duct to each zone from the multizone air handler (Figure 77-10). The air handler is a blow-through design with a hot and a cold deck. The airflow from each of these decks is fed through a set of dampers, 90 degree opposed to each other on the same shaft. When the cooling damper is fully closed, the heating damper is fully open, and vice versa. Most of the time, the dampers are modulating in response to a temperature signal from the zone. The overall system is low to medium

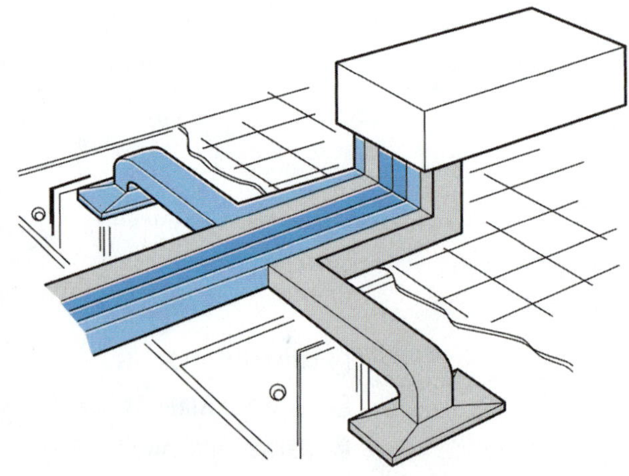

Figure 77-10 Constant-volume multizone heating and cooling unit.

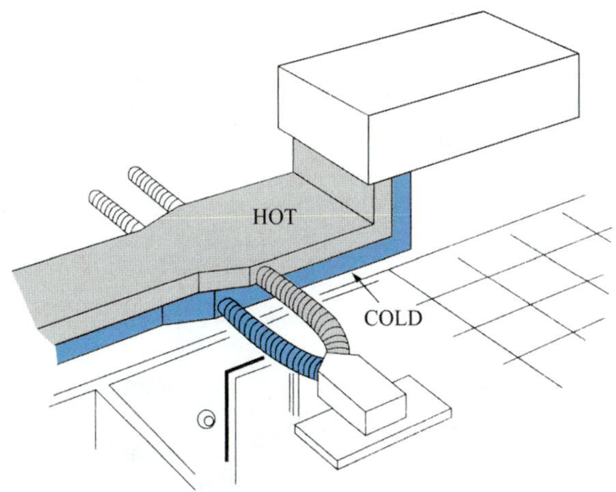

Figure 77-11 Double-duct system.

pressure and constant volume. The multizone's limitations, from a design standpoint, are that only a limited number of zones are available on the air handler. Consequently, adjustments are always needed to get an even balance. The smallest zone is typically 8–10 percent of the full load.

Double-duct systems are the ultimate in design flexibility but are very high in initial cost. Double-duct systems are high-velocity, high-pressure systems. Some are variable volume, but most are constant volume. Two full-sized supply ducts are required for the system, one for cooling and one for heating (Figure 77-11). The major benefit of these systems is that the two ducts each serve one terminal mixing box, which controls the air temperature for each zone. The mixing boxes are available in very small sizes, down to 200 CFM, allowing for zones as small as ½ ton. Zoning is limited by this small increment and the overall size of the system. The high-pressure air in the main supply ducts (4–5 in wc) is reduced in the mixing boxes, so that the distribution ducting is normally low-pressure design, with low noise levels.

77.8 MIXING DAMPERS

Dampers installed on AHUs control the flow of air through various parts of the system. The function of dampers depends on the design of the system. There are two general types of multiple-leaf dampers: parallel blade and opposing blade. Parallel-blade dampers tend to direct the air as they open, whereas opposing-blade dampers are usually constructed to more readily offer a positive close-off.

TECH TIP

The proper adjustment of mixing dampers is critical to maintaining an efficiently operating system that provides the proper air distribution within a building. Mixing dampers blend the air so that you have the proper temperature and are able to control humidity within the building. If these dampers are not set properly, the building operating cost will go up dramatically.

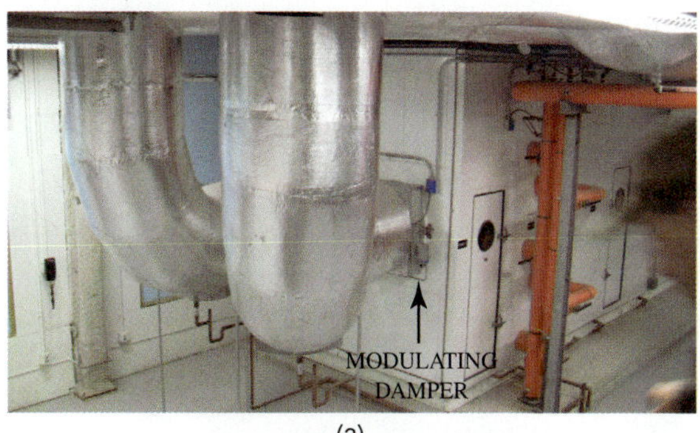

(a)

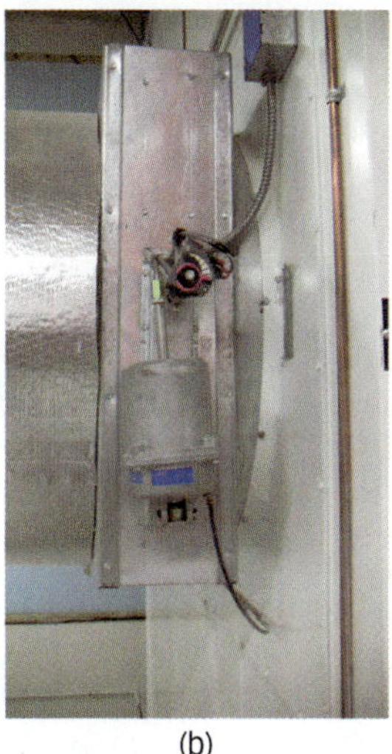

(b)

Figure 77-12 (a) Large air handler for central system showing location of a modulating damper; (b) modulating damper.

Dampers can operate either in two positions, open or closed, or they can be modulated so that they can be positioned anywhere between fully open and fully closed, depending on the requirements of the control system. Most dampers on central station air handlers are of the modulating type (Figure 77-12).

Dampers can be linked together, so that when one opens the other closes. For example, in a mixing box such as shown in Figure 77-13, where the return air is mixed with outside air, the dampers can be linked and controlled so that when the outside air is increased, the return air is decreased (and vice versa). This provides the control of the airflow needed for the wide variety of conditioning requirements.

77.9 MIXED-AIR CONTROL

To illustrate some of the many uses of dampers, refer to Figure 77-14. Of primary importance is the mixing of return air and outside air. During normal operation, only the return air and the minimum outside air dampers are involved. The minimum outside air damper is a two-position damper. Normally it is fully open to provide ventilation air to meet code requirements. The only time it closes is when the building is unoccupied or the system is shut down.

Provision is made in the damper system and the control system to use additional outside air for free cooling (economizer cycle) when practical. A maximum outside-air

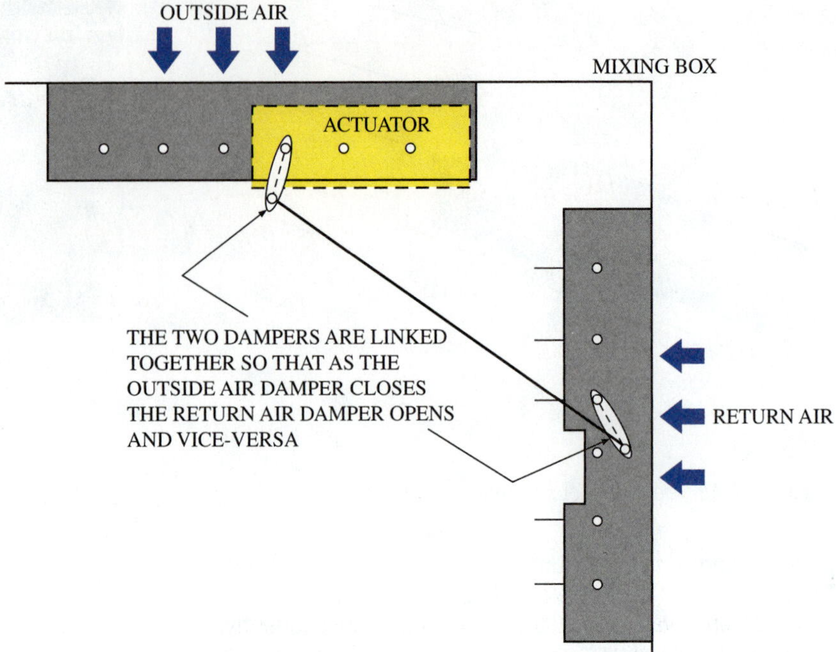

Figure 77-13 Mixing box parallel-blade dampers.

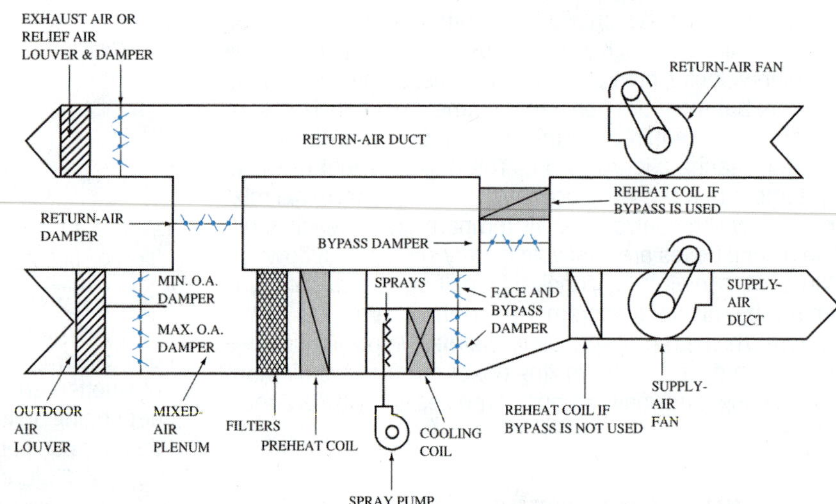

Figure 77-14 Schematic diagram of a typical AHU showing the various types of damper arrangements.

damper is modulated along with the return-air and exhaust-air damper to maintain a slightly positive pressure in the building. The total air quantity through the air handler does not change, but the proportions of outside air to return air do change to meet the control requirements.

77.10 FACE AND BYPASS CONTROL

One way to control the amount of cooling or heating is to use a face damper in series with the coil surface and bypass dampers around the coil and in a duct connection for the return air, as shown in Figure 77-14. With this arrangement, the total air volume remains constant even though the space load demand is changing. The air entering the fan can be adjusted for different load conditions by modifying the air mixture.

As an example, outside and return air are normally mixed in the air plenum chamber and passed through the filters, preheater, and cooling coil, and then to the fan. If the outside air temperature is low enough, the face dampers on the cooling coil can begin to close while the bypass dampers around the coil begin to open. This reduces the load on the cooling coil and utilizes the outside air to assist in cooling. The damper control modulates the air quantity from each source to match the load. The two sets of dampers are linked together either mechanically or electrically so that when one modulates toward the open position, the other modulates toward the closed position (and vice versa). Because the dampers are proportioned evenly, the flow of air through the fan remains unchanged. Some systems will include bypass fans as well as dampers (Figure 77-15).

77.11 ECONOMIZER

An economizer is a mixing box assembly that will allow 0–100 percent of the supply air to be drawn in from the outside. Typically the outdoor and return-air supply vents are located in one assembly (Figure 77-16). The outside

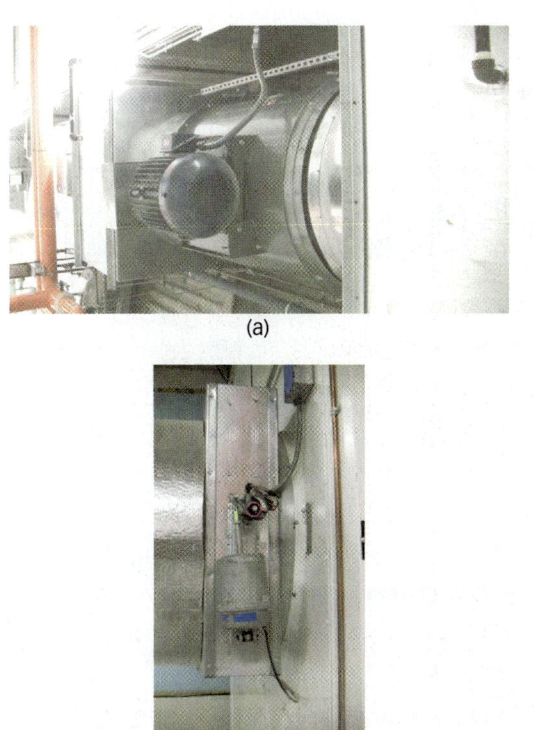

(a)

(b)

Figure 77-15 (a) Separate bypass fan and duct; (b) bypass fan.

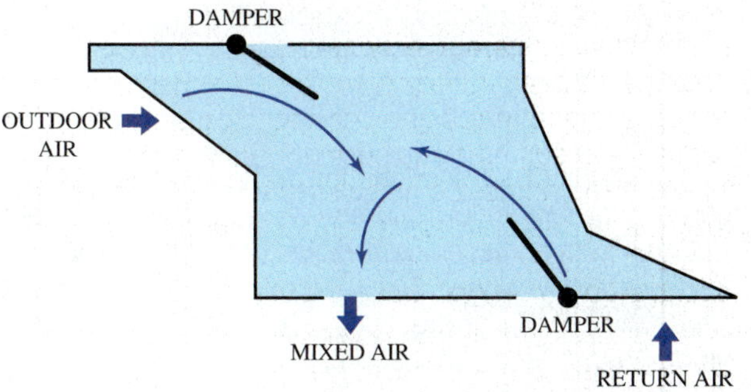

Figure 77-16 Economizer with outdoor and return-air vents.

temperature is monitored by a control system that will position the mixing damper for the most efficient operation. As an example, in the cooling mode, if the outside air is cooler than the inside air temperature, the damper will allow for more outside air to enter the building, thus reducing the total cooling load on the air-conditioning system. If the mixing damper positions allow for a high flow rate of outside air and very little return air, pressure will increase inside the building. To accommodate for this, an equivalent amount of air will need to be removed from the building by the exhaust fan.

77.12 ENERGY-RECOVERY VENTILATION

Energy-recovery ventilation is a more elaborate system for utilizing outdoor air as compared to an economizer. Energy-recovery sections often include an economizer function in their operation. However, energy-recovery ventilation not only increases efficiency but also allows for an increased amount of outdoor air to be admitted to a building. This increase in the amount of outdoor air reduces the level of indoor pollutants, as too much recirculated air can lead to an unhealthy environment. The additional load that an increased outdoor air supply places on the system can be very expensive. Energy-recovery systems allow for increased outdoor air supply without adding any considerable additional load. Systems are being designed to allow for increases in outdoor air supplies from current levels of 5 CFM up to 15 CFM per person without substantially increasing the operating costs.

The principle component for this system is often a rotating desiccant wheel that is designed to transfer water vapor and heat from one airstream to another (Figure 77-17). An energy-recovery system precools and dehumidifies the outdoor air by using the exhaust air during the summer. As shown in Figure 77-18, the return air will pass across a filter and then through the recovery wheel and out the exhaust. The outdoor air enters through a filter and then passes across the recovery wheel and into

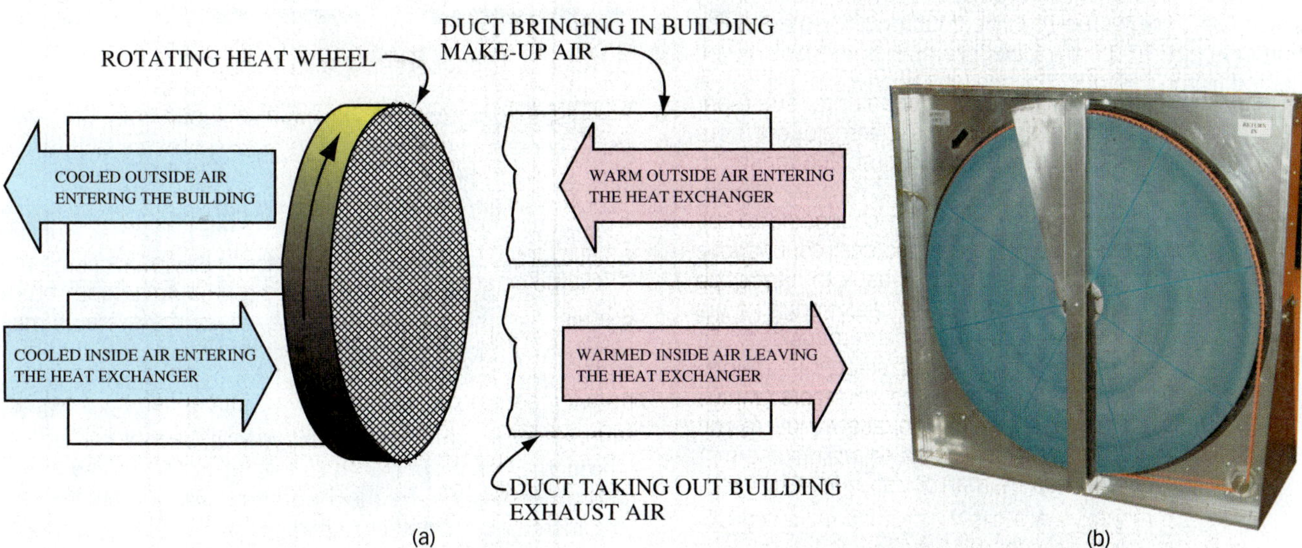

(a)

(b)

Figure 77-17 Energy-recovery wheel.

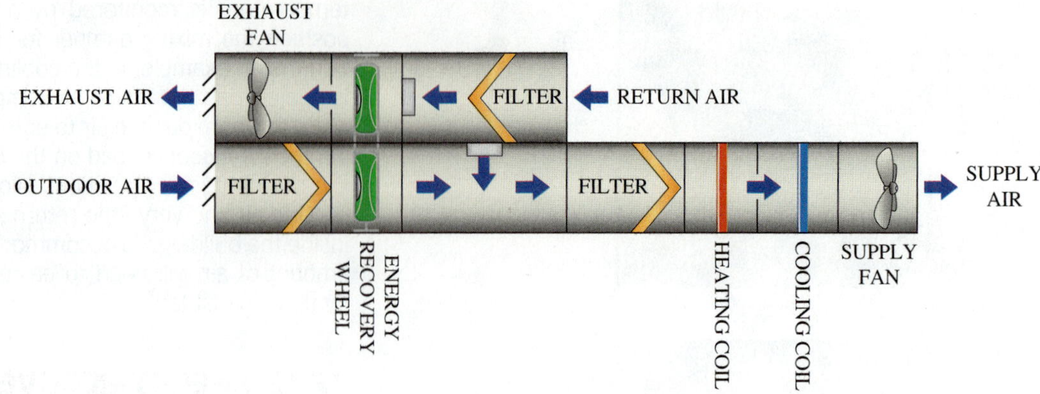

Figure 77-18 Draw-through air-handler energy-recovery section.

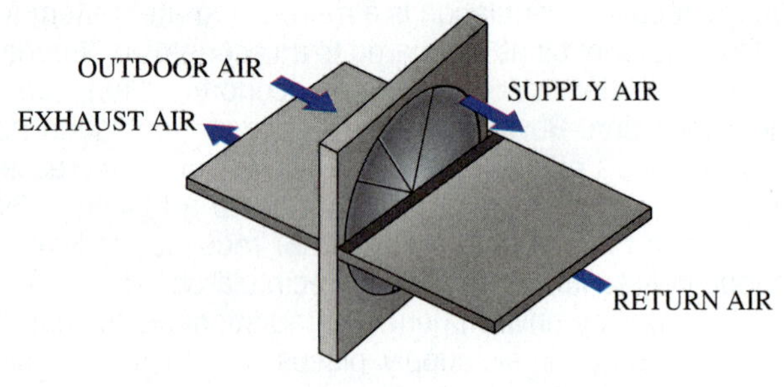

Figure 77-19 Flow-through energy-recovery wheel.

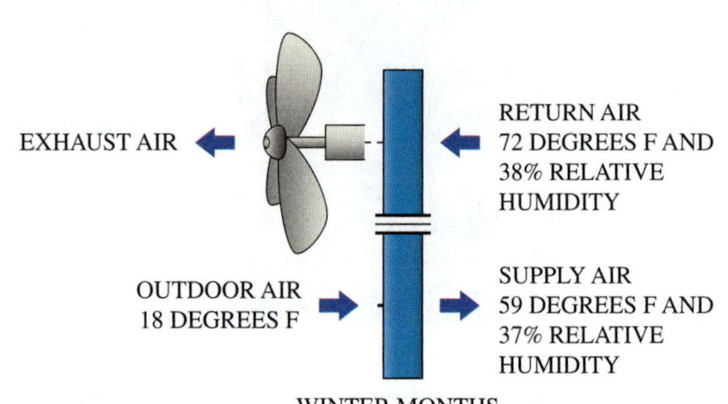

Figure 77-21 Energy wheel winter condition.

the mixing chamber. Here some portion of the return air can be mixed with the outdoor air before passing on through the next set of filters and then on to the heating and cooling coils.

The principal of operation is shown in Figure 77-19. The energy-recovery wheel rotates and outdoor air passes across the top portion of the wheel while exhaust air passes across its bottom portion. During the summer months, the heat and moisture from the outdoor air will be transferred to the exhaust air. This is shown in Figure 77-20. For this particular application, the outdoor air temperature drops from 92°F to 79°F while the relative humidity still remains low. During the winter months, as shown for the application in Figure 77-21, the cold incoming outdoor air is preheated from 18°F to 59°F along with an increase in relative humidity.

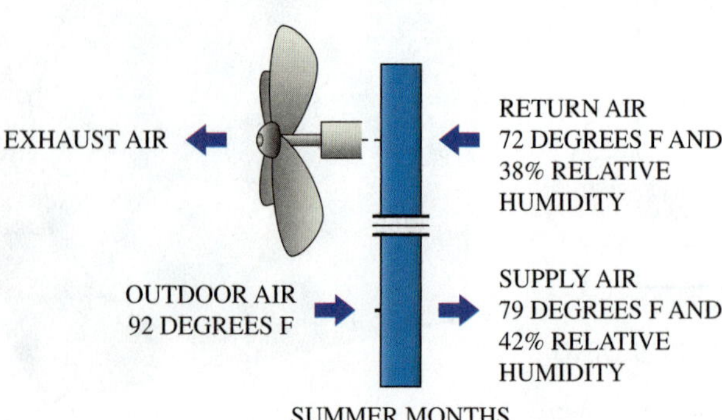

Figure 77-20 Energy wheel summer condition.

Energy-recovery sections for air handlers can be configured in a number of different ways. In colder climates, frost may form and eventually block the airflow through the energy-recovery wheel. The methods for frost protection include the use of a preheat coil, a bypass for the wheel, on and off control, and exhaust only.

UNIT 77—SUMMARY

Air-distribution control systems use input from temperature, condition, and static air-pressure sensors to position dampers for directing and mixing the airflow and/or control the supply fan. Some of these inputs are also used to regulate the temperature control of the preheater, cooling coil, heating coil, and reheater. Mixing dampers can be positioned to provide for the best mix of outside and return air through the handler. Variable-air-volume systems may also adjust the fan air supply. Inlet guide vanes are sometimes used on large supply fans that can be closed to reduce the air supply or can be opened to increase the air supply. This is a more efficient way to control the airflow through a fan as compared to regulating the airflow at the fan outlet. Variable-frequency drive supply fans are becoming increasingly popular in variable-air-volume supply systems because they can be controlled to operate at the most efficient speed and load.

WORK ORDERS

Service Ticket 7701

Customer Complaint: Supply-Air Pressure Low

Equipment Type: Packaged AHU with Centrifugal Fan

The service technician answers a call for a small office building with a recently installed packaged air handler. The owner complains that the space is too hot and that there does not seem to be very much air blowing from the supply registers in each space. The system was only installed a couple of weeks ago and seemed to work properly, but the past few days the outside temperature has been very hot. The owner is wondering if maybe the unit is undersized.

The air handler is located in the back of the building, and the technician shuts off the power and then removes the access covers for the fan section. The fan seems to be mounted correctly and the drive belts appear to be tensioned properly. Turning the fan on, it runs smoothly, but it has minimal airflow. With the fan running, the technician can see the problem is in the fan rotation. It is rotating backward. The technician shuts off the power supply and reverses two of the electrical leads to the fan motor. Once the unit is restarted, the air supply is more than adequate for the system.

Service Ticket 7702

Customer Complaint: Space Too Hot

Equipment Type: Packaged AHU Unit with Direct Expansion Cooling Coil

The service technician answers a service call for a veterinary building and speaks to the office receptionist. The receptionist explains that the waiting room and the lab areas are much warmer than they should be, and this started a few hours ago. The system had been running fine prior to that. The technician is led to the back of the building where the air handler is located. The unit is rumbling and producing considerable vibration. The technician opens the access doors and immediately sees that the fan motor is loose on its mounting bolts. The unit is shut down and the drive belt tension adjusted as the motor mounting bolts are tightened. The technician replaces the original mounting nuts with lock nuts.

The unit is restarted and the vibration is eliminated and the unit is running smoothly. However, the mixing dampers do not seem to be operating. They are frozen in one position with the outside air supply almost wide open, placing a considerable load on the direct expansion cooling coil. The technician again shuts the unit down and then tries to manually stroke the dampers and finds they are not frozen into position. The locknut in the linkage from the actuator to the dampers has come off due to all of the vibration. The linkage is reconnected and the actuator stroked back and forth to confirm the damper movement. Once everything seems to be all clear, the technician restarts the unit and the mixing dampers immediately modulate to their proper position, minimizing the amount of outside air through the handler.

Service Ticket 7703

Customer Complaint: No Air Supply—Space Too Hot

Equipment Type: Packaged AHU Unit with Variable-Frequency Drive Fan

The technician answers a service call for a multistory building with an air handler located on the roof. There is no air coming through the registers, so evidently the fan has tripped off. The control panel has an alarm light indicating a dirty filter. The technician finds that it is almost plugged. The filter is replaced and after resetting the fan motor overload, the unit is restarted. The measured current draw of the fan motor is high. The technician checks the fan speed with a strobe tachometer and finds that it is operating at its maximum speed setting.

The technician shuts the unit off again and locates the static air-pressure sensor. It is removed and found to be partially plugged at the inlet. The sensor is falsely reading a low static pressure, and this led to the increase in the fan speed. After cleaning, the sensor is reinstalled and the unit restarted. The fan is still running fast because of the high cooling demand on the system. As the conditioned space temperature begins to return to normal, the fan speed also begins to drop. After a few hours, the system begins to stabilize and return to normal operation.

UNIT 77—REVIEW QUESTIONS

1. Why is outside air mixed with return air?
2. How much outside air should be admitted to a space?
3. Can outside air ever be completely shut off from a space?
4. When will a preheater turn on?
5. What is the purpose of a filter status differential pressure switch?
6. Why are steam or electric reheaters inefficient?
7. What is an advantage of a high-velocity air-distribution system?
8. How does duct static air pressure affect the supply-fan load?
9. How does a perimeter heating or cooling load differ from a core load?
10. What type of an air-distribution system is a core zone most likely to have?

11. Can constant-volume fans be used in variable-air-volume systems?
12. Why are variable inlet guide vanes sometimes used for supply fans?
13. What is meant by a modulating damper?
14. What is an advantage of face and bypass control dampers?
15. What is the difference between a draw-through air system and a blow-through air system?
16. How do carbon dioxide levels relate to outside air supply requirements?
17. What are the common types of preheaters?
18. What are the three common classifications for commercial air-distribution systems?
19. What is the principle of operation for an economizer?
20. How do direct expansion cooling coils differ from indirect expansion coils?
21. How is the total amount of cool air in CFM required for a particular space determined?

UNIT 78

Commercial Control Systems

OBJECTIVES

After completing this unit, you will be able to:

1. explain how a chilled-water system operates.
2. list the common temperature input signals for a packaged air handler.
3. describe the function of air-quality and air-condition sensors.
4. describe how water or steam coil control valves operate.
5. explain the differences and advantages of different types of control systems.

78.1 INTRODUCTION

Air-handling units (AHUs) range greatly in size and application. Residential air handlers typically have a direct expansion coil and airflow rates in the range of 400 CFM per ton of air-conditioning capacity. In comparison, large commercial air handlers may have airflow rates of 55,000 CFM or greater. These units often use cooling coils supplied with chilled water from a central air-conditioning unit that may supply one or more air handlers.

Air handlers are often used for both cooling and heating. For the heating mode, smaller residential units may use electric resistance heat. Large air handlers have heating coils supplied with hot water or steam delivered from a boiler. In addition to heating and cooling, air handlers may be specially equipped to improve dehumidification of the air in the cooling mode and to humidify the air in the heating mode. Many air handlers today also can be configured with energy-recovery devices to increase the unit's efficiency without compromising comfort.

78.2 CONTROL SYSTEMS

The control system must direct the operation of all elements automatically. In large installations, the control system is usually separate from the air-conditioning equipment. Controls may be electric, electronic, pneumatic (air), or a combination of all three. The controls must be included in the initial construction stages to provide the total integration of the system.

CAUTION

Because these systems can be started and stopped automatically, you must make sure the power to the unit is off before beginning service to prevent the unit from starting while you are working on it.

Many air handlers will have an electronic control module installed in a control box located on the unit. The control module will continuously monitor inputs such as supply-air temperature, return-air temperature, air quality, space temperature, outdoor air temperature, duct static pressure, and space humidity. Dependent on the input, the control module is programmed to control outputs such as the supply fan, cooling and heating coil valves, mixed air dampers, electric heat, and direct expansion cooling. Multiple air handlers can be configured to be linked together.

Supply-Fan Control

The supply fan can be started or stopped based upon a number of different programmable values: the time the space is occupied, the temperature of the space, an unoccupied cool-down or warmup period, and economizer cooling, dependent on outside air conditions. Fans can be a constant speed, constant volume; two-speed variable volume; or variable-frequency drive (VFD) variable volume.

Heating and Cooling Coil Control Valves

The heating or cooling coil control adjusts the steam, hot water, or chilled-water valve. Control valves are typically two way for steam and two way or three way for water. Control valves can come as normally open or normally closed. The correct sizing of proportional control valves is critical to proper system performance. An undersized valve will result in insufficient heating or cooling and damage from cavitation due to the excessive velocity through it. An oversized valve will also experience rapid wear of the seat and disk due to the increased velocity through a mostly closed valve. There will also be rapidly fluctuating temperatures and cycling (hunting), especially at light loads for an oversized valve.

A typical installation for a two-way control valve for a cooling coil chilled-water or a heating coil hot-water loop is shown in Figure 78-1. The design pressure drop across the control valve is measured with the valve fully open. The recommended maximum pressure drop for two-way and three-way valves is 35 psig. Pressure drops in excess of this value can result in cavitation, erosion of disks, or wire drawing off the seat on a steam valve. Refer to the manufacturer's recommendations or valve sizing charts when selecting new valves or replacing existing valves in the system.

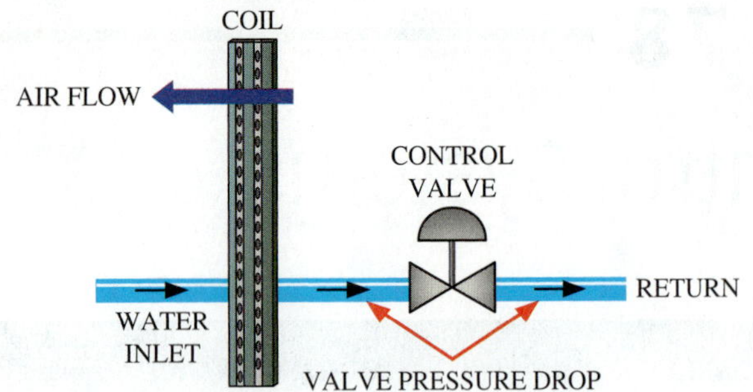

Figure 78-1 Coil two-way valve.

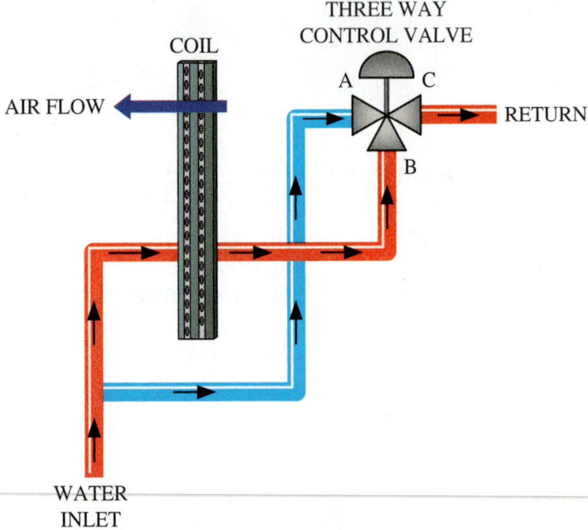

Figure 78-2 Coil three-way valve—fails open as indicated in red.

A three-way valve installation is shown in Figure 78-2. The outlet from the coil is connected to the B port of the valve so that in this arrangement the valve will fail open, allowing for full flow. Figure 78-3 shows an opposite arrangement, where the control valve will fail closed, thereby stopping all flow through the coil.

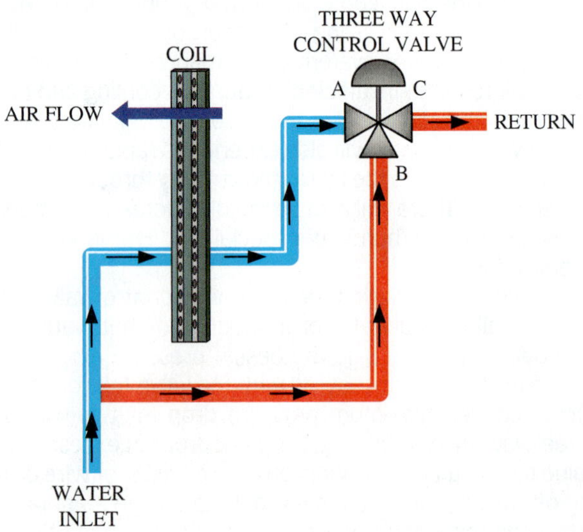

Figure 78-3 Coil three-way valve—fails closed as indicated in red.

Figure 78-4 Heating coil steam-trap condensate-return lines.

A two-way steam control valve modulates the flow of steam through the coil based on the signal generated by the control. The pressure leaving the steam trap (system return pressure) is assumed to be zero (Figure 78-4). Low-pressure steam applications are 15 psig or less. Medium steam pressure applications operate with steam pressures higher than 15 psig. Most steam control valves are designed to operate with steam pressure not exceeding 35 psig.

78.3 PNEUMATIC CONTROL SYSTEMS

New systems are being designed today that utilize solid-state controls. However, many older systems still in operation continue to rely on pneumatic controls. System upgrades may sometimes combine both solid-state and pneumatic controls, so a fundamental understanding of pneumatic control systems is important.

In a basic control system for cooling, a thermostat sensing the chilled-water outlet temperature operates the chiller. Each air handler will have a control valve regulating the flow of chilled water through the cooling coil. The position of inlet, mixing, and bypass dampers will need to be controlled. The condenser-cooling water pump will always be operating when the chiller is running. The cooling tower fan for the condenser-cooling water may cycle, and often there may be a water bypass control.

For the heating system, boiler firing is activated by the hot-water thermostat, often referred to as an aquastat. Individual space thermostats and humidistats control the functions of the air handler.

Pneumatic control systems use compressed air to supply energy for the operation of valves, motors, relays, and other pneumatic control equipment. Consequently, the circuits consist of air piping, valves, orifices, and similar mechanical devices.

Pneumatic control systems offer some distinct advantages:

- They provide an excellent means of modulating control operation.
- They provide a wide variety of control sequences with relatively simple equipment.
- They are relatively free of operational problems.
- They cost less than electrical controls if the codes require electrical conduit.

Pneumatic controls are made up of the following elements:

- A constant supply of clean, dry, compressed air
- Air lines consisting of mains and branches, usually copper or plastic, to connect the control devices
- A series of controllers, including thermostats, humidistats, humidity controllers, relays, and switches
- A series of controlled devices, including motors and valves called operators or actuators

The air source is usually an electrically driven compressor (Figure 78-5), which is connected to a storage tank. The air pressure is maintained between fixed limits (usually between 20 and 35 psi for low-pressure systems). Air leaving the tank is filtered to remove the oil and dust. Many installations use a small refrigeration system to dehumidify the air. Pressure-reducing valves control the air pressure.

Air Dryers

Specialty compact refrigeration systems are often used as air dryers for pneumatic control systems. Figure 78-6 shows the operating display panel for an air dryer. The air is cooled as it passes through one side of a tube-in-tube evaporator coil. Then water is drained off from the air as it leaves the coil. The cooled dry air picks up heat as it leaves the unit by the warm air entering. The refrigerant cycle is typical of most systems and utilizes a compressor, air-cooled condenser, refrigerant expansion device such as a capillary tube or TEV, and a tube-in-tube evaporator coil (Figure 78-7).

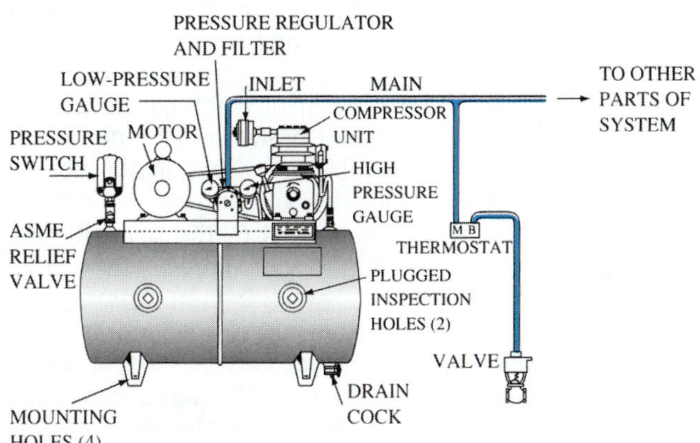

Figure 78-5 Air compressor for a pneumatic control system.

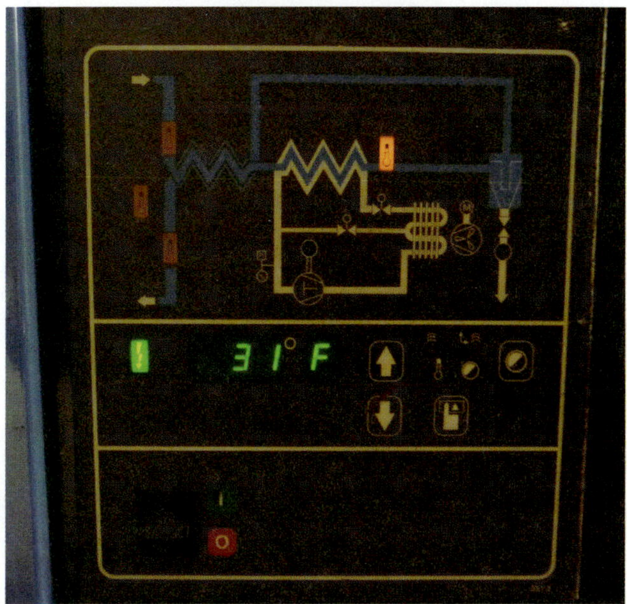

Figure 78-6 Air dryer operating display panel.

Figure 78-7 Air dryer with the cover off to show the insulated evaporator coil, hermetic compressor, and air-cooled condenser.

78.4 HOW THE PNEUMATIC CONTROLS OPERATE

The controller function is to regulate the position of the controlled device. It does this by taking air from the supply main at a constant pressure and adjusting the delivered pressure according to the measured conditions.

One type of thermostat is the bleed type, shown in Figure 78-8. The bimetal element reacts to the temperature and controls the branch line bleed-off pressure. These thermostats do not have a wide range of control, therefore the

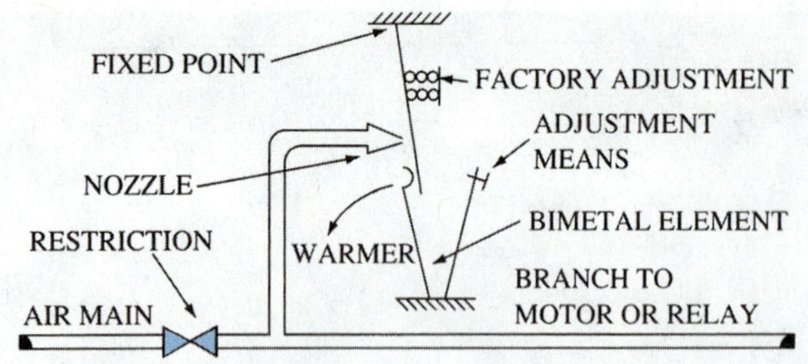

Figure 78-8 Diagram of a bleed pneumatic thermostat.

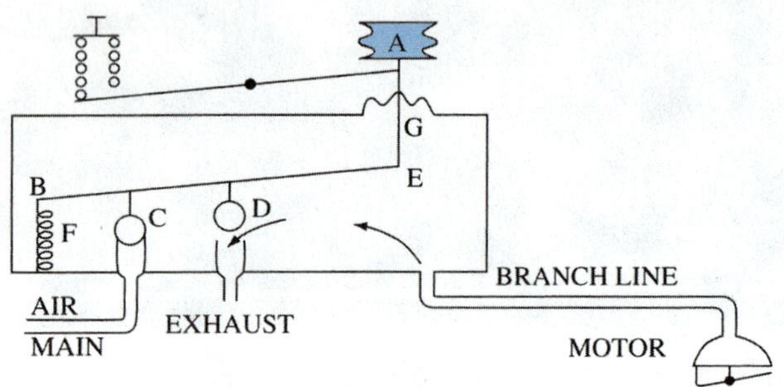

Figure 78-9 Diagram of a nonbleed pneumatic thermostat.

branch line is often run to a relay that controls the action. Bleed controls cause a constant drain on the compressed air source.

Nonbleed controllers use air only when the branch line pressure is being increased. The air pressure is regulated by a system of valves (Figure 78-9), which eliminates the constant bleeding characteristic of the bleed-type unit. Valves C and D are controlled by the action of the bellows (A) resulting from the changes in room temperature. Although the exhaust is a bleeding action, it is relatively small and occurs only on a pressure increase.

Controlled devices, operators or actuators, are mostly pneumatic damper motors or valves. The principle of operation is the same for both. Figure 78-10 is a diagram of a typical motor. The movement of the bellows as the branch line changes activates the lever arm or valve stem. The spring exerts an opposing force so that a balanced, controlled position can be stabilized. The motor arm L can be linked to a number of functions.

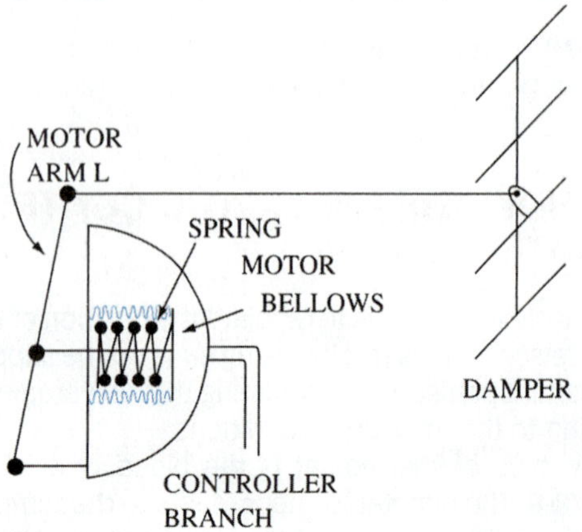

Figure 78-10 Pneumatic actuator with normally open damper.

Figure 78-11 shows a pictorial review of some of the functions in a pneumatic control system. There is always some crossover between the air devices and the electrical system. The device most widely used is the pneumatic/electric relay.

TECH TIP

Unlike electronic and electrical control systems, pneumatic systems do not require any energy when they are in a "hold position." Some electrical and electronic devices consume power even when they are off. This power consumption is referred to as the background power. Although the background power for any single device can be very small, the total power consumed by a large building's control system can be significant. The federal government feels that background power consumption has become so large nationally that the EPA has established guidelines for manufacturers regarding background power consumption. Because pneumatic systems do not have any background power, they are exempt from these regulations.

78.5 ELECTRONIC CONTROL SYSTEMS

Electronic control may also be used effectively for central station equipment. Due to a number of advantages, it is rapidly gaining in popularity. There are no moving parts, the response is fast, and the regulatory element can easily be a reasonable distance from the sensing element. This will allow for adjustments to be made at a central location. There are also cleaner conditions at a central location than at the location of the sensing element.

Only simple low-voltage connections are needed between the sensing element and the electric circuit. Flexibility is important since electronic circuits can be combined with both electric and pneumatic circuits to provide results that could not usually be achieved separately. Electronic circuits can coordinate temperature changes from several sources, such as room, outdoor air, and fan discharge air, and program action accordingly.

Electronic controls are based on the principle of the Wheatstone bridge (Figure 78-12), which is composed of two sets of series resistors (R_1 and R_2, R_3 and R_4), connected in parallel across a DC voltage source. A galvanometer G (a sensitive indicator of electrical current) is connected across the parallel branches at junctions C and D between the series resistors. If switch S is closed, voltage E (DC battery) energizes both branches. If the potential at C equals the potential at D, the net potential difference is zero. When this condition exists, the bridge is in balance.

If the resistance of any one leg is changed, the galvanometer will register a flow of current. The bridge is now unbalanced. If that resistance is changed as a result of a temperature reaction, we now have an electronic method of measuring current in relation to temperature change. With a few changes we can create an electronic

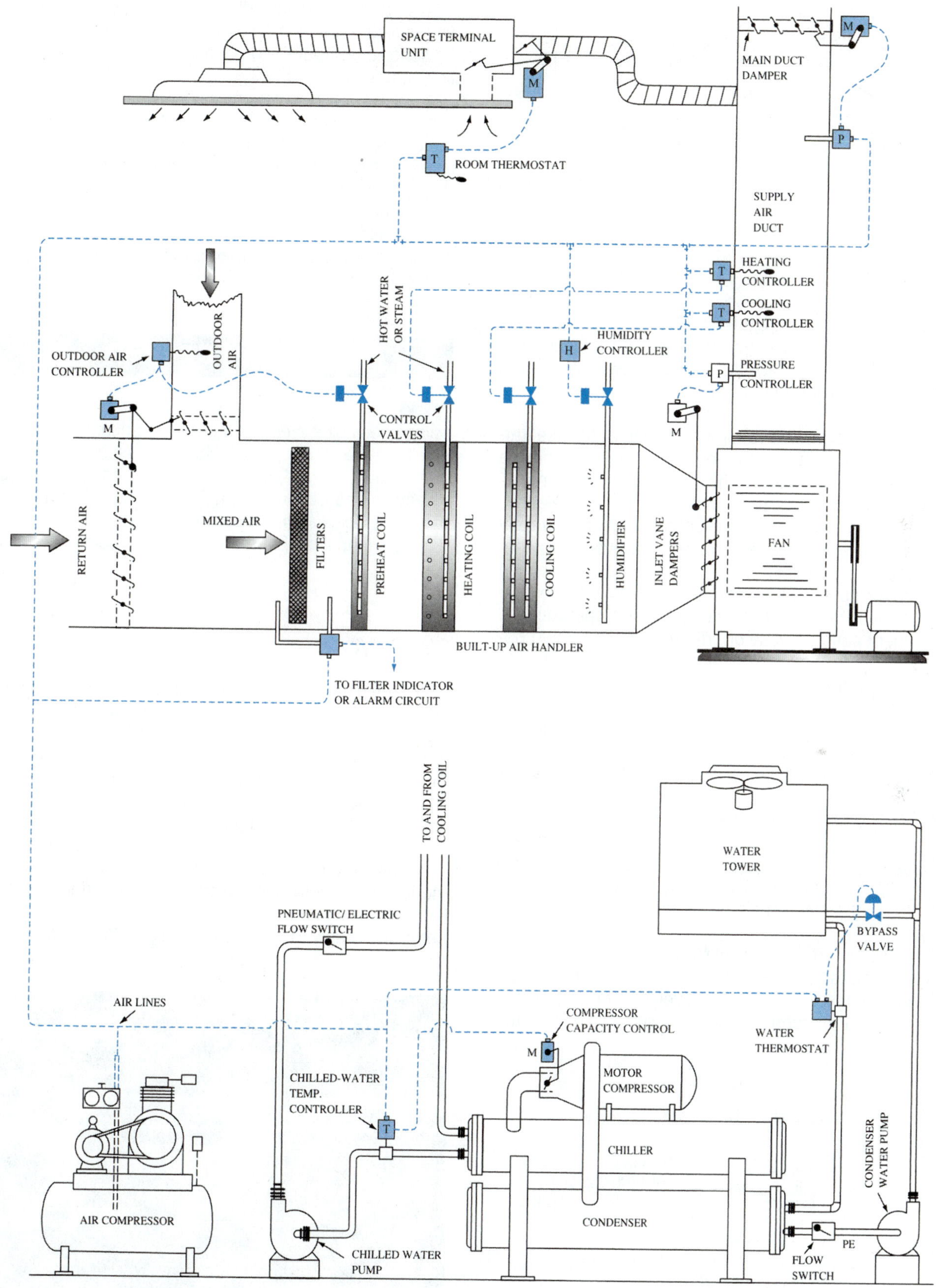

Figure 78-11 System diagram for a central station installation showing pneumatic controls.

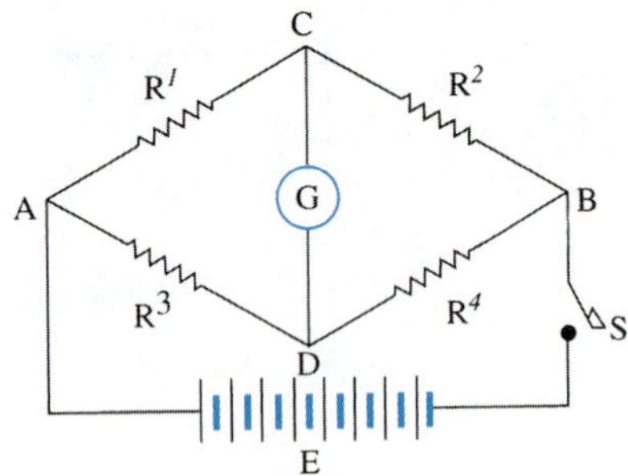

Figure 78-12 Wheatstone bridge circuit.

main circuit, such as in Figure 78-13. A 15 V AC circuit replaces the DC battery. The galvanometer is replaced by a voltage amplifier and phase discriminator switching relay. Resistor R_2 is replaced by sensing element T of an electronic controller.

The purpose of the voltage amplifier is to take the small voltage from the bridge and increase its magnitude by stage amplification to do the work. Phase discrimination means determining the sensor action. In an electric bimetal thermostat, the mechanical movement is directly related to the temperature changes; however, the electronic sensing element is a nonmoving part and the phase discriminator determines whether the signal will indicate a rise or fall in temperature. The relay then operates the final element action. Phase discrimination can be two position or, with certain modifications, can be converted to a modular system.

The crossover point from the electronic to electric occurs at the output of the amplifier and relay signal. The motor that it operates is a conventional ON/OFF motor or a proportional (modulating) electric motorized valve or damper actuator.

Electronic temperature-sensing elements are room thermostats, outdoor thermostats, insertion thermostats for ducts, or insertion thermostats for liquids. The typical room thermostat is a coil of wire wound on a bobbin. The resistance of the wire varies directly with the temperature changes.

A sensor is any device that converts a nonelectrical impulse such as sound, heat, light, or pressure into an electrical signal. Sensors have been developed to provide the necessary input for controlling the pressure in a duct or the relative humidity in a conditioned area. The sensor provides an input to a solid-state controller. The logic in the controller sends an output to some mechanical device to produce the required action.

For example, to control humidity, a pair of electronic sensors read wet- and dry-bulb temperatures. The logic in the controller converts these readings to a relative humidity value. Based on the limits set up, the connected mechanical device is programmed to add or deduct moisture from the air to meet the requirements. The use of electronic equipment makes possible more accurate control of the space conditions than can be accomplished by pneumatic or electromechanical control equipment.

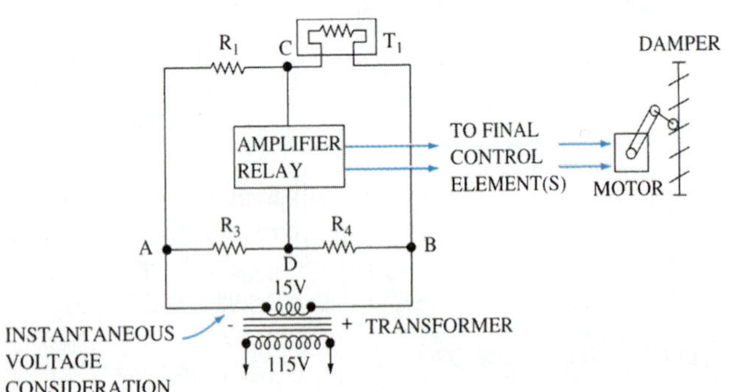

Figure 78-13 Electronic main bridge circuit.

SERVICE TIP

Make sure you follow the manufacturer's procedures when replacing electronic sensors, because some can be damaged by improper handling. Because they may be easily damaged, some supply houses will not allow electronic sensors to be returned once they are removed from their packaging. Make sure you are ready to use the sensor right away before you open the package.

78.6 PACKAGED AIR-HANDLER CONTROL SYSTEM

Packaged air handlers can be factory assembled as one unit or field installed in multiple configurations and modules. Field installation of multiple packaged modules allows for the best match of components for a specific application. Once installed, the packaged air handler will have some type of control panel. Controllers are typically configured to accept hundreds of configuration settings and setpoints. A control module contains the software and microprocessor that control the operation of the unit (Figure 78-14). The module will receive input from all of the different sensors located throughout the air handler that measure air temperature, quality, and condition (Figure 78-15). It will then

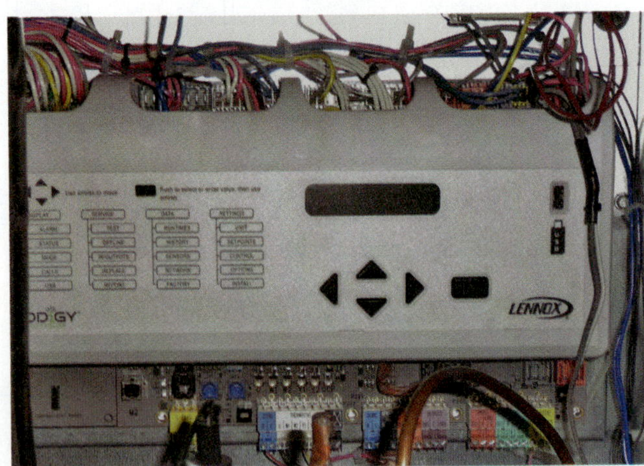

Figure 78-14 Air handler control module.

Figure 78-15 Inputs and outputs for temperature, indoor air quality, humidity, and smoke.

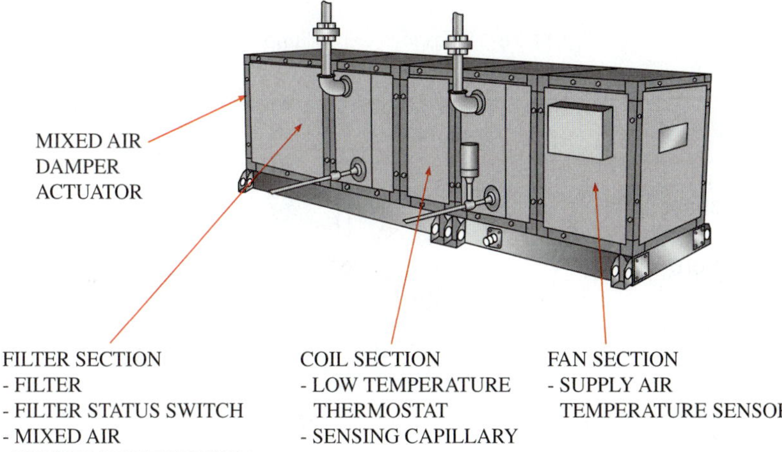

Figure 78-16 Typical air-handler actuator and sensor installation locations.

regulate the supply fan, mixing dampers, and heating and cooling coils as necessary to achieve the programmed condition setpoints.

SERVICE TIP

Most control panels will have a display for fault codes. When troubleshooting a unit, always check this first. The manufacturer's literature will identify the fault code so that you can immediately begin correcting the problem.

78.7 CONSTANT-AIR-VOLUME CONTROL SYSTEM

Depending on the system configuration, there will be a variety of temperature, condition, and alarm sensors. These include supply-air temperature, low-temperature thermostat, filter differential pressure switch, mixed air temperature, airflow switch, and preheat temperature. Typical sensor locations are shown on the air handler in Figure 78-16.

These measured input signals are compared to the setpoints programmed into the controller, and the appropriate actions will be taken on the supply-fan relay (start/stop), mixed-air damper actuator, heating coil valve/electric heater, cooling coil control valve or direct expansion control (depending on the system), and the exhaust damper position.

Temperature Input Signals

A number of the input signals are temperature measurements. The supply-air temperature sensor is installed on the discharge side of the air handler for measuring the supply-air temperature for the conditioned space (Figure 78-17). The low-temperature thermostat measures the temperature of the air as it passes across the cooling coil. The thermostat consists of a long length of serpentine capillary tubing that is installed on the inlet side of the coil. It will be set to shut down the fan and set off an alarm if the air

Figure 78-17 Temperature sensor located in the supply-air plenum.

Figure 78-19 Spring-return damper actuator.

temperature drops below 35°F. This is essentially a backup for the preheater, which receives its signal from the preheat temperature sensor. This preheat sensor is also serpentine capillary tubing, but it is located on the outlet of the preheater coil rather than at the inlet (Figure 78-18).

The mixed-air temperature sensor is mounted in the mixing box and has direct input for the spring-return damper actuator (Figure 78-19). The damper control assembly will modulate the damper position through a linkage (Figure 78-20), based upon the input signal from the mixed-air sensor.

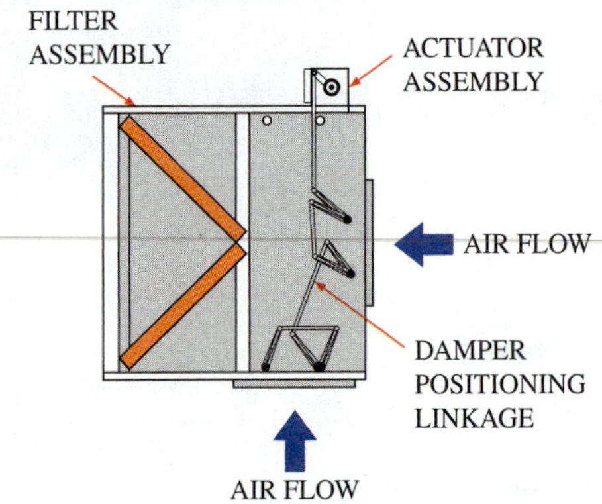

Figure 78-20 Assembled actuator and linkage.

SERVICE TIP

A malfunctioning sensor can sometimes be difficult to diagnose, but by logically working through a step-by-step sequence of operation, the different possible problems can be eliminated one by one. More difficult to troubleshoot is an improperly placed sensor. Always follow the manufacturer's recommendations when reinstalling sensors.

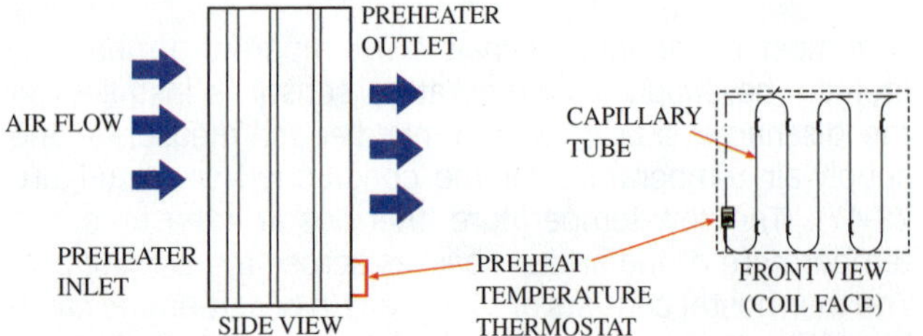

Figure 78-18 Capillary tube routing for low-temperature thermostat on heating application only.

Airflow Input Signals

There are also airflow input signals. The airflow switch consists of a probe mounted in the fan discharge duct. This snap action switch verifies that the fan is either on or off. The filter status switch is set to alarm as the differential pressure increases once the filters become dirty (Figure 78-21).

78.8 VARIABLE-AIR-VOLUME CONTROL SYSTEM

This system differs from the constant-volume/variable-temperature system since basically it maintains a nearly constant air temperature but matches the load by changing the air volume. This system became popular when attention was given to conserving energy without decreasing the comfort level during partial-load conditions.

One of the features of these systems is the use of a variable-frequency fan-speed controller that is applied to the air-handler fan (Figure 78-22). This electrical device, along

Figure 78-21 Filter status switch.

with the necessary sensors, controls the speed of the central fan to match the total air volume required by the individual zone terminal units. As an example, a static pressure sensor can be installed in the airflow to reduce the speed of the fan as the air supply demand decreases and the static air pressure rises (Figure 78-23). Since the power required for driving the fan is proportional to the cube root of the fan speed, tremendous power savings can result from slowing down the fan when the extra air is not needed.

Many of the input signals and controls are similar to the constant-air-volume system, with some exceptions. The fan motor is a variable-frequency drive unit (VFD). Because of this, a static pressure transducer and probe are installed rather than an airflow switch. The fan speed can be varied according to the static pressure in the system to supply more or less air as the system requires. There is also a temperature sensor located in the return-air supply.

Also included in a VAV system is a high-pressure switch. This switch will shut down the fan if the duct pressure exceeds the setpoint. Additional fan shutdowns are the low-temperature thermostat and the fire shutdown relay.

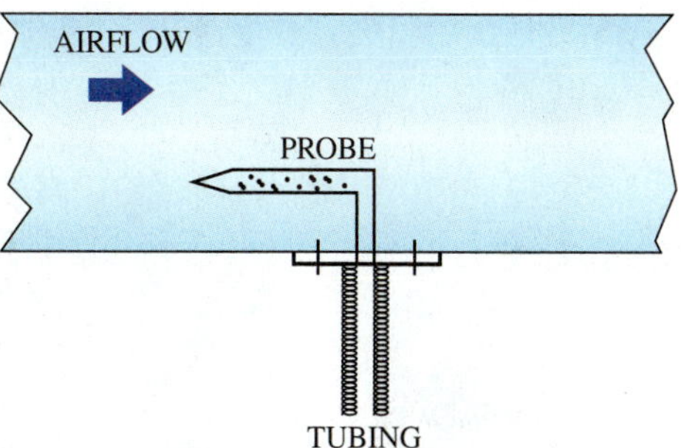

Figure 78-23 Duct static pressure probe.

78.9 AIR-QUALITY AND AIR-CONDITION SENSORS

Sensors can be used to measure the quality and condition of the air, such as CO_2 levels, enthalpy, and humidity throughout different locations within the air-distribution system. They can be used for simple data acquisition where the measured output—as an example, the CO_2 level—is read from a display. A more significant advantage to sensors is that many electronic control systems will use their output to automatically control actions such as damper positioning and fan modulation.

Air Quality

CO_2 levels are monitored in the conditioned air space and can also be monitored in the return-air duct. These sensors use infrared technology with a range of 0–2,000 ppm. The amount of outside air admitted will be adjusted to control the setpoint level.

Differential Enthalpy Control

Two individual sensors are used, with one sensing the outside air enthalpy and the other sensing the return air enthalpy (Figure 78-24). This is used for the economizer

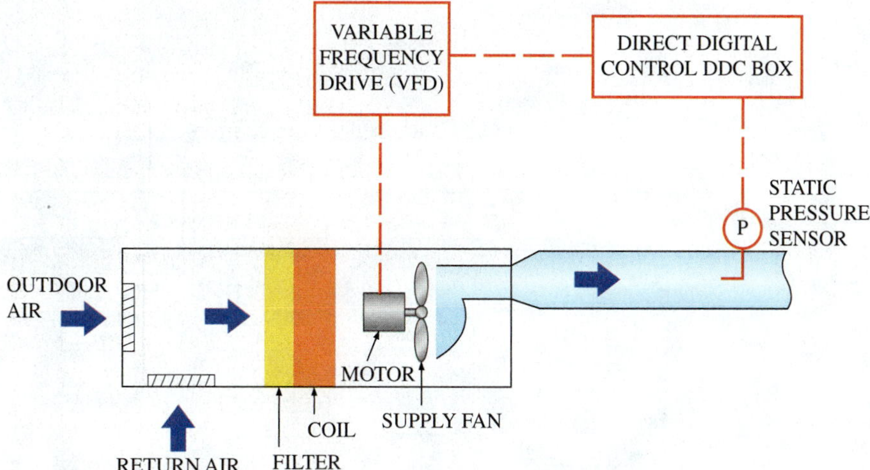

Figure 78-22 Supply-fan control.

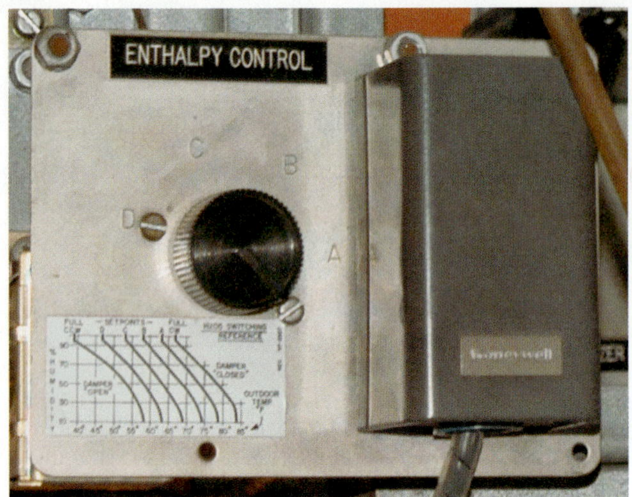

Figure 78-24 Differential enthalpy control sensor.

section. The airstream with the lowest enthalpy will be directed by dampers through the economizer section.

Space Humidity

The humidity level is monitored in the conditioned space and is mounted in a location that is representative of the entire zone. This sensor can be used for adjusting the cooling coil face and bypass dampers to allow for sufficient cooling of the air for dehumidification.

78.10 DIRECT DIGITAL CONTROL SYSTEMS

In a direct digital control (DDC) system, the computer acts as the primary control for all HVACR functions. Valves, dampers, fan speeds, and so on are all controlled by the computer without the use of conventional control devices such as thermostats, humidistats, or timers.

The computer directly senses the building environmental conditions and, based on a user-defined programmed set of instructions, initiates the proper control actions in the HVACR system. Direct digital control of HVACR components gives more accurate control and greater flexibility than other commonly used mechanical and electrical control devices. It also has the capability of coordinating inputs from a number of sensing devices and arriving at an output that takes into consideration numerous influencing factors. The following are some examples of the capabilities of the DDC control systems:

- The DDC system can control a VAV terminal box to discharge the proper air supply based on a variety of inputs, such as dry-bulb temperature, relative humidity, and mean radiant temperature. In this way, considering the total environmental conditions, a greater feeling of comfort is produced for the occupants in the space. Figure 78-25 shows a computer-generated control screen for an exhaust-air handler.
- In a central station system with a large supply fan, the microprocessor can regulate the speed of the fan to

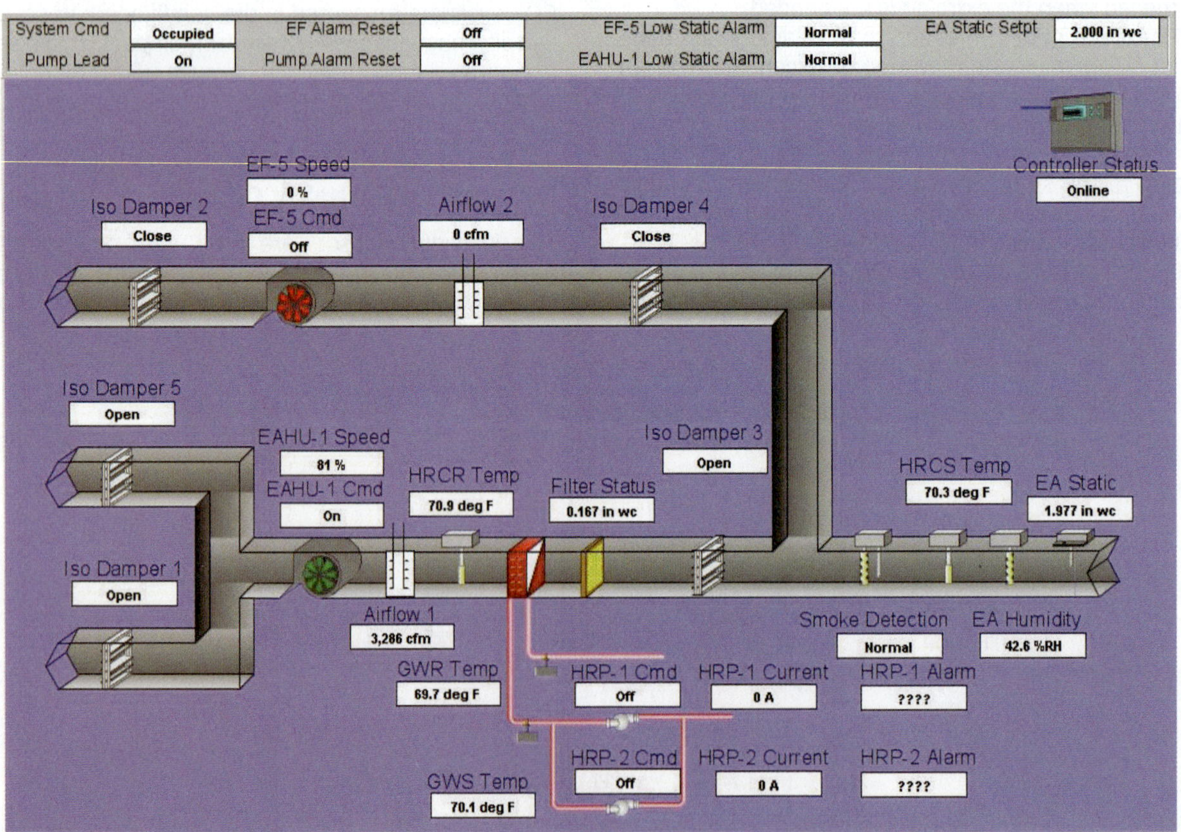

Figure 78-25 Computer-generated control screen for an AHU. *(Courtesy of the Jackson Laboratory)*

produce the required airflow using a minimum amount of power.

- Computers are currently used to turn on chillers and boilers at an optimum time to recover from a period when the building is unconditioned. Figure 78-26 shows a computer-generated control screen for a bank of chillers, the chilled water pumps, and the operating temperatures and pressures for the system.
- Control from a remote location. As an example, a building in California could be monitored and controlled from a location in Georgia.

These systems operate with a central stand-alone controller connected to a series of remote-control units.

The individual remote-control units are located in the building near the equipment being controlled. These units have a series of input and output wiring connections that go to sensors and controls on the HVACR system that permit control of the operations of the system. Both the input and output functions are of two types: analog and digital. The analog functions supply or deliver modulated information. For example, an analog temperature sensor may be capable of reading temperatures between 0°F and 100°F. This input could be converted by the computer to produce an analog output signal to control a damper to any position between fully open and fully closed.

The digital function is a binary, or two-position, function. For an input, the signal could monitor whether a switch is open or closed. For output, the binary signal could position the switch in either an ON or OFF position.

The DDC systems are often called energy-management systems since one of their main functions, and usually justification for their adoption, is saving energy. Special provision has been made in the selection of sensors to make possible continuous monitoring of the energy usage. The control system is set up to energize loads only when necessary and to use such features as free cooling (economizer operation) whenever possible.

The control center collects key operating data from the HVACR system and incorporates remote-control devices to supervise the system's operation. The elaborateness of the data center is related to the type and size of the system and to economic considerations. Some control centers have continuous scanners with alarm indicators to monitor refrigeration machines, oil and refrigerant pressures, chilled-water temperatures, air filter conditions, low water conditions in the boiler, and conventional space temperature and humidity conditions in each zone.

An example of a computer-generated control screen for the heat-reclaim section of an air handler is shown in Figure 78-27. This includes fan operation, all the damper positions, and the valve positions for coil control of glycol, steam, and chilled water. The screen shown in Figure 78-28 controls the cooling water tower loop for the chillers.

Screens are designed to help the operator visualize the process with a graphical display that best represents the

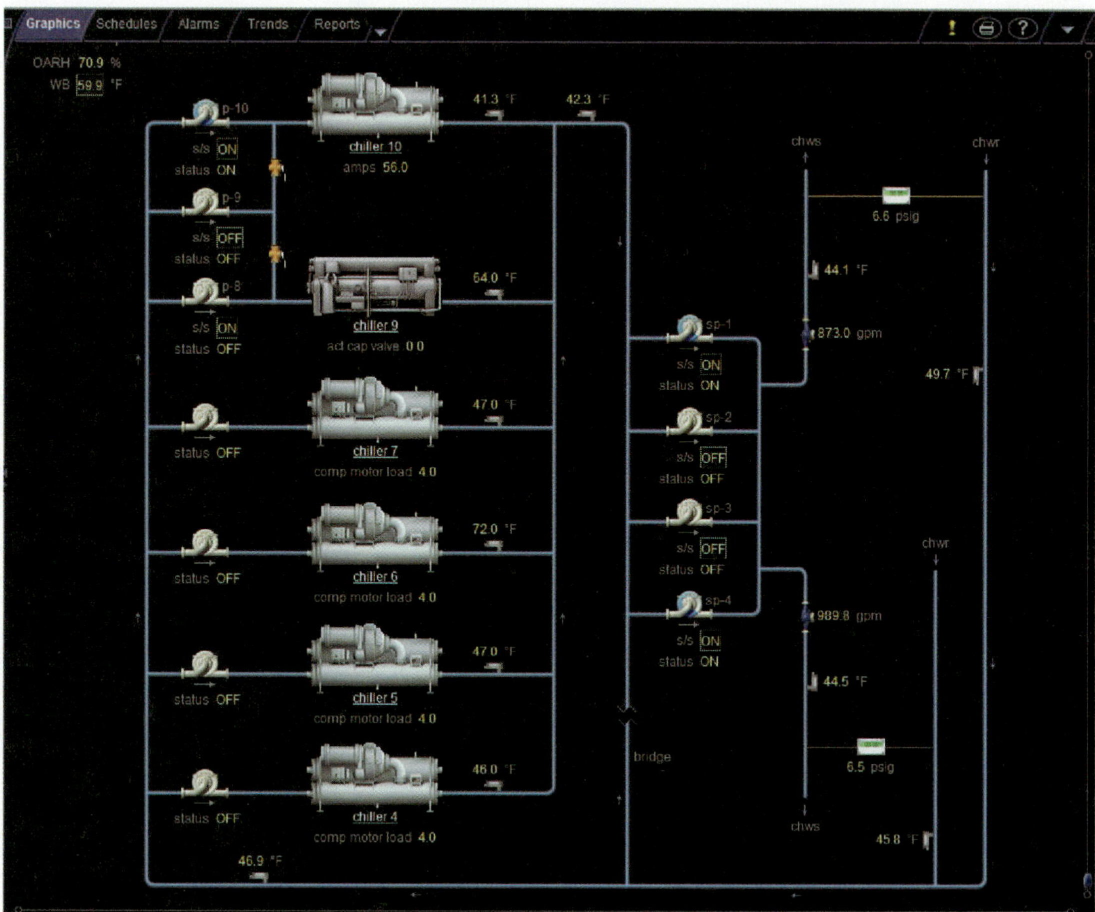

Figure 78-26 Computer-generated control screen for a bank of chillers and chilled-water pumps. *(Courtesy of the Jackson Laboratory)*

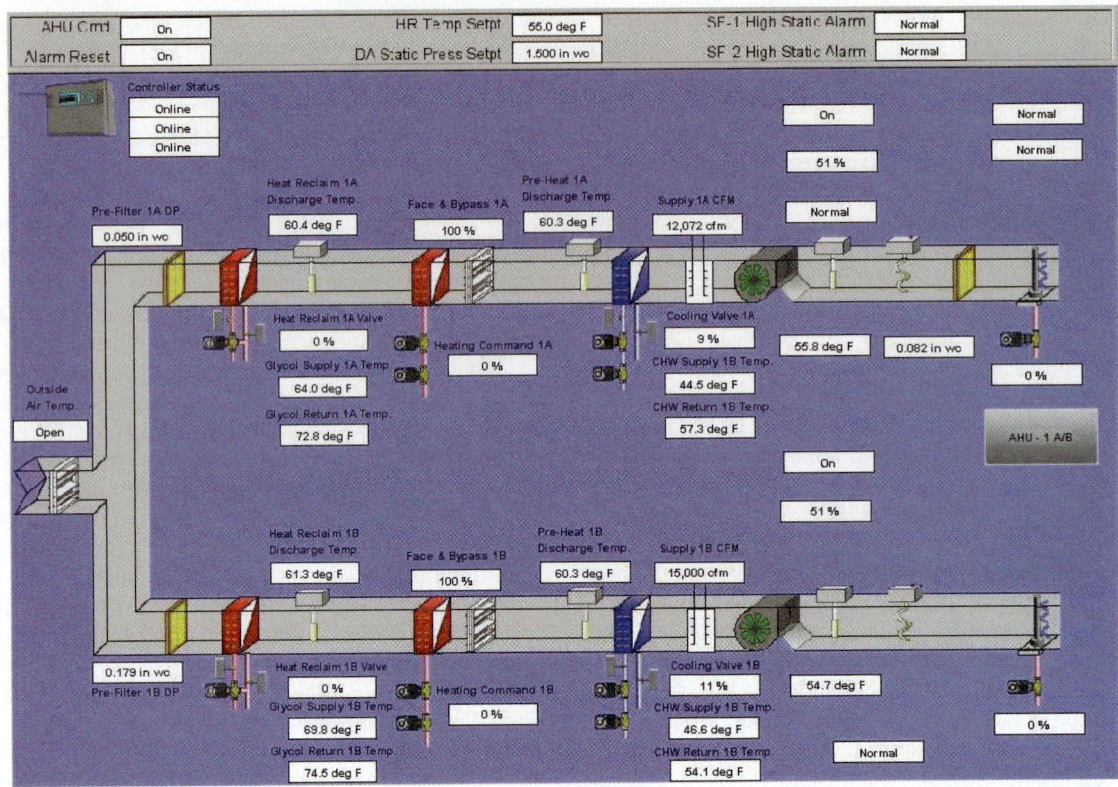

Figure 78-27 Computer-generated control screen for heat-reclaim section of an air handler. *(Courtesy of the Jackson Laboratory)*

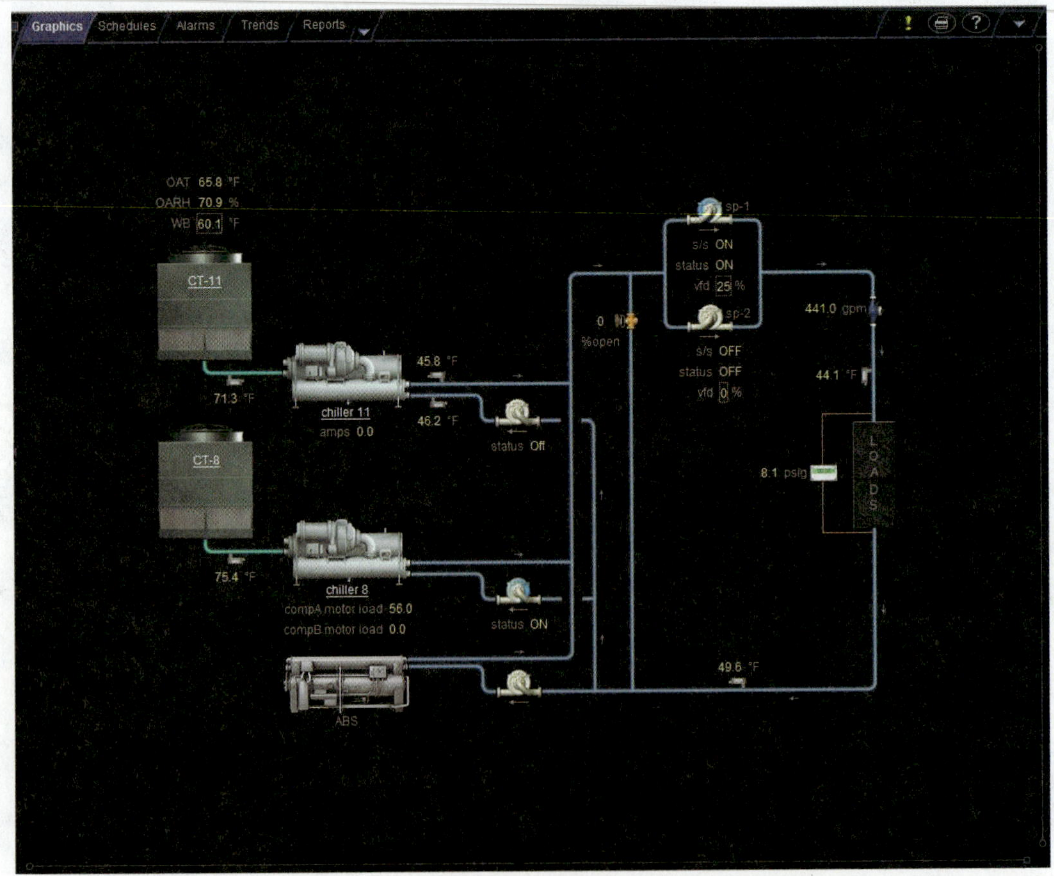

Figure 78-28 Computer-generated control screen for cooling water towers. *(Courtesy of the Jackson Laboratory)*

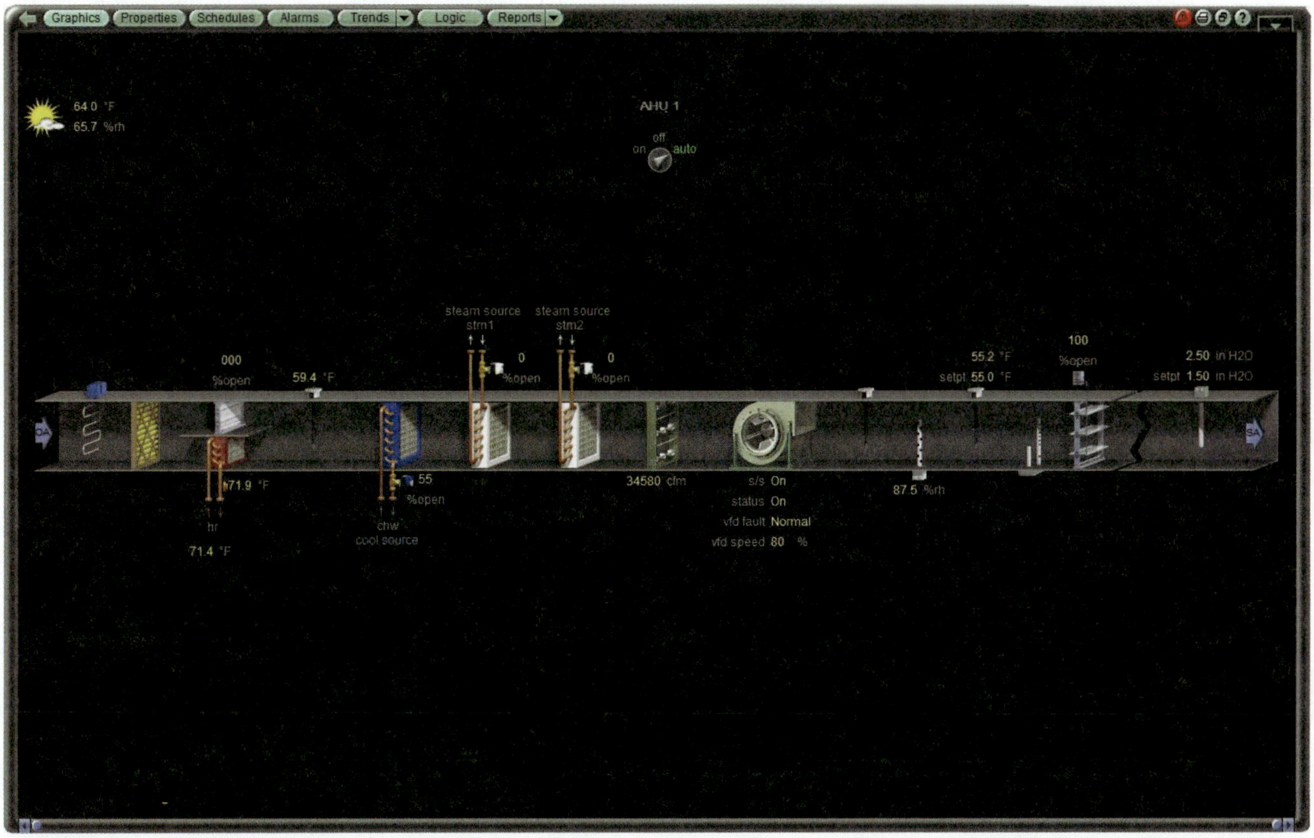

Figure 78-29 Computer-generated control screen for an AHU with enhanced graphics. *(Courtesy of the Jackson Laboratory)*

system, as shown in Figure 78-29. This screen has state-of-the-art graphics that represent coils and valves and an air flow sequence that can be easily interpreted. The detail is such that even the steam humidifiers and sensing probes are shown.

Vibration Monitoring

Vibration-monitoring systems are part of what is called "predictive" maintenance. Rotating equipment such as fans can be continuously monitored for excessive vibration levels. To reduce vibration, large fans may be mounted on resilient mounts, such as springs (Figure 78-30). Even so, over time the fan bearings will wear and vibration levels will increase. Vibration can also result from worn belts, loose pulleys, and loose foundation bolts.

For continuous monitoring, sensors need to be installed on the equipment (Figure 78-31). These vibration sensors will measure the "amplitude" (acceleration) of the vibration and trigger an alarm if levels exceed set limits. This vibration signature can be tracked over time to help determine when a piece of equipment may need to have bearings replaced or be in need of other types of overhaul. The life of

Figure 78-30 Large fan mounted on springs to reduce the effects from vibration.

(a)

Figure 78-31 (a) Rotating equipment such as fans can be fitted with vibration sensors; (b) (see next page) vibration sensor wiring.

Figure 78-31 *(Continued)*

the equipment can be predicted in a sense, so that components are not replaced prematurely. Continuously monitoring the equipment also reduces the chance that the unit will fail unexpectedly and without warning.

The frequency of the signature helps to determine the cause of the vibration. Different problems occur at different frequencies. Vibration from a bad bearing will provide a different frequency signature as compared to an unbalanced fan blade or loose hold-down bolt. An example of a computer-generated screen showing the frequency and amplitude signature for a fan motor is shown in Figure 78-32.

UNIT 78—SUMMARY

Many control systems today are electronic; however, a combination of electric, electronic, and pneumatic controls can be used on large commercial systems.

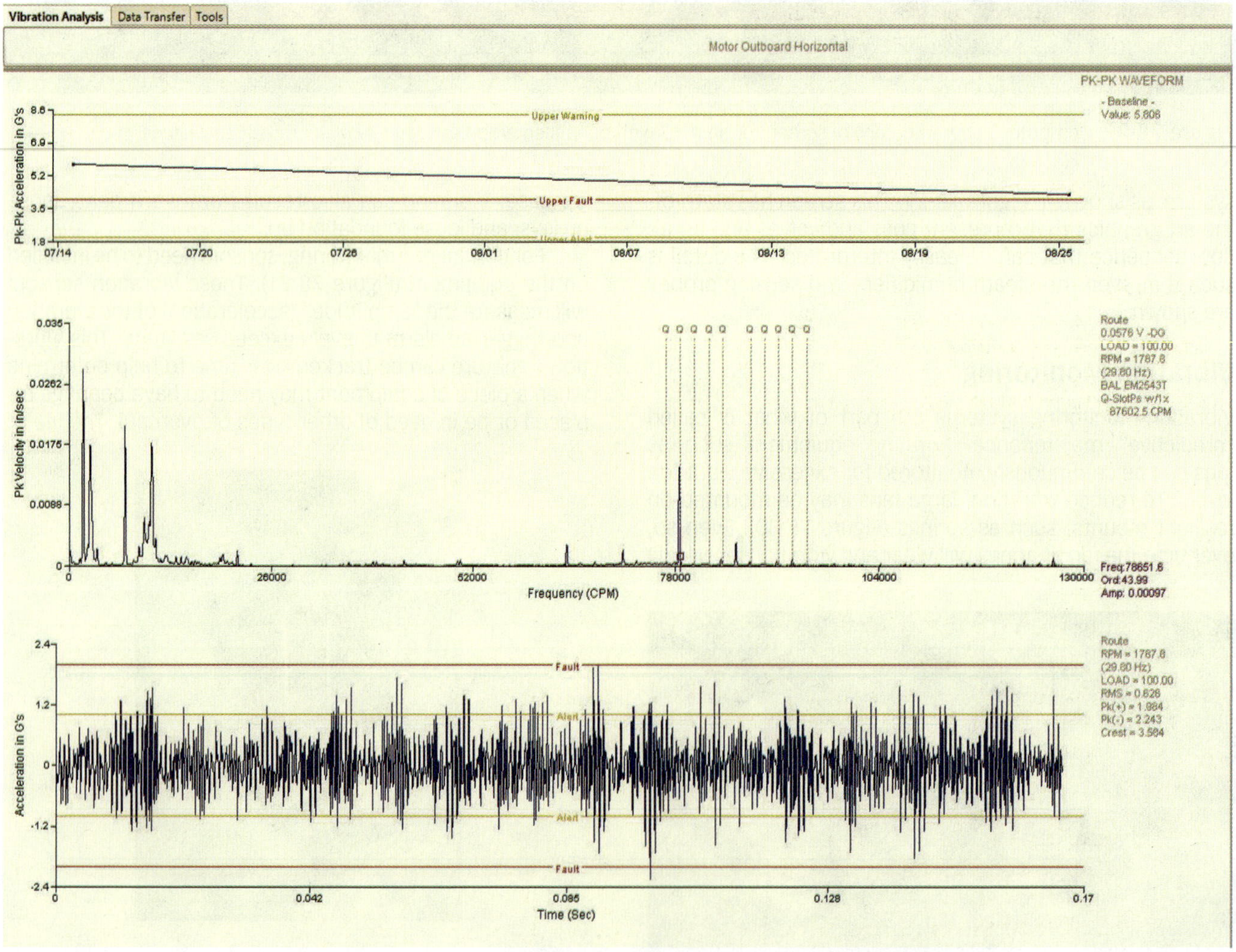

Figure 78-32 Continuously monitored vibration signature for a motor showing the amplitude and frequency of vibration. *(Courtesy of the Jackson Laboratory)*

WORK ORDERS

Service Ticket 7801

Customer Complaint: Compressors Run Constantly Even When the Outside Air Is Cool Enough for System to Be in Economizer Mode

Equipment Type: Light Commercial Air Handler with DX Coil and Economizer Package

The technician operates the system while doing a visual check. Even though the outside temperature is low enough for economizer operation, the compressor is not turning off and the outside air dampers (OA) are not opening. After examining the operation sequence and electrical diagram, the technician suspects a faulty economizer control thermostat. After testing, the technician proves that the control is defective. The technician replaces and sets the thermostat. The technician then tests the system by placing the sensing bulb in ice water. The compressor cycles off and the OA damper opens per the manufacturer's specifications. The system is now operating as designed.

Service Ticket 7802

Customer Complaint: System Has Shut Down Completely

Equipment Type: 100-Ton Central Station Unit with Pneumatic Control

The technician talks to the maintenance department and realizes there must be a complete shutdown of either the electrical power or the control system. The technician checks the electrical power and all the systems are energized. The technician checks the control system and finds the control system air compressor is not running and the pressure is down to zero. Troubleshooting the air compressor reveals a bad pressure controller that would not allow the control air compressor to operate. After replacing the switch, the technician turns the compressor on and the system operates as designed.

Service Ticket 7803

Customer Complaint: The Third Floor of a Multistory Building Has Reported a "Cold Call"

Equipment Type: Air Handler with Heating and Cooling Operated from a Large Central Station Unit

The technician is taken to the area by a maintenance technician. The AHU is located in a third-floor fan room. The technician connects the laptop computer to the ports furnished in the DDS controller located on the unit. The technician tries to cycle the unit from cool to heat and diagnoses a bad controller that will not cycle the damper, bring on the heating fan, or open the HW control valve. The technician replaces the DDS controller inside the control panel, and the system now heats and cools as designed.

UNIT 78—REVIEW QUESTIONS

1. What is the function of a low-temperature thermostat on a packaged air handler?
2. What is the function of an airflow switch on a packaged air handler?
3. What input signal would control the speed of a variable-frequency drive supply fan?
4. What is the purpose of a high-pressure switch in a packaged air handler?
5. What will an air-quality sensor typically measure?
6. What is the purpose of a differential enthalpy sensor?
7. How does a refrigeration air dryer operate?
8. How would a humidity control operate?
9. What is an important consideration to take into account when replacing sensors?
10. Why are direct digital control (DDC) systems sometimes referred to as energy-management systems?
11. What could cause vibration in a large air-handler fan?
12. What do vibration monitors specifically measure?
13. What types of control valves are used for heating and cooling coils?
14. What can happen if the pressure drop through a heating or cooling coil control valve is excessive?
15. List some of the advantages of pneumatic control systems.
16. What is background power?
17. What are some of the advantages of electronic control systems?
18. List some electronic temperature-sensing elements.
19. What acts as the primary control for all HVACR functions in a DDC system?
20. What is a sensor?

UNIT 79

Chilled Water Systems

OBJECTIVES

After completing this unit, you will be able to:

1. describe the difference between a flooded and a dry (direct expansion) type chiller.
2. explain why low-pressure chillers need purge recovery.
3. explain why some chillers are better suited for air-cooled condensers while others operate with water-cooled condensers.
4. list the different types of compressor arrangements found on chillers.
5. describe the function of a chiller economizer.
6. describe the operation for an oil-return system for a chiller.
7. transfer refrigerant to and from a chiller.
8. explain how an absorption chiller operates.

79.1 INTRODUCTION

In a large multistory building or complex, it is impractical to deliver refrigerant long distances from a central air-conditioning compressor and condensing unit. Excessive energy would be required for the compressor to deliver the refrigerant to different air handlers located on different levels of the building. Any leaks in the system could lead to a substantial loss of refrigerant before they could be detected.

The more common way to deliver refrigerant to individual air handlers is to use a secondary refrigerant such as chilled water. The chilled water is cooled in a central station chiller by a primary refrigerant. The chilled water is then delivered throughout the building to individual air handlers by circulating pumps (Figure 79-1). Any leaks in this system are easily detected. The primary refrigerant does not leave the central air-conditioning unit.

79.2 FLOODED AND DRY-TYPE WATER CHILLERS

The water in a chiller is cooled to approximately 43–45°F. The chilled-water pump circulates the chilled-water supply (CHWS) to the cooling coil in the air handler. The chilled-water loop will have an expansion tank and fill connections for makeup water, and it will also have air vents, balancing valves, and cooling coil flow controls, as shown in Figure 79-2. The heat absorbed from the air by the cooling coil warms the water about 10°F at full load, with a chilled-water return (CHWR) temperature of 53–55°F.

The cooler section is the evaporator where the primary refrigerant changes state and absorbs heat from the chilled water. The cooler section, often referred to as the chiller, is often cylindrical in shape and of the shell-and-tube design. A "dry" direct expansion (DX) cooler will

(a)

(b)

Figure 79-1 (a) Chilled-water cooling loop from central unit to multiple zones; (b) glycol and chilled-water piping connections to the air handler.

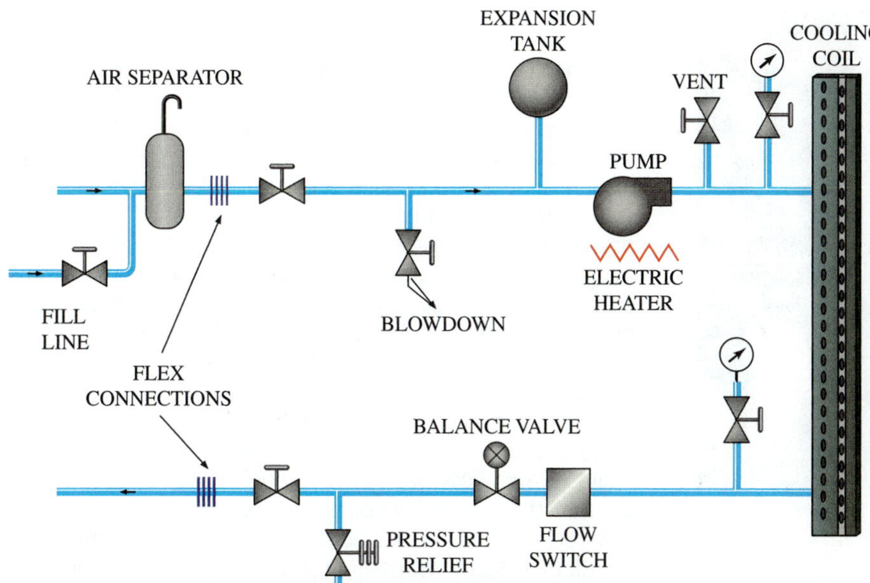

Figure 79-2 Chilled-water cooling loop for cooling coil.

Figure 79-3 Centrifugal compressor chiller.

circulate the refrigerant through the inside of the tubes, and the chilled water will surround the tubes. For larger tonnage units, flooded coolers are more commonly used. In a flooded cooler, the refrigerant surrounds the tubes and the chilled water is circulated through the inside of the tubes. Flooded coolers have greater cooling capacities, but they require more refrigerant and have a greater potential for freeze-up as compared to DX coolers. Figure 79-3 shows a centrifugal compressor chiller with a flooded cooler and subcooler section.

79.3 LOW-PRESSURE CHILLERS

The chilled-water temperature for an air-conditioning system should not fall much below 40°F. A temperature sensor located in the chilled-water circuit provides an input signal to the compressor operation and capacity controls. Many systems will automatically shut down if the chilled-water temperature drops to 35°F. Low-pressure refrigerants are those that have boiling points of 50°F or above. This high boiling temperature would not be suitable for low-temperature applications; however, it is suitable for chiller applications. To achieve the desired refrigerant temperature, many chillers operate with suction pressures below atmospheric pressure.

For large air-conditioning loads, a considerable amount of refrigerant must be circulated by the compressor. Centrifugal compressors are able to pump large volumes of refrigerant as compared to reciprocating compressors. However, they do not develop the high-side discharge pressures of a reciprocating compressor. Since low-pressure refrigerants do not require such high discharge pressures, they are ideally suited for centrifugal chiller applications.

79.4 PURGE RECOVERY UNITS

A low-pressure chiller will operate with a negative suction pressure that is below atmospheric. Any leaks in the system will draw in air. This air typically collects in the top section of the condenser. The air will increase the discharge pressure of the unit and lead to higher compressor power consumption. The air will also contain water that will oxidize the oil in the system and lead to the formation of sludge and acids. Chillers of this type generally have some type of purge recovery unit (Figure 79-4). This unit will purge the air from the top of the condenser and recover the refrigerant on a continuous basis while the chiller is running.

79.5 LEAK TESTING LOW-PRESSURE CHILLERS

Leak testing a low-pressure chiller is difficult because air will be leaking in rather than refrigerant leaking out. One way to check for refrigerant leaks in a low-pressure chiller is to raise the temperature of the chilled water. The unit is shut down and the chilled water is heated and circulated through the cooler. This will raise the pressure in the unit above atmospheric pressure so that a conventional leak detector can be used to find any refrigerant leaks.

Figure 79-4 Purge recovery unit.

Figure 79-5 Centrifugal compressor suction-side rupture disk.

CAUTION

Never allow a low-pressure chiller to reach or exceed the rupture disk setting when testing for leaks. The rupture disk setting for low-pressure chillers is normally set at 15 psig, but it is advisable not to exceed 10 psig. If this happens, then the chiller rupture disk will release and all of the refrigerant in the unit will be lost (Figure 79-5).

79.6 RECIPROCATING COMPRESSOR CHILLERS

Reciprocating compressors have been commonly used with small to moderate size chillers. This has been changing, since many newer units are now utilizing more efficient scroll compressors. The total refrigerant capacity for reciprocating compressors is somewhat limited, so large chillers will use centrifugal and screw-type compressors.

Air-Cooled Condensers

Starting at the lower capacities, packaged water chillers use one or more reciprocating compressors. Figure 79-6 shows an example of a small air-cooled chiller in the 30-ton range. This type of unit would be mounted directly outside or on a rooftop with supply and return connections for the chilled water piped into the building. For head (discharge) pressure control, the condenser fan control receives input from a pressure transducer (Figure 79-7). The fans are maintained at the lowest condensing pressure and temperature possible to maintain the highest unit efficiency.

The coolers for this type of unit are typically the "dry" DX type. Refrigerant flows through the tubes and will generally make two passes for standard operation, which provides a counterflow arrangement, opposite to the typical chilled-water flow pattern. The cooler shell is suspended beneath the condenser coil and fan section. The cooler shell (evaporator) must be protected from freezing. Electric heating elements are wrapped around the shell and then covered with a thick layer of insulation. Some manufacturers also add a final protective metal jacket that doubles as a good vapor barrier.

The larger package chillers with reciprocating compressors can be air cooled or water cooled and range upward from 40 to 200 tons. The cooler and compressor can be located indoors and the air-cooled condenser located on the rooftop, as shown in Figure 79-8a, or outside the building, as shown in Figure 79-8b. Another option is to locate a water-cooled condenser alongside the compressor. If a ready supply of water for the condenser is not available, then a cooling tower will be required. The cooler shell and suction lines must be properly insulated to prevent sweating. This is done at the factory with a layer of closed-cell foam insulation prior to painting.

Water-Cooled Condensers

Most water-cooled condensers are shell-and-tube construction. Water flows through the tubes, and refrigerant vapor fills the shell. The refrigerant vapor condenses to a liquid as heat is transferred from the refrigerant to the water. The liquid collects in the bottom, where it is subcooled an additional 10–15°F for greater cooling capacity. In the water circuit, the coldest condenser water enters the lower part of the shell and circulates through the tubes. The water will make two or three passes through the shell before it is discharged. This is arranged by circuit baffles in the condenser heads. The higher the number of passes, the greater the pressure drop and pressure required to produce the required condenser cooling-water flow rate. Condenser capacity is normally based on 85°F entering water temperature. There is normally a 10°F rise between the exiting and entering water temperatures.

TECH TIP

A special seawater condenser is used for marine duty where saltwater is the cooling medium. These tubes are made of cupronickel steel to withstand the salt corrosion effects.

Figure 79-6 Package chiller with reciprocating compressor.

Figure 79-7 Control sensor: pressure transducer.

79.7 RECIPROCATING COMPRESSOR CAPACITY CONTROL

The larger tonnage units will have multiple compressors that permit close control of capacity. They are also generally equipped with cylinder unloaders and hot-gas bypass. As an example, consider a two-compressor unit with one compressor set as the lead and the other as the lag. During peak periods, both compressors would be running fully loaded. In the early evening as the load begins to decrease, the chilled-water temperature will begin to drop. This will signal the lag compressor to begin unloading cylinders sequentially and run at a reduced capacity. At some point the load will decrease to the point that the lag compressor stops. If the load continues to decrease, the lead compressor will begin unloading cylinders. Eventually, when the lead compressor is fully unloaded, the hot-gas bypass will open.

79.8 SCROLL COMPRESSOR CHILLERS

In addition to reciprocating compressors, the scroll-type compressor is being utilized increasingly in packaged chillers. These compressors use a pair of mating scroll-shaped

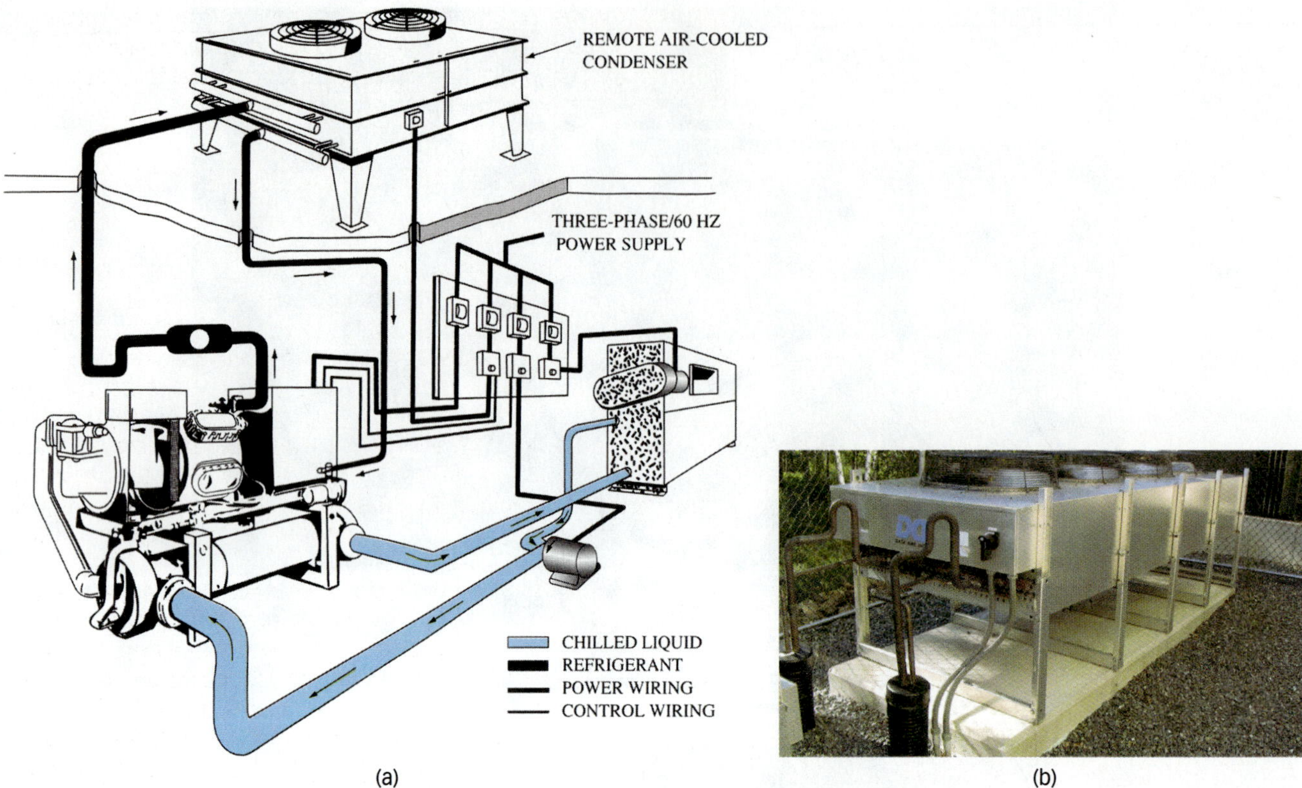

Figure 79-8 (a) Air-cooled condenser installation with a package chiller; (b) remote air-cooled condenser.

surfaces to compress the refrigerant with pure rotary motion, as shown in the cutaway diagram in Figure 79-9. The individual scroll compressors are very efficient and are used in sizes from 5 to 20 tons. Since they have limited capacity control, they are usually used in multiples of two, three, or four compressors that cycle on and off dependent on the load demand.

79.9 CENTRIFUGAL COMPRESSOR CHILLERS

For very large installations, the industry offers a range of hermetic centrifugal compressor water chillers of up to 2,000 tons in a single assembly. Very large buildings or complexes such as college and school campuses, sports arenas, airports, and high-rise office buildings may use more than one chiller to meet their cooling needs.

Hermetic centrifugal compressors (Figure 79-10) vary in design and refrigerant type. This unit has a hermetically sealed motor that will drive the compressor through a set of transmission-speed-increasing gears. The compressor speed is typically much higher than the motor speed. The refrigerant vapor is drawn from the cooler into the suction eye of the rotating impeller (Figure 79-11). The pressure of the refrigerant is increased as it passes through the discharge of the spiral compressor casing (volute) and into the condenser (Figure 79-12). The high-pressure refrigerant vapor is condensed and passes through a fixed orifice or float chamber to return to the cooler. There is also a much smaller additional cooling circuit fed by expansion valves

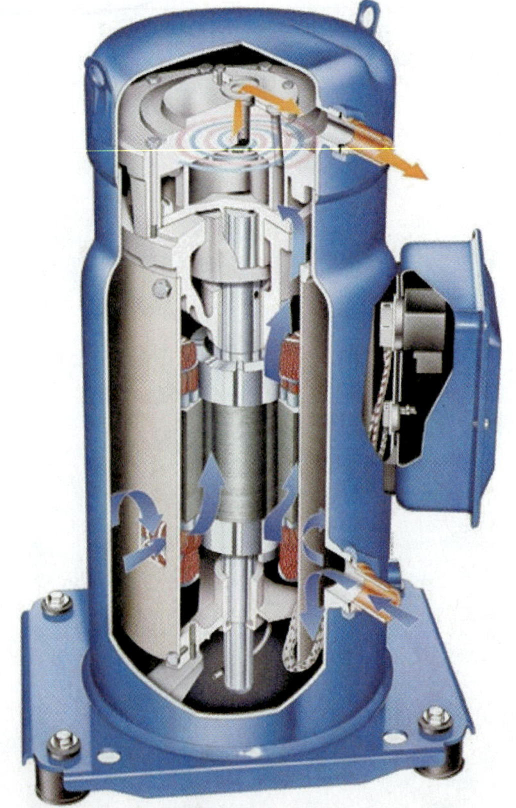

Figure 79-9 Cutaway view of a scroll compressor. *(Courtesy of Danfoss)*

Figure 79-10 Hermetic centrifugal compressor.

Figure 79-11 Centrifugal compressor impeller.

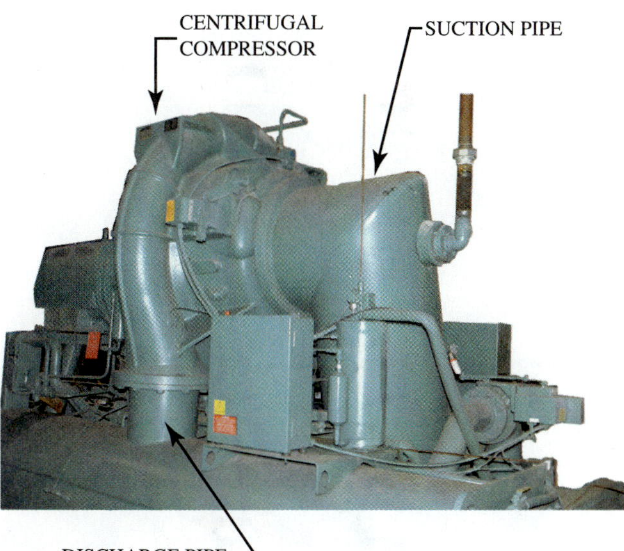

Figure 79-12 Compressor discharge through casing volute to condenser.

for the oil, hermetic motor, and the variable-frequency drive. The typical configuration is for the compressor, condenser, and cooler to all be located together to circulate the refrigerant a minimal distance.

79.10 CENTRIFUGAL COMPRESSOR STAGING

Single-stage centrifugal compressors will have only one impeller. A two-stage unit will have two impellers. The first impeller will discharge into the suction of the second impeller and then to the condenser. A three-stage unit will have three impellers, and so on. More stages results in higher discharge pressures, which allows for the use of high-pressure refrigerants such as R-134a. Low-pressure chillers operating with low-pressure refrigerants such as R-123 can be simple single-stage units that operate at high speeds in the 10,000 rpm or higher range. Some gear-driven impellers reach speeds of 20,000–25,000 rpm.

Typically, more stages and higher pressures require more power input and the compressor arrangement is somewhat more complex. However, multiple stages generally require lower speeds, which can offset the power requirements. Lower speed units are also less susceptible to the vibration problems that can be encountered with very-high-speed units. High-pressure chillers do not require continuous purge recovery systems.

79.11 CENTRIFUGAL COMPRESSOR DRIVES

The most efficient form of capacity control for a centrifugal compressor is to regulate the speed. At low loads the speed of the unit will be decreased. This requires a variable-speed prime mover. Depending on the location and application for the unit, there are a number of possible choices for prime movers. Variable-frequency drive (VFD) AC motors are becoming more popular. The price for this type of drive has been coming down, and the savings from efficient operation help to defray the initial cost.

Steam turbines provide for very reliable speed control, but a ready supply of steam is required. Gas turbines only need a connection to a natural gas source, which is commonly available in many urban areas. Simple cycle gas turbine efficiency is rather low, however, so this type of system when utilized is better suited to a combined cycle application. In a combined-cycle system, the energy from the heat of the gas turbine exhaust is recovered to generate steam or to augment some other heating requirement.

TECH TIP

Some chiller systems use heat-recovery units to provide hot water for the building. Heat-recovery units pick up waste heat from the condenser to provide "free" hot water.

Diesel engines have much higher efficiencies than gas turbines, but they are not suited for operating at low loads. The newest diesel engines designed to specifically operate chillers use natural gas or propane as a fuel rather than diesel fuel. These engines are referred to as "spark ignited" diesel engines. Many of the inherent problems encountered from operating a traditional diesel engine at low loads are significantly reduced. If there is ready access to a gas line, this type of prime mover can be a reasonable choice.

79.12 CENTRIFUGAL COMPRESSOR CAPACITY CONTROL

The most efficient method of capacity control is to reduce the compressor speed at low loads. This cannot be the only method because many older centrifugal units are driven by constant-speed AC motors. Variable-speed prime movers may also have limitations in speed reduction and only operate within a specific range. To accommodate this, most centrifugal compressors will have variable inlet guide vanes located at the compressor suction. These vanes will be closed at low load and at compressor start-up, as shown in Figure 79-13a, and open at high load, as shown in Figure 79-13b.

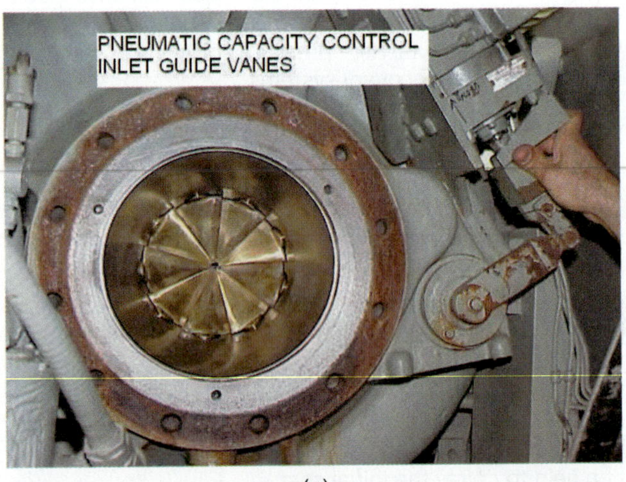

(a)

(b)

Figure 79-13a (a) Inlet guide vanes in the closed position; (b) inlet guide vanes in the open position.

As the load on the centrifugal compressor decreases and the inlet guide vanes begin to close, the pressure in the condenser may become unbalanced. When this occurs, the compressor will surge. To reduce compressor surging at low loads, hot-gas bypass is frequently utilized. A portion of the discharge will be bypassed through a solenoid-operated valve back into the suction side. When the compressor is operating at normal loads, the hot-gas bypass valve should be closed.

79.13 SCREW COMPRESSOR CHILLERS

An important type of packaged water chiller uses the helical rotary compressor (Figure 79-14), commonly known as a screw compressor because of the appearance of the rotors. Screw compressors and their chiller packages were originally developed as effective units for the refrigeration needs of the food and chemical industries. They have been successfully adapted to the needs of the comfort air-conditioning market, with industrial-based technology.

The twin rotor screw compressor, illustrated in Figure 79-15, uses a mating pair of rotors with lobes that rotate

Figure 79-14 Screw compressor, end view.

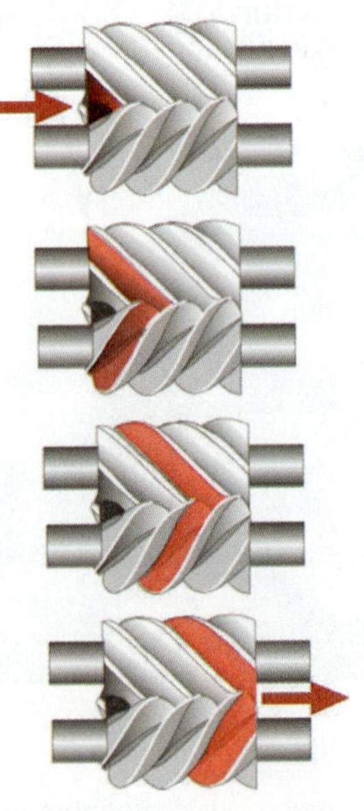

Figure 79-15 Diagram of the rotors of a screw compressor.

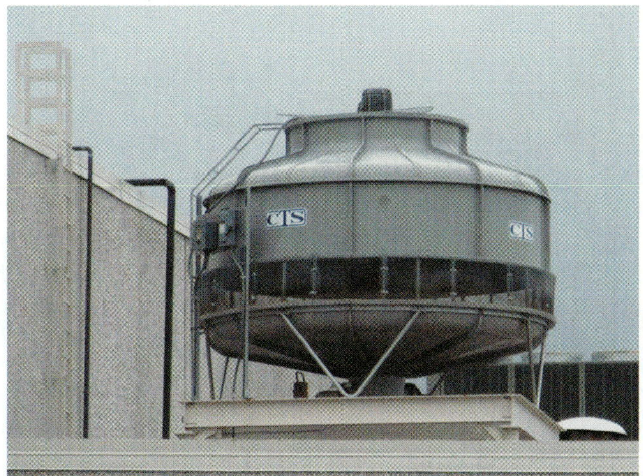

Figure 79-16 Cooling water tower for closed-loop water-cooled condenser.

much like a pair of gears. During rotation, the space or mesh between the lobes first expands to draw in the suction gas. At a point where the interlobe space is at maximum, the lobes seal off the inlet port. As the lobes rotate, the interlobe space becomes smaller as the gas is carried to the discharge end of the compressor. The refrigerant gas is internally compressed by this positive-displacement compressor until the rotors uncover the discharge port, where the compressed gas is discharged from the compressor.

Screw compressors are used because of their high capacity for a small unit and their continuously variable (stepless) capacity control, typically modulating from 100 to 10 percent of full capacity. Being positive displacement, screw compressors have piping flexibility to accommodate remotely located condensers and can be used with high-pressure refrigerants. They can operate with both water- and air-cooled condensers. Cooling water towers may be used for closed-loop water-cooled condensers (Figure 79-16).

Screw compressors come in a variety of configurations, depending on the manufacturer and application. One common type is the horizontal open-drive unit, typically driven at 3,500 rpm by an external motor. Many packaged chiller units use semihermetic construction on the compressor, eliminating the problems of shaft alignment and mechanical shaft seal leakage. Some newer designs utilize three screws. This shortens the overall length and increases the compression efficiency.

79.14 SCREW COMPRESSOR CHILLER OPERATION

The cycle for a screw compressor chiller with a water-cooled condenser is similar to the centrifugal compressor chiller. Figure 79-17 shows a hermetic screw compressor drawing suction from a flooded cooler. The refrigerant vapor is compressed and discharged into the condenser. A float valve will control refrigerant flow and allow for the pressure drop required for the refrigerant to expand in the cooler. A hot-gas bypass solenoid-operated valve is used to recirculate refrigerant at low loads. Also shown are the cooling circuits for the hermetic motor and the variable-frequency drive. Capacity for this unit is controlled by adjusting the motor speed.

79.15 CHILLER ECONOMIZER OPERATION

Many large commercial centrifugal and screw-type chillers utilize economizer sections. Their purpose is to subcool the liquid refrigerant before it enters the cooler. An economizer section is shown in Figure 79-18. The refrigerant leaving the condenser as a liquid passes through an orifice into the economizer section. The pressure in the economizer is between the condenser and the cooler operating pressures. Due to the drop in pressure in the economizer, some of the liquid refrigerant flashes into a vapor and is returned back to the compressor. This flashing in the economizer subcools the liquid that then passes on to the cooler. Due to this subcooling, less flash gas is produced when the liquid refrigerant passes through the fixed orifice and enters the cooler. This will provide more available liquid refrigerant in the cooler.

The refrigerant that flashes in the economizer is not piped directly back to the compressor suction. Since the economizer is at a higher pressure than the suction pressure, the vapor can enter the compressor at an intermediate point, or in the case of a staged compressor, in a later stage. If no economizer section is used, this vapor will still be formed as flash gas in the cooler but will require more energy to compress it from suction pressure to discharge pressure as compared to compressing the vapor from economizer pressure to discharge pressure.

79.16 CHILLER OIL RETURN

The oil in many large commercial centrifugal and screw-type chillers travels through the system and collects in the bottom of the cooler. The compressor normally draws suction from the top of the cooler to reduce any possibility of a liquid slug. Without some type of oil-return system, the oil will not be able to be reintroduced into the lubrication circuit.

Figure 79-19 illustrates an example of a typical oil-return system for a hermetic screw-type compressor chiller. A metered amount of oil collecting in the bottom of the cooler is allowed to pass to the vaporizer section. Hot gas from the condenser passes through a coil located in the vaporizer. This will vaporize any liquid refrigerant that has been carried over with the oil. The oil then passes from the vaporizer to the oil sump, where an electric heater is utilized to further vaporize any refrigerant liquid entrained in the oil. Mist eliminators are used to remove any entrained oil in the refrigerant vapor before returning to the compressor suction. This reduces the possibility of any oil carryover that would slug the compressor.

79.17 CHILLER REFRIGERANT RECOVERY AND TRANSFER

Periodically, chillers may need to be pumped down for servicing. Unlike some commercial systems that utilize receivers, there is no place to transfer the refrigerant other than an external storage tank. The external storage tank

Figure 79-17 Screw compressor refrigerant flow without economizer.

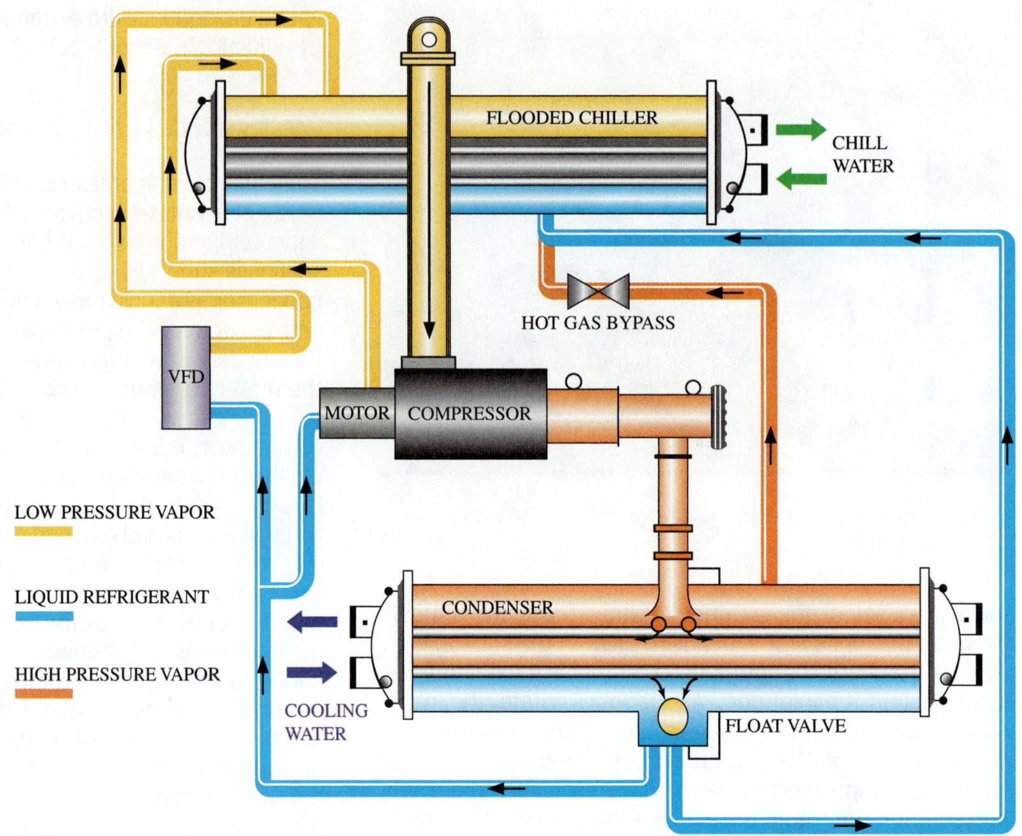

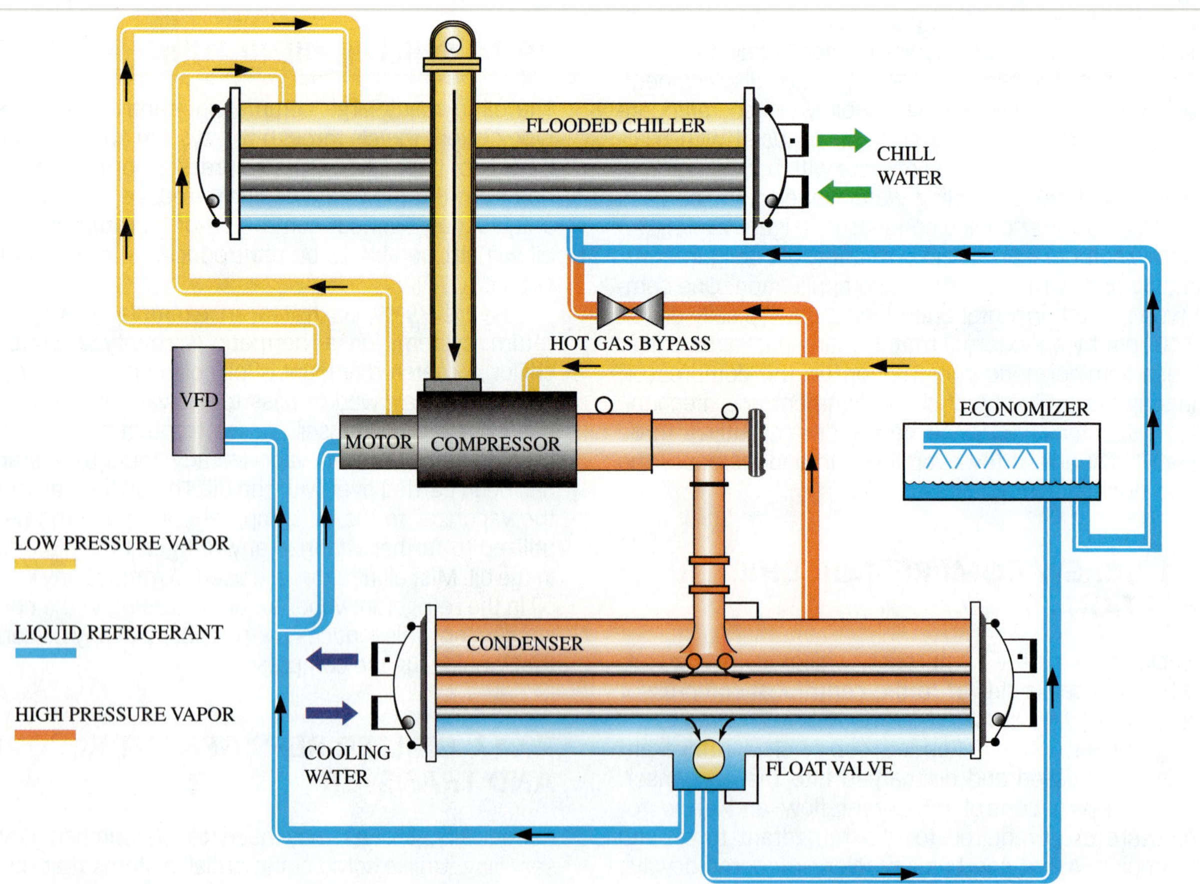

Figure 79-18 Screw compressor refrigerant flow with economizer.

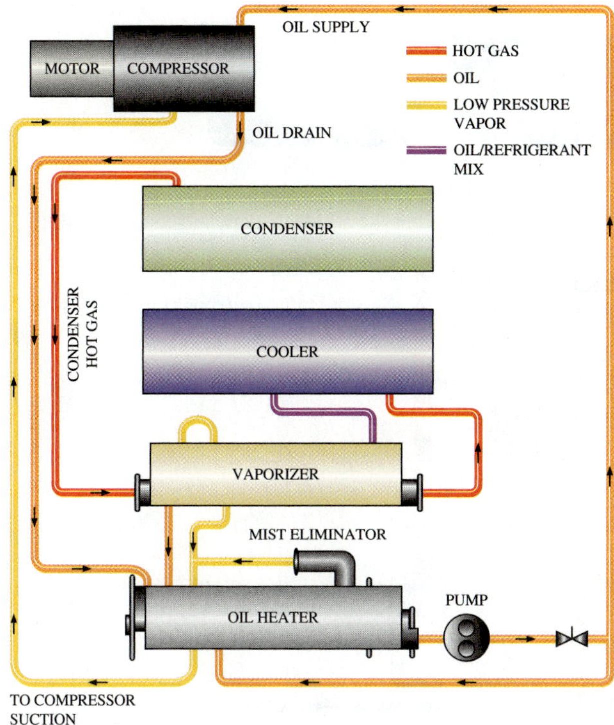

Figure 79-19 Screw compressor oil flow.

or receiver should be of sufficient size to hold the entire charge of refrigerant, and it should be rated for the same or a higher maximum working pressure as the system. Some systems have external receivers that can be used for multiple chiller circuits.

CAUTION

Always begin the refrigerant transfer from a chiller unit using liquid-recovery methods. If vapor is transferred from the chiller while liquid still remains, the chiller will freeze up and the tubes will crack to cause extensive damage to the chiller. Circulate chilled water and condenser cooling water through the unit during the recovery process to reduce the possibility of freeze-up.

An example of a transfer arrangement is shown in Figure 79-20. This system is permanently installed and can be used to transfer refrigerant from multiple chiller circuits. Chillers that do not have permanently installed transfer systems will require a portable unit. This is different from conventional portable recovery units because of the valve arrangement.

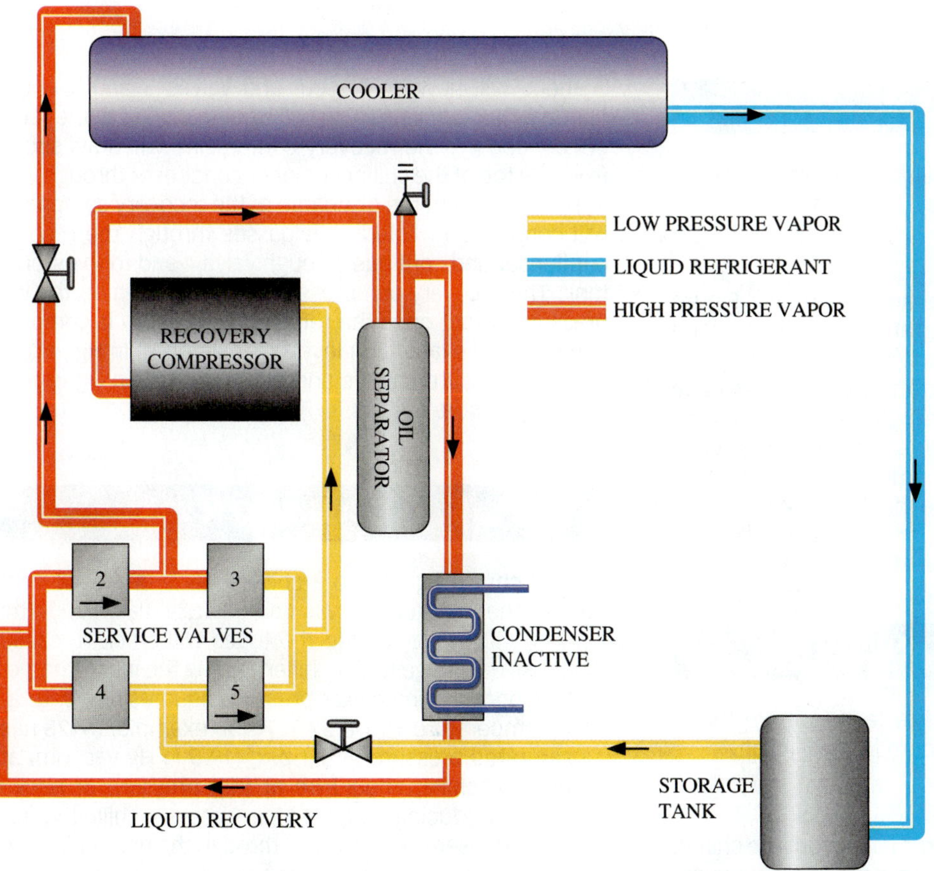

Figure 79-20a Liquid recovery first

Figure 79-20b Vapor recovery follows liquid recovery.

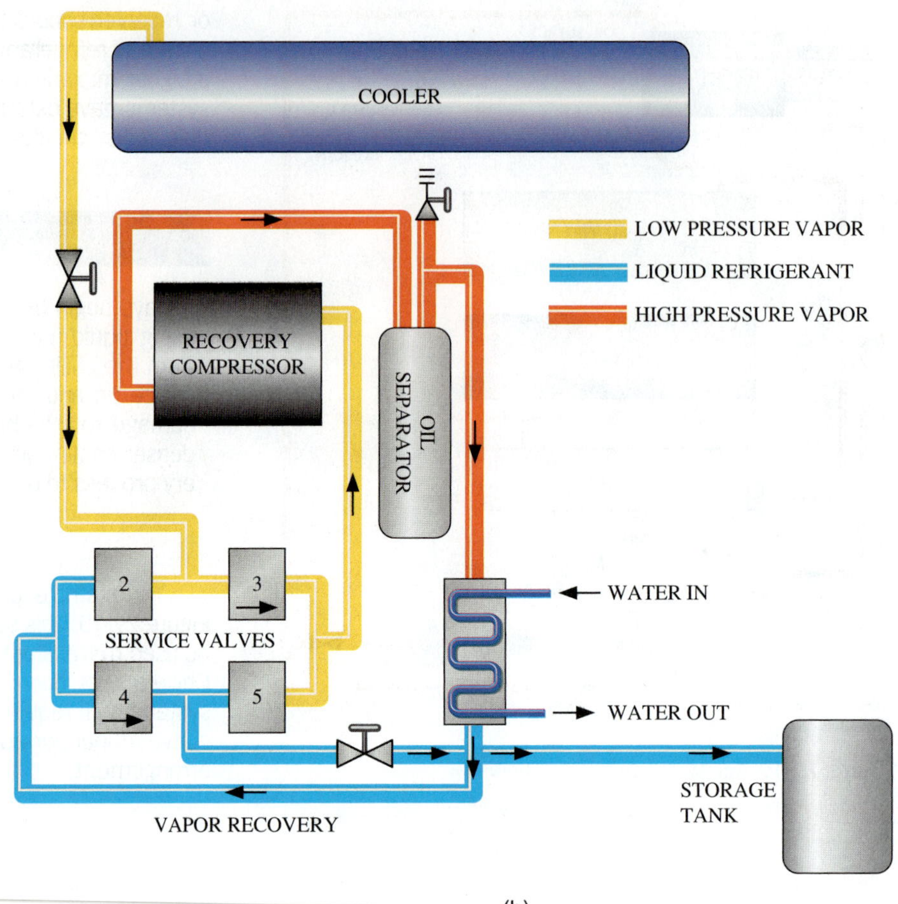

(b)

CAUTION

Do not mix refrigerants from chillers that use different compressor oils.

Always transfer liquid refrigerant first. Using Figure 79-20a as an example, the valve arrangement will be lined up with valves 2 and 5 open. This will allow the recovery compressor to draw suction from the top of the storage tank through valve 5 and deliver high-pressure vapor to the top of the chiller cooler or condenser through valve 2. Valves 3 and 4 will remain closed. During this liquid transfer, the recovery condenser will be inactive. The vapor entering the top of the chiller cooler or condenser will force liquid out of the bottom and to the storage tank.

CAUTION

During transfer, never fill a storage tank to more than 80 percent of capacity to allow room for expansion.

After all of the liquid has been removed from the chiller, the valve alignment will be changed for the vapor-recovery portion of the transfer, as shown in Figure 79-20b. Valves 2 and 5 will now be closed, and valves 3 and 4 will be opened. Cooling water flow will be established to the recovery condenser. The recovery compressor will draw suction from the top of the chiller cooler or condenser through valve 3. The vapor from the discharge of the recovery compressor will condense to liquid as it passes through the recovery condenser and continue through valve 4 and to the storage tank. The recovery compressor will shut down on the low-pressure cutout once the refrigerant recovery is complete. The required evacuation level is 7 psia or 15 in Hg vacuum for high-pressure chillers and 0.5 psia or 29 in Hg vacuum for low-pressure chillers.

CAUTION

Never charge liquid into a chiller that is in a deep vacuum. The refrigerant will immediately flash and the chiller will freeze up and the tubes will crack to cause extensive damage to the chiller. During the initial charge, add vapor until the chiller pressure equates to a saturation temperature above 32°F. As an example, R-123 has a saturated pressure of 5.1 psia (19.5 in Hg vacuum) at 35°F. The chiller should be at that pressure or above before introducing liquid R-123. Circulate chilled water and condenser-cooling water through the unit during the charging process to reduce the possibility of freeze-up.

79.18 ABSORPTION CHILLERS

Unlike the conventional mechanical compression refrigeration cycle used in all the other equipment discussed, an absorption chiller uses steam, hot water, or direct firing by natural gas as an energy source to produce a pressure differential in a generator section. Some absorption units generate both chilled water and hot water from the same unit. Since they can operate on waste steam or direct-fired natural gas, the actual costs of operation may be less than electrically driven equipment, depending on the relative energy costs.

SERVICE TIP

Most power companies use a building peak load as the basis for establishing the billing rate for electrical power. The peak load for electrical billing rates is based on the highest electricity usage during the prime electrical usage time of the day. Because chillers require larger amounts of energy during start-up than during operation, they must be started before the beginning of the prime cooling time. In most parts of the country, the prime electrical usage time is between 9 AM and 6 PM.

Once the peak load rate is established, it will remain in effect for all electrical usage for the next 12 months. Starting a chiller late and establishing a new higher peak rate can cost a building operator tens of thousands of dollars in higher utility bills.

Absorption chillers operate most efficiently when they run at a steady state for long periods of time. For this reason they may be one unit as part of a large building's multiple-chiller system. In a multiple-chiller installation, the absorption chiller can provide the base cooling needed to maintain the building over 24 hours, and other chillers can be started and stopped during the day as needed to meet peak cooling loads.

79.19 ABSORPTION CHILLER OPERATION

The absorption-cycle components and operational cycle are shown in Figure 79-21. The refrigerant used is simply water, and the absorbent is lithium bromide (LiBr). The evaporator shell pressure is reduced to 0.12 psia (29.7 in Hg vacuum). At this low pressure, water will evaporate at 41°F.

The refrigerant (water, colored blue) is sprayed over the outside of the chilled-water tubes in the evaporator and evaporates to remove heat from the chilled water. A strong solution of LiBr (colored orange) is sprayed into the absorber section and absorbs the water vapor. The heat given off by this absorption process is transferred to the cooling water circuit. The solution of LiBr diluted with water (colored yellow) falls to the bottom of the vessel.

The diluted LiBr solution passes through a heat exchanger and into the generator section. Heat is supplied from a heat source (hot water), through a series of coils within the generator. The water entrained in the LiBr-diluted solution is released by the heat and becomes water vapor. The water vapor passes into the condenser section,

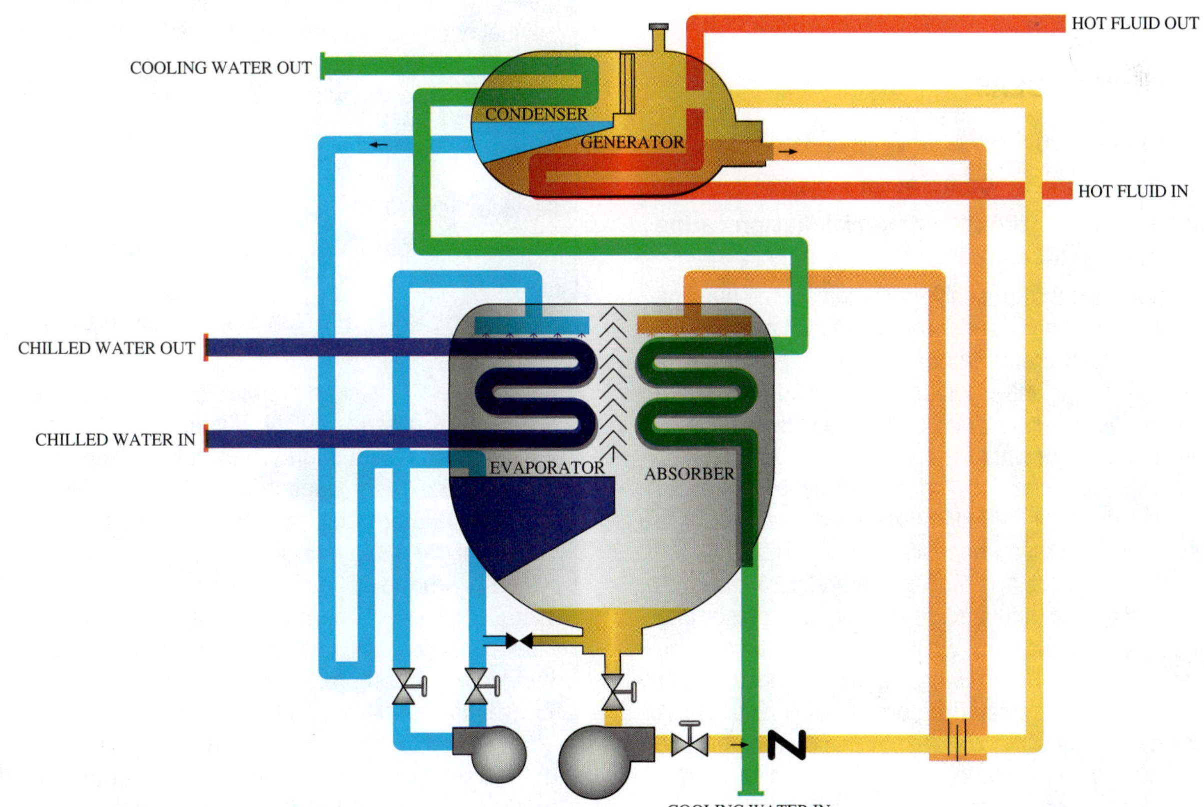

Figure 79-21 Absorption-cycle operation.

condenses back into liquid, and returns to the evaporator section. Due to the release of the absorbed water, the LiBr once again becomes a strong solution (orange) and returns via the heat exchanger back to the absorber.

UNIT 79—SUMMARY

Air-conditioning chillers are typically designed for indirect expansion. Chilled water is circulated to individual air handlers or fan rooms rather than the primary refrigerant. Chillers come in various sizes, with smaller tonnage units typically utilizing reciprocating or scroll compressors with electric motor drives. Larger tonnage units require more refrigerant flow and therefore are generally configured with centrifugal and screw-type compressors. These can also be driven by electric motors with the trend being toward variable-frequency (adjustable speed) AC drive units to increase efficiency. Other suitable drives can be diesel, steam, or gas turbine.

It is important to understand the refrigerant cycle and operation of a chiller before transferring refrigerant. Always remove liquid refrigerant from a chiller first before recovering the vapor. Failure to do this could result in chiller freeze-up and considerable damage. This also applies to charging a unit. Never charge liquid refrigerant to a chiller that is under a deep vacuum. Never raise the pressure in a chiller above the rupture disk burst pressure. Follow all manufacturer's recommendations and procedures whenever servicing a chiller.

WORK ORDERS

Service Ticket 7901

Customer Complaint: Compressor Fails to Load

Equipment Type: Light Commercial Reciprocating Chiller with Dual Compressors

The complaint is that the chiller fails to cool the building during periods of high loads. The technician arrives during the middle of the afternoon during the peak cooling period. The chiller unit is located just outside the building. The chiller operates with two four-cylinder reciprocating compressors with one in lead and the other in lag. Both compressors have cylinder head bypass solenoid unloaders. The solenoid coil on the lead compressor is warm to the touch and appears to be energized. However, the solenoid coil on the lag compressor feels cold.

The technician shuts down the lag compressor and secures its power supply while leaving the lead compressor running. The circuit is tested to ensure that the power is disconnected before the solenoid coil is removed. The coil is then tested with an ohmmeter. The coil shows infinite resistance, which indicates an open in the coil. A new coil is wired into place and the compressor starts and now loads normally.

Service Ticket 7902

Customer Complaint: Compressor Fails to Load

Equipment Type: 150-Ton Centrifugal Water-Cooled Chiller

The technician meets with the supervisor of the maintenance department and is told that the centrifugal chiller is not loading up and the chilled-water temperature leaving the unit is very high. The technician checks the temperatures and pressures registered on the unit and finds the chilled-water return temperature to be 65°F and the chilled-water outlet temperature to be 60°F. The normal chilled-water temperature outlet should be 45°F.

The compressor is a constant-speed unit operating at 12,000 rpm with variable inlet guide vanes for capacity control. The guide vanes are in the fully closed position or unloaded position. The control circuit for the inlet damper position is controlled by the chilled-water leaving temperature, so the guide vanes should be fully open, as the chilled-water temperature is high. The technician places the inlet guide vanes into manual control and slowly strokes the vanes open and closed to verify they are working correctly.

The technician follows the capillary tube from the control cabinet to the thermal bulb located in the cooler shell. The sensing line is kinked perhaps from someone stepping or dropping something on it. The sensing bulb could be easily removed without draining the chilled water from the cooler. A new sensing line and bulb are installed. The technician slowly loads the compressor with manual control of the inlet guide vanes. Once the unit is nearly fully loaded, the technician switches the control back to automatic and the guide vanes are now being controlled by the chilled-water leaving temperature.

Service Ticket 7903

Customer Complaint: Chiller Tripping on High Pressure

Equipment Type: 150-Ton Low-Pressure Centrifugal Water-Cooled Chiller

The technician checked the unit which was tripped out. The display message on the digital controller indicates a high-pressure situation. The cooling water circuit is checked and there appears to be a sufficient supply of cooling water going to the unit. Another possibility for the high-pressure condition is excessive air in the system. With the system down, the chiller pressure has risen above atmospheric. The technician uses an electronic leak detector and finds a leak at a flared fitting leading to the compressor suction pressure gauge. The fitting is tightened, and this stops the leak.

Next, the technician checks the purge recovery unit. This is a self-contained unit with a small compressor that continually removes gas from the top of the condenser, which is then condensed and returned to the unit. The noncondensable gases collected are purged to the atmosphere. The compressor for the purge recovery unit has also tripped out. The technician resets the purge recovery unit and it begins operating normally. The technician recommends the unit to be checked regularly for leaks. The purge recovery unit also needs to be monitored because it is an older unit and could be in need of replacement soon.

UNIT 79—REVIEW QUESTIONS

1. Describe the difference between a flooded and a dry (direct expansion) type cooler.
2. What are some advantages and disadvantages of flooded coolers?
3. Why are low-pressure refrigerants well suited for centrifugal compressor chillers?
4. What is the function of a purge recovery unit?
5. What precaution should be taken whenever leak testing a low-pressure chiller?
6. What precaution should be taken for coolers that are located outside?
7. Why is a cooler shell and suction line insulated on a unit that is located indoors?
8. Explain the operation of lead and lag compressors.
9. How can the centrifugal compressor speed be higher than the motor speed?
10. What is the difference between a single-stage and two-stage centrifugal compressor?
11. What is the most efficient method of capacity control for a centrifugal compressor?
12. What types of prime movers can be used for centrifugal chillers?
13. What type of capacity control is used for a centrifugal compressor that is driven by a constant-speed motor?
14. What can be considered an advantage of a screw-type chiller as compared to a centrifugal chiller?
15. What is the advantage of a screw-type compressor that utilizes three screws?
16. What is the purpose of a chiller economizer?
17. What happens to the flash gas formed in a chiller economizer?
18. What is the purpose of an oil-return system for a chiller?
19. How should refrigerant be transferred from a chiller?
20. What could happen if you charge liquid refrigerant into a chiller that is under a deep vacuum?

UNIT 80

Hydronic Heating Systems

OBJECTIVES

After completing this unit, you will be able to:

1. list the common types of terminal units used for hydronic heating systems.
2. list the different types of piping systems that can be used for hydronic heating.
3. explain how a multiple-zone heating system is configured.
4. describe the operation of a motorized valve used for hydronic heating systems.
5. list the advantages of radiant floor heating.
6. describe the operation of mixing valves.
7. select a proper circulator for a heating system.

80.1 INTRODUCTION

Hydronics can be defined as a science that utilizes water or steam to transfer heat from the source where it is produced to an area where it can be used through a closed system of piping. Depending on the application, it may be desirable to circulate hot water to radiant heaters, fan coil units, or air handlers located throughout a building. Water-to-water and air-to-water heat pumps are also examples of hydronic heating systems. Another method growing in popularity is radiant floor heating where hot water is piped through the floor. To service and install hydronic systems, it is important to understand how these systems are configured and what advantages and disadvantages these have over other alternative heating methods.

80.2 HYDRONIC SYSTEM SELECTION

The decision to install a hydronic heating system should be weighed against all other available alternatives. Although water can be circulated by gravity, modern heating systems use forced circulation, which requires an electrically driven pump or series of pumps. Therefore, the installation of a hydronic system will generally require an extensive network of piping. The common choice for piping in the past was copper, which required the fitting and soldering of varied lengths, and this was a time-consuming process. New materials that are much easier to work with, such as cross-linked polyethylene tubing, have sped up the installation process.

The advantages of hot-water heating systems are based on overall comfort for the following reasons:

- Heat is supplied at the base of the outside wall, warming the interior wall surface by convection and conduction to near room temperature. This prevents the uncomfortable feeling of losing body heat to cold outside walls.
- The natural circulation of warm air rising from the radiant heaters supplies heat to the room by convection with a gentle movement of air. This prevents uncomfortable drafts and air velocities that absorb moisture from the skin, creating a cooling effect.
- The hot water heats and cools slowly. This prevents sudden changes in temperature.

The disadvantages of hot water heating are as follows:

- It costs more than an equivalent warm-air installation.
- If humidification and air filtration are desired, a separate air system needs to be installed.
- If cooling is desired as well as heating, dual systems are required.

80.3 WATER AS A HEAT-TRANSFER MEDIUM

Water expands and contracts as its temperature changes. In a liquid state, the change in volume with temperature is relatively small, but this can still be significant in a closed-loop system. Therefore, in closed hot-water heating systems, to allow for expansion, certain safety measures are provided as follows:

- Both temperature- and pressure-limiting controls are used to maintain the water within safe conditions.
- A pressure-relief device is supplied in the piping to relieve excess pressure should it occur.
- An expansion tank is provided in the piping, partly filled with air, to permit normal expansion and contraction of the water.

Water will freeze into ice at temperatures around 32°F, depending on the pressure. As water freezes, it expands and can burst pipes or containers. Provisions need to be made to do the following:

- Prevent the water from reaching the freezing point.
- If the water cannot be prevented from reaching the freezing point, use an antifreeze solution in the system in place of pure water, to lower the freezing point to a safe value.

Water has considerable weight (62.3 lb/ft^3), and allowances often need to be made in the supporting structure for equipment containing water. Water has friction as it flows through piping, fittings, and equipment. This frictional force must be considered when selecting the flow of a circulating pump.

CAUTION

It is not easy to drain all of the fluid out of hydronic system pipes for service. Sometimes water can be trapped and then be released suddenly as the result of vibration from shaking the pipe or when air breaks the siphon. It is important when working on electrical components such as pump motors that you make certain that the power to the pump is off and all the water in the system has been removed. A sudden gush of water can create a major safety hazard from electric shock if these precautions are not taken.

80.4 HYDRONIC SYSTEM RESIDENTIAL APPLICATIONS

A hydronic heating system for residential use often includes a gas- or oil-fired boiler. The boiler will be piped with the proper accessories to allow for expansion, water makeup, and mixing valves for different types of zone heating applications. This typical arrangement is shown in Figure 80-1. The hot water leaves the boiler through a 2-in supply valve and is circulated throughout the building. The water returns to the boiler from a 2-in return line and through a circulating pump. In this example, the water is pressure fed into the boiler by the circulator pump. There are a number of different design types, with another common example being a straight gravity return line without a circulating pump. Also note the expansion tank and an air vent located in the return line. The cold-water inlet is for the makeup water. This will be connected to the house water supply plumbing. A pressure-reducing valve is used to regulate the water admitted into the boiler. The boiler is equipped with a pressure-relief valve. There are shutoff valves to isolate different parts of the system and to drain lines for use when servicing or repairing the system.

TECH TIP

The term *radiator* has been a commonly used term for heating systems. Although this term is commonly accepted, it is misleading because it implies that the heat transfer is through radiation, which is incorrect. Heat transfer from terminal units is through conduction (heat transfer through a solid substance) and convection (heat transfer through a fluid). This means that air circulation for convective heat transfer is critical for proper operation. Radiant floor heating also has this same type of misleading terminology; and heat transfer for this type of installation likewise depends on convection and conduction. A radiant floor heating system placed beneath a thick carpet will be less efficient than a radiant heating system placed beneath a tile floor due to the conductive heat transfer properties of the different materials.

80.5 TERMINAL UNITS

Terminal units transfer heat from the hot water to the various building areas. The heat supplied by these units is controlled to provide comfortable conditions. Normally this

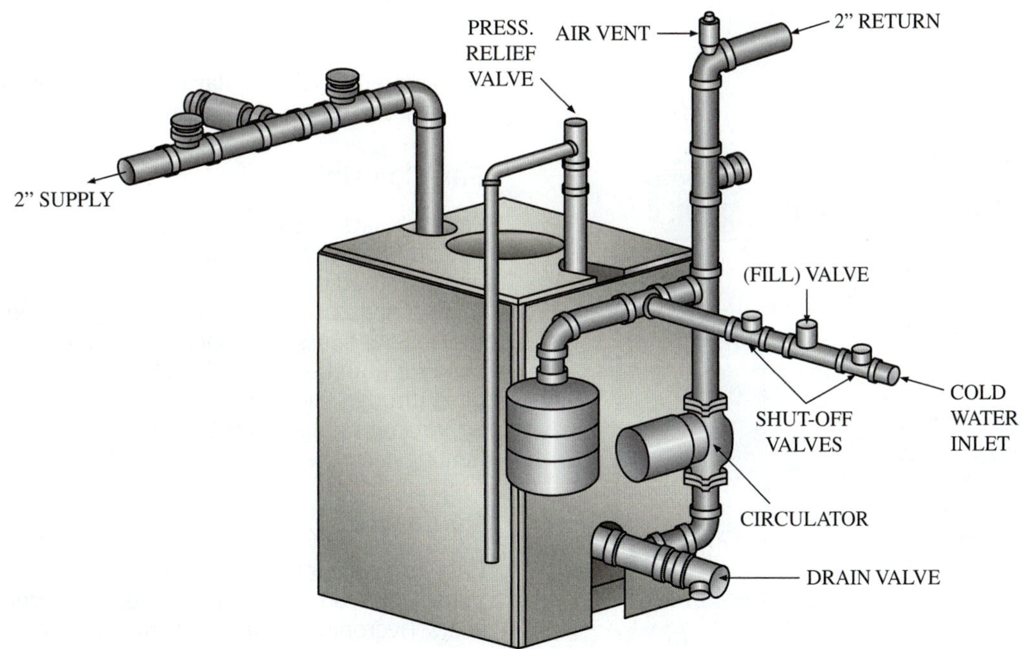

Figure 80-1 Hydronic boiler piping arrangement.

Figure 80-2 Copper tubing and aluminum fins.

equipment is rated in BTU/hr or MBTU/hr; however, these ratings can be stated in terms of equivalent direct radiation (EDR). For hot-water systems, 1 ft^2 EDR = 150 BTU, based on an average water temperature of 170°F. Hot-water units are rated based on the water temperature drop through the unit, which can be 10°F –30°F. A number of types of terminal heating units are available, including baseboard units, convectors, fan coil units, and radiators.

Baseboard Units

Baseboard units are installed near the floor on the outside walls of each room. Heat is supplied by conduction and convection. The heating element is finned-tube construction or cast iron. The most common units have elements using copper tube and aluminum fins, as shown in Figure 80-2. A typical finned tube along with accessory pieces for corners and ends is shown in Figure 80-3.

Convector Units

Convector units such as the one shown in Figure 80-4 can be freestanding or recessed. These units heat the room mainly by convection, with the air entering the bottom and leaving from the upper front. The freestanding model is supplied either with a flat or slanting top. An adjustable damper can be supplied if desired.

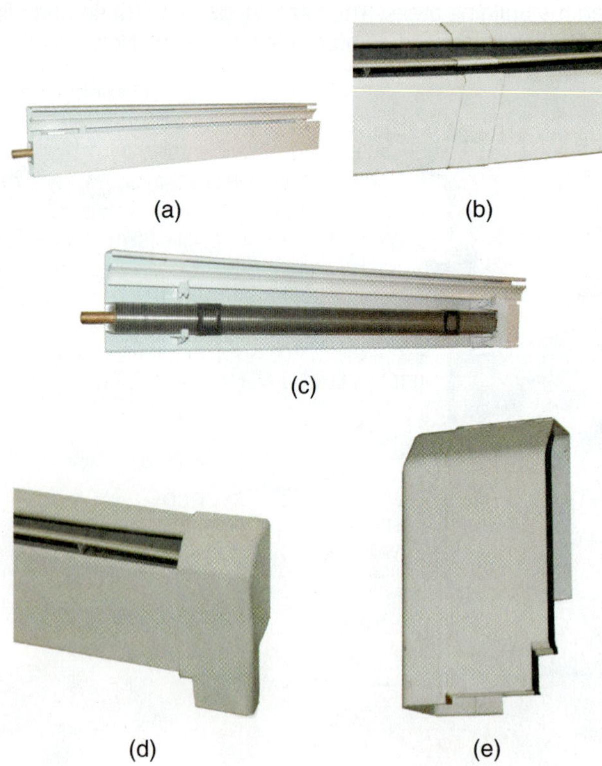

Figure 80-3 Trim pieces for baseboard enclosures.

Figure 80-4 Typical convector.

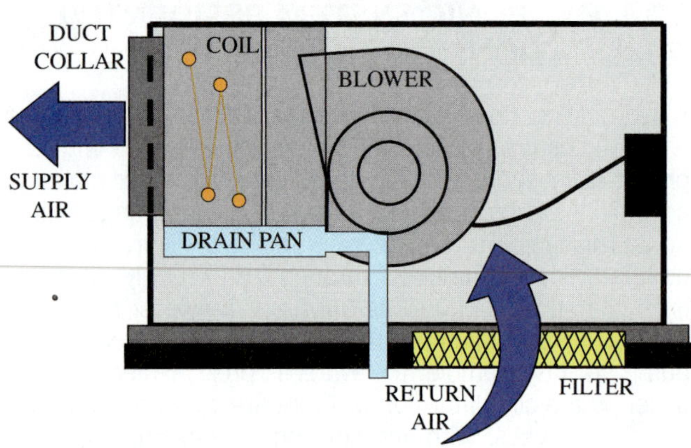

Figure 80-5 Horizontal fan coil.

Fan Coil Units

Fan coil units as shown in Figure 80-5 can serve the same function as convectors, but they have additional features and increased efficiency. The unit has a centrifugal fan with manual speed control for circulating the room air through the unit. Air filters are standard equipment. The unit is often used for both heating and cooling. Outside air can be supplied through the back of the unit. Units are available with air volume ratings from 200 to 1,200 CFM and hot-water heating capacities from 7.6 to 112 MBTU/hr.

Radiators

Cast-iron radiators are primarily used for steam installations (Figure 80-6). They were used more frequently in past years. Hydronic systems are sometimes configured to work with existing cast-iron radiators. These typically operate with water temperatures of 125°F to 200°F.

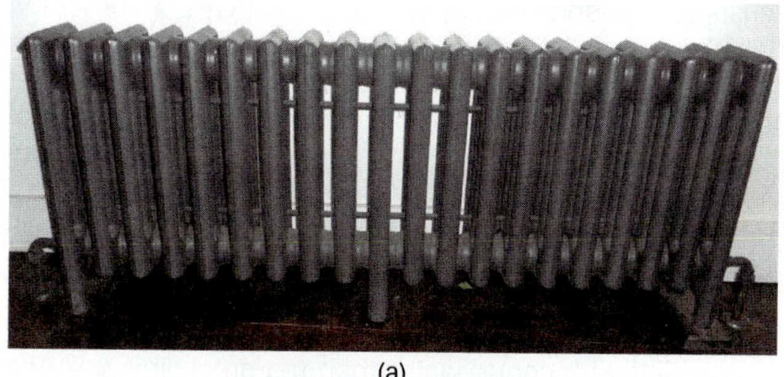

(a)

(b)

Figure 80-6 (a) Cast-iron radiator front view; (b) side view.

(a)

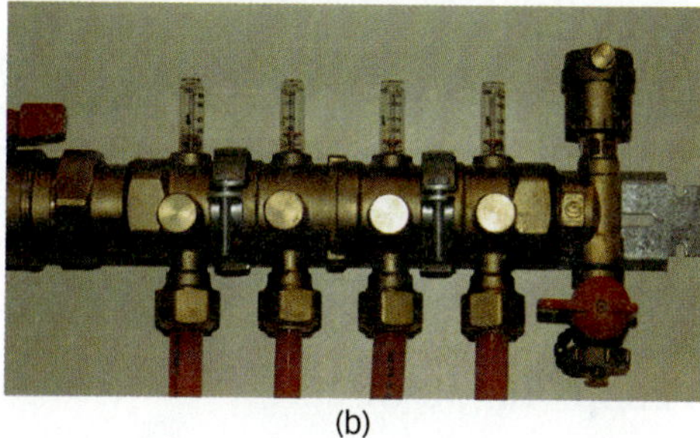

(b)

Figure 80-8 (a) Radiant floor manifolds with individual flow control; (b) manifold flow meters.

80.6 RADIANT FLOOR HEATING

Radiant floor heating has been recently growing in popularity, although this type of heating has a long history. The Romans were the first to use this type of heating. American architect Frank Lloyd Wright used radiant floor heating in many of his home designs from the 1930s. Radiant floor heating is more comfortable than conventional heating because it directs the heat to the occupant level and less heat is wasted up near the ceiling level, as shown in Figure 80-7. There are also no baseboards or radiator panels offering obstruction to furniture. This allows for more effective use of wall space within the room. Radiant floor heating is also quiet when compared to the fan noise levels found with a forced-hot-air system.

Radiant heat piping will have a number of heating loops distributed through a manifold, as shown in Figure 80-8. Often cross-linked polyethylene pipe is used because it is flexible and relatively easy to connect. The piping can be placed into the poured concrete slab for the building. This works well, as the entire slab heats evenly to provide for a continuous and prolonged heat sink. For upper stories

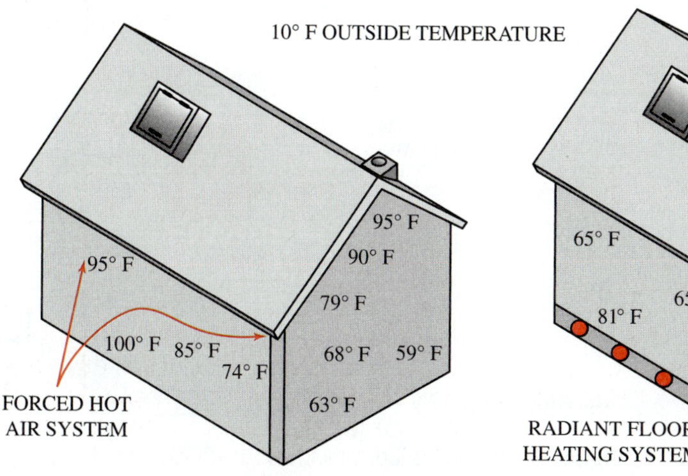

Figure 80-7 Radiant floor temperature comparison.

of the building, the piping can be installed under the floor through the floor joists. This type of installation can be performed after the building is complete by drilling through the floor joists and installing the piping. This can also be installed during the initial construction of the building.

TECH TIP

Radiant floor heating can heat effectively with water temperatures as low as 90°F to 105°F. This allows lower boiler temperatures and increased operating efficiency.

80.7 TYPES OF PIPING SYSTEMS

Hot-water heating systems can be installed with a variety of piping arrangements. Each of these arrangements controls the supply of hot water from the boiler to the terminal units (and the return) in such a way that comfortable conditions are produced. The common systems for hot-water heating applications in residential and small commercial installations are each briefly explained.

Series Loop, Single Circuit

The series-loop system is probably the simplest of all the piping arrangements. A schematic diagram of the single-circuit piping layout is shown in Figure 80-9. In this piping system, the main supply pipe from the boiler enters the first baseboard unit. The outlet from the first unit is connected directly to the inlet of the second unit. This arrangement is continued until all baseboard sections are connected in series. The outlet from the last section is connected to the circulator pump, which connects to the boiler. Thus, all the water flowing through the system passes through each unit.

In many cases, the bathroom or kitchen does not have enough wall space for the proper length of baseboard heating, so a convector is used in place of the baseboard. The convector can be tied to the system using a venturi fitting (see Figure 80-9). The convector is the first terminal unit after the boiler. The venturi is simply a restriction in the line that ensures that some water will flow through the convector, which is in parallel with the other terminal units in the circuit.

When these systems are installed, the entire perimeter of the building has baseboard enclosures. The amount of finned-tube element placed in these enclosures is dependent on heat-loss requirements. Where the main pipe passes a doorway, an offset is made in the piping to run it down below the floor until it passes the door opening to come up again on the other side.

With this piping arrangement, each downstream unit is supplied with cooler water than the preceding one. The actual hot-water temperature drop between terminal units is about 2°F. To compensate for this, each successive unit is oversized by about 2 percent. The use of a series-loop system reduces the cost of the installation to a minimum.

Series Loop, Double Circuit

On larger systems, better performance can be obtained by dividing the system into two (or more) approximately equal circuits (Figure 80-10). When this is done, balancing valves are installed in each branch to permit adjustment of the flow.

One-Pipe Venturi System, Single Circuit

The one-pipe venturi fitting system has a single main line extending from the supply to the return connection of the boiler. The terminal units are fed by a supply and return branch connected to the main, as shown in Figure 80-11. The supply connection uses a standard tee, and the return connection uses a special venturi fitting. The venturi creates a negative pressure at the main connection, drawing the necessary flow through the branch.

This system differs from the series loop in that only a portion of the total flow enters the terminal unit. There is still a temperature drop in the main line as it reaches the location of successive units. The reduced flow, however, improves the performance of the terminal units.

One-Pipe Venturi Unit, Double Circuit

On larger systems using this piping arrangement, better performance can be obtained by dividing the system into

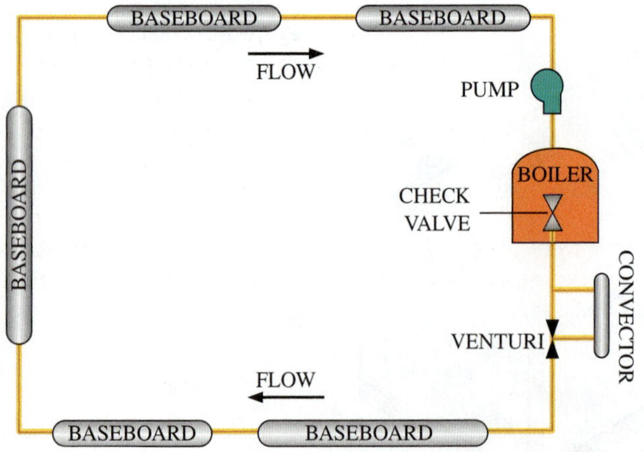

Figure 80-9 Series loop piping layout, single circuit.

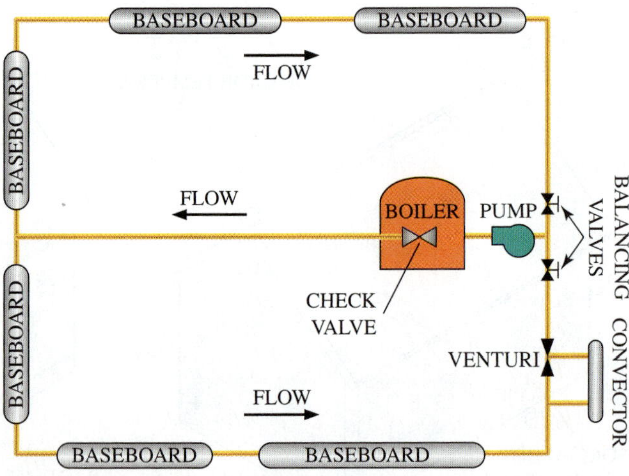

Figure 80-10 Series loop piping layout, double circuit.

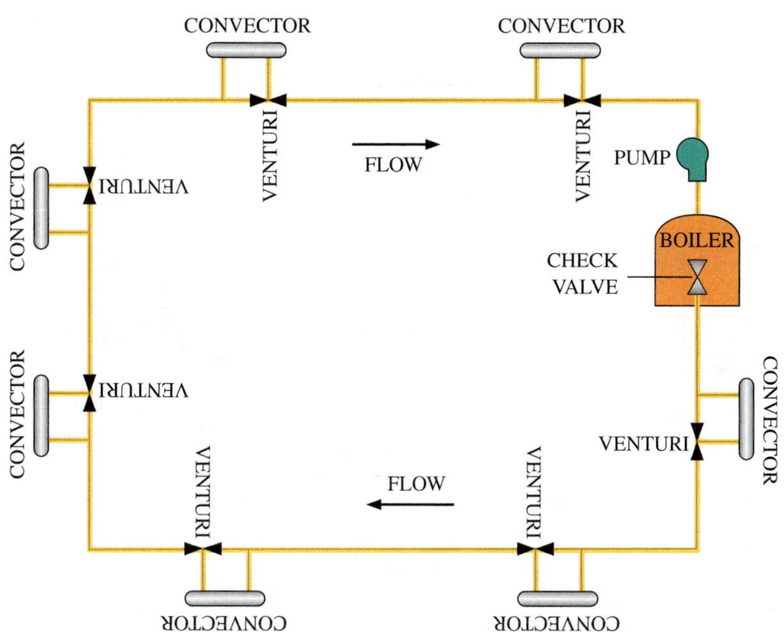

Figure 80-11 One-pipe venturi system, single circuit.

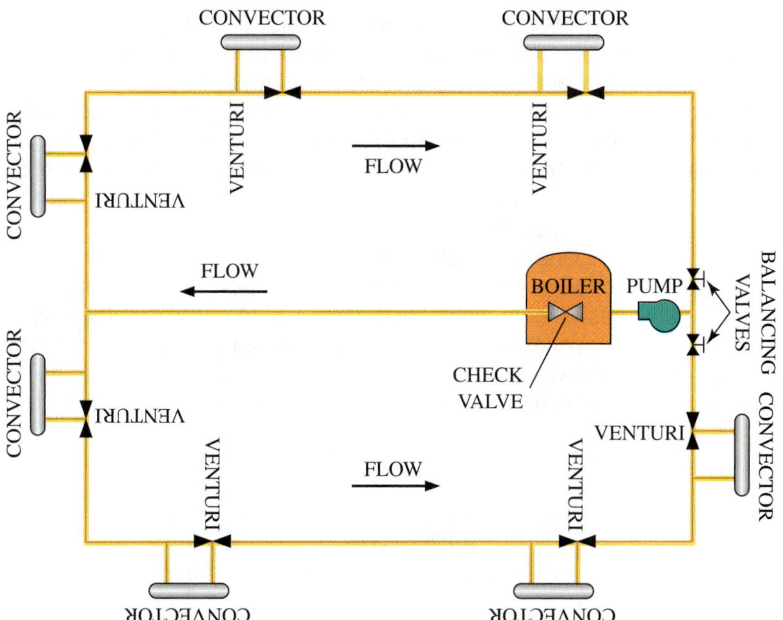

Figure 80-12 One-pipe venturi system, double circuit.

two (or more) approximately equal circuits (Figure 80-12). When this is done, balancing valves are installed in each branch to permit adjusting flow.

Two-Pipe System, Direct Return

In the two-pipe system, separate supply and return mains are used. The return water from the terminal units is collected and returned to the boiler. With this piping arrangement, each unit will receive the same supply-water temperature.

With the direct return, the two mains are run side by side with a supply branch run to each unit from the supply main and a return branch run from each unit to the return main. The first unit taken off the supply main is the first unit on the return main before it reaches the boiler (Figure 80-13). Obviously, the shortest piping is used on the unit nearest the boiler. The longest piping is used on the unit farthest from the boiler. This difference in piping pressure drop must be equalized by using a square head cock in either the supply or return branch to each unit.

The two-pipe direct system uses more pipe than the one-pipe system; however, it is considered to be more efficient by providing better water distribution and better control. This system can also be designed using two or more circuits.

Two-Pipe System, Reverse Return

The two-pipe reverse-return system is similar to the direct return system except that the piping is arranged so that the same length of pipe (supply + return) is used for each

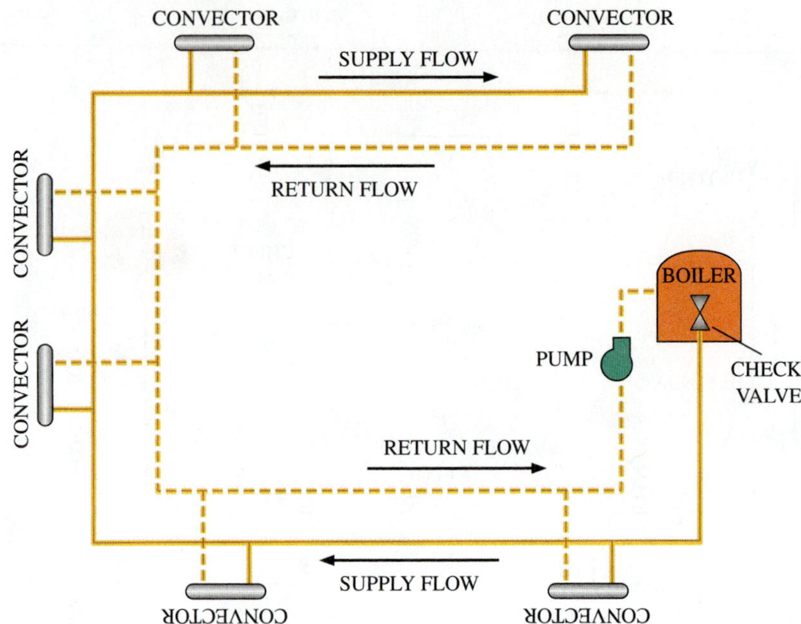

Figure 80-13 Two-pipe direct-return system.

terminal unit. This is accomplished, as shown in Figure 80-14, by taking the unit nearest the boiler off the supply main first, connecting it to the return main last, and proceeding on this basis to equalize the piping lengths for all units. This equalization of the piping loss to each unit provides a balanced condition so that each unit receives its proper share of water.

80.8 ZONED SYSTEMS

Zoning is provided where it is desirable to control the supply of heat (hot water) to different areas of the building by separate thermostats. The areas selected can be determined by usage or building orientation. For example, to zone by usage, the living areas of a residence can be placed in one zone and the sleeping areas in another. To illustrate orientation zoning, a small commercial building can place all the rooms with southern exposure in one zone and the rooms with northern exposure in another. Whatever separation is made, allowance must be made in the piping and controls to provide this feature.

One advantage of hydronic heat is that installing a zoned heating system is relatively easy. Each heating zone has its own piping loop. One way to control water flow to the zones is by a motorized zone valve (Figure 80-15). The zone valve is a normally closed electrically operated valve. A zone thermostat controls the zone valve, opening it when

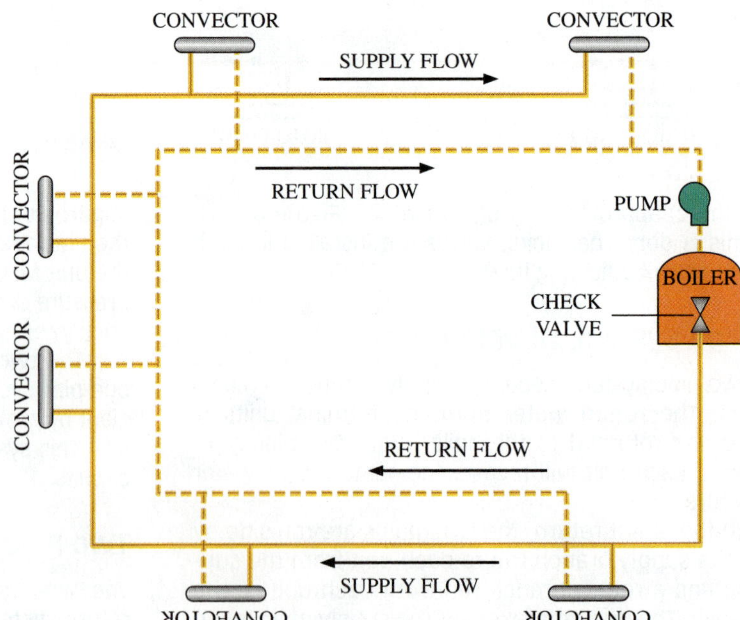

Figure 80-14 Two-pipe reverse-return system.

Figure 80-15 Motorized zone valve.

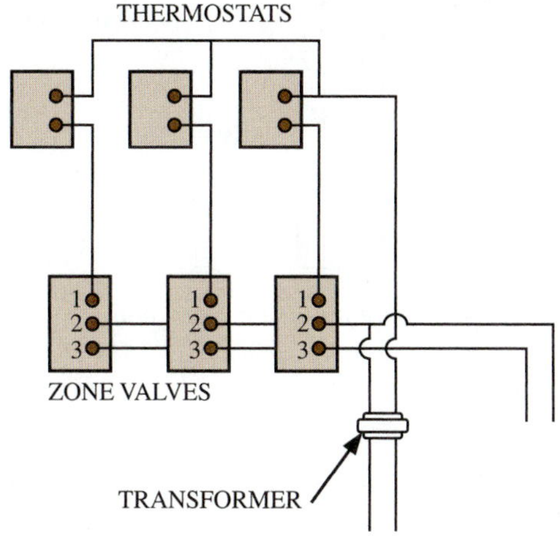

Figure 80-16 Zone valve wiring.

there is a call for heat. The zone valve has a set of contacts that close to energize the circulator pump, providing heat for the zone. The contacts for the valves are wired in parallel with each other so that any of the zone valves can energize the pump (Figure 80-16). This arrangement allows individual temperature control of multiple zones without complex controls and with relatively low installation cost.

There are many ways to pipe a zoned hydronic system. A major design challenge is to ensure steady water flow for even heat distribution. A simple method of achieving this is called reverse return. In reverse-return piping, the zone with the shortest distance to the hot-water supply has the longest distance to the return.

Figure 80-17 shows a typical piping system using zone valves. Balancing valves are used to compensate for different pressure drops and different load requirements. A balancing valve is placed in series with each zone, allowing the pressure drop though each zone to be adjusted.

Another way to control the flow of hot water in each zone is to provide a separate circulator for each zone, as shown in Figure 80-18. Any one of the systems previously described can be provided with zoning. The circulator in each zone responds to the requirement of the thermostat in that zone.

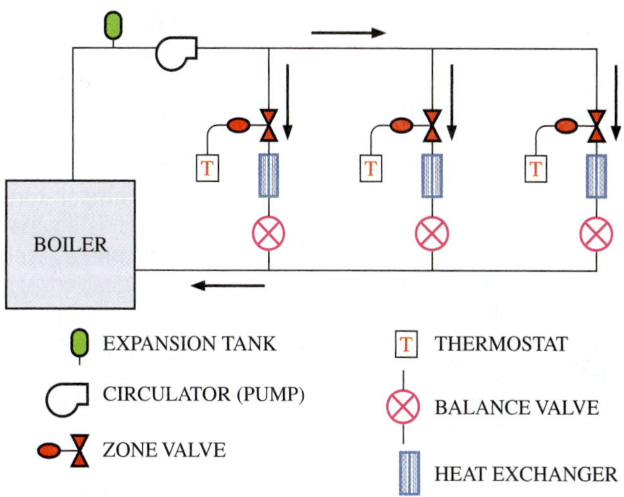

Figure 80-17 Zoned hydronic system piping.

Figure 80-18 Zoned system using separate circulators for each zone.

TECH TIP

A zoning hydronic system can have a significant energy savings in two ways. It provides heat only where it is needed, and the thermal mass of the water within the system will allow the heating unit to operate for longer periods of time, even during light loads. As the water in the system is heated and the heat is used at a terminal unit located in a zoned area, the boiler may cut off and remain off for a period of time while the circulating pump still provides hot water to the zone. The boiler will not relight until the water temperature has dropped below the upper setpoint. This allows for longer boiler run times, which increases the overall operating efficiency.

80.9 RESIDENTIAL ZONE HEATING

Figure 80-19 shows the configuration for a multiple-zone system. As an example, zone 1 could be the basement, zone 2 the first story of the house, and zone 3 the second floor. Each zone would have its own thermostat. The upstairs bedrooms could be easily maintained at a lower temperature than the downstairs living spaces. This system has a separate circulator pump supplying water to each zone as well as a circulator on the return circuit to the hot-water boiler.

The boiler shown in Figure 80-20 has a circulator pump that delivers water from the boiler to the zone circulator pumps. When the zone calls for heat, the circulator pump for the zone will run. In this way, if zone 1 is running and then zone 2 comes on, there should be little reduction in the supply to zone 1.

The boiler in Figure 80-21 uses zone control valves instead of individual circulator pumps for the zones. One circulator pump will supply both zones. This type of system utilizes fewer pumps, thereby reducing installation costs and electricity usage. When the zone calls for heat, the zone control valve will open and the circulator pump will run. One apparent disadvantage should be evident. If zone 1 is running and then zone 2 comes on, there could be a reduction in the supply to zone 1.

The determination for the number of zones and the placement of circulator pumps depends on a number of factors. One is the personal preference of the homeowner. Every room could be a separate zone if desired. However,

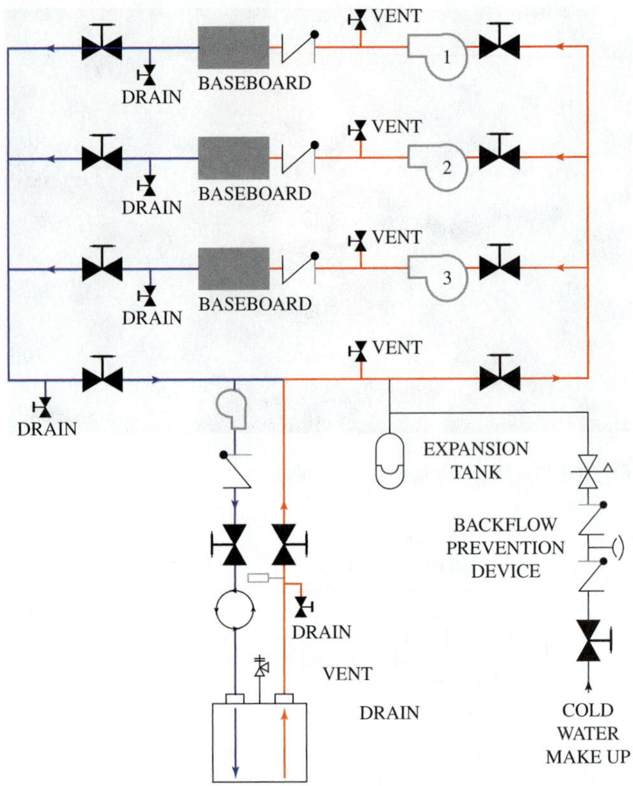

Figure 80-19 Multiple zones using circulators for supply and return.

Figure 80-20 Multiple zones with circulator pumps.

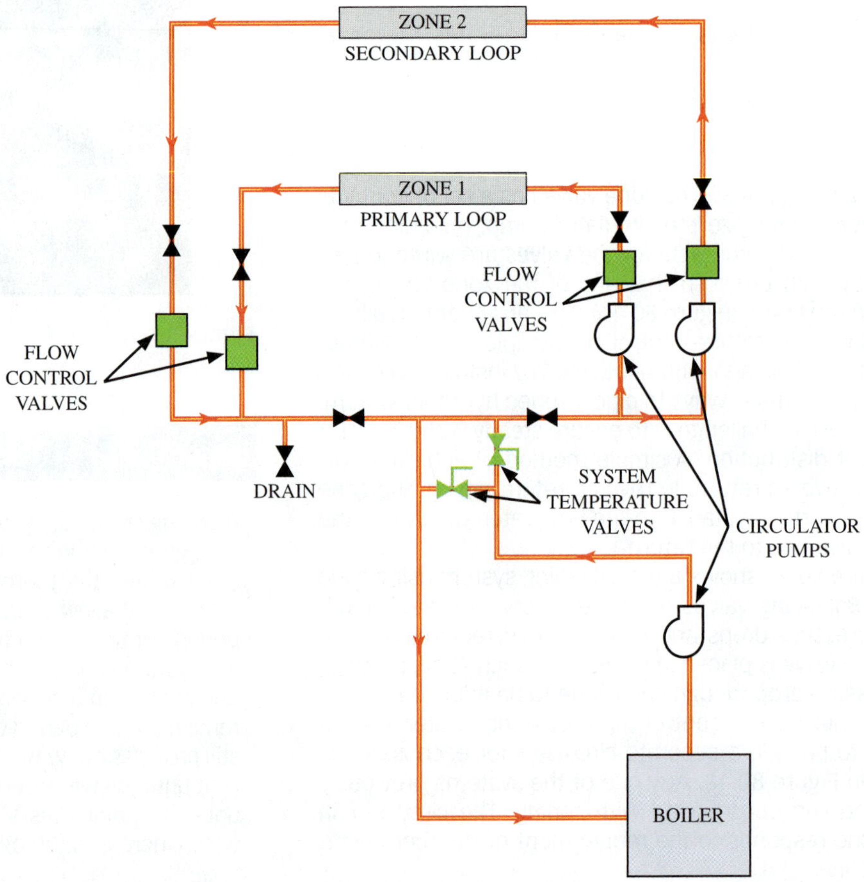

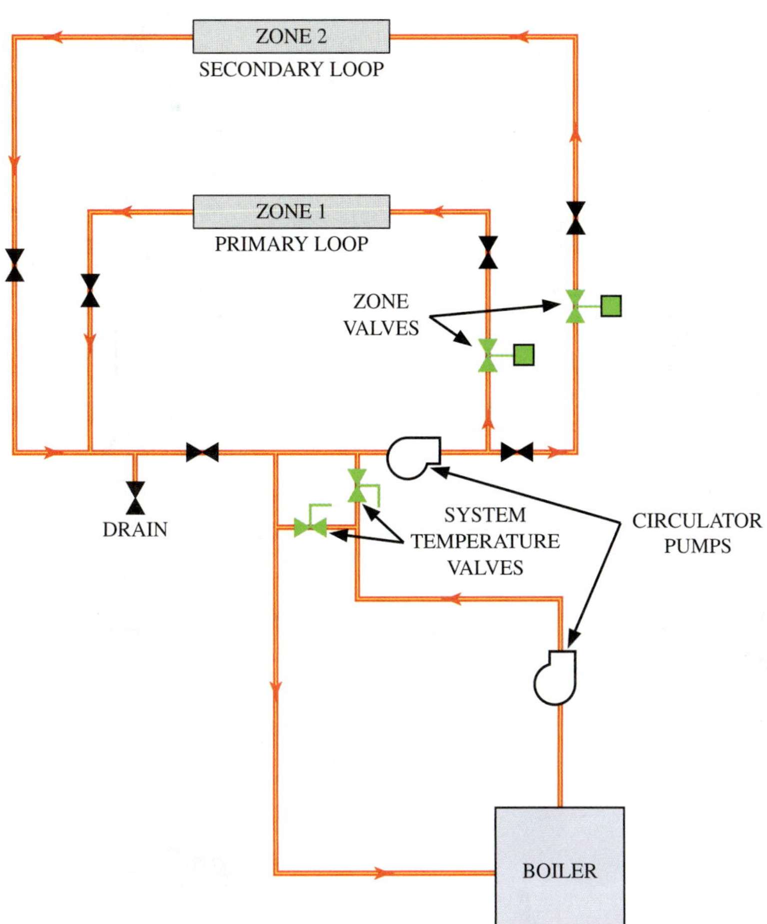

Figure 80-21 Multiple zones with zone control valves.

this would involve a complex series of piping and zone valves or circulator pumps. It is not unusual for a house to have two zones. The first floor and second floor are often the two zones. One circulator pump with zone control valves may not be sufficient to deliver the water to both the first and second floor. In this case, individual zone pumps may be required. If the house is a single-story building with two zones on the first floor, then one circulator pump with zone control valves may prove to be sufficient.

80.10 ZONED SYSTEM, USING MOTORIZED VALVES

The zoning arrangement using motorized valves is similar to the system with pump zoning, with the exception that the motorized valves control the flow in each zone. Each supply circuit has a motorized valve controlled by the thermostat in that zone. A motorized valve can also be used as a bypass around the circulator pump to relieve the pressure when only one zone valve is open.

The cross-sectional view of a motorized zone valve, as shown in Figure 80-22, can be used to describe its operation. When the thermostat contacts close on a call for heat, the 120/24 V transformer is energized. The heating element wrapped around the wax-filled element in the power head heats up. The heated expanding wax pushes down on

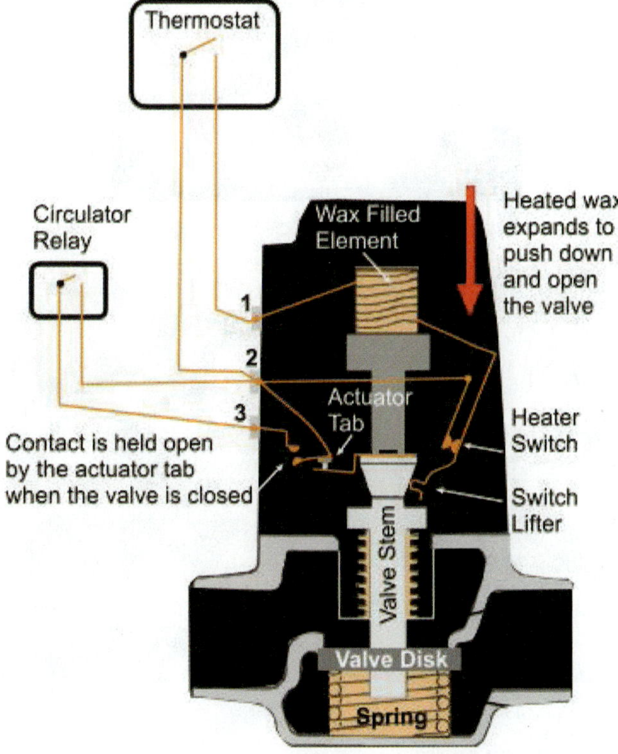

Figure 80-22 Heat motor valve cross-section.

the valve stem and water begins to flow through the valve to the zone. The stroking action of the valve stem piston closes electrical contacts to complete the circuit through terminals 2 and 3. This triggers a relay to start the circulator pump. The opposite sequence occurs when the thermostat contacts open.

80.11 MIXING VALVES

Mixing valves are used to mix the hot boiler water with the cooler return water (Figure 80-23). These can be used for different purposes. As the outdoor temperature drops, the supply-water temperature can be increased due to the greater heating demand. The higher the water temperature, the higher the heat output at the terminal unit. At higher outside temperatures, less heating is required and the mixed water temperature can be lowered, providing energy savings. A three-way mixing valve with outside temperature control is shown in Figure 80-24.

Another application of a three-way mixing valve is for radiant floor heating. The temperature setpoint of the hot-water boiler may be higher than the temperature required for the radiant loop. Generally, radiant floor heating loops operate at much lower temperatures than baseboard or convectors. The mixing valve will control the final temperature to the radiant loop, as shown in Figure 80-25.

(a)

(b)

Figure 80-23 (a) Cutaway view of mixing valve; (b) mixing valve with a metering scale marked on the outside.

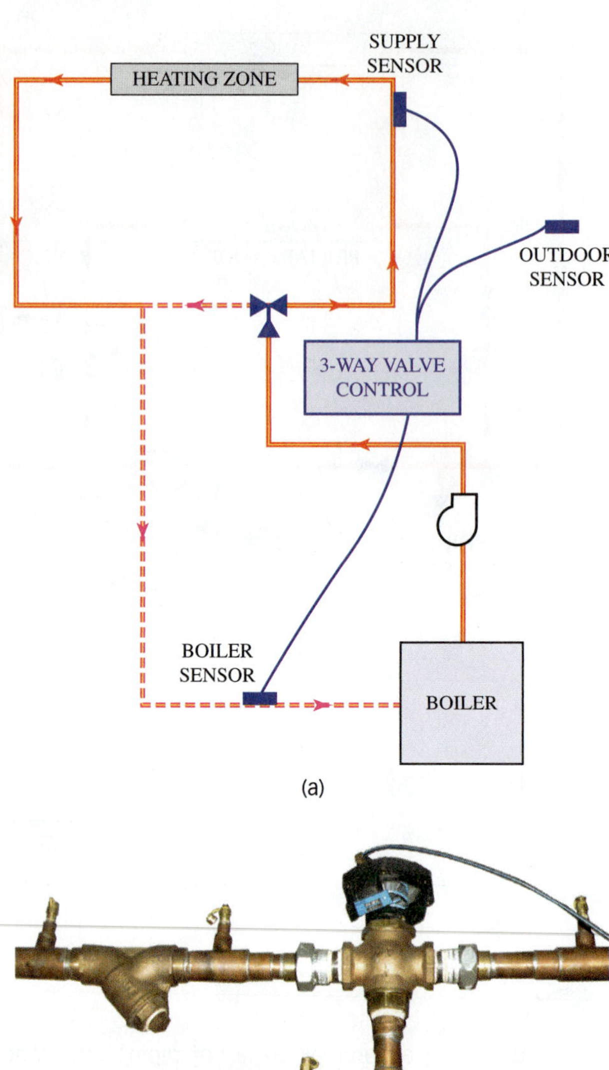

(a)

(b)

Figure 80-24 (a) Three-way mixing valve circuit; (b) three-way mixing valve with outside temperature control.

80.12 CIRCULATOR PUMPS

The pumps used for hydronic heating systems are usually the centrifugal type (Figure 80-26). These are referred to as non–positive displacement pumps. Unlike a gear-type pump, a centrifugal pump can be operated with the discharge valve

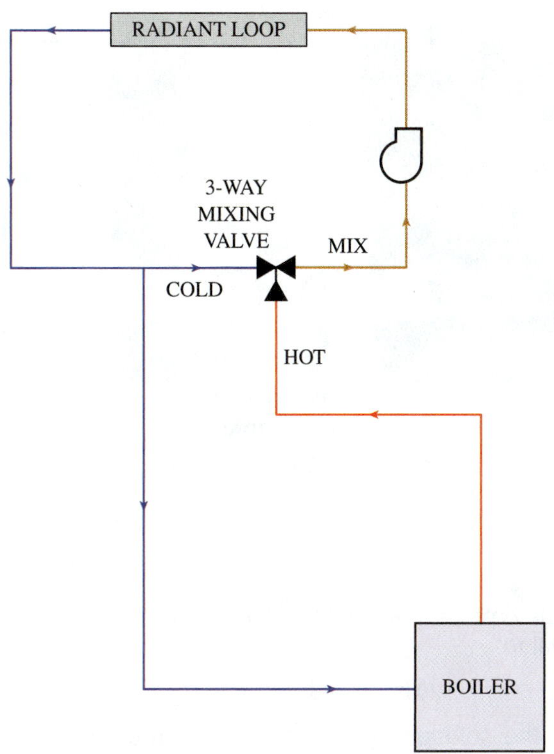

Figure 80-25 Three-way mixing valve operation for radiant loop.

Figure 80-26 Horizontal centrifugal pump.

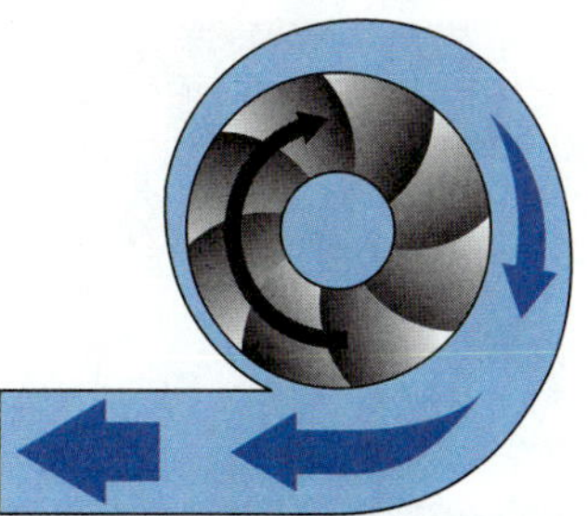

Figure 80-27 The spinning impeller will increase the velocity of the water through the pump.

(a)

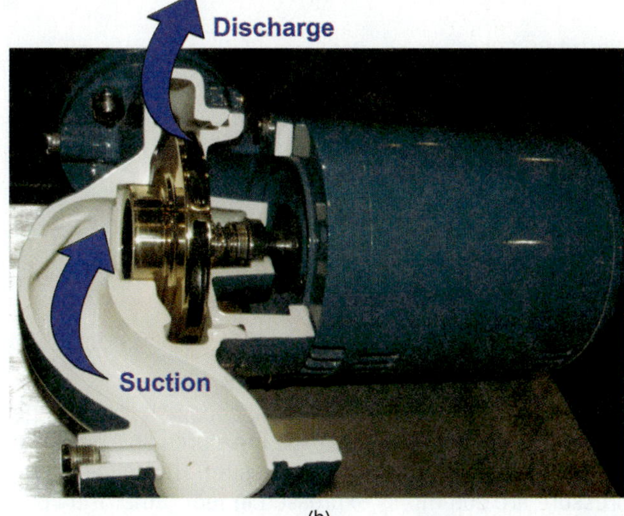

(b)

Figure 80-28 (a) An impeller for a small circulator pump; (b) closed impeller centrifugal pumps showing direction of water flow (continued next page);

closed. This is called the pump shutoff head. However, if the centrifugal pump operates for an extended period of time with the discharge valve closed, it will overheat due to the fluid friction developed.

Flow Rate

The principle of operation for a centrifugal pump is shown in Figure 80-27. The pump motor spins an impeller, which can be similar to a flat disk with vanes (Figure 80-28a). This is called an open impeller. There are many different sizes and types of impellers, as shown in Figures 80-28b–d. The spinning impeller increases the water velocity, thereby making

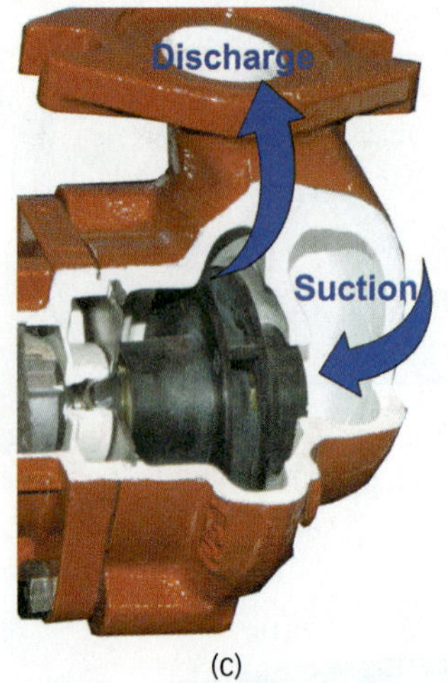

(c)

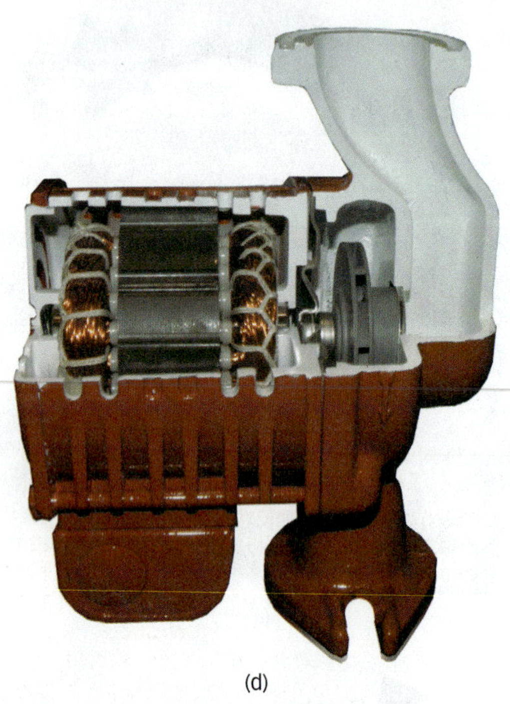

(d)

Figure 80-28 (c) another closed impeller centrifugal pumps showing direction of water flow; (d) cross-section of circulator to show impeller and motor windings.

it flow out of the pump and through the piping. The flow of water is measured in gallons per minute (gpm).

Head Pressure

The weight of water in a pipe or tank exerts a force. As the height of water in a pipe or tank becomes greater, the force increases. This force can be measured as a pressure and is often referred to as head pressure. The units of head pressure are commonly expressed in feet rather than psi. The height of the water column is used to determine the head pressure. One foot of water exerts a pressure equivalent to 0.433 psi. The pressure as measured by a gauge

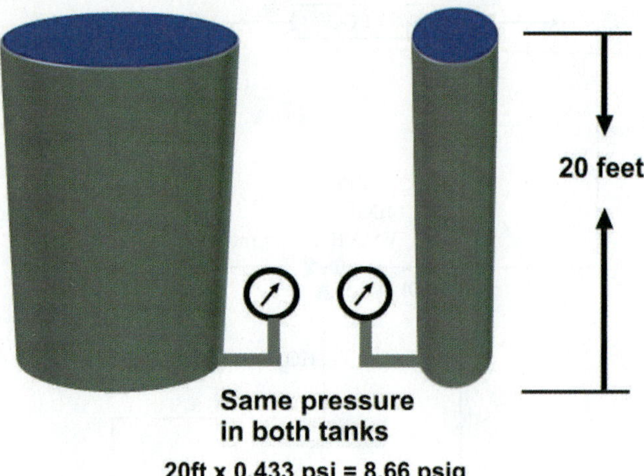

Figure 80-29 Relationship between tank size and head pressure.

at the bottom of an open water tank 20 ft high would be equal to:

$$20\text{ ft.} \times 0.433\text{ psi/ft} = 8.66\text{ psi}$$

The solution shows that a pressure gauge at the bottom of an open water tank 20 ft high would read 8.66 psig. The diameter of the tank does not matter, just the height (Figure 80-29). This is because the pressure measured is per square inch, and 1 in^2 of area in a 10-ft diameter tank is the same as 1 in^2 area of a 5-ft diameter tank. It must be understood then, that the head pressure of a 1-in diameter pipe is the same as the head pressure in a ½-in-diameter pipe if both are 20 ft high. For quick estimates, the relationship between water height and pressure is almost two to one. 100 feet of water would have a pressure at the bottom of about 50 psig. However realize that the more accurate measurement would be $100 \times 0.433 = 43.3$ psig.

Head Pressure and Flow

Head pressure is measured in feet of head or psi, and flow is measured in gpm. When a pump is required to pump water higher and higher, the head pressure will increase. The head pressure required to deliver water from the basement to the third floor will be much higher than the head pressure required for delivering water from the basement to the first floor.

As the head pressure increases, the total flow (gpm) through the pump decreases, as shown on the pump performance curve in Figure 80-30. This pump with would be able to deliver approximately 18 to 19 gpm with 8 ft of total head. Some pumps such as this are designed for high flow rates and low head pressures. These are ideal for applications where a lot of water is required that does not have to travel too far or too high. Other pumps are designed with high head pressures but lower flow rates, as shown on the pump performance curve in Figure 80-31. These are ideal for applications where less flow is required but the water has to be delivered to the second or third floor of the building.

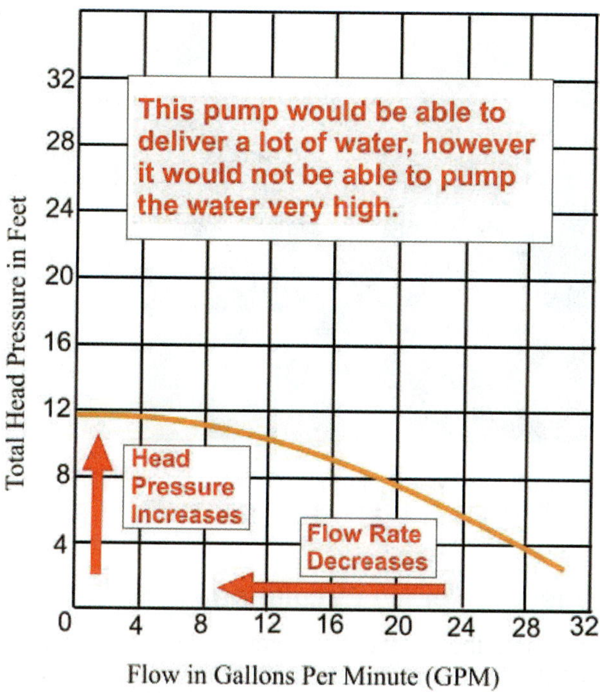

Figure 80-30 As head pressure increases, the pump flow rate decreases.

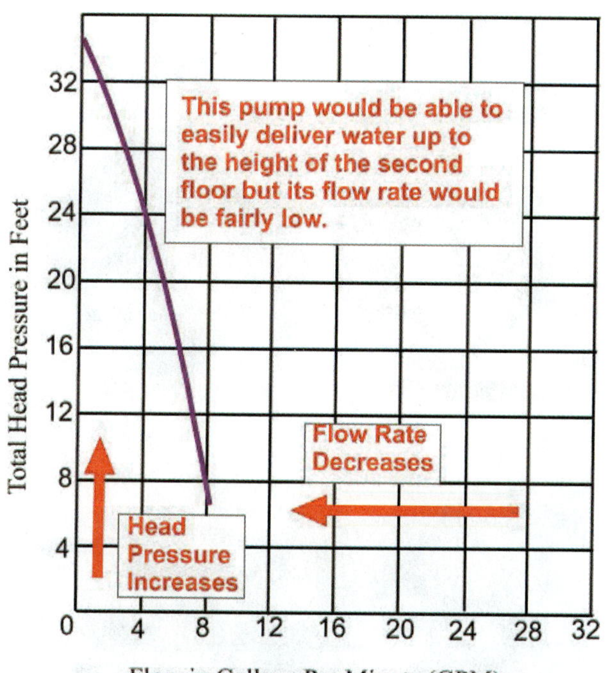

Figure 80-31 Pump designed for high head pressure but relatively low flow rate.

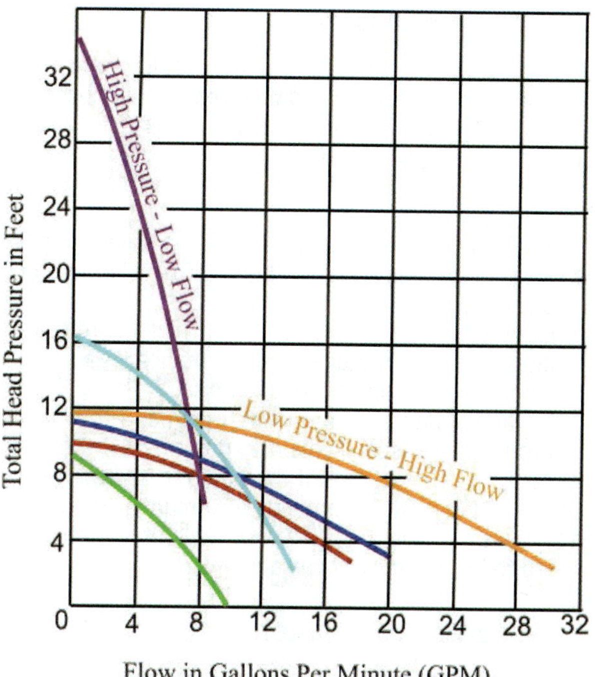

Figure 80-32 Typical performance curves for centrifugal pumps.

Pumps are selected to perform a specific purpose, providing the proper flow at a pressure to overcome the resistance of the circuit in which they are placed. Manufacturers provide pump design data, and it is important to select the proper pump depending on the application. An example of a performance chart for an assortment of circulator pumps is shown in Figure 80-32. This allows for matching the most suitable pump to meet the system requirements.

Pump Performance Curves

It is also important to note that pump efficiency will change at any point along the performance curve. The total head pressure and flow rates developed are determined to some extent by the impeller diameter and the brake horsepower (bhp) rating of the pump. A larger-diameter impeller requires a greater bhp and will produce more flow and higher head pressures as compared to a smaller-diameter impeller and lower bhp pump. It is important to note that the best efficiency is not at the maximum head pressure or the maximum flow rate but usually somewhere between the two. As an example, one model of pump rated at 7.5-bhp with an 8.25-in diameter impeller would have a maximum efficiency of over 80 percent at a total head pressure of approximately 46 ft and a flow rate of 400 gpm. If the total head pressure increases to 50 ft and the flow rate drops to 325 gpm, then the efficiency for this pump would drop to less than 77 percent. The same is true if the total head decreases and the flow rate increases.

80.13 TROUBLESHOOTING PUMPS

A common troubleshooting problem is pump noise and vibration. This is a nuisance and the vibration can cause the piping to fatigue and break over time. If the pump loses its prime, its impeller can overheat and become so damaged that the pump will not work. Common causes of pump noise and vibration are misalignment, air entrainment, and clogged strainers.

Pump or System Noise

Noise could be the result of a shaft misalignment which would need to be checked and then realigned if necessary. A worn coupling can lead to noise and if this is the cause it will need to be replaced and the shaft realigned. The same is true for worn pump or motor bearings which may require lubrication or replacement.

Improper pump foundation installation or loose bolting will cause noise. Piping strain due to expansion and contraction can lead to loose foundation bolts. In this case additional hangers may be required or the piping may need alteration. Improper motor speed or rotation or a clogged strainer or impeller can also lead to noise.

The actual pump performance may need to be compared to its design specification to determine if the pump impeller is the correct size. If the pump is operating close to or beyond the end point of its performance curve, noise can result. Excessive throttling of balance or control valves will lead to high water velocity and noise.

Cartridge-Type Circulator Pumps

Some small circulator pumps used in hot-water systems have replaceable cartridges, such as the one shown in Figure 80-33. The replacement cartridge contains all the moving parts and allows the pump to be serviced instead of replacing the entire unit (Figure 80-34). It is self-lubricating and contains no mechanical seal. When troubleshooting this type of pump, the motor windings can be easily inspected. Notice the charred burnt windings of the motor and the discoloration of the capacitor (Figure 80-35).

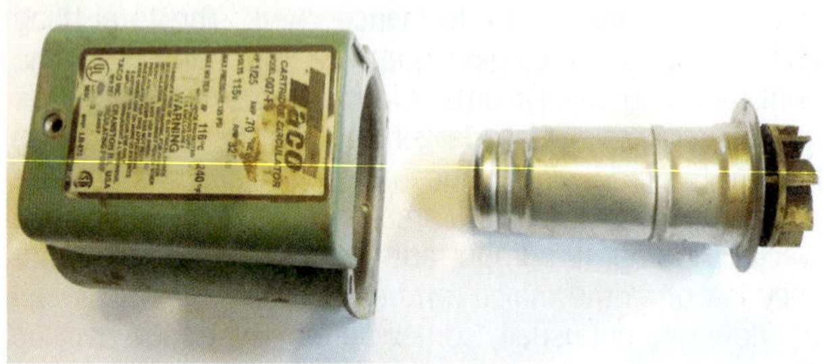

Figure 80-33 Cartridge-type circulator pump.

Figure 80-34a Pump cartridge and motor.

Figure 80-34b Motor and pump cartridge being removed from the pump housing.

Inadequate or No Circulation

This is often the result of an improperly filled system that is air bound. Typically the vent piping, make up feed regulator, or expansion tank may be at fault. It is important to get all of the air out of the piping system before leaving the job.

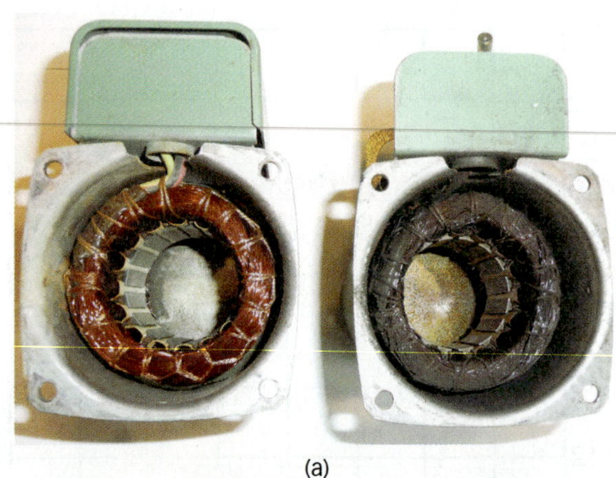

(a)

(b)

Figure 80-35 (a) Burnt-out motor shown on the right; (b) discoloration of the capacitor shown on the right.

Figure 80-36 Automatic air-purge valves are located at the high points of the radiant heat piping loop.

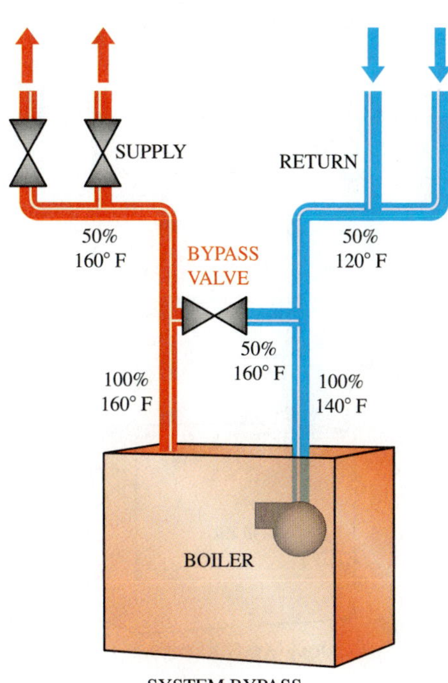

Figure 80-37 System bypass arrangement.

Air trapped in the system can cause noise and may result in the water pump losing its prime. Radiant piping systems will often have automatic venting valves located at the high points of the circulating water loop (Figure 80-36). If it is determined that air is not the problem, then another possible cause is a clogged suction strainer.

TECH TIP

Circulating pumps on closed-loop systems only work against the vertical lift head when the system is initially being filled. Once the system is completely filled with water, the pump effective head (discharge head – suction head) is only the flow resistance of the piping system itself. The suction head developed from a height of 10 ft of water on the return side of the pump will balance the discharge head developed from a height of 10 ft of water on the outlet side of the pump.

80.14 SPECIAL PIPING CONSIDERATONS

Most gas furnaces are limited to return-water temperatures of 140°F or greater. This is because the low water temperature would bring the boiler flue gas temperature below its dew point. Condensation in the boiler flue gas would lead to corrosion in the heat exchanger. The return water is mixed with the supply water to achieve the desired return-water temperature.

System Bypass Piping

This type of system is used on single-zone systems or multiple-zone systems using zone control valves. Notice the circulator is located on the return piping (Figure 80-37). The bypass valve will mix supply water with the return water to ensure that the return-water temperature does not drop too low. Typically, gas furnaces should not operate with return-water temperatures of less than 140°F. Plug valves offer good control to regulate the amount of flow that will be diverted to the return.

Pump-Away Bypass

The pump-away bypass is another method for controlling return-water temperature for a single-circulator pump system that is using zone control valves. The bypass is installed on the discharge side of the circulator, as shown in Figure 80-38. Hot-water flow regulated through this bypass is used to raise the temperature of the return water. The cost of installing this type of bypass is low, and the temperature can be easily set with a contact thermometer.

Pumped Bypass

For systems with multiple zones and circulators, a dedicated bypass circulator provides a strong blending flow without reducing the flow to any other heating zone. A typical common circulator can be used for this purpose (Figure 80-39).

80.15 HOT-WATER HEATING APPLICATIONS

It is common for oil-fired boilers to be installed to accommodate both heating and domestic hot-water service. Coil-type heat exchangers can be located directly within the boiler, or a separate circuit is installed for an indirect-fired hot-water heater. More recently, solar heating systems are being used to supplement the hot-water demand and are not only becoming more common but are required for new construction in some localities. These circulate either water or glycol from a hot-water holding tank through outdoor solar panels.

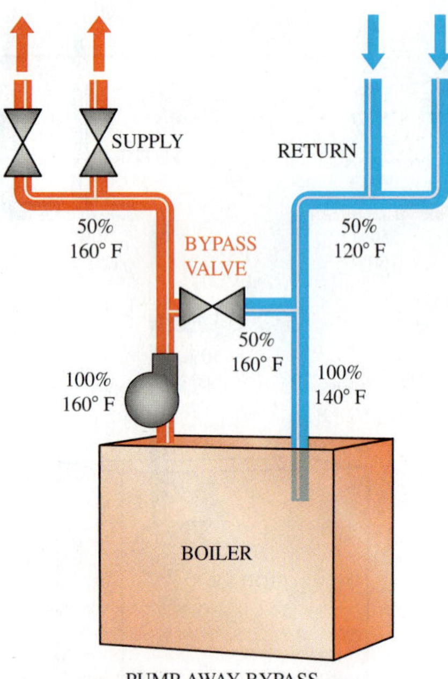

Figure 80-38 Pump-away bypass arrangement.

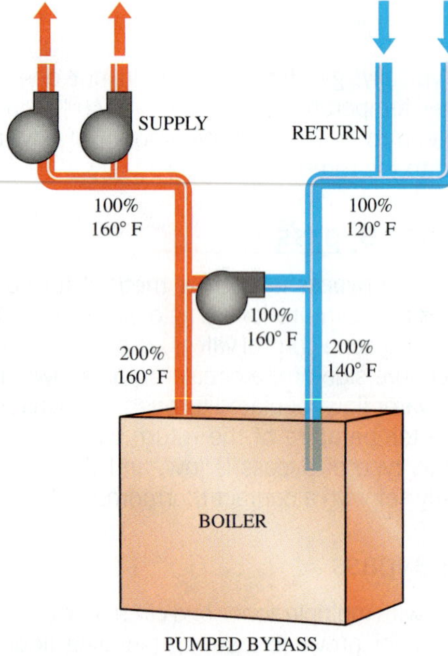

Figure 80-39 Pumped bypass arrangement.

Figure 80-40 Heating coil located inside an indirect-fired hot-water heater.

Figure 80-41 Indirect-fired hot-water heater showing the temperature control and the relief valve.

Indirect-Fired Hot-Water Heaters

Hot-water heating systems can be configured to also supply domestic hot water. A separate hot circulating loop is delivered to an indirect-fired hot-water heater. The heater acts as a storage tank and will have a coil-type heat exchanger located inside that is supplied with hot water from the boiler (Figure 80-40). This type of hot-water heater will have separate controls to maintain the domestic water at the setpoint temperature and a relief valve (Figure 80-41). The tank will need to be large enough to deliver an average supply of hot water based upon the household size. A drawback with this type of system is the amount time required to replenish the hot water. Hot water being used for multiple uses at the same time, such as washing clothes while taking showers, may lead to a temporary drop in water temperature.

UNIT 80—SUMMARY

Hydronic heating systems require circulator pumps and piping loops for delivering hot water from a boiler to a terminal unit. The different types of terminal units include baseboard heaters, convectors, fan coil units, and radiators. Radiant floor heating is another popular application of a hydronic system. There are a number of different types of piping arrangements, such as series loops, one-pipe venturi systems, two-pipe systems, and zoned systems. These systems also require expansion tanks, and many systems also require balancing valves, mixing valves, and motorized control valves. Most hot-water heating systems do not come as packages and must be installed individually based upon the building floor plan and heating requirements.

WORK ORDERS

Service Ticket 8001

Customer Complaint: No Heat on Second Floor

Equipment Type: Gas-Fired Boiler with Forced Hot Water—Two Zones

The technician arrives at a no-heat call and speaks with the customer. The boiler appears to be running fine, but there is no heat on the second floor. The technician first checks the thermostat and finds that it is calling for heat as it should, but the circulator for the zone is not operating. The motor circuit is tested with a multimeter for voltage, and there is 120 V. The power is disconnected from the circuit, and the circulator motor electrical leads measure infinite resistance, indicating an open in the winding. The circulator is replaced as a unit and electrically wired. When the power is turned back on, the circulator begins running as normal.

Service Ticket 8002

Customer Complaint: One Room Too Cold

Equipment Type: Radiant Floor Heating System

The technician meets with the customer, who explains that a newly installed radiant floor heating system works well except for in the bedroom. The system is an underfloor system, and all of the floors are tile except for the bedroom, which has a thick wall-to-wall carpet. The technician locates the supply manifold for the individual circuits and notes that the water temperature is mixed to be about 110°F. The technician adjusts the mixing valve for the bedroom to elevate the mixed temperature to 140°F. The technician tells the customer that it may take a few hours before any difference is noticed, but eventually the room should be warmer than it has been. The radiant heat temperature for the bedroom needed to be set slightly higher than the rest of the rooms because the wall-to-wall carpeting does not transfer the heat as well as the tile.

Service Ticket 8003

Customer Complaint: Banging Noise from Pipes

Equipment Type: Gas-Fired Boiler—Three-Zone Hot-Water Heating System

The customer tells the technician that the system was fine until the heat was recently turned on to the upstairs bedrooms. Ever since that time, there has been an occasional banging in the pipes. The upstairs heat had been turned off to conserve energy, and now that company has arrived it is back on again. The technician suspects that the water from the second floor has drained back to the boiler through a faulty check valve, allowing air to enter the system. The technician turns up all the thermostats so that all of the circulators will be running and then goes to each baseboard unit and vents air from each vent. After all the air is vented from the system, there seems to be no more knocking noise.

The technician tells the customer that the system will need to be shut off and drained to replace the check valve. The check valve may have just had something stuck underneath its seat and may not need replacement. The technician recommends leaving the check valve alone for now and replacing it when the heating season is over if the banging noise returns.

UNIT 80—REVIEW QUESTIONS

1. Define *hydronics*.
2. What was the common choice for hydronic piping in the past, and how is that changing?
3. What are the advantages of hot-water heating systems based on overall comfort?
4. What are the disadvantages of hot-water heating systems?
5. What safety measures are used to allow for expansion in a hot-water heating system?
6. What provisions should be made to prevent water from freezing during low-temperature periods of operation for a hydronic heating system?
7. Why is it important when working on electrical components such as pump motors to make sure that all of the water in the system has been removed?
8. List the different types of terminal units used for hot-water heating systems.
9. How does a fan coil unit differ from a convector?
10. What are the advantages of radiant floor heating?
11. Describe a series loop, single-circuit hot-water piping system.
12. What is the purpose of the venturi on a one-pipe venturi system, single circuit?
13. What is the advantage to a two-pipe system, direct return?
14. How does a two-pipe system, reverse return, differ from a two-pipe system, direct return?
15. How is hot water controlled to a zone?
16. How is the water returned from a zone to the boiler?
17. How does a motorized zone control valve work?
18. What are mixing valves used for?
19. Why might mixing valves be used for radiant floor heating?
20. How does a radiant floor heating system transfer heat?
21. How does an open piping system differ from a closed-type piping system?

UNIT 81
Boilers and Related Equipment

OBJECTIVES

After completing this unit, you will be able to:

1. explain the difference between steam boilers and hot-water boilers.
2. list the working pressures and temperatures for low-pressure, medium-pressure, and high-pressure boilers.
3. list the accessories required for a hot-water boiler.
4. vent air from a hot-water boiler.

81.1 INTRODUCTION

All boilers are used to heat water. Where hot water is used to transfer heat, the output of the boiler maintains a temperature below the steaming temperature. At sea level atmospheric pressure, water boils at 212°F; at higher pressures the boiling point increases. Where large amounts of heat need to be transferred, the boiler is used to produce steam. This is because about 970 BTU/lb of heat can be added to water at 212°F and atmospheric pressure to change it into steam at the same temperature. This same amount of heat will be transferred to the air passing across a heating coil as the steam turns back into condensate.

In general, for residential and small commercial installations, hot-water boilers are used rather than steam due to the size of the installations and accuracy of control that can be provided.

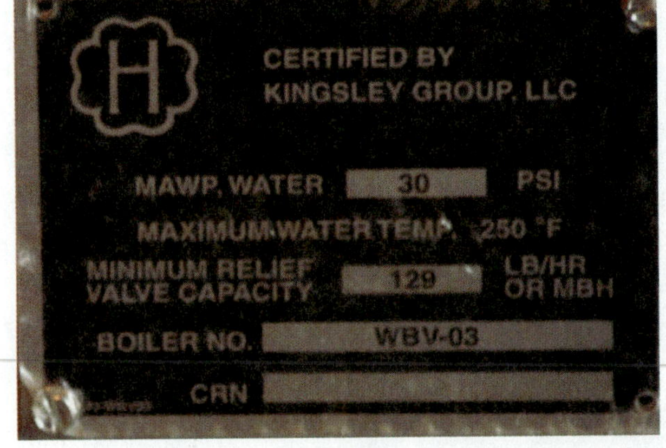

Figure 81-1 Low-pressure residential hot-water boiler nameplate.

81.2 BOILER CLASSIFICATIONS

Boilers are classified three ways: (1) working temperature/pressure, (2) fuel used, and (3) materials of construction.

Low-Pressure Boilers

Low-pressure boilers have a working pressure of up to 15 psi steam and/or up to 160 psi water pressure at a maximum of 250°F operating temperature. Low-pressure boilers are most commonly used for heating systems. Some are equipped with internal or external heat exchangers to supply domestic hot water. Residential heating boilers are gas- and oil-fired boilers with inputs less than 300,000 BTU/hr. Commercial heating boilers are those with heating inputs of 300,000 BTU/hr and larger. Water boilers are available from outputs of 50,000 BTU to 50,000,000 BTU (50 MBH). Every steam and hot-water boiler is rated at a maximum working pressure determined by the ASME code under which it is constructed and tested. A typical nameplate rating for a low-pressure hot-water boiler is shown in Figure 81-1. The boiler is rated for a maximum allowable working pressure (MAWP) of 30 psi, a maximum water temperature of 250°F, and a relief valve capacity of 129 lb/hr.

Medium- and High-Pressure Boilers

Medium- and high-pressure boilers operate above 15 psi steam and/or over 160 psi water pressure at temperatures above 250°F (Figure 81-2). Steam boilers are available for up to 50,000 lb/hr of steam. Many of them are used in central

Figure 81-2 Steam pressure gauge for a high-pressure boiler.

Figure 81-3 Central station high-pressure boilers.

station systems for heating medium and large commercial buildings that are beyond the range of a hot-water boiler (Figure 81-3). Often industrial boilers are used for the dual purposes of generating process steam and heating steam. As an example, the process steam may be used in a hospital for sterilizing surgical equipment, while the heating steam that is reduced to a lower pressure is used in the heating coils.

CAUTION

Boiler operators are licensed and regulated by local, state, and federal agencies. This typically applies to commercial boilers and boilers located in public buildings. Only individuals holding appropriate licenses may be operating engineers in charge of boilers. Domestic hot-water boilers are typically excluded from this category.

SAFETY TIP

All boilers must be routinely inspected and certified for operation. The boiler certification must be posted in plain view in the boiler room. This typically applies to commercial boilers and boilers located in public buildings. Domestic hot-water boilers are typically excluded from this category.

81.3 BOILER FUELS

Boilers may be designed to burn coal, wood, various grades of fuel oil, or various types of fuel gas or operate as electric boilers. Heat-recovery steam generators utilize the exhaust gas from a gas turbine to generate steam. Each fuel has its own special firing arrangement depending on the type of application.

By far the most common types of fuel for residential and light commercial applications are gas and oil. Natural gas is clean burning and convenient if it is readily available through a pipeline connection (Figure 81-4). The burner arrangement for a gas-fired hot-water boiler is very similar to a gas-fired hot-air furnace. Typically, the return-water temperature for a gas-fired hot-water boiler should not be below 140°F.

Figure 81-4 Gas-fired boiler.

In rural areas where there is no gas pipeline connection, propane may be used, but this requires a storage tank and regular delivery. The propane storage tank will be located outside the building and will be subject to proper installation to meet local codes. A common alternative is the use of #2 heating oil (Figure 81-5). The oil storage tank may

Figure 81-5 #2 oil-fired high-pressure boiler with gas for initial igniter start.

Figure 81-6 Typical cast iron gas boiler.

Figure 81-7 Typical packaged cast iron oil boiler with burner assembly not yet installed.

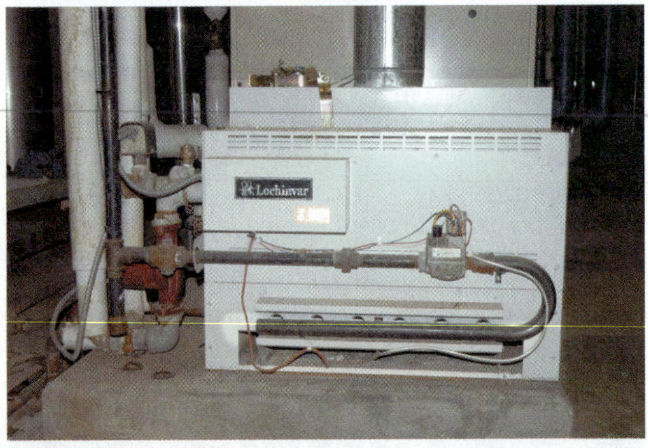

Figure 81-8 Steel boiler, gas fired.

be located within the building, typically in the basement. Similar to an oil-fired hot-air furnace, an oil-fired hot-water boiler will have a burner arrangement with an oil pump assembly.

81.4 MATERIALS OF CONSTRUCTION

Boilers must be designed to withstand high operating pressures and temperatures. It is important that the materials and methods of construction meet local and national codes. Many boilers are required to be constructed of materials in accordance with ASME (American Society of Mechanical Engineers) standards.

Cast Iron Boilers

Hot-water boilers are constructed using cast iron, steel, or stainless steel. A typical gas-fired cast iron boiler is shown in Figure 81-6. An oil-fired cast iron boiler is shown in Figure 81-7. Both of these boilers are assembled and shipped as packaged units. The package usually includes the circulator pump and complete controls. The wall thermostat is shipped separately for field mounting and wiring. Some packages also include a diaphragm-type expansion tank and a check valve piped into the unit. Cast iron boilers are constructed of individually cast sections assembled in groups or sections. Push or screw nipples or an external header join the sections together by applied pressure to seal them tight. This type of construction provides passages for water, steam, and products of combustion.

Steel Boilers

A steel boiler is shown in Figure 81-8. This boiler is also shipped as a package with accessories and controls. Newer designs with condensing type heat exchangers produce efficiencies in the AFUE 90+ percent range. Condensing boilers have recently been developed to improve the efficiency of boilers. Previously, the flue gases were not allowed to condense in the boiler due to the corrosion that they could cause. These new condensing boilers are constructed with heat exchangers made of materials to resist corrosion.

Steel boilers are fabricated into one assembly of a given size, usually by welding. The heat exchangers for steel boilers can be fire tube or water tube design. The fire tube construction, which is most common, has flue gas passage space between the water holding sections. The water tube construction uses water-filled tubes for the heat exchanger with the flue gases in contact with the external surface of the tubes.

The surface of steel boilers can be electroplated or clad with nickel or other corrosion-resistant material. Although this adds significant cost to the boiler, it significantly extends the unit operational life. Corrosion protection can also be provided by placing a sacrificial zinc rod in the tank. Zinc is more reactive than steel, so it corrodes away first and can be easily replaced when it is gone.

For larger hot-water installations, a high-efficiency pulse-type steel boiler is available, such as the examples shown in Figure 81-9. These boilers produce efficiencies as high as AFUE 96 percent. Both combustion air and gas metering valves are standard equipment. A small fan is used to deliver combustion air on start-up but shuts off after ignition. The unit is primarily controlled by a microprocessor.

Other Boiler Types

Copper boilers are usually some variation of the water tube type. The two most common tube types are parallel-finned copper tube coils with headers and serpentine copper tube units. Some are offered as residential wall-hung boilers. Electric boilers are in a separate class. No combustion takes place and no flue passages are needed. Electric elements can be immersed in the boiler water.

81.5 RESIDENTIAL HOT-WATER BOILERS

A boiler is a pressure vessel designed to transfer heat (produced by combustion or by electrical resistance) to a fluid, usually water. If the fluid being heated is air, the unit is called a furnace. If the fluid is water, then it is called a boiler. The heating surface of a boiler is the surface area with water on one side and the fire and exhaust gases on the other side. Boiler design provides for connections to a piping system that delivers heated fluid to the point of use and returns the fluid to the boiler.

Residential hot-water boilers are classified as low-pressure boilers. A hot-water boiler typically heats water to a final temperature of 180°F to 200°F. The hot-water pump delivers this hot-water supply to terminal units in the heating circuit loop or to heating coils in an air handler. Water temperatures for most heating systems have a range of 130°F to 180°F, while radiant floor heating systems are much lower, at 85°F to 140°F. Too low a temperature will result in inadequate heating, while too high a temperature could be damaging to the piping or materials used in the heating loop. Because of this, there are different methods of mixing the supply and return water to achieve the desired supply temperatures.

The water will transfer its sensible heat in the terminal unit. The temperature difference from the supply to the return line is normally 20°F–40°F. In many boilers, the return water is mixed with the supply water to maintain a temperature of at least 140°F returning to the boiler to reduce the chance of reaching the dew point in the boiler flue gases.

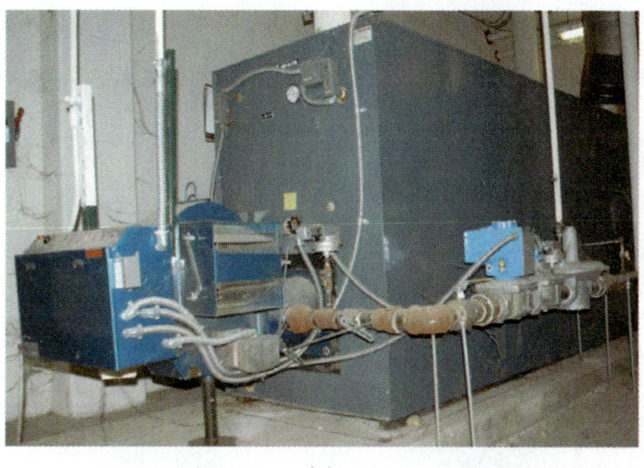

(a)

(b)

Figure 81-9 (a) Forced-draft large high-efficiency steel boiler; (b) pulse high-efficiency boiler.

81.6 RESIDENTIAL HOT-WATER BOILER PIPING DETAILS

A typical hot-water heating system diagram is shown in Figure 81-10. Key components include an oil- or gas-fired boiler, a circulation pump, an air separator, an expansion tank, radiant floor piping, a makeup water valve, a temperature/pressure-relief valve, and a backflow-prevention device required by code in many states. The piping arrangement for a wall-mounted gas-fired boiler hot-water heating system is shown in Figure 81-11.

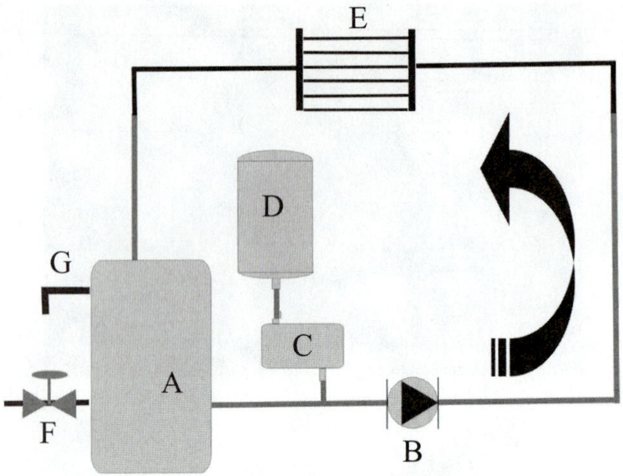

Figure 81-10 Basic hydronic system: (a) oil-fired boiler; (b) circulation pump; (c) air separator; (d) expansion tank; (e) radiant floor piping; (F) makeup water valve; (G) temperature-pressure relief.

Figure 81-11 Piping arrangement for a hot-water heating system.

Expansion Tank

An expansion tank is placed in the piping to permit the expansion and contraction of the water volume. The tank holds both air and water. The air is compressed when the water volume expands, and the air expands when the water volume decreases. An open type of expansion tank is located above the highest terminal unit, usually in the attic, and has a gauge glass and overflow pipe to a drain. These are not very common today.

SERVICE TIP

After installing or working on an expansion tank, watch the boiler pressure as the boiler heats up. If the pressure rises too high, this could be due to an expansion tank sizing or operational problem.

Closed-type expansion tanks are welded gas tight and located above the boiler. The tank is partially filled with water to leave a cushion of air for expansion. Do not use an automatic air vent on this type of tank. The air will escape from the system rather than returning to the tank. Eventually the tank will waterlog. This type of tank should utilize a fitting that permits the air bubbles to rise directly into the tank but restricts the flow of water back into the tank, as shown in Figure 81-12.

Diaphragm-type expansion tanks are also welded gas tight but have a flexible rubber membrane between the air space and the water space that moves during operation to allow for the changing volume of water. The air is precharged (usually 12 psig for residential units) into the expansion tank and unlike the closed type does not come in contact with the water in the system. An automatic air vent should be installed on top of a diaphragm expansion tank, as shown in Figure 81-13. A recommended expansion tank location is shown in Figure 81-14.

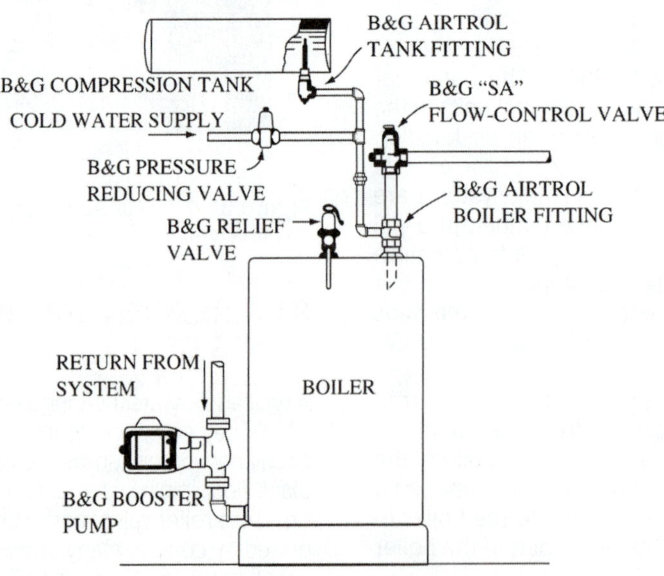

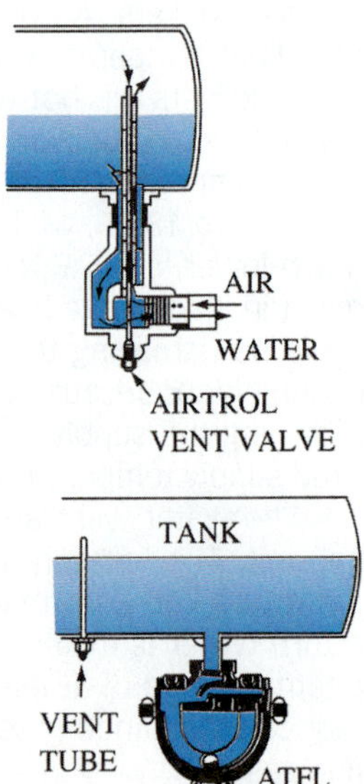

Figure 81-12 Compression-type expansion tank.

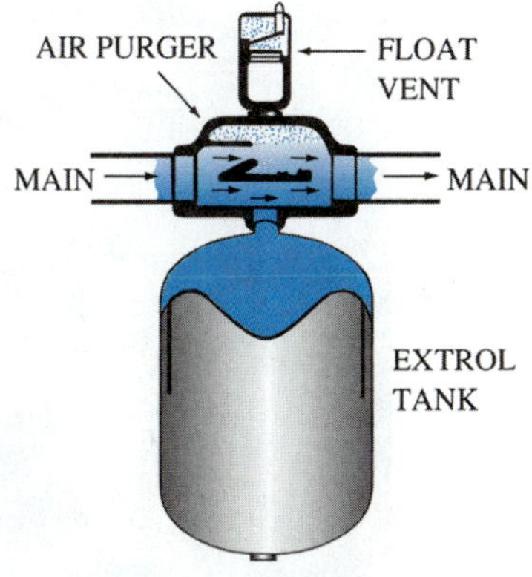

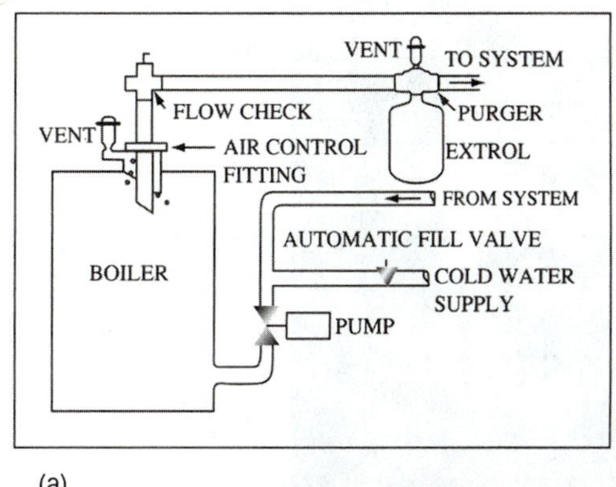

(a)

(b)

(c)

Figure 81-13 (a) Diagram of a diaphragm-type expansion tank; (b) diaphragm expansion tank and automatic air vent; (c) cross-section of automatic air vent.

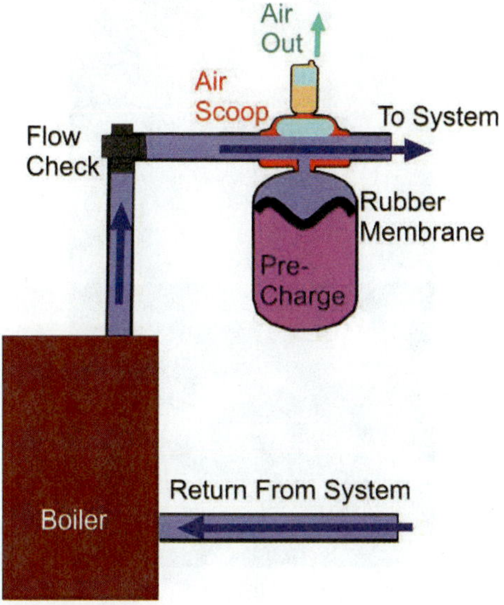

Figure 81-14 Recommended expansion tank hookup connected to discharge side of boiler.

SERVICE TIP

A weeping relief valve may be an indication of an expansion tank problem. If the tank is a closed type, it may be waterlogged. If the tank is a diaphragm type, the membrane may be damaged.

TECH TIP

To reduce the expense in piping for large water systems, steel pipe is used. However, steel pipe has the potential problem of rusting. To reduce the rust or corrosion problem, steel pipes are treated with a "pickling" solution. This treatment forms a barrier to rust formation in the pipe. In addition, chemicals are added to the water circulating through the system that further retard rust. Check with the chemical manufacturer for the proper mixing ratio.

Figure 81-15 Automatic purge valve located on top of residential boiler.

Figure 81-16 Water hammer arrester.

Air Removal and Water Hammer

Air must be eliminated from the piping system. Air vents need to be placed at high points of the system and on terminal units (Figure 81-15). Even if the system is free from air when filled, air can enter during normal operation. Air in the system can lead to noise and vibration similar to water hammer in the lines. Depending on the selection of accessories, air that collects during operation can be vented at the expansion tank. Some air remains in the expansion tank to act as a cushion for the expansion and contraction of water during temperature change.

Water hammer is a pressure surge or wave that is created when the water is forced to stop or change direction suddenly, such as quickly closing a valve. To reduce this effect, hammer arresters can be installed in lines where water hammer becomes a problem, particularly where there are sharp 90-degree bends (Figure 81-16).

Pressure-Relief Valves

The location of the pressure-relief valve is shown in Figure 81-17. The pressure-relief valve is usually set for 30 psig on a residential or small commercial system, and the piping

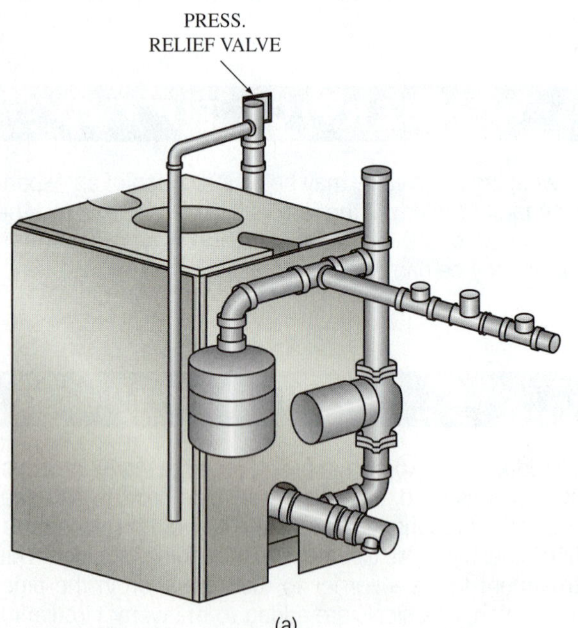

(a)

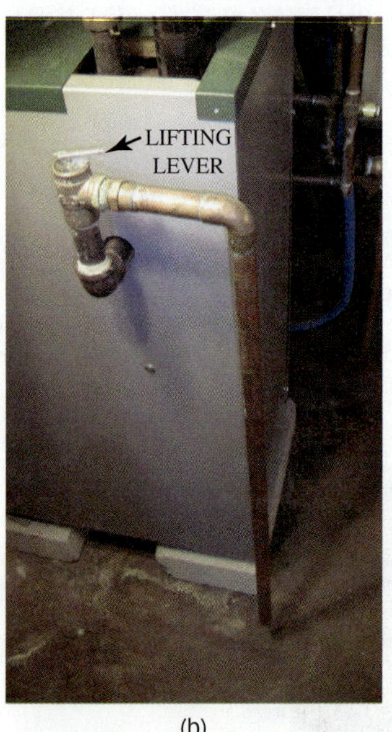

(b)

Figure 81-17 (a) Pressure-relief valve location; (b) manual lifting lever for pressure-relief valve.

Figure 81-18 Residential pressure-regulating valve for makeup water.

Figure 81-19 Energy manager control module.

is arranged to discharge into an approved drain. The outlet must run to a safe place to eliminate the possibility of severe burns if the valve should lift. No shutoff valve can be installed between the relief valve and the boiler. The valve may be tested by lifting the lever located on its top. The valve lever should be operated at least once each year. Take precautions when lifting the valve to drain away and protect against burns from the hot water. If no water comes out, then the valve must be replaced. It is recommended that the valve be physically removed and inspected thoroughly at least once every 3 years.

Pressure-Regulating Valve

The boiler will need some makeup water from time to time. Typically, an automatic pressure valve supplied from the main building water line will be connected to the piping circuit (Figure 81-18). This will keep the system topped off with water. It is important to note, however, that minerals in water can build up on heat-transfer surfaces and cause overheating and eventual failure of cast iron sections. The water should fall between a 7.0 pH and an 8.5 pH with less than 7 grains hardness. Water treatment is recommended in areas where water quality is a problem.

Freeze Protection

If the conditions warrant, antifreeze can be circulated through the unit for layup periods. This should be thoroughly flushed prior to recommissioning the boiler. Use only approved solutions. Never use an automotive-type or RV-type antifreeze. An alternative to using antifreeze is to thoroughly drain the entire system of water. A drain valve will be located on the bottom of the boiler.

81.7 RESIDENTIAL HOT-WATER BOILER CONTROLS

If a package boiler is used, the temperature- and pressure-limiting devices are supplied with the package; otherwise they must be applied during installation. There are a number of ways that residential or small commercial hot-water heating systems can be controlled, such as by thermostat, aquastat, manual reset high limit, and low-water cutout. Some residential boilers now come with an electronic control module for supervising most boiler functions (Figures 81-19 & 81-20).

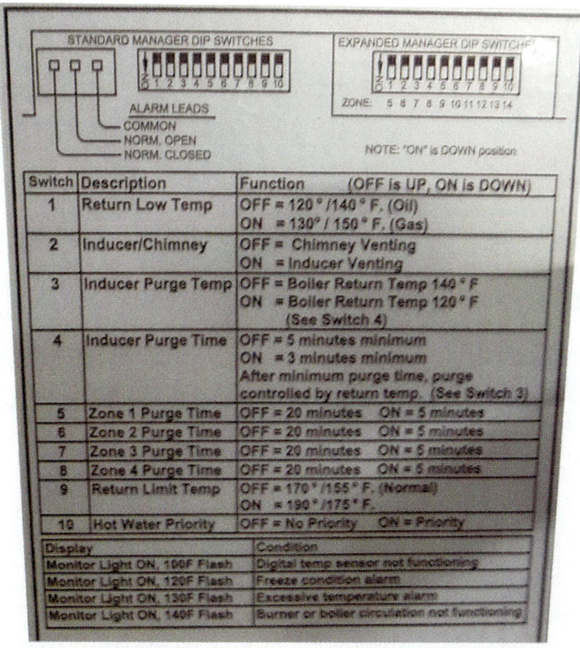

STANDARD MANAGER DIP SWITCHES

ALARM LEADS
COMMON
NORM. OPEN
NORM. CLOSED

EXPANDED MANAGER DIP SWITCHES

ZONE: 5 6 7 8 9 10 11 12 13 14

NOTE: "ON" is DOWN position

Switch	Description	Function (OFF is UP, ON is DOWN)
1	Return Low Temp	OFF = 120°/140° F. (Oil) ON = 130°/150° F. (Gas)
2	Inducer/Chimney	OFF = Chimney Venting ON = Inducer Venting
3	Inducer Purge Temp	OFF = Boiler Return Temp 140° F ON = Boiler Return Temp 120° F (See Switch 4)
4	Inducer Purge Time	OFF = 5 minutes minimum ON = 3 minutes minimum After minimum purge time, purge controlled by return temp. (See Switch 3)
5	Zone 1 Purge Time	OFF = 20 minutes ON = 5 minutes
6	Zone 2 Purge Time	OFF = 20 minutes ON = 5 minutes
7	Zone 3 Purge Time	OFF = 20 minutes ON = 5 minutes
8	Zone 4 Purge Time	OFF = 20 minutes ON = 5 minutes
9	Return Limit Temp	OFF = 170°/155° F. (Normal) ON = 190°/175° F.
10	Hot Water Priority	OFF = No Priority ON = Priority

Display	Condition
Monitor Light ON, 100F Flash	Digital temp sensor not functioning
Monitor Light ON, 120F Flash	Freeze condition alarm
Monitor Light ON, 130F Flash	Excessive temperature alarm
Monitor Light ON, 140F Flash	Burner or boiler circulation not functioning

Figure 81-20 Electronic control system settings for residential boiler.

Thermostat

The thermostat signal is typically used to start the circulator pump. As the water begins to circulate, the boiler water temperature begins to drop and an aquastat signal fires the boiler based on boiler water temperature. Thermostat signals can also be used to open zone control valves. Motorized zone control valves open slowly while sending a signal to start the circulator pump. Some thermostat arrangements are configured to control both the fuel-burning device and the circulating pump. The advantage of this arrangement is that in mild weather the boiler water temperature is lower due to the shorter running time, producing more efficient operation.

Aquastat

An aquastat can be used to maintain boiler water temperature by controlling boiler firing. A single aquastat has one sensor typically located in the return-water line. A dual aquastat has sensors in both the supply and return lines and is often used for step firing control. A high-limit aquastat is used for a safety control only. A reverse-acting aquastat is sometimes used on a single-zone system (Figure 81-21). It will not allow the circulator to start until the boiler has reached its operating temperature.

High Temperature Limit

This is a safety control to shut the boiler down if the temperature becomes too high (Figure 81-22). This is usually above 200°F. The limit is normally reset manually.

Figure 81-22 High temperature limit.

CAUTION

It is extremely important that boilers not be allowed to run low or out of water, as they can shut down or implode as violently as if it exploded. Implosion comes from the rapid condensing of steam into water, creating a negative pressure in the boiler.

Low-Water Cutout

If the water level in the boiler is low, a serious condition could result. If a boiler runs low on water, it can overheat, causing damage to the boiler. If a boiler runs out of water, it can explode as a result of steam pressure. The low-water cutout will automatically shut off the boiler on a low water level.

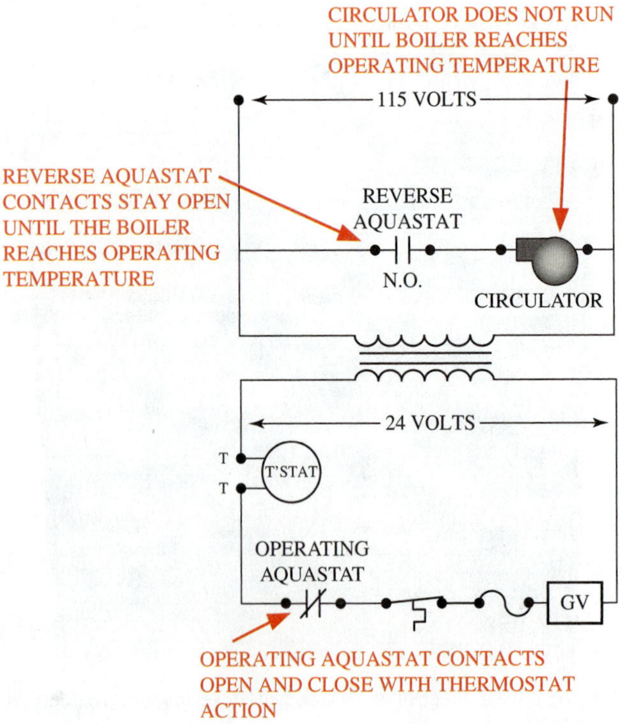

CIRCULATOR DOES NOT RUN UNTIL BOILER REACHES OPERATING TEMPERATURE

115 VOLTS

REVERSE AQUASTAT CONTACTS STAY OPEN UNTIL THE BOILER REACHES OPERATING TEMPERATURE

REVERSE AQUASTAT

N.O.

CIRCULATOR

24 VOLTS

T

T

T'STAT

OPERATING AQUASTAT

GV

OPERATING AQUASTAT CONTACTS OPEN AND CLOSE WITH THERMOSTAT ACTION

Figure 81-21 Reverse-acting aquastat.

81.8 STEAM-HEATING BOILERS

Steam-heating boilers are most often used for commercial and industrial applications rather than residential use. Large buildings may use steam heat rather than hot water or a combination of the two. Hospitals, universities, and manufacturing facilities often have steam-heating boilers (Figure 81-23).

Figure 81-23 Oil-fired high-pressure steam-heating boiler.

Figure 81-24 Boiler steam safety valves vent to the outside of the building.

Figure 81-25 Boiler steam safety valve with manual release lever.

Steam heat reduces the total number of circulating pumps required. The steam will flow from the boiler to the heating coils located in the air handlers without needing to be pumped. However, a steam boiler will require different piping materials, relief valves, and control systems as compared to a hot-water boiler. The requirements are more stringent, and these types of boilers often require operators to obtain a boiler license.

Steam Boiler Safety Valves

Steam released from a ruptured pipe will expand rapidly depending on the operating pressure. If a high-pressure steam pipe on a boiler were to burst, the entire space would quickly fill with steam and possibly suffocate anyone present. This is why steam boilers utilize safety valves that vent outside of the building (Figure 81-24). The boiler safety valve will be set to open and relieve the steam pressure well before the burst pressure of any of the piping materials or boiler internals is reached (Figure 81-25).

Steam Boiler Controls

Instead of controlling water temperature, steam boiler controls regulate the steam pressure. The burner is either fired on and off to maintain a constant steam pressure or modulated between higher and lower firing rates. Most automated steam boilers have modulating burners that automatically adjust the air and fuel flow to meet the steam demand. The water level in the boiler remains constant and is controlled by a feed-water regulator. Commercial and industrial steam-heating boilers will have a water level indicator, steam pressure gauges, switches, and modulating controls (Figure 81-26).

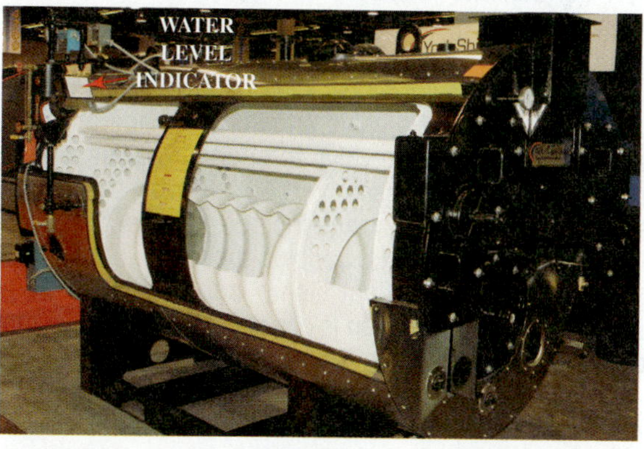

(a)

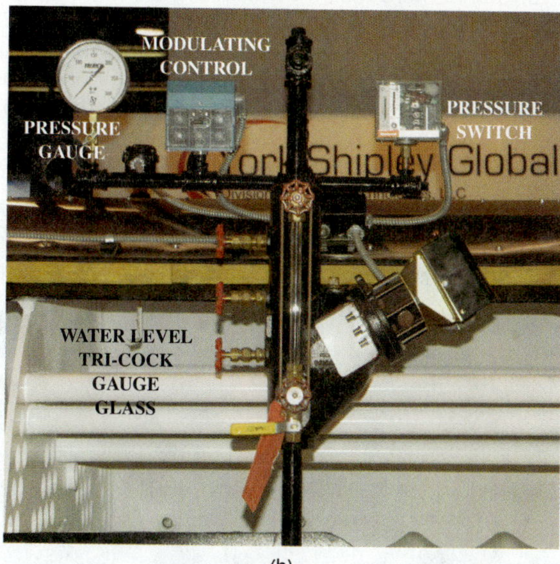

(b)

Figure 81-26 (a) Commercial steam heating boiler showing location of water-level-gauge glass; (b) tri-cock water-level-gauge glass and steam pressure-sensing instruments.

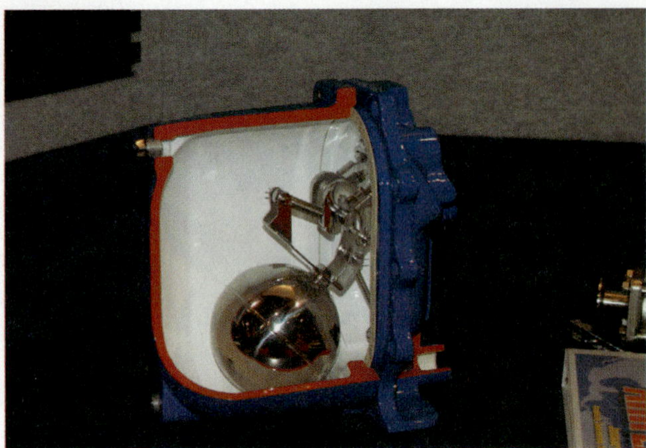

Figure 81-27 Ball-type mechanical float trap.

Steam Traps

Steam traps located in steam-return lines drain condensate from the lines without allowing the steam to escape. Maximum efficiency of a steam-heating coil occurs only when condensate leaves the coil, not steam. If steam leaves the heating coil, then its useful latent heat will be wasted.

There are many different kinds of steam traps. They all consist essentially of a valve and some device or arrangement that will cause the valve to open and close as necessary to drain the condensate without allowing the escape of steam. Common steam traps include the mechanical ball-float and the thermostatic bellows types.

A mechanical ball-float steam trap is shown in Figure 81-27. The valve of the trap is connected in such a way that the valve opens when the float rises. When the steam leaving the heating coil begins to condense, the resulting water level rises and lifts the float, allowing the valve to open. The condensate drains out and the float lowers to close the valve until more steam condenses. The float does not necessarily need to have the shape of a ball (Figure 81-28).

A thermostatic bellows steam trap is shown in Figure 81-29. It has fewer moving parts as compared to a mechanical-type steam trap. The valve for the trap is attached to the bellows in such a way that the valve closes when the bellows expands. Steam enters the trap body and heats the volatile liquid contained inside the sealed bellows. This causes the bellows to expand and close the valve. As the steam cools and condenses, the bellows contracts to open the valve, and this action allows the condensate to drain.

Y-type strainers are located in steam systems to collect scale and rust that travels along with the steam (Figure 81-30). This protects valves and traps from becoming plugged up. Periodically the y-strainers should be isolated, the pressure allowed to drain, and the strainer removed and cleaned.

Figure 81-28 Mechanical float trap.

Figure 81-29 Cutaway of a thermostatic bellows steam trap.

Figure 81-30a Cutaway of a y-type steam strainer.

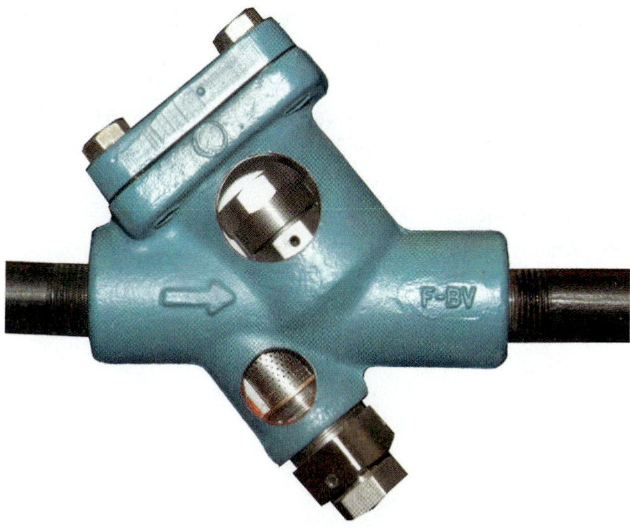

Figure 81-30b Cutaway of a y-type steam strainer and thermostatic trap.

UNIT 81—SUMMARY

Most boilers used for residential and light commercial applications are low-pressure hot-water systems. The firing arrangements are typically oil or gas and the burner assemblies are similar to those used in hot-air furnaces. The firebox, however, is surrounded by water and therefore takes on a different construction. The boiler can be directly controlled by a thermostat. More commonly, aquastats are used, which fire the furnace dependent on water temperature. Routine maintenance should include a test of the low-water cutout and the boiler pressure-relief valve.

WORK ORDERS

Service Ticket 8101

Customer Complaint: Routine Service

Equipment Type: Oil-Fired Residential Hot-Water Boiler

The technician is performing an annual maintenance call on an oil-fired hot-water boiler. The fuel filters are replaced and also the burner nozzle. The furnace flue gas passages are cleaned out with a vacuum. Before the boiler is started for a test run, the technician connects a short piece of hose from the outlet of the relief valve piping to a drain and lifts the relief valve manually. No water comes from the relief valve. The technician closes the isolation valves on the boiler, shuts off the automatic water supply, and opens the drain valve to release some of the water from the boiler. The relief valve is removed and replaced with a new one. Water is added back to the boiler while air is vented out and the automatic makeup valve is placed back into service. Before starting the boiler, the new relief valve is tested and it releases water. The boiler is started and comes up to temperature normally. The relief valve is once again manually lifted and it drains water and closes tight when the lever is released.

Service Ticket 8102

Customer Complaint: Routine Service

Equipment Type: Gas-Fired Residential Hot-Water Boiler

The technician is conducting an annual maintenance call on a gas-fired hot-water boiler. The boiler pressure-relief valve drain piping has a trickle of water slowly coming from it. The technician checks the air pressure in the diaphragm expansion tank and it is zero. A small hand pump is connected to the fitting on the expansion tank, but it will not hold any pressure. The technician isolates that section of the piping system and replaces the expansion tank. After the installation, the system is vented for air. The technician lifts the pressure-relief valve by hand to see whether it will close tightly, and it does.

Service Ticket 8103

Customer Complaint: Not Enough Heat

Equipment Type: Gas-Fired Residential Hot-Water Boiler

The customer complains that the thermostat setpoint temperature is not being reached. The circulator pumps are running a majority of the time, but the baseboards still seem cold to the touch. The technician checks the pressure and temperature gauge on the furnace, and both are too low. The technician starts the boiler manually by placing a jumper around the aquastat. The boiler starts and the water temperature begins to rise. The combustion flame looks good. The technician shuts the boiler down and replaces the aquastat. The boiler is restarted, and now it runs normally.

UNIT 81—REVIEW QUESTIONS

1. When would a steam boiler be preferable over a hot-water-type boiler?
2. What water temperatures do most hot-water boilers heat up to?
3. What water temperature do radiant floor heating systems typically operate at?
4. What is the typical return-water temperature for a gas-fired boiler?
5. What is the working pressure for a low-pressure boiler?
6. What is the heating input for a low-pressure residential boiler?
7. What is the working pressure for medium- and high-pressure boilers?

8. What materials are boilers commonly made from?
9. What are some of the accessories required for a hot-water boiler?
10. What can be done to treat steel piping to help keep it from rusting?
11. Where should air vents be located in a hot-water heating system?
12. Why is it recommended to never put an air vent on a closed-type expansion tank?
13. What is a diaphragm type of expansion tank?
14. What could be the cause of a weeping relief valve on a hot-water boiler?
15. What can happen if a boiler runs out of water?
16. What pressure is the relief valve typically set at on a residential or small commercial system?
17. Can a hot-water boiler pressure-relief valve be tested?
18. What is the function of an aquastat?
19. What is a low-water cutout?
20. Where do high-pressure steam safety valves vent to?

UNIT 82

Cooling Towers

OBJECTIVES

After completing this unit, you will be able to:

1. explain the difference between mechanical- and atmospheric-draft cooling towers.
2. list the different types of mechanical-draft cooling towers.
3. explain how cooling towers lower the water temperature below ambient.
4. list the common terms associated with cooling tower operation.
5. describe the piping arrangements used for cooling towers.
6. perform routine maintenance on a cooling tower.
7. explain how an evaporative condenser operates.

82.1 INTRODUCTION

The cost and scarcity of water (unless drawn from a lake or wells and returned) has become prohibitive and is even outlawed for refrigeration and air conditioning use by some local city codes. Ordinances often restrict the use of water to the point where such installations are forced to use all air-cooled equipment, water-saving devices such as the evaporative condenser, or a water tower.

Cooling towers are an essential part of most central station air-conditioning systems. For the smaller systems, even up to 100 tons, there has been a trend toward the use of remote air-cooled condensers. Their use is limited, however, to installations where the length of the refrigerant piping can be short enough to be practical. Almost all large condensing units and large packaged chiller units use water-cooled condensers with cooling towers.

A cooling tower is a large evaporative cooler that is used to reject heat from water. They are used with water-cooled condensers to cool the warm water leaving the condenser and return it to the condenser ready to absorb more heat. Cooling towers can be either open loop or closed loop. The water circulating in an open-loop system is also the water circulating through the condenser. Closed-loop systems use a heat exchanger to separate the water in the condenser loop from the water in the tower. That way only the tower water is exposed to the air. Closed-loop systems reduce the amount of water maintenance required for the water in the condenser loop. Closing the loop reduces the opportunity for contaminants to be introduced into the condensing loop.

82.2 COOLING TOWERS

The function of the water tower is to pick up the heat rejected by the water flowing through the condenser and discharge it into the air passing through the tower. The cooling tower delivers the condenser water supply at about 85°F, pumped by the condenser pump, to the water-cooled condenser on the water chiller. The condenser water is warmed by the rejected heat from the chiller to about 95°F and returned through the condenser water return to the cooling tower, where it is cooled by heat transfer to the airstream and evaporation.

Figure 82-1 is a schematic drawing showing the principle of operation used by a cooling tower. The water from the heat source is distributed over the wet deck surface (fill media) by spray nozzles. Air is simultaneously blown upward over the wet deck surface to absorb sensible heat from the water. In addition, a small portion of the water will evaporate, which removes latent heat from the remaining water. The cooled water is collected in the tower sump and pumped back to the condenser.

82.3 MECHANICAL-DRAFT COOLING TOWERS

A mechanical-draft tower utilizes a motor-driven fan to move air through the tower, the fan being an integral part of the tower. A typical mechanical-draft cooling water tower is shown in Figure 82-2. They are available in many configurations and arrangements. The smallest cooling towers are structured for only a few gallons of water per minute, while the largest cooling towers may handle upwards of thousands of gallons per minute.

Induced-Draft and Forced-Draft Cooling Towers

Mechanical-draft towers are categorized as either being induced draft or forced draft. An induced-draft tower has the fan located in the airstream leaving the tower and draws the air through it (Figure 82-3). A forced-draft tower has the fan located in the airstream entering the tower and blows the air through the tower.

The type of tower installed is dependent on location and application. The fan static discharge pressure will be higher for a forced-draft tower as compared to an induced-draft tower. A crossflow tower can be configured so that

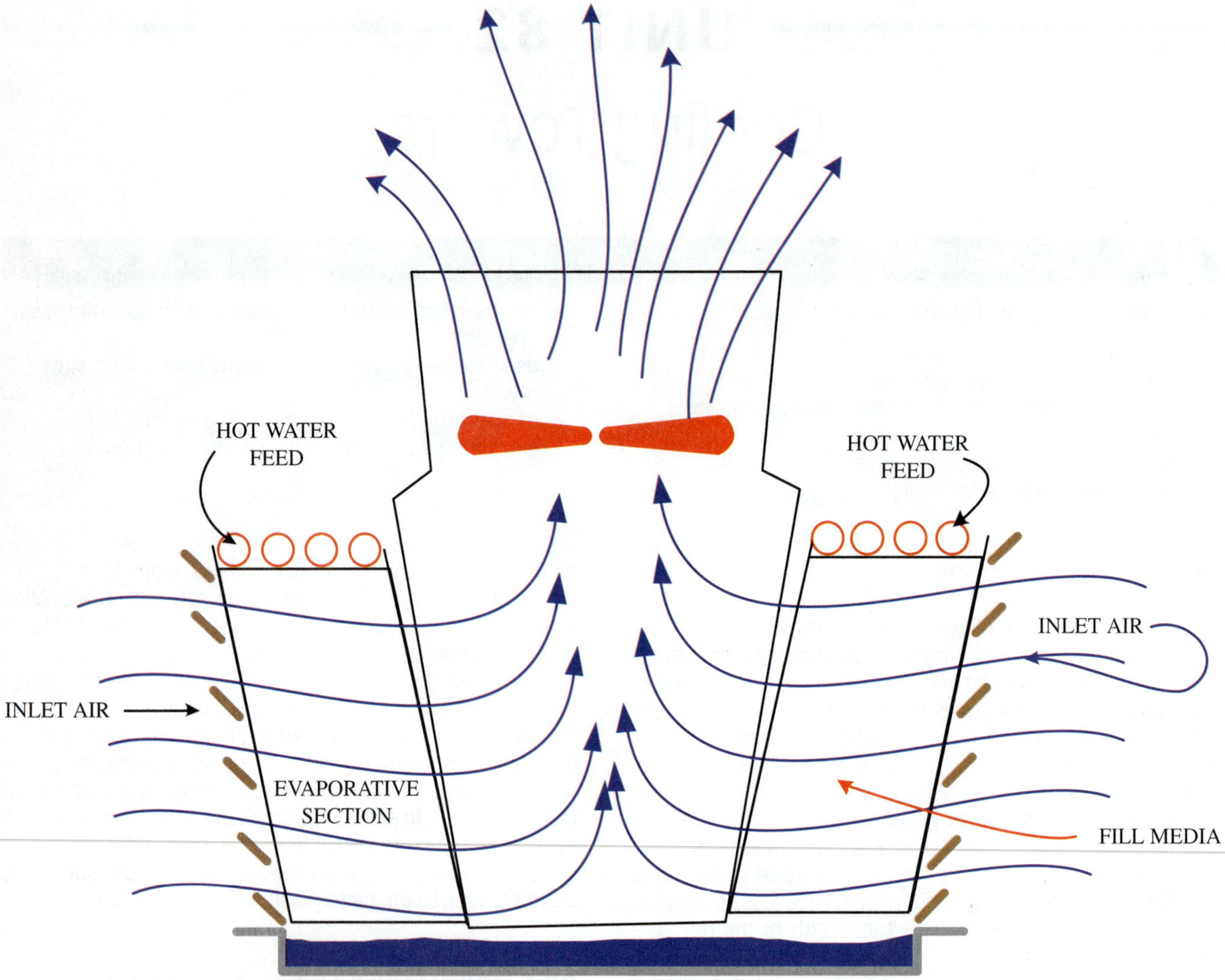

Figure 82-1 Schematic view of cooling tower construction.

the height is reduced compared to a counterflow cooling tower. It will be shorter but wider. Crossflow cooling towers are also sometimes used for indoor applications. Two connections to the outside are required. One connection is for drawing outdoor air into the tower and another is for sending it back outside again.

Counterflow Cooling Towers

An induced-draft counterflow cooling tower is shown in Figure 82-4. The air and water flow in opposite directions in a counterflow design. Typically, the water enters the tower

(a)

(b)

Figure 82-2 (a) Air inlet on a mechanical-induced-draft crossflow cooling water tower; (b) water inlet and outlet piping.

Figure 82-3 Fan blades and grille located on top of induced-draft cooling tower.

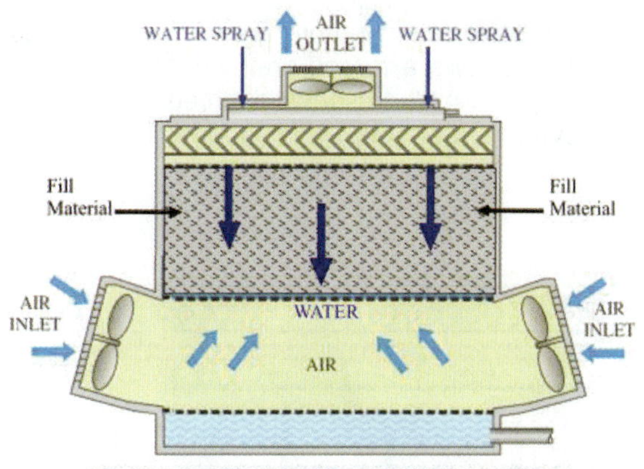

Figure 82-4 Induced-draft counterflow cooling tower.

at the top through a distribution header (Figure 82-5). The water is sprayed through nozzles in the form of a shower either by pump pressure or by gravity pressure from the water-filled headers (Figure 82-6). The hot water flows from the header into the tower fill material. This material is usually constructed of plastic or wood shells. The fill is much like a honeycomb found in a bee's nest and slows the fall of the water and increases its surface exposure (Figure 82-7). The fan pulls air through the fill. This air passes over and directly contacts the water to transfer heat from the warm water into the air. The cooled water falls to collect in the lower basin of the tower. A wet sump has the basin located

Figure 82-5 Water is pumped to the top of the tower.

Figure 82-6 Water-distribution spray nozzles located on the top of the tower.

(a)

(b)

Figure 82-7 (a) Plastic honeycomb fill; (b) support bracing for fill looking down from the top of the tower.

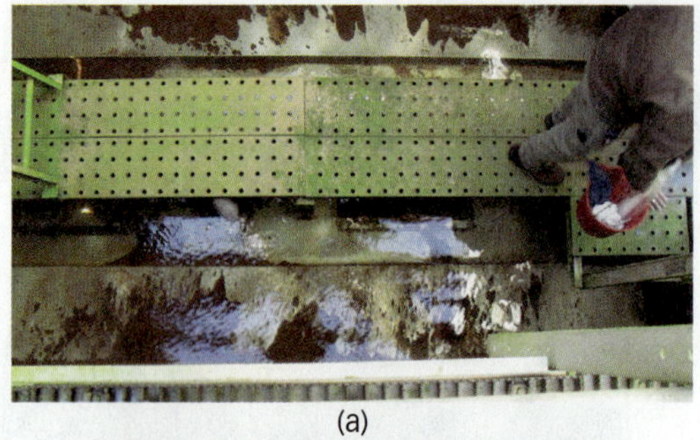

(a)

(b)

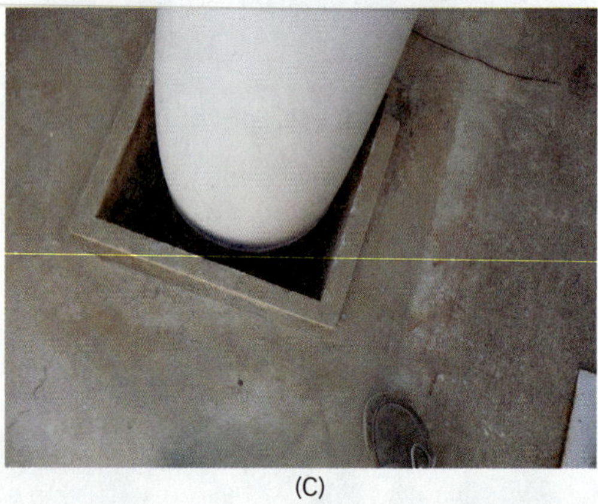

(C)

Figure 82-8 (a) Wet cooling tower sump; (b) dry cooling water sump; (c) cooling water piping to the dry basin sump.

inside the tower, while a dry sump has the basin located underneath (Figure 82-8). The water is then pumped back to the water-cooled condenser to absorb more heat from the refrigerant.

Crossflow Cooling Towers

A slightly different arrangement is shown in the induced-draft crossflow cooling tower shown in Figure 82-9. The air is drawn across the cooling water rather than up through it. Forced-draft towers are arranged so that the air is blown through rather than drawn through the fill material. Whether the air is drawn through or blown through, the basic configurations are still similar in design. Figure 82-10 shows a forced-draft counterflow cooling tower, while Figure 82-11 shows a forced-draft crossflow cooling tower.

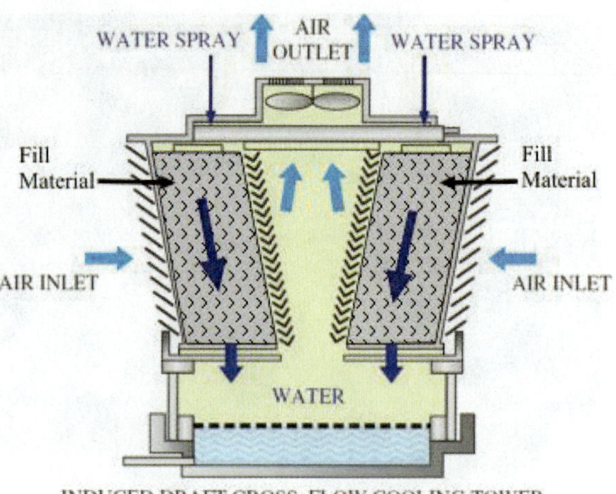

Figure 82-9 Induced-draft crossflow cooling tower.

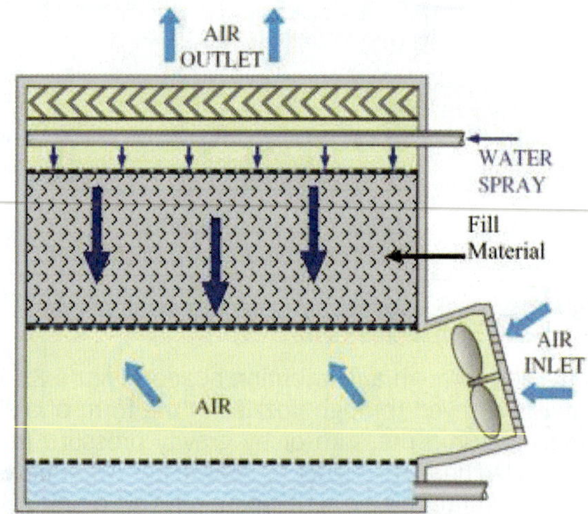

Figure 82-10 Forced-draft counterflow cooling tower.

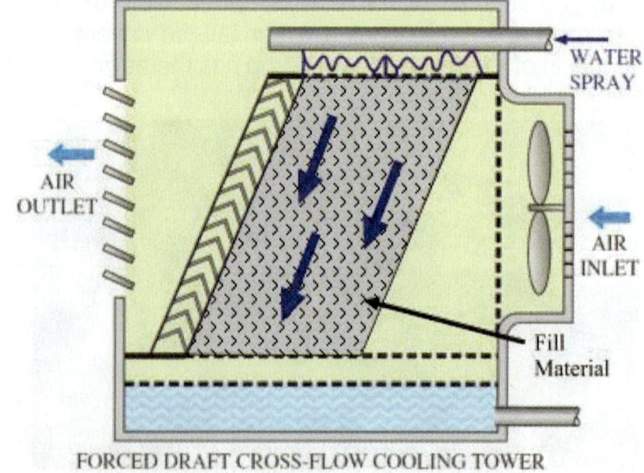

Figure 82-11 Forced-draft crossflow cooling tower.

TECH TIP

The visible cloud, called the plume, that exits cooling towers during cool, damp weather can be a problem in some circumstances. Airports, for example, cannot have large plumes of water vapor rising that might interfere with control tower visibility at the airfield. Plume-suppression kits are available. These kits draw in additional air and disperse the plume in a turbulent draft caused by the suppression fans. Even though some plume may be visible a few feet above the tower, the plume does not extend hundreds of feet in the air, as is the case with an unsuppressed plume.

82.4 ATMOSPHERIC (NATURAL) DRAFT COOLING TOWERS

An atmospheric-draft cooling tower is also referred to as a hyperbolic "natural"-draft cooling tower (Figure 82-12). Natural-draft cooling towers rely on prevailing winds and convection currents to move air across the wetted media in the tower. Natural-draft towers are normally larger than mechanical-draft towers for the same capacity. A properly sized natural-draft tower normally has a low operating cost since no power is used to move air.

This type of tower has no fill material or fan and depends on the spray nozzles to break up the water into a spray. Airflow through this tower is produced by natural circulation due to the density difference between the heated air in the tower and the cooler outside air. In comparison to mechanical-draft towers, the size, weight, and location requirements of an atmospheric-draft cooling tower reduce its use considerably. They are more often used for large electrical power stations and are seldom encountered in refrigeration service work.

Figure 82-12 Atmospheric-draft cooling tower.

82.5 EFFECT OF WET-BULB TEMPERATURE ON TOWER OPERATION

A cooling tower evaporates a portion of the water stream to cool down the remaining water. Drier air will allow more water to evaporate and increase the cooling tower efficiency as compared to humid air. Therefore, one of the most significant factors in cooling tower performance is the outdoor wet-bulb temperature, which provides an indication of how humid or dry the air is.

Sensible Heat Transfer

When the air in the cooling tower comes into contact with the water, the air absorbs sensible heat and becomes hotter and the water rejects sensible heat and becomes cooler. When the air temperature and the water temperature are equal, no more sensible heat transfer takes place. On a day when the outside dry-bulb temperature is higher and the air is hotter, less sensible heat transfer will take place. To adjust for higher outside dry-bulb temperatures, the water flow, the airflow, or a combination of both will need to be increased.

Latent Heat Transfer

The primary cooling of the circulating water is provided by latent heat of evaporation. Each pound of water evaporated by the cooling tower removes 970 BTU of latent heat. At that rate, it takes about 12 lb of water being evaporated to remove each ton of heat from the circulating water: 12,000 BTU/ton ÷ 970 BTU/lb = 12.3 lb of water. This amounts to approximately 1.5 gallons per hour. Compared to 1.5 gallons per minute for wastewater systems, it is easy to see that cooling towers save water.

Another outside air condition other than dry-bulb temperature that affects cooling tower performance is the outside humidity. If the relative humidity is low, then some of the water passing through the cooling tower will evaporate. Any water that evaporates (phase change, which may be called a change of state) will require the addition of latent heat. Some of this heat will come from the water that does not evaporate. As some of the water evaporates, the remaining water becomes cooler. In this way the water is not only cooled by the sensible transfer of heat to the air but also by the latent heat transfer due to evaporation.

Wet-Bulb Temperature

Cooling towers cool the circulated water through evaporation by spraying water through a moving airstream. In a typical tower, water "rains" down while air is pulled through the water spray using a fan. The middle of the tower is filled with wetted media to increase the surface area of the water exposed to the air. The fill increases the exposure of the water to the air and increases evaporation.

Cooling tower performance is governed by the entering-air wet-bulb temperature. Dry air has a lower wet-bulb temperature than wet air because the rapid evaporation of water cools the sock on the wet-bulb thermometer. Dry air can more easily evaporate water, so cooling tower capacity is highest with dry air. Nominal tower capacities are sized to cool the incoming water 10°F, from 95°F to 85°F with an outdoor wet-bulb temperature of 78°F.

A wet-bulb thermometer is an ordinary dry-bulb thermometer with a wetted cotton wick. As water from the wick evaporates, latent heat is removed from the thermometer, lowering its temperature. At a lower relative humidity, more water evaporates from the bulb and gives a greater difference between wet-bulb and dry-bulb readings. This means the drier the air, the more water will evaporate and the wet-bulb temperature will be lower.

It follows, then, that cooling tower operation does not depend solely on the dry-bulb temperature. The ability of a cooling tower to cool water is a measure of how close the tower can bring the water temperature to the wet-bulb temperature of the surrounding air. The lower the ambient wet-bulb temperature, the lower the tower can cool the water. It is important to remember that no cooling tower can ever cool water below the wet-bulb temperature of the incoming air. In actual practice, the final water temperature will always be at least a few degrees above the wet-bulb design temperature.

82.6 COOLING RANGE, APPROACH, AND HEAT LOAD

The following terms and definitions apply to all cooling towers and are illustrated in Figure 82-13.

- **Cooling range** The number of degrees Fahrenheit through which the water is cooled in the tower. It is the temperature difference between the hot water entering the tower and the cold water leaving the tower.
- **Approach** The difference in degrees Fahrenheit between the temperature of the cold water leaving the cooling tower and the wet-bulb temperature of the air entering the tower.
- **Heat load** The amount of heat "thrown away" by the cooling tower in BTU per hour. It is equal to the pounds of water circulated multiplied by the cooling range.

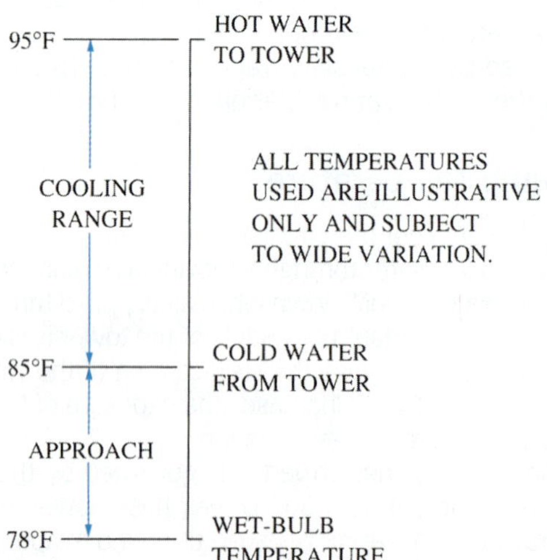

Figure 82-13 Range and approach as applied to cooling tower selection.

EXAMPLE 82-1: SOLVING FOR COOLING TOWER CAPACITY

Given a tower circulating 18 gpm with a 10°F cooling range, what would be the capacity?
Use the following formula:

$$Q = \text{gpm} \times 500 \times \text{TD}$$

where

Q = heat rejection in BTU/hr
gpm = flow in gallons per minute
500 = factor derived from specific heat of water (1) × 60 min/hr × 8.33 lb/gal
TD = temperature difference in °F

Therefore,

$$Q = 18 \text{ gpm} \times 500 \times 10 \text{ F}$$
$$= 90{,}000 \text{ BTU/hr}$$

82.7 DRIFT, EVAPORATION, BLOWDOWN, AND MAKEUP

Some of the water that evaporates from a cooling tower will be carried away by the air, and this is called drift. The amount of water lost will vary with the size of the cooling tower. Under certain ambient conditions, large cooling towers can produce a large white plume of water vapor that looks similar to fog.

Drift

Drift is the small amount of water lost in the form of fine droplets carried away by the circulating air, as shown in Figure 82-14. It is independent of, and in addition to, evaporation loss. For mechanical cooling towers, this will range from 0.1 to 0.3 percent of the total water being cooled.

Figure 82-14 Drift can be seen rising out of this cooling tower on a hot, humid summer day.

Evaporation

Theoretically, the evaporation rate is 1 percent of the water supplied to the tower for every 10°F the water is cooled.

Blowdown

The drift and evaporation of the water in the cooling circuit will leave solids behind, depending on the quality of the water. Blowdown (sometimes called bleed-off) is the continuous or intermittent draining of a small fraction of circulating water to prevent the buildup and concentration of scale-forming minerals and other nonvolatile impurities in the water.

Makeup

Makeup is the water required to replace water lost by evaporation, drift, and blowdown.

TECH TIP

Algae and slime are also present in the cooling water circuit and must be controlled chemically. It is also important that cooling towers be treated with the appropriate biocides to minimize the risk of *Legionella pneumophila* bacteria contaminating the tower.

82.8 CONTROLLING WATER-COOLED CONDENSER CAPACITY

The capacity of water-cooled systems using water towers can be affected by the outdoor temperature and humidity. It is possible for the cooling tower to cool so well that a minimum desired head pressure is not maintained. On water-cooled condenser systems, a temperature-controlled water valve that bypasses the tower and mixes cooling tower water with condenser water can be used. With this arrangement, the desired condenser water temperature can be maintained (Figure 82-15). The temperature-sensing element for the bypass valve is placed at the water entrance to the condenser.

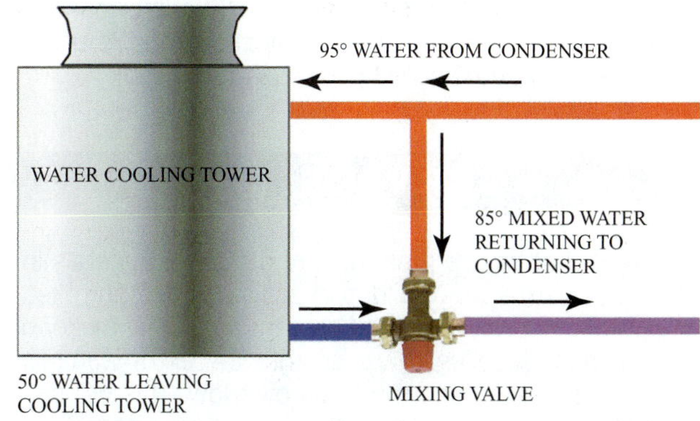

Figure 82-15 A mixing valve mixes hot and cold water from the condenser with cold water from the cooling tower to keep the returning water from being too cold.

82.9 COOLING TOWER PIPING CONNECTIONS

Figure 82-16 represents a typical mechanical-draft tower piping arrangement to the condenser of a packaged refrigeration or air-conditioning unit. Water supply lines should be as short as conditions permit. Standard-weight steel pipe (galvanized), type L copper tubing, and CPVC plastic pipe are among the satisfactory materials, subject to job conditions and local codes. The entire piping circuit should be analyzed to establish the proper location of valves for operation and maintenance of the system. A means of adjusting water flow is desirable. Shutoff valves should be placed so that each piece of equipment can be isolated for maintenance.

Cooling tower pump head is the pressure required to lift the returning warm water from the cold-water basin operating level to the top of the tower and force it through

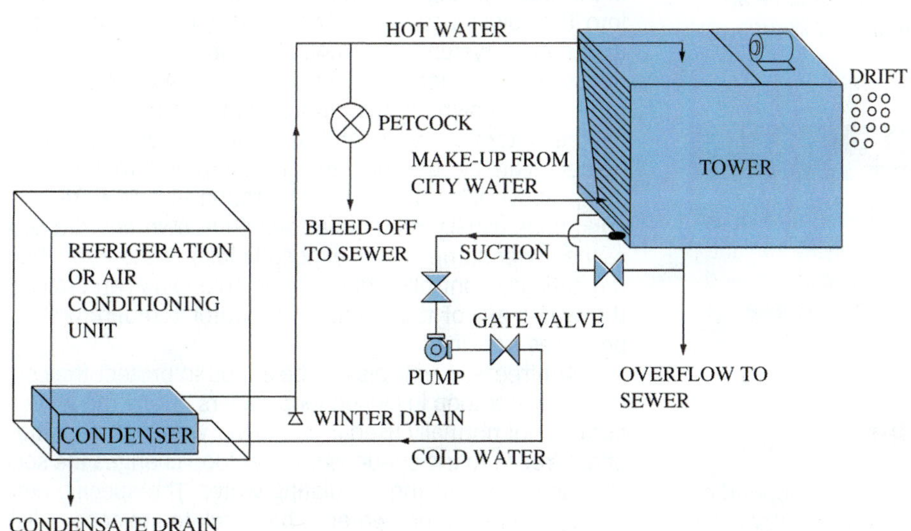

Figure 82-16 Piping diagram for a system using water-cooled condenser and cooling tower.

the distribution system spray nozzles. This information is found in the manufacturer's specifications and is usually expressed in feet of head (1 psi = 2.31 ft of head).

SERVICE TIP

In areas that have a great deal of dissolved minerals in the water, extensive buildup of scale can be controlled by putting approximately 10 percent additional makeup water in the cooling tower so that the extra water is flushed down through the basin overflow. This will significantly increase the time intervals between descaling.

There are many types of pumps from which to choose. For most air-conditioning applications, an iron-body, bronze-fitted, end-suction, centrifugal pump with mechanical seals will do the job. Close-coupled 3,500-rpm pumps are economical and do not have to be aligned. If continuity of service is important, install a standby pump. By locating the condenser pump outdoors below the tower, a leaking seal will be less of a problem. For some applications, 1,750-rpm base-mounted pumps are specified. Motor replacements are easier, and the motors run quieter.

The installation of the pumps includes the following:

- The pump should be located between the tower and the refrigeration or air-conditioning unit so that the water is "pulled" from the tower and "pushed" through the condenser. It is good practice to place a flow control valve in the pump discharge line.
- The pump should be installed so that the pump suction level is lower than the water level in the cold-water basin of the tower. This ensures pump priming.
- If the pump is located indoors, consideration should be given to noise levels and water leakage should the seal fail.
- If an open, drip-proof motor is used outdoors, a rain cover will provide additional protection. Make sure ventilation is adequate so the motor does not overheat.
- The pump should be accessible for maintenance and installed to permit complete drainage for winter shutdown.

SERVICE TIP

When replacing motors that are frame mounted, it is very important that the alignment between motor and pump be accurate. Misalignment will cause noise and excessive bearing wear. Shims are available to make the alignment process easier.

82.10 COOLING TOWER WIRING

The wiring and control arrangement will vary depending on the size of the equipment being installed. In every case, the objective is to provide the specified results with optimum operating economy and protection to the equipment involved.

For small refrigeration and air-conditioning systems, the ideal arrangement is based on a start-up sequence beginning with the cooling tower pump. The starter controlling the fan and pump will activate the compressor motor starter through an interlock. This method, illustrated in Figure 82-17, ensures sufficient condenser water flow so that compressor short-cycling is eliminated in the event of pump motor failure. In other words, the compressor cannot run unless the tower is operating. The tower fan is wired to allow cycling by a tower thermostat to maintain the proper condenser water supply temperature.

There are other, more economical methods based on using the compressor starter to activate the pump and fan, but water-temperature or flow-sensing devices should be incorporated as protection for the compressor. Where multiple refrigeration units are used on a common tower, the first unit that is turned on activates the cooling tower.

82.11 PROTECTIVE MEASURES FOR COOLING TOWERS IN COLD WEATHER

Winter operation or low ambient operation of a cooling tower is subject to special treatment for temperatures near freezing. Excessively cold water exiting the tower may cause thermal shock to the condenser and result in a very low condensing temperature. Cooling towers can also freeze during very cold weather. With enough heat load, the water in the tower will not freeze. Freezing is more likely to occur when operating at reduced load in cold weather. Ice accumulation on towers can add considerable weight to the tower, resulting in collapse of overloaded components.

One method of preventing the temperature of the water leaving the tower from falling too low is to turn off or reduce the speed of the tower fan. This will reduce the tower's thermal capability, causing the water temperature to rise. If this is not sufficient, a bypass valve can be placed in the tower piping to permit dumping warm water directly into the tower base, thus bypassing the spray or water-distribution system. A combination of both fan and total bypass may be needed to ensure leaving-water temperatures remain high enough during cold weather, but reducing the water flow will contribute to tower freeze-up. Sump heaters can be thermostatically controlled to keep the sump above 40°F. Prolonged operation at below-freezing conditions usually requires a total shutdown and draining of the tower sump and all exposed piping. Automatic thermostatically controlled drain-down solenoids can drain all the water out of the sump if the water temperature approaches freezing.

Antifreeze chemicals can be added to protect the condenser water loop in closed-loop towers during the winter, but it is not normally used in the tower water. The addition of antifreeze to the condenser water loop changes the specific heat value of the circulating water. The specific heat drops, requiring more water to be circulated to accomplish the same amount of heat removal.

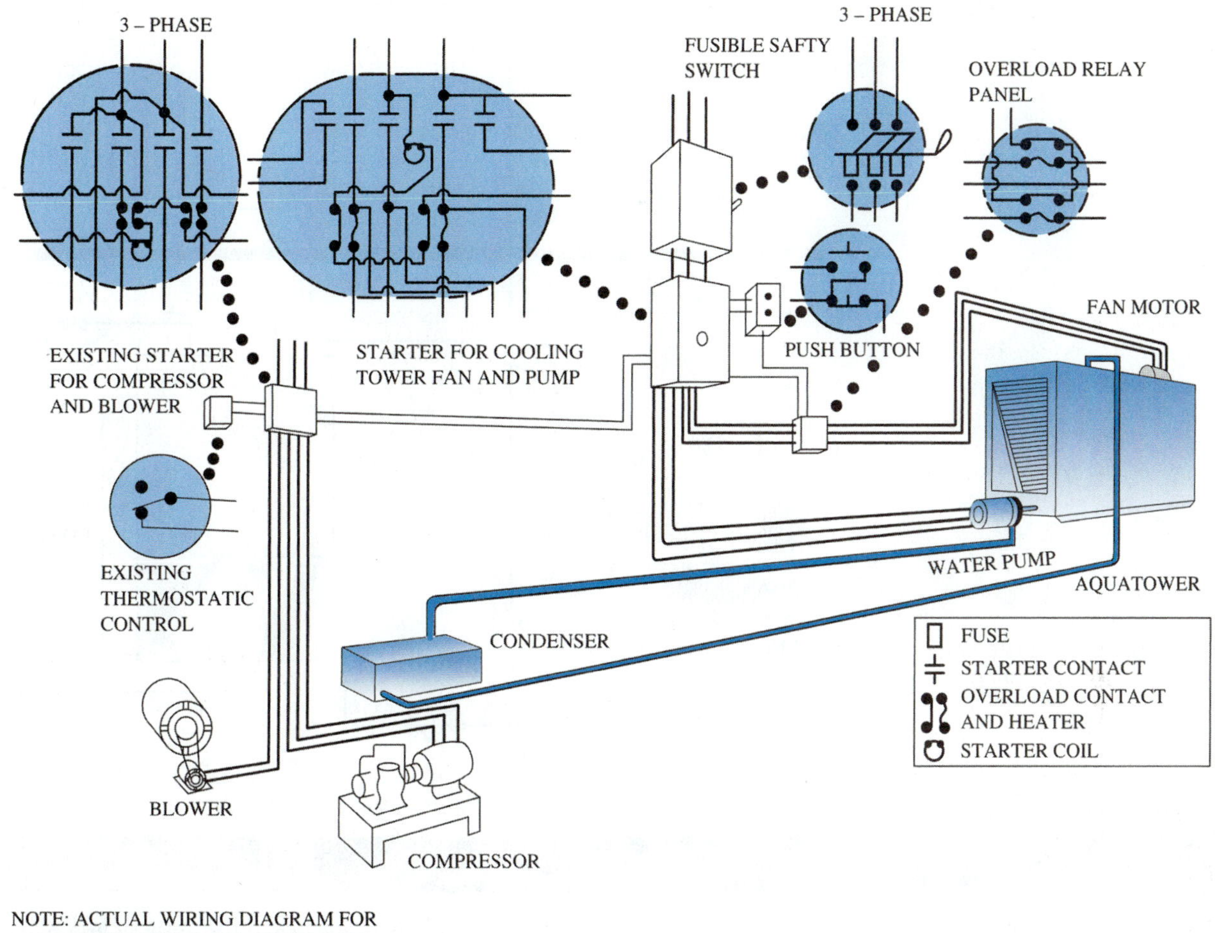

Figure 82-17 Typical wiring diagram for water-cooled condensing unit and cooling tower.

CAUTION

For a number of reasons, including environmental, expense, and maintenance, some cooling tower operators do not put antifreeze in their cooling towers. These towers are typically protected by either draining the towers in prolonged cold spells or running the circulator pumps. A significant disadvantage to this method is that a sudden cold spell may freeze the system before it can be drained. Also, power failure resulting from cold weather can stop the circulating pumps, which can result in a freeze-up.

A piping and control arrangement to provide winter operation is shown in Figure 82-18. This arrangement provides an inside sump. During normal operation, when the ambient temperature is above freezing, the water flows from the inside sump, through the condenser into the tower, and back to the sump. When the thermostat senses near-freezing water temperatures, the three-way valve is repositioned to direct the flow of water from the condenser directly to the inside sump, bypassing the tower. When there is no flow to the tower, the tower fan is cycled off.

82.12 COOLING TOWER MAINTENANCE

Proper maintenance is extremely important to keep a cooling tower operating at peak efficiency. Maintenance should include routine cleaning and chemical treatment to reduce mineral scale deposits, corrosion, bacteria growth, and fouling of the water.

Fans

Cooling tower fans are often belt driven (Figure 82-19). It is important to regularly check belt tension and wear. Vibration sensors will help to continuously monitor the fan condition and provide an early indication of problems. Existing fans can be equipped with vibration sensors that can be wired back to a continuous monitoring system (Figure 82-20).

Cooling Tower Cleaning

Peak performance of a water-cooled condenser and cooling tower system depends heavily on regular maintenance. Growths of slime or algae, which reduce heat transfer and clog the system, should be prevented. The success of any water treatment lies in starting it early and using it regularly. Once scale deposits have formed, it can be costly to remove them.

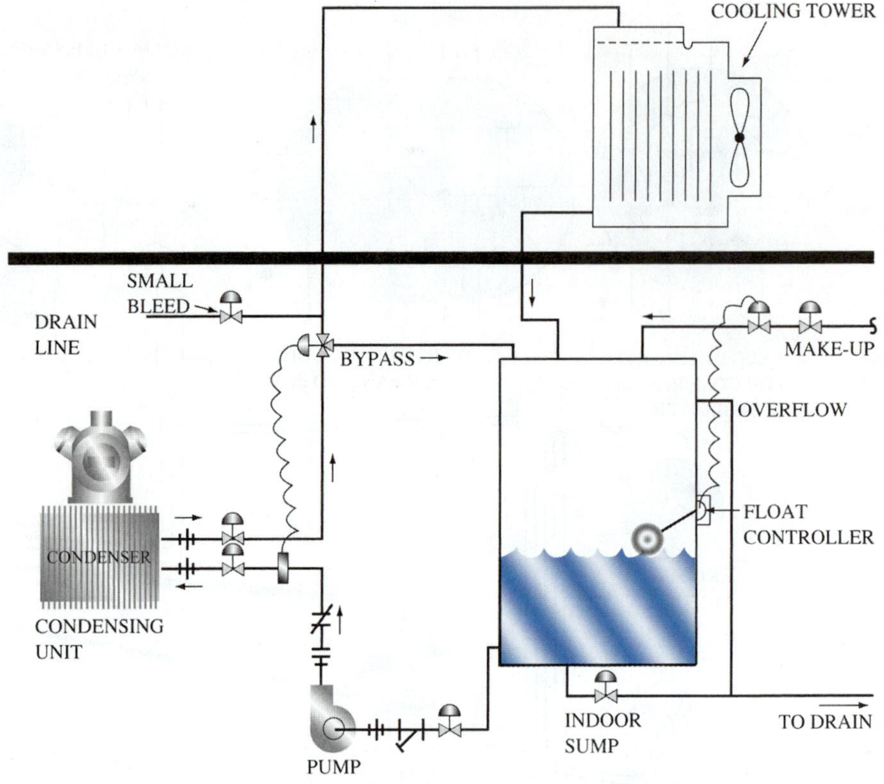

Figure 82-18 Winter operation using a cooling tower.

SERVICE TIP

The buildup of algae or other growth on water-cooled condensers can have the same adverse effect as debris on air-cooled condensers. It will reduce the system's operating efficiency. You must use a manufacturer-approved cleaning solution to remove these deposits. If the deposits are recurring, it may be necessary to use a water treatment to retard their growth.

In selecting condensers, application engineers will usually allow for the results of fouling so that the condenser will have sufficient excess tube surface to maintain satisfactory performance during normal operation, with a reasonable period of service between cleanings. For conditions of extreme fouling and poor maintenance, higher fouling factors are used. Proper maintenance depends on the type of condenser; mechanical or chemical cleaning—or both—may be needed or employed to remove scale deposits.

(a)

(b)

Figure 82-19 (a) Belt-driven fan; (b) fan motor pulley.

(a)

(b)

(c)

Figure 82-20 (a) Vibration sensor attached to the top of the fan motor; (b) vibration sensor attached to the bottom of the fan motor; (c) the vibration sensors are wired to an interface unit .

The technician should become familiar with the companies, chemicals, and cleaning techniques recommended locally. Water treatment and maintenance can be expensive and time-consuming. Once commonly used compounds, chromates are no longer used. Chemicals used for treatment are increasingly becoming regulated. A treatment must be selected that complies with local regulations.

Cooling Tower Water Treatment

In addition to preventing scale, water treatment should protect the system components against corrosion. This is critical on systems using steel pipe. The water tower continuously aerates the water, adding oxygen. Open-circuit cooling tower loops are much more subject to corrosion than closed-circuit chilled- or hot-water systems.

In addition to chemical treatment, regular draining, cleaning, and flushing of the tower basin is recommended. Also, a blowdown valve should be adjusted to cause a small amount of overflow that trickles down the drain. This blowdown is the continuous or intermittent removal of a small amount of water (1 percent or less) from the system. Dissolved concentrates are continually diluted and flushed away.

SERVICE TIP

Cooling towers are protected with wire screens to prevent birds such as pigeons from roosting in the towers (Figure 82-21). However, even the best screening allows some birds access to the tower. Often towers become roosts for large flocks of pigeons during evening off hours. Unfortunately, when the tower is brought online in the morning, many of these birds drown. Their carcasses are washed into the basin. This organic material can cause the mineral deposits normally found in the basin during cleanup to become extremely nauseating. Better screening and more frequent basin cleanout will help reduce this problem.

82.13 EVAPORATIVE CONDENSERS

In an evaporative condenser, the hot-gas piping from the compressor discharge passes through the condenser like it does on an air-cooled condenser, except the tubes are

Figure 82-21 Wire screen for keeping birds out of the tower.

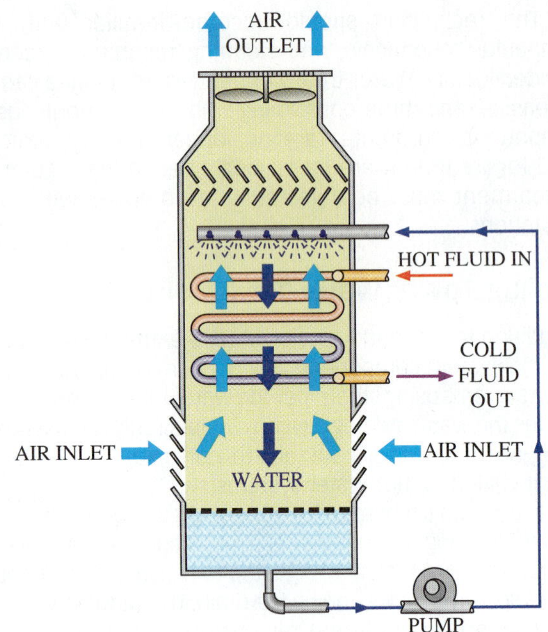

Figure 82-22 Evaporative condenser.

bare, and no fins are needed (Figure 82-22). In addition, the evaporative condenser has water nozzles that spray water over the tubes. As water flows over the hot tubes, it evaporates and picks up heat from the refrigerant, causing it to condense. At the same time, the blower in the evaporative condenser exhausts the humid, heat-laden air to the outside.

The evaporative condenser has a sump in the bottom with a float valve. As the water is evaporated, makeup water is added to the sump to replace it. Water from the sump is pumped to the spray nozzles at the top of the condenser and recirculated to provide continuous flow. When the water evaporates, it leaves a mineral residue in the sump that can accumulate. To prevent this, a continuous blowdown is provided. This residue also will adhere to the tubes and needs to be periodically cleaned off. If it accumulates on the tubes it will reduce the transfer rate. Water treatment is also recommended.

Evaporative condensers range in capacity from about 10 to 1,000 tons. They can be located inside or outside the building. If they are located inside the building, then outside air must be supplied. The heat-laden air they exhaust needs to be ducted to the outside of the building. They are more efficient than either the air-cooled or water-cooled condenser. This is because the capacity of cooling towers and evaporative condensers is governed by the lower wet-bulb temperatures, whereas air-cooled condensers are related to the higher ambient air dry-bulb temperatures. But of the three types, evaporative condensers are the least popular. Internal corrosion causes them to have the highest maintenance requirements. Their use is more frequent on industrial HVACR systems.

An interesting application of the evaporative condenser principle is included in the design of a room cooler. The condensed water vapor from the evaporator is directed to a small sump in the bottom of the condenser fan housing. A slinger ring on the fan splashes this condensate onto the surface of the condenser coil. When this moisture evaporates from the surface of the hot condenser coil, it removes heat, similar to the action that takes place using an evaporative condenser. When the condensation is light or nonexistent, the condenser acts strictly as an air-cooled condenser.

UNIT 82—SUMMARY

Cooling towers allow the use of water-cooled condensers. They can also cool the water below the outside ambient dry-bulb temperature. Most cooling towers for HVACR systems are of the mechanical-draft type. The water used in a cooling water tower is circulated in a loop. It is important that it be treated to reduce slime, algae, *Legionella* bacteria, and scale buildup. A small measure of water is continually drained from the loop to reduce the buildup and concentration of scale-forming minerals. Makeup water is added to replace the water lost through evaporation, drift, and blowdown. Normal maintenance includes draining and cleaning the water basin, adding water treatment chemicals, and adjusting blowdown and makeup. System leaks or pumping problems should also be corrected. Proper cooling tower operation will result in cleaner condenser tubes and higher efficiency.

WORK ORDERS

Service Ticket 8201

Customer Complaint: Routine Service

Equipment Type: Rooftop Induced-Draft Cooling Tower

The technician finds the cooling tower on the rooftop and discovers that the water basin is very cloudy. The unit is shut down and the fans secured. The water basin is drained of water, and it is discovered that birds had found a way in and died in the basin. The technician cleans the basin and adds fresh water. The water is tested for pH and hardness, and it is within the limits. The correct amount of water treatment additive is put into the basin. The mesh screen surrounding the tower is repaired so that birds will not be able to find their way into the tower. The tower is then put back into service.

Service Ticket 8202

Customer Complaint: Compressor Tripped

Equipment Type: Central Air Conditioning with Mechanical Cooling Tower

The technician is told that the screw compressor for a water-cooled chiller was tripping on high-pressure cutout. The technician checks the condenser cooling water inlet and outlet temperatures and finds them to be high. The condenser circulating pump pressure appears adequate and the pump seems to be operating normally. The technician next goes to the mechanical cooling tower. Only one of the two induced fans is operating. The power is secured and the fan motor circuit is tested and appears normal. Next the fan motor itself is tested and found to have a short. The fan motor is replaced and the tower put back into service.

Service Ticket 8203

Customer Complaint: Condenser Pressure High

Equipment Type: Central Air Conditioning with Mechanical Cooling Tower

The building superintendant tells the technician that the unit operating pressure has risen slowly over the past 6 months. This is also reflected in higher electricity bills for the complex. The cooling water to the condenser is wide open and therefore can be increased no more. The technician takes a water sample from the cooling tower and has it checked for hardness. It is above the recommended limits. The unit is shut down and drained. The condenser heads are taken off from each end to expose the tubes. A mechanical cleaner is used to pass though each tube and remove the scale buildup. After cleaning is complete, the condenser is put back together and the cooling water loop is refilled with the proper quality water. The unit is started. Now the condenser water regulating valve is only halfway open while maintaining a proper condenser discharge pressure.

UNIT 82—REVIEW QUESTIONS

1. What are the two common mechanical-draft cooling tower arrangements?
2. How does an induced-draft cooling tower differ from a forced-draft tower?
3. What is cooling tower fill material?
4. What is an advantage of a crossflow tower as compared to a counterflow cooling tower?
5. What is a cooling tower plume?
6. What is the working principle for an atmospheric-draft cooling tower?
7. What type of heat is required to evaporate water?
8. Where does the heat come from to evaporate water in a cooling tower?
9. Can a cooling tower cool below the outside wet-bulb temperature?
10. What is meant by cooling range?
11. What is meant by the approach?
12. How is the heat load for a cooling tower calculated?
13. What is meant by drift?
14. Theoretically, how much water evaporates in a cooling water tower?
15. What is blowdown?
16. How much makeup water is required for a cooling tower?
17. 1 psi is equal to how many feet of head?
18. Not including friction, what is the static discharge head for a water line 30 ft high?
19. How should the circulating pump be located between the condenser and the tower?
20. In addition to preventing scale, water treatment should protect the system components against what other deteriorating condition?

UNIT 83

Thermal Storage Systems

OBJECTIVES

After completing this unit, you will be able to:

1. list the different types of ice thermal storage systems.
2. describe the operation of ice thermal storage systems.
3. explain why thermal storage systems are used.
4. list the benefits of thermal storage systems.
5. explain the modes of operation for thermal storage systems.

83.1 INTRODUCTION

Thermal storage is the temporary storage or removal of heat for later use. An example of thermal storage is the storage of solar heat energy during the day to be used at a later time for heating at night. In the HVACR field, this type of application using thermal storage for heating is less common than using thermal storage for cooling. An example of the storage of "cold" heat removal for later use is ice made during the cooler night hours for use during the hot daylight hours (Figure 83-1). This ice storage is produced when electrical utility rates are lower. This is often referred to as "off-peak" cooling.

When used for the proper application with the appropriate design, off-peak cooling systems can lower energy costs. The U.S. Green Building Council, as described in Unit 68 has developed the Leadership in Energy Efficiency and Environmental Design (LEED) program to encourage the design of high-performance buildings that will help protect our environment. The increased levels of energy performance by utilizing off-peak cooling may qualify for credits toward LEED certification.

Figure 83-1 Ice-storage system. *(Courtesy of CALMAC Manufacturing Corporation)*

A stratified water thermal storage system capable of providing 24,628 ton-hours of thermal storage is shown in Figure 83-2. Specially engineered diffusers or nozzles distribute warm water at the top of the tank and cool water at the bottom. This provides for a separation between the cool supply water and the warm return water. This 3.3-million-gallon storage tank is filled with chilled water every night.

Figure 83-2 Stratified water thermal storage tank. *(Courtesy of Office of Energy Efficiency and Renewable Energy, U.S. Department of Energy)*

83.2 THERMAL STORAGE SYSTEMS

The advantages of thermal storage are the following:

- Commercial electrical rates are lower at night.
- It takes less energy to make ice when it is cool at night.
- A smaller, more efficient system can do the job of a much larger unit by running for more hours.

Most thermal storage applications involve a 24-hour storage cycle, although weekly and seasonal cycles are also used. In the context of the energy-efficient system applications, storage of cooling is the most popular. For HVACR purposes, water and phase-change materials (PCMs)—particularly ice—constitute the principal storage media. Water has the advantage of universal availability, low cost, and transportability through other system components. Ice, however, will absorb more heat than water and therefore has the advantage of taking up less space.

Then every day from noon until 8 PM the chilled water is used to meet the cooling load. Thermal storage water systems can also serve to provide large volumes of water onsite for fire protection systems.

83.3 MODES OF OPERATION

There are five modes of operation of a cooling storage system that stores and releases thermal energy.

Charging Storage

In charging storage, the refrigeration system is extracting heat from the storage vessel. This would be preferably done during an off-peak period. If this is a water-storage system, then the water would be cooled. For an ice-storage system, ice would be formed.

Simultaneous Recharging Storage and Live Load Chilling

In simultaneous recharging storage and live load chilling, some of the refrigeration capacity is being used by the load at the same time the storage is being recharged. This situation occurs when there is extra capacity available. Rather than operating the refrigeration system at partial load, it is operated at full load with the additional cooling applied to the storage tank. This is preferred during off-peak periods when the cost of operation is low.

Live Load Chilling

Live load chilling is identical to normal water chiller operation, providing cooling as needed. In this situation, no cooling of the storage tank is taking place. All of the cooling capacity is directed to the building load.

Discharging and Live Load Chilling

In discharging and live load chilling, the refrigeration unit operates and the storage vessel is being discharged at the same time to satisfy the cooling load requirement. In this case, the cooling demand may be so high that additional cooling is required. Another scenario is that the peak load cost is high. The refrigeration system is operated at reduced load during this high-cost period, with the storage system making up the difference.

Discharging

In discharging, cooling needs are met only from the storage, with no refrigeration system operating. The storage tank could be used during peak loads when costs are high. This could also be used during low-demand periods when operation of the refrigeration system at low loads is undesirable. Another scenario is during a power outage when the emergency power supply is limited. The uninterrupted power supply for large computer systems will come on automatically, and these can generate a considerable amount of heat. The emergency power may not be sufficient to run the chiller unit itself, but the auxiliaries can be run and cooling supplied from the cold storage.

83.4 SELECTING THE METHOD OF OPERATION

Thermal storage, particularly ice storage, has its roots in applications involving short-duration loads, with relatively long times between loads. The classic examples are churches, sports facilities, and older movie theaters. The peak cooling load would only occur during scheduled events when the spaces are occupied. Current systems take into account these peak demand periods, but they are also now being primarily designed to offset increasing electric utility rates, day rate schedules, and large demand charges for electric power.

Thermal storage systems are typically classified as being either full storage or partial storage. The essential purpose of ice storage is to make ice while the electrical power rates are low. Full thermal storage systems will generate ice when the costs are lowest during the off-peak periods, typically at night. This ice is then melted to cool the building during the daytime hours while the large-capacity refrigeration chiller is shut down. Partial thermal storage systems differ in that they utilize a reduced-size refrigeration chiller that is operating along with the melting ice to cool the building. The optimum system type is derived from an analysis of all the elements involved, including the following:

- Total cooling load
- Total daily ton-hours of load (the sum of each of the hourly loads in a day's use)
- Available ice recharging time
- Steady loads and temporary loads
- Energy cost, peak and off-peak cost, and hours of billing
- Space for storage
- Relative costs of type of systems

Water-based thermal storage has certain specialized applications and has been successfully applied many times. However, there are inherent problems with this type of system. It is difficult to provide stratified storage temperatures, so that the recirculated, warmer water does not disturb the stored chilled water. Due to the large volume of water required, there are space and structural limitations. Open systems will require water treatment.

Using ice for storage systems has some distinct benefits. These systems can vary in size. Ice-storage systems are practical on systems as low as 5 tons. Applications utilizing ice-storage systems require modifications of conventional system designs. Ice provides cold (33°F –34°F) water for cooling and has the following advantages:

- Colder water provides a large temperature differential, which means less flow will be required for a given cooling load.
- Reduced flow rates call for smaller pipes and lower pumping costs.
- Colder water provides for lower leaving-air temperatures from cooling coils.
- Colder supply air will absorb more heat; therefore, less air needs to be delivered by the fan.
- Reduced fan airflow (CFM) calls for smaller fans and ductwork.

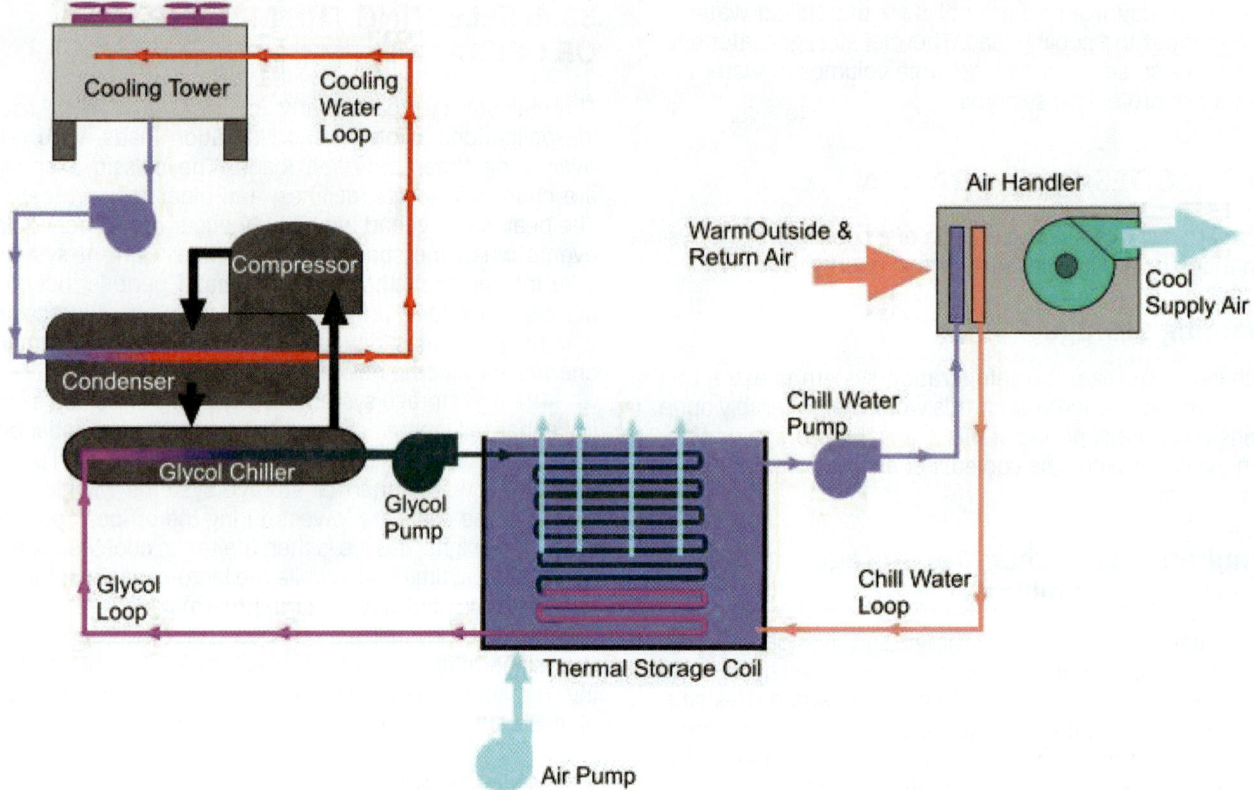

Figure 83-3 Thermal storage system schematic.

> **TECH TIP**
>
> Because of the high energy efficiency obtainable through thermal storage systems, many manufacturers are working on applying this concept to smaller residential systems. As these become more commonly available, they can provide significant energy savings to homeowners.

83.5 ICE-ON-COIL (EXTERNAL MELT) SYSTEMS

In these types of systems, multiple banks of hot-dipped galvanized steel coils are submerged underwater in field-constructed concrete tanks. Refrigerant is circulated through the inside of the coils to generate ice on the outside. The refrigerant circulated may be a direct-expansion type, but the most common kind is a system-circulating glycol supplied from a chiller (Figure 83-3). This type of system is commonly used for district cooling or for systems where cold temperatures are required.

Full thermal-storage systems have two modes of operation, referred to as "ice-build" and "melt-out." During the ice-build phase, low-temperature glycol is circulated through the coils. For most air-conditioning applications, a solution of water and 25 to 30 percent ethylene glycol is used. The glycol warms as it passes through the coil, which forms an ice taper that is thicker near the inlet and thinner near the outlet. This can lead to wasted volume in a thermal storage tank (Figure 83-4a). To increase efficiency, countercurrent flow circuits are utilized (Figure 83-4b). Elliptical-tube coil design also allows for better ice-packing efficiency (Figure 83-5). This is the ratio of the volume actually formed and stored in comparison to the available space for ice around the coil assembly, excluding the necessary clearance spaces.

The amount of ice formed in the storage system can be determined by measuring the water level in the tank. It will increase and rise as the ice is being formed because the ice is less dense than the water. Ice thickness controllers using conductivity measurement are also used to sense the thickness of the ice. When the ice-build is complete, the controllers shut off the glycol flow to the ice coils. In addition, an air-agitation system consisting of perforated PVC pipes located beneath the coils and supplied with low-pressure air from a blower is used during both the ice-build and ice-melt-out phases.

The melt-out phase for an external melt system begins as the 32°F tank water is circulated to the building to provide the required cooling. The warm water returning to the system will begin to melt the ice on the tubes from the outside in. During the melt, the ice surface area is reduced. With about 50 percent of the ice remaining on the tubes, the water temperature will begin to increase. Because the water temperature slowly begins to rise during the melt-out phase, this type of application is suitable for cooling loads that are high early in the cycle and slowly go down later in the cycle.

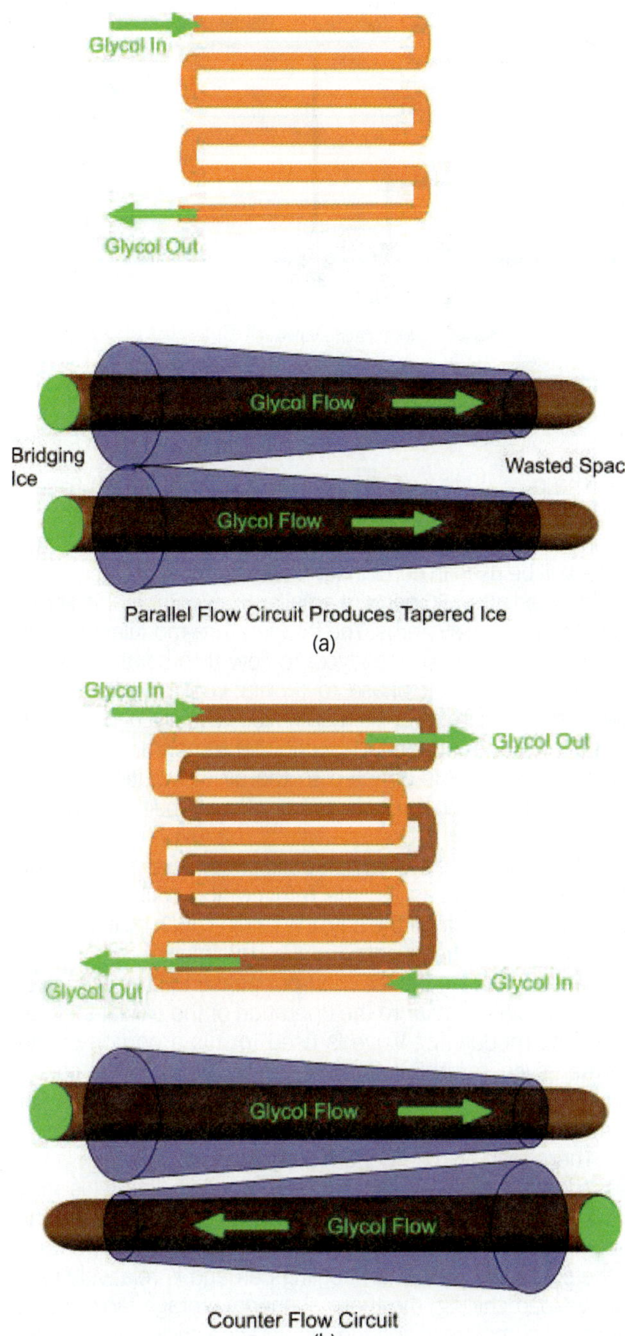

Figure 83-4 (a) Parallel flow circuits; (b) countercurrent flow circuits.

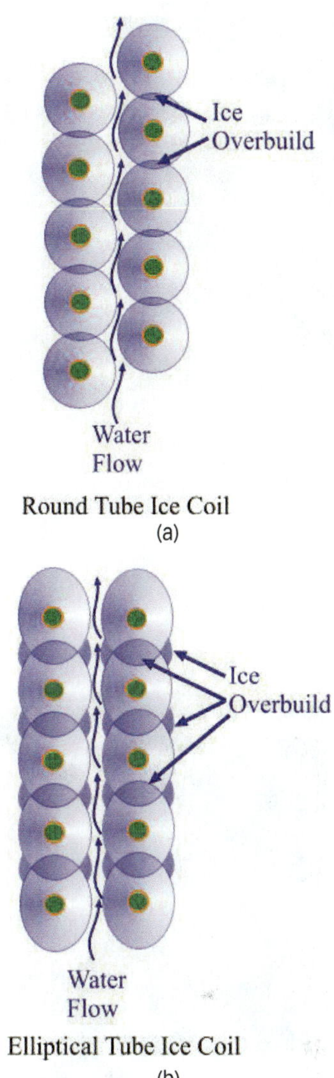

Figure 83-5 (a) Round-tube ice coil—reduced packing efficiency; (b) elliptical-tube ice coil—increased packing efficiency.

83.6 ICE-ON-COIL (INTERNAL MELT) SYSTEMS

Internal melt systems also have an ice-build phase. However, during the melt-out phase, warm glycol is circulated from the building load back to the storage tank through the inside of the heat exchanger in such a manner as to melt the ice from the inside out. Because the glycol temperature slowly begins to lower during the melt-out phase, this type of application is suitable for cooling loads that are low early in the cycle and slowly increase later in the cycle.

The storage tanks utilized for internal melt systems are manufactured in a variety of sizes, from 45 to over 500 ton-hours (Figure 83-6). They are well-insulated, seamless, one-piece tanks of polyethylene containing a spiral-wound polyethylene-tube heat exchanger (Figure 83-7). Water containing 25 percent ethylene glycol is circulated through the heat exchanger at approximately 25°F during the ice-build phase. About 95 percent of the water in the tank is frozen solid. Since the ice is built uniformly throughout the tank, the water does not become surrounded by the ice during the freezing process, allowing it to freely move and prevent damage to the tank.

This type of system can be configured for either full- or partial-storage conditions. The charge cycle for a partial-storage system is shown in Figure 83-8. During the off-peak utility hours, the glycol leaving the chiller is set for a temperature of 25°F. In this cycle, the glycol is used only for building ice and is diverted back through the chiller until approximately 95 percent of the water inside the tank has been frozen solid.

Figure 83-6 Fossil Ridge High School. *(Courtesy of CALMAC Manufacturing Corporation)*

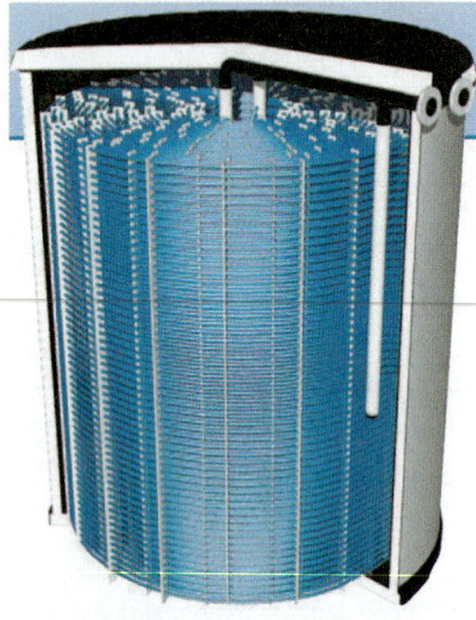

Figure 83-7 Thermal energy-storage tank IceBank. *(Courtesy of CALMAC Manufacturing Corporation)*

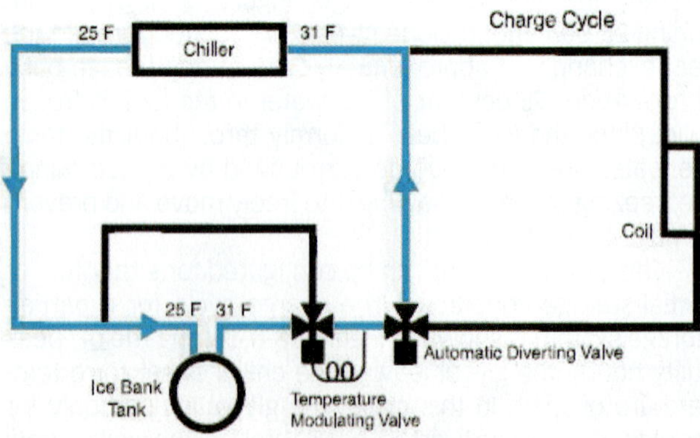

Figure 83-8 Partial thermal energy-storage charge cycle. *(Courtesy of CALMAC Manufacturing Corporation)*

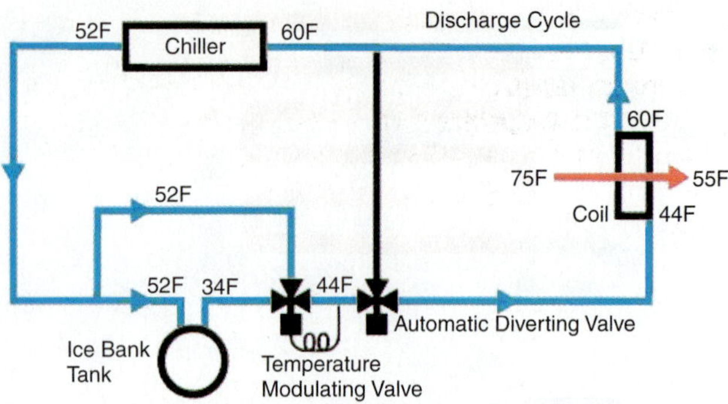

Figure 83-9 Partial thermal energy-storage discharge cycle. *(Courtesy of CALMAC Manufacturing Corporation)*

The partial-storage discharge cycle is shown in Figure 83-9. During the heat of the day, the glycol temperature leaving the chiller is set to a temperature of 52°F. Because the chiller will be used in combination with the ice-storage tank, it can be sized 40 to 50 percent smaller than comparable HVAC non-storage-type systems. The temperature-modulating valve will allow a portion of the glycol to flow through the storage tank for the ice-melt phase to further cool the glycol. This lower-temperature glycol is then mixed with glycol bypassing the tank to control the final glycol outlet temperature at 44°F. The glycol at 44°F is directed to the air-distribution system cooling coil to remove the heat from the air being circulated within the building.

The full-storage system ice-build phase occurs during the off-peak utility hours with the refrigeration chiller in operation. During the heat of the day, the refrigeration chiller is shut down. However, the glycol continues to circulate through the storage tank during the ice-melt-out phase to supply cooling for the building. Similar to the operation of the partial-storage cycle, the modulating valve is used to mix a portion of the warmer glycol returning directly from the coil with the colder glycol passing through the storage tank to maintain the set temperature of the glycol supplied to the coil.

There may also be periods when the storage tank is bypassed altogether, referred to as live-load chilling. This may occur when electrical rates are low and the ice-storage tank is being conserved for later use or for mild-temperature days in the spring and fall when cooling demand is relatively low. For live-load chilling, the thermal energy storage tank will be bypassed completely, and the refrigeration chiller will be used to meet the cooling demand, as shown in Figure 83-10.

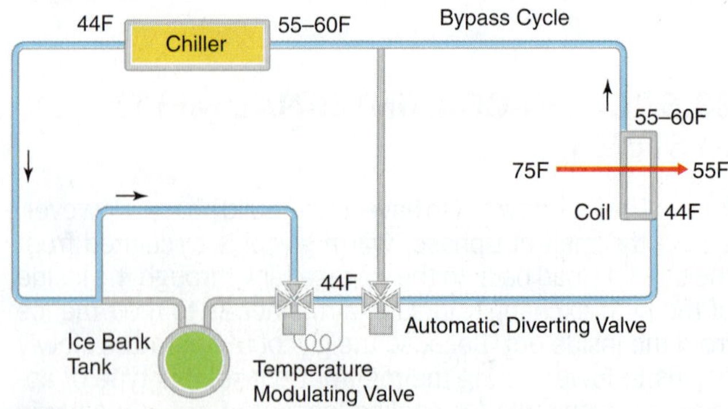

Figure 83-10 Bypass cycle (chiller-only cooling).

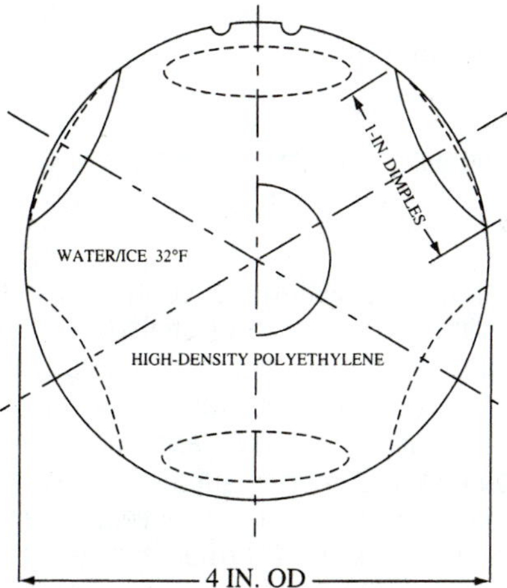

Figure 83-11 Spherical container for ice thermal storage.

The polyethylene construction and seamless tank design result in a low-maintenance demand for this type of system. The water level of the storage tank is checked annually. A biocide is added to the water every 2 years. The concentration of the glycol solution is checked annually.

83.7 ENCAPSULATED ICE-STORAGE SYSTEMS

In this type of system, a number of water-filled containers are placed inside a tank. A glycol solution fills the tank and circulates over and around the containers. Some containers are spherical, with dimples to allow for expansion on freezing, as shown in the diagram in Figure 83-11. Other containers are flat and rectangular like a giant hot-water bottle or annular, shaped like a donut. These containers are stacked in storage tanks that can vary in shape and design. In this type of system, the ice is formed within each spherical container from contact with a cold glycol solution, as shown in Figure 83-12.

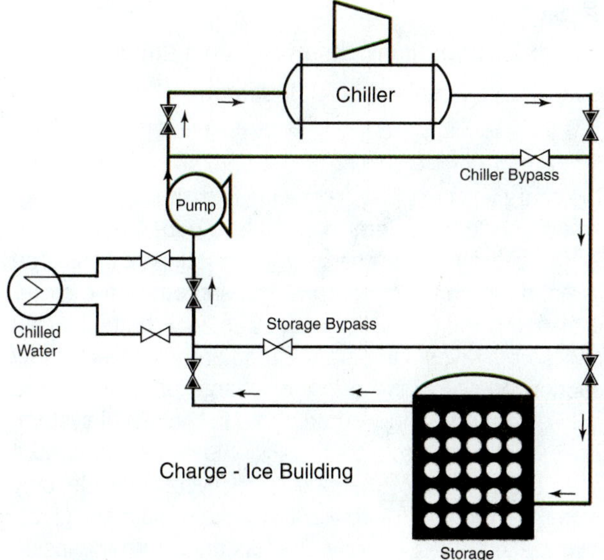

Figure 83-12 Encapsulated ice-storage system—ice building.

The piping in and out of the tank must create a flow path that is effective for both charging and discharging. The control system monitors the liquid level of the tanks to determine the amount of ice formed. As ice is formed, the containers expand, which raises the glycol level in the tank. The tank level is monitored to determine when the system is fully charged.

The economic factors for this type of system include storage tank shape, configuration, and installation. The non-managed flow path of encapsulated ice-storage systems does not offer repeatable performance. Because of this, there is a tendency for designers to oversize them. Also because of the design, encapsulated ice-storage systems typically will use more glycol as compared to ice-on-coil systems.

83.8 ICE-HARVESTING SYSTEMS

Ice-harvesting systems are another application of industrial ice makers/water chillers. These types of systems require large and heavy equipment with limited configurations and so are becoming less common. These units build ice in sheets on the surface of vertical refrigerated plates. The ice is harvested by slightly warming the plates with refrigerant discharge gas, which causes the ice sheets to break away from the plates and fall into a water-filled storage tank. The sheets of ice break up when they hit the water, providing many exposed ice surfaces for heat transfer to the water. The thin sheets of ice provide a very even discharge cycle, with consistently low water temperatures. Monitoring of the tank level determines refrigeration operation.

The storage tank must be constructed with weirs (a weir is like a dam) to prevent the ice from getting into the circulation system. Since the ice floats and distributes itself in the tank, water flow distribution is less critical than with some of the other systems.

SERVICE TIP

Ice-harvesting equipment is the most complex of the various thermal storage systems. However, it is often the most efficient method of ice production when properly set up. A major factor in the efficiency of ice harvesters is the actual harvesting cycle. Harvesting is typically performed by diverting the hot condenser gas back into the evaporator plate where the ice has formed. This momentary heating causes the ice to release, clearing the evaporator surface for more ice to form. Manufacturer guidelines list recommended harvest times to optimize ice production.

83.9 THERMAL STORAGE FOR HEATING

Applications of thermal storage for heating in the HVACR field are less common than cooling. However, there are applications that have been becoming increasingly popular. These systems usually require some type of a thermal storage tank or reservoir.

Heat Pump Applications

Most ground- and water-source heat pumps take advantage of sources of heat without necessarily requiring a thermal storage tank. However, heat pumps can be used in combination with thermal storage systems. As an example, rooftop solar collector panels can be used during the day to heat water that is stored in a thermal storage tank. During the cooler evening hours, the heat pump can be used to remove heat from the storage tank and deliver it to the building.

Thermal Storage Under-Floor Heating

Some homes are designed with electric cables or heating panels buried 4–12 inches deep in a bed of sand. The cement slab for the building is then poured on top. During the low-cost, off-peak hours, the sand is heated to create a thermal reservoir beneath the floor. During the cooler evening hours, the heat from the floor will be continually radiated into the building.

Coil-type mats are used for permanent and/or temporary ice applications, such as skating rinks (Figure 83-13). Similar mats are also used for subfloor heating and other applications, such as turf warming for year-round plant growing at stadiums and greenhouses, snow melting for athletic fields and dog tracks, and permafrost prevention for refrigerated warehouses and ice-skating rinks. A heated liquid is circulated through a series of polyethylene tubes. Lake-coupled geothermal heating systems involve laying a series of plastic tubes near the bottom of a lake to be used in combination with a heat pump system. Standard-size preassembled mats are commonly available, as are custom designs to meet almost any application.

Boreholes

For this application, an array of bored holes resembling standard drilled wells is spaced in a grid. A plastic pipe with a U bend is then inserted into each borehole. Solar heaters and a pumping system provide for a flow of hot water down into the boreholes to heat the ground during the summer months. This heat is then recovered and sent to the building during the winter months.

Figure 83-13 IceMat® being unrolled for installation. *(Courtesy of CALMAC Manufacturing Corporation)*

UNIT 83—SUMMARY

Most thermal storage systems encountered in HVACR are used for cooling. The storage of chilled water is popular, but this requires a large storage tank that is typically of the stratified design. Ice-storage systems require less space and are the most popular type. There are a number of different types of ice-storage systems, such as ice-on-coil internal melt, ice-on-coil external melt, ice harvesting, and encapsulated ice-storage systems. Thermal storage systems allow for charging during periods of low-cost electrical off-peak rates. This stored cooling capacity is then discharged during high-cost peak periods, which reduces or even eliminates the need to run the chiller units. Thermal storage systems therefore will reduce total operating costs.

WORK ORDERS

Service Ticket 8301

Customer Complaint: Ice Too Thin

Equipment Type: Ice-on-Coil System

The ice builder is charging very quickly and the ice formation is not very thick. During discharge periods, the total cooling period is short. The technician first checks the refrigeration chiller, and it was found to be operating normally. The ice formation is uniform across the coil, which indicates good refrigerant flow and water coverage. However, the layer of ice is thin. The technician locates the ice sensors used to monitor the ice thickness and replaces them. Now the unit charges to the proper ice thickness and the discharge period is considerably longer.

Service Ticket 8302

Customer Complaint: Insufficient Ice Buildup

Equipment Type: Ice-on-Coil System

The glycol coil is not freezing the ice solid, and the refrigeration unit is running for prolonged periods during the charging cycle. The technician first checks the refrigeration chiller and determines that the load on the unit is low, although the charging cycle is in progress. The glycol temperature difference into and out of the chiller is also low, indicating little heat transfer. The technician checks the glycol circulating pump and discovers a low suction pressure and a low discharge pressure. There is a restriction to flow somewhere in the glycol system. The technician is told that a glycol leak was repaired a few days back. Because there is no glycol available, only water is added back into the system to replace the glycol that leaked out. Evidently, ice has been forming inside the glycol circuit and restricting flow.

The technician recommends that the system be shut down to allow for the ice inside the circuit to melt. A glycol sample will be taken and tested for specific gravity. This will determine how much of the existing glycol solution needs to be drained and replaced with new glycol. When this is completed, the system should operate normally.

Service Ticket 8303

Customer Complaint: Ice Harvest Not Uniform

Equipment Type: Ice-Harvesting System

The ice-harvesting machine when working properly provides for uniform sheets of ice that drop from the coil into the water-filled storage tank. But now the ice is becoming too thick. To get the ice to drop off, the unit needs to be shut down until the sheet melts and falls. The technician checks the hot-gas solenoid valve. This valve is set on a timer to automatically direct hot gas from the compressor through the evaporator coil so that the ice melts and falls. The gas solenoid valve electromagnet solenoid coil is tested for electrical resistance and is found to have infinite resistance, which indicates that it needs to be replaced. The replacement is a fairly simple process of wiring in a new solenoid. The ice harvester is now working properly.

UNIT 83—REVIEW QUESTIONS

1. What are the advantages of thermal storage systems?
2. What common types of phase-change materials are used for thermal storage systems?
3. What is an advantage of ice-storage systems as compared to water-storage systems?
4. What is meant by a stratified thermal water-storage tank?
5. What is meant by charging storage?
6. What is meant by discharging storage?
7. Would a system ever be charged and discharged at the same time?
8. What is meant by live-load chilling?
9. What are the determining factors for when to charge and when discharge from thermal storage?
10. What is the difference between full and partial thermal storage systems?
11. What can happen if the ice sensor for an ice-on-coil system is faulty?
12. What type of regular maintenance is required for an ice-on-coil internal melt thermal energy system?
13. What percentage of ethylene glycol is commonly used for an ice-on-coil (external melt) system?
14. Why is clear ice formation on an ice-on-coil thermal storage system desired?
15. Why do spherical encapsulated ice containers have dimples on their exterior surface?
16. How does an encapsulated ice energy-storage system compare to an ice-on-coil energy storage system?
17. How is the ice harvested from the evaporator plates in an ice-harvesting system?
18. Of all the thermal storage systems for ice, which is the most complex?
19. What are the advantages of water for use in a thermal energy-storage system?
20. What melts the ice on an ice-on-coil external melt thermal energy storage system?

UNIT 84

Food Preservation

OBJECTIVES

After completing this unit, you will be able to:

1. explain how various refrigeration processes are used to preserve perishable food.
2. explain the purpose and methods of flash freezing.
3. list the design consideration for a long-term storage building.
4. compare short- and long-term storage of food products.

84.1 INTRODUCTION

The perishable food industry is one of the largest industries in the country. An industry of this size is extremely important, and proper refrigeration is an important factor in the success of this business.

Perishable foods can be classified as follows:

- Meats
- Poultry
- Seafood
- Fruits
- Vegetables
- Dairy products

Figure 84-1 Blueberries are a fruit, picked when ripe, and can be preserved temporarily with refrigeration; but for long-term storage, they are flash frozen.

84.2 PRESERVATION OF PERISHABLE FOODS

Perishable foods can be divided into three groups: animal, vegetable, and dairy products. Each group requires separate treatment to preserve the products and to keep them palatable. The storage of animal products requires the prevention of deterioration of the nonliving products. Fruits and vegetables, however, are as much alive while they are being transported as they were while growing and require an entirely different set of preservation conditions (Figure 84-1). The principal causes of food spoilage are as follows:

- **Microbiological** These include bacteria, molds, and fungi.
- **Enzymes** These are chemical in nature and do not deteriorate.
- **Oxidation changes** These are caused by atmospheric oxygen coming in contact with the food, producing discoloration and rancidity.
- **Surface dehydration** In freezing, this is called freezer burn.
- **Wilting** This applies to vegetables that lose their crispness.
- **Suffocation** Certain fresh vegetables must have air. When sealed in cellophane bags, the bags must have holes.

SERVICE TIP

It is very important that the equipment used to process, store, or serve food products be cleaned. Foodborne contaminants are a major health issue, and only stringent cleaning of equipment can control this problem.

84.3 MEATS, POULTRY, AND SEAFOOD

Although poultry is a meat, it is generally classified in a separate category by the United States Department of Agriculture (USDA). In regard to preservation, all of these products deteriorate through the action of bacteria. Sanitation is the most important factor in controlling bacteria. Air has many forms of bacteria present. One of the best ways of controlling bacterial infection is through the use of germicidal or ultraviolet lamps.

Meats

Common types of meat include beef, pork, and lamb. The method of handling, preservation, and storage will depend on the type of meat product. As an example, a side of beef is aged, a while a pork ham and bacon are normally smoked and sausage is cured in a dry room. The aging process for beef allows its natural enzymes to act as a tenderizer. This utilizes the good effects of the enzymes without the harmful effects of bacteria. Oxidation is detrimental to meats, causing undesirable appearance and deterioration of the flavor. Dehydration can be controlled to a large extent by maintaining high humidity in the storage room. High humidity also protects against moisture loss, which lowers the weight of the meat.

Pork should be rapidly cooled after it is cut. This prevents destructive enzymatic action that causes discoloration, rancidity, and poor flavor. The keeping qualities of a variety of dry sausages produced depend on curing ingredients, spices, and removal of moisture from the product by drying. Dry rooms are used to remove about 30 percent of the moisture, to a point where the sausage will keep for a long time virtually without refrigeration. This process is used as an alternative to the smoking process. The U.S. Department of Agriculture (USDA) requires that a dry room for sausage be maintained at temperatures above 45°F, and the length of time in the room depends on the diameter of the sausage after stuffing and method of preparation.

Poultry

Problems associated with the preservation of poultry are similar to those of meat in many respects, except that poultry spoils much faster. Poultry, however, can be precooled by the use of cold water without detrimental effects. This is a relatively simple and effective process and therefore quite generally used. Bacteria and enzyme action is useful only in preserving game birds, as such action has a tendency to enhance the "game flavor."

CAUTION

All meat products have some level of bacteria. However, poultry has a particularly high level of *Salmonella*. If you cut yourself while working on equipment that processes or stores any raw meat, especially poultry, you must take particular precautions to ensure that the cut does not become infected. If it does become infected, you should seek medical advice.

Seafood

This product is the most perishable of all the animal foods, yet there is a vast difference in the keeping quality of different kinds of fish. For example, swordfish can be kept refrigerated for 24 days and be in a more palatable condition than mackerel refrigerated for 24 hours. Commercial fish are usually refrigerated with ice.

84.4 FRUITS AND VEGETABLES

The unique situation with fruits and vegetables is that they are still alive after they are picked. They grow, breathe, and ripen. Most fruits and vegetables are picked in an unripe condition. The best tasting products are ripened before they are picked. The purpose of refrigeration is to slow down the ripening process so that these products can reach consumers before spoiling.

Vegetables quickly lose their vitamin content when surface drying takes place. It is interesting to note that products shipped from California to Chicago that have been properly iced after harvest will be fresher than produce supplied from Illinois farms and shipped to a Chicago market without being iced.

SERVICE TIP

All fruits and vegetables give off some quantity of moisture. This moisture release is called respiration. The ability to cool a product can be significantly affected by its rate of respiration due to the added latent load it represents.

Another way to improve the product when it reaches the user is to package it. This cuts down on surface drying. Packages are usually made of cellophane or some similar plastic product. These containers must have holes so that the product can breathe (exchange oxygen and CO_2). Otherwise, the product will die, and a dead product will spoil rapidly.

A number of products require special treatment—bananas, for example. These are picked green and must be ripened for marketing. Banana ripening is initiated by the introduction of ethylene gas. For this to be effective, banana rooms must be airtight. Refrigeration is provided by using a refrigerant other than ammonia, because leaks will damage the fruit. Banana rooms are cooled using 45°F–65°F air. Keeping a design temperature difference of 15°F and a refrigerant temperature of 40°F is considered good practice.

Freezing Vegetables

All prepared vegetables are precooked and cooled before freezing, as discussed in the previous section. Refrigeration is used for every step in the preservation process. This includes the raw product cooling and storage facility, cooling the product after blanching, freezing the product, and finally storing the product in a warehouse. Loads vary widely depending on the product.

In vegetable facilities that operate only for short periods at peak capacity—1,500–2,500 hr/yr—spare equipment cannot be economically justified, and good maintenance is important to avoid downtime losses.

Freezing Potatoes

Prepared potato products in various forms dominate the frozen ready-to-use vegetable group and are processed year round. Products include french fries, hash browns, twice-baked potatoes, potato skins, and boiled potatoes. French fries are probably the most popular.

Raw potatoes for fries are steam peeled and trimmed and then cut into desired shapes. The slivers are graded out for use as puffs, tots, and wedges. The fries are blanched and then partially dried and oil fried. They are frozen on a straight-belt freezer system with three separate conveyors for precooling and totally freezing the fries to 5°F–10°F. Sorting is done at 15°F and packaged in an air-conditioned area.

84.5 DAIRY PRODUCTS

Sanitation is extremely important in all stages of handling milk. The bacteria content of milk must be controlled. Mechanical refrigeration begins to cool it even during milking, from 90°F down to 50°F within the first hour and from 50°F down to 40°F within the next hour. As more milk is added, the blended liquid must not rise above 45°F. Limits are set for the number of bacteria (the bacteria count) for milk supplied by the producer.

Milk is stored in insulated or refrigerated silo-type tanks that maintain a 40°F temperature. After milk is pasteurized and homogenized, it is again cooled in a heat exchanger (a plate or tubular unit) to 40°F or lower and packaged.

Butter

Butter is manufactured from 30 to 40 percent cream obtained from the separation of warm, acidified milk. It is cooled to 46°F–55°F and then churned to remove excess water. Butter keeps better if stored in bulk. If kept for several months, the temperature should not be above 0°F and preferably below –20°F. For short periods, 32°F–40°F is satisfactory. If stored improperly, the quality of butter deteriorates from absorption of atmospheric odors, loss of weight through evaporation, surface oxidation, growth of microorganisms and resulting activity of enzymes, and low pH (high acid) of salted butter. Low temperatures; a clean environment; use of a good quality cream; avoidance of light, copper, and iron; and adjustment of the pH to 6.8 to 7.0 eliminates most of these problems.

Cheeses

Cheeses are refrigerated to prevent too rapid mold growth. The surface must be kept moist or the cheese will become hard and brittle. Moisture, meanwhile, facilitates mold. While some mold enhances the flavor, too much mold creates waste because it must be removed before sale. The ideal storage temperature for various types of cheese is in the range of 30°F to 34°F for natural cheeses and 45°F for processed cheeses. Maximum temperatures range from 45°F to 60°F for the natural cheeses, while the processed cheeses may be kept on open shelves at 75°F.

TECH TIP

Many cheeses contain active organisms that provide the cheese with its unique flavor. Improper storage of these cheeses can damage these organisms and the flavors that they provide the cheese. It is very important that the specific temperature and humidity range for a particular cheese be obtained from the customer before a storage system is established. Some cheeses have very broad temperature and humidity tolerances, while others are extremely narrow and sensitive. The active organisms in cheese can also continue to generate heat after the cheese is in the cooler. This contributes additional load and extends the time required to cool the product (pull-down time) to its final storage temperature.

Eggs

Eggs should be refrigerated at all stages of handling and storage. Shell eggs account for about 75 percent of all eggs used. Table 84-1 lists the temperatures, humidity, and time they can be stored.

Research has shown that microbial growth associated with *Salmonella* can be controlled by holding eggs at less than 40°F. This has led to major changes in storage and display areas not refrigerated or inefficiently refrigerated. All egg storage areas should maintain ambient temperatures of 45°F; however, with mechanized processing and packaging procedures that insulate the eggs within cartons, it may require up to 1 week of storage before the eggs reach the temperature of the storage room. If shipped earlier to sell a "fresh" product, eggs are only partially cooled. Methods are being developed to improve cooling in the processing plant.

Shipment in refrigerated trucks is mandatory. Problems that arise are associated with the truck design, manner of loading, size of the shipment, and distance.

84.6 FROZEN FOODS

The freezing of food, although highly product specific, is basically a time/temperature-related process of three phases: (1) cooling to freezing point, (2) changing the water in the product to ice, and (3) lowering the freezing temperature to optimum frozen storage temperature. Product differences as well as quality relate to the specific values and to the rates of these stages.

Table 84-1 Storage Period of Eggs at Various Temperatures and Humidity Conditions

Temperature (°F)	Relative Humidity (%)	Storage Period
51–60	75–80	2–3 weeks
45	75–80	2–4 weeks
29–31	85–92	5–6 months

The following factors are considered in selecting a freezing system for a specific product:

- Special handling requirements
- Capacity
- Freezing times
- Quality consideration
- Yield
- Appearance
- First cost
- Operating costs
- Automation
- Space availability
- Upstream/downstream processes

84.7 FLASH FREEZING

Quick freezing was first undertaken by Clarence Birdseye in 1924. This essential process produces small ice crystals that are less damaging to the product. Ideally, the ripe produce should be frozen immediately after harvest—the sooner the better. Small packages are better to freeze than large packages because the interior freezes more quickly.

Flash freezing is the term most commonly used today. Flash freezing generally denotes quick freezing at cryogenic temperatures (below −238°F). Many quick freezing applications such as blast freezers operate at low temperatures but still well above cryogenic. Even so, the term flash freezing is generally preferred to describe most quick freezing processes.

Small packages may be frozen on or between refrigerated plates or in a "blast" freezer. Foods are frozen at temperatures between −20°F and −5°F. Freezer burn should be avoided. This is the condition of surface oxidation that causes discoloration of the product. It is prevented by packaging in airtight containers or by waxing or glazing the product. Ice glazing is used to prevent surface drying of fish. Fruits are often glazed with a sugar syrup to prevent oxidation.

TECH TIP

As the water in products begins to freeze, it forms very long, sharp crystals. The slower the freezing process, the larger these crystals become. If the crystal size is large enough, it will puncture the cell walls of the product being frozen. When the product thaws and the ice crystals melt, fluids in the individual cells of the product can be lost. This loss of fluid results in a significant decrease in the quality of the product. To reduce this problem, many products are flash frozen. This process produces the smallest possible crystal form, thus reducing the loss of special product fluids and maintaining product quality.

Vegetables must be blanched before freezing. This consists of placing the product in boiling water or steam to kill bacteria and to stop enzyme action. Air is removed from citrus juice before freezing.

Commercial freezing systems can be divided into four groups:

- Air-blast freezers
- Contact freezers
- Immersion freezers
- Cryogenic freezers

Air-Blast Freezers

Air-blast freezing can best be described as a convection system, where cold air at high velocities is circulated over the product. The air removes heat from the product and releases it to an air refrigerant heat exchanger before circulation occurs again.

The air-blast freezer using a stationary tunnel produces satisfactory results for practically all products, in or out of packages. Products are placed in trays that are held in racks, placed so that air bypass is minimized. Air-blast freezers can also be mechanized to provide a continuous process. Conveyer-belt-type freezers use vertical airflow and greatly improve the contact between air and product.

An example of a single-conveyer-belt blast freezer for a blueberry-processing plant is shown in Figure 84-2. The blast freezer is fed a continuous supply of blueberries, as shown in Figure 84-3. The blueberries are first cleaned to remove twigs and stones in a float tank (Figure 84-4). After cleaning, the belt delivers the blueberries to the blast freezer (Figure 84-5).

This method is commonly used for freezing blueberries. Once frozen, blueberries should not stick together in clumps but be separate from one another. A frozen container of blueberries should rattle when shaken. The blueberries will enter the tunnel shortly after harvesting, so they will be laden with moisture. If frozen in a container, the blueberries will all stick together. On the conveyor, they are spread out to allow the low-temperature air to circulate up through them as they pass through the tunnel. After flash freezing, the blueberries are placed in large walk-in coolers for storage.

A slightly different method uses air to move the product rather than a belt. Solid particulate products such as

Figure 84-2 Inside one of the air-blast freezers at a blueberry-processing plant.

Figure 84-3 This blast freezer is a tunnel type that flash freezes the blueberries as they pass through the unit from one end to the other.

Figure 84-4 Twigs and stones are removed from the blueberries prior to freezing.

Figure 84-5 A series of conveyor belts continuously deliver blueberries to the blast freezer.

peas, sliced and diced carrots, and shredded potatoes or cheese are floated upward and through the freezer by streams of air. The product is frozen in 3–11 minutes by refrigerant temperatures of −40°F. These freezers are packaged, factory-assembled units.

Contact Freezers

Contact freezers are conduction-type freezers. Products are placed on or between horizontal or vertical refrigerated plates that provide efficient heat transfer and short freezing time. They can be arranged as racks with the product placed directly on top of the refrigerated plate. This method is sometimes used for seafood such as fillets of fish. Product that is 2 to 3 in thick such as this is used with contact freezer applications.

Immersion Freezers

Immersion freezing is the fastest method available for flash freezing. The product is passed through a shallow liquid nitrogen bath, which forms an instant crust, which in turn freeze-locks in the natural flavors and moisture of the product.

Some products, such as shrimp, are frozen by immersion in a boiling, highly purified refrigerant. The surface of a sticky or delicate product is "set" by this rapid freezing, reducing dehydration and improving the handling characteristics of the product. The product is then removed and the freezing process is completed in a mechanical freezer. In these systems, the refrigerant is recovered by condensing on the surface of refrigerated coils.

Cryogenic Freezers

Cryogenic freezers utilize both convection and/or conduction by exposing food to temperatures below −76°F in the presence of liquid nitrogen or liquid carbon dioxide refrigerants. Liquid nitrogen boils at −320°F, and carbon dioxide boils at approximately −110°F. The freezers may be cabinets, straight-belt freezers, spiral conveyors, or liquid immersion freezers. The boiling liquid comes in direct contact with the product. After use, the refrigerant is wasted to the atmosphere. Spiral freezers can be used for flash freezing on trays. Internal freezer temperatures can reach −150°F. While operating costs are high, the small initial investment makes this economical for certain foods, such as delicate and difficult to freeze products such as bakery products and entrees.

84.8 REFREEZING

When a vegetable product is in a frozen state, for all practical purposes it is considered "dead"; however, microbes and enzymes may remain there in an inactive state. When thawed, a large amount of water is present from the ruptured tissues to provide a favorable environment for the growth of microbes and deterioration due to the enzymes. All of these processes serve to lower the quality of the product and could continue after refreezing. The product should be heated sufficiently to kill these destructive agents before refreezing. Some canneries freeze products to prevent spoilage until they can schedule the final canning process.

84.9 FOOD PROCESSING

The production of precooked and prepared foods developed into an important industry during the last half of the twentieth century. These foods, which include ready-to-use foods for main dishes and meals, vegetables, and potato production, require refrigeration and air-conditioning facilities.

Main dishes, which constitute the largest group of products in this area, include complete dinners, lunches and breakfasts, soups/chowders, low-calorie/diet specialties, and ethnic meals. They are characterized by having a large number of ingredients, several unit operations, an assembly-type packaging line, and final refrigerating or freezing of individual packages or cartons. Production falls into the following operations in the processing plant:

1. Preparation, processing, and unit operations. This involves the initial preparation of all ingredients to be assembled, including refrigeration and/or freezing needs. These require specific attention to individual requirements for selecting processes, equipment, space, controls, and safeguards.
2. Assembly, filling, and packaging. This includes all handling of components for putting into containers or packages, packaging, and placing the containers and packages in refrigerators or freezers. It is considered good practice to air-condition filling and packaging bins to control bacteria and increase worker productivity.
3. Cooling, freezing, and casing. There is a constant effort to improve the economy and efficiency of production, often altering original design conditions. Space and equipment capacity allowances should be 25–50 percent for increasing requirements. Maintenance should include checking temperatures and other specific conditions of the particular product involved. Defrosting on conveyor systems should be checked when there are hang-ups or stoppages. When the position or place of packages is changed in the storage processes, the quality of the product should be closely monitored.
4. Finishing: storage and shipping. Infiltration and product pulldown loads occur when there is negative air pressure due to exhausting more air than is supplied by ventilation. This causes a serious load on the refrigeration, making it difficult to maintain proper storage temperatures.

(a)

(b)

Figure 84-6 (a) Equipment room with two screw-type ammonia compressors; (b) ammonia screw-type compressor suction line.

SERVICE TIP

Many plants have requirements for hairnets and protective clothing while working around food-processing equipment. You are required to wear the same type and level of personal hygienic materials as other workers. These items may include hairnets, beard nets, lab coats, and shoe covers. It is also important that you keep your tools very clean and disinfected.

84.10 REFRIGERATION EQUIPMENT AND LOADS

Records need to be kept of operating conditions to identify poor performance and to provide guidance for new systems. These records should show conditions for time of day, season, on/off shift production, evaporator temperature, and equipment type/function.

The refrigerant used for many large refrigeration systems is ammonia. To avoid the hazard of a potential ammonia spill to workers in the plant, glycol chillers are used by some plants to circulate propylene glycol to evaporators located in the production areas.

Reciprocating compressors are common, but for large loads, screw-type compressors will provide for greater capacity (Figure 84-6). Unlike many other common refrigerants, ammonia is lighter than air. Therefore, ammonia condensers will not have air purge lines on the top of the condenser. An air-cooled ammonia condenser and receiver are shown in Figure 84-7.

Figure 84-7 Ammonia air-cooled condenser and receiver.

Due to its toxicity, precaution must always be taken with ammonia systems. Ammonia-detection systems are available that will continuously monitor ammonia levels for an enclosed space or equipment room (Figure 84-8).

(a)

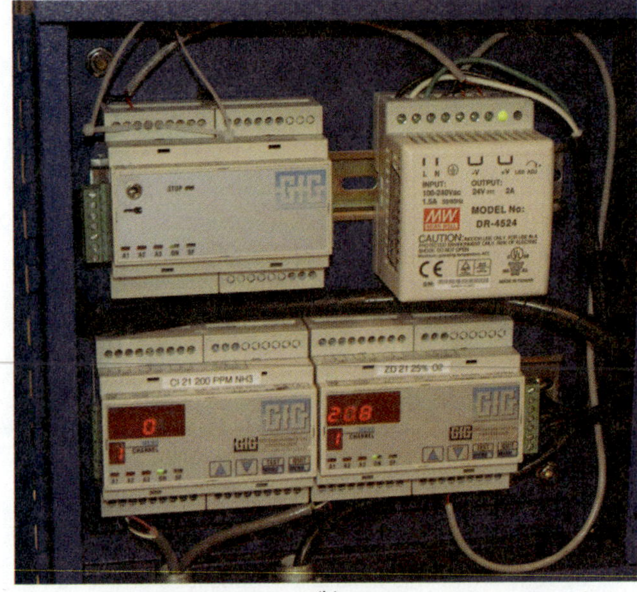

(b)

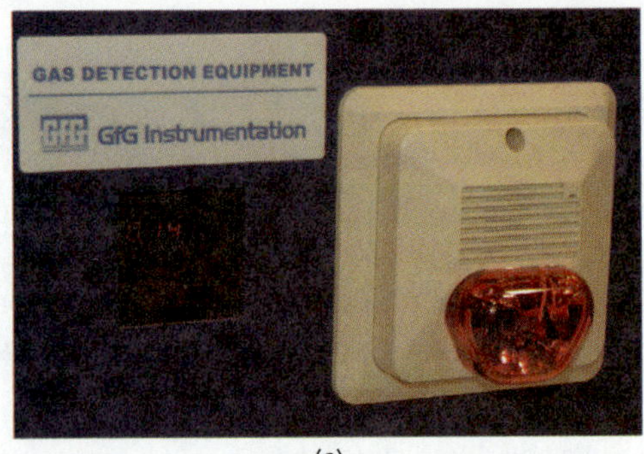

(c)

Figure 84-8 (a) Ammonia sensor; (b) ammonia and oxygen levels for the space; (c) alarm light.

84.11 LONG-TERM STORAGE

Freezing and thawing temperatures of animal, vegetable, dairy, and egg products vary widely. Temperatures must be maintained that preserve the quality and safety of products over the time periods required. As a result of ongoing research, lower temperatures are being recommended.

Most modern refrigerated warehouses are one-story structures. The building is designed for easy loading and unloading of tractor trailer trucks (Figure 84-9), and the inside design allows enough room to maneuver a forklift (Figure 84-10). Steel racks allow for the product to be stacked and yet still have enough air circulation to maintain the proper temperature (Figure 84-11). Even ice-making facilities will require warehouse storage (Figure 84-12). An ice and water slurry will be compressed into block ice and then bagged and stored (Figure 84-13).

Older storage buildings will have the evaporator coil located directly in the space (Figure 84-14). Today, it is usually more convenient to use penthouse refrigeration equipment rooms, as shown in (Figure 84-15). In this way, the air can be blown down into the space from the penthouse (Figure 84-16). The penthouse will contain all of the refrigeration equipment and an air handler.

Building Design Considerations

Building design consideration for long-term storage must be given to the following elements:

- Entering temperatures
- Duration of storage
- Required product temperature for maximum/minimum protection
- Uniformity of temperatures
- Air movement and ventilation
- Humidity
- Traffic into and out of storage space
- Sanitation
- Light

Figure 84-9 Loading dock for tractor trailer trucks.

Figure 84-10 There must be sufficient clearance for forklift operation.

Figure 84-11 The product is stacked on racks to allow for sufficient air flow.

Figure 84-12 Long-term storage for ice.

(a)

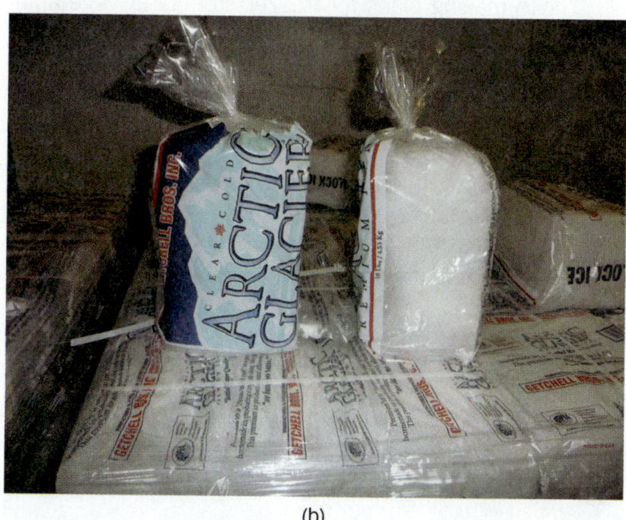

(b)

Figure 84-13 (a) Ice-bagging machine; (b) bags of crushed and block ice in storage.

(a)

(b)

Figure 84-14 (a) Older buildings may have the evaporator coil located in the storage space; (b) evaporator in the defrost mode with its panel closed.

Those making plans for freezer storage at vegetable-processing plants must also consider the following potential additional loads:

- Extra reserve capacity needed for product pulldown during peak processing
- Negative pressure that can increase infiltration by direct flow-through
- The process machinery load (particularly pneumatic conveyors) associated with repack operations

Figure 84-15 Penthouse located on top of the storage building.

Figure 84-16 Cold air is blown into the storage space by ducts connected to the penthouse, which is located above the building.

Regulations and guidelines for the refrigerated storage of foods have been established by the following agencies and should be familiar to the serviceperson working in these areas:

- The Association of Food and Drug Officials (AFDO)
- Occupational Safety and Health Act (OSHA)
- U.S. Department of Agriculture (USDA)
- U.S. Environmental Protection Agency (EPA)

UNIT 84—SUMMARY

In 2006, the United States consumed more heat-and-eat and thaw-and-eat food than ever in history. This trend was brought about by the advancement in food preservation, which has allowed these products to retain their appearance, flavor, and nutritional value. Without the skills and expertise of HVACR technicians, these advances would not have been possible. As more and better processes are developed, it is hoped one day that hunger worldwide could be eliminated. It is an exciting time and hopefully a profitable time to be involved in refrigeration service work.

WORK ORDERS

Service Ticket 8401

Customer Complaint: Milk Is Going Sour Too Quickly

Equipment Type: Walk-in Cooler

The technician arrives at the facility and finds that the temperature in the cooler is properly set, and the system temperature log indicates that the box is maintaining the temperature.

Upon investigation of the cooler, the technician finds that cases of eggs have been stacked too high and are blocking the airflow from the coil. This is allowing the milk, which is stored in the front of the cooler, to become too warm.

The technician shows the owner the problem, and the cases are moved. The box temperature shows an initial rise of a few degrees but quickly responds. The rise in box temperature was a result of the door being open and the full load of the box being seen by the coil. With the cases of eggs in front of the coil, the unit was really only cooling the small space behind the cases.

UNIT 84—REVIEW QUESTIONS

1. List the six categories of perishable foods.
2. What are the principal causes of spoilage in foods?
3. What is the most important factor in controlling bacteria?
4. What problems are associated with the preservation of poultry?
5. What is the most perishable of all the animal foods?
6. What is the purpose of refrigerating fruits and vegetables?
7. Describe the special treatment required for bananas.
8. Describe the special treatment required for milk.
9. What happens if butter is stored improperly?
10. What is the ideal temperature for storing cheese?
11. Using Table 84-1, determine the conditions needed to properly store eggs for 2 to 4 weeks.
12. What factors must be considered when selecting a freezing system for a specific product?
13. How can freezer burn be avoided?
14. What is the blanching process, and why should it be used before freezing vegetables?
15. Describe the process of air-blast freezing.
16. How are products frozen in conduction-type freezers?
17. Describe the cryogenic freezing process.
18. Describe the immersion process of freezing.
19. What should be done before refreezing a vegetable product?
20. What four operations are performed in a processing plant for precooked and prepared foods?
21. What operating conditions should be recorded for refrigeration loads?
22. What are some factors that must be considered when designing a refrigerated warehouse?
23. What organizations establish regulations and guidelines for the refrigerated storage of foods?

UNIT 85

Commercial Refrigeration Systems

OBJECTIVES

After completing this unit, you will be able to:

1. describe the differences between high-temperature, medium-temperature, and low-temperature refrigeration systems.
2. explain how a multiple-compressor system operates.
3. explain the operation of a defrost system for a walk-in freezer.
4. explain how to size a walk-in cooler or freezer.

85.1 INTRODUCTION

Commercial refrigeration systems serve a wide variety of purposes. The most common systems are those used for food-storage applications, such as a walk-in cooler at a grocery store or a food display case, and that is what is covered in this unit. However, with the exception of air conditioning, refrigeration is used for almost everything else that needs to be cooled. There are large commercial and industrial refrigeration systems that are designed for specialty applications such as making ice, quick freezing and storing blueberries, potatoes, fish, and pizza (see Unit 84). Refrigeration is also used in manufacturing, where it is referred to as process cooling.

There are many types of refrigeration systems, and each type is suitable for specific applications. The equipment design must ensure that the quality of the product being cooled remains satisfactory. In addition, a minimum amount of energy must be used to perform the operation. Finally, the process must be operated to comply with the laws relating to protecting the environment.

85.2 SYSTEMS

Refrigeration is divided into four broad areas based on the evaporator coil temperature. High-temperature refrigeration coil temperatures range from +20°F to +55°F. Medium-temperature refrigeration coil temperatures range from −10°F to +30°F. Low-temperature refrigeration coils range from −40°F to +10°F. Cryogenics involves a very-low-temperature process with temperatures of −250°F and below.

High-temperature refrigeration is used for product storage above freezing. When we think of high-temperature refrigeration, we usually think of food storage. Food items like milk, cheese, and fresh fruit and vegetables are stored at temperatures below 40°F and above 32°F. Storage temperatures within this range are also used for things like flowers or medicine.

Medium-temperature refrigeration is used for storing frozen foods. The most common example of medium-temperature refrigeration is a home refrigerator's freezer compartment.

Low-temperature refrigeration is used to quickly freeze and store foods. Temperatures below 0°F freeze foods quickly to preserve their quality. Foods at these temperatures can be stored for long periods of time as compared to foods stored in the medium-temperature range. The most common example of low-temperature refrigeration is a deep freeze.

Cryogenic temperatures are used for the preservation of materials for an extremely long period of time. Materials stored at these temperatures are almost suspended in time. Maintaining cryogenic temperatures requires a lot of refrigeration and is very expensive, so this type of storage is mainly used to store medical and scientific materials.

TECH TIP

Generally, the colder a food product is stored, the longer its shelf life. For example, milk stored at 40°F in a refrigerator may only last a few days before souring, as compared to milk stored at 32°F, which may last two or more weeks before souring. Even foods that are frozen will remain at a higher quality longer the lower the storage temperature. Food in a refrigerator/freezer can be stored and used longer when the temperature in the freezer is closest to 0°F. A deep-freeze temperature is below 0°F, and foods at these temperatures can be stored for much longer periods of time while maintaining their quality.

85.3 REFRIGERATION SYSTEM COMPONENTS

The essential components required for a basic refrigeration system are shown in Figure 85-1. They include the compressor, condenser, metering device, and evaporator. Also shown are some of the important accessories common to commercial refrigeration systems: a high/low pressure control, condenser fan, receiver tank, suction to liquid-line heat exchanger, filter drier, evaporator fan, thermostat, and suction-line accumulator. Each of these accessories helps maintain an efficient system.

Refrigeration systems require the addition of components not normally found on air-conditioning systems.

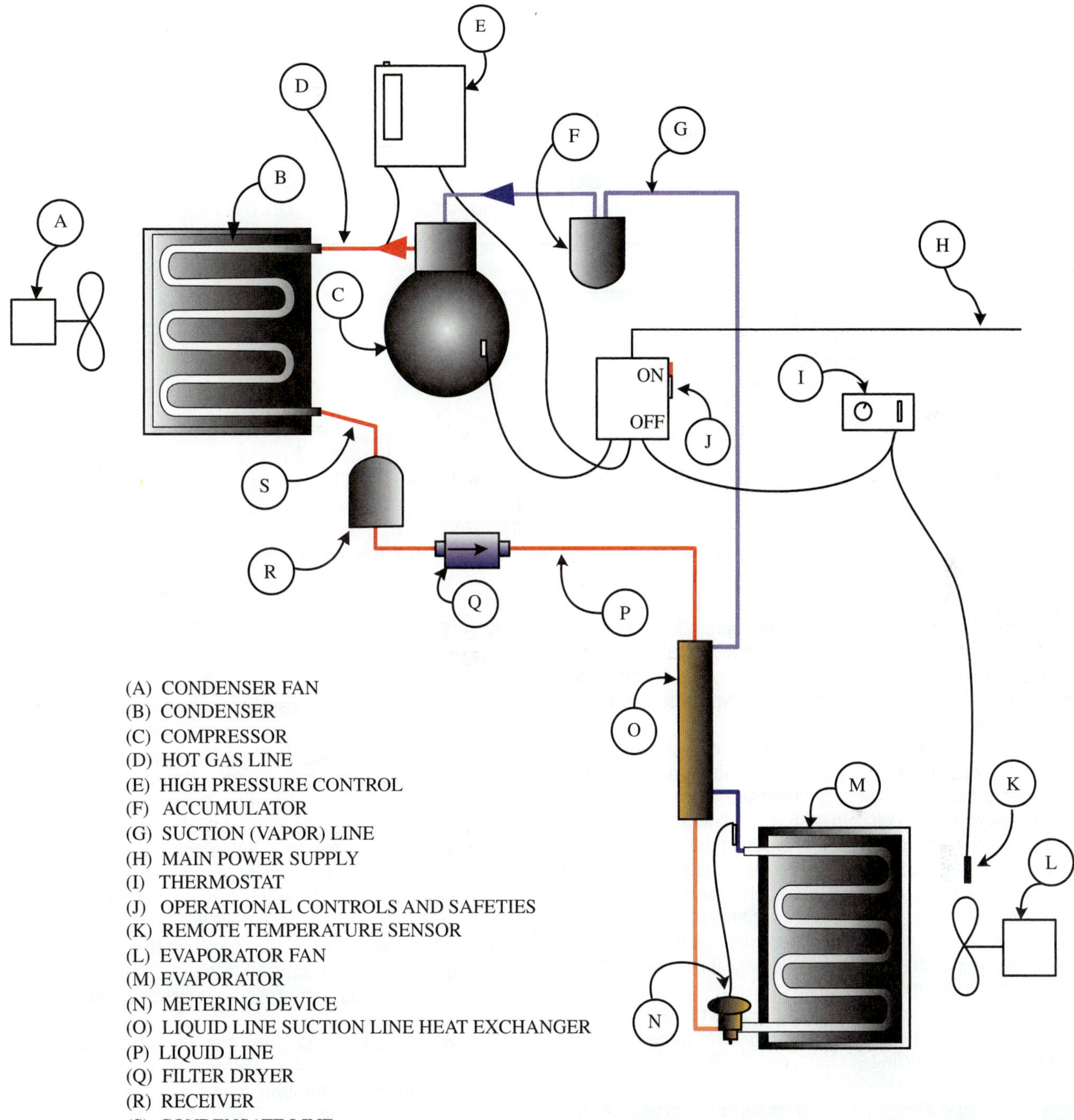

Figure 85-1 A simple refrigeration system with a single compressor and air-cooled condenser.

Each of these devices that are added will create resistance to the flow of liquid or vapor refrigerant. Such resistance reduces the overall operating efficiency of a system. For that reason, technicians must exercise good judgment when deciding whether to add additional accessories to an operating refrigeration system. They must weigh the advantages and necessity for such accessories to the proper operation of the system as compared to any detrimental effect it has on system capacity. This is of prime concern with systems that are marginally meeting current demand.

85.4 MULTIPLE-EVAPORATOR SYSTEMS

Multiple-evaporator systems are common in supermarkets. Multiple evaporators make it possible to use a single compressor to control a number of different case or fixture temperatures. Figure 85-2 shows a three-temperature system.

Evaporator pressure regulators (EPRs) are placed in the suction lines to the two higher temperature evaporators. These are adjusted to maintain the desired evaporator temperature.

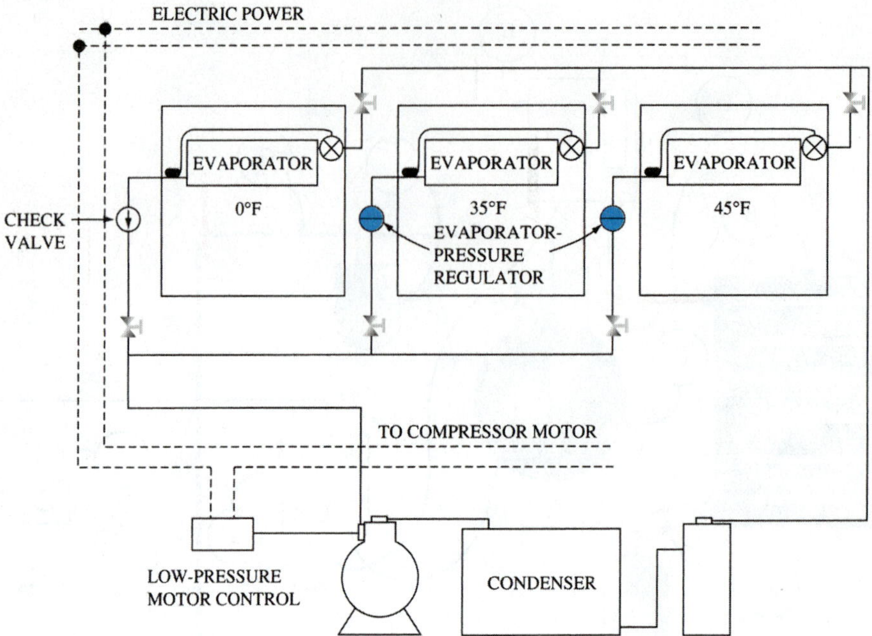

Figure 85-2 Multiple-evaporator system using evaporator pressure regulators to maintain different conditions in each box.

A check valve is installed in the suction line from the lowest-temperature coil. This prevents migration of the refrigerant from the higher-temperature coils to the low-temperature coil when the compressor is producing cooling for the higher-temperature coils.

When all of the coils require refrigeration, such as on start-up, the compressor will operate at the suction pressure required to cool the highest-temperature coil first, then the middle-temperature coil, and, last, the low-temperature coil. The compressor must be sized to produce the entire cooling load at the evaporator pressure of the lowest-temperature coil. The low-temperature coil will receive very little refrigeration until the higher-temperature coils are satisfied. For this reason, the load on the low-temperature coil must account for 60 percent of the total load, otherwise it may not receive adequate refrigeration.

On a small system it is general practice to install a surge tank in the suction line of the low-temperature evaporator to prevent short-cycling.

SERVICE TIP

Restaurants come and go, but the refrigeration equipment stays behind. Frequently, refrigeration equipment designed for the original restaurant may not meet the needs of the new operation. When the primary loads change, it is not possible simply to make adjustments to multiple-evaporator systems to accommodate these changes. It is not possible to simply turn down or up the low-temperature side of a system without adversely affecting the medium- or high-temperature side of a system. Restaurant managers are not always aware of this issue. It is in your best interest to take the time to explain to the manager what the difficulties can be for changing the load for refrigeration systems.

85.5 EVAPORATOR TEMPERATURE CONTROLS

Evaporator temperature can be controlled with a conventional thermostat, a suction pressure cutout control, or an evaporator pressure regulator. If the evaporator coil temperature is excessively low, the food stored in the space will be robbed of moisture and dry out. This is particularly true of vegetables and fruits. This moisture collects on the evaporator coil where it will freeze and lead to a buildup of ice.

An EPR maintains a steady refrigerant pressure in the evaporator coil. Since both liquid and vapor refrigerant fill the evaporator coil, the refrigerant is in its saturated state. For any given preset refrigerant pressure there will be a corresponding refrigerant saturation temperature. Regulating for a constant evaporator pressure provides a stable evaporator temperature.

Conventional thermostats and suction pressure cutout controls do not always maintain a constant evaporator pressure like an EPR. With these types of controls, the evaporator pressure is most likely to fall with a decrease in load. This can result in a lowering of the coil temperature to undesirable levels.

85.6 MULTIPLE-COMPRESSOR SYSTEM

Where the refrigeration load varies over a wide range, such as in supermarkets, it is desirable to use multiple compressors connected in parallel, as shown in Figure 85-3. Suction pressure control is used to turn on and off individual compressors as required to match the load (Figure 85-4). These controls also have the ability to change the lead compressor to obtain approximately equal running time on each compressor. A single condenser and receiver are used for all units.

The refrigerant piping must be done properly, as shown in Figures 85-5 and 85-6. Referring to Figure 85-5,

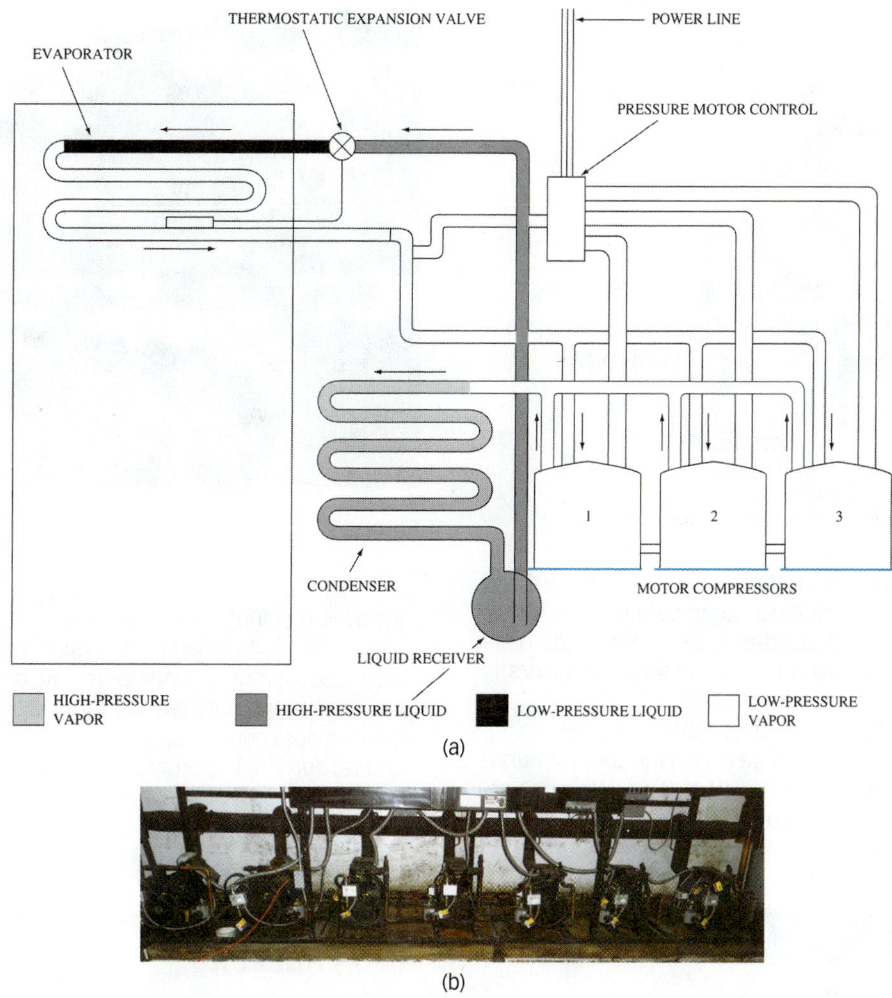

Figure 85-3 (a) Diagram of a multiple-compressor system using three compressors connected in parallel; (b) multiple-compressor rack.

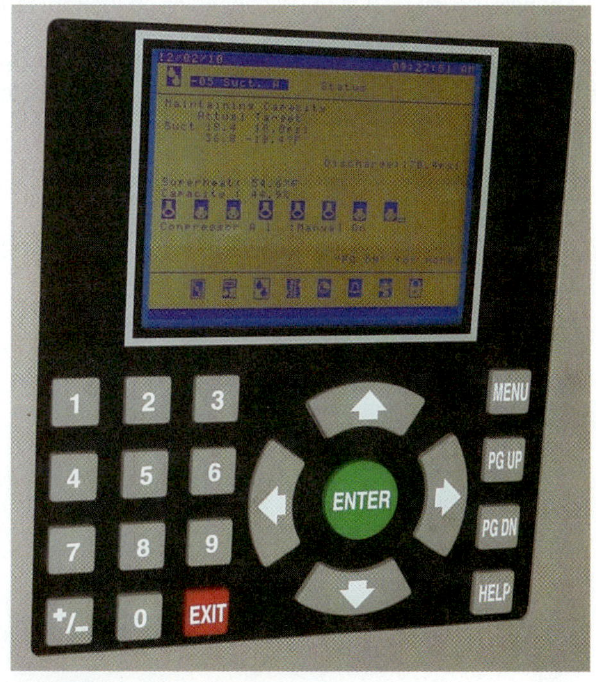

Figure 85-4 Multiple-compressor control panel.

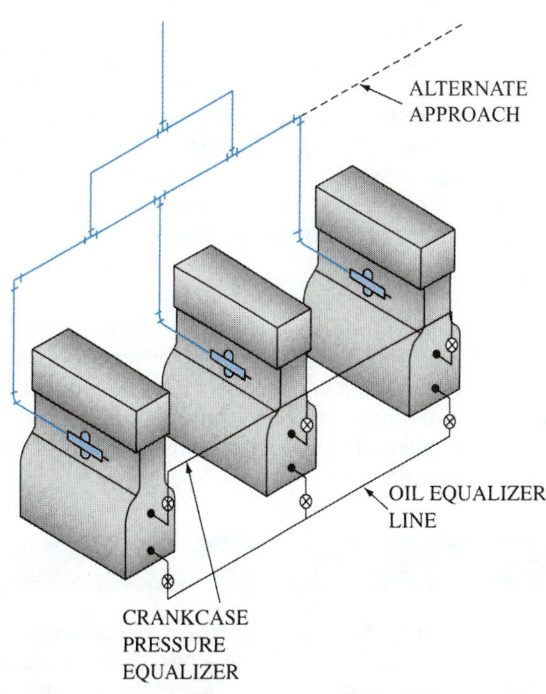

Figure 85-5 Suction-line piping for parallel compressors.

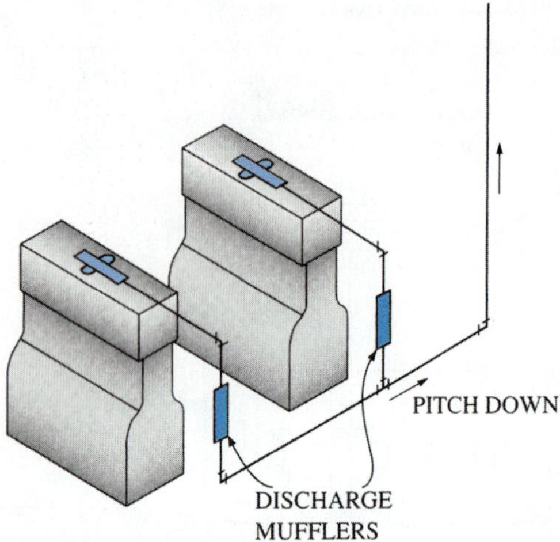

Figure 85-6 Discharge-line piping for parallel compressors.

the suction piping should be brought in above the level of the compressor. With multiple compressors, a common suction header should be used and the piping should be designed so that the oil return to several compressors is as nearly equal as possible.

If an oil-level control is not used, the discharge piping, as shown in Figure 85-6, should be used with the discharge piping running to a header near to floor level. With this arrangement, a discharge line trap is not required since the header serves this purpose.

CAUTION

Maintaining the proper oil level between multiple compressors is critical. If the compressor oil level drops, the system can be severely damaged or destroyed. Compressors have oil safety switches that are designed to trip and take the compressor offline when the oil pressure drops below a critical level. However, system owners and managers frequently see oil trips as a "nuisance" call and will push the reset button themselves. It is important that you stress to your customers that you must be notified each and every time a system shuts down as a result of an oil problem so that you are able to identify the root cause of the problem and correct it.

Pressure in the reservoir is reduced by boiling the refrigerant contained in the oil and relieving the pressure above the oil through a vent line to the suction header. An oil-level control meters the oil to the compressors equal to the pumping rate and thereby maintains the oil level specified by the manufacturer.

TROUBLESHOOTING

Oil is returned to a compressor as a result of the velocity of refrigerant vapor in the suction line. If an evaporator defrost system is not working correctly, then the evaporator will ice over, resulting in a slower flow of refrigerant vapor. This can result in oil slugging of the evaporator. If the oil level in an entire system appears to be low, do not simply add oil. Look for the primary cause of the oil shortage. Refrigerant oil does not simply evaporate. It stays in the system somewhere, and it is important that you locate it and determine what is necessary to have it return to the compressor during system operation.

Figure 85-7 Walk-in freezer.

85.7 WALK-IN REFRIGERATORS AND FREEZERS

Walk-in coolers and freezers are one of the most common applications for commercial refrigeration (Figure 85-7). Their cabinets are normally made with modular insulated metal-clad panels, usually 1–4 in thick, depending on the temperature requirement inside the cabinet. Using corner sections and standard wall and ceiling panels can make various-sized boxes, as shown in Figure 85-8. Panels can be assembled

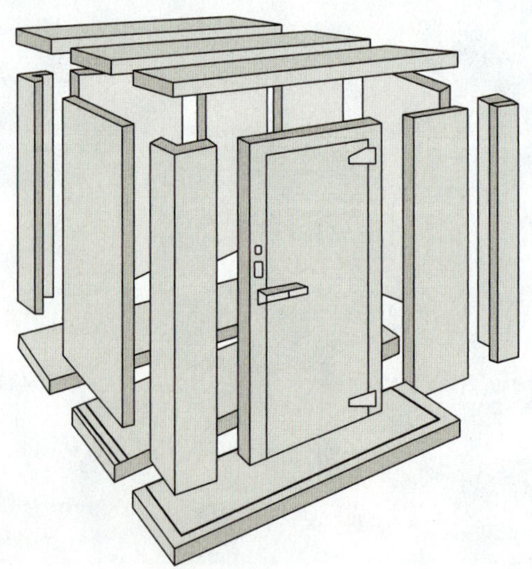

Figure 85-8 Modular walk-in cooler/freezer.

(a)

(b)

Figure 85-9 (a) Aluminum cabinet exterior; (b) stainless steel cabinet exterior.

on the job using special eccentric cam fasteners. Insulated panels are used for the floors. The interior surfaces of wall and ceiling panels are usually clad with aluminum and floors with galvanized iron. Stainless steel is also commonly used both outside and inside the cabinet (Figure 85-9).

Occasionally one structure will contain both a cooler and a freezer. Normally the freezer door is located in the cooler compartment. This makes the overall system more economical, since a common wall is used. Also it allows less heat to enter the freezer when its door is opened.

CAUTION

Local, state, and federal laws require that all walk-in refrigerators and coolers have safety latches that can be opened from the inside. It is a good practice when working alone to confirm that the latch is operating properly so that the door can be opened easily from the inside. There may also be a safety indicator visible from the outside (Figure 85-10).

Figure 85-10 Safety indicator light located on the outside of a walk-in freezer.

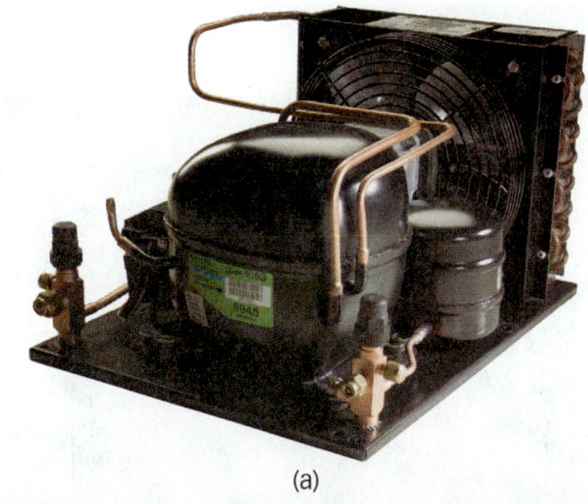
(a)

(b)

Figure 85-11 (a) Drop in unit sits on top of cabinet *(Courtesy of Danfoss)*; (b) controls located outside the walk-in for a refrigeration system located in an equipment room.

The refrigeration system for these cases can be either remote built-up systems or packaged systems that drop in place (Figure 85-11). Remote systems can have their condensing unit located in the building either near the walk-in unit or in an equipment room. The condensing unit can also be installed outdoors. When located outdoors, special controls are added to allow it to operate in low ambient temperatures.

Each time the door is opened, warm air enters the case. To prevent some of this warm air from entering the case, some systems will have a door switch,(Figure 85-12),

Figure 85-12 A door switch used on a walk-in freezer.

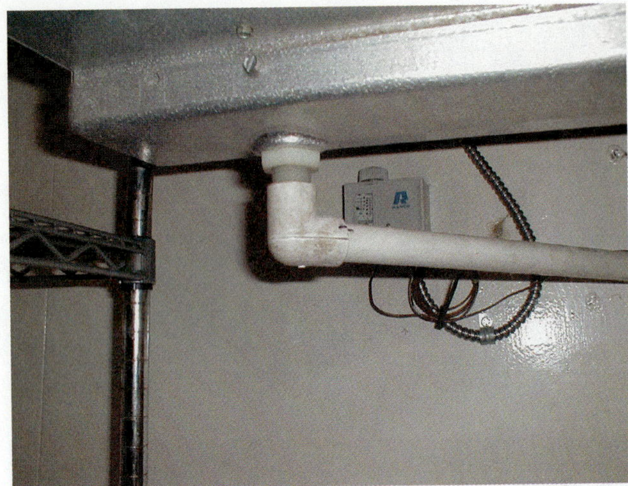

Figure 85-13 A condensate drain line connected to an evaporator.

which will shut off the fans when the door is opened. This door switch may also control a liquid-line solenoid, which stops the flow of refrigerant to the evaporator each time the door is opened. Some cases may also install a door curtain to help inhibit this warm air from entering the case.

Evaporator condensate lines normally connect to a drain outside of the cabinet. Slope the line ¼ in per foot for proper drainage. Install a trap so outside air does not enter the cabinet. Figure 85-13 shows a typical condensate installation. Heat the lines if they are subject to freezing temperatures, such as inside a freezer. Drain-line heaters inside the cabinet are always energized. Drain-line heaters outside the cabinet are thermostatically controlled and only energized when needed.

85.8 DEFROST SYSTEMS

Since most refrigeration systems operate at evaporating temperatures below 32°F, frost will accumulate on the evaporator's surface. This frost must be removed on a regular basis, otherwise it will begin to insulate the surface of the evaporator and prohibit the transmission of heat from the product to the refrigerant within the evaporator.

There are different methods by which an evaporator can be successfully defrosted. The method used will depend on the system application. For most medium-temperature applications, the air within the case can be used to successfully defrost an evaporator. At regular intervals, the compressor is shut down and the evaporator fan(s) will continue to draw air across the evaporator, warming the coil and melting any frost. The number of times the compressor is cycled off and the length of time the compressor stays off will vary. The condition of the environment surrounding the case will determine if more or less time is needed. If the case is located in a very humid location, it will require more time for defrosting.

For low-temperature applications, the case air is too cold and cannot be used to defrost the evaporator, so a supplemental heat source must be added. There are two methods commonly used. One method de-energizes the compressor and energizes a resistive-type heater, which is attached directly to the evaporator coil. The electric heaters stay energized and the compressor stays de-energized until the frost is removed from the coil. Another method is to introduce discharge vapor from the compressor directly into the evaporator and use this heat to melt the frost from the evaporator. The discharge vapor from the compressor will continue to directly enter the evaporator until the frost is melted.

When electric resistive-type heaters are used (Figure 85-14), they are mounted on the outer surface of the evaporator's coils. These resistive heaters are sized to provide

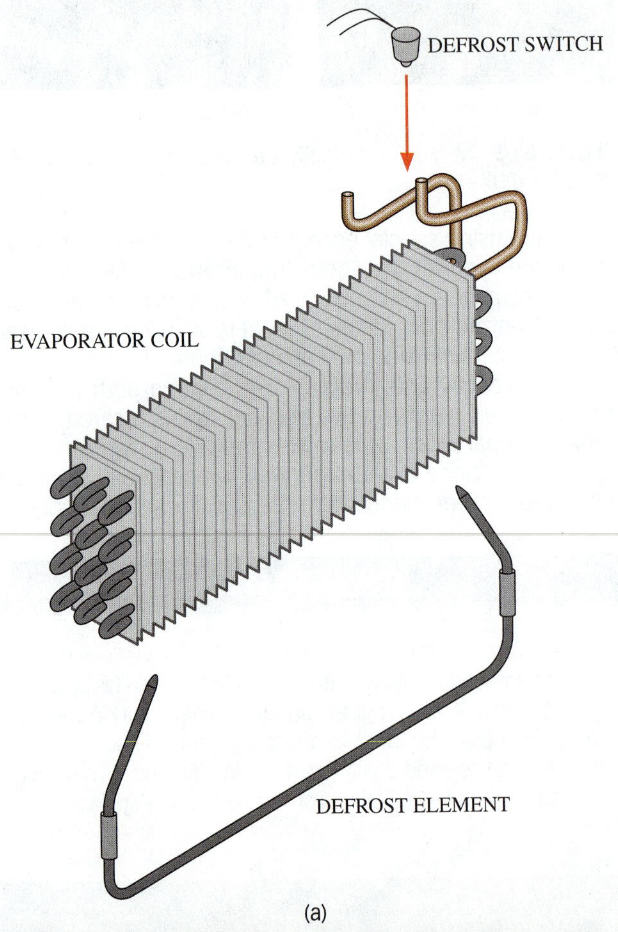

(a)

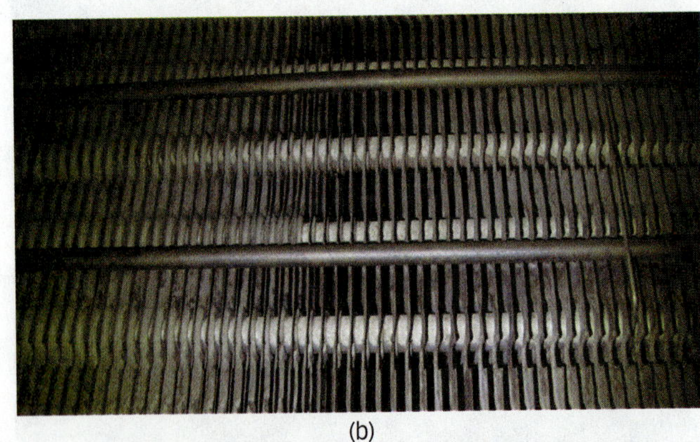

(b)

Figure 85-14 (a) Electric defrost heater diagram; (b) evaporator fins and electric heater strips; (c) (see next page) electric heater strips held in place by wire clips.

(c)

Figure 85-14 *(Continued)*

sufficient heat to effectively defrost the coil. Their capacity is normally rated in watts per foot. They are shaped to fit snugly onto the coil surface, creating an efficient heat transfer. Most heaters are manufactured for a specific coil, and when replacing these heaters it is best to obtain the original equipment manufacturer (OEM) replacement. There are universal defrost heaters available, but matching their wattage and shape may be very difficult at times.

A defrost timer is used to control the defrost operation. A typical mechanical defrost timer is shown in Figure 85-15. It initiates the defrost cycle, controls the operation of the compressor and defrost heaters, and is part of the defrost termination. Defrost timers can be adjusted to defrost once a day or several times a day. Screw-fastened pins are located at positions around the 24-hour clock dial. As shown in Figure 85-16, the pins are located at 6 AM and 6 PM. The inner knob adjusts the total defrost time anywhere from 2 to 110 minutes.

The number of defrosts per day depends on case location. Freezers are usually designed to defrost once or twice a day. The more humid and warm a location is, the more defrosts it will need. If a system needs to be defrosted more frequently, add only one additional defrost period at a time and monitor the results. Avoid adding too many defrost periods—it may not be beneficial to the system or to the customer.

Figure 85-15 A typical defrost timer for a walk-in freezer.

Figure 85-16 Mechanical defrost timer can be set for number of defrost cycles, time of cycle, and duration of cycle.

Figure 85-17 illustrates a common wiring diagram and the instructions for an electric defrost system. As the diagram indicates, the time motor is energized continuously. The normally closed contacts 2–4 of the defrost timer are wired in series with the compressor and the evaporator fan motor. The normally open contacts 1–3 are wired in series with the electric defrost heaters and the timer-release solenoid (TRS). The timer motor controls the operation of contacts 2–4 and 1–3. They work opposite each other. When contacts 2–4 are closed, 1–3 are opened. When contacts 2–4 are opened, 1–3 are closed. When the timer motor initiates a defrost, contacts 2–4 will open and 1–3 will close. This will stop the compressor and the evaporator fan motor and energize the defrost heaters (Figure 85-18).

A defrost cycle is typically terminated based on either coil temperature or low-side pressure. Most defrost timers will also have a failsafe time that can be set to terminate defrost based on time. On systems being terminated by a pressure or temperature switch, if the termination switch fails, the defrost will be terminated by time. The failsafe time should be set long enough to allow the system to terminate by the temperature or pressure switch and short enough to prevent the system from over defrosting or creating a hazardous condition. Usually the failsafe time is set to 35–45 minutes.

Terminating a defrost cycle by temperature is the most popular method. A temperature switch, such as one shown in Figure 85-19, is used to terminate a defrost cycle. It is installed on the evaporator at a location where the design engineers feel that frost will leave the coil last. The temperature switch may have a range and differential setting. The settings shown in Figure 85-20 are for a walk-in freezer that operates at 0 °F. The range and differential are set at approximately 30°F and 24°F, respectively. With these settings, the defrost cycle would terminate when the evaporator temperature at the switch location reaches approximately 30°F.

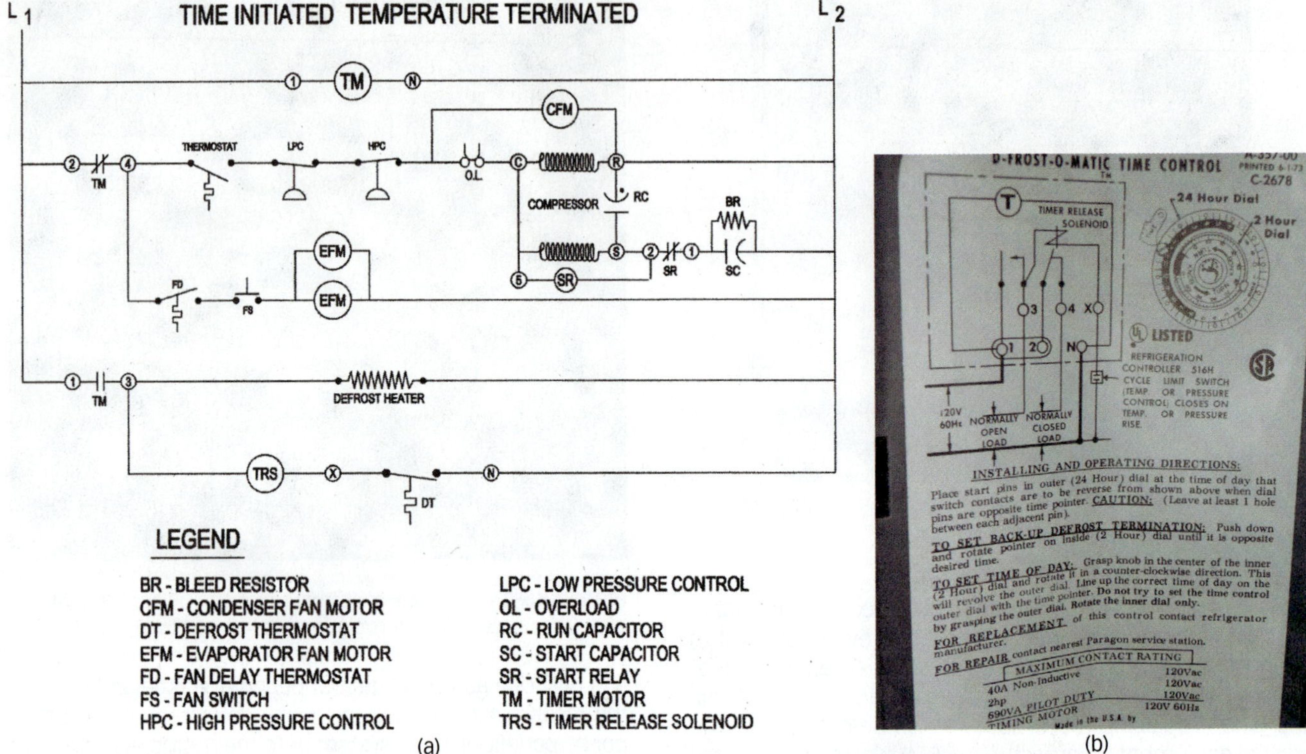

Figure 85-17 (a) A wire diagram for an electric defrost system; (b) instructions for setting the defrost timer.

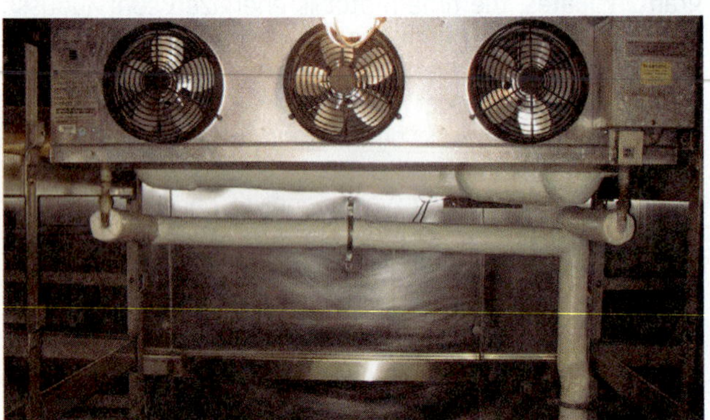

Figure 85-18 Evaporator fans stop during the defrost cycle.

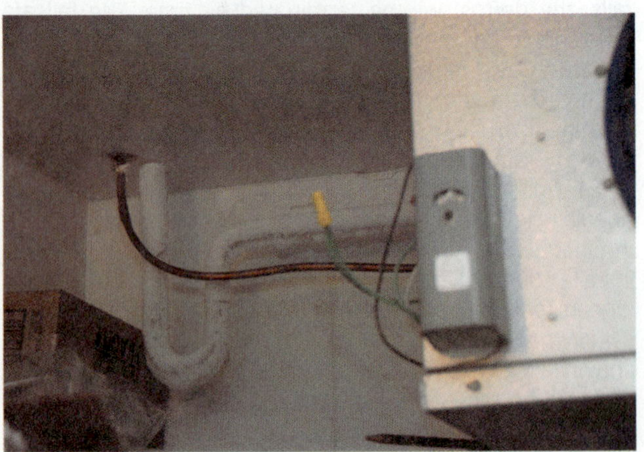

Figure 85-19 Adjustable temperature termination switch.

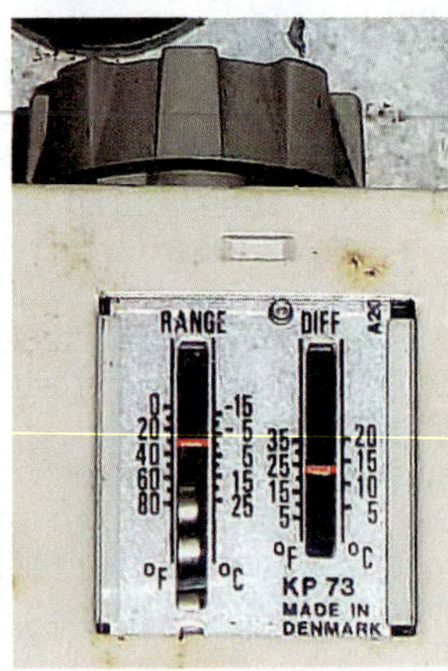

Figure 85-20 Range and differential settings for a temperature termination switch on a freezer.

At this specified temperature the defrost termination switch will close and energize the timer release solenoid (TRS), switching the system back into the refrigerating mode. The TRS is an electric solenoid located in the defrost timer. When the TRS is energized, it will mechanically switch the timer contacts: 2–4 will close and 1–3 will open. The defrost cycle will not initiate again until the temperature reaches the range setting minus the differential setting. In this case, this is 6°F: 30°F − 24°F = 6°F. Many temperature termination switches are not adjustable, and the temperature at which the switch closes may vary from design to design. It is best to check with the manufacturer of the system to determine the correct temperature setting. Some defrost termination switches will have the setting stamped onto the body of the device.

When the defrost cycle is terminated by pressure, a pressure control is used as the defrost termination switch. When the control senses a low-side pressure that will ensure that all of the ice is removed from the evaporator, it will close and terminate the defrost cycle. Many times the pressure control and the defrost timer are combined into one unit. The pressure at which the switch closes will depend on the type of refrigerant used in the system. Check with the manufacturer for the recommended pressure setting.

The defrost timer can also be used to terminate defrost by time, although this method is not very popular. The time required to defrost an evaporator will vary depending on how much frost has developed on the coil. The ambient humidity level and the usage of the case will be factors in how much frost develops on the coil. If the case is in a very humid location with heavy usage, a heavy accumulation of frost will develop on the coil. If the case is installed in a location where the humidity level is low and has very little usage, the frost that develops on the coil will be less for the same timespan.

Defrost systems can run into two problems: the system will be either over-defrosted or under-defrosted. Over-defrosting is when the defrost heaters stay energized too long, causing the case temperature to rise too high. The product may begin to melt and then refreeze. This problem can be identified by monitoring the box temperature during defrost or by examining the product in the case. Ice crystals forming on frozen product may be a sign of over-defrosting. Over-defrosting is normally caused by a malfunction in the termination of the defrost cycle. For time termination defrost systems, either the defrost time is set too long or the time clock is defective. On pressure-terminated systems, either the terminating pressure control is set too high, the pressure control is defective, or the solenoid coil in the defrost timer is incorrectly wired or defective. On temperature-terminated systems, either the temperature control is set too high, the control is defective, or the solenoid in the defrost timer is defective.

Systems that are under-defrosting will result in a frozen evaporator. The frost that normally develops on the evaporator coil will continue to build until the entire surface of the evaporator is iced over. This can be caused by a defective or incorrectly set time clock that does not initiate a defrost cycle. Or it can be caused by an open heater element or a defective defrost termination switch that continually terminates the defrost on each initiation of a new cycle.

Sizing Walk-in Coolers and Freezers

Sizing walk in coolers and freezers can be complicated because of the many factors involved. To simplify this process, there are many reference charts available that list the required refrigeration loads for different size walk-ins. Generally they provide two load estimates—one for average usage and one for heavy usage. It is important to remember that these load calculations are only estimates and the actual load requirements for a particular walk-in cooler or freezer may be different.

Heat-load calculations for any walk-in cooler or freezer are based on four separate loads:

- Heat transmission, which is the heat gained through the walls, floors, and ceiling of the walk-in structure
- Air infiltration, which is heat gained due to the air entering the structure
- Product load, which is the heat removed from a product to cool it from one temperature to a lower temperature; also includes the heat of respiration from fruits and vegetables
- Supplemental loads, which are made up of the heat dissipated by people or mechanical equipment in the walk-in

When sizing a walk-in cooler or freezer, a technician and/or salesperson must understand the application and environment in which the walk-in will be used. There are several situations where a person relying on the sizing chart alone could miscalculate the actual load for the walk-in.

If, for example, the walk-in were to be located outdoors rather than indoors, the heat gain through the walls would be much greater in the hot days of the summer than normally accounted for in many sizing charts. The size of the refrigeration equipment may need to be increased to remove this additional load. Two important questions a person sizing a walk-in cooler or freezer should ask are, Where is the walk-in going to be installed? and What will the maximum ambient temperature outside of the box be?

Other important questions to ask concern the product going into the walk-in: How much and what type of product will be stored? What is the initial temperature of the product entering the walk-in? At what temperature does it need to be stored? Again, most charts are based on averages. If the temperature of the product entering the walk-in is higher than normal or if an unusual amount of product will be stored in the walk-in, the capacity of the refrigeration equipment may need to be increased. It is not unusual for the load from the product to exceed all other loads added together.

Another potential problem with using generic sizing charts is with the supplemental loads. If the supplemental load is higher than the averages, then the capacity as listed on a generic chart may need to be increased. For example, if many people will be working in the walk-in cooler, or if some type of major equipment will be running in the walk-in on a consistent basis, the load may need to be increased to handle this additional load.

There are other instances that may cause the actual load requirements to differ from the load as shown on a basic chart. Any person sizing walk-in coolers and freezers should know the process of calculating the actual load on a walk-in box. This will allow them to determine if a particular application falls outside the loads as stated on the charts and if a complete load calculation needs to be done.

85.9 REFRIGERATION EQUIPMENT SELECTION

Many field-assembled refrigeration system installations require the installing contractor or sales engineer to select the proper system components. This involves selecting the correct condensing unit (compressor), unit cooler (evaporator), metering device, refrigerant, accessories, and piping. Before any of the system components can be selected, determine the required system BTU capacity by referring to a sizing chart or performing a heat-load calculation. Knowing the required BTU capacity for the project is the baseline for selecting the proper components for any installation. Many of the major system components are selected based on the BTU capacity of the system.

Normally, the first major system component selected is the condensing unit, or compressor if the system will have a remote condenser. The BTU capacity of any compressor or condensing unit is based on the suction pressure of the refrigerant entering the compressor. Because of this, most manufacturers will provide the BTU capacity of their compressors or condensing units at a specific entering suction pressure or, more correctly, at a specific saturation temperature.

Determine the refrigerant saturation temperature by using a standard pressure/temperature (PT) chart and convert the suction pressure to its equivalent saturation temperature. For example, if HFC-134a is the refrigerant and the entering suction pressure is 18.5 psig, the corresponding saturation temperature entering the compressor would be 20°F. Next determine the design evaporator temperature to determine the required saturation temperature entering a compressor. Many standard medium-temperature applications use an evaporator temperature of 20°F–25°F, and many low-temperature applications use an evaporating temperature of –10°F. Once the design evaporator temperature is determined, subtracting any suction line pressure losses will allow the approximated saturation temperature at the inlet of the compressor to be determined. For most applications, a pressure loss equivalent to a 2°F reduction in saturation temperature is used. For example, if the design evaporating temperature were 22°F, the assumed saturation temperature entering the compressor would be 20°F.

The next component selected is the unit cooler or evaporator. The unit cooler must closely match the design BTU capacity of the compressor or condensing unit. Manufacturers of unit coolers will provide their BTU capacity at a specific temperature difference (TD), which is the difference between the air entering the unit cooler and the evaporating temperature of the refrigerant leaving the unit cooler. The desired TD of the unit cooler is selected to match the humidity requirements of the product being stored. For storage of general packaged products at a relative humidity of 85 percent, a TD of 10°F is normally used. Other types of products may require different relative humidity conditions, which will change the design TD. See Unit 18 for more information on evaporator temperature drop and humidity control.

Once the unit cooler has been selected, the metering device is next. Most field-installed systems will use a thermostatic or electronic expansion valve. Capillary tubes and automatic expansion valves are normally used on self-contained systems that are completely assembled at the factory. The thermostatic expansion valve (TEV) is also selected based on the BTU capacity of the system. It is chosen to closely match the BTU capacity of both the condensing unit and unit cooler. The capacity of a TEV is based on several system characteristics. Always follow the selection guidelines of the TEV manufacturer for their recommendations. See Unit 17 for more information on TEV sizing.

Once all the major system components have been chosen, select the ACR tubing size. Size the tubing based on the BTU capacity of the system as well as the length of run. It is important to use correctly sized suction, discharge, and liquid lines on any application. Using an undersized suction line reduces overall system capacity. Using an oversized suction line leads to oil return problems. Always follow the equipment component manufacturer's guidelines for their recommendations. See Unit 26 for more information on refrigerant piping.

85.10 LOCATING AIR-COOLED CONDENSING UNITS

When installing walk-in coolers and freezers, the installing contractor will need to select the proper location of the condensing unit, such as shown in Figure 85-21. Location is an important consideration for the overall success of the installation. There are several items to consider when selecting the location of this type of condensing unit.

First, the location should be discussed with the customer. Be sure the customer is in agreement with the placement of the condensing unit. The customer may have a plan for the location different from where the installing contractor decides to place the unit. As retail space becomes more expensive, customers will want to take advantage of all the indoor space and may want the condensing unit located outdoors or in an indoor location that will not interfere with the retail operation.

If the condensing unit is to be placed indoors, make sure the location has adequate ventilation. Low-temperature condensing units will require approximately 200 CFM per 1,000 BTU, and medium-temperature units will require approximately 165 CFM per 1,000 BTU. Always check with the condensing unit manufacturer for ventilation requirements, as they may differ from these approximate values. Another consideration with air-cooled units located indoors is the heat they may add to the space. This additional heat may be objectionable to the equipment owner.

An alternative to locating the condensing unit indoors is to place it outdoors. When installing outdoors, there are several additional components that should be installed on the condensing unit. The condensing unit should have a crankcase heater installed. This will help to prevent refrigerant migration during the off cycle when the condensing unit is exposed to temperatures that are colder than the evaporator. The condensing unit will also need some means of keeping the high-side pressure above a minimum value during periods of low outdoor temperatures. This is normally accomplished with either a fan cycling control or a head pressure controller.

Figure 85-21 Air-cooled condensing unit.

Figure 85-22 Outdoor condensing unit with enclosure.

Figure 85-23 An installation of an evaporator coil in a walk-in cooler.

The method of controlling the case temperature may also need to be modified. The system should use a pump-down method to control the operation of the condensing unit. A standard air-sensing temperature controller should be used to control the operation of a liquid-line solenoid. When the temperature controller is satisfied, it will close the liquid-line solenoid, causing the refrigerant to be trapped in the condenser and receiver. This will cause the low-side pressure to drop. The system's low-pressure control will then shut down the condensing unit when the low-side pressure drops to an appropriate value.

The outdoor unit should also be covered to protect it from the outdoor environment, as shown in Figure 85-22. If the condensing unit is ordered for an outdoor application, it will normally come with some type of enclosure. When moving an indoor condensing unit to the outdoors, make sure an adequate enclosure is constructed to properly protect the unit.

85.11 INSTALLING EVAPORATOR COILS IN A WALK-IN COOLER OR FREEZER

When installing a walk-in cooler or freezer, the installation of the evaporator coil is vital to the proper operation of the entire refrigeration system. A typical installation of an evaporator coil is shown in Figure 85-23. There are several general guidelines that apply to most installations; however, it is always best to check with the manufacturer of the evaporator for specific instructions. Below is a list of several guidelines:

- The air pattern of the evaporator must envelop the entire case. This is important to provide a more uniform temperature throughout the entire box. Do not install the coil where product could potentially be stacked that would block the airflow. Always allow for sufficient space between the rear and sides of the evaporator to permit free return air.
- Never install an evaporator above a door. This can cause the evaporator to draw warm, humid air through the coil each time the door is opened. This will cause the evaporator to frost up more quickly and may cause an icing problem.
- Do not install an evaporator coil in an area where it will interfere with the cooler's aisles or storage racks. This may cause a problem for the storeowner, which could result in the evaporator needing to be relocated. The location of the evaporator should allow for easy storage and removal of the refrigerated product.
- The evaporator should be installed in a location that will provide the shortest possible distance between it and the condensing unit. Excessive piping can increase the pressure drop between the evaporator and the condensing unit. An excessive pressure drop will decrease the system's overall refrigeration capacity.
- The location of the condensate drains should provide for minimum pipe length. This will allow for proper drainage. In addition, the unit should be kept level for proper drainage, and it should be trapped. When draining two or more evaporators into a common drain, always trap drain lines individually to prevent vapor migration. On low-temperature evaporators, the traps should be located outside the case. If freezing of the traps is still a potential problem, they should be heated and wrapped.
- Supports for hanging the unit should be sufficient to hold its weight. For coils up to 250 lb, 5/16-in bolts or threaded rods should be sufficient. For coils up to 500 lb, 3/8-in bolts or threaded rods should be used. Coils over 500 lb should use 5/8-in bolts or threaded rods. When using rod hangers, allow adequate space between the top of the unit and the ceiling for cleaning. If fastening the evaporator flush to the top of the cooler, seal the joint between the top and ceiling with an approved sealant.

Other considerations include mounting of the TEV sensing bulb and the liquid-line solenoid thermostat sensing bulb. The expansion valve bulb will be attached at an appropriate location and in direct contact with the evaporator coil. The bulb will be wrapped with insulation. Figure 85-24 shows the pressure and bulb sensing lines for an externally equalized TEV.

The liquid-line solenoid thermostat sensing bulb is often installed alongside the sensing bulb used to display the walk-in temperature (Figure 85-25). They should be located

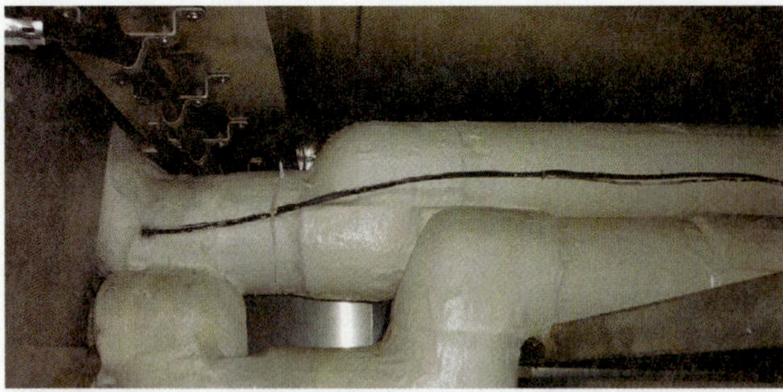

Figure 85-24 Pressure- and temperature-sensing lines for an externally equalized TEV.

Figure 85-25 Walk-in temperature display located outside the cabinet.

at a position in the walk-in to measure the average temperature of the cabinet. They should not be located near the door. It is good practice to shield the sensing bulbs so they do not accidently become damaged and to seal the points of entry into the cabinet (Figure 85-26).

Figure 85-26 Sensing bulbs should be protected from accidental damage and points of entry; sensing lines penetrating through the walk-in wall should be sealed.

UNIT 85—SUMMARY

There are many types of commercial refrigeration systems designed for meeting the requirements of specific applications to maintain the quality of the product at minimal expense while meeting current laws and regulations regarding the environment. One way to classify these different systems is to categorize them by the four broad based areas of high, medium, low, and cryogenic evaporator coil temperatures.

Single-component systems will cool just one space, while multiple-evaporator systems commonly found in supermarkets will utilize a single compressor to control the temperature for any number of different spaces. In many of these systems, regulator control valves for evaporator pressure are used to maintain the desired evaporator coil temperatures. Multiple-compressor arrangements can be utilized for conditions where the load may vary considerably. The compressors can cycle on and off as the load dictates.

WORK ORDERS

Service Ticket 8501

Customer Complaint: Ice Buildup on Evaporator

Equipment Type: Supermarket Walk-in Cooler

The service technician meets with the produce manager of the supermarket who describes a problem with one of the walk-in coolers. The evaporator coil exterior is covered in ice and the temperature of the space has begun to gradually increase. The technician asks what type of produce was being stored in the space, realizing that some foods give off considerable amounts of moisture that collects on the evaporator coil.

At the door to the walk-in cooler, the technician carefully checks the door seals to make sure that air is not being drawn into the space. The technician also asks the produce manager if the door has been left open for long periods of time. The technician then checks the automatic timer defrost controls, which are operating properly. The coils are normally defrosted by electric strip heater elements that are now completely covered in ice. The drain line out of the space from the drip pan is protected with electric heat tape that feels warm to the touch.

The technician then decides to check the refrigerant controls. The space has a thermostatic expansion valve, a box solenoid valve, and an evaporator pressure regulator (EPR). The box solenoid valve is energized and open as it should be due to the high space temperature. The technician then notices that the pressure gauge attached to the EPR is reading very low. Looking closer, the technician finds that the bypass valve around the evaporator pressure regulator has been left open. The technician closes the bypass valve and the evaporator coil pressure immediately begins to rise. Evidently, the last time the system was pumped down, the EPR bypass was left open.

Service Ticket 8502

Customer Complaint: Compressor Tripping Out

Equipment Type: Supermarket Multiple-Compressor System

The service technician meets with the produce manager of the supermarket, who describes the problem. One of two compressors has tripped out on low lube oil pressure, and its oil level appears low. The compressors are being operated in parallel to cool six different spaces. The one operating compressor is having difficulty maintaining the load by itself, and the space temperatures are beginning to slowly rise.

The first thing that the technician does is check to see if the compressors are equipped with oil separators on the discharge side. Upon investigation, the technician suspects that the float inside the oil separator on the idle compressor has failed and the oil is not returning to the crankcase. The technician will have to order a new separator and expects this will take a few days to arrive.

In the meantime, the technician needs to bring the space temperatures back down to normal. This is accomplished by bypassing the faulty oil separator and then adding oil to the compressor. However, now the compressor can no longer be operated in parallel without an oil separator. The technician changes over the valve configuration so that the two compressors are split. Each compressor will be lined up to separately cool three of the six spaces. Once the new separator arrives, the faulty one will be replaced and the system will be once again lined up for parallel operation.

Service Ticket 8503

Customer Complaint: System Is Not Defrosting

Equipment Type: Walk-in Freezer

A technician is called out to service a walk-in freezer operating at a temperature of 28°F and climbing. The walk-in freezer uses R-404a refrigerant, a semihermetic compressor, a thermostatic expansion valve, an air-cooled condenser, and a time-initiated, temperature-terminated defrost method.

When the technician arrives on the job, the evaporator's coils are completely covered in ice and the evaporator fans appear to be running normally. After further investigation, the technician observes the following conditions:

- An operating suction pressure of 20 psig
- An operating discharge pressure of 229 psig
- No bubbles in the liquid-line sight glass
- A 95°F air temperature leaving the condenser

A manual defrost is initiated and the following conditions are observed:

- The evaporator fans shut down.
- Current draw of 5 A is measured through the defrost heaters.
- Defrost heaters stay energized for 35 minutes and then the defrost cycle terminates.
- The evaporator fans turn back on after a delay.
- After running for 20 minutes, the case temperature starts to drop below 20°F.

Based on these observations, the technician determines that the defrost timer must be replaced.

UNIT 85—REVIEW QUESTIONS

1. What are three important items to consider in the design of commercial refrigeration systems?
2. Refrigeration is divided into four broad areas based on the evaporator coil temperature. What are these?
3. Food items like milk, cheese, and fresh fruit and vegetables are usually stored at what temperatures?
4. Medium-temperature refrigeration coil temperatures ranging from 32°F down to 0°F are generally used for what applications?
5. A deep freeze would be an example of what coil temperature application?
7. List the essential components required for a basic refrigeration system.
8. List the important accessories common to commercial refrigeration systems.
9. What type of system makes possible using a single compressor to control a number of different case or fixture temperatures?
10. What would be installed in the suction line from the lowest-temperature coil in a multiple-evaporator system?
11. List the three ways that evaporator temperature can be controlled.
12. What type of system would be used when the refrigeration load varies over a wide range, such as in a supermarket?
13. Explain a method used on walk-in coolers and freezers to help prevent warm air from entering the case when the door is opened.
14. Why do commercial refrigeration systems need a defrost system?
15. List the ways an evaporator can be defrosted.
16. What are the three methods by which a defrost cycle can be terminated?
17. List some of the loads that must be considered when sizing the BTU requirements of a walk-in cooler or freezer.
18. What are the four separate loads used to calculate the heat load for walk-in coolers or freezers?
19. What is the first step in selecting commercial refrigeration equipment for new installations?
20. What are the approximate ventilation requirements for air-cooled condensing units located in an indoor environment?
21. When locating an air-cooled condenser outdoors, why must the system be modified?
22. List at least three requirements when installing an evaporator coil in a walk-in cooler or freezer.

UNIT 86

Supermarket Equipment

OBJECTIVES

After completing this unit, you will be able to:

1. explain the difference between multiplex, distributed, and secondary-loop systems.
2. explain the purpose, arrangement, and components of compressor racks.
3. sketch a simple hot-gas defrost system layout.
4. troubleshoot open refrigerated display cases.
5. explain the difference between series, parallel, and hydronic heat-reclaim systems.
6. Describe how basic CO_2 cascade refrigeration systems operate.

86.1 INTRODUCTION

Energy-efficient designs are at the forefront of today's advanced supermarket refrigeration systems. Traditional multiplex direct-expansion (DX) systems require very large refrigerant charges and consume substantial amounts of electrical power. In addition, these systems require long lengths of refrigerant piping, which can lead to significant refrigerant losses. New designs such as distributed, secondary loop, low-charge multiplex, and advanced self-contained refrigeration systems are replacing the traditional multiplex DX systems. Many of these newer designs use less energy and have reduced amounts of refrigerant. New types of condenser and evaporator multichannel- and microplate-type heat exchangers are being designed to reduce the total refrigerant charge required by the system.

There is some movement toward using more nontraditional refrigerants such as CO_2. This is because many of the newer ozone-friendly refrigerants have relatively high global warming potentials. Refrigerant such as CO_2 has drawbacks because of the high operating pressures it requires, but CO_2 has the ability pound per pound to absorb a lot of heat and has no impact on the ozone. There are over 500 supermarkets in Europe that are currently using CO_2 refrigeration systems.

Waste-heat-recovery improvement is also one of the main priorities for new supermarket design. Heat-reclaim systems have been in use for some time now, but today's advanced systems are utilizing secondary loops and water-source heat pumps to gain maximum efficiency. The use of heat pumps allows for a large amount of the rejected heat to be reclaimed without increasing the discharge (head) pressure like traditional systems. Lower head pressure results in reduced power consumption. Advanced supermarket refrigeration systems with waste-heat recovery can substantially reduce the combined refrigeration and HVAC energy use and are becoming increasingly more commonplace. These systems are being designed to minimize the total equivalent warming impact.

86.2 "TRADITIONAL" SUPERMARKET LAYOUT

A typical supermarket layout is shown in Figure 86-1. The most common refrigeration system used in "traditional" supermarkets is the multiple DX (multiplex) system. The compressors are located in a machine room near the back of the store or on the roof. Heat rejection is usually accomplished with air-cooled condensers because they are the easiest to maintain. Refrigerant is piped from the discharge of the compressor to DX coils located within the cases and then back again (Figure 86-2).

Some supermarket systems will be designed to use a single compressor for each refrigerated case. Others are designed to use multiple refrigerated cases connected to a single compressor. Still other supermarkets may use a rack of multiple compressors (rack systems) piped together and connected to multiple refrigerated cases. This rack-type system is the most common configuration for multiplex DX systems.

Compressor Racks

Most often there are multiple racks that serve a number of cases that are operating at different temperatures. As an example, there may be a 15°F rack, a 20°F rack, and a –25°F rack (Figure 86-3). The use of individual racks allows the refrigerant temperature in the case evaporator coil to be close to the case operating temperature. This is accomplished with a rack suction pressure controller, which will indirectly control the refrigerant temperature for the rack.

Most racks utilize more than one compressor, usually three to five (Figure 86-4. There is an efficiency advantage to having multiple parallel compressors, sometimes of different capacities (called an uneven rack system). This is because supermarket systems must be designed for full-load operation even though the actual average load is generally below the peak load. This design peak load might only be reached on a few very hot days in the summer. Therefore, a rack with just one large compressor would be required to operate unloaded or cycle on and off frequently. Producing

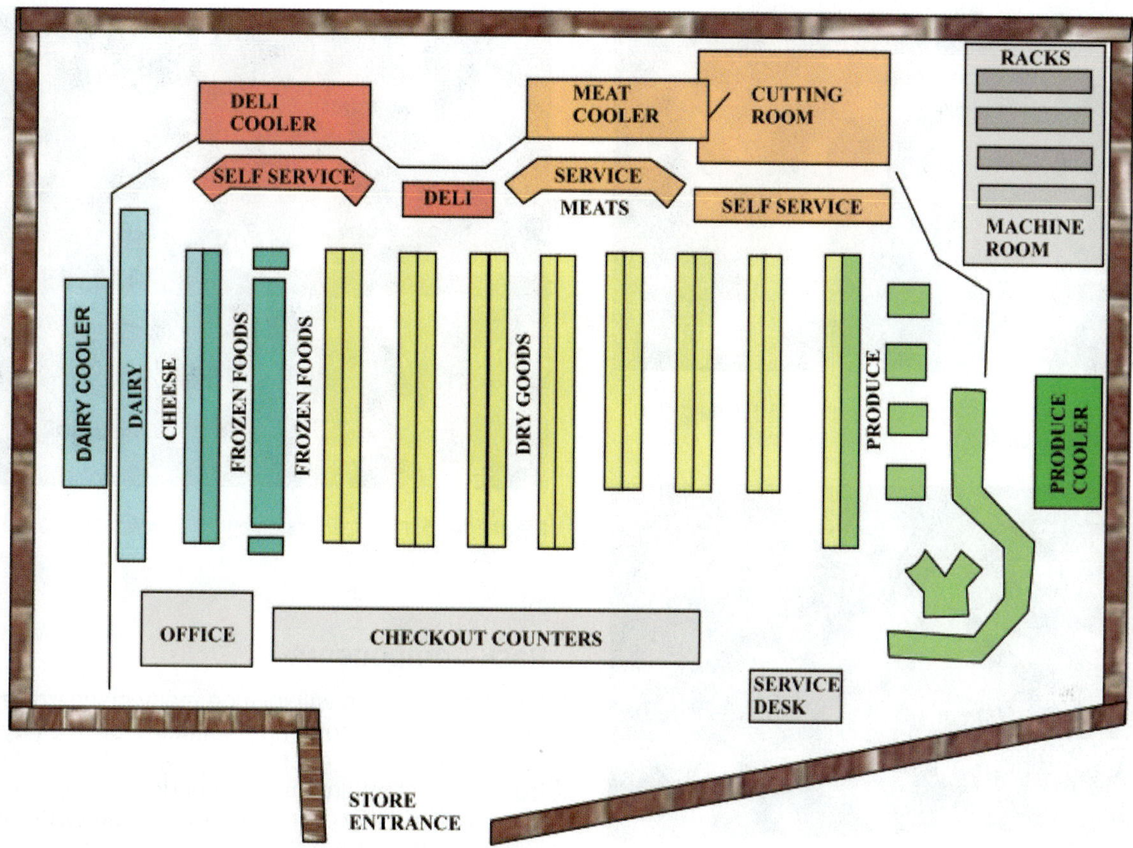

Figure 86-1 Typical supermarket layout.

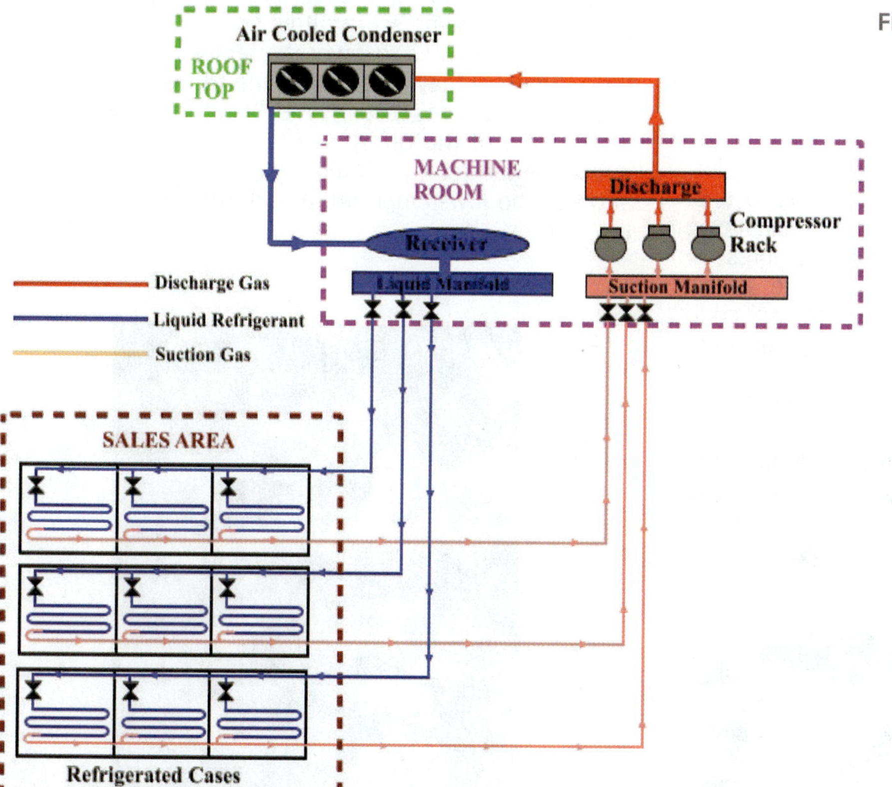

Figure 86-2 Multiple DX system (multiplex).

Figure 86-3 Compressor rack C operates at 15°F, and rack B can be seen directly behind it.

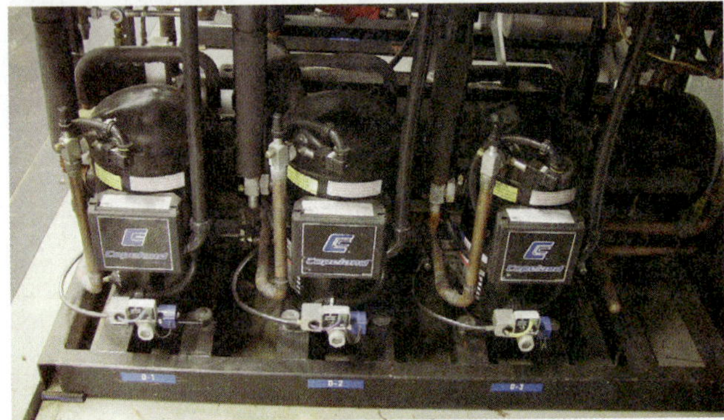

Figure 86-4 Rack with three scroll-type compressors.

Figure 86-5 Piping arrangement of a rack system to several refrigerated cases.

more capacity than needed wastes considerable costly energy, and frequent cycling on and off accelerates wear and shortens the useful lifetime of motors, contactors, and other components. A rack that utilizes multiple parallel compressors is better designed to match the load under varying conditions.

As an example, three 10-horsepower (hp) compressors used in an even parallel rack will provide the four capacities of 0, 10, 20, or 30 hp. An uneven parallel rack of one 10- and two 20-hp compressors will provide more combinations that include 0, 10, 20, 30, 40, or 50 hp. Four compressor racks can provide ten different capacities or more. These types of uneven rack systems rely on microprocessor controllers and control system software provided with temperature- and pressure-sensor information for accurate and repeatable control. The increased number of capacity steps better matches the load and increases system efficiency.

Piping from a rack system to the individual cases is carried in trenches underneath the floor or overhead using hangers. Piping must be properly insulated and isolated to prevent the possibility of electrolytic action, as shown in Figure 86-5.

TECH TIP

One disadvantage of parallel systems is that a single leak can shut down a sizable number of cases. When possible, a refrigerant monitor should be used to alert the owner or service company of a refrigerant leak.

Rack Components

Compressor racks will include many standard refrigeration system components depending on their type and design. Liquid-line driers are installed to remove water from the system and must be periodically replaced as a preventative maintenance item (Figure 86-6). Compressor oil is separated from the discharge gas and returned to the crankcase. An oil separator is shown in Figure 86-7, and an electronic oil level control is shown in Figure 86-8. The liquid receiver will have a level sensor (Figure 86-9). Compressor safety and cycling controls include high-pressure and low-pressure cutout switches (Figure 86-10). Subcoolers are used to lower the temperature of the liquid refrigerant leaving the machine room to reduce the amount of flash gas through the expansion valve located at the inlet of the refrigerated case evaporator coil. A subcooler using the high-temperature rack refrigerant to subcool the low-temperature rack refrigerant is shown in Figure 86-11.

Figure 86-6 Liquid-line drier located on compressor rack.

Figure 86-7 Oil separator.

(a)

(b)

Figure 86-8 (a) Electronic oil-level control. (b) Electronic oil-level control located on the compressor.

(a)

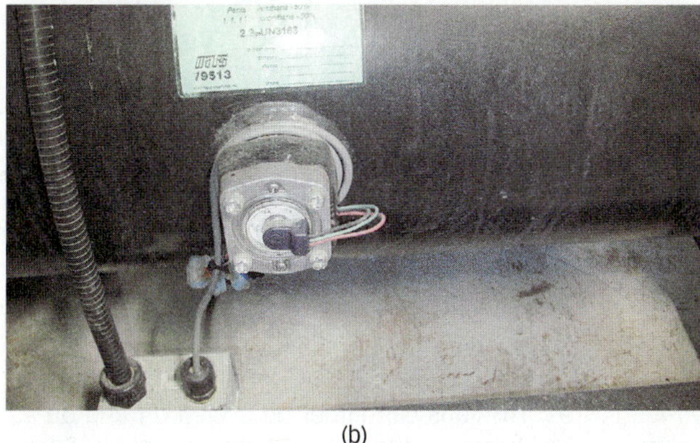

(b)

Figure 86-9 (a) Compressor rack liquid receiver. (b) Level indicator for liquid receiver.

Figure 86-10 Compressor rack high-pressure and low-pressure cutout switches.

Condensers

Air-cooled condensers are very common due to their low cost and maintenance. However, the condensing temperature is directly affected by the outside ambient air temperature and will vary considerably. Higher discharge pressures use more energy, and lower discharge pressures can affect the pressure drop across the expansion valve. Therefore, some type of condenser pressure controller will be required, such as cycling fans, condenser bypass, or some combination. Evaporative condensers are more efficient and will reduce condensing temperatures, but they are more difficult to maintain.

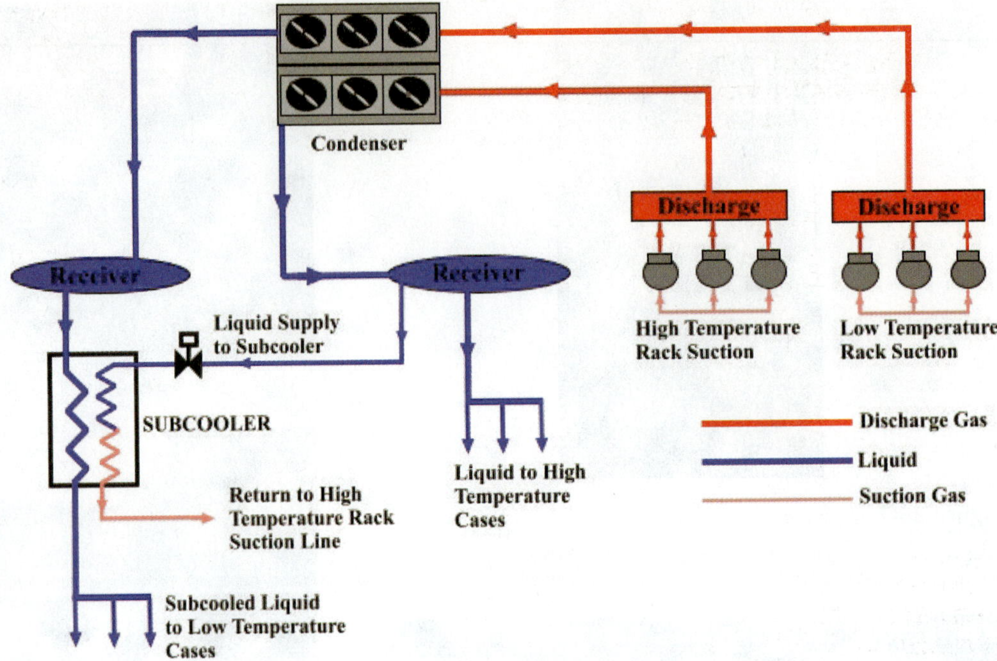

Figure 86-11 Subcooler arrangement.

Typically the operation of the fans can be controlled to maintain a minimum high-side pressure during low outdoor temperatures. The fans can either be cycled off and on as needed or the speed of the fans can be controlled. Some designs will use a split condenser, which is a condenser split into halves and piped in parallel. At a certain outdoor temperature, the cycling of the fans may not be sufficient to maintain a minimum high-side pressure. In this design, with the use of solenoid-controlled three-way valve and a check valve, half of the condenser can be shut off from the system (Figure 86-12). This reduces the condensing area and will cause the high-side pressure to increase.

Figure 86-12 A three-way valve used on a split condenser.

Direct Expansion Coils

The refrigerant must travel from where the compressors are located in the machine room throughout the store to the individual refrigerated cases. This results in the need for a very large amount of refrigerant, as much as 3,000 to 5,500 lbs. Because of the large quantity of refrigerant, it is not unusual for refrigerant leakage rates to be as high as 30 percent.

The case coil will have temperature and refrigerant superheat controllers. Older systems may have typical solenoid valves that open and close to control the case temperature and diaphragm-operated TEVs for superheat control. New multiplex systems will have electronic valves and controls.

Electronically Controlled Expansion Valves

Electronically controlled expansion valves can be pulse-width modulated, heat motor valves, or step motor valves. They do not rely on diaphragm operation like traditional expansion valves. A separate case controller that receives temperature, and pressure inputs from the coil will electronically control the valve.

The pulse-width modulated valve shown in Figure 86-13 is located inside the case under the meat trays (Figure 86-14). This type of valve opens and closes quickly to control refrigerant flow to the coil. Within a period of 6 seconds, a voltage signal from the controller will be transmitted to and removed from the valve coil. If there is a demand for refrigerant, the valve will remain open for almost all of the 6 seconds. If there is little demand, the valve will only stay open for a fraction of the 6-second period. This type of valve includes a tight shut-off function, which eliminates the need for an additional solenoid valve.

A step motor expansion valve will deliver more precise control and excellent repeatability. These valves are further

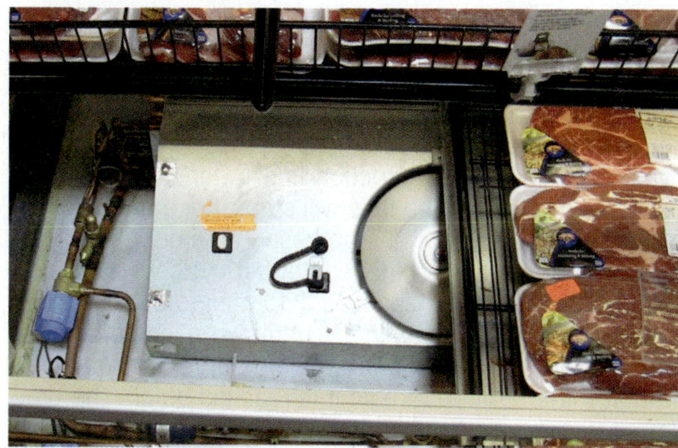

Figure 86-13 Pulse-width modulated expansion valve located in refrigerated meat case.

Figure 86-14 Expansion valve is located under the left side of the meat case.

explained in Unit 17, Metering Devices. Power pulses will rotate the stepper motor in either direction, which connected to the valve spindle will either open or close the valve. They can also be designed for a solenoid tight-shutoff function as well as bi-flow operation.

Case Controllers

Case control modules receive inputs from the coil sensors measuring temperatures and pressures. These electronic controllers are able to regulate all of the case functions, such as day and night thermostats, defrost, fan control, rail heat control, alarm functions, lighting, etc. They are designed to operate the case in the most efficient manner and can also provide signals to other controllers tied into a network.

Where electronically controlled valves are used, the typical case configuration is not very different from that found in a traditional refrigeration system. There will be a case temperature sensor acting as a thermostat to modulate the flow of refrigerant on or off. There will be a temperature sensor and a pressure sensor to measure the condition of the refrigerant leaving the coil for superheat control (Figure 86-15). These sensors send their information to the controller, which then regulates the valve accordingly.

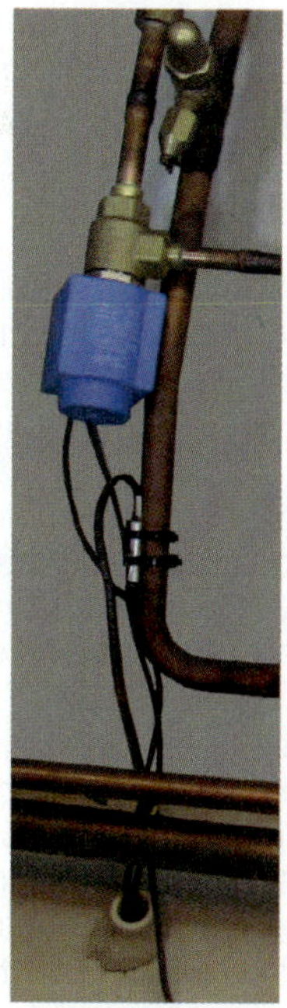

Figure 86-15 Valve sensors wired to case controller.

These inputs will also be used to determine if alarm conditions have been met. Some alarm limits include the following:

- Too low a temperature
- Too high a temperature during normal operation
- Too high a temperature during cool-down after:
  - Internal or external stop
  - Defrost cycle
  - Power failure
  - Appliance cleaning

System Controllers

System controllers can be networked to the individual case controllers for monitoring all functions at one convenient location (Figure 86-16). In addition, the refrigeration rack controls and condenser controls can be networked to the system controller. System controllers can also be used for the stores HVAC systems. This allows the entire refrigeration and HVAC system to be monitored and controlled

Figure 86-16 System controllers for both the refrigeration and HVAC systems located in the machine room.

through a supervisory system. System controllers can also be tied to external communications to allow off-site monitoring or control of the systems.

Defrost

There are a number of different defrost methods that can be used, such as natural defrost, electric heat, or hot gas. There are also different ways to begin a defrost cycle. There can be manual defrost, regular interval defrost, and an automatic (adaptive) defrost. A manual defrost is initiated when the controller is set to begin a defrost sequence. An interval defrost can be programmed to begin on a regular basis, such as every 8 hours. An automatic defrost occurs when sensors indicate there is a need to begin the defrost cycle.

Adaptive defrost systems initiate the defrost system based upon measured parameters rather than a timed cycle. As an example, if the registered air flow through the coil is reduced due to ice buildup, then a defrost cycle will be initiated. The case controller does not necessarily require an actual air flow measurement. Electronically controlled expansion valves can act as mass flow meters. This can be accomplished by comparing the energy admission on the refrigerant side of the valve to the energy admission on the air side of the valve.

TECH TIP

Electronic systems are "intelligent." Multiple inputs from a variety of sensors can be used to determine system operating conditions. Take your automobile as an example. There may be an indication of the transmission temperature, but there is no sensor that measures this directly. Other sensors that are measuring engine torque and speed are used to calculate what the expected transmission temperature should be.

A natural defrost cycle begins by shutting off the refrigerant flow and the coil is allowed to empty any remaining refrigerant. The fans remain on during the defrost cycle to help defrost the coil using air circulation. This method will take a considerable amount of time if there is substantial ice buildup. This leads to increased case temperatures during the defrost cycle

Electric heat is faster than natural defrost and is also fairly simple, but it will consume a considerable amount of power. During the defrost cycle, the refrigerant flow is stopped and the coil is allowed to empty any remaining refrigerant, just as in the natural defrost cycle. However, at this point the fans will be shut off and the electric strip heater will turn on. When the defrost cycle is concluded, the heater will turn off and the refrigerant will begin to flow again. The fans will only start after a brief delay so that any remaining drops of water will freeze on the coil rather than being splattered throughout the case.

Hot-gas defrost is more complex but it is fast and requires less overall energy. The cycle will progress in the same manner as the electric defrost, but instead of an electric heater turning on, a hot-gas valve will open to allow high pressure and temperature refrigerant gas to flow through the evaporator coil. The refrigerant flows backward through the coil and out through a check valve into the liquid line (Figure 86-17). Upon completion of the defrost cycle, a small drain line is used to equalize pressure between the evaporator coil and the suction line. After the refrigerant begins to flow again, the fans will start after a brief delay.

The defrost cycle will stop based on time, a temperature limit, or manually. A timed defrost cycle will last for a specific period of time such, as a 10-minute defrost cycle. A temperature defrost will last until a set temperature is reached. Timed and temperature defrosts can be combined so that if the temperature limit is reached before the time limit, the defrost cycle will end. Or, conversely, the defrost cycle can be set so that if the time limit is reached before the temperature limit, the defrost cycle will end.

86.3 SECONDARY-LOOP REFRIGERATION SYSTEMS

Secondary-loop systems reduce the total amount of refrigerant charge required. In traditional multiplex DX systems, a large amount of refrigerant is contained in both the condenser and the case evaporator coils. One type of secondary loop uses chilled fluid (often a propylene glycol solution) to cool the cases rather than a primary expanding refrigerant. Another type of loop circulates glycol through the condenser located next to the compressor rack to minimize the length of refrigerant piping. This condenser secondary loop can be integrated with a heat recovery system that utilizes air-source or water-source heat pumps. These two types of secondary loops can be used separately or combined. A case coil secondary loop combined with a condenser secondary loop would have the lowest total refrigerant charge.

- Case coil DX/air-cooled condenser (largest charge)
- Case coil DX/condenser secondary loop
- Case coil secondary loop/air-cooled condenser
- Case coil secondary loop/condenser secondary loop (least charge)

Case Coil Secondary Loops

For systems that use a case coil secondary loop, the chilled fluid such as glycol does not evaporate and change phase but instead is pumped through the cooling case coil as liquid (Figure 86-18a). The glycol is cooled when it returns from the case to the evaporator, which is located close to the compressor rack, which in this configuration is often referred to as a "chiller." The primary refrigerant is used for cooling the glycol in the chiller and is not delivered to the individual cases. This eliminates the need for long pipe runs of refrigerant as required in multiplex DX systems, which considerably reduces the amount of charge required. Long refrigerant runs of multiplex systems incur large pressure drops and result in high-suction-gas superheat. These conditions reduce the mass flow rate of the compressors, which results in loss of refrigeration capacity for multiplex systems. Even so, secondary loops also have their own drawbacks. The total energy demand for secondary-loop systems is slightly higher as compared to multiplex DX

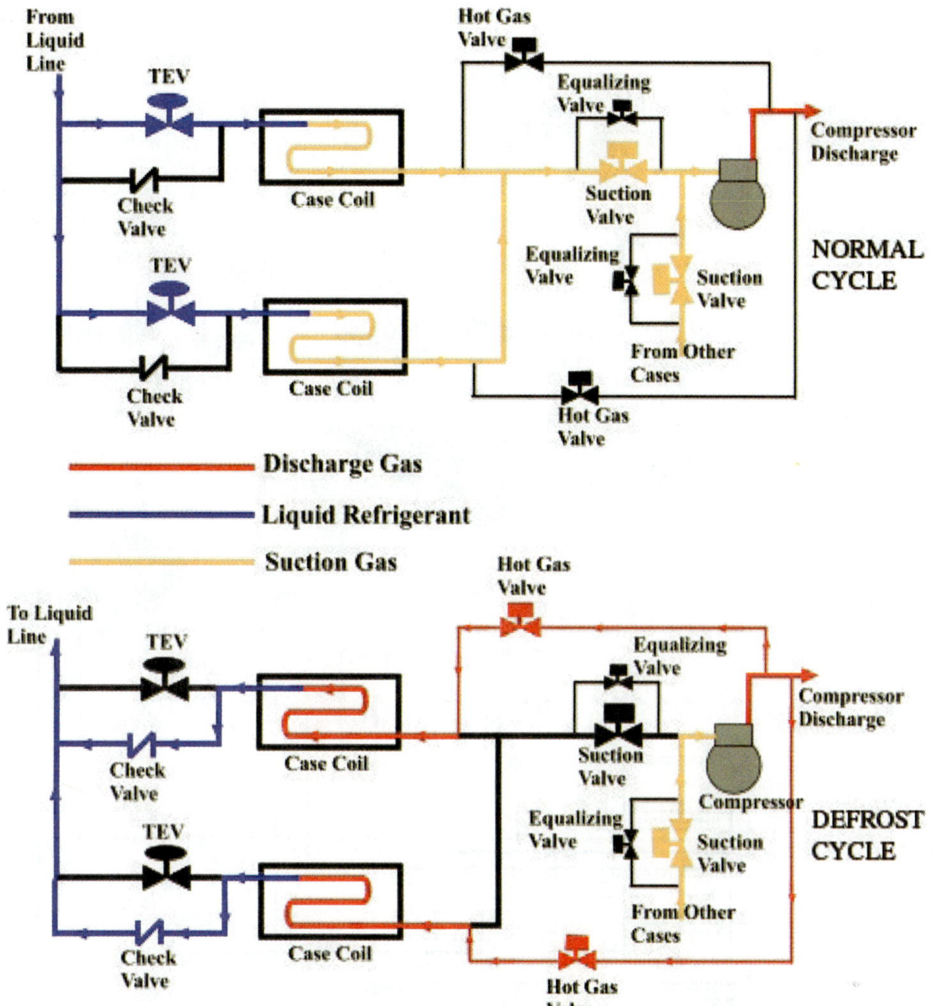

Figure 86-17 Hot-gas defrost arrangement for refrigerated case evaporator coil.

systems. This is partly due to the additional power required to pump the secondary refrigerant.

The cooling cases require coils and temperature control valves, and fluid flow is regulated to maintain the air temperature setpoint. Some secondary fluids such as propylene glycol can operate at temperatures below −20°F and are low enough in toxicity to be used in food applications. The piping used to circulate the glycol can be steel, copper, or high-density polyethylene. The glycol is dyed so that leaks can be easily detected.

Since the glycol remains a liquid, only sensible heat is transferred in the case coil, so it must enter the case at a low temperature. To keep the glycol temperature low enough for good heat transfer, the piping requires insulation. The typical glycol temperature change across the coil is 7°F to 10°F, and because of this the glycol flow rate is fairly high. This high flow rate, along with a higher viscosity of the glycol at low temperatures and almost continuous circulation, requires considerable pumping energy. A single stand-alone glycol-circulating pump must bypass excess fluid back to the suction side to avoid throttling losses at low loads. A better alternative is to incorporate multiple, parallel pumps that can be cycled on and off. This will better match the cooling load and use less energy. Pumps are added or subtracted based upon the pressure difference between the supply and return flow of the glycol.

There are generally at least two separate secondary loops and chillers. One for the low-temperature frozen food cases (see Figure 86-18b) and a second for the medium-temperature chilled food cases (see Figure 86-18c). However, depending on the store layout and case operating temperatures, more than two different temperature loops may be installed. This will raise the effective average evaporator temperature of the system and improve energy efficiency, although initial installation costs will be considerably higher.

Secondary-Loop Defrost

The defrost cycle for the case coil can be designed to utilize the glycol that has been heated rather than chilled. A heat exchanger located just after the receiver in the chiller liquid line will warm the glycol and also subcool the refrigerant. This is desirable because subcooling the refrigerant will increase the capacity of the system. A second heat exchanger located at the compressor discharge raises the glycol to a final temperature of approximately 65°F–80°F.

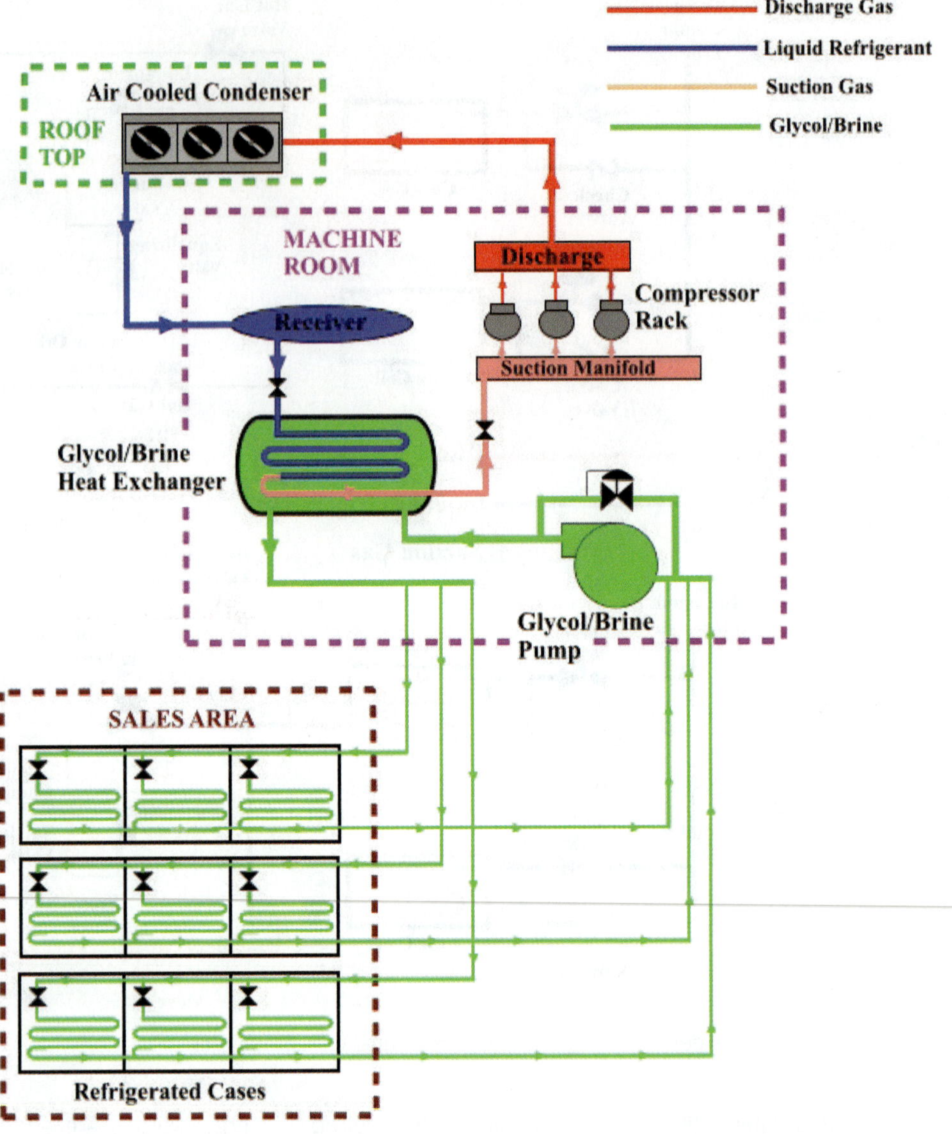

Figure 86-18a Secondary glycol/brine loop for a refrigerated case coil.

During defrost, the warm glycol flows to the case coil through a separate pipe used exclusively for this purpose. At this time, the chilled glycol supply to the coils is shut off to allow the warm glycol to pass through the case coil. The warm glycol leaving the coil is returned to the chiller located in the machine room through the common return pipe system. The warm glycol is cooled as it melts the frost and therefore returns to the system at close to its normal temperature, adding little excess heat. This defrost method is faster than electric defrost.

Condenser-Cooling Secondary Loops

Secondary loops can also be used for heat rejection from the rack condenser, which can be located in the machine room rather than on the roof (Figure 86-19). This eliminates the long refrigerant lines required to deliver the refrigerant to rooftop air-cooled condensers. The secondary loop can be designed to utilize cooling water that passes through the machine room condenser and pumped to a cooling water tower located on the roof. An advantage of this type of system is that the heat is rejected through both latent and sensible heat exchange in the cooling water tower, thereby increasing the efficiency. A disadvantage of this type of system is that some of the water will continually evaporate, leaving behind deposits that will need to be cleaned. In addition, this water will need to be continually replaced with makeup. Another problem is that this type of system is difficult to operate in cold climates because of the possibility of freeze-up.

Instead, a closed secondary loop can be designed so that a fluid with a lower freezing temperature, such as propylene glycol, can be used. Rather than using evaporative cooling, the glycol rejects only sensible heat to an air-cooled heat exchanger located on the roof. This can also be used in combination with a heat-reclaim coil to deliver heat back into the store during the colder winter months. Air-source, ground-source, and deep-well water-source heat pumps installed for the HVAC system can also be connected to the secondary loop for the most efficient heat reclaim.

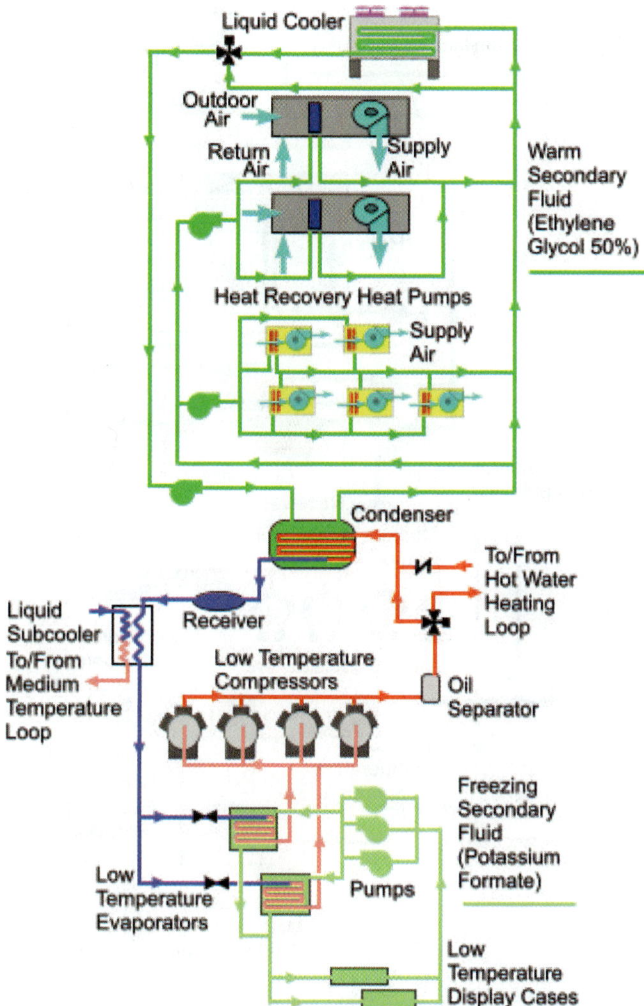

Figure 86-18b Configuration of the low-temperature secondary freezing zone.

Secondary-Loop Fluid

The loops that freeze at the lowest temperatures often use a water-based inhibited potassium formate salt (brine) solution, which can operate at temperatures down to –58°F. It is nontoxic and has a high specific heat for higher thermal efficiencies. It also has a fairly low viscosity at these reduced temperatures as compared to glycol. Because of this, it requires less energy to pump through the system. Potassium formate is compatible with many piping materials, such as aluminum, brass, bronze, stainless steel, cast iron, and copper, when it is not exposed to oxygen.

The two common glycols used for secondary loops are propylene glycol and ethylene glycol. Propylene glycol is nontoxic and therefore safer than ethylene glycol. It is rated for operating temperatures down to –20°F, with some solutions rated as low as –50°F, depending on the glycol concentration. Propylene glycol is used in the secondary loop for food display cases using medium-temperature refrigeration because it is nontoxic.

Ethylene glycol is the primary ingredient in automotive antifreeze and is toxic. It can operate at temperatures down to –40°F. Condenser-cooling secondary loops often use ethylene glycol because it has better heat-transfer properties as compared to propylene glycol. Both types of glycol are compatible with steel, copper, or high-density polyethylene piping.

86.4 DISTRIBUTED AND SELF-CONTAINED SYSTEMS

Distributed systems utilize DX case coils similar to the multiplex DX design. However, to minimize the long runs of refrigerant lines, the compressors are distributed throughout the store near the cases being cooled (Figure 86-20). This reduces the total refrigerant charge required and the potential for refrigerant leaks. A typical store might have as many as ten to fifteen distributed systems located throughout the entire store.

Each compressor cabinet in a distributed system is similar to a multiplex rack only smaller. Multiple compressors of several sizes, usually three to five compressors for each cabinet, are piped in parallel and are cycled to match refrigeration capacity and load. Because the compressors

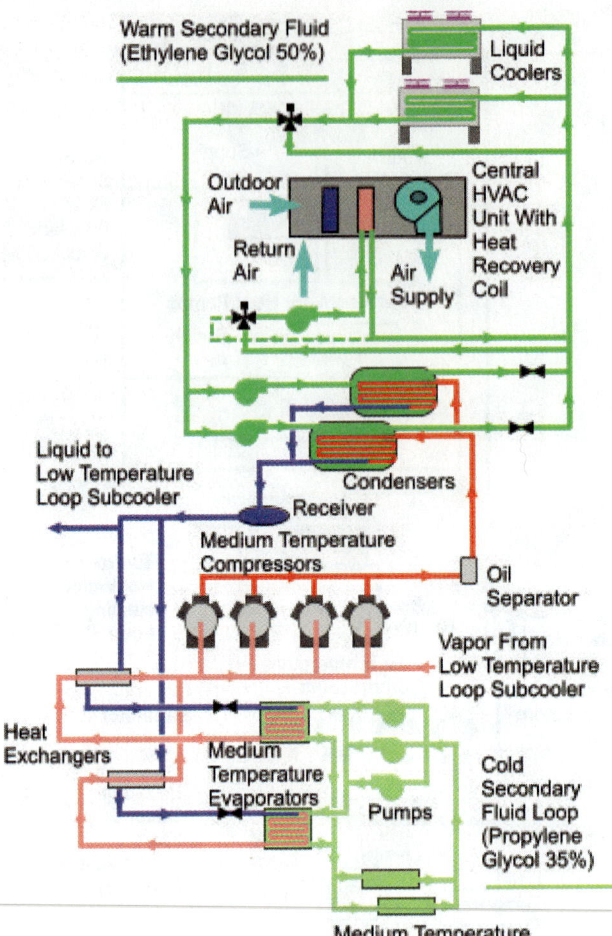

Figure 86-18c Configuration of the medium-temperature secondary fluid refrigeration zone.

Figure 86-19 Plate-type condenser located in the machinery room.

are located in the store sales area, scroll compressors that provide quiet operation and low vibration levels are most often used. A compact plate condenser is part of each package and is cooled from the closed-loop fluid cooler system similar to the condenser-cooling secondary loop described

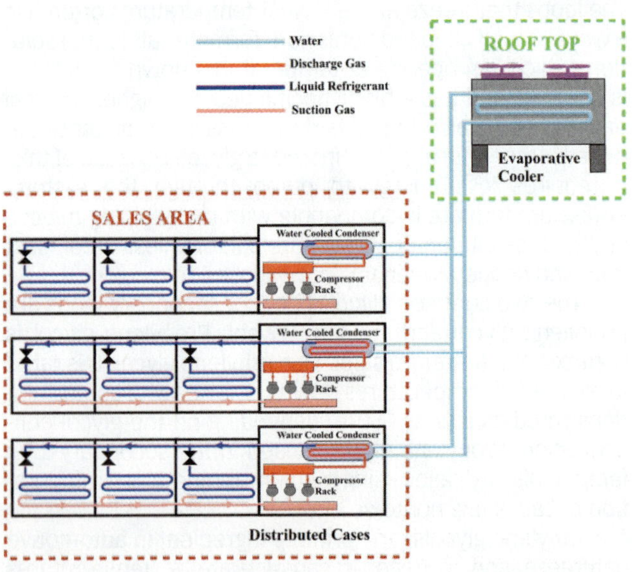

Figure 86-20 Distributed system layout.

Figure 86-21 An air-cooled condensing unit connected to a single refrigerated case.

in section 86.3. Glycol is circulated to each distributive unit plate condenser and its heat rejected to a rooftop air-cooled or ground-well heat exchanger.

Each unit includes an electronic controller that manages both compressor cycling and scheduled defrosts. Compressors are cycled to match the load requirements. A central pedestal-mounted power-distribution panel furnishes electrical power to each unit by means of a four-wire drop cord that supplies power for the compressors, fans, lights, and anti-sweat heaters.

Self-Contained Systems

Self-contained refrigeration systems are generally used in only a limited amount of display cases where the cases are inaccessible to the central refrigeration piping. An example is a refrigerated soda case located right next to the checkout lane. These units have their own air-cooled condensing units and reject heat directly into the store. An air-cooled self-contained unit used for cooling one individual case is shown in Figure 86-21.

86.5 VARIABLE-FREQUENCY DRIVES

Variable-frequency drives (VFD) to control compressor and fan motors are being used more frequently as a way to reduce energy costs. Most of these motors are three-phase AC synchronous induction motors. Their speed is directly related to the number of motor poles and the frequency of the applied current. Low frequency and therefore low motor speed at start-up reduces the high current surge encountered with conventional motors. The frequency can then also be adjusted to operate the motor at a speed that is best matched for the demand.

With conventional rack systems, parallel compressors are turned on and off by a control system, and this cycling leads to a high current inrush each time a compressor starts. This starting and stopping will also lead to wear and tear on the compressors. In comparison, rack systems that use a VFD for motor speed control can adjust the compressor capacity throughout a wide range of demand, which limits cycling, decreases power consumption, and increases the useful life of the compressor. This same application can be applied to air-cooled condenser and cooling tower fans and also for case coil cooling fans.

86.6 OPEN DISPLAY CASES

To maintain a more appealing shopping environment, many supermarkets will display some of their refrigerated merchandise in open display cases (Figure 86-22). This is due to the belief that open display cases allow the product to be more sellable. However, current studies have shown that this may not necessarily be true, and new supermarket design is moving toward more energy-efficient closed display cases (Figure 86-23). This is partly because open display cases can absorb heat from the inside of the store, which adds to the total refrigeration load and increased energy consumption. Because of this, they have a considerable impact on the inside store temperature, which is kept relatively low. This low inside temperature may prove uncomfortable for customers and also places additional demands on the HVAC system.

Open display cases are designed and engineered to hold product at the temperature at which it enters the case (Figure 86-24a,b). They are not designed to reduce a product's temperature. They are designed to primarily handle heat leakage into the case. They also are not designed to

Figure 86-22 A typical multideck open meat case.

Figure 86-23 A closed meat case.

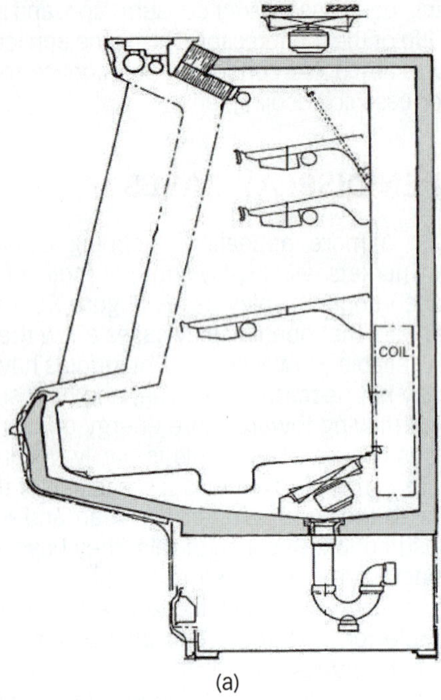

(a)

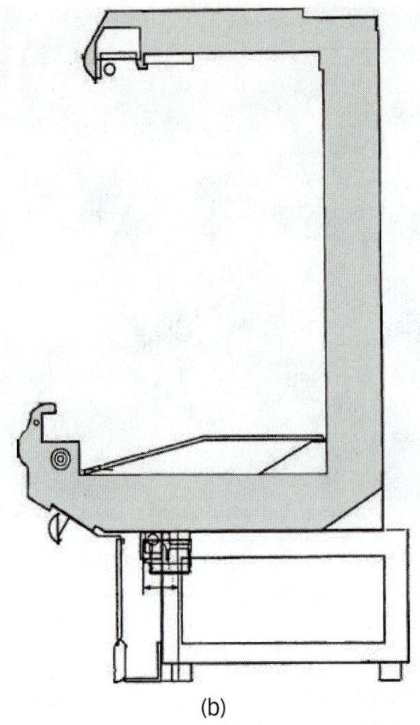

(b)

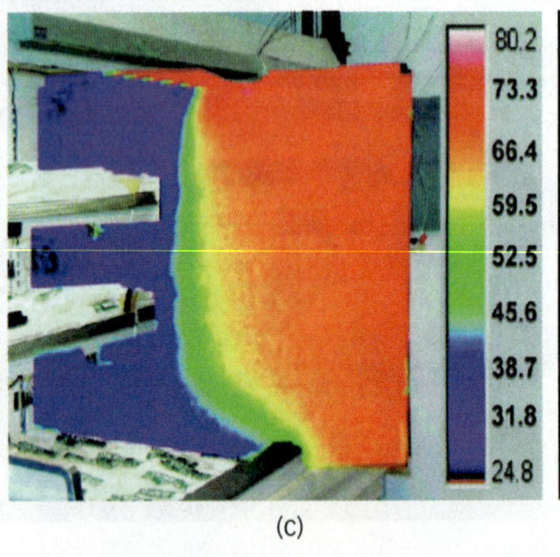

(c)

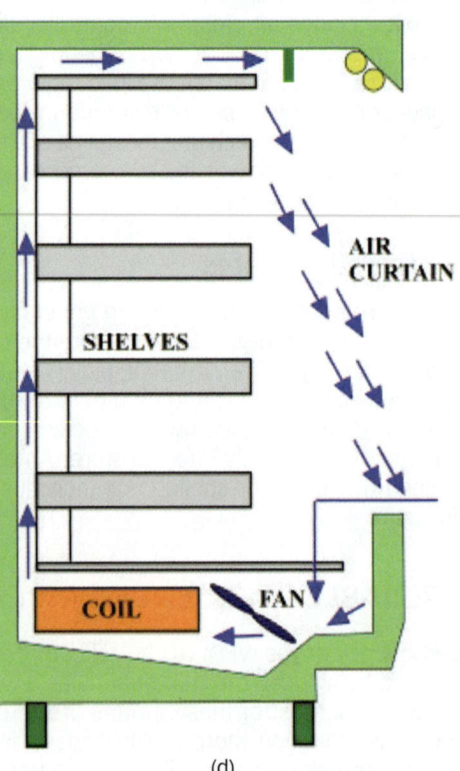

(d)

Figure 86-24 (a) Older-type open display case; (b) newer-type open display case; (c) air curtain temperature profile of an actual display case captured using infrared photography; (d) air flow through an open multideck case.

store product for a long period of time. Because of this, most case manufacturers recommend rotating product on a regular basis. There are several types of open display cases, which can be categorized into two general types: single deck and multideck. Since there are no doors on open cases, they rely on an air curtain to provide a barrier between the store's environment and the product (Figure 86-24c). The evaporator coil and its fans are located at the bottom of the cabinet with the fans located in front of the coil (Figure 86-24d). In a multideck case, the fans force air through the evaporator coil, up a channel at the rear and top of the case, through a set of honeycombs located on the top front of the case, and discharge the air down in front of the case's shelving. The honeycombs are designed to even out the airflow and develop a uniform air pattern down the front or across the top of the case. In a single-deck case, the air is discharged across the top.

Single-Deck Cases

Single-deck display cases, sometimes referred to as coffin cases, are refrigerated cabinets resembling a coffin in which

Figure 86-25 A closed single-deck case.

the product is merchandised horizontally. Open single-deck cases are designed to provide an air curtain across the top of the case. These cases are less prone to problems as compared to multideck cases since the colder air, which is heavier than the warmer store air, seems to stay in the case and not spill out. Both low- and medium-temperature single-deck cases will usually have a single air curtain. An energy-efficient closed single-deck display case is shown in Figure 86-25.

Multideck Cases

Multideck cases are different than single-deck cases in that they merchandise the product in a layered vertical fashion. This allows the product to be at eye level, making it more appealing to the customer. Produce is often displayed in multideck cases. A commonly used produce case is shown in Figure 86-26. It is sometimes referred to as a vision-type case because mirrors are used to enhance the appearance of the product. Some of these cases will also have a misting system to wet down the produce as needed, as some produce products need to be kept moist. This increases the shelf life of the product and increases its appeal to the customer. A basic misting system consists of a water tube with multiple spray heads, a timer, a water filter, and a solenoid valve.

Figure 86-26 A typical produce case.

Figure 86-27 A lineup of multideck dairy cases.

Dairy products can also be displayed in multideck cases. A typical dairy case is shown in Figure 86-27. These cases normally operate at a discharge-air temperature of approximately 36°F. Single-deck or multideck cases can be used to display meat products and frozen food products. These cases normally operate at a discharge-air temperature of approximately 26°F.

Multideck Open-Case Air Curtains

The multideck open-case air curtain is directed down the front of the case opening with the discharge air on the top of the opening and the return on the bottom of the opening. As compared to single-deck cases, these are more prone to problems since it is easy for the air curtain to become disturbed. Therefore, most service problem relating to air curtains will be found on the multideck cases.

Medium-temperature multideck cases usually have a single air curtain, while low-temperature multideck cases will usually have three separate air curtains. Since low-temperature multideck cases are more prone to problems with heat leakage into the case, three air curtains are needed to provide a sufficient barrier: a primary air curtain, a secondary air curtains, and an ambient air curtain. The primary air curtain is the coldest air curtain and is closest to the product. The secondary air curtain is slightly warmer than the primary and is in front of it. The ambient air curtain is actually a blanket of store air that is directed straight down the front of the case. This air curtain is not circulated within the case.

Air Curtain Velocity

Since open cases rely on an air curtain to seal the product from the store's environment, it is important to make sure it is sufficiently maintained. Otherwise, warmer and more humid store air will migrate into the case, causing the case temperature to rise and the evaporator coil to frost up more quickly. To test an air curtain, the air discharge velocity at the supply outlet is measured. This velocity can then be compared to the manufacturer's guidelines for the case. It is important to obtain the correct velocity from the manufacturer for a particular case. As a general guideline, medium-temperature single-deck cases will have an approximate discharge velocity of 120 to 140 feet per minute (fpm). Single-deck low-temperature cases will have a discharge velocity of approximately 200–250 fpm. Multideck medium-temperature cases will have a discharge velocity of approximately 170–400 fpm. (This depends on the application and manufacturer.) The primary air curtain of a multideck low-temperature case will have a discharge velocity of approximately 550 fpm; the secondary air curtain will have a velocity of approximately 400 fpm; and the ambient air curtain 250 fpm.

Air Curtain Troubleshooting

An air curtain can be disturbed by several factors. A technician needs to identify potential problems and correct them. A heating or air-conditioning diffuser can disturb an air curtain. If a supply diffuser is directed toward a case, it could disrupt its air curtain. Generally, velocities of over 20 fpm by the case are considered excessive.

Another way an air curtain can be disturbed is by how the product is loaded in the case. Each case is normally marked with a load-limit line. The product should not be loaded beyond this line; otherwise the air curtain will be disturbed and cause a problem.

If an air curtain is suspected of being disturbed, a technician can usually identify this in one of two methods. One method is to place a thin plastic bag in front of the case and its air curtain. (The bags used for fresh produce in a supermarket work very well.) If the plastic bag is drawn into the case, the store air will be, too; the cause will need to be identified. Another method to test the integrity of an air curtain is to smoke the case, as shown in Figure 86-28. By using a smoke candle to introduce smoke into the return-air grille, a technician can easily see if the air curtain is intact or is being disturbed. Smoke candles can be found at most local HVACR supply houses. These candles can be purchased to burn from 30 to 60 seconds. A 30-second burn is usually sufficient to test an air curtain.

Open-Case Discharge-Air Temperature

Another area of importance when troubleshooting open cases involves checking the discharge-air temperature. It is standard practice for an open case to be controlled and monitored by its discharge-air temperature, shown in Figure 86-29, and *not* the interior case temperature or the return-air temperature. This is different from other types of refrigeration equipment, which may be controlled and monitored by either the return-air temperature or box temperature. Since there are major differences between the discharge-air temperature, the box temperature, and the return-air temperature, a technician should not confuse the three when servicing these cases.

Figure 86-28 Smoking a refrigerated meat case.

Figure 86-29 Temperature sensor located at the discharge-air honeycombs of an open display case.

The control system for open display cases can also be configured to set off an alarm if the discharge air temperature exceeds a predetermined temperature for a specific delay period. For example, an alarm temperature of 45°F with a 60-minute delay could easily be used to signal a problem with the case. The alarm can be either a local or remote alarm. A local alarm will set off some type of light or buzzer in the store to notify personnel of a problem. A remote alarm will notify someone outside the store that there is a problem, such as a facilities manager, refrigeration contractor, or alarm company.

TECH TIP

Measuring various types of temperatures is a common task for HVACR technicians. The thermometers used must be accurate for a technician to be able to properly troubleshoot a system, so technicians should occasionally test the accuracy of their thermometers.

Place a thermometer in a solution of ice and water to do this. Fill a bucket with crushed ice and add water so that ¾ of the ice is immersed in the water. It is best to have a solution that has more ice than water. Let the temperature of the solution stabilize—it should stabilize close to 32°F.

Once the temperature of the solution has stabilized, place the thermometer in the ice/water solution. Let the temperature measured on the thermometer stabilize, and observe its reading. If the thermometer measures a temperature relatively close to 32°F, then the thermometer is relatively accurate. If not, the thermometer needs to be adjusted, repaired, or discarded.

Store Humidity and Radiation Effects

Air curtains cannot totally eliminate all of the store air from migrating into a case. Small amounts of air will always migrate through the air curtains into the case. It is important to limit the store humidity level to 55 percent to keep these cases working properly. High store humidity will cause the case's evaporator to frost up more quickly, resulting in frost accumulating on the product and the system consuming more energy. A 10°F increase in the store wet-bulb temperature will increase the compressor's power usage by 25 percent.

Another source of potential problems with these cases is the introduction of radiant heat into the case. Radiant heat can be introduced into a case from its own case lighting, the store's lighting, or, depending on its location, sunlight. This can be a major source of a load on the case and should be minimized when possible.

Open-Case Defrost and Cleaning

Verifying the correct defrost times is also an important check on open cases. When possible, always check with the manufacturer for the recommended settings. General defrost norms are as follows: two to three defrosts a day for medium-temperature single-deck cases; three to four defrosts a day for medium-temperature multideck cases; one to two defrosts per day for low-temperature single-deck cases; and two to four defrosts per day for low-temperature multideck cases. Defrost times will depend greatly on the store's humidity level. The higher the store's humidity, the more defrosts the case might need.

An open case should not be installed tight to a wall, as it may cause condensation to form on the back or bottom of the case. Always allow an air space at the back of the case. When joining cases together, be sure to follow the manufacturer's guidelines and procedures. A lineup of cases that defrost together should be sealed off from any adjoining cases that defrost at a different time.

Regularly cleaning open cases is a very important maintenance task. They should be periodically shut down and cleaned out entirely to help prevent loss of product due to equipment failure. Without regular cleaning, drain lines can become clogged, causing ice to form in the drain pan and eventually rise up to stop the fans and shut the case down. Even with properly tuned refrigeration systems and proper defrost time settings, ice will develop on the ends of the evaporator coil and eventually grow large enough to possibly affect the operation of the case. Regular cleaning will prevent this from becoming a problem.

86.7 ENCLOSED DISPLAY CASES AND WALK-IN COOLERS

The refrigeration load in an open display case includes heat gains from radiation, conduction, lighting, fans, defrost, and anti-sweat heaters. And while this heat gain is significant, it accounts for only 30 percent of the total load, with the remaining 70 percent caused by infiltration. Reducing case infiltration saves energy because the refrigeration equipment accounts for approximately 50 percent of the total electrical energy consumption in a supermarket.

Figure 86-30 Closed frozen case.

This is a concern because the national focus on reducing annual energy use is beginning to affect supermarkets because they are such large energy users. Therefore, new supermarket design and retrofit have been focused on reducing energy consumption in all ways possible. Because of this, many open display cases are starting to be replaced, because it is obvious that glass-doored enclosed display cases will not be subject to such high-infiltration heat gains and therefore energy losses (Figure 86-30). Even so, there is a concern that customers find enclosed cases inconvenient and stores fear the loss of sales if open display cases are replaced. However, recent studies have shown that the enclosed display cases have little impact on product sales.

Enclosed display cases can reduce compressor rack size by as much as 15 percent. Not only that, but food safety is improved, as wide variation in product temperatures is reduced. Indoor store temperatures are warmer due to less cold air spillage, and this is more comfortable to customers. This also reduces the load on the HVAC system, which will not need to be adjusted to meet the demands of partial cooling and dehumidification by the refrigeration system.

Even with these advantages, open display cases still remain very popular. This is because even though enclosed cases reduce the total refrigeration load, they have additional energy requirements as compared to open cases. Anti-sweat heaters located in the glass doors require a substantial amount of energy. Enclosed cases also require more lighting as compared to open cases. Because of this there is very little total efficiency advantage between enclosed and open cases and not enough to warrant a retrofit change-out. However, new enclosed cases are being designed to use less energy by using "no heat" doors and LED lighting operated by motion sensors. As these improvements are standardized, the use of enclosed cases will become increasingly more cost effective.

Although enclosed cases are less prone to problems with fluctuations in the store's environment, they are still designed to operate within a desired temperature and humidity range. The store's environment should not exceed a dry-bulb temperature of 75°F and a relative humidity of 55 percent. Cases subjected to higher conditions may develop frosting and icing problems in the case, on its doors, and on the product. Even though these cases have glass doors to seal the case from the store's environment, there is still a definite air pattern set up within them just as with open display cases. The discharge temperature usually operates in a range of −10°F to 0°F (−15°F to −5°F for ice cream applications).

Figure 86-31 Walk-in cooler.

Figure 86-32 Evaporator coil for walk-in cooler.

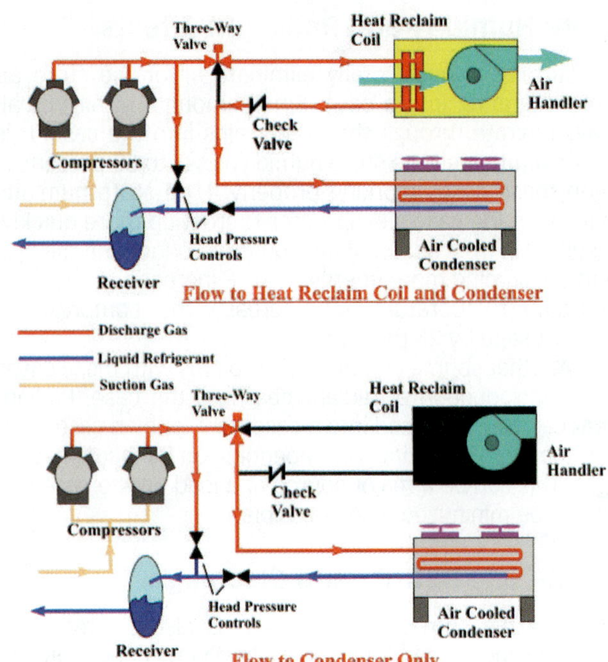

Figure 86-33 Traditional series heat-reclaim system.

Walk-in Coolers

There may be one or more walk-in coolers that are used to store food products until they are placed on the refrigerated cases (Figure 86-31). Walk-ins will have fan coil units located in the space (Figure 86-32). Electric defrost or hot-gas defrost is commonly used. The product is sometimes stored in the walk-in before being placed into the refrigerated display cases.

86.8 HEAT-RECLAIM SYSTEMS

Supermarket heat-reclaim systems have been in use for many years. The older series heat-reclaim systems were not very efficient, but today's systems utilizing heat pumps are designed to reduce operating costs, making them popular.

Series Heat-Reclaim Systems

Early conventional series heat-reclaim methods utilize a heat-reclaim coil located in the air handler. Heat from the refrigerant in the compressor discharge line is used to preheat the air for the store's HVAC system. When heating is desired, a three-way valve is used to direct the refrigerant in series through the heat-reclaim coil and then to the air-cooled condenser (Figure 86-33). The heat transferred by the refrigerant to the air handler is mainly in the form of sensible heat only. This system was originally desirable as a method to capture "free" heat that would normally be rejected to the outside through the air-cooled condenser.

The actual efficiency of series heat-reclaim systems is generally poor. This is because the refrigerant discharge temperature needs to be high enough to recover an adequate amount of sensible heat. The system relies on head pressure control valves (holdback valves) to maintain a high enough condensing temperature sufficient for heat recovery. If the refrigerant condensing temperature is too low, then the compressor's compression ratio decreases, which results in lower mass flow rates and less total heat available. A major disadvantage is that most often the electrical energy consumed by the compressors required by this floating head pressure control offsets the gains made by heat recovery. In addition, this type of system requires long refrigerant line lengths to and from the air handler. For these reasons, series heat-reclaim systems are seldom used today.

Parallel Heat-Reclaim Systems

A parallel heat-reclaim system recovers both the sensible and latent heat from the refrigerant. As compared to a series heat-reclaim system that only transfers sensible heat, the refrigerant is condensed in a parallel heat reclaim system. The reclaim coil is not in series but rather in parallel with the air-cooled condenser (Figure 86-34). This design allows for a number of different options, such as locating the heat-reclaim coil in the air handler. A disadvantage to this location is the heat rejected from the reclaim coil needs to satisfy both the compressor head pressure requirements and the air-handler demands. This requires properly designed coil sizing and is not easily controlled like a typical HVAC system. As an example, the reclaim coil could reject more heat than the HVAC system requires, which would raise the building temperature to an undesirable level.

An alternative method is to include a secondary-loop heat exchanger to condense the refrigerant. After absorbing heat from the condensing refrigerant, the secondary fluid is pumped to a sensible heat-reclaim coil located in the air handler (Figure 86-35). A disadvantage to this system is

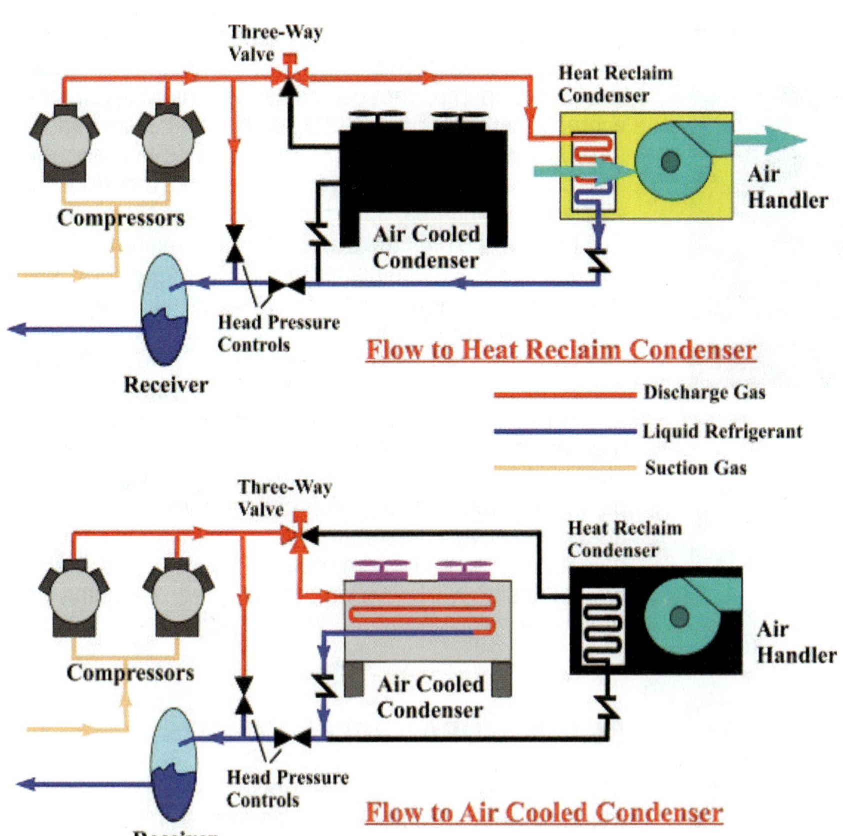

Figure 86-34 Parallel heat-reclaim system.

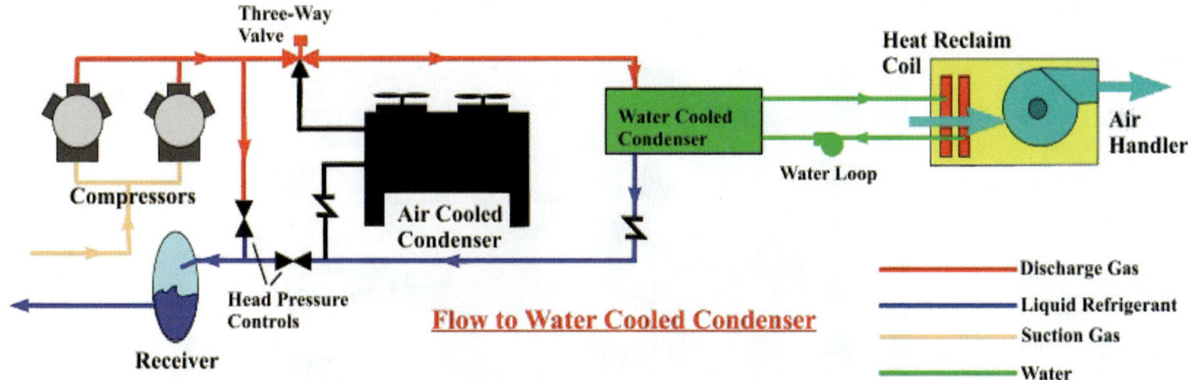

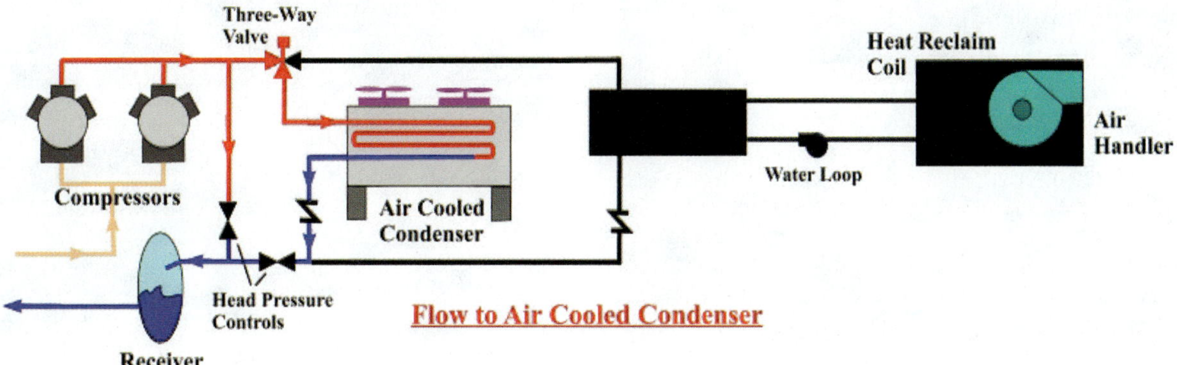

Figure 86-35 Secondary-loop heat-reclaim system.

Figure 86-36 Glycol piping.

that although both sensible and latent heat are recovered from the refrigerant, only sensible heat is transferred in the air handler. Also, additional pumps for the secondary fluid will increase electrical energy consumption.

Hydronic Heat-Reclaim Systems with Heat Pumps

Current refrigeration systems are being designed to reduce energy costs and minimize the total refrigerant charge. Hydronic heat-reclaim systems that incorporate water-source heat pumps are becoming increasingly popular. The refrigeration system uses a condenser-cooling secondary loop, and in this way the refrigerant condenser can be located next to the compressor rack in the machine room. The secondary fluid, typically ethylene glycol (Figure 86-36), is pumped to air-cooled heat exchangers located on the roof (Figure 86-37). This same loop can incorporate water-source heat pumps that can be used for heating during the winter months (Figure 86-38). With outside air temperatures below freezing, air-source heat pumps will have a very low COP (coefficient of performance) and therefore poor efficiency. In comparison, the water-source heat pumps that recover the rejected heat from the refrigeration system will operate at a much higher COP.

To gain further efficiency, geothermal heat pumps are used for the secondary loop. Geothermal heat pumps will cool the store in the summer while the rooftop air-cooled heat exchangers reject heat from the refrigeration system. During the winter months when heat is required, the heat pumps will be used in the loop for heat reclaim. This type of system requires extensive secondary loop piping and pumps along with change-over valves (Figure 86-39). However, the initial cost for installation should be paid for over time through reduced energy costs, and this type of design will better meet new "green" building code requirements.

86.9 NATURAL REFRIGERANTS

CFC and HCFC refrigerants are being replaced by HFC refrigerants that are safer for the environment because they have a lower ozone-depletion potential. However, many of these new environmentally acceptable refrigerants have been found to increase energy consumption and therefore actually contribute to the global warming potential (GWP)

(a)

(b)

(c)

Figure 86-37 (a) Directional control valves for heat-reclaim loop; (b,c) rooftop heat exchangers.

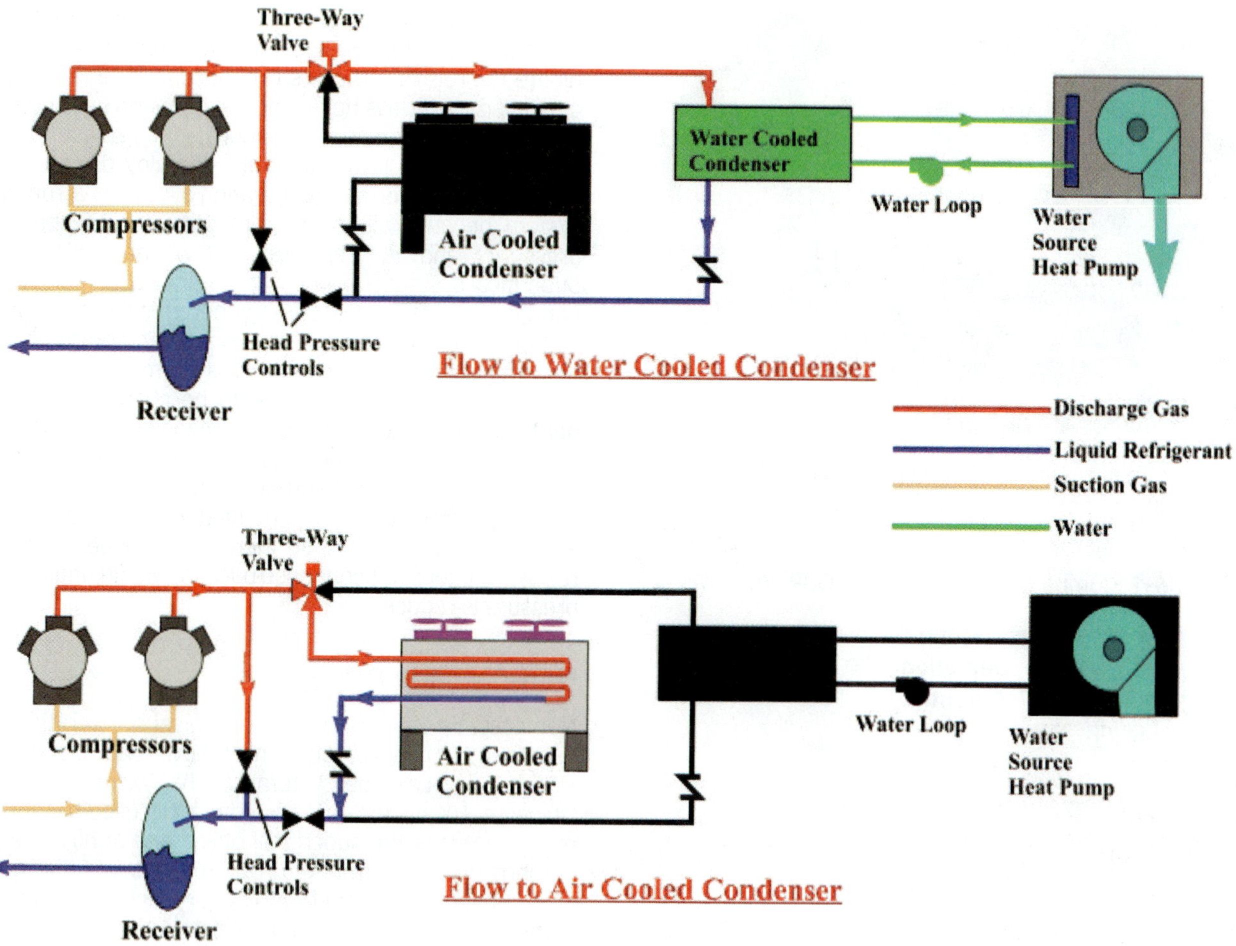

Figure 86-38 Secondary heat-reclaim system with heat pumps.

(Table 86-1). The current "green" building climate is moving toward carbon neutrality and a reduction in worldwide greenhouse emissions. Current alternative refrigerants for commercial use in large supermarkets include ammonia, CO_2, and hydrofluoro olefins (HFO). HFOs are newly developed types of refrigerant that have a low global-warming potential (GWP).

Carbon Dioxide (CO_2) Systems

Ammonia is both toxic and flammable, and HFOs are still in the early stages of development. This has led new supermarket

Figure 86-39a Heat-reclaim piping manifold.

Figure 86-39b Heat-reclaim directional control valve.

Figure 86-39c Heat-reclaim piping from the machine room to the roof top.

Table 86-1 Comparison of refrigerant ODP and GWP.

Refrigerant	Ozone Depletion Potential	Global Warming Potential (100 yr)
R-12 (CFC)	1.0	4750
R-22 (HCFC)	0.05	1810
R-134a (HFC)	0	1430
R-404A (HFC)	0	3922
R-410A (HFC)	0	2088
R-290 (Propane)	0	20
R-717 (Ammonia)	0	<1
R-744 (CO_2)	0	1
HFO-1234yf	0	4

design to consider the use of CO_2 as a refrigerant. CO_2 has a low GWP, is nontoxic (within a ventilated space), nonflammable, and doesn't deplete the ozone.

CO_2 has a very high volumetric cooling capacity. It can absorb 5 or 6 times as much heat as HFCs at medium temperatures. However, CO_2 systems operate at considerably higher pressures. At room temperature, the pressure of CO_2 is approximately 913 psi. CO_2 can operate in the subcritical, transcritical, and supercritical phase. In the supercritical range at temperatures above 88°F, the vapor and liquid CO_2 are identical, and vapor CO_2 cannot condense into a liquid. Supercritical systems operate best in the 1,600 to 2,500 psi range. Systems are designed as non-cascade and cascade.

Subcritical, Supercritical, and Transcritical Systems

Subcritical, supercritical, and transcritical all describe a refrigeration system's operating pressures and temperatures relative to the critical point of the refrigerant in the system. The critical point is the highest pressure and temperature where the refrigerant can still condense. At and above the critical point there is no distinction between gas and liquid, so no condensation or evaporation can take place. The very top of the hump in a refrigerant enthalpy diagram is the critical point. A normal refrigeration system is a subcritical system because all the system components operate at pressures and temperatures below the critical point. Subcritical systems use an evaporator to absorb heat and a condenser to reject heat. In a supercritical system, all the components operate at pressures and temperatures above the critical point. A supercritical system operates on gas compression and expansion with no change of state. All heat transfer occurs by the gas changing temperature. A transcritical system operates both above and below the critical point. Heat is absorbed in an evaporator where liquid evaporates to a gas, but heat rejection takes place above the critical point, so there is no condensation. The refrigerant does not condense back to a liquid until after the pressure is reduced.

Non-Cascade CO_2 Systems

Non-cascade CO_2 systems use a typical refrigeration cycle loop with compressor, condenser, flow control, and evaporator. A non-cascade system can be DX or transcritical. However, DX systems are not commonly used because the refrigerant is in the subcritical phase and at high pressure at normal room temperatures. Because of this, an air-cooled condenser would require very high operating pressures and therefore not be practical.

A transcritical system can operate in both the subcritical and supercritical ranges. In temperatures above 88°F, the CO_2 would be supercritical and the condenser would become a gas cooler. At temperatures below this, the system would operate as a DX subcritical system. This type of system is much more efficient when operating in the subcritical range. Because of this, the system has poor operation and efficiency in warm climates.

Cascade Systems

A cascade CO_2 system uses a secondary loop with a different refrigerant (Figure 86-40). In a subcritical DX cascade system, both loops use compressors to circulate the refrigerants. HFC refrigerant is circulated to the medium-temperature refrigerated cases and also to a heat exchanger for cooling the CO_2 (Figure 86-41). This lower-temperature cooling with the HFC allows the CO_2 to remain in the subcritical range. The CO_2 is used for cooling the low-temperature refrigerated cases. This type of system reduces the total refrigerant charge and maintains operating pressures fairly low, at approximately 450 psi.

A cascade secondary loop with liquid recirculation uses both pumps and compressors to circulate the refrigerant. HFC refrigerant is circulated by a compressor only for cooling the CO_2. This CO_2 heat exchanger is a flooded coil that has both liquid and vapor CO_2 leaving (Figure 86-42). The vapor CO_2 is delivered by compressors for cooling the low-temperature refrigerated cases. The liquid CO_2 is pumped

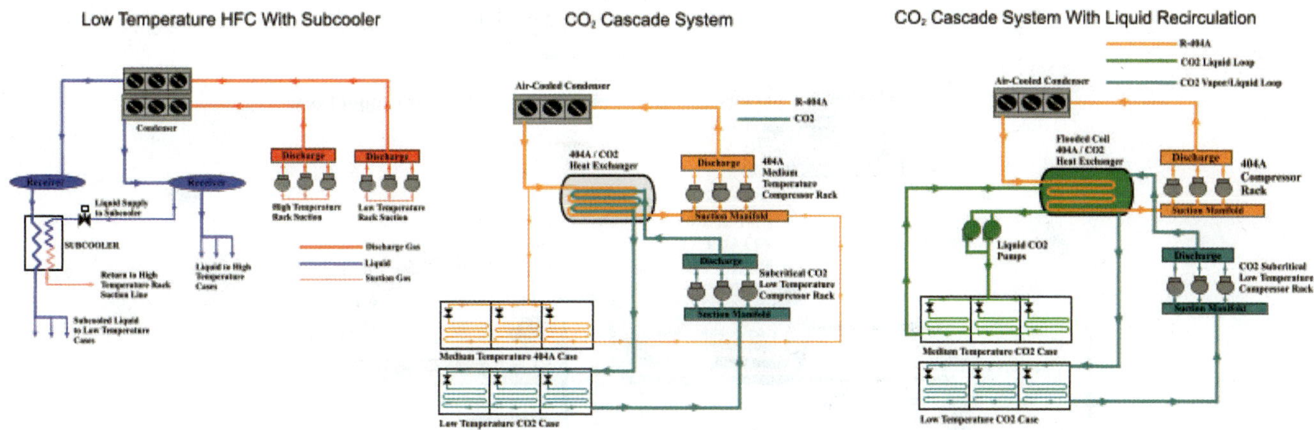

Figure 86-40 Three different types of low-temperature systems.

to the medium-temperature refrigerated cases. Similar to a cascade DX system, this reduces the total refrigerant charge and maintains operating pressures fairly low, at approximately 450 psi.

Future of CO_2

Over a dozen CO_2-based supermarket installations have been completed successfully in the United States. Over five hundred supermarkets are currently using CO_2 in Europe. Since this technology is relatively new, there are concerns about having properly trained people to operate and service these systems. Just a small increase in temperature results in a significant rise in pressure. Systems need to be kept below 88°F at all times in the event of a power outage. Systems must be designed to vent the CO_2 directly outside to prevent personal injury in occupied spaces. However, even with these operational concerns, if the trend continues toward reducing high GWPs and no significant developments are made in developing hydrofluoro olefins, then more CO_2 systems will continue to be installed.

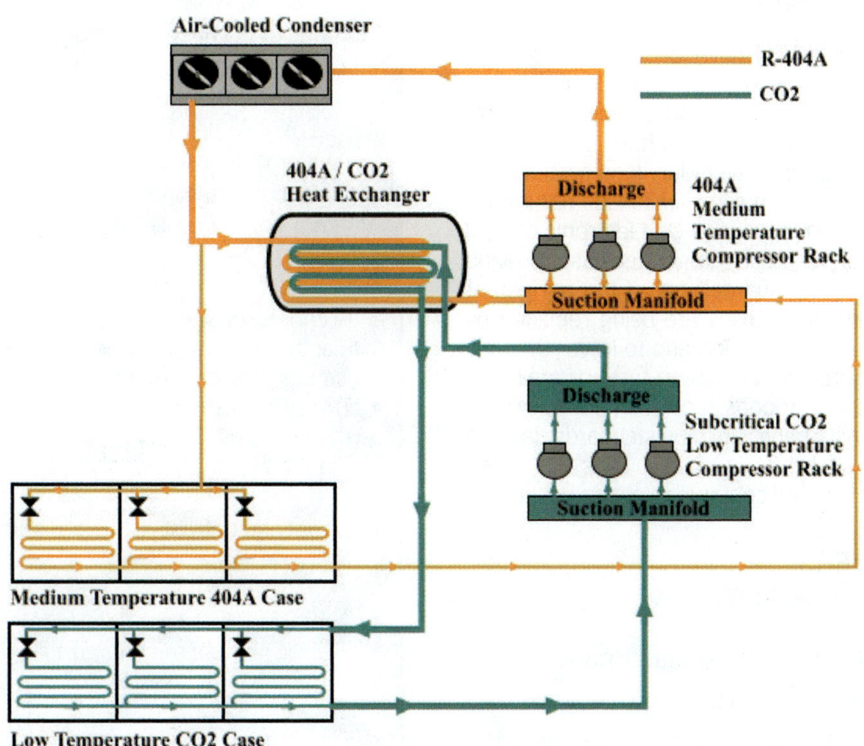

Figure 86-41 CO_2 cascade system.

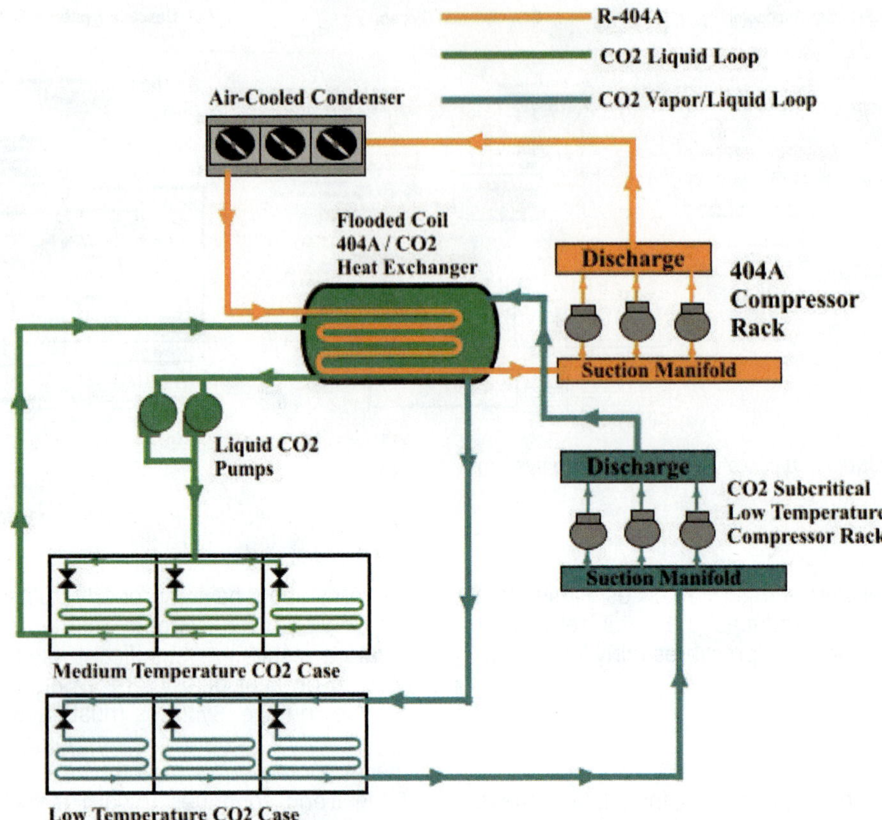

Figure 86-42 CO_2 cascade system with liquid recirculation.

UNIT 86—SUMMARY

The impact that supermarkets have on the total equivalent warming impact has been considered to be significant. Because of this, supermarket design is focusing on becoming more energy efficient. New systems have considerably lower total refrigerant charges. Secondary-loop systems are becoming more popular, and in some instances natural refrigerants such as CO_2 are replacing HFCs. Electronic control systems for compressor racks and refrigerated cases provide for more accurate metering and delivery of the refrigerant and are now common. Open cases are being replaced by closed cases to reduce air infiltration and to lower the load on the store's HVAC system. Heat-reclaim systems are being designed to utilize secondary loops in combination with geothermal heat pumps. As "green" building standards develop and new refrigerants such as hydrofluro olefins are developed, supermarket design will continue to change.

WORK ORDERS

Service Ticket 8601

Customer Complaint: Ice on Evaporator Coils

Equipment Type: Walk-in Freezer

A technician is called out to service a walk-in freezer with an air-cooled condenser, R-404A refrigerant, and an external equalized TEV operating at a 30°F case temperature.

When the technician arrives on the job, the evaporator has an excessive amount of ice on its coils, and the compressor, condenser, and evaporator fans are running. The system incorporates electric defrost heaters and a time-temperature defrost arrangement with three defrost periods per day and a failsafe time of 35 minutes. The system also uses a standard temperature controller to control case temperature. The technician decides to initiate a defrost cycle and then observes the following conditions:

- Compressor and evaporator fans shut down.
- Electric heaters are energized.
- Ice starts to melt off the coils of the evaporator.
- The system stays in defrost for 35 minutes and then the compressor cycles back on with the evaporator fan coming on after a slight delay.

A second defrost is initiated and the following conditions are observed:

- Compressor and evaporator fans shut down.
- Electric heaters are energized.
- The remaining ice melts off the coils of the evaporator.
- The system stays in defrost for 30 minutes and then the compressor cycles back on with the evaporator fan coming on after a normal delay.

After running for approximately 45 minutes after the second defrost, the following conditions are observed:

- The operation of the time clock is observed and the time advances normally.
- The case temperature is down to 10°F.

Based on these observations, the technician determines that the system is operating normally. After speaking with some employees working in the kitchen, it is discovered that the door was accidentally left slightly open for several hours, which most likely caused the elevated case temperature and the excessive ice on the evaporator.

Service Ticket 8602

Customer Complaint: Ice on Evaporator Coils

Equipment Type: Walk-in Freezer

A technician is called out to service a walk-in freezer using R-404A refrigerant, an air-cooled condenser, TEV metering device, and a time-initiated temperature-terminated electric defrost strategy and operating at a case temperature of 28°F and climbing.

When the technician arrives on the job, the compressor, condenser, and evaporator fans are running, and the evaporator is completely covered in ice. After further investigation, the technician observes the following conditions:

- The condenser coil is clean.
- The ambient temperature is 78°F.
- The operating suction pressure is 20 psig.
- The operating discharge pressure is 229 psig.
- The resistances of the defrost heaters are measured, and each heater has a measurable resistance.
- The defrost time clock is observed for 30 minutes, and it seems to advance normally.
- A manual defrost is initiated, and the defrost time clock immediately jumps back into the refrigeration mode.

The technician then decides to replace the termination temperature switch.

UNIT 86—REVIEW QUESTIONS

1. What is the most common type of system used in "traditional" supermarkets?
2. How many compressors are typically found on a compressor rack?
3. An uneven parallel rack of one 10- and two 20-hp compressors will provide for what horsepower combinations?
4. What type of pressure control can be used for rooftop air-cooled condensers?
5. How does a pulse-width modulated expansion valve operate?
6. What types of defrost methods can be used for refrigerated case coils?
7. Explain how an adaptive defrost cycle operates.
8. What is the advantage of a case coil secondary-loop system?
9. What types of fluids are used for secondary-loop systems?
10. How is a distributed system arranged in a supermarket?
11. What is the advantage to variable-frequency drive motors?
12. Secondary-loop fluid potassium formate is compatible with which piping materials?
13. Do enclosed display cases use more or less energy than open cases?
14. How does a series heat-reclaim system operate?
15. What is the disadvantage of a parallel heat-reclaim system that has the reclaim coil located in the air handler?
16. How do open display cases maintain the temperature within the case without any doors?
17. The temperature is monitored in an open display case at what location?
18. How many air curtains are used in a multideck open display freezer?
19. What are the advantages of CO_2 as a refrigerant?
20. What are the disadvantages of CO_2 as a refrigerant?

UNIT 87

Ice Machines

OBJECTIVES

After completing this unit, you will be able to:

1. estimate the size of the ice machine required for a particular application.
2. explain the differences between cubes, crushed ice, flake ice, shaved ice, and block ice and what applications are bested suited for these ice types.
3. determine if an ice machine needs to be cleaned.
4. describe the sequence of the freeze and harvest cycles of a cube ice machine.
5. describe the operation of a flake ice machine.
6. troubleshoot cube and flake ice machines.

87.1 INTRODUCTION

Ice has always been a valuable commodity. Before the advent and commercialization of mechanical refrigeration systems, ice was harvested from frozen lakes and rivers and then cut into large blocks with hand saws (Figure 87-1). Besides being stored and used locally, some of this ice was packed in sawdust for insulation and transported by sailing ships to places such as the islands of the Caribbean. Later, mechanical refrigeration systems allowed ice to be made year round, making it available for more people. Much of this ice was produced in large industrial ice-making facilities as large blocks. These large blocks were sawed into smaller blocks and cubes. The ice shavings left behind from the electric saws were collected and used in stores to keep vegetables and seafood displays cold. This is similar to the flake ice that is made specifically for such purposes today.

Industrial ice-making facilities still provide for much of the bagged ice that can be purchased in supermarkets and convenience stores. However, most commercial institutions such as restaurants and cafeterias rely on the operation of packaged ice-making machines. These commercial ice makers come in many different sizes and are available for a wide range of applications.

Even though there are many different types of ice makers, they all have the same major refrigeration components, which include a compressor, condenser, metering device, and an evaporator (Figure 87-2). This is why it is important to have a good understanding of the basic refrigeration cycle before working on commercial ice makers.

Figure 87-1 Ice harvesting in 1940. The sawed ice blocks from the river are being channeled to the conveyor and into the icehouse. *(Collections of the Bangor Museum and History Center)*

TECH TIP

Most ice machines work on the principle of the basic mechanical refrigeration cycle. The unit will have a compressor, condenser, metering device, and evaporator. Troubleshooting these units requires a good understanding of refrigeration systems. Although there are many different types of ice machines, most manufacturers provide detailed operating and troubleshooting guidelines that are unit specific.

87.2 COMMERCIAL ICE MACHINE APPLICATIONS

Commercial ice machines are often used for food service applications, such as restaurants, institutional kitchens, cafeterias, and correctional foodservice. They can also be found in such places as hotels, hospitals, theme parks, resorts, casinos, and cruise ships. Some commercial ice machines can be compact for small-volume ice production. These are used when there are limited ice needs, and they do not take up much space. They can be found in places such as small kitchens, hotel hallways, and office lunchrooms. Larger machines are required when a lot of ice is needed on a fairly continuous basis (Figure 87-3). There are some machines that also automatically bag the ice, which adds to the mechanical complexity of the unit.

Production of ice will vary, with small ice machines making 50 lb of ice per day or less and large machines making

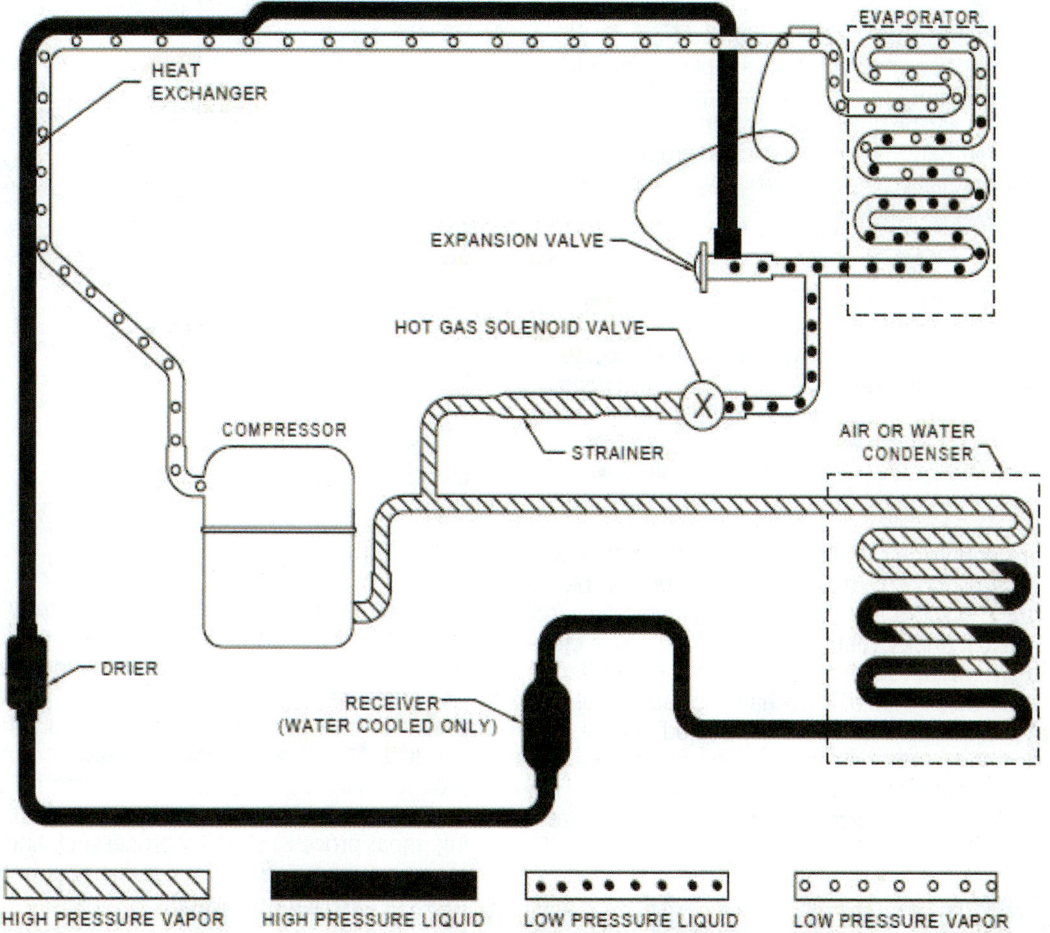

Figure 87-2 Mechanical refrigeration cycle for an ice machine. *(Courtesy of Manitowoc)*

Figure 87-3 Filling buckets with ice from an industrial ice machine. *(Edward. J. Westmacott/Alamy)*

over 3,000 lb per day. The total amount of ice required is dependent on the application. An example of estimated ice usage based on application is provided in Table 87-1.

Table 87-1 Estimated Ice Usage

Application	Ice Type	Daily Approximation
Restaurant	Cube	3 lb per person
Cafeteria	Cube	3 lb per seat
Cafeteria	Crushed	6 lb per seat
Hotel room service	Cube	5 lb per person
Hotel food service	Cube	1.5 lb per person
Hospital	Cube	5 lb per bed
Hospital	Crushed	10 lb per bed
Salad bar	Crushed	30 lb per cubic foot
12-oz drinking cup	Cube	6 oz of ice

TECH TIP

Some small soda fountains have self-contained ice makers, while others must be loaded with ice from a large centrally located ice machine that supplies the entire cafeteria. The advantage to a unit that must be loaded with ice is that it can be quickly replenished if it becomes empty. If the centrally located unit is sized properly, then it should not normally run out of ice.

87.3 ICE TYPES

Very often we think of ice as always being cubed. However, there are many other different types of ice, such as crushed, shaved, flake, and block. Each type is made differently and has specific uses for which it is best suited.

Cubes

Ice cubes are popular for drinks because they melt slowly and cause less dilution. A full-size square cube can be as large as 1-1/4 × 1-1/4 × 1-1/4 in. This size is typically used for bagged ice, bar drinks, room service, and airlines. A dice-size cube is 7/8 × 7/8 × 7/8 in. There would be about forty-eight dice size cubes in 1 lb of ice. A half-cube size would be about 1-1/4 × 1-1/4 × 5/8 in. These are a good size for fast food establishments and restaurants. The size of a quarter cube would be about 1-1/4 × 5/8 × 5/8 in. These are good for beverage service and are a very versatile and popular size.

TECH TIP

In 1914, Fred Wolf invented a refrigerating machine called the DOMELRE (Domestic Electric Refrigerator), which had a simple ice cube tray. The first flexible stainless steel, all-metal ice cube tray was created by Guy L. Tinkham and sold for 50 cents in 1933. The idea for the first rubber ice cube tray was thought of by Lloyd Groff Copeman in 1928. Sales from this invention earned him approximately $500,000, which is equivalent to about $10 million today.

Crushed Ice

Crushed ice is often used in cold drinks and is easy to chew. However, it melts faster than cubes and adds more water to the beverage. It is generally crushed from cubes, and this can be done by hand but is very time-consuming. It is known as "cracked" ice when left in larger fragments. Crushed ice has a finer texture and may need to be hand scooped rather than dispensed by a machine. This is why many machines incorporate a built-in ice crusher to allow for a choice between cube ice and crushed ice (Figure 87-4). The crusher does not add to the complexity of the refrigeration components but does add a mechanical feature that will require maintenance.

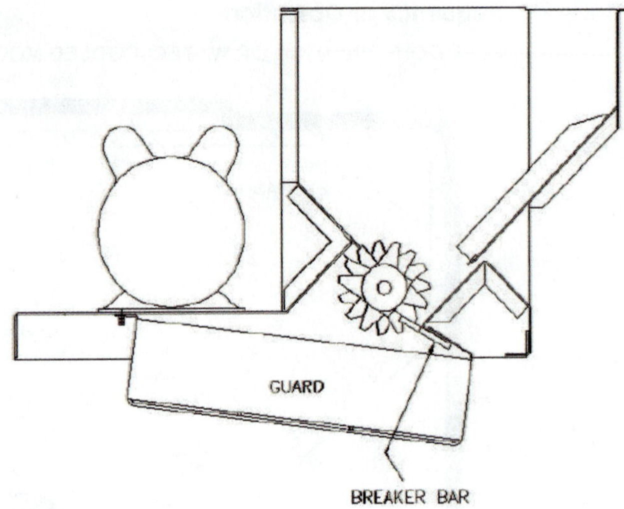

Figure 87-4 Ice crusher located in the ice chute to make crushed ice from cubes. *(Courtesy of KOLD-DRAFT)*

Flake Ice

Flake ice is small, hard bits of ice ideally suited for any "contact" presentation, such as displaying fish at the supermarket. It forms completely around the contact item and cools quickly. Because of its thin, flat shape, flake ice provides more surface area per ton than any other type of ice. It is often used in the processing of perishable food products when quick, efficient cooling is important. Such applications include seafood processing, poultry processing, meat processing, dairy processing, and even chemical manufacturing. The seafood industry is one of the largest users of flake ice worldwide.

Shaved Ice

This is a very fine ice that looks almost like snow and is often used for a specialty ice-based dessert product that is flavored. The ice is shaved from a block using an ice-block shaving machine. The popular vendor item known as the "snow cone" can be made from shaved ice but is more often made from crushed ice.

Block Ice

Block ice can vary in sizes from 10 lb (6 × 8 × 10 in) to as much as 300 lb (10 × 20 × 40 in). Blocks will last longer than cubes, so they are often used by campers and picnickers for coolers. For the same reason and because blocks take up less space, they are sometimes preferred for food processing. However, for the ice to be used effectively, it will still need to be crushed or ground into small pieces to make good contact with the food product. Another popular application for blocks is to make shaved ice.

87.4 ICE QUALITY

Generally, the ice machine will not freeze water like the simple ice cube tray you place in your freezer. The ice cube tray method will produce cloudy cubes as a result of air

Figure 87-5 Cube ice machine with cover removed to show evaporator and freezing cells.

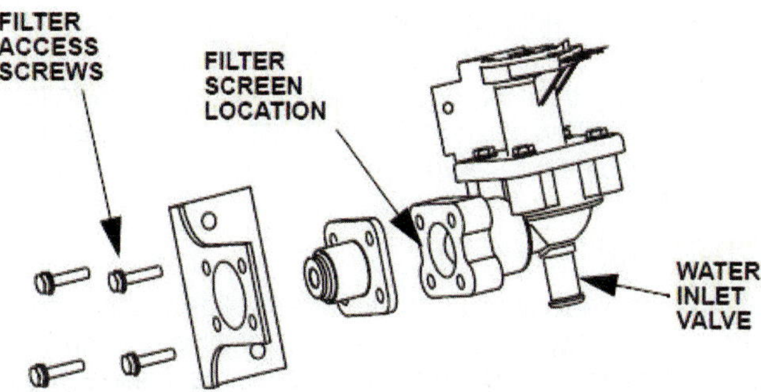

Figure 87-6 Location of the filter screen for a water inlet valve. *(Courtesy of Manitowoc)*

bubbles trapped inside the ice. These bubbles come from the dissolved gases that can no longer stay in solution as the ice freezes. As the ice forms, it floats to the top, and this ice layer traps the bubbles within the cube. Commercial ice machines utilize systems in which water is circulated continuously through the freezing cell (Figure 87-5). This helps to eliminate air and flushes out any impurities. The cube should be completely clear.

TECH TIP

The speed at which water flows down the evaporator plate will directly affect ice purity, clarity, and hardness.

Temperature is critical because cold, pure, hard ice is denser, melts slower, and therefore lasts longer than softer ice. The amount of refrigeration effect also depends on the quality of the ice. Ice quality is measured in percentage of hardness, which is a measurement that represents thermal cooling capacity. Ice that has a higher percentage of hardness will have more cooling ability. Ice hardness is calculated by conducting a calorimeter test as specified in ANSI/ASHRAE Standard 29-2009.

Ice cubes and block ice will generally be within the 95 to 100 percent hardness range. Flake ice has more minerals carried over from the water during its formation, and more air is present. This reduces the overall hardness and cooling effect of flake ice to about 70 percent.

CODES AND STANDARDS

ANSI/ASHRAE Standard 29-2009, Methods of Testing Automatic Ice Makers, prescribes a method for testing automatic ice makers and includes information about testing procedures, instrumentation and apparatus required for testing, and a uniform method for calculating results.

87.5 WATER QUALITY

Water conditions vary widely from location to location. Cleaner water produces cleaner ice. It is important to have a good clean water supply that has filters or strainers that are regularly changed or cleaned. The filter screen location for a water inlet valve is shown in Figure 87-6. Ice machine problems are often a result of poor water quality. It goes without saying that poor-quality water will ruin the taste of any drink.

TECH TIP

Filtering systems that include sediment removal, activated carbon, and reverse-osmosis (R/O) units can be considered where water quality is questionable. Ultraviolet (UV) lamps can be used in the final stage to destroy any microbes that may remain. Although these systems provide very pure water, they can be expensive and should not normally be used where city tap water is available.

Testing Water

Water quality should be tested to determine if treatment is needed and what type of treatment would work best. Water treatment facilities often provide sample bottles and sampling instructions. The sample is taken and then sent to the facility to be analyzed. Kits are also available to test the water in the field, but the analysis is less comprehensive.

Suspended Solids

Suspended solids can usually be removed with a simple filter system. Mostly composed of minerals, these solids will lead to cloudy ice cubes and a scaled-up evaporator surface. This scale reduces water flow through the system, reduces heat transfer, and will not allow the cubes to harvest from the evaporator plates. This is a common reason for ice machine freeze-up.

Dissolved Solids

During the freezing process, dissolved solids separate from the water and will lead to the same problems encountered with suspended solids. The difference is that dissolved solids are not removed with simple filtration systems. If dissolved solids are present, treatment is required to remove, reduce, or neutralize the minerals present. Hard water (water hardness) has greater concentrations of solids than soft water. Water softeners are used to produce an ion exchange that reduces the hardness of the water, making it less likely to produce scale.

TECH TIP

The filter flow rate is critical for proper sizing of a treatment system. Low water flow can lead to ice machine freeze-up. A properly selected filter should last from 4 to 6 months or longer before cartridges need to be changed.

Corrosion

Often pH is used to determine the acidity or alkalinity of the water. The pH scale ranges from 0 to 14. A neutral pH of 7 is desired. Readings below 7 are acidic. Acidic water can cause corrosion, while alkaline water usually has more scale-forming minerals.

Local municipalities have been adding higher levels of chlorine, chloramines, and sometimes chlorine dioxides to reduce bacteria in the water. Stainless steel ice-maker surfaces are susceptible to corrosion from the exposure to chlorine gas. These gases stick to the wet surfaces to form hydrochloric acid. This results in a rust-colored corrosion and pitting of the stainless steel.

Slime Buildup

Slime buildup results from the growth of bacteria and fungi and occurs partly due to the moist conditions present in an ice machine. When the bin door is opened for ice removal, airborne bacteria will enter the bin cavity. The bacteria will adhere to the wet surfaces on the evaporator. This buildup can lead to odors and is a health concern. Chlorine sanitizing solutions and tablets should be used on a regular basis to keep bacteria and fungi growth in check.

87.6 CUBE ICE MACHINE OPERATION

Cube ice machines all have the four basic components of the refrigeration cycle: compressor, condenser, metering device, and evaporator. A cross-sectional view of an ice machine is shown in Figure 87-7. The evaporator has individual cells built into it, similar to an ice cube tray (Figure 87-8). A water pump is used to distribute the water evenly across the cells to allow the freezing process to begin (freeze cycle) (Figure 87-9). When the ice is formed to the proper thickness, a harvest cycle is initiated to release the cubes from the cells and to be collected in the ice bin (harvest cycle). When the ice bin is full, the ice machine will stop producing ice.

Freeze Cycle

Evaporators are made of copper to allow for good heat transfer. They are tin plated or sandwiched between stainless steel plates for protection against corrosion (Figure 87-10). Evaporators can be either vertically or horizontally positioned. The operation allows for a fixed amount of water supplied from a reservoir to circulate continuously through the freezing cells to form ice smoothly on the cold surfaces. The water is directed evenly to the cells by a water-distribution tube (Figure 87-11). The pure water freezes first and impurities are washed down and away from the cells. This circulation of water that allows the ice to grow from the freezing surfaces outward also eliminates the formation of air bubbles.

There is often a short period for chilling the evaporator coil at the beginning of the freeze cycle (approximately 30 seconds) before the water begins to circulate. This allows the ice to immediately begin forming once the water contacts the cold surfaces. The reservoir will have a float switch or water-level probe that terminates the freeze cycle when the water level is low, indicating that the ice cubes have been formed.

An ice-thickness sensor can also be used to determine when the ice has reached its maximum thickness (Figure 87-12). This will be more accurate than relying merely on the water level in the reservoir. When the ice reaches the proper thickness, the freeze cycle is terminated. There is typically a slight delay (several seconds) before the freeze cycle is terminated to ensure that the ice-thickness sensor is accurate and is not being prematurely activated by splashing water.

Harvest Cycle

The term *harvest* is a throwback to the days when ice was "harvested" from lakes and rivers. When the ice is properly formed, it needs to be released from the freezing cells. The harvest cycle is initiated either when the reservoir float switch indicates a low water level or when the ice sensor determines the proper ice thickness has been reached. A hot-gas solenoid valve will direct the hot discharge gas from the compressor directly into the evaporator coil rather than to the condenser (Figure 87-13). This method causes a rapid heating of the freezing cells, allowing the cubes to be released to drop and collect in the ice bin located directly below. Some ice machines have water assist. A separate loop of water circulation aids in releasing the ice from the freezing cells. The harvest cycle will continue until a thermistor (evaporator temperature probe) measuring evaporator temperature is satisfied that the proper harvesting temperature has been reached.

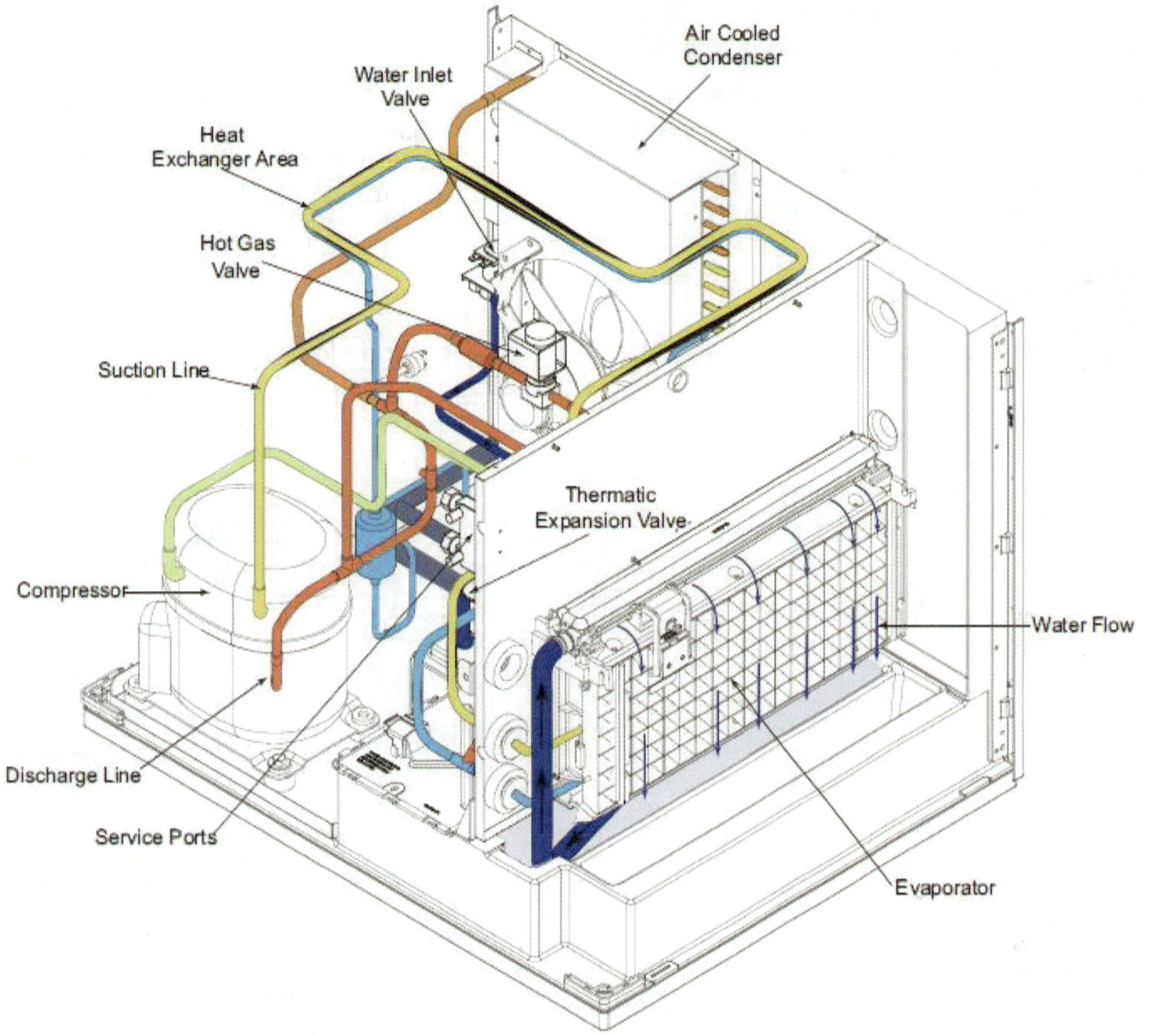

Air Cooled Condenser
Water Inlet Valve
Heat Exchanger Area
Hot Gas Valve
Suction Line
Thermatic Expansion Valve
Compressor
Water Flow
Discharge Line
Service Ports
Evaporator

Figure 87-7 Cross-section of a cube ice machine. *(Courtesy of Scotsman Ice Systems)*

Figure 87-8 Evaporator and freezing cells.

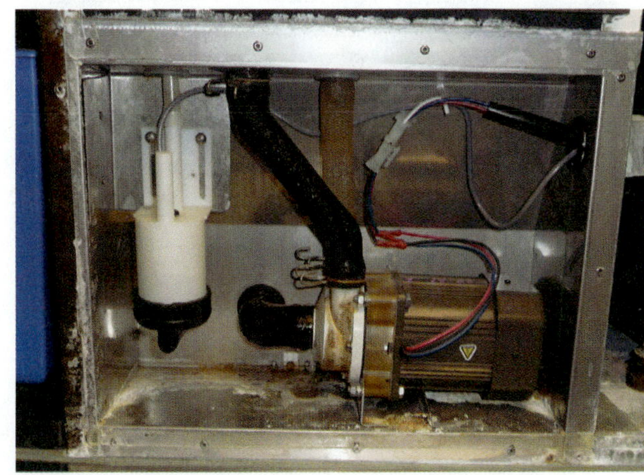

Figure 87-9 Water-circulation pump.

Figure 87-10 Ice machine evaporator.

TECH TIP

Have you ever had trouble at home with breaking, fragmented cubes that are difficult to remove from an ice cube tray? Place the entire tray under running warm water for several seconds and you will find the cubes much easier to remove. In a similar manner, some harvest cycles use water assist. The circulating water will help to release the cubes from the freezing cells.

Reservoir and Float Switch

The reservoir is filled by a water inlet valve. A water pump drawing from this reservoir will circulate the water across

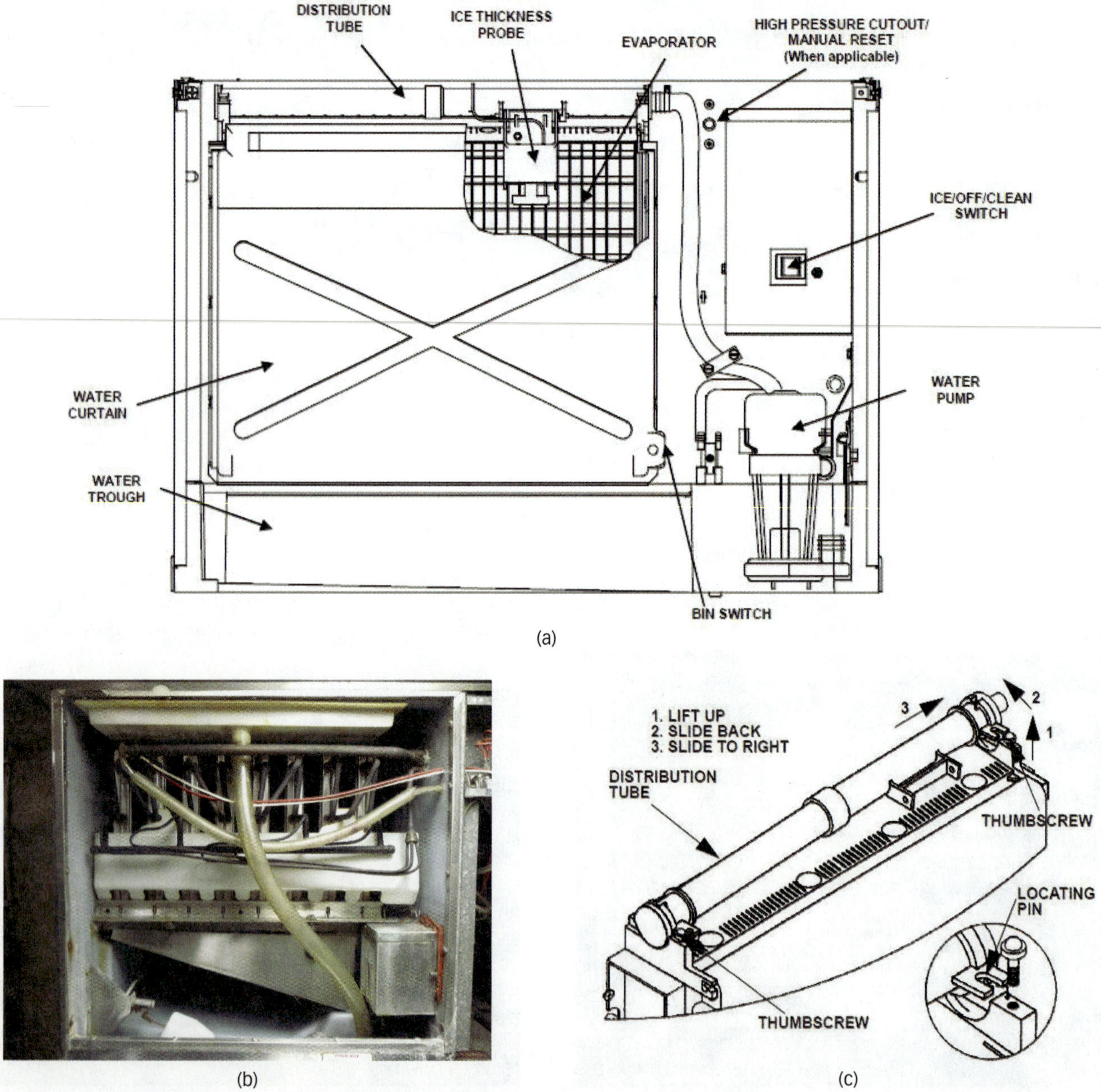

(a)

(b)

(c)

Figure 87-11 (a) Ice machine with location of water-distribution tube shown; (b) water distribution tube; (c) removing a water-distribution tube for cleaning. *(Courtesy of Manitowoc)*

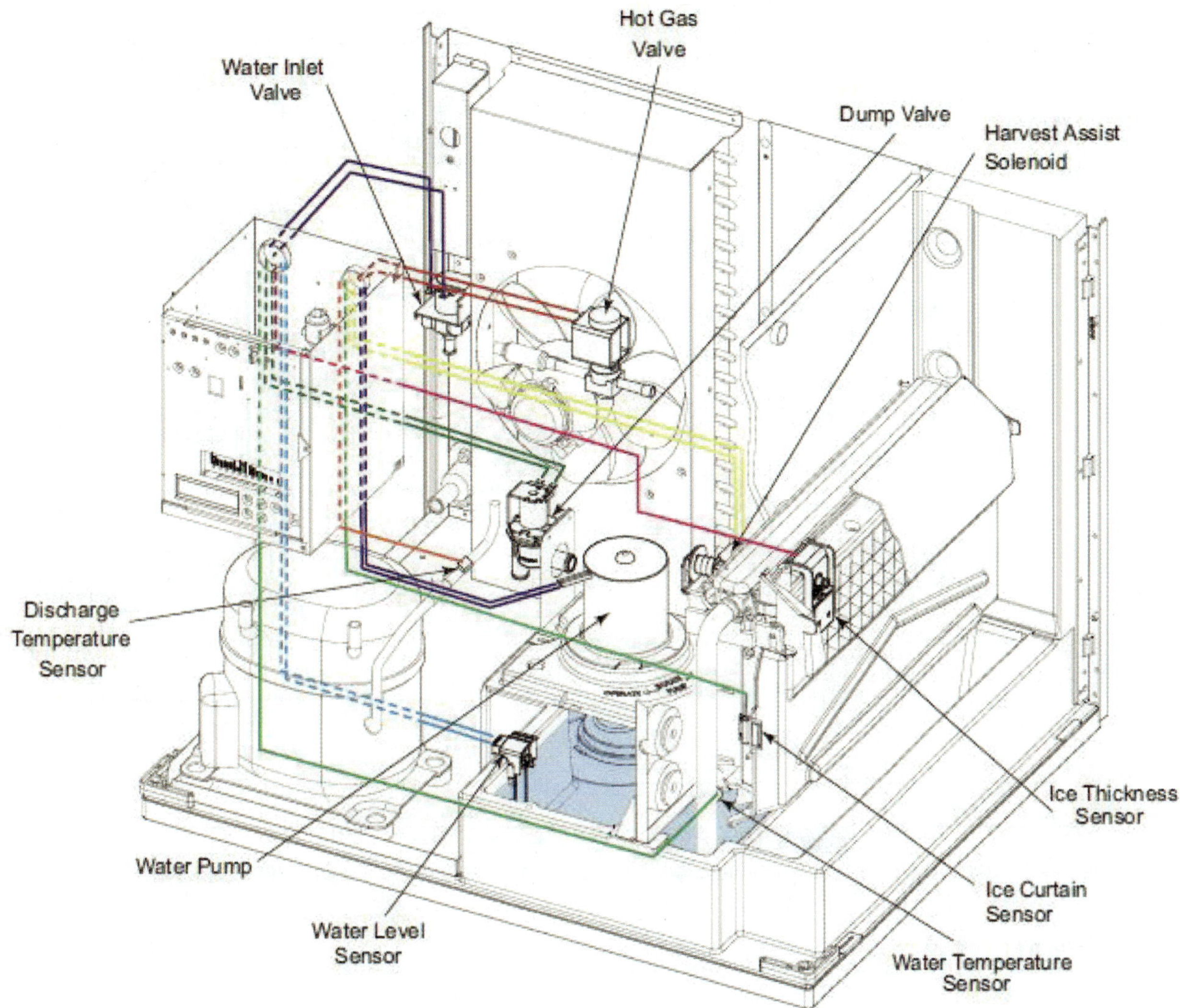

Figure 87-12 Cross-section of ice machine showing location of ice thickness sensor. *(Courtesy of Scotsman Ice Systems)*

the freezing cells. Typically, when the reservoir is full, the freezing cycle will begin. The level in the reservoir is determined by a float switch or a level probe. The evaporator is generally allowed to chill for a brief period before the water begins to circulate.

Since there is a fixed amount of water in the reservoir, the cubes should be fully formed when the water level becomes low. A common float that rides up and down with the water level is attached to a switch (Figure 87-14). This will determine if the level is at maximum, minimum, or somewhere in between. A water-level probe does not rely on a float but instead on the conductivity of the water (Figure 87-15). With either type of level indicator, the cycle process is still similar.

When the reservoir is low, the ice cubes should be formed, and the harvest cycle is initiated. The harvest cycle will be terminated when the evaporator has reached a sufficient temperature to ensure that the formed ice has been released. The end of the harvest cycle will allow for the reservoir to be refilled by the water inlet valve. An excess amount of water is supplied and overflows a standpipe located in the reservoir. This excess water "flushes" out the impurities that have collected in the water from the previous freeze cycle. When the reservoir is full, the evaporator will pre-chill, and then the freezing cycle will begin once again. This cycle of freezing and harvesting will continue until the ice bin is full.

Ice Bin Controls

The ice bin is where the ice is collected and stored for later use (Figure 87-16). The ice drops from the freezing cells through the ice chute into the ice bin (Figure 87-17). There are different types of controls to determine when the ice bin is full. A thermostatic capillary tube (ice-level probe) attached to a pressure switch can be mounted in the ice drop zone (Figure 87-18). When the bin is full, ice touches the thermostatic bulb and the pressure inside the bulb drops. This opens the contacts in the pressure switch to shut the ice machine down.

A mechanical-type bin control may consist of a proximity switch and a paddle. When the bin is full, ice pushes the paddle away from the proximity switch and the unit shuts down. Another type of mechanical bin control utilizes a curtain that opens to allow the newly formed ice to pass through to the bin during the harvest. At the edge of the curtain is a switch (Figure 87-19). If the curtain opens and closes during

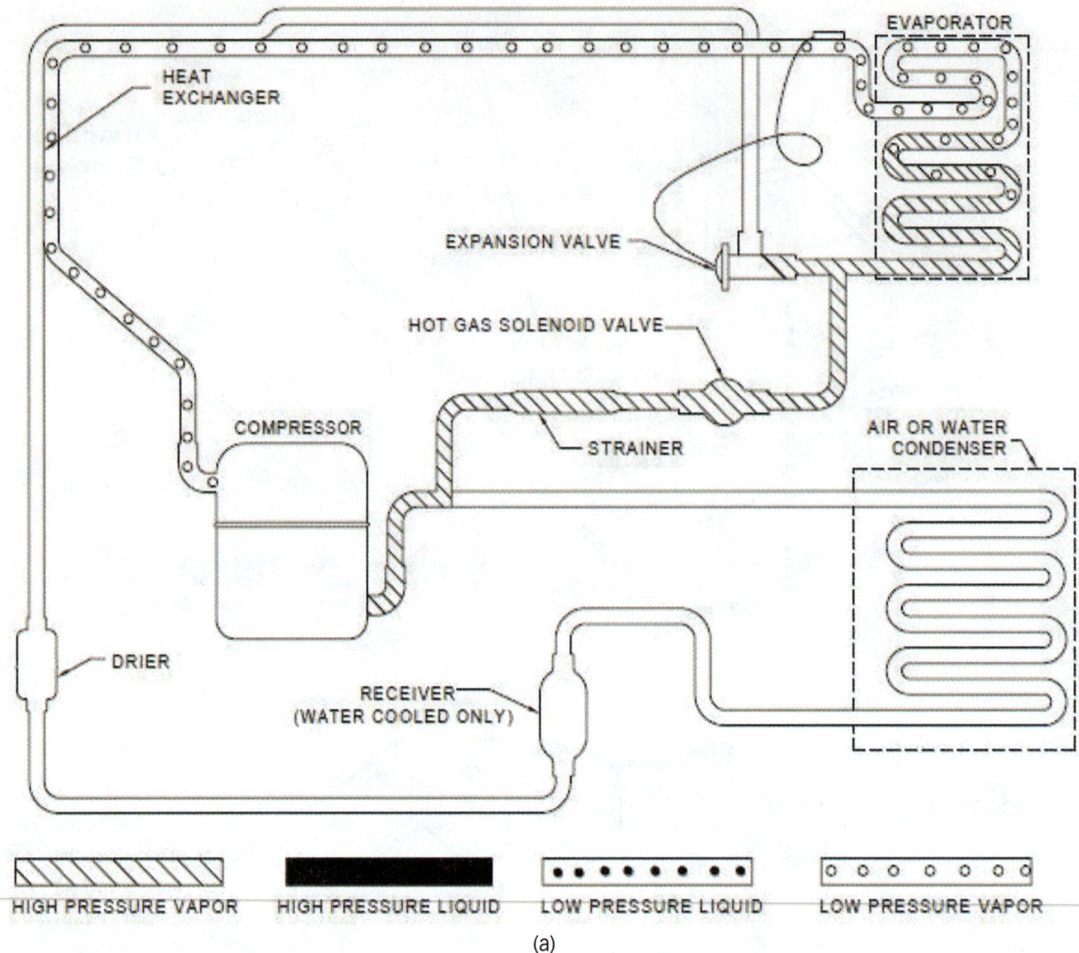

(a)

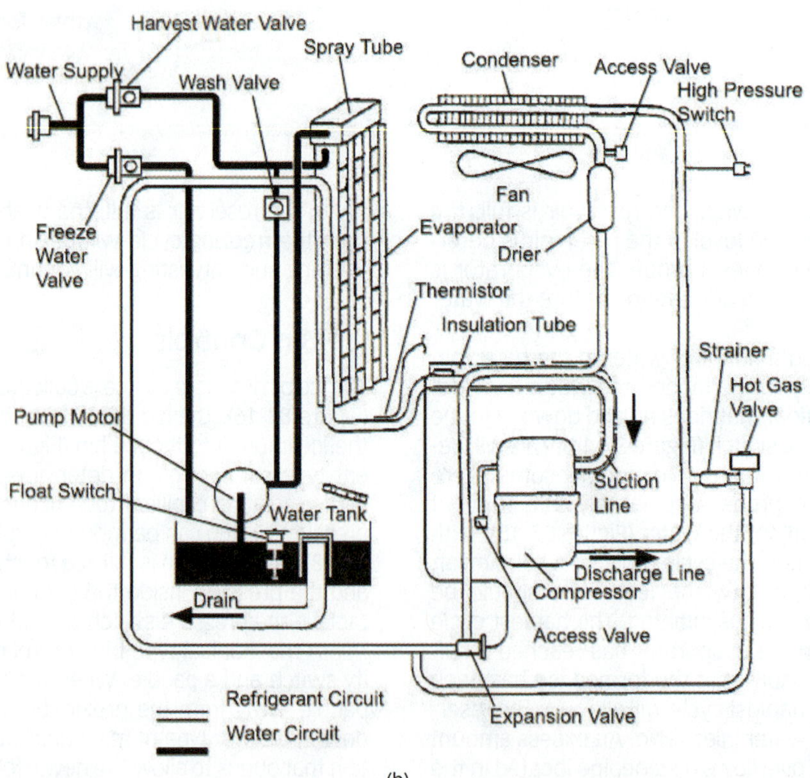

(b)

Figure 87-13 (a) Harvest cycle *(Courtesy of Manitowoc)*; (b) ice machine schematic showing location of hot-gas valve. *(Courtesy of Hoshizaki Technical Support)*

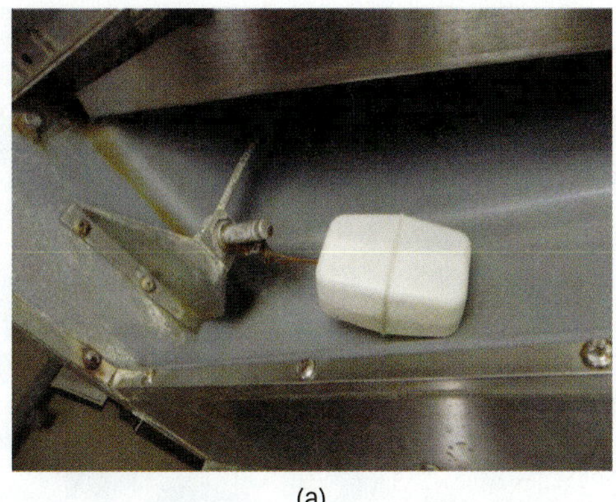

(a)

(b)

Figure 87-14 (a) Simple float in water reservoir is connected to a float switch; (b) float located below pump.

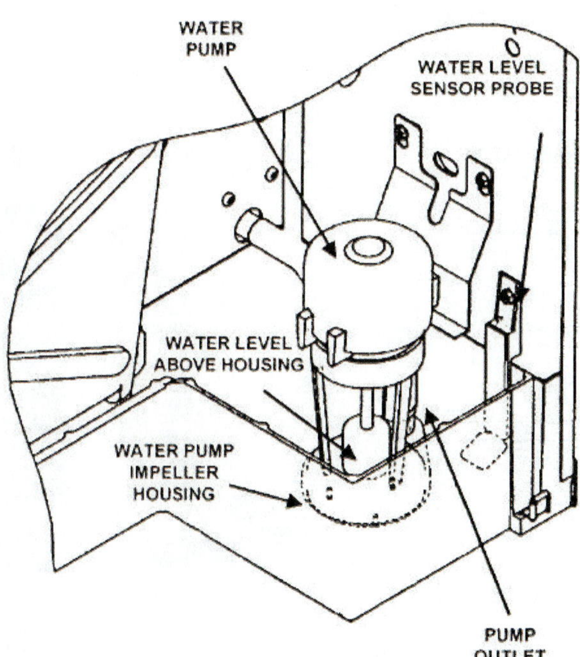

Figure 87-15 Water-level sensor probe. *(Courtesy of Manitowoc)*

Figure 87-16 Ice bin.

Figure 87-17 Ice chute directing cubes to ice bin.

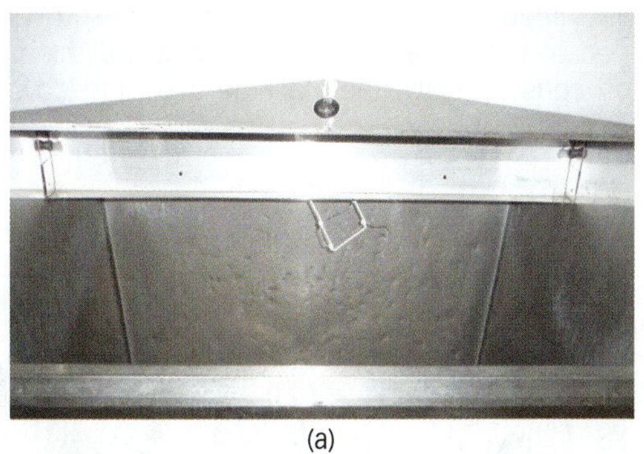

(a)

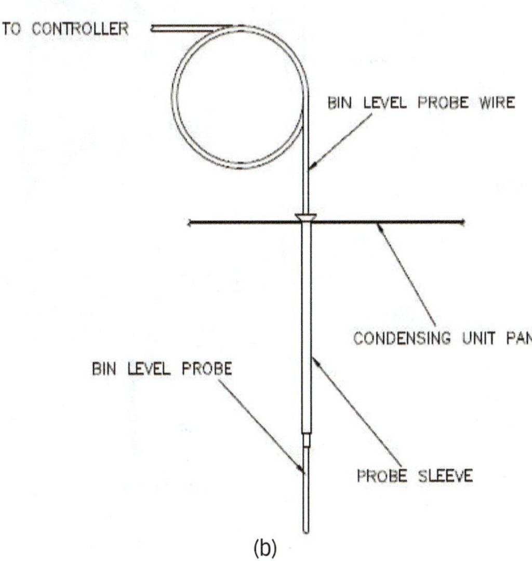

(b)

Figure 87-18 (a) Ice-level probe location in ice bin; (b) ice-level probe. *(Courtesy of KOLD-DRAFT)*

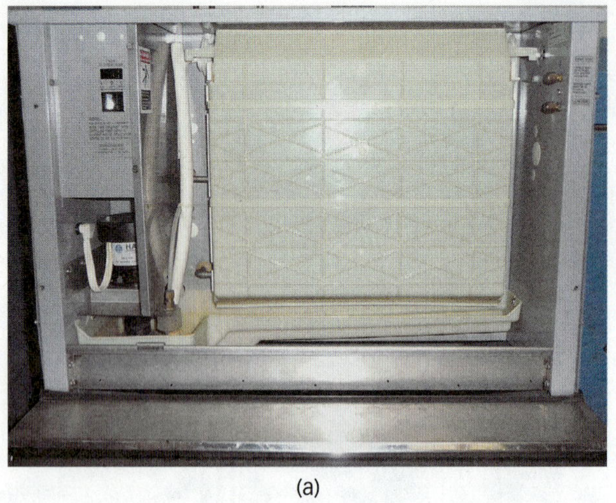

(a)

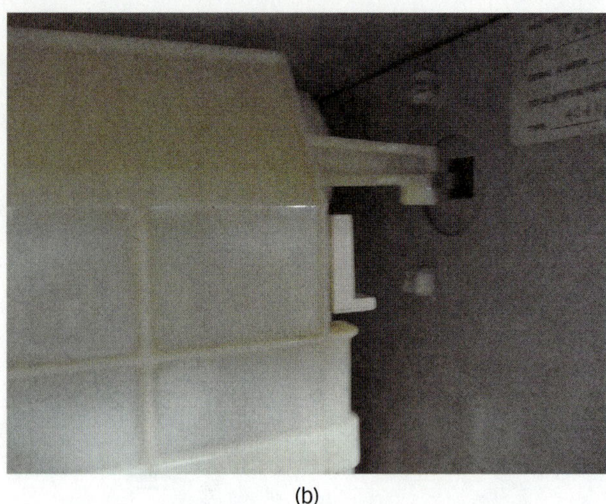

(b)

Figure 87-19 (a) Ice curtain; (b) ice curtain switch.

the harvest cycle, the controller will start a fresh freeze cycle. If the curtain remains open during harvest, indicating the bin is full of ice, the controller will shut the machine off.

Photo-Electric Eyes

Bin level control does not protect against ice machine malfunction but only shuts the unit down when the bin is full of ice. Some ice machines use photo-electric eyes to detect ice as it is made. If no ice production is sensed within a set period of time, the unit will shut down. If this fault is repeated consecutively, the machine may shut down and need to be manually reset. There is a delay programmed into the controller to allow time for the machine to begin making ice when it is first started. This type of system protects the ice machine from damage that could occur from freeze-up and other such problems.

87.7 FLAKE ICE MACHINES

Unlike cube ice machines that go through freezing and harvesting cycles, flake ice making is a continuous process. Most flake ice machines utilize a vertical cylindrical, tube-shaped stainless steel evaporator (Figure 87-20).

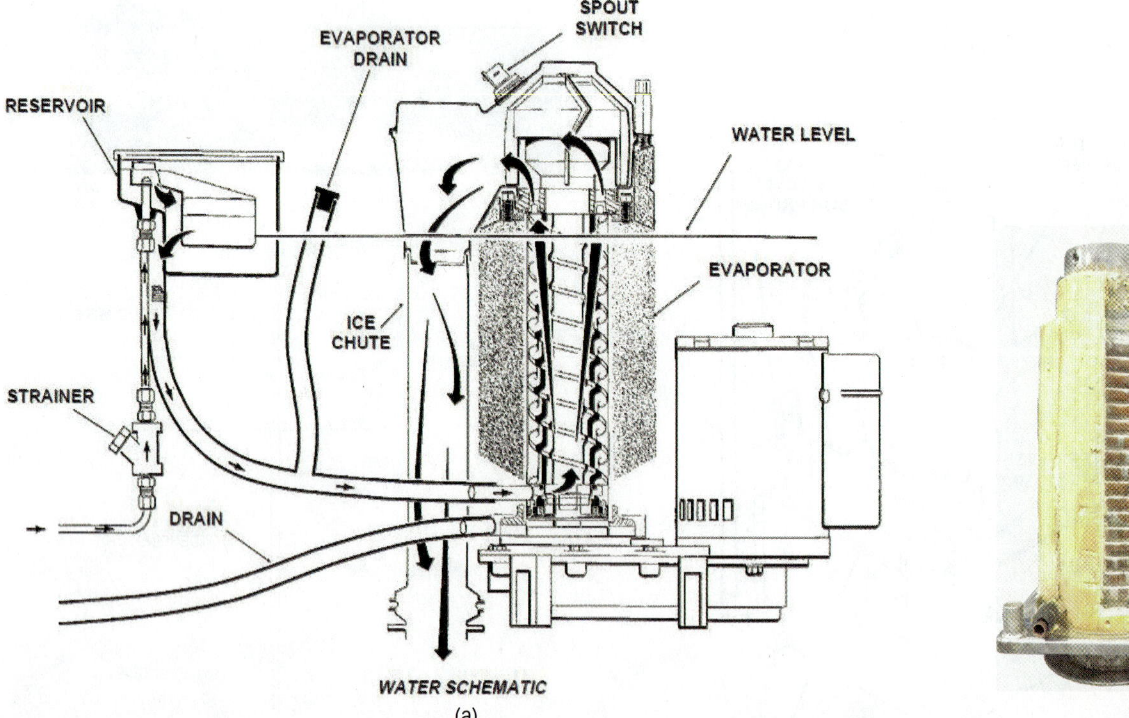

(a)

(b)

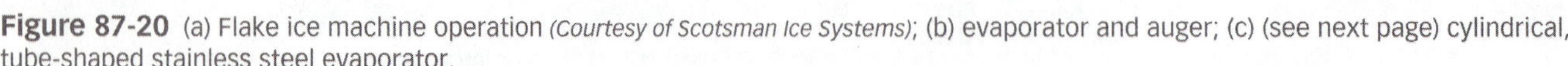

Figure 87-20 (a) Flake ice machine operation *(Courtesy of Scotsman Ice Systems)*; (b) evaporator and auger; (c) (see next page) cylindrical, tube-shaped stainless steel evaporator.

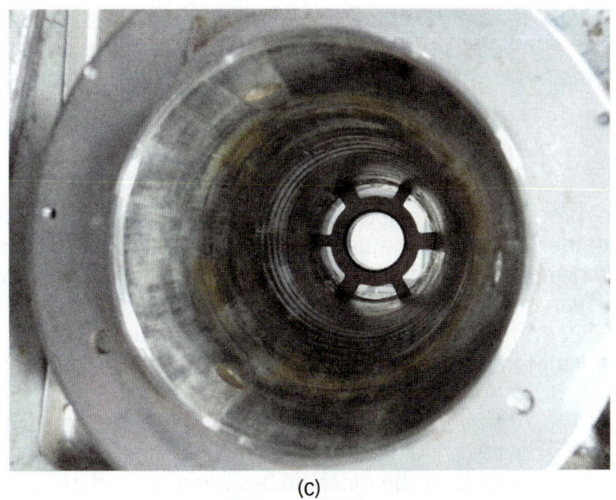

(c)

Figure 87-20 *(Continued)*

Water flows from a reservoir to the evaporator, where it comes into contact with the stainless steel freezing surfaces and freezes immediately on contact. A float switch in the reservoir maintains the water level in the evaporator. A slow-turning spiral-fluted auger located within the evaporator forces the crystalline ice formed on the inside of the evaporator to crack and break away. The flake ice moves upward and out through the top of the evaporator following the rotation of the auger.

Small holes sometimes called breakers at the top of the evaporator tube help to squeeze out any remaining water before it passes to the ice chute and on to the ice bin. An ice sweep revolves with the auger at its top to help move the ice into the ice chute (Figure 87-21). In addition to the typical bin controls described with cube ice machines, a photo-electric eye may be located in the base of the ice chute to detect the production of ice.

Auger Seal

The auger will be driven by an electric motor through a gearbox arrangement. A positive seal prevents water from entering the gear box and oil from entering the ice. It is important for the auger to be perfectly straight inside the cylinder so that it does not wobble. The auger flute should not contact the inner wall of the stainless steel evaporator. An auger seal and its removal are shown in Figure 87-22.

Safety Devices

Due to their design and application, flake ice machines have some additional safety devices. A low-water safety protects against dry operation or freeze-up in the evaporator due to low water flow. A safety in the gear motor circuit will not allow the refrigeration system to operate unless the auger

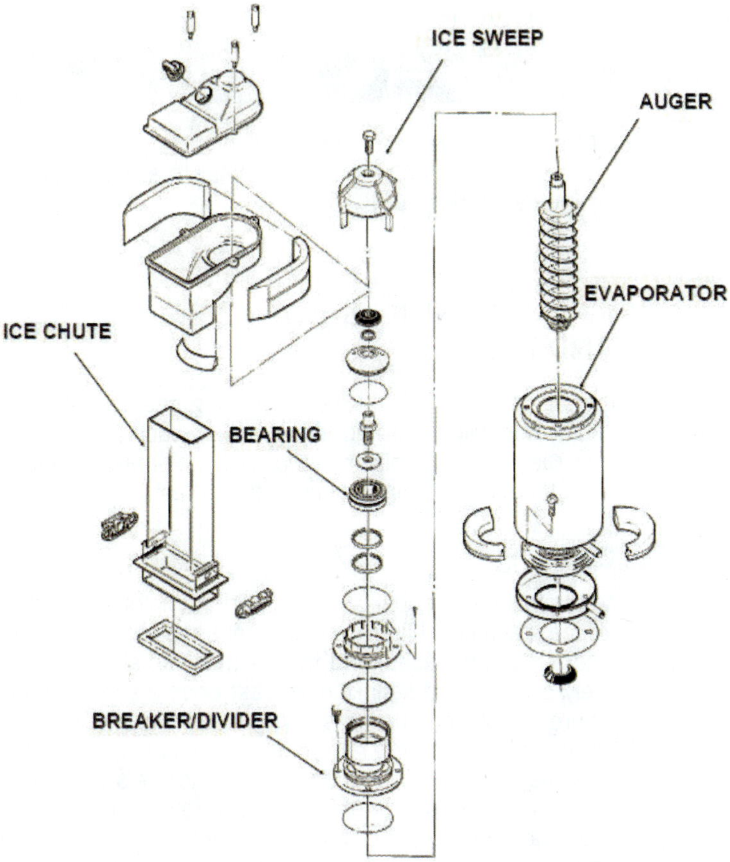

ICE SWEEP
AUGER
EVAPORATOR
ICE CHUTE
BEARING
BREAKER/DIVIDER

(a)

(b)

Figure 87-21 (a) Flake ice machine components showing ice sweep *(Courtesy of Scotsman Ice Systems)*; (b) ice sweep.

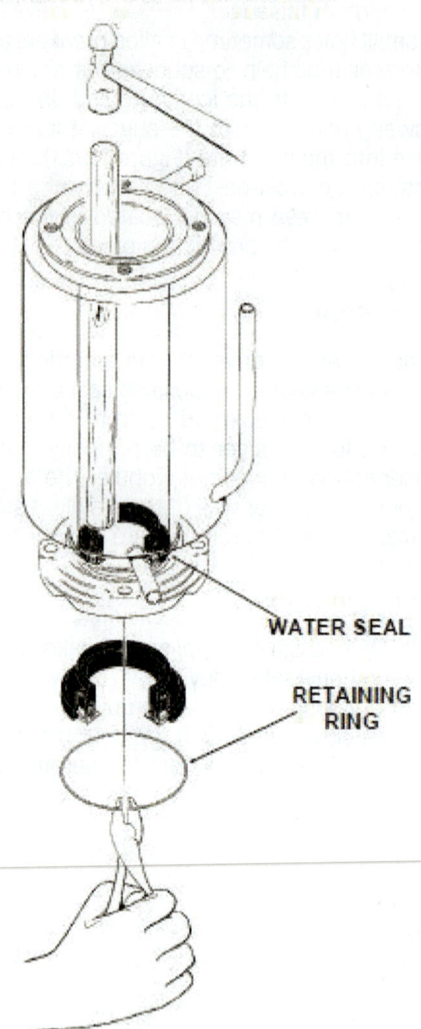

(a)

(b)

Figure 87-22 (a) Removing seal for flake ice machine auger *(Courtesy of Scotsman Ice Systems);* (b) flake ice machine auger seal.

is rotating. A safety in the gear motor circuit will not allow the refrigeration system to operate if there is excessive load on the auger, typically indicated by a high amperage draw.

87.8 ICE PRODUCTION

Ice production will decrease during periods of warmer weather when outside and incoming water temperatures are high. Because of this, the actual ice production per day will vary and must be cross-referenced to the manufacturer's performance data chart. The measured ice production, ambient temperature, and inlet water temperature must be applied to the chart, and ice production should fall within +/– 10% of rated capacity. For the best accuracy, ice production checks should be performed during a normal freeze cycle, not just after the machine has been opened for service or shut down.

For cube ice machines, the entire freeze cycle must be timed (total amount of minutes) from the beginning of one cycle to the beginning of the next. To determine the number of cycles per day, divide 1,440 (the total minutes in a day) by the total minutes that were timed for the cycle. All the ice made during that cycle must be collected and weighed. Multiply the weight of the collected ice by the number of cycles per day to calculate the total pounds of ice produced per day.

For flake ice machines, collect all of the ice made for a 10-minute period. The ice collected must be weighed. Multiply the weight of the ice collected by 144 to calculate the total pounds of ice produced per day.

TECH TIP

To calculate the total number of minutes per day, multiply 24 hours by 60 minutes per hour to arrive at 1,440 minutes per day.

87.9 CLEANING, SANITIZING, AND FLUSHING

Proper maintenance includes keeping the ice machine clean. This will ensure good ice quality, such as taste and size. Cleaning also helps to keep up the ice-making capability of the machine for maximum ice harvest.

Cleaning

Maintaining water quality is an important maintenance item for ice machines. Filters and strainers need to be regularly cleaned and replaced, and the ice machine should be entirely cleaned and sanitized at least once a year. In areas where the water quality is poor, more frequent cleaning will be needed. If parts fail due to calcium or scale buildup, they may not be covered under the machine's warranty if proper cleaning procedures were not followed. There are a number of different cleaning solutions available. It is important to review the MSDS for the cleaner you are using. These should be available when purchasing the product or downloaded online.

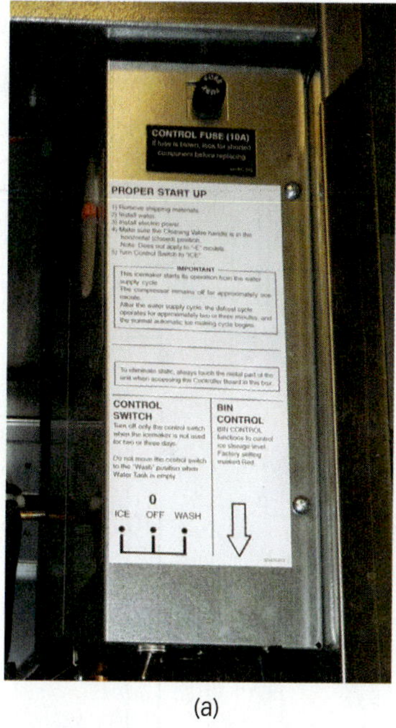

(a)

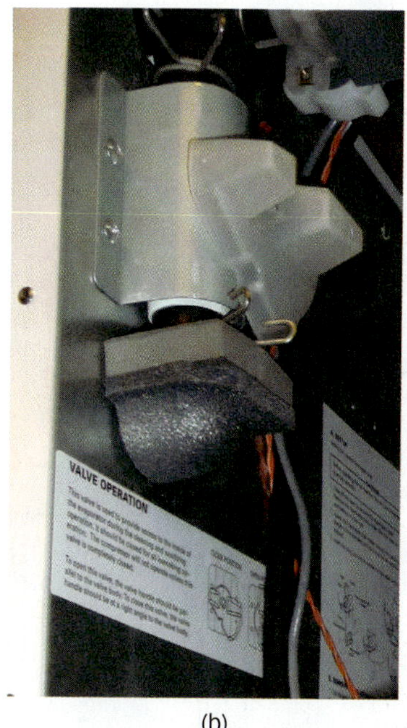

(b)

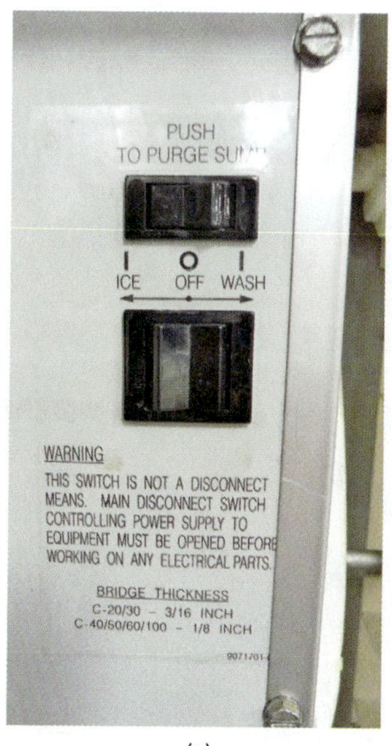

(c)

Figure 87-23 (a) Ice machine instructions for cleaning cycle; (b) switch for cleaning ice machine; (c) valves must be manually repositioned when cleaning this ice machine.

SAFETY TIP

Cleaning solutions used for ice machines must be used correctly, as they are mostly acid solutions. Check the MSDS for the solution that you are using. Whenever you handle solutions, wear chemical splash goggles and chemical-resistant impervious gloves. Use the solutions in a well-ventilated area.

Many ice machines have a cleaning cycle (Figure 87-23). To clean an ice machine, all the ice must be removed from the bin so that the cleaning solution will not contaminate the ice. The water should be turned off and the sump tank drained. The cleaning solution (descaling solution) is mixed with warm water in the correct ratio and poured into the sump tank. When the unit is turned on to the cleaning cycle, the cleaning solution will flow through the system and should be circulated until the unit is clean.

A nylon bristle brush that fits tightly down the channels on the evaporator may be used when the evaporator is extremely dirty and simple flushing is insufficient to remove the built-up scale. Be careful not to damage the evaporator plates. Individual components such as spray tubes and float switches can be taken apart and allowed to soak in solution. The inlet water valve should be taken apart and cleaned along with its strainer.

SAFETY TIP

Augers used for flake ice machines may need to be removed and cleaned. After the auger is removed and dried, it should appear bright and shiny. Be careful when handling the auger because it will have sharp edges.

Sanitizing

Most cleaning solutions will not kill the algae and bacteria that can develop on the ice machine. A separate sanitizing solution or sanitizing tablets should be used. The sanitizing solution will be circulated in the same manner as the cleaning solution. Do not mix the sanitizer with the descaling solution. Sanitizing should be performed regularly and after every repair or replacement of parts. After cleaning and sanitizing, the machine should be thoroughly rinsed and allowed to air-dry before renewed operation.

SAFETY TIP

Never mix cleaning solution (descaling solution) and sanitizing solution together. Mixing the two solutions can create a strong chlorine gas that is difficult to breathe.

Flushing

With cube ice machine design, the water flows across the evaporator plates and the minerals and contaminants are washed away so as not to collect on the evaporator surfaces. This leads to higher and higher concentrations of solids for the water that returns to the reservoir (sump). Periodic flushing of the reservoir reduces the buildup of scale-forming solids. To reduce this buildup, a standpipe is located in the reservoir that regularly overflows, thus flushing out the contaminants.

In addition to a standpipe, a periodic flush cycle controlled by a timer will cycle the unit down and open a flush valve to allow the complete water system to drain (Figure 87-24). This is important for flake ice machines that utilize an auger rather than evaporator plates. Flake ice machines will have greater concentrations of scale buildup as compared to cube

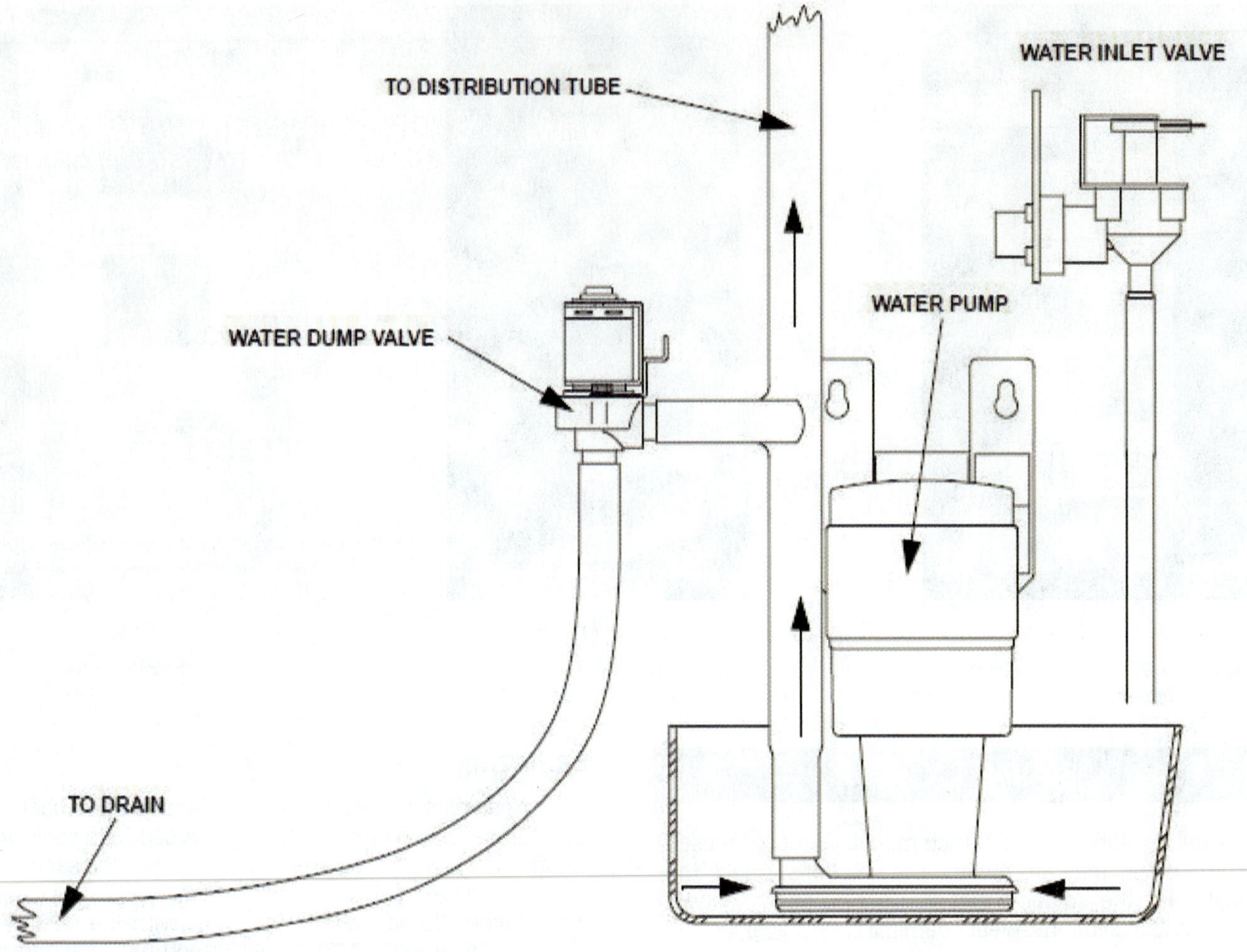

Figure 87-24 Water drain valve. *(Courtesy of Manitowoc)*

ice machines. Since solids collect near the bottom of the auger, a complete draining and flush of the system is required.

SERVICE TIP

A strange sound coming from a flake ice machine as a high-pitched squeak or squeal in combination with low production and soft, mushy ice is an indication of a dirty evaporator.

Some machines have built-in features that provide for a regular automatic cleaning or sanitizing cycle. The number of ice-making cycles is tracked by the machine, and the appropriate cycle is initiated when the limit of ice-making cycles is reached. This reduces the need to manually clean the machine as frequently.

87.10 COMPRESSOR DISCHARGE (HEAD) PRESSURE

The importance of operating with a correct discharge pressure for an ice machine is not very different from any other refrigeration system. Proper refrigerant liquid-line pressure is required to maintain the necessary pressure drop across the metering device. Just as in any typical refrigeration system, low discharge pressure can result from a low refrigerant charge, too much condenser cooling, and a restriction in the liquid line. High discharge pressure is often the result of inadequate condenser cooling. Troubleshooting these conditions follows the normal progression for diagnosis. However, cube ice machines are different in that a hot-gas harvest cycle is employed to release the newly formed ice.

Hot-Gas Harvest Cycle

Low discharge pressure will severely affect the harvest cycle for cube ice machines. This can occur at low ambient temperatures when the condenser has too much cooling. Cube ice machines will have a hot-gas valve open during the harvest cycle to allow the discharge from the compressor to directly enter the evaporator. It is the heat from the gas that melts the cubes just enough to cause them to release from the freezing cells and fall into the ice bin. If the discharge pressure is too low, the hot gas will not properly flow through the evaporator and the harvest cycle will not complete.

To artificially increase the discharge pressure during low ambient conditions, headmaster controls are often used. The headmaster is a valve with three different ports. One connects to the condenser, one to the receiver, and

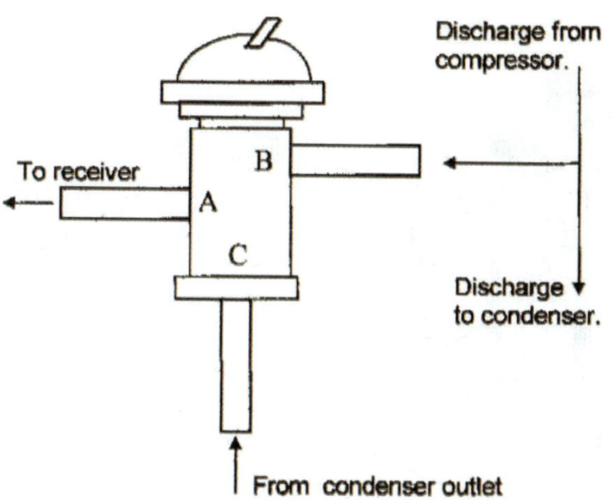

Figure 87-25 Head pressure control valve. *(Courtesy of Hoshizaki Technical Support)*

one directly to the discharge line (Figure 87-25). Normally the refrigerant flows from the compressor to the condenser to the receiver. If the discharge pressure begins to drop, some refrigerant will by bypass the condenser and go directly to the receiver as controlled by the headmaster valve (Figure 87-26). Other methods for controlling discharge pressure include regulating the condenser fan speed for air-cooled condensers and regulating the condenser-cooling water flow for water-cooled condensers.

> **TECH TIP**
>
> The evaporators on cube ice machines continually switch back and forth between freezing cycles and harvest cycles. The evaporators are specifically designed to withstand the resulting effects of contraction and expansion. However, if an ice maker's controls are malfunctioning and extended periods of harvesting (hot gas) result, the evaporator can be damaged.

Remote Condensers

The main reason for high discharge pressure is reduced heat transfer in the condenser. This can be a particular problem for ice machines that are located in high-temperature areas. Sometimes restaurants and cafeterias locate the ice machine near the cooking area for easy access to the ice. This can result in excessive ambient temperatures, and the ice machine condenser will operate poorly. Under these conditions, the ice production will decrease. Even if the unit works

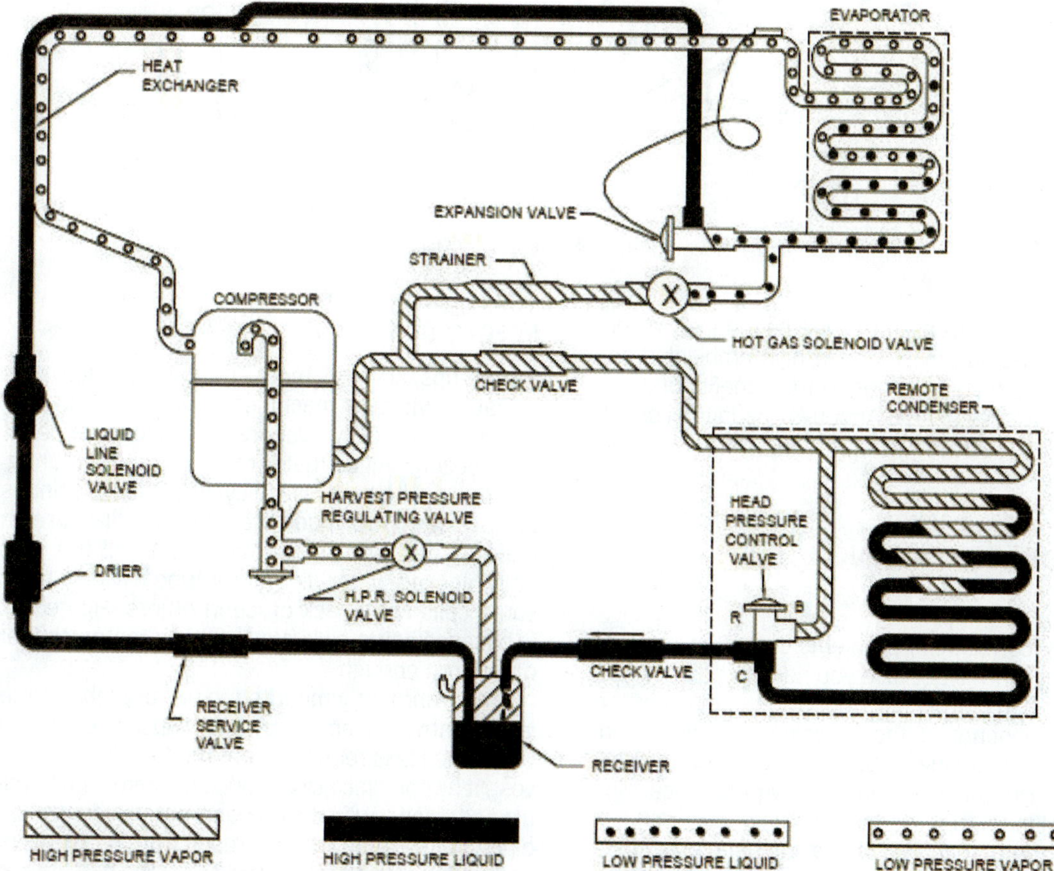

Figure 87-26 Operation of head pressure control valve. *(Courtesy of Manitowoc)*

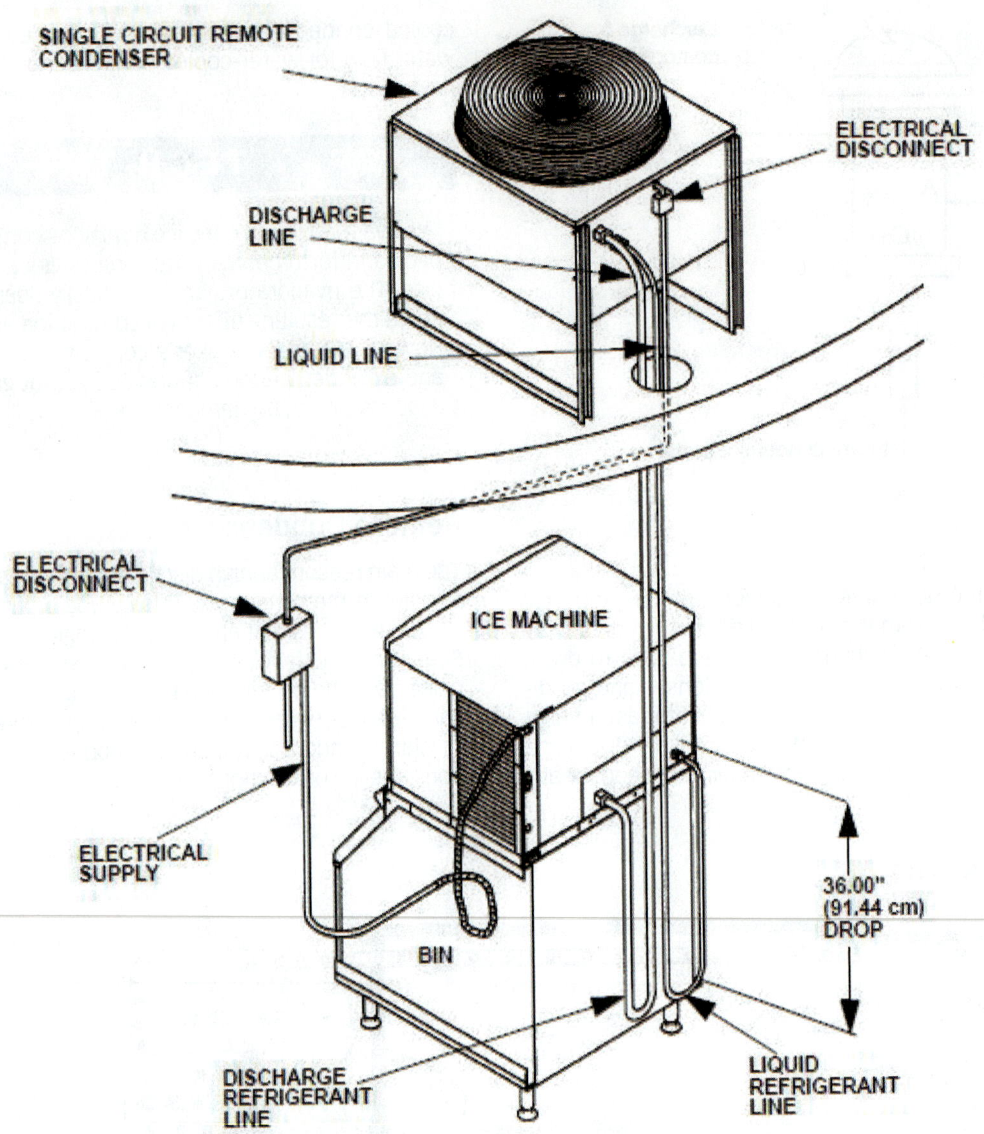

Figure 87-27 Ice machine with remote condenser located outside. *(Courtesy of Manitowoc)*

somewhat satisfactorily, its useful life will be reduced. When installing an ice machine located in a high ambient area, you should advise the customer as to the benefits of a remote condenser (Figure 87-27). Although the initial cost is greater, the unit will perform better, produce more ice, and last longer.

87.11 TROUBLESHOOTING ICE MACHINES

Troubleshooting ice machine problems can often be particularly annoying. This is because you need to address not only the refrigeration components but also the many other mechanical components of the machine. The first step is to gain an understanding of basic refrigeration system troubleshooting. Next, become familiar with how ice machines operate. Finally, refer to the manufacturer's service information for the particular unit that requires attention. Most ice-machine manufacturers provide very good troubleshooting guides and service manual information.

Freeze-ups

Freeze-ups are the most frequent service problem associated with ice machines. Always check the simplest things first. A dirty evaporator causes the ice to stick to the freezing cell so that it may not release during harvest, and during the next freeze cycle the ice continues to build up (bridges). Too much water flow will cause the ice to build up and bridge downward over all the cells. A partially blocked water-distributor tube will lead to bridging of some cells and a lack of ice in others (Figure 87-28). This is because some cells are receiving too much water while others not enough.

An improperly mounted ice-level probe or mechanical level control for an ice bin will cause the ice to back up. When the ice is released from the freezing cells during harvest, it has no place to go. Since every freezing cycle distributes additional water, the evaporator will freeze into a block of ice. A freeze-up will also result from problems pertaining to the flow of hot gas that will affect the harvest cycle. Table 87-2 provides a list of some possible common problems.

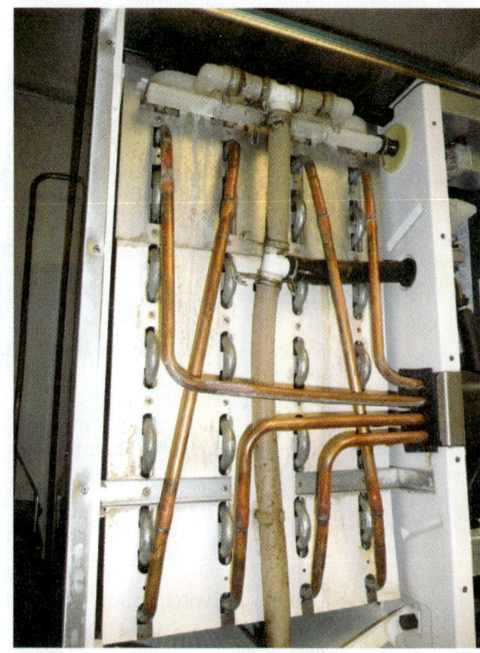

Figure 87-28 Water-distribution tube may need to be cleaned periodically.

Table 87-2 Possible Causes of Ice Machine Freeze-up

Condition	Possible Cause
Ice sticks and does not release during harvest and builds up	Dirty evaporator
Too much water bridges cubes together	Inlet water valve leaking
Bridging on some cells with lack of ice in others	Blocked distribution tube
Ice backs up into evaporator section	Ice bin control faulty
No hot gas through evaporator for harvest cycle	Hot-gas valve stuck closed
Not enough hot gas through evaporator for harvest cycle	TEV leaking during harvest
Not enough hot gas through evaporator for harvest cycle	Low discharge pressure

Poor-Quality Ice and Production

Cloudy ice is often the result of poor-water quality and dirty evaporator surfaces. A decrease in ice production can result from a low water flow. Inlet control valve screens and inlet filters should be checked. A lack of ice production can result from a refrigerant undercharge, which should coincide with a low suction and discharge pressure. Improper TEV superheat can affect cube quality. High superheat can affect the cubes located toward the outlet of the evaporator, making them smaller. Low ice production can also result from high discharge pressure or a partially blocked metering device. A faulty evaporator thermistor can lead to extended harvest times and potentially cause damage to the evaporator. Table 87-3 provides a list of some possible common problems.

Table 87-3 Possible Causes of Poor Ice Quality and Low Production

Condition	Possible Cause
Cloudy ice	Poor water quality
Small cubes and lack of production	Low inlet water flow
Small cubes and lack of production	Refrigerant undercharge
Uneven ice formation	TEV superheat
Small cubes and lack of production, frequent trips	High discharge pressure
Small cubes and lack of production	Blocked metering device
Lack of production	Faulty evaporator thermistor

Control Systems

Electronic control systems consist of a central controller that receives inputs from assorted sensors that monitor the status of such levels as ice thickness, water level, water temperature, and discharge temperature (Figure 87-29). Controllers operate from a low-voltage power supply and use the input from sensors to switch relays on and off. Most mechanical relays found on older units have been replaced by solid-state relays. The relays operate motors and solenoids that require line voltage.

Control boards utilize multilayered circuitry to improve safety protection and help with service diagnosis. LED lights help to troubleshoot a problem. These lights come on during sequences of each cycle, such as fill, harvest, freeze, and pump out (Figure 87-30). When troubleshooting, the LEDs should light in the proper sequence of operation.

Faults can be identified by a lit fault LED. Some units have alarm codes or beeps to indicate such faults as high temperature, long harvest cycle, long freeze cycle, shorted connection, open connection, low voltage, and high voltage. Some units store fault codes that can be later recalled to provide a brief history of the machine's condition to help troubleshoot problems.

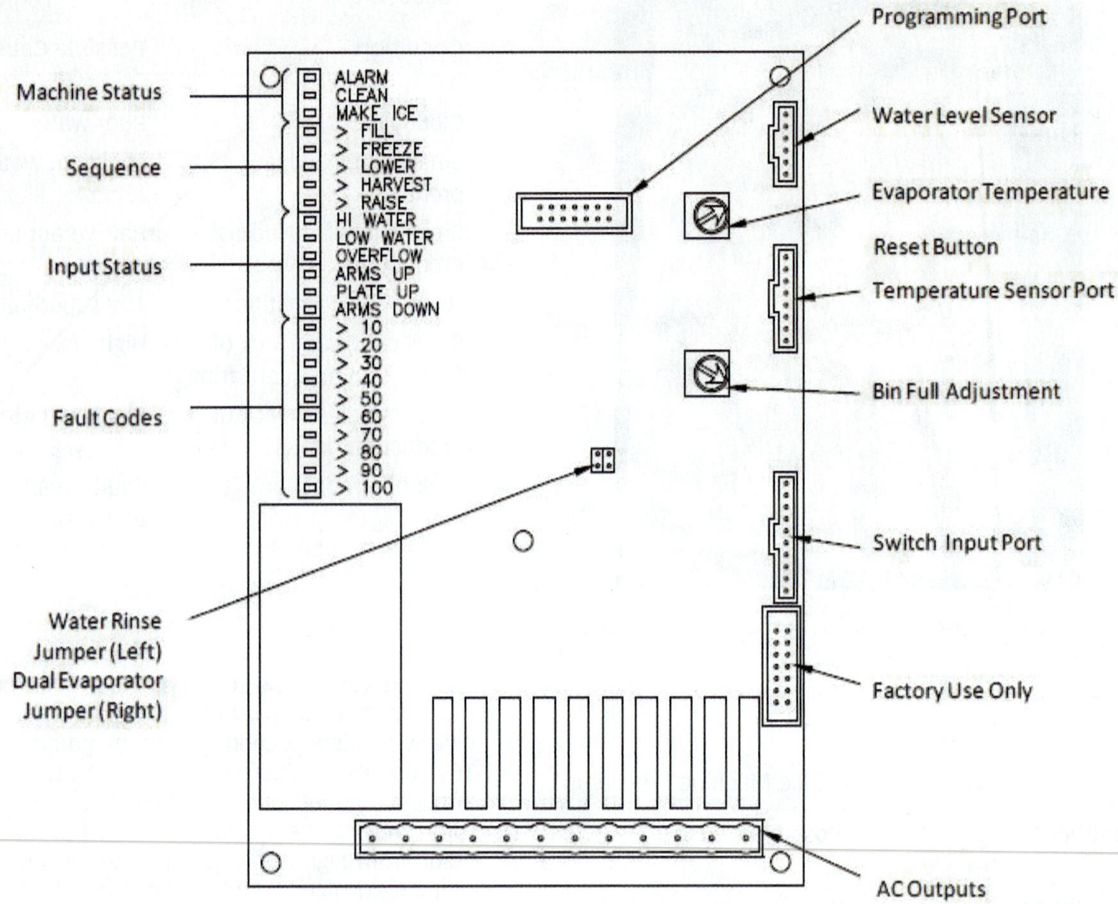

Figure 87-29 Ice machine controller showing inputs and fault codes. *(Courtesy of KOLD-DRAFT)*

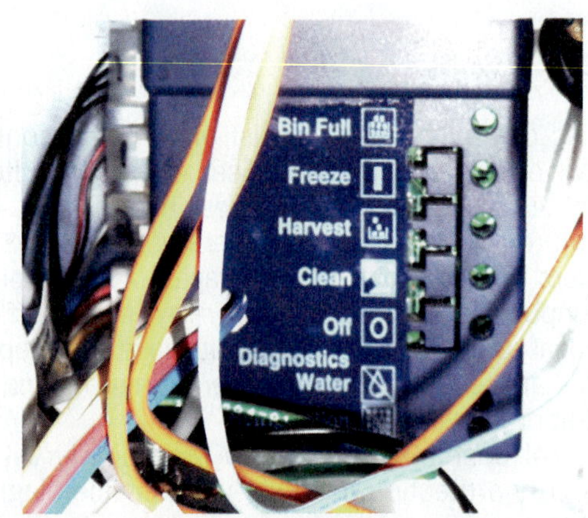

Figure 87-30 LED lights on an ice machine controller.

UNIT 87—SUMMARY

There are many different models and types of ice machines. The most common types are cube ice and flake ice machines. Both types utilize the standard refrigeration cycle. Cube ice machines have a harvest cycle that requires hot gas, while flake ice machines continuously produce ice and utilize a rotating auger. Whenever troubleshooting these machines, try to determine if the malfunction is refrigerant related. Most manufacturers provide detailed service manuals and troubleshooting guides that are unit specific. Controller LED fault code indicators will help to diagnose problems. Take the troubleshooting sequence step by step. Look at the symptoms first before jumping to conclusions, because there are no such things as "phantom" freeze-ups. There is always a logical reason for an operating problem.

WORK ORDERS

Service Ticket 8701

Customer Complaint: Ice Machine Freeze-up

The service technician is told by the customer that the ice machine has stopped producing ice. Upon inspecting the machine, it is found that the evaporator had a freeze-up. The LED display on the controller indicates the unit shut down because the freeze cycle lasted less than 5 minutes for three consecutive cycles. This is because the ice was not releasing from the freezing cells and the ice-thickness sensor was terminating the freeze cycle almost immediately at the beginning of each cycle. The first step taken is to allow the evaporator to thaw out completely before restarting. Once started, the machine seems to operate through a number of cycles as normal. The technician suspects that something has stopped the ice from releasing normally. The evaporator surfaces appear clean and the hot-gas solenoid and water-level probes seem to be working normally. Checking the ice bin probe, the technician finds that its capillary tube has become kinked. The technician replaces the ice bin probe. Evidently, the kinked capillary tube provided a false signal, allowing the ice to back up from the ice bin, which led to the freeze-up.

Service Ticket 8702

Customer Complaint: New Ice Machine

The restaurant manager tells the technician that she wants to replace her ice machine. It is only a few years old but it had problems right from the start. It never produced enough ice and the machine tripped out frequently. She would never buy that brand again. The technician is very familiar with the brand and knows that it is a reliable machine. The machine is located in a corner of the kitchen, and there is little air flow for the condenser. However, this is about the only place the machine could be located due to the limitation of space. Because the manager is familiar with the operation of that unit, the technician suggests a similar model of the same brand ice machine but one equipped with an outside condenser. This will cost more but will provide for better operation and longer life. The manager is at first skeptical but decides to take the technician's recommendation. A month after the installation, the technician checks with the manager to find that she is very happy with the new machine and always has a good supply of quality ice.

Service Ticket 8703

Customer Complaint: No Ice

The bartender tells the technician that the ice machine shuts down and is not making any more ice. Checking the controller, the technician finds that the LED fault indicator for the harvest cycle is lit. For protection, the unit automatically shuts down when the harvest cycle exceeds 20 minutes. The long harvest could be due to ice sticking to the freezing cells and not releasing because the evaporator is dirty. However, if the cycle goes on for an extended period, the evaporator could reach a temperature over 120 degrees, and this would also lead to a trip. But the evaporator high-temperature LED fault indicator is not lit. Checking the maintenance records, the technician discovers that the machine was recently cleaned and the water filter changed. So instead of cleaning the machine, the technician manually resets it and places it into the harvest cycle. The hot-gas solenoid does not seem to be functioning because the line between it and the evaporator is not hot. The machine is shut down and the power to the unit secured. The solenoid coil is removed from the hot-gas valve and checked with a multimeter. The reading shows an infinite resistance, indicating an open coil. The technician replaces the coil and restarts the machine. It now begins to harvest and cycle normally.

UNIT 87—REVIEW QUESTIONS

1. List eight applications for commercial ice machines.
2. How much ice will a 12-oz drinking cup approximately hold?
3. About how many dice ice cubes would it take to equal 1 lb of ice?
4. How is crushed ice made?
5. Why is flake ice ideally suited for any "contact" presentation, such as displaying fish at the supermarket?
6. Can ice machine cleaning solution be mixed with sanitizing solution?
7. Why is the flushing cycle important for flake ice machines?
8. What is the purpose of the auger seal on a flake ice machine?
9. List three safety devices used for flake ice machines.
10. What might cause a high-pitched squeak or squeal in combination with low production and soft, mushy ice in a flake ice machine?
11. Why would it be desired to increase the head pressure for a cube ice machine?
12. How can discharge pressure for a cube ice machine be artificially increased?
13. What could cause frequent tripping on high-discharge pressure and poor ice production?
14. What would happen if the ice machine had a partially blocked water distributor tube?
15. What could cause ice to build up and bridge downward over the freezing cells?
16. What could happen if there were a burned-out solenoid on a hot-gas valve?
17. What could cause ice to build up in the ice bin?
18. List two things that might lead to a decrease in ice production.
19. Why might an ice machine produce cloudy ice?
20. What type of fault could lead to extended harvest times?

UNIT 88

Troubleshooting Refrigeration Systems

OBJECTIVES

After completing this unit, you will be able to:

1. explain the different methods to electrically troubleshoot a refrigeration system.
2. explain how to troubleshoot a defective compressor.
3. explain how to troubleshoot a defective start relay.
4. explain how to troubleshoot a defective capacitor.
5. explain how to troubleshoot metering devices.
6. explain how to troubleshoot common evaporator problems.
7. explain how to troubleshoot common condenser problems.
8. explain how to search for leaks in a refrigeration system.

88.1 INTRODUCTION

Patience is a requirement when troubleshooting refrigeration systems. It is important to take the time necessary to properly diagnose a system problem, as rushing to a quick diagnosis often leads to misdiagnosing the root cause of the problem.

A good service practice is to spend some time speaking to the customer before looking at the system. Listen to the customer's complaints and then ask some pertinent questions about the system, such as the following:

- How old is the unit?
- When was it last repaired, and what was done?
- Has the unit been working OK up until the time of the breakdown?
- Have you noticed any strange sounds or erratic operation lately?

After speaking with the customer, it is time to gather some information about the system. Conduct a visual inspection of the system and look for any obvious system problems. This is another important part of the troubleshooting process. During the initial visual inspection try to determine the following:

- Unit model and serial numbers
- Required supply voltage and phase
- Current ratings of motors
- Full load amps for most motors
- Rated load amps (RLA) for compressor motors
- Type of metering device used
- Type of refrigerant in the system
- Condition of the evaporator and condenser coils
- Overall condition of the equipment

It may be necessary to remove panels on the equipment to see everything. A crucial part of the initial inspection is the examination of the evaporator and condenser coils. An iced-up evaporator coil and dirty condenser coil are two common problems that can easily be identified during a visual inspection. The cause may not be as apparent, but the symptom is easy to discover.

Identifying the refrigerant type helps to determine the correct operating pressure when it comes time to install a set of service gauges on the system to read its working pressures. Finding the correct supply voltage and whether it is a single- or three-phase system will also help when testing the circuit.

Another part of a visual inspection is to look for any oily pipes or parts, which is usually a good indication of a refrigerant leak. Also carefully (it may be hot) feel the head of the compressor to see if it is cold, warm, or hot. This helps determine if the compressor is the cause of the problem. For example, if the compressor is extremely hot and not running, it may be off due to an overload. You may want to start troubleshooting there to find out why this has happened.

Visually inspecting all the major components of a system takes a little extra time, but it is time well spent and is necessary to efficiently troubleshoot any refrigeration system.

SAFETY TIP

Taking shortcuts on a job may save time but may cause a technician to work in an unsafe manner. This is not a wise tradeoff. Do not take any shortcuts that may cause you to work in an unsafe manner. It is simply not worth the few extra minutes you may have gained.

88.2 ELECTRICAL TROUBLESHOOTING

Electrical troubleshooting can be divided into two areas: (1) electrically diagnosing controls (switches, relays, and contactors) and (2) electrically diagnosing loads (coils, fan motors, and compressors).

There are three major tools commonly used to troubleshoot the various electrical components of a refrigeration system: (1) voltmeter, (2) ammeter, and (3) ohmmeter. Each measures a different electrical characteristic in a circuit and is needed to properly diagnose an electrical fault within a system. Many technicians will use a multimeter, as shown in (Figure 88-1), to measure all of these electrical characteristics.

Figure 88-1 A digital multimeter used to measure various electrical characteristics of a circuit.

SAFETY TIP

When replacing or repairing any electrical components, always verify that the voltage source is truly disconnected from the circuit. Test the circuit for the presence of voltage with some type of voltmeter or voltage indicator. Do not solely rely on the electrical disconnect to ensure the voltage is disengaged. Always verify this yourself.

A voltmeter measures electrical potential (voltage) at a switch or a load (such as a motor or the coil of a contactor). For a load to operate properly, the correct voltage must be applied. Most units will have the required applied voltage stamped on a clearly marked nameplate. If a load is not energized, one of the first steps in the troubleshooting process is to determine if the proper voltage is applied to the nonenergized load. For most loads, the applied voltage must be within ± 10 percent of the nameplate voltage. If no voltage or the incorrect voltage (outside the ± 10 percent tolerance) is present at a load, then the problem lies in the voltage supply, controls, or wiring leading to that load. When checking the applied voltage to a load, always measure the voltage at the load and not at an electrical junction before the load. There may be an issue with the wiring leading to the load, so it is more accurate to measure the applied voltage directly there.

A voltmeter can also be used to check the operation of the various electrical controls in a circuit. Electrical controls fail in one of two ways: they fail to close or they fail to open. A voltmeter can be used to determine whether a control is electrically open or closed by measuring the line voltage across the control's contacts. A measured voltage indicates that the contacts are electrically open. If the contacts should be closed, then the circuit problem lies in the control.

SAFETY TIP

When working on electrical circuits, safety should always be first on your mind. Electrical circuits should always be de-energized before repairing or replacing any electrical component or wiring. De-energize at the system's disconnect and then follow standard lockout/tag-out procedures.

A zero measured voltage would seem to indicate the control is closed. However, this is not always true. The absence of a line voltage across a contact does not always indicate the contact is closed. If two or more controls wired in series are electrically open in a circuit, then there will not be a line voltage drop across their contacts. Therefore, this method of electrically troubleshooting a control is not always reliable.

A more reliable method of electrically troubleshooting controls is a procedure referred to by many technicians as hopscotching. This allows a technician to easily isolate an open electrical component. First, attach one probe of a voltmeter to a common point in the circuit (Figure 88-2). The blue probe is attached to the L2 leg of the circuit and will stay attached at this point during the process. This is an electrically common point for all the loads in the circuit. Now the red probe can be moved around the circuit to help determine which switch is electrically open or closed.

Here is an example of the process. As shown in Figure 88-2, the red probe of the voltmeter is initially connected before the CR contact and should read line voltage, which is 115 V in our example. When the red probe is moved to point 6 on the diagram, 115 V should again be read on the voltmeter if the CR contact is electrically closed. If the voltmeter reads 0 V, the CR contact is electrically open. This process can be repeated for all of the switches in the circuit. If line voltage is read before and after the switch, the switch is

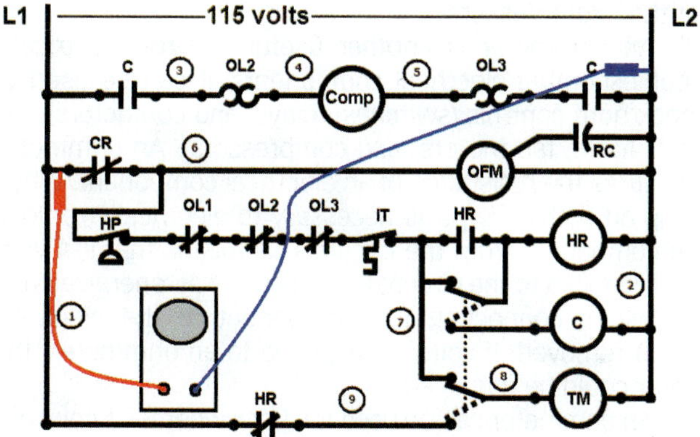

Figure 88-2 Hopscotching an electrical circuit.

Figure 88-3 A digital clamp-on ammeter used to measure the amperage drawn of the load in a circuit.

electrically closed. If line voltage is read before but not after the switch, the switch is electrically open.

An ammeter is another effective troubleshooting tool. A clamp-on ammeter is shown in Figure 88-3. It is used to measure how much current a load is drawing. This enables a technician to verify that a load is electrically energized. If a load is drawing current, it is electrically energized. However, the total amount of current being drawn will indicate whether the unit is operating properly. If a load is drawing current outside the specifications of the manufacturer of the component, then there is a problem with either its operation or application.

Always check with the manufacturer of the component to determine its correct operating current range. For many (but not all) fan motors, the current drawn by the fan motor should be within a range of –25%/+10 percent of its nameplate full load amperage rating. The amperage drawn by a compressor varies depending on its applied suction, discharge pressure, and applied voltage. Compressor manufacturers will publish the correct operating current draws for their compressors.

An ohmmeter is another useful tool for electrically troubleshooting electrical components. It can be used to check both controls (switches, relays, and contactors) and loads (coils, fan motors, and compressors). An ohmmeter measures the resistance of an electrical component. When using an ohmmeter, it is necessary to electrically remove the component from the circuit. This means that if a voltage is applied to the circuit, it needs to be de-energized and the wiring connecting the component to the electrical circuit removed. If voltage is applied to an ohmmeter, the meter could be damaged.

An ohmmeter can be used to check whether a control's contact is electrically open or closed. Once the component is disconnected from the circuit, the leads of an ohmmeter can be placed across the control's electrical contacts. If the ohmmeter measures a resistance of approximately 0 Ω, the contacts of the control are closed. If the ohmmeter measures an infinite resistance (OL) across the control's contacts, the contacts are open.

When using an ohmmeter to check the condition of a load, the leads are placed across the windings of the load. If the measured resistance agrees with the manufacturer's specifications, the windings of the load are OK. If a 0 Ω resistance is measured across the windings of the load, the windings are shorted. If an OL value is measured across the windings, the windings are open.

88.3 MECHANICAL REFRIGERATION TROUBLESHOOTING

Troubleshooting the mechanical side of a system involves analyzing or investigating the refrigerant flow either throughout the entire system or through any of the components. This requires being able to determine the refrigerant's pressure and temperature at various locations throughout a system.

When analyzing the entire system, the actual refrigerant conditions at various locations in a system are compared to the design conditions of a properly operating system. This will normally allow a technician to determine a system's defect. For example, if a system is discovered to be operating with both lower than normal suction and discharge pressures, and the superheat value of the refrigerant leaving the evaporator is higher than normal with a lower than normal subcooling value leaving the condenser, a technician can determine that the system has a low refrigerant charge.

Many times a technician will need to analyze the process through a component such as the compressor, condenser, or evaporator. This requires an understanding of the component and how it is designed to function in a system. For example, a refrigeration compressor is designed to pressurize a low-pressure refrigerant vapor to a higher-pressure refrigerant vapor. If a compressor is electrically operating but fails to sufficiently increase the pressure of the refrigerant from its inlet to its outlet, the compressor is mechanically defective, and more than likely the compressor will need to be replaced.

In general, a technician can analyze the operation of the evaporator as well. If the superheat value of the refrigerant leaving an evaporator is higher than normal, the evaporator is starved for refrigerant. If the superheat value of the refrigerant is lower than normal, the evaporator is flooded with refrigerant. Finding the reason why the evaporator is either being starved or flooded, however, will require looking at the entire system and analyzing other conditions.

When analyzing the condition of the refrigerant throughout a system or through a system component, a technician must allow the pressures and temperatures of the refrigerant to stabilize upon restarting if the system has been off. When troubleshooting, it is a good service practice to let a system run for about 10 minutes before recording any of the system's pressures or temperatures.

88.4 TROUBLESHOOTING A DEFECTIVE COMPRESSOR

Replacing a compressor on a refrigeration system is never an easy or inexpensive task. If a compressor is found to be defective, every effort should be made to verify that a correct diagnosis is made. Sometimes compressors that have been changed out in the field are later found to be operational.

Compressor problems can be divided into two groups: mechanical and electrical. Mechanical defects are problems that affect the operation inside the compressor, such as blown valve plates, a broken crankshaft, or worn pistons. These defects will either cause a compressor not to pump any refrigerant or pump to below its rated capacity.

It is generally easy to determine if a compressor is not pumping adequately. At start-up, verify that the compressor is electrically energized, then measure its amperage draw. It will be lower than normal. Next, monitor the compressor's suction and discharge pressures. Neither the suction nor the discharge pressure will change dramatically. There may be a very slight change in pressure, but the suction pressure will stay very high and the discharge will stay low.

To determine whether a compressor is pumping to its rated capacity is not quite as simple. A technician must measure the compressor's actual amperage draw and compare this to the amperage draw as stated by the manufacturer. The amperage should be within ±5 percent of the manufacturer's stated value. If it is drawing less than 5 percent of its rated amperage draw, it is not pumping to its rated capacity.

Compressors can fail to start as a result of either an electrical defect or a mechanical defect. Mechanically, one or more of the pistons within the compressor can become locked and fail to move. The compressor will attempt to start, but since the pistons will not move, the compressor draws high amperage, causing it to shut down on overload.

There are three major electrical defects that will cause a compressor not to operate. One possibility is the motor windings of the compressor are open, shorted, or grounded. Another possibility is that the starting relay or capacitor as shown in Figure 88-4 are defective. The third possibility is that the incorrect voltage is applied to a compressor.

Figure 88-4 Starting components (relay and capacitor) used on a single-phase compressor.

Figure 88-5 Measuring the resistance of a compressor.

To check the condition of the compressor's motor windings, a technician will need to measure the resistance of the windings using a standard ohmmeter, as shown in Figure 88-5. For single-phase compressors, measure the resistance value of both the run and start windings and measure the resistance of each winding to ground. For three-phase compressors, each of the three windings will need to be checked. If the correct resistance value is measured across its windings, as shown in Figure 88-6, and an OL is measured from the windings to ground, as shown in Figure 88-7, then the windings are satisfactory.

TECH TIP

The voltage applied to a three-phase motor must be balanced, meaning the applied voltage to each leg must be relatively the same. The applied voltage must not deviate more than 2 percent from the average supplied voltage. A voltage imbalance of greater than 2 percent will cause the windings within a motor to generate heat beyond safe levels, leading to premature motor failure.

Figure 88-6 Measuring the resistance of a good compressor.

Figure 88-7 Measuring the resistance of an open winding on a compressor.

There is a common scenario in which a compressor may appear to be defective when in fact it is not. A compressor with an internal overload that has overheated will shut down and will not restart until the compressor has cooled. If a technician performs a resistance check on the compressor, the first step will be to measure the resistance from common to run and common to start, which will be OL. The technician may interpret this to be an open winding in the compressor. However, if the compressor is allowed to cool down and the internal overload to reset, the technician may find that the compressor will start normally and will not need to be replaced. The reason why the compressor overheated will need to be identified and resolved.

A compressor with a defective run capacitor will draw higher than normal amperage and cycle off on its overload or not start at all. If a compressor incorporates a start relay and a start capacitor and either is defective, the compressor will most likely not start.

Incorrect voltage applied to a compressor may cause it to run for a brief time then cycle off on its overload or not start at all. A service technician must first verify the correct voltage for the compressor and then measure to see what actual voltage is applied. Most compressors are rated with a tolerance of ±10 percent. If the applied voltage is outside these limits or those stated by the manufacturer, it must be corrected before the compressor can be properly diagnosed.

Using a Compressor Analyzer

One option when troubleshooting a single-phase hermetic compressor that does not start is to use a compressor analyzer. This tool will allow a technician to start a single-phase compressor directly without using any of the system's starting components. It will connect directly to the terminals of the compressor and use its own starting components to start the compressor. This will allow a technician to determine if the fault is with the compressor or with its starting components. Some compressor analyzers will also allow a technician to check the integrity of the compressor's windings and momentarily reverse the rotation of the rotor within the compressor.

Compressor Amperage Ratings

Compressor manufacturers will publish a rating chart for each compressor they manufacture. These charts are most often in a table, or in graphic format. The compressor chart will list the correct amperage draw for the compressor under various operating conditions. To use the chart, a technician must know the evaporating temperature, condensing temperature, and voltage applied to the compressor. By using the compressor chart, a technician can determine the correct amperage draw for the compressor and use that information to accurately troubleshoot the compressor and the system.

Obtaining these charts can be time-consuming, but the extra time involved in finding the correct amperage is worthwhile. With the use of the Internet, this information is now easier to obtain.

Most manufacturers will stamp the rated load amperage (RLA) on their compressors. However, a service technician cannot use this value to determine the correct operating amperage. RLA is a mathematical calculation required to meet Underwriters Laboratories (UL) approval. The compressor manufacturer must run a series of tests to determine the maximum continuous amperage before the overload trips. Once that has been determined, UL divides the MCC by 1.56 to determine the RLA. The primary value of the RLA is to determine at what amperage draw the compressor overload will trip. This will be used to determine the fuse/circuit breaker size and the wire size. The RLA is not a reliable indicator of what the compressor amp draw should be. The amp draw of a properly operating compressor can be considerably lower than its RLA rating. A better indicator would be current ratings for specific operating conditions.

When measuring the amperage draw of a compressor, make sure to read only the amperage drawn by the compressor. Choose a location on the wiring diagram that will isolate the compressor from any other loads, such as the condensing fan motor. It is important to measure amperage draw of the *compressor only* when trying to determine if it is operating properly.

Compressor Overheating

A major cause of premature compressor failure is the compressor operating at elevated temperatures. Temperatures in the compressor head and cylinder become so hot that the oil thins and loses its ability to lubricate. This may cause rings, pistons, and cylinders to wear, resulting in blow-by, leaking valves, and metal debris in the oil.

Cylinder temperatures exceeding 300°F will begin the breakdown of the oil, and at 350°F it will vaporize. When servicing a system, it is a good practice to check the operating temperature of the cylinders. To measure the cylinder temperature of a compressor, place a temperature gauge no more than 6 in out on the discharge line from the compressor and add 50°F–75°F to the measured pipe temperature. This will give an approximate temperature of the cylinders. For most applications, the discharge line temperature should be 225°F and below. A temperature of 250°F indicates a dangerous level. A temperature of 275°F and above indicates certain failure of the compressor.

Figure 88-8 This compressor has been changed out four times.

Preventing Repeat Compressor Failure

When a compressor fails, its troubleshooting and replacement are relatively simple and common procedures. However, discovering *why* the original compressor failed may not be as simple. But it is extremely important for a technician to find this answer. Replacing the defective compressor without finding the cause will most likely result in the replacement compressor failing as well. Replacing the compressor a second, third, or fourth time is not advantageous to the service contractor or to the customer. It can cause the service contractor to lose money and quite possibly the customer. Figure 88-8 shows a compressor that has been changed four times within an 18-month period.

In most cases, upon original inspection of a defective compressor it is difficult to determine what caused the compressor to fail. The compressor is not normally running and the system is not refrigerating. After replacing the compressor, the technician should spend time discovering why the original compressor failed. With a detailed inspection, the technician should be able to find the cause. The inspection should include checking the operating suction and discharge pressures, the amount of superheat at the compressor inlet (Figure 88-9), the return-gas temperature,

Figure 88-9 Checking the superheat at the inlet of a compressor.

the discharge-gas temperature, the amperage draw, and the applied voltage to the compressor. By comparing these readings to the manufacturer's specifications, the cause should be identifiable.

Some typical causes of a compressor failure are liquid returning to the compressor, high return-gas temperature, high discharge-gas temperature, or incorrectly applied voltage to the compressor. Most of these causes can be found once the compressor is up and running.

There are times, however, when the apparent cause cannot be found upon start-up. The technician should monitor the system's operation for a period of time to find the cause. The problem may develop after the system has been running for some time—for example, during or after a defrost cycle—or as the result of an iced evaporator coil.

Another helpful troubleshooting step is to disassemble the defective compressor and examine the valve plates, pistons, crankshaft, bearing surfaces, and windings. Although this is time-consuming, it can significantly help to identify or confirm the cause of the original failure. Unit 15 has more details on specific types of compressor failures and how to recognize them.

88.5 STARTING RELAYS

Many commercial refrigeration systems use single-phase compressors. Single-phase compressor motors have a run winding and a start winding. The main purpose of the start winding is to add an additional electrical circuit that is out of phase with the main circuit. The phase displacement between the start and run windings creates the torque required to start the compressor. An external relay is used to remove the starting components from the circuit after the compressor starts. Depending upon the type of compressor motor used, the electrical components that must be dropped out can be just the start winding, just a start capacitor, or the start winding and start capacitor. Split-phase motors drop out the start winding, capacitor start motors drop out both the start capacitor and the start winding, and capacitor start-run motors drop out just the start capacitor.

There are three types of starting relays used on single-phase compressors:

- Current relay (Figure 88-10)
- PTC relay (Figure 88-11)
- Potential relay (Figure 88-12)

Current Relays

Current relays are unique. They operate based on the amp draw through the coil, not applied voltage. Current relay coils are wired in series with the run winding of the motor they are controlling (Figure 88-13). The normally open contacts of the current relay are wired in series with the start winding (Figure 88-13). When the compressor is first energized, there is a circuit through the current relay coil to the run winding but not to the start winding, because the current relay contacts are normally open. The compressor cannot start and draws locked rotor amps. The high amp draw causes the normally open contacts to close, energizing the starting winding. The compressor starts, the amp

Figure 88-10 Current relay.

Figure 88-11 PTC relay.

Figure 88-12 Potential relays: RBM on left, GE on right.

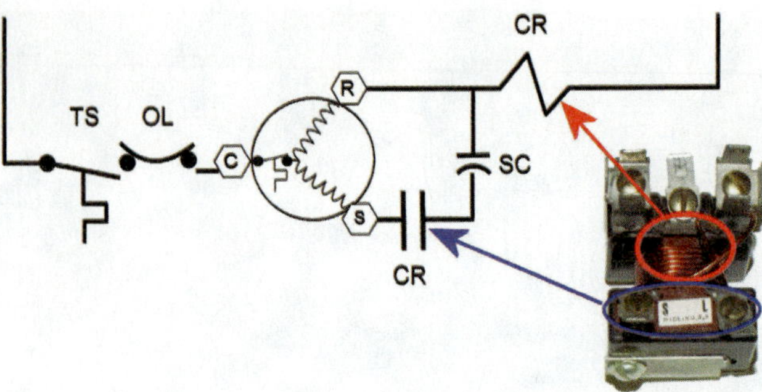

Figure 88-13 Current relay wiring diagram.

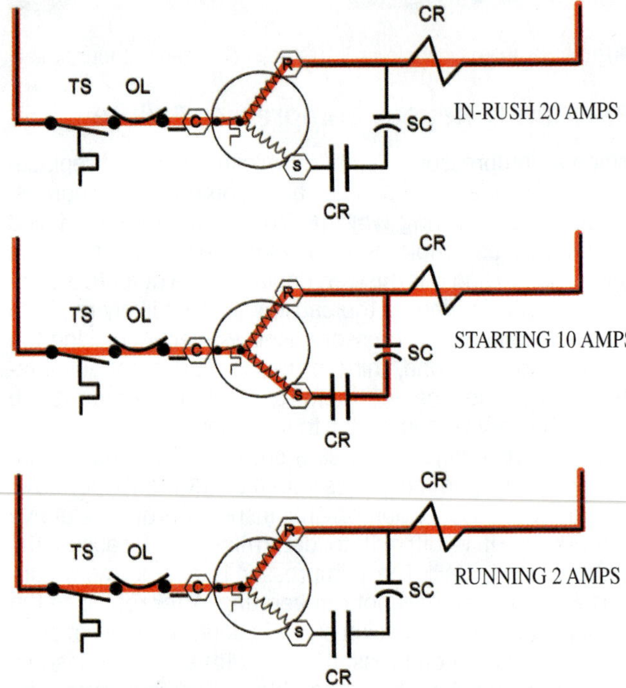

Figure 88-14 Diagram showing current relay operation.

draw drops, and the current relay contacts open back up. Thus, the contacts start out open, they close during start-up, and they open again during operation (Figure 88-14). Current relays must be carefully matched to the motors they control since the motor operating characteristics determine how the relay reacts.

Current relays are only used on fractional-horsepower compressors because the compressor locked rotor current must pass through the relay coil. Current relays typically plug directly onto the compressor terminals for small compressors (less than 1/4 hp) (Figure 88-15). Current relays for larger compressors (1/2 hp) are mounted separately (Figure 88-16).

PTC Relays

PTC stands for "positive temperature coefficient" thermistor. The "contacts" are actually a resistor that increases in resistance as it heats up. Unlike actual contacts, the thermistor never really opens. Its resistance just gets so high that the amount of current passing through is negligible. PTC relays are now more common on small compressors

Figure 88-15 The current relay plugs directly onto the compressor terminals on small compressors.

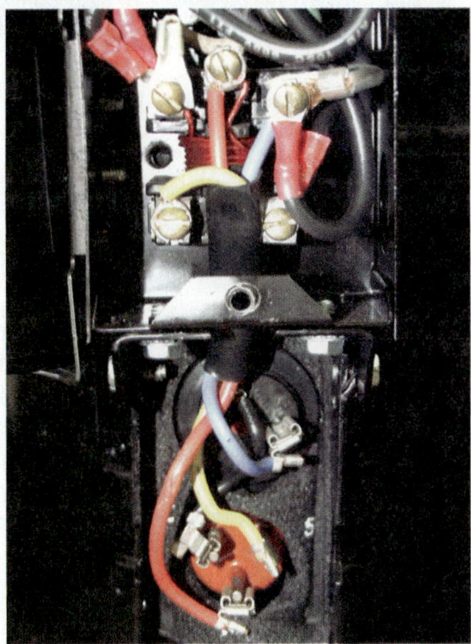

Figure 88-16 The current relay is mounted away from the terminals on larger compressors.

than current relays. Like small current relays, PTC relays typically mount right to the compressor terminals (Figure 88-17). PTC relays are wired in series with the start winding (Figure 88-18).

Potential Relays

Potential relays are used for larger single-phase motors. They have a coil that wires in parallel to the compressor start winding and a normally closed set of contacts that is wired in series with the start winding (Figure 88-19). Potential relays operate at a voltage that is higher than the line voltage supplied to the compressor because the back electromotive force (back emf) produced across the start winding is usually higher than the line voltage. The back emf increases as the compressor speed increases. The potential relay is calibrated to open its contacts when the compressor reaches approximately 75 percent of its full speed.

Figure 88-17 PTC relays mount right to the compressor terminals.

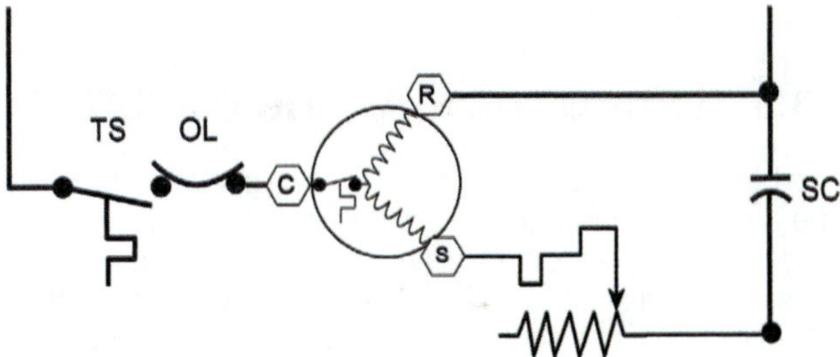

Figure 88-18 Diagram of a PTC starting relay.

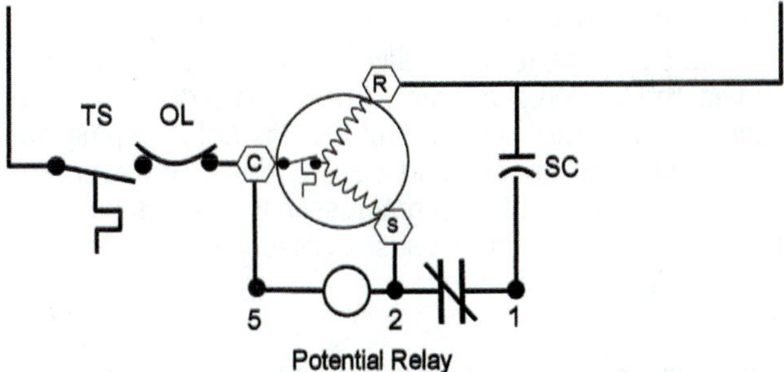

Figure 88-19 Diagram of a potential relay.

TECH TIP

Counter electromotive force, or back emf, is a voltage produced in AC magnetic devices as a result of the changing current flow through the device. It is called counter, or back electromotive force because it opposes the original voltage. Back emf is the primary opposition to current flow in most AC magnetic devices, including motors. The faster the motor turns, the higher the generated back emf.

Figure 88-20 When checking the amp draw through the start capacitor, you should see a reading when the compressor starts, then it should quickly drop to 0.

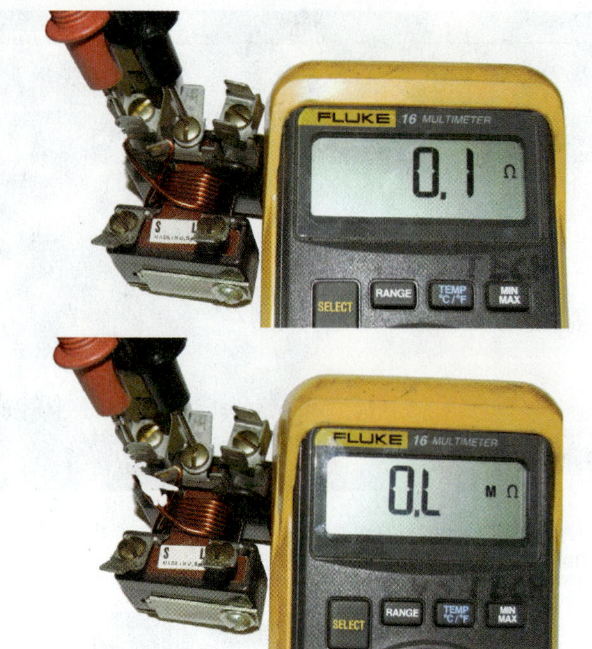

Figure 88-21 The current relay coil on top is good; the one on the bottom is bad.

88.5 TESTING THE STARTING CIRCUIT

The best way to test any of these starting circuits is to check the amp draw through the starting circuit. It should spike when the compressor starts and then drop to zero after the compressor is started. This all happens very quickly. Figure 88-20 illustrates how this works. If any of the starting components are open, the amp reading on the starting circuit will be zero. The amp draw on the run winding will likely be locked rotor amps because the compressor will not start. In this case you need to check the starting relay, capacitor, and start winding to see if any are open. If the amp draw spikes, the compressor starts, but the amp draw does not drop to zero, then the starting relay is not dropping out. This usually indicates a bad starting relay. However, this can also be caused by a compressor that starts but does not get up to full speed because of mechanical or refrigeration problems.

88.6 TESTING START RELAYS

Many times a technician will need to determine if the capacitor or starting relay used on a compressor is the cause of the compressor not starting or operating properly. To properly inspect these components, they must be electrically removed from the system. Then, using a standard multimeter or capacitance meter, the component can be tested to see if it is electrically acceptable.

Current Relay

To test a current relay using an ohmmeter, first remove the relay from the circuit. Hold the relay in the upright position and check the resistance between the L and M terminals of the relay. If the ohmmeter reads approximately 0.1 Ω, the coil of the relay is good. If the meter reads an infinite reading (OL), the coil is open and the relay needs to be replaced (Figure 88-21).

Next, place the ohmmeter leads on the M and S terminals of the relay. If the ohmmeter reads an OL, the contacts are open—as they should be. If the meter reads 0 or a measureable resistance, the contacts are stuck closed. The relay will need to be replaced (Figure 88-22). Next, try turning the relay upside down and shaking it. The contacts should be

Figure 88-22 The contacts should be open (OL) with relay upright and closed (0) with the relay upside down.

Figure 88-23 The PTC relay on the top is good; the relay on the bottom is bad.

closed, showing a low or zero resistance on the ohmmeter (Figure 88-22). However, turning the relay upside down will not close the contacts on all current relays.

PTC Relay

To check a PTC relay, check the resistance through the relay. The relay should have a low resistance when it is cold. Make sure the relay is cool. If it is still hot, the thermistor will still have a high resistance. If you read a high resistance or infinity with a cold PTC relay, the relay is bad (Figure 88-23).

Potential Relay

Remove the relay from the circuit and read the resistance between the 2 and 5 terminals of the relay. If the ohmmeter reads a high resistance (approximately 6,000–14,000 Ω) the coil is good. If the coil reads OL, it is open and the relay needs to be replaced. Next, read the resistance between the 1 and 2 terminals on the relay. If the ohmmeter reads zero, the contacts are closed—as they should be. If the ohmmeter reads a high resistance or an OL, the contacts are defective and the relay needs to be replaced.

88.7 TESTING CAPACITORS

Capacitors will normally fail open or shorted. They can be tested with a multimeter or a capacitor tester. Normally a visual inspection of a capacitor will indicate a defect. On run capacitors, any sign of bulging of its body is a sign of a defect (Figure 88-24). Start capacitors will have a rubber membrane located on the top portion of the capacitor. If this membrane blows out, it is a sign of a defect (Figure 88-25).

Figure 88-24 The bulge in this run capacitor indicates that it is bad.

SAFETY TIP

Capacitors have the ability to store an electrical charge. Because of this, a technician should always discharge a capacitor before handling or checking it. A capacitor can be safely discharged using a 20,000 Ω, 5 W resistor with an insulated pair of pliers. Some start capacitors have a bleed resistor across the terminals designed to discharge the capacitor. It is still a good practice to discharge these capacitors with your own resistor. The resistor on the capacitor may be electrically open and may not have discharged the capacitor.

Figure 88-25 The ruptured membrane on the top of the capacitor and the white residue on the bottom capacitor indicate that they have been overheated and are bad.

Testing with an Analog Ohm Meter

The following procedure can be used to test a capacitor using an analog ohmmeter:

1. Remove the capacitor from the circuit.
2. Discharge the capacitor with a bleed resistor or a voltmeter.
3. Set an ohmmeter to its highest scale. Zero the ohmmeter.
4. Place the ohmmeter leads on the capacitor terminals.
5. Watch for one of the following indications of the conditions of the capacitor:
   a. Good—needle will swing toward 0 and then slowly return to infinity.
   b. Shorted—needle will swing toward 0 and remain there.
   c. Open—needle will stay at infinity.
6. Make a second test by reversing the leads of the ohmmeter.
7. For run capacitors with metal cases, place one of the ohmmeter leads on a terminal of the capacitor and one lead on the body. If the ohmmeter reads a resistance, the capacitor is grounded and needs to be replaced. Next, test the other lead of the capacitor.

Testing with a Digital Multimeter or Capacitor Analyzer

Most digital multimeters intended for HVACR use have a scale specifically for testing capacitors (Figure 88-26). These meters typically will discharge the capacitor and then display its capacity in microfarads. To test capacitors using a capacitance meter, simply discharge the capacitor and place the meter leads across the terminals of the capacitor. If the capacitor is OK, the actual capacitance (within 10 percent of its original value) will be displayed on the screen. If the capacitor is shorted, then the display will indicate a value close to zero. If the capacitor is open, it will normally show an OL on its display.

Figure 88-26 Capacitor test setting on digital multimeter.

SERVICE TIP

The capacitor testing function on many multimeters will not read correctly if the capacitor has a bleed resistor. The bleed resistor must be removed from one side of the capacitor to use the capacitor test function on these meters.

88.8 TROUBLESHOOTING EVAPORATOR PROBLEMS

An evaporator is a heat exchanger used to absorb heat from the refrigerated product. Heat is transferred from the product through the tubing and fins of the evaporator to the refrigerant. If the evaporator's surface becomes insulated as the result of either being dirty or covered with ice, the amount of heat transfer will be dramatically reduced.

Ice buildup can occur when the heat load on the evaporator is reduced. As a result, the temperature of the evaporator drops and an excessive amount of frost and ice accumulates on its surface. For example, a blower-type evaporator needs the correct amount of airflow across its surface. If this airflow is restricted as a result of one or more defective fan motors, the surface of the evaporator will develop an excessive amount of ice and the system will not be able to absorb enough heat from the product to maintain the desired case temperature.

With an evaporator temperature below 32°F, frost will naturally develop on the surface of the evaporator even if the correct heat load is applied. The design of the refrigeration system must incorporate some means of defrosting the evaporator on a regular basis to prevent an excessive amount of frost from developing on the surface of the evaporator. Medium-temperature refrigeration systems will use the case air to defrost the evaporator. When the refrigeration system shuts down, the evaporator fan continues to operate and will continue to draw case air across the evaporator to melt any frost that has accumulated. However, with low-temperature systems the case temperature is well below 32°F and therefore the case air cannot be used to defrost the evaporator. Low-temperature systems require some type of supplemental heat to defrost the evaporator surface.

Troubleshooting Defrost Problems on Freezers

A frozen evaporator coil is a common problem found when troubleshooting freezers. Although there are several possible causes for this problem, one common cause is the defrost system. The system is not properly defrosting the evaporator's coil on a regular basis. To effectively troubleshoot this problem, a technician must understand the design and operation of the defrost systems typically used.

Defrost system problems fall into two categories. The system may be over-defrosted or under-defrosted. Over-defrosting is when the defrost heater stays energized too long, causing the box temperature to rise too high during defrost. The product may begin to melt and then refreeze. This problem can be identified by monitoring the box temperature during defrost or by examining the product in the case. Ice crystals forming on frozen product may be an indication of melting and refreezing. Over-defrosting is normally caused by a malfunction in the defrost cycle termination. On time defrost systems, either the defrost time is set too long or the time clock is defective. On time-pressure systems, either the terminating pressure control is set too high, the pressure control is defective, or the solenoid coil in the defrost timer is incorrectly wired or defective. On time-temperature systems, either the temperature control is set too high, the control is defective, or the solenoid in the defrost timer is defective.

Systems that are under-defrosting will result in a frozen evaporator. The frost that normally develops on the evaporator coil will continue to build until the entire surface of the evaporator is iced over. This can be caused by a defective or incorrectly set time clock that does not initiate a defrost cycle. It can also be caused by an open heater element or a defective defrost termination switch that continually terminates defrost on each initiation of a new cycle.

When an evaporator ices over, a technician will need to defrost the coil. Sometimes this can be done by manually initiating a defrost cycle. This can be done on a mechanical defrost timer by rotating the inner knob counter-clockwise (Figure 88-27). Many times, however, this cannot be done. The technician will have to manually defrost the coil. Extreme care must be taken when doing this. Do not use any sharp objects to chip away ice from the coil—this could easily puncture the coil and cause a refrigerant leak to develop in the evaporator. Using water is the best method but is not always practical, since the water will need to be drained away. If water cannot be used easily, a heat gun usually works well. Defrosting a coil manually is time-consuming. Do not rush this procedure, as making a careless mistake can be costly.

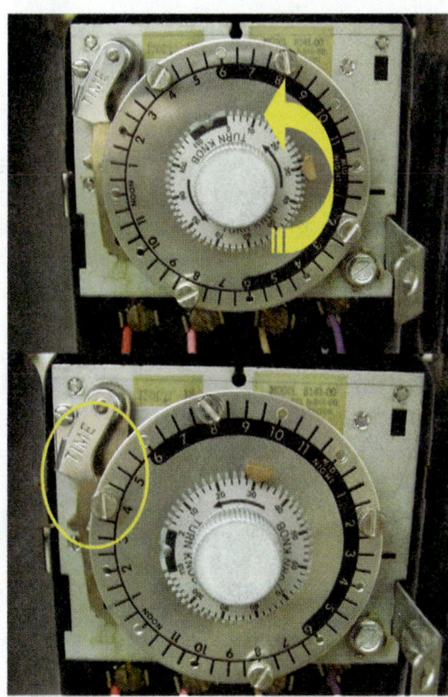

Figure 88-27 To manually put this unit in defrost, turn the dial until one of the initiation pins reaches the arrow and the points click.

88.9 TROUBLESHOOTING CONDENSER PROBLEMS

A condenser is a heat-exchange device used to reject heat energy from the refrigerant to its cooling medium. Heat energy from the refrigerant is rejected through the tubing and fins of the condenser and absorbed by the condensing medium. There are several problems that can occur that can impede this process. If the condenser's surface becomes insulated as a result of being dirty, it will reduce the ability of the cooling medium to absorb heat from the refrigerant. This will cause the discharge pressure of the system to rise beyond its normal operating range and cause the system to malfunction.

Determining Discharge Pressures

When troubleshooting commercial refrigeration systems, many times it is necessary to measure the discharge pressure of a system and determine if the pressure is normal, higher than normal, or lower than normal. To make this determination, the correct operating discharge pressure must be known. For systems utilizing an air-cooled condenser, this can be determined by measuring the system's suction pressure and the temperature of the air entering the condenser.

Below are the general procedures for determining the correct operating discharge pressure of a commercial refrigeration system utilizing an air-cooled condenser.

Step 1. Measure the dry-bulb temperature of the air entering the condenser. (We will refer to this as the EAT—entering air temperature.)

Step 2. Measure the system's operating suction pressure.

Step 3. Using a pressure/temperature (PT) chart, convert the suction pressure to its equivalent saturation temperature.

Step 4. Determine the system's application for this procedure: low temperature, medium temperature, or high temperature. A low-temperature system is one that operates at an evaporating temperature of −40°F to +10°F. A medium-temperature system is one that operates at an evaporating temperature of between −10°F and +30°F. A high-temperature system is one that operates at an evaporating temperature of between +20°F and +55°F.

Step 5. Using the chart in Table 88-1, determine the appropriate temperature rise (TR). The TR is the difference between the EAT and the condensing temperature of the refrigerant in the condenser.

Step 6. The condensing temperature (CT) can then be determined by adding the EAT and the TR. CT = EAT + TR.

Step 7. Once the CT is known, its equivalent saturation pressure can be determined using a PT chart. This will be the correct operating discharge pressure of the system.

For example, suppose we are working on a walk-in cooler using R-134a refrigerant with an operating suction pressure of 18.4 psig and 76°F air temperature entering the condenser.

Step 1. The EAT in our example is 76°F.

Step 2. The system's operating suction pressure is 18.4 psig.

Step 3. Using a PT, chart we convert our suction pressure of 18.4 psig to a saturation temperature of 20°F.

Table 88-1 Condenser Temperature Rise for Different Evaporator Temperatures

Condenser Temperature Rise (TR) °F			
	Evaporator Temperature		
Condenser Temp Rise (TR) °F	Low Temp	Medium Temp	High Temp
20	−25	0	—
23	−20	5	25
26	−15	10	30
29	−10	15	35
32	−5	20	40
35	0	25	45

Step 4. We determine the system's application to be medium temperature.
Step 5. Using Table 88-1, we determine the appropriate TR to be 32°F.
Step 6. The CT of our system will be 108°F (108°F = 76°F + 32°F).
Step 7. Again using a PT chart, we can conclude that our operating discharge pressure should be 142.8 psig.

Using this procedure should help determine if the discharge pressure of a commercial refrigeration system is normal, higher than normal, or lower than normal.

Checking for Noncondensable Gases

The only fluids circulating within a refrigeration system should be refrigerant and oil. Any other fluids contained in the system may reduce its capacity and cause harm to the system. One common contaminant that can enter a system is air. Air contains nitrogen and oxygen, both of which are harmful to a refrigeration system. When a system has been open to the atmosphere for servicing, it must be completely evacuated to remove the unwanted air before adding refrigerant.

One problem when air is contained in a system is that it will become trapped in the condenser and will not condense. It will take up space, causing less surface area for the refrigerant to condense. This will cause the high-side pressure to elevate, causing further problems, such as higher discharge temperatures and reduced system capacity.

Occasionally a technician may need to check whether noncondensables such as air are contained in a system. Use the following procedure to determine this in a refrigeration system with an air-cooled condenser.

- Determine the type of refrigerant in the system.
- Electrically disable the compressor and allow the condenser fan to operate.
- Attach a temperature probe to both the discharge line and liquid line, shown as in Figure 88-28.
- Place a third temperature probe to measure the temperature of the air entering the condenser.
- Connect a pressure gauge to the system to measure the pressure of the refrigerant in the condenser.

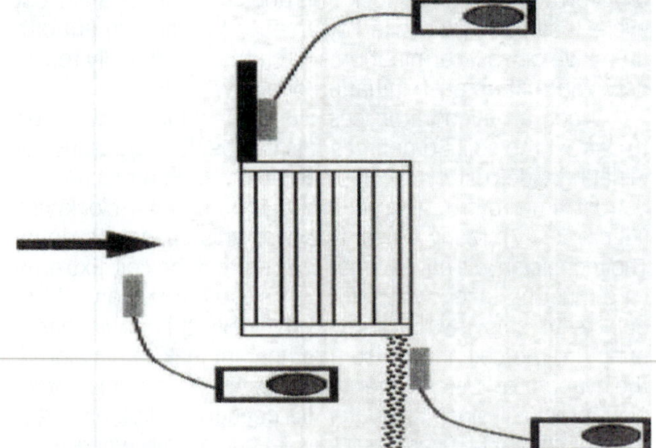

Figure 88-28 When checking for air in a refrigeration system, take temperatures at three locations: the discharge line at the inlet of the condenser, the liquid line at the outlet of the condenser, and the temperature of the air entering the condenser.

- When all three temperature probes (discharge line, liquid line, and air entering the condenser) read the same temperature, record the pressure of the refrigerant in the condenser.
- Using a PT chart, convert the measured pressure to its saturation temperature.
- The converted temperature should be within a few degrees of the measured discharge, liquid, and air entering temperatures.
- If the converted temperature is higher than the measured temperatures by more than a few degrees, there are noncondensables in the system that need to be removed.

88.10 TROUBLESHOOTING METERING DEVICES

Metering devices are refrigerant flow-control devices that feed the proper amount of refrigerant into an evaporator. If a metering device is defective it will either feed too much or too little refrigerant into the evaporator. To determine if too much or too little is being fed into an evaporator, a technician must measure the superheat level of the refrigerant leaving the evaporator. If the superheat value is higher than recommended by the equipment manufacturer, too little refrigerant is being fed into the evaporator. If the superheat value is lower than recommended by the equipment manufacturer, too much refrigerant is being fed into the evaporator. Both of these conditions could be the result of a defective metering device. However, before deciding that the problem is the metering device, make sure all other possible causes are investigated and eliminated. Many times it may appear the metering device is not functioning properly but another system problem is the actual cause.

88.11 TROUBLESHOOTING THERMOSTATIC EXPANSION VALVES

A thermostatic expansion valve (TEV) is designed to maintain a specific amount of superheat at the outlet of the evaporator. If the superheat value is too high or too low at the outlet of the evaporator, the cause may be the TEV. However, as mentioned previously, make sure all other possible causes are investigated first before determining that it is the TEV.

If the TEV is the cause of the system problem, try adjusting the valve first. To increase the superheat setting of the valve, turn the adjustment stem clockwise. To decrease the superheat setting of the valve, turn the adjustment stem counterclockwise. If adjusting the valve does not resolve the problem, then the valve is defective and will need to be replaced.

Some common causes of an evaporator with a low superheat and a high suction pressure are as follows:

- Oversized valve
- TEV seat is leaking
- Low superheat adjustment
- Sensing bulb making poor thermal contact
- Wrong thermostatic charge
- Incorrectly located external equalizer
- Restricted or capped external equalizer

Some common causes of an evaporator with a high superheat and a low suction pressure are as follows:

- Restricted filter drier
- Restricted TEV inlet screen
- Undercharged system
- Undersized valve
- High superheat adjustment
- Gas charge condensation
- Dead thermostatic element charge
- Wrong thermostatic charge
- Evaporator pressure drop—no external equalizer
- External equalizer location

88.12 TROUBLESHOOTING CAPILLARY TUBES

A capillary tube is simply a section of tubing with a very small inside diameter. A problem that does occur with these metering devices is that they can become restricted. Either they can be totally restricted where no refrigerant will pass through or partially restricted where some refrigerant will pass through but not enough to properly feed the evaporator. A system with a restricted capillary tube will have a low suction pressure and a discharge pressure that is slightly lower than normal.

Occasionally it may be difficult to determine the difference between a system with a low refrigerant charge, a restricted capillary tube, or a restricted filter drier (or strainer). A system with a low refrigerant charge will have both a low suction pressure and a low discharge pressure. A system with a restricted capillary tube will have a low suction pressure and a discharge pressure that is lower than normal but not quite as low as a system with a low refrigerant charge. It may be difficult to determine the difference between a low discharge pressure and a slightly lower than normal discharge pressure. One method is to recover the refrigerant from the system, evacuate the system, and then recharge with the correct weighed amount of refrigerant. If the system operates properly, the problem was a low refrigerant charge and the system leak will need to be found and repaired. If weighing in the charge does not resolve the problem, then the capillary tube or the filter drier is restricted. Repeat the process, but now replace the filter drier. If the system operates properly, the filter-drier was the cause of the problem. If the problem remains, the cause is a restricted capillary tube.

88.13 USING A SHORT REFRIGERANT GAUGE

Small-capacity refrigeration systems hold relatively small amounts of refrigerant. When troubleshooting these systems, avoid installing a gauge manifold to read the system pressures when possible. The small amount of refrigerant released from the process of installing and removing refrigeration hoses may cause a system problem. Enough refrigerant could be lost to cause the system to operate with a low refrigerant charge. This is especially true if 6-ft hoses are used or if a gauge manifold set is installed on and off several times.

If a technician needs to read the system's pressure, use a short gauge setup (Figure 88-29). This will diminish the amount of refrigerant released while reading the system's pressures.

88.14 REFRIGERANT LEAK DETECTION

Loss of refrigerant from a system due to a leak is a very common service call. This problem is normally quite simple to diagnose, but locating the source of the leak can be difficult. There are several methods of finding leaks, and each has its own advantages and disadvantages. Some methods may work well to locate a leak on one system and not as well on another. A technician should be able to use more than one leak-detection method. If one method is not working well, an alternate one can be tried. Also, many times it is beneficial for a technician to employ two methods while searching for a leak: one to locate the area of the suspected leak and the other to pinpoint its location.

Figure 88-29 Use a short refrigeration gauge on systems containing small amounts of refrigerant.

Figure 88-30 Soap bubble solution used to identify refrigerant leaks. *(Courtesy Highside Chemicals, Inc.)*

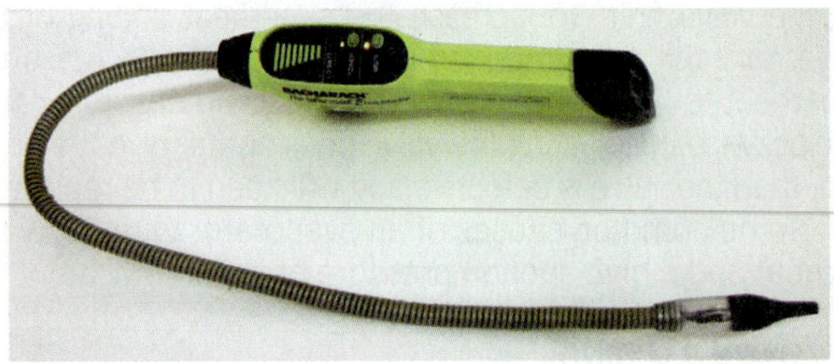

Figure 88-31 An electronic leak detector used to search for refrigerant leaks.

Below are various ways to search for a refrigerant leak:

- **Visual inspection.** As refrigerant leaks from a system, so does a portion of the refrigeration oil. The oil will normally stain the area surrounding the location. An easy way to search for a leak is to visually inspect the lines and fittings of the system for signs of oil. Although this may not pinpoint the location, it gives a technician the general area where the leak is located. This method works well for those leaks where a technician can visually inspect all of the piping easily. This is not always possible—many times it is very difficult, or even impossible, to visually inspect all of the areas of a system.
- **Soap bubbles.** Using an approved soapy solution (Figure 88-30) is a very common method to locate leaks. Simply saturate the suspected area with the soapy solution. If a leak is present, the solution will bubble. This method works well if the area of the leak is known and if the system is adequately pressurized. This method also works well in conjunction with the other methods used to verify the location of a leak. It is normally best to use a soapy solution purchased from a supply house, but some technicians make their own solution. Homemade solutions work well as long as they do not corrode or freeze onto the tubing or fittings.
- **Electronic leak detectors.** Electronic leak detectors (Figure 88-31) are very common today. Most currently produced electronic leak detectors will do an adequate job of locating leaks. A technician should always verify the operation of the electronic leak detector before relying on it. The sensors on most detectors will need to be changed after a period of usage. By testing the operation of these leak detectors, a technician can be assured that time is not being wasted searching for a leak with a faulty detector.
- **Refrigerant dyes.** Refrigerant leaks can be identified by injecting a dye into the system. A refrigerant dye kit is shown in Figure 88-32. The dye mixes with the refrigerant oil, and the point where this mixture leaks from the system will become stained with dye. A technician can then easily locate the leak by looking for the stained area. Two basic types of dyes used are red dye and fluorescent dye.
- **Ultrasonic leak detectors.** As refrigerant is escaping from a system, it will generate a sound at a higher than normal frequency. An ultrasonic leak detector (Figure 88-33), can be used to pick up these sound waves. This method works well as long as there are no other sources within the system area that produce sound waves at the same frequency.

SAFETY TIP

It is a common practice for a technician to warm a refrigerant cylinder while adding refrigerant to keep the vapor pressure within the cylinder high enough to allow the refrigerant to efficiently transfer from the cylinder to the system. The best method for warming a refrigerant cylinder is to place it in a bucket of warm water. Never use a torch or an open flame to warm a refrigerant cylinder. A water temperature of approximately 90°F is ideal. Never use water at a temperature higher than 125°F as it may overpressurize the refrigerant cylinder.

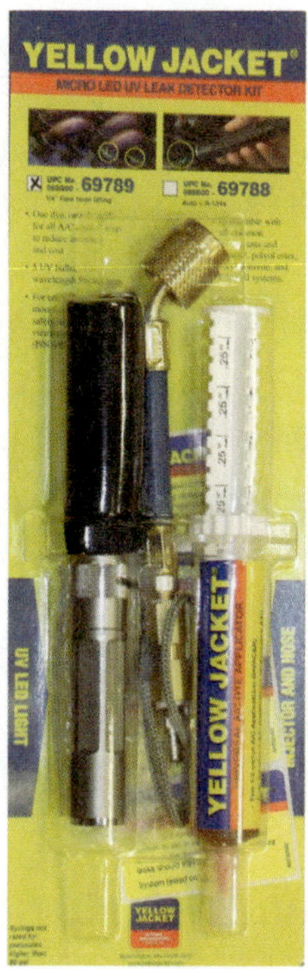

Figure 88-32 A refrigerant dye kit (HVAC fluorescent leak detection system).

Figure 88-33 An ultrasonic leak detector.

One basic rule a technician *must* follow while searching for a leak is *have patience*. At times this process can be very tedious, and a technician must have patience and be diligent; otherwise it can be a very frustrating procedure.

UNIT 88—SUMMARY

Patience is a requirement when troubleshooting refrigeration systems. A good service practice is to spend some time speaking to the customer before looking at the system and then conduct a visual inspection of the system to look for any obvious system problems.

Electrical troubleshooting can be divided into two types: (1) electrically diagnosing controls (switches, relays, and contactors) and (2) electrically diagnosing loads (coils, fan motors, and compressors). A voltmeter measures whether an electrical potential (voltage) is present at a switch or a load. An ammeter is used to measure how much current a load is consuming. An ohmmeter measures the resistance of an electrical component. Capacitors will normally fail open or shorted. This can be tested with an ohmmeter or a capacitor tester.

Compressor problems can be divided into two groups: mechanical or electrical. Mechanical defects are problems that affect the operation of the mechanical pump inside the compressor, such as blown valve plates, a broken crankshaft, or worn pistons. Compressor manufacturers will publish a rating chart for each compressor they manufacture. The compressor chart will list the correct amperage draw for the compressor under its various operating conditions. A major cause of premature compressor failure is the compressor operating at elevated temperatures.

A technician manually defrosting an evaporator must use extreme care. Do not use any sharp objects to chip away ice from the coil, as this could easily puncture the coil and cause a refrigerant leak to develop in the evaporator.

A thermostatic expansion valve (TEV) is designed to maintain a specific amount of superheat at the outlet of the evaporator. If the superheat value is too high or too low at the outlet of the evaporator, the cause may be the TEV. However, make sure all other possible causes are investigated first before determining that it is the TEV.

Small-capacity refrigeration systems hold relatively small amounts of refrigerant. When troubleshooting these systems, avoid installing a gauge manifold to read the system pressures when possible.

WORK ORDERS

Service Ticket 8801

Customer Complaint: Compressor Not Running

Equipment Type: Walk-in Cooler

A technician is called out to service a walk-in cooler using R-134a refrigerant, an air-cooled condenser, TEV metering device, and a constant cut-in temperature control and operating at a case temperature of 60°F.

When the technician arrives on the job, the condenser and evaporator fans are running, and the compressor is not running and cycles on overload when voltage is applied to the compressor. After further investigation, the technician observes the following conditions:

- The condenser coil is clean.
- The low pressure is set for a cut-in of 20 psig and a differential of 15 psig.
- The high-pressure control is set for a cut-out pressure of 250 psig.
- 232 V are measured across the common and run terminals of the compressor.
- The following resistances are measured across the terminals of the compressor.
- Common to run—3 Ω

- Common to start—10 Ω
- Run to start—13 Ω
- The run capacitor reads a capacitance of 35 mfd.
- The start capacitor reads a capacitance of 198 mfd.
- A resistance reading of 7,800 Ω is measured across terminals 2 and 5 of the potential relay.
- A resistance reading of infinity is measured across terminals 1 and 2 of the potential relay.

The technician decides to replace the potential relay.

Service Ticket 8802

Customer Complaint: Freezer Not Cooling Properly

Equipment Type: Walk-in Freezer

A technician is called out to service an 8 ft × 10 ft × 7ft, 6-in walk-in freezer that is operating at a 20°F case temperature. The system has an air-cooled condensing unit with a semihermetic compressor located indoors on top of the freezer. The system uses R-404A refrigerant and an externally equalized TEV. The system is controlled by a temperature controller and a pump-down solenoid. The system is using an electric defrost heater and a time/temperature defrost method.

When the technician arrives on the job, the compressor, condenser, and evaporator fans are running. After further investigation, the following conditions are observed:

- Ambient temperature is 75°F.
- The condenser and evaporator coils are relatively clean and there is no obstruction to airflow through them.
- Operating suction pressure is 16 psig.
- Operating discharge pressure is 240 psig.
- The suction line temperature is 20°F at the outlet of the evaporator.
- The liquid line temperature is 85°F at the outlet of the condenser.
- The sight glass is clear.
- There is no temperature drop across the liquid-line filter drier.

Based on these observations, the technician determines to replace the TEV.

Service Ticket 8803

Customer Complaint: Reach-in Cooler Not Cooling Properly

Equipment Type: Reach-in Cooler

A technician is called out to service a reach-in cooler using R-134a refrigerant, an air-cooled condenser, capillary-tube metering device, and constant cut-in temperature control and operating at a case temperature of 60°F.

When the technician arrives on the job, the condenser and evaporator fans are running, but the compressor is hot and not running. After further investigation, the technician observes the following conditions:

- The evaporator coil is clean.
- The condenser coil is relatively clean.
- The temperature control is set on its #5 position.
- 115 V is measured at the C and R terminals of the compressor.
- With the voltage removed and the relay disconnected from the compressor, an infinite resistance (OL) is measured between the C and R terminals of the compressor.
- With the voltage removed and the relay disconnected from the compressor, 12 Ω is measured between the S and R terminals of the compressor.
- With the start capacitor removed from the system, 189 mfd is measured across its terminals.
- With the current coil relay removed from the system, approximately 1 Ω is measured across its coil.
- With the current coil relay removed from the system, 0 Ω is measured across its contacts.

The technician then decides to replace the current coil relay.

UNIT 88—REVIEW QUESTIONS

1. What are the two areas into which electrical troubleshooting can be divided?
2. What are the three major tools commonly used to troubleshoot the various electrical components of a refrigeration system?
3. Which tool measures whether an electrical potential is present at a switch or a load?
4. Which tool measures how much current a load is drawing?
5. Explain the groups into which compressor problems can be divided.
6. How can a technician determine whether a compressor is not pumping to its rated capacity?
7. What are the three major electrical defects that can cause a compressor not to operate?
8. What is the maximum voltage imbalance of a three-phase compressor?
9. What does RLA stand for when stamped on a compressor?
10. At what cylinder temperature will oil break down inside a compressor?
11. What should a technician do after changing out a defective compressor?
12. Explain how to test whether a current relay is defective using an ohmmeter.
13. Explain how to test whether a potential relay is defective using an ohmmeter.
14. Explain how to test whether a capacitor is defective using an ohmmeter.
15. Explain how a technician should manually defrost a frozen evaporator coil.
16. Explain a method used to determine the correct operating discharge pressure of a commercial refrigeration system.
17. Explain a method used to determine whether noncondensables are in a refrigeration system.
18. List some common causes of an evaporator operating with a low superheat and a high suction pressure.
19. A technician should use a short refrigeration gauge on what type of refrigeration systems?
20. List some methods of searching for refrigerant leaks.

UNIT 89

Installation Techniques

OBJECTIVES

After completing this unit, you will be able to:

1. list the industry organizations that set equipment and installation standards.
2. list and describe the major phases of system installation.
3. discuss the factors affecting equipment placement.
4. discuss the procedure for selecting and sizing the power wiring, disconnect, and overcurrent protection.
5. discuss correct refrigeration piping procedures.
6. determine the correct charge for a new air-conditioning installation.

89.1 INTRODUCTION

The first step in a successful installation is reading the manufacturer's instructions. The secret to avoiding unpleasant surprises is taking some time to research what the manufacturer and all applicable national and local codes require. The time spent preparing will make the installation run smoothly and save time overall. Paradoxically, the key to speeding up is slowing down.

Proper installation is critical to every system. An 18-SEER air conditioner incorrectly installed could perform like a 10-SEER system. In most cases, a correctly installed lower-efficiency system will outperform an incorrectly installed higher-efficiency system. A good installation is the first step toward providing long equipment life, reliable operation, and energy efficiency.

89.2 STANDARDS, CODES, ORDINANCES, AND LAWS

Equipment installation instructions typically include a statement such as "All work should be performed in accordance with applicable national and local codes and standards." These regulations may apply to the design and performance of a product, its application, its installation, or safety considerations. Understanding what the rules are that you must follow is important. Standards, codes, ordinances, and laws are all essentially sets of rules.

Standards are generally not intended to be legal documents but rather a description of best practices. They are intended to inform and guide professionals. Standards are normally offered by industry groups such as ACCA, AHRI, ASHRAE, ASME, CGA, or UL.

Codes also describe how things should be done, but they generally describe the minimum acceptable installation. Codes focus heavily on safety. Codes are intended to be adopted as law. They are written by industry entities that specialize in writing code, including ICC, NFPA, and IAPMO. New code versions are published every 3 years. Although codes are intended to be legal documents, the organizations that write them are not part of the government and have no legal authority to enforce their codes. Codes have no legal effect until a government agency adopts them.

Code names such as the National Electric Code, the International Residential Code, or the Uniform Mechanical Code sound as if they are part of national law. Despite their names, there is no federal or national code-enforcement agency. Code enforcement is done at the state or local level. When states adopt a code, they often add a few changes, or amendments, to the existing code. Application of the same basic code can vary slightly from state to state. The amended code is usually referred to in that state as the state code. For example, the Georgia State Minimum Standard Mechanical Code is the International Mechanical Code with Georgia Amendments. Local codes are usually divided into the following sections:

- Electrical
- Plumbing
- Mechanical (air conditioning and refrigeration)
- Other (such as sound control)

Local inspectors enforce these codes based on the permits issued.

89.3 STANDARDS ORGANIZATIONS

There are many different industry organizations that publish standards. For most of these organizations, writing standards is only part of what they do. Each has a different core constituency, but all of them contribute to the body of standards we use in the HVACR industry.

ACCA (Air Conditioning Contractors of America)

ACCA is a nonprofit association serving more than 60,000 professionals and 4,000 businesses in the HVACR community. ACCA works to promote professional contracting, energy efficiency, and healthy, comfortable indoor living for all Americans. They produce many publications on air-conditioning system design and installation, including Manual J, Manual D, and the Quality Installation Standard.

AHRI (Air-Conditioning Heating and Refrigeration Institute)

AHRI is an association of manufacturers of refrigeration, air conditioning, and heating equipment and allied products. Although it is a public relations and information center for industry data, one of the institute's most important functions is establishing product or application standards by which the associated members can design, rate, and apply their hardware. In some cases, the products are submitted for test and are subject to certification and are listed in nationally published directories. The intent is to provide the user with equipment that meets a recognized standard. The standards used for rating HVACR equipment are published by AHRI.

ANSI (American National Standards Institute)

ANSI was formed in 1918. The Institute oversees the creation, promulgation, and use of thousands of norms and guidelines that directly affect businesses in nearly every sector, including HVACR. The ANSI mission is to enhance both the global competitiveness of U.S. business and the U.S. quality of life by promoting and facilitating voluntary consensus standards and conformity-assessment systems and safeguarding their integrity. ANSI often strengthens the standards of HVACR organizations by adopting their standards as an ANSI standard.

ASHRAE (American Society of Heating, Refrigeration, and Air-Conditioning Engineers)

ASHRAE started in 1904 as the American Society of Refrigeration Engineers. Today its membership includes thousands of engineers and technicians from all areas of the industry. ASHRAE creates standards for system design and application. ASHRAE Standard 90, Energy Standard for Buildings, is very popular because of the increased emphasis on energy-efficiency. ASHRAE publishes a series of handbooks that have become the reference bibles of the HVACR industry. These include *Fundamentals, Applications, HVAC Systems and Equipment*, and *Refrigeration*.

ASME (American Society of Mechanical Engineers)

ASME is concerned primarily with codes and standards related to the safety aspects of pressure vessels. Around the turn of the twentieth century, steam power was being widely used. Boilers and boiler accidents were common. Thousands of people were killed each year because of boiler explosions. To counter this, the American Society of Mechanical Engineers started listing safe materials and methods of construction used in boilers in an effort to improve boiler safety. Today most mechanical codes require that all large pressure vessels holding a pressure of over 15 psig be ASME rated. This includes refrigerant receivers and high-pressure chillers as well as boilers.

ASTM International

Formerly known as the American Society for Testing and Materials, ASTM is a globally recognized leader in the development and delivery of international voluntary consensus standards. Today, some 12,000 ASTM standards are used around the world to improve product quality, enhance safety, facilitate market access and trade, and build consumer confidence. ASTM primarily tests material properties. ASTM testing is often referred to in other standards and codes. For example, the ASTM testing procedures for flame spread and smoke are referenced in standards and codes applying to duct materials.

CGA (Compressed Gas Association)

The mission of CGA is to promote the safe manufacture, transportation, storage, transfilling, and disposal of industrial and medical gases and their containers. CGA standards are often quoted in codes dealing with refrigeration cylinders, propone cylinders, acetylene cylinders, and oxygen cylinders.

CSA (Canadian Standards Association)

The CSA is a not-for-profit membership-based association serving business, industry, government, and consumers in Canada and the global marketplace. As a solutions-oriented organization, CSA works in Canada and around the world to develop standards that address real needs, such as enhancing public safety and health, advancing the quality of life, helping to preserve the environment, and facilitating trade. CSA standards applicable to HVACR include standards covering construction, electrical, mechanical systems, and occupational safety.

International Organization for Standardization (ISO)

ISO is the world's largest developer and publisher of international standards. ISO is a network of the national standards institutes of 162 countries, one member per country, with a Central Secretariat in Geneva, Switzerland, that coordinates the system. ANSI is a member body of ISO.

NEMA (National Electrical Manufacturers Association)

NEMA is the trade association of choice for the electrical manufacturing industry. Founded in 1926 and headquartered near Washington, DC, its approximately 450 member companies manufacture products used in the generation, transmission and distribution, control, and end-use of electricity. NEMA standards affecting HVACR include standards covering electric motors, motor controls, disconnect switches, and electrical protection equipment.

UL (Underwriters Laboratories)

UL is a testing and code agency that specializes in the safety aspects of electrical products, while also including an overall review of some products. The UL seal on household appliances is a familiar one, and it is also applied to the approval of refrigeration and air-conditioning equipment. Its scope of activity has expanded into large centrifugal refrigeration machines.

UL approval for certain types of refrigeration and air-conditioning products may be mandatory by local electrical inspectors. Compliance with UL is the responsibility of the manufacturer, and the approved products are listed in a directory. Installation in accordance with the approved standards is the responsibility of the installer. Violation of these standards can cause a safety hazard as well as possibly voiding the user's insurance coverage should an accident or fire result. Installation procedures should therefore always conform to UL approval standards.

89.4 CERTIFYING AND TESTING ORGANIZATIONS

Independent laboratory testing is used to ensure that products actually meet the standards they are designed to meet. Many organizations that write standards also perform certification testing to ensure that products meet specific standards. Common certifying and testing organizations include the Air-Conditioning Heating and Refrigeration Institute (AHRI), the Canadian Standards Association (CSA), and Intertek (ETL). Codes often reference these independent industry certifications or listings. For example, electrical codes typically require appliances to be listed by a testing organization such as UL, CSA, or ETL. The capacity of HVACR equipment is rated by AHRI. This is especially important for split systems. An AHRI listing of the two components assures the installer and customer that the two pieces will work properly together and perform as listed by AHRI. When a unit is listed by a certifying agency, their certifying mark is applied to the unit. Equipment is often listed by more than one certifying agency (Figure 89-1).

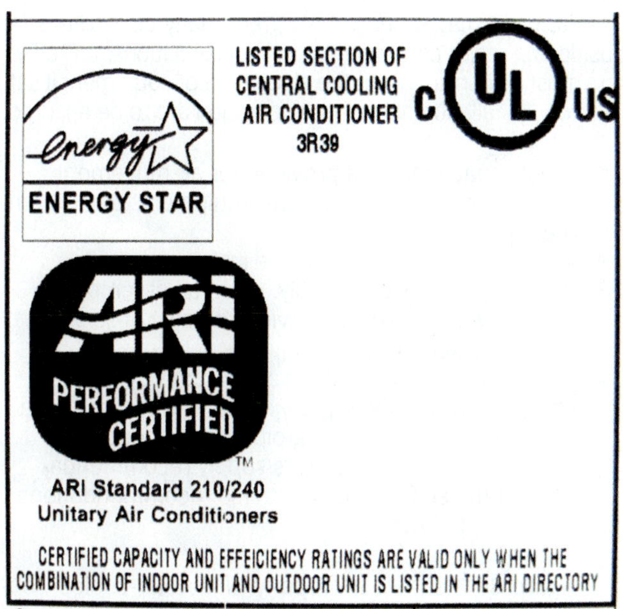

Figure 89-1 This unit's performance has been certified by AHRI; it also meets Energy Star efficiency standards and is listed by UL.

89.5 CODE ORGANIZATIONS

The three main code-writing organizations are the National Fire Protection Association (NFPA), the International Code Council (ICC), and the International Association of Plumbing and Mechanical Officials (IAPMO). Each organization is responsible for several codes. The NFPA's focus is on fire prevention. NFPA codes are typically branded as "national" codes. For example, the NFPA writes the National Electrical Code and the National Fuel Gas Code. The ICC's focus is toward construction and development. ICC codes are branded as "international" codes. ICC codes used in HVACR include the International Building Code, the International Residential Code, the International Mechanical Code, the International Fuel Gas Code, and the International Energy conservation Code. The IAPMO focuses on plumbing and mechanical systems. Their codes are branded as "uniform" codes. Their HVACR-related codes include the Uniform Mechanical Code and the Uniform Solar Energy Code.

NFPA (National Fire Protection Association)

The mission of the international nonprofit NFPA, established in 1896, is to reduce the worldwide burden of fire and other hazards on the quality of life by providing and advocating consensus codes and standards, research, training, and education. The world's leading advocate of fire prevention and an authoritative source on public safety, NFPA develops, publishes, and disseminates more than three hundred consensus codes and standards intended to minimize the possibility and effects of fire and other risks. Two codes, the National Fuel Gas Safety Code and the National Electric Code have become mainstays in the HVACR industry. The National Electric Code is truly nationally recognized, adopted, specified, and quoted.

IAPMO (International Association of Plumbing and Mechanical Officials)

The IAPMO has been protecting the public's health and safety for more than eighty years by working in concert with government and industry to implement comprehensive plumbing and mechanical systems around the world. IAPMO publishes the Uniform Mechanical Code, which is used in many locations to govern the installation of HVACR equipment. They also produce the Uniform Solar Energy Code and the Green Plumbing and Mechanical Code Supplement.

ICC (International Code Council)

The ICC is a membership association dedicated to building safety and fire prevention. ICC develops the codes and standards used to construct residential and commercial buildings, including homes and schools. The ICC's International Codes, or I-Codes, provide minimum safeguards for people at home, at school, and in the workplace. The I-Codes are a complete set of comprehensive, coordinated building safety and fire prevention codes. Building codes benefit public safety and support the industry's need for one set of codes without regional limitations. Fifty states and the District of Columbia have adopted the I-Codes at the state or jurisdictional level. The most commonly referenced I-Codes in the HVACR industry are the International Residential Code, the International Mechanical Code, and the International Fuel Gas Code.

89.6 MECHANICAL AND GAS CODES

For many years, the codes that specified the installation of HVACR equipment have been the Mechanical and Fuel Gas Codes. Aspects such as equipment location, duct installation, and refrigerant piping are covered in the Mechanical Codes. Gas piping, combustion air, and venting are covered in the Fuel Gas Codes. The two mechanical codes most often used are the Uniform Mechanical Code by IAPMO and the International Mechanical code by ICC. The two prevalent Fuel Gas Codes are the National Fuel Gas code by NFPA and the International Fuel Gas Code by ICC.

TECH TIP

Although there are many up-to-date codes available only those codes approved by the local or state governing agents are in effect. As a result, the most current code may not have been adopted in your area. In these cases, you must follow the earlier adopted code unless you receive a waiver from the city or state allowing you to use the more current codes. Failure to get prior approval, even for a newer code, can result in your job failing an inspection and receiving a red tag.

89.7 NATIONAL ELECTRIC CODE

Closely associated with the work of UL in terms of electrical safety is the National Electrical Code, sponsored by the National Fire Protection Association (NFPA). The original code was developed in 1897 as a united effort of various insurance, electrical, architectural, and allied interests. Although it is called the National Electrical Code, its intent is to guide local parties in the proper application and installation of electrical devices. It is the backbone of most state or local electrical codes and ordinances.

The most current code is 2011. The code specifies wiring methods and procedures necessary for the safe installation and service of electrical systems and equipment. Sections of the NEC that are particularly appropriate for air-conditioning installation are sections 310, Conductors for General Wiring; 430, Motors Motor Circuits and Controllers; and 440, Air Conditioning and Refrigeration Equipment.

89.8 INTERNATIONAL RESIDENTIAL CODE

The International Residential Building Code is written by ICC. It is a comprehensive, stand-alone residential construction code that has been adopted by many states and municipalities. It is often used instead of separate plumbing, mechanical, gas, and electric codes for one- and two-family dwellings. This code encompasses all aspects of residential construction, including all aspects of HVACR mechanical systems.

89.9 MAJOR PHASES OF INSTALLATION

Installations can be broken down into phases, based on the required skills and the work performed. The major phases of installations are as follows:

- Equipment placement
- Ductwork installation
- Piping installation (refrigerant, water, and gas)
- Electrical connection
- Evacuation and charging
- Start and check

In larger companies, these phases are often performed by separate crews. The equipment placement and duct installation is typically done by the same crew. The erection of piping is sometimes performed by installers who specialize in piping. Electrical connections are frequently done by crews who primarily handle electrical wiring and installation. Frequently the wires are pulled in place by the equipment installers, but the connections are made by the wiring and controls crew. Evacuation, charging, and start-up are usually performed by crews who specialize in refrigeration.

Any number of these phases can be performed by the same technicians in some companies. In smaller companies, a single installation crew is responsible for all phases of installation.

89.10 EQUIPMENT PLACEMENT

Despite what may seem to be a great many possibilities for positioning major components during installations, three factors must be considered in the placement of equipment if satisfactory installation and proper operation are to be ensured:

- Ample space must be provided for air movement around air-cooled condensing equipment to and from the condenser.
- All major components must be installed so that they may be serviced readily. When an assembly is not easily accessible for service, the cost of service becomes excessive.
- Vibration isolation must always be considered, not only in regard to the equipment itself but also in relation to the interconnecting piping and sheet metal ductwork. All manufacturers supply recommendations of the space required; these recommendations should be followed.

Figure 89-2 shows a properly placed air-cooled condensing unit. Sufficient room has been left for the passage of incoming air around and over the unit.

An example of a common error in positioning major system components is shown in Figure 89-3. A shell-and-tube

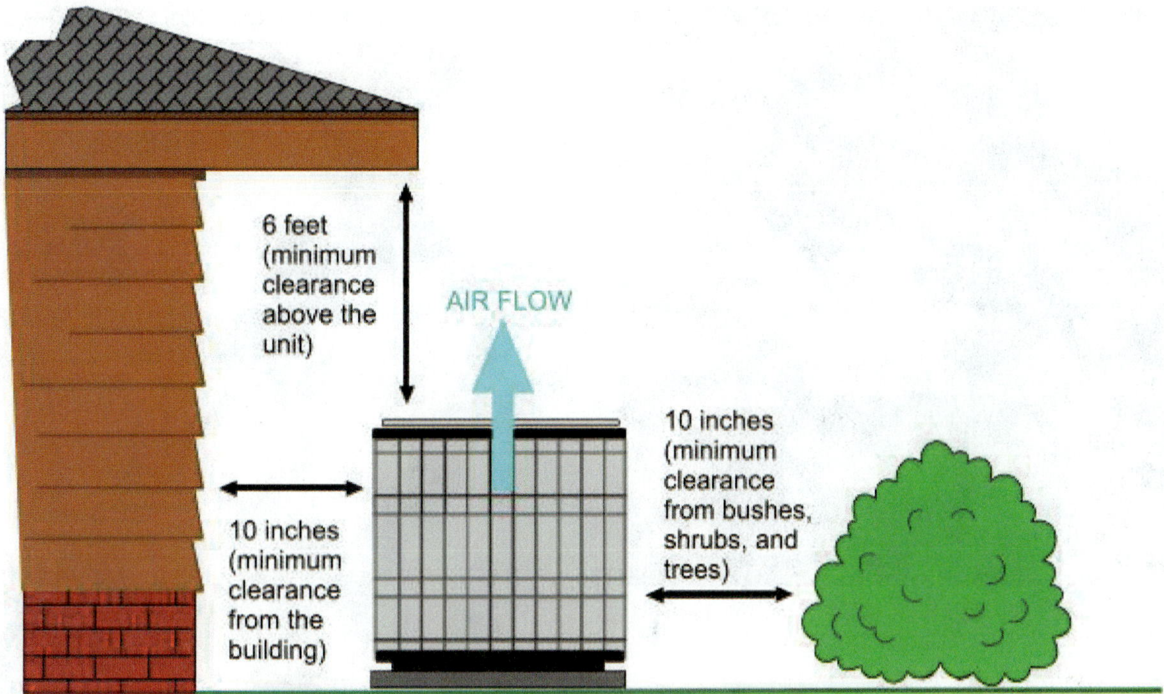

Figure 89-2 Air-cooled condensing units must be placed so that their airflow is not restricted.

Figure 89-3 Service accessibility is important in positioning major system components.

Figure 89-4 Internal isolation springs inside hermetic compressor.

water-cooled condenser has been placed in such a position that the entire condensing unit must be moved to replace a single condenser tube. Be sure to allow room for replacement of items such as compressors, fan motors, fans, and filters. High service costs are often attributable to the poor placement of system components.

Noise is also an important factor in the placement of air-cooled condensing equipment. The discharge air transmits noise generated within the unit. It is poor practice to "aim" the condenser discharge air in a direction where noise may be disturbing, such as a neighboring window.

Vibrations set up by rotating assemblies such as compressors, fans, and fan motors can break refrigerant lines, cause structural building damage, and create noise. Vibration isolation is required on all refrigeration and air-conditioning equipment where noise or vibration may be disturbing. Almost all manufacturers use some form of vibration isolation in the production of their equipment. The compressor is the chief source of system vibration. Hermetic reciprocating compressors are usually mounted on springs inside the compressor shell (Figure 89-4). Most compressors are also externally mounted on some form of vibration-dampening material, such as the rubber bushings shown in Figure 89-5.

Figure 89-5 Hermetic compressor mount on rubber isolation bushings.

Figure 89-6 Hanging spring suspension mounts.

TECH TIP

Some metropolitan areas have maximum-noise-level ordinances affecting HVACR equipment. These ordinances may affect the equipment selected and equipment placement to keep the sound level within the ordinance guidelines. Check with your local building office regarding noise level ordinances.

When air-conditioning units with compressors are suspended from the roof or in the upper stories of a multistory building, vibration isolators may be used (Figure 89-6). These spring suspensions isolate unit vibration from the surface from which the unit is suspended.

Isolation pads designed specifically for vibration dampening (Figure 89-7) are placed under the equipment to reduce the transmission of vibrations to the surface on which the unit is resting. This type of material is designed to dampen the vibration from a given amount of weight per square inch of area. Consult the manufacturer's installation recommendations when selecting vibration isolation materials.

Figure 89-7 Isolation pads in various sizes.

Vibration eliminators are installed in the discharge and suction lines of the compressor to isolate compressor movement and vibration from the refrigerant lines (Figure 89-8). The eliminator consists of a flexible corrugated metal hose core with an overall metal braid. The braid will allow some sideways movement of the flexible material, but no expansion or contraction. This type of eliminator is normally installed in the compressor refrigerant lines as close to the compressor as practical. It is particularly effective on installations where the compressor and condenser are on different bases yet quite close together. The vibration isolator should be placed in the line so that the movement it absorbs is in a plane at right angles to the device. Do not place this unit in a position that will put tension on it, as its useful life will be shortened.

(a)

(b)

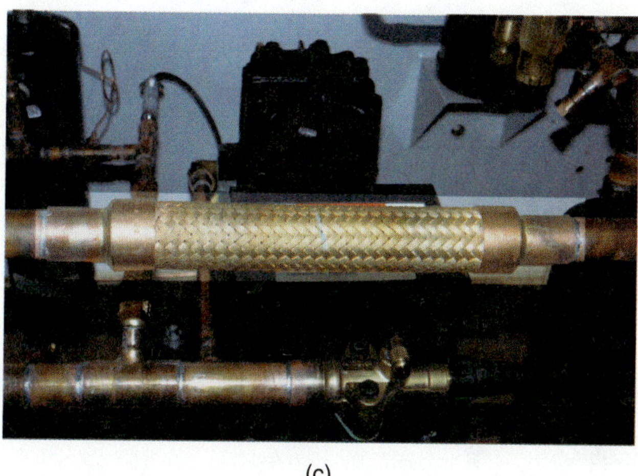

(c)

Figure 89-8 (a) Vibration eliminator; (b) internal ribs that allow a vibration damper to flex; (c) typical installation of a vibration damper as it is installed on a refrigeration system.

SERVICE TIP

When hanging an air handler from the attic rafters, it can be set on wooden blocks first so it can be leveled before attaching the hanging straps. This will make leveling the unit much easier. In addition to the common sheet metal straps, there are other hanging systems, including cable and threaded rods, which make hanging and leveling much easier and faster.

Figure 89-9 Blower motors are mounted to the blower housing using rubber bushings.

The blower assembly in furnaces and air handlers is a possible source of vibration. Most manufacturers isolate the fan and fan motor inside the unit with rubber mounts (Figure 89-9). When furnaces or air handlers are installed in the attic, suspending them (Figure 89-10) will isolate them from the ceiling and reduce sound transmission. The unit may also rest directly on the ceiling joists if vibration isolation pads are used (see Figure 89-7).

SERVICE TIP

Never rest a unit directly on the ceiling joists or a platform fastened to the ceiling joists without using vibration-isolation pads. Placing the unit directly on the ceiling joists turns the entire ceiling into a sounding board that amplifies the unit noise.

89.11 BASEMENT AND CLOSET INSTALLATION CHECKLIST

The following checklist highlights important considerations when installing equipment in basements, closets and alcoves. Figure 89-11 illustrates the listed specifications. Code numbers follow each item in parentheses.

- Appliances shall not be installed in storage closets. (IRC G2406.2)
- There must be a minimum of 3 in clearance along the sides, back, and top of equipment. (IRC M1305.1.1)
- The space must be at least 12 in wider than the equipment. (IRC M1305.1.1)
- Furnaces must have a minimum 6-in clearance in front of the firebox. (IRC M1305.1.1)

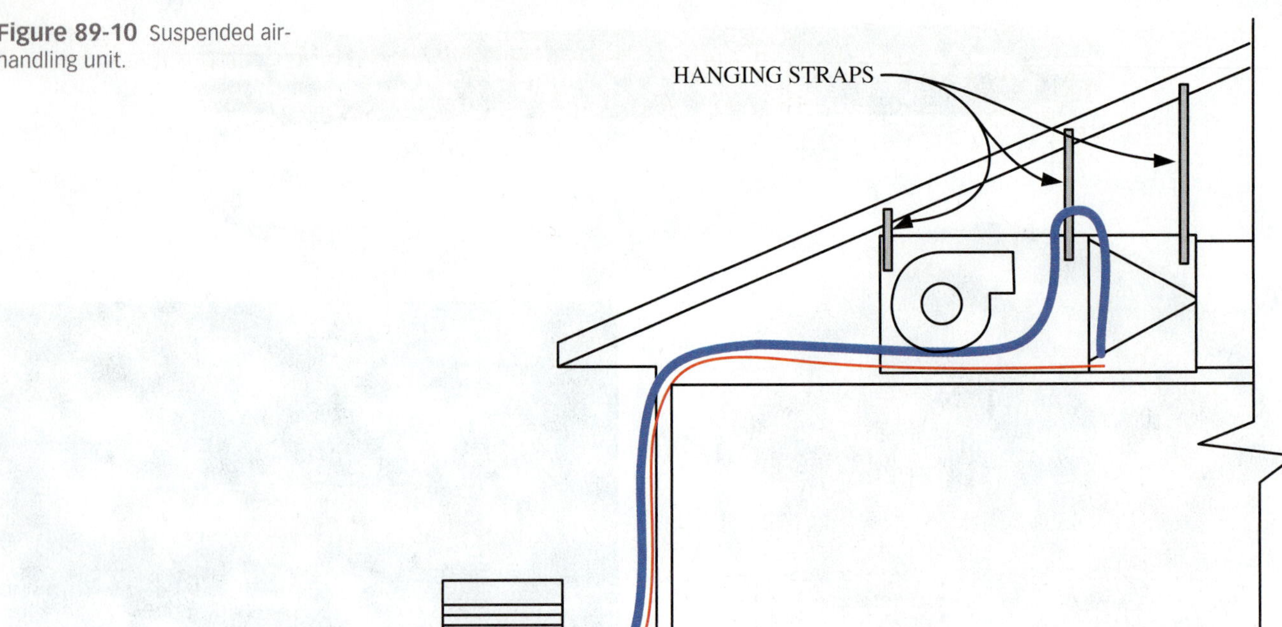

Figure 89-10 Suspended air-handling unit.

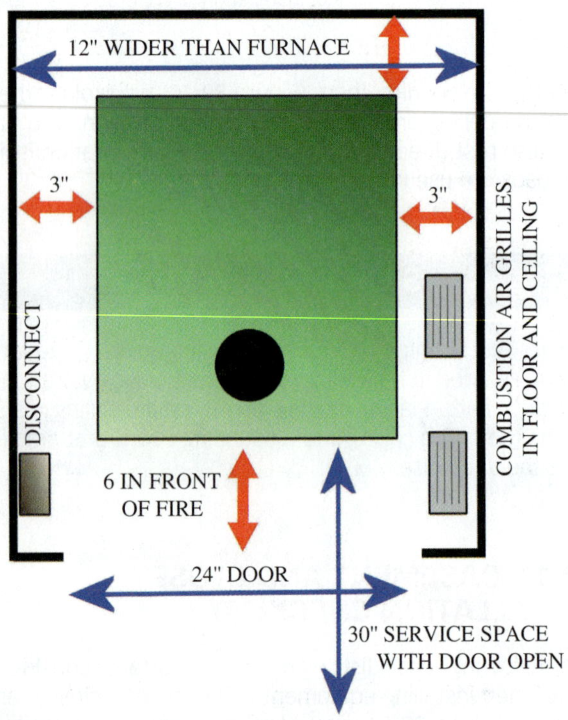

Figure 89-11 Specifications for equipment installed in a closet.

- The door to a closet must be at least 24 in wide. (IRC M1305.1.2)
- Equipment must fit through the door. (IRC M1305.1.2)
- A minimum service space is required. It should be located by the service access panel, 30 in deep and the height of the appliance. but not less than 30 in high. (IRC M1305.1.2)
- Where damage to building components will occur from water overflow, an auxiliary drain pan is required for condensing equipment, including air conditioners, heat pumps, and 90 percent furnaces. This provision is normally applied to attic installations but can also be applied to any installation where water can damage the floor or walls. (IRC M1411.3.1)
- Electrical disconnect must be within sight of the unit. (IRC Table E4101.5)
- For combustion appliances, ensure that adequate combustion air is provided. (IRC G2407) (For more details, see Unit 51, Gas Furnace Installation.)

89.12 ATTIC INSTALLATION CHECKLIST

The following checklist highlights important considerations when installing equipment in attic spaces. Figure 89-12 illustrates the listed specifications. Code numbers follow each item in parentheses.

- The minimum clear opening is 20 in x 30 in. (IRC M1305.1.3)
- The equipment must fit through the opening. (IRC M1305.1.3)
- There must be a solid floor at least 24 in wide from the entrance to the unit. (IRC M1305.1.3)
- There must be a light switch at the entryway. (IRC M1305.1.4.3)
- There must be a light and receptacle near the unit. (IRC M1305.1.4.3)
- There must be an electrical disconnect within sight of the unit. (IRC E4101.5)
- Maximum distance must be 20 ft from the entry to the unit. However, if the passageway is at least 6 ft high

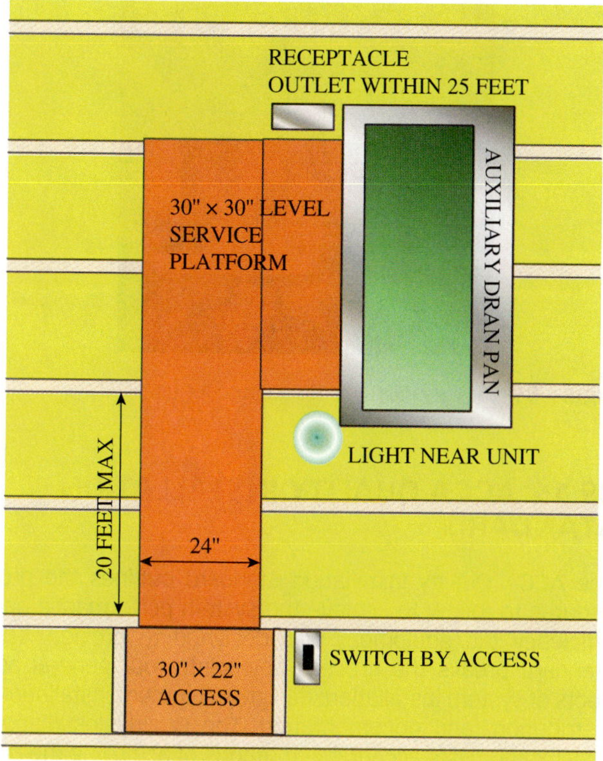

Figure 89-12 Specifications for equipment installed in an attic.

and 22 in wide, the distance may be up to 50 ft. (IRC M1305.1.3)

- A minimum 30 in x 30 in level service platform is required by the service panel of the equipment. (IRC M1305.1.3)
- An auxiliary drain pan is required for condensing equipment, including air conditioners, heat pumps, and 90 percent furnaces. Although there are several approved methods for protection, the most common is to use a pan with a separate drain installed underneath the equipment. The pan must be at least 1.5 in deep and 3 in larger in all dimensions than the equipment. (IRC M1411.3.1)
- The appliance should be supported above the flood rim of the auxiliary drain pan. (IRC M1411.3.3)
- For combustion appliances, ensure that adequate combustion air is provided.(IRC G2407) (For more details, see Unit 51, Gas Furnace Installation.)

89.13 CRAWLSPACE INSTALLATION CHECKLIST

The following checklist highlights important considerations when installing equipment in crawlspaces. Figure 89-13 illustrates the listed specifications. Code numbers follow each item in parentheses.

- The minimum rough framed opening for the access door is 22 in x 30 in. (IRC M1305.1.4)
- The equipment must fit through the opening. (IRC M1305.1.4)

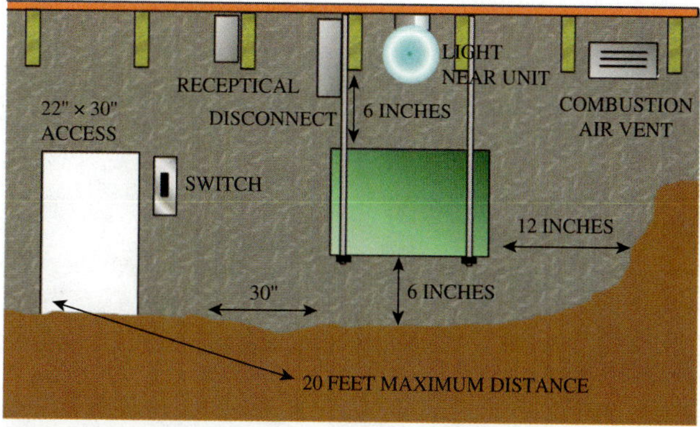

Figure 89-13 Specifications for equipment installed in a crawlspace.

- The passageway to the equipment must be at least 22 in wide and 30 in high. (IRC M1305.1.4)
- If the passageway is 12 in or more below grade, the walls must be lined with masonry or concrete extending 4 in above grade. (IRC M1305.1.4)
- Maximum distance must be 20 ft from the entry to the unit. However, if the passageway is at least 6 ft high and 22 in wide, the distance may be unlimited. (IRC M1305.1.4)
- If excavation is required, there should be a 6 in clearance under the unit, 12 in clearance on all sides, and a 30 in service clearance at the appliance. (IRC M1305.1.4.2)
- A minimum 30 in x 30 in level service space is required by the service panel of the equipment. (IRC M1305.1.4)
- Equipment supported from the ground must be on a concrete slab at least 3 in above the adjoining ground. (IRC M1305.1.4)
- Equipment suspended from the floor shall have a ground clearance of at least 6 in. (IRC M1305.1.4)
- A light switch must be located at the entryway. (IRC M1305.1.3)
- A light and receptacle must be located near the unit. (IRC M1305.1.4.3)
- The electrical disconnect must be located within sight of the unit. (IRC E4101.5)
- For combustion appliances, ensure that adequate combustion air is provided. (IRC G2407) (For more details, see Unit 51, Gas Furnace Installation.)

89.14 GARAGE INSTALLATION CHECKLIST

SAFETY TIP

Although code does allow installation of HVACR equipment in a garage under certain conditions, there are safety considerations that are best addressed by not locating HVACR equipment in a garage. The first is the possibility of igniting gasoline vapors. The second is the possibility of circulating carbon monoxide produced by vehicles through the system duct work. Simply not locating the equipment in the garage is the best safeguard to both of these potentially fatal safety concerns.

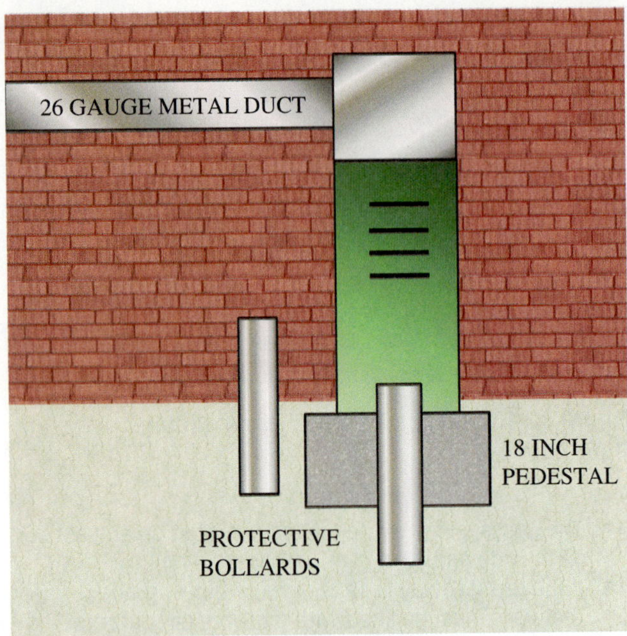

Figure 89-14 Specifications for a unit installed in a garage.

Garage installation of HVACR equipment is not recommended. If it cannot be avoided, the following checklist will reduce the associated hazards. Figure 89-14 illustrates the listed specifications. Code numbers follow each item in parentheses.

- Equipment should be located at least 18 in above the garage floor. (IRC M1307.3)
- Equipment shall be protected from impact. (IRC M1307.3.1)
- Systems supplying the living spaces shall not draw air from or supply air to the garage. (IRC M1601.6)
- Ductwork should be constructed of 26-gauge sheet metal. (IRC R302.5.2)
- Ductwork shall have no opening to the garage. (IRC R302.5.2)

89.15 OUTDOOR UNIT INSTALLATION CHECKLIST

The following checklist highlights important considerations when installing a condensing unit outside. Figure 89-15 illustrates the listed specifications. Code numbers follow each item in parentheses.

- Outdoor unit shall be raised at least 3 in above the ground. (IRC M1403.2)
- Refrigerant vapor lines shall be insulated with R-4 insulation. (IRC M1411.5)
- Outdoor refrigerant access ports shall have locking caps. (IRC M1411.6)
- Electrical disconnect must be located within sight of the unit. (IRC E4101.5)
- Refrigerant lines running inside the wall should be protected by steel shield plates. (IRC M1308.2)

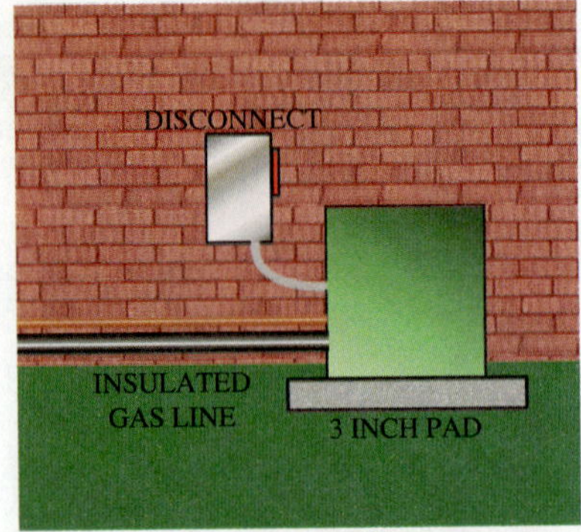

Figure 89-15 Specifications for an outdoor unit.

89.16 ACCA QUALITY INSTALLATION STANDARD

The ACCA Quality Installation Standard outlines the procedures to follow to achieve the system performance and efficiency the equipment was designed to provide. The standard breaks the process down into four general aspects of system installation: design, equipment installation, distribution, and documentation. The Quality Installation Specification lists acceptable methods of achieving the desired results and the type of documentation the contractor should provide to comply with it. The specific areas addressed by the standard and the units in this text that cover those areas are listed in Table 89-1.

89.17 DUCTWORK

Ductwork is an often overlooked part of a system installation. Customers often focus on the equipment ratings and don't pay enough attention to the workmanship of the installation. The equipment cannot deliver its rated capacity and efficiency if it is connected to a bad duct system. Since the ducts distribute the conditioned air to the house, if they are improperly sized, leaky, or poorly insulated, the house will not be properly conditioned. Further, the energy cost of poorly designed, leaky, and poorly insulated ducts is usually much higher than the difference between the operating cost of an entry-level unit and a high-efficiency unit. A good duct system must be properly designed, installed, sealed, insulated, tested for leaks, and balanced. Duct systems should be sized based on the heat load of the house. Ducts should be sealed with products made expressly for that purpose. They should be UL 181 rated for the specific type of duct material being used. All ductwork should be insulated and have a vapor barrier to prevent condensation. Duct insulation should have a minimum R value of 6. Ducts located in attics should be insulated with a minimum R value of 8. After installation, ducts should be pressurized and leak tested. Finally, the airflow delivery to the different rooms of the house should be checked and the system balanced to deliver the correct airflow.

The details of each of these steps are covered in the text. Unit 70, Residential Load Calculations, covers heat load calculations and equipment sizing. Unit 71, Duct System Design, covers duct design. Unit 46, Duct Installation, covers duct installation, insulation, and leak testing. Unit 73,

Table 89-1 ACCA Quality Installation Specification

Design	
Ventilation	43 Ventilation and Dehumidification
Building Load Calculations	70 Residential Load Calculations
Equipment Selection	70 Residential Load Calculations
Geothermal Ground Heat Exchanger	64 Geothermal Heat Pumps
Matching System Components.	18 Evaporators
Equipment Installation	
Airflow Through Indoor Heat Exchangers	41 Psychrometrics and Airflow 60 Testing and Balancing
Water Flow Through Indoor Heat Exchangers	64 Geothermal Heat Pumps
Refrigerant Charge	31 Charging
Electrical Requirements	35 Electric Meters and Test Instruments
On-Rate for Fuel-Fired Equipment	51 Gas Furnace Installation 54 Residential Oil Heating Installation
Combustion Venting System	51 Gas Furnace Installation 54 Residential Oil Heating Installation
System Controls	39 Control Systems
Distribution	
Duct Leakage	46 Duct Installation
Airflow Balance,	73 Testing and Balancing
Hydronic Balance	64 Geothermal Heat Pumps 80 Hydronic Heating Systems
Documentation	
Proper System Documentation to the Owner	89 Installation
Owner/Operator Education	89 Installation

Testing and Balancing Air Systems, covers the procedures for balancing the airflow through the duct system.

89.18 PIPING

The art of making flared, soldered, and brazed connections in copper tubing has been discussed in Unit 24, Piping and Tubing, and Unit 25, Soldering and Brazing. The procedures for sizing refrigerant lines were discussed in Unit 26, Refrigerant System Piping. Several important points should be remembered while actually erecting the system piping.

When hard copper tubing is selected for refrigerant piping, it is recommended that silver alloy brazing materials be used. These alloys have flow points in the range of 1,100°F–1,400°F. To attain these temperatures, oxyacetylene equipment is required (Figure 89-16).

The use of nitrogen is recommended to keep the interior of the pipe clean during the brazing process. Nitrogen brazing requires a nitrogen regulator and pressure-relief valve (Figure 89-17). During brazing, the surface of the copper will reach a temperature at which the metal will react with oxygen in the air to form a copper oxide scale. If this scale forms

Figure 89-16 Oxyacetylene torch used for brazing.

Figure 89-17 Nitrogen regulator with pressure-relief valve.

on the interior surface of the pipe, it might be washed off by the refrigerant and oil circulating in the system. The scale can clog strainers or capillary tubes and will plug orifices.

This scaling can be prevented by replacing the air in the pipe with nitrogen or carbon dioxide. Flowing nitrogen or carbon dioxide through the pipe prevents scaling because the nitrogen and carbon dioxide are inert gases and will not combine with copper, even under high-temperature conditions. The pipe interior will remain clean during brazing and no scale is formed, though discoloration may occur with overheating.

TECH TIP

When using an inert gas as a purge to prevent oxide formation, only a very slight flow of gas is required. Excessively high flow rates may create enough internal pressure to cause pinholes during the brazing process as the gas is forced out through the joint.

Figure 89-18 shows the connections for introducing nitrogen into the system during the brazing operation. Connect the nitrogen bottle to the liquid-line service Schrader valve with a refrigeration hose. A slight pressure is admitted into the tubing using the pressure-regulating valve on the nitrogen bottle. This is just enough pressure to ensure that air will be forced from the pipe. If the nitrogen flow can be felt on the palm of the hand, the flow is sufficient. This nitrogen pressure is kept in the tubing throughout the entire brazing operation.

Figure 89-18 System connections for nitrogen brazing.

CAUTION

Never use oxygen or acetylene to develop pressure when checking for leaks. Oxygen will cause an explosion in the presence of oil. Acetylene will decompose and explode if the regulator pressure is over 15–30 psig (30–45 psia or 210–310 kPa).

Clean pipe is essential in refrigeration installation. The use of nitrogen or carbon dioxide is extremely important in the brazing operation to ensure scale-free interior pipe walls.

The temperatures during brazing operations can warp metals and burn or distort internal elastomeric or plastic components. Many valves have rubber O-ring seals that can be damaged by excess heat. After they are melted, often the only fix is to replace the valve. The valves can be protected with a wet rag wrapped around the body (Figure 89-19). The water absorbs the heat that flows to the valve body during the brazing operation. By keeping the rag wet, the valve and its component parts are protected from heat damage. Heat-absorbing pastes and sprays are available to control heat and protect components (Figure 89-20). Using a heat-absorbing paste or spray is often necessary in places where wet rags will not go.

After all interconnecting tubing has been assembled, nitrogen is introduced into the system for leak testing. By using nitrogen, the pressure in the system may be built up to approximately 100 psig for leak testing.

CAUTION

Never pressurize the system above the low-side system test pressure (Figure 89-21) or above 150 psig for systems with hermetic compressors! Overpressurizing a compressor shell can cause catastrophic shell failure, turning the compressor shell into a bomb!

When using nitrogen to pressurize the system, leaks may be detected with soap bubbles and ultrasonic leak detectors. Halide torches and electronic leak detectors will not work with nitrogen. A small amount of R-22 or a non-ozone-depleting refrigerant may be used with the nitrogen as a trace gas if you wish to use an electronic leak detector. Since nitrogen is an inert gas, the nitrogen must be vented and the system evacuated after it has been determined to be leak free.

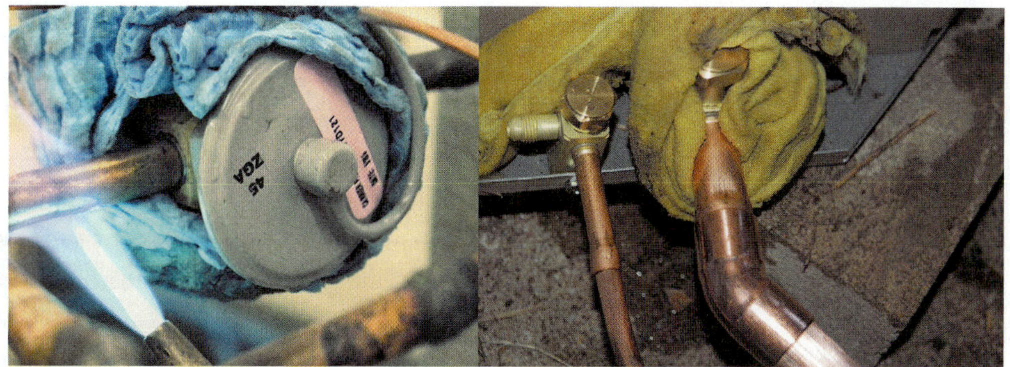

Figure 89-19 Valves can be protected from damage during brazing with a wet rag.

Figure 89-20 Heat-absorbing paste or spray gel is easier to use in many situations than a wet rag.

CONTAINS HCFC – 22	DESIGN PRESSURE
FACTORY CHARGE	278 HI PSIG
12 LBS 8 OZS	144 LO PSIG
ELECTRICAL RATING	NOMINAL 208/230 VOLTS
1 PH 60 HZ	MIN 197 MAX 253
COMPRESSOR(S):(1)	FAN MOTOR(S): (1)
PH 1	PH 1
RLA 23.8	FLA 1.7
LRA 129	HP 1/4
MIN. CKT AMPACITY AMPERAGE MINIMUM 31.5	MAX FUSE OR CKT.BKR. FUSIBLE/COUPE CIRCUIT (HACR PER NEC) 50
FOR OUTDOOR USE	
VERIFIED	VERIFIE

Figure 89-21 This system should not be pressurized past 144 psig when leak testing.

Figure 89-22 Suction lines should be insulated to avoid condensation and loss of system efficiency.

TECH TIP

Venting mixtures of R-22 and nitrogen used as a holding charge or a leak test gas does not violate the prohibition on venting ozone-depleting substances. However, other ozone-depleting refrigerants may not be used as trace gases.

The suction line must be insulated to avoid condensation and excessive heat from being absorbed by the refrigerant (Figure 89-22).The outside surface temperature of the suction line is frequently below the dew point of the surrounding air, causing moisture in the air to condense on the exterior of the pipe. This will create problems where this continuous moisture drips and can be annoying.

TECH TIP

Moisture has been associated with the formation of mold and mildew in buildings. Condensate can be a significant source of moisture. It is very important that tubing insulation on refrigerant lines be sealed at the joints so that condensate cannot form.

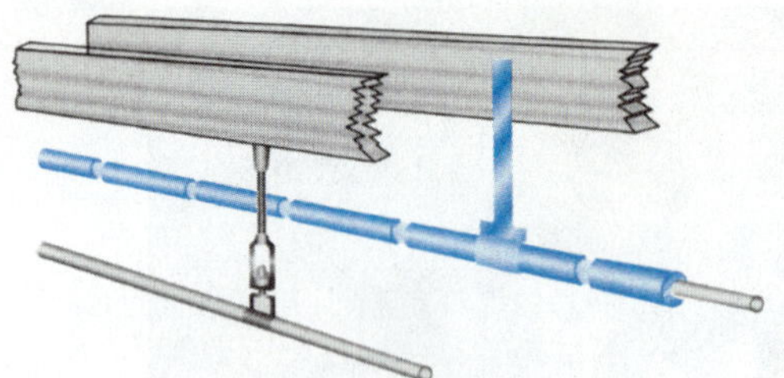

Figure 89-23 Pipe hangers used to support insulated pipes.

The insulation must be of good quality so that the temperature of its exterior surface will never drop below the dew point of the surrounding area. If the system is a heat pump, make sure the insulation is approved for use with heat pumps. It must also be well sealed, so that air and the moisture it contains cannot reach the pipe, thus causing condensation underneath the insulation.

An example of a typical pipe hanger that might be used on a small commercial refrigeration installation is shown on the right in Figure 89-23. This hanger serves as a vibration isolator as well as a pipe supporter. A short length of light-gauge rustproof metal is used to form a cradle to support the insulated pipe. A metal strap has been attached to this length of metal. In some cases it is merely wrapped around the length of sheet metal. The free end of the strap is then fastened to the joist or ceiling. The insulation in such a hanger will act as the vibration isolator. The purpose of the short length of metal is to prevent the thin strap from cutting the insulation.

The hanger shown on the left in Figure 89-23 has a height-adjustment feature for leveling or pitching the pipe if required for oil return. This type of hanger can be used with or without insulation.

ELECTRICAL RATING		NOMINAL 208/230 VOLTS	
1 PH	60 HZ	MIN 197	MAX 253
COMPRESSOR(S):(1)		FAN MOTOR(S): (1)	
PH	1	PH	1
RLA	23.8	FLA	1.7
LRA	129	HP	1/4
MIN. CKT AMPACITY AMPERAGE MINIMUM	31.5	MAX FUSE OR CKT.BKR. FUSIBLE/COUPE CIRCUIT (HACR PER NEC)	50
FOR OUTDOOR USE			
VERIFIED ENERGY PERFORMANCE	CSA	VERIFIE RENDEMENT ENERGETIQUE	
UL	LISTED 29EF SECTION OF HEAT PUMP		C UL

Figure 89-24 The minimum circuit ampacity is in yellow; the maximum fuse size is in green. The wire should be sized for no less than 31.5 amps; the fuse should be no larger than 50 amps.

89.19 ELECTRICAL CONNECTIONS

The service technician who installs the refrigeration equipment is sometimes responsible for the final wiring connections between the installed unit and the fused disconnect switch. All electrical power to the refrigeration unit must pass through this switch. When this switch is pulled, or opened, all electrical power to the unit must be disconnected. This same disconnect switch also contains fuses, which will interrupt the flow of current whenever a severe electrical overload occurs. This mechanism is a protection against fires and explosions and also against electrical shocks to people.

Three components of the power wiring to the unit that must be properly sized are the conductor, the disconnect, and overload protection devices.

Conductors are sized by current carrying capacity and voltage drop. Voltage-drop calculations are normally not necessary for runs under 100 ft if the minimum wire sizing charts in Section 310 of the NEC are followed. A conductor should be selected from the wire sizing charts whose current rating is equal to or greater than the minimum circuit ampacity listed on the unit data plate (Figure 89-24).

Air-conditioning units must have a readily accessible disconnect within sight of the unit (Figure 89-25). The disconnect current rating should be 115 percent of the minimum circuit ampacity. The overload protection should be no larger than the maximum fuse size shown on the data plate (see Figure 89-24). Standard fuses and circuit breakers are subject to nuisance trips from the momentary high inrush current when compressors start. The overcurrent protection for systems with compressors should be either time-delay fuses or an HVACR-rated circuit breaker because they allow the compressor to start without tripping.

Figure 89-25 All equipment should have a disconnect switch within sight of the unit and readily accessible.

CAUTION

Although codes require that all electricity to equipment be able to be disconnected at a single disconnect point, this is not always the case. Occasionally, host installation wiring may bring voltage into equipment that does not go through the disconnect. Before you begin working inside electrical panels on equipment, make certain that there is no voltage. Use your voltmeter to test all connections between each power connection and each power connection and a known ground.

89.20 SYSTEM EVACUATION

The refrigerant system needs to be leak tested and evacuated before it is charged. Nitrogen is commonly used as a leak test gas after making all the refrigerant piping connections. Any nitrogen left in the system from leak testing should be released before proceeding to evacuate the system. Connecting a vacuum pump to a system with a nitrogen charge will blow the oil out of the vacuum pump, making a mess and possibly damaging the vacuum pump.

TECH TIP

Since the air is 78 percent nitrogen, it is perfectly legal to vent the nitrogen holding charge. Venting de minimus mixtures of R-22 and nitrogen used for leak testing is also permitted by the EPA. However, nitrogen mixed with other ozone-depleting refrigerants may NOT be vented.

Be careful to release only nitrogen charges. Some systems, such as commercial refrigeration systems, come with only a nitrogen holding charge, which must be released before charging (Figure 89-26). The service valves on these units must be cracked to release the nitrogen. On the other

Figure 89-26 This equipment has a nitrogen holding charge.

CONTAINS HCFC-22		DESIGN PRESSURE	
FACTORY CHARGE		278	HI PSIG
12 LBS	8 OZS	144	LO PSIG
ELECTRICAL RATING		NOMINAL 208/230	VOLTS
1 PH	60 HZ	MIN 197	MAX 253

Figure 89-27 This unit holds 12 lb, 8 oz of HCFC 22 refrigerant.

hand, air-conditioning systems usually have a factory refrigerant charge that should not be released (Figure 89-27). These valves are typically shipped front-seated. They should not be turned until after the lines and coil have been evacuated. Read the installation instructions carefully to determine which type of system you are installing.

The system should be evacuated from both the high and low sides of the system to the vacuum level specified by the manufacturer, typically 500 microns or lower. The only way to ensure the proper level of evacuation is to use a vacuum gauge to verify the results. The amount of time required varies depending upon the size of the system and the length of refrigerant piping. However, clean systems rarely require longer than an hour unless they are very large.

SERVICE TIP

Many technicians save time by leak testing and evacuating the system before making the electrical connections. The system will typically be fully evacuated by the time they finish making all the electrical connections, so no additional installation time is required to properly evacuate the system.

89.21 CHARGING THE SYSTEM WITH REFRIGERANT

Charging the system with refrigerant is one of the most commonly performed tasks in the field. Unfortunately, it is often done incorrectly. Improper charging is most often due to the technician's failure to read and follow the manufacturer's instructions.

Packaged Units

Packaged units normally do not require charging for new installations because they are charged at the factory. In the event a packaged unit does require charging, you should first determine if the unit was shipped without a charge or if the charge leaked out. If the charge leaked out then the first step is to locate and repair all leaks. After the system is leak-tight, it should be evacuated and charged. The amount and type of refrigerant will be on the data plate (see Figure 89-27). The unit should not be operating while the refrigerant is weighed in. For units that hold more than 1 lb of refrigerant, the charge should be weighed in as a liquid into the king valve or liquid-line service valve. For units that hold

Figure 89-28 The unit arrives with the valves front-seated, as shown on the left; turning the valves counterclockwise opens them and releases the refrigerant.

less than 1 lb, the refrigerant should be weighed in as a gas into both sides of the system at once.

Split Systems

Split systems usually do require some adjustment to the refrigerant charge to account for their refrigerant lines. Most manufacturers ship their condensing units with enough charge for the outdoor unit, the indoor coil, and a set length of refrigerant line. This charge is held in the condensing unit with split-system shutoff valves (Figure 89-28).

These valves are front-seated when the unit is shipped. The Schrader access fittings on these valves are always open to the refrigerant line regardless of the valve's position. The shutoff valves are opened only after installing, leak testing, and evacuating the refrigerant lines and indoor coil.

CAUTION

Do not overtighten the installation valves when opening them. Stop turning the valve as soon as you feel increased resistance or see the top of the plug. A lock ring is all that keeps the valve plug in. Overtightening it in the counterclockwise direction will result in the valve plug shooting out at a pressure exceeding 100 psig, and all the refrigerant will come with it!

Opening the valves allows the refrigerant that is trapped in the condenser to travel throughout the system. If the actual line length is exactly the same as the manufacturer's assumed line length, you are done. If the lines are longer than the assumed length, you will need to add refrigerant. If the lines are shorter, you will need to remove refrigerant. The amount of assumed line length varies from 15 ft to 30 ft, depending upon the manufacturer. Many systems now include information on the unit that lists the assumed line length and gives the required per foot charge adjustment (Figure 89-29).

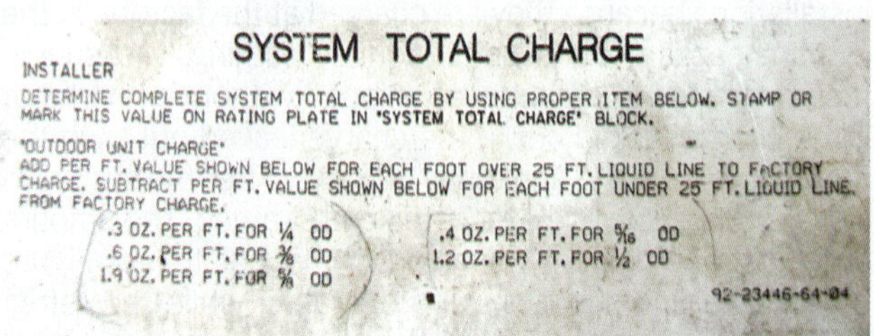

SYSTEM TOTAL CHARGE

INSTALLER

DETERMINE COMPLETE SYSTEM TOTAL CHARGE BY USING PROPER ITEM BELOW. STAMP OR MARK THIS VALUE ON RATING PLATE IN "SYSTEM TOTAL CHARGE" BLOCK.

"OUTDOOR UNIT CHARGE"
ADD PER FT. VALUE SHOWN BELOW FOR EACH FOOT OVER 25 FT. LIQUID LINE TO FACTORY CHARGE. SUBTRACT PER FT. VALUE SHOWN BELOW FOR EACH FOOT UNDER 25 FT. LIQUID LINE FROM FACTORY CHARGE.

.3 OZ. PER FT. FOR ¼ OD
.6 OZ. PER FT. FOR ⅜ OD
1.9 OZ. PER FT. FOR ⅝ OD
.4 OZ. PER FT. FOR 5/16 OD
1.2 OZ. PER FT. FOR ½ OD

92-23446-64-04

Figure 89-29 The assumed line length of 25 ft is included in the charging instruction that are included on this unit's service panel.

Table 89-2 Liquid Line Charge Allowance

Liquid Line Diameter (in)	Ounces per Foot of Liquid Line
¼	0.3
5/16	0.4
⅜	0.6
½	1.2
⅝	1.9

To calculate the amount of refrigerant needed:

1. Find the extra line length by subtracting the assumed length from the actual length.
2. Look up the per-foot charge for the size liquid line used.
3. Multiply the extra line length by the charge per foot to find the amount of additional charge required.

Table 89-2 shows the per foot charge required for liquid lines of various sizes. Note that an allowance is made only for the liquid line. There is not enough refrigerant in the suction line to calculate.

Under normal charging conditions, the refrigerant cylinder will be at ambient temperature and corresponding pressure. When the unit is operating, the refrigerant cylinder pressure will usually be below the head pressure of the condenser and above the back pressure of the evaporator. The easiest way to add the small amount of refrigerant needed for the line allowance is to add vapor to the suction side with the unit operating. For all refrigerants besides zeotropes, the refrigerant should be leaving the cylinder as a vapor with the cylinder in the upright position (Figure 89-30).

For zeotropes, the refrigerant must leave the cylinder as a liquid (Figure 89-31). This poses a problem because liquid should not be charged into the suction side of a system. The liquid refrigerant must be restricted between the cylinder and the system to safely add a zeotropic refrigerant with the unit operating. This can be accomplished with a charging device that is put in line with the cylinder (Figure 89-32) or by throttling the refrigerant using the hand valve on your gauges. Caution must be taken when charging a system with liquid refrigerant so that the compressor is not overloaded or slugged with liquid. Allowing too much liquid to enter the suction line is a fast way to kill a compressor.

Figure 89-30 The refrigerant leaving this cylinder will be a gas when the cylinder is upright.

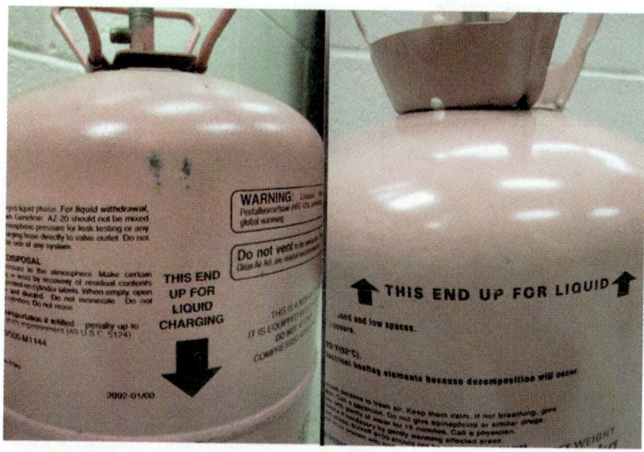

Figure 89-31 The cylinder on the left must be inverted to deliver liquid; the cylinder on the right delivers liquid while in the upright position.

Figure 89-32 This device flashes off liquid to gas, allowing the refrigerant to leave the cylinder as a liquid even when charging into the suction line. *(Courtesy of HVAC Department, Terra Community College, Fremont, Ohio)*

TECH TIP

Many of the systems that are in service today are overcharged. There is a feeling among some technicians that a little extra refrigerant should be added to the system just in case there is a leak. If you suspect a leak, fix the leak; do not overcharge the system. Excessive refrigerant charges can be as detrimental to the equipment and to the equipment performance as undercharged equipment.

89.22 PRESTART CHECKS

The initial start-up is a critical time in a unit's life. Making certain that the installation meets all the manufacturer's specifications is a good way to prolong the life of the equipment. Installations that are not up to spec often lead to premature equipment failures that could have been avoided by taking a little extra time before starting the unit for the first time.

It is a good idea to have a routine to follow for initial start-up. Using a routine saves time and ensures that all necessary items are covered. Figure 89-33 is a general-purpose initial pre-start checklist. If the manufacturer has supplied a suggested start-up procedure, follow it! The installation and start-up directions are normally inside the unit or shipping carton.

Clearances

Check to see that an adequate amount of space has been left on all sides of the unit. The specific distances for a particular unit can be found in the installation instructions. Clearances on air-conditioning condensing units are required for proper airflow and adequate service access. Heat pump condensing units also require clearance from the ground in certain locations to elevate them above ice and snow. Also, all outdoor units should be protected from roof water runoff. This can be accomplished by locating the unit outside of the roof overhang, by using gutters to divert the water flow, or by using a strip of angled flashing installed on the edge of the roof above the unit. Furnaces require clearances from combustible materials. Tables 89-3 and 89-4 list typical clearances for air-cooled condensers.

Table 89-3 Vertical Discharge Air-Cooled Condensing Units

Type of Clearance	Distance (in)
Clearance above the top of the unit for airflow	48
Clearance from any side of the unit where air is drawn in	18
Clearance from any service access panel	30

System Pre-Start Check List		
FURNACES	All Sides and Back 3" clearance Top 6" clearance Gas line has shut off, drip tee, and union Combustion Air Grilles Vent Connected Type B Vent 1" Clearance	
ALL INDOOR UNITS	Unit Is Level Evaporator Fan Spins Freely Clean Air Filter in Place Evaporator Drain Pan Drains Secondary Drain Pan Drains Correct Voltage at Disconnect Correct Voltage at Unit Unit is Grounded Thermostat is Level Overcurrent Protection is Correct Size	
OUTDOOR CONDENSING UNITS	Sides 18"Clearance Service Panel 30 " Clearance Vertical Discharge 48" Clearance Horizontal Discharge 48" Clearance Pad is at least 3" Above Ground Heat Pumps Elevated Above Snow Line Unit Is Level Condenser Fan Spins Freely Installation Valves Open (backseated) *Note: If valves are NOT open, the lines and coil must be evacuated before opening them* Unit Has Refrigerant (Check by Installing Gauges) Disconnect Within Sight and Accessible Overcurrent Protection is Correct Size Unit is Grounded Correct Voltage at Disconnect Correct Voltage at Unit	

Figure 89-33 General-purpose initial pre-start checklist.

Table 89-4 Horizontal Discharge Air-Cooled Condensing Units

Type of Clearance	Distance (in)
Clearance from all sides where air is discharged	48
Clearance from any side of the unit where air is drawn in	18
Clearance from any service access panel	30

Mechanical Precheck

Prestart mechanical checks should include the following:

- Visually look for and remove all packing and shipping supports and bolts.
- Make sure that a clean air filter is in place inside.
- Pour water in the evaporator drain pan to confirm that it drains correctly.
- Pour water in the secondary drain pan to confirm that it drains correctly.
- Check the thermostat with a small level.

- Use a level to check the level of the unit.
- Spin the condenser fan by hand to make sure it rotates freely.
- Check to see that the charging valves on the condensing unit are back-seated.
- If they are not, then you will need to evacuate the lines and evaporator coil *before* opening them.
- Connect refrigerant gauges to ensure that the unit has a charge. If there is no charge, you will need to determine why before proceeding.
- Check all line connections for leaks.

Electrical Precheck

Electrical checks to be completed before starting the unit should include the following:

- The power wire should be *copper* (*not* aluminum).
- The power wire connections should be tight.
- Make certain that the unit is grounded.
- There should be a disconnect switch for each piece of equipment *in sight* of the equipment.
- Check that the overcurrent protection is not more than the maximum size stated on the unit's nameplate.
- Use the wiring diagram to check the low-voltage connections. With the thermostat and disconnect switch turned off, check the voltage going into the disconnect switch. It should be within ± 10 percent of the nominal voltage rating on the nameplate. If it is not, do not turn the disconnect switch on.
- Turn the disconnect switch on and check the voltage available to the unit. It should be within the minimum and maximum voltages stated on the unit data plate.

Gas Precheck for Combustion Appliances

- Check that gas line has a shutoff valve, drip tee, and union.
- Check to see that all gas piping is properly supported.
- Check for combustion air provisions.
- Check to see proper vent materials and connections.

89.23 INITIAL START-UP

A quick check of unit operation should be done immediately after start-up to verify that it is operating safely. For all types of systems, check the operating voltage and compare it to the system data plate. If the voltage was correct before start-up but drops below the unit's minimum voltage rating after starting, either the wire is too small or there is a bad connection. In either case, the unit must be shut off and the problem must be corrected. If the voltage is within the manufacturer's specifications, check the operating current against the data plate and make sure that it is not too high.

Do a quick check of the refrigerant pressures. At this point, you just need to verify that they are within safe operating limits. If the suction pressure starts to pull down to 0 psig, there is a refrigerant restriction that must be corrected. Shut down the unit and look for possible restrictions. Start by double-checking the position of the line installation valves. Check to see that the high-side pressure is not too high. As a general rule, R-22 pressures over 300 psig or R-410A pressures over 450 psig indicate a problem. Do not let a unit operate for an extended period of time above these pressures. Excessive head pressures could be the result of poor airflow over the condenser or an overcharge. On a new unit, poor airflow would most likely be caused by incorrect condenser fan rotation, the condenser fan not operating, or coils blocked by shipping plastic or cardboard. In any event, the high side must be brought down to a safe level. On a three-phase unit with a scroll compressor, if the compressor starts but the high-side and low-side pressures remain the same, the compressor is most likely turning backwards. Shut off the power and reverse two of the power legs at the compressor contactor.

Once the unit has started and you have verified that is operating safely, you can measure the system operational data and check it against the manufacturer's specifications. Before checking the system operating pressures and temperatures, the unit should operate long enough to bring the house close to normal AHRI inside design conditions: 80° dry bulb, 67° wet bulb. While the unit is operating, check the total system airflow by checking the static pressure across the blower and comparing it to the manufacturer's specifications. Next, check the airflow at the registers and compare it to the design airflow. Now check the system operational data, including high- and low-side pressures, superheat, subcooling, temperature drop across the evaporator, and temperature rise across the condenser. Compare these data to the manufacturer's specifications. These data are usually found in the installation instructions and on the equipment service panels. Units with expansion valves often have a subcooling specified right on the condenser nameplate (Figure 89-34). Figure 89-35 is a general-purpose initial prestart and start-up checklist.

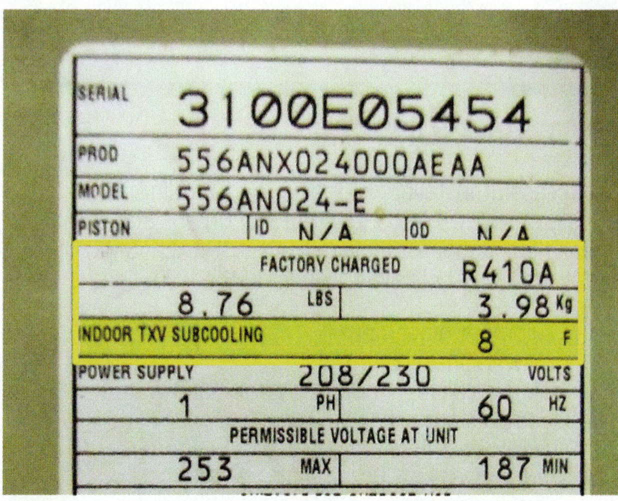

Figure 89-34 The data plate on this unit indicates that it should operate with 8° subcooling.

System Initial Start Up Check List		
Furnaces	Gas Manifold Pressure	
	Vent Draft Pressure	
	Firing Rate (Btuh Gas Consumption)	
	Temperature Rise	
	Static Pressure Across Indoor Blower	
	Indoor Airflow (based on static reading)	
Air Cond	Indoor Wet Bulb Temperature	
	Indoor Dry Bulb Temperature	
	Static Pressure Across Indoor Blower	
	Indoor Airflow (based on static reading)	
	Evaporator Temperature Drop	
Outdoor Condensing Units	Outdoor Ambient Temperature	
	Type of Refrigerant	
	Equalized Refrigerant Pressure	
	Voltage at Unit After Startup	
	Unit Operating Current (Amps)	
	Initial Suction Pressure After Startup	
	Initial Condenser Pressure After Startup	
	Suction Pressure	
	Suction Line Temperature	
	Suction superheat	
	Liquid Pressure	
	Liquid Line Temperature	
	Subcooling	
	Condenser Temperature Rise	

Figure 89-35 General-purpose initial start-up checklist.

UNIT 89—SUMMARY

A successful installation begins with reading the instructions and applicable codes. National organizations that set installation standards include AHRI, ASHRAE, ASME, UL, and NFPA. Phases of installation include the following:

- Equipment placement
- Ductwork installation
- Piping installation (both refrigerant and water)
- Electrical connection
- Evacuation and charging
- Start and check

Equipment should be located where it can be serviced. Air-cooled condensers should be placed so that the airflow to them is not obstructed. All piping should be supported. Refrigerant lines should be purged with nitrogen during brazing to prevent oxidation. All equipment should have an electrical disconnect within sight of the equipment. Power wiring should be sized to carry the minimum circuit ampacity marked on the unit data plate. The circuit overcurrent protection should not exceed the maximum fuse size listed on the unit data plate. Refrigeration systems should be evacuated to 500 microns before charging. A careful prestart-up inspection and initial system performance check can help avoid future problems. Using a checklist to document the steps taken during the prestart inspection and initial start-up can ensure that all steps have been taken. Regardless of equipment type, a good installation is the best way to ensure long equipment life, reliable operation, and energy efficiency.

WORK ORDERS

Service Ticket 8901

Initial System Start-up

Equipment: Small Split-System Air-Conditioning Unit in Commercial Setting

A technician is performing a prestart-up inspection and notices that the system's minimum circuit ampacity is listed as 15 A and the unit is wired with 14-gauge wire. The NEC wire sizing chart shows that 15 A is the most that wire will carry. The technician also notes that the power wire has been run a long way because the unit is over 200 ft from the power panel. The system nominal voltage rating is 208–230 V with a minimum of 190 V and a maximum of 250 V. The voltage check with the unit off shows 208 V available. The technician is wary because the voltage is in the lower end of the acceptable range. When the unit is started, the line voltage drops to 188 V and the technician shuts the system off. The technician recommends that the power wire be replaced with a larger gauge wire to reduce the voltage drop and keep the unit operating consistently above the minimum voltage rating.

Service Ticket 8902

Initial System Start-up

Equipment: Split-System Residential Air-Conditioning Unit

A technician is sent to perform an initial system start-up. During the prestart check the technician notices that the refrigerant line installation valves are still front-seated. The technician decides to evacuate the lines and coil before opening the valves in case they were not evacuated properly. The lines and coil are blanked off after they are evacuated to 500 microns. The vacuum holds, so the technician opens the refrigeration line installation valves.

Service Ticket 8903

Customer Request: Packaged Unit Replacement

Equipment Type: Residential Packaged Air Conditioner with Vertical Condenser Air Discharge

An estimator is sent to collect information to bid a job replacing a packaged air-conditioning unit. Upon arrival, the estimator notices that the old system is located under a porch, has an over-under duct configuration, and has a horizontal discharge condenser. The systems their company normally sells have side-by-side duct connections and a vertical discharge condenser. The estimator returns and explains that either a different system must be used or the system must be relocated to allow for proper condenser airflow and major duct modifications.

Service Ticket 8904

Customer Request: Heat Pump Installation

Equipment Type: Residential Split-System Heat Pump in Northern Climate

A technician is sent to install a split-system heat pump and notices that no provision has been made to keep the outdoor unit above the snow level in the winter. The company has never installed a heat pump before and was planning on setting the outdoor unit on the concrete pad like an air-conditioning condensing unit. The technician shows the foreman the installation instructions that show a metal frame elevating the unit above the top of the snow. The technician fabricates a galvanized metal frame that holds the outdoor unit 3 ft off the ground.

Service Ticket 8905

Customer Request: Gas Furnace Replacement

Equipment Type: 80 Percent Induced-Draft Furnace Replacing Existing Natural-Draft Furnace

A technician is sent to replace an existing natural-draft furnace with a new 80 percent induced-draft furnace. The technician notices that the existing furnace uses a 7-in single-wall metal vent connector leading to a masonry chimney. The new furnace has a vent collar of 5 in. The new furnace installation instructions clearly say that single-wall pipe should not be used and that masonry chimneys should not be used unless a metal liner is used. The installer recommends lining the masonry chimney with a metal liner and replacing the old 7-in single-wall vent connector with 5-in double-wall pipe.

UNIT 89—REVIEW QUESTIONS

1. List the associations that establish national standards for the HVACR industry.
2. What is the function of AHRI?
3. What does the UL specialize in?
4. List all major phases of system installation.
5. What factors must be considered in equipment placement?
6. Outline the correct brazing procedure for hard copper refrigeration piping.
7. What electrical rating on a unit data plate is useful for sizing the power wire for the unit?
8. What electrical rating on the unit data plate is useful for sizing the circuit breaker for the unit?
9. Use the information on the data plate in Figure 89-24 to determine the minimum current rating for sizing the power wire to the unit.
10. Use the information on the data plate in Figure 89-24 to determine the minimum current rating for the unit disconnect.
11. Use the information on the data plate in Figure 89-24 to determine the maximum circuit breaker size for the unit.

12. What type of overcurrent protection is used with air-conditioning equipment?
13. How should ductwork be sealed against air leaks?
14. How can vibrations and sound transmission in HVACR equipment be minimized?
15. Use Figures 89-27 and Table 89-2 to determine the correct amount of additional charge for a new split system with a 40-ft line set and a ⅜-in liquid line. The factory charge includes enough refrigerant for 15 ft of line.
16. What position are the unit installation valves in when the unit is shipped?
17. When installing a split system, what should be done prior to opening the installation valves?

UNIT 90

Planned Maintenance

OBJECTIVES

After completing this unit, you will be able to:

1. describe the four phases of a planned maintenance service call.
2. discuss the advantages of planned maintenance.
3. describe the planned maintenance process for refrigeration and air-conditioning systems.
4. describe the planned maintenance process for heating systems.
5. list system operational safety concerns that should be addressed on a planned maintenance service call.
6. discuss aspects of planned maintenance that are common to most systems.

90.1 INTRODUCTION

An often repeated bit of folk wisdom is "If it ain't broke, don't fix it." The problem with this philosophy is that it limits service work to emergency operations that are performed under duress after the worst has already happened. It makes life easier to keep things from breaking in the first place.

Planned maintenance enables efficient use of service labor resources. Regular maintenance is scheduled during slow periods, freeing service labor for more emergency service during peak seasons. Scheduling maintenance regularly helps catch small problems before they become big ones, saving both time and money. Keeping systems operating according to manufacturer's specifications saves both energy and money as well.

90.2 PLANNED MAINTENANCE

Scheduled maintenance calls include four important elements:

1. Routine maintenance on parts that need regular checking, cleaning, or replacing
2. Prestart inspection of the equipment
3. Operation check of the equipment
4. Troubleshooting and correction when necessary

Routine Maintenance

Maintenance schedules are developed according to the season and the system requirements. Seasonal maintenance on heating systems is typically done in the fall, and seasonal maintenance on air-conditioning systems is usually done in the spring. These seasonal checkups are intended to ensure that the systems perform when needed. They are the customer's insurance policy against system malfunctions during the peak season. However, some system components may need attention more than once a year. Air filters are a good example. Standard 1-in light-duty filters need checking monthly in most houses. The maintenance schedule and services performed will vary depending upon the type of system. What remains constant is that neglecting the required maintenance will lead to increased operating costs and shorter equipment life.

Prestart Check

There are a few things that need to be checked before starting the system. The best time to discover that the voltage is incorrect is before you start the unit! First find the unit data plate (Figure 90-1) and note all information about the equipment operational requirements. These could include the following:

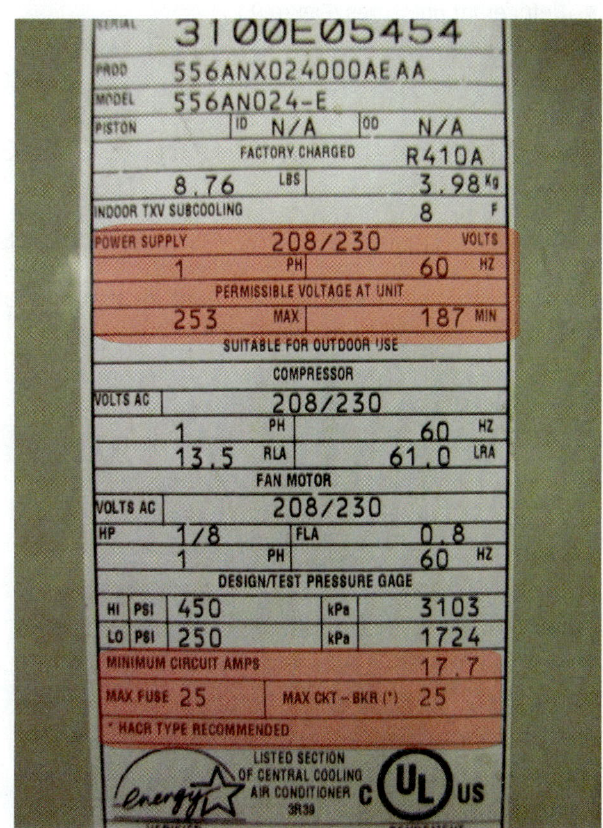

Figure 90-1 Unit data plate.

Figure 90-2 Manufacturer's refrigerant pressure chart.

			Outdoor Ambient Temperature																							
			65°F				75°F				85°F				95°F				105°F				115°F			
			Entering Indoor Wet Bulb Temperature																							
IDB	Airflow		59	63	67	71	59	63	67	71	59	63	67	71	59	63	67	71	59	63	67	71	59	63	67	71
70	1519	MBh	39.6	41.1	45.0	-	38.7	40.1	43.9	-	37.8	39.2	42.9	-	36.9	38.2	41.9	-	35.0	36.3	39.8	-	32.4	33.6	36.8	-
		S/T	0.73	0.61	0.42	-	0.76	0.63	0.44	-	0.78	0.65	0.45	-	0.80	0.67	0.47	-	0.83	0.70	0.48	-	0.84	0.70	0.49	-
		ΔT	18	15	12	-	18	15	12	-	18	15	12	-	18	16	12	-	18	15	12	-	17	14	11	-
		kW	2.79	2.85	2.93	-	3.00	3.06	3.16	-	3.18	3.25	3.35	-	3.34	3.42	3.53	-	3.48	3.56	3.67	-	3.60	3.68	3.80	-
		Amps	12.3	12.5	12.9	-	13.1	13.4	13.8	-	14.1	14.4	14.9	-	15.0	15.3	15.8	-	15.8	16.2	16.7	-	16.7	17.1	17.6	-
		Hi PR	222	239	252	-	249	268	283	-	283	305	322	-	322	347	366	-	363	390	412	-	401	431	455	-
		Lo PR	109	116	127	-	116	123	134	-	120	128	139	-	126	134	146	-	132	141	154	-	137	145	159	-
	1350	MBh	38.5	39.9	43.7	-	37.6	38.9	42.7	-	36.7	38.0	41.6	-	35.8	37.1	40.6	-	34.0	35.2	38.6	-	31.5	32.6	35.8	-
		S/T	0.70	0.58	0.40	-	0.72	0.61	0.42	-	0.74	0.62	0.43	-	0.77	0.64	0.44	-	0.80	0.66	0.46	-	0.80	0.67	0.46	-
		ΔT	18	16	12	-	19	16	12	-	19	16	12	-	19	16	12	-	18	16	12	-	17	15	11	-
		kW	2.77	2.82	2.91	-	2.97	3.04	3.13	-	3.16	3.22	3.33	-	3.32	3.39	3.50	-	3.45	3.53	3.64	-	3.57	3.65	3.77	-
		Amps	12.2	12.5	12.8	-	13.0	13.3	13.7	-	14.0	14.3	14.7	-	14.9	15.2	15.6	-	15.7	16.1	16.5	-	16.5	16.9	17.4	-
		Hi PR	220	236	250	-	246	265	280	-	280	302	318	-	319	343	363	-	359	386	408	-	397	427	451	-
		Lo PR	108	115	126	-	114	122	133	-	119	126	138	-	125	133	145	-	131	139	152	-	135	144	157	-
	1181	MBh	35.5	36.8	40.3	-	34.7	35.9	39.4	-	33.9	35.1	38.4	-	33.0	34.2	37.5	-	31.4	32.5	35.6	-	29.1	30.1	33.0	-
		S/T	0.67	0.56	0.39	-	0.70	0.58	0.40	-	0.72	0.60	0.41	-	0.74	0.62	0.43	-	0.77	0.64	0.44	-	0.77	0.65	0.45	-
		ΔT	19	16	12	-	19	16	12	-	19	16	12	-	19	16	12	-	19	16	12	-	18	15	12	-
		kW	2.70	2.76	2.84	-	2.90	2.96	3.06	-	3.08	3.15	3.25	-	3.24	3.31	3.41	-	3.37	3.44	3.55	-	3.48	3.56	3.68	-
		Amps	11.9	12.2	12.5	-	12.7	13.0	13.4	-	13.7	14.0	14.4	-	14.5	14.8	15.3	-	15.3	15.7	16.1	-	16.1	16.5	17.0	-
		Hi PR	213	229	242	-	239	257	272	-	272	293	309	-	310	333	352	-	348	375	396	-	385	414	437	-
		Lo PR	105	112	122	-	111	118	129	-	115	123	134	-	121	129	141	-	127	135	147	-	131	140	153	-

- The minimum and maximum voltage rating for the equipment
- The current draw of all motors
- The fuel source and fuel pressure for combustion appliances
- The temperature rise for furnaces

Operational Checks

The operation of the equipment should be checked against the manufacturer's specifications. Items to be checked could include the following:

- Operating voltage
- Operating current
- Refrigerant pressures (Figure 90-2)
- Gas manifold pressures
- Temperature rise (Figure 90-3)

90.3 REFRIGERATION/AIR-CONDITIONING MAINTENANCE

Although different types of systems share certain aspects of routine maintenance, each type of system will have areas of maintenance that are vitally important for that type of system. For air conditioning, all aspects of affecting airflow are crucial to proper system operation.

Routine Maintenance

Air-conditioning and refrigeration equipment should receive an annual checkup and cleaning. This is normally done in the spring. The first order of business is to inspect and clean the coils and filters. The condenser and evaporator should be checked and cleaned if necessary.

Lubrication and Adjustment

Now is a good time to lubricate all bearings and tighten any belts. Most residential blowers use sleeve bearings Some sleeve bearings require semi-annual lubrication. Many motors have lubrication instructions on the motor (Figure 90-4).

You should only apply oil to oiler tubes. Do not attempt to oil a motor that has no place for adding oil. Do not over-oil; five to ten drops are normally sufficient. Most residential equipment no longer uses belts. If you are working on a piece of equipment with a belt, you should check the belt, the belt alignment, and the belt tension.

Figure 90-3 The temperature rise of gas-fired heating equipment can be found on the unit data plate.

Figure 90-4 The lubricating instructions for this motor are on its data plate.

Figure 90-5 Overheating causes the terminal to discolor and could cause wire insulation to melt.

Figure 90-6 Burned and pitted contactor points, like the ones on the bottom, will create a voltage drop and cause motor overheating.

Prestart Inspection

Check the power supply to the unit with the unit off. It should fall between the minimum and maximum voltage ratings. Visually inspect the wiring for connections that appear discolored. Loose connections create heat, which makes the wires and insulation brittle and burned looking (Figure 90-5).

Inspect the contactor. Its armature should move freely without sticking. The contactor points should be smooth, not pitted. Some discoloration of the points is normal (Figure 90-6).

SERVICE TIP

Do not try to sand or file the points, as you will most likely do more harm than good. Filing removes the silver-cadmium outer plating, which is much more resistant to oxidation than the copper underneath.

Install gauges on the equipment to make sure it has refrigerant. You should not operate equipment without a charge. Check for refrigerant leaks at all mechanical connections and visible braze joints. Particularly look for signs of oil on the lines or the pipe insulation (Figure 90-7).

If you find leaks, they should be repaired before proceeding.

Operational Checks

Start the system and check the incoming voltage with the unit operating. A significant drop in voltage indicates problems in the power supply. If the voltage drops below the minimum voltage, shut down the system. The contactor contacts should also be checked for voltage drop. The ideal voltage reading from one side of the contact to the other with current going though is 0 V. A reading of more than a few tenths of a volt indicates bad contacts on the contactor. If there is a significant voltage drop across the contacts, shut the system down and change the contactor. A voltage drop across the contactor contacts can kill the compressor.

With the unit operating, check all the following operational characteristics:

- Amp draw of all motors
- Airflow
- Low-side and high-side pressures
- Superheat and subcooling
- Temperature drop across the evaporator

Compare your readings with the manufacturer's specifications.

SERVICE TIP

Most equipment has some operating specifications on the inside of the service panel, including refrigerant charging charts. The electrical specifications can be found on the unit nameplate.

Figure 90-7 Oil is a natural leak indicator; this Vibrasorber has a leak at one end where it flexes. *(Courtesy of HVAC Department, Terra Community College, Fremont, Ohio)*

90.4 HEATING SYSTEMS MAINTENANCE

An annual heating system check should include checking the fuel source, performing prestart checks, and checking the air temperature rise. The combustion should be checked on gas and oil furnaces and boilers. See Unit 51 Gas Furnace Installation, and Unit 54 Oil Furnace installation for more details on combustion checks.

90.5 ELECTRIC HEATING

The most important aspects of maintenance for electric furnaces relate to the large amount of electric current they require. Electrical connections are vitally important. Poor connections quickly cause major problems.

Routine Maintenance

The only routine maintenance necessary on an electric furnace would be changing or cleaning the air filter and lubricating the blower. In the rare cases of an electric furnace with a belt-drive blower, you should also check the belt, belt alignment, and belt tension.

Prestart Inspection

The incoming voltage should be checked and compared to the nameplate rating. Perform a visual inspection on all wires and wire connections. Overheating and degradation of wires, wire lugs, and fuse blocks are common problems in electric furnaces. The large current draw turns any small resistance into a small heater that degrades and discolors electrical connections over time. Any visibly discolored wires or wiring components should be replaced.

Operational Check

Checking the power source and all electrical connections is essential when performing annual maintenance. Because electric furnaces use large amounts of power, any resistance in any of the power-handling components leads to heat and failure of that component. More seriously, heated wires, wiring lugs, or fuse blocks can be dangerous.

The voltage to the furnace should be checked before and after starting the furnace. A voltage drop indicates a problem with the power wiring to the furnace. The voltage at the heating elements should be checked and compared to the voltage entering the furnace; they should be the same. A voltage drop indicates that there is a problem in the furnace. This can be located by measuring the voltage across all the power-handling components. An infrared thermometer can also help. The point of voltage drop will be hotter than the rest of the wires and components. The current draw of each heating element should be checked and compared to the nameplate rating. A heat strip that does not draw any current indicates a problem. First check to make sure it has voltage. If no voltage is being applied to the heat strip, the problem is in the controls to that strip. If voltage is being applied and it still has no current draw, then either the heat strip or its fusible link is open. After all heating elements are operating, check the temperature rise of the furnace. Measure the temperature of the air entering and leaving the furnace. The difference is the temperature rise. The temperature rise should fall within the range listed on the furnace nameplate.

Figure 90-8 These burners are no longer serviceable and should be replaced.

90.6 GAS HEATING

Possibly the most important checks you can perform with a gas furnace is to check for adequate combustion air and proper venting. These are major safety concerns that are easily overlooked if you are focusing solely on the equipment.

Routine Maintenance

As is the case for all forced-air systems, the air filters should be cleaned or changed and the blower motor should be lubricated. For belt-drive systems, the belt, belt alignment, and belt tension should all be checked. Ribbon, slotted, or drilled port burners should be removed so they can be inspected and cleaned. Burners with more than light surface rust should be replaced pending the outcome of the prestart inspection (Figure 90-8).

SERVICE TIP

Loose rust on the burners may not be from the burners but from the heat exchanger. If the burners have large flakes of rust on them, the heat exchanger needs to be carefully checked.

In-shot burners generally do not require cleaning but should be inspected for rust. This can usually be done with an inspection mirror without removing the burners. Extremely rusty burners could indicate a more serious problem that might require replacing the furnace. If you have removed the burners to clean them, leave them out for the prestart check.

On low-NOX furnaces, check the NOX rods or screens for rust, corrosion, or warping. Ribbon burners use rods

Figure 90-9 Deteriorating NOX screen on in-shot burner.

installed over the burner, whereas in-shot burners use screens (Figure 90-9).

SAFETY TIP

Some AQMD (air-quality management districts) require the removal of nitrogen and oxygen compounds (NOX) formed as a byproduct of combustion. Metal rods or screens are placed in the flame to reduce the NOX emissions from the furnace. These rods get red hot and cause a complete burn of the combustion byproducts. Replacing NOX rods that are rusted or corroded is an important safety issue. NOX rods tend to deteriorate, and rust and parts of the rod can fall into ribbon burner slots. Rods have broken, fallen down, and caused fires. They also warp out of their retention brackets and bump into firebox walls. When they go unchecked and warp or rust, they can cause hotspots on the firebox walls, causing cracks and warping of the firebox. They can also restrict gas flow from burners, causing uneven burns or hotspots or premature firebox failure. Routine inspections can ward off issues with these NOX rods.

Prestart Check

The heat exchanger should be visually inspected for holes, cracks, and rust. Some surface rust is normal, but flaky rust that comes off in chunks should be carefully inspected. The heat exchanger may be viewed from the burner end and from the draft diverter end for older-style natural-draft furnaces. An inspection mirror can be helpful, but a flexible fiberoptic or video inspection tool is preferable (Figure 90-10).

SERVICE TIP

On noncondensing furnaces, a better view of the heat exchanger can be gained by removing the blower and looking through the furnace from the blower end. This will not work on condensing furnaces because the recuperative heat exchanger is in the way.

Figure 90-10 This Testo 318 inspection tool can be used to inspect heat exchangers for cracks. *(Courtesy of Testo, Inc.)*

The heat exchanger will have a separate tube or clamshell for each burner. Each section needs to be inspected individually.

The vent system should be inspected for loose connections, signs of corrosion, or rust. A rusty or corroded vent indicates condensation in the vent (Figure 90-11).

The cause of the condensation should be determined and corrected. It could be using a single-wall vent in a cold ambient temperature, an oversized vent, a vent with an excessively long horizontal run, a restricted vent, or lack of combustion air. A malfunctioning vent is dangerous; the cause should be found and corrected!

The combustion and venting processes cannot work correctly without adequate combustion and ventilation air. The room the furnace is in should normally be provided with combustion air somewhere near the floor and also near the ceiling. This can be a hole or grille to a ventilated crawlspace or attic or through ducts that go outside. Locate the combustion air openings and verify that they are not blocked (Figure 90-12).

Find the nameplate and write down the following information:

- The type of gas
- The minimum and maximum input gas pressure
- The manifold gas pressure
- The temperature rise
- The operating voltage

Check the incoming gas pressure to see if it falls between the minimum and maximum input pressures. Check the incoming voltage to see if it falls between the minimum and

Figure 90-11 The water in the flue has been condensing in this vent, causing the rust at the joints of the elbow. *(Courtesy of HVAC Department, Terra Community College, Fremont, Ohio)*

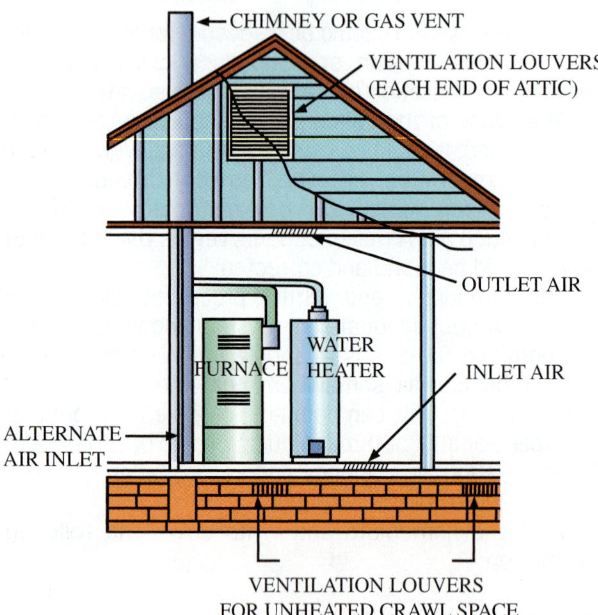

Figure 90-12 Combustion appliances should be supplied with combustion air openings both high and low.

maximum voltage. Install the manometer on the manifold gas pressure port before turning on the system. Any reading other than 0 in wc indicates a leaking gas valve. This is a very rare but dangerous situation. If the gas valve leaks, it must be replaced. Turn off the gas supply and leave the area to let the accumulated gas dissipate before proceeding. Leave the manometer on for use during the operational check.

SAFETY TIP

Do not put your face directly in front of the burner area when the furnace first lights. Flame rollout and flashback can occur, causing flames to come out from the furnace.

Operational Check

For category I and II furnaces, place a draft gauge in the vent and check the draft with the burners operating. You should have a draft of at least –0.02 in wc. Watch the draft reading when the indoor blower comes on. A reduction in draft or an increase to a positive pressure indicates a leak in the heat exchanger. This is a serious safety issue that generally warrants replacing the furnace.

Check the manifold gas pressure after the burners light. The correct manifold gas pressure at sea level is indicated on the furnace nameplate. The manifold pressure is usually slightly less at higher altitudes. The manufacturer's installation instructions are the best source of information for de-rating gas appliances.

Look at the color of the flames. The flames should be light blue in the center with a darker blue outer area. There should be no yellow. (Figure 90-13).

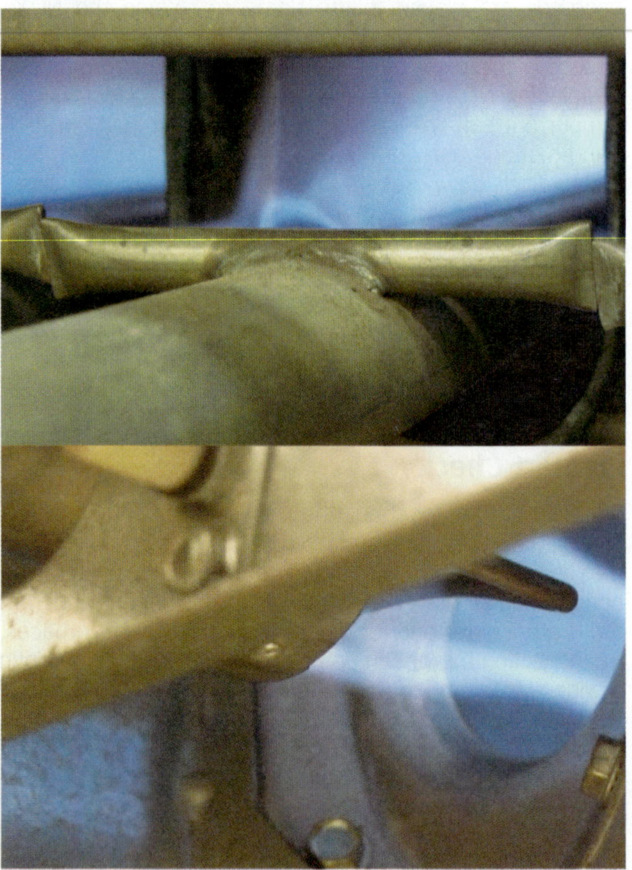

Figure 90-13 The flame at the top shows a correctly adjusted ribbon burner. The flame on the bottom shows a correctly adjusted in-shot burner.

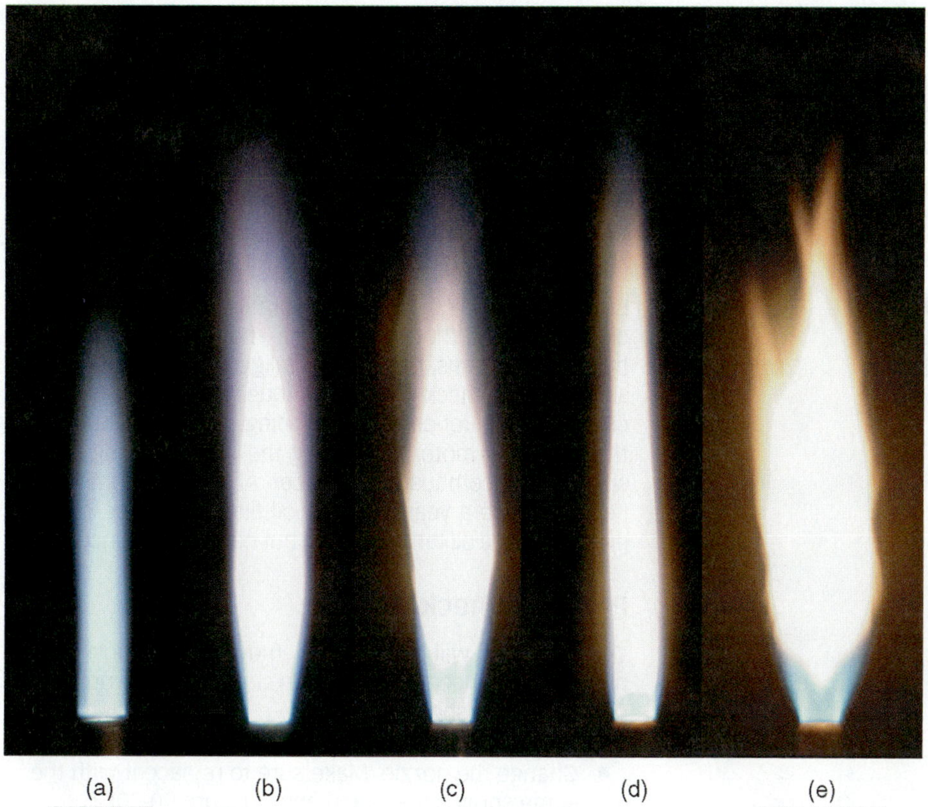

Figure 90-14 Natural gas flames: (a) correct mixture of fuel and air; (b)-(e) show the effects of decreased primary air. As the primary air is decreased the flame becomes wider and more irregular with a bright yellow color.

Yellow flames indicate incomplete combustion and the presence of carbon monoxide (Figure 90-14).

Yellow flames are the result of too much gas, too little air, or faulty burners. Older gas furnaces have adjustable primary air intakes (Figure 90-15).

The primary air can be adjusted until the yellow disappears. However, opening the shutters too far can cause delayed ignition, especially on cold days. Too much primary air also causes the flames to lift away from the burner face. In-shot burners, like those in most modern furnaces, do not have any primary air adjustment. Yellow flames on these furnaces indicate a lack of combustion air in the room or too much gas. Check the orifice size against the manufacturer's specifications or an orifice chart to make certain the furnace has the correct size of gas orifices (Figure 90-16 and 90-17).

The combustion gases in the vent should be checked for carbon monoxide. Ideally, the carbon monoxide level

Figure 90-15 Some gas burners have primary air shutters for adjusting the amount of primary air to the burner.

Figure 90-16 The orifice size is stamped on it; this is a #43.

ORIFICE SIZE FOR 23,000 BTUH GAS FIRED BURNER					
Altitude	0 – 7,000	7,001 – 11,000	0 – 7,000	7,001 – 9,000	9,001 – 11,000
Type of Gas	LP	LP	Natural	Natural	Natural
Orifice Size	#55	#56	#43	#44	#45

Figure 90-17 The same furnace may require four different orifices depending upon the type of gas and altitude.

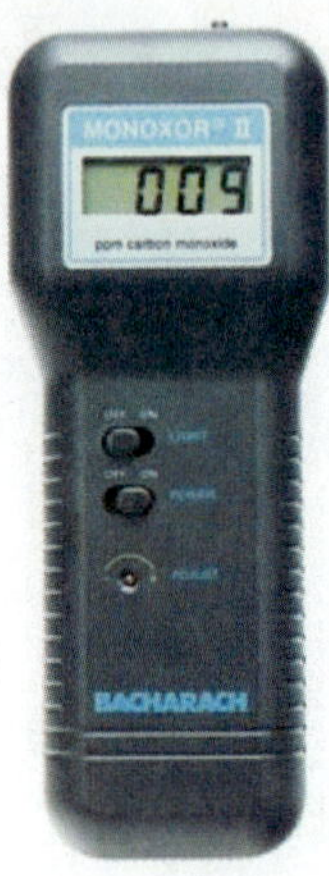

Figure 90-18 Carbon monoxide testers can read the level of CO in the flue gas.

should be 0 ppm for complete combustion. Realistically, it is common for some low level of CO to be present in the combustion gases. Levels of CO in the vent gases exceeding 50 parts per million warrant a closer look at the combustion process (Figure 90-18).

After the blower has been operating for a few minutes, check the temperature of the air entering and leaving the furnace. The difference is the temperature rise (Figure 90-19).

Be sure to check the leaving-air temperature in a location that is not directly in the line of sight of the heat exchanger. If the thermometer can "see" the heat exchanger, it may read higher due to the heat exchanger's radiant heat. If the temperature rise is too high, the airflow needs to be increased. If the temperature rise is too low, the airflow needs to be reduced.

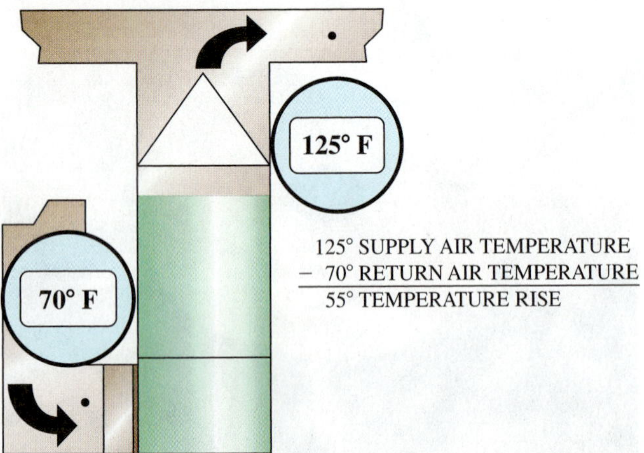

Figure 90-19 Temperature rise is the difference between supply- and return-air temperatures.

90.7 OIL-FIRED HEATING

As with gas furnaces, combustion air and venting are major safety concerns. Annual maintenance is a requirement to keep oil-fired equipment operating properly. Measuring and adjusting the burner to produce proper flame and combustion gas characteristics should be performed annually.

Routine Maintenance

Like all forced-air systems, the air filters should be cleaned or changed and the blower motor should be lubricated. For belt-drive systems, the belt, belt alignment, and belt tension should all be checked. Items unique to oil furnaces include changing the fuel oil filter, replacing the nozzle, lubricating the oil burner motor, inspecting the drive coupling, and inspecting the combustion chamber. All of these items should be done once a year. The fuel oil filter should be changed every time the oil tank is refilled but no less than once a year.

Prestart Check

The oil burner will have to be removed to perform the annual maintenance and preseason check. After removing the oil burner, you should:

- Change the nozzle. Make sure to replace it with the same spray pattern and angle (Figure 90-20).
- Inspect the electrodes for cracks in the ceramic insulators (Figure 90-21).

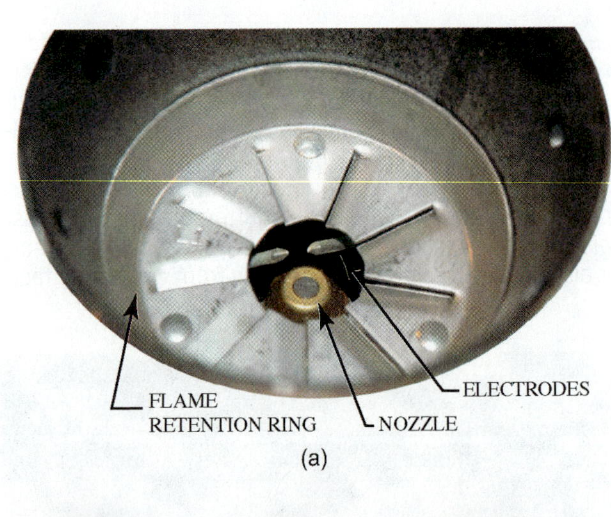

(a)

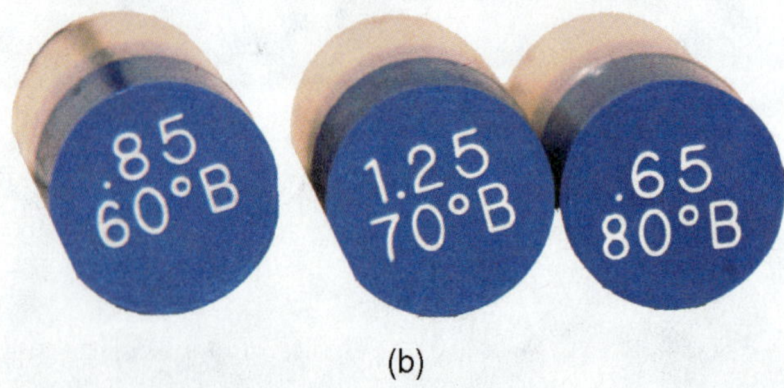

(b)

Figure 90-20 (a) Nozzle, electrodes, and air; (b) various sizes and types of nozzles; (see next page) (c) nozzle size is stamped on the nozzle; (d) nozzle angle is stamped on the nozzle.

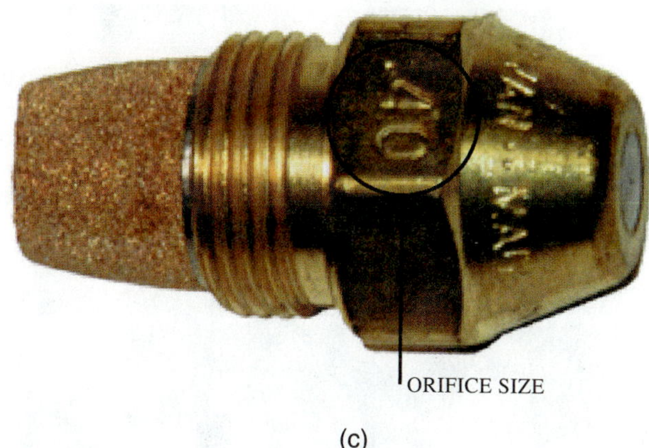

(c)

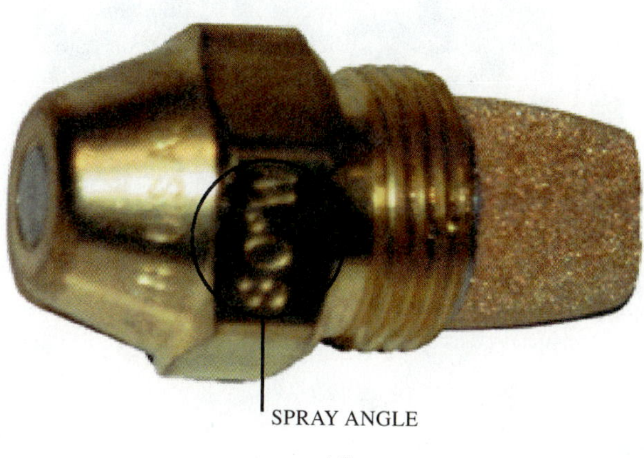

(d)

Figure 90-20 *(Continued)*

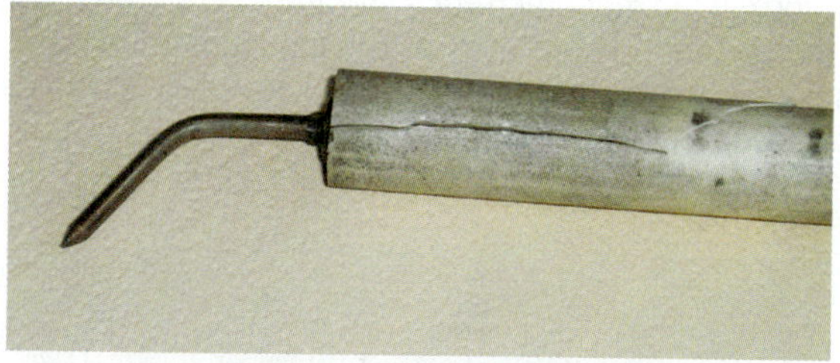

Figure 90-21 Cracks in the electrode porcelain can allow spark to arc at unwanted places and reduce the arc available at the electrode tip. *(Courtesy of HVAC Department, Terra Community College, Fremont, Ohio)*

- Clean the electrodes and set them to the correct angle and gap (Figure 90-22).
- Adjust the *Z* dimension of the burner tube (Figure 90-23).
- Inspect the drive coupling (Figure 90-24).
- Clean the cad cell sensor (Figure 90-25).

While the burner is removed, clean any accumulated debris out of the combustion chamber and inspect the combustion chamber liner. If it has deteriorated, it should be replaced. Replace the burner after these steps are completed.

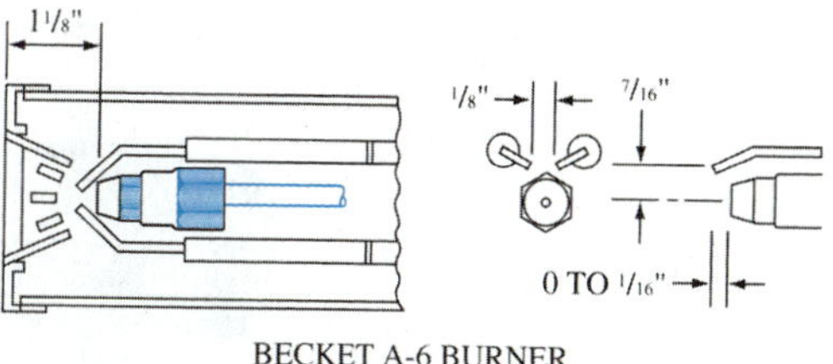

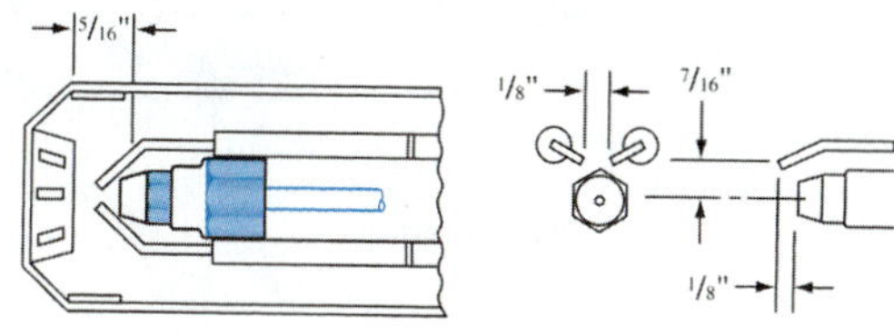

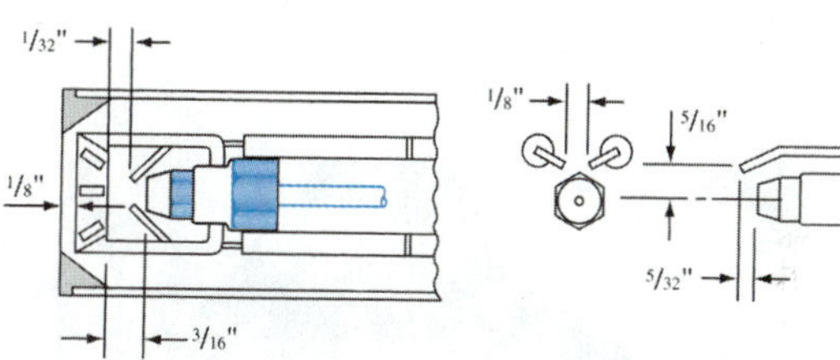

Figure 90-22 Electrode-spacing dimensions for an oil burner.

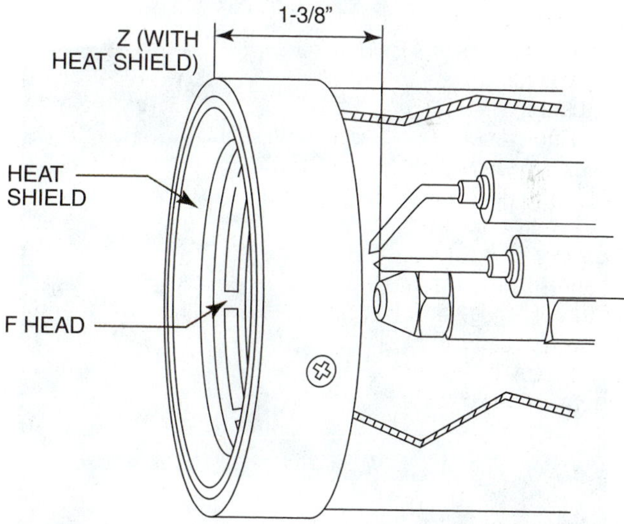

Figure 90-23 The "*Z* dimension" is the distance from the tip of the nozzle to the face on the burner head; an incorrect setting can cause incomplete combustion.

SAFETY TIP

If the burner has not been firing, a dangerous accumulation of oil in the combustion chamber can result from pushing the reset button several times. The customer may have tried resetting the burner several times before calling for service. This oil needs to be cleaned from the combustion chamber as thoroughly as possible. Lighting the furnace with a soaked combustion chamber may lead to an explosion!

Figure 90-24 The drive coupling makes the connection between the oil pump motor and the oil pump. *(Courtesy of HVAC Department, Terra Community College, Fremont, Ohio)*

Figure 90-25 Accumulated dirt and soot on the cad cell can keep it from seeing the flames. *(Courtesy of HVAC Department, Terra Community College, Fremont, Ohio)*

The vent system should be inspected for loose connections, signs of corrosion, and rust. The barometric damper should move freely (Figure 90-26).

Look to make sure adequate combustion air is provided. Find the nameplate and write down the temperature rise and the minimum and maximum operating voltages. Check the incoming voltage to see if it falls between the minimum and maximum voltage.

Figure 90-26 The barometric damper maintains the correct over-fire draft in an oil furnace. *(Courtesy of HVAC Department, Terra Community College, Fremont, Ohio)*

Operational Check

To correctly adjust the burner, it will be necessary to measure the oil pump pressure, stack temperature, and carbon dioxide flue gas percentage. Install a pressure gauge on the pressure port of the oil pump (Figure 90-27).

Look in the manufacturer's literature or on the oil burner for the correct pump pressure. The most common pressure has been 100 psig for many years. However, most recent oil burners operate at higher pressures—140–170 psig. Start

Figure 90-27 This fuel oil test kit can be used to check the operation oil pump pressure. *(Courtesy of Ritchie Engineering Company, Inc., Yellow Jacket Products Division)*

Figure 90-28 The oil pressure is increased by turning the adjusting screw clockwise and decreased by turning it counterclockwise. *(Courtesy of HVAC Department, Terra Community College, Fremont, Ohio)*

Figure 90-29 The temperature- and pressure-relief valve is an extremely important safety item; it opens to relieve excessive pressure, preventing boiler explosion.

the burner and compare the oil pump pressure to the recommended pressure. If the pressure is not correct, turn the adjusting screw clockwise to increase pressure or counterclockwise to decrease pressure (Figure 90-28).

The operation of the barometric damper should be checked. The manufacturer's instructions should list the recommended draft pressure between the damper and the furnace: –0.02 in wc is common. Adjust the weight on the barometric damper to achieve the correct draft.

Measure the smoke spot and the CO_2 percentage. The correct flame should have no smoke and a CO_2 level between 10 and 12.5 percent. If there is smoke in the flame or if the CO_2 level is outside of the 10 to 12.5 percent range, you need to adjust the air. To do this, follow these steps:

1. Adjust the air inlet vane until you see a trace of smoke in the flame.
2. Measure the CO_2 content of the flue gas.
3. Open the air until you have lowered the CO_2 by 1 or 2 percent.
4. The correct flame should have no smoke and CO_2 between 10 and 12.5 percent.

Ideally, the combustion efficiency should be measured and the flue gas should be checked for carbon monoxide. Measure the stack temperature and subtract the equipment ambient temperature from it to get the net stack temperature. The combustion efficiency can be calculated using the CO_2 percentage, the net stack temperature, and a combustion efficiency slide rule. The flue gas should have no carbon monoxide, but trace amounts of up to 50 ppm are acceptable. Check the temperature rise and blower current draw once the indoor blower comes on.

90.8 HYDRONIC HEATING

One advantage of hydronic heating systems is their low maintenance requirements. Maintenance can be divided into three general areas:

- Boiler
- Pump and controls
- Piping, radiators, and convectors

Hydronic boilers do not really boil the water; they are more like efficient hot-water heaters for nonpotable water. The temperature-pressure relief valve should be checked annually (Figure 90-29).

Pull the lever on top of the valve and water should come out the discharge pipe connected to the temperature-pressure valve. Take caution, because this releases scalding hot water under pressure. Keep away from the discharge outlet and do not stand directly in front of the valve. The remaining maintenance requirements vary depending upon the fuel the boiler uses. They can operate on electricity, fuel oil, or gas.

Electric Boilers

Electric boilers require the least amount of maintenance because there is no combustion process. Because they use large amounts of current, the power source, all controls, and connections should be checked annually. Visually check for overheated wires and wire connections. Measure the element voltage and compare it to the nameplate voltage and the actual incoming voltage. Any voltage drop in the circuitry or controls should be located and eliminated. Finally, the current draw of each element should be checked against the nameplate rating.

Gas-Fired and Oil-Fired Boilers

Gas-fired and oil-fired boilers should have their combustion air, combustion process, and vent system checked. These checks are similar to the checks used on gas and oil furnaces.

Water Pump

A water pump is used in hydronic heating systems to circulate the hot water. The vast majority of the pumps used today in residential systems are maintenance free. They

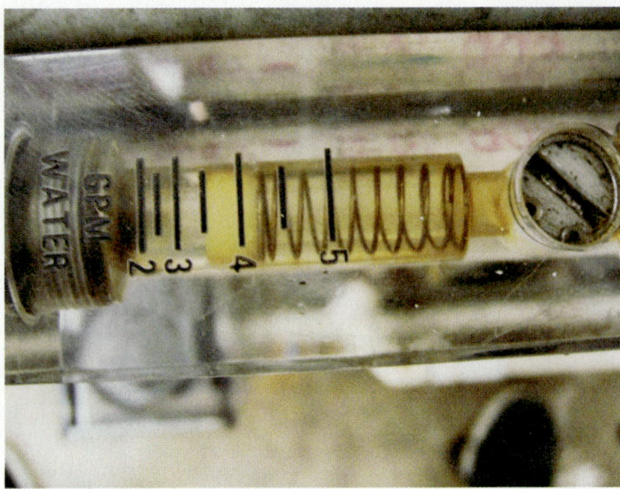

Figure 90-30 This system's water flow can easily be checked by looking at this water flow meter.

have no seals to leak and they are typically cooled and lubricated by the water they pump. Checking the water pump consists of checking the amount of water it is pumping. This is done two ways. Some systems have water flow indicators that visually show the rate of water flow (Figure 90-30). Unfortunately, many systems do not have a water flow indicator. The other common method of checking the water flow is to measure the pressure drop across the pump and compare it to the pump specifications. This is quite accurate but requires the pump specifications .

Figure 90-31 This pressure gauge shows the pressure in the water loop.

Piping and Fluid

The system should be inspected for water leaks. Systems that use a lot of makeup water have leaks. The system water pressure should also be checked (Figure 90-31).

Most hydronic systems should be kept at a pressure of 10 to 20 psig. Below 10 psig, systems experience more problems with air entrainment and corrosion. The pressure-relief valve on most residential boilers opens at 30 psig, so the loop pressure should be kept enough below 30 psig to prevent opening the pressure-relief valve.

Most hydronic systems are closed loops and use glycols and corrosion inhibitors in the water. The water pH level should be checked to ensure it is not corrosive to the system piping and components. This can be done with a pH meter (Figure 90-32).

Generally, the pH should be between 8.5 and 10.5. The fluid's pH drops as the inhibitors wear. Adding inhibitor increases the water pH level. Check with the boiler manufacturer for the correct glycol and inhibitor solutions for the system.

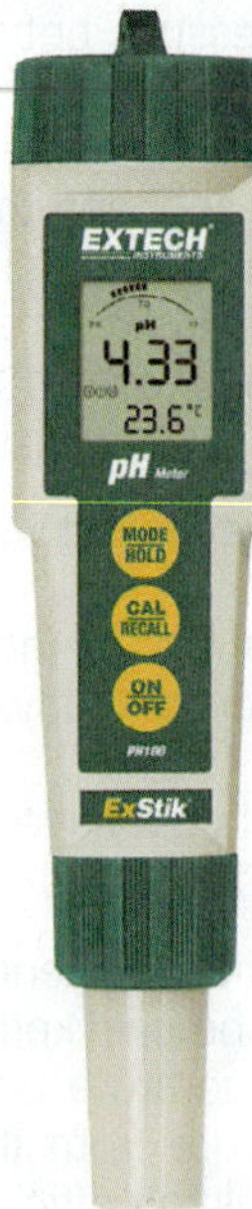

Figure 90-32 A pH meter like this one can be used to check the water pH. *(Courtesy of Extech Instruments, A FLIR Company)*

TECH TIP

Generally speaking, it is not a good idea to use automotive antifreeze. The inhibitors added are specific to the types of materials in the system. Because automobile engines contain different materials than hydronic heating systems, the antifreeze formulation is different.

90.9 HEAT PUMPS

During a heat pump planned service call, the cooling, heating, defrost, auxiliary heat, and emergency heat cycles should all be checked whenever possible. However, most air source heat pump systems should not be operated in the cooling cycle at temperatures below 50°F or in the heating and defrost cycles at temperatures above 70°F.

Cooling Cycle

In cooling, heat pumps are functionally very similar to air-conditioning systems. Checking a heat pump in the cooling cycle is essentially the same as checking an air-conditioning system.

Heating Cycle

The routine maintenance for the heating cycle is also similar to air conditioning:

- Clean or change the air filter.
- Clean the outdoor coil.
- Check the indoor coil.
- Lubricate the motors.

The operational check should include the following:

- Amp draw of all motors
- Airflow
- Low-side and high-side pressures
- Superheat and subcooling
- Temperature rise across the indoor coil

Determining the correct charge in the heating cycle can be tricky because the system capacity varies with the operating conditions. Typically the heating capacity of most air-source heat pumps at 45°F is similar to their cooling capacity at 95°F. But their heating capacity at 17°F is just a little more than half that. There is also about half the amount of refrigerant circulating. Manufacturers have different ways of handling this. It is essential that the equipment manufacturer's charge-checking method be used for accurate results. The most common methods of checking the heating cycle operation are as follows:

- Temperature rise across the indoor coil
- Pressure-temperature charts (Figure 90-33)

HEATING CHECK CHART FOR R-22 AIR SOURCE HEAT PUMP								
INDOOR TEMPERATURE	PRESSURE	OUTDOOR TEMPERATURE						
		60°F	50°F	40°F	30°F	20°F	10°F	0°F
60°F	HIGH	217	201	186	170	154	139	123
	LOW	68	50	41	53	35	27	17
70°F	HIGH	248	230	213	195	178	161	143
	LOW	71	62	53	44	36	27	17
80°F	HIGH	279	260	241	222	203	185	166
	LOW	74	65	55	46	36	27	17

USE OF CHART
Table indicates whether a correct relationship exists between system operating pressure and air temperature entering indoor and outdoor units. If pressure and temperature do not match on chart, system refrigerant charge may not be correct or other system abnormalities may exist. Do not use table to adjust refrigerant charge.

CHARGING
When charging is necessary during heating season, remove any refrigerant remaining in system, evacuate and recharge by weight. Weigh in total charge as indicated on unit rating plate. Rating plate charge is for systems with 15 ft. line-set. Adjust charge at rate of 0.6 oz/ft of 3/8" liquid line over 15 ft.

Figure 90-33 A check chart is used to check the heating operation, but it is not used for adjusting the charge.

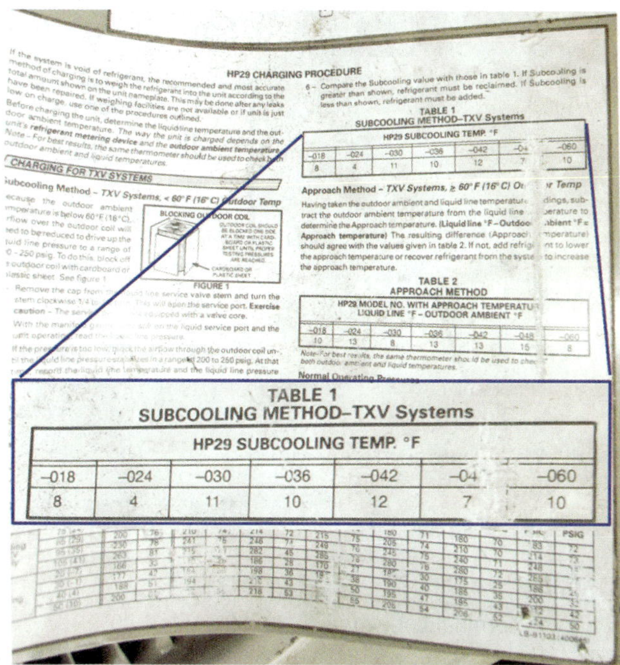

TABLE 1
SUBCOOLING METHOD–TXV Systems

HP29 SUBCOOLING TEMP. °F						
–018	–024	–030	–036	–042	–04	–060
8	4	11	10	12	7	10

Figure 90-34 The charge may be checked using subcooling when the temperature is below 60°F; operate the unit in cooling and block the condenser to raise the liquid pressure to 200–250 psig (R-22).

- Subcooling (Figure 90-34)
- Ambient temperature approach (Figure 90-35)
- Suction superheat (Figure 90-36)
- Discharge line temperature

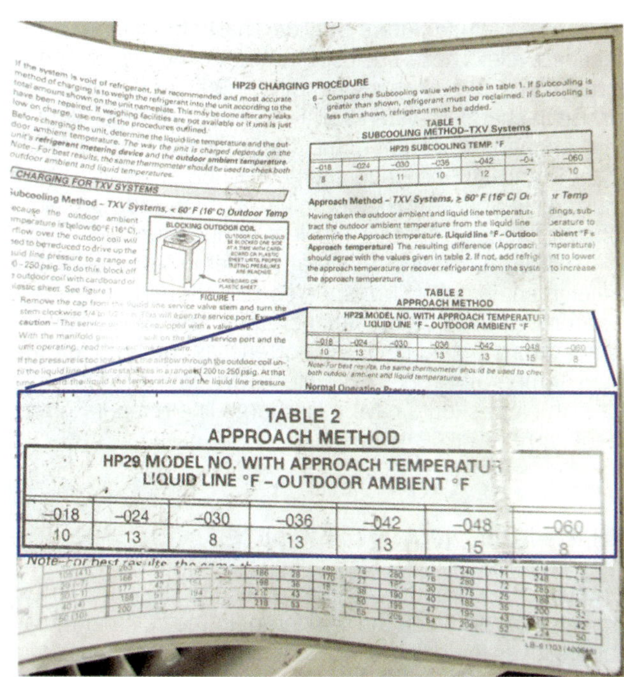

TABLE 2
APPROACH METHOD

HP29 MODEL NO. WITH APPROACH TEMPERATU LIQUID LINE °F – OUTDOOR AMBIENT °F						
–018	–024	–030	–036	–042	–048	–060
10	13	8	13	13	15	8

Figure 90-35 The approach method compares the outdoor temperature to the liquid-line temperature while operating in cooling; the ambient temperature must be above 60°F.

Figure 90-36 This chart gives the system's suction superheat in the heating cycle.

Superheat at Compressor Suction in Heating Operation

Indoor Temperature	Outdoor Ambient Temperature							
	-10°F	0°F	10°F	20°F	30°F	40°F	50°F	60°F
55°F	NR	6	8	10	13	20	27	37
70°F	NR	NR	NR	NR	5	12	20	30
80°F	NR	NR	NR	NR	NR	8	16	26

Note: Chart values assume design airflow over both indoor and outdoor units

Table 90-1 Outdoor Temperature

	0°F	10°F	20°F	30°F	40°F	50°F
Temperature rise	9	11	14	18	23	28
Percent nominal capacity	33%	40%	50%	65%	82%	100%

The temperature-rise method can be applied to any unit if you know the indoor airflow and the specified system BTU/hr capacity at the current outdoor temperature.

The expected temperature rise can be calculated using the formula

$$\text{Temp rise} = \text{Btu}/(\text{CFM} \times 1.08)$$

The chart in Table 90-1 shows a calculated temperature rise for single-speed systems whose airflow is 400 CFM per ton. The percentage capacity used for the calculations is shown in Table 90-1.

Some manufacturers provide charts for checking the heating cycle. These are used to check the system pressures at specific operating conditions but are not intended for use in charging systems. This is because it is possible to have enough refrigerant in the system to operate correctly at 25°F and still be undercharged at 45°. Typically these charts give you an expected high- and low-side pressure range for specific outdoor ambient temperatures.

The discharge-line temperature method is used by one major manufacturer. It compares the discharge-line temperature to the outdoor temperature. The discharge-line temperature should be 110°F hotter than the outdoor ambient temperature. For example, a system operating in a 40°F ambient should have a discharge-line temperature of 150°F: 40°F + 110°F = 150°F.

90.10 COIL MAINTENANCE

Condenser coils and evaporator coils are designed to exchange heat between the refrigerant in the coil and the air passing over the coil. Dirty coils cannot perform their function because the dirt adds resistance to airflow, reducing the amount of air passing over the coil. The dirt also acts as an insulator, slowing the transfer of heat between the refrigerant and the air.

Figure 90-37 Dirty air-cooled condensers are a common problem.

Air-Cooled Condensers

Dirty condenser coils restrict airflow, raise the condenser temperature and pressure, reduce efficiency, and shorten equipment life (Figure 90-37). Air-cooled condensers should be cleaned annually regardless of how clean they appear. A coil that appears clean can have dirt packed inside the fins where you cannot see.

Always follow the equipment manufacturer's recommendations when selecting cleaning agents. Cleaning with the wrong chemicals or applying the chemicals incorrectly can do more harm than good. In particular, check that you have properly diluted the coil cleaner (Figure 90-38).

Before cleaning the coil, turn the power to the unit off and check it with a meter to ensure that it is off. Remove matted material with a stiff brush. Apply coil cleaner according

Figure 90-38 Make sure to dilute coil cleaner according to the manufacturer's directions.

to the chemical manufacturer's instructions. Rinse going in the opposite direction of air flow to help with the removal of pet hair and material caught between the fins. Material can get caught between the coils on some double-row coils. If possible, separate the two coils to remove the trapped material. Debris can also get caught between the coil and the metal louvered coil guards on some units. Removal of the metal panels may be necessary to get this material out.

Many types of coil cleaner are sold concentrated and must be diluted with water before using. Care should be taken to rinse the coils thoroughly so that no cleaning chemicals remain on the coil. For spiney-fin coils that have the fins wrapped around the tubes, the amount of pressure used to apply both the cleaner and the water needs to be very low. These fins are quite thin and will crush easily. Flattening the fins essentially does the same thing as having them matted with dirt; the airflow is restricted (Figure 90-39). These coils cannot be straightened with a fin comb, so damaged coils are nearly impossible to repair.

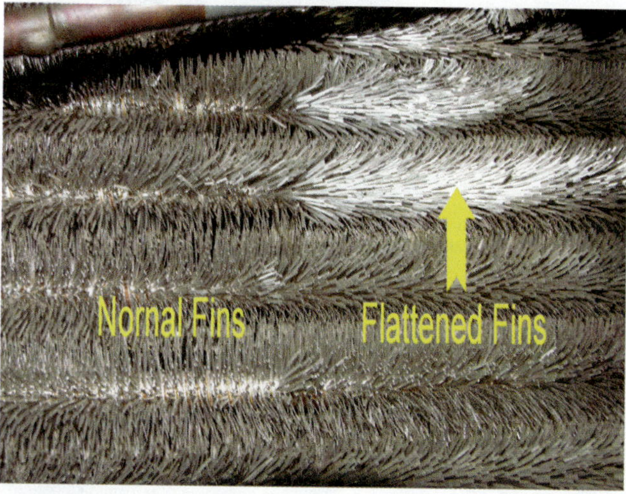

Figure 90-39 Be careful when cleaning spiney fins; they can be easily crushed.

Figure 90-40 Dirty evaporators are a common problem that is often misdiagnosed as a refrigerant undercharge. *(Courtesy of Nu-Calgon)*

Air-Conditioning Evaporator Coils

Air-conditioning evaporator coils should be protected by an air filter. You should always clean or replace the indoor air filter when servicing an air conditioner. If this filter is cleaned or changed regularly, then the evaporator will usually stay clean. If possible, you should visually inspect the evaporator to make certain that it is clean, but it is not necessary to clean the evaporator every year. However, if the filters are not kept clean, the evaporator can become matted with dirt (Figure 90-40).

Since the evaporator is usually wet when it runs, dirt sticks to it. In severe cases, this can reduce the airflow to the point that the evaporator freezes up. A frozen coil can be spotted by the ice around the suction line leaving the coil. If the unit runs long enough with a frozen evaporator, the suction line will freeze all the way back to the compressor (Figure 90-41).

The side of the coil that needs cleaning is where the air enters the coil. After removing the access door on an evaporator casing, what is usually visible is the side of the coil where the air leaves. You may have to remove interior sheet metal, cut into ductwork, or even entirely remove the evaporator coil to gain access to the side that needs cleaning (Figure 90-42).

Obviously it is easier to keep the coil clean in the first place! Another complication when cleaning evaporator coils is controlling the mess that is made. Condensers are usually outside, so the resulting mess is not as critical. But evaporators are normally inside, and it just is not feasible to wash it down with a garden hose. Some cleaners are designed to be rinsed by the water that condenses on the evaporator as it operates. However, you should check with the equipment manufacturer before using them. A hand pump chemical sprayer is a practical way to apply coil cleaner and control the amount and direction of the water spray when rinsing the coil.

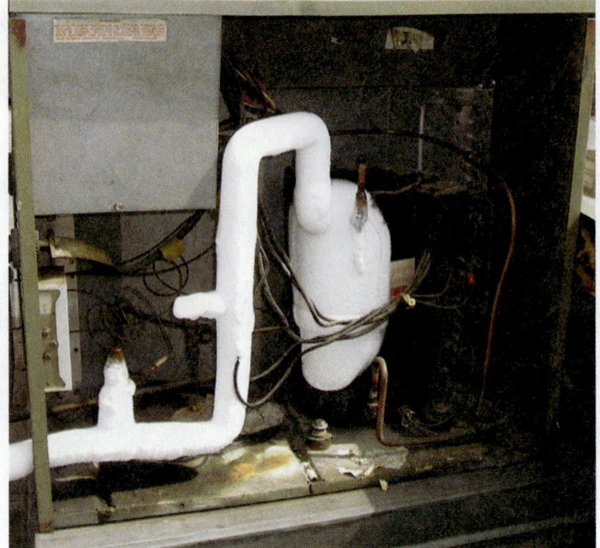

Figure 90-41 If a unit continues to operate after the evaporator coil freezes, the ice will travel down the suction line and even up onto the compressor. *(Courtesy of HVAC Department, Terra Community College, Fremont, Ohio)*

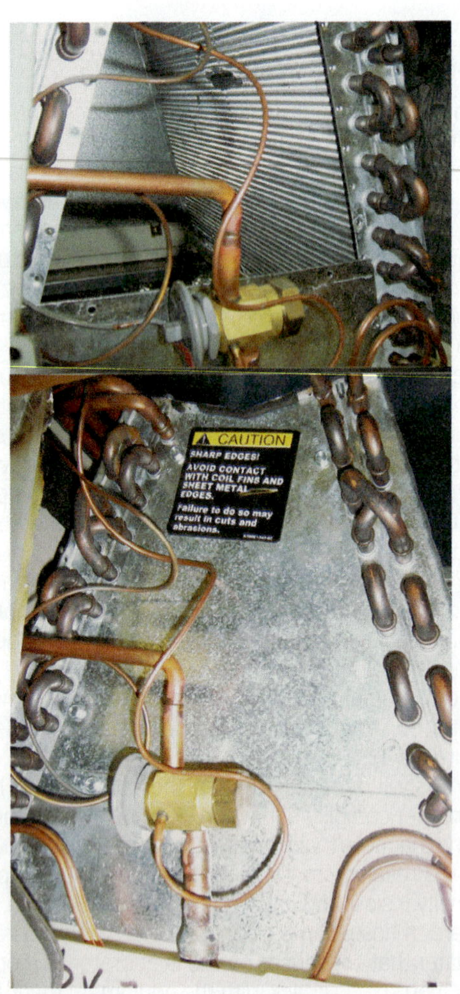

Figure 90-42 The triangle endplate must be removed from the coil to gain access to the underside of the coil where dirt collects.

To clean the evaporator coil, you should do the following:

1. Remove any visible dirt accumulation with a stiff brush.
2. Clean out the drain and drain pan. A free-flowing drain will make the rest of the job easier.
3. Apply a properly diluted coil cleaner to both sides of the coil.
4. Thoroughly rinse the coil starting from the leaving-air side. This will help push wedged debris out.

There are methods for cleaning the coil from the leaving-air side. But without seeing the results, you cannot be certain you have properly completed the job. These methods include high-pressure air to blow backwards through the coil and chemical foams that are applied to the leaving-air side. One problem with blowing the dirt off and leaving it is that the dirt is still there and could become trapped in the coil again. The foam cleaner works by expanding and pushing the dirt out the other way. The water condensing on the coil when the unit operates is then supposed to rinse off the foam.

Even though the evaporator coil does not always need cleaning, the drain pan and drain line usually do. The quickest and most effective way to clean a drain pan and drain is with a tool that uses air pressure or nitrogen pressure to create suction. This lets you suck the water and slime out of the drain pan and drain (Figure 90-43).

Figure 90-43 A sludge sucker can be used to clear a drain line from the outside. *(Courtesy of Uniweld Products, Inc.)*

Figure 90-44 Example of mineral deposits on an ice machine water float.

Water Coils

Systems that have water circulating in open loops should have the water checked weekly. Any water that is open to the air will grow things.

From ice machines to high-temperature water-jacketed heat exchangers, if the water is open to the air, something will grow in the water. Evaporation of water in open-loop systems leads to increased mineral concentration in the water. This leads to salting or calcification of all the water-handling parts (Figure 90-44).

This can be controlled with regular maintenance. Regular maintenance on the water side of systems typically consists of testing the water and adding chemicals as needed to prevent biological fouling and mineral scaling. Regular cleaning is necessary for systems such as ice machines where chemicals cannot be present in the water during normal operation. Most ice machines should be cleaned monthly.

90.11 MAINTENANCE OF ELECTRONIC AIR FILTERS

Service on electronic air cleaners consists of the following steps:

1. Inspecting the air cleaner cells
2. Checking the operation of the air cleaner
3. Cleaning the prefilter
4. Cleaning the collector plates

Inspection

Turn off the power to the air cleaner. Normally there is a switch on the door or on top of the air cleaner. Most air cleaners have power indicator lights that will show when power is on. Remove the access door on the side of the air cleaner. The cells will slide out. Typically there will be one or two cells. Pay careful attention to how the cells are oriented so that you can reinsert them correctly. Many have arrows showing the direction of airflow or keyways to prevent misalignment (Figure 90-45).

Figure 90-45 The arrow on the side of the air cleaner cell indicates the airflow direction.

Visually check the fine wires on the front of the cells. There should be no loose wires. Reinsert the cells for the operational check.

Operational Check

Many electronic air cleaners have test buttons and/or test lights to indicate that the air cleaner is operating correctly. The test lights are wired to a transformer and tuned circuit in the air cleaner. They light when the air cleaner power supply is operating correctly (Figure 90-46).

Figure 90-46 The light at the top right is lit to indicate that the air cleaner is energized. *(Courtesy of HVAC Department, Terra Community College, Fremont, Ohio)*

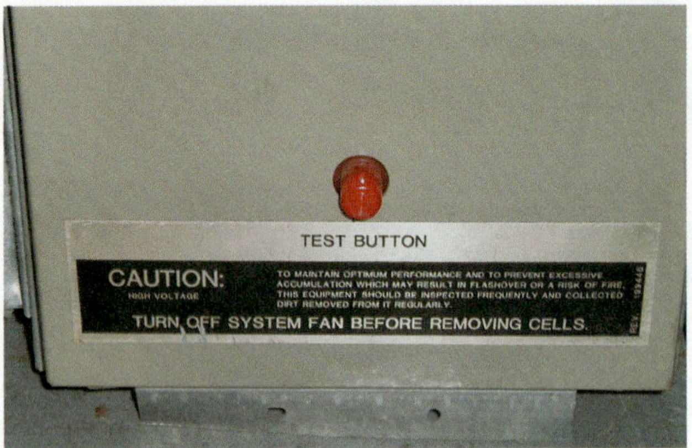

Figure 90-47 Pushing the test button at the bottom produces a snapping sound to indicate that the air cleaner is working. *(Courtesy of HVAC Department, Terra Community College, Fremont, Ohio)*

The test button temporarily shorts across the positive and negative plates. This causes a loud zapping noise that indicates the air cleaner is working (Figure 90-47).

Do not try to short across the plates with anything else. The plates have 8,000 V DC on them, and touching them during operation would literally be a shocking experience!

Some air cleaners have circuits that detect airflow and automatically energize the cells when airflow is detected. These usually have lights indicating when airflow is detected (Figure 90-48).

Make sure that the light comes on when the fan comes on and goes off when the fan goes off. These circuits are a common source of problems. If the circuit does not work, check to see that the air sensor tube is not plugged with debris (Figure 90-49).

Prefilter Cleaning

The prefilter is similar to a cleanable 1-in low-efficiency air filter. Its job is to collect large dust particles and keep them out of the collector plates. When a large particle does occasionally make it into the air cleaner, you will hear a loud zapping noise similar to a bug zapper. The prefilter should be cleaned monthly.

Figure 90-48 The green light on this air cleaner lights up whenever the air cleaner detects airflow and turns itself on.

Figure 90-49 This sensor energizes the air cleaner whenever it senses airflow.

Cleaning the Air-Cleaner Cells

The air-cleaner cells should be cleaned once a year. They can be cleaned in a dishwasher, or they can be cleaned with coil cleaner and rinsed. They should be allowed to dry before being placed back into the air cleaner. It would not be good to apply 8,000 V to a dripping-wet metal appliance.

90.12 AIR FILTERS

Air filters for residential systems are available in sizes from 1–5 in, in efficiencies from MERV 1 to MERV 12, and can be disposable or cleanable. The least expensive low-efficiency filters are intended to protect your equipment, while the higher-efficiency filters are intended to protect the equipment and clean the air.

Disposable 1-in Filters

Most 1-in filters should be changed monthly, including the standard 1-in disposable air filter. Disposable 1-in filters come in several efficiencies. As consumers become more concerned with indoor air quality, high-efficiency pleated 1-in filters are growing in popularity. Unfortunately, the higher efficiency 1-in disposable filters have a very high pressure drop, even when clean. The system airflow should be checked if lower-efficiency 1-in filters are replaced with higher-efficiency 1-in filters. It is essential that these filters be changed monthly. The cost of changing the filter monthly is far less than the cost of the extra energy consumed as a result of operating a system with dirty filters. In a worst-case scenario, dirty air filters can lead to blocked indoor air coils and dead compressors. The cost of replacing a dead compressor will exceed the cost of replacing filters monthly for several years.

Pay attention to the directional arrows on the filter when inserting one. The arrows point in the direction of airflow. This should be toward the unit (Figure 90-50).

Figure 90-50 Filters have arrows on the sides to indicate the direction of airflow.

Cleanable 1-in Filters

Most cleanable filters are 1 in thick. Cleanable low-efficiency filters include hog's hair, expanded metal, and foam. These should be cleaned monthly. They may require more frequent cleaning in abnormally dirty conditions. Rinsing with water and mild detergent works well for these. Electrostatic air filters are a type of cleanable medium-efficiency 1-in filters. These use materials that build up a static charge as air passes over them. Keeping these filters clean is critical. They have a higher pressure drop than standard air filters when they are clean. When they are dirty, they restrict airflow to the point that most of them will not operate correctly. They are also very difficult to clean when they have become packed with dirt. These should be cleaned monthly.

High-Efficiency Pleated Filters

Pleated disposable filters are available in 4-in and 5-in widths (Figure 90-51). These filters have MERV ratings from 8 to 12. The pleats provide more surface area for the air to pass through, so the filters can use a denser material. This allows higher-efficiency filtering without imposing large pressure drops. However, they do have a higher pressure drop than low-efficiency 1-in filters. Large pleated filters have the added benefit of lasting longer between filter changes. Pleated high-efficiency 5-in filters normally only need to be changed every 6 months. Some may go as long as a year depending upon the amount of material in the air. Two types are available: accordion-fold filters that are assembled in the field and box filters that have a cardboard support around the pleats. The field-assembled filters typically have combs that separate and support the pleats. Assembling these can be tricky if you have not done it before. Pay attention to the way the parts fit together when disassembling the old filter so that you will know how to reassemble the new one. Make certain the arrows indicating the airflow are facing the correct direction.

Figure 90-51 The combs in this filter separate the pleats; they must be removed and reinstalled when changing the filter media.

90.13 BELT-DRIVE ASSEMBLY CHECKING

Belt-drive assemblies are used to drive fans and compressors in commercial equipment. Some older residential equipment used belt-drive fans, but residential equipment built after 1970 is all direct drive. When checking a belt-drive assembly, you should check the following:

- Shaft alignment
- Pulley alignment
- Pulley wear
- Shaft bearings
- Belt tension
- Belt condition

Safety

All work on belt-drive systems should be done with the system off and locked out. You should never try to adjust a belt-drive system with the power on. An unexpected system start-up could cause severe personal injury.

Shaft Alignment

The drive shaft and the driven shaft should be parallel in two axes. This can be checked by viewing the shafts from above and from the side. The shaft centers should be the same distance apart at the front and back of the shaft. Typically, the motor mounts are adjustable, and any shaft misalignment is corrected through adjustments to the motor mount (Figure 90-52).

Pulley Alignment

The two pulleys should be in line so that a straight edge will lay flat across the outside face of both pulleys (Figure 90-53). The pulley alignment can also be checked by looking at how the belt leaves the pulleys. If the belt appears to veer to the right or left after leaving the pulley, the pulleys are not properly aligned (Figure 90-54). This can be corrected by loosening the setscrew on the pulley and moving it until the belt comes off the pulley in a straight line.

Pulley Wear

Improper belt tension and misaligned belts both can cause excessive wear on aluminum pulleys. Pulleys that are cupped or grooved on the inside should be replaced.

Shaft Bearings

Belt-driven devices have bearings that should be checked and lubricated. Belt-drive fans typically use pillow-block

Figure 90-52 These shafts are not in alignment; the straight edge does not rest straight across the faces of the pulleys.

Figure 90-53 These shafts are in alignment; the straight edge rests flat against the faces of the pulleys.

Figure 90-54 These pulleys are out of alignment; the belt is pulling to the left.

bearings that are lubricated with a small amount of grease added through the grease fittings using a grease gun.

Belt Tension

The correct tension for any belt-drive application is the minimum tension required to keep the belt from slipping at its maximum torque. For blowers, this is at start-up. Belt manufacturers publish specifications for the design load of their belts, but a belt may not be operating at its maximum design load on any particular application. Tensioning a belt past the required torque can actually shorten its life. If a fan belt does not slip on start-up, it is tight enough.

Belt-drive blowers typically use V-belts. New V-belts need to operate a while to settle in. They need to be re-tensioned after the first several hours of operation. This requires going back to the job to re-tension the belt. Some belt manufacturers specify different tensions for new belts and old belts. The new belt tension is higher to allow for the belt settling in. The intent is to avoid going back to re-tension the belt. Belt tensions should be checked every 6 months . There is no single tension specification that is accurate for all belts. The most common method is to check deflection of the belt while applying force to the center of the belt span (Figure 90-55).

The amount of force and amount of deflection differ from belt to belt. Belt manufacturers publish charts showing the proper amount of force and deflection for any given belt. In general, longer belts will have more deflection than shorter belts.

Belt Condition

Belts wear out. They wear out more quickly if the alignment and tension are not correct, but even if all drive components are properly aligned and the belt is properly tensioned, it will still wear out. Worn belts will have one or more of the following characteristics:

- Hard and glazed sides (Figure 90-56).
- Dried and cracked, not soft and pliable
- Deformed sides
- Frayed

Belt Replacement

Most belts will last for several years. When it is time to install a new belt it is important to do it correctly. A common but incorrect method of installing a belt is to roll it over the outside edge of the pulley (Figure 90-57). This can damage the belt and shorten its life. The belt tension adjustment should

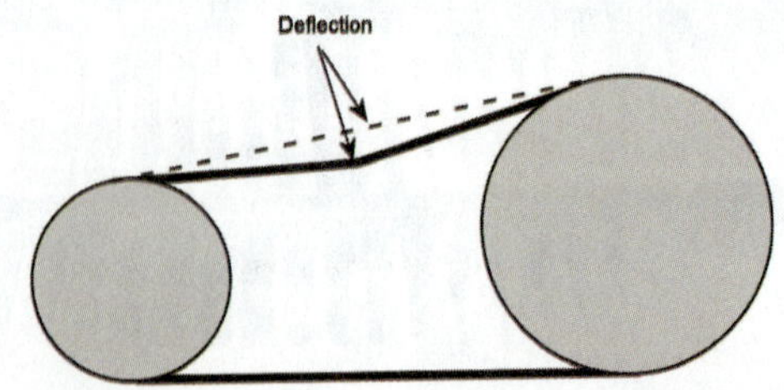

Figure 90-55 Belt-tension gauges measure the amount of force required to produce belt deflection.

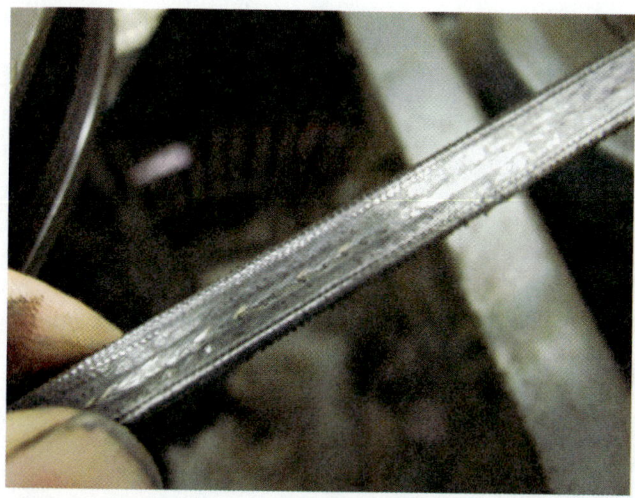

Figure 90-56 The sides of this belt are shiny and glazed; this belt needs replacing.

Figure 90-57 Belts should *not* be installed by rolling them over the pulley.

be loosened until the belt fits easily over the pulley and then retightened after installing the belt. Be sure to check the pulley alignment and belt tension after installing a new belt.

Common Problems and Misconceptions

A sure sign of a belt-drive problem is vibration. Belt vibrations are caused by alignment problems or hardware problems. If the pulleys are misaligned, the belt has to jump over the side of the pulley, causing vibrations. Warped pulleys or shafts can do the same thing. This problem is frequently mistaken as a sign of a loose belt. Tightening the belt reduces the size of the arc in the belt vibration but does not solve the problem. In fact, tightening a belt that is oscillating may actually increase belt wear.

Loose belts slip, causing a squealing noise. The proper fix is to ensure proper alignment and tighten the belt. Unfortunately, a common "remedy" is lubricating the belt or treating it with belt dressing sprays. These only fix the sound. The lubricated belt slips even more, but it does so quietly—and usually just long enough for the technician to leave. Belt dressings are designed to make the belt tacky and reduce slippage. However, that is also short lived. If the belt is loose it should be tightened; if it has glazed walls it should be replaced.

90.14 SOUND-LEVEL MONITORING

Sound-level monitoring is one technique used for predictive maintenance. The goal of predictive maintenance is to monitor system behavior, document changes in behavior, and repair or replace components that are on a predictable path toward failure. Machines all have their own sounds. When these sounds change, something has changed about the machine. Ultrasonic sound-level meters can be used to hear sounds outside of human hearing. Increased friction leads to increased sound. When a machine changes its tune, it is trying to tell you something.

Dosimeters are used to measure ultrasonic sound levels. They measure ultrasound in decibels relative to 1 millivolt, abbreviated dBmv (Figure 90-58). The ultrasonic output

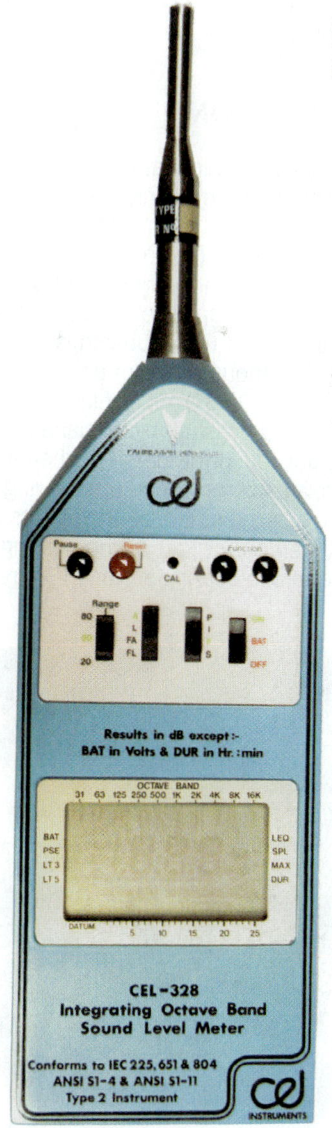

Figure 90-58 Noise dosimeter with data logger.

Figure 90-59 The sleeve bearing is shown highlighted in blue.

of a bearing should be relatively constant at ±3–4 dBmv. Small increases in ultrasonic output indicate the need for bearing lubrication. Larger increases warn of the earliest stages of bearing failure.

90.15 LUBRICATION

There is really not a lot to lubricate on most modern heating and air-conditioning systems. Only the fan motors need lubricating on most systems. Even many fan motors are now using sealed bearings that cannot be lubricated. Two types of bearings are used: ball bearings and sleeve bearings. Most residential applications use sleeve bearings.

Sleeve bearings are metal bushings whose inside diameter is just a few thousandths of an inch larger than the shaft diameter (Figure 90-59). An oil film separates the shaft from the bearing, and the shaft hydroplanes on the bearing. The shaft does not touch the bearing when it is operating. In small motors, a packing around the bearing holds oil like a sponge and releases it as the bearing heats up.

A slice is taken out of part of the bearing to allow the oil to reach the inside bearing surface (Figure 90-60). This type of bearing is oiled through the oiler ports on the motor (Figure 90-61). Ideally, these should be pointing up, but they are sometimes on the side coming off at an angle. They should never be pointing down. Many motors have lubrication instructions on them. Typically they use 5–10 drops of SAE 10 to 20 weight oil. Never use penetrating oils or spray lubricants. They are too thin and the solvents in them can attack the materials in the motor. The time period before the first lubrication is normally longer than subsequent lubrications. The initial operating period before lubrication is often as long as 2 years. The time between normal lubrications varies from 3 months to a year.

Figure 90-60 The slice at the top of the sleeve bearing allows oil to feed into the area between the shaft and the bearing.

Figure 90-61 These oiler tubes carry oil to the packing around the sleeve bearing.

TECH TIP

Do not attempt to oil any part of an electric motor other than at the oiler holes. Spraying oil all around the motor will just accumulate dust faster and contribute to winding breakdown.

Larger motors typically use ball bearings or roller bearings. These are lubricated with grease. The grease should create a film over all the bearing components. It should not fill up all the space in the bearing (Figure 90-62). If the bearing is completely filled with grease, the balls have to push through it. This actually increases the bearing temperature.

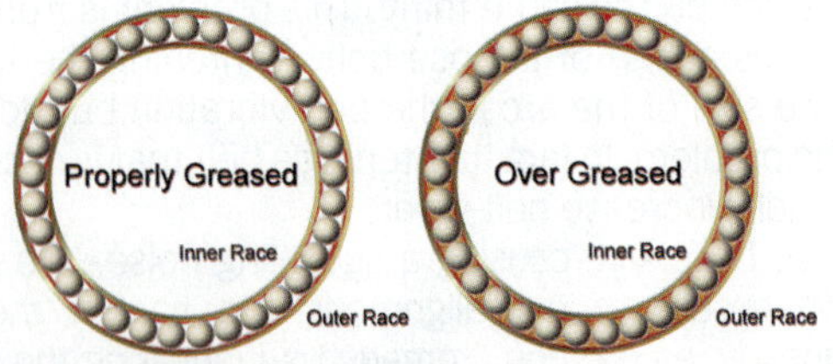

Figure 90-62 A properly lubricated ball bearing should not have grease packed in every space between the ball bearings but just around the races.

Figure 90-63 Many large motors have grease ports for greasing the bearings.

Further, the grease is whipped up and loses some of its ability to lubricate. Expansion of excess grease will force open the seals in the bearings and destroy the bearing. To lubricate a bearing with grease fittings you should follow these steps:

1. Open both the fill and drain ports (Figure 90-63).
2. Operate the motor until the bearing is warm.
3. Add the manufacturer recommended amount of grease in the fill port.
4. Operate the motor with the ports still open until the bearing is warm.
5. Replace the port plugs.

If the manufacturer's recommendations are not available, use an ultrasonic tester to listen to the bearing and add small amounts until the bearing noise abates.

SERVICE TIP

A common misconception is that grease should be added until all the old grease is purged and new grease starts to exit the drain port. This will certainly fill all the bearing cavities with grease; thus this procedure is guaranteed to over-grease the bearing and shorten its life.

90.16 VISUAL INSPECTION

Your eyes are one of the most powerful tools at your disposal when performing routine inspections. Many system problems give visual clues about a system's operation. Take some time at the beginning to look the system over carefully. Oil spots on refrigerant lines indicate leaks. Stains on gas vents indicate venting problems. Darkened or discolored wires have been overheated. Corroded electrical connections indicate loose connections. System installation errors can often be readily seen. A few items to look for would be the following:

- Combustion air grilles in the case of gas and oil appliances
- Mechanical components properly supported
- Ductwork insulated
- Suction line insulated
- Service clearance in front of unit
- Drains running downhill

90.17 CALIBRATION

To calibrate a control, compare its output to an accepted standard and adjust it if necessary to conform to that standard. In air conditioners, the only parts that occasionally need calibrating are thermostats. Most mechanical thermostats use a bimetal to measure the temperature with a mercury bulb mounted on it to do the switching. These are position sensitive. The tilt of the bimetal controls the calibration of the thermostat.

To check the calibration you should follow these steps:

1. Place an accurate thermometer next to the thermostat and wait for its reading to stabilize.
2. Move the thermostat to its lowest setting.
3. Slowly raise the thermostat setting and note the temperature where the mercury rolls from one end of the bulb to the other.
4. Slowly lower the thermostat setting and note the temperature where the mercury rolls back to the other end of the bulb.
5. The bulb should turn off and on at either side of the actual temperature.

For example, if the actual temperature is 75°, the bulb should switch just above and below 75°. The most common cause of thermostat calibration problems is installation. If a mercury bulb thermostat is not mounted level on the wall, it will be out of calibration. The fix is to level the thermostat. Some mercury bulb thermostats can be calibrated using a small wrench to adjust the tilt of the mercury bulb. However, you should check to see if the thermostat is level before adjusting it. A few electronic thermostats can be calibrated by turning a small trim pot. When calibrating either type of thermostat you should make sure the thermostat is at room temperature, that the thermometer you are using is accurate, and that you avoid breathing on and handing the thermostat. Your body heat will cause it to read a few degrees higher.

TECH TIP

Many instruments can be ordered with certified NIST calibration. These instruments have been sent to a laboratory where the instrument was calibrated according to NIST standards. Many jobs in testing and balancing or combustion efficiency testing require certified calibrated instruments.

90.18 SCHEDULING PARTS CHANGES

All systems have parts that wear out. Scheduling replacement of these parts near the end of their expected service life can save system shutdowns and emergency calls. The advantage to the customer is reliable operation. A gas furnace's hot surface igniter is an example of a part that might be considered for a replacement schedule.

90.19 SAFETY CHECKS

A system that is not safe should never be left in operation. Systems should be inspected for possible safety hazards before work begins and before leaving then job site. Potential hazards should be corrected before proceeding with any work. Before operating the equipment, ensure the following:

- All equipment is properly grounded (Figure 90-64).
- Wire size and type are appropriate for the application.
- All electrical wiring passes through strain relief when entering equipment.
- All equipment, ductwork, wiring, and piping are properly supported.
- All moving parts are covered or protected by a safety guard.
- Combustion appliances have adequate combustion air and venting.
- Volatile fluids and other combustibles are not stored near the equipment.

Operational safety checks include the following:

- Current draw and temperature of power wiring during operation
- Pressure in vent system of combustion appliances
- CO level in combustion gases for combustion appliances
- CO level in equipment room for combustion appliances

Figure 90-64 All electrical equipment should be properly grounded.

UNIT 90—SUMMARY

The aim of planned maintenance is to extend the life of the equipment and keep it running efficiently. Planned maintenance also reduces emergency maintenance. All maintenance calls should start with a visual inspection, paying particular attention to safety issues. A planned maintenance call typically consists of routine maintenance, a prestart inspection, an operational check, and troubleshooting if needed. Many important operational characteristics can be found on the equipment's data plate or inside the service panels. During a planned maintenance service call, the technician has an opportunity to make the job easier in the future.

WORK ORDERS

Service Ticket 9001

Customer Request: Spring Air-Conditioning Checkup

Equipment Type: R-22 Split-System Air Conditioner, Horizontal Blower Coil in Crawlspace

During the initial prestart visual check, the technician notices water around the indoor unit and a very dirty filter. The technician changes the air filter, checks the evaporator and sees no ice, and operates the system. The airflow is noticeably low and the suction line starts to frost. System readings are:

Outside ambient	80°F
Return air	78°F
Supply air	50°F
Suction pressure	55 psig
Liquid-line pressure	175 psig
Suction-line temperature	30°F
Liquid-line temperature	84°F

The low suction pressure and low superheat point to an airflow problem. The water is a result of a plugged drain, which is complicated by ice forming on the evaporator. The technician concludes that the evaporator coil and drain pan both need cleaning.

Service Ticket 9002

Customer Request: Electric Furnace Fall Check

Equipment Type: 15 kW Electric Furnace Installed in a Closet

A technician is sent to do a fall check on an electric furnace. The voltage to the unit with the unit off is 240 V, the nameplate voltage. The technician checks the voltage at the fuse block with the unit operating and finds 240 V. When checking the voltage to the electric strips, there is 240 V on two strips, but only 210 V on the third strip. There is a 30 V drop across the sequencer contacts, and the sequencer temperature is noticeably higher than the rest of the control panel. The technician concludes that the sequencer contacts are defective and changes the sequencer.

Service Ticket 9003

Customer Request: Gas Furnace Fall Check

Equipment Type: 80 Percent Upflow Furnace in Basement

A technician is called to perform a fall preseason check on a gas furnace. This customer has the furnace serviced every year and there is no record of any problems. The customer finished the basement over the summer and the furnace that was in an open basement is now in a small closet. The technician performs the furnace check with the closet door open and finds no problems. However, the technician wonders if the furnace will perform as well with the closet door closed because there are no visible combustion air openings. The furnace is operated with the door closed, and 15 minutes after lighting, the flames start to turn yellow and lazy. The technician's CO monitor indicates high levels of CO in the closet. The technician shuts down the furnace and concludes that the closet needs to have combustion air openings added for safe operation of the furnace. The open basement provided enough combustion air, but the unventilated closet does not.

Service Ticket 9004

Customer Request: Heat Pump Fall Check

Equipment Type: Split-System Air-Source Heat Pump

A technician is called to perform a fall check on the heat pump. The outside temperature is 50°F. While operating the unit on first-stage heat, the technician notices that the temperature rise across the indoor coil is only 10°F. The airflow is measured and found to be within specification for that unit. Further checks indicate that the refrigerant pressures are lower than expected using the heating performance check chart. The technician concludes that the unit is undercharged. The technician looks for leaks and finds a small leak at a flare joint. Tightening the flare stops the leak. The refrigerant is recovered, the system evacuated, and a full charge of refrigerant is weighed in according to the manufacturer's specifications.

Service Ticket 9005

Customer Request: Spring Air-Conditioning Checkup

Equipment Type: Split-System Air Conditioner with an Upflow Furnace in a Mechanical Closet

A technician is called to perform a spring check on a split-system air conditioner. The technician has to move cardboard boxes of Christmas ornaments stacked between the gas furnace and gas water heater to get to the air filter. After concluding the checkup, the technician moves the boxes out of the furnace room and informs the homeowner that storing flammable items next to the furnace and water heater is dangerous.

Service Ticket 9006

Customer Request: Spring Heat Pump Checkup

Equipment: Packaged Air-Source Heat Pump, Filter Grille in Hall

A technician is called to do a spring check on an air-source heat pump. The technician finds a new high-efficiency electrostatic filter from DIY Depot in the filter grille. The airflow is noticeably low and the indoor coil starts to freeze over. The technician checks the indoor coil and finds it clean. When the high-efficiency filter is replaced with a standard filter, the airflow improves and the coil does not freeze. The technician informs the homeowner that the filter she was using had too much pressure drop for the system to operate correctly and offers to have a company representative discuss ways of improving the filtration without choking off the airflow.

UNIT 90—REVIEW QUESTIONS

1. List the four phases of a planned maintenance service call.
2. What is the purpose of checking the supply voltage both before and after operating the system?
3. Name key safety aspects of combustion appliances that should be checked on a planned maintenance service call.
4. List operational checks to make during an air-conditioning planned maintenance service call.
5. Why should caution be exercised when replacing a low-efficiency air filter with a higher-efficiency air filter?
6. How can performing planned maintenance save the customer money?
7. How does planned maintenance improve system reliability?
8. How can the technician determine what the system voltage parameters are?
9. How can the heating operation of a heat pump be checked without gauges?
10. The supply voltage to a system drops below the minimum voltage when the unit is operated. What could be causing this?
11. What color should the flames be in a correctly adjusted gas burner?
12. List the items to be checked when servicing a belt-drive system.
13. How can a worn belt be identified?
14. How is a temperature-pressure–relief valve on a hot water boiler tested?
15. Describe how sound is used in predictive maintenance.
16. Discuss the difference between predictive maintenance and planned maintenance.
17. What side of an air-conditioning A coil generally needs cleaning?
18. How does the technician know which way to install an air filter?
19. List some maintenance items that can be inspected visually.
20. How are water pumps in hydronic heating systems checked?

UNIT 91

Refrigeration System Cleanup

OBJECTIVES

After completing this unit, you will be able to:

1. list the data that should be recorded for baseline measurements.
2. explain the need for system cleanup after a compressor burnout.
3. explain how to clean up a system after a compressor burnout.
4. describe methods of performing a compressor-oil acid test.
5. discuss the safety issues in debrazing refrigeration components.
6. describe the procedure for flushing a refrigeration system.

91.1 INTRODUCTION

Refrigeration system component failures are often accompanied by system contamination. Compressor failures are often accompanied by system contamination. The contamination can be caused by a compressor failure. Contamination can also be a contributing factor in causing the compressor failure. Either way, leaving the contamination is certain to shorten the life of any replacement compressor and reduce system efficiency. Servicing a refrigeration system after a major component failure should always include system cleanup. One of the best ways to determine that a cleanup was successful is to compare the final operating conditions to the baseline performance data that were collected when the system was first installed.

91.2 BASELINE PERFORMANCE DATA

Baseline performance data comprise a record of all the system performance specifications when the system is operating as designed. Baseline performance shows what the system actually does in that particular installation, so it provides a benchmark to real operating conditions that are specific to that unit and application. Baseline information should include operating conditions, system pressures and temperatures, system superheat and subcooling, air or water temperature difference across both the evaporator and condenser, and amp draw of all motors—especially the compressor. Figure 91-1 shows a list of readings that should be collected for baseline data measurements.

It is a very good idea to collect baseline performance data on all new systems to allow comparison later in the life of the unit. System baseline performance can provide a valuable set of data for comparison when servicing a system. The effectiveness of a system cleanup can be determined by comparing the system operating characteristics after the cleanup to the baseline data collected when the system was first installed. System operating characteristics should be checked after any major repair and compared to the baseline data. System baseline data can also be measured just prior to any major change in the system. Baseline performance measurements are often taken when replacing an ozone-depleting refrigerant with a newer, non-ozone-depleting replacement. The baseline performance data allow the system operating characteristics with the new refrigerant to be compared to the refrigerant the system was designed to use.

REFRIGERATION SYSTEM BASELINE DATA	
Compressor Data	
suction pressure	
suction temperature	
discharge pressure	
discharge temperature	
amp draw	
Condenser Data	
temperature of air or water in	
temperature of air or water out	
condenser flow in CFM or GPM	
liquid line temperature	
Evaporator Data (Including Metering Device)	
entering air wet bulb temperature (for air coils)	
temperature of air or water in	
temperature of air or water out	
evaporator flow in CFM or GPM	
liquid line temperature	
suction line temperature	

Figure 91-1 List of readings for baseline data measurements.

TECH TIP

Compressor manufacturers and refrigerant suppliers both recommend taking baseline measurements prior to changing the refrigerant in a system to a newer replacement refrigerant. You should get the system operating as close to the manufacturer's specifications using the original refrigerant before proceeding with any refrigerant conversion.

91.3 CAUSES OF REFRIGERATION SYSTEM CONTAMINATION

System contamination is often the result of improper installation. Common installation-related contaminants include carbon, oxides, air, and water. Additional contaminants can form later as a result of these, including acids and hydrolyzed compressor lubricant. Oxides are formed during brazing without using an inert gas. Oxygen in the air combines with the heated metal parts to form metal oxides. These oxides can come loose and collect in bearings and in the metering device. Figure 91-2 shows a screen that is stopped up by copper oxide deposits as result of brazing without nitrogen.

TECH TIP

Newer systems using HFC refrigerants and polyol ester lubricants are less tolerant of oxides than older HCFC systems, which use mineral oil. HFC refrigerants and ester-based lubricants are strong solvents. They will scrub oxides off the walls of oxidized tubing and deposit them in metering devices, screens, and orifices.

Figure 91-2 Screen that is stopped up by copper oxide deposits.

Air and moisture are left in the system when the system is not properly evacuated. Noncondensables, such as air and nitrogen, take up space in the condenser and raise the system high-side pressure. This increases the compression ratio, decreases the system capacity, and increases the compressor operating temperature. The higher operating temperatures encourage acids to form by chemical reactions between the refrigerant, air, and water. The acids then attack the system components from the inside out. Moisture left in the system can freeze up at the metering device, restricting the flow of refrigerant and severely decreasing system capacity.

TECH TIP

Newer systems using HFC refrigerants and polyol ester lubricants are more adversely affected by moisture than older HCFC systems, which use mineral oil. Ester based lubricants are cousins to soap. Polyol ester is a hygroscopic lubricant—water is attracted to it. They break down when mixed with water in the presence of metals and heat to form soapy deposits that stop up metering devices, screens, and orifices throughout the system.

System contamination can also be the result of a major compressor burnout. Hermetic compressor motors are surrounded and cooled by the refrigerant that they are pumping (Figure 91-3). When the compressor motor burns out, carbon compounds contaminate the refrigerant and oil. Contaminants can be pumped throughout the system if the compressor motor burns out slowly. Figure 91-4 shows a burned-out hermetic compressor motor. If

Figure 91-3 Hermetic motor in shell.

Figure 91-4 Burned hermetic compressor motor.

the compressor has operated at elevated temperatures for a period of time, the compressor oil will be carbonized. If this carbon is left in the system, it can ground out the new compressor by providing a conductive path between the motor terminals and the shell because the electrical terminals are exposed inside the shell (Figure 91-5). Acids left in the system can attack system components, including the compressor windings.

91.4 EFFECTS OF SYSTEM CONTAMINATION

Problems caused by system contamination range from loss of operating efficiency and higher operating cost to failure of system components and shorter system life. All these problems cost the customer money, so it is in the customer's best financial interest to clean up contaminated systems. It will help technicians working on chronically poor-performing systems to recognize the effects of some common contaminants.

Figure 91-5 Compressor terminals grounded by carbon.

Carbon

Carbon particles circulating with the refrigerant and oil can plug small holes in refrigeration components, such as screens, metering devices, control capillary tubes, and suction-accumulator oil-return holes. Carbon can also cause internal shorts and grounds on hermetic compressors (see Figure 91-5). To clean up a system contaminated with carbon: recover the refrigerant, flush the system, install filter driers, evacuate the system, and charge the system with clean refrigerant.

Noncondensables

Noncondensables, such as air or nitrogen, take up space in the condenser and raise the system high-side pressure. This increases the compression ratio, which increases the compressor amp draw and decreases the compressor capacity. The end result is decreased system capacity, increased compressor operating temperature, and increased system operating costs. The only way to remove noncondensables once they are in the system is to recover the entire charge, evacuate the system, and recharge it with the correct refrigerant charge, being careful not to reintroduce noncondensables during the charging process.

Moisture

Moisture is normally introduced when air is let into the system, because nearly all air contains moisture. Moisture attacks the metals inside the system, causing corrosion and oxidation. A drop of water can freeze up at the metering device, causing a restriction. The drop in system capacity can be dramatic. A restriction caused by moisture will often clear if the system is turned off and allowed to warm up. When the system is restarted, the problem appears to have gone away. But the problems reappear when the moisture makes its way back to the metering device. To remove moisture from a system, recover the refrigerant, install a filter drier, evacuate the system down to 500 microns, and recharge with clean refrigerant.

Acid

Acid is rarely introduced into the system; instead, it is formed by the cumulative effect of other contaminants. The combination of air, water, and refrigerant at elevated temperatures can react chemically to form acids that are concentrated in the compressor oil. Most of the materials inside the system can be attacked by acid, literally eating the system away from the inside out. Another effect of acid is copper plating. Acidic oil dissolves copper from the inside of the lines and deposits it in other places, like compressor valves and bearing surfaces. A system with a coppery sheen on internal steel parts has been operating under acidic conditions (Figure 91-6).

91.5 OIL ACID TEST

The compressor oil is a good indicator of the overall system cleanliness because oil acts like a scavenger, collecting contaminants from all over the system. After installing

Figure 91-6 Copper plating inside compressor.

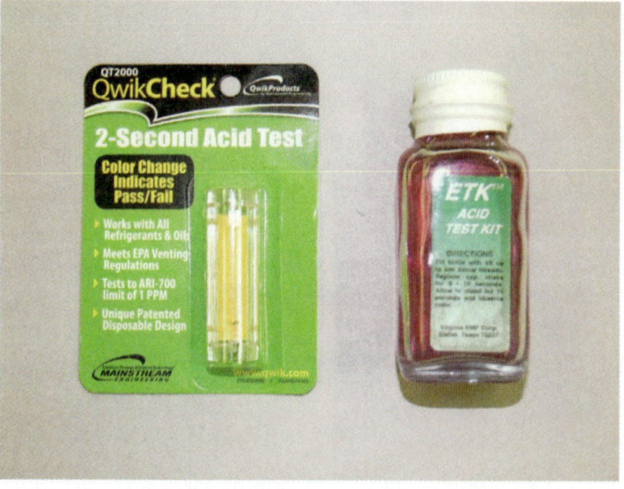

Figure 91-7 Oil acid test strip and test kit.

Figure 91-8 Color change indicating acid.

TECH TIP

New high-efficiency systems only perform at their rated capacity and efficiency if they are properly matched with a listed AHRI coil and installed according the manufacturer's specifications. System contamination can adversely affect the system performance and shorten the equipment life. A customer that has purchased an expensive new high-efficiency system will not be pleased when it performs no better than the older unit being replaced.

a replacement compressor, the compressor oil should be regularly monitored to ensure that the system stays clean. Ideally, the compressor oil should be checked during the first week of operation after the cleanup. Acid is the primary contaminant to look for. Oil can appear clean but still have a high acid content. Two types of oil acid tests are available: kits that use a physical oil sample and indicators that allow a small amount of refrigerant to pass through them without getting an oil sample (Figure 91-7). In both cases, an indicator chemical changes color when acid is present.

The liquid oil acid test kits can only be used with compressors that have an oil drain plug, such as open compressors and semihermetic compressors. Remember that the compressor crankcase is under pressure. You must valve off the compressor and recover the refrigerant in the crankcase before taking the oil sample. Leave a slight pressure in the crankcase of 1 to 2 psig to help push the oil out without letting air in. A specific quantity of oil is collected and mixed with the indicator chemical. Acid is indicated by a change in color (Figure 91-8). Follow the manufacturer's instructions for specific details.

It is not practical to get an oil sample with hermetic compressors because they must be physically removed to change the oil or take an oil sample. Indicators are available that get a trace amount of oil by purging a small amount of refrigerant through them while the system is operating, making it unnecessary to collecting an oil sample. These indicators typically connect to the suction line with the system operating, reducing the refrigerant pressure on them. Refrigerant vapor is purged through them for a few seconds. The indicator changes color if there is acid in the refrigerant vapor flowing through the indicator tube (Figure 91-9).

If the test indicates that the compressor oil is acidic, the compressor oil and system filter driers should be changed. In the case of a hermetic compressor, the oil is normally not changed due to the impracticality of removing the compressor to change the oil. It is possible to clean up a system through repeated filter drier changes; it just takes longer.

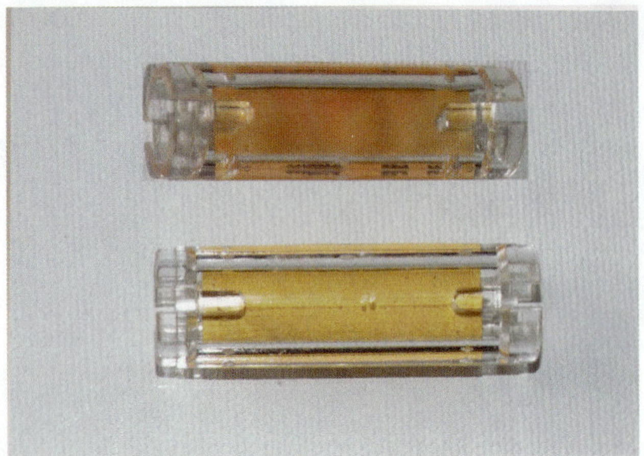

Figure 91-9 Strip color change showing acid.

91.6 STAGES OF REFRIGERATION SYSTEM CLEANUP

The stages in cleaning up a system are as follows:

1. Recovering the contaminated refrigerant
2. Changing the contaminated compressor oil
3. Flushing the lines and coils
4. Performing any system modifications or repairs
5. Installing liquid-line and suction-line driers
6. Evacuating and charging the system
7. Operating the system and checking performance data

91.7 RECOVERING CONTAMINATED REFRIGERANT

Contaminated refrigerant should be recovered into an empty recovery cylinder marked "Contaminated" to keep it from contaminating other relatively clean recovered refrigerant (Figure 91-10). Charges for destroying contaminated refrigerant can become expensive.

SAFETY TIP

If you are recovering R-410a, make sure all your equipment is rated for R-410a, including the gauges, hoses, recovery machine, filter, and recovery cylinder. Using equipment that is not specifically designed for R-410A can be dangerous. Equipment not rated for R-410A can rupture due to its higher pressures.

The refrigerant should pass through a filter before entering the recovery unit to reduce the amount of contaminants entering the unit, thereby prolong its life (Figure 91-11). The refrigerant should not be reused. Contaminants that remain in the refrigerant will shorten the life of any new components and reduce system efficiency. The cost of new refrigerant is minor compared to the cost of a new system. Instead, the contaminated refrigerant should be returned to

Figure 91-10 Recovery cylinder marked "Contaminated."

Figure 91-11 Filter inline to recovery unit.

a refrigerant processor, who can either reclaim or destroy it. The refrigerant processor may charge for disposing of the contaminated refrigerant depending upon the level of contamination and the capabilities of the refrigerant processor. Keeping severely contaminated refrigerant separated from other recovered refrigerant reduces your refrigerant processing costs.

Ideally, your gauges should be connected to both the suction line and the liquid line. Figure 91-12 shows recovery connections using a standard two-valve manifold. Recovering liquid first speeds the recovery process and

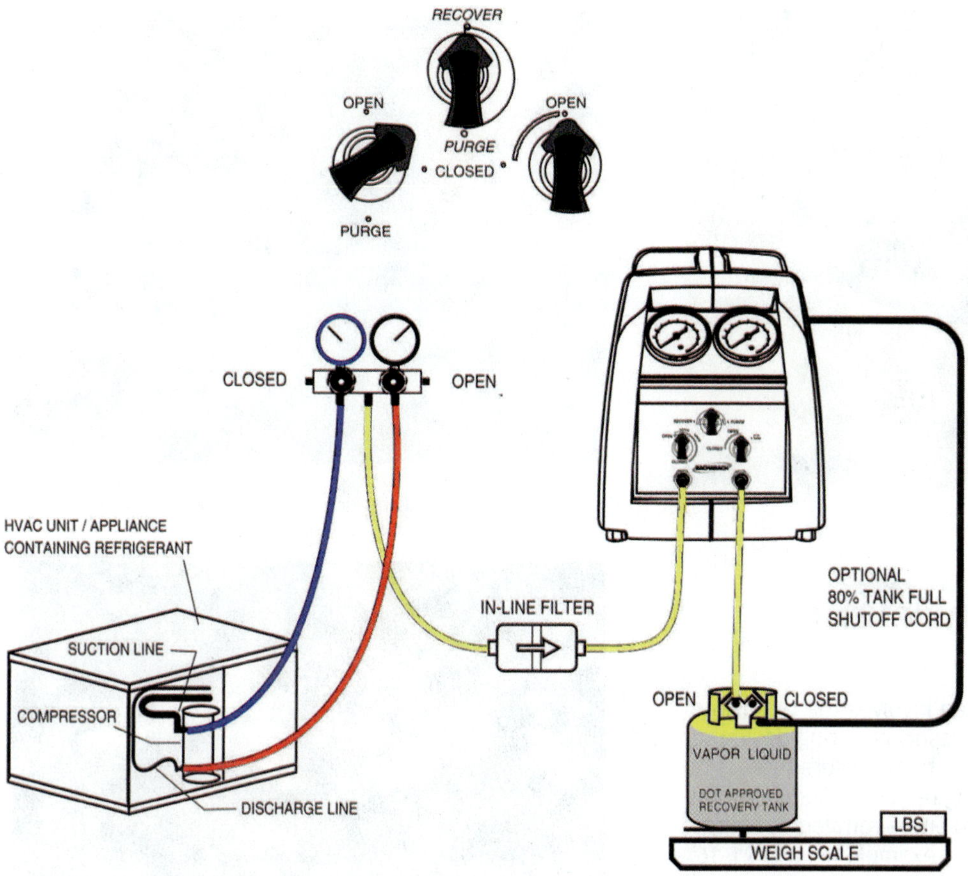

Figure 91-12 Recovery unit connections to manifold and unit.

maximizes contaminant removal. For liquid recovery, only the high-side manifold hand wheel will be opened and the recovery unit valve(s) will be set for liquid (Figure 91-13). Some gauges have a sight glass that can be used to monitor the refrigerant, allowing you to see when liquid is no longer being recovered (Figure 91-14). The temperature of the middle gauge port is also an indicator. Liquid refrigerant flashes as it passes through the gauges, causing the port to become cold. When only vapor refrigerant is flowing through the gauges, the port warms back up.

Figure 91-13 Recovery unit valve set to liquid recovery.

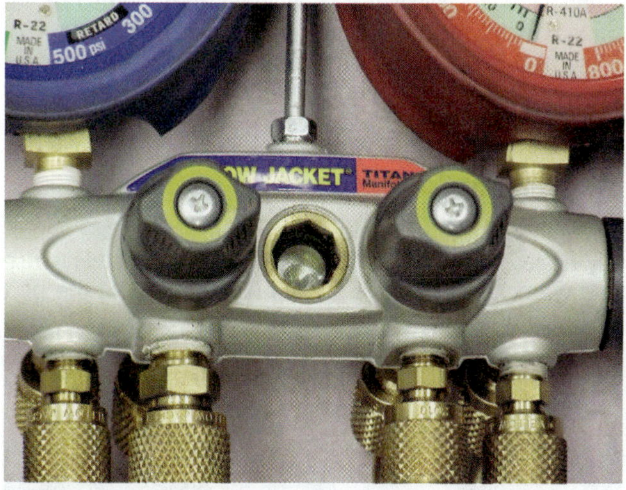

Figure 91-14 Sight glass in gauges.

After removing as much liquid as possible, switch to vapor recovery. Vapor should be recovered from both sides of the system to avoid pressure drop through the compressor and metering device. During vapor recovery, both manifold hand wheels will be opened and the recovery unit valve(s) will be set for vapor (Figure 91-15). You should recover vapor down to the recovery level specified by the EPA. Recovery may be stopped before reaching the specified level on systems with leaks, but in no case should the final system pressure be above 0 psig.

Figure 91-15 Recovery unit set for vapor recovery.

Figure 91-17 Semihermetic compressor with service valves.

91.8 REFRIGERANT COMPONENT REMOVAL

Refrigeration system service often involves the removal of components such as the compressor, metering device, or filter driers. The first step is always to make sure there is no pressure on the component being removed. Some components will require recovery of the entire refrigerant charge: a hermetic compressor is a good example (Figure 91-16). Some components can be isolated, minimizing the amount of refrigerant recovery. A semihermetic compressor with manual service valves can be isolated from the rest of the system by front-seating both service valves (Figure 91-17). Systems with king valves or liquid-line service valves can often be pumped down before removing metering devices or filter driers (Figure 91-18).

Figure 91-18 King valve for pumping down.

Mechanical Connections

Components that have mechanical connections are generally easier to remove and replace. You should always use two wrenches: one for a backup wrench to hold the part that should not turn and the other to do the turning. Always use correct fitting, properly sized wrenches (Figure 91-19). Fixed wrenches are preferable to adjustable wrenches. Pliers, vise grips, Robogrips, and channel locks should not even be considered (Figure 91-20). As a precaution, loosen

Figure 91-16 Hermetic compressor in system.

Figure 91-19 Two wrenches removing part.

Figure 91-20 Channel locks should not be used on flare nuts or valves.

the connection and listen for escaping refrigerant before completely removing the connection. Most mechanical connections will turn easily by hand after the initial loosening. A connection that remains tight after the first partial turn may still have pressure behind it. Do not stand directly in line with the connection being removed in case it is still pressurized. You do not want to be in the path in a worst-case scenario with a large accidental release of refrigerant. Ideally, you should have the replacement part ready to go so that it can be installed immediately after removing the old component. If this is not possible, the opening should be capped or plugged to prevent system contamination from air and moisture (Figure 91-21).

Figure 91-21 Line is capped to prevent contamination.

Brazed Connections

There are two general methods for removing components that are brazed in: debrazing and cutting. Debrazing has the advantage of leaving the copper tubing in the original configuration. This makes exact replacement components easier to fit and reduces the number of brazed connections necessary. Debrazing has the disadvantage of being inherently more dangerous. Debrazing can also create more system contamination. Filter driers should not be debrazed because heating them can drive contaminants captured in the driers back into the system.

It is crucial to ensure that the component is not pressurized before debrazing. Although most of the refrigerants we work with are nonflammable, the refrigerant oil is flammable. A sudden release of refrigerant and oil can ignite with a flamethrower effect, even with an A1 refrigerant. When debrazing a compressor, cut open the process tube before proceeding to relieve any pressure left in the compressor (Figure 91-22). Even with the refrigerant recovered, a short puff of flames often comes out of the connection when it is pulled loose (Figure 91-23). Another hazard is created by the

Figure 91-22 Cut process tube on hermetic compressor.

Figure 91-23 Flames coming from de-brazed connection.

Figure 91-24 Tubing broken off from bad de-brazing.

decomposition of halogenated refrigerants. Even after recovery, there is still enough refrigerant left inside the system to cause a problem during debrazing. The refrigerant breaks down to form poisonous gases in the presence of flames. It can be very difficult to breathe if you are positioned over the opening. Debrazing is more difficult than brazing. The entire joint must be molten before you can pull the tubing out. Tubing can break off in the joint if you try to pull it out before all the braze material is molten (Figure 91-24). For safety reasons, some manufacturers recommend cutting compressors out instead of debrazing them.

Figure 91-25 Oil drain plug on compressor.

Figure 91-26 Oil fill plug.

91.9 CHANGING CONTAMINATED COMPRESSOR OIL

The refrigerant oil acts like a scavenger, absorbing contaminants as it circulates through the system. The compressor can hold large quantities of contaminants suspended in the oil and settled at the bottom of the crankcase. If the compressor is being changed, this contaminated oil will be removed with the compressor. However, if the compressor is not being replaced, the compressor oil should be changed whenever practical. The fastest way to clean a contaminated system is to keep changing the oil until it is clean. The oil can often be changed in semihermetic and open compressors, but it is not practical to change the oil in a hermetic compressor.

SAFETY TIP

Contaminated compressor oil can sometimes be very acidic. It should always be handled as if it is acidic. Severe burns can occur from contacting acidic contaminated compressor oil.

Most open compressors and semihermetic compressors have an oil drain plug (Figure 91-25). Remember that the compressor crankcase is under pressure. You must valve off the compressor by front seating the suction and discharge service valves, and recover the refrigerant in the crankcase before removing the drain plug. Drain the oil into a clean, empty container so that the level of contamination can be checked. Measure the amount of oil removed, and replace just that much oil. There will be more oil out in the system. If you fill the crankcase, you will most likely have too much oil in the system. Remove the fill plug and pour in the new oil (Figure 91-26). Because the compressor was opened to the atmosphere, it must be evacuated. However, only the compressor needs to be evacuated, because it was isolated from the rest of the system when you front-seated the service valves. This should not take long. After the compressor is evacuated, you can backseat the compressor service valves and put the system back into operation. If the compressor has a sight glass, observe the oil level after starting the compressor to make certain it is correct.

91.10 SYSTEM FLUSHING SOLVENTS

The lines and coils will be coated with refrigerant oil. If the system is contaminated, the oil in the lines and coils may also be contaminated. A nitrogen purge can blow out some of these contaminants, but many contaminants will remain in the system. The lines and coils can be flushed with a solvent that is manufactured specifically for cleaning refrigeration systems to remove the residual contaminants. Do not use general-purpose solvents, especially not flammable petroleum products.

Figure 91-27 Oil-return hole in accumulator.

SAFETY TIP

An inert gas such as nitrogen or CO_2 may be used to purge the system of contaminants and residual solvent. NEVER use oxygen or compressed air because it can create an explosive mixture with the refrigerant oil in the system.

You should never flush through refrigeration components that act as natural traps, such as liquid receivers, accumulators, or compressors. Contaminants will flush in but not out. If a receiver or accumulator is suspected of holding a large amount of contaminants, it should be replaced. Accumulators in particular have a small oil-return hole. If this becomes stopped up, oil will be trapped in the accumulator when the system operates and the compressor is sure to fail eventually from lack of lubrication (Figure 91-27). Filter driers should be replaced, not cleaned. Do not flush through the metering device; the restriction in the metering device can trap small particles and stop it up. Expansion valves and orifices should be removed prior to flushing the evaporator coil.

TECH TIP

Most system flush products are low-pressure chemicals that remain a liquid while flushing but will easily evaporate from the system. You do not want to use a product that will remain in the system. Ideally, any flush product that remains in the system will evaporate when you pull a vacuum on the system before recharging it.

91.11 FLUSH PROCEDURE

Make certain the system refrigerant has been recovered before proceeding. You cannot flush systems that still have a refrigerant charge. Components that cannot be flushed should either be removed or bypassed. You should only be flushing lines and coils—not compressors, accumulators, receivers, metering devices, or filter driers.

Figure 91-28 Front-seated split-system shutoff valves.

With split systems that have installation valves, front-seating the valve will isolate the line from the unit (Figure 91-28). This will direct the flushing agent through the valve and into the lines, not into the system. The end of the lines at the blower coil should be disconnected or cut and placed in a container to catch the contaminated oil and debris. The lines should first be purged with nitrogen to blow out as much oil and debris as possible. Nitrogen is much cheaper than flushing solvent. Adjust the nitrogen regulator to between 100 and 120 psig and purge with nitrogen for 5 minutes. Never exceed the system's low-side design pressure.

SERVICE TIP

For best results, remove to cores from Schrader valves before purging and flushing through the valve and use a connecting hose with no core depressor. The pressure drop across a Schrader valve is dramatic. Removing the core and core depressor increases the flow of the purging gas and flushing solutions, improving the effectiveness of the purge and flush operation.

After purging with nitrogen, close the nitrogen cylinder, back out the regulator, and disconnect the nitrogen hose from the system. You are now ready to flush the lines with flushing solvent. The specific procedure may vary depending upon the solvent manufacturer. Always read and follow the solvent manufacturer's recommendations.

SAFETY TIP

Always use a regulator and pressure-relief valve when using nitrogen cylinders. Never attempt to connect the nitrogen cylinder directly to the system without going through a regulator.

Typically, the solvent comes in a disposable metal cylinder containing both the solvent and propellant. The cylinder must be inverted to deliver the solvent (Figure 91-29). Flushing time varies depending upon the type of solvent, the size of the line, and the length of the line being flushed. This can be from 30 seconds to several minutes. Ideally, you

Figure 91-29 Inverted flushing solvent cylinder.

Figure 91-30 Liquid-line and suction-line filters.

want to flush until no oil or debris comes out of the line. Often you may not be able to see the other end of the line. In these cases, use multiple flushes of 1 to 2 minutes and examine the contents of the container between flushes. Finally, reconnect the nitrogen and purge at 100 to 120 psig until no solvent comes out.

SAFETY TIP

Many flushing solvents come in disposable cylinders. These cylinders should never be exposed to system pressure or pressure from a nitrogen cylinder because they could explode under these higher pressures. To avoid any chance of accidentally pressurizing the disposable cylinder, the cylinder of flushing solvent should never be connected to the system at the same time as a nitrogen or refrigerant cylinder.

91.12 USING FILTER DRIERS FOR SYSTEM CLEANUP

Some of the particles loosened by the flushing procedure are not removed from the system. Filter driers should be added to remove the remaining contaminants and protect the refrigeration system components. A liquid-line filter should be added to the liquid line to protect the metering device; a suction-line filter should be added to protect the compressor (Figure 91-30). Specially designed cleanup filter driers should be used following a flushing procedure. Cleanup filter driers are designed to maximize acid and particle removal; they can remove particles down to 3 microns in size (Figure 91-31).

Drier sizing should take into account both the volume of refrigerant in the system and the pressure drop across the drier. In general, driers should be sized for a pressure drop of 1 psig or less. Filter drier manufacturers provide sizing guidelines on the box and in the instructions that come with the drier (Figure 91-32). Suction filters are sized for a

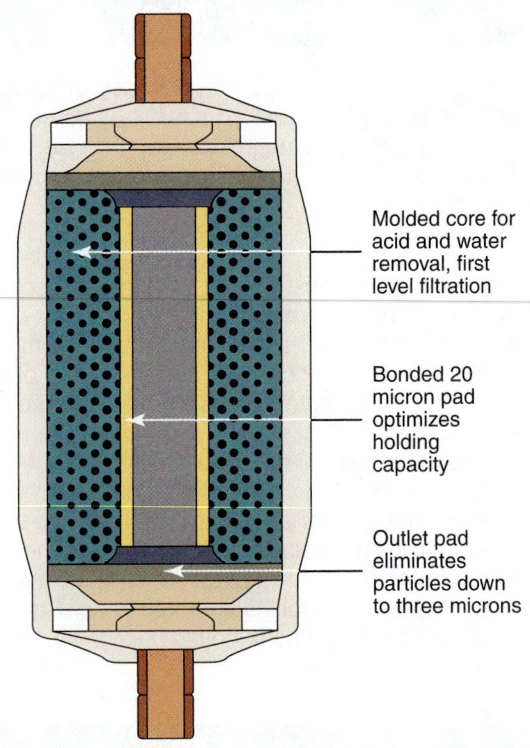

Figure 91-31 Special cleanup filter.

pressure drop equivalent to a 2°F saturation temperature drop. A pressure drop equivalent to a 5°F saturation temperature drop is acceptable for temporary application.

Although regular filter driers can also remove acid and particles, their primary purpose is moisture removal. The two desiccants used most often in filter driers are molecular sieve and activated alumina. Molecular sieve is superior at moisture removal but is not very effective on inorganic acids. Activated alumina is effective on both organic and inorganic acids but is less effective at moisture removal. Activated charcoal is often added to cleanup driers to remove hydrocarbon residues and wax. Driers that are designed primarily for moisture removal will contain high

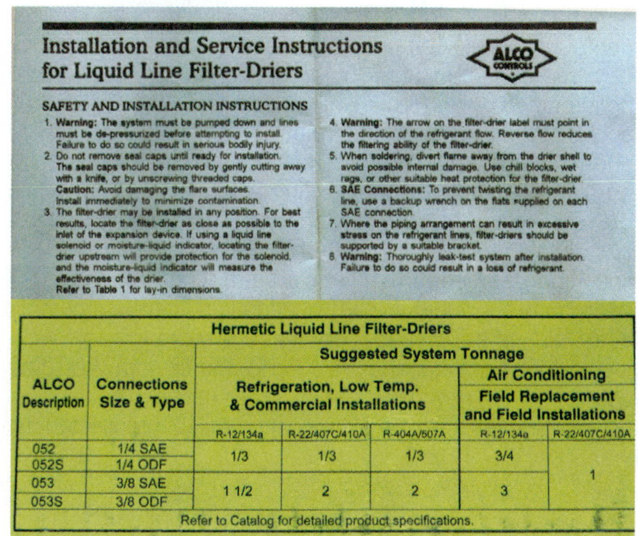

Installation and Service Instructions for Liquid Line Filter-Driers

ALCO CONTROLS

SAFETY AND INSTALLATION INSTRUCTIONS

1. **Warning:** The system must be pumped down and lines must be de-pressurized before attempting to install. Failure to do so could result in serious bodily injury.
2. Do not remove seal caps until ready for installation. The seal caps should be removed by gently cutting away with a knife, or by unscrewing threaded caps. **Caution:** Avoid damaging the flare surfaces. Install immediately to minimize contamination.
3. The filter-drier may be installed in any position. For best results, locate the filter-drier as close as possible to the inlet of the expansion device. If using a liquid line solenoid or moisture-liquid indicator, locating the filter-drier upstream will provide protection for the solenoid, and the moisture-liquid indicator will measure the effectiveness of the drier. Refer to Table 1 for lay-in dimensions.
4. **Warning:** The arrow on the filter-drier label must point in the direction of the refrigerant flow. Reverse flow reduces the filtering ability of the filter-drier.
5. When soldering, divert flame away from the drier shell to avoid possible internal damage. Use chill blocks, wet rags, or other suitable heat protection for the filter-drier.
6. **SAE Connections:** To prevent twisting the refrigerant line, use a backup wrench on the flats supplied on each SAE connection.
7. Where the piping arrangement can result in excessive stress on the refrigerant lines, filter-driers should be supported by a suitable bracket.
8. **Warning:** Thoroughly leak-test system after installation. Failure to do so could result in a loss of refrigerant.

Hermetic Liquid Line Filter-Driers						
ALCO Description	Connections Size & Type	Suggested System Tonnage				
		Refrigeration, Low Temp. & Commercial Installations			Air Conditioning Field Replacement and Field Installations	
		R-12/134a	R-22/407C/410A	R-404A/507A	R-12/134a	R-22/407C/410A
052	1/4 SAE	1/3	1/3	1/3	3/4	1
052S	1/4 ODF					
053	3/8 SAE	1 1/2	2	2	3	
053S	3/8 ODF					
Refer to Catalog for detailed product specifications.						

Figure 91-32 Filter drier application notes on box and sheet.

Figure 91-33 Gauge ports on filter drier.

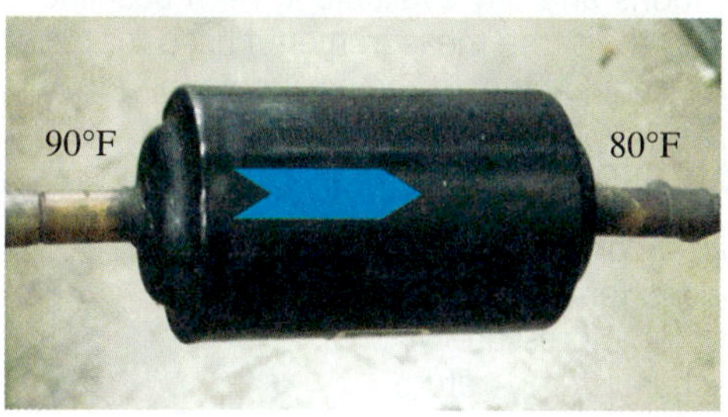

Figure 91-34 Temperature drop across filter.

levels of molecular sieve but may have little or no activated alumina for acid removal. Cleanup driers contain high levels of activated alumina and activated charcoal to handle acids and hydrocarbon residues. Typically, cleanup driers also offer high levels of filtration for particles down to 3 microns compared to 30 microns for standard driers.

Another difference is that regular driers are designed to remain in the system. Cleanup driers are removed after they have performed their job. Logically, driers should be removed after collecting contaminants simply to remove the contaminants from the system. Suction-line filters in particular are designed to be removed from the system after the system is clean. This not only removes the contaminants from the system but also minimizes pressure drop in the suction line.

TECH TIP

Some manufacturers will void the compressor warranty if a suction-line filter drier is left on the system indefinitely.

There is a slight refrigerant pressure drop across any filter drier, even a clean one. The refrigerant pressure drop increases when the drier has absorbed all the contaminants it can hold. Most filter driers should be replaced whenever the pressure drop across the drier exceeds 2 psig, because this pressure drop causes a loss in system efficiency. Pressure drop across a liquid-line filter can cause flash gas in liquid lines, reducing overall system capacity by reducing the amount of liquid delivered to the metering device. Suction-line filter pressure drop reduces the suction pressure to the compressor, increasing compression ratio. Many filters come with Schrader valves to make checking the pressure drop across the filter easy (Figure 91-33). Another way to check for a restricted filter drier is to measure the temperature on either side of the filter drier with the system operating. A measureable difference between the inlet and the outlet indicates a restriction (Figure 91-34).

91.13 EVACUATE AND CHARGE

The system should be reassembled, leak tested, evacuated, and charged according to the manufacturer's specifications. Failure to properly evacuate the system will introduce new contaminants. There is no point spending time and money cleaning up a system and then undoing all your work by not properly evacuating and charging the system. A leak test with nitrogen should be the first step after reassembly. Finding and repairing any leaks will save time in the long run. Refer to Unit 29 for more detail on refrigerant leak testing.

Next, release the nitrogen and evacuate the system to under 500 microns. It is especially important to get a good evacuation on a system after a cleanup. Refer to Unit 30 for more detail on system evacuation.

Finally, weigh in the manufacturer's specified charge. You may need to add additional charge for the refrigerant capacity of filters, which is typically stated in the instructions (Figure 91-35). If the refrigerant is being changed to a more environmentally friendly refrigerant, remember that the weight of the correct amount of the new refrigerant will be different from the older refrigerant. Older CFC and HCFC refrigerants are denser than most replacement HFC refrigerants. It does not take as much weight of the newer refrigerants to fill up the same space, so less refrigerant is required. Refer to the equipment manufacturer and the refrigerant producer for technical specifications and conversion directions. For more detailed discussion of refrigerant charging, refer to Unit 29.

Weight of Refrigerant in Ounces - EK Series						
Unit Size	R12 Liquid temp.		R22 Liquid temp.		R502 Liquid temp.	
	75°F	125°F	75°F	125°F	75°F	125°F
03	2.9	2.6	2.6	2.3	2.7	2.4
05	6.5	5.9	5.9	5.3	6.1	5.4
08	8.3	7.6	7.5	6.8	7.8	7.0
16	10.2	9.4	9.3	8.4	9.7	8.6
30	28.7	26.1	26.1	23.5	27.1	24.0
41	40.0	36.4	36.4	32.5	37.8	33.4
75	72.4	65.8	65.8	59.2	68.4	60.5

Figure 91-35 Refrigerant capacity of drier on instruction sheet.

91.14 OPERATE AND ADJUST

System performance should be verified by measuring and recording key system operating performance specifications. These should then be compared to the manufacturer's specifications and any available system baseline performance data. The same measurements used for baseline performance should be checked, including the following.

Compressor Data

- Suction pressure
- Suction temperature
- Discharge pressure
- Discharge temperature
- Amp draw

Condenser Data

- Temperature of air or water in
- Temperature of air or water out
- Condenser flow in CFM for air or GPM for water
- Liquid-line temperature

Evaporator Data (Including Metering Device)

- Entering-air wet-bulb temperature (for air coils)
- Temperature of air or water in
- Temperature of air or water out
- Evaporator flow in CFM for air or GPM for water
- Liquid-line temperature
- Suction-line temperature

UNIT 91—SUMMARY

System contamination can cause restricted metering devices and compressor failures. Contamination can be caused by improper installation and service or a catastrophic compressor failure. Typical system contaminants include moisture, noncondensables, carbon, acid, and metal shavings. Systems should be cleaned before installing a new compressor to prevent the contaminants from killing the new compressor. Cleaning methods include nitrogen purging, flushing with solvents, filter installation, changing the compressor oil, and system evacuation. Technicians should only purge lines and coils. Do not purge through components such as the compressor or accumulator, because they can trap contaminants. Contaminated compressor oil can be identified by performing an oil acid test. Two types of oil acid tests are available: kits that use a physical oil sample and indicators that allow a small amount of refrigerant to pass through them without getting an oil sample. The stages of system cleanup are recovering the contaminated refrigerant, changing the contaminated compressor oil, flushing the lines and coils, performing any system modifications or repairs, installing liquid-line and suction-line driers, evacuating and charging the system, operating the system, and checking performance data. Baseline performance data should be collected on all new installations. These data can be used for comparison later after major service or cleanup operations.

UNIT 91—REVIEW QUESTIONS

1. What are the most common types of refrigeration system contaminants introduced because of improper installation?
2. How does carbon affect a system?
3. What effect does the presence of noncondensables have on a system?
4. What system condition is indicated by copper plating?
5. What are the most common sources of refrigeration system contamination?
6. Explain how contaminants from a previous compressor failure can cause a replacement compressor to fail.
7. What is the purpose of collecting system baseline performance data?
8. List the system characteristics that are checked when taking system baseline performance.
9. List the stages of system cleanup.
10. Why does changing the compressor oil help clean up a contaminated system?
11. What is the difference between a regular filter drier and a cleanup filter drier?
12. Why should a cleanup filter drier always be used after performing a system flush?
13. Which refrigeration system components should not be flushed?
14. List the steps in flushing a system.
15. What are the two general types of oil acid tests available?
16. Why is it safer to cut out refrigeration components for removal rather than debraze them?
17. Why are suction-line accumulators particularly vulnerable to system contamination?
18. When retrofitting a CFC or HCFC system with a newer HFC refrigerant, why is the HFC refrigerant charge usually less than the original factory charge?
19. Why is a good evacuation critical after performing a system cleanup?
20. Describe the correct way to loosen a mechanical fitting when removing a refrigeration component.
21. Why should refrigerant from a compressor burnout be recovered into a separate cylinder marked "Contaminated"?
22. What factors are taken into consideration when sizing filter driers?
23. Why should suction-line filter driers be removed after the system is clean?

UNIT 92

Troubleshooting

OBJECTIVES

Upon completion of this unit, you will be able to:

1. discuss common questions customers ask service technicians.
2. outline the basic phases of troubleshooting.
3. outline an electrical troubleshooting procedure that uses the process of elimination.
4. discuss the use of common refrigeration system performance indicators in troubleshooting refrigeration problems.
5. explain how to use manufacturer-provided troubleshooting aids.

92.1 INTRODUCTION

Service technicians earn their living with their troubleshooting skills. Troubleshooting is the process of identifying problems, determining their cause, and correcting the cause of the problem. Effective troubleshooting requires problem solving, and the technician needs to do more than try to match symptoms with causes because most symptoms have many possible causes. Rather, technicians must understand exactly how equipment operates so they can analyze the cause of improper operation. Troubleshooting often requires performing tests on equipment, but taking a lot of readings is not troubleshooting. The data are worthless if they are not understood. Each test should take the technician a step closer to solving the problem. Most important, troubleshooting is not replacing parts until the system operates. Not only is this approach expensive, but it frequently fails to repair the problem. It is possible to replace all the parts and still have the original problem if the cause is not a defective component. A methodical approach to troubleshooting that is based on knowledge of fundamental system operation will result in less time spent overall, less cost in unnecessary replacement parts, and fewer mistakes. Technicians who practice systematic troubleshooting techniques can often correct problems in equipment that they have never seen before.

92.2 SIX CRUCIAL QUESTIONS ALL SERVICE TECHNICIANS MUST BE ABLE TO ANSWER

An important part of providing service is answering the customer's questions. Being prepared to answer questions completely and confidently builds the customer's confidence in the service technician. At the conclusion of a troubleshooting call, technicians should be able to answer six general questions:

- How is the equipment supposed to operate?
- What is the equipment doing wrong?
- What is the root cause of the problem?
- What specific tests or procedures were performed to determine the cause of the problem?
- How can this be corrected?
- How much will this cost?

How Is the Equipment Supposed to Operate?

Being able to explain how the equipment is supposed to operate makes explaining what is wrong easier. If the customer perceives that the technician does not understand the equipment, he or she will not have much faith in the technician's proposed solution.

What Is the Equipment Doing Wrong?

The technician should be able to identify what the equipment is doing wrong. What is the equipment doing that it should not do, or what is it not doing that it should do? Technicians should be able to clearly explain to their supervisors and to their customers what is incorrect about the operation of the equipment.

What Is the Cause of the Problem?

Of course, the customer wants to know exactly what is causing the problem. If possible, explain why this component or situation is causing the problem. The customer will have more confidence in a cause that can be logically explained. This is where being able to explain the system operation comes in handy.

What Tests Were Performed to Determine the Cause of the Problem?

Technicians should be able to explain to their supervisor or the customer what specific tests they did. Condemning parts or systems without testing is guessing. Technicians should keep an accurate log of all work performed. Repairing a system by replacing parts is expensive and may not solve the problem! Systems can malfunction for reasons other than bad parts. An air-conditioning system with a dirty condenser may shut down because of high pressure. This does not mean that the high-pressure switch needs to be replaced or that the system is overcharged. In fact, replacing the high-pressure switch would have no effect on system operation.

How Can This Be Corrected?

Most customers will not be satisfied with simply finding out why their unit does not work; they want the problem solved. The service technician should be prepared to offer solutions. Remember that replacing a failed component is not necessarily the solution. Replacing a compressor that has experienced liquid slugging without addressing the cause of the slugging has not solved the customer's problem. The cause of the liquid slugging must also be found and corrected.

How Much Will This Cost?

A practical reason for recording all tests performed is that this information is generally used to generate bills. Many companies now use service pricing books to calculate the cost of a service call. Most price books are based on the specific services performed. It is very difficult to produce an accurate bill without a specific list of services performed.

The cost is the foremost question in the customer's mind. The service technician needs to be able to address this question, which is certain to be asked. Many companies now require their technicians to generate and present the bill. Even technicians working for companies that do not generate bills in the field should be prepared to explain the company's pricing policy.

SERVICE TIP

One way of avoiding conflict is to make sure the customer understands the company's pricing policies up front. Also, allow the customer to accept your proposed solution before implementing it. You should clearly explain the problem, the cause of the problem, and your proposed solution to the customer. Let the customer agree with the solution before implementing it.

92.3 SYSTEMS APPROACH TO TROUBLESHOOTING

The major processes involved in operating air-conditioning systems include the electrical, refrigeration, airflow, and combustion processes. To make the material manageable, the different processes are often discussed separately. However, technicians are called to work on the entire system, not an isolated part of it. All system components depend upon each other. In truth, it is difficult to examine any single aspect of system performance while ignoring the rest of the system.

Service technicians must understand the interrelationship between different system components and be aware of the effect one process has on the rest of the system. For example, low evaporator airflow in an air conditioner will cause low suction pressures, causing the compressor to trip out on the low-pressure switch. What may initially look like an electrical problem, or a refrigeration problem, may in fact be an airflow problem. Similarly, refrigeration system problems can have an electrical cause. An inoperative evaporator fan motor can reduce the low-side pressure, but the solution is certainly not more refrigerant.

SERVICE TIP

The low-side pressure is closely tied to airflow or water flow over the evaporator. Reduced airflow over the evaporator reduces the amount of heat available for evaporating the liquid refrigerant. The refrigerant temperature starts dropping due to reduced heat load, and so the pressure also starts to drop. Airflow or water flow is the first possible cause that should be investigated when examining a system experiencing a low evaporator pressure.

Low evaporator airflow can also cause many other problems that are not as immediately obvious as low suction pressure. The low evaporator pressure that results from poor evaporator airflow can cause high compression ratio and loss of capacity. Low suction-gas superheat that results from poor evaporator airflow can damage the compressor by allowing liquid in the return-gas line.

SERVICE TIP

The high-side pressure is closely tied to airflow or water flow over the condenser. Reduced airflow over the condenser reduces the rate of heat rejection from the condenser. This causes increased refrigerant temperature and pressure. Airflow or water flow is the first possible cause that should be investigated when examining a system experiencing high condenser pressure.

Low condenser airflow can also cause other problems that are not as immediately obvious as high discharge pressure. The high discharge pressure that results from poor condenser airflow can cause high compression ratio and loss of capacity. High discharge pressure and reduced refrigerant flow can lead to compressor overheating because many compressors are cooled by the refrigerant they pump.

92.4 STAGES OF TROUBLESHOOTING

Troubleshooting can be broken down into several distinct steps, each step logically following the other:

- Understanding system operational sequence
- Preliminary system inspection
- Collecting operational data
- Recognizing what is operating incorrectly
- Testing to isolate the cause
- Recommending corrective action

92.5 UNDERSTANDING THE SYSTEM OPERATIONAL SEQUENCE

It is necessary to understand how a machine is supposed to operate to recognize when it is operating incorrectly. In most cases, this step should be accomplished before arriving at the job. It is possible to troubleshoot an unfamiliar system, but the first step is to understand how it operates. Sources of information on the system sequence of operation include the manufacturer's installation and operation and manuals, unit wiring diagrams, unit charging charts, and general textbooks. Some manufacturers provide detailed information about the unit operation on equipment panels. Increasingly, equipment and component manufacturers are providing specific information about current products on their Web sites. The Web site is often the best place to find the latest service bulletins and updates. The key point is that you must learn what the system is supposed to do and how it is supposed to do it before you can effectively troubleshoot it.

92.6 TYPICAL OPERATING SEQUENCES

Technicians should have an understanding of the basic sequence of operation for the most common types of systems they work on. Although all induced-draft furnaces do not use the same parts or work in exactly the same manner, there are some basic operations that are common to most of them. The same can be said of air conditioners and heat pumps. Figures 92-1 through 92-5 show typical operating sequences for common residential systems.

The charts are read from the top down. Events that take place first are at the top; events that take place last are at the bottom. The items in a column occur in sequence, meaning that all the actions listed above an entry must occur before that action can take place. The events are in bold; details of each event follow.

92.7 PRELIMINARY SYSTEM INSPECTION

Do the easy things first! Many systems have simple problems like loss of power, loss of control voltage, dirty air filters, or dirty coils. It is never wrong to check the air filter or check to see how dirty a condenser coil is. There are several items that should be routinely checked on any service call:

- Is the thermostat or system control set correctly?
- Is the air filter clean?
- Is there enough airflow or water flow (depending on the type of unit)?
- Is the system operating voltage within the nameplate rating?
- Is the system control voltage correct?

AIR CONDITIONER OPERATING SEQUENCE		
Thermostat terminal R, Control Power, R must receive 24 volts for unit to work. The thermostat connects the control voltage at R to Y and G to control unit operation.		
Thermostat terminal Y, Cooling Y is energized anytime there is a call for cooling.		**Thermostat terminal G,** Indoor Fan In **AUTO** position, G is energized on a call for cooling. In **ON** position, G is energized all the time.
Evaporator drain float switch Breaks Y signal if evaporator drain is full.		**Indoor fan relay** Closes contacts when coil is energized.
Low-pressure switch Breaks Y signal if suction pressure is low.		**Run capacitor** Required for fan operation.
High-pressure switch Breaks Y signal if discharge pressure is high.		**Indoor fan motor**
Anti–short-cycle timer Prevents immediate restarting.		
Compressor contactor Closes contact when coil is energized.		
Run capacitor Required for fan compressor. Frequently the fan and compressor run capacitors are both in one dual capacitor.	**Run capacitor** Required for fan operation.	
Start capacitor Required by some compressors to start.	**Condenser fan motor**	
Potential relay Required if compressor has a start capacitor.		
Compressor		

Figure 92-1 Air conditioning operating sequence.

AIR SOURCE HEAT PUMP SEQUENCE OF OPERATION

R control power thermostat receives 24 volts at R terminal and sends it to the other terminals to control unit operation.

Y compressor Y is energized in both first-stage heating and cooling.	**G Indoor Fan** **AUTO** energized on call for cooling. **ON** position energized all the time.	**O Reversing Valve** *** B Reversing Valve** O is energized in cooling to energize the reversing valve on most heat pumps. B is energized in heating to energize the reversing valve on a few heat pumps.	**W2 Second-Stage Heat Defrost Control** W2 is energized to energize the second stage heat. The defrost control energizes the second-stage heat during defrost.
Evaporator float switch Breaks Y signal if evaporator drain pan is full.	**Indoor Fan Relay** Closes contacts when coil is energized.	**Defrost Control** Defrost control energizes reversing valve in defrost for O-controlled valves. Defrost control de-energizes reversing valve in defrost for B controlled valves.	**Heat Sequencer** Close their contacts after a time delay when the coil is energized.
Liquid low-pressure switch Breaks Y signal if liquid pressure is low	**Run Capacitor** Required for fan operation.	**Reversing Valve** Swaps functions of the indoor and outdoor coils. Energized in cool for most heat pumps. Energized in heat on a few heat pumps.	**Thermal Overload** The thermal overloads are really safety devices in series with the heat strips.
High-pressure switch Breaks Y signal if discharge pressure is high.	**Indoor Fan Motor** Circulates air inside.		**Strip Heaters** Come on for supplemen tal heat.

Anti–short-cycle timer
Prevents immediate restarting.

Compressor contactor
Closed contacts when coil is energized.

Defrost control Opens contacts to outdoor fan during defrost cycle. **Run capacitor** Required for fan operation.	**Run capacitor** Frequently the fan and compressor run capacitors are both in one dual capacitor. Required for compressor operation.
Outdoor fan motor Blows air over outdoor coil.	**Start capacitor** Required by some compressors to start.
	Potential relay Required if compress or has a start capacitor. **Compressor**

Figure 92-2 Air-source heat pump sequence of operation.

WATER-SOURCE HEAT PUMP SEQUENCE OF OPERATION			
R control power thermostat receives 24 volts at R terminal and sends it to the other terminals to control unit operation.			
Y compressor Y is energized in both first-stage heating and cooling.	**G Indoor Fan** **AUTO** energized on call for cooling. **ON** position energized all the time.	**O Reversing Valve** *** B Reversing Valve** O is energized in cooling to energize the reversing valve on most heat pumps. B is energized in heating to energize the reversing valve on a few heat pumps.	**W2 Second-Stage Heat** W2 is energized to energize the second stage heat. Not all water-source heat pumps have second stage heat.
Evaporator float switch Breaks Y signal if evaporator drain pan is full.	**Indoor Fan Relay** Closes contacts when coil is energized.	**Reversing Valve** Swaps functions of the in door and outdoor coils. Energized in cool for most heat pumps. Energized in heat on a few heat pumps.	**Heat Sequencer** Close their contacts after a time delay when the coil is energized.
Liquid low-pressure switch Breaks Y signal if liquid pressure is low.	**Run Capacitor** Required for fan operation.		**Thermal Overload** The thermal overloads are really safety devices in series with the heat strips.
High pressure switch Breaks Y signal if discharge pressure is high.	**Indoor fan motor** Circulates air inside.		**Strip heaters** Come on for supplemental heat.
Anti–short-cycle timer Prevents immediate restarting.			
Compressor contactor Closed contacts when coil is energized.			
Run capacitor Frequently the fan and compressor run capacitors are both in one dual capacitor. Required for compressor operation.			
Start capacitor Required by some compressors to start.			
Potential relay Required if compressor has a start capacitor.			
Compressor			

Figure 92-3 Water-source heat pump sequence of operation.

<table>
<tr><th colspan="3">INDUCED-DRAFT GAS FURNACE OPERATING SEQUENCE</th></tr>
<tr><td colspan="3">Manual reset rollout switches
These shut down the system if the flames come outside of the normal combustion area. They may be after the R terminal on some furnaces.</td></tr>
<tr><td colspan="2">Thermostat terminal R, Control Power, R must receive 24 volts for unit to work.
R connects 24 volts to W to start a call for heat.</td><td></td></tr>
<tr><td colspan="2">Thermostat terminal W, Heating
W is energized when there is a call for heating.</td><td>Thermostat terminal G, Indoor Fan
In AUTO position, G is NOT energized for heating.
In ON position, G is energized all the time.</td></tr>
<tr><td colspan="2">Manual reset auxiliary limit
Auxiliary limits prevent the fan compartment from overheating. They are not typically found on upflow only furnaces.</td><td>Indoor fan relay
Closes contacts when coil is energized.</td></tr>
<tr><td colspan="2">Auto reset primary limit
The primary limit shuts down the flames if the heat exchanger area overheats.</td><td>Run capacitor
Required for fan operation</td></tr>
<tr><td colspan="2">Induced-draft fan motor relay
This is usually part of the furnace control board it energizes to operate the induced draft fan motor.</td><td>Indoor fan motor</td></tr>
<tr><td colspan="2">Indoor-draft fan motor
Creates the draft in the combustion area to pull combustion air and combustion gases through the heat exchanger.
Typically, the induced-draft fan operates for a few minutes in a prepurge cycle to clear out old combustion gasses before the ignition process is attempted.</td><td></td></tr>
<tr><td colspan="2">Draft pressure switch, normally open
Must close to indicate correct draft. The ignition process does not proceed if this switch does not close.
These are usually normally open and must close to indicate draft.</td><td></td></tr>
<tr><td colspan="2">Hot surface igniter relay
This is actually part of the furnace control board. It energizes to send voltage to the hot-surface igniter.</td><td></td></tr>
<tr><td colspan="2">Hot-surface igniter
The hot-surface igniter is energized long enough for it to get hot, typically 30 seconds to a minute.</td><td></td></tr>
<tr><td colspan="2">Ignition trial: gas valve energizes
The gas valve is energized for a short time, usually no longer than 3 seconds.</td><td></td></tr>
<tr><td colspan="2">Ignition trial flame rectification circuit
The flame safety looks for a microamp signal on the flame sensor, indicating the presence of flames.
If the signal is not seen in 2 to 3 seconds, the gas valve is shut down.
Most furnaces will cycle through the prepurge to ignition trial several times attempting to light.
If they don’t light after four or five attempts, the heat circuit is locked out for a period of several hours.</td><td></td></tr>
<tr><td colspan="2">Indoor blower relay
This is actually part of the furnace board. The furnace control board energizes the indoor blower after a time delay of 1.5 to 3 minutes to allow the heat exchanger to heat up.</td><td></td></tr>
<tr><td colspan="2">Indoor blower motor
The indoor blower motor is energized to circulate warm air throughout the house</td><td></td></tr>
<tr><td colspan="2">W terminal de-energized, all for heat satisfied
The thermostat quits sending 24 V to the W terminal once the temperature is satisfied</td><td></td></tr>
<tr><td>Post-purge
The induced-draft fan motor continues to operate for 1 to 2 minutes to purge the combustion chamber of any remaining combustion gases.</td><td>Heat exchanger cool down
The indoor blower remains to get the remaining heat in the heat exchanger into the house.</td><td></td></tr>
</table>

Figure 92-4 Induced-draft gas furnace operating sequence.

<table>
<tr><th colspan="3">OIL FURNACE OPERATING SEQUENCE</th></tr>
<tr><td colspan="3">Manual reset primary control safety
Shuts down primary control if ignition is not achieved in time.</td></tr>
<tr><td colspan="3">Thermostat terminal R, Control Power, R must receive 24 V for unit to work.
R connects 24 volts to W to start a call for heat.</td></tr>
<tr><td colspan="2">Thermostat terminal W, Heating
W is energized when there is a call for heating.</td><td>Thermostat Terminal G, Indoor Fan
In AUTO position G is NOT energized for heating.
In ON position, G is energized all the time.</td></tr>
<tr><td colspan="2">Auto reset primary limit
The primary limit shuts down the flames if the heat exchanger area overheats.</td><td>Indoor fan relay
Closes contacts when coil is energized.</td></tr>
<tr><td colspan="2">Burner motor relay
This is part of the furnace primary control. It energizes to operate the burner motor.</td><td>Run capacitor
Required for fan operation.</td></tr>
<tr><td>Oil burner motor</td><td>Ignition transformer</td><td rowspan="11">Indoor blower motor</td></tr>
<tr><td>Turns on both the oil pump and the combustion air blower.</td><td>Creates the high-voltage spark to ignite the oil.</td></tr>
<tr><td colspan="2">Cad cell ignition check
Cad cell sees the light of the flames and signals to primary control that ignition has occurred.
If flames are not seen in 90 seconds, the primary control safety trips and shuts down the furnace.</td></tr>
<tr><td colspan="2">Indoor blower relay
This is usually part of the primary control. The furnace primary control energizes the indoor blower after a time delay of 1.5 to 3 minutes to allow the heat exchanger to heat up.</td></tr>
<tr><td colspan="2">Indoor blower motor</td></tr>
<tr><td colspan="2">The indoor blower motor is energized to circulate warm air throughout the house.</td></tr>
<tr><td colspan="2">W terminal de-energized when call for heat satisfied
The thermostat quits sending 24 V to the W terminal once the temperature is satisfied.</td></tr>
<tr><td colspan="2">Oil burner motor stops</td></tr>
<tr><td colspan="2">The burner motor stops.</td></tr>
<tr><td colspan="2">Indoor blower motor stops</td></tr>
<tr><td colspan="2">The indoor blower continues to operate for 2 to 3 minutes to remove the heat remaining in the heat exchanger.</td></tr>
</table>

Figure 92-5 Oil-fired furnace operating sequence.

It takes very little time to check these items, and the most common problems can be traced to one of them. Further, if any of these are not correct, system performance tests will not produce accurate results.

92.8 COLLECTING OPERATIONAL DATA

The next step is to collect data that tell how the system is actually operating. Sources for the current system operational data include the homeowner, general observation, diagnostic boards, and system performance tests. All of these are not necessary in every situation, but service technicians typically use more than one of these methods to collect data about system operation. Remember that troubleshooting is a lot like solving a puzzle: the more clues you have, the easier it is to solve.

Homeowner Information

The homeowner can be a valuable source of information, but information gained from talking to homeowners should be verified because they may not always be able to make accurate statements based on their observations. Information from homeowners should be general, not specific. Do not ask the homeowner if the compressor is operating, as he or she may not know what a compressor is or where it is located.

Diagnostic Codes

Many systems now have electronic control boards with built-in diagnostics. The diagnostics typically consist of LEDs that flash a one- or two-digit code. The codes are typically listed in the manufacturer's service literature and frequently can also be found somewhere on the unit. The

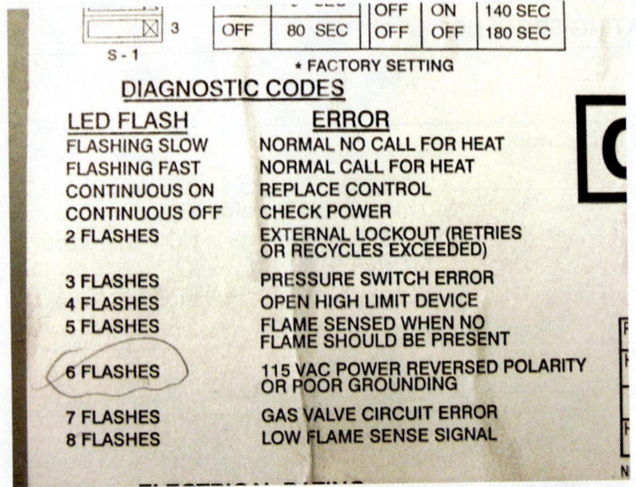

Figure 92-6 Typical ignition board fault code.

number of LED flashes corresponds to the code number. In Figure 92-6, three flashes (fault code 3), indicate "pressure switch error." For two-digit codes there is a pause between the flashes representing the first digit and the flashes representing the second digit. The code then repeats. Some boards differentiate between the first and second digits by the length of the flashes; long flashes representing one digit and short flashes representing the other. For example, a series of two long flashes and four short flashes would represent a code 24.

System Performance Tests

System performance tests provide more detailed information about how a system is operating. The specific tests will vary depending on the type of equipment being checked. Some common system performance tests include checks for refrigerant pressure, superheat, subcooling, temperature change, and electrical current draw. The purpose of these tests is to allow comparison between actual performance and factory-specified performance. Figure 92-7 shows system performance data for a water-source heat pump. System performance checks provide a more accurate picture of how well a system is performing than casual observation does. Rather than simply feeling the supply air to see if it is cool, measuring its temperature provides a quantifiable measurement. For systems that are checked regularly, recording this information provides a baseline to compare current and past system performance.

92.9 RECOGNIZING INCORRECT OPERATION

Troubleshooting often appears to begin at this step. Experienced technicians may already understand the system operation thoroughly, so the first step has essentially already been done ahead of time. An experienced

R-410A WATER SOURCE HEAT PUMP OPERATING DATA

ENTERING WATER	FLOW GPM	COOLING PERFORMANCE DATA				HEATING PERFORMANCE DATA			
		WATER TEMP RISE	SUCTION PRESSURE	DISCHARGE PRESSURE	AIR TEMP DROP	WATER TEMP DROP	SUCTION PRESSURE	DISCHARGE PRESSURE	AIR TEMP RISE
30°F	5	*NR*	*NR*	*NR*	*NR*	5-6	72-87	296-361	21-25
	7	*NR*	*NR*	*NR*	*NR*	3-4	75-92	301-368	22-26
40°F	5	14-17	114-139	155-190	22-27	6-7	88-107	314-384	24-29
	7	10-12	108-132	147-180	23-28	4-5	92-112	321-392	25-30
50°F	5	13-16	116-142	192-234	21-26	7-9	104-127	333-407	27-33
	7	9-12	111-135	182-222	22-27	5-6	109-133	340-415	28-34
60°F	5	13-16	119-146	228-279	21-26	8-10	120-146	352-430	30-37
	7	9-11	113-138	217-265	22-27	6-7	125-153	359-439	32-39
70°F	5	13-15	122-149	262-323	20-25	9-12	136-166	371-453	33-41
	7	9-11	116-142	251-307	21-26	7-8	142-174	378-462	35-43
80°F	5	12-15	125-152	301-368	20-24	11-13	152-185	389-476	36-44
	7	9-11	118-145	286-349	21-26	8-9	159-194	397-485	38-47
90°F	5	12-15	127-156	337-412	19-24	12-15	168-205	408-499	39-48
	7	9-10	121-148	320-392	20-25	8-10	176-215	416-509	41-51

NOTE: CHART VALUES ASSUME 80°F DRY BULB/ 67°F WET BULB ENTERING AIR IN COOLING, 70°F DRY BULB ENTERING AIR IN HEATING.

Figure 92-7 Water-source heat pump performance data.

technician can often quickly recognize incorrect operation in a system that they are already familiar with. However, comparing the actual operation to factory specifications is still the most accurate way to recognize incorrect operation. Collecting operational data is not always necessary initially. If a system is not operating at all, there is not a lot of useful data to collect. However, system performance data should be taken after getting the unit operating. Remember that the component or condition preventing unit operation may be another symptom, not the root cause of the problem.

92.10 TESTING TO ISOLATE THE CAUSE

Many people think of this part of the process as troubleshooting. Indeed, the majority of a technician's troubleshooting time is spent tracking down the cause of the problem. However, technicians should avoid the temptation to jump straight into solving a problem that is not clearly defined. Jumping straight to determining the cause can waste time by looking for the cause of nonexistent problems.

Proceed from General to Specific

Try to resist the temptation of starting immediately to diagnose specific components. The slowest way to solve a problem is by jumping from one component to the next looking for the specific cause before you have a good idea of exactly what the system is doing. First make sure you understand what the system is supposed to do. Next, determine what the system is actually doing, and then determine what it is doing incorrectly. Once you know what the system is doing wrong, start with more general investigations like system line voltage or control voltage.

The Process of Elimination

The key to determining the root cause in a reasonable amount of time is to follow a logical pattern of investigation and use the process of elimination. You should have a purpose for each test performed. For example, if a system is not operating at all, a logical question is whether it has voltage available. Taking a voltage reading at the power source will help decide where to look next. If no voltage is available, then the reason for the loss of power must be investigated. On the other hand, if voltage is present, then it would make sense to look at the controls. Each test should eliminate some avenues of investigation and help narrow the focus of further tests.

92.11 LOGICAL ELECTRICAL TROUBLESHOOTING PROCEDURE

Electrical problems are by far the most common problems encountered in the field. A general procedure for quickly isolating the cause of an electrical failure is to check the following:

1. Line voltage to the unit
2. Control voltage
3. Voltage to the component that is not functioning
4. Circuit to the component that is not functioning

Line Voltage to the Unit

A significant number of problems are solved at this point. If the system does not operate at all, check to see that it has the correct voltage. If a fuse is blown or a breaker is tripped, look for the reason. Overload devices can open because of reasons other than problems in the unit, like power interruptions. However, anytime a fuse is replaced or a breaker is reset, the operation of the system should be checked carefully. Before operating the system, check all major loads for shorts or grounds and check the main power wire connections. Run the system and check the amp draw on the circuit to the unit and at each major load.

Control Voltage

If a system has no control voltage, the controls cannot operate the system. Control voltage can be out because the part of the system that houses the transformer has no line voltage, or it can be an issue with the transformer that supplies the control voltage. If a low-voltage fuse is blown, or if the transformer needs replacing, be sure to look for shorts and grounds in the low-voltage components. Check the secondary voltage immediately after the fuse or transformer is replaced. If a short is causing the transformer to overheat, the secondary voltage will drop very low, possibly close to 0 V. The transformer cannot take this for more than a few seconds, so it must be recognized immediately. If there are no problems at power-up, monitor the secondary voltage while running the system through all its operating cycles.

Voltage to Nonfunctional Component(s)

If there are nonfunctional components and the system has both line voltage and control voltage, the next step is to determine if these components are receiving voltage. At this point, you are trying to determine if the problem is in the component or in the circuit to the component.

If the component is receiving the correct voltage and is not operating, the problem is most likely that component. However, be careful with single-phase motors. Most single-phase motors have some sort of starting components: run capacitor, start capacitor, or starting relay. Failure of the starting components will keep a single-phase motor from operating. It is more common for these starting components to fail than for the actual motors to fail.

Circuit to Nonfunctional Component(s)

If the nonfunctional component is not receiving voltage, the circuit to the component needs to be checked. Since you now know that the unit has both line voltage and control voltage and that the voltage is not getting to the inoperative component, it is logical to conclude that the problem is somewhere in the circuit to the component. You need to locate the break in the circuit. The most effective method of locating a break in a circuit is a technique called hopscotching.

92.12 HOPSCOTCHING

This technique works by starting at a point where there is correct voltage and moving one lead at a time from point to point in the circuit toward the nonfunctional load until the voltage is lost. Most circuits put all the controls on one side of a load. Use the schematic diagram to identify all the controls in series with the load and the path that voltage takes to get to the load.

Place the leads at a point in the circuit before the controls, where the correct voltage will be read. Follow the schematic and move the meter lead on the control side of the circuit from one control to the next, checking the voltage in and out of the control. The voltage will drop to 0 immediately after the control that is opening the circuit.

Next, you need to determine why the control is open. For example, if the control is a pressure switch, check the system pressures. If the control is a relay, check to see if the relay coil is receiving voltage. If it is not receiving voltage, then you need to investigate the circuit to the coil using the same hopscotching technique. If the control is receiving the correct input and is not closing, it needs to be replaced.

For example, Figure 92-8 shows the low-voltage schematic for an air-conditioning unit with a contactor that is not closing. The contactor coil is shown in blue. Three controls are in series between the thermostat and the contactor

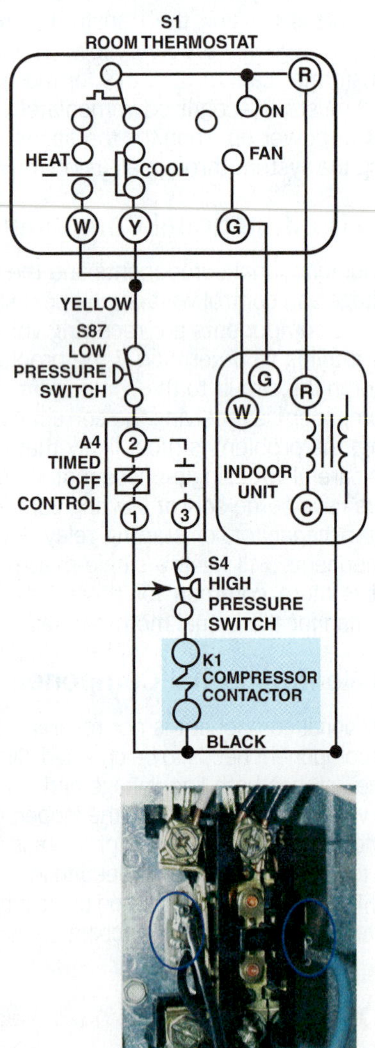

Figure 92-8 Schematic of low-voltage circuits in air-conditioning condensing unit.

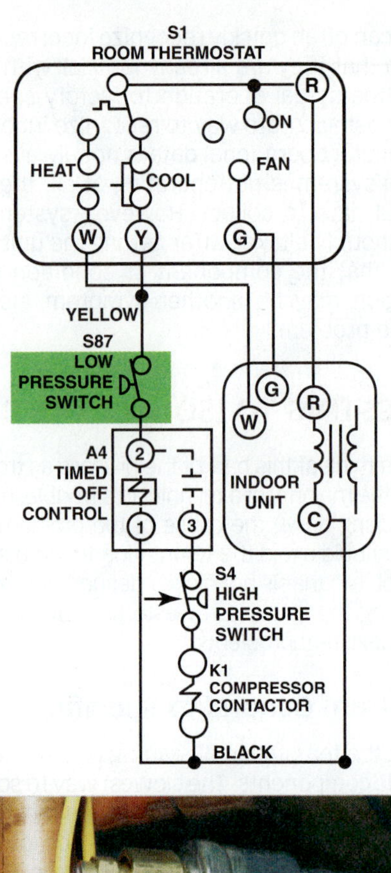

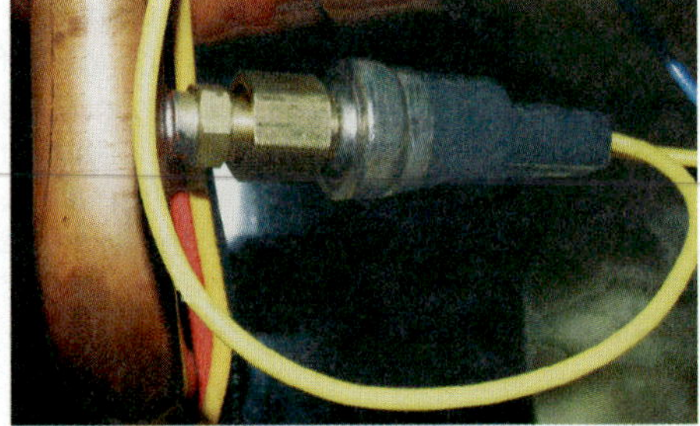

Figure 92-9 The low-pressure switch is in series with the contactor coil.

coil: a low-pressure switch (Figure 92-9), a timed off control (Figure 92-10), and a high-pressure switch (Figure 92-11). Start by checking the voltage to the circuit (Figure 92-12). The meter shows that the unit is receiving a control voltage of 25 V. Place one meter lead on the side of the load that is not fed through controls. In this case, the side of the coil wired with black wires (Figure 92-13). The schematic shows that the first control in the circuit is the low-pressure switch (see Figure 92-9). We know it is receiving voltage. The low-pressure switch feeds the time-delay circuit board. Checking voltage at the input of the time-delay board will check the voltage leaving the low-pressure switch (Figure 92-14). The meter shows that there is 25 V between the low-pressure switch and common, indicating that the low-pressure switch is not breaking the circuit. Next, check the voltage leaving the timer (Figure 92-15). The meter shows that there are 25 V at the timer out terminal, indicating that the timer is not breaking the circuit. The diagram shows that the high-pressure switch is next in the circuit (see Figure 92-11). The leads for the high-pressure switch are wired directly to the timer and the contactor coil. The meter reading drops to near

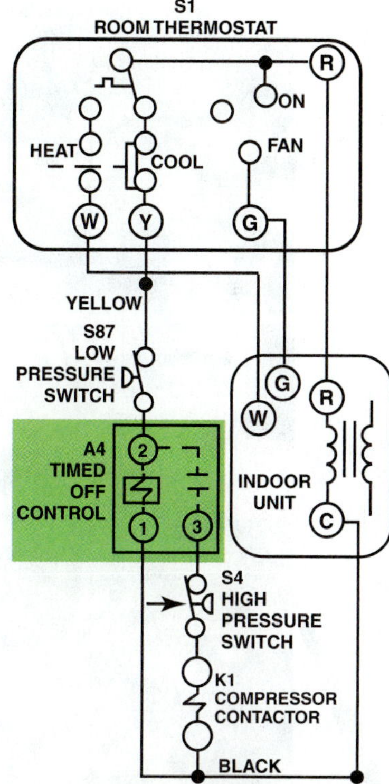

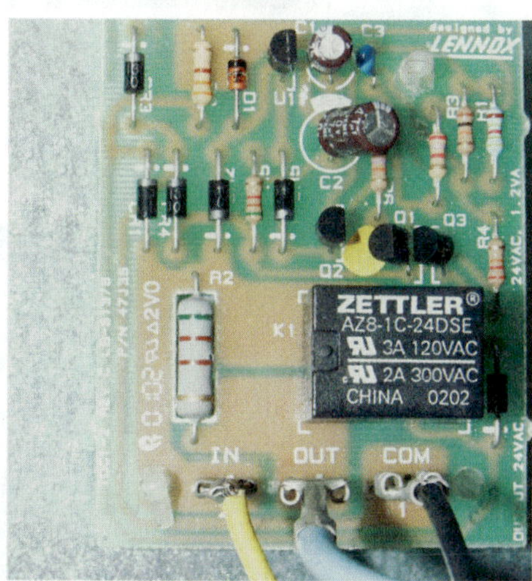

Figure 92-10 The timed off circuit board is in series with the contactor coil.

0 V when checking the voltage at the contactor coil, indicating that the high-pressure switch is open (Figure 92-16). The next step would be to check the system high-side pressure to determine whether the pressure switch should be open. If the system pressure is high, then the technician should look for causes of high discharge pressure.

92.13 REFRIGERATION CYCLE TROUBLESHOOTING

The first thing to check with any refrigeration problem is the flow across the evaporator and condenser. There is no point in taking pressure and temperature readings on a system that does not have the correct airflow or water flow across one of its coils. The majority of refrigeration problems can

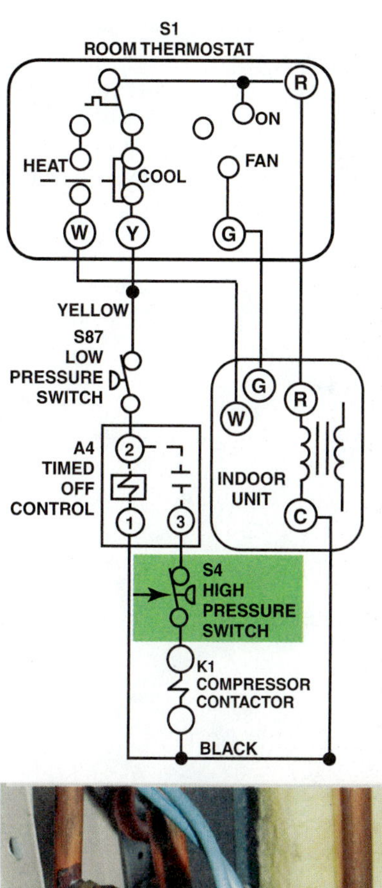

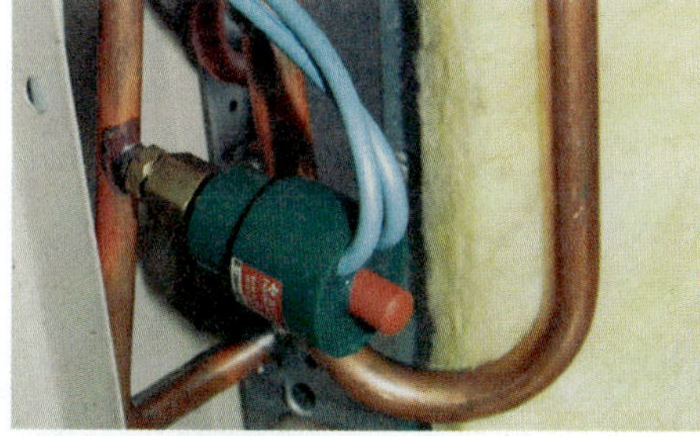

Figure 92-11 The high-pressure switch is in series with the contactor coil.

be solved by establishing a proper flow across both the condenser and evaporator.

Once proper flow is established, system operating data can be collected. As always, the best benchmarks are those provided by the manufacturer. However, technicians should be able to recognize many common refrigeration cycle problems even without the manufacturer's data. Many problems can be identified by looking at these system performance indicators:

- Low-side pressure
- High-side pressures
- Suction-line temperature
- Discharge-line temperature
- Liquid-line temperature
- Air or water temperature rise across the condenser
- Air or water temperature drop across the evaporator

Four of the most important indicators require a comparison of data:

- Suction-line superheat
- Liquid-line subcooling

Figure 92-12 Checking the low-voltage control signal coming into the unit.

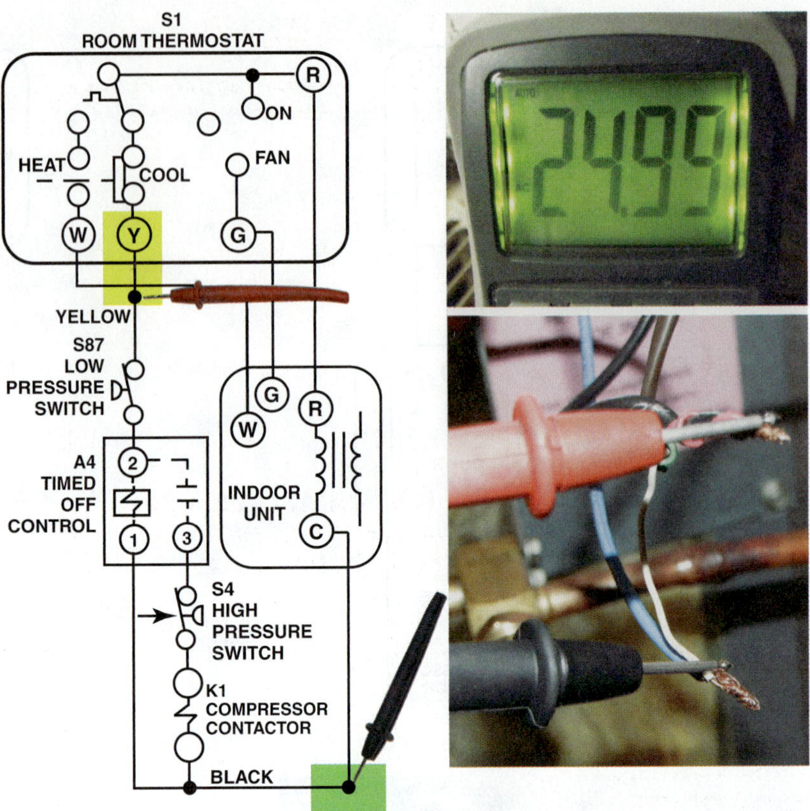

Figure 92-13 The common side of low voltage can be read at the common side of the contactor coil.

- Condenser temperature difference (delta T)
- Evaporator temperature difference (delta T)

Suction-line superheat is the difference between the actual suction-line temperature and the refrigerant saturation temperature. The saturation temperature is obtained from the suction pressure using a refrigerant pressure-temperature chart. Liquid-line subcooling is the difference between the saturation temperature obtained from the discharge pressure using a pressure-temperature chart and the actual liquid-line temperature.

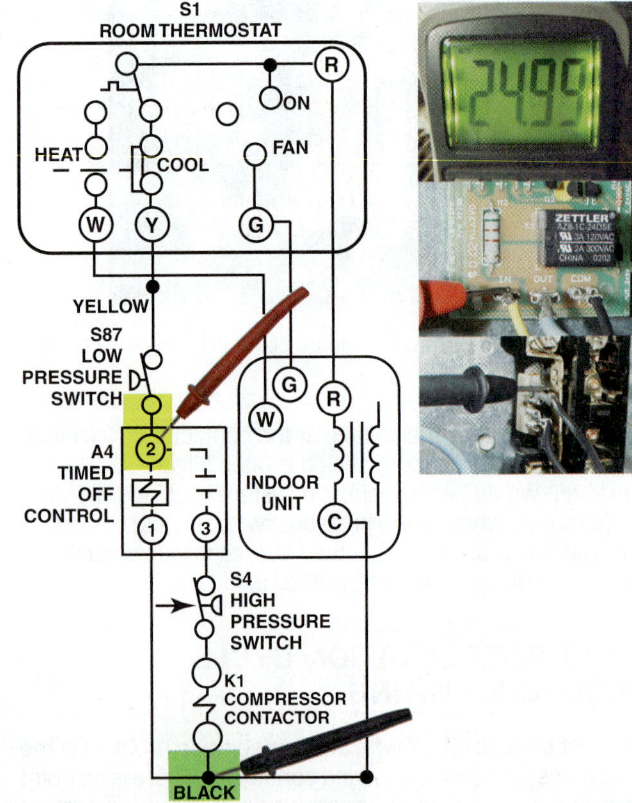

Figure 92-14 Checking the voltage leaving the low-pressure switch and entering the timed off board.

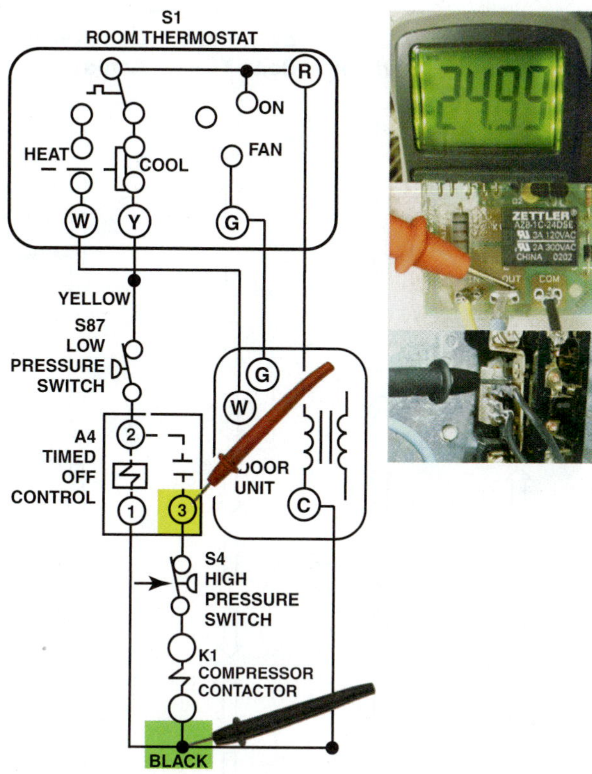

Figure 92-15 Checking the voltage leaving the timed off board and centering the high-pressure switch.

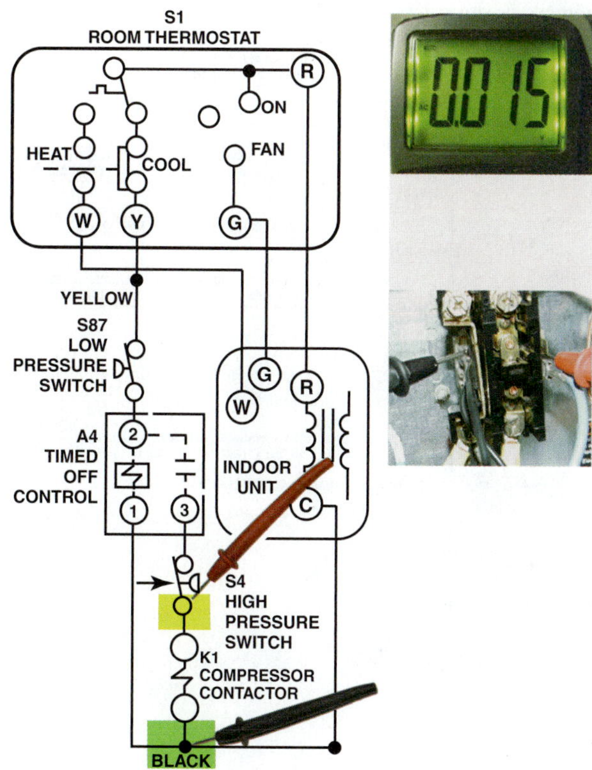

Figure 92-16 Checking the voltage leaving the high-pressure switch and entering the contactor coil.

TECH TIP

Many technical manuals use the word *delta* when talking about the difference in a condition. For example, delta-T is used for temperature difference and delta-P is used for pressure difference. Sometimes the Greek letter delta is used, which looks like a triangle: △. So temperature difference is abbreviated as △T and pressure difference is abbreviated as △P. These are simply shorthand ways of writing temperature difference or pressure difference.

The condenser temperature difference is simply the difference between the temperature of the air or water entering the condenser and the temperature of the air or water leaving the condenser. The evaporator temperature difference is the difference between the temperature of the air or water entering the evaporator and the temperature of the air or water leaving the evaporator.

The chart in Figure 92-17 summarizes the effect common problems will have on the important refrigeration cycle indicators. There are no numbers on the chart, only arrows. The arrows show if a system performance indicator will increase or decrease from normal as a result of a particular problem. The technician still must know what normal is to use the chart.

92.14 RECOMMENDING CORRECTIVE ACTION

The technician's job is to solve problems, so the customer should be offered at least one solution to the problem at hand. Sometimes the corrective action is obvious. If a part has failed, it must be replaced. However, make sure that you solve the problem and not the symptoms. If a part has failed because of abuse or lack of maintenance, be sure to include correcting the cause of the failure as part of the solution.

When possible, it is best to keep customers informed as you proceed. They will have a much easier time accepting a large repair bill if they had an opportunity to choose not to implement the repair. Again, understanding how the system works and being able to logically explain your solution are crucial at this point. If the solution being proposed involves replacing parts, be prepared to explain why each of the parts is necessary. If the solution involves customer education, be patient. It is never helpful to be condescending toward customers.

92.15 USING MANUFACTURERS' TROUBLESHOOTING CHARTS

Most manufacturers supply some form of troubleshooting chart in the service manuals for their equipment. These tend to take three general forms: logical troubleshooting flowcharts, lists of common symptoms and causes, and references to fault codes provided by diagnostic LEDs on the system controls.

Logic Flowcharts

A logic flowchart provides pass/fail tests to help guide the technician in using the process of elimination to locate the cause of a system malfunction (Figure 92-18). These charts

Problem	Discharge Pressure	Subcooling	Condenser Delta T	Suction Pressure	Superheat	Evaporator Delta T	Compressor Amps
Low flow of evaporator air or water	⇩	⇧	⇩	⇩	⇩	⇧	⇩
Dirty evaporator	⇩	⇧	⇩	⇩	⇩	⇧	⇩
liquid line restriction	⇩	⇧	⇩	⇩	⇧	⇩	⇩
Underfeeding expansion device	⇩	⇧	⇩	⇩	⇧	⇩	⇩
Undercharge	⇩	⇩	⇩	⇩	⇧	⇩	⇩
Low-capacity compressor	⇩	⇧	⇩	⇧	⇧	⇩	⇩
Low flow of condenser air or water	⇧	⇩	⇧	⇧	⇩	⇩	⇧
Dirty air-cooled condenser	⇧	⇩	⇧	⇧	⇩	⇩	⇧
Scaled water cooled condenser	⇧	⇩	⇩	⇧	⇩	⇩	⇧
Overcharge	⇧	⇧	⇩	⇧	⇩	⇩	⇧
Overfeeding expansion device	⇧	⇩	⇩	⇧	⇩	⇩	⇧

Figure 92-17 Refrigeration cycle troubleshooting chart.

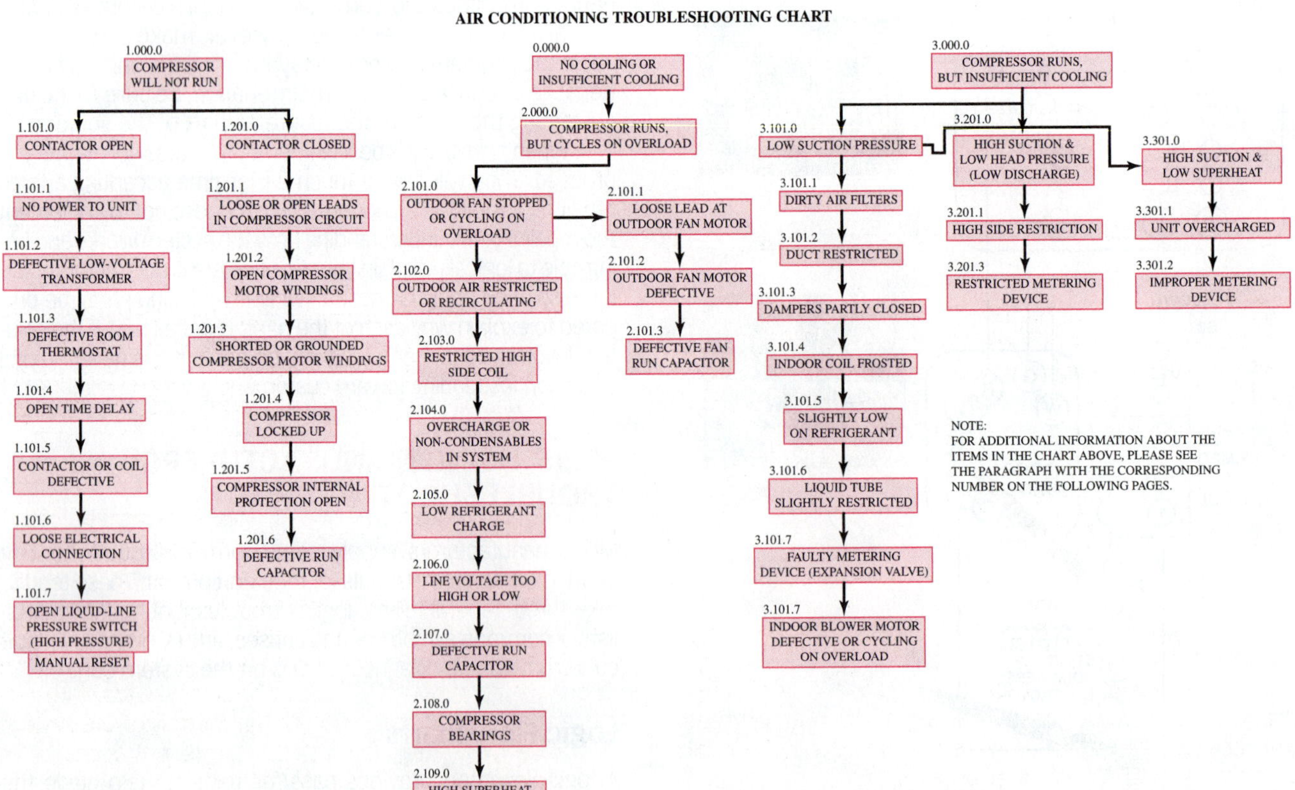

Figure 92-18 Example of a troubleshooting logic flowchart.

WATER SOURCE HEAT PUMP TROUBLESHOOTING CHART		
PROBLEM	**CAUSE**	**CORRECTION**
Nothing operates in any thermostat setting	No power to unit	Check circuit breaker or fuses
	Defective control transformer	Replace defective transformer
	Loose connection	Repair connection
	Defective thermostat	Replace thermostat
Unit short cycles	Thermostat location	Move thermostat
Blower operates but compressor does not operate	Defective compressor overload	Replace (if external)
	Defective compressor contactor	Replace
	Supply voltage too low	Check power wire size
	Defective compressor capacitor	Replace capacitor
	Defective windings	Replace compressor
	Limit switches open	Check cause and replace
Insufficient system capacity	Dirty air filter	Replace air filter
	Blower RPM too low	Increase blower RPM
	Leaky duct work	Repair and/or seal ductwork
	Low refrigerant charge	Locate leak, recover charge, repair leak, evacuate and recharge system
	Restricted expansion valve	Replace expansion valve
	Defective reversing valve	Replace reversing valve
	Thermostat located in supply air	Move thermostat
	Undersized system	Perform load calculation
	Insufficient water flow	Increase water flow
	Scaled or fouled water coil	Clean or replace water coil
	Water temperature outside of manufacturer's specified operating limits	Correct water temperature
High pressure switch open or high head pressure	Inadequate water flow	Increase water flow
	Water too hot	Decrease water temperature
	Dirty or fouled water coil	Clean or replace water coil
	Inadequate airflow	Increase airflow
	Dirty air filter	Replace air filter
	Overcharged with refrigerant	Recover charge
	Defective pressure switch	Replace pressure switch
	Non-condensables in system	Recover refrigerant, evacuate and recharge
Low pressure switch open or low suction pressure	Inadequate airflow	Increase airflow
	Dirty air filter	Replace air filter
	Inadequate water flow	Increase water flow
	Water too cold	Increase water temperature
	Undercharged with refrigerant	Add refrigerant charge
	Defective pressure switch	Replace pressure switch
	Restricted expansion valve	Replace expansion valve

Note: Items in blue apply only to cooling cycle, items in red apply only to heating cycle.

Figure 92-19 Example of a troubleshooting systems list.

can be especially useful for determining if proprietary control boards are functioning properly. The danger of these troubleshooting aids is that answering one question incorrectly can send you off in an entirely incorrect direction. Technicians who have too little knowledge of how a system actually works may not understand why tests are being performed or even exactly how to perform them. Incorrect readings cause misinterpretation of the chart.

Symptom Lists

Many manufacturers provide a list of common symptoms and their most common causes (Figure 92-19). These troubleshooting aids can help speed up diagnosis of common problems. These are generally easier for people to follow and require less testing. However, it is not really possible to make a complete list of all possible symptoms, or even a complete list of all the possible causes of any particular symptom. Another issue is the temptation to simply go down the list of causes for a particular symptom. If more than one component can cause the same symptom, multiple components may be replaced by someone who is just running the list and does not really know how to check each component.

Fault Code Reference List

Most units today employ some type of electronic control. Many of these have built-in diagnostics. Typically, these diagnostics will tell the technician what the control is "thinking." A list of fault conditions is typically located in the service manual (Figure 92-20) and also frequently on the unit (Figure 92-21). The explanation can be a short description, such as "open pressure switch," or it can be a more detailed explanation of the problem including a fault history of the system.

Technicians need to understand the limitations of on board diagnostics. On-board diagnostics can only report issues for which there are inputs. All the board knows is that an input is correct or incorrect. The accompanying chart can

Figure 92-20 Fault code from a service manual.

FURNACE FAULT CODES

Red LED	Green LED	Indication
Simultaneous Slow Flash		Power on Normal operation. No call for heat. Also signaled during cooling and continuous fan.
Simultaneous Fast Flash		Normal operation with a call for heat.
Alternating Slow Flash		Flame not sensed after ignition attempt.
Alternating Fast Flash		Broken igniter or igniter circuit.
Fast Flash	Slow Flash	Main power polarity reversed or improper main ground.
Slow Flash	Fast Flash	Low flame signal.
Slow Flash	On	Primary or auxiliary limit switch open.
Off	Slow Flash	Draft pressure switch fault.
Slow Flash	Off	Flame sensed without gas valve energized.
On	Slow Flash	Roll out switch open.
On	On	Control board wired incorrectly or defective.
Off	On	Control board wired incorrectly or defective.
On	Off	Control board wired incorrectly or defective.

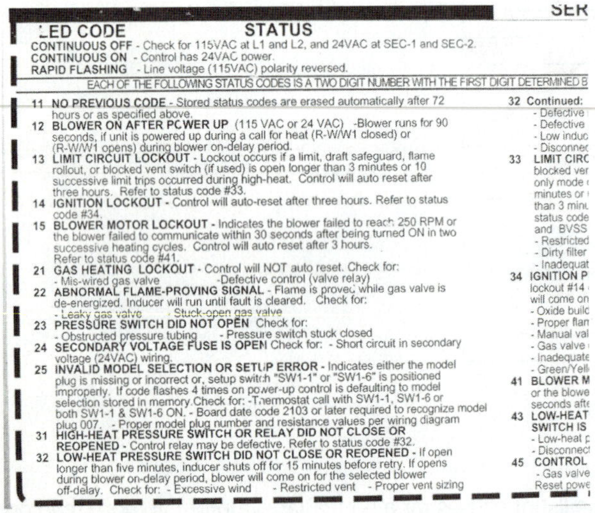

LED CODE STATUS

CONTINUOUS OFF - Check for 115VAC at L1 and L2, and 24VAC at SEC-1 and SEC-2.
CONTINUOUS ON - Control has 24VAC power.
RAPID FLASHING - Line voltage (115VAC) polarity reversed.

EACH OF THE FOLLOWING STATUS CODES IS A TWO DIGIT NUMBER WITH THE FIRST DIGIT DETERMINED B

11 NO PREVIOUS CODE - Stored status codes are erased automatically after 72 hours or as specified above.
12 BLOWER ON AFTER POWER UP (115 VAC or 24 VAC) -Blower runs for 90 seconds, if unit is powered up during a call for heat (R-W/W1 closed) or (R-W/W1 opens) during blower on-delay period.
13 LIMIT CIRCUIT LOCKOUT - Lockout occurs if a limit, draft safeguard, flame rollout, or blocked vent switch (if used) is open longer than 3 minutes or 10 successive limit trips occurred during high-heat. Control will auto reset after three hours. Refer to status code #33.
14 IGNITION LOCKOUT - Control will auto-reset after three hours. Refer to status code #34.
15 BLOWER MOTOR LOCKOUT - Indicates the blower failed to reach 250 RPM or the blower failed to communicate within 30 seconds after being turned ON in two successive heating cycles. Control will auto reset after 3 hours. Refer to status code #41.
21 GAS HEATING LOCKOUT - Control will NOT auto reset. Check for: - Mis-wired gas valve -Defective control (valve relay)
22 ABNORMAL FLAME-PROVING SIGNAL - Flame is proved while gas valve is de-energized. Inducer will run until fault is cleared. Check for: - Leaky gas valve - Stuck-open gas valve
23 PRESSURE SWITCH DID NOT OPEN Check for: - Obstructed pressure tubing - Pressure switch stuck closed
24 SECONDARY VOLTAGE FUSE IS OPEN Check for: - Short circuit in secondary voltage (24VAC) wiring.
25 INVALID MODEL SELECTION OR SETUP ERROR - Indicates either the model plug is missing or incorrect or, setup switch "SW1-1" or "SW1-6" is positioned improperly. If code flashes 4 times on power-up control is defaulting to model selection stored in memory.Check for: -Thermostat call with SW1-1, SW1-6 or both SW1-1 & SW1-6 ON. - Board date code 2103 or later required to recognize model plug 007. - Proper model plug number and resistance values per wiring diagram
31 HIGH-HEAT PRESSURE SWITCH OR RELAY DID NOT CLOSE OR REOPENED - Control relay may be defective. Refer to status code #32.
32 LOW-HEAT PRESSURE SWITCH DID NOT CLOSE OR REOPENED - If open longer than five minutes, inducer shuts off for 15 minutes before retry. If opens during blower on-delay period, blower will come on for the selected blower off-delay. Check for: - Excessive wind - Restricted vent - Proper vent sizing

SER
32 Continued: - Defective - Defective - Low induc - Disconnec
33 LIMIT CIRC blocked ver only mode minutes or than 3 minu status code and BVSS - Restricted - Dirty filter - Inadequat
34 IGNITION P lockout #14 will come on - Oxide build - Proper flan - Manual val - Gas valve - Inadequate - Green/Yell
41 BLOWER M or the blowe seconds afte
43 LOW-HEAT SWITCH IS - Low-heat p - Disconnect
45 CONTROL - Gas valve Reset powe

Figure 92-21 Fault code located on an equipment panel.

suggest what the cause is, but it does not really know. Verifying the cause is still the service technician's job. Although diagnostics greatly simplify locating system malfunctions, they do not replace understanding how a system operates or knowing how to diagnose faulty components.

UNIT 92—SUMMARY

Troubleshooting is the process of identifying the problems, determining their cause, and correcting the cause of the problem. Technicians should view units as systems, keeping in mind the interrelationship between the electrical, refrigeration, and airflow system components. Troubleshooting can be broken down into the following steps:

- Understanding system operational sequence
- Preliminary system inspection
- Collecting operational data
- Recognizing what is operating incorrectly
- Testing to isolate the cause
- Recommending corrective action

Technicians must understand exactly how equipment operates so they can analyze the cause of improper operation. Understanding the operating sequence of the most common types of systems is important. Sources of information on system operation include the manufacturer's installation and operation and manuals, unit wiring diagrams, unit charging charts, and general textbooks. Preliminary system checks should include checking the thermostat setting, air filter, airflow and/or water flow, line voltage, and control voltage. Operational data can come from the homeowner, general observation, diagnostic boards, and system performance tests. Troubleshooting logic flowcharts, common symptom lists, and diagnostic codes can all be helpful in identifying the cause of system malfunction. The hopscotching technique is useful for identifying the problem in a malfunctioning electrical circuit. When troubleshooting the refrigeration cycle in a system, key performance indicators include the following:

- Low-side pressure
- High-side pressures
- Suction-line temperature
- Discharge-line temperature
- Liquid-line temperature

- Air or water temperature rise across the condenser
- Air or water temperature drop across the evaporator
- Suction-line superheat
- Liquid-line subcooling
- Condenser delta T
- Evaporator delta T

The technician should be able to explain to the customer exactly what is causing the problem and recommend a solution.

WORK ORDERS

Service tickets 9201–9204 all start with the same symptoms, but they each have a different root cause. The beginning of each call starts out as follows:

A customer calls because his air-conditioning system is not cooling properly. The system is operating when the technician arrives, and the technician notices that the condensing fan motor outside is running but that the compressor is not running. The compressor line-voltage wires on the contactor have the correct voltage. The compressor shell is very hot. The technician turns off the power to the unit and measures the resistance of the compressor terminals. There is no continuity between common and either run or start, but there is continuity between start and run. This indicates that the internal overload is open.

Each service ticket picks up from this point.

Service Ticket 9201

Customer Complaint: Poor Cooling

Equipment Type: R-410a Residential Split-System Air Conditioner

... The condenser coil appears clean. The technician checks the compressor starting components and finds that the compressor run capacitor is open. The run capacitor is replaced. The compressor starts and runs correctly after it cools down and the internal overload resets. The technician checks the system performance against the manufacturer's specifications on the service panel to verify that it is operating correctly.

Service Ticket 9202

Customer Complaint: Poor Cooling

Equipment Type: R-410a Residential Split-System Air Conditioner

... The condenser coil appears clean. The technician checks the compressor starting components and finds that they are all good. Next, the technician checks the system pressure and finds the equalized pressure is 100 psig. The technician concludes that the system is undercharged, leading the compressor to overheat. A leak inspection reveals a leak at the mechanical fitting on the suction line. The technician tightens this fitting, stopping the leak. The unit is charged according to the manufacturer's performance chart after the compressor cools and is able to operate.

Service Ticket 9203

Customer Complaint: Poor Cooling

Equipment Type: R-410a Residential Split-System Air Conditioner

... The condenser coil appears to be clogged with debris. The technician concludes that poor condenser airflow has caused the compressor to overheat, turns off the power to the unit, and cleans the condenser coil. The compressor starts and runs correctly after it cools down and the internal overload resets. The technician checks the system performance against the manufacturer's specifications on the service panel to verify that it is operating correctly.

Service Ticket 9204

Customer Complaint: Poor Cooling

Equipment Type: R-410a Residential Split-System Air Conditioner

... The condenser coil appears clean. The technician checks the compressor starting components and finds that they are all good. Next, the technician checks the system pressure and finds the equalized pressure is 200 psig. The technician decides that the system will have to operate to make any further diagnosis. The compressor starts and runs after it cools down and the internal overload resets. The high-side pressure climbs quickly to close to the manufacturer's specification. The low side, however, drops to around 50 psig, much lower than the manufacturer's specification. The superheat and subcooling are both high, and the air temperature exchange across the condenser is lower than normal. The technician begins looking for a temperature drop in the liquid line that would indicate a restriction. None is found. The expansion device is an expansion valve. The technician removes the thermal bulb and holds it, looking for a change in the low-side pressure or superheat. There is no change. The technician concludes that the thermostatic expansion valve is underfeeding, causing the compressor to overheat. The refrigerant is recovered, the expansion valve changed, and the system is evacuated and recharged according to the manufacturer's specifications. The technician operates the system and compares the system performance to the manufacturer's specifications to verify that the system is performing correctly.

Service Ticket 9205

Customer Complaint: No Cooling

Equipment Type: Residential Packaged Heat Pump

... A customer complains that he has no cooling. The technician arrives, makes sure the thermostat is set to Cool, and checks the line voltage to the unit. The line voltage is 0 V. The technician finds a blown fuse in the outdoor disconnect switch. Next, all major loads are checked for shorts and grounds. None are found. The technician notes that the fuses in the disconnect box are one-time fuses and replaces both fuses with time-delay fuses. The unit comes on and runs after the fuses are replaced. The technician checks the system amp draw and the amp draw of each component and finds them at or below the data plate specifications. Next, the system performance is checked against the manufacturer's specifications to verify correct operation.

Service Ticket 9206

Customer Complaint: No Cooling

Equipment Type: Residential Packaged Heat Pump

... A customer complains that they have no cooling. The technician arrives, makes sure the thermostat is set to cool, and checks the line voltage to the unit. The line voltage is 240 V. The technician checks the control voltage and finds 0 V. The technician finds 24 V fuse to the R terminal is blown. The fuse is replaced and it immediately blows again. The technician starts checking the low voltage controls for shorts and finds that the compressor contactor coil has a resistance reading of ½ ohm. The contactor is replaced and the system is turned on again. This time the system starts and operates. Next, the system performance is checked against the manufacturer's specifications to verify correct operation.

Service Ticket 9207

Customer Complaint: Noisy Unit

Equipment Type: R-22 Residential Split System Air Conditioner

... A customer calls complaining about his condensing unit making a lot of noise. The customer states that all his neighbors have the same systems, and they are all noisy. However, he cannot stand the noise any longer and is ready to change out the unit if necessary. The customer also asks for the "extra cooling" option, explaining that he paid the original contractor $100 to add extra refrigerant to improve the system's performance. The technician explains that adding extra charge does not help the unit cool better. If anything, it makes it cool worse. The technician checks the system performance characteristics and finds that both pressures are high, the superheat is essentially 0°F, and the subcooling is 30°F. The technician suspects that the system charge was not done correctly at installation and elects to start from scratch. So all the charge is recovered, the system evacuated, and a full charge weighed in according to the manufacturer's data plate information, including line allowances. The system starts and runs much more quietly, and the pressures fall to within the manufacturer's specifications.

UNIT 92—REVIEW QUESTIONS

1. List some common questions customers ask service technicians.
2. Explain what is meant by a systems approach to troubleshooting.
3. Give an example of an airflow problem that can look like a refrigerant charge problem if a service technician is only looking at system pressures.
4. List the stages of troubleshooting.
5. What checks are typically performed during the preliminary inspection?
6. What should a service technician do if there is a blown fuse in the unit disconnect switch during the preliminary inspection?
7. What should a service technician do if there is a blown fuse in series with the low-voltage side of the transformer?
8. Outline a method of isolating an electrical fault using the process of elimination.
9. Describe the process of hopscotching.
10. Using the logic flowchart in Figure 92-18, what is the first condition listed to check for a compressor that will not run when the contactor is closed?
11. Using the symptoms list in Figure 92-19, what possible causes are listed for a unit that experiences high-pressure-switch tripping only during the cooling cycle?
12. Using the chart in Figure 92-20, what conditions cause both lights to display an alternating fast flash?
13. Using the air-source heat pump sequence chart in Figure 92-2, list the devices in order that control the outdoor fan.
14. Using the chart in Figure 92-17, what problems can cause both a high superheat and a high subcooling?
15. Using the induced-draft sequence chart in Figure 92-4, what should be checked if the induced-draft fan comes on but the igniter is never energized?
16. List some resources the air-conditioning technician can use to learn about the system operational sequence.
17. List some sources the air-conditioning technician can use to get system operational data.
18. Using Figure 92-7, what should the approximate suction pressure be for a system operating in heating with an entering water temperature of 50°F and a water flow of 5 GPM?
19. Why are diagnostic codes not always able to identify the system problem?
20. Why should service technicians be familiar with company pricing policies?

Glossary

A Spanish version of the Glossary is available in MyHVACLab.

A Coil Evaporator consisting of two coils joined together forming the shape of the letter A.

Absolute Humidity The actual amount of moisture in a given volume of air.

Absolute Pressure Gauge pressure plus atmospheric pressure.

Absolute Temperature Temperature measured from absolute zero.

Absolute Zero Temperature at which all molecular motion ceases (–460°F and –273°C).

Absorbent Substance that has the ability to take up or absorb another substance.

Absorber A device containing liquid for absorbing refrigerant vapor or other vapors.

Absorption Diffusion of one substance into the inner composition of another substance.

Absorption Chiller Device used to produce chill water using the absorption cycle.

Absorption Cycle Refrigeration cycle based upon the absorption of the refrigerant by a chemical substance rather than by mechanical compression.

Absorption Refrigerator A refrigerator that utilizes the principles of the absorption cycle.

Accessible Hermetic A hermetically sealed unit that can be opened for repair and maintenance, also referred to as semi-hermetic.

Accumulator A storage vessel located in the suction line before the compressor. Used to limit liquid refrigerant return to the compressor and store excess refrigerant in the heating mode.

Acetylene Gas commonly used with oxygen for brazing and soldering.

ACR Tubing Air conditioning and refrigeration tubing sealed on each end to keep the tubing clean and dry.

Activated Alumina An aluminum oxide desiccant used in refrigeration system driers to absorb moisture.

Activated Carbon Specially processed carbon commonly used to clean air.

Active Refrigerant Recovery Device A self contained recovery unit that is not an integral part of the refrigeration system.

Active Solar System A system designed to utilize solar energy.

Actuator A device used to convert fluid power, thermal energy, and electrical energy into mechanical motion.

Adiabatic Compression Compressing refrigerant gas without removing or adding heat.

Adsorbent Substance that has the property to hold molecules of fluids without causing a chemical or physical change.

Air The fluid surrounding us composed mainly of oxygen and nitrogen.

Air-Acetylene Brazing torch that draws in the surrounding air rather than requiring a separate oxygen supply.

Air Binding A condition in which a bubble or other pocket of air is present in a pipeline that prevents the desired flow in the pipeline.

Air Break An inverted opening placed in a chimney to prevent a downdraft from affecting the furnace flame.

Air Changes The amount of air leakage through a building in terms of the number of building volumes exchanged.

Air Cleaner A filtration system used to remove airborne impurities.

Air Coil A refrigeration heat exchanger that has air passing across its outer surface.

Air Conditioner Device used to control temperature, humidity, cleanliness, and movement of air in conditioned space.

Air Conditioning The process to control temperature, humidity, cleanliness, and movement of air in conditioned space.

Air Conditioning Heating and Refrigeration Institute (AHRI) A national trade association representing manufacturers of over 90% of U.S.-produced central air conditioning, gas appliances, and commercial refrigeration equipment.

Air Cooled Condenser A heat exchanger that uses air passing across it to condense refrigerant from vapor to liquid.

Air Core Solenoid A solenoid that has a hollow center.

Air Curtain Forced air used to shield the food product in an uncovered display case.

Air Cushion Tank (see Expansion Tank)

Air Defrosting A defrost method of shutting the unit off while allowing the warm air to circulate.

Air Diffuser An air register designed to distribute air in a specific arrangement.

Air Friction Chart A chart used for sizing round ducts.

Air Gap The air space between the rotating and stationary components of a motor or generator.

Air Handler A unit housing a blower and heating/cooling coils that is used for delivering air to a space.

Air Heat Exchanger A heat exchanger that allows for air to be drawn across its exterior surface.

Air Loop The ducted air side on a heat pump system.

Air Pressure Switch Measures air pressure drop across an outdoor heat pump coil to determine frost build-up.

Air Sensor Installed in air handlers to measure conditions such as temperature, pressure, velocity, and humidity.

Air Shutter An adjustable shutter on the primary air openings of a burner that is used to control the amount of combustion air.

Air Source Heat Pump Heat pump that transfers heat from outdoor air to an indoor air circulation system.

Air Standard The ASHRAE standard designation for dry air at 70°F and atmospheric pressure.

Air-to-Air Heat Pump (see Air Source Heat Pump)

Air Vent A valve installed at the high points in a hot water system to remove air from the system.

Air Washer Filtration of air using water or other liquid.

Aldehyde A class of compounds that can be produced during incomplete combustion of a fuel gas.

Algae A light to dark green plant growth that can develop in water systems.

Alkylbenzene A synthetic oil compatible with CFC and HCFC refrigerants.

Allen Head Screw Screw with recessed head designed to be turned with a hex shaped wrench.

Alternating Current (AC) Current that reverses polarity or direction with regular frequency. It can be single phase, two phase, three phase, or polyphase.

Alternative Refrigerant A replacement refrigerant that is different from the existing system refrigerant.

Alternator A machine that converts mechanical energy into alternating current.

Altitude The measured distance above sea level.

Altitude Adjustment Thermostat adjustment to account for changes in elevation dependent on geographical location.

Ambient Compensator An electronic device that provides enough heat to cycle the refrigeration unit when ambient temperatures are low.

Ambient Temperature Temperature of the air that surrounds an object.

American National Standards Institute (ANSI) Institute that determines refrigeration standards along with ASHRAE.

American Society of Heating, Refrigeration, and Air Conditioning Engineers (ASHRAE) Its membership is composed of thousands of professional engineers and technicians from all phases of the HVACR industry. ASHRAE also creates equipment standards for the industry.

American Society for Testing and Materials Engineers (ASTME) Society that, in addition to materials testing, determines wind and rain infiltration standards.

American Standard Pipe Thread Identifies the standard pipe thread size encountered in the field for installation purposes.

American Wire Gauge (AWG) A system of numbers that designate cross-sectional area of wire. As the diameter gets smaller, the number gets larger, for example, AWG #14 = 0.0641 in, AWG #12 = 0.0808 in.

Ammeter An electric meter used to measure current.

Ammonia A widely used refrigerant for large commercial systems due to its high latent heat value.

Amperage The measured flow of electrical current.

Ampere Unit of electric current equivalent to the flow of one coulomb per second.

Ampere Turn (AT) (NI) Unit of magnetizing force produced by a current flow of one ampere through one turn of wire in a coil.

Amplitude The maximum instantaneous value of alternating current or voltage. It can be in either a positive or negative direction.

Analog Electronic Device A device that delivers a modulated signal.

Analog Signal A modulated signal as compared to a digital signal, which is binary.

Analog VOM A volt-ohm meter using a needle measurement scale rather than a digital display.

Anemometer Instrument for measuring the flow rate of air.

Aneroid Barometer An instrument used for measuring barometric pressure.

Angle Valve A valve that is constructed with a 90° bend.

Angle of Lag or Lead Phase angle difference between two sinusoidal waveforms having the same frequency.

Annealing Process of heat treating metal to obtain desired properties of softness and ductility.

Annual Fuel Utilization Efficiency (AFUE) A furnace efficiency comparison for consumers.

Anode The positive electrode in an electrolytic cell.

Anticipator A heater used to adjust thermostat operation to produce a closer temperature differential than the mechanical capability of the control.

Antifreeze Solution A mixture of glycol and water with a freezing point below 32°F.

Approach Term used for cooling water towers to determine effectiveness. A measure of the cooling air ambient wet bulb compared to the temperature of the cooling water.

Approach Temperature Chilled water leaving temperature compared to refrigerant temperature.

Arcing Sparking discharge between two electrodes or contacts.

Armature The moving or rotating component of a motor, generator, relay, or other electromagnetic device.

ASME American Society of Mechanical Engineers.

Aspect Ratio Ratio of length to width of a rectangular duct.

Aspirating Psychrometer A device that draws in a sample of air to measure for humidity.

Aspiration To draw a fluid by creating a suction.

Atmospheric Dust Spot Efficiency A measure of the ability for a filter to remove atmospheric dust from a test sample of air.

Atmospheric Pressure The envelope of atmospheric gases creating a pressure equal to approximately 14.7 psi at a temperature of 70°F at sea level.

Atom Smallest particle of an element.

Atomic Weight The number of protons in an atom of a material.

Atomize Process of changing a liquid to a fine spray.

Atomization (see Atomize)

Atomizing Humidifier A humidifier that supplies finely atomized water into the airstream.

Attenuate Decrease or lessen in intensity.

Auger Rotating device resembling a huge corkscrew used in ice machines.

Authorized Dealer A dealer authorized to install heat pumps by the manufacturer.

Automatic Changeover Thermostat A thermostat that automatically changes modes dependent on the room temperature.

Automatic Defrost System of removing ice and frost from evaporators automatically.

Automatic Expansion Valve (AEV) Pressure controlled valve used as a metering device.

Automatic Pumpdown System A system designed to pump down refrigerant to the receiver prior to shutting down.

Autotransformer A transformer in which both primary and secondary coils have turns in common. Stepup or stepdown of voltage is accomplished by taps on a common winding.

Auxiliary Drain Pan Located beneath the evaporator coil to collect and drain away water that condenses from the air.

Azeotrope Two or more refrigerants mixed together by the refrigerant manufacturer that have the combined properties of a single refrigerant.

B Vent Double walled vent pipe used for gas furnace applications.

BW Vent Double walled vent pipe used with wall type gas furnaces.

Back Electromotive Force (BEMF) Voltage generating effect of an electrical motor's rotor.

Back Pressure Pressure in the low side of the refrigeration system. Also called low side pressure or suction pressure.

Back Seat The position of a service valve that is all the way counterclockwise or fully open.

Bacteria Microscopic organisms found almost everywhere; while some types can be useful, other types can cause disease.

Baffle Similar to a damper used to regulate the flow of a fluid.

Balance Fitting A pipe fitting or valve designed so that its resistance to flow may be varied. This type of fitting is used to obtain the desired flow rate through parallel circuits.

Balance Point The outdoor temperature at which the heating capacity of a heat pump in a particular installation is equal to the heat loss of the conditioned area. Below the balance point, supplementary heat is needed to maintain indoor comfort.

Balance Point, Initial The outdoor ambient temperature at which the heating capacity of the heat pump only balances the heat loss of the conditioned area.

Balance Point, Second The outdoor temperature at which the heating capacity of the heat pump plus the auxiliary heat capacity balances the heat loss of the conditioned area.

Balanced-Port TEV A thermostatic expansion valve that compensates for wide variations in high to low side pressure and evaporator loads.

Ball Check Valve A check valve that uses a ball against a seat as a shut-off means.

Ball Valve A valve in which modulation or shut-off is accomplished by a one quarter turn of a ball that has an opening through it.

Bar One bar is 100 kilopascals (kPa) with 101.3 kPa equal to standard atmospheric pressure.

Barometer Instrument for measuring atmospheric pressure.

Base Semiconductor terminal.

Baseboard A terminal unit resembling the base trim of a house.

Bath A container holding a liquid where a component can be washed or heated.

Battery Two or more primary or secondary electrically interconnected cells.

Baudelot Cooler Heat exchanger in which water flows by gravity over the outside of the tubes or plates.

Bearing Ball bearings are used in pumps and motors to hold the rotating shaft in alignment with the casing.

Bellows Corrugated cylindrical container that moves as pressures change.

Bellows Seal A crankshaft seal made tight by a bellows arrangement.

Belly-Band-Mount Motor An electric motor held in place with a metal strap.

Belt A flexible material used to connect rotating shafts through a pulley arrangement, such as a motor driving a fan.

Bending Spring Coil spring that is used to keep a tube from collapsing while being bent.

Bio-aerosol An airborne microbial contaminant, such as a virus, bacteria, fungus, or algae.

Bimetal Strip Two dissimilar metals with unequal expansion rates fused together.

Blast Freezer A unit that deep freezes the product faster than conventional freezers.

Bleeding Slowly draining pressure from a system.

Bleed Off The continuous or intermittent wasting of a small fraction of the water in the basin of a cooling tower to prevent the buildup of scale-forming minerals.

Bleed Valve Valve that permits a minimum fluid flow when valve is closed.

Blend A mixture of two or more refrigerants.

Blocked Suction A method for unloading a reciprocating compressor.

Blowdown Draining a controlled amount of water from a cooling water tower to reduce the level of dissolved solids.

Blown Building insulation can be applied to a structure using this method. A melted fuse can be considered blown.

Blower A centrifugal fan.

Boiler A furnace that generates either hot water or steam.

Boiler, High Pressure A boiler that operates at a pressure higher than 15 psig.

Boiler, Low Pressure A boiler that operates at a pressure below 15 psig.

Boiler Horsepower The equivalent evaporation of 34.5 lb of water per hour at 212°F. This is equal to a heat output of $970.3 \times 34.5 = 33{,}475$ Btu/hr.

Boiling Point Temperature at a given pressure at which a fluid changes from a liquid to a gas.

Boiling Temperature Temperature at which a fluid changes from a liquid to a gas.

Bonnet The sheet metal chamber in a furnace where heat collects.

Booster Pump A second pump used to raise the pressure in the system higher than created by the first pump.

Boot Connection between a floor register and a branch duct.

Bore Inside diameter of a cylindrical hole.

Bourdon Tube Thin walled tube of circular shape that tends to straighten as pressure inside is increased.

Boyle's Law Law stating the pressure-volume relationship of gases.

Brazing Soldering with a filler material whose melting temperature is higher than 840°F.

Break Electrical discontinuity in the circuit generally resulting from the operation of a switch or circuit breaker.

Breaker (see Circuit Breaker)

Breaker Strip Strip of plastic used to cover the joint between the outside case and the inside liner of a refrigerator.

Breeching Transition piece located between furnace and stack.

Bubble Point The formation of bubbles in a saturated liquid refrigerant zoetrope will begin at this temperature.

Building Related Illness (BRI) Health disorders linked to the environment of modern airtight buildings.

Built Up System Unit is not packaged but instead the individual components (condenser, compressor, etc.) are installed together at the job site.

Brine Water saturated with a chemical such as salt.

British Thermal Unit (Btu) The amount of heat required to raise the temperature of 1 lb of water (about one pint) by 1°F.

Bulb Used as a sensing element for valves and controls.

Bull Head The installation of a pipe tee in such a way that water enters (or leaves) the tee at both ends of the run (the straight through section of the tee) and leaves (or enters) through the side connection only.

Bunsen Burner A gas burner in which combustion air is premixed with the gas supply within the burner body before the gas burns at the burner port.

Burner A device for the final conveyance of gas, or a mixture of gas and air, to the combustion zone.

Burnout Accidental passage of high voltage through an electrical circuit or device that causes damage.

Burr Raised edges left behind from cutting tubing or a pipe. This burr must be removed before installation.

Butane A highly flammable, colorless, odorless, easily liquefied gas.

Butterfly Valve A valve with a handle that can be positioned 90° to be either parallel (open) or perpendicular (closed) to the fluid flow.

Bypass A connection that diverts the fluid flow around a valve or component.

Cadmium Cell A device whose resistance changes according to the amount of light sensed.

Calibrate To adjust an indicator so that it correctly indicates the variable sensed.

Calorie Heat required to raise the temperature of 1g of water 1°C.

Calorimeter A device for measuring the heat of chemical reactions.

Cam A projection on a rotating disc that can be used to operate defrost timers as the cam slowly turns.

Capacitance A measure of the amount of energy stored by a capacitor.

Capacitive Reactance The capacitive electrical impedance opposing the alternating current in a circuit.

Capacitor Type of electrical storage device used in starting and/or running circuits on many electric motors.

Capacitor Start Motor (CSR) A single-phase motor that has a capacitor in series with the start winding.

Capacity The rating of a unit generally based on size, power, and refrigerant amount.

Capillary Attraction The ability of one substance to draw another substance into it.

Capillary Tube A fixed restriction pressure reducing device. Usually consists of lengths of small inside diameter tubing. The flow restriction produces the necessary reduction in pressure and boiling point of the liquid refrigerant before entering the evaporator.

Capillary Tube System A refrigeration unit that utilizes a capillary tube type of flow control device.

Carbon Dioxide (CO_2) A nontoxic product of combustion.

Carbon Dioxide Indicator Device used to measure carbon dioxide levels.

Carbon Filter Activated charcoal filter used for adsorption purposes.

Carbon Footprint The total set of greenhouse gas emissions caused by an organization, event, product or person.

Carbon Monoxide (CO) A chemical resulting from incomplete combustion. It is odorless, colorless, and toxic.

Carbon Neutrality Refers to achieving net zero carbon emissions by balancing a measured amount of carbon released with an equivalent amount.

Cascade System One having two or more refrigerant circuits, each with a compressor, condenser, and evaporator, where the evaporator of one circuit cools the condenser of the other (lower temperature) circuit.

Cathode Electrode through which positive electric current flows.

Cavitation Vapor bubbles forming in a liquid region due to a drop in pressure that creates turbulence.

Celsius The metric system temperature scale.

Cellulose Used as an environmentally preferable material for building insulation.

Centigrade An earlier name for the Celsius scale.

Centimeter Metric unit of linear measurement, equals 0.3937 in.

Central Air Conditioning Air is treated at a central station and then distributed to the conditioned spaces by a series of ducts.

Central Station The equipment required for treating the air is all located together at one location.

Centrifugal Compressor Compressor that compresses gaseous refrigerants by centrifugal force.

Centrifugal Force The outward force created by the circular motion of an object.

Centrifugal Pump A pump with a rotating impeller that utilizes centrifugal force to pump a liquid.

Centrifugal Switch A switch that opens or closes due to the action of centrifugal force.

Ceramic Ignitor Ignitor used for gas stoves and heaters.

Change of State (see Phase Change)

CFM Cubic feet per minute (ft^3/min).

Charge The amount of refrigerant in a system.

Charging Cylinder A device for charging a predetermined weight of refrigerant into a system.

Charging Scale Digital scale that measures the weight of refrigerant in a cylinder.

Charging the Loop Filling the earth loop with the correct mixture and purging all the air from the loop.

Charles' Law The ideal gas law.

Check Valve A flow control valve that permits flow in one direction only.

Chemical Refrigeration (see Absorption Cycle)

Chill Factor (see Wind Chill Factor)

Chiller Refrigeration system designed to cool water, glycol, or brine.

Chilled Water System The series of pumps and heat exchangers used to deliver chilled water to a conditioned space.

Chimney Stack used for venting hot flue gases from a boiler or furnace.

Chimney Connector Connecting pipe to connect the furnace to the chimney.
Chimney Effect The tendency of air or gas in a duct or other vertical passage to rise when heated.
Chimney Flue Chimney stack.
Chlorinated Polyvinyl Chloride (CPVC) A thermoplastic used for hot and cold water pipes.
Chlorofluorocarbon (CFC) Refrigerant composed of chlorine, fluorine, and carbon.
Circuit A tubing, piping, or electrical wire installation that permits flow from the energy source back to energy source.
Circuit Breaker A device that senses current flow and opens when its rated current flow is exceeded.
Circuit, Parallel (see Parallel Circuit)
Circuit, Pilot A device used to initiate a circuit.
Circuit, Series (see Series)
Circulator A pump used for hydronic heating systems.
Clamp-on Ammeter Current measuring device that is not connected directly to the circuit but instead temporarily clamps around the electrical wire.
Clearance Space Amount of access space required on all sides and above and below an installed unit.
Clean Room Found in semiconductor manufacturing plants to maintain a perfectly clean environment.
Clearance Space in cylinder not occupied by the piston at the end of the compression stroke.
Climate Weather conditions for a specific geographical region established over long periods of time.
Climate Control HVAC is often referred to as a method of climate control.
Closed Circuit Completed path for current flow with no open switches.
Closed Circuit Cooling Tower The process cooling fluid is completely isolated from the atmosphere.
Closed Loop Any cycle in which the primary medium is always enclosed and repeats the same sequence of events.
Closed-Loop Control Configuration A sensor monitors the output and provides feedback to maintain setpoint conditions.
Closed-Loop Heat Pump A geothermal heat pump that circulates a glycol solution through a ground coil.
Clutch, Magnetic Transmits torque and speed without contact.
Clutch, Centrifugal The driving shaft nested inside the driven shaft uses centrifugal force for engagement.
Coaxial Fitting A fitting that connects to a single port of a hot water storage tank. Or, tube in a fitting. The hot water goes through the tube into the tank, and the cold water comes out of the fitting surrounding the tube.
Coaxial Heat Exchanger A spiral shaped heat exchanger consisting of an inner and outer tube for counterflow applications.
Code Local, state, and national rules and regulations.
Coefficient of Conductivity Thermal conductivity of a substance (*k*).
Coefficient of Expansion Change in length per degree of temperature change of a substance.
Coefficient of Performance (COP) A ratio calculated by dividing the total heating capacity provided by the refrigeration system, including circulating fan heat but excluding supplementary resistance heat (Btu/hr), by the total electrical input (watts) $\times$ 3.412.
Cogeneration Combined heat and power.
Coil A wound conductor that creates a strong magnetic field when current passes through it.
Coil, Device Subcooler A section in the outdoor evaporator coil used to increase the liquid subcooling during the cooling mode as well as act as an extra defrost surface during the defrost mode.
Coil, Inside The coil located in the inside portion of the heat pump system. Performs as the evaporator in the cooling mode and the condenser in the heating mode.
Coil, Outside The coil located in the outside portion of an air-to-air heat pump system. Performs as an evaporator in the heating mode and as the condenser in the cooling mode.
Cold The absence of heat.
Cold Anticipator (see Cooling Anticipator)
Cold Junction The part of the thermoelectric system that absorbs heat.
Cold Trap Used with vacuum pumps to collect condensation.
Collector Semiconductor terminal.
Combined Annual Efficiency (CAE) Annual performance of combination appliances such as space and water heaters.
Combustible Liquids Class II and Class III liquids with closed cup flash points at or above 100°F.
Combustion The rapid oxidation of fuel gases accompanied by the production of heat.
Combustion Air Air supplied for the combustion of a fuel.
Combustion Analyzer Device used for analysis of flue gases for optimizing combustion efficiency.
Combustion Products Constituents resulting from the combustion of a fuel gas with the oxygen in air, including the inerts, but excluding excess air.
Comfort Those conditions that are most suitable for human well being.
Comfort Chart Chart used in air conditioning that shows the dry bulb temperature and humidity for human comfort.
Comfort Zone The temperature and humidity range within which one is comfortable.
Commercial Buildings Such buildings as stores, shops, restaurants, motels, and large apartment buildings.
Commercial System Systems designed for nonmanufacturing business establishments.
Commutator A ring of copper segments insulated from each other and connecting the armature and brushes of a motor or generator. It passes power into or from the brushes.
Compound Gauge Pressure gauge that has scales both above and below atmospheric pressure.
Compound Pump A two stage pump.
Compound Refrigerating System System that has several compressors in series.
Compression A reduction of gas volume resulting in an increase in pressure.
Compression Gauge A gauge used to measure compression pressure.
Compression Ratio The absolute discharge pressure divided by the absolute suction pressure for a compressor.

Compression Ring Piston ring closest to the cylinder head on a reciprocating compressor.
Compression Tank (see Air Cushion Tank)
Compressor The pump of a refrigerating mechanism that draws a vacuum or low pressure on the cooling side of refrigerant cycle and squeezes or compresses the gas into the high pressure or condensing side of the cycle.
Compressor Crankcase Location for the crankshaft and oil sump.
Compressor Displacement The volume discharged by a compressor in one rotation of the crankshaft.
Compressor, External Drive A compressor that requires a crankshaft seal because it is externally coupled to a motor.
Compressor Head (see Cylinder Head)
Compressor, Hermetic A sealed unit with the compressor and motor located within the same casing.
Compressor, Multiple Stage Compressor with more than one stage of compression.
Compressor Oil Cooler Heat exchanger used to cool the oil in a compressor crankcase.
Compressor, Open (see Compressor, External Drive)
Compressor, Reciprocating Compressor that uses pistons driven by a crankshaft.
Compressor, Rotary Screw Compressor that utilizes two meshed rotating positive-displacement helical screws.
Compressor Seal Leakproof seal between crankshaft and compressor body.
Compressor Shaft Seal (see Shaft Seal)
Compressor, Single Stage Compressor with only one stage of compression.
Condensate Water that has condensed on the surface of a cooling coil.
Condensate Drain Trap A pipe arrangement that provides a water seal in the drain line to prevent airflow through the drain line.
Condensate Pan A pan located under an evaporator to collect condensate from the coil and carry it to the drain line.
Condensate Pump Device used to remove fluid condensate.
Condensation Liquid that forms when a vapor is cooled below its condensing temperature or dew point.
Condense The action of changing a saturated vapor to a saturated liquid.
Condenser The part of the refrigeration system that receives the high pressure superheated vapor from the compressor and extracts the heat in the vapor reducing it to a high pressure saturated liquid.
Condenser, Air Cooled A heat exchanger that transfers heat energy from the refrigerant vapor to air.
Condenser Fan A fan used to deliver air across the surface of an air cooled condenser.
Condenser Flooding Condenser containing excessive liquid refrigerant.
Condenser, Liquid Cooled A heat exchanger that transfers heat energy from the refrigerant vapor to the liquid heat sink.
Condenser Water Pump A device used to supply cooling water to the condenser.
Condensible A gas that can be easily converted to liquid form.
Condensing Boiler Boiler where the flue gas is allowed to drop below the dew point temperature and condense.
Condensing Furnace (see Condensing Boiler)
Condensing Pressure The refrigerant pressure inside the condenser coil.
Condensing Temperature The temperature at which a vapor changes to a liquid at a given pressure.
Condensing Unit The portion of the refrigeration system that converts high pressure superheated vapor to high pressure saturated liquid. Commonly called the high side.
Conductance, Thermal (C) The time rate of heat flow per unit area through a material.
Conduction, Thermal Particle to particle transmission of heat.
Conductivity The ability of a substance to allow the flow of heat or electricity.
Conductor Material or substance that readily passes electricity.
Connected Load The sum of the capacities or continuous ratings of the load-consuming apparatus connected to a supplying system.
Connecting Rod The part of a compressor that connects the piston to the crankshaft.
Console An interface to operate computers and control systems.
Constant A known and unchanging value.
Constrictor Tube or orifice used to restrict flow.
Contact The part of a switch or relay that carries current.
Contactor A device for making or breaking load carrying contacts by a pilot circuit through a magnetic coil.
Contaminant A substance (dirt, moisture, etc.) foreign to refrigerant or refrigerant oil in a system.
Continuous Absorption Cycle An absorption cycle that uses a limited amount of heat with no moving parts.
Continuous Operation Unit that is always running.
Contractual Agreement Agreed to by both parties and legally binding.
Control Switches and other devices to run HVAC/R equipment.
Control Motor Motor used to control an action such as the positioning of a damper.
Control, High Pressure Shuts down the unit if the high pressure setpoint is reached.
Control Loop Control system with sensors and feedback.
Control, Low Pressure A pressure operated control connected to the suction side of the refrigeration system to prevent unit operation below a set pressure or coil operating boiling point.
Control Module Component that contains the control system inputs and outputs.
Control Point The actual value as compared to the setpoint.
Control, Refrigerant A device used to provide the necessary pressure reduction of the liquid refrigerant to obtain the proper boiling point of the refrigerant in the evaporator.
Control System System used to manage, command, direct, or regulate other devices.
Control, Temperature A device that uses changes in temperature to operate contacts in an electrical circuit.
Control Valve Valve used for directing flow.

Controlled Environment (see Climate Control)
Controlled Medium The conditioned air or water distributed through an HVAC/R system.
Controller Measures the difference between sensed output and desired output and initiates a response to correct the difference.
Convection Transfer of heat by means of movement of a fluid, either liquid or air.
Convection, Forced Transfer of convective heat assisted by a pump or fan rather than natural circulation.
Convection, Natural Transfer of convective heat by natural circulation with warm air rising and cold air falling.
Conversion Factor Constant value used to change from one scale to another such as from Celsius to Fahrenheit.
Cooler A walk-in or reach-in refrigerated compartment.
Cooling Anticipator A resistor in a room thermostat that causes the cooling cycle to begin prematurely.
Cooling Coil An evaporator coil.
Cooling Tower Device that cools water by evaporation in air. Water is cooled to the wet bulb temperature of air.
Copper Plating Electrolysis process that leads to copper deposits on the compressor internals.
Core, Air (see Air Core Solenoid)
Core, Magnetic (see Solenoid)
Core Valve (see Schrader Valve)
Corrosion The breakdown of material such as the weakening of iron due to oxidation known as rust.
Cotter Pin Split type fastener also called a cotter key or split pin.
Coulomb An electrical unit of charge; one coulomb per second (C/s) equals one ampere (A) or 6.25×10^{23} electrons past a given point in 1s.
Counter EMF Counterelectromotive force. The EMF induced in a coil that opposes applied voltage.
Counterflow Two liquids and/or vapors flowing in directions opposite to each other.
Coupling Device used to connect two shafts together.
Cracking a Valve Opening a valve a small amount.
Cradle-Mounted Motor A motor cradled on each end and held down with a bracket.
Crankcase Heater A heating device fastened to the crankcase or lower portion of the compressor housing intended to keep the oil in the compressor at a higher temperature than the rest of the system to reduce the migration of the refrigerant.
Crankcase Pressure Regulating Valve (CPR) Device to limit the compressor suction pressure to reduce the load at startup.
Crankshaft Converts rotating motion to reciprocating motion through the connecting rod to the piston.
Crankshaft Seal Leakproof joint between crankshaft and compressor body.
Crank Throw Location where the connecting rod is attached to the crankshaft.
Creosote The buildup of carbon deposits in chimneys from wood-burning fires.
Crisper The enclosed vegetable compartment in a refrigerator.
Critical Pressure The vapor pressure at the critical temperature.
Critical Temperature The temperature above which liquid and gas phases do not exist.
Cross-Charged A TEV sensing bulb charged with a fluid that is different from the refrigerant used in the system.
Cryogenic Temperatures typically below –238°F.
Cryogenic Fluid Liquid nitrogen and liquid helium are commonly used in cryogenic applications.
Cryogenic Food Freezing Liquid nitrogen is used in blast freezing or immersion freezing systems.
Cryogenics The branches of physics and engineering that involve the study of very low temperatures.
Crystallization A highly concentrated brine solution that precipitates as salt.
Cupronickel An alloy of copper and nickel that does not corrode in seawater.
Current The flow of electrons through a conductor.
Current Relay Opens or closes contacts with a change in current flow through it.
Current Limiting Fuse Opens the circuit if the current is above the setpoint.
Current Transformer Provides current in its secondary winding proportional to the current flow through its primary winding.
Customer Relations The evaluation of the technician by the customer as a result of the job performance and attitude.
Cut In The temperature or pressure at which an automatic control switch closes.
Cut Out The temperature or pressure at which an automatic control switch opens.
Cycle A repeated series of events from start to end, such as the refrigeration cycle.
Cylinder, Compressor Compression chamber that houses the piston in a reciprocating compressor.
Cylinder Head Part that encloses compression end of compressor cylinder.
Cylinder, Refrigerant Rigid container holding refrigerant that may be of the disposable type or reusable type, such as a recovery cylinder.
Cylinder Unloading Method of reducing compressor load by cutting out individual cylinders.
Cylindrical Commutator Armature for a motor.

Dalton's Law Law of partial pressures.
Damper Movable plate or louver for controlling airflow.
Damper, Return Air Positioned to control air flow through the return air duct.
Damper, Outdoor Air Positioned to control airflow through the outdoor air duct.
DC Motor Direct current motor.
Dead Band A range of temperature in a heating/cooling system in which no heating or cooling is supplied.
Decibel Unit used for measuring relative loudness of sounds.
Deck, Cold Air directed over a cooling coil located in separate cool air duct for a large zone system.
Deck, Hot Air directed over a heating coil located in separate hot air duct for a large zone system.
Dedicated Geothermal Well Water is drawn from the top of the well and returned to the bottom of the well. The well supplies only the heat pump.
Deep Vacuum A vacuum of 500 microns or less.
Defrost Control A control system used to detect frost or ice buildup on the outside coil of an air-to-air heat pump

during the heating mode and causes the system to reverse in order to remove the frost or ice from the coil.

Defrost Cycle Refrigerating cycle in which evaporator frost and ice accumulation is melted.

Defrosting Method of removing frost and ice from an evaporator coil.

Defrosting Evaporator An evaporator that utilizes hot gas defrost.

Defrost Mode The portion of system operation when the evaporator frost or ice is removed.

Defrost Timer A device connected into the electronic control system that controls the frequency and duration of the defrost operating mode.

Degreaser Cleaning solvent used to remove grease.

Degreasing Using degreaser to remove grease buildup.

Degree Day Unit that represents one degree of difference from a standard average outdoor temperature of one day.

Dehumidify Lowering air temperature below its dew point to remove moisture.

Dehumidifier Device such as a cooling coil used for dehumidification.

Dehydrate To remove water in all forms from a material or system.

Dehydrated Oil Moisture free oil.

Dehydrator Removes moisture in a refrigeration system and is often referred to as a drier.

Deice To remove ice such as in defrosting.

Deice Control (see Defrost Control)

Delta Connection The connection in a three phase system in which terminal connections are triangular, similar to the Greek letter delta (△).

Delta-*P* Change in pressure.

Delta-*T* Change in temperature.

Delta Connected Motor (see Delta Connection)

Demand The size of any load generally averaged over a specified interval of time.

Demand (Billing) The demand upon which billing to a customer is based.

Demand Meter An instrument used to measure peak kilowatt hour consumption.

Density Weight per unit volume.

Deodorizer Substance used to inhibit the growth of bacteria that causes odors.

Department of Energy (DOE) Name for the government agency in North America devoted to national energy.

Department of Transportation Name for the government agency in North America devoted to transportation.

Desert Bag Watertight bag that may be used for storage. A bag used to keep water cool in the desert.

Desiccant Substance used to collect and hold moisture.

Desiccant Drier Usually located in the liquid line, a drier charged with desiccant.

Design Load The amount of heating or cooling required to maintain inside conditions when the outdoor conditions are at design temperature.

Design Pressure Rated pressure the system is normally designed to operate at.

Design Temperature Difference The difference between the design indoor and outdoor temperatures.

Design Water Temperature Difference The difference between the temperature of the water leaving the unit and entering the unit when the system is operating at design conditions.

Desuperheater A heat exchanger in the hot gas line between the compressor discharge and reversing valve. Used to heat domestic water by removing the superheat from the hot refrigerant vapor.

Detector, Leak A test instrument used to detect and locate refrigerant leaks.

Detent Allows for a controlled hesitation in movement.

Dew Water that has condensed out of air.

Dew Point Temperature at which water vapor begins to condense out of air. Air at the dew point—100% relative humidity—is referred to as saturated air.

Diagnostics Methodical identification of symptoms to identify problems with operating equipment.

Diagnostic Thermostat A thermostat that can receive input to diagnose problems.

Diaphragm Flexible membrane.

Diaphragm Valve The valve stem acts on a diaphragm to open or close the valve.

Dielectric An insulator or nonconductor.

Dies (Thread) Tool used to cut external threads.

Differential As applied to controls, the differential between the cut in and cut out temperature or pressure setpoints of a control.

Diffuser Air distribution outlet designed to direct airflow into a room.

Diffuse Radiation Solar radiation reaching the earth's surface after having been scattered by molecules in the atmosphere.

Digital Electronic Devices Devices that send or receive bit stream digital signals.

Digital Electronic Signal A waveform that switches between two voltage levels representing two states (0 and 1).

Digital Volt Ohm Milliammeter (VOM) A multimeter with a digital output.

Dilution Air Air that enters a draft hood and mixes with the flue gases.

Diode A device that will carry current in one direction but not the reverse direction.

Dual Inline Pair Switch (DIP) A small low amperage single-pole, double-throw switch used in electronic circuits.

Direct Current (DC) Electric current that flows only in one direction.

Direct Digital Control (DDC) The ability to control HVAC devices via microprocessors.

Direct Drive Compressor Compressor directly connected and driven at motor speed.

Direct Drive Motor Motor directly connected to device it is driving.

Direct Expansion Evaporator The primary refrigerant expands and absorbs heat from the air as opposed to an indirect expansion system that uses a secondary refrigerant such as chill water.

Direct Return A two pipe system in which the first terminal unit taken off the supply main is the first unit connected to the return main.

Direct Spark Ignition (DSI) The gas burner is lit by a direct spark rather than a pilot.

Direct Vent Flue gases can be vented directly outside the building without the use of a chimney.

Discharge Pressure Refrigerant pressure on the high side at the compressor outlet of a system.

Discharge Valve A valve located directly at the outlet of the compressor.

Disconnecting Switch A knife switch that opens a circuit.

Discus Compressor A compressor with discus valve design to allow for capacity modulation to reduce cycling.

Discus Valve A valve designed to allow for less clearance volume vapors to be trapped when the piston is at top dead center.

Dispensing Freezers Units that process and freeze previously pasteurized product, such as soft ice cream.

Displacement The total volume of refrigerant that a compressor can draw in during the intake stroke.

Distributed Controls The controls are not in a central location but rather distributed throughout the system.

Distributor Allows for distribution of refrigerant to individual refrigerant circuits in an evaporator coil.

District Heating and Cooling A system for distributing heat generated in a centralized location for residential and commercial use.

Disturbed Earth Effect Denotes the heat content change that affects the earth's temperature by the addition or removal of heat energy by the heat pump system.

Diverter Tee Fitting used for a hydronic system to direct water through a terminal unit branch circuit.

Dome Hat Compressor housing for a hermetically sealed refrigeration unit.

Domestic Hot Water Heated water used for cooking, washing, and so on.

Domestic Geothermal Well Water is drawn from the top of the well. Only the water from the heat pump is returned to the bottom of the well. The difference is used for building requirements.

Door Heater A heater located around the opening of a refrigerator door to prevent the door from freezing closed.

Downflow Furnace Furnace where the return air enters at the top, flows down through the heat exchanger, and exits out the bottom.

Double Pole Switch Simultaneously opens and closes two connections of a circuit.

Double Thickness Flare Tubing end that has been formed into a two-wall thickness.

Dowel Pin Pin pressed through two assembled parts to ensure accurate alignment.

Downdraft Downward flow of flue gas.

Draft Gauge Instrument used to measure air pressure.

Draft Indicator Instrument used to measure chimney draft.

Draft Regulator Device used to automatically control chimney draft.

Drier A substance or a device used to absorb moisture within a refrigeration system.

Drift Entrained water carried from a cooling tower by wind.

Drilled Port Burner A burner in which the ports have been formed by drilled holes.

Drilled Well A water supply for a geothermal water source heat pump.

Drip Pan Pan shaped panel or trough used to collect condensate from evaporator coil.

Drip Proof Motor A motor typically designed for outdoor use or a wet environment.

Drive Clip Steel fastener for connecting ductwork.

Dry Bulb Temperature The actual (physical) temperature of a substance. Usually refers to air.

Dry Cell Battery Common standard battery with a paste low-moisture electrolyte as compared to a wet cell type.

Dry Ice Solid carbon dioxide.

Dry System Refrigerant is metered into the evaporator coil rather than flooding the coil.

Dry Well A water return for a geothermal water source heat pump open loop.

Dual Pressure Regulator A high and low pressure control mounted together.

Duct Round or rectangular sheet metal or fiberglass pipe that carries the air between the conditioning unit and the conditioned area.

Duct Sweeper A cleaning tool used to remove dirt and debris from the inside of air ducts.

Dust Mites Microscopic bug often found in bedding and one of the most common allergens that trigger asthma.

Dynamometer Device for measuring power.

E Symbol for volts. From the term "electromotive force."

Earth Coupled Heat Pump Geothermal heat pump.

Earth Temperature, Minimum (*T*) The lowest temperature the earth will become in the winter season. Used for heat pump design. Will vary with location and weather conditions.

Eccentric A circle or disk mounted off center on a shaft.

Economizer, Air Conditioning Multiple flash chamber for air conditioning chiller designed to gain efficiency.

Economizer, Air Supply Adjusts for optimum outside air supply designed to maintain maximum efficiency.

Economizer, Evaporator Heat exchanger used to reduce flash gas through the metering device.

Eddy Currents Circulating flow of current within a conductor that can generate heat.

Effective Area The actual opening in a grill or register through which air can pass. The effective area is the gross (overall) area of the grill face minus the area of the deflector vanes or bars.

Effective Latent Heat (see Net Refrigeration)

Effective Temperature Calculated nonmeasured temperature based upon temperature, humidity, and air movement.

Effectiveness, Absorption Systems Method to evaluate absorption systems using heat input and cooling effect.

Efficiency, Electrical A percentage value denoting the ratio of power output to power input.

Ejector A venturi arrangement used as a pump with no moving parts.

Electric Defrosting An electric resistance strip heater attached to the evaporator coil used for defrosting.

Electric Field A magnetic region in space.

Electric Forced Air Furnace Return air is blown across electric resistance strip heaters.

Electric Heating A heating system that uses electric resistance elements as the heat energy source.

Electric Heating Element A unit consisting of resistance wire, insulated supports, and connection terminals for connecting wire to a source of electrical power.

Electric Hydronic Boiler Electric resistance heating elements used to heat water.

Electric Water Valve Electrically operated valve to control water flow.

Electrical Circuit Electrical wire installation that permits flow from the energy source back to energy source.

Electrical Erasable Programmable Read-Only Memory (EEPROM) Memory used in computers and electronic devices to store small amounts of data.

Electrical Power Measured in units of watts, equal to voltage × amperage.

Electrical Resistance The degree to which an object opposes the flow of electric current through it.

Electrical Shock Contact with a voltage source allowing current to flow through the body.

Electricity The presence and flow of an electric charge.

Electrodes Used with high voltage applications for lighting oil and gas burners.

Electrolyte A solution of a substance (liquid or paste) that is capable of conducting electricity.

Electrolytic Capacitor Capacitor using a paper spacer soaked in electrolyte in one of its plates.

Electromagnet A magnet created by the flow of electricity through a coil of wire.

Electromagnetic Energy Energy derived from an electromagnet.

Electromechanical Controls Mechanical movement controlled through electrical input or motors.

Electromotive Force (EMF) The difference in potential or voltage of electrical energy between two points.

Electron Subatomic particle that carries a negative electric charge.

Electronic Charging Scale (see Charging Scale)

Electronic Circuit Board Printed board used to support and connect electronic solid state components.

Electronic Controls Control systems utilizing electronic circuit boards with solid state components.

Electronic Expansion Valve Electronically controlled, electrically operated type of valve.

Electronic Leak Detector An electronic instrument that measures the changes in electron flow across an electrically charged gap that indicates the presence of refrigerant vapor molecules.

Electronic Relay Electronic switch such as a Triac.

Electronics Field of science dealing with electronic devices and their uses.

Electronic Sight Glass Signals when system is low on refrigerant.

Electronically Commutated Motor (ECM) Brushless motor relying on an external power drive for commutation.

Electrostatic Filter Type of filter that gives particles of dust electric charge.

Embrittlement The loss of elasticity of a material, making it brittle.

End Bell End structure of electric motor that usually holds motor bearings.

End-Mount Motor Motor held in place at the motor housing end.

Endplay Slight movement of shaft along a center line.

Endothermal Chemical reaction where heat is converted to chemical bond energy.

Energized Having current flow.

Energy The ability to do work.

Energy Efficiency Ratio (EER) The comparison of the heat transfer ability of a refrigeration system and the electrical energy used as expressed in Btu/hr per watt. A ratio calculated by dividing the cooling capacity in Btu per hour (Btu/hr) by the power input in watts at any given set of rating conditions, expressed in Btu/hr per watt.

Energy Management System of computer-aided tools to monitor and control HVAC functions most efficiently.

Energy Management Control System (see Energy Management)

Energy Recovery Ventilator (ERV) Air-to-air heat exchanger installed in a duct system to gain efficiency.

Energy Star Government sponsored program designed to save consumers money and protect the environment through energy-efficient products and practices.

Energy Utilization Index (EUI) Comparison of energy usage within different areas of a building.

Enthalpy Heat content of a refrigerant, usually with respect to a reference point (Btu/lb).

Entropy Energy that cannot be used for external work.

Environment The surrounding conditions.

Environmental Protection Agency (EPA) Agency of the federal government responsible for safeguarding the natural environment.

Epoxy A synthetic plastic adhesive usually composed of two parts, the resin and the hardener.

Equalizer Tube Device used to maintain equal pressure or equal liquid levels between two containers.

Equivalent Length Calculated nonmeasured piping length taking into account friction losses.

Error Codes Code numbers used with control system to indicate faults.

Ester Lubricant Synthetic refrigeration oil.

Ethane Colorless and odorless gas used as a refrigerant in cryogenic refrigeration systems.

Eutectic Solution used for cold storage within evaporator coils.

Eutectic Point Temperature at which eutectic changes from liquid to solid.

Eutectic Solution Mixture of two solids so that the melting point is as low as possible.

Evacuate To remove water in all forms from a material or system.

Evacuation Removal of air, noncondensable gases, and moisture from a refrigeration system.

Evaporation A term used to describe the changing of a liquid to a vapor by the addition of heat energy. Change from liquid to vapor at a relatively slow rate.

Evaporative Condenser A device that uses a water spray to cool a condenser.

Evaporative Cooling The use of a water mist or spray to cool the air.

Evaporative Humidifier Addition of moisture to the airstream using a water mist or spray.

Evaporator Part of a refrigerating mechanism in which the refrigerant vaporizes and absorbs heat.

Evaporator Coil (see Evaporator)

Evaporator, Dry Type An evaporator into which refrigerant is fed from a pressure reducing device.

Evaporator Fan Fan that moves air through the evaporator.

Evaporator, Flooded An evaporator containing liquid refrigerant at all times.

Evaporator Pressure Regulator (EPR) Device used to regulate and control evaporator pressure.

Excess Air Air that is in excess of the required amount for complete combustion of a fuel gas.

Exfiltration Outward flow of air from an occupied area through openings in the building structure.

Exhaust Opening Any opening through which air is removed from a space.

Exhaust Valve Discharge valve for a positive displacement compressor.

Exothermal Chemical reaction where heat is released.

Expansion Joint A fitting or piping arrangement designed to relieve stress caused by expansion of piping.

Expansion Tank A closed tank that allows for water expansion without creating excessive pressure.

Expansion Valve A device in a refrigerating system that maintains a pressure difference between the high side and the low side.

Expendable Refrigerant System System that discards the refrigerant after it has evaporated.

Explosion Proof Motor Nonsparking motor designed to operate in explosive locations.

Extended Surface Heat transfer surface, one side of which is increased in area by the use of fins.

External Drive (see Compressor, External Drive)

External Equalizer A pressure connection from an area below the diaphragm of a thermostatic expansion valve and the outlet of an evaporator.

External Motor Protection Overload mounted on the outside of a motor.

Fahrenheit The common scale of temperature measurement in the English system of units.

Fail-Safe Designed to ensure safety of the unit in the event a component fails to operate.

False Defrost The system goes into the defrost mode even if the outdoor has no frost or ice buildup.

Fan A radial or axial flow device using rotating blades for moving air.

Fan Coil A terminal unit consisting of a finned tube coil and a fan in a single enclosure.

Fan Relay Relay used to start and stop a fan.

Farad The base unit of measurement for capacitance. Since a farad is a very large unit, a much smaller unit is usually used. (See Microfarad)

Faraday Experiment An experiment on electromagnetic induction.

Fast Acting Fuse A fuse that opens the circuit immediately if the rated current is exceeded.

Feedback The transfer of energy output back to input.

Feedback Control System Closed loop system with sensors that provide feedback based upon the setpoint.

Feet of Head (ft HD) The term used to designate the flow resistance the pump must overcome to deliver the required amount of liquid. 2.307 ft HD = 1 psig.

Female Thread A thread on the inside of a pipe or fitting.

Ferrous Objects made of iron or steel.

Fiberduct Duct made from fiberglass.

Fiberglass Fine glass fiber and resin composite.

Field The space involving the magnetic lines of force.

Field Pole Located as part of the motor stator winding.

Fill Honeycomb material for cooling water towers to slow the fall of water and increase surface exposure.

Film Factor Value representing resistance to heat transfer through the surface film of a material.

Filter-Drier A combination refrigerant drier and filter in one component.

Filter A device for removing foreign particles from vapor or liquid.

Fin Comb Comblike device used to straighten the metal fins on coils.

Finned-Tube Evaporator Tube type evaporator with attached fins to increase surface area and heat transfer.

Firepot Refractory-lined combustion chamber.

Fixed-Bore Device Refrigerant metering device with fixed orifice openings.

Fixed Resistor A resistor in which the value does not change.

Fire Tube Boiler A steel boiler in which the hot gases from combustion are circulated through tubes that are surrounded by water.

Flake Ice Small hard bits of ice ideally suited for any "contact" presentation, such as displaying fish at the supermarket.

Flame Detector The components of a flame detection system that detects the presence or absence of a flame.

Flame Impingement Excessively wide or jagged burner flame making contact with the side walls of a furnace.

Flame Rectification AC is converted to DC by the flame of a gas furnace.

Flame Sensing Element A device that uses heat, infrared, optical, rectification, rectifying photo cell, or ultraviolet light to detect a flame.

Flammability The ease at which a substance will ignite.

Flapper Valve A lightweight reed type valve hinged on one end allowing it to flap open and closed.

Flare An angle formed at the end of a tube.

Flare Nut Fitting used to clamp the tubing flare against another fitting.

Flashback The movement of the gas flame down through the burner port upon shutdown of the gas supply.

Flash Gas Vapor that is produced by the boiling of liquid refrigerant as it passes through the metering device.

Flash Point Temperature at which an oil will give off sufficient vapor to support a flame.

Flash Weld A resistance or spot weld where metal sections under pressure are fused together by electrical current.

Flexible Duct Designed to allow for some movement and bending to accommodate installation.

Flexible Lines Lines designed to accommodate bending.

Float Valve Type of valve that is operated by a sphere that floats on the liquid surface and controls the level of liquid.

Floating Head Pressure (see Head Pressure, Control)

Floc Point The temperature at which the wax in an oil will start to separate from the oil.

Flooded System Type of refrigerating system in which liquid refrigerant fills the evaporator.

Flooding Condition used to describe excessive liquid refrigerant in the evaporator or condenser.

Fluorescent Capable of emitting fluorescence such as in lighting.

Flow Check Piston Fixed orifice metering device with a piston assembly serving as a check control for reverse flow.
Flow Hood A direct reading instrument that measures air quantity in cubic feet per minute (CFM) out of or into a duct system.
Flow Meter Instrument used to measure the volume of a liquid flowing through a pipe or tube in quantity per time (gallons per minute, pounds per hour, liters per minute, etc.).
Flow Switch An automatic switch that senses fluid flow.
Flue A passage in the chimney to carry flue gas.
Flue Gases Products of combustion and excess air.
Fluid Substance in a liquid or gaseous state.
Flushing the Loop Forcing liquid through the loop at sufficient velocity to remove all forms of foreign material and air.
Flux Brazing, Soldering Substance applied to surfaces to be joined to free them from oxides.
Flux, Electrical The electric or magnetic lines of force in a region.
Flux, Magnetic Lines of force of a magnet.
Foaming Formation of foam in an oil refrigerant mixture due to rapid evaporation of liquid refrigerant entrained in the oil.
Foot of Water A measure of pressure. One foot of water is the pressure created by a column of water 1 ft high. This is equal to 0.433 psig.
Foot Pound Unit of work.
Force A push or pull which will cause an object to begin moving.
Force Feed Oiling Oil is fed to lubrication points by a pump.
Forced Air Air movement produced by a fan.
Forced Circulation Evaporator Evaporator with forced air passing across it.
Forced Convection Heat transfer by the movement of a liquid or vapor by a mechanical means such as a pump or fan.
Forced Draft Burner A burner in which combustion air is supplied by a blower.
Forced Draft Cooling Tower A cooling tower utilizing a fan to draw air through rather than natural draft.
Forced Draft Evaporator (see forced circulation evaporator)
Fossil Fuels Mineral fuels; mainly hydrocarbons, such as petroleum found within the top layer of the earth's crust.
Four-Way Valve Multi-ported valve used for reverse cycle heat pumps.
Fractionation Tendency for refrigerants contained in a zeotropic blend to vaporize and separate.
Free Area The total area of an opening in a grill or register through which air passes.
Free Wheeling Allowed to move or rotate without obstruction.
Freeze Drying Dehydration process that involves freezing the material first.
Freeze-up Frost or ice formation on an evaporator that restricts airflow or liquid flow through the evaporator.
Freezer Burn Damage to improperly packaged food product due to moisture being drawn out.
Freezing Change of state from liquid to solid.
Freezing Point The temperature at which a liquid will solidify upon removal of heat.
Freezing Point Depression The lowered freezing point of a liquid resulting from mixing in an additive.
Freon Trade name for a family of synthetic chemical refrigerants manufactured by DuPont, Inc.
Frequency The number of complete cycles in a unit of time.
Friction Force resisting the relative motion of two surfaces in contact, solid or liquid.
Friction Head In a hydronic system, the loss in pressure resulting from the flow of water in the piping system.
Front Seated A closed refrigeration valve is considered to be front seated.
Frost Back A condition in which liquid refrigerant boils in the suction line and produces a frosted surface on the line.
Frostbite Damage caused to skin and tissue due to extreme cold.
Frost Control, Automatic A control that automatically reverses a refrigeration system to remove frost/ice from the evaporator.
Frost Free Refrigerator A refrigerated cabinet that operates with an automatic defrost.
Frost Line The depth to which the ground freezes over a winter season. The frost line is usually set at 12 in below the deepest recorded frost penetration.
Frozen When a liquid has turned to a solid.
Fuel Oil Liquid petroleum fossil fuel used for an oil fired furnace.
Full Load Amperage (FLA) The amount of amperage an inductive load (motor) will draw at its full design load.
Furnace That part of a warm air heating system in which combustion takes place.
Fuse Electrical safety device consisting of a strip of fusible metal in a circuit that melts when the circuit is overloaded.
Fusible Link Link within an electrical circuit that will melt and open due to excessive temperature.
Fusible Plug A plug with a low melting temperature metal to release pressure in the event of a fire.

Gain Increase in the electronic amplification of a signal.
Galvanic Chemically generated current resulting from two dissimilar conductors immersed in an electrolyte.
Galvanic Action The destruction of one metal by electrolysis when two different metals are joined.
Gang Mechanical connection of two or more switches.
Gas Valve Valve used to control the flow of gas.
Gasket A resilient material used between mating surfaces to provide a leakproof seal.
Gas A substance in the vapor state.
Gas, Noncondensible A gas that will not form into a liquid under typical pressure-temperature conditions.
Gate Valve Valve with a vertical sliding gate normally fully open or fully closed, not throttled.
Gauge, Compound A pressure indicating instrument that measures pressures both above and below atmospheric pressure.
Gauge, Gas Manifold Pressure A direct reading gauge for measuring the gas manifold pressure.
Gauge, High Pressure An instrument used to measure pressures above atmospheric pressure in the 0–500 psig range.

Gauge, Low Pressure An instrument used to measure pressures above atmospheric up to 50 psig.
Gauge, Manifold A device that contains a combination of gauges and control valves to control the flow of refrigerant for service applications.
Gauge Port Access connection to attach gauges.
Gauge, Standard An instrument designed to measure pressures above atmospheric.
Gauge, Vacuum An instrument used to measure pressures below atmospheric.
Generator A machine that converts mechanical energy into electrical energy (also see Alternator).
Geothermal Heat that comes from within the earth.
Geothermal Heat Pump (see Ground Coupled Heat Pump)
Geothermal Well A drilled well in which the returned water is returned to the same well from which the supply water is taken.
Germanium Semiconductor material used for the manufacture of transistors.
Global Warming Increase in the average measured earth temperature.
Global Warming Potential (GWP) Relative scale used to determine the increase in global warming attributed to a piece of operating equipment.
Globe Valve Valve with a spherical disk and seat that may be used for throttling.
Glow Coil Automatically relights the pilot for a gas burner.
Glycol Antifreeze.
Grade Level The top surface of the earth.
Graduated Cylinder (see Charging Cylinder)
Grain Unit of weight measure. 7,000 grains equal 1lb.
Gram Metric unit of weight measure.
Gravity Flow (see Gravity System)
Gravity System A heating system in which the distribution of the warm air or water relies upon natural convective currents. There is no fan or pump.
Green Construction Practices used today that efficiently use energy, reduce waste and pollution, and create a healthy environment for the occupants.
Grill An ornamental or louvered opening through which air enters or leaves a duct system or area.
Grommet A round or oval shaped rubber, metal, or plastic shield to protect wire and pipe cabinet penetrations.
Gross Capacity, Heating The gross capacity of a heat pump in the heating mode is the total amount of heat energy transferred to the inside air. This is made up of the heat energy picked up from a heat source (air or liquid) plus the heat energy of the electrical power required to operate the heat pump.
Gross Output A rating applied to boilers. It is the total quantity of heat that the boiler will deliver and at the same time meet all limitations of applicable testing and rating codes.
Ground Connection between an electrical circuit and the earth.
Ground Coil, Earth Coil A heat exchanger coil buried in the earth that is used to either remove heat energy from the earth or add heat energy to the earth.
Ground Coupled Heat Pump A heat pump that uses a ground coil to exchange heat.
Ground Fault An electrical leak from the circuit to ground.
Ground Fault Circuit Interrupter (GFCI) Breaker specifically designed to open the circuit in the event of a ground fault.
Ground Wire An electrical wire that will safely conduct electricity from a structure into the ground.
Guide Vanes, Inlet, Prerotation Located at the inlet of centrifugal compressors and fans to vary capacity.
Gun Burner Oil forced through the burner orifice is atomized as it enters the furnace.

Halide Refrigerants Family of refrigerants containing halogen chemicals.
Halide Torch Type of torch used to detect halogen refrigerant leaks.
Halogens Compounds composed of fluorine, chlorine, bromine, iodine, and astatine.
Hand Truck Two wheeled dolly used for moving heavy objects.
Hanger Device used to support tubing, piping, ductwork, and so on.
Harvest Cycle Term used when ice is released from the freezing cells in an ice machine.
Head Pressure difference.
Head, Pressure Control A pressure operated control that opens an electrical circuit if the pressure exceeds the cutout point of the control.
Head, Static Pressure of a liquid or vapor in terms of the height of a column of water or mercury.
Head, Total In flowing fluid, the sum of the static and velocity pressures at the point of measurement.
Head, Velocity In flowing fluid, height of fluid equivalent to its velocity pressure.
Head Pressure The pressure in the condensing or high side of the refrigeration system.
Header A piping arrangement for interconnecting two or more supply or return taps of a boiler.
Heat A form of energy. Heat affects the molecular activity of a substance. This is reflected in the temperature of the substance. The addition of heat energy increases the molecular activity and temperature. The removal of heat energy lowers the molecular activity and temperature.
Heat Anticipator A resistor in a room thermostat that shuts the heating cycle off prematurely.
Heat Coil (see Heating Coil)
Heat Exchanger A device used to transfer heat energy from a higher temperature to a lower temperature. Evaporators, condensers, and earth coils are examples.
Heat Gain Total amount of heat absorbed by a building used to determine HVAC loads.
Heating Coil A heat transfer device designed to add heat to liquid or vapor.
Heating Control Device to control heating function.
Heating Mode The operating phase of a system that is adding heat energy to an occupied area.
Heating Seasonal Performance Factor (HSPF) The total heating output of a heat pump during its normal annual usage period for heating divided by the total electric power input in watt-hours during the same period.
Heating Value The number of Btu produced by the combustion of 1ft^3 of gas.
Heat Lag When a substance is heated on one side, it takes time for the heat to travel through the substance.

Heat, Latent Heat energy used to change the state of a substance; solid to liquid and liquid to vapor by heat addition, or vapor to liquid and liquid to solid by heat removal, without a change in the temperature.

Heat Load The amount of heat energy, measured in Btu/hr, that is required to maintain a given temperature in a conditioned area at design conditions both inside and outside the conditioned area.

Heat Loss The rate of heat transfer from a heated building to the outdoors.

Heat Motor An electrical device that produces motion by means of the temperature change of bimetal elements. Used to produce time delay in the sequence of operation of a device (also see Sequencer).

Heat of Compression The heat added to the refrigerant by the work being done by the compressor.

Heat of Fusion The heat energy needed to accomplish the change in state of a material between a liquid and solid; addition for solid to liquid, removal for liquid to solid.

Heat Pipe Liquid filled pipes used in high efficiency gas furnaces.

Heat of Respiration When carbon dioxide and water are given off by foods in storage.

Heat Pump A name given to an air conditioning system that is capable of either heating or cooling an area on demand.

Heat Pump, Air-to-Air A device that transfers heat between two different air quantities in either direction on demand.

Heat Pump, Air-to-Liquid A device that transfers heat from an air source to a liquid by means of a refrigeration system.

Heat Pump, Liquid-to-Air A device that transfers heat between liquid and air in either direction on demand.

Heat Pump, Liquid-to-Liquid A device that transfers heat between two different liquids in either direction on demand.

Heat Pump, Water Heater An air-to-liquid refrigeration system used to heat domestic water with air as the heat source.

Heat Pump; Water Heater, Remove The refrigeration system is a complete unit designed to be connected to an existing water tank.

Heat Pump; Water Heater, Self Contained The component parts of the refrigeration system (condenser) are an integral part of the hot water tank.

Heat Reclaim Utilizing the heat rejected by the condenser for other heating purposes.

Heat Recovery System System utilizing an energy recovery ventilator.

Heat Recovery Ventilator (see Energy Recovery Ventilator)

Heat, Sensible The heat energy used to change the temperature of a material—solid, liquid, or vapor—without a change in state.

Heat Sink A place or material into which heat energy is placed. In a heat pump, the air outside the house is used as a heat sink during the cooling cycle.

Heat Source A place or material from which heat energy is obtained. In an air source heat pump, the air outside the house is used as a heat source during the heating cycle.

Heat, Total The sum of both the sensible and latent heat energy in air. Expressed as Btu/lb of air.

Heat Tape Electrical resistance heating wire that is wrapped around drain piping to prevent freezing.

Heat Transfer Movement of heat energy from one body or substance to another. Heat may be transferred by any combination of or all of the three methods radiation, conduction, and convection.

Heat Transfer Coefficient (U) Value assigned to designate heat transfer through a combination of given materials.

Heat Transfer Rate (Q) Amount of heat transferred per unit of time.

Helix Coil Bimetal coil used for dial thermometers.

Henry (H) The unit of inductance.

Hermetic Completely sealed.

Hermetic Motor Compressor drive motor sealed within the same casing that contains the compressor.

Hertz (Hz) Measure of frequency.

Hg Symbol for mercury.

High Glide Blend The range from the bubble point to the dew point is greater than 1°F at constant pressure.

High Efficiency Gas Furnace Furnace rated at 90% AFUE or better.

High Efficiency Particulate Air Filter (HEPA) Removes at least 99.97% of airborne particles 0.3 micrometers in diameter.

High Limit Control Safety that shuts off the furnace burner if the bonnet temperature is too high.

High Pressure Cutout An electrical pressure control switch operated by the high side pressure that is set to stop the compressor if the pressure safety limit is reached.

High Side Parts of a refrigerating system that are under condensing or discharge pressure.

High Side Float Refrigerant flow control located in the condenser.

High Temperature Refrigeration System with evaporator temperatures above 35°F.

High Vacuum Pump (see Deep Vacuum)

High Velocity System Usually large commercial or industrial air distribution systems designed to operate with static pressures of 6–9 in WC.

Horizontal Furnace The air flows horizontally through the furnace.

Horsepower A unit of power equal to 33,000 ft-lb of work per minute.

Hot Gas Compressor discharge vapor.

Hot Gas Bypass Method of compressor capacity control.

Hot Gas Defrost A defrosting system in which hot refrigerant gas from the compressor is directed through the evaporator to remove frost or ice from the evaporator.

Hot Gas Line The refrigerant tube that carries the high pressure, high temperature refrigerant vapor from the compressor to the condenser.

Hot Junction That part of a thermoelectric circuit that releases heat.

Hot Pulldown Lowering the temperature of a warm space.

Hot Surface Ignition A silicon carbide element is heated for the purpose of lighting the main burner.

Hot Water Heating System System that distributes hot water with pumps for heating a building.

Hot Wire Live wire with voltage potential to ground.
Humidifiers Device that directs water spray or mist into an airstream for humidification.
Humidistat A control that is affected by changes in the relative humidity in air. It is often used to control the operation of a humidifier.
Humidity Related to the amount of water vapor in the air.
Hunting Cycling above and below a setpoint such as a valve continuously opening and closing.
Hydraulics The use of fluid properties for the purpose of fluid power.
Hydrocarbon Any of a number of compounds composed of carbon and hydrogen.
Hydrochlorofluorocarbon (HCFC) Refrigerant composed of hydrogen, chlorine, fluorine, and carbon.
Hydrofluorocarbon (HFC) Refrigerant composed of hydrogen, fluorine, and carbon.
Hydrogen A colorless, odorless, highly flammable gas. It is the lightest of all elements.
Hydrofluoro Olefins (HFO) Newly developed types of refrigerant that have a low global warming potential.
Hydrometer Floating instrument used to measure specific gravity of a liquid.
Hydronics Pertaining to heating or cooling with water.
Hygrometer An instrument used to measure the percentage of moisture (relative humidity) in air.
Hygroscopic Ability of a substance to absorb and retain moisture.

Ice Bank A thermal accumulator in which ice is formed or charged during off peak periods of refrigeration demand, and used during peak demand for refrigeration to supplement the compressor capacity by melting the ice.
Ice Bin Location where ice is collected and stored for later use by an ice machine.
Ice Harvest Switch A device located at the end of the water plate. It resets the timer on the status indicator control card after each ice harvest.
Ice Maker A cyclic-type automatic ice making machine that has separate and sequential water fill, freezing, and ice harvesting phases for the production of ice.
Ice Ring An accumulation of ice at the bottom of the outdoor coil due to incomplete defrost operation.
Identification Plate Equipment data or name plate.
Idler A pulley used on some belt drives to provide the proper belt tension.
Ignition System Method of lighting a furnace.
Ignition Temperature The minimum temperature at which combustion can be started.
Ignition Transformer A transformer designed to provide a high voltage current for furnace electrodes.
Impedance A type of electrical resistance that is only present in alternating current circuits.
Impeller Rotating part of a centrifugal pump.
Impingement (see Flame Impingement)
Incandescent Heat driven light emissions allowing an object to glow from the heat.
Inches Hg Pressure Measurement of pressure in inches of mercury (in Hg) absolute.
Inches Hg Vacuum Measurement of vacuum in inches of mercury (in Hg) vacuum.
Inclined Water Manometer (see Manometer, Inclined)
Incomplete Combustion A condition where a portion of the fuel passing through a furnace remains unburned.
Indoor Air Quality A measurement of the factors attributing to the air quality condition within a building.
Indoor Coil The portion of a heat pump that is located in the house and functions as the heat transfer point for warming or cooling indoor air.
Induced Draft Burner A burner that depends on draft induced by a fan or blower at the flue outlet to draw in combustion air and vent flue gases.
Induced Magnetism Ability of a magnetic field to induce magnetism in a metal.
Inductance The characteristic of an alternating current circuit to oppose a change in current flow.
Induction The act that produces induced voltage in an object by exposure to a magnetic field.
Induction Motor An AC motor that operates on the principle of the rotating magnetic field. The rotor has no electrical connection, but receives electrical energy by transformer action from field windings.
Inductive Circuit Electrical circuit where the current lags the voltage.
Inductive Load A device that uses electrical energy to produce motion.
Inductive Reactance Opposition, measured in ohms, to an alternating or pulsating current.
Inerts Noncombustible substances in a fuel, or in flue gases, such as nitrogen or carbon dioxide.
Infiltration The air that leaks into the refrigerated or air conditioned space.
Infrared Unnoticeable rays just below red in the visible spectrum that transfer heat by radiation.
Infrared Humidifier A humidifier that utilizes infrared lamps to evaporate the water into the airstream.
Infrared Lamp An electrical device that emits infrared rays.
Inherent Motor Protection Built in motor protection such as a thermal overload located in the winding.
Inhibitor Substance that prevents a chemical reaction.
Injection Term used for a hydronic system circulating pump forcing water into a secondary heating loop.
In Phase The condition existing when two waves of the same frequency have their maximum and minimum values of like polarity at the same instant.
Input/Output Board Circuit board with measured inputs and controlled outputs.
Inspection To examine equipment and installation and compare to established guidelines and codes.
Instrument The addition of sensors with feedback and control loops to an existing system.
Insulation, Electric A substance that is a poor conductor of electricity with few free electrons.
Insulation, Thermal Substance used to retard or slow the flow of heat through a wall or partition.
Insulator Nonconductive material such as glass, rubber, or plastic used to shield bare wires and electrical connections.
Integrated Circuit A miniaturized electronic circuit consisting of mainly semiconductor devices.
Integrated Circuit Board Electronic circuits are manufactured on a thin surface of semiconductor material that is also referred to as microchip or just chip.

Interlock A safety device that allows power to a circuit only after a predetermined function has taken place.
Intermittent Absorption Cycle Closed system charged with ammonia and water and activated by a gas burner.
Intermittent Ignition Furnace ignition system that operates only as required.
Internal Motor Overload (see Inherent Motor Protection)
International Green Construction Code (IGCC) Comprehensive set of requirements intended to reduce the negative impact of buildings on the natural environment.
Interstate Commerce Commission (ICC) Federal regulatory body to oversee common carriers such as railroads and trucking.
Inverter An electrical device for converting DC to AC.
Ion An ion is an atom that is either negatively (anion) or positively (cation) charged.
Ion Generator A device that charges particles.
IR Drop Voltage drop resulting from current flow through a resistor.
Isolation Relays A relay designed to prevent undesirable electrical feedback.
Isothermal At constant temperature.

Joints, Brazed A solder type connection or joint obtained by the joining of the metal parts with metallic mixtures or alloys that have a melting temperature above 1,000°F and up to 1,500°F.
Joint, Soldered A solder type connection or joint obtained by the joining of the metal parts with metallic mixtures or alloys that have a melting temperature below 1,000°F.
Joint, Welded A solder type pipe connection or joint obtained by the joining of the metal parts with metallic mixtures or alloys that have a melting temperature above 1,500°F.
Joule A measure of heat in the metric system.
Joule-Thomson Effect A drop in gas temperature due to a sudden increase in its volume.
Journal, Crankshaft Part of the shaft that contacts the bearing.
Junction Box Group of electrical terminals housed in protective box or container.

Kelvin Scale (K) Thermometer scale on which the unit of measurement equals the centigrade degree and according to which absolute zero is 0°K, the equivalent of –273.16°C.
Kilocalorie (C) Measure of heat equal to 1,000 calories.
Kilometer (km) Measure of distance equal to 1,000 m.
Kilopascal (kPa) A measure of pressure in the metric system.
Kilovoltampere (kVA) Measure of electrical work equal to 1,000 voltamperes.
Kilowatt (kW) A unit of electrical energy equal to 1,000 W.
Kilowatt Hour (kWh) A unit for measuring electrical energy equal to 3,413 Btu.
Kinetic Energy Energy in motion.
King Valve The valve located at the outlet of the receiver.

Lacquer Clear coating on wire that provides insulation for electric windings.
Ladder Diagram Electrical diagram that shows a circuit by circuit arrangement. The power lines are the "rails" of the ladder, and each individual circuit is a "rung" of the ladder.
Lagging Thermal insulation around pipes and boiler jackets.
Latent Heat Heat change that occurs when a substance changes state.
Latent Heat of Condensation Heat released when a gas condenses to a liquid.
Latent Heat of Deposition Heat released when a gas freezes to a solid.
Latent Heat of Fusion Heat released to change a liquid to a solid.
Latent Heat of Sublimation Heat absorbed when a solid changes directly to a gas without first becoming a liquid.
Latent Heat of Vaporization Heat absorbed when a liquid changes to a vapor.
Leadership in Energy and Environmental Design (LEED) Green Construction rating system created by the U.S. Green Building Council.
Leads, Meter The wires that are used to connect an electric meter to the circuit being tested.
Leads, Motor The connection wires on a motor.
Leak Detector A device or instrument used to detect and locate leaks of gases or vapors.
Light Emitting Diode (LED) A diode that lights up when electrical current flows through it.
Legionella The bacteria that can thrive in cooling towers and cause Legionnaire's disease.
Legionnaire's Disease A bacterial pneumonia named after an outbreak at the July 1976 American Legion convention in Philadelphia.
Limit Control A device used to open or close electrical circuits if temperature or pressure reaches preset limits.
Limit Switch A normally closed switch on a furnace that opens if the furnace temperature exceeds the limit setpoint.
Line Diagram Another name for a ladder or schematic diagram.
Line Set Copper tubing used for the liquid and vapor lines to connect the two parts of split system air conditioners and heat pumps.
Line Tap Valve (see Piercing Valve).
Line Voltage Voltage existing at wall outlets or terminals of a power line system above 30 V.
Line Voltage Thermostat Thermostat designed for use with voltages of 120 V or higher.
Liquefied Natural Gas (LNG) Natural gas that has been cooled until it becomes a liquid.
Liquefied Petroleum Gases (LP, LPG) Fuel that is stored as a liquid under pressure, but used as a gas. Includes any fuel gas that is composed predominantly of any of the following hydrocarbons: propane, propylene, normal butane or isobutene, and butylenes.
Liquid A fluid form of matter that takes on the shape of its container but whose volume is not determined by the volume of the container.
Liquid Charge Bulb A temperature sensing bulb with enough refrigerant to ensure that there is always some quantity of liquid in the bulb.
Liquid Floodback Liquid refrigerant returning to the suction side of a compressor during operation.
Liquid Hammer A pounding sound made by the sudden increase in static pressure in the pipes when flowing liquid is stopped suddenly by the closing of a valve.

Liquid Injection Cooling low temperature compressors by introducing small amounts of liquid refrigerant into the shell of the compressor.

Liquid Line The tube that carries liquid refrigerant from the condenser or liquid receiver to the pressure reducing device.

Liquid Nitrogen (LN_2) Nitrogen in liquid form.

Liquid Receiver Cylinder connected to condenser outlet for storage of liquid refrigerant in a system.

Liquid Refrigerant Charging Adding refrigerant to a refrigeration system in liquid form.

Liquid Refrigerant Recovery Removing refrigerant from a system in liquid form.

Liquid Slugging Hydraulic pressures created by liquid that is trapped in the clearance pocket of a reciprocating compressor during operation.

Liquid-Vapor Valve The valve on refillable refrigerant cylinders that allows either liquid or vapor to be removed from the cylinder.

Liter (L) Metric unit of volume.

Lithium Bromide The absorbent in large absorption chillers that use water as the refrigerant.

Load The amount of heat per hour that the refrigeration system is required to supply at design conditions. The amount of electrical energy expected to be connected to an electrical power supply.

Load Matching Matching the system capacity to the heat load.

Load Shedding Removing electrical loads in a building for periods of time as part of an energy management strategy.

Locked Rotor Amperage (LRA) The amount of energy a motor will draw under stalled conditions.

Lock Out Relay A control scheme that prevents the restarting of a compressor after a safety control opens, even after the safety control resets itself.

Louvers Sloping, overlapping boards or metal plates intended to permit ventilation and shed falling water.

Low Ambient Control A device used to allow proper condensing operation at low outdoor ambient temperatures.

Lowboy Furnace A furnace with the blower and heat exchanger side by side having both the return and supply air connections at the top of the furnace.

Low Glide Blend A zeotropic refrigerant with a typical temperature difference between the bubble point and the dew point of less than 3°F.

Low-Loss Fitting A fitting on the end of a refrigerant hose designed to prevent refrigerant from leaving the hose when it is disconnected from the system.

Low Pressure Control A device used to start or stop the system operation when the low pressure (low side pressure, suction pressure) drops to preset limits.

Low Pressure Cutout Device used to keep the low side evaporating pressure from dropping below a certain pressure.

Low Side The portion of the refrigeration system that operates at the evaporator pressure. Consists of the pressure reducing device, evaporator, and suction line.

Low Side Float Controls flow into the evaporator using a mechanical float located on the low side of the system.

Low Side Pressure Pressure in the low side of the refrigeration system. Also called low side pressure or suction pressure.

Low-Temperature Refrigeration a commercial refrigeration system with an evaporator saturation temperature between–30°F and 10°F and a nominal compressor rating point of–10°F.

Low Voltage An AC voltage system no higher than 30 V with a maximum rating of 100 VA. Typically 24 V in air conditioning control systems.

Low Voltage Control Controls designed to operate at voltages of 20–30 V.

Low-Voltage Thermostat A thermostat designed for a low voltage control system.

MBH Abbreviation for 1,000 Btu/hr.

Machine Room A room in a building dedicated to mechanical equipment.

Machine Screw Fastener with a head, a cylindrical body, and fine threads.

McLeod Gauge An instrument for accurate measurement of very low pressures.

Magnetic Clutch Assembly that transfers power from an outer pulley to an open compressor when energized.

Magnetic Field The space in which a magnetic force exists.

Magnetic Flux The lines of force around a magnetic field.

Magnetic Gasket A door seal on a refrigerator or freezer that uses a magnet to ensure tight closure.

Magnetic Overload Protection Breaks the circuit to an electric motor when high current draw produces a strong magnetic field.

Magnetism A force that attracts certain metals like iron and alloys of nickel and cobalt

Makeup Air Fresh air. The air supplied to a building to replace air exhausted from the building.

Makeup Water Line The line connected to the loop system for filling or adding liquid as necessary.

Male Thread A pipe or fitting with a thread on the outside.

Manifold A pipe with several outlets for distributing fluid flow.

Manifold Gauge A device constructed to hold compound and high pressure gauges and valves to control flow of fluids through it.

Manometer An instrument used to measure the pressure of gases or vapors. The gas pressure is balanced against a column of liquid, such as water or mercury, in a U-shaped tube open to atmospheric pressure.

Manometer, Inclined A manometer on which the liquid tube is inclined from horizontal to produce wider, more accurate readings over a smaller range of readings.

Manual Reset A safety device with a button or flag that must be pushed in order to restore normal operation.

Mapp Gas Dow Chemical trade name for methylacetylene-propadiene, a liquefied petroleum gas that burns hotter than propane and is sometimes used for soldering and brazing.

Marine Water Box The water manifolding portion of a chiller or boiler with a removable cover.

Mass A quantity of matter cohering together to make one body.

Matched Belts A set of belts that are precisely sized to the same circumference for use on a multiple belt drive.

Matched System A split system air conditioner or heat pump whose indoor and outdoor components are designed for use with each other.

Matter The substance that physical objects are made of. Matter has weight and takes up space.

Maximum Operating Pressure (MOP) The maximum pressure a pressure limiting thermostatic expansion valve will allow in the evaporator.

Mean Annual Earth Temperature The mean average temperature of the earth as it changes throughout the year. The largest factor for earth temperature change is sunshine.

Mean Temperature Difference The average temperature between the temperature before a process begins and the temperature after the process is completed.

Mechanical Controls Controls that operate by purely mechanical means, such as pressure regulating valves.

Mechanical Cycle Cycle that is a repetitive series of mechanical events.

Medium-Temperature Refrigeration A commercial refrigeration system with a saturated evaporator temperature between–10°F and 30°F and a nominal compressor rating point of 20°F.

Mega (M) SI prefix for 1,000,000.

Megohm (MΩ) 1,000,000 Ω of electrical resistance.

Megohmmeter A meter that can read millions of ohms of resistance.

Melting Point Temperature at atmospheric pressure, at which a substance will melt.

Mercaptan An odorant additive that gives natural gas its characteristic odor.

Mercury Bulb An electrical circuit switch that uses a small quantity of mercury in a sealed glass tube to make or break electrical contact with terminals within the tube.

Meter Metric unit of linear measurement.

Metering Device Controls the flow of refrigerant into the evaporator of a refrigeration system.

Methane A hydrocarbon gas with the formula CH_4, the principal component of natural gas.

Metric System The measurement system used worldwide based on units of ten.

Micro A combining form denoting one millionth.

Microfarad (MFD) The unit of measurement for capacitance, equal to a millionth of a farad.

Micrometer A device for making measurements accurate to 1/1,000th of an inch.

Micron (μm) A metric unit of length equal to one millionth (1/1,000,000) of a meter used for measuring vacuums.

Micron Gauge Measures deep vacuums in microns of mercury absolute pressure.

Microprocessor A single integrated chip that combines all the functions of a computer central processing unit.

Midposition Description of the position of a refrigeration service valve when it is in between the front seat and back seat of the valve.

Migration of Refrigerant Refrigerant moving from one part of a refrigeration system to another during the off cycle.

Milli (m) A term used to denote one thousandth (1/1,000) of a unit. For example, a milliampere is 1/1,000 of an ampere.

Mineral Oil (MO) Lubricant made by refining crude oil. It is used in CFC and HCFC systems.

Minimum Efficiency Reporting Value (MERV) System used to compare the relative efficiency of air filters.

Miscibility The ability of a refrigerant and a lubricant to mix.

Modulating A type of device or control that tends to adjust by increments (minute changes) rather than by either full on or full off operation.

Module A group of components or circuits in a single housing. Usually electronic.

Moisture Indicator Instrument used to measure the moisture content of a refrigerant.

Mold A fungus that thrives in wet environments and can contribute to indoor air quality problems.

Molecular Motion Vibration and movement of individual molecules of a substance that increases as the level of molecular energy increases.

Molecule The smallest portion of a compound that retains the identity and characteristics of the compound.

Mollier Diagram Graph of refrigerant pressure, heat, and temperature properties.

Montreal Protocol An international agreement to control the production and release of ozone depleting substances.

Motor Converts electrical energy into rotating mechanical motion using electromagnetism.

Motor, Permanent Split Capacitor A single phase induction motor that uses both run and start windings in the running mode. The start winding is connected in series with a capacitor to change the electrical characteristics of the winding.

Motor Burnout A condition in a motor where the insulation has deteriorated due to overheating.

Motor Control A control to start and stop a motor at preset pressures or temperatures.

Motor Starter A high capacity electrical contact operated by an electromagnetic coil and containing properly sized overload cutouts.

Muffler A refrigeration component located immediately after the compressor that reduces sound and vibration set up by the pulsing flow of refrigerant.

Mullion Stationary structure between two doors in a refrigerator or freezer.

Mullion Heater A heater that keeps the mullion temperature above the dew point.

Multimeter A meter that reads multiple electrical properties, typically including volts and ohms.

Multiple Circuit Coil A coil that has more than one path for simultaneous fluid flow.

Multiple Evacuation Evacuating a system, charging it with dry nitrogen, releasing the nitrogen, and then evacuating the system again.

Multistage System A system that delivers heating or cooling in two or more steps, or stages.

Naphthenic Oil Mineral oil that has a low wax content.

National Electrical Code (NEC) A code of electrical rules based on fire underwriters' requirements for interior electric wiring.

National Environmental Balancing Bureau (NEBB) An organization that certifies companies and technicians in building science services, including building commissioning, testing, adjusting, and balancing, and clean rooms.

National Fuel Gas Code (NFPA 54) A safety code for the installation and utilization of fuel gas piping systems and equipment.

National Fire Protection Association (NFPA) Organization dedicated to fire prevention that writes the National Electric Code and the National Fuel Gas Code.

National Green Building Standard Developed by the National Association of Home Builders that identifies green building practices in six different categories.

National Institute of Standards and Technology (NIST) A nonregulatory federal agency within the U.S. Department of Commerce that sets standards for measurement and technology.

National Pipe Taper (NPT) The standard specification for pipe threads.

Natural Convection Circulation of a gas or liquid due to difference in density resulting from temperature differences.

Natural Draft Airflow created by natural convection.

Natural Draft Cooling Tower A cooling tower that depends on natural convection instead of a fan.

Natural Gas Any gas found in the earth, as opposed to gases that are manufactured.

Needle Valve A valve used to accurately control very low flow rates of fluids.

Negative Electrical Charge An electrical charge resulting from an atom or object having more electrons than protons.

Negative Temperature Coefficient Thermistor (NTC) A semiconductor whose electrical resistance increases as its temperature decreases. Commonly used to measure temperature of outdoor coils on heat pumps to initiate and terminate defrost cycles.

Neoprene A synthetic rubber that is resistant to hydrocarbon oil and gas.

Net Oil Pressure The difference between the pressure at the outlet of the oil pump and the crankcase pressure in a compressor.

Net Refrigeration A measure of the amount of cooling that takes place in an evaporator. It is the enthalpy of the gas leaving the evaporator minus enthalpy of the gas entering the evaporator. The units are Btu/lb.

Net Stack Temperature The difference between the temperature of the flue gases and the temperature of the combustion air.

Neutralizer A chemical added to a system to counteract the effect of acids.

Neutron A subatomic particle located in the nucleus of an atom that does not have an electrical charge.

Newton (N) A measure of force in the SI system. One Newton accelerates one kilogram one meter per second squared.

Nichrome An alloy of nickel and chromium used in the construction of electrical heating elements.

Nitrogen A common, inert gas. The air is approximately 78% nitrogen.

No-Frost Freezer A freezer with an automatic defrost cycle.

Noise Dosimeter An instrument used to measure exposure to sound levels.

Nominal The named rating, or standard rating, for quantities and dimensions whose actual value is variable.

Nominal Size Measurement for tubing and pipe equal to the approximate inside diameter.

Nominal Voltage The name plate voltage for an appliance. AHRI standard 110–2002 sets the nominal voltages for air conditioning equipment.

Noncondensible Gas Gas that does not change into a liquid at operating temperatures and pressures.

Nonferrous Group of metals and metal alloys that contain no iron; also metals other than iron or steel. In heating systems, the principal nonferrous metals are copper and aluminum.

Non-Frosting Evaporator A commercial refrigeration evaporator that operates above freezing.

Noninductive Load An electrical load that does not have any electromagnetic components.

Normally Closed (NC) Switch contacts closed with the circuit deenergized.

Normally Open (NO) Switch contacts open with the circuit deenergized.

North Pole The pole on a magnet that the magnetic lines of force flow out of.

Nozzle The device on the end of the oil fuel pipe used to form the oil into fine droplets by forcing the oil through a small hole to cause the oil to break up.

NOx A term for various compounds of nitrogen and oxygen that are byproducts of combustion.

Nut Driver A tool with a handle, shaft, and socket on the end for tightening and loosening hexagonal nuts and screws.

Odorant A substance added to an otherwise odorless, colorless, and tasteless gas to give warning of gas leakage and to aid in leak detection.

Off-Cycle Defrost A naturally occurring defrost in commercial refrigeration equipment with an evaporator temperature below freezing and a box temperature above freezing.

Off Mode Cycle That part of the operating mode when the system has been shut down by the controls.

Offset The difference between the control setpoint and the actual maintained condition.

Ohm (Ω) A unit of resistance.

Ohmmeter A meter for measuring electrical resistance.

Ohm's Law States the relationship between voltage, current, and resistance in a purely resistive circuit.

Oil Binding Physical condition when an oil layer on top of refrigerant liquid hinders it from evaporating at its normal pressure-temperature condition.

Oil Burner The part of an oil furnace or boiler that produces heat by burning fuel oil.

Oil Level Regulator Controls the level of oil in a compressor in a multiple compressor system.

Oil Pressure, Compressor The pressure created by the oil pump of a compressor with a positive pressure lubrication system.

Oil Pressure, Oil Burner The pressure created by the oil pump in an oil burner.

Oil Pressure Safety Switch A timed pressure differential switch that opens the control circuit to a compressor if the oil pressure drops below safe levels.

Oil Reservoir The vessel that holds reserve oil in an oil level control system used for a system with multiple compressors.

Oil Rings Expanding rings mounted in grooves and piston; designed to prevent oil from moving into compression chamber.

Oil Separator A device used to separate refrigerant oil from refrigerant gas and return the oil to crankcase of compressor.

Oil Slugging Hydraulic pressures created by oil that is trapped in the clearance pocket of a reciprocating compressor during operation.

One Pipe Fitting A specially designed tee for use in a one-pipe system to connect the supply or return branch into a circuit.

One Pipe System A forced hot water system using one continuous pipe or main from the boiler supply to the boiler return.

One Time Relief Valve A pressure relief device that must be replaced after it has opened.

One Time Fuse A fuse with little time delay in opening and little tolerance for surge current.

Open Circuit An interrupted electrical circuit that prevents electrical flow.

Open Compressor One with a separate motor or drive.

Open Display Case A commercial refrigerator or freezer that is designed to remain open on the top or front.

Open-Loop, Control System A control system where the controller output does not affect the condition the controller is sensing.

Open-Loop, Water System A recirculating water system that is open to the atmosphere somewhere in the system.

Open-Loop, Heat Pump A water source heat pump that uses groundwater for its water source.

Open Refrigeration System A refrigeration system that uses an open compressor.

Open Winding A motor winding with a break in it.

Operating Control Cycles system components during normal operation. Commonly used to refer to the control that cycles the compressor in a refrigeration system.

Operating Pressures The low and high side pressures in an operating refrigeration system.

Orifice An accurately sized opening that controls the flow of vapor or liquid at given pressures across the opening.

Orifice, Oil Metering A small hole in the pickup tube in the accumulator used to ensure oil return to the compressor in small easily handled quantities.

Orifice Spud A removable plug containing an orifice that determines the quantity of gas that will flow.

O-Ring An elastic seal shaped in a circle used for sealing mechanical parts.

Oscilloscope A fluorescent coated tube that visually shows an electrical wave.

Outdoor Coil The portion of a heat pump that is located outside the home and functions as a heat transfer point for collecting heat from or dispelling heat to the outside air.

Outside Air Atmosphere exterior to the conditioned area.

Overload Load greater than the load for which the system or mechanism was intended.

Overload Protector A device that will stop operation of the unit if dangerous conditions arise.

Oxidation A chemical reaction between oxygen and other materials that forms new compounds. Rust is iron oxidation.

Oxygen An element normally found in gas form that makes up approximately 21% of the atmosphere. It is required for combustion.

Ozone A gaseous form of oxygen (O_3).

Ozone Depletion Reduction of the ozone concentration in the stratosphere.

Ozone Depletion Potential (ODP) A rating of a chemical's potential to deplete stratospheric ozone. By definition, the ozone depletion potential of CFC 11 is 1.

Package Boiler A boiler having all components assembled as a unit.

Packaged Terminal Air Conditioner (PTAC) A small capacity, through the wall packaged air conditioner designed for use in motels, dorm rooms, and efficiencies.

Packaged Unit A complete refrigeration or air conditioning system in which all the components are factory assembled into a single unit.

Packing The material that is packed around the stem of a manual valve to seal it.

Paraffinic Oil Mineral oil with a high wax content.

Parallel Circuit connected so current has two or more paths to follow.

Parallel Circuit An electrical circuit in which there is more than one different path across the power supply.

Partial Pressures Condition where two or more gases occupy a space and each one creates part of the total pressure.

Parts per Million (ppm) A measurement of the concentration of gases and contaminants at relatively low levels.

Part Winding Start Starting a dual voltage, three phase motor using only half of the motor windings when used on low voltage to reduce the motor inrush current.

Pascal (Pa) The SI measurement of pressure equal to 1 N of force applied over a square meter of area.

Pascal's Law A pressure imposed upon a fluid is transmitted equally in all directions.

Passive Refrigerant Recovery Device A refrigerant recovery device that has no operating components.

Passive Solar Heating Solar heating that depends on architectural design rather than mechanical and electrical components.

Peltier Effect The principle that makes thermoelectric refrigeration work. When current flows through a thermocouple, one end heats up and the other cools down.

Permeance The ratio of water vapor flow to the vapor pressure difference.

pH A term based on the hydrogen ion concentration in water, which denotes whether the water is acid, alkaline, or neutral.

Permanent Magnet A magnet that retains in magnetic effect and produces its own magnetic field.

Permanent Split Capacitor Motor (PSC) (see Motor, Permanent Split Capacitor).

Phase, Electrical The number of separate sine wave forms providing power to an alternating current device.

Phase, Physics The physical state of a substance: solid, liquid, gas, or plasma.

Phase Change A substance changing from one physical state to another.

Phase Loss Monitor A safety device that monitors three phase current and shuts down a system or component if one of the three phases drops out.

Phase Reversal A rearranging of the three power legs in a three phase circuit that results in three phase motors reversing their rotation.

Photoelectricity A physical action wherein an electrical flow is generated by light waves.

Photovoltaic Cell A semiconductor device that converts sunlight into electricity.

Pictorial Wiring Diagram Shows the electrical components as they appear on the equipment and shows the actual routing and color of all interconnecting wiring.

Piercing valve A valve that bolts or brazes onto the outside of a pipe and pierces it to gain access to the system.

Piezoelectric A property of some crystals that produces electric potential when they are struck.

Pilot A small flame that is used to ignite the gas at the main burner.

Pilot Duty Relay A relay used for switching other controls. Only rated for switching small amounts of current.

Pilot Generator A device used to generate voltage in a millivolt heating system.

Pilot Safety Valve A valve that will shut off the main gas if the pilot flame is not proved.

Pilot Switch A control used in conjunction with gas burners. Its function is to prevent operation of the burner in the event of pilot failure.

Pinch Off Tool Device used to press the walls of tubing together until fluid flow ceases.

Pipe Used to convey fluids. Pipe is measured by its approximate inside diameter and has relatively thick walls compared with tubing.

Piston Close fitting part that moves up and down in a cylinder.

Piston Displacement Volume obtained by multiplying area of cylinder bore by length of piston stroke.

Pitch Pipe slope.

Pitot Tube A device that senses static and total pressure.

Plenum A cube shaped duct or box on the supply or return side of an air handling unit to connect the supply or return duct system to the unit.

Pneumatic Controls Controls that are operated by air pressure.

Point of Vaporization The location of the position in the evaporator coils where the last bit of liquid refrigerant is vaporized.

Polarity The condition denoting the direction of current flow.

Pollen Count An index that shows the number of pollen grains in a cubic meter of air.

Polyalkylene Glycol (PAG) A synthetic lubricant used with HFC refrigerants.

Polybutylene (PB) A plastic used for pipe material that has excellent creep resistance as well as high resistance to stress cracking. Recommended for earth loops.

Polychlorinated Biphenyl (PCB) A dielectric oil that was used in transformers and capacitors. It is no longer used because it is a carcinogen.

Polyethylene (PE) A plastic used for tubing for ground loop systems, cold water lines, and heat pump piping.

Polyethylene, Cross Linked (PEX) A derivative of polyethylene that is stronger, more elastic, and can withstand higher temperatures.

Polyol Ester (POE) A synthetic lubricant that can be used with most CFC, HCFC, and HFC based refrigerants.

Polyphase An electrical device or system that is designed to operate on more than one phase of alternating current.

Polyphase Motor An electric motor that is designed to operate on more than one phase of alternating current.

Polystyrene Plastic used as an insulation in some refrigerator cabinet structures.

Polyvinyl Chloride (PVC) A thermoplastic material used for making cold water piping and drain lines.

Ponded Roof Flat roof designed to hold a quantity of water.

Positive Displacement A compressor that compresses gas by filling the space occupied by the gas with a physical part, such as a piston.

Positive Electrical Charge An electrical charge resulting from an atom or object having more protons than electrons.

Positive Pressure A pressure above atmospheric pressure.

Positive Temperature Coefficient Thermistor (PTC) A semiconductor whose electrical resistance increases as its temperature increases. Sometimes used as an electronic start relay in the start circuits for compressors.

Potable Water Water that is suitable for human consumption.

Potential The amount of voltage or electrical pressure between two points of an electrical circuit.

Potential Energy Stored energy.

Potential Relay A relay that senses the back emf of the start winding and drops out the starting components of a compressor.

Potentiometer A variable resistor.

Pounds per Square Inch (psi) Unit of pressure measurement.

Pounds per Square Inch Absolute (psia) Pressure measured in pounds per square inch absolute.

Pounds per Square Inch Gauge (psig) A symbol or initials used to indicate pressure in pounds per square inch gauge.

Pour Point Lowest temperature at which a particular oil will pour.

Powder Actuated Tool (PAT) A tool that drives nails and anchors into concrete and steel using gunpowder.

Power Time rate at which work is done or energy consumed. Source or means of supplying energy.

Power Burner A burner that uses a blower to force in combustion air and control the mixture of fuel and air.

Power Element Sensitive element of a temperature operated control.

Power Factor The rate of actual power as measured by a wattmeter in an alternating circuit to the apparent power determined by multiplying amperes by volts.

Predictive Maintenance Performing calibrated tests on operating equipment and comparing them to base line measurements so that equipment can be repaired or replaced as it wears, before it actually fails.

Pressure Force per unit area.

Pressure, Absolute (psia) (see Absolute Pressure)

Pressure, Atmospheric Pressure exerted by the atmosphere.

Pressure, Back The pressure the pump must overcome to be able to force the liquid into the pressure tank. The pressure in the tank.

Pressure, Gauge (psig) Pressure measured above atmospheric pressure.

Pressure, Static (P_s) The pressure in the system when the pump is idle.

Pressure, Suction The pressure in the low pressure or evaporator section of the refrigeration system.

Pressure Drop (ΔP) The pressure difference between two locations in a circuit, the difference being a result of flow resistance in the circuit.

Pressure-Enthalpy Diagram (PH) (see Mollier Diagram)

Pressure Gauge Instrument for measuring pressure in pounds per square inch gauge (psig).

Pressure Limiter Device that remains closed until a certain pressure is reached and then opens and releases fluid to another part of the system.

Pressure Motor Control A device which opens and closes on electrical circuit as pressures change to desired pressures.

Pressure Limiting Thermostatic Expansion Valve A thermostatic expansion valve that will not feed more refrigerant once it reaches its maximum operating pressure.

Pressure Operated Altitude Valve (POA) Device that maintains a constant low side pressure independent of altitude of operation.

Pressure Reducing Device The device used to reduce the pressure on the liquid refrigerant and thus the boiling point before entering the evaporator.

Pressure Reducing Valve A diaphragm operated valve installed in the makeup water line of a hot water heating system.

Pressure Regulator A device for controlling a uniform outlet gas pressure.

Pressure Relief Valve A device for protecting a tank from excessive pressure by opening at a predetermined pressure.

Pressure Switch An electrical switch that is opened and closed by changes in pressure.

Pressure-Temperature Chart (PT Chart) Shows refrigerant saturation pressures and temperatures.

Pressure-Temperature Plugs (P/T) Allow thermometers and pressure gauges entrance into a water system without draining or removing the pressure of the system.

Pressure-Temperature Relationship The predictable relationship between the pressure and temperature of a saturated mixture.

Pressure Transducer Produces a variable electrical signal based on the applied pressure.

Pressure Vessel An enclosure designed to hold pressure.

Pressure Water Valve A valve that controls the flow of water through a water cooled condenser based on the condenser pressure.

Preventative Maintenance Performing regularly scheduled maintenance in order to prevent equipment failure and shutdown.

Primary Air The combustion air introduced into a burner that mixes with the gas before it reaches the port.

Primary Control One type of operating controller for an oil burner.

Primary Voltage The voltage of the circuit supplying power to a transformer.

Primary Winding The input winding on a transformer.

Process Tube Length of tubing fastened to hermetic unit dome, used for servicing unit.

Product The material being cooled in a commercial refrigeration system.

Product Heat Load The heat that is added to the refrigeration system from the product being cooled.

Programmable Thermostat A thermostat that can be programmed to automatically change setpoints based on the time.

Propane A hydrocarbon fuel.

Propeller Fan A fan with blades mounted perpendicular to the shaft arranged in a circle. Airflow is parallel to the shaft.

Proportional Controller A control that can blend two streams of air or water in varying proportions.

Propylene Glycol A nontoxic chemical used for antifreeze solutions in HVACR applications.

Proton A positively charged subatomic particle found in the nucleus of atoms.

Psychrometer An instrument used to measure the dry bulb and wet bulb temperatures of air.

Psychrometric Chart A chart that graphs the relationship of the temperature, pressure, and moisture content of air.

Psychrometrics The study of the thermodynamic and physical properties of air.

Puffback An explosion in the firebox of an oil burner accompanied by a puff of smoke created by delayed ignition. Potentially very hazardous.

Pulley A wheel with a groove around its circumference used for transmitting mechanical power using a belt designed to fit in the groove.

Pulse Combustion A series of small explosions that produces extremely efficient combustion.

Pulse Furnace A furnace that uses pulse combustion for high operating efficiency.

Pulse Width Modulation (PWM) A pulsing electronic signal that communicates based on the length of the pulses and the time intervals in between the pulses. Used in some types of electronic control systems.

Pump A motor driven device used to mechanically circulate water in the system.

Pump, Centrifugal A pump that works by taking water into the center of a high-rpm impellar and throwing it to the outside of the wheel using centrifugal force.

Pump, Screw Moves liquid using two intermeshing, auger shaped screws turning in opposite directions.

Pumpdown A service procedure where the refrigerant is pumped into the receiver.

Pump Head The difference in pressure on the supply and intake sides of the pump created by the operation of the pump.

Purging Releasing compressed gas to the atmosphere through some part or parts for the purpose of removing contaminants from that part or parts.

Pyrometer Instrument for measuring temperatures.

Quenching Submerging hot solid object in cooling fluid.

Quench Valve A valve used on transport refrigeration units to meter small amounts of liquid refrigerant into the suction side of the compressor to reduce the compressor discharge temperature.

Quick Connect Coupling A device that permits an easy and fast means of connecting two fluid lines or fittings together without the use of solder.

R Value, SI The thermal resistance of a building material in units of $K{\cdot}m^2/W$.

R Value, U.S. The thermal resistance of a building material in units of $ft^2 \cdot °F \cdot h/Btu$.

Rack System A commercial refrigeration system with multiple compressors manifolded together on a rack.

Radiant Heating A heating system in which only the heat radiated from panels is effective.

Radiation Transfer of heat by heat rays.

Radiator A heating unit exposed to view within the room or space to be heated.

Radiator Valve A valve installed on a terminal unit to manually control the flow of water through the unit.

Radon A naturally occurring radioactive gas sometimes found in basements and crawl spaces.

Range Pressure or temperature settings of a control.

Rankine (R) Absolute Fahrenheit scale.

Rated-Load Amperage (RLA) The current that a motor draws when it is operating at its rating condition.

Reactance Opposition to alternating current by either inductance or capacitance, or both.

Reamer A tool for removing the burr left by a tubing cutter.

Receiver A refrigeration component located after the condenser that stores high pressure liquid refrigerant.

Receiver-drier A refrigerant receiver that incorporates a desiccant bag for dehydrating the refrigerant.

Reciprocal The result of dividing any number into 1 is its reciprocal.

Reciprocating Action in which the motion is back and forth in a straight line.

Reciprocating Compressor A compressor that uses pistons, cylinders, and valves to compress gas.

Recirculated Air Return air passed through the conditioner before being again supplied to the conditioned space.

Recirculated Water System A heating or cooling system that reuses the same water by circulating the water through a closed loop.

Reclaim To reprocess refrigerant so that it meets the AHRI 700–2006 standard for new refrigerant.

Recording Ammeter An ammeter that can record the amp draw over a period of time either on paper or in the meter's memory.

Recording Thermometer A thermometer that can record the temperature over a period of time either on paper or in its memory.

Recording Voltmeter A voltmeter that can record voltage over a period of time either on paper or in the meter's memory.

Recovery Removing refrigerant from a system and storing it in an external container without cleaning or processing the refrigerant.

Recovery Cylinder A DOT approved cylinder for storing recovered refrigerant.

Rectifier, Electric An electrical device for converting AC into DC.

Rectifier, Refrigeration A component in an absorption system that condenses water vapor and returns it to the generator to improve ammonia concentration to the condenser.

Recuperative Heat Exchanger A secondary heat exchanger on high efficiency gas and oil furnaces that recovers heat from the combustion gases through condensation.

Recycling Removing refrigerant from a system, processing it using filters and oil separators, and storing it in an external container.

Reducing Fitting A pipe fitting designed to change from one pipe size to another.

Reed Valve Thin flat tempered steel plate fastened at one end.

Reference Point A specific known condition used for calibrating instruments.

Reference Pressure The pressure that data in a chart or table are based on.

Reference Temperature The temperature that the data in a chart or table are based on.

Refractory A material that can withstand very high temperature that is used to line the combustion chamber of furnaces.

Refrigerant Substance used in a refrigerating mechanism to absorb heat in an evaporator coil and to release its heat in a condenser.

Refrigerant Blend A mixture of two or more refrigerants that do not chemically combine.

Refrigerant Charge Quantity of refrigerant in a system.

Refrigerant Control Another name for the refrigerant expansion device or metering device in a refrigeration system.

Refrigerant Quality The percentage of liquid and vapor in a saturated mixture.

Refrigerant Reclaim (see Reclaim)

Refrigerant Recovery (see Recovery)

Refrigerant Recycling (see Recycling)

Refrigerating Effect The amount of heat in Btu/hr the system is capable of transferring.

Refrigerant Expansion Device Another name for the metering device or refrigerant control in a refrigeration system.

Refrigerated Air Drier Reduces the dew point of compressed air using a refrigeration system.

Refrigeration The process of transferring heat from one place to another.

Refrigeration Cycle A process during which a refrigerant absorbs heat at a relatively low temperature and reflects heat at a relatively higher temperature.

Refrigeration Oil Specially prepared oil used in refrigerator mechanism. It circulates with refrigerant.

Refrigeration System The combination of interconnecting devices, tubes, and/or pipes in which the refrigerant is circulated for the purpose of exchanging heat to produce cooling.

Register Combination grill and damper assembly on an air opening or at the end of an air duct.

Regulator A device that controls the flow through it to control the downstream pressure.

Relative Humidity (RH) The ratio of the weight of moisture that air actually contains at a certain temperature as compared to the amount that it could contain if it were saturated.

Relay Electrical mechanism that uses small current in the control circuit to operate a valve switch in the operating circuit.

Relief Opening The opening in a draft hood to permit ready escape to the atmosphere of flue products.

Relief Valve Safety device designed to open before a dangerous pressure is reached.

Reluctance The opposition to magnetic lines of force passing through a magnetic material.

Remote Bulb A temperature sensing element that works by pressure change that is transmitted through a capillary tube to the control. It is located in a different location from the component it controls.

Remote Condenser A condenser that is located by itself in a different location from the compressor.

Repulsion Start Induction Motor Type of motor that has an electrical winding on the rotor for starting purposes.

Resilient Motor Mount A motor mount that contains flexible material such as rubber to reduce transmission of vibration and noise.

Refrigerant Control Another name for the refrigerant expansion device or metering device in a refrigeration system.

Refrigerant Dye Used to locate difficult to find leaks in refrigeration systems. It actually dyes the refrigerant oil, not the refrigerant.

Refrigerant Management System An accounting system to document refrigerant use.

Resistance The opposition to current flow by a physical conductor.

Resistor An electronic component with a fixed electrical resistance that opposes current flow.

Resonance The pipe organ effect produced by a gas furnace when the frequency of the burner flame combustion and the pressure wave distance in the burner pouch are in exact synchronization.

Restrictor A refrigerant metering device with a small orifice that creates a pressure drop because of its restriction.

Retrofit, Mechanical Reworking existing equipment to meet current efficiency or code requirements.

Retrofit, Refrigerant Reworking an existing system to replace an older ozone depleting refrigerant with a newer environmentally friendly refrigerant.

Return Air Air returned from conditioned or refrigerated space.

Return Well The well that water from an open loop water source heat pump is returned to.

Return Piping That portion of the piping system that carries water from the terminal units back to the boiler.

Reverse Acting Control (RA) A switch controlled by temperature and designed to open on temperature drop and close on temperature rise.

Reverse Cycle Defrost A method of defrosting the evaporator by means of flow valve(s) to move hot vapor from the compressor into the evaporator.

Reverse Cycle Refrigeration System Commonly called a heat pump. A refrigeration system capable of reversing its operation and direction of heat transfer.

Reverse Return A two-pipe system in which the return connections from the terminal units into the return main are made in the reverse order from that in which the supply connections are made.

Reversing Valve A device used to change the direction of refrigerant vapor flow between the evaporator to compressor and from the compressor to the condenser depending on the heating or cooling effect desired.

Rheostat An adjustable or variable resistor.

Rich Mixture A mixture of gas and air containing too much fuel or too little air for complete combustion of the gas.

Riser A vertical pipe carrying gas or fluid up against gravity.

Rod and Tube Consists of a rod with a low expansion rate inside a tube with a high expansion rate. Thermal change produces linear movement for operating switches.

Rollout A condition where flame rolls out of a combustion chamber when the burner is turned on.

Roof Mounted The unit is mounted on a platform designed to distribute the weight of the unit over as wide an area of the roof as possible.

Root Mean Square Voltage (RMS) An accurate method of determining the effective voltage of an alternating current waveform.

Rotary Compressor Mechanism that compresses by trapping gas in a cylinder between an offset disc and one or more blades.

Rotating Blade Rotary Compressor A rotary compressor with blades housed in the offset disc that rotate with the compressor shaft.

Rotor Rotating part of a mechanism.

Run Capacitor A capacitor designed to stay in the circuit all the time in a permanent split capacitor motor.

Running Winding Electrical winding of motor that has current flowing through it during normal operation of the motor.

Runout This term generally applies to the horizontal portion of branch duct between the main trunk and the diffuser.

Rupture Disk A disk on a low pressure chiller that breaks to keep the pressure of the chiller from exceeding 15 psig.

S Lock An S shaped metal strip used to hold the long sides of rectangular metal duct together.

SI System The international system of units based on the metric system.

Saddle Valve Valve body shaped so it may be silver brazed to refrigerant tubing surface.

Safety Control Device that will stop the refrigerating unit if unsafe pressures and/or temperatures are reached.

Safety Interlock Switch A switch that prevents the operation of a machine at an unsafe condition, such as with a panel removed.

Safety Plug Device that will release the contents of a container above normal pressure conditions and before rupture pressures are reached.

Sail Switch A switch that detects airflow in ductwork using a "sail."

Satellite Compressor A compressor on a rack system that is dedicated to the lowest temperature and pressure evaporators.

Saturated Vapor A vapor condition that will result in condensation into droplets of liquid as vapor temperature is reduced.

Saturation Condition when both liquid and vapor are present or either phase is just about to appear or disappear.

Saybolt Universal Seconds (SUS, SSU) Used to measure oil viscosity. It is the time required for 60 ml of oil to flow through the calibrated orifice.

Scaling The formation of lime and other deposits on the water side surfaces of heat exchangers.

Schematic Wiring Diagram Another name for ladder or line diagram.

Schrader Valve Spring loaded device that permits fluid flow when a center pin is depressed.
Scotch Yoke Mechanism used to change reciprocating motion into rotary motion or vice versa.
Screw Compressor Compresses gas using two intermeshing, auger shaped screws turning in opposite directions.
Scroll Compressor Reduces the gas volume using a fixed and an orbiting scroll (spiral) to move gas from the outside of the scrolls to the inside.
Sealed Combustion Provides combustion air for a furnace, water heater, or boiler through a pipe, sealing off the combustion process from the air surrounding the equipment.
Seasonal Energy Efficiency Ratio (SEER) The total cooling of a central air conditioner in Btu during its normal annual usage period for cooling divided by the total electric input in watt-hours during the same period.
Seat The part of a valve body that the moveable portion of the valve rests against when sealing off.
Secondary Air Combustion air externally supplied to a burner flame at the point of combustion.
Secondary Refrigerant The water or brine that is circulated in a chiller system from the chiller to the heat load.
Secondary Refrigeration System Uses a primary refrigerant in a compression refrigeration system to cool a fluid that is circulated to the actual heat load. The secondary refrigerant uses all sensible cooling and does not change state.
Secondary Voltage The output, or load supply voltage, of a transformer.
Second Law of Thermodynamics Heat will flow only from material at a certain temperature to material at a lower temperature.
Seebeck Effect The generation of electrical current when a temperature difference exists between the two ends of a thermocouple.
Semiconductor A material that conducts electricity better than an insulator but not as well as a conductor.
Semihermetic Compressor A serviceable hermetic compressor.
Sensible Heat Heat energy added to or removed from a material that causes a change in temperature of the material without a change in state of the material.
Sensor A material or device that changes characteristics (electrical or mechanical) with a change in temperature and pressure.
Sequencer A control device used to control electrical circuits that use a heat motor for time delay of the operation of the device (see Heat Motor).
Series A circuit with one continuous path for current flow.
Series Wound Motor A brush-type motor with a wound armature that is wired in series with the field windings. It can operate on either DC or AC current.
Serviceable Hermetic Hermetic unit housing containing motor and compressor assembled by use of bolts or threads.
Service Factor (SF) A decimal factor that produces the maximum safe motor output when multiplied by the motor's nominal capacity rating.
Service Valve Device used by service technicians to check pressures and change refrigerating units.
Setpoint The temperature setting on a thermostat.
Shaded Pole Motor A small AC motor used for light start loads.
Shaft Seal A device used to prevent leakage between shaft and housing.
Shell and Coil Condenser A water cooled condenser consisting of a coil of tubing inside a shell. Water runs through the coil, and the refrigerant is in the shell.
Shell and Tube A water cooled condenser consisting of straight tubes passing through a shell. Water runs through the tubes, and the refrigerant is in the shell.
Short Circuit A low resistance connection (usually accidental and undesirable) between two parts of an electrical circuit.
Short Cycling Refrigerating system that starts and stops more frequently than it should.
Shroud A baffle around a propeller fan that separates the intake air from the exhaust air.
Sick Building Syndrome (SBS) A combination of ailments associated with an individual's exposure to a particular building.
Sight Glass Glass tube or glass window in refrigerating mechanism that shows the amount of refrigerant or oil in the system.
Silica Gel Chemical compound used as a drier that has ability to absorb moisture.
Silicon A semiconductor material used in the manufacture of electronic components.
Silver Brazing Brazing process in which brazing alloy contains some silver as part of the joining alloy.
Sine Wave The shape created when the voltage of an alternating current is graphed. It appears like a regular set of repeating peaks and valleys connected by a single smooth curve.
Single Package Heat Pump A system that has all components completely contained in one unit.
Single Phase An alternating current with a single sine waveform supplied by two wires.
Single-Phase Motor An electric motor designed to operate on single phase alternating current.
Single Phasing Occurs when a three phase motor is suddenly operating on a single phase of current instead of three because one of the three legs of power drops out. The current in the remaining two legs increases and the motor windings rapidly overheat.
Single Pipe System, Steam Heat A method of piping that allows a single pipe to carry both the steam to the radiators and the condensate back to the boiler.
Single Pipe System, Oil Heat Using a single pipe to carry the oil from the tank to the oil burner.
Single Pole Double Throw Switch (SPDT) A switch that can connect a center post to one of two circuits.
Single Pole Single Throw Switch (SPST) A switch that can open or close a single circuit.
Skin Condenser A condenser commonly found on chest type freezers that uses the outer surface of the cabinet to radiate heat.
Slinger Ring A ring around the outside of the condenser fan on window units that throws evaporator condensate water onto the condenser, causing the water to evaporate.
Sling Psychrometer Humidity measuring device with wet and dry bulb thermometers.

Slinky Loop A type of loop used in ground source heat pump installations in which a coil of tubing is stretched out in a ditch a flattened slinky.

Slip The difference between the synchronous speed and the actual speed of an induction motor.

Slug A quantity of liquid in the compressor clearance space that causes a hydraulic hammer.

Slugging A condition where liquid refrigerant is entering an operating compressor.

Smoke Test Test made to determine completeness of combustion.

Snap Action A means of ensuring positive closure of a mechanically operated electrical switch.

Snap-Disc A warped bimetal disc that snaps at a particular temperature to operate switches.

Soft Flame A flame partially deprived of primary air such that the combustion zone is extended and the inner cone is ill defined.

Solar Cell (see Photovoltaic Cell)

Solar Collector Typically, an insulated box that converts solar energy into thermal energy using a black surface.

Solar Heat Heat from visible and invisible energy waves from the sun.

Solder An alloy with a low melting temperature used for joining metals.

Soldering Joining two metals by adhesion of a low melting temperature metal (less than 840°F).

Solderless Terminals Wire terminals that connect to the wires using mechanical crimping rather than solder.

Solenoid A movable plunger activated by an electromagnetic coil.

Solenoid Valve Valve actuated by magnetic action by means of an electrically energized coil.

Solid The physical state of matter that has both a definite shape and volume.

Solid Fuel Combustible material such as wood, coal, or corn pellets.

Solubility The ability of one substance to be dissolved in another.

Solution, Chemical A liquid with another solid, liquid, or gas dissolved in it.

Solution, Refrigeration A mixture of absorbent and refrigerant in an absorption system.

Solution, Strong The solution leaving the absorber with a high percentage of refrigerant.

Solution, Weak The solution leaving the generator with a low percentage of refrigerant.

Sone A sound rating used for rating ventilation fans.

Soot A black substance, mostly consisting of small particles of carbon, that can result from incomplete combustion.

South Pole The pole on a magnet that magnetic lines of force flow into.

Space Heater A small heater designed to sit in the area that it heats.

Specific Gravity (SG) For a liquid or solid, the ratio of its density compared to water. For a vapor, the ratio of its density to air.

Specific Heat The amount of heat energy needed to change the temperature of a material at a given pressure as compared to an equal quantity of water or air at the same pressure.

Specific Volume The volume of a substance per unit mass; the reciprocal of density units; cubic feet per pound, cubic centimeters per gram, and so on.

Splash Lubrication Lubrication based on compressor components dipping into the oil in the crankcase.

Split Phase Motor Motor with two stator windings.

Split System Refrigeration or air conditioning installation that places condensing unit remote from evaporator.

Split System Heat Pump A heat pump with components located both inside and outside of a building—the most common type of heat pump installed in a home.

Spray Pond Used for cooling condenser water through evaporation by spraying water into the air.

Squirrel Cage Fan that has blades parallel to fan axis and moves air at right angle to fan axis.

Stack Switch A type of primary control and safety for oil furnaces that senses the temperature of the flue for operation.

Standard Air Air at standard conditions.

Standard Atmosphere Air at a pressure of 14.696 psia, 59°F, with a relative humidity of 30%.

Standard Conditions 68°F and 14.696 psia atmospheric pressure.

Standard Rating Conditions The conditions used for AHRI performance ratings. For cooling, 80°F dry bulb, 67°F wet bulb return air, and 95°F dry bulb outside air.

Standing Pilot A small flame that burns continuously that is used to light the main burners of a gas fired appliance.

Start Capacitor A capacitor that is used only during startup to increase the starting torque of a single phase motor.

Start Relay An electrical device that connects and/or disconnects the start winding of electric motor.

Starting Winding Winding in electric motor used for only the brief period when the motor is starting.

Starved Evaporator A term used to describe a condition where less refrigerant flows into the evaporator than is required for the heat load.

Static Condenser A natural draft condenser that depends more on radiation than convection for its cooling effect.

Static Electricity Electrical potential that results from the physical displacement of electrons from one object to another.

Static Head The amount of pressure difference a pump must overcome before it can move any water.

Static Pressure (P_s) The pressure exerted against the inside surfaces of a container or duct. Sometimes defined as burst pressure.

Stationary Blade Rotary Compressor A rotary compressor with a single blade housed in the compressor body that does not rotate with the compressor shaft.

Stator Stationary part of electric motor.

Steady State Condition The voltage and current levels of the controls on a gas furnace after ignition has been achieved and the indoor blower is operating.

Steam Jet Refrigeration Refrigerating system that uses a steam venturi to create high vacuum (low pressure) on a water container causing water to evaporate at low temperature.

Steam Water in a vapor state.

Steam Heating A heating system in which water absorbs heat in a boiler where it boils and gives off heat from radiators where it condenses.

Steam Trap A device that will prevent the flow of steam, but will allow the flow of condensate.

Step Motor A motor used for precisely positioning valves that moves in small repeatable steps.

Strainer A screen used to retain solid particles while liquid passes through.

Stratification Condition in which air lies in temperature layers.

Stratosphere The portion of the atmosphere approximately 12–30 mi above the earth. The ozone layer is in the stratosphere.

Stroke, Actuator The distance that an actuator for a damper or valve can move.

Stroke, Compressor The distance that a piston travels in the cylinder of a reciprocating compressor.

Subbase The mounting base for a low voltage thermostat where electrical connections are made.

Subcooled Liquid Liquid at a temperature lower than is possible when it is in equilibrium with its vapor. The pressure is higher than the vapor pressure.

Subcooling The reduction of the temperature of a liquid below its condensing temperature.

Subcritical A refrigeration process that take place below the refrigerant critical point.

Sublimation Condition where a substance changes from a solid to a gas without becoming a liquid.

Suction Line Tubing or pipe used to carry refrigerant vapor from the evaporator to the compressor in single function systems or from the reversing valve to the compressor in dual function systems.

Suction Line Accumulator (see Accumulator)

Suction Pressure Pressure in the low side of the refrigeration system. Also called low side pressure or suction pressure.

Suction Service Valve A two way manually operated valve located at the inlet to the compressor.

Suction Valves The valves on the underside of the valve plate of a reciprocating compressor that allow gas to enter the cylinder on the downstroke.

Sump A reservoir at the bottom of a piece of equipment that circulates water, such as a cooling tower.

Supercritical A refrigeration process that takes place above the refrigerant critical point.

Superheat Heat energy added to a gas so that the enthalpy is higher than for saturated vapor at the same pressure. The heat added to a vapor to raise the sensible temperature of the vapor above its boiling point.

Supplementary Heat The auxiliary heat provided at temperatures below the heat pump balance point. In most cases this is done with electric heating elements that are part of the heat pump system installation. A gas or oil furnace also can be used to provide supplementary heat when a heat pump is added to an existing fossil fuel heating system.

Surge Tank Container connected to a refrigerating system that increases gas volume and reduces the rate of pressure change.

Sustainability A building designed in such a way that can continue indefinitely into the future without damaging or depleting natural resources.

Swaging Enlarging a tube end so that another tube of the same size will fit inside it.

Swaging, Tool A tool for swaging tubing.

Swamp Cooler An air conditioner that uses water evaporation to reduce the temperature of the air.

Swash Plate Wobble plate. Device used to change rotary motion to reciprocating motion.

Sweating This term has two definitions in air conditioning or heat pump work: (1) formation of moisture on the outside of cold pipes or ducts; and (2) joining of two metals by the adhesive action of a third metal.

Sweet Water Tap water.

Synchronous Speed The theoretical speed that an alternating current induction motor would turn if the rotor turned at the same speed as the magnetic pulses in the motor windings.

Synthetic Not naturally occurring, manmade.

Synthetic Lubricant A synthetically derived chemical, such as polyol ester, that is used for lubrication.

Synthyetic Dust Weight Arrestance A method of testing air cleaner efficiency using manufactured "dust" with its size carefully controlled.

System Charge The type and amount of refrigerant in a system, usually stated on the data plate.

System Lag Time delay between when a condition is sensed and when a system begins to act on the condition.

System Overshoot Space temperature exceeding the thermostat setpoint in a heating system.

Tankless Water Heater An indirect water heater designed to operate without a hot water storage tank.

Tap (Screw Thread) Tool used to cut internal threads.

Temperature Degree of hotness or coldness as measured by a thermometer.

Temperature Glide The temperature difference between the dew point and the bubble point of a zeotropic refrigerant.

Temperature-Humidity Index An index that combines air temperature and humidity to describe levels of comfort.

Temperature-Pressure Relationship (see Pressure-Temperature Relationship)

Temperature-Pressure Chart (TP Chart) (see Pressure-Temperature Chart)

Temperature-Pressure Relief Valve (TPR) A valve that opens to relieve the pressure on a closed water system if the pressure or temperature exceeds a safe level.

Temperature Sensing Bulb A bulb filled with a volatile fluid or gas with a capillary tube that transmits pressure from the bulb to the controlled device. Used with thermostatic expansion valves or thermostats.

Temperature Swing The difference between the low temperature and the high temperature in a controlled space.

Terminal Units Radiators, convectors, baseboard, unit heaters, finned tube, and so on.

Terminal Velocity The speed of the airstream leaving a diffuser when it has slowed to a point that it no longer effectively mixes with room air. Typically between 50–150 fpm.

Test Light An electrical test instrument that indicates the presence of voltage by illuminating a light.

Testing, Adjusting, and Balancing (TAB) Testing system performance and making adjustments in the system to make actual system operation match design specification. The term is often used to refer to airflow adjustments, but all aspects of system performance should be evaluated.

Testing, Adjusting and Balancing Bureau (TABB) An organization affiliated with the sheet metal industry that certifies TAB technicians.

Therm A unit of heat having a value of 100,000 Btu.

Thermal Conductivity The ability of a material to transmit heat.

Thermal Conductivity, Earth The rate at which heat energy flows through an earth material. Expressed in Btu/ft^2 of the material surface times the temperature difference per thickness in feet of the material.

Thermal Cutout An overcurrent protection device that contains a heater element that affects a bimetal element designed to open a circuit in the event of electrical current flow above the rated amount of the device.

Thermal Resistance The resistance a material offers to the transmission of heat.

Thermistor An electrical device that changes electrical resistance with a change in temperature of the device.

Thermocouple Device that generates electricity using the principle that if two dissimilar metals are welded together and the junction is heated, a voltage will develop across the open ends.

Thermodynamics The study of heat and heat flow.

Thermoelectric Refrigeration A refrigerator mechanism that depends on the Peltier effect.

Thermometer Device for measuring temperatures.

Thermopile A pilot generator.

Thermostat A control device that responds to surrounding air temperatures.

Thermostat, Outdoor Ambient A control used to limit the amount of auxiliary electric heat according to the outdoor ambient temperature to reduce electrical surge and cost.

Thermostat, Termination A thermostat mounted on the outdoor coil that interrupts the defrost mode when the temperature of the coil reaches the cutout setpoint of the control.

Thermostat Droop Maintaining temperature lower than the setpoint.

Thermostatic Control A device that controls the operation of equipment according to the temperature of the air surrounding the control.

Thermal Expansion Valve (see Thermostatic Expansion Valve)

Thermostatic Expansion Valve (TEV, TXV) A valve operated by temperature and pressure within the evaporator coil.

Thermostatic Valve Valve controlled by thermostatic elements.

Thermostatic Water Valve A water regulating valve controlled by temperature used on water cooled systems.

Three Phase An alternating current with three sine waveforms separated by 120°, supplied by three wires.

Three Phase Motor An electric motor designed to operate on three phase alternating current.

Three Way Valve, Control A valve that controls flow from one inlet to two outlets.

Three Way Valve, Manual A manual valve with three ports and two seats often used as a manual refrigeration service valve.

Throttling Expansion of gas through an orifice.

Throw The distance that air travels from an air diffuser or register before reaching terminal velocity.

Thrust Bearing A bearing designed to withstand force pushing against the bearing in parallel with the shaft. Vertical mount fans require thrust bearings.

Timed Off Control A timer that begins timing when the unit cycles off to ensure a minimum off cycle time before allowing the unit to restart.

Time Delay A delay between when a control is energized and when it acts.

Time Delay Fuse A fuse that can withstand momentary overload. Commonly used with motors.

Time Delay Relay A relay whose coil is energized for a time period before its contacts open or close.

Timers Mechanism used to control the time cycling of an electrical circuit.

Timer, Defrost A timer that operates at the same time as the refrigeration system. After a set period of operating time, the timer trips to initiate a defrost operation if the termination thermostat is closed.

Ton Refrigerating effect equal to the melting of one ton of ice in 24 hr.

Torque Turning or twisting force.

Torque, Full Load The maximum torque that a motor can provide at its rated output.

Torque, Stall The amount of torque that a motor produces when energized and mechanically locked. This is the highest torque that a motor can produce.

Torque, Starting The amount of torque a motor produces when starting, typically much higher than full load torque.

Torque Wrenches Wrenches that may be used to measure the torque applied.

Torr A non-SI unit of vacuum equal to 1/760 of standard atmospheric pressure. A perfect vacuum is 760 Torr.

Total Energy Management (TEM) An energy management strategy that looks at the total energy use of a building rather than concentrating on specific components or systems.

Total Equivalent Warming Impact (TEWI) An index of global warming that takes into account both the direct and indirect global warming impact of a refrigerant.

Total Heat The sum of sensible heat and latent heat, referred to as enthalpy.

Total Pressure (P_t) The sum of the static pressure and the velocity pressure at the point of measurement.

Toxicity The degree to which something is poisonous. Refrigerants that are harmful in concentrations of 400 ppm or less are listed as higher toxicity.

Transcritical A refrigeration system which has part of it's processes operating above the refrigerant critical point and part of it's processes operating below the refrigerant critical point.

Transducer A device that produces an electronic signal based on a physical property such as temperature or pressure.

Transformer A device designed to change voltage.

Transmission, Heat Heat loss or gain through building components.

Triple Point The temperature and pressure where a substance can exist as a solid, liquid, and gas simultaneously.

Troposphere The lowest portion of the atmosphere, close to the earth's surface.

True RMS Meter Uses frequent value sampling and the root mean square formula to determine alternating current values.

Tube in a Tube Coaxial A heat exchanger constructed of a tube inside a tube sealed off from each other. Usually liquid is in the inner tube and refrigerant is in the outer tube.

Tubing Used for conveying fluids. Tubing has a consistent outside diameter and relatively thin walls compared to pipe.

Turbulent Flow The movement of a liquid or vapor in a pipe in a constantly churning and mixing fashion.

Turndown The ratio of maximum to minimum input rates.

Twinning Installing and wiring two furnaces to operate simultaneously as one larger unit.

Two Pipe System A hot water heating system using one pipe to supply heated water to the terminal units, and a second pipe to return the water from the terminal units.

Two Position Control A control that is either on or off.

Two Stage Air Conditioner Can operate at two separate capacities to better match unit capacity to load.

Two Stage Compressor Can operate at two separate capacities to better match the compressor capacity to the load. Used in two stage air conditioners and heat pumps.

Two Stage Vacuum Pump Produces a lower pressure and deeper vacuum by having two vacuum pumps in series; the discharge of the low stage feeds the suction of the high stage.

Two Stage Furnace Has a low firing rate and a high firing rate that can better match furnace capacity to the load.

Two Stage Heat Pump Can operate at two separate capacities to better match unit capacity to load. Allows increased heating capacity without oversizing the air conditioning.

Two Temperature Valve Another name for evaporator pressure regulator.

Two Way Valve A valve with one inlet and one outlet that has two positions: opened and closed.

U Factor Unit of measure of thermal conductivity.

U Value Another name for U factor.

Ultrasonic Leak Detector Detects gas leaks in refrigeration systems by hearing the ultrasonic whistle of the gas escaping.

Ultraviolet (UV) Invisible radiation waves with frequencies shorter than wavelengths of visible light and longer than X-ray.

Ultraviolet A (UVA) Long wavelength ultraviolet light with wavelengths of 400–315 nm.

Ultraviolet B (UVB) Medium wavelength ultraviolet light with wavelengths of 315–280 nm.

Ultraviolet C (UVC) Short wavelength ultraviolet light with wavelengths of 280–100 nm.

Underwriters Laboratories (UL) Independent standards and testing laboratory for the examination and testing of devices.

Unit Heater A fan and motor, a heating element, and an enclosure hung from a ceiling or wall.

Unit Ventilator A terminal unit in which a fan is used to mechanically circulate air over the heating coil.

Universal Motor A series wound DC motor that can operate on either AC or DC current.

Unvented Space Heater A gas fired space heater that releases all the products of combustion into the space it is heating.

Universal Motor A series wound motor used for small electric appliances and power tools. Universal motors can operate on either AC or DC current.

Unloader Device that reduces the capacity of a compressor by preventing some cylinders from compressing.

Upflow Furnaces and air handlers with airflow that is up vertically, entering at the bottom of the unit and leaving at the top.

Urethane Foam Type of insulation that is foamed in between inner and outer walls of a display case.

U.S. Green Building Council (USGBC) A nonprofit organization committed to a prosperous and sustainable future through cost-efficient and energy-saving green buildings

U-Tube Manometer Measures pressure by the height of liquid column that it will support in a U-shaped tube. Low pressure manometers are filled with water; higher pressure manometers are filled with mercury.

V Belt Drive belt with a V-shaped cross section.

Vacuum Reduction in pressure below atmospheric.

Vacuum Gauge Measures the level of a deep vacuum, usually in microns.

Vacuum Pump A high efficiency vapor pump used for creating deep vacuum in refrigeration systems for testing and/or drying purposes.

Valve Device used for controlling fluid flow.

Valve, Check A valve that will permit fluid flow in only one direction. Sometimes called a one way valve.

Valve, Expansion A modulating refrigerant metering device.

Valve Plate Part of compressor located between the top of the compressor body and the head that contains the compressor valves.

Valve, Reversing A valve used to change the direction of refrigerant flow in a heat pump system. Because there are four pipe connections, it is also called a four way valve.

Valve Seat (see Seat)

Valve, Service A device used by service technicians to connect pressure gauges into the refrigeration system.

Valve, Slide The slide valve portion of the reversing valve that shifts the refrigerant flow.

Valve, Solenoid A flow control valve controlled by an electromagnetic coil actuating a plunger off its seat.

Valve, Suction (see Suction Valves)

Valve, TXV, Bi-Flow A thermostatic expansion valve that is designed to provide pressure reduction and refrigerant flow control in either direction.

Valve Stem Depressor The part on the end of a refrigeration hose that pushes in the valve core when connecting to a Schrader valve.

Vapor The gas form of a substance. It takes the shape and volume of its container.

Vapor Barrier Thin plastic or metal foil sheet used to prevent water vapor from penetrating insulating material.

Vapor Charge Bulb Also called gas charge or limited charge. A temperature sensing bulb that has a limited amount of liquid so that all the liquid will be vaporized at the upper end of the bulb's operating temperature range.

Vapor Line Found only in dual action heat pumps. It is the suction line in the cooling mode and the hot gas line in the heating mode.

Vapor Lock The restriction or blockage of flow in a liquid line due to the presence of vapor in the line.

Vapor, Saturated A vapor whose temperature has been reduced to the point of condensation but condensation has not started.

Vapor Pressure The pressure exerted by a vapor on its surroundings.

Vapor Refrigerant Charging Adding refrigerant to a system in gas form.

Vapor Refrigerant Recovery Removing refrigerant from a system in vapor form.

Vaporization Changing physical state from a liquid to a gas.

Variable Air Volume (VAV) A method of zone control that varies the volume of air to the zone depending upon the load.

Variable Frequency Drive (VFD) A control for an AC electric motor that changes the speed of the motor by varying the frequency of the current to the motor.

Variable Pitch Pulley Pulley that can be adjusted to provide different pulley ratios.

Variable Resistor A resistor whose value can be adjusted, typically between 0 Ω and its maximum value.

Variable-Speed Motor A motor whose speed can be adjusted to meet varying load requirements.

Velocimeter Instrument used to measure air velocities.

Velocity (v) Speed.

Velocity Pressure (P_v) The pressure exerted in the direction of flow.

Vent-Free A gas fired appliance that is designed to operate without a vent whose combustion products are released into the area it is in.

Vented Space Heater A gas fired space heater whose combustion products are vented through a flue to the outside.

Vent Gases Products of combustion.

Ventilation The introduction of outdoor air into a building by mechanical means.

Venturi A section in a pipe or a burner body that narrows down and then flares out again.

Vibrasorber A refrigeration component for reducing vibration in refrigerant piping consisting of a corrugated inner tube and a woven metal outer casing.

Viscosity Measure of a fluid's ability to flow. Typically measured in Saybolt universal seconds.

Volatile A fluid that evaporates readily.

Volt (V) The unit of electrical potential or pressure.

Voltage The potential difference between two points.

Voltage, Effective A measure of AC voltage that states its effect in a resistive circuit.

Voltage, RMS Root mean square voltage is used to calculate the effective AC voltage. The instantaneous voltage is measured many times each cycle. Each of these values are squared, the average of the squares is found, and finally the square root of the average is the effective voltage.

Voltage, Minimum The minimum supply voltage to a piece of equipment.

Voltage, Maximum The maximum supply voltage to a piece of equipment.

Voltage, Nominal The standard AHRI voltage rating for a piece of equipment.

Voltage, Peak The voltage at the peak of an AC sine waveform.

Voltage Relay One that functions at a predetermined voltage value.

Voltmeter A meter that reads voltage.

Volt Ohm Milliammeter (VOM) A meter that measures voltage, resistance (ohms), and milliamps.

Volumetric Efficiency Ratio of the actual performance of a compressor and calculated performance.

Walk-in Cooler Large commercial refrigerated room.

Wastewater System A water cooled refrigeration system that does not recirculate the condenser cooling water.

Water Box The water manifolding portion of a chiller or boiler.

Water Coil A heat transfer coil designed for water.

Water Column (wc) A unit of pressure comparing the height of water column it will support.

Water Cooled Condenser A heat exchanger that uses water to remove the heat from the high temperature compressor discharge vapor.

Water Cooled Condensing Unit A condensing unit (high side) that is cooled by the use of water.

Water Cooler An appliance that cools drinking water.

Water Hammer A pounding sound made by the sudden increase in static pressure in the pipes when flowing water is stopped suddenly by the closing of a valve.

Water Loop The piping and components in a closed circuit.

Water Regulating Valve A valve that regulates the water flow through a water cooled condenser in response to condenser pressure.

Water Tube Boiler A hot water boiler in which the water is circulated through the tubes and the hot gases from combustion of the fuel are circulated around the tubes.

Watt (W) A unit of electrical power.

Watt-Hour (Wh) Using electrical power at the rate of 1 W for a period of 1 hr.

Welded Hermetic Compressor A compressor with both the compressor and motor inside a welded shell.

Well Water Source, Closed Loop A system that removes water from the earth by means of a drilled well and returns it to the earth by means of a separate drilled well.

Well Water Source, Open Loop A system that removes water from the earth by means of a drilled or bored well and returns the water to the earth through a separate disposal system.

Wet Bulb Temperature (wbt) A temperature taken with a wet bulb thermometer.

Wet Bulb Depression The difference between the dry bulb and wet bulb temperatures.

Wet Bulb Thermometer A thermometer that uses a wet sac on the bulb to measure the evaporation rate of the air sample.

Wet Cell Battery A battery with a liquid electrolyte solution.

Wet Heat Heating system using hot water or steam.

Wind Chill Factor The reduction in perceived temperature because of wind.

Wind Effect The increase in evaporation rate due to air travel over a water surface.

Winding Thermostat A temperature sensor that opens if the motor winding temperature exceeds a safe level.

Windmilling Backward fan rotation caused by wind blowing through outdoor fans when they are off.

Window Unit An air conditioner designed to be installed in a window.

Wire Connectors Electrical terminals that are fastened to wires by soldering or crimping that facilitate wire connection to components.

Wobble Plate Compressor A compressor with pistons that are parallel to the crankshaft that uses an offset disc to create reciprocating motion.

Work The transfer of energy.

Work Hardening Metal becoming brittle as a result of being repeatedly flexed.

Wye Connected Motor A three phase motor whose windings are joined at the center to form a wye.

Wye Transformer A three phase transformer whose windings are joined at the center to form a wye.

Zeotropic A blend that changes in composition as it boils.

Zone That portion of a building that has its temperature controlled by a single thermostat.

Zone Control The controller that coordinates operation of the dampers or valves in a zone control system.

Zone Damper A damper that controls the airflow to a particular zone in a zoned system.

Zone Control System A control system that maintains the condition of separate zones within a building.

Zone Valve A valve that controls water flow to a zone in a zoned hydronic heating system.

Index

J

K

L

R

S

U

V

W

Z